Dinichthys terrelli

Cooksonia

Horse-tails

Cycad

Paleozoic

| Silurian | Devonian | Carboniferous | Permian |

408 360 286 245

Major extinction:
70% of all species

Major extinction:
90% of all species

FOR GEORGE DUSHECK AND EVE TOBIN
AND IN LOVING MEMORY OF NINA DUSHECK AND MAURICE TOBIN

Asking About Life

Allan J. Tobin
*University of California,
Los Angeles*

Jennie Dusheck
Santa Cruz, California

Saunders College Publishing
Harcourt Brace College Publishers

Fort Worth   Philadelphia   San Diego   New York   Orlando   Austin
San Antonio   Toronto   Montreal   London   Sydney   Tokyo

Publisher: Emily Barrosse
Executive Editor: Edith Beard Brady
Product Manager: Erik Fahlgren
Developmental Editor: Lee Marcott
Project Editor: Elizabeth Ahrens
Production Manager: Charlene Catlett Squibb
Art Director: Caroline McGowan
Text and Cover Designer: Ruth A. Hoover
Art Development: Elizabeth Morales

ABOUT THE COVER
Like humans, many animals are curious about the world around them. These giraffes are obviously intrigued by the activities of the photographer and seem to be studying the reader. Three hang back hesitantly, but the fourth has moved forward, fearless and determined to find out what's up. That attitude typifies the best science and exemplifies the questioning theme of *Asking About Life*.

The giraffes also remind us of Lamarck's theory of evolution and how human is the scientific enterprise. Like many modern scientists, Lamarck suffered humiliating rejection by his peers. Charles Darwin's efforts to avoid a similar fate inspired him to build an irrefutable and monolithic theory of evolution that modern biologists accept nearly as Darwin wrote it. (*Mitch Reardon/Tony Stone Images*)

Frontispiece Credit: *Yoshio Otsuka/Photonica*
Chapter 3 Opener: *The LEGO® Brick is a Trademark of the LEGO Group. © Models protected by Copyright owned by the LEGO Group. Both used here with special permission.*
Chapter 14 Opener: *David Hanover/Tony Stone Images (woman); Douglas Struthers/Tony Stone Images (DNA). Concept and design by Mycoff Advertising Inc. for New England Biolabs Inc. Used with permission.*
Chapter 22 Opener: *© 1997 The Georgia O'Keeffe Foundation/Artists Rights Society (ARS), New York. Photo courtesy Colorado Springs Fine Arts.*

Printed in the United States of America

ASKING ABOUT LIFE
0-03-072046-X

Library of Congress Catalog Card Number: 97-68078

890123456   032   10 98765432

PREFACE

When Charles Darwin published *The Origin of Species* in 1859, the first edition sold out overnight. His revolutionary theory of evolution was of interest not only to his fellow scientists, but to great numbers of ordinary people, who avidly read *The Origin* and argued over the details of the theory.

Today, public interest in biology is even greater. When a previously unknown Scottish biologist cloned the first mammal, a sheep, in February of 1997, the story made front-page headlines all over the world and sharply increased the value of biotechnology stocks overnight.

Biology is a discipline in full flower. Biologists now have the capacity to understand the workings and the interactions of organisms and—at the cellular level—to alter them, almost at will. Biologists have bred new crops, discovered the frailties and strengths of precious ecosystems, developed new treatments for diseases, and begun to solve many puzzles of the human mind.

Spectacular as these developments are, they are a mixed blessing for instructors and their students. The sheer numbers of facts can be overwhelming. How, for example, can we remember the difference between a missense mutation and a nonsense mutation? Between a plasmodial slime mold and a cellular slime mold? And why should we care?

Yet, to make the simplest decisions in the 21st century, you will need to understand how science works and, at least, the basics of biology. If you are sick, should you take an antibiotic? If one of your parents has a genetic disease, should you be tested for the disease-causing allele? If the vacant lot down the road is to be turned into a playing field, should the creek that runs through it be preserved? Is there any harm in running the creek through an underground, concrete culvert? The answers to all of these questions and hundreds of others depend on the ability to understand and evaluate scientific arguments.

The greatest barrier to understanding science is the common perception that science is inaccessible to ordinary folks. We know that it is accessible. It's about asking questions, getting a partial answer, then asking another question.

What Is *Asking About Life* About?

Asking About Life is about curiosity. Our very title stresses our conviction that questions are what drive the process of science. The history of science reveals that while the answers to questions change over time, the questions themselves, if they are good ones, remain the same. One good question has been "How do new species form?" Biologists have been trying to answer this question for 150 years. But it is a question with many answers. And the answers that seemed correct in 1880 or 1960 continue to be refined, expanded, or even overthrown.

Asking About Life consistently emphasizes the importance of questions and the process of finding answers. To remind ourselves and our readers of this, headings and subheadings are more often questions than statements.

Stephen J. Krasemann/Photo Researchers

The task of student and scientist alike is to ask questions as insistently as possible and to answer them as cleverly as possible. While we understand that the "facts" of biology are important, science progresses only by discovering new relationships of facts to questions: the difference between a nonsense mutation and a missense mutation takes on pressing significance if a relative has a disease caused by a genetic mutation. Suddenly, we find it much easier to remember why a missense mutation often does no harm, while a nonsense mutation nearly always does harm. Just hearing a story about someone with a genetic disease would help us remember why this distinction is important. In learning, story and context (whether emotional or historical) are everything.

Throughout our book, we have emphasized the story behind scientific discoveries. *Asking About Life* illustrates the passion of the scientific enterprise by beginning each chapter with a story about a piece of research. Each story illustrates how an individual biologist pursued a scientific question, often in the face of intellectual or social adversity—how, for example, Barry Marshall convinced first himself and then others that bacteria, not stress, cause ulcers; and how Rosalind Franklin struggled in deep social isolation to elucidate the structure of DNA. Our anecdotes are about real people—their triumphs, their frustrations, their genius, and their persistence.

The story of how biologists came to ask a question and how they went about answering that question helps us remember not just one fact, but all the facts and ideas associated with that question. All the stories illustrate that science is both an intellectual endeavor and a social one. Scientists must not only ask and answer questions, they must also persuade their colleagues that their answers are correct and interesting. The most successful scientists are often the most sociable ones—though not always.

We have tried to make our stories lively and engaging. Asking and answering questions about life is a job for the most lively and engaged of humans. Biology is not a mysterious activity performed by those who are remote, calculating, or cold. It is an exquisitely human endeavor in which enthusiastic beginning students can participate as fully as veteran scientists. In fact, the fresh view of the engaged beginner sometimes unlocks puzzles that have stumped seasoned scientists for years, as illustrated in the story about Walter Sutton in Chapter 9.

The flowering of biology in the late 20th century has already produced new seeds for the 21st. It is our hope that student readers of *Asking About Life* will be able to nurture and appreciate the growth of new ideas that we feel will dominate the intellectual and economic landscape of the next generation.

What Is the Philosophy Behind *Asking About Life?*

In some respects, the philosophy of our text runs counter to current trends in textbook construction. Many textbooks now break information up into self-contained units. Each bit of information is presented as if unrelated to the information in the rest of the book. Much information is tucked into separate illustrations, with no text to weave disparate facts together. The rationale for this is that today's students belong to the "visual-information" generation and are incapable of sustained reading.

We have more faith in students. We know they can read, and even enjoy reading, provided the reading is interesting and rewarding. Our philosophy is that biology is a story, and, as such, it must be presented as continuously as possible. Each section of the book includes background information, usually historical, that provides a context for current research.

For example, in the introductory chapter to the section on Diversity, where we discuss different systems of classification, we show students that biologists have been arguing about the definition of a species for nearly 200 years. In the previous section, we have emphasized Darwin's frustration in defining species of barnacles when he could see that many species blended one into the next. In that context, the opening story in the Diversity section about the current and highly politicized debate over whether the red wolf is a species or a hybrid makes sense. At the same time, the red wolf story brings to life what would otherwise be an abstract discussion of classification.

In this context of questions and process, we naturally emphasize the experimental nature of science. Relying on headings that are phrased as questions, we lead students from one critical experiment to the next. Students learn not only the results of experiments, but why biologists asked certain questions and how they discovered the answers. In Chapter 10, for example, we do not content ourselves with describing the structure of DNA, but lay out for the student all the clues that Rosalind Franklin, James Watson, and Francis Crick used to deduce DNA's structure. The reader thus experiences anew the excitement of the original discovery.

Throughout the book we emphasize current research wherever appropriate. As a result, instructors have the opportunity to discuss with their students topics appearing in the daily news. The chapter on human genetics, for example, focuses heavily on current debates about policy and ethics in the application of genetic techniques to human disease and reproduction.

Naturally, we also try to minimize abstract discussions through the use of metaphors and analogies. For example, in our discussion of the movements of chromosomes during mitosis and meiosis, we compare the pairs of chromosomes to pairs of socks going through the laundry. In our discussion of osmosis, we compare the swelling of cells in a hypotonic solution to the swelling of raisins in a pot of water. And in Chapter 3, we compare functional groups—the small chemical groups that give molecules their characteristic chemistries—to the different parts of a Swiss Army knife.

More than in other texts, the art program in *Asking About Life* makes such metaphors visual—leaving the student with memorable and often amusing images. Vivid illustrations help fix otherwise abstract ideas in our minds in the same way a mordant fixes a dye. For example, many books compare the structure of tRNA to a clover leaf. Only *Asking About Life* actually shows the clover leaf—complete with stem and leaflets—folded into the classic "L" shape. In the same chapter, Chapter 11, the ribosome is depicted as a sewing machine that stitches together amino acids to form polypeptides.

In Chapter 34, a drawing of a mouse with an 8-inch fur coat helps students remember that the metabolic rates of small animals must be much higher than those of large ones. Wherever possible, we have used visual metaphors and other striking images that act as icons for ideas.

Each chapter begins with an image, either a photograph or drawing, that was chosen to complement the opening story or to sum up a theme of the chapter. These are not captioned and are intended to be brain-teasers for the reader, who may find it fun and interesting to guess the connection being made.

For instance, the opening image for Chapter 1—the dinosaur in the egg—represents evolution and the continuity of life. The dinosaur, not unlike a modern barnyard chick, illustrates how life continues from generation to generation. Life changes and yet stays the same.

In Chapter 3, the thousands of Lego™ pieces in the Legoland elephant symbolize the building blocks of life. Similarly, just a few molecules combine to make the large molecules that make up all organisms.

In Chapter 5, the rusting car symbolizes entropy, the concept that ordered systems and structures tend to become disordered over time.

The chapter-opening images for all 44 chapters are identified and their significance is explained in the Teaching Suggestions section of each chapter in the *Instructor's Manual*.

We regard illustration as a teaching tool in its own right, not just a backup for the text. In this, we have been privileged to work closely with two outstanding illustrators—Elizabeth Morales and Elizabeth McClelland—whose skill, attention, and insight have contributed not only to the art manuscript but to the text as well. Morales, the book's art developmental editor, created a friendly style that perfectly complements our informal text. Her clean designs and simple illustrations greatly clarify sometimes difficult material. In addition, Elizabeth McClelland contributed dozens of beautifully rendered illustrations of animals and plants, as well as many excellent diagrams throughout the book.

What Kinds of Pedagogy Does *Asking About Life* Employ?

Asking About Life has a variety of features designed to engage the reader and to aid student learning:

Each chapter begins with a story about a piece of research that draws students into the subject of the chapter and also introduces the key questions and ideas that are discussed.

Within or after each chapter story, readers will find a list of **Key Concepts**, which are some of the most important and basic ideas covered in the chapter.

Headings are often posed in the form of questions throughout each chapter. These question headings focus the reader's attention on the most significant question to be explored in that section.

Subsections are followed by **Summary Statements**—brief summaries of the take-home message. These provide students with a reality check. If the student doesn't understand the summary statement, that is a cue to study the preceding material more closely.

Drawings and photographs support concepts covered in the text and help students visualize the structures of objects as diverse as molecules and ecological communities. Photographs of structures that are too small to be seen with the naked eye are accompanied by size bars to give a sense of scale.

Figure legends are designed to stand alone, so that even a student flipping through the chapter for the first time, glancing at the diagrams and reading the captions, will come away with some important information.

Visual metaphors in the illustrations drive home key points introduced in the text.

Boldface terms throughout the text help students to locate key terms and their definitions.

Tables and **graphs** summarize key facts and additional material.

Most chapters feature one or more **Boxes**, which discuss some topics in greater depth or bring into sharper focus a piece of especially important research. In Chapter 12, which focuses on viruses and jumping genes, a box discusses current interest in the old idea of using bacteriophages as a therapeutic treatment for bacterial infections in humans. Chapter 26, on ecological succession, features a box on forensic ecology—the technique by which coroners determine when a person died by using the stages of development of different insect larvae found on a corpse.

The end of each chapter features a **Study Outline with Key Terms.** All of the boldface terms found in the chapter are used again in a highly compressed summary. This provides students with another opportunity to check their understanding of the chapter. If they encounter terms they don't remember or ideas that seem unfamiliar, they can return to the main text and illustrations.

The Study Outline is followed by a set of **Review and Thought Questions**. Our Thought Questions are especially engaging, frequently bringing ideas in the chapter into the everyday world.

Selected Readings emphasize readings in science that are accessible to a general audience and available in any good library. The majority of the readings are books, most of which the authors of *Asking About Life* have themselves enjoyed. The readings are intended as thought-provoking pleasure reading, not as grist for term papers. Class papers can always be researched in the conventional way—at the library, or increasingly, online.

An appendix on standard weights and measures is provided for student reference (Appendix B).

A **Glossary** provides a complete list of Key Terms and their definitions.

Supplements

To further facilitate learning and teaching, a supplements package has been carefully designed for the student and instructor.

The **Study Guide** by Lori Garrett of Danville Area Community College includes Chapter Objectives that restate the Key Concepts of the text as material to be mastered, Key Concepts, an Extended Chapter Outline that gives an overview of the most important topics covered in the chapter, Vocabulary Building exercises, and Chapter Tests. Each Chapter Test has five parts: Multiple Choice; True/False; Matching; Short Answer; and Essay/Thought Questions. All answers are provided, with the exeption of the Essay/Thought Questions.

The **Instructor's Manual** by Michael Ulrich of Elon College includes an introductory section of classroom teaching suggestions, research paper grading criteria, and evaluation procedures. Each chapter has Lecture Outlines; all the answers to the Review and Thought Questions from the text; and Teaching Suggestions, which include how to use the opening stories to motivate student interest and the significance of the chapter-opening images.

The **Test Bank** by Frederick Peabody of University of South Dakota comprises 2000 questions of assorted type (multiple choice, fill-in-the-blank, and short-answer essay questions) that are organized by the main chapter headings as well as keyed to the Key Concepts as they appear at the start of chapters. The **Computerized Test Bank** is available for Windows™ and Macintosh platforms.

Other important components of the supplements package for *Asking About Life* include a set of 200 **Overhead Transparencies** based on the drawings from the book and **Bio-Art**, which is a set of 100 black-and-white unlabeled line drawings from the text.

Thinking Toward Solutions: Problem-Based Learning Activities in General Biology by Deborah Allen and Barbara Duch, of the University of Delaware, presents complex, open-ended problems for introductory biology covering all aspects of the discipline from cells to the environment. Problems address real-world applications of biology to questions of ethics, economy, and daily living. Problems foster critical thinking, cooperative learning, and problem-solving skills. A detailed Instructor's Manual provides practical suggestions on how to use the problems in any size course and in a variety of teaching styles. *Thinking Toward Solutions* also has an Internet component, which provides links to resources to aid students in solving each problem.

The Process of Science: Discovering Biology™ CD-ROM has been developed to reflect the spirit of inquiry that characterizes *Asking About Life*. It allows students to explore the discoveries of some of the most important concepts in biology. In the *Interactive Investigations*, students retrace the steps of scientists' experiments and discoveries using the scientific process as their road map. An *Investigator's Notepad* allows the students to track their progress through each investigation, to pose new questions for themselves to pursue, or to initiate a discussion with the instructor or other students via an Internet connec-

tion. *Concept Tutorials* provide students with essential background information in general biology for the course and the investigations. The CD will be available for use with classes in the fall of 1998; a demonstration disk on the topic of transmission genetics will be available to preview in January 1998.

The **Biology Survival Kit CD-ROM** is available for both IBM and Macintosh formats. The Biology Survivial Kit includes **BioXL+™** and **Biology A₂Z™ The Dynamic Glossary.** The BioXL+™ CD-ROM program presents a series of interactive animations that illustrate essential biological concepts. These simulations ensure a better retention of important concepts because students are actively engaged in a visual and dynamic way. BioXL+™ focuses on the basics, while guiding students through the concepts and providing real-world applications. Biology A₂Z™ The Dynamic Glossary CD-ROM provides text definitions, audio pronunciations, graphics, and animations for approximately 800 key terms traditionally presented in introductory biology courses. In addition to the definitions and audio, terms are supported by the text explanations, figures, and animations that put the key term in a conceptual context. Students will access the audio glossary from an extensive table of contents organized by concepts.

Biology MediaActive, Version III (Version III 1998). The CD-ROM Biology Media Bank contains imagery from Tobin/Dusheck: *Asking About Life,* Goodenough/Wallace/McGuire: *Human Biology*, Karleskint: *Marine Biology,* and Raven/Berg/Johnson: *Environment,* Second Edition. This CD-ROM is available as a presentation tool to be used in conjunction with commercial presentation packages, such as PowerPoint™ and Persuasion™, and will be available on the Biology MediaActive CD-ROM. Available for both Windows and Macintosh platforms.

Please visit our *Asking About Life* **Website** at

http://www.saunderscollege.com/lifesci/

Click on Tobin/Dusheck: *Asking About Life.* It offers an on-line study section for students and a continually updated resource for instructor materials.

A Textbook Is Born

Our developmental editor, Lee Marcott, has been a true mid-wife to this book, and both authors are extremely grateful to her. In addition to editing the manuscript and choosing excellent reviewers, artists, and photo researcher, Lee has also tactfully managed the two authors—pushing us to meet deadlines, calming us down when we panicked, organizing us, and giving us pep talks.

We also thank photo researcher Amy Ellis Dunleavy, who consistently found just the right photo, or had it shot, and project editor Beth Ahrens, who managed the endless details during the production of this first-edition book. We deeply appreciate, as well, the work of all the other people at Saunders College Publishing who made this book happen—including Donald Jackson, Michael Brown, and Edward Murphy, who nursed the manuscript in its infancy; and Elizabeth Widdi-combe, Julie Levin Alexander, Edith Beard Brady, and Emily Barrosse, who supported and nurtured this project as it developed.

We are also grateful to the art and production people at Saunders College Publishing: Beth Ahrens, of course, and Carol Bleistine, Joanne Cassetti, Ruth Hoover, Sue Kinney, Sally Kusch, Caroline McGowan, and Charlene Squibb, to name only some of the dozens of people who helped to shepherd this book into being. We thank Erik Fahlgren for his enthusiastic support of the book and his creative marketing strategy. We also thank our copy editors, Toni Wrighton and Sue Nelson; proofreader, Beth Morel; layout artists, Claudia Durrell and Julie Anderson; and indexer, Kathi Unger. We thank Don Lovett for calculating the scale bars found throughout the book and for fact checking many of our captions. Special thanks also to Leslie Sweeney, Jonathan Knight, Mari Jensen, Nili Kirschner, Don Ellis, Lydia Okelberry, and Chelsea Forbes.

We thank all the reviewers who took the time to read and comment on this manuscript—correcting our errors, asking thought-provoking questions, and suggesting examples, alternative wordings, or new ways of thinking. Although we never met any of our reviewers in person, working with them has been a rewarding intellectual experience. Both the process of writing this book and the resulting book itself would not have been the same without the reviewers.

Finally, we thank the following individuals, who took the time to talk to us or to answer our letters: Seymour Benzer, of Caltech; Harry Greene, of UC Berkeley; Ross Koning, of Eastern Connecticut State University; Gail Martin, of UC San Francisco; James Patton, of UC Berkeley; Peter Radetsky, of UC Santa Cruz; Gunther Stent, of UC Berkeley; and Robert Wayne, of UCLA.

We dedicate this book to our mentors in the art of communicating: Janet Hadda, David Tobin, Adam Tobin, and Eve Tobin; and to our mentors in the art of communicating science: Nina Dusheck, George Dusheck, John Dobson, and John Wilkes.

Allan J. Tobin
Jennie Dusheck
November 1997

REVIEWERS

Juan Aninao, *Dominican College of San Rafael*
Edwin A. Arnfield, *Macomb Community College*
Linda W. Barham, *Meridian Community College*
George W. Barlow, *University of California, Berkeley*
Brenda Blackwelder, *Central Piedmont Community College*
Mildred Brammer, *Ithaca College*
Richard B. Brugam, *Southern Illinois University, Edwardsville*
William Bowen, *Jacksonville State University*
Bradford Boyer, *Suffolk County Community College*
Hara Dracon Charlier, *Miami University*
H. Tak Cheung, *Illinois State University*
Karen Crombie, *Fresno City College*
Donald Cronkite, *Hope College*
Tom Daniel, *University of Washington*
Darleen A. DeMason, *University of California, Riverside*
Leah Devlin, *Pennsylvania State University, Abington College*
Ernest F. DuBrul, *University of Toledo*
Peter Ducey, *State University of New York at Cortland*
Steven H. Everhart, *Campbell University*
Lynn Fancher, *College of DuPage*
Cynthia Fitch, *Seattle Pacific University*
Dietrich Foerstel, *Champlain Regional College & Bishop's University*
Sally Frost-Mason, *University of Kansas*
Jack Gallagher, *William Rainey Harper College*
Lori K. Garrett, *Danville Area Community College*
Ben R. Golden, *Kennesaw State College*
Glenn A. Gorelick, *Citrus College*
Nels H. Granholm, *South Dakota State University*
Herbert H. Grossman, *The Pennsylvania State University*

Lonnie J. Guralnick, *Western Oregon State College*
Ross Hamilton, *Okaloosa-Walton Community College*
Richard Harrison, *Cornell University*
Wiley Henderson, *Alabama A&M University*
Bob Highley, *Bergen Community College*
Kathleen L. Hornberger, *Widener University*
Linda Hsu, *Seton Hall University*
David Inouye, *University of Maryland*
William A. Jensen, *The Ohio State University*
J. Morris Johnson, *West Oregon State University*
Peter Kareiva, *University of Washington*
James Karr, *University of Washington*
Tanseem Khaleel, *Montana State University, Billings*
Robert Kitchin, *University of Wyoming*
Ross Koning, *Eastern Connecticut State University*
Dan Krane, *Wright State College*
James W. Langdon, *University of South Alabama*
Anton Lawson, *Arizona State College*
Charles Leavell, *Fullerton Community College*
Kathleen Lively, *Marquette University*
Melanie Loo, *California State University, Sacramento*
Linda A. Malmgren, *Franklin Pierce College*
Nilo Marin, *Broward Community College*
Theresa Martin, *College of San Mateo*
Dorrie Matthews, *Sage Jr. College of Albany*
Gary F. McCracken, *University of Tennessee, Knoxville*
Robert J. McDonough, *DeKalb College*
Joseph McGrellis, *Atlantic Community College*
John Mertz, *Delaware Valley College*
Debbie Meuler, *Cardinal Stritch College*

Robert Morris, *Widener University*

Alison M. Mostrom, *Philadelphia College of Pharmacy and Science*

David M. Ogilvie, *The University of Western Ontario*

Bruce Parker, *Utah Valley State College*

Lee R. Parker, *California Polytechnic State University, San Luis Obispo*

Frederick Peabody, *University of South Dakota*

Ed Perry, *Faulkner State Community College*

Gary Pettibone, *State College of New York at Buffalo*

Jay Pitocchelli, *St. Anselm College*

David M. Polcyn, *California State University, San Bernadino*

Shirley Porteous-Gafford, *Fresno City College*

Paul F. Ramp, *University of Tennessee, Knoxville*

Franklin Robinson, *Mountain Empire Community College*

Lyndell Robinson, *Lincoln Land Community College*

Earle Rowe, *Walters State Community College*

Donna Rowell, *Holmes Community College*

Andrew Scala, *Dutchess Community College*

Shirley Seagle, *Jacksonville State University*

Phillip R. Shelp, *Brookhaven College*

John Simpson, *Gadsden State College*

Linda Simpson, *University of North Carolina, Charlotte*

Michael E. Smith, *Valdosta State University*

Fred L. Spangler, *University of Wisconsin, Oshkosh*

Steven Spilatro, *Marietta College*

Herbert Stewart, *Florida Atlantic University*

Cindy Stokes, *Kennesaw State College*

Jeffrey Thompson, *California State University, San Bernadino*

Janice Toyoshima, *Bakersfield College*

John Tramontano, *Orange County Community College*

Michael J. Ulrich, *Elon College*

Kristin Vessay, *Bowling Green University*

James A. Winsor, *The Pennsylvania State University, Altoona Campus*

Daniel Wivagg, *Baylor University*

Tom Worcester, *Mt. Hood Community College*

Allan J. Tobin

is Director of the UCLA Brain Research Institute. He holds the Eleanor Leslie Chair in Neuroscience at UCLA, where he is both Professor of Neurology and Professor of Physiological Science. Tobin is also Scientific Director of the Hereditary Disease Foundation (HDF), where he helped organize the consortium that identified the gene responsible for Huntington's disease. Both at UCLA and at HDF, he has encouraged the application of cell biology and molecular genetics to disorders of the brain.

Tobin's undergraduate degree is in literature and biology (MIT, 1963), and his doctoral degree is in biophysics, with an emphasis on physical biochemistry (Harvard, 1969). Tobin did postdoctoral work at the Weizmann Institute of Science, in Israel, and at MIT. At UCLA, his active research laboratory studies the production and action of GABA, the major inhibitory signal in the brain. These studies may eventually lead to new therapeutic approaches to epilepsy, Huntington's disease, and juvenile diabetes. Tobin is the recipient of a Jacob Javits Neuroscience Investigator Award from the National Institute of Neurological Disorders and Stroke.

For more than 25 years, Tobin has taught introductory courses in cell biology, molecular biology, developmental biology, and neuroscience. He is the recipient of a Faculty Teaching and Service Award and is regarded as an excellent and highly interactive teacher.

Jennie Dusheck

is a writer living in Santa Cruz, California, and a member of the National Association of Science Writers. Her undergraduate degree is in zoology (University of California, Berkeley, 1978) and her master's degree (by thesis) is in zoology (UC Davis, 1983). She also holds a certificate in science writing from UC Santa Cruz (1985).

Dusheck has done field and laboratory research on bird song; social behavior in field mice; food preferences of deer, cattle, and skipper butterflies (the subject of her thesis research at UC Davis); as well as axis formation in *Xenopus laevis*. While working for the Department of Molecular Biology at UC Berkeley, under a contract with NASA, she designed and wrote a protocol for an experiment that sent live frog embryos into space on the Fall 1992 Space Shuttle flight. She has taught university lab classes in introductory zoology, embryology, and comparative anatomy.

Dusheck has written for *Science News, Science* magazine, and other publications. From 1985 to 1993, she worked as a Principal Editor at the University of California, Santa Cruz. She has received several national awards, including the Gold Medal for Best-in-Category from the National Council for the Advancement and Support of Education.

CONTENTS OVERVIEW

CONTENTS

Runk/Schoenberger from Grant Heilman

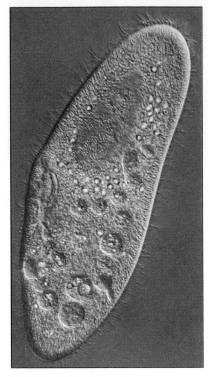

M.I. Walker/Photo Researchers

4   WHY ARE ALL ORGANISMS MADE OF CELLS?   80

5   DIRECTIONS AND RATES OF BIOCHEMICAL PROCESSES   114

Zig Leszczynski/Animals Animals

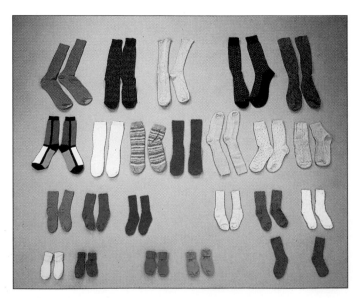

Paraskevas Photography

© 1990 Peter Menzel/Stock Boston

Runk/Schoenberger from Grant Heilman

III   Evolution   355

Daniel J. Cox/Natural Selection

IV   Diversity   465

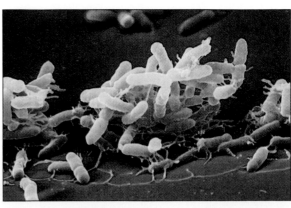

Meckes/Ottawa/Photo Researchers

S.E. Georgia/Animals Animals

NASA

Art Wolfe/Tony Stone Images

VI Structural and Physiological Adaptations of Flowering Plants 683

© 1981 by Darwin Dale/Photo Researchers

---

37   BLOOD, CIRCULATION, AND
THE HEART   *804*

*William Harvey, Miguel Serveto, and the
Pulmonary Circulation*   *804*

Francois Gohier/Photo Researchers

---

38   HOW DO ANIMALS OBTAIN AND
DISTRIBUTE OXYGEN?   *822*

Stanton Glantz and the Tobacco Industry   *822*

Francois Gohier/Photo Researchers

Art Wolfe/Tony Stone Images

Tom Bledsoe/DRK Photo

Bacteria: Enough To Give You an Ulcer

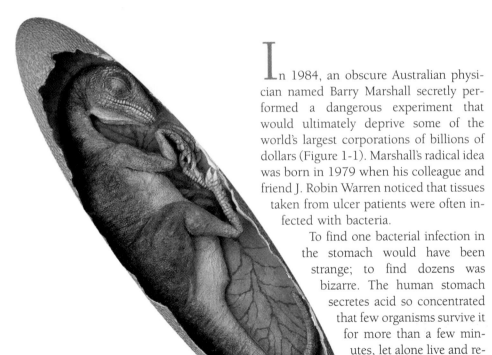

In 1984, an obscure Australian physician named Barry Marshall secretly performed a dangerous experiment that would ultimately deprive some of the world's largest corporations of billions of dollars (Figure 1-1). Marshall's radical idea was born in 1979 when his colleague and friend J. Robin Warren noticed that tissues taken from ulcer patients were often infected with bacteria.

To find one bacterial infection in the stomach would have been strange; to find dozens was bizarre. The human stomach secretes acid so concentrated that few organisms survive it for more than a few minutes, let alone live and reproduce in it. Yet Warren found bacteria flourishing there.

Warren's discovery suggested an alternative to doctors' long-standing belief that ulcers are caused by excess stomach acid. Ulcers, every medical textbook reported, were caused by the oversecretion of stomach acid in people with overanxious, frustrated personalities. Such personality problems were thought to be aggravated by the stressful pace of modern life. But if a bacterium could infect the stomach, Marshall and Warren realized, maybe it could cause ulcers. Intrigued by this idea, Marshall began ordering biopsies for all his patients who had stomach problems. He found that nearly every patient with ulcers was infected with the same bacterium.

The most common kind of ulcer is a peptic ulcer, an open wound located where the stomach joins the small intestine, at the bottom of the stomach. The word peptic comes from the Greek word *peptein,* to digest. Nearly one in ten adults has a peptic ulcer. Some people with ulcers feel no discomfort, but most experience at least mild pain, and many suffer excruciating pain for weeks at a time throughout their adult lives. In rare cases, blood may pour from the wound so freely that the victim bleeds to death. The standard treatment for ulcers had always been a high-fat, bland diet, tranquilizers, psychotherapy, and, in severe cases, surgery. Mainly, however, doctors prescribed antacids—lots of antacids.

Antacids are the biggest-selling prescription drugs in the world. In 1992, Americans bought $4.4 billion worth of the drugs. Prescription antacids are remarkably effective at controlling the secretion of stomach acid, but remarkably ineffective at controlling ulcers. Ninety-five percent of ulcer patients have a new ulcer within 2 years of treatment. That means people who have ulcers take the $100-a-month antacids almost continuously. In a lifetime, an ulcer-sufferer can spend tens of thousands of dollars on antacids.

Yet, if ulcers were caused by a bacterium, a 2-week course of antibiotics might cure millions of people permanently of what would otherwise be a lifetime of suffering. If Marshall's hunch was right, he had very good news, although not for the companies selling antacids.

Marshall, however, had insignificant credentials as a doctor and none as a researcher. In 1980, he seemed to have no more chance of selling his idea to the biomedical community than his bacteria had of flourishing in the corrosive environment of the stomach. Nevertheless, in 1983, Marshall presented his hypothesis at a scientific conference in Brussels. His presentation was a disaster. He was unknown; he was young, inexperienced, and overexcited; and he had a seemingly screwball idea. "He didn't have the demeanor of a scientist," recalled Martin Blaser, professor of medicine at Vanderbilt University, "He was strutting around the stage. I thought, this guy is nuts." When Marshall's presentation was over, his audience of eminent medical researchers shifted uneasily in their seats, embarrassed. A few laughed. They couldn't believe he was serious. Most bacteria can barely survive a brief passage through the stomach. How could they flourish there for months or years?

Besides, Marshall had no scientific evidence to back up his claim. Maybe, his audience told him, the bacteria had contaminated the stomach samples *after* the stomach tissues had been removed. Or maybe the bacteria were harmless and unrelated to the ulcers. Or maybe the bacteria were able to colonize the stomach *as a result* of the ulcer.

Marshall realized that the only way to settle these questions was to study the bacterium in an experimental animal. He needed to find an animal whose stomach could be infected with the bacterium. After returning to Australia, he began feeding the bacteria to rats. The bacteria died in the rats' stomachs without having any effect. He fed the bacteria to pigs, with the same result. Now he began to wonder, could the bacteria really infect a stomach? Maybe the researchers in Brussels had been right to laugh at him.

Desperate to prove that he was no nut, Marshall did something highly unusual (and controversial). First, he had a stomach exam and biopsy to make sure his stomach was healthy. Then he made an "ulcer bug" cocktail containing at least a billion bacteria, and, in a few swift gulps, drank it down. The cocktail was enough, he hoped, to infect his stomach. He told no one ahead of time—not the medical ethics board at the hospital, not his wife. They wouldn't have approved, he knew.

At first, nothing happened. Then, 8 days later, nausea woke him early and he vomited. For another week he was tired, irritable, and hoarse. He had headaches and foul breath. A second stomach exam and biopsy showed that his stomach was inflamed and swarming with bacteria.

By the third week, Marshall was lucky enough to have recovered completely. He had not proved that the bacterium could cause ulcers, or even that it could infect the stomach for years at a time. But he had done something that strongly suggested that the bacterium, still unnamed at the time, could at least infect a healthy human stomach—one that didn't already have an ulcer.

In time, other, more-established researchers began to take an interest. Mainly they were interested in proving Marshall wrong. But by the end of the 1980s, the

If a bacterium could infect the stomach, Marshall and Warren realized, maybe it could cause ulcers.

Figure 1-1 Barry Marshall showed that ulcers are caused not by psychological problems, but by bacteria. *(Courtesy, Dr. Barry Marshall)*

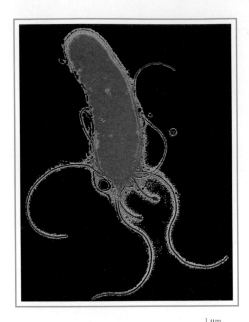

1 μm

Figure 1-2 **The culprit.** *Helicobacter pylori* infects the lining of the stomach. *(P. Hawtin, University of Southampton/Science Photo Library/Photo Researchers)*

evidence that the bacterium could infect the stomach was unassailable. In 1989, the bacterium was named *Helicobacter pylori* because of the bacterium's helical shape and because it was known to colonize the pylorus, near the stomach's exit (Figure 1-2). By 1993, definitive studies by other researchers had shown that some 80 percent of ulcers were caused by *H. pylori*, and were treatable with antibiotics.

Drug companies, initially somewhat negative about Marshall's idea, began to see a silver lining in the cloud the Australian had created. It was true that people treated with only $20 worth of antibiotics had a relapse rate of only about five to ten percent—which meant an ulcer sufferer would spend thousands of dollars less on drugs. On the other hand, infection by *H. pylori* turned out to be one of the most common bacterial infections in humans in the world—infecting up to half of all people worldwide. And mounting evidence suggested that chronic infection by *H. pylori* not only caused ulcers but also increased the risk of developing stomach cancer. Here was a market for antibiotics consisting of billions of people. Marshall's cloud definitely had a silver lining.

By 1993, drug companies were hastily developing new diagnostic tests for *H. pylori* and new antibiotics to treat the infection. Biotechnology companies were trying to develop a vaccine, to be given in childhood, that would protect against *H. pylori*, ulcers, and maybe even stomach cancer, a leading cause of death in Asia. In early 1994, experts at the National Institutes of Health declared antibiotics the official treatment for most ulcers. For the first time, doctors began treating their patients with antibiotics that would offer a permanent cure. Change came slowly. By 1996, only one-third of doctors were prescribing antibiotics for ulcer patients. The rest continued to prescribe antacids only.

Marshall ultimately won acceptance as a scientist, earning a position on the faculty at the University of Virginia Medical School. He succeeded in part because he possessed many of the attributes of a good scientist. He had curiosity, intelligence, vision, and the dogged determination to pursue an idea—even when his stubbornness made him appear foolish. Perhaps most important, Marshall displayed an unusual independence of thought that allowed him to pursue an idea unimaginable to more dogmatic thinkers.

In this book we will meet many scientists who are as curious, independent, and stubborn as Marshall. Some are impulsive, like Marshall. Others take years to reach their conclusions. Most work long hours for years on end. A very few seem merely to play at science, reaping brilliant discoveries from a few hours work. We'll see some of them risk their reputations to defend the ideas they believe in. A few even lose their lives. Right or wrong, all are fascinated by questions about what makes living things tick. Good biologists, like other scientists, are people who are intensely engaged with life.

In this first chapter of *Asking About Life*, we will examine two important aspects of **biology**—the science of life. We will see how biologists try to answer questions about life, and we will try to define life itself: What is life? How do biologists study life?

KEY CONCEPTS

1. A model is a simplified view of a particular process, generally based on observations. Good models lead to hypotheses. Hypotheses are valuable only if they can be tested and disproved through experiment.

2. All living organisms consist of well-ordered parts, obtain energy from their surroundings, perform chemical reactions, change with time, respond to their environment, reproduce, and share parts of a common history.

3. Living organisms consist of one or more cells.

4. Organisms maintain a relatively constant internal environment.

5. Natural selection leads to the accumulation of changes in an organism— a process called evolution.

6. Some of these changes are adaptations, features that increase the chances that an individual will live and reproduce.

7. Genes, which are made of DNA (deoxyribonucleic acid), ensure continuity between generations.

8. The characteristics of an organism depend on both the genes and the environment of the organism.

HOW DO BIOLOGISTS ASK QUESTIONS?

From early childhood, everyone asks questions about life: What makes me alive? What goes wrong when I am sick? Where did I come from? How am I like other living things?

The beginning of all scientific inquiry is curiosity. Something in the world commands our attention, and we begin to ask questions. Sometimes our attention is captured by pure wonderment or by arresting beauty. In other cases, we are stimulated by a practical need. We would like to see a sick person get well or the hungry provided for.

Natural curiosity such as this is the beginning of science. The physicist Albert Einstein attributed his scientific achievements to his childlike curiosity. Einstein asked the kinds of questions that most people ask only as children. But he was still asking them when he had the intellectual power to answer his questions.

Many scientists pride themselves on such childlike curiosity. But scientific inquiry is more than asking questions. Science is curiosity that is controlled and channeled. For example, asking whether a bacterium causes ulcers was a question that could be answered. All good scientists try to limit themselves to thinking about questions that can be answered.

Do Scientists Use the Scientific Method?

In the first part of this chapter, we will discuss how scientists choose questions to ask, and how scientists answer those questions. Because science has been so successful in increasing knowledge, many philosophers and historians have tried to understand how scientists work. One explanation for science's success is the **scientific method,** a set of rules for formulating, testing, and eliminating ideas. In reality, most scientists do not think of themselves as following the rules of this or any other single "method." Rather, they pursue knowledge in a variety of individual and creative ways. Nonetheless, we can give a general description of how most scientists go about learning about the world.

The scientific method refers to a formal set of rules for forming and testing hypotheses. Most scientists use the scientific method only loosely and often unconsciously.

How Do Observations Lead to Models?

A scientist's first step in understanding something is to focus on a single question or small set of questions. Because science, especially biology, works in tiny steps, scientists rarely answer big, general questions all at once. A scientist does not ask, How do animals move? and expect to come up with a simple answer to that question.

When we ask how something works, whether it be an automobile or an organism, we generally want to know how particular parts are used. For example, consider two familiar exercises—the chin-up and the push-up. From experience we know that a chin-up requires the forceful bending of the arm at the elbow, while a push-up requires the forceful *unbending* of the arm (Figure 1-3). But what allows such forceful movements? What structures change as the arms bend and unbend? These are specific questions that can be answered.

The second step in understanding is *observation.* Looking, hearing, smelling, and touching may all contribute to our observations. For example, during a chin-up, one can feel the bulging of the biceps, which is the major muscle on the front side of the upper arm. During a push-up, the triceps, which is the long muscle on the back side of the upper arm, hardens and bulges. When the triceps bulges, the biceps flattens out. When the biceps bulges, the triceps flattens.

If we examine (observe) the internal structure of the arm, we see that the skeleton of the upper arm consists of a single, long bone connecting the shoulder and the forearm. The biceps connects the front of the shoulder to the forearm, and the triceps connects the back of the shoulder to the forearm.

Now we are ready for the next step: coming up with a **model**—a simplified view of how the components of a structure operate that is consistent with previous scientific knowledge and also offers new insight. A model often comes from comparing a process that is not understood to one that is. The forearm, for example, resembles a seesaw, which is a kind of lever. When the biceps contracts, the forearm is pulled up and the tip of the elbow rotates down. When the triceps contracts, the forearm is pulled down and the elbow rotates up (Figure 1-3).

A scientist begins by asking a narrowly focused question about a particular phenomenon, then by observing that phenomenon. Then, comparing the phenomenon to some already understood process, the scientist constructs a model that suggests how the phenomenon works.

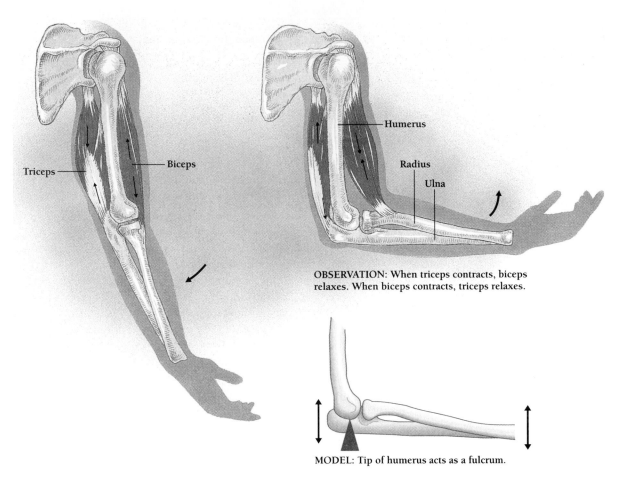

OBSERVATION: When triceps contracts, biceps relaxes. When biceps contracts, triceps relaxes.

MODEL: Tip of humerus acts as a fulcrum.

Figure 1-3   Like a children's seesaw, the biceps and triceps operate the main movements of the arm. The ulna and radius bones (in the lower arm) are a lever, which rocks back and forth, and the tip of the humerus (in the upper arm) is the fulcrum. When the triceps contracts, the elbow end of the lever tips up and the hand drops down. When the biceps contracts, the hand tips up and the elbow drops down.

What Is a Hypothesis?

Once we have a model, we can formulate a **hypothesis**—an informed guess about the way a process works or a structure is organized that allows us to predict what will happen in different situations. If the predictions of a hypothesis are incorrect, then the hypothesis and its underlying model are probably also incorrect. If, however, the predictions of a hypothesis turn out to be correct, then scientists gain confidence in the hypothesis and the underlying model.

A good model suggests how other similar processes work. For example, the model that we develop for the movements of the arm applies to the legs, the hips, and other muscle groups. Our model of the forearm accomplishes the two important purposes of a model: it suggests both why muscles bulge and how muscles work; that is, for every movement some muscles contract, and others relax.

Where do hypotheses come from?

Often, a hypothesis comes from other hypotheses. Indeed, most scientific progress depends on the continual refinement and ex-

pansion of accepted theory. Such work represents the most fascinating kind of puzzle solving: the scientist, already having a good idea of what he or she is looking for, proposes and tests hypotheses to reveal, piece by piece, a picture that no one else has seen. In this way, scientific knowledge continually expands. The 17th century physicist Isaac Newton best summarized the process by which scientists continuously build on one another's work in talking about his own great work: "If I have seen farther than others it is because I have stood on the shoulders of giants."

A hypothesis is an informed guess about the way a process works that enables a scientist to predict what will happen in different situations. It is a product of logic, previous knowledge, insight, and creativity.

How Do Scientists Decide If a Hypothesis Is Correct?

A scientist can sometimes prove that a hypothesis is *incorrect*, but a scientist can never prove that a hypothesis is *correct*. This is because it is always possible to obtain the predicted result—even when the model is incorrect. We might, for example, hypothesize that Runner A can run the 100 m faster than anyone else in the world. If we find someone else who can run the 100 m faster than Runner A, we have disproved our hypothesis. That part is easy. But what if we cannot find anyone who can run the 100 m faster? How can we be sure that there isn't some undiscovered individual who can run it faster? The answer is that we can never be sure. We cannot prove our hypothesis right. An experiment can only *disprove* a hypothesis.

Even disproving a hypothesis is not always easy. When the predicted result of an experiment is wrong, the hypothesis still may not be wrong. For example, we might hypothesize that bears eat berries. We predict that we will find berry skins in bears' stools. Yet, when we examine stool samples from 25 bears, we find grass, insect parts, and even small plastic bags, but no sign of berries. Does this mean the hypothesis was wrong? Not necessarily. Our results might only signify that berries were not in season, and that we should repeat our study in the fall. Or we might wonder if berry skins are digested completely and leave no trace in the stool. In that case, we could feed berries to captive bears to test that hypothesis.

Any experiment may fail to give the predicted result for reasons that have nothing to do with the hypothesis being tested. The prediction might have been wrong because of faulty logic, or the experiment might have been poorly designed or executed. It is the scientist's job to figure out why the model's prediction was incorrect, and then to come up with either a better model or a better experiment.

When a hypothesis is tested, it can sometimes be proved false, but it can never be proved true.

What Makes a Good Experiment?

Generally, an experiment is any procedure carried out under conditions controlled by the experimenter either to test a hypothesis or to discover some unknown effect. A good scientific experiment, however, requires much more.

In a good experiment, the outcome of the experiment must depend only on the proposed cause. Suppose, for example, that we inject baby mice with extract of cinnamon to see if the cinnamon slows their growth. We find that their growth seems slow compared with that of mice in other studies. Can we conclude that cinnamon slowed their growth? We cannot. Maybe we were not feeding the mice enough food. Maybe the room they are in is too hot. Maybe chemicals in the extract other than the cinnamon are slowing their growth. The outcome of this experiment could depend on many causes besides the cinnamon.

To evaluate an experiment properly a researcher must compare experimental results with the results of a **control**—a version of the experiment in which everything is the same except for the single factor being tested. To test our cinnamon hypothesis properly, we need *at least* two controls. We need a matching set of mice that receive no injection so that we know how fast mice normally grow under the conditions in our laboratory, not someone else's. We need, as well, a set of mice that receive an injection containing everything in the extract except the cinnamon. This second control shows whether something else in the extract slows the growth of baby mice.

The design of a control is critically important and not always obvious. For example, the trauma of being handled might be enough to slow the mice's growth. To make sure that is not the case, we might want to include a third control in which researchers merely pretend to give baby mice injections—sticking them with a needle, but not injecting anything. This might sound silly, but it is good science. Because an experiment must distinguish causes from irrelevant factors, an ideal experiment allows a researcher to evaluate the effects of one factor at a time.

When Barry Marshall swallowed his mixture of stomach bacteria, he was performing an experiment—but only just barely. It was not a good, controlled experiment. His symptoms—nausea, tiredness, and bad breath—might have been caused by the flu, a sinus infection, or any number of things. Even the inflammation of his stomach shortly after he drank the ulcer bug cocktail might have been a coincidence.

Experiments on a single individual may be suggestive, as Marshall's was. But such experiments do not provide results that mean anything scientifically. To do the experiment properly, Marshall would have had to persuade at least three groups of people to participate. The first group, the "experimental" group, would have gotten the ulcer bug cocktail. The second group would have gotten a different bacterium—one not believed to cause ulcers—to rule out the possibility that any concentrated mixture of bacteria can make someone sick. The third group would have gotten the same mixture except without any bacteria, to rule out the possibility that the cocktail—even without any bacteria—is itself a sickening mixture.

An experiment must include one or more carefully designed controls.

How Do Scientists Design Experiments?

The design of experiments that test a hypothesis is both the greatest challenge to a scientist's ingenuity and the greatest source of excitement. One of the most famous stories concerns Otto Loewi's elegant test of the hypothesis that nerves control muscle contractions by producing a chemical messenger.

Loewi, an early 20th century German physiologist, reasoned that a muscle ought to respond to the right chemical even if the nerve itself were not present.

Loewi had been working with the vagus nerve, which, among other things, controls the rate at which the heart beats.

(Text continued on page 8.)

BOX 1-1

Reductionism

One way to study a thing is to look at its parts. If we look at the wheels of a bicycle, for example, we see they turn and are attached to the bicycle's frame by means of intricately designed hubs. Much of biology involves studying the parts of organisms, whether their organs, their cells, or their molecules. For every level of organization, biologists seek to explain structure and function in terms of the next finer level of structure. We can, for example, understand something about the action of arm muscles by studying the anatomy of the bones and muscles. The effort to understand the whole in terms of the parts is called **reductionism.**

We can understand more about how muscles work by looking at the structure of the muscles. A microscope allows us to see that muscles consist of bundles of thin fibers, each up to several centimeters long.

Regular stripes, called *striations,* cross each fiber (Figure A). When a muscle contracts, the distance between the striations decreases. Closer examination reveals still smaller structures—minute fibers called *filaments* (Figure B). As a muscle contracts, the filaments slide past each other in the same way that the two parts of an extendible ladder do (Figure C).

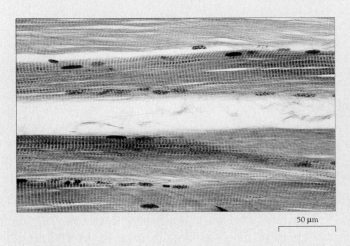

50 μm

0.5 μm

Figure A   Skeletal muscle shows obvious striations. Shown are parts of four muscle fibers.   *(Bruce Iverson)*

Figure B   Filaments. Each muscle fiber consists of many "myofibrils." Here, five myofibrils run diagonally from lower left to upper right.   *(Dwight Kuhn)*

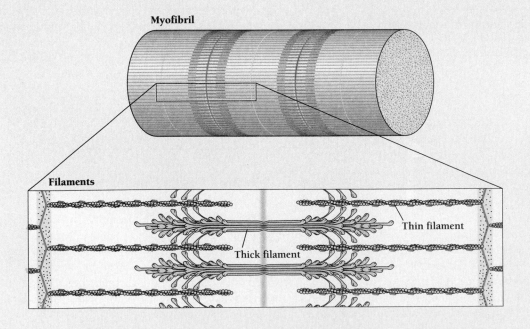

Myofibril

Filaments

Thin filament

Thick filament

Figure C   The value of the reductionist approach to science. Understanding specific parts helps explain larger, more complex biological processes. By studying the movement of muscle filaments, for example, we understand how muscles contract. Each myofibril is composed of thick and thin filaments, which slide past one another in the same way as the two parts of an extendible ladder.

By understanding the movements of the filaments we understand more about how muscles contract, and we see how understanding specific parts helps explain larger, more complex biological processes. Many of the questions that we ask in this book concern the relationships of structures and processes. The questions are often the same: What are the involved structures? What are they made of? How do they capture energy? What controls their operation? The answers to our questions may lie at the level of the molecules, cells, whole organisms, populations, or ecosystems.

The reductionist approach to science has been and will continue to be immensely successful. The most successful and enthusiastic reductionists are molecular biologists, who seek to answer an enormous range of questions—from the inner workings of cells to the evolution of whole families of organisms over millions of years—solely by looking at the behavior and structure of molecules.

However, reductionism has definite limits. All of the properties of an object are not explainable in terms of the object's parts. A whole is greater than the sum of its parts. The painting in Figure D illustrates this principle beautifully. When we examine the painting under high magnification, we see only an abstract pattern of colored dots. If we use a lower magnification, the situation is not much improved. We see patches of color and vague shapes. It is only when we step back and look at the whole picture at once that we see a landscape.

(b)

The landscape is an **emergent property,** a characteristic that arises only at complex levels of organization. Living organisms are no different. To understand how bees pollinate flowers, for example, we need to study whole populations of bees and flowers, not just molecules or genes. Just as a painting is much more than a collection of brush strokes, an organism is much more than a collection of organs, and a group of organisms is more than a collection of individuals. A detailed knowledge of the behavior of solitary humans, for example, would not allow us to predict such bizarre group behaviors as war and Tupperware parties. Complex structures interact in ways that cannot be predicted only from a knowledge of the component parts.

Just as in the case of the painting, the most interesting and beautiful aspects of biology involve emergent properties. While it is certainly true that organisms and cells must obey the laws of physics and chemistry, physics and chemistry by themselves provide little insight into disease, sexuality, fear, and the complex relations among the organisms of an ecosystem. Nevertheless, as we will see throughout this book, physics and chemistry explain much about how cells and organisms operate. Biologists therefore accept both the power and the limitations of reductionism.

(a)

Figure D The limits of the reductionist approach to science. A landscape, like many of the characteristics of organisms, is an emergent property—a characteristic that arises only at complex levels of organization. (a) When we look at the individual brushstrokes, we see only an abstract pattern of colored dots. (b) Even if we look at groups of brushstrokes, we see only patches of color and vague shapes. (c) It is only when we step back and look at the whole picture at once that we see a landscape—"La Seine à Herblay," by Maximilien Luce (Musée d'Orsay, Paris). *(Erich Lessing/Art Resource)*

(c)

The heart is a hollow muscle that contracts with each beat. When Loewi electrically stimulated a frog's vagus nerve, the heart rate slowed. But, he asked, did the slowing result from a direct connection of the nerve to the heart or from a chemical produced and secreted by the vagus nerve? Loewi had no idea what such a chemical might be. Designing an experiment to test his hypothesis seemed impossible.

In fact, conceiving the experimental design took him 17 years. Loewi himself described how the experiment finally came to him:

> The night before Easter Sunday of 1920 I awoke, turned on the light, and jotted down a few notes on a tiny slip of thin paper. Then I fell asleep again. It occurred to me at six o'clock in the morning that during the night I had written down something most important, but I was unable to decipher the scrawl. The next night, at three o'clock the idea returned. It was the design of an experiment to determine whether or not the hypothesis of chemical transmission that I had uttered seventeen years ago was correct. I got up immediately, went to the laboratory, and performed a simple experiment on a frog heart according to the nocturnal design.

In hindsight, Loewi's 3 a.m. experiment seems straightforward, almost routine. Loewi isolated two frog hearts, one with the vagus nerve still attached, the other without the nerve. He stimulated the vagus nerve of the first heart, collected the fluid surrounding it, and transferred the fluid to the second heart, the one with no nerve connections. The second heart slowed. This result showed that the slowing was a response to some substance produced by the vagus nerve. Loewi called this substance "vagus stuff," and we now know it as a chemical signal called acetylcholine.

Loewi eventually received a Nobel Prize for his work, which brought to biology a new understanding of how nerves communicate. The story of his discovery is a dramatic illustration of the importance of inspiration in experimental design.

Designing experiments to test a hypothesis requires a mind that is both creative and logical.

What Is a Testable Hypothesis?

We can quickly generate lots of hypotheses to explain any observation. If we note that ants in the kitchen are crowding around the sink, we might hypothesize: (1) that they are looking for water, (2) that they are looking for food particles, (3) that they are looking for a new nest site, or (4) that they want to clean the dishes for us. Clearly, some hypotheses are better than others. Scientists need ways of deciding which hypotheses to explore.

In general, a hypothesis is valuable only when it is **testable,** meaning that someone can devise an experiment that would disprove the hypothesis if it were incorrect. The hypothesis that

bears ear berries is testable. The hypothesis that bears eat rocks is also testable. Testable (or *disprovable*) means that we can design an experiment in which the result depends only on whether the hypothesis is correct or incorrect.

The hypothesis that extraterrestrial beings influence the movement of ants in your kitchen is untestable—*not* because the hypothesis is wrong, but because there is no way to prove that it is wrong. The reason that this hypothesis is untestable—and therefore of no interest to science—is that designing a control experiment in which the influence of the alleged extraterrestrials is ruled out is impossible. We can't prevent their alleged influence because we have no information about what kinds of influence extraterrestrials might exert. We have no reliable observations of extraterrestrials. The same arguments apply to fairies, gnomes, elves, and deities.

A hypothesis is valuable only if it is testable.

How Do Scientists Evaluate Experiments That Give Varying Results?

Until now, we have talked as if an experiment yields the same results each time it is performed. But what happens when the results are not the same every time? Not all the individuals within an experimental group necessarily respond the same way. For example, if the mice injected with cinnamon extract grew 10 percent more slowly, on average, than control mice, that could mean that some grew normally and some grew much more slowly. Because individuals vary, biological experiments are always done on groups of organisms, rather than on single individuals.

The results must be evaluated carefully using **statistics**—the science of collecting and analyzing numerical data. Statistics applies a branch of mathematics called "probability theory" to practical problems in such diverse areas as experimental science, economics, census taking, and opinion polling. Statistics is a powerful tool that can tell a scientist whether the results of an experiment are accurate. Statistics can also suggest new questions and new hypotheses.

Most importantly, statistics tell a scientist whether an experiment has produced a result that is really different from the control. If the mice who received cinnamon extract grew 10 percent more slowly than the control mice, how do we know that this small effect is not the result of chance variation? Maybe the group of mice that ended up in the experimental group just happened to be slow growers compared to those that ended up in the control group. Statistics can tell us how likely it is that a difference of that size could occur by chance.

A good experiment includes statistical analysis as well as appropriate control experiments.

What Is a Theory?

When a hypothesis makes correct predictions about experimental results, scientists naturally gain confidence both in the hypothesis and in the model that suggested it. Furthermore, the more a model explains, the more important it becomes to other scientists. For example, the model of arm flexion and extension suggests that pairs of muscles working in opposition to each other might be responsible for the movements of legs, fingers, and toes. For each moving part, biologists have another hypothesis to test. If experiments on each body part produce the same result, scientists may accept a generalized model that muscular movement occurs by means of "antagonistic" pairs of muscles.

Once scientists have worked sufficiently hard to test general models for a group of related structures or processes, they may decide that they can accept a fairly broad set of hypotheses. They may refer to this set of related hypotheses as a **theory**—a system of statements and ideas that explains a group of facts or phenomena, such as "atomic theory" or "evolutionary theory."

Scientists and nonscientists use the word "theory" differently. To a nonscientist, a theory is an untested idea or speculation. To a scientist, it is a set of interconnected, rigorously tested hypotheses. Sometimes this difference in word usage leads nonscientists to think that scientists are less certain about their theories than they actually are.

A set of related hypotheses that consistently resist scientists' attempts to disprove them may become recognized as a theory.

Biologists Ask Many Different Kinds of Questions

The scope of biology is enormous, so that the task of understanding biology sometimes seems overwhelming to students and professional biologists alike. Every biologist has a unique view of the field as a whole. For example, one biologist may see biology as being primarily concerned with the inner workings of living organisms. Another is more interested in the history, or evolution, of life. Equally important, different fields of biology demand different work styles. A biochemist analyzes changes in chemical structure that often take place in minutes or fractions of seconds, while some ecologists study changes in forest ecosystems that take decades or centuries. Biochemists look for immediate answers to their questions. Ecologists are more likely to take the long view—patiently waiting for the answer to emerge from thousands of hours of work and reams of mathematical analysis.

Students may also find different ways of appreciating biology. Some may enjoy learning about how biology can improve health, nutrition, and food production. Others may be interested in their ancient biological origins and the enormous diversity of species now living on Earth. Still others want to learn how to slow the steady destruction of life on Earth. In the next 20 years, the Earth's booming human population and dwindling resources will force scientists and nonscientists to make many tough decisions. Educated men and women need to understand not just the "facts" of biology, but also the ways that biologists go about learning and integrating knowledge.

This book tries to address the concerns of all biologists and interested students. We will see that even the most complicated process can be understood in relatively simple terms. This understanding will require work. But the inevitable rewards include both a deeper understanding of life and a finer appreciation of our place in the exquisite living world.

Finally, throughout the book we will see again and again that it is people who ask questions and people who have insights. These insights often depend on the peculiarities of a given observation or the cleverness of an experiment. First and foremost, biology is a human endeavor.

WHAT DO WE KNOW ABOUT LIFE?

Everywhere on Earth, there is life: from the depths of the Pacific Ocean to the top of Mt. Everest, from the frozen wastes of the Arctic Ocean to the tropical rain forests of the Amazon. For each living thing, the same questions arise: How is it put together? How does it work? How did it get here? In answering these questions, biologists have accumulated millions of individual facts (only a tiny fraction of which we will study). Tying these disconnected facts together, however, are a few basic themes. Once we grasp these few generalities, the study of life becomes much easier.

How Are Organisms Different from One Another?

Organisms come in many sizes, shapes, and colors. They range in size from the smallest species of bacteria to California's coast redwoods, the most massive organisms in the world. Organisms range in shape from streamlined dolphins to prickly cacti, and in color from the softest brown moth to the brightest blue peacock. The more we look, the more different forms of life we see.

Biologists now divide these many organisms into three large categories called domains (Figure 1-4). Two domains, the **Eubacteria** and the **Archaea**, comprise distinct groups of bacteria. The third domain, the **Eukarya**, encompasses four *kingdoms*. The most familiar of these kingdoms include the organisms that are always multicellular—**Animalia** (animals) and **Plantae** (plants). Plants range from mosses to ferns, giant redwoods, and mahogany trees. Biologists have named some 248,000 species of plants.

Animals are even more diverse, with more than 1 million known species. Only 42,000 animal species are what many people consider "real" animals—fish, amphibians, reptiles, birds, and mammals. Nearly 200,000 animals are sponges, jellyfish, worms, spiders, snails, and sea urchins and their kin. But the vast majority of animals are insects: some 800,000 described animal

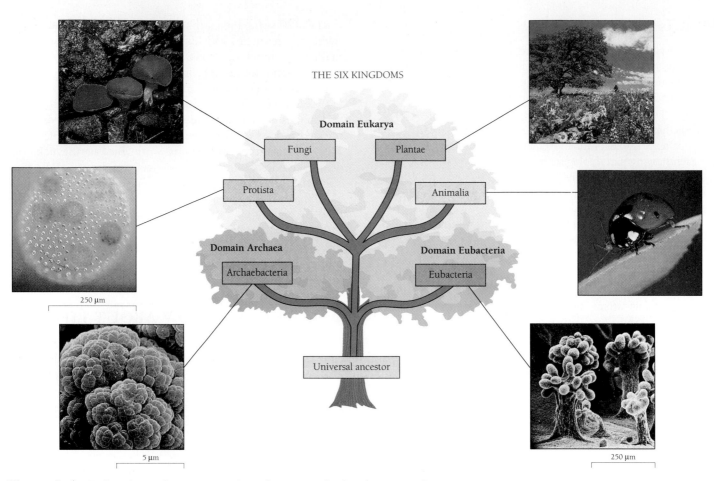

THE SIX KINGDOMS

Domain Eukarya

Fungi

Plantae

Protista

Animalia

Domain Archaea

Archaebacteria

Domain Eubacteria

Eubacteria

Universal ancestor

250 μm

5 μm

250 μm

Figure 1-4 **Six kingdoms of organisms in three domains.** In the first domain are the four kingdoms of eukaryotes. The seven spotted lady beetle represents the **Animalia.** An oak tree surrounded by wild flowers (arrowleaf balsam root, paintbrush, and lupine) represents the **Plantae.** The stalked scarlet cup, *Sarcoscypha occidentalis,* represents the **Fungi.** *Volvox* represents the **Protista.** The second domain includes just one kingdom, the **Eubacteria,** represented here by aggregating myxobacteria. The third domain includes one last kingdom, the Archaebacteria, or **Archaea,** represented by this colony of methanobacteria, *Methanosarcina.* *(animal, Runk/Schoenberger from Grant Heilman; plant, F. Stuart Westmorland/Photo Researchers; fungi, R.M. Meadows/Peter Arnold, Inc.; protista, Biological Photo Service; eubacteria, Patricia L. Grilione/Phototake; archaea, R. Robinson/Visuals Unlimited)*

species are insects. Beetles alone, with some 300,000 to 350,000 species, constitute about 30 percent of all known animal species.

The **Fungi,** with just under 70,000 named species, include both multicellular organisms such as mushrooms and single-celled organisms such as yeast. A fourth kingdom, the **Protista,** consists largely of single-celled organisms. The more than 30,000 protists include the amebas such as those that live in pond water (and the kind that give humans dysentery); the dinoflagellates, some of which cause poisonous red tides in coastal waters; as well as many kinds of algae, water molds, and slime molds.

In the other two domains, the Archaea include one to three kingdoms and the Eubacteria include one to several. The exact numbers depend on who is counting. Altogether, these simple, single-celled organisms include 4800 eubacteria and archaebacteria, of which the majority are free living. Only a small minority of bacteria cause infections in humans and other organisms.

Within these three domains, named species of organisms total about 1.4 million. Biologists believe, however, that most living species have yet to be named. The world may hold 10 million or even 100 million species. No one really knows how many species the world holds. Whatever the total, it is only the tip of the iceberg when thought of in terms of the history of life. Less than one percent of the organisms that have ever lived are alive today. The vast majority of species—in all their diversity—went extinct millions of years ago.

Life is enormously diverse. More than 1 million living species have names, but millions more have no names or are already extinct.

BOX 1-2

Homeostasis and negative feedback

Homeostasis, the process by which organisms maintain a relatively stable internal environment, always involves two stages: (1) detecting changes from the stable state, and (2) counteracting such changes.

Once an organism detects changes, it often uses a strategy, called **negative feedback,** to return to the stable state. The household thermostat provides a familiar nonliving example of negative feedback. When the temperature of a room falls below the set temperature, the thermostat activates a heater that warms the room. When the room reaches the set temperature, the

thermostat turns off the heater. The heater stays off until the temperature again falls below the thermostat setting. Although the temperature of the room varies slightly as the heater cycles on and off, the negative feedback mechanism keeps the room temperature nearly constant.

Humans and other mammals also maintain a remarkably constant internal temperature. A nude person, for example, can be exposed to temperatures as low as 55°F and as high as 140°F and still maintain a body temperature of between 98°F and 99°F. When the body temperature rises

too high, humans perspire, drink water, and look for a cool place to rest (Figure A). When the body temperature drops, humans shut down the flow of blood to the skin, stop perspiring, and increase the production of heat (by increasing the rate at which they burn calories and by shivering). Humans can also put on warm clothes or turn up the thermostat. All of these responses, whether physiological or behavioral, are homeostatic responses designed to keep the temperature of the body at between 98°F and 99°F.

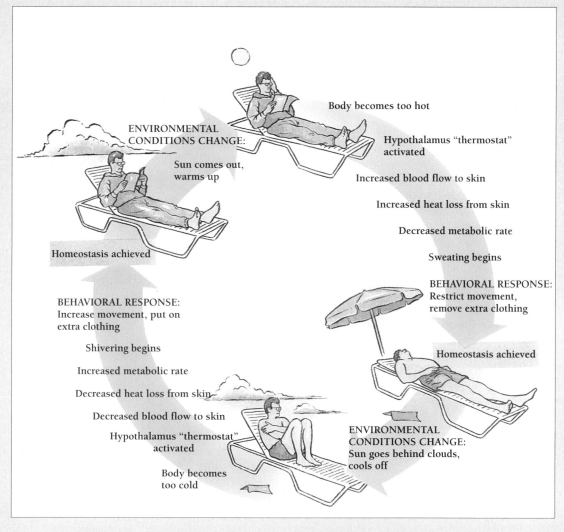

Figure A Homeostasis is the process by which organisms maintain a relatively stable internal environment. Here Jim tries to stay comfortable on a day when it's hot in the sun but cool when the fog comes in.

How Are Organisms Alike?

Since **biology** [Greek, *bios* = life + *logo* = word] is the study of life, the first general question we must answer is, What is life? Most of the time we can distinguish the living from the nonliving as easily as we can tell a live bear from a teddy bear. Although arriving at a definition of life that is both precise and general is considerably harder, we can agree on some common characteristics of all living things (Table 1-1):

1. *Living things consist of* **organized parts.** Organization is an essential property of life. Indeed, biologists call living things "organisms," a word with the same root as "organization." We easily see organization in the shell of the chambered nautilus, a marine mollusk related to the squid and octopus (Table 1-1A). All organisms must be highly organized to live.

2. *Organisms* **perform chemical reactions.** Organisms modify molecules in their environment. The honeybee, for example, converts flower nectar into honey (Table 1-1B). Some of these chemical changes produce the materials that make up their bodies. For example, humans change plant proteins into animal proteins. Other chemical changes produce energy. Reactions that produce energy may drive movement such as the contraction of muscle, or they may generate heat. **Metabolism** [Greek, *metabole* = change] refers to all the chemical reactions occurring within an organism.

3. *Living organisms* **obtain energy from their surroundings.** Organisms use energy to maintain their organization. For almost all organisms, sunlight is the ultimate source of energy. Plants obtain energy directly from sunlight (Table 1-1C). They may use this energy to build or to replace new parts, or they may store it away in the form of carbohydrates (for example, potatoes or grains). Animals obtain energy by breaking down energy-rich molecules from plants (or from other animals that have eaten plants).

4. *Organisms* **change** *with time.* Even as organisms such as the tadpole in Table 1-1D maintain their organization, they also develop, change form, and grow.

5. *Organisms* **respond to their environments.** The plants in Table 1-1E are bending toward a window to obtain more light energy. All organisms behave in reaction to their surroundings. When a dog gets too hot, it drinks water, finds shade to lie in, and pants. These behaviors help keep the dog's body at a constant temperature. They· are a form of **homeostasis** [Greek, *homeo* = like, similar + *stasis* = standing], the tendency of organisms to maintain a stable internal environment in the presence of a changing external environment. Homeostasis is one way that organisms respond to their environments.

6. *Organisms* **reproduce.** Organisms produce more organisms of the same kind. Sometimes a single organism reproduces itself. Often, however, reproduction is sexual and requires the collaboration of two parents. Every individual organism comes from another organism or pair of organisms (Table 1-1F).

7. *Organisms share a* **common evolutionary history.** All organisms use the same chemical building blocks, put them together in the same ways, and pass on information to their offspring in the same way. The simplest explanation for these similarities is that all organisms on Earth are related to one another and share a single common ancestor. This explanation is supported both by the fact that the remains of ancient organisms strikingly resemble modern organisms as well as by many other lines of evidence (Table 1-1G).

All living things are organized into parts, perform chemical reactions, obtain energy from their surroundings, change with time, respond to their environments, reproduce, and share a common evolutionary history.

How Can We Tell If Something Is Alive?

None of the first six characteristics of life by itself distinguishes every living thing from every nonliving thing. Spores and seeds, for example, are certainly "alive," but do not perform most of the processes on the list. Computers exhibit many of the characteristics of life. Certainly computers are well ordered; obtain energy—electrical energy—from their environments; change with time—as they acquire new programs; and respond to their environments—according to their programmed instructions. Although computers do not perform chemical reactions or reproduce, we can easily imagine a computer managing robots that perform chemical reactions or build other computers.

The one requirement that computers can never fulfill, however, is that they do not share a common origin with organisms. The shared history of life on our planet is a major theme both in this book and among contemporary biologists. We will see the astonishing similarities in the ways all organisms extract energy to fuel their own activities and in the molecules of which they are composed.

Living organisms share many individual traits with nonliving things. However, only living organisms share a common ancestry.

Organisms Are Made of Cells

Organisms come in great variety, but they are all made of cells. The cells themselves are wonderfully diverse and specialized. We can see some examples of such specialization in the microscopic picture of human skin cells in Figure 1-5: nerve cells that carry touch information, surface cells that serve as a barrier to the outside world, and "follicle" cells that produce hairs. Plant cells are similarly specialized. Some transform light energy into energy-rich sugar molecules, while others distribute these molecules to the rest of the plant. There is no easy way

Table 1-1 Characteristics of Life

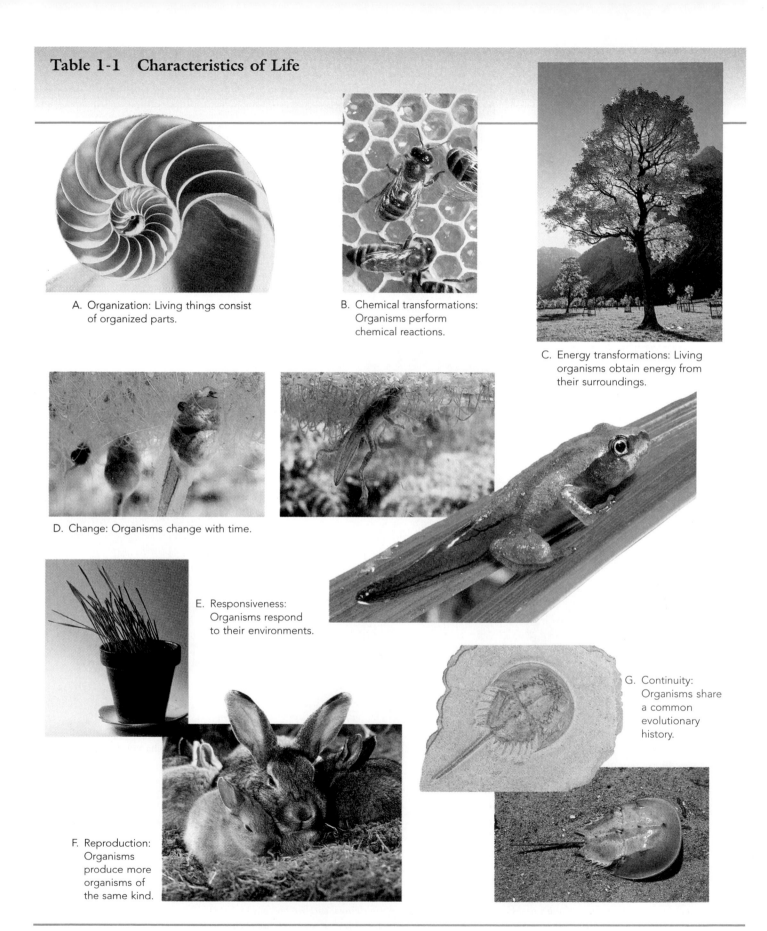

A. Organization: Living things consist of organized parts.

B. Chemical transformations: Organisms perform chemical reactions.

C. Energy transformations: Living organisms obtain energy from their surroundings.

D. Change: Organisms change with time.

E. Responsiveness: Organisms respond to their environments.

F. Reproduction: Organisms produce more organisms of the same kind.

G. Continuity: Organisms share a common evolutionary history.

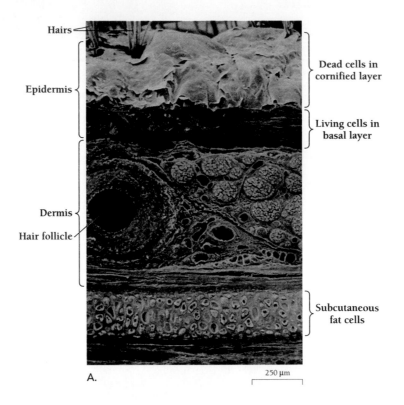

Hairs

Epidermis

Dead cells in cornified layer

Living cells in basal layer

Dermis

Hair follicle

Subcutaneous fat cells

A.   250 μm

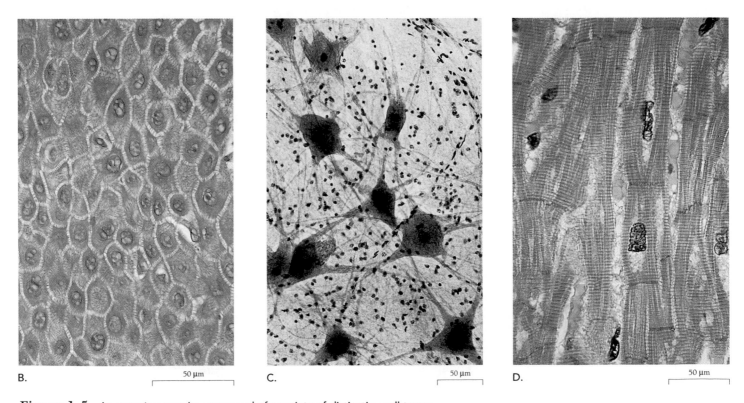

B.   50 μm   C.   50 μm   D.   50 μm

Figure 1-5 An organism may be composed of a variety of distinctive cell types.
A. Skin cross section. Even a single tissue can include many kinds of cells. Notice, for example, dead (corni-fied) skin cells at the surface, cells of the basal layer of epidermis, cells of the dermis, and the fat cells in the subcutaneous layer. B. Human squamous epithelial cells from the epidermis. C. Motor neurons in spinal cord. D. Cells of cardiac muscle.   *(A, C.Y. Shih & R.G. Kessel/Visuals Unlimited; B, G.W. Willis, M.D./Biological Photo Service; C, Manfred Kage/Peter Arnold, Inc.; D, Ed Reschke/Peter Arnold, Inc.)*

to count all the different kinds of cells. Mammals alone have at least 200 different kinds of cells.

Nonetheless, cells are amazingly alike. With some interesting exceptions, such as the enormous single cells that make up birds' eggs, cells are all about the same size. As one biologist wrote, "Nature keeps to the same bricks whether she builds a great house or small."

All cells conform to just two basic plans. **Eukaryotic** cells [Greek, *eu* = true + *karyon* = nucleus] contain a central **nucleus**, which is surrounded by a membrane. Inside the nucleus is the genetic material. Eukaryotic cells also contain other internal membranes, some of which enclose specific types of organelles. In contrast, the **prokaryotic** cells [Greek, *pro* = before] of the Archaea and Eubacteria contain neither nuclei nor membrane-bounded organelles.

All eukaryotic cells are made of the same internal parts. Every cell is enclosed by a fairly standardized membrane. Inside the cell are about 20 kinds of compartments, each of which performs a different task. The largest compartments are called **organelles** [little organs]. These include, for example, the numerous **mitochondria**, which supply usable energy to the cell, and the nucleus, which contains the genetic material.

All plants and animals are made of eukaryotic cells. Indeed, the Animalia, Plantae, Fungi, and Protista are all eukaryotes. The prokaryotes—including both archaebacteria and eubacteria—number only 4800 named species. Yet if these tiny organisms are few in kind, they are great in number. More bacterial cells live in our intestines, for example, than we have cells in our bodies. And it is from the prokaryotes that all eukaryotes—including humans—are descended. As we shall see in Chapter 18, scientists believe that eukaryotic cells evolved from prokaryotic ancestors.

The number of kinds of cells that compose organisms is limited. Two basic plans for cells exist. Eukaryotic cells have a membrane-bounded nucleus and membrane-bounded organelles. Most eukaryotic cells are about the same size, and contain the same organelles. Prokaryotic cells are smaller and lack membrane-bounded organelles.

How do living organisms replace worn-out parts?

A living organism, unlike a nonliving thing, constantly replaces and remolds itself, exchanging old atoms and molecules for new ones. All the while, the organism's form and internal properties stay almost constant. In organisms that are **multicellular** (made of many cells), cells live and die independently of the whole organism. For example, in humans and other animals, the cells of the skin, blood, and intestines are continuously replaced during life. The advantage of such replacement is that the life of a multicellular organism can extend far beyond the life of an individual cell.

A multicellular organism can lose and replace individual cells.

How much do organisms resemble one another chemically?

If we look at the tissues of different organisms under the microscope, we find great similarities. Cells are more alike than the organisms they make up. Millions of species are made of just a few thousand kinds of cells. And these relatively few cell types, in turn, contain just a few kinds of subcellular organelles.

This theme extends even further. Cells are made almost entirely from water and just four other kinds of molecules—**sugars** (which link together to form polysaccharides), **amino acids** (which link to form proteins), **lipids** (which link to form fats and other molecules), and **nucleotides** (which link to form polynucleotides such as DNA and RNA). At the molecular level, then, cells are even more alike than they appear under the microscope. The diversity of whole organisms is far greater than the diversity of their molecular components.

Organisms are even more alike at the molecular level than at the structural level.

Why Do Children Look Like Their Parents?

A visitor to a museum may be surprised to find that a 6000-year-old Egyptian cat looks just like a house cat from Houston. Species change very little from one generation to the next. Just as today's children look much like those of 30 years ago, they also resemble children born 30 or 40 generations ago in 1200 A.D.

The maintenance of these similarities depends on the way that the parents transmit "plans" to their offspring. Every individual inherits a set of **genes** [Greek, *gen* = to produce], which are detailed instructions for making the molecular constituents of an organism.

Genes ensure continuity from generation to generation.

What are genes made of?

As we will discuss in detail in Chapters 10 through 14, biologists have learned that organisms store and transmit genetic information in long thin molecules called **deoxyribonucleic acid (DNA)** (Figure 1-6). Each DNA molecule consists of long chains of **nucleotides**—relatively small molecules composed of carbon, hydrogen, oxygen, nitrogen, and phosphorus atoms.

DNA contains only four kinds of nucleotides. But these four simple molecules play the same role as the 26 letters of the alphabet. Just as each word on this page consists of a sequence of letters, each gene consists of a sequence of nucleotides. The sequence of nucleotide letters in each gene is

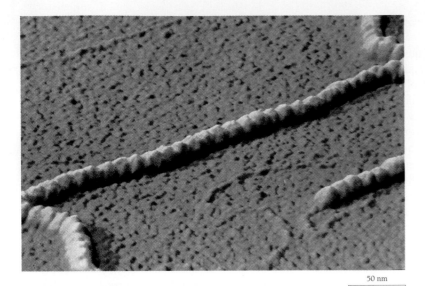

50 nm

Figure 1-6   DNA, or deoxyribonucleic acid, is a long, thin molecule consisting of a string of simple molecules called nucleotides. Just as the letters in this sentence carry information, the nucleotides in the DNA carry information.   *(Science VU/BMRL/Visuals Unlimited)*

written in a code that can be translated into a set of instructions. Most genes provide the information that cells need to build particular proteins. Thus, a set of genes is in many ways analogous to the list of ingredients in a recipe. The genes are a recipe for building an organism.

Genes are the means by which organisms transmit the information needed to build and maintain organisms.

What creates variation from one generation to the next?

The sequence of nucleotides in DNA is extremely stable from generation to generation. This is why offspring resemble their parents. However, when parents reproduce sexually, the offspring receive one-half of their genes from each parent and then mix the two sets of genes together. Sexual reproduction allows new combinations of genes, so that each new individual is unique. Suppose, for example, a female cat with short, brown hair mated with a tom cat with long, black hair. She might give birth to kittens with short, brown hair or long, black hair, but also to kittens with long, brown hair or short, black hair—combinations of traits different from those of the parents (Figure 1-7A).

None of the offspring would likely have curly hair or no hair. Even though each individual is unique, it carries the same genes as its parents. This ensures continuity from generation to generation, and the characteristics of a species remain the same—sometimes for millions of years.

Nonetheless, mistakes in the copying of DNA, or other alterations in DNA, called **mutations,** can create variations that can be inherited. For example, a mutation that eliminated color in the skin, hair, and eyes would result in a kitten with white hair and pink eyes—an albino.

Mutations create new genes and sexual reproduction creates new combinations of genes in each generation.

What determines the characteristics of an individual?

Just as the taste and texture of a meal depends on both the quality of the ingredients and the temperature of the oven, the form of a mature plant or animal depends on both its genetic plans and the environment in which the plans are executed. For example, it takes many years for a fertilized human egg, smaller than the dot in 0.1 mm, to grow into an adult. The results of this amazing development depend both on the instructions coded in the DNA and on such environmental factors as breathable air, protection from the cold, adequate food and water, and significant mental stimulation. Environmental insults as different as poor nutrition, a chemical that causes birth defects, or a gunshot wound can severely alter the course of development.

All the properties of an organism constitute its **phenotype** [Greek, *phainein* = to show + *typos* = impression]. The phenotype includes the appearance, the chemical makeup, and the behavior of an organism throughout its life history. Height, for example, is a particularly visible aspect of a human's phenotype. Height has a genetic component: tall parents are likely to have taller children than short parents. Yet height also depends on diet: a malnourished child of tall parents may not grow as tall as the well-fed child of short parents. Variations in both genes and environment can lead to diversity among individuals within a single species (Figure 1-7B). Genes and environment interact in complex ways to produce an individual's phenotype.

Each of the characteristics that make up the phenotype of an individual depends on both genes and environment.

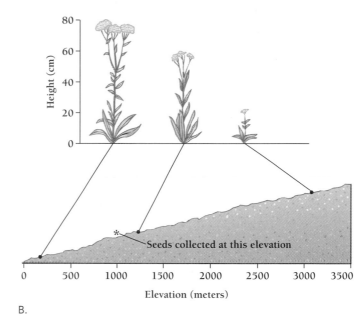

A.

B.

Figure 1-7 What creates variation from one generation to the next? A. When organisms reproduce, the genetic information in the DNA allows the offspring to be both like their parents and unlike their parents. Here, the offspring of a cat show a diversity of phenotypes because each has a unique set of genes inherited from mother and father. B. But it is not only genes that effect the phenotype of an organism. Here, the flower *Achillea millefolium* (yarrow) grows tall or short depending on the altitude at which it grows. All three plants are genetically similar, yet their phenotypes differ according to the environment in which they grow. *(A, Francois Gohier/Photo Researchers, Inc.)*

How Do Organisms Become Different from One Another?

We have seen that even though organisms are enormously diverse, they all share the same principles of cellular and subcellular organization, and all share the same set of chemical building blocks. These common characteristics (and others as well) have persuaded biologists that all organisms are genetically related and are descended from a common ancestor. How, then, can we understand the enormous diversity of life?

Species Differ in Their Adaptations to Distinct Environments

Each species has its characteristic **adaptations**, special structures or behaviors that fit it for life in its particular environment. Adaptations increase the chance that an individual will live and reproduce. The thick fur of the fur seal, for example, helps it to stay warm in cold waters. The *Amanita* mushroom has two adaptations designed to keep animals from eating it: a poison that sickens but does not kill and striking red and white colors that remind forgetful mushroom eaters to look elsewhere.

Size, form, color, internal structure, and special chemical properties all contribute to an organism's ability to survive in a particular environment, to produce energy in a characteristic way, and to reproduce more of its own kind. Despite the differences among individuals within a species, all the members of a given species are similarly adapted for a particular way of life. Given the opportunity, all cows stand around eating grass, all polar bears hunt seals, all kittens pounce on small objects, and all human children laugh and chase one another.

Part of the diversity of life results from the different sets of adaptations in each species. Among the favorite questions of biologists is, How does a particular structure (or behavior pattern or biochemical process) help this organism to survive and reproduce? Put more concretely, What good is an elephant's trunk, a mallard's mating dance, or a cactus' thick stems? Of ourselves we might ask, Of what use are a large brain and hairless skin? These are the same general questions that we might ask about the parts of a sewing machine or a car engine: What does it do? How else could it be done?

Because every species has so many adaptations, the same general questions arise again and again. After a while, biology students might begin to wonder how such a variety of adaptations came to be in the first place. As we mentioned earlier, all organisms appear to share a common origin. Biologists must therefore ask how individual species became so different from one another and why there are so many different species. The question, Why are organisms so diverse? becomes, How did so many different species derive from a single kind of living being?

Every species has a unique set of adaptations.

17

How Do We Know That the Diversity of Life Resulted from Evolution?

Until the middle of the 19th century, most Europeans believed that the Earth was only about 6000 years old, and that modern organisms were no different from those present at the beginning. After all, ancient Egyptian paintings and mummies revealed creatures and plants indistinguishable from those of modern times (Figure 1-8).

However, the discovery of **fossils**—objects dug from the ground that contain remains or traces of past life—strongly suggested that organisms very different from the ones we know today had once lived on Earth (Figure 1-9). In the early 19th century, geologists began to recognize that the Earth was far older than 6000 years. They understood for the first time that over millions of years, huge parts of the Earth's surface had moved about, new mountain ranges had arisen, and wind and water had shifted land from one place to another. Fossil organisms that had been near the surface became covered over with new layers of rock. The successive layers produced by these processes, scientists realized, contained the history of the Earth and its inhabitants (Figure 1-4G).

We now know that during the last 4.5 billion years, our planet has changed from a lifeless sphere of rock and water to a rich, green world alive with fabulously diverse organisms. This transformation is the result of **evolution**—the process by which species have arisen and changed as they descended from common ancestors.

In the 19th century, scientists began to realize that fossils represented extinct forms of life very different from anything living today.

Figure 1-8 Organisms, to a large extent, remain the same from one generation to the next. This hippopotamus, carved more than 2000 years ago, closely resembles hippopotami that live today. *(Larry Tackett/Tom Stack & Associates)*

Charles Darwin provided evidence for evolution

By the early 19th century, many scientists believed that during the Earth's history some species had disappeared and new ones had appeared. No one understood how these changes could have occurred, however, until the 19th century English biologist Charles Darwin (1809–1882) proposed his theory of *evolution through natural selection* (Figure 1-10).

Nothing in Darwin's youth suggested that he would revolutionize the world, becoming perhaps the most influential biologist in history. As we will see in Chapter 15, Darwin was a miserable student. Nonetheless, opportunity came his way. After he graduated from Cambridge University, he was invited to serve as an unpaid naturalist on the *H.M.S. Beagle,* a British surveying ship. Darwin accepted and, during the 5-year voyage around the world, collected and studied thousands of plants and animals. For the first time, he began to think like a scientist. He was struck by the enormous diversity of life he saw in his travels through South America. Everywhere he went he saw novel creatures and plants. Some resembled their familiar European cousins, others were astonishingly different. Darwin asked why there were so many species and why some species closely resembled others.

When Darwin returned to England, a marriage to a wealthy heiress, along with his share of the family estate, freed him from having to work, and he spent the rest of his life at home with his wife and children, studying his specimens, experimenting, thinking, and writing. In 1859, 23 years after his voyage, he published the most influential scientific book in history—*The Origin of Species by Means of Natural Selection, or the preservation of favoured races in the struggle for life.* In *The Origin of Species,* as we now call his book, Darwin presented two ideas. First, he presented convincing evidence for the idea that organisms have evolved, and, second, he proposed a possible explanation for how evolution occurs.

Darwin's model for evolution was the techniques that plant and animal breeders have used for thousands of years to produce useful new varieties of organisms, whether wheat, corn, dogs, or horses. The method is simple: the breeder selects certain individuals to breed—the most productive wheat plants, the fiercest dogs, the fastest horses. Breeders depend on the naturally occurring variation in any population of individual plants or animals. Individuals with less desirable characteristics are simply not allowed to reproduce. Figure 1-11 illustrates the power of this strategy, showing the results of artificial selection in wild sea cabbage to produce broccoli, cabbage, Brussels sprouts, and more. Using such artificial selection, breeders have produced incredible varieties of plants and animals.

Darwin saw that a similar kind of selective breeding, which he called natural selection, could operate in nature. **Natural selection** is the differential survival and reproduction of individuals with certain inherited traits. While artificial selection depends on the conscious decisions of a breeder, natural selection depends on the demands of an organism's environment.

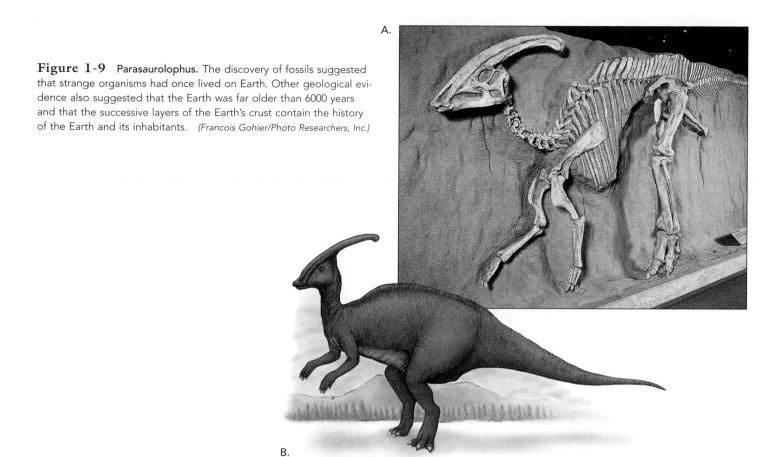

Figure 1-9 Parasaurolophus. The discovery of fossils suggested that strange organisms had once lived on Earth. Other geological evidence also suggested that the Earth was far older than 6000 years and that the successive layers of the Earth's crust contain the history of the Earth and its inhabitants. *(Francois Gohier/Photo Researchers, Inc.)*

A.

B.

Organisms that possess traits that improve their ability to escape predators; to withstand environmental changes; or to compete for food, sunlight, or other limited resources generally live longer and produce more offspring.

Charles Darwin supplied extensive evidence for the idea that species have evolved from common ancestors. He also suggested that species have evolved by means of natural selection.

How are traits passed from one generation to the next?
Darwin understood how evolution might work, but he had no idea how living organisms pass traits to their offspring or why the offspring are different from their parents and different from each other. Darwin knew neither why we look like our parents nor why we look slightly different from our parents. The question of how heredity works puzzled and frustrated him for the rest of his life. Ironically, the answer lay close at hand.

In 1856, the Austrian naturalist monk Gregor Johann Mendel (1822–1884) began a series of experiments on pea plants in the gardens of the monastery at Brno, Austria. In 1866, just 7 years after Darwin published *The Origin of Species*, Mendel published a long, difficult paper describing the results of his 10 years of experiments. He formulated the laws of heredity that laid the groundwork for the modern science of genetics. His paper was virtually ignored until 1900, when his work was rediscovered and recognized (Chapter 9).

Figure 1-10 Charles Darwin persuaded scientists that all life on Earth evolved from simple forms. Darwin suggested that the mechanism for evolution is natural selection, which he compared to the artificial selection used by plant breeders. *(The Granger Collection, New York)*

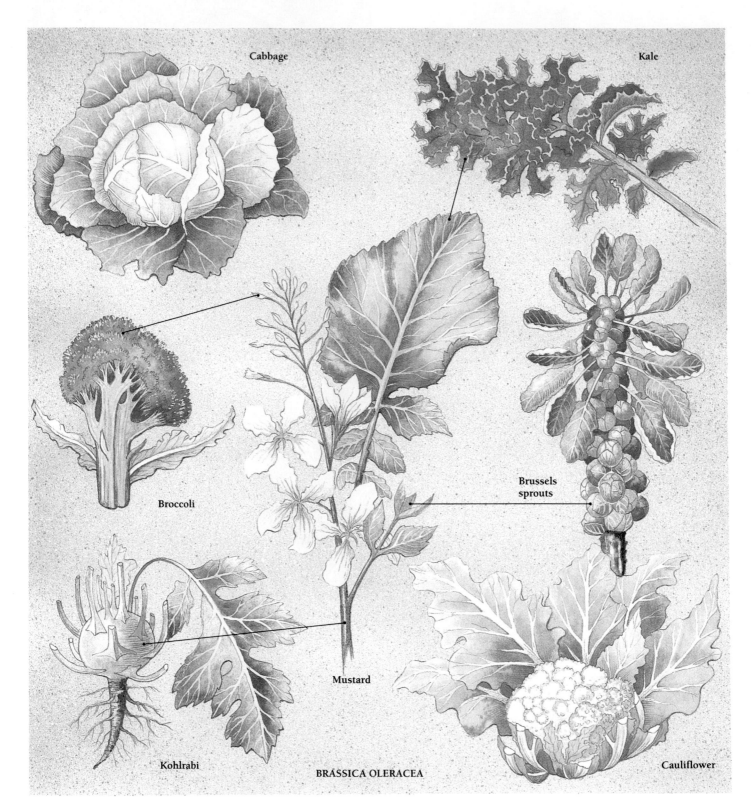

Cabbage

Kale

Broccoli

Brussels sprouts

Mustard

Kohlrabi

Cauliflower

BRASSICA OLERACEA

Figure 1-11   European plant breeders developed a great variety of vegetables from a single species of plant, the sea cabbage *Brassica oleracea.* Above are cabbage, cauliflower, broccoli, Brussels sprouts, kale, and kohlrabi. In the case of kale and cabbage, farmers bred only those plants with large, edible leaves. To create broccoli and cauliflower, they selected plants with the most flower buds. For Brussels sprouts, they chose those with the largest, most edible leaf buds. In kohlrabi, breeders selected plants with the biggest, most edible stems.

According to Mendel's principle of segregation, each sexually reproducing organism has for each characteristic two versions of each gene. These two versions segregate (or separate) randomly during the production of eggs or sperm, so-called gametes. As a result, half the gametes carry one version of the gene and half carry the other version.

According to Mendel's principle of independent assortment, genes for different traits are distributed to the gametes (eggs and sperm) independently of one another. For example, as we mentioned earlier, a female cat with short, brown hair mated to a tom cat with long, black hair might give birth to kittens with short, brown hair or long, black hair, but also kittens with long, brown hair or short, black hair. The genes for fur color and the genes for fur length assort independently when they go to the gametes.

Gregor Mendel discovered the genetic rules that underlie natural variation.

We have seen how biologists try to answer questions about life and how they define life. We have outlined, very briefly and superficially, how living organisms reproduce themselves and evolve, subjects that we will study in far more detail in the later chapters on genetics and evolution. In the next few chapters, we will introduce ourselves to the chemistry of life. We will look at the chemistry of water; the building blocks of proteins, carbohydrates, and fats; the internal structure of cells; and the elaborate chemical reactions that allow organisms to extract energy from either sunlight or molecules.

STUDY OUTLINE WITH KEY TERMS

Scientists answer questions using the **scientific method.** However, most scientists do not stick to any rigid set of rules. Thinking up both questions to ask and ways to answer those questions requires inspiration and creativity, as well as hard work, patience, and a methodical approach. In general, a scientist begins by narrowing an area of interest to a specific question, then **observing** a phenomenon. A scientist then compares the phenomenon to an already understood process, which serves as a **model.** From the model, the scientist derives a **hypothesis** about how a process or structure works. The hypothesis allows the scientist to make a **testable** prediction that the process or structure will act in a certain way under certain specified circumstances. The hypothesis can be proven wrong, but not right, by an experiment. Essential elements of most scientific experiments are the **control** experiment and **statistical** analysis. A set of hypotheses that consistently resist scientists' attempts to disprove them may become accepted as a **theory.**

All living things are organized into parts, obtain energy from their surroundings, perform chemical reactions (or **metabolize**), change with time, reproduce, share a common evolutionary history, and respond to their environments. Mechanisms that maintain a constant internal environment are a form of **homeostasis.**

The millions of species of organisms are organized by biologists into three domains: the **Archaea,** the **Eubacteria,** and the **Eukarya.** The Eukarya comprise just four kingdoms: **Animalia, Plantae,** **Fungi,** and **Protista.** Biologists do not agree whether the Archaea and Eubacteria each comprise one kingdom or several.

According to the **cell theory,** all organisms are composed of one or more cells, which are themselves alive, and which all come from other cells. **Multicellular** organisms can lose and replace individual cells. **Eukaryotic** cells have a **nucleus** enclosed by a membrane and are organized into recognizable **organelles** such as **mitochondria.** **Prokaryotic** cells have no membrane-bounded nucleus. All organisms are composed of the same basic kinds of building block molecules: **sugars, amino acids, lipids,** and **nucleotides.**

Living things all pass information from one generation to the next by means of **genes,** which are made of **deoxyribonucleic acid,** or **DNA.** DNA consists of long sequences of **nucleotides** that carry information in the same way that sequences of letters carry information. When a cell copies a sequence of nucleotides incorrectly, we call it a **mutation.** The **phenotype** of an organism develops in response to complex interactions between the genes and the environment.

Traits that are inherited and help a species thrive are considered **adaptive.** The discovery of **fossils** suggested to 18th- and 19th-century scientists that organisms had changed, or evolved, over long periods. Charles Darwin assembled compelling evidence for the idea of **evolution** and suggested a mechanism, called **natural selection.** Gregor Mendel discovered and articulated the first rules of inheritance.

REVIEW AND THOUGHT QUESTIONS

Review Questions

1. Explain the differences between a hypothesis, a model, and a theory.
2. Explain why a hypothesis can be proved wrong, but can never be proved right.
3. Give an example of an untestable hypothesis. Why is it untestable? What would you need to know or do to test it?
4. Suppose you decide to test the hypothesis that your favorite soft

drink contains substances that cause cancer. You plan to feed the drink to 100 mice and then count the number of tumors they develop. Describe an appropriate control experiment.

5. Describe the seven distinguishing characteristics of living things. Give an example of each one.
6. Define homeostasis. List two examples of homeostasis in an animal.
7. What do the genes do? What molecule makes up the genes?
8. Explain what is meant by natural selection. What is "selected"?

Thought Questions

9. Contrast the opinion of reductionists with that of their critics. Does each position have strong and weak points? How does each position influence the way in which scientists study biology?

10. The ants in your kitchen are driving you nuts. Using the "scientific method," how would you find out where they were entering your house? How would you determine why they were entering your house? Think of several alternative hypotheses. Now design a series of experiments whose results will either support or eliminate the various alternative explanations.

11. A virus is a bit of DNA surrounded by a protein coat. A virus can attach to the wall of a cell and inject the DNA into the cell. Inside the cell, the viral DNA forces the cell to make new viruses. The cell makes so many new viruses that the cell eventually bursts open, releasing thousands of new viruses, which go on to infect other cells. Is the virus alive? Using the definitions of life given at the beginning of this chapter, explain why a virus is alive or why a virus is not alive.

12. From your own experience, describe a moment of inspiration—intellectual or otherwise. How did it feel? How did it change you?

SELECTED READINGS

Eiseley, Loren, *The Immense Journey,* Vintage Books, New York, 1946. A literary essay by an anthropologist and naturalist about our place in nature.

Gould, Stephen J., *Ever Since Darwin, Reflections in Natural History,* W. W. Norton & Co., 1977. A collection of essays on evolution, natural selection, and other topics.

Monmaney, Terence, "Marshall's Hunch," in *The New Yorker* magazine, September 20, 1993.

Thomas, Lewis, *The Lives of a Cell,* Bantam Books, 1974. A collection of short essays that are simultaneously both personal and scientific describing our strange relations with other living organisms. The author was an eminent biologist.

▶ On-line materials relating to this chapter are on the World Wide Web at http://www.saunderscollege.com/lifesci/ Click on Tobin/Dusheck: *Asking About Life.*

PART I

Chemistry and Cell Biology

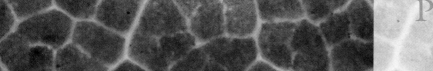

Micrograph of a sugar maple leaf turning red. *(© John Eastcott/Yva Momatiuk, DRK Photo)*

Alchemy and Chemistry

In the spring of 1992, Texas A & M University Professor of Chemistry John Bockris received a friendly phone call from a man named Joe Champion. Champion said he was a Tennessee inventor who had developed a method for turning silver into gold. Gold is worth 20 times more than silver, so Champion's discovery, if legitimate, would be better than owning a gold mine. Besides offering to share this valuable secret with Bockris, Champion also offered $200,000 for Bockris to investigate and test the potentially lucrative new technique. It was an offer that Bockris couldn't resist, and he soon had the members of his laboratory testing Champion's technique.

Champion came to Texas to show Bockris and his assistants how to mix the right chemicals (which included one of the components of gun powder) to create gold. As long as Champion was helping out in the lab, the technique seemed to work. But when Champion left town, the technique mysteriously failed.

Bockris' colleagues at Texas A & M were acutely embarrassed. What Bockris was attempting was plain old-fashioned **alchemy**—an ancient pseudoscience one

Figure 2-1 *The Alchemist* painted by David Teniers the Younger in 1648. Alchemists of the medieval period and later attempted to turn lead and other cheap metals into gold and silver. *(Scala/Art Resource, NY)*

of whose aims was to turn "base" metals, such as lead or copper, into silver or gold (Figure 2-1). Alchemy reached the height of its popularity during the Middle Ages (about 700 years ago) and was completely discredited in the 19th century. Silver, gold, copper, and lead are all **elements**—substances that cannot be reduced to simpler substances by chemical means. In this chapter we will see why no element can be transformed into another element by chemical reactions.

What happened to Champion and Bockris? Champion spent a couple of years in jail, charged with criminal fraud

in another case. A third man, who provided the $200,000 to Bockris, was

---

Champion's discovery, if legitimate, would be better than owning a gold mine.

---

charged by the federal Securities and Exchange Commission with fraudulently selling unregistered stocks to innocent in-

vestors. Texas A & M froze Bockris' research account, refusing to give him any more of the probably ill-gotten $200,000. And 11 of Bockris' 38 colleagues in the chemistry department formally demanded that Bockris resign from the university, charging that he was ruining their reputations as chemists. Bockris kept his job, however, while still arguing that elements could be changed one into another by chemical means. Ultimately, he abandoned his researches in alchemy and turned to more conventional studies of the chemical bond.

WHAT IS MATTER?

Anyone whose car has ever run out of gas knows that when the gauge says empty, the gas is gone. Where does it go? Chemistry tells us that when the chemicals in gasoline come into contact with the oxygen in air, a spark (from the spark plug) triggers a chemical reaction in which the gas is burned, or "oxidized." Most of the gasoline is transformed into water and carbon dioxide (the same stuff that makes the bubbles in soft drinks).

Chemistry is the science dealing with the properties and the transformations of all forms of matter—whether gasoline, air, protons, people, or planets. **Matter** is any substance, whether it is a solid, a liquid, a gas, or even the plasma that makes up the stars. One form of matter can change into other forms in the course of **chemical reactions**—transformations in which different forms of matter combine or break down.

Understanding the chemical properties of matter depends upon first identifying what forms matter takes. The kinds of matter that cannot be converted into simpler substances in chemical reactions are called elements. Hydrogen, helium, oxygen, and carbon are elements. So are silver and gold, the carbon in a pencil, the copper in a wire, and the mercury in a thermometer (Figure 2-2). Altogether, scientists know of 92 naturally occurring elements. Elements are composed of tiny particles called **atoms**—the smallest units that still have the properties of an element. Atoms themselves are made of smaller "subatomic" particles, including protons, neutrons, and electrons.

Each element has a standard one- or two-letter abbreviation. These usually correspond to the beginning of the element's English name, though sometimes it's the Latin or Greek name. "C" stands for carbon and "O" for oxygen, while "Na" stands for sodium [Latin, *natrium*]. Four elements—carbon (C), oxygen (O), nitrogen (N), and hydrogen (H)—make up more than 99 percent of the matter in organisms. Table 2-1 shows the main elements of the human body.

Chemical reactions cannot change or destroy elements. Carbon cannot be converted to oxygen. Silver cannot be converted to gold. However, elements can be created or changed by means of **nuclear reactions,** the high-energy transformation of individual atoms that occur in the bowels of stars such as our sun or in nuclear reactors.

Although oxygen is an element, we do not breathe oxygen in its elemental form. Instead, we breathe "molecular" oxygen, which is two atoms of oxygen bound together by a "chemical bond." A **molecule** is a stable assembly of two or more atoms.

Table 2-1 Percentage of the Human Body Composed of Each Element, by Weight

Humans, like other organisms, are primarily water. The commonest two atoms in our bodies, then, are hydrogen and oxygen (H_2O). Hydrogen is a much lighter atom than oxygen, so most of our mass is made up of oxygen atoms. Next comes carbon, which, like hydrogen and oxygen, is in virtually every other molecule in our bodies.

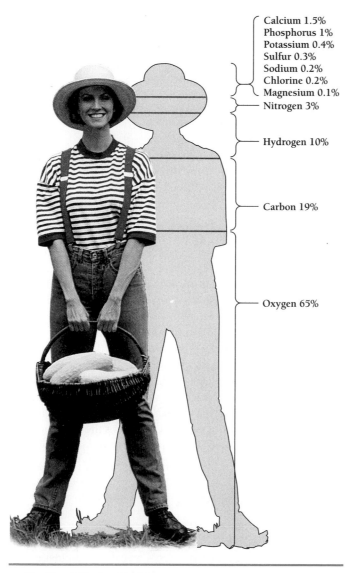

Calcium 1.5%
Phosphorus 1%
Potassium 0.4%
Sulfur 0.3%
Sodium 0.2%
Chlorine 0.2%
Magnesium 0.1%
Nitrogen 3%

Hydrogen 10%

Carbon 19%

Oxygen 65%

(Roy Morsch/The Stock Market)

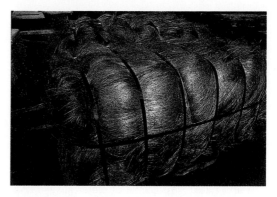

Figure 2-2 **Copper wire in bundles.** Copper is the 29th element in the Periodic Table and an unusually good conductor of electricity and heat. Historically, copper was used in bronze tools and weapons, which replaced stone tools and weapons. Today, copper is primarily used in electrical wiring. *(John Cancalosi/DRK Photo)*

alone or oxygen alone, and the properties of carbon dioxide are different from those of carbon and oxygen.

A compound consists of a single kind of molecule. Chemists designate the composition of a molecule by writing the number of atoms of each type as subscripts to the right of the symbol for each element. Water is always H_2O, carbon dioxide is CO_2, and ordinary table sugar (sucrose) is $C_{12}H_{22}O_{11}$.

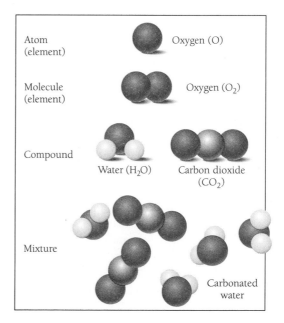

Atom (element) Oxygen (O)

Molecule (element) Oxygen (O_2)

Compound Water (H_2O) Carbon dioxide (CO_2)

Mixture Carbonated water

Figure 2-3 **Atom, molecule, compound, and mixture.** Elements are substances that cannot be reduced to simpler substances by chemical means. The smallest unit of matter with the properties of an element is an atom. A stable assembly of two or more atoms is a molecule. Two atoms of the same element, such as two oxygen atoms, may form a molecule. But most molecules are compounds forged from two or more different elements. Water is a compound of hydrogen and oxygen, and carbon dioxide is a compound of carbon and oxygen. Most substances are mixtures of different compounds. Sparkling water, for example, is a mixture of carbon dioxide and water.

Most molecules are **compounds**—substances forged in chemical reactions from two or more *different* elements. Water, for example, is a compound of the elements hydrogen and oxygen (Figure 2-3). Carbon dioxide is a compound of carbon and oxygen.

A compound is unlike the elements that compose it. The properties of water are different from those of either hydrogen

In our day-to-day lives, we rarely encounter either compounds or elements in pure form. Most matter is a **mixture** of different compounds (Figure 2-3). Sand and salt poured together, for example, is a mixture. One component of a mixture does not affect the properties of the other components. If we pour salt into boiling water to make spaghetti, the salt is still salt and the water is still water. If we boil away all of the water, the salt forms crystals at the bottom of the pot, and we can retrieve the same salt we poured into the pot.

Unlike the components of a compound, which are always present in the same fixed ratio, the components of a mixture may be present in any ratio. A molecule of water always contains twice as much hydrogen as oxygen. But you can mix a little sugar into your cup of coffee, or a lot.

Elements are pure substances that cannot be converted into simpler substances. An atom is the fundamental unit of an element, and a molecule is the fundamental unit of a compound.

What Is Chemistry?

Until the 16th century, chemistry as we know it did not exist at all. Alchemists combined and recombined various compounds and elements, hoping to make silver, gold, and other valuable elements. Alchemists did not yet recognize that the elements could not be changed one into another by ordinary means. Some alchemy was science of a sort, but much of it was simple superstition. None of the alchemists in the Middle Ages possessed a modern understanding of what matter was and why it acted as it did.

The only other kind of chemistry in the Middle Ages was *metallurgy*—the study of the properties and transformations of metals. Long before the Middle Ages, bronze tools replaced stone tools and, later, iron tools replaced bronze tools. Steel tools have, in turn, replaced iron ones. Metallurgy was, and is, so important that its mastery remains a measure of the progress of a civilization.

In the 1500s, the chemistry of medicines emerged as a serious discipline. Then, bit by bit, Renaissance scientists and philosophers began reinterpreting the chemical facts discovered by the alchemists. Not until the 19th century, however, did scientists begin to understand the chemical nature of matter. Today, on the verge of the 21st century, chemistry extends beyond metals and medicines to all matter, including the substances that make up organisms. One of the great achievements of 20th century biology has been to explain the chemistry that underlies all cellular processes—from the workings of a gene to the contraction of a muscle.

One of the great questions in biology was whether living matter was in any way different from nonliving matter. Are a horse and a plow made of the same basic stuff? In this chapter, we will see that the molecules of living organisms indeed obey the same chemical laws as the molecules of nonliving matter.

From a chemist's point of view, just two aspects of living matter distinguish it from nonliving matter—the elaborate organization of the molecules and the limited number of kinds of molecules that comprise living matter. All organisms consist of the same few kinds of small organic molecules, which we may think of as "building blocks." Just as a good cook can make a dazzling array of cookies, cakes, and pies from just a few ingredients, so organisms can produce extraordinary structures from a small number of standard molecules.

Chemistry deals with the properties and transformations of both living and nonliving matter.

WHAT DETERMINES THE PROPERTIES OF AN ATOM?

By the early 19th century, chemists were able to distinguish compounds—which could be broken down into simpler substances—from elements—which could not. They then began to ask: Does the way in which a substance breaks down reveal something about its basic structure? Are there units that cannot break down further?

To answer these questions, the English chemist John Dalton revived an idea first developed by the Greek philosopher Democritus in the 5th century B.C.: Matter is built from small particles. Each element, Dalton proposed, was composed of tiny particles called atoms.

The atoms of each element have a characteristic weight. For example, an atom of carbon weighs 12 times as much as an atom of hydrogen. Chemists and physicists prefer to talk of the "mass" of an atom, rather than its weight. **Mass** is the amount of matter in something. Weight results from the action of gravity on mass. You may weigh 125 pounds on Earth and zero pounds in space, but your mass remains the same no matter where you are. Mass is closely related to weight, but the mass of an object, unlike its weight, is constant.

What Are Atoms Made Of?

In the late 18th and early 19th centuries chemists discovered that they could measure the relative mass of each element in any compound. Such measurements revealed a fascinating relationship: The mass of almost every kind of atom is a multiple of the mass of a hydrogen atom. An atom of helium weighs twice as much as an atom of hydrogen. An atom of lithium weighs three times as much as hydrogen. Carbon weighs 12 times as much; oxygen, 16 times; and nitrogen, 14 times.

This pattern suggested the possibility that the larger atoms were composed of groups of hydrogen atoms. That is, an atom of carbon was composed of 12 hydrogen atoms; an atom of oxygen, 16 hydrogen atoms; and so forth. Ultimately, chemists concluded that atoms are not actually made of hydrogen atoms, but of subatomic particles, each of which weighs about as much as a hydrogen atom.

The masses of atoms are extremely small. An oxygen atom, for example, weighs less than 3×10^{-23} grams: that's 0.00000000000000000000003 grams. (A gram is about as heavy as one-third of an animal cracker.) For convenience, chemists have defined a measuring unit for the mass of an atom or a molecule that is much smaller than a gram. A dalton, named after the chemist John Dalton, is approximately the mass of a hydrogen atom. An atom of hydrogen has a mass of 1 dalton, and an atom of oxygen a mass of 16 daltons. Since a molecule consists of a fixed number of atoms, every molecule has a characteristic molecular mass, usually called its **molecular weight**. A molecule of water (H_2O), for example, has a molecular weight of 18:

$$18 = 16 \text{ (oxygen)} + 1 \text{ (hydrogen)} + 1 \text{ (hydrogen)}$$

Atoms are made of subatomic particles similar in size to one hydrogen atom.

What Is the Internal Structure of an Atom?

Once physicists knew that atoms were made up of hydrogen-sized particles, they began to wonder about those particles. What were the hydrogen-sized particles, and how did they fit together to make an atom? The first important finding about atomic structure was that atoms are almost entirely empty space. In 1910, a New Zealand physicist working in England devised an ingenious way to probe the structure of the atom. Ernest Rutherford's approach, diagrammed in Figure 2-4, was like searching for a rock in a haystack by shooting bullets into the haystack. If a bullet goes straight through the haystack, it means that it has failed to hit anything solid. If it comes out at an angle, then it must have bounced off something. The "bullets" Rutherford used were small, positively charged α particles, produced from the radioactive decay of radium. The "haystack" was a thin piece of gold foil.

Rutherford found that the proton bullets passed right through most parts of a gold atom, only bouncing off a tiny, central part of each atom. He proposed that the atom is mostly empty space, with almost all of its mass concentrated in a small heavy core, or **nucleus**, at its center.

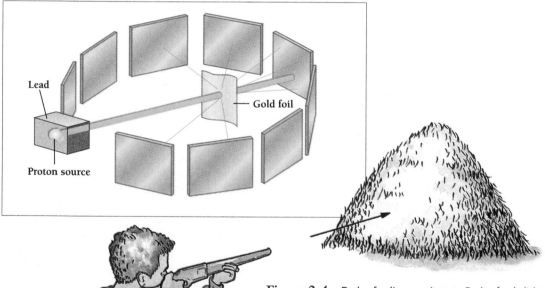

Lead

Gold foil

Proton source

Figure 2-4 Rutherford's experiment. Rutherford did not perform his now famous experiment himself. He delegated it to a student in his lab, to whom he initially did not even speak. Rutherford later wrote, "One day Geiger came to me and said, 'don't you think that young Marsden, whom I am training in radioactive methods, ought to begin a small research?' . . . I said, 'Why not let him see if any alpha particles can be scattered through a large angle?' I may tell you in confidence that I did not believe that there would be, since we knew that the alpha particle was a very fast massive particle, with a great deal of energy, and you could show that . . . the chance of an alpha particle being scattered backwards was very small. Then I remember two or three days later Geiger coming to me in great excitement and saying, 'We have been able to get some of the alpha particles coming backwards . . .' It was quite the most incredible event that has ever happened to me in my life. It was almost as incredible as if you fired a 15-inch shell at a piece of tissue paper and it came back and hit you." *(Rutherford and the Nature of the Atom, by E.N. da Costa Andrade, Doubleday, Garden City, NY, 1964; quoted by S. Weinberg,* The Discovery of Subatomic Particles, *Scientific American Library, New York, 1983, p. 124.)*

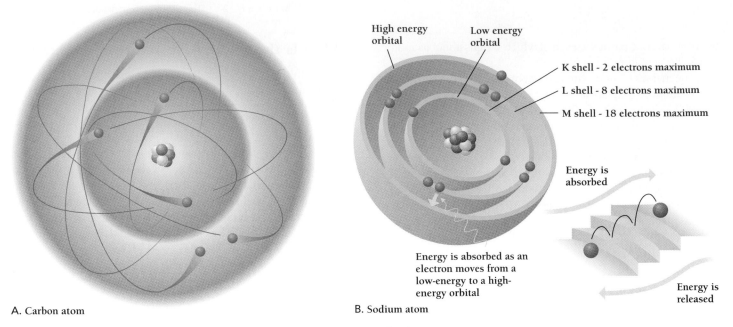

High energy orbital   Low energy orbital

K shell - 2 electrons maximum
L shell - 8 electrons maximum
M shell - 18 electrons maximum

Energy is absorbed

Energy is absorbed as an electron moves from a low-energy to a high-energy orbital

Energy is released

A. Carbon atom

B. Sodium atom

Figure 2-5 **The structure of an atom.** A. Carbon atom. By convention, orbitals are drawn as flat rings on the page as in B, but in reality, electron orbitals are diffuse three-dimensional spheres (or other shapes), as shown here. B. Sodium atom. Orbitals are grouped into shells, each of which has a different energy. Closest to the nucleus is the K shell. It has the lowest energy and holds a single orbital with a capacity for two electrons. Farther out is the L shell, which contains four orbitals with a capacity for eight electrons. Still farther out is the M shell, which has a total capacity of 18 electrons in 9 separate orbitals.

The diameter of the nucleus is only 1/10,000 that of the whole atom. Imagine a pingpong ball (the nucleus) at the center of two vacant city blocks (the atom), and you will have an idea of the proportions of an atom. Because the tiny nucleus reflected Rutherford's positively charged particles, he reasoned that the nucleus must itself have a positive charge that repelled his "bullets." The positive charge of the nucleus is balanced by negatively charged particles, called **electrons,** which surround it (Figure 2-5). Rutherford imagined that the negative electrons orbit the positive nucleus in the same way that the planets orbit the sun.

In fact, later researchers confirmed that the nucleus contains particles called **protons.** The positive charge of a proton exactly balances the negative charge of an electron. A neutral (uncharged) atom has the same number of protons and electrons. The nucleus of an atom also contains **neutrons,** which have no electric charge.

The mass of an atom depends almost entirely on the total number of protons and neutrons in the nucleus. The tiny electrons have little mass. Each has just 1/2000th the mass of a proton.

The atoms of each element always have the same number of protons. Oxygen, for example, always has 8 protons. However, some forms of an element, called **isotopes,** may have different numbers of neutrons. Carbon, for example, has three common isotopes: carbon 12, carbon 13, and carbon 14. The atoms of all three isotopes have six protons and six electrons and the same chemical properties. However, each has a differ-

ent number of neutrons and, therefore, different masses: carbon 12 has six neutrons, carbon 13 has seven, and carbon 14 has eight. The three isotopes of carbon are also written ^{12}C, ^{13}C, and ^{14}C.

Some isotopes are unstable and tend to break down or decay. The nucleus of ^{14}C, for example, which consists of six protons and eight neutrons, is unstable. It is more likely to come apart than a ^{12}C nucleus, which consists of six protons and six neutrons. Every second, the nuclei of about four ^{14}C atoms in a trillion (10^{12}) break up. As a result of this decay, an electron speeds away from the nucleus (Figure 2-6). An electron released in this manner is called a beta particle, and is easily detected and counted.

Isotopes that release beta particles are called **radioactive isotopes.** The decay process occurs at a constant rate that is characteristic of each radioactive isotope. The time required for half the atoms of a radioactive isotope to decay is called that isotope's **half-life.** Half-lives for different isotopes range from as little as 0.3 microseconds to as long as 2,000,000,000,000,000 (10^{15}) years. The half-life of carbon 14 (^{14}C) is about 5760 years, while that of uranium 235 (^{235}U) is about 704 million years.

Atoms are mostly empty space. They consist of a tiny central nucleus orbited by even tinier electrons. The nucleus has a positive charge, balanced by the negative charge of the electrons.

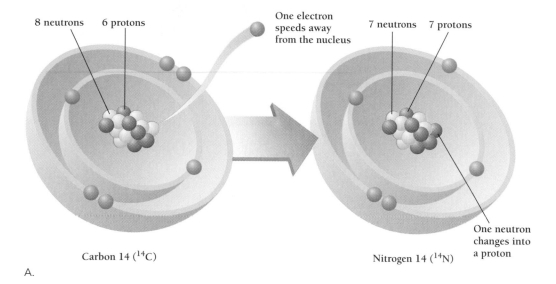

8 neutrons   6 protons

One electron speeds away from the nucleus

7 neutrons   7 protons

Carbon 14 (^{14}C)

Nitrogen 14 (^{14}N)

One neutron changes into a proton

A.

B.

Figure 2-6   **Half-life.** A. Breakdown of ^{14}C isotope. When the neutron releases its electron (the beta particle), the neutron becomes a proton. As a result, the altered atom has one less neutron and one more proton, for a total of seven neutrons and seven protons. An atom with seven protons is no longer carbon, however, but nitrogen. B. Half-life of chocolate. (B, Paraskevas Photography)

What Does Radioactivity Do to Living Organisms?

Beta particles released by radioactive isotopes are a form of radiation that damages and kills cells. Alpha particles, beta particles, and x rays, for example, are all different forms of "ionizing radiation." Ionizing radiation causes the water molecules in tissues and cells to break down into *free radicals.* Free radicals are chemically active, oxygen-containing molecules that easily react with and damage other molecules, such as proteins and DNA. Free radicals can cripple or kill cells.

The cells most vulnerable to radiation are those that divide frequently. Ionizing radiation quickly kills cells lining the mouth, the stomach, and the intestines; the cells in the skin that produce hair; the cells in the marrow of the bones that produce red blood cells; and more. As a result, people and

other mammals exposed to intense doses of ionizing radiation—whether by accident or as therapy for cancer—suffer nausea, vomiting, diarrhea, hair loss, and anemia. Ironically, because ionizing radiation can damage the DNA in cells, it can also cause subsequent bouts of cancer.

Where Are the Electrons in an Atom?

Although Rutherford envisioned the electrons neatly orbiting the nucleus like tiny planets, physicists have found that electrons actually move around the nucleus randomly, like a cloud of gnats. Although we still speak of electron **orbitals,** the distributions of electrons more closely resemble clouds. Like clouds, orbitals have fuzzy boundaries but definite shapes (Figure 2-5).

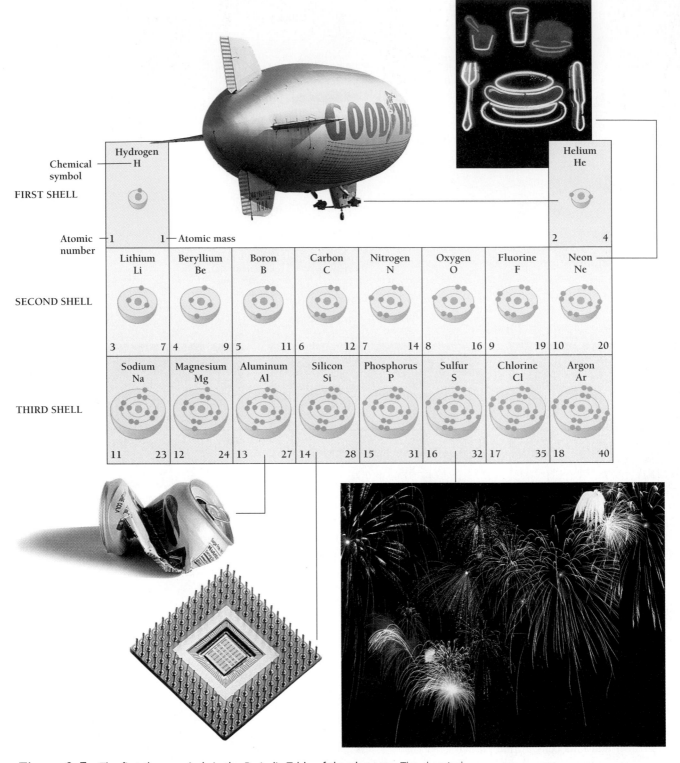

								Helium He
Hydrogen H								

Chemical symbol

FIRST SHELL

Atomic number — 1 ‖ 1 — Atomic mass ‖ 2 ‖ 4

Lithium Li	Beryllium Be	Boron B	Carbon C	Nitrogen N	Oxygen O	Fluorine F	Neon Ne
3 ‖ 7	4 ‖ 9	5 ‖ 11	6 ‖ 12	7 ‖ 14	8 ‖ 16	9 ‖ 19	10 ‖ 20

SECOND SHELL

Sodium Na	Magnesium Mg	Aluminum Al	Silicon Si	Phosphorus P	Sulfur S	Chlorine Cl	Argon Ar
11 ‖ 23	12 ‖ 24	13 ‖ 27	14 ‖ 28	15 ‖ 31	16 ‖ 32	17 ‖ 35	18 ‖ 40

THIRD SHELL

Figure 2-7 **The first three periods in the Periodic Table of the elements.** The chemical and physical properties of the 92 elements depend on the atomic numbers (*left*) and the atomic mass (*right*) of the nucleus. The atomic number is the number of protons in the nucleus. The atomic mass is the total number of protons plus neutrons. On the far right are elements such as helium and argon, whose electron orbitals are full. These "noble gases" remain aloof from other elements and rarely form chemical compounds. On the far left are hydrogen, sodium, and other highly reactive elements. These elements are missing all but one electron in their outer orbital and are "eager" to bond with other atoms. In the middle, near the top, are carbon, nitrogen, and oxygen, all fairly reactive, but not explosively so. These three elements, together with hydrogen, form the basis of all organic molecules. See if you can find phosphorus and sulfur, two other important elements to life. *(silicon chip, Alfred Pasleka/Peter Arnold, Inc.; neon sign, Allan Kaye/DRK Photo; blimp, Gianni Tortoli, Photo Researchers; aluminum can, Paraskevas Photography; fireworks, AlaskaStock)*

32

An electron in a particular orbital has a characteristic **energy**—the capacity to perform work. Each orbital has a different amount of energy. When an electron moves from a high-energy orbital to a low-energy orbital, an atom may release energy, often in the form of light. Conversely, when an atom absorbs energy, one or more electrons may move from a low-energy orbital to a high-energy one. We can visualize the energy of electrons as steps on a staircase. A ball on a staircase tends to bounce down the stairs. Just so, electrons tend to fall into orbitals with the lowest energy.

Two rules determine the distribution of electrons in an atom: (1) electrons occupy orbitals with the lowest possible energy, and (2) each orbital may contain only two electrons.

The atoms that are most common in organisms contain fewer than 18 electrons. The electrons of each of these atoms occupy only the nine orbitals with the lowest energies. Chemists group electron orbitals into three **shells**—groups of orbitals whose electrons have nearly equal energy. The first shell has just one orbital containing two places for electrons. The second shell includes four orbitals, with room for eight electrons each. The third shell includes 9 orbitals, with room for 18 electrons.

The arrangement of electrons in the outermost shells is the basis for the Periodic Table (Figure 2-7). The elements in each column have the same number of electrons in their outermost shell. These outer electrons are those that interact with other atoms and determine the chemical properties of an atom.

Electrons occupy orbitals with the lowest possible energy, and each orbital may contain only two electrons.

WHAT HOLDS MOLECULES TOGETHER?

An atom is most stable when its outermost shell is full, meaning (usually) that it contains eight electrons. The atoms of some elements already contain a full outer shell. These elements, neon or argon, for example, are chemically unreactive. Most atoms, however, do not have full outer shells. Atoms can acquire enough electrons to make a full outer shell by sharing them with other atoms or by borrowing them. Such arrangements are the basis of chemical bonds. Bonds may be weak or strong. Different kinds of bonds play different roles in the chemistry of life.

Covalent and Ionic Bonds Are the Strong Interactions Among Atoms

Two kinds of bonds—covalent bonds and ionic bonds—hold atoms together especially strongly and are difficult to break. Such "strong" bonds hold together all the major molecules of life or biochemicals—from the carbohydrates in a loaf of bread to the DNA in your cells.

Two or more atoms that share one or more electrons are said to be held together by **covalent bonds**. Other atoms gain or lose electrons to form an **ion**—an atom or a molecule with a net electrical charge. Ions with opposite charges may be drawn together to form an **ionic bond.**

Atoms Share Electrons in Covalent Bonds

The molecules of life are mainly built from just six kinds of atoms: hydrogen (H), carbon (C), nitrogen (N), oxygen (O), phosphorus (P), and sulfur (S). One way to remember this list is to arrange the symbols for each element into the nonsense word "SPONCH." The outer shells of these six atoms determine their ability to form molecules. Carbon, for example, has only six electrons—two in the first shell and four in the second (Figure 2-8A). To fill its outer shell with eight electrons it must find other atoms with which to share an additional four electrons. These four shared electrons can form four covalent bonds, as in the case of methane (CH_4), the simplest organic compound (Figure 2-8B).

Two atoms may share more than one pair of electrons. For example, carbon may share two pairs of electrons with a single other carbon atom in a **double covalent bond.** The molecule ethylene (C_2H_4), important in the ripening of apples, bananas, and other fruits, has a double bond between its two carbon atoms. Some other molecules, including the gaseous nitrogen (N_2) in our atmosphere, contain **triple covalent bonds,** in which two carbon atoms share three pairs of electrons.

Most biochemicals are made from just six kinds of atoms, bound together by covalent bonds.

What Determines the Shape of a Molecule?

The covalent bonds within a molecule have fixed directions and lengths. For example, the distance between a carbon atom and a hydrogen atom is the same whether in a complex molecule or a simple one such as methane (Figure 2-8). Similarly, the angles between bonds are always the same.

Because bond lengths and angles are constant, every molecule has a definite size and shape. A water molecule, for example, always has the same boomerang shape. We can calculate the shape of a molecule by making a three-dimensional model, based on what we know about the positions of the atoms and the bond lengths and angles. Chemists and biologists represent the structures of molecules in a variety of ways—on a two-dimensional page, in several types of three-dimensional models, or on a computer screen (Figure 2-8).

Because bond lengths and angles are constant, every molecule has a definite size and shape.

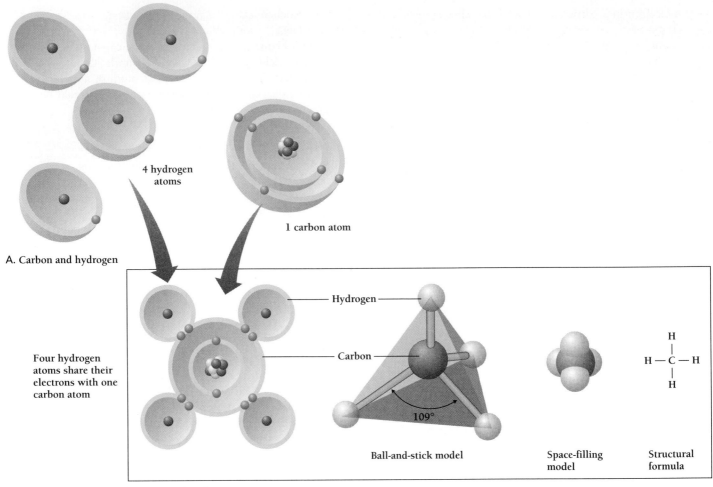

4 hydrogen atoms

1 carbon atom

A. Carbon and hydrogen

Four hydrogen atoms share their electrons with one carbon atom

Hydrogen

Carbon

109°

Ball-and-stick model

Space-filling model

Structural formula

B. Methane (CH₄)

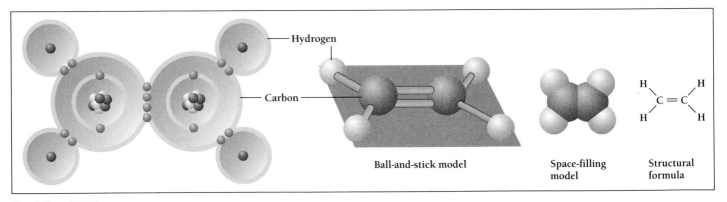

Hydrogen

Carbon

Ball-and-stick model

Space-filling model

Structural formula

C. Ethylene (C₂H₄)

Figure 2-8 Covalent bonding in hydrogen and carbon. A. Each hydrogen atom can share one electron and each carbon atom can share four. B. The four electrons in carbon's outer shell form covalent bonds with the hydrogen atoms to form a methane molecule. Molecules may be depicted in several ways. Simple structural formulas *(far right)* are easy to write, but they do not clearly show the angles at which the atoms connect to one another—the "bond angles." Space-filling models *(second from right)* suggest the arrangements of the atoms. But ball-and-stick models *(third from right)* show bond angles most clearly. In methane (CH₄), each bond points to one corner of a tetrahedron. The angle between any two bonds is 109.5°. In the structural and space-filling models, bonds drawn vertically are understood to point away from the viewer toward the back of the page; bonds drawn horizontally point out at the viewer. C. In ethylene, carbon atoms make double covalent bonds by sharing two pairs of electrons.

BOX 2-1

How are radioisotopes useful?

Although some radioactive isotopes are dangerous, many others are safe enough to use in medicine and scientific research. Isotopes can be used, for example, to determine the age of rocks and fossils; to track individual molecules in viruses, cells, or organisms; and even to identify the origins of products ranging from brand-name beverages to the explosive materials used in bombs.

Most of the isotopes found in nature are stable forms. Many of the isotopes of the heaviest elements, however, are unstable, or radioactive. Radioisotopes break down spontaneously into entirely different elements. As they break down, or decay, they release high-energy radiation.

The radiation from radioisotopes can be used to kill cancer cells. Radiation is now a standard therapy for many forms of cancer. Radioisotopes can also be used as tracers. For example, because the radioactive form of carbon (carbon 14) has the same properties as the regular form (carbon 12), organisms process both forms in the same way. Researchers can use sensitive instruments or photographic films to trace the movements of carbon 14 and answer many questions about molecular processes.

Does the carbon in the sugar manufactured by plants come from the carbon dioxide in air? By growing the plant in air containing radioactive carbon 14, researchers can tell how the carbon 14 was used. The applications of this technique are nearly limitless, and researchers all over the world daily trace different molecular processes using radioisotopes.

In a recent, especially inventive application of the technique, for example, researchers first marked bacteria with radioactive tracers, then introduced the marked bacteria into laboratory mice. As the bacteria multiplied and infected the mice, the researchers could visualize the path of the bacteria by using photographic film sensitive to the radiation. The mice could be treated with different antibiotics and the results again visualized with photographs taken from outside the body hour by hour.

Figure A   **Marie Curie and her two daughters, Irène and Eva.** The Polish-born French chemist Marie Curie did some of the earliest and best work on radioisotopes. Marie Curie and her husband Pierre shared the 1903 Nobel Prize for Chemistry with H. Becquerel for the discovery of radioactivity. After Pierre's death in 1906, Marie Curie assumed his professorship at the Sorbonne and went on to win a second Nobel Prize, in 1911, for the isolation of pure radium. Her subsequent work laid the groundwork for the discovery of the neutron and the synthesis of artificial radioactive elements. This latter work was carried out by Curie's daughter Irène Joliot-Curie and her husband Frèdèric Joliot. In 1934, Marie Curie died of leukemia at age 67, a result of her lifelong exposure to radiation. The following year, Irène and Frèdèric won the Nobel Prize for Chemistry. Irène Curie also died of leukemia, in 1956, at age 59, and her husband Frèdèric assumed her professorship at the University of Paris. *(Culver Pictures)*

Atoms Lose and Gain Electrons in Ionic Bonds

Atoms form other kinds of bonds different from covalent bonds. The most familiar are those in salts such as ordinary table salt—sodium chloride. The sodium atoms and the chlorine atoms in sodium chloride have stable electron configurations, with eight electrons in their outer shells (Figure 2-9). Instead of sharing electrons, however, the sodium atom gives away an electron and the chlorine atom takes one. The result is positively and negatively charged ions. The positively charged sodium ions and the negatively charged chloride ions (abbreviated Na^+ and

Cl^-) are held together by simple electrical attraction, in what is called an **ionic bond.** Electrostatic attraction is the same thing that makes your socks cling to your shirts when they come out of the dryer.

Electrical forces hold together ions of opposite charge, forming a complex of positive and negative ions rather than a true molecule. In a salt crystal, for example, each Na^+ ion is surrounded by Cl^- ions, and each Cl^- ion by Na^+ ions (Figure 2-9). If salt is dissolved in water, however, each ion is surrounded by water molecules, as we will discuss later.

Figure 2-9 Table salt. A. Ionic bonding between sodium and chloride ions in a crystal of sodium chloride. B. Reaction of sodium and chloride to form salt. C. Salt crystals. *(C, Bruce Iverson)*

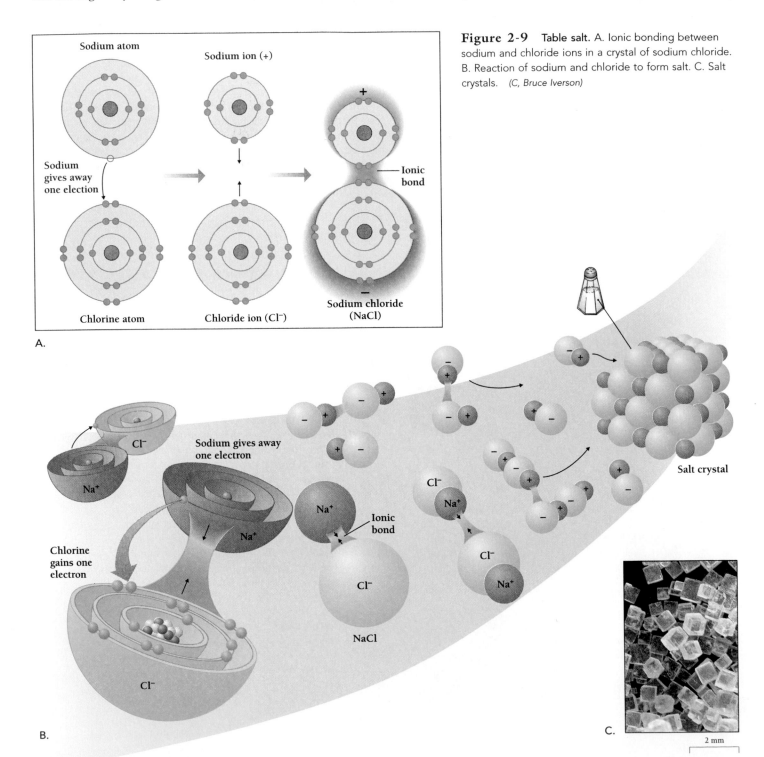

Sodium atom

Sodium ion (+)

Sodium gives away one electron

Ionic bond

Chlorine atom

Chloride ion (Cl⁻)

Sodium chloride (NaCl)

A.

Cl⁻

Na⁺

Sodium gives away one electron

Chlorine gains one electron

Na⁺

Na⁺

Ionic bond

Cl⁻

Cl⁻

Na⁺

Cl⁻

Na⁺

NaCl

Salt crystal

B.

C.

2 mm

An ionic bond is a simple electrical attraction between two ions.

Atoms Have Different Tendencies To Gain or Lose Electrons

A chlorine atom has a high tendency to gain an electron to form a chloride ion, while a sodium atom has a high tendency to lose an electron to form a sodium ion. A carbon atom, on the other hand, tends to share its outer electrons to form covalent bonds. The American chemist Linus Pauling realized that the atoms of each element have a characteristic **electronegativity**— the tendency to gain electrons. Oxygen and chlorine, for example, have a great tendency to gain electrons and are among the most electronegative elements. In contrast, sodium, potassium, and lithium are among the least electronegative elements—they have the greatest tendency to lose electrons. In between these are elements, such as carbon, that have no great tendency either to lose or to gain electrons. (We can think of electronegativity as a measure of electron greediness.)

Knowing the electronegativity of two atoms allows one to predict whether a bond between them will be covalent or ionic. For example, because chlorine has high electronegativity and sodium has low electronegativity, we can predict that they are likely to react together. The larger the difference in the electronegativities of two atoms, the more likely they are to form an ionic rather than a covalent bond. Sodium and chloride, for example, have a large difference in electronegativities and there-fore form ionic bonds. Carbon and nitrogen, on the other hand, have similar, moderate electronegativities and usually form covalent bonds.

Even in a covalent bond, however, atoms may not share electrons equally. When atoms differ in electronegativity, they do not share electrons equally. Instead, the diffuse clouds of shared electrons tilt toward the more highly charged atoms. In the water molecule, for example, oxygen shares electrons with two hydrogens. But the shared electrons are more concentrated around the oxygen nucleus than around the two hydrogen nuclei (Figure 2-10). As a result, the oxygen atom has a slight negative charge, and the two hydrogen atoms have a slight positive charge.

Molecules that have uneven distributions of electrical charge are said to be **polar**, since they have positive and negative poles in the same way that a magnet has two poles (Figure 2-10). When a polar molecule such as water comes close to an ion or to another polar molecule, its negative pole points towards the other molecule's positive pole, and its positive pole toward a nearby negative pole. Molecules with approximately uniform charge distributions are said to be **nonpolar.**

Atoms with great differences in electronegativity tend to form ionic bonds. Atoms with only slight differences in electronegativity tend to form covalent bonds. Molecules with uneven charge distributions are said to be polar.

Water molecule (H$_2$O)

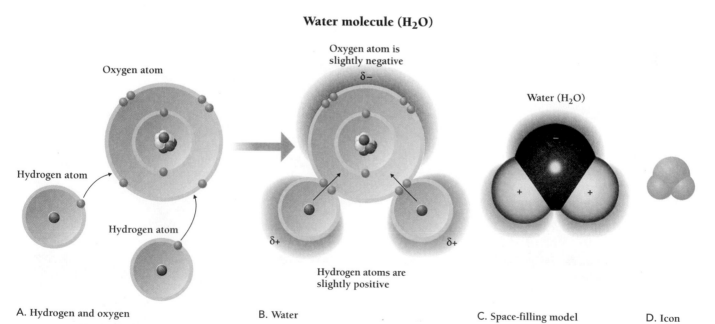

A. Hydrogen and oxygen   B. Water   C. Space-filling model   D. Icon

Figure 2-10   The electronegativity of water. A. Two atoms of hydrogen and one atom of oxygen share electrons in two covalent bonds to form a molecule of water. B. Because the oxygen atom is more electronegative than the hydrogen atoms, the electrons spend more time hovering around the oxygen end of the water molecule. As a result, the oxygen end has a slightly negative charge, while the hydrogens have a slight positive charge. Such a partial charge, less than one full electron, is symbolized by δ. C. Space-filling model of water molecule. D. Icon for water.   *(C, Tom Pantages/Phototake)*

Weak Interactions Also Hold Atoms Together

The forces that hold atoms together in covalent bonds are strong. Breaking them requires lots of energy. Electrostatic interactions in salt crystals (ionic bonds) are also strong. Most of the chemistry that concerns biologists, however, happens in or around water. In such an **aqueous** (watery) environment, a variety of weak interactions bind molecules together. Although individually each interaction is weak, in sum, weak interactions are responsible for many of the complex structures of organisms.

Hydrophobic Molecules Cling Together in an Aqueous Solution

When a bartender mixes an alcoholic drink such as brandy and soda water, the result is a smooth mixture, uniform throughout. Yet, when a cook mixes oil and vinegar to make salad dressing, the oil runs together in globs and floats to the top of the vinegar (Figure 2-11A). Why is this? The answer is that polar molecules, such as alcohol, mix well with water and are said to be **hydrophilic** [Greek, *hydro* = water + *phili* = love], or water-loving. But because water molecules attract and surround themselves with other polar molecules, nonpolar molecules

such as oil seem to be repelled by water. We say that nonpolar molecules are **hydrophobic** [Greek, *hydro* = water + *phobos* = fear], or water-fearing. In water, hydrophobic molecules are driven together by their mutual tendency to avoid water. The aggregation of nonpolar molecules is called a **hydrophobic interaction** (Figure 2-11B).

The properties of polar and nonpolar molecules play a key role in life. The tendency of nonpolar molecules to separate from water underlies the formation of the membranes that separate the inside of a cell from the outside (Chapter 4).

Oils and other hydrophobic molecules cling together when confronted with an aqueous environment.

Van der Waals Attractions Reinforce Hydrophobic Interactions

Because electrons move around rapidly, the charges on the different parts of a molecule change constantly. At a given instant, such fluctuations may temporarily produce positive and negative poles even in a nonpolar molecule. A temporarily polar molecule may, at the instant of polarity, induce a change in an adjacent molecule, so that it too becomes temporarily polar.

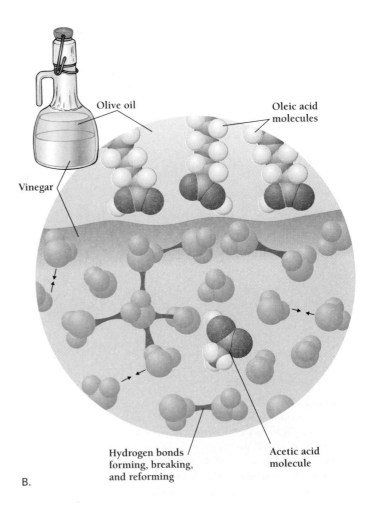

A.

Figure 2-11 **Fear of water.** A. Separation of oil and vinegar in a salad dressing. Vinegar, a dilute solution of acetic acid, is mostly water. B. Hydrophobic interactions. Water molecules attract and surround themselves with other polar molecules (including water) but repel nonpolar molecules such as oils. Nonpolar molecules are driven together by their mutual tendency to avoid water. *(A, Barry Runk/Grant Heilman Photography)*

Olive oil

Oleic acid molecules

Vinegar

Hydrogen bonds forming, breaking, and reforming

Acetic acid molecule

B.

For a moment, then, the two molecules are attracted to each other. Such **van der Waals interactions** operate only over very short distances and tend to reinforce the hydrophobic interactions among nonpolar molecules in water.

If two molecules are already very close, van der Waals attractions can pull them closer together.

Hydrogen Bonds Are Weak But Important

When a hydrogen atom attaches to a highly electronegative atom such as oxygen or nitrogen, the resulting covalent bond is polar. In this case, the hydrogen atom acquires a slight positive charge. Such a hydrogen can then participate in a **hydrogen bond**—a weak attraction to a negatively charged atom in another molecule (Figure 2-12). The most common hydrogen bonds are those between water molecules, but other hydrogen bonds also play a critical role in the structure of proteins and DNA.

The polarity (or nonpolarity) of molecules underlies their noncovalent interactions.

HOW IS WATER ESPECIALLY WELL-SUITED FOR ITS ROLE IN LIFE?

Life and water are intimately connected (Figure 2-13). As we will discuss in Chapter 18, life probably originated in water. Wherever life is found there is water, and wherever liquid water is found there is life. Water makes up more than 70 percent of the material of living organisms themselves and covers more than 75 percent of the Earth's surface. It is the medium

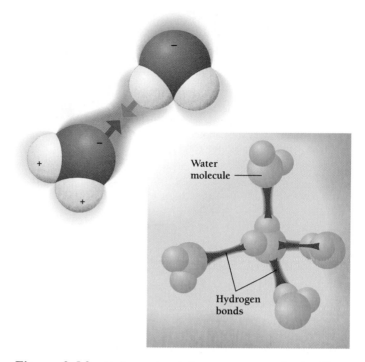

Figure 2-12 Hydrogen bonds between water molecules. The hydrogen atoms of one water molecule are attracted to the oxygen atom of another water molecule.

in which most cells are constantly bathed and the major component of cells themselves. Not only do most biochemical reactions occur in water, but water itself participates in many biochemical reactions. Although water is common—at least on Earth—its properties are highly unusual. Water's special attributes make it uniquely fit for its important role in life.

Water Is Denser as a Liquid Than as a Solid

Most solids sink into their corresponding liquids, but ice floats on water (Table 2-2A). Because ice floats, lakes and ponds freeze from the top down, instead of from the bottom up. The surface ice of a wintry pond, such as that in Figure 2-14 (p. 42),

Figure 2-13 Water is essential to every organism. Killer whales (*Orcinus orca*) spend their entire lives in the ocean. *(Francois Gohier/Photo Researchers)*

Table 2-2 How Does Water Compare with Ethanol and Oleic Acid?

	Water	Ethanol	Oleic acid
A. Density. In water, the liquid phase is more dense than the solid phase (ice), which is why ice floats. In most substances, including ethanol and oleic acid, the solid phase is denser than the liquid phase.			
B. Cohesion. In water, the attraction of the water molecules for one another creates a web of molecules that resists the sinking of the spoon. Neither ethanol nor oleic acid is as cohesive as water, and the spoon sinks.			
C. Adhesion. Water molecules' polarity enables them to form stronger interactions with the molecules in paper, so water rises faster than does ethanol or oleic acid.			
D. Ability to dissolve other substances. Water dissolves table sugar (sucrose) more quickly than does ethanol or oleic acid.			
E. If you were trapped on a desert island which of these would you choose?			

(A–D, Charles D. Winters; E: © 1994 Zefa Germany/The Stock Market; © 1997 Paraskevas Photography; © Camerique/The Picture Cube)

40

insulates the liquid water below the surface from the freezing air above. As a result, ponds, lakes, and rivers remain liquid at the bottom, allowing fish and other organisms to survive through long winters.

Ice has a lower density than liquid water because in ice the water molecules form a rigid lattice or crystal that holds them somewhat apart from one another: each water molecule is hydrogen-bonded to four other water molecules. In liquid water, the network of water molecules constantly changes, and there is less open space between the molecules.

Ice floats on water because ice has a lower density than water.

Water Absorbs More Heat Than Most Substances

In our daily lives, we know that temperature is a measure of how warm or cold something is. But to a scientist, temperature is a measure of the motion of molecules.

Molecules move constantly. In solids, molecules merely jiggle. In gases, molecules whiz around at mind-boggling speeds, bumping into other molecules. At room temperature, for example, the average speed of an oxygen molecule in the air is about 500 meters per second (more than 1000 miles per hour). The higher the temperature, the more rapidly molecules move about. Heat is energy that increases the random movement of molecules.

For most substances, any heat is converted into the more rapid motion of whole molecules, that is, into increased temperature. In water, however, much of the heat is converted into a more rapid exchange of hydrogen ions between adjacent molecules. As a result, for a given amount of heat, the temperature of water increases much less than would be expected for other substances. Water is said to have a high **heat capacity.**

One consequence of water's heat capacity is that the oceans heat up and cool down very slowly, moderating the climates of nearby land. In winter, the interiors of the world's continents disappear under a layer of snow, while temperatures in coastal areas remain relatively mild. Conversely, in summer, the continental interiors reach blazing temperatures, while coastal areas remain cool. Every large body of water has a similar influence on the land surrounding it.

Water not only cools the coasts of continents, it also cools the human body. To keep our body temperature within the narrow range that many biochemical reactions demand, we must rid ourselves of excess heat. We do so largely by perspiring. As the water in sweat evaporates, it carries the heat away. Dogs accomplish the same thing by panting (Figure 2-15). When a substance changes from a liquid to a gas it absorbs heat. As water evaporates, it absorbs a lot of heat—more than twice as much as the same amount of ethanol, for example. (Ethanol is also called ethyl alcohol or just plain "alcohol" when in wine, beer, or spirits.) Water absorbs a lot of heat when it turns into

water vapor because each of the hydrogen bonds that hold the water molecules together must be broken.

Water's enormous heat capacity stabilizes temperatures in oceans, lakes, coastal areas, and even within individuals.

Water Molecules Cling to One Another

An attraction between molecules of the same substance is called **cohesion.** Chewing gum and water are both highly cohesive. Under tension, they stretch but do not break. In water, cohesion also creates high **surface tension**—the tendency of a substance to form a smooth round surface. Because water molecules cohere, a water surface resists pressure. For example, as illustrated in Table 2-2B, the bowl of a teaspoon balanced on the edge of a teacup will rest on the surface of the tea as though floating. But a cup of alcohol or olive oil will not support the same spoon.

Water's cohesiveness is caused by hydrogen bonds among its molecules. The attraction among water molecules is so great that water has the highest surface tension of any liquid except for liquid mercury. The surface of water acts like a thin skin. Some insects, such as the water strider, take advantage of water's surface tension and walk easily over the surface of a pond.

Hydrogen bonding makes water molecules cohere, creating surface tension.

Water Molecules Cling to Many Other Substances

An attraction of one substance to another is called **adhesion.** Water clings to and "wets" surfaces such as glass that are composed of polar or charged molecules. Water's adhesiveness allows a close interaction between it and many other biological surfaces.

Because water both clings to itself (cohesion) and clings to other surfaces (adhesion), it readily enters and climbs tiny tubes, in a process called **capillary action.** If we place fine glass tubes (capillaries) into beakers of water, ethanol, and olive oil, the water moves upward a fair distance, the ethanol moves less, and the olive oil not at all. Such capillary action is important in the movement of water through soil, and up the narrow tubes inside of trees and other plants. Water's cohesiveness and adhesiveness help trees raise water from roots deep in the ground to leaves hundreds of feet in the air (Figure 2-16).

Water's adhesiveness results from the tendency of water molecules to form hydrogen bonds with other polar or charged molecules. Water molecules stick to glass capillaries because glass contains charges on its surface. Similarly, paper towels, which are made of long polar molecules called cellulose, absorb water very well. In contrast, olive oil, which is mostly non-

Water ⟶ **Ice lattice**

Ice layer

Air space

Algae

Figure 2-14 **Pond life under winter ice.** Because water is denser than ice, the ice floats and covers lakes and ponds in the winter. This layer of ice insulates the water below from freezing air temperatures. Fish, frogs, insects, copepods, protistans, and other life forms survive the winter, all because water is denser than ice.

polar, is relatively poorly absorbed by the cellulose in paper towels. That's why cleaning up water is so much easier than cleaning up olive oil (Table 2-2C).

Both a three-hundred-foot redwood tree and a child cleaning up spilled milk depend on water's adhesiveness.

Water Is a Powerful Solvent

Most people think of a solvent as the fluid that the dry cleaner uses to clean a wool coat, or maybe the gasoline you use to get grease off the wheels of your car. In fact, a **solvent** is any fluid in which other substances dissolve. Anyone who has poured salt into boiling water or stirred sugar into a cup of hot coffee will have noticed how rapidly salt and sugar dissolve in water. Water is a powerful solvent. It is an especially good solvent for ions, such as table salt, and polar molecules, such as sugar. Salt and sugar, which dissolve so well in water, dissolve poorly in ethanol and hardly at all in olive oil, which is nonpolar (Table 2-2D).

The substances that dissolve within a solvent such as water are called **solutes**. All salts are composed of ions, and all are extremely soluble in water. Sodium chloride (table salt), for example, breaks into sodium ions (Na^+) and chloride ions (Cl^-) when the highly polar water molecules surround the two halves of the salt molecule, like devoted fans crowding around a sports star (Figure 2-17). This shell of water molecules shields the positive and negative ions from each other, allowing the ions to remain farther away from each other than they can in

Figure 2-15 **Why are dogs' noses wet?** Water's enormous heat capacity makes it a useful coolant. As the water in a dog's nose evaporates, it carries heat away, cooling the inside of the nose. On a hot day, overheated blood from the body flows through the cold nose before reaching the brain, so that the dog's brain tissues remain cool. *(John Cancalosi/DRK Photo)*

Figure 2-16 Only because water is so cohesive and adhesive can trees raise it hundreds of feet off the ground. Individual California coast redwoods (*Sequoia sempervirens*) may grow to heights of 360 feet, taller than a 25-story building. *(Larry Ulrich/DRK Photo)*

their solid form (crystals), where they have a regular arrangement.

Sugars and other polar molecules also dissolve well in water, but for a slightly different reason. When water molecules surround a polar molecule such as sugar, hydrogen bonds form between the water molecules and the electronegative oxygen atoms of the sugar, isolating the molecules of sugar from one another.

The polar properties of water explain why some molecules are hydrophilic and others are hydrophobic. Hydrophilic molecules interact strongly with water molecules, while nonpolar, hydrophobic, molecules do not. Oil, a mixture of nonpolar molecules, does not readily mix with water.

Many biologically important molecules contain both hydrophilic and hydrophobic regions and are said to be **amphipathic** [Greek, *amphi* = both + *pathos* = feeling]. When an amphipathic molecule is placed in water, its hydrophobic regions avoid the water, and its hydrophilic regions bond with water (Figure 2-18). Mayonnaise, for example, consists of tiny oil droplets suspended in a watery solution.

All **detergents** are made of amphipathic molecules that interact both with water and with hydrophobic molecules such as oil. Detergents remove oil from dirty dishes and clothes by making the oil soluble in water.

Water's polar properties make it an excellent solvent for ions and polar molecules.

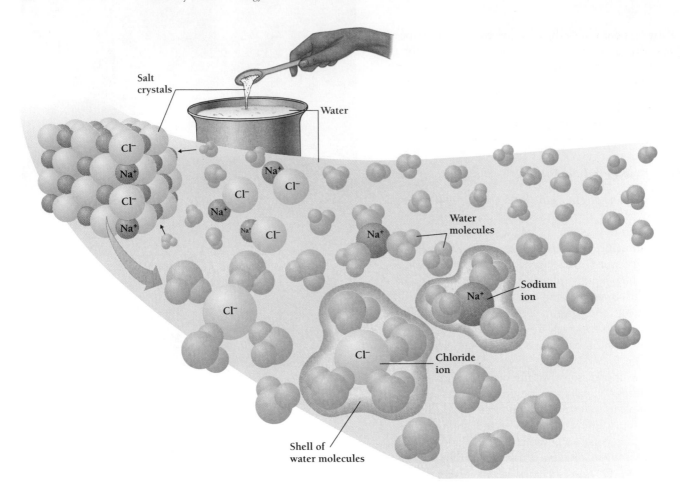

Figure 2-17   Dissolving salt. When sodium and chloride ions dissolve in water, water molecules surround each ion, forming "hydration shells," which prevent sodium ions from bonding with chloride ions.

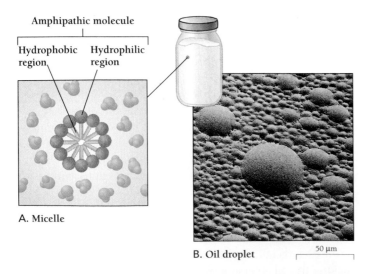

A. Micelle

B. Oil droplet   50 μm

Figure 2-18   Amphipathic molecules contain a hydrophobic tail and hydrophilic head. A ball of these molecules, called a micelle, forms when water molecules cluster around the hydrophilic heads and the hydrophobic tails hide together in the middle.   *(B, Dr. Jeremy Burgess/Science Photo Library/Photo Researchers, Inc.)*

Water Participates in Many Biochemical Reactions

Water is present wherever there is life, and many metabolic processes produce or use water. Besides serving as the medium in which most biochemical reactions occur, water itself participates in many reactions. For example, when we break down the large carbohydrate molecules in spaghetti into the little molecules of sugar that we need for fuel, we do so by adding water molecules. The negatively charged oxygen atom in water tends to approach positively charged atoms in other molecules. The oxygen may then form a covalent bond with the other molecule by simultaneously breaking its covalent bond with one of its hydrogen atoms.

Water is the most important polar molecule both within cells and in their external environments. Water molecules interact strongly with each other and with other charged and polar molecules. These interactions, and the lack of interactions with nonpolar molecules, are the most important factors in establishing biological structures.

Water Molecules Continually Split into Hydrogen Ions and Hydroxide Ions

Recall that the electrons that form the chemical bonds in a water molecule are closer to the oxygen atom than to the hydrogen atoms. Sometimes the electrons get so close to the oxygen atom that one of the hydrogen atoms has no electron at all and becomes a naked nucleus, or **hydrogen ion** (H^+). A hydrogen ion without an electron is just a proton, and it can jump back and forth between the oxygen atoms of two different water molecules (Figure 2-19). When that happens, one water molecule can end up with an extra proton and the other water molecule ends up missing a proton. The deprived water molecule is called a **hydroxide ion** (OH^-) and has a charge of -1. The water molecule with the extra hydrogen ion (H^+) is called a **hydronium ion** (H_3O^+) and has a charge of $+1$. We can represent this exchange of hydrogen ions by the following chemical equation:

$$H_2O + H_2O \longrightarrow OH^- + H_3O^+$$

water + water $\longrightarrow$ hydroxide ion + hydronium ion

Notice that each side of the arrow has equal numbers of hydrogen atoms and oxygen atoms, and that the charges are balanced. For the sake of brevity, we can leave one of the water molecules out of the equation. We can represent the shuttling of hydrogen ions as the splitting of water into a hydroxide ion and a naked hydrogen ion:

$$H_2O \longleftrightarrow H^+ + OH^-$$

water $\longleftrightarrow$ hydrogen ion + hydroxide ion

The opposite charges of hydrogen ions and hydroxide ions ensure that they do not get far. They quickly recombine with other hydroxide ions and hydrogen ions to form uncharged water molecules again. Water, then, constantly splits into hydrogen ions and hydroxide ions, and the ions constantly recombine to form water.

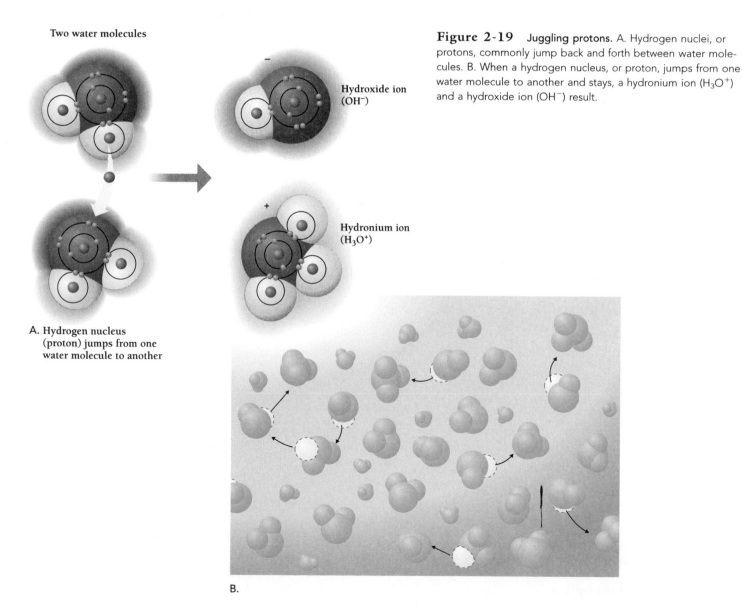

Two water molecules

Hydroxide ion (OH⁻)

Hydronium ion (H₃O⁺)

A. Hydrogen nucleus (proton) jumps from one water molecule to another

B.

Figure 2-19 **Juggling protons.** A. Hydrogen nuclei, or protons, commonly jump back and forth between water molecules. B. When a hydrogen nucleus, or proton, jumps from one water molecule to another and stays, a hydronium ion (H_3O^+) and a hydroxide ion (OH^-) result.

When the splitting of water into hydrogen ions and hydroxide ions exactly balances the recombining of hydrogen ions and hydroxide ions back into water, the number of ions at any instant is always the same, and we say that the solution is "at equilibrium." Such a balance between forward and reverse reactions is called **chemical equilibrium** [Latin, *aequus* = equal + *libra* = balance].

In pure water at equilibrium, the number of hydrogen ions exactly equals the number of hydroxide ions. However, in most solutions the number of hydrogen ions differs from the number of hydroxide ions. Vinegar has a million times more hydrogen ions than hydroxide ions. Stomach acid, as we saw in Chapter 1, is even stronger. It contains about one trillion times more hydrogen ions than hydroxide ions.

The **concentration** of hydrogen ions—the number of hydrogen ions in a given amount of liquid—varies so widely in different solutions that chemists have defined a scale, called **pH**, to measure the concentration of hydrogen ions in a solution (Figure 2-20). The pH scale is based on the logarithm (to the base 10) of the concentration of hydrogen ion. We can think of pH as standing for "power of hydrogen." But the logarithm is actually a negative number. So a pH of 7, for example, represents a concentration of hydrogen ions equal to 10^{-7} moles

per liter (a *mole* is a measure of numbers of molecules, so moles per liter is a unit of concentration). The middle of the pH scale is 7, which represents a neutral solution such as water, in which H^+ and OH^- concentrations are the same. Acid solutions have a lower pH. For example, the pH of vinegar is 4 (10^{-4} moles per liter). The pH of the hydrochloric acid secreted by the stomach is less than 1 (10^{-1} moles per liter), or 1/10 mole per liter—very concentrated. In short, the greater the concentration of hydrogen ions, the lower the pH value. A tenfold increase in concentration means a decrease of one pH unit.

A molecule that can give up a hydrogen ion is called an **acid.** A molecule that can accept a hydrogen ion from an acid is called a **base.** An acid dissolved in water contributes hydrogen ions to water molecules and raises the concentration of hydrogen ions, which decreases the pH. Conversely, bases dissolved in water result in a decreased concentration of hydrogen ions and increased pH. Bases, then, have pH values higher than 7.

Water molecules split into hydrogen ions and hydroxide ions. The exact balance between the two kinds of ions determines the pH (the relative acidity) of the solution.

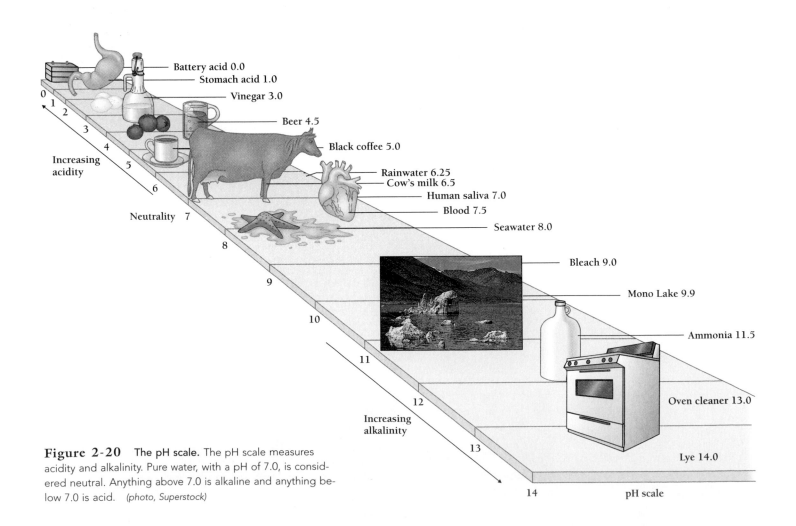

Figure 2-20   The pH scale. The pH scale measures acidity and alkalinity. Pure water, with a pH of 7.0, is considered neutral. Anything above 7.0 is alkaline and anything below 7.0 is acid. *(photo, Superstock)*

Battery acid 0.0
Stomach acid 1.0
Vinegar 3.0
Beer 4.5
Black coffee 5.0
Rainwater 6.25
Cow's milk 6.5
Human saliva 7.0
Blood 7.5
Seawater 8.0
Bleach 9.0
Mono Lake 9.9
Ammonia 11.5
Oven cleaner 13.0
Lye 14.0

Increasing acidity

Neutrality   7

Increasing alkalinity

pH scale

BOX 2-2

Life at low pH

Organisms that live in water usually require a pH near 7. For thousands of years, people have killed unwanted organisms in their food by pickling—by soaking meats or vegetables in vinegar, a weak acid. Some microorganisms, however, thrive in acidic environments. As lactobacilli bacteria grow in milk, they reduce the pH of their environment. This acidity not only prevents the growth of most other organisms, it also induces the coagulation of the milk into buttermilk, sour cream, cheese, or yogurt (depending on the species of lactobacilli). The characteristic flavors of these milk products all develop from the action of the bacteria.

Other microorganisms can tolerate still lower pH. Bacteria growing at pH below 2 have been isolated from the acid hot springs of Yellowstone National Park, from the refuse piles of coal mines, and from rivers and ponds polluted with industrial wastes. These bacteria appear to have special adaptations that maintain their internal pH near 7 in spite of their acidic external environment.

The organisms that can withstand extremes of pH are mostly small. This means that when mining or industrial pollution reduces the pH of a lake or river, only microorganisms survive. The fish, the frogs, and the multicellular plants begin to die off.

Less extreme acidity also affects organisms. When soil water becomes acidic, plants and animals are less able to survive. The land areas surrounding acid-polluted waters—like the waters themselves—have reduced vegetation and animal life.

As illustrated in Figure A, acidity in soils, lakes, and rivers has become increas-

Figure A Acid rain. Trees along the Blue Ridge Parkway in North Carolina are dying as a result of acid rain. *(Clyde H. Smith/Peter Arnold, Inc.)*

ingly widespread in the last 50 years. The burning of sulfur-containing coal and of nitrogen-containing gasoline releases sulfur and nitrogen oxides into the atmosphere. Winds can carry these oxides hundreds or even thousands of miles, where they can dissolve in rain water and produce strong acids—sulfuric, sulfurous, nitric, and nitrous. The resulting "acid rain" reduces the pH of lakes, ponds, rivers, and soils.

Pollution by acid rain is both a national and an international problem. The industrial wastes of England and Germany have changed the pH of the lakes and rivers of Sweden, and the factories of the American midwest have a direct effect on the wildlife and agriculture of the northeastern United States and Canada.

Why Is pH Important to Organisms?

Changing the pH of a solution affects the properties of other molecules. For example, the characteristic sour taste of acids results from the interaction of hydrogen ions with molecules on the tongue that report taste to the brain. Similarly, basic solutions feel slippery because they change the characteristics of molecules on the skin.

When a molecule donates or accepts a hydrogen ion, its net charge changes. Such changes alter a molecule's interactions with other molecules and ions. As a result, pH dramatically af-fects both the structure and the chemical reactivity of most biologically important molecules. In animals, the maintenance of a blood pH close to neutral is crucial for life.

In humans, the pH of the blood is normally 7.4. If the pH of the blood decreases until it is even slightly acidic—to pH 6.95—the nervous system becomes unresponsive, and coma and death soon follow. Alternatively, if the blood pH increases to as little as pH 7.7, the nervous system becomes overreactive, and muscle spasms and convulsions begin. Normally, such disasters are prevented by a number of physiological adaptations, which include the regulation of the heart rate, changes in the rate and depth of breathing, and the secretion of hydrogen ions by the kidneys.

However, a wide variety of illnesses, injuries, and drugs can cause acid or alkaline blood. One of the most common causes of acid blood, for example, is severe diarrhea. Basic blood can result when a physician gives a patient excessive amounts of diuretics (drugs that cause the body to eliminate large amounts of water and hydrogen ions) or if a person takes too much bicarbonate of soda for an upset stomach.

Changes in pH do not have to be dangerous. Many organisms use pH to control normal processes. For example, a sperm in the testes must remain motionless in order to save energy for the time when it is near an egg. Yet something must activate the sperm when the moment is right. In sea urchins, that something is a decrease in pH. The pH of the sea urchin's testes and sperm is about 7.2, very slightly basic. When the sea urchin releases its sperm into seawater, which has a pH of about 8.0, the sperm become more basic (about 7.6), which activates the sperm. Off they swim.

Both inside and outside of cells, pH powerfully influences the chemistry of life.

Buffers: How Do Organisms Resist Changes in pH?

Because pH dramatically affects both the structure and the chemical reactivity of most biologically important molecules, cells have mechanisms that maintain nearly constant pH. The interior contents of cells almost always have a pH of about 7.

Organisms use both chemical and physiological devices to maintain constant pH. The chemical strategy is to employ **buffers**—molecules that easily interconvert between acidic and basic forms by donating or accepting hydrogen ions. In human blood, the main buffer is carbonic acid and bicarbonate.

Many chemical reactions in organisms involve exchanges of hydrogen ions. For example, when a person exercises, the muscles produce carbon dioxide (CO_2), which the blood carries to the lungs. Most of the carbon dioxide in the blood, however, combines with water to form carbonic acid (H_2CO_3),

which tends to break down into a hydrogen ion and a bicarbonate ion (HCO_3^-):

$$CO_2 + H_2O \longrightarrow H_2CO_3 \longrightarrow H^+ + HCO_3^-$$

carbon dioxide + water $\longrightarrow$ carbonic acid $\longrightarrow$ hydrogen ion + bicarbonate ion

When the bicarbonate ions reach the lungs, they reacquire a hydrogen ion and reform carbon dioxide and water:

$$H^+ + HCO_3^- \longrightarrow H_2CO_3 \longrightarrow CO_2 + H_2O$$

hydrogen ion + bicarbonate ion $\longrightarrow$ carbonic acid $\longrightarrow$ carbon dioxide + water

The equilibrium between carbonic acid and bicarbonate helps maintain a constant pH. When the digestion of food produces extra hydrogen ions, bicarbonate ions can absorb them by becoming carbonic acid, and the pH of the blood changes little. Similarly, hydrogen ions provided by carbonic acid can "soak up" hydroxide ions added to the blood. In both cases, the presence of the buffer minimizes changes in the total concentration of hydrogen ions. Buffers such as carbonic acid/bicarbonate reduce, but do not eliminate, the pH changes of organisms and cells.

To maintain the constant internal environment necessary for life, cells and organisms rely on physiological adaptations. For example, as mentioned earlier, blood pH is maintained within narrow limits by the regulation of the heart rate, changes in the rate and depth of breathing, and the secretion of hydrogen ions by the kidneys.

Organisms resist changes in pH with chemical buffers and other physiological mechanisms.

The biological functions of molecules are determined by their structures. Since water is the universal medium of life, the attraction or avoidance of water by molecules or parts of molecules is the basis of the organization of cells and organisms. The interactions of molecules with water and with each other, in turn, depend on the properties and the arrangements of the atoms that compose them.

STUDY OUTLINE WITH KEY TERMS

All matter is composed of **atoms.** An atom is the smallest particle of an element that has the properties of that element. **Elements** are forms of matter that cannot be broken down to simpler substances. **Compounds** can be broken down into elements. Compounds consist of **molecules,** which are arrangements of two or more atoms. The atoms of each element have characteristic properties that determine what kinds of molecules can be formed.

Atoms are made of still smaller particles. Almost all of an atom's **mass,** or weight, is contained in its **nucleus.** Typical atoms are elec-

trically neutral; the positive charge of the **protons** in the nucleus is balanced by an equal number of negatively charged **electrons.** Electrons are outside the nucleus and occupy most of the space of the atom. The number of protons in the nucleus of an atom is called the **atomic number.** The **mass number** of an atom is the total number of protons and **neutrons** in the atom's nucleus.

The atoms of an element all have the same atomic number, but they may have different mass numbers, that is, different numbers of neutrons. The different kinds of atoms of a single element are called

isotopes. Some isotopes spontaneously disintegrate, giving off radioactive particles that can be detected photographically or electronically.

Electrons are arranged in **orbitals,** each of which can hold a different number of electrons. Groups of orbitals, in turn, form **shells,** which differ in energy. The chemical properties of an atom are largely determined by the number of electrons in the outermost shell.

In forming chemical bonds, electrons may be rearranged in two distinct ways: (1) by sharing with other atoms to form **covalent bonds,** or (2) by donating or accepting electrons to form **ionic bonds.** Individual atoms have different tendencies to gain or to lose electrons. Atoms that are most apt to gain electrons are said to be the most **electronegative.** Differences in the electronegativity of atoms in a molecule determine the extent to which the electrons in a chemical bond are equally distributed. Molecules with uneven distributions of electrons are said to be **polar;** those with uniform charge distributions are said to be **nonpolar.** Individual molecules may interact via several types of relatively weak forces—**hydrogen bonds, hydrophobic interactions,** and **van der Waals interactions.**

Water is the most abundant molecule in all organisms, and its unusual properties are especially suited for its role in life. Among these properties are its **cohesiveness,** its higher density as a liquid than as a solid, its high **heat capacity,** its **adhesiveness,** its ability to serve as a powerful **solvent,** its ability to absorb heat, and its participation in many biochemical reactions.

The properties of water result from its polarity. The polarity of water permits interaction with **ions** and polar molecules. These **hydrophilic** substances readily dissolve in water. Nonpolar molecules do not dissolve readily in water, and they are said to be **hydrophobic.** Many biologically important molecules are **amphipathic,** containing both hydrophilic and hydrophobic regions.

The hydrogen nuclei that participate in hydrogen bonds can also jump between water molecules. This ability leads to water's high heat capacity.

A molecule that can donate a hydrogen ion to another molecule is called an **acid.** A molecule that can accept a hydrogen ion from an acid is called a **base.** Some of the time, the jumping hydrogen ions create **hydroxide ions** and **hydronium ions.** Water is thus both an acid and a base. The **pH scale** measures the concentration of H^+ and OH^- ions in a solution. Changes in pH alter the charge on molecules and affect their chemical and physical properties. The fluids of organisms contain **buffers,** which resist changes in pH. Cells and organisms also have physiological adaptations that maintain nearly constant pH.

REVIEW AND THOUGHT QUESTIONS

Review Questions

1. Define the words element, atom, compound, and molecule.
2. What is an isotope? What is meant by the term "radioactive isotope"? Why are only some isotopes radioactive?
3. What is an electron orbital?
4. Explain the difference between a covalent bond and an ionic bond. How is each formed? Name a common compound that exhibits each type of bond.
5. Explain how covalent bonds can be polar or nonpolar. Give examples of each type of bond. Name a polar molecule that is of great importance to organisms and their environment.
6. Describe each of the following: hydrogen bonds, hydrophobic interactions, and van der Waals interactions. Which of these reactions are likely to occur between (1) polar compounds or (2) nonpolar compounds?
7. Define cohesion and surface tension. What is the relationship between these two properties of water?
8. How does adhesion differ from cohesion? What is a capillary?
9. Give examples of polar and ionic compounds that dissolve easily in water as well as nonpolar ones that do not. Which type of molecule would you use to make waterproof bags?

10. What type of bond is formed between adjacent water molecules? Why do they cohere? Is this more similar to a covalent or to an ionic bond?
11. What types of bonds tend to produce (1) hydrophilic substances or (2) hydrophobic ones?
12. What is a hydrogen ion? If a hydrogen ion is formed from a water molecule, what other ion is also formed? If a hydrogen ion is formed from HCl, what other ion is also formed? How is the pH of water defined?

Thought Questions

13. Explain how the results of Rutherford's experiment caused him to propose that the nucleus of an atom was very small and dense, and that most of the atom was nearly empty space.
14. The tendency of living things to remain in a narrow range of temperatures is part of homeostasis—one of the characteristics of living things. How do the properties of water help organisms regulate temperature? Consider organisms that live in water, then organisms such as humans that live outside water.
15. Explain why acid rain is so damaging to ecosystems. Would basic rain also be damaging? Why?

SELECTED READINGS

Henderson, L.J., *The Fitness of the Environment,* Macmillan, New York, 1913 (republished 1970, P. Smith, Gloucester, MA). A classic discussion of the properties of water and the significance for life.

Stryer, L., *Biochemistry,* 4th ed., W.H. Freeman, New York, 1995. A beautifully written text that is advanced but accessible to interested readers.

▶On-line materials relating to this chapter are on the World Wide Web at http://www.saunderscollege.com/lifesci/
Click on Tobin/Dusheck: *Asking About Life.*

Linus Pauling and the Alpha Helix

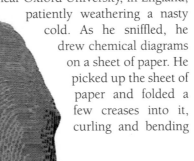

One day in April of 1948, the American physical chemist Linus Pauling lay in a bed near Oxford University, in England, patiently weathering a nasty cold. As he sniffled, he drew chemical diagrams on a sheet of paper. He picked up the sheet of paper and folded a few creases into it, curling and bending the paper this way and that. Then he tried refolding the creases so that they bent at different angles. Suddenly, he had it.

For 11 years Pauling had worked off and on trying to find a general structure for proteins, and now he was almost certain he knew the answer (Figure 3-1). He was not certain enough to tell anyone, but he was certain enough to return to his laboratory at Caltech, in Pasadena, California, and perfect the idea. It would be more than two and a half years before he made his revolutionary discovery public.

The story behind Pauling's discovery began at the beginning of the 20th century, when the German physicist Max von Laue invented a technique for studying the internal structure of crystals. Crystals are solids whose regular geometric shapes reflect the regular arrangement of their atoms. Ice, sugar, salt, and glass, for example, all form crystals. When light passes through a crystal, the light bends and spreads out into a rainbow. This is the same effect that produces the rainbow of colors in a polished diamond.

X rays also bend as they pass through crystals. In 1912, von Laue was able to show that a crystal can scatter—or diffract—x rays in such a way that they form a pattern of spots that can be recorded on photographic film. Von Laue found that the pattern of spots reflected the arrangement of the atoms and molecules inside

Legoland, Denmark

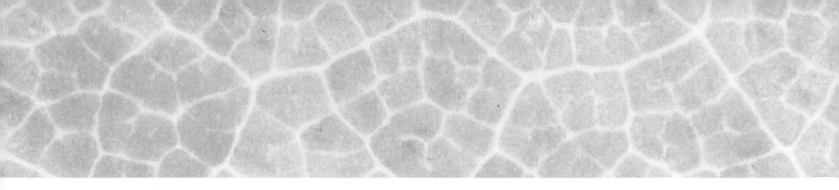

the crystal. His new technique, called **x-ray crystallography**—or x-ray diffraction—consisted of analyzing the scattering or "diffraction" of x rays by crystals.

Soon after von Laue's discovery, two British physicists, William Henry Bragg and his son Lawrence Bragg (Figure 3-2), showed that analysis of x-ray diffraction patterns could reveal the precise atomic structure of the molecules in a crystal. Because each atom in a crystal contributes to the pattern of spots, crystallographers can calculate the structure of the individual molecules of the crystal from the intensities of these spots (Figure 3-3). The Braggs' discovery allowed the determination of all the molecular structures that we will discuss in the rest of this chapter.

In 1927, young Linus Pauling went to visit the Braggs' laboratory in England and came away with enough of an understanding of their new technique to use it himself. He formalized a set of six rules describing how relations among atoms determine their position in a molecule. For example, he described how sizes and electrical charges of atoms determine their arrangement in a molecule. Pauling's rules influenced generations of chemists.

In 1934, the English chemist J. D. Bernal showed that not only small molecules could be studied with x-ray crystallography, but also proteins and other giant molecules. Bernal's 1934 discovery opened the door to the possibility of untangling the structures of proteins. Proteins were known to be composed of smaller units called amino acids. But no one knew how the amino acids were arranged within the protein. Some scientists envisioned proteins as long chains of amino acids. Others thought they were more like cages of linked amino acid rings.

In 1937, Pauling devised a spiraling chain of amino acids that conformed to his notion of what a protein molecule might

Figure 3-1 Linus Pauling (1901–1994). Pauling made major contributions to chemistry, which he summarized in his influential book, *The Nature of the Chemical Bond*. He was largely responsible for the concept of electronegativity, and he won two Nobel Prizes: one for his scientific work, and one for his efforts on behalf of world peace, which led to the 1964 treaty outlawing atmospheric tests of nuclear weapons. *(Thomas Hollyman/Photo Researchers)*

Figure 3-2 William Henry Bragg (1862–1942) and Lawrence Bragg (1890–1971). Father and son shared the Nobel Prize in 1915. The younger Bragg (*inset*) left the front during the First World War to accept the prize. Lawrence became the director of the Cavendish Laboratory in Cambridge, where scientists later used x-ray diffraction to determine the first three-dimensional structures of proteins. *(AP/Wide World Photos; UPI/Corbis-Bettmann)*

look like. However, his model conflicted with results from x-ray crystallography studies. To try to come up with a better model, Pauling and his colleague Robert Corey began a 12-year study of the structures of the individual amino acids. By 1948, the two men knew more about the structures of amino acids than anyone else in the world, and they could accurately predict what kinds of bonds each amino acid would form with any other—including the exact lengths and angles of each of those bonds. Pauling had every reason to be confident that he and Corey would eventually determine the structure of protein.

So when, in 1948, Pauling found himself bored and sick in bed at Oxford, he couldn't help thinking about the problem. With his now incomparable knowledge of the chemistry of amino acids, he soon realized that the dimensions of the spiraling chain he had conceived in 1937 were, in fact, probably correct. (Research published in 1950 showed what Pauling had long suspected: The x-ray crystallography study that had conflicted with his original ideas about proteins had been wrong. The photos were accurate enough, but the researcher had misunderstood what they meant.) In 1948, Pauling could see that if he adjusted the steepness of the spiral just so, hydrogen bonds connected the turns, holding the spiral in place.

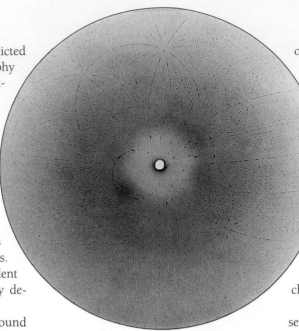

Figure 3-3 X-ray diffraction pattern from a crystal of myoglobin. *(Joel Berendzen/Los Alamos National Lab, NM)*

an article that, for the rest of their lives, they would regret publishing.

Back at Caltech, Pauling and Corey had begun building large and detailed models of various proteins—including keratin, the protein found in hair, hooves, and horn. The two scientists adhered strictly to the rules that they had learned over the years, confirming that Pauling's spiral model was possible and likely. In the fall of 1950, the two scientists announced their discovery in a short letter to a respected journal of chemistry.

He picked up the sheet of paper and folded a few creases into it, curling and bending the paper this way and that.

Meanwhile, Lawrence Bragg and several other researchers at Cambridge University had also become fascinated by the structure of proteins. They threw themselves at the problem, and, in the spring of 1950, published a scientific paper that suggested that the amino acids in a protein called alpha keratin were arranged in a ribbonlike chain that folded back and forth, like a chain of paper dolls. It was

Soon afterward, Pauling gave a memorable lecture describing the results of the 15 years of work. Scientists jammed into the lecture hall to find out what Pauling had discovered, and the charismatic chemist entertained his audience with pointed references to the Cambridge group's mistake. Bragg and his colleagues had failed because, as Pauling later put it, they were ignorant

of the "principles of chemistry, of structural chemistry."

The following spring, Pauling and Corey published a long series of papers describing the structures of several important proteins, including the blood protein hemoglobin and various proteins found in hair, feathers, muscle, silk, and gelatin. They described in detail several structural forms, including the spiral amino acid chain envisioned by the bedridden Pauling. Pauling called the spiral structure an α (alpha) helix (Figure 3-4). The biochemical community was bedazzled.

Bragg was furious, mainly with himself. In his rush to publish something about the structure of proteins, he had gone completely astray. His colleague Max Perutz, whose name was also on the notorious Cambridge paper, was so beside himself that he immediately thought of a way to test Pauling's idea. If keratin really had an α-helix structure, Perutz reasoned, then x-ray crystallography should show a spot where no one had ever reported one before (Figure 3-5).

Perutz knew that there were only two possible reasons that no one had seen the spot. Either the spot didn't exist, and Pauling and Corey were wrong, or else no one had set up a keratin molecule at the right angle. It was a Saturday, but only an hour or two after he had finished reading Pauling and Corey's series of papers, he went to his lab. He found a horsehair—a rich source of keratin—set the hair at the angle he had calculated was necessary to show the spot, then took a single x-ray diffraction photo. There was the spot.

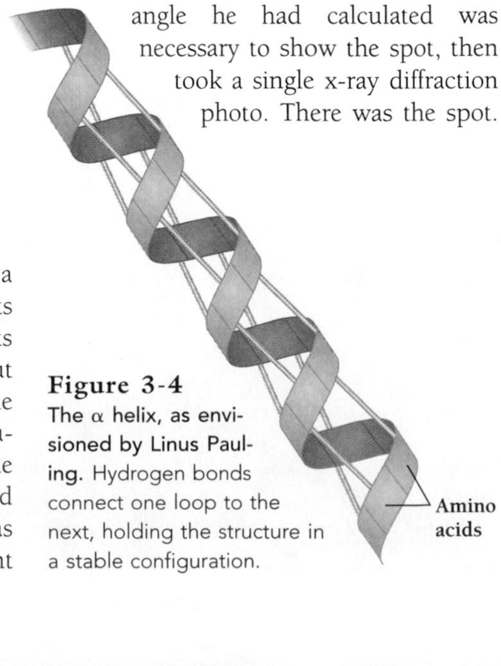

Figure 3-4
The α helix, as envisioned by Linus Pauling. Hydrogen bonds connect one loop to the next, holding the structure in a stable configuration.

Amino acids

Perutz's spot was the first independent experimental confirmation of Pauling's hypothesis. Perutz later showed that the α helix could be found in a variety of proteins, just as Pauling had predicted. In 1954, Pauling received a Nobel prize for his work with Corey on the structures of large molecules. Eight years later, in 1962, Perutz also received a Nobel prize for his contributions.

Pauling and Corey's work was some of the finest work on the structure of large molecules that has ever been done. In this chapter we will explore a few of the chemical principles that determine how biological molecules, both small and large, fit together. Many of these are the same principles that made Pauling's work possible.

Figure 3-5 **Three Nobel Prize winners.** Max Perutz *(center)* in a Cambridge laboratory in 1962, with Francis Crick *(left)* and John Kendrew *(right)*. All three scientists used x-ray crystallography to study the structures of proteins. Crick received his Nobel Prize for his work on the structure of DNA, a polynucleotide. *(Express Newsphotos/Archive Photos)*

KEY CONCEPTS

1. Organisms build almost all their structures from four types of building blocks: lipids, sugars, amino acids, and nucleotides.

2. Each type of building block has characteristic functional groups that give it its chemical properties.

3. Macromolecules are formed from their component building blocks by condensation reactions that remove the equivalent of one molecule of water.

4. Each kind of protein molecule has a distinct sequence and characteristic properties.

5. The three-dimensional structure of a protein depends on interactions among its amino acids.

HOW DO ORGANISMS USE BIOLOGICAL MOLECULES TO BUILD?

If we analyze the sizes of all the molecules in a cell, we discover a surprising thing. Cells have many small molecules, with molecular weights less than 300, and many large molecules, with molecular weights greater than 10,000. But cells have very few molecules with intermediate sizes.

How Big Are Biological Molecules?

Biologists call large molecules **macromolecules** [Greek, *macro* = large]. Macromolecules consist of small molecules linked together in long chains called **polymers** [Greek, *poly* = many + *meros* = part]. Each of the component parts of a poly-

mer is called a **monomer** [Greek, *mono* = single]. In this chapter we will discuss both small molecules and polymers.

Nearly all molecules found in living organisms share one important trait. They are made of chains of carbon atoms. In fact, living organisms are associated so closely with carbon that all carbon-containing compounds are called **organic** molecules (even when they have nothing to do with organisms). **Organic chemistry** is the study of the structures and reactions of carbon compounds. **Biochemistry,** or biological chemistry, is the study of the chemistry of living organisms.

The carbon-based building blocks of life are both simple and universal. Even though there are a nearly infinite variety of carbon-containing molecules, organisms use only a few basic building blocks. In this chapter we will discuss the proportions of 35 small molecules that together form almost all biological structures (Table 3-1).

Table 3-1   The Building Blocks of Life

Living organisms are largely built from just 35 small molecules. These include 20 amino acids, 6 lipids or phospholipids, glucose, and the 8 molecules that make up the nucleic acids DNA and RNA (ribose, deoxyribose, phosphate, and 5 bases).

Amino acid

Amino acids with basic side chains

Lysine (Lys, K) Arginine (Arg, R) Histidine (His, H)

Amino acids with acidic side chains

Aspartic acid (Asp, D) Glutamic acid (Glu, E)

Amino acids with uncharged polar side chains

Asparagine (Asn, N) Glutamine (Gln, Q) Serine (Ser, S) Threonine (Thr, T) Tyrosine (Tyr, Y)

Amino acids with nonpolar side chains

Glycine (Gly, G) Alanine (Ala, A) Valine (Val, V) Leucine (Leu, L) Isoleucine (Ile, I)

Phenylalanine (Phe, F) Methionine (Met, M) Cysteine (Cys, C) Proline (Pro, P) Tryptophan (Trp, W)

Glucose

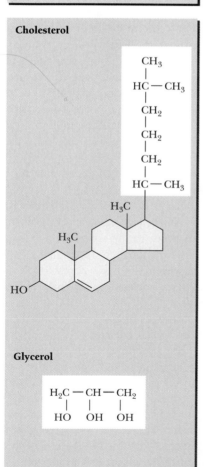

Ribose

Deoxyribose

Cholesterol

Glycerol

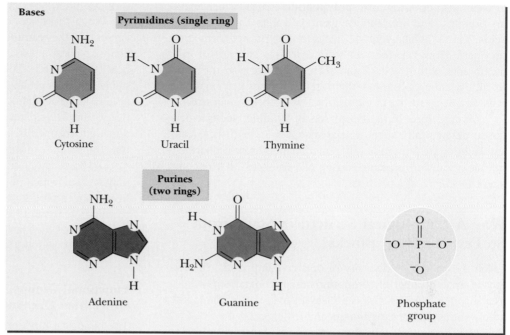

Bases

Pyrimidines (single ring)

Cytosine   Uracil   Thymine

Purines (two rings)

Adenine   Guanine   Phosphate group

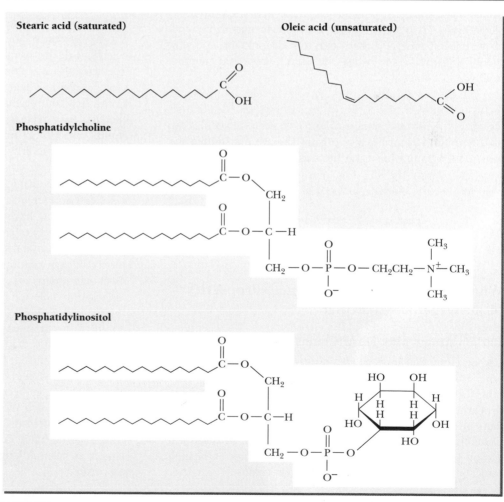

Stearic acid (saturated)

Oleic acid (unsaturated)

Phosphatidylcholine

Phosphatidylinositol

55

These small molecules fall into four classes: sugars, amino acids, nucleotides, and lipids. Sugars form polymers called **polysaccharides,** such as the starch in bread or the cellulose in paper. Amino acids form **polypeptide** chains, which make up proteins, and nucleotides form polymers called **nucleic acids.** Lipids, which include fats and oils, do not form long polymers. However, they do form extensive galleries of thin membranes within cells. Lipid membranes are also the basis for all plasma membranes—the thin membranes that surround cells and regulate their interactions with their environment.

We may think of these four sets of building blocks as letters in different alphabets. Just as sentences in English, Greek, and Hebrew are written in different alphabets, polysaccharides, nucleic acids, and proteins consist of "sentences" that use different biochemical alphabets.

Why Are Biological Structures Made from So Few Building Blocks?

Given the enormous diversity of organic compounds, the relatively small number of compounds used by organisms comes as a surprise. The surprise is a pleasant one, both to biologists who want to design experiments and to students who must familiarize themselves with the parts of life. But it should also make us wonder: Just why should there be so few building blocks? Why, for example, should every single organism on this planet use the same 20 amino acids to make proteins, and why should so many organisms use the sugar glucose as their principal energy source?

The extremely limited number of molecules that form all the wonderfully diverse forms of life supports the idea that all life on Earth had a common origin. The same molecules are used again and again because organisms have transmitted the genetic information for using the same molecules to their descendants. Thirty-five building blocks are enough to accomplish most of the essential tasks of life.

All living organisms are made of the same few small molecules.

What Determines the Biological Properties of an Organic Molecule?

Carbon Atoms May Form Chains of Any Length

Carbon atoms can form single covalent bonds with up to four other atoms. Carbon can also form double or triple bonds with other carbon atoms. A common biological structure is a **hydrocarbon chain**—a chain of connected carbon atoms, with pairs of hydrogen atoms bonded to each carbon in the chain. Octane, a component of gasoline, for example, consists of 8 carbon and 18 hydrogen atoms (Figure 3-6A). Many of the molecules in gasoline are simple hydrocarbon chains.

Hydrocarbon chains can also form rings. Because carbon can form covalent bonds only at certain angles, most carbon-containing rings have either five or six atoms. A ring sometimes includes other types of atoms, especially oxygen and nitrogen.

When the atoms of a ring are connected by single covalent bonds, they cannot lie in the same plane. Typically, the ring bends to form a structure that looks rather like a chair or a boat (Figure 3-6B). Some ring compounds are flat, unusually stable, and unreactive (Figure 3-6C and D). These belong to a special group of organic molecules called **aromatic** compounds. Acetylsalicylic acid (aspirin), benzene (an industrial solvent), dimethylpyrazine (a component of chocolate), and phenylalanine (one of the building blocks of cellular proteins) are all aromatics. Some aromatic compounds consist of carbon atoms only, and some include one or two nitrogen atoms in their rings. Many are quite literally aromatic, including such strong-smelling compounds as the oil that makes almonds bitter, the quinine in tonic water, and the benzene, toluene, and xylene that together give high-performance gasoline its zing.

Functional Groups: How Do the Small Molecules Differ from One Another?

The biological functions of molecules are determined by their structures. Just as the parts of a car or a bicycle must fit together for the machine to work, the molecules in a cell must also fit together or interact for the cell to work. Every molecule has a distinctive shape, charge distribution, and ability to interact with water and other kinds of molecules. The properties of a molecule depend on the particular arrangements of its atoms. In analyzing the structures of many organic molecules, chemists have repeatedly found standard small groupings of atoms, called **functional groups,** which help determine the properties of a molecule.

Functional groups are comparable to the standardized parts of a Swiss Army knife, which may include, for example, a knife, a corkscrew, and a can opener (Figure 3-7). A Swiss Army knife may have only one or two such attachments or many, in a variety of combinations. In the same way, a biological molecule may have a few functional groups or many, in a variety of combinations. However, the exact combination of functional groups on a biological molecule profoundly affects the way the molecule behaves.

Table 3-2 shows the properties and structures of half a dozen of the most important functional groups. Many more functional groups exist that are not shown. Here we will discuss the different effects of just three functional groups: hydroxyl, carboxyl, and amino. Figure 3-7 shows the structures of four simple molecules: ethane, ethanol, acetic acid, and aminoethane. Each molecule contains two carbon atoms and different numbers of hydrogen atoms. Of these molecules, ethane has no functional groups, while each of the other three molecules has a different functional group that determines its properties.

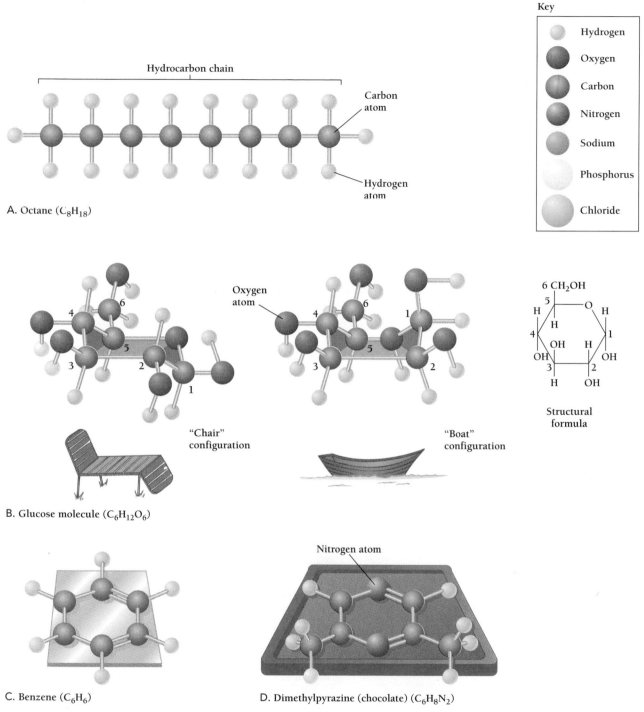

Key

Hydrogen

Oxygen

Carbon

Nitrogen

Sodium

Phosphorus

Chloride

Hydrocarbon chain

Carbon atom

Hydrogen atom

A. Octane (C_8H_{18})

Oxygen atom

6 CH_2OH

"Chair" configuration

"Boat" configuration

Structural formula

B. Glucose molecule ($C_6H_{12}O_6$)

Nitrogen atom

C. Benzene (C_6H_6)

D. Dimethylpyrazine (chocolate) ($C_6H_8N_2$)

Figure 3-6 Carbon atoms frequently form single covalent bonds with other carbon atoms in chains or rings. A. Octane has 8 carbons, with 18 hydrogen atoms hanging off the sides. During combustion inside an automobile engine, each of the carbon-carbon bonds in octane is broken to release the stored potential energy. Many of the other molecules in gasoline are also simple hydrocarbon chains. B. The glucose ring bends into a "chair" structure or a "boat" structure. C. In contrast, a benzene ring lies flat. D. Dimethylpyrazine (an important component of chocolate) also lies flat.

Table 3-2   Functional Groups

Functional Group	Ball and Stick Model	Structural Formula
Hydroxyl group		R— OH
Carbonyl group		$R-\overset{\overset{O}{\parallel}}{C}-H$ (or R)
Carboxyl group		$R-C{\overset{O}{\underset{OH}{}}}$
Amino group		$R-N{\overset{H}{\underset{H}{}}}$
Sulfhydryl group		R— SH
Phosphate group		$R-O-\overset{\overset{O}{\vert}}{\underset{\underset{O^-}{\vert}}{P}}-O^-$

Ethane has the simplest structure: two carbon atoms, each with three hydrogens attached. It is an odorless gas that forms about nine percent of natural gas, the kind used for heating homes and cooking (Figure 3-7A). In high concentrations, ethane has a narcotic effect, but it is not particularly toxic.

Ethanol, the kind of alcohol found in beer, wine, and liquor, looks like ethane except that in place of one hydrogen atom is a **hydroxyl,** or OH group (Figure 3-7B). That single oxygen atom makes ethanol very different from ethane. Ethanol (or ethyl alcohol) is quite toxic. It is used as an antiseptic to kill both prokaryotic and eukaryotic cells. Ethanol is also used as an industrial solvent, as an additive in gasoline, and, in veterinary medicine, to kill nerve tissue. Its most well-known use is in beverages such as beer, wine, and hard liquor. If ingested in sufficient quantity, ethanol causes nausea, vomiting, mental excitement or depression, incoordination, stupor, coma, and death. Chronic use damages the liver and circulatory system.

The hydroxyl group, which is an extremely common functional group, allows molecules that contain it to form hydrogen bonds. Since the electrons of ethanol's hydroxl group tend to spend more time near the electronegative oxygen, the hydrogen in this functional group tends to form hydrogen bonds with other molecules such as water. As a result, ethanol is highly hydrophilic and dissolves readily in water. In contrast, ethane bonds are all nonpolar, and ethane molecules do not interact much, even with each other.

Ethane and ethanol differ in other ways as well. At room temperature, for example, ethane is an odorless gas, while ethanol is a sweet-smelling liquid. Other molecules that contain an O–H group attached to a carbon atom, including sugars and other alcohols, have properties similar to ethanol's.

If we substitute a **carboxyl** group (–COOH) in place of the carbon and three hydrogens in ethane, we get acetic acid, the acid that gives vinegar its familiar sour smell and sharp bite (Figure 3-7C). Just as the O–H group characterizes alcohols and sugars, the carboxyl group characterizes organic acids, since it tends to give up its hydrogen easily. Both of the two oxygens in a carboxyl group pull electrons toward them so strongly that the hydrogen atom actually loses the electron that binds it to the carboxyl group. The hydrogen atom breaks away, leaving behind a COO^- ion.

Now suppose that instead of a hydrogen or hydroxyl or carboxyl group we substitute an **amino** group, a single atom of nitrogen bonded to two atoms of hydrogen (NH_2). The amino group tends to make molecules basic because it takes up hydrogen ions, forming NH_3^+ ions. Substituting an amino group for one of the hydrogen atoms in ethane creates aminoethane (Figure 3-7D). Like ethanol, aminoethane is a liquid, but it is otherwise very different. Highly alkaline, aminoethane's powerful vapors smell intensely of ammonia and can severely irritate the skin, eyes, and mucous membranes. Aminoethane is 20 times more toxic than ethanol.

Aminoethane belongs to a class of compounds called "amines." Amines all have amino functional groups and nearly all smell terrible or at least unpleasant. Two of the most offensive amines are putrescene and cadaverine, which together give memorable odors to rotting fish, aging corpses, urine, semen, and bad breath.

A small number of functional groups determine the properties of a diverse array of biological molecules.

Cells Build Complex Molecules from Four Types of Building Blocks

We are now in a position to understand the properties of many organic molecules. Attached to the carbon skeleton of each organic molecule are hydrogen atoms or functional groups that determine the character of the molecule.

The most important biological properties of a molecule depend on its interactions with water. For example, the boundaries between and within cells consist of hydrophobic molecules that repel water. In contrast, molecular interactions within the watery interior of a cell mostly involve hydrophilic molecules.

Recall from Chapter 2 that whether a molecule is hydrophobic, hydrophilic, or amphipathic depends on the polarity of individual bonds within the molecule. Polar bonds between atoms have a charge at one end, which, in some cases, allows the molecule to form hydrogen bonds with water molecules. For this reason, polar molecules are more water soluble than nonpolar molecules. The polarity of each bond depends on the electronegativity of the participating atoms. The biological functions of a molecule therefore depend not only on its shape and the number and kind of atoms, but also on the specific arrangements of atoms.

As we mentioned at the beginning of this chapter, all the structures of a cell are built from combinations and polymers of the 35 small-molecule building blocks listed in Table 3-1. These small molecules fall into four classes: (1) **lipids**, compounds that are less soluble in water than in nonpolar solvents such as olive oil; (2) **sugars**, molecules that have the equivalent of one molecule of water (that is, two hydrogen and one oxygen atom) for every atom of carbon; (3) **amino acids**, molecules that contain both amino and carboxyl groups; and (4) **nucleotides**, molecules that each consist of a nitrogen-containing aromatic ring compound, a sugar, and a molecule of phosphoric acid.

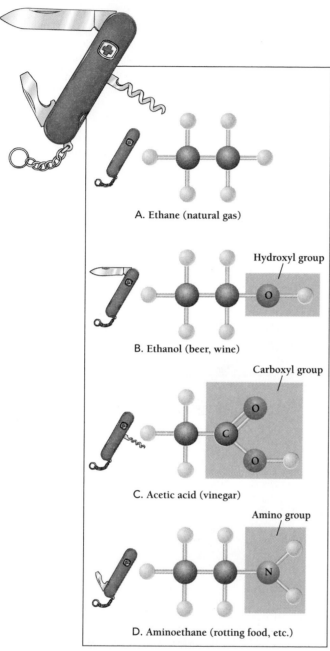

A. Ethane (natural gas)

Hydroxyl group

B. Ethanol (beer, wine)

Carboxyl group

C. Acetic acid (vinegar)

Amino group

D. Aminoethane (rotting food, etc.)

Functional groups

Figure 3-7 Functional groups are like the different attachments on a Swiss Army knife. Which groups a molecule has determine its structure and function. A. Ethane, a nontoxic, flammable gas, has only hydrogen atoms attached to its two-carbon chain. B. The ethanol in beer, wine, and spirits is, like all alcohols, characterized by a hydroxyl (OH^-) group. C. Acetic acid, in vinegar, is a single carbon with an attached carboxyl (–COOH) group—a functional group that characterizes many organic acids. D. Aminoethane is two carbons with an attached amino group (NH_2), which tends to make molecules alkaline, or basic. Like many other amines, aminoethane is strong smelling and toxic.

SMALL BIOLOGICAL MOLECULES ARE THE BUILDING BLOCKS OF LIFE

Lipids Include a Variety of Nonpolar Compounds

Many familiar compounds from the kitchen are lipids, including oils, fats, and candle wax, for example. Because lipids contain more chemical energy per gram than other biological molecules, lipids often serve as energy stores (Figure 3-8).

One characteristic of lipids is that they dissolve in nonpolar solvents, such as gasoline or olive oil, rather than in water. The reason that lipids do not dissolve in water is that they have few functional groups that are polar. The limited solubility of

Figure 3-8   Fat cells, or adipocytes, store energy in the form of fats. *(CNRI/Phototake)*

25 μm

lipids in water explains why removing a butter or gravy stain often requires dry cleaning, that is, treatment with nonpolar organic solvents such as benzene.

However, not all lipids are hydrophobic. Some are amphipathic: they contain a polar functional group as well as nonpolar chains of carbon and hydrogen atoms. One class of amphipathic lipids is the **fatty acids**—a major component of the cell membrane that separates cells from their usually watery external environments.

Fatty acids can differ both in the number of carbon atoms they contain and in the relative numbers of single and double bonds. The hydrocarbon chain in a fatty acid consists of carbon atoms linked either to one or two other carbon atoms and to hydrogen atoms. When all the bonds between carbon atoms are single bonds, the hydrocarbon chain contains the maximum number of hydrogen atoms and is said to be **saturated** with hydrogen atoms. Stearic acid, an important component in beef fat, is a highly saturated fatty acid (Figure 3-9A).

BOX 3-1

The biochemistry of cholesterol

A quick scan of any health magazine would leave most people with the impression that cholesterol is always bad for you. In fact, you can't live without it. The outer membranes of your cells are rich in this lipid, having as much as one cholesterol molecule for every phospholipid molecule. Cholesterol's bulky shape contributes to membrane fluidity, in the same way the double bonds in phospholipids' fatty acid chains do. Unlike phospholipids, however, cholesterol is hydrophobic, which means that, by itself, it doesn't dissolve well in blood.

Cholesterol also serves as a chemical precursor to the sex hormones testosterone and estradiol (a form of estrogen), as well as two other important steroid hormones. All steroids contain the large, nonpolar ring structure of cholesterol and are fat soluble (hydrophobic).

Cholesterol comes from two sources: the liver and the diet. The liver makes cholesterol as a component of bile, a substance secreted into the upper intestine to aid digestion. In addition, most Americans consume about 550 mg of cholesterol a day, all

of which comes from animal products. Some of this cholesterol passes through the digestive tract into the feces, and some gets absorbed into the blood. The blood plasma of a typical college-age American contains about 180 mg/dl (1 deciliter, or dl, = 100 ml) of cholesterol. Total plasma cholesterol rises slowly with age to peak at about 230 mg/dl in men and 250 mg/dl in women by age 55.

Cholesterol doesn't travel alone in the blood. If it did, its hydrophobicity would cause it to form big clumps, like oil in a bottle of salad dressing. Instead it travels in lipoprotein complexes. The most common are called low-density lipoproteins, or LDLs. These are responsible for delivering cholesterol to cells and tissues where it is needed. High-density lipoprotein, HDLs, remove cholesterol from cells and transport it to the liver, where it is secreted into the bile.

Sometimes people call LDL cholesterol "bad" and HDL cholesterol "good." This is because if not enough HDLs are present to remove cholesterol, the excess builds up

along the walls of arteries, contributing to deposits called "plaques." Plaques cause hardening of the arteries, or atherosclerosis. Atherosclerosis reduces blood flow to the organs, including the heart, and damages the arteries. Tiny bits of plaque may break off and lodge in an artery, blocking blood flow to the heart and causing a heart attack.

While limiting cholesterol intake is sound dietary advice, the best indicator of the chance of developing heart disease from cholesterol deposits in the arteries is not total cholesterol, but the ratio of bad to good cholesterol. The average American has an LDL to HDL ratio of 5:1. Reducing that ratio to 3.5:1 by increasing plasma HDL cuts the risk of heart disease in half.

How can people increase HDLs? The following items all tend to raise HDLs: female sex hormones; regular aerobic exercise; monosaturated fats such as olive oil—instead of, not in addition to, saturated fats; weight loss; and certain forms of fiber. On the other hand, tobacco smoke *lowers* plasma HDL (and also contributes to heart disease in other ways).

LIPIDS

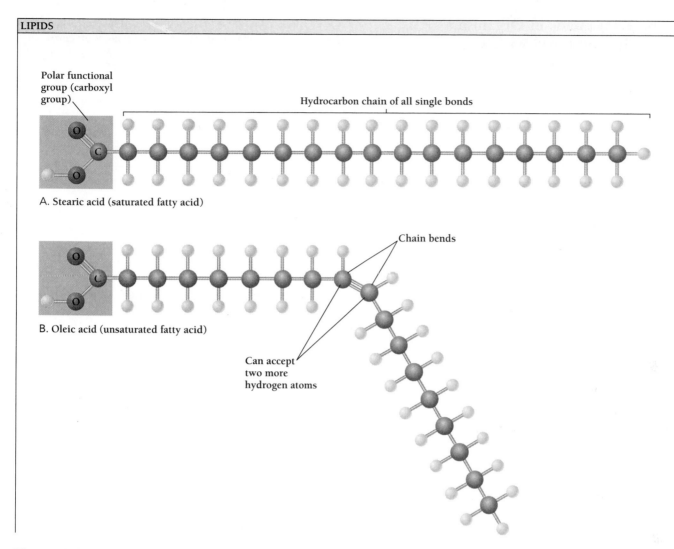

Polar functional group (carboxyl group)

Hydrocarbon chain of all single bonds

A. Stearic acid (saturated fatty acid)

Chain bends

Can accept two more hydrogen atoms

B. Oleic acid (unsaturated fatty acid)

Figure 3-9   Saturated fatty acids, such as the stearic acid in beef fat, pack together more closely than unsaturated fatty acids, such as the oleic acid in olive oil. Because saturated fatty acids pack closely, they tend to form solids such as lard and butter. Unsaturated fatty acids tend to form liquid oils. In saturated fatty acids, carbon atoms form single covalent bonds with carbon or hydrogen atoms. In unsaturated fatty acids, carbon atoms form one or more double bonds with other carbon atoms. Unsaturated fatty acids can be synthetically "hydrogenated," or saturated. "Hydrogenated vegetable oils"—found on virtually every list of ingredients on packages of snack food—give cookies, crackers, and chips a crispier texture that most people prefer.

When a fatty acid contains one or more carbon-to-carbon double bonds, and is therefore missing two or more hydrogen atoms, the fatty acid is said to be **unsaturated**. Oleic acid, an important component of vegetable oils, is unsaturated because it can accept two more hydrogen atoms per double bond (Figure 3-9B). A fatty acid with many double bonds is said to be **polyunsaturated.**

Fats, waxes, and other lipids that have saturated hydrocarbon chains, are solid at room temperature. Lipids that are polyunsaturated tend to be liquid at room temperature, and we call them **oils.** Because single C–C bonds tend to be straight,

while double C=C bonds tend to be bent, saturated fats are linear and pack more tightly together than unsaturated fats, which are kinky (Figure 3-9). This is why saturated fats are solid at room temperature and polyunsaturated fats (or oils) are liquid. In general, animals contain more fats, and plants more oils.

Lipids, which include both fats and oils, are usually oily to the touch, insoluble in water, and high in energy. One class of lipids, the fatty acids, may be more or less "saturated."

Most of the Fatty Acids in Organisms Are Chemically Combined with a Molecule Called Glycerol

Glycerol is a simple three-carbon molecule with three hydroxyl groups (Figure 3-10A). Each hydroxyl group can attach to the carboxyl carbon of a fatty acid. A glycerol with all three hy-

droxyl groups attached to fatty acids, so that the glycerol has three, long hydrocarbon "streamers," is a **triacyl glycerol** (Figure 3-10B). When a triacyl glycerol (or triglyceride) forms, the hydrophilic property of the carboxyl group is lost, so triacyl glycerols are even less soluble in water than fatty acids. Most of the fats that humans and other animals store, as well as the oils of plants, consist of triacyl glycerols.

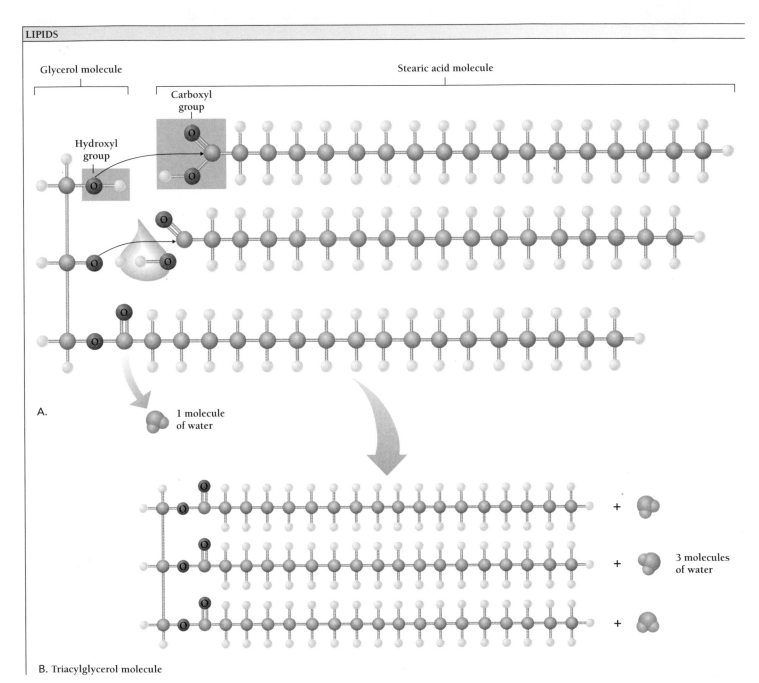

LIPIDS

Glycerol molecule

Stearic acid molecule

Carboxyl group

Hydroxyl group

A.

1 molecule of water

B. Triacylglycerol molecule

+
+ 3 molecules of water
+

Figure 3-10   A triacyl glycerol (or triglyceride) forms when three molecules of water condense from glycerol and three fatty acid chains, as in A and B. C. A phospholipid forms when just two fatty acids attach to glycerol. D. Phospholipids tend to form membranes in which the hydrophilic heads face out into the aqueous environment and the hydrophobic tails hide together in the middle.

A glycerol molecule with a phosphate group attached in place of the third hydroxyl group, and just two fatty acid "streamers," instead of three, is a **phospholipid** (Figure 3-10C). Phospholipids are highly amphipathic. Each has a hydrophobic "tail," which consists of the hydrocarbon chains of two fatty acids, and a hydrophilic "head," which contains the charged phosphate group.

In water, the hydrophobic tails of phospholipids clump together to form an oily interior, while the hydrophilic heads point outward into the surrounding water (Figure 3-10D). The membrane boundaries between cells and cell compartments are composed primarily of phospholipids. All of these membranes have a hydrophobic interior and a hydrophilic exterior.

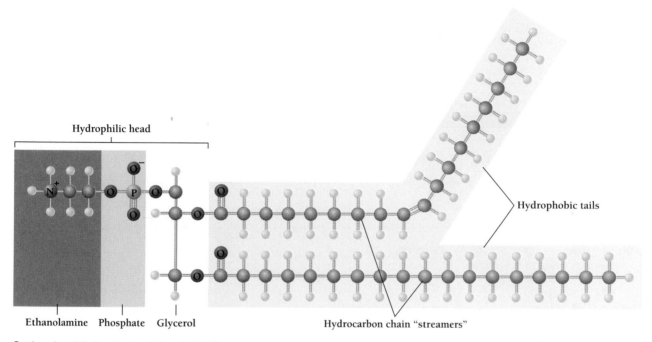

Hydrophilic head

Hydrophobic tails

Ethanolamine   Phosphate   Glycerol

Hydrocarbon chain "streamers"

C. Phosphatidylethanolamine (phospholipid)

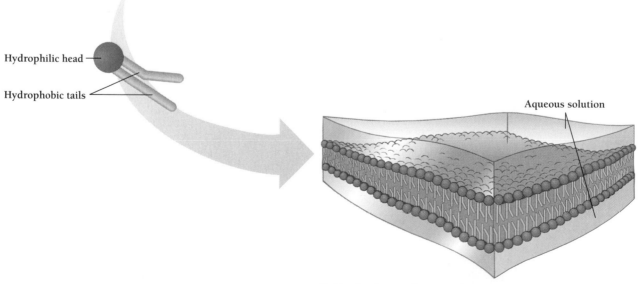

Hydrophilic head

Hydrophobic tails

Aqueous solution

D. Bimolecular membrane

Another class of lipids is **steroids**—hydrocarbon chains with four interconnected rings (Figure 3-11). Many of the hormones responsible for sexual development and reproductive functions in animals are steroids. Small differences in the attached functional groups can make enormous differences in a steroid's biological properties. For example, testosterone—the hormone most responsible for inducing the physical characteristics of males—differs by only a few atoms from the female hormones progesterone and estrogen (Figure 3-11C and D). Both sexes have both "male" and "female" hormones, but in different proportions. Figure 3-12 shows the striking effects of testosterone and estrogen on sexual development. The starting point for the synthesis of all other steroids is **cholesterol**—a hydrocarbon with the same pattern of four interconnected rings (Figure 3-11A).

A glycerol molecule with three, long carbon chains is a triacyl glycerol, the principal component of fats and oils. A glycerol molecule with two carbon chains and a phosphate group is a phospholipid, a principal component of membranes. Steroid molecules have four interconnected hydrocarbon rings.

Sugars Contain Many Hydroxyl Groups

Sugars and polysaccharides are important energy storage molecules in cells. Animals and plants use simple sugars for short-term energy storage. Animals, for example, store energy for immediate use in glucose, while plants usually use disaccharides,

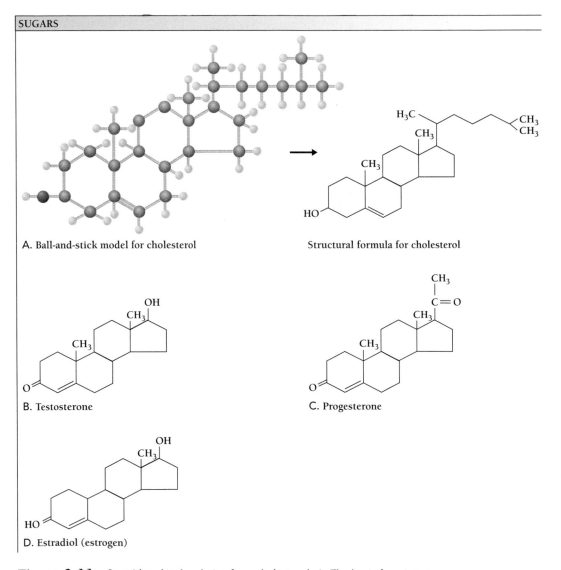

SUGARS

A. Ball-and-stick model for cholesterol

Structural formula for cholesterol

B. Testosterone

C. Progesterone

D. Estradiol (estrogen)

Figure 3-11   Steroid molecules derive from cholesterol. A. The basic four-ring structure of cholesterol. B. Testosterone, a male sex hormone. C. Progesterone, a female sex hormone. D. Estradiol, one of the more active forms of estrogen, another female sex hormone.

Figure 3-12 In males, the testes (male gonads) produce large amounts of the steroid hormone testosterone. At puberty, testosterone stimulates the development of secondary sexual characteristics. Here, a hen that has been treated with testosterone develops a red comb like that of a rooster. ("Hormones," Scientific American, *March, 1957. Reprinted in* Hormones and Reproductive Behavior, *W.H. Freeman and Company, San Francisco, 1979.)*

especially sucrose. Virtually all the interrelated biochemical reactions that produce energy are in some way connected to the synthesis and breakdown of glucose. Glucose is so important that all cells and organisms use special adaptations to keep its concentration constant (Figure 3-13).

Simple sugars or **monosaccharides** [Greek, *mono* = one + *saccharine* = sugar) may contain from three to nine carbon atoms. Figure 3-13 shows glucose and ribose, two monosaccharides that are important components of biological molecules. The most common is **glucose,** which contains six carbon atoms and is therefore called a **hexose** [Greek, *hex* = six]. Ribose and deoxyribose, which are components of nucleotides, contain five carbons and are therefore called **pentoses** [Greek, *pente* = five]. Table sugar, or **sucrose,** consists of two monosaccharides, glucose and fructose, linked together.

All sugars and polysaccharides are classified as **carbohydrates**—compounds that contain the equivalent of one water molecule (one oxygen atom and two hydrogen atoms) for every carbon atom. Most of the carbon atoms in carbohydrates are attached to both a hydrogen atom and a hydroxyl group. Each of these hydroxyl groups can form a hydrogen bond with a water molecule, which is why sugars and other carbohydrates dissolve in water.

A Nucleotide Has Three Parts, Each with Different Functional Groups

Nucleotides are the building blocks of the two kinds of information-carrying molecules, RNA (ribonucleic acid) and DNA (deoxyribonucleic acid). A nucleotide consists of three parts: a sugar (ribose or deoxyribose), one or more phosphate groups, and a **nitrogenous base.** Each nitrogenous base contains nitrogen, accepts hydrogen ions (as all bases do), and includes one or two rings of carbon and nitrogen atoms. Attached to these rings are hydrogen atoms and various polar functional groups. Cells commonly use just five nitrogenous bases (Figure 3-14).

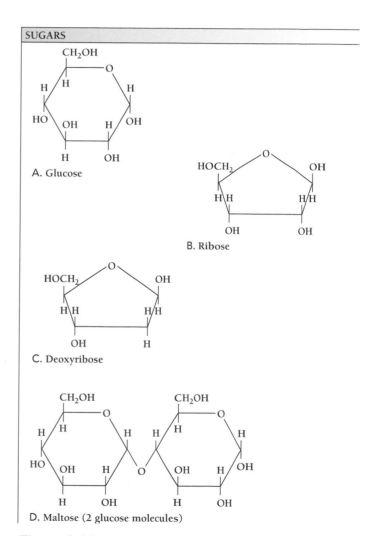

Figure 3-13 Three monosaccharides (simple sugars) and a disaccharide. A. Glucose is a six-carbon sugar. B. Ribose is a five-carbon sugar and a component of ATP and RNA. C. Deoxyribose is a five-carbon sugar and a component of DNA. D. Maltose is a two-ring sugar composed of two molecules of glucose linked together. Table sugar, or sucrose, is a disaccharide of the simple sugars glucose and fructose.

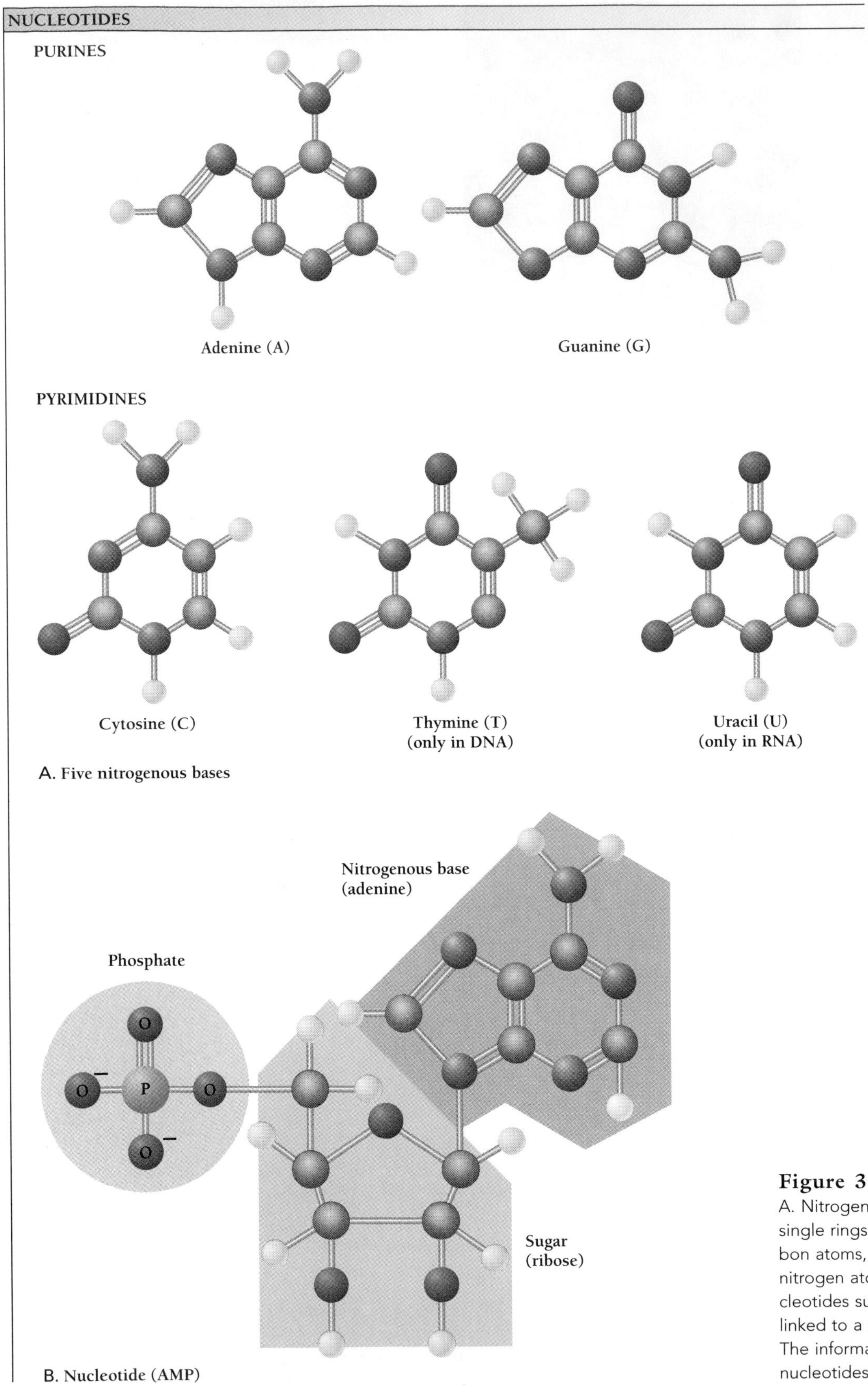

NUCLEOTIDES

PURINES

Adenine (A)   Guanine (G)

PYRIMIDINES

Cytosine (C)   Thymine (T)   Uracil (U)
                     (only in DNA)   (only in RNA)

A. Five nitrogenous bases

Nitrogenous base
(adenine)

Phosphate

Sugar
(ribose)

B. Nucleotide (AMP)

Figure 3-14 The five nitrogenous bases.
A. Nitrogenous bases are either pyrimidines, with single rings of two nitrogen atoms and four carbon atoms, or purines, with double rings of four nitrogen atoms and five carbon atoms. B. Nucleotides such as this are nitrogenous bases linked to a sugar linked to a phosphate group. The information molecule DNA is composed of nucleotides linked together in long chains.

NUCLEOTIDES

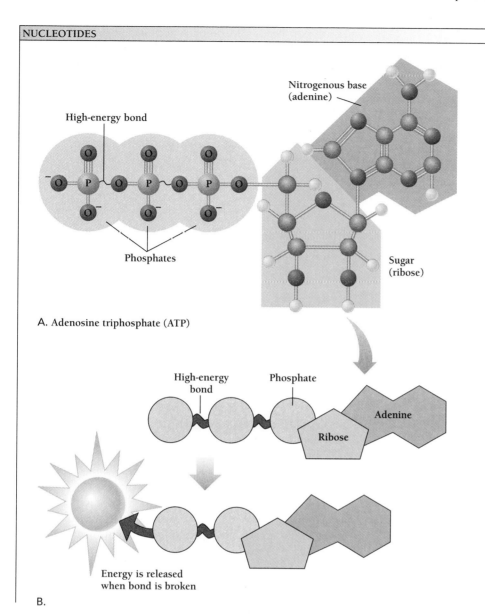

High-energy bond

**Nitrogenous base
(adenine)**

Phosphates

**Sugar
(ribose)**

A. Adenosine triphosphate (ATP)

**High-energy
bond**

Phosphate

Adenine

Ribose

**Energy is released
when bond is broken**

B.

Figure 3-15 ATP is the common currency of energy for all organisms. ATP is a nucleotide consisting of the nitrogenous base adenine, the sugar ribose, and the three phosphate groups joined by high-energy bonds.

These five bases are divided into two types: **pyrimidines**, which contain one ring of atoms, and **purines**, which contain two interlocking rings (Figure 3-14A). Cells contain two purines—adenine and guanine—and three pyrimidines—cytosine, uracil, and thymine. Thymine and uracil are almost identical, differing only in the presence of an extra methyl ($–CH_3$) group in thymine.

Like sugars and amino acids, nucleotides themselves serve important biological roles, particularly in the production and packaging of energy. For example, free nucleotides (those that are not part of a nucleic acid) may contain one, two, or three phosphate groups linked to one another, as in the case of **ATP (adenosine triphosphate)** (Figure 3-15A). ATP molecules, which store energy, are the immediate source of energy for all

biological processes in all cells. By breaking the phosphate-phosphate bonds in ATP, cells release a bit of energy that can then be used to drive other chemical reactions (Figure 3-15B).

Nucleotides are essential to life both as the building blocks DNA and RNA, and as molecules for energy storage.

Amino Acids Contain Both Carboxyl and Amino Groups

All proteins in all living organisms are composed of combinations of just 20 amino acid building blocks. These 20 amino

AMINO ACIDS

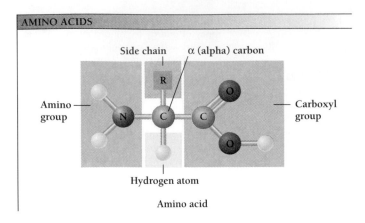

Amino acid

Figure 3-16 An amino acid includes a carboxyl group, an amino group, a hydrogen atom, and a side chain (R). All four are attached to a single, central carbon atom, called the α carbon.

acids are all variations on the same basic structure. Each amino acid contains a carboxyl group, which makes it an acid, and an amino group, which also makes it a base (Figure 3-16). Both the carboxyl group and the amino group are attached to the same carbon atom, called the α **(alpha) carbon.**

Each amino acid has a characteristic group of atoms, called a side chain, or **R group,** also attached to the α-carbon atom. Table 3-1 shows the R side chains of the 20 amino acids found in proteins. Side chains may contain hydrocarbon chains of different lengths, as well as different functional groups. Some side chains are hydrophobic, some hydrophilic. Some normally have a positive charge, some a negative charge, and some are neutral. The properties of the side chains establish the individual character of each amino acid.

Although amino acids' greatest role is as building blocks for proteins, several amino acids and their derivatives themselves serve important biological functions. For example, some are **neurotransmitters**—chemicals secreted by nerve cells to convey information to other nerve cells or to muscle cells.

Amino acids, the building blocks of proteins, have a carboxyl group, an amino group, and an R group.

HOW DO SMALL MOLECULES LINK TOGETHER TO FORM MACROMOLECULES?

Recall that sugars form polysaccharides, nucleotides form nucleic acids, and amino acids form proteins, but that lipids do not form polymers. Polysaccharides, nucleotides, and proteins are chains of building blocks linked end-to-end, like a child's

plastic "pop beads." In proteins and nucleic acids the chains are simple, linear sequences of small molecules (Figure 3-17A). The building blocks themselves may have branches of atoms, but the "backbone chain" remains unbranched. In contrast, polysaccharide chains may be highly branched (Figure 3-17B). Even though proteins and nucleic acids are unbranched, however, they form much more complex molecules than polysaccharides. As Linus Pauling showed, the chains may coil and fold to form complex three-dimensional shapes that determine their biological behavior.

What Holds Building Blocks Together in a Macromolecule?

When we digest an apple, we break down its carbohydrates into sugars and its proteins into amino acids. Likewise, we continually break down and remake the proteins of our skin, muscles, and bones. Like all animals, we break down and reuse the building blocks from the macromolecules we eat, and recycle the building blocks of our own macromolecules as we grow and change.

All organisms need to assemble and disassemble macromolecules easily. The bonds that hold macromolecules together must be strong enough so that the macromolecules will not easily come apart. On the other hand, the bonds must be weak enough so that organisms can disassemble the macromolecules easily. It turns out that all biological building blocks are held together by similar kinds of chemical bonds.

When cells link small molecules to make macromolecules, they form each link by using enzymes to eliminate two hydrogen atoms and one oxygen atom—the equivalent of a molecule of water (Figure 3-17C). The loss of the water molecule is called a "dehydration." The linking of two building blocks is also called a "condensation," in the same sense that milk and books are condensed. All are made more compact by taking something away. The linking of two small molecules is thus called a **dehydration condensation reaction.** The whole process of linking chains of building blocks into a macromolecule is called **polymerization.** Sugars, proteins, and nucleic acids are all formed by removing a water molecule from between each two building blocks in the chain.

To take macromolecules apart, cells reverse the dehydration process. To detach each small molecule from the macromolecule, one molecule of water is added. This breaking down process is called **hydrolysis** [Greek, *hydro* = water + *lysis* = breaking] (Figure 3-17D).

Cells link small molecules into chains by taking away the equivalent of one water molecule from each two small molecules—a process called a dehydration condensation reaction. The reverse process, hydrolysis, breaks the links between the molecules in a chain.

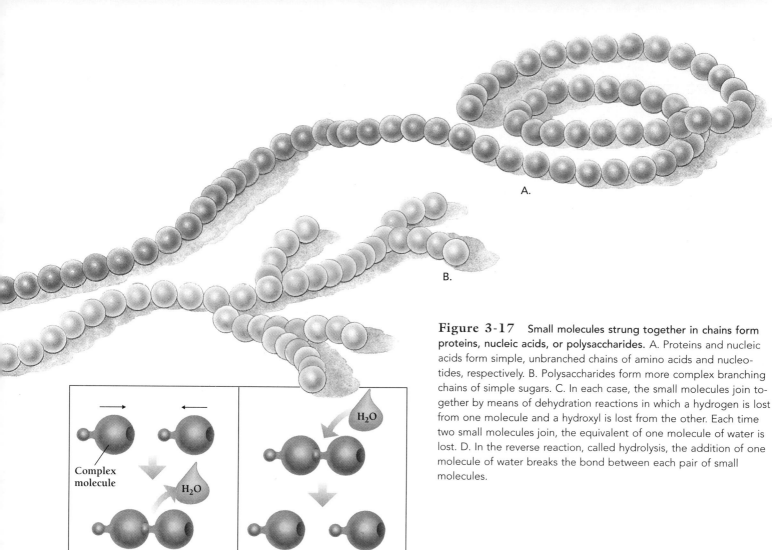

A.

B.

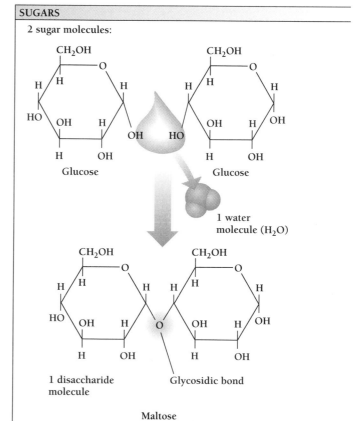

Complex molecule

H_2O

H_2O

C. Dehydration condensation

D. Hydrolysis

Figure 3-17 Small molecules strung together in chains form proteins, nucleic acids, or polysaccharides. A. Proteins and nucleic acids form simple, unbranched chains of amino acids and nucleotides, respectively. B. Polysaccharides form more complex branching chains of simple sugars. C. In each case, the small molecules join together by means of dehydration reactions in which a hydrogen is lost from one molecule and a hydroxyl is lost from the other. Each time two small molecules join, the equivalent of one molecule of water is lost. D. In the reverse reaction, called hydrolysis, the addition of one molecule of water breaks the bond between each pair of small molecules.

How Are Sugars Held Together To Form Polysaccharides?

Two sugar molecules can join together to form a two-sugar "disaccharide." Two glucose molecules, for example, can link to form maltose (malt sugar), while a glucose and fructose form sucrose (table sugar). The two sugars are linked by a **glycosidic bond,** in which an oxygen atom forms a bridge between carbon atoms on two sugar molecules (Figure 3-18). This process eliminates the equivalent of one water molecule.

Organisms can also link sugars into enormous polysaccharides containing hundreds or thousands of sugar building blocks. Organisms use polysaccharides—**starch** in plants and **glycogen** in animals—for long-term energy storage. The starches in potatoes and pasta, for example, are energy-rich

SUGARS

2 sugar molecules:

Glucose   Glucose

1 water molecule (H_2O)

1 disaccharide molecule   Glycosidic bond

Maltose

Figure 3-18 The loss of a molecule of water from two molecules of glucose creates a glycosidic bond between them and a new disaccharide molecule—maltose.

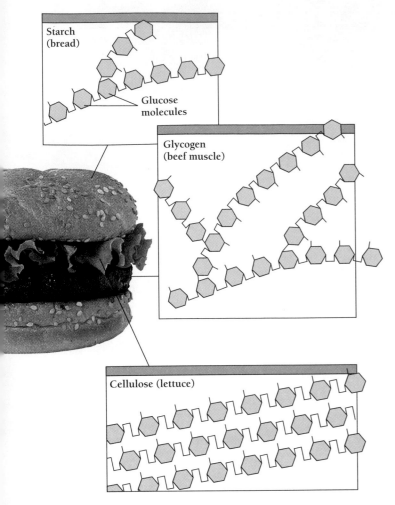

Figure 3-19 A typical hamburger contains three important polysaccharides: starch (in the bun), glycogen (in the beef), and cellulose (in the lettuce). The bonds in cellulose are different from those in starch and glycogen and we do not have the enzymes necessary to break them. Our bodies can break down the starch and glycogen thoroughly, but much of the lettuce is indigestible. *(Paraskevas Photography)*

polysaccharides manufactured by plants (Figure 3-19). **Cellulose**—the major structural material of wood, cotton, and paper—is another polysaccharide made by plants.

The three-dimensional structure and the biological properties of polysaccharides depend both on the kinds of sugars they are made of and on the way in which the sugars are joined. Cells use different sugars to make polysaccharides, but the most common polysaccharides—starch, cellulose, and glycogen—consist of glucose alone. Glycogen tends to be highly branched. In contrast, starch and cellulose tend to be linear molecules with only occasional branching.

The most important difference between cellulose and starch is that—due to the arrangement of the glycosidic bonds between the glucose molecules—one molecule is digestible and the other is not (Figure 3-19). In cellulose some of the links between glucose are oriented upside down, as compared to those in starch. As a result, the enzymes that work so well on

starches have no effect on cellulose, and special enzymes are needed to digest cellulose.

Because most cells do not have the special enzymes necessary to break the reversed links in cellulose, most organisms cannot digest cellulose. Only a very few microbes, such as those found in the guts of cows and termites, have the necessary enzymes. For the rest of us, the cellulose in bran flakes and prunes is just "roughage" or fiber—a water-absorbing material that keeps things moving in the digestive tract.

Indeed, cellulose is one of those no-calorie foods that the food industry uses to thicken milk shakes or to make low-calorie breads. Because we cannot digest it, these foods are said to be "low in calories." One useful result of the geometry of the glycosidic bonds of cellulose is that wood, books, and clothes do not easily rot: few organisms can break them down.

How Are Nucleotides Held Together To Form Nucleic Acids?

The nucleic acids DNA and RNA are composed of long chains of nucleotides, called **polynucleotides.** Polynucleotides are always unbranched. Just as in polysaccharides, the bond that joins the building blocks forms with the elimination of the equivalent of one molecule of water (Figure 3-20).

The long polynucleotide chains of DNA and RNA both tend to twist, like the curly cord on a telephone. A molecule of DNA usually consists of two chains, while a molecule of RNA usually consists of just one chain. The three-dimensional structures of both RNA and DNA help determine the way they interact with specific proteins. In Chapter 10 we discuss the structure and function of DNA in detail.

DNA and RNA are both composed of long chains of nucleotide bases linked by dehydration condensation reactions.

How Are Amino Acids Held Together To Form Proteins?

Every kind of organism has its own unique set of proteins. The human body, for example, contains 50,000 to 100,000 different proteins. Proteins are some of the most important molecules in an organism. They hold the body together, work as enzymes, and regulate the expression of our genes.

Collagen, the commonest protein in the human body, gives strength to connective tissues in structures such as tendons, ligaments, bones, muscle, and skin. Dissolved in sugar water, collagen gives Jell-O its jiggle. Yet, in the body, collagen is as strong as steel. The protein elastin makes skin stretchy, like rubber, while keratin strengthens hair, horns, nails, and claws. Some proteins transport materials. Hemoglobin, for example, takes up oxygen from the lungs and transports it, by way of the blood, to the cells in the rest of the body. Other proteins

NUCLEOTIDES

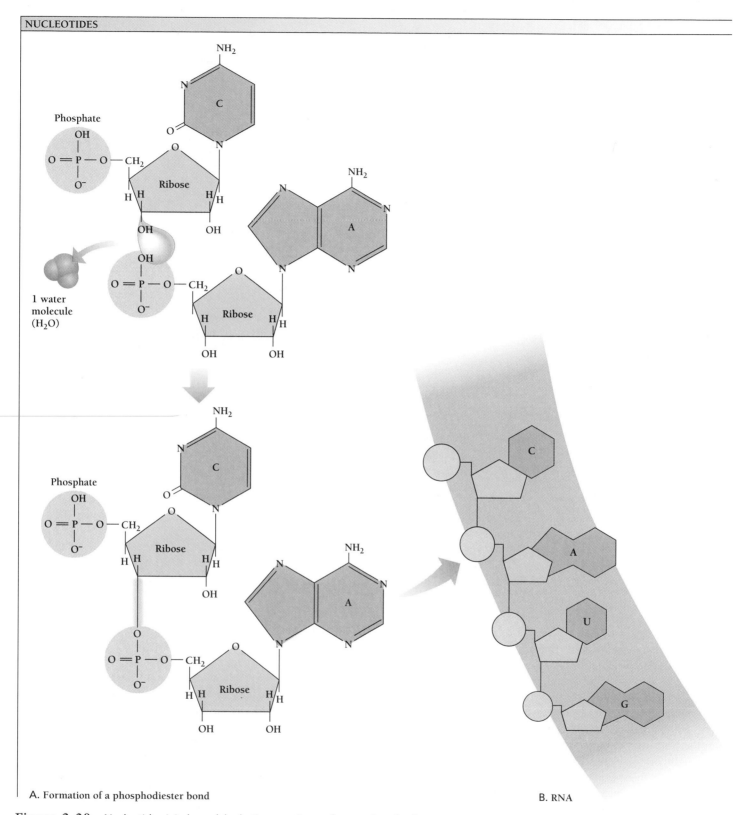

A. Formation of a phosphodiester bond

B. RNA

Figure 3-20 Nucleotides join by a dehydration reaction to form a phosphodiester bond. A chain of such nucleotides form RNA, which, like DNA, carries genetic information.

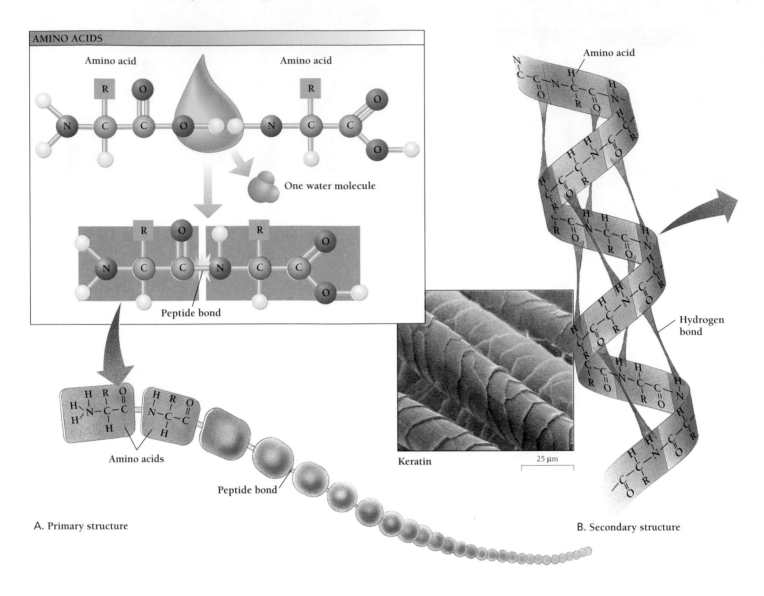

AMINO ACIDS

Amino acid   Amino acid

One water molecule

Peptide bond

Amino acids

Peptide bond

A. Primary structure

Keratin   25 μm

Amino acid

Hydrogen bond

B. Secondary structure

act as hormones, antibodies, and poisons (for example, rattlesnake venom). Still others transport molecules across the membranes of cells.

One of the most important roles for proteins is as **enzymes**—biological molecules that speed up reactions between other molecules. Enzymes are indispensable to life. Without them, many chemical reactions would occur too slowly to sustain life. Enzymes replicate and repair DNA, allow cells to extract energy from sugars, digest our food, and much more. For example, enzymes in the mouth help break down the polysaccharides in bread into simple sugars. That is why bread gets sweeter and sweeter as you chew it. We will discuss enzymes in greater detail in Chapter 5.

Like polysaccharides, proteins are composed of chains of building blocks joined together by dehydration reactions. The building blocks in proteins are amino acids, however, and the bond that links two amino acids together is called a **peptide bond** (Figure 3-21). Chains of amino acids are called **polypep-**

tides. Polypeptides are always linear, never branched like those in starch or cellulose. However, polypeptides fold into balls and other shapes. A protein may consist of a single, folded polypeptide chain or of several chains folded together. In proteins with more than one chain, the individual chains are held together tightly, so that the protein behaves like a single molecule.

Proteins play many roles. They hold together the parts of an organism, they act as enzymes, and they act as identification tags on the surfaces of cells. The biological role of a protein depends exquisitely upon its three-dimensional shape. A protein's shape depends, in turn, upon the sequence of amino acids in each of its polypeptides.

Peptide bonds link amino acids together into polypeptide chains. One or more polypeptides together folded into a unique shape make a protein.

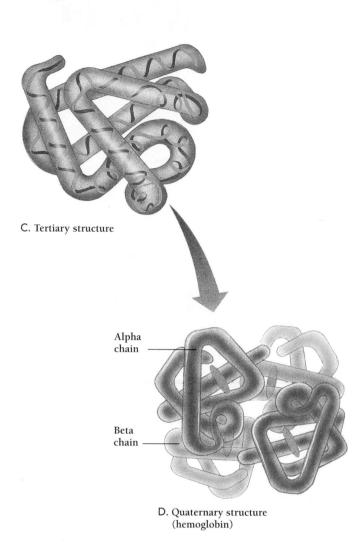

C. Tertiary structure

Alpha
chain

Beta
chain

D. Quaternary structure
(hemoglobin)

Figure 3-21 Amino acids join by a dehydration reaction to form a peptide bond. A chain of amino acids forms a polypeptide chain. A. The primary structure of a protein is simply the sequence of amino acids in its polypeptide(s). B. The secondary structure is the curling of each polypeptide into a three-dimensional form, such as the α helix shown here. C. The tertiary structure is the folding of the curled polypeptide into a more complex three-dimensional shape, such as that of the myoglobin shown here. D. The quaternary structure is the fitting together of two or more folded polypeptides, as in the hemoglobin molecule shown here. *(Andrew Syred/Science Photo Library/Photo Researchers)*

By making different versions of a polypeptide, biochemists can see how tiny changes in its structure affect its biochemical behavior. In growth hormone, for example, changing one amino acid may have little effect, while changing another may completely alter its structure and abolish its biological activity.

When biochemists compare the structures of proteins from different organisms they often find such small differences. A muscle protein isolated from a cow may have a sequence of amino acids slightly different from the same protein isolated from a human. In such cases, a slight difference does not change the protein's three-dimensional structure or the way it functions.

In general, the changes that make the biggest differences in function are those that alter the chemical behavior of side chains. In the blood protein hemoglobin, for example, a single substitution of a positively charged amino acid for a negatively charged one results in a hemoglobin molecule that causes sickle cell anemia. On the other hand, substitution of one negatively charged side chain for another would probably have little effect.

In some cases, a single altered amino acid can make the difference between life and death. In the fatal childhood disease Tay-Sachs, a single altered amino acid in the enzyme hexosaminidase A leads to an accumulation of fat in the nerve cells, mental retardation, blindness, loss of muscular control, and early death.

Small changes in the structure of proteins can enormously
alter the way the protein behaves or have no effect at all,
depending on whether the change alters
the chemistry of the protein molecule.

The Structure of a Polypeptide Determines Its Function

With the advent of bioengineering, biochemists have learned to manufacture useful proteins such as **growth hormone**, which stimulates growth, milk production, and other processes in humans, cows, and other mammals. Growth hormone has the same structure and biological effects, whether it comes from the body or from the laboratory. Biochemists have learned that the molecules they make behave the same as the ones made by a living cell, *as long as the structure is the same* (Box 3-2).

The chains that make up proteins and nucleic acids must fold in precisely the right way to form biologically active molecules. The way a polypeptide folds depends on the sequence of building blocks in its chains. In contrast, the exact number and order of sugars in a polysaccharide has little effect on its function. The number of glucoses in a molecule of starch, for example, does not affect its ability to store energy.

What Determines the Three-Dimensional Structure of a Protein?

What Shapes Do Proteins Assume?

Since the function of a protein molecule depends on its structure, we would like to know what individual proteins look like. Biochemists have now purified and studied several thousand proteins, each of which has a characteristic structure and function.

BOX 3-2

Molecules with similar shapes can mimic one another

The general arrangement of atoms in a molecule determines many of its properties, including, for example, whether it is polar or nonpolar. But the specific arrangement of atoms determines the molecule's size and shape. In many cases, the shape and charge distribution of a molecule—much more than its chemical properties—determines its biological effects.

Synthetic molecules made in the laboratory, for example, may have shapes and charge distributions so similar to those of natural compounds that the synthetic versions mimic biological effects. A particularly dramatic example of such mimicry is the drug amphetamine, whose three-dimensional structure closely resembles epinephrine, a powerful chemical signal (Figure A). Epinephrine, secreted by the adrenal glands, helps coordinate the "fight or flight" reaction, by which animals respond to threats. It is a stimulant that increases heart rate and blood pressure, redistributes blood to the muscles, and raises blood sugar.

During World War II, flyers and soldiers on all sides used amphetamine to keep alert during night missions. Amphetamine may also cause a host of side effects, including restlessness, euphoria, headaches, dry mouth, constipation, and weight loss. After 1945, returning soldiers continued to use the drug and amphetamine became a widely abused drug in both Japan and the United States. Today amphetamine is still widely prescribed by physicians as a treatment for obesity, hyperactivity in children, and narcolepsy, the overwhelming tendency to fall asleep.

Because the brain cannot regulate the effects of amphetamine as it regulates those of epinephrine, amphetamine has severe disorganizing effects on the brain. Fatigue and depression inevitably follow the use of amphetamines. Chronic amphetamine intoxication tends to result in insomnia, irritability, hyperactivity, and personality changes. Overdose produces tremors, confusion, assaultive behavior, hallucinations, and panic, as well as nausea, vomiting, and abdominal cramps. Fatal overdose is preceded by convulsions and coma.

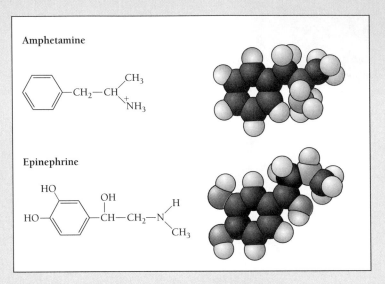

Figure A Amphetamine and epinephrine.

Because proteins are made of hundreds or thousands of amino acids, proteins are far larger than the small molecules discussed earlier in this chapter. A typical globular protein such as hemoglobin, for example, is 400 times as big as a molecule of glucose.

X-ray diffraction studies yield so much detailed information about protein structure that it is hard to understand the data for a even single protein. The complete structure of a protein requires more than 25,000 measurements and equations, with more than a billion terms! Computer technology has greatly accelerated the analysis of crystallographic data, and crystallographers have now determined the complete structures of more than 100 proteins. But until recently such computing power was nonexistent.

Proteins are also extraordinarily diverse. There are, for example, 10^{130} ways (that is 1 with 130 zeros) to make a sequence of just 100 amino acids, and 100 amino acids is a very short polypeptide. Longer sequences of, say 10,000 amino acids, allow seemingly unlimited arrangements. Cells, then, can potentially manufacture any of a nearly infinite number of different proteins. Recall that the human body, for example, manufactures 50,000 to 100,000 different proteins.

Despite their diversity, we can nonetheless describe proteins in general terms. Most proteins are compact, roundish molecules called **globular proteins.** A classic example of a globular protein is hemoglobin, the iron-rich protein that carries oxygen in the blood. Distinct from the globular proteins are the **fibrous proteins,** which provide structural support within organisms. Fibrous proteins, such as the keratin in hair and silk and the collagen in skin and bone, have an elongated shape (Figure 3-21B).

Biochemists have defined four levels of protein structure. The **primary structure** of a protein is the linear sequence of amino acids in each chain (Figure 3-21A). The **secondary structure** of a protein is the three-dimensional arrangement of the polypeptide chains that compose the protein (Figure 3-21B). One kind of secondary structure, Pauling's **α helix**, found in keratin—the fibrous proteins in hair, wool, feathers, nails, hooves, and horns—looks like a ribbon wrapped around a cylinder.

The **tertiary structure**, or **conformation**, of a protein is the three-dimensional folding of an entire polypeptide chain (Figure 3-21C). If you curl a ribbon into a spiral (or helix) by scraping it with a blade, you have produced secondary structure. If you then take the curled ribbon and tie it

around a birthday present, you have given the ribbon a tertiary structure.

Many proteins also have **quaternary structure**—the fitting together of two or more folded chains (Figure 3-21D). This is analogous to piecing together several different ribbons into an elaborate bow. Just as the ribbons can be either all the same color or a mixture of different colors, proteins can be made of several identical polypeptide chains or of several different ones. Hemoglobin, for example, is made of four polypeptide chains, two of one kind of chain and two of another.

Besides the **α** helix, two other common kinds of secondary structure deserve special mention. The **beta (β) structure**, shown in Figure 3-22, is the main feature of the fibrous proteins in silk. Globular proteins also contain β structures, with two to five parallel sections of a single chain forming a **β sheet** (Figure 3-22). The fibrous protein collagen—found in the skin and throughout the body—contains a third type of secondary

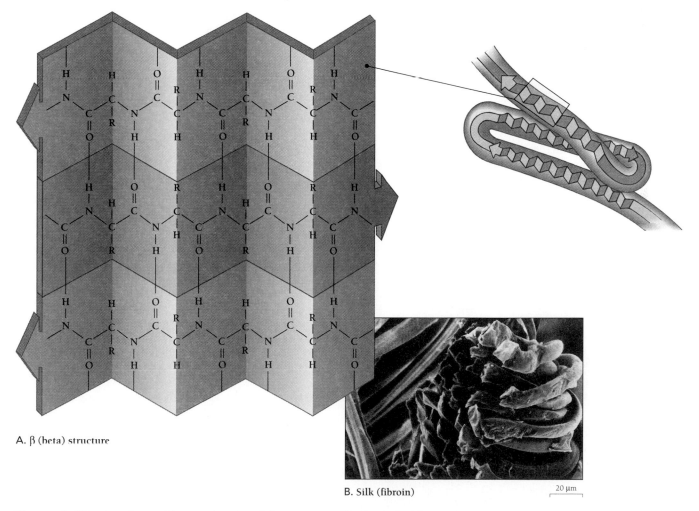

A. β (beta) structure

B. Silk (fibroin)

20 μm

Figure 3-22   A β sheet, with several antiparallel segments and hydrogen bonding. Linus Pauling correctly predicted two types of β structures: A. the parallel β structure, in which the hydrogen-bonded polypeptides run in the same direction; and B. the antiparallel β structure, in which the chains run in opposite directions.   (B, K. Jakes, The Ohio State University)

structure called a **collagen helix** (Figure 3-23). A collagen helix consists of three polypeptide chains wound around each other.

The Structure of a Protein Depends on Interactions Among Its Amino Acids

The α helix, the β structure, and the collagen helix all share an important trait. As regular repeating forms, they are strong. Their simple shapes allow them to function as standardized building materials, comparable to the wooden beams and bricks with which a house is built. All fibrous proteins, whose role is to provide mechanical support, are made of these simple repeating forms.

The three secondary structures discussed each result from an extremely regular sequence of amino acids in the polypeptide chain. In the β structures of silk protein, for example, every other amino acid is glycine, and in the collagen helix, every third amino acid is proline.

In contrast to the fibrous proteins, globular proteins have nonrepeating sequences of amino acids. As a result, globular proteins fold into intricate shapes that are unique to each protein (Figure 3-24). The many unique shapes of globular proteins allow them to play more diverse roles than the fibrous proteins. Globular proteins act as hormones, as antibodies in the immune system, as blood proteins such as hemoglobin, and as enzymes.

The tertiary folding of polypeptide chains results from interactions of the side chains on each amino acid. Each side chain has its particular characteristics. Some are polar, some nonpolar. Some can serve as proton donors in hydrogen bonds. Others serve as proton acceptors. Either a hydrogen bond or the attraction between positively and negatively charged side chains can stabilize

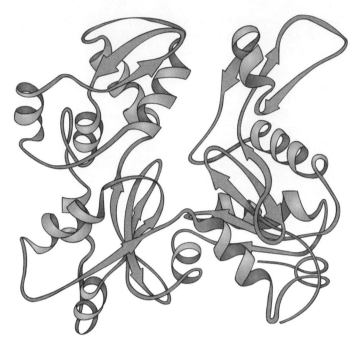

Figure 3-24   The globular protein insulin, shown here as a ribbon model, has a unique shape.

a particular shape. Conversely, some interactions, such as the repulsion between two like-charged side chains, destabilize a conformation.

Singly, any of these interactions contributes only weakly to the stability of a given molecular structure. Together, however, the various interactions establish a three-dimensional structure that can be extremely stable in certain chemical environments. Because the internal environment of a cell is relatively constant, each protein usually maintains its characteristic conformation. Subtle changes in the cellular environment, however, can change the shape of a protein and change the way it behaves. The sensitivity of protein form and function to environmental factors is the key to the regulation of biological processes.

Structural proteins tend to have regular sequences of amino acids and regular secondary structures such as the α helix, the β structure, and the collagen helix. Globular proteins have nonrepeating amino acid sequences and irregular and unique shapes that facilitate, for example, their roles as antibodies and enzymes.

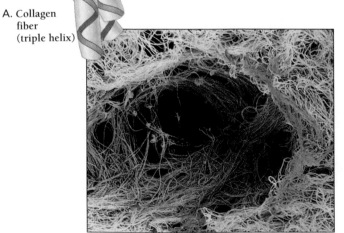

A. Collagen fiber (triple helix)

B. Collagen fibers   5 μm

Figure 3-23   Collagen is the major constituent of tendons, ligaments, bones, muscle, and skin. A. A three-stranded collagen helix. B. Collagen fibers.   *(B, Professor P.M. Motta & E. Vizza/Science Photo Library/Photo Researchers)*

Are the Interactions Among Amino Acids Enough To Determine Protein Structure?

We have seen that the interactions among the functional groups in the polypeptide backbone and among the amino acid side chains stabilize the conformation of a protein. But biochemists wondered if these interactions were enough to determine the

BOX 3-3

Dorothy Crowfoot Hodgkin

When Dorothy Crowfoot Hodgkin was ten years old, she watched her first crystals form on a string dangling in a glass of salt water. Many children before and since have done the same, but in Hodgkin's case, the sparkling geometric shapes kindled a fascination that would lead her to world fame.

In 1964, nearly half a century later, Hodgkin received a Nobel Prize in Chemistry for teasing out the structures of penicillin and vitamin B12 from photographic images of their crystals. She made the images with a technique called x-ray crystallography, which involved firing x rays through a crystal to determine the arrangement of the atoms in it. It was a bit like determining the shape of a jungle gym from its shadow.

Born in 1910, Hodgkin spent the first few years of her life in Cairo, where her father was an official in the British colonial government. Most of her education had been at home, but once back at school in England, her keen interest in crystals won the attention of her schoolteacher. Hodgkin and a friend got special permission to join the boys studying chemistry. By age 12, she was doing chemistry experiments on rocks she found in her garden to see what they contained.

That summer, while visiting her father in Khartoum, Sudan, she met Dr. A.F. Joseph, a friend of her father's and a well-known soil chemist. Joseph took her on a tour of his laboratory. Pleased by her intense interest, he put together a small chemistry set for her, which she took back to England and set up in her mother's attic. It was her first laboratory.

Hodgkin enrolled at Oxford University, where she eventually specialized in x-ray crystallography. At the time, the analysis of the structures of even the sim-plest chemicals by x-ray crystallography required at least 30 sets of calculations, all done by hand. The work demanded perseverance and diligence, and a good head for math. Under these conditions, Hodgkin flourished.

Seeking a greater challenge after college, Hodgkin went to Cambridge to study with a young crystallographer named J.D. Bernal. Together they solved some of the most complex chemical structures ever attempted, including those of several vitamins and sex hormones. They took the first x-ray photographs of a protein—the stomach enzyme pepsin—showing that proteins form regular crystals.

In 1937, Hodgkin received her doctorate. Within a few months, she also married historian Thomas Hodgkin, taking his name. The Hodgkins were a two-career family, working in different towns and commuting on alternate weekends to see each other. Dorothy Hodgkin remained at Oxford, where she continued her research, taught university classes, and raised three children.

When the demand for penicillin soared during World War II, chemists all over the world raced to determine its structure. Experimental chemists used chemical reactions. Structural chemists, such as Hodgkin, used crystallography. Despite daunting calculations, Hodgkin and her students at Oxford completed the structure in 1949, beating the experimental chemists and establishing x-ray crystallography as an indispensable tool in biochemistry. Even as Hodgkin was finishing her analysis of penicillin, however, she had already begun a study of vitamin B12, widely used to treat pernicious anemia. In 1957, she published the structure of this 180-atom molecule.

When she was awarded the Nobel Prize in 1964, she told a group of students

Figure A Dorothy Crowfoot Hodgkin. *(Express Newsphotos/ Archive Photos)*

at the ceremonies in Stockholm, Sweden, that she hoped her position as the only woman to receive the prize that year "will not be so very uncommon in the future. . . as more and more women carry out research in the same way as men."

But what was perhaps Hodgkin's greatest success came after the Nobel Prize, when she tackled the biggest molecule of her career. Insulin, a protein that regulates the body's sugar storage, contains over 1000 atoms. A deficiency in or insensitivity to insulin causes diabetes, a complex disease that causes suffering in several hundred million people worldwide. Hodgkin solved the structure of insulin in only five years. Her achievement proved that proteins have regular shapes, and it spawned research that ultimately led to effective treatments for diabetes.

---

precise three-dimensional structure of a protein. Might not some outside molecule somehow direct the shaping of a polypeptide? In that case, the shape of a polypeptide would depend not only on its amino acid sequence but also on the presence of some other "instructional" molecule.

To answer this question, the biochemist Christian Anfinsen performed an experiment with the enzyme ribonuclease, which hydrolyzes, or breaks, the bonds in RNA molecules (Figure 3-25). Anfinsen treated ribonuclease with a solution that interfered both with hydrogen bonds and hydrophobic interactions. This treatment changed the enzyme's shape and destroyed its ability to function. Yet, when the solution was removed, the ribonuclease spontaneously reverted back to its "native," functional shape.

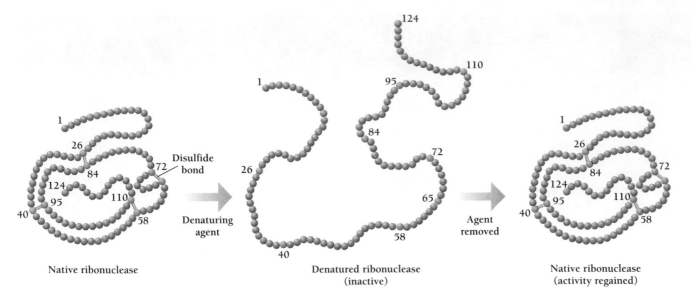

Figure 3-25 Christian Anfinsen denatured, then renatured, the protein ribonuclease to show that this polypeptide could regain its "native," functional shape without help.

This recovery required no other molecule to guide the protein's folding. The amino acid sequence of the protein was sufficient to determine its three-dimensional structure and biological activity. Anfinsen's experiment suggested that proteins spontaneously fold into their biologically active forms. For many years, molecular biologists accepted the idea that the three-dimensional structure of a protein is implied by its amino acid sequence. Biologists guessed that sequences that led to multiple structures had probably been selected out over millions of years of evolution.

More recently, as molecular biologists have begun reprogramming cells to produce proteins, they have discovered that many energy-requiring events contribute to the folding of a protein. A pure polypeptide in a dilute solution will ultimately adopt a single, stable conformation. But newly made proteins left to themselves usually do not fold properly—whether in test tubes of protein chemists or in the interiors of cells.

At the high-protein concentrations within the cell, for example, individual peptide chains tend to form bonds with other peptides, rather than with themselves. Molecular biologists call such intermolecular bonding **promiscuous interactions.** Special proteins, called **chaperones,** prevent such promiscuous interactions by binding to unfolded polypeptides and catalyzing correct folding.

The interactions among the amino acids in a polypeptide were once thought to be sufficient to determine the three-dimensional structure of a protein. Today researchers are discovering that other molecules often help catalyze the correct folding of a protein.

In this chapter we have seen how the structure of a molecule—whether at the level of individual functional groups or at the level of amino acid sequences and tertiary structure—profoundly affects the chemistry of the molecule. The structure of a cell determines its behavior and capabilities just as profoundly, as we will see in the next chapter.

STUDY OUTLINE WITH KEY TERMS

After water, nearly all molecules found in living organisms are made of chains of carbon atoms. In fact, all **organic** molecules contain carbon. **Organic chemistry** is the study of the structures and reactions of carbon compounds, while **biochemistry** is the study of the chemistry of living organisms.

The properties of small molecules depend on **functional groups,** such as the **hydroxyl, carboxyl,** and **amino** groups. The properties of macromolecules depend on the nature and sequence in which the component building blocks assemble.

Some biological molecules form rings of carbon. Flat, five- or six-carbon rings are called **aromatic** rings. Most macromolecules are linear, however. The small building blocks, or **monomers,** all assemble into different kinds of macromolecules, or **polymers,** by the same kind of general reaction. This chemical reaction is called the **dehydration condensation reaction** because the equivalent of one molecule of water is eliminated each time two building blocks join together. Conversely, the breaking down of a macromolecule into its constituent building blocks is called **hydrolysis** because the equivalent of one molecule of water is added as each small molecule is removed from the chain.

The building blocks of cells fall into four classes—**sugars, amino acids, nucleotides,** and **lipids.** Sugars form polymers called **polysaccharides.** Amino acids form **polypeptides,** and nucleotides form **nucleic acids.**

Although lipids do not form polymers, some, called **fatty acids,** do contain long carbon chains. Fatty acids may be **saturated, unsaturated,** or **polyunsaturated.** A **glycerol** molecule with two fatty acid chains and a phosphoric acid group is a **phospholipid,** a highly amphipathic kind of molecule that makes up cell membranes. A glycerol molecule with three fatty acid chains is a **triacyl glycerol.** Different kinds of triacyl glycerols make up the **fats** (solid at room temperature) and **oils** (liquid at room temperature) of animals and plants. A third class of lipids consisting of four interconnected rings is the **steroids,** including both **cholesterol** and **steroid hormones** such as **progesterone** and **testosterone.**

Simple sugars, or **monosaccharides,** are the building blocks of **disaccharides** and **polysaccharides.** Sugars, such as **glucose** and **sucrose,** and polysaccharides, such as **starch** and **cellulose,** are all classed as **carbohydrates.** A five-carbon sugar is a **pentose,** while a six-carbon sugar is a **hexose.**

Every kind of protein has a unique sequence of amino acids and characteristic properties. The folding of a **polypeptide** into a precise three-dimensional structure generally requires energy. This energy comes from noncovalent interactions among the atoms of the polypeptide backbone and those of the amino acid side chains, as well as between the protein and its watery environment.

X-ray crystallography reveals the three-dimensional structures of proteins and other molecules. Polypeptides form both regular and irregular structures. The regular structures within protein molecules include the α helix, β structures, and the **collagen helix.**

REVIEW AND THOUGHT QUESTIONS

Review Questions

1. Distinguish between the terms polymer and monomer.
2. Where would you find a large concentration of lipids in a cell? Why?
3. What building blocks make up proteins? How many kinds of these building blocks are there?
4. What three components make up a nucleotide?
5. What small molecule is removed during a dehydration condensation reaction? Why are these types of reactions also called polymerization reactions? What is the opposite reaction called? What happens in that case?
6. What is the relationship between polysaccharides, sugars, and carbohydrates? What three roles do carbohydrates play in cells?
7. Define peptide bond, polypeptide, protein, and enzyme.
8. Define nucleic acid.
9. Describe the four levels of structure that occur in protein folding.
10. Distinguish between the α helix, the β structure, and the collagen helix, all of which occur in fibrous proteins. Who discovered the α helix?

Thought Questions

11. What is it about the shape of a protein that determines its properties and therefore its functions? Describe the shape of a globular protein and of a fibrous protein. How are these two shapes usually put to different purposes in a cell?
12. The 35 small molecules listed in Table 3-1 are called building blocks. How can the same building blocks be used to construct billions upon billions of different macromolecules?

SELECTED READINGS

Judson, Horace Freeland, *The Eighth Day of Creation,* Simon and Schuster, New York, 1979. A comprehensive history of molecular biology for a lay audience.

Needham, Joseph (ed.), *The Chemistry of Life,* Cambridge University Press, Cambridge, 1970. Eight essays on the history of biochemistry.

Stryer, L., *Biochemistry,* 4th ed. W. H. Freeman, New York, 1995. A lucid beautifully written introduction to biochemistry.

▶ On-line materials relating to this chapter are on the World Wide Web at http://www.saunderscollege.com/lifesci/
Click on Tobin/Dusheck: *Asking About Life.*

Very Little Animalcules

More than three hundred years ago, an uneducated Dutch cloth merchant named Antonie van Leeuwenhoek (1632–1723) discovered something so interesting that the Czar of Russia, King James II of England, and Frederick the II of Prussia all came to visit the merchant. Like many young men, Leeuwenhoek had a consuming interest that distracted him from his work (Figure 4-1). He liked to make glass lenses by grinding and polishing bits of glass and then mounting the lenses between gold or copper plates. He used these simple, single-lens microscopes to examine anything he encountered.

To Leeuwenhoek's delight, his lenses—some no larger than a pinhead—revealed hundreds of tiny living beings never before seen by human eyes. Leeuwenhoek astonished his friends and acquaintances with descriptions of red blood cells in blood, sperm in semen, and "very little animalcules" in rainwater. He even saw bacteria just 1/10th the size of a red blood cell. With his lenses, Leeuwenhoek also demonstrated that fleas, ants, weevils, and other pests do not arise spontaneously from dust or wheat, as was commonly believed, but developed from larvae that hatch from tiny eggs laid by the adults. Leeuwenhoek's discoveries dazzled educated people throughout Europe. Aristocratic men and women came from all over Europe to peer through his lenses. Amazingly, Leeuwenhoek's microscopic techniques were so advanced, and his methods so secret, that no one was able to repeat or confirm some of his observations of bacteria for 200 years.

In 1673, when Leeuwenhoek was 41, he began writing of his discoveries to the recently created Royal Society of England, one of the few scientific organizations in the world at that time. Over the next 50 years, until Leeuwenhoek's death in 1723, the Society received and published 375 letters from him, the last when the Dutch lens grinder was 90, not long before he died.

Over the years Leeuwenhoek described how everywhere he looked—in pond water, in the human mouth and intestine—he found tiny organisms. In 1677, he discovered one-celled organisms, and also confirmed the presence of sperm in semen, correctly guessing that sperm are the agent of reproduction. He found that blood is full of cells, and, in 1683, he became the first person to see bacteria. Leeuwenhoek's great contribution to science was to demonstrate that life is not limited to organisms visible with the naked eye.

In England, the Royal Society welcomed Leeuwenhoek's fascinating observations through the microscope. They had already hired Robert Hooke to be their "curator of instruments." Hooke's job was to demonstrate the microscope and other new technologies for the entertainment of the gentlemen at the Royal Society's

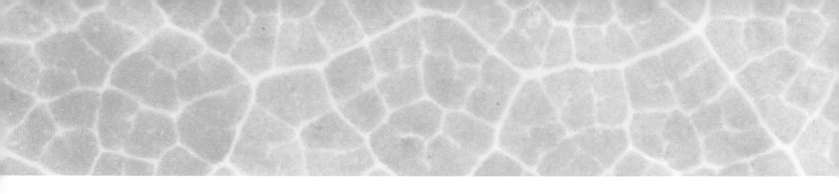

meetings. In 1665, Hooke published his book *Micrographia,* a series of illustrations and descriptions of his observations using new and improved microscopes. Hooke showed, for example, the structure of feathers, the stinger of a bee, and the foot of a fly.

Hooke also coined the term **"cellulae"** [Latin, *cella* = small room] to describe the boxlike cavities (in other words, cells) he saw when he examined a slice of cork under a microscope. Cellulae, in those days, referred to the small rooms of a monastery, nunnery, or prison. Hooke later published detailed drawings of cells in other plant tissues, noting that they contained "juices."

Leeuwenhoek, Hooke, and the few other microscopists of the day described a whole new world of tiny living organisms, and their research revealed an unsuspected level of structural complexity in larger organisms. With these discoveries, it might have seemed that the history of cell biology was about to take off. Yet, oddly, these discoveries lay dormant for nearly 200 years.

Figure 4-1 Antonie van Leeuwenhoek built primitive microscopes so effective that he saw bacteria 200 years before anyone else. *(Science Photo Library/Photo Researchers)*

Leeuwenhoek showed that life is not limited to organisms visible with the naked eye.

Neither Leeuwenhoek's nor Hooke's studies led directly to any unifying biological principles. Most scientists dismissed Leeuwenhoek's animalcules as little more than amusing natural curiosities. Hooke's "cellulae" were simply bubbly stuff in plants. No one guessed that cells come from other cells by the process of cell division, nor that cells are the fundamental unit of life—that all tissues in all organisms are made of cells. No one guessed that some microscopic organisms cause disease. Leeuwenhoek's microorganisms and Hooke's cells seemed no more important to most people than the specks of dust in a ray of sunshine.

One reason that neither Hooke's nor Leeuwenhoek's observations were taken seriously by the important thinkers of the day was the two men's social class. Leeuwenhoek was an uneducated amateur, and Hooke was an employee of the Royal Society, not a full member. Neither man was, in other words, a "gentleman," so neither commanded respect.

Equally important, 17th-century scientists and philosophers carried a bias, left over from the Middle Ages, that valued theory over experiments and observations. Hooke's and Leeuwenhoek's microscope work was entirely descriptive.

Leeuwenhoek described thousands of objects but, probably because of his lack of education, he did not attempt to present any unifying explanation for *why* things were as they were.

The microscope was viewed as more of a toy than a scientific instrument. By the end of Leeuwenhoek's life at the beginning of the 18th century, few people were even making lenses anymore. Those simple microscopes already in existence were playthings for ladies, of whose observations there is little record.

Another important reason that study of the microscopic details of life languished was the poor quality of 17th- and 18th-century microscopes. Leeuwenhoek's handmade lenses were far better than any others in existence. Some of the 400 lenses he left behind when he died could magnify objects 300 times, which is powerful enough to show the largest bacteria. And his mysterious techniques—probably sophisticated lighting—which he never revealed, allowed him to see details that neither Hooke nor any of the other prominent microscopists of the time could match. Yet, even Leeuwenhoek's microscopes were crude by today's standards (Box 4-1). His finely polished lenses distorted both the shapes and the colors of the objects under the microscope. It was not until the late 19th century, a hundred years after the beginning of the Industrial Revolution, that technology provided microscopes adequate for the study of the insides of cells.

BOX 4-1

How do microscopes help biologists study cells?

An important way to learn about cells is to look at them. Each major improvement in microscopy has led to a burst of new knowledge and a host of new questions. In the 17th century, the development of the microscope made it possible to see cells for the first time. In the 19th century, successive improvements in microscope design revealed that all organisms are made of cells and that their elaborate adaptations allow specific functions. By the mid-20th century, the development of new types of microscopes had further revolutionized the study of cells—revealing the rich internal structure of cytoplasm, nuclei, and membranes.

How can we see such details? We can see an object only when it is large, clear, and stands out from its background. In order to see an object as small as a cell, first we must use lenses to magnify its image. The ratio between the size of the image and the size of the object itself is called the **magnification.**

Even when we can magnify an object, however, we may not be able to see it clearly enough to learn anything about it. When small objects are too close together, we cannot tell whether we are looking at one or several objects. The **resolution** of a microscope or other optical instrument is the minimum distance between two objects that allows them to form separate images. We can say that resolution is a measure of image clarity. Our eyes have a resolution of about 0.1 mm. A microscope resolves images less than 1/100th that size (0.001 mm or 1 μm). Such resolving power is enough to distinguish individual prokaryotic cells and some internal structure in eukaryotic cells.

Visibility depends not only on magnification and resolution but also on **contrast**—how well an object stands out from its background. Contrast depends on an object's tendency to absorb more or less light than its surroundings. A transparent object is invisible. So is an object that is the same shade as its background.

Microscopists have devised ingenious ways of increasing contrast. Nineteenth-century biologists, for example, found that different dyes would bind to different parts of the cell—one dye to the nucleus, another to the membrane, another to parts of the cytoplasm. Dyes that bind differently to cell components are called stains.

But staining cells usually kills them, so early biologists could study living cells only with difficulty. Beginning in the 1930s, biologists developed techniques for studying living cells. In conventional light microscopy, contrast depends on how different cellular structures absorb light. Most materials in unstained cells absorb little visible light. However, substances that do not absorb or reflect much light can still change the speed at which light passes through. For example, light passes through the cell nucleus more slowly than it does through the watery cytoplasm. The light passing through the nucleus becomes "out of phase" with light waves that pass through the cytoplasm. Special optical arrangements such as "phase contrast" and "interference contrast" microscopy can heighten this effect, revealing phase differences among different cell components without the use of stains. These techniques permit the study of living cells.

Figure A(a) shows the workings of an ordinary light microscope. Because a light microscope contains several lenses, it is also called a **compound microscope.** It consists of a tube with lenses at each end, a stage that holds a specimen, and a light source. The light source has its own lens, which allows the user to control the light that passes through the specimen. The lens near the eye is called the **eyepiece,** and the lens at the other end of the tube, near the object being examined, is called the **objective.**

The objective lens magnifies the image of the specimen and projects the image to the eyepiece. The eyepiece magnifies the image again. The final magnification is the product of the magnification of the objec-

tive and that of the eyepiece. For example, an objective that magnifies 40 times (40×) and an eyepiece that magnifies 10 times (10×) would produce an image 400 times (400×) as large as the original specimen.

A light microscope often has three objective lenses (for example, with magnifications of 10×, 40×, and 100×) mounted on a rotating turret, so that the viewer can change magnifications. Because eyepieces usually contribute another 10-fold magnification, the final image usually ranges from 100 to 1000 times the size of the specimen.

Even the best lenses, however, have a resolution limited by the nature of light. Light is composed of waves, and lenses cannot resolve structures smaller than the wavelength of light. The wavelength of visible light ranges from abut 0.4 μm (for blue light) to about 0.7 μm (for red light). Light microscopes reveal no details of subcellular structures smaller than about 0.5 μm.

One solution is to use wavelengths far smaller than those of visible light. Electron microscopes, which see objects as small as 0.002 μm, depend on the wave properties of moving electrons. In a 100,000-volt electron microscope, the wavelength of an electron is 0.004 nanometers (nm), or 0.000004 μm, less than the diameter of an individual atom.

The lenses of an electron microscope are electromagnets that bend the path of electrons just as a glass lens bends the path of light. In reality, the resolution for looking at cells is only about 2 nm (0.002 μm), but still about 100 times better than the best light microscope (Figure A(b) and (c)).

Biologists commonly use two kinds of electron microscopes—the transmission electron microscope (TEM) and the scanning electron microscope (SEM). In each case the electrons form an image on a phosphorescent screen, like that of a television set. Figure A compares a *Paramecium* seen with a light microscope, a transmission electron microscope, and a scanning electron microscope.

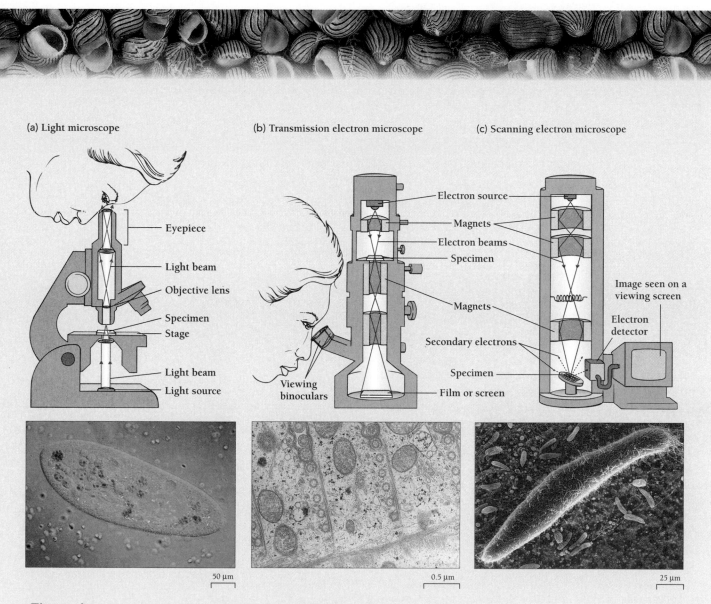

(a) Light microscope

- Eyepiece
- Light beam
- Objective lens
- Specimen
- Stage
- Light beam
- Light source

(b) Transmission electron microscope

- Electron source
- Magnets
- Electron beams
- Specimen
- Magnets
- Secondary electrons
- Specimen
- Viewing binoculars
- Film or screen

(c) Scanning electron microscope

- Image seen on a viewing screen
- Electron detector

50 µm   0.5 µm   25 µm

Figure A   Views of *Paramecium* with different types of microscopes.
(a) With a light microscope, we see the internal anatomy of this one-celled organism and a fringe of hairlike cilia, which these creatures use to propel themselves through the water. (b) With a transmission electron microscope, we see a cross section through three rows of cilia at much greater magnification and in great detail. (c) With a scanning electron microscope, set at low power, we see the cilia protruding from the surface in greater detail than with the light microscope but nothing of the *Paramecium's* internal anatomy.   (a, Eric Grave/Phototake; b, David Phillips/Visuals Unlimited; c, Dr. Dennis Kunkel/Phototake)

In TEM, a beam of electrons passes through the specimen in the same way that a beam of light would in a light microscope. SEM is different. In SEM, a narrow beam of electrons moves back and forth across the surface of the specimen. Wherever the beams hits the surface of the specimen, it causes the surface to emit more electrons, called secondary electrons. These secondary electrons produce a detailed, almost three-dimensional, picture of the specimen's surface.

KEY CONCEPTS

1. All organisms are made of one or more cells—the basic living units of function and organization.

2. Every cell has a boundary, a set of genes, and a cell body. Different cells types are distinguished by different combinations of organelles and other subcellular structures.

3. Internal membranes divide eukaryotic cells into distinct compartments, each with a characteristic structure and function. Prokaryotic cells have no internal membranes.

4. Cells of all six kingdoms have both common and distinctive internal features.

5. Membranes are organized arrangements of lipids and proteins.

6. Membranes help regulate the chemical composition of the spaces they enclose.

7. The outer surfaces of cells are responsible for their interactions with other cells and with other molecules in their environments.

8. The cytoskeleton contributes to the internal organization of eukaryotic cells.

WHY ARE ALL ORGANISMS MADE OF CELLS?

In the 1820s, better microscopes paved the way for rapid advances in cell biology. Within 10 years, biologists recognized for the first time two important parts of cells. They discovered that both animal and plant cells contain a nucleus, a dark mass that we now know contains the genetic material. Researchers also finally realized that the "juice" observed by Robert Hooke was a living substance, which they named **protoplasm** [Greek, *proto* = first + Latin, *plasma* = a thing molded or formed].

All Organisms Are Made of Cells

In 1838, the German botanist Matthias Schleiden proposed that all plants consist of cells—an idea seconded for animal cells in 1839 by German zoologist Theodor Schwann. Cells, they wrote, are the elementary particles of all living organisms. In addition, Schwann and Schleiden argued that all cells are alive—independent of the organisms to which they belong. In other words, a liver cell is alive in its own right even though it is only a small part of a living organism. Remove the cell from the liver and the cell is still alive.

Recall from Chapter 1 that animals and plants are all multicellular, that the Fungi and Protista include both multicellular organisms and single-celled organisms, and that the Archaea and Eubacteria are all single-celled organisms. Noncellular organisms do not exist. Further, all cells—from Leeuwenhoek's one-celled "animalcules" to the individual cells of the liver—conform to the definitions of life. Cells are highly organized, obtain energy and materials from their food, change with time, respond to their environments, and reproduce.

Biologists now recognized the importance of cells. Cells, they saw, were everywhere. But where did cells come from? Where did the new cells that heal a wound come from? Where did the cells that build a child's growing body come from? Where did the burgeoning masses of cells in cancers come from? And where did the one-celled organisms living in a pond come from?

Schleiden and Schwann proposed that cells crystallized spontaneously out of shapeless matter. However, ever-improving microscopes allowed more and more scientists to see that individual cells can divide, each giving rise to what biologists call "daughter" cells. In time, biologists accepted the idea that all new cells come from the division of preexisting cells.

In 1858, the widely respected physician and biologist Rudolf Virchow formalized this understanding with the phrase *omnis cellula e cellula,* "all cells from cells." In a book written for physicians, Virchow summarized his theory about the role that cells play in disease. In doing so, he simultaneously revolutionized both biology and medicine. Virchow argued (1) that cells never arise from noncellular material and (2) that diseases result from changes in specific kinds of cells.

For the first time, biologists universally accepted the principle that cells always come from other cells. Cells healing a wound result when preexisting skin cells begin dividing. A child's growth results from the rapid multiplication of cells. Cancers result from the uncontrolled division of an organism's cells. Single-celled organisms in ponds descend from other single-celled organisms.

Today, we summarize the work of Schleiden, Schwann, and Virchow in the **cell theory,** which states that (1) all organisms are composed of one or more cells; (2) cells, themselves alive, are the basic living unit of organization of all organisms; and (3) all cells come from other cells.

The cell theory says that all organisms are composed of one or more cells, cells are the basic living unit of organization of all organisms, and all cells come from other cells.

Every Cell Consists of a Boundary, a Set of Genes, and a Cell Body

A look at Table 4-1 shows that the cells of organisms from the six kingdoms can differ greatly. Even within a single organism, cells come in wildly different types (Figure 4-2). Nonetheless, all cells have three common features: (1) a boundary that separates the inside of the cell from the rest of the world, (2) a set of genetic instructions, and (3) a cell body.

The **plasma membrane** is a cell's boundary, a highly organized and responsive structure. Like the outer wall of a house, with its windows and doors, the plasma membrane not only defines the limits of a cell, but also helps to regulate the cell's internal environment by selectively admitting and excreting specific molecules.

The genetic instructions for each cell are contained in one or more molecules of DNA. In eukaryotes (animals, plants, fungi, and protists), each cell's DNA is confined within a membrane-enclosed structure called the **nucleus** (Figure 4-3A). The presence of a true nucleus characterizes the eukaryotes and distinguishes them from the prokaryotes (eubacteria and archaebacteria). In the prokaryotes, the DNA occupies a limited region of the cell, called the **nucleoid**, which has no membrane (Figure 4-3B and Table 4-1).

Throughout the 19th century and well into the 20th century, biologists continued to think of the protoplasm as a more or less homogeneous jellylike substance. Improved techniques for looking at cells, however, revealed within the protoplasm of eukaryotic cells more and more tiny structures, called **organelles**, each of which we now know performs a specialized task. Prokaryotes, by contrast, have no membrane-enclosed organelles.

Biologists discarded the word protoplasm and renamed the cell body—that part of the cell outside the nucleus but inside the membrane—the **cytoplasm** [Greek, *cyto* = a hollow container (a cell) + Latin, *plasma* = a thing molded or formed].

Table 4-1   Cells of the Six Kingdoms

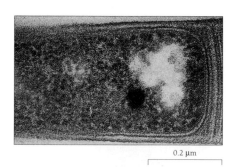

0.2 μm

A. Archaea: *Methanospirillum hungatei.*

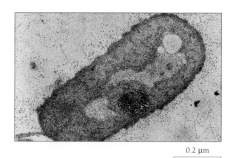

0.2 μm

B. Eubacteria: *Bacillus megaterium.*

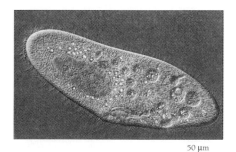

50 μm

C. Protista: *Paramecium.*

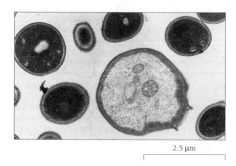

2.5 μm

D. Fungi: yeast.

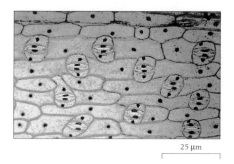

25 μm

E. Plantae: onion leaf epidermis.

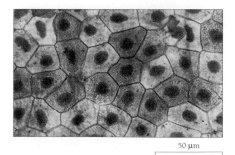

50 μm

F. Animalia: Salamander squamous epithelium.

(A, T.J. Beveridge/Visuals Unlimited; B, Ralph A. Slepecky/Visuals Unlimited; C, M.I. Walker/Photo Researchers; D, E.H. White/Science VU/Visuals Unlimited; E,F, © Dwight Kuhn)

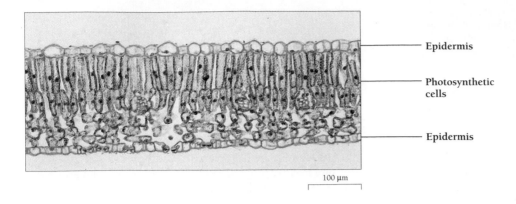

Epidermis

Photosynthetic
cells

Epidermis

100 μm

Figure 4-2 An organism may be made
of many kinds of cells. Cross section of lilac
leaf (*Syringa* sp.) shows epidermal cells and, inside, photosynthetic cells.   *(Runk/Schoenberger
from Grant Heilman)*

The part of the cytoplasm not contained within membrane-bounded organelles was named the **cytosol.** Most of a cell's biochemical work occurs within the cytosol. In addition, running through the cytosol is a complicated network of protein fibers, called the **cytoskeleton,** which gives the cell its shape, holds organelles in place, and participates in cell movement.

Every cell is bounded by a plasma membrane and keeps its DNA
in a nucleus (eukaryotes) or nucleoid (prokaryotes). In
eukaryotes, other organelles also occupy the cytoplasm.

What Are the Advantages of Cellular Organization?

Single-celled organisms are nearly all very small. Plants, animals, and other multicellular organisms can be as large as trees and elephants, but a free-living cell never grows that big. In fact, with some remarkable exceptions, cells are nearly all about the same size. Most eukaryotic cells range from 10 to 100 micrometers (μm), while prokaryotic cells are smaller, ranging

from 0.4 to 5 μm. No cells are smaller than 0.4 μm, and almost none are larger than 100 μm. Exceptional cells may be enormous. Some plant fiber cells are a meter long (1 million μm), while the nerve cells in the legs of a giraffe may be several meters long.

One reason that cells are nearly all so small is that subdivision into tiny cells offers organisms many advantages. To understand these advantages, we can start by asking a simple question: What problems would you face if you were one very large cell?

The Cell's Need To Regulate Its Internal Environment Limits Its Size

For the biochemical machinery of a cell to function, the cell must maintain a relatively constant internal environment. Otherwise, enzymes and other proteins will not adopt the shapes needed for biological activity. To maintain a constant internal environment, the cell must keep both the concentration of salts and the pH from changing. At the same time, the cell must also be able to take in useful molecules and dispose of waste molecules.

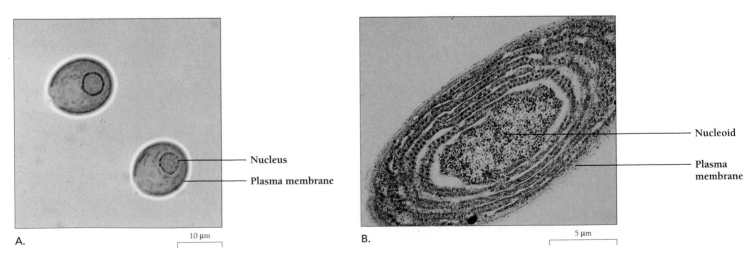

Nucleus

Plasma membrane

Nucleoid

Plasma
membrane

A.   10 μm

B.   5 μm

Figure 4-3 Eukaryotes have a nucleus. A. A eukaryotic cell, such as the green alga
Chlamydomonas, has both a plasma membrane and a nucleus. B. A prokaryotic cell such as
a cyanobacterium has a plasma membrane, but no nucleus.   *(A, Philip Sze/Visuals Unlimited; B,
Elizabeth Gentt/Visuals Unlimited)*

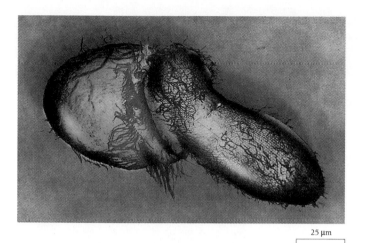
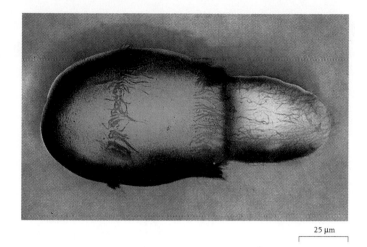

25 µm

25 µm

Figure 4-4 *Didinium* eating a *Paramecium*. The *Didinium* has a mouth with which it can consume other protists. *(Biophoto Associates/Science Source/Photo Researchers)*

If you were one big cell, you would have to maintain just the right environment for each of the different molecular activities in your body. The powerful acid that you use to digest food would have to somehow coexist with the delicate pH balance needed for protein synthesis.

All the materials that come into a cell or leave a cell must go through the thin membrane that envelopes the cell. If you were one big cell, your plasma membrane would have to admit or exclude specific molecules selectively. Somehow you'd have to absorb enough pizza to get you through the day.

Protists, such as the *Didinium* pictured in Figure 4-4, do all of this and more. In fact, unlike most protists, the *Didinium* even has a sort of mouth, called a **cytostome** [Greek, *cyto* = cell + *stoma* = mouth], with which it is eating another protist (a *Paramecium*). Protists, however, are all small. Accomplishing all that a protist does wouldn't be so easy for a one- or two-hundred-pound cell like yourself. The membrane's ability to admit or to excrete molecules is limited by its size. A big cell certainly has more external membrane than a small one. As a cell grows, however, its volume increases faster than its surface area.

Let's think, for a minute, about the relationship between the surface area and the volume of any object. Suppose we have a tiny cardboard cube, one centimeter on each side. If we double the linear dimensions of the cube, so that each side is 2 cm by 2 cm, the surface area of the cube increases by a factor of 4. But its *volume* increases by a factor of 8 (Figure 4-5).

The larger cube has proportionately less surface area than the smaller cube. We can see this if we look at the **surface-to-volume ratio**—the amount of surface area for each bit of volume. For the small cube, the surface-to-volume ratio is 6:1. For each cm^3 of volume, the smaller cube has 6 cm^2 of surface. But the surface-to-volume ratio for the larger cube is 24:8 (or 3:1). For each cm^3 of volume, the larger cube has only 3 cm^2 of surface.

Because larger objects have smaller surface-to-volume ratios, larger cells have proportionately less membrane with which to regulate their internal space. As a result, larger cells have a harder time obtaining nutrients, getting rid of wastes, and regulating the internal concentrations of ions and molecules. If you were one big cell, you could bury yourself in pizza and you'd still go to bed hungry. You could not absorb enough pizza molecules through your membrane to survive.

Knowing how a cell's size affects the regulation of its internal environment, we might wonder how eukaryotic cells, which range from 10 to 100 µm, can be so much larger than prokaryotic cells, which range from 0.4 to 5 µm. The answer is that eukaryotic cells have special adaptations to increase their surface areas. For example, many eukaryotic cells have convoluted surface membranes and almost all have elaborate internal membrane systems, which we will discuss later in this chap-

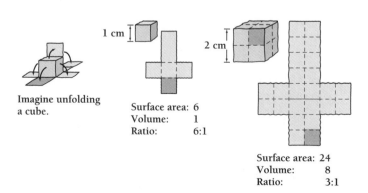

1 cm

2 cm

Imagine unfolding a cube.

Surface area: 6
Volume: 1
Ratio: 6:1

Surface area: 24
Volume: 8
Ratio: 3:1

Figure 4-5 **The relationship between surface area and volume determines many of the properties of cells.** In particular, smaller cells have more surface area per volume than do larger cells. Here we can see that a cube 1 cm on a side has a surface area of 1 square centimeter on each of its 6 sides, so that it has a total surface area of 6 square centimeters (6 cm^2). Its volume is 1 cm^3. When we double the linear dimensions of the cube, we have a cube 2 cm by 2 cm on each side, which gives a surface area of 24 cm^2. Its volume is 8 cm^3. We can see that when the linear dimensions of an object double, surface area increases 4 times, while volume increases 8 times.

BOX 4-2

Caulerpa, the world's largest single-celled organism?

Many cells vie for the title "World's Largest Single Cell." Near the head of the line are the enormous nerve cells of some animals—those in giraffes' legs may be several meters long. Yet these cells are not self-perpetuating organisms, but the pampered parts of a body, cared for by a complex system of blood vessels and immune cells. The ostrich egg might be a contender, to say nothing of the still larger eggs of Cretaceous dinosaurs. But an egg cell is as much the creation of a complex multicellular animal as a nerve cell. It is an individual, but it is not a mature, functioning individual.

Some researchers give a score of 10 to the green alga *Caulerpa*—without question, one of the world's largest single-celled organisms. *Caulerpa* flourishes in warm, shallow, tropical and subtropical seas throughout the world (Figure A). Some species grow to be a meter long—10,000 times larger than most other cells. If most cells are limited in size to less than 100 microns across, how does *Caulerpa* manage so well?

First, *Caulerpa* divides into compartments. Far from being a huge, undifferen-

tiated blob, *Caulerpa* consists of a slender filament that grows along the ocean floor and thin, leaflike projections that extend up into the water. The green fronds in your dentist's saltwater aquarium may, in fact, be *Caulerpa*.

This giant cell is bounded not only by its plasma membrane, but by a cell wall, a rigid structure enclosing the membrane. The cell wall provides support. In addition, rods, acting as struts, protrude from the cell wall into the cytoplasm, spanning the cell and further stiffening it. Running both perpendicular and parallel to the long axis of the cell, the rods form a dense lattice throughout the cell and act as an internal skeleton.

Why isn't acquiring nutrients and excreting wastes an insurmountable problem for this giant cell? One answer is shape: being long and thin, not short and fat, helps *Caulerpa* by increasing its surface-to-volume ratio. Even so, it would appear that, to transport nutrients and wastes in and out of the cell, *Caulerpa* has to move them further than does a typical cell.

The answer may be *Caulerpa*'s vacuoles, fluid-filled sacs inside the cell. The

Figure A   The plant-sized algal cell *Caulerpa* can grow up to 1 m in length. *(Philip Sze/Visuals Unlimited)*

vacuoles take up most of the space inside the cell, squeezing the cytoplasm into a thin layer between the vacuole and the cell wall. In *Caulerpa*, that layer of cytoplasm is only five microns thick. As a result, whether a *Caulerpa* cell needs to absorb nutrients or excrete wastes, the distance from the cytosol to the external environment—the warm, tropical ocean—is similar to that of other, smaller cells.

ter. Indeed, some eukaryotic cells, such as the white blood cell, have 50 times more membrane inside than outside.

body. You wouldn't have lungs to extract oxygen from the air. You'd just have 125 pounds of cytoplasm.

The larger a cell becomes, the smaller its surface-to-volume ratio. A low surface-to-volume ratio limits a cell's ability to absorb nutrients and expel wastes.

Cellular organization allows organisms to make a division of labor among specialized cells.

Individual Cells May Specialize for Different Tasks

In multicellular organisms, individual cells often perform separate tasks. For example, red blood cells carry oxygen, while muscle cells contract and move. Cells in the roots of a tree absorb nutrients from the soil, while cells in the leaves harvest the energy of sunlight. Just as a division of labor makes human societies more productive, the specialization of cells allows an organism to function efficiently.

If you were one big cell, organizing your body to perform all its different jobs would be difficult. You wouldn't have a heart or arteries to move oxygen and nutrients around your

A Multicellular Organism Can Lose and Replace Individual Cells

Subdividing the body into cells has another advantage. Cells often live and die independently of the whole organism. In fact, the death of some cells is a normal part of development. The fingers of the hand, for example, develop from paddlelike appendages as a result of the death of the cells between the fingers. The lenses of the eyes, the red blood cells that carry oxygen to all the tissues, and the outer layer of the skin all develop because of the death of specific cells. Equally important, the cells of the skin, blood, and intestines continuously replace themselves. The advantage of such replacement is that the life of a multicellular organism can extend far beyond the life of an individual cell.

Cellular organization allows organisms to outlive the cells that compose them.

WHAT'S IN A CELL?

Eukaryotic cells contain so many different organelles, each with a special set of tasks, that we can imagine each cell as a tiny walled city (Figure 4-6). Inside the wall of the city are power stations, a central library of genetic information, warehouses for packaging proteins, and much more.

When scientists first began using the transmission electron microscope in the 1940s and 1950s, one of the great surprises was the huge amounts of internal membrane in eukaryotic cells (Box 4-1). These membranes divide cells into at least seven kinds of compartments: the nucleus, the cytosol, the endoplasmic reticulum, the Golgi complex, the lysosomes, the peroxisomes, and the mitochondria. In addition, a plant cell contains one more kind of compartment—the plastids.

Eukaryotic cells also contain a variety of membrane-bounded sacs without any obvious internal structure. The **vesicles** of plant and animal cells are small, generally about 100 nm in diameter. Plant cells also contain **vacuoles**—large sacs that may occupy as much as 95 percent of the volume of the cell. Table 4-2 summarizes the characteristics of cells from the six kingdoms of organisms.

Table 4-2   The Cells of All Six Kingdoms Have Both Common and Distinctive Features

Cell Component	Function	Eubacteria Archaebacteria	Protist	Fungus	Plant	Animal
Cell wall	Protect and support cell	Yes	Yes	Yes	Yes	None
Plasma membrane	Regulate communication with other cells and movement of materials in and out of cell	Yes	Yes	Yes	Yes	Yes
Membrane-bounded nucleus	Isolate DNA from cytoplasm	None	Yes	Yes	Yes	Yes
DNA	Encode information for of proteins	Single molecule	Many molecules			
RNA	Construction of proteins from information in DNA	Yes	Yes	Yes	Yes	Yes
Nucleolus	Synthesize ribosomes	None	Yes	Yes	Yes	Yes
Ribosome	Synthesize proteins	Yes	Yes	Yes	Yes	Yes
Endoplasmic reticulum	Modify proteins, synthesize lipids	None	Yes	Yes	Yes	Yes
Golgi complex	Tag proteins and lipids, package them for export outside cell	None	Yes	Yes	Yes	Yes
Lysosome	Contain digestive enzymes	None	Yes	Yes	Yes	Yes
Mitochondrion	Synthesize ATP	None	Yes	Yes	Yes	Yes
Chloroplast	Photosynthesize	None	Some	None	Yes	None
Nonchloroplast plastids	Store food, pigments	None	Some	None	Yes	None
Central vacuole	Provide turgor pressure; contain water and wastes	None	None	Yes	Yes	None
Cytoskeleton	Shape, strengthen, and move cell	None	Yes	Yes	Yes	Yes
Flagellum, cilium	Move cell through fluid or fluid past cell surface	None	Yes	Yes	Yes	Yes
Bacterial flagellum	Move cell through fluid or fluid past cell surface	Yes	None	None	None	None

Figure 4-6   Two eukaryotic cells, an animal cell and a plant cell.
A eukaryotic cell, with its plasma membrane and membrane-bounded
organelles, is like a walled medieval city. The plasma membrane regulates the passage of molecules in and out of the cell, just as the city
wall regulates the passage of people. Inside, the mitochondria supply
power to the cell in the form of ATP. The nucleus is a library, housing
all of the information a cell needs to produce proteins. The nuclear
membrane determines which molecules have access to the information inside and which bits of information (books) may be checked out
for use in the cytoplasm. The ribosomes act as workbenches for the
production of proteins. The vacuole serves as a depository for waste
materials, much like a toxic waste dump. The Golgi complex is like a
mail room. Here glycoproteins are packaged and labeled for export
or intracellular transport. See if you can think of analogies for some of
the other structures in these cells. To what would you compare a
chloroplast?

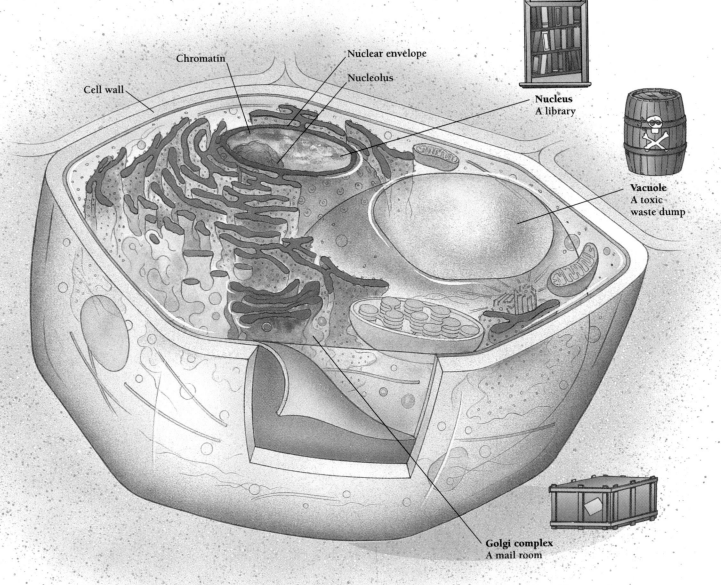

Cell wall

Chromatin

Nuclear envelope

Nucleolus

Nucleus
A library

Vacuole
A toxic
waste dump

Golgi complex
A mail room

PLANT CELL

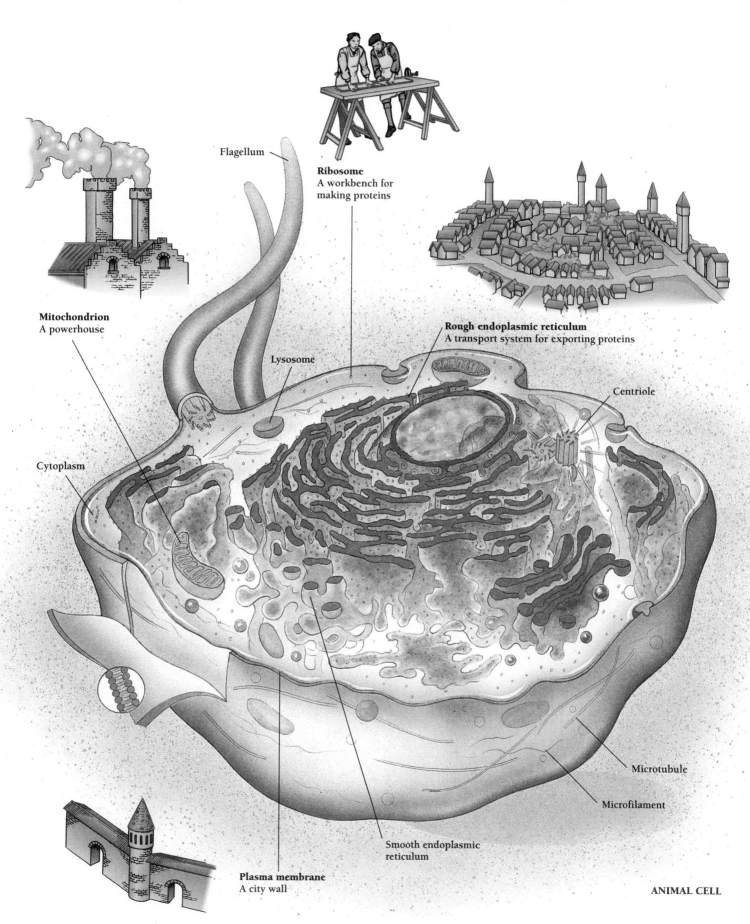

Flagellum

Ribosome
A workbench for
making proteins

Mitochondrion
A powerhouse

Rough endoplasmic reticulum
A transport system for exporting proteins

Lysosome

Centriole

Cytoplasm

Microtubule

Microfilament

Smooth endoplasmic
reticulum

Plasma membrane
A city wall

ANIMAL CELL

What Role Does the Nucleus Play in the Life of a Cell?

The nucleus usually makes up 5 to 10 percent of the volume of the cell. In cells that are not dividing, the nucleus stains as a diffuse mass. This pattern results from the binding of stains to **chromatin** [Greek, *chroma* = color], a complex of DNA and protein. In cells that are dividing, the chromatin forms distinct pieces called **chromosomes** [Greek, *chroma* = color + *soma* = body] (Figure 4-7).

An electron microscope shows that the nucleus' boundary is a double membrane called the **nuclear envelope** (Figure 4-8). The two membranes are separated by a space about 20 to 40 nm wide. In many places, however, the two membranes are fused to form interruptions called **nuclear pores**, which form channels between the nucleoplasm (the contents of the nucleus) and the cytoplasm. These pores resemble the holes in a tea ball that allow water to enter the tea ball and tea to leak out into the surrounding water. However, nuclear pores play a more active role than mere holes. They control the movement of materials in and out of the nucleus in much the same way that a turnstile controls the movements of people in and out of a subway.

The nucleus serves as the library for the rest of the cell. It contains the genetic information (coded in DNA) that is passed from one generation to the next. When a cell divides, each daughter cell gets a complete copy of the genetic instructions in the nucleus. The DNA in each cell nucleus includes instructions for building every polypeptide the body will ever use. In Part II of this book, we will consider how DNA carries

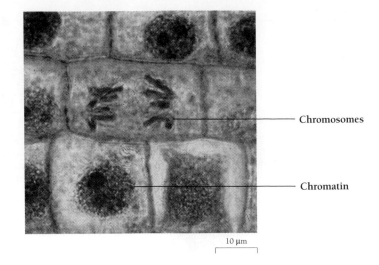

Figure 4-7   Chromatin in the nucleus of an onion cell. Chromatin condenses in dividing cells, making the chromosomes visible. (*Runk/Schoenberger from Grant Heilman*)

genetic information, how cells use this information, and how they pass it to their daughter cells.

In the nucleus of eukaryotic cells lie the chromosomes, which are made of chromatin, a complex of DNA and protein. The DNA acts as a library of information, which cells pass on to their daughter cells.

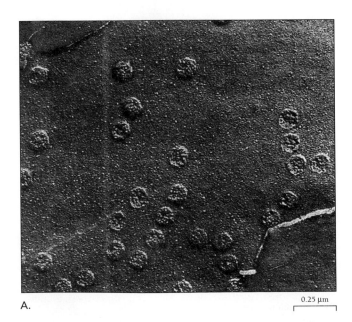

A.   0.25 μm

Figure 4-8   The nuclear membrane and nuclear pores. The structure of the pores is revealed by a technique called freeze fracture that reveals surface structure by splitting the membrane. (*A, R. Kessel-G.Shih/Visuals Unlimited*)

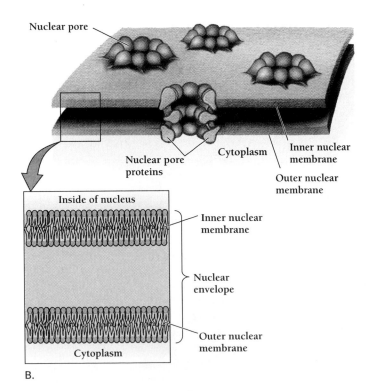

B.

The Cytosol Is the Cytoplasm That Lies Outside the Organelles

The cytosol makes up about half of a cell's volume. Cell biologists usually isolate the cytosol from the organelles suspended in it by breaking open cells with detergents and then spinning the mixture in a centrifuge. A centrifuge is a machine that spins a set of tubes (Figure 4-9). The tubes are arranged so that when they are spinning their contents come under a force greater than gravity. This force—centrifugal force—is the same force that presses clothes in a washing machine to the outside wall of the tub during the spin cycle. In the centrifuge, a tube spins so fast that the contents of the tube sink to the bottom, with bigger particles and molecules sinking faster than smaller ones. When cell biologists centrifuge cytoplasm from cells, the organelles, which are heavier than the cytosol, drop soonest to the bottoms of the tubes.

The cytosol contains thousands of different kinds of enzymes responsible for producing building blocks, degrading small molecules to yield energy, and synthesizing proteins. Although the cytosol is aqueous (mostly water), about 20 percent of its weight is protein, giving it the consistency of Jell-O. Embedded in the cytosol are organelles, vesicles, and vacuoles, as well as smaller structures that are not enclosed by membranes. These include granules of energy-rich glycogen, droplets of stored fat, and *ribosomes*—tiny, round organelles, about 15 to 30 nm in diameter.

Ribosomes are complexes of RNA and protein that are indispensable to protein synthesis. Ribosomes [Latin, *soma* = body + "ribo" because they contain ribonucleic acid (RNA)] act as workbenches on which protein molecules are stitched together with peptide bonds. Ribosomes in the cytosol are called "free ribosomes." Ribosomes associated with a cell's internal membranes are called "membrane-bound ribosomes."

The jellylike cytosol, which surrounds the cell's organelles, contains ribosomes, glycogen, fat, thousands of enzymes, and much more.

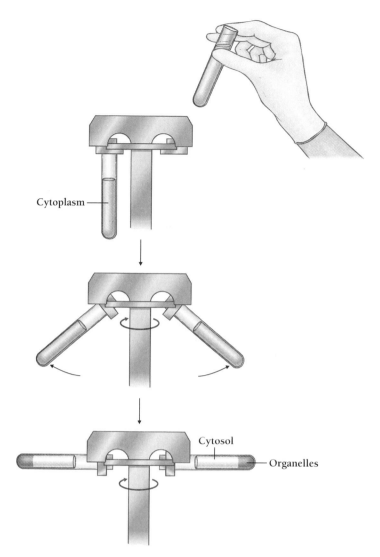

Cytoplasm

Cytosol

Organelles

Figure 4-9   The centrifuge spins a tube of macromolecules up to 80,000 revolutions per minute (rpm). Such speed places the contents of the tube under forces more than 100,000 times that of gravity. Large molecules travel to the bottom of the tube fastest.

The Endoplasmic Reticulum Is a Folded Membrane

The **endoplasmic reticulum (ER)** [Greek, *endon* = within + *plasmein* = to mold + Latin, *reticulum* = network] is an elaborate structure roughly similar in appearance to the marbling in marble cake. When biologists cut a cell into uniform slices, as we would slice a loaf of marble cake, the ER in each slice, or "section," resembles an extensive and convoluted network. Careful study of successive sections, like those in Figure 4-10, show that the ER is actually a single sheet of membrane enclosing a network of interconnected cavities and channels referred to as the **lumen** [Latin, *lumen* = light, an opening]. The ER lumen is like a three-dimensional maze. The ER membrane is continuous with the outer membrane of the nucleus. In many eukaryotic cells, the lumen of the ER makes up about 15 percent of the cell's volume, and the membranes of the ER more than half the total membrane.

The ER consists of two parts that are distinct in both appearance and function. The **rough ER** is studded with ribosomes on the cytoplasmic side of the membrane and is mostly devoted to modifying recently synthesized proteins, especially those destined for export to organelles or out of the cell (Figure 4-10B). The **smooth ER** has no ribosomes and is mostly devoted to the synthesis and metabolism of lipids (Figure 4-10C). Another important function of the smooth ER is detoxification of substances such as alcohol. Both rough and smooth ER are present in most eukaryotic cells. But since the two kinds of ER serve different purposes, some specialized cells may have especially large amounts of one or the other. Cells specialized for the production of secreted proteins, such as those in the

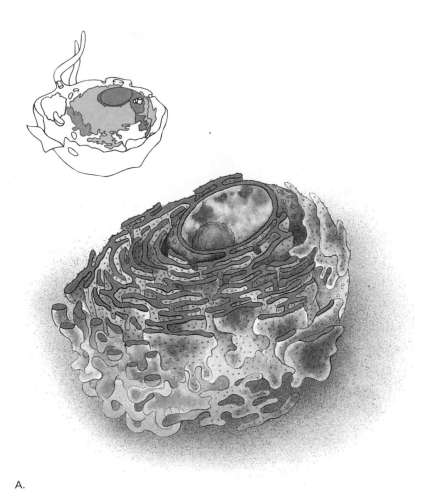

A.

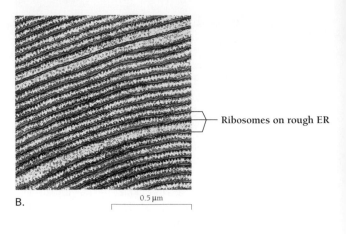

Ribosomes on rough ER

B.   0.5 μm

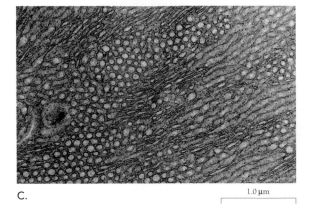

C.   1.0 μm

Figure 4-10 The two parts of the endoplasmic reticulum, the rough ER and the smooth ER, have different functions. A. The rough ER is continuous with the smooth ER. B. The rough ER is studded with ribosomes and is mostly devoted to synthesis and modification of proteins, especially those destined for export to organelles or out of the cell. C. The smooth ER has no ribosomes and is mostly devoted to the synthesis and metabolism of lipids and the detoxification of substances such as alcohol.   *(B, Don Fawcett/Science Source/Photo Researchers; C, David M. Phillips/Visuals Unlimited)*

pancreas, have large amounts of rough ER. Cells specialized for the production of lipids, such as the cells that produce steroid hormones, contain large amounts of smooth ER.

The endoplasmic reticulum (ER), with its mazelike internal lumen, consists of rough ER and smooth ER.

The Golgi Complex Directs the Flow of Newly Made Proteins

The **Golgi complex** functions as both a packaging center and as a traffic director. It helps form structures that stay in the cell, such as lysosomes, and it also helps shuttle materials outside the cell.

The Golgi complex, usually found near the nucleus, consists of sets of flattened discs, like stacked dinner plates (Figure 4-11). Each stack is about 1 μm in diameter and typically consists of about 6 discs. Surrounding the stacks of flat discs are some smaller, spherical vesicles.

Every eukaryotic cell has a Golgi complex, but the number of stacks varies among different cell types. The cells with the most Golgi stacks are those that secrete large quantities of **glycoproteins,** which are proteins, such as the albumin in egg white, that have sugars or carbohydrates attached. Mucus is also made of glycoproteins, and the mucus-secreting goblet cells of the intestines have particularly large numbers of Golgi stacks (Figure 4-11). Most of the proteins that are exported to the outside of the cell are glycoproteins. Biologists like to say, "All exported proteins are sugar-coated."

To see what the Golgi complex does, George Palade and Marilyn Farquhar, at Rockefeller University, labeled newly made glycoproteins with radioactive tracers. Proteins destined for ex-

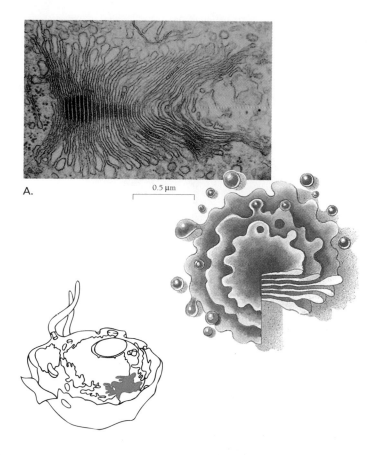

A.

0.5 μm

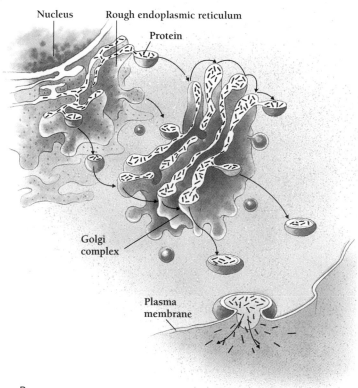

Nucleus   Rough endoplasmic reticulum

Protein

Golgi
complex

Plasma
membrane

B.

Figure 4-11 Newly made proteins pass through the ER and then the Golgi complex. Vesicles on the surface of the Golgi complex shuttle glycoproteins from the rough ER to the Golgi stack, which "labels" the glycoproteins and packages them into vesicles for secretion. *(A, Cabisco/Visuals Unlimited)*

port appeared first in the rough ER and then in the Golgi complex (Figure 4-11). The Golgi complex modifies the glycoproteins, then packages them into membrane-bounded secretory vesicles. The vesicles eventually fuse with the plasma membrane and discharge their contents to the outside of the cell.

The Golgi complex also manages the flow of proteins to different destinations. The Golgi complex "addresses" glycoproteins by modifying the carbohydrates (which were originally attached to the glycoproteins in the lumen of the ER). Proteins acquire specific "tags" that direct them to the mitochondria, the lysosomes, or other places in the cell. Other proteins are labeled specifically for export outside of the cell.

The Golgi complex packages and labels glycoproteins for destinations both inside and outside the cell.

The Lysosomes Function as Digestion Vats

Lysosomes are small, membrane-bounded vesicles found in all cells (Figure 4-12A). Lysosomes vary widely in size, but all contain enzymes that break down proteins, nucleic acids, sugars, lipids, and other complex molecules. The large vacuole of a plant cell contains similar enzymes, and we may consider it a giant lysosome. Lysosomes are particularly numerous in cells that actively perform **phagocytosis** [Greek, *phagein* = to eat + *kytos* = hollow vessel, or cell], the process by which cells take in and consume large particles of solid food. Free-living amebas and many other single-celled eukaryotes feed primarily by phagocytosis (Figure 4-12B and C). Many of our own white blood cells regularly engulf and digest bacteria, viruses, and assorted cellular debris, for example, and these white cells are packed with lysosomes. Once a cell has engulfed material within a vesicle, the vesicle fuses with a lysosome, whose enzymes then digest the engulfed material.

The membrane of a lysosome keeps its enzymes from digesting the cytosol's own proteins. If the lysosome membrane breaks down, however, the cell begins to digest itself and dies.

Lysosomes are made by other organelles in the cell. Lysosomes first form from membranes derived from the Golgi complex. Their digestive enzymes are made on the ribosomes of the rough ER. They are packaged into lysosomes after passing through the lumen of the ER, where each lysosomal enzyme acquires a special attached sugar. This sugar provides the lysosomal proteins with an address tag, so that the Golgi complex directs them to lysosomes rather than exporting them to the outside of the cell.

Lysosomes contain enzymes that the cell uses to digest food, wastes, and foreign invaders. Lysosome membranes normally isolate these dangerous enzymes from the rest of the cell.

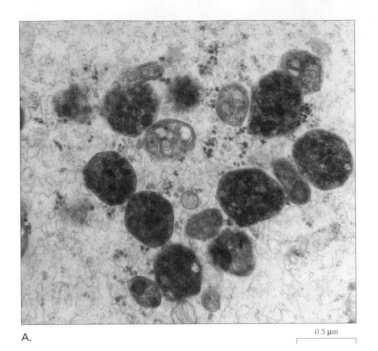

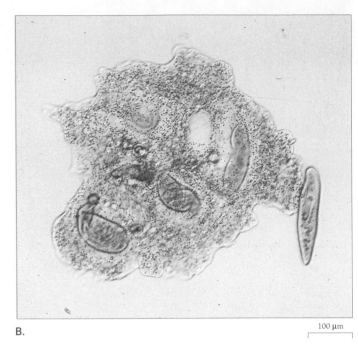

A.   0.5 μm   B.   100 μm

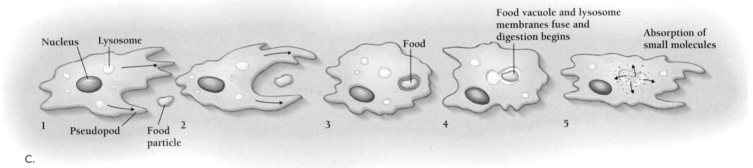

C.

Figure 4-12   Phagocytosis. A. Lysosomes, which contain digestive enzymes, are abundant in cells that perform phagocytosis. B. An ameba engulfs its food. C. The cell membrane fuses around the particle to form a large vesicle, which travels to the interior of the cell. Finally, the vesicle fuses with a lysome.   *(A, Don Fawcett/Visuals Unlimited; B, Runk/Schoenberger from Grant Heilman)*

The Peroxisomes Produce Peroxide and Metabolize Small Organic Molecules

Peroxisomes are another kind of membrane-bounded vesicle widespread in all kinds of eukaryotes (Figure 4-13). **Peroxisomes** contain enzymes that use oxygen to break down various molecules by means of biochemical reactions that produce hydrogen peroxide (H_2O_2), hence their name. Because hydrogen peroxide is highly reactive and can easily damage cells, peroxisomes also contain a protective enzyme, called *catalase*, that breaks down hydrogen peroxide into water and oxygen.

Peroxisomes, whose diameters can range from 0.15 μm to more than 1 μm, can occupy as much as half the volume of a cell. They are especially abundant in cells that synthesize, store, or break down lipids. For example, peroxisomes in the seeds of some plants contain an enzyme that breaks down stored fats and therefore provides energy for the growing plant embryo.

Peroxisomes contain a chemical dangerous to the cell itself—hydrogen peroxide. Cells use enzymes that give off hydrogen peroxide to break down fats and oils.

The Mitochondria Capture the Energy from Small Organic Molecules in the Form of ATP

Mitochondria [singular, mitochondrion; Greek, *mitos* = thread + *chondrion* = a grain] are the powerhouses of all eukaryotic cells. In cells that do not directly harvest sunlight, mitochondria produce nearly all of the ATP the cell needs to power the chemical reactions of the cell. The production of ATP by mitochondria depends on the organization of the internal and external membranes of mitochondria.

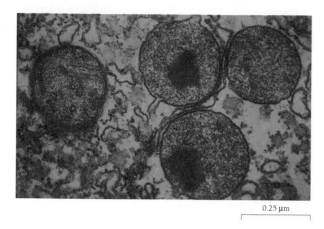

0.25 μm

Figure 4-13 Peroxisomes contain enzymes that degrade molecules in reactions that produce hydrogen peroxide. *(D. Friend & D. Fawcett/Visuals Unlimited)*

Even under the best light microscope, mitochondria, often less than 1 μm in diameter, can barely be seen. However, the transmission electron microscope (TEM) reveals that mitochondria are some of the most numerous organelles in most eukaryotic cells (Figure 4-14). Different kinds of cells have different numbers of mitochondria. A liver cell, for example, may contain several thousand mitochondria, which together make up about a quarter of the cell volume. Cells that use particularly large amounts of energy, such as those of the heart, may contain still more.

TEM also reveals the mitochondria's characteristic structure. Two membranes, 6 to 10 nm apart, separate the mitochondrion's interior from the cytosol. The outer membrane of a mitochondrion is smooth, but the inner membrane forms elaborate folds, called *cristae*. Mitochondria come in a variety of sizes and shapes. Time-lapse photography with a phase contrast microscope reveals that mitochondria may be constantly in motion, changing shape, fusing, and dividing.

Mitochondria synthesize energy-rich
ATP in their mazelike interior.

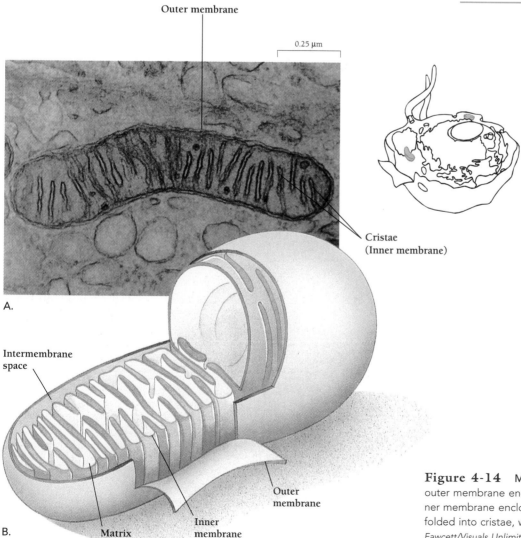

Outer membrane

0.25 μm

Cristae
(Inner membrane)

A.

Intermembrane
space

B. Matrix

Inner
membrane

Outer
membrane

Figure 4-14 Mitochondria manufacture ATP. The outer membrane encloses the whole organelle, and the inner membrane encloses the matrix. The inner membrane is folded into cristae, which extend into the matrix. *(A, Don Fawcett/Visuals Unlimited)*

A Plant Cell's Chloroplast Is Just One Kind of Plastid

Plastids are organelles that, like mitochondria and nuclei, are surrounded by a double membrane. The most important plastid is the chloroplast, which is responsible for turning sunlight into chemical energy. Most organisms on Earth ultimately depend on chloroplasts for food. Plants also have other kinds of plastids, which serve as storage depots for various kinds of molecules (Figure 4-15).

A **chloroplast** is an organelle, found only in plants, that uses the energy of sunlight to build energy-rich sugar molecules in the process of **photosynthesis** [Greek, *photo* = light + *suntithenai* = to put together]. During photosynthesis, sugars are synthesized within the chloroplast and temporarily stored there. The size, shape, and number of chloroplasts per cell all vary from species to species and from cell type to cell type. A leaf cell in a flowering plant, for example, typically contains several dozen chloroplasts. A stem cell contains only a few, and a flower cell may contain none at all.

A typical chloroplast is a large, round, green organelle. The distinctive shade of green comes from the pigment *chlorophyll*. It is 5 to 10 μm in diameter, but only 2 to 4 μm high, a thick disc. Like a mitochondrion, a chloroplast has an intricate internal structure of folded membranes. But the chloroplast's membranes are not continuous with the inner membrane of the chloroplast envelope. Instead, the internal membranes lie in flattened disclike sacs, called **thylakoids** [Greek, *thylakos* = sac + *oides* = like] (Figure 4-15A). The thylakoids are piled up like plates. Each stack, of 10 or more discs, is called a **granum** [Latin, *granum* = grain] (plural, grana) because the stacks resemble the granaries used by farmers to store grain. Each chloroplast contains many grana.

Two other kinds of plastids are chromoplasts and amyloplasts. **Chromoplasts** contain the color pigments that give yellow, orange, and red color to certain flowers, fruits, vegetables, and autumn leaves (Figure 4-15B). Chromoplasts arise from chloroplasts through the reshaping of the inner membrane and the breakdown of chlorophyll. **Amyloplasts** do not derive directly from chloroplasts, but from the common precursor of all plastids, which is called a *proplastid*. Amyloplasts, which store starches, are abundant in roots such as sweet potatoes and turnips, as well as in wheat, rice, and other seeds (Figure 4-15C).

Chloroplasts turn solar energy into chemical energy in the process of photosynthesis. Other kinds of plastids store pigments or starches.

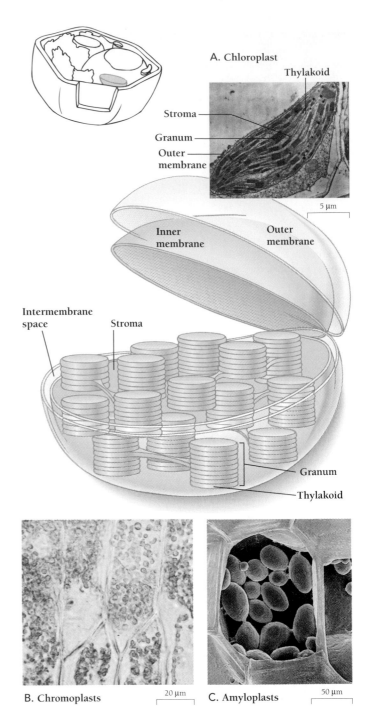

Figure 4-15 Plastids. A. A chloroplast uses sunlight to build sugar molecules, which are temporarily stored within the chloroplast. A chloroplast's green color comes from the pigment *chlorophyll*—a central player in photosynthesis. Like mitochondria, chloroplasts have an outer membrane, which encloses the whole organelle, and an inner membrane, which encloses a matrix. In chloroplasts, the inner membrane is even more highly folded than the inner membrane of mitochondria, forming pockets called thylakoids, arranged in stacks called grana. The intermembrane space is called the stroma. B. Chromoplasts in the cells of a yellow petal (from a tulip). C. Amyloplasts in raw potato. *(A, Grant Heilman Photography; B, Biophoto Associates/Science Source/Photo Researchers; C, Dr. Jeremy Burgess/Science Photo Library/Photo Researchers)*

What Does the Cytoskeleton Do?

Eukaryotic cells have a complex network of protein filaments, called the **cytoskeleton,** which is visible in electron micrographs (Figure 4-16A). Like our bony skeleton, the cytoskeleton gives structure and support to the cytoplasm and its component organelles.

The name cytoskeleton is slightly misleading, however, since the cytoskeleton does more than give mechanical support. Like muscle, it also provides the force for most of the movements of cells, including the transport of materials and changes in the overall shape of a cell. In fact, some of the specialized proteins of the cytoskeleton are closely related to the proteins that make up muscles. While muscle proteins are essentially permanent structures, however, the cytoskeleton is constantly dissolving and reforming itself. At least three types of filaments make up the cytoskeleton: microtubules, actin filaments, and intermediate filaments (Figure 4-16).

Microtubules are hollow cylinders about 25 nm in diameter, which vary in length (Figure 4-16B). Microtubules play a critical role in cell division. When a cell divides, microtubules distribute the DNA and other materials to the two daughter cells.

The cylinders of microtubules consist of two globular protein molecules called **tubulins.** Pure tubulin in a test tube can form microtubular structures all by itself. In living cells, how-

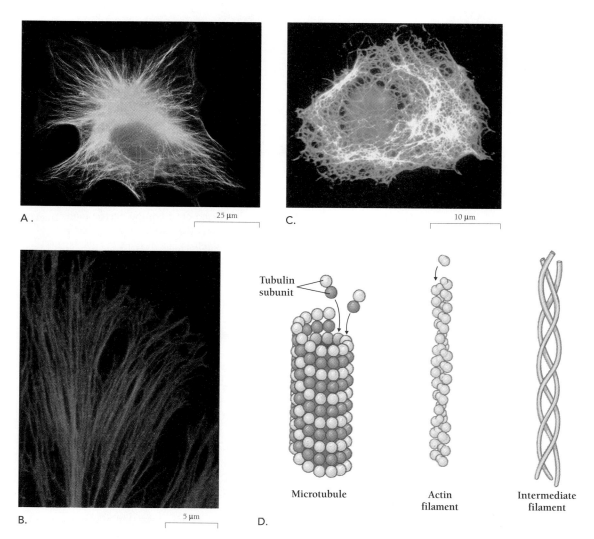

A.   25 μm

C.   10 μm

B.   5 μm

D.

Tubulin subunit

Microtubule

Actin filament

Intermediate filament

Figure 4-16   **The cytoskeleton of a cell is a complex network of protein filaments.**
The cytoskeleton's three types of protein filaments have distinctive appearances and distributions. A. In this cell, microtubules are labeled green and actin microfilaments are labeled red. B. Microtubules. C. Intermediate filaments. D. The tubulin molecules are arranged in a helix to form a hollow microtubule. Actin filaments are finer solid filaments made of actin. Intermediate filaments, thicker than actin filaments but finer than microtubules, are made of fibrous proteins such as keratin.   *(A, B, M. Schliwa/Visuals Unlimited; C, K.G. Murti/Visuals Unlimited)*

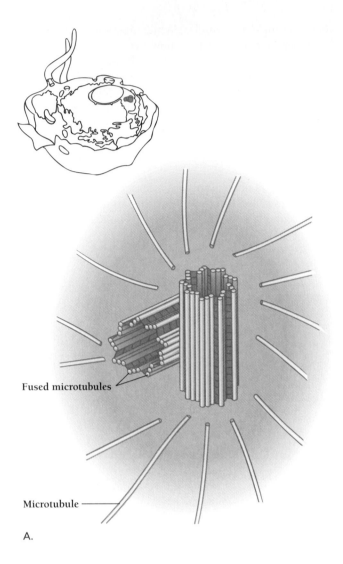

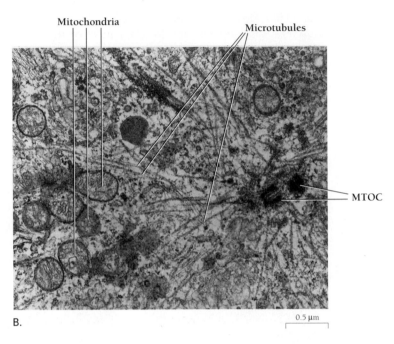

Figure 4-17 Microtubules radiating out from the microtubule organizing center. Most MTOCs are near the nucleus in a zone called the centrosome, which plays a central role in cell division. (B, K.G. Murti/Visuals Unlimited)

ever, up to 50 other proteins may associate with microtubules, influencing their distribution and stability.

Although microtubules appear throughout the cytoplasm, and even in the nucleus, their assembly always seems to originate in structures called **microtubule organizing centers (MTOCs)** (Figure 4-17). In the cytoplasm of most eukaryotic cells, the major MTOCs are near the nucleus in a zone called the **centrosome**—a region that (in many cells) also contains a distinct organelle called the **centriole**. The MTOCs of the centrosome play an important role in cell division, as we will discuss in Chapter 8.

Actin filaments, also called **microfilaments,** are much finer than microtubules—only about 7 nm in diameter (Figure 4-16A). Actin filaments are often anchored to the cell surface and provide the force for movement and shape changes. Actin filaments consist of a single kind of protein, called actin, which is also present in muscle fibers.

In muscle, actin filaments are permanent structures that interact with other proteins to produce the force leading to contraction. In nonmuscle cells, actin filaments are constantly being built and destroyed. Even in the test tube, actin molecules can self-assemble to form a two-stranded helix apparently iden-

tical to the actin filaments in intact cells. In cells, they form cross-linked cables that provide mechanical support within cells. They also associate with other proteins to furnish musclelike contractile forces. The proteins that associate with actin filaments in generating a pulling force have been compared to the winches that pull cables.

Intermediate filaments, usually 8 to 10 nm in diameter, are fibrous proteins such as keratin (Figure 4-16C). Recall from our discussion of Linus Pauling's work in Chapter 3 that keratin strengthens wool, hair, fingernails, the outer layers of the skin, and other epithelial sheets. All intermediate filaments seem to be common in parts of cells that are subject to mechanical stress, suggesting that these filaments strengthen cells and tissues.

The cytoskeleton is both skeleton and muscle system for the cell. Hollow microtubules, along with the microtubule organizing center near the centrosome or centriole, provide the cell with a shape. Microfilaments supply the dynamic contractile force, and intermediate filaments provide strength.

BOX 4-3

The 9+2 structure of flagella

Many prokaryotes, including *E. coli*, have long fibers on their surfaces. Each of these structures is about 10 to 20 nm thick and up to 10 μm long and is called a **flagellum** [Latin, = whip; plural flagella]. The flagellum's proportions are similar to those of a half-inch rope 40 feet long. Flagella act as propellers, driving bacteria at speeds of tens of micrometers per second. In terms of the bacteria's own dimensions, this would be equivalent to a 15-foot-long car moving at 100 miles per hour.

The flagella of eukaryotes are different from those of prokaryotes. When these structures are long and few in number, they are called flagella (Figure A). When they are short and numerous, they are called cilia. However, both cilia and flagella have the same characteristic cross section. Each consists of a ring of nine sets of microtubules, surrounding two more sets in the center. The microtubule organizing center of a cilium or flagellum also has a nine-membered ring of microtubular structures and is called a basal body. Protists and some cells of animals (including sperm) use flagella or cilia to propel themselves. Cilia on the surfaces of the digestive, respiratory, and reproductive tracts push fluids along the surface.

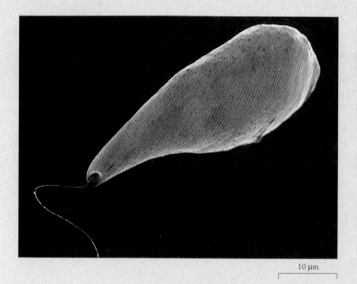

10 µm

1 µm

Figure A   Flagella. The protist *Euglena* (*left*) has a long flagellum (*right*), which it whips around like a propeller.   *(left, David M. Phillips/Science Source/Photo Researchers; right, Manfred Kage/Peter Arnold, Inc.)*

WHAT ENCLOSES A CELL AND ITS COMPARTMENTS?

We saw earlier in this chapter that cell size is partly limited by the amount of membrane, which controls the transfer of materials in and out of the cell, just as a system of freeways and rail lines carries materials in and out of a city. Membrane, then, is as important to cells as a transportation system is to a city.

But membranes do more than just move materials in and out of the cell. In the last 20 years, cell biologists and biochemists have found four major roles for cellular membranes:

1. Membranes are essential boundaries that separate the inside from the outside.

2. Membranes regulate the contents of the spaces they enclose.
3. Membranes serve as a "workbench" for a variety of biochemical reactions, especially those involving the metabolism of lipids and the secretion of proteins.
4. Membranes participate intimately in energy conversions (as we will discuss in Chapters 6 and 7).

What Kinds of Molecules Do Membranes Contain?

The plasma membranes of all six kingdoms look the same, appearing in cross section as two dark lines, 7 to 10 nm apart, with a lighter zone between them (Table 4-1). The membrane

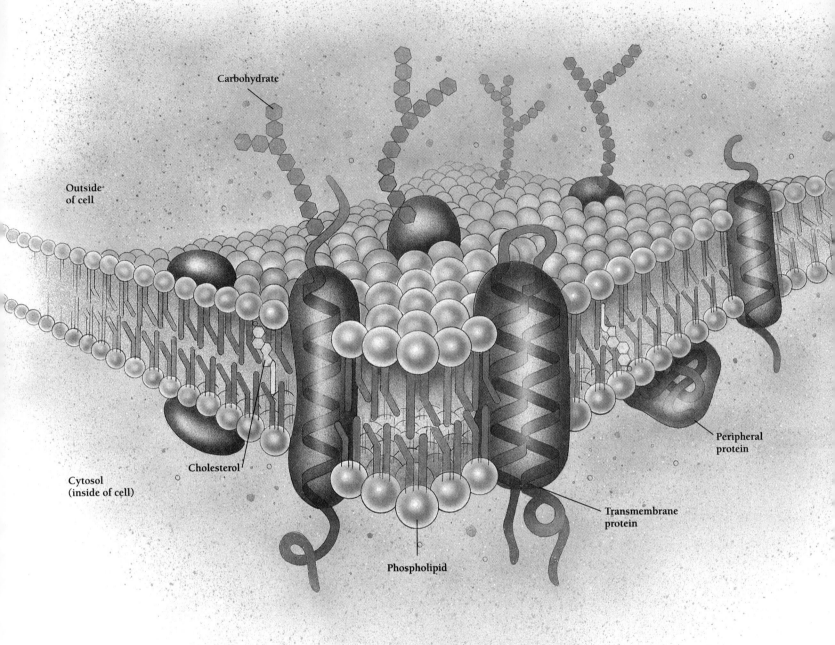

Outside of cell

Carbohydrate

Cholesterol

Cytosol (inside of cell)

Phospholipid

Peripheral protein

Transmembrane protein

consists, then, of at least two parallel layers separated by a slight space. Our first question concerns the structure of these layers. What molecules make up the layers and what holds them into the same characteristic structure in all organisms?

The first understanding of the molecular structure of membranes came from studies of red blood cells. The mammalian red blood cell is ideally suited for membrane studies because it lacks nearly all cell structure. Red blood cells have no internal membranes, no organelles, and few types of macromolecules other than hemoglobin, the red oxygen-binding protein. Red blood cells do not even have nuclei. Like empty plastic bags, they can be squashed into many shapes. This flexibility allows them to pass more easily through the tiny vessels of the circulatory system.

Biochemists can obtain essentially pure plasma membrane by breaking open the red blood cells to release the hemoglo-

bin and then using a centrifuge to separate the empty plasma membranes, called red blood cell "ghosts." Biochemical analysis has shown that these membranes consist mostly of lipids and proteins, as well as some carbohydrates. In the 1920s, biochemists calculated that a red cell ghost contained just enough lipid to enclose the cell with a continuous layer two molecules thick. This result suggested that the ghosts might indeed consist of a lipid "bilayer."

The most important lipids in the membrane are phospholipids. Recall from Chapter 3 that phospholipids are highly amphipathic; that is, they have a hydrophilic (water-loving) head and a hydrophobic (water-fearing) tail. When pure phospholipids are placed in water, they can spontaneously organize into sheets that are two layers thick. The oily tails of the lipids point inward towards one another, while their hydrophilic heads point outward. Once biologists understood this, they wondered

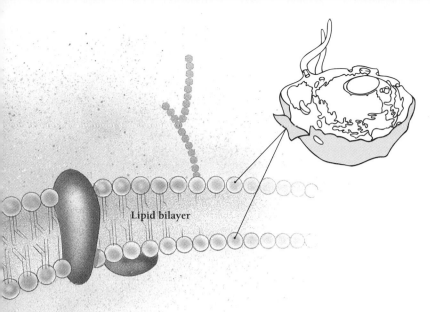

Figure 4-18 **Figure 4-18** A plasma membrane. Carbohydrates stud the outer surface of the membane and transmembrane proteins pass through the lipid bilayer.

Lipid bilayer

The Fluid Mosaic Model Summarizes Current Understanding of the Workings of Biological Membranes

The plasma membrane has a consistency similar to that of thin oil, such as you might use to lubricate a sewing machine or a bicycle. Water and other polar (hydrophilic) molecules can diffuse rapidly from one side of the membrane to the other, and also within the membrane.

What about the molecules of the membrane itself? Are they also able to move about? To answer this question, biochemists marked certain lipids known to occur in the membrane with a special chemical group that could be detected, but which would not much alter the lipids' chemistry. After a cell had taken up the marked lipids, the biochemists could study the movements of the lipids. These experiments showed that membrane lipids move freely within the plane of the membrane, but that the lipids in one layer do not usually move to the other layer.

Proteins also move freely within each layer of the membrane. To study the movement of proteins, biochemists marked membrane proteins with a fluorescent (glowing) dye, then used a laser to bleach the dye in a small region of the membrane. They then watched the membrane through a microscope to see if new fluorescent molecules would appear in the bleached area. Indeed, glowing proteins entered from adjacent areas. These experiments showed that about half of the proteins in a membrane move freely, and about half remain tightly bound in a particular position within the membrane.

We can see that both lipid and protein molecules move about with relative ease within each layer of the membrane. The molecules move easily within each layer because the layers are **fluid.** On the basis of this fluidity, two biochemists at the University of California San Diego, Jonathan Singer and Garth Nicholson, proposed a new model of membrane structure, called the **fluid mosaic model.** Singer and Nicholson's model emphasizes that proteins and lipid molecules can usually move freely within the lipid bilayer. Since this model was first advanced in 1972, cell biologists have used it as a basis for understanding all cellular membranes. Today we summarize the fluid mosaic model—somewhat modified since Singer and Nicholson first proposed it—as follows:

if a biological membrane was in fact just a phospholipid bilayer. If so, where would the membrane proteins lie?

By the mid-1930s, biologists were convinced that biological membranes were indeed formed from lipid bilayers. Since biologists knew that proteins studded both the inner and outer surface of the cell membrane, they hypothesized that the membrane proteins were scattered on the inner and outer surface of the lipid bilayer, in contact with the polar groups of the phospholipid. But this is not quite correct.

Researchers have now determined that proteins not only occupy the two layers of the plasma membrane, but they also occupy the space *in between* the two layers. Some membrane proteins are located only on the outside of the membrane; others (called transmembrane proteins) span the entire lipid bilayer (Figure 4-18).

Membrane proteins are amphipathic, with both hydrophilic and hydrophobic regions. The hydrophilic regions interact with the aqueous solutions of the cytosol and the cell exterior, as well as with the hydrophilic heads of the lipids. The hydrophobic regions interact with the hydrophobic interior of the membrane.

The two layers of the lipid bilayer differ slightly from one another (Figure 4-18). One sign that the two layers are different is that certain enzymes can break down the lipids in one layer, but not the other. Another difference is that carbohydrates appear only on the outer face. The inner and outer layers also differ in the way they interact with membrane proteins. Proteins that span the entire thickness of the membrane, for example, consistently orient themselves in a certain direction (Figure 4-18). We will see later in this chapter that the orientation of a protein is crucial, because some membrane proteins work like turnstiles, letting molecules pass through the membrane in only one direction.

The plasma membrane consists of an asymmetric phospholipid bilayer. Proteins occupy both the inner and outer surfaces and the interior of this bilayer. Carbohydrates occupy the outer surface only.

---

1. The basic structure of the membrane is a lipid bilayer, with two sheets of phospholipids arranged tail to tail.

2. Proteins are dispersed through the membrane, like the individual pieces of stone or tile in a mosaic. Proteins contribute to membrane structure and function. Some proteins span the membrane, while others are confined to the inner or outer surface.

3. The membrane is fluid; both protein and lipid molecules move freely within the plane of the membrane.

4. The lipid bilayer serves as a hydrophobic barrier, confining hydrophilic molecules to the inside or the outside of a cell (or organelle).

5. Some membrane proteins help transport specific molecules across the membrane.

HOW DO MEMBRANES REGULATE THE SPACES THEY ENCLOSE?

Throw a handful of raisins into a pot of boiling water and the raisins gradually fill with water, swelling until they resemble the grapes from which they were made (Figure 4-19). The raisin skin is a kind of membrane. Water and other small molecules easily pass through the skin, but larger molecules such as cellulose cannot pass through the skin. Large molecules inside the raisin are trapped, those outside are excluded. Such a membrane is said to be **selectively permeable,** meaning that it allows the passage of some molecules (especially of water), but not of others.

Plasma membranes are also selectively permeable. Proteins, for example, cannot directly move through plasma membranes at all. Yet, as we shall see, cells and organelles need to move large molecules—digestive enzymes, for example—across their membranes and also to regulate the movements of small molecules into and out of cells and organelles.

Such regulation is accomplished by molecules within the plasma membrane that actively draw specific small molecules through the membrane. Before we can discuss the role of membranes in regulating a cell's internal environment through such maneuvers, however, we must first discuss the rules that govern the movement of water and dissolved substances. These rules are the same physical laws that govern the movement of water into and out of raisins.

Why Does Water Move Across Membranes?

What Is Diffusion?

If you pour a teaspoon of salt into a pot of water, most of the salt will at first drop to the bottom of the pot in a little pile. Gradually, the salt will dissolve into sodium ions and chloride ions. At first, the concentration of ions will be very high near the spot where you first poured the salt. The **concentration** is simply the number of molecules in a given volume. However, as time passes, the salt ions will move about randomly in amongst the water molecules until the ions are distributed evenly throughout the pot of water. Even after the ions are dis-

tributed evenly, they continue to move about randomly. Since as many ions move out of a given area as into it, however, the net charge is still zero. The random movement of like molecules or ions from an area of high concentration to an area of low concentration is called **simple diffusion** [Latin, *diffundere* = to pour out]. Stirring speeds this process up, but diffusion will happen all by itself if you wait long enough.

The presence of an area of high concentration adjacent to an area of low concentration is called a **concentration gradient** (Figure 4-19). Concentration gradients are virtually everywhere. Cigarette smoke moving from the smoking section of a

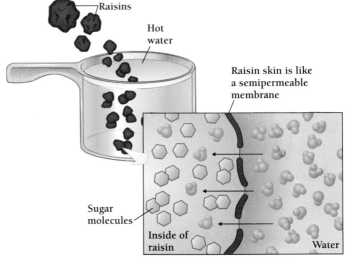

A. Concentration gradient

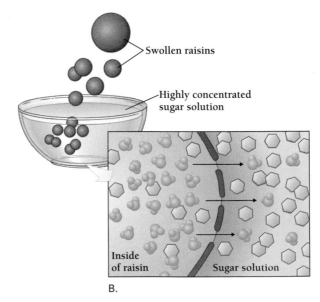

B.

Figure 4-19   Concentration gradient. A. A raisin skin is like a semipermeable membrane. The high concentration of sugar molecules inside the raisin creates a concentration gradient. Water flows inside. B. In a concentrated sugar solution the water in the swollen raisin flows outward. Because raisins are so sweet, such a sugar solution would have to be very concentrated.

restaurant to the nonsmoking section creates a concentration gradient from an area of dirtier air to an area of cleaner air. A teaspoon of sugar poured into a cup of coffee creates a concentration gradient—for a while.

Concentration gradients are always temporary, since simple diffusion, as well as ordinary mixing, or turbulence, tends to eliminate differences. How fast diffusion occurs depends on a variety of factors, including temperature, the sizes of the molecules, and the steepness of the gradient. At higher temperatures, molecules move faster and so diffuse faster. Small molecules generally diffuse faster than large ones. Finally, the greater the difference in concentration at the two ends of a gradient, the faster diffusion occurs.

In simple diffusion, solutes move from an area of high concentration to an area of low concentration.

What Is Osmosis?

Diffusion occurs across membranes as well as through open solutions. For example, raisins immersed in a pot of plain water fill with water because the raisins have a high concentration of sugar and other molecules. The sugar molecules cannot pass easily through the skin of the raisin and out into the water. However, water molecules move in and out of the raisin easily. From the perspective of a water molecule, then, a raisin is an area of low water concentration. The difference between the sugary inside of the raisin and the watery outside constitutes a concentration gradient, along which the water molecules move.

The movement of the water molecules along this concentration gradient into the raisin is called osmosis. **Osmosis** [Greek, *osmos* = push, thrust] is the movement of water across any selectively permeable membrane in response to a concentration gradient. Osmosis can exert a force—called **osmotic pressure**—that is comparable to physical pressure.

The influx of water into a raisin stretches the skin until it begins to resist. If the skin is weak, it may break. Otherwise, it pushes back, meeting osmotic pressure with mechanical pressure. The physical pressure exerted by the raisin's skin counterbalances the osmotic pressure. The difference between osmotic pressure and physical pressure on either side of a selectively permeable membrane is called the **water potential.**

Osmosis is the movement of water across any selectively permeable membrane in response to a concentration gradient. The pressure exerted by osmosis, called osmotic pressure, is comparable to mechanical pressure. The balance between osmotic pressure and mechanical pressure is the water potential.

What Is Tonicity?

We have seen what happens when a closed membrane (the raisin) with a high concentration of solutes (in this case, various sugars) is dropped into a solution with a low concentra-

tion of solutes (plain water). What happens if we take the water-filled raisins and put them into a highly concentrated solution of sugar?

The answer depends on how concentrated the sugar is. Look inside a can of fruit cocktail and you'll see what happens when the concentrations of sugar in a grape and the solution are about the same. The grape neither shrinks nor expands, but retains its normal shape. Put the grape into a highly concentrated sugar solution, however, and you will see it actually shrink as the water molecules inside move out through the skin, following the concentration gradient (Figure 4-19).

In the raisin example, the osmotic pressure forcing water into the raisin is greater than the osmotic pressure forcing water out. The water in which the raisin sits is said to be **hypotonic** [Greek, *hypo* = under + *tonos* = tension] compared to the inside of the raisin. In ordinary canned fruit cocktail, the osmotic pressure inside and outside of a grape is balanced perfectly. In that case, the sugary solution in which the grape is sitting is said to be **isotonic** [Greek, *isos* = equal] with the inside of the grape. If we put grapes into too heavy a syrup, however, the concentration of the syrup solution is so high that osmotic pressure inside the grape forces water out. In that case, the solution outside the grape is said to be **hypertonic** [Greek, *hyper* = above] compared to the inside of the grape.

Cells, it turns out, behave in the same way. A red blood cell immersed in an isotonic solution, such as blood plasma, or a suitable salt solution, remains intact (Figure 4-20A). But

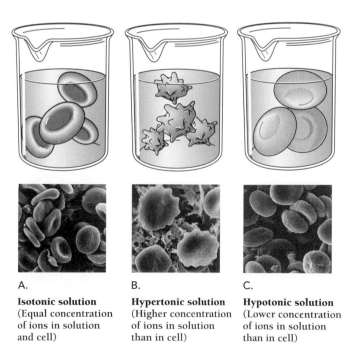

A.
Isotonic solution
(Equal concentration of ions in solution and cell)

B.
Hypertonic solution
(Higher concentration of ions in solution than in cell)

C.
Hypotonic solution
(Lower concentration of ions in solution than in cell)

Figure 4-20 Iso, hyper, hypo. A. A red blood cell in an isotonic salt solution remains intact. B. A red blood cell in a hypertonic salt solution shrivels. C. A red blood cell in a hypotonic salt solution swells with water. In a still more hypotonic solution, such as plain water, the cells in C would fill up and burst open. *(Dennis Kunkel/Photo-take NYC)*

if we immerse the red blood cell in a highly concentrated (hypertonic) solution, the cell shrivels (Figure 4-20B). The hypertonic solution has a lower water potential than the cell's interior, and so the selectively permeable membrane allows water to pass freely out of the cell. When enough water flows out so that the new concentration of molecules within the cell is again the same as outside the cell, no more water flows.

Conversely, a red blood cell immersed in a hypotonic solution such as plain water will tend to fill with water and burst. When we decrease the solute concentration outside the cell, we raise the water potential above that of the cell's interior. Water then flows into the cell, causing it to expand (Figure 4-20C). Theoretically, water will flow in until the water potential is equal on both sides of the membrane.

However, in practice there may be too few solute molecules on the outside for this equalization to occur, as, for example, when the cells are suspended in pure water. In this case, the water will continue to flow into the cell until the cell bursts and releases its contents. (This is how cell biologists prepare red cell ghosts.)

When a selectively permeable membrane separates two solutions with different water potentials, water flows in the direction that tends to equalize the concentrations of solutes on the two sides of the membrane, a process called osmosis. A solution may be hypotonic, isotonic, or hypertonic, relative to the concentration of solutes on the other side of the membrane.

How Does Turgor Pressure Support Plants?

When plant cells, such as those in the onion skin shown in Figure 4-21, are exposed to a hypotonic solution, water moves into the cell, but does not accumulate in the cytoplasm. In-stead, the inflowing water passes from the cytoplasm to a vacuole. The expansion of the vacuole pushes the surrounding cytoplasm against the **cell wall**—the rigid boxlike structure that encloses all plant cells. Conversely, when an onion cell is placed in a hypertonic solution, water flows out of the vacuole and the cell separates from the cell wall, a phenomenon called **plasmolysis** [Greek, *plasma* = form + *lysis* = loosening].

In a hypotonic solution, the mechanical pressure exerted on the cell by the cell wall prevents the further flow of water into the cell. (Water actually continues to flow equally in both directions through the plasma membrane, but there is no *net* inflow of water.) The mechanical pressure thus balances the osmotic pressure. As a result, the total water potential of the cell's interior is equal to that of its environment. The pressure of a plant cell's contents against its cell wall is called **turgor pressure** [Latin, *turgere* =to swell].

Turgor pressure supports nonwoody plants in the same way that water supports the shape of a water balloon. This is why plants (and water balloons) begin to droop, or wilt, when they do not have enough water. The same phenomenon accounts for the way lettuce wilts soon after it is covered with salad dressing. Because the dressing is hypertonic, it draws water out of the cells of the lettuce leaves, causing them to wilt.

What Determines the Movement of Molecules Through a Selectively Permeable Membrane?

We can ask two questions about the passage of molecules through a membrane: "Which way?" and "How fast?" In addressing the first question, membrane researchers distinguish between "passive transport" and "active transport." **Passive transport** occurs spontaneously, without the expenditure of en-

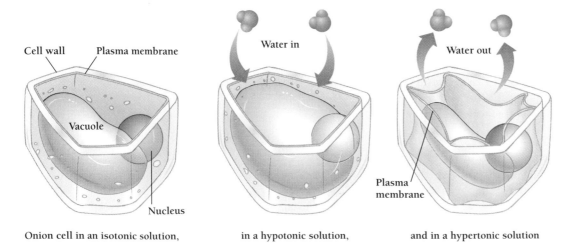

Cell wall   Plasma membrane

Water in

Water out

Vacuole

Nucleus

Plasma membrane

Onion cell in an isotonic solution,   in a hypotonic solution,   and in a hypertonic solution

Figure 4-21   Onion cell. Plant cells behave like red blood cells (and raisins). The water content of the vacuole increases or decreases according to whether the cell is in an isotonic, hypertonic, or hypotonic solution. In a hypotonic solution, the vacuole fills with water and exerts (turgor) pressure against the cell wall and helps support the plant. In a hypertonic solution, the vacuole shrinks and the plant wilts.

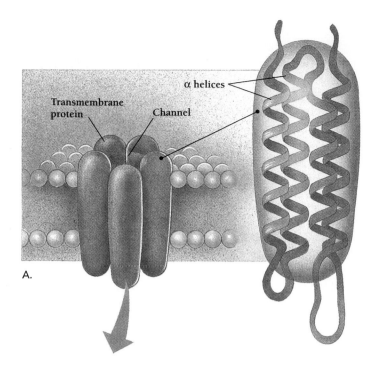

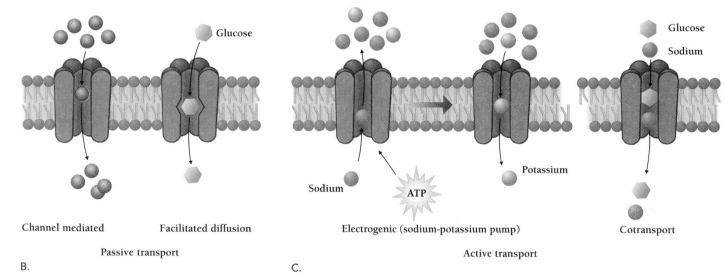

Figure 4-22 **Passive and active transport.** A. Transmembrane proteins help transport molecules and ions across the cell membrane. B. In passive transport, no energy is required. In channel-mediated transport, ions or small molecules pass through the transmembrane channels without regulation or help. In facilitated diffusion, molecules move faster than they ordinarily would thanks to the action of transporter proteins, which carry molecules across the membrane. Transporters bind only to certain molecules, such as glucose. C. Active transport requires energy, as well as a carrier protein that couples transport with energy expenditure. (1) The sodium-potassium pump binds to ATP and Na^+ ions on the inside of the plasma membrane and to K^+ ions on the outside of the membrane. It then transports Na^+ and K^+ in opposite directions, using the energy of ATP. (2) In cotransport, a cell membrane transports molecules using the energy from a concentration difference across a membrane (instead of using ATP). Cotransporter proteins transport glucose using a concentration difference in Na^+ ions, linking the "uphill" transport of one molecule or ion to the "downhill" transport of another.

ergy (Figure 4-22A). **Active transport** moves molecules against a concentration gradient and requires energy (Figure 4-22B).

The answer to the question "Which way?" is different for passive transport and active transport. Passive transport results in equal concentrations of a molecule on the two sides of a membrane, at least for uncharged molecules. For charged molecules or ions, passive transport results in unequal distribution. This is because the excess negative charge from other ions repels negative ions and attracts positive ions. Biologists can calculate how unequal the distribution of ions should be, and any deviation from the predicted distribution indicates that active transport has occurred.

Active transport comes in several forms. In one form, cells move ions across the membrane so that a net charge accumulates on one side. For example, an important active transporter, the **sodium-potassium pump,** simultaneously transports

sodium ions (Na^+) out of cells and potassium ions (K^+) into cells, using the energy of ATP.

Facilitated Diffusion Accelerates Passive Transport

Some molecules move across membranes much faster than would be expected on the basis of diffusion alone, even though their final concentrations are consistent with passive transport. Such **facilitated diffusion**—an increased rate of passive transport—depends on the action of specific molecules within the membrane that help, or "facilitate," transport. Glucose, for example, usually moves across cell membranes by means of facilitated diffusion.

The transmembrane protein glucose permease is responsible for glucose transport in red blood cells. To show that it causes the facilitated diffusion of glucose, researchers studied the rates of entry of radioactive glucose into liposomes—tiny bubbles of artificial phospholipid bilayers that behave like membranes. Pure liposomes allow the entry of scarcely any glucose. But when glucose permease is added, the rate of glucose entry increases dramatically, showing that permease facilitates the entry of glucose into the liposome.

Transport proteins such as glucose permease bind to the molecules they transport, then carry the molecules across the membrane in some manner (Figure 4-22). Like enzymes, transporters bind only to certain molecules. Glucose permease, for example, only transports glucose and other similar sugars.

Active Transport Uses Energy To Move Ions

How can a membrane accomplish active transport? To transport a substance actively across a membrane requires a special "pump" mechanism that couples transport to an energy source (such as ATP). For example, the sodium-potassium pump binds to ATP and Na$^+$ ions on the inside of the plasma membrane and to K$^+$ ions on the outside of the membrane (Figure 4-22). When inserted into liposomes, the pure protein (which consists of two polypeptide chains) transports Na$^+$ and K$^+$ in opposite directions, using the energy obtained by hydrolyzing ATP.

Somewhat surprisingly, other examples of active transport do not directly depend on ATP. Instead, the cell harnesses energy from an already existing concentration difference across a membrane, usually of Na$^+$ or H$^+$ ions.

The coupled transport of two substances is called **cotransport** and depends on specific transmembrane protein molecules, called cotransporters. Cotransporters couple the "uphill" transport of one molecule or ion to the "downhill" transport of another. This mechanism is comparable to the action of a children's seesaw. As gravity pulls one child down, the child at the other end is raised up against the force of gravity. In a cotransporter used by the cells that line the small intestine, for example, the concentration difference of Na$^+$ ions supplies the energy for the cells to take up glucose (Figure 4-22).

We can see that membranes are important regulators of a cell's contents. They allow water to pass freely, with the net movement dependent on the relative water potential inside and outside the cell. They also admit larger molecules, but selectively. In some cases, they speed the passive transport of specific molecules with transport proteins such as glucose permease. In other cases, they harness the energy of ATP or of existing concentration gradients to accumulate or exclude particular molecules or ions.

While we have spoken almost entirely about plasma membranes, the internal membranes of eukaryotic organelles are equally important in regulating the contents of the spaces that they enclose. In Chapters 6 and 7, we will discover how crucial the membranes of some organelles are to the production of ATP.

Molecules pass through plasma membranes by means of passive transport, facilitated diffusion, and active transport. Active transport mechanisms use energy from ATP or from a concentration gradient of an ion or molecule different from the one being transported.

HOW DO MEMBRANES INTERACT WITH THE EXTERNAL ENVIRONMENT?

The plasma membrane functions in its environment like the outer surface of an organism. In multicellular organisms, the immediate environment of the cell may be created by the cell itself. For example, the plasma membranes of plant and fungus cells, most prokaryotes, and many protist cells are surrounded by a rigid external structure made of carbohydrate, called the cell wall (Figure 4-23A). The cell wall may continue

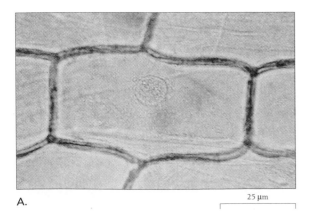

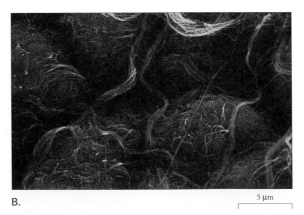

A.   25 μm   B.   5 μm

Figure 4-23   The cell wall. A. A rigid, external cell wall made of carbohydrate surrounds the plasma membranes of plant and fungus cells, most prokaryotes, and many protist cells. B. An extracellular matrix of carbohydrates and proteins encloses many animal cells.
(A, Kevin Collins/Visuals Unlimited; B, Fred Hossler/Visuals Unlimited)

functioning after the death of the cell inside. In fact, some plant cells function *only* after death. Many of the vessels that conduct fluids in vascular plants, as well as many structural fibers, for example, are composed of the walls of dead cells. Animal cells may be enclosed in a network of extracellular molecules, both carbohydrates and proteins, called the **extracellular matrix** (Figure 4-23B). But cells, like organisms, must both communicate and also consume and excrete materials. Membranes facilitate both of these activities.

Membrane Fusion Allows the Uptake of Macromolecules and Particles

The tendency of membranes to avoid water, together with their fluidity, means that biological membranes can change shape rapidly. The ruptured membranes of red cells, for example, can reform as intact "ghosts" of cells that do a remarkably good job of regulating their internal environments by pumping ions and molecules.

The reorganization of membranes is an important part of many cellular functions. In cell division, for example, the plasma membrane of the parent cell divides and reseals itself around each daughter cell. In phagocytosis, cells engulf rela-

tively large particles, such as microorganisms or cell debris. First, part of the plasma membrane extends outward and surrounds the particle (Figure 4-12). The membrane then fuses to form a large vesicle, which travels to the interior of the cell. The vesicle later fuses with the membranes that surround the lysosomes, which contain the digestive enzymes.

In contrast, in **endocytosis,** the cell takes in tiny amounts of materials in vesicles that arise by the inward folding (or invagination) of the plasma membrane. The pouchlike vesicles formed in endocytosis are tiny, about one-tenth the size of those formed in phagocytosis.

Cell biologists now recognize two types of endocytosis. In **pinocytosis** [Greek, *pinein* = to drink], the cell takes up any bits of liquid and dissolved molecules. In **receptor-mediated endocytosis,** the cell takes up only specific substances, which it recognizes by means of special proteins on the cell's surface, called receptors (Figure 4-24A).

Exocytosis is the export of molecules from a cell by a process that is approximately the reverse of pinocytosis. Molecules to be exported are surrounded by membranes that move to the cell surface (Figure 4-24B). They then fuse with the plasma membrane, releasing their contents to the outside of the cell. Molecules exported by exocytosis include proteins that

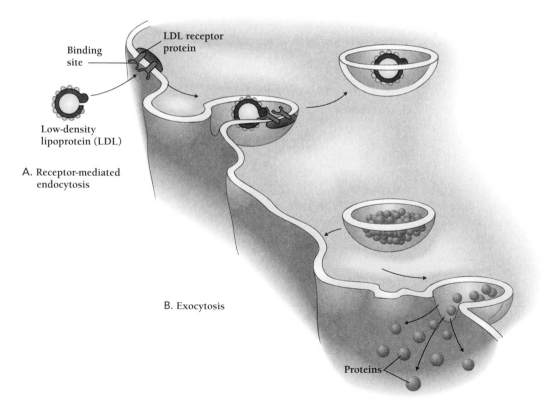

Figure 4-24 **Endocytosis and exocytosis.** A. Endocytosis plays a central role in the regulation of low-density lipoprotein (LDL) and cholesterol levels in the blood, an important predictor of heart disease. In one form of receptor-mediated endocytosis, LDL particles full of cholesterol bind to LDL receptors on the cell membrane. The cell membrane folds around the LDL particle and engulfs it, then the cell digests it. B. In exocytosis, the reverse happens: vesicles full of proteins, hormones, or neurotransmitters made inside the cell fuse with the cell membrane and diffuse into the outside environment of the cell.

function outside of cells, such as the digestive enzymes of the gut, and chemical signals, such as hormones and neurotransmitters.

Plasma membranes fuse easily and can spontaneously form vesicles. Cells use membrane fusion to take up particles, large molecules, and small quantities of fluid, as well as to expel specific molecules.

How Do Cells in Multicellular Organisms Communicate?

Because the individual cells in all multicellular organisms must cooperate, communication among the cells is paramount. In general, cells communicate by means of chemical signals such as hormones. Such signals allow the activities of one cell to influence those of another.

How Do Cells Communicate with Distant Cells?

Chemical signals all act by binding to specific **receptors**, specialized proteins that can lie either on the surface of a cell or within the cell. A particular receptor normally only binds with one signaling molecule. Some chemical signals stay outside the cell, influencing what is going on inside by means of cell surface receptors, which influence the cell in many different ways. Surface receptors can, for example, open channels in the plasma membrane to allow the passage of new materials into or out of the cell. Or they can regulate the levels of an influential chemical inside of the cell.

Other chemical signals enter the cell by passing through the plasma membrane and then binding with receptors within the cytoplasm. The steroid hormone testosterone, for example, acts in this way, actually entering cells and altering the expression of individual genes. At puberty, a young man's hypothalamus (a part of the brain) sends a chemical signal to the pituitary gland, at the base of the brain. The pituitary responds by secreting a hormone that is carried throughout the body by the bloodstream. When the hormone reaches the cells of the testes, the male reproductive organs, it binds to cell surface receptors, changing the cells' properties. The cells of the testes then begin to secrete the hormone testosterone, which enters the bloodstream and binds to cytoplasmic receptors in different cells all over the body. In each cell, the receptor-hormone complex moves into the cell nucleus and influences the production of various proteins. The results include a gradual deepening of the voice, a rapid spurt in body growth, and the development of a sex drive.

Cells respond to chemical signals by means of specialized proteins called receptors.

How Do Cells Communicate with Their Immediate Neighbors?

We have seen that the cells of plants are both cemented together and separated from each other by rigid cell walls. All the living cells of flowering and other higher plants are nonetheless in intimate contact through fine intercellular channels called **plasmodesmata** (singular, plasmodesma) [Greek, *plassein* = to mold + *desmos* = to bond]. Plasmodesmata, which are between 20 and 60 nm in diameter (Figure 4-25), allow molecules to pass directly from cell to cell. Some kinds of cells, such as those in the tips of a plant's roots, have large numbers of plasmodesmata. The more plasmodesmata that connect neighboring cells, the more rapidly materials pass from one cell to the next.

Animal cells are not separated from each other by cell walls, and they communicate with each other in a variety of ways. Some intercellular communication, such as the transmission of information from one nerve cell to another or from a nerve cell to a muscle cell, depends on chemical signaling. Molecules made by one cell bind to specific receptors in other cells, triggering a series of specific events.

Some animal cells easily pass ions and molecules, even proteins, through specialized regions of their membranes called **gap junctions.** In gap junctions, the membranes of adjacent cells are only 2 to 4 nm apart, instead of the usual 10 to 20 nm (Figure 4-26A). In addition, gap junctions connect cells with a protein called connexin that allows the cells to exchange chemical signals. Heart muscle cells, for example, are connected

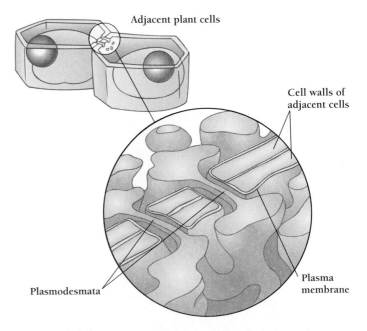

Figure 4-25 Fine intercellular channels called plasmodesmata connect all the living cells of higher plants.

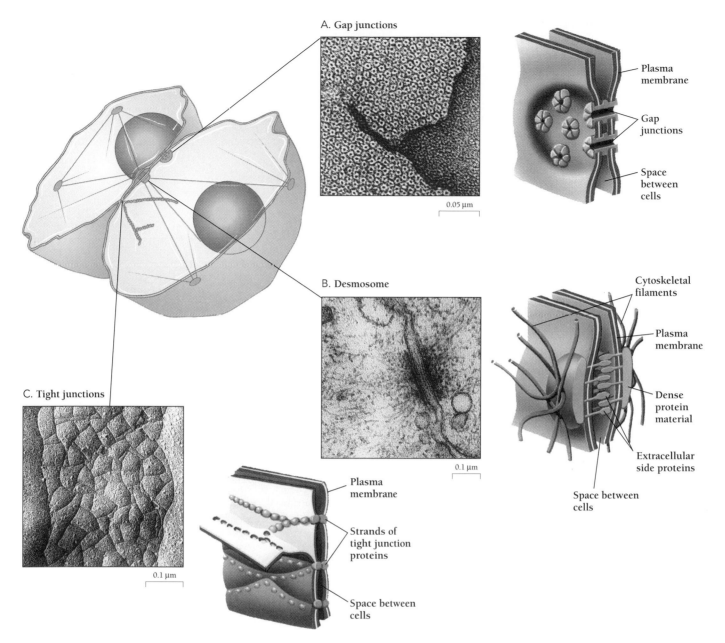

Figure 4-26 **Cell junctions.** A. Gap junctions are protein assemblies that form channels between adjacent cells. B. Like rivets, desmosomes anchor cells together, without interfering with the passage of molecules from one cell to the next. C. Tight junctions seal cells together and prevent molecules from leaking between them. *(A, Gilulap, Fawcett/Visuals Unlimited; B, Farquhar, Palade, Fawcett/Visuals Unlimited; C, Hull, Staehelin, Fawcett/Visuals Unlimited)*

A. Gap junctions

0.05 µm

Plasma membrane

Gap junctions

Space between cells

B. Desmosome

0.1 µm

Cytoskeletal filaments

Plasma membrane

Dense protein material

Extracellular side proteins

Space between cells

C. Tight junctions

0.1 µm

Plasma membrane

Strands of tight junction proteins

Space between cells

by gap junctions so that all of the muscle cells can contract at once to pump the blood.

Other animal cells communicate using "adhering junctions." **Adhering junctions** help provide strength to resist mechanical stress without affecting the passage of molecules between them. Adhering junctions usually consist of **desmosomes** [Greek, *desmos* = to bond + *soma* = body], buttonlike structures that hold two membranes together (Figure 4-26B). Desmosomes may be localized in spots, like the rivets on a pair of jeans, or they may be spread in bands or belts.

Several types of specialized junctions connect **epithelial cells** [Greek, *epi* = upon + *thele* = nipple], the cells that make up the skin and intestines. Epithelial cells are linked together tightly to form a sheet, called an **epithelium** [plural, epithelia]. Many epithelial sheets enclose fluid-containing cavities, such as the blood vessels and the intestines. One task of these epithelia is to keep the fluids from escaping. Impermeable junctions not only weld cells together, but also prevent any molecules from leaking between them. In mammals and other vertebrates, the main kind of impermeable junction is called a

tight junction, in which the membranes of adjacent cells actually fuse (Figure 4-26C).

Plant cells communicate with their neighbors by means of plasmodesmata. Animal cells communicate by means of narrow gap junctions, adhering junctions, and tight junctions.

In this chapter we have seen how biologists first recognized the importance of cells and made an acquaintance with the cell theory. We have examined the fine internal structure of cells and discovered a little about how membranes work. In the next three chapters, we will use some of this knowledge to try to understand the biochemical processes that take place inside of cells.

STUDY OUTLINE WITH KEY TERMS

Cells are not only the units of biological structure, they are also the units of function. Individual cells are specialized to perform distinct tasks. Cells are living entities that are highly organized, obtain energy and materials, change with time, respond to their environments, and (usually) reproduce.

The **cell theory** states that (1) all organisms are composed of one or more cells; (2) cells, themselves alive, are the basic living units of organization of all organisms; and (3) all cells come from other cells.

The subdivision of organisms into cells and of cells into organelles allows an efficient division of labor and makes it easier to maintain a constant environment within each cell (because of a higher **surface-to-volume ratio**). Multicellular organisms can survive the deaths of individual cells because of cell replacement.

Cells from all six kingdoms contain three common features: a **plasma membrane** that separates the inside of the cell from the rest of the world, a set of genetic instructions encoded in DNA, and a cell body (or cytoplasm). In the cells of all **eukaryotes,** the genetic material is enclosed in a membrane-bounded **nucleus.** The nucleus contains the cell's genetic instructions in the form of **chromosomes,** which are made of **chromatin,** a complex of DNA and protein. The two membranes of the **nuclear envelope** are fused in many places to form **nuclear pores.**

The Archaea and Eubacteria, or prokaryotes, lack a membrane-bounded nucleus. They keep their DNA in a region called the **nucleoid.** The **cytoplasm** of most eukaryotic cells (formerly referred to as the **protoplasm**) is divided into well-defined compartments, or **organelles,** by internal membranes. The jellylike cytosol, which surrounds all the cell's organelles, contains **ribosomes,** glycogen, fat, and thousands of enzymes.

The **cytosol** is the part of the cytoplasm outside the membrane-enclosed organelles. It contains most of the enzymes that carry out the chemical reactions of the cell. Within the cytoplasm are several well-defined kinds of membrane-bounded organelles: the **endoplasmic reticulum, Golgi complex, lysosomes, peroxisomes, mitochondria,** and such **plastids** as **chloroplasts, chromoplasts,** and **amyloplasts.** In addition, eukaryotic cells contain **vesicles** and **vacuoles,** which are also bounded by membranes but whose structures and functions vary widely.

The endoplasmic reticulum is an extensive network of membranes that encloses a space called the **endoplasmic reticulum lumen.** One part of the endoplasmic reticulum—the **rough endoplasmic reticulum**—is studded with ribosomes and makes proteins mainly destined for export to the outside or the outer surface of the cell. The other part of the endoplasmic reticulum—the **smooth endoplasmic reticulum**—contains no ribosomes and appears to function in the synthesis of lipids.

The Golgi complex is a set of discs, each of which is enclosed by a membrane. It serves as the traffic director in the cell, helping to package some proteins for export and others for inclusion in other membrane-enclosed organelles.

Lysosomes and peroxisomes are small membrane-bounded organelles that are responsible for digestion. They dispose of both material taken in from outside the cell and the cell's own debris.

Mitochondria and chloroplasts are the largest organelles outside the nucleus. Each is bound by a double membrane, and each contains complicated internal membranes. Mitochondria are the powerhouses of cells, making most of the cell's ATP by breaking down small, organic molecules in the presence of oxygen. Chloroplasts are found only in photosynthetic cells. They are the sites of **photosynthesis.**

All cells are surrounded by membranes. Membranes consist mainly of lipids and proteins. The lipids, mostly phospholipids, are arranged in two layers, a "bilayer." Their hydrophobic tails make up the interior of the membrane and their hydrophilic heads point toward the cytoplasm and toward the outside of the cell. Some proteins are located on the external and internal surfaces of the bilayer, in contact with both the hydrophobic interior of the membrane and the hydrophilic cytoplasm or cell exterior. Other proteins pass completely through the membrane. The membrane is **fluid,** both protein and lipid molecules move freely within it. This picture of membrane structure is called the **fluid mosaic model.**

Membranes are **selectively permeable:** they allow some molecules to pass in and out freely and confine others to the inside or the outside. In **simple diffusion,** solutes move from an area of high **concentration** to an area of low concentration. That is, solutes move along a **concentration gradient. Osmosis** is the movement of water across any selectively permeable membrane in response to a concentration gradient. The pressure exerted by osmosis, called **osmotic pressure,** is comparable to mechanical pressure. The balance between osmotic pressure and mechanical pressure is the **water potential.**

Another way to define osmosis, then, is to say that when a selectively permeable membrane separates two solutions with different water potentials, water flows in the direction that tends to equalize the concentrations of solutes on the two sides of the membrane. A solution may be **hypotonic, isotonic,** or **hypertonic,** relative to the concentration of solutes on the other side of the membrane.

In plants, the central vacuole fills with water, which presses against the **cell wall,** creating **turgor pressure**—the principal mechanical support of nonwoody plants. When a plant cell is immersed in a hypertonic solution, water flows out of its vacuole and the cell separates from the cell wall in **plasmolysis.**

Molecules pass through plasma membranes by means of **passive transport** (which requires no energy), **facilitated diffusion,** and

active transport (which requires energy from ATP or **cotransport**). The **sodium-potassium pump** is one active transport mechanism.

Membrane fusions allow the uptake of macromolecules and even larger particles. Certain cells engulf large particles by extending their plasma membranes in a process called **phagocytosis.** Most cells are also capable of **endocytosis, pinocytosis,** and **exocytosis.** Endocytosis may be nonspecific or may depend on specific receptors on the membrane.

The outer surfaces of cells are responsible for their interactions with other cells and with other molecules in their environments. Cell walls enclose most prokaryotic and many eukaryotic cells (though not animal cells). Animal cells often are surrounded with **extracellular matrix,** a complex of carbohydrates and proteins.

Multicellular organisms depend on coordination of activities among many cells. Plants and animals have a variety of specialized structures by which cells interact. Some of these structures—**plasmodesmata** in plants and **gap junctions** in animals—allow the free passage of some molecules between adjoining cells. Other specialized junctions between cells—**adhering junctions**—rely on **desmosomes** and provide mechanical strength to sheets of **epithelial cells. Tight junctions** weld some cells together and prevent the passage of molecules through them.

Together with internal membranes, the **cytoskeleton** provides internal organization for eukaryotic cells. It consists of three sets of filaments: **microtubules** (which are made of two kinds of **tubulin**), **actin filaments,** and **intermediate filaments.** In most cells, the cytoskeleton consists of structures that are assembled and disassembled continually. Their assembly seems always to originate in structures called **microtubule organizing centers (MTOC).** The major MTOCs are near the nucleus in a zone called the **centrosome,** a region that (in many cells) also contains a distinct organelle called the **centriole.** Some cells have permanent structures, such as **cilia** and **flagella,** which are composed of elements of the cytoskeleton.

REVIEW AND THOUGHT QUESTIONS

Review Questions

1. State the cell theory. Be sure your statement includes the concepts of what a cell is, where cells occur, where they come from, and what they do.
2. Describe three features that all cells have in common.
3. Describe the differences between prokaryotic and eukaryotic cells.
4. Explain how the surface-to-volume ratio of a cell is related to its ability to take in resources and to rid itself of wastes. Why are most cells so small? How can a chicken egg, which is one cell, be so big?
5. Sketch and label a mitochondrion and a chloroplast. What is the principal function of each? Where does each obtain fuel?
6. Describe the fluid mosaic model of membrane structure.
7. A plasma membrane functions to keep the material outside a cell separate from the material inside. What is one property of the membrane that allows this?
8. Explain how membrane-bounded vesicles move materials into, out of, and within cells.

9. What is the role of the cytoskeleton in eukaryotic cells?
10. Compare the sizes of microtubules, actin filaments, and intermediate filaments. Describe one activity or structure in which each participates.

Thought Questions

11. Between conception and death, most of your cells die while new ones are produced by the division of other cells. What makes you really "you"? What is it that is constant about your structure?
12. Explain how the endoplasmic reticulum might increase the effective surface-to-volume ratio for a cell, even though it is located inside the outer perimeter of the cell. Can it really assist in bringing resources in and getting rid of wastes?
13. Chloroplasts are also called the "powerhouses" of cells. What do they produce to earn this description? What types of cells contain chloroplasts?

SELECTED READINGS

De Duve, C., *A Guided Tour of the Living Cell,* Scientific American Books, New York, 1984. A masterful view of the insides of cells, by one of the pioneers of modern cell biology.

Thomas, Lewis, *The Lives of a Cell,* Viking Press, 1975. A collection of beautiful but brief essays on cells and biology.

▶ On-line materials relating to this chapter are on the World Wide Web at http://www.saunderscollege.com/lifesci/
Click on Tobin/Dusheck: *Asking About Life.*

Ludwig Boltzmann: A Man Left Behind or a Man Ahead of His Time?

In September 1906, the eminent Austrian physicist Ludwig Boltzmann fell into the last of many deep depressions. Hoping to lift his spirits, he took his wife and daughters on a trip to Duino, Italy, a spectacular resort on the Adriatic Sea.

The mercurial Boltzmann had always been emotional (Figure 5-1). He could be moved to tears of joy by a well-played sonata, the color of the sea, or a piece of poetry. At other times he lapsed into long, grief-stricken silences. He wept at the railroad station whenever he left his wife for any extended period.

Although Boltzmann was unstable, his growing despondency had roots in the scientific milieu of his time. At 60, the physicist was convinced that much of his life's work had been wasted. For years, he had championed the idea that matter is composed of atoms and that the behavior of atoms accounts for all chemical transformations. Although throughout the 19th century chemists and physicists had discussed atoms extensively, they had not actually accepted atoms. By the end of Boltzmann's life, most scientists argued that atoms were merely scientific abstractions.

The theory that matter was made of tiny particles called atoms was first popularized by the Greek philosopher Democritus in the fifth century B.C. Not until the 17th century, however, did "atomism" begin to develop as a scientific idea. Ironically, the same philosophical mood that gave birth to atomism also gave birth to the philosophy of "positivism," a school of thought whose proponents would help drive Boltzmann to despair.

Nineteenth-century positivists asserted that science is the only true knowledge, and that facts are the only way of coming to that knowledge. Because atoms and molecules were, for 19th-century scientists, completely undetectable, there was, they argued, no reason to believe in them. Boltzmann's peers, all eminent scientists and heartfelt positivists, insisted that assuming that atoms were real—actually basing scientific theories on their existence—was little better than mysticism. Indeed, one of Boltzmann's stoutest critics, the physicist Ernst Mach, used to harass Boltzmann as Boltzmann lectured on atoms by calling out from the audience, "Have you ever seen one?" Of course, Boltzmann had not.

Boltzmann could be moved to tears of joy by a well-played sonata, the color of the sea, or a piece of poetry.

Nonetheless, Boltzmann felt certain that atoms were real and, for years, had written scathing critiques of even his closest friends' work. In his personal life, Boltzmann was charming, modest, and earnest, but in debates and in the lecture hall, he was a bulldog, relentlessly attacking arguments he believed were wrong and pushing his own ideas.

For almost 40 years, Boltzmann worked to show how atoms accounted for the behavior of all forms of matter. His greatest achievement was to find a way to understand the behavior of matter in terms of large collections of molecules and atoms. He saw that the statistical laws of large numbers could be applied to collections of atoms or molecules.

But Boltzmann's work could be correct only if atoms were real. And, after 40 years of arguing, fewer and fewer people seemed to think that atoms existed. As Boltzmann wrote to a friend, "I am conscious of being only one individual struggling weakly against the stream of time," and later, "I present myself to you therefore as a man left behind."

Boltzmann did not know it, but by 1906, the year he took his family to Italy, the positivist view had already peaked. Already, the darkest days for the atom—and Boltzmann's ideas—were over. Bright young scientists were beginning to take a second look at atoms. For example, in 1905, twenty-six-year-old Albert Einstein had published a paper showing that the constant jiggling of particles that scientists noticed when looking down a microscope (called Brownian motion) was most likely due to bombardment by atoms—real, physical atoms, not abstract statistical averages. In just a few years, Ernest Rutherford would locate the tiny, central nucleus of the atom. And soon after that, the physicist Niels Bohr would calculate the movements of electrons.

But Boltzmann had given up. On a sunny day in September 1906, while his wife and daughters went swimming, Boltzmann ended his life by hanging himself in his hotel room. Had he endured just a few more years of frustration, he would have lived to see the triumph of his own ideas and the humiliation of his intellectual enemies.

Indeed, Boltzmann's intellectual victory was so complete that much of the material in this chapter—written nearly 100 years after his death—rests heavily on his work. The most important applications of Boltzmann's idea were in the field of

thermodynamics [Greek, *thermo* = heat + *dynamis* = power, force], the study of the relationships among different forms of energy. For example, as a car converts the chemical energy of gasoline into the energy of motion, the car moves forward. Every event, every reaction, every process is accompanied by a "transformation of energy."

The direction a transformation takes depends on the rules of thermodynamics. A scientist can predict the direction of a reaction without knowing exactly how the reaction takes place. We know, for example, that water always flows downhill, even if we do not know whether it gets there through a pipe, a stream, or a waterfall.

Before Boltzmann, thermodynamics was understood as a pair of abstract mathematical laws. These two laws could pre-

Figure 5-1 Ludwig Boltzmann (1844–1906). *(Science Photo Library/Photo Researchers)*

dict which way a reaction would go, but they said nothing about the actual process of the reaction.

Boltzmann showed that the reason for thermodynamics was the behavior of large numbers of individual atoms and molecules. He showed *how* chemical reactions occur, what heat actually is, and why some chemical reactions produce heat while others absorb heat. Most important, Boltzmann's work paved the way for the theory of **kinetics** [Greek, *kinetikos* = moving], the study of the rates of reactions. Before Boltzmann, chemists could only predict whether a reaction should take place. Kinetics, by explaining why a reaction takes place, could predict how long the reaction would take to occur.

Understanding a biological or chemical process (or any process, for that matter) requires more than noticing changes in structures. For example, we may observe that most of the gasoline in a car's engine is converted to water and carbon dioxide, with a tremendous release of heat and energy of motion. But we cannot understand what is happening unless we know the answers to two general questions: (1) Which way will the reaction occur? (2) How fast will it occur? We must ask, Why does the conversion of gasoline to water and carbon dioxide occur in one direction only? Why don't the carbon dioxide and water present in the air we breathe combine spontaneously to form gasoline? And why does the conversion from gasoline to carbon dioxide and water occur in fractions of a second? What if converting a tank of gasoline to water and carbon dioxide took several years? In that case, gasoline would be a poor fuel. Why does the reaction occur as rapidly as it does? What factors determine how fast any reaction goes?

Predicting the speed of a process requires that we know something about the process itself. For example, we know that water plunges quickly downhill in a waterfall and meanders slowly in a gentle stream. The rate at which the water moves depends on the steepness of its descent. Thermodynamics predicts that water will flow downhill. But only kinetics can tell us how fast the water will flow. In this chapter we will see that thermodynamics tells us which way a reaction will go and kinetics tells how fast.

Thermodynamics predicts the direction of an energy transformation. Kinetics predicts the rate of an energy transformation.

KEY CONCEPTS

1. The Second Law of Thermodynamics predicts the direction of any process. Energy becomes less available for work as a reaction or process occurs.

2. Chemical reactions depend on the random movements of molecules. The rate of a chemical reaction depends on the activation energy needed to rearrange chemical bonds.

3. Enzymes speed up biochemical reactions by lowering their activation energies.

4. Cells and organisms regulate their own metabolism by changing the shapes and activities of enzymes.

WHAT DETERMINES WHICH WAY A REACTION PROCEEDS?

Everyone can tell when a movie runs backwards: a waterfall goes up instead of down; a balloon spontaneously assembles from scattered pieces; a rubber ball bounces up a flight of stairs (Figure 5-2). We all know the direction of time's arrow. Biological processes also have direction: a child matures into an adult; a bud unfolds into a leaf; dead organisms decay.

The key to predicting direction lies in understanding the conversions of energy. In any process—whether the flow of water, the growth of a tree, or the maturation of a child—energy changes form. The study of thermodynamics has provided rules that govern all energy changes, both biological and nonbiological. These rules establish the direction of time's arrow.

Figure 5-2   We all know the direction of time's arrow. An egg may break, but it can never reassemble.   *(Peter Parks/Oxford Scientific Films/Earth Scenes)*

How May Work Be Converted to Kinetic or Potential Energy?

Most of us think of work as any process that requires energy. There is, however, a more precise definition of work: **work** is the movement of an object against a force. For example, lifting a sack of concrete against the force of gravity is work. So is winding the spring in a child's toy (Figure 5-3).

Work, in this sense, can be stored for later use, the way energy is stored by a battery. The wound-up spring in a toy beetle can propel the toy by operating its legs. When the spring is wound, we say that the work of winding the spring is stored as **potential energy**. When the spring unwinds and the beetle leaps forward, we say that the toy has **kinetic energy**—the energy of moving objects. A boulder at the edge of a cliff has potential energy. A moving car, a running athlete, and a contracting muscle all have kinetic energy. Moving molecules also have kinetic energy, which we call heat. Like kinetic energy, potential energy exists in many forms, including gravitational, electrical, and chemical.

Rearranging atoms in a chemical reaction is also a form of work. Recall, for example, that the energy-rich molecule ATP consists of a molecule of adenosine linked to three negatively charged phosphate groups (Figure 5-3A). Making a molecule of ATP requires the squeezing together of three negatively charged phosphate groups. This squeezing is work against an electrical force. Once the atoms are squeezed together into a molecule, the work is stored as potential energy. Thus ATP has potential energy. This energy can later be released to do work. The energy can be released, for example, in the form of kinetic energy by making a muscle move.

Work is the movement of an object against a force such as gravity or an electrical attraction or repulsion. Making a molecule of ATP is a form of work because it involves squeezing atoms together.

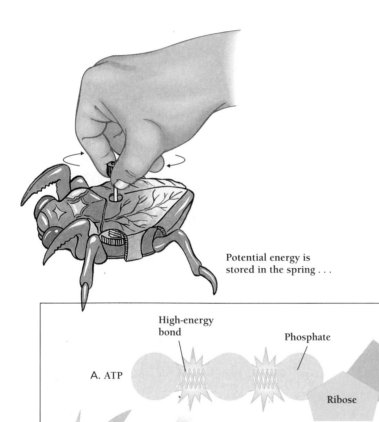

Potential energy is stored in the spring . . .

High-energy bond

Phosphate

A. ATP

Adenine

Ribose

B.

. . . and is released as kinetic energy.

Figure 5-3 Energy may be stored in a spring or in the high-energy bonds of ATP. Tightening a spring takes work. Releasing the spring discharges the potential energy stored in the spring. A. Creating a high-energy phosphate bond takes work. B. Breaking the bond discharges the potential energy stored in the bond.

How Does Thermodynamics Predict the Direction of a Reaction?

Scientists and engineers first began to understand the relationship among different forms of energy in the late 18th century. The major stimulus for the development of the rules of thermodynamics was the increasing use of the steam engine, which converted the potential energy of coal into kinetic energy capable of powering a locomotive or a factory. As we shall see, the same rules that apply to steam engines also apply to the energy conversions in organisms.

One of the first and most important conclusions of the new field of thermodynamics was that, in any process, energy is neither created nor destroyed—it only changes form. The **First Law of Thermodynamics** states that the total amount of energy in any process stays constant (Figure 5-4A). In this view, energy may switch from potential energy to kinetic energy and back again, but it is neither created nor destroyed. Any heat generated during a process is merely another form of kinetic energy. At 22, Boltzmann pointed out that atoms nicely account for the First Law of Thermodynamics. If atoms and molecules actually exist, he argued, then heat could be merely the rapid movement of atoms and molecules. Molecules and atoms—though invisible to us—have kinetic energy in the same way that a ball moving through the air has.

The second major conclusion of thermodynamics was that, in any process, energy becomes less and less available to do work (Figure 5-4B). More succinctly, the **Second Law of Thermodynamics** states that, in the course of any process, the energy available for work decreases. The Second Law explains why we can tell when a film is running backwards, why, in short, time has direction (Figure 5-2). At 27, Boltzmann showed—by means of difficult mathematics—that the existence of atoms accounted for the Second Law, as well. Oddly, though Boltzmann was widely regarded as a genius throughout his life, few people understood his arguments. And Boltzmann's peers, for the most part, ignored his conclusions about atoms and thermodynamics, even though everyone accepted the laws of thermodynamics themselves.

We can see how the two laws of thermodynamics work together to explain the direction of a reaction by applying them to the contraction of a muscle in a chin-up. When the biceps contract, they perform work—the lifting of the rest of the body. The energy for this work comes from the chemical energy in ATP. The ATP molecule gives up part of its stored energy by splitting off one of its three phosphate groups. The remaining molecule has two phosphate groups, and is then called adenosine *di*phosphate or ADP [Greek, *di* = two]. ADP still has potential energy, but less than ATP.

According to the Second Law, muscle contraction converts only a fraction of the energy from ATP into useful work. The rest of the energy becomes heat—the random motion of molecules. From the organism's point of view, heat is largely wasted energy. Even though some energy is dissipated as heat, we know that no energy is actually destroyed. The energy released by

ATP exactly equals the work done by the contracting muscle plus any heat released in the process.

The First Law implies that all forms of energy—heat and work, whether chemical, electrical, or mechanical—can be measured in the same units. The unit of energy that we will use in this book is the **calorie**—the amount of energy needed to raise the temperature of one gram of water by 1°C. The energy contained in food is often reported in Calories (with a capital C). One Calorie is equal to 1000 calories or 1 kcal ("kilocalorie"). The daily diet of a college student usually contains from 2000 to 3000 kcal (Calories) of energy.

Every time energy changes form—whether from chemical to electrical or from electrical to light energy—a little more of the energy turns into heat. (Think of the chemical energy in a flashlight battery transformed to electricity and then to light.) Over time, more and more energy turns into heat and less and less energy becomes available for work. For many processes, then, the answer to "which way?" is the direction that releases, rather than absorbs heat. A ball will bounce down rather than up a staircase because the impact of each bounce converts part of the ball's kinetic energy into heat. With each bounce, less energy is available for the next bounce, so the ball can never bounce back up the previous stair. It can only go down.

For a chemical reaction, the predictions of the Second Law are more complicated than for a mechanical system such as a ball on a staircase. In general, the formation of a chemical bond between any two atoms releases energy. Likewise, breaking a chemical bond usually consumes energy. However, the amount of energy released by forming a chemical bond varies. The overall change in energy in a reaction depends on which bonds are broken and which bonds are formed.

The First Law of Thermodynamics states that energy changes form, but is neither created nor destroyed. The Second Law of Thermodynamics states that in any process, the energy available to do work decreases. The loss of heat often establishes the direction of the process, preventing the process from reversing itself without an input of energy.

Free Energy Changes Predict the Direction of a Reaction

To predict the direction of a chemical reaction, we need an overall measure of the availability of energy. **Free energy** is the energy in a system available for doing work. Free energy is abbreviated G, after the American chemist J. Willard Gibbs, who, in the 1870s, single-handedly worked out the theory of chemical thermodynamics.

Every chemical has a certain amount of free energy. And every chemical reaction involves a "change in free energy." When two molecules (or **reactants**) react to form one or more new molecules (or **products**), the change in free energy is equal to the free energy of the products minus the free energy of the

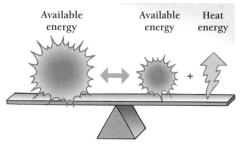

A. First Law of Thermodynamics: In a closed system, energy is neither lost nor gained, only transformed from one kind of energy to another.

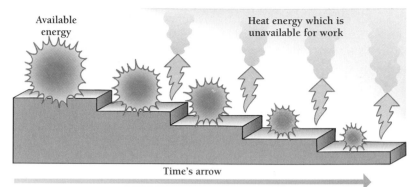

B. Second Law of Thermodynamics: The energy available for work decreases as energy is lost to heat.

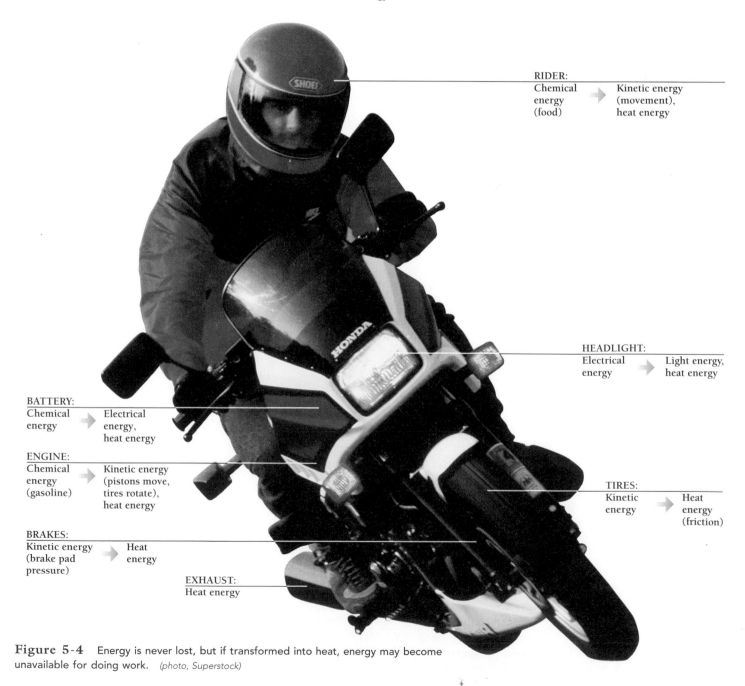

RIDER:
Chemical energy (food) → Kinetic energy (movement), heat energy

HEADLIGHT:
Electrical energy → Light energy, heat energy

BATTERY:
Chemical energy → Electrical energy, heat energy

ENGINE:
Chemical energy (gasoline) → Kinetic energy (pistons move, tires rotate), heat energy

TIRES:
Kinetic energy → Heat energy (friction)

BRAKES:
Kinetic energy (brake pad pressure) → Heat energy

EXHAUST:
Heat energy

Figure 5-4 Energy is never lost, but if transformed into heat, energy may become unavailable for doing work. *(photo, Superstock)*

reactants. The change in free energy is abbreviated as ΔG. (Δ, pronounced "delta," is the mathematical symbol for a change in a quantity.) Chemists abbreviate this statement by writing:

$$\Delta G = G_{products} - G_{reactants}$$

This equation might look intimidating, but it is no different from saying that the amount of weight you have gained or lost in the last week is equal to the amount you have today minus the amount you had last week:

$$\Delta lbs = lbs_{today} - lbs_{last\ week}$$

(where Δlbs equals the change in your weight, the amount of weight you lost or gained).

In any process, the change in free energy depends only on the initial and final states. The change does not depend on the way the process occurs or on how long it takes. A ball on a staircase might fall down the whole staircase in one big bounce or in many smaller bounces. The ball may sit unmoving on the top step for years on end until it is jostled by an earthquake, a passing truck, or a curious child.

Thermodynamics cannot predict how long it will take for a ball to fall down the stairs, or for a muscle cell to break down ATP into ADP and phosphate. All that thermodynamics tells us is that, if we wait long enough, the ball will eventually roll down the stairs (not up) and that ATP will eventually form ADP and phosphate. In both cases, the process goes "downhill," from higher to lower free energy.

Both a ball falling and ATP changing to ADP are **exergonic**, or energy-releasing, processes. The reaction between the octane in gasoline and the oxygen in air is highly exergonic: a lot of energy is released. On the other hand, carrying a coconut up a tree and converting ADP to ATP are **endergonic**, or energy-consuming, processes (Figure 5-5).

In exergonic processes, free energy decreases, and ΔG is negative. As a result, exergonic processes, such as ATP changing to ADP, occur spontaneously (given enough time). In endergonic processes, free energy increases and ΔG is positive. Endergonic reactions will not occur spontaneously, but require energy from some other source.

The change in free energy—the amount of energy available to do work—that occurs during the course of a reaction determines whether the reaction can occur spontaneously.

Where Does the Free Energy Released or Consumed During a Reaction Come From?

When two isolated atoms form a chemical bond, energy is released. The tighter the bond, the more energy is released. During a chemical reaction, some of the bonds in the reactants break, and other bonds in the products form. Let us consider the conversion of ATP to ADP:

$$ATP + H_2O \longrightarrow ADP + Phosphate$$
$$(Reactants) \longrightarrow (Products)$$

In this reaction, the oxygen atom in the water molecule breaks off from the water molecule and then displaces the oxygen atom in the ATP's third phosphate group, so that a new P–O (phosphate-oxygen) bond forms. As a result, two bonds break—an O–H bond in water and a P–O bond in ATP—and two new bonds form—a P–O bond in the ADP, and an O–H bond in phosphate (Figure 5-6). Because of electrical repulsion by the other phosphate groups, however, the P–O bond that breaks in ATP is looser than the new one that forms in ADP.

A.   B.

Figure 5-5 **Work and play.** A. Carrying a coconut up a tree is an endergonic (energy-consuming) process. B. Dropping a coconut is an exergonic (energy-releasing) process.

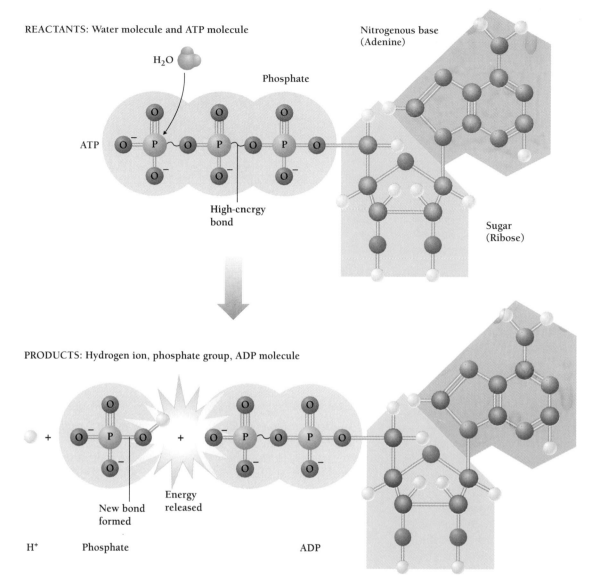

REACTANTS: Water molecule and ATP molecule

H_2O

Phosphate

Nitrogenous base
(Adenine)

ATP

High-energy
bond

Sugar
(Ribose)

PRODUCTS: Hydrogen ion, phosphate group, ADP molecule

New bond
formed

Energy
released

H^+ Phosphate   ADP

Figure 5-6 Splitting a high-energy phosphate bond in ATP releases energy that can be used in other reactions. A water molecule attacks and hydrolyzes the phosphate bond, releasing a phosphate group from the ADP molecule.

As a result, the free energy of the products (ADP and phosphate) is less than the free energy of the reactants (ATP and water). ADP has less energy available to perform work than ATP.

The ATP $\longrightarrow$ ADP reaction favors the breaking of less stable bonds and the forming of more stable bonds. Biochemists speak of **high-energy bonds**—unstable chemical bonds that give up energy to form more stable, low-energy bonds. The bonds that hold the second and third phosphates of ATP are high-energy bonds. Biochemists sometimes represent such a high-energy bond as a squiggle (~).

Chemists can determine the direction of a chemical reaction by measuring the free energy changes associated with the making and breaking of chemical bonds. For example, the free energy change for the conversion of ATP into ADP and phosphate is -7.3 kcal/mole. (A mole is a fixed, huge number of molecules—6×10^{23}.) The minus sign shows that the free energy decreases in this reaction. Such a reaction is therefore exergonic and occurs spontaneously.

The bonds between the atoms of a molecule have characteristic energies that change in the course of a reaction. Rearrangements of these chemical bonds may consume or release energy.

How Can One Process Provide the Energy for Another?

We saw earlier that an energy-consuming, or endergonic, reaction needs a source of energy. That energy can come from an exergonic process. We can see how this happens by studying the chimpanzees and coconuts in Figure 5-7. When one process contributes energy to a second process, as when a coconut falls onto a seesaw and lifts a small chimpanzee, the two processes are said to be **coupled.**

The Second Law requires only that the coupled reaction be exergonic. Changes in free energy of coupled reactions also determine the direction of biological processes. In each case, the coupled biochemical reactions must—together—be exergonic. For almost all the biological processes that we discuss in this book, the driving exergonic reaction is the splitting of ATP.

How does a cell accomplish the coupling of endergonic and exergonic reactions? The answer lies in the ability of cells to speed up particular chemical reactions selectively. Just as the arrangement of a rope and pulley determines whether a falling ball will lift another, cellular mechanisms can couple two chemical reactions.

The agents by which cells accomplish this coupling are large molecules—usually proteins—called **enzymes.** Enzymes do not affect the change in free energy of a reaction, but they do speed up chemical reactions. Enzymes can also couple two reactions together by allowing the product of an endergonic reaction to participate in an exergonic reaction, so the two reactions together are exergonic. In muscle contraction, for example, an enzyme couples the hydrolysis of ATP to the movement of tiny muscle filaments. That is, the energy that comes from breaking the bonds in ATP is used by the muscle filaments to contract. The energy in the ATP comes, in turn, from the energy in sugars and other energy-rich food molecules. The energy in food molecules comes, ultimately, from sunlight, as we will discuss in Chapter 7.

With the help of enzymes, an exergonic reaction such as the splitting of ATP can drive an endergonic reaction.

Disorder (Entropy) Always Increases in Spontaneous Processes

We have seen that the Second Law of Thermodynamics says that uncoordinated motion is more probable than coordinated motion. It is, for example, more likely that the pieces of a burst balloon will fly apart than it is that they will spontaneously reassemble. It is more likely that the books in your room will become disorganized than that they will remain neatly shelved in alphabetical order. The Second Law is a formal statement that

Figure 5-7   A coupled reaction. When the adult chimpanzee drops the coconut, the baby chimp is thrown up into the air. The falling coconut is *coupled* to the flying chimp.

disorder is more probable than order. The amount of disorder in the universe tends to increase.

Physicists have defined a formal measure of disorder, called **entropy.** Entropy has a high value when objects are disordered or distributed at random, and a low value when they are ordered. Because disorder and random motion are more probable, the tendency of entropy to increase can be used to produce work (Figure 5-8).

How Does the Concentration of a Substance Affect Its Free Energy?

Any process that converts an orderly arrangement to a less orderly one can perform work. For example, when we burn wood (converting well-ordered molecules of cellulose to disordered molecules of carbon dioxide and water) we can use the energy from this exergonic reaction to warm a house, to grill a hamburger, or to power a steam engine.

On the other hand, any process that converts a disorderly arrangement to an orderly one consumes energy. The oak tree uses energy from the sun and CO_2 from the air to build the cellulose in its massive trunk and branches (Figure 5-9). Likewise, a cell uses energy to maintain concentrations of molecules that are different from those outside the cell. Cells must perform work to keep their internal environments constant. When no energy supply is available to perform this work, the differences between the inside and the outside of the cell disappear, and the cell dies.

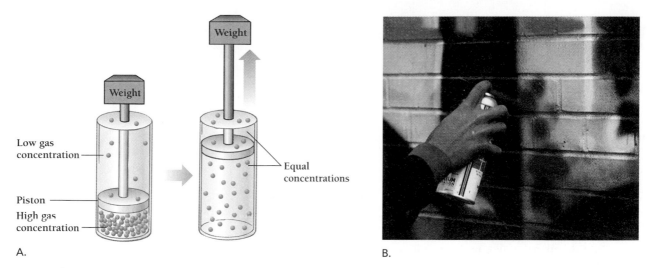

Low gas
concentration

Piston

High gas
concentration

Weight

Weight

Equal
concentrations

A.

B.

Figure 5-8   A difference in concentration constitutes potential energy. A. The higher concentration of gas molecules in the bottom of the chamber pushes the piston up, lifting a weight. B. The high concentration of gas (relative to the outside air) in a can of spray paint is a source of potential energy.   *(B, Martha Cooper/Peter Arnold, Inc.)*

Figure 5-9   Disorder into order—order into disorder. The sun supplies energy for the tree to do the work of building its massive trunk, its branches, and its leaves. When the leaves die, they fall to the ground, where bacteria and fungi break them down and use the energy that is stored within them.   *(W. Banaszewski/Visuals Unlimited)*

Since the movement of molecules from a more concentrated to a less concentrated state can perform work, we can conclude that free energy—the energy available to perform work—depends on the concentration of the molecules (Figure 5-8). The greater the difference in concentration between two sides of a cell membrane or other barrier, the more work that can be done. This means that whether a reaction occurs depends not only on the energy in the individual bonds but also on the concentrations of both chemical reactants and products.

Imagine a chemical reaction in which reactant A is converted to a product B. If the reaction is exergonic (energy-releasing), then the reaction will proceed, and product B will begin to accumulate. As the amount of B increases, however, so does B's free energy (the energy available to do work). Gradually, the difference in energy between the reactants and the products decreases. When the free energy of B equals the free energy of A (and $\Delta G = 0$), the reaction is no longer "downhill," and an **equilibrium** exists; no further net change will take place.

Thermodynamics predicts where the equilibrium lies, but gives no information about how long a reaction will take to get to equilibrium. In the next section we will find out how the rules of kinetics predict the rates of chemical reactions.

The end, or equilibrium, point of a reaction depends on the concentration of the reactants and products, as well as on the free energy in the bonds between the atoms of each molecule.

WHAT DETERMINES THE RATE OF A CHEMICAL REACTION?

In the 19th century, archaeologists opened the tomb of an Egyptian pharaoh and found, among other artifacts, a perfectly preserved breakfast laid out for him. If the pharaoh had lived long enough to eat this meal, the food would have been oxidized in a matter of minutes. Yet, even after sitting for several thousand years in an oxygen-filled room, the pharaoh's breakfast was chemically unchanged. What kept the pharaoh's breakfast from being oxidized? In other words, How can cells initiate and speed up chemical reactions that run slowly or not at all outside the body?

We have seen that thermodynamics predicts both whether a reaction will occur and the direction of a reaction, independent of the precise mechanism. But to find out how *rapidly* a reaction occurs, we must understand its mechanism. We need to know what actually happens—what "path" a process takes from start to finish. For chemical reactions, this means going beyond discussing changes in bond energies and concentrations. Predicting the rates of chemical reactions requires an understanding of how molecules behave.

How Does Molecular Motion Help Explain Reaction Rates?

Molecules and atoms are constantly in motion. When a substance absorbs heat, its temperature increases and the atoms and molecules bounce and jostle about more rapidly. This simple fact is the basis for **kinetic theory,** or **kinetics.** Kinetics allows us to understand the rates of chemical reactions and how temperature and other environmental factors influence these rates.

For a moment, picture molecules as tiny frogs constantly jumping about at random (Figure 5-10A). Imagine individual frogs (molecules) jumping around at random in an area called "Level A," and in an adjacent area called "Level B." Some frogs will naturally jump into Level B. If Level B is lower than Level A, then the frogs will have a harder time going from B to A than in going from A to B. As a result, once the frogs arrive at the lower level, fewer will be able to jump back to Level A—only those with higher kinetic energy.

According to kinetics, molecules move about randomly and with varying energies.

What Stops a Chemical Reaction?

As we saw at the beginning of this chapter, the idea that molecules are constantly in motion and that their behavior explains many of the properties of matter was developed in the 1870s

Level A

Level B

High activation energy barrier

A.

Level A

Level B

Decreased activation energy barrier

B.

Figure 5-10   The random movements of molecules can be compared to the random leaps of frogs.
A. Even if all the frogs begin at level A, more will end up at Level B because it is harder for them to jump back to A.
B. With a decreased activation energy barrier, frogs can move more easily between the two levels, and it will take less time to reach equilibrium.

by Ludwig Boltzmann. Boltzmann's work led to the realization that temperature was a measure of the average kinetic energy of the moving molecules. The higher the temperature is, the faster the molecules move. However, we have to recognize that since temperature measures only the average energy of the molecules, they may have a range of kinetic energies. In our frog model, we would say that some frogs jump higher than others at any given temperature. At higher temperatures, more frogs will be able to jump over a fixed barrier, such as that between Level A and Level B (Figure 5-11).

Low temperature

Level A

Level B

A.

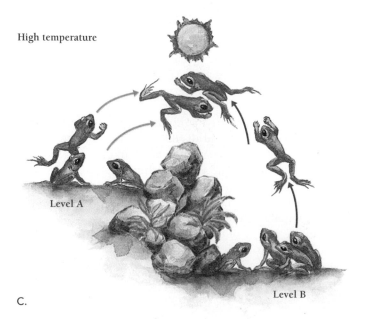

Increased temperature

Level A

Level B

B.

High temperature

Level A

Level B

C.

As time passes, the number of frogs (molecules) jumping from A to B and from B to A reaches equilibrium (that is, each minute just as many frogs will jump from A into B as jump from B into A). For this to happen, there must be more frogs in B than in A, since a smaller fraction of the frogs in B will be able to make the jump. The lower energy of Level B means that more frogs will end up in Level B than in Level A. We can see that this model leads to the same conclusion as thermodynamics—that molecules move in the direction of lower free energy. However, the jumping frog model also allows us to see once more that the concentrations of reactants and products influence chemical equilibrium.

The random bouncing of molecules helps explain chemical equilibrium.

What Starts a Chemical Reaction?

A ball sitting at the top of a flight of stairs will not bounce down the stairs until something starts it moving. Just so, every process—no matter how much it is favored thermodynamically—needs to be started in the right direction. In chemical reactions, we must ask, where does this push come from? The answer is that the energy for the reactions between molecules comes from the random movements of molecules.

To understand the factors that start chemical reactions, let us return to the jumping frog model. The difficulty in starting a chemical reaction is comparable to the height of the barrier between the frogs. To get from Level A to Level B, the frogs first must jump over the barrier (Figure 5-10). The molecules or frogs must have a minimum energy to do this. The minimum energy needed for a process to occur is called the **activation energy**.

The number of the molecules in a space that are able to react depends both on the temperature and on the height of the activation energy barrier (Figure 5-10). The higher the temperature, the greater the number of molecules with enough energy to jump the barrier. The lower the activation energy, the greater the number of molecules that can jump the barriers and the more rapidly equilibrium will be achieved.

Laboratory and industrial chemists can speed up a specific chemical reaction by using a **catalyst**—a substance that lowers the activation energy of a reaction but does not participate

Figure 5-11 **The rate of equilibration depends on the temperature of the frogs.**
A. At a low temperature, the frogs do not have the activation energy to get to level B. At equilibrium, all the frogs remain at level A.
B. At higher temperatures, half the frogs have enough energy to get to level B.
C. At still higher temperatures, all the frogs have enough energy to get to level B, but only some can leap back. The reaction equilibrates with 7 at level B and 3 at level A.

directly in the reaction. That is, a catalyst, by definition, is not changed by the reaction it speeds up. A catalyst, in essence, lowers the barrier between the two groups of frogs.

An enzyme, which we first encountered in Chapter 3, is no more than a biological catalyst. Almost all enzymes are proteins. Enzymes allow organisms to lower the activation energy for thousands of specific chemical transformations, and so shorten the time required to attain equilibrium. Hastening reactions is essential to life. We all need to be able to digest our breakfast in something less than 10,000 years.

Kinetic theory helps us understand why some thermodynamically favored reactions take place while others do not. Spontaneous reactions will actually occur only when a relatively large fraction of the molecules possess the needed activation energy. The carbohydrates in a piece of bread can be oxidized either by increasing the temperature (who hasn't burned toast?) or by exposing them to enzymes such as those in the human digestive tract. Likewise, an organism can convert the energy stored in ATP to muscular movement only if its cells possess the needed catalytic enzyme.

The presence of an active enzyme allows a cell to use a molecule such as ATP in specific ways. Depending on how particular enzymes couple the splitting of ATP to other reactions, the result may be the contraction of muscle, the pumping of ions into or out of cells, or the synthesis of other molecules. Cells control their chemistry by regulating the production and activity of individual enzymes.

The movement of molecules initiates chemical reactions. Temperature, a measure of the kinetic energy of molecules, is one factor that determines whether a reaction will occur. In organisms and cells, enzymes hasten chemical reactions by lowering the activation energy of the reaction.

HOW DO ENZYMES WORK?

In the last section we saw that enzymes speed up biological reactions by lowering activation energies. In this section we will see how enzymes accomplish this feat.

How Does an Enzyme Bind to a Reactant?

Enzymes accelerate specific chemical reactions by binding to the reacting molecules, which are called **substrates.** The interactions between enzyme and substrate are the basis for the remarkable properties of enzymes. Enzymes have enormous *catalytic power.* The right enzyme makes a chemical reaction go a million times faster than it would without the enzyme. A single molecule of the enzyme catalase, for example, can break down 5 million molecules of hydrogen peroxide per minute at 0°C. Enzymes are highly *specific.* Each enzyme usually only catalyzes a single chemical reaction.

Every enzyme has an **active site**—a groove or cleft on the enzyme's surface by which it binds to the small substrate molecule (Figure 5-12). The binding of substrate and enzyme lowers the activation energy for a particular chemical reaction. Different kinds of enzymes may be able to bind to the same substrate, *but each enzyme lowers the activation energy for a different reaction.*

The substrate temporarily binds to the enzyme by weak noncovalent bonds—including hydrogen bonds, ionic bonds, hydrophobic interactions, and van der Waals attractions. Often, as the two molecules bind, the shape of the enzyme changes (Figure 5-12). When this occurs, the interaction of enzyme and substrate is called an **induced fit.**

The binding of the enzyme and the substrate is reversible, and the reactant or product molecules are quickly released back into the solution. Some enzymes are capable of binding, processing, and releasing individual substrate molecules more than 100,000 times a second.

Even so, the interactions between the enzyme and the substrate occur one molecule at a time. So although the rate of any biochemical reaction depends on temperature and the concentration of the reactants (as in the case of the jumping frogs or molecules), it also depends on how many molecules of enzyme are present.

The active site on an enzyme binds to a substrate by means of weak noncovalent bonds. Each enzyme lowers the activation energy for only one reaction.

How Does an Enzyme Lower the Activation Energy of a Chemical Reaction?

The formation of noncovalent bonds between enzyme and substrate provides the free energy needed to form a more ordered state. This new arrangement of substrate lowers the activation energy for a reaction. In all cases, the enzyme itself remains unchanged after the completion of the reaction.

First, enzymes operate by bringing substrates together in a position that favors a reaction. Without the enzyme, substrate molecules in solution bump into each other at random, but only a few of these collisions lead to a reaction (Figure 5-13).

Second, when an enzyme binds to a substrate molecule, the enzyme may strain and distort the covalent bonds of the substrate. Bonds under such strain break more easily. The distorted form of a substrate is a **transition state**—a molecular form intermediate between the starting reactant and the final product. The transition state exists only for a brief moment—as little as a billionth of a second.

The free energy of the transition state molecule is higher than that of either the reactants or the products. The extra free energy comes from interactions with the enzyme. The net effect of binding is to lower the activation energy barrier.

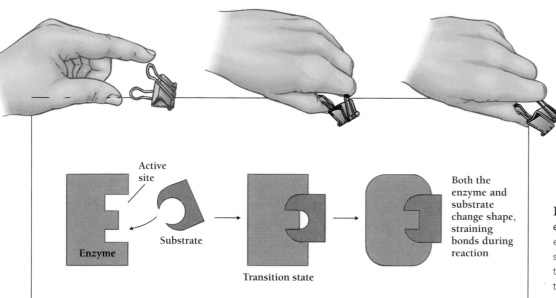

Figure 5-12 Induced fit model of enzyme-substrate interaction. Both the enzyme and the substrate change shape, straining bonds during the reaction. The strained form of a substrate—the transition state—exists for as little as a billionth of a second.

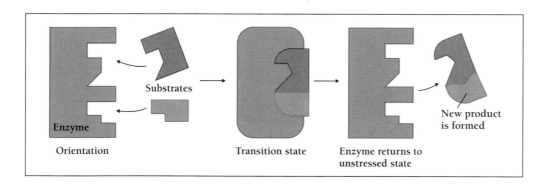

Figure 5-13 How does an enzyme catalyze a reaction between two other molecules? The enzyme brings both substrates close together in the proper orientation and also strains the covalent bonds of the substrates, lowering the activation energy of the reaction.

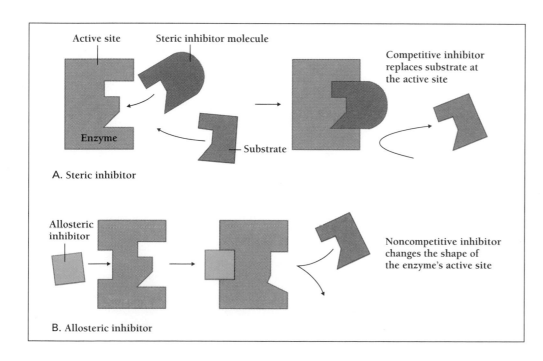

Figure 5-14 What can prevent an enzyme from catalyzing a reaction? A. Steric (competitive) inhibition. A steric inhibitor is a molecule that binds to the enzyme's active site, preventing the substrate from binding. A steric inhibitor competes with the substrate for the enzyme's active site. B. Allosteric (noncompetitive) inhibition. An allosteric inhibitor does not compete with the enzyme, but binds to another site on the enzyme in such a way that the active site no longer functions.

Finally, enzymes also may contribute directly to the chemical reaction—for example, by lending or temporarily accepting an atom or an ion. This change strains the substrate molecule, bringing it closer to its transition state. Again, the net effect is to lower the activation energy barrier to the reaction.

Enzymes lower the activation energy for a reaction by orienting two or more substrate molecules to increase the chances of their interaction, by straining the covalent bonds of the substrate molecules, or by participating directly in the chemical reaction.

How Do Environmental Conditions Affect the Rates of Enzymatic Reactions?

Enzymes work most efficiently at certain temperatures, pH values, and salt concentrations. Most human enzymes, for example, work best at body temperature. In our previous discussion of kinetics, we saw that all chemical reactions occur more rapidly at higher temperatures because more molecules possess enough kinetic energy to get over the activation energy barrier. But increasing the temperature of an enzymatic reaction increases the rate only to a point. Once the temperature reaches a certain point, the bonds that maintain the enzyme's structure begin to break. The enzyme **denatures**—it loses its activity. When an organism's enzymes begin to denature, it dies. Most organisms therefore cannot survive temperatures much higher than 40°C. A few unusual organisms live at higher temperatures. Bacteria that live in hot springs, for example, have evolved enzymes that are unusually insensitive to temperature.

Most enzymes also work best in only a narrow range of pH values. The pH of the environment can, for example, alter the shape of an enzyme and add or remove hydrogen ions from the enzyme, changing the way the enzyme binds to the substrate. Most human enzymes work best at a pH of between 6 and 8. However, pepsin, the protein-digesting enzyme secreted by the stomach, works best at a pH of about 2. The fuller the stomach becomes, the more the stomach acid is diluted and the higher the pH rises, though usually no higher than about 3.5. If the pH of the stomach contents were to rise above 5, the pepsin would no longer digest.

Many enzymes work by acting as acids or bases, donating or accepting hydrogen ions to certain reactions. For example, the amino acid histidine, when part of the active site of certain enzymes, can act as a base or an acid. If the pH of the environment is about 6.5 (slightly acid), histidine is an acid. At a pH of about 7.5 (slightly alkaline), however, histidine loses a hydrogen ion and becomes a base. An enzyme with an active histidine side chain can donate a hydrogen ion and so break a particular chemical bond, but only if the pH is right.

Most enzymes do not work in highly acid environments. This is why vinegar is such an effective preservative. The acetic acid in vinegar disables any enzymes present either in food itself or in any microorganisms that have slipped in. Without enzymes, reactions take place so slowly that pickled cucumbers, onions, peppers, herring, cabbage, or plums may keep for decades or longer.

The rates of enzymatic reactions increase with temperature, but only up to a point. Most enzymes work in a narrow range of pH values.

HOW DOES A CELL OR ORGANISM REGULATE ITS OWN METABOLISM?

Organisms and cells are enormously flexible in the way they process, or metabolize, food. They function in lean as well as in fat times. They can obtain energy when food is available or break down their own stores when it is not. They can make building blocks not obtainable from the diet, or they can stop making a particular building block when it becomes available. An organism with a diet rich in protein, for example, need not waste hard-won ATP making its own amino acids.

Organisms adjust their metabolism by regulating the amounts and activities of enzymes. In doing so, they both adjust the overall flow of energy through the cells and channel energy to perform individual tasks, such as the production of specific building blocks.

Enzymatic Reactions Often Occur in Small Steps

When organisms and cells transform one molecule into another, they frequently do so by means of a series of small steps or subreactions, each one mediated by a different enzyme. For example, if a chemical reacts to form product B, which reacts to form C, which reacts to form D,

$$A \longrightarrow B \longrightarrow C \longrightarrow D$$

each step is catalyzed by a different enzyme. In humans, separate reactions number in the thousands. Ten different reaction steps and ten different enzymes are necessary in the initial extraction of energy from the sugar glucose. Humans, plants, and bacteria all use exactly the same enzymes to digest this sugar (Chapter 6).

The network of biochemical reactions is complex for at least four reasons. First, some of the chemical transformations of cells are complex, and can only be accomplished in several steps. Second, many transformations are endergonic. For these reactions to proceed, they must be coupled to exergonic reactions, such as the breakdown of ATP to ADP. Such coupled endergonic reactions are especially common in reactions of **anabolism**—the synthesis of large molecules from smaller ones. Third, many exergonic transformations produce more energy than the cell can store all at once. Other biochemical reactions, which release smaller amounts of energy, allow the cell to capture the same energy in smaller steps. Smaller reactions allow the cell to store the energy for later use and also minimize the buildup of heat. Transformations of this kind are especially common in reactions of **catabolism**—the breakdown of food molecules to smaller molecules.

Fourth, the intermediate products in some reaction sequences serve as starting materials for the synthesis of needed building blocks. For example, some of the intermediate steps in the breakdown of sugars to carbon dioxide and water are important precursors of the building blocks from which proteins are made.

Many chemical transformations performed by organisms consist of numerous subreactions.

What Inhibits an Enzyme?

Because an enzyme recognizes the shape of its substrate, enzymes can be "fooled" into binding to molecules that are similar in size and shape. Such impersonators may block the active site and prevent the enzyme from acting (Figure 5-14 on page 127). The impersonator molecule is a kind of **inhibitor.** Specifically it is a **steric inhibitor** [Greek, *steros* = solid] because the inhibition has to do with shape. Steric inhibition can be overcome, however. When the concentration of the substrate is much greater than the inhibitor molecule, the enzyme still binds to the substrate. In this case, the inhibitor is called a competitive inhibitor.

Many inhibitors, called **allosteric inhibitors,** bind to another site on the enzyme besides the active site. Allosteric inhibitors cannot be overcome by increasing the concentration of the molecule that normally binds to the active site. Such **noncompetitive inhibitors,** including, for example, lead and other heavy metals, work by changing the overall shape and chemistry of an enzyme. Some noncompetitive enzyme inhibitors act only temporarily. But other inhibitors act irreversibly by forming a bond with the enzyme. The bond permanently alters the enzyme's shape. The antibiotic penicillin, for example, irreversibly inhibits a bacterial enzyme that bacteria use to build their cell wall.

Enzyme inhibitors prevent an enzyme from catalyzing a reaction. Inhibitors may either (1) compete with the substrate for the active site or (2) temporarily or permanently alter the shape of the enzyme.

How Does a Cell Regulate Enzyme Function?

Cells must themselves control the activity of enzymes by selectively inhibiting them. To make unnecessary building blocks would waste energy. Organisms that conserve energy by regulating the activity of enzymes therefore have an advantage over organisms that must spend more energy. An example of such regulation is the synthesis of the amino acid isoleucine. When isoleucine is not available, cells make it from another amino acid, threonine, in a sequence of five enzymatic reactions. When enough isoleucine has accumulated, the pathway from threonine shuts down. Isoleucine, the end product of the pathway, specifically inhibits the first step of this pathway (Figure 5-15).

This feedback inhibition is one of many examples of homeostasis in metabolism. Other molecules besides end products may also change the activity of enzymes—sometimes inhibiting enzyme action, sometimes increasing enzymatic action. A molecule or ion that changes the activity of an enzyme is called an **effector.** Inhibitors are just one kind of effector. Other effectors increase the activity of enzymes. When the effector binds to the enzyme at a place other than the active site, the effector is called an **allosteric effector.**

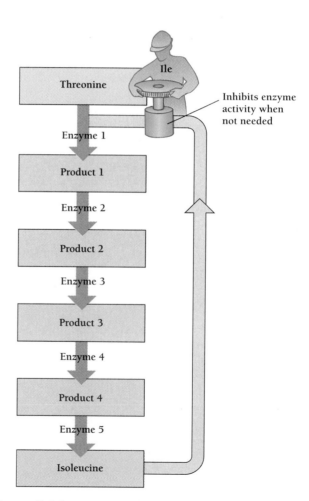

Figure 5-15   **How can cells regulate enzyme function?** The production of the amino acid isoleucine self-regulates by means of negative feedback. In the absence of isoleucine, a sequence of five enzymatic reactions synthesizes isoleucine from threonine. However, when isoleucine molecules accumulate, they inhibit the enzyme that catalyzes the first of the five reactions, shutting down the isoleucine pathway.

What is the mechanism of allosteric regulation? An enzyme can adopt a variety of shapes. Most enzymes have evolved so that their most biologically active shape is also their most stable shape. The binding of an allosteric effector changes the enzyme's shape to one that is either more or less active. Thus, an enzyme subject to an allosteric effector has two alternative shapes. The allosteric effector switches the enzyme from one form to the other, "pulling" or "pushing" the enzyme into a more or less active shape.

The various molecules within a cell could undergo countless thermodynamically favored chemical reactions. The only reactions that happen rapidly enough to contribute to the cell's economy, however, are those catalyzed by enzymes. Such reactions include those that provide cells with the energy needed to accomplish the tasks of living. Not only do enzymes serve as catalysts for specific reactions, but they also regulate the rates of reactions in response to environmental changes.

One way cells regulate enzyme function is through effectors, which increase or decrease the activity of enzymes by altering their shape.

STUDY OUTLINE WITH KEY TERMS

All processes—biological and nonbiological—are accompanied by changes in the form of energy, which are subject to the laws of **thermodynamics.** The **First Law of Thermodynamics** states that the total amount of energy does not change. The **Second Law of Thermodynamics** states that the total amount of energy in a closed system becomes less and less available for **work;** more and more of it is converted to heat. In short, **entropy** increases.

The laws of thermodynamics allow scientists to predict whether any given process occurs spontaneously. In biochemical reactions, the making and breaking of chemical bonds between atoms leads to changes in available energy. A process will occur spontaneously if the energy available to do work—the **free energy**—decreases. A reaction is **exergonic** if free energy (ΔG) decreases and **endergonic** if free energy increases. Exergonic reactions occur spontaneously, but thermodynamics provides no information about how quickly they occur. By **coupling** endergonic processes to exergonic reactions, organisms can accomplish "uphill reactions." Cells achieve such coupling by using **catalysts** called **enzymes,** which are almost always protein molecules. In many cases, cells derive energy from the **high-energy bonds** of ATP. During chemical reactions and other processes, **kinetic energy** (for example, heat) may be converted to **potential energy** (for example, a high-energy bond in ATP).

The free energy change in a chemical reaction depends on the concentrations of **reactants** and **products,** as well as on the number of chemical bonds that are made and broken. Reactions tend to go in the direction of increased disorder, or entropy, in which the concentrations of molecules become more uniform.

To determine how long a process will take to occur requires knowledge of how it will happen. The study of **kinetics** allows scientists to predict the rates of chemical reactions from the properties of molecules. The basis of kinetics is that molecules are constantly in motion, with the average energy of a group of molecules determined by the temperature, measured in **calories.**

At chemical **equilibrium** the direction of a reaction is balanced by the opposite reaction. **Enzymes** and other **catalysts** speed up specific chemical reactions but do not affect the free energy changes of chemical reactions. They increase the speed at which equilibrium is approached, but they do not affect the character of the equilibrium.

Enzymes accelerate particular chemical reactions by lowering the **activation energy** of the reaction. Enzymes accomplish this by binding to the reacting molecules by means of noncovalent interactions. An enzyme may join two or more **substrate** molecules. Alternatively, it may strain the bonds of a single substrate molecule in an **induced fit,** forcing the substrate into a **transition state.** Some enzymes actually participate in the chemical reaction by temporarily giving or receiving atoms or ions. The activity of an enzyme is highly sensitive to environmental conditions that can affect its activity. Temperature, pH, and the presence of other molecules may **denature** an enzyme.

Cells regulate the activity of many enzymes with **effectors,** including **allosteric effectors, steric inhibitors,** and **allosteric inhibitors.** Steric inhibitors block the enzyme's **active site,** while allosteric inhibitors and effectors change the enzyme's shape and activity by binding to a nonactive site.

Many chemical transformations performed by organisms in **anabolism** and **catabolism** consist of numerous small reactions. Such small reactions more easily allow the cell to capture the energy for later use.

REVIEW AND THOUGHT QUESTIONS

Review Questions

1. State the First and Second Laws of Thermodynamics and explain their relationships to an example of work.
2. What is free energy? What is its relationship to the Second Law of Thermodynamics? In what direction must it change in all reactions (or processes of work)? Compare this to the meaning of entropy.
3. Define endergonic and exergonic in terms of the free energy change associated with reactions. Which reaction involving ATP is endergonic? Exergonic? Which reaction occurs spontaneously?

4. What is the activation energy for a reaction? How might it influence the speed with which a reaction proceeds?

5. Enzymes are biological catalysts. How do they perform the task of speeding up biological reactions?

6. What is meant by the active site of an enzyme? How is it related to the substrate of that enzyme? What is meant by induced fit? How does this make the substrate more likely to undergo a reaction?

7. Describe four benefits that cells derive from organizing their reactions into pathways.

Thought Questions

8. Where does the energy come from when you perform the work of riding your bicycle? Did ALL of the energy in the food you

used for this task actually help turn the wheels of your bicycle? Explain using the First and Second Laws of Thermodynamics.

9. Explain why unraked leaves might form a neat pile in a specific spot in your backyard. Does this mean that the Second Law of Thermodynamics is not working? What is the source of energy for this organization of the leaves?

10. Why do you think that some exergonic reactions will NOT go at all in the absence of a catalyst? If the change in free energy is no obstruction, what is?

SELECTED READINGS

Blum, H.F., *Time's Arrow and Evolution,* Princeton University Press, Princeton, New Jersey, 1968. A close but accessible look at the operation of thermodynamics in biology.

Monod, Jacob, *Chance and Necessity,* Alfred Knopf, New York, 1971. A philosophical discussion of the role of chance in biology, by an eminent molecular biologist. *Chance and Necessity* argues for scientific objectivity as an ethical choice and requires a background in modern molecular biology.

▶ On-line materials relating to this chapter are on the World Wide Web at
http://www.saunderscollege.com/lifesci/
Click on Tobin/Dusheck: *Asking About Life.*

Louis Pasteur and Vitalism

In 1835, the French scientist Charles Cagniard de la Tour observed that the yeast in fermenting beer consisted of tiny living cellular organisms. Cagniard de la Tour noted that these tiny organisms multiplied by budding, and he suggested that they were somehow linked to the process of alcoholic fermentation—the conversion of sugar to ethanol and carbon dioxide. In the same year, two other scientific papers described similar observations. One was by Theodor Schwann, the German botanist whose work helped establish the cell theory described in Chapter 4.

Schwann also showed that if grape juice was boiled to kill the yeast, the juice would not ferment to wine unless yeast cells were added back to the juice. Schwann correctly concluded that yeast plays an essential role in fermentation. He then showed that fermentation began at the same time that the yeast cells first appeared, progressed with their multiplication, and stopped as soon as the yeast cells stopped multiplying.

Amazingly, the scientific community rejected these important discoveries out of hand, thanks to a handful of powerful personalities and the scientific and philosophical temper of the time. The most influential scientists insisted that yeast was not a living organism at all, but a substance, and probably an unimportant one.

No less than the father of modern chemistry, Antoine Lavoisier (1743–1794), had already argued that fermentation could be understood as a simple chemical reaction. Fermentation, he and others had written, consisted of a reaction in which the six-carbon sugar glucose broke down into two molecules of ethanol and two molecules of carbon dioxide:

$$C_6H_{12}O_6 \longrightarrow 2\ C_2H_5OH + 2\ CO_2$$

$$\text{glucose} \longrightarrow \text{ethanol} + \text{carbon dioxide}$$

This nicely balanced formulation seemed so simple and so right that chemists, riding high on the successes of 18th-century physics and chemistry, could not imagine why yeast would be necessary. Even if yeast did turn out to be necessary, where could it possibly go in the equation? They accepted that yeast was present during fermentation. But there was no question of yeast actually taking part in the reaction. At most, they said, dead and dying yeast cells might release some substance that made the reaction possible but did not itself take part in the reaction. Nowadays, we would call such a substance a "catalyst."

Nineteenth-century chemists firmly believed that all biological processes were chemical in nature. To believe otherwise,

Paraskevas Photography

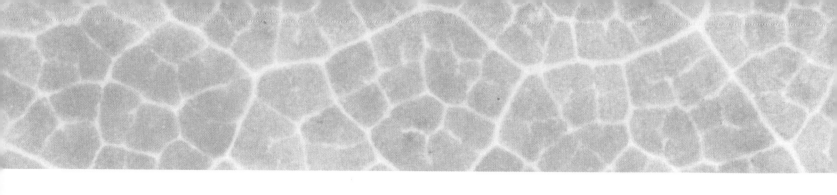

to insist on some mysterious role for living organisms that was not purely chemical in nature, was **vitalism**—the belief that living systems have powers beyond those of nonliving systems.

In a review of all the yeast fermentation papers, the chemist Jons Jacob Berzelius angrily dismissed the work of Schwann and the others. The idea that a simple chemical reaction needed tiny living organisms to occur seemed absurd. In 1839, two of the most prominent early chemists, Friedrich Wöhler and Justus von Liebig, composed and published a lampoon designed to humiliate Schwann and the other yeast researchers. Wöhler and Liebig pretended to give a scientific description of the anatomy and behavior of yeast cells. A yeast cell, they said, looks exactly like a miniature still (the apparatus used to distill whiskey). Yeast cells suck in sugar and excrete ethanol and carbon dioxide, Wöhler and Liebig joked, writing, "The urine bladder in the full condition is shaped like a champagne bottle."

The lampoon must have worked, for Schwann retired from the debate, and Liebig and Wöhler's opinions prevailed unquestioned for 20 years. However, in 1854, young Louis Pasteur, a bright new star in the scientific world, was appointed professor of chemistry at Lille, France.

Lille was a major center for the manufacture of ethanol, through the fermentation of beet juice, and it was not long before one of Pasteur's students appealed to him to help solve problems at the family distillery. Pasteur immediately became intrigued by the process of fermentation and just as quickly realized that fermentation depended entirely on the presence of yeast.

Pasteur was ambitious, and he must have known that proving the eminent scientist Liebig wrong would be a magnificent coup. In September of 1855, Pasteur began a series of studies in fermentation. As his wife, Marie Pasteur, wrote to her father-in-law, "Louis is up to his neck in beet juice. He spends all his days in the distillery. He has probably told you that he teaches only one lecture a week; this leaves him much free time which, I assure you, he uses and abuses."[1]

The scientific community rejected these important discoveries out of hand, thanks to a handful of powerful personalities.

By 1860, Pasteur had succeeded in growing the yeast in an inorganic medium seeded with tiny numbers of yeast cells. The ethanol produced was proportional to the multiplication of the yeast. Normally, vast numbers of yeast cells are present in fermenting solutions. And Liebig had long insisted that some chemical released by the decomposing bodies of yeast cells initiated fermentation. But Pasteur's yeast cells were multiplying, not dying, and the scientific community soon rallied to Pasteur's side. Liebig stubbornly rejected Pasteur's results to the end of his life. Fermentation, he said, was a chemical process, and vitalism was not science.

Ironically, although the scientific evidence was all on Pasteur's side, Liebig was largely right. In fact, fermentation *is* a chemical process—though far more complicated than that envisioned by early chemists. And fermentation can take place outside of living cells—provided the right enzymes are available.

In 1897, Hans and Eduard Büchner decided to make an extract of yeast, called "zymase," to be used as a patent medicine. However, the extract kept rotting, so the brothers added sugar, for the same reason that fruit canners add sugar to canned peaches and berries. In high concentration, sugar is an excellent preservative. To their amazement, the yeast extract began bubbling furiously, as carbon dioxide gas burst from the surface of the extract. Hans Büchner immediately recognized that the sugar was fermenting in the absence of living yeast cells. The Büchners' zymase contained essential enzymes that catalyzed the alcoholic fermentation reaction—just as Liebig had predicted.

Nonetheless, Pasteur was also right. Fermentation normally does not occur outside of living organisms. It is a highly specialized process by which microorganisms extract energy from sugar. And while Liebig's idea that life would ultimately be understood in terms of chemistry is once more accepted, organisms are more than mere bags of enzymes. Their organization and complex structures guide and control the chemical reactions that contribute to life. In this chapter, we will see how cells of all kinds extract energy from biological molecules.

[1]From Rene Dubos, *Louis Pasteur, Free Lance of Science,* Little, Brown, 1950.

KEY CONCEPTS

1. Cellular respiration converts the chemical energy of food molecules into the chemical energy of ATP.

2. Oxidative phosphorylation couples electron transport to the production of high-energy phosphate bonds in ATP in a process that depends on the intact organization of mitochondria.

3. Glycolysis derives energy by rearranging the molecules of glucose to form two molecules of a three-carbon compound called pyruvate.

4. Respiring cells convert two of the three carbon atoms of pyruvate into an activated form of acetic acid, called acetyl-CoA.

5. The citric acid cycle removes eight electrons from the two carbon atoms of acetyl-CoA. These electrons reduce two types of electron carriers—NAD^+ and FAD—in eight enzymatic steps.

6. Heterotrophs digest the macromolecules of food into smaller molecules that can enter cells and participate in energy metabolism.

HOW DO ORGANISMS SUPPLY THEMSELVES WITH ENERGY?

All organisms need energy, and the ultimate source of energy for most organisms is sunlight. Plants convert light energy from the sun into chemical energy through photosynthesis. Plants and all other organisms that obtain energy and synthesize organic molecules from inorganic material are **autotrophs** [Greek, *auto* = self, same + *trophe* = to nourish]. The vast majority of autotrophs, including most plants and many algae and bacteria, photosynthesize. Nonetheless, not all autotrophs photosynthesize. Autotrophs include a few bacteria that obtain energy by oxidizing inorganic substances such as sulfur and ammonia.

Animals, fungi, and other organisms that obtain chemical energy from other organisms are **heterotrophs** [Greek, *hetero* = other + trophe = to nourish]. Both heterotrophs and autotrophs obtain energy by breaking down organic molecules in the process called **cellular respiration**. The difference between heterotrophs and autotrophs is that autotrophs make these organic molecules themselves, while heterotrophs take them from others—either from autotrophs or other heterotrophs. Ultimately, nearly all the energy that powers living organisms comes from sunlight.

What Is the Common Currency of Energy for Organisms?

In all cells, the common currency of energy is ATP (adenosine triphosphate). Plants produce ATP during photosynthesis, as we will see in Chapter 7. But both plants and animals can produce ATP in another way, by capturing energy from the breakdown of carbohydrates, lipids, and proteins.

The main way that organisms break down organic molecules is **cellular respiration,** the oxygen-dependent process by which cells extract energy from food molecules. Cellular respiration perfectly illustrates how chemical pathways occur step by step, how some pathways occur in cycles, and how eukaryotes all share the same metabolic pathways. Animals and other heterotrophs obtain almost all of their energy through cellular respiration. Plants also depend on cellular respiration for ATP at night or at other times when they cannot photosynthesize.

Cellular respiration is distinct from ordinary respiration, or breathing. Breathing—inhaling oxygen and exhaling carbon dioxide—is nonetheless closely related to cellular respiration. Breathing supplies our cells with oxygen for cellular respiration and disposes of the carbon dioxide that is a waste product of cellular respiration.

Most eukaryotic cells (and many prokaryotic cells) are **aerobic,** that is, they depend on oxygen for life. All aerobic organisms, from earthworms to arctic terns, produce the bulk of their ATP by means of cellular respiration (Figure 6-1).

Many prokaryotes and a few eukaryotes can live **anaerobically,** without oxygen. The microbe *Clostridium botulinum,* which grows in canned meats and other foods and causes botulism (food poisoning), grows anaerobically (Figure 6-2A). Even our own cells, which are mainly aerobic, can produce energy anaerobically for short periods by a process similar to alcoholic fermentation (Figure 6-2B). Towards the end of this chapter, we will see how organisms produce ATP anaerobically. Until then, our discussion will center on cellular respiration, which is always aerobic. In particular, we will see how nearly all organisms use the same set of oxygen-dependent chemical reactions to extract energy from the simple sugar glucose.

In all organisms, the common currency of energy is ATP. Whether organisms produce energy through photosynthesis, anaerobically, or by means of cellular respiration, all store and use energy as ATP.

Figure 6-1 When it comes to sustained aerobic activity, the arctic tern, *Sterna paradisaea*, is the champion. This arctic tern nests each summer near the Arctic Circle in North America, then migrates across the Atlantic Ocean to Europe, then south to South Africa, then across the South Atlantic to Antarctica, a distance of 18,000 km (11,000 miles). In the spring, the bird flies all the way back around the world to the Arctic Circle to nest once more. *(Dwight Kuhn)*

How Do Heterotrophs Extract Energy from Macromolecules?

Cellular respiration begins with small molecules such as glucose—a simple six-carbon sugar. Cells cannot extract energy directly from complex carbohydrates, proteins, or fats. Therefore, large molecules must undergo **digestion** into smaller units—proteins to amino acids, polysaccharides to glucose and other simple sugars, and fats to fatty acids and glycerol. Digestion occurs through the process of hydrolysis—breaking each link in a polymer through the addition of a molecule of water.

Digestion takes place outside of the cytosol of the cell. In animals and fungi, most digestion takes place outside of the cell through the action of secreted enzymes, in the intestinal tract of animals, or in the space surrounding a fungus (Figure 6-3). Even in protists and in some animal cells that digest inside the cell, digestion occurs inside lysosomes separated from the cytosol. The lysosomal membrane separates digestive enzymes from the rest of the cell (Figure 4-12).

Digestion does not provide energy, and it does not occur in a fixed sequence. Digestion generates small, energy-rich molecules that move across the cell membrane into the cytosol. One of the most useful of these molecules is glucose. Inside the cell, enzymes transfer the energy from glucose into the chemical bonds of ATP.

Heterotrophs prepare for cellular respiration by capturing macromolecules and by digesting them with enzymes into their component building blocks. Digestion itself does not produce energy.

What Are the Four Stages of Cellular Respiration?

Cells extract energy from glucose by breaking the bonds between each of its six carbon atoms and moving the energy from those bonds to the high-energy bonds of ATP. During cellular respiration, glucose is completely broken down to carbon diox-

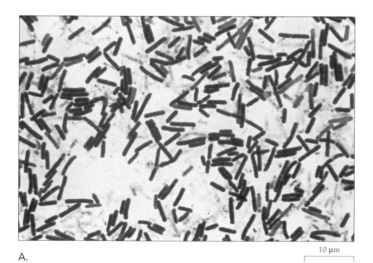

A.   10 μm

B.

Figure 6-2 Anaerobic respiration. A. The obligate anaerobe *Clostridium botulinum*, which causes a serious form of bacterial food poisoning, cannot reproduce in the presence of oxygen. It is known, however, for its ability to multiply inside of sealed canned goods. B. Among vertebrates, the red-eared turtle, *Chrysemys scripta elegans*, is unusual in its ability to live without oxygen. It can stay under water for 2 weeks at a time, relying on glycolysis for energy production. *(A, CNRI/Phototake; B, Zig Leszczynski/Animals Animals)*

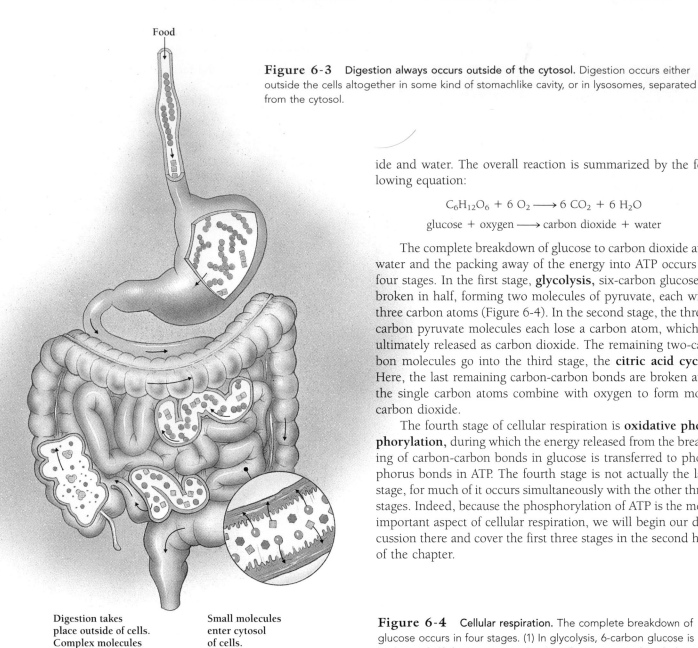

Figure 6-3 **Digestion always occurs outside of the cytosol.** Digestion occurs either outside the cells altogether in some kind of stomachlike cavity, or in lysosomes, separated from the cytosol.

Food

Digestion takes place outside of cells. Complex molecules are broken down.

Small molecules enter cytosol of cells.

Small molecules, such as glucose, enter the cytosol of the cell

ide and water. The overall reaction is summarized by the following equation:

$$C_6H_{12}O_6 + 6\ O_2 \longrightarrow 6\ CO_2 + 6\ H_2O$$

glucose + oxygen $\longrightarrow$ carbon dioxide + water

The complete breakdown of glucose to carbon dioxide and water and the packing away of the energy into ATP occurs in four stages. In the first stage, **glycolysis**, six-carbon glucose is broken in half, forming two molecules of pyruvate, each with three carbon atoms (Figure 6-4). In the second stage, the three-carbon pyruvate molecules each lose a carbon atom, which is ultimately released as carbon dioxide. The remaining two-carbon molecules go into the third stage, the **citric acid cycle.** Here, the last remaining carbon-carbon bonds are broken and the single carbon atoms combine with oxygen to form more carbon dioxide.

The fourth stage of cellular respiration is **oxidative phosphorylation,** during which the energy released from the breaking of carbon-carbon bonds in glucose is transferred to phosphorus bonds in ATP. The fourth stage is not actually the last stage, for much of it occurs simultaneously with the other three stages. Indeed, because the phosphorylation of ATP is the most important aspect of cellular respiration, we will begin our discussion there and cover the first three stages in the second half of the chapter.

Figure 6-4 **Cellular respiration.** The complete breakdown of glucose occurs in four stages. (1) In glycolysis, 6-carbon glucose is broken in half, forming two molecules of pyruvate, each with three carbon atoms. (2) In the second stage, the two pyruvate molecules each lose a carbon atom, ultimately released as carbon dioxide. (3) In the citric acid cycle, the bonds of the remaining 2-carbon molecules are broken and the single carbon atoms combine with oxygen to form more carbon dioxide. (4) During oxidative phosphorylation, which occurs during the other stages, the energy from the broken carbon-carbon bonds in glucose is transferred to ATP.

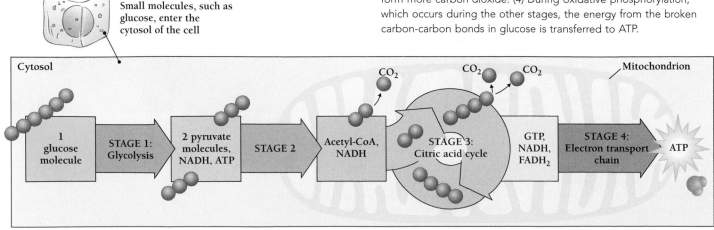

ELECTRON TRANSPORT: HOW DOES THE ENERGY IN GLUCOSE REACH ATP?

If you spend all your time today studying biology, you will use the energy from about 10^{25} molecules of ATP in the next 24 hours—some 90 lbs of ATP. If you were to run a marathon, you would use the same amount in just two hours. Almost every time one of your cells performs an energy-consuming reaction, it derives the energy it needs by splitting ATP, the universal currency of biological energy. Fortunately, however, cells recycle ATP.

A major goal of biologists and biochemists has been to learn how cells convert the energy of sunlight or food into the high-energy bonds of ATP. Until the early 1960s, biochemists believed that they would one day come to understand how organisms make ATP simply by studying the rearrangements of chemical bonds in reactions catalyzed by enzymes in test tubes. Indeed, decades of ingenious research produced detailed maps of such enzyme-catalyzed transformations, many of which made or used ATP. Yet researchers gradually realized that in living cells, most ATP was made in some other way. But how?

The first clue came not from the properties of enzyme reactions, however, but from the properties of membranes and subcellular organelles. In producing ATP, cells rely on their internal structure in a way that eluded most biochemists.

It was biochemists' rejection of vitalism that misled them. Because biochemists are more interested in the chemistry of cells than in the structure of cells, biochemists tend to take a "grind-and-find" approach to cell biology. Ignoring the complex structure of living organisms, they simply grind up cells to find out what chemicals are in them. This approach has worked very well. Decades of research have shown that the majority of cellular processes depend on simple reactions that can be reproduced in a test tube. The grind-and-find approach has been so successful, in fact, that, until the 1970s, biochemists were blind to the importance of intact cellular structures.

As it turns out, however, the steps of cellular respiration that release the most energy occur in functioning intact mitochondria. Cellular respiration so depends on mitochondria that we can pick out the cells that are most active in cellular respiration by counting the number of mitochondria in them.

Biochemists' "grind-and-find" approach to cell biology prevented them from recognizing the importance of intact mitochondria in cellular respiration.

What Is Oxidation?

Most energy-producing processes in cells involve the transfer of electrons from one molecule to another. A molecule that accepts an electron is called an electron acceptor, or **oxidizing agent**. A molecule that donates an electron is called an electron donor, or **reducing agent**. Oxidation and reduction reactions always go hand in hand. The oxidation of one atom, molecule, or ion provides the electrons for the reduction of another atom, molecule, or ion. Such reactions are called **reduction-oxidation reactions** or **redox reactions**.

In cellular respiration, electrons from the chemical bonds of glucose molecules combine with oxygen (and hydrogen ions) to form water. As we have seen, the six carbons in glucose ultimately break down into six molecules of carbon dioxide. Carbon atoms donate electrons, oxygen atoms accept electrons, and hydrogen ions join the reduced oxygen to form water.

How Does the Flow of Electrons from Electron Donors to Oxygen Release Energy to the Phosphate Bonds in ATP?

In Chapter 2 we saw that different atoms have different affinities for electrons. Oxygen is the most electron-hungry atom in the environment. Organisms have evolved mechanisms for using oxygen's enormous electronegativity to obtain energy from reactions in which oxygen is the ultimate electron acceptor. As electrons move from glucose and other energy-rich molecules to oxygen, however, they must pass through many intermediate electron acceptors.

The two most important electron acceptors in energy metabolism are NAD^+ and **FAD**, which take electrons from glucose and transfer them to oxygen in a series of steps. Each NAD^+ can accept two electrons and a hydrogen ion, which converts it to its reduced form, **NADH**. Each FAD can accept two electrons (and two hydrogen ions) to form $FADH_2$. In eukaryotes, the transfer occurs in the mitochondrial membrane. In prokaryotes, these reactions occur in association with the cell membrane.

NADH and $FADH_2$ are examples of **coenzymes**—organic molecules that are necessary participants in certain enzyme reactions, often as electron donors or acceptors. Most of the B vitamins, found in whole grains, for example, are coenzymes. Unlike enzymes, coenzymes can be changed in the course of a reaction. NAD^+ and NADH participate in many enzyme reactions. NADH donates electrons and NAD^+ accepts electrons.

Altogether, the process of cellular respiration transfers 24 electrons from glucose to oxygen. The first electron acceptor for 20 of the 24 electrons is NAD^+. The other four electrons of glucose are transferred to **FAD**. Altogether about 90 percent of the energy stored in the chemical bonds of a glucose molecule is converted to chemical energy in the form of NADH and $FADH_2$.

The two electrons received by each molecule of NADH or $FADH_2$ ultimately go to reduce a single oxygen atom. The reduced atom of oxygen, together with two hydrogen ions from the surrounding water, results in the formation of one molecule of water.

The energy from the oxidation of NADH or $FADH_2$ is used to make high-energy phosphate bonds in ATP by a process that releases the energy in a series of small steps. This is similar to breaking a big waterfall into a gentle cascade.

Electrons flow to oxygen as spontaneously as water flows downhill, and for the same reason. The flow of electrons to oxygen is a "downhill," energy-releasing, process. However, the making of a high-energy phosphate bond to form ATP from ADP, called **oxidative phosphorylation**, is definitely an uphill, energy-consuming, process. One important question, then, is how cells couple the downhill flow of electrons to the uphill production of high-energy bonds in ATP. The answer is that the coupling occurs in stage four of cellular respiration—electron transport phosphorylation.

The electrons from NADH and FADH$_2$ make their way to oxygen by way of a series of other molecules, called **electron carriers**. The pathway of the electrons (from one carrier to another) is called the **electron transport chain**. Biochemists like to compare the electron transport chain that carries electrons to oxygen to the "bucket brigade" that old-time firefighters used to carry water to a fire (Figure 6-5D). Each electron carrier passes its electrons to the next carrier in the line. A reduced carrier becomes oxidized as it gives up its electrons, and the next carrier becomes reduced as it receives electrons.

Cellular respiration transfers electrons from glucose to oxygen by way of a series of intermediaries. The first electron acceptor is either NAD$^+$ or FAD. NADH or FADH$_2$ then release their electrons to the electron transport chain, which transfers electrons to oxygen through a series of small steps. Each step releases a little bit of energy.

Which Molecules Serve as Electron Carriers?

In 1925, the British biochemist David Keilin discovered that cellular respiration depends completely on a set of specific proteins that change color as they accept or donate electrons. These proteins, called **cytochromes** [Greek, *kytos* = hollow vessel + *chroma* = color], have a reddish color that derives from **heme**, the same iron-containing ring structure that gives hemoglobin its color. Heme, like NAD$^+$ and FAD, may both donate and accept electrons. Keilin and others showed that cytochromes could accept electrons from NADH and also from each other. By comparing the flow of electrons both in isolated cytochromes and in whole cells, biochemists worked out the path of electrons from NADH to oxygen in the electron transport chain.

Iron-containing cytochromes contribute to the transfer of electrons through the electron transport chain.

How Do Cells Harvest the Energy of Electron Transport?

After the discovery of the electron transport chain, the big question still remained: How do cells harvest the energy released by the flow of electrons? The experts were convinced that intermediate molecules that would couple electron flow to ATP production must exist. But no one was able to find the intermediates. Biochemists concluded that the intermediate molecules were so unstable that they fell apart before they could be isolated in an experiment.

The situation was this: an electron transport chain in a test tube could pass electrons, but it would generate no ATP. Isolated mitochondria could perform both electron transport and ATP synthesis, but as soon as the mitochondria were broken open, ATP synthesis stopped (even though electron transport continued). These experiments clearly showed the importance of intact mitochondrial organization for the production of ATP. But what did the mitochondria do?

The puzzle was solved in the early 1960s by Peter Mitchell, an English biochemist. Mitchell had been studying the way bacteria pump hydrogen ions across their membranes. The inside of a bacterium has a higher pH (a lower hydrogen ion concentration) than its environment. The difference in hydrogen ion concentration across a membrane is called a **proton gradient.**

Mitchell showed that bacteria could use the energy in this proton gradient to transport other substances into the cell. In 1961, he suggested that mitochondria, like bacteria, might be able to make use of the energy stored in a proton gradient. The idea that a proton gradient could be used to make ATP was, at the time, thoroughly unconventional. Mitchell was unorthodox in other respects, as well. For example, he performed his research not in a university laboratory, but in a manor house in the English countryside where he lived.

Mitchell proposed a two-part hypothesis. He suggested first that the flow of electrons through the electron transport chain establishes a proton gradient across the mitochondrial membrane. Second, he argued, some special molecular machinery in the membrane captures the energy stored in the proton gradient by making ATP. Mitchell's hypothesis showed why biochemists had not found the intermediate molecules that supposedly coupled electron flow to ATP production. They did not exist. Mitchell argued instead that the formation of ATP was the result of **chemiosmosis** [Greek, *osmos,* to push]—the combined electrostatic and osmotic gradients generated by the electron transport chain.

In the 1960s and 1970s, Mitchell, his colleague Jennifer Moyle, and others showed that the coupling of a proton gradient to the synthesis of ATP helps explain not only cellular respiration in mitochondria but also photosynthesis in chloroplasts. As we shall see, Mitchell's radical suggestion, called the **chemiosmotic hypothesis,** has now become the basis for understanding the harvesting of energy both in photosynthesis and cellular respiration.

The English biochemist Peter Mitchell proposed that the energy from the electron transport chain drives ATP production by means of chemiosmosis, not by means of intermediate molecules.

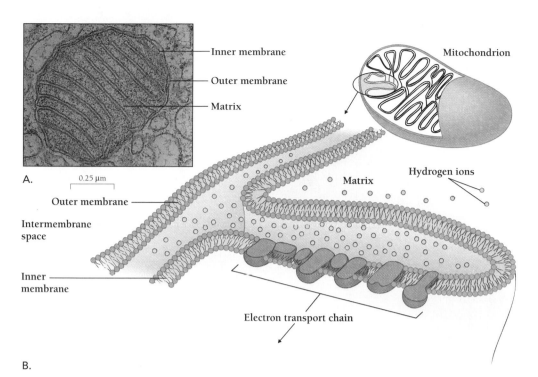

A.   0.25 µm

Outer membrane

Intermembrane space

Inner membrane

Matrix

Hydrogen ions

Mitochondrion

Electron transport chain

B.

Figure 6-5   **The electron transport chain is like a bucket brigade.** A. TEM of a single mitochondrion. B. The electron transport chain is located in the inner membrane. The intermembrane space, between inner and outer membranes, has a high concentration of hydrogen ions. In contrast, the mitochondrial matrix, inside the inner membrane, has a low concentration of hydrogen ions. C. The electron transport chain drives hydrogen ions out of the matrix, creating an electrochemical gradient. D. The electron transport chain consists of a series of electron carriers that pass electrons from one to the next like the members of an old-fashioned bucket brigade trying to put a fire out. Instead of passing buckets of water, however, electron carriers pass electrons, derived from the breakdown of glucose. The electrons ultimately go to oxygen molecules.   *(A, D. Friend, D. Fawcett/Visuals Unlimited)*

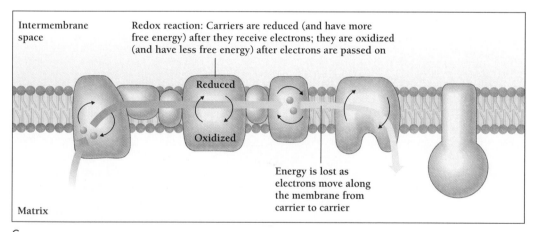

Intermembrane space

Redox reaction: Carriers are reduced (and have more free energy) after they receive electrons; they are oxidized (and have less free energy) after electrons are passed on

Reduced

Oxidized

Energy is lost as electrons move along the membrane from carrier to carrier

Matrix

C.

Glucose

D.

How Do Mitochondria Generate a Proton Gradient?

Recall that a mitochondrion has two membranes. A mitochondrion, therefore, provides four separate locations for the different parts of cellular respiration: the outer membrane; the inner membrane; the space between them, called the intermembrane space; and the space within the inner mitochondrial membrane, called the **mitochondrial matrix** (Figure 6-5 A and B).

Most of a cell's NADH and $FADH_2$ is produced within the mitochondrial matrix, which makes up about two-thirds of the volume of a typical mitochondrion. The electron transport chain lies within the inner membrane. The electron carriers are oriented within the inner membrane so that, as they pass electrons from one to another, hydrogen ions move from the matrix out to the intermembrane space by crossing the inner mitochondrial membrane (Figures 6-5C and 6-6). This movement of hydrogen ions (or protons) out of the matrix establishes a proton gradient across the inner membrane, which lies between the matrix and the intermembrane space. Careful measurements with intact mitochondria confirm that a proton gradient really exists across the inner mitochondrial membrane, with the pH lower outside the membrane than inside. The movement of electrons through the electron transport chain in the inner membrane causes protons to move *out* of the matrix.

Experiments such as that diagrammed in Figure 6-7 have demonstrated that even fragments of inner membranes can pump protons. In these studies, mitochondria were broken open and their membranes allowed to reassemble spontaneously into empty vesicles. As long as electrons flowed in one direction, protons moved in the opposite direction. When the

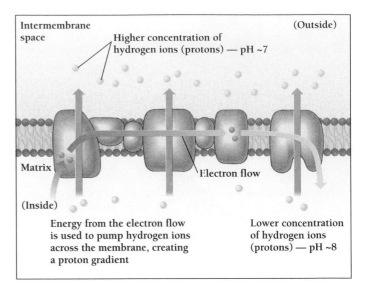

Figure 6-6   More than 40 proteins participate in the electron transport chain. The most important are five cytochromes and two other molecules pictured here.

experimenter stopped electron flow with a poison, however, proton pumping stopped also. These experiments established the ability of the electron transport chain to pump protons.

The electron transport chain within the inner membrane of the mitochondrion generates a proton gradient that pumps protons out of the matrix.

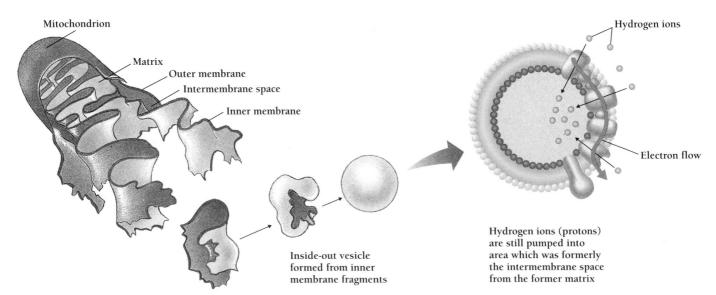

Figure 6-7   Even fragments of inner membranes can pump protons. In one experiment, the inner membranes of mitochondria were broken open and allowed to reassemble into inside-out vesicles. As long as the vesicles were supplied with electrons, the mitochondrial membrane pumped protons into the inside of these vesicles.

What Pumps Protons out of the Mitochondrial Matrix?

By taking apart the inner membrane still further, biochemists found four distinct electron transport complexes in the inner mitochondrial membrane. Three of these complexes use the energy derived from transporting electrons part of the way down the electron transport chain to pump protons across the inner membrane (Figure 6-8). (The fourth complex transfers electrons but does not move protons.)

Each of the three complexes has a fixed orientation in the membrane, so that it pumps protons in a single direction only. For example, one of the respiratory complexes, called cytochrome oxidase, passes electrons from the last carrier in the electron transport chain (cytochrome *c*) to oxygen. Purified cytochrome oxidase complex will transfer electrons from reduced cytochrome *c* to oxygen and simultaneously pump protons, even in a vesicle made from artificial membrane.

The outer mitochondrial membrane allows the easy passage of small molecules and ions, so the intermembrane space has the same neutral pH as the bulk of the cytosol, about pH 7. In contrast, the pH of the matrix increases to about pH 8 as the electron transport chain pumps protons out of the matrix. The inner membrane is not permeable to protons (except through special channels), so the pH difference between the matrix and the intermembrane space is stable as long as electron flow continues (Figure 6-8).

Proton pumping establishes a charge gradient as well as a pH gradient. This is because pumping the positive protons (hydrogen ions) out of the matrix leaves behind an excess of negative ions. The proton pump generates an **electrochemical gradient,** a double gradient composed of a *chemical* gradient (the difference in hydrogen ion concentration, or pH) and an *electrical* gradient (the difference in charge). Both components of the electrochemical gradient favor the return of protons to the matrix.

Three electron transport complexes in the inner membrane of the mitochondrion pump protons out of the matrix, generating an electrochemical gradient that favors the return of protons to the matrix.

How Does the Flow of Protons Back into the Matrix Cause the Synthesis of ATP?

Once hydrogen ions are pumped out of the matrix, there is only one way back in—through special channels in the otherwise impermeable inner membrane. As the protons flow "downhill" back into the matrix—from an area of high concentration to an area of low concentration and from an area of positive charge to one with a negative charge— a protein complex called **ATP synthase** uses the flow of protons to drive the synthesis of ATP (Figure 6-9). **ATP synthase** works like the turbine that converts the kinetic energy of a waterfall to electrical energy.

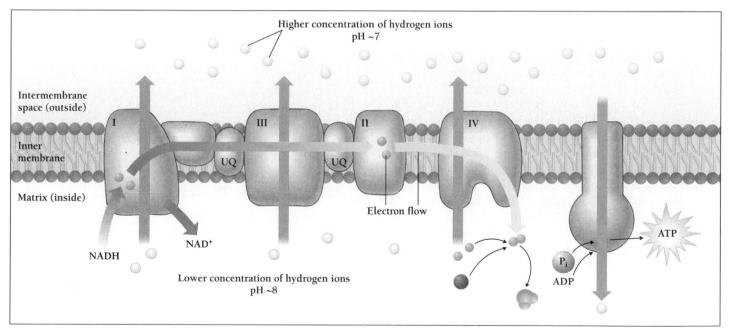

STAGE 4: ELECTRON TRANSPORT CHAIN

Figure 6-8   Each electron carrier is alternately reduced and oxidized as it gains and loses each electron passing down the chain. As the electrons pass, they lose energy to proton pumps that push hydrogen ions, or protons, out of the matrix.

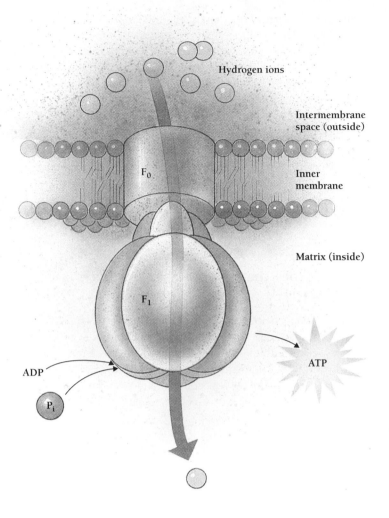

The lollipop-shaped F_0-F_1 complex, or ATP synthase channels the flow of protons back into the mitochondrial matrix and uses the free energy from that flow to synthesize ATP.

We can actually see ATP synthase protein complexes as components of the inner mitochondrial membrane. Each looks like a miniature lollipop protruding from the inner mitochondrial membrane into the mitochondrial matrix (Figure 6-9). Biochemists have found that each contains two major components: the stick, which spans the inner membrane, called F_0, and the top, called F_1.

Under certain conditions, fragments of inner mitochondrial membrane form inside-out vesicles from which F_0-F_1 complexes protrude (Figure 6-10). Gentle shaking removes F_1 from the membrane, leaving F_0 behind. After this treatment, the inside-out vesicles still allow protons to flow down an electrochemical gradient, suggesting that F_0 channels protons back into the matrix. But the removal of F_1 abolishes the ability of the vesicles to make ATP. Replacing the F_1 fragment causes ATP synthesis to resume (Figure 6-10). So F_1 appears to be responsible for making ATP and is called the **coupling factor.** We still do not fully understand, however, exactly how the F_0-F_1 complex harnesses the proton flow to create new high-energy phosphate bonds in ATP. We do know, however, that F_1 spins like a motor as it cranks out ATP.

Are Proton Pumping and ATP Synthesis Really Separate Processes?

Even before the discovery that the F_0-F_1 complex was responsible for the coupling of proton flow to ATP synthesis, biochemists became convinced of the chemiosmotic hypothesis because of experiments with photosynthetic bacteria from the salt flats near San Francisco Bay. These bacteria provided a particularly elegant demonstration of the separateness of proton pumping and ATP synthesis. A purple protein called *bacteriorhodopsin* pumps protons across the bacterial membrane in response to light, as part of the process of photosynthesis. Even when researchers put this protein into artificial membrane vesicles, light stimulates the purple protein to pump protons

F₀-F₁ complexes

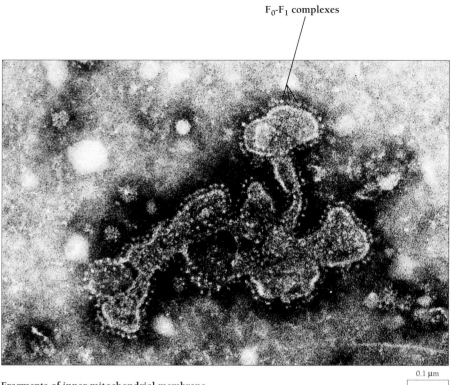

Fragments of inner mitochondrial membrane

0.1 μm

A.

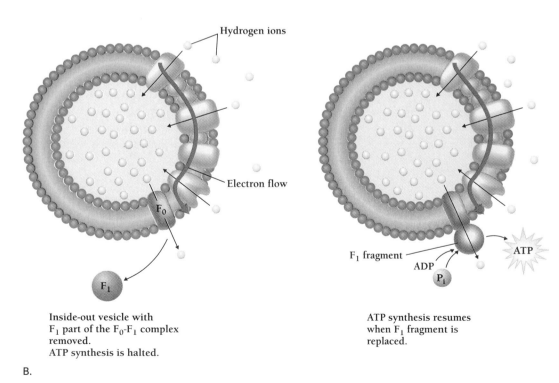

Hydrogen ions

Electron flow

F₀

F₁

Inside-out vesicle with
F₁ part of the F₀-F₁ complex
removed.
ATP synthesis is halted.

F₁ fragment

ADP

Pᵢ

ATP

ATP synthesis resumes
when F₁ fragment is
replaced.

B.

Figure 6-10   The F₀-F₁ complex. **A.** This micrograph shows hundreds of F₀-F₁ complexes dotting the surface of an inside-out inner mitochondrial membrane. **B.** When the F₁ fragment (the coupling factor) is replaced, ATP synthesis resumes.   *(A, R. Bhatnagar/Visuals Unlimited)*

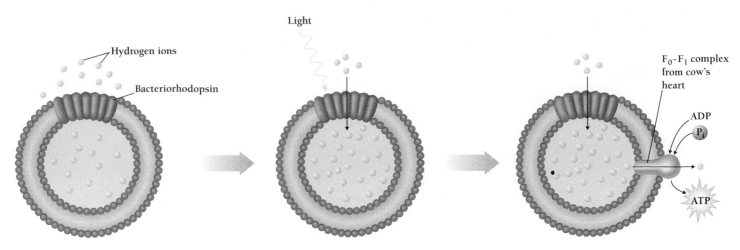

Figure 6-11 Even an isolated F_0-F_1 protein complex can generate ATP, as long as it is embedded in a membrane spanning a proton gradient. In this experiment, researchers took an F_0-F_1 protein complex isolated from the heart muscle of a cow and placed the F_0-F_1 complex into the membrane of a photosynthetic bacterium. A purple protein called *bacteriorhodopsin* pumped protons across the bacterial membrane in response to light. But the F_0-F_1 complex allowed protons to flow back out of the vesicle and generate ATP.

into the vesicles. This proton pumping establishes a pH gradient between the inside and the outside (Figure 6-11).

To these artificial vesicles researchers then added the F_0-F_1 protein complex isolated from the mitochondria of a cow's heart (Figure 6-11). The F_0-F_1 complex allowed protons to flow back out of the vesicles and generate ATP. This experiment demonstrated that the F_0-F_1 complex can harness the energy of proton flow even in an artificial situation, with separate components from two extraordinarily different sources—a rare microscopic bacterium and a familiar large animal.

The chemiosmotic hypothesis explains how cells transfer the energy from the reduced coenzymes NADH and $FADH_2$ to ATP: (1) the electron transport chain creates an electrochemical gradient across the inner mitochondrial membrane, and (2) the F_0-F_1 complex couples the flow of protons down the gradient to the synthesis of ATP.

The main point is that the creation of the proton gradient and the generation of high-energy bonds in ATP are separate processes. The experiments with *bacteriorhodopsin* and cow F_0-F_1 complex support this idea and further suggest that this strategy for ATP synthesis is nearly universal. In Chapter 7, we will see that proton gradients not only produce ATP in mitochondria but also produce ATP in chloroplasts during photosynthesis.

Proton pumping and ATP synthesis are separate processes.

In the rest of this chapter we will see how cells produce NADH and FAD—the starting materials for oxidative phos-

phorylation—from glucose and other food molecules. The reactions fall into three stages (Figure 6-4):

1. The splitting of one 6-carbon glucose into two 3-carbon molecules each.
2. The conversion of each 3-carbon compound into a 2-carbon compound.
3. The conversion of the 2-carbon compounds to 1-carbon carbon dioxide.

Each of these stages reduces (or donates electrons to) the cofactors NAD^+ and $FADH_2$. The first and third stages also include reactions that directly produce high-energy bonds. (Oxidative phosphorylation, which we have just discussed, is the fourth stage.)

HOW DO CELLS EXTRACT ENERGY FROM GLUCOSE?

The first of the four stages in the metabolism of glucose is **glycolysis** [Greek, *glykys* = sweet (referring to sugar) + *lyein* = to loosen], a set of ten chemical reactions (or steps) that convert glucose, a sugar with six carbon atoms, into two identical smaller molecules, called **pyruvate**, with three carbon atoms each (Figure 6-12). The ten reactions also convert some of the energy of glucose into two high-energy ATP bonds, and some into two high-energy molecules of NADH. All ten reactions take place in the cytosol, outside the mitochondria.

The most remarkable fact about glycolysis is that **all** organisms accomplish it in exactly the same way. This universality suggests that the common ancestors of all present-day

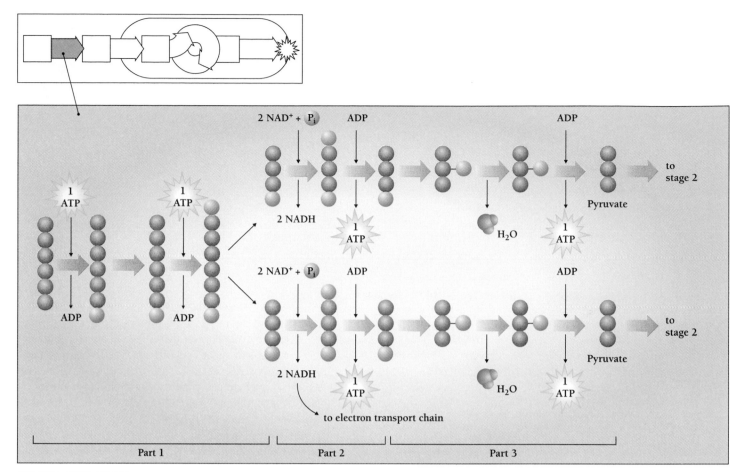

STAGE 1: GLYCOLYSIS

Figure 6-12 Stage 1: The first stage of metabolism is glycolysis. Glycolysis is a set of ten chemical reactions that break each 6-carbon glucose into two molecules of 3-carbon pyruvate. The ten reactions—only nine of which are shown here—also convert some of the energy of glucose into high-energy ATP and NADH. All of the reactions take place in the cytosol, outside the mitochondria.

organisms performed glycolysis in the same way that we do today. In that case, the enzymes that catalyze the ten reactions must have evolved at least 3.8 billion years ago, at the very dawn of life on Earth.

Glycolysis: How Do Cells Capture Energy in ATP and NADH?

As glycolysis breaks glucose into two molecules of pyruvate, it also converts two molecules of NAD^+ to NADH and two molecules of ADP to ATP. The major question about glycolysis is how cells capture energy in ATP and NADH. Like other biochemical pathways, glycolysis proceeds in small steps. We can divide the ten chemical reactions of glycolysis into just three parts:

Part 1. Splitting of the 6-carbon glucose into two 3-carbon molecules.

Part 2. Oxidation of the 3-carbon molecules and reduction of NAD^+.

Part 3. Further oxidation of the new 3-carbon molecules to pyruvate.

Part 1, the conversion of the 6-carbon glucose molecule into two 3-carbon molecules, requires five separate reactions (or steps), each catalyzed by a different enzyme. This process transfers two phosphate groups from two ATP molecules to glucose and splits glucose in half. The energy realized from the splitting of the high-energy phosphates pushes the reactions in the direction of the 3-carbon product, glyceraldehyde-3-phosphate (PGAL). At the end of part 1, glycolysis has consumed rather than harvested energy.

In part 2 of glycolysis, NAD^+ oxidizes the two molecules of PGAL to phosphoglycerate, each molecule of PGAL generating one ATP and one NADH. In addition, a phosphate ion from the cytosol is attached first to PGAL, then to ADP to form a new molecule of ATP. For each starting glucose molecule, then, part 2 yields two ATPs and two NADHs, thus recovering the ATP investment of part 1.

The reaction steps of part 2 require both phosphate ions and NAD^+. If either is missing, the reaction simply cannot

145

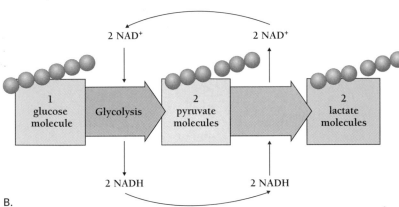

A.

B.

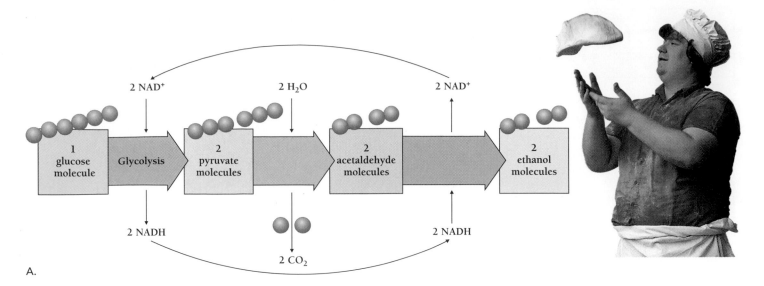

Figure 6-13 **Two kinds of fermentation.** In each case, cells regenerate NAD^+ for further glycolysis. A. In yeast cells, pyruvate forms acetaldehyde and ethanol. B. In muscle cells, pyruvate forms lactic acid (lactate). *(pizza maker, Jim Erickson/The Stock Market; sprinters, David Madison)*

occur. Because these two ions are so essential and cannot be in short supply for the cell to survive, cells recycle both of these. Phosphate is usually not a problem, since it is present at fairly high concentrations in most cells and is continuously released as ATP is split. In aerobic organisms, the regeneration of NAD^+ generally requires transfer of NADH's electrons to the electron transport chain and therefore usually depends on the availability of oxygen. In the absence of oxygen (and in anaerobes), however, the cell must regenerate NAD^+ by other means, which we will discuss later.

Part 3 converts phosphoglycerate to pyruvate and transfers its phosphate to ADP, forming one more ATP molecule for each phosphoglycerate (that is, two ATP molecules for each glucose). Since the reactions in part 2 had already recovered the two ATP molecules invested in part 1, the net gain of ATP during glycolysis is two ATPs per molecule of glucose, as well as two molecules of NADH. But this is only a small bite out of a very large apple. The pyruvate molecules still contain lots of energy. This energy is extracted only in the second stages of glucose metabolism, which we will discuss later in this chapter. First, however, we will see how cells can use glycolysis to obtain energy even in the absence of oxygen.

Glycolysis, the first stage in the extraction of energy from glucose, breaks glucose into two molecules of pyruvate, and produces two molecules each of ATP and NADH.

How Can Glycolysis Continue Without Oxygen?

To sustain glycolysis, a cell must regenerate more NAD^+ from NADH. Just how it does this depends on the species of organism and on whether oxygen is present.

In the presence of oxygen, most cells can oxidize NADH to NAD^+ by the reactions of oxidative phosphorylation. In the absence of oxygen, however, cells must regenerate NAD^+ by using NADH to reduce pyruvate. Their manner of doing so depends on what enzymes they have.

Many microorganisms live by **fermentation** [Latin, *fervere* = to boil]—the anaerobic extraction of energy from organic compounds (Figure 6-13). Yeast cells, for example, ferment the sugars in beer, wine, and bread, converting glucose into carbon dioxide and ethanol. The carbon dioxide makes beer bubble and bread rise. The ethanol gives beer its punch and freshly baked bread its sweet aroma. To perform this feat, yeasts employ all the reactions of glycolysis that we have discussed. In addition, they use the accumulated NADH to reduce pyruvate to form ethanol. In the process, NADH is oxidized back to NAD^+. Yeast is a *facultative anaerobe*—an organism that can live either anaerobically, by fermentation, or aerobically, by oxidative phosphorylation. Other microorganisms, such as *Clostridium botulinum,* are *obligate anaerobes*, meaning that they can grow only in the absence of oxygen (Figure 6-2A).

Animals, including humans, often suffer a temporary lack of oxygen in some of their cells. For example, during sprinting or other vigorous exercise, the muscle cells need more oxygen (to oxidize NADH) than the blood and lungs can deliver. In this case, and for short periods only, the cells can regenerate NAD^+ in the absence of oxygen. Instead of producing carbon dioxide and ethanol (as a yeast cell does), however, muscle cells make lactate, a compound with three carbon atoms (Figure 6-13B). During a vigorous sprint, lactate accumulates in the muscles, sometimes causing soreness. Once oxygen becomes available again, however, much of the lactic acid reverts to pyruvic acid. Heart muscle is especially efficient at extracting energy from lactic acid by first converting it to pyruvic acid.

Glycolysis can continue without oxygen if cells use pyruvate to oxidize NADH to NAD^+.

What Regulates the Overall Rate of Glycolysis?

In the cell, the overall process of glycolysis is highly exergonic. This means that glycolysis proceeds spontaneously and irreversibly. Since all the needed enzymes are present in cells, the reactions should also proceed quickly. How, then, does the cell keep all its glucose from being immediately converted to pyruvate?

We can see that glycolysis must be regulated. The question is where and how? In general, biochemical pathways are regulated at the first step that is unique to the pathway. This strategy ensures that molecules are not processed needlessly when they could be used for alternate purposes. In glycolysis, regulation takes place in reaction 3, by means of the enzyme phosphofructokinase (Figure 6-14).

We would expect glycolysis to proceed rapidly when the cell needs ATP and to proceed slowly or not at all when the cell already has lots of ATP. This is indeed the case. ATP inhibits the action of phosphofructokinase, so when ATP is present, it slows glycolysis at reaction 3. As a result, glycolysis stops producing ATP when it is not needed.

Because ATP inhibits reaction 3 of glycolysis, cells with plenty of ATP stop extracting energy from glucose. Cells with little ATP can proceed with glycolysis.

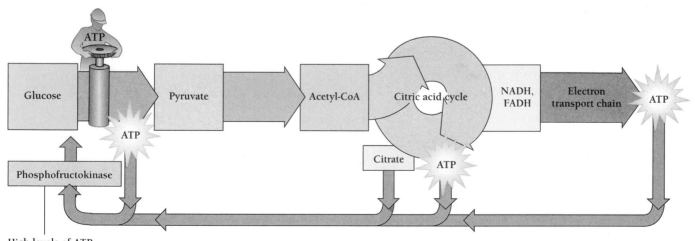

High levels of ATP
and citrate bind to
the allosteric site
of this enzyme. The
enzyme becomes inactive
and glycolysis stops.

Figure 6-14 **Negative feedback regulates glycolysis.** When ATP levels rise, the ATP and citrate produced by glycolysis bind to phosphofructokinase, inactivating it and slowing glycolysis.

BOX 6-1

Sprints, dives, and marathons

Respiration yields at least 15 times more ATP than glycolysis. It's no wonder that animals—especially warm-blooded ones—depend on respiration for most of their energy. However, because oxygen is the ultimate electron acceptor in respiration, available energy depends on a steady supply of oxygen. For vertebrates, continuous energy production requires efficient lungs (to take in oxygen-rich air) and hearts (to pump oxygen-rich blood from the lungs to the muscles).

Training can improve the performance of the systems that deliver oxygen, but humans and other animals often need to spend energy more quickly than our oxygen supply will allow. We all need to sprint occasionally, whether we are a pedestrian dashing to avoid an inattentive driver, an outfielder swooping down on a wayward baseball, a cheetah chasing a gazelle, or a gazelle eluding a cheetah. Several energy reservoirs allow such bursts of energy: (1) muscles have stores of ATP and other compounds with similar high-energy phosphate bonds, enough, in any case, to allow several minutes of work without any new ATP production; (2) muscles can use blood glucose and stored glycogen to produce ATP by anaerobic glycolysis, producing lactate as the end product.

The length of time that an organism can depend on anaerobic glycolysis for ATP production is limited both by training and by genetic capacity. Humans can manage only a few minutes. But some animals, such as whales, seals, and other diving animals, can function by means of glycolysis alone for much longer periods. The freshwater red-eared turtle shown in Figure 6-2B, for example, can stay underwater for as long as two weeks.

Anaerobic glycolysis provides animals with a way of borrowing energy. But the loan is only short term. After anaerobic exertion, all animals must get rid of the accumulated lactate and restore their supplies of glucose and glycogen. These processes require oxygen. In most animals, the "oxygen debt" must be repaid almost immediately—through heavy breathing and increased circulation. Animals extract energy from the accumulated lactate by converting it to pyruvate, which can be used in aerobic respiration. In some vertebrates, some of the lactate travels via the blood to the liver, where it can be recycled into glucose. The liver then resupplies the glucose reservoir of the blood.

Anaerobic glycolysis allows untrained school children to sprint up to 17 miles per hour in a 100-meter dash. But the necessity for oxygen prevents even champion athletes from averaging more than about 11.5 miles per hour in a 26-mile marathon (Figure A).

Nonmammalian vertebrates (fish, amphibians, and reptiles), which live their lives at a slower metabolic pace than we do, may function without air for long periods. Instead of using oxygen to recycle the accumulated lactate, fish, amphibians, and reptiles may excrete the lactate in their urine. This represents a huge waste of energy, but it allows these animals to feed and function in environments in which they could not otherwise live.

Figure A   1995 Worlds women's marathon. The speed that a marathoner can attain is limited by the need for oxygen.   *(David Madison)*

The Formation of Acetyl-CoA Is the Second Stage in the Extraction of Energy from Glucose

In eukaryotic cells, the steps of energy extraction after glycolysis take place in the mitochondria. The pyruvate produced by glycolysis must enter the mitochondria for cellular respiration to proceed. After pyruvate arrives in the mitochondria, it is oxidized through a series of steps to a compound called acetyl-CoA whose acetyl group contains two carbons (Figure 6-15).

Acetyl-CoA forms from a molecule of acetate (the same as the acetic acid that makes vinegar) linked by a high-energy bond to a molecule called **coenzyme A (CoA).** During the formation of acetyl-CoA, the reaction removes from pyruvate one molecule of carbon dioxide, as well as two electrons and two protons. NAD^+ serves as the electron acceptor.

Acetyl-CoA has only two principal fates: (1) it can enter the third stage of energy metabolism and generate more ATP, as discussed in the following section, or (2) it can be used to synthesize new lipids, either fatty acids or cholesterol. When a cell's ATP levels are high, the cell redirects acetyl-CoA into a set of reactions that make fat. This redirection of acetyl-CoA allows the body to store energy for later use. It is the main reason that we accumulate fat when we consume too many molecules of any substance, whether the carbohydrates in an apple or the stearic acid in hamburger fat.

The pyruvate produced by glycolysis enters a cell's mitochondria, where it is converted to acetyl-CoA. Acetyl-CoA can generate more ATP or more fat.

How Does a Cell Generate ATP from Acetyl-CoA?

In nearly all organisms, acetyl-CoA enters directly into a set of reactions that converts the carbon atoms of acetyl-CoA into carbon dioxide. These reactions are called the **citric acid cycle** or the **Krebs cycle,** after Hans Krebs, who began working out the details of the cycle in the 1930s.

In eukaryotes, the citric acid cycle takes place within the mitochondrial matrix. In its barest outlines, the citric acid cycle takes molecules of the 2-carbon compound acetyl-CoA and processes it through the following cycle. The 2-carbon molecule is added to a 4-carbon molecule, called **oxaloacetate,** to form a 6-carbon molecule (citric acid), from which is removed first one carbon, then another, to form another 4-carbon molecule, which reenters the cycle once more.

The citric acid cycle removes eight electrons from the two carbons of acetyl-CoA. Three NAD^+ molecules accept six of these electrons, and one molecule of FAD accepts the other two. The citric acid cycle also forms one high-energy phosphate bond, but, unlike the other reactions that we have discussed, the phosphate acceptor is guanosine diphosphate (GDP), rather than ADP. The resulting GTP (guanosine triphosphate) carries the same amount of energy as ATP.

Oxygen is the ultimate acceptor of electrons and is therefore absolutely essential for the citric acid cycle to proceed. Only when oxygen is available to accept electrons from NADH and $FADH_2$ can the cell regenerate more NAD^+ and FAD and allow the cycle to continue.

The citric acid cycle not only provides energy, but it also serves as the major source of building blocks for many amino

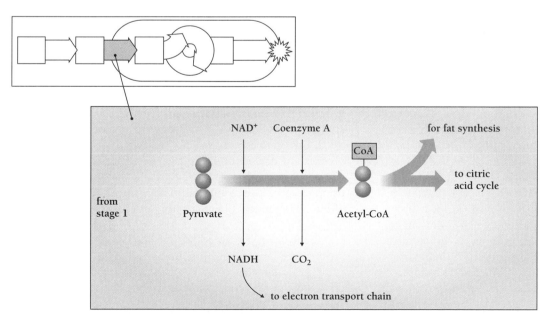

STAGE 2: FORMATION OF ACETYL-CoA

Figure 6-15 Stage 2: After glycolysis, pyruvate enters the mitochondria for the second stage of energy extraction. In the mitochondria, pyruvate is oxidized to form acetyl-CoA. Acetyl-CoA can either enter the third stage of energy metabolism and generate more ATP or it can be used to synthesize body fat.

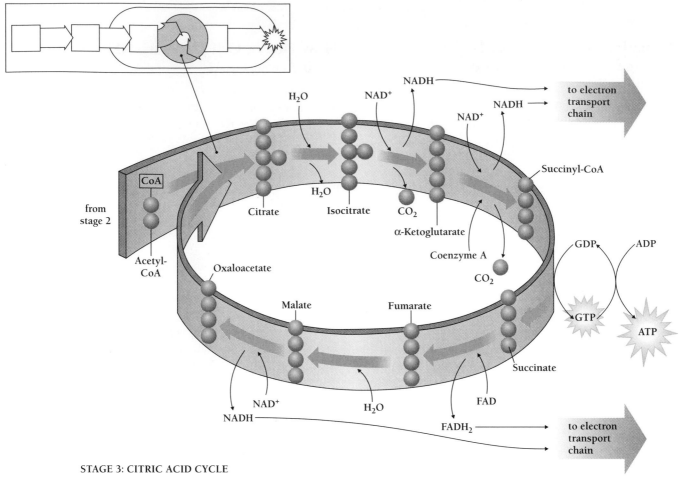

STAGE 3: CITRIC ACID CYCLE

Figure 6-16   Stage 3: During the citric acid cycle, a single molecule of acetyl-CoA generates 3 NADH, 1 $FADH_2$, 1 GTP, and 2 CO_2.

acids and other small molecules. The flow of molecules through the cycle then responds not only to the energy needs of a cell, but also to the availability of amino acids.

Figure 6-16 summarizes the steps of the citric acid cycle. The central events of the citric acid cycle are its four oxidation-reduction reactions, which employ two kinds of electron acceptors—NAD^+ and FAD. The reactions of the cycle fall into three parts: the first steps form 6-carbon citrate (citric acid) through the addition of 2-carbon acetyl-CoA to 4-carbon oxaloacetate, and then rearrange the molecule citrate into isocitrate; the next steps convert isocitrate into a 4-carbon compound called succinate, which is linked to CoA; and the last steps produce a 4-carbon compound, called oxaloacetate, which can combine with acetyl-CoA to regenerate citrate.

To review, each turn of the citric acid cycle begins with the condensation of oxaloacetate and acetyl-CoA to form citrate. With each turn, citrate loses two carbon atoms as carbon dioxide and regenerates the 4-carbon oxaloacetate. Notice, however, that the carbon atoms of the carbon dioxide from each turn are not those that entered the cycle as acetyl-CoA. Nor is the oxaloacetate regenerated by the cycle the same as the one that started the cycle. In each turn of the cycle, citrate loses a total of eight electrons.

The citric acid cycle converts acetyl-CoA to carbon dioxide, transferring, in the process, eight electrons from acetyl-CoA to NADH and $FADH_2$.

How Was the Citric Acid Cycle Discovered?

Biochemists tried for years to understand how cells use oxygen to obtain energy from organic molecules. But it was the Hungarian biochemist Albert Szent-Györgyi, a man fascinated by all oxidative processes, who solved the problem. The first observation that Szent-Györgyi made was that extracts of pigeon breast muscle were strikingly efficient at taking up and using oxygen. This is not surprising, since the flight muscles of birds require lots of energy and are packed with mitochondria.

Szent-Györgyi then added various compounds to his extracts to see how these compounds were oxidized. To his surprise, he found that certain 4-carbon compounds stimulated the uptake of oxygen far more than they should have. He had calculated the amount of oxygen required for the complete oxidation of each added compound, but the extracts were using more oxygen than complete oxidation would require.

The reason for this discrepancy, we now realize, is that the 4-carbon compounds that Szent-Györgyi added were intermediates in the citric acid cycle—they increase the rate at which acetyl-CoA enters the cycle. This, in turn, increases the amount of NADH and $FADH_2$ produced and therefore the amount of oxygen needed to regenerate NAD^+ and FAD (Figure 6-16).

By 1937, biochemists knew that pigeon muscle extracts could convert citric acid to these same 4-carbon compounds. The missing link and the final interpretation were provided in 1938 by Hans Krebs, a refugee from Nazi Germany working in England. Krebs showed that Szent-Györgyi's muscle extracts could make citric acid from the 4-carbon compound oxaloacetate and acetyl-CoA.

Krebs's crucial contribution was the idea that the conversion of acetate to carbon dioxide involved a cycle. Krebs showed how the particular 4-, 5-, and 6-carbon compounds fit together in a single cycle. He also showed that the same chemical transformations occurred both in cells and in test tubes.

The first clue to the existence of the citric acid cycle was Szent-Györgyi's observation that 4-carbon oxaloacetate added to muscle tissue stimulated the uptake of more oxygen than was required to oxidize the oxaloacetate.

The Citric Acid Cycle Intersects Other Biochemical Pathways

The citric acid cycle lies at the center of the complex network of biochemical conversions that are collectively called metabolism. Metabolism is divided into two classes of pathways: **catabolism** [Greek, *cata* = down + *ballein* = to throw]—the breakdown of complex molecules, such as those in food; and **anabolism** [Greek, *ana* = up + *ballein* = to throw]—the synthesis of complex molecules (Figure 6-17). In general, catabolism produces energy and usually involves oxidation—the removal of electrons and the production of NADH. Because anabolism usually consumes energy and involves reduction, it uses NADH and other reducing agents.

Figure 6-18 shows that different molecules enter the cellular respiration pathway at different points. Carbohydrates in such foods as bread are broken down into simple sugars and take the same path as glucose. Fats and oils are broken down into glycerol, which may enter glycolysis, or fatty acids, which enter the pathway with acetyl-CoA. When we eat protein, it is digested into its component amino acids, which are converted to pyruvate, acetyl-CoA, or other intermediates that enter later in the cellular respiration pathway.

Cells regulate the rates at which pathways or sets of pathways operate. High levels of ATP favor anabolism, and low levels favor catabolism. When ATP levels are high but the products of anabolism are abundant, the energy in food is stored for future use—as carbohydrates, proteins, or fats. Despite the importance of glucose and other carbohydrates in energy metabolism, the preferred form of energy storage in animals is fat.

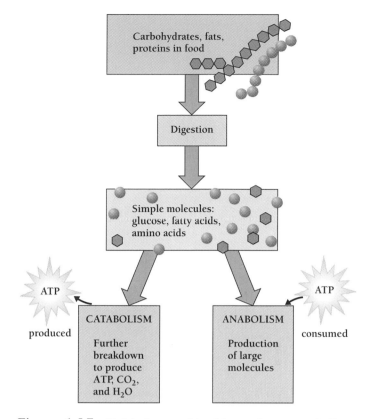

Figure 6-17   Metabolism consists of two pathways: catabolism and anabolism. Catabolism produces energy through the oxidation of molecules. Anabolism usually consumes energy in the process of building large molecules.

The end product of the catabolism of fats is acetyl-CoA. Fats are more reduced compounds than sugars; that is, for each carbon atom, fats provide more electrons for oxidative phosphorylation. Fats, therefore, contain more free energy—about 9.4 kcal per gram of fat, instead of the 4.1 kcal per gram of carbohydrate and 5.6 kcal per gram of protein. In addition, fats are less polar and less hydrophilic than carbohydrates, which may bind as much as twice their weight in water. As a result of these factors, each gram of stored fat contains about six times as much energy as each gram of stored glycogen. A 50 kg (110 lb) woman, for example, has fuel reserves of about 70,000 kcal in fats, about 18,000 kcal in proteins (mostly in muscle), about 400 kcal in glycogen, and about 30 kcal in glucose. Her fat stores, however, weigh only about 8 kg (18 lb). To store the same amount of energy as carbohydrates (with all their bound water), she would need to weigh at least 90 kg (200 lb)!

The citric acid cycle lies at the intersection of the catabolic and anabolic pathways. High levels of ATP favor anabolism, and low levels favor catabolism. When ATP levels are high, the most efficient way to store energy is in the form of fat.

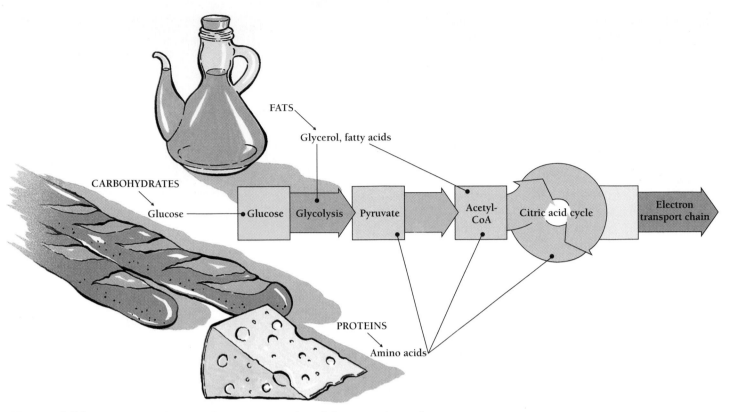

Figure 6-18   Fat, carbohydrates, and proteins enter the cellular respiration pathway at different points. The busiest crossroads is at acetyl-CoA, which then enters the citric acid cycle. All fatty acids and many amino acids are converted into acetyl-CoA.

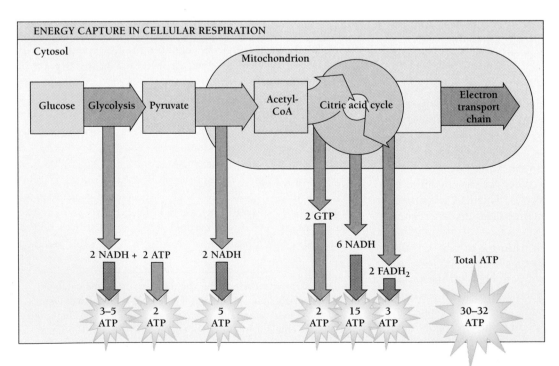

Figure 6-19   Energy capture in cellular respiration. For each glucose molecule, glycolysis produces 2 ATP and 2 NADH, and the production of acetyl-CoA yields another 2 NADH. The citric acid cycle then yields 2 GTP, 6 NADH, and 2 FADH$_2$. The electron transport chain captures the energy of NADH and FADH$_2$ in 26–28 ATP.

BOX 6-2

The metabolism of alcohol

For thousands of years and in many societies, ethanol has provided a popular means of changing mood and behavior. In heavy drinkers, however, ethanol can cause permanent damage to cells in the liver, nerves, brain, and heart. The metabolic pathways described in this chapter help us to understand how this happens.

In humans, the only cells in the body that can metabolize ethanol are cells in the liver. In these cells, NAD^+ oxidizes ethanol, first to acetaldehyde and then to acetic acid, generating abundant amounts of NADH. The NADH then donates its electrons to the electron transport chain. Oxidative phosphorylation then occurs, but without the participation of the citric acid cycle. Sustained use of large amounts of alcohol leads to strange-looking mitochondria, the result of active oxidative phosphorylation and the shut-off of the citric cycle. In addition, carbohydrates that would ordinarily enter the citric acid cycle are instead converted to fat. The fat is secreted by the liver into the blood in little vesicles (produced by the endoplasmic reticulum).

Alcohol accumulates in the blood because it moves from the intestines into the blood more rapidly than the body can eliminate it. The blood carries the alcohol to cells throughout the body. Nerve and brain cells are particularly vulnerable to damage by alcohol. Heavy alcohol use irreparably damages heart muscle after about ten years in men, sooner in women.

However, the most common organic disease among heavy drinkers is liver failure. After a few years, the liver cells begin to fill with fat and to cease functioning. Eventually, the liver becomes heavily scarred—a condition called **cirrhosis** (Figure A). Once the damaged liver loses its ability to metabolize alcohol and other toxins, damage to the body accelerates. Cirrhosis of the liver is the seventh leading cause of death in the United States.

Our ability to metabolize alcohol demonstrates how adaptable animals are. Alcohol is just one of hundreds of chemicals from which we can obtain energy. No matter what we eat, biochemical homeostasis keeps the cells supplied with ATP and reducing power. Sources of energy are rarely wasted; they are used or stored. But animals also have evolved within a relatively narrow set of conditions. We cannot tolerate too large a variation of diet for an extended period without damage to the cells of our body.

Figure A　**Cirrhosis of the liver.** In the numerous dark regions, the small blood vessels of the liver have become blocked. *(Martin Rotker/Photo Researchers)*

How Much Usable Energy Can a Cell Harvest from a Molecule of Glucose?

Actual measurements in isolated mitochondria show that the oxidation of each molecule of NADH leads to the movement of 10 hydrogen ions across the inner membrane and to the synthesis of 2.5 molecules of ATP. Similar experiments show that each $FADH_2$ can yield 1.5 molecules of ATP. For each molecule of glucose, then, the NADH produced in the mitochondria yield 20 molecules of ATP and the $FADH_2$ yields 3 molecules of ATP (Figure 6-19).

Glycolysis produces two more molecules of NADH. These are made in the cytoplasm and do not readily cross the inner mitochondrial membrane. The two molecules of NADH made during glycolysis produce either three or five molecules of ATP, depending on the tissue.

Finally, two molecules of ATP are produced in glycolysis and two molecules of GTP in the mitochondria. Since the high-energy phosphates of the GTPs readily produce ATP, we may say that we have an additional four molecules of ATP per molecule of glucose.

Cellular respiration thus produces 30 or 32 molecules of ATP for each molecule of glucose. (Until recently, biochemists estimated that each molecule of glucose yielded 36 to 38 molecules of ATP, because it was believed that each NADH yielded three ATP, rather than 2.5, and each $FADH_2$ yielded two ATP, instead of 1.5.) All but four of the ATP molecules produced from a glucose molecule are produced by oxidative phosphorylation. Yet, in the absence of oxygen, glycolysis alone can produce only two molecules of ATP.

The ATP molecules produced in the mitochondria travel across the mitochondrial membranes to the cytoplasm, where they can be used. Once their high-energy phosphate bond is broken, ADP molecules return to the mitochondria to be oxidized once more. Within the mitochondria, the rates of the citric acid cycle and oxidative phosphorylation depend on the relative amounts of ATP and ADP. Because of such regulation, the body catabolizes sugars and fats five to ten times more rapidly during exercise than when at rest.

The process by which energy is extracted from glucose under cellular conditions is highly efficient. A gasoline engine has an efficiency of only 10 to 20 percent. That is, only 10 to 20 percent of the energy in the gasoline is converted to forward motion. In contrast, cells extract energy from glucose with a phenomenal efficiency of 50 percent. It is hard to think of another process with such high efficiency.

Cellular respiration is a highly efficient process for extracting energy—in the form of 30 or 32 molecules of ATP—from glucose.

STUDY OUTLINE WITH KEY TERMS

Modern biochemists reject **vitalism,** the belief that living systems have powers beyond those of nonliving systems. But the production of ATP by the mitochondria presents an excellent example of a living system with powers that biochemists so far cannot reproduce artificially.

Plants and other photosynthesizing organisms that produce their own food from sunlight are called **autotrophs.** Organisms that get their energy from other organisms are **heterotrophs.** Heterotrophs prepare for cellular respiration by capturing macromolecules and **digesting** them into their component building blocks. Digestion itself does not produce energy.

Almost all eukaryotic cells (and many prokaryotic cells as well) are **aerobic,** producing most of their ATP by **cellular respiration.** Most energy-producing reactions in cells are **reduction-oxidation (redox) reactions,** which transfer electrons from the **reducing agent** to the **oxidizing agent.**

Cellular respiration transfers 24 electrons from glucose to oxygen. For 20 of these electrons, the first acceptor is NAD^+, which can accept two electrons and a hydrogen atom to give its reduced form **NADH.** The other four electrons of glucose are transferred to **FAD,** whose reduced form is **FADH$_2$.** These reduced **coenzymes**—the major energy storage molecules in energy metabolism—ultimately transfer their electrons to oxygen in a series of steps. In eukaryotes, the steps occur within the inner mitochondrial membrane. In prokaryotes, these reactions occur in association with the plasma membrane.

Oxidative phosphorylation couples the oxidation of NADH and FADH$_2$ to the production of high-energy phosphate bonds in ATP. The electrons from NADH and FADH$_2$ reach oxygen, their ultimate acceptor, through the **electron transport chain** (or respiratory chain), which consists of a series of **electron carriers.** Five of the electron carriers are **cytochromes** whose reddish color derives from **heme.**

Oxidative phosphorylation operates by a mechanism (called **chemiosmosis**) summarized in the **chemiosmotic hypothesis:** (1) the flow of electrons through the electron transport chain pumps protons out of the **mitochondrial matrix,** establishing an **electrochemical gradient** across the mitochondrial membrane; and (2) some special molecular machinery in the membrane captures the energy of this gradient in ATP.

The orientation of the electron carrier complexes establishes the direction of the **proton gradient.** As the protons flow down the electrochemical gradient through channels in the inner membrane, their free energy decreases. A protein complex, the **F$_0$-F$_1$ complex,** or **coupling factor,** uses the free energy from the flow of protons to drive the synthesis of ATP. The F$_1$ protein, also called **ATP synthase,** synthesizes ATP.

Glycolysis is the first stage in the metabolism of glucose. Glycolysis is a set of ten chemical reactions that breaks glucose into two molecules of **pyruvate.** These ten reactions also convert some of the energy of glucose into two high-energy ATP bonds.

Many microorganisms obtain energy from **anaerobic** glycolysis, or **fermentation.** Yeast cells, for example, avoid the need for oxygen by using NADH to reduce pyruvate to form ethanol.

In eukaryotic cells, energy metabolism (other than glycolysis) takes place in the mitochondria. Pyruvate enters the mitochondria, where cellular respiration proceeds. Pyruvate then undergoes an oxidation reaction, in which NAD^+ serves as the electron acceptor. In acetyl-CoA, acetate is linked, via a high-energy bond, to a molecule called **coenzyme A.**

The **citric acid cycle,** or **Krebs cycle,** converts the carbon atoms of pyruvate to acetyl-CoA and then carbon dioxide. The citric acid cycle is a nearly universal set of reactions. The citric acid cycle adds the 4-carbon molecule **oxaloacetate** to 2-carbon acetyl-CoA to form a 6-carbon molecule, from which two carbons are removed in succession. Oxaloacetate then reenters the cycle once more.

The citric acid cycle, which not only provides energy but also serves as the major route for the synthesis of many small molecules, lies at the center of the complex network of biochemical conversions that collectively are called metabolism. Metabolism is divided into two classes of pathways: **catabolism** and **anabolism.**

Cellular respiration produces 30 (or 32) molecules of ATP for each molecule of glucose. All but four of these ATP molecules are produced by oxidative phosphorylation. In the absence of oxygen, glycolysis alone produces only two molecules of ATP.

REVIEW AND THOUGHT QUESTIONS

Review Questions

1. Write down a reaction that describes the following situation: substance A carries a net charge of -1, that is, it has one extra electron, while substance B is electrically neutral. During a reaction, the extra electron spontaneously moves from A to B, so that A becomes neutral and B acquires a negative charge. Indicate each of the following: the oxidized substance, the reduced substance, the oxidizing agent, the reducing agent, the more electronegative substance, the substance with lower affinity for electrons.

2. Name the process in which most of the energy from glucose is converted to ATP. What role does oxygen play in this process? What gas is produced as a waste product? Where in a cell do these reactions occur?

3. Why is the electron transport pathway divided into so many small reactions?

4. Across what eukaryotic membrane is a gradient established during cellular respiration? What is unequally distributed across these membranes? What caused this unequal distribution?

5. Describe and name three stages that aerobic organisms use to extract energy from glucose.

6. How many carbon atoms does a glucose molecule have? Name the end product molecule that is produced during glycolysis. How many carbon atoms does it have?

7. How many ATPs are produced in the citric acid cycle? How does this number compare with that produced in glycolysis?

8. Define metabolism, anabolism, and catabolism, and give examples of catabolic and anabolic pathways. Which type of metabolism is most favored when a cell (a) has abundant ATP? (b) contains little ATP?

Thought Questions

9. Why might you lose weight (a lot of weight) if your mitochondria suddenly lost the ability to couple electron transport to the production of ATP?

10. When you exercise heavily, you consume oxygen faster than you can take it in through your lungs and circulatory system, so your cells have to rely at least partly on anaerobic processes to extract energy from food molecules. Which of the three stages in energy extraction can your cells not perform under these conditions? How much ATP can you form from food molecules under these conditions?

SELECTED READINGS

Dubos, Rene, *Louis Pasteur, Free Lance of Science,* Little, Brown and Company, Boston, 1950. A highly readable biography of the great microbiologist.

Mass, Ralph W., *Free Radical, Albert Szent-Györgi and the Battle over Vitamin C,* Paragon House Publishers, New York, 1988. A readable but somewhat breathless biography of Szent-Györgi, the firebrand of oxidation chemistry.

▶ On-line materials relating to this chapter are on the World Wide Web at http://www.saunderscollege.com/lifesci/
Click on Tobin/Dusheck: *Asking About Life.*

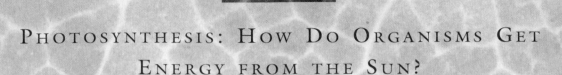

7

PHOTOSYNTHESIS: HOW DO ORGANISMS GET
ENERGY FROM THE SUN?

The Chemical Evangelist

In the early 17th century, the Belgian physician and alchemist Johann Baptista van Helmont [1579–1646] laid the first stone in the foundation for the modern understanding of **photosynthesis** [Greek, *photos* = light + *syntithenai* = to put together]—the process that changes light energy into the energy of chemical bonds. Van Helmont's contribution was a simple and dramatic experiment that scientists have admired for nearly 400 years.

Van Helmont was a man who, in our time, would be considered highly unusual. He was deeply religious, nearly fanatical in his beliefs. Yet he was a born skeptic, always questioning the beliefs of those around him. It was an attitude that would bring upon him the fury of the Spanish Inquisition.

As a teenager, van Helmont found his education at the Catholic University of Louvain so disappointing that he refused the title "Master of Arts" because he said he had learned nothing substantial or true. On leaving the university, he then refused a well-paid position as a canon, or clergyman, because he opposed "living on the sins of the people." He tried living the austere life of a Christian Stoic, until the hardship made him ill.

Perhaps illness inspired the young van Helmont, for he then began to study medicine, with particular emphasis on herbal remedies. Again he was disappointed,

Paraskevas Photography

156

however. The herbalists, he complained, had discovered nothing new about medicine since ancient Greek times, and argued endlessly over details that had nothing to do with making people well. When van Helmont was 20 years old, he accepted a medical degree from Louvain. Yet he never practiced medicine, believing that he had learned nothing that would actually help anyone who was sick.

rogance. Then in the 1620s, one of his many enemies published one of van Helmont's tirades against Jesuits and physicians. In 1623, the faculty at the Louvain medical school denounced the pamphlet, which van Helmont had not intended to publish, as "monstrous." Two years later, the Spanish Inquisition declared that 27 statements in this manuscript were almost certainly heretical. Between 1624 and

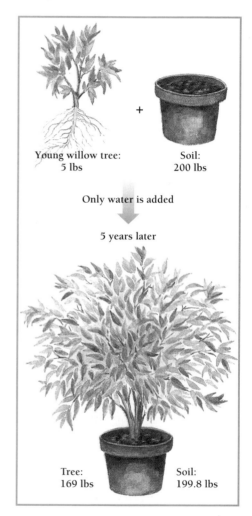

But van Helmont knew Aristotle was wrong. Anyone could see that a potted plant didn't use much soil; it used water.

His new interest was alchemy, a result, he later wrote, of a call from God. (He was called the "chemical evangelist.") Although he had frequently claimed to detest money, at 28, he married a woman so wealthy that he never needed to work for a living again. For 15 years, he happily pursued his own researches into alchemy, religious philosophy, medicine, and natural science.

He determined, for example, that digestion is accomplished by acid and bile. He recognized gas as a form of matter, defining it as the volatile part of a substance, distinct from the mixture of gases we call air. He recognized that the lung is an organ for gas exchange. He used a simple scale to show that matter is conserved during chemical reactions.

He wrote profusely of his discoveries and investigations, never hesitating to criticize the thinking of others. He especially attacked practicing physicians—for their ignorance and superstition—and the Jesuits, a powerful order of Catholics—for what he viewed as their ignorance and ar-

1642, the Spanish Inquisition imprisoned and interrogated van Helmont, then locked him up in a convent and, finally, confined him to house arrest for many years. During all of this time, van Helmont published nothing. Yet, he continued to work and, equally important, to write.

Much of van Helmont's work was directed at undermining the teachings of the Jesuits. The Jesuits based their teachings on Aristotle, who was the most influential of the Greek philosophers. Aristotle had argued that plants derive their sustenance from the soil. A plant's roots, Aristotle said, extract materials from soil in much the same way that an animal's stomach extracts materials from its food. This view seemed logical: farmers had long known that even the richest soils eventually lose the ability to support luxuriant plant growth.

But van Helmont knew Aristotle was wrong. Anyone could see that a potted plant didn't use much soil; it used water. Besides van Helmont believed that all natural bodies come from water. To knock

Figure 7-1 Van Helmont's willow tree experiment showed the importance of water and unimportance of soil in providing the material that makes up a plant. After 5 years, the willow tree had gained 164 pounds, but the soil had lost only a few ounces.

Aristotle from his pedestal and bolster his own theory, van Helmont performed a now famous experiment (Figure 7-1). He planted a 5-pound willow sapling in 200 pounds of soil, which he had carefully dried in a furnace, weighed, and then placed in a large pot. He covered the pot with an iron plate to keep dust from falling into it and watered the growing tree as needed. He did not weigh the leaves that fell from the tree each autumn. After 5 years, he again weighed both the tree and the soil. The willow tree now weighed 169 pounds, while the soil weighed only 2 ounces less than the starting 200 pounds. Van Helmont concluded that "164 pounds of wood, bark, and roots arose out of water alone."

Van Helmont's experiment was excellent, but his interpretation was partly mistaken. His experiment did prove that Aristotle (and the Jesuits) were wrong: the tree's mass could not have come from the soil. However, van Helmont himself was also wrong in assuming that his experiment showed that the tree came entirely from water. As the discoverer of gas, it was ironic that van Helmont did not consider that something in the air might also have contributed to the growth of his willow tree.

KEY CONCEPTS

1. Carbon dioxide and water are the raw materials of photosynthesis.

2. Specialized photosynthetic machinery absorbs light and transforms light energy into chemical energy.

3. Photosynthesis requires two distinct but interacting sets of light-dependent reactions, called photosystem I and photosystem II.

4. The energy of sunlight, first captured in the chemical bonds of NADPH and ATP, is stored in the chemical bonds of carbohydrates.

5. The efficiency of photosynthesis in a plant depends both on its genetic capacity and on environmental conditions.

HOW DO WE KNOW HOW PLANTS OBTAIN CARBON AND OXYGEN?

The first suggestion that the atmosphere contributed to the stuff of plants came not from chemists or alchemists, but from microscopists. With the new microscopes developed in the late 17th century, an Englishman and an Italian almost simultaneously discovered minute pores in the leaves of plants called **stomata,** [Greek, *stoma* = mouth]. Stomata appeared to allow air to pass to the interior of the leaves (Figure 7-2). The activities of plants, they speculated, must somehow depend on interactions with air. By the mid-18th century, scientists had concluded that "vegetation is planted in the air as well as in the earth."

What Do Plants and Air Do for Each Other?

The next big clue to the interaction between plants and air came from experiments in which a mouse or a candle was placed in an airtight container. After some time the mouse would die or the candle would go out. This was because the mouse, or the burning candle, consumed the oxygen in the container. When the oxygen was gone, the mouse died and the candle went out.

In the 1770s, the English clergyman Joseph Priestley performed an experiment to show that a sprig of mint could restore the air in the jar and keep the mouse alive (Figure 7-3). Priestley showed that plants put back what living animals (or burning candles) take out (Figure 7-4). This discovery was nothing less than the discovery of oxygen, whose existence Antoine Lavoisier independently found at about the same time. Credit for the discovery usually goes to both men.

Scientists soon learned that the material that animals and candles remove from the air, and that plants put back, is oxygen. But plants remove something different—carbon dioxide.

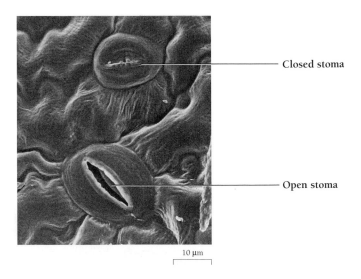

Closed stoma

Open stoma

10 μm

Figure 7-2   **Stomata open and close depending on whether plants need to conserve water or collect carbon dioxide.** The discovery of stomata in the surfaces of leaves suggested that plants use gases from the air as well as water from the ground. Scientists concluded that "vegetation is planted in the air as well as in the earth." *(© Dwight Kuhn)*

Figure 7-3 Joseph Priestley. Even as modern chemistry took its first breath, its greatest exponents—Joseph Priestley and Antoine Lavoisier—were caught up in the turmoil surrounding the American and French revolutions. In 1791, Priestley's sympathies for the French Revolution inspired an angry mob to destroy his home, library, and laboratory. Within 3 years, he fled to the United States. In the same year, on the other side of the English Channel, the nobleman Lavoisier, who had served King Louis XVI, was guillotined. His executioners proclaimed, "The Republic has no need for scholars," and threw his body into a common grave. *(The Granger Collection, New York)*

In light, plants take carbon dioxide from the air, and release oxygen as a waste product (Figure 7-5A). However, plants also respire, taking in oxygen and releasing carbon dioxide (Figure 7-5B). When plants are in the dark, respiration provides their only energy source.

Today we know that, during photosynthesis, plants take in water and carbon dioxide, with which they synthesize glucose, a simple sugar made of the elements carbon, hydrogen, and oxygen. To a large extent, then, plants are made of water and air. However, as we saw in Chapter 3, proteins and most other biological molecules also include atoms of nitrogen, phosphorus, and other elements. These come, in minute quantities, from the soil. The two ounces of soil that van Helmont measured may have included nutrients essential to the growth of the willow tree.

In this chapter, we will ask how cells accomplish photosynthesis. How does photosynthesis capture light energy? How does it store the captured energy? What molecules and subcellular structures participate?

Plants take carbon dioxide from the air and water from the soil,
and they release oxygen back into the air.

Where Do the Atoms Go in Photosynthesis?

Now that we know the raw materials of photosynthesis, we can summarize the overall chemical transformation in the following equation:

A. Dead mouse

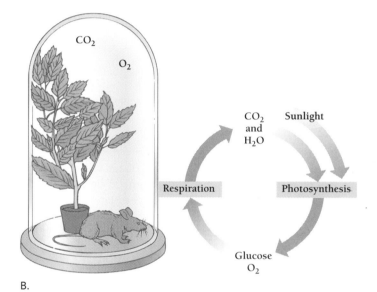

B.

Figure 7-4 Priestley's experiment showed that plants put something into air that mice need. A. The mouse in the jar uses up the oxygen and fills the jar with carbon dioxide, which suffocates the mouse. B. A photosynthesizing plant placed in the jar takes up the carbon dioxide released by the mouse and replaces it with oxygen. The mouse is fine.

$$6\ CO_2 + 6\ H_2O \longrightarrow C_6H_{12}O_6 + 6\ O_2$$

carbon dioxide + water $\xrightarrow{\text{light}}$ glucose + oxygen

Notice that this is the exact reverse of the overall equation for respiration given in Chapter 6 on page 136. Another way of writing the equation for photosynthesis is:

$$CO_2 + H_2O \longrightarrow C(H_2O) + O_2$$

carbon dioxide + water $\longrightarrow$ carbohydrate + oxygen

Here, $C(H_2O)$ is the general formula for a carbohydrate, with each carbon atom associated with the equivalent of one molecule of water.

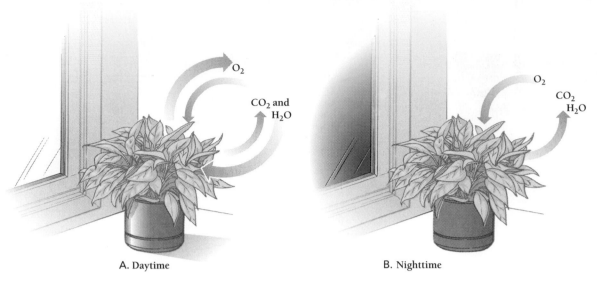

A. Daytime   B. Nighttime

Figure 7-5   In light, plants perform photosynthesis. They use carbon dioxide and water to make carbohydrates. In both the dark and the light, they derive energy by respiration, starting with the carbohydrates they produced themselves.

To early 20th-century chemists, this equation suggested a hypothesis for how photosynthesis might occur: light splits the carbon dioxide into oxygen and carbon. Then oxygen is expelled into the atmosphere, and the carbon combines with water to form carbohydrate. Although this idea is appealing, it is wrong.

The first realization that this view of photosynthesis was wrong came from studies of photosynthetic bacteria. In the early 1930s, a graduate student at Stanford University named Cornelius van Niel was examining photosynthesis in bacteria that used hydrogen sulfide (H_2S) instead of water as a raw material.

$$CO_2 + H_2S \longrightarrow C(H_2O) + S + O_2$$

carbon dioxide + hydrogen sulfide $\longrightarrow$ carbohydrate + sulfur + oxygen

Instead of producing only oxygen as a by-product, these bacteria produce both oxygen and sulfur (Figure 7-6). Van Niel reasoned that these bacteria probably used the same biochemical pathways as other photosynthesizing organisms. Water (H_2O) and hydrogen sulfide (H_2S) probably served the same biochemical role. Yet, the sulfur produced by photosynthetic bacteria clearly comes from hydrogen sulfide. It stood to reason, then, that the oxygen produced by plants came from splitting water, not carbon dioxide.

However, neither van Niel nor anyone else had any direct evidence to support this hypothesis. No one could be certain which oxygen atoms came from which molecules. More than 10 years passed before scientists found a way to distinguish between the oxygen molecules in water and those in carbon dioxide. The key was the isotope ^{18}O, an oxygen atom that is slightly heavier than usual. Plants provided with normal carbon dioxide and so-called heavy water, $H_2{}^{18}O$, produce heavy oxygen, but *not* heavy carbohydrate:

$$CO_2 + H_2{}^{18}O \longrightarrow C(H_2O) + {}^{18}O_2$$

carbon dioxide + water $\longrightarrow$ carbohydrate + oxygen

This experiment showed conclusively that water, not CO_2, is split by sunlight (Figure 7-6C).

The oxygen released by plants comes from water, not carbon dioxide.

Biochemists now had a clear understanding of what raw materials plants used to synthesize glucose. But two difficult questions still remained: How do plants channel light energy into this chemical reaction? And how do plants use that chemical energy to synthesize carbohydrates?

Photosynthesis consists of two processes. The so-called light reactions use light to generate energy-transfer molecules such as ATP. The light reactions are called **photophosphorylation**, because they use photons to "phosphorylate" ADP to ATP. The second part of photosynthesis consists of the so-called dark reactions, which use the energy packaged in the light reactions to make glucose from carbon dioxide. Actually, the dark reactions can occur in light, but light is unnecessary. They are more correctly called light-independent reactions. In the rest of this chapter, we will examine the light reactions and the light-independent reactions in detail (Figure 7-7).

HOW DO PLANTS COLLECT ENERGY FROM THE SUN?

Julius Mayer, one of the discoverers of the First Law of Thermodynamics, elegantly summarized the business of photosynthesis: "Nature has put itself the problem of how to catch in flight light streaming to the earth and to store the most elusive of all powers in rigid form."

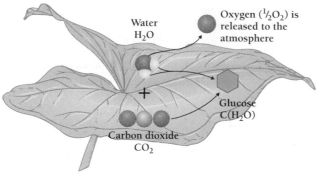

A. Photosynthesis in plants

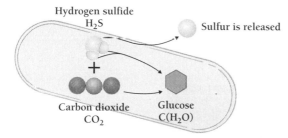

B. Photosynthesis in bacteria

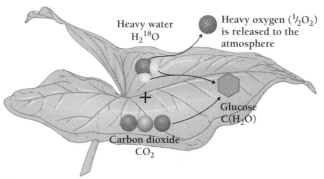

C. Photosynthesis in plants using heavy water

Figure 7-6 **Where do the oxygen atoms come from?** A. The oxygen released by plants during photosynthesis comes from water, not from carbon dioxide. Two experiments demonstrate this: B. First, sulfur released by bacteria during photosynthesis comes from hydrogen sulfide, which is chemically analogous to water. C. Second, in plants supplied with "heavy water," the oxygen released during photosynthesis comes from water, not carbon dioxide.

What Is Light?

Light is always moving. Light stops moving only when some object absorbs it, and then it is no longer light but some other kind of energy. The rug, curtains, and furniture of your room absorb the morning sunlight and transform light energy into heat. In photosynthesis, specialized cellular machinery absorbs light and transforms light energy into chemical energy.

Ordinary rules cannot describe light, which behaves both like a particle and like a wave. Like a particle, light travels in straight lines (Figure 7-8A). Like the ripples in a rain puddle, light has troughs and crests that interfere with one another (Figure 7-8B). Nothing in our ordinary experience, however, provides a good model for the behavior of light. We are left talk-

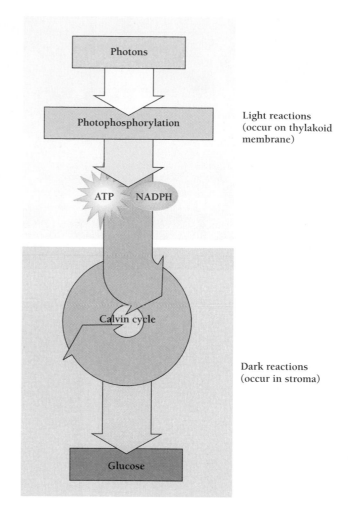

Figure 7-7 **The two parts of photosynthesis.** The light-dependent reactions, called photophosphorylation, generate energy-transfer molecules such as ATP and occur in the thylakoid membrane of the chloroplast. The light-independent, or "dark," reactions use the energy packaged in the light reactions to make glucose from carbon dioxide. These reactions occur outside the thylakoid membrane in the stroma.

ing about separate properties, some "wavelike" and some "particlelike" properties.

Light is a form of electromagnetic radiation, a form of energy transmitted through space as periodically changing electrical and magnetic forces. Light waves are fluctuations of force, distinct from the fluctuations of matter we usually call waves. The most important wavelike property of light is its **wavelength**—the distance between the crests of two successive waves (Figure 7-8C). Wavelengths vary in length from kilometers to nanometers. A nanometer, which is abbreviated as nm, is a billionth of a meter.

The light of our everyday experience is a mixture of many wavelengths. Our eyes and brains allow us to see only those wavelengths ranging from about 390 nm to about 760 nm—what we call **visible light**. We see different wavelengths of visible light as different colors (Figure 7-9). For example, we see 400 nm light as violet, 500 nm as blue-green, and 600 nm as orange-red. What we see as white light consists of a mixture of many wavelengths. In fact, the light we see is only a tiny

A.

Figure 7-8 Light is both a particle and a wave.
A. Light is made of particles that move in straight lines and bounce off surfaces in the same way as marbles or billiard balls. B. The circular ripples created by raindrops on a pond make a pattern of troughs and crests. Light also forms waves with crests and troughs that interfere with one another. C. Wavelength is the distance from the crest of one wave to the next. *(A, Leonard Lessin/Peter Arnold, Inc.; B, (left) Runk/Schoenberger from Grant Heilman, (right) Yoav Levy/Phototake)*

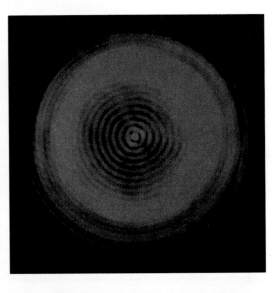

B.

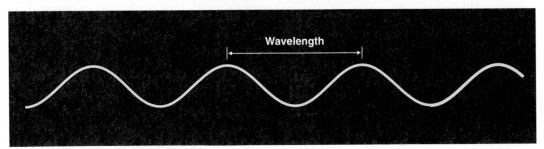

Wavelength

C.

part of all possible wavelengths that make up the electromagnetic spectrum (Figure 7-10). Other kinds of electromagnetic radiation range from gamma rays (with wavelengths less than 1 nm) to radio waves (with wavelengths measured in meters and even kilometers).

To understand photosynthesis, we must see how light interacts with matter. For this, we must talk of light's particlelike and wavelike properties at the same time. A light particle, or **photon,** is a package of energy. The amount of energy in each photon depends entirely on the wavelength of light. Photons

of shorter wavelengths (toward the blue end of the visible spectrum) have more energy than photons of longer wavelengths (toward the red end of the visible spectrum). For example, blue light, which has a short wavelength, has about twice as much energy per photon as red light, which has a long wavelength. An atom or a molecule can absorb a photon of light if the energy of the photon matches the energy needed to boost one of its electrons from one energy level to a higher one.

The energy of individual photons is unrelated to the brightness of the light. Brighter light consists of more photons than

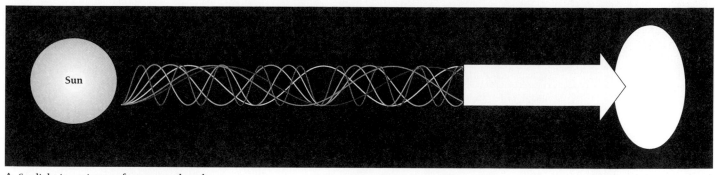

A. Sunlight is a mixture of many wavelengths

All colors absorbed except purple

Light

All colors absorbed except orange

C.

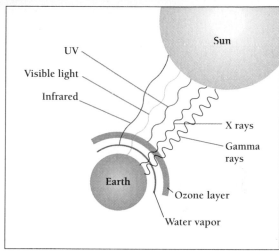

B.

Figure 7-9   Sunlight. A. Sunlight is a mixture of many wavelengths of light. B. Most sunlight that reaches the Earth's surface is visible light because the Earth's atmosphere reflects most ultraviolet light and absorbs most infrared light. C. Tiger lilies, irises, and other living organisms have evolved pigments and other molecules that can use the energy available in visible light.

a dimmer light. A 150-watt bulb, for example, emits more photons than a 50-watt bulb. Brighter light can excite more atoms or molecules, but only if it has the proper wavelength.

We see visible light because it excites electrons in specialized molecules in our eyes. Visible light also alters the chemicals in photographic film. In some situations, a photon can induce an electrical signal, for example, in a video camera or a solar collector. In photosynthesis, light excites electrons in par-

ticular molecules and thereby increases their free energy. The photosynthetic machinery captures some of this free energy and converts it into high-energy chemical bonds.

Light has both wavelike and particlelike properties. Light energy comes packaged in photons, which can excite molecules.

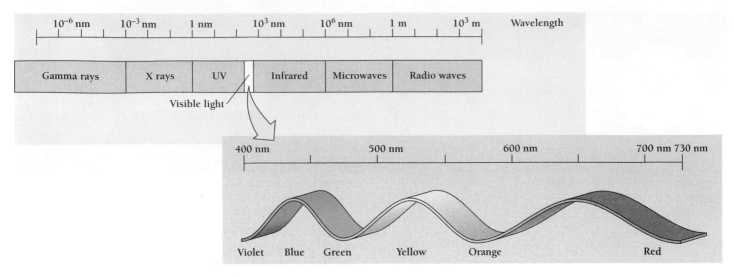

Figure 7-10   The wavelengths of visible light range from about 390 nanometers (nm) to about 760 nm. We see 400 nm light as violet, 500 nm as blue-green, and 600 nm as orange-red. Other kinds of electromagnetic radiation range from gamma rays (with wavelengths less than 1 nm) to radio waves (with wavelengths measured in meters and even kilometers).

How Much Energy Does a Photon Contain?

Most of the photons that reach the Earth's surface have wavelengths in the visible range (Figure 7-9). Most of the more energetic photons of ultraviolet ("beyond violet") light are absorbed by ozone in the upper atmosphere. And most of the less energetic photons of infrared ("below red") light are absorbed by water vapor and carbon dioxide in the atmosphere.

Not surprisingly, life has evolved molecules that make use of the energy in visible light—the light that best penetrates the atmosphere. Life takes full advantage of available energy. The energy contained in a visible photon is enough to produce the chemical reactions of vision and photosynthesis.

Short wavelength photons have high energy, and long wavelength photons have low energy.

The Effectiveness of Different Wavelengths Can Reveal Which Molecules Are Responsible for a Light-Dependent Process

Different molecules absorb different wavelengths of light. For example, a red flower absorbs most wavelengths in the visible range except for those that we see as red, which are reflected. The green pigment in plants absorbs every color but green. The particular set of wavelengths that a molecule absorbs is called its **absorption spectrum.**

A molecule's absorption spectrum tells us how much energy the molecule needs to move an electron from one orbital to another (Figure 7-11). Most of the molecules in a cell absorb ultraviolet light only. This is because the energy required to move an electron to a higher-energy orbital is usually quite high. Some molecules, however, can change their electronic arrangements with relatively little energy. These molecules, called **pigments,** absorb visible light (Figure 7-12).

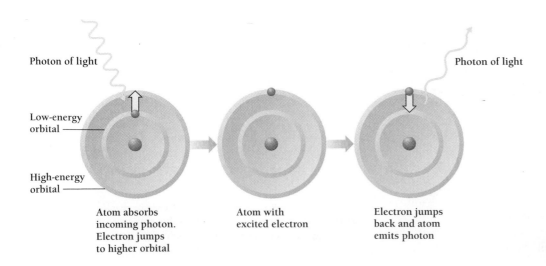

Photon of light

Photon of light

Low-energy orbital

High-energy orbital

Atom absorbs incoming photon. Electron jumps to higher orbital

Atom with excited electron

Electron jumps back and atom emits photon

Figure 7-11   When an atom *absorbs* light, a photon propels an electron to a higher-energy orbital. This temporary state of excitement ends when the electron falls back to a lower energy orbital, releasing the photon. Combinations of atoms and molecules absorb characteristic wavelengths of light.

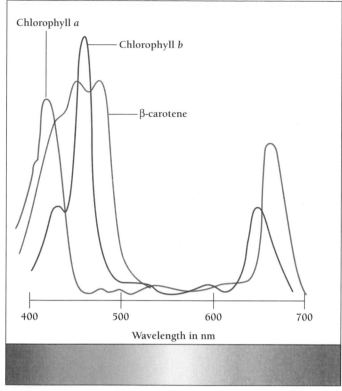

B.

Figure 7-12 Plant pigments. A. Absorption spectra of plant pigments. A purple iris absorbs most wavelengths in the visible range except those that we see as purple, which it reflects. Chlorophyll, the photosynthetic pigment in plants, absorbs red and blue photons but allows green ones to pass or to be reflected. The particular set of wavelengths that a molecule absorbs is its absorption spectrum. B. Autumn leaves along the Kancamagus Highway in New Hampshire. Carotenoids are pigments responsible for the beautiful colors of autumn leaves. *(B, Stephen J. Krasemann/Photo Researchers)*

For example, a solution of isolated **chlorophyll**—the green pigment in plants—absorbs red and blue photons and allows photons of intermediate (green) wavelengths to pass or to be reflected. These photons (with green wavelength) excite pigment molecules in our eyes, and we perceive a green color. Similarly, the red color of hemoglobin in blood results from the selective absorption of blue and green photons. The remaining long-wavelength light excites other pigment molecules in our eyes, and we see red.

For any light-dependent process, we also can determine an **action spectrum**—the relative effectiveness of different wavelengths in promoting the process. For example, we can measure the rate of photosynthesis under lights of different colors. The action spectrum of photosynthesis roughly follows the absorption spectrum of chlorophyll (Figure 7-13). This correspondence suggests that the absorption of light by chlorophyll may be responsible for photosynthesis.

More careful measurements of the action spectrum of photosynthesis show that the action spectrum does not *exactly* correspond to the absorption spectrum of chlorophyll. Light with a wavelength of 500 nm (blue-green light) can stimulate photosynthesis even though blue-green light is not absorbed by chlorophyll. From this, it is clear that pigments other than chlorophyll must play a role in photosynthesis. These pigments are the yellow or orange **carotenoids**, which absorb blue-green

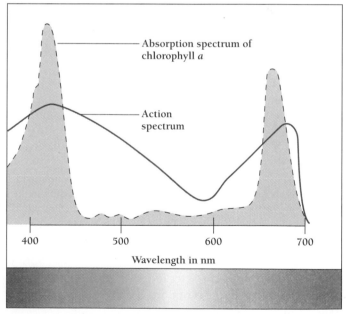

Figure 7-13 An action spectrum is the range of wavelengths that effectively promotes a light-dependent process. The action spectrum of photosynthesis, for example, roughly parallels the absorption spectrum of the pigment chlorophyll a.

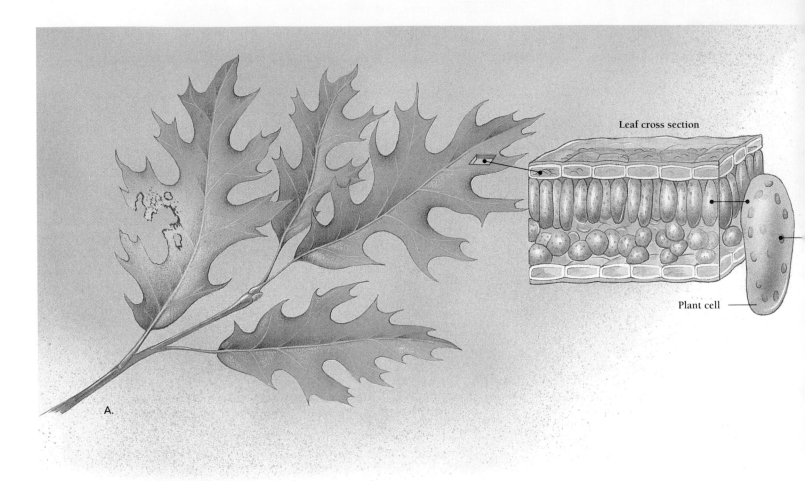

Leaf cross section

Plant cell

light. Carotenoids work by transferring the energy of their excited electrons to chlorophyll.

In fact, most chlorophyll works the same way. Researchers have found that most of the chlorophyll molecules in a leaf do not actually photosynthesize. Instead, hundreds of chlorophyll molecules associate with a few carotenoids to form a weblike **antenna complex,** which traps light and transfers energy to a **photochemical reaction center,** where the captured light energy is actually converted to chemical energy (Figure 7-14). In the photochemical reaction center are the small number of chlorophyll molecules that actually accept electrons.

The efficiency of photosynthesis (the action spectrum) in plants roughly corresponds to the absorption spectrum of the pigment chlorophyll, but not exactly. Only a small number of chlorophyll molecules transfer electrons to $NADP^+$. Most chlorophyll molecules trap and transfer light energy.

Light Excites Electrons in Two Types of Reaction Centers

We can now describe a fairly simple model of photosynthesis: (1) chlorophyll and carotenoids absorb light; (2) these pigments transfer the captured energy to the reaction center; and (3) the energy splits oxygen from water and somehow forms chemical bonds.

If this simple model were correct, each absorbed photon would contribute equally to photosynthesis, which is indeed the case in bacterial photosynthesis. Photosynthesis in plants, however, appears to operate by a more complicated mechanism. Some of the wavelengths that are readily absorbed by plant pigments are inefficient at promoting photosynthesis. When absorbed light has a wavelength longer than 680 nm, efficiency drops abruptly. Deep red light (with a wavelength of 700 nm or longer) will promote photosynthesis only if shorter wavelength light (680 nm) also illuminates the plant. Yet when both 680 nm (red) and 700 nm (deep red) light shine on a plant, photosynthesis is more efficient than with 680 nm light alone. In other words, photosynthesis works better with light of two wavelengths than with either alone.

Plants' need for more than one wavelength of light suggests that photosynthesis involves two distinct, but interacting, sets of reactions. Each set of reactions depends on light of different wavelengths. One set of reactions, called **photosystem I,** best absorbs light with wavelengths of about 700 nm (deep red). The other set, called **photosystem II,** requires light with wavelengths of about 680 nm (red light). (In contrast, photosynthetic bacteria have only a single photosystem.) The existence of the two photosystems in plants explains why indoor plants grow better when lit with both a bluish fluorescent bulb

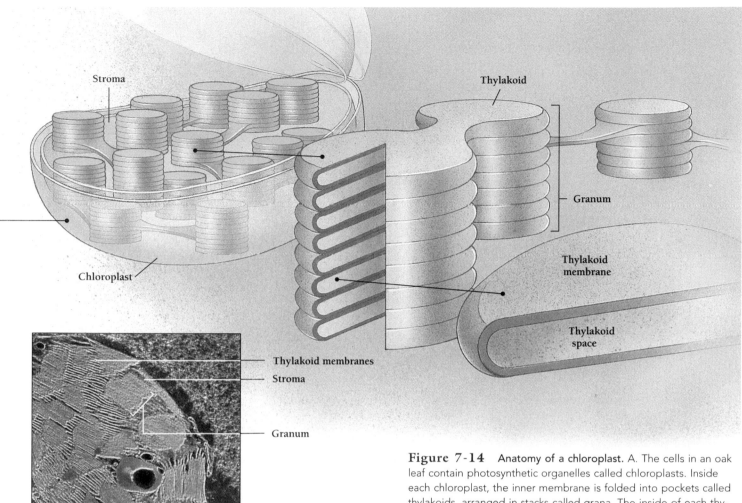

Figure 7-14   **Anatomy of a chloroplast.** A. The cells in an oak leaf contain photosynthetic organelles called chloroplasts. Inside each chloroplast, the inner membrane is folded into pockets called thylakoids, arranged in stacks called grana. The inside of each thylakoid pocket is called the thylakoid space. The space outside the stacks of thylakoids (but inside the chloroplast) is called the stroma. Light-*ind*ependent reactions occur in the stroma. Light-*d*ependent reactions occur in the photochemical reaction center in the antenna complex of the thylakoid membrane. B. TEM of a chloroplast. *(B, James Dennis/CNRI/Phototake)*

(which has little 700 nm light) and a yellowish incandescent bulb (which has little 680 nm light) than with either kind of bulb alone. Later, we will see how the two photosystems each contribute to harnessing light energy.

The two photosystems lie in separate subcellular particles. Both particles are inside of chloroplasts. There they associate with the thylakoid membranes, which, recall from Chapter 4, enclose the stacked vesicles called grana (Figure 7-14). Each photosystem particle consists of chlorophyll molecules together with carotenoids, specific proteins, and other molecules and ions (Figure 7-15A). Each photosystem performs its own biochemical reactions, and each is required for plant photosynthesis.

The chlorophyll in photosystem I absorbs light most efficiently at 700 nm and is called P_{700} (P stands for pigment). Similarly, the chlorophyll molecules of photosystem II absorb light best at 680 nm and are called P_{680}.

Plant photosynthesis depends on two separate photosystems, each of which absorbs a different wavelength of light most efficiently.

The Excited Electrons of Photosystem I Can Move in a Circle Back to Chlorophyll and Produce ATP

When P_{700} absorbs light from the antenna complex, its free energy increases, and it becomes a strong reducing agent (electron donor). As shown in Figures 7-16 and 7-17, the excited P_{700} readily gives up a high-energy electron to one of two electron transport chains. When P_{700} loses its excited electron, it also loses most of the free energy it gained from the absorbed photon. Because P_{700} has now given up an electron, it is in its oxidized state.

One of the two electron transport chains associated with photosystem I leads the excited electron back to oxidized P_{700}

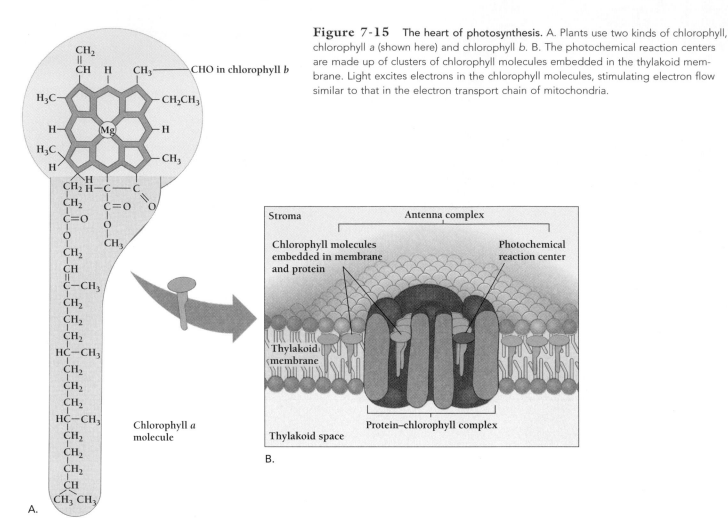

Figure 7-15   **The heart of photosynthesis.** A. Plants use two kinds of chlorophyll, chlorophyll *a* (shown here) and chlorophyll *b*. B. The photochemical reaction centers are made up of clusters of chlorophyll molecules embedded in the thylakoid membrane. Light excites electrons in the chlorophyll molecules, stimulating electron flow similar to that in the electron transport chain of mitochondria.

(Figure 7-16). This cycle restores P_{700} to its original state and makes it ready to absorb another photon, a process called **cyclic photophosphorylation.**

The electron acceptor here is a cytochrome complex, which passes the electron on to another acceptor, which reduces oxidized P_{700}, completing the cycle. As in mitochondria, the passage of electrons through the cytochrome complex results in a proton gradient across a membrane. Protons move from a space just outside the thylakoids, called the **stroma**, into the lumen of the thylakoids (Figure 7-16). The stroma corresponds to the mitochondrial matrix. In mitochondria, protons move out of the matrix to the intermembrane space. In chloroplasts, protons move out of the stroma into the lumen, or intermembrane space of the thylakoids.

Like the inner mitochondrial membrane, the thylakoid membrane contains turbinelike ATP synthetase molecules, which produce ATP as the protons flow back down their gradient (Figure 7-16). The ATP is made in the stroma, just as ATP is made in the mitochondrial matrix. The net effect of the electron route just described is the production of ATP from the energy of absorbed light. The ATP made in the stroma is used to synthesize glucose.

> In cyclic photophosphorylation, excited electrons from P_{700} pass through an electron transport chain and back to P_{700}. This cyclic passage of electrons creates a proton gradient across a membrane in the chloroplast capable of generating ATP.

The Excited Electrons of Photosystem I Can Produce the Strong Reducing Agent NADPH

Photosystem I also includes an alternate, noncyclic electron path, which does not produce ATP (Figure 7-17). Instead, the energy of the excited electron of P_{700} is used to make NADPH, a derivative of NADH. NADPH is the electron-carrying coenzyme that chloroplasts use ($NADP^+$ in its oxidized form). While NADH and NAD^+ are used in pathways such as glycolysis and respiration (catabolism), NADPH and $NADP^+$ are used in the building of high-energy molecules such as glucose (anabolism).

The noncyclic route of photosystem I supplies most of the reducing power for glucose synthesis. But after P_{700} has given up its excited electron to form NADPH, P_{700} needs another electron before it can again serve as an electron donor. That is, some reducing agent must contribute electrons to oxidized P_{700}.

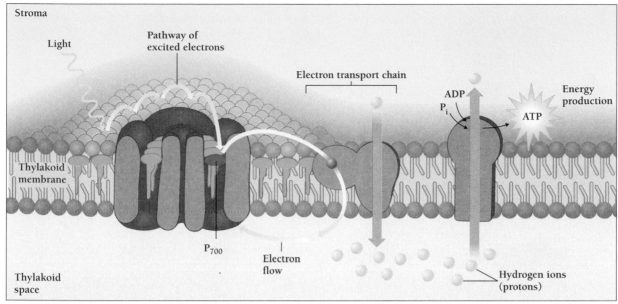

Stroma

Light

Pathway of
excited electrons

Electron transport chain

ADP
P_i

ATP

Energy
production

Thylakoid
membrane

P_{700}

Electron
flow

Thylakoid
space

Hydrogen ions
(protons)

PHOTOSYSTEM I: Cyclic photophosphorylation

Figure 7-16 Cyclic photophosphorylation in photosystem I. Excited electrons from P_{700} pass through an electron transport chain and back to P_{700}. This cycle of electrons creates a proton gradient across the thylakoid membrane that can generate ATP.

> In the noncyclic route of photosystem I, the excited electron from P_{700} goes to help form NADPH. P_{700} is then missing one of its electrons.

Photosystem II Provides Electrons To Reduce P_{700}

The chlorophyll molecules of photosystem II absorb light best at 680 nm and are called P_{680}. Just as excited P_{700} is a powerful reducing agent that can donate a high-energy electron to an electron acceptor, excited P_{680} can also serve as an electron donor (Figure 7-18A). Electrons move from excited P_{680}, through the electron transport chain of photosystem II, to oxidized P_{700} (which is produced by photosystem I). By regenerating reduced P_{700}, photosystem II supplies electrons to photosystem I even when there is no cyclic flow of electrons.

Photosystem II also produces ATP as protons flow down the gradient generated by its electron transport chain. This ATP production is called **noncyclic photophosphorylation,** to distinguish it from the ATP production in the cyclic path of photosystem I.

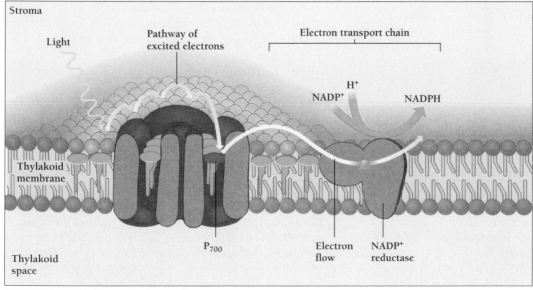

Stroma

Light

Pathway of
excited electrons

Electron transport chain

H^+
$NADP^+$ NADPH

Thylakoid
membrane

P_{700}

Electron
flow

$NADP^+$
reductase

Thylakoid
space

PHOTOSYSTEM I: Noncyclic electron flow

Figure 7-17 Noncyclic photophosphorylation in photosystem I. An excited electron from P_{700} goes to help form NADPH. P_{700} then is missing an electron, giving it a positive charge.

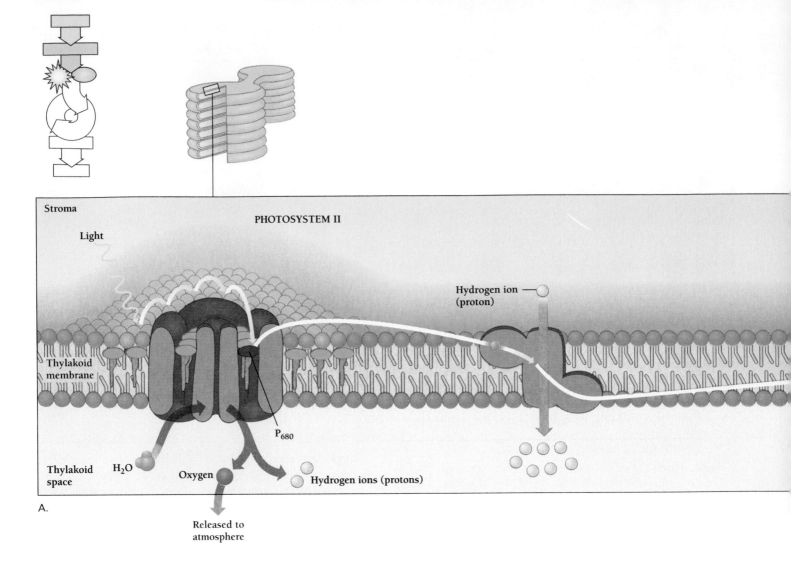

PHOTOSYSTEM II

Stroma

Light

Hydrogen ion (proton)

Thylakoid membrane

P_{680}

Thylakoid space

H_2O

Oxygen

Hydrogen ions (protons)

A.

Released to atmosphere

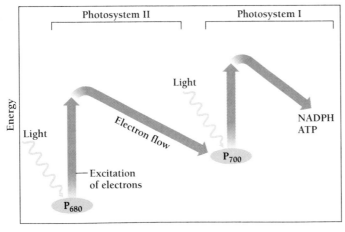

Photosystem II   Photosystem I

Energy

Light

Electron flow

Light

NADPH
ATP

P_{700}

Excitation of electrons

P_{680}

B. Z scheme of photosynthesis

But we have the same question all over again: If the electron from excited P_{680} is transferred to oxidized P_{700}, how is the reduced form of P_{680} regenerated? How does it get an electron back again? The answer is that the electrons for regenerating P_{680} come from water. This fact is essential to our own existence. When photosystem II in all plants strips electrons from water molecules, it generates the free oxygen that we breathe (Figure 7-18).

Until the photosystem II pathway evolved, at least 3 billion years ago, the Earth's atmosphere contained virtually no oxygen. Only when photosystem II evolved did oxygen begin to accumulate in our atmosphere and allow the evolution of animals and other organisms unable to make their own nutrients.

In noncyclic photophosphorylation, photosystem II's P_{680} supplies electrons to P_{700} by way of an electron transport chain. The flow of electrons generates a proton gradient that allows ATP synthesis. P_{680} gets new electrons from water. The evolution of photosystem II allowed the accumulation of oxygen in our atmosphere, which triggered the evolution of animals and other heterotrophs.

Figure 7-18 Photosystems I and II. A. Photosystem II's P_{680} molecule acquires electrons from water, then supplies these electrons to P_{700} by way of an electron transport chain. The flow of electrons generates a proton gradient that allows ATP synthesis. The breakdown of water into oxygen and hydrogen by photosystem II is the ultimate source of the oxygen we breathe. The evolution of photosystem II underlies the evolution of animals and other heterotrophs. B. The Z scheme of photosynthesis summarizes the flow of electrons in photosystems I and II.

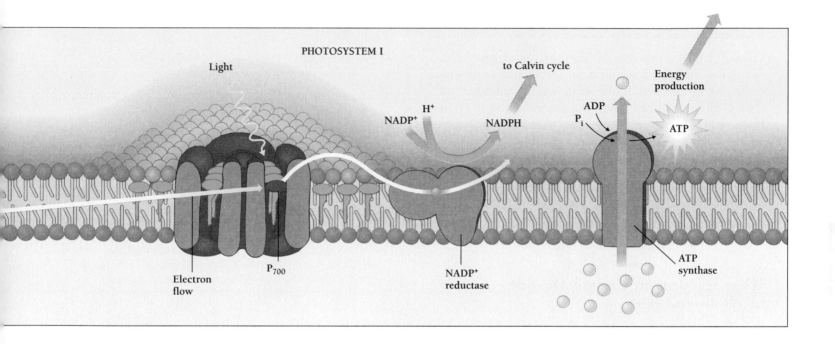

The Light-Dependent Reactions of Photosynthesis Transform Light Energy into the Chemical Bonds of NADPH and ATP

Figure 7-18B shows how photosystems I and II together make up the light-dependent reactions of photosynthesis. The zigzag path of the electron, which is apparent in the figure, is often described as the "Z scheme of photosynthesis." The absorption of light by the two photosystems causes (1) the splitting of water to oxygen and hydrogen ions, (2) the production of ATP, and (3) the production of NADPH. Oxygen is a waste product released into the atmosphere through the stomata. ATP and NADPH remain mostly in the chloroplast stroma, where they contribute to the light-independent reactions of photosynthesis—the synthesis of glucose.

The mid-20th century biochemist Albert Szent-Györgyi characterized this efficient process as "a little electric current driven by the sunshine." The reduction of each molecule of $NADP^+$ to NADPH requires two electrons. To provide each excited electron for this electron path, photosystems I and II must each absorb one photon. To provide the two electrons, then, requires the absorption of four photons, two by each photosystem. Therefore, the production of a molecule of NADPH requires four photons of light.

As each electron flows from photosystem II to photosystem I, it pumps enough protons to produce a molecule of ATP. So the energy of four photons is transformed into the chemical energy of one molecule of NADPH and of two molecules of ATP.

The energy stored in the chemical bonds of NADPH and ATP is used to make glucose, a more stable form in which to store energy. Plants then store glucose in the form of starch, sucrose, or some other carbohydrate. Starch forms a solid deposit in chloroplasts. In contrast, sucrose, or table sugar, remains in solution and can move throughout the plant. Just as animals transport sugar from cell to cell in the form of glucose, plants transport sugar from cell to cell in the form of sucrose.

The process by which plants transform the energy in ATP and NADPH into more stable forms such as glucose is the subject of the next section. The synthesis of sugar is the second part of photosynthesis.

HOW DO PLANTS MAKE GLUCOSE?

Glucose provides plants with their most important source of energy, as well as building blocks that contain carbon, hydrogen, and oxygen. We've seen that the hydrogen for glucose comes from water and the carbon and oxygen come from carbon dioxide in the atmosphere.

The overall reaction of photosynthesis is the reverse of respiration (though the individual steps are not). So we can guess, from what we learned in Chapter 6, that the synthesis of glucose from carbon dioxide and water must consume a lot of energy:

$$6\ CO_2 + 12\ H_2O \longrightarrow C_6H_{12}O_6 + 6\ H_2O + 6\ O_2$$

carbon dioxide + water $\longrightarrow$ glucose + water + oxygen

Recall that the individual reactions in the breakdown of glucose divide the huge downhill process into small steps whose energy can be captured. In the same way, the individual reactions of glucose synthesis allow the energy of light to be harnessed a little bit at a time. The energy "hill" is climbed in a set of small steps.

We know, from the last section, that the H_2O in this equation provides the electrons for reducing the carbon atoms in carbon dioxide. But we still need to find out what happens to the carbon dioxide in photosynthesis. That is, what is the chemical process by which the single carbon atoms in carbon dioxide are combined into 6-carbon glucose?

What Happens to Carbon Dioxide in Photosynthesis?

The discovery of the radioactive isotope ^{14}C provided a way to answer this question. In 1940, Martin Kamen and Samuel Ruben, then working at the University of California, Berkeley, discovered ^{14}C and immediately showed that $^{14}CO_2$ could be used to investigate the chemical transformations of carbon dioxide during photosynthesis. In 1945, Melvin Calvin, another researcher at Berkeley, began the crucial experiments (Figure 7-19).

Calvin injected a small amount of $^{14}CO_2$ into a flask of algae that were actively growing in light. Quickly, sometimes within 5 seconds, he killed the algae by adding alcohol. Then, to discover the fate of the radioactive carbon atoms, he analyzed the compounds made by the algae using a technique called paper chromatography (Figure 7-19). In paper chromatography, compounds dotted on a piece of paper are carefully washed with solvents. As the solvents move across the paper, the various compounds move at different rates, separating out from one another like runners in a foot race.

Afterwards Calvin detected the radioactive compounds by pressing the paper against a photographic film. Wherever ^{14}C was present, the film became black (Figure 7-19). Calvin later reflected that the primary data of his experiments was contained "in the number, position, and intensity—that is, radioactivity—of the blackened areas. The [film] ordinarily does not print out the names of these compounds, unfortunately, and our principal chore for the succeeding ten years was to properly label those blackened areas on the film."

The first detectable product was a 3-carbon compound. When Calvin collected his algae within 5 seconds after the injection of $^{14}CO_2$, all the radioactivity was in a single spot. He and his colleagues later identified that spot as 3-phosphoglycerate, which is also an intermediate in glycolysis. At successively later times, the pattern of spots became much more complicated, and it took a while to figure out just which compounds formed when (Figure 7-19). Calvin and his colleagues worked out the set of reactions that converted the carbon atoms of carbon dioxide into the carbon atoms of glucose. This set of reactions forms a cycle that regenerates the molecule that initially combines with carbon dioxide. This cycle is called the **Calvin cycle,** after its principal discoverer (Figure 7-20).

The Calvin cycle consists of three parts:

1. The production of phosphoglycerate from the combining of carbon dioxide with an acceptor molecule.
2. The conversion of phosphoglycerate into glyceraldehyde phosphate in two reactions that use ATP and NADPH produced in the light reactions.
3. The regeneration of the initial carbon dioxide acceptor.

Melvin Calvin used paper chromatography to obtain an ordered list of the chemical products of photosynthesis. Calvin was then able to suggest a reaction sequence that explained the order in which the various chemicals appeared in time. This reaction sequence came to be known as the Calvin cycle.

In Part 1 of the Calvin Cycle, Carbon Dioxide Combines with a 5-Carbon Molecule and Forms Two 3-Carbon Molecules

The first reaction of the Calvin cycle is the production of phosphoglycerate (Figure 7-20). The carbon dioxide acceptor in this reaction is a 5-carbon compound called ribulose bisphosphate. Together they form an unstable 6-carbon compound, which immediately breaks down into two molecules of phosphoglycerate.

The enzyme that catalyzes this reaction, ribulose bisphosphate carboxylase (Rubisco), has a challenging task: it is almost solely responsible for the recycling of carbon dioxide back into the biosphere and supplying carbon for all living things. The

Figure 7-19 Melvin Calvin deduced the products of photosynthesis by using paper chromatography. Calvin then suggested a reaction sequence that explained the order in which the various chemicals appeared—a reaction sequence that came to be known as the Calvin cycle. A. Radioactive $^{14}CO_2$ is added to algae growing in bright light; the reactions are stopped and the algae extracted after different amounts of time have passed. B. Compounds in the extracts are separated from one another by two-dimensional paper chromatography. C. The radioactive products show up as black spots on paper sensitive to radiation. D. After 5 seconds, most of the radioactivity is in a single spot, corresponding to 3-phosphoglycerate. E. Later, the pattern becomes more complicated, reflecting the subsequent conversion of 3-phosphoglycerate to other compounds. *(A, Melvin Calvin, U. C. Berkeley)*

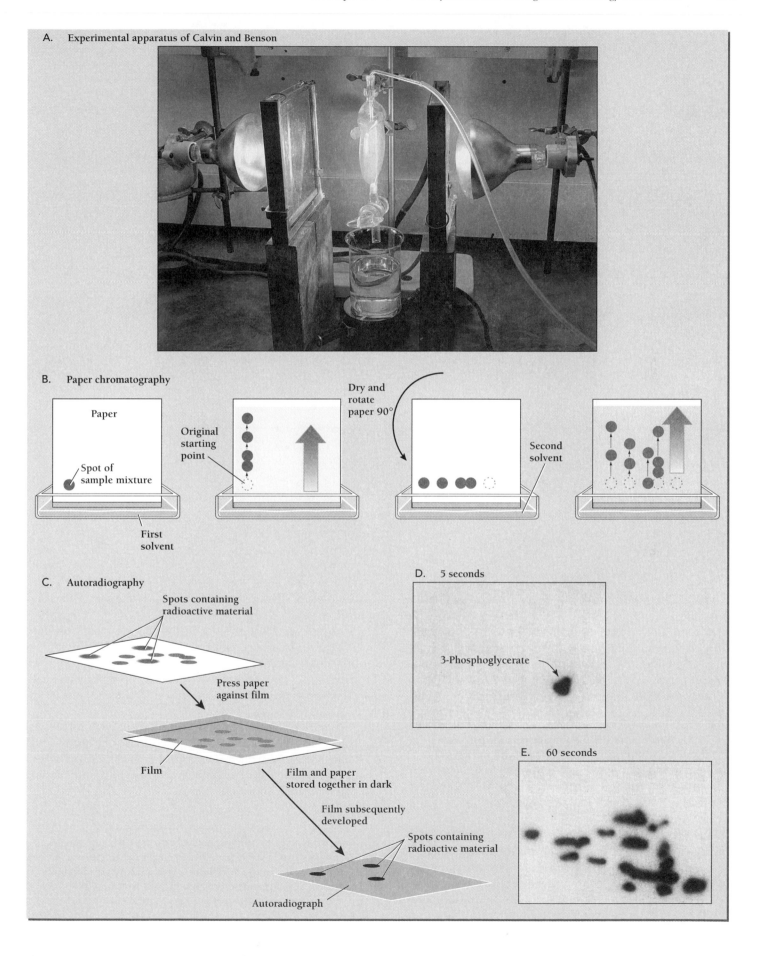

A. Experimental apparatus of Calvin and Benson

B. Paper chromatography

Paper

Spot of sample mixture

First solvent

Original starting point

Dry and rotate paper 90°

Second solvent

C. Autoradiography

Spots containing radioactive material

Press paper against film

Film

Film and paper stored together in dark

Film subsequently developed

Spots containing radioactive material

Autoradiograph

D. 5 seconds

3-Phosphoglycerate

E. 60 seconds

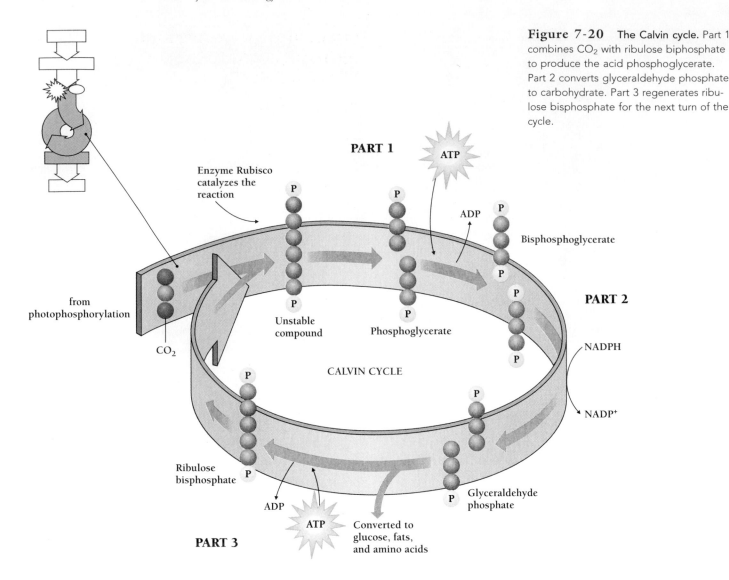

Figure 7-20 The Calvin cycle. Part 1 combines CO_2 with ribulose biphosphate to produce the acid phosphoglycerate. Part 2 converts glyceraldehyde phosphate to carbohydrate. Part 3 regenerates ribulose bisphosphate for the next turn of the cycle.

enzyme performs this job slowly, processing only about three molecules per second (Figure 7-20). This compares to 1000 per second for a typical enzyme and 10,000 per second for some champions. To compensate for this sluggishness, plants typically produce lots of Rubisco, which may make up more than 50 percent of the total protein in chloroplasts. Rubisco is probably the most abundant protein on the planet.

In the first reaction of the Calvin cycle, carbon dioxide combines with ribulose bisphosphate with the aid of the enzyme Rubisco. The resulting six-carbon molecule immediately breaks down into two molecules of phosphoglycerate.

Part 2 of the Calvin Cycle Produces Carbohydrate

The next two steps of the Calvin cycle convert phosphoglycerate into glyceraldehyde phosphate, a 3-carbon sugar and a true carbohydrate. The two reactions of part 2 are similar to those in glycolysis (though they run in reverse directions). In chloroplasts, however, the electron-carrying coenzyme is not NADH, but NADPH.

The reactions of part 2 of the Calvin cycle require one molecule of ATP and one molecule of NADPH for each molecule of phosphoglycerate. Since one molecule of carbon dioxide entering the cycle produces two molecules of phosphoglycerate, each turn of the cycle requires two molecules each of ATP and NADPH.

Some of the glyceraldehyde phosphate enters the cytoplasm, where it is converted to glucose, fats, or amino acids. Some of the glyceraldehyde phosphate remains in the chloroplast, where it regenerates ribulose bisphosphate, the task of part 3 of the Calvin cycle.

The two molecules of phosphoglycerate are each converted into a 3-carbon sugar. This sugar either is converted to glucose, fats, or amino acids (in the cytoplasm), or is used to regenerate the ribulose bisphosphate consumed in part 1 of the Calvin cycle.

Part 3 of the Calvin Cycle Regenerates Ribulose Bisphosphate

The regeneration of the carbon dioxide acceptor ribulose bisphosphate is the messy part of the Calvin cycle, involving seven different enzymatic steps. The end result is the production of the 5-carbon ribulose bisphosphate from the 3-carbon glyceraldehyde phosphate. In the last step of part 3, ATP contributes a second phosphate group to form ribulose bisphosphate.

The Calvin cycle is the heart of the light-independent reactions of photosynthesis. The Calvin cycle uses the enzyme Rubisco to synthesize 3-carbon sugars from ordinary carbon dioxide. These sugars then can be converted to glucose, fats, or amino acids or else used to regenerate ribulose bisphosphate.

How Much ATP and NADPH Is Required To Make a Molecule of Glucose?

For each molecule of glucose produced, six molecules of carbon dioxide must enter the Calvin cycle. As the cycle turns 6 times, it produces 12 molecules of 3-carbon glyceraldehyde phosphate. Of these, 2 form glucose, and the remaining 10 (which together contain 30 carbon atoms) regenerate 6 molecules of 5-carbon ribulose bisphosphate (containing the same 30 carbon atoms).

The cost of making a molecule of glucose is considerable. Each turn of the Calvin cycle requires one molecule of ribulose bisphosphate and two molecules each of ATP and NADPH. The 6 turns needed to make a molecule of glucose use 12 molecules of ATP and 12 molecules of NADPH. The production of the six molecules of ribulose bisphosphate requires an additional 6 molecules of ATP. The total cost of making a molecule of glucose from 6 molecules of carbon dioxide, then, is 18 molecules of ATP and 12 molecules of NADPH. We can also express the cost as three molecules of ATP and two of NADPH per carbon atom turned to glucose.

The energy driving the Calvin cycle comes from the high-energy bonds of NADPH and ATP, which formed during the light-dependent reactions. To make the 12 molecules of NADPH, a cell must absorb 48 photons of light by means of photosystems I and II. The same 48 photons provide energy for 12 of the needed molecules of ATP. The additional 6 molecules of ATP, generated by cyclic photophosphorylation, require another 24 photons (approximately).

For each molecule of glucose synthesized, photosynthesis consumes 18 molecules of ATP and 12 molecules of NADPH, or the energy from approximately 72 photons of light.

WHAT DETERMINES THE PRODUCTIVITY OF PHOTOSYNTHESIS?

The factors that influence plant productivity are not only scientifically interesting, but also crucially important to feeding our entire planet. We have already seen that the wavelength of light influences the efficiency of photosynthesis: light that cannot be absorbed cannot contribute to photosynthesis. (The overall efficiency of photosynthesis is about 20 percent.) The intensity of light is also important, since plants do not produce as much carbohydrate in the shade as in the light. The length of the day, the length of the growing season, and the angle at which sunlight enters the atmosphere all influence how many photons reach the surface of a leaf and, therefore, how much light energy is available to a plant. Other environmental conditions influence the amount of light reaching a leaf as well. In London, for example, pollution and fog reduce by half the total sunlight reaching the ground. In a tropical rain forest, 15 to 20 layers of vegetation may absorb and reflect more than 95 percent of the light before it reaches the ground. Only a small proportion of that light drives photosynthesis. Most is reflected back up into the sky. The rest passes through leaves or heats up the forest.

Photosynthesis also requires water and carbon dioxide. Land plants take water from the soil and transport it to their leaves. Some of this water is used in photosynthesis, but most of the water evaporates into the surrounding air, passing through the stomata in the leaves. This continuous flow of water from the roots to the leaves provides both a means of transporting nutrients from the soil to other parts of the plant and a way of regulating the plant's temperature through evaporation.

Besides acting as escape valves for water, the stomata also allow oxygen to leave the interior of the plant and carbon dioxide to enter. The various purposes of the stomata conflict with one another, however. When the weather is hot and dry, the stomata close to save water. As photosynthesis continues, oxygen levels inside the plant quickly rise and carbon dioxide levels drop. For desert plants, which must keep their stomata closed as much as possible to avoid drying out, maintaining carbon dioxide levels high enough for photosynthesis is especially challenging.

Plummeting carbon dioxide levels not only slow or halt photosynthesis, they also initiate a dangerous chemical pathway called **photorespiration,** in which plants waste much of their hard-won carbohydrate. The problem occurs because Rubisco requires a relatively high concentration of carbon dioxide to work properly. Normally in the first step of the Calvin cycle, Rubisco converts ribulose bisphosphate into two molecules of phosphoglycerate. However, when carbon dioxide levels are low compared to oxygen levels, Rubisco converts ribulose bisphosphate into one molecule each of phosphoglycerate and phosphoglycolate. As a result, plants waste as much as half

BOX 7-1

How do some herbicides kill plants?

When we think of farmers and gardeners, we imagine them nursing tender seedlings into healthy plants. But growing plants also involves killing other plants, weeds that compete with whatever a farmer or gardener is trying to grow. A weed is simply a plant that is unwelcome for any reason. The easiest way to kill a weed, or any plant, is to spray it with an herbicide, a chemical that kills plants. Farmers began using modern herbicides long before anyone understood how these chemicals hurt plants. In many cases, however, research has revealed precisely the metabolic processes in which herbicides act.

Some herbicides selectively kill different types of plants. For example, 2,4-D—a synthetic compound that mimics a plant hormone—kills C_3 plants more effectively than it kills C_4 plants. Many weeds are C_3 plants, while corn, a major U.S. crop, is a C_4 plant. Thus, farmers can apply 2,4-D to corn fields, killing the weeds, but not the corn.

Most herbicides kill all plants indiscriminately. Such herbicides act by interfering with photosynthesis or other basic processes. Two herbicides, atrazine and diuron, do so by interfering directly with electron transport in photosystem II (Figure A). Both compounds prevent the transfer of electrons from excited P_{680} to one of the cytochromes. The site of interference is a thylakoid protein, called QB, that helps transfer electrons. Atrazine and diuron bind to QB and prevent electron transfer.

Corn farmers have used atrazine and related compounds heavily, since corn contains enzymes that make it resistant to atrazine. In contrast, diuron is as toxic to corn as it is to weeds. So farmers can only use it when there is no risk of killing their crop. In recent decades, however, many species of weeds have evolved tolerance to atrazine. In other words, these weeds thrive despite being sprayed with massive amounts of atrazine.

Researchers have been able to identify the exact cause of this resistance. Atrazine-resistant weeds have an altered QB protein. In nonresistant weeds, QB becomes disabled when it binds to atrazine. But a single change in one of QB's 300 amino acids can prevent atrazine from binding, rendering the weed safe from atrazine.

Another widely used herbicide that inhibits photosynthesis is paraquat. Paraquat and related compounds accept electrons from photosystem I and transfer them directly to O_2 to produce "free radicals," highly reactive ions that destroy delicate cellular machinery. These radicals rapidly attack thylakoid membranes, destroying their ability to sustain photophosphorylation and leading to plant death. Because paraquat serves as an electron shuttle—taking electrons from photosystem I and transferring them to O_2—it continues to do damage for as long as photosystem I is operating. Unfortunately, paraquat is also toxic to humans and must be used with great caution. In humans and other animals, paraquat can potentially damage the digestive tract, the kidneys, and the lungs—thereby causing diarrhea, vomiting, and sometimes death.

Some herbicides act on other basic metabolic pathways that are unique to plants. The widely used herbicide glyphosate (Roundup), for example, interferes with the synthesis of aromatic amino acids (phenylalanine, tyrosine, and tryptophan). Unlike paraquat, however, glyphosate is not particularly toxic to animals. In animals, the synthetic pathways for the aromatic amino acids are completely different from those in plants.

A number of herbicide-producing companies are now using genetic engineering techniques to produce herbicide-resistant crop plants. For example, some crops now carry the atrazine-resistant QB protein. Chemical companies argue that such herbicide-resistant crops allow farmers to destroy weeds selectively by applying a particular herbicide. As a result, farmers do not have to plow their fields to get rid of weeds and less topsoil is lost to erosion. However, most environmentalists doubt the wisdom of this strategy, since it will ultimately increase herbicide use.

Hundreds of herbicides are used everywhere today: in fields where fruits and vegetables are grown; on crops we feed to animals, themselves bound for our dinner table; along roadsides; and on our lawns and gardens. We find herbicides useful, indeed, hard to live without. Some break down soon after they are applied and are nearly harmless to the environment in the long run. Most do not harm humans. A few are hazardous in high doses, such as those farm workers might be exposed to. Some herbicides remain in the environment for months or years, continuing to poison plants and, in some cases, animals. All plants regularly exposed to herbicides ultimately evolve resistance, forcing researchers to develop new toxins.

Figure A   Structures of some herbicides. (a, b) Atrazine and diuron block the transfer of electrons in photosystem II. (c) Paraquat transfers electrons from photosystem I to O_2 molecules, creating free radicals that destroy other delicate enzymes and other molecules. (d) Glyphosate (Roundup) blocks the synthesis of aromatic amino acids in plants but not in animals.

of the carbohydrates they produce. Photorespiration seems especially useless, since it produces no ATP or NADH. Its function is still obscure.

However, carbon dioxide inhibits photorespiration, and at high enough carbon dioxide concentrations Rubisco produces only phosphoglycerate. Some plants have evolved a special biochemical pathway that allows them to maintain high levels of carbon dioxide inside their cells, even when their stomata are closed much of the time. These plants have a unique set of enzymes that provide special carbohydrate-producing cells with adequate carbon dioxide.

In these plants, the first reaction of carbon dioxide does not produce the usual 3-carbon molecule phosphoglycerate, but the 4-carbon compound oxaloacetate (which is also part of the citric acid cycle of Chapter 6). One of the special enzymes converts oxaloacetate to another 4-carbon compound, which moves into the carbohydrate-producing cells. There it breaks down to release carbon dioxide, which immediately enters the Calvin cycle via the Rubisco reaction.

Plants that use the 3-carbon phosphoglycerate, which include most plants, are called **C₃ plants.** Plants that use the 4-carbon oxaloacetate pathway are called **C₄ plants.** C₄ plants, such as corn and sugar cane, have a distinctive anatomy and grow well in hot, dry climates. The C₄ pathway and anatomy are adaptations that allow plants to minimize water loss through the stomata, while still providing enough carbon dioxide to maximize photosynthesis and minimize photorespiration. C₄ plants waste far less sugar than C₃ plants, but C₄ plants need more ATP—and therefore more light—than C₃ plants.

Desert-adapted plants, such as cacti, use a modified C₄ pathway called *crassulacean acid metabolism,* or *CAM,* a type of photosynthesis that conserves water. CAM plants open their stomata to collect CO_2 at night only, when the air is cool and water loss is minimal. They then store the CO_2 in a 4-carbon molecule until the next day, when photosynthesis can proceed. The energy cost of CAM plants' nighttime respiration keeps their growth rates extremely low.

In hot, dry climates with plenty of light, C₄ plants predominate. In temperate zones with less light and more water, C₃ plants have the advantage. In Kentucky, for example, the humid summers allow plants to leave their stomata open much of the time, and C₃ plants (such as Kentucky bluegrass) thrive. However, many West Coast admirers of Kentucky bluegrass, which is the most popular landscaping plant in the United States, are frustrated by the selective advantage of C₄ plants. As the hot, dry summer wears on, their beautiful green lawns are taken over by ugly, yellow-green crabgrass, a vigorous C₄ plant.

The C₄ pathway represents an important adaptation both to naturally growing plants and to crops, such as sugar cane and corn. C₄ crops are especially efficient at capturing solar energy, converting as much as 8 percent of the energy falling on a field to the chemical bonds in carbohydrates. As oil and coal reserves diminish, some countries have begun to fuel their cars and trucks with alcohol produced from such efficient C₄ plants as corn and sugar cane.

Compared to C₃ plants, C₄ plants minimize water loss and photorespiration, while maximizing photosynthesis, by maintaining high concentrations of carbon dioxide inside special cells. C₄ plants are more efficient than C₃ plants, but need more light than C₃ plants.

In these first few chapters of *Asking About Life,* we have examined some of the basic chemical processes that enable cells to live. In the next few chapters, we will see how cells multiply and grow. Then we will see how sexual reproduction allows cells to recombine their genetic material in novel arrangements. We will learn how the structure of DNA was discovered and find out how genes are expressed. Finally, we will explore the ways that molecular biologists have learned to control the expression of genes and how this dizzying array of new knowledge applies to humans.

STUDY OUTLINE WITH KEY TERMS

Green plants and other photosynthetic organisms harvest the sun's energy and convert it into the chemical energy of carbohydrates. The raw materials for photosynthesis are water and carbon dioxide. The products are carbohydrates and oxygen.

Photosynthesis uses light energy to split water, generating oxygen and high-energy electrons. These electrons provide the reducing power to convert carbon dioxide to carbohydrates. Light has both wavelike and particlelike properties. Each particle of light, called a **photon,** carries a fixed amount of energy, which corresponds to its **wavelength. Visible light,** which ranges from about 400 nm to about 750 nm, is the rainbow of colors (wavelengths of light) that humans can see. Plants contain two sets of **pigments** that absorb light, the **chlorophylls** and the **carotenoids.** In plants, energy absorbed by a large number of chlorophyll and carotenoid molecules, which together form **antenna complexes,** is transferred to one of two kinds of **pho-**

tochemical reaction centers, each of which absorbs light with characteristic wavelengths. Every molecule has an **absorption spectrum** and every light-dependent process has an **action spectrum.**

Each reaction center participates in a separate photosystem, located on the thylakoid membrane of the chloroplast. P_{700} is part of **photosystem I,** and P_{680} part of **photosystem II.** The excited electrons of each photosystem are transferred to special electron acceptors, leaving the reaction center in an oxidized state. The excited electrons of each photosystem make different contributions to the harnessing of energy.

Excited electrons of photosystem I can flow back to reduce the oxidized P_{700}, via an electron transport chain that generates a proton gradient, which in turn can generate ATP. This looping flow of electrons is called **cyclic photophosphorylation.** Its end result is the regeneration of P_{700} and the production of ATP.

Alternatively, the excited electrons of photosystem I can flow down a different electron transport chain to produce NADPH, which is used directly in the synthesis of carbohydrates. This noncyclic path requires that P_{700} be reduced by an excited electron of photosystem II.

The electron transport chain of photosystem II carries an excited electron from excited P_{680} to the oxidized form of P_{700}. As the electron moves down the chain, it generates a proton gradient (between the **stroma** and the thylakoids), which generates ATP. The net result is **noncyclic photophosphorylation.**

The electron needed to regenerate reduced P_{680} comes from water. The reduction of each molecule of $NADP^+$ to NADPH requires two electrons and the absorption of four photons (two by each photosystem). At the same time, two molecules of ATP are formed.

The light-independent reactions of photosynthesis are those that produce carbohydrates by using the ATP and NADPH formed in the light-dependent reactions. The set of reactions that produce glucose from carbon dioxide, NADPH, and ATP is called the **Calvin cycle.** The production of a molecule of glucose from 6 carbon dioxide molecules requires 72 photons of light.

Several environmental factors limit the efficiency of photosynthesis. There must be sufficient supplies of light, water, and carbon dioxide. In hot, dry weather, plants must close their **stomata** to limit water loss. With the stomata closed, plants may begin **photorespiration. C_4 plants** avoid photorespiration by maintaining high levels of carbon dioxide inside the cells that make glucose. As long as carbon dioxide is limiting and there is enough light to provide the extra energy needed, C_4 plants are more efficient producers of carbohydrates than ordinary **C_3 plants.** The efficiency of photosynthesis depends on both the genetic capacity of the plant and its environmental conditions.

REVIEW AND THOUGHT QUESTIONS

Review Questions

1. What is photosynthesis? What material does it synthesize? Where in a cell does this process occur?
2. Write the overall reaction for photosynthesis. How do the reactants get into the leaf and into the cells? How does oxygen get out of the leaf cells and back into the air outside the plant?
3. Which parts of the electromagnetic spectrum do plants use in photosynthesis? How does a comparison of the absorption spectrum of chlorophyll and the action spectrum of photosynthesis lead to a conclusion about the molecular basis of photosynthesis? How did such studies lead to the discovery of the role of carotenoids?
4. Describe the cyclic photophosphorylation associated with photosystem I. What is the relationship between the absorption of light and the excitation of electrons? How does the transfer of electrons lead to the synthesis of ATP?
5. Describe the reactions that make up the Calvin cycle. What enzyme captures carbon dioxide from the atmosphere? What are the products of this reaction? How is the energy from light introduced in the Calvin cycle? Why must the cycle produce more ribulose bisphosphate?
6. Under what environmental conditions might a plant close its stomata? Compare the rates at which plants can carry out photosynthesis when their stomata are open and when they are closed. Why should the opening of stomata affect the rate?

7. Explain how photorespiration causes C_3 plants to waste some of their photosynthetic efforts under conditions of low carbon dioxide concentration.
8. Describe Priestley's experiment that showed that a plant could restore the air in a bell jar. Make a sketch of a bell jar containing only a mouse, and explain why the mouse dies. Then sketch a bell jar containing both a mouse and a plant, and show by arrows how oxygen is produced and consumed.
9. What advantages do C_4 plants have over C_3 plants? Why don't C_4 plants take over everywhere? What limits them?

Thought Questions

10. What prevents the roots of a plant from carrying out photosynthesis? By what metabolic process do you suppose roots extract energy from their available resources?
11. Describe in a few sentences the concept of how light energy is used to drive the reaction of glucose synthesis. (Be careful to give the outline of the story, so that a nonbiologist could understand what you mean. Don't just list a lot of details.)
12. Why is the process of photosynthesis important to feeding our entire planet? Can you name any living things that do not depend on plants (directly or indirectly)?

SELECTED READINGS

Daviss, B., "Going for the Green," *Discover,* 13:20 February 1992.
Galston, Arthur W., *Life Processes of Plants,* Scientific American Library Series, 49, distributed by W.H. Freeman, New York, 1994.
Youvan, D., and B. Marrs, "Molecular Mechanisms of Photosynthesis," *Scientific American,* 256:42–50, 1987.

▶ On-line materials relating to this chapter are on the World Wide Web at http://www.saunderscollege.com/lifesci
Click on Tobin/Dusheck: *Asking About Life.*

PART II

Genetics:
The Continuity
of Life

Ponderosa pine cones. *(Art Wolfe APR94)*

8

CELL REPRODUCTION

The Unfortunate Henrietta Lacks

Early in 1951, an anxious 31-year-old woman checked into the medical clinic at Johns Hopkins University in Baltimore (Figure 8-1). Her anxiety was confirmed when her doctor found a one-inch purple growth on her cervix, the neck-shaped entrance to the womb. He told her he would have to do a biopsy, and he cut out a small piece of the growth and sent it to a pathologist, a medical specialist who can tell, among other things, whether a human cell is cancerous or not. The report came back positive. Henrietta Lacks's growth was dangerously malignant. Its cells were dividing rapidly, and, without treatment, the tumor would kill her.

The news came as a blow. She and her husband had two children to raise. She was too young, she thought, for so serious a disease, one that does not usually afflict women until their 40s or 50s. Within days, the doctors at Johns Hopkins began treating her with high doses of radiation, in what was to be a futile effort to save her life. In the autumn of 1951, the best part of Henrietta Lacks died. Ironically, her tumor cells survived and live today in laboratories all over the world.

The day of her biopsy, Lacks's doctor took a piece of her tumor to a group of researchers who were struggling to grow human cells in plastic dishes containing a mixture of nutrients and blood products. They tried adding different nutrients or changing the pH of the mixtures. No one had yet succeeded. Most human cells died immediately. A few divided several times, then died. Still, the researchers hoped that any day they would find the right cells or the right conditions for growing human cells in the laboratory.

When Henrietta Lacks's cells arrived in the lab, they were different from anything anyone had ever seen before. They grew without restraint, dividing and redividing every 24 hours. The same virulence that would finally kill Henrietta Lacks made these cells unstoppable in the laboratory. Before Lacks's life was over, her cells took up an independent existence, thanks to the work of medical researchers.

Although Henrietta Lacks's cells, named HeLa cells, were the first human cells to be successfully cultured, hundreds of others soon followed. Unlike her cells, most of these were difficult to culture. In general, human cells die easily outside the body. A few stray cells on the side of a glass flask, or in the fluid left in a pipette, expire almost immediately. Beginning a

culture of human cells in a plastic culture dish requires care and attention. It isn't something that normally happens accidentally.

thought—but on Henrietta Lacks's cervical cells. The story was the same for research on prostate cells, kidney cells, liver cells, and more.

The same virulence that would finally kill Henrietta Lacks made these cells unstoppable in the laboratory.

Still, in the 1950s and 1960s, growing human cell cultures was exciting work. For the first time, scientists could study human biology without experimenting on humans themselves. They could study the way human cells divide, the way human cells move, the way cancer cells differ from normal cells. The possibilities seemed endless. Every week a lab somewhere developed a new kind of human cell to culture. These new cell lines were rapidly distributed to other labs—some down the hall, others on the other side of the globe.

In the early 1970s, odd things began to happen. The National Institutes of Health (NIH), in Washington DC, sent five kinds of human cells, which had come from five different research labs in the Soviet Union, to a lab in Berkeley. There, an alert researcher named Walter Nelson-Rees discovered that the five supposedly different cell lines were all HeLa cells.

Soon HeLa cells began to turn up everywhere. Nelson-Rees began checking every cell line he knew of. He published lists of cell cultures that were not what they were supposed to be. By 1981, he had found 90 different cell lines, each of which was HeLa cells. Researchers studying the biology of breast cells, both normal and cancerous, were horrified to find that their years of research had not been done on breast cells—as they had

The repercussions were enormous. Years of research by scores of researchers had become meaningless. Much of the biology of cancer had to be rethought. For years researchers had believed that all cancer cells have the same nutritional needs and all cancer cells have abnormal chromosomes, the so-called mark of cancer. Then, as one cell line after another proved to be not prostate cells, not kidney cells, not this and not that, but HeLa cells, biologists came to realize that, in fact, cancer cells are not all alike. The only cancer cells that were all alike were HeLa cells. The abnormal chromosomes were typical of HeLa cells, but not of all cancer cells.

But how had it happened? The answer was that Henrietta Lacks's cancer cells were unbelievably vigorous. If so much as a single one of her cells fell by accident onto a culture dish, her cells could take over, no matter what other cells were living in the dish. One group of researchers found that HeLa cells could float through the air inside the tiny bits of spray created when a rubber stopper was pulled from a bottle containing HeLa cells. The floating cells could then drop down into nearby culture dishes. Within a week, a culture of prostate cells was taken over by cervical cells. It was a scenario that scientists had never imagined.

Figure 8-1 Young Henrietta Lacks died of cancer in 1951, but biologists have kept her cancer cells alive ever since. In the 1950s, her cells, called HeLa cells, were used to develop a vaccine against polio. Today, genetic engineers insert genes for valuable human proteins into HeLa cells, which then synthesize the proteins. *(Science Photo Library/Photo Researchers)*

Today, cell biologists have learned to be more cautious. They work with cell cultures whose ancestry they've checked. And, for the most part, researchers work with just one cell line at a time.

But HeLa cells are still used, and Lacks's cells live on. All over the world her cells continue to divide relentlessly day after day. In the 1950s, her cells were used as hosts for growing polio virus, enabling researchers to develop the vaccine against polio. Today, genetic engineers insert genes for valuable human proteins into HeLa cells, which then synthesize the proteins. Henrietta Lacks died before she could finish raising her own two children, yet her cells have been instrumental in basic research and have helped saved the lives of millions of children all over the world.

BOX 8-1

Cervical cancer and the Pap smear

In the 1930s, more women died of cervical cancer in the United States than of any other form of cancer. Today, the rate of cervical cancer is less than one-tenth that of breast cancer and declining rapidly. Even more encouraging, the overall survival rate (more than two-thirds) is one of the highest of all cancers, thanks to a diagnostic test invented by George N. Papanicolaou in 1943.

In the last 50 years, the widespread use of the Pap test, or Pap smear (named after its inventor), has transformed a diagnosis of cervical cancer from a virtual death sentence to a highly survivable disease. The survival rate for cervical cancer would be even higher if more women were tested regularly.

The test consists of a simple office procedure in which a doctor or nurse scrapes a few living cells from the surface of the cervix. The scraping may hurt very slightly, but the life-saving test is well worth the second or two of discomfort. Once the cells have been collected from the cervix, they are sent to a laboratory pathologist for analysis.

In most forms of cervical cancer, cells undergo changes in appearance years before they become cancerous. Such "precancerous" cells have distinctive characteristics that a good pathologist can easily recognize (Figure A). The cervical cells are usually classified into five grades: normal, atypical, low-grade precancerous lesion, high-grade precancerous lesion, and invasive cancer. In the great majority of cases, all the cells collected will be normal. Occasionally, some of the cells will be "atypical," usually because of some temporary infection of the cervix or vagina.

Women with a low-grade or high-grade lesion are immediately tested for any sort of growth. Any precancerous growths that are found are removed from the cervix, using cryosurgery (freezing) or lasers. After treatment, a woman is normally cancer-free and able to bear children if she wishes.

If the Pap smear reveals cancerous cells, as opposed to precancerous ones, the recommended treatment is a hysterectomy (the removal of the uterus). If the cancer has spread, the surgeon will also remove the upper part of the vagina. If the cancer is caught early, when it is still localized on the cervix, the survival rate is 91 percent. If the cancer is diagnosed late, after it has spread beyond the reproductive organs, the survival rate is less than 9 percent.

Cervical cancer strikes women of all ages and ethnic groups. African American women, however, are twice as likely as white women to die of the disease. Researchers attribute the difference to lack of good medical care, especially regular Pap tests. Fortunately, increasing numbers of African American women are receiving the Pap test and the difference in survival is therefore decreasing.

Because most cervical cancer develops slowly, it is quite easy to diagnose the disease while it is still in one of the precancerous stages. Public health experts recommend that all women have annual Pap smears starting at age 18 or at the onset of sexual activity, whichever occurs first. After three or more negative tests, a women is safe in having an exam every other year. Women who are at risk for cervical cancer should continue to have the exam every year.

The main risk factor for cervical cancer is infection with human papilloma virus (HPV), also known as genital warts. In 93 percent of all cases of both cervical cancer and precancerous lesions, HPV DNA is present. Indeed, the HP virus incorporates two genes into the genomes of previously healthy cervical cells. These two genes encode proteins that cause the cells to form tumors. In Chapter 12, we will learn more about how viral DNA infects cells and how tumor viruses cause these cells to form tumors.

Because HPV is a sexually transmitted disease (STD), the best ways to avoid infection are the same as for AIDS and other STDs: abstinence or else barrier methods of birth control (condom or diaphragm) and fewer than five sexual partners over a lifetime.

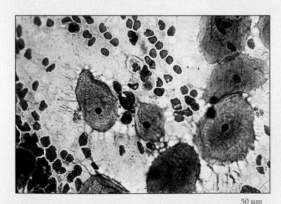
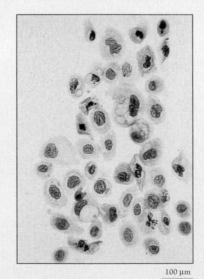
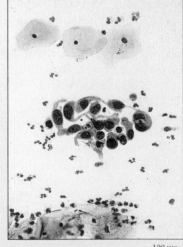

50 μm

100 μm

100 μm

Figure A Normal cervical cells (*left*) are distinctly different from precancerous (*middle*) and cancerous (*right*) cervical cells. Note the sperm-shaped cells that characterize invasive cancer cells. (© *Biology Media/ Photo Researchers; Science Photo Library/Photo Researchers; M. Rotker/Photo Researchers*)

KEY CONCEPTS

1. Cell reproduction occurs in an orderly sequence of events called the cell cycle.

2. Cell reproduction involves three distinct but interconnected processes: the duplication of genetic information in the nucleus, its equal distribution to each of two daughter cells, and the division of the cytoplasm.

3. In eukaryotic cells, DNA forms compact complexes with proteins. During mitosis, these complexes become even more compact and are visible as distinct chromosomes.

4. During mitosis, microtubules and other subcellular components bring about chromosome movements that distribute one copy of each chromosome to each daughter cell.

5. Mitosis ensures that each new cell inherits a complete set of chromosomes.

6. Eukaryotic cells, ranging from yeast to human cells, use similar molecular mechanisms to regulate passage through the cell cycle.

7. Interactions with other cells and extracellular molecules regulate cell division in multicellular organisms.

8. Cancer results when the molecular mechanisms that regulate cell division go awry.

CELL DIVISION

Cells are the fundamental living units of all organisms. An organism may consist of a single cell or many cells. Bacteria such as *Lactobacillus,* yeasts such as *Candida albicans,* and protists such as *Paramecium* are all single-celled organisms. An adult human contains about ten trillion (10^{13}) cells. Because all the cells in a multicellular organism originate from a single cell, the continuation of life ultimately depends on cell reproduction.

How Do Cells Divide?

In the five minutes or so that it will take you to read this page, your body will produce a billion (10^9) new cells. Most of these replace battered and dying cells of your skin, intestines, and blood. Each newly made cell contains both a copy of the genetic information you inherited from your parents and the molecular machinery and materials needed to interpret that information, including membranes, organelles, macromolecules, and small molecules.

Cell reproduction is so common that it is easy to forget how extraordinary it is. Our wonder at this copying process may be heightened by our everyday experience with nonbiological copying—the recording of music, the filming and taping of movies and television programs, and the printing of books. In each case, complicated machinery carries out a complex but ordered process. But cell reproduction is even more remarkable than these other processes, since the cell is both the thing being copied and its own copying machine.

Cell reproduction has different results in single-celled and multicelled organisms. Single-celled organisms divide to create new individuals. Cell division in yeast, for example, leads directly to an increase in the number of individuals. In a multicellular organism such as a frog the same kind of cell reproduction allows one individual to grow from a single cell into a whole organism. Cell reproduction also allows a multicellular organism to replace dying cells. A dandelion root can regenerate its leaves and flowers after they are cut away by a lawn mower. In the same way, children quickly replace the skin cells that they scrape and tear off in their inevitable bangs and falls (Figure 8-2). Before cells divide, they must first duplicate their contents, including their DNA. Eukaryotic cells must also duplicate their internal membranes and organelles.

An altogether different kind of cell division also occurs in sexually reproducing organisms. Plants, animals, and other sexually reproducing organisms make sperm and eggs (or their equivalents) through an elaborate series of chromosomal divisions and fusions. We discuss this process, called *meiosis,* in Chapter 9. In this chapter, we examine simple cell division, the kind that bacteria use to reproduce. Humans and other multicelled organisms use the same process to grow and to replace the cells of tissues such as leaves, bark, muscle, and hair. In simple cell division, the cell duplicates its DNA, then divides it in a process called mitosis.

Both single-celled organisms and multicellular organisms make more cells by dividing. Just before they divide, cells double their DNA and then divide it in a process called mitosis.

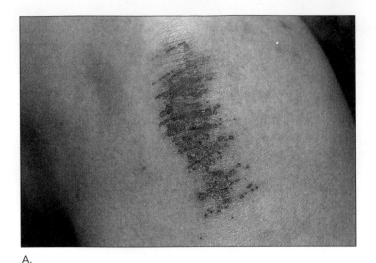

A.

B.

Figure 8-2   Cell reproduction enables organisms to replace cells. A. When cells are destroyed in a wound, such as this scrape on a child's knee, nearby cells divide in two and then divide again and again until the gap is filled. B. Within days, the wound is completely healed, and the child is back at play. *(A, Christian Grzimek/OKAPIA/Photo Researchers; B, Arthur Tilley/FPG)*

What Did Early Biologists Discover About Chromosomes?

In the 19th century, the cell theory established that all organisms are made of cells and all cells come from other cells. But the cell theory suggested a problem. How does a cell know how to make another cell just like itself? How does the cell pass on the knowledge of what kind of cell it is? How, for example, does a cell from Koko the gorilla know how to make another Koko cell, a cell unlike that found in any other gorilla and unlike that found in other primates such as your classmates?

The answer, we now know, is in the **chromosomes** [Greek, *chroma* = color + *soma* = body]—the threads of DNA and protein that carry the genes. Yet, 19th-century microscopists saw chromosomes and examined them in great detail without recognizing that they were the means by which organisms inherit traits. Early biologists held the key to the problem of inheritance without knowing it.

In the late 19th century, improvements in the lenses of microscopes for the first time allowed microscopists to see the tiny structures within cells. Some of the most important advances, however, came from the use of natural dyes developed for the burgeoning 19th-century textile industry. The two most useful dyes, still in use today, were the purple-and-blue dye hematoxylin (which comes from the logwood tree of Central America) and the synthetic red dye eosin.

Thanks to these dyes, microscopists of the 1870s were able to describe in detail the threadlike chromosomes in a variety of animal and plant cells. In fact, during one 4-year period, researchers published over 200 papers on the complex dance of the chromosomes during cell division. By the 1880s, biologists had discovered that cells reproduce themselves in a process called **cell division**, in which a parent cell gives rise to two **daughter cells**. Cells stained with hematoxylin and eosin showed the cell's nucleus to be full of a grainy material called **chromatin** [Greek, *chroma* = color] (Figure 8-3). Biologists also knew that when a cell divides, its nucleus divides, providing one-half of its material to each daughter nucleus.

The German physician and microscopist Walther Flemming pieced together a detailed description of the process by which the chromosomes are equally divided and distributed to

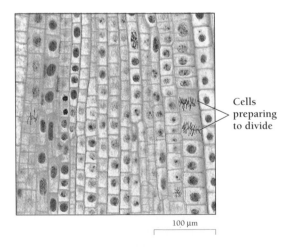

Cells preparing to divide

100 µm

Figure 8-3   Onion cells reproduce by dividing in two. Cells stained with a dye that binds to DNA show the DNA in the cell's nucleus. In some cells, the DNA stains as a diffuse grainy material called chromatin. In dividing cells, the chromatin condenses into distinct chromosomes. Chromatin is present in nearly all eukaryotic cells, while chromosomes are visible only in dividing cells. *(Dwight Kuhn)*

the two new cells during cell division. Flemming named this elaborate division of the chromosomes **mitosis** [Greek, *mitos* = thread].

One of the first events that Flemming noted was the gathering of the chromatin into visible threads (later called chromosomes). He also observed in salamander cells the duplication of an object that he called an **aster** [Latin, = star]—the center of a finer set of threads just outside the nucleus. During mitosis, the two asters migrate to opposite ends of the cell. While the asters are migrating, the chromosomes line up in a row in the middle of the cell, and then each one splits into two. Each of the two half-chromosomes then migrates toward one of the asters. Finally, the chromosomes disappear, and the nuclei regain their grainy appearance.

Late 19th-century microscopy revealed all the major details of cell division in eukaryotes without explaining the role of the nucleus or chromosomes in the life of the cell.

How Did Biologists Discover the Function of the Chromosomes?

But what was the point of this careful division? For the moment, no one knew. The nucleus of the cell had been recognized as a standard component of cells as early as the 1830s, but its function was elusive. During cell division, the nuclear membrane breaks down, and the nucleus seems to disappear. Because the nucleus disappears, biologists did not at first consider that it might play an important role in cell division. Many biologists assumed that the cell constructs a new nucleus after each cell division.

The first hint that the nucleus might be important came with studies of fertilization, the process by which a sperm and egg fuse to form a *zygote*. In the 1870s, several microscopists noticed that the sperm is little more than a nucleus with a tail to propel it. Further studies showed that a zygote forms from exactly one sperm and one egg and that the sperm nucleus and the egg nucleus fuse to form a single nucleus. In time, the zygote develops into an embryo by dividing into more and more cells.

By 1880, biologists knew that all the nuclei of a developing organism are descended from a single nucleus created by the fusion of the nuclei of the egg and the sperm. This fact did not, however, suggest to biologists that the nucleus contained the hereditary material. First, 19th-century biologists were not necessarily looking for a hereditary material. Many biologists of the time believed that the process of fertilization did not transmit a material substance, but merely provided some form of energy or force that triggered the development of the embryo.

In addition, biologists of the time were unaware that the material of the chromosomes carries information. Biologists assumed that the individual chromosomes in a cell are all alike, no more different from one another than are individual pret-

zels. Although they might look different from one another, dividing them would be similar to dividing a bag of pretzels between two people. It wouldn't matter who got which pretzel, as long as each person got half.

Flemming and others had demonstrated the presence of individual chromosomes. For example, we now know that every human cell has the same set of 23 pairs of chromosomes. Each member of a pair has a distinctive and unique shape and size. Flemming saw the uniqueness of chromosomes in other organisms and described how each chromosome splits down its length, one-half "predestined for one new cell and [one-half] for the other." If we divided a bag of pretzels in half, however, we would not cut each pretzel down its whole length. We would just give each person half the pretzels in the bag.

It was the great embryologist Wilhelm Roux who first observed that the complex way the chromosomes split during mitosis strongly suggested that the chromosomes were not uniform at all. Why, he asked, would the cell use such a complex process to divide the nucleus, dividing each chromosome neatly in half, if simply cutting the nucleus in two would work? The chromosomes in the nucleus, argued Roux, must contain material of different kinds and the cell must be carefully dividing the material so that each daughter cell receives some of each.

In 1889, August Weismann insisted that the nucleus contained the hereditary material, which he called the *germ plasm*. "Heredity," Weismann wrote, "is brought about by the transference from one generation to another, of a substance with a definite . . . molecular constitution." But Weismann was known for his radical ideas, and most biologists of the time dismissed Weismann's claim.

In the same year, the German cell biologist Theodor Boveri performed an unusual experiment that lent firm support to the idea that the nucleus can determine the characteristics of an organism. Boveri discovered that, by vigorously shaking the eggs of a sea urchin until they broke open, he could knock the nucleus right out of the eggs. He then fertilized these damaged "enucleated" eggs with sperm from a different species of sea urchin. Unlike normal zygotes, these had no genetic material from the mother (Figure 8-4).

Miraculously, these zygotes developed into normal sea urchins. But the resulting adult sea urchins did not look like the species from which the eggs came. Instead they looked like the species from which the sperm came. In a control experiment, Boveri used sperm that belonged to the same species as the eggs. The resulting sea urchins looked like the species from which the eggs and sperm both came. Since the sperm contributes virtually nothing but a nucleus, here was dramatic evidence that the nucleus by itself can determine the characteristics of the organism. In 1891, Boveri wrote, "in all cells derived . . . from the fertilized egg, one half of the chromosomes are of strictly paternal origin, the other half of maternal."

But Boveri's experiment left hardly a ripple in the scientific community. As we will see in the next chapter, evidence that was even more persuasive failed to persuade many biologists that the chromosomes are the hereditary material. And, as recently as the 1930s, a few respected biologists still rejected the notion that the chromosome is the vehicle of inheritance.

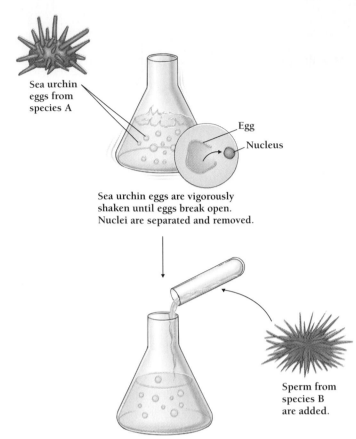

Sea urchin eggs from species A

Egg

Nucleus

Sea urchin eggs are vigorously shaken until eggs break open. Nuclei are separated and removed.

Sperm from species B are added.

Larva

Zygotes develop into normal sea urchins resembling species B.

Figure 8-4 The nucleus determines the characteristics of an organism. In an experiment in 1889, Theodor Boveri shook sea urchin eggs with such force that the nuclei fell out. "Enucleated" eggs fertilized with sperm from a different species of sea urchin developed into individuals like the species from which the sperm came. Boveri concluded that a nucleus (in this case, the sperm's) can determine the characteristics of the organism.

Today we have much more information about both chromosomes and the nucleus. In later chapters, we will see how biologists learned more about chromosomes, nuclei, and heredity. For now, it is important to understand that heredity does indeed have a material basis, and that the material is DNA. We will see that each daughter cell carries genetic information inherited from the parent cell, and that the genetic material de-

termines how the organism interacts with its environment. In all organisms (and all cells) the genetic material is DNA; the sequence of nucleotides in DNA contains coded information about cellular molecules.

Despite detailed descriptions of the division of chromosomes during mitosis, 19th-century biologists did not realize that the chromosomes carry hereditary information.

HOW DOES A DIVIDING CELL ENSURE THAT EACH DAUGHTER CELL RECEIVES AN EXACT COPY OF THE PARENT CELL'S DNA?

One of the main tasks of cell reproduction is to **replicate**, or copy, the cell's DNA (Chapter 10). After the DNA replicates, the cell's machinery distributes one copy to each of the two daughter cells. When a single molecule of DNA is involved, as in most bacteria, equal distribution is relatively simple. In eukaryotes, where several DNA molecules are involved, the distribution process is more complex. Cell division also splits the rest of the cell's contents between the two daughter cells. Cell reproduction, then, accomplishes three tasks: the replication of DNA, the equal distribution of the DNA to the two daughter cells, and the division of the other cell components.

We can describe the process of cell reproduction in terms of three M's—materials (the small molecules that carry energy or serve as building blocks), machinery (the organelles and macromolecular structures needed to carry on cellular processes), and memory (the information, contained in DNA, for building cellular machinery from the building blocks). Before a cell divides, it usually doubles its materials, machinery, and memory. In most cases, each daughter cell then receives an equal share of each.

Each of the three M's is essential to cell reproduction. DNA, for example, contains the information needed to produce the next generation of cells and organisms. Without the memory encoded in DNA, the cell's materials and machinery are useless. On the other hand, without the cell's materials and machinery, the DNA has no environment in which to express itself.

The information in the DNA comes from both parents. But the materials and machinery come exclusively from the mother. The precise doubling of DNA employs some of the cell's most elaborate molecular mechanisms. We discuss these mechanisms at length in Chapter 10. In this chapter, we see how dividing cells coordinate the duplication and distribution of DNA and cytoplasm.

Each DNA molecule consists of two polynucleotide chains. After the DNA replicates, each daughter DNA molecule contains a newly synthesized chain and one of the parent chains. The cell's machinery then must distribute one copy to each of the two daughter cells. It is this process of mitosis that drew so much attention to the chromosomes in the 1870s and 1880s.

How Do Prokaryotic Cells Divide?

Cell division in prokaryotes is fairly straightforward. Prokaryotes divide by **binary fission**, in which a cell pinches in two, distributing its materials and molecular machinery more or less evenly to the two daughter cells. Since prokaryotes have no membrane-enclosed nucleus, binary fission is simpler than mitosis. In most prokaryotes, genetic information is stored in a single molecule of DNA, which usually forms a closed loop (Figure 8-5A). This loop is referred to as a "circle" of DNA, even though it looks like a true circle only in a diagram. Before the prokaryotic cell divides, it duplicates its DNA. Each of the two resulting daughter DNA molecules then attaches to the

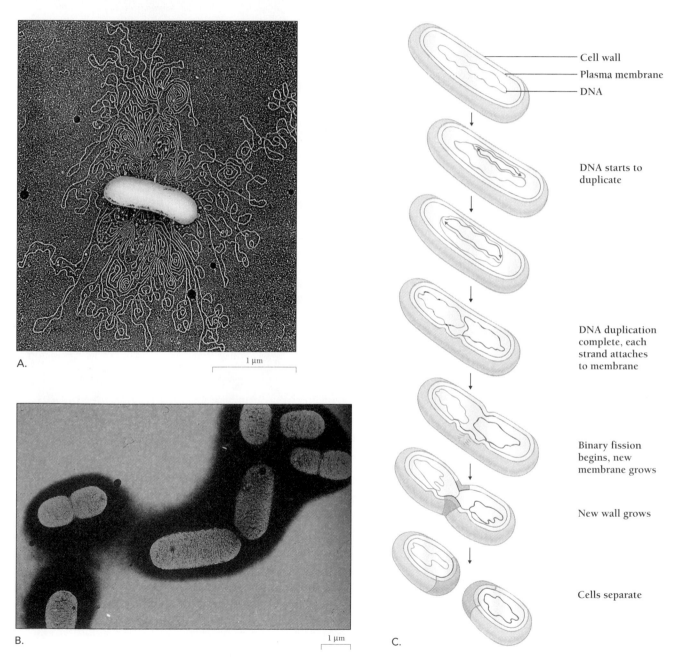

A.

1 μm

B.

1 μm

C.

Cell wall
Plasma membrane
DNA

DNA starts to duplicate

DNA duplication complete, each strand attaches to membrane

Binary fission begins, new membrane grows

New wall grows

Cells separate

Figure 8-5   Prokaryote cells divide by pinching in two. A. In *E. coli* and most other prokaryotes, genetic information is stored in a circular loop of DNA. Here circular DNA is spilling out in loops from a broken *E. coli* bacterium. B and C. The cell copies its DNA, forming two loops, each of which attaches to folds in the cell membrane. As the cell divides, the two DNA molecules are pulled apart by their attachments to the membrane, so that each daughter cell receives a single molecule of DNA.   *(A, Dr. Gopal Murti/Science Photo Library/Photo Researchers; B, K.G. Murti/Visuals Unlimited)*

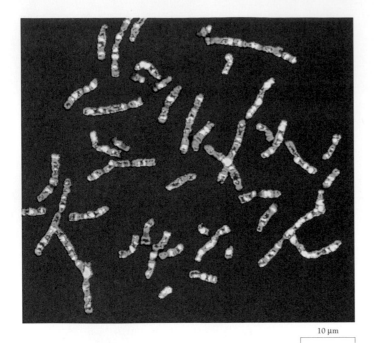

10 µm

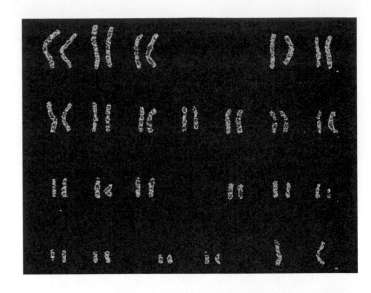

A.

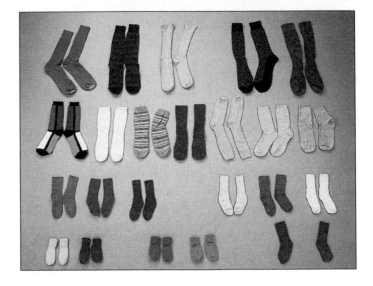

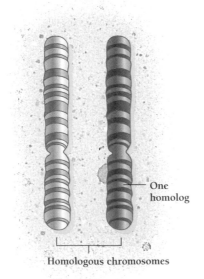

One
homolog

Homologous chromosomes

B.

Figure 8-6 The 46 chromosomes in the human karyotype come in pairs, like socks. A. If we take a photograph of a set of chromosomes, then cut out the individual chromosomes and arrange them in pairs of decreasing size, we create a tidy picture of all 46 chromosomes called a karyotype. B. Each member of a pair is similar, or homologous, but not identical to its mate. In each chromosome, the DNA loops and clusters in characteristic ways to form a pattern of bands that becomes visible in a microscope when the bands are stained. Biologists can detect abnormal chromosomes by comparing the banding patterns of abnormal chromosomes with those of normal chromosomes. *(A, (top) Custom Medical Stock, (bottom) Paraskevas Photography)*

membrane fold at which binary fission begins (Figure 8-5B and C). As the cell divides, the two DNA molecules are pulled apart by their attachments to the plasma membrane, so that each daughter cell receives a single molecule of DNA.

In prokaryotes, division of the genetic material is distinct from mitosis in eukaryotes.

How Do Eukaryotic Cells Divide?

The internal membranes and compartments of eukaryotic cells make the process of cell reproduction far more complicated than in prokaryotes. Just the existence of a nucleus poses a problem: DNA cannot be easily distributed to daughter cells as long as it is surrounded by the nuclear membrane. Further, a eukaryotic cell may contain more than 1000 times as much DNA as a prokaryotic cell—far more than any single circle of DNA could include.

We can observe the doubling and distribution of DNA if we stain dividing cells with a dye that binds to DNA. Figure 8-3 shows individual cells from an onion root in various stages of division. We can see that, in some cells, the material of the nucleus has become organized into chromosomes. The cells of each species contain a characteristic number of chromosomes: onions have 16, humans have 46, and the fruit fly *Drosophila melanogaster* has 8. (Don't conclude, however, that 46 chromosomes make you smarter than a fruit fly or an onion. Your baked potato had 48 chromosomes in each cell before it went into the oven.)

Nearly every cell in the human body has copies of the same 46 chromosomes. (One exception is the red blood cell; as a red cell develops, it sheds its nucleus and ends up with no chromosomes at all.) The chromosomes of most sexually reproducing eukaryotes come in pairs, like socks (Figure 8-6). Humans, for example, have 23 pairs of chromosomes, whereas fruit flies have 4 pairs. Each member of the pair is similar but not identical to its mate. Pairs of matching chromosomes are called **homologous chromosomes** [Latin, *homo* = same], and each member of the pair is called a **homolog**.

Everything that happens to a cell from the time that it first forms until it divides is part of the **cell cycle**—the orderly sequence of events that accomplish cell reproduction. The time from one division to the next is one cycle. Biologists distinguish three major stages in the cell cycle:

1. **Mitosis,** the division of the nucleus.
2. **Cytokinesis,** the division of the cytoplasm and formation of two separate plasma membranes.
3. **Interphase,** the time when the DNA replicates and when the chromosomes are not condensed and the cytoplasm is not dividing.

The amount of time that a single cell cycle takes depends on the organism and on its circumstances. Most plant and animal cells are capable of dividing in about one day, with mito-

sis and cytokinesis occupying one or two hours. But every cell is different. For example, a neuron spends most of its life in interphase, while rapidly dividing cancer cells spend very little time in interphase. The replication of DNA during interphase is invisible, and nothing appears to be happening in the cell. Only when biologists recognized the importance of DNA replication did they realize that what happens during interphase is just as important as what happens during mitosis and cytokinesis.

The amount of DNA in a cell provides a useful way of dividing the cell cycle into different phases (Figure 8-7). The most visually dramatic phase is **M**, for *m*itosis, which includes both mitosis and cytokinesis, (the actual division of the cell). After cytokinesis come the three parts of interphase: G_1, the gap, or growth phase, between the completion of M and the beginning of DNA synthesis; **S**, the period of DNA synthesis, when a cell doubles its DNA content; and G_2, the gap between the completion of DNA synthesis and the beginning of M (of the next cell cycle). During G_1, a cell accumulates the materials and machinery needed for the DNA synthesis that will occur during S. In G_2, the cell prepares for mitosis and cytokinesis, assembling the molecular machinery needed to sort the chromosomes and divide the cell.

With each new phase of the cell cycle, the characteristics of the chromosomes change. At the beginning of mitosis, when the chromosomes first become visible, each chromosome consists of two separate but connected bodies, called **sister chromatids** (Figure 8-8). The two chromatids are joined at a single point called the **centromere.** Both the chromatids and the centromere consist of molecules of DNA joined to various proteins. The two sister chromatids that make up a single chromosome are duplicate copies with exactly the same genetic information. In contrast, homologs are similar but different. During mitosis, the sister chromatids within each chromosome separate and go to opposite poles of the cell.

During cytokinesis each chromosome consists of a single chromatid, which gradually unfolds into chromatin and disappears. All the chromosomes remain unchanged throughout G_1, and we cannot see them in a microscope. During S, each chromosome duplicates, a process that takes 8 to 10 hours in most eukaryotic cells. The amount of DNA also doubles, so that each chromosome again consists of two (still invisible) sister chromatids. During G_2, the paired chromatids remain invisible. Only at the beginning of mitosis do the paired sister chromatids again become visible.

Researchers cannot tell whether a cell is in S, G_1, or G_2 by looking. They can sometimes tell what phase a cell is in, however, by measuring the amount of DNA in each cell. One way to measure the DNA content of individual cells is to treat cells with a fluorescent dye that binds to DNA. The amount of fluorescence from each cell is a measure of its DNA content and so of its phase in the cell cycle (Figure 8-8). To see if a cell is in S, researchers supply radioactively labeled precursors called nucleosides. Cells in S take up these radioactive nucleosides (to make new DNA), while cells in G_1 or G_2 do not.

(*Text continued on page 192.*)

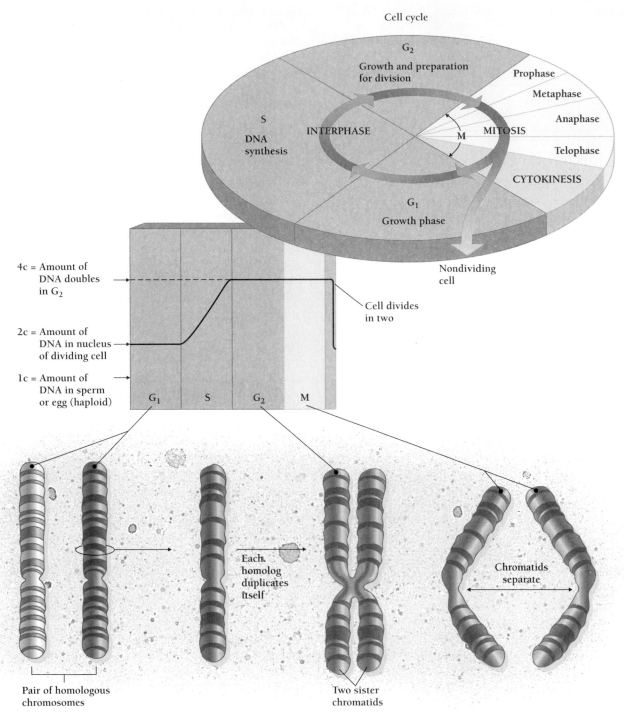

Cell cycle

G_2

Growth and preparation
for division

Prophase

Metaphase

Anaphase

S

DNA
synthesis

INTERPHASE

M

MITOSIS

Telophase

CYTOKINESIS

G_1

Growth phase

Nondividing
cell

4c = Amount of
DNA doubles
in G_2

2c = Amount of
DNA in nucleus
of dividing cell

1c = Amount of
DNA in sperm
or egg (haploid)

Cell divides
in two

G_1 S G_2 M

Each
homolog
duplicates
itself

Chromatids
separate

Pair of homologous
chromosomes

Two sister
chromatids

Figure 8-7   The cell cycle consists of G_1, S, G_2, mitosis, and cytokinesis. After the first
growth phase (G_1), the DNA doubles (S). The cell then increases materials (G_2) in preparation
for the separation of identical sister chromatids (mitosis) and the separation of the two
daughter cells (cytokinesis).

Figure 8-8   The stages of mitosis appear clearly in microscopic ▶
images of cells whose chromosomes have been stained. Look for
the mitotic spindle, sister chromatids, centriole, kinetochores, centromere, kinetochore microtubules, polar microtubules, metaphase
plate, and asters.   *(B, M. Abbey/Photo Researchers)*

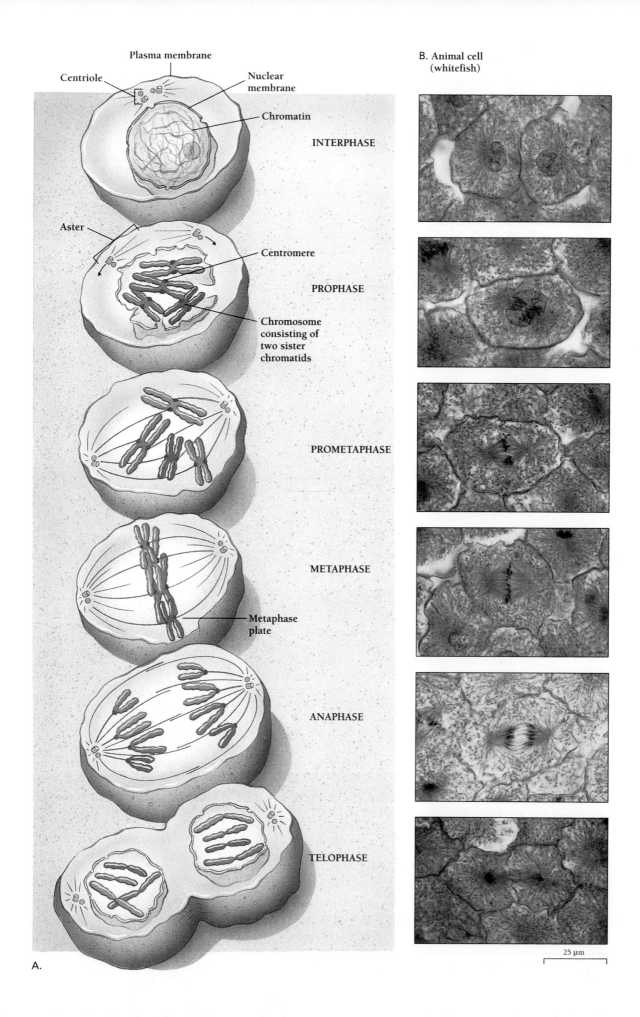

Plasma membrane

Centriole

Nuclear membrane

Chromatin

INTERPHASE

Aster

Centromere

PROPHASE

Chromosome consisting of two sister chromatids

PROMETAPHASE

METAPHASE

Metaphase plate

ANAPHASE

TELOPHASE

A.

B. Animal cell (whitefish)

25 µm

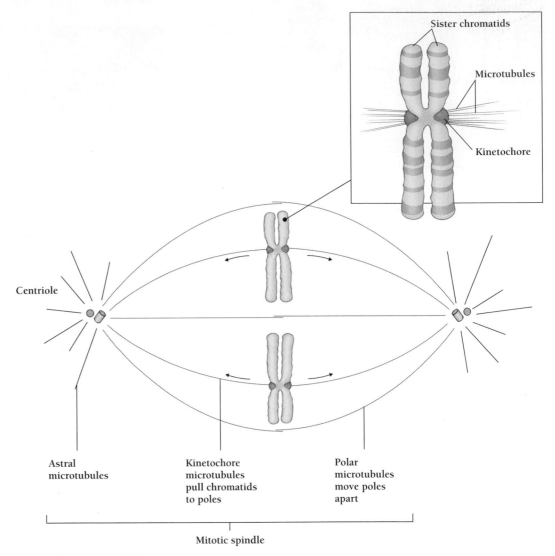

Figure 8-9 Each chromatid develops a disc-shaped kinetochore that attaches the mitotic spindle to the chromosome's centromere. Polar microtubules attach to the kinetochores (and so to chromosomes) and become "kinetochore microtubules." By the end of prometaphase, animal cells have three sets of microtubules originating at a spindle pole: kinetochore microtubules, polar microtubules, and astral microtubules.

During G_1, the cell and the DNA appear to do nothing. Yet this is the time when the cell does most of the work of reproduction. During G_1, the cell doubles nearly all of its materials and machinery (other than DNA). If the cell runs out of nutrients, it stops the cycle in G_1.

The cell cycle consists of the three parts of interphase (G_1, S, and G_2), mitosis, and cytokinesis. During interphase, a cell doubles all of its materials, including the DNA. During mitosis, the sister chromatids separate. During cytokinesis, the cell divides into two daughter cells.

HOW DOES MITOSIS DISTRIBUTE ONE COPY OF EACH CHROMOSOME TO EACH DAUGHTER CELL?

The major task of mitosis is to distribute one chromatid from each chromosome to each daughter cell. Mitosis normally ensures that each offspring cell inherits two sets of chromosomes. Remarkably, all eukaryotic cells accomplish mitosis with the same chromosomal ballet that Flemming first described more than 100 years ago. Modern optical techniques, however, allow biologists to watch the whole process in living cells far more conveniently than could Walther Flemming.

Mitosis Is a Continuous Process, But Biologists Distinguish Four Phases

Cell biologists divide mitosis into four phases. These are prophase, metaphase, anaphase, and telophase. The boundaries between these phases are arbitrary since mitosis is a continuous process. Nonetheless, the events described in these four phases always occur in exactly the same order (Figure 8-8).

During Prophase, Chromosomes Condense and the Mitotic Spindle Forms

During the first phase of mitosis, **prophase** [Greek, *pro* = before], the diffuse chromatin condenses into discrete chromosomes, each consisting of two chromatids, joined together at the centromere. As the chromosomes become visible within the nucleus, the nucleoli disappear. This process takes 10 to 15 minutes in mammalian cells growing in culture.

At the same time, a new structure, called a **mitotic spindle,** develops outside the nucleus. The mitotic spindle, shaped like an American football, consists of prominent bands of **microtubules,** hollow tubes constructed of the protein tubulin. Microtubules form the machinery that moves the chromatids apart. The microtubules of the mitotic spindle are identical in appearance to those found in cilia, flagella, and the cellular cytoskeleton, as discussed in Chapter 4.

In animal cells (and in some other eukaryotes as well), each pole of the mitotic spindle contains a pair of small cylindrical **centrioles,** which lie at right angles to one another. Microtubules originate from each pole of the spindle. Some of the microtubules, the "polar microtubules," run toward the equator. Others, called "astral microtubules," extend outward from each centriole to form an aster, as first noticed by Flemming. By the end of prophase, the spindle has started to elongate, and the two poles (with or without asters) begin to move apart. (The cells of most plants usually have neither centrioles nor asters, and mitosis in these cells is said to be "anastral.")

During the transition from prophase to metaphase, so much happens that researchers give this period its own name, "prometaphase." The nuclear membrane disappears, and the chromosomes are free to attach to the mitotic spindle. Throughout mitosis, the homologs in each pair of chromosomes behave independently of one another. Like socks in the washing machine, they do not pair up.

In mammalian cells, the chromosomes take 10 to 20 minutes to attach to the mitotic spindle. Each chromatid develops a **kinetochore,** a specialized disc-shaped structure that attaches the mitotic spindle to the centromere (Figure 8-9). Some of the polar microtubules attach to the kinetochores (and so to chromosomes) and become "kinetochore microtubules." By the end of prometaphase, animal cells have three sets of microtubules, all of which originate at a spindle pole: kinetochore microtubules, polar microtubules, and astral microtubules. Most plant cells contain kinetochore and polar microtubules, but no astral microtubules. After the spindle apparatus has formed, the attached chromosomes are moved along the spindle toward the cell's equator.

During Metaphase, Chromosomes Align

By the end of **metaphase,** the second and longest stage of mitosis, the chromosomes have moved halfway between the two poles of the spindle, where they form a disc, called the **metaphase plate.** For about one hour, all the chromosomes then lie in a single plane at the equator (at right angles to the spindle fibers). At this stage, the individual chromosomes look different from one another, in size and in the positions of the centromeres.

During Anaphase, Chromatids Separate

The most dramatic stage of mitosis is **anaphase,** the time of chromatid separation. The centromere of each chromosome splits so that each chromatid contains its own centromere connected to the kinetochore microtubules. Then, all at once, the two chromatids that make up each chromosome begin to move away from the metaphase plate. The paired chromatids separate and move in opposite directions. For 5 to 10 minutes, all the chromatids move at the same speed (about 1 μm per minute) toward the two poles of the spindle. Each centromere leads the way, with the rest of the chromatid dragging behind, as if the kinetochore microtubules are pulling the chromatid toward the poles.

Even as the chromosomes move toward the spindle poles, the poles themselves are moving further apart, often doubling the distance between them. By the end of anaphase, the two processes, acting simultaneously, have separated the two sets of chromatids.

During Telophase, the Mitotic Apparatus Disappears

During **telophase** [Greek, *telos* = end], the last phase of mitosis, the mitotic apparatus (including kinetochore, polar, and astral microtubules) disperses. The chromosomes once more assume the diffuse appearance of chromatin, losing their distinct identities, and the nucleoli again become visible. The nuclear membranes form and enclose the two new sets of daughter chromosomes. Each daughter nucleus has the same number of chromosomes as the parent nucleus, but now each chromosome consists of a single chromatid. (Each chromosome will replicate during the next S phase.) After the completion of telophase, the cell usually completes cytokinesis, which we will discuss later.

What Propels the Chromosomes During Mitosis?

We have described three sets of movements whose molecular mechanisms we would like to understand: (1) the movement of the chromosomes toward the equator to form the metaphase plate; (2) the movement of the separated chromatids toward

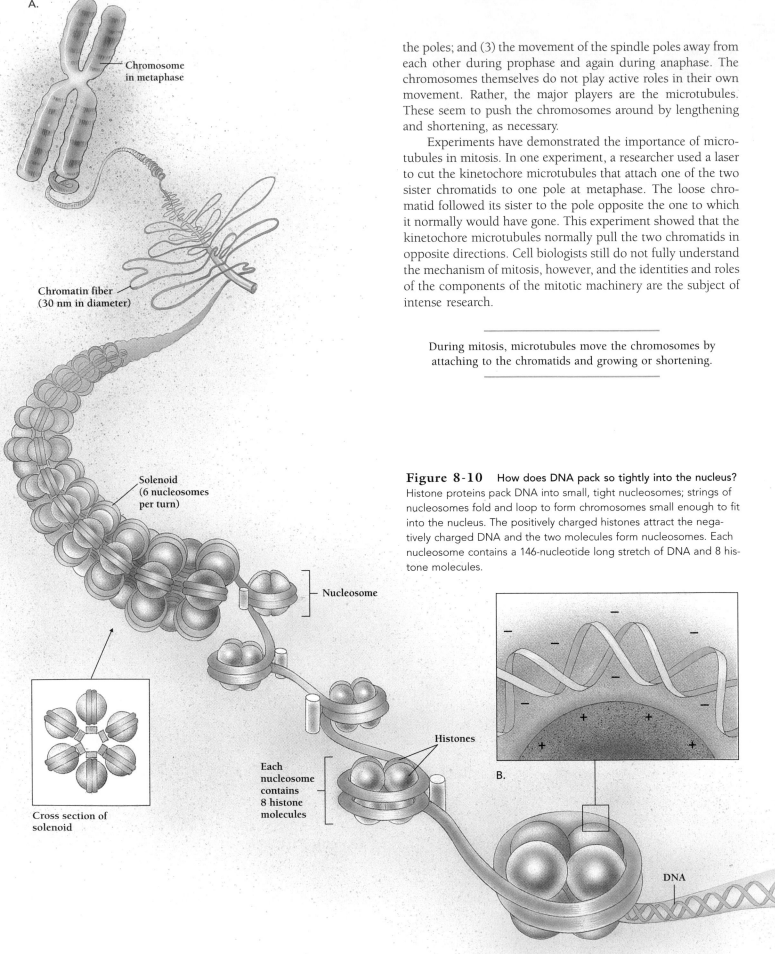

A.

Chromosome
in metaphase

Chromatin fiber
(30 nm in diameter)

Solenoid
(6 nucleosomes
per turn)

Cross section of
solenoid

Nucleosome

Each
nucleosome
contains
8 histone
molecules

Histones

B.

DNA

the poles; and (3) the movement of the spindle poles away from each other during prophase and again during anaphase. The chromosomes themselves do not play active roles in their own movement. Rather, the major players are the microtubules. These seem to push the chromosomes around by lengthening and shortening, as necessary.

Experiments have demonstrated the importance of microtubules in mitosis. In one experiment, a researcher used a laser to cut the kinetochore microtubules that attach one of the two sister chromatids to one pole at metaphase. The loose chromatid followed its sister to the pole opposite the one to which it normally would have gone. This experiment showed that the kinetochore microtubules normally pull the two chromatids in opposite directions. Cell biologists still do not fully understand the mechanism of mitosis, however, and the identities and roles of the components of the mitotic machinery are the subject of intense research.

During mitosis, microtubules move the chromosomes by attaching to the chromatids and growing or shortening.

Figure 8-10   How does DNA pack so tightly into the nucleus? Histone proteins pack DNA into small, tight nucleosomes; strings of nucleosomes fold and loop to form chromosomes small enough to fit into the nucleus. The positively charged histones attract the negatively charged DNA and the two molecules form nucleosomes. Each nucleosome contains a 146-nucleotide long stretch of DNA and 8 histone molecules.

HOW DOES A CELL FIT ALL ITS DNA INTO A NUCLEUS?

The DNA in a cell carries information that the cell uses to construct cell machinery and to carry out cell processes. When a cell divides, the cell must duplicate this information and pass it on to its two daughter cells.

Each molecule of DNA contains two polynucleotide strands, wrapped around each other in a double-helix. This ropelike structure is so fine that a DNA molecule long enough to circle the earth at the equator would weigh less than a grain of fine sand. In contrast, the same length of fine cotton thread would weigh about 275 kg (over 600 lbs) The DNA in a human cell weighs about 6×10^{-12} g (6 trillionths of a gram) and has a total length of about 2 m. Yet a chromosome is only a millionth of a meter long, and the nucleus is only 5 μm (millionths of a meter) in diameter. A cell must therefore solve a major problem: how to fit 2 m of DNA into the tiny nucleus without the long threads of DNA getting tangled.

How Does the DNA Fold Up?

Collapsing DNA into so small a structure depends on folding and bundling the DNA with the aid of specialized proteins, the most abundant of which are the **histones** [Greek, *histos* = web], a set of small, positively charged proteins. Most cells have five types of histones (called H1, H2A, H2B, H3, and H4). A cell may contain some 60 million copies of each type of histone.

The positive charges of histones come from their unusually high content of the basic amino acids (arginine and lysine), which attract hydrogen ions and acquire positively charged side chains. Because DNA has a negative charge, the positively charged histones bind to it tightly (Figure 8-10B). In a test tube, purified DNA and four of the five kinds of histones associate with each other to form what look like beads on a string (Figure 8-10A). Each bead, along with its strand of DNA, is called a **nucleosome** and is a DNA-histone complex about 11 nm in diameter. Each nucleosome contains a 146-nucleotide long stretch of DNA and 8 histone molecules. The formation of nucleosomes is the first stage in the packing of DNA into a nucleus.

Histones from all eukaryotes resemble one another. For example, histone H4 of cows differs by only 2 amino acids (out of 102) from that of peas. This extraordinary similarity suggests that the structure of histone H4 has remained almost unchanged for more than a billion years. We can conclude that this method of packing DNA evolved long ago.

Proteins called histones pack DNA into small, tight bundles called nucleosomes that resemble beads on a string.

How Do DNA, Histones, and Other Proteins Form Such Compact Structures?

Histone H1 appears to be responsible for holding together groups of nucleosomes to form a fiber that is 30 nm in diameter. As a result of this packing, the DNA of a typical human chromosome could be condensed from an average length of 5 cm to about 1 mm—a terrific achievement, but still not small enough for it to fit into a 5 μm nucleus. We still do not know exactly how DNA is made more compact. In some specialized cells, microscopists can see loops extending from chromosomes (Figure 8-11). Many biologists now think that such loops are common to the chromatin of all cells.

In chromosomes, the DNA is even more condensed than in the diffuse chromatin. Still more loops form, and these loops cluster in characteristic ways in each chromosome. The specific pattern of DNA clustering results in **chromosome banding**, the

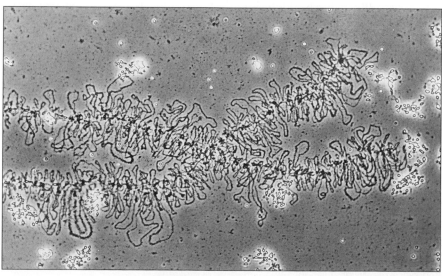

Figure 8-11 In some cells, the DNA forms loops, obvious in this micrograph. *(Joseph Gall, Carnegie Institution)*

25 μm

pattern (visible in a microscope) that results from the selective binding of certain dyes (Figure 8-6A (top)). Biologists can detect abnormal chromosomes by comparing the banding patterns of abnormal chromosomes with those of normal chromosomes.

Strings of nucleosomes fold and loop to form the chromosomes, the highly compact arrangement of chromatin that fits into the nucleus. No one know how this occurs.

HOW DOES A CELL DIVIDE ITS CYTOPLASM?

So far we have discussed how a cell distributes the two copies of its genetic information. But we still need to describe cytokinesis—the process by which a dividing cell partitions its cytoplasm, including the two new nuclei that form during mitosis.

Cytokinesis and mitosis are separate processes, and they depend on different molecular machinery. In some organisms, for example, mitosis sometimes occurs without cytokinesis, leading to two or more nuclei within a single cell, as in the early fruit fly embryo and in the specialized tissues of a plant seed. This shows that cytokinesis and mitosis are truly separate processes. Nonetheless, cytokinesis almost always accompanies mitosis, usually beginning during anaphase and finishing shortly after the end of telophase.

During cytokinesis, the plane of cell division is always perpendicular to the axis of the mitotic spindle (Figure 8-12A). In early anaphase, dividing animal cells form a beltlike **contractile ring,** a bundle of actin filaments that surrounds the dividing cell. The contractile ring works like a purse string, pinching the cell into two parts through the spindle's equator. The force for this movement comes from a mechanism similar to that which causes muscle contraction. In contrast to the permanent contractile structures of muscle cells, however, the contractile ring is a temporary structure like the mitotic spindle.

As the beltlike contractile ring tightens, the cell membrane is pulled into a deepening groove, called the **cleavage furrow** (Figure 8-12B). As telophase proceeds, the two daughter cells become almost, but not quite, completely separated. A thin connection between the daughter nuclei, called the midbody, persists until the end of mitosis. The midbody is packed with microtubules from the spindle apparatus. Finally, at the end of telophase, the microtubules disassemble, and the midbody breaks, leaving the daughter cells completely separated.

As the two daughter cells separate, they must enlarge their cell membranes (since two cells have more surface area than a single cell of the same volume). The additional membrane material comes from a supply of extra membrane made by the parent cell during interphase.

A plant cell, which has a rigid cell wall, cannot simply pinch itself in two the way an animal cell does. Cytokinesis in a plant cell requires the building of a new cell wall between the daughter cells. The new wall begins in telophase as a small, flattened disc, called the early **cell plate,** in the space between the two daughter nuclei (Figure 8-13). The disc grows to become a cell plate, which, except for connecting plasmodesmata, completely seals off the two daughters from one another.

During cytokinesis in animal cells, actin filaments form a contractile ring that pinches the cell in two. In plant cells, telophase includes the building of a new cell wall between the two daughter cells.

HOW DOES A CELL REGULATE PASSAGE THROUGH THE CELL CYCLE?

In this chapter, we have seen how normal cells accomplish the equal distribution of chromosomes by mitosis and the division of cytoplasmic components by cytokinesis. But how do cells know when to begin the cell cycle? How do they coordinate

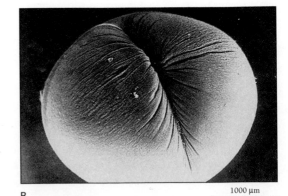

Figure 8-12 During cytokinesis, actin filaments inside the cell form a contractile ring, or cleavage furrow, and pinch the dividing cell into two. *(B, David M. Phillips/Visuals Unlimited)*

A.

B.

1000 μm

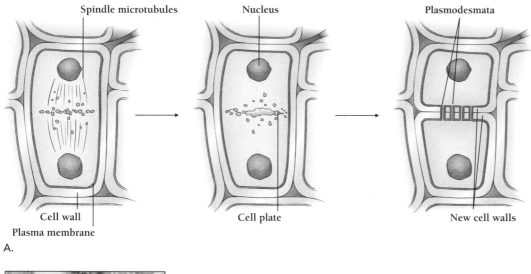

Spindle microtubules   Nucleus   Plasmodesmata

Cell wall   Cell plate   New cell walls

Plasma membrane

A.

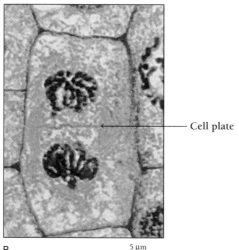

— Cell plate

B.   5 μm

Figure 8-13   In plants, a new cell wall forms between daughter cells. The cell wall begins during telophase as a small, flattened disc, called the early cell plate, then grows into a full-sized cell wall, complete with interconnecting plasmodesmata.   *(B, R. Calentine/ Visuals Unlimited)*

its separate parts? And how do they know when to stop dividing?

If cell division continued without stopping, single-celled organisms would soon run out of space and food. For example, after just four days of undisturbed growth, a single yeast cell could have divided 48 times and produced 2^{48} (about 10 million billion) descendants, as many cells as are in 100 people.

In multicellular organisms, cells normally subordinate such growth and division to the needs of the body's tissues and organs. Cells divide rapidly during periods of growth and more slowly (or not at all) in mature organisms. The only exception to this rule is the uncontrolled growth and division of cancer cells, such as those of Henrietta Lacks. These multiply at the expense of all the other cells in the body.

How Do Normal Cells Determine When To Stop Dividing?

What normally prevents such runaway cell division? Two general mechanisms appear to operate—cell senescence (aging) and growth control. **Cell senescence** limits the number of

times a cell can divide: the more times a cell divides (at least under conditions of laboratory culture), the more likely it is to withdraw from the cell cycle. An average cell taken from a newborn baby, for example, will divide about 50 times in a standard culture medium. But cells taken from an 80-year-old stop cycling after about 30 divisions.

A second set of regulatory mechanisms—those involved in **growth control**—prevent uncontrolled division by allowing the cell cycle to proceed under some conditions and to stop under others. As an example of growth control, think about what happens when you cut yourself. The skin cells on the edges of the cut begin to divide and fill the space left by the wound. The cells divide until the two edges again touch. When the cells on each side of the cut meet, and the wound is healed, cell division ceases.

In a culture dish, cells regulate their division in much the same way (Figure 8-14). Cells attach to the bottom of a plastic dish and divide until they form a single, continuous layer of cells that occupies the whole surface. Then they stop dividing. If we now make a "wound" in this monolayer by scraping away a swath of cells, the cells on the margins move into the

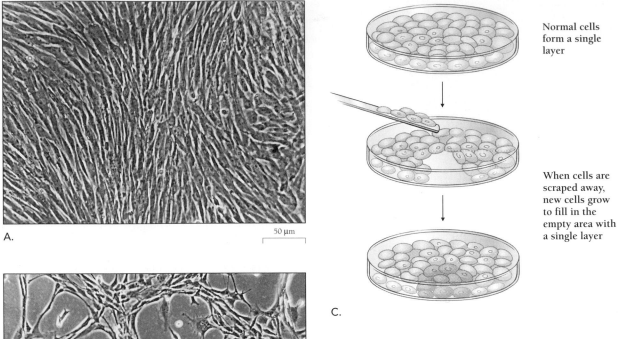

A.

50 μm

B.

50 μm

C.

Normal cells form a single layer

When cells are scraped away, new cells grow to fill in the empty area with a single layer

Figure 8-14   Cells regulate their division by means of contact inhibition. A. Normal cells stop dividing when they come into contact with other cells. B. Cancer cells continue to divide even after they come into contact with other cells. C. In cell cultures, cells divide until they form a single, continuous layer of cells, then stop. If we scrape a "wound" in this monolayer of cells, cells from the edge move into the open space and divide until the wound is filled.   *(A,B, L.B. Chen)*

open space and begin to divide. Division stops when the space is filled.

We can summarize the growth control of normal cells in culture with the following rules: (1) cells stop dividing during G_1 when they run out of free space on which to spread—that is, when neighboring cells all touch each other; and (2) cells that have proceeded beyond G_1 begin to divide when contact with their neighbors ceases. This kind of growth control is called **contact inhibition** of cell division.

Cancer cells provide an excellent example of what happens when contact inhibition fails. Cancer cells do not exhibit contact inhibition in cell culture (Figure 8-14B). Instead of forming a continuous monolayer, they pile up on top of one another and grow to much higher densities than normal cells.

Growth control mechanisms such as contact inhibition prevent normal cells from growing out of control the way cancer cells do.

How Do Normal Cells Determine When It Is Time To Divide?

Most cells divide only after they have first doubled their mass. Otherwise, daughter cells would get smaller with every generation. Somehow cells sense when they have reached a critical size and have enough nutrients to supply two daughter cells. We do not know how eukaryotic cells accomplish this sensing, but once they do, they become irreversibly committed to cell division. Once a cell proceeds beyond a certain point in G_1, the cell proceeds through the rest of the cycle, including mitosis and cytokinesis. This "point of no return" is called **Start.** Arrival at Start depends heavily on the environment of the cell—including the availability of nutrients and signals from other cells. Once the cell has gone through Start, however, the rest of the cycle proceeds, independent of the extracellular environment.

One important exception to the general rule that cells double their mass before dividing are egg cells, which grow to enormous sizes without dividing. For example, the ostrich egg is a single cell. Another important exception are the early embryos of many animal species. Embryos derived from large eggs begin life after fertilization with a series of rapid cleavage divisions, which divide the embryo into many cells without increasing its mass. A frog zygote, for example, is an enormous single cell about 1 mm in diameter. Cleavage divisions rapidly divide it into cells that are more nearly the size of cells in the adult, about 10 to 20 μm in diameter. Even a tiny human zygote divides into 50 or 60 cells in the week before it connects to its mother's blood supply. Only after the human zygote connects can it increase the size of its cells.

The cells of mature organisms only divide when they have doubled their mass unless they are egg cells or the cells of an early embryo.

What Triggers the Main Events of Mitosis?

Cell reproduction requires the coordination of three cycles:

1. The duplication and packaging of DNA, which we discuss in Chapter 10.
2. The duplication of the centrosomes and the operation of the mitotic spindle apparatus.
3. Cytokinesis, which depends on the contractile ring.

Recent research has shown that all eukaryotes, from yeasts to humans, use almost identical mechanisms to coordinate and initiate the events of these three processes.

Fluctuating amounts of various proteins seem to regulate the cell's passage through mitosis. One way researchers study cell cycles is to create **synchronous cell populations,** groups of cells that are all at the same stage of the cell cycle. An easy way of producing such populations is to block DNA synthesis with a chemical inhibitor that stops the cell cycle just before S. When the inhibitor is removed, all the cells begin S at the same time and continue in lock step through the rest of the cell cycle.

Synchronous cell populations have allowed researchers to show that cells undergoing mitosis contain a molecule called **MPF (M-phase-promoting factor),** which triggers mitosis. In one experiment researchers fused pairs of cells that were in different stages in mitosis to form a single cell with cytoplasm in one stage and a nucleus in another stage (Figure 8-15). For example, in a cell made by fusing a cell in mitosis (M) with a nucleus in G_1, the G_1 chromatin prematurely condensed into chromosomes (with only one chromatid each). From this, the researchers concluded that MPF triggers mitosis even in an unprepared cell.

A new and attractive view of cell cycle regulation is that changing levels of MPF and one of its component proteins, called **cyclin,** entirely determine a cell's passage through the cycle. In one kind of experiment, researchers prevent an ex-

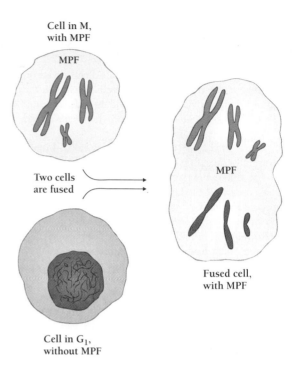

Cell in M, with MPF

MPF

Two cells are fused

MPF

Fused cell, with MPF

Cell in G_1, without MPF

Figure 8-15   **MPF triggers chromosome condensation even in an unprepared nucleus.** When a dividing cell (in M) is fused with a nondividing G_1 nucleus, the G_1 chromatin condenses into chromosomes even though it has not yet doubled. The fused cell contains chromosomes consisting of only one chromatid instead of two.

tract of cytoplasm from synthesizing any proteins except cyclin. Nuclei in the extract enter mitosis, showing that cyclin by itself is enough to stimulate mitosis. Other experiments have shown that the nuclei cannot get beyond metaphase unless the cyclin is destroyed. The periodic increase and decrease of cyclin is both necessary and sufficient for cell cycling.

Various events regulate a cell's passage through the cell cycle, including the completion of DNA synthesis and the fluctuation in the amount of MPF in the cell.

In this chapter we have seen how eukaryotic cells with two sets of chromosomes divide their genetic material to create two daughter cells, each with two sets of chromosomes. Eukaryotes, such as ourselves, employ such mitotic division to create genetically identical copies of cells. In the next chapter, we will see how eukaryotic cells divide a cell with two sets of chromosomes to create gametes (for example sperm and eggs)—cells with just one set of chromosomes. This process, called meiosis, creates cells that are genetically different from the parent cells and also genetically different from one another. Along the way, we will see that sex and meiosis are two sides of the same coin.

STUDY OUTLINE WITH KEY TERMS

Cells reproduce themselves in a process called **cell division,** in which a parent cell gives rise to two **daughter cells,** each with the same genetic information as the parent cell. Dividing cells replicate DNA and distribute it during an orderly sequence of events called the **cell cycle.**

Prokaryotes divide by a simple process called **binary fission,** which distributes one copy of the parent cell's DNA to each daughter cell. Cell reproduction is more complicated in eukaryotic cells, whose DNA is enclosed in a membrane-bounded nucleus. Eukaryotes employ a spindle apparatus, but prokaryotes do not. **Chromosomes** are highly condensed and visible in dividing eukaryotic cells and more diffuse (**chromatin**) in nondividing cells. Chromosomes come in nonidentical pairs called **homologous chromosomes.** Each member of a pair is a **homolog.** When the DNA in a cell has doubled, each chromosome consists of two identical **sister chromatids.**

Cell biologists divide the cell cycle into five phases: mitosis, cytokinesis, G_1, S, and G_2. The process of evenly distributing chromosomes to daughter cells is called **mitosis.** The process of dividing a cell's cytoplasm is called **cytokinesis.** Mitosis and cytokinesis together make up **M.** During G_1, a cell accumulates the materials and machinery necessary to **replicate** its DNA; during **S,** it synthesizes DNA, making two copies of the parent cell's DNA; during G_2, it prepares for mitosis and cytokinesis. G_1, S, and G_2 together are called **interphase.** The point of no return is **Start.**

Just before a cell begins mitosis, each chromosome consists of two identical chromatids. Mitosis is a continuous process, but biologists distinguish four major phases: prophase, metaphase, anaphase, and telophase. During **prophase,** chromosomes condense and the **mitotic spindle** forms. As the **centrioles** of animal cells move apart, **microtubules** radiating outward from the centrioles form **asters.** Between prophase and metaphase (sometimes called prometaphase), the nuclear membrane disappears and the chromosomes attach to the spindle fibers. During **metaphase,** the chromosomes align in a disk called a **metaphase plate,** then bind to the **kinetochores** that form on the **centromere** of each chromatid. During **anaphase,** the two chromatids that make up each chromosome separate. During **telophase,** the mitotic apparatus disappears.

During interphase the cell duplicates its single centriole. During prophase, a mitotic spindle assembles with microtubules emerging from each centriole to form a mitotic spindle. Microtubules connect to each chromatid in a structure called a kinetochore. The mitotic apparatus is responsible for chromosome movements during mitosis.

The DNA of eukaryotic cells is highly folded. Protein molecules called **histones** bind to DNA and form particles called **nucleosomes** that resemble beads on a string. DNA, histones, and other proteins fold to form still more compact structures. The orderly packing of DNA produces a characteristic pattern of **chromosome banding** in each metaphase chromosome.

During cytokinesis, the plane of cell division is always perpendicular to the axis of the mitotic spindle. The pinching movement of cytokinesis results from the action of a **contractile ring,** which surrounds the dividing cell. As the contractile ring contracts, it pulls the membrane inward to form a **cleavage furrow,** which deepens and separates the two daughter cells. Cytokinesis in plant cells requires a special mechanism for building a new cell wall. The new wall begins in telophase as a small, flattened disc that grows to become a **cell plate.**

Cells actively regulate passage through the cell cycle. Regulated cell division in multicellular organisms depends on **cell senescence** and **growth control.** Cell senescence limits the number of times a cell can divide. Growth control regulates cell reproduction according to external conditions. For example, many cells exhibit **contact inhibition** of cell division, in which cells divide only when they are not in contact with their neighbors. Growth control may also involve growth factors—extracellular proteins that stimulate cell division. Cancer cells fail to show normal growth control. Studies of **synchronous cell populations** show that the regulatory protein **MPF** stimulates cells to enter mitosis. Changing levels of MPF and one of its component proteins, **cyclin,** may entirely determine a cell's passage through the cell cycle.

REVIEW AND THOUGHT QUESTIONS

Review Questions

1. Why must a cell double its materials, machinery, and memory before it can divide?
2. Why did the discovery that DNA was the genetic material change the way that cell biologists regarded interphase?
3. Describe a method for measuring the amount of DNA in a cell. How does the amount of DNA in a cell change during the cell cycle?
4. Describe and draw the events of the four stages of mitosis. When does the actual division of the chromosomes occur?
5. Using diagrams, compare the events in prophase to those in telophase. Why do you think they are so nearly opposite?
6. Compare the role of the centriole in mitosis with that of a kinetochore.
7. What factors regulate a cell's ability to proceed to the next round of cell division?

8. Describe the role of histones in the structure of eukaryotic chromosomes. How do histones help DNA fit into a nucleus?
9. How does cytokinesis in a plant cell differ from that in an animal cell?

Thought Questions

10. How would prevention of DNA synthesis affect mitosis?
11. What is the relationship between the materials accumulated in G_1 and G_2 and the processes that follow?
12. How might cancer be associated with improper regulation of the cell cycle?

SELECTED READINGS

Gold, Michael, *A Conspiracy of Cells: One Woman's Immortal Legacy and the Medical Scandal It Caused,* State University of New York Press, Albany, 1986. A highly readable account of the HeLa cell disaster.

Moore, John A., *Science as a Way of Knowing: The Foundations of Modern Biology,* Harvard University Press, Cambridge, Massachusetts, 1993. Presents fascinating accounts of the twisty routes by which biologists arrived at our present understanding of biology.

▶ On-line materials relating to this chapter are on the World Wide Web at http://www.saunderscollege.com/lifesci/
Click on Tobin/Dusheck: *Asking About Life*.

9

FROM MEIOSIS TO MENDEL

Why Is the Yellow Dog Yellow?

Early in the spring of 1902, a young graduate student at Columbia University informed his professor with barely suppressed excitement that he had solved the problem of heredity. Walter Stanborough Sutton (1877–1916), a six-foot tall, 215-pound Kansas farm boy who towered over everyone around him gleefully announced, "I know why the yellow dog is yellow." At 25, Sutton knew with certainty that the chromosomes carry the hereditary material (Figure 9-1).

Sutton had discovered that the chromosomes are the mechanism by which organisms inherit particular traits and that he held the key to the then-infant science of genetics. To Sutton's frustration, his professor, Edmund B. Wilson (1856–1939), the most distinguished cell biologist in the United States, merely nodded indulgently. No one else took Sutton any more seriously.

That summer, Wilson invited Sutton to the seaside at Beaufort, North Carolina, to study marine animals, to sail, to swim, and to talk. For the first time, Wilson began to pay attention to what Sutton was telling him. "It was only then," Wilson later wrote, "that I first saw the full sweep, and the fundamental significance of his discovery."

Ripsaw, Inc.

Figure 9-1 At age 25, Walter Sutton showed that the genes lay on the chromosomes. *(University of Kansas Archives)*

Sutton had come to biology partly by chance and he hadn't decided yet if he would stay. In 1896, he had begun studying engineering at the University of Kansas. When he returned home for the summer in 1897, he took with him a case of typhoid fever that felled his entire family. When it was over, Walter Sutton's beloved younger brother John was dead at 17. In his grief, Sutton decided to abandon engineering for medicine. When he returned to the University of Kansas at Lawrence in the fall, he began to study biology, and he quickly became friends with a young professor of zoology named Clarence McClung.

On vacations at home, Sutton collected grasshoppers from the wheat fields on the family farm to send to McClung, who admired the grasshoppers' enormous cells. The cells of *Brachystola magna*, the lubber grasshopper, were so large, in fact,

that they were perfect for studying the fine structure of the cell. McClung was especially interested in the fact that some of the sperm cells appeared to have an extra chromosome. At this time, no one knew what chromosomes were. But McClung called this chromosome the "accessory chromosome" and theorized that it determined sex. He was largely correct. McClung's idea that the accessory chromosome could give grasshopper individuals specific traits suggested to Sutton that the other chromosomes might do the same.

In April of 1900, Sutton published a paper describing in great detail the sex cells of the lubber grasshopper, including 41 drawings and 10 photographs. A year later, Sutton received his master's degree based on his work on the cell divisions in the sperm cells of the grasshopper. In the fall of 1901, encouraged by McClung, he transferred to Columbia University to work with Wilson.

Within 6 months, Sutton was hanging on Wilson's sleeve trying to persuade

1910 were a breathtaking period in biology, and Bateson's talk electrified his American audience.

As Sutton listened to Bateson, any doubts he had about the importance of his discovery evaporated. Everything that Bateson said made perfect sense in the light of his knowledge of chromosomes. Within a month, Sutton had published a short paper describing his idea that the chromosomes are the physical basis for heredity. Within 4 months he had published a full-length paper describing all of the theoretical implications of his idea. Sutton's ideas are now so well accepted that we must be reminded of why more-experienced biologists did not recognize the role that chromosomes play in inheritance.

All later research has confirmed Sutton's theory in nearly every respect. As is often the case in science, however, other biologists did not embrace this idea at once. It would be nearly 40 years before Sutton's theory was fully accepted.

At 25, Sutton knew with certainty that the chromosomes carry the hereditary material.

him that he had solved the mystery of heredity. That fall, the famous British biologist William Bateson came to Columbia to talk about the new science of genetics. Two years before, Bateson reminded his audience, three botanists had rediscovered the laws of inheritance first described and published by the Austrian monk Gregor Mendel in 1866. These laws laid a baseline for all future work on inheritance. The years between 1900 and

Perhaps it was partly the cool response of his elders. Perhaps it was Sutton's idea that he could do more immediate good in the world as a doctor. Whatever the reason, Sutton pursued genetics no further. In the summer of 1903, he dropped out of graduate school. Two years passed before, pressured by his father, he returned to Columbia, where he resumed, not his research in biology, but his long-delayed medical training. He was as outstanding a medical student as he had been a biologist. In 1907, he received his medical degree, and, returning to Kansas in 1909, he practiced as a surgeon for the rest of his short life. Still a bachelor, he died of a burst appendix at 39.

Sutton was a good surgeon, but it was his research in biology that changed the world. At 25, he discovered a fundamental fact of biology that had eluded older scientists for decades. Sutton revealed the basic mechanism by which organisms inherit traits—why, as he put it, "the yellow dog is yellow." His work is the foundation for all of modern genetics and molecular biology. In the next section, we will see why it took a young biologist such as Sutton to recognize what now seems so obvious. In the rest of the chapter, and in succeeding chapters, we will try to find out exactly what the chromosome does.

KEY CONCEPTS

1. The relationship between the phenotype of an individual and its genotype is mediated by the environment from the moment the individual comes into being.

2. Sexually reproducing eukaryotes have *pairs* of homologous chromosomes, which meiosis distributes—one chromosome from each pair—to each daughter cell.

3. At fertilization the chromosomes of the two haploid gametes combine so that the resulting diploid zygote has pairs of chromosomes again—the same number of chromosomes as the parents.

4. The behavior of chromosomes underlies the rules of inheritance.

5. The inheritance of genes parallels the inheritance of chromosomes.

6. Gregor Mendel established the principles of genetics even before biologists had discovered chromosome movements.

7. A chromosome contains many genes.

WHY WAS THE CHROMOSOMAL THEORY OF INHERITANCE SO HARD TO ACCEPT?

Historically, biologists had no trouble posing theories to explain how organisms stay the same between generations. Before the 17th century, a favorite theory was that miniature versions of future organisms were already "preformed" in sperm or eggs, so that the properties of all future generations were already determined. Most philosophers believed that the egg contributed the materials for an individual, and the sperm contributed some life force that determined the form or shape of the individual. Philosophers (male philosophers) compared the creative action of the sperm to the divine creation of the universe from formless matter.

Blending Inheritance: A Wrong Turn

In contrast, plant breeders had long known that when two varieties of plants are crossed, both parent plants contributed equally to the form of the next generation, no matter which variety of plant provided the sperm (pollen) and which variety provided the egg. In the face of such evidence, most biologists reasoned that the different types of offspring somehow result from the "blending" of the genes of the parents. We are not surprised, for example, when a black mare and white stallion produce a gray foal or when a red snapdragon crossed with a white snapdragon produces a plant with pink flowers (Figure 9-2). The color of the offspring appears to be the result of blending, as if two cans of paint have been poured together and mixed.

The **blending** model of inheritance seemed to work well for traits that vary continuously, such as height, weight, and sometimes color. The blending model did not, however, account for the reappearance of unblended characteristics in later generations. Pink flowers can give rise to red or white offspring, and brown-eyed parents can have blue-eyed children.

The blending model also created a serious problem in understanding the mechanism of evolution. If blending occurred all the time, variation should eventually disappear in each species, and natural selection would have no inherited variations on which to act. Blending inheritance and Darwin's theory of natural selection, then, were incompatible theories.

Most biologists of the 19th century initially assumed that blending inheritance was right and Darwin's theory of natural selection was wrong. They accepted evolution, but not natural selection. That left them without a mechanism for evolution and without a theory of inheritance that could account for unblended traits. The problem of inheritance was so intriguing that, in time, nearly every important biologist of the second half of the 19th century had a different hypothesis about how inheritance worked.

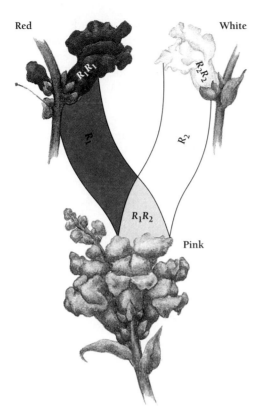

Red   White

R_1R_1 R_2R_2

R_1 R_2

R_1R_2

Pink

Figure 9-2   Crossing red and white snapdragons produces pink ones. Blending inheritance seemed reasonable to early biologists.

Most 19th-century biologists mistakenly accepted blending inheritance and rejected natural selection.

Chromosomes Are Individually Unique and Exist Continuously from Generation to Generation

Recall from the last chapter that a wealth of clues suggested that the chromosomes carry the hereditary material. Walther Flemming had shown that during mitosis, the chromosomes divide down the middle, so that half of each chromosome goes to each cell. Wilhelm Roux had argued that this complex way of dividing the chromosomes at mitosis suggested that each chromosome was unique. Theodor Boveri had shown that if the nucleus of an egg is removed, the sperm nucleus can determine by itself the characteristics of an organism. Boveri also showed that in all cells descended from the single cell of a zygote, one-half of the chromosomes come from the father and one-half come from the mother.

Nonetheless, the idea that the chromosomes were responsible for heredity was no more than a possibility. Most 19th-century biologists did not take the idea seriously. One exception was the German biologist August Weismann, who argued that organisms inherit their traits by means of some information-carrying chemical that is in the nucleus of the cell. He fur-

ther suggested that since sex involves the fusion of hereditary material from the mother and the father, logically, the egg and sperm must each contain only half the normal amount of hereditary material. Otherwise, argued Weismann, there would be a doubling of the hereditary material in each generation.

Within 3 years, cell biologists confirmed Weismann's prediction. First, they showed that sperm and eggs contain only half the usual number of chromosomes. Second, they observed the special process of **meiosis** by which cells with the normal number of chromosomes divide to produce daughter cells (eggs and sperm) with only half the normal number of chromosomes. By the late 19th century, most cell biologists accepted that the chromosomes were central to the process of both mitosis and meiosis and probably somehow essential for the normal development of an embryo into an adult. But they did not know what the chromosomes did. No one yet knew that the chromosomes were the hereditary material.

In 1900, however, an intellectual earthquake transformed the scientific landscape. In that year, three scientists independently rediscovered the work of Gregor Mendel, a visionary scientist who had described several important principles of heredity in 1866 (Chapter 1). Mendel's work on patterns of inheritance in pea plants showed that every organism has two sets of genes. In just a few months, the entire field of genetics came into full flower. As biologists leapt to consider Mendel's ideas, they realized that the behavior of the chromosomes during meiosis might explain Mendel's principles. Chromosomes suddenly became of intense interest to a great many biologists.

Two important problems, however, stood in the way of recognizing that the chromosomes carry the genes. First, because the chromosomes disappear in between cell divisions, biologists had no evidence that the chromosomes were passed intact from generation to generation. The chromosomes might well be created anew after each cell division. If so, biologists reasoned, chromosomes could no more carry information from parent cell to daughter cell than yesterday's news could be reconstructed from a recycled newspaper.

Second, biologists had no evidence that the individual chromosomes were different from one another in any important way. True, the chromosomes seemed to come in a variety of shapes. But they all behaved in the same way, and they all seemed to be made of the same material. There was nothing to suggest that each one might be carrying unique information. The arguments of Weismann and others that each chromosome always maintains its individuality and integrity were good guesses only. Walter Sutton was the first to state explicitly that Mendel's genes were on the chromosomes and to give good reasons why this must be so.

Biologists at first failed to recognize that the chromosomes were the genetic material because the chromosomes seemed to disappear during cell division and because biologists had no reason to think that each chromosome carried unique information.

HOW DO ORGANISMS PASS GENETIC INFORMATION TO THEIR OFFSPRING?

Genetics, the study of inheritance, today faces the same contradictory challenges it did 100 years ago. Genetics must explain both how like begets like—why children resemble their parents—and also the origin and maintenance of variation—why children differ from their parents. In this chapter, we will discuss **transmission genetics,** the study of how variation is passed from one generation to the next. We will see that the behavior of chromosomes during meiosis and fertilization does much to explain the inheritance of variation.

Distinct from transmission genetics is **molecular genetics,** the study of how DNA carries genetic instructions and how cells carry out these instructions. Molecular genetics helps explain how a single-celled zygote divides and eventually develops into liver cells, skin cells, nerve cells and more; and also how the DNA in these various kinds of cells helps direct day-to-day operations in the cell. We discuss molecular genetics in Chapters 10 through 13. In this chapter, we begin our discussion of transmission genetics by briefly defining how the zygote uses the genetic information it receives from its parents to become a unique individual. No two organisms are identical, and the unique collection of traits that define an individual is called the "phenotype."

What Is Phenotype?

Everything about an organism, from its size and chemistry to its behavior constitute its **phenotype** [Greek, *phenein* = to show]. Phenotype encompasses both physical and behavioral characteristics. In humans, height and skin color are easily observed phenotypic traits, the individual aspects of phenotype in which individuals may vary. In contrast, the **genotype** is the genetic constitution of a cell or organism. A genotype is a collection of genes. Biologists use the word "genotype" to refer both to all the genes in an organism, its **genome,** or, alternatively, to the subset of genes (or even one gene) that influences a particular trait, such as eye color.

A **gene** is a region of the DNA that either specifies a particular protein or that helps regulate the expression of other genes. Most genes specify proteins and are called **structural genes.** Other genes, called **regulatory genes,** are regions of DNA that regulate the expression of structural genes. Under the influence of an organism's environment, the genes of the genotype help determine the characteristics of the organism.

An organism's phenotype reflects both its genotype, what genes it has, and the environment in which it has developed. Even identical twins raised together exhibit different phenotypes (Figure 9-3). Identical twins result when a zygote divides into two separate cells, which then develop separately. Identical twins share the same genome, the total genetic information carried by a cell or organism. (Fraternal twins, which develop from two eggs fertilized by two sperm, are no more related than ordinary brothers and sisters.) Most of the differences between identical twins are the result of differences in their environments: in the womb, as they grow up, and as adults.

One of the most difficult tasks of genetics is to separate the environmental and genetic influences on phenotype. Genes and environment affect the phenotype separately, but also together. In an extremely simple example, a child born with the genetic defect "PKU" (phenylketonuria) becomes severely retarded if raised on a normal diet. The same child raised on a special diet that is low in the amino acid phenylalanine may seem perfectly normal. In this case, we cannot say that either the PKU genotype by itself or the environment determines the phenotype. How the environment and the genotype *interact* determines the phenotype. The interaction of genes and environment is usually more subtle, depending on the trait. In addition, genes interact with one another in complex ways.

Figure 9-3 There is more to phenotype than genes. Even identical twins, with exactly the same genes, often look and behave differently. *(Kathryn Abbe)*

The word "genotype" can refer not only to an organism's entire set of genes, but also to a restricted set of genes that contributes to a particular phenotypic trait. For example, we might say that a flower has the "blue-petal" genotype, by which we mean that the plant has a gene that specifies blue petals. The genome, the set of all genes, of an individual is a relatively passive entity within the cell. Like a cookbook, it provides lists of ingredients (proteins) and directions for how much of each protein to use.

Although geneticists can study the inheritance of genes themselves, the most interesting genetic questions concern the phenotype of the resulting organism and how genes influence that phenotype, not the genetic plans themselves. An organism's phenotype is not only more interesting to us because we are fellow organisms, but also because it determines the way that the organism gets on in the world. It is the phenotype that is subject to natural selection. The crucial biological questions—how well does a given individual compete with other organisms and how much will it contribute to the next generation—depend on the phenotype.

Phenotype results from an interaction of genes and environment.

The Same Laws of Inheritance Apply to All Sexually Reproducing Organisms

All sexually reproducing eukaryotes—from the most primitive algae to the most complex primates—follow the genetic rules that we describe in this chapter. There are some notable exceptions to these rules, including the inheritance of genes carried in the DNA of mitochondria and chloroplasts, which we will discuss in Chapter 12. But the rules of genetics are otherwise nearly universal. The reasons for this unity are

1. All organisms use DNA as the genetic material.
2. The DNA of all eukaryotic organisms is organized into chromosomes.
3. Almost all chromosomes exist in pairs at some time during a sexual life cycle.
4. These pairs of chromosomes behave in the same ways during meiosis and at fertilization in all eukaryotes.

HOW DO SEXUALLY REPRODUCING ORGANISMS KEEP THE SAME NUMBER OF CHROMOSOMES FROM GENERATION TO GENERATION?

In Chapter 8, we saw how the individual cells within an organism manage to distribute identical sets of chromosomes to their daughter cells. In this chapter, we will see how sexually reproducing organisms pass on half-sets of chromosomes to their sperm and egg cells. When a sperm and egg fuse, the re-

sulting individual has, once again, a full set of chromosomes. This is the essence of sexual reproduction.

Mitosis is an easy way for single-celled eukaryotes to reproduce themselves. Multicelled organisms can likewise reproduce through mitosis, by splitting off a single cell or group of cells capable of developing into a whole individual. In both cases, reproduction through mitosis, called **asexual reproduction**, produces offspring with genes from just one parent. All the offspring have the same genes, those of the parent. The offspring that result from asexual reproduction constitute a **clone**, a set of genetically identical individuals.

In contrast, **sexual reproduction** produces offspring that inherit genetic information from two parents, rather than one. Since the genetic information from each parent is unique, sexual reproduction combines already unique combinations of genes in new ways, so that, in every generation, a new set of unique individuals is created. Sexual reproduction creates enormous diversity.

But sexual reproduction creates a problem: how can each new organism receive chromosomes from both parents and still maintain the same total number of chromosomes?

The answer to this question lies in the fact that chromosomes, like socks, come in pairs. Figure 9-4 shows the 46 chromosomes of a human being. Notice that for each chromosome, with a characteristic size and banding pattern, there is another that looks just like it. Pairs of matching chromosomes are called **homologous chromosomes,** and each member of the pair is called a **homolog.** In other words, the 46 chromosomes are actually 23 pairs of chromosomes. Cells that contain two sets of chromosomes are said to be **diploid** [Greek, *di* = double + *ploion* = vessel]. Like the two socks in a pair, homologous chromosomes are basically alike, but they are not perfect copies.

Sexually reproducing organisms maintain the diploid number of chromosomes in each generation by supplying their offspring with just half of the needed chromosomes. Each parent provides one of the chromosomes in a pair. In humans, for example, each parent contributes 23 chromosomes. (This is analogous to receiving for your birthday 23 unmatched socks from your mother and 23 unmatched socks from your father. You put them together and discover that, miraculously, you have 23 matching pairs.)

A sexually reproducing organism passes its single, unmatched set of chromosomes to its offspring in specialized reproductive cells called **gametes** [Greek, *gamos* = marriage]. Each gamete is **haploid** [Greek, *haploos* = single], meaning that it carries a single set of chromosomes, in humans 23. (Sometimes the word haploid is confusing, since it sounds like "half." But haploid means single, not half.)

In all animals and plants, the gametes are distinct from one another. One kind, called the **egg** or **ovum** [Latin, = egg; plural, *ova*], is large and usually cannot move itself, while the other, called the **sperm,** or **spermatozoon** [Greek, *sperma* = seed + *zoos* = living], is small and motile (able to move under its own power). No matter how small or strange an organism is to us, if it produces eggs we say it is female. Likewise, anything that produces sperm is defined as male.

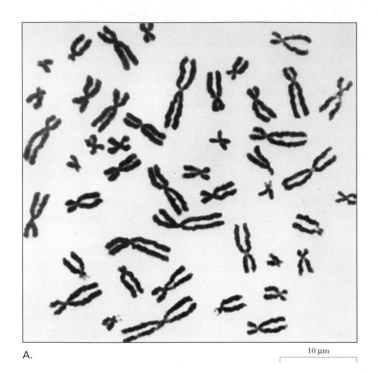

A.

10 μm

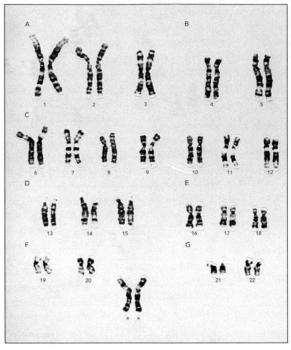

B.

Figure 9-4   The 46 mitotic chromosomes from a skin cell of a female human. Chromosomes, like socks, come in pairs. A. The chromosomes in a disorganized pile. B. The same chromosomes in 23 matched pairs and ordered by size to make a karyotype.   *(A, Biophoto Associates/Photo Researchers; B, Leonard Lessin/Peter Arnold)*

Gametes are the genetic link between generations. But where do these haploid cells come from? The answer is that haploid gametes arise from diploid cells by **meiosis,** a process that allots one haploid set of chromosomes to each of four daughter cells. Gametes and the special cells from which they arise are called **germ cells** or the **germ line** (Figure 9-5A). Gametes derive only from germ cells. In adult animals, the germ cells are found in special gamete-producing organs, called **gonads** [Greek, *gonos* = seed]. The gonads are called **ovaries** in females and **testes** in males.

All of the rest of the cells in a multicelled organism are called **somatic cells** [Greek, *soma* = body]. Mature somatic cells are capable of undergoing mitosis only, not meiosis. In all sexually reproducing organisms, genetic instructions pass exclusively through the germ line.

***

The germ cells of sexually reproducing organisms undergo meiosis to produce haploid gametes, which combine to form a diploid individual.

***

The First Cell of the New Generation Has Two Sets of Chromosomes

Every new generation begins with **fertilization,** the union of the two haploid gametes (one from each parent) to form a diploid cell, called the zygote (Figure 9-5B). In humans, fertil-

ization is also called conception. Some biologists prefer the term **syngamy,** since "fertilization" wrongly suggests that the sperm makes the egg fertile. In fact, the egg and sperm are both equally fertile before they fuse to form the zygote. Nonetheless, this book uses the older term, fertilization.

Fertilization depends heavily on adaptations of the gametes. For example, the spermatozoon's whiplike tail specializes it for movement. The egg, in turn, is specialized for physically moving the sperm nucleus and its own nucleus together, for fusing the two haploid sets of chromosomes into one diploid set, and for dividing rapidly after fertilization. In addition, the egg is packed with molecules and organelles that support the early development of the organism.

In mammals, birds, reptiles, insects, and many other animals, fertilization is "internal," meaning that it takes place within the female's reproductive system and therefore requires copulation. Animals that practice internal fertilization often rely on elaborate behavioral and structural adaptations to accomplish fertilization, as shown in Figure 9-6. Putting the egg and sperm together is only the beginning, of course. A host of factors determine whether fertilization occurs, including pH, temperature, immune responses, sperm competition, and more.

Despite the many different ways that species may achieve fertilization, it always serves the same purpose: to combine two haploid sets of chromosomes into one diploid set of chromo-

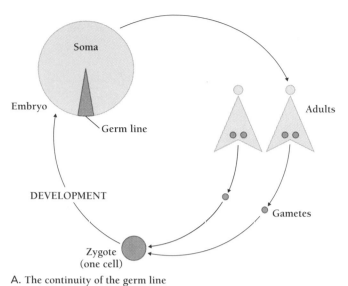

A. The continuity of the germ line

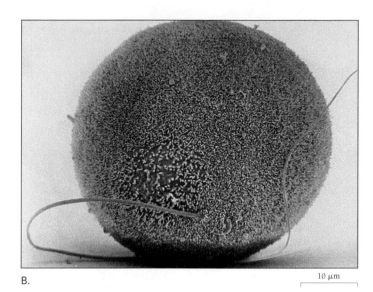

B.

10 μm

Figure 9-5   The germ line consists of the gametes and the tissues of the body that are capable of undergoing meiosis to produce more gametes. A. The soma, or body cells, specialize to form skin, nerves, and muscle. Cells of the soma are not part of the germ line. B. Union of hamster egg and sperm.   *(B, David Phillips/Visuals Unlimited)*

A.

Figure 9-6   Internal fertilization requires elaborate behavioral and structural adaptations. A. Blue-footed boobies and other animals engage in complex dances and other rituals before they mate. B. Penis bones from various kinds of mice. The penis bone, or baculum, is found in a variety of male mammals, including insectivores (shrews and moles), bats, rodents, carnivores (lions and wolves) and primates. Among primates only humans lack a penis bone.   *(A, Hal Harrison/Grant Heilman Photography)*

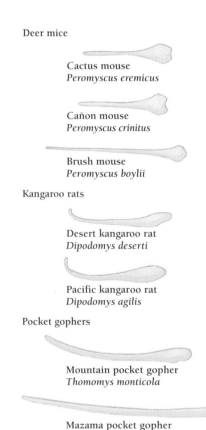

Deer mice

Cactus mouse
Peromyscus eremicus

Cañon mouse
Peromyscus crinitus

Brush mouse
Peromyscus boylii

Kangaroo rats

Desert kangaroo rat
Dipodomys deserti

Pacific kangaroo rat
Dipodomys agilis

Pocket gophers

Mountain pocket gopher
Thomomys monticola

Mazama pocket gopher
Thomomys mazama

5 mm

B.

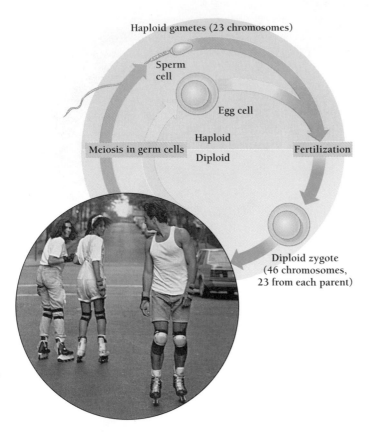

Figure 9-7 **Meiosis and fertilization.** Meiosis produces haploid gametes. Fertilization combines two haploid sets of chromosomes into one diploid set of chromosomes. *(photo, Rob Lang)*

somes (Figure 9-7). Meiosis serves to produce gametes with those haploid sets of chromosomes. All sexually reproducing eukaryotes accomplish meiosis in essentially the same way, in an elaborate chromosomal ballet that both resembles and differs from mitosis.

How Does Meiosis Distribute Chromosomes to the Gametes?

Meiosis consists of two cell divisions, called **meiosis I** and **meiosis II.** These occur only in cells of the germ line during the production of gametes. Like mitosis, meiosis is a continuous process, which biologists divide into a series of steps only for purposes of discussion. An important point to remember is that DNA replication occurs before the first division (meiosis I), but *not* before the second division (meiosis II).

Biologists divide each of the two meiotic divisions into four phases (Figure 9-8). These are: prophase, metaphase, anaphase, and telophase. In general outline, each phase of meiosis resembles the corresponding stage of mitosis. During each **prophase,** chromosomes condense, and the nuclear membrane breaks down. During each **metaphase,** the chromosomes move to the equator of the spindle apparatus. During each **anaphase,** spindle fibers move the chromosomes toward the poles. And, during each **telophase,** the nuclear membrane reforms. But the

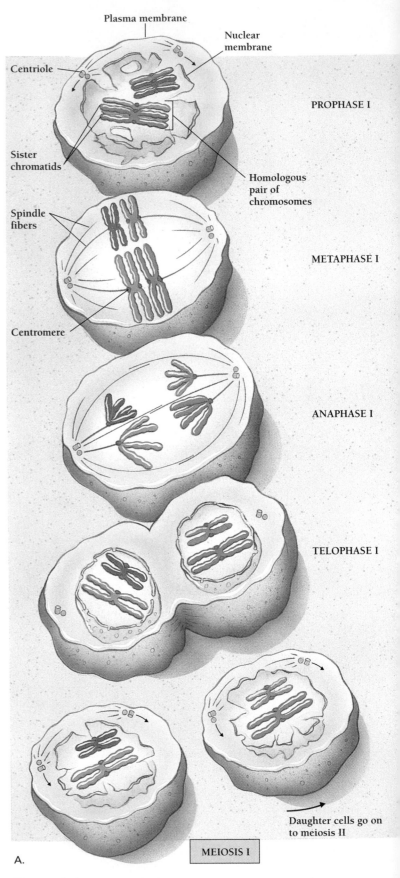

Figure 9-8 **Meiosis is a continuous process consisting of two sequential divisions: meiosis I and meiosis II.** Each of the four daughter cells receives one chromosome from each homologous pair.

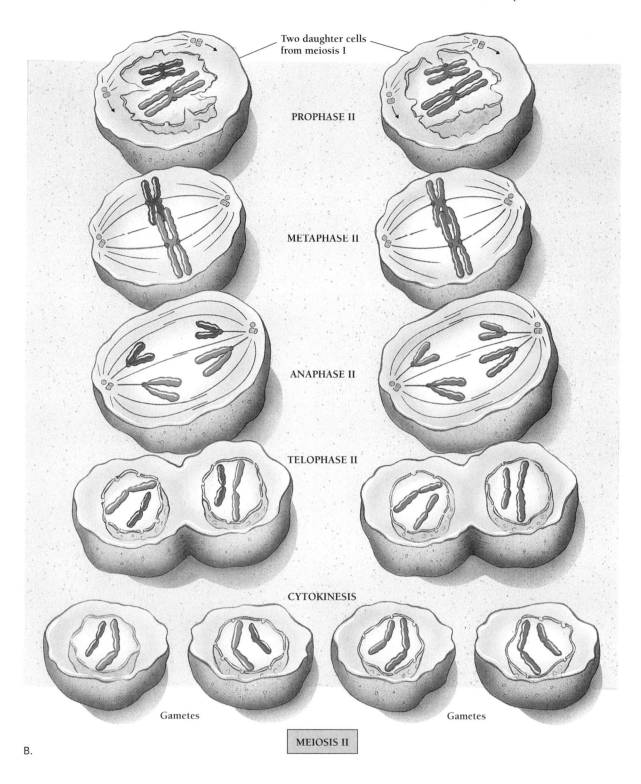

Two daughter cells from meiosis I

PROPHASE II

METAPHASE II

ANAPHASE II

TELOPHASE II

CYTOKINESIS

Gametes   Gametes

MEIOSIS II

B.

end results of meiosis and mitosis are crucially different. Every mitotic division produces two diploid daughter cells, with two of each chromosome. In contrast, the two divisions of meiosis produce four haploid daughter cells, with only one of each chromosome.

For mitosis, a cell replicates its DNA once and divides once, resulting in two diploid daughter cells. For meiosis, a cell replicates its DNA once but divides twice, resulting in four haploid daughter cells.

During Prophase I, Homologous Chromosomes Form Pairs

Prophase I occupies more than 90 percent of the total time of meiosis. As in mitotic prophase, the chromosomes condense, the nucleoli and the nuclear membrane disappear, and the spindle apparatus begins to form.

But meiotic prophase differs from mitotic prophase in one essential way. After the chromosomes begin to condense, the homologous pairs come together in the process of **synapsis** [Greek, = union], in which homologous chromosomes align exactly. Each chromosome itself consists of two sister **chromatids,** so that the four chromatids, form a **tetrad** (Figure 9-9). Synapsis resembles the process of pairing socks before putting them away. From a disorganized pile of 46 socks, we can put together 23 pairs.

Synapsis is the central event of meiosis. It is the process that allows the sorting of the homologous chromosomes into two haploid sets. Following synapsis, the chromo*tids* of each chromosome stay together throughout meiosis I. In contrast, the homologous chromo*somes* go their separate ways. (In mitosis and in meiosis II, it is the *sister chromatids* that separate, not the chromosomes.)

Another important event that occurs during synapsis is **crossing over,** in which homologous chromatids break and exchange equivalent pieces, resulting in new combinations of genes from the two parents (Figure 9-9). The visible result of crossing over is a crosslike configuration called a **chiasma**

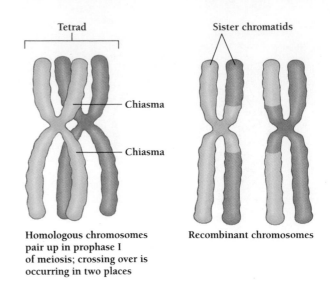

Homologous chromosomes pair up in prophase I of meiosis; crossing over is occurring in two places

Recombinant chromosomes

Figure 9-9 **Crossing over at two chiasma.** During synapsis, homologous chromatids can break and exchange equivalent pieces, resulting in new combinations of genes from the two parents.

[Greek, = cross; plural, **chiasmata**]. At each chiasma, homologous chromosomes may exchange chromatid segments. Crossing over is another way that sexually reproducing organisms recombine genes. We will discuss such recombination later.

The extended duration of prophase I provides the time needed to build up materials and machinery in the egg cell. In the females of some species, meiosis may actually stop for extended times during prophase of meiosis I. In frogs, for exam-

Figure 9-10 **The separation of the chromatids.** During meiosis I, sister chromatids go to the same pole. During meiosis II, sister chromatids go to opposite poles (as in mitosis).

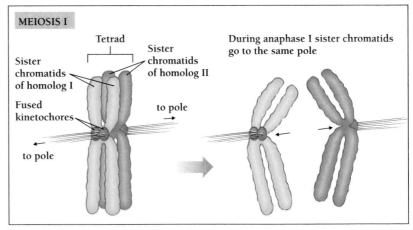

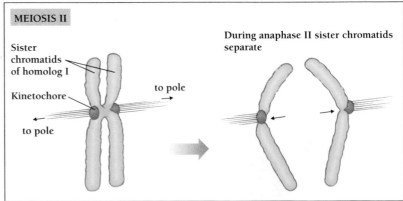

212

ple, prophase I may last for several years. In humans, meiosis begins before birth, pauses in prophase I, and does not resume until after puberty. Starting at around age 12, one or two eggs complete meiosis each month. Since female humans may continue to ovulate into their 50s, some individual cells may not resume meiosis for more than 50 years.

During the Rest of Meiosis I, One Chromosome from Each Homologous Pair Goes to Each Daughter Cell

In the next movement of the meiotic dance, metaphase I, each tetrad migrates to the equator (Figure 9-10). The two centromeres of each tetrad are attached to spindle fibers from opposite poles.

Then, during anaphase I, something happens that is very different from what occurs in mitosis. The sister chromatids go to the *same* pole (whereas, in mitosis, the two chromatids go to *opposite* poles). Telophase I and cytokinesis rapidly follow. The chromosomes may remain partly condensed, and the daughter cells of meiosis I move right into meiosis II without any more DNA synthesis.

We may ask how the tetrads align themselves during metaphase I. Do all the maternally derived chromosomes go to one pole, and the paternally derived chromosomes to the other? If so, meiosis would recreate the same chromosome sets generation after generation.

In fact, the assortment of chromosomes is random. Each pair of homologs aligns independently. As a result, meiosis is a major source of genetic diversity. Sutton showed that just two pairs of chromosomes can form 4 (2^2) kinds of gametes (Figure 9-11). Similarly, three pairs of chromosomes can form 8 (2^3) kinds. A single human germ cell, with 23 pairs of chromosomes, can therefore form 2^{23} (about 8 million) kinds of gametes, each one unique. A single couple, then, could theoretically produce more than 64 trillion genetically unique offspring

(8 million eggs × 8 million sperm). All this diversity comes from the random alignment and subsequent separation of tetrads in metaphase I. No wonder sexually reproducing species (including our own) are so wonderfully diverse!

During anaphase I of meiosis, sister chromatids go to the same pole in the cell. Homologs, carrying different information, go to opposite poles. In mitosis, sister chromatids go to opposite poles. Telophase and cytokinesis I resemble their counterparts in mitosis.

Meiosis II Distributes Sister Chromatids to Daughter Cells

Meiosis II resembles mitosis in a haploid cell. The cell at the end of meiosis I is haploid; and after it divides it is still haploid. Meiosis II separates the sister chromatids, so that each chromosome of the daughter cell contains a single chromatid.

Prophase II (unlike prophase I) is brief, since the chromosomes are still mostly condensed from meiosis I. If a nuclear membrane has reappeared during telophase I, it breaks down during prophase II. The chromosomes attach to newly assembled spindle fibers. Then, during metaphase II, the chromosomes line up across the equator of the spindle apparatus, with each centromere connected to spindle fibers from both poles. During anaphase II, the chromosomes split—one chromatid moves to each pole. In telophase II, the nuclear membrane forms once more. The daughter cells each have a complete haploid set of chromosomes, each consisting of a single chromatid.

Meiosis II resembles mitosis except that each resulting cell is haploid, rather than diploid.

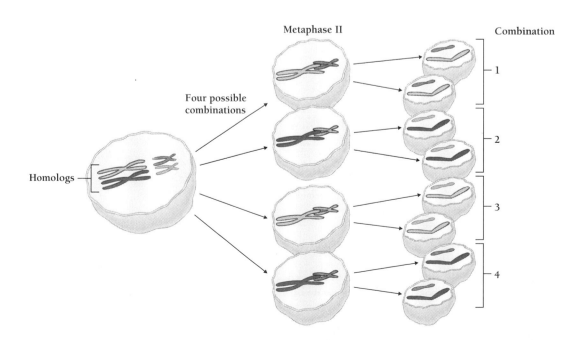

Figure 9-11 Two pairs of chromosomes can form four kinds of gametes. If the chromosomes arrange themselves as in the first cell in this figure, the resulting gametes have chromosome combinations 1 and 2. If, in other germ cells, the big chromosomes on the left swap places (which occurs at random), then the resulting gametes have combinations 3 and 4.

Metaphase II

Combination

Four possible combinations

Homologs

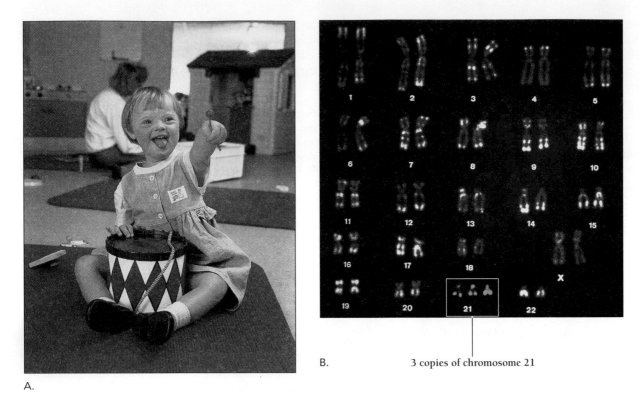

A.

B.   3 copies of chromosome 21

Figure 9-12   Children with Down syndrome display many distinctive characteristics. These include, for example, mental retardation, a flat facial profile, excess skin on the neck, double-jointedness, an unusual crease in the palm of each hand, and heart defects. B. Down syndrome results from the presence of three copies of chromosome 21.   *(A, Hattie Young/Science Photo Library/Photo Researchers; B, Kunkel/Phototake NYC)*

Disjunction and Nondisjunction

The moving apart of each pair of homologous chromosomes in anaphase I and of each pair of sister chromatids in anaphase II is called **disjunction.** Occasionally, separation fails to proceed normally, an event called **nondisjunction.** This results in a gamete having too many or too few chromosomes. In humans, for example, a sperm or egg might have 22 or 24 chromosomes instead of 23. After fertilization, the resulting zygote likewise has the wrong number of chromosomes. A cell or individual with the correct number of chromosomes is said to be **euploid** [Greek, *eu* = good, true + *ploion* = vessel], while one with an abnormal number of chromosomes is said to be **aneuploid** [Greek, *an* = not + euploid].

One of the most common results of nondisjunction in humans is **Down syndrome,** a disorder that leads to mental retardation and the abnormal development of the face, heart, and other parts of the body (Figure 9-12). Down syndrome is associated with a chromosomal abnormality called **trisomy 21** [Greek, *tri* = three + *soma* = body], the presence of three, rather than two, copies of chromosome 21.

Trisomy 21 usually results from nondisjunction during meiosis I. If, for example, the two chromosomes 21 fail to separate during anaphase I, then two of the four possible gametes

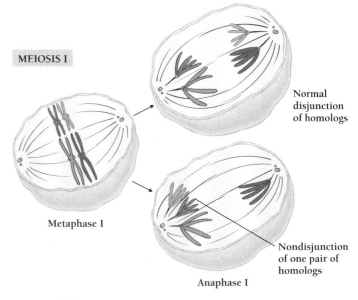

MEIOSIS I

Metaphase I

Anaphase I

Normal disjunction of homologs

Nondisjunction of one pair of homologs

Figure 9-13   In nondisjunction, homologous chromosomes fail to separate during anaphase I. The resulting gametes have either three chromosomes or one, but not the usual two.

BOX 9-1

What happens when meiosis goes wrong?

Among human embryos with trisomy 21, only about 20 percent survive until birth, although those that survive to birth may live a long time. The other 80 percent die as embryos in **spontaneous abortions,** abor-

Table A Incidence of Trisomy 21

Age of Mother	Chance of Giving Birth to Child with Trisomy 21
under 25	1/1400
under 30	1/1000
at age 35	1/350
over 40	1/100

tions that occur because the embryo is not viable. For comparison, about 70 percent of all human zygotes survive to birth. But three chromosomes are not as lethal as one. Almost none of the embryos with only one copy of chromosome 21 survive.

About one baby in every 700 has Down syndrome. Although the extra chromosome usually comes from the mother, about one-fifth of the time, the extra chromosome 21 comes from the father. Older women are much more likely than younger women to give birth to a child who has Down syndrome, but the risk increases gradually (Table A). As a rule, physicians test the embryos of women older than 35

for Down syndrome, but there is nothing magical about age 35.

Cell biologists believe that the chances of nondisjunction (at least for chromosome 21) increase with maternal age because of the pattern of meiosis in humans. The cells that form human eggs begin meiosis while the mother is herself still an embryo in the womb. But meiosis stops in the middle of prophase I, well before birth, and does not resume until after the egg is released from the ovaries, anywhere from 13 to 55 years later. Nondisjunction of chromosome 21 may be the result of interrupting the complicated process of meiosis for so long a time.

produced during meiosis will each have two copies instead of one, and the other two will have no copies of chromosome 21 (Figure 9-13). When these gametes unite with normal gametes at fertilization, they produce abnormal zygotes—some with three copies of chromosome 21, and some with only one copy (Box 9-1).

If, during anaphase I, homologous chromosomes fail to separate, the gametes will have too many or too few chromosomes.

WHY SEX?

Many organisms can reproduce either sexually or asexually. Plants, for example, may reproduce sexually, as when pollen finds its way to a flower, or asexually, as when the roots of an aspen tree sprout new trees. Animals, such as the hydra, for example, can also reproduce asexually, by budding or by regenerating a whole organism from a small part. Some animals are capable of reproducing through **parthenogenesis,** the process by which fully formed offspring develop from haploid egg cells. In populations of parthogenic lizards, for example, all individuals are females, which reproduce without mating. All the offspring are necessarily female as well, for there are no males to pass on a Y chromosome or its equivalent.

Asexual reproduction is the quickest and energetically least expensive way to reproduce. In contrast, sexually reproducing organisms must make hundreds, thousands, or even millions of gametes to ensure that a few meet for fertilization. A codfish produces some 9 million eggs per year, of which only a small portion are fertilized. A typical male human may produce 20 to 30 billion sperm over the course of the weeks or months it takes him to successfully fertilize a single egg. The female human is born with 2 million egg cells (oocytes). In addition, humans and other animals invest huge amounts of energy into finding, courting, and mating with other individuals. Many plants construct elaborate and showy flowers stocked with sweet nectar, adaptations for attracting insects and other animals that carry pollen from one plant to another. What makes sexual reproduction worth such expense?

One answer is genetic variation in one's offspring. We have seen that, because of the independent assortment of chromosomes during metaphase I, a single pair of humans can produce more than 64 trillion genetically unique zygotes. In contrast, asexual reproduction produces offspring that are genetically identical to the parent.

Sexual reproduction does far more than just replace individuals. It increases genetic diversity (and so phenotypic diversity) in populations. The genetic diversity that results from sexual reproduction makes it worth an enormous expense in energy.

Of What Value Is Genetic Variation?

In an unchanging environment, a single, perfectly adapted genotype might, theoretically, persist unchanged for millions of years. Most environments, however, change constantly, and parents that produce a diversity of offspring are more likely to have some of them survive and reproduce than parents that produce offspring that are all identical to the parents. Indeed, evolutionary biologists generally find that the less predictable an environment is, the more likely that the organisms living in it will have evolved methods for increasing diversity.

We have some hint of the importance of genetic diversity from studies of species that actually change their mode of reproduction in different environments. For example, when food and space are abundant, the water flea shown in Figure 9-14 can proliferate rapidly by asexual means, producing large numbers of genetically identical individuals. When food becomes limited or uncertain, however, the same fleas begin to reproduce sexually. The resulting phenotypic diversity increases the chance that at least some individuals will survive.

In general, a genetically diverse population is far more likely to survive a changing environment than a genetically uniform population. Genetic diversity is the raw material for natural selection and evolution. Any mechanisms that help produce variants will be favored. We must concede, however, that many biologists doubt that even the great value of genetic variability, by itself, makes sexual reproduction worth the trouble. The reasons for sex remain controversial.

Ultimately, all genetic variation arises first from **mutations**—random, spontaneous changes in the genetic information coded in the sequence of nucleotides of DNA. However, it is sexual reproduction—mixing new and old mutations into unique combinations—that increases genetic diversity most.

Genetic diversity increases the chances that some individuals will survive, no matter how the environment changes.

What Are Sex Chromosomes?

In most organisms, both males and females have complete sets of homologous (matching) chromosomes. In mammals, birds, and some other organisms, however, either the male or the female may have a pair of unmatched chromosomes. In mammals (including humans), for example, the female has two homologous "X" chromosomes, while the male has one "X" chromosome and one "Y" chromosome (Figure 9-15). The **X** and **Y** chromosomes are called the **sex chromosomes**; the rest of the chromosomes are called the **autosomes**. A human cell has 22 pairs of autosomes and one pair of sex chromosomes. (Protists, fungi, plants, and many animals lack sex chromosomes. In these organisms, sexual phenotype is determined by different mechanisms.)

The X chromosome in the male human is the same as the X chromosome in the female. The Y chromosome, however, is much smaller and bears far fewer genes than the X chromosome. In both females and males, the two sex chromosomes (XX in females and XY in males) pair and segregate during meiosis

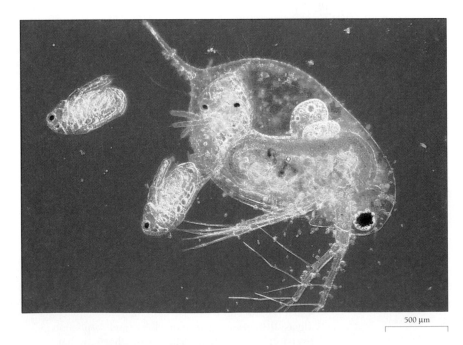

500 µm

Figure 9-14   When food is abundant, the water flea, *Daphnia*, reproduces asexually.
(Oxford Scientific Films/Animals Animals)

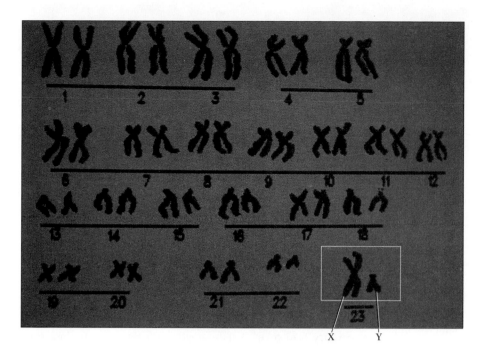

Figure 9-15   Karyotype of a male human. In certain organisms, one sex has a pair of unmatched sex chromosomes. Cells from a human male have 22 matched chromosomes and 2 sex chromosomes called X and Y.   *(Don Kelly/Grant Heilman Photography)*

in the same way as homologous pairs of autosomes. How can X and Y pair in synapsis if they differ so much in size and form? The answer is that they share a region that is homologous. Despite their striking differences in size and form, then, they still behave as homologs during meiosis. In Chapter 14, we will see that the differences between the X and Y chromosomes have important consequences for human genetics.

In mammals, the female has two homologous X chromosomes, and the male has one X chromosome and one Y chromosome.

THE INHERITANCE OF GENES PARALLELS THE INHERITANCE OF CHROMOSOMES

A Diploid Cell Has Two Copies of Every Gene

We can see that a cell that has two copies of each chromosome also has two copies of each gene. This is the key to understanding genetics in higher organisms. In addition, each copy of a gene may differ slightly from its counterpart on the match-

ing chromosome. Alternative versions of the same gene are called **alleles.** For example, in humans both chromosomes in a pair may have a certain gene that influences eye color, but one homologous chromosome might have an allele that results in brown eyes, while the other homolog might result in blue eyes.

Most genes are simply instructions for the production of a particular protein. Alleles are variations in those instructions. Just as your aunt June makes chocolate brownies one way and your cousin Ralph makes them another, different alleles specify different proteins. Sometimes the differences are subtle; sometimes the instructions for a protein (the allele) are garbled so badly that the protein does not function at all.

Let us consider the inheritance of flower color in the snapdragon (Figure 9-2). Suppose we obtain two varieties of snapdragons, one of which produces white flowers, and the other red. We can show that each of these varieties is **true-breeding** for flower color, that is, all the progeny (offspring) of the red-flowered variety produce red flowers, generation after generation, and all the progeny of the white-flowered variety produce only white flowers. In such true-breeding varieties, every chromosome carries the same allele, which specifies the same protein.

Flower color in snapdragons depends on the two alleles of a single gene. One allele (which we will call R_1) results in red flowers, and the other allele (which we will call R_2) results in white flowers. One might imagine, for example, that the R_1 gene allows the plant to make a red protein pigment, while the

R_2 gene allows the plant to make a white protein pigment (or no functioning pigment at all).

We now understand that genes are on chromosomes and that each diploid cell has two copies of each chromosome. So each cell in a snapdragon plant (apart from the gametes) should have two copies of the color gene. True-breeding varieties contain two copies of the same allele. An organism is said to be **homozygous** for an allele when it has two copies of the same allele. An organism is **heterozygous** when it contains two different alleles for a single gene. We would say, then, that the genotype of a true-breeding, red snapdragon is R_1R_1 (homozygous) and that the genotype of the true-breeding, white snapdragon is R_2R_2 (homozygous). The genotype of the pink heterozygote would be R_1R_2.

Ordinarily, snapdragons "self-fertilize." Each snapdragon flower produces both pollen (which produces sperm) and egg cells. The sperm from the pollen fertilizes eggs within the same flower. A plant breeder can prevent self-pollination by snipping off the pollen-carrying structures and then artificially pollinating the flower with pollen from another plant. This method of breeding two genetically distinct organisms is called **cross breeding** (or **crossing**), and the progeny of such a cross are called **hybrids.** In snapdragons, the results of crossing a true-breeding (homozygous), red snapdragon with a true-breeding (homozygous), white snapdragon are heterozygous hybrids with pink flowers.

The original parents in a cross are called the **parental,** or **P,** generation, and the progeny are the "first filial" [Latin, *filia* = daughter, *filius* = son], or **F1,** generation. The **F2** generation are the progeny of the F1 generation. In the snapdragon cross, then, the parents are R_1R_1 and R_2R_2 homozygotes, and the F1 generation are R_1R_2 heterozygous hybrids. The genotypes of the heterozygous hybrids are all R_1R_2, and their phenotypes are pink flowers.

The key to understanding genetics in higher organisms is the realization that there are two copies of each gene, one copy on each homolog in a chromosome pair.

What Are the Genotypes and Phenotypes of the F2 Generation?

A typical question that a geneticist would ask is, "What color flowers will a snapdragon grower obtain by crossing these F1 hybrids with one another?" We know that meiosis produces gametes that have just one of each pair of homologous chromosomes. The genotype of the F1 hybrids is R_1R_2, so each snapdragon gamete receives either R_1 or R_2. Half of all the gametes contain the R_1 allele, and half the R_2 allele. We can now make some guesses about the distribution of genotypes in the F2.

A convenient way to make these calculations relies on a checkerboard, shown in Figure 9-16, invented by an early 20th

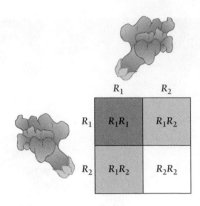

Figure 9-16 Inheritance of flower color in snapdragons presented as a Punnett square. A cross between two R_1R_2 heterozygotes produces three genotypes: snapdragons homozygous for R_1, snapdragons homozygous for R_2, and R_1R_2 heterozygous snapdragons.

century British geneticist named Reginald Punnett. In a **Punnett square,** we represent each kind of gamete produced by one parent along the top of the square and each kind of gamete produced by the other parent along the left side of the square. Within the small squares, we can then write the genotypes that would be produced by each combination. Note, however, that the R_1R_2 genotype is the same as the R_2R_1 genotype. That is, the flowers are pink no matter which allele comes from which parent.

The Punnett square shows that the three possible genotypes of the F2 generation of our snapdragon experiment should be in the ratio $1:2:1$. Because the pink phenotype of the heterozygote is different from the phenotypes of either homozygote (red or white), we can directly count the ratios of the genotypes in the F2 generation. These experiments always produce a ratio fairly close to the expected $1:2:1$.

If we cross two R_1R_2 heterozygotes, we get three kinds of genotypes. One-quarter of the offspring are homozygous for R_1, one-quarter are homozygous for gene R_2, and one-half are heterozygous R_1R_2.

GREGOR MENDEL ESTABLISHED THE PRINCIPLES OF GENETICS

Today, the inheritance of flower color in snapdragons makes sense to us because we understand the relationship between the chromosomes and the genes. But the explanation would not seem at all simple if we did not already have a model in mind. Nineteenth-century biologists knew little about chromosomes and nothing about genes. For them, Gregor Mendel's principles of inheritance were difficult to fit into a larger understanding of inheritance. Our hard-won understanding of the behavior of chromosomes and genes gives us a great advantage.

Mendel Chose Traits and Plant Stocks Carefully

Gregor Mendel was the first person to realize the significance of the ratios of different phenotypes in genetic crosses (Figure 9-17). In his monastery's garden in what is now the Czech Republic, Mendel studied seven different pairs of traits (Figure 9-18): round versus wrinkled seeds, yellow versus green seeds, purple versus white flowers, inflated versus constricted pods, green versus yellow pods, axial versus terminal flowers, and tall versus dwarf stems. Each of these pairs of traits showed the same pattern of inheritance. Mendel chose pairs of traits in which each variant was distinct (in other words, tall could not be confused with short).

In each case, Mendel showed that the parental stocks were true-breeding. For example, self-pollination of plants with

Figure 9-17   Gregor Mendel in his monastery garden. Before Mendel's death, and just 16 years before the dramatic rediscovery of his work, he remarked, "Though I have suffered some bitter moments in my life, ... My scientific work has brought me a great deal of satisfaction, and I am convinced that it will not be long before the whole world acknowledges it." *(The Bettmann Archive)*

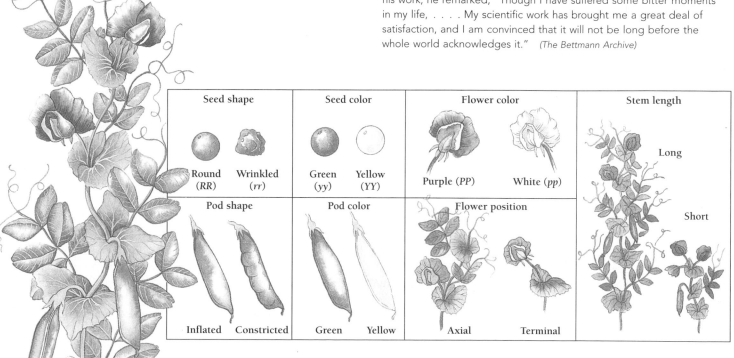

Figure 9-18   Mendel's seven traits in pea plants. A pea plant has just seven chromosomes, and each of the genes for the seven traits that Mendel chose lies on a different chromosome.

Figure 9-19   Mendel's studies of inheritance in peas required that he prevent self-fertilization. Here, the pollen-producing anthers are snipped from a flower and the flower is fertilized with pollen from another flower.

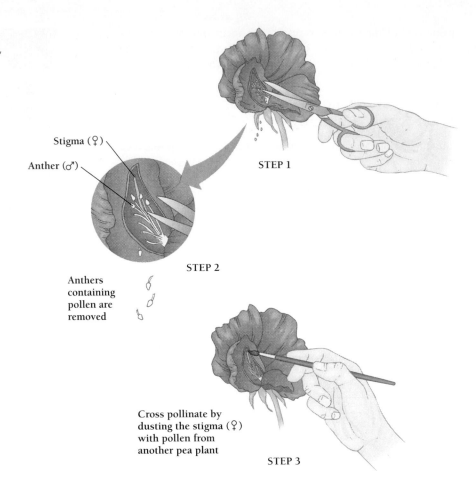

Stigma (♀)

Anther (♂)

STEP 1

STEP 2

Anthers containing pollen are removed

Cross pollinate by dusting the stigma (♀) with pollen from another pea plant

STEP 3

round seeds always produced progeny with round seeds. Mendel then prevented self-pollination and cross pollinated flowers with different traits, a purple-flowered pea with a white-flowered pea, for example (Figure 9-19).

F1 hybrids do not always have an intermediate phenotype like the pink flowers of hybrid snapdragons. Instead, a heterozygous hybrid may have the same phenotype as that of the parents. For example, when Mendel crossed true-breeding, purple-flowered peas with true-breeding, white-flowered peas, all the hybrids had purple flowers (Figure 9-20A). The F1 hybrids are all heterozygotes, but only the allele for purple flowers is expressed. We say that an allele is **dominant** when it alone determines the phenotype of a heterozygote, as in the case of the purple allele in peas. An allele is **recessive** when it contributes nothing to the phenotype of a heterozygote, as is the case of the white allele of peas.

Today we know that a dominant allele usually contains the instructions for a functional protein, while a recessive allele often lacks this capacity. Many recessive alleles, such as one that causes type O blood, are extremely common. Others, such as the allele for the genetic disease Tay-Sachs, which kills children at 3 or 4 years of age, are quite rare. Lethal recessives such as Tay-Sachs are rare because people who are homozygous for this allele die before passing on the gene in reproduction. This is an extreme form of natural selection.

When two alleles each contribute to the phenotype, as in the case of the R_1 and R_2 alleles of snapdragons (which together gave a color different from either homozygote), the alleles are said to lack dominance. In **partial dominance**, for example, both alleles are expressed. In the 19th century, partial dominance seemed to provide evidence for blending inheritance. But blending inheritance cannot explain the reappearance of red and white snapdragons after crossing the hybrids. By a convention invented by Mendel, we begin the names of dominant and codominant alleles with a capital letter, and those of recessive alleles with a small letter.

In each of Mendel's seven crosses, all of the F1 hybrids had the same phenotype: plants that produced round peas crossed with wrinkled pea stocks gave F1 hybrids with round peas; tall pea stocks crossed with short pea stocks gave tall F1 hybrids; and so on. We say that the purple flower allele is dominant to the white flower allele, the round allele is dominant to the wrinkled allele, and the tall allele is dominant to the short allele.

Mendel began his experiments with true-breeding stocks, choosing pairs of traits in which the two alternate forms were distinct. Each trait was determined by just two alleles, and each allele in a pair was clearly dominant or recessive.

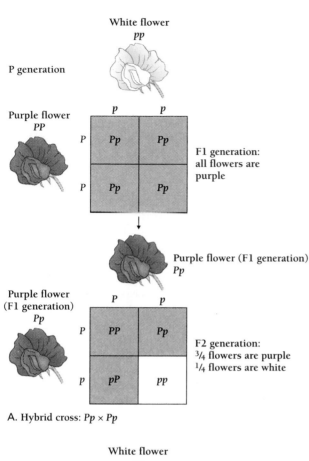

Figure 9-20   A backcross of purple- and white-flowered peas.
A. First, a homozygous white flower is crossed with a homozygous purple flower to produce a 1:2:1 mix of homozygotes and heterozygotes. B. Then, in the backcross, a white homozygote is crossed with a purple heterozygote; half the offspring are purple heterozygotes and half are white homozygotes.

Mendel's Insights Came from Careful Counting

Mendel allowed the F1 plants to self-pollinate, then counted the F2 progeny. In every case, the dominant allele phenotype made up about three-fourths of the progeny, and the recessive allele phenotype made up the remaining one-fourth of the progeny.

We can understand this 3:1 ratio in terms of the behavior of chromosomes in meiosis and fertilization. Just as in the snapdragon cross, one-half of the progeny were homozygotes, and one-half were heterozygotes.

In each of Mendel's crosses, one allele was dominant and one recessive, so the F2 heterozygotes had the same phenotype as the homozygotes. (They had the same phenotype, despite their different genotypes.) In the case of the round versus wrinkled cross, for example, $\frac{1}{4}$ were homozygous for the round allele, $\frac{1}{4}$ were homozygous for the wrinkled allele, and $\frac{1}{2}$ were heterozygous. Since the round allele is dominant, the heterozygotes had the same phenotype as the homozygote for the round allele. So $\frac{3}{4}$ ($\frac{1}{4} + \frac{1}{2}$) were round, and $\frac{1}{4}$ were wrinkled.

The Principle of Segregation

Mendel had to make sense of his data without the knowledge of genes and chromosomes that we now possess. He did this just by considering the possible meanings of the ratios of phenotypes in his breeding experiments.

Mendel's first conclusion is called the **Principle of Segregation**: Each sexually reproducing organism has two "determinants" (or genes, in modern terms) for each characteristic; these two determinants *segregate* (or separate) during the production of gametes. From the rules of probability, Mendel could predict the ratios of progeny with different characteristics.

To test the Principle of Segregation, Mendel performed a new kind of cross, called a **backcross**. Instead of allowing the F1 heterozygotes to self-pollinate, he crossed them with the homozygous parental stock, those which contained only the recessive allele (Figure 9-20B). A backcross to a homozygous recessive that is intended to reveal the genotype of the other parent is called a **testcross**. The parents in this cross would look just like those in the original cross—a purple-flowered plant with a white-flowered plant. But Mendel had guessed that the purple-flowered F1 hybrids included two different genotypes and should therefore produce two kinds of gametes, one with the purple allele, and the other with the white allele (whereas the original purple parents produced only purple alleles). The plants with white flowers, on the other hand, would only produce gametes with the white allele. He correctly predicted that half the progeny should have purple flowers and half white flowers.

Mendel's hybrid crosses and backcrosses produced offspring with ratios of phenotypes that suggested the Principle of Segregation; each sexually reproducing organism has two genes for each characteristic, which segregate during the production of gametes.

The Principle of Independent Assortment

In one of his next experiments, Mendel took two true-breeding stocks that differed in both the color and the form of the peas they produced. One parent produced round, yellow peas;

the other wrinkled, green peas. From previous crosses, he knew that round was dominant to wrinkled and yellow was dominant to green. As he expected, all the F1 peas were yellow and round. He then determined how these traits were inherited in the F2 generation. The allele specifying round peas is *R*, and the one for wrinkled peas is *r*; the allele specifying yellow peas is called *Y*, and the one for green peas is *y* (Figure 9-21). The genotypes of the parents are therefore *RRYY* and *rryy*.

Homozygous parents can produce only a single kind of gamete: *RRYY* plants produce *RY* gametes, and *rryy* plants produce *ry* gametes. The genotype of the F1 plants therefore must be *RrYy*.

What happens in the F2 generation? Mendel allowed the F1 to self-pollinate and then counted 556 peas in the F2. Of these, 315 were yellow and round (representing the action of the dominant alleles), and only 32 were green and wrinkled (Figure 9-21). The rest represented new phenotypes, unlike either of the parental stocks: 101 were yellow and wrinkled, and 108 were green and round. The two crosses had produced new

combinations of genes, illustrating again how sexual reproduction provides diversity in a population. But how did Mendel make sense of the numbers?

First, he noted that the traits associated with each gene were in a ratio of about 3 : 1, just as in his experiments with only one trait. In the F2 generation there were three times as many yellow as green peas (416 : 140 = 3 : 1), and three times as many round as wrinkled (423 : 133 = 3.2 : 1). Mendel immediately saw that the two traits were behaving independently; whether a pea was yellow or green had no effect on whether it was wrinkled or round.

Mendel could explain the data if he assumed that each F1 plant produces four kinds of gametes: *ry*, *Ry*, *rY*, and *RY*. The Punnett square of Figure 9-21 predicts the distribution of genotypes and phenotypes of the F2 generation of this cross. It shows that 9 out of the 16 combinations of gametes produce plants with round, yellow peas, but only one combination produces plants with wrinkled, green peas. The overall ratio of phenotypes predicted by this analysis is 9 : 3 : 3 : 1.

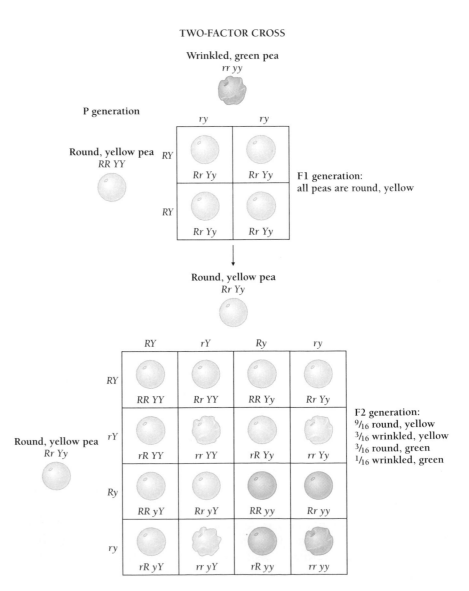

Figure 9-21   A two-factor cross tests two traits at once. Mendel crossed round, yellow peas with wrinkled, green peas. In the first (F1) generation, all the peas were round, yellow heterozygotes. When he crossed the F1 generation, he got an assortment of different peas.

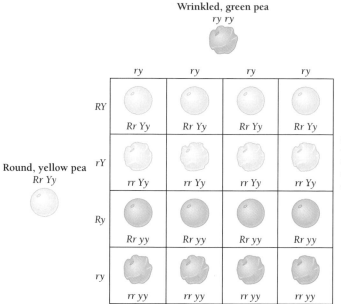

TWO-FACTOR BACKCROSS

Wrinkled, green pea
ry ry

Round, yellow pea
Rr Yy

Offspring:
4/16 (1/4) round, yellow
4/16 (1/4) wrinkled, yellow
4/16 (1/4) round, green
4/16 (1/4) wrinkled, green

Figure 9-22 A two-factor backcross allowed Mendel to distinguish genotypes of the offspring of his two-factor cross.

The correspondence of Mendel's experiments to this prediction is called the **Principle of Independent Assortment.** This principle states that each pair of genes is distributed independently during the formation of the gametes. The independent assortment of homologous chromosomes during meiosis underlies the independent assortment of genes and thereby results in genetic recombination.

Mendel verified his understanding of the genotypes of the gametes produced by the heterozygous F1 hybrids by doing another testcross with plants that produce green, wrinkled peas (Figure 9-22). Because both *r* (for wrinkled peas) and *y* (for green peas) are recessive, the genotype of the plants used in the backcross must be *rryy*. The gametes produced by these plants all must be *ry*.

If the genotype of the F1 plants is really *RrYy* and the chromosomes bearing the two genes assort independently, then the four kinds of gametes (*RY, rY, Ry,* and *ry*) should be made in equal numbers (Figure 9-22). When these gametes form zygotes with *ry* gametes, we expect equal numbers of the four phenotypes, in contrast to the 9:3:3:1 ratio expected in the case of self-fertilization of the F1 plants. Again, Mendel obtained the expected ratios, confirming his understanding of the process.

Mendel's Principle of Independent Assortment states that
each gene in a pair is distributed independently during
the formation of the gametes.

SUPPORT FOR THE CHROMOSOMAL THEORY OF INHERITANCE

Mendel presented his work to the Brünn Society for the Study of Natural Sciences in 1865 and published a formal report in the society's journal *Transactions* the following year. His experiments and analysis are among the most elegant and important in the whole history of science. In the succeeding years, other scientists frequently mentioned his work in their own papers. In fact, a 19th-century edition of the *Encyclopedia Britannica* cited Mendel's paper. Yet, partly because biologists did not understand Mendel's mathematics, his discoveries made absolutely no impact until 16 years after his death in 1884.

Then, in 1900, three botanists, Hugo de Vries in the Netherlands, Karl Correns in Germany, and Erich Tschermak in Austria, independently announced the rediscovery of Mendel's laws. These biologists repeated Mendel's work and confirmed his conclusions. In 1902, Walter Sutton and the German biologist Theodor Boveri independently published papers that showed that the behavior of chromosomes during meiosis accounted for the Mendelian principles of segregation and independent assortment. Sutton's argument consisted of six separate points that together formed an almost fortresslike argument for what came to be known as the **Sutton-Boveri chromosomal theory of inheritance.** Sutton's first three points merely confirm the ideas of others; the second three are his own ideas.

First, said Sutton, the chromosomes are actually pairs of chromosomes, one of which comes from the mother and one from the father. Second, synapsis consists of the pairing of homologous maternal and paternal chromosomes. Because the chromosomes of the lubber grasshopper had distinctive sizes and shapes, Sutton was able to see, as had others, that the chromosomes occurred in pairs, which separated during the course of meiosis.

Third, the chromosomes retain their individual forms throughout the cell cycle. Sutton argued that the constant differences in size and shape of the chromosomes suggested that the chromosomes are not recreated every cell cycle. The best evidence for this idea came from Boveri, who had studied sea urchin cells with reduced numbers of chromosomes. Only cells with the correct number of chromosomes developed normally. Boveri concluded that since every chromosome had to be present for normal development, each chromosome must be unique and must make a unique contribution to development.

Fourth, said Sutton, the chromosomes contain Mendel's genes. Fifth, meiosis creates new combinations of genes in each generation. During meiosis, when the diploid cell is reduced to two haploid cells, each homolog in a pair orients randomly with respect to the two poles of the dividing cell. As a result, either homolog may end up in either new cell, regardless of the way all the other chromosomes divide. This, said Sutton, accounted for Mendel's principle of independent assortment. In addition, because the maternal and paternal homologs of each chromosome migrate to each new cell independently of the other homologs, each parent can produce 2^n kinds of gametes, where n is the number of chromosome pairs. As we saw earlier, and as Sutton himself argued, an organism with two pairs of chromosomes, for example, would have 2^2, or 4, types of gametes, while a human could produce 2^{23}, or more than 8 million kinds of gametes. (In fact, the process of crossing over, about which Sutton knew nothing, allows organisms to make even more kinds of gametes than Sutton realized.)

Sixth, argued Sutton, each chromosome carries a different set of genes, and all of the genes on one chromosome are inherited together. That different chromosomes are inherited independently accounts for Mendel's principle of independent assortment. We now see Sutton's insights as the basis for understanding the chromosomal basis of the laws of inheritance. Table 9-1 summarizes the rules of inheritance that we can take both from Sutton's work and from subsequently gained knowledge. Because of Sutton's synthesis and because of vast amounts of subsequent work in genetics, we are in a better position to understand genetics than were the 19th-century biologists who first discovered the physical basis of heredity.

How Did Thomas Hunt Morgan Bolster the Theory of Chromosomal Inheritance?

The fruit fly *Drosophila melanogaster* has been a favorite organism for genetic research since 1909, when the embryologist and geneticist Thomas Hunt Morgan first began breeding the tiny flies. As Morgan discovered, fruit flies have many advantages

Table 9-1   Walter Sutton's Six Rules of Inheritance*

1. The chromosomes are actually pairs of chromosomes, one of which comes from the mother and one from the father.
2. Synapsis is the pairing of homologous maternal and paternal chromosomes. These pairs separate during meiosis.
3. The chromosomes are not broken down and created anew during each cell cycle.
4. The chromosomes carry Mendel's genes.
5. Meiosis creates new combinations of genes in each generation. During meiosis, either homolog may end up in either new cell, regardless of the way all the other homologous chromosomes divide. This accounts for Mendel's principle of segregation.
6. Each chromosome carries a different set of genes, and all of the genes on one chromosome are inherited together. That different chromosomes are inherited independently accounts for Mendel's principle of independent assortment.

*The first three rules confirmed the ideas of others; the second three were Sutton's own ideas.

for genetic research. Among them: (1) they breed rapidly, growing from eggs to sexually mature adults in less than two weeks (Figure 9-23); (2) they are prolific, with each female laying hundreds of eggs; (3) they are small (about 3 mm long) and easy to maintain; (4) they are intricately and precisely built organisms, with dozens of easily identifiable traits under genetic control.

Morgan collected flies by leaving a piece of ripe pineapple on his windowsill. He kept the flies in small milk bottles "borrowed" from his home milk delivery man. Morgan's laboratory at Columbia University, in New York, called the "fly room," was the center of genetic research for over 25 years.

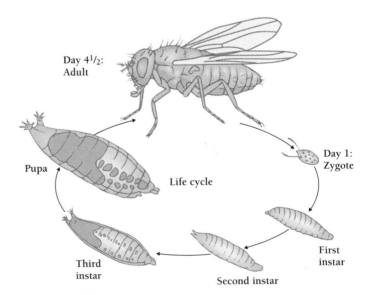

Day 4½: Adult

Pupa

Life cycle

Day 1: Zygote

Third instar

Second instar

First instar

Figure 9-23   The short life cycle of the fruit fly *Drosophila melanogaster.*

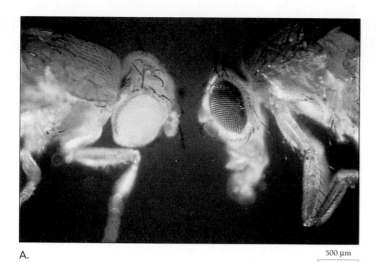

A.
500 µm

Figure 9-24 Geneticists have isolated scores of mutant types in the fruit fly *Drosophila melanogaster.* A. Some have white eyes instead of red eyes. B. Some have legs growing where their antennae should be. *(A, Dr. Jeremy Burgess/Science Photo Library/Photo Researchers; B, Thomas Kaufman)*

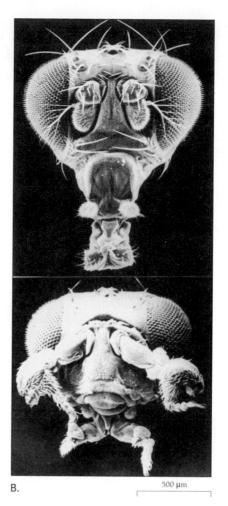

B.
500 µm

Since 1909, thousands of geneticists and hundreds of thousands of undergraduate biology students have studied inheritance in *Drosophila,* showing again and again that traits such as eye color, body color, and wing shape are all inherited in accordance with Mendel's laws (Figure 9-24).

Genes on the X Chromosome Are Inherited Differently in Males and Females

Nearly 90 years of detailed genetic analysis of *Drosophila* has allowed geneticists to understand much about the relationships between genes and chromosomes. One of Morgan's first discoveries concerned a gene for eye color, which, he showed, lies on the X chromosome.

Drosophila has four pairs of chromosomes: males have three pairs of autosomes plus one pair of unmatched sex chromosomes, called X and Y; *Drosophila* females have three pairs of autosomes plus one pair of X chromosomes. Many genes that lie on the X chromosome, then, will be present in two copies in females, but in only one copy in males. Males are therefore said to be *hemizygous* for certain genes on the X chromosome. That is, like male humans, male flies have only one (instead of two) copies of certain X-chromosome genes.

Most wild fruit flies have red eyes. One day, however, Calvin Bridges, a Columbia undergraduate hired to wash the dirty milk bottles in Morgan's laboratory, noticed a single male fly with white eyes. Morgan crossed the white-eyed male with red-eyed females. All the F1 progeny had red eyes, showing that the white-eye allele must be recessive and the corresponding red-eye allele dominant.

Morgan then crossed the F1 males and females and obtained an overall ratio of 3:1 of red-eyed flies to white-eyed flies. But all of the white-eyed flies were male! In addition, half of the F2 males had red eyes, but all of the females had red eyes. What was going on?

Morgan hypothesized that a single gene determined eye color in this case and that that gene was on the X chromosome. We can diagram Morgan's experiment with Punnett squares, designating the dominant red allele as X^W (because it is on the X chromosome) and the recessive white allele as X^w (because it is on the X chromosome, as well).

We see that the males and females in the F1 generation have different genotypes. The females have the white-eye allele (*w*) on one X chromosome and the dominant red-eye allele (*W*) on the other X chromosome. The males have the red eye allele on their single X chromosome and a Y chromosome with no allele at all. The F1 males are hemizygous for the red allele on the X chromosome because they only get the one allele (from their mother). The F1 females are heterozygous: they receive one *W* and one *w*.

In the F2 generation, all the males receive their Y chromosomes from their fathers and their X chromosomes from their mothers. Because the Y chromosome has no *white* allele,

(Text continued on page 228.)

BOX 9-2

Sex linkage in humans

Some 10,000 genes—about 10 percent of all human genes—are located on the X chromosome. Because the Y chromosome lacks many of these genes, human males have only one copy of such **sex-linked genes.** A woman may carry an abnormal recessive allele on one of her X chromosomes without any effect, since she nearly always carries a normal allele on the other X chromosome. The normal gene may produce all the protein she needs. A man who inherits the same abnormal recessive, however, may have no other allele to compensate for the nonfunctioning gene, and he is left missing a crucial protein or with an altered form of the protein that behaves differently from the usual version.

A number of abnormal phenotypes result from such sex-linked traits. For example, about 8 percent of American men are color blind, because they have only one of

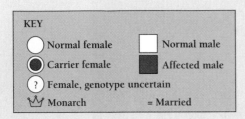

KEY

- ◯ Normal female
- ◉ Carrier female
- ⑦ Female, genotype uncertain
- ♔ Monarch
- ☐ Normal male
- ■ Affected male
- = Married

Figure A Hemophilia in the family of **Queen Victoria.** One of Victoria's sons (Leopold) was a hemophiliac, and two of her daughters (Alice and Beatrice) were carriers. Alice and Beatrice had four carrier daughters and three hemophiliac sons. One of these carriers was Alexandra, who married the Czar of Russia and bore Alexis (bottom, center).

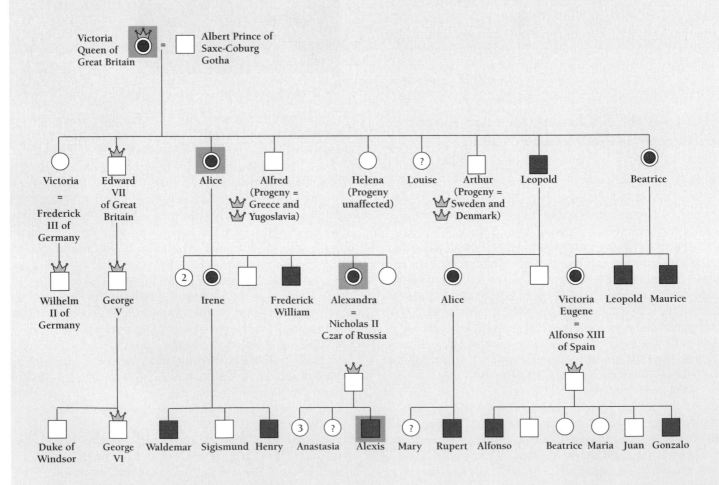

two genes on the X chromosome that specify the eye pigments responsible for color vision. Still other alleles for genes on the X chromosome cause severe diseases of muscles, brain, and blood. The most famous sex-linked disease is *hemophilia*, which is caused by a defective allele for a protein needed for normal clotting of the blood (Figure A). In the absence of this protein, the blood will not clot, and, until recently, a hemophiliac who suffered even a minor injury invariably bled to death (Figure B).

Until now, the major treatment for hemophilia has been transfusion, both to replace lost blood and to provide the protein that is missing in hemophiliacs. Patients who receive frequent transfusions, however, are susceptible to infection by viruses carried in the transfused blood, including hepatitis and HIV (the virus that causes AIDS). Today testing programs make it possible to identify and discard most virus-contaminated blood, but many people who have AIDS are hemophiliacs who received

blood before such tests were available. For many hemophiliacs, an even safer solution came from molecular biology—the isolation of the gene for the clotting protein, called Factor VIII, that is missing in one type of hemophilia. Recombinant DNA techniques, such as those discussed in Chapter 13, can now provide a cheap, safe source of Factor VIII.

(a)

(b)

Figure B   The hemophilia allele played an important role in the Russian Revolution of 1917. (a) The Russian Royal Family just before the Revolution. Nikolas II, the last Czar of Russia, married Queen Victoria's granddaughter Alexandra, who carried the allele for hemophilia. (b) Alexandra eventually fell under the influence of the "mad monk" Rasputin, who convinced her that only he could save Alexis. Rasputin's power over the royal family intensified Nikolas's growing unpopularity with the Russian people and led, ultimately, to the family's downfall. *(Culver Pictures)*

they receive no allele from their fathers. From their mothers, half of the male F2 flies receive X^W and have red eyes, and half receive X^w and have white eyes.

The females receive one X chromosome from each parent. All the females receive the dominant X^W from their fathers and have red eyes. Of these, however, half should also receive the dominant X^W from their mothers and be homozygous, and half should receive the recessive X^w and be heterozygous.

Morgan checked this prediction by performing a backcross, mating the presumably heterozygous F1 females with the original white-eyed male. His results were exactly as he predicted. Half of the males and half of the females had white eyes. This result established that the white-eye gene indeed lay on the X chromosome.

The gene studied in this experiment is one of almost 1000 genes in *Drosophila* that are sex-linked. In flies and other organisms with unmatched sex chromosomes, **sex-linked genes** have different patterns of inheritance in males and females because they lie on chromosomes that also determine the gender of the offspring. In mammals and flies, for example, the X chromosome contains many more genes than the Y chromosome, and sex linkage therefore nearly always refers to genes on the X chromosome. This experiment and others performed by Morgan and his students helped establish that genes were indeed on chromosomes, just as Sutton had proposed a decade earlier. (Sex-linked genes are not the same as genes that determine what each sex looks like. Many genes for sex determination do not lie on the sex chromosomes.)

Experiments with sex-linked genes by Thomas Hunt Morgan provided support for Sutton's theory that the genes lie on the chromosomes.

Genes Lie on the Chromosomes

When Gregor Mendel published his work on peas in the mid-19th century, he described independent assortment in seven pairs of traits. What he did not know, and what modern geneticists at first found astounding, is that the pea has only seven chromosomes. For Mendel's seven traits to assort independently, each gene would have to be on a separate chromosome. At first glance, it looked as if Mendel had been remarkably lucky in choosing exactly seven traits on exactly seven different chromosomes. In fact, it is now clear that Mendel sifted through scores of traits in some 34 varieties of pea plant for the traits that would confirm his hypothesis.

Some scientists have objected, in retrospect, to Mendel's approach because it is not how science is properly done. Others take a more generous view. They interpret Mendel's published work as a carefully constructed demonstration of what he did understand, in which he excluded the many aspects of genetics that were then unexplainable. He was merely pre-

senting the simplest possible cases, much as molecular biologists began by studying the simple genomes of viruses and bacteria instead of mice. Now we will see what happens to Mendel's rules when two genes lie on the same chromosome.

Morgan's work on sex linkage helped establish that genes were indeed on chromosomes. The door was now open for a new kind of question: just *where* on the chromosome does a particular gene lie? There are many more genes than there are chromosomes. At least 1000 genes lie on the *Drosophila* X chromosome, and the human X chromosome may contain 10,000 or more. The principle of independent assortment depends on the independent assortment of chromosomes, but it cannot hold for genes that are physically attached to one another on the same chromosome. That is, if Mendel had chosen an eighth trait, it would not have assorted independently. Sutton recognized this problem in 1903, but Morgan and his students were the first geneticists to design experiments around this question. In so doing, they further established the chromosomal basis of inheritance.

Morgan crossed two types of flies. One kind had wings of a more or less standard length and red eyes—characteristics usually found in wild populations. These flies were called the *wild type*. The second kind of fly had short, stubby "vestigial" wings and purple eyes (Figure 9-25). Morgan had previously shown that purple-eyed flies with vestigial wings bred true, and he concluded that the purple, vestigial phenotype results from the homozygous condition of two genes. Morgan named the purple allele *pr* and the vestigial allele *vg*; he called the corresponding wild-type alleles pr^+ and vg^+. So the genotype of the purple vestigial flies was *prpr vgvg*. Each such fly could produce eggs or sperm carrying both the *pr* and the *vg* allele.

Purple vestigial flies crossed with true-breeding, wild-type flies produced phenotypically normal (wild-type) F1 flies (Figure 9-25A). This showed that *pr* and *vg* were recessive alleles, and pr^+ and vg^+ were dominant.

To assess the independent assortment of the *pr* and *vg* genes, Morgan performed a backcross. He mated the heterozygous F1 flies with purple vestigial flies. If the two genes assorted independently, this cross should have yielded four possible phenotypes, in equal proportions: $\frac{1}{4}$ purple eyes, vestigial wings; $\frac{1}{4}$ purple eyes, normal wings; $\frac{1}{4}$ red eyes, vestigial wings; and $\frac{1}{4}$ red eyes, normal wings (Figure 9-25B).

Instead, almost 90 percent of the flies had the same phenotypes as the two original parents—either purple eyes and vestigial wings or red eyes and normal wings. The remaining 10 percent of the flies were divided into flies with purple eyes and normal wings and flies with red eyes and vestigial wings.

Morgan saw that the two genes whose alleles produced purple eyes and vestigial wings did not assort independently. The *vg* and *pr* alleles stay together, and the vg^+ and pr^+ alleles stay together. Morgan guessed that these alleles stayed together because they were on the same chromosome, and that genes assort independently only when they are on separate chromosomes (or, as we will see later, quite far apart on the same chromosome).

A. ORIGINAL CROSS

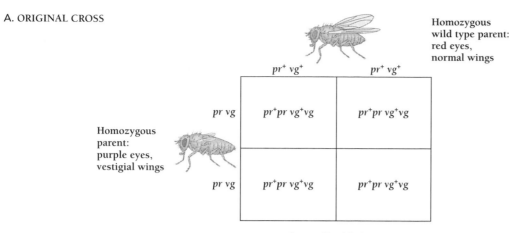

Homozygous
wild type parent:
red eyes,
normal wings

	$pr^+ vg^+$	$pr^+ vg^+$
$pr\ vg$	$pr^+pr\ vg^+vg$	$pr^+pr\ vg^+vg$
$pr\ vg$	$pr^+pr\ vg^+vg$	$pr^+pr\ vg^+vg$

Homozygous
parent:
purple eyes,
vestigial wings

Result: F1 all wild phenotype

B. TEST CROSS

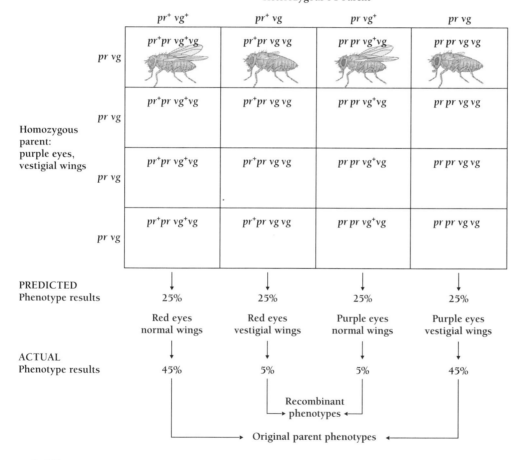

Heterozygous F1 Parent

	$pr^+ vg^+$	$pr^+ vg$	$pr\ vg^+$	$pr\ vg$
$pr\ vg$	$pr^+pr\ vg^+vg$	$pr^+pr\ vg\ vg$	$pr\ pr\ vg^+vg$	$pr\ pr\ vg\ vg$
$pr\ vg$	$pr^+pr\ vg^+vg$	$pr^+pr\ vg\ vg$	$pr\ pr\ vg^+vg$	$pr\ pr\ vg\ vg$
$pr\ vg$	$pr^+pr\ vg^+vg$	$pr^+pr\ vg\ vg$	$pr\ pr\ vg^+vg$	$pr\ pr\ vg\ vg$
$pr\ vg$	$pr^+pr\ vg^+vg$	$pr^+pr\ vg\ vg$	$pr\ pr\ vg^+vg$	$pr\ pr\ vg\ vg$

Homozygous
parent:
purple eyes,
vestigial wings

PREDICTED
Phenotype results

25%	25%	25%	25%
Red eyes normal wings	Red eyes vestigial wings	Purple eyes normal wings	Purple eyes vestigial wings

ACTUAL
Phenotype results

45%	5%	5%	45%

Recombinant phenotypes

Original parent phenotypes

Figure 9-25   Genetic linkage. The plus signs indicate the dominant alleles: red eyes and normal wings. A. In the original cross between a red-eyed parent with normal wings and a purple-eyed parent with vestigial wings, all the offspring are heterozygotes with red eyes and normal wings (the "wild type"). B. In a backcross among these heterozygotes, the ratio among the different genotypes differs from what the Law of Independent Assortment predicts. These two genes do not assort independently because they are on the same chromosome.

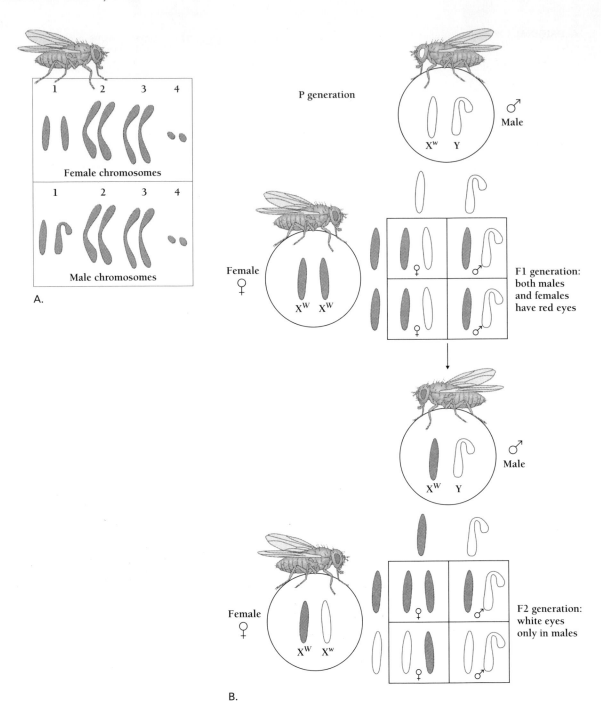

Figure 9-26   The four linkage groups in fruit flies correspond to the four pairs of chromosomes. A. As in humans, the male has an unmatched set of sex chromosomes. B. The white gene is on the sex chromosome and we can trace its descent through two crosses by looking at the way the sex chromosomes assort.

The tendency of genes on the same chromosome to stay together is called **genetic linkage**. Morgan and his students found that all of the fly alleles they studied fell into just four distinct **linkage groups**, sets of genes that do not assort independently. The four linkage groups, Morgan realized, corresponded to the four pairs of chromosomes in *Drosophila* (Figure 9-26). One linkage group consisted of sex-linked traits, indicating that it included all of the genes on the X chromosome. The fact that genes lie on chromosomes explains not only Mendel's laws, but, also the existence of linkage groups.

Morgan's discovery of linkage groups further bolstered the chromosomal theory of inheritance.

Genetic Recombination Can Arise from Independent Assortment or from Crossing Over

Genetic linkage explains why 90 percent of the flies in the cross shown in Figure 9-25B resemble one or the other parental phenotype. But why do some flies in the backcross have purple eyes and normal wings?

The answer is that homologous chromosomes frequently exchange material with each other during crossing over in meiosis. Recall from earlier in this chapter that crossing over occurs when homologous chromosomes pair up during meiosis I. Single chromatids in homologous chromosomes break, exchange material, and rejoin.

After crossing over, a new chromosome contains some genes from each of the two homologs. These new combinations of genes then stay together through the rest of meiosis. The chromosomes of the resulting gametes thus differ from the chromosomes in the gametes of the previous generation.

Crossing over, like independent assortment, recombines alleles.

Studies of Corn Showed That the Chromosomes Literally Cross Over

The physical basis of crossing over was first demonstrated in 1931 by Harriet Creighton and Barbara McClintock at Cornell University. Creighton and McClintock, who were studying the genetics of corn, took advantage of a pair of homologous chromosomes that differed in appearance (Figure 9-27). One homolog was relatively plain, while the other had a knob at one end and a long tail at the other end. Creighton and McClintock compared the pattern of inheritance of alleles in corn with the appearance of the chromosomes. Corn plants whose linked traits clearly indicated crossing over had homologs with a knob or a tail at one end, but not both. Creighton and McClintock showed that genetic recombination involves actual physical exchanges between paired homologous chromosomes.

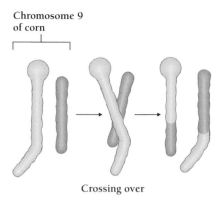

Chromosome 9 of corn

Crossing over

Figure 9 27   Crossing over in corn chromosomes. Creighton and McClintock easily showed crossing over in chromosome 9 because this pair of chromosomes is unmatched.

Harriet Creighton and Barbara McClintock showed that crossing over is a physical exchange of material between homologous chromosomes. Their work was immediately accepted as evidence that confirmed the chromosomal theory of inheritance.

The Frequency of Crossing Over Between Two Genes Reflects the Physical Distance Between Them

The farther apart two genes are on the chromosome, the greater the likelihood that they will cross over. The discoverer of this insight was Alfred Sturtevant, who was, like Bridges, an undergraduate in Morgan's laboratory. Sturtevant realized that he could express the genetic distance between two genes as the chance that recombination would occur between them. He defined a **map unit** as the distance between two genes that would produce 1 percent recombinant gametes (and 99 percent parental gametes). He then constructed a **genetic map** (or **linkage map**) that summarized the distances between genes. Geneticists refer to the position of a gene on a chromosome as its locus [plural, **loci**; Latin, = place]. The distance between the *vg* locus and the *pr* locus, for example, is about 10 map units.

The distance between any two genes on a chromosome is proportional to the likelihood that they will recombine.

In this chapter we have seen how cells divide to create haploid gametes by means of meiosis and how the twin processes of syngamy and meiosis complement each other. We have also seen that the behavior of the chromosomes during meiosis explains the patterns in which genes are inherited. In the next chapter, we learn what a gene does and what it is made of. We will also see that the structure of DNA explains how it can copy itself during the S phase of the cell cycle.

STUDY OUTLINE WITH KEY TERMS

Genetics is divided into **transmission genetics,** the study of how variation is passed from one generation to the next, and **molecular genetics,** the study of how DNA carries genetic instructions and how cells carry out these instructions. For many years, biologists thought that the characteristics of offspring represented a **blending** of the characteristics of their parents. But most genes are present in two copies, one from each parent, and—most of the time—these genes do not change as they pass from generation to generation. The **Sutton-Boveri chromosomal theory of inheritance** states that the units of heredity are genes carried on chromosomes.

Phenotype encompasses all physical and behavioral characteristics of an organism. **Genotype**—the genetic constitution of a cell or organism—is a collection of genes. The word genotype can refer to the entire **genome,** or to a subset of genes. A **gene** is a region of DNA. A **structural gene** specifies a particular protein, while a **regulatory gene** helps regulate the expression of other genes.

Asexual reproduction, through mitosis, produces offspring with genes from just one parent, a **clone.** In **parthenogenesis,** some animals reproduce fully formed offspring developed from haploid egg cells. **Sexual reproduction** produces offspring that inherit genetic information from two parents. This genetic information is contained in two sets of chromosomes, one set from each parent.

The **somatic cells** of an organism all have two copies of each chromosome and are said to be **diploid.** The **sperm** and **ova (eggs),** or **gametes,** each carry only one copy of each chromosome and are said to be **haploid.** Gametes arise from the **germ line** or **germ cells** in the **gonads**—**ovaries** in females and **testes** in males.

Male and female gametes come together at fertilization (or syngamy) to form a **zygote. Fertilization** (or **syngamy**) restores the diploid chromosome number. The genetic information in the zygote, however, is a mixture of information from the parents. Adult organisms produce new haploid gametes by means of **meiosis,** a process that allots one haploid set of chromosomes to each of four daughter cells. Because the distribution of chromosomes occurs randomly, meiosis generates combinations of genes that are different from those present in the parents and thus accounts for much of the genetic diversity of individuals within a species.

Meiosis consists of two cell divisions (**meiosis I** and **meiosis II**), each of which consists of **prophase, metaphase, anaphase,** and **telophase.** During prophase I homologous chromosomes form precisely aligned pairs in a process called **synapsis.** Each chromosome consists of a **tetrad.** During the rest of meiosis I, one chromosome from each pair (with two **chromatids**) goes to each daughter cell. Meiosis II distributes the sister chromatids of each chromosome to daughter cells.

Each member of a matching pair of chromosomes, or **homologous chromosomes,** is called a **homolog.** Where homologous chromosomes pair and **cross over** during synapsis, **chiasmata** form. **Dis-**junction, the separation of homologous chromosomes (resulting in **euploid** gametes), sometimes fails. **Nondisjunction** leads to the production of gametes with an abnormal number of chromosomes, or **aneuploidy.** One example is **Down syndrome,** or **trisomy 21.** Most aneuploid embryos die long before birth in **spontaneous abortions.**

Alternative versions of the same gene are called **alleles.** All variant alleles ultimately result from **mutations,** changes in the sequences of nucleotides in DNA. An organism with two copies of the same allele is **homozygous,** one with two different alleles is **heterozygous,** and one with only a single allele is **hemizygous.**

An allele is **dominant** when it alone determines the phenotype of a heterozygote and **recessive** when it contributes nothing to the phenotype of a heterozygote. In **partial dominance** both alleles are expressed.

Mendel made use of **true-breeding** varieties of peas to show his **Principles of Segregation** and **Independent Assortment.** Breeding two genetically distinct organisms is called **cross breeding** (or **crossing**), and the progeny of such a cross are called **hybrids.** To test the Principle of Segregation, Mendel performed a new kind of cross, called a **backcross.** A backcross to a homozygous recessive that is intended to reveal the genotype of the other parent is called a **testcross.** The original parents in a cross are called the **parental,** or **P,** generation, and the progeny are the "first filial," or **F1** generation, **F2** generation, and so on. All kinds of crosses can be represented in a **Punnett square,** with each kind of gamete produced by one parent along the top of the square and each kind of gamete produced by the other parent along the left side of the square.

The position of a gene on a chromosome is its **locus.** The tendency of genes on the same chromosome to stay together is called **genetic linkage.** Geneticists can construct a **genetic map,** (or **linkage map**) that summarizes the distances between genes in **map units. Linkage groups** are sets of genes that do not assort independently. Compelling evidence that genes are on chromosomes came from the analysis of genetic linkage in *Drosophila.* Genetic maps of *Drosophila* and other organisms identified the location of the genes on the chromosomes and explained deviations from the Principle of Independent Assortment.

In mammals and fruit flies, males have two unmatched **sex chromosomes,** called **X** and **Y,** as well as homologous **autosomes.** Females have the same autosomes plus a pair of matched X chromosomes. The X and Y chromosomes have homologous regions and pair during synapsis. Because males have only one X chromosome, they are susceptible to **sex-linked** genetic defects such as hemophilia and color-blindness that result from being hemizygous.

Harriet Creighton and Barbara McClintock showed that crossing over is a physical exchange of material between homologous chromosomes. Their work was immediately accepted as confirming the chromosomal theory of inheritance.

REVIEW AND THOUGHT QUESTIONS

Review Questions

1. How are homologous chromosomes similar to each other? How are they different?
2. Describe and draw the steps of meiosis I and explain what they accomplish. Then do the same for meiosis II. Which part of meiosis actually accomplishes the reduction in chromosome number?
3. Explain how crossing over of chromatids in prophase I leads to genetic recombination, so that the gametes produced from one parent are virtually all unique.
4. Describe the main ways in which meiosis differs from mitosis.
5. Define the Principles of Independent Assortment and Segregation. What aspects of meiosis explain these principles?
6. Explain the difference between autosomes and sex chromosomes. How many of each do you have? How many of each does a potato have?
7. What causes Down syndrome and other forms of aneuploidy? Why should the frequency of aneuploid eggs increase as a mother ages?

Thought Questions

8. In what way is fertilization the opposite of meiosis? Why is it important for a complete sexual life cycle to contain both processes?
9. If you found a new species in which the sexes look different, what criteria would you use to decide which to call "female" and which to call "male"?
10. Textbooks make a point of contrasting the genotype and the phenotype as mutually exclusive. If the phenotype is every measurable aspect of an organism, including its structure, chemical makeup and behavior, could an organism's genotype be considered a part of the phenotype? Why or why not?

SELECTED READINGS

Orel, Vitezslav, *Mendel,* Oxford University Press, New York, 1984. A short (100 pages), pithy biography of the first geneticist.

Robinson, Gloria, *A Prelude to Genetics,* Coronado Press, 1979. An engaging account of 19th-century theories of inheritance.

▶ On-line materials relating to this chapter are on the World Wide Web at http://www.saunderscollege.com/lifesci/ Click on Tobin/Dusheck: *Asking About Life.*

Rosalind Franklin and the Double Helix

On February 28, 1953, James D. Watson and Francis Crick discovered the structure of DNA, the hereditary material of the genes (Figure 10-1). Their breathtaking discovery laid a foundation for all of the most spectacular advances in biology until the present time. It was "the secret of life," as they announced at the time. The two men were helped in this momentous accomplishment by conversations with many other scientists. Yet, a young scientist named Rosalind Franklin, working in near total isolation, almost beat them to the punch.

In 1951, Watson was a 23-year-old former radio Quiz Kid, cherished by his elders as an up-and-coming genius. He had just received his Ph.D., studying the genetics of bacteria and the viruses that infect them, and his professors had arranged for him to go to Copenhagen to study the chemistry of nucleic acids. But the chemistry of nucleic acids bored him, and he began casting about for something else to work on.

While on a trip to Naples, Italy, Watson found that something—a question so fundamental that its answer seemed guaranteed to bring a Nobel Prize. At a lecture by an English

Figure 10-1 James D. Watson and Francis Crick. *(A. Barrington Brown/Science Source/Photo Researchers)*

physicist turned biochemist named Maurice Wilkins, Watson saw a slide of an x-ray diffraction pattern from what Wilkins said was a crystalline form of DNA. Wilkins argued that his photos, taken with the help of graduate student Raymond Gosling, were the first step in deducing the structure of DNA. The molecular structure of DNA would, of course, shed light on how genes work.

Watson was transfixed. He understood that if DNA could form crystals, it must therefore have a regular and relatively

Paraskevas Photography

simple shape. When, within just a few days, he read Linus Pauling's sensational papers on the alpha helical structures within proteins, Watson was galvanized. He longed to do something equally spectacular. With DNA and helices swirling in his head, Watson persuaded his former professors back in the United States to get him out of Copenhagen and into a lab in England where he could study DNA.

In a few months, Watson arrived at the Cavendish Laboratories in Cambridge, in the same building as Max Perutz and John Kendrew, pioneers of the application of x-ray diffraction to macromolecules. Also in the lab was Francis Crick, a 31-year-old graduate student working on his Ph.D. project. Working under Perutz, Crick was studying the x-ray diffraction of polypeptides and proteins, especially the blood protein hemoglobin.

Crick immediately liked Watson, and the two began to do an enormous amount of talking, so much so, in fact, that Perutz and Kendrew finally banished the two younger scientists to a room of their own. Watson knew nothing about x-ray crystallography except that he needed it to solve the structure of DNA. He nagged Crick relentlessly to help him work on DNA.

Crick liked to talk about DNA, but he wasn't about to start working on its structure. With a wife and daughter to support, he had to finish his Ph.D. and get a job. Besides, etiquette forbade Crick from working on the DNA problem. Crick and Wilkins were old friends going back to their joint work for the British Admiralty during World War II. Just as important, the Cavendish and the King's College Laboratory in London, where Wilkins worked, were sister facilities—and the DNA problem formally belonged to King's.

Still, it didn't hurt to talk. The first step to solving the three-dimensional structure of DNA was to study its density, size, and other properties using x-ray crystallography. Because DNA is a long, thin molecule, it would be impossible to obtain true DNA crystals until the 1960s, when researchers learned to produce crystals from short pieces of DNA synthesized in the laboratory. However, Wilkins had acquired DNA *fibers*, in which hundreds of millions of DNA molecules lie parallel to one another (Figure 10-2).

In 1951, the best x-ray diffraction photos of DNA fibers were those prepared

---

Franklin displayed none of the deference that Wilkins expected. Instead, she treated Wilkins as a colleague, vigorously arguing with his ideas about DNA, in the spirited style she had picked up in Paris.

---

by Rosalind Franklin (Figure 10-3), who was also working at the King's Laboratory. In November 1951, Watson took the train down to London to see Maurice Wilkins and to hear Rosalind Franklin give a lecture on the structure of DNA.

Franklin had arrived at King's at the beginning of 1951. A talented and recognized authority in industrial physical chemistry, she had just completed 4 successful years working on the structure of coals at a lab in Paris. Her work laid the foundation for modern industrial carbon-fiber technology. Today, carbon-fiber technology is used in making strong, lightweight (and expensive) bicycle frames, sailboats, and more. For her analysis of coals, Franklin learned and improved on known x-ray diffraction techniques, developed new mathematical techniques for interpreting the resulting photos, and constructed scale models like those that Linus Pauling had used in his work. She returned to her native England captivated by the prospect of using x-ray diffraction to work out the structure of an important biological molecule.

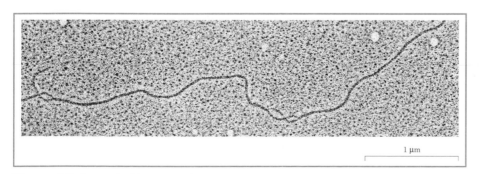

Figure 10-2 A double strand of DNA is only 2 nanometers thick but may reach lengths of many centimeters. DNA fibers consist of many parallel molecules. *(K.G. Murti/Visuals Unlimited)*

John Randall, the head of King's Laboratory, asked Franklin to set up and head an x-ray crystallography lab, with all the newest, most powerful equipment available. Once the lab was set up, her assigned task was to work out the structure of DNA. He assigned a graduate student to assist Franklin. It was Raymond Gosling, the same young man who had taken pictures for Wilkins.

What Randall did not explain to Franklin was that Maurice Wilkins, just down the hall, had already begun work on DNA and considered the DNA problem his own. Randall, in fact, had never even mentioned Wilkins in his letter offering her the job. Worse, at her first meeting with Randall and Gosling to discuss the project, Wilkins was not present. With that peculiar oversight, Randall set the stage for one of the most disastrous personality clashes in 20th-century science.

Wilkins was equally in the dark about Franklin. He viewed her more as a technician than a scientist, and he appears to have believed that she was hired specifically to do his x-ray crystallography work, in which he had no expertise. He handed over his best DNA fibers, hoping she

would produce for him the photos that would reveal their structure. But Franklin displayed none of the deference that Wilkins expected. Instead, she treated Wilkins as a colleague, vigorously arguing with his ideas about DNA, in the spirited style she had picked up in Paris. It was her way of trying to make friends. Offended by what he perceived as combativeness, he struggled to put her in her place. Franklin found him unaccountably touchy; he was given to turning away in the middle of conversations he didn't like.

Wilkins spitefully attacked her behind her back, undermining her relationships with other scientists at the lab. This was all too easy, for Franklin was never around when the other scientists gathered to talk at lunchtime or teatime: women were not allowed in the King's dining room.

Over the next 6 months, Franklin set up the x-ray crystallography lab, and then, with Gosling's assistance, began working out the structure of DNA. Though she liked her work, the social atmosphere was poisonous, unlike anything she had ever experienced in other labs. Franklin was used to making warm friendships with her

Figure 10-3   Rosalind Franklin. Franklin produced most of the data needed to discover the structure of DNA.   *(Vittorio Luzzati)*

colleagues. But between Wilkins' hostility and her banishment from the dining room, this seemed impossible. She longed for her happy days in Paris. Wilkins resentfully moved on to other projects. Randall, who might have reconciled the two researchers, did nothing.

KEY CONCEPTS

1. DNA consists of two polynucleotide strands, which wind together to form a double helix.

2. DNA from different species may have different proportions of nucleotides, but the amount of adenine (A) always equals the amount of thymine (T), while the amount of guanine (G) always equals the amount of cytosine (C).

3. The two strands of DNA contain complementary versions of the same information. During DNA replication, each strand directs the synthesis of a complementary strand.

4. Experiments with bacteria and bacterial viruses showed that genes are made of DNA.

5. Each gene (with some important exceptions) specifies the structure of a single polypeptide.

6. Each allele of a gene specifies a polypeptide with a different amino acid sequence.

7. A mutation may result in an organism that lacks a particular enzyme.

8. Enzyme systems in cells correct most, but not all, mistakes in DNA.

WHAT IS THE STRUCTURE OF DNA?

When Franklin began her work in 1951, the chemical makeup of DNA had been known for about 30 years. Franklin knew that DNA consisted of long chains of nucleotides linked together into polynucleotides. She knew that each nucleotide consisted of a sugar, a phosphate, and a base, and that the sugar and phosphate units were linked together alternately (sugar-phosphate-sugar-phosphate, etc.) into a sugar-phosphate "backbone" (Figure 10-4A).

From the sugar-phosphate backbone hung bases of four kinds. Two of the four bases are **pyrimidines,** which consist of rings made up of four carbon atoms and two nitrogen atoms. The other two bases are **purines,** which contain a double ring of six carbon atoms and four nitrogen atoms (Figure 10-4B). The purine bases are **adenine** and **guanine,** and the pyrimidine bases are **cytosine** and **thymine.** Table 10-1 gives the full names of the nucleotides that contain each base, but we shall generally refer to each nucleotide by a single letter abbreviation: A, G, C, or T. This convention not only saves space, it

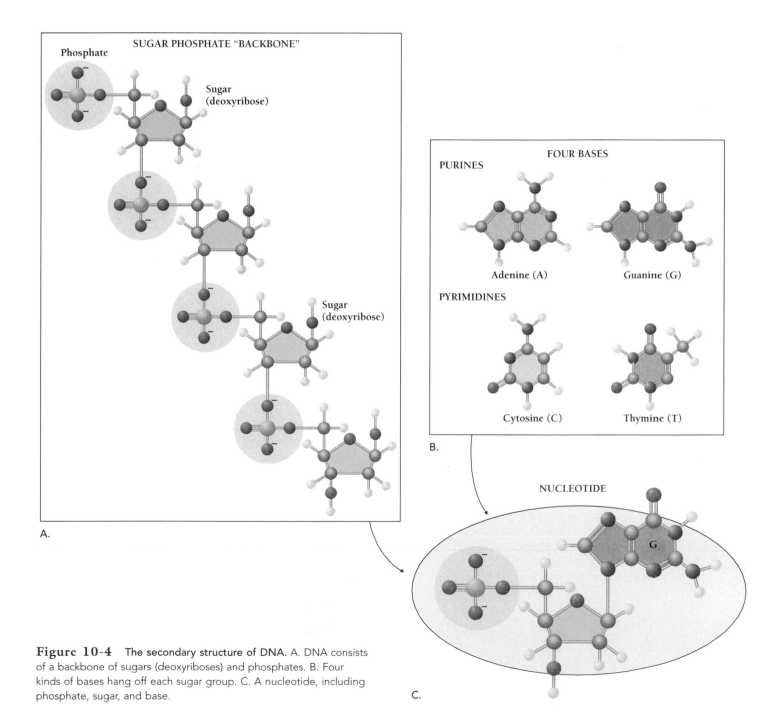

Figure 10-4   The secondary structure of DNA. A. DNA consists of a backbone of sugars (deoxyriboses) and phosphates. B. Four kinds of bases hang off each sugar group. C. A nucleotide, including phosphate, sugar, and base.

Table 10-1 The Nucleotides of DNA and RNA

Bases	DNA Deoxynucleotide	RNA Nucleotide
Purines		
A Adenine	Deoxyadenosine monophosphate (dAMP)	Adenosine monophosphate (AMP)
G Guanine	Deoxyguanosine monophosphate (dGMP)	Guanosine monophosphate (GMP)
Pyrimidines		
C Cytosine	Deoxycytidine monophosphate (dCMP)	Cytidine monophosphate (CMP)
U Uracil	none	Uridine monophosphate (UMP)
T Thymine	Thymidine monophosphate (TMP)	None

also emphasizes that the nucleotides are indeed the four letters in the DNA alphabet (Figure 10-4C).

What was *not* known by anyone at the time that Franklin set to work was that a molecule of DNA resembles a ladder twisted around its long axis (Figure 10-5). The two uprights of the ladder are the two sugar-phosphate chains. Each rung of the ladder consists of a pair of bases (A always goes with T, and G always goes with C). Each twist of the ladder contains 10 rungs, which are 0.34 nm apart. A single molecule of DNA may be either short or long. In humans, a molecule of DNA may be very long—about 9 cm (about $3\frac{1}{2}$ inches) long—and may consist of millions of nucleotide pairs. The diameter of the ladder, however, is always 2 nm.

Significantly, the two sugar-phosphate chains run in opposite directions. Each polynucleotide chain has a direction. In the sugar-phosphate backbone, the third carbon of one sugar is attached via a phosphate group to the fifth carbon of the next sugar. The third carbon is called the 3′ carbon, pronounced "three-prime carbon," while the fifth carbon is called the 5′ carbon, or "five-prime carbon." One end of each polynucleotide has a free 5′ carbon, called the **5′ end,** while the other end has a free 3′ carbon, called the **3′ end.** If we draw an arrow from the 5′ end to the 3′ end of each strand, one arrow points up the helix, the other down.

Each nucleotide in DNA consists of a sugar, a phosphate, and a base. The sugar and phosphate units are linked together alternately (sugar-phosphate-sugar-phosphate, etc.) into a sugar-phosphate "backbone." The two parallel backbones resemble a ladder twisted around its long axis, with each rung consisting of a pair of bases (A-T or G-C).

How Much Did Franklin Discover on Her Own?

By the time Franklin gave her talk in November of 1951, her 5 months of work had revealed some essential facts about the structure of DNA. She believed that DNA was probably a big

helix with two, three, or four parallel chains, with the phosphates wound around the outside of the helix, like the uprights of a ladder. She had also accurately measured the density of DNA and the number of water molecules associated with each nucleotide (eight), and discovered that DNA fibers have two slightly different structures, depending on whether they are wet or dry. Finally, she had identified the "symmetry" of DNA. Crystallographers classify crystals into 230 types according to their symmetry. Which of these symmetries DNA had might seem like an obscure point, but it would be crucial to Watson and Crick's solution to the structure of DNA.

Only the crystallographers working at the Cavendish, particularly Crick, could fully appreciate the meaning of that particular symmetry. Had Crick heard Franklin speak, he would have understood that Franklin's parallel chains were symmetrical in a particular way. A chain that ran in one direction only would look different when rotated 180°, while a pair of symmetrical chains running opposite to each would look the same when rotated. Franklin's symmetry implied that there were two chains, one running up, the other running down. The chains could have been four (two up and two down), but Franklin's density measurements ruled that out.

Franklin did not have the extensive training of a full-fledged crystallographer, and unlike Crick, she had never before worked with a large biological molecule, so she couldn't yet deduce the shape of the DNA molecule. Nonetheless, she was clearly on her way.

Watson watched Franklin talk, but he understood little of what she said and took no notes. Afterwards, Wilkins told him that Franklin didn't know what she was doing. When Crick interrogated Watson the next day, he had almost nothing to say about Franklin's talk. Crick was annoyed, but it would be 14 months before he learned how much he had missed.

What Watson did bring back was one wrong fact, the amount of water in the molecule, and the mistaken impression—gained from Wilkins—that neither Franklin nor Wilkins was capable of solving the structure of DNA. Clearly, the two at King's were not working together, and Wilkins seemed to be going nowhere by himself. Watson does not appear to have considered that Franklin might solve the structure. It seemed

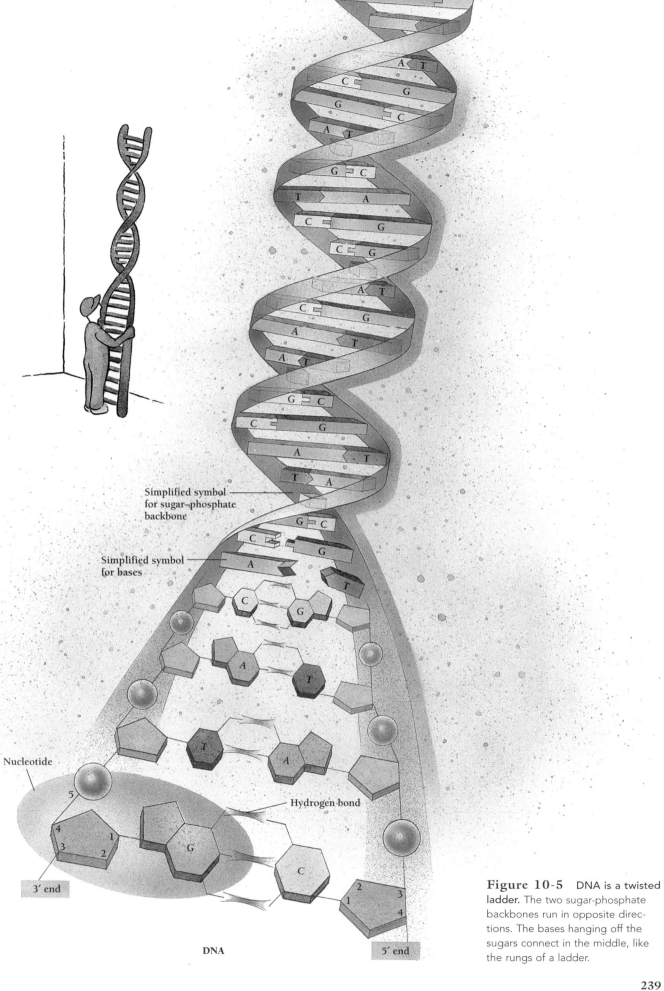

Simplified symbol
for sugar–phosphate
backbone

Simplified symbol
for bases

Nucleotide

5

4
3
1
2

3' end

G

C

Hydrogen bond

2
1
3
4

DNA

5' end

Figure 10-5 DNA is a twisted ladder. The two sugar-phosphate backbones run in opposite directions. The bases hanging off the sugars connect in the middle, like the rungs of a ladder.

239

to him ridiculous to allow Franklin to bumble away, not solving this important problem, when he and Crick could be deciphering DNA themselves.

Back in Cambridge, Watson finally persuaded Crick to help him attack the problem. Within days, the two produced a model of DNA consisting of three intertwined helices, sugar-phosphate backbones on the inside, and far too little water. They called Wilkins and Franklin to come up to Cambridge the next day and look over their model. Franklin glanced at the model, knew at once that it was completely wrong, and said so. Crick was humiliated, for he knew instantly that she was right.

Watson had missed everything important that she had said. Worse, Watson and Crick had worked on a problem that belonged to King's, and they had been wrong. The head of the Cavendish Lab, Sir William Lawrence Bragg, sternly reminded them that the DNA problem was not theirs to work on. Reluctantly, they agreed to stop trying to model its structure, but they continued to talk about DNA. They couldn't help it. They loved to talk.

For a year, Franklin studied DNA in near solitude, talking only to Gosling. For a while she thought that maybe the structure couldn't be a helix after all. The dry form of the DNA fiber gave pictures that seemed to suggest DNA was not a helix. There was, in fact, no certain evidence for a DNA helix early on. Pauling had argued that a repeating chain of any kind, whether a molecule or a telephone cord, will tend to twist about its axis into a helix. A helix seemed likely. But the pictures of the dry form did not lend support to a helix.

Franklin played with other possibilities, including repeating figure eights. In the summer she took an excellent picture of wet DNA fibers, showing a pattern that fairly shouted "helix." She labeled it photo number 51. She liked it a lot, and spent several days thinking about it. She had been working on the dry form for so long, however, that she soon returned to that. If she had had a collaborator, someone to argue with, she might have given more thought to photo number 51. But the atmosphere at King's was as bad as ever. She was already making plans to go to another lab.

Back at the Cavendish, Watson frittered away his time, flirting with women and fretting vaguely about DNA. He was, in his own words, "totally underemployed." Then, in the fall of 1952, two young researchers arrived at the Cavendish—Linus Pauling's son Peter and one of Pauling's graduate students. In mid-December, Peter announced that his father had discovered the structure of DNA. It was more than Watson could stand. In January, he begged Peter to get his father to send the manuscript describing his structure for DNA.

Franklin, who had also heard the news, wrote to Pauling's lab with the same request. Meanwhile, she began calculating bond lengths and angles for the different building blocks of DNA. Her months of work had yielded the main dimensions of the molecule. She suspected that DNA was a double helix. She knew for certain that it was 2 nm in diameter and that if it was a helix, each complete turn of the helix would be 3.4 nm long. Now, with these and other hard-won clues at hand, she quietly prepared to build a model of DNA, the same ap-

proach that had worked so well for Pauling in deducing the α helix of proteins.

On January 30, 1953, Watson went to King's College and stopped to visit Wilkins. Finding Wilkins busy, Watson dropped in on Franklin to taunt her with his copy of Pauling's manuscript. When she looked at the manuscript she could see as well as Watson that Pauling's three-chain helix was hopelessly wrong.

She also discovered that Pauling and his co-author, Robert Corey, had snubbed her. The previous spring, Corey had come to King's and asked to see her pictures, and she had graciously shown him her work. Wilkins and Gosling had taken earlier pictures of DNA also, but these were vastly inferior to Franklin's. Now Pauling and Corey had not only ignored her letter requesting their manuscript, they had also cited Wilkins' inferior x-ray diffraction pictures of DNA—which they had never seen—and not her own, which they had. Watson later recalled that he then began to goad her again, implying that she was incompetent to interpret her own x-ray diffraction photos. Although few people believe him, Watson claimed that the petite Franklin was so angry that she chased him from her lab.

According to Watson, he then found Wilkins, and the two men then took turns complaining about Franklin. Wilkins reminded Watson of Franklin's ideas on DNA, the same ideas she had presented 14 months earlier, then showed Watson her striking picture of the wet form of DNA, number 51. Crick and another scientist had already worked out a mathematical understanding of the way that a helix would interact with x rays. That work, which Crick had published and which Franklin was fully aware of, had shown that a helix should generate a cross-shaped pattern. Now, with that knowledge and without Franklin's scores of other pictures to confuse him, Watson leapt to the conclusion that DNA must be a helix. It was obvious to everyone who had seen it—Franklin, Wilkins, and Watson—that this photo was of a helix. Nonetheless, Wilkins told Watson that Franklin stubbornly insisted that DNA was *not* a helix.

Watson could only conclude that Franklin really was incompetent. He was tired of the British laboratory etiquette that prevented him from working on DNA himself. He would build a model of the wet form of DNA, whose picture looked so simple. Instinctively, he decided to try two backbones. He had a feeling that for DNA to duplicate itself most easily, it should have two chains. Four chains seemed needlessly complex, and three chains seemed illogical. He was still convinced, however, that the sugar-phosphate backbones must be on the inside of the helix, with the bases hanging off the outside, despite Franklin's assertion that the sugar-phosphate backbones were on the outside. Watson rushed back to the Cavendish and at last succeeded in persuading Bragg to let him work on DNA. Feverishly, Watson began working to build a model. Crick, still writing his Ph.D. thesis, watched from across the room with amusement and growing interest.

Watson made little progress until, on February 5, 1953, Crick finally persuaded him to try putting the two sugar-phosphate backbones on the outside, as Franklin had suggested. For the first time, Watson had the beginnings of a model. The following weekend, Wilkins came up to Cambridge and reluctantly gave Watson and Crick permission to work on the DNA problem. Wilkins promised Crick in a letter to "tell you all I can remember & scribble down from Rosie."

No one asked Franklin's permission. Unaware that Wilkins and Bragg had now given away her research problem, she restarted her work on a structure for the easier, wet form of DNA. She was trying to decide if it was one chain or two.

For the first time, Crick felt free to work on DNA. The best source of information about DNA was Franklin, but Crick had no more intention of talking to her than had Watson or Wilkins. Crick soon learned, however, that Franklin and Gosling had summarized their unpublished work in an annual report on research at King's. Crick tracked down a copy of the report and read it carefully.

Now Crick knew almost everything that Franklin and Gosling had so laboriously worked out. For the first time, he began to take the whole enterprise rather seriously. When he read Franklin's report describing DNA—its density, its water content, and, critically, its symmetry—it gradually dawned on Crick that DNA must have two chains and that they must run in opposite directions.

Franklin and Gosling also reported that the distance along the wet form of the molecule through one complete turn was 3.4 nm. Previous work had shown that the distance between the bases was 0.34 nm, so it was easy to see that there were probably ten bases per 360° turn. Wilkins and Gosling had earlier shown that the diameter of DNA was about 2 nm, and Franklin and Gosling had confirmed this. Things were beginning to come together.

Crick tried to persuade Watson to build a model with the chains running in opposite directions. But Watson couldn't figure out how to make it come out right. So, while Watson was off playing tennis, Crick abandoned his Ph.D. dissertation for an afternoon and took over the model. He built a more open double helix than Watson's, with sugar-phosphate chains running in opposite directions as required by the symmetry, and 10 bases per 360° turn of the helix. The building blocks, with their bond angles and lengths, all fit nicely. Each complete turn was 3.4 nm, as Franklin had measured, and the distance from one base to the next was 0.34 nm. The only problem now was to find a way to fit the bases together in the middle.

It wasn't immediately obvious how this could be done. The four bases were of very different sizes and shapes. Some pair combinations would be wider than 2 nm and would make the molecule bulge and strain. Other combinations would have been too narrow, and would have left a gap in the middle or pinched the two backbones in. Franklin's x-ray data showed, however, that the helix had a constant diameter. It neither bulged nor pinched in (Figure 10-6).

Watson began reading all the biochemistry of bases that he could find. If DNA was to carry genetic information, it should be able to accommodate any sequence of nucleotides, just as each line of print on this page must be able to accommodate any sequence of letters. Two weeks after he first put the sugar-phosphate backbone on the outside, Watson theo-

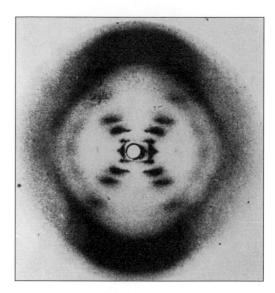

Figure 10-6   One of Rosalind Franklin's x-ray diffraction photos of DNA.   *(Cold Spring Harbor Laboratory Archives)*

rized that like bases paired (guanine with guanine and so on). But within days, Jerry Donohue, the young chemist from Pauling's lab, completely demolished Watson's theory, explaining that the molecular forms that Watson had used, from standard textbooks, were all wrong. Besides, added Crick, Watson's like-with-like base pairing did not square with Chargaff's rules.

The Double Helix Explains Chargaff's Rules

In the 1920s and 1930s, chemists believed that DNA contained almost equal amounts of each base. Then, in the 1940s, Erwin Chargaff developed a new method for analyzing the base composition of DNA. Chargaff found that although the proportions of the bases can vary widely from species to species, the amount of A is always equal to the amount of T and the amount of G is always equal to the amount of C. As a result, the proportion of A plus T to G plus C is constant within each species. Neither Chargaff nor anyone else at the time understood the significance of this finding, summarized in what came to be called "**Chargaff's rules**."

In the spring of 1952, nine months before Watson and Crick discovered the correct structure of DNA, Chargaff had come to Cambridge. Watson and Crick were introduced to Chargaff, and the older scientist took an immediate dislike to the two young men. "It struck me as a typical British intellectual atmosphere, little work and lots of talk," Chargaff later complained to the writer Horace Judson. Crick knew almost nothing about Chargaff's work, but the older man brusquely set him straight.

After a little thought, Crick saw that Chargaff's rules might imply complementary base pairing. That is, A always pairs with T and G always pairs with C. Such a relationship, he reasoned, could provide a means for DNA to copy, or replicate, itself. Nine months later, however, Crick had practically forgotten his own insight.

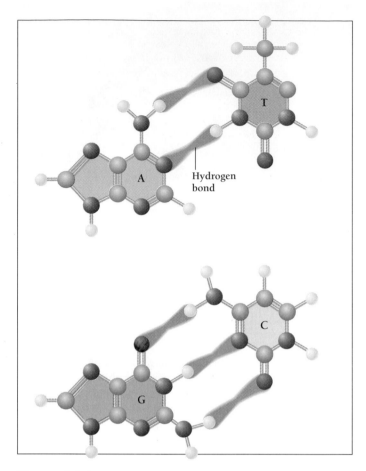

Figure 10-7 The bases connect, or "pair," in the middle of the twisted ladder by means of hydrogen bonds. The nucleotides A and T can form two stable bonds with each other, and G and C can form three. When matched in this way, the width of a G-C pair is the same as an A-T pair, as predicted by Franklin's x-ray diffraction data.

By Saturday, February 28, Watson had constructed cardboard cutouts representing the shapes of the four bases. He fiddled with them, fitting them together this way and that. He struggled to come up with pair combinations that would have the same symmetry and diameter, and still allow any order of bases on the backbone. Finally, he tried pairing A with T and G with C. The fit was unbelievably good. An AT pair had the exact shape of a GC pair (Figure 10-7).

Donohue quickly assured Watson that the two sets of bases could in fact be bonded together that way. The arrangement of atoms in the bases is such that A is perfectly arranged to make two hydrogen bonds with T, while G can form three hydrogen bonds with C. The chemistry worked, and for the first time, Watson and Crick understood the entire structure. It would be another week before they had the materials to build a complete model, but their work was done. Euphoric, they announced to everyone they met that they had "discovered the secret of life."

On Saturday, March 7, Watson and Crick eagerly took model bases made for them in the Cavendish machine shop and built an accurate model of the DNA double helix. On the same day, Wilkins wrote a note to Crick, including the following:

. . . I think you will be interested to know that our dark lady leaves us next week and much of the . . . data is already in our hands. I . . . have started up a general offensive on Nature's secret strongholds on all fronts: models, a theoretical chemistry and interpretation of data, crystalline and comparative. At last the decks are clear and we can put all hands to the pumps!

It won't be long now.
Regards to all,
Yours ever,
M.

Franklin Comes Within Two Steps of the Correct Model

But the race was over. Wilkins little knew how close Franklin had come to solving the problem on her own, or how late he was in restarting his own work. Nearly 2 weeks earlier, on February 23, Franklin had begun remeasuring photo number 51. By the end of the next day, she had correctly concluded that DNA was a double helix, with a diameter of 2 nm. She was now certain that the sugar-phosphate backbones were on the outside. She guessed that the backbones had some peculiar relationship to each other, but what? She hadn't yet realized that the two backbones ran in opposite directions. She knew about Chargaff's rules, and she had begun playing with the problem of how the bases would fit together in the middle. She was two steps from the answer.

But Franklin's work on DNA was at an end. Randall had told her she could leave King's for another lab if she wanted, but she could not take DNA with her. Accordingly, she left King's and, with Gosling, wrote up the results of their 18 months of work on the structure of DNA. Their work was not complete, but it was substantial.

The potent combination of Franklin's experimental work, Crick's theoretical work, and Watson's infectious enthusiasm had provided everything needed to solve the problem. Yet, amazingly, when Watson and Crick wrote up their results for the prestigious British journal *Nature,* Franklin was the one person whose work was not formally cited. This deficiency was apparently no oversight. According to writer Horace Judson, Wilkins asked Watson and Crick to delete a sentence from their manuscript that said, "It is known that there is much unpublished experimental material." Instead, Watson and Crick explicitly asserted that they were unaware of Franklin's work, writing that

. . . We were not aware of the details of the results presented [in separately published accompanying papers by Franklin and Wilkins] when we devised our structure, which rests mainly though not entirely on published experimental data and stereochemical arguments.[1]

[1]From Watson, J.D., and Crick, F.H.C., "Molecular Structure of Nucleic Acids: A Structure for Deoxyribose Nucleic Acid," *Nature,* 25 April 1953, vol. 171, pp. 737–738.

In fact, according to later accounts, Watson and Crick's structure rested heavily on the unpublished data of Rosalind Franklin and Raymond Gosling, details of which Watson and Crick were well aware.

Not until March 18, when Wilkins first received Watson and Crick's manuscript, did Franklin learn of their achievement. She took the train up to Cambridge to see their model. Watson was astonished at how quickly and graciously she agreed that the model was correct. He had been sure that she would find some grounds on which to criticize it, for he little appreciated how much she knew and understood. She, for her part, had no reason not to be gracious, for she little suspected how much Watson and Crick had relied on her own work for their success. The elegant beauty of the structure delighted her. She was filled with admiration for Crick's genius.

Franklin was glad, in any case, to be hard at work in her new lab. She was now studying the structure of tobacco mosaic virus. Over the next 5 years, Franklin directed a highly productive research team. Later, she collaborated with researchers at Berkeley and Yale in the United States, and at Tübingen in Germany. When she discovered she had cancer, in 1956, she continued to work cheerfully and productively until the very end, so devoted was she to the research she loved. She died in 1958, only 37 years old, without ever learning of her own crucial contribution to science.

When a couple of years later it appeared that Watson and Crick might get a Nobel Prize for their work, Sir Lawrence Bragg, who had headed the Cavendish, made sure that Wilkins received the prize with them. Recounting the story to writer Horace Judson, Bragg said that "when it came to the Nobel Prize, I put every ounce of weight I could behind Wilkins getting it along with them. It was just frightfully bad luck, really." Wilkins was lucky enough, however, to receive in 1963 a third share of a Nobel Prize for the discovery of the structure of DNA. Franklin herself was ineligible, for only the living are honored.

In 1968, ten years after Franklin's death, James Watson wrote a personal account of the discovery of the structure of DNA. In *The Double Helix,* Watson for the first time made clear how essential Franklin's experimental results had been to their discovery. He also falsely depicted her as Wilkins' rebellious and shrewish assistant, who seemed to deserve any bit of ill-treatment that came her way. Watson's book was attacked and reviled by nearly everyone who had been present during the discovery. It has never been considered an accurate account. Yet, as a work of literature and as a window into the mind of a successful molecular biologist, *The Double Helix* is a masterpiece.

Modern Science Is a Social Endeavor

The discovery of DNA's structure is the best-documented story in the history of science. Watson, Crick, and others have all written extensively and passionately about both the science and the scientists. It is clear that Franklin's failure to discover the structure of DNA before Watson and Crick was a direct result of her social isolation. Watson himself said in 1993, "[Franklin] would be famous for having found DNA if she'd just talked to Francis [Crick] for an hour."

Watson and Crick had each other to talk to. They had Max Perutz, John Kendrew, Jerry Donohue, Peter Pauling, and even, grudgingly, Erwin Chargaff. All told, at least a half dozen scientists provided essential clues that guided Watson and Crick toward the correct answer. Franklin, isolated by her feud with Wilkins, as well as her banishment from the King's dining room, had no one to talk to but Gosling, who had little useful knowledge.

Watson and Crick's discovery of the structure of DNA was a borrowed victory that rested heavily on Franklin's discoveries. Their real contribution to biology was their early recognition that DNA's structure explains how it works. In fact, Watson and Crick's insight into the implications of base-pairing stands as one of the most influential advances in the entire history of biology. It has provided the sole basis for the twin fields of molecular genetics and genetic engineering, and transformed every other area of biology.

What was so important about their model? We've seen that the structure was like a twisted ladder with the uprights running in opposite directions. Each rung of the ladder was a pair of bases held together by hydrogen bonds. Watson and Crick's rules of complementary base-pairing meant that the sequence of one polynucleotide strand exactly specified the sequence of the second strand. That is, one strand could give directions for building the other. This explained how a chromosome could replicate itself during cell division. Further, it was clear that the bases could be arranged in a nearly infinite variety of different sequences. That meant that the sequence of bases in DNA, like the sequence of letters on a page, could contain coded information.

In the rest of this chapter, we will examine the structure of DNA in more detail. Along the way, we will find out how biologists found out that the genetic material was DNA, what a gene is, and how DNA replicates and repairs itself. In the next chapter, we will see how cells actually use the information encoded in the DNA.

WHAT IS A GENE?

Defining a gene remains a problem not only to students taking biology examinations but also to scientists searching for biological molecules. To Mendel in 1865, a gene was an abstraction that somehow determined the traits of individuals. To Sutton and Morgan in the early years of the 20th century, genes were parts of physical structures, the chromosomes.

But biologists still did not know what genes actually were or how cells executed and replicated the information that genes contain. That knowledge came only when geneticists tried to understand the function and the structure of genes in chemical terms. In the rest of this chapter, we will see how researchers established (1) what genes do, (2) what they are made of, and (3) how they replicate. We will learn that genes specify the

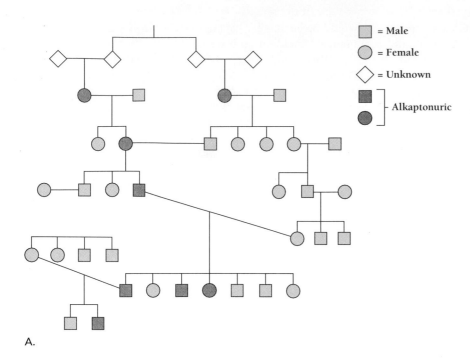

□ = Male

○ = Female

◇ = Unknown

■
● } = Alkaptonuric

Figure 10-8 Inheritance of alkaptonuria. A. A pedigree of alkaptonuria from a Lebanese family. The marriage of cousins and other closely related family members increases the incidence of homozygous recessive individuals with alkaptonuria. B. Cells can break down phenylalanine into homogentisic acid (HA), but the lack of a crucial enzyme prevents degradation of HA and the toxic material accumulates in the body.

A.

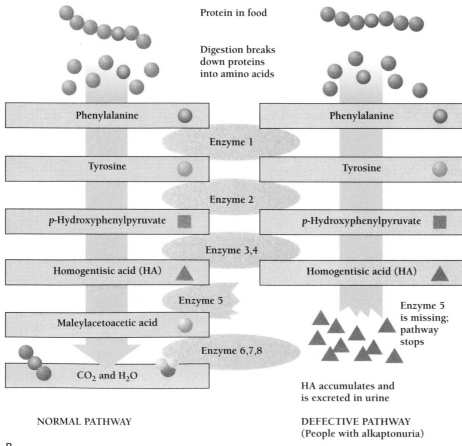

B.

sequences of amino acids in polypeptides, that they are made of DNA, and that the three-dimensional structure described by Watson and Crick is especially favorable for their replication and repair.

In this and later chapters we will also see that genes are not only physical entities or places on the chromosomes, but information. In this sense, they are indeed abstractions, as Mendel conceived them. In the same way that information about biology, for example, can be communicated as pictures, as printed text, or as electric currents, genetic information can be relayed in many forms. The physical expression of that information can be viewed as separate from the information itself.

How Do Genes Affect Biochemical Processes?

The first hint of the biochemical importance of genes came in 1902 from the work of an English physician, Archibald Garrod. Among Garrod's patients at the Hospital for Sick Children in London was a baby, whose diapers developed black urine stains. The baby's disease, alkaptonuria, had first been described in 1649 and named for the dark "alkapton bodies," which formed after the urine was exposed to air. The precursor of the alkapton bodies was identified as homogentisic acid (HA), which oxidized to form the alkapton bodies.

Garrod, working just 2 years after the rediscovery of Mendel's paper, guessed that alkaptonuria might be an inherited disease, since the baby's parents were first cousins. After studying the way the disease was inherited, he realized that alkaptonuria behaved as if it were caused by a recessive allele. Each parent had contributed the same recessive allele (inherited from one of their common grandparents) to the baby, which was homozygous (Figure 10-8A).

Like the amino acids phenylalanine and tyrosine, HA contains a benzene ring. Garrod suggested that HA was an intermediate in the metabolism of phenylalanine and tyrosine. Most people have an enzyme called HA oxidase, which catalyzes the oxidation of HA, so that the HA is excreted as carbon dioxide and water (Figure 10-8B). Alkaptonuria patients, however, cannot oxidize HA, and it turns to alkapton bodies that accumulate in the urine. Garrod realized that the absence of HA oxidase might cause alkaptonuria. He hypothesized that the recessive allele whose inheritance he had charted could explain the missing enzyme. If an alkaptonuria patient had two copies of a defective allele of the gene for HA oxidase, he or she would not be able to make the enzyme.

The first clue that genes influence phenotype through biochemistry came from Garrod's work in 1902 on infants with the genetic disease alkaptonuria.

One Gene—One Enzyme

Garrod's insight—that genes control the activity of enzymes—at first seemed to apply to only a few rare diseases: alkaptonuria and albinism, for example. In all of these conditions, the failure to make a particular enzyme explains the observed phenotypic abnormality. But how general is this picture of what genes do?

The answer did not come until 1941, when two Stanford University geneticists, George Beadle and Edward Tatum, began looking for other genetic defects in metabolism. Beadle and Tatum chose to study a simple fungus, the pink bread mold, *Neurospora* (Figure 10-9). Beadle and Tatum induced mutations in *Neurospora* with x rays and then looked for biochemical defects. Such biochemical defects were easy to detect because *Neurospora* is a haploid organism—it has only a single copy of each gene. When that single copy is altered, the change immediately shows up in the phenotype. (In diploid organisms such as humans, two copies of a recessive allele must be present for the defect to be expressed.)

Wild-type *Neurospora* can grow on an extremely simple medium (called a minimal medium), which contains glucose, ammonia, and salts. But Beadle and Tatum identified a large number of mutant *Neurospora* molds that could not grow on such a minimal medium (Figure 10-9). These mutants required the addition of various compounds, which the molds were incapable of synthesizing themselves. One mutant strain, for instance, would grow only if the minimal medium was enriched with vitamin B_6. Another mutant strain needed the amino acid arginine.

Beadle and Tatum then mapped the chromosomal location of each of their mutants by studying genetic linkage, as described in the last chapter. The mutants that needed arginine, for example, fell into three groups. Beadle and Tatum established that the three groups corresponded to mutations in three separate genes, on three separate chromosomes. Each gene coded for a separate enzyme needed to make arginine.

Figure 10-9 shows the growth properties of the three groups of mutants. One group of mutants (*arg*-1) would grow not only when arginine was added but also in the presence of two related compounds, ornithine and citrulline. The second group of mutants (*arg*-2) would grow in the presence of arginine or citrulline, but not ornithine alone. The third group (*arg*-3) required arginine; neither citrulline nor ornithine would do.

Beadle and Tatum realized they could explain their results with the following model of arginine synthesis:

$$\text{precursor} \xrightarrow{\text{enzyme 1}} \text{ornithine} \xrightarrow{\text{enzyme 2}} \text{citrulline} \xrightarrow{\text{enzyme 3}} \text{arginine}$$

Each step in the synthesis of arginine, they reasoned, is catalyzed by a different enzyme. The *arg*-1 mutations, they argued, must lie in the gene for enzyme 1. The absence of enzyme 1, then, could be overcome by ornithine, citrulline, or

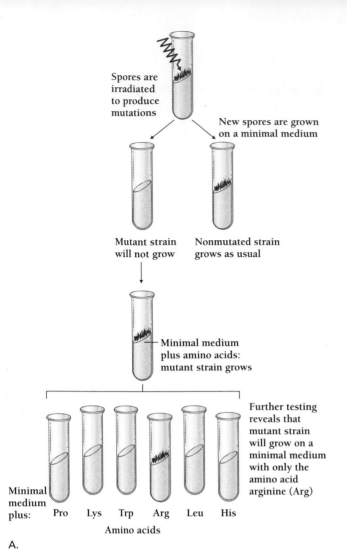

Spores are irradiated to produce mutations

New spores are grown on a minimal medium

Mutant strain will not grow

Nonmutated strain grows as usual

Minimal medium plus amino acids: mutant strain grows

Further testing reveals that mutant strain will grow on a minimal medium with only the amino acid arginine (Arg)

Minimal medium plus:

Pro   Lys   Trp   Arg   Leu   His

Amino acids

A.

Figure 10-9 Mutant *arg* genes in the bread mold *Neurospora crassa.* A. Different mutant strains of *Neurospora* need different amino acid supplements to survive. B. Each mutated *arg* gene is responsible for the absence of a different enzyme in the pathway of arginine synthesis. C. *Neurospora crassa.* *(C, James W. Richardson/Visuals Unlimited)*

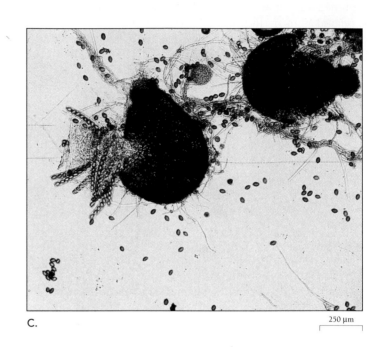

C.

250 μm

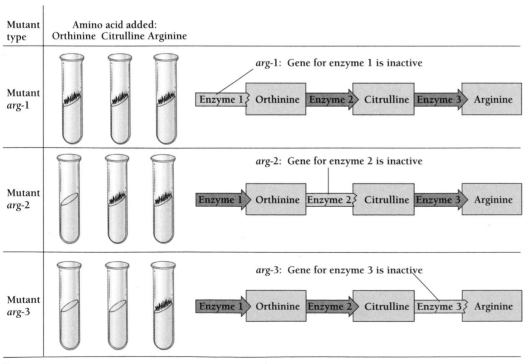

Mutant type	Amino acid added: Orthinine Citrulline Arginine		

Mutant arg-1

arg-1: Gene for enzyme 1 is inactive

Enzyme 1 → Orthinine → Enzyme 2 → Citrulline → Enzyme 3 → Arginine

Mutant arg-2

arg-2: Gene for enzyme 2 is inactive

Enzyme 1 → Orthinine → Enzyme 2 → Citrulline → Enzyme 3 → Arginine

Mutant arg-3

arg-3: Gene for enzyme 3 is inactive

Enzyme 1 → Orthinine → Enzyme 2 → Citrulline → Enzyme 3 → Arginine

B.

arginine. Similarly, defects in the gene for enzyme 2 can be overcome by the addition of citrulline or arginine, but not of ornithine, since the organism cannot process it. Finally, defects in the gene for enzyme 3 can only be overcome by the addition of arginine itself. Beadle and Tatum's work helped unravel previously unknown biochemical pathways, since the mutations blocked the pathways at different steps and allowed biochemical analysis. Still more importantly, however, their work suggested that every gene somehow affects the function of a single enzyme. Beadle and Tatum captured this concept in a single phrase: "**One gene—one enzyme.**"

So do genes only contain information for making enzymes? Are enzymes the only link between genotype and phenotype? Or can genes code for other proteins besides enzymes?

Beadle and Tatum showed that one gene could specify one enzyme.

One Gene—One Polypeptide

The answer to this question came with the unraveling of the molecular basis of another genetic disease, sickle cell anemia. The name "sickle cell" comes from the curled appearance of the red blood cells (Figure 10-10). The distorted blood cells clump and interfere with the circulation, causing excruciating pain, anemia, heart trouble, and brain damage. The body's defense mechanisms eliminate the defective cells as rapidly as possible, bringing about a severe anemia. People with sickle cell disease almost always die young.

Sickle cell disease, like alkaptonuria, results from the action of a recessive allele. Unlike alkaptonuria, however, it is both serious and common—afflicting 1 of every 625 African-Americans. And 1 in 13 carry the allele. In 1949, Linus Pauling and his collaborators showed that in sickle cell patients, the red oxygen-binding protein of the blood, **hemoglobin,** was different from the usual adult hemoglobin, called HbA. Sickle

2.5 μm

Figure 10-10   Sickled and standard red blood cells. The distortion of red blood cells that causes the symptoms of sickle cell disease results from altered hemoglobin molecules.   *(Stan Flegler/Visuals Unlimited)*

cell hemoglobin is called HbS. The difference between HbS and HbA can easily be demonstrated by a technique called electrophoresis, which detects differences in the sizes and charges of molecules by the way they move in an electric field. (Figure 10-11).

The parents of sickle cell patients have both HbS and HbA, but the children lack HbA entirely. A simple explanation of the disease emerged:

1. A recessive allele codes for an abnormal protein, HbS.
2. Heterozygotes (such as the parents) have genes for both normal and abnormal proteins.
3. In the presence of HbA, HbS does not cause the disease; so heterozygotes are healthy.
4. Homozygotes for the recessive allele make only HbS.
5. In the absence of the normal protein (HbA), HbS causes sickle cell disease.

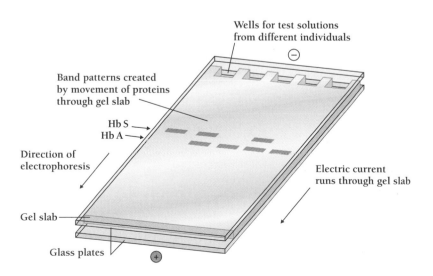

Figure 10-11   Electrophoresis. Molecular biologists separate macromolecules according to charge, shape, and size. DNA, RNA, or polypeptides with an electrical charge are pulled through a thick gel by an electric current. Smaller molecules move through the gel more quickly than the larger ones. Here, the change in polypeptide sequence in sickle cell hemoglobin (HbS) reduces its negative charge and causes it to move through the gel more slowly than normal hemoglobin (HbA).

Since differences in the gene had caused differences in the hemoglobin protein, not in any enzyme, Pauling's work suggested that genes controlled other proteins besides enzymes. Pauling extended "one gene—one enzyme" to "one gene—one protein."

Exactly how does an abnormal gene change a protein? To answer this question Vernon Ingram, a young biochemist working at Cambridge University in 1956, used new methods of protein chemistry to analyze the structural differences between HbS and HbA.

Hemoglobin consists of two types of polypeptide chains, called α (alpha)-globin and β (beta)-globin. Each hemoglobin molecule has two α-globin polypeptides and two β-globin polypeptides, for a total of four polypeptide chains. Ingram showed that only the β-globin polypeptides were abnormal in HbS. In the β chain's entire amino acid sequence of 146 amino acids, a single amino acid change produced an altered protein and, in homozygotes, a deadly disease.

Amino acid 6, ordinarily glutamic acid in HbA, is valine in HbS. Since glutamic acid is negatively charged and valine is neutral, this substitution changes the charge of the hemoglobin molecule, which alters the way that hemoglobin molecules pack into the crowded red cell. The result is that the HbS tends to crystallize within the cell, leading to the distortion that gives the disease its name. It is the same charge difference that changes the movement of HbS during electrophoresis. Ingram's work suggested that genes somehow control the amino acid sequence of proteins.

Ingram found that the sickle cell gene produced abnormalities only in the β-globin polypeptide chain, never in the α-globin chain. In fact, further research revealed that the gene for another hereditary blood disease, which produced abnormalities in the α-globin chain, was completely separate from the sickle cell gene. These findings suggested that each gene was responsible for the production not of an entire protein, but of a single polypeptide chain. "One gene—one protein" became **"one gene—one polypeptide."**

But even this rule has important exceptions. For example, the sequences of a few kinds of polypeptides involved in the immune system are determined by more than one gene.

Nonetheless, we can say that, generally, genes influence phenotype by carrying the information necessary for the synthesis of polypeptides. And each polypeptide chain—whether it is part of an enzyme like HA oxidase, a protein of the cytoskeleton, or a regulatory protein—plays a part in determining an organism's physical characteristics. But exactly how does a gene code for a polypeptide? To answer this question, we must first look more closely at what a gene is.

Genes carry information necessary for the
synthesis of polypeptides.

HOW DID BIOLOGISTS LEARN WHAT GENES ARE MADE OF?

Research on an amazing variety of organisms—including humans, peas, flies, and bread mold—has contributed to our understanding of what genes do. To find what genes are made of, biologists turned to less complex, smaller organisms: bacteria and their viruses.

Extracts of Bacteria Can Change the Genetic Properties of Other Bacteria

Before general improvements in hygiene and the development of antibiotics, bacterial pneumonia was an often fatal disease. Research directed at its prevention led directly to the discovery of the nature of genes.

One approach to preventing bacterial pneumonia was to develop a vaccine against the bacterium *Streptococcus pneumoniae,* also called pneumococcus. In 1928 Frederick Griffith, an English bacteriologist, was trying to provoke the immune defenses of mice against pneumococcus with injections of the bacterium.

Griffith worked with two strains of pneumococcus. One strain, called the S (or smooth) strain, produced pneumonia, while the other, called the R (or rough) strain, did not. When strains of pneumococcus with S and R phenotypes grow in artificial media, the virulent (disease-causing) S strain grows into

Injections of pneumococcus bacteria

Live S   Live R   Heat-killed S   Live R *plus* heat-killed S

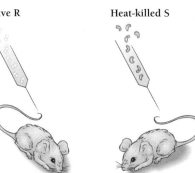

Figure 10-12 Griffith's experiment. DNA from a virulent strain of the bacterium *Streptococcus pneumoniae* can transform a harmless strain into a deadly one.

glistening smooth colonies, while the nonvirulent R strain forms rough-looking colonies. We now know that the virulent S bacteria form smooth colonies because they produce a polysaccharide capsule. This capsule prevents their digestion by white blood cells in the blood and makes them effective in causing pneumonia. The R bacteria are mutants—they lack an enzyme needed to produce the polysaccharide capsule. They are much more susceptible to attacks from cells of the immune system and therefore do not produce pneumonia.

Griffith tried to produce immunity to pneumonia in mice by injecting them with different combinations of live and heat-killed pneumococcus. As he expected, the mice injected with live S strain bacteria got pneumonia and died, while the mice injected with either live R strain bacteria or with heat-killed S bacteria survived. Griffith was amazed, however, by the results of injecting a mouse with a mixture of heat-killed S strain bacteria and live R strain bacteria (Figure 10-12). The mice died! Equally astonishing, the blood of the sick mice contained live S bacteria. Something in the dead S bacteria had transformed the harmless R bacteria into virulent S bacteria.

In 1928, Frederick Griffith discovered that dead virulent bacteria could transform harmless bacteria into virulent ones.

What Is the Transforming Factor?

Other scientists soon found that they could convert R strains into S strains in a test tube as well as in a mouse. Extracts of S strains, dead or alive, could transform R strains. Once the R strains were transformed, moreover, their descendants inherited the properties of S strains. **Transformation,** as we now understand it, is the transfer of one or more genes from one organism to another. While experiments using transformation now go on in thousands of laboratories, Griffith's work was the first example and provided the first opportunity to learn what genes were actually made of.

In the early 1930s, Oswald Avery at the Rockefeller Institute in New York set out to determine what molecules caused transformation in pneumococcus. With his coworkers Colin MacLeod and Maclyn McCarty, Avery carefully isolated different kinds of molecules from heat-killed S bacteria and tested to see which ones were responsible for transformation.

At the time, almost everyone expected that the transforming factor—and genes in general—would turn out to be made of protein, since proteins were already known to perform so many other cellular functions. Yet, surprisingly, transformation had occurred even when the bacteria had been heat-killed. Heat denatures (unfolds) proteins and so destroys them.

Avery and his colleagues worked almost 10 years on this problem, isolating substance after substance from heat-killed S strain bacteria and then testing each substance to see if it could transform R strain bacteria. They treated each isolated substance with a variety of enzymes that would hydrolyze proteins

into amino acids, polysaccharides into sugars, and nucleic acids into nucleotides. Finally, in 1944, they were forced to conclude that the transforming factor was DNA. Only the DNA fraction of the cells produced transformation, and the transforming activity was destroyed by an enzyme (DNAase) known to break down DNA, but not proteins.

Despite the care with which Avery and his colleagues had carried out their work, the prejudice against DNA continued, and most biologists did not accept Avery's conclusion for another 10 years. Biologists believed that DNA was a simple repeating molecule, no more capable of carrying information than a string of identical beads. During the 1940s, however, another group of researchers began work with an even more unusual set of genes, those of a bacterial virus. This work finally convinced the scientific world of DNA's importance.

By 1944, Avery and his colleagues had shown that bacteria are transformed by DNA, not proteins. Nonetheless, biologists continued to believe that DNA could not be the genetic material.

What Is a Virus?

In 1915, even before Griffith's work on pneumococcus, the bacteriologist Felix d'Herelle discovered that bacteria themselves could be killed by a tiny infectious agent called a **bacteriophage** [Greek, *bakterion* = little rod + *phagein* = to eat], or **phage.**

D'Herelle, a young French microbiologist working in Mexico, first became aware of bacteriophages while trying to halt a plague of locusts then sweeping the Yucatan. D'Herelle collected bacteria from the diarrhea of dying locusts and cultured it. Then, as hoards of the locusts advanced across the Mexican countryside, d'Herelle dusted the vegetation before them with his bacterium. As the locusts ate, they sickened and died. The plan worked magnificently.

D'Herelle was puzzled, however, by the appearance of clear spots in his cultures, places where the bacteria seemed to have died. What was killing his bacteria?

The answer came to d'Herelle in a moment of inspiration some 5 years later, while studying an epidemic of dysentery in a squadron of French cavalry stationed near Paris. D'Herelle had been culturing the Shiga dysentery bacillus from sick soldiers when the bacteriophages once more made their presence unmistakably obvious to him.

> The next morning . . . I experienced one of those rare moments of intense emotion which reward the research worker for all his pains: at the first glance I saw that the broth culture, which the night before had been very turbid, was perfectly clear: all the bacteria had vanished, they had dissolved away like sugar in water . . . in a flash I . . . understood: what caused my clear spots was in fact an invisible microbe, . . . a virus parasitic on bacteria. Another thought came to me also: "If this is true, the

same thing has probably occurred during the night in the sick man, who yesterday was in a serious condition. In his intestine, as in my test tube, the dysentery bacilli will have dissolved away under the action of [the virus]. He should now be cured."

And, indeed, d'Herelle dashed to the hospital to find the sick man greatly recovered.

D'Herelle knew that bacteriophages were small enough to pass through the filters he used to remove bacteria. Also, these "filterable agents" could reproduce themselves within bacteria and, therefore, seemed to be alive.

Bacteriophages are not actually alive, however. They are **viruses,** molecular assemblies of protein and either DNA or RNA. Viruses are not cells. They cannot reproduce outside of a host cell, and they themselves cannot obtain energy from their environments. A single virus can, however, infect a cell and subvert the cell's molecular machinery to make more viruses. A virus, then, is a molecular assemblage of nucleic acid and protein that parasitizes cells. Viruses are not confined to infecting bacteria. They may infect plant cells, animal cells, and other kinds of eukaryotic cells.

Viruses are parasitic assemblies of protein and DNA or RNA capable of forcing host cells to make more viruses. Viruses that specialize in infecting bacterial cells are called bacteriophages, or phages.

Bacterial Viruses Provided More Evidence That Genes Are Made of DNA

In the 1940s, Max Delbruck and Salvador Luria began to study the genetics of the bacteriophages that infect the common intestinal bacterium (Figure 10-13A). Because bacteriophages reproduce so rapidly, they were, Delbruck and Luria decided, the ideal material for genetic studies. A bacterium such as *E. coli* infected with a single bacteriophage breaks open after about 20 minutes and releases about 200 new phage particles (Figure 10-13B). Each of these new phages can then infect another bacterium. If the bacteria are growing on a solid culture medium, the phages create a clear spot, or "plaque," on the otherwise continuous bacterial "lawn." As the phages multiply, their descendants may eventually consume almost the entire lawn.

Delbruck, Luria, and other members of the "phage group" catalogued many inherited variants of the bacteriophage. The phage group produced mutant phages with distinct phenotypes. For example, some phages produced clear plaques and others produced cloudy plaques. Still others infected one strain of *E. coli* but not another.

Such variant phenotypes of each mutant were inherited: phage particles isolated from cloudy plaques themselves produced cloudy plaques. Bacteriophages, then, like organisms, have genes. Alleles of bacteriophage genes specify alternative phenotypes such as differences that produce clear or cloudy plaques. The phage researchers knew that bacteriophages consisted of complexes of proteins surrounding a molecule of nucleic acid, usually DNA. Their genes therefore had to be either protein or DNA.

The nature of phage genes was settled in 1952 by Alfred Hershey and Martha Chase. They performed an experiment specifically designed to discover if DNA was also the hereditary material in bacteriophages. Hershey and Chase knew from

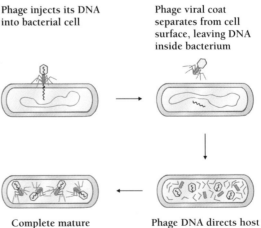

Phage attaches to surface of bacterium

Phage injects its DNA into bacterial cell

Phage viral coat separates from cell surface, leaving DNA inside bacterium

Cell wall bursts, releasing 100 to 200 new phages

Complete mature phages assemble

Phage DNA directs host cell to produce new viral proteins and DNA

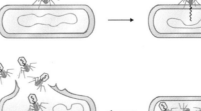

A.   B.

Figure 10-13   **The phage group.** A. Max Delbruck and Salvador Luria. B. Life cycle of the bacteriophage.   *(A, Cold Spring Harbor Laboratory Archives)*

A.

Figure 10-14 **The genetic material is DNA (or RNA), not pro-tein.** A. Martha Chase and Albert Hershey showed that when a phage infects a bacterium, it is the DNA that enters and infects the bac-terium, not the protein. B. In the Hershey-Chase experiment, radioac-tive DNA from a phage shows up inside the infected bacterium. Ra-dioactive protein stays outside.   *(A, Cold Spring Harbor Laboratory Archives)*

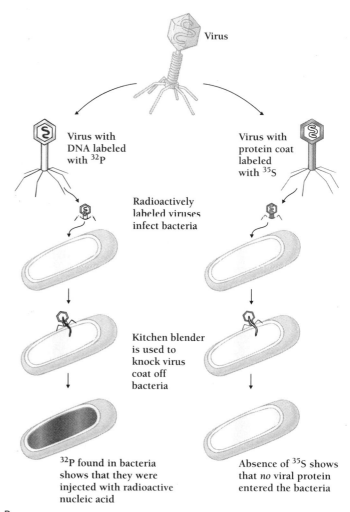

B.

microscope studies that during a phage infection only part of the virus enters the host cell. The rest remains outside. They reasoned that the material that actually enters the cell must carry the virus's genetic information. So the question was, What was that material?

Hershey and Chase were among the first geneticists to use radioisotopes in their experiments (Figure 10-14A). The avail-ability of radioisotopes had increased greatly after World War II, partly as a result of the work done to develop the atomic bomb. Hershey and Chase grew bacteria in ^{32}P-labeled phos-phate or in ^{35}S-labeled sulfate (Figure 10-14B). Recall that the amino acids that make up proteins contain sulfur but no phos-phorus, and that nucleic acids have phosphorus but no sulfur. The bacteria (and any phages that infected them) would there-fore have ^{35}S incorporated in their proteins and ^{32}P in the sugar-phosphate backbone of their DNA (and RNA). Hershey and Chase then infected these radioactive bacteria with bacterio-phages. The descendants of the infecting phages became ra-dioactively labeled. The first group of phages acquired ^{32}P-labeled DNA; the second got ^{35}S-labeled proteins.

Hershey and Chase then used the radioactive phage to in-fect new, nonradioactive bacteria. They waited just long enough for the infection to begin and then interrupted the process by dumping the bacteria into a kitchen blender. The force of the blender's blades knocked the empty phage particles off the out-side of the bacteria and separated these particles from the bac-teria. Hershey and Chase then examined the infected bacteria to see what part of the phage had entered the bacteria. Was it the DNA or the protein?

Hershey and Chase found that the phage's DNA entered the infected cell, while the phage's protein could be knocked off the cell's surface with the blender. They also showed that the phages released by the infected cell contained ^{32}P, but never ^{35}S. Hershey and Chase therefore concluded that the genes of the bacteriophage were made of DNA, not protein.

The phage group had convinced many geneticists that the genes of bacteriophages must be similar to the genes of organ-isms. The Hershey-Chase experiment therefore immediately convinced a large group of influential biologists in a way that the Avery paper, 8 years earlier, had not. Watson and Crick's 1953 paper on how DNA's structure might allow it to specify proteins convinced any remaining skeptics. After 1953, nearly everyone agreed that DNA was the genetic material.

Radioactively labeled phages showed that when a phage infects a bacterium, it is the DNA that enters and infects the bacterium, not the protein.

Why Was Hershey and Chase's Experiment More Persuasive Than Avery's?

Avery's 1944 paper showing that DNA was responsible for the transformation of pneumonia-causing bacteria should have convinced scientists that DNA was the genetic material. Yet, most biologists continued to regard DNA as merely a structural component of the chromosomes. Biologists had two important reasons for being skeptical. First, if DNA were the genetic material, the DNA of bacteria should differ from the DNA of butterflies and bananas. Yet, all DNA seemed to be chemically identical. Second, most biochemists accepted an idea proposed in the 1920s that DNA was a "tetranucleotide," the same sequence of four nucleotides repeated over and over. In that case, DNA would be unable to carry information.

Chargaff's 1950 measurements of base ratios, however, broke down both these barriers to DNA's acceptance. First, he had shown that species differences did exist—that, indeed, the DNA of butterflies and bacteria had different proportions of bases. Second, without equal proportions of each base, DNA could not be a regular repeating tetranucleotide after all. By the time Hershey and Chase published their famous experiment in 1952, scientists were ready to accept DNA as the genetic material.

It took from 1903 to 1952 to persuade biologists that DNA was the genetic material. The next questions were, What is the structure of DNA? and How does it work? As we have seen, Watson and Crick substantially answered both questions within a month of Hershey and Chase's paper.

HOW DOES ONE DNA STRAND DIRECT THE SYNTHESIS OF ANOTHER?

Watson and Crick's complementary base-pairing rule established the basis of DNA's ability to replicate itself. Since a base on one strand can match with only one other base, the sequence of bases on each strand implies the sequence of bases in the complementary strand. In their first paper describing their model, the two coyly commented on this property: "It has not escaped our notice that the specific pairing we have postulated immediately suggests a possible copying mechanism for the genetic material." (Years later, they confessed that they wanted to claim the idea without saying so much that they sounded foolish if they turned out to be wrong.) In the same way that a photographer can make a positive print from a negative, or a negative from a positive print, each strand of DNA has the capacity to direct the synthesis of the other.

This complementarity is the basis for understanding both how a chromosome replicates itself during cell division and how genes actually work. The structure of DNA explains both how cells copy their DNA and how genes specify proteins. Here we will examine how dividing cells copy the DNA.

During DNA Replication, Each Strand of DNA Remains Intact as It Directs the Synthesis of a Complementary Strand

The rules of base pairing in DNA mean that the sequence of one polynucleotide strand specifies the sequence of the second strand. The two strands of DNA are complementary, meaning that they fit together to make a unified whole—the double helix.

The basis for DNA replication is that the sequence of bases in one chain can determine the sequence of bases in the other. That is, the two strands of the DNA double helix contain the same information in two different forms. Picture a molecule of DNA coming apart into two complementary single strands. If the right bases and the proper machinery were available, each strand could act as a template for the assembly of a new complementary strand. The result would be the formation of two daughter DNA molecules, each a replica of the parent DNA.

If this model of DNA replication is correct, each daughter DNA molecule should contain one old strand and one new strand. According to this model, DNA replication is **semi-conservative**, that is, half of each parent molecule would be present in each daughter molecule (Figure 10-15A and B). Each new DNA molecule, then, is really half old and half new.

DNA replication is semi-conservative: When the DNA replicates, each half of the double helix builds a new mate.

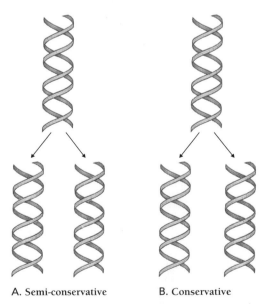

A. Semi-conservative   B. Conservative

Figure 10-15   Semi-conservative replication. At one time biologists theorized that DNA replication was conservative; both halves of the old molecule stayed together, while a whole new molecule was created (B). In fact, when DNA replicates, each half of the old double helix acquires a new mate (A).

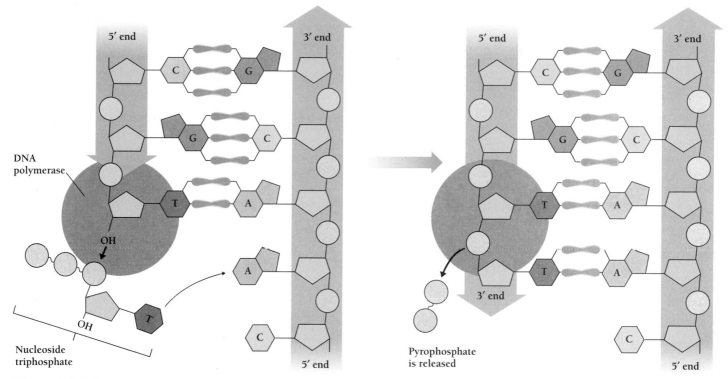

Figure 10-16 **DNA polymerization.** The enzyme DNA polymerase makes a copy of each strand of DNA by adding nucleotides to the 3' end of the polynucleotide chain.

Our discussion has so far focused on how DNA passes information to daughter DNA molecules. But what is the mechanism? How do cells actually produce new DNA? As in the case of other biochemical reactions, DNA replication depends on enzymes. The full process of DNA replication requires at least a dozen enzymes. Even now, more than 40 years after molecular biologists first synthesized DNA in a test tube, they still do not know all the details of DNA replication. The most basic steps of the process, however, are well understood.

DNA Polymerase Catalyzes the Ordered Addition of Nucleotides to Each DNA Strand

The assembly of DNA involves putting together many small molecules (nucleotides) into a larger one, a process that is called a **polymerization** [Greek, *polys* = many + *meros* = part]. The enzyme that strings together the nucleotides is called **DNA polymerase.** In each strand, the nucleotides must polymerize in a fixed order, guided by the order of bases in the complementary strand (Figure 10-16).

The Watson-Crick model for DNA replication suggests how each strand of DNA serves as a **template,** or guide, for the assembly of a complementary sequence (Figure 10-17A). Our understanding of just how the template works and how the new chains are assembled depends heavily on the work of Arthur Kornberg and his colleagues. By 1957, Kornberg had demonstrated that extracts of bacterial enzymes could catalyze DNA

synthesis. He and his colleagues then isolated DNA polymerase and studied the reactions it catalyzed.

Kornberg found that DNA polymerase can only add nucleotides to an already existing polynucleotide called a **primer** (Figure 10-17B). DNA polymerase adds nucleotides to the 3' end of the polynucleotide primer, so that the polynucleotide grows in a single direction—from its 5' end toward its 3' end.

The particular nucleotide added at each step depends upon the base sequence of the template strand. Each added nucleotide forms hydrogen bonds with the next base on the template strand. So each growing strand extends in a single direction, starting with the sequence closest to the 5' end of the primer and extending from its 3' end.

Once a free 3' end exists, we can see how the process continues. But if DNA can only add nucleotides to the 3' end of existing polynucleotides, how can the replication process ever begin? The resolution of this problem is that DNA replication does not begin with a DNA primer, but with a temporary RNA primer.

The RNA primer arises from the action of **primase,** an RNA polymerase that copies short stretches (fewer than 10 nucleotides) of DNA into RNA, without requiring a primer. DNA polymerase can then use this RNA primer's 3' end to begin synthesizing the new strand of DNA, adding nucleotides, step-by-step, to match the template and to create a DNA primer.

RNA polymerase creates an RNA primer that enables DNA polymerase to begin making a copy of each strand of DNA.

How Does a Cell Copy Both Strands of DNA Simultaneously?

Researchers can actually see cellular DNA in the process of replication with the electron microscope. The DNA in a dividing tissue culture cell looks like a long, thin molecule punctuated by bubbles. As DNA replicates, these bubbles expand. Outside of the bubble are the two strands of parental DNA, while the interior of the bubble consists of the four strands of replicated DNA, two parental strands and two newly made strands.

Each bubble consists of two **replication forks,** Y-shaped regions of DNA where the two strands of the helix have come unzipped (Figure 10-18). At the replication fork, each strand begins to direct the assembly of a new complementary strand. As replication proceeds, the forks move away from each other. In bacteria, the replication forks move away from each other at about 500 nucleotides per second, and, in mammals, at about 50 nucleotides per second.

> DNA replication occurs in replication forks, Y-shaped regions of DNA where the two strands of the helix have come apart.

How Does DNA Replicate in the 3′ to 5′ Direction?

As the replication fork moves along both strands of DNA, it moves from the 5′ end to the 3′ end along one strand, but from the 3′ end to the 5′ end on the other strand. We have seen how simply and continuously the first strand is replicated. But the second strand cannot be replicated in this way because nucleotides must be added to its 5′ end, a task that DNA polymerase cannot perform. How is this strand made?

The answer is that the second strand is produced discontinuously. After the moving fork has exposed about 1000 nucleotides of template for the lagging strand, primase produces

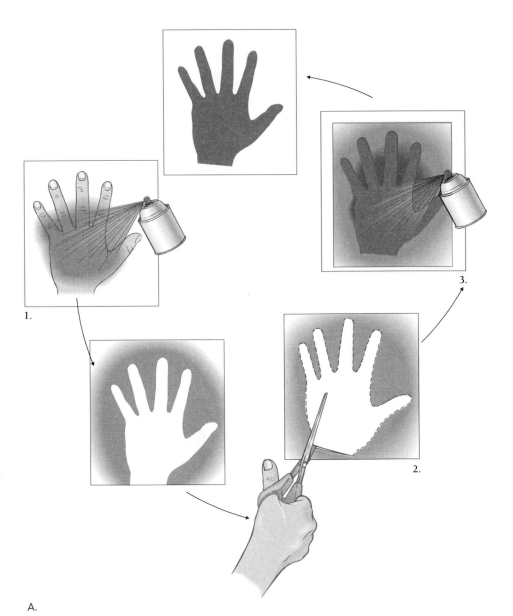

Figure 10-17   A template. A. A template is a negative that can be used to create a new version of the positive. Cookie cutters, stencils such as this one, and photographic negatives are all templates. B. The blue strand of DNA acts as a template for the creation of a new complementary strand of DNA (rust). Helicase unwinds the strands. At left, DNA polymerase adds nucleotides in a continuous string, working from the 3′ end of the template to the 5′ end. At right, DNA polymerase works in the opposite direction, starting with RNA primers to create Okazaki fragments that are later linked together by DNA ligase.   A.

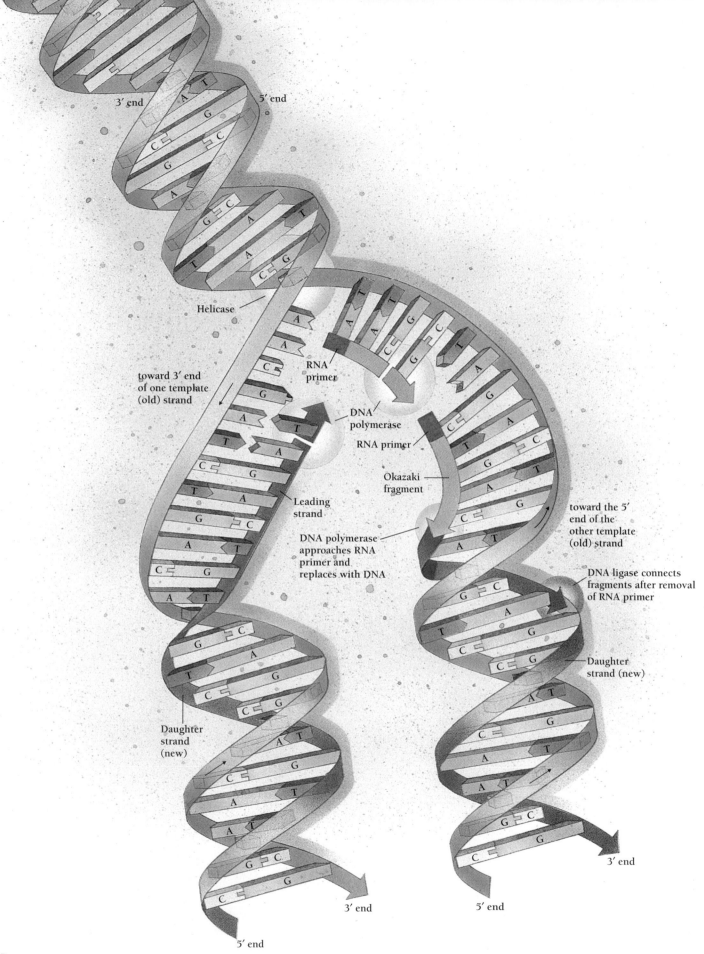

3′ end

5′ end

Helicase

toward 3′ end
of one template
(old) strand

RNA
primer

DNA
polymerase

RNA primer

Okazaki
fragment

Leading
strand

DNA polymerase
approaches RNA
primer and
replaces with DNA

toward the 5′
end of the
other template
(old) strand

DNA ligase connects
fragments after removal
of RNA primer

Daughter
strand (new)

Daughter
strand
(new)

3′ end

3′ end

5′ end

5′ end

B.

255

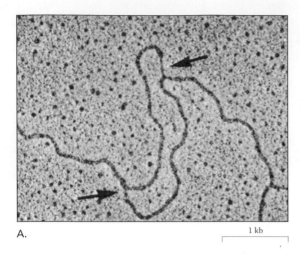

A.

1 kb

Figure 10-18   Replicating DNA. A. TEM of a replication fork.
B. A simplified view of a DNA replication fork.   *(A, David Hogness and
Henry Kriegstein, Stanford University School of Medicine,* Proceedings of the
National Academy of Science, *71 (#1), January 1974, pp. 135–139)*

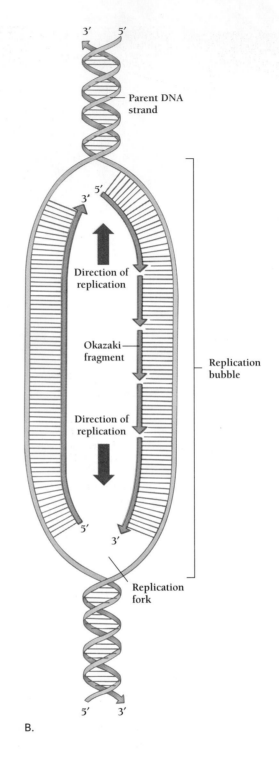

B.

a short piece of complementary RNA. This RNA serves as a
primer for DNA polymerase, which attaches nucleotides to its
3' end. This process produces fragments, called **Okazaki fragments,** after their discoverer Reiji Okazaki. Each Okazaki fragment contains a short stretch of RNA connected to about 1000
nucleotides of a DNA strand. DNA polymerase then removes
the RNA primer, and yet another enzyme, called **DNA ligase,**
stitches the fragments together (Figure 10-18).

In the leading strand of a replication fork, nucleotides add
directly to the 3' end. The other strand, however, is produced
in fragments. Each fragment is begun with a short
RNA primer and then is joined to the other
fragments by the enzyme DNA ligase.

How Does DNA Synthesis Begin?

DNA always begins replication at the **origin of replication**.
E. coli, the organism whose DNA synthesis has been most studied, has a single piece of circular DNA with a single origin of
replication (Figure 10-19). Starting from this single site, two
replication forks move in opposite directions away from the
origin of replication, and DNA synthesis stops when the fork
that is moving clockwise meets the counterclockwise fork.

Several proteins participate in DNA replication. An enzyme
called helicase uses energy from ATP to untwist the helix, while
another enzyme called DNA gyrase prevents the accumulation
of kinks at the stems of the replication forks. Still another protein binds to the unwound DNA and keeps open the jaws of
the replication fork.

In eukaryotic cells, DNA replication is more involved. Each
cell contains thousands of origins of replication. Groups of 20
to 50 origins, called **replication units,** form replication forks

at the same time. Within a replication unit, the forks move in
opposite directions from each origin until they encounter a fork
from an adjacent origin.

DNA synthesis always begins at special sequences in the DNA
called origins of replication. Proteins bind to these origins of
replication and open and untwist the double helix.

Figure 10-19 Replication of bacterial DNA. Replication begins at the origin of replication (*arrows*), then works in both directions around the circle.

MUTATIONS ARE THE ULTIMATE SOURCE OF ALL GENETIC DIVERSITY

Every living organism shares the same chemistry of life. Your DNA is made of the same stuff and follows the same rules as the DNA of the roses nodding in your neighbor's yard. Yet you are different from a rose. If, for example, you want lunch, you must get up and make yourself a sandwich, while the rose simply waits passively for the sun to emerge from behind a cloud. What makes us so different from a rose?

Each species has its own characteristic DNA. But essentially all biologists think that life began on Earth just once. If that is the case, then at one time all organisms were of the same kind and all of their genetic material was the same. How, over billions of years of evolution, did we become so different from one another? The short answer is **mutations,** permanent changes in the sequence of DNA. The different alleles of a gene are created by mutations in the original sequence. The original sequence itself is a mutation of some other gene. All genetic differences ultimately depend on the accumulation of mutations in the genome.

Although the evolution of organisms depends on the accumulation of changes in the DNA, the vast majority of mutations are either harmful or meaningless. If we opened up the

back of a television set and switched two wires at random, the chances that this would improve the picture or the general performance of the television are almost zero. Mutations in highly evolved genomes are the same. Most are a disaster for the phenotype. Every now and then, a new mutation is useful under certain circumstances. In that case, natural selection may increase its representation in a population of organisms.

We have already seen in Chapter 9 how the recombination (crossing over) and the mixing of gametes during sexual reproduction create unique combinations of alleles. But, ultimately, all new alleles result from mutations. Our focus in the last part of this chapter is the origin and consequences of mutations: How do mutations arise? How do they influence the phenotype of the organism?

What Causes Mutations?

Mutations occur spontaneously as random events in the cell. The structure of any molecule—including DNA—can change as a result of a random collision with another molecule or ion. For example, the bonds between the A and G (purine) bases in DNA and the backbone sugar are slightly unstable. As a result, each cell in the human body loses some 5000 to 10,000 A and G nucleotides every day.

Occasionally, mutations occur because DNA polymerase places the wrong nucleotide in a sequence during the synthesis (S) phase of the cell cycle (Chapter 8). In addition, certain agents, called **mutagens,** increase the rate of mutation. Ultraviolet light, high-energy radiation from x rays or radioactive decay, and many chemicals are all mutagens. Each mutagen has a characteristic spectrum of effects. Ultraviolet light, for example, may link together adjacent Ts in the same DNA strand. X rays are particularly effective producers of chromosomal mutations, causing DNA to break and rejoin in strange new arrangements. Chemical mutagens, such as alcohol and dioxin, for example, also increase mutation rates.

Mutations occur spontaneously. But they occur more frequently when the DNA is exposed to ultraviolet light, x rays, chemicals, and other mutagens.

How Often Do Mutations Occur?

Changes in DNA sequence occur spontaneously all the time. Mutations can initially result from random collisions between molecules, from mistakes in replication (for example, from a failure to form a correct base pair), or from changes in the structure of a nucleotide.

The **mutation rate** is how often mutations occur. More specifically, the mutation rate is the number of changes per nucleotide per generation. The mutation rate depends both on how often a sequence mutates and how efficiently cells repair these mutations. Some sequences within a gene, called

(Text continued on page 260.)

BOX 10-1

Determining the sequence of nucleotides in DNA

Current molecular biological techniques allow scientists to prepare large amounts of individual pieces of DNA and to study both their chemical structure and their biological function. One of the first questions in such studies is, What is the sequence of nucleotides in a given piece of DNA?

The most widely used method for DNA sequencing, developed in England, by Fred Sanger in the 1960s, depends on DNA polymerase. Sanger's method depends on five "rules" of DNA synthesis:

1. DNA synthesis is semi-conservative.
2. A newly made DNA strand grows in a single direction, adding one nucleotide at a time to the 3′ end.
3. The nucleotide sequence of the template strand determines the sequence of nucleotides in the growing strand.
4. DNA polymerase requires a primer polynucleotide.
5. The copying of a single strand into a new complementary strand is catalyzed by a single enzyme, DNA polymerase.

In addition, DNA sequencing requires that a researcher be able to separate DNA fragments of different lengths.

Sanger began by heating the DNA until it separated into two single template strands. He then added a specific complementary primer, DNA polymerase, and plenty of A, C, G, and T. The DNA polymerase automatically set to work extending the primer until it produced a complete copy of the template strand.

Figure A   To sequence DNA, biologists use electrophoresis to order DNA fragments by length. Biologists then "read" the last letter of each fragment, starting with the shortest fragment. Reading from shortest to longest, the nucleotide letters are A-T-C-G-T-T-G-A.

In a laboratory, gels stand vertically, with the DNA samples at the top. Gels are then read from the bottom (shortest) to the top (longest). We have reversed this to increase the clarity of the illustration.

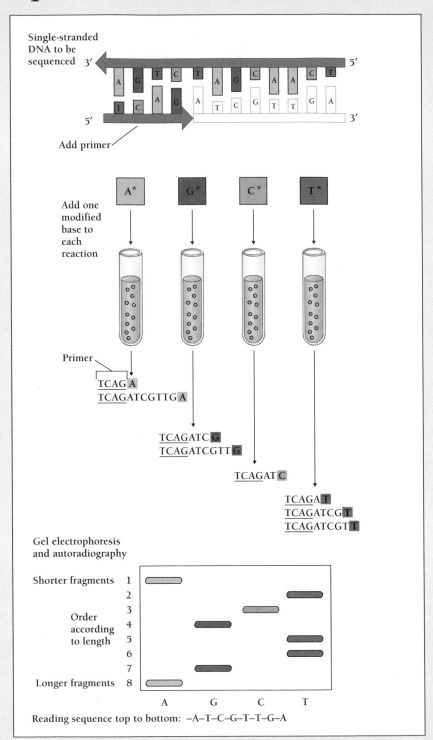

Sanger then devised a way of stopping the polymerase reaction at different places, specifically at A, C, G, or T. He did this by replacing part of a base with one that would not allow further polymerization. DNA polymerase adds each nucleotide to the 3′-hydroxyl group of the previous nucleotide. But Sanger could replace an A, for example, with a similar molecule that lacked a 3′-hydroxyl group. Therefore, once this modified A "base" was added to the growing strand, no more bases could be added (Figure A).

Each time the DNA polymerase sees a T in the template strand, it places an A in the new strand. But if the selected A has no 3′-hydroxyl group, the chain stops growing. Some DNA molecules stop after the first A in the new strand, some after the second, some after the third, and so on. The result is a collection of DNA fragments, all of which end in A. Electrophoresis allows us to distinguish the sizes of each of these fragments (Figure B), even ones that differ in length by no more than one nucleotide.

Sanger could stop polymerization after C, G, or T in the same way. In each case, we can separate the resulting collections of fragments by electrophoresis, which reveals the sizes of the fragments that end with each type of nucleotide. From these four sets of bands, researchers can work out the complete sequence of the newly made DNA (Figure C).

Today, commercial machines do most of this work automatically. The power of these automatic machines is such that we will soon be able to determine the entire sequence of all 3 billion nucleotides in human DNA.

The sequencing of DNA provides important information. It allows researchers to detect genetic mutations associated with diseases, such as sickle cell disease. In addition, new methods allow the selective alteration of nucleotide sequences in genes and the placing of genes into cells to study the effects of these changes. By altering the sequence of a gene, researchers alter the structure of the resulting polypeptide protein and they then can study how changes in the structure of a protein affect its function.

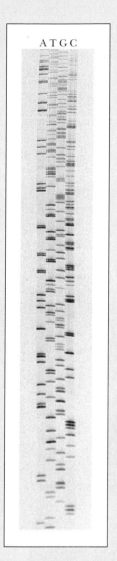

Figure C Actual data from a DNA sequencing gel. Reading from the 5′ end (bottom of the gel), the sequence is TATC-CCGTTGGAAGGTCGTCTGCTCCCTG-GAAGTAG. *(James D. Colandene, University of Virginia)*

Figure B Electrophoresis separates DNA fragments of different sizes. Here a scientist studies a DNA electrophoresis gel. The fragments have been treated with a stain that fluoresces under ultraviolet light. *(Geoff Tompkinson/Science Photo Library/Photo Researchers)*

mutation hot spots, are more likely to mutate than others. In bacteria, for example, some sequences mutate 25 times more rapidly than others. Even the highest rates of mutation, however, are far less than the error levels to which we are accustomed in everyday life. A first-class typist, for example, makes only 1 mistake in 10,000 characters. Yet, the average rate of mutation in bacteria during ordinary DNA replication is only 1 mutation in 10 million. DNA replication is at least 1000 times more accurate than the best human typist.

Mutations in bacterial DNA occur at the rate of about 1 mutation in 10 million base pairs each generation.

What Kinds of Mutations Are There?

We can categorize mutations on the basis of the sizes of the changes in DNA. Geneticists often distinguish between **point mutations,** which change one or several nucleotide pairs, and **chromosomal mutations,** those which change large regions of chromosomes. A point mutation may be (1) a **base substitution**—the replacement of one base (nucleotide) by another, (2) an **insertion**—the addition of one or more nucleotides, or (3) a **deletion**—the removal of one or more nucleotides.

Chromosomal mutations are larger than point mutations and may affect large regions of chromosomes, or even whole chromosomes. Chromosomal mutations include (1) **deficiencies**—deletions that are larger than a few nucleotides, (2) **translocations**—in which part of one chromosome is moved to another chromosome, (3) **inversions**—in which a segment of a chromosome is flipped over 180° from its normal orientation, and (4) **duplications**—in which part of a chromosome appears twice. **Aneuploidy,** abnormal chromosome number, is regarded as another type of chromosomal mutation. Aneuploidy includes **trisomy,** the presence of three copies of a certain chromosome, and **monosomy,** the presence of only a single copy in a cell that normally has pairs of chromosomes.

A mutation is any change in the sequence or number of nucleotides, ranging from small changes of one or a few nucleotides to chromosomal inversions or aneuploidy.

HOW DO CELLS HANDLE MISTAKES IN THE NUCLEOTIDE SEQUENCE OF DNA?

Because DNA mutates, cells must be able to repair damage to DNA if they are to maintain the continuity of life. This is especially important for the germ line cells that pass information

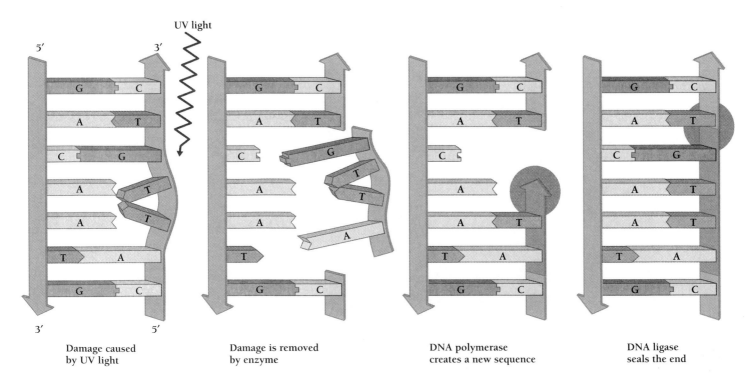

Damage caused by UV light

Damage is removed by enzyme

DNA polymerase creates a new sequence

DNA ligase seals the end

Figure 10-20 Healthy cells are quick to repair damaged DNA. Ultraviolet light can damage DNA by linking together adjacent Ts in the same strand. Repair enzymes excise the abnormal nucleotides and restore the original sequence, then DNA polymerase and DNA ligase repair the gap as in normal replication.

to the next generation. However, mutations in somatic cells are dangerous as well. In cancer, for example, a cell with a defective gene for growth control multiplies so fast that immature daughter cells take over the area.

Both prokaryotic and eukaryotic cells have sophisticated mechanisms to repair damage to DNA and to correct mistakes in replication. These mechanisms are crucial for the stability of genetic information. They depend upon the duplex structure of DNA and on the presence of enzymes that recognize commonly occurring mistakes.

In each case, an enzyme detects something wrong in one strand of the DNA and removes the error. For example, cytosine may be accidentally converted to uracil, a base that is ordinarily not present in DNA. A special enzyme called uracil-DNA glycosidase can recognize the strange base and remove it. In another example, two T nucleotides may bond to one another instead of to their respective A nucleotides (Figure 10-20).

Once correction enzymes remove the offending nucleotides, DNA polymerase goes to work. It copies the information in the intact second strand and creates a new stretch of DNA that is properly matched. **DNA ligase** then seals the gap, and the original sequence is restored. Over 50 enzymes "proofread" DNA for errors—unusual bases, gaps, and bulges.

Nonetheless, errors still can slip through this elaborate system. For example, the genes for error correction enzymes may themselves be damaged. In that case, large numbers of mutations can accumulate (Figure 10-21). If these uncorrected mutations occur in genes that help control cell division, the cells become cancerous.

Cells have sophisticated mechanisms for repairing damage and for correcting mistakes made during DNA replication.

In this chapter, we have seen how we know that genes are made of DNA and that DNA has a special structure that enables it to replicate. We also saw that genes are both informa-

Figure 10-21   Sunburns damage cells and lead to skin cancer. DNA repair mechanisms break down with age or exposure to the ultraviolet light in sunlight. In the absence of DNA repair, many genes become damaged, including those that control the cell cycle. In early rounds of sunburn, most cells with damaged DNA self-destruct, and the skin peels. One theory holds that a few cells with damaged DNA remain but do not reproduce because they are surrounded by healthy cells. During subsequent sunburns, however, more cells with damaged DNA self-destruct, clearing the field for cancerous cells to begin growing.   *(© 1990 Peter Menzel/Stock Boston)*

tion and physical regions in the DNA. In the next chapter, we will see how DNA's structure allows the "expression" of genes as polypeptides and how cells control the expression of genes—so that a given polypeptide is synthesized at the right time and in the right amount. We will also further explore how mutations in the structure of DNA change the information in genes.

STUDY OUTLINE WITH KEY TERMS

This chapter discusses two ways of defining a gene. One describes the effects of altered genes on phenotype; the other describes the chemical structure of genes.

Many genes contain the information for the structure of enzymes. When an organism cannot make a functional enzyme, it cannot carry out the biochemical conversions catalyzed by the enzyme. The result is often the failure to produce a needed compound and the buildup of some intermediate in the biochemical pathway leading to that compound. Thus patients who have alkaptonuria lack the capacity to

process tyrosine, and they accumulate a compound that turns the urine black.

By studying similar interruptions in biochemical pathways in the bread mold *Neurospora,* Beadle and Tatum reached the conclusion that each gene determined the structure of a single enzyme, shortening this insight to **"one gene—one enzyme."** However, many genes do not determine the structure of enzymes but of other proteins, such as hemoglobin. A better statement of gene function is **"one gene—one polypeptide."**

Different alleles of a gene produce different versions of the same protein. Thus, an alteration in the gene coding for the β-globin polypeptide leads to hemoglobins with different structures and properties. Sickle cell disease results from the homozygous occurrence of an allele that alters the structure of β-globin at a single amino acid.

The idea that genes are made of DNA came from work on bacteria and bacterial **viruses,** called **bacteriophages,** or simply **phages.** A virulent strain of bacteria can transfer the genetic instructions for virulence to a nonvirulent strain by means of DNA, in a process called **transformation.**

DNA consists of nucleotides held together by bonds between the third (3′) carbon of the sugar of one nucleotide and the fifth (5′) carbon of the sugar of the next nucleotide. Each polynucleotide chain has a direction with a 5′ end and a 3′ end. DNA contains four kinds of nucleotides, which differ in the nitrogen-containing bases attached to their sugar components. The bases are two purines (adenine and guanine, abbreviated A and G) and two pyrimidines (cytosine and thymine, abbreviated C and T). According to **Chargaff's rules,** the amount of C always equals the amount of G, and the amount of A equals the amount of T in the DNA of a given species.

In cells, DNA consists of two polynucleotide strands twisted around each other in a double helix. In the early 1950s, James Watson and Francis Crick used the results of x-ray diffraction studies of DNA by Rosalind Franklin to build a molecular model of the double helix. The two strands of DNA are held together by specific hydrogen bonds that pair G with C and A with T. These specific pairings, argued Watson and Crick, explain the equal concentrations of G and C, and of A and T. The specific pairings also mean that the information contained in the sequence of nucleotides was equivalent in the two strands. Each strand can direct the synthesis of a complementary strand in DNA replication.

DNA replication is **semi-conservative:** when DNA replicates, each strand of the double helix acquires a new mate. The main enzyme that catalyzes the **polymerization** of DNA is called **DNA polymerase.** DNA polymerase assembles nucleotides into sequences spec-

ified by one strand of DNA, the **template** strand. DNA polymerase can add nucleotides only to a preexisting RNA **primer,** a molecule that arises from the action of **primase,** an RNA polymerase that copies short stretches (fewer than 10 nucleotides) of DNA into RNA, without a primer. DNA polymerase adds nucleotides only to the 3′ end of the primer and continues to add nucleotides to the 3′ end of the growing polynucleotide.

DNA replication occurs in **replication forks,** Y-shaped regions of DNA where the two strands of the helix have come apart. Nucleotides add directly to the 3′ end of one strand in the fork, called the leading strand. The other strand, however, is produced discontinuously, in **Okazaki fragments** of about 1000 nucleotides. Each of these fragments starts with a short stretch of RNA, which serves as a primer. Another enzyme, called **DNA ligase,** joins the fragments together as DNA replication proceeds. Special sequences in DNA serve as **origins of replication,** where DNA synthesis always begins. The sequences bind to proteins that open and untwist the double helix. Groups of 20 to 50 origins, called **replication units,** form replication forks at the same time.

DNA molecules continuously suffer inevitable damage in the form of mutations. A **mutation** is any change in the sequence or number of nucleotides. **Point mutations** include multiple **base substitutions, insertions,** and **deletions. Chromosomal mutations** include **deficiencies, translocations, inversions,** and **duplications,** as well as **aneuploidies** such as **trisomy** and **monosomy.** Mutations are the ultimate source of all genetic diversity.

Mutation rates are usually low—occuring in bacterial DNA, for example, at the rate of about 1 mutation in 10 million base pairs each generation. **Mutation hot spots** mutate more often than other parts of the genome. Mutations can occur spontaneously, but they are hastened by ultraviolet light, x rays, and many chemicals called **mutagens.**

Cells have sophisticated mechanisms to repair damage to DNA and to correct mistakes made during DNA replication. One result of uncorrected mistakes is cancer.

REVIEW AND THOUGHT QUESTIONS

Review Questions

1. What facts about the structure of DNA did Franklin contribute? What did Watson and Crick contribute?
2. How does Watson and Crick's theory of base pairing explain how DNA replicates itself?
3. Do genes specify enzymes, proteins, or polypeptides? Why did biologists worry about this distinction?
4. What does it mean for a bacterial cell to be *transformed*? What does the transforming?
5. How is semi-conservative replication different from conservative replication?
6. What serves as a template for the replication of a single strand of DNA?
7. What enzyme polymerizes nucleotides into a new strand of DNA?
8. Why does the cell need an RNA primer for DNA replication to begin? Where on the DNA does DNA polymerase begin replication?

9. What is the ultimate source of all genetic variation?
10. Why do cells need mechanisms for repairing DNA?

Thought Questions

11. D'Herelle and others thought that bacteriophages might provide a way to kill disease-causing bacteria—an idea that was especially exciting in the days before the discovery of penicillin and other antibiotics. Yet, in more than 80 years of research, no one has ever found a way to use bacteriophages to destroy bacteria that have already infected the body. What problems might prevent medical researchers from making bacteriophages kill bacteria in the body?
12. Why does the dentist place a lead shield over your lap before x-raying your teeth? What might happen if the dentist did not take this precaution?

SELECTED READINGS

Judson, Horace Freeland, *The Eighth Day of Creation,* Simon and Schuster, New York, 1979. A highly detailed 700-page history of molecular biology, including fascinating anecdotes and clear descriptions of experiments. Considering the level of detail, this book is surprisingly accessible to lay readers.

Sayre, Anne, *Rosalind Franklin and DNA,* Norton, New York, 1975. A detailed biography and impassioned defense of Franklin by a friend of Franklin's.

Watson, James, *The Double Helix,* Penguin, New York, 1968. Watson's riveting account of the discovery of the structure of DNA.

▶ On-line materials relating to this chapter are on the World Wide Web at http://www.saunderscollege.com/lifesci/
Click on Tobin/Dusheck: *Asking About Life.*

How Was the Genetic Code Discovered?

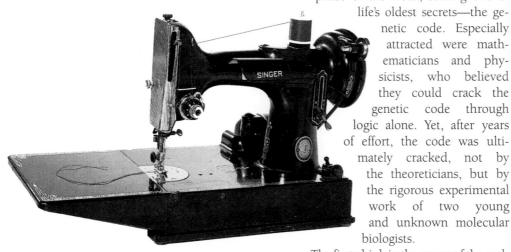

As soon as Watson and Crick discovered the structure of DNA, scientists all over the world wanted to become involved in the obvious next phase of the work, solving one of life's oldest secrets—the genetic code. Especially attracted were mathematicians and physicists, who believed they could crack the genetic code through logic alone. Yet, after years of effort, the code was ultimately cracked, not by the theoreticians, but by the rigorous experimental work of two young and unknown molecular biologists.

The first chink in the armor of the coding problem came in 1955, when Marianne Grunberg-Manago, working at New York University, discovered an enzyme that could link nucleotides together into long polymers. Grunberg-Manago took a year to isolate the enzyme. Once she had the enzyme, however, she began making artificial RNA polymers. She could, for example, link together long chains of A (adenine) nucleotides, which she called "poly A," or long chains of C (cytosine) nucleotides, called "poly C." In addition, she could make poly U, poly G, poly AU, and other combinations.

For the moment, however, no one knew what combination of nucleotides could actually be translated by a cell into a polypeptide. It was 6 more years before the next big break came. Then, in 1960, 31-year-old Johann Heinrich Matthei arrived at the National Institutes of Health (NIH) near Washington, DC. Matthei had come from Germany earlier that year looking for interesting scientific work. He had not found anything satisfactory yet, but he was especially interested in working on protein synthesis. At NIH, he got the names of three researchers who were trying to synthesize proteins artificially outside of a cell. Thirty-three-year-old Marshall Nirenberg stood out from the other two (Figure 11-1). His interests and instincts matched Matthei's own, and in November 1960, Matthei and Nirenberg embarked on a productive, if sometimes strained, collaboration.

Although, technically, Nirenberg was Matthei's boss, Matthei had come to NIH for collaboration, not direction. He was fully able to work on his own. Nirenberg's goal was to synthesize the polypeptides that make up proteins. He wanted to make them in a test tube. To do this, he planned to add ATP (for energy) and free-floating amino acids to a mixture of ribosomes, nu-

Figure 11-1 Marshall Nirenberg. Nirenberg and Johann Matthei cracked the genetic code in 1960. *(UPI/Corbis-Bettmann)*

cleic acids, and enzymes extracted from cells. Other researchers had used this technique and had shown that the amino acids were incorporated into polypeptides. No one, however, knew what substances in the cell extracts were making the polypeptides and no one knew what kind of polypeptides were being synthesized.

The obvious next step was to try to make a particular polypeptide by putting genetic information into the test tube. Several labs in the United States were hotly pursuing this idea. Before Matthei arrived, Nirenberg had tried to synthesize penicillinase, the enzyme that antibiotic-resistant bacteria use to break down the antibiotic penicillin. Nirenberg added bacterial DNA and RNA to his test tube system. In ex-

periment after experiment, however, he got no penicillinase. But without knowing how the code worked, no one would ever be able to synthesize any polypeptide, let alone a particularly valuable one such as penicillinase. Nirenberg needed to ask a simpler question.

After Matthei arrived, the two researchers asked a basic question: What kinds of RNA stimulate the synthesis of polypeptides? To answer this question, they took an extremely methodical approach. Before trying to synthesize any protein, they would make sure that their test tube system could respond to RNA, any RNA. The two spent months perfecting their system. In the end, they could detect even minute amounts of protein synthesis. Their test tube system was now highly sensitive to RNA.

Nirenberg and Matthei knew that protein synthesis required ribosomes and that ribosomes contained RNA. Like most other scientists of the period, they believed that RNA must carry a sort of message from the DNA in the nucleus to the ribosomes in the cytoplasm, where protein synthesis normally happened. But although Nirenberg and Matthei's test tubes were packed with ribosomal RNA, they were seeing very little protein synthesis. Instinctively, they felt that, besides ribosomal RNA, there must be another kind of RNA—a kind that carried information that could specify a polypeptide and

would make protein in their test tubes. But where was it? What was it?

They drew up a list of some 200 kinds of RNA to test. One of the first possibilities was RNA from tobacco mosaic virus. (In some viruses, RNA carries the genetic information.) The results were spectacu-

On Friday night, Matthei stayed up all night again. Early on Saturday morning, bleary-eyed but elated, Matthei had his answer.

lar. Nirenberg and Matthei's test tube system incorporated huge amounts of amino acids into some mystery protein.

Nirenberg instantly recognized a golden opportunity. Another molecular biologist, Heinz Fraenkel-Conrat, at the University of California, Berkeley, had recently worked out the sequence of the tobacco mosaic virus protein. Nirenberg called Fraenkel-Conrat, forged an instant collaboration, and in May of 1961 left to spend a month in California to try to synthesize tobacco mosaic virus proteins. If he succeeded, he would gain scientific celebrity.

Matthei stayed behind at NIH patiently testing the other possibilities on the list of 200 different kinds of RNA. High on the list were the artificial RNAs made using the methods invented by Marianne Grunberg-Manago. A week after Nirenberg left, Matthei set up the test tubes, the different enzymes, the ribosomes, the ATP, and the 16 amino acids he had on hand. Then to the different test tubes he added poly U, poly A, and poly AU. He knew that if he got any synthesis at all, poly U would make a polypeptide consisting en-

Figure 11-2 A simplified view of Nirenberg and Matthei's experiments. A. Matthei added 4 different radioactively labeled amino acids to each of 5 test tubes containing enzymes, ribosomes, and ATP, as well as poly U. One test tube synthesized large amounts of an unknown polypeptide. B. To find out which amino acid had been joined into polypeptides, Matthei added just one amino acid to each of 5 more test tubes. The test tube that synthesized protein was the one that contained the amino acid phenylalanine. The synthesized polypeptide consisted entirely of phenylalanine.

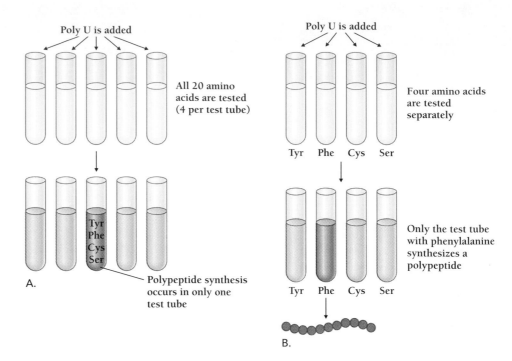

Poly U is added

All 20 amino acids are tested (4 per test tube)

Tyr
Phe
Cys
Ser

A.

Polypeptide synthesis occurs in only one test tube

Poly U is added

Four amino acids are tested separately

Tyr   Phe   Cys   Ser

Only the test tube with phenylalanine synthesizes a polypeptide

Tyr   Phe   Cys   Ser

B.

tirely of one kind of amino acid. Poly A would too, but it would be a different amino acid.

The results again were dramatic (Figure 11-2). Poly A and poly AU did almost nothing, but poly U stimulated 12 times the protein synthesis Matthei and Nirenberg got with nothing else added. The only question now was, What polypeptide had been made? In other words, Which of the amino acids had been polymerized? Matthei attacked the problem aggressively. For 5 days, he worked almost round-the-clock, testing one amino acid after another.

On Friday night, Matthei stayed up all night again. Early on Saturday morning, bleary-eyed but elated, Matthei had his answer. Poly U coded for a polypeptide made exclusively of the amino acid phenylalanine. He now knew that U nucleotides coded for phenylalanine, but how many nucleotides it took—whether one, two, three, or even more nucleotides—was another question. The number of nucleotides required to specify one amino acid was still unknown. Nonetheless, as Matthei stood in the lab, he realized that he alone in all the world knew the first word of the genetic code. He told Nirenberg's boss at NIH, but Matthei did not call Berkeley.

Nirenberg did not hear the news until he returned from California nearly 3 weeks later. His experiments at Berkeley had been inconclusive. The protein that tobacco mosaic virus RNA seemed to be synthesizing could not be identified. He must have heard Matthei's news with a mixture of emotions—delight at Matthei's success, disappointment that he had not been there to share in it.

Everyone at NIH knew what Matthei and Nirenberg had done. But the rest of the world either had not heard or did not believe it. Nirenberg was quiet, soft-spoken. As one well-placed molecular biologist later said apologetically in reference to Nirenberg, "either a person . . . is someone who's in the club and you know him, or else his results are unlikely to be correct, because he [isn't] in the club." Nirenberg was definitely not in the club. And, as far as the fast-moving world of molecular biology was concerned, Matthei might as well have not existed.

Together Nirenberg and Matthei wrote up the results of all their work so far and sent it to a scientific journal. A week later, NIH sent Nirenberg to Moscow for the Fifth International Congress of Biochemistry. In a small, nearly empty room, Nirenberg presented the results of his and Matthei's months of work. Of the handful

of scientists present, only molecular biologist Matthew Meselson really listened.

As Meselson later recalled, "I was bowled over by the results, and I went and chased down Francis [Crick], and told him that he must have a private talk with the man." Crick talked to Nirenberg, then arranged for him to give his talk once more in front of an audience of hundreds of biochemists and molecular biologists. Nirenberg's scientific fortune was made. He might not belong to the club, but he had been heard by nearly everyone who did.

While Nirenberg was in Moscow, Matthei nailed down still another "word" in the genetic code. Poly C, he found, made a polypeptide composed entirely of the amino acid proline.

Nirenberg returned to NIH in triumph. Yet within days of his announcement in Moscow, another research lab began using Nirenberg and Matthei's refined technique to crack the rest of the genetic code. Although initially taken aback by the aggressiveness of this move, Nirenberg soon threw himself into the competition, determined to be first.

In his excitement, Nirenberg began making all the decisions, and Matthei began to feel like a fifth wheel. His friendship and collaboration with Nirenberg

A.

First letter (5′ end)	Second letter				Third letter (3′ end)
	U	**C**	**A**	**G**	
U	UUU Phe UUC UUA Leu UUG	UCU UCC Ser UCA UCG	UAU Tyr UAC UAA Stop UAG Stop	UGU Cys UGC UGA Stop UGG Trp	U C A G
C	CUU CUC Leu CUA CUG	CCU CCC Pro CCA CCG	CAU His CAC CAA Gln CAG	CGU CGC Arg CGA CGG	U C A G
A	AUU AUC Ile AUA AUG Met	ACU ACC Thr ACA ACG	AAU Asn AAC AAA Lys AAG	AGU Ser AGC AGA Arg AGG	U C A G
G	GUU GUC Val GUA GUG	GCU GCC Ala GCA GCG	GAU Asp GAC GAA Glu GAG	GGU GGC Gly GGA GGG	U C A G

Nucleotide base   Codon   Amino acid

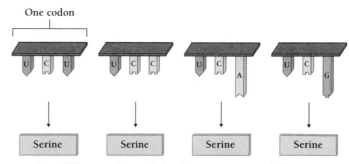

Four different codons code for the same amino acid. Note difference is in last base.

B.

Figure 11-3   **The genetic code.** A. This table shows which amino acid corresponds to each RNA triplet. All the DNA "T"s are replaced by "U"s, since this is RNA. B. Some amino acids have more than one codon. Usually the first two bases are the same, but the third base is different.

soured, and 18 months after they began to work together, Matthei was back in Germany working by himself once more.

Nirenberg never missed a step. He began working with another collaborator, Philip Leder, and the two soon showed that a length of RNA just three nucleotides long could specify an amino acid. They were able to define a **codon** as a group of three nucleotides that specifies a single amino acid within a polypeptide.

Then another researcher stepped into the picture. Organic chemist Gobind Khorana found a way to link nucleotides together in any order he wanted. In one experiment, he put together long chains of poly AC, which contained the sequence ACACACACAC and so on. Khorana found that poly AC RNA stimulated the production of a polypeptide consisting of two alternating amino acids—threonine and histidine.

Khorana showed that ACA specified threonine and CAC specified histidine. Thus ACA CAC ACA CAC coded for threonine-histidine-threonine-histidine. From a series of such experiments, Khorana and his colleagues established a dictionary of codons. It took 5 years, but by 1966, Nirenberg, Khorana, and others had written a complete dictionary for the genetic code (Figure 11-3). There were 64 different codons. Sydney Brenner and Francis Crick showed that 3 of the 64 possible

codons coded for no amino acids at all. They were "nonsense" codons whose role was to signal the end of a polypeptide chain, just like the period at the end of a sentence.

In 1968, Nirenberg and Khorana won a Nobel Prize for their elucidation of the genetic code. In the same year, James Watson's bestselling book *The Double Helix* was published. In the world of biology, the first era in the history of molecular biology had come to an end.

The genetic code was deciphered by means of careful laboratory experiments.

KEY CONCEPTS

1. Protein synthesis in test tubes, using cell extracts, revealed how information flows from messenger RNA (mRNA) to polypeptide.

2. Genetic information flows from DNA to RNA to polypeptides.

3. DNA and RNA carry information in a 4-letter alphabet, while proteins carry information in a 20-letter alphabet.

4. RNA polymerase produces molecules of RNA by copying one strand of DNA starting at a promoter sequence and ending at a terminator sequence.

5. The RNA products of eukaryotic transcription often undergo extensive modification.

6. Translation of messenger RNA requires correct initiation, elongation, and termination of polypeptide synthesis.

7. In eukaryotic cells, special amino acid sequences help specify the destinations of newly made polypeptides.

8. In both prokaryotes and eukaryotes, several factors may influence the expression of a single gene.

9. Energy organelles—mitochondria and chloroplasts—contain independent genetic systems and probably evolved from ancient associations of prokaryotes.

What Is the Genetic Code Like?

The nucleotide sequences of DNA and RNA are written in a different language from the amino acid sequences of polypeptides. The DNA/RNA language is written with an alphabet of just 4 nucleotides, while the polypeptide alphabet has 20 amino acid letters (Figure 11-4).

Although the DNA and protein languages are different, the relationship between them is simple. This is because the **genetic code** clearly specifies which nucleotide sequences correspond to which amino acid sequences. Each set of three nucleotides, called a codon, specifies an amino acid.

Since different combinations of any 3 nucleotides can code for 64 different amino acids, and there are only 20 amino acids, what do the extra 44 codons do? The answer is that most of these extra codons also code for the same 20 amino acids. The genetic code has several ways of specifying each amino acid. Because several codons may have the same meaning, the genetic code is redundant.

Of the 64 triplets, only 61 specify amino acids. The other three codons (UAA, UAG, and UGA) are called **nonsense**, or **stop, codons** because rather than specifying an amino acid, they normally signal the end of a polypeptide chain.

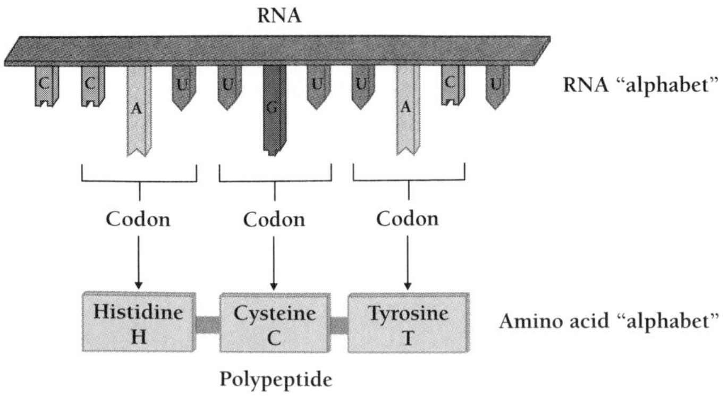

Because the same code is used by all species, the genetic code is said to be **universal.** One demonstration of this universality comes from studies of hemoglobin, the oxygen-carrying molecule in red blood cells. Before red blood cells enter the bloodstream, they synthesize large amounts of hemoglobin from two polypeptides, called globins, made from two kinds of RNA. When this RNA is purified from rabbit red blood cells and then added to wheat germ cells, the wheat cells make rabbit globin (Figure 11-5). This demonstrates that wheat cells and rabbit cells use the same code to transcribe RNA. Indeed, almost all eukaryotes use the same code and the same molecular machinery.

Nonetheless, the genetic code is not perfectly universal. Mitochondria, whose DNA is passed on separately from the DNA in the nucleus, use a slightly different code from the one in Figure 11-3. And yeast mitochondria use a different genetic code from human mitochondria.

We have seen how molecular biologists solved the genetic code, which helps explain how genetic information flows from DNA to RNA to proteins. But we must also ask about the mechanism by which this information flows. What happens, step by step?

The genetic code translated the language of DNA and RNA into the language of polypeptides. Three nucleotides—a codon—specify one amino acid. The genetic code is redundant and nearly universal, and 3 of the 64 codons are nonsense, or stop, codons.

RNA "alphabet"

Codon   Codon   Codon

Histidine   Cysteine   Tyrosine
H   C   T   Amino acid "alphabet"

Polypeptide

Figure 11-4   Three nucleotides code for one amino acid. The DNA/RNA language is written with an alphabet of nucleotides, while the polypeptide language is written with an alphabet of amino acids.

GENETIC INFORMATION FLOWS FROM DNA TO RNA TO POLYPEPTIDES

By the mid-1950s, biologists knew that genes were sequences of nucleotides that specify the sequences of amino acids in polypeptides. The next question was, How does a sequence of

BOX 11-1

How do we know that a codon is three nucleotides long?

Experimental confirmation that each codon consists of a triplet of nucleotides came in 1961 from studies with **acridine dyes**—chemical mutagens that cause mutations by inserting themselves into the backbone of DNA double helix. Each inserted dye molecule leads to the insertion of a single nucleotide. Working in Cambridge, England, Francis Crick and Sydney Brenner showed that a single acridine-induced insertion would cause a mutation of the bacteriophage gene that they were studying. A second such insertion would still be nonfunctional, but a third single-base insertion in the same gene restored nearly normal function. The beauty of this experiment was that it was done by observing phenotype alone.

The results clearly supported the hypothesis that each codon consists of three nucleotides. Each insertion, Crick and Brenner reasoned, caused a shift in reading frame, the grouping of nucleotides into codons that specify an amino acid sequence. A shift of one or two nucleotides in a triplet code led to a complete garbling of the message. In contrast, the insertion of three nucleotides restored most of the message's original sense, even though it was slightly different from the original. Crick and Brenner reasoned that a code unit must be a triplet because three nearby insertions seemed to restore the reading frame and confer a nearly normal phenotype.

On the basis of these experiments, molecular biologists concluded that a triplet of nucleotides in DNA specified each amino acid in a protein. Further confirmation came from later studies that showed that three insertions resulted in a polypeptide chain with one extra amino acid, while three deletions resulted in a polypeptide chain missing one amino acid.

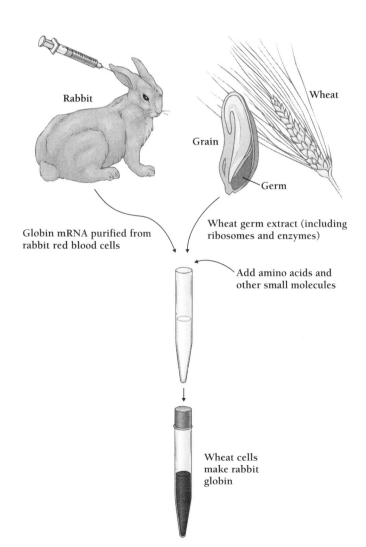

Globin mRNA purified from rabbit red blood cells

Wheat germ extract (including ribosomes and enzymes)

Add amino acids and other small molecules

Wheat cells make rabbit globin

nucleotides specify a sequence of amino acids? Molecular biologists such as Nirenberg and Khorana confirmed what Watson and others had long suspected: RNA is always an intermediate in the production of proteins.

Francis Crick summarized cellular information flow in a statement he called the **central dogma** of molecular biology (Figure 11-6): "DNA specifies RNA, which specifies proteins." According to Crick's central dogma, information can flow *only* from DNA to RNA to proteins. It cannot flow from proteins to RNA or DNA. Crick put this idea another way, writing "Once 'information' has passed into protein, *it cannot get out again.*" Proteins cannot change the information in the genes.

Crick used the word dogma ironically because "dogma" means an opinion, belief, or a tenet of faith. It is not the word to use to describe scientific facts or well-supported theories. But, in fact, molecular biologists had no good evidence supporting the central dogma. Crick and others just thought it must be true. Crick's joke was lost on most other researchers, however, and the "central dogma" has now been presented in textbooks *as dogma* for decades. As we will see in the next chapter, Crick had good reason for his twinge of doubt. The central dogma is mostly true, but not completely: information in RNA and proteins can influence the expression of the DNA's genes.

Nevertheless, the idea that DNA specifies RNA, which specifies proteins, is a good starting point for understanding how genes are expressed through the two processes of tran-

Figure 11-5 Nearly all organisms use the same genetic code. The protein-making machinery of a wheat plant can synthesize blood proteins normally found only in rabbits.

THE CENTRAL DOGMA

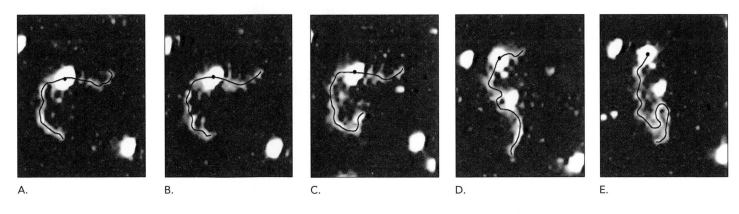

Figure 11-6 **The central dogma.** DNA is transcribed into RNA, which is translated into protein. In eukaryotes, transcription occurs in the nucleus and translation occurs outside the nucleus, in the cytoplasm. In prokaryotes, which lack a nucleus, both transcription and translation occur in the cytoplasm.

scription and translation. The process by which DNA transmits its information to RNA is called **transcription.** During transcription, the information in the DNA is rewritten, or transcribed, to RNA. It is from the information in the RNA that the cell makes polypeptides. Using RNA as a secondary information source is analogous to working from a photocopy of an important document, while keeping the original in a safe place, in this case the nucleus.

The RNA copy of the information from DNA is composed of **messenger RNA (mRNA).** Molecules of mRNA have the same information as the DNA original. RNA, like DNA, is a polynucleotide. It differs from DNA in just three ways:

1. The backbone sugar is ribose instead of deoxyribose.
2. Uracil (U) replaces thymine (T) as one of the pyrimidine bases.
3. RNA is usually single stranded, whereas DNA is usually double stranded.

Despite these differences, a message in the RNA alphabet (A,G,C, and U) is the same as in the DNA alphabet (A,G,C, and T).

The conversion of the message carried by the RNA into strings of amino acids—actual polypeptides—is called **translation** (Figure 11-7A–E). Translation takes the mRNA text and translates it into the language of amino acids. This is the essence of the central dogma.

$$DNA \xrightarrow{\text{Transcription}} RNA \xrightarrow{\text{Translation}} Polypeptide$$

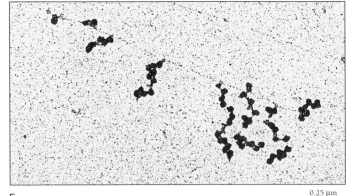

A.   B.   C.   D.   E.

Figure 11-7 **Transcription and translation.** A–E. In this sequence of micrographs, RNA polymerase transcribes a strand of DNA into mRNA (which is not visible). RNA polymerase, the large, round blob, moves from the middle of the strand (A) to the end (E) in less than 5 minutes. F. In prokaryotes, which lack a nuclear membrane, transcription and translation can occur simultaneously. In this electron micrograph, the DNA runs diagonally from the upper left to the lower right. Transcription began at the upper left end of the DNA, so the longest mRNAs (with the most ribosomes) are at the lower right. Clusters of up to 13 ribosomes read individual strands of mRNA, even as they are being transcribed from the DNA. *(A–E, M. Guthold, X. Zhu & C. Bustamante, University of Oregon; F, Dr. Barbara Hamkalo, University of California, Irvine)*

F.   0.25 μm

BOX 11-2

Which strand of the DNA is transcribed?

At the beginning of this chapter, we saw that protein synthesis in a test tube helped solve the genetic code. Such test-tube protein synthesis also allowed researchers to discover how RNA polymerase transcribes the DNA.

Just as DNA polymerase assembles new molecules of DNA one nucleotide at a time, RNA polymerase assembles RNA by adding one nucleotide at a time. However, DNA polymerase copies both strands of the DNA double helix, while RNA polymerase copies only one of the two strands of DNA. The strand that is copied is called the **template strand**. Although the two strands are chemically the same, the information in the template strand is completely different from that in the nontemplate strand of DNA. The template strand, also called the **antisense strand**, is a reverse copy of its mate, the **sense strand**. If this sentence were the sense strand, we might think the template strand would look something like this: :siht ekil gnihtemos kool dluow dnarts etalpmet eht kniht thgim ew ,dnarts esnes eht erew ecnetnes siht fI

Actually, however, the template strand is not only a reverse copy of the sense strand. It is also composed of the Watson-Crick base pairs, as if we had substituted different letters into our reversed sentence. For example, if the sense strand of DNA read CATTAG, then the antisense strand, read forward by RNA polymerase, would be CTAATG.

sense strand →

CATTAG

GTAATC

← antisense strand

The most important result of this reversal and substitution is that the antisense strand cannot code for a protein. The antisense strand is a chaotic sequence of codons that includes so many stop codons that any strange polypeptides made following its instructions could not be more than a few amino acids long. The sense strand, in contrast, has stop codons only at the end of each gene.

The antisense strand is useful, however, because the messenger RNA created from the antisense strand is not nonsense. The mRNA is an RNA version of the sense strand, the reverse of a reverse. The antisense strand at left, for example, would be transcribed as CAUUAG, since wherever DNA has a T, the mRNA has a U.

RNA polymerase copies the antisense DNA strand by holding in place an RNA nucleoside triphosphate base, which pairs with the next nucleotide in the template. The mRNA strand grows from its 5′ end to its 3′ end, as it copies the template DNA strand from its 3′ end to its 5′ end. The new mRNA molecule has the same sequence of bases as the sense strand of DNA, except that wherever DNA has a T the RNA has a U.

RNA polymerase transcribes from the template, or antisense, strand of DNA, which cannot itself code for a protein. The resulting mRNA is a reverse of a reverse—a copy of the sense strand.

In eukaryotes, transcription occurs in the nucleus and translation occurs in the cytoplasm outside the nucleus. Messenger RNA passes through the nuclear membrane out to the cytoplasm. There, ribosomes, acting like molecular sewing machines, stitch together amino acids into polypeptides.

In the next part of this chapter, we will discuss the details of the molecular machinery responsible for the flow of information from DNA to RNA to polypeptide. The process of building a cell from the information in DNA results from chemical reactions that biologists can now describe in detail. These reactions are subject to the same thermodynamic laws as other chemical processes. The same kinds of forces drive them, and no special force is needed to build the machinery of life.

The message in the DNA is transcribed into mRNA, which is translated into polypeptides. The central dogma says that information flows from DNA to RNA to protein only. Information in protein is not supposed to be able to influence the information in the DNA.

RNA POLYMERASE TRANSCRIBES DNA INTO RNA

When a cell makes a polypeptide we say that the gene for that polypeptide has been "expressed." Gene expression begins when the enzyme **RNA polymerase** transcribes the DNA gene into mRNA. Actually, there are several kinds of RNA polymerases, but since they all make RNA in the same way, we will discuss them as if they were a single enzyme.

How Does RNA Polymerase Begin Transcription?

In the last chapter we saw that DNA polymerase can build DNA in two directions and that the enzyme begins by adding nucleotides to an existing polynucleotide called a primer. In contrast, RNA polymerase transcribes DNA into RNA in only one direction and does not require a preexisting RNA primer (Figure 11-8). The starting signal for mRNA synthesis is a special sequence of DNA called a **promoter**. RNA polymerase (working with other proteins) recognizes the promoter, unwinds and

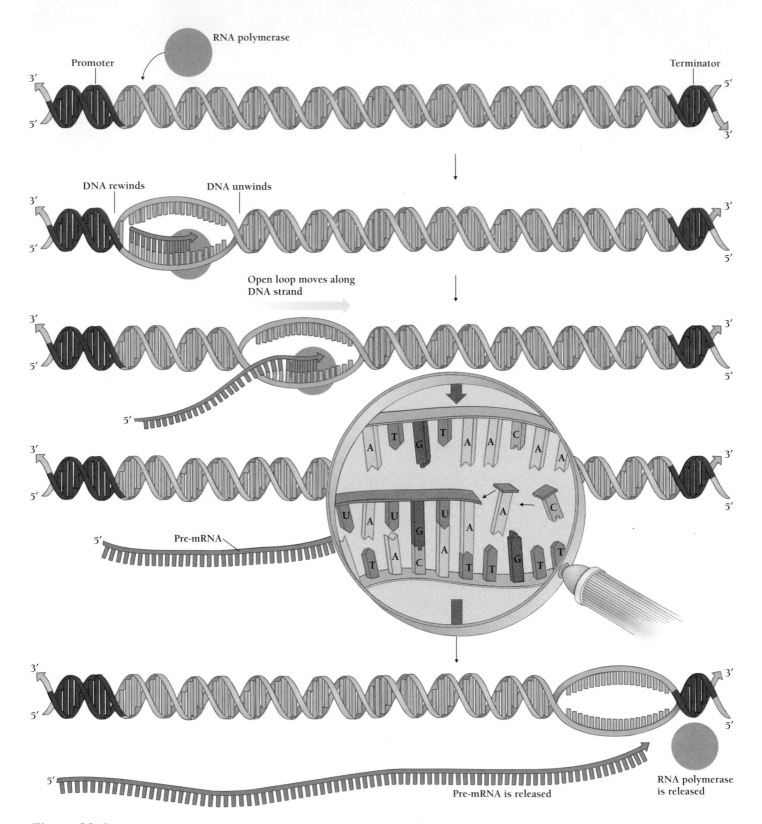

Figure 11-8 How does RNA polymerase begin transcription? Starting at the DNA promoter, RNA polymerase transcribes DNA into mRNA until it reaches the terminator.

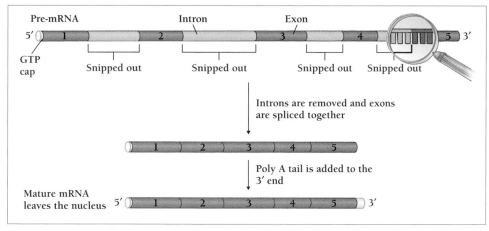

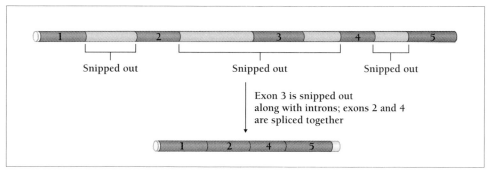

B.

Figure 11-9 RNA editing. In this example, introns are snipped from the mRNA, the five exons are spliced together, and each end of the edited mRNA is punctuated. Like a capital letter, the GTP cap indicates the beginning of the gene sentence. Like a period, the poly A tail indicates the end. B. The same pre-mRNA can be snipped and spliced in different ways to produce mRNAs that code for different polypeptides.

separates the two strands of DNA near the promoter site, and then attaches to the promoter.

In bacteria, the promoter consists of two characteristic sequences each about 6 nucleotides long, with about 25 other nucleotides between them. RNA polymerase binds to the promoter and begins to copy the template strand.

In eukaryotes, the production of RNA involves an extra step. Unlike prokaryotic RNA polymerases, eukaryotic RNA polymerases do not themselves recognize the promoter. Instead, they depend on the assistance of other proteins, called **transcription factors**, which lead the RNA polymerase to the right promoter. One transcription factor, for example, always takes RNA polymerase to a DNA sequence called a "TATA box," since it usually contains the sequence TATA.

Both eukaryotic and prokaryotic RNA polymerases stop transcription at special sequences, called **terminators**, which stop the enzyme from transcribing any more of the DNA.

With the help of transcription factors, RNA polymerase produces molecules of RNA by copying one strand of DNA starting at a promoter sequence.

How Does the Cell Alter the mRNA Transcribed from the DNA?

In prokaryotes, the mRNA binds to the ribosomes and begins directing protein synthesis as soon as the mRNA is made. Sometimes translation begins even as the DNA is being transcribed (Figure 11-7F). In eukaryotes, however, transcription and translation occur separately. Transcription occurs in the nucleus and translation occurs, hours later, in the cytoplasm.

Transcription and translation in eukaryotes are separated by much more than just the nuclear membrane. After the mRNA is transcribed from the DNA, in the nucleus, it is called pre-mRNA. While still in the nucleus, pre-mRNA undergoes a number of modifications that transform it into a molecule of "mature" mRNA that can be translated into a polypeptide. These modifications include: (1) capping the 5′ end of the RNA with a GTP (a relative of ATP), (2) adding a "tail" of 150 to 200 A's to the 3′ end of most mRNAs, and (3) removing large pieces of the RNA and splicing the remaining pieces together. In addition, using a recently discovered process called RNA editing, scientists can add, remove, or change the sequence of a few selected nucleotides in some regions of the mRNA (Figure 11-9A).

The cutting and splicing of pre-mRNAs was a surprising discovery since it demonstrated that genes are interrupted by nucleotide sequences that do not code for proteins. These interruptions in pre-mRNA (and the corresponding sequences in DNA) are called **introns**, for intervening sequences. The RNA sequences that directly code for proteins are spliced together to form the mature mRNA molecule. These sequences (and the corresponding sequences in DNA) are called **exons**, for expressed sequences. In a typical gene, most of the sequence consists of introns. The exons make up only a tiny fraction of the gene. A single pre-mRNA molecule may contain as many as 50 introns, each one ranging in size from 80 to 10,000 nucleotides. Yet all the exons in a single mRNA are, together, rarely more than 3000 nucleotides long.

Most of the DNA in a gene, then, consists of introns and does not code for a functioning polypeptide. In fact, more than 98 percent of human DNA does not directly specify polypeptide sequences, and a large part of this apparently useless DNA is introns. No one knows what introns do. Some scientists believe that the DNA in the introns is basically "junk DNA," which has accumulated in the genome in the same way that old keys and odd parts of things accumulate in the back of a kitchen drawer. Others believe that all DNA is in the genome for a purpose.

Pre-mRNA includes long stretches of noncoding DNA sequences, called introns, that must be removed. The remaining exons are spliced together to form mature mRNA.

One Gene—Heaven Only Knows How Many Polypeptides

We have seen that most eukaryotic genes consist of exons interspersed with introns. Transcription, we have said, then yields a single kind of mRNA containing all the introns and exons. Molecular machinery within the nucleus cuts and splices the pre-mRNA to produce mature mRNAs without introns.

As it turns out, transcription can be far more complicated than this simple picture suggests. In some cases, cells can actually transcribe many different mRNAs from the same gene. For example, the mRNA transcript may start or end at different sites on the same gene. Or a single pre-mRNA may be broken up and spliced together in several different ways, so that one mRNA's exon becomes another mRNA's intron (Figure 11-9B). As a result, a single gene may encode several polypeptides. What is more, the cell may express all of these various polypeptides at once.

Geneticists spent decades arriving at our definition from Chapter 10, "one gene—one polypeptide." But studies of gene expression beginning in the late 1970s have conclusively shown that one gene can actually encode several polypeptide chains, occasionally with very different functions. Most molecular biologists now say that a gene is a DNA sequence that is transcribed as a single unit and encodes either a single polypeptide or a set of related polypeptides.

A single gene can encode several different polypeptides.

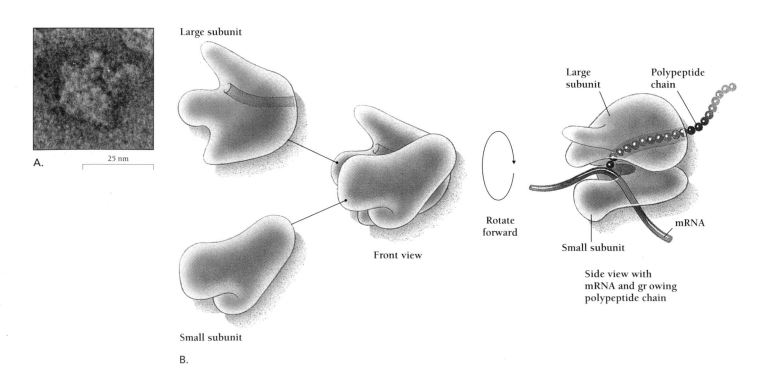

Figure 11-10   How do ribosomes work? A. Electron micrograph of whole ribosome. B. The two parts of a ribosome fit together around the mRNA molecule and the growing polypeptide chain.   *(A, courtesy of James Lake)*

HOW DOES A CELL TRANSLATE mRNA INTO A POLYPEPTIDE?

The synthesis of proteins requires a multitude of molecular tools. Among them are three distinct kinds of RNA: (1) Messenger RNA carries genetic information from DNA that specifies the amino acid sequence of a polypeptide. (2) Transfer RNAs carry each amino acid to a codon in the mRNA. (3) Ribosomal RNA, complexed with proteins to form ribosomes, serves as the site for polypeptide synthesis. In addition, an array of enzymes and other proteins assist in the synthesis and folding of new proteins.

What Do Ribosomes Do?

Ribosomes are like subcellular sewing machines that link amino acids into polypeptides. All cells contain ribosomes. Each ribosome consists of a small and a large subunit (Figure 11-10). Each subunit contains both RNA and proteins. The *E. coli* ribosome, for example, consists of three molecules of RNA, called **ribosomal RNAs**, or **rRNAs**, and 55 kinds of protein molecules. Eukaryotic ribosomes are even more complicated, with four molecules of rRNA and some 82 kinds of protein.

Ribosomal RNA is very different from messenger RNA. The sequence and length of messenger RNA varies, depending on the kind of polypeptide it encodes. But the RNA that makes up the ribosome is always the same, no matter what kind of protein the ribosome is producing.

Ribosomal RNAs from organisms of all kingdoms are highly similar to one another. Such similarity suggests that the exact shape of the ribosome is important to the way it works. We can guess that any individual with even minor changes in the standard ribosome design would be unable to make proteins and would therefore die immediately.

Ribosomes help catalyze the formation of peptide bonds between amino acids, in an order determined by mRNA. Ribosomes' main tasks are to read the information in mRNA and to translate it into a sequence of amino acids. In addition, ribosomes must recognize where to start and where to stop reading the mRNA for the production of each polypeptide.

For example, consider the translation of the mRNA sequence AUGCAUGCA. If a ribosome started reading at the first nucleotide, it would assemble the peptide methionine-histidine-arginine. If, however, the ribosome started with the second nucleotide, it would assemble cysteine-methionine. If the ribosomes started with the third nucleotide, it would assemble alanine-cysteine. The grouping of nucleotide triplets is called the **reading frame** for translation (Figure 11-11).

Because starting at the right place is crucial to reading mRNA correctly, mRNAs must contain signals that say "start" and "stop." Start and stop signals are like punctuation that tells where a sentence begins and ends. In prokaryotes, the rRNA of the smaller ribosomal subunit finds the right place to start by forming complementary base pairs with a start site in the mRNA.

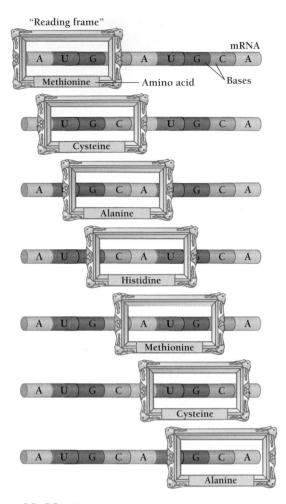

Figure 11-11 **The reading frame.** The exact base at which the ribosome begins translating the mRNA determines how it interprets all of the three-base codons from there onward.

Eukaryotic ribosomes seem to work in the same way as prokaryotic ribosomes. Each ribosome contains two grooves. On the small subunit is a groove for reading the mRNA. On the large subunit is a groove for the growing polypeptide chain. Protein synthesis requires the participation of both subunits. The small subunit is responsible for starting synthesis at the right place, while the large subunit contains the machinery for making the peptide bonds that hold together the polypeptide.

Ribosomes, which are made of rRNA and proteins, link amino acids together into polypeptides. The small ribosomal subunit reads the mRNA, and the large ribosomal subunit makes the peptide bonds between each pair of amino acids in the growing chain.

How Do tRNAs Serve as Adapters Between mRNAs and Amino Acids?

Once molecular biologists began to suspect that RNA dictated the sequences of amino acids in polypeptides, they tried to imagine how this could occur. An early idea was that each

BOX 11-3

How do the ribosomes read different kinds of mutations?

A base substitution in a gene may result in a **missense mutation,** which changes the codon for one amino acid into the codon for another amino acid. The resulting polypeptide may be either nonfunctional or altered in the way it functions. For example, some altered polypeptides result in **conditional mutations,** in which the resulting protein may be able to function under some conditions, but not under others. In one *Drosophila* mutant, for example, the fly functions normally at 20°C, but becomes paralyzed at 30°C. This is due to a protein that regulates ion transport across nerve cell membranes, which fails at 30°. Such temperature sensitivity results when an amino acid substitution reduces a protein's stability at higher temperatures, so that it denatures, or changes shape, more readily than the normal protein.

A base substitution can also result in a **nonsense mutation,** which changes the codon for an amino acid into a nonsense or stop codon, leading to a shortened polypeptide (Figure A). Such a shortened polypeptide almost certainly will not function at all in its former role.

Mutations that have no effect on phenotype are called **silent mutations.** Some are simply single-base substitutions that result in a codon that codes for the same amino acid (because of the redundancy of the genetic code). In that case, the polypeptide is unchanged. Most silent mutations, however, lie in DNA sequences that do not actually code for a polypeptide. Other supposedly silent mutations may merely produce changes in the phenotype so subtle that researchers cannot tell the difference.

Insertions and deletions in structural genes produce **frameshift mutations,** which alter the groupings of nucleotides into codons. The addition or subtraction of a single base can wreak havoc in the ribosomes' reading of mRNA downstream from the mutation. The result is often a string of missense changes, followed by a stop codon.

Sometimes point mutations can reverse themselves in a process called **reversion.** Reversion mutations, sometimes called **suppressor** mutations, consist of a second mutation that reverses the effects of the first mutation. A reversion might be an exact restoration of the original nucleotide sequence. Alternatively, the DNA might have a nucleotide sequence that is not identical to the original, but which specifies the same amino acid—a silent mutation.

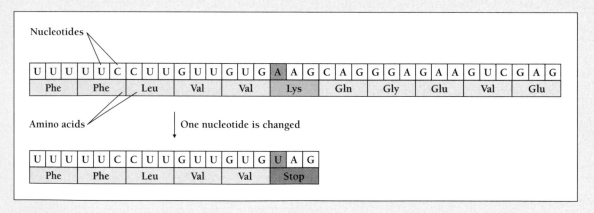

Figure A **Nonsense mutation.** A nonsense, or stop, codon shortens the polypeptide.

codon in an mRNA molecule bound itself to a specific amino acid. No one, however, could find any such binding. Besides, it was hard to imagine how a sequence of just three nucleotides could distinguish among amino acids as similar as leucine, isoleucine, and valine, to name a few.

The actual explanation, proposed by Francis Crick, is that the amino acids are first linked to special adapter molecules, called **transfer RNAs,** or **tRNAs.** The mRNA codons bind to the tRNA, not to the amino acid itself. Each tRNA consists of

75 to 85 nucleotides and folds to form a secondary structure shaped like a cloverleaf (Figure 11-12). The "stems" are double strands of tRNA held together by base pairing. The "leaves" are loops of tRNA with unpaired bases. X-ray diffraction studies of tRNA's tertiary structure show that the whole cloverleaf bends into a shape resembling the letter **L**.

One loop of each tRNA attaches to the correct amino acid (Figure 11-12). The opposite side of each tRNA molecule binds to a specific codon in the messenger RNA. In one of the loops

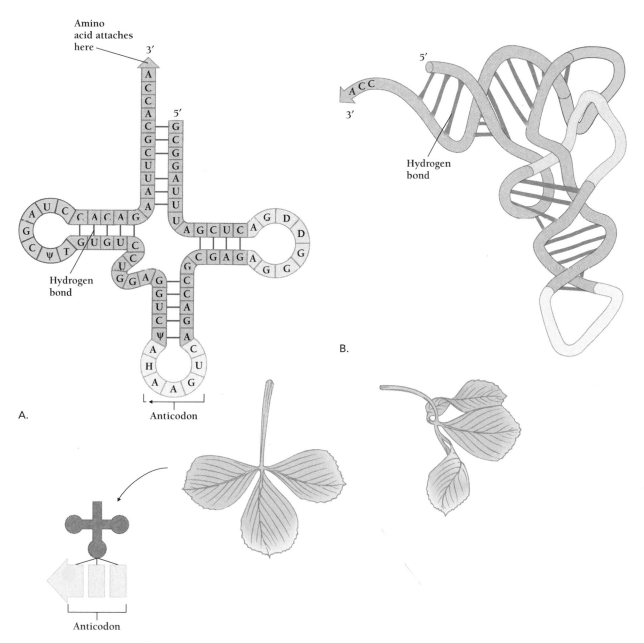

Figure 11-12   Cloverleaf-shaped tRNAs translate codons into amino acids. A. The anticodon is at the tip of the middle "leaflet." B. Like many macromolecules, tRNAs have a complex folded structure. Like so many hair pins, hydrogen bonds hold this complex shape together. C. The anticodon binds to the codon.

is a sequence of three nucleotides, called the **anticodon,** that forms base pairs with the matching codon sequence in the mRNA (Figure 11-12). The tRNAs are highly specialized. A given tRNA has only one kind of anticodon and can pick up only one kind of amino acid.

We may wonder how many tRNAs a cell needs—one for each amino acid or one for each codon? A cell must have at least 1 tRNA for each of the 20 amino acids. Yet we might expect 61—one for every codon except the 3 codons that do not specify any amino acid. In fact, different species have different

numbers of tRNAs. However, every species has more than 20 and fewer than 61 kinds of tRNA. Organisms can have fewer than 61 tRNAs because some tRNA anticodons recognize more than one codon.

How do we know that mRNA recognizes the tRNA, rather than the amino acid itself? To find out, researchers "tricked" the messenger RNA. They took the tRNA for cysteine, attached the cysteine, then chemically converted the attached cysteine into alanine so that the tRNA for cysteine actually carried an alanine. Finally, researchers added this incorrectly loaded tRNA to a cell extract that was synthesizing protein. The extract made incorrect polypeptides. Wherever the mRNA specified cysteine, the tRNA inserted alanine instead. This result shows that the mRNA recognizes the tRNA, not the amino acid itself.

Transfer RNA helps assemble amino acids into polypeptides by binding first to an amino acid and then to the correct codon in the mRNA.

How Do tRNAs Recognize the Right Amino Acid?

We have seen how the tRNA anticodon recognizes the correct codon. But how does tRNA recognize and load the correct amino acid? The task is to match the unique shape and charge distribution of each amino acid with the shape and charge distribution of the appropriate tRNA. The linking of tRNAs to their corresponding amino acids to form "charged" tRNAs—tRNAs joined to their amino acids by high-energy bonds—depends upon a set of enzymes. The tRNAs and amino acids recognize each other because the different tRNAs have different three-dimensional structures recognized by the enzymes. Each charging enzyme has two binding sites: one holds a certain kind of tRNA, and the other site holds the corresponding amino acid (Figure 11-13).

Special charging enzymes help tRNAs hook up with the correct amino acid.

How Do Ribosomes Begin Translation?

We have seen how each molecule of mRNA contains signals that tell the ribosomes where to start and where to stop. We have also seen that starting translation at the right place is crucial since a mistake of only a single nucleotide would lead to a complete misreading of the information.

Figure 11-13 Special enzymes charge each tRNA with the right amino acid. Here, the enzyme serine tRNA synthase joins a tRNA to serine. Different tRNAs join different amino acids.

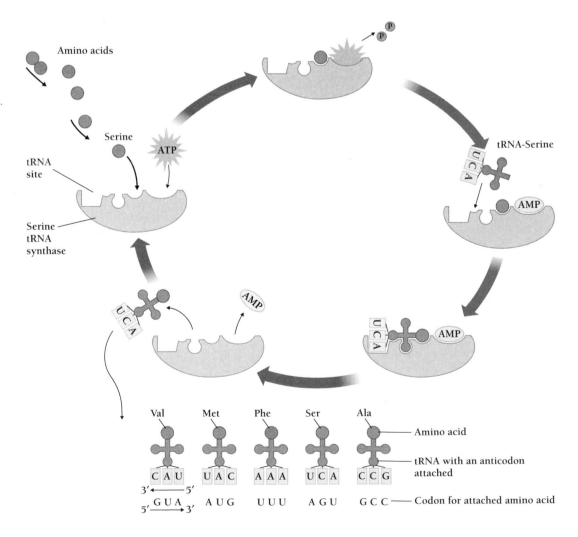

BOX 11-4

Antibiotics and protein synthesis

Antibiotics are usually complex organic molecules that allow bacteria and fungi to dispose of (and often to eat) their competitors. The medical uses of antibiotics began at least 2500 years ago, when the Chinese began to use the moldy curd of soybeans to treat boils and similar infections. In the 20th century, the pursuit of antibiotics as a treatment for human disease started with Alexander Fleming. In 1928 Fleming discovered that the growth of bacteria in one of his experiments had been stopped by the presence of a green mold.

Fleming identified the mold as a species of *Penicillium*. Later, other scientists isolated the active substance and named it penicillin. Penicillin and its many derivatives have remained the most frequently used antibiotics. But dozens of other antibiotics have been isolated from molds or bacteria.

Penicillin acts by interfering with cell wall synthesis in many bacteria. Other antibiotics act by interfering with protein synthesis. This characteristic has made them extremely important tools for dissecting the steps of protein synthesis. Each of the steps of elongation, for example, is subject to inhibition by a different set of antibiotics: in prokaryotic cells (bacteria), tetracycline interferes with positioning of the ribosome, chloramphenicol with peptide bond formation, and erythromycin with the movement of the ribosome on the mRNA (Figure A).

Some of the most useful antibiotics interfere with prokaryotic protein synthesis without affecting that in eukaryotic (human) cells. For example, while erythromycin blocks translocation of the growing peptide chain in bacteria, it does not inhibit this step during protein synthesis on

eukaryotic ribosomes. This specificity reflects differences in the proteins of prokaryotic and eukaryotic ribosomes.

Antibiotics have dramatically decreased the numbers of deaths caused by infectious disease. They not only can cure an initial infection, but they also can prevent long-term consequences such as the rheumatic fever which, until this century, often resulted from strep throat. In addition, killing the bacteria that make one person sick prevents the spread of an infection. Finally, the ability to deal with infections has made surgery safer, and allowed the development of operations ranging from the removal of cancers to the replacement of vital organs.

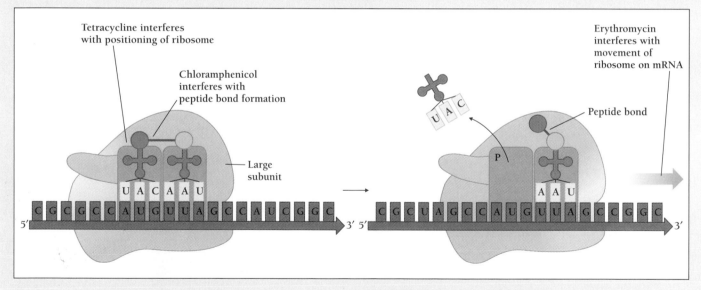

Figure A Each of the steps of elongation is inhibited by a different set of antibiotics. In prokaryotic cells, tetracycline interferes with positioning, chloramphenicol with peptide bond formation, and erythromycin with the movement of the ribosome along the mRNA.

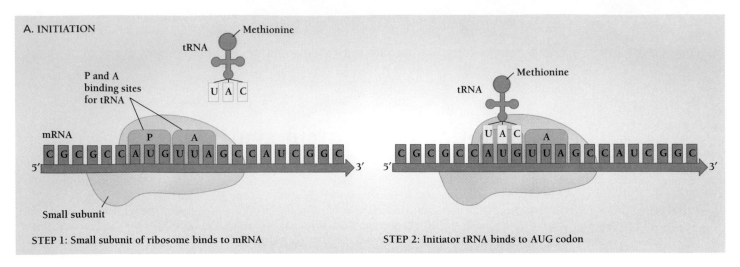

A. INITIATION

Methionine

tRNA

P and A binding sites for tRNA

mRNA

P | A

CGCGCCAUGUUAGCCAUCGGC

5'   3'

Small subunit

STEP 1: Small subunit of ribosome binds to mRNA

Methionine

tRNA

UAC

A

CGCGCCAUGUUAGCCAUCGGC

5'   3'

STEP 2: Initiator tRNA binds to AUG codon

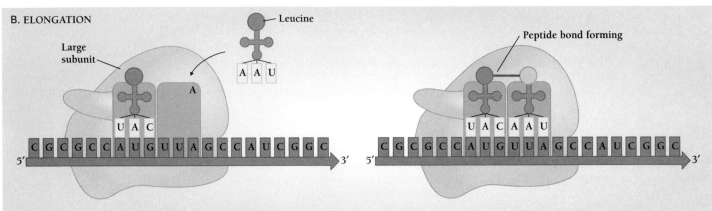

B. ELONGATION

Leucine

Large subunit

A

AAU

UAC

CGCGCCAUGUUAGCCAUCGGC

5'   3'

Peptide bond forming

UACAAU

CGCGCCAUGUUAGCCAUCGGC

5'   3'

STEP 1: Large subunit of ribosome binds to mRNA; next tRNA with amino acid leucine binds to UUA codon

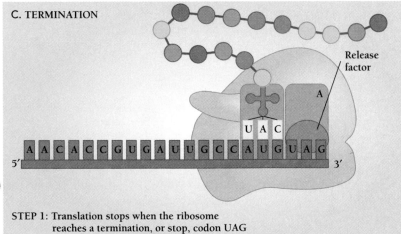

C. TERMINATION

Release factor

A

UAC

AACACCGUGAUUGCCAUGUAG

5'   3'

STEP 1: Translation stops when the ribosome reaches a termination, or stop, codon UAG

The first step of translation is **initiation,** which begins with the attachment of the ribosome's small subunit to the mRNA. The actual initiation site, or **start site,** in mRNA is always the codon AUG, which specifies methionine. This means that *all* recently translated polypeptides begin with methionine. In most cases, however, a special enzyme later clips away the methionine.

But how can the ribosome distinguish between the AUG meaning "start" and the AUGs that specify methionines in the

Figure 11-14   Translation: initiation, elongation, and termination. A. During initiation, the small subunit of the ribosome binds to AUG (the start site) of the mRNA and moves the amino acid methionine into place. B. During elongation, a second charged tRNA arrives and the large subunit of the ribosome—holding the two charged tRNAs side by side in the P and A sites—binds the two amino acids together. The ribosome and its attached polypeptide chain then move forward on the mRNA and another charged tRNA moves into place. C. When the ribosome reaches a stop codon, a protein called release factor binds to the stop codon and an enzyme cuts the newly made polypeptide chain from the last tRNA. The ribosomal subunits then separate from each other and become available for another round of initiation and elongation.

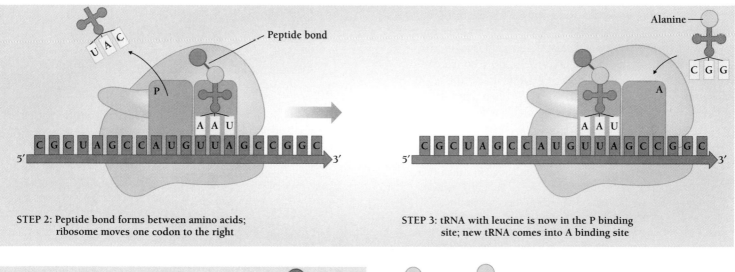

STEP 2: Peptide bond forms between amino acids; ribosome moves one codon to the right

STEP 3: tRNA with leucine is now in the P binding site; new tRNA comes into A binding site

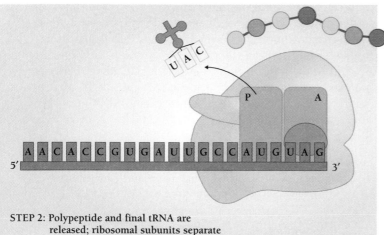

STEP 2: Polypeptide and final tRNA are released; ribosomal subunits separate

and its attached methionine are in place, elongation of the polypeptide chain can begin.

In the first step of translation—initiation—the ribosome attaches to the start site—the codon AUG. In prokaryotes, a sequence of six nucleotides distinguishes the start AUG from the AUG codon for methionine. In eukaryotes, the start AUG is simply the first one.

---

middle of a polypeptide? The answer to this question differs in prokaryotes and eukaryotes. In prokaryotes, a sequence of about six nucleotides, slightly upstream from the initiation AUG, distinguishes between initiation sites and internal methionine. In eukaryotes, each ribosome simply attaches to the 5' end of the mRNA and moves along until it encounters the first AUG, where translation then begins (Figure 11-14).

The initiation process aligns the mRNA for proper reading and moves the first amino acid into place. After the first tRNA

How Do Ribosomes Make Peptide Bonds?

The process of polypeptide-chain **elongation** consists of three steps (Figure 11-14): (1) putting the next amino acid into position; (2) forming a peptide bond between the growing polypeptide and the next amino acid; and (3) moving the ribosome to the next codon. Every amino acid is joined to its neighbor in exactly the same way.

Elongation, unlike initiation, requires the participation of the large ribosomal subunit (which binds to the initiator tRNA) as well as the mRNA and the small ribosomal subunit. The large subunit contains two special pockets for tRNAs, called the P site [for *p*eptide] and the A site [for *a*mino acid]. The P and A sites bind two tRNAs to adjacent codons in the mRNA. This is the first step of elongation. When both sites are filled, an enzyme within the ribosome links the attached amino acids with a peptide bond. This is the second step of elongation. The tRNA in the P site then falls away from the ribosome, leaving the tRNA in the A site, and its attached amino acid, which is now linked to the entire polypeptide chain. In the third step, the ribosome moves forward so that the tRNA in the A site with its attached polypeptide chain ends up in the P site. The repositioning of the tRNA and the attached polypeptide to the P site completes the three steps of the elongation cycle.

The cycle now repeats over and over until all the codons in the mRNA are read. In a bacterium, each elongation cycle (the addition of one amino acid) takes about 1/20th of a second, so the synthesis of an average-sized polypeptide (including 400 to 1200 amino acids) takes 20 to 60 seconds. We can see now how the A and P sites get their names: the A site is where an *a*mino acid is first added to the chain, and the P site is where the rest of the *p*olypeptide chain is anchored, ready to receive the next amino acid.

During elongation, amino acids are joined one after the other in exactly the same way. A tRNA first moves an amino acid into position next to the last amino acid. Once two amino acids are in place, an enzyme in the ribosome joins them together with a peptide bond. The ribosome then moves forward on the mRNA, repositioning the tRNA and its attached polypeptide chain.

Several Ribosomes Can Read a Single Strand of mRNA Simultaneously

As soon as a ribosome has moved far enough away from the AUG start site, the start site is free for another ribosome to attach and begin translation. As a result, several ribosomes, spaced as little as 80 nucleotides apart, may be attached to a single mRNA at one time, each one overseeing the production of identical polypeptides. This complex is called a **polysome,** short for polyribosome (Figure 11-15). In this way, the cell can synthesize many copies of a polypeptide at once. Such mass production is the only way to quickly produce the enzymes and other materials needed by a cell. We will see later in this chapter, for example, that an *E. coli* bacterium can begin producing enzymes to break down milk sugar (lactose) as soon as the bacterium detects the milk sugar.

How Does Polypeptide Synthesis Stop?

The appearance of a stop codon in the mRNA signals the end of a polypeptide chain, a process called **termination.** When the ribosome reaches one of the three stop codons (UAA, UAG, or UGA), there will be no corresponding tRNA. Instead a pro-

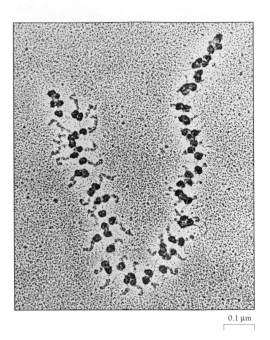

0.1 μm

Figure 11-15   A large polysome. Individual ribosomes move along a strand of mRNA, translating as they go. The ribosomes on the far left are just beginning translation and the polypeptide strands are not yet visible. The ribosomes on the right and in the middle have produced visible polypeptide strands.   *(Oscar L. Miller, University of Virginia)*

tein called a **release factor** binds to the stop codon (Figure 11-14). The enzyme that makes peptide bonds during elongation now cuts the polypeptide from the last tRNA, releasing the completed polypeptide into the cytoplasm. The ribosomal subunits then separate from each other and become available for another round of initiation and elongation.

All of the steps of protein synthesis, including transcription and translation, are summarized in Figure 11-16. We can see that once a polypeptide chain is assembled, it must then fold into a functioning protein.

When the ribosome reaches a stop codon, a protein called release factor binds to the stop codon and an enzyme cuts the newly made polypeptide chain from the last rRNA.

How Do Cells Direct Newly Made Proteins to Different Destinations?

Newly made polypeptides fold into three-dimensional structures according to their amino acid sequences. As we saw in Chapter 3, protein molecules may contain a single polypeptide or several. Most of the proteins produced on the ribosomes stay in the cytosol, but many must find their ways to various cytoplasmic organelles, to the plasma membrane, or to the outsides of cells. But how do newly made proteins find their way to the right destinations?

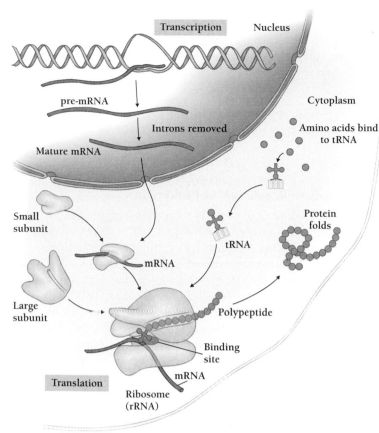

Transcription

Nucleus

pre-mRNA

Introns removed

Mature mRNA

Cytoplasm

Amino acids bind to tRNA

Small subunit

Protein folds

tRNA

mRNA

Large subunit

Polypeptide

Binding site

Translation

mRNA

Ribosome (rRNA)

Figure 11-16 Transcription and translation in a eukaryotic cell.

The answer is that many newly made polypeptides have "address labels." The amino acid sequence of a polypeptide contains not only the information needed to specify its three-dimensional structure, but also an address tag that specifies where the protein should go. Clear evidence for this idea came in 1980 from the work of Gunther Blobel, a cell biologist at Rockefeller University in New York City. Blobel prepared mRNA for a protein secreted from cells. He then translated the mRNA in a test tube. The resulting polypeptide was longer by several amino acids than the polypeptide as it is normally secreted by cells. The extra amino acids, Blobel showed, directed the newly made polypeptide to a particular organelle.

The extra block of 4 to 12 amino acids is called a **signal peptide.** In proteins destined for the plasma membrane or for secretion outside of the cell, for example, the signal peptide directs the newly made polypeptide into the space (or "lumen") that is surrounded by the endoplasmic reticulum (ER) during protein synthesis. In one experiment, molecular biologists attached a signal peptide to a protein not normally sent to the ER. The protein then entered the ER, demonstrating that the signal peptide indeed acts as an address label. Other polypeptides contain signals that direct them to the mitochondria, the chloroplasts, or the nucleus. In the case of secreted proteins—such as those studied by Blobel—the signal peptide is removed. In other cases, the address tag remains part of the protein.

Cells direct proteins to different destinations by addressing the protein with a signal peptide.

HOW DO CELLS REGULATE GENE EXPRESSION?

We have seen now how cells express the information in the DNA. But cells contain thousands of genes. Some of these are expressed as polypeptides all the time, some only occasionally, and many are never expressed. Prokaryotes transcribe almost all of their genes into RNA. But eukaryotes regulate gene expression more closely. Cells, especially eukaryotic cells, do not make all of their polypeptides at once for the same reason that we do not leave all of the water taps in our homes flowing at once. Just as we turn on hot or cold water according to our needs, a eukaryotic cell produces different proteins according to its needs. Further, just as a water tap can be turned on full blast or at a trickle, a cell can turn out thousands of copies of a polypeptide, or only two or three copies.

Eukaryotes, most of which are multicellular, face an additional problem. The dramatic cell specialization that characterizes most multicellular organisms requires the differential expression of genes in cells that have identical genomes. A human being, for example, consists of more than 200 distinguishable types of cells. Nearly every one of these cells contains the same set of about 100,000 genes. Yet each kind of cell expresses only a fraction of its 100,000 genes.

A typical eukaryotic cell transcribes only about 20 percent of its DNA into RNA. In a human skin cell, for example, this 20 percent would include information for basic life-support proteins that all human cells need—such as membrane proteins, ribosomal proteins, and metabolic enzymes—as well as information for specialized proteins, such as the stretchy collagen that makes skin so tough. In a skin cell, the 80 percent of the DNA not transcribed would include, for example, the genes for α and β globin, various muscle proteins, and a host of digestive enzymes.

Cells can regulate the expression of genes at many levels. The most efficient and most usual way is so-called **transcriptional control,** in which cells increase or decrease the amount of mRNA transcribed from the DNA. Every cell in the body has the gene for the globin polypeptides, for example, yet only the developing red blood cells transcribe them. In every other kind of cell in the body, the globin genes sit unused.

Cells can regulate the expression of genes in other ways. In **posttranscriptional control,** the cell transcribes the mRNA for a gene, but modifies the pre-mRNA before it reaches the cytoplasm. In **translational control,** the cell finishes making the mRNA for a given polypeptide, but regulates the rate at which the mRNA is translated. In **posttranslation control,** the cell goes so far as to make the polypeptide, but then modifies its structure and behavior. Even after a protein has been synthesized, the cell can regulate protein activity.

Cells regulate the expression of genes in many ways, but transcriptional control is the most efficient and most common way.

How Do Prokaryotes Regulate Gene Expression?

Bacteria provide the best understood examples of gene regulation. Most of the time, bacteria regulate by controlling transcription. Like other organisms, bacteria must respond to their environment. They have evolved sophisticated and sensitive molecular mechanisms that ensure the production of enzymes when they need them. When a bacterium's environment contains high concentrations of amino acids, it devotes most of its energy to rapid division and does not squander energy making more amino acids. If no amino acids are available, however, the bacterium rapidly synthesizes amino acids from simple molecules, usually taking carbon atoms from sugar and nitrogen atoms from ammonia, for example.

To do all this, the bacteria must make the right enzymes at the right time. For example, the synthesis of just one amino acid, histidine, from simple carbon- and nitrogen-containing compounds requires the participation of seven different enzymes. Conveniently, the genes for these seven enzymes lie right next to each other on the bacteria's DNA.

Molecular biologists were surprised to find that the cell transcribes the seven genes into a single huge molecule of mRNA instead of into seven separate lengths of mRNA. Transcribing the seven genes together ensures that all the enzymes needed to make the final product (histidine) are turned on and off at the same time. A stretch of DNA that includes several genes under coordinated control is called an "operon."

Each of the seven genes for the enzymes needed to make histidine is an example of a "structural gene." A **structural gene** is one that codes for a polypeptide that plays some role in the organism outside of the nucleus (in eukaryotes) or apart from the nucleoid region (in prokaryotes). But cells also use **regulatory genes**—genes that help govern the *amount* of a polypeptide made by a cell. One class of regulatory genes encodes small polypeptides that influence the expression of other genes. Such regulatory proteins are called **transcription factors.** The other kind of regulatory gene, a **regulatory site,** is a DNA sequence to which a transcription factor binds. A regulatory site does not itself code for a polypeptide. In the rest of this chapter, we will see how these two kinds of regulatory genes control the amount of polypeptide made.

Regulatory genes, which include both the regulatory site and genes for transcription factors, regulate the amount of polypeptide made by a cell.

Negative Regulation in Prokaryotes

The best-studied operon in prokaryotes regulates the use of lactose (the main sugar in milk) by *E. coli,* a common bacterium in the intestines of humans and other mammals. This elegant system enables bacteria to produce costly enzymes only when they are needed. As long as glucose is available in the intestines, *E. coli* will not digest the milk sugar lactose. When glucose is absent, however, *and* the host organism (that's you) drinks milk, then *E. coli* begins synthesizing three enzymes that digest lactose. These are the enzyme β-galactosidase (which splits lactose into the sugars galactose and glucose) and two other enzymes.

The three genes encoding these three enzymes lie next to each other on the DNA of *E. coli* and are part of the lactose operon, or **lac operon.** *E. coli* transcribes the DNA of the *lac* operon into an mRNA that encodes all three proteins.

In the absence of lactose, each bacterium contains only about three molecules of β-galactosidase. However, when glucose is absent and lactose becomes available, the lactose stimulates, or **induces,** the expression of the *lac* operon, and the bacterium synthesizes about 3000 molecules of β-galactosidase. Lactose is called the "inducer" because it induces gene expression.

In 1960, two French molecular biologists—François Jacob and Jacques Monod—discovered how *E. coli* control these dramatic changes. Jacob and Monod discovered two genes that regulate the expression of the *lac* operon. One gene, which lies a short distance from the structural gene for β-galactosidase, is the structural gene that encodes the **lac repressor,** a protein that prevents the expression of the *lac* operon (Figure 11-17A). The other gene, a sequence of 21 nucleotides called the **operator,** lies immediately next to the gene for β-galactosidase and is part of the operon. The operator does not encode a protein. It is a regulatory site to which the *lac* repressor can bind.

The operator overlaps with the operon's *promoter,* which, recall, is the base sequence that signals the start of a gene. When the *lac* repressor binds to the operator, it prevents RNA polymerase from binding to the promoter and transcribing the three enzyme genes in the operon (Figure 11-17B). The lac repressor protein was the first example of a protein that regulates gene expression. The operator was the first example of a regulatory site.

How do the repressor and operator allow lactose to control gene expression in the *lac* operon? The key to understanding the induction of the *lac* operon is the realization that the *lac* repressor has two binding sites: it can bind to the operator or it can bind to lactose. But it cannot bind to both at

Figure 11-17 Negative feedback in the *lac* operon. A. Peanuts entice an elephant from the tracks so that a locomotive can pass. In the same way, lactose binds to the repressor so that RNA polymerase can transcribe the DNA. B. The repressor sits on the operator and blocks RNA polymerase. C. When lactose is present, the repressor binds to lactose instead of the operator and the RNA polymerase transcribes the mRNA for three enzymes that break down lactose.

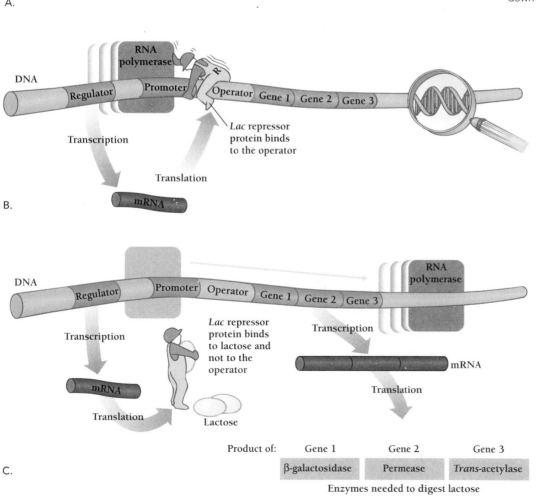

Product of:	Gene 1	Gene 2	Gene 3
	β-galactosidase	Permease	*Trans*-acetylase

Enzymes needed to digest lactose

the same time. When lactose binds to the repressor, the repressor can no longer bind to the operator and prevent transcription (Figure 11-17C). RNA polymerase can then bind to the *lac* operon's promoter and transcribe the three enzymes needed to digest lactose. The *lac* repressor is an example of a **negative regulator,** a molecule that inhibits transcription.

When lactose binds to the *lac* repressor (a protein), the *lac* operon's promoter becomes unblocked, and RNA polymerase can transcribe the three genes that encode enzymes for digesting lactose.

Positive Regulation in Prokaryotes

Bacteria also use **positive regulators,** molecules that increase the rate of transcription. One such positive regulator is the **catabolite activator protein** (CAP) (Figure 11-18). Like the *lac* repressor, CAP is a transcription factor with two binding sites—one for a particular sequence in DNA and one for cyclic AMP, a signaling molecule. Cyclic AMP is a nucleotide generated from ATP in response to certain hormones. High concentrations of cyclic AMP tell the cell that glucose is unavailable and that the cell needs to get energy from another source, such as lactose.

In the presence of lactose, CAP can stimulate the expression of the *lac* operon by binding to the *lac* operon's promoter.

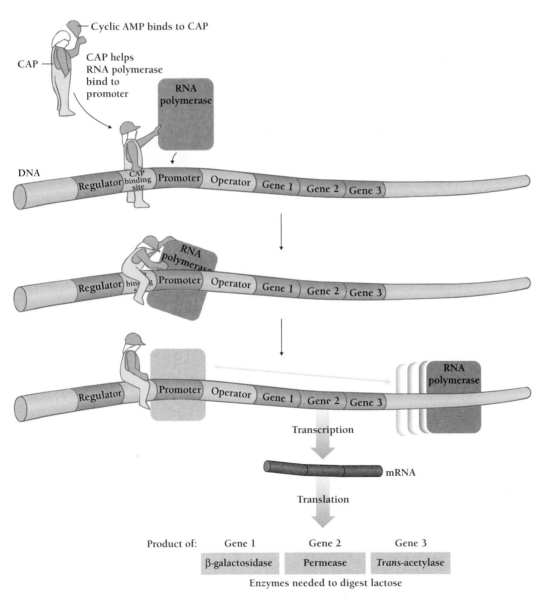

Figure 11-18   Positive feedback in the *lac* operon. In the presence of glucose, cyclic AMP activates CAP, which binds to the promoter and stimulates transcription of the three genes that code for lactose-digesting enzymes.

This increases the rate at which RNA polymerase binds, and so stimulates the synthesis of *lac* operon mRNAs. When the cell has lots of glucose and little cyclic AMP, however, the *lac* operon is not active even in the presence of lactose.

The combination of negative control and positive control allows bacteria to respond quickly to changes in their environment. When lactose is present but glucose is not, *E. coli* quickly begin to make the enzyme β-galactosidase. Similarly, when glucose becomes available or when lactose is exhausted, *E. coli* almost immediately stop production of the enzyme. This system saves the bacteria from expending energy making polypeptides they do not need.

CAP regulates the transcription not only of the *lac* operon but also of other genes. All of the genes code for enzymes needed to obtain energy from molecules other than glucose. CAP provides an example of how a single regulatory gene can affect the transcription of genes that are physically separate from one another but functionally related.

The positive regulator CAP increases the transcription rate for certain mRNAs, provided that other regulatory mechanisms allow these mRNAs to be transcribed at all.

How Do Eukaryotes Regulate Gene Expression?

Until the late 1980s, molecular biologists knew far more about regulation in prokaryotes than in eukaryotes. However, new techniques have resulted in an explosion of information about gene expression in eukaryotic cells. We now know that eukaryotes control gene expression at many more levels than prokaryotes. Nonetheless, as in prokaryotes, most regulation occurs at the level of transcription.

The major differences between regulation in prokaryotes and eukaryotes result from the presence, in eukaryotes, of a nuclear membrane and of a much larger amount of DNA. The presence of the nuclear membrane has several important consequences. One is that only proteins in the nucleus can contribute to regulation. Another consequence of the nuclear membrane is that the mRNA can be modified before it goes out into the cytoplasm to be translated. By interfering with any of the steps involved in modifying the pre-mRNA, the cell can regulate the expression of the gene. This is "posttranscriptional regulation."

Eukaryotes also have much more DNA than prokaryotes. Each human cell, for example, contains about 1000 times as much DNA as is in an *E. coli* bacterium. Recall from Chapter 10 that to fit the DNA of a eukaryotic cell into its nucleus, the cell packages the DNA into tightly folded nucleosomes. Such packaging makes the DNA less accessible to RNA polymerase than in prokaryotes. As a result, eukaryotic transcription requires a more elaborate set of transcription factors. This gives the cell yet another way to regulate which genes are expressed.

Eukaryotes also have many intracellular compartments and elaborate mechanisms for the distribution of newly made proteins. Sending the right proteins to the nucleus, for example, is especially important for building the chromatin and regulating gene expression. Holding up the delivery of one of these essential proteins is another way to regulate gene expression.

The protein tubulin, from which the cell builds its cytoskeleton, provides a nice example of regulation at the level of translation. The microtubules of the cytoskeleton form through the assembly of α and β tubulin polypeptides into αβ tubulin "dimers" [Greek, *dimeres* = have two parts]. When enough dimers accumulate in the cytoplasm, they activate an enzyme that blocks the mRNA on ribosomes that are producing tubulin polypeptides. Once the ribosomes stop producing α and β tubulin polypeptides, the concentration of dimers in the cytoplasm drops, and the enzyme stops interfering with the production of α and β tubulin.

Eukaryotes can regulate gene expression in more ways than prokaryotes because of the presence of the nuclear membrane, because of the way the DNA is folded and packaged, and because of the many compartments in eukaryotic cells.

In Transcriptional Regulation, Proteins in the Nucleus Bind to Special Sites on the DNA

Just as in prokaryotes, the best-studied cases of transcriptional regulation in eukaryotes depend on the interaction of proteins with regulatory sites on the DNA. The large amount of DNA in eukaryotic cells means that the expression of a single gene may depend on many DNA-protein interactions.

A transcription factor can affect transcription in two ways. One, the transcription factor can change the rate of transcription. For example, it can increase or decrease initiation by RNA polymerase—in the same way that the repressor and activator proteins turn the *lac* operon on and off. And two, the transcription factor can produce a permanent change in chromatin structure that makes a particular gene more or less accessible to the transcriptional machinery. (The gene itself is not altered.) Once the chromatin is changed, however, the gene may be expressed differently long after the transcription factor is gone.

Among the DNA sequences identified so far are: (1) *promoters,* which specify the starting point and direction of transcription; (2) *response elements,* also called *enhancers,* because they usually stimulate the transcription of neighboring DNA without themselves serving as promoters; (3) *silencers,* which help inhibit transcription; and (4) *termination signals.*

How Complex Is the Problem of Gene Regulation in Eukaryotes?

The regulation of the 100,000 or so genes in the human genome is an enormously complex problem. Researchers have estimated that as many as 10 percent of the approximately 1000 genes in *E. coli* are regulatory. Gene regulation in the mammalian nervous system is certain to be far more complex. Biologists estimate that mammals have about 5000 transcription factors and a corresponding number of response elements (silencers and enhancers). Each transcription factor may interact with more than one response element, and each response element can respond to more than one transcription factor. The number of possible interactions at the transcription level alone is huge. Yet the sets of 10,000 to 100,000 kinds of protein molecules in each of the trillion or so nerve cells in the human brain probably requires even more complex regulation than that.

In eukaryotes, the huge numbers of transcription factors and regulatory sites allow nearly unlimited varieties of interactions. The problem of gene expression is enormously complex.

A Single Transcription Factor May Affect Many Genes

The hormone testosterone, which works by binding to a transcription factor, shows the diverse effects of a transcription factor. In human and other mammalian embryos, testosterone is one of a group of hormones called androgens that stimulate the formation of the male genitalia (penis, scrotum, and the internal ducts that carry sperm). Later, at puberty, androgens stimulate the development of male secondary sexual characteristics—male distribution of fat, muscle, and body hair; deepening of the voice; and other patterns of growth and behavior.

We know that androgens accomplish all this by binding to an **androgen receptor.** This androgen receptor is actually a transcription factor that binds both to testosterone and to response elements in the DNA, thereby influencing the expression of genes related to the secondary sexual characteristics.

In a condition called **androgen insensitivity syndrome,** a chromosomal (XY) male produces testosterone, but no androgen receptors. Despite the presence of the Y chromosome and plenty of testosterone, such an individual develops into a woman, with breasts and female external genitalia. However, she has no uterus, or only a rudimentary one, and no ovaries. She has testes, but these are high in the body cavity, rather than descended. This XY individual is female in nearly every other respect. Androgen insensitivity syndrome shows how the actions of testosterone and its receptor increase the rate of transcription of a host of important genes.

The absence of a single transcription factor—the androgen receptor, for example—can greatly alter gene expression and the resulting phenotype.

In this chapter we have seen how the coded information in the gene flows from DNA to RNA to polypeptides and how cells regulate that flow. In the next chapter we will see that genetic information (in the form of individual genes) is not confined to the DNA in the nucleus, but can move about within the cell or from cell to cell. The consequences of this genetic mobility are enormous.

STUDY OUTLINE WITH KEY TERMS

DNA and RNA carry information in the sequences of nucleotides. Protein synthesis translates this information into a sequence of amino acids. The **genetic code** is the **universal** dictionary that specifies the relationship of the nucleotide sequence to the amino acid sequence. A **codon** is a group of three nucleotides that specifies a single amino acid. There are 64 codons. Of these, 61 specify the 20 amino acids and 3—**nonsense,** or **stop, codons**—signal the end of a polypeptide.

According to Francis Crick's **central dogma:** "DNA specifies RNA, which specifies proteins." Cells **transcribe** information from DNA into RNA and **translate** information from **messenger RNA (mRNA)** into protein. **RNA polymerase** copies information from one strand of DNA into RNA. The starting point for transcription is called a **promoter.** In eukaryotes, **transcription factors** lead the RNA polymerase to the right promoter. Eukaryotic and prokaryotic RNA polymerases stop transcription at **terminators.**

In prokaryotes, RNA polymerase produces mature RNAs directly. In eukaryotes, cells modify the initially transcribed pre-mRNAs into functional messenger RNAs. Such modifications include cutting out **introns** and splicing together the remaining **exons** to form a patchwork mRNA.

Ribosomes, made of **ribosomal RNAs (rRNAs)** and proteins, are the molecular factories that assemble amino acids into polypeptides. Several ribosomes may associate with a single mRNA molecule to form a **polysome.** Before ribosomes can arrange the sequence of amino acids according to the information encoded in mRNA, however, the amino acids are first linked to adapter molecules called **transfer RNAs (tRNAs).** Each tRNA contains an **anticodon,** a specific nucleotide sequence that forms base pairs with a codon in mRNA. The pairing of specific tRNAs with their appropriate amino acids is the task of enzymes that recognize one amino acid and the corresponding tRNA and link them together.

The translation of information from mRNA into the sequence of amino acids in a polypeptide occurs in three distinct steps, called initiation, elongation, and termination. **Initiation** positions the mRNA in the correct **reading frame.** Initiation always starts with AUG, which is both a signal for methionine and the **start site. Elongation** consists of three steps: (1) putting the next amino acid in place, (2) forming a peptide bond, and (3) moving the ribosome to read the next codon. **Termination** occurs when the ribosome encounters a stop codon to which a protein called a **release factor** has bound.

Eukaryotic cells sort newly made polypeptides to different addresses such as the cytosol, organelles, membranes, or locations outside of the cell. The destination of a polypeptide depends on the presence of **signal peptides.**

Most molecular biologists now say that a gene is a DNA sequence that is transcribed as a single unit and encodes either a single polypeptide or a set of related polypeptides. Cells regulate the expression of genes in many ways, including **posttranscriptional control, translational control, posttranslation control,** and **protein activity.**

But **transcriptional control** is the most efficient and the most common way. **Regulatory genes,** which include both the **regulatory site** and genes for **transcription factors,** regulate the amount of polypeptide made by a cell. In contrast, **structural genes** code for polypeptides that generally function outside of the nucleus.

Two types of transcriptional control in prokaryotes are positive regulation and negative regulation. The *lac* operon is an example of **negative regulation.** When lactose binds to the *lac* repressor (a protein), the lac operon's **promoter** becomes unblocked, and RNA polymerase can transcribe the three genes that encode enzymes for digesting lactose. We say lactose **induces** the expression of the *lac*

operon. When lactose is absent, the lac repressor binds to a regulatory site called the **operator** and prevents transcription.

CAP is an example of a **positive regulator.** CAP increases the transcription rate for certain mRNAs, provided that other regulatory mechanisms will allow these mRNAs to be transcribed at all. Regulatory proteins are called **transcription factors.**

Eukaryotes can regulate gene expression in more ways than prokaryotes because of the presence of the nuclear membrane, because of the way the DNA is folded and packaged, and because of the many compartments in eukaryotic cells. In eukaryotes, the huge numbers of transcription factors and regulatory sites allow nearly unlimited varieties of interactions. The problem of gene expression is enormously complex. In **androgen insensitivity syndrome,** the absence of a single transcription factor—the **androgen receptor,** for example—can greatly alter gene expression and the resulting phenotype.

REVIEW AND THOUGHT QUESTIONS

Review Questions

1. What is the central dogma?
2. What is the difference between the template, or antisense, strand of DNA and its mate, the sense strand? Which one is transcribed into RNA? What would happen if the sense strand were transcribed?
3. What kinds of proteins help RNA polymerase recognize the right promoter? What is a promoter?
4. How do tRNAs mediate between the mRNA and the amino acids? What does "t" stand for? What did Francis Crick call tRNAs before they were discovered?
5. What is the difference between positive regulation and negative regulation of gene expression? Give an example of each.
6. Why do bacteria need to regulate the expression of their genes?
7. What is a regulatory site? Give an example of one.
8. Why do eukaryotes need to regulate the expression of their genes? Why do multicellular organisms need to regulate gene expression?

Thought Questions

9. Why is it hard to clone humans and other mammals? If you could clone someone, whom would you choose to clone? How many copies of the person would you make? Would you expect them to be completely identical in every respect? What would you hope to accomplish by this? Would you send them all to college?
10. Suppose that you spent half your girlhood and all of your teenage years training to be a sprinter. You have qualified for the Olympic team and will compete in the 2000 Olympics. The night before the first heat, your coach tells you that your chromosome test has revealed that you have a Y chromosome and you will not be allowed to compete. You have never thought of yourself as anything except a girl, now a young woman. How do you feel? How will this change your life?
11. If you are a heterosexual man, what would you do if you discovered that your wife of 5 years has a Y chromosome?

SELECTED READINGS

Crick, Francis, *What Mad Pursuit: A Personal View of Scientific Discovery,* HarperCollins, 1988. An engaging autobiography with the emphasis on science.

Judson, Horace Freeland, *The Eighth Day of Creation,* Simon and Schuster, New York, 1979. A highly detailed 700-page history of molecular biology, including fascinating anecdotes and clear de-

scriptions of experiments. Considering the level of detail, this book is surprisingly accessible to lay readers.

▶ On-line materials relating to this chapter are on the World Wide Web at
http://www.saunderscollege.com/lifesci/
Click on Tobin/Dusheck: *Asking About Life.*

12

JUMPING GENES AND OTHER UNCONVENTIONAL
GENETIC SYSTEMS

Jumping Genes and Indian Corn

In the 1940s, a geneticist named Barbara McClintock made a discovery that would eventually revolutionize molecular biology, a science that did not even exist yet (Figure 12-1). Like Gregor Mendel, McClintock was so much ahead of her time that her research made no impact on the world of biology for 30 years.

McClintock began her career in 1918 at Cornell University, where she was first an undergraduate student, then a graduate student, and finally a researcher. In the 1920s and 1930s, geneticists there noticed that the normal purple color of corn kernels switched to pale yellow or white as a result of a mutation in a pigment gene. Multicolored

"Indian" corn is normal corn: the pale corn we buy at the store has been bred to prevent the expression of these purple pigment genes. Oddly, however, the mutation that turns purple kernels yellow or white can reverse itself, or partly reverse itself, from generation to generation.

Each kernel of corn on a cob contains a separate, unique individual—the result of a separate pairing of sperm and egg, a separate pairing of maternal and paternal chromosomes. As a result, some kernels on a cob might have the mutation and some might have it reversed: some kernels are white and some are purple. Still others may be streaked or spotted—the result of partial expression of the pigment gene. When McClintock was at Cornell, no one understood how these "unstable mutations" could flip back and forth so easily.

In 1941, McClintock moved her own research on corn to Cold Spring Harbor Laboratory, New York. She was especially interested in broken chromosomes. She had noticed a corn plant in which all of the chromosomes always broke in the same place, an oddity, since chromosomes usually break randomly. McClintock called the place where the chromosome always broke the *Ds* gene.

Ds was short for "dissociator," the gene that took apart (dissociated) the chromosome. She soon discovered that the *Ds* gene could move around on the chromosome, or "transpose," from generation to generation. Then came an even more startling revelation. McClintock noticed that the same corn plants in which *Ds* had moved simultaneously acquired the unstable pigment mutation.

That connection suggested to McClintock that the unstable pigment mutation might be caused by the movement of the *Ds* gene into the middle of the pig-

ment gene. Taking another step, she saw that if the *Ds* gene jumped back out of the pigment gene, the pigment gene would function again, coloring the kernel purple. This would explain how the kernels seem to gain and lose the mutation, switching

operon, and most geneticists did not know what to make of her results.

McClintock was nonetheless a highly respected geneticist. In 1931, recall, she and Harriet Creighton had demonstrated that chromosomes cross over, exchanging

---

McClintock might as well have told them that the streets they drove everyday moved from place to place.

---

back and forth between a normal phenotype and an abnormal one, from generation to generation. McClintock's hypothesis not only explained how unstable mutations worked, but introduced a completely new idea to the science of genetics: the idea that genes could move. She coined the word **transposon** to describe these genes that jumped around the genome. Others would come to call them "**jumping genes.**"

In time, McClintock found more complex examples of transposons. She also realized that just as the *Ds* gene could influence the expression of the pigment gene, regulatory genes could finely regulate the expression of structural genes. McClintock was the first person to recognize that one gene could regulate the expression of another. It was McClintock who first discovered promoter and suppressor genes. Unfortunately, she discovered them in corn, an organism far more complex than the bacteria in which Jacob and Monod discovered the *lac* operon. The system of regulation that McClintock discovered was far more difficult—both to explain and to understand—than the *lac*

genes during meiosis (Chapter 9). The two women had provided the conclusive evidence for the theory of chromosomal inheritance. Yet, in 1951, when McClintock presented her work on transposons to a meeting of geneticists specifically interested in mutations, her talk "met with stony silence," as McClintock's biographer Evelyn Fox Keller wrote in her biography of McClintock, *A Feeling for the Organism*. "With one or two exceptions, no one understood. Afterward, there was mumbling—even some snickering—and outright complaints. . . . It was impossible to understand." McClintock knew her work was complex, but she had not anticipated such rejection. "It was such a surprise that I couldn't communicate," she told Keller. "It was a surprise that I was being ridiculed, or being told I was really mad."

The idea that genes can move from one part of the genome to another is not, in fact, extraordinarily difficult to understand. But it was extraordinarily difficult for geneticists to accept. McClintock's colleagues just assumed that—except for crossing over and other changes that oc-

Figure 12-1 Barbara McClintock in the early 1920s, at her parents' home in Brooklyn, New York. *(Cold Spring Harbor Laboratory Archives. Permission of Marjorie M. Bhavnani.)*

cur during meiosis—the genome is a rigid, stable structure. McClintock might as well have told them that the streets they drove everyday moved from place to place. She was viewed as something of a mad genius. Her explanations were extraordinarily complicated; her writing lacked the spare elegance of Jacob and Monod's discussions of the *lac* operon. For 30 years, most geneticists did not even try to understand McClintock's work.

Not until the early 1970s, when molecular biologists began to discover the same things that McClintock had discovered in the 1940s and 50s, did biologists begin to recognize the importance of her work. They realized that McClintock's transposons, or jumping genes, are just one example of a whole class of **mobile genes**—genes that move around. Finally, biologists began to see how broadly Mc-

Clintock's ideas applied to all sorts of mobile genes, not just jumping genes and not just in corn, but to mobile genes in all living organisms.

By the 1980s, young molecular biologists had begun to use mobile genes as tools for the study of how genomes work. And, as we will see in the next chapter, mobile genes also play a pivotal role in the development of a great range of genetic engineering techniques. In 1983, when McClintock was 81, her great contributions to biology were finally recognized with a Nobel Prize.

Since the beginning of the 20th century, geneticists have carefully studied both the information in a gene and the gene's place in the chromosome. While the information in a gene is important, it is a gene's location that determines how the gene will be inherited. We have already seen, for example, that genes that are far apart on the chromosome are more likely to recombine than genes that lie next to one another. Barbara McClintock's work showed that individual genes on the chromosome can actually move from place to place, with profound consequences.

In this chapter we will also examine an odd collection of **unconventional genetic systems.** Unconventional genes are either not in the regular genome of the cell nucleus or they are in the genome but move about the genome or from one place to another. Mitochondria and chloroplasts have their own DNA. This DNA does not move, but it constitutes an unconventional genetic system because it is passed down from organelle to organelle separately from the DNA in the nucleus and without any kind of sexual recombination.

The other unconventional genes we will discuss are all mobile genes. Jumping genes, or transposons, are the simplest mobile genes. Transposons replicate as part of a host cell's DNA. In contrast, **plasmids** usually exist as simple loops of DNA separate from the cell's own DNA. Plasmids, which exist in prokaryotic cells (and yeast cells), can pass from cell to cell, and may sometimes integrate into the cell's DNA. **Viruses,** familiar to us as the purveyors of colds and flu, are another variety of mobile gene. Viruses move genetic material from cell to cell and can insert genetic information into the genomes of both prokaryotic cells and eukaryotic cells.

As we survey this odd collection of unconventional genes, however, we will see that the replication and expression of all genes, unconventional or conventional, can occur only within a cell. The study of unconventional genes has fundamentally changed cell biology. It has given biologists new insights into diseases ranging from the common cold to cancer and AIDS; it has suggested new views about the origin of eukaryotic cells and of life itself; and it has led directly to the extraordinary growth of the biotechnology industry.

KEY CONCEPTS

1. Unconventional genetic systems include all genes that are not part of a cell's nuclear genome, including mitochondrial and chloroplast DNA and all mobile genes.

2. Mobile genes include viruses, plasmids, and transposons.

3. A virus is an assemblage of protein and nucleic acid that can replicate within a living cell.

4. Many plasmids and viruses have an origin of replication that allows them to replicate separately from other types of DNA. Other mobile genes can only be replicated if they are first integrated into other DNA.

5. Transposons replicate only when integrated into genomic DNA or mobile DNA that can replicate autonomously.

6. The genomes of both prokaryotes and eukaryotes contain DNA that originally derived from mobile genes.

7. Energy organelles—mitochondria and chloroplasts—contain independent genetic systems and probably evolved from ancient associations of prokaryotes.

WHAT IS A VIRUS?

Viruses are our first example of an unconventional genetic system. A **virus** is an assemblage of nucleic acid (DNA or RNA) and proteins (and occasionally other components, such as lipids or carbohydrates). Viruses have many properties of life: they are highly organized, they reproduce, they respond to their environments, and they evolve. However, viruses cannot obtain energy from their surroundings, and cannot form the building blocks needed for the synthesis of their components. Because a virus cannot perform metabolic reactions, it can replicate its genetic material and build its proteins only by using the biochemical machinery of a host cell. In short, a virus is not alive.

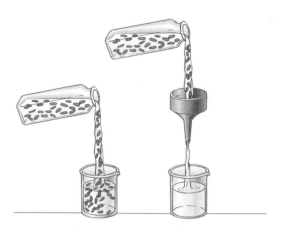

Figure 12-2 Viruses are much smaller than bacteria. A. Fine filters remove the bacteria that cause milk to sour. B. Fine filters do not remove the virus that causes tobacco mosaic disease.

A. Porcelain filter removes bacteria from sour milk

B. Tobacco mosaic virus passes through porcelain filter

Most viruses that infect bacteria, plants, and animals do not cause disease. Many are relatively harmless and some introduce useful genes into the genomes of other organisms. In fact, in the next chapter, we will see how bioengineers use this attribute of viruses to create new strains of bacteria, plants, and animals.

Nonetheless, the study of viruses has, historically, been the study of disease-causing viruses. In the last century, the word "virus" [Latin, *virus* = slimy liquid or poison] came to mean any material associated with death and disease. By the late 19th century, Louis Pasteur and others had shown that bacteria and other small cells cause many infectious diseases and also spoil food. Pasteur's colleague Charles Chamberland developed a way of removing even the smallest cells from liquids by forcing them through a fine filter (Figure 12-2A). But biologists soon discovered that some infectious agents managed to get through the bacterial filters. These mysterious entities were called "filterable viruses." One by one the bacteria specifically responsible for many diseases were identified and named. The only infectious agents left unnamed were the mysterious filterable viruses, which microbiologists simply call "viruses."

In the 1890s, just a few years after Chamberland's filter became available, two men—the Russian pathologist Dmitri Iwanowski and the Dutch microbiologist Martinus Beijerinck—independently demonstrated that a filterable virus was responsible for a disease of tobacco leaves (Figure 12-2B). Because the diseased leaves have a patchy, or mosaic, appearance, the disease was called tobacco mosaic disease, and the virus that causes it is now called tobacco mosaic virus.

Other microbiologists soon discovered that viruses cause diseases in animals as well, including yellow fever (a deadly human disease carried by mosquitoes) and hoof-and-mouth disease (a disease of cattle). In 1915, the English bacteriologist Frederick Twort discovered that viruses even infect bacteria. Twort called his bacteria-infecting virus a **bacteriophage** [Greek, *phagein* = to eat].

Discoveries such as these demonstrated that disease-causing agents could be smaller than the 0.5 μm pores in a Chamberland bacterial filter. No one, however, knew what viruses actually were. Most people thought that they were just cells too small to be stopped by a filter and too small to be seen through a microscope.

In 1935, this view fell apart, however, when Wendel Stanley, a biochemist working at the Rockefeller Institute, produced crystals of tobacco mosaic virus (Figure 12-3). This was confusing indeed. A crystal forms only when identical molecules arrange themselves in a regular lattice (as in the case of the familiar salt and sugar crystals on the breakfast table). How could

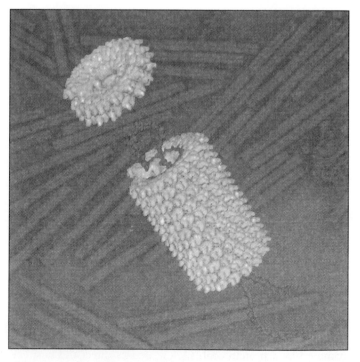

Figure 12-3 Three-dimensional model of tobacco mosaic virus. *(Dr. Gerald Stubbs & Dr. Hong Wang, Vanderbilt University)*

BOX 12-1

Can viruses make you well?

Imagine one day going to the doctor with a case of strep throat and getting a spoonful of virus to cure it. As more and more bacterial infections become resistant to antibiotic treatments, that day may not be so far off. In some parts of the world, bacteriophages, the viruses that infect bacteria, are already in use to combat disease.

Many infections that were once easily cured have become resistant to all but a few antibiotics. The list includes some kinds of pneumonia, ear infections, strep throat, urinary-tract infections, cholera, tuberculosis, and others. The heavy use of antibiotics during and after surgery has made hospital recovery rooms into breeding grounds for antibiotic-resistant strains of bacteria. In recent years, some hospitals have reported outbreaks of hospital-acquired pneumonia that are resistant to every known antibiotic.

Bacteriophages are natural parasites of bacteria. Once a phage invades a bacterial cell, it reproduces rapidly, and within an hour the cell bursts open to release millions of phage progeny, which then hunt down more bacteria.

Phages have several advantages over antibiotics. Because antibiotics are toxic, they often cause side effects, such as intestinal disorders and hearing problems. Some people are violently allergic to antibiotics. And many people don't take the complete course of antibiotic. In such cases, all of the bacteria may not be eliminated—either by the antibiotic or by the immune system—and the surviving bacteria in such a lingering infection tend to be those that are resistant to the antibiotic.

Phages have none of these problems. They function regardless of whether the bacteria are antibiotic resistant or not. They produce no side effects and tend not to be allergenic. And, since they multiply themselves in the process of slaughtering the bacteria, forgetting to take a dose or stopping treatment too early are never problems.

But phages are no panacea. Each strain is so specific for a particular type of bacteria that doctors have to know exactly what kind of bacterium is causing an infection. That could mean a delay of several days while the species of bacterium is identified. Also, some harmful bacteria have no known

phage predators. And, unfortunately, bacteria may evolve resistance to infection by phage, just as they do to antibiotics. Finally, little is known about the efficacy of phage therapy.

In the 1930s and 1970s, researchers made some attempts to treat infections with phages and many hoped that they would be the ultimate treatment for bacterial infections. In Sinclair Lewis's novel *Arrowsmith,* written in 1924, the hero tried to combat a devastating epidemic with a bacteriophage treatment. Nonetheless, the method was all but abandoned with the advent of antibiotics, but only in the West.

In parts of the former Soviet Union, however, phages have been used routinely to treat everything from burn infections to food poisoning. Doctors at a hospital in Tbilisi, Georgia, report using phages successfully to disinfect operating rooms and surgical instruments, as well as to reduce the spread of antibiotic-resistant pneumonia and strep throat. While bacteriophage therapy may be low-tech, a few researchers hope that this therapy will be the wave of the future.

---

an infectious agent—something that was assumed to be alive—form the precisely ordered arrays of a crystal? One certainly couldn't imagine a crystal of bacteria or of mice.

Stanley's discovery seemed to imply that a virus was just a large molecule, and that viral diseases were similar to chemical poisoning. But a poisonous substance would be diluted and broken down if it were passed from individual to individual, while tobacco mosaic virus from a dissolved crystal could be passed indefinitely from plant to plant. Clearly, tobacco mosaic virus was replicating itself—something ordinary molecules are not known to do.

We now understand that viruses are relatively large, regular molecular complexes, some of which can form crystals. By analyzing such crystals with x-ray diffraction techniques, researchers have been able to determine the structure of some viruses in great detail. Most viruses consist of a nucleic acid core (either DNA or RNA) that carries the virus's genes, all surrounded by a protein coat. (A few viruses direct the host cell to make a viral coat out of the host cell's own plasma membrane. This trick helps an infecting virus conceal its identity.)

Some of a virus's genes direct the synthesis of proteins and nucleic acids that make up the virus, while other genes specify proteins that regulate the host cell's DNA. Such regulatory genes force the host cell to make new virus particles.

A virus is a large complex of nucleic acid and protein capable of infecting a cell. Infection may be helpful, harmless, or fatal to the cell.

What Is the Form of a Virus?

With the development of the electron microscope in the 1940s, scientists could actually see viruses for the first time. They saw that viruses are large particles, much smaller than cells, but larger than single protein molecules. The biggest viruses are the pox viruses such as smallpox virus, which range from 250 to 400 nm in diameter. The smallest ones, including, for example, poliovirus, are 20 to 30 nm in diameter (Figure 12-4).

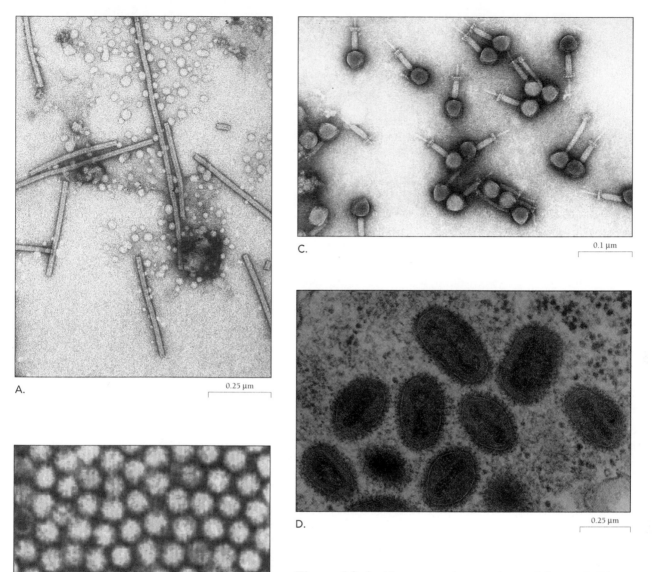

Figure 12-4 **Viruses come in many sizes and shapes.** A. Tobacco mosaic virus. B. Like the poliovirus, tomato bushy stunt virus is a 20-sided icosahedron. C. T4 bacteriophage. D. HIV, which causes AIDS. *(A, Jack D. Griffith, University of North Carolina; B, Andrew O. Jackson, University of California, Berkeley; C, Dr. Thomas Broker/Phototake NYC; D, Hans Gelderblom/Visuals Unlimited)*

Viruses have many shapes, but most have one of two forms: a long helix, such as tobacco mosaic virus (Figure 12-4A); or a compact, nearly spherical particle, such as poliovirus (Figure 12-4B). The poliovirus is not actually spherical. It is a 20-sided object, called an **icosahedron** [Greek, *eikosi* = twenty, *hedra* = base], a little like a soccer ball. Unlike the sides of a soccer ball, however, each facet is shaped like a triangle. Other viruses have more complex forms. The bacteriophage T4 is among the most complex of viruses (Figure 12-4C). With an icosahedral head and an intricate tail structure, it looks and functions like a tiny medical syringe, injecting the bacteriophage's DNA into its bacterial host.

The outside surface of a virus is either a coat of protein—a **capsid**—or a **membrane envelope**—a membrane like that which surrounds a cell. Many viruses that infect animals, such as influenza virus and human immunodeficiency virus (HIV, the AIDS virus), direct the host cell to make a viral membrane from the host's own cell membrane, together with proteins specified by the virus's genes (Figure 12-4D). Otherwise, viruses consist almost entirely of proteins and nucleic acid. The nucleic acid may be either DNA or RNA, depending on the virus. Just as in humans and other living organisms, the nucleic acid serves as the virus's genetic material and contains the information for most, if not all, of the virus's proteins.

Figure 12-5   The reproductive cycle of a bacterio-phage.

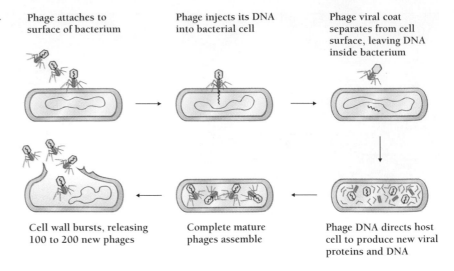

Phage attaches to surface of bacterium

Phage injects its DNA into bacterial cell

Phage viral coat separates from cell surface, leaving DNA inside bacterium

Cell wall bursts, releasing 100 to 200 new phages

Complete mature phages assemble

Phage DNA directs host cell to produce new viral proteins and DNA

Viruses have regular shapes, such as an icosahedron or a helix. They consist of a core of genetic material—DNA or RNA—enclosed in a protein capsid or a membrane envelope.

How Do Viruses Make More Viruses?

Virologists' studies of how viruses infect cells in the laboratory have revealed much about how viruses replicate. Although viruses can make more viruses in many different ways, their histories always include the following stages: (1) the virus attaches to the host cell; (2) the viral nucleic acid enters the cell; (3) the cell synthesizes proteins specified by the virus's genes; (4) the cell replicates the virus's DNA or RNA; (5) the new viral protein and DNA or RNA made by the host cell assembles into new viruses, also called virus particles; and (6) the new viruses are released from the cell (Figure 12-5).

Each stage in a virus's multiplication cycle depends on the interaction between the host cell and the virus. A given kind of virus can only infect certain kinds of cells. We say that viruses are adapted to specific hosts. Most cold viruses and the HIV virus, for example, are specific to humans. In this section, we will briefly discuss four of these six stages of viral infection. After that, we will discuss the replication of the viral DNA or RNA in greater detail.

How Does a Virus Attach to a Cell?

The attachment of the virus to a cell depends on an interaction between a viral protein and receptor molecules on the host cell's membrane (Figure 12-6). Specific proteins on the surface of the virus bind to molecules on the surface of the host cell. The attachment site may be on all of the host's cells or just on a special set of cells. Human cold and flu viruses, for example, specifically infect cells of the respiratory tract. Others, such as chicken pox virus and other varieties of herpes viruses, infect skin and nerve cells. Poliovirus initially attacks cells in the digestive tract, then moves into the bloodstream where it can gain access to all the cells in the body. In 1 percent of polio cases, poliovirus attaches to and destroys nerve cells, causing paralysis or even death. In comparison, HIV primarily infects cells of the immune system and the brain.

Viruses attach to protein receptors on the surfaces of cells.

How Does the Viral DNA or RNA Enter the Host Cell?

The mechanism by which the viral nucleic acid enters the host cell depends on the kind of virus and the kind of host cell. The T4 bacteriophage, for example, contains an enzyme that digests away part of a bacterial cell wall, allowing the phage to inject its DNA into its host. Other viruses, such as the flu virus, are taken into cells by endocytosis, the process by which the cell's membrane enfolds particles (Figure 12-7). Still other viruses, including HIV, are enclosed in bits of the cell's own membrane, which the virus particles capture as they leave an infected cell. Such viruses easily enter their next host cell because the virus's stolen membrane envelope fuses with the host cell's membrane (Figure 12-7).

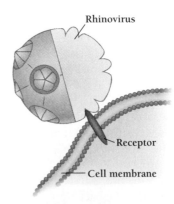

Rhinovirus

Receptor

Cell membrane

Figure 12-6   Viruses latch onto cell surface receptors.

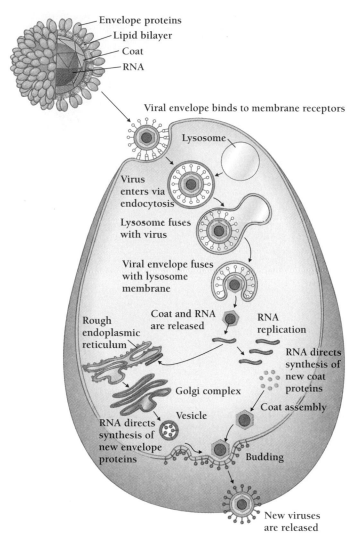

Figure 12-7 **Like the Trojan horse, an influenza virus tricks a cell into taking up the virus.** The virus attaches to cell surface receptors, which initiate endocytosis of the virus. Once inside, the virus releases RNAs, which the cell duplicates and packs in viral protein coats for release to the outside.

Viruses enter cells either by cutting a hole in the cell membrane, or by inducing the cell to engulf the virus particle, or by fusing with the cell's membrane.

How Are Virus Particles Manufactured and Released?

Once inside the host cell, the virus's genes direct the synthesis of viral proteins in a variety of ways. Some direct the host's protein-synthesizing machinery to make an enzyme that destroys the host's own DNA. Other viruses rely on promoters and other regulatory sequences that ensure rapid replication, transcription, and translation. In most cases, a host cell is forced to make viral proteins instead of its own.

For a bacteriophage, the whole cycle, from attachment and entry to assembly of new virus particles, may take only 20 min-

utes and may produce 100 or more new bacteriophages from each original invading bacteriophage inside the cell (Figure 12-7). After these new viruses are produced, a viral enzyme causes the host cell membrane to break open in a process called **lysis** [Greek, *lysis* = loosening]. The ruptured cell then releases the newly made viruses, which are ready to infect more host cells. Viruses that destroy their host cells in this way are called **lytic viruses.**

Not all viruses cause cell lysis, however. Killing the host—whether the host is a single cell or a large multicellular organism—is not the only strategy by which a virus can successfully replicate. After all, the host the virus kills may be the last one it encounters for a while, so killing the host could spell the end of the line for a virus. Some viruses have a multiplication cycle similar to that of a lytic virus, but exit the cell without destroying the host. Other viruses stay inside the cell for long periods, as we will see.

Viruses force the host cell to make viral proteins—sometimes by destroying the host's DNA, sometimes through subtler forms of gene regulation. Many viruses escape the host cell by breaking open the cell membrane in a process called lysis.

HOW DO VIRUSES AND OTHER MOBILE GENES REPLICATE THEIR GENES?

"A hen is only an egg's way of making another egg," wrote the English novelist and essayist Samuel Butler. Just so, a virus is only a gene's way of making another gene. This statement might sound extreme, but the idea behind it helps us to understand the strategies that viruses and other mobile genes use to replicate.

A virus is the most elaborate of the mobile genes, but there are many other kinds of mobile genes that consist only of "DNA or RNA." Such mobile genes dispense with any kind of exterior coat or membrane. Like viruses, however, they include sequences that induce the host cell to make copies of their genes.

Mobile genes, including viruses, plasmids, and transposons, have two reproductive strategies. In **autonomous replication,** the genes are separate from the DNA of the host cell and can be replicated even when the host DNA is not replicating. In **integrated replication,** the genes become integrated into the host's DNA and are replicated along with those of the host. Integrated genes can *only* replicate when the host DNA replicates. Some mobile genes can switch back and forth from autonomous to integrated replication depending on circumstances.

In the next few sections, we will discuss **plasmids,** mobile genes that usually replicate only autonomously; **transposons,** mobile genes that can replicate only when integrated into the host's DNA or into a plasmid; and **episomes,** viruses and plas-

Figure 12-8 Conjugation in bacteria. During a brief encounter, bacteria exchange genes carried on the bacterial DNA or on plasmids. The bacteria may, for example, transfer genes on plasmids that confer antibiotic resistance. By means of conjugation, bacteria can spread genes for antibiotic resistance among different strains and even different species of bacteria. *(A, Dr. Dennis Kunkel/Phototake NYC)*

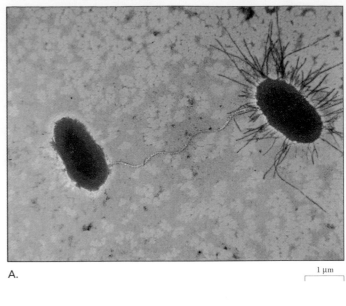

A.

1 μm

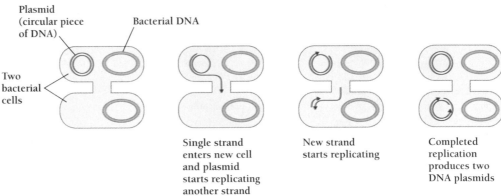

Plasmid (circular piece of DNA)

Bacterial DNA

Two bacterial cells

Single strand enters new cell and plasmid starts replicating another strand

New strand starts replicating

Completed replication produces two DNA plasmids

B.

mids that can switch back and forth between integrated and autonomous replication.

How Do Plasmids Replicate Autonomously?

A **plasmid** is a small piece of circular DNA or RNA in a bacterial or yeast cell that includes from 3 to 300 genes. Most plasmids do not integrate themselves into the host cell's DNA, but exist as independent entities. Usually, plasmids replicate only when the DNA of their hosts does. However, many plasmids can replicate even when host DNA synthesis has stopped. Such plasmids are said to replicate **autonomously.** A plasmid can do this because it contains an origin of replication, the DNA sequence where DNA replication begins. As a result of this autonomous replication, a cell may contain up to 1000 copies of each plasmid. And the cell may carry many different kinds of plasmids.

Plasmids, unlike viruses, do not produce protein coats that would allow them to move easily from one cell to another. However, plasmids have another way to get around. Bacteria sometimes engage in a sexlike process called **conjugation,** during which two bacterial cells temporarily attach to one another and

exchange genetic material (Figure 12-8). During conjugation, plasmids can move from one bacterium to another.

Plasmids often contain genes that bring their hosts some advantage. For example, some plasmids—called **resistance factors** or **R factors**—contain genes for enzymes that make the cell resistant to antibiotics (Figure 12-9). Other plasmids con-

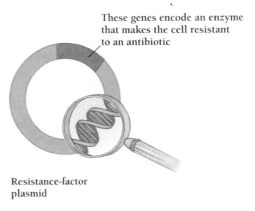

These genes encode an enzyme that makes the cell resistant to an antibiotic

Resistance-factor plasmid

Figure 12-9 A plasmid carrying a gene for antibiotic resistance.

tain genes for resistance to heavy metal poisoning, for the metabolism of unusual compounds (such as those found in oil spills), or for the production of toxins that kill other bacteria. In addition, molecular biologists have synthesized thousands of **recombinant plasmids,** plasmids that contain combinations of genes never found in nature. We will discuss the production and uses of such recombinant DNAs in Chapter 13.

A plasmid is a circular piece of DNA or RNA capable of autonomous replication that carries genes, such as antibiotic resistance, from one bacterial cell to another.

Many Mobile Genes Can Replicate Only When Integrated into a Host Cell's DNA

Unlike a plasmid, a **transposon** is a mobile DNA sequence that can replicate only when it is part of the host cell's DNA. Transposons, or jumping genes, are genetic sequences that move, either by moving themselves to new locations or by duplicating themselves for repeated insertions elsewhere. A transposon, by definition, includes a gene for an enzyme called **transposase,** which catalyzes insertion into new sites (Figure 12-10). While plasmids are found mostly in bacteria and in some yeast cells, transposons are common in all cells, including McClintock's corn.

Eukaryotic transposons mostly reproduce only in the integrated form. They are surprisingly mobile, however. In eukaryotic organisms from corn to humans, they can jump from place to place within the genome, duplicating themselves, disrupting other genes, deleting and repeatedly duplicating whole stretches of DNA, breaking chromosomes, and sometimes even

attaching part of one chromosome to another. The mapping of eukaryotic genomes has revealed that virtually all genomes are littered with DNA sequences that appear to be relics of ancient transpositions by jumping genes.

A transposon may be copied many times, and each copy moved about the host's genome separately. For example, at least six different transposons are present in the genome of *E. coli.* Several of these transposons have been copied 5 to 10 times within each *E. coli* genome.

A **simple transposon** specifies transposase and nothing else. As far as biologists know, a simple transposon's only function is to make more simple transposons. Simple transposons may perhaps help bacteria in a very indirect way. Because transposons rearrange the genome and interrupt genes, they increase genetic variability in their hosts. Greater genetic variability can theoretically allow species to adapt to new conditions more quickly and more effectively.

When two simple transposons lie near each other on the host genome, they can jump together, carrying with them any part of the host's genome that lies between them. Such combinations are called **complex transposons.** Some complex transposons carry antibiotic resistance genes, for example (Figure 12-10). Transposons can move these genes into plasmids or into the host cell's chromosomal DNA. In this way, transposons can help a plasmid accumulate genes for resistance to multiple antibiotics (tetracycline and penicillin, for example), which the plasmid then carries from one bacterium to another.

As we will see in Chapter 20, transposons' ability to move genes ultimately explains the awful mystery of how antibiotic resistance spreads so rapidly among bacteria. Strains of the bacteria that cause tuberculosis (TB), for example, now exist that are resistant to every known antibiotic. The medical treatment for people infected with such bacteria is essentially the same

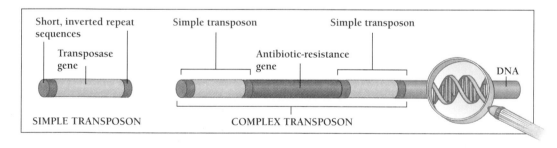

Short, inverted repeat sequences

Transposase gene

Simple transposon

Antibiotic-resistance gene

Simple transposon

DNA

SIMPLE TRANSPOSON

COMPLEX TRANSPOSON

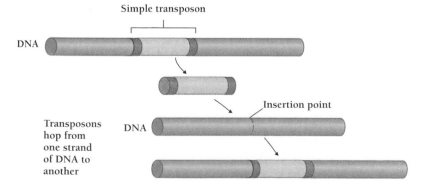

Simple transposon

DNA

Insertion point

DNA

Transposons hop from one strand of DNA to another

Figure 12-10 Transposons. A simple transposon carries the gene for transposase and little else. A complex transposon is two simple transposons with some other genes carried between them.

Figure 12-11   **Virus life cycles.** A. How do DNA tumor viruses cause tumors? DNA tumor viruses integrate their DNA into the DNA of eukaryotic cells, then disrupt the cell cycle, forcing infected cells into excessive growth. B. The lysogenic cycle of the bacteriophage lambda. Lambda can either integrate into the DNA of its host or initiate a lytic cycle.

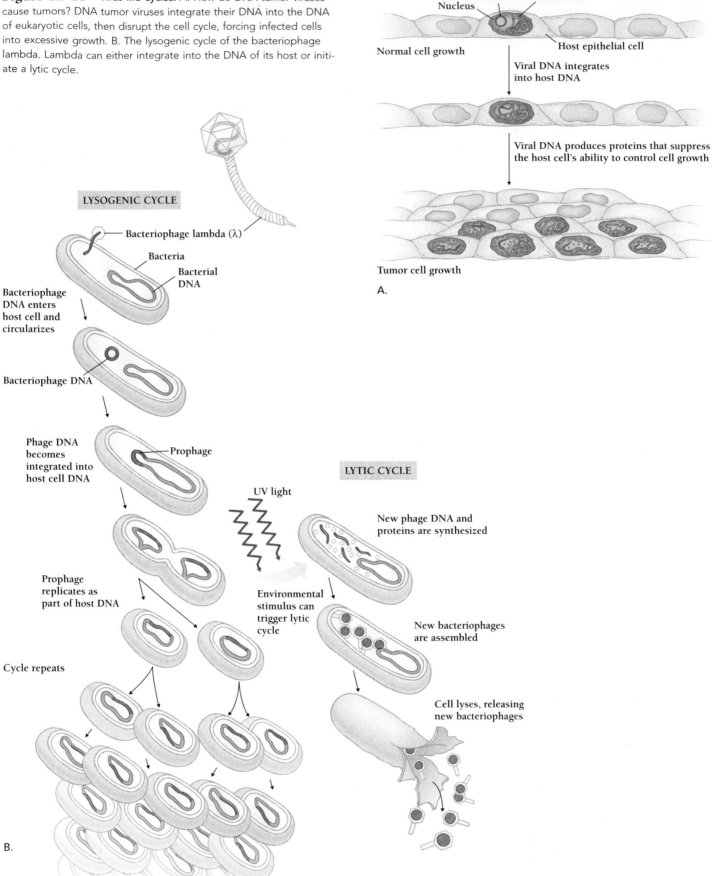

LYSOGENIC CYCLE

Bacteriophage lambda (λ)

Bacteria

Bacterial DNA

Bacteriophage DNA enters host cell and circularizes

Bacteriophage DNA

Phage DNA becomes integrated into host cell DNA

Prophage

Prophage replicates as part of host DNA

Cycle repeats

B.

Nucleus

Viral DNA   Host DNA

Host epithelial cell

Normal cell growth

Viral DNA integrates into host DNA

Viral DNA produces proteins that suppress the host cell's ability to control cell growth

Tumor cell growth

A.

LYTIC CYCLE

UV light

New phage DNA and proteins are synthesized

Environmental stimulus can trigger lytic cycle

New bacteriophages are assembled

Cell lyses, releasing new bacteriophages

as that used 50 years ago, before the availability of antibiotics. TB patients with multiple-antibiotic-resistant TB are kept isolated, to prevent them from passing this dangerous form of TB to others. But there is no specific medical treatment that will help them get well.

A simple transposon codes for nothing but transposase, which is all the transposon needs to insert itself into a new site in the host's DNA. Complex transposons carry other genes as well. Both eukaryotes and prokaryotes contain mobile genes, which can move within the genome of the host cell and somehow affect the expression of other genes.

Tumor Viruses Also Insert Themselves into Chromosomal DNA

Many viruses replicate autonomously, lyse the cell, and escape to infect more cells. But viruses that cause tumors are a special case. Rather than multiplying as rapidly as possible and killing their host cells in a lytic cycle, tumor viruses multiply only when their eukaryotic hosts do (Figure 12-11A).

Tumor viruses exist independently of the host's DNA only during a brief, independent, mobile phase when the tumor virus finds the eukaryotic cell and infects it. After infection, the genes of most tumor viruses become integrated into the DNA of their hosts and stay there. The virus uses the host's enzymes to express one or more viral genes. The resulting viral proteins ruin the host's ability to control cell division, forcing the cell to divide rapidly, which causes a tumor to form. With each round of cell division, a new generation of virus genes is created.

The viral genes whose expressed proteins interfere with the host cell's ability to control cell division are called **oncogenes** [Greek, *onkos* = mass, swelling], a word that means "genes that cause tumors." Most oncogenes are genes of our own that have been damaged. But some viruses also contain oncogenes. Viral oncogenes resemble eukaryotic genes that normally control growth and differentiation, called **proto-oncogenes** or **cellular oncogenes.** Viral oncogenes almost certainly evolved from the proto-oncogenes of cells. Biologists now suspect that the oncogenes of tumor viruses merely provide an overdose of normal cellular proteins that ordinarily stimulate normal cell division.

Tumor viruses integrate their DNA into that of a eukaryotic host, then force the host to replicate both viral and genomic DNA more often than normal. The result is many copies of the viral DNA.

Episomes Can Replicate Both Autonomously and Through Integration

Episomes are viruses or plasmids that switch hit between autonomous replication and integrated replication. Episomes can reproduce either autonomously or as part of the host's chromosome, as the occasion requires.

The bacteriophage λ (lambda), for example, is an episome with two alternative reproductive cycles in its *E. coli* hosts. In the **lytic cycle,** λ's DNA is separate from that of its host, and λ lyses its host. Occasionally, however, λ integrates its DNA into *E. coli's* DNA. The integrated form of a virus is called a **provirus** or, in the case of a bacteriophage, a **prophage** [Greek, *pro* = before]. The provirus (or prophage) and the lytic cycle together make up the viral **lysogenic cycle** (Figure 12-11B).

When the λ bacteriophage is in the prophage stage, the host cell does not make viral proteins. The reason for this is that the λ DNA directs the synthesis of a powerful repressor, called the *lambda repressor,* which prevents the transcription of any other λ genes. Ultraviolet light and other environmental insults that threaten to destroy the host cell trigger an "escape" mechanism in the prophage. In escape mode, the λ DNA directs the synthesis of a protein that represses the lambda repressor. Once the lambda repressor is no longer preventing the expression of the λ genes, the virus enters a lytic cycle, destroys the host cell, and new λ bacteriophages escape the dying cell.

Because a virus in the provirus stage does not lyse its host cell, it cannot infect other cells. Although a provirus or prophage does not force the cell to make new virus particles, the nucleic acid of the provirus is copied along with the cell's own DNA each time the cell divides or otherwise copies its DNA. As a result, all of the daughter cells of an infected cell are also infected with proviruses, and an infected eukaryotic organism cannot rid itself of the virus.

If the host cell is exposed to ultraviolet light or certain DNA-damaging chemicals, however, the viral DNA then escapes and begins a lytic cycle (Figure 12-11B). This may explain why intense sunlight can activate the herpes virus that infects the cells of the mouth and lips. When the virus becomes active, it enters a lytic cycle, lysing cells and causing a painful cold sore.

Episomes are viruses or plasmids that switch back and forth between integrated and autonomous replication as the occasion demands.

How Do RNA Viruses Reproduce and Express Their Genes?

Many mobile genes use genetic material other than double-stranded DNA. These genetic entities must use special tricks to replicate themselves. For example, the virus φX-174 uses single-stranded DNA. This virus simply copies the single strand into a complementary strand of DNA to make a double-stranded "replicative form" (Figure 12-12A).

Other viruses and plasmids have RNA genes. These include the viruses that cause polio, influenza, AIDS, and many animal tumors. How do these genes function? The short answer is that they use special enzymes, encoded in the virus's genome. The

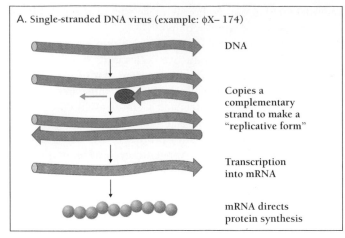

A. Single-stranded DNA virus (example: φX–174)

DNA

Copies a complementary strand to make a "replicative form"

Transcription into mRNA

mRNA directs protein synthesis

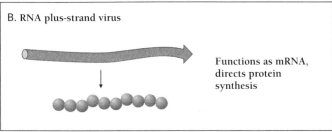

B. RNA plus-strand virus

Functions as mRNA, directs protein synthesis

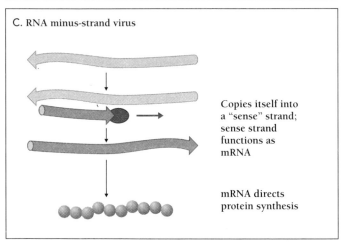

C. RNA minus-strand virus

Copies itself into a "sense" strand; sense strand functions as mRNA

mRNA directs protein synthesis

Figure 12-12   How do viruses express their genes? A. Single-stranded DNA virus. B. RNA plus-strand virus. C. RNA minus-strand virus.

particular function of the enzyme varies with the virus, but RNA viruses all depend on standard base-pairing to make a complementary strand.

Figure 12-12 shows several different ways that RNA viruses express their genes. Some viruses carry their genes on a strand of RNA that, like messenger RNA, actually codes for various viral proteins. Such RNA is expressed directly by the host cell's machinery as soon as the virus enters the cell. Such viruses are called "plus-strand" RNA viruses (Figure 12-12B). Other RNA viruses, called minus-strand viruses, use an "antisense" RNA strand (Chapter 11). Such viruses contain a viral enzyme that

first copies the RNA into a complementary (sense) strand that functions as mRNA and codes for proteins (Figure 12-12C).

Plus-strand viral RNA works like mRNA. Minus-strand RNA is antisense RNA, which is transcribed into RNA that works like mRNA.

RNA Tumor Viruses Present Special Problems

By the mid-1960s virologists had discovered many of the ways that RNA viruses could reproduce. But the reproductive cycle of RNA tumor viruses, such as the Rous sarcoma virus, mystified them. DNA tumor viruses inserted their genes directly into the DNA of their hosts, permanently changing their genetic makeup. But where could an RNA tumor virus leave a permanent copy of its genes? Certainly single-stranded RNA could not insert directly into chromosomal DNA.

The answer was first suggested in the 1960s by Howard Temin, a virologist at the University of Wisconsin. Perhaps, Temin argued, the RNA is recopied into DNA, and the DNA is then inserted into the host cell's genome. Many scientists found this suggestion heretical, since it violated Francis Crick's central dogma: "DNA specifies RNA, which specifies proteins."

In 1970, however, Temin and David Baltimore, a molecular biologist at MIT, independently discovered that RNA tumor viruses contain an enzyme, called **reverse transcriptase,** which indeed copies RNA into DNA. The enzyme received its name because the direction in which information flows is backwards from that found in normal transcription (where information goes from DNA to RNA). As we will see in the next chapter, reverse transcriptase has become an essential tool for molecular biologists. Viruses that use reverse transcriptase in this way are called **retroviruses.** The discovery of reverse transcriptase supported Temin's hypothesis.

Subsequent work showed that Temin was exactly right. In most viruses, only the virus's nucleic acid actually enters the host. In retroviruses, however, the reverse transcriptase enzyme is actually packaged *with* the viral RNA, so the enzyme (not just the gene for the enzyme) enters the cell along with the virus's RNA (Figure 12-13). The reverse transcriptase then copies the RNA first into single-stranded and then into double-stranded DNA. The double-stranded DNA then integrates into the host cell's DNA, so that the virus's genes are replicated along with those of the host. Like DNA-containing tumor viruses, many (but not all) retroviruses contain an oncogene, whose expression leads to tumor formation. Retroviruses are known to cause cancer in birds, mice, monkeys, and humans. HIV is a retrovirus that infects cells of the human immune system, but does not contain an oncogene.

Retroviruses, which use reverse transcriptase to transfer information from RNA to DNA, violate the central dogma.

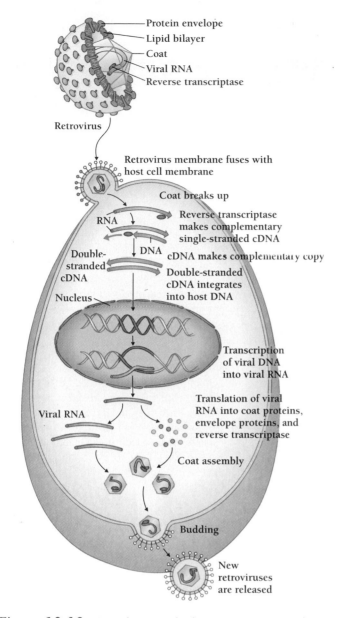

Labels on figure:
- Protein envelope
- Lipid bilayer
- Coat
- Viral RNA
- Reverse transcriptase

Retrovirus

Retrovirus membrane fuses with host cell membrane

Coat breaks up

RNA

Reverse transcriptase makes complementary single-stranded cDNA

DNA

cDNA makes complementary copy

Double-stranded cDNA

Double-stranded cDNA integrates into host DNA

Nucleus

Transcription of viral DNA into viral RNA

Viral RNA

Translation of viral RNA into coat proteins, envelope proteins, and reverse transcriptase

Coat assembly

Budding

New retroviruses are released

Figure 12-13 How do HIV and other retroviruses reproduce?

HOW DO MOBILE GENES EVOLVE?

Viruses and other mobile genes live so closely with cells that we might not realize at first that they are influenced by evolutionary forces independent of those that influence their hosts. Even though mobile genes are not alive, they are still independent entities whose replication is subject to natural selection—the mechanism by which Darwin explained the evolution of organisms.

Although Viruses Are Not Alive, They Are Still Subject to Natural Selection

Because a virus uses energy-expensive materials from its host cell, the cell's ability to grow and reproduce is impaired. As a result, hosts that evolve ways to resist viral infection are more

likely to survive and reproduce. This is simply natural selection. Cells have therefore evolved general mechanisms for resisting most kinds of viruses and also resistance mechanisms aimed at specific viruses. Many strains of *E. coli,* for example, guard against viruses by producing **restriction enzymes,** enzymes that cut all foreign DNA into little pieces. Most restriction enzymes do not cut the DNA at random, but at special "target sites"—particular sequences of nucleotides that occur commonly in all genomes. Although a restriction enzyme can protect *E. coli* (or another bacterial strain) from infection by a bacteriophage, such an enzyme would be self-destructive if it also digested *E. coli*'s own DNA. To prevent such molecular suicide, *E. coli* modifies its DNA by adding methyl groups that protect the bacterial DNA from digestion. Since the discovery of restriction enzymes, geneticists have put them to work in genetic engineering, as we will see in the next chapter.

As hosts evolve mechanisms for resisting a virus, the virus, in turn, evolves mechanisms for getting past the cell's defenses. The kinds of viruses that are able to infect *E. coli,* for example, all have DNA that is missing the sequence targeted by *E. coli* restriction enzymes. Because the viral DNA lacks this target sequence, the cell's restriction enzymes cannot destroy the viral DNA.

Viral multiplication cycles—from infection to release—have evolved to become highly efficient. Cycles are generally rapid, with lots of new viruses produced in each infected cell. Viral regulatory genes are so powerful that the host's transcription and translation machinery recognizes them in preference to the host's own genes.

One way that viruses speed up their cycle times is to keep their genetic material to a minimum. Unlike eukaryotic genomes, which are loaded with "junk DNA," every gene in a virus matters. One kind of cold virus, for example, specifies several different proteins from a single gene merely by arranging to have the mRNA spliced in different ways.

How May Viruses Have Evolved?

Some scientists believe that viruses are descended from free-living cells that became parasites on other cells. According to this view, over eons the viruses lost most of their cellular components, letting the host cells do all the work.

Others believe viruses may have evolved from plasmids, which evolved from transposons. Knowing what we now know about mobile genes, we can imagine that mobile genes first appeared as simple transposons that could move from place to place in the genome. Transposons that also included genes that benefited the host cell—such as genes that allow the host to break down poisons or antibiotics—would have allowed the host cells that had them to reproduce faster than cells without them. As a result, transposons with beneficial genes would have replicated much faster than those without beneficial genes. This is simply natural selection.

Later, beneficial genes could have acquired the ability to replicate independently as plasmids do. Finally, some au-

BOX 12-2

How can researchers design drugs that will combat AIDS?

The greatest epidemic in recent history is acquired immune deficiency syndrome (AIDS). Most biologists now agree that AIDS is caused by human immunodeficiency virus (HIV). HIV infects specialized white blood cells, called T lymphocytes, that normally help find and destroy invading pathogens. Over several years, the virus reduces the number and effectiveness of these crucial immune cells, making the body susceptible to infections and cancers that a healthy immune system normally fends off. These secondary illnesses, rather than the HIV virus itself, are what kill most people infected with AIDS.

Because HIV grows and replicates in the body for years before producing any symptoms, a person can test positive for HIV antibodies, indicating the presence of the virus, long before the onset of any illness. During this early stage of infection, bodily fluids such as blood and semen harbor active viruses that can transmit the infection to a new host. Most people contract the virus by sharing contaminated hypodermic needles or by having unprotected sex with an infected partner.

Like all retroviruses, HIV carries its genetic material in a single RNA strand accompanied by a molecule of reverse transcriptase. Once inside the cell, the reverse transcriptase generates a DNA strand copied from the RNA template. A second round of synthesis produces a double-stranded DNA copy of the viral RNA, which inserts itself into a chromosome in the host cell's nucleus. Without reverse transcriptase, HIV could still get into the cell, but it would never be able to replicate.

Since human cells do not depend on reverse transcriptase, the virus's enzyme makes a perfect target for anti-AIDS drugs. One type of antiviral therapy deceives the enzyme with synthetic molecules that resemble ordinary nucleotides, the raw materials for DNA synthesis. Reverse transcriptase incorporates these nucleotide impostors into the growing strand of viral DNA. But since the false nucleotides are subtly different from real ones, replication stops. AZT and 3TC, two drugs now commonly used to slow viral replication in patients with AIDS, are examples of false nucleotides. Human DNA polymerase is more picky than reverse transcriptase in selecting nucleotides and rarely incorporates AZT or 3TC. As a result, human cells exposed to the drugs continue to grow and divide normally. Unfortunately, HIV mutates rapidly, and strains that no longer use the false nucleotides quickly emerge. These resistant strains thrive in patients treated with these drugs, rendering the drugs useless.

Another class of HIV drugs works by interfering with a different viral enzyme—one that cuts viral mRNAs. After viral DNA integrates into the host's chromosome, host proteins transcribe the viral genes. Rather than producing a separate mRNA for each viral protein, however, the messages for several proteins form a single, long mRNA molecule. The resulting mRNA gets translated into a long polypeptide that must be cut into separate proteins before they will function. One of the proteins in the viral polypeptide chain is the enzyme that does the cutting. This enzyme, called a protease, begins cutting even while it is still part of the "polyprotein." Without the protease, the polyprotein remains unclipped and inactive, and the virus cannot go on to infect more cells.

In an attempt to stop HIV by knocking out the protease, researchers devised tiny molecules that nestle into the active site of the protease enzyme and prevent it from doing its job. The first of these protease inhibitors, called saquinavir, was released in December 1995—with dramatic results. A drug cocktail containing saquinavir, AZT, and 3TC can reduce the amount of virus in a patient's body to undetectable levels. For the first time since the discovery of HIV, this "triple therapy" has transformed AIDS from a death sentence to a potentially survivable disease.

But triple therapy has its problems. The cost of these AIDS drugs, particularly protease inhibitors, is still extremely high, placing an economic barrier between many sufferers of AIDS and good treatment. In the United States, not all insurance covers AIDS therapies, and in Africa, where most of the world's AIDS cases occur, the price is out of reach of almost everyone who has the disease. Also, triple therapy doesn't work well for everyone, and it will be several years before researchers know whether the drug cocktail can permanently clear the virus out of the body. Indeed, researchers worry that HIV may evolve resistance to protease inhibitors, as it has to other drugs. In that case, it could still return in full force, insensitive once more to all drug treatments.

To help stave off that possibility, research is underway on still another class of drugs that may close the door on the virus, preventing it from getting into T cells in the first place. To enter T cells, an HIV virus must bind to a pair of proteins on the cell surface, one of which is the recently identified chemokine receptor CCR5. Chemokines are chemical signals that induce inflammation at a site of injury or infection. Researchers noticed in 1995 that chemokines block HIV from infecting cultured cells. Not long after, more work revealed that CCR5 helps HIV enter cells and that the chemokine was interfering with that process. Drug companies are now racing to develop a drug that binds to CCR5 without setting off the inflammatory pathway. One reason to believe that this approach may work is that studies of people with a defective CCR5 gene have shown that the defect causes no obvious health problems, yet provides genetic resistance to many strains of HIV.

The battle against the AIDS epidemic may finally be succeeding. In 1996, the number of patients dying of AIDS in the United States declined for the first time. While this is a significant landmark, the epidemic is far from over. As has been true of many other infectious diseaeses, the AIDS virus will continue to evolve and researchers may always be just one step ahead.

tonomously reproducing plasmids could have acquired genes for protein coats that allowed them to move efficiently from cell to cell. Such a protein-wrapped plasmid would look just like a virus.

Molecular biologists theorize that over evolutionary time, transposons acquired the ability to replicate autonomously, making them plasmids. Some plasmids then acquired the genes for making a protein coat, allowing them to evolve into a modern virus.

MITOCHONDRIA AND CHLOROPLASTS CONTAIN THEIR OWN DNA

Chloroplasts and mitochondria contain their own DNA. Since all eukaryotes have mitochondria and many have chloroplasts as well, almost every eukaryotic cell therefore carries at least one genetic system independent of the nucleus. In a mammalian cell, each mitochondrion contains 5 to 10 identical DNA molecules, each about 16,000 nucleotides long. Since a single cell may have hundreds of mitochondria, up to 1 percent of the total DNA in the cell may consist of mitochondrial DNA. Each piece of mitochondrial DNA is circular, meaning that it forms a continuous loop, like the DNA of bacteria and many viruses (Figure 12-14A).

Plant mitochondria also contain circular DNA molecules, but usually of several different kinds. The total length and the total information content of mitochondrial DNA in plants is 30 to 100 times more than in animal mitochondria. Chloroplasts contain DNA as well (Figure 12-14B). Each species has a single kind of chloroplast DNA. The chloroplast DNA is circular and usually about 150,000 nucleotides long. In higher plants, chloroplast DNA may constitute up to 15 percent of the total DNA in a cell.

The reproduction of mitochondria and chloroplasts depends on cellular machinery that is specified both by the DNA in the nucleus and by the DNA in the organelle itself. The replication and expression of mitochondrial or chloroplast DNA, for example, requires about 90 proteins, far more than the organelle's DNA can usually encode. These proteins are specified by the DNA in the nucleus and made on cytoplasmic ribosomes. The DNA of mitochondria and chloroplasts nonetheless contain important information. For example, both mitochondria and chloroplasts contain the proton-pumping complex F_0-F_1 that harnesses the energy of proton gradients to make ATP. Each complex contains nine different polypeptides, only some of which are specified by the DNA in the nucleus. The other genes are in the mitochondria or chloroplasts.

Mitochondria and chloroplasts contain complete systems for transcription and translation, including not only DNA, but ribosomes, transfer RNAs (tRNAs), and an RNA polymerase similar to that of bacteria. Some components of the machinery of protein synthesis are made in the cytoplasm and other components are made within the organelles.

Mitochondria and chloroplasts rely on both their own genetic information and that in the nucleus.

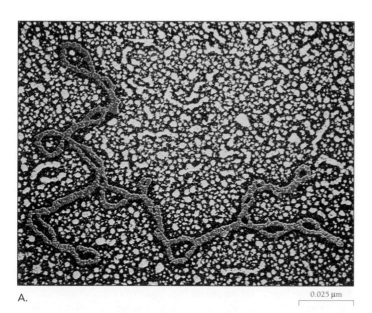

A.   0.025 µm

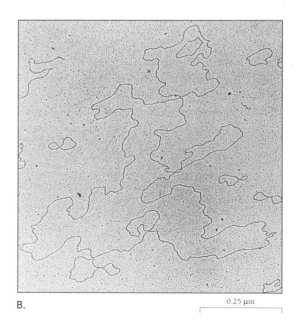

B.   0.25 µm

Figure 12-14   **Circular DNA in energy organelles.** A. From a mitochondrion. B. From a chloroplast. Chloroplast DNA is 10 times longer than that in animal mitochondria. The smaller circular DNA shown is from a virus.   *(A, CNRI/Phototake NYC; B, Dr. Richard Kolodner, Ludwig Institute for Cancer Research.)*

Studies of the DNA and RNA of Energy Organelles Support the Theory That They Derive from Ancient Prokaryotes

Like prokaryotic DNA, the DNA of both mitochondria and chloroplasts is circular. The ribosomes of mitochondria and chloroplasts are also more like bacterial ribosomes than like the ribosomes in the cytoplasm of eukaryotic cells. These and many other similarities between the energy organelles and prokaryotes support the **endosymbiotic theory** [Greek, *endon* = within + *syn* = together + *bios* = life]. This theory holds that present-day mitochondria and chloroplasts are descended from prokaryotes that came to live inside ancient eukaryotic cells. There are many modern examples of cells that host endosymbionts (Figure 12-15).

The genetic code used in mitochondria is slightly different from that used in prokaryotes or in eukaryotic nuclei, with four codon assignments that are different. If mitochondria are descended from free-living prokaryotes, their different genetic code suggests an origin so ancient that it predates the establishment of a universal genetic code. In Chapter 18, we will discuss the origin of mitochondria and chloroplasts in more detail.

The genetic systems of mitochondria and chloroplasts differ from those in both eukaryotic nuclear DNA and in prokaryotes, suggesting that these organelles have an ancient and separate origin.

In this chapter, we have seen that not all genes reside in the chromosomes, nor do all genes pass neatly from generation to generation by means of meiotic divisions. Genes exist in mitochondria and chloroplasts, suggesting that these organelles were once independent organisms. Even more intriguing, however, genes also exist in mobile forms, such as viruses, plasmids, and jumping genes. In the next chapter, we see that mobile genes are the power behind genetic engineering, a new field that may have profound effects on society in the next century.

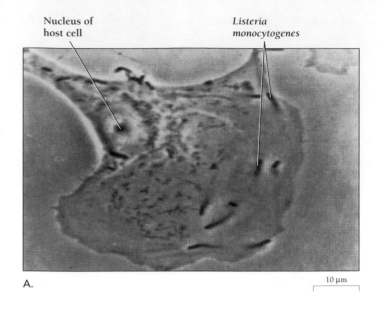

Nucleus of host cell

Listeria monocytogenes

A.   10 μm

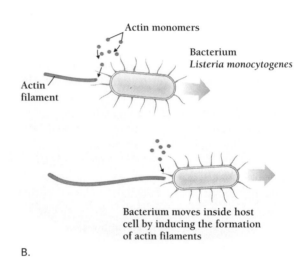

Actin monomers

Bacterium
Listeria monocytogenes

Actin filament

Bacterium moves inside host cell by inducing the formation of actin filaments

B.

Figure 12-15   An intracellular parasite. *Listeria monocytogenes* is a bacterium that infects human cells, including cells in the nervous system and the genital area and the white blood cells of the immune system. It most frequently causes disease in pregnant women and their babies. *Listeria monocytogenes* is remarkable for its ability to move about inside cells with the aid of one to four actin filaments, which work like flagella.   *(A, J.A. Theriot)*

STUDY OUTLINE WITH KEY TERMS

Barbara McClintock showed that some genes can move from place to place. She showed that **jumping genes,** or transposons, cause the curious dappled colors of Indian corn. **Unconventional genetic systems** are DNA or RNA sequences that can replicate and function apart from chromosomal DNA, including the DNA of mitochondria and chloroplasts and mobile genes. **Mobile genes,** which include **viruses, plasmids,** and **transposons,** can move from cell to cell or from place to place within a chromosome. All of these unconventional genes depend on cellular machinery specified by chromosomal DNA.

Viruses, including **bacteriophages,** are assemblages of nucleic acid and proteins that can reproduce within a living cell. Whether shaped like an icosahedron or a long helix, viruses consist of an inner core of nucleic acid surrounded by a protein **capsid** or **membrane envelope.** Viruses were first discovered as disease-causing agents that could pass through filters fine enough to remove bacteria. Viruses have many properties of life: they reproduce, they respond to their environments, and they evolve. However, viruses cannot obtain energy from their surroundings and cannot form the building blocks needed for the synthesis of their components.

Although viruses are not alive, they evolve through natural selection. Viruses are able to take over the machinery of their host cells and to produce more viruses. The growth of viruses in laboratory cell cultures has allowed researchers to study their multiplication cycles. These cycles consist of six steps: (1) attachment to the host cell, (2) entry of viral nucleic acid into the cell, (3) synthesis of viral proteins, (4) replication of viral nucleic acid, (5) assembly of new virus particles, and (6) release of new viruses.

Many viruses kill their hosts in the process of replicating themselves. These viruses, which break open the host membrane in a process called **lysis**, are called **lytic viruses.** Sometimes, viral genes can integrate into the chromosomal DNA of their hosts. The integrated virus, called a **provirus**, does not immediately produce more virus. However, when the host cell replicates, it replicates the viral genes as well, a process called **integrated replication.** Other viruses make new viruses through **autonomous replication.** Actual virus particles are produced only when something induces the provirus to enter a lytic cycle. In a **lysogenic cycle,** a virus or bacteriophage switches back and forth between a **lytic cycle** and an integrated **provirus,** or **prophage** stage.

RNA serves as the genetic material in some viruses. In plus-strand viruses, the viral RNA directs polypeptide synthesis. In minus-strand viruses, an enzyme copies the viral RNA into a complementary strand that directs polypeptide synthesis. In AIDS and other **retroviruses,** **reverse transcriptase** copies viral RNA into DNA, which then integrates into the host cell's chromosomal DNA.

Mobile genes include **plasmids,** which can replicate only autonomously; **transposons,** which can replicate only in integrated form; and **episomes,** which can replicate either autonomously or in integrated form. **Simple transposons** carry nothing but the gene for **transposase,** while **complex transposons** carry genes that benefit their bacterial hosts such as those for antibiotic resistance. Plasmids artificially constructed in the lab are called **recombinant plasmids.**

Plasmids called **resistance factors** or **R factors** contain genes for enzymes that make the cell resistant to antibiotics. Bacteria engaging in **conjugation** exchange genetic material, including both bacterial DNA and plasmids. In eukaryotic cells, some transposons, or jumping genes, move within the chromosomal DNA, disrupting the genome. Bacteria defend themselves against bacteriophages using **restriction enzymes,** which cut DNA at special "target sites."

Viral **oncogenes,** which give cells an overdose of proteins that stimulate cell division, resemble, and almost certainly evolved from, **proto-oncogenes** or **cellular oncogenes** in the eukaryotic host cell.

Energy organelles—chloroplasts and mitochondria—have their own DNA. The circular form of these DNAs and the genes they contain suggest that they derived from ancient prokaryotes that lived within ancient eukaryotic cells, a theory called the **endosymbiont theory.**

REVIEW AND THOUGHT QUESTIONS

Review Questions

1. What characteristics of a virus makes it nonliving?
2. How can a virus or other nonliving genetic entity evolve through natural selection? Give two examples.
3. How does a virus attach to a cell and transfer its genetic material into the cell?
4. What advantages does integrated replication have over autonomous replication?
5. What is a retrovirus? How does it differ from other RNA viruses?
6. What is the difference between a plasmid and a transposon?
7. What evidence suggests that mitochondria and chloroplasts were once prokaryotes that lived independently of other cells?

Thought Questions

8. If you take antibiotics frequently, you will select for *E. coli* that carry genes for antibiotic resistance. Suppose that you contracted tuberculosis (TB) from your roommate. Could plasmids then transmit antibiotic resistance from the *E. coli* in your intestines to the TB bacteria? Why or why not?
9. Many biologists suspect that transposons move genes among different species of eukaryotes. For example, some evidence suggests that within the last 100 years, a parasitic mite has carried transposons called P elements from one species of fruit fly to another. What evolutionary consequences might result from such transfers?

SELECTED READINGS

Garrett, Laurie, *The Coming Plague: Newly Emerging Diseases in a World Out of Balance,* Farrar, Straus & Giroux, New York, 1994. A carefully considered discussion of the deadly new viruses in Africa and the rest of the world. Garrett—a journalist with a graduate degree in immunology—won a Pulitzer Prize in 1996 for her reporting of the Ebola virus.

Keller, Evelyn Fox, *A Feeling for the Organism: The Life and Work of Barbara McClintock,* W. H. Freeman, New York, 1983. A fascinating portrait of a quiet genius.

Preston, Richard, *The Hot Zone,* Random House, New York, 1994. A popular "thriller" account of deadly viruses emerging in Africa.

Radetsky, Peter, *The Invisible Invaders: The Story of the Emerging Age of Viruses,* Little Brown, Boston, 1991. An engaging and highly readable history of viruses and virologists, beginning with Edward Jenner's discovery of a vaccine for smallpox up through modern research into HIV and other viruses.

▶ On-line materials relating to this chapter are on the World Wide Web at http://www.saunderscollege.com/lifesci/
Click on Tobin/Dusheck: *Asking About Life.*

13

The Maverick

Early in the morning of October 13, 1993, reporters and a photographer showed up at the door of 50-year-old unemployed surfer and molecular biologist Kary Mullis (1944—). The Nobel Prize winners for 1993 had been announced in the middle of the night, and all the winners were being interviewed. The reporters gathered around Mullis's beachfront apartment in Southern California and caught him on his way out for a morning of surfing (Figure 13-1). They asked him, Did he know he had won? Was he happy? Surprised? How did he feel when the phone call came from Stockholm, Sweden? Mullis knew—he had already got his call in the small hours of the morning. Naturally, he was happy. "I realized I did something significant, the kind of thing you get the Nobel Prize for," he told reporters, "but never in my wildest dreams did I ever think I'd win."

He wasn't the only one surprised. Explained one colleague to *Esquire* magazine, "He doesn't fit the normal mold of a Nobel-prize winner. . . . He's a wild man. There's a certain amount of politicking involved, and I was not sure his personality and his lifestyle would have allowed him to win." Mullis is indeed a maverick. Eventually, he offends nearly everyone, and he seems to enjoy doing it. At one scientific conference, he began laughing when someone else was speaking. When the speaker chided Mullis for laughing at a serious matter, he responded "I'm not laughing at that. I'm laughing at *you*." Mullis is a far cry from the serious, dedicated researcher who still goes to the lab seven days a week long after retirement. In fact, he hasn't worked in a lab in years.

Mullis, whose personality is equal parts boyish charm and astonishing boorishness, grew up in South Carolina, where, like many a Southern boy, he learned about life by roaming the woods and building rockets from drugstore supplies. At Georgia Tech, he earned a degree and got married for the first time.

In 1966, he arrived at the University of California, Berkeley, with a young wife and baby daughter, to work on a Ph.D. in biochemistry. Berkeley was all he had hoped it would be, and then some. "Six years in the biochemistry department didn't change my mind about DNA," he told UC Berkeley's alumni magazine, "but six years of Berkeley changed my mind about almost everything else." Besides psychedelics such as LSD, Mullis also discovered free-love and divorce. "It wasn't a good place to be married."

By 1981, he was on his third marriage, with two young sons. For a while he worked as the night manager at the

Paraskevas Photography

Buttercup Bakery in Berkeley. When he took a job as a chemist for one of the earliest biotechnology companies, Cetus Corporation, it was as a low-level technician. At Cetus he fell in love with another biochemist and, at 36, ended his third marriage.

foolproof—he couldn't find anything wrong with it. In theory, he should be able to take a test tube filled with thousands of genes, single out a stretch of nucleotides, and make millions of copies within a day. For a molecular biologist, it was undreamed-of power.

"Never in my wildest dreams did I ever think I'd win."

One Friday night in the spring of 1983, the two biochemists headed for Mullis' weekend cabin in the coastal mountains of Northern California. Late that night, as Mullis drove along a twisting mountain road through vineyards and redwoods, he filled the silence by thinking about DNA. Specifically, he was interested in making his job at Cetus—building short pieces of synthetic DNA—a little easier.

As the winding road unfolded before him, he thought about the way the two strands of DNA untwist and separate so that they can be copied. He designed experiments in his head, wondering how he might single out one particular nucleotide in the DNA and make copies of it. He saw ways his imaginary experiments might go wrong, and twisting his thoughts this way and that, fixed them. His idea seemed

He realized with a shock that his idea was almost good enough to win him a Nobel Prize. There had to be something wrong with his idea, he thought. It was too obvious, too slick, too easy. The idea was this: Mullis could make an oligonucleotide, a short piece of DNA, that binds to a complementary sequence in a far longer strand of DNA (Figure 13-2). Once the oligonucleotide was bound to the DNA, it could serve as a primer for DNA polymerase. The DNA polymerase would copy only the DNA immediately adjacent to the bound oligonucleotide. In the next step, the newly made DNA could itself be used as a template to make still another strand (as we discuss later in this chapter). The result would be a chain reaction: one copy of a specific DNA sequence yielding 2 copies, then 4, then 8, then 16, and so on.

Figure 13-1 Kary Mullis, inventor of PCR. *(© 1997 Darryl Estrine Photography)*

Back at Cetus, Mullis's boss told him to take all the time he needed to develop his idea. Colleagues pitched in and helped him make the idea work. By mid-December, Mullis had worked out all the bugs in a new technique called the **polymerase chain reaction (PCR)**. PCR specifically and repetitively copies any segment of DNA between any two defined nucleotide sequences. Today, specialized machines about the size of a toaster oven perform this process automatically, so that a single copy of a chosen sequence can be multiplied into millions or billions of copies in less than a day.

Finding a short sequence of nucleotides in a cell's DNA is like searching for a needle in a haystack. PCR finds the right sequence, then copies it relentlessly until, after 30 rounds of replication, the original haystack is dwarfed by a haystack of needles. Not only does PCR help find particular sequences, it also makes them in usable amounts. For molecular biologists to study a particular nucleotide sequence, they need more than one copy of a sequence; they need thousands of identical copies. Mullis's idea was a singularly valuable technique that would allow hundreds, eventually thousands, of other molecular biologists to make important discoveries.

The uses for PCR seem infinite. PCR enables crime experts to use DNA fingerprinting on mere specks of blood and other tissues to identify who has been present at the scene of a crime. PCR allows researchers to pick out trace amounts of DNA in blood from HIV and other viruses, as well as pick up early signs of cancer and genetic defects in human eggs and embryos. PCR allows archaeologists and paleontologists to study minute amounts of DNA from ancient animals and plants—from the 130-year-old tissues of Abraham Lincoln to an 18-million-year-old magnolia leaf preserved in a peat bog.

Mullis's discovery might have launched his career, even made him happy. But it

did not. Almost as soon as Mullis had perfected the technique, other researchers at Cetus began using the technique to do experiments. Mullis thought his name would be on those papers—as a coauthor—but his colleagues didn't agree. A simple acknowledgment of his invention near the end of the paper seemed like enough to them. Tempers flared, and Mullis was asked to leave if he couldn't get along with his colleagues. In the end, Mullis left Cetus, taking with him a $10,000 bonus for his invention. It was little enough. Cetus made millions of dollars from his invention and ultimately sold the rights to PCR for a mind-boggling $300 million. But even that was a

Figure 13-2   PCR is a simple, powerful technique for multiplying specific sequences of DNA. A. When DNA is heated, the two strands uncoil. They are then cooled and replicated. The cycle of heating, cooling, replicating, and then heating again is repeated until millions or billions of copies of the sequence are obtained. B. Short segments of single-stranded DNA called oligonucleotides act as primers and allow researchers to replicate a particular sequence, not just any DNA. The 20 or so bases of the oligonucleotide pair with the correct segment of the DNA and initiate replication.

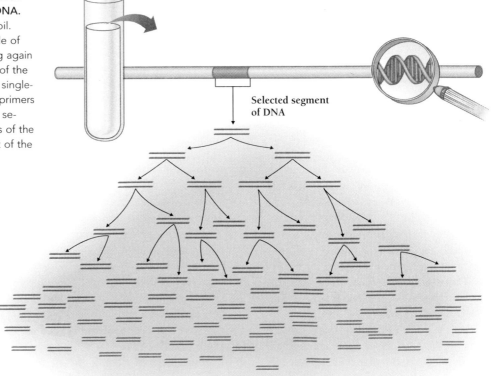

Selected segment of DNA

From one small piece of DNA a million copies can be duplicated in about one hour

A.

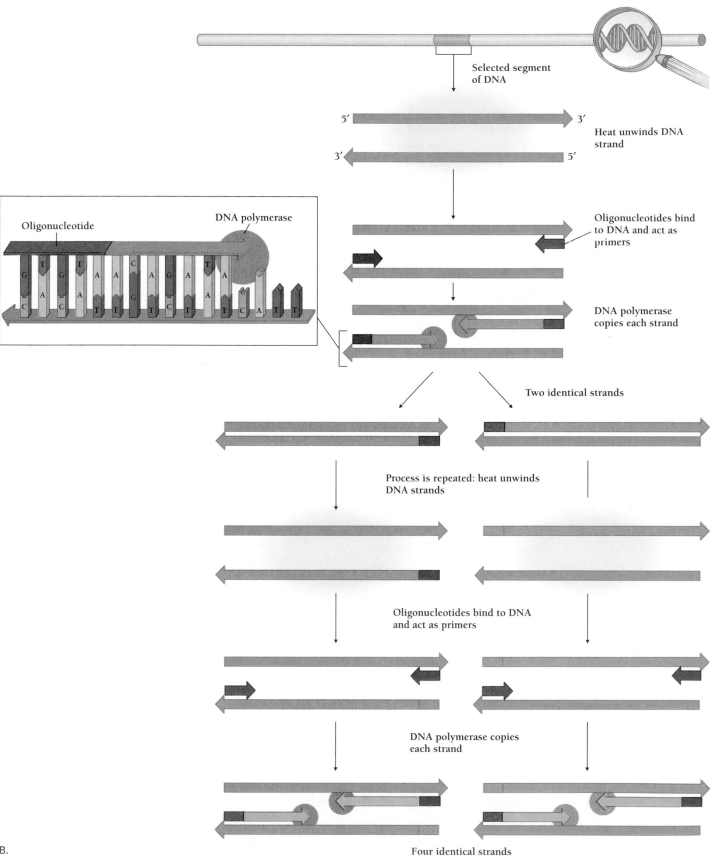

Selected segment of DNA

5′ ——————→ 3′
3′ ←—————— 5′

Heat unwinds DNA strand

Oligonucleotides bind to DNA and act as primers

Oligonucleotide

DNA polymerase

DNA polymerase copies each strand

Two identical strands

Process is repeated: heat unwinds DNA strands

Oligonucleotides bind to DNA and act as primers

DNA polymerase copies each strand

B.

Four identical strands

bargain. Today, PCR technology is a $1-billion-a-year industry.

Except for a brief stint as a consultant, Mullis has mostly abandoned conventional science. Friends and colleagues suggest that his talents are being wasted. Modern molecular biology, however, is as much a social endeavor as an intellectual one, and Mullis' intellectual talents are of little value in an environment where getting along with people is paramount. Doing science, especially doing science that others notice, means fitting in, belonging to one or another "club." A few researchers—loners like Barbara McClintock—do not entirely mind if no one appreciates their discoveries. They enjoy their work so much that they continue in a virtual social vacuum.

Mullis will be starved for neither money nor attention. He eventually won $450,000 from the prestigious Japan Prize and a $412,500 Nobel Prize for his invention of PCR. And although he may not have a desk at a university or his name on a door at a biotechnology company, his Nobel Prize has indirectly provided him with many opportunities for fame and fortune.

KEY CONCEPTS

1. Genetic manipulation depends on choosing organisms with desirable genetic properties from a population that contains natural or induced variation.

2. Recombinant DNA techniques allow the production of relatively large amounts of pure genes and gene products.

3. Recombinant DNAs can reprogram both prokaryotes and eukaryotes to have new properties.

WHAT IS GENETIC ENGINEERING?

The Origin of Genetic Engineering

Thousands of years ago, before cities or farms existed, nomadic hunter-gatherers collected grain and seeds from the plant varieties that yielded the most grain. As they worked with the grain, carrying it or storing it, they spilled some of the grain. In the spring it sprouted and people had only to harvest their crop. In time, they learned to deliberately plant some of their seed in places where it would grow well. They became farmers. Populations that had been nomadic settled in one place to farm.

In time, these first farmers learned to observe closely the individual plants in their fields. Grains from the plants that produced the most seed, the largest seeds, or the best-tasting seeds were saved for the next year's crop. Ten thousand years ago, the first farmers began practicing artificial selection, breeding both plants and animals.

The success of agriculture allowed human populations in the Middle East, northwestern China, and southern Mexico to flourish. As these populations became larger and denser, they built villages and cities. By 5000 years ago, the peoples of the Middle East had mastered enough microbiology to ferment grain into beer and to culture milk into cheese and yogurt.

The early domestication of plants and animals depended primarily on selecting which individuals to breed. Once breeders had chosen desirable varieties of plants or animals, the trick was to preserve and enhance their genetic traits. They did this by allowing individuals with the desired qualities to breed only with each other. Sheep with the thickest wool, for example, were bred to one another. This ancient practice is depicted in the Bible, where Jacob keeps his spotted goats apart from the nonspotted goats of his father-in-law. Brewers and cheese makers likewise learned to select the right strains of microorganisms by starting each batch with the right "starter."

Beer brewing, cheese making, and agriculture itself are early examples of **biotechnology**—the use of living organisms for practical purposes. Modern biotechnology depends as heavily on the selection of organisms for their desirable genetic qualities as the first biotechnology did 10,000 years ago. Genetic engineers can alter the genetic traits of any organism, whether sheep, goats, wheat, corn, yeast or mold (Figure 13-3). The greatest difference between ancient methods and modern ones is the precision and rapidity with which organisms can be altered. But modern genetic engineering also transgresses the natural boundaries among species. Genetic engineers now happily move genes among all kinds of organisms, including humans, mice, tomatoes, yeasts, and bacteria.

Biotechnology first arose 10,000 years ago when humans began selecting and breeding useful plants, animals, fungi, and microorganisms.

A.

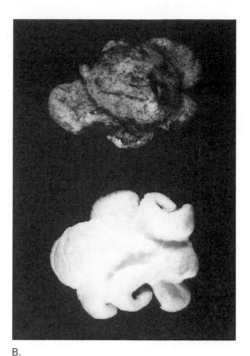

B.

Figure 13-3 Prehistoric corn. A. Cob from a cave in Mexico. Humans have been breeding bigger and bigger corn cobs for thousands of years. Some biologists believe that early corn was bred specifically to be popped. B. A recently popped prehistoric corn kernel compares favorably with the modern version. *(A, Science VU/Visuals Unlimited; B, courtesy of C.P. Mangelsdorf)*

Genetic Engineering Makes Use of Both Natural and Induced Variation

In 1929, Alexander Fleming discovered penicillin as the result of a chance contamination of one of his bacterial cultures. Fleming isolated the mold responsible for killing his bacteria and grew the mold in pure cultures. In 1938, nine years after Fleming's discovery, two Oxford University scientists—Ernst Chain, a refugee from Nazi Germany, and his professor, Howard Florey, an Australian pathologist—realized that penicillin might be able to play a role in controlling diseases caused by bacteria. Within two years, Florey and Chain announced that penicillin, even in very small amounts, could protect mice from an otherwise fatal injection of bacteria.

But humans need far larger amounts of penicillin than mice. To treat just one adult for a bacterial infection, a doctor needs about 7.5 grams of penicillin (the weight of two and a half animal crackers). But to obtain even a millionth of a gram of penicillin, Florey's team had to process enough mold to fill nearly 30 bathtubs. Penicillin was so rare that, in some cases, the precious drug was extracted from the urine of patients and given to them once again in subsequent doses. It was clear that to make penicillin in the amounts needed to control human disease, scientists had to find a way to make much more penicillin. They did this partly by finding better ways to grow the mold, and partly by choosing strains of mold that could produce more penicillin.

To find molds that produced more penicillin, microbiologists searched through, or screened, soil samples from all over the world. The soil sample that produced the most penicillin came from a U.S. Department of Agriculture Laboratory in Peoria, Illinois. Microbiologists, in government laboratories, uni-

versities, and pharmaceutical companies, then increased the genetic variation in the Peoria strain of mold by subjecting it to mutagens—x rays, ultraviolet light, and any chemicals that would cause its DNA to mutate. The changes in the DNA were random and most were useless, but because of the great number of mutations, a few turned out to increase the production of penicillin. After each round of mutagenesis, researchers chose the strains with the highest yield of penicillin. By the end of this selection process, researchers had found a mold that produced nearly 100 times as much penicillin as the original Peoria strain.

Biotechnology selects useful traits from a range of variation. This variation can be natural or induced by means of mutagens.

Knowing Biochemical Pathways Helps Molecular Biologists Design Useful Organisms

To find the strain of mold that could produce the most penicillin, Florey and Chain and the other researchers depended on the same general approach that breeders have always used, screening large numbers of individuals for those with the best expression of a particular trait—in this case penicillin production. They improved their chances by treating the molds with mutagens. However, they had no idea why the Peoria strain produced more penicillin than other strains. In the 1940s, researchers could not predict what mutations would increase the yield of penicillin.

Today, however, molecular biologists know so much about the biochemical pathways in some organisms that they can predict what kind of mutation will produce a desired trait. The genetic engineering of tomatoes provides one example. In order for tomato farmers to get tomatoes safely to the supermarket, they must pick the tomatoes when they are green and hard. If the tomatoes were picked when they were ripe and soft, they would be crushed to a pulp by the weight of the other tomatoes in the box.

Normally, green tomatoes ripen into sweet, flavorful red tomatoes before we eat them. Once the tomatoes are ready to be delivered to a store, they are treated with the plant hormone ethylene gas, which triggers their ripening. The tomatoes also produce ethylene themselves, and, regardless of whether the gas is natural or factory made, the tomatoes ripen just the same. Unfortunately, about 40 percent of tomatoes are picked too soon and can never ripen properly. These are the tomatoes we all know so well that turn red but taste like wet cardboard. Only tomatoes that have developed different "flavor components," chemicals that make them taste good, are actually ready to be picked. But it is impossible for the grower to tell the green tomatoes that are ready to be picked from those that are not. All green tomatoes look the same.

One biotechnology company solved this problem by deliberately damaging a gene that controls the production of ethylene in the tomato. Such tomatoes can be left on the vine until they turn a pinkish orange, which signals that they have developed their flavor components. These tomatoes will never soften (or turn red and sweet) on their own, for they cannot make ethylene. Only after the tomato grower has harvested the tomatoes and the tomato distributor has treated the tomatoes with ethylene will they ripen. Since the tomatoes do not need to be picked until they have both color and flavor components, growers won't pick them too soon.

Another solution, from a different biotech company, is to suppress the expression of a gene for an enzyme called polygalacturonase that softens ripening tomatoes. Tomatoes in which polygalacturonase is suppressed can also be picked when pink, because although they ripen normally otherwise, they do not soften. These tomatoes stay hard long enough to survive the trip to the supermarket.

Biotech companies have also used specific knowledge of biochemical pathways to select hundreds of commercially important strains of bacteria and fungi. Some of the molecules produced by these strains include amino acids such as glutamic acid (used as the flavoring monosodium glutamate or MSG), vitamins such as riboflavin, and even industrial chemicals such as ethanol and acetone. Microorganisms also produce commercially important enzymes, such as those used to convert starch into fructose for use as a sweetener and the protein-digesting enzymes used in laundry detergents.

Molecular biologists can design useful organisms by inserting or destroying genes that code for proteins involved in specific biochemical pathways.

WHAT IS RECOMBINANT DNA AND HOW IS IT USEFUL?

The discovery, in the early 1950s, that genes are nothing more than chains of nucleotides led directly to our present ability to both alter individual genes and move them around wherever we want them. By the early 1970s, molecular biologists had learned to cut and paste pieces of DNA from different organisms (as well as from various mobile genes), and by the early 1980s, they had learned how to alter the individual nucleotides in a gene. Earlier work had already revealed how viruses and plasmids shuttle genes around the genomes of bacteria and other host cells. Molecular biologists found it easy to use these tiny molecular parasites to carry along and reproduce pieces of DNA that had been patched together in a test tube.

How Do Molecular Biologists Use Recombinant DNA?

Recombinant DNA is a DNA molecule consisting of two or more DNA segments that are not found together in nature. For example, researchers can join a gene for antibiotic resistance with a plasmid, then infect a cell with the plasmid, giving the cell a gene for antibiotic resistance. Because molecular biologists can now design and make recombinant DNA that fulfills particular requirements—scientific, medical, or commercial—the resulting recombinant DNAs are sometimes called "designer genes."

Recombinant DNA technology is so powerful that it is difficult to summarize all the ways in which it has proved useful. Recombinant DNA has provided scientists with an irreplaceable tool for studying the structure, regulation, and function of individual genes; for unraveling the molecular bases of genetic diseases; and for turning plants, animals, and microorganisms into chemical factories that can churn out vast quantities of proteins and other substances that these organisms would never make on their own (Figure 13-4).

Recombinant DNA promises purer vaccines (which have fewer risks), new drugs, gene therapy for treating genetic diseases such as cystic fibrosis, and genetic tests for hereditary diseases such as Huntington's disease or partly hereditary diseases such as breast cancer. Recombinant DNA means brand new proteins, oils, and carbohydrates—designed by molecular biologists and manufactured by bioengineered organisms. Finally, recombinant DNA presents a thousand moral and ethical questions that cannot be answered solely by the scientists who raise them. These questions must be understood and answered by all educated citizens.

The development of recombinant DNA technology depended on the union of many different lines of research, including work on (1) restriction enzymes in bacteria, (2) DNA replication and repair, (3) replication of viruses and plasmids, and (4) the chemical synthesis of specific nucleotide sequences. We will see how information and techniques from these different fields all contributed to the ability to make recombinant DNAs.

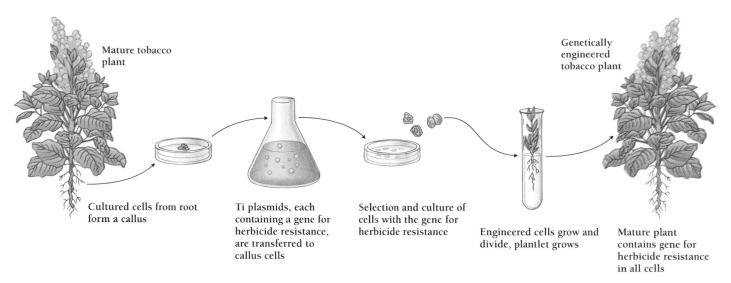

Mature tobacco plant

Genetically engineered tobacco plant

Cultured cells from root form a callus

Ti plasmids, each containing a gene for herbicide resistance, are transferred to callus cells

Selection and culture of cells with the gene for herbicide resistance

Engineered cells grow and divide, plantlet grows

Mature plant contains gene for herbicide resistance in all cells

Figure 13-4 Recombinant DNA technology in action. Cells from a tobacco plant are grown in culture, then infected with a Ti plasmid carrying a gene for herbicide resistance and other genes. The growing cells are treated with herbicide and only the cells expressing the new genes survive. These herbicide-resistant cells can then be grown into mature plants, which will bear seeds carrying the new genes.

Recombinant DNA techniques join together pieces of DNA in combinations that do not occur in nature.

How Do Restriction Enzymes Cut Up a Genome?

DNAase and RNAase are enzymes that cut the links between the nucleotides in DNA or RNA, respectively. DNAase usually cuts DNA into short fragments at random (Figure 13-5A). In a test tube filled with DNA, each of the many copies of a genome will be cut differently. The resulting fragments are all cut into different sizes and the pieces from one genome overlap those in another. In such a mess, there is little hope of finding a specific sequence. To isolate a specific piece of DNA, a researcher must instead use **restriction enzymes**, special DNAases that cut DNA only at particular sequences (Figure 13-5B).

Happily, such enzymes exist. Bacteria use restriction enzymes to defend themselves against viruses and other mobile genes. When a bit of alien DNA enters a bacterium, the restriction enzymes cut it to bits. Bacteria have many kinds of restriction enzymes, each of which recognizes a different nucleotide sequence. The sequence recognized by a given restriction enzyme is called a **restriction site** and typically consists of four or six nucleotides, though some restriction sites are longer. Some restriction enzymes cut each of the two strands of DNA in the same place, so that the resulting fragment of DNA has "blunt" ends (Figure 13-5B). Other restriction enzymes make staggered cuts, which result in single-stranded, "sticky" ends that easily join with other such fragments that have complementary sequences (Figure 13-5C).

Researchers have prepared restriction enzymes from over 200 different bacterial strains. Each enzyme recognizes a specific sequence, with over 90 such restriction sites now described. One strain of *E. coli,* for example, uses a restriction enzyme, called *EcoRI,* that binds to and cuts DNA wherever it finds the sequence GAATTC (Figure 13-5C). A restriction enzyme such as *EcoRI* will cut multiple copies of DNA everywhere a particular sequence appears. The result is a matching set of **restriction fragments**—pieces of DNA that begin and end with a restriction site.

By comparing the sizes of restriction fragments researchers can establish a **restriction map**, which shows how the restriction sites are placed within a piece of DNA. More important, genetic engineers can join these fragments together in novel ways. A restriction enzyme that makes fragments with sticky ends can be used to make human and mouse fragments that stick together (Figure 13-6A). They can, for example, join a restriction fragment from a human being to a fragment from a mouse or a bacterium.

Restriction enzymes, which bacteria use to defend themselves against viruses, cut DNA molecules at specific sequences. Restriction enzymes allow the preparation of DNA fragments with defined lengths and sequences.

How Do Molecular Biologists Join Restriction Fragments Together?

Recall from Chapter 10 that during ordinary DNA replication, the enzyme DNA ligase links together the separate pieces of DNA into one continuous strand. Bioengineers depend on DNA

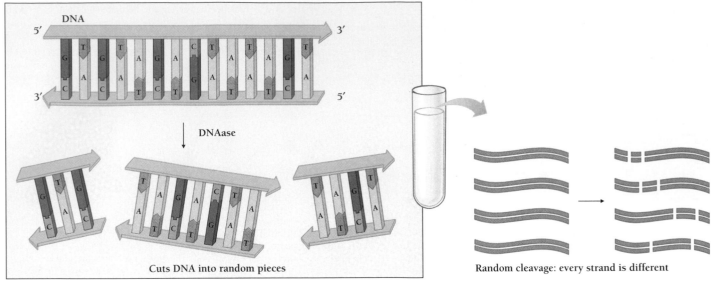

A.

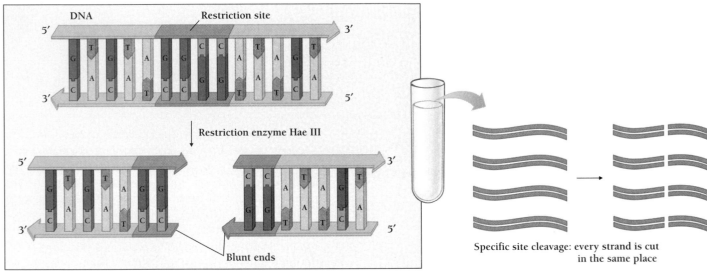

B.

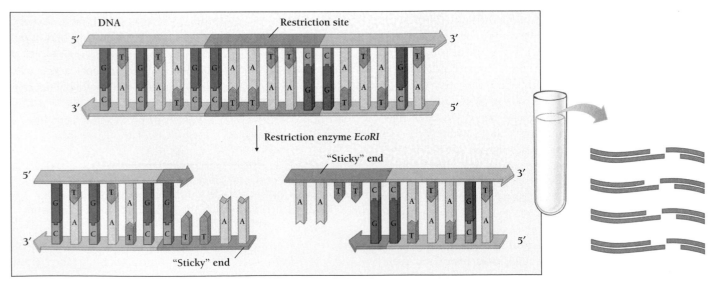

C.

Figure 13-5 Cutting up DNA. A. DNAases cut DNA at random. Each copy of DNA is cut differently. B. Restriction enzymes cut each copy of DNA at the same places. Matching fragments have the same length and sequence. Restriction enzymes can cut the DNA with blunt ends, as in B, or with sticky ends, as in C.

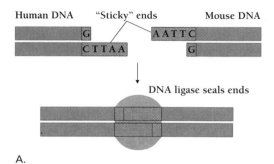

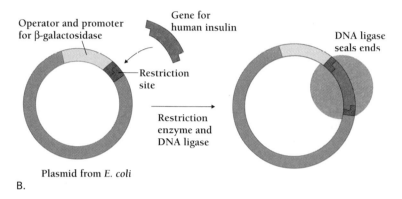

Figure 13-6 **Genetic engineering.** A. Sticky ends enable researchers to join pieces of unrelated DNA. B. DNA from a mouse and human can be joined, as long as they have been cut at the same sequence by the same restriction enzyme.

ligase to link restriction fragments together. For example, to make the hormone insulin, researchers initially inserted a sequence coding for human insulin next to the operator and promoter for β-galactosidase from *E. coli* (Figure 13-6B). The manufactured recombinant DNA then contained one piece of a bacterial regulatory gene and one piece of a mammalian structural gene. This recombinant DNA, put back into *E. coli*, produces insulin when the bacteria are exposed to the milk sugar lactose. Ordinarily, the presence of lactose would signal the bacteria to prepare to digest lactose. But bacteria carrying the recombinant DNA respond to lactose by producing human insulin.

DNA ligase can link together two pieces of DNA from different sources to produce recombinant DNA.

How Do Molecular Biologists Express Recombinant DNA in Bacteria and Other Hosts?

Once scientists taught themselves to make recombinant DNA using DNA ligase and restriction fragments, they faced two serious challenges. The first challenge was to produce large numbers of particular genes. The second was to induce bacteria or other host cells to express these recombinant genes as usable proteins.

How Can Bacteria Be Induced To Make Great Numbers of Copies of a Gene?

The answer to the first problem (production of large numbers of specific genes) was plasmids, which are capable of forcing a bacterial cell to make large numbers of copies of a single gene. Recall from Chapter 12 that R (resistance) factors are naturally occurring plasmids containing genes for antibiotic resistance. By the early 1970s, researchers had isolated and studied many such R factors. Using restriction enzymes and DNA ligase, they made streamlined plasmids that contained only two parts: (1) an origin of replication—the DNA sequence needed to start DNA synthesis; and (2) one or more genes for antibiotic resistance. The antibiotic resistance acted as a marker, allowing researchers to distinguish and select bacteria that carried the plasmid from those without the plasmid.

Using DNA ligase, researchers can hook any gene to such a plasmid. The recombinant DNA plasmid ensures that the gene gets copied and transcribed inside a bacterial cell. Bacteria will copy essentially any DNA sequence as long as it is not too long. Researchers then grow the bacteria in an antibiotic. Because the plasmid's antibiotic resistance gene protects the host bacteria against being killed by the antibiotic, only the genetically engineered bacteria survive.

Microbiologists have long used the word **vector** [Latin = carrier, rider] for any organism or virus that spreads disease from organism to organism. Bioengineers therefore use the word vector to describe anything that spreads genes from organism to organism. Plasmids are not the only vectors. Researchers have also used specially modified bacteriophages, animal and plant viruses, and transposable elements.

Vectors are DNA sequences from viruses, plasmids, and other mobile genes that are capable of carrying recombinant DNA into cells.

How Can Bacteria Be Induced To Make Eukaryotic Genes?

If a researcher wants to clone a gene from a virus or a bacterium, pulling out a particular gene for cloning is not too hard. Often enough, the entire sequence is already known. Extracting genes from the chromosomes of plants and animals, however, is not only more difficult, it is often useless. Recall that a eukaryotic gene is interrupted by long sequences of introns. Bacteria, which must express the gene, do not have any machinery for recognizing and removing introns from the resulting pre-messenger RNA. (Prokaryotes, recall, have no introns.)

If the bacteria cannot remove the introns, they cannot make a translatable mRNA. So how can researchers acquire a clean, intronless copy of a eukaryotic gene that can be cloned using bacteria?

The solution comes from retroviruses, the viruses that use RNA as their nucleic acid, yet insert themselves into the DNA genomes of eukaryotes. Recall that retroviruses use the enzyme **reverse transcriptase** to copy RNA into DNA. Reverse transcriptase copies almost any RNA into DNA. Researchers who want a particular gene can start with the mature mRNA for that gene, which has already had all the introns removed. (The mature mRNA comes from eukaryotic tissues that express the desired gene.) Reverse transcriptase then copies that mRNA back into DNA. The result is a DNA molecule, called **complementary DNA** or **cDNA,** which is complementary to the mRNA from which it was copied. Unlike the DNA in the genome, cDNA has no introns as it is a copy of the mature mRNA.

Another enzyme copies the single-stranded cDNA into double-stranded cDNA, the second strand of which is identical in sequence to the original mRNA. The cDNA "gene" is now ready to be joined to a vector and cloned.

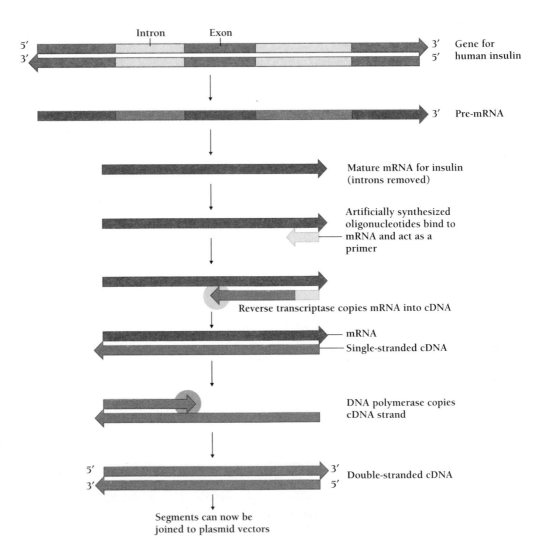

Figure 13-7 How do researchers derive human genes without introns? In a human cell, the gene for human insulin is transcribed into pre-mRNA. The cell edits out the introns to form mature mRNA. Researchers then isolate the mature insulin mRNA and copy it into DNA, which can be joined to a plasmid vector and expressed by a bacterium.

Intron   Exon
5′   3′   Gene for
3′   5′   human insulin

3′   Pre-mRNA

Mature mRNA for insulin
(introns removed)

Artificially synthesized
oligonucleotides bind to
mRNA and act as a
primer

Reverse transcriptase copies mRNA into cDNA

mRNA
Single-stranded cDNA

DNA polymerase copies
cDNA strand

5′   3′   Double-stranded cDNA
3′   5′

Segments can now be
joined to plasmid vectors

Actual genes from eukaryotes cannot be expressed in bacteria. Molecular biologists must make a DNA copy (called cDNA) of a eukaryote's mRNA, using reverse transcriptase. The cDNA gene can then be cloned using ordinary methods.

Can Host Cells Be Induced To Express Polypeptides in a Usable Form?

Unfortunately, not all eukaryotic genes can be expressed in bacteria. As we saw in Chapter 11, many genes require extensive modification of the polypeptide itself in order to function. Bacteria cannot perform these modifications. Most membrane proteins, for example, require modifications that can only be made in eukaryotic hosts (Figure 13-7). Recombinant *E. coli* designed to produce insulin, for example, cannot produce insulin that will function properly in the human body. Nonetheless, hundreds of functional proteins have been expressed from recombinant DNAs in bacteria.

Proteins that need to be modified after translation cannot be expressed in bacteria.

How Do Researchers Make Multiple Copies of Recombinant DNA?

As we mentioned above, researchers often like to have many copies of the gene itself, not just its product. They need enormous numbers of copies of a gene in order to work out the sequence of a gene, to detect mutations, or to study the way proteins interact with genes to influence gene expression.

For researchers to sequence a gene, for example, they need at least a few milligrams of the gene in an extremely pure state. Yet, the entire human body only contains about 30 millionths of a gram of any given gene. So even if researchers could extract and purify a particular gene from a human body—itself an insurmountable problem—they would need a thousand human bodies for each gene. In short, until the advent of recombinant DNA technology researchers had no source for the quantities of DNA they needed.

Throughout the 1970s and early 1980s, the only practical way to make enough copies of a particular DNA sequence to study it was to introduce a single recombinant DNA molecule (gene plus plasmid vector) into a bacterial host cell. The plasmid can induce the host cell to make up to a thousand copies of the gene it carries. In addition, researchers induce the bacterial host cell to divide rapidly. As the bacteria multiply, so does the recombinant DNA. Each colony of bacteria may grow into 10^{12} or more bacteria. If each one has 1000 (10^3) copies of the plasmid, such a colony may contain 10^{15} identical copies of the recombinant DNA. Researchers call this process "cloning the recombinant DNA."

The word "clone" comes from gardening. Gardeners and fruit growers often grow whole new plants from live twigs cut from mature trees and bushes. All of the plants derived in this way are genetically identical and constitute a **clone** [Greek, *klon* = twig]—a group of genetically identical cells or organisms. In humans, identical twins are natural clones. Any group of genetically identical individuals constitutes a clone. Similarly, the offspring of a single bacterial cell are called a clone, and the many copies of a plasmid are called a clone. By extension, molecular biologists also refer to the many copies of recombinant DNA as a clone.

Cloning recombinant DNA depends on the DNA replication machinery of bacteria or other cells to copy recombinant DNA. Since Kary Mullis' invention of PCR, however, researchers have been able to make millions, or even billions, of copies of DNA sequences in a test tube, using a DNA polymerase purified from a bacteria. PCR can make huge numbers of individual DNA sequences very rapidly. The sequences must be short, usually no more than 20,000 nucleotide pairs, but the process is much faster than cloning.

PCR depends on the ability to make two specific polynucleotides that correspond to sequences at each end of the sequence to be copied. These two synthetic polynucleotides are typically each about 20 nucleotides long and are examples of **oligonucleotides** [Greek, *oligos* = few]. Each oligonucleotide serves as a primer for DNA polymerase. No enzyme is needed to unwind the DNA. Instead, in PCR, the DNA is simply heated until the two strands come apart. The DNA is cooled, and the two primers then bind to the complementary sequences—one on each strand (Figure 13-2). DNA polymerase then copies each strand until the experimenter (or the machine) stops the reaction by again raising the temperature.

The increased temperature also separates the template strand and the newly made complementary copy. Lowering the temperature then allows DNA polymerase to go back to work, again starting with the oligonucleotide primers. The only region that will be copied—again and again with further cycles—is the segment flanked by the two primers. The result is the rapid and exclusive multiplication of one specific segment of DNA.

Cloning and PCR technologies allow researchers to make thousands or millions of copies of individual genes.

How Do Biologists Find the Right DNA Sequence in a Recombinant DNA Library?

The genome of any organism—whether bacterium, plant, or animal—can be cut into pieces with restriction enzymes. The collection of thousands or millions of restriction fragments from a single genome is called a **gene library.** These fragments can be combined with plasmid vectors and introduced into populations of bacteria, which can then replicate each fragment in the library indefinitely. If the library contains at least one representative of every sequence in a genome, it is called "a complete genomic library." Alternatively, a recombinant DNA library

BOX 13-1

Can biologists clone dinosaurs?

Scary as *Tyrannosaurus rex* must have been, few people would pass up a chance to see the great flesh-eating dinosaur alive. Is it possible that DNA technology will someday bring dinosaurs back from extinction to populate theme parks everywhere? Can we scrape together bits of prehistoric DNA from fossils, stuff them in a big egg, and watch a baby brontosaurus poke through the shell?

The answer is no. Researchers cloned frog embryos in the early 1950s, and cloning adult animals became reality in 1997 with the birth of Dolly, a lamb whose genetic makeup was taken from the mammary tissue of an adult ewe. But the creation of a living organism from scraps of ancient DNA is an entirely different problem.

The first challenge for would-be dinosaur cloners is to find the genetic material of a race of animals that died out 65 million years ago. Little more remains of the ancient beasts than fossilized bones, in which minerals have replaced the organic material, essentially turning the bone into rock. Fossils therefore contain little or no DNA. Blood from ancient mosquitoes preserved in amber may be a more likely source of Jurassic genes. But tapping into that resource raises other questions: What animal provided the insect's last meal? Did mosquitoes even bite dinosaurs? And if they feasted on several different species, which DNA in the insect's belly came from whom?

Even if scientists found a reliable source of dino DNA, a complete dinosaur genome would be nearly impossible to recover. For one thing, many scientists believe that DNA is not likely to survive intact for so long. One study has shown that

DNA in a water solution will break down almost entirely into individual nucleotides after about 50,000 years. Still, under the right conditions, DNA can last longer. In 1990, geneticists successfully extracted DNA from a 17-million-year-old fossilized magnolia leaf dug up in Idaho. Two years later, researchers used PCR to amplify short gene fragments from a termite and a stingless bee, both encased in amber for 30 million years.

Attempts to recover dinosaur genes have been less successful. In 1993, scientists in Montana found blood cells in a *T. rex* bone that had only partly fossilized, thus protecting the central cavity from the elements. Unfortunately, the only sequences isolated from the sample later turned out to be human DNA contaminating the specimen. Other labs have had similar contamination problems in trying to squeeze dino DNA from ancient bones. Nevertheless, some researchers remain confident that PCR will eventually produce authentic dinosaur gene fragments. But even then, those few hundred bases will be a far cry from the millions and possibly billions of bases of sequence that once resided in dinosaur chromosomes.

Suppose by some miracle researchers got their hands on the complete *Tyrannosaurus rex* sequence. A much bigger hurdle would still await them: how to bring the DNA to life. As we saw in Chapter 11, chromosomes are coated with proteins that regulate the expression of their genes. Many of those proteins bind to very specific DNA sequences in key locations in the code. Without the right proteins, the chromosome is effectively dead.

In addition, numerous other proteins that control genes, modify and transport RNA, replicate DNA, initiate cell division, and regulate cell growth reside in the nucleus separate from the chromosomes. Although many of these proteins look very similar in organisms as different as yeast and humans, most have subtle and important differences even in closely related species. Unfortunately, the proteins that once gave life to dinosaur DNA are long gone.

The lack of an appropriate egg is yet another nail in *T. rex*'s coffin. Even in the unlikely event that scientists discovered a cache of perfectly preserved dinosaur chromosomes, they would need to place them in an egg from some suitable modern animal. Whose egg would they try?

If dinosaurs lie on the evolutionary pathway to modern birds, as many biologists theorize, then the ostrich egg might not be a bad one to start off with. But an egg is more than just a protective coating for DNA. Eggs contain information about how cell division should proceed, or how the embryo will be oriented (which side is up, for example). It seems improbable that an ostrich egg, or any modern egg, would resemble a dinosaur egg closely enough to support the growth of a baby dinosaur, even with the proper genetic instructions.

Cloning living animals is at least technologically feasible. And now that successful clones of sheep have been made, successes in other animals, including humans, may not be far off. But for better or worse, the great dinosaurs are gone for good.

may be constructed from mRNA alone (using reverse transcriptase), in which case it is called "a cDNA library." In either case, a gene library is useful only if researchers can find the gene they want. Until recently, isolating the right gene was like trying to find a needle in a haystack.

Molecular geneticists have developed techniques for finding and cataloging the tens of thousands of sequences in a gene

library. Just as a magnet can help find a needle even in a haystack, molecular biologists can use two kinds of molecular "magnets" to find particular pieces of DNA. One kind of magnet is a **probe,** a short bit of single-stranded DNA whose sequence is complementary to a small part of the sequence in the gene researchers are looking for, much like the DNA primers used in PCR. The other magnets are **antibodies,** proteins from

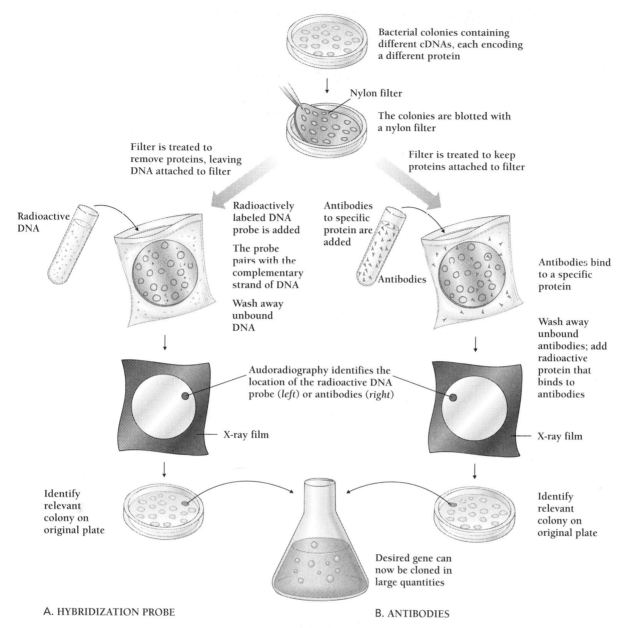

Bacterial colonies containing different cDNAs, each encoding a different protein

Nylon filter

The colonies are blotted with a nylon filter

Filter is treated to remove proteins, leaving DNA attached to filter

Filter is treated to keep proteins attached to filter

Radioactive DNA

Radioactively labeled DNA probe is added

The probe pairs with the complementary strand of DNA

Wash away unbound DNA

Antibodies to specific protein are added

Antibodies

Antibodies bind to a specific protein

Wash away unbound antibodies; add radioactive protein that binds to antibodies

Audoradiography identifies the location of the radioactive DNA probe (*left*) or antibodies (*right*)

X-ray film

X-ray film

Identify relevant colony on original plate

Identify relevant colony on original plate

Desired gene can now be cloned in large quantities

A. HYBRIDIZATION PROBE

B. ANTIBODIES

Figure 13-8 Two techniques for locating a gene. A. A hybridization probe locates a specific DNA sequence. B. Antibodies locate the protein product of the same sequence.

the immune system that recognize and bind to other proteins. As we will see, antibodies do not detect the recombinant DNA itself, but only the proteins specified by the DNA.

A probe, also called a **hybridization probe,** can hybridize by base-pairing with DNA in the library to form **hybrid DNA**— DNA whose two complementary strands come from different sources (Figure 13-8A). Such a probe, dropped into a test tube filled with thousands of restriction fragments, can locate and pair with the one fragment whose sequence complements its own as easily as a bloodhound can track down a lost child. Once the probe binds to the right piece of DNA, it is easy to locate the probe if it has been chemically or radioactively la-

beled. The technique of hybridizing a labeled DNA probe in order to find the right piece of recombinant DNA in a gene library is called molecular hybridization.

A probe is easy to make, since molecular biologists usually already know either some part of the sequence of the gene they want, or else they can guess the sequence from part of the amino acid sequence of the protein. As little as 9 bases are all that are required to identify 95 percent of human genes. To screen an entire library requires thousands of copies of a probe, but PCR makes replicating probes easy. Probes consist of hundreds or thousands of nucleotides. In general, the longer the probe, the more accurate it is.

Researchers who know what protein they are looking for can also use antibodies to find the gene for that protein (Figure 13-8B). Antibodies are proteins with surfaces that recognize the shapes of foreign molecules, thereby giving animals an important defense against infection. When an animal is exposed to a foreign protein, it responds by making antibodies that bind specifically to that protein. Researchers start with a complete cDNA library laid out in a half million colonies of bacteria, each colony carrying and expressing a different cDNA "gene" from the human genome. Over a few days, the 500,000 colonies can be treated with labeled antibody, which binds to and identifies the colony that is synthesizing the protein and therefore carries the gene the researchers are looking for. From the bacterial colony carrying the correct gene, researchers can clone the gene for study.

Two kinds of molecular probes allow biologists to locate specific genes. Hybridization probes detect genes in recombinant DNA clones, in cell extracts, and in cells. Antibodies detect the synthesis of specific proteins in colonies of bacteria containing recombinant DNA.

RECOMBINANT DNA CAN REPROGRAM CELLS TO MAKE NEW PRODUCTS

The ability to manipulate genes has already had important practical consequences. Hundreds of biotechnology, chemical, and pharmaceutical companies have begun to use recombinant DNA to make proteins useful to medicine, to agriculture, and to the chemical industry. Recent products include newly engineered forms of well-known proteins (such as an improved version of an enzyme used in laundry detergents), as well as previously unknown proteins—such as the protein missing in the lungs of cystic fibrosis patients or the protein growth factors that may prevent nerve death in various neurological diseases. In addition, genetic engineers are breeding plants and animals with new characteristics—plants that are resistant to insects or to particular herbicides, sheep that produce a clotting factor in their milk, and pigs that produce human hemoglobin. Such gene products are then purified and used in medicine.

Genetically Engineered Bacteria and Eukaryotic Cells Can Make Useful Proteins

Recombinant DNA technology allows the reprogramming of cells to make an extraordinary number of products. These products include insulin, growth hormone, and other protein hormones, as well as ingredients for processed foods and enzymes needed to produce valuable small molecules or to destroy pollutants (Figure 13-9).

Vaccines are another protein product that can now be produced by means of recombinant DNA technology. Recall that viruses can only grow inside host cells. In the first half of this century, virologists learned to grow infectious viruses in tissue culture cells or in chick embryos. But to produce a vaccine, they had to inactivate (or "attenuate") the virus so that it would not cause disease but would produce an immune response, the production of antibodies to the virus. Even such inactivated vaccines sometimes cause illness, however, either because of the presence of a small amount of virus that remained active or from impurities in the vaccine. In addition, proteins from the cells in which the virus grew could cause an allergic reaction.

Figure 13-9 Effects of human growth hormone (HGH). HGH cures certain forms of dwarfism. Formerly HGH had to be isolated from human cadavers and recipients of the growth hormone too often contacted hepatitis and other viral diseases. Today, HGH is produced by genetically engineering bacteria and is free of viruses that infect humans. *(© Jerry Cooke/Photo Researchers)*

Recombinant DNA techniques allow the preparation of un-contaminated vaccines because the immunizing molecules are made from viral DNA rather than from active viruses. Instead of growing the virus and then inactivating it, virologists isolate a piece of viral DNA that encodes a single viral protein. A single protein cannot cause a viral disease, but it can provoke an immune response that will guard against infection by the actual virus. Recombinant DNA technology has produced safer vaccines for many kinds of disease-causing viruses. Researchers hope that one day such techniques may also produce vaccines for diseases such as herpes and AIDS, for which no effective vaccines now exist.

Some of the protein products of recombinant DNA are so rare and so hard to isolate that little was known of them before they could be produced from recombinant DNAs. **Growth factors,** for example, are rare proteins that keep specific kinds of animal cells alive, stimulate cell division, or trigger particular types of cell specialization. Until recently, researchers knew little about the structure of growth factors or how they work. Recombinant DNA techniques, however, have produced relatively large amounts of pure growth factors, allowing researchers to study their structure and mode of action. Many academic and commercial laboratories are now investigating the possibility that growth factors may be able to prevent nerve degeneration in such devastating diseases as Alzheimer's disease, Parkinson's disease, amyotrophic lateral sclerosis (ALS, also called Lou Gehrig's disease), and Huntington's disease.

Genetically engineered bacteria and eukaryotic cells can make useful proteins, including enzymes, vaccines, drugs, and human proteins for treating genetic diseases.

Recombinant DNA Technology Allows the Production of Novel Proteins

Not all of the new products of recombinant DNA technology are familiar proteins. Some are novel proteins designed by researchers to accomplish new tasks. These range from enzymes with altered properties to specific antibodies to signaling molecules designed to act on particular cells. An especially promising line of research is the attempt to modify antibodies in such a way that they can serve as enzymes. These engineered enzymes, called "abzymes," are as specific as antibodies, but may serve as enzymes. For example, abzymes may be designed to destroy whatever molecule they attach to. Researchers dream of making abzymes that will break down the many synthetic chemicals that now pollute our environment because nothing in nature can break them down.

Recombinant DNA technology can produce new proteins that are not present in nature.

Gene Therapy: Products of Recombinant DNAs Can Be Released Directly into the Body from Engineered Somatic Cells

Usually, the products of recombinant DNAs are made in laboratories or factories, then purified, packaged, and distributed in the same way as other products. A number of laboratories and companies, however, are now experimenting with a new way of delivering recombinant DNA products. They are engineering the cells of a living person to make and deliver needed proteins. This infant technology is called **gene therapy.**

One strategy is to take cells out of the body, engineer them to produce the desired product, and then to implant them back into the body. One example comes from efforts to "cure" a form of hemophilia, the genetic disease in which the blood fails to clot because of the absence of a protein called a clotting factor. Currently, this clotting factor is isolated from human blood and injected into the patients who need it. Treatment costs as much as $100,000 a year, however, and the patient runs the risk of becoming infected by hepatitis and other viruses present in the blood from which the clotting factor is prepared. Researchers are experimenting with inserting the gene for clotting factor into cells that can then be implanted in the person with hemophilia. That way the body makes its own clotting factor. In one version, researchers extract the cells that line the blood vessels, insert a gene for clotting factor, then reintroduce the engineered cells into the lining of the blood vessels.

Other approaches include attaching a needed gene to a virus and engineering the virus to infect either the whole body or certain organs (Figure 13-10). To treat cystic fibrosis, the genetic disease that disrupts the flow of fluids in and out of cells, for example, researchers are attempting to introduce a recombinant gene into the body using cold viruses as vectors that carry the missing protein into the lungs.

These are just two of numerous approaches that researchers are now experimenting with. Gene therapy in humans remains controversial, however. Skeptics suggest that much of the hype associated with gene therapy, as reported in newspapers and magazines, is a result of the financial interests of biotechnology companies, which are strongly motivated to make experimental technologies seem more immediately practical than they actually are. Nonetheless, the techniques being developed for gene therapy are potentially valuable. As a way of treating genetic diseases, gene therapy is theoretically unrivaled. The question is, When will gene therapy be practical and for what kinds of diseases? Equally important, the techniques now being developed will be valuable tools for biologists studying how organisms work.

Gene therapy as a treatment for genetic diseases is still highly experimental, but the techniques being developed may be invaluable.

Figure 13-10 Engineering cells that line blood vessels to make a clotting factor.

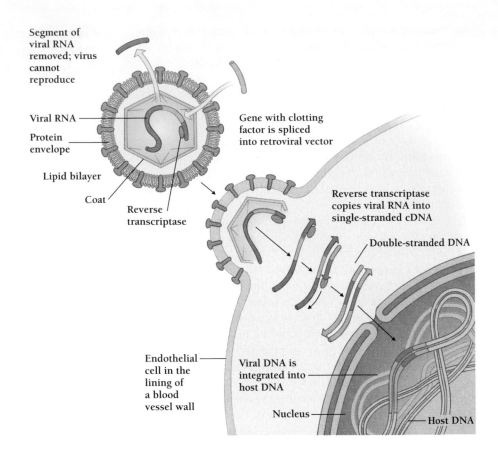

RECOMBINANT DNA CAN GENETICALLY ALTER ANIMALS AND PLANTS

Recombinant DNA can be inserted into the cells of whole plants and animals. Organisms that carry recombinant DNA in their genomes are called **transgenic organisms.** The added DNA is called a **transgene.**

To be expressed in a higher organism, a transgene must contain the appropriate control signals. To identify such regulatory regions, researchers usually join together a segment of DNA that is suspected of being regulatory with a protein coding gene whose expression can be easily determined.

How Do Researchers Produce a Transgenic Mammal?

For a gene to be expressed in all the appropriate cells of an animal, researchers must put the transgene into the zygote before the beginning of embryonic development. Then all the cells of the organism will contain the engineered DNA.

In the early 1980s, several research groups succeeded in producing transgenic mice. Perhaps the most dramatic of these early experiments produced mice carrying the gene for human growth hormone. At birth, such transgenic mice were already twice the size of their litter mates (Figure 13-11A).

To produce a transgenic mammal, researchers must surgically remove eggs from the ovaries in the abdomen of the female. As a result, not very many eggs are available at one time. These few eggs are put into a dish and fertilized *in vitro* with sperm from a male. The resulting zygote is then injected with the engineered gene. This technique has produced transgenic pigs, goats, and sheep, and it seems likely to yield similar results with other species (Figure 13-11B). So far, however, only one transgenic animal is born for every 100 injected zygotes. Despite tabloid claims to the contrary, no one is known to have made any transgenic human embryos, though it is no more difficult to do in humans than in mice or goats.

The engineering of transgenic animals faces serious obstacles. First, few eggs actually incorporate the recombinant DNA. Even genes that initially seem to have been incorporated turn out to be not in the chromosomes, but in some mobile DNA outside the nucleus. Such genes can be lost in subsequent cell division. Perhaps more troubling, current genetic engineering techniques put the recombinant gene into the animal's chromosomes at random, both adding to the host's own genes and, often, disrupting other genes. The disruption of key genes creates mutant animals that die young, which is one reason why it would be unethical to attempt this technique in human zygotes. In addition, a transgenic mouse also has two alleles of its own (one on each chromosome).

Researchers would like to replace the animal's own genes with the recombinant version instead of merely adding to them.

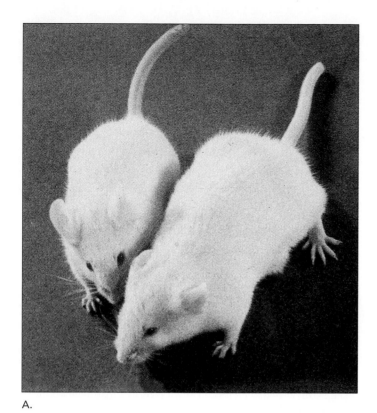

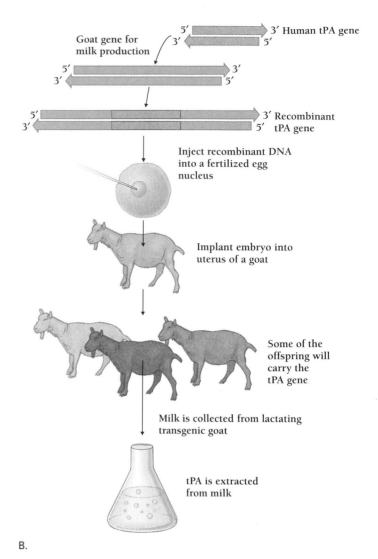

Figure 13-11 Transgenic animals carry foreign genes that they can pass on to their own offspring. A. The first transgenic mammal was a mouse carrying the gene for human growth hormone. B. Today, transgenic goats secrete valuable proteins in their milk. The proteins are isolated from the milk for use as pharmaceuticals. This goat carries the gene for tissue plasminogen activator (tPA), a protein that activates the anticlotting system and is used to save the lives of heart attack patients. *(A, Dr. R.L. Brinster/Peter Arnold, Inc.)*

This is relatively easy in prokaryotes and yeast, but more difficult in animal cells. One area of success is the production of genetic "knockouts"—animals in which a particular gene has been inactivated (Figure 13-12). Knockouts are a useful way of creating research animals with specific genetic defects, such as those found in humans. Usually, knockouts are used as a way of studying what role a gene plays in determining phenotype. The phenotypes that result from knockouts can surprise biologists. Figure 13-12, for example, shows a mouse in which researchers knocked out a gene specifying a growth factor believed to influence early embryonic development. The researchers expected the knockouts to stop developing long before birth. Instead, normal-looking baby mice were born, but with hair that grew so long that the knockouts looked a bit like Persian cats.

Recently, researchers have succeeded in creating "knockin" mice, in which an allele has been *replaced* rather than merely added. Like knockouts, knockins are difficult to make. In a knockin, however, the allele goes exactly where it is supposed to and the old allele is eliminated.

The knockout approach has several drawbacks. Quite often, all of the individual knockouts die as early embryos and little can be learned about the adults or even the developing embryos. Even more discouraging for researchers, many knockouts show no effects at all from the loss of a functioning gene. Other genes apparently compensate when a particular protein is absent. In one recent example, researchers created knockout mice unable to make oxytocin, a hormone produced during the birth process and commonly given to women to induce labor. Yet, knockout mice unable to produce oxytocin delivered their pups with no problems.

Interpreting the phenotypic results of knockout organisms is also problematic. Knocking out genes to see what they do is very like trying to find out what the different objects in a house do by removing them one at a time. If we remove the washing machine, the laundry no longer gets washed. But the same thing occurs if we remove the laundry detergent, or the clothes, or the electrical outlet into which the washing machine is normally plugged. Each of them is essential to getting the clothes done, but no single one of them is solely responsible for clean

Figure 13-12   Knockout mouse. Researcher Gail Martin decided to knock out a growth factor gene, called FGF5, believed to play a role in the very early formation of an embryo. Martin and her colleagues expected that none of the homozygous embryos would finish developing. Instead, the knockouts all developed normally, but with hair nearly 50 percent longer than normal. Although FGF5 is normally expressed in muscle cells, brain cells, and motor neurons, all of these tissues were normal.   *(Gail Martin, University of California, San Francisco)*

clothes. Similarly, when molecular biologists claim to have found a "master gene" for a given trait, the gene could equally well be described as the "weak link" in the intricate chain of events and materials that make up the phenotype. Teasing apart that chain of events is a major challenge for current molecular biology.

Creating transgenic mammals presents many problems. First, they must be made one at a time. Second, in knockouts the recombinant genes are inserted into the genome at random, which means that the genes may not function as researchers hope and may also disrupt other genes. Genetic knockouts and knockins can nonetheless provide clues about how previously mysterious proteins function in the body.

The Genetic Engineering of Plants Is Easier Than That of Animals

Perhaps the most rapidly expanding area of genetic engineering is in agriculture. Plant cells are generally easier to clone than animal cells. Equally important, plants can be grown in vast fields, so that a human protein, such as hemoglobin, produced by a plant, can be grown in hundreds of pounds per acre, instead of in grams per tub of *E. coli.* Tobacco plants, for example, can synthesize mouse antibodies that seem to behave just like mouse-made antibodies. Tomato plants can synthesize human serum albumin, a protein used to treat burn victims.

Genetic engineering of plants also has the potential to be enormously lucrative. If one of the engineered tomatoes men-

tioned earlier in this chapter becomes popular, for example, the inventors stand to gain a virtual monopoly in the tomato market. (Americans eat more tomatoes than any other vegetable, by far.)

One way that researchers carry genes into plant cells involves a species of soil bacterium called *Agrobacterium tumefaciens. A. tumefaciens* helps cause plant tumors, called crown galls. The actual tumor-causing agent is not *A. tumefaciens,* however, but a plasmid called the **Ti plasmid** [for *tumor-inducing*], which resides within the bacterium. A tumor results when a piece of plasmid DNA moves into the genome of the host plant. Researchers can engineer the Ti plasmid to carry specific genes into the genome of a plant. As in other forms of genetic engineering, the recombinant DNA usually includes an antibiotic marker that allows researchers to quickly distinguish the cells that have taken up the recombinant DNA from those that have not.

In another technique, sometimes called the **gene gun,** tiny particles of gold or tungsten are coated with fragments of recombinant DNA, then fired into a young plant with a miniature "shotgun" powered by gunpowder or compressed gas. The young plant is then chopped up and its cells grown on a culture containing an antibiotic. The antibiotic kills all of the cells except those that have taken up the recombinant DNA introduced with the gene gun.

Genetic engineering has produced new varieties of flowers and remarkable crops (Figure 13-13). For example, researchers are working to produce glow-in-the-dark Christmas trees by inserting into a Monterey pine the *lux* gene from a deep-sea fish. The *lux* gene could also be inserted into oleanders, the shrubs often planted down highway medians, to create a wall of light.

Many experimental products are controversial, however. For example, in 1994, nearly one-third of all the genetically engineered plants being tested were designed to resist an herbicide. Weeding a crop of such plants is simple. The farmer sprays the herbicide on the field, killing the weeds but not the genetically resistant crop. Herbicide manufacturers, who are themselves engineering many of these new crops, argue that this practice will reduce the use of herbicides. But critics argue that such engineered crops will *increase* the use of herbicides and profoundly affect the ecology of surrounding areas.

Many genetic engineers hope to design plants that will reduce the need for insecticides, which are more dangerous to humans and other animals than herbicides. One of the earliest attempts was a variety of potato that expressed high levels of a natural insecticide. This potato was highly resistant to damage by pests, and did not need to be treated with pesticides. Unfortunately, it was also too toxic to eat.

A more successful example is a squash genetically engineered to resist a damaging virus. Researchers discovered that by inserting a gene for a viral coat protein into the squash genome, they could prevent the virus from infecting the squash plants. The new squash plants can theoretically reduce pesticide use since several pests damage the squash only by spreading the virus. The pests themselves do not harm the crop. If

Figure 13-13   **Transgenic cotton.** Both the normal cotton plant (*left*) and the transgenic cotton plant (*right*) have been sprayed with a potent herbicide. Only the transgenic plant flourishes, however, for it alone possesses a gene for herbicide-resistance.   *(Calgene)*

the squash plants cannot be infected, squash farmers will not have to spray. Strangely, how the viral coat gene vaccinates the plant is not understood.

> Using either the Ti plasmid or the gene gun, molecular biologists can genetically engineer plants that can, for example, synthesize animal or plant proteins, resist herbicides, or resist infection by plant viruses.

What Are the Environmental Risks of Recombinant DNA?

The incredible power of recombinant DNA technology lies partly in how precisely and quickly changes can be made and partly in the ability of molecular biologists to transfer genes from one organism to another. The long-term consequences of moving genes from one organism to another are unknown. Some critics argue that the ecological effects, if any, can be global and irreversible. Genetically engineered plants can, for example, exchange their new genes with wild cousins, either through conventional interbreeding or by way of plasmids and other vectors. Will genes inserted into a crop plant insert themselves into unrelated pest plants, creating, for example, herbicide-resistant weeds? Can a wild plant acquire traits that allow it to take over a local ecology or become a serious agricultural pest? Can the genes for antibiotic resistance present in engineered tomatoes and other vegetables insert themselves into the bacteria in our intestines? Will engineered plants and animals interact with their environments in unforeseen, potentially destructive ways? Biologists and others can only speculate. No one knows if any of these scenarios are real threats. Biologists have not yet encountered any problems.

The Application of Recombinant DNA Technology Poses Moral Questions for Society

At the other end of the spectrum are the moral and ethical issues that come from recombinant DNA's precision. The diagnosis of genetic diseases, for example, is far in advance of treatment. Thus, in the next century, people may know that they are at higher-than-average risk for certain forms of cancer or mental illness, for example, but they will not necessarily be able to do anything about their fate. This is already true for genes that predispose people to certain forms of cancer of the breast, colon, and prostate. People at risk for Huntington's disease can now find out if they carry the gene. Those with the Huntington's gene have no way, however, of avoiding the disease, so the test is of very limited value. Yet, potential employers and insurance companies want to know what people's genetic flaws are so that they can avoid hiring or insuring people likely to get sick.

Many people worry about applying genetic engineering to humans. Researchers are already working to treat genetic diseases by inserting genes in human cells. Some researchers may try to eliminate certain diseases altogether by replacing defective genes even before conception. Such a heritable genetic cure has already been achieved in mice for a neurological disease characterized by abnormal brain development.

However, while few people object to curing deadly diseases, what constitutes a "genetic disease" may become controversial. Societies have often designated people with certain characteristics as "undesirable," and readily embraced arguments that these characteristics are genetic. If biologists develop the capacity to replace genes in humans, will genetic engineers also try to modify genes that affect characteristics other than those responsible for disease? Will future societies attempt to produce more (or less) intelligent citizens? or less (or more) aggressive citizens? Already, many physicians treat shorter-than-average children with growth hormone, so that the children will "fit in." There is no reason to think that either parents or physicians would hesitate to "treat" such children with gene therapy.

While many aspects of physical appearance and behavior are strongly genetic, it seems unlikely in principle that gene replacement could ever produce a uniform society. Possibly, the wealthiest members of society could afford to have their children engineered for such traits as good teeth and other aspects of health and appearance. Meanwhile, the rest of society would go on as usual. Conventional medicine already suffers from this defect: the wealthiest members of society get better medical care than the poorest. It is clear that all the people in a society need to discuss the issues raised by the future possibilities of genetic engineering. As in the case of any new advance (such as nuclear weapons and nuclear power), the use of technology is a concern for the society as a whole, not just for scientists.

We do not yet know all the limitations of the new technology. Society must carefully evaluate the risks, benefits, and moral implications that arise from the ability to produce new or previously unavailable products, especially engineered plants and animals. In the next chapter, the last chapter of this section on genetics, we will explore some of these issues in more detail.

STUDY OUTLINE WITH KEY TERMS

Until the development of **recombinant DNA** techniques, **biotechnology** consisted mostly of the selection of varieties of organisms with desired traits.

Recombinant DNAs consist of two or more DNA segments that are not found together in nature. For example, by combining DNA pieces from humans and E. coli, researchers made a recombinant DNA molecule with the regulatory region of β-galactosidase and the structural gene for insulin. Bacteria that contain this DNA make insulin in response to the presence of lactose, instead of β-galactosidase.

Making recombinant DNA depends on the use of enzymes that can cut, link, and modify DNA. **Restriction enzymes** cut DNA at **restriction sites,** where specific sequences occur, allowing molecular biologists to isolate individual DNA fragments, called **restriction fragments. DNA ligase,** an enzyme used in DNA replication, links pieces of DNA from different sources. A **restriction map** shows the positions of the restriction sites in a piece of DNA.

DNAs derived from viruses and plasmids can serve as **vectors** for the propagation of recombinant DNAs within host cells. Some of the most widely used vectors derive from R-factors, plasmids that carry antibiotic resistance genes. Other vectors include bacteriophages, as well as plant and animal viruses.

Polymerase chain reaction (PCR) is now a basic tool for finding and multiplying, or **cloning,** fragments of DNA. **Oligonucleotides,** short pieces of single-stranded DNA, act as primers for DNA polymerase. Alternating heating and cooling do much of the rest.

The starting materials for recombinant DNA may come from the genomes of viruses or organisms or from DNA copies of mRNA. Most plant and animal genes contain introns, which would interfere with their expression in bacteria. So many recombinant DNAs coding for plant and animal proteins are derived from **complementary DNAs (cDNAs),** copied from mRNA by the **reverse transcriptase** of a retrovirus.

Collections of recombinant DNAs, called **gene libraries,** may contain thousands or millions of different DNAs. Finding a desired DNA sequence usually depends on the use of a **probe.** A **hybridization probe**—often a piece of DNA made in the laboratory—bonds with one strand of the DNA to form **hybrid DNA.** By labeling the probe with radioactive nucleotides or a chemical marker, the experimenter can identify recombinant DNA. **Antibodies,** which have surfaces that recognize the shapes of specific molecules, can help identify bacteria that are producing a particular protein.

In many cases, researchers do not know the function of a particular protein until recombinant DNA techniques make enough of the protein available to study. Recombinant DNAs are likely to be useful in the preparation of safe vaccines for use in disease prevention.

Recombinant DNAs can reprogram organisms to make new products. Products such as insulin and **growth factor** that are made by genetically engineered bacteria are already commercially important as pharmaceuticals. Treating humans with genetically engineered human cells is called **gene therapy.** Recombinant DNA can be used to alter animals and plants genetically. **Transgenic organisms** carry deliberately added **transgenes** in their genomes. The present method for making transgenic mammals involves the injection of an engineered gene into a zygote produced by in vitro fertilization. Molecular biologists can also damage selected genes to make genetic knockouts.

Making transgenic plants is easier than making transgenic animals because experimenters can easily add or replace genes in single cells in culture. The engineered cells can then be grown into whole plants. Genes are introduced either by means of a tumor-inducing (or **Ti**) **plasmid,** which normally resides in bacteria that cause plant tumors, or by means of a **gene gun.**

REVIEW AND THOUGHT QUESTIONS

Review Questions

1. How does modern genetic engineering using recombinant DNA differ from breeding techniques practiced over the last 10,000 years? How are these two kinds of genetic engineering the same?
2. How can knowing a biochemical pathway help molecular engineers design useful organisms?
3. What is recombinant DNA?

4. What two enzymes are basic to assembling recombinant DNA? What do the two enzymes do, and what is their role in nature?
5. Why do molecular biologists need multiple copies of a gene?
6. What two techniques allow molecular biologists to make many copies of a gene? What are the strengths and weaknesses of each technique?
7. Why do molecular biologists want to engineer cells to express

large amounts of the gene product, or protein? How do they accomplish this?

8. What prevents bacteria from expressing a gene cut directly from a human genome? How do molecular biologists get around this problem?

9. How do molecular biologists find the right DNA sequence in a recombinant DNA library?

Thought Questions

10. What is your opinion about creating transgenic humans? Suppose that you knew that you carried a gene that would predispose your children to a mental illness. Would you opt to conceive by means of *in vitro* fertilization so that your zygotes could be screened and the gene replaced? Suppose large numbers of other people chose not to use this service. Some of their children would continue to develop severe mental illness, eventually becoming rather expensive wards of the state. Would you feel that everyone should be required to screen their zygotes? What if you knew that the families of people carrying this gene tend to be more intelligent and more creative than average? Would that make a difference in your opinion? Why or why not?

11. Would you mind eating a genetically engineered tomato? Why or why not?

SELECTED READINGS

Fussell, Betty, *The Story of Corn: The Myths and History, the Culture and Agriculture, the Art and Science of America's Quintessential Crop,* Knopf, New York, 1992. The subject of this book sounds dull, but *The Story of Corn* is actually well written, engaging, and amusing. With both humor and respect for her subject, Fussel covers corn from archaeology and art to *Zea mays* and Zuni culture. Many wonderful illustrations. Includes a good explanation of how corn may have been domesticated thousands of years ago.

Hall, Stephen S., *Invisible Frontiers: The Race to Synthesize a Human Gene,* Tempus Books of Microsoft Corporation, Redmond, Washington, 1987. A compelling and detailed history of how the gene for human insulin was first sequenced and then synthesized.

Hubbard, Ruth, and Wald, Elijah, *Exploding the Gene Myth: How Genetic Information Is Produced and Manipulated by Scientists, Physicians, Employers, Insurance Companies, Educators, and Law Enforcers,* Beacon Press, Boston, 1993. A highly readable argument against "geneticization," the tendency to link all of our health and other problems to genes. Hubbard is a professor of biology at Harvard University, and Wald is her son.

▶ On-line materials relating to this chapter are on the World Wide Web at http://www.saunderscollege.com/lifesci/
Click on Tobin/Dusheck: *Asking About Life.*

To Map the Genome or Hunt the Huntington's Gene?

David Botstein, burly and bombastic, jumped up from the table and paced the small conference room, loudly lecturing before the other 12 scientists still seated at the table. It was October 1979, and the 13 scientists had met at the National Institutes of Health near Washington, DC, to discuss the impact of molecular biology on inherited diseases, especially Huntington's disease. But the discussion had come to a critical point of disagreement, and Botstein, a brilliant and temperamental molecular biologist from the Massachusetts Institute of Technology (MIT), could no longer contain himself (Figure 14-1).

Beginning a search for the Huntington's gene now, argued Botstein, was premature and a waste of time. New technology would soon revolutionize human genetics, allowing scientists to map and identify every human gene and find the molecular basis of every genetic disease. Once a general map was established, it would be child's play to find the molecular defects that cause Huntington's disease, cystic fibrosis, muscular dystrophy, and any of the other 3000 inherited diseases.

At the table, still seated, was David Housman, another molecular biologist from MIT. Housman watched Botstein impassively, then quietly disagreed. Yes, eventually, all would be known, he said, but wasn't it better to start first—right now with a single disease? And why not Huntington's disease? Researchers need not wait for a complete genetic map to find the Huntington's gene. There was no reason not to jump ahead, applying knowledge and techniques that were already available. Housman himself had already shown how the abnormal regulation of hemoglobin genes could cause certain inherited anemias called thalassemias.

Botstein scoffed. It was short-sighted to jump into problems that might be personally pressing but were not yet scientifically "ripe." Better, said Botstein, for scientists to wait for basic knowledge. Indeed, despite the efforts of dozens of researchers, the Huntington's gene was not sequenced for 14 more years. The early mapping and sequencing of individual disease genes moved excruciatingly slowly. Only when basic research by Botstein and others provided new techniques for faster advances did the mapping of defective alleles begin to proceed quickly.

Yet it was Housman who prevailed at the 1979 conference. The conference had been sponsored by a small foundation called the Hereditary Disease Foundation (HDF), which is based in Los Angeles and dedicated to promoting research on Huntington's disease. The HDF had been

Figure 14-1 David Botstein. *(Courtesy, Stanford University Medical Center)*

started by Milton Wexler, a prominent Beverly Hills psychoanalyst. Wexler's two daughters, Alice and Nancy, were at risk for developing Huntington's; in 1968 their mother Leonore had been diagnosed with the fatal disease.

Huntington's disease is a devastating neurological disease that causes jerky movements, mood swings, personality changes, and premature death. The most obvious early symptom of the disease is the constant motion of the arms and legs, giving the disease its previous name, "Huntington's chorea" [Greek, *chorea* = dance, as in "choreography"]. Huntington's usually begins in middle age. Over the course of 10 to 20 years, patients gradually lose control of movements, memory, and mood. The bodies of Huntington's patients gradually waste away, and for the last few months of life patients stop moving altogether.

Huntington's disease is a classic genetic disease. An affected parent passes the disease to half of his or her offspring, indicating that the disease is caused by a single dominant allele of some gene. Nearly everyone who inherits the Huntington's disease allele eventually develops the

symptoms. This high level of expressivity makes the disease particularly suitable for genetic analysis. In 1979, however, researchers and physicians knew nothing about the gene and almost nothing about the damage it caused at a biochemical and cellular level.

In 1969, Milton Wexler had begun engaging young scientists to brainstorm new approaches to understanding Huntington's disease. He attracted these biologists both by inviting them to parties that included some of the most famous artists and entertainers in the Los Angeles area and by encouraging these young biologists to think freely and creatively. Few young biologists could resist these brushes with Hollywood glamour. But the real purpose of the gatherings was the scientific workshops, in which Wexler brought together

small groups of biologists who did not ordinarily meet—researchers in fields ranging from genetics to neurology. These informal workshops were intellectual incubators for new ideas not only about Huntington's disease but about the application of new techniques to the study of all inherited diseases, especially disorders of the brain. Among the young scientists stimulated by these workshops was one of this book's authors, Allan Tobin. In 1979,

Wexler appointed Tobin Scientific Director of the Hereditary Disease Foundation. It was Tobin who organized the 1979 workshop on molecular genetics, with help from his former Cambridge neighbor, David Housman.

Botstein's fears that attempting to map the Huntington's gene would be difficult turned out to be well-founded. To find a particular gene among the 100,000 genes on the 46 chromosomes, geneticists had to look for markers, obvious phenotypic characteristics such as eye color, that are always inherited with the defective allele, and are therefore near the defective gene. Geneticists had constructed linkage maps in *Drosophila* through the painstaking analysis of the selective mating of flies with identifiable phenotypes—such as white eyes or notched wings. But geneticists can-

The Hereditary Disease Foundation's first-ever gene search—undertaken despite Botstein's fierce arguments for patience—dramatically demonstrated the power of molecular biology to help understand human disease.

not set up matings of humans; human genetics instead depends on the analysis of matings that, to a geneticist, seem arbitrary and senseless. A few markers in humans, such as those for blood proteins, work well, but none of these were inherited with the Huntington's allele.

The picture changed, however, when molecular biologists began to use recombinant DNA techniques to clone and sequence specific genes. Researchers then

Figure 14-2   The fight against Huntington's disease. The families of the famous singer Woody Guthrie and little-known biology teacher Leonore Wexler used their considerable influence to initiate and support the search for the Huntington's gene. *(Guthrie, Culver Pictures; Wexlers, courtesy of Alice Wexler)*

discovered numerous small differences among individuals at the DNA level. Even though two people may make exactly the same hemoglobin molecules, for example, their hemoglobin genes are likely to differ from one another. These "silent" differences lie in DNA sequences that are not transcribed and translated into protein—for example in introns or in flanking sequences. By the late 1970s, researchers were convinced that (except for identical twins) every person is genetically distinct from every other. At the DNA level, the homologous genes inherited from mother and father differ at about 1 nucleotide in 500.

The technology that so excited David Botstein was a method for identifying such differences, which he called **"restriction fragment length polymorphisms"** or **RFLPs** (pronounced "RIFFlips"), and which we discuss below. Within a few years, differences in DNA sequences would not only allow geneticists to produce a map of the human genome, it would also allow police to match blood and tissue samples with a previously unimaginable accuracy—changing forever the identification of criminal suspects. By 1980, Botstein, his collaborators, and his competitors had established the feasibility of making a map of the human genome based on differences in DNA sequence, rather than on differences in phenotypic

traits. By 1989, researchers had succeeded in making a relatively complete RFLP map of the human genome.

But enthusiasm for making a map of the human genome was not universal, even among molecular biologists. The research required was repetitive and dull. Indeed, the excitement for learning about human genes came from the public's interest in finding the alleles that underlie diseases ranging from Huntington's to cancer. The first success in the application of the new molecular genetics to human disease was, in fact, Huntington's disease.

Huntington's disease provided the key and the momentum for several reasons. First, its inheritance was particularly clear. Second, motivated families, especially those of Leonore Wexler and of the great folksinger Woody Guthrie, who also died of Huntington's disease, encouraged young researchers both psychologically and financially through their support of several foundations concerned with Huntington's disease (Figure 14-2). In particular, Nancy Wexler, the younger daughter of Milton and Leonore, has devoted her professional life to advancing the search for the gene for Huntington's disease. As a researcher herself, she documented the inheritance of Huntington's disease within an extended family that lives in three villages on the shores of Lake Maracaibo in Venezuela.

Within this group of about 10,000 people, several hundred have already developed Huntington's disease, and hundreds more carry the gene. By tracking the simultaneous inheritance of Huntington's disease and of DNA markers within this family, James Gusella (a former student of David Housman), Nancy Wexler, and their colleagues were able—in 1983, years before the construction of a complete human genetic map—to determine the approximate chromosomal location of the Huntington's disease gene. This accomplishment depended directly on the RFLP method advocated by Botstein. The Huntington's allele was the first disease gene ever found with that method.

It took an international consortium organized by the Hereditary Disease Foundation another 10 years to identify the actual gene for Huntington's disease. By then (1993), recombinant DNA techniques had already led to the precise identification of many other disease-causing alleles, including that for cystic fibrosis, the most common genetic disease among Caucasians and a good target for the development of gene-based therapies. But the Hereditary Disease Foundation's first-ever gene search—undertaken despite Botstein's fierce arguments for patience—dramatically demonstrated the power of molecular biology to help understand human disease.

Because of the efforts of the HDF, dozens of researchers focused on applying known techniques to the specific problem of Huntington's disease. Choosing a single disease on which to focus their efforts gave biologists a sense of mission and a single, clear-cut goal that was emotionally appealing. They could imagine that their day-to-day research might one day save the Wexler daughters and other sufferers of Huntington's disease. In addition, the HDF's intimate workshops promoted many fruitful scientific friendships and collaborations. And finally, Gusella's location of the Huntington's gene in 1983 provided the proof of concept that may well have convinced both scientists and scientific administrators that mapping the human genome was feasible. In this chapter

we will see, however, that it was the basic research carried out by Botstein and others that invested the science of molecular genetics with the promise it now holds.

The genetics of humans is no different from the genetics of other organisms. For this reason, many readers of this text in its manuscript form objected to the very idea of having a chapter on human genetics. The *study* of human genetics, however, does differ in one very significant way from the study of the genetics of fruit flies, pea plants, or corn. Humans cannot be crossed at the whim of geneticists. As a result, the study of genetics in humans (as well as in other undomesticated animals) is much harder than in laboratory organisms in which any two individuals can be crossed.

Another difference between the genetics of humans and that of other organisms is that our own traits are, to us at least, infinitely more interesting and subtle. The degree to which a trait is the product of genetic influences or the product of environmental influences rouses people to intense political debate that makes objective discussion nearly impossible. In this chapter we will try to present enough information for readers to make intelligent judgments about what they read and hear. We will learn about genetic diversity in humans, the complex relationship between genotype and phenotype, the Human Genome Project and, finally, both the promise and limitations of applying knowledge about human genetics.

KEY CONCEPTS

1. Even though humans are more than 99 percent genetically identical, all humans, except identical twins, are genetically unique.

2. Phenotype is always a product of the interaction between genotype and environment.

3. A particular gene is expressed in both a physical environment and a developmental environment. In addition, every gene is expressed in the context of the other genes in the genome.

4. The relationship between genotype and phenotype is complex and therefore loose.

5. The Human Genome Project promises to map and sequence the entire human genome by about 2002.

6. Many aspects of human genetics, including, for example, diagnostic testing and gene therapy, raise serious ethical issues.

GENETIC DIVERSITY IN HUMAN BEINGS

On a Saturday evening in the winter of 1983, five-year-old Alice Brown was abducted just steps from her own front door. She and her mother, Sarah Brown, were about to sit down to dinner with a neighbor in their 200-unit apartment complex in northern California when Alice's three-year-old brother Brad begged for his boots. Alice took Brad home, just two doors down, but when Brad returned moments later he was alone. Alice's mother never saw her daughter alive again.*

*We have changed the names of those involved.

When Alice's body was found a week later, police discovered that she had been raped and then suffocated. Sarah said that Alice would never have gone with any man except one: a neighbor in the same apartment complex with whom Sarah had gone out several times. Police considered Harry Gordon, then 31, a prime suspect, but they had no concrete evidence to link him to the case. They had to have "probable cause."

All of the evidence was circumstantial. He'd given contradictory stories about his whereabouts on the night of the murder. Police had arrested him in 1981 for exposing himself to a little girl and again, in 1984, on suspicion of molesting a third girl. He was eventually convicted of a misdemeanor for exposing himself while drunk. Gordon's ex-wife said he was a likely suspect and police pursued dozens of leads. But Gordon denied any involvement in Alice's death and nothing gave the police the evidence they needed to arrest Gordon.

Sarah and Brad moved away and changed their names. Thirteen years passed. Then, in the spring of 1996, when Alice would have been 18 years old, police finally arrested Gordon for the rape and murder of Alice Brown. It turned out that police had had the concrete evidence they needed all along. Subtle changes in California law and advances in molecular biology had finally made it possible for prosecutors to use that evidence to link Gordon to the crime and to know that they could probably convict him.

Just as every individual has a unique set of fingerprints, so every individual also has a unique set of gene fragments called a **DNA fingerprint.** DNA for a DNA fingerprint can be extracted from any cell that has a nucleus. Such cells can be found in blood stains, in saliva on an envelope, in tears on a cheek, in skin cells, under fingernails, in hair, or in semen. Thanks to the invention of PCR, researchers can now analyze DNA from the barest traces of cells. When police found Alice's body in 1983, they had been able to collect samples of semen from her body and clothes. As evidence, these samples had been carefully stored. But in 1983 they were of little value, for DNA fingerprinting had not yet been invented.

By the early 1990s, prosecutors knew that the DNA in the semen could tell them who had murdered Alice Brown. A legal problem, however, prevented prosecutors from acting. Unlike most states, the California Supreme Court had not issued a ruling on the admissibility of DNA fingerprinting as evidence. As a result, lower courts had issued conflicting rulings and a conviction based on DNA fingerprinting stood a good chance of being overturned on appeal. In 1992, a northern California judge had ruled that DNA fingerprinting was not as reliable as a real fingerprint. If prosecutors tried to convict Gordon and the court rejected the DNA fingerprinting, prosecutors knew they would have lost their only chance to convict Gordon, and at great cost. Just arguing the admissibility of DNA fingerprinting in court could cost $50,000—the price of the expert witnesses.

How Reliable Is DNA Fingerprinting?

The haploid human genome contains some 3 billion base pairs of DNA. Only 5 percent of this DNA (150 million base pairs) codes for polypeptides. The polypeptide-coding regions are almost identical from one individual to the next because natural selection eliminates most mutations.

The 95 percent of the DNA that does not code for polypeptides is much less subject to natural selection. As far as biologists can tell, such noncoding DNA has no central role in the phenotype, and so it can mutate without impairing an individual's ability to survive and reproduce. Noncoding DNA accumulates large numbers of mutations and contains far more variation than coding DNA. Estimates suggest that each pair of chromosomes differ by about 1 base in 500, which on a single chromosome could add up to some 500,000 base-pair differ-

ences. As a result, except for identical twins, every individual is genetically unique and DNA sequencing can reliably distinguish every individual in a population.

Researchers create a DNA fingerprint by using restriction enzymes to break up an individual's DNA into restriction fragments (Figure 14-3). Fragments from some regions of the DNA vary in length from individual to individual. The length of a fragment depends on whether the sequence includes a restriction site for the restriction enzyme in use. (Other parts of the DNA are consistently the same length from one person to another.) Fragments that vary in length are said to be polymorphic and are called restriction fragment length polymorphisms, or RFLPs. This long name merely means that for a given RFLP sequence, the presence or absence of a restriction site determines whether the fragment is long or short.

In the next step, electrophoresis, the fragments of DNA are exposed to an electric current that moves them through a jelly-like medium, called a "gel." The DNA fragments, which are negatively charged, move through the gel toward the positive electrode. Some move quickly, some slowly, depending on their size. The shortest fragments travel through the gel rapidly, while the longer fragments are left behind. After a while, the fragments are spread out across the gel like runners in a race. When stained, the fragments produce a pattern of bands.

Every genome yields a unique pattern of dark bands. In theory, fragments taken from the whole genome in different individuals would reveal a DNA fingerprint as unique as a real fingerprint. In practice, however, fingerprinting laboratories do not test all of the DNA in the genome. Instead they take a sample of noncoding DNA known to have many variants. Researchers then calculate the probability that someone else in the general population has that exact set of base changes. Researchers can come up with a probability of uniqueness ranging from 1 in 500 to one in several billion.

Two factors affect this number. The first is the number of different markers examined. The greater the number of markers surveyed, the more specific the DNA fingerprint will be, and the more other people in the world can be excluded from consideration. The second factor is how "the general population" is defined. As we will see below, different populations are more or less likely to have certain genotypes. The chances of someone having the same DNA fingerprint as Harry Gordon might be one number for the population of San Francisco Bay Area and another number for the population of all of California. In any case, geneticists do not necessarily know what those probabilities are for different populations. They have estimates for large populations that have been studied, such as those of the United States, Argentina, or Spain, but not for individual cities such as San Francisco or Los Angeles.

Why Did Police Finally Arrest Gordon?

In the fall of 1995, prosecutors in the Harry Gordon case began to sense that DNA fingerprinting was coming to be accepted in California. They no longer believed that if they took

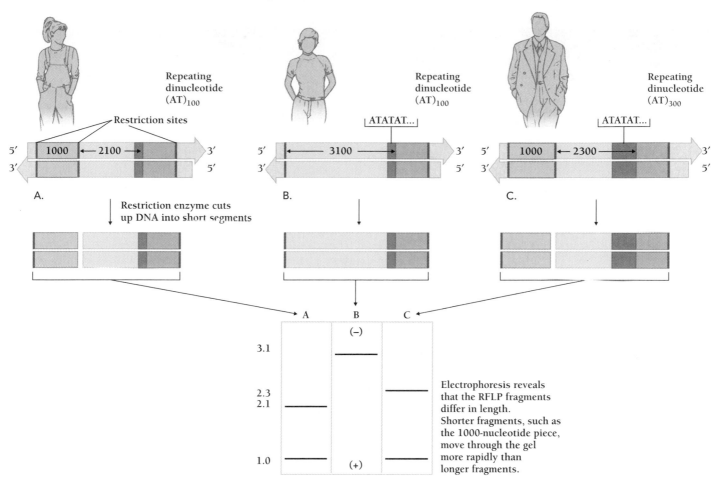

Same segment of DNA from three different individuals:

Figure 14-3 RFLPs (restriction fragment length polymorphisms) in three individuals. Restriction enzymes break the DNA of each individual into fragments. For individual A, two fragments result that are 1 kilobase and 2.1 kilobases, respectively. The shorter one moves across the electrophoretic gel faster than the longer one. In B, the middle restriction site is absent and a single long fragment results that is 3.1 kilobases long (1 + 2.1 = 3.1). This long fragment moves across the gel very slowly. In C, the 1 kilobase fragment is the same as the one in A. But a repeating sequence has made the 2.1 kilobase fragment in A 200 bases longer. C's longer fragment moves more slowly than the one in A, but faster than the one in B.

Gordon (or any other suspect) to court that they would have to prove that a DNA fingerprint is unique. In the fall of 1995, police obtained a court-ordered search warrant and forced Gordon to give a blood sample. They sent the blood sample and the sample of the semen from Alice Brown's body to a laboratory in Maryland. There technicians determined that the two samples matched. In RFLP after RFLP, the pattern of dark bands was the same.

To minimize conflict in court, the lab tested so much DNA that technicians were able to estimate that the DNA in the semen from Alice could be matched by only one Caucasian male in 5.7 billion, some 10 times the number of Caucasian males in the world. Prosecutors finally had the tangible evidence for which they had waited so long. In the spring of 1996, they arrested Gordon and charged him with Alice's murder. His case is not expected to go to trial until some time in 1998. Many

people are hoping that his arrest will bring to an end a string of similar unsolved murders in the area.

What Might Gordon's Defense Attorneys Argue?

Even if the DNA fingerprinting work done on Gordon's blood samples included dozens of RFLPs and even if the probability of someone else in the world having Gordon's DNA fingerprint is virtually zero, his defense attorneys will likely still argue that the semen samples are so old that the evidence is unreliable. Is this a legitimate argument?

The use of DNA-based evidence has been much debated in U.S. courts and in the pages of scientific journals. A variety of arguments have been advanced against the reliability of DNA fingerprinting. Gordon's attorneys are likely to argue that the semen samples are very old and were improperly stored. But

BOX 14-1

Southern, northern, and western blotting

Stop by a molecular biologist's lab bench today and chances are you will find a plastic box filled with gel and wired up to run an electric current through the slab of clear gel. Using a technique called **gel electrophoresis**, researchers can separate large molecules on the basis of their size. Gel electrophoresis is one of the most useful techniques available for studying DNA, RNA, and proteins.

In a gel containing sequences of DNA, the current draws the negatively charged DNA through the gel—a porous matrix that slows down the larger molecules as they snake and bump their way toward the positive electrode. A special stain reveals bands of DNA molecules, which group at different positions in the gel according to their size. The smaller fragments have moved farther, and the larger fragments remain closer to the rectangular well in the gel where they were first "loaded."

Sometimes, however, the bands all blur together and it can be hard to tell one fragment from another. For such complex samples, molecular biologist Ed Southern in 1975 developed a technique that lets the experimenter choose which fragments to look at. Usually, only a particular part of the genome is of interest—perhaps a specific gene or a polymorphic region used for DNA fingerprinting. In **Southern blotting and hybridization**, the DNA is separated by means of electrophoresis (Figure A). Then, for more convenient handling, the DNA is transferred from the thick, watery gel to special paper. The paper, placed between the gel and a stack of paper towels, wicks the DNA and all the liquid out of the gel. The result is a Southern blot.

Next, the researcher applies a radioactive DNA probe that base pairs, or hybridizes, with the region of interest. (One or more of the phosphorus atoms in the DNA probe is radioactive.) Since the DNA is normally already double-stranded, heat is applied to melt the strands apart, allowing the probe to find its target. When the temperature is again lowered, some of the single-stranded probe molecules hybridize with DNA fragments in the blot.

The result of this Southern hybridization is a sheet of paper the size of the original gel with radioactive bands of DNA wherever there were fragments of the gene of interest. The Southern blot is then exposed to a sheet of film, which reveals only the bands of DNA lit up by the probe. The DNA sequences of interest light up, while irrelevant bits of DNA remain invisible.

Several years after Southern developed his blotting technique for DNA, scientists began studying complex samples of RNA with a similar procedure and dubbed the technique **northern blotting**. Other scientists blotted proteins out of protein gels and called it **western blotting**. There is no such thing as an eastern blot, but the other three points of the compass come up in laboratory discussions every day.

Figure A **Blotting techniques.** Specific nucleotide sequences are detected by hybridization to a blot of fragments separated by electrophoresis. If the fragments are DNA, the technique is a Southern blot. If the fragments are RNA, the technique is a northern blot. ▶

how does improper storage affect the results? The answer is that if the DNA were to break down, for example, as a result of exposure to heat, sunlight, or chemicals, the changes in sequence would be random and extremely unlikely to exactly match Gordon's own DNA. In other words, damaged DNA will not convict an innocent person. In any case, DNA is amazingly durable stuff. Researchers have succeeded in sequencing DNA taken from fossils that are millions of years old, and certainly not stored under ideal conditions. That doesn't mean that DNA cannot break down. It does. But age or poor storage is no guarantee that it will deteriorate.

In some cases, defense attorneys have argued that the tissue samples from the site of a murder have been contaminated by DNA from other sources. Because PCR-based DNA finger-

printing is so sensitive, a few cells from a different person could change the results, they argue. In reality, unless the contaminating cells came from the suspect, they would not match the suspect's DNA. The only way for the suspect's cells to be added or substituted would be if someone deliberately set out to frame the suspect. This is possible, but no more likely with DNA fingerprinting than with other forms of evidence.

DNA fingerprinting identifies variations in sequences of DNA taken from blood or other tissues and provides an estimate of how common that particular pattern of variation is in the general population.

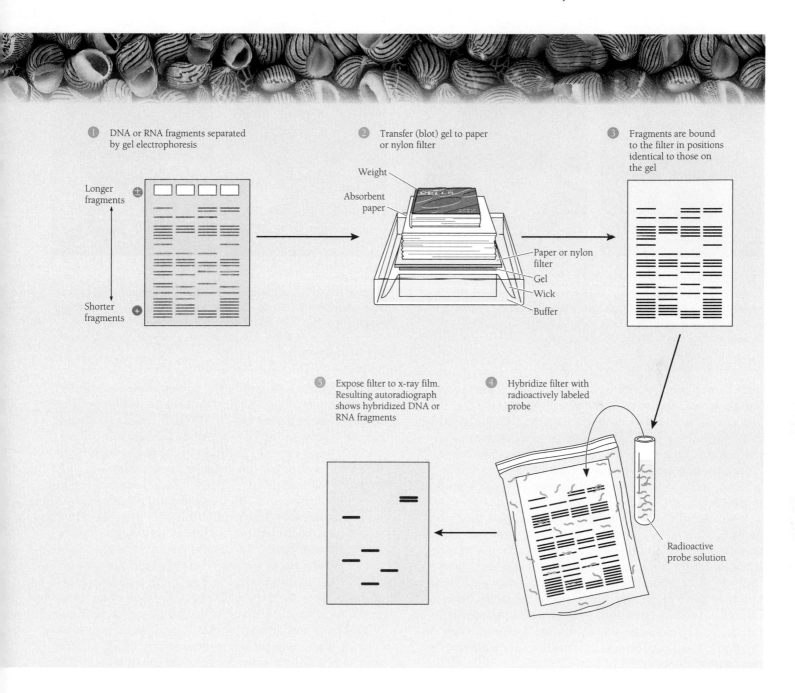

① DNA or RNA fragments separated by gel electrophoresis

Longer fragments

Shorter fragments

② Transfer (blot) gel to paper or nylon filter

Weight

Absorbent paper

Paper or nylon filter
Gel
Wick
Buffer

③ Fragments are bound to the filter in positions identical to those on the gel

⑤ Expose filter to x-ray film. Resulting autoradiograph shows hybridized DNA or RNA fragments

④ Hybridize filter with radioactively labeled probe

Radioactive probe solution

How Much Do Individuals Vary?

How different is one person from another? Even if we look only at the coding regions of our DNA, we are still all unique. We share some 99 percent of the same genes. Yet the 1 percent of genes that differ comes to some 1000 genes, each with so many alleles that everyone is genetically unique. If we look at the ABO blood types, for example, we see that about 45 percent of Europeans have type O blood, 35 percent have type A, 15 percent have type B, and 5 percent have type AB. That means that nearly half the population of Caucasians have the same blood type, O. But blood contains many more sets of proteins than just the ABO group. At least 12 groups of proteins have been identified, each with 2 to 7 different possible phenotypes.

The commonest single set of all these different proteins would make up only 0.2 percent of the population. If we include all of the proteins in the body, then each particular phenotype becomes so rare as to be effectively unique.

Every individual who is not an identical twin has a unique genotype.

What Are Human "Races"?

We all know that the human species varies in appearance and in physiology geographically. People of Indian descent, for ex-

A.   B.

Figure 14-4 People from different geographic regions vary in such traits as height, blood type, skin color, and facial features. A. An Indian woman. B. An ethnic minority woman from China.
(A, Superstock; B, Paul Grebliunas/Tony Stone Images)

A.   B.

Figure 14-5 Classifying humans into a few discrete races has been unsuccessful. Named races frequently include markedly distinct peoples. A. A Bushman from Namibia. B. A rubber plantation foreman from the Ivory Coast. *(A, M.P. Kahl/Photo Researchers; B, Charles O. Cecil/Visuals Unlimited)*

ample, are recognizably different from those of Chinese descent. We can look at the shape of the nose, the eyes, and the ears, for example, and we see differences (Figure 14-4). We can look at the distribution of individual alleles and see differences. Fifty-one percent of Nigerians have type O blood, compared to only 30 percent of Japanese. Twenty percent of Russians have type B blood, while the Amerindians of Lima, Peru, have no detectable levels of type B blood at all.

Nineteenth-century anthropologists struggled to classify human groups into a few major races. Some systems identified only 12 races, while other systems listed 30 or more. One problem was that no matter how anthropologists divided humans, there always seemed to be tribes or nations that would not fit into any known group. The Basques, who live in the Pyrenees mountains between France and Spain, for example, appear European. Yet their language and culture are unlike any other in the world, and some researchers argue that they are direct descendants of stone-age Europeans. Similarly, the Bushmen are unique among African groups in both appearance and physiology (Figure 14-5).

A more serious problem with the grouping of humans into races is that most groups do *not* stand out from those around them; they blend. Because groups of humans inevitably mix, through migration, warfare, and trade, human "races" are never pure. Both the Japanese and British, for example, take pride in the purity of their island races. Yet the Japanese are a graded mixture of Korean and Ainu north islanders (a people of possibly European descent). This mix shows up in the distribution of blood types from one end of Japan to the other (Figure 14-6).

The British are even more of a melting pot than the Japanese. The Bronze Age Beaker Folk mixed with the Indo-European Celts in the first thousand years B.C. In the next thousand

years, the Angles, the Saxons, the Jutes, and the Picts arrived, followed by the Vikings and their descendants, the French Norman invaders. In the last century, Africans, Indians, Pakistanis, and others have lent spice to this already heady mix.

If the Japanese and the British, seemingly isolated by geography, are well mixed, most mainland groups represent a true continuum of traits. In fact, studies of the distributions of different alleles have convinced researchers that human races—in the biological sense—do not exist. This is because, in humans, genetic variation within populations is greater than that between nations or races. Of all human genetic variation, 85 percent is variation among the individuals of a nation or tribe. Another 8 percent is variation among larger populations that belong to the same "race." Only about 7 percent of all human genetic variation comes from differences among races (Figure 14-7). The greatest genetic variation among humans is found in Africa. If a disaster killed everyone on Earth except for those living in Africa, the human species would still retain at least 93 percent of its genetic diversity. Less than 7 percent of all genetic diversity would be lost.

In practice, we focus on the small numbers of traits (for example, skin color and eye shape) that vary geographically—a subset of the 7 percent—and we use just those few traits to identify groups of people who differ from us more in their cultural practices than in their genetic constituencies.

Human races blur into one another, so that defining consistent races is impossible. Of all human genetic variation, only 7 percent is among so-called races. The remaining 93 percent is variation among individuals or populations.

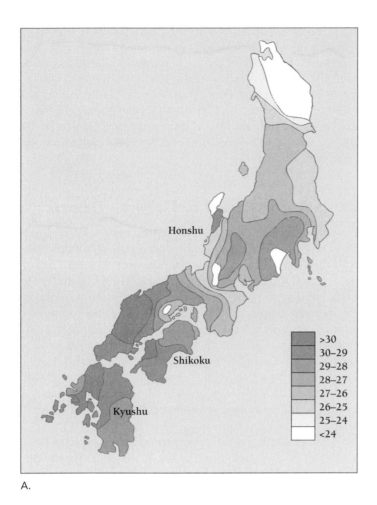

A.

B.

Figure 14-6 Variation in blood group A in modern Japan. A. Even the Japanese, a relatively homogeneous island race, show variation that demonstrates intermarriage with other groups. Populations closest to mainland Korea have the greatest concentration of A alleles. Populations closest to Hokkaido have the lowest concentrations of A alleles. B. In Hokkaido lives a remnant population of Ainu, sometimes called the "hairy Ainu." The Ainu, who were once a white minority in Japan, are considered primitive by other Japanese. The Ainu had white skin, abundant body hair, and round eyes. Descendants of Causasoid peoples from northern Asia, the Ainu once lived on all four of Japan's major islands. Today they live only in the north. Intermarriage and cultural assimilation by the Japanese have made the traditional Ainu virtually extinct.   *(A, from Human Diversity by Lewontin © 1982 by Scientific American Library. Used with permission of W.H. Freeman and Company; B, Superstock)*

Population A

Population B

Figure 14-7 Differences among the members of a population of humans are usually greater than the **differences between populations.** If we consider only eye color, we would never confuse a member of population A with one from population B. The two "races" are distinct. Yet, when we consider other attributes, the average difference between the two populations is minuscule compared to the large differences among the individuals.

THE RELATIONSHIP BETWEEN GENOTYPE AND PHENOTYPE IS COMPLEX

Genes provide the continuity between the generations, but they do not by themselves define individuals. Every individual—as expressed in the phenotype—results from complex interactions between genes and environment. The environment includes both the physical and chemical environment of the developing organism, and the ongoing environment of the mature organism. In humans, environmental influences include the quality of the cytoplasm in the egg, hormonal and nutritional influences in the womb, upbringing, and the physical and social atmosphere of the home or workplace. It is already known that in other animals gene expression and even brain structure are altered by the social environment of the adult (Figure 14-8). The tiny Japanese goby *Trimma okinawae*, for example, can change sex repeatedly, from female to male and back again. Schools of female fish breed with a single male. If a local breeding male is replaced by a larger male, a juvenile male in the

school becomes female and remains to breed with the new male. If the breeding male leaves, however, one of the females becomes male and takes his place. These fish can change the structure of the gonads, genitals, and brain, as well as their behavior, in as little as four days. Their anatomy and behavior are the same as that of fish that have always been one sex. In short, the same genotype can impart more than one phenotype. In this section, we will explore the complex relationship between genotype and phenotype.

Genes Are Expressed to Different Degrees

Genes vary enormously in their **expressivity**—the range of phenotypes associated with a given genotype. Two individuals with the same allele, for example, may express it very differently. The human disease neurofibromatosis, for example, is caused by a dominant allele whose expressivity varies so much that in some people it causes blindness or horribly disfiguring growths, whereas in others the only symptom is a coffee-colored spot (Figure 14-9). Still others with the same neurofibromatosis allele have no symptoms at all. Huntington's disease varies, too, in how strongly it is expressed and when. Although most people show no symptoms until middle age or after, a few children may exhibit severe symptoms as early as 2 years of age.

Genes also vary in their **penetrance.** While expressivity is a measure of how strongly a trait is expressed, penetrance describes the likelihood that an individual with a dominant allele will show the phenotype usually associated with that genotype. A dominant allele, such as the allele for purple flowers in Mendel's peas, has "complete penetrance," meaning that it will always be expressed.

A dominant allele with "incomplete penetrance" will, in certain physical or genetic environments, not be expressed. In flowers, this would simply mean that only a certain proportion (say, 10 percent or 75 percent) of individuals with a given genotype would show the phenotype. But penetrance can be a slippery concept. The allele for Huntington's disease, for example, is said to be completely penetrant. Theoretically, those who have the allele always get the disease. Because the allele may be expressed late in life, however, some people with the allele die of other causes before they experience any symptoms.

Why a gene varies in its expressivity and penetrance is usually not known. In well-studied species such as *E. coli* and *Drosophila,* specific genes, called "suppressor genes," suppress the effects of dominant harmful alleles. Both the environment (temperature and nutrition, for example) and interactions with other genes play crucial roles in determining the expression of individual genes. But the possession of a particular allele usually does not by itself predict what an individual's phenotype will be.

Figure 14-8 **The phenotype of an animal can change with its social environment.** The scalefinned anthias, shown here, lives in the Red Sea. It is one of dozens of fish that can change sex in response to a change in its social environment. If a school of females loses a breeding male, the dominant female changes into a male and begins breeding. Over a period of a few days, a female's ovaries convert into testes, while "her" coloring changes from golden orange to a pale background with magenta head and fins. Four days after the fish's transformation, he begins courting females. His anatomy, coloring, and behavior are indistinguishable from those of any other male's. *(David Hall/Photo Researchers)*

Because alleles vary in their expressivity and penetrance, the genotype does not define the phenotype.

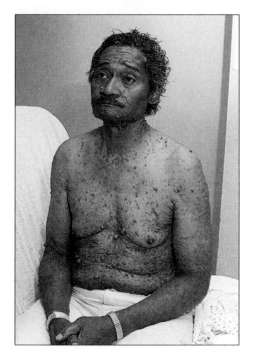

Figure 14-9 The symptoms of neurofibromatosis are distinctive. *(L. Steinmark/Custom Medical Stock Photo)*

A Single Gene Can Affect Many Traits

When a baby is born with the genetic disease cystic fibrosis, the baby's doctors know that it is destined for a life with chronic respiratory infections. Cystic fibrosis sufferers tend to have unusually salty perspiration and thick, sticky mucus that clogs the lungs and leads to repeated bouts of pneumonia, bronchitis, and other respiratory infections (Figure 14-10). In addition, many have blocked pancreatic ducts, which prevents the normal secretion of digestive enzymes. Males are frequently sterile, as the *vas deferens,* the tube that carries sperm from the testes, may also be blocked.

Cystic fibrosis can be caused by more than one kind of mutation. In about 70 percent of cases, however, all of the symptoms of cystic fibrosis are caused by the deletion of just three base pairs in a gene for a polypeptide called cystic fibrosis transmembrane conductance regulator, or CFTR for short. CFTR is 1480 amino acids long. The deletion of three base pairs results in the deletion of the 508th amino acid (phenylalanine) from the CFTR protein (Figure 14-11).

The CFTR polypeptide is a plasma membrane protein in *epithelial cells* that pumps chloride ions out of the cell. Epithelial cells are the cells that line the inner and outer surfaces of cavities. They can be found in every kind of duct—in the intestines, the lungs, the nose, and the mouth. Skin cells are also epithelial cells. The altered version of the CFTR polypeptide functions poorly, and epithelial cell fluids do not go where they are supposed to. As a result, sweat secreted by the skin is too concentrated, and the mucus secreted by the lungs, digestive tract, and sex organs is so thick and sticky that the ducts that secrete it frequently become clogged and infected. We can see,

then, that one mutation can have wide-ranging effects on the phenotype.

Not all mutated genes produce polypeptides that function as poorly as the altered CFTR polypeptide. Some altered proteins may be able to function reasonably well (and the phenotype will be normal, or nearly so). In between are cases such as sickle cell anemia, in which the mutation produces a protein that functions much of the time, but not always.

The allele that causes cystic fibrosis affects many aspects of phenotype. The capacity of one gene to have diverse effects is called **pleiotropy** [Greek, *pleios* = more + *trope* = turning]. Every gene that has been studied has been found to be pleiotropic to some degree. We can say with some assurance, then, that all genes in humans and in other organisms are pleiotropic.

A small mutation in a single gene, such as the one for the CFTR polypeptide, can affect many aspects of phenotype, a phenomenon called pleiotropy. All genes are pleiotropic.

A Single Trait Can Be Influenced by Many Genes

Phenotype depends on the interactions of many genes, as well as the effects of environment. Most traits are **polygenic,** meaning that they are influenced by more than one gene. Usually,

Figure 14-10 A child who has cystic fibrosis. Children who have cystic fibrosis must be thumped on the chest several times a day to loosen accumulated mucus in the lungs. *(© 1997 Abraham Menashe)*

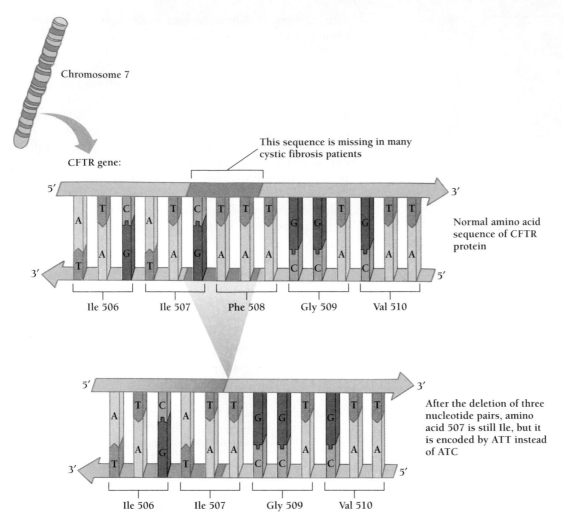

Figure 14-11   Translation of the gene for CFTR. Most cases of cystic fibrosis result from the deletion of three base pairs, one in the 507th codon and two in the 508th codon. The 507th amino acid is unaffected, but the 508th amino acid is missing and CFTR functions poorly.

we know little about the identities and effects of each of the genes. In a few cases, however, we know a great deal. In mice, for example, fur color depends on the action (and interaction) of at least five genes: one gene determines the distribution of pigment within individual hairs; a second gene determines the color of the pigment; a third gene allows or prevents the production of the pigment molecules; a fourth gene determines the intensity of the pigmentation; and a fifth gene controls the presence or absence of spots (Figure 14-12). The gene that regulates the intensity of pigmentation is called a modifier gene, because it regulates the action or "expression" of another, separate gene.

Many genes are now thought to be back-up systems for other genes. That is, the genome has several ways of accomplishing the same thing. More and more researchers are finding that knocking out seemingly important genes may have no effect on phenotype.

The long-haired mice mentioned in the last chapter are an example of such redundancy in genetic systems. In 1993, Gail Martin, at the University of California, San Francisco, found that knocking out a growth factor gene believed to influence early development in mice actually had no effect on development. The gene FGF5 was believed to play a role in the very early formation of an embryo. Martin expected that knocking out the gene would prevent early development and that none of the homozygous embryos would finish developing. The researchers were quite surprised and disappointed to see healthy mice born, but astonished when the mice all grew hair that was nearly 50 percent longer than normal (Figure 13-12). FGF5 normally is expressed in early embryo tissues, in muscle cells, in the hippocampus of the brain (the seat of learning), and in motor neurons. Yet, only the hair follicles were affected by the absence of the gene. Every other aspect of phenotype appeared normal.

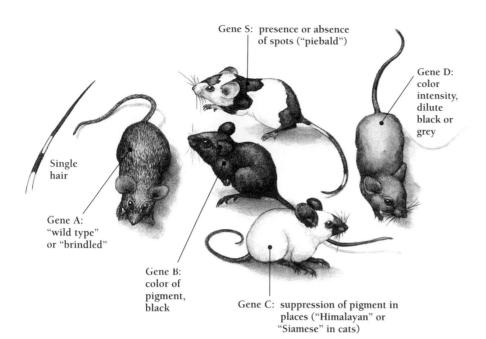

Gene S: presence or absence of spots ("piebald")

Gene D: color intensity, dilute black or grey

Single hair

Gene A: "wild type" or "brindled"

Gene B: color of pigment, black

Gene C: suppression of pigment in places ("Himalayan" or "Siamese" in cats)

Figure 14-12 **Several interacting genes determine coat color in mice.** Gene A determines where on the hair the pigments go. In wild mice, a band of yellow in the middle of the hair creates an "agouti" or "mousy" coat color. One allele confers a narrow yellow band, which gives a dark coat. Another allele confers a broader yellow band, which results in a lighter coat. In mice with a coat of a solid color, the yellow band is not present at all. The normal allele of gene B encodes a black pigment. Gene C suppresses the expression of pigment in the warmest parts of the body. Only the cooler ears, nose, tail, and paws are dark. Gene D regulates how much pigment each hair receives. Gene S determines the presence or absence of pigment in patches.

Martin and other molecular biologists now believe that nerve, muscle, and brain function survive knocking out the FGF5 gene, not because FGF5 has no function in these tissues, but because other genes provide backup systems. If this is true, it is reasonable to ask why other genes do not seem to provide backup for hair length. Possibly, short hair is not as crucial to survival and long hair is occasionally useful, unlike a defect in the development of the brain or the nervous system. For the moment, however, no one really knows.

In humans, as in mice and other organisms, the vast majority of traits are polygenic and probably redundant as well. Eye color, for example, is controlled by at least two genes. Studying polygenic traits is extremely difficult, since with each added set of alleles, the number of possible genotypes increases geometrically. If these are difficult to study in research animals or plants, they are nearly impossible to study in humans. As a result, almost nothing is known about polygenic traits (most traits) in humans.

The expression of each trait is the result of many genes acting together.

A Single Gene May Have Multiple Alleles

Most of the time we have spoken as if each gene can have only two alleles—a dominant allele and a recessive allele. But a gene may exist in many alternate forms, each of which contains a distinct sequence of nucleotides that specifies a distinct polypeptide. Some of the polypeptides cannot function at all, while others may function differently.

One familiar example of a gene locus with multiple alleles is the one responsible for the ABO blood types. This locus has three alleles, called I^A, I^B, and i, which specify different versions of a surface marker on red blood cells. (Although three alleles exist in most human populations, each individual has only two of the three alleles, one on each homologous chromosome.) Alleles I^A and I^B each specify an enzyme that helps to synthesize a glycoprotein marker on the surface of red blood cells. These two glycoproteins are called, respectively, glycoproteins A and B. Allele i specifies no enzyme and so neither cell surface glycoprotein is synthesized. An individual human may have one of six genotypes (Table 14-1): I^AI^A, I^BI^B, I^AI^B, I^Ai, I^Bi, or ii. Because alleles I^A and I^B are dominant to i, individuals with I^AI^A or I^Ai genotypes have type A blood, individuals with I^BI^B or I^Bi have type B blood, and individuals with ii genes lack these surface glycoproteins, so-called type O blood. Alleles I^A and I^B are codominant, meaning that in individuals with both alleles, both alleles are fully expressed. I^AI^B individuals have type AB blood.

Table 14-1 Human ABO Blood Groups

Genotype	Blood Type	Red Blood Cell Surface	Plasma Antibodies
I^AI^A	A	A glycoprotein	anti-B
I^Ai	A	A glycoprotein	anti-B
I^BI^B	B	B glycoprotein	anti-A
I^Bi	B	B glycoprotein	anti-A
I^AI^B	AB	A and B glycoproteins	no antibodies
ii	O	no glycoproteins	anti-A and anti-B

A single gene may have many functioning alleles. While some amino acid changes result in proteins that do not function at all, other amino acid changes alter the activity of the resulting protein. When both alleles are expressed, they are said to be codominant.

Are Recessive Alleles Rare?

Biologists often talk as if a dominant allele always contains the instructions for a functional protein, and a recessive allele always lacks this capacity. Often enough, this is true, and the dominant allele is the common, functional form. But these are only generalizations. The allele that gives people black hair, for example, is both common and recessive. And type O blood, the most common blood type in the world, results from a double dose of the recessive allele (ii). Further, a dominant allele is not necessarily either functional or common. In Huntington's disease, for example, the disease-causing allele is dominant. In addition, although this allele "functions" in a new way, it does not function in a way that extends the life of the individual.

Recessive alleles are not necessarily rare, and dominant alleles are not necessarily common. It is possible for a dominant allele to encode a nonfunctioning form of a protein.

THE HUMAN GENOME PROJECT

If each of the six billion human beings on Earth is genetically unique, we can conclude that human beings harbor enormous genetic diversity. At the same time, we must remember that we are still more than 99 percent identical. For this reason, it makes sense to most people to study "the human genome" as if it were a definable entity.

What Is the Human Genome Project?

In 1988, the National Institutes of Health (NIH), the federal agency that oversees and funds research in the health sciences, launched the largest and most expensive single project in the history of biology. The Human Genome Project is a 15-year, $3 billion project jointly funded by the Department of Energy and by NIH's National Center for Human Genome Research. The project is designed to map the human genome in four major ways:

1. The project has already made linkage maps of more than 3000 genes, restriction enzyme cut sites, or other markers. (Recall from Chapter 9 that the closer two genes are on a chromosome, the less likely they are to cross over during meiosis and assort independently.) Using family histories, geneticists can estimate the distance between any two genes—for which traits are known—by how often they cross over.

2. The project will construct a physical map of each human chromosome by first cutting the chromosomes into restriction fragments, then identifying unique sequences that can act as landmarks, and determining the distances to the next landmark.

3. The project also aims to sequence all 3 billion base pairs on one set (23 pairs) of chromosomes for two individuals (one man and one woman), then load this information onto a computer to create an electronic version of the genome.

4. The selection of *which* man and *which* woman was, naturally, extremely controversial, but the two individuals are of European descent. The project will therefore also collect additional information on alleles common in other parts of the world.

The earliest objection to the Human Genome Project was its cost. With a projected budget of $50 to $100 million a year, many biologists and biomedical researchers felt that the money could be better spent on hundreds of smaller, more focused projects. But supporters argued that the Human Genome Project would increase understanding of normal human physiology, human genetic diseases, and human evolution. And big science projects, such as the space shuttle, hold a lot of appeal in Washington, DC.

In 1990, work on the project was begun in laboratories all over the world under the direction of Nobel Laureate James Watson, the codiscoverer, with Francis Crick, of the structure of DNA. In spite of the large number of researchers involved, the task of sequencing the entire human genome seemed insurmountable. At the time, researchers had sequenced fewer than 5 percent of all genes.

Then, just as the project was getting under way, a molecular biologist at NIH named Craig Venter found a way to simplify one part of the job. Since only 3 to 5 percent of the 3 billion bases in the human genome actually code for proteins, he would extract mRNA from human tissues, copy it into cDNA with reverse transcriptase, clone the cDNA, then sequence it. That way, he would sequence only the 150 million nucleotides (100,000 genes × 1500 nucleotides) that actually code for proteins, leaving out all the introns and other junk DNA. Not only that, Venter used an automated apparatus recently invented by molecular biologist Leroy Hood to sequence the cDNA at a phenomenal rate. Venter's "genes," however, were only the exons of genes.

Hundreds of researchers all over the world are working on the rest of the Human Genome Project's goals. Advances speeded the work, and by the end of 1995, researchers had already completed the second part of the project, mapping more than 15,000 markers (such as restriction enzyme cut sites), which covered some 95 percent of the genome. Recently, researchers at the University of Washington have proposed that further research should be concentrated on sequencing only. Echoing the debate between Botstein and Housman described at the beginning of this chapter, the Seattle researchers argue that current DNA sequencing techniques are more than adequate to finish the project if the program's directors just focus efforts on sequencing.

If the plan to focus on sequencing only works, it would put the project 5 years ahead of schedule and, possibly, under budget. By 2001 or 2002, researchers hope to have finished sequencing all human DNA, as well as that of several other organisms, such as the fruit fly *Drosophila melanogaster,* the nematode worm *Caenorhabditis elegans,* the domestic dog *Canis familiaris,* and the plant *Arabidopsis thaliana.* Researchers have already finished sequencing the DNA of several species, including the bacterium *E. coli* and the yeast *Saccharomyces cerevisiae.*

The Human Genome Project offers to sequence an entire human genome, as well as the genomes of a few other organisms, and to construct linkage maps and physical maps of each human chromosome. By 2002, scientists will probably have sequenced the entire human genome, as well as the genomes of at least a half dozen other organisms.

How Do Researchers Locate Genes for Specific Diseases?

The drive to sequence the human genome has rapidly provided a gene library of actual gene fragments stored in numbered test tubes, as well as a computer database listing the nucleotide sequence of each human chromosome. Researchers express many of these genes (from mRNA) to find out what sorts of proteins they make. But these libraries have no catalogue to help researchers identify and locate individual genes. The contents of a gene library are therefore still something of a mystery. What genes does the library contain? How do they function in the body? Which chromosomes contain which genes? How can researchers find individual genes among the 80,000 to 100,000 human genes in the genome?

Pedigree Analysis

The time-honored way to locate a gene in the genome is by means of experimental crosses. Recall from Chapter 9 that experimental crosses reveal how frequently genes cross over, providing a measure of how far apart they are on a chromosome or whether they are on separate chromosomes. Although researchers cannot arrange experimental crosses in humans, they can obtain helpful information from many families about physical traits or genetic diseases extending back for generations.

Such records, arranged in the form of family pedigrees, show patterns of inheritance. Family pedigrees are useful for analyzing the inheritance of obvious traits that are passed in a very simple manner (Figure 14-13). Traits such as fatal diseases and whether the earlobe is "attached" are easy to identify.

A more sophisticated approach is to look for genetic markers that seem to travel with a disease from generation to generation, a technique called **pedigree analysis.** Researchers first collect blood samples from individuals in families that carry a given disease, then use DNA fingerprinting to look for RFLP alleles that are consistently associated with the disease. Recall that over long stretches of DNA, a restriction enzyme will cut fragments of varying lengths, depending on the number of nucleotides between cutting sites. Inevitably, some of these fragments will be near the gene in question.

In the search for the Huntington's disease allele, for example, James Gusella, at Massachusetts General Hospital, began looking for a particular RFLP allele that traveled with the Huntington's disease allele in some families. Because the human genome is enormous, and because no one knew even on what chromosome the Huntington's gene might lie, the task appeared Herculean. Yet, miraculously, Gusella immediately happened on a polymorphic marker that, in certain families, including the huge family at Lake Maracaibo, in Venezuela, was always inherited with Huntington's disease. This RFLP site came from the short arm of chromosome 4. In 1983, for the first time, the approximate location of the Huntington's gene was known.

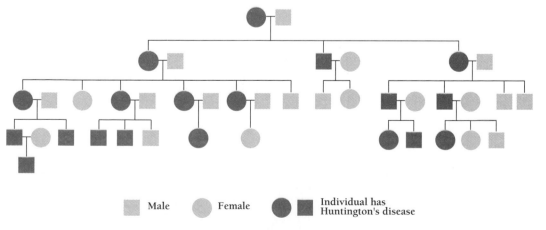

Male   **Female**   **Individual has Huntington's disease**

Figure 14-13   Pedigree for a family with Huntington's disease. The marriages and offspring of unaffected individuals have been omitted.

Pedigree analysis allows researchers to find a genetic marker associated with a particular trait or disease that shows approximately where on the genome a gene lies.

Positional Cloning

Once the approximate position of a gene has been located between two RFLP markers some million base pairs apart, researchers can use **positional cloning** to pinpoint the gene. The first step in positional cloning is the assembly of a library of clones that cover the entire region between the two RFLP markers on either side of the gene (Figure 14-14). Researchers clone numerous pieces of DNA that are known to overlap, then string them together into a continuous string that extends from one marker to the other, a so-called "contig."

Next, researchers narrow their search down to DNA sequences that are likely to encode a protein in an odd procedure called a *zoo blot* (Figure 14-15). They do this by hybridizing single strands of human DNA with single strands of DNA from rats, mice, or even flies or worms. DNA from a mouse, for example, can hybridize with human DNA to form a double-stranded molecule, one strand mouse and one strand human. Only sequences that have stayed more or less the same since mice and humans diverged millions of years ago, however, will hybridize. Sequences that do not hybridize are very different from one another and most likely represent junk DNA or introns that do not encode functional proteins. In contrast, the hybridized sequences mark genes whose exact sequences are so important that they have been conserved in the course of evolution. Mutations in these genes are far more likely to cause death or disease.

In the next step, molecular biologists look for mRNA in the tissues where the gene is expressed. A diseased gene that affects the brain, for example, would presumably be expressed in brain tissue. Researchers then compare the conserved sequences from diseased tissues with similar sequences from healthy individuals. Any substantial insertion or deletion is easy to spot as a difference in the length of the gene's restriction fragment.

Positional cloning is a time-consuming, tedious procedure, taking anywhere from 10 to 100 person-years to locate one gene. But many genes have now been isolated, including the gene for Duchenne muscular dystrophy, the gene for neurofibromatosis, and the gene for Huntington's disease.

Positional cloning allows researchers to locate a particular gene precisely on a chromosome.

Of What Use Is a Sequenced Gene?

Once a gene has been sequenced, geneticists can test individuals to see if they have a normal allele or a disease-causing allele of the gene. In practice, this may be neither easy nor necessarily useful. For example, in recent years, researchers cloned two genes, called BRCA1 and BRCA2, some alleles of which predispose women to breast cancer. Some of the alleles that cause cancer give a woman an 85 percent chance of developing breast cancer. The disease alleles of these two genes, however, can result from any of hundreds of different mutations, each of which may carry a different risk of causing breast can-

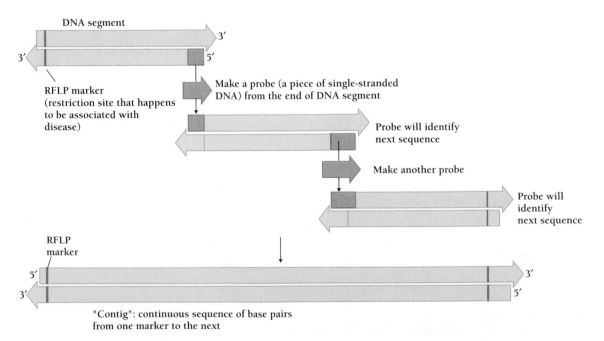

Figure 14-14   Assembly of a contig. To pinpoint a gene, researchers assemble a single long sequence of DNA between two RFLP markers on either side of the gene.

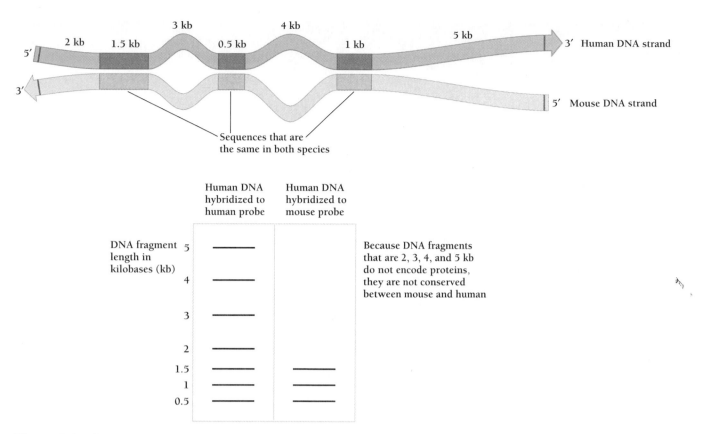

Figure 14-15   **Zoo blot hybridization.** Single strands of DNA from two different organisms will match up (and base pair) in some places, but not in others.

cer. In addition, other genes, both known and unknown, can alter that risk as well.

In any case, the vast majority of women who get breast cancer do not have an abnormal allele of either BRCA1 or BRCA2. It only makes sense to test women who come from families where breast cancer is strongly inherited and common. Even for that minority of women, however, the test is of limited value. Molecular biologists' knowledge of the sequences of BRCA1 and BRCA2 does not enable them to either prevent or cure breast cancer. Finally, once tests reveal that a person has a gene that predisposes him or her to a particular disease, insurance companies will not want to provide health insurance to that person for that disease. Most geneticists argue that no genetic information should be made available to insurance companies. But insurance companies have a powerful economic incentive to overturn or sidestep any rules limiting their access to genetic information. For the average person, then, a genetic test can potentially do more harm than good.

On the other hand, researchers who know the sequence of a disease allele are in a good position to begin to understand how the gene does its damage. Huntington's disease provides a good example of this. In 1995, researchers found that the gene involved in Huntington's disease codes for a polypeptide called huntingtin that is expressed in every cell in the body. A second protein, HAP-1, which is expressed only in brain tis-

sue, binds to huntingtin in brain cells. Researchers hope to discover whether either of these interactions is responsible for the gradual destruction of brain cells that leads to the dementia and loss of motor control in Huntington's disease. Already they have found that still another protein—a common metabolic enzyme found in all cells—also binds to the disease-causing version of huntingtin.

If researchers can understand how an alteration in huntingtin leads to cell destruction, then someday they may be able to devise a treatment for this disease. Thus, sequencing human genes remains a field of bright, but as yet unfulfilled promises.

Can Human Genes Be Patented?

In 1991, Craig Venter suggested that NIH patent the protein-encoding cDNA, so that any commercial uses for the genes would earn royalties for the federal agency. In June 1991, NIH filed applications on 315 gene fragments. But the scientific community protested. The human genome was viewed as a common heritage, not something that could be patented so that one person or one business could make money from it. Even religious coalitions have made formal objections to the idea of patenting human genes, and whether the patents will be upheld if they are challenged legally is still in doubt.

BOX 14-2

Human cloning: A multiplicity of sticky options

When the *Wall Street Journal* asked business leaders whether they would clone themselves if the technology became available, most said yes. Albert Dunlap, CEO of Sunbeam Corporation, said he would like to have 12 copies of himself, because he could think of 12 companies that could use his help to restructure. "You could have 12 fantastic turnarounds," he said.

Is Dunlap right in his assumption that his clones would possess his business savvy? If you could drop a few skin cells into a cloning machine and have your spitting image stroll out the other side, would you do it? Would your clone walk, talk, or behave like you?

If you have ever known a pair of identical twins, you already have part of the answer. Identical twins share their entire genetic endowment, just as you would with your clone. In fact, identical twins are real life clones. While twins generally look alike and may share many habits and traits, they also have different hopes and aspirations and follow different paths.

But your clone would be even more different from you than a twin sister or brother. The *Journal*'s question assumed that technology will someday make it possible to produce photocopies of ourselves, pre-aged and suffused with a life history and a mature consciousness. In fact, this will never be possible.

The cells in the body of a typical college student, for example, have taken nearly two decades to reach their current state. Stem cells (undifferentiated precursor cells that generate your skin and other tissues) have divided thousands of times, making adult skin tougher and coarser than baby skin. Your immune system has evolved to

resist thousands of diseases, and your brain and nervous system have grown and developed in response to a lifetime of experience. None of that was in your DNA when you were born. If technology existed that could clone you, your genetic double would have to begin at the beginning, as an embryo. Your twin would be 20 years younger than you, growing up in a different home, a different school, with different friends, fashions, music, and customs. In 20 years, you might not even recognize yourself.

So far, no one has cloned a human being, but scientists have been cloning animals since the 1950s. The reason for early cloning attempts was to find out whether all cells retain their complete genetic potential as they divide and specialize to form tissues such as skin, blood, and brain. Beginning with amphibians, scientists replaced the nuclei of a frog egg with nuclei from the cells of an embryo, tadpole, or adult frog. Eggs transplanted with nuclei from young embryos nearly always developed into normal tadpoles. But nuclei taken from older embryos and from tadpoles produced progressively fewer tadpole clones. In some cases the egg did not even divide once. Scientists eventually succeeded in generating swimming tadpoles with nuclei taken from adult frog skin, but these tadpoles always died before metamorphosis. It appeared that nuclei from specialized cells were somehow different from the nucleus of an egg.

Cloning adult animals finally became a reality in 1997 when Ian Wilmut, a biologist in Scotland, announced the birth of Dolly, a lamb whose genetic constitution had come from cells of her "mother's" udder. Cloning itself was not new. For years,

agricultural biologists had been cloning sheep, pigs, and cattle—but only from the cells of very early embryos that had not yet implanted in the mother's uterus.

In order to clone an adult, Wilmut and his colleagues found that they had to trick the donor cell into behaving as if it were part of an early embryo. By growing the mother's mammary cells in a dish with a limited supply of nutrients, the researchers found they could push the cells back to a less specialized state. Wilmut's cloning success proved that differentiated cells have not necessarily lost their genetic potential.

But is animal cloning useful? Some biotechnology companies think so. These companies already genetically alter farm animals to produce proteins for treating cancer and other diseases, or to create organs for human transplant. Now they have taken an interest in cloning as a way to make copies of mature animals with laboriously constructed genotypes. But others are waiting to see if the technology can be perfected. The Scottish team had to transplant some 300 nuclei to get just one lamb.

Should cloning ever be tried in humans? Wilmut has said he finds the idea morally repugnant and sees no reason even to try. Others object based on the possibility of selecting a nucleus with damaged DNA, which might produce a clone with a genetic disease. Nonetheless, nightmare visions of hordes of mutant drones manufactured to perform menial tasks for a leisure class are probably exaggerated. Cloning human beings would merely produce more human beings with the same natural revulsion to enslavement we all possess.

Ultimately, the applications were withdrawn, but Venter was determined. He started a nonprofit organization called The Institute for Genomic Research (TIGR) that would continue his work at NIH on a massive scale, but funded by private companies. By 1995, TIGR had sequenced pieces of nearly 90 percent of the genes expressed in human tissues, that is, the cDNA version of all the different kinds of mRNA. A second, for-profit company, Human Genome Sciences, Inc. (HGS), patents and sells whatever interesting genes TIGR clones. Pharmaceutical companies have lined up to give TIGR millions of dollars in exchange for a stake in its discoveries. Drug companies are gambling that human proteins made from cDNA will turn out to be useful and lucrative drugs that will, for example, boost the immune system or dull the appetite. Meanwhile, researchers

BOX 14-3

CAG repeats in huntingtin

When the Huntington's Disease Collaborative Research Group cloned the HD gene in 1993, they got a surprise. Instead of just two alleles, normal and abnormal, they found dozens, all differing by a single factor: the number of glutamines in the protein product.

The length of huntingtin, the protein encoded by the HD gene, varies depending on the number of CAG codons in the gene. CAG is the codon that specifies the amino acid glutamine. Normal alleles can contain as few as 10 and as many as 35 repeats of CAG, while alleles associated with Huntington's disease may have anywhere from 37 to 121 CAGs.

These abnormal alleles are unstable, meaning that more CAG repeats accumulate as cells divide. Not only do different cells in the body of a Huntington's patient have different numbers of CAG repeats, but the offspring of Huntington's sufferers also tend to end up with even more of the triplets than their parents. Generally, the more CAG repeats beyond the normal range in a huntingtin allele, the earlier the onset of the disease.

No one knows what the polyglutamine tract does, but it may help huntingtin bind to other proteins. So far, scientists have discovered two proteins that bind to huntingtin. One has no known function, but the other is a fuel-burning protein called glyceraldehyde-3-phosphate dehydrogenase, or GAPDH.

GAPDH, found in every cell, is a crucial enzyme that participates in glycolysis, a reaction that mobilizes the energy stored in glucose. Since the brain gets nearly all of its energy from glucose, brain cells rely heavily on GAPDH. Huntingtin binds this protein via the polyglutamine stretch, suggesting that perhaps the expanded glutamine repeats found in Huntington's disease inhibit the normal function of GAPDH. The resulting decrease in energy production would gradually kill off the brain and spinal cord cells where huntingtin is expressed.

Expanded glutamine repeats occur in four other degenerative diseases of the nervous system besides Huntington's, and at least one of the proteins involved also binds to GAPDH. Although not all the evidence is in, GAPDH (or some other shared protein partner) may be the common factor in all of these diseases. In that case, a single therapy might treat them all.

at many universities and several pharmaceutical companies are freely publishing the full sequences of as many genes as they can, so that at least some human gene sequences will be public property and not patentable. Other universities are themselves patenting DNA sequences.

Businesses and universities are rapidly patenting cDNAs that code for human proteins. Many people, however, doubt that all these patents will withstand legal challenges.

PREVENTING GENETIC DISEASE

Genetic research has greatly advanced our understanding of how certain genetic diseases are inherited and how they make people ill. Medical researchers have used this information to try to help people avoid having children with serious defects, or, in a few rare cases, to actually treat people with genetic diseases through "gene therapy." In this section we will see how this is done.

Only about 3 percent of all human diseases are caused by defects in a single gene. The most common genetic disease in the United States is cystic fibrosis, which affects about 1 person in 10,000. None of these single-gene diseases—several of which we have discussed in this chapter—are the ones that kill most

adults. For comparison, cardiovascular disease kills more than one person in three. About half that number die of cancer. People fall prey to one disease or another—lung cancer versus stroke, for example—because of lifelong behaviors and experiences and because a constellation of different genes slightly predisposes them to one weakness or another. Every person carries at least 5 to 10 genes that could make them seriously ill under certain circumstances. As geneticist Michael Kaback puts it, "We are all mutants. Everybody is genetically defective."

Genetic Counseling: How Can Parents Decide?

The first kind of genetic testing is simply to test couples who are considering having a baby to see if they carry alleles for genetic diseases, such as sickle cell anemia, Tay-Sachs, cystic fibrosis, or Huntington's disease. Geneticists can now test for more than 100 different genetic defects. Usually, however, geneticists test only for the alleles that are relatively common or for those that the parents seem most likely to carry. A genetic counselor can calculate the likelihood that a child will have a certain disease and help the couple decide whether to risk having children or not.

Tay-Sachs, for example, is a genetic disease that results from a single defective gene for an enzyme normally synthesized by nerve cells. In healthy children, this enzyme breaks down lipids. But in children homozygous for the Tay-Sachs allele, lipids accumulate in the brain, causing mental retardation, blindness, and failure to develop control of the muscles. All Tay-Sachs children die in early childhood, usually by age 4. Tay-Sachs has a simple pattern of inheritance because all children homozygous for the single Tay-Sachs allele become ill. Heterozygous individuals, who have one defective allele and one normal allele, are called "carriers." Carriers can be identified by their low levels of the lipid-digesting enzyme, but they are otherwise healthy.

The Tay-Sachs allele is most common among Eastern European Jews and their descendants. If both parents are of Eastern European Jewish descent, a genetic counselor will likely recommend that they be tested for the allele. If only one parent is Jewish, however, the counselor will probably not recommend the test, because a baby can only become ill if it receives the allele from both parents. This population of people has taken such care to avoid having Tay-Sachs children, however, that Tay-Sachs is now quite rare and, at least in America, the disease may now be more common in the population at large than among those descended from Eastern European Jews. Tay-Sachs is one of the great successes of genetic testing.

Genetic counselors can tell couples whether they carry certain alleles and the likelihood that their offspring will express certain traits.

Prenatal Testing: Can Parents Tell Before?

Once a woman becomes pregnant, physicians can obtain cells from the fetus in order to search for genetic defects. Besides looking for chromosomal abnormalities by examining the chromosomes, genetics clinics can now detect genetic diseases in the cells of an early embryo. Once the parents know that their baby has a genetic disease, however, they have only two options. They may decide to have the mother carry the fetus to term (birth) and cope as best they can with the baby's condition, or they can decide to abort the fetus.

Amniocentesis

By examining chromosomes from a fetus early in pregnancy, geneticists can tell if the parents have produced a zygote with a chromosomal abnormality. The most common chromosomal abnormality found in humans is trisomy 21, or Down syndrome. Geneticists usually obtain the fetal chromosomes by taking a sample of fetal cells from the **amniotic fluid** [Greek, *amnion* = membrane around a fetus], the watery fluid that surrounds the fetus. This procedure, called **amniocentesis** [Greek, *centes* = puncture], is usually recommended to women over 35 years old and to women with a family history of chromosomal abnormalities or other diseases (Figure 14-16A). Amniocentesis can detect more than 100 genetic abnormalities besides Down syndrome.

Figure 14-16   Prenatal testing.
A. In amniocentesis, a physician extracts amniotic fluid through a needle inserted through the mother's abdominal wall.
B. In chorionic villi sampling, a physician extracts cells from the placenta through the mother's vagina and cervix.

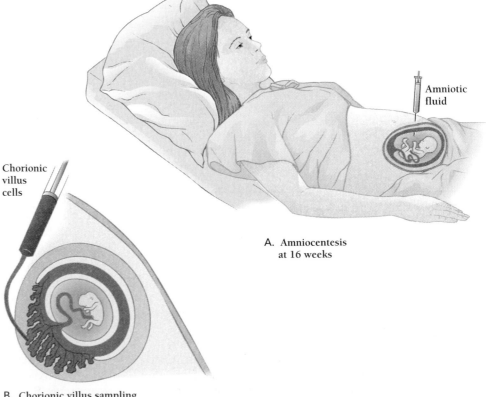

Amniotic fluid

A. Amniocentesis
at 16 weeks

Chorionic villus cells

B. Chorionic villus sampling
at 6–12 weeks

At 15 or 16 weeks after conception, a physician inserts a needle through the mother's abdominal wall, just below her navel, and into her uterus. The needle is used to remove from the womb amniotic fluid that contains cells from the fetus. This procedure is virtually painless, but can be emotionally upsetting for some women. Amniocentesis is very safe, but not perfectly so. Approximately 0.5 percent of fetuses exposed to amniocentesis later die because of spontaneous abortion, commonly known as miscarriage. (Some 10 to 15 percent of known pregnancies abort spontaneously before the 20th week, usually because of chromosomal or other abnormalities in the fetus. After 20 weeks, an early delivery is called a "preterm birth.")

To prepare the chromosomes for examination takes 3 to 4 weeks, during which a geneticist grows the fetal cells in culture. During this time, the fetus continues to mature and, indeed, approaches the age when it becomes viable outside of the womb, which is about 22 weeks. (A full-term baby is 32 to 40 weeks.) Parents meanwhile think about what they will decide if the fetus turns out to have Down syndrome or some other chromosomal defect.

Amniocentesis allows a geneticist to look for chromosomal or other genetic abnormalities in a four-month-old fetus.

Chorionic Villus Sampling

A second procedure, called **chorionic villus sampling (CVS)**, uses fetal cells present in the placenta—the tissue that carries nutrients and oxygen from the mother's blood to the fetus and that carries wastes from the fetus back to the mother (Figure 14-16B). Chorionic villus sampling may be done as early as 6 to 12 weeks after conception, much earlier in pregnancy than amniocentesis. Because the results of the test become available so much sooner, a decision to abort becomes more acceptable to some people. Alternatively, news that the fetus is healthy means a shorter period of worry for the parents. The increased risk of spontaneous abortion due to CVS is about the same as in amniocentesis, a half a percent.

Chorionic villus sampling allows geneticists to test for the same genetic defects as amniocentesis, but earlier in pregnancy.

Aborting a Fetus

In the next section, we see that researchers are working hard to develop techniques for actually curing genetic diseases, or, at least, mitigating their effects. For now, however, prenatal testing implies only one decision: to abort or not. For many people abortion is unacceptable under any circumstances. For them, prenatal testing provides time to plan for the care of a child with unusual needs.

For most people in the United States, however, abortion is acceptable under certain circumstances. Each family's cir-

cumstances are unique, however, and what is acceptable varies enormously from individual to individual and from situation to situation. For example, more people find it acceptable to abort a fetus with Tay-Sachs than a fetus with Down syndrome or sickle cell anemia. A child with Tay-Sachs always dies young and cruelly. The effects of Down syndrome, however, vary enormously, and many of its symptoms can be treated. A child with Down syndrome can experience a happy childhood and may live well into adulthood (Figure 14-17).

Likewise, a person with sickle cell anemia may have few or no symptoms much of the time and live well into middle age. New treatments may prevent any significant ill effects. Even now, when a sickle cell crisis does occur, and such a crisis can be both life-threatening and agonizingly painful, it usually can be controlled with prompt medical attention. Many symptoms of cystic fibrosis can also be effectively treated with drugs.

Many parents might decide to abort a fetus with Down syndrome, sickle cell disease, or cystic fibrosis, but many others would choose to let the fetus develop into a baby. In neither case is the decision a light one. Parents must weigh their feelings about abortion against their family's ability to marshal both the emotional and the financial resources necessary to provide adequate care for the child (as well as the adult the child becomes) who has special needs. The ethical values of the family, the developmental age of the fetus, the financial status of the family, and the stability of the parents' relationship may all be considerations.

Figure 14-17 A child who has Down syndrome *(right)*. *(Gary Parker/Science Photo Library/Custom Medical Stock Photo)*

Test-Tube Babies from Selected Gametes

Recently, medical researchers have come up with a way for parents to avoid having a baby with a specific genetic defect without confronting the question of abortion. Very simply, a doctor takes unfertilized eggs from a mother who is heterozygous for a particular genetic defect and selects the eggs that lack the allele in question.

The techniques for doing this are rather complex. In one example, the woman must first be treated with hormones that induce her to ovulate several eggs at once. A surgeon then removes these freshly ovulated eggs from the mother's abdomen and takes them to a laboratory. Freshly ovulated eggs have barely completed meiosis, and they are still attached to a sister cell. Because the two daughter cells do not divide evenly, however, the egg cell is very large, while the other cell, called the *polar body,* is very tiny.

The polar body is normally sloughed off, but geneticists have found a use for it. Because homologous chromosomes separate during meiosis, only one of the two cells gets the defective gene. If researchers find that the polar body has the defective gene, then they know that the egg has the normal copy. The normal eggs can then be combined with sperm and implanted in the mother's uterus.

The number of genes that researchers can examine by means of this method will increase 10 times in the next decade, since the Human Genome Project is producing potential new genetic tests at the rate of about one a week. Because this method of making babies is rather expensive, only the most motivated couples—such as those carrying alleles for Tay-Sachs—are likely to take advantage of it.

Medical researchers have invented ways to select human eggs that are free of certain genetic defects, but this technology is cumbersome and expensive.

Gene Therapy: Can We Fix It Later?

Gene therapy is a technology researchers are developing to deliver normal genes to the tissues of people whose own genes are somehow defective. Researchers have attempted to treat some 18 different diseases, using dozens of different methods. So far, however, not one variety of gene therapy has resulted in a complete cure of even one person. Neither the genotype nor the phenotype has been "fixed." In the best cases, researchers can effect a partial, temporary cure that must be renewed every few months, often in combination with conventional drug therapy (Figure 14-18).

Gene therapy may, theoretically, apply to somatic or germ cells. Recall that the germ cells are the cells that, through meiosis, form the gametes. Genetically engineered germ cells can pass on their altered genes to future generations. Researchers can currently create transgenic mice, pigs, and goats, for example, in which a gene is inserted into the genome of an animal when it is still a zygote, right after fertilization. These an-

Figure 14-18   The most successful example of gene therapy. In 1990, researchers at the National Institutes of Health treated Ashanti DeSilva for the genetic disease SCID (severe combined immunodeficiency disease), which almost completely disables the immune system. Researchers removed immune cells from the 4-year-old DeSilva, added a functioning allele for the enzyme adenosine deaminase, then injected the genetically engineered cells back into her body. With repeated treatments, DeSilva has been transformed from a chronically sick child with a short expected life span to an evidently healthy child.   *(Ted Thai/Time Magazine/© Time Inc.)*

imals can pass the gene they receive from researchers on to their own offspring. Human germ-cell therapy would require that researchers successfully replace nonfunctioning alleles with functional alleles in either zygotes or in germ cells of more mature individuals. Although knockin technology now makes this theoretically possible, the technique is still too new for use in humans.

For the foreseeable future, gene-therapy research will be confined to treating only somatic cells. Research on gene therapy for cystic fibrosis constitutes some of the most promising work in the field of somatic-cell gene therapy. Cystic fibrosis is the most common serious genetic defect in the United States, afflicting some 30,000 people, or about 0.01 percent of the population. Recall that the many symptoms of cystic fibrosis, from chronic respiratory infections to blocked pancreatic ducts, all seem to result from a defective ion transport molecule in the cell membrane.

In many genetic diseases, the body's cells normally need to express large amounts of the missing protein. In cystic fibrosis, however, even very low levels of expression of the CFTR gene—say 10 to 100 molecules per cell—reduce symptoms dramatically. This fact makes cystic fibrosis more amenable to treatment through gene therapy than other diseases.

One partly successful approach to treating cystic fibrosis through gene therapy has been to spray genetically engineered cold viruses into the noses of patients who have the disease. The engineered *adenovirus* carries a normal version of the CFTR gene into the cells of the inner surface of the nose. The virus is deliberately damaged so that it cannot replicate in the body and make the patient ill. It can only deliver its cargo—the CFTR gene—into the infected cells.

The treated cells express the normal CFTR gene, and the patient's symptoms are alleviated, but only partly. Treating the back of the nose does not, for example, unblock the pancreatic ducts or help alleviate symptoms in any other part of the body. Further, the cells that the virus infects live only a few weeks before they are shed and replaced by new cells. These new cells must be infected with another treatment of virus. Patients must therefore return again and again for more treatments.

Why isn't the gene passed on to daughter cells? Recall that a gene can be integrated into the genome (into one of the chromosomes) or it can simply be deposited into the cell along with some regulatory DNA that ensures that it will be transcribed. Researchers could insert the CFTR gene into the genome of a human cell if they wanted to and then the CFTR gene would be passed to daughter cells. But a gene randomly inserted by a viral vector can affect nearby genes. It can, for example, turn off an adjacent tumor suppressor gene and trigger cancerous growth. So far, researchers cannot insert genes into chromosomes in cells in the human body without the risk of damaging other genes.

Currently, researchers in gene therapy almost always put the gene into the cell nucleus, but not into the chromosomes. Such genes are not replicated when the cells divide. In some cases, however, researchers can insert genes into the chromosomes of dividing cells, so that the cells' progeny also carry the new gene. Although such engineered cells are passed on within a particular tissue, other tissues of the body do not have the engineered genes. In particular, sperm and eggs will not carry the engineered gene, so it cannot be passed to any children.

Researchers have also recently created artificial chromosomes, which undergo mitosis in cultured human cells and are passed on to progeny cells. Artificial chromosomes may some-day be used to carry functioning genes. However, the simplest solution to the technical challenges of gene therapy may be the perfection of the technology for making "knockins"—alleles that are inserted in place of a damaged allele in exactly the right place.

Until then, gene therapy will be only a temporary cure that will need to be renewed every few weeks or months. And, as mentioned earlier, an individual with a genetic disease will always harbor the defective allele in the germ cells and can continue to pass it on to offspring. For now, gene therapy remains an area of basic research, not of practical results.

As a panel convened by NIH reported in late 1995, gene therapy is still in its infancy despite expenditures of more than $400 million per year and more than 100 clinical trials in humans. Many of these trials, said the report, were poorly designed, and none has demonstrated even one technique that actually cures genetic disease. The NIH report further accused researchers of both "overselling" gene therapy and of neglecting basic questions about gene regulation, vectors, stem cell function, and other areas. In short, people with genetic diseases should not expect effective gene therapy any time soon.

Using new, gene-therapy techniques, researchers can cause human cells to express a missing gene. Expression is limited, so far, to a relatively few molecules of the desired protein. Because the cells usually cannot pass the gene on to new daughter cells in the body, treatments must be repeated. Also, the treated patient cannot pass on the "good" allele acquired in this way to any offspring.

In this last chapter on genetics, we have used the knowledge gained from previous chapters to examine genetic diversity in humans, the complex relationship between genotype and phenotype, the Human Genome Project, and the promise and limitations of applied human genetics. Now that we understand inheritance, we are ready to tackle evolution, the most important idea that biology has ever produced.

STUDY OUTLINE WITH KEY TERMS

DNA fingerprinting identifies variations in DNA, called **RFLPs (restriction fragment length polymorphisms),** derived from blood or other tissues. It provides an estimate of how common that particular pattern of variation is in the population. Humans are 99 percent genetically identical. Yet every individual who is not an identical twin has a unique genotype. Human races blur into one another, so that defining consistent races is essentially impossible. The differences among the members of one race are greater than the average differences between races.

A small mutation in a single gene, such as the one that causes most cases of cystic fibrosis, can affect many aspects of phenotype. But the relationship between genotype and phenotype is generally extremely loose. One reason is that alleles vary in their **expressivity** and **penetrance.** In addition, each gene affects many aspects of phenotype—a phenomenon called **pleiotropy.** The expression of each trait is **polygenic,** the result of many genes acting together. Finally, a single gene may have many functioning alleles. Because of all this complexity, the expression of any gene is heavily influenced by both the physical and developmental environment and also by the genomic environment of the gene.

The Human Genome Project offers to sequence an entire human genome, as well as the genomes of a few other organisms, and to construct linkage maps and physical maps of each human chromosome. By 2002, the Human Genome Project will probably have sequenced the entire human genome. Meanwhile, businesses and universities are patenting DNA fragments that encode human proteins.

Pedigree analysis allows researchers to find a genetic marker associated with a particular trait or disease that show approximately where on the genome a gene lies. **Positional cloning** allows researchers to obtain a particular gene on the basis of its precise location.

Genetic counselors can tell couples both whether they carry certain alleles and the likelihood that their offspring will express certain traits. In **amniocentesis,** geneticist sample a 4-month-old fetus's **amniotic fluid** and look for chromosomal or other genetic abnormali-ties. **Chorionic villus sampling (CVS)** allows geneticists to test for the same genetic defects as amniocentesis, but earlier in pregnancy.

Using new **gene-therapy** techniques, researchers can induce human cells to express a gene that the cells were previously unable to express. Expression is presently limited, however, to relatively few molecules of the desired protein. Because the cells cannot pass the gene on to new daughter cells in the body, treatments must be repeated. Also, the treated patient cannot pass on the "good" allele acquired in this way to any offspring.

REVIEW AND THOUGHT QUESTIONS

Review Questions

1. What factors ensure that the relationship between genotype and phenotype is a loose one? What is pleiotropy?
2. What are the limitations and strengths of DNA fingerprinting? What is a RFLP?
3. What is the purpose of the Human Genome Project?
4. How do researchers locate genes for specific diseases such as Huntington's or cystic fibrosis?
5. How is prenatal testing different from genetic counseling?
6. If a prenatal test reveals a genetic defect in a fetus, what choices do the parents have?

Thought Questions

7. If you and your spouse both carried the gene for cystic fibrosis, would you get prenatal testing? If you decided to have a test, would you prefer amniocentesis or chorionic villus sampling? Why?
8. How would your current knowledge about research on gene therapy for cystic fibrosis influence your decision?
9. If you knew you had an allele that slightly predisposed you to not get heart disease, would that influence your eating and exercise habits? If you had an allele that slightly predisposed you to get liver cancer, would you care if your insurance company and your employer knew?

SELECTED READINGS

Cranor, Carl F., ed., *Are Genes Us? The Social Consequences of the New Genetics,* Rutgers University Press, New Brunswick, New Jersey, 1994. A collection of essays on ethical and other social consequences of research on genetics, especially human genetics.

Kevles, Daniel J., and Leroy Hood, eds., *The Code of Codes: Scientific and Social Issues in the Human Genome Project,* Harvard University Press, Cambridge, Massachusetts, 1992. Another collection of essays about the new genetics.

Watson, James D., Michael Gilman, Jan Witkowski, and Mark Zoller, *Recombinant DNA,* 2nd ed., Scientific American Books, W.H. Freeman, New York, 1992. The last five chapters of this book discuss the impact of recombinant DNA on human genetics. Watson is past director of the Human Genome Project.

Wexler, Alice, *Mapping Fate: A Memoir of Family, Risk, and Genetic Research,* Times Books, Random House, 1995. A revealing and emotional account of the Wexler family's trials with Huntington's disease, as well as a lucid account of the nearly 20 years of research on the Huntington's gene.

Wilkie, Tom, *Perilous Knowledge: The Human Genome Project and Its Implications,* University of California Press, Berkeley, 1993. An accessible discussion of the flowering of human genetics by an experienced science writer.

▶ On-line materials relating to this chapter are on the World Wide Web at http://www.saunderscollege.com/lifesci/
Click on Tobin/Dusheck: *Asking About Life.*

PART III

Evolution

Grand Canyon National Park seen from Mohave Point at sunset.
(© 1982 Gary Withey/Bruce Coleman, Inc.)

The Scopes Trial

"Nothing in biology makes sense except in the light of evolution."
—Theodosius Dobzhansky

In Dayton, Tennessee, on the morning of July 10, 1925, high-school teacher John T. Scopes went on trial for violating a state law that prohibited the teaching of evolution in public schools. In the hot courtroom, spectators mopped their brows and swatted at flies. On a table in a back corner stood a dozen journalists trying to get a better view. Midway through the testimony, the table collapsed with a groan. Through the open windows, drowning out the judge's calls for order, came the cheers and boos of a rambunctious audience who were listening from the wide lawn outside. In quiet moments, the trumpeting voices of a score of evangelists outside filtered through the testimony.

The town of Dayton was bedecked as if for a carnival, with banners and posters of monkeys. Newly constructed stands selling hot dogs, lemonade, and sandwiches lined the sidewalks, and a circus man ran a sideshow with two chimpanzees, brought to town to "testify for the prosecution," as the man put it. Two hundred and twenty-five newspapermen had arrived from New York, Baltimore, Philadelphia, and points beyond. They would put the tiny southern town on the map forever.

The law prohibiting the teaching of evolution—passed by the Tennessee legislature in a moment of boredom—was never intended to be enforced. It nonetheless attracted the attention of the American Civil Liberties Union (ACLU), which believed the law violated the First Amendment right to free speech. Eager to test the law in a federal court, the ACLU advertised in the *Chattanooga News* for a teacher willing to be arrested. Dayton boosters jumped at the chance to liven up their sleepy town with some national attention.

Scopes agreed to be arrested. An affable high-school physics teacher and athletic coach, Scopes had never actually taught evolution, but he'd once substituted for a biology teacher who was ill and assigned the textbook pages that covered evolution.

The trial—loudly and forcefully argued by two nationally famous lawyers, William Jennings Bryan and Clarence Darrow, and covered by every major newspaper in the country—was, nevertheless, a rather dull affair. The attorneys represent-

The Granger Collection, New York

ing each side wanted the same outcome—Scopes's conviction. The ACLU wanted a clean conviction so they could appeal the case in a higher court, and, perhaps, have the antievolution law overturned.

Eager to test the law in a federal court, the ACLU advertised in the Chattanooga News *for a teacher willing to be arrested.*

But it was not to be. In an ironic twist, the ACLU won the battle it was trying to lose and so lost the war. Scopes was initially convicted, but then acquitted on a technicality. The judge technically violated the law when he mistakenly levied Scopes's $100 fine himself, instead of letting the jury do it. The ACLU never got to appeal the case, and the Tennessee antievolution law stayed in effect for 42 years, until 1967.

The Tennessee antievolution law was never again applied. But American textbook publishers—ever shy of controversy—nevertheless took the hint, reducing or even eliminating discussions of evolution in most high-school textbooks. By 1942, fewer than half of all high-school science teachers were covering evolution at all. Only with the Soviet Union's 1957 launching of the first satellite, Sputnik, did the United States begin emphasizing better science education and textbooks. But the increasing appearance of evolution in textbooks sparked a new reaction among antievolutionists, who realized that their fight against science was far from over.

In 1973, antievolutionists in Tennessee passed a bill calling for equal time for teaching evolution and creation, the biblical myth that the universe was created by the Judeo–Christian God in seven days. **Creationism** says, in particular, that the Earth was created very recently and that each species was created individually. The "equal-time" bill was later declared unconstitutional on the grounds that it violated the separation of church and state. Then, in 1981, a creationist filed suit against the state of California, charging that the teaching of evolution in the schools violated his children's religious freedom. The judge rejected the suit.

In the same year, Arkansas and Louisiana passed identical bills, mandating that teachers present the scientific evidence for the biblical story of creation whenever they taught evolution. Creationists argued that the account given in the Bible's book of Genesis was the basis for "Creation science." The bills stated:

"Creation-science" includes the scientific evidences and related inferences that indicate: (1) Sudden creation of the universe, energy and life from nothing; (2) The insufficiency of mutation and natural selection in bringing about the development of all living kinds from a single organism; (3) Changes only within fixed limits of originally created kinds of plants and animals; (4) Separate ancestry for man and apes; (5) Explanation of the earth's geology by catastrophism, including the occurrence of a worldwide flood; and (6) A relatively recent inception of the earth and somewhat later of life."

In 1982, the Arkansas law was overturned by Federal Judge William R. Overton, who wrote, ". . . the evidence is overwhelming that both the purpose and effect of Act 590 is the advancement of religion in the public schools." He also noted, in reference to creation science, that "While anybody is free to approach a scientific inquiry in any fashion they choose, they cannot properly describe the method used as scientific if they start with a conclusion and refuse to change it regardless of the evidence developed during the course of the investigation."

On the basis of Overton's decision, another judge struck down the Louisiana law as well. And in June 1987, the United States Supreme Court affirmed this decision in a 7–2 vote, effectively ending a dispute that had begun 128 years before.

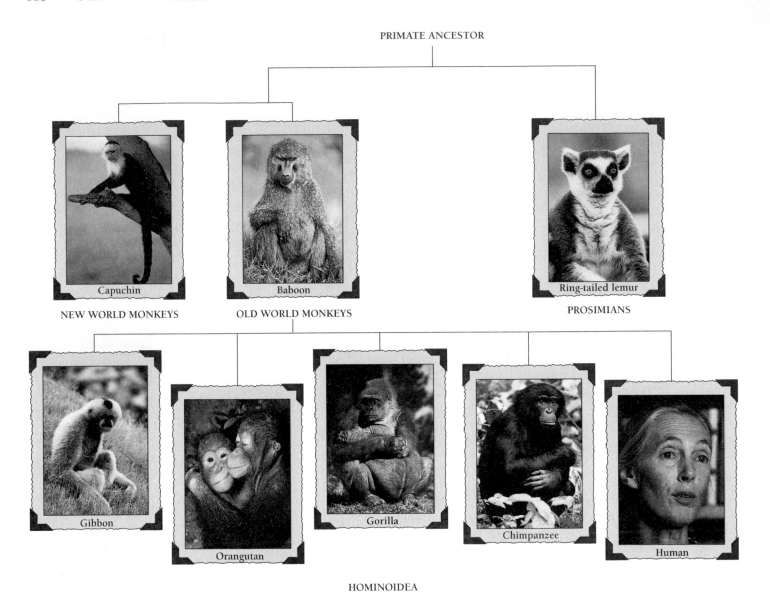

Figure 15-1 **Our family tree.** Biologists classify humans as primates not only because we look like other primates, but because we are descended from the same stock as they are. We are most closely related to chimpanzees, gorillas, and orangutans. *(capuchin, Richard Gibson/Natural Selection; baboon, Stephen G. Maka/DRK Photo; lemur, Kevin Schafer; gibbon, Joe McDonald/Bruce Coleman, Inc.; orangutan, Daniel J. Cox/Natural Selection; gorilla, Mimi Forsyth/Bruce Coleman, Inc.; chimpanzee, Frans Lanting/Minden Pictures; Jane Goodall, Jack Dykinga, Bruce Coleman, Inc.)*

Modern debate about evolution began in 1859, when Charles Darwin published his arguments for the evolution of species in *On the Origin of Species by Means of Natural Selection, or the Preservation of Favoured Races in the Struggle for Life.* In that book, Darwin rejected the idea that each species had been specially created. The book disturbed a great many people. Three of Darwin's ideas offended most often: (1) that the Earth was very old, (2) that one species could change into another, and (3) that humans and apes were related (Figure 15-1).

In this chapter we will discuss the main ideas of evolution, trace the history of evolution, and present some of the evidence that supports this revolutionary idea. Along the way, we will see why creation "science" has no scientific basis, and why some people find the theory of evolution so disturbing that they would prefer that their children not know about it. We will also note that Darwin's theory of evolution does not actually address the origin of life. Evolution only accounts for the development of new species from previous ones.

KEY CONCEPTS

1. Evolution is an idea with a long, varied, and controversial history.

2. Some biologists before Darwin understood that evolution had occurred, but they could not come up with a persuasive model for how evolution could happen.

3. Darwin's theory of evolution was a synthesis of several ideas, which can be roughly divided into two parts. One group of ideas described evolution, and the other group of ideas described how and why evolution works.

4. A single structure such as a limb or a molecule may evolve different uses in different species. Such structures are said to be homologous when inherited from a common ancestor. The degree of difference between homologous structures in any two species generally reflects the effects of evolution and, therefore, the degree of relatedness.

5. Such comparisons, made at both anatomical and molecular levels, allow biologists to group species into taxonomic groups that reflect actual relatedness.

6. The story of evolution told by such comparisons is confirmed by biogeography, the fossil record, and embryology.

WHAT IS EVOLUTION?

Biological evolution is the process that has taken life on Earth from perhaps a few mere flecks of RNA to the breathtaking variety of more than a million species. What is popularly understood to be Darwin's theory of evolution is really a group of several ideas that can be roughly divided into two parts.

The first part is evolution itself, the idea that species change over time. Darwin called evolution "**descent with modification.**" A concise definition of evolution says it is heritable change in organisms over time. Some of this change is *adaptive,* which means that the change increases an organism's chances of perpetuating its genes in future generations. If the average beak size of a population of finches increases, for example, the finches have evolved. Whether any trait, such as increased beak size, is adaptive depends on the environment of the organism. When many large, hard seeds are present, a large beak is adaptive. When few large, hard seeds are available, but plenty of small soft seeds are available, the large beak may be maladaptive.

Not all evolutionary change is adaptive. Some change can occur through random processes. Evolutionary biologists argue vigorously over how much of evolutionary change is adaptive and how much is not.

The second part of Darwin's theory is (1) his specific mechanism for evolution, natural selection, and (2) a description of the way he thought evolution occurred, gradualism. Darwin's **gradualism** was the idea that species evolve gradually and continuously through the steady accumulation of changes, rather than through sudden changes.

For Darwin, natural selection was the heart of his theory. It was a mechanism that set his theory of evolution apart from similar theories. Briefly, **natural selection** is the differential reproductive success of individuals due to genetically inherited traits. Darwin described natural selection as analogous to artificial selection. **Artificial selection** is the process by which animal breeders and farmers create new lines of dogs, pigeons and corn, for example, by selecting the most desirable individuals for breeding, generation after generation. At the end of this chapter, and in later chapters, we will examine natural selection and other mechanisms for evolution in more detail.

Most 19th-century scientists, even some of Darwin's most faithful supporters, rejected natural selection and gradualism. Even modern biologists, who do not contest the idea that natural selection and gradualism operate in nature, debate the importance of these mechanisms in the evolution of new species. In sharp contrast, modern biologists consider evolution—the principle that the heritable traits of organisms change over time—itself an incontestable fact. Likewise, 19th-century scientists nearly wholly accepted evolution within just a few years of the publication of *The Origin of Species*.

Biologists have always debated the importance of natural selection and gradualism in evolution. In contrast, few biologists since Darwin have doubted that evolution has occurred. The evidence he presented was nearly unassailable.

Four Revolutionary Ideas

Darwin's theory of descent with modification through natural selection rested on four ideas. First, he recognized that the world is ever-changing and very old (rather than constant and recently created). Second, he recognized that species change.

(As we shall see, neither of these ideas was original with Darwin.) Third, Darwin argued that species are composed of populations of individuals. And fourth, he argued that every species and group of species is descended from a common ancestral species. Ultimately, Darwin said, all organisms—including humans—may derive from a single origin of life on Earth.

Every one of these ideas conflicted not only with the biblical account of creation but also with a major strand of Western philosophy, called **essentialism.** According to the Greek philosopher Plato (427–347 B.C.), individuals, whether chairs, daisies, or people, are only distorted shadows of an ideal, or essential, form. The variations that make every daisy unique, said Plato, are distortions of the ideal or essential daisy, which alone is real.

Because of essentialism, most scientists in Darwin's time viewed species in terms of their ideal or essential forms. These were perfect, unchangeable, sharply defined "natural kinds." In contrast, Darwin viewed every organism as an individual, not a distortion of some ideal form. As evolutionary biologist Ernst Mayr has written, "It was Darwin's genius to see that this uniqueness of each individual is not limited to the human species but is equally true for every sexually reproducing species of animal and plant." Darwin recognized that species are not ideal forms, but groups of individuals. Thus, the flaws that make a daisy unique are just as "real" as the features common to all daisies.

Darwin asserted (1) that the world is ever-changing and very old; (2) that species are made up of individuals; (3) that species are plastic; and (4) that all organisms are related by descent from a common ancestor. His theory conflicted with both the biblical account of creation and with essentialism.

Evolution Before Darwin

Although Darwin was the first to persuade the world that plants and animals had evolved, he was hardly the first to believe in evolution. His own grandfather, Erasmus Darwin (1731–1802), argued in 1794 that all species originated from the same primitive "filament." However, Erasmus Darwin propounded so many outlandish ideas on so many topics, and supported it all with so little fact, that no scientist took his ideas on evolution seriously. In that era, to be accused of "Darwinizing," that is, engaging in wild speculation, was an acute professional embarrassment.

Anaximander, Plato

Evolution as an idea actually predates even Plato. One hundred years earlier, in the 6th century B.C., the Greek philosopher Anaximander argued that the world was not created, as most ancient religions say. He believed, instead, that animals evolved, and that humans, like all other vertebrate animals, were descended from fishes. It was 2500 years before Anaximander's ideas found scientific support and general acceptance.

In large part, this was because Christian beliefs thoroughly incorporated Plato's essentialism. And essentialism, along with creationism, established a nearly unbreachable intellectual barrier to the idea of the evolution of species.

Another serious barrier to the idea of evolution was the notion—accepted by nearly everyone in 17th- and 18th-century Europe—that the Earth had been created in 4004 B.C. This estimate had been worked out in 1650 by Archbishop James Ussher, who added together the life spans of all the patriarchs named in the Bible, beginning with Adam.

Even so, by the end of the 18th century, the biblical account of creation and the notion that species do not change faced serious challenges from three increasingly disturbing lines of evidence. First was the discovery of a host of species not mentioned in the Bible. Explorers returning to Europe from Africa, the Americas, and the Pacific, brought back hundreds of thousands of plants and animals never seen before. Equally disturbing was the increasing number of discoveries in Europe itself of fossil plants and animals that were likewise unknown. In addition, by the middle of the 19th century, geologists presented compelling evidence that the Earth was far older than Ussher's estimated 6000 years.

Five Scientists Lay the Groundwork for Evolution

The difficulty of reconciling these new facts with biblical doctrine is apparent in the thinking of five of the greatest scientists of the 18th and early 19th centuries. All of them discovered evidence for evolution, yet because of creationism and essentialism all of them interpreted the evidence in nonevolutionary terms.

Linnaeus introduces modern taxonomy

The first of the five is Carolus Linnaeus (1707–1778), a Swedish physician who single-handedly invented the science of **taxonomy** [Greek, *taxis* = arrangement]—the naming and grouping of organisms (Figure 15-2). Linnaeus's most familiar contribution to biology was the two-part Latin names—genus and species—that biologists assign to all organisms. The scientific name for a human being, for example, is *Homo sapiens*. We belong to the **genus** *Homo* ("human"), and our **species** name is *sapiens* ("wise"). Using this system, Linnaeus named more than 4000 plants and animals. Biologists since Linnaeus have used his system to name some 1.4 million species, while millions more species remain to be discovered and named.

Just as enduring as Linnaeus's naming system, however, and historically even more important, is his hierarchical system of classification—the heart of taxonomy. For example, the genus *Homo* belongs to the **family** Hominidae. And the family Hominidae belongs to the **order** Primates, one of 16 orders in the **class** Mammalia. Thus, we group species into genera, genera into families, families into orders, and orders into classes. Each of these groupings is a **taxon.** Together they are **taxa.**

Linnaeus firmly believed that each species was individually and separately created by God, yet he arranged them into groups that unambiguously reflected their similarities to one

THE SIX KINGDOMS

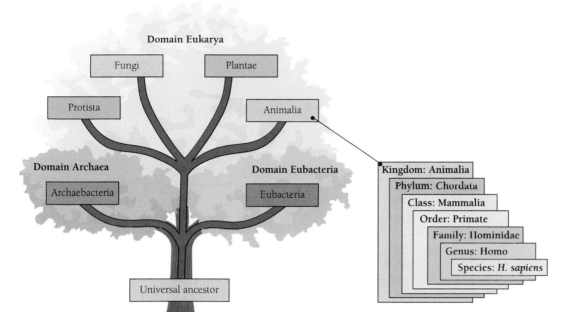

Figure 15-2   The Tree of Life. Carolus Linnaeus named all organisms known in 18th-century Europe and grouped them into categories according to similarities. Linnaeus designed a hierarchy of classification extending from kingdoms to genera and species. Recently, biologists have added a new layer of classification, the domain. In this book, we classify organisms into three domains, which together include six kingdoms.

another. He could see, for example, that mice were more closely related to rats than to dogs or horses, and his formal system for recognizing such natural relations was fundamental to Darwin's arguments 100 years later.

Linnaeus avoided asking *why* organisms fall so naturally into families, piously writing, *"Deus creavit, Linnaeus disposuit"*—God creates, Linnaeus arranges. But one of his contemporaries, the great French biologist Count George–Louis Leclerc Buffon (1707–1788), couldn't leave the question alone.

Buffon discovers evolution, but is afraid to admit it

Buffon discovered evolution, but was afraid to admit it. In his early manhood, Buffon boldly stated that the Earth must be far older than 6000 years. Church elders threatened him with excommunication for this heresy, however, and he quickly recanted.

In later years, he prudently recanted each heresy as he committed it. In a 36-volume natural history, in which he interspersed descriptions of plants and animals with philosophical discourses on nature, Buffon noted the remarkable similarity of the horse and the ass. He wrote, in 1753, "But if we once admit that there are families of plants and animals, so that the ass may be of the family of the horse, and that the one may only differ from the other through degeneration from a common ancestor, we might be driven to admit that the ape is of the family of man." And, as if that were not bad enough, he added, ". . . then there is no further limit to be set to the power of nature, and we should not be wrong in supposing that with sufficient time she could have evolved all other organized forms from one primordial type."

To placate the church, he quickly contradicted this insight, "It is certain through Revelation that all animals have shared equally in the grace of creation and that each emerged from the hands of the creator as it appears today."

Hutton shows that the Earth may be older than 6000 years

Beginning in the mid-18th century, the Industrial Revolution's insatiable appetite for coal and ore had turned geology into a practical and hugely popular science. For the first time, scientists began to seriously explore and understand the surface of the Earth.

They knew, for example, that mountains were eroded by wind, frost, and running water, and they noticed how quickly soil and rock eroded in some areas. Indeed, in some places, whole mountain ranges seemed to have eroded completely away. But could all this have happened, geologists wondered, in just 6000 years?

In 1788, the Scottish geologist James Hutton (1726–1797) suggested that it could not. The Earth, he said, was eternal. The perpetual erosion was repaired by the uplift of sediments and the expulsion of magma from the Earth's depths. At the time, most geologists rejected the idea that forces within the Earth were capable of renewing its surface by raising continents and mountains. But Hutton insisted that the Earth was not a static, passive body but a "beautiful machine," powered by the heat of its interior. In a theory he called **gradualism** (distinct from Darwin's much later theory), he proposed that cycles of erosion, sedimentation, and uplift molded the Earth's surface. Our world, he said, ran through endless cycles

A.

B.

Figure 15-3 Catastrophism and uniformitarianism. A. The Grand Canyon. James Hutton wrote, "Thus . . . from the top of the mountain to the shore of the sea . . . everything is in a state of change; the rock and solid strata slowly dissolving, breaking and decomposing, for the purpose of becoming soil; the soil traveling along the surface of the earth on its way to the shore; and the shore itself wearing and wasting by the agitation of the sea, an agitation which is essential to the purposes of a living world. Without those operations which wear and waste the coast, there would not be wind and rain; and without those operations which wear and waste the solid land, the surface of the earth would become sterile." B. Noah's Flood. The deluge as portrayed in the first edition of the Luther Bible, published in 1534.   *(A, Jack Dermid/Bruce Coleman, Inc.)*

without intervention from God, and the Earth could be immeasurably old. In his now-famous conclusion, he wrote, "The result, therefore, of our present inquiry is, that we find no vestige of a beginning—no prospect of an end."

It was, of course, heresy. But if Hutton was a thorn in the side of theologians, Baron Georges Cuvier (1769–1832) was their savior. In 1812, the great French anatomist and paleontologist found an ingenious way to reconcile geology and Genesis—a way to reconcile the obviously great age of the Earth with theological doctrine (Figure 15-3).

Catastrophism: Cuvier tries to reconcile Genesis and geology

Cuvier was a gifted anatomist, and the first to study fossils systematically. He invented both comparative anatomy and **paleontology**, the study of fossils [Greek, *palaios* = ancient + *on* = being + *logos* = discourse].

Fossils had puzzled thinking people for thousands of years. Fossils looked like plants and animals or parts of plants and animals. But they rarely looked like *familiar* plants and animals. And they were made of stone. Fossils were thought to be God's first experiments with creation or, alternatively, the remains of plants and animals that had literally missed the boat. Failing to board Noah's ark, they had perished in the great flood. But Cuvier knew that fossil organisms could not all have died in the same biblical flood.

Geologists already suspected that fossil organisms had lived and died at different times in the Earth's history. By 1760, geologists had begun to recognize that rocks could be arranged in a sequence of layers, or strata. William "Strata" Smith, a humble canal engineer and geologist, had shown that fossils were arranged within these layers "like slices of bread and butter on a breakfast plate," appearing and disappearing from the fossil record according to their kind.

Cuvier argued that the organisms from the different layers had perished in a series of catastrophes, of which Noah's flood was merely the most recent and dramatic. Each of the ages within the fossil record, said Cuvier, had been terminated by some new catastrophe, after which life arose once more in a new creation. Cuvier believed that species were unchangeable, that the organisms of previous ages were not precursors to modern ones, and the older layers had nothing at all to do with present life.

The previous catastrophes were not recorded in Genesis, he suggested, because God didn't think Moses needed to know about them. **Catastrophism,** as his theory came to be called, allowed for the great age of the Earth, without implying that Genesis was wrong (Figure 15-3B).

Catastrophism stumbles

Catastrophism remained immensely popular for decades. In fact, the 1981 Arkansas equal-time bill (mentioned earlier) called for a revival of this very theory. But more and more facts began to undercut catastrophism. For example, although Cuvier had originally proposed only four or five catastrophes before Noah's flood, fossil hunters continued to find so many distinct layers of fossils that more and more catastrophes had to be invented to account for them all. A student of Cuvier's, the Swiss–American biologist Louis Agassiz (1807–1873), maintained, for example, that the fossil record revealed 50 to 80 separate catastrophes, requiring an equal number of separate creations.

And, while some organisms dutifully appeared once in the fossil record and then disappeared forever, others kept reappearing, virtually unchanged, in creation after creation (Figure 15-4). Archaeologists had even begun to find human artifacts and bones, which suggested that humans predated not only the biblical flood, but the postulated creation of the world in 4004 B.C.

Uniformitarianism: Lyell rejects catastrophism

In 1830, catastrophism received its most damaging blow when the Scottish geologist Charles Lyell (1797–1875) published the first volume of his three-volume *Principles of Geology.* Lyell explicitly rejected catastrophism, instead expanding on Hutton's gradualism and renaming the idea **uniformitarianism.** Lyell argued that the processes that now mold the Earth's surface—erosion, sedimentation, and upheaval—are the ones that have always molded it. Geologic change, he said, is slow, gradual, and steady, not catastrophic. Given enough time, mountains could erode down to mere hillocks, sea beds rise to great heights, and rivers cut deep canyons. Gradual processes alone could create the world we know. Catastrophism, argued Lyell, need not be invoked to explain geologic history.

Lyell, who had been trained as a lawyer, provided copious evidence to back up his (and Hutton's) claims and produced an immensely influential work. Darwin later wrote that Lyell was one of the only two great influences in his intellectual life. "I have always thought that the great merit of the *Principles* was that it altered the whole tone of one's mind, and therefore that, when seeing a thing never seen by Lyell one yet saw it partially through his eyes." Lyell's *Principles* provided fertile soil for Darwin's ideas. But Lyell, like Linnaeus, Hutton, and Cuvier, believed firmly that species were fixed. Not until Darwin published *The Origin of Species,* 30 years later, did Lyell change his mind.

Clearly, neither the great age of the Earth nor the fossil record was enough to convince scientists that species evolve. The idea was in the air for a hundred years before Darwin, but the idea of evolution had a very bad name among reputable scientists.

The Industrial Revolution's demand for fossil fuels created the science of geology. The insights offered by geology made clear the great age of the Earth.

 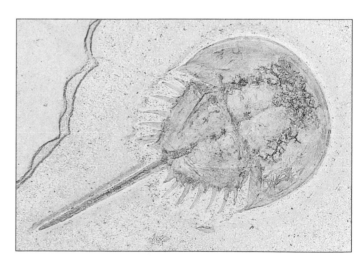

Figure 15-4 **Horseshoe crab.** Many groups of organisms have changed hardly at all over millions of years. The modern horseshoe crab remarkably resembles its Jurassic ancestor, which lived nearly 200 million years ago. *(left, Alan Desbonnet/Visuals Unlimited; right, John Cancalosi/Peter Arnold, Inc.)*

LAMARCKISM

DARWINISM

Ancient giraffes had short necks.

Ancient giraffes had varying neck lengths.

With climate change, giraffes stretched necks to reach tall food trees.

With climate change, long-necked giraffes could feed on tall trees; short-necked ones could not.

Giraffes acquired long necks from stretching for food and passed this trait to offspring.

Short-necked giraffes died; long-necked giraffes survived to reproduce.

Figure 15-5   **Lamarckism and Darwinism.** Lamarck suggested that evolution occurs when organisms pass useful *acquired* traits to their offspring. Darwin suggested that every individual is unique and that evolution occurs when some individuals reproduce more successfully than others.

Lamarck Proposes Evolution, But No One Listens

Buffon and Erasmus Darwin had proposed that species evolve. But neither man was taken seriously. In 1809, the great French biologist Jean Baptiste Lamarck (1744–1829) again proposed the idea, this time in great detail.

Lamarck came from an impoverished noble family, and served as a tutor to Buffon's son. He studied medicine and botany in Paris, and published two works on botany before being appointed professor of zoology at the Jardin des Plants in 1793. An avid taxonomist, he distinguished himself from Linnaeus by arguing that plants and animals are related, coining the word "biology" to express this union. He named hundreds of species and five different groups of animals, including the Annelida (segmented worms), the Crustacea (crabs, shrimp, and lobsters), and the Arachnida (spiders). He wrote more than a dozen volumes on fossil invertebrates (a word he also coined).

In his systematic studies of both living and fossil organisms, Lamarck recognized a clear line of descent from older fossils to more recent fossils to modern species. The oldest fossils looked the least like modern organisms. Later fossils included more and more familiar organisms.

Like Darwin, Lamarck believed that the accumulated changes were adaptive—that is, they made organisms better able to survive under new environmental conditions. However, Lamarck argued that the changes resulted from an inherent drive toward perfection and increased complexity, and from the "felt needs" of the organism.

A giraffe, for example, reaching ever higher in the trees for leaves to eat, would feel the need to gradually stretch its neck, and would thus acquire a permanently longer neck. The giraffe's offspring would then inherit longer necks. Lamarck said that adaptive characteristics acquired by the parent are inherited by the offspring (Figure 15-5).

We now know that only characteristics with a genetic basis can be inherited. Characteristics acquired during the life of the parents cannot be inherited by the offspring. No matter how much the giraffe stretches its neck, its offspring will not be born with longer necks.

Today, biologists accept evolution as such but reject Lamarck's mechanism, the theory of **the inheritance of acquired characteristics.** Ironically, Lamarck's contemporaries did just the opposite. The inheritance of acquired characteristics, Lamarck's mechanism for evolution, was nearly universally accepted. Even Darwin rather reluctantly accepted it as a source of variation on which natural selection could work. Some biologists accepted the inheritance of acquired characteristics well into the 1930s (Chapter 16). Nevertheless, Lamarck's peers vehemently rejected evolution itself, and he was vilified by virtually the entire scientific community.

The fiercest opposition came from Cuvier, who firmly believed in the fixity of species. As a French cabinet minister and an eminent scientist, his opinion carried considerable weight. Cuvier criticized Lamarck all of his life, and he continued to do so even after Lamarck was dead. In 1832, three years after Lamarck had died—poor and almost forgotten—Cuvier delivered an "elegy" to Lamarck at the prestigious French Academy, in which he firmly dismissed Lamarck's theory of evolution and even his qualifications as a biologist.

When Darwin began writing his theory of evolution, he did everything he could to distance himself from Lamarck, and to anticipate every possible objection to his own ideas.

Darwin Develops an Unassailable Theory of Evolution

By the middle of the 19th century, scientists and others had begun to accept the Earth's great age, to recognize that previous ages had contained plants and animals that no longer existed, and to appreciate the constantly changing nature of the Earth's surface.

At the same time, natural science had become enormously popular with the lay public. Every sector of English society was obsessed with the exotic and the unusual in nature. Most middle-class homes housed an aquarium, a fern-case, a butterfly cabinet, or a shell collection. Mothers routinely taught their children the names of ferns and fungi. Aristocrats turned their estates into parks for exotic animals such as elands, beavers, and kangaroos. Books on natural history were so popular that one—*Common Objects of the Country*—sold 100,000 copies in a week, a total that today would earn the book a place on the *New York Times* best-sellers list.

Darwin's Education

Charles Robert Darwin (1809–1882) came of age in the middle of this national obsession, and was himself entirely caught up by it. He spent his boyhood tramping through the woods and fields of Shrewsbury, hunting, fishing, and collecting insects. As a young man his interests narrowed down to hunting. His father despaired of Charles ever making anything of himself, writing, "You care for nothing but shooting, dogs and rat-catching, and you will be a disgrace to yourself and all your family."

Darwin's utter lack of ambition forced his father to pick a profession for him, and in 1825 Robert Darwin sent young Charles to Edinburgh University to study medicine. In the course of his studies, Darwin attended lectures on geology and zoology and thought they were the most boring thing he'd ever had to endure in his life. His only interest, outside of shooting birds, was the meetings of a natural history club. As for medicine, if he had ever considered becoming a doctor, he changed his mind after watching two horrifying surgical operations done without anesthesia, one on a child. His resistance to studying medicine, or anything else, only hardened when he realized that his wealthy father would probably support him for the rest of his life, whatever he chose to do (or not to do).

Dr. Darwin then decided that Charles should study for the clergy and in 1828 sent him to Cambridge University. Fortunately for modern biology, the clergy was then a haven for naturalists, and 19-year-old Charles Darwin soon shook off his

A.

B.

Figure 15-6   Darwin and Wallace. A. Charles Darwin and Emma Wedgwood Darwin. Even in his most productive years, Charles Darwin was sickly and worked only for short periods. B. Alfred Russel Wallace's ideas about evolution generally matched Darwin's. However, although Wallace believed that human beings had evolved from ancestral primates, he did not believe that our capacity for thought and spiritual passion could have evolved in the same way.   *(A, Bridgeman/Art Resource, NY; B, The Granger Collection, New York)*

perpetual boredom and rediscovered his boyhood passion for nature.

In three years, he managed to complete his B.A., and when he graduated from Cambridge in 1831, one of his professors recommended him for a position as ship naturalist on board the H.M.S. *Beagle*. The *Beagle* was a survey ship about to embark on a trip around the world, its five-year mission to map the coast of South America. Darwin's duties were to collect specimens of the flora, fauna, and rocks from this unexplored new world. In December 1831, just before the *Beagle* sailed, a professor sent Darwin the first volume of Lyell's *Principles of Geology*. His course was set.

During the long trip across the Atlantic, he immersed himself in Lyell's arguments and thought of nothing but geology for months. Lyell formalized two important ideas: first, that the Earth is very old, and second, that geologic processes occur very slowly and gradually. Darwin's theory of evolution, as he would later conceive it, likewise stated that change (evolutionary, in this case) was gradual and required huge spans of time.

As Darwin explored South America over the next five years, he began to suspect that all animals and plants had descended from a common ancestor. He took notes; he collected specimens. He noticed what animals ate, where they lived, what kind of soil this plant or that seemed to prefer.

Almost as soon as he returned to England in 1836, Darwin began to classify his specimens and to organize his notes.

But it was six more years before he "allowed himself the satisfaction" of even outlining his theory (Figure 15-6). In the meantime he read the work of the second great intellectual influence after Lyell—Thomas Malthus.

Malthus: Too Many Children

In 1798, the English economist and clergyman Thomas Robert Malthus (1766–1834) wrote a highly controversial paper, *An Essay on the Principle of Population*. Malthus argued that human populations tend to increase geometrically, while food supplies and other resources increase only arithmetically, if at all. For example, if a couple has three children, and each of their children has three children, and the nine grandchildren each have three children, the family farm cannot possibly support the 27 great-grandchildren. The family may be able to increase the amount of food they can produce on their farm, but they cannot possibly triple production every generation. Poverty, famine, overcrowding, disease, and war are all the inevitable consequences of this excessive population growth, argued Malthus. He predicted that without some check on population growth, human populations would face a continuing, ferocious struggle for existence in the face of limited resources.

When Darwin read Malthus in 1838, he saw for the first time how evolution could work. Malthus's arguments applied to all species, reasoned Darwin, not just humans. Many more individuals of each species are born than can possibly survive, and, consequently, there is a constant struggle for existence.

"It follows," Darwin later wrote, "that any being, if it vary however slightly in any manner profitable to itself, under the complex and sometimes varying conditions of life, will have a better chance of surviving, and thus be *naturally selected*." In short, the differential survival of the fittest could generate the kind of gradual change that would drive evolution. His theory of evolution through natural selection was beginning to take shape.

Wallace Scoops Darwin

Still, Darwin delayed publishing. Month by month, year by year, he added new facts to bolster his theory. Lyell, who was by then a friend, still didn't accept evolution, but he nevertheless urged Darwin to publish his idea before someone else thought of it and published first. Then, in 1858, Alfred Russel Wallace (1823–1913), a young land surveyor and naturalist working in the Malay Archipelago (in what is now Malaysia) independently hit upon natural selection as a mechanism for evolution.

Wallace had written to Darwin the year before asking for information about selective breeding. Darwin wrote back with the information and added tactfully that he had been working on similar ideas for some 20 years. Again Wallace wrote, this time outlining his theory of evolution—but without natural selection. Darwin responded politely, but hinted that he knew more. Wallace was obviously hot on his heels, yet Darwin did nothing.

Then, a few months later, in 1858, Wallace wrote once more. In the interim, he had read Malthus and experienced exactly the same flash of insight Darwin had. Eagerly, Wallace wrote to Darwin, outlining his theory of evolution by natural selection.

Darwin, utterly crushed, was ready to give all of the credit to Wallace. In a letter to Lyell, he wrote:

> I never saw a more striking coincidence; if Wallace had my [manuscript] sketch written out in 1842, he could not have made a better abstract! Even his terms now stand as heads of my chapters. Please return me the [manuscript], which he does not say he wishes me to publish, but I shall, of course, at once write and offer to send it to any journal. So all my originality, whatever it may amount to, will be smashed.

Then more misfortune struck. Scarlet fever swept Darwin's family, killing his 19-month-old son Charles in a matter of days. With the whole family in a turmoil, Darwin threw up his hands and turned everything over to Lyell. Lyell rescued Darwin, presenting to the Linnaean Society both Wallace's paper and an extract from Darwin's unpublished outline. The following year Darwin finally published *On the Origin of Species by Means of Natural Selection, or the Preservation of Favoured Races in the Struggle for Life*.

His book is probably the single most influential scientific book ever written. When it appeared in 1859, all 1250 copies of the first printing sold out in a single day. Its arguments persuaded even those, such as Lyell, who had previously rejected the idea that species evolved. What was so convincing?

WHAT IS THE EVIDENCE FOR EVOLUTION?

The book's strength was Darwin's gluttony for evidence. Darwin provided evidence for evolution from the fossil record, from the distribution of plants and animals (biogeography), from taxonomy, from comparative anatomy, from comparative embryology, and from domestic breeding. Since Darwin's time, other biologists have further buttressed his arguments with evidence from fields as diverse as comparative molecular biology, classical genetics, population ecology, developmental biology, and animal behavior. In the rest of this chapter we will briefly examine some of this evidence.

The Fossil Record

How Does the Fossil Record Establish the Time Scale of Evolution?

In the 18th and 19th centuries, scientists recognized that the fossils in the top layers of sediment were the most recent, while the fossils in the deepest layers were the oldest. For the first time, scientists were able to read the layers of rock as if they were the pages of a history book. They discovered a startling history of the last 570 million years.

Geologists divided this history into chapters. Each chapter, or **period,** is 30 to 75 million years long and contains distinctive forms of life in its fossil record (Figure 15-7). Periods are grouped into four long **eras,** of from 65 million to several billion years each. The most recent chapters are subdivided into relatively short **epochs.** These distinctive layers appeared in rock formations around the world.

Nineteenth-century geologists could not give exact ages to each layer. However, they could tell the order in which the layers had been deposited. And they estimated the ages of the various periods and eras recorded by these fossil layers remarkably well. The modern development of radioactive dating techniques, in the 1940s and 1950s, allowed paleontologists to assign actual dates to evolutionary events and confirm many of the old estimates.

The Fossil Record Is Not a Complete History of Life on Earth

What are fossils?

Fossils [Latin, *fossilis* = dug up] are the remains or imprints of past life. The most impressive fossils are the intact skeletons of extinct animals, such as dinosaurs or even entire woolly mammoths retrieved from Arctic glaciers. Such spectacular finds, however, are relatively rare compared to the hundreds of thousands of small fossilized shells and individual bones dug up in the last several hundred years. Fossils include not only

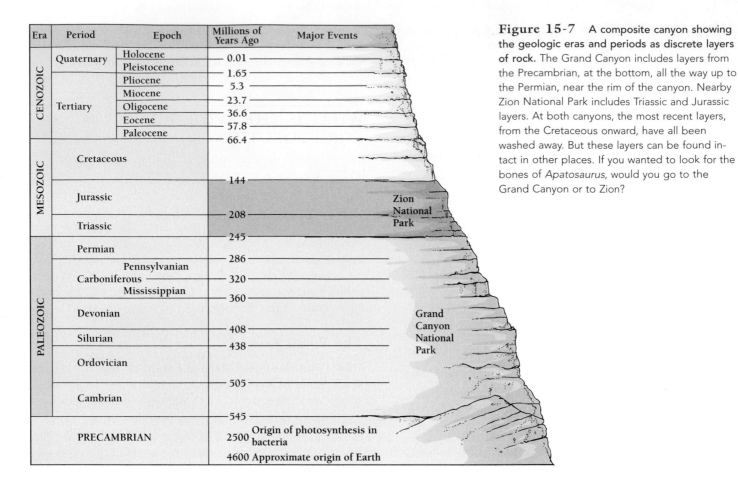

Era	Period	Epoch	Millions of Years Ago	Major Events
CENOZOIC	Quaternary	Holocene	— 0.01	
		Pleistocene	— 1.65	
	Tertiary	Pliocene	— 5.3	
		Miocene	— 23.7	
		Oligocene	— 36.6	
		Eocene	— 57.8	
		Paleocene	— 66.4	
MESOZOIC	Cretaceous			
			— 144	
	Jurassic			
			— 208	
	Triassic		— 245	
PALEOZOIC	Permian		— 286	
	Carboniferous	Pennsylvanian	— 320	
		Mississippian	— 360	
	Devonian		— 408	
	Silurian		— 438	
	Ordovician		— 505	
	Cambrian			
			— 545	
	PRECAMBRIAN		2500	Origin of photosynthesis in bacteria
			4600	Approximate origin of Earth

Figure 15-7 A composite canyon showing the geologic eras and periods as discrete layers of rock. The Grand Canyon includes layers from the Precambrian, at the bottom, all the way up to the Permian, near the rim of the canyon. Nearby Zion National Park includes Triassic and Jurassic layers. At both canyons, the most recent layers, from the Cretaceous onward, have all been washed away. But these layers can be found intact in other places. If you wanted to look for the bones of *Apatosaurus*, would you go to the Grand Canyon or to Zion?

preserved bones and teeth but also footprints, impressions of the surfaces of ferns, petrified trees, bits of fossilized dung, as well as ants and other insects preserved in hardened tree resin (amber).

Why are many organisms missing from the fossil record?

An animal or plant becomes a fossil only if it dies under the right conditions for preservation. The vast majority of plants and animals are consumed by other organisms and leave no trace of their existence. Dead things persist only if their bodies are rapidly buried in a bog or the bottom of a sea or lake. There, protected from scavengers, erosion, and decay, organisms may in time become more deeply buried, as successive layers of mud and sand settle over them.

Ultimately, pressure from succeeding layers turns these layers of mud, sand, and other sediment into **sedimentary rock.** In time, geologic forces may raise the rock up into mountains. Rivers and streams begin to flow down the sides of the mountains, eroding canyons through the layers of old sediment, revealing the ancient fossils.

Because organisms fossilize most often when buried in sediment at the bottom of a sea, lake, or pond, most fossils are marine or freshwater organisms. Terrestrial plants and animals are usually only fossilized if their remains happen to be carried to such a site. Furthermore, only the hard, inedible parts of an-

imals—shells, teeth, and bones—and the woody parts of plants are likely to survive long enough to form fossils. The possibility of a soft-bodied terrestrial organism such as a banana slug becoming fossilized is remote.

The chanciness of fossilization means that the fossil record is incomplete. For example, most insects are too small and delicate to be preserved, and so the fossil record of insects is rather poor. However, where they are preserved, they often occur in great numbers and in great variety. Large numbers of insects may appear in one layer, disappear—leaving a gap—then reappear millions of years later.

Gaps in the fossil record: gaps in evolution?

If we compare the story of evolution, as rendered in the fossil record, to a movie, we must say that most of the individual picture frames are missing. The result is a jerky, discontinuous image. Gaps in a film, however, can result either from missing frames or from deliberate interruptions in the story. In the slow pace of evolutionary time, what appears to be a sudden jerk in the fossil record may correspond to tens of thousands of years of gradual change. One of the questions that modern evolutionary biologists discuss most intensely is whether the jerkiness of the evolutionary story is only apparent, reflecting the incompleteness of the fossil record, or real, reflecting discontinuities in the rate of the evolutionary process itself.

BOX 15-1

Radioactive dating

The fossil record provides a useful chronicle of the sequence of organisms that have lived on Earth. Fossils also provide an accurate way to determine the relative ages of different layers. Knowing the **relative age** means knowing which layers came first and which came later.

Still, even if scientists know the relative ages of every single layer, they still want to know the **absolute age**—the age in years—of each of the layers. Scientists had no way of estimating the absolute ages of rocks until the discovery of radioactivity by Marie Sklodowska Curie and Pierre Curie in 1898.

As we discussed in Chapter 2, radioactive atoms, such as uranium, break down into other materials by giving off alpha, beta, and gamma rays. The breakdown, or **decay**, occurs at a steady rate that is unaffected by temperature, pressure, or other environmental variables. Every form of an atom, called an **isotope**, breaks down into a series of other isotopes. Half of the isotope carbon-14, for example, decays to nitrogen-14 every 5600 years. We say that the **half-life** of carbon-14 is 5600 years.

By comparing the relative proportions of different isotopes in a rock sample, scientists can tell the age of the rock. Different radioactive isotopes have different half-lives. Carbon-14, denoted as ^{14}C, is useful for dating objects that are only a few thousand years old. In contrast, rubidium-87 has a half-life of about 5×10^{10} years (50 billion years), so it is useful for dating rocks that originated at the same time as the solar system (about 5 billion years ago) or even earlier.

Radioactive isotope dating is not a perfect method. The age provided is only accurate to within about 10 percent of the actual age of the rock. However, relative dating of the sedimentary layers allows scientists to refine these estimates. Radioactive isotope dating places these divisions in a rough time frame, while fossils allow finer divisions into periods.

Despite its limitations, the fossil record clearly shows a history of life on our planet that has evolved according to the sequence summarized below.

What the Fossil Record Shows

The Precambrian Era: life appears

The **Precambrian Era** comprises most of the history of the Earth, beginning with the planet's formation 4.6 billion years ago. Before the discovery of the first Precambrian fossils in the 1950s, paleontologists viewed the Precambrian as one long lifeless era. Modern paleontology has subdivided the Precambrian, however, and in Chapter 18 we will discuss some of the major events of the Precambrian. For now, we will just note a few highlights. The first fossils, some as old as 3.8 billion years, resemble modern prokaryotes, single cells lacking any membrane-bounded organelles (Figure 15-8). These ancient fossils vividly illustrate the breathtaking antiquity of life on Earth. Fossils of the first eukaryotes (cells with organelles) appear in rocks about 2 billion years old.

Only near the end of the Precambrian, about 800 million years ago, do large numbers of multicellular organisms appear in the record. Thus, nearly 3 billion years elapsed between the appearance of the first cells and that of the first multicelled organisms. But once they evolved, they appeared in amazing variety, and included animals resembling modern jellyfishes, corals, and segmented worms, as well as creatures unlike anything alive today (Figure 15-8). Because these animals were soft-bodied, their fossils are rare.

The Paleozoic Era: invertebrates and fishes flourish; plants, insects, and vertebrates colonize the land

The **Paleozoic Era** [Greek, *palaios* = ancient + *zoos* = life] began about 570 million years ago. All of the major groups of animals appeared before the end of this era. The Paleozoic Era consists of six periods—each one from 30 to 75 million years long.

The Cambrian Period [Latin, *Cambria* = Wales] marks the beginning of the abundant fossil record. Most numerous and striking are the fungi, the algae, and the trilobites, which are relatives of modern crustaceans. By the end of the Cambrian, some 500 million years ago, the first vertebrates—the armor-plated fishes—had already appeared.

During the next two periods of the Paleozoic Era, the Ordovician and the Silurian, increasing numbers of fishes with bony skeletons appeared. The first land plants—simple stalks with scalelike leaves—appeared, accompanied by fungi and arthropods such as scorpions.

The Devonian Period witnessed a dramatic receding of the oceans and a corresponding increase in the amount of dry land. The fishes continued to diversify strikingly, however, and the Devonian is sometimes called the Age of Fishes.

The Carboniferous Period, together with the Permian Period, is sometimes called the Age of Amphibia. During the Carboniferous, the seas rose and fell repeatedly, and the Amphibia, the first class of vertebrates to establish themselves permanently on the land, diversified and proliferated. The first reptiles appeared. Flying insects, including the cockroaches and dragonflies, appeared. Some dragonflies had wingspans of 75 cm (2.5

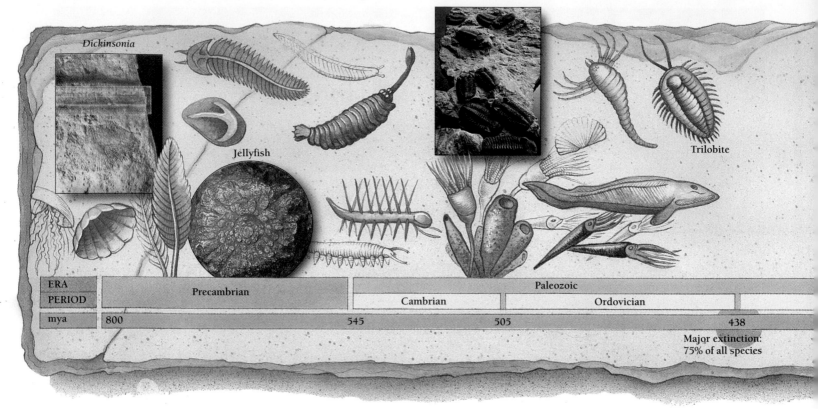

Dickinsonia

Jellyfish

Trilobite

ERA	Precambrian		Paleozoic		
PERIOD			Cambrian	Ordovician	
mya	800	545	505	438	

Major extinction:
75% of all species

Figure 15-8 (part 1)   Distinctive fossils from the geologic past.   *(Photos from left to right: Dickinsonia, Biological Photo Service; fossil jellyfish, William E. Ferguson; trilobites, James L. Amos/Photo Researchers; Devonian fish Dinichthys terrelli, James L. Amos/Photo Researchers; calamites, Ted Clutter/Photo Researchers; fern fossil, J.C. Carton/Bruce Coleman, Inc.)*

ft). Swamps filled with ferns, horsetails, and cone-bearing trees covered huge areas of the Earth. These fossilized swamps persist today as the coal deposits from which we still get coal, oil, and natural gas.

The extraordinary diversification of plants and animals that occurred during the Carboniferous ended in the Permian. During this period, the continents coalesced into a single supercontinent, called Pangaea, the seas contracted, and the climate became arid. About half of all marine species went extinct, including the once-abundant trilobites.

The Mesozoic Era: land plants and animals diversify

The **Mesozoic Era** [Greek, *mesos* = middle + *zoos* = life], often called the Age of Reptiles, began about 248 million years ago and ended 65 million years ago. The Mesozoic comprises just three periods—the Triassic, Jurassic, and Cretaceous.

During the Triassic Period, reptiles diversified enormously. More kinds of reptiles lived then than ever before or since (Figure 15-8). They were everywhere. In the seas, long-necked aquatic plesiosaurs fed on Mesozoic fish. In the air flapped enormous pterosaurs, some with wingspans of as much as 10 meters (over 30 feet). On land, one group of primitive reptiles, the thecodonts [Greek, *theke* = socket + *odon* = tooth], gave rise to the crocodiles, the birds, and the dinosaurs (Figure 15-8). With them came the newly evolved lizards and turtles. Di-

nosaurs first appeared in the late Triassic, proliferated throughout the Jurassic, then disappeared at the end of the Cretaceous.

The first mammals, small shrewlike animals, also appeared in the Triassic. They did not diversify much, but they hung on for 200 million years until the dinosaurs were gone.

The Mesozoic Era also saw the appearance of many new forms of marine invertebrates. Especially abundant were the ammonites, a now-extinct group of mollusks that resembled the modern chambered nautilus (Figure 15-8). The ammonites disappeared at the same time as the dinosaurs, at the end of the Cretaceous, accompanied by the plesiosaurs and the pterosaurs.

Many organisms survived the mass extinction at the end of the Cretaceous. These included the insects, flowering plants, birds, mammals, and many reptiles. The insects and flowering plants, in particular, diversified enormously during the Cretaceous (Figure 15-8). But the birds and mammals, which had first appeared in the Jurassic, awaited the arrival of the Cenozoic Era to flourish.

The Cenozoic Era: mammals, birds, insects, and flowering plants diversify

The **Cenozoic Era** [Greek, *cainos* = recent], the shortest era of geological time, extends from about 65 million years ago to the

Dinichthys terrelli

Cooksonia

Horse-tails

Cycad

Paleozoic				
Silurian	Devonian	Carboniferous		Permian
408	360	286		245

Major extinction:
70% of all species

Major extinction:
90% of all species

present. During the Cenozoic, the insects and flowering plants have continued to flourish and diversify, as have the modern birds and mammals.

Because so little time has elapsed since the beginning of the Cenozoic—a mere 65 million years—the fossil record for this period has been relatively undisturbed. Many beds of sedimentary layers have not been uplifted, crushed, and twisted by geologic events, and fossils from this era are relatively common and well preserved. Fossils of mammals are especially abundant, and paleontologists have been able to piece together fairly continuous histories for many modern species, including those of humans and horses.

The Fossil Record Tells a Story of Evolution

The sequence of organisms in the fossil record tells a story with an obvious meaning. At the bottom, where sediments are oldest, lie the primitive prokaryotes and eukaryotes. Above them come simple multicellular organisms, then trilobites and other invertebrates, and then the first vertebrates, the fishes. Many more fish follow. Then, in rapid succession, amphibians and reptiles arrive. Newer, and higher in the sedimentary strata, are the dinosaurs, crocodiles, birds, and mammals.

Three facts stand out. First, fossils are distributed consistently. Rocks of the same age contain approximately the same groups of organisms. Cambrian rocks, which generally contain trilobites, never contain dinosaurs or horses, only trilobites. Recent Pleistocene rocks contain no trilobites. Such a pattern is consistent with the idea that once organisms go extinct they stay extinct. Extinct species do not suddenly reappear millions of years later as a new creation.

Second, the order in which organisms are laid down in the fossil record suggests a sequence of evolution that is, as we shall see, independently confirmed by other fields of biology. For example, the first true mammals do not appear in the fossil record until *after* the appearance of the mammal-like reptiles, which are intermediate between early reptiles and mammals. This pattern is consistent with the idea that mammals evolved from these mammal-like reptiles. Similar patterns appear over and over in the fossil record.

Finally, recent fossils look most like modern organisms, while the oldest fossils generally look least like modern organisms, a pattern that suggests the accumulation of change—or evolution.

The order in which organisms appear in the fossil record is consistent with the theory of evolution.

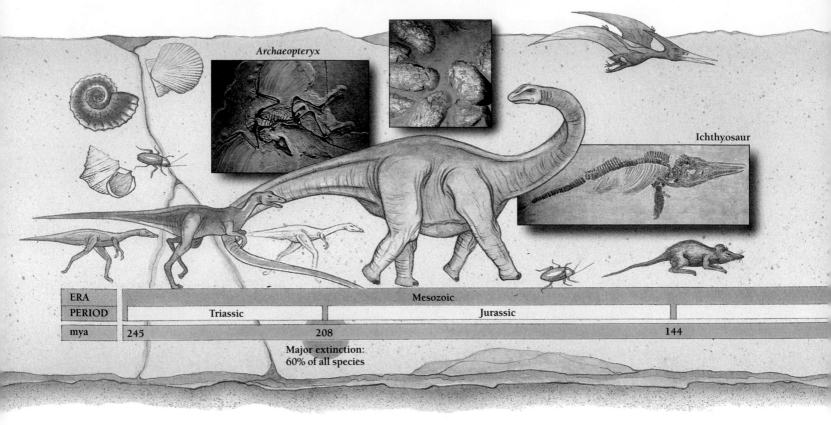

ERA	Mesozoic		
PERIOD	Triassic	Jurassic	
mya	245 208	144	

Major extinction:
60% of all species

Figure 15-8 (part 2)   Distinctive fossils from the geologic past.   *(Photos from left to right: Archaeopteryx fossil, James L. Amos/Photo Researchers; fossil dinosaur eggs (Protoceratops andrewsi), Ken Lucas/Visuals Unlimited; ichthyosaur fossil, E.R. Degginger/Bruce Coleman, Inc.; fossil flower, William E. Ferguson; plesiosaur fossil, E.R. Degginger/Earth Scenes; orangutan and baby, Tim Davis/Photo Researchers)*

Biogeography

Biogeography is the study of past and present distributions of plant and animal species. It considers not only where species live and have lived, but also how they are related to one another.

One of Darwin's most persuasive arguments for evolution was the close resemblance of species on islands to those on nearby continents (Chapter 1). Island species are often **endemic,** that is, found nowhere else in the world. But endemic species do have relatives, species that resemble them most closely.

If species arose as described in the Bible, at divine command, then similar species should live in similar habitats, since each species would have been perfected for a particular life. A 19th-century biologist might well have expected that the closest relative of a species of finch from the Galápagos Islands, off the coast of Equador in South America, would be found in a similar group of islands, matched for climate and geology. But that was not what Darwin found. He found that the closest relatives of endemic island species were generally those on the

nearest continent, not species on nearly identical islands halfway around the world.

For example, recall from Chapter 1 that the Galápagos Islands are home to several endemic species of finches. These birds, though distinct species, more closely resemble finches on the nearby South American mainland than they resemble any other finches in the world. This resemblance suggests that the Galápagos finches are related to the South American finches. The ancestors of the Galápagos finches must have come from South America.

Similarly, the closest relatives of continental species are found in other, very different habitats on the same continent—not in similar habitats on different continents. Species in the South American tropics, for example, are most closely related to other nontropical South American species than to tropical species of Africa or Asia.

Species all over the world are most closely related to those that live nearby. Darwin's theory of common descent accounts for this pattern.

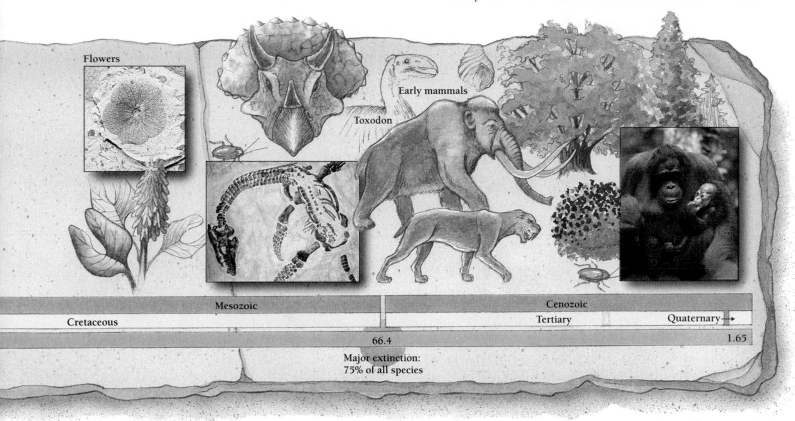

Flowers

Toxodon

Early mammals

| | Mesozoic | | Cenozoic | |
| Cretaceous | | | Tertiary | Quaternary → |

66.4

1.65

Major extinction:
75% of all species

Taxonomy

Linnaeus was a great proponent of the fixity of species, yet he provided a powerful theoretical framework for considering the evolution of species through common descent. As discussed at the beginning of this chapter, his hierarchical classification looks exactly like a family tree.

For example, all the dogs, wolves, coyotes, and other species in the genus *Canis* are like siblings. The members of related genera—*Canis* and the foxes *Vulpes* and *Fennecus,* for example—are like cousins. All the dogs, foxes, and wolves belong to the same family, the Canidae, and are therefore more closely related to one another than to the bears, cats, and other families in the order Carnivora (Figure 15-9).

The members of the genus *Canis* look alike, just as members of the same family should. The wolf, the coyote, and the jackal also look remarkably similar. The other genera in the family—various kinds of foxes—look a little less doglike, but they still look much more like dogs than do raccoons, bears, or cats.

Nonetheless, notice that dogs themselves vary enormously. The variation among the different breeds of dogs is due to in-tense artificial selection. The English bulldog, for example, is the result of selection for a developmental deformity in which the upper jaw fails to develop at the same rate as the lower jaw. The lesson here is that nearly all species possess tremendous untapped genetic variation, a point to which we will return.

Linnaeus did not intend for his classification to be interpreted literally as a family tree. We now understand, however, that the careful anatomical comparisons that Linnaeus and his successors made reflect actual relatedness.

Comparative Anatomy

Homologous Structures

Species descended from a common ancestor may evolve in quite different directions and yet retain many of the same character-istics. For example, consider the forelimbs of humans, horses,

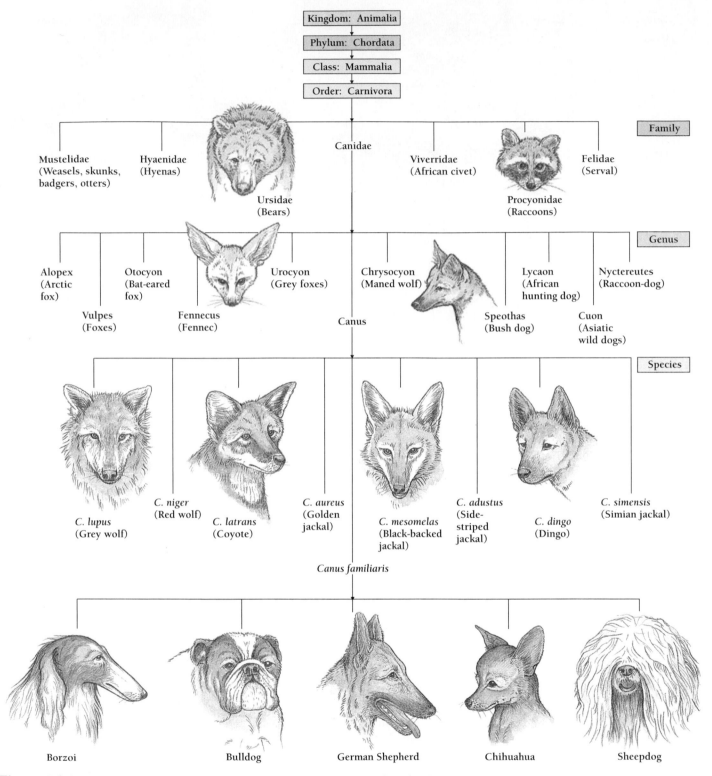

Figure 15-9 The canids demonstrate hidden variation. All the members of the family Canidae are distinctive—recognizably different from bears, raccoons, cats, and other carnivores. In their natural forms, all the canids look similarly doglike. Yet, artificial selection in dogs (*Canis familiaris*) has exaggerated certain traits and deformities to such an extent that the physical variation expressed by just five breeds surpasses that shown by all the other species in the genus *Canis* and even in the whole family Canidae. There are limits to the variation present in a single species, however. No dog could be mistaken for a bear, a cat, a hyena, or any of the other noncanid families in the order Carnivora.

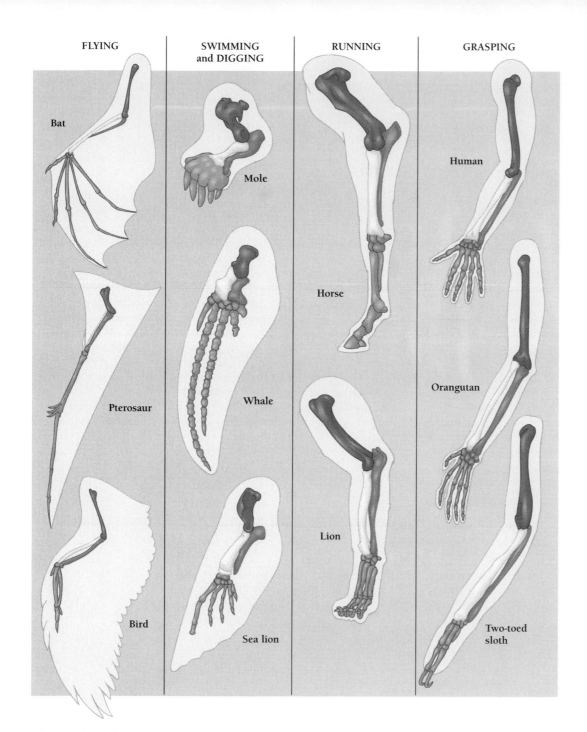

FLYING

Bat

Pterosaur

Bird

SWIMMING and DIGGING

Mole

Whale

Sea lion

RUNNING

Horse

Lion

GRASPING

Human

Orangutan

Two-toed sloth

Figure 15-10 Homology in vertebrate limbs. The limbs of bats, moles, horses, and humans contain the same set of bones, modified, respectively, for flying, digging, running, and throwing. This limb homology extends to birds, reptiles, and amphibians, and even to crossopterygian fishes.

bats, moles, and whales (Figure 15-10). Each of the different forelimbs possesses the same bones arranged in the same way. These bones are said to be **homologous,** for they all arose from the same structures in a common ancestor.

Homology in forelimbs extends even beyond the mammals. The forelimbs of toads, turtles, and turkeys are likewise homologous (Figure 15-10). The general appearance of these forelimbs differs more widely than in the mammals, but all are homologous.

Homologous structures often perform vastly different functions. Humans write, horses stride, bats fly, moles dig, and whales swim. Each such use represents a specific adaptation to a different way of life.

Homologous structures may perform the same function. For example, the bird and the bat have evolved different arrangements of the same bones to perform the same function. The surface of a bird's wing covers the entire forelimb, while

(Text continued on page 378.)

BOX 15-2

Continental drift

Studies in biogeography reveal that species resemble nearby species more than species that live far away. However, similar organisms can occur in places that are remote from one another. For example, rheas, ostriches, and emus, all large flightless birds, occur in South America, Africa, and Australia, respectively. We may wonder if these birds are related to one another and if so, how they came to be so widely dispersed.

The fossil record shows many examples of a similar pattern of dispersal, with closely related species in the southern continents. The distinctive fossil plant *Glossopteris,* for example, appears in similar rock formations from the late Paleozoic Era in both Brazil and South Africa. How could

organisms that are so closely related live so far from one another?

In 1915, the German geologist Alfred Wegener, observing the jigsaw-puzzle fit of the African and South American coastlines (Figure A), proposed the theory of continental drift. Wegener argued that Africa and South America, as well as Australia, Antarctica, and India, were once a single supercontinent, which fragmented into separate continents and drifted apart. Most geologists rejected Wegener's theory, however, and biologists and paleontologists continued to puzzle over the similarities between the species on the southern continents.

Then, in the early 1960s, geophysicists discovered that the Earth's surface consists of large blocks or **plates,** which move with

respect to one another at the rate of a few centimeters a year. The discovery that both the continents and the sea floor were moving about on the surface of the Earth immediately rejuvenated Wegener's theory.

Geologists now understand that this movement comes from the continuous formation of new crust at ridges under the ocean. As lava emerges from below the surface, the new rock pushes the older rock of the plates on either side. When two plates collide, one may dive beneath the other, pushing up mountains and precipitating earthquakes and volcanic activity. The Himalayas, for example, rise a little higher each year as the Indian–Australian plate crashes—ever so slowly—into the Eurasian plate. Similarly, the Pacific plate, beneath

Figure A   **Continental drift.** Groups of plants and animals common to two or more distant continents supported the idea that the continents were formerly united.

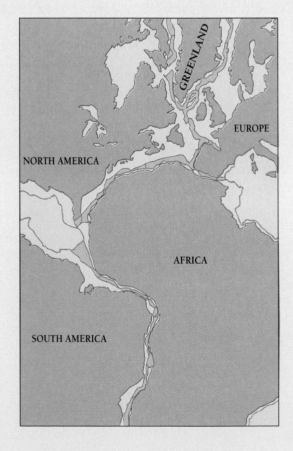

the Pacific Ocean, scrapes northward past the North American plate creating California's earthquake-prone San Andreas Fault (Figure B).

Geologists have been able to puzzle out many of the past movements of the continents by studying the magnetic orientation of rocks. The outline of this history explains several features of the fossil record and of the distribution of modern species. About 250 to 200 million years ago, all of the continents were united into one supercontinent called Pangaea. Pangaea was surrounded by Pantalassa, the ancestral Pacific Ocean (Figure C). In time, Pangaea began to break up, allowing independent evolution to occur on two great continents—Laurasia (Eurasia, Greenland, and North America) and Gondwanaland (India, Africa, South America, Madagascar, Antarctica, and Australia).

By the beginning of the Cretaceous, 135 million years ago, Gondwanaland had also begun to break up, separating South American and African species from those on Australia and Antarctica. By about 65 million years ago, South America had split from Africa and Australia had separated from Antarctica, after which time Australian species evolved in isolation. The land connection between North and South America was formed only about 3 million years ago, so there has been only limited flow of species between these two continents.

Figure B Tectonic plates. San Francisco Bay. At the San Andreas Fault, the Pacific plate slides north past the North American plate. Tectonic plates that make up the Earth's crust float on a sea of molten rock. These plates are in constant motion, bumping into one another, sliding past one another, and diving under one another. *(Terranova International/Photo Researchers)*

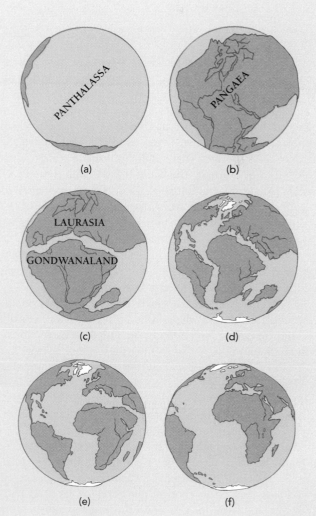

Figure C The history of the continents. (a,b) Alfred Wegener's view of the Earth as it existed 200 million years ago. At this time there was one ocean (Panthalassa) and one continent (Pangaea). (c) By about 20 million years later, the supercontinent had begun to split into northern Laurasia and southern Gondwanaland. (d) Separation continued, and (e) looked this way about 65 million years ago. (f) Further widening of the Atlantic and northward migration of India brings the Earth to its present state.

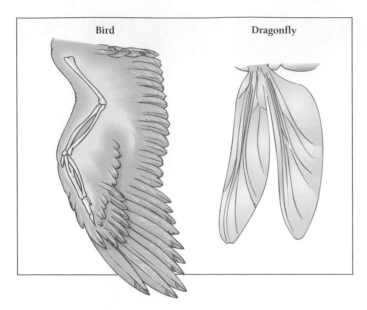

Figure 15-11 Analogy in the wings of bird and dragonfly. Not all similarities in structure are due to homology. Birds and insects have evolved wings separately. The two kinds of wings look similar because flying exerts similar selective pressures on all flying organisms.

that of a bat consists of a web that covers just four of the five digits. And, in an extraordinary 150-million-year-old fossil of the pterosaur *Rhamphorhynchus,* we find still another kind of wing, with a leathery membrane extending from one long digit to the trunk of the body. All these wings evolved separately from homologous structures, and each, in its own way, allows flight.

Not all structures that fulfill the same function are homologous. For example, the wings of birds and of insects (Figure 15-11) are not homologous. Even though they evolved to serve the same function, they derive from different ancestral structures, and so they are said to be *analogous* structures. Analogous structures may have many physical similarities, since they serve the same function. For example, all wings must be light and have broad surfaces. But a dragonfly's wing and a sandpiper's wing have little else in common.

Evolution exhibits a diversity of solutions to the same design problems. Bats, birds, and butterflies all have very different kinds of wings. Fish swim by arching their backs from side to side. Dolphins swim by arching their backs up and down.

All of these examples show that evolution is opportunistic (Figure 15-12). Just as a handyman can use a bit of wire here, a bit of glue there, to fix a bicycle or a light fixture, so evolution may employ a humerus here, a radius there, to allow an organism to write, gallop, fly, dig, or swim.

Paleontologists can trace changes in homologous structures through the fossil record. For example, the first recognizable horse lived about 50 million years ago. The size of a large dog, it had four toes on its forefeet and three on its hind feet. Its teeth were too weak for it to chew the tough grasses that modern horses live on, so paleontologists believe that these ances-

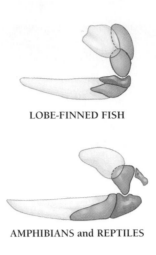

LOBE-FINNED FISH

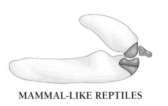

AMPHIBIANS and REPTILES

MAMMAL-LIKE REPTILES

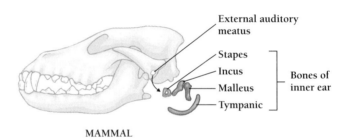

External auditory meatus

Stapes
Incus
Malleus
Tympanic

Bones of inner ear

MAMMAL

Figure 15-12 Evolution is opportunistic. The four tiny bones of the mammalian inner ear evolved from the jaw bones of early reptiles. The jaw bones evolved, in turn, from the branchial arches (structures that carry gills) of our jawless ancestors.

tral horses lived on leaves, which are softer and more easily digested. Subsequent horse fossils, however, show a progressive lengthening of the legs, strengthening of the teeth, and reduction in the number of toes. In the foreleg of a modern horse, the lower bones (radius and ulna) are fused and long. The last segment of a horse's leg is actually a long middle finger, the only one remaining of the five digits found in primitive ancestral mammals. These successive changes in the homologous bones give the modern horse longer legs and extraordinary fleetness.

Vestigial Structures

Some homologous structures have no apparent use. The modern horse has a single, useless side toe, called a splint. The splint is an example of a **vestigial structure** [Latin, *vestigium* = footprint], a part of an organism with little or no function but which had a function in an ancestral species. A vestigial

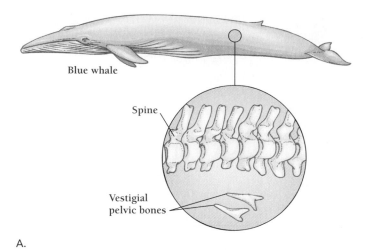

A.   B.

Figure 15-13   Vestigial structures. A. The blue whale has vestigial pelvic bones, remnants from the whale's ancestors' life as land animals with legs. B. The South African python's short claw is a vestigial leg.   *(B, Anthony Bannister Photo Library)*

structure is a remnant. Our own tailbone, invisible without an x-ray image, is a vestige of another way of life. Similarly, some snakes and whales have vestigial pelvic and leg bones, left over from ancestors that walked (Figure 15-13). Vestigial organs, like other homologous structures, are indicators of biological history and thus offer further evidence for evolution.

Homologous structures and vestigial structures suggest an
evolutionary process whereby ancient structures are
redesigned or abandoned.

Comparative Embryology

Not only are the early embryos of all mammals amazingly alike, they also resemble to a remarkable degree the embryos of birds, reptiles, and even amphibians and fishes (Figure 15-14).

For example, all vertebrate embryos, including humans, have tails and **gill slits.** In fishes and amphibians, these gill slits develop into gills for breathing under water. Tadpoles, like most amphibians, lose their gills only when they change into adult frogs. Reptiles, birds, and mammals, however, dispense with most of the gill slits earlier in development. One gill slit, however, is obvious even in humans. It develops into the ear hole, which leads from the outside of the ear inward to the eardrum. The other gill slits normally disappear into the developing neck. In rare cases, however, human babies are born with vestigial gill slits (Figure 15-15).

In fishes and amphibians, the tissues between the gill slits, called the **branchial arches,** develop into gills proper, absorbing oxygen and releasing carbon dioxide as our lungs do. But in reptiles, birds, and mammals, the branchial arches develop into completely different structures in the adult ears, mouth, and respiratory tract. In humans, for example, these include

the bones of the ears and throat, some of the muscles of the face and neck, the parathyroid and thymus glands, as well as a major blood vessel of the lungs. All of these structures can be traced to the embryonic gill slits and branchial arches. They unambiguously indicate our descent from fishes.

Developing organisms frequently pass through stages that
resemble the organisms from which they evolved, a fact
consistent with the theory of evolution.

Comparative Molecular Biology

Just as the morphology and embryology of organisms provide clues about their relatedness, so too does their molecular makeup. An organism's genes and the products of those genes— the proteins—are a clear record of its heredity.

At the most fundamental level, all cells rely on the same molecular machinery. All cells use DNA to carry the genetic information; RNA, ribosomes, and (approximately) the same genetic code to translate that information into proteins; the same 20 amino acids to build proteins; and ATP to carry energy. The universality of this system suggests that it is a common heritage to all of life, passed down from some truly ancient common ancestor.

Widely divergent organisms use similar or identical versions of the same proteins as well. Modern biologists' ability to precisely determine the amino acid sequences of proteins and the nucleotide sequences of DNA allows them to compare these different proteins directly. From such comparisons, biologists can construct hierarchies of relatedness much like those derived from comparative anatomy. The more closely related two species are, the greater the similarity between the amino acid sequences of all their proteins.

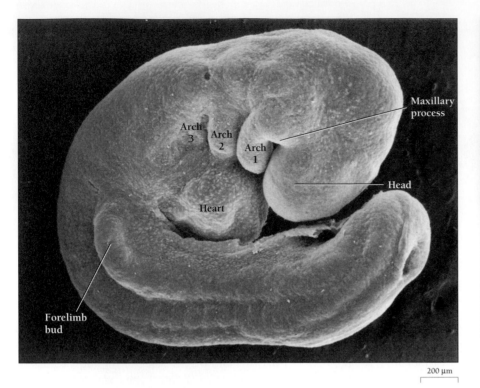
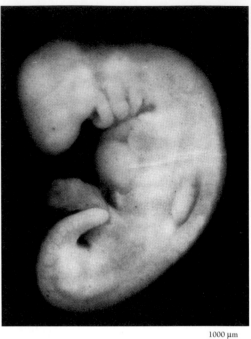

Figure 15-14   Early embryos of a mouse (*left*) and a human (*right*). Early embryos of vertebrates are remarkably alike, displaying gill arches and gill slits, paddlelike hands and feet and tails.   *(left, courtesy of K. Sulik; right, Professor Hideo Nishimura)*

Amino Acid Sequencing: Cytochrome *c*

One of the first proteins whose structure biochemists analyzed in this way was cytochrome *c*, a participant in the electron transport chain of all aerobic organisms. Cytochrome *c* is a single polypeptide chain of 104 amino acids. The exact amino acid sequence has been determined in dozens of species. In all of the species examined, 35 of the 104 amino acids are always identical. The remaining 69 vary from species to species.

More sequence differences appear between species whose anatomy suggests that they are only distantly related. Fewer differences appear in species whose anatomy suggests that they are closely related. For example, anatomical comparisons suggest that humans and chimpanzees are closely related, while humans are only very distantly related to tuna. Cytochrome-*c* analysis confirms this relationship. The resulting **phylogenetic** tree is essentially identical to a tree made on the basis of other comparative studies (Figure 15-16). The correspondence is ample evidence of the power of molecular comparisons.

The similarities and differences among amino acid sequences also reflect evolutionary history as recorded in the fossil record. For example, the family tree derived from analyzing cytochrome *c* sequences suggests that the common ancestor of humans and monkeys lived after the common ancestor of mammals and reptiles. The fossil record shows that, indeed, the first

anthropoid fossils date from about 30 million years ago, while the first mammal-like reptiles date from at least 180 million years ago.

DNA–DNA Hybridization

Molecular biologists can also derive phylogenetic trees by going directly to the DNA and sequencing individual genes. Individual genes, however, evolve at different rates depending on selective pressures. A phylogenetic tree based on just one gene or protein might not be reliable. A faster, more comprehensive technique, called *DNA–DNA hybridization,* allows researchers to compare hundreds of genes at once. In this technique, the single strands of DNA from two different species are allowed to base pair (hybridize) into a two-species double helix. The two strands of DNA are not completely complementary. Only the paired bases will hold the two strands together, however. The rest are like the broken rungs on a ladder. The more different the two strands of DNA, the more rungs will be "broken" and the more easily the hybrid will separate if it is heated. The temperature at which the two strands separate (or "melt") is a measure of the genetic distance between the two species. Thus, any pair of species has a characteristic *melting temperature,* from which molecular biologists can construct a phylogenetic tree.

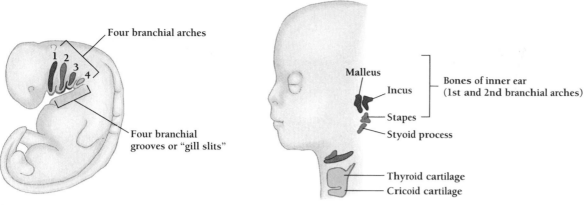

A. 4-week-old human embryo

Four branchial arches

Four branchial grooves or "gill slits"

B. 24-week-old human fetus

Malleus
Incus
Stapes
Styoid process

Bones of inner ear (1st and 2nd branchial arches)

Thyroid cartilage
Cricoid cartilage

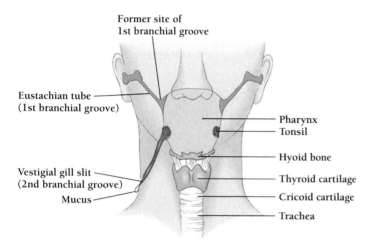

Former site of 1st branchial groove

Eustachian tube (1st branchial groove)

Pharynx
Tonsil
Hyoid bone
Thyroid cartilage
Cricoid cartilage
Trachea

Vestigial gill slit (2nd branchial groove)
Mucus

C. Adult

Figure 15-15   Gills in adult humans? Not quite. In some rare individuals, however, the second gill slit remains open even in adulthood. This opening from the throat to the outside of the neck appears in C. In all adult humans, the opening into the outer ear and the Eustachian tube derive from the first gill slit. In addition, the bones of the inner ear derive from the first and second branchial arches of the gills, as shown in A and B. Bones in the throat derive from the 2nd, 3rd, and 4th branchial arches.

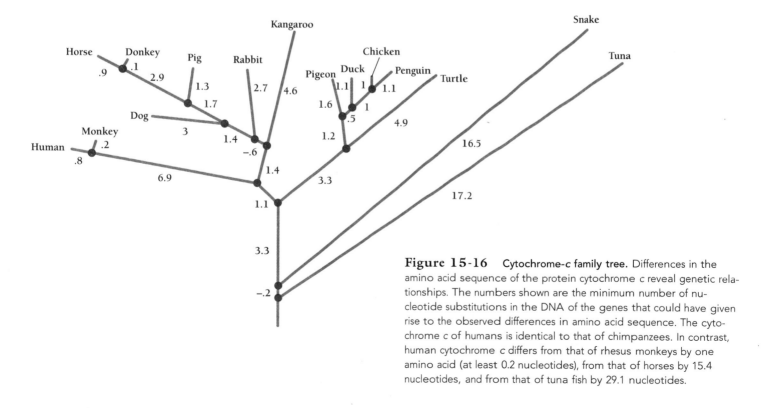

Figure 15-16   Cytochrome-c family tree. Differences in the amino acid sequence of the protein cytochrome c reveal genetic relationships. The numbers shown are the minimum number of nucleotide substitutions in the DNA of the genes that could have given rise to the observed differences in amino acid sequence. The cytochrome c of humans is identical to that of chimpanzees. In contrast, human cytochrome c differs from that of rhesus monkeys by one amino acid (at least 0.2 nucleotides), from that of horses by 15.4 nucleotides, and from that of tuna fish by 29.1 nucleotides.

The Molecular Clock

By comparing family trees derived from molecular analysis with dates derived from the fossil record, evolutionary biologists have discovered an intriguing and useful relationship. In any two species, the number of molecular differences in a shared molecule is proportional to the amount of time that has elapsed since the two species diverged. Molecular changes seem to accumulate in each type of protein, or in the corresponding piece of DNA, like the ticks of a molecular clock. Because each gene or protein changes at a different rate, each clock needs a separate calibration. Further, the same gene may, for example, evolve faster in rodents than in primates. Nonetheless, these clocks allow researchers to estimate when two species diverged, even when no common ancestor exists in the fossil record.

Comparative molecular biology confirms and elaborates the story told by comparative anatomy and the fossil record. The molecules of closely related organisms are more similar than the molecules of distantly related organisms. Molecular clocks also allow researchers to estimate the amount of time that has elapsed since two related species diverged.

The Evidence for Evolution: Summary

The fossil record shows a sequence of organisms that suggests that modern organisms have evolved from those that came before. The particular order of appearance of thousands of species is consistent with the idea of evolution. No other theory accounts for the specific arrangement of organisms in the fossil record. Studies in biogeography, comparative anatomy, embryology, and comparative molecular biology, as well as genetics, ecology, and animal behavior, all independently support the theory of evolution through common descent.

Darwin made a strong case for the theory of evolution. And the evidence that scientists have accumulated since then has strengthened this theory. We can accept the idea that organisms have changed over time, and we can accept the idea that modern organisms are descended from ancient ones. But what of Darwin's proposed mechanism, natural selection? Darwin believed that natural selection, acting gradually over immense spans of time, created not only new species, but new genera, new families, and all the higher taxa. Why did he think that was possible?

NATURAL SELECTION: DARWIN'S MECHANISM FOR EVOLUTION

Once Darwin conceived the idea of evolution by natural selection, he developed an obsessive interest in the work of gardeners, farmers, and stock breeders. It was perfectly apparent to him that many of the different breeds of dogs and pigeons,

for example, would, if found in nature, be identified as separate species (Figure 15-10).

Artificial Selection

Breeders derive such varied forms from a single species by carefully selecting and breeding only individuals that show certain desired traits most strongly. Individuals that do not show the trait, or show it only weakly, are not bred. Under this selective regimen, each generation will show the selected trait more clearly than the last.

Darwin's particular interest was the work of Victorian pigeon fanciers. He collected many varieties of pigeons, studied the differences in their skeletons, and bombarded fanciers with questions (Figure 15-17). Ultimately, he concluded that all breeds of pigeons were descended from a single species—the European rock pigeon. Yet, the more than 20 different breeds were so different, Darwin argued, that an ornithologist [Greek, *ornitho* = bird] who thought that they were wild birds would certainly classify them into a dozen or more species and quite possibly into half a dozen genera, as well.

Clearly, artificial selection by breeders could transform the characteristics of a breed. Success required only (1) that individual organisms vary in their characteristics, (2) that these variations be heritable, and (3) that the breeder consistently select certain traits in each generation.

Artificial selection does not result in the production of new species. At least new breeds are not recognized as species, no matter how morphologically distinct they are. A mammologist from Mars wouldn't hesitate to classify a bulldog as a species separate from a chihuahua or a German shepherd. Morphologically they are distinct. But we know their history and that they can interbreed, so we classify them all as one species. Artificial selection demonstrates how much genetic variation exists in populations of organisms and how much change can occur in just a few years. More importantly, however, artificial selection hints at the much greater degree of change that might occur under the constant pressure of natural selection over millions of years.

Artificial selection provides a model for natural selection, a possible mechanism for evolution.

Darwin's Argument for Natural Selection

How can nature make selections in the same way as human breeders, generation after generation? In Malthus, Darwin found the answer. Malthus, recall, had argued that the production of food and other resources can never keep up with the enormous increases that human populations are capable of, that early death is the fate of a large proportion of individuals born in every generation.

Figure 15-17 Artificial selection. Charles Darwin collected different breeds of pigeons to study the variation possible within a single species. Shown here are a pair of baldheads, a carrier cock, two white fantails, two white pouters, and several other varieties. *(The Illustrated London News Picture Library)*

Darwin saw that Malthus's arguments applied to populations of any species. Populations of organisms, although capable of enormous increases, do not continue to grow exponentially, but instead are limited by resources, predation, and disease. Natural selection must occur, Darwin argued, because nature cannot sustain unlimited exponential growth. To illustrate this point, he considered the descendants of a single pair of elephants, among the slowest breeders of all organisms. Even elephants, however, have the potential to breed geometrically. Over the course of 750 years, two elephants could have some 19 million descendants. Yet, historically, the elephant population has been fairly stable. So it is likely that two elephants in the year 1000 had only two descendants alive in the year 1750. The question Darwin posed was, Which two?

According to Darwin's theory, the survivors would be the descendants of whichever elephants were best adapted to the environment over the course of those 750 years. In artificial selection, the decisions of the breeder determine which individuals produce the next generation. In nature, the individuals that manage to survive and reproduce produce the next generation.

We can easily see how this argument might apply to the evolution of horses, for example. Individuals with long legs and teeth capable of chewing the tough grasses so common on the plains apparently survived and reproduced, whereas those with short legs and small teeth died out.

Darwin argued that if large changes can occur in domestic plants and animals over just a few generations, then even larger changes can certainly occur over the millions of years available for evolution.

The rate at which changes accumulate depends both on the intensity of selection and on the extent of inherited variation. For example, if only a few young horses from each gen-

eration survive, because most cannot escape a fast predator or cannot digest grass well enough, selection is strong and advantageous traits would accumulate rapidly. However, if most of the young horses from each generation survive and reproduce, then selection is weak and change occurs slowly or not at all.

The rate of change also depends on the extent of variation within the population. To take an extreme case, if all the individual horses of a generation were identical, then the next generation of horses could not differ from their parents. But, in fact, individuals vary enormously.

Natural Selection Analyzed

Modern evolutionary biologists have broken down natural selection into a series of basic facts and inferences.

1. Fact: **Superfecundity.** All species of organisms are so fertile that their populations would increase exponentially if all offspring survived.
2. Fact: **Not all offspring reproduce.** Populations of organisms do not attain these staggering proportions because many offspring succumb to disease, predation, and other fates before they reproduce. Those that do reach reproductive age find that the world does not contain enough food, territory, and other resources for all of them to reproduce.

   Inference: **Struggle for existence.** Whenever extinction does not occur, some individuals will escape death and disease and avail themselves of the limited resources more successfully than others. In the "struggle for existence," as Darwin called it, some individuals will survive to reproduce, while others will not.

3. Fact: **Individual variation.** Among sexually reproducing species, the individuals of a population differ. Most individuals are unique.

4. Fact: **Heredity.** Many of the differences between individuals are heritable—that is, individual differences can be passed on to offspring. Darwin knew nothing about genetics, but he could see, for example, that the short legs of the dachshund were inherited from its parents.

Inference: **Adaptive traits may be passed on differentially.** Darwin argued that in the struggle for existence, individuals whose traits best suit them to their environment survive and reproduce more successfully than those less well suited to the environment. Adaptive traits that are heritable are more likely to be passed on to the next generation. For example, we might say that cheetahs that can run 70 miles per hour catch more gazelles than cheetahs that can run only 65 miles per hour. And cheetahs that can catch more gazelles are more likely to live long, healthy lives themselves, and more likely to rear many kittens free from starvation and disease. *If* the ability to run faster (because of longer legs and a more flexible backbone, for example) is heritable, the kittens will tend to inherit those traits.

Inference: **Adaptive traits accumulate within a population.** As a result of the process of natural selection, various traits become more or less common in a population, and the overall character of the individuals within that population changes. For example, if the speed at which an animal can run is both heritable and adaptive, then, over time,

more and more cheetahs will be able to run 70 miles per hour. In this example, the genetic traits that allow a cheetah to run faster are *adaptive* traits. They are traits that allow it to survive long enough to reproduce.

Biologists accept the reality of both evolution and natural selection. The evidence that natural selection operates in natural populations is now irrefutable, yet some modern biologists question whether natural selection is the primary means by which new species and higher taxa such as genera and families are created. Over the years, biologists have proposed a host of alternatives to natural selection. A few biologists have argued that mechanisms other than natural selection may be responsible for macroevolution, the evolution of species, genera, and higher taxa of organisms. In the next chapter, we will examine natural selection more closely and also consider alternative mechanisms for evolution.

In addition, we will explore the genetic basis of variation. We can see that for natural selection to work, organisms must vary, and that the differences that distinguish individuals must be heritable. But Darwin himself knew nothing about *how* variation was inherited. The details of inheritance would not become clear until after 1900, with the rediscovery of Mendelian genetics and the advent of population genetics. Only in this century did biologists begin to understand how genes, carrying the hereditary information, provide the basis for natural selection.

STUDY OUTLINE WITH KEY TERMS

When Darwin published *On the Origin of Species* in 1859, he changed the way we view the world as much as did Copernicus when he showed that the Earth was not the center of the universe. The theory of evolution contradicted **creationism,** the Bible's version of the story of creation, the philosophy of **essentialism,** and the Christian doctrine that the Earth was only 6000 years old.

By the middle of the 19th century, Hutton's **gradualism** and Lyell's **uniformitarianism** had persuaded the scientific community that the Earth was very old, and that change occurs slowly and gradually, not catastrophically. Examination of the fossil record by Cuvier and others suggested a very long history of life on Earth—a history far longer than the 6000 years implied by the Bible. In a theory that came to be called **catastrophism,** Cuvier suggested that fossil species were organisms that had perished in periodic catastrophes, each followed by a creation.

Linnaeus had created a system of naming organisms and a hierarchy that resembled a family tree. Biologists such as Buffon and Lamarck believed in evolution, and Lamarck formally introduced the idea. But Lamarck thought that species evolved in response to "felt needs," and proposed that evolution depended on the **inheritance of acquired characteristics**. Many European scientists accepted the inheritance of acquired characteristics, but rejected evolution.

In 1859, Charles Darwin published the influential *On the Origin of Species,* which defined evolution, which he called **descent with**

modification, and described a mechanism by which evolution could work, **natural selection.** Natural selection is the differential survival and reproduction of individuals with certain inherited traits. Darwin also believed that evolution occurs gradually through the slow accumulation of changes, an idea called **gradualism.**

Darwin argued that natural selection worked like **artificial selection.** In artificial selection the breeder selects the varieties that will produce the next generation. In natural selection the individuals who best survive and reproduce are those that will produce the next generation. Natural selection, like artificial selection, favors some forms over others.

Artificial selection suggested a mechanism for evolution, but other lines of evidence suggested evolution itself. One of the most persuasive was the fossil record. **Fossils** are the remains or imprints of past life. Most come from layers of **sedimentary rocks,** which form from the compression of settling mud, sand, and debris. The order of the layers (or strata) of sedimentary rock establishes the sequences in which ancient life evolved. The simplest organisms lie at the bottom of the record and become both increasingly complex and increasingly more like modern organisms, a pattern that suggests evolution. Paleontologists, scientists who study fossils, usually divide the history of life into four long **eras**—the **Precambrian,** the **Paleozoic,** the **Mesozoic,** and the **Cenozoic** eras. Each era in turn consists of several shorter divisions called **periods,** and, more recently, **epochs.**

Another line of evidence for evolution comes from **biogeography. Endemic** species tend to resemble neighboring species in different habitats more than they resemble species that are in similar habitats but far away. This pattern of geographical distribution suggests that species evolved from one another.

Linnaeus's **taxonomy,** a hierarchical grouping of organisms by **species, genus, family, order, class,** and other **taxa,** suggests patterns of relatedness among different species, not separate creations. The relationships implied by taxonomy—so-called **phylogenies**—are derived from studies in comparative anatomy. Species that closely resemble one another are considered to be more closely related than species that do not resemble one another. **Homologous** structures, such as limb bones, provide additional evidence for evolution. **Vestigial structures,** such as the horse's splint, indicate evolutionary history.

Embryology provides another highly persuasive line of evidence for evolution. Embryos repeat the development of their ancestors to a limited but nonetheless remarkable degree. The early embryos of mammals, birds, and reptiles, for example, all have the **gill slits** and **branchial arches** typical of the fishes from which they evolved.

Comparative molecular biology confirms the lines of descent suggested by both comparative anatomy and the fossil record. Moreover, molecular biologists have discovered that the number of differences between the sequences of amino acids in a particular protein (or of nucleotides in a particular gene) in any two species is roughly proportional to the time elapsed since the two species being compared branched from their common ancestor.

Natural selection, Darwin's mechanism for evolution, rests on several facts and inferences. All species are **superfecund,** capable of producing far more offspring than the world can support. Not all of these offspring survive long enough to reproduce. In each generation, the offspring of sexually reproducing organisms vary. That is, each individual is unique, and this uniqueness has a distinct heritable component. Those individuals that survive to reproduce determine which traits will predominate in the next generation. Such individuals can be said to be naturally selected.

REVIEW AND THOUGHT QUESTIONS

Review Questions

1. What three ideas kept most scientists from seriously considering the idea of evolution until the 19th century?
2. Which of the above three ideas did Cuvier's theory of catastrophism address? Describe the theory of catastrophism.
3. Describe the theory of uniformitarianism. In what two ways did it influence Darwin?
4. Who wrote the essay that inspired both Darwin and Wallace to invent the theory of natural selection? What did the essay say?
5. Explain why the fossil record is not complete. Describe two possible interpretations for the gaps in the fossil record.
6. During which era did mammals flourish? Were dinosaurs present at the same time?
7. How did Darwin's observations on the distribution of species suggest that species have evolved?
8. If you were a bat, what bones (that you have as a human) would you use to fly with?
9. If you were a fish, instead of a human, how would you use the hole that in your human form lets sound into your ear?
10. Describe the molecular clock. What is its use?
11. Darwin's theory of natural selection can be divided into four facts and three inferences. What are they?

Thought Questions

12. The world population is currently about 6 billion. If every couple in the world had 20 surviving children, what would the world population be in 2 generations?
13. If the individuals of a species did not vary, but were all identical genetically, what would be the consequences of natural selection?

SELECTED READINGS

Barber, Lynn, *The Heyday of Natural History 1820–1870,* Jonathan Cape, London, 1980. A highly accessible account of the times in which Darwin lived.

Burkhardt, Frederick, *Charles Darwin's Letters: A Selection 1825–1859,* Cambridge University Press, 1996. A wonderfully revealing selection of letters by Darwin to friends, family, and colleagues.

Darwin, Charles, *On The Origin of Species by Means of Natural Selection, or the Preservation of Favoured Races in the Struggle for Life,* Avnel Books, New York, 1859. Darwin's original explanation of evolution by natural selection.

Darwin, Charles, *The Voyage of the Beagle,* Bantam Books, New York, 1972. Darwin's own account of his adventures.

Futuyma, Douglas, *Science on Trial,* Pantheon Books, New York, 1983. A history of science and creationism in the United States.

Mayr, Ernst, *One Long Argument, Charles Darwin and Genesis of Modern Evolutionary Thought,* Harvard University Press, Cambridge, Massachusetts, 1991. A sophisticated examination of Darwin's ideas.

▶ On-line materials relating to this chapter are on the World Wide Web at http://www.saunderscollege.com/lifesci/ Click on Tobin/Dusheck: *Asking About Life.*

A Fatal Disagreement

On January 26, 1943, an internationally famous and much beloved plant geneticist died of starvation in a Soviet prison hospital. He must have wondered at the irony of his own death. Nikolai Vavilov, Director of the prestigious Institute of Plant Breeding, in Leningrad, had devoted his life to the study and improvement of wheat, corn, and other cereal crops, in order to eliminate hunger in the vast, new Soviet Union. His "crime" had been to become involved in a long-running dispute about how evolution works. He, quite literally, contradicted the party line and paid for it with his life.

A cheerful, good-natured scholar, Vavilov traveled all over the world in the 1920s collecting seeds from different kinds of crops. He was the world's expert on the biogeography of wheat and a friend to eminent biologists in every part of the world. In the United States, he picked up the expression "Keep smiling," and liked to use it on anxious Soviet colleagues.

Vavilov's own troubles began in the mid-1920s when he befriended a hard-working and charismatic young scientist, named Trofim Lysenko. Lysenko, the son of a peasant, worked from sunup until sundown, talking to his experimental plants as if they were people. He continually imagined new ways of "training" crops to yield more grain or grow in different climates. Lysenko passionately wanted to improve the lot of the Soviet people. However, Lysenko had little education in biology and disdained what he did not know. He especially disdained genetics, which he considered to be "harmful nonsense." Even his friends in those days joked, "Lysenko is sure that it is possible to produce a camel from a cotton seed."

Vavilov ignored the young worker's ignorance, and, hoping to win him over to genetics, began inviting him to scholarly meetings, although Lysenko was not well received by other biologists. Vavilov had reasons to promote the young man. The Soviet government was pressuring Soviet scientists to abandon basic science and come up with "practical" results. Moreover, the government preferred uneducated workers of peasant stock to educated bourgeois elitists. Peasant scientists were especially hard to come by, and Vavilov leaped at the chance to bring along one who also worked entirely in the pursuit of practical results.

In the early 1930s, Lysenko joined forces with Isai Prezent, a philosophy student. Prezent could see that Lysenko needed a theoretical framework to make his shoot-from-the-hip approach to biology seem more studied, more scholarly. Prezent knew no more about genetics than Lysenko, but he had studied the theory of the inheritance of acquired characteristics, by now in its death throes in the West. The theory perfectly supported Lysenko's approach to plant breeding, stating as it did that traits developed during the life of an organism in response to the envi-

Frans Lanting/Minden Pictures

ronment—long roots grown during a dry season, for example—could be passed on to the offspring. By then, modern genetics had supplanted Lamarckism, and few biologists in the world believed that acquired characteristics could be inherited. Yet, surprisingly, no one had actually disproved Lamarckism.

Imagine! Eight-hundred thousand tweezers!

For this reason, and because Lysenko's peasant background and pragmatic philosophy were above criticism, government officials granted him more and more privileges and opportunities to carry out his research. He promised spectacular results with his techniques. Lysenko claimed he would double or triple the wheat crop by "training" wheat seedlings to grow in the far North. Grain would be superabundant. Soviet bureaucrats loved it, and newspaper reporters fell all over themselves to interview the young genius.

Lysenko's techniques consistently failed to increase harvest yields, but no one seemed to notice. When one idea failed, he came up with another. Each one's promise eclipsed the failures of the last. In the late 1930s, Lysenko was put in charge of all of Soviet agriculture. The whole nation became his "experimental garden," and every peasant farmer became a scientist. Once, Lysenko ordered tens of thousands of collective farms to test a theory about cross-pollinating different varieties of wheat. He requisitioned 800,000 pairs of tweezers for transferring pollen, one pair for each of the peasant farmers who would implement his "experiment."

Joseph Stalin, the USSR's totalitarian leader, was entranced by the scale of Lysenko's work. Imagine! Eight-hundred thousand tweezers! Vavilov and his fellow biologists were horrified. Cross-pollinating most of the wheat crop in the Soviet Union would surely eliminate many of the country's best varieties of wheat. And it did.

Nonetheless, Lysenko and Prezent were more outspoken than ever in their disdain for genetics. Under their influence, Stalin agreed that genetics would no longer be taught at the universities, medical schools, or high schools. Increasingly, seminars at Vavilov's Institute of Plant Breeding, a world-class research center for genetics, deteriorated into confrontations between geneticists and Lysenko's student supporters. Lysenko took Vavilov's persistent advocacy of genetics as a personal affront. In an era when millions of Soviet citizens were executed for far less than irritating a friend of Stalin's, Vavilov's defense of genetics was courageous but fatal.

In the summer of 1940, Vavilov was arrested for "wrecking" Soviet agriculture and sentenced to be shot. That sentence was later commuted to life in prison. But in the two and a half years he spent in prison, he was never moved to a work camp. Instead, along with thousands of other Soviet intellectuals, he died a slow and humiliating death in prison. In his last year, underfed and wasted by dysentery, Vavilov and two other professors passed the time by giving a series of lectures on history, biology, and forestry. Vavilov delivered more than a hundred hours of lectures. But his efforts to civilize the dirty, crowded cells could not stave off the hungry prisoners who stole his bread and the slow death that finally took him.

Vavilov knew that by continually contradicting Lysenko he had put his life in danger. Yet he never renounced genetics. Why did the truth not prevail? If Lysenko was wrong in his belief in Lamarckian inheritance, why couldn't Vavilov and his fellow geneticists prove that inheritance was genetic, clearly and unambiguously, for Lysenko's supporters?

The short answer to this question is that although geneticists of the 1920s and 1930s knew that genes were somehow passed from generation to generation, they could not prove beyond doubt that genes were real entities rather than just theoretical constructs. Nor could they prove that the genes in the eggs and sperm were free from the influence of the adult organism. In short, they could not prove that ac- quired characteristics did not somehow influence the genes.

The final proof that the genes had a physical reality in DNA, whose structure was unlikely to be influenced by the "needs" of the organism, had to wait un- til molecular genetics blossomed in the 1950s—too late for Vavilov.

KEY CONCEPTS

1. The genetic material in the sperm and eggs is not affected by the experiences of the parent.

2. Genetic variation is the fuel of evolution.

3. Mutations in genes and chromosomes are the ultimate source of all genetic variation.

4. Sexual reproduction spreads new alleles through a population and recombines different mutations into new genotypes.

5. Sexual reproduction does not by itself change the genetic characteristics of a population.

6. Natural selection can bring about evolution by changing the genetic characteristics of a population.

7. Natural selection is only one of several forces that can change the frequency of alleles within a population. Other forces are mating patterns, random drift, mutation, and breeding between populations.

HOW ARE VARIANTS CREATED AND MAINTAINED?

We saw in Chapter 15 that evolution as an idea has a long and venerable history, dating back to the ancient Greeks. Yet even when biologists finally accepted that evolution must have oc- curred, they couldn't agree on how it could work. Darwin had proposed natural selection. But without an understanding of inheritance, Darwin's mechanism was incomplete. We know now that Darwin's theory of natural selection was, in fact, largely correct. But it took biologists more than 70 years of pa- tient research and bitter dispute to appreciate Darwin's inge- nious mechanism for evolution—natural selection.

In this chapter we will see how biologists gradually came to understand how organisms inherit traits. We will study vari- ation in populations, the basis for Darwin's theory of natural selection, the sources of variation, and the means by which that variation is passed from generation to generation. We will ex- amine, as well, some of the theories that have been proposed to account for the evolution of populations.

Why Did Charles Darwin Accept Blending Inheritance?

Darwin wrote the *Origin of Species* primarily to show that evo- lution had occurred and to discredit the idea that species were specially and individually created by God. His theory of nat- ural selection was a separate idea—advanced to account for how evolution might have happened. Darwin thoroughly un- derstood selection—the differential survival and reproduction of genetic variants. But he did not know where the variation came from.

The gene, as concept or fact, was completely unknown in 1859. Most scientists, including Darwin, believed that inheri- tance was "blended"—much as yellow and blue watercolors blend to make green. Biologists could see that when two races or species were crossed, the resulting offspring generally ap- peared to possess a smooth blending of the characters of each of the parents. For example, a cross between a horse and a ze- bra produces an intermediate animal, called a "zebroid," not simply a horse with stripes or a zebra without stripes (Figure 16-1). Many 19th-century scientists assumed that the heredi- tary material, whatever it was, must blend smoothly also.

Yet Darwin intuitively suspected the true particulate na- ture of inheritance. As he wrote to Thomas Huxley in 1856, "I have lately been inclined to speculate, very crudely and indis- tinctly, that propagation by true fertilisation will turn out to be a sort of mixture, and not true fusion, of two distinct individ- uals, or rather of innumerable individuals, as each parent has its parents and ancestors." However, Darwin had no evidence for this and did not dispute the theory of **blending inheritance** publicly (Figure 16-2).

In 1866, just a few years after Darwin's publication of the *Origin of Species,* Gregor Mendel published his groundbreaking

Figure 16-1 **Zebroid and horse.** A zebroid is a hybrid of a horse and a zebra with some of the traits of each. How are its stripes not quite like those of a zebra? *(Carl Purcell/Photo Researchers)*

research, which showed, among other things, the particulate nature of inheritance (Chapter 9). For evolutionary theory, Mendel's most important conclusion was that the individual units of inheritance, which we now call genes, remained intact and did not blend or fuse with one another. But as we saw in Chapter 9, neither Darwin nor any of his contemporaries ever read Mendel's paper.

Darwin's public acceptance of blending inheritance was unfortunate, for it left him open to a criticism for which he had no answer and led him into a second fundamental misunderstanding about the nature of inheritance.

How Did an Engineer Influence Biology?

In 1867, Fleeming Jenkin, an enterprising Scottish engineer, announced that natural selection could never work in sexually reproducing organisms. Favorable variants, Jenkin argued, would not be able to pass on their advantages to the next generation. Instead, interbreeding would quickly "dilute" any new traits. His idea was that when a faster cheetah bred with a slower one, for example, the offspring would be only half as much faster as the faster parent. And *their* offspring would be only a quarter as much faster as the fast grandparent. Eventually, the increased speed would disappear altogether. How, he asked, could natural selection operate on variants that faded away as soon as they appeared? Assuming blending inheritance, he had a point.

Thanks to Mendel's work, however, we now know that inheritance is particulate, not blending, and that sexual reproduction does not necessarily dilute traits, but instead creates an assortment of different traits. In the cheetah example, the offspring would not all be exactly half intermediate in speed between the faster parent and the slower parent. Each young cheetah would be different, ranging from fast to slow, and each one's ability to reproduce would be correspondingly different.

Jenkin's logic was fine, but his grasp of biology was weak, and he was mistaken in nearly everything he said. Darwin could easily have demolished the engineer's argument on a variety of grounds. Yet Darwin was so uncertain about the nature of inheritance that, rather than counter the particulars of Jenkin's arguments, or question the basic assumption behind it—blending inheritance—Darwin made a second mistake. He took refuge in the inheritance of acquired characteristics, an idea that had been universally accepted for hundreds of years.

Darwin knew very well that individuals are unique. He knew that natural populations contain enormous reservoirs of

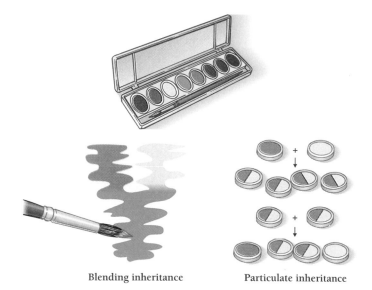

Blending inheritance   Particulate inheritance

Figure 16-2 **Blending inheritance or particulate inheritance?** Nineteenth-century biologists believed that the hereditary material from each parent blended together in the offspring much as watercolors blend. Darwin suspected the truth: each parent's genetic contribution remains separate.

variability on which natural selection can act. If Jenkin was right that variability was constantly blending away, where were new variants coming from? The **inheritance of acquired characteristics,** which Darwin accepted in principle, seemed to provide the only way for new variants to be continuously generated and passed on. According to the theory

1. Changes in the environment create changes in the needs of organisms.
2. Changes in needs occasion changes in behavior to satisfy these needs.
3. Changes in behavior result in the increased use and development of certain parts, or the use and development of new parts.
4. These changes are passed on to the offspring.

After Jenkin's critique, Darwin began to concede a greater role to the inheritance of acquired characteristics. Darwin died in 1882, one year short of seeing the problem of inheritance solved by one of his most ardent followers.

Fleeming Jenkin demonstrated that blending inheritance and evolution by natural selection were incompatible. Darwin's acceptance of blending inheritance, as opposed to particulate inheritance, forced him to accept the theory of acquired characteristics.

How Did Biologists Come To Reject the Theory of Acquired Characteristics?

Of all the world's biologists, the German biologists were the most enthusiastic about Darwinism. Among them was a man named August Weismann (1834–1914), who became one of the first biologists to unequivocally reject the inheritance of acquired characteristics and the first to suggest a coherent theory of molecular genetics.

As a boy, Weismann enthusiastically collected butterflies, beetles, and plants. Like Darwin, he studied medicine, but then returned to natural science, devoting himself to the embryology of insects. Ultimately, an eye disease forced him to give up these studies, which required long hours at the microscope, and he turned increasingly to theoretical work.

Weismann soon realized—as Darwin apparently had not—that the controversy about natural selection could not be resolved until inheritance was fully understood. Like other biologists of the time, Weismann at first accepted the inheritance of acquired characteristics. But in the 1870s he began testing the theory in a series of experiments. By the time he was 49 years old he had completely changed his mind, and, in 1883, he published a paper rejecting Lamarckism entirely.

It was a radical proposal, and biologists of the day nearly universally repudiated it. Only Wallace, Darwin's codiscoverer of evolution through natural selection, publicly supported Weismann. Wallace and Weismann alone believed that natural selection—without the inheritance of acquired characteris-

Figure 16-3 Four-week-old human embryo. At four weeks, the germ plasm, from which the ova or testes will form, is already separate from the other tissues of the developing body. *(Petit Format/Nestle/Science Source/Photo Researchers)*

tics—was the major mechanism for evolution. Other biologists had no reason to abandon the inheritance of acquired characteristics. Lamarckism was not, in fact, inconsistent with natural selection; it had not been disproved; and it constituted a likely source of variation. On the other hand, as Weismann saw it, no one had any idea how the inheritance of acquired characteristics could work, and Lamarckism was not necessary for natural selection to work.

Furthermore, Weismann had a better idea—another mechanism for inheritance. In 1882, a German cell biologist had discovered the cell's nucleus and noticed the way the chromosomes divide at mitosis. Weismann theorized that the genetic material was all in the nucleus. His theory, he said, was "founded upon the idea that heredity is brought about by the transmission from one generation to another of a substance with a definite chemical, and above all, molecular constitution." He was, of course, exactly right. But no one knew that, and not even he could be sure.

Weismann went on to argue that the hereditary material, once it is formed, is essentially isolated from the rest of the body. Embryologists had already noted that in vertebrate animals, at least, the tissue that eventually forms the ova or testes—the **germ plasm**—separates from the rest of the body early in development. In humans, for example, the cells that form the ova or testes differentiate from other embryonic cells early in the fourth week of development. The ova and testes become distinctly formed by 20 weeks (Figure 16-3).

Weismann concluded that the germ plasm is separate from the rest of the body from the very beginning, and virtually nothing in the development or environment of the body can influence the germ cells or the hereditary material. The genetic material, he said, is passed from generation to generation intact, unblended, and unchanged. This theory came to be known as the "continuity of the germ plasm." Nothing could influence the germ plasm, the location of the hereditary material. And only the germ plasm was passed on to the offspring. Therefore, the inheritance of acquired characteristics must be impossible.

Modern molecular biology has confirmed Weismann's conclusion. There is no way for the child of a woman weight lifter to inherit her large muscles. Lifting weights does not influence the molecular structure of the DNA in the mother's ova, which formed months before she was born. Genetic information flows almost entirely from the DNA to RNA to protein.

August Weismann proposed that the hereditary material was in the nucleus and that it was molecular. He argued that the hereditary material was not susceptible to influence by developments in the rest of the organism and that, therefore, the inheritance of acquired characteristics was impossible.

How Did Weismann Explain Variation in Individuals?

Weismann's theory of inheritance ruled out Lamarckism. However, he still needed to explain the origin of the variation on which natural selection acts. Remarkably, he had an answer that was partly correct—sexual reproduction.

By 1883, cell biologists had observed that maternal and paternal chromosomes do not fuse during fertilization, but merely restore the two sets of chromosomes normally found in most cells. From this observation, Weismann leapt to the conclusion that the effect of sexual reproduction was not to blend parental characters (as blending inheritance implied) but to recombine separate "hereditary tendencies."

Weismann rightly concluded that sexual reproduction, far from homogenizing parental traits, as Jenkin had argued, brings together old traits into new combinations. Even though Mendel's work was still buried in obscurity and the word "gene" had yet to be coined, Weismann had, to a remarkable degree, solved the problem of heredity.

Weismann's view remained a minority opinion throughout the last part of the 19th century and into the beginning of the 20th century. But he and Wallace kept alive the idea of natural selection without Lamarckism. Without these two scientists, the idea might have been forgotten. Not until the 1940s did biologists appreciate Weismann's enormous contribution to evolutionary theory.

In the meantime, one major piece of the heredity puzzle remained to be discovered. While sexual reproduction results in an almost unlimited supply of new variants, the mixing of parental genes into new combinations does not account for the origin of new traits. With the rediscovery of Mendelian genetics in 1900, another piece of the puzzle fell into place.

Weismann showed how sexual reproduction, by creating new combinations of traits, could act as a source of new variants for natural selection.

How Did the Rediscovery of Mendel's Work Transform Biology?

In 1900, three biologists working in three different countries independently rediscovered Mendel's laws. In the Netherlands, Hugo Marie de Vries (1848–1935) published a paper showing that if he crossed even numbers of hairy and smooth primroses he got three-to-one ratios of hairy and smooth plants. From this he concluded that the alternative traits—smooth and hairy—were controlled by a single pair of what we now call genes. Each parent, he explained, passes on only one of each of its pairs.

As soon as de Vries's paper came out, he immediately received copies of papers from two younger biologists—Karl Franz Joseph Correns (1864–1935), in Germany, and Erich Tschermak (1871–1962), in Austria—each claiming that they too had discovered the three-to-one ratios, each offering the same genetic explanation as de Vries, and each pointing out to him that Gregor Mendel had anticipated them all by 35 years.

Historians believe that both Correns and Tschermak probably read Mendel's paper before they had their respective "flashes of insight." De Vries himself received a copy of Mendel's paper from a fourth biologist sometime before he published his first paper. Whether there were actually any "independent discoveries" besides Mendel's we may never know. One thing is clear: Mendel's ideas made a splash.

In 1900, as the British biologist William Bateson (1861–1926) was traveling to London to deliver a paper to the British Horticultural Society, he passed the time on the train by reading de Vries's paper. He was thunderstruck. In London, he cast aside his prepared notes and proclaimed the death of Darwinism (by which he meant gradual evolution) and "the birth of a new science" (later called "genetics").

Bateson, de Vries, and others, who came to be known as the Mendelians, argued that evolution occurred not gradually, through natural selection, but in leaps and bounds, or saltations, by means of genetic mutations. These earliest geneticists firmly rejected the idea that evolution occurred through the slow selection of variants in a population. The Mendelians rejected natural selection and gradualism and any suggestion that evolution occurred among groups of individuals. They believed that evolution occurred suddenly, from one generation to the next, when genetic mutations transformed an individual organism into a radically new type. They argued that genetic mutations in an individual were the cause of new species, not to mention new genera, families, and other higher groupings of plants and animals.

The Mendelians contended that every evolutionary change, large or small, was the result of a separate individual mutation, and that natural selection, individual variation, and sexual reproduction played no important role in evolution. Normal continuous individual variation, such as adult height in humans—the very foundation for Darwin's theory of natural selection—was not even heritable, argued the Mendelians.

Naturalists of the era were horrified. They knew, from their close observations of wild populations of plants and animals, that most variation was continuous and that continuous variation was indeed heritable. But they, in turn, overreacted. The naturalists claimed that although the rules of Mendelian inheritance might apply to discontinuously inherited traits such as smooth or hairy plants or blue or brown eyes, such traits were unusual and irrelevant to evolution. Continuous variation, the stuff of evolution, they argued, was not governed by the rules of Mendelian inheritance. Rejecting all of genetics, they clung to a variety of other theories, including Lamarckism. The polarized debate between the Mendelians and the naturalists raged for nearly 40 years.

The rediscovery of Mendelian genetics precipitated an intense debate. Early geneticists believed that evolution occurred from one generation to the next as a result of radical mutations in individuals. Biologists who studied natural populations of organisms insisted that Mendelian genetics did not apply to the kind of variation that was the basis of evolution. Both sides were wrong.

The Modern Synthesis

In the late 1930s, a remarkable marriage of genetics and evolutionary theory occurred, which, for the first time, elucidated the genetic basis of variation and natural selection. This marriage, called the **modern synthesis**, did much to end the international feud between geneticists and naturalists. The modern synthesis fully embraced Weismann's theories and added to them all that was known about genetics. Much of what we will consider in the rest of this chapter comes from a field of genetics that contributed substantially to the modern synthesis—population genetics. Population genetics combined Mendelian genetics with the recognition that evolution can occur at the population level.

Population genetics explains the processes by which variation is generated and passed on within populations of organisms in precise mathematical terms. These processes are sometimes called **microevolution**. Population geneticists define microevolution narrowly as *changes in the frequencies of alleles of genes in a population*. For example, consider a wildflower population that includes some individuals with a gene for red flowers and other individuals with a gene for white flowers. If the relative proportions of red and white flowers change over the course of several generations, the population of flowers is said to evolve. Microevolution is distinct from **macroevolution**—the processes by which species and higher groupings (taxa) of organisms originate, change, and go extinct, a topic we will discuss in the next chapter.

Population genetics explains variation at the population level and provides a basis for natural selection and other evolutionary forces. In the next section, we will discuss what genetic variation is, where it comes from, and how it spreads and persists in populations of organisms.

WHAT IS THE SOURCE OF THE VARIATION THAT FUELS EVOLUTION?

A Review of Genetics

Until the 1930s, few biologists believed that genes were real entities. No one had ever seen a gene, and there was no other evidence for their existence. Geneticists carefully stated that genes were purely theoretical constructs. Gradually, however, geneticists began to notice more and more cases where one genetic trait, such as eye color, was consistently linked to some other genetic trait, such as size. It began to look as if the genes themselves were physically linked together when they were passed from parent to offspring.

As we saw in Chapter 10, genes are very real, consisting of strings of nucleic acids on the DNA in the chromosomes. Each set of three nucleic acids codes for one amino acid. Strings of amino acids, constructed on the ribosomes, fold into proteins.

Recall also that most sexually reproducing organisms have two of each chromosome. The two chromosomes in a pair are said to be **homologous.** One homologous chromosome possesses the same genes in the same positions as its homologue (unless they are unmatched sex chromosomes). Eggs and sperm have only one of each chromosome. When the egg and sperm fuse into a zygote, the two sets of chromosomes pair up. Thus each parent supplies one chromosome in a chromosome pair.

Genetic variation originates when a gene mutates into different forms, called **alleles.** If each chromosome in a pair has the same allele at a particular gene position, the organism is said to be **homozygous** for that gene. If each chromosome has a different allele at a particular gene position, the organism is said to be **heterozygous** for that gene.

What Is Genetic Variation?

Most genetic variability comes from gene mutations. Such mutations change only one or a few nucleotides in a gene and, thus, alter only a single protein. However, chromosomal mutations, which affect the number of chromosomes or the number or arrangement of genes on a chromosome, can change a large number of proteins at once. Chromosomal mutations tend to be lethal or extremely damaging in animals, though less so in plants. However, even rearrangements of chromosomes can, on occasion, benefit the organism. For example, the duplication of a chromosome segment, if harmless, can be passed on. In time, gene mutations can allow the genes on the redundant segment to take on new functions.

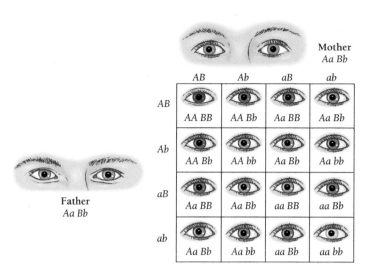

Figure 16-4 Polymorphism. In many animals, every individual has approximately the same color eyes. Human eye color varies from pale blue to nearly black. Most of this variation derives from just two alleles of two genes. The 4 alleles can combine in 9 ways.

In a population at large, a given gene may have just one form, or allele. In that case, every member of the population would have to be homozygous for that gene. On the other hand, a given gene may have two or even hundreds of alleles. If a gene had several major alleles, members of the population could be either homozygous or heterozygous in one of several different ways. For example, members of a population possessing a gene with three alleles *A, a,* and *a'* could be homozygous in three ways—*AA, aa,* or *a'a'*—or heterozygous in three ways—*Aa, Aa',* or *aa'*. If a population has two or more alleles of a given gene, it is said to be **polymorphic** [Greek, *poly* = many + *morph* = form]. The word "polymorphic" also has a second, related meaning. If the members of a population come in two or more forms, the population is also said to be polymorphic. Humans, for example, are polymorphic for eye color (Figure 16-4).

The original source of variation is gene mutations or chromosomal mutations. A population is polymorphic for a given gene if that gene has more than one allele.

Most Traits Are Polygenic

Most genetic research deals with alleles that specify discrete characters such as the presence or absence of a specific enzyme. Most traits, however, are **polygenic**, governed by many genes (Figure 16-5). Such traits—height, weight, and skin color, for example—vary smoothly and continuously within a population. We may portray such continuous variation by means of a graph (Figure 16-6). The curves describing such distributions often have the same bell shape. Two aspects of such curves de-

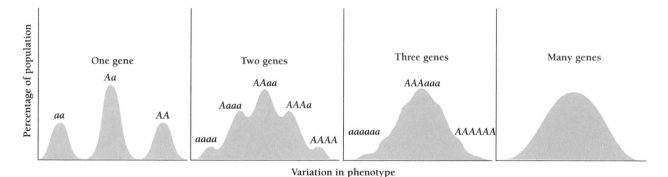

Figure 16-5 Most traits are polygenic. The more genes involved that influence a trait, the more continuously the trait will vary in a population. If only one gene with two alleles is involved, then no more than three basic types will appear. Two genes with two alleles each can result in five phenotypes (as in eye color above). However, even those individuals with the same genotype differ from one another because of environmental influences. Most traits are influenced by many genes and display a smooth bell-shaped curve.

Figure 16-6 A bell curve.
Because adult height in humans is controlled by many genes, the distribution of heights in adults is a smooth curve, as illustrated by this photo of students. The students are arranged according to height with the shortest on the left and the tallest on the right. (*photo, courtesy, Joiner Associates, Inc.*)

Number of individuals

Height in inches

serve special attention, the average value and the breadth of the curve. The breadth of the curve is a measure of the variability of the trait. We all know that the average value of a trait may vary from population to population. African Pygmies are, on average, shorter than people living in Boston (Figure 16-7). Populations may also vary in the variability of a trait. The population of African Dinkas pictured in Figure 16-7, for example, shows a greater range of heights than that of either the Pygmies or the Bostonians.

Most traits are influenced by many genes.

What Determines the Genetic Variability of a Population?

Populations that contain many alleles vary more than populations that have few alleles per gene. In a changing environment, highly variable populations can evolve relatively rapidly because such populations present more variation for natural selection to select from than less variable populations.

Three factors determine the genetic variability of a population: (1) the rate at which mutations accumulate in the DNA; (2) the rate at which changes spread through a population (principally by sexual reproduction); and (3) the rate at which deleterious mutations are eliminated from a population by natural selection, a topic we will discuss in more detail at the end of this chapter.

How Often Do Mutations Occur?

Mutations cause random changes in protein structure and, thus, most mutations damage the organism. Occasionally, however, random changes may cause no harm or may even give an advantage to the organism that bears them. Such rare advantageous mutations are the fuel of evolution. Without a constant supply, evolution would stop. Instead, mistakes and mutations

persist and occasionally provide every species with adaptive variants. For example, on average, each new human baby carries one or two new mutations.

Mutations are not the result of directed molecular surgery, but arise constantly as the result of random processes. Because they occur constantly and randomly in all genes, geneticists have been able to estimate the probability that a given nucleotide will mutate in each generation. The allele for eyelessness in the fruit fly *Drosophila melanogaster* occurs in 6 of every 100,000 fly gametes. Because mutation occurs more or less constantly in a species, and because the total number of genes is great (about 100,000 for humans, for example), the total number of variations within a species can be enormous.

How Do Mutations Spread Through a Population?

The rate at which a population evolves depends on how rapidly new mutations spread through the population. In fact, although evolution could not long continue without new mutations, the spread of those mutations, rather than their original occurrence, is the most important factor in increasing variability in a population. The main way for genetic variation to spread is through sexual reproduction and recombination. Recombination, which occurs during sexual reproduction, mixes mutations that have arisen independently into endless new combinations.

Genetic variation originates when genes or chromosomes mutate. Genetic variation spreads when sexual reproduction and recombination mix these changes into endless new combinations.

How Much Genetic Variation Is There?

How Do Geneticists Measure Variation?

Since variability is the raw material for selection, one of the first questions population geneticists ask is "How much variation is there in single genes within a population?"

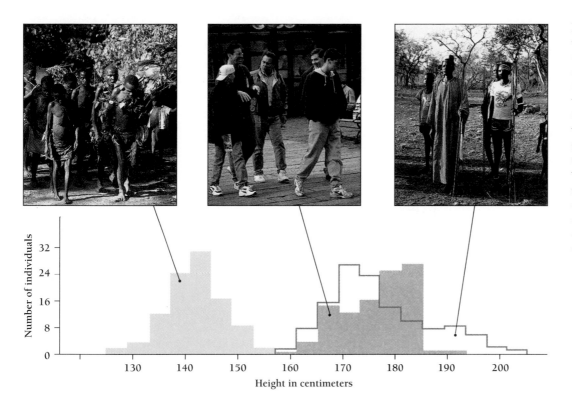

Figure 16-7   **Even variation varies.** Heights for male African Pygmies, male American Bostonians, and male African Dinkas. Nearly all Pygmies are shorter than Bostonians and Dinkas. Dinkas are, on average, taller than Bostonians, but they have a great range of heights, so that many Dinkas are actually shorter than many Bostonians. Which group varies most? *(Pygmies, Klaus Payson/Peter Arnold, Inc.; Bostonians, Frank Siteman/Stock, Boston; Dinkas, F. Jackson/Bruce Coleman, Inc.)*

To examine the extent of variation of single genes, we can look at the polypeptides and proteins that are the products of genes or at the genes (DNA) themselves. One way of estimating the level of variation within proteins depends on the technique of electrophoresis, which separates molecules according to their charge, size, and shape. By studying the enzyme activity or structure of proteins through electrophoresis, researchers can detect genetic variations in the structures of enzyme molecules. Electrophoresis has been widely used to estimate the degree of polymorphism in many populations.

In humans, about 25 percent of all proteins have an alternate form present in at least 5 percent of the population. Other species examined show similar degrees of polymorphism. In populations of *Drosophila* flies, for example, about 53 percent of genes are polymorphic. In a survey of invertebrate animals, nearly half of all proteins examined had more than one form in the population at large.

Another way of expressing the level of genetic variation in a population is to estimate the likelihood that any individual will be heterozygous for a given gene (Figure 16-8). In humans about 7 percent of the genes that code for proteins are heterozygous. That is, we have received a different version of a gene from each parent in about 7 percent of our genes. In the study of invertebrates, individuals were heterozygous for an average of 13 percent of genes examined.

Most natural populations vary enormously.

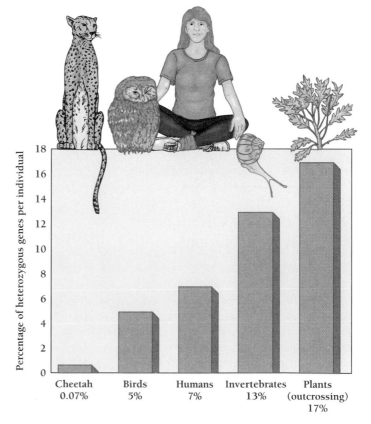

Figure 16-8   Genetic variation in some major taxonomic groups. Most organisms have far more genetic variation than cheetahs. Humans, for example, are heterozygous at about 7 percent of their 100,000 genes. By contrast, cheetahs are heterozygous at only 0.07 percent of gene positions.

Is All Genetic Variation Subject to Natural Selection?

The only variants important for natural selection are those that contribute to changes in the whole individual in ways that affect the individual's ability to survive and reproduce. The expressed traits of a whole individual organism are called the **phenotype** [Greek, *phainein* = show]. The phenotype, which includes the appearance, the physiology, and the biochemical makeup, is a result of both the genetic makeup of the individual—its **genotype**—and the effects of the environment on the development of the individual.

Selection acts on the phenotype. For example, an individual homozygous for some lethal, recessive trait, such as cystic fibrosis, expresses the trait and dies. The gene has expressed itself in the phenotype, and selection has occurred. Another individual, heterozygous for the trait, will have only one copy of the deadly allele and will have a phenotype that is perfectly healthy. In the heterozygote, the allele is not expressed in the phenotype, and so selection against the allele cannot operate even though the allele is present in the genotype. Most lethal mutations are recessive and are usually not subject to selection unless they are homozygous.

With the development of rapid methods for determining DNA sequences, evolutionary biologists have begun to accumulate information on the extent of variation in the DNA. Humans appear to be heterozygous for virtually every gene. On average, about 1 nucleotide in every 500 is different in the 2 homologous chromosomes. Most of these differences, however, lie in nucleotides that do not code for proteins. They contribute to high variability at the DNA level, but do not alter the structure of a protein or change the way the organism functions. Such mutations do not affect phenotype and are, therefore, not subject to natural selection.

Mutations that do not affect the phenotype are not subject to selection.

Why Is Genetic Variation Essential to Evolution?

If mutations are the fuel of evolution and recombination is its motor, then the driver—the guiding force—of evolution is natural selection. Because some variants survive and reproduce more successfully than others, the mutations that best equip their possessors to survive and reproduce will increase in frequency in each generation.

When biologists study genetic variation within a species, they almost always find lots (Figure 16-8). In some exceptional cases, however, genetic variation is limited. For example, most thoroughbred horses are direct descendants of a few dozen horses that lived in the late 18th century. Although breeders have allowed some outbreeding—breeding with nonthoroughbreds—the relative lack of genetic variability has led to stagnation. Despite the best efforts of highly skilled breeders, the winning times for the Kentucky Derby have not changed significantly in 50 years. The lack of change in winning Derby times could be considered an example of microevolution that has come to a standstill. Later in this chapter we will discuss some examples of natural populations that lack genetic diversity.

Evolution by natural selection depends on the continued existence of variation. Yet selection tends to reduce variation by eliminating organisms less able to survive and reproduce. We might expect, then, that species would evolve so that all individuals will have the same "best" genotype. If this were to happen, evolution of that species could not continue. (However, environments change constantly and so do selection pressures. What is "selected" changes over time.)

In the absence of genetic variation, evolution would cease.

In the rest of this chapter, we will see how populations maintain variability, even as their gene distributions change during evolution. Such variability allows populations to adapt to continuing changes in both the physical and the biological environment.

How Do Populations Maintain Genetic Equilibrium?

Hardy–Weinberg Equilibrium

In 1908, G. Hardy, an English mathematician, and G. Weinberg, a German physician, independently reconsidered Fleeming Jenkin's idea that sexual reproduction eliminates variation in light of the fresh knowledge that heredity is nonblending. Recall that Mendel's Principle of Segregation says that alleles do not blend with other alleles; they stay intact. Hardy and Weinberg each reached the same conclusion, now called the Hardy–Weinberg Principle: *Sexual reproduction by itself does not change the frequencies of alleles within a population.* Mendelian genetics, argued Hardy and Weinberg, overturned Jenkin's case against natural selection.

To see what this means, we will need to define some terms. We speak of all the genes of all the individuals in a population as the **gene pool.** And we calculate the *frequency* of a given allele within the gene pool by dividing the number of times the allele is present in a population by the number of times it could be present if every organism in the population had the allele on both of the paired homologous chromosomes. In that case, the allele would be present on every single chromosome on which the gene appears. *The allele frequency is the number of times the allele is present in the population divided by the total number of chromosomes on which the gene appears.*

To understand the Hardy–Weinberg Principle, let us consider a gene that has two alleles, *A* and *a*. For example, the color of mussels on the north coast of California depends on a single gene with two alleles. One allele, *A*, gives the mussels a blue color and is *dominant.* The other allele, *a,* makes the

BOX 16-1

Does selection act on preexisting variants?

Many 19th-century biologists believed, as Lamarck did, that variation arose in response to the needs of the organism. Thus even after mutations and genetics were understood, an important question arose: Do mutations arise spontaneously without reference to the needs of the organism or selective pressure or do they arise in response to selective pressure? In short, are the variants that selection acts on preexisting ones, or not?

In the early 1950s, the American geneticists Joshua and Esther Lederberg decided to test the idea that variants preexist. They used bacteria, because they grow so fast, and an elegant technique they had devised called replica plating. Replica plating allowed them to study the descendants of thousands of individual cells in different environments.

After growing bacteria in a nutritious broth, the Lederbergs spread the bacteria onto a petri dish containing the Jell-O–like medium agar. Each single bacterium on the agar divided to produce a visible colony composed of thousands of cells. The Lederbergs then touched the surface of the agar to a piece of velvet. The velvet picked up a sample of each colony. By touching the velvet to another agar plate, they made a replica of the original plate. Every colony of the replica had the same position as the "parent" colony on the original plate. The corresponding colonies all derived from the same cell in the liquid broth and so had the same genotypes.

The Lederbergs used their replica plating technique to study antibiotic-resistant variants of the human gut bacterium *E. coli*. In plates containing the antibiotic streptomycin most of the bacteria died, but a few survived and grew into colonies. All of the cells in each of the colonies themselves produced streptomycin-resistant progeny, showing that the resistance to the drug is a genetic property.

But were individuals with genetic resistance present in the original population or did they arise only as a result of exposure to the drug? Replica plating provided an easy way to answer this question. The Lederbergs simply examined the replicate colonies that had not yet been exposed to streptomycin. Because they knew the identity of each colony from its position on the plate, they could pick out the ones whose brethren, on other plates, had been able to flourish on streptomycin-spiked agar. Here were colonies of the same genotype that had not yet been exposed to streptomycin. Were they resistant to the drug, too? The answer was yes. The Lederbergs proved that the variation preexisted in the original bacterial population, even though the bacteria had no need for such resistance. This result supports the idea that adaptation to changed environment can result from selection among preexisting variants.

mussels brown and is *recessive*. That is, whether the mussel is *AA* or *Aa*, it will be blue. Only when the mussel is *aa* will it be brown.

The fraction of sperm and ova in the population that carries the *A* allele we call "p" and the fraction that carries the *a* allele we call "q." The fraction of the blue allele (*A*) in a particular mussel population is p = 0.6, while the fraction of the brown allele (*a*) is q = 0.4. Since there are only two alleles for this gene, p + q = 1. Geneticists call p and q "frequencies." Thus the *A* allele has a frequency of 0.6 and the *a* allele has a frequency of 0.4.

But the frequencies of the alleles are not the same as the frequencies of the genotypes. Suppose the mussels release their eggs and sperm into the seawater. We know that 60 percent will carry the *A* allele and 40 percent will carry the *a* allele. Now the sperm and eggs begin joining up at random and fusing into zygotes with genotypes *AA*, *Aa*, and *aa*. What are the frequencies of those genotypes? To calculate the frequencies of the three genotypes among the offspring, or F1 generation, we use a Punnett square, as described in Chapter 9.

If the frequency with which a sperm or egg carries the *A* allele is 0.6, then the frequency of an *AA* homozygote will be the probability that an *A* sperm will fertilize an *A* egg. That probability is p × p = p² = 0.36. Similarly, the frequency of an *aa* homozygote is q² = 0.16. The frequency of a heterozygote is the probability that an *A* sperm will fertilize an *a* egg plus the probability that an *a* sperm will fertilize an *A* egg, that is

$$(pq) + (qp) = 2pq = 2(0.4)(0.6) = 0.48$$

The frequencies of *AA*, *aa*, and *Aa* are 0.36, 0.16, and 0.48, respectively. Notice that the sum of the frequencies is p² + 2pq + q² = (p + q)² = 1² = 1. Likewise, 0.36 + 0.48 + 0.16 = 1. In other words, we have accounted for 100 percent of the population.

From the genotype frequencies we can derive the allele frequencies in the F2 generation just now conceived and see if they are the same as in the previous generation, in short, p = 0.6 and q = 0.4.

First, let's assume that we have 100 individuals, each with 2 alleles, for a total of 200 alleles in the gene pool. What fraction of these 200 are the *A* allele? Each of the *AA* homozygotes carries two copies of *A*, so they contribute 2p² of the *A* alleles. That's 2(0.36) = 0.72. So, among the 100 mussel offspring are 72 *A* alleles from homozygotes. In addition, each of the *Aa* heterozygotes also carries a copy of an *A* allele, so they contribute 2pq, or 0.48, *A* alleles. That's 48 more *A* alleles. In the accompanying table, notice that we can derive the frequency for

Table 16-1   The Essence of Hardy–Weinberg

The frequency of the *A* allele remains p and the frequency of the *a* allele remains q. Generally, then, sexual reproduction does not change allele frequencies. The total number of *A* alleles is $2p^2 + 2pq$. Because there are 2 chromosomes, the *frequency* of the *A* alleles is this number divided by two, or $(2p^2 + 2pq)/2 = p^2 + pq = p(p + q)$. Remember, however, that $p + q = 1$, so $p(p + q) = p$. So if a population starts with gene frequencies of $p = 0.6$ and $q = 0.4$, the next generation will have the same frequencies.

Genotype	*AA*	*Aa*	*aa*
Frequency	p^2	$2pq$	q^2
	0.36	0.48	0.16
Number of individuals with that genotype in the population	36	48	16
Total alleles	72A	48A 48a	32a

A alleles = 72 + 48 = 120
a alleles = 32 + 48 = 80
Number of individuals is 100.
Total alleles is 200.
So *A* alleles have a frequency of 120/200 = 0.6
And *a* alleles have a frequency of 80/200 = 0.4

the *a* allele in much the same way. We see that in spite of sexual reproduction, the allele frequencies have not changed. In successive generations, the frequencies for the genotypes *AA*, *Aa*, and *aa*—$p^2 + 2pq + q^2$—will also remain the same (Table 16-1).

So far, we have considered only two alleles for each gene, but we know that mutations are possible at every nucleotide of DNA. The number of possible alleles for each gene is large indeed. We could easily accommodate more alleles in our discussion. We might, for example, designate the frequencies of three alleles (*A*, *a*, and *a'*) as p, q, and r, so that $p + q + r = 1$. We would find that the Hardy–Weinberg Principle remains the same. The frequencies of the individual alleles in a population do not change as a result of sexual reproduction.

The frequencies of each allele therefore stay constant in successive generations, and sexual reproduction does not lead to the dilution of variation, as Fleeming Jenkin suggested. Equally important, genotype frequencies stay at the same **Hardy–Weinberg equilibrium,** a stable distribution of genotype frequencies maintained by a population from generation to generation.

The attainment of such an equilibrium, however, occurs only when the population meets all of the following five conditions: random mating, large population size, no mutations, no breeding with other populations, and no selection. In reality, these conditions are rarely met. Consequently, gene frequencies change and evolution occurs. If a population is not at equilibrium—that is, if the genotype frequencies are changing from generation to generation—then we know that one or more of the five conditions are not being met.

The Hardy–Weinberg Principle shows that sexual reproduction does not lead to the dilution of variation but maintains a stable distribution of genotype frequencies from generation to generation, called the Hardy–Weinberg equilibrium. The Hardy–Weinberg equilibrium is a baseline against which the evolution of populations can be measured and is the foundation for the genetic theory of evolution.

WHAT CAUSES GENE FREQUENCIES TO CHANGE?

Of all the causes of changing allele and genotype frequencies—nonrandom mating, small population size, mutations, breeding between populations (gene flow), and natural selection—only natural selection consistently leads to *adaptive* changes in allele frequencies (Figure 16-9). The others tend to cause changes in gene frequencies that are not directly related to the ecology of the population and may or may not be adaptive. These effects are said to be "nonadaptive."

Nonrandom Mating

In deriving the Hardy–Weinberg Principle, we assumed that individuals of one genotype do not mate preferentially, either with their own genotype or with others. Our calculation would be invalid, for example, if blue mussels mated only with other blue mussels, and not with brown mussels. In fact, mussels mate by releasing into seawater gametes that then come together essentially at random. The same is true of most plants and many other animals.

But mammals and birds usually choose their mates very carefully. Such nonrandom mating, called **assortative mating,** occurs when individuals choose mates on the basis of the mate's genotype (as reflected in the phenotype). Such choices may be made independent of the individual's own genotype. For example, regardless of their own appearance, humans generally show a preference for mates with average facial structure. Birds and mammals prefer mates with a high degree of symmetry.

Individuals may also choose mates on the basis of similarities (or dissimilarities) to the individual's own genotype. Humans show positive assortative mating with respect to skin color and height. That is, on average, they mate preferentially with those of similar height and skin color. Positive assortative mating tends to increase the proportion of homozygotes.

An extreme form of positive assortative mating is **inbreeding,** mating among close relatives, which greatly increases the proportion of homozygotes. Figure 16-10 shows a baby suffering from Ellis–van Creveld syndrome, a genetic defect whose relatively frequent occurrence among the Amish is a consequence of 200 years of inbreeding. Recessive genes may have negative effects that are only expressed in homozygotes. In sexual species, at least, inbred individuals who are homozygous

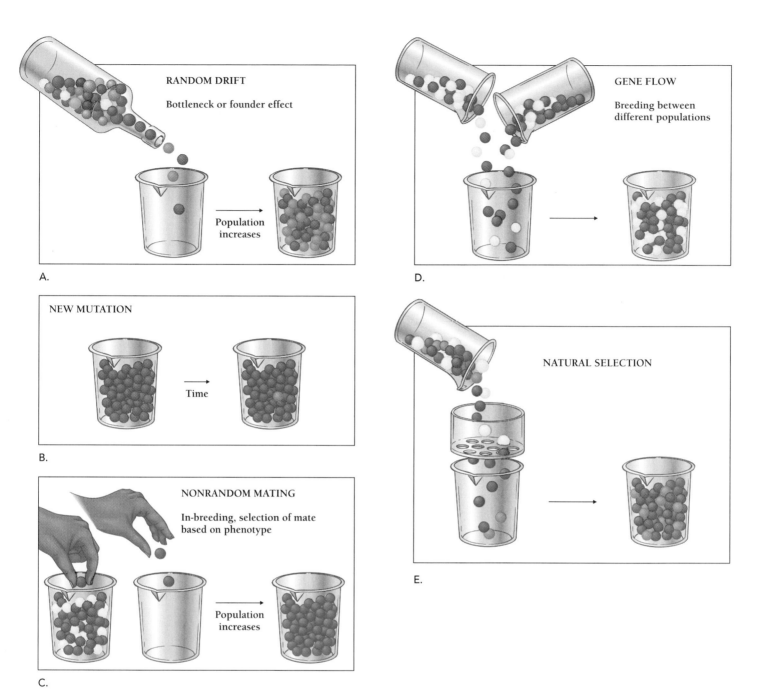

Figure 16-9 **What causes changes in gene frequency?** A. Random drift. B. New mutation.
C. Nonrandom mating. D. Gene flow. E. Natural selection.

for many genes are, on average, less healthy than individuals with greater heterozygosity.

Neither assortative mating nor inbreeding changes overall allele frequencies. However, any increase in the number of homozygotes facilitates change by natural selection by exposing alleles to selection.

Most species are **outbreeders,** with complex physical or behavioral adaptations that promote crossbreeding with usually not-too-related individuals. That is, individuals choose mates somewhat unlike themselves, a process called negative assortative mating.

Random Drift

Flip a coin 1000 times and your chance of getting all heads is virtually zero. You can reasonably expect that the number of heads will be close to 500, the "expected" number. However, if you flip the coin only four times, your chance of getting all heads is rather high (1 in 16), and you're not likely to get heads exactly one-half of the time. In fact, your chances of getting two heads is less than 40 percent. Such chance fluctuations, which affect small samples much more than large ones, are known as "sampling errors." In small populations of organisms,

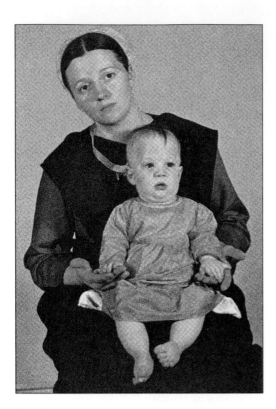

Figure 16-10 Inbreeding. A modern Amish woman with her baby, who suffers from a genetic defect known as Ellis–van Creveld syndrome. In the 18th century, a young Amish couple from Switzerland joined a colony of 200 Amish in Pennsylvania. The couple carried with them a rare allele for Ellis–van Creveld syndrome, which produces short arms and legs, extra fingers, and sometimes heart defects. Over the years, the tiny colony of 200 Amish grew to about 8000, but the people remained reproductively isolated. In 1964, a survey showed that this population had a frequency of the allele for Ellis–van Creveld syndrome of 7 percent, more than 70 times that of the general population — a consequence of 200 years of inbreeding. *(Courtesy, Dr. Victor A. McKusick)*

sampling errors can cause allele frequencies to change randomly from generation to generation.

For example, in a small population of mussels, the actual proportion of eggs and sperm with the *A* allele may differ widely from 0.6. The smaller the population, the greater the likelihood that the frequency of various alleles will deviate from what they are "supposed" to be. The effects of chance in small populations can lead to **random drift**—changes in gene frequency not due to selection, mutation, or immigration, but due to random events. Such random events can lead to changes in gene frequency within a small population.

The Founder Effect and the Bottleneck Effect

In small populations, random drift can have pronounced effects. Two closely related kinds of random drift are the founder effect and the bottleneck effect. In both cases, a whole population descends from a small number of individuals.

In the **founder effect,** a small subset of a larger population founds a new population. In the **bottleneck effect,** a large

population is reduced to a few surviving individuals by some sort of random disaster or harsh selection pressure. For example, in the early 19th century, elephant seals flourished along the California coast from Point Reyes, just north of San Francisco Bay, south to Baja California. By 1890, hunters had virtually eliminated the entire population (Figure 16-11). Best estimates are that the population was reduced to between 20 and 100 individuals. Under federal protection, the elephant seal population now numbers some 100,000 animals. All of these animals have descended from the same few individuals.

The founder effect and the bottleneck effect each produce a population whose gene pool is a tiny sample of the original. In this new population, the frequency of an allele is likely to be quite different from that in the original population. Many alleles will not be represented at all. Since the allele frequencies have changed, the population can be said to have evolved in the absence of natural selection.

Founder and bottleneck effects in the human population—together with the inbreeding and increased homozygosity that occur in small populations—are thought to be responsible for the high incidence of genetic diseases in certain human populations. For example, the founder effect is responsible for the high frequency of otherwise rare alleles among Afrikaaners of South Africa (descended from about 30 Dutch families in the 17th century), the Old Order Amish in the United States (descended from only a few individuals in the 18th century), and Swedes in Lapland (descended from three families in the 18th century). The high incidence of several genetic diseases in Ashkenazi Jews probably results from a bottleneck in the Middle Ages in Eastern Europe. Among present-day Ashkenazi Jews, for example, the incidence of the allele for Tay Sachs disease—invariably fatal to homozygous babies—is almost 1 in 30.

New Mutations

Random drift can play an important role in the spread of new mutations. Mathematicians have shown that, even in the absence of selection, a new mutation will either disappear completely or spread through the population. The smaller the population, the less time it takes for either to happen. If a new mutation enters the gene pool, it will at first exist in only a few individuals and will have a significant chance of being eliminated. For example, imagine a population of 25 elephants, 3 of which have a new allele for hairy ears. A volcano that erupts every 10,000 years happens to eliminate the three hairy-eared elephants. Their elimination is not selection, since smooth ears do not protect elephants from volcanoes. It's just chance. Yet, the result is the same as if natural selection had eliminated the hairy-eared elephants—a population of smooth-eared elephants.

Random Drift in Large Populations

The role of random drift in speciation in large populations is not so clear cut. However, even in large populations, random

Figure 16-11 Bottleneck effect. In the 19th century, heavy hunting reduced the number of elephant seals to as few as 20 individuals. Legally protected since 1884, the huge seals are now flourishing, but have become highly inbred. *(Frank S. Balthis Photography)*

drift will *tend* to cause alleles to disappear. In the absence of natural selection, new mutations, recombination, and migration, random drift would cause the frequencies of all alleles in even large populations to go to either 0 percent or 100 percent. It would happen slowly, but it would happen eventually, leading, ultimately, to a population that was completely homozygous. In such a population, evolution would cease. We can see that new mutations "restock" genetic variation, and the constant mixing of such mutations, within populations and between populations, is key to evolution.

In small, isolated populations, random drift can cause significant changes in gene frequencies. In large populations, drift operates so slowly that its importance may be small compared to the effects of mutation, gene flow, and selection.

Mutation

Slight changes in allele frequencies also occur as the result of mutation. As we saw in the last section, mutation can generate every possible change in the amino acid sequence of each protein during the lifetime of a species. Indeed, mutations provide the raw material of evolution. How do such mutations affect the genetic equilibrium of a population?

The answer is, very little. Random drift—given enough time—can lead to either the disappearance of a new mutation or its spread through the population. However, the number of new mutations in each generation does not much change allele frequencies. Although the rate of mutation is high enough that each new human baby carries one or two new mutations, the change in the frequency of any single allele is low.

Mutation barely alters the overall frequency of any given allele.

Gene Flow: Breeding Between Populations

Hardy–Weinberg equilibrium requires that a population not interbreed with other populations. However, such isolation rarely occurs. Among animals, individuals immigrating into a population bring new alleles with them. Among plants, pollen carried into a population, by wind or insects, brings new alleles. In either case, the resulting **gene flow** causes changes in allele frequencies. In our own species, we have many obvious examples of gene flow as people move to new countries. Despite a general tendency to find a mate within one's own group, gene flow among different groups is common (Figure 16-12). Gene flow has two major effects on allele frequencies. Within a given local population, gene flow from outside populations increases genetic variation. However, the same gene flow makes adjacent populations more like one another.

Gene flow increases variation within a local population, but makes adjacent populations more alike.

Natural Selection

A group inept
Might better opt
To be adept
And so adopt
Ways more apt
To wit, adapt.

—*John M. Burns*

How Does Natural Selection Change Allele Frequencies?

A large, isolated population whose members mated randomly and whose mutation rate was low would maintain constant allele frequencies—except for one thing: **natural selection**.

Figure 16-12   Gene flow. Humans travel singly or in groups, spreading their customs and genes. In November 1996, hundreds of thousands of Rwandan refugees began returning home after living in Zaire for months.   *(Christopher Morris/Black Star)*

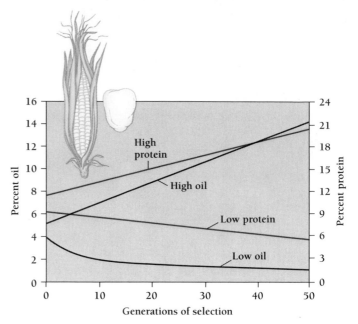

Figure 16-13   Selection is a powerful way to change gene frequencies. In four, long-running experiments, researchers select for corn with more protein or less protein and for corn with more oil or less oil. After 50 generations, researchers have continued to produce plants with greater and greater amounts of protein and oil. However, in the experiments selecting for less protein or less oil, the amounts of oil and protein reach a minimum and microevolution stops. Apparently, plants need a minimum amount of oil and protein to survive.

Because natural selection reduces the frequency of deleterious alleles, it is a powerful way to change gene frequencies (Figure 16-13). We can understand this action most easily in the case of individuals who carry a lethal genotype, such as that for Tay Sachs disease. Children with Tay Sachs die before they are 4 or 5 years old. Having Tay Sachs therefore eliminates the possibility of reproducing. Although the allele can survive in heterozygotes, homozygotes *never* pass the allele to offspring. As a result, the allele survives only at low frequencies. Recessive alleles such as Tay Sachs tend to survive at low frequencies because they "hide" from natural selection in the form of heterozygotes. In contrast, natural selection nearly eliminates dominant alleles that kill before the age of reproduction. As the frequency of any harmful allele approaches zero, however, the rate of mutation by the good allele into the deleterious one may come to balance the selection against the deleterious allele.

To calculate the rate at which recessive lethals will be eliminated, imagine that all *aa* homozygotes die before they reproduce. Recall that, according to the Hardy–Weinberg Principle, the frequencies of the *AA, Aa,* and *aa* genotypes are p^2, $2pq$, and q^2, respectively. We can say, then, that the frequency of the *a* allele in the population will decrease by q^2 in each genera-

tion. When the lethal *a* allele is present at relatively high frequencies, many individuals are eliminated in each generation. When the allele is present at low frequencies, however, selection eliminates very few individuals. For example, if the frequency (q) of *a* is 0.5, then 25 percent of the population will die young each generation. But if q = 0.01, then only 1 infant in 10,000 will die from having the *aa* genotype, and only $2(0.01)(0.99) = 2$ percent of the population will carry the gene. Thus, while selection can dramatically change allele frequencies, even the most lethal recessive alleles can still persist indefinitely as heterozygotes.

Nevertheless, even a small difference in the probability of reproduction can lead, over time, to the effective elimination of a deleterious allele. Suppose, for example, that individuals with the homozygous *aa* genotype are slightly less well adapted than individuals with either *AA* or *Aa* genotypes. Say that each *aa* individual leaves only 999 progeny for every 1000 of the *AA* or *Aa* individuals. Even if we start with a gene frequency for the *A* allele of only 0.00001, we can predict a steady increase in the frequency of the *A* allele, until after only 23,400 generations its frequency becomes 0.99. For the fruit fly *Drosophila*, this could occur in less than 500 years. For species with generation times as long as those of humans it would take about half a million years—certainly a long time, but only a tick in the history of life. Even subtle selection pressures thus have large effects on allele frequencies.

Natural selection cannot, by itself, eliminate alleles, but natural selection is, nevertheless, a powerful way to change allele frequencies.

What Kinds of Selection Occur in Nature?

Theoretically, then, natural selection seems to be a powerful force for change. But, we may ask, does selection occur in nature? One example is the development of resistance to toxins by natural populations of "pests." Most people are now aware that populations of bacteria, parasites, and pests have changed as a result of society's waging chemical warfare against them. After World War II, countries around the world began to use increasing amounts of the pesticide DDT to destroy disease-carrying insects. The heavy spraying of DDT resulted in strong selection against the mosquitos, house flies, and other pests

that had no resistance to the pesticide. Only the individuals that happened to carry alleles that protected them from DDT survived and reproduced. Such examples are common. Increasing numbers of bacteria are resistant to common antibiotics, and household cockroaches have even evolved an aversion to the glucose (sugar) used to bait roach traps (Figure 16-14).

A classic example of selection in nature concerns the peppered moth (*Biston betularia*). Because of a long British tradition of collecting moths and butterflies, biologists in Great Britain have been able to trace a change in the moth population over more than 100 years. In the early 19th century, all the peppered moths collected in Britain were pale and mottled. Collectors netted the pale moths as they fluttered through the air. Occasionally, a collector found a dark variant of the peppered moth. These were rare and therefore highly prized (Figure 16-15).

The industrial revolution dramatically changed this scene, however. In the 19th century, thousands of coal furnaces filled the air with dark soot, which blackened the landscape. The light peppered moth became rare, and collectors found more and more of the dark variants. While lighter moths continued to prevail in unpolluted rural areas, the frequency of dark moths around factory cities increased to about 98 percent.

Looking at the historical data (from old collections of moths and butterflies), the British evolutionary biologist J.B.S. Haldane calculated that, in industrial areas, the darker variant was twice as likely to survive and reproduce as its lighter coun-

Two apparently identical cockroaches touch/taste glucose-baited poison with their antennae.

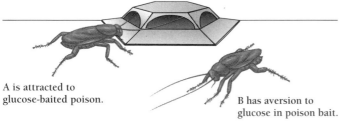

A is attracted to glucose-baited poison.

B has aversion to glucose in poison bait.

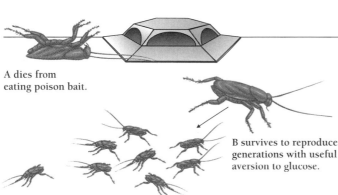

A dies from eating poison bait.

B survives to reproduce generations with useful aversion to glucose.

Figure 16-14 Natural selection in action. Because cockroach traps and poisons are commonly baited with the sugar glucose, intense selection pressure favors cockroaches that dislike the flavor of glucose. Researchers have discovered populations of German cockroaches from Florida to California that no longer take such bait because selection has allowed only the glucose-haters to survive.

Figure 16-15 Directional selection. The English peppered moth *Biston betularia* comes in two forms, light and dark. The light ones, whose colors blend well with the pale lichens found on trees, formerly predominated throughout England. The dark form predominates in areas were industrial pollution has blackened the landscape. *(Michael Tweedie/Photo Researchers)*

terpart. A reasonable explanation for this was that, against the now dark background of trees and rocks, the dark variant had a selective advantage because it was less likely to be seen and eaten by birds. A fine theory, but the naturalists of the time said that they had never seen a bird eating a peppered moth. Indeed, no one had actually seen a peppered moth resting on a tree or rock. The moths were always caught while flying. The question remained: Was the change in frequency of dark moths the result of selection by birds?

To answer this question H.B.D. Kettlewell, a biologist at Oxford University, undertook experiments that now provide among the best evidence concerning the operation of natural selection. In the 1950s, Kettlewell captured several hundred dark and light moths and marked them with little spots of paint. He released some of the dark and light marked moths in a smoky industrial area near Birmingham and the rest in an un-polluted woodland near Dorset. He then recaptured as many moths as he could and calculated the percentage of recovered dark and light moths (Figure 16-15). Near blackened Birmingham, the fraction of dark moths recovered was twice that of the light. Near rural Dorset the situation was reversed; the percent of light survivors was twice that of the dark. This experiment supported the idea that the change in frequency was related to the color of the tree trunks.

But were birds the agents of selection? Kettlewell showed that the moths really were being eaten by birds and that the color of the moth does, indeed, change the chance of being eaten. After placing moths on tree trunks, Kettlewell recorded the menu decisions of the birds in the area with hidden cam-

eras. As expected, the birds preferentially chose the dark moths on the lighter trees of Dorset and the light moths on the darker trees near Birmingham.

Kettlewell's work seemed to demonstrate both the operation and the actual agents of natural selection. The darkening of moths and butterflies to give better protective coloration in sooty environments is now called "industrial melanism." It has occurred in dozens of species not only in Britain, but also near coal-burning industries in the United States and Germany. Beginning in 1956, clean air laws returned the forests to their normal color and the fraction of dark moths decreased. In Michigan, black peppered moths increased throughout the 1950s, and then, after the United States instituted its first clean air laws in 1965, the white form came to predominate.

But there is a flaw in this story. At one site where the moths were studied, the pale moths began to return long before the lichens began coming back. And, despite intensive searching, no researcher has ever found a peppered moth resting on a lichen-covered tree (unless the researcher put it there). Some researchers now postulate that the moths roost in the topmost branches of trees and that selection occurs there.

Kettlewell showed that in a blackened environment, dark forms of the peppered moth have a selective advantage over lighter forms. Researchers do not yet know, however, whether selection by birds accounts for such color changes in response to pollution.

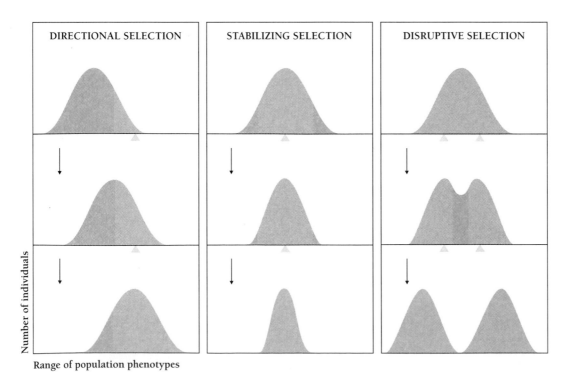

| DIRECTIONAL SELECTION | STABILIZING SELECTION | DISRUPTIVE SELECTION |

Number of individuals

Range of population phenotypes

Figure 16-16   Directional, stabilizing, and disruptive selection.

Directional Selection

Directional selection shifts the frequency of one or more traits in a particular direction (Figure 16-16). Industrial melanism is a classic example of directional selection. Selection for resistance, whether to a pesticide or an antibiotic, is also a form of directional selection. Likewise, agricultural breeders have long practiced directional artificial selection to produce such things as chickens that lay more eggs and cows that give more milk.

In nature, directional selection is typical of a changing environment and is a likely explanation for the directional changes that occur during the evolution of new species. The fossil record shows, for example, that the descendants of early humans and horses have become progressively larger. These changes presumably reflect the action of various selective pressures. In the case of horses, paleontologists suggest that the major selective force was the presence of predators. As horses became larger, they could run faster and were better able to avoid being caught and eaten.

Other forms of natural selection also operate. These include stabilizing selection, disruptive selection, and sexual selection.

Stabilizing Selection

Stabilizing selection tends to act against extremes in the phenotype, so that the average is favored (Figure 16-16). For example, variability in wasps drops dramatically during the winter when their mortality is high. An average wasp stands a better chance of surviving a harsh winter than those with unusual phenotypes, which die in great numbers.

Stabilizing selection is characteristic of stable environments because such environments consistently favor the average and rarely or never favor extreme variants. Stabilizing selection is one explanation for the almost unchanging morphological characteristics of such "living fossils" as the horseshoe crab, the coelacanth, and the ginkgo tree (Figure 16-17). If the environment for such organisms were sufficiently stable, the same phenotype could be adaptive for hundreds of millions of years.

Disruptive Selection

Disruptive selection is the opposite of stabilizing selection; it increases the frequency of extreme types in a population, at the expense of intermediate forms (Figure 16-16). Like directional selection, it is a likely cause of the evolution of new species. A nice example of disruptive selection is the African swallowtail butterfly (*Papilio dardanus*), which comes in three distinct variants, each of which lives in a different part of the butterfly's range (Figure 16-18). All three variants appear unappetizing to their predators because of their resemblance to another particularly foul-tasting butterfly. However, each variant resembles a different foul-tasting butterfly that lives in the same area. Predators in each area avoid the three extreme forms, but eat any in-

Ginkgo tree

Figure 16-17 They haven't changed a bit. Stabilizing selection probably accounts for organisms that hardly change over millions of years of evolution. Shown here are the ginkgo tree (modern and 250 million years old) and the coelacanth (modern and 350 million years old). *(fossil ginkgo, Kjell B. Sandved/Photo Researchers; fossil coelacanth, John Cancalosi/DRK Photo; modern coelacanth, Tom McHugh/Photo Researchers)*

Figure 16-18   **Disruptive selection.** Female African swallowtail butterflies, *Papilio dardanus*, (*top row*) taste good to birds. The butterflies avoid bird predators by mimicking one of three foul-tasting species of butterflies that also live in the area (*bottom row*). Intermediate forms of the African swallowtail are snapped up by birds, and only the swallowtails that most closely resemble the bad-tasting species survive and reproduce.   *(Hope Entomology Collection, Oxford/Biological Photo Service)*

termediate variants because they resemble none of the three foul-tasting butterflies.

Sexual Selection

Physical or behavioral differences between the sexes are extremely common. Among birds, for example, the male is often showy, with brilliant plumage, a long tail or crest, and a lively song, while the female is drab and inconspicuous. Among mammals, the male tends to be larger and more aggressive, as in humans. Many male mammals have horns, antlers, or other rather theatrical weapons. These are used in displays or fights with other males in competitions for territory or access to females. Such differences between the sexes, called **sexual dimorphism,** are sometimes the result of **sexual selection,** the differential ability of individuals with different genotypes to acquire mates.

Biologists generally recognize two kinds of sexual selection, both of which revolve around sexual reproduction. These are **female choice** (or, rarely, male choice) and **male competition.** In male competition, a male competes with other males for territory or access to females (Figure 16-19). He is more likely to leave numerous descendants if he can win the contest. Anything that gives him an edge in a contest with other males, whether larger horns or greater aggressiveness, will give him a selective advantage.

The second variety of sexual selection, female choice, involves courtship behavior. The purpose of courtship is to allow the female to assess the male (Figure 16-20). In most cases, females have fewer opportunities to breed than males. Females also invest more time and energy into each of their offspring, on average, than males do. It is important, therefore, for females to be more selective in choosing a mate than males are. Females will, theoretically, try to mix their own genes with the highest quality genes they can. As a result, most species have evolved complex rituals that enable the female to decide if her prospective mate is genetically "good enough" and choose accordingly.

For example, males of outbred and inbred populations of the fruit fly *Drosophila subobscura* differ in their ability to court female flies. To gain a female's favor, a male must perform a series of complex maneuvers, including tapping the female on

the head with his front legs and moving around to approach her head on while extending his proboscis, the tonguelike structure with which insects feed and taste. The female quickly sidesteps back and forth, forcing the male to keep adjusting his position. This she does, apparently, to prolong the courtship. The longer the courtship, the more easily she can assess his vigor and skill. Outbred males, which show far greater athletic ability in performing these fly dances, are more successful in their courtship. They have a distinct selective advantage over inbred males.

Figure 16-19   **Sexual selection.** Male competition leads to selection for traits that increase the reproductive success of males but which are of no value to females. In most animals that have horns, for example, only the males have elaborate horns.   *(Fred Bruemmer/ DRK Photo)*

Figure 16-20   **Female choice, in humans.** Among the Wodaabe people of Niger, the men dress up in elaborate costumes and makeup for the Geerewal dance, where women observe and judge, selecting the most beautiful men. In America and in many other cultures, it is women who compete in beauty contests and men who do the choosing.   *(© Carol Beckwith/Robert Estall Photo Library)*

HOW CAN NATURAL SELECTION PROMOTE GENETIC VARIATION?

Given the power of natural selection to change allele frequencies, we may well wonder why significant polymorphism persists at all. The maintenance of genetic heterogeneity in a population is not particularly surprising if different alleles of the same gene are equally useful. For example, within the range of a single interbreeding population, local environments may select for different alleles. The climate of a mountainside may select for short trees that are less likely to be toppled by winter winds, while a milder valley climate may favor larger trees that produce more seeds. Or one allele may be better adapted at one time of year than another. Among the members of one species of ladybug, for example, black ladybird beetles are more frequent in the fall, while red ladybird beetles predominate in the spring. Such a situation creates a balance of different alle-les in a population, called a **balanced polymorphism.** The most famous case of a balanced polymorphism in nonhumans is, of course, the current balance between dark and light peppered moths.

Populations may maintain high frequencies of alleles that are deleterious or even lethal when homozygous. In such cases, polymorphism results from the selective superiority of the heterozygote. Sickle cell anemia is an excellent example of a case where heterozygote superiority creates a balance of different alleles.

As we discussed in Chapter 10, people with two copies of the β^S-globin allele suffer from sickle cell disease. Recall that hemoglobin consists of four polypeptide chains, two βs and two αs. In people of African and Middle Eastern descent, the normal allele for the beta chain is sometimes replaced by an allele known as β^S. In β^S homozygotes, low oxygen levels, such as are found in the capillaries, cause the abnormal hemoglobin to precipitate in the red blood cells, leading to distortion and stiffening of the cells. The sickled cells are so stiff that they can block capillaries and so fragile that they break open in the strong currents of larger arteries. In the United States, victims of sickle cell disease rarely live beyond age 40, dying of infections, kidney failure, or the blockage of blood vessels in vital areas such as the lungs. Most suffer periodic bouts of pain, increasing weakness, and constant infections.

In Central and West Africa, homozygotes often die in their teens and 20s and rarely reproduce. Nevertheless, there, the β^S allele has strong selective value—not in homozygotes, clearly, but in β^A/β^S heterozygotes. Heterozygotes, whose blood contains both normal and abnormal hemoglobins, are not anemic yet they are resistant to infection by the malaria parasite, *Plasmodium falciparum*. Since this parasite is a common source of disease and death in Africa, the heterozygotes have a distinct selective advantage over β^A/β^A homozygotes, who are susceptible to the *Plasmodium* parasite. Thus the environment provides a strong selective pressure for the maintenance of the β^S-globin allele.

In some regions of Central and West Africa, the frequency of the β^S allele is about 20 percent (Figure 16-21). Using the Hardy–Weinberg Principle, we can calculate that the frequency of heterozygotes (2 pq) is then 2(0.2)(0.8) = 0.32, or 32 percent. The expected frequency of homozygotes would be $(0.2)^2$ = 0.04, or 4 percent. The homozygotes usually die before reproducing, and thus represent the "cost" of maintaining the β^S allele in the population. There is a trade-off between the deaths caused by the homozygous condition and the lives saved by the heterozygote condition.

In the United States, where malaria is nearly nonexistent, the β^S allele probably gives people no selective advantage, either in the homozygote or the heterozygote individual. In the United States, the frequency of homozygotes among African Americans is less than one-hundredth the frequency in Central and West Africa, while the frequency of heterozygotes is only about one-third that in Africa.

The reduction in the frequency of the β^S allele is due not only to its selective disadvantage but also to the effects of gene

Figure 16-21   A balanced polymorphism. A. The distribution of the sickle cell allele (β^S) closely parallels that of malaria. Blue shows areas of southern Europe, Africa, and the Middle East where malaria is or was once common. Red illustrates areas where sickle cell anemia occurs. Considerable overlap (purple) suggests selection for the sickle cell allele (β^S) in areas with malaria. B. Sickled red blood cell. C. The allele for sickle cell alters the shape of a polypeptide in hemoglobin. The altered hemoglobin changes the shape of individual red blood cells. *(B, Roseman/Custom Medical Stock)*

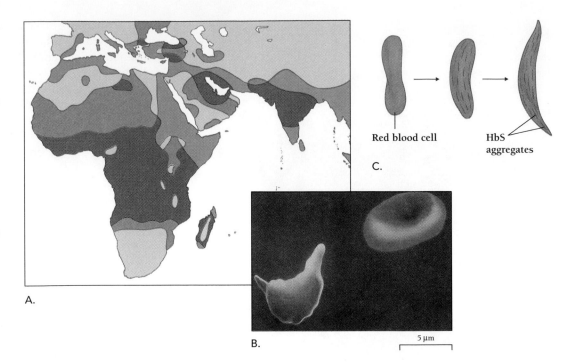

flow. On average, one-third of the genes of contemporary African Americans come from their European ancestry, so the frequency of the β^S allele would be reduced by dilution as well as by selection.

A balanced polymorphism may result from heterozygote superiority or from different selective pressures in time and space within the same population.

In this chapter we have seen where variation comes from, how it is inherited, and some of the ways that gene frequencies can change. Changes in gene frequency in a population, called microevolution, can occur through natural selection, or, in small, isolated populations, through random drift. However, explanations for the larger patterns of evolution—the multiplication of species and higher taxa, as well as patterns of extinction, for example—remain controversial. In the next chapter we will consider why it is so hard to define a species, how new species form, and why species multiply and go extinct.

STUDY OUTLINE WITH KEY TERMS

Natural selection was controversial because heredity was not understood. Darwin believed in **blending inheritance,** which led him to wrongly embrace the theory of **the inheritance of acquired characteristics.**

Weismann knew nothing of genes or mutations, so he could not know the ultimate source of genetic variation. Nevertheless, he was able to correctly guess that inheritance is not blending but particulate, that the genetic material is in the **germ plasm** and is unaffected by the life of the parent, and that sexual reproduction is a means of creating new variants by mixing and recombining old traits.

The **modern synthesis** wedded natural selection and genetics into an essentially modern theory of evolution. Population geneticists attempt to understand evolution in terms of the changes in the frequency of alleles within populations and species. **Microevolution** is shifts in gene frequencies in a population. **Macroevolution,** which we will study in the next chapter, deals with the processes by which species and higher groupings of organisms originate, change, and go extinct.

Genetic variation is the fuel of evolution. Gene and chromosome mutations are the ultimate source of all genetic variation. Genetic recombination, enhanced by sexual reproduction, spreads new **alleles** through a population by recombining different alleles into new genotypes.

Most organisms have two sets of **homologous** chromosomes. When the same allele is on each chromosome in a pair, the organism is said to be **homozygous** for that gene. When a different allele is on each chromosome in a pair, the organism is said to be **heterozygous** for that gene. Traits determined by two or more genes are said to be **polygenic.** Organisms displaying two or more alternative forms are said to be **polymorphic.**

Studies of DNA and proteins have revealed that most natural populations contain significant genetic variation. In humans, for example, studies show that about 25 percent of all proteins have an alternative form present in at least 5 percent of the population.

The **Hardy–Weinberg Principle** shows that under certain conditions, a single generation of random mating establishes a distribution of genotype frequencies $p^2 + 2pq + q^2$, and neither these frequencies nor the allele frequencies p and q will change in future generations. The Hardy–Weinberg Principle disproves Fleeming Jenkin's assertion that sexual reproduction by itself would change the genetic characteristics of a population. The Hardy–Weinberg Principle applies only if the following conditions are met: random mating, large population size, no mutations, no gene flow between populations, and no selection. **Natural selection** is only one of several forces that can change the frequency of alleles within a population. Others are **nonrandom mating** (including **assortative mating, inbreeding,** and **outbreeding**), **random drift, mutation,** and **gene flow.** Both the **bottleneck effect** and the **founder effect** are the result of chance effects within small populations that are tiny subsamples of a larger population. A **gene pool** is all the alleles of all the individuals in a population.

Natural selection can bring about evolution by changing gene frequencies in a population. Natural selection can lead to changes in allele frequencies even when individuals with different genotypes have only slight differences in reproductive success. Natural selection acts on the **phenotype** directly, and the **genotype** only indirectly. Although natural selection reduces the frequency of deleterious alleles, deleterious recessive alleles—even when they are lethal in homozygotes—persist at low frequencies for many generations. Natural selection can also act to decrease the genetic variability in a population (**stabilizing selection**), to change the quantitative characteristics of a population in a particular direction (**directional selection**), to increase the frequency of extreme types in a population (**disruptive selection**), or to cause **sexual dimorphism** (**sexual selection**). Biologists believe that sexual selection results most often from **female choice** and **male competition.** Selection for heterozygotes and simultaneous selection for different alleles maintains genetic variation in a population (**balanced polymorphism**).

REVIEW AND THOUGHT QUESTIONS

Review Questions

1. Darwin understood how natural selection works in the presence of a population with many variant types. What does that variation consist of? How does the variation arise? How does it spread through the population?
2. How much genetic variation do human populations exhibit?
3. Do the eye colors exhibited by your classmates fall neatly into the five colors shown in Figure 16-4? If not, why might they be different?
4. Why doesn't sexual reproduction, by itself, change allele frequencies in a population?
5. Why do the frequencies of the alleles in a gene pool add up to one?
6. What conditions must be met for a population to be at a Hardy–Weinberg equilibrium?
7. How does chance lead to changes in allele frequencies?
8. Why are certain genetic diseases more common in certain human groups than in others?

9. Give an example of gene flow in a human population and in another animal population.
10. Why would a dominant lethal allele die out in one generation while a recessive lethal might persist for hundreds of generations?
11. Why is the lethal recessive allele for sickle cell anemia so common in some human populations?

Thought Questions

12. Why doesn't natural selection eliminate variation in a population?
13. Why did Darwin think that sexual reproduction must eliminate variation in a population? What was his mistake?
14. What facts tended to suggest that the theory of the inheritance of acquired characteristics was wrong? What facts were needed to prove that it was wrong?
15. Is it possible that the apparent changes in the proportions of light and dark peppered moths are due to the moths being invisible to collectors, rather than to birds? Discuss your reasoning.

SELECTED READINGS

Gould, Stephen J., *Ever Since Darwin,* W.W. Norton and Company, New York, 1977. A collection of original and accessible essays about Darwin and the consequences of Darwin's ideas.

Koestler, Arthur, *The Case of the Midwife Toad,* Vintage Books, New York, 1973. A novelistic account of an unsolved case of scientific fraud.

Lewontin, Richard, *Human Diversity,* Scientific American Books, W.H. Freeman and Company, New York, 1982. The science and politics of human genetics are presented clearly and engagingly. This is a short, beautifully designed and illustrated book.

Popovsky, Mark, *The Vavilov Affair,* Archon Books, Hamden, Connecticut, 1984. An engaging account of Vavilov's life.

Quammen, David, *The Flight of the Iguana: A Sidelong View of Science and Nature,* Delacorte Press, New York, 1988. A collection of charming essays about evolution and natural history.

▶ On-line materials relating to this chapter are on the World Wide Web at
http://www.saunderscollege.com/lifesci/
Click on Tobin/Dusheck: *Asking About Life.*

MACROEVOLUTION: HOW DO SPECIES EVOLVE?

What Darwin Saw at the Galápagos Islands

When Charles Darwin arrived at the Galápagos Islands in the summer of 1835, he found a group of dreary volcanic islands, consisting of hundreds of craters, the largest some three or four thousand feet high. Five main islands dominated the little group of 14 islands, 600 miles off the coast of Ecuador. In spite of their position on the Equator, the Galápagos were not the lush, tropical islands typical of other parts of the Pacific, but barren desert islands populated by a small number of cacti, reptiles, and small, brown birds. It was the dry season.

"Nothing could be less inviting than the first appearance," Darwin wrote in his diary. "A broken field of black basaltic lava, thrown into the most rugged waves, and crossed by great fissures, is everywhere covered by stunted, sunburnt brushwood, which shows little signs of life. The dry and parched surface, being heated by the noonday sun, gave to the air a close and sultry feeling, like that from a stove: we fancied even that the bushes smelt unpleasantly." The most striking animals were the reptiles, large marine iguanas and immense Galápagos tortoises (*Geochelone elephantopus*), some of them $5\frac{1}{2}$ feet tall and weighing 500 pounds (Figure 17-1).

Heather Angel/Biofotos

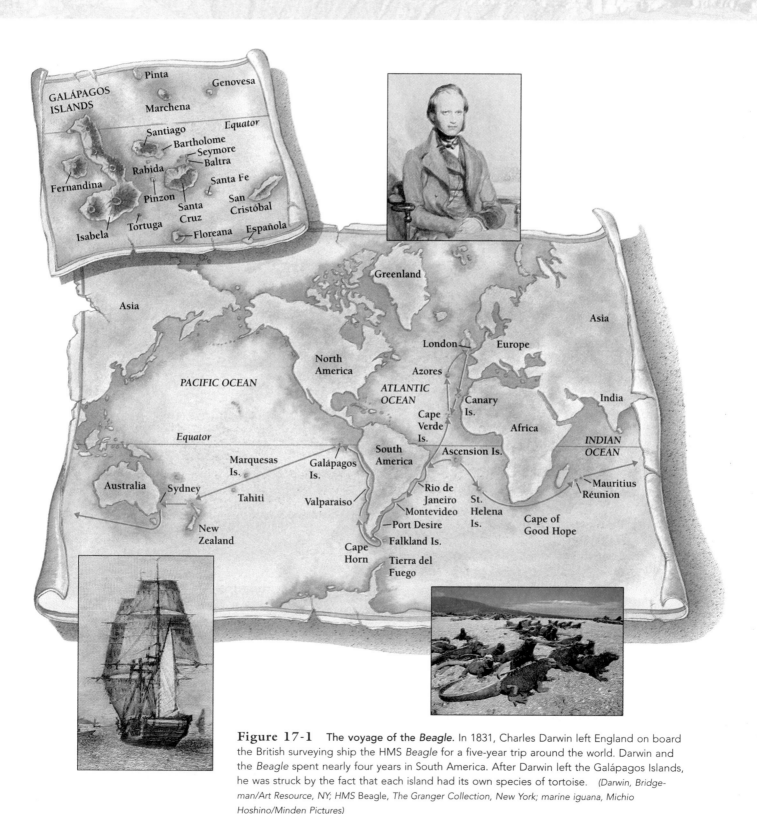

Figure 17-1 The voyage of the *Beagle*. In 1831, Charles Darwin left England on board the British surveying ship the HMS *Beagle* for a five-year trip around the world. Darwin and the *Beagle* spent nearly four years in South America. After Darwin left the Galápagos Islands, he was struck by the fact that each island had its own species of tortoise. *(Darwin, Bridgeman/Art Resource, NY; HMS Beagle, The Granger Collection, New York; marine iguana, Michio Hoshino/Minden Pictures)*

411

Darwin spent a month in the islands, exploring and collecting rocks, plants, and animals. He discovered that almost all of the terrestrial animals and at least half of the plants were endemic species, that is, found nowhere else in the world. Yet, all of the Galapagan species resembled species found on the nearby South American mainland—all, that is, but a European rat, probably brought by ship to the islands.

to James and Chatham Islands. Darwin's plant collection, more carefully labeled, was a gold mine. For example, of 71 species found on James Island, 38 were endemic to the Galápagos and 30 of those were endemic to James Island alone. Every island had the same story to tell.

"One is astonished," he wrote in his diary, "at the amount of creative force, if such an expression may be used, displayed on these small, barren, and rocky

would have been surprising. According to the 19th-century theological concept of special creation, similar habitats should support identical species. Every dark forest should have its bear and every clear stream its trout. And, since each species was thought to have been specially created by the Judeo–Christian God, species were not supposed to be related to one another. Most biologists, like Linnaeus, could see that animals and plants group naturally into rough family–treelike hierarchies. Yet they carefully ignored the suggestion of genealogy suggested by these hierarchies.

Frantically, he went back through his specimens to see if different species came from different islands.

The island-bound plants and animals were related to those on the mainland. But how were they related, and why were they different?

Darwin's observation that so many Galapagan species were unique to the islands did not at first strike him as meaningful. Even when the Vice-Governor of the islands casually mentioned that tortoises from different islands were so distinctive that he could easily tell which island a particular tortoise had come from, Darwin paid no attention. Then, just days before the *Beagle* was to set sail for Tahiti, the Vice-Governor's remark struck Darwin with sudden force.

Maybe, Darwin thought, each island supported a different set of organisms. Frantically, he went back through his specimens to see if different species came from different islands. But he had labeled most of them "Galápagos." He had hopelessly mixed up the different kinds of tortoises and the finches. There was no way to tell which islands they had come from.

But one group of birds, the mocking thrushes, "thoroughly aroused" his attention. Of three species, one was unique to Charles Island, one to Albemarle, and one

islands; and still more so, at its diverse yet analogous action on points so near each other."

Two years after his visit to the Galápagos, Darwin began a notebook on the "transmutation of Species." It would be years before he realized the full implications of all that he had seen in the Galápagos. It was no accident, he ultimately realized, that Galápagos species were so closely related to each other and to those on the South American mainland. Each endemic species had evolved on one of the islands, having descended from migrants from other islands or the mainland.

Nowadays, every child knows that different sorts of animals and plants populate each of the different continents. Africa has its lions and camels, South America its jaguars and llamas. Lions and jaguars, although descended from the same ancestral cat, have evolved independently since the continents began to separate 135 million years ago. The same is true of camels and llamas.

In Darwin's day, however, the presence of similar animals in very different habitats, or of completely different assemblages of animals in similar habitats,

But Darwin—scion of a distinguished family—knew the importance of genealogy. He realized that 5 million years ago, the volcanic Galápagos Islands had burst steaming and erupting from the sea, devoid of life. Gradually, plants and animals began to colonize the islands. Sea birds roosted on the barren rocks, leaving seeds that had stuck to their wings or passed through their stomachs. Some of these seeds took root and managed to survive. Ocean currents brought more plant seeds and an occasional reptile. Winds also brought seeds, as well as finches and sparrows and other land birds. Some plants and animals survived and reproduced, others did not. Some moved from island to island, spreading and interbreeding with other populations. Others remained confined to one island. These slowly evolved away from their cousins on the other islands and the parent population on the mainland. In this way, one species became many.

In this chapter we will study **macroevolution,** the origin and multiplication of species. We will see what a species is, and explore how it comes into being. We will ask what the patterns in the fossil record tell us about evolution. And we will discover why, after over 150 years, evolutionary biologists are still arguing about the questions raised by Darwin after his visit to the Galápagos Islands. Near the end of the chapter, we will briefly discuss the evolution of humans.

KEY CONCEPTS

1. A species is a group of organisms that actually (or can potentially) interbreed to produce viable offspring but usually do not interbreed with members of other such populations.

2. A population may form a separate species when it becomes reproductively isolated. Populations may become reproductively isolated in many ways. Geographic isolation is the most easily understood way. However, barriers to reproduction can arise in a variety of ways, either before fertilization or afterwards.

3. Many kinds of adaptations contribute to the reproductive isolation of new species.

4. Adaptive radiation is the generation of diverse new species from a single ancestral species. It can occur under a variety of circumstances, but is most obvious in oceanic island chains, such as the ones that Darwin and Wallace explored.

5. Species probably evolve at different rates at different times. But evolutionary biologists do not agree on whether species stop evolving for long periods of time.

WHAT IS A SPECIES?

How Many Species?

In the next section of this book we will explore the enormous biological diversity of life on Earth. For now, we will merely note that biologists have so far identified about 1.4 million species of living organisms—give or take a hundred thousand. Newly discovered species pour into the museums monthly, so this number is probably less than one tenth of the total number of species currently living. The actual number could be anywhere from 10 million to 100 million. Of the known species, more than half are insects (Figure 17-2). Nearly half of all insects—290,000 species—are beetles, some 20 percent of all species of organisms combined. The biologist J.B.S. Haldane once ironically remarked that God "must have had an inordinate fondness for beetles."

Why Is It So Hard To Define a Species?

The average person, faced with groups of familiar organisms such as mammals or birds, easily recognizes one species from another. Even when presented with a set of unfamiliar mammals, the average person can generally distinguish most species. In 1927, for example, a young biologist named Ernst Mayr, who later became a leading figure in evolutionary biology, led an expedition into the Arafak Mountains of New Guinea to catalogue new species of animals. After he had identified 138 species of birds, he was surprised to discover that the local Papua natives themselves recognized 137. Two of the species were so similar that the Papuans considered them a single species.

What Is a Species?

Unfortunately for biologists, such similar-looking species are common. Two species may even look identical. Some species of frogs, for example, are distinguishable only by their calls. Adult forms of the malaria-carrying mosquito *Anopheles maculipennis* cannot be distinguished from those of several other species that do not carry malaria. A species that is extremely similar to, but reproductively isolated from, another is called a **sibling species.** We can conclude that appearance alone fails to define a species.

Defining a species is awkward. When biologists believed that organisms were all variants of "ideal types," as 19th-century essentialists believed, the problem was much easier. The definition of a species was a description of the physical features of the ideal type, a sometimes arbitrarily chosen "type specimen." Although type specimens are still useful for identifying species collected in the wild, we now realize that every individual in a species is unique—ideal types do not exist. Even an individual that happened to be completely average in every respect would not represent its species better than any other individual in the population.

The definition of species must take account of the enormous variation among individuals and recognize that two species can be distinct even if they are very much alike. Darwin found himself exasperated by the problem of classifying species. Writing to his friend J.D. Hooker in 1853, Darwin wrote:

> After describing a set of forms as distinct species; tearing up my [manuscript], & making them one species; tearing that up & making them separate, & then making them one again (which has happened to me) I have gnashed my teeth, cursed species, & asked what sin I committed to be so punished.

TOTAL NUMBER OF NAMED SPECIES: 1.4 MILLION

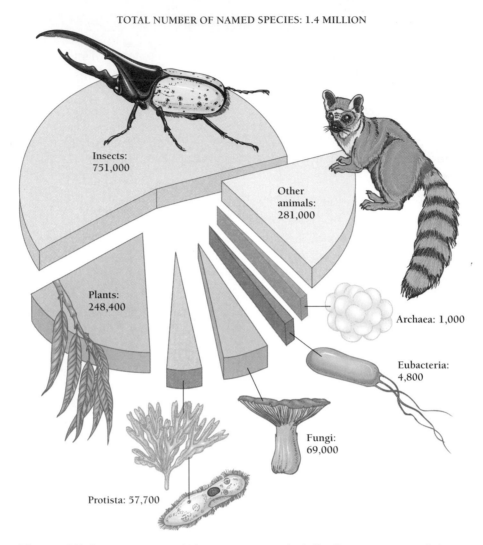

Insects:
751,000

Other
animals:
281,000

Plants:
248,400

Archaea: 1,000

Eubacteria:
4,800

Fungi:
69,000

Protista: 57,700

Figure 17-2   **The diversity of life.** At least one and a half million species now inhabit the Earth.

Darwin concluded that since species gradually evolve one from another, the term "species" is a convenient but arbitrary designation that has no biological meaning. He found this a great "relief," for he no longer felt he had to worry about whether organisms were truly separate species or not.

Modern biology takes a different view from Darwin. We now think that species are, for the most part, discrete and very real entities. Further, biologists have found that in order to discuss evolution, they need to define what a species is. In the 1930s and 1940s, the makers of the modern synthesis produced the **Biological Species Concept,** one modern definition of a species. In 1942, Ernst Mayr defined a **biological species** as follows: *"Species are groups of actually or potentially interbreeding populations, which are reproductively isolated from other such groups."* A biological species, then, is the largest unit of a population of similar organisms in which gene flow is possible.

Most species conform nicely both to the biological species concept and to our sense that species should look different from one another. Bullfrogs (*Rana catesbeiana*) and wood frogs (*Rana sylvatica*), for example, look different from each other and do not interbreed (Figure 17-3). They are members of different species.

But the biological species concept is a tricky definition. First, it does not apply to asexual organisms such as prokaryotes, certain fungi, and even some plants. For example, *E. coli,* the distinctive bacterium that inhabits the human digestive tract, is almost always asexual. Even though it does not satisfy the definition of the biological species, taxonomists have nonetheless assigned it to a genus and species.

Second, the biological species concept does not apply easily to extinct organisms. Paleontologists naming fossils have no way of knowing if two so-called species interbred. Many fossil species gradually give way to increasingly different forms. Pa-

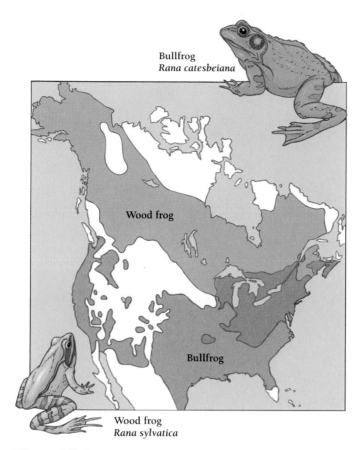

Bullfrog
Rana catesbeiana

Wood frog

Bullfrog

Wood frog
Rana sylvatica

Figure 17-3 **Separate species.** The wood frog *Rana sylvatica* and the bullfrog *Rana catesbeiana* look different and occupy different, though overlapping, habitats. They probably do not interbreed; they are true species. *(Two range maps from A Field Guide to Western Reptiles and Amphibians, 2e. Copyright © 1985 Robert C. Stebbins. Reprinted by permission of Houghton Mifflin Company. All rights reserved.)*

appearance—may or may not be capable of interbreeding. Lions and tigers interbreed if put together in zoos, which should make them the same species. But in the wild, they never interbreed, so they are effectively separate. But then again, since their habitats do not overlap, there is no way to know if they would interbreed if they lived closer to one another. In garter snakes of the genus *Thamnophis* the two species *hydrophila* and *elegans* live in the same area in northern California and southern Oregon without interbreeding. They seem to be what taxonomists call "good" biological species. Yet, in a small area just south of the Oregon–California border, these two species interbreed.

"Species are groups of actually or potentially interbreeding populations, which are reproductively isolated from other such groups." However, for practical purposes, taxonomists give species names to organisms that do not necessarily conform to this definition.

Other Taxonomic Categories

Many species have **races** or **subspecies,** morphologically distinct subpopulations, which can interbreed. For example, along the western coast of North America, the song sparrow has been divided into as many as 34 subspecies (Figure 17-5). However, it is not always clear that such races are specifically adapted to different environments. For example, six races of carpenter bee inhabit a single island in Indonesia, with no apparent differences in the way they live.

In fact, almost any subpopulation can be distinguished from others on the basis of some cluster of characteristics. For example, the people living in Minneapolis might differ from those living in St. Paul in having, on average, longer noses and a slightly higher frequency of ingrown toenails. But we would not necessarily conclude that the populations of the Twin Cities were separate races. Many evolutionary biologists therefore consider the concept of race or subspecies to be just a normal manifestation of natural variation and without any real biological meaning. The differences among human "races" constitute only about 15 percent of the genetic diversity in the human species. The other 85 percent is individual variation.

Nevertheless, even though most races or subspecies of organisms interbreed easily, a few do not. Subspecies that do not interbreed may be in the process of separating into species and are sometimes called **incipient species.** Distinguishing ordinary subspecies from incipient species is often a judgment call that only a biologist who has thoroughly studied the population in question can make.

As we will see in Chapter 19, genera and higher taxonomic categories are still harder to define. The hierarchical system of classification that taxonomists use to organize biological species into higher groupings generally represents degrees of anatomical difference and so, to some extent, evolutionary distance.

leontologists arbitrarily give an organism a new species name when it looks sufficiently different from its ancestors. But no one knows whether the modern horse (*Equus*), for example, could interbreed with its predecessor *Hippidium* (Figure 17-4).

Furthermore, paleontologists must give taxonomic species names to fossil organisms on the basis of morphological differences in the hard parts alone. Muscles and other soft parts, as well as clues about physiology and behavior, are almost never fossilized. The two mosquito species mentioned earlier, for example, would never be recognized as two species if they appeared only as fossils. Paleontological species are therefore subjective designations. A group of fossils may be lumped together as a single species or divided into several. But these same organisms might be judged differently if they were alive today. This point will be important to keep in mind when we discuss the evolution of humans at the end of this chapter.

The biological species concept does not even apply well to all living sexually reproducing organisms. It applies poorly to organisms that do not normally live together. Geographically isolated organisms—whether very different or very similar in

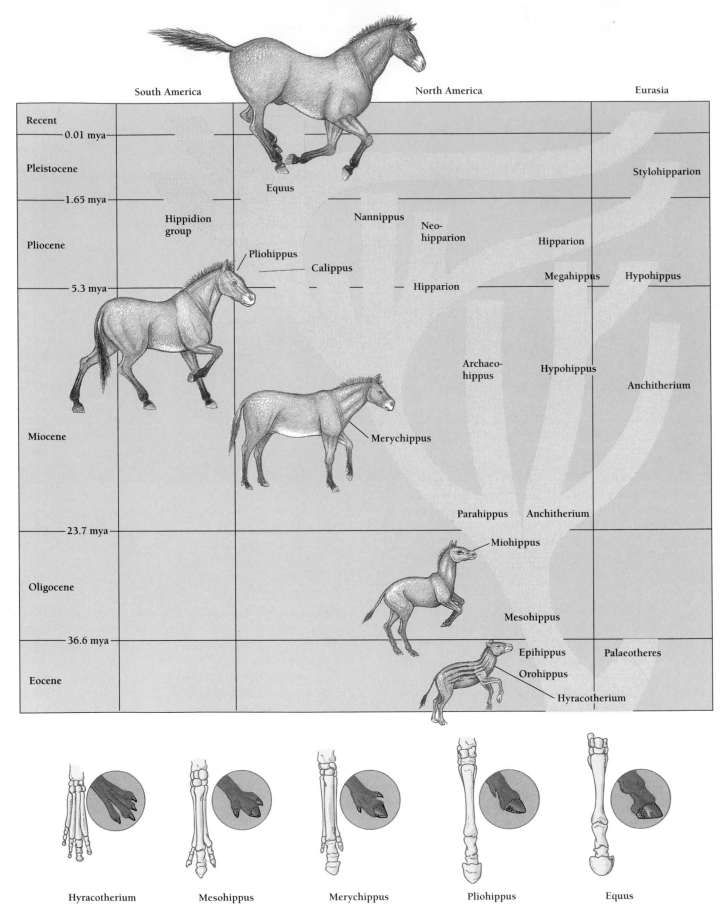

South America   North America   Eurasia

Recent

—0.01 mya—

Pleistocene   Stylohipparion

Equus

—1.65 mya—

Hippidion group   Nannippus   Neo-hipparion

Pliocene   Hipparion

Pliohippus

Calippus   Megahippus   Hypohippus

—5.3 mya—   Hipparion

Archaeo-hippus   Hypohippus

Anchitherium

Miocene

Merychippus

Parahippus   Anchitherium

—23.7 mya—

Miohippus

Oligocene

Mesohippus

—36.6 mya—

Epihippus   Palaeotheres

Orohippus

Eocene

Hyracotherium

Hyracotherium   Mesohippus   Merychippus   Pliohippus   Equus

Figure 17-4 **Evolution of the horse.** Great numbers of fossil horses have given paleontologists a good picture of how horses gradually evolved from smaller, many-toed ancestors. If *Pliohippus* were alive today, could it breed with *Equus*? No one knows.

Central California
(*M. m. heermani*)

San Francisco Bay
(*M. m. samuelis*)

Desert form
(*M. m. saltonis*)

Alaskan form
(*M. m. inexpectata*)

Subspecies of the song sparrow, *Melospiza melodia*

Figure 17-5 **Subspecies.** Here are just 4 of the 34 known subspecies of Western song sparrow.

A higher taxon is usually defined by a few characters that consistently distinguish its members from the species in other taxa. Mammals, for example, are distinguished by hair, two sets of teeth, and mammary glands. If we find an animal with these characters, we can be sure it is a mammal.

However, the division of a group of species into separate genera, families, or higher taxa can be very subjective. For example, in 1947, a botanist decided that a recently discovered daisylike plant whose flowers had no petals was so novel that it should be put into its own genus. Only later did botanists discover that the plant was a simple variant of a well-known species that normally had petals. The "new genus" was not even a new species.

Species are biological entities apart from human definitions—even though they are sometimes hard to define. Higher taxa, as well as subspecies and races, are still more subjective groupings.

HOW DO SPECIES FORM?

What Barriers Reproductively Isolate Populations and Species?

We have seen that for two populations to become separate species they must become **reproductively isolated,** unable to interbreed. Often, the process by which one species becomes two begins when one population becomes geographically iso-

lated from the rest of the species. If that happens, the physically isolated population has a chance of becoming reproductively isolated from the parent species, and with that a chance of becoming a separate species. Some populations merely happen to become reproductively isolated and then evolve apart.

Other populations, however, may first evolve separate adaptations and then develop reproductive barriers that allow them to retain their respective adaptations. Many of the most remarkable features of organisms function to maintain reproductive isolation. Such features often include the diverse structures of flowers and the elaborate rituals of animal mating.

The larkspur (*Delphinium*) provides one example of instant reproductive isolation. Most species of larkspur have blue or purple flowers (Figure 17-6). Bees visit blue larkspur (and other blue and purple flowers), carrying pollen from flower to flower. But one northern California larkspur has a mutation that gives

Figure 17-6 **Reproductive isolation.** Some time in the past, the scarlet larkspur shown here became isolated from the rest of its species. A mutation produced a generation with red flowers instead of the usual blue. But red flowers tend to attract hummingbirds instead of the bees and other insects that pollinate the blue flowers. Because the hummingbirds do not carry pollen between the red and blue flowers, the scarlet larkspur has become reproductively isolated from its blue relatives.

it scarlet flowers—a color bees cannot see. Fortunately, hummingbirds see red flowers very well, and they pollinate the red larkspurs. The bees and hummingbirds do not intrude on one another's flowers, and the blue and scarlet larkspurs are reproductively isolated.

For populations to become reproductively isolated, gene flow between them must cease. Barriers to gene flow fall into two broad categories: (1) **prezygotic barriers,** which prevent syngamy, the fusion of the sperm and egg to form a zygote, and (2) **postzygotic barriers,** which make a zygote either inviable (certain to die) or sterile. The red and blue larkspurs are an example of a prezygotic barrier, since the flowers' different colors prevent cross fertilization. Prezygotic barriers also include differences in the ways that species live (**ecological isolation**), in the times at which they reproduce (**temporal isolation**), in their mating behaviors (**behavioral isolation**), in the complementarity of male and female reproductive organs (**mechanical isolation**), and in the compatibility of their gametes (**gametic isolation**).

Postzygotic barriers operate when fertilization has already occurred, but the resulting zygote either dies or fails to reproduce successfully. An individual that results from cross fertilization between two different species, or, sometimes, between two distinct populations, is called a hybrid. For example, a cross between a horse and a donkey results in a mule, a hybrid that is nearly always sterile. The mule is an example of **hybrid sterility.** However, hybrids between related species often die during development or early in life because of chromosomal and genetic incompatibilities, a barrier referred to as **hybrid inviability**. In some cases, the hybrids of two species may be viable and fertile, but *their* offspring are weak or sterile (*hybrid breakdown*).

In plants, changes in chromosome number can simultaneously create new variants and cause their immediate reproductive isolation. For example, if homologous chromosomes fail to separate during meiosis, then each gamete will have the diploid number (2n) of chromosomes instead of the haploid number (n). Fertilization can then produce a tetraploid (or 4n) zygote, with four rather than two sets of homologous chromosomes (Figure 17-7). Such tetraploid plants will produce gametes with 2n chromosomes. These gametes cannot form zygotes with gametes of the parental stock, since such zygotes would contain 3n chromosomes, which could not pair at meiosis. Tetraploid plants thus become immediately isolated from the main population. Other multiples of the haploid number of chromosomes also occur. Plants that contain two or more complete sets of chromosomes are said to be **polyploid.** Polyploidy has played an important role in the evolution of grasses, ferns, and other vascular plants.

Another common postzygotic barrier in plants results when the zygote combines the chromosome sets of two species. Organisms formed from the gametes of two different species are ordinarily sterile, if not inviable. Such hybrids rarely produce gametes because the maternal and paternal chromosomes are not homologous, and so they cannot pair during meiosis. However, pairing can occur if the hybrid doubles its chromosomes. Such a doubling forms an **allopolyploid.** Each set of chromosomes is present twice, and can engage in proper meiotic pairing and gamete formation. But the gametes of the hybrid can no longer form viable zygotes with the gametes of either parental species, so they become reproductively isolated from both parental populations. Since many plant species can self-fertilize, however, a single plant can, potentially, establish a new species.

Populations become reproductively isolated by prezygotic or by postzygotic barriers.

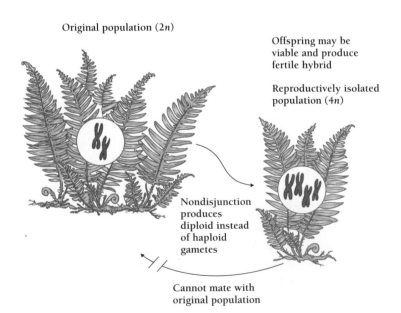

Original population (2n)

Offspring may be viable and produce fertile hybrid

Reproductively isolated population (4n)

Nondisjunction produces diploid instead of haploid gametes

Cannot mate with original population

Figure 17-7 Reproductive isolation by polyploidy. An error during meiosis produces diploid gametes, which form tetraploid offspring. The offspring can breed only with one another, however, because if the diploid gametes fuse with normal haploid gametes, they form triploid zygotes, which cannot survive.

Speciation as a Function of Geography and Gene Flow

Allopatric Speciation

In most cases, populations must be geographically isolated from the parent population before other barriers to reproduction arise (Figure 17-8). Oceans, mountain ranges, and rivers, for example, are all capable of sharply reducing or eliminating interbreeding with a parent population. Once a population is isolated, it can follow its own evolutionary course. Its gene frequencies can change independently, according to the forces of selection and genetic drift. In time, the isolated population may acquire new chromosomal arrangements and reproductive strategies that prevent interbreeding with the parent population if the two should again come into contact. These isolating mechanisms preserve the isolated population's differences and allow it to **speciate**—form a new species. Species formation by geographical isolation is called **allopatric speciation** [Greek, *allos* = other + *patra* = country].

Allopatric speciation is speciation that proceeds from geographical isolation. Once two populations are geographically isolated, independent selective pressures and genetic drift can further differentiate the two populations.

Sympatric Speciation

The formation of separate species without geographical isolation is called **sympatric speciation** [Greek, *syn* = together + *patra* = country]. For a single population to spontaneously break into two reproductively isolated subpopulations, something must inhibit gene flow between the two subpopulations (Figure 17-9). Some barriers to gene flow, such as polyploidy in plants, can instantaneously isolate subpopulations from their parent populations. However, most barriers to gene flow would have to evolve relatively gradually. Whether this can happen within a single interbreeding population is a question of great debate among evolutionary biologists.

Most models of sympatric speciation invoke disruptive selection as a mechanism. The hawthorn fly may present a good example of the first steps in sympatric speciation by disruptive selection. In 1864, farmers in New York State noticed that the small hawthorn fly *Rhagoletis pomonella,* which normally fed on hawthorn fruits, had begun to feast on apples as well. Because the apples at once became both food source and mating site for some of the hawthorn flies, it has been suggested that the apple-eating flies rapidly became reproductively isolated from the rest of the population. In 1960, some of the flies switched to cherries, possibly creating yet another isolated, but sympatric, race.

The flies move back and forth from one fruit to another, however, and researchers are not certain how different or how well isolated these three populations are. Although the three subpopulations have divided the same environment into three

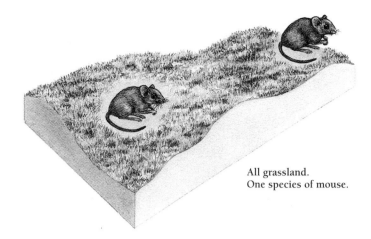

All grassland.
One species of mouse.

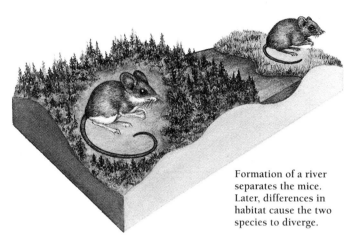

Formation of a river separates the mice. Later, differences in habitat cause the two species to diverge.

Figure 17-8 **Allopatric speciation.** When a single population becomes divided—by a river, for example—the two subpopulations evolve separately and may form separate species.

potentially separate habitats, so far biologists still recognize only a single species. Whether these are incipient species remains to be seen.

Sympatric speciation, the formation of species without geographical isolation, is controversial.

Parapatric Speciation

A third type of speciation, called **parapatric speciation**, occurs when a species splits into two populations that are geographically separated, but still have some contact. Parapatric speciation is thus intermediate between allopatric and sympatric speciation. Reproductive barriers can isolate parapatric populations if the environments of the two populations impose different selective pressures on each, and if gene flow between the two populations is small (lest it overwhelm the divergent changes).

Figure 17-9 **Sympatric speciation.** The California newt *Ensatina eschscholtzi* has diverged into a series of overlapping subspecies along the Coast Range and the Sierra Nevada. Little or no habitat exists in the Great Central Valley, which runs down the center of the state between the two mountain ranges. Adjacent subspecies of the newt interbreed. The two subspecies at either end of the newt's horseshoe-shaped range are linked by a continuum of subspecies, but do not themselves interbreed. They appear to be reproductively isolated. Are they separate species? *(Excerpted from* Life on the Edge, *copyright © 1994 by Bio Systems Analysis, Inc., with permission from Ten Speed Press, P.O. Box 7123, Berkeley, CA 94707)*

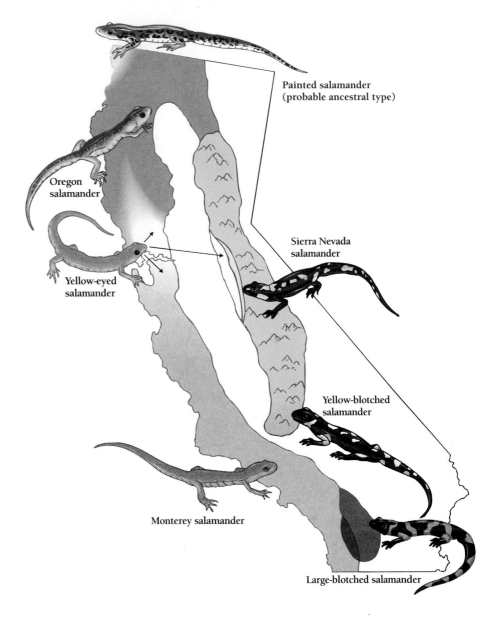

The growth of grasses around the entrance to a Welsh lead mine provides a good example of parapatric isolation and the probable beginning of parapatric speciation. The soil immediately around the mine has a high concentration of lead—high enough to kill most of the species of grass that flourish in the adjacent pasture. But a variety of one species of grass, *Agrostis tenuis,* tolerates lead and grows well on the mine tailings. This variety has become reproductively isolated from the neighboring parental pasture grass because it happens to flower at a different time.

If we could be sure that the separation were complete—that gene flow had completely ceased—then we would call the two populations separate species. At this point, however, we can only note the different selective pressures of tailings and pasture and the limited gene flow between them. Such a situation could easily lead to true reproductive isolation and speciation.

Parapatric speciation may occur if gene flow between adjacent, nonoverlapping populations is sufficiently limited.

Patterns of Descent

Paleontologists have identified several patterns of evolutionary change in the fossil record. The separation of one species of organisms into two (or more) species is also called **divergent evolution** (Figure 17-10). **Convergent evolution,** the independent development of similar features in separate groups of organisms, is not the opposite of divergent evolution, but a separate, unrelated idea. Recall from Chapter 15, for example, the convergent evolution of wings in the unrelated birds, bats, and insects. The best understood example of divergent evolution is **adaptive radiation,** the generation of diversely adapted new

Warbler finch

Needle-nose pliers

Medium tree finch

High-leverage diagonal pliers

Large ground finch

Heavy-duty linesman's pliers

Figure 17-10 **Adaptive radiation.** Less than 5 million years ago, the Galápagos Islands began erupting from the floor of the Pacific Ocean. Sometime after the islands had been colonized by plants, a single species of finch colonized the islands. This finch then radiated into 13 separate species, which taxonomists have grouped into 4 genera. Here are just three, representing three genera. Each finch is specialized for a different way of life. The finch with the largest beak, for example, breaks open large, hard seeds, impossible for the other two finches to crack.

species from a single ancestral species. For example, biologists believe Darwin's 14 species of Galápagos finches all descended from a single mainland species.

Patterns of Diversification

Adaptive radiation is the key to understanding why there are so many species. We will first discuss examples of adaptive radiation in chains of islands and then try to understand how diversity increases worldwide.

Adaptive Radiation on Chains of Islands

The most dramatic cases of adaptive radiation occur on chains of islands (or archipelagos), such as the Galápagos. We have seen that the formation of new species generally requires either reproductive isolation (so that descendant species do not interbreed with one another) and one or both of two other conditions: either distinct selective pressures (so that different descendant species acquire different adaptations) or else small population size (so that genetic drift can operate).

Archipelagos sometimes provide all of these conditions (Figure 17-11). Geographical isolation from the mainland ensures reproductive isolation from mainland species. The same isolation also means that relatively few individuals arrive at each island. Plants may arrive as seeds carried by wind or birds. Animals arrive only if they can fly, swim, or float on a piece of driftwood or other natural raft. In short, founder populations

tend to be so small in numbers that founder effects alone cause gene frequencies in the island populations to differ from those in their mainland parent populations. In some cases, the environments of different islands differ enough to exert different selective pressures (Figure 17-12).

When Darwin visited the Galápagos Islands in the Pacific and the Cape Verde Islands in the Atlantic, he noted three surprising facts (Figure 17-1):

1. The species in each chain of islands resembled those on the closest mainland, rather than of other islands. Despite the similar climate and terrain of the Cape Verde Islands and the Galápagos Islands, the birds on the Cape Verde Islands were similar to those of Africa and those of the Galápagos to those of South America. This suggested that the islands were colonized by migrants from the mainland.
2. The archipelagos contained a much smaller number of species than the mainland. The islands' biological poverty supported Darwin's idea that only a small number of ancestral organisms had colonized the islands.
3. The individual islands in a chain had distinct species. Tortoises from the drier islands had longer necks and peaked shells that allowed them to reach the cactus and tree branches, while those from wetter islands had shorter necks, better adapted to feeding on ground vegetation.

Darwin recognized that the distribution of species in chains of islands provides a record of the arrival of immigrants from the mainland. For example, of the hundreds of bird species in

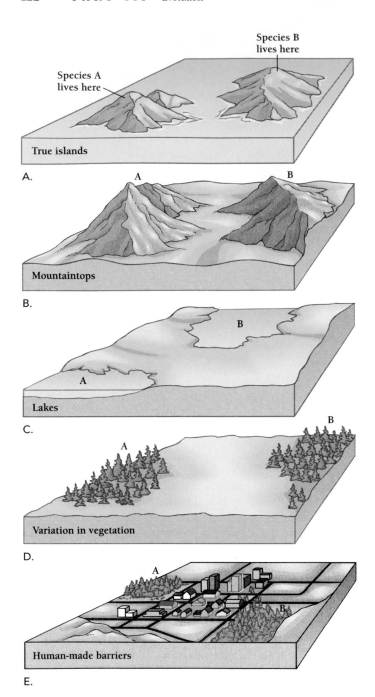

Figure 17-11   Geographical isolation comes in many shapes.
Although oceanic archipelagos feature many classic cases of adaptive radiation, other kinds of "islands" also facilitate adaptive radiations. To a species of fish, a series of lakes or ponds is an archipelago. To a mouse, the individual tops of mountains or plateaus can function as islands. A. True islands. B. Mountaintops. C. Lakes, rivers, and oceans. D. Variation in vegetation. E. Human-made barriers such as freeways or suburban sprawl isolate small patches of forest or meadow.

South America, only eight succeeded in reaching and colonizing the Galápagos. Yet 1 of those 8 gave rise to no less than 14 species of finches. The Galápagos finches offer the most famous example of adaptive radiation.

Figure 17-12   Galápagos tortoises. On each island, the tortoises evolved separately from those on the other islands and so differ with respect to the shape of the shell, the length of the neck, and other traits. The first written account of the differences among Galápagos tortoises came from the captain of the United States frigate *Essex*, which visited the Galápagos during the War of 1812 and, by itself, nearly destroyed the British whaling fleet stationed there. Before the *Essex* was finally captured by the British in 1814, Captain David Porter found time to describe the tortoises and various lava lizards.

The Galápagos finches are all distinct from any on the continent of South America. Yet they were enough alike to suggest to Darwin that they all stemmed from a common ancestor. Contemplating the finches, Darwin wrote his first statement on adaptive radiation: "One might really fancy that . . . one species had been taken and modified for different ends."

The probable mechanism of adaptive radiation on islands such as the Galápagos is relatively easy to understand. Over the course of the 5 million years since the islands first appeared—and possibly as recently as half a million years ago—a few finches (at least one male and one female) managed to fly the 1000 km (600 miles) from the South American mainland—no easy task, since finches are poor long-distance flyers. The founders were probably ground birds, with short bills useful for crushing seeds. Descendants of these birds managed to colonize the 13 islands in the Galápagos chain, as well as Cocos Island, about 1000 km away. However, because of the finches' limited flying abilities, flights among the islands must have been rare. So the finches on each island mated only with other finches from the same island. In short, the finches on each island became reproductively isolated.

Because each island differs slightly in geology, terrain, vegetation, and insect life, the finch population on each island experienced different selective pressures. Acted upon by both natural selection and genetic drift, the populations came to differ in many ways: size and form of beaks, nesting patterns, songs, plumage, and courtship behavior. Separate species formed through allopatric speciation. Over time, some of the new species managed to colonize the other islands, so that some islands came to sustain several species of finches.

Detailed studies of the finches have shown that pairs of species that share an island are more different from one another—in beak size and other measures—than are species that live on different islands. In short, where several finch species live together, individual species tend to evolve away from each other and become more specialized. Like all organisms on Earth, these birds feel the tug and pull of natural selection and will evolve (or go extinct). Only in these finches and a few other organisms, however, have researchers documented the effects of natural selection so clearly.

Yet, it is the absence of more specialized birds in the Galápagos that has allowed the finches to radiate. For example, because the Galápagos have no true woodpeckers, one species of Galápagos finch has evolved into a sort of imitation woodpecker. The Galápagos woodpecker finch does not have the special tongue and beak and other adaptations of a true woodpecker (Figure 17-13). Yet the woodpecker finch manages to feed on the insects that live in tree bark by using a cactus spine or a twig clamped in its beak to dig the insects from the bark.

Similarly, on the island chain of Hawaii, which also lacks true woodpeckers, a species of honeycreeper digs insects from tree bark by chiseling the bark with its lower jaw and picking out insects with its curved upper bill. Neither the Hawaiian honeycreeper nor the Galápagos finch seems as well adapted as a true woodpecker, but each manages to flourish in the "woodpecker" niche in the absence of direct competition from a true mainland woodpecker.

Plants also undergo adaptive radiation. The sunflower family (Asteraceae) provides two especially dramatic examples. In the Hawaiian Islands, 28 species of the most spectacular plants in the islands have all derived from a single small and unremarkable California tarweed. The amazing Hawaiian tarweeds, some of which are shown in Figure 17-14, include the Haleakala silversword, which grows within the cone of a volcano on the island of Maui, as well as the yuccalike iliau, which grows among the scrub vegetation of Kauai. Most of these plants, like their relatives in California, grow in dry areas, but a few have adapted to wet forests and bogs.

An equally dramatic radiation within the sunflower family has occurred on the remote Atlantic island of St. Helena, halfway between Africa and South America. The usually low-growing members of this family have evolved into large "trees." Reproductive isolation and the absence of competitors have evidently allowed the evolution of these distinctive species, which are found nowhere else on Earth.

Adaptive radiation, the evolutionary divergence of a single line of organisms, is the key to biological diversity.

Diversification on the Continents

The multiplication of species is not confined to archipelagos. On continents, two ways that species diversity increases are **dispersal,** the spread of a taxon through a large area, and

North American hairy woodpecker

Galápagos woodpecker finch

Hawaiian honeycreeper (akiapolaau)

Figure 17-13 **Convergent evolution.** In the absence of mainland woodpeckers, island birds can evolve adaptations for taking grubs from tree bark.

vicariance, the fragmentation of an already dispersed species, or group of species.

Birds and bats and even insects are, of course, very good at dispersing all over the world. So are plants, whose tough seeds hitch rides on animals and winds. Ants, horses, and other terrestrial organisms may not be able to cross oceans, but they can disperse enormous distances by simply walking a little farther, generation after generation.

In vicariance, a widely dispersed population can become fragmented either through the extinction of populations in between or through the creation of geographic barriers. As we saw in Chapter 15, tectonic plates move around, alternately drifting apart and bumping into one another over millions of years. Such movement not only separates and joins whole con-

Figure 17-14   Adaptive radiation in Hawaii. In the isolation of the Hawaiian Islands, an unremarkable California tarweed of the genus *Madia* has radiated into 28 species. A. The California tarweed *Madia madiodes*. B. Hawaiian iliau *Wilkesia gymnoxiphium*. C. The Haleakala silversword *Argyroxiphium sandwicense*.   *(A, N.H. (Dan) Cheatham/Photo Researchers; B, Gerald D. Carr; C, C.K. Lorenz/Photo Researchers)*

A.

B.

C.

tinents, it also raises mountain ranges and lifts old sea beds that are then cut by massive canyons. Once populations are isolated by such geologic changes, they can speciate allopatrically.

The fossil record makes sense in terms of the movements of the continents. For example, the giant ground birds of Australia, Africa, and South America are all related and were separated only by the Mesozoic breakup of Gondwanaland. In contrast, the camel family evolved after the continents were already separated. The Camelidae originated in North America during the Eocene. They later dispersed into Eurasia by way of the Bering Land Bridge (long since disappeared) that connected Alaska and Siberia and into South America by way of the Central American isthmus. Today the Camelidae are extinct everywhere except Asia, North Africa, and South America.

Likewise, the modern horse evolved in North America, then migrated to Eurasia and to South America in periods when those continents were connected with North America. Because the current land connection between North and South America was formed only about 3 million years ago, no horse fossils in South America date back before that time.

Continental diversity increases through the dispersal of organisms to new parts of the world and through vicariance, the fragmentation of dispersed groups into populations that can evolve separately.

Adaptive Radiation on the Continents

The Bering and Central American land bridges provided a route for the dispersal of many species between Eurasia and North and South America—including, for example, horses, camels, and humans (Figure 17-15). But there were no such bridges to

evolution of the marsupials in Australia parallels that of the placental mammals elsewhere, in what is an equally extraordinary display of large-scale convergent evolution. The present-day marsupials include animals that resemble the placental mammals of other continents, both in appearance and in habit. Although kangaroos do not look much like horses or cows, Australian marsupials include species that look like rabbits, mice, moles, anteaters, cats, wolves, squirrels, and woodchucks. Distinctly wolflike creatures have separately evolved from both a marsupial and a placental ancestor.

The parallel radiations of marsupials and placental mammals show the opportunism of evolution and help account for some of the enormous diversity we see around us.

Australian marsupials illustrate both adaptive radiation and convergent evolution.

THE CAMBRIAN EXPLOSION

Up until the Cambrian Period, almost 600 million years ago, life on Earth seems to have been limited to a few soft, one-celled organisms, such as algae and bacteria. Until quite late in the Precambrian, the only complex organisms that paleontologists find are a few jellyfishlike creatures.

Then, at the beginning of the Cambrian, life bloomed. In a burst of diversification unmatched in the history of the world, all of the modern animal phyla that have fossilizable skeletons appeared, though not necessarily in their current forms. For example, the arthropods, which today include the insects, spiders, crustaceans, and barnacles, were represented by the trilobites. Also present in the Cambrian were the brachiopods (lamp shells), the mollusks (snails, clams, and squid), the sponges, the echinoderms (starfish and sea urchins), the cnidarians (jellyfish and corals), not to mention the first chordates (today including amphibians and mammals).

Taxonomists have based the classification of animal phyla on basic body plans—the two-layered radial symmetry of the jellyfish; the three-layered bilateral symmetry of flatworms; the three-layered bilateral symmetry, plus notochord, of chordates, such as ourselves, to name a few. Although enormous evolutionary change has occurred since the early Cambrian, and the level of diversity is arguably as great today as it ever was, almost no new body plans have evolved in the half a billion years since these basic forms first appeared.

Evolutionary biologists refer to this one-time evolutionary radiation as the **Cambrian explosion.** Paleontologists remain divided over the exact time span during which this explosion occurred. Some evidence suggests that the major animal phyla may have diverged as much as 900 million years ago, hundreds of millions of years before the Cambrian. Other evidence suggests that all the phyla may have appeared at once at the very end of the Precambrian. Either way, evolutionary biologists are

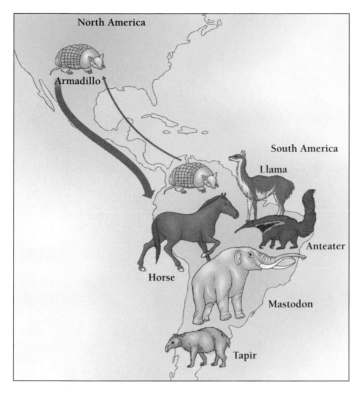

Figure 17-15 Land bridge. The North and South American continents have been isolated and connected at different times in geologic history, depending on the level of the oceans. Land bridges allow many organisms to migrate to new habitats. When sea levels rise and again isolate the continents, separate species arise through allopatric speciation.

Australia. Australia became separated from all the other continents, near the end of the Cretaceous Period, some 65 million years ago, after which Australian species evolved in isolation. As no horses or other grass-eaters reached Australia, other grass-eaters evolved in Australia and populated the plains—the kangaroos.

Kangaroos, like almost all other Australian mammals, are marsupials, mammals who bear young as fetuses and carry them in a pouch until the young are able to fend for themselves. In contrast, humans and other placental mammals bear more-developed young. Placental mammal mothers nurture their young in the uterus with the aid of a specialized tissue, called the *placenta* [Latin, *placenta* = flat cake], which passes nutrients, oxygen, and wastes between the mother and the developing embryo.

Marsupials and placental mammals diverged from each other in the early Cretaceous, and the marsupials happened to reach Australia first, in the late Cretaceous. With the exception of bats, which flew there, and rats, which must have rafted from island to island in the East Indies, all the native mammals of Australia are marsupials.

The marsupials of Australia are an extraordinary example of adaptive radiation (Figure 17-16). All appear to have evolved from a primitive, opossumlike marsupial of the Cretaceous. The

Figure 17-16   Adaptive radiations in parallel. The marsupial mammals of Australia belong to an ancient and very distinct lineage of mammals. Yet, marsupial mammals (*left in each pair*) and placental mammals (*right*) have separately evolved many of the same body types. The sugar glider and the flying squirrel, for example, both have long bushy tails and webbed limbs—and for the same reason. Both animals quickly cover great distances by sailing from the tops of tall trees down to the ground.

MARSUPIALS   PLACENTALS

Koala   Tree sloth

Wombat   Woodchuck

Marsupial mole   Common mole

Dunnart   Shrew

left with a major puzzle. What has prevented major new body plans from evolving since the Cambrian?

One argument is that evolution has somehow slowed since then, that changes simply do not occur as rapidly as they did in the Precambrian and the Cambrian. However, modern populations are capable of evolving quite rapidly. For example, as we saw in Chapter 16, both bacteria and insects evolve resistance to antibiotics and pesticides within a few years or even months. Vertebrates, such as fish and birds, are also capable of rapid changes. Research on Darwin's finches in the Galápagos has shown that natural selection can change their average beak size, according to what size seeds are available, within a matter of months. All of Darwin's finches probably evolved from a

single species within the last half a million years, a short period in geologic time.

Another argument is that once the major body plans formed, **developmental constraints**—the rules of embryological development that determine the general form of an organism—prevented new body plans from evolving. Many biologists believe that in most organisms, development has become so finely tuned since the Cambrian explosion that any dramatic change results in organisms too defective to flourish. The idea is that modern organisms are so highly evolved that there is less and less room for improvement. An organism with an entirely new form could not compete with organisms that are relatively "perfected."

MARSUPIALS   PLACENTALS

Banded anteater   Giant anteater

Sugar glider   Flying squirrel

Tiger cat (Quoll)   Ocelot

Tasmanian wolf   Gray wolf

But, the ancient body plans are not necessarily optimal designs that could never be improved on. Indeed, very likely they are just accidents of history, like the homologous vertebrate limbs illustrated in Chapter 15. Animal body plans, such as the particular arrangement of bones in the limbs of horses, whales, and humans, may simply be a case of organisms "making do" with what they have.

Evolution at the species level occurs as rapidly as ever, but the evolution of new phyla seems to have ceased. The Cambrian explosion may have been a one-time phenomenon.

EXTINCTION AND THE RATE OF EVOLUTION

Evolutionary biologists estimate that more than 99 percent of all the species that have ever lived are now extinct and that the vast majority of these went extinct millions of years ago. But why species go extinct is, in most instances, something of a mystery. By looking at species that have gone extinct in historical times, we can say that some species go extinct because of competition from other species; some because their habitat is drastically changed; and some because of a new predator. Human beings, for example, have driven a variety of species to

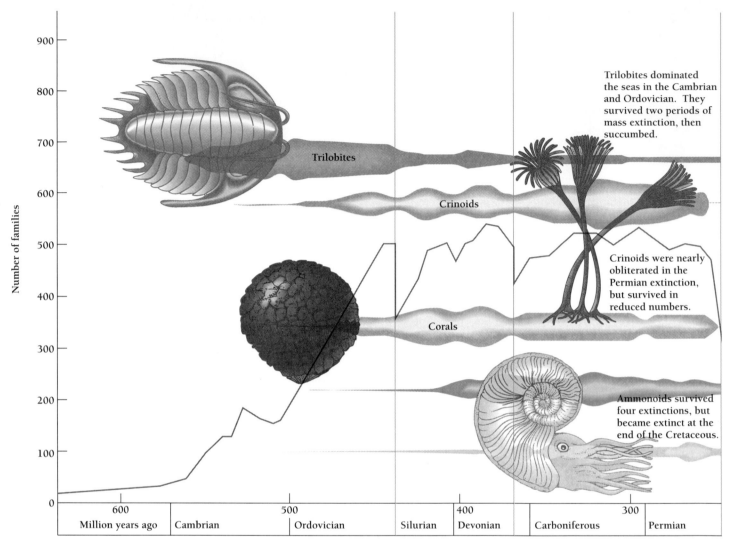

Trilobites dominated the seas in the Cambrian and Ordovician. They survived two periods of mass extinction, then succumbed.

Crinoids were nearly obliterated in the Permian extinction, but survived in reduced numbers.

Ammonoids survived four extinctions, but became extinct at the end of the Cretaceous.

Number of families

Trilobites

Crinoids

Corals

| 600 | | 500 | | | 400 | | 300 |
| Million years ago | Cambrian | | Ordovician | Silurian | Devonian | Carboniferous | Permian |

Geologic time (10^6 years)

extinction—the American passenger pigeon, for example—through heavy hunting. But in most extinctions, probably no single factor determines whether a species will go extinct. Some species survive such onslaughts, while others do not.

The Mass Extinctions

The fossil record shows that, in the long run, species go extinct at a more or less regular rate. However, superimposed on this ongoing extinction rate are five intriguing mass extinctions, including the death of the dinosaurs (Figure 17-17). Near the end of five geologic periods—the Ordovician, the Devonian,

the Permian, the Triassic, and the Cretaceous—the number of extinctions increased by as much as 2.5 times the background extinction rate.

Why these mass extinctions occurred is one of the most controversial questions in paleontology. Some researchers deny that the mass extinctions are qualitatively different from the background extinction rate. They argue that these mass extinctions are just random fluctuations in the usual loss of species over time. Other researchers argue that the five extinctions are the consequence of major worldwide catastrophes. The exact nature of the catastrophes is hotly debated.

The most controversial mass extinction is the one at the end of the Cretaceous. Sixty-five million years ago, all the di-

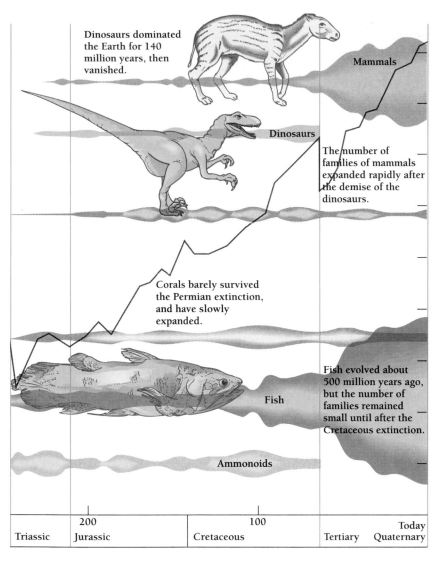

Dinosaurs dominated the Earth for 140 million years, then vanished.

Mammals

Dinosaurs

The number of families of mammals expanded rapidly after the demise of the dinosaurs.

Corals barely survived the Permian extinction, and have slowly expanded.

Fish evolved about 500 million years ago, but the number of families remained small until after the Cretaceous extinction.

Fish

Ammonoids

200	100	Today
Triassic	Jurassic	Cretaceous	Tertiary	Quaternary

Geologic time (10^6 years)

Figure 17-17 Increasing diversity and mass extinctions in a selection of animals. The number of species or organisms tends to increase over time, except when disaster strikes in the form of a mass extinction. Recovery from a mass extinction like the one the Earth is experiencing today takes 5 million years or longer. Many groups of species, such as the dinosaurs and the trilobites, never recover.

nosaurs and half of all species of plants and animals went extinct. Every conceivable catastrophe has been suggested to explain this extinction. But the majority of researchers now seem to believe that this most recent extinction occurred when a giant asteroid hit the Earth with enough force to blast a 180-km crater and plunge the world into almost total darkness for months.

We are now in the midst of a sixth mass extinction. The full extent of this extinction is somewhat controversial. No one knows precisely how many families of organisms have gone extinct, nor how many will go extinct before this extinction event is over. But the cause of this mass extinction is not at all controversial. It is entirely the result of human activity.

After each major extinction in the past, diversity increased dramatically (Figure 17-17). For example, the mammals radiated spectacularly after the Cretaceous extinction of the dinosaurs and other reptiles 65 million years ago. We might guess that the only circumstances under which really new organisms might flourish are where a mass extinction has eliminated much of the competition.

All species go extinct eventually. The rate of extinctions since the Cambrian has been relatively constant except for five mass extinctions. Evolutionary biologists have yet to agree on what caused these mass extinctions.

BOX 17-1

Did it come from outer space?

It has been less than 20 years since scientists began to suspect that the Age of Dinosaurs ended with a bang, not a whimper.

In 1979, Walter Alvarez was studying rocks in Gubbio, Italy, at what geologists call the K/T boundary—the point where rocks laid down 65 million years ago, at the end of the Cretaceous (K), are overlaid by rocks laid down during the following geologic age, the Tertiary (T).

At Gubbio, the boundary is clearly marked by a 2- to 3-cm layer of clay (Figure A). To determine how much of Earth's history is represented by that inch-deep clay layer, Alvarez sampled the layer for iridium, a metal rare on the Earth's surface. A constant hail of micrometeorites continually introduces small amounts of iridium into the atmosphere. The iridium then falls onto the Earth's surface. Like the sand in an hourglass, the buildup of iridium can be used to determine how much time has passed.

Alvarez expected to find less than 1 part per billion (ppb) of iridium. Instead, he found 10 ppb, but only in the clay layer itself. Both the Cretaceous rock below the clay and the Tertiary rock above had the expected amount—1 ppb or less. What had caused the spike in iridium in the clay?

Collaborating with his father, Nobel Prize–winning physicist Luis Alvarez, and geochemist colleagues Frank Asaro and Helen Michel, the Alvarez team went on to find similar iridium spikes at two other K/T sites—in Denmark and in New Zealand. In 1980, the team published their groundbreaking paper, "Extraterrestrial Cause for the Cretaceous–Tertiary Extinction," in the

journal *Science.* In it, the researchers argued that the iridium spike, or *iridium anomaly,* as it is often called, was the result of a gigantic asteroid, 10 km across, that slammed into the Earth and caused the extinction of all of the dinosaurs and many other forms of life, both terrestrial and marine.

Although extraterrestrial causes of the K/T extinction, such as supernovas or radiation from outer space, had been suggested before, the iridium anomaly was the first strong supporting evidence for any of these scenarios. The Alvarez team's paper, controversial from the beginning, stimulated other scientists to investigate the team's findings. So far, iridium anomalies have been found at more than 100 different K/T boundary sites, both on land and in ocean sediments.

In addition, other evidence for the asteroid has been discovered at iridium-rich K/T sites in the western United States. If an asteroid 10 km across smashed into the Earth, the pressure generated would liquefy the meteorite and the rock it smashed into, spewing molten meteorite and rock into the air—just as water droplets fly into the air when a boulder is heaved into a pond. As the molten meteorite droplets cooled, they would solidify into *microtektites* and fall back to Earth. Such microtektites, as well as *shocked quartz*—deformed pieces of the mineral quartz—have been found in association with known meteor sites, such as Meteor Crater in northern Arizona.

Perhaps the most dramatic evidence would be a huge hole in the ground. Until the late 1980s, no one had found the 150- to 300-km crater that scientists predicted

such an asteroid would create. If the crater was deep in the ocean and now filled with sediment, it would not be easy to find.

Yet, off the Gulf Coast of Texas, researchers found evidence that, just at the K/T boundary, huge boulders had been moved around as if by a tidal wave. Then another researcher reported a subterranean bowl-shaped structure, 180 km across, located in the northwestern corner of the Yucatan Peninsula near the town of Chicxulub (pronounced CHICKS-uh-loob). In 1991, others reported that samples from the Chicxulub bowl contained shocked quartz. The layer of melted rock underneath the huge bowl has now been dated at 65 million years old—the age of the Cretaceous–Tertiary (K/T) boundary.

Few people now doubt that the structure at Chicxulub is the crater from an enormous asteroid that slammed into Earth 65 million years ago. That this impact caused global catastrophe is also indisputable. The impact would have incinerated plants and animals across the North American continent and spewed debris into the atmosphere, blotting out the sun worldwide for months or even years.

Today, the only debate is over the exact consequences of that global catastrophe. Not everybody agrees that the asteroid alone caused the mass extinction at the end of the Cretaceous. Some lineages of organisms disappear in the Cretaceous fossil record before the K/T boundary clay layer. For some paleontologists, that means there was a more gradual increase in extinctions during the late Cretaceous, not a sudden catastrophic reduction in species. The as-

The Pace of Evolution

In 1859, Charles Darwin was nearly alone in his belief that evolution was a gradual process. He held that not only individual species, but also genera, orders, and other higher taxa all arose through the slow transformation of species. His closest friends tried to persuade him that evolution might very well occur in jumps, or **saltations.** On the eve of the publication of the *Origin of Species,* in 1859, Darwin's friend Thomas H. Huxley wrote

to him, "You have loaded yourself with an unnecessary difficulty in adopting *natura non facit saltum* [Nature makes no jumps] so unreservedly."

Gradualism versus Saltationism

For Darwin, the conflict was between what he saw everywhere around him in nature and what the fossil record showed. As we noted in Chapter 15, gaps in the fossil record pose a seri-

teroid's impact may have come long after many species were already extinct.

But why might they have gone extinct? The best alternative explanation is that massive volcanic activity in western and central India, combined with volcanic activity in the Pacific Rim, released enormous amounts of volatile gases into the atmosphere. Some researchers think the large release of volcanic gases and dust could have altered the global climate, caused acid rain, and even damaged the ozone layer. Such large-scale climatic alteration could have caused extinctions.

Debate continues about whether the late Cretaceous extinctions were rapid or gradual. If the extinctions were sudden, the asteroid is the most likely culprit. If the extinctions were more gradual, the asteroid may have acted merely as a *coup de grâce*.

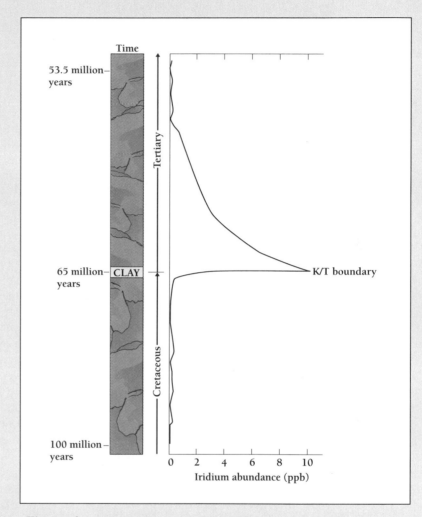

Figure A   **An asteroid impact?** At the end of the Cretaceous, the amount of iridium present in deposits increases dramatically from 1 part per billion (ppb) to 10 ppb, then gradually decreases again over a period of about 10 million years. Iridium is a metal rare on Earth but common in asteroids and meteors.

ous problem in understanding the evolution of species. To demonstrate that evolution happened gradually, we would like to see a fossil representing each small step in an evolutionary line. But, in general, the fossil record does not fulfill this wish. Transitional forms between distinct species are disappointingly unusual.

Paleontologists are working with a random sample of organisms through time. They may find a well-preserved fossil of species A in one layer, then no more of that species until 50,000 years later (in the fossil beds). If the next fossil they find is different enough from the first, they may say it is a second species, B, but they won't know how B is related to A. Is B descended directly from A? Are A and B instead descended from a common ancestor C? If so, is C much older than A and B, or just a little older? Did B (or C) gradually turn into A, leaving no unevolved remnants? Or could A, B, and C have all lived at the same time?

The lack of intermediate forms may merely represent the incomplete preservation of ancient organisms. Just as most individual human beings are not named in the pages of recorded history, so most species are not preserved in the layers of rock that make up the fossil record. However, another interpretation for the gaps in the fossil record is that species mostly do not change at all, evolving only in short bursts, or saltations. The intermediate forms are not preserved, either because they never existed, or because they existed only momentarily in geological time (less than 10,000 years).

Darwin firmly rejected this latter line of reasoning. He argued that the gaps in the fossil record were the result of the incomplete preservation of ancient organisms.

The living species Darwin had studied all his life seemed to argue against saltationism. First, species were so subtly different that it was hard to decide if they were true species. Three Galápagos Island mockingbird species, each of which came from a different island, were so alike that Darwin assumed they were merely races of a South American species until he took his specimens back to England. There, however, an ornithologist pronounced each a separate species. Later in his life, Darwin spent eight long years struggling to work out the taxonomy of barnacles. Each barnacle species graded so smoothly into the others that he complained bitterly about how hard it was to tell one from another. In short, everywhere Darwin looked, species seemed to grade into one another. Even Linnaeus had commented on this difficulty.

Gradual selection was clearly capable of producing enormous changes. The dramatically different races of domestic pigeon, without question the result of gradual selection, would be considered separate species, if in nature (Figure 15-17). And nowhere did Darwin see the living members of populations abruptly change into new species. (If he had, he would have been able to make a much stronger case for evolution.)

Nevertheless, the discontinuous nature of the fossil record convinced many 19th-century biologists that evolution did indeed occur by fits and starts, a view known as **saltationism.** After the rediscovery of Mendel's work in 1900, Mendelians argued strenuously that species arose abruptly from one generation to the next because of so-called macromutations—mutations that caused large changes in individual organisms. By the 1940s, however, Darwin's gradualism was finally in vogue and saltationism was out. By the 1960s, most biologists subscribed to **gradualism**—the view that most species evolved gradually and continuously in direct response to selective pressures from the environment. Paleontologists could object that the fossil record did not support this idea, but geneticists dominated biology at that time, and few of them were interested in accounting for the discrepancies in the fossil record.

Nonetheless, a few renegades held out for saltationism. The most famous of these was the German–American biologist Richard Goldschmidt (1878–1958). He argued that although macromutations generally result in deformed organisms, some of these monstrosities have potential if they happen to arise in the right environment. He called these potential species **hopeful monsters.** "A fish undergoing a mutation which made for a distortion of the skull carrying both eyes to one side of the body is a monster," he argued. "The same mu-

Figure 17-18 Hopeful monster? The flounder begins its life like other fish, with one eye on each side of its head. As the embryonic fish develops, one eye migrates over the top of the head to join the other eye. The adult fish lies on its side in the sand and the mud at the bottom of the ocean and looks up.

tant in a much compressed form of fish living near the bottom of the sea produced a hopeful monster, as it enabled the species to take to the life upon the sandy bottom of the ocean, as exemplified by the flounders" (Figure 17-18). Goldschmidt's term "hopeful monsters" was both memorable and unfortunate. Evolutionary biologist Ernst Mayr's equally memorable term of derision for Goldschmidt's idea—"hopeless monsters"—was quickly picked up by other biologists.

Nevertheless, by 1950, the fossil record, which Darwin had found so imperfect, had improved little. A hundred years of research still revealed few cases with all the intermediate stages in the evolution of a lineage. When new species appeared, they did so suddenly, the intermediate stages leaving no trace. And many groups hardly seemed to change at all over long periods of geological time. For example, garfishes and sturgeons show only minimal morphological change over 80 million years. Horseshoe crabs have barely changed at all in 230 million years. Such stability over long periods is called **stasis.**

The periods of apparent stasis in the fossil record may not be what they seem, however. Only hard tissues such as bones and shells are preserved. Even extensive changes in soft anatomy, physiology, and behavior of a species would not appear in the fossil record.

In addition, some scientists argue, paleontologists often ignore intermediate forms. Faced with a series of fossils, paleontologists tend to classify individual fossils in a layer as one species or another, not as "something in between." An intensive study of 15,000 trilobite fossils from Wales provided a good example of what happens when a scientist looks instead for the in-between species. The trilobites, a massively successful group of arthropods that flourished through the Permian, showed so many intermediate forms that classifying individual fossils into species became an exercise in arbitrariness.

Nonetheless, even if we accept that species may not actually evolve quite as abruptly as paleontology suggests, we can still try to explain why speciation should seem sudden at all. Species' abrupt appearances suggest two alternatives. In one scenario, new species appear almost instantaneously, as, for example, in a single generation through saltations. In the other scenario, new species evolve quite rapidly—by geological standards—over just millions or even mere thousands of years at the time of speciation. In either case, the intermediate forms would not appear in the fossil record. The Mendelians had believed that new species appeared almost instantaneously.

The latter idea—rapid, but not instantaneous, speciation—was not inconsistent with what population genetics had suggested about speciation in small, isolated populations. The founder effect, the bottleneck effect, as well as allopatric speciation and even sympatric speciation are all consistent with the idea of rapid speciation in small, isolated populations. Mayr had himself suggested this idea in 1942. In the case of rapid speciation, intermediate forms would exist in such small numbers and for such a short time geologically that the chances of their being preserved in the fossil record would be almost nil.

Until the 1970s, biologists paid little attention to the connection between rapid speciation and the gaps in the fossil record. Most biologists were interested in microevolution and population studies, not macroevolution. And paleontologists, interested primarily in macroevolution, wondered how biologists could go on believing in gradualism when there were so few examples of it in the fossil record.

Punctuated Equilibrium

In the 1970s, two paleontologists—Niles Eldredge at the University of California, Berkeley, and Stephen J. Gould at Har-vard—argued that the fossil record is exactly what we would expect it to be if the formation of new species was very rapid compared to the accumulation of changes within a species.

This view, called **punctuated equilibrium**, proposes two main ideas. First, *species change very little most of the time* (stasis). Second, *most anatomical or other evolutionary change in individual species occurs during a geologically brief period at the time of speciation.* In this view, speciation events punctuate an otherwise stable equilibrium, and the most important events in evolution are those that lead to reproductive isolation. Discontinuities (even catastrophes) are paramount, and gradual changes within species are mere fine-tuning.

That most species show little change over long periods is not controversial. But not all evolutionary biologists agree that significant change can occur only when a species splits into two or more species.

Punctuated equilibrium is an extreme alternative to extreme gradualism. The two ideas are truly abstractions, not alternative theories that can be tested against one another. No biologist believes that *all* evolutionary change occurs very gradually over hundreds of thousands or millions of years. And only a very few biologists say that all or most evolution occurs only when a species splits into two or more. Punctuated equilibrium and gradualism are extreme ends of a continuum of possible scenarios (Figure 17-19). The history of every species has occurred in a more or less punctuated way.

Even extreme punctuated equilibrium does not contradict the findings of population geneticists or Darwin's view that evolutionary change occurs gradually. The brief periods during which populations are said to speciate—5000 to 50,000 years—are invisible in the fossil record (and therefore "instan-

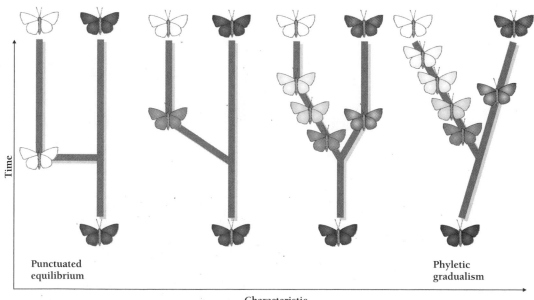

Time

Characteristic
(wing color)

Punctuated
equilibrium

Phyletic
gradualism

Figure 17-19 Is evolution gradual or abrupt? The answer probably depends on the organism. Punctuated equilibrium and gradualism are the ends of a continuum.

BOX 17-2

Modern tropical rain forests:
mass extinction, or not?

The fossil record shows five natural mass extinctions during the past 600 million years. Many biologists believe that a sixth, human-induced mass extinction is now underway.

The most species-rich regions of the world are tropical environments, such as the Amazon Basin. For any group of organisms—whether birds, insects, trees, fungi, or ants—the number of species is greatest in the tropics. Although tropical rain forests cover less than 7 percent of the world's land area, one-half to two-thirds of the Earth's 10 million to 100 million species inhabit these rich forests.

Few of these species are known to science—only 1.4 million of the world's species have been named. Researchers are struggling to keep ahead of the chainsaws and forest fires that daily destroy the habitats of thousands of tropical forest species, half of which are insects. Tropical rain forests are dense with species. A 1-hectare (2.5 acres) plot of Brazilian forest may contain 200 species of trees, whereas the same size plot in Canada would support fewer than 10. And each species of tree may support 5 to 10 species of insects that specialize only on that one species of tree. Eliminating that tree species causes all the insect species to vanish also. Such cascades of extinctions are not limited to plant–insect interactions. When peccaries (a piglike mammal) were eliminated in one Brazilian forest, so were three species of frogs. All the frogs, it seemed, lived in the mud wallows the peccaries created.

As of today, less than one-half of the original tropical forests remain. And logging,

Figure A **Slash and burn agriculture.** In Brazil, farmers burn small areas of tropical rain forest. They farm for a while and then move on to burn another plot of land, as young trees grow up in the last plot. As human population size increases, however, rain forests are being burned faster than they are growing back. (Earth Scenes/© 1996 Dr. Nigel Smith)

taneous"), but these brief periods in evolutionary history provide ample time for gradual change to occur.

Early biologists argued about whether most evolution occurred gradually or by means of saltations. Punctuated equilibrium, as it is now understood, represents a compromise view that is consistent with the gaps in the fossil record, stasis, and the small genetic changes that occur in most natural populations from generation to generation.

THE EVOLUTION OF HUMANS

When we visit the zoo, we are often drawn to the animals that most resemble ourselves—the monkeys and apes. Taxonomists say that apes, monkeys, and lemurs all belong to the same order of mammals. This group, called the **primates,** are remarkably unspecialized. Like the earliest mammals, we have five digits on each hand or foot and unspecialized teeth. All primates lack, for example, the sharp cutting teeth and broad grinding surfaces that characterize the teeth of horses, rabbits, and other

agriculture, cattle ranching, and firewood harvesting continue to degrade and destroy tropical forests (Figure A). In some cases, regions of burning forests are so vast that the fires are visible from space. Harvard biologist E.O. Wilson estimates that if tropical forests continue to be destroyed at current rates, less than 10 percent of the original forests will remain by the year 2050, and between 2.5 and 25 million species will go extinct.

How does this loss of species compare with the five prehistoric mass extinctions? By examining the fossil record, scientists estimate that the normal, or background, extinction rate may be as high as one extinction per year. In a mass extinction, the number of extinctions increases to at least 2.5 times the background rate. Since 1600, more than 1000 species of plants and animals have gone extinct—yielding a minimum extinction rate of more than 2.5 species per year (Figure B).

However, there were probably at least 10 times as many extinctions during that time period. Recall that scientists have, at best, named only about one-tenth of the 10 million to 100 million species thought to exist. Just as the named species are a small fraction of the total species, the tally of documented extinctions includes a similarly small fraction of the total number of extinctions. For the 400 years since 1600, the minimum extinction rate may well approach 25 species per year. In addition, as the remaining tropical forests grow smaller, species are lost at a faster rate. The current loss of tropical forest species is comparable in magnitude to the prehistoric mass extinctions.

Although evolution will continue, the world will not recover its former biological diversity in our lifetime—or in our great-great-great-grandchildren's lifetime. If we can look to the fossil record for an answer, it may be 5 million years before the Earth once more supports the diversity of life that existed when Columbus sailed to the New World.

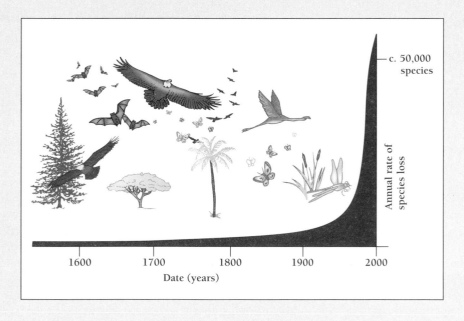

— c. 50,000 species

Annual rate of species loss

1600 1700 1800 1900 2000

Date (years)

Figure B The rate of species loss has increased precipitously in the last 100 years.

herbivores. We also lack the long canines and shearing teeth of the carnivores such as dogs.

Nonetheless, despite our lack of specialization, we primates have a set of common characteristics that together distinguish us from other mammals. These include large brains and binocular vision, flexible shoulder joints, and hands and feet with five grasping digits, a grasping thumb, and flat fingernails rather than claws.

Most of the 166 species of living primates are tree dwellers. Our flexible shoulder joints and grasping hands and feet en- able us to climb trees and swing from one branch to another. As a group, primates are basically herbivores that live on leaves and fruits. Nonetheless, many primates are opportunistic omnivores who will occasionally eat grains, grubs, and even flesh. Most 20th-century humans eat grains and flesh regularly.

The Classification of the Apes

Primates are one of the oldest orders of mammals, dating back some 65 million years. Early primates, which more closely re-

Figure 17-20   Phylogeny of primates as revealed by DNA hybridization. By measuring base-pairing between matching sequences of DNA from different species, biologists can determine relative differences and estimate the time in years since two species diverged. This phylogenetic tree, for example, suggests that humans diverged from pygmy chimps 5 to 10 million years ago and that we are more closely related to pygmy chimps than to common chimps.

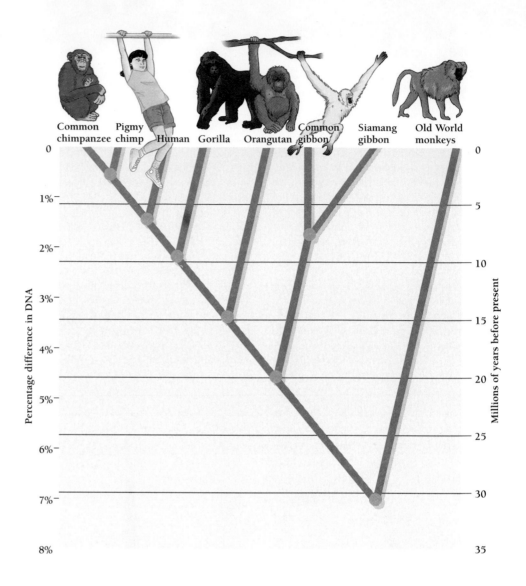

semble modern lemurs than monkeys or apes, had smaller bodies and longer snouts than modern monkeys and apes (Figure 17-20).

Monkeys and apes consist of two large groups. The more primitive group are the **platyrrhines** [Greek, *platyus* = flat + *rine* = nose], monkeys whose flat noses have widely separated nostrils that point sideways (Figure 17-21A). Many platyrrhines also have prehensile (grasping) tails. All platyrrhines live in South and Central America. Based on fossil and anatomical evidence, zoologists have concluded that they first evolved about 26 million years ago.

The second group, called **catarrhines** [Greek, *kata* = down + *rina* = nose], evolved in Africa and Asia, somewhat later. The catarrhines have downward-pointing nostrils and often no tail at all (Figure 17-21B). Gibbons and other primitive catarrhines first appeared in East Africa 20 to 17 million years ago (Figure 17-22A). All of the catarrhines except for the humans still live only in Africa or Asia.

Taxonomists group the catarrhines into the catarrhine monkeys, on one hand, and the **hominoids** [Latin, *homo* = hu-

man], on the other. Hominoids have large skulls and long arms, and they tend to walk at least partially erect (Figure 17-22B). The hominoids include the gibbons, which are relatively primitive, as well as gorillas, baboons, orangutans, humans, and chimpanzees.

Exactly how all the different apes are related to each other and to humans is a subject of great debate among all kinds of biologists. Molecular biologists compare protein or DNA sequences and draw one conclusion. Zoologists and **physical anthropologists** [Greek, *anthropo* = human], researchers who study humans, compare anatomy and behavior and draw slightly different conclusions. Paleontologists focus on the fossil record and see a subtly different picture. All these researchers agree, however, that humans share a common ancestry with other apes.

Were it not for our tendency to distance ourselves from other animals, humans would probably be lumped into the same family as the gorillas, chimpanzees, and orangutans—together called the **Pongidae.** We are closely related to both gorillas and chimpanzees. Our own DNA differs from that of

Figure 17-21 Platyrrhines and catarrhines. A. Howler monkey. The platyrrhine monkeys (from South and Central America) have nostrils that point sideways. B. Green monkey. The catarrhines—the Old World monkeys and apes—diverged from the platyrrhine monkeys from 17 to 20 million years ago. The catarrhines' nostrils point downward. *(A, Tom & Pat Leeson/Photo Researchers; B, Glenn Vanstrum/Animals Animals)*

A.

B.

chimpanzees and their cousins the bonobos ("pygmy chimps"), for example, by less than one percent.

Despite the similarities between humans and pongids, we humans have given ourselves a separate family, the **Hominidae.** Fossil and molecular data suggest that the common ancestors of orangutans diverged from those of the other pongids and the hominids about 12 million years ago. Hominids diverged from the chimpanzees and bonobos only 5 to 6 million years ago.

Homo sapiens is not the only hominid. We count as relatives a number of different species of hominids, some of which lived more than 4 million years ago. Two characteristics distinguish all hominids from other apes: (1) hominids are **bipedal,** that is, they consistently walk on two rather than four feet, and (2) hominids have rounded, rather than rectangular jaws (Figure 17-23).

All hominid species except ours are now extinct, but anthropologists have discovered the buried bones of hominids

Figure 17-22 Two hominoids. The hominoids include the primitive gibbon and the more evolved bonobos and humans. A. White-handed gibbon. As shown in Figure 17-20, the gibbons are the most primitive of the hominoids. B. Bonobo and her baby. Bonobos, or pygmy chimps, are our closest relatives. *(A, Gerard Lacz/Peter Arnold, Inc.: B, M. Long/Visuals Unlimited)*

A.

B.

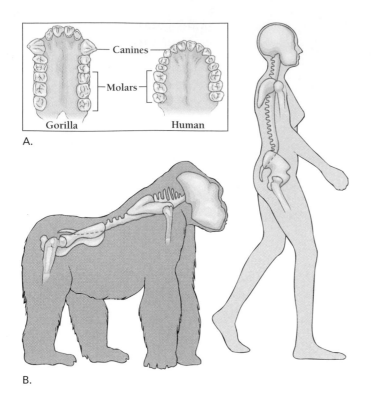

A.

B.

Figure 17-23 Hominid characteristics. Several characteristics distinguish *hominids* from other *hominoids*. A. Hominids tend to have more rounded jaws than chimps and gorillas. B. In addition, because hominids walk upright, we have arched rather than straight spines.

throughout Africa, Asia, and Europe. In 1996, researchers digging near the shores of Lake Turkana, in East Africa, discovered the oldest known hominid. *Australopithecus anamensis,* as its discoverers named it, is 4.2 million years old and the earliest example yet of an **australopithecine** [Latin, *australis* = south + Greek, *pithekos* = ape]. *A. anamensis* had a small brain, a squarish jaw, and long apelike arms. But it walked upright.

How are we different from *Australopithecus?* Later hominids such as ourselves have large brain cases. Our backs arch, and our thigh bones attach differently to the pelvis and angle inward, so that we are "knock-kneed" compared to the pongids (Figure 17-24A). We have lost the opposable toe that characterizes other primates. Instead, we have the reduced toes and long feet seen in other running animals, such as horses and dogs (Figure 17-24B).

In addition, *Homo sapiens* combine extremely sparse body hair (compared to other apes) with an extravagant growth of hair on the head and, in the case of males, on the face as well. When illustrating hominids, artists are forced to decide whether to draw our ancestors naked like us, or hairy like pongids. Most artists compromise and draw our ancestors as semihairy, but with short hair on the head. In reality, the degree to which other species of *Homo* shared our strange distribution of hair will probably never be known, as hair is rarely fossilized.

Primates are divided into the lemurs, the platyrrhine monkeys, and the catarrhine monkeys and apes. The catarrhine apes, including every ape from gibbons to humans, are called hominoids. Humans and their relatives are hominids, while chimpanzees, gorillas, and orangutans are pongids.

Why Did *Australopithecus* Stand Up?

The question of how walking hominids evolved from a more primitive ancestor has haunted anthropologists for more than a century. Some researchers have argued that the evolution of an upright stance and a hairless body was an adaptation for remaining cool on a hot African savanna millions of years ago. The idea is that a person (or ape) standing up exposes less surface area to the sun. Other researchers have suggested that the first hominids stood up because they could run faster that way—either from predators, or, alternatively, in pursuit of prey. Still others have argued that hominids stood up so that they could carry and wield weapons and other tools. Some now argue that hominids became walking apes when they began scavenging over large territories.

Most of these ideas lack specific evidence in their favor. The only certainty is that an ancestral ape did evolve an upright stance and the long legs of a long-distance walker.

Nearly all of the hypotheses about why hominids evolved an upright stance depend on the idea that hominids left the forest to live in an open grassland, or savanna. One of the most persistent theories is that our upright stance evolved in association with a global cooling 2.5 to 2 million years ago. At that time, the world was a colder and drier place. So much water was frozen into glaciers in the far north and south that ocean levels worldwide were much lower than today.

Researchers have theorized for many years that the dry climate caused African forests to turn into savanna. Our herbivorous, tree-dwelling forebears then abandoned both the remaining forest and their vegetarian habits and began hunting other animals out on the savanna, or so goes this widely accepted story. Out on the savanna, hominids evolved an erect posture, stone tools, large brains, and hairless skin. This story has been repeated so many times and with such assurance that most nonscientists are under the impression that it is an accepted scientific fact.

Unfortunately, it is just a story. Recent studies of the remains of plants in 2-million-year-old soils in East Africa suggest no dramatic change from forest to grassland between 2.5 and 2 million years ago. Researchers have concluded that hominids of the area most likely never encountered an open grassland. Other research, on the dust in ocean sediments, has shown only a single *dramatic* change in climate in Africa in the last 5 million years. This dry period came about 2.8 million years ago, more than a million years too late to account for the evolution of the first known bipedal hominid, 4.2 million years ago (*A. anamensis*). This arid period, therefore, could not have been responsible for the evolution of bipedalism.

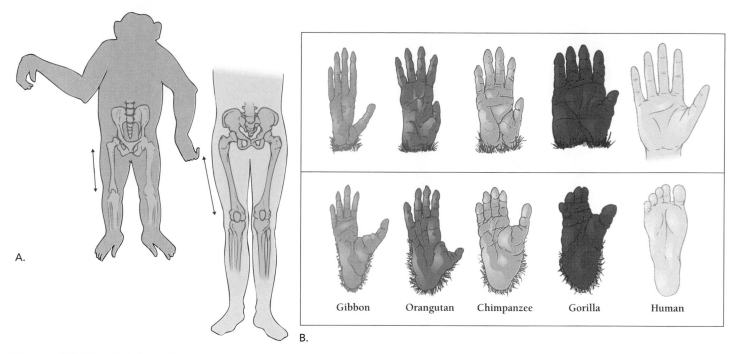

Figure 17-24 Tools for walking. A. Humans and other hominids have evolved the long lower limbs that characterize dogs, cheetahs, horses, and other animals that walk or run. Our legs bend in at the knee in a "knock-kneed" position. B. Even our feet are longer. Our big toe is no longer very good at grasping and all our toes are much reduced compared to those of other apes. In this respect, we resemble horses, which also have reduced digits.

Gibbon   Orangutan   Chimpanzee   Gorilla   Human

Later hominids did not necessarily live on the savanna either. The 3.2-million-year-old hominid *Australopithecus afarensis*, until recently one of the oldest hominids known, was found at two sites in East Africa. One site was a dry, open habitat, as anthropologists had expected, but the other site was a woodland divided by ancient rivers.

Anthropologists trying to save the "savanna hypothesis" argue that bipedalism evolved in a dry, open habitat. It does not matter, they say, where other, later hominids lived or what the climate was like worldwide. They are pinning their hopes for validation on the idea that *A. anamensis*, the oldest known bipedal hominid, lived in an arid, open habitat. Other researchers argue that huge rivers fed Lake Turkana, where *A. anamensis* lived. These rivers and the lake itself, these researchers assert, must have been rimmed with forests. Anthropologists are now vigorously debating whether the habitat around Lake Turkana was arid or wet 4.2 million years ago.

The hypothesis that the first hominids stood up as a result of moving from the dark forest to a bright, open savanna carries great appeal. So far, however, no one has provided much evidence for it. Other related theories are similarly weak. The idea that standing up and losing the fur were adaptations to the heat of the savanna has little support, as most other animals on the African savanna walk and run on four legs and retain their fur. The ostrich, which is bipedal and runs well, has a broad horizontal black back, well suited to absorbing heat. It seems unlikely that the ostrich's bipedal gait is an adaptation for staying cool. Excluding naked mole rats, which live underground, the few African mammals that lack fur are large herbivores—elephants, rhinoceroses, and hippopotamuses, for example.

Did the first hominids stand up in order to carry something? Possibly, but that something was emphatically not stone weapons. The earliest stone tools, which experts say were most likely used to chop and mash vegetables (not to kill zebras), do not appear in the fossil record until about 2.5 million years ago, long after our ancestors stood up. Absent stone tools, it's difficult to imagine our australopithecine ancestors fashioning any other tools or objects important enough to confer a selective advantage on an upright posture.

The only object an ape or australopithecine might have found worth carrying around all the time would be a baby. The babies of other monkeys and apes cling to their mother's fur, however. So researchers would have to hypothesize that hominids lost their fur before they evolved an upright posture, an idea for which there is no evidence either.

With luck, researchers may someday agree about what selective pressures caused hominids to evolve an upright stance. For the moment, we can only say with certainty that they did. From the earliest australopithecines to the later ones, the skeleton increasingly became specialized for walking and running while upright.

No one knows exactly why hominids evolved the back, hips, and legs of a bipedal walker and runner. Most anthropologists assume that our ancestors left a life in the trees to live in an open, probably grassy, habitat. Researchers assume that the first hominids had to walk long distances and that this new lifestyle created selective pressures for an upright stance. But we may never know exactly where our ancestors were walking or why.

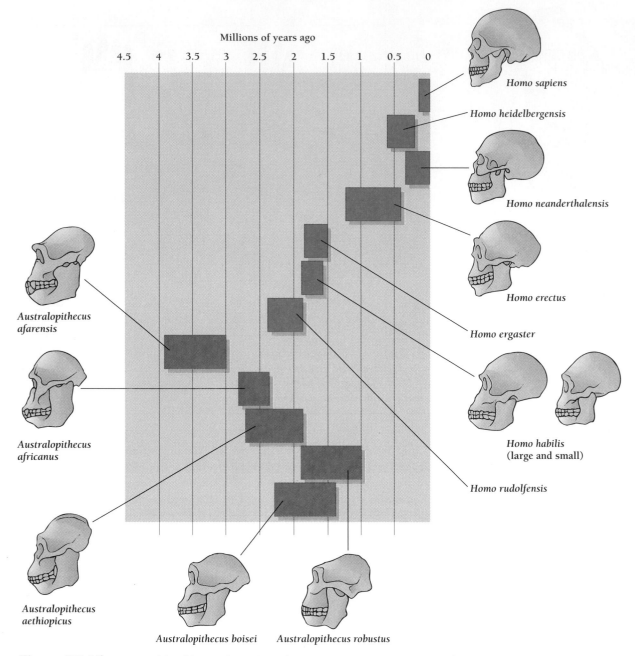

Figure 17-25 A timetable of hominids. Anthropologists may not necessarily know who descended from whom. But they do know which kinds of hominid species existed when. At least a dozen other hominids preceded us.

Who Were the Earliest Known Hominids?

The australopithecines fall into three to six distinct species, which lived at slightly different but overlapping times (Figure 17-25). All of them come from Africa, south of the Sahara desert. In fact, all hominids older than about 2 million years come from Africa, and most anthropologists agree that the first hominids evolved in Africa. The different species of australopithecines appear to have lived between 4.2 and 2.5 million years ago. For example, *Australopithecus afarensis* has turned up in Chad and in East Africa, where a nearly complete skeleton was unearthed (Figure 17-26).

This specimen of *A. afarensis,* named "Lucy" by her discoverer, was small, only about 1.1 meters (3 feet 7 inches) tall and weighing less than 30 kg (about 60 pounds). Lucy's adult body was the size of a 9-year-old *Homo sapiens,* but her brain was far smaller. Brain size for most normal adult humans ranges from 1000 to 1600 cubic centimeters (cm^3) (although the full range is 900 to 2000 cm^3). In contrast, the brain of *A. afarensis* was about 400 cm^3, about the size of a chimpanzee's brain (300 to 500 cm^3).

Other australopithecine species, taller and with larger brains, appeared during the next 2 million years. But the first fossils to be recognized as members of the same genus as our-

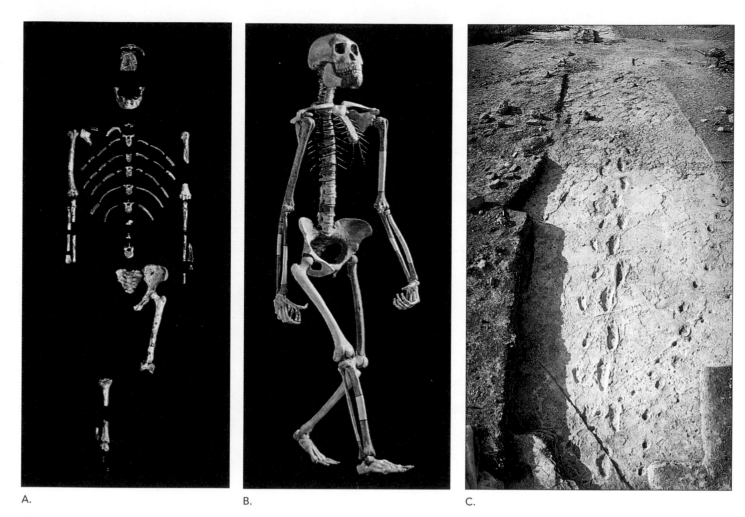

A.   B.   C.

Figure 17-26 *Australopithecus afarensis.* One of the earliest australopithecines,
A. afarensis is represented both in the nearly complete skeleton of "Lucy"—discovered in
Ethiopia in the 1970s—and in other, more fragmented, fossils from Tanzania. Evidence that
this australopithecine was bipedal comes not only from the pelvis, but from the extraordinary
footprints found by Mary Leakey and her colleagues. A. The pieces of Lucy. B. Reconstruc-
tion of Lucy. C. Footprints of *A. afarensis.*   *(A, National Museum of Ethiopia/Photo © 1985 David L.
Brill; B, Dr. Owen Lovejoy and students, Kent State University/Photo © 1985 David L. Brill; C, John
Reader/Science Photo Library/Photo Researchers)*

selves—*Homo*—date from 2 to 3 million years ago. These hominids all used crude stone tools, which is one measure that
sets them apart from the australopithecines. Members of the
genus *Homo* include *H. ergaster, H. habilis, H. erectus, H. heidelbergensis,* and *H. sapiens.*

The earliest fossils classified in the genus *Homo* have been
discovered in Ethiopia. Estimated dates for these fossils range
from 2.7 to 2.2 million years. No one is yet sure exactly how
old they are. *Homo ergaster* appears in different parts of Africa
from as early as 2.35 million years ago to as late as 1.4 million
years ago (Figure 17-27). Towards the end of their existence,
H. ergaster peoples used a more advanced method of chipping

stones to form tools. Two or three million years ago, they
chipped only one side of a stone to form a sharp surface for
cutting and chopping—probably plant material. After about a
million years, they learned to chip both sides to make a sharper
edge.

Homo habilis [Latin, *habilis* = able], the "handy man," used
stone tools and also apparently used them to butcher large animals, which *H. habilis* either hunted or scavenged (Figure 17-
25). *Homo habilis* stood about 1.7 meters (5 feet) tall and had
a 590- to 700-cm^3 brain, considerably larger than that of
A. afarensis. Living side by side with *H. habilis* in at least two
sites in East Africa was a hominid that is not included in the

Figure 17-27 *Homo ergaster.* The earliest complete skeleton of *Homo ergaster* is a boy from Turkana, Kenya. Not yet full grown, "Turkana boy" would have reached about 6 feet if he had lived. He had a high, domed cranium and light facial bones like those of *Homo sapiens.* *(Alan Walker © National Museums of Kenya)*

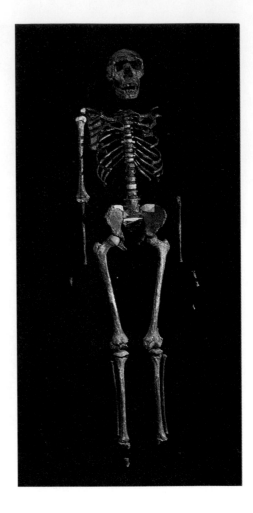

genus *Homo.* Researchers now believe that *Paranthropus*, which lived from 2 million until 1.5 million years ago, was another branch of the hominid family. Researchers dispute vigorously over which of the various australopithecines and hominids are our true ancestors (Figure 17-28).

The early hominids, or australopithecines, were short and small-brained compared to modern humans and other members of the genus *Homo.*

Out of Africa

Nearly 2 million years ago, *Homo* left Africa and migrated to Asia. Asian fossils of *Homo erectus* exist from Java in Eastern Asia to northern China. *Homo erectus* fossils have also been found in Africa, so it's reasonable to guess that the *H. erectus* in Asia came from Africa.

Homo erectus was by far the longest-lived species of *Homo.* The earliest fossils date from 1.8 million years ago and the last

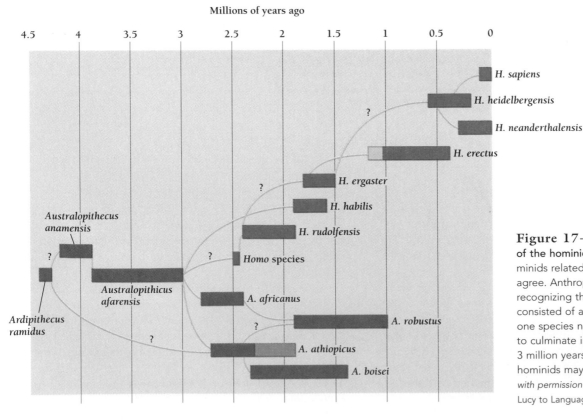

Figure 17-28 **Branching evolution of the hominids.** How are the various hominids related? Anthropologists do not agree. Anthropologists are increasingly recognizing that human evolution has not consisted of a single line of hominids, with one species neatly evolving into the next to culminate in *H. sapiens*. Between 2 and 3 million years ago, as many as 6 different hominids may have coexisted. *(Reprinted with permission from Johanson/Edgar/Brill: From Lucy to Language, Simon & Schuster Publishers)*

BOX 17-3

What is the nature of anthropological evidence?

The kinds of evidence that molecular biologists use when studying living organisms and their relationships to one another are different from the evidence that paleontologists and anthropologists must use. The essence of most scientific evidence is reproducibility, and studies of living organisms tend to yield reproducible results, provided the studies are done carefully. For example, if a molecular biologist finds that chimpanzees share a certain proportion of the same alleles with humans on Tuesday in Los Angeles, a different researcher, using the same material and methods, should be able to get the same result on Friday in Boston.

Anthropologists, however, have limited material with which to work, and much of their evidence is not reproducible. For example, the number of fossils of early apes and humans is minuscule. A species of ape may be represented by a single jaw bone or a few teeth.

One of the most complete skeletons of an early hominid is "Lucy," a 3.2-million-year-old specimen of *Australopithecus afarensis*. Researchers strongly suspect that Lucy was female. Based on that assumption, they have hypothesized what a male of the same species would look like. They have further assumed that the male would be larger by a certain percent than Lucy, and some textbooks report the height and weight of this theoretical male as if it were a fact. But other researchers have argued that there is no certainty that Lucy is female. Maybe "she" was a male. Further, without hundreds of complete skeletons of the same species, there is no way to know if Lucy was in any way typical of males *or* females. Lucy may have been unusually tall or unusually short.

Nonetheless, the scarcity of evidence in no way detracts from the fact that a long line of species has existed, and that those species are morphologically intermediate between modern apes and modern humans. We may never know either the true average dimensions of skulls and limbs of each species or whether males and females were mostly the same size or different sizes. Nor may we ever know with certainty how each ancient hominid species is related to any other. But we do know that we are closely related genetically to modern chimpanzees and other apes. And we do know that the 3.2 million-year-old fossil Lucy walked upright, just as we do. There may be only one Lucy, but Lucy lived.

---

date from just 100,000 years ago, for a total species life span of 1.7 million years. *Homo erectus* existed nearly 17 times as long as our own species has.

Homo erectus skeletons are larger than those of *H. habilis* and the brain case is also proportionately larger, ranging to more than 1000 cubic centimeters in the earliest specimens. In Java, where *H. erectus* lived until just 100,000 years ago, the brain cases are as large as 1300 cm^3, well within the limits of our own brain size (1000 to 1600 cm^3).

Some anthropologists and paleontologists have suggested that *H. erectus* migrated from Africa, starting about 1.5 million years ago, with subsequent waves of migration evolving separately into varieties, or races, that differ from one another. Some of these, they argue, evolved into *H. sapiens*.

Others argue that *H. erectus* living in different parts of the world were replaced first by *Homo heidelbergensis* and later by *H. sapiens,* each of which migrated separately from Africa. In this view, *H. erectus* was a side branch in the evolution of humans, and we are descended from another line of *Homo* that left Africa after *H. erectus*. The most likely candidate seems to be *H. heidelbergensis,* found in Ethiopia and dated at about 130,000 years old. Several skulls that seem transitional between *H. heidelbergensis* and *H. sapiens* have been found in Ethiopia.

Alternatively, *H. sapiens* may have evolved in Africa 200,000 to 100,000 years ago, separately and at the same time as *H. heidelbergensis,* then emigrated to Asia and Europe. The earliest known representatives of *H. sapiens* were found at two sites in Israel and are dated at about 90,000 years old.

The early *H. sapiens* are commonly divided into two kinds, the **Cro-Magnons** and the **Neanderthals.** The Neanderthals, who looked much like us, but with a heavier build, occupied parts of Europe until just 35,000 years ago. Their thick arms and legs and heavy brows provide cartoonists with the classic "caveman" image; and, because of differences in Neanderthal throat bones, some researchers doubt that they spoke as well as we do, if at all. Nonetheless, Neanderthals, who first evolved between 200,000 and 100,000 years ago, probably had a fair amount of mental horsepower. Their brains were as large or larger than those of the Cro-Magnons, and some Neanderthal may have played musical instruments (Figure 17-29).

Researchers do not know why the Neanderthals became extinct or if they interbred with Cro-Magnons. At one site, in Israel, the Cro-Magnons and Neanderthals appear to have lived within a few miles of each other at the same time. If we knew that they interbred, we could apply the biological species concept and conclude that they were one species. But no one knows, and researchers do not agree whether the Neanderthals were a separate species (*H. neanderthalensis*), a subspecies, or just a now-extinct race of *H. sapiens*. Recent analysis of DNA extracted from a Neanderthal bone suggests that they were genetically distinct from modern humans. But we still do not know if they were equally different from our Cro-Magnon forebears.

Even if we ultimately conclude that the Neanderthals were *H. sapiens,* we probably had other *Homo* cousins in the early days of our existence (Figure 17-25). Both *H. heidelbergensis* and *H. erectus* lived up to 100,000 years ago.

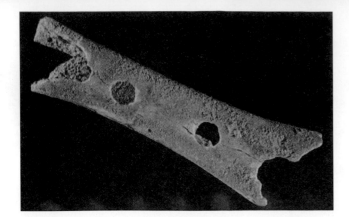

Figure 17-29 Neanderthal flute? In 1996, a team of researchers discovered what appeared to be a bone flute in a 43,000-year-old Neanderthal cave in Slovenia, in Eastern Europe. Apparently fashioned from the thigh bone of a young bear, the hollow bone has four holes spaced as if for a diatonic scale (do, re, mi...). Similar flutes have been found in newer, Cro-Magnon caves, and researchers around the world reacted with delight to the idea that Neanderthals might have played music. Some researchers, however, flatly reject the idea. They argue that the holes were made by wolves' canine teeth and that the spacing is purely accidental. *(Blackwell, Queens College)*

Anthropologists do not agree whether *Homo sapiens* evolved in Africa. In the last 2 million years, several species of *Homo* appear to have been living on Earth at the same time.

In the first two chapters of this section, we reviewed the evidence for evolution and studied the ways that changes in gene frequency allow evolution to occur. In this chapter we discussed how species change and multiply and examined some of the larger patterns of evolution that are evident in the fossil record. In the next chapter, the last in this section, we will explore a topic that is related, but in many ways quite different—the question of how and where life first originated.

The evolution of life is, like gravity, a simple fact. The origin of life, however, is truly mysterious. We know that life appeared on Earth within a few million years of the cooling of the Earth's crust, but we can only surmise how it got there.

STUDY OUTLINE WITH KEY TERMS

Macroevolution is the origin and multiplication of species. According to the commonly accepted **Biological Species Concept,** a **biological species** is a group of organisms that actually or potentially interbreed to produce viable offspring, but cannot interbreed with members of other such populations. Populations that do not interbreed with other populations are said to be **reproductively isolated.** **Sibling species** are species that are extremely similar, yet reproductively isolated. **Races** or **subspecies** are subpopulations of a species that are distinct, yet not reproductively isolated. In other words, races can interbreed with other races of the same species. Subspecies that interbreed only rarely are sometimes called **incipient species.**

Populations become reproductively isolated from one another through two kinds of barriers to gene flow: **prezygotic barriers,** which prevent fertilization, and **postzygotic barriers,** which either make the zygote inviable or make the resulting adult sterile. Prezygotic barriers include **ecological, temporal, behavioral, mechanical,** and **gametic isolation.** Once fertilization occurs, postzygotic barriers can come into play. These include chromosomal and genetic incompatibilities that cause the hybrid zygote to die young (**hybrid inviability**); or, if the individual reaches maturity, to produce no gametes (**hybrid sterility**); or, in some cases, if the hybrid produces offspring, to have *its* offspring be weak or sterile (**hybrid breakdown**).

Two other postzygotic barriers, peculiar mostly to plants, reproductively isolate hybrid offspring from the parent population in just one generation. These include the multiplication of chromosomes, or **polyploidy;** and the combining and doubling of nonhomologous chromosomes from different species, or **allopolyploidy.**

Geography frequently isolates populations from one another well before other isolating mechanisms develop. Populations can become isolated through **dispersal** and through **vicariance.** When reproductively isolated populations diverge in morphology or behavior, they are said to **speciate.** Geographic speciation follows several patterns. In **allopatric speciation,** geographic isolation halts nearly all gene flow. Genetic drift or different selective pressures complete the speciation process. In **parapatric speciation,** adjacent populations with limited gene flow manage to speciate. In **sympatric speciation,** populations that occupy the same geographic area become reproduc-

tively isolated through barriers to gene flow—temporal, ecological, or other.

Divergent evolution is the separation of a lineage into two or more species or lineages. **Convergent evolution** is a separate idea, referring to the acquisition of similar characters in unrelated lineages. An important kind of divergent evolution is **adaptive radiation,** the generation of diverse new species from a single ancestral species. Because no new body plans evolved after the **Cambrian explosion,** some biologists have suggested that **developmental constraints** limit evolutionary pathways.

Despite the gaps in the fossil record, Darwin believed in **gradualism,** the view that evolution was a gradual process. Early biologists rejected Darwin's gradualism in favor of **saltationism,** the view that evolution occurred in sudden leaps, or saltations. The last well-known saltationist suggested that genetic monstrosities, which he called **hopeful monsters,** began new evolutionary lineages.

Although gradualism gained favor from the 1930s through the 1960s, the fossil record seemed to show that many species remained unchanged for millions of years (**stasis**) and then abruptly gave rise to new species. A full record of intermediate stages was usually lacking. Thus, evolution, according to the fossil record, was neither gradual nor continuous. It seemed to happen in fits and starts.

In the 1970s, two paleontologists, Niles Eldredge and Stephen J. Gould, argued that the formation of new species is rapid compared to the rate of accumulation of changes within a species. This view, called **punctuated equilibrium,** rests on two main ideas. The first is that species change very little most of the time. The second is that most anatomical or other evolutionary change in individual species occurs during a geologically brief period at the time of speciation.

Anatomical and molecular evidence suggests that humans are **primates** most closely related to chimpanzees, gorillas, and other pongids. The **Pongidae** are a family of **hominoids,** which are the ape members of the catarrhines. The **catarrhines,** which include Old World monkeys and apes, are a more recently evolved group of primates than the **platyrrhines,** which are tailed monkeys that live in South America and Central America. Humans and their ancestors all belong in the family **Hominidae.**

Hominids are **bipedal** and have rounded jaws. **Physical anthropologists** call the earliest hominids **australopithecines.** Later hominids also have large brains, arched backs, and reduced toes. Two genera of hominids succeed the australopithecines: *Paranthropus* and *Homo.* All species of *Paranthropus* and *Australopithecus* are now extinct, and only one species of *Homo* remains, *H. sapiens.* No one knows exactly why hominids evolved a bipedal gait. Most anthropologists assume that the first hominids left a life in the trees to live in an open, probably grassy, habitat where they had to walk long distances.

Anthropologists do not agree whether *H. sapiens* evolved in Africa, as did all or most other species of *Homo,* or if *H. sapiens* evolved from an intermediate ancestor, which evolved in Africa but then dispersed to Europe and Asia. In the last 2 million years, several species of *Homo* appear to have been living on Earth at the same time. The longest lived species was *H. erectus,* which lived in parts of Africa and Asia for nearly 2 million years and became extinct only about 100,000 years ago. Early *H. sapiens* are divided into the heavy-boned **Neanderthals** and the more delicate **Cro-Magnons.** Modern humans more closely resemble the Cro-Magnons.

REVIEW AND THOUGHT QUESTIONS

Review Questions

1. Define the word "species," then explain why that definition is tricky. What are sibling species? What is a subspecies? Why is a subspecies not a separate species?
2. What conditions are required for speciation?
3. Distinguish between prezygotic and postzygotic barriers to gene flow. Give several examples of each. Is polyploidy a prezygotic barrier or a postzygotic barrier?
4. Distinguish among sympatric, allopatric, and parapatric speciation.
5. Distinguish between phyletic evolution and cladogenesis.
6. What clues from the Galápagos Islands suggested to Darwin the idea of adaptive radiation?
7. Describe the Cambrian explosion. What happened 600 to 900 million years ago that has not happened since? Why hasn't it happened again?
8. Name a postzygotic isolating barrier in plants that is consistent with the saltationist view of evolution.
9. Define gradualism and punctuated equilibrium. What characteristics of the fossil record favor the theory of punctuated equilibrium? Explain why these two ideas are or are not in conflict with each other.
10. What are some weaknesses in the argument for stasis?

Thought Questions

11. Consider an oceanic archipelago consisting of only 10 islands. Describe a pattern of colonizations whereby 20 or more species could arise from a colonization by a single mainland species.
12. With Goldschmidt's "hopeful monsters" in mind, make up a science-fiction scenario in which a human baby born with two hearts would have some selective advantage.
13. Do you feel that there is a conflict between what you know about the evolution of life on Earth and religious belief? Why or why not?
14. The current mass extinction will likely result in an adaptive radiation over several million years. What organisms do you think will survive and what niches will they evolve to fill 5 million years from now?

SELECTED READINGS

Dennett, Daniel C., *Darwin's Dangerous Idea: The Evolution and the Meanings of Life,* Simon & Schuster, New York, 1995. With the enthusiasm of a convert, Dennett, a philosopher at Tufts University, explores the implications of evolution.

Fortey, Richard, *Fossils: The Key to the Past,* Harvard University Press, Cambridge, Massachusetts, 1991. A gentle introduction to paleontology, with fascinating photos.

Gould, Stephen J., *Wonderful Life: The Burgess Shale and the Nature of History,* W.W. Norton and Company, New York, 1989. Gould at his most engaged, discussing what he knows best—fossils and scientists.

Stumpke, Harald, *The Snouters: Form and Life of the Rhinogrades,* The Natural History Press, New York, 1967. A tongue-in-cheek natural history of some imaginary creatures that says a lot about taxonomy.

Weiner, Jonathan, *The Beak of the Finch,* Vintage Books, New York, 1994. Weiner chronicles the research of Peter and Rosemary Grant, who have demonstrated beyond doubt that Darwin's finches continue to evolve from month to month and from year to year. This enjoyable book, which won a Pulitzer Prize and The Los Angeles Times Book Prize, is highly worthwhile.

Wilson, Edward O., *The Diversity of Life,* Harvard University Press, Cambridge, Massachusetts, 1992. In this beautifully illustrated book, Wilson introduces the great diversity of species and explains how we know that a mass extinction is occurring.

▶On-line materials relating to this chapter are on the World Wide Web at http://www.saunderscollege.com/lifesci/
Click on Tobin/Dusheck: *Asking About Life.*

Can Organisms Arise Spontaneously from Nonliving Matter?

Until the last few hundred years, most people believed that new life arose spontaneously and daily from nonliving materials. Anyone could see that flies and maggots arose from rotting meat, frogs from swamps, mice from moist grain. Nearly everyone believed that spontaneous generation of life was common among the simple—so-called "lower"—plants and animals. Some people even argued that, under carefully controlled conditions, it might be possible to generate a man from a corpse.

Of course, every farmer knew that wheat and many other plants came from seeds, and that the cows and sheep were born of their mothers. Here and there, a few skeptics also insisted that insects and other creatures come from the eggs or sperm of parent organisms. Not until 1668, however, was this idea put to the test.

The skeptic was Francesco Redi (1626–1697), an Italian physician, poet, and naturalist, and one of the first modern experimental biologists. Heavily influenced by Galileo's belief that the natural world could be understood best through the senses, Redi strove to perform controlled experiments to test his ideas. Among these was a series of experiments challenging the idea of spontaneous generation.

In his own account of these experiments Redi first states his hypotheses:

> I shall express my belief that the Earth, after having brought forth the first plants and animals at the beginning by order of the . . . Creator, has never since produced any kinds of plants or animals . . . and everything which we know in past or present times that she has produced came solely from the true seeds of the plants and animals themselves, which thus . . . preserve their species. And, although it be a matter of daily observation that infinite numbers of worms are produced in dead bodies and decayed plants, I feel . . . inclined to believe that . . . the putrefied matter in which they are found has no other office than that of serving as a place . . . where animals deposit their eggs at the breeding season, and in which they also find nourishment; otherwise, I assert that nothing is ever generated therein.

NASA

Redi then describes how, after preliminary studies of the development of different kinds of maggots into various species of flies, he put several kinds of fresh meat into two sets of jars. One set of jars was open, the other closed. Those that were open all developed maggots and, later, flies. Those that were closed developed no maggots, though, Redi notes, "Outside on the paper cover there was now and then a deposit, or a maggot that eagerly sought some crevice by which to enter and obtain nourishment."

Finally, Redi wanted to prove that the absence of air was not keeping the maggots from developing on the meat. He placed fresh meat into large vases that were either open or covered with very fine mesh. To prevent maggots from crawling through the holes in the netting, he put the covered vases inside a large net-covered frame and carefully removed any maggots that managed to penetrate the outer netting (Figure 18-1). The results were the same. "I never saw any worms in the meat, though many were to be seen moving about on the net-covered frame." The meat in the open vases was, of course, riddled with maggots, which, Redi observed, eventually turned into various

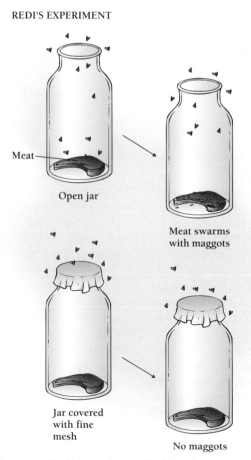

REDI'S EXPERIMENT

Meat

Open jar

Meat swarms with maggots

Jar covered with fine mesh

No maggots

Figure 18-1 Francesco Redi's experiment. Do flies arise spontaneously from rotten meat? Or do they develop from eggs laid on the meat by adult flies?

Figure 18-2 Lazzaro Spallanzani's experiment. Do microorganisms arise spontaneously in a rich broth?

SPALLANZANI'S EXPERIMENT

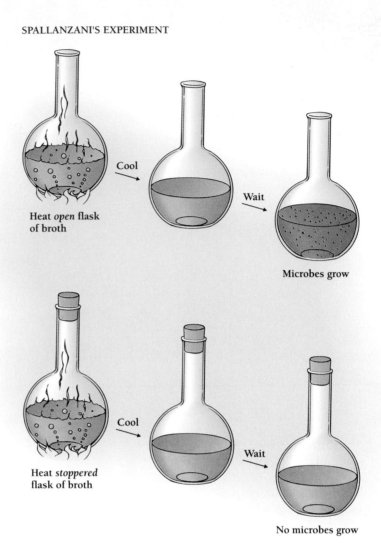

Heat *open* flask of broth

Cool

Wait

Microbes grow

Heat *stoppered* flask of broth

Cool

Wait

No microbes grow

kinds of flies. Redi showed that maggots did not appear spontaneously.

Redi's 1668 experiments convinced most scientists that the common conception of spontaneous generation was incorrect. However, within 10 years, van Leeuwenhoek used his new microscope to find little "animalcules" (microorganisms) everywhere. Most people felt that even if maggots and flies could not arise by spontaneous generation, very likely these microorganisms could.

It was not until a hundred years later that anyone showed that they could not. In the 18th century, Lazzaro Spallanzani (1729–1899), an Italian physicist, microscopist, and physiologist, demonstrated that no organisms grew in a rich broth if it were first heated and allowed to cool in a stoppered flask (Figure 18-2). Other ex-

perimenters less careful than Spallanzani contaminated their broth and so obtained the opposite result, however. Moreover, some scientists argued that, by heating the broth and air, Spallanzani may have destroyed some substance needed for spontaneous generation.

For another hundred years, these objections stood. Finally, in 1860, the controversy over spontaneous generation became so intense that the prestigious French Academy of Sciences offered a prize to anyone who could resolve the matter. In 1864, Louis Pasteur, the founder of microbiology, designed a simple but irrefutable experiment that won the prize.

Pasteur repeated Spallanzani's experiment, but with one essential difference. Instead of stoppering the flask to prevent

organisms or spores from falling into the broth, Pasteur used a flask with a long neck like that of a swan (Figure 18-3). Air could pass freely in and out of such a flask, but microbes could not get through the long, curved neck. Only when Pasteur broke the neck of the flask was the broth "seeded" by bacteria and fungi falling from the air. Some of Pasteur's open flasks—still sterile after more than 130 years—are on display in Paris.

"All life from life," Pasteur emphatically concluded. "Never," he said, "will the doctrine of spontaneous generation recover from the mortal blow of this simple experiment."

Organisms do not arise spontaneously from nonliving matter.

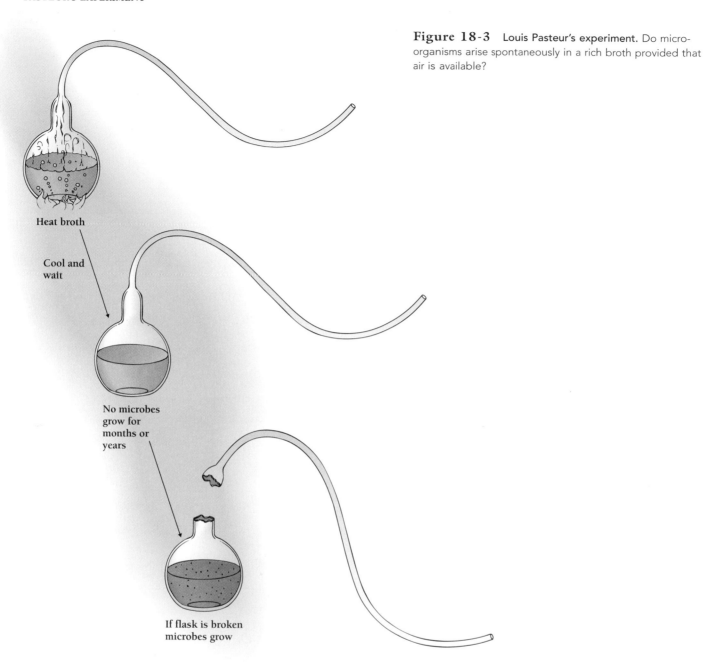

Figure 18-3 Louis Pasteur's experiment. Do microorganisms arise spontaneously in a rich broth provided that air is available?

Heat broth

Cool and wait

No microbes grow for months or years

If flask is broken microbes grow

KEY CONCEPTS

1. Life appeared just a few hundred million years after the Earth's crust hardened.

2. Scientists can construct a reasonable scenario for the evolution of the first biological molecules.

3. Scientists can construct a reasonable scenario for the generation of abiotic cell-like structures.

4. The evolution of the first cells required a means of coupling protein synthesis to genetic instructions.

5. Eukaryotic cells probably arose through a symbiotic association of prokaryotic cells.

WHERE DOES LIFE COME FROM?

If Spontaneous Generation Is Impossible, How Did Life Originate?

Darwin and biologists since then have shown that all modern species have evolved from previous ones. But we must ask, From what did the millions of species on Earth evolve? If Pasteur was correct in saying that all life comes from other life, we must account for the presence of evolved life on a planet that itself came into being a finite number of years ago. Where did life first come from?

Every culture on Earth—ancient, primitive, or modern—has an answer to this question. Many invoke an intelligent, creative force. However appealing many of these myths may be, they do not constitute scientific explanations, and they tell us little more than that "life happened."

Scientists cannot say with certainty how life arose on Earth, either. However, they can speculate, based on a modern understanding of astronomy, planetary science, geophysics, and biology, what might reasonably have happened. Scientists see essentially two alternative answers to the question, Where did life on Earth come from? Either life on Earth arose spontaneously from nonliving organic molecules in the primitive environment or else it arrived from outer space by means of a meteor, comet, or asteroid.

In this chapter, we will see why neither of these ideas is quite as outlandish as it might at first seem. The bulk of our discussion will be devoted to exploring how life might have originated on Earth.

Evolutionary biologists have inferred part of the story from the shared characteristics of modern organisms. But much of the story must be guesswork. Because the origin of life is not an ongoing, testable process, as evolution is, biologists can only surmise how life might have arisen. The point is to show that life could have arisen by some series of steps, each of which is reasonable in terms of our current understanding of the laws of physics, chemistry, and biology. By showing that they can tell such a reasonable story, evolutionary biologists have attempted to refute the idea that there is an uncrossable gulf between living and nonliving matter. We will also see why, despite all this, Pasteur was after all correct: spontaneous generation could not happen today.

When and Where Did Life First Appear?

Life Appeared Soon After the Earth's Origin

Before 1950 no one had seen a fossil older than about 570 million years (the beginning of the Cambrian Era). Older fossils went unnoticed, it turns out, because they are extremely rare, usually quite small, and found in a kind of rock different from that of later fossils.

Cambrian and later fossils are usually hard bones and shells preserved in shales, sandstones, limestones, and other sedi-

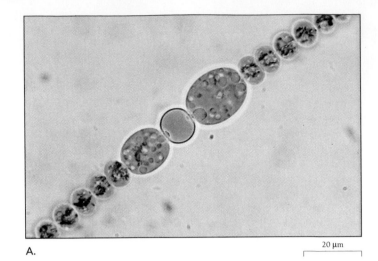

A.   20 µm

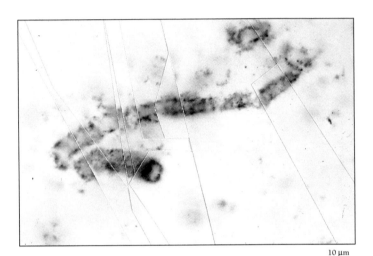

10 µm

B.

Figure 18-4 Filamentous cyanobacteria. Some filamentous fossils closely resemble their modern counterparts. A. Modern filamentous cyanobacterium. B. 3.5-billion-year-old filamentous cyanobacterium from the Warrawoona rocks of western Australia. *(A, Biological Photo Service; B, Stanley Awramik/Biological Photo Service)*

mentary rocks. But most Precambrian organisms had few or no hard parts, and their soft remains usually dissolved or decayed. Their soft, tiny bodies persisted only when caught in an extremely fine-grained rock called chert. Such rocks are now generally far below the Earth's surface, covered by later deposits. Even cherts that have returned to the surface may be hopelessly deformed by heat and pressure. In only a few places on

Figure 18-5 Living stromatolites like these once filled vast shallow seas. These cushions of cyanobacteria at Shark Bay, western Australia, resemble 3.5-billion-year-old fossil stromatolites. *(John Reader/Science Photo Library/Photo Researchers)*

Earth are the cherts that bear Precambrian fossils both accessible and intact.

But for scientists interested in Precambrian life, finding fossil-bearing cherts was only half the battle. In order to see the fossils, paleontologists had to slice the chert into thin sections, polish them until they were transparent, and then examine the sections under a microscope.

The oldest and most primitive fossils known resemble modern prokaryotes—simple rods, spheres, and filaments with no nucleus (Figure 18-4). These ancient fossils, embedded in cherts from southern Africa and western Australia, are about 3.5 billion years old. More recently, researchers have dated fossil organisms that are even older, some 3.8 billion years.

Another important group of ancient fossils are the stromatolites, banded columns of limestone or chert, as old as 3 billion years (Figure 18-5). For many years, stromatolites were not positively identified as the remains of living organisms. This is because the organisms that make them are not preserved, only an unusually orderly arrangement of sediments. But living stromatolites, discovered in a few remote intertidal areas, solved the mystery. Stromatolites are formed when huge, mounded colonies of cyanobacteria and other microorganisms trap fine sediment. The sediment settles in thin layers within each mound. The cyanobacteria in modern stromatolites are photosynthetic. Paleontologists assume that fossil stromatolites were also photosynthetic. In other words, photosynthesis probably evolved at least 3 billion years ago (Figure 18-6). Later in this chapter, we will see how the photosynthetic abilities of early cyanobacteria profoundly changed the world, allowing the evolution of all higher organisms.

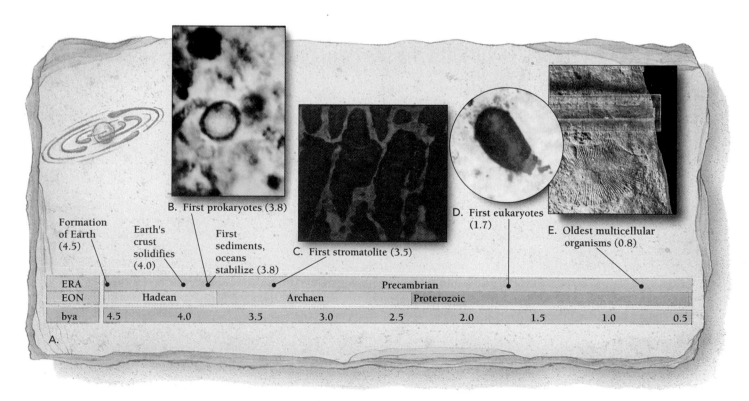

Figure 18-6 A calendar of Earth's early history. Life appeared on Earth almost as soon as conditions would allow it. The first stromatolites (photosynthetic cyanobacteria) appeared 3.5 billion years ago. Eukaryotes appeared nearly 2 billion years after the first prokaryotes. *(B, Science VU-USM/Visuals Unlimited; C, William E. Ferguson; D, Andrew H. Knoll, Knoll and Calder, Palaeontology 26:467, 1983; E, Biological Photo Service)*

BOX 18-1

Is there life elsewhere in the universe?

In 1959, a young astronomer assigned to the Greenbank Observatory in rural West Virginia began using the big observatory radio telescope to listen for signals from extraterrestrial beings. Frank Drake and his colleagues tried to keep it a secret, but word got out quickly. Newspaper reporters besieged the observatory with calls, but Drake could not honestly report any definite communications from outer space.

Yet, for nearly 40 years, Drake hasn't stopped listening. In 1974, he used Cornell University's Arecibo radio telescope, a giant, 300-meter telescope in Puerto Rico, to send the first deliberate message from Earth. The message described Earth's position in the Milky Way galaxy, what humans look like, and how many of us there were.

Today, Drake and others are still listening for signs of intelligent life in the universe. This project, called the Search for Extraterrestrial Intelligence (SETI) asks, Is anyone out there? (SETI's work is depicted melodramatically in the recent Hollywood film "Contact.") Biologists tend to be conservative about such things. Many believe that the emergence of life on Earth is so remarkable, that it would be astonishing if it occurred elsewhere in the universe. Indeed, the biologist and writer Jared Diamond has argued that the development of the radio on Earth was extremely unlikely. (He adds that, given the way humans treat less advanced civilizations and less intelligent creatures on our own planet, we will be fortunate indeed if we are alone in the universe.) Biologists are also eager to distance themselves from people who claim encounters with intelligent extraterrestrials on Earth.

Astronomers tend to take a different view. They argue that to insist that life could have arisen only on Earth is arrogant and excessively geocentric, or Earth-centered. Of course, they dismiss flying saucers—so-called UFOs (Unidentified Flying Objects).

Even the closest stars are too far away for spaceships to get here, let alone home again. And if intelligent life exists, the chances are that it is far away, indeed. Any signals that SETI picks up will be thousands or millions of years old.

But we are almost certainly not alone. Drake has suggested that the number of civilizations in the galaxy can be calculated using the following equation, called the Drake Equation:

$$N = R f_p \, n_e \, f_l f_i f_c f_L$$

In English, this equation says, the number of technological (detectable) civilizations in the galaxy (N) is equal to the rate (R) of star formation, times the fraction of stars with planets (f_p), times the number of planets ecologically suitable for life (n_e), times the fraction of those planets where life actually evolves (f_l), times the fraction that develops intelligent life (f_i), times the fraction that communicates (f_c), times the fraction of the planets' life occupied by the communicating civilization (f_L).

Clearly, the answer depends on what values are plugged into this equation. However, many of these values can be estimated with some confidence. For example, astronomers estimate that the Milky Way contains about 400 billion stars. Conservative estimates for most of the rest of the variables give predictions such as 4 billion planets with some form of life and 4000 existing technological civilizations in the Milky Way. And that is in our galaxy only. From astronomers' perspective, the evolution of life was practically inevitable.

But if extraterrestrial life is out there, we haven't found it yet. Soil samples from other planets have revealed no life, and SETI has so far turned up nothing concrete. The only tantalizing hint so far are some complex organic compounds from a meteorite believed to have come from Mars.

Figure A **Are we alone?** Probably not. Frank Drake, president of the SETI Institute and emeritus professor of astronomy at the University of California, Santa Cruz, first began searching for intelligent extraterrestrials in 1959. *(Bill Lovejoy)*

With luck, NASA's Pathfinder mission will supply evidence supporting this discovery.

Organic molecules seem to abound in the universe. Astronomers have detected the presence of a number of carbon- and nitrogen-containing compounds in interstellar gas clouds. These compounds include those that we think were present in the Earth's primitive atmosphere—hydrogen, hydrogen cyanide, carbon monoxide, and ammonia.

Soil samples brought back from the moon contained several amino acids. Meteors, such as the one that fell in Murchison, Australia, in 1969 (Figure 18-10), contain the amino acids found in organisms as well as other amino acids and more complex molecules. The fragments from the Murchison meteorite contained 1 or 2 percent organic material. If such an endowment is typical of comets, asteroids, and meteors in the rest of the galaxy, it is hard to imagine how life could not have evolved elsewhere.

Where Did Life First Come From?

As we have noted, there are two general explanations for the relatively sudden appearance of prokaryotic organisms soon after the Earth had cooled: (1) life arose spontaneously from nonliving organic molecules in the primitive environment; or (2) life arrived from elsewhere in the universe. It is almost impossible to imagine how we could prove or disprove either of these explanations. We can, however, examine the reasons for thinking either of them reasonable.

The second hypothesis at first appears to be the simpler explanation for the appearance of life. According to this view, the present diversity of life all derives from a colonization of

Earth followed by a huge adaptive radiation. An appealing feature of this hypothesis is that the universe is thought to have originated at least 12 to 16 billion years ago, 8 to 12 billion years before the cooling of the Earth. This far longer history would give life more time to develop. Furthermore, an extraterrestrial origin of life explains some otherwise unexpected chemical properties of organisms. It would explain, for example, the requirement for molybdenum—an extremely rare element on this planet—in key reactions of photosynthesis and nitrogen fixation.

Until recently, no evidence existed for life on the other planets in the solar system. In August of 1996, however, researchers discovered signs of life in a meteorite composed of 3.6-billion-year-old Martian rock. The rock, which crashed into Antarctica 13,000 years ago, contains what look like tiny fossils and also complex organic molecules that, on Earth, are produced when organisms decay or burn. These molecules are most concentrated in the interior of the rock, suggesting to some researchers that they did not infiltrate the rock from the outside after it landed on Earth.

It is not at all clear if these fossils represent life from Mars. Even if Mars did once harbor living organisms, there is no evidence that life on Earth was seeded by Martian organisms. Earth may have seeded Mars. Or life may have arisen independently on each planet. Another possibility is that life arrived on both planets from some source outside the solar system.

For life to colonize Earth (and Mars), it would have to have come from at least as far away as the nearest star, some 4 light years away—38 trillion km (23 trillion miles). Living organisms would have had to survive extremely violent events, for nothing else would have propelled them into interstellar space. There they would have to survive intense radiation for untold millions of years. Recent research has suggested, however, that bacteria might survive the intense radiation if they were encased in ice.

If such a trip were possible, we would still want to understand how (and where) those organisms first arose. The problem of the origin of life would be virtually insurmountable, since we would not know where in the universe such organisms might have come from.

For now, the colonization-from-space hypothesis is only remotely likely. Perhaps more important, the hypothesis is so inaccessible to inquiry that most scientists interested in such things have focused their attention on the possible origin of life on Earth. Most biologists now believe that life must have developed from nonliving molecules on the early Earth's surface. They have accepted the challenge of showing that the development of organisms from nonliving matter—called **prebiotic evolution**—is possible.

Most biologists believe that life originated on Earth. No one knows enough about conditions in the rest of the universe to guess how extraterrestrial life might have evolved or safely arrived on Earth.

PREBIOTIC EVOLUTION: HOW COULD COMPLEX MOLECULES EVOLVE?

By the 1930s biochemists realized that all organisms are made from the same building blocks—the sugars, lipids, amino acids, and nucleotides that make up the "biochemical alphabets" described in Chapter 3. Two influential scientists—Alexander Ivanovich Oparin, a Russian biochemist, and J.B.S. Haldane, the English evolutionary biologist—proposed that these building blocks had been formed in the primitive environment. The original synthesis of these materials, they argued, was **prebiotic** and therefore **abiotic**—before life and therefore without life.

Important differences separate spontaneous generation from theories of prebiotic evolution. First, while spontaneous generation had to occur nearly instantly, the origin of life may have taken some 300 million years to unfold. Second, the conditions during which life must have first appeared were dramatically different from those that exist today. One of the most important differences was that the atmosphere contained virtually no free oxygen. Had oxygen been present in any quantity, it would certainly have oxidized the starting materials for life. A second important difference was the absence of life itself. Today, life could not arise on Earth because a plethora of hungry organisms find and consume organic molecules.

Darwin himself grasped this distinction, writing to a friend in 1871:

> It has often been said that all of the conditions for the first production of a living organism are now present which could ever have been present. But if (and oh! what a big if!) we could conceive in some warm little pond, with all sorts of ammonia and phosphoric salts, light, heat, electricity, etc., present, that a protein compound was chemically formed ready to undergo still more complex changes, at the present day such matter would be instantly devoured or absorbed, which would not have been the case before living creatures were formed.

Was Darwin right? In the absence of devouring organisms and destructive oxygen, can the building blocks of life manage to assemble themselves into living organisms? To decide we must ask whether we can imagine a reasonable scenario for the "spontaneous" production of the building blocks of life and whether laboratory experiments demonstrate that the steps in the presumed process can actually occur. To discuss these questions, however, we must first understand the conditions on Earth shortly after it formed.

The conditions available for the origin of life differ from those for spontaneous generation in three dramatic ways—the long period of time available, the absence of life, and the absence of oxygen.

What Was the Earth Like When It Was Young?

The Formation of the Solar System

The Earth formed about 4.6 billion years ago, when a spinning cloud of rocks, dust, and gas formed the sun and its planets. Astronomers believe that the spinning cloud gradually spread out and flattened as it rotated, like a lump of spinning pizza dough (Figure 18-7). The bulk of the material, at the center, condensed into an enormous compact mass that became the sun. The rest of the spinning cloud gradually formed the planets, comets, meteors, and other objects that make up the solar system.

The debris orbiting closest to the protosun [Greek, *protos* = primitive] moved fastest, and there the friction of constant collisions generated the most heat. Close to this hot center, pieces of silicon, metals, metal oxides, and other materials that remain solid even at high temperatures fell together, drawn to one another by their own gravities. These hot lumps of solid material gradually formed the inner protoplanets—Mercury, Venus, Earth, and Mars. Water, methane, and other compounds that solidify only at low temperatures formed the cold, outer protoplanets, such as Jupiter, far from the center of the spinning disk. The huge, outer planets attracted and held cold atmospheres of hydrogen, helium, and other gases. But the tiny, inner planets had gravities too weak to hold the hot gases, which escaped into space.

The early protosun was much cooler than today's sun. But Earth was far hotter. Gravity compacted the Earth, and, for a billion years, huge meteors, many of them hundreds of miles in diameter, bombarded the planet. The compaction and the bombardment kept the molten Earth from solidifying for another half a billion years. The heaviest elements sank to the center of the Earth, while the lightest ones floated to the top. Iron and nickel sank to the core, the lightest silicates (granite, for example) rose to the surface, while denser silicates of iron and magnesium floated in between.

After 500 to 600 million years, the surface cooled enough to form a thin crust. Dense gases escaped from the interior through cracks and volcanoes in the new crust to form a primitive atmosphere. Among these gases was water vapor, immense quantities of which then condensed into clouds. Torrential rains pummeled the still hot Earth for millions of years, creating the oceans and eroding the Earth's rocky surface. Although the oceans formed relatively early, constant bombardment by huge meteors, some hundreds of miles in diameter, may have repeatedly vaporized the oceans until about 3.8 billion years ago. About that time, the first sediments and the first life appeared.

The Evolution of the Atmosphere

The exact composition of the gases in the primitive atmosphere has been a matter of much debate. The modern atmosphere is 78 percent nitrogen, 21 percent oxygen, and about 1 percent inert argon gas, with varying amounts of water vapor close to the Earth. Another 0.1 percent of the atmosphere consists of

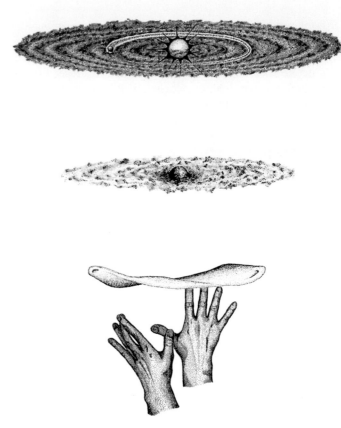

Figure 18-7   **Formation of the solar system.** Like a spinning lump of pizza dough, a cloud of dust and rocks flattened, and then condensed into a central mass that became the sun. Material further out condensed into rotating balls of rock and gas that we now know as the planets.   *(Source: Kristin Levier, Science Notes, University of California, Santa Cruz)*

traces of carbon dioxide, carbon monoxide, sulfur oxides, nitric oxides, and other gases.

Scientists long assumed that the early atmosphere contained hydrogen (H_2), ammonia (NH_3), and methane (CH_4), with traces of carbon dioxide (CO_2) and hydrogen sulfide (H_2S). Recent research has cast doubt on these assumptions, however.

About one important gas there has been little argument. The early atmosphere had almost no oxygen. The strongest evidence for this is the near total absence of oxidized iron (rust) in the earliest sediments. Had oxygen been present in any quantity, it would certainly have oxidized the starting materials for life. Oxygen, we will see later in this chapter, came much later, probably as a by-product of photosynthesis.

The absence of oxygen had other important consequences. Chief among these was the absence of an ozone layer. The ozone (O_3) layer, about 40 km (25 miles) up, today shields the Earth from intense ultraviolet (UV) light from the sun. Modern organisms need that protection because UV light, extremely good at making and breaking chemical bonds, interferes with normal life functions.

However, 4 billion years ago, any chemical synthesis would have required considerable energy. Recall from Chapter 5 that synthetic reactions are almost all thermodynamically uphill. So UV radiation from the sun would have helped to promote novel chemical reactions. Heat from the hot interior of the Earth provided more energy. But the most dramatic energy source was undoubtedly the lightning accompanying the storms that filled the oceans.

The Earth's crust solidified about 4 billion years ago, but the oceans were not stable until about 3.8 billion years ago. Since the first fossil organisms are about 3.8 billion years old, we can conclude that life on Earth originated relatively quickly—within a 300-million year window.

Is Prebiotic Synthesis Plausible?

By the early 1950s, scientists had some idea of what the early Earth was like: they believed it had the raw materials for organic synthesis—water, methane, ammonia, and hydrogen—and it had sources of energy—UV radiation, heat, and lightning. But scientists did not know whether the conditions of the Earth's early history would allow the abiotic synthesis of biologically important molecules.

Then, in 1953, a 23-year-old graduate student working in the laboratory of Harold Urey, at the University of Chicago, undertook to simulate the primitive conditions of the Earth. Stanley Miller simulated the early Earth inside a simple, glass apparatus built from the standard equipment found in a sophomore organic chemistry laboratory (Figure 18-8). The simulated Earth consisted of water in a boiler (the ocean) and a chamber fitted with electrodes (the atmosphere, with lightning). In one experiment, Miller mixed together an atmosphere of hydrogen (H_2), methane (CH_4), and ammonia (NH_3). Miller let the gases circulate past the flashing electrodes for a week and then analyzed the contents.

He found that more than 10 percent of the carbon from the methane was now in organic molecules. These included at least five amino acids, as well as compounds that are precursors of other amino acids and of the bases of nucleic acids. Among these precursors was hydrogen cyanide (HCN), five molecules of which can directly form adenine (Figure 18-9).

Miller and others repeated this basic experiment, with different atmospheres and different sources of energy. As long as free oxygen (O_2) was absent, the same building blocks appeared. Other sources of energy also work—most notably UV light, the principal energy source available on the Earth's surface before the Earth developed its protective ozone layer.

As a result of these experiments, it seemed likely that—within a few million years—the primitive oceans have loaded up with organic molecules capable of becoming the building blocks of life. This primordial soup may have contained as much as 1 percent organic molecules in some places, about the same as a watery chicken broth. Remarkably, the amino acids most easily produced in these abiotic experiments are the ones most common in proteins today. And the most easily synthe-

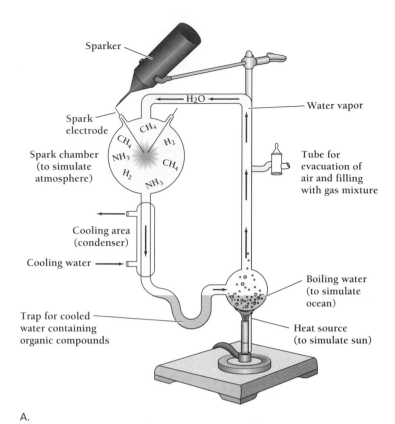

A.

B.

Figure 18-8 **Stanley Miller's experiment.** Can conditions similar to those on Earth 4 billion years ago give rise to the building blocks of life? Miller treated a mixture of water and gases with flashes of artificial "lightning." Among the complex molecules produced after one week were five amino acids. *(B, © 1988 Roger Ressmeyer/Starlight)*

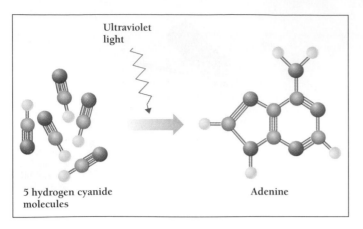

Figure 18-9 First steps in the formation of life? Miller's apparatus also produced hydrogen cyanide. In the presence of ultraviolet light, 5 molecules of hydrogen cyanide form adenine, the nitrogenous base in ATP.

sized base—adenine—is biologically the most important of all the nitrogenous bases.

A distinct problem with this scenario, however, is that recent research by geoscientists suggests that the early atmosphere contained neither methane nor ammonia. Heavy doses of UV light from the sun would have destroyed such hydrogen-based molecules, and even free hydrogen is so light that it would have escaped into space. Geoscientists say that the major components of the atmosphere were more likely carbon dioxide and nitrogen, with water and traces of carbon monoxide and free hydrogen, the same gases that are ejected from modern volcanoes.

Such an atmosphere would not have allowed the kind of amino acid synthesis suggested by Miller's experiment. Formaldehyde (H_2CO), essential to the formation of sugars, would have formed easily enough. But a likely biochemical pathway for the synthesis of hydrogen cyanide, the building block for adenine and other amino acids, has yet to be found.

Miller, now a professor of chemistry at the University of California, San Diego, defends his early study, saying that smoke and clouds might have shielded the fragile methane and ammonia molecules from UV light. And several Japanese researchers, supporting Miller, argue that solar particles and cosmic rays would have countered the effect of UV light by releasing free hydrogen from water molecules and so promoting the synthesis of methane and ammonia.

Prebiotic synthesis is plausible, but deriving the likely biochemical pathways depends on knowing the composition of the early atmosphere.

Was Prebiotic Synthesis Necessary?

Some astronomers have suggested that the debate over the exact composition of the early atmosphere may be moot. They believe that another source of simple organic molecules is pos-

sible. Researchers have discovered that asteroids, meteors, and comets are rich in complex organic compounds created during the formation of the solar system (Figure 18-10). Further, based on the calculated number and size of comets and meteors likely to have entered the Earth's atmosphere, astronomers estimate that as much as 10 million kg of complex organic molecules might have showered the Earth, each year. At the end of a billion years, they calculate, the Earth could have accumulated as much as a million billion (10^{15}) kg of organic molecules, more than the total mass of all life on Earth today.

The first organic molecules could have come from asteroids, meteors, or comets.

How Can Biological Building Blocks Assemble Outside of Cells?

If we assume that the Earth's early environment either allowed the production of amino acids and other building blocks or received them passively from space, we must still account for the assembly of those building blocks into other amino acids, nucleic acids, ATP, and other macromolecules. Can this process be simulated in the laboratory?

In the Miller atmosphere, adenine easily forms from hydrogen cyanide (Figure 18-9). The other nucleic acid bases require more complex reactions, so adenine could have been the first nitrogenous base to form. If so, it would not be surprising for its activated product, ATP, to have become the universal energy currency of life.

ATP itself may have formed rather easily. Laboratory experiments show that mixtures of adenine, ribose, and phosphate can form ATP when exposed to UV light. Other triphosphates may have arisen in the same way.

Even prebiotic replication of nucleic acids is not at all far-fetched. Biochemists have discovered that nucleotides can combine in the test tube to form short RNAs. These RNAs, moreover, form double-stranded molecules that follow the Watson–Crick rules for base pairing.

Simulating the polymerization of amino acids into polypeptides is more of a challenge. Many researchers, however, think that ATP in the primordial broth may have somehow contributed to the synthesis of activated amino acids. Solutions of such activated amino acids can spontaneously give polypeptide chains up to 50 amino acids long. Polypeptides also form when researchers heat dry mixtures of amino acids. Sidney Fox, a biochemist at Southern Illinois University, has suggested that such polypeptides could have formed on volcanic cinder cones and washed into the sea.

The earliest polypeptides almost certainly would not have had specific sequences of amino acids. In fact, Fox has dubbed the early polypeptides **proteinoids,** to distinguish them from the true proteins, which are the products of living organisms and whose sequences are defined by the genes in DNA.

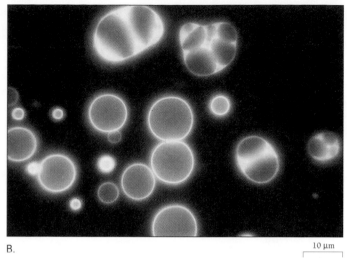

A.

B.

10 μm

Figure 18-10   **Remains of extraterrestrial life?** A. This fragment of a meteorite hurtled through the atmosphere and landed in Murchison, Australia, in 1969. Scattered through the rock are particles of organic compounds. B. When extracted, these compounds self-assembled into vesicles. The yellow-green color is the fluorescence of complex organic compounds called polycyclic aromatic hydrocarbons.   *(D.W. Deamer, University of California, Santa Cruz)*

The abiotic synthesis of RNA, ATP, and even
polypeptides seems plausible.

Some variant of the chemical reactions described above may have filled the open seas, or even small pools and puddles, with the building blocks of life, perhaps as long ago as 4 billion years. But the materials of life are not the same as life. Two important steps would have to have occurred—the evolution of reproduction and the evolution of metabolism. The evolution of reproductive ability depended on establishing a highly specific relationship between the self-replicating nucleic acids and the sequence of amino acids. This meant the development both of the genetic code and of a mechanism of translation.

Before this could happen, however, polypeptides and other molecules probably began to come together into organized associations. As we will see, such primitive cells, or **protobionts,** could concentrate organic molecules and maintain an internal environment different from that of their surroundings. They could, in short, begin to develop the first suggestion of metabolism.

Life required the evolution of cells, reproduction,
and metabolism.

How Can Prebiotic "Cells" Form?

Polypeptides, nucleic acids, and polysaccharides in solution will spontaneously concentrate themselves into discrete tiny, balloonlike droplets, which the Russian biochemist Oparin

called **coacervates**. Coacervates have a discrete inside and outside separated by a membranelike boundary.

If enzymes are present in the solution, these will be bound in the coacervates, which then act like little cells. They absorb molecules from the solution and release the products of the reactions catalyzed by the enzymes. For example, Oparin showed that coacervates containing the enzyme phosphorylase would take up glucose-1-phosphate from the surrounding medium, convert it to starch, and concentrate it inside the droplet (Figure 18-11).

Other mixtures can also give rise to isolated, cell-like structures. For example, Sidney Fox found that the proteinoids formed from dry amino acids would form tiny spheres. These structures, called **proteinoid microspheres,** could absorb more polymers from solution, forming buds or even a new generation of spheres (Figure 18-12). Microspheres dropped in wa-

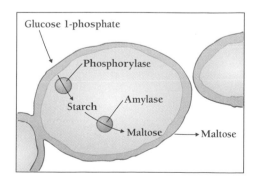

Figure 18-11   **Prebiotic metabolism?** Macromolecules in solution can form droplets, which Alexander Oparin called coacervates. In one experiment, coacervates that contained the enzymes phosphorylase and amylase could convert glucose-1-phosphate to starch and then to maltose.

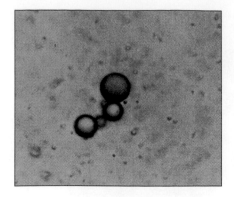

Figure 18-12 Proteinoid microspheres. These nonliving polymer bubbles can reproduce themselves by growing and budding. *(Steven Brooke and Richard LeDuc)*

ter are differentially permeable. They swell or shrink according to the concentration of salts in the water. Some will even store energy, in the form of a differential electric charge across the membrane. Such microspheres can release a charge in much the same way a nerve cell does. A mixture of phospholipids and proteins can form yet another kind of enclosed structure, called a **liposome**. Some liposome membranes have two layers, much like those in modern cells.

None of these self-organizing structures are cells. Yet they look so much like cells that experienced biologists have occasionally mistaken them for bacteria and even tried to classify

them. And protobionts show some of the characteristics of life, including growth, metabolism, and responsiveness to their environment. The existence of protobionts shows that cell-like structures could have formed spontaneously in the primordial soup.

Further, we can envision the evolution of primitive metabolic pathways. As coacervates used up a given compound in the environment, those coacervates able to make the compound from another component of the soup would persist. Once the precursor was exhausted, further growth of protobionts would depend on the ability to produce the needed product from still another precursor. According to this scheme, a number of pathways would evolve—one step at a time (Figure 18-13).

EARLY BIOTIC EVOLUTION: HOW DID THE FIRST GENETIC SYSTEM EVOLVE?

Could RNA Have Served as the Original Genetic Material?

Even with the evolution of primitive metabolic pathways, protobiont "life" would have been severely limited. Despite proto-

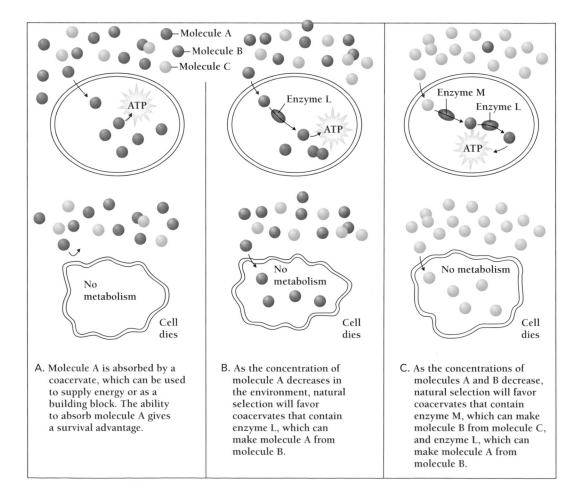

Figure 18-13 How might metabolic pathways have evolved?

A. Molecule A is absorbed by a coacervate, which can be used to supply energy or as a building block. The ability to absorb molecule A gives a survival advantage.

B. As the concentration of molecule A decreases in the environment, natural selection will favor coacervates that contain enzyme L, which can make molecule A from molecule B.

C. As the concentrations of molecules A and B decrease, natural selection will favor coacervates that contain enzyme M, which can make molecule B from molecule C, and enzyme L, which can make molecule A from molecule B.

bionts' ability to grow and to bud, they could not reproduce. Even if they had accumulated catalysts for specific chemical reactions, protobionts could not make more of these specific catalysts. As unique molecular catalysts were passed on to increasing numbers of "offspring" protobionts, the catalysts became increasingly diluted. To invest the new protobionts with equal concentrations of functional molecules required the invention of a way of "teaching" the next generation to make its own catalysts. Life would need a way of storing, replicating, and translating the information for the building of enzyme catalysts—in short, a genetic system.

A modern cell stores its genetic information as DNA, transcribes the information into RNA, and then translates the message into specific polypeptides. When the cell divides into two cells, it gives a copy of the DNA to each new cell, and so passes on the recipe for each polypeptide.

The DNA-to-RNA-to-polypeptide scheme is too complicated to have arisen all at once in the first cells. So biologists have tried to guess which of these three kinds of molecules could have first stored information, directed the synthesis of proteins, and also replicated itself. But DNA cannot form without catalytic proteins, and proteins cannot form without DNA.

That leaves RNA. In modern cells, RNA plays a central role in translation, supporting the idea that RNA could have been the primitive genetic material. Still, if RNA acted as the original genetic material, some sort of mechanism had to exist for aligning amino acids along sequenced strands of genetic RNA. Such a mechanism would have to allow RNA to store information, direct the synthesis of proteins, and also replicate itself. Considerable evidence now supports the idea that RNA can do this.

First, scientists have found that some RNA can spontaneously create itself and replicate itself in the test tube. Short strands of nonordered abiotic RNA can self-assemble from individual ribonucleotides. And, in the presence of an RNA template, ordered sequences of five to ten nucleotides will polymerize according to the standard base-pairing rules.

In the early 1980s, molecular biologists Thomas Cech and Sidney Altman showed that certain kinds of RNA could act as enzymes, snipping off a bit at an end or cutting out a bit and splicing the remaining pieces together. It was a dramatic discovery that earned the two researchers a Nobel prize in 1989 and suggested that perhaps RNA could replicate itself without the help of protein enzymes (polymerases).

Indeed, in 1989, Jennifer A. Doudna and Jack Szostak created an artificial RNA enzyme, or **ribozyme**, that could make new RNA molecules. However, although the molecule can splice preassembled strands of RNA along a template, it cannot add individual bases to a growing chain.

We know that RNA is capable of storing information, and it might, perhaps, be able to replicate itself. But the hardest question to answer is how translation could have evolved. How a correlation between nucleotide sequences and amino acid sequences could have arisen remains a mystery.

How Did Translation Evolve?

Since the genetic code plays such a pivotal role in translation, we may look to it for clues about the evolution of translation. The first clue is the fact that all organisms use the same triplet genetic code. This suggests that the system originated with life itself. Second, most of the information for each codon is in the first two nucleotides: for 7 of the 20 amino acids, the third "letter" is irrelevant. This does not mean that the code was originally a doublet, however. If the code had started out as a doublet, then all previously evolved genes would have become useless when the code became a triplet. Instead, the present characteristics of the code were almost certainly fixed after its initial success.

As molecular biologist Francis Crick put it, the code is a "frozen accident." Information in messenger RNA (mRNA) was probably read three bases at a time from the beginning. But initially, the first two letters of each codon contained all the meaning.

Because the code is universal among both prokaryotes and eukaryotes, it must have completed its refinement before the time of the common origin of all modern organisms.

How Did Ribosomes Evolve?

Modern ribosomes contain more than 50 different proteins as well as ribosomal RNAs (rRNAs). Since each protein has a defined amino acid sequence, ribosomal proteins could not have been part of the original translational machinery. It seems more likely that the primitive translational machinery depended primarily on RNA. Indeed, the mRNAs of many prokaryotic and eukaryotic bacteria can form specific base pairs with part of the smaller rRNA.

The anatomy of rRNAs also supports the idea that rRNAs directly participated in protein synthesis. For example, researchers have determined the nucleotide sequences of the smaller rRNAs of many widely divergent organisms, ranging from bacteria to vertebrates, as well as of ribosomes from mitochondria and chloroplasts. For each rRNA researchers predicted the pattern of stems and loops that could be formed by the formation of duplexes of complementary sequences. While the sequences of the rRNAs vary widely, the patterns of stems and loops are much the same (Figure 18-14). Such conservation suggests an essential function for the pattern of stems and loops, probably as a catalyst for directed protein synthesis.

How Did tRNAs Evolve?

All modern cells attach amino acids to transfer RNAs (tRNAs) before assembling a polypeptide. Transfer RNAs play a critical role in translation, allowing unlike molecules—nucleotides and amino acids—to be connected. All modern tRNAs have similar stem-and-loop structures, and all attach to an amino acid at their 3′ ends. In addition, all tRNAs contain an anticodon

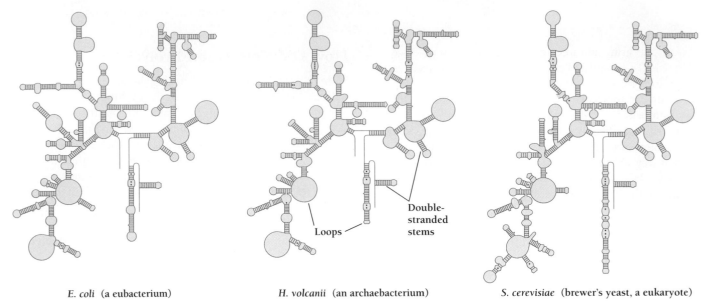

E. coli (a eubacterium)   H. volcanii (an archaebacterium)   S. cerevisiae (brewer's yeast, a eukaryote)

Figure 18-14   Ribosomal RNA. All modern ribosomal RNAs have a similar "stem-and-loop" structure. Notice that the pattern of stems and loops in eukaryote (yeast) RNA is more like the RNA of an archaebacterium than like that of a eubacterium. This supports the theory that we are descended from archaebacteria rather than eubacteria.

with common surrounding elements. Thus, like the genetic code, tRNAs probably diverged from a single tRNA early in the history of life.

How Did the First Cells Use Genetic Information?

The first true cell would have had to have the following properties:

1. A boundary, such as that in artificial protobionts.
2. Enzymes for extracting energy from chemicals in the environment.
3. Energy storage in ATP.
4. RNA that specified the structure of specific enzymes.
5. RNA replication, using the Watson–Crick rules of base pairing.
6. Specification of each amino acid by a triplet codon, with most of the information conveyed by one or two bases.
7. Primitive tRNAs to connect amino acids to nucleotides.

The first enzymes probably arose by the random assembly of amino acids into short polypeptide segments. Such primitive enzymes were probably neither as active nor as stable as modern enzymes. But if a polypeptide segment could form the active site of a catalyst, then its possessor would have enormous advantages over its competitors.

Similarly, the first genetically useful RNAs probably had only short stretches of information. Once the first genetic system had evolved, however, natural selection would begin to work. Natural selection would favor cells with increased capacity to reproduce, as well as cells better able to extract energy from chemicals in the environment. Reproduction, protein synthesis, and metabolism must therefore have evolved

together. The most successful organisms would be those that best coupled their genetic and metabolic systems—those best able to pass on useful catalysts to their descendants.

HOW DID MODERN CELLS EVOLVE?

The first fossils that are clearly eukaryotic are about 1.7 billion years old. More than 2 billion years elapsed between the first appearance of life 3.8 billion years ago and the first eukaryotes. What, we may ask, happened during that period to allow more complex forms to evolve? What biochemical pathways did the earliest organisms develop, and, equally important, how did these developments fundamentally change the Earth's environment?

How Did Early Life Affect the Environment?

The Evolution of Autotrophy

The first cells were **heterotrophs** [Greek, *heteros* = other + *trophe* = nourishment], organisms that obtain energy only by degrading existing organic molecules. Human beings and other modern animals are all heterotrophs, too. We obtain both energy and building blocks, such as amino acids, by eating other organisms. We can extract energy in the process of respiration by oxidizing preexisting organic molecules from other organisms.

The earliest organisms also obtained both energy and building blocks from preexisting organic molecules in the primordial soup. Like all heterotrophs, they consumed what was in the soup, but contributed no energy. But unlike animals, they could only break down molecules by means of glycolysis, or fermentation. They were incapable of true respiration, since there was no oxygen. And they were incapable of obtaining energy from the sun by photosynthesis or from inorganic molecules, as plants and other modern **autotrophs** (Greek, *auto* = self + *trophe* = nourishment) do.

Of the three ways to extract energy from molecules—photosynthesis, respiration, and fermentation—only fermentation is common to all bacteria and eukaryotes. It seems likely, then, that this metabolic pathway is the oldest and had already evolved in the earliest cells—at least those whose descendants have survived.

However, the source of nourishment for early heterotrophs, the organic molecules in the primordial soup, were in limited supply. Early organisms would have gradually exhausted them of their nutritive value, since no autotrophs existed to renew them. Early organisms probably forestalled the exhaustion of the Earth's first resources by developing elaborate chemical pathways for exploiting more and more of the molecules in the soup. Eventually, though, all the original molecules, and all the energy in them, would have been exhausted. One alternative was to begin extracting energy from inorganic chemicals by oxidizing them and to use the energy to build new organic molecules from carbon dioxide. The present-day *Sulfolobus* bacterium, for example, gets its energy by converting sulfur to sulfur-dioxide.

The Evolution of Photosynthesis

The most abundant energy source was not inorganic molecules, however, but sunlight. We may guess at how photosynthesis gradually developed, at least 2.7 billion years ago, by examining the variety of ways in which modern organisms harness solar energy.

The first step in the direct harnessing of solar energy by organisms was probably the development of a light-driven proton pump, such as that in the modern purple bacterium. This bacterium pumps protons across its cell membrane, which allows the bacterium to store energy in ATP, as well as actively transport other substances across the membrane.

More advanced modern bacteria use light to generate reducing power (hydrogen atoms or electrons) to synthesize carbohydrates from carbon dioxide. Some photosynthetic bacteria obtain the needed electrons from such donors as hydrogen gas or hydrogen sulfide.

The largest potential source of electrons, however, is water. Water is difficult to break down, but it is the electron source for nearly all modern photosynthetic organisms. The first ancient organism to synthesize organic molecules from sunlight, water, and carbon dioxide won for its descendants a freedom from want that has lasted more than 2 billion years. Cells that could photosynthesize would no longer have to compete for energy-bearing organic molecules. They could simply make their own from the abundant water, carbon dioxide, and sunlight.

The First Global Toxic Waste Problem

However, breaking down water created a serious global toxic waste problem. The very steps needed to separate the hydrogen atoms from water also released huge quantities of molecular oxygen. Oxygen is a powerful oxidant that easily rips apart many molecules. Sensitive organic molecules would have been destroyed by oxygen. Many of the original heterotrophs, which depended on organic molecules, probably starved, their food supply destroyed by the wastes of their own autotrophic cousins.

The development of photosynthesis drastically altered the Earth's early environment. Major contributors to this change were the cyanobacteria making up the stromatolites, mentioned at the beginning of this chapter. The wide distribution of stromatolite fossils suggests that they covered vast areas of the world's shallow waters from about 3 billion years ago until the end of the Precambrian. In the absence of modern grazing sea animals, photosynthetic cyanobacteria flourished nearly unchecked for millions upon millions of years. For perhaps as long as 2.5 billion years, they pumped oxygen into the oceans and the atmosphere. The cyanobacteria began to decline about a billion years ago. However, single-celled algae began to appear throughout the open oceans, their photosynthetic activity further boosting the oxygen content of the seas and the atmosphere.

The availability of oxygen allowed the evolution of a new class of heterotrophic organisms that could use oxygen to break down organic molecules. The complete oxidation of organic compounds to water and carbon supplies more energy than glycolysis. The advantage of exploiting oxygen was so great that today anaerobic organisms persist in only a few isolated environments.

The vast flats of cyanobacteria and oceans of algae must have presented another undeniable opportunity. Some of the new organisms that had developed the ability to breathe oxygen now developed the ability to eat the very organisms that had created this caustic waste. The first autotrophs thus provided the basis of a food chain that still survives today. Most animals that live in the ocean eat smaller animals that eat planktonic animals that eat single-celled algae.

How Did Mitochondria and Chloroplasts Arise?

The First Eukaryotes Probably Possessed Mitochondria

Almost all eukaryotes—protists, fungi, plants, and animals—have mitochondria, the organelles that produce most of a cell's energy. Only plants and some protists, however, have chloroplasts. Biologists conclude from this that the first eukaryotes probably had mitochondria, but no chloroplasts.

Because mitochondria depend on oxygen to generate useful energy, they could have evolved only after the photosynthetic prokaryotes had filled the world with oxygen. Mito-

Codon	UGA	AUA	CUA	AGA/AGG
Universal code	STOP	Ile	Leu	Arg
Human mitochondria	Trp	Met	Leu	STOP
Drosophila mitochondria	Trp	Met	Leu	Ser
Yeast mitochondria	Trp	Met	Thr	Arg
Plant mitochondria	STOP	Ile	Leu	Arg

*(Rows from "Human mitochondria" through "Plant mitochondria" are bracketed as **Mitochondrial codes**.)*

Figure 18-15   Genetic code in mitochondria. Mitochondrial DNA encodes polypeptides using a slightly different code than that used by nuclear DNA. The top codons translate to stop, isoleucine, leucine, and arginine, respectively, in the universal code. The mitochondria of humans, fruit flies, and yeast each translate one or more of these codons differently. In contrast, plant mitochondria use the universal code. What does that suggest about plant mitochondria?

chondria are probably extremely ancient, however, as they use a genetic code that differs from the universal code used by all modern eukaryotes and most prokaryotes. In fact, the mitochondria of different organisms use slightly different codes (Figure 18-15).

———

Mitochondria are probably older than chloroplasts.

———

Are Mitochondria and Chloroplasts Endosymbionts?

In both size and internal organization chloroplasts resemble cyanobacteria (Figure 18-16). These similarities have long suggested that chloroplasts derived from free-living organisms. Indeed, there are many examples today of photosynthetic organisms living within animals or protists. Such a close association of two organisms, one of which lives inside the other, is called **endosymbiosis** [Greek, *endon* = within + *syn* = together + *bios* = life]. Endosymbiosis is mutually beneficial since it provides an internal source of food for the host and mobility and protection for the inner organism.

Most biologists now favor the view that both chloroplasts and mitochondria are the endosymbiotic descendants of free-living prokaryotes (Figure 18-17). One of the most forceful proponents of this view has been Lynn Margulis, an evolutionary biologist at the University of Massachusetts. Margulis has gone one step further than many other biologists. She speculates that the microtubules of the eukaryotic flagellum arose from a symbiosis with a spirochete bacterium. It would be but a short step, she argues, to construct the mitotic spindle apparatus, once cells acquired the ability to produce a flagellum.

Some biologists still consider the endosymbiosis theory too speculative. Many reject Margulis's argument for the endosymbiotic origin of flagella and the mitotic apparatus. Others feel that chloroplasts may well have originated as free-living cyanobacteria, but that the case for mitochondria is much weaker. And most biologists think that there is as yet no convincing argument for the origin of eukaryotic nuclei, though a strong candidate has recently emerged. Most prokaryotes lack the proteins histone and actin. But the archaebacterium *Thermoplasma acidophilum*, a heat-loving prokaryote, does make histones and actin. The ancestors of this organism, which lacks a cell wall, is a popular best guess for the origin of the nucleus.

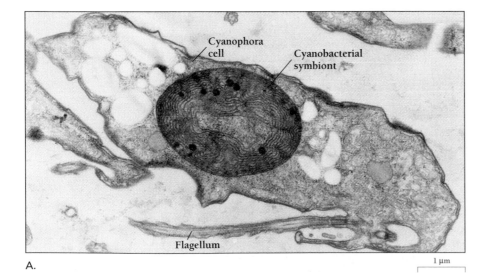

Cyanophora cell
Cyanobacterial symbiont
Flagellum
A.   1 μm

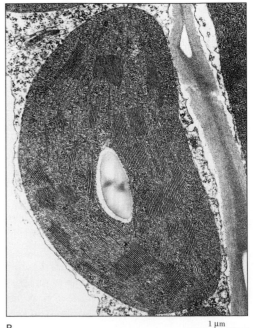

B.   1 μm

Figure 18-16   Endosymbiosis. A. Living inside the photosynthetic protist *Cyanophora* is a photosynthetic cyanobacterium. B. A chloroplast resembles a cyanobacterium.   *(A, Biophoto Associates; B, Omikron/Photo Researchers, Inc.)*

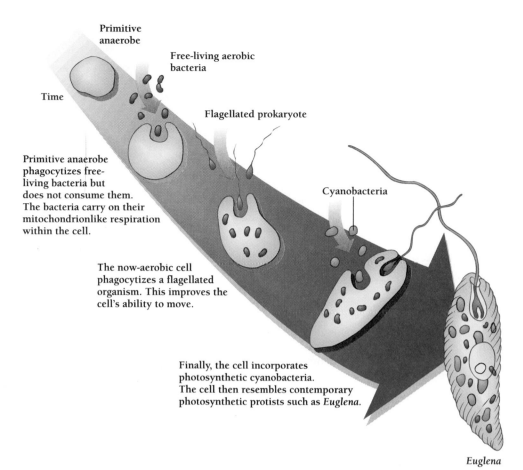

Primitive
anaerobe

Free-living aerobic
bacteria

Time

Flagellated prokaryote

**Primitive anaerobe
phagocytizes free-
living bacteria but
does not consume them.
The bacteria carry on their
mitochondrionlike respiration
within the cell.**

Cyanobacteria

**The now-aerobic cell
phagocytizes a flagellated
organism. This improves the
cell's ability to move.**

**Finally, the cell incorporates
photosynthetic cyanobacteria.
The cell then resembles contemporary
photosynthetic protists such as *Euglena*.**

Euglena

Figure 18-17 How did the subcellular organelles of eukaryotes evolve? One theory holds that each organelle represents an endosymbiotic event—a sort of housing cooperative. Mitochondria may be bacteria that took up residence in another cell. Chloroplasts may be photosynthetic cyanobacteria that likewise moved in. Even flagella may be the remnants of flagellated prokaryotes that flourished inside another cell. In each case, biologists speculate, the host cell attempted to phagocytize a smaller bacterium and succeeded in enveloping it, but did not digest it.

The actual origin of eukaryotic cells can never be proven, but Margulis's arguments—some of which remain controversial—have aroused considerable interest in this subject.

Did Eukaryotes Arise Through the Proliferation of Internal Membranes?

Opponents of the endosymbiotic theory argue that eukaryotes arose instead by the proliferation of internal membranes to form new cellular compartments. Lysosomes and the endoplasmic reticulum, for example, form from cellular membranes—not from free-living forms. Perhaps mitochondria and chloroplasts formed when membranes surrounded pieces of mobile DNA (plasmids or transposons) that detached from nuclear genes.

The endosymbiotic theory produces a more complicated evolutionary tree than the traditional view. With this complexity, however, comes a consistency with current views of prebi-

otic evolution. For the moment, at least, most biologists seem to find endosymbiosis the most convincing general explanation for the origin of eukaryotic cells.

Mitochondria and chloroplasts, which resemble prokaryotes, probably began their association with eukaryote cells as endosymbionts.

In this chapter, we have explored scientists' hypotheses about how life might have evolved on Earth billions of years ago. Our discussion of such long-ago events has been, necessarily, speculative. In the next section of *Asking About Life* we return to the present with a review of the great diversity of modern life. We will examine how biologists classify organisms and then survey each of the six kingdoms of organisms.

STUDY OUTLINE WITH KEY TERMS

By the middle of the 19th century, the work of Redi, Spallanzani, and Pasteur had convinced scientists that life could not arise spontaneously. Yet all life must have some ultimate origin. While some scientists have argued that life may have arrived from elsewhere in the universe, most researchers assume that life evolved from nonliving, or **abiotic,** organic molecules here on Earth.

The theory of the origin of life on Earth is different from the theory of spontaneous generation. Life could not evolve on Earth today

because oxygen would oxidize the components and other organisms would eat the components before enough time elapsed for anything to happen. Ideas about the origin of life rest on the demonstration that plausible **prebiotic** chemical reactions could have produced molecules and organized structures like those in modern organisms.

After the Earth formed some 4.6 billion years ago, and its crust had hardened half a billion years later, torrential rains began creating the first sediments. Precambrian rocks contain fossils as old as 3.8 billion years, only 300 million years after the formation of the oldest known sedimentary rocks. The Earth's early atmosphere contained no free oxygen. Consequently, no ozone layer protected the atmosphere from ultraviolet (UV) light. UV light and lightning probably contributed the intense energy needed for the formation of the first organic molecules. The early atmosphere may have contained hydrogen, ammonia, and methane, as well as other gases and water. In the laboratory, such a mixture, exposed to electrical discharges, produces many organic molecules, including amino acids and precursors of nucleotides. However, recent research by geoscientists suggests that the early atmosphere was more likely composed of carbon dioxide and nitrogen. Such an atmosphere could have easily given rise to sugars, but to amino acids only with difficulty.

Biologists assume that the primitive oceans became loaded with organic molecules. These organic molecules can form polymers, called **proteinoids.** Mixtures of different kinds of molecules can assemble into organized cell-like structures, called **protobionts.** Polypeptides, nucleic acids, and polysaccharides form **coacervates.** Proteinoids will form tiny spheres called **proteinoid microspheres.** A mixture of phospholipids and proteins form **liposomes,** which have a bilayered membrane similar to that of a true cell.

Although protobionts resemble cells and are capable of cell-like behavior and **prebiotic evolution,** they are not alive because they cannot reproduce themselves. Living cells must be able to store, translate, and replicate genetic instructions. By comparing the characteristics of protein synthesis in contemporary organisms, scientists have inferred that the first organisms used RNA as the genetic material, with an early version of the triplet genetic code. RNA can store information and replicate itself. It can also snip and splice itself like an enzyme. RNA catalysts, called **ribozymes,** can catalyze the synthesis of tRNA, rRNA, and mRNA, the major components of translation.

The first organisms were **heterotrophs.** Photosynthetic and other forms of **autotrophic** organisms probably evolved after early life had depleted the primitive oceans of energy-rich organic molecules. The photosynthetic organisms pumped the oceans and the atmosphere full of deadly free oxygen.

The presence of oxygen allowed the evolution of a new class of heterotrophs that could respire. The presence of autotrophs also allowed the development of herbivores and the beginnings of a modern food chain.

Eukaryotes evolved long after prokaryotes, probably after the appearance of oxygen. Most biologists believe that mitochondria and chloroplasts derived from previously free-living prokaryotes that became **endosymbionts.**

REVIEW AND THOUGHT QUESTIONS

Review Questions

1. Give two reasons why it took paleontologists so long to find Precambrian fossils.
2. Describe the two possible alternative compositions of the Earth's primitive atmosphere, including the disagreements scientists have about them. List two major sources of prebiotic organic molecules.
3. Describe Stanley Miller's landmark experiment to test the idea of prebiotic synthesis. What is the "primordial soup?"
4. What is a coacervate made of, how is it formed, and what can it do? What is a proteinoid microsphere made of, how is it formed, and what can it do? What is a liposome made of, how is it formed, and what can it do?

5. What would a protobiont need to become a living cell?
6. Why do scientists think that RNA was the first genetic material?
7. In what way did the early heterotrophs change the environment? What were some of the consequences of this change?

Thought Questions

8. Can natural selection operate on nonliving structures such as coacervates? Explain.
9. Pasteur's sterile broths contain no organisms that could consume any life that might arise spontaneously. Why have these flasks not given rise to new life in 130 years?

SELECTED READINGS

Drake, Frank, and Dava Sobel, *Is Anyone Out There? The Scientific Search for Extraterrestrial Intelligence,* Delacorte Press, New York, 1992. An exciting history of astronomers' long-running search for extraterrestrial intelligence.

Fortey, Richard, *Fossils: The Key to the Past,* Harvard University Press, Cambridge, Massachusetts, 1991. A gentle introduction to paleontology, with many fascinating photos.

Laporte, Leo F., ed., *Evolution and the Fossil Record,* Readings from *Scientific American,* W.H. Freeman and Company, San Francisco, 1978. A collection of *Scientific American* offprints on the topic of evolution.

Redi, Francesco, *Experiments on the Generation of Insects,* Kraus Reprint Co., New York, 1969. A charming discussion of the details of all of Redi's experiments.

▶On-line materials relating to this chapter are on the World Wide Web at http://www.saunderscollege.com/lifesci/
Click on Tobin/Dusheck: *Asking About Life.*

PART IV

Diversity

Coral from the Flores Sea of Indonesia. *(© Mickey Gibson/Animals Animals)*

CLASSIFICATION: WHAT'S IN A NAME?

Is the Red Wolf a Species?

In the 18th century, when naturalists first began mapping and naming the animals and plants of North America, the destruction of the Eastern forests had already driven countless species from the landscape. One of the first to go was the gray wolf, which had, before the time of Columbus, occupied nearly the whole of North America. By the 1970s, the gray wolf had been driven out of virtually all of the lower 48 states, its range reduced to Alaska, Canada, and tiny bits of northern Michigan and Minnesota (Figure 19-1). In the gray wolf's wake were left a rapidly shrinking population of red wolves, which had once lived throughout the American South, from Florida to Texas; and a rapidly expanding population of coyotes, which gradually spread from the Rockies and the Great Plains—north, east, and west—into the once-forested areas vacated by the gray and red wolves.

In 1967, the federal government designated the red wolf an endangered species, and biologists working for the U.S. Fish and Wildlife Service began surveying the South for red wolves (Figure 19-2). The red wolf, they discovered, was all but extinct. By 1972, the last few red wolves clung to the southern edge of North America, where Texas and Louisiana drop into the Gulf of Mexico. These last wolves were actively breeding with coyotes.

Government biologists realized with a shock that the only way to save the red wolf from complete extinction was to capture all the remaining wolves and breed them in captivity. From 1974 to 1980, government biologists trapped more than 400 animals. In 1980, in a marsh edging an industrial area of Galveston, Texas, the last wild red wolves in America walked into traps. Of the 400 captured animals, only 43 actually looked like wolves, however, and only 14 produced descendants that looked like wolves rather than coyotes.

Under the watchful eyes of biologists, this group of 14 animals has produced hundreds of offspring. Many of these captive-bred wolves have been released into places such as the Great Smoky Mountains of Tennessee and the Alligator River National Wildlife Refuge, in North Carolina. The rest of the captive wolves continue to produce offspring to be released in years to come. The red wolf reintroduction program, as this project is called, is the most successful captive-breeding program in existence. By the early 1990s, the red wolf had been brought back from the very brink of extinction.

Not everyone has applauded the program's success, however. Rural residents worry about the safety of their children and pets. Others question the price of saving the red wolf. By 1995, the red wolf reintroduction program had already cost millions of dollars, and, to maintain the

Historical

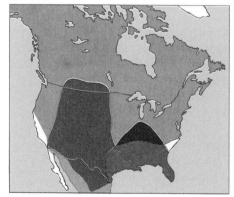

Modern

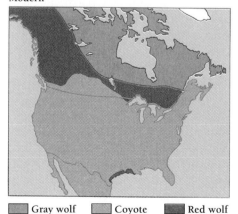

Gray wolf   Coyote   Red wolf

Figure 19-1 The disappearance of the gray wolf. As settlers moved across North America in the 1900s, they used guns, traps, and poison to virtually eliminate the gray wolf from the United States. But, the coyote, which has been subjected to similar organized massacres, has flourished. *(Michael Goodman)*

program, government biologists expected to spend nearly a million dollars a year between 1995 and 2000. Yet the program's most vocal opponents were not taxpayers, but sheep and cattle ranchers. For them, the reintroduction of wolves to areas of the country now occupied by cattle and sheep realized the romantic notions of city-dwelling environmentalists at the risk of valuable livestock.

Into this heated conflict stepped David Mech, one of the world's leading wolf experts. In 1989, at an Atlanta meeting of experts on wolf biology, Mech challenged his fellow researchers to tell him how they could justify spending so much money rescuing the red wolf when it might not even be a species.

Since the 18th century, biologists had argued over the status of the red wolf. In 1791, naturalists designated the red wolf a subspecies of the gray wolf. Later biologists decided, however, that the red wolf was a separate species. In 1979, U.S. Fish and Wildlife Service biologist Ronald Nowak carefully compared the skulls of gray wolves, red wolves, and coyotes, and noticed that the size and shape of the red

In 1980, the last wild red wolves in America walked into traps.

wolf skull fell midway between that of the coyote and the gray wolf. Nowak's interpretation of the fossil record further suggested to him that intermediate skulls like that of the red wolf skull first appeared in North America more than a million years ago, well before the first wolves or coyotes. Nowak concluded that the red wolf was not only a unique species, but the ancient ancestor of both the gray wolf and the coyote.

Additional paleontological evidence suggested that the gray wolves of Europe and Asia originated in North America. Therefore, argued Nowak, the red wolf must be a direct descendant of the first wolf. The rescue of the red wolf from the brink of extinction, then, could be seen as the salvation of a southern species of nearly royal lineage. It was a compelling idea, one that persisted almost unchallenged for 10 years, throughout the early years of the red wolf reintroduction program.

But David Mech had a different theory about red wolves. In a 1970 book, Mech had proposed that the red wolf was neither species nor subspecies, but a hy-

467

Figure 19-2   The red wolf *(center)* is intermediate between the gray wolf *(left)* and the coyote *(right)*—in size, conformation, and behavior.   *(gray wolf, Tom & Pat Leeson/DRK Photo; red wolf; Jane McAlonan/Visuals Unlimited; coyote, C.K. Lorenz/Photo Researchers)*

brid produced by interbreeding between the gray wolf and the coyote.

Mech's hypothesis accounted for Nowak's data showing that the size and shape of the red wolf skull was midway between that of the gray wolf and the coyote. Furthermore, northern gray wolves were known to interbreed with coyotes in some areas, and the offspring had skull measurements similar to those of red wolves. Nowak's skull measurements could be interpreted in two ways, Mech argued: either the red wolf species had split over evolutionary time into three species, the red wolf, the gray wolf, and the coyote; or red wolves were hybrids of gray wolves and coyotes.

In 1989, two University of California biologists, Robert Wayne (of UCLA) and Susan Jenks (of UC Berkeley), approached the U.S. Fish and Wildlife Service and offered to settle the matter once and for all. Like Nowak, Wayne was an expert on the morphology and taxonomy of wolves and other canids (members of the dog family). But unlike Nowak, Wayne had added molecular genetics to his repertoire of research techniques. Wayne and Jenks offered to search the red wolf genome for some unique stretch of DNA that neither the coyote nor the gray wolf had, something that would clearly identify the red wolf as a separate species.

The government agreed to fund the study, and the two biologists began examining DNA from red wolves, gray wolves, and coyotes. Wayne and Jenks began with mitochondrial DNA because, in general, it accumulates mutations more rapidly than DNA from the nucleus. This rapid evolution makes mitochondrial DNA particularly useful for comparing groups of closely related organisms. The base sequence of the mitochondrial DNA of gray wolves (*Canis lupus*) and coyotes (*Canis latrans*), for example, differs by about four percent.

The mitochondrial DNA of the gray wolf, Wayne and Jenks found, included sequences that the coyote did not have. Similarly, the DNA of the coyote included sequences that the gray wolf did not have. The gray wolf and the coyote were clearly separate species. But, to Wayne and Jenks's surprise, the mitochondrial DNA of the red wolf (*Canis rufus*), included *no sequences unique to red wolves*. Red wolf mitochondrial DNA only included sequences identical to those of gray wolves and sequences identical to those of coyotes. The two biologists tentatively and somewhat reluctantly concluded that the red wolf was most likely a hybrid of the gray wolf and the coyote.

Nowak and the other biologists at the U.S. Fish and Wildlife Service could not

believe what they were being told. Maybe, argued the government biologists, Wayne and Jenks had simply missed the DNA sequences that distinguished the red wolf. Maybe they had not looked at enough DNA.

To put to rest any lingering doubts, Wayne and other colleagues turned to special repetitive regions of the nuclear DNA called microsatellites. Wayne checked 10 microsatellite regions in blood samples taken from hundreds of living animals and skin samples from 16 pre-1930s skins stored in a vast collection of furs at the Smithsonian Institution, in Washington, DC.

The results were the same. Neither the samples of blood from living red wolves, nor the samples from the skins of pre-1930s red wolves showed any unique sequences. By 1994, Wayne had found no evidence that the red wolf had ever been reproductively isolated from either gray wolves or coyotes. The red wolf had to be a hybrid of the gray wolf and the coyote.

Wayne's genetic data proved to be an embarrassment to the U.S. Fish and Wildlife Service, which had poured millions of dollars into the reintroduction program in the belief that the red wolf was a unique and endangered species. Yet, the agency had acted in good faith. Until Wayne and his colleagues finished their research, the U.S. Fish and Wildlife Service had no way of knowing that the red wolf was not a species.

Now the government agency was faced with a terrible dilemma. Wayne's results threatened to discredit the reintroduction program, strip the red wolf of its endangered status, and further undermine the increasingly battered public image of the federal Endangered Species Act itself. The act had been under attack from its very inception by real estate developers, ranchers, miners, and other industries that find species preservation inconvenient and expensive.

For now, the red wolf hybrid has not lost protection under the Endangered Species Act, but its hybrid status certainly weakened its claim to protection. The

American Sheep Industry Association, for example, formally petitioned the U.S. Department of the Interior to take the animal off the endangered list. The Department of the Interior turned down this petition in 1992, arguing that "limited genetic exchange" between groups does not undermine species status. In short, the agency did not acknowledge that the red wolf was probably not a species.

To protect the red wolf, the U.S. Fish and Wildlife Service began pressuring Wayne to avoid the word "hybrid" in his research papers and to substitute the term "intergrade species" and other similar phrases. By 1994, it was clear, however, that the red wolf was not a species engaging in "limited genetic exchange" with other species.

In 1995, the U.S. Department of the Interior issued a legal opinion that said that hybrids would be protected under the Endangered Species Act if *morphological* evidence showed that the hybrids were similar to the endangered "pure" form. In essence, if they looked like red wolves, they would be protected. But the genetic data did not support the idea that a "pure" form of the red wolf had ever existed, certainly not in the last 100 years. In issuing this opinion, the agency excluded all the genetic evidence regarding the red wolf's species status. The only question was whether the red wolf looked different from the coyote and the gray wolf. It did, and, therefore, until such time as the government acknowledges the genetic data, the red wolf will be considered a species.

KEY CONCEPTS

1. Biologists classify organisms into both species and larger, more inclusive groups, such as genera, families, and orders.

2. Biologists use both biological and typological concepts to classify species.

3. Naming organisms allows biologists to distinguish one from another and to assign organisms to groups of similar organisms. Linnaeus standardized both the naming of organisms and the classification of organisms.

4. Some biologists try to classify organisms only on the basis of their observable characteristics, while others insist that classification must reflect evolutionary relationships.

5. Natural selection shapes organisms (and their genes) in ways that may obscure patterns of relatedness. Molecular approaches to classification are extremely useful, but they are not necessarily more objective than traditional methods.

WHY IS IT HARD TO NAME AND CLASSIFY ORGANISMS?

The story of the red wolf illustrates that deciding how to classify organisms into tidy groups is no easy task. Surprisingly, recognizing species is usually the *least* controversial task facing a **taxonomist,** a biologist who classifies organisms. In contrast, determining whether a species should be subdivided into subspecies (and, if so, how many subspecies) nearly always raises an argument. Likewise, grouping species into higher taxa— such as families, orders, or classes—immediately stirs up a host of controversies, which may be scientific, personal, philosophical, or political. In this chapter, we will briefly discuss how biologists classify organisms. Along the way, we will see why classification is nearly always a controversial topic.

Why Do We Name Things?

"What's in a name?" asks Shakespeare's Juliet, "That which we call a rose by any other name would smell as sweet."

Why do we name things? By naming the rose, we distinguish it from other flowers—both in our own minds and in the minds of others with whom we speak. The name allows us to talk with others about the rose's significance—romantic, evolutionary, or otherwise. To the extent that everyone agrees on a name, we can each talk of roses with confidence that we will be understood. Naming things also allows us to communicate what we know about something, whether the something is an individual object or organism or a class of objects or organisms. We can distinguish palatable plants from poisonous ones, and dangerous animals from docile ones.

Figure 19-3 Classification forces attention to detail. A. A child's drawing of mint *(Mentha spicata)*. B. An illustrator's rendering of the same herb. *(A, Daniel Dusheck Wilkes; B, © Carolyn Cohen)*

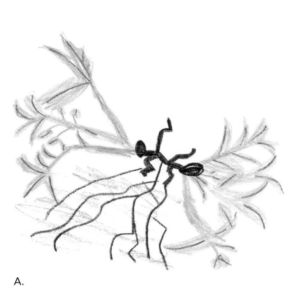

A.

B.

Further, by naming things, we force ourselves to make distinctions that we might not otherwise have noticed. In thinking about a rose, most people picture a red flower, yet we would not confuse a red rose with a red poppy or red carnation. Nor would we mistake a yellow rose for a buttercup or a dandelion. Naming an object, like drawing it, forces us to know it (Figure 19-3).

Another reason for naming an object is to classify it, or assign it to a group. A couple might name their children David and Nina simply to distinguish them from others, for example. But parents also give their children a last name to identify the children as part of a particular group, in this case a family.

Naming objects allows us to both distinguish one object from another and assign objects to groups of similar objects.

What Did Linnaeus's System of Classification Accomplish?

To classify objects, we usually establish different categories, or levels, of classification, and place these categories in a **hierarchy**, an arrangement in which larger groups contain smaller groups, which contain still smaller groups. We can illustrate a hierarchy of classification with a familiar example. Your address probably falls into each of the following categories: apartment number, street number, street, zip code, city, county, state, country, continent, planet, etc. Notice that the categories form a hierarchy, with each category often (but not necessarily) including more than one example of the lower category (Figure 19-4). Each state, for example, includes many counties, and most counties include several cities and towns.

Recall from Chapter 15 that biologists use a hierarchical taxonomy established by the 18th-century Swedish physician and biologist Carolus Linnaeus. Before Linnaeus's work, biologists had already used a category called a **genus** [plural = genera; Greek, *genos* = to be born], a group of species that share many morphological characteristics. Linnaeus and his later followers established higher categories, so that today we traditionally recognize the following hierarchy of classification: species are grouped into genera, genera into families, families into orders, orders into classes, classes into phyla, and phyla into kingdoms. (Some people remember these categories—in the reverse order—using the mnemonic: "Kings Play Chess On Fine-Grained Sand." But you can invent your own mnemonic.) Recently, biologists have further arranged the kingdoms into three domains. Figure 19-4 shows the hierarchical classification of several species including our own. We are members of the domain Eukarya, the kingdom Animalia, the phylum Chordata, the class Mammalia, the order Primates, the family Hominidae, the genus *Homo,* and the species *Homo sapiens.*

Biologists use the word **taxon** [plural, taxa; Greek, *taxis* = arrangement] to mean a group of organisms at any level of the classification hierarchy. So we may speak of all the primates as one taxon and all the bats as another. At the same time, we may speak of all the individuals of a species such as *Homo sapiens* as another taxon. As we move from the higher taxa to the lower taxa, we become more and more specific about the characteristics of each taxon.

Linnaeus not only introduced a system of classification, but also standardized the naming of organisms. The official name of a species is a **binomial**, a two-part name that contains both the genus (*Homo,* for example) and species (*sapiens,* for example). It is not correct to leave out the genus name or the species name when referring to a species. The fruit fly (or, more properly, the vinegar fly) most often used in genetics experiments is *Drosophila melanogaster.* We may, after writing the name out fully once, abbreviate the genus to its initial letter, as *D. melanogaster.* But scientists and laypersons alike often refer

TAXON	Human	Barn owl	Box turtle	Maize
Domain	Eukarya	Eukarya	Eukarya	Eukarya
Kingdom	Animalia	Animalia	Animalia	Plantae
Phylum	Chordata	Chordata	Chordata	Anthophyta
Class	Mammalia	Aves	Reptilia	Monocotyledones
Order	Primates	Strigiformes	Chelonia	Commelinales
Family	Hominidae	Tytonidae	Emydidae	Poaceae
Genus	Homo	Tyto	Terrapene	Zea
Species	*Homo sapiens*	*Tyto alba*	*Terrapene carolina*	*Zea mays*

Your address	
Galaxy	Milky Way
Star	Sun
Planet	Earth
Continent	North America
Country	United States
State	California
City	San Francisco
Street	Haight St.
Number	555

Figure 19-4 Hierarchical classification means that each level is included within the levels above.

to *Drosophila melanogaster* as simply *Drosophila*. This is informal and technically incorrect. North America alone has nearly 200 different species that all belong to the genus *Drosophila*. We cannot call all of them *Drosophila* and know what we are talking about.

Binomial names are always Latin or Greek (or at least they are meant to be). Latin was, until this century, regarded as the "universal language" by educated Europeans. The use of Latin names eliminates uncertainty in translation from one language to another (as the Latin name need never be translated). Latin also prevents confusion that results when people in different regions give the same common name to different species. For example, the English robin (*Erithacus rubecula*) bears only passing resemblance and relation to the robin we know in the United States (*Turdus migratorius*) (Figure 19-5).

Linnaeus standardized both the naming of organisms (with a Latin binomial) and the classification of organisms (into a strict hierarchy).

English robin

American robin

Figure 19-5 Although unrelated, both the English robin and its American namesake are called "robins." Scientific names are intended to minimize such confusion.

BOX 19-1

Foucault's Chinese encyclopedia

"This book first arose out of a passage in Borges, out of the laughter that shattered, as I read the passage, all the familiar landmarks of my thought—our thought, the thought that bears the stamp of our age and our geography—breaking up all the ordered surfaces and all the planes with which we are accustomed to tame the wild profusion of existing things, and continuing long afterwards to disturb and threaten with collapse our age-old distinction between the Same and the Other.

"This passage quotes a 'certain Chinese encyclopaedia' in which it is written that 'animals are divided into: (a) belonging to the Emperor, (b) embalmed, (c) tame, (d) sucking pigs, (e) sirens, (f) fabulous, (g) stray dogs, (h) included in the present classification, (i) frenzied, (j) innumerable, (k) drawn with a very fine camelhair brush, (l) et cetera, (m) having just broken the water pitcher, (n) that from a long way off look like flies.'

"In the wonderment of this taxonomy, the thing we apprehend in one great leap, the thing that, by means of the fable, is demonstrated as the exotic charm of another system of thought, is the limitation of our own, the stark impossibility of thinking that."

—From the introduction to *The Order of Things, An Archaeology of the Human Sciences,* by Michel Foucault, Vintage Books, New York, 1970.

Figure A   A Chinese hanging scroll depicting hens and roosters in a barnyard. *(Werner Forman Archive, National Gallery, Prague/Art Resource)*

Who Names a Species?

The first person to describe a new species assigns it its name. Sometimes biologists discover that two people have described and named the same species. The name given first is always considered the correct name and has "priority." Sometimes this is too bad. One hundred years ago, the Yale paleontologist Charles Marsh assigned the charming name Eohippus ("dawn horse") to the 55-million-year-old ancestor of the modern horse. Marsh did not know that the tiny fossil horse had already been described and named by someone who thought it was a hyrax (an animal that looks like a guinea pig, only bigger). Eohippus, which is both poetic and accurate, lost out to *Hyracotherium,* a mouthful that means hyraxlike mammal. Because of the rules of priority in naming, we are stuck with this unfortunate name—rules are rules.

Taxonomists use their naming privileges for different purposes. Some name species after colleagues they admire. Some try to describe the species: *Rosa alba,* for example, is a white rose (alba, as in albion or albumen, means white), while *Rosa chinensis* is a Chinese rose. Other biologists just want to have fun: the names for some genera of stink bugs (family Pentatomidae), for example, end with the suffix -chisme, pronounced "kiss me."

Species are generally named by the first person to recognize the species as new.

How Do We Recognize Species?

Long ago, the Greek philosopher Plato described how we make ordinary classifications. Plato argued that the individual objects of the world are all variations of universal "types," forms that share an ideal underlying plan (Chapter 15). A chair, for example, always has a seat large enough to support a sitter and, usually, has four legs and a back.

Aristotle extended Plato's concept of types to organisms. In this view, the **typological species concept,** each species consists of individuals that are variants of a fixed underlying plan. The key word here is "fixed." A typological species cannot evolve and become something different. Aristotle would have argued that a robin and a cat each have a fixed underlying plan.

Eighteenth- and nineteenth-century biologists applied the philosophy of types in the following way. When they first recognized a new species, they killed and collected several examples of the organism. One individual was then selected as the "type specimen" for each species and kept in a museum for reference. All other organisms believed to belong to that species were then compared to the type specimen to verify that they were the same. Because of Darwin, however, biologists have recognized that the typological species concept has two important limitations:

1. All species change over time.
2. The individual members of species vary.

Twentieth-century biologists have mostly defined **biological species** as groups of actually or potentially interbreeding populations that are reproductively isolated from other such groups (Chapter 17). In this view, called the **biological species concept,** individuals are members of a species only in relation to other species. In the same way, "brother" is not a category that can be applied to a single individual. Someone is a brother only in relation to someone else.

To decide if a group of organisms constitutes a separate species, most biologists would prefer to use the biological species concept whenever possible. For huge numbers of organisms, however, the biological species concept is virtually impossible to apply. Certainly we can obtain no information about breeding populations for fossil organisms or for organisms that reproduce asexually, including great numbers of plants, fungi, protists, and bacteria.

Even living, well-studied, sexually reproducing organisms pose problems. No one was certain, for example, how frequently gray wolves and coyotes interbred until biologists tested DNA in blood samples drawn from hundreds of wild wolves and coyotes. Such efforts are expensive and unusual. Many sexually reproducing plants and animals readily form hybrids with groups traditionally classified as separate species.

We can see two objections, then, to the biological species concept: (1) the frequent lack of information about whether two groups can interbreed; and (2) the fact that many recognized species can and do interbreed.

Biologists cope with the practical problems presented by the biological species concept by recognizing that a biological species is usually a reproductive unit, a genetic unit, and an ecological unit. A species is a **reproductive unit** in the sense that the species is the group of potential mates. Among organisms that do not reproduce sexually, that definition of a species does not apply. A species—whether sexually reproducing or not—is also a **genetic unit,** because a species can be defined as a gene pool (usually isolated from other "species" by reproductive barriers).

Finally, a species is an **ecological unit:** within a given habitat, a species interacts in a predictable fashion with other species in the environment. Gray wolves, for example, hunt in packs of 7 to 20 if large animals such as deer, moose, and caribou are available. In contrast, coyotes do not hunt in packs and prefer to prey on smaller animals such as mice and hares. Because of their size and behavior, coyotes interact with their environment differently from wolves. Biologists now know, for example, that although under certain circumstances wolves and coyotes interbreed, they also know that the two species remain distinct, genetically and ecologically.

Even these rules are not enough to help classify all organisms. As we will see in the next chapter, bacteria frequently exchange genes with other unrelated species. Thus bacteria are not generally reproductive units or even genetic units, but only ecological ones, and the way that bacteria interact with their environment can change radically with the sudden acquisition of a single gene.

In reality, biologists usually classify species by the way they look, relying on morphology—the form or visual appearance—of organisms for the provisional definition of a species. As we saw at the beginning of this chapter, biologists name even living, sexually reproducing species such as the red wolf, on the basis of their morphology simply because it is easier.

Fortunately, the morphologically defined species usually corresponds exactly to the biologically defined species. The gray wolf and the coyote, for example, look different and behave differently. In large part, they occupy different habitats. In short, they look like true species and act like true species. Genetic data confirm that the two species have unique sequences of DNA, showing that they have been reproductively isolated from each other. That these two species can interbreed extensively does not make them the same species. In general, the morphologically defined species corresponds to the biologically defined species because reproductive isolation tends to lead to the evolution of divergent morphological characters. Despite most biologists' noisy objections to typological species, then, they frequently begin defining a species on a typological basis.

Biologists use both biological and typological concepts to name and classify species.

HOW DO BIOLOGISTS GROUP SPECIES?

Although biologists with different agendas argue whether particular groups deserve the species label, modern biologists all consider species to be real entities, not abstract or arbitrary classifications. Because species are natural entities, recognizing what constitutes a species is obvious most of the time. Biologists disagree surprisingly rarely on the boundaries that separate individual species. The same researchers may bicker constantly, however, about how to subdivide species into subspecies and how to place species into larger groups such as genera, families, orders, phyla, kingdoms, and now domains.

Classification of species according to their evolutionary relationships is the main business of systematics. **Systematics** is the scientific study of the kinds and diversity of organisms and of any and all relationships among them. In contrast, **taxonomy** is the theoretical study of classification, including the study of the principles, procedures, and rules of classification.

BOX 19-2

Are guinea pigs rodents?

One group of animals whose classification has come under close scrutiny recently is the guinea pigs and their relatives. One of the defining characteristics of rodents is a single pair of razor-sharp incisors with which they can gnaw through the husks of nuts and seeds. In fact, the word rodent means "to gnaw." Because guinea pigs possess these distinctive gnawing incisors, as well as other adaptations for gnawing, taxonomists have traditionally classified them in one of three suborders of rodents.

The three suborders—the squirrels, the mice and rats, and the guinea pigs and their relatives—are classified according to the anatomy of their gnawing muscles. The principle jaw muscle is the masseter muscle, which closes the jaws and also pulls the lower jaw forward, a movement critical to gnawing. In mice and rats, the two sections of the masseter—the lateral masseter and the deep masseter—both attach well forward on the jaw, and these animals' ability to gnaw is unparalleled. In squirrels, only the lateral masseter extends forward. In guinea pigs, only the deep masseter extends forward. Consequently, neither squirrels nor guinea pigs gnaw as well as mice and rats.

The suborder to which the guinea pigs belong, called the caviomorphs, differs from other rodents in numerous ways. Caviomorphs are more diverse in their morphology and habits than either the squir-

Figure A   Mother guinea pig and her baby. Are they in the right class? *(Gerard Lacz/Peter Arnold, Inc.)*

rels or the rats and mice. They include porcupines, chinchillas, various guinea pigs (called cavys), the capybara, the pacas, and the agoutis. The capybara, for example, is a large, social herbivore. Individuals may reach 150 pounds. Caviomorphs all have proportionately larger heads than other rodents and they reproduce differently (Figure A). While a rat may bear litters of 11 every three weeks, a guinea pig bears 2 or 3 young every few months.

Like mice and rats, guinea pigs are frequent subjects of scientific experiment. As a result, quite a bit is known about guinea pig physiology and genetics. In 1991, researchers suggested that the protein sequences of guinea pigs are sufficiently different from those of rats and mice to justify

These two words have different meanings, but, for simplicity, we use the word taxonomy only in this chapter.

Biologists would like to classify organisms according to *relatedness,* but they must do so from measurements of *similarity.* In classifying any kind of objects—organisms, books, or buildings—we naturally group things according to their similarities. We can understand the problems this presents if we imagine attending a big family reunion of 200 people and trying to decide who is related to whom. A group of redheads might appear to be related. We might find, however, that one of them is simply married to someone else in the family and is

not related to anyone else at the reunion. Similarity by itself does not always signify relatedness, and which similarities are important ones is often a matter of much dispute.

For example, in Papua New Guinea, the giant tropical island just north of Australia, biologists have compared their own classification of plants and animals with that used by the tribespeople who live there. The biologists and the tribespeople agree nearly exactly on the number and the identity of species. They disagree, however, on the higher classifications. The biologists group organisms according to their biological similarities—lumping all the birds together in the class Aves, for example.

giving the guinea pigs and their relatives a separate order.

A series of studies in the early 1990s on both protein and nucleotide sequences gave contradictory answers, however, and the guinea pigs remained classified as rodents. Then, in 1997, a group of Italian and Swedish researchers carefully analyzed differences in the mitochondrial DNA (mtDNA) of 16 animals, whose complete mtDNA sequences were known (Figure B).

The results were dramatic. Three different methods of classification all put the guinea pigs in a separate taxon from the mice and rats. Indeed, the guinea pigs appeared to be more closely related to rabbits and primates than to rodents. Although the five researchers called for further research—"the phylogeny will be clearer when a larger number of species can be considered in phylogenetic analysis"—they titled their research report in the journal *Nature* decisively "The guinea-pig is not a rodent."

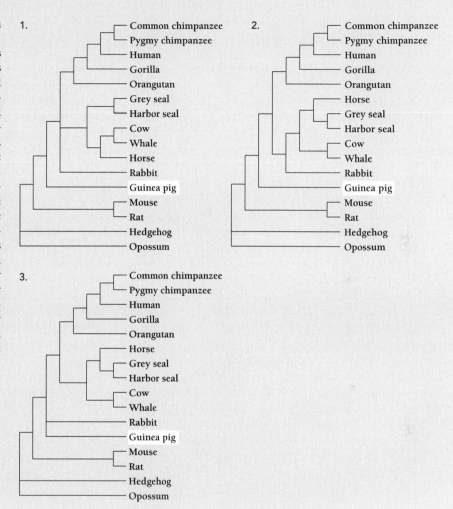

Figure B Three methods of analyzing the genetic differences among 16 species of mammals produced similar classifications of the guinea pig. In each case, guinea pigs appear to be at least as different from rodents as rabbits are.

The tribespeople classify according to use—lumping together all the species that are edible, all that are poisonous, or the ones that provide materials for clothing. Both systems of classification serve legitimate purposes.

Taxonomy invites dissent. As a result, biologists now use at least four methods for classifying organisms. The first is classical evolutionary taxonomy, in use for the last 150 years. More recently, two groups of biologists have each proposed a new way to classify organisms. The first of these approaches—phenetics—focuses on the practical problem of taxonomy and proposes a seemingly objective step-by-step method for classify-

ing organisms. The second approach—cladistics—proposes to classify organisms according to their evolutionary history. In addition, molecular biologists have begun classifying organisms according to their molecular similarities.

Classical Evolutionary Taxonomy

Classification by means of classical taxonomy relies explicitly on the judgment and experience of the taxonomist. In fact, in one definition of a species, biologists like to say, "A species is what a good taxonomist says it is. And a good taxonomist is one whose work has stood the test of time." A good taxonomist examines the similarities and differences in as many traits as possible. The most commonly used traits in living animals are morphological ones—including external form, internal anatomy, tissue types, and chromosomal morphology. Mammals, for example, are classified into three broad groups according to their skeletons. Adult monotremes such as the duck-billed platypus and the spiny anteaters have no teeth and lack certain bones of the skull that other mammals have. Their limbs stick out to the sides in a fashion that resembles that of lizards and other reptiles. Because of these and other traits, monotremes are considered primitive mammals.

Classical taxonomists also rely on embryological evidence. For example, the early embryos of all mammals, including humans, include a stage with gill pouches (Figure 15-15). This fact, with others, suggests that we are descended from aquatic organisms that needed gills to extract oxygen from water.

Not all similarities between organisms mean that they are related. Bats and birds have both evolved wings for flight, for example, but bats are no more related to birds than are humans. Similarities that are inherited from a common ancestor are said to be **homologous.** Similarities that perform the same function but that evolved separately are said to be **analogous.** The legs of insects are analogous to those of humans and other vertebrates. They are jointed and muscled in a similar manner because they perform the same job, not because they are related. In contrast, the leg of a horse and the leg of a human are homologous.

When classifying organisms, a classical taxonomist must take into account which characteristics are likely to be homologous and which only analogous, giving weight to homologous characters and disregarding analogous. Although the wings of bats and birds seem obviously different to us, distinguishing analogous structures from homologous ones can sometimes be difficult. Intuitive judgments may appear subjective or even arbitrary.

Homologous characters indicate relatedness, while analogous characters indicate similar adaptations in unrelated organisms. Taxonomists try hard to distinguish homologous characters from analogous ones.

How Do Pheneticists Attempt To Produce an "Objective" Classification?

Phenetics is a method of classification that attempts to avoid the subjective choices of classical taxonomy. **Phenetics** is based on all observable characteristics. The word "phenetics" comes from the same Greek root as "phenotype," *phainein* = to show. The phenetic taxonomist characterizes the phenotype of an organism by a series of yes or no questions. Each of these questions defines a *unit character,* a characteristic of an organism that cannot be split into a set of finer descriptions. For an animal, some unit characters are the answers to questions such

as: (1) Does it have five digits on its front limb? (2) Does it have four digits on its front limb? (3) Does it have a backbone? (4) Does it have wings? The pheneticist records the answers to these questions in a binary (1 or 0) code: if the answer is yes, he or she records a 1; if the answer is no, a 0. Phenetics assigns equal weight to each unit character. That is, every character is considered to be equally important. Whether an animal has a tail would be considered as important as whether it has a backbone.

By compiling as many unit characters as possible for each organism, the pheneticist assembles a scorecard consisting of a string of 1s and 0s for each organism. Using a computer, the pheneticist can sort organisms according to their degree of similarity. Pheneticists (who sometimes call themselves "numerical taxonomists") argue that the resulting groupings are objective measures of the similarities among organisms and do not depend on observation of reproductive behavior.

But does phenetics really represent an "objective" taxonomy? Most traditional biologists think not. To begin with, the choice of unit characters is, in the final analysis, subjective. Second, characters that arose independently in different groups will suggest groupings unrelated to evolution. For example, the wings and other adaptations to flight in bats and birds would suggest that these two groups are more closely related than they are. Traditional taxonomists believe that some characters reveal more than others: the presence or absence of a backbone, for example, is considered more important than the number of digits on a limb. (This saves the taxonomist from finding relatedness between unrelated organisms such as the horse and the mosquito, each of which has just one toe per limb.) Furthermore, environmental conditions can affect some characters

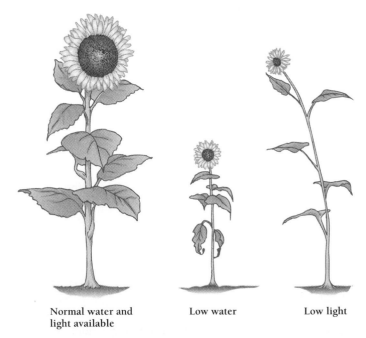

Figure 19-6 A sunflower's characteristics depend on how much water and light it receives.

Normal water and light available

Low water

Low light

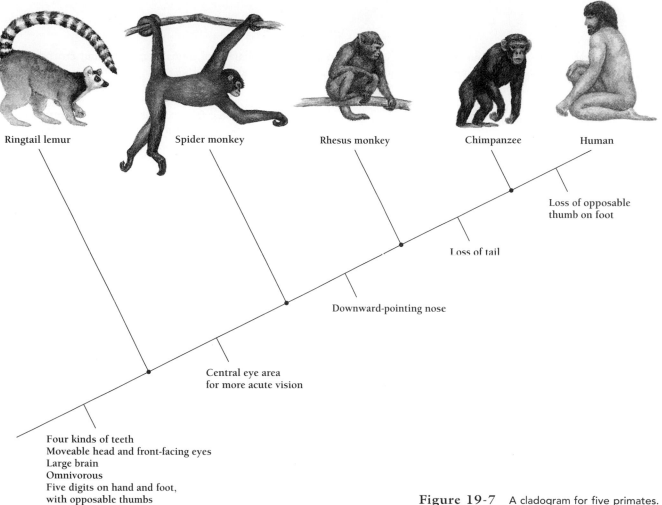

Ringtail lemur

Spider monkey

Rhesus monkey

Chimpanzee

Human

Loss of opposable
thumb on foot

Loss of tail

Downward-pointing nose

Central eye area
for more acute vision

Four kinds of teeth
Moveable head and front-facing eyes
Large brain
Omnivorous
Five digits on hand and foot,
with opposable thumbs

Figure 19-7 A cladogram for five primates.

(such as size) more than others (such as the number of digits on a limb or the number of lobes on a leaf) (Figure 19-6).

Such judgments are difficult to make in groups such as the bacteria. In fact, bacteria are commonly classified phenetically, as they are difficult or impossible to classify by traditional methods. Nonetheless, most biologists prefer to make their own considered judgments about which characters of an organism are important for classification, rather than to average together all possible characteristics.

Pheneticists classify species and higher groupings according to a "numerical taxonomy," in which they assign equal value to every character that they list.

How Do Cladists Attempt To Produce an Evolutionary Classification?

Another approach to classification is **cladistics** [Greek, *clados* = branch]. A cladistic taxonomist's first goal is to describe the groupings of organisms in a way that shows their **phylogeny** [Greek, *phylon* = race, tribe + *geneia* = birth, origin], or evolutionary history. Cladists therefore present their groupings of organisms in evolutionary trees, called cladograms.

According to cladists, every taxon above the species level—whether it is a genus, a family, or a higher category—should be **monophyletic**, that is, each taxon should include the ancestral species and all its descendants. Birds, for example, form a monophyletic taxon: all birds are descended from a common ancestor, which was itself a bird, and all the descendants of this ancestor are birds.

Cladists reject any taxon that is **polyphyletic**—one that includes descendants of more than one ancestor. As we will see in the following chapters, a number of higher taxa, including the Animals, appear to be polyphyletic. Although the sponges have long been classified as animals, for example, it now appears that the sponges evolved from separate ancestors. For this and other reasons, cladists reject the traditional grouping of animals into the kingdom Animalia, which is, nonetheless, still in use.

To see how cladists classify organisms, we can construct a cladogram for our own order, the primates. Primates are one of 20 orders of mammals. All primates have four limbs with five digits each, four kinds of teeth, moveable heads with front-facing eyes, good eyesight, a poor sense of smell, large brains, and a diet that includes both plants and animals.

Figure 19-7 shows a cladogram for five primates: a ringtail lemur, a spider monkey, a rhesus monkey, a chimpanzee, and a human. Constructing the cladogram requires that we note characters in which these species differ. Four of these species (all except the lemur) possess a specialized area in the center of the eye that allows more acute vision. Three species (rhesus monkey, chimp, and human) have nostrils that point downwards. Humans and chimps differ from the other three species in that they lack tails.

The cladogram in Figure 19-7 is a classification hierarchy that represents a series of hypotheses about the evolutionary history of the primates. Biologists can test these hypotheses by examining other features of these five organisms and by comparing these organisms with the fossil record. They can also compare the sequences of specific proteins and genes.

In making the cladogram, we choose characters that distinguish subgroups of the five species. As it happens, the cladistic groupings correspond to the classical taxonomic groupings. For contrast, let us look at an example of a conflict between traditional classification and cladistics.

Cladists group organisms according to a system of logic designed to clarify evolutionary relationships between groups.

How Does the Classification of Reptiles Illustrate the Conflict Between Cladistics and Traditional Taxonomy?

The cladists' argument that phylogeny must be the basis of classification leads them to reject the entire taxon of what we know as reptiles. At first glance, reptiles appear to be a monophyletic group. That is, all reptiles appear to have descended from a common ancestor that was itself a reptile. But, in fact, reptiles are **paraphyletic**—meaning that some of the descendants of reptiles have evolved into organisms not considered reptiles. Specifically, birds and mammals are descended from early reptiles, yet are not classified as reptiles.

The first reptiles arose about 300 million years ago. Among their descendants were the stem reptiles, which split into many separate lines, including the turtles, lizards, snakes, primitive mammals, and archosaurs. The archosaurs, in turn, split into various dinosaurs, primitive birds, and crocodiles.

From the cladogram in Figure 19-8, we can construct a nested set of taxa for birds, mammals, and reptiles. According to this scheme, reptiles are not a legitimate taxon because they are not truly monophyletic—unless we were to include the birds and mammals as reptiles. In such a scheme, birds and mammals would be reptiles, classified at the same level as crocodiles, for example.

Traditional biologists tend to dislike this scheme, however, because such a cladistic classification obscures major similarities among all the reptiles. From a physiological and an ecological standpoint, a lizard and a crocodile resemble one another more than either resembles a dog or an eagle. Cladistics has nonetheless contributed a great deal to current understanding of evolutionary relationships. For example, the common nesting behavior of dinosaurs, birds, and crocodiles reveals their familial ties.

The rules of cladistics force cladists to ignore similarities in physiology and lifestyle when grouping species into higher taxa.

Do DNA and Protein Sequences Provide "Objective" Criteria for Classification?

As we saw in Chapter 15, the sequence of a protein evolves in much the same way as a morphological feature. For this reason, comparisons of amino acid sequences (or nucleotide sequences) suggest relationships between species. For example, all other things being equal, two species with identical genes for hemoglobin are more likely to be closely related than two species whose genes for hemoglobin differ by several bases.

Because base sequences are linear and unambiguous, it is easy to imagine ranking all genetic differences equally, as pheneticists suggest we do with morphological differences. Indeed, evolutionary biologists have embraced the molecular approach enthusiastically, partly for this reason. The gene sequences that specify proteins, however, evolve in the same way as larger-scale characters, and molecular biologists may not always know whether similar sequences are homologous or analogous. Base sequences, then, are not automatically more objective measures of relatedness than morphological characters.

Molecular biologists must also make subjective choices about which characters are important. These choices are necessary because characters do not all evolve at the same rate. We can see that a structure such as the backbone may persist for 100 million years, while another character such as fur color may change dramatically in just a few generations. It therefore makes more sense to group animals according to whether they have backbones than according to their external coloring.

Genes, and the polypeptides they specify, also evolve at different rates. Histone 4, a protein that binds tightly to nuclear DNA, appears to have remained nearly unchanged for over 500 million years, while other protein sequences diverge strongly over relatively short periods. Proteins, like morphological characters, evolve at different rates according to how much selective pressure they experience. The divergence of sequences, then, reflects natural selection. It makes sense when classifying organisms, then, to give priority to the proteins that have evolved most slowly.

Even within a single protein, some amino acids change more than others. For example, in cytochrome *c*—which transfers electrons in the electron transport chain—27 amino acids are the same in all organisms from yeast to humans. Other positions in the polypeptide chain vary widely, while still others vary only slightly. Some positions are occupied only by uncharged amino acids, others only by charged amino acids. At the level of DNA, base substitutions that change the meaning of individual codons, and so change the amino acid sequence

Cladogram

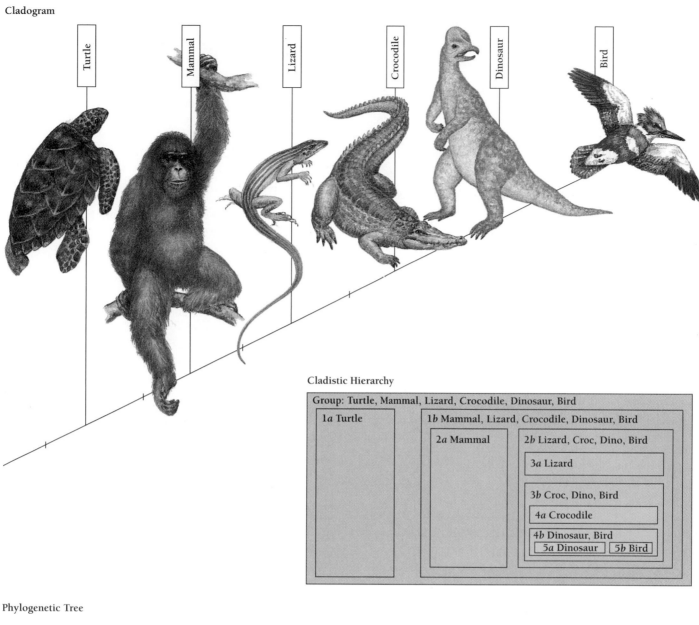

Cladistic Hierarchy

Group: Turtle, Mammal, Lizard, Crocodile, Dinosaur, Bird

1a Turtle	1b Mammal, Lizard, Crocodile, Dinosaur, Bird

Within 1b:

2a Mammal	2b Lizard, Croc, Dino, Bird

Within 2b:

3a Lizard

3b Croc, Dino, Bird

4a Crocodile

4b Dinosaur, Bird

| 5a Dinosaur | 5b Bird |

Phylogenetic Tree

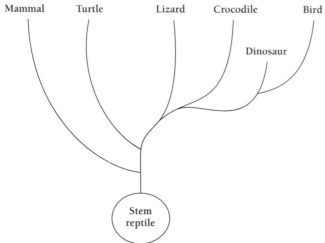

Mammal   Turtle   Lizard   Crocodile   Bird

Dinosaur

Stem reptile

Figure 19-8 Cladistic classification of birds, mammals, and reptiles. In the cladistic view, reptiles are a legitimate taxon only if they include birds and mammals.

of a protein, diverge more slowly than those that merely substitute a redundant codon. It seems that no matter how we look at phylogeny, we must ultimately accept that some characters are more useful indicators than others.

But what about DNA not subject to natural selection? Molecular biologists argue that DNA sequences that do not code for proteins may mutate at a constant rate that is not influenced by the environment of the genomes. If so, changes in such sequences would represent the tick of a molecular clock.

As with other traits, some nucleotides and amino acid sequences are more important and more stable than others. Molecular approaches to classification are extremely useful, but they are not necessarily more objective than traditional methods.

Why Do Cladists, Pheneticists, and Classical Taxonomists Disagree?

Cladists insist that classification must reflect evolutionary history. They accept the biological species concept but argue about higher categories. Pheneticists, on the other hand, rely solely on observed differences in modern organisms. Pheneticists take issue with the biological species concept, saying that it relies on subjective judgments nearly always made in the absence of sufficient information.

If all genes and morphological features changed at constant rates, then cladists, pheneticists, and traditionalist taxonomists would nearly always agree on classification. Similarly, paleontologists, systematists, and molecular biologists would be likely to agree on phylogeny. But evolution does not conform to the desires of biologists. Instead, the characters of organisms change in response to the chance appearance of new mutations, the changing geography of our planet, and the many forces of natural selection.

In the following chapters, we follow a traditional classification. The view taken in this book is that (1) biologists and students must be aware of the difficulties of classification; and (2) we must be able to communicate with one another. We therefore will follow the most widely accepted conventions for classification, while occasionally noting the difficulties of conventional wisdom.

Classification is made difficult by the fact that the forces of natural selection shape organisms (and their genes) in ways that may obscure patterns of relatedness.

HOW MANY KINGDOMS OF ORGANISMS ARE THERE?

Linnaeus divided all organisms into two kingdoms, the plants and the animals. As biologists learned more about life on Earth,

however, they became dissatisfied with this grouping. The fungi might not move about as animals do, for example, but they were clearly not photosynthetic like plants. In 1969, Robert H. Whittaker of Cornell University proposed a five-kingdom classification that most biologists accepted. Whittaker put all the prokaryotes into the kingdom Monera, then divided the eukaryotes into four kingdoms, the protists, fungi, plants, and animals.

In recent years, the five-kingdom classification has come under increasing attack. Cladists object to it because none of the five kingdoms is monophyletic. Eukaryotes are all descendants of prokaryotes and therefore clearly members of other kingdoms, and the four eukaryotic kingdoms each include lines of organisms with different origins. For example, among the protists, the green algae and the protozoa each arose from different prokaryotic ancestors. Among the animals, the sponges probably have a separate origin from other animal phyla.

Researchers have suggested several other classifications including one with eight kingdoms. One group of organisms that taxonomists now recognize as distinct are the archaebacteria, or Archaea, a group of prokaryotes that have a separate phylogeny from that of other bacteria. The Archaea are anatomically and metabolically distinct from all other bacteria and so diverse that some taxonomists argue that the Archaea comprise three kingdoms of organisms. Whether the Archaea are three kingdoms or one, most biologists now accept the idea that they must be in a domain separate from either the Eukarya or the Bacteria.

In this book, we give them just one kingdom, the Archaebacteria, within the domain Archaea. That is, we use a six-kingdom classification. Many biologists, however, still employ the five-kingdom classification, as it nicely reflects similarities and differences among very large groups of organisms.

The five- and six-kingdom classifications distinguish among organisms in two ways—cellular organization and source of energy. The cells of the prokaryotes differ from those of the other four kingdoms: prokaryote cells lack a nucleus or other membrane-enclosed organelles. The protists, fungi, plants, and animals are all eukaryotic, meaning that they have nuclei and membrane-enclosed organelles. The five- and six-kingdom schemes distinguish among plants, animals, and fungi by the manner in which their members derive energy—plants from photosynthesis, animals from eating other organisms, and fungi from absorbing chemicals from their surroundings. The protists are usually, although not always, single-celled eukaryotes that do not fit into any of the other three kingdoms of eukaryotes.

In the following five chapters, we will survey some of the major groups within each kingdom. In doing so, we will examine the diverse ways by which species survive and flourish.

The six-kingdom classification we will study reflects differences and similarities in cellular organization and lifestyle, rather than strict evolutionary relationships.

STUDY OUTLINE WITH KEY TERMS

A **taxonomist** is a biologist who classifies organisms. Each classification of organisms is a **taxon.** Many biologists now use the word **systematics** for the classification of organisms and **taxonomy** for the study of *how* organisms should be classified. Until this century, biologists usually viewed the members of a species as variants from a fixed underlying plan, in what is now called the **typological species concept.** Today, biologists usually define a species according to the **biological species concept,** which says that a species is a group of actually or potentially interbreeding organisms that do not interbreed with members of other such groups. Besides being **reproductive units, biological species** are also **genetic units** and **ecological units.**

Most of the time, morphologically defined species correspond to the biologically defined species. Classification of species into larger groups, however, may be difficult. Biologists use a **hierarchical** scheme, in which species are grouped into **genera,** genera into **families,** families into **orders,** orders into **classes,** classes into **phyla** (or divisions), and phyla into **kingdoms.** Kingdoms are further arranged into three **domains**—the Archaea, the Bacteria, and the Eukarya. Each species has a two-word designation, called a **binomial,** which names both the **genus** and the species.

The biological species concept is not always useful. Classifiers may know nothing about breeding patterns, and many species interbreed with other species. **Phenetics** gives equal weight to every character, grouping organisms with similar clusters of characteristics. But not all common characters indicate relatedness. Similarities that are inherited from a common ancestor are said to be **homologous** (and are given more weight by taxonomists and systematists). Similarities that perform the same function but that evolved separately are said to be **analogous.**

Cladistics insists that the only meaningful classification scheme is one that reflects **phylogeny,** or evolutionary history The goal of these biologists is to construct evolutionary trees, or cladograms, that show the relationship of groups of species. Cladists insist that every recognized grouping in their classifications be **monophyletic,** that is, each group should include an ancestral species and all its descendants. Many traditionally accepted groupings, including the five kingdoms, then, are **paraphyletic** or **polyphyletic** and therefore unacceptable to cladists. The six-kingdom classification reflects differences and similarities in cellular organization and lifestyle, rather than strict evolutionary relationships.

REVIEW AND THOUGHT QUESTIONS

Review Questions

1. Define a species according to the biological species concept. What two kinds of problems often make this definition impractical?
2. What two purposes are accomplished by Linnaeus' system of classification? Name the seven levels of taxa commonly used by taxonomists.
3. Explain how a numerical taxonomist would decide if a group of organisms constituted a species. Define phenetics.
4. What is cladistics? Why do cladists require that a lineage be monophyletic to be considered part of a single taxon?
5. Why do cladists insist that the reptiles are not a legitimate taxon?
6. Define paraphyletic.
7. Does DNA sequencing provide a more objective method for classifying organisms?

Thought Questions

8. Do you think that the red wolf should be saved from extinction? Do you think that the decision to save it should be influenced by whether it is a species, a subspecies, or a hybrid?
9. If you found a new species of bird, what would you name the bird? If you were in charge of classifying this bird, what system of classification would you use? Why? Now describe the process, step-by-step, by which you would classify your bird. You may make up characters.

SELECTED READINGS

Hull, David L., *Science as a Process: An Evolutionary Account of the Social and Conceptual Development of Science,* University of Chicago Press, Chicago, 1988. A scholarly sociological and historical treatise on the evolution of scientific thought. Hull focuses especially on taxonomy and systematics.

Steinhart, Peter, *The Company of Wolves,* Alfred A. Knopf, New York, 1995. A thorough and readable history of biological research on wolves. Steinhart interviewed dozens of opinionated biologists for this fascinating account.

Wayne, Robert K., and Gittleman, John L., "The Problematic Red Wolf," *Scientific American,* July 1995. A review of the genetic evidence bearing on whether the red wolf is a species or a hybrid.

Weiner, Jonathan, *The Beak of the Finch,* Vintage Books, New York, 1995. Highly readable. This Pulitzer Prize–winning book explores 20 years of research on natural selection in Galápagos finches by Peter and Rosemary Grant and their colleagues. Variation and speciation are discussed clearly and concretely and interspersed with amusing anecdotes about the birds themselves and about research on a desert island.

▶ On-line materials relating to this chapter are on the World Wide Web at
http://www.saunderscollege.com/lifesci/
Click on Tobin/Dusheck: *Asking About Life.*

Tracking Down a Killer

In 1983, a young scientist named Scott A. Holmberg showed for the first time that feeding antibiotics to farm animals can sicken and even kill people. Most of us have experienced food poisoning, the severe nausea and diarrhea that results from consuming meat, milk, or eggs contaminated with salmonella bacteria. Farm animals—whether cattle, pigs, or chickens—can carry salmonella, but animals fed antibiotics, Holmberg showed, carry especially deadly strains of the bacteria.

In February 1983, Holmberg was working at the Centers for Disease Control (CDC), the nation's premier center for the study of all public health problems, from AIDS to smoking among teenagers. His boss, Mitchell Cohen, had received a report that 11 Minnesotans had been hospitalized with salmonella poisoning. Eleven people with salmonella in Minnesota was hardly an epidemic: after all, 40,000 Americans are hospitalized with salmonella every year, and another 5 million suffer the severe diarrhea and vomiting associated with salmonella poisoning but don't need hospitalization. What was both disturbing and exciting about the Minnesota cases was, first, that most of the patients had been taking antibiotics when they came down with food poisoning and, second, that each patient was extremely sick. One man

had died. Cohen strongly suspected that antibiotics were somehow causing or helping to cause the food poisoning. But how?

Cohen called Holmberg on a Saturday, and by the next morning, Holmberg was on a flight to Minnesota to find out. In Minnesota, Holmberg met with other **epidemiologists** [Greek, *epidemia* = how widespread a disease is]—researchers who study the incidence and transmission of diseases in populations. By that afternoon, Holmberg and state epidemiologist Michael Osterholm had two facts in hand.

First, all of the victims were infected with the same strain of *Salmonella newport*. This strain of bacteria, resistant to several common antibiotics, including penicillin and tetracycline, was unusual. The fact that all the victims had been made ill by the same rare form of *S. newport* suggested that they had all acquired the infection from a common source. In short, they must have all eaten the same thing. But what?

Second, 8 of the 11 victims had been taking antibiotics manufactured by the same drug company. At first glance, it looked like the antibiotics themselves were contaminated with salmonella bacteria. However, because 3 of the 11 patients had not been taking antibiotics at all, Holmberg and Osterholm ruled out that possibility.

Nonetheless, another clue suggested that antibiotics played an important role. One woman had been hospitalized with salmonella poisoning the day after she had begun taking amoxicillin for a sore throat. Her husband then developed a sore throat, too, and took some of his wife's amoxicillin. Within 48 hours, he was in the hospital with salmonella poisoning.

Holmberg and Osterholm hypothesized that the antibiotic had actually triggered the couple's salmonella poisoning (Figure 20-1). The couple must have been infected with antibiotic-resistant salmonella that suddenly flourished when the amoxicillin killed off all other competing bacteria. Without competition, the salmonella bacteria multiplied from a few thousand cells into a full-scale infection of millions upon millions of cells in just a few hours. But why were the couple harboring antibiotic-resistant salmonella bacteria in the first place? Where had they come from?

Holmberg spent the following two weeks searching for the common source of the food poisoning in the homes of salmonella victims. He looked through refrigerators and cupboards, searching for some sort of food they had all eaten—some common source for the antibiotic-resistant bacteria.

Finally, with still no answer, he returned to the CDC frustrated and discouraged. He and his boss, Mitchell Cohen, knew that the most likely scenario was that all of the patients had consumed meat, milk, or eggs infected with resistant *S. newport*. In the United States, large numbers of farm animals are fed low doses of antibiotics to increase their growth rate.

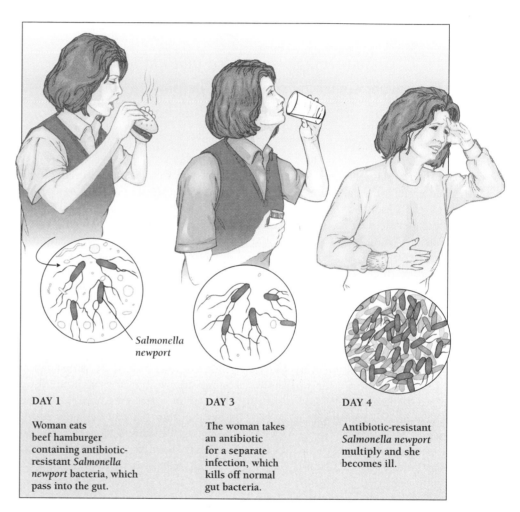

Figure 20-1   How antibiotics can trigger an infection.

Salmonella newport

DAY 1

Woman eats beef hamburger containing antibiotic-resistant *Salmonella newport* bacteria, which pass into the gut.

DAY 3

The woman takes an antibiotic for a separate infection, which kills off normal gut bacteria.

DAY 4

Antibiotic-resistant *Salmonella newport* multiply and she becomes ill.

This practice kills bacteria that are sensitive to antibiotics, encouraging the proliferation of antibiotic-resistant bacteria. In 1983, Cohen and other biologists strongly suspected that farm animals raised on antibiotics were making people sick, but they had not yet been able to prove it.

Cohen knew of a new technique, however, that might help Holmberg make a direct link between individual farm animals and individual salmonella victims. Molecular biologists had recently devel-oped a way to "fingerprint" bacteria by looking at the genetic profile of plasmids, the simple loops of DNA that inhabit bacteria (Chapter 12). With this new tool, every strain of bacteria could be distinguished from all others according to the kinds of plasmids it carried.

Cohen suggested that Holmberg find out where else in the United States *S. newport* existed. Holmberg wrote to epidemiologists all over the United States, and, within a month, he had a clue that would solve the case. The state epidemiologist in South Dakota wrote to say that antibiotic-resistant *S. newport* had recently sickened four people.

From Atlanta, Holmberg interviewed the South Dakota victims by phone. One of the victims, a dairy farmer, told Holmberg that his dairy cows had all had diarrhea the previous November, and that one dead calf had been autopsied and found

that between this point here in South Dakota and that point there in Minnesota, there was a connection. Nature does not hand you coincidences like this. I was pretty sure that the beef herd was the source [of the *S. newport*]. . . . Now all I wanted to do was connect the two points . . ."

Connecting the points, it turned out, was not so easy. In mid-January, just days

the source of the antibiotic-resistant salmonella in Minnesota.

The new genetic fingerprinting technique had given epidemiologists their first clear evidence that feeding food animals antibiotics can make humans sick. Antibiotic-resistant bacteria are, it turns out, unusually dangerous. Holmberg showed that people are 21 times more likely to die from an infection by antibiotic-resistant salmonella than from an infection by non-resistant salmonella. Similarly, a recent study showed that 25 percent of burn patients carrying antibiotic-resistant *Staphylococcus aureus* end up with a potentially fatal staph infection, compared with only 4 percent of burn patients carrying ordinary *S. aureus*.

Cohen strongly suspected that antibiotics were somehow causing or helping to cause the food poisoning. But how?

---

to have died of salmonella. Researchers at the state agricultural laboratory that had analyzed the dead calf's salmonella told Holmberg that it had the same plasmid profile as the salmonella bacteria in the Minnesota patients. Holmberg soon discovered that three of the four South Dakota victims had eaten beef from cattle raised on a farm adjacent to the dairy farm. The beef farmer said that his cattle had been perfectly healthy, adding that he regularly laced their feed with the antibiotic tetracycline, a handful to every ton of grain. Even if the beef cattle had shown no symptoms of salmonella, however, they could have been carrying salmonella just the same.

"I knew what the story was," Holmberg later told *Science* magazine. "I knew

before the salmonella outbreak in Minnesota, the beef farmer had sent all 105 head of cattle to slaughter. Now the beef cattle were gone, and Holmberg had no way to test the cattle for salmonella.

Nonetheless, Holmberg was able to follow the path the tainted meat had traveled to supermarkets in Minnesota and Iowa, where 10 of the salmonella victims had shopped. In addition, Holmberg was able to show that the strain of *S. newport* from the dead dairy calf was truly distinctive. He examined 91 samples of different strains of *S. newport* from all over the United States, yet only the bacteria from the dead calf had the same plasmid profile as the salmonella in the sick Minnesotans. Holmberg no longer doubted that the beef cattle in South Dakota were

Amazingly, the practice of feeding antibiotics to farm animals continues today. Despite the clear hazards to public health, no laws regulate the practice. As a result of public pressure, only 20 percent of poultry growers now use antibiotics, the poultry industry reports. The beef industry recommends that its members only use antibiotics not used in humans. The pork industry continues to use human antibiotics without restraint.

How bacteria develop resistance to antibiotics, and how they pass that resistance among themselves, is a topic that reveals much about the lives of bacteria. In the rest of this chapter, we will learn a little about how bacteria are different from the rest of us and what kinds of bacteria populate the Earth. Along the way, we will see why bacteria are so good at becoming resistant to antibiotics.

KEY CONCEPTS

1. Plasmids enable prokaryotes to accumulate resistance to several antibiotics.

2. The structural characteristics of prokaryotes distinguish them from the eukaryotes.

3. Prokaryotes are successful because of their rapid growth rates and their metabolic versatility.

4. Prokaryotes are metabolically far more diverse than eukaryotes.

5. The metabolic abilities of prokaryotes provide a basis for their classification.

PROKARYOTES AND PEOPLE

The number of bacteria in your mouth at this moment exceeds the total number of people that have ever lived. You carry many more bacteria on your skin and in your gut than you have cells in your body. From a bacterium's point of view, your body is just another island to be colonized.

But bacteria are not limited to living in or on animals. They live in the bodies of every kind of organism, even inside other cells. They live in soil, ice, and water. A pinch of ordinary soil, for example, contains some 10 billion bacteria. And bacteria can live almost anywhere on Earth, from the ice fields of Antarctica to the steaming cauldron of a Yellowstone hot spring, from the bottom of the ocean to the driest desert soils. In the rest of this chapter, we will see that bacteria are the most numerous, the most diverse, and the most ancient of organisms. In short, the bacteria—the Archaebacteria together with the Eubacteria—are two of the most successful of the six kingdoms of organisms. Remarkably, bacteria, all of which are prokaryotes, are also the simplest organisms. In fact, whether we call these simple organisms prokaryotes or bacteria, we mean the same thing. In this chapter, however, we will refer to both the Archaea and the Eubacteria as prokaryotes or bacteria.

How Do Bacteria Make Us Ill?

Most people think of bacteria as lowly pests—carriers of disease and spoilers of food. In fact, most bacteria are very useful to us in one way or another. The trillions of bacteria in a handful of soil, for example, help recycle carbon and other nutrients. All the organisms on Earth depend on bacteria to pull nitrogen from the air so that we can use it to build proteins. Even the bacteria inside our bodies are useful. Some help us digest various kinds of foods. Others produce vitamins that we would otherwise have to get from special foods. In short, bacteria are a normal and mostly benign part of our lives.

But bacteria can also be **pathogens,** agents that cause disease. To cause a disease a bacterium usually must accomplish two things:

1. It must invade the host, attaching and multiplying on or in the host's tissues and cells.
2. It must produce a toxin—a chemical that destroys or interferes with the host's normal processes.

In some cases, disease results just from the excessive growth of pathogenic bacteria. But in most cases, disease results from toxins made by the pathogen. These toxins may be **endotoxins,** toxic substances that are components of the bacterial cells themselves or **exotoxins,** molecules that are secreted by the bacteria.

In the late 19th century, Robert Koch, a German physician, established rigorous criteria for identifying the pathogen for a given disease. The four rules, now known as **Koch's postulates,** are:

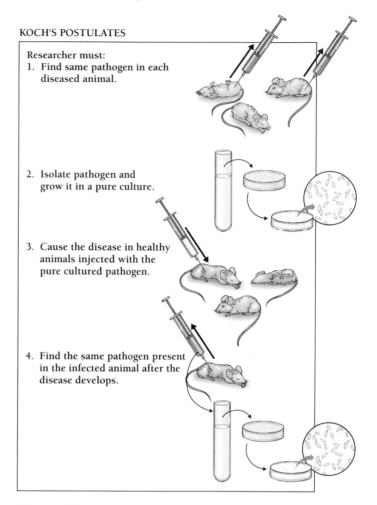

Figure 20-2 Koch's Postulates. Pathogens such as viruses that can only be grown inside of cells cannot be isolated in a pure culture.

1. The same pathogen must be present in every person with the disease.
2. The pathogen must be capable of being grown in a pure culture—one uncontaminated with other organisms.
3. The pure cultured pathogen, introduced into an experimental animal, must cause the disease.
4. The same pathogen must be present in the infected animal after the disease develops (Figure 20-2).

Although Koch's postulates are now the accepted guidelines for identifying pathogenic organisms, the postulates cannot always be fulfilled. For example, the bacterium, *Treponema pallidum,* which causes the sexually transmitted disease syphilis, has yet to be cultured in the lab. In addition, bacteria that live inside other cells (such as the sexually transmitted bacterium *Chlamydia*) cannot grow in pure cultures because these bacteria always require the presence of host cells. Likewise, all viruses must grow inside other cells (Chapter 12), so pathogenic viruses such as HIV (human immunodeficiency virus) and influenza cannot be grown in a "pure culture."

Not all pathogenic bacteria are strangers. Many are normal residents of our bodies that are out of place or have multiplied

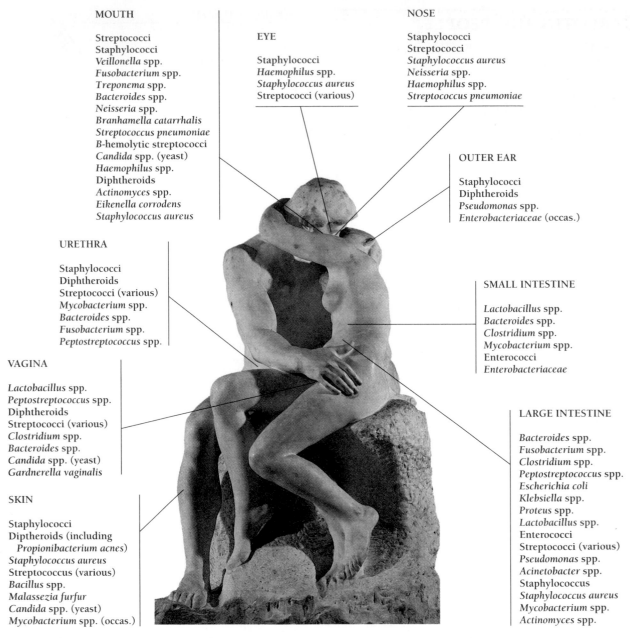

MOUTH

Streptococci
Staphylococci
Veillonella spp.
Fusobacterium spp.
Treponema spp.
Bacteroides spp.
Neisseria spp.
Branhamella catarrhalis
Streptococcus pneumoniae
B-hemolytic streptococci
Candida spp. (yeast)
Haemophilus spp.
Diphtheroids
Actinomyces spp.
Eikenella corrodens
Staphylococcus aureus

EYE

Staphylococci
Haemophilus spp.
Staphylococcus aureus
Streptococci (various)

NOSE

Staphylococci
Streptococci
Staphylococcus aureus
Neisseria spp.
Haemophilus spp.
Streptococcus pneumoniae

OUTER EAR

Staphylococci
Diphtheroids
Pseudomonas spp.
Enterobacteriaceae (occas.)

URETHRA

Staphylococci
Diphtheroids
Streptococci (various)
Mycobacterium spp.
Bacteroides spp.
Fusobacterium spp.
Peptostreptococcus spp.

SMALL INTESTINE

Lactobacillus spp.
Bacteroides spp.
Clostridium spp.
Mycobacterium spp.
Enterococci
Enterobacteriaceae

VAGINA

Lactobacillus spp.
Peptostreptococcus spp.
Diphtheroids
Streptococci (various)
Clostridium spp.
Bacteroides spp.
Candida spp. (yeast)
Gardnerella vaginalis

LARGE INTESTINE

Bacteroides spp.
Fusobacterium spp.
Clostridium spp.
Peptostreptococcus spp.
Escherichia coli
Klebsiella spp.
Proteus spp.
Lactobacillus spp.
Enterococci
Streptococci (various)
Pseudomonas spp.
Acinetobacter spp.
Staphylococcus
Staphylococcus aureus
Mycobacterium spp.
Actinomyces spp.

SKIN

Staphylococci
Diptheroids (including
  Propionibacterium acnes)
Staphylococcus aureus
Streptococcus (various)
Bacillus spp.
Malassezia furfur
Candida spp. (yeast)
Mycobacterium spp. (occas.)

Figure 20-3 Microorganisms that normally inhabit the human body. All of the micro-organisms listed here live—usually harmoniously—on the surfaces and in the interiors of human bodies. *Candida albicans*, commonly known as yeast, is a fungus that lives on the skin and in the mouth and vagina. *Candida* is, of course, a eukaryote, not a prokaryote. *(Erich Lessing/Art Resource)*

when the body's defenses weaken (Figure 20-3). *Escherichia coli* in our intestines, for example, are normal and useful members of the community of organisms in the body. But the very same organism in the bloodstream can kill in a matter of hours.

Most pathogenic bacteria make us ill by releasing toxic chemicals that kill our own cells. Koch's postulates are guidelines for determining whether a microorganism causes a specific disease.

How Do Bacteria Acquire Resistance to Antibiotics?

Whenever we are sick from a bacterial infection, we and our doctors want to eliminate the pathogenic bacteria as quickly as possible. The easiest way to do that is with an antibiotic. **Antibiotics** are chemicals that kill bacterial cells without harming our own eukaryotic cells. Since the first antibiotics came into use in the 1940s, medical researchers have discovered or synthesized more than 150 different antibiotics.

Figure 20-4 A TB isolation ward. In the 1920s, sick children were isolated on this ferry boat in New York's harbor. With the advent of antibiotics, hospitals no longer needed isolation wards for patients with TB. Today, because of strains of TB that are resistant to all known antibiotics, such wards have returned. *(UPI/Corbis-Bettmann)*

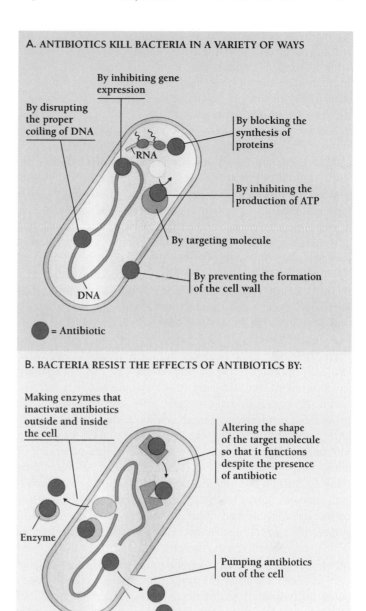

Figure 20-5 How does it work? A. Every antibiotic has a distinctive way of preventing bacteria from reproducing. B. Bacteria have different ways of resisting the effects of antibiotics.

Amazingly, pathogenic bacteria now exist that are resistant to all antibiotics. In New York City, patients with certain strains of highly resistant tuberculosis, for example, are now treated the same way tuberculosis patients were treated 50 years ago, by isolating them. There is no cure for such people, and health workers try to prevent them from passing their resistant strains of tuberculosis bacteria to other patients or to hospital staff (Figure 20-4).

Where do bacteria acquire resistance to antibiotics? First we must understand how antibiotics work and where that resistance comes from. Most antibiotics are naturally synthesized by soil bacteria or fungi. In fact, soil bacteria produce so many chemicals that are toxic to one another that pathogenic bacteria usually die when exposed to ordinary dirt.

The genes for resistance to these antibiotics come from the very organisms—mostly soil organisms—that make the antibiotics. A soil bacterium needs to be immune to its own toxins, after all. The evidence for this is persuasive: researchers have even isolated genes for antibiotic resistance from commercial antibiotics. These genes have the potential to transform bacteria in our bodies from nonresistant forms to resistant forms. Thus, even as we take an antibiotic, we may be handing the infecting bacteria the tools they need to fend off the antibiotic in the future.

What are these tools? That depends on the antibiotic. Some antibiotics, such as penicillin, kill bacteria by preventing the bacteria from synthesizing a cell wall. Other antibiotics prevent bacteria from expressing their genes; interfere with DNA folding (which can disrupt gene expression and DNA replication);

inhibit the synthesis of folic acid, which is important in nucleotide synthesis; or block protein synthesis.

Resistance to antibiotics can be classified into two main varieties (Figure 20-5A). Some resistance genes merely enable the bacteria to survive exposure to an antibiotic. Such genes may alter the shapes of bacterial molecules in such a way that the antibiotics can no longer bind to them, pump the antibiotic out of the cell, or destroy the antibiotic as it enters the cell.

Other resistance genes enable the bacteria to release enzymes that actually destroy antibiotics in the environment (Figure 20-5B). This kind of resistance is the most dangerous. If, for example, a college student regularly took the antibiotic tetracycline to treat acne, he could gradually accumulate intestinal and other bacteria able to destroy tetracycline. If he then became infected by nonresistant *Chlamydia trachomatis,* for example, and his doctor prescribed tetracycline, his intestinal bacteria would destroy the tetracycline before it could kill the invading *Chlamydia* bacteria.

Resistance accumulates in populations of bacteria in several ways. In all populations of bacteria, a few individuals carry genes for resistance to antibiotics. Even baboons living in the wild far from any humans carry small numbers of intestinal bacteria that are resistant to some antibiotics. Humans have far higher numbers of resistant bacteria. One study of resistance in a group of students showed that 1 in 10 students carried large numbers of *E. coli* that were resistant to a combination of two antibiotics first introduced in the 1970s.

When humans or other animals are exposed to antibiotics, the resistant bacteria flourish in the absence of competing bacteria, multiplying until they are the majority. This is simply natural selection. The longer the antibiotic is present, the more likely the antibiotic-resistant bacteria are to increase in numbers. Thus, beef cattle regularly fed the antibiotic tetracycline, gradually accumulate populations of bacteria that are resistant to tetracycline. Unfortunately, the story does not end there.

If antibiotic-resistant bacteria only accumulated in the presence of antibiotics, through selection, then society's problems with antibiotic resistance would be relatively minor. Strains of bacteria without resistance to a specific antibiotic could never acquire resistance except through chance mutations. Unfortunately, bacteria can pass fully functioning antibiotic resistance genes to completely different kinds of bacteria, so that resistance spreads rapidly. Earlier in this book, we mentioned that bacteria engage in a process called conjugation, in which one bacterium can pass plasmids and other DNA to another bacterium (Figure 12-8). Plasmids, simple loops of DNA carried separately from the bacterium's DNA, generally carry genes for traits such as antibiotic resistance that are beneficial to the bacteria that they inhabit. Because plasmids themselves pass genes around to one another by means of jumping genes, plasmids accumulate genes for resistance the way a snowball accumulates snow as it rolls downhill. Plasmids can carry resistance to as many as eight classes of antibiotics. Bacteria then pass such plasmids, through conjugation, not only to other bacteria of the same species, but also to bacteria that are unrelated. Unlike most species of organisms, bacteria are not discrete genetic entities, but a vast integrated microbial world.

As a result, multiple antibiotic resistance is the rule, rather than the exception. In 1995, hospital workers in Atlanta, Georgia, found that 25 percent of white children fighting ordinary ear and respiratory infections by *Streptococcus pneumoniae* were infected with strains resistant to two or more antibiotics.

The tendency of plasmids to carry genes for resistance to several antibiotics has important consequences. Cattle treated with tetracycline, for example, can acquire bacteria that are resistant not only to tetracycline but to several other antibiotics as well. This is why the South Dakota beef cattle, which had been treated with tetracycline, carried *Salmonella* resistant to both amoxicillin and tetracycline. The plasmid that protected the bacteria from tetracycline also happened to carry genes for resistance to other antibiotics.

Resistance may allow a bacterium simply to survive exposure to an antibiotic or, alternatively, actively to destroy any antibiotic in its environment. Resistance genes accumulate in populations of bacteria, first, through natural selection, and second, by moving from one bacterial strain to another by means of plasmids and other mobile genes.

HOW DO WE KNOW THAT AN ORGANISM IS A PROKARYOTE?

The single characteristic of prokaryotes that sets them apart from all other organisms is the absence of a true nucleus (Chapter 4). Instead of neatly packing its DNA in a membrane-enclosed nucleus (as a eukaryote does), a prokaryote tucks its DNA into a concentrated mass called a **nucleoid.** The DNA in the nucleoid consists of a single molecule of double-stranded DNA that forms a closed loop. A prokaryote may also contain one or more smaller loops of DNA called plasmids (Chapter 12). Because a prokaryote's DNA is circular, it can replicate continuously, unlike eukaryotic DNA, which must replicate in pieces. Also unlike the eukaryotes, prokaryotes undergo neither mitosis nor meiosis. Prokaryotes do not require sex to reproduce. However, as we have seen, they exchange genetic information so freely that reproductive isolation is meaningless in bacteria. Unlike other species of organisms, then, a "species" of bacterium is neither a reproductive unit nor a genetic unit, and classifying bacteria into species is something of an art.

Prokaryotes lack not only a nucleus, but also chloroplasts and mitochondria and all other membrane-bounded organelles. Some prokaryotes do have extensively folded plasma membranes, however, and use these large surfaces for respiration, photosynthesis, and DNA replication.

Prokaryotic cells come in a variety of shapes (Figure 20-6). The most common are the rod-shaped **bacilli** [singular, *bacillus,* Latin, = little rod]. In addition are the spiral-shaped **spirilla** [singular, spirillum; Latin, *spira* = coil], and the spherical **cocci** [singular, coccus; Greek, *kokkos* = berry]. Most bacteria are single-celled organisms, though some aggregate to form colonies of unspecialized cells (Figure 20-7).

The defining characteristic of prokaryotes is the absence of a nucleus or other membrane-enclosed organelles. The DNA of prokaryotes forms a single circle of double-stranded DNA.

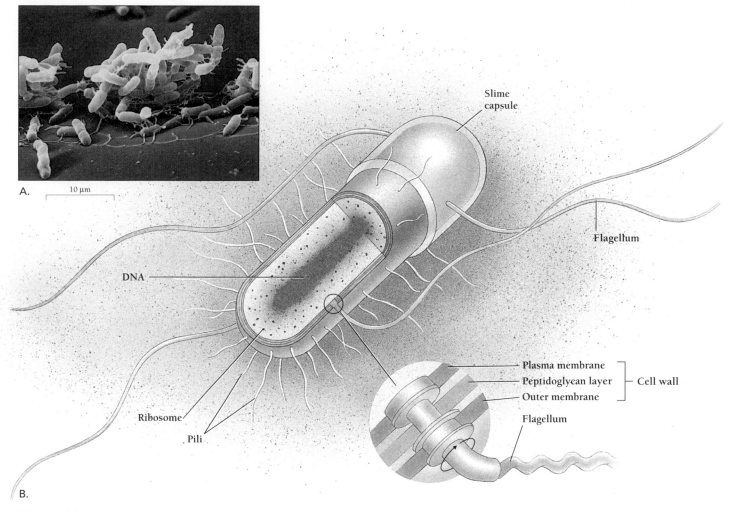

A.

10 μm

Slime capsule

Flagellum

DNA

Plasma membrane
Peptidoglycan layer — Cell wall
Outer membrane

Flagellum

Ribosome

Pili

B.

Figure 20-6 The familiar shape of the bacillus typifies both *Salmonella* and *E. coli*. A. *Salmonella* sp. B. *Escherichia coli*. *(A, Meckes/Ottawa/Photo Researchers)*

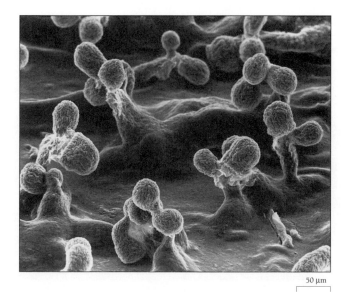

50 μm

Figure 20-7 Some bacteria, such as this myxobacterium, are colonial. Cells of this species form a "fruiting body" to reproduce. *(Karen Stephens/Biological Photo Service)*

Most Prokaryotes Have Complex Cell Walls

Like plant cells, most bacterial cells have rigid cell walls outside their plasma membranes. Like the wall of a plant cell, the wall of a bacterial cell gives the cell a shape and protects it from injury. However, while plant cell walls are made of cellulose, prokaryotic cell walls are made of a mixture of polysaccharides and polypeptides called **peptidoglycan**. The cell wall is a single, giant, bag-shaped macromolecule composed of a network of cross-linked peptidoglycan. The peptidoglycan cell wall is important to biologists for two reasons. It helps distinguish two major groups of bacteria, and it makes some bacteria susceptible to antibiotics.

Biologists divide all of the bacteria into two major groups according to the way their cell walls soak up various dyes. In the 1880s, the Danish bacteriologist Hans Christian Gram found that when he treated bacteria with a particular purple dye, some kinds of bacteria became permanently stained while others could be washed clean. Those that became permanently stained he called "stain-positive." The others were "stain-negative." Today bacteriologists call these two categories of bacteria

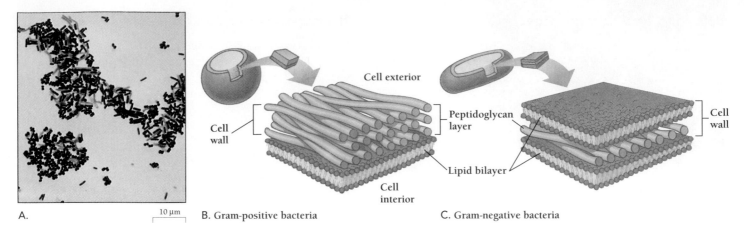

A.   10 μm

B. Gram-positive bacteria   C. Gram-negative bacteria

Figure 20-8   It's all in the wall. Bacteria absorb stain or not depending on what kind of cell wall they have. A. A mixture of gram-negative (light) and gram-positive (dark) bacteria. B. The gram-positive coccus bacterium has a two-layered, peptidoglycan-rich cell wall. C. The gram-negative bacillus has a three-layered, peptidoglycan-poor cell wall.   *(A, G.W. Willis/Biological Photo Service)*

gram-positive and **gram-negative,** after Gram's stain (Figure 20–8).

What sets gram-negative and gram-positive bacteria apart from one another is the structure of the cell wall. Gram-negative bacteria—such as *E. coli* or the gonococcus bacteria that cause the sexually transmitted disease gonorrhea—have a thin, three-layered cell wall, whose outermost layer does not bind the gram stain (Figure 20-8C). In contrast, gram-positive bacteria—such as the streptococci (which can cause strep throat) and the staphylococci (which can cause boils and other infections)—have a thick, single-layered cell wall containing at least 20 times as much peptidoglycan as the gram-negative bacteria. The peptidoglycan, itself arranged in up to 40 layers, binds the gram stain so that it cannot be washed away.

In gram-negative bacteria, the same outer layer of the cell wall that prevents the gram stain from penetrating also prevents antibiotics such as penicillin from penetrating. As a result, penicillin and other "narrow-spectrum" antibiotics kill gram-positive bacteria only. "Broad-spectrum" antibiotics such as tetracycline kill both gram-negative bacteria and many gram-positive bacteria.

Outside of the cell wall, some prokaryotes have other materials. The outer surfaces of some prokaryotes possess *pili* [singular, pilus; Latin, = hair], hairlike structures that attach the bacteria to surfaces and to other cells. The F-pilus, for example, provides a bridge for the transfer of DNA during bacterial conjugation.

The bacterium that causes pneumonia has a gelatinous layer of polysaccharides and proteins outside its cell wall called a **capsule.** Capsules protect disease-causing bacteria from attack by the body's white blood cells, enabling the bacteria to overwhelm the body's immune system. Such bacteria are more deadly, or virulent.

Other prokaryotes possess a **slime layer,** a tangled web of polysaccharides which, like a capsule, lies outside the cell wall.

A sticky slime layer protects the bacterial cell against dehydration, helps keep food from diffusing away, and also allows cells to attach to and move about on the surfaces on which they grow. Dental plaque is a slime layer made by the bacteria that inhabit the mouth. Plaque helps bacteria grow smoothly over the surfaces of our teeth (Figure 20-9).

The presence of large amounts of peptidoglycan distinguishes the prokaryote cell wall from the cell walls of eukaryotes. Gram-positive bacteria have a thick, peptidoglycan-rich, cell wall that leaves them susceptible to penicillin and other narrow-spectrum antibiotics.

5 μm

Figure 20-9   Dental plaque bacteria, which coat our teeth in the morning, are one kind of bacteria that have a slime layer. *(S. Fiegler/Visuals Unlimited)*

How Do Prokaryotes Move?

Some prokaryotes move by secreting slime and gliding on it. Others use a propeller—the prokaryotic **flagellum** (Chapter 4). Prokaryotic and eukaryotic flagella are entirely different. The eukaryotic flagellum is an extension of the cytoplasm surrounded by membrane that beats back and forth. In contrast, the prokaryotic flagellum is a thin, rotating protein filament attached to the outside of the cell.

Many bacteria use their flagella for directed movement, or **taxis** [Greek, = to put in order]. (A taxicab performs directed movements.) The direction of movement depends on the operation of the flagella. For example, when the flagella of an *E. coli* rotate in a counterclockwise direction, they twist around each other and propel the bacterium forward. When the flagella rotate in a clockwise direction, however, they separate and become uncoordinated. As a result, the bacterium tumbles aimlessly. When the flagella resume a counterclockwise spin, the bacterium moves in a new direction. By controlling the timing of clockwise and counterclockwise rotations, bacteria can move toward or away from food, oxygen, heat, light, and other stimuli. Some bacteria even use the Earth's magnetic field to guide them to the lower depths of the ocean. **Chemotaxis** is taxis in response to varying concentrations of specific chemicals.

Prokaryotes move in directed fashion, by gliding on slime or by rotating propellerlike flagella.

How Do Prokaryotes Grow So Rapidly?

Because bacteria keep their DNA in a nucleoid instead of a nucleus, bacteria can replicate their DNA continuously if necessary and divide very rapidly. In fact, the cells in well-nourished colonies of *E. coli* can divide every 20 minutes. Every time the bacteria divide, the colony doubles in size. That is, after one generation, one cell becomes two. After another generation the 2 cells become 4, then 8, 16, and so forth. After 20 such doublings (in less than 7 hours), a single cell would yield 2^{20}, or about a million, new cells; and after 30 doublings (10 hours), 2^{30}, or about a billion cells. Biologists describe such growth as exponential or logarithmic (Chapter 29).

If a single *E. coli* bacterium could be supplied with enough food for it to actually divide every 20 minutes continuously, the combined mass of the bacterium and all its descendants would exceed that of the entire planet after only two days. We can see that, in reality, exponential growth can last for only short times. Nonetheless, the rapid proliferation of prokaryotes means that a favorable new mutation such as antibiotic resistance can spread quickly to a huge number of progeny.

A streamlined metabolism and a simple means of replicating DNA together allow bacteria to grow and divide more rapidly than eukaryotes.

Prokaryotes Are Metabolically Diverse

We eukaryotes are monotonous from a biochemical point of view. Nearly all of us use glycolysis and respiration to produce energy. Nearly all of us need oxygen to live. Likewise, all photosynthetic eukaryotes photosynthesize in the same way. In fact, the relentless sameness of all eukaryotes' metabolic pathways strongly supports the idea that we are all descended from a common ancestor.

In contrast, prokaryotes use a great variety of biochemical pathways. Some, for example, perform glycolysis, but with end products other than ethanol or lactic acid (Chapter 6). Many of these end products contribute to the pleasant (or not so pleasant) odors of wines and cheeses. Some prokaryotes do not use glycolysis at all. Instead, they produce ATP by capturing the chemical energy of molecules in their environment. Methane-producing bacteria, for example, capture energy released when carbon dioxide reacts with hydrogen gas (H_2) to produce methane (CH_4).

Because prokaryotes are so diverse, they flourish in places where eukaryotes could never survive. Over billions of years, prokaryotes have evolved an amazing variety of adaptations that contribute to their continuing success. Some of these adaptations are (1) spore formation, which allows bacteria to survive, multiply, and disperse no matter how hostile the environment; (2) the production of antibiotics that can kill competitors; (3) the use of resistance transfer factors and plasmids to spread resistance to antibiotics from one bacterium to another; and (4) the ability to obtain energy from the waste products of other species (for example, methane).

The metabolic diversity of prokaryotes contributes to their enormous success.

HOW DO BIOLOGISTS CLASSIFY PROKARYOTES?

Because prokaryotes do not reproduce sexually, taxonomists cannot use the biological species concept, which defines a species as an interbreeding population, to classify prokaryotes (Chapter 19). Reconstructing the phylogeny of bacteria is equally difficult, as the fossil record for bacteria is nearly nonexistent. Instead, taxonomists rely on morphological, biochemical, and molecular differences and similarities. Even this approach is fraught with difficulties, however. Two strains of a single species may look very different. One may possess a distinctive feature such as an external capsule, while the other lacks the capsule. In addition, genes carrying distinctive traits such as antibiotic resistance can spread rapidly—through viruses, plasmids, or transfer factors—among many unrelated bacteria. So far, some 1700 species of prokaryotes have been named, although estimates of the number of species vary widely—from 5000 up into the millions.

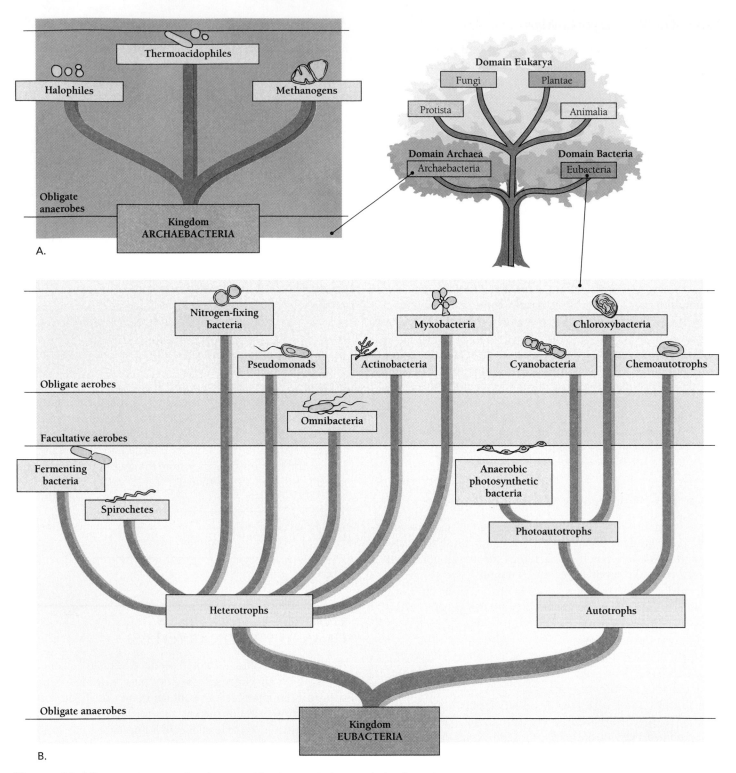

Figure 20-10   Classification of prokaryotes. The domain Archaea includes three distinct groups, which may be kingdoms or phyla. The domain Eubacteria includes far more groups, a dozen of which are shown here. The Eubacteria are divided into heterotrophs and autotrophs. The autotrophs are further divided into photoautotrophs and chemoautotrophs.

How Do Biologists Classify the Archaea?

Although many biologists still put all of the bacteria into a single kingdom, the consensus has changed in recent years. The three phyla of **Archaea** [Greek, *archaio* = ancient] have been given a separate kingdom and a separate domain, as well (Figure 20-10). Whether that domain contains one kingdom with three phyla or, alternatively, three kingdoms (corresponding to the three phyla) is controversial. The Archaea differ from both all other bacteria and from all eukaryotes in the composition of their ribosomes, in the construction of their cell wall, and in the kinds of lipids in their cell membranes. The Archaea entirely lack peptidoglycan in their cell walls.

The three phyla (or kingdoms) of Archaea are the **halophiles,** the salt-loving bacteria (Figure 20-11); the **thermacidophiles,** the bacteria that inhabit hot sulfur springs such as those in Yellowstone National Park (Figure 20-12); and the **methanogens,** the methane producers (Figure 20-13).

The methanogens' distinctive property is their ability to live by making methane (CH_4) from carbon dioxide (CO_2) and hydrogen gas (H_2). Methanogens, all obligate anaerobes, live principally in two environments: (1) swamps and sewage treatment plants, and (2) the guts of animals, especially of cows, termites, and other animals that digest cellulose. Even the human gut harbors methanogens that produce methane, which, when released, is called flatulence.

Methanogens play an essential role in the ecology of the Earth. Where methanogens flourish in swamps made anaerobic by the growth of other bacteria, they release methane, or "marsh gas," into the atmosphere. There, the methane is quickly oxidized to carbon dioxide, which is then available for photosynthesis. If methanogens did not recycle carbon from organic molecules back into the atmosphere, other organisms would gradually run out of carbon. Natural gas, which we use to heat

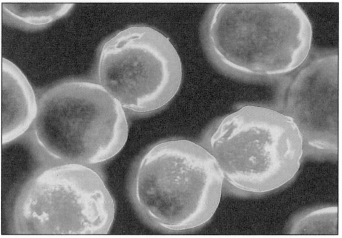

1 μm

Figure 20-11   Halophiles. Most bacteria cannot grow in high concentrations of salt. Because salt prevents the growth of most bacteria, it acts as a preservative in ham, beef jerky, and other salty foods. Unlike any other bacteria, halophiles such as these *Halococcus* bacteria flourish in briny places the world over—from the Dead Sea to Utah's Great Salt Lake and California's Mono Lake. Halophiles, usually flagellated rods or cocci (like these), form a lavender-pink scum over their salty homes. The pink color comes from a purple pigment in the plasma membrane that captures light energy.   *(Alfred Pasteka/ Science Photo Library/Photo Researchers)*

our homes and cook our meals, is about 98 percent methane—all provided by the methanogens.

The Archaea consist of three groups, the most widespread and ecologically important of which are the methanogens.

Figure 20-12   Thermoacidophiles. Thermoacidophiles may be chemoautotrophs (reaping energy from sulfur compounds or methane) or heterotrophs (reaping energy and carbon from small organic compounds made by other organisms). The best-known genus, *Sulfolobus,* lives in the hot sulfur springs of Yellowstone National Park. *Sulfolobus* can survive a temperature as high as 88°C (190°F) and an acidity as low as pH = 1. Although *Sulfolobus* thrives in heat, it dies of cold at temperatures below 55°C (130°F). *(Richard Thom/Visuals Unlimited)*

BOX 20-1

Carl Woese and the archaea

In the summer of 1996, a researcher working on an obscure microbe at The Institute for Genomic Research in Rockville, Maryland, announced a finding that immediately hit the television nightly news and the front pages of newspapers all over the United States. In sequencing the microbe's genome, the researcher, Carol Bult, had provided irrefutable evidence that all organisms, historically classified as either eukaryotes or bacteria, actually fall into three basic categories.

The microbe, named *Methanococcus jannaschii*, was only the fourth organism for which researchers had completed such a sequence, but the implications were profound. Bult's work not only supported the theory that the archaea are distinct from both bacteria and eukaryotes, it also provided independent support for two other ideas—first, that we eukaryotes are more closely related to the archaea than to the bacteria and, second, that both eukaryotes and bacteria are descended from ancient organisms that resembled archaea.

Yet, despite the drama with which these important results were revealed, it was an oddly anticlimatic end to a very long story. Thirty years before, Carl Woese, a young molecular biologist at the University of Illinois, Urbana, began the painstaking task of working out the evolutionary relationships of the bacteria. Earlier researchers had made the attempt and failed. "The ultimate scientific goal of biological classification cannot be achieved in the case of bacteria," wrote the eminent microbiologist Roger Stanier.

Woese admired Stanier's own considerable efforts to classify the bacteria but did not accept that classifying the bacteria was impossible. Indeed, he had an ingenious plan for discovering their relations to one another. He knew that ribosomal RNAs are among the most conserved parts of all organisms. Because the work of translating RNA into proteins is fundamental, organisms with mutations in their ribosomal RNA nearly always die and ribosomes change hardly at all over millions of years.

Ribosomes are, in short, some of the best sources of information about long-term evolutionary relationships among organisms.

For ten years Woese (pronounced woes) catalogued short strings of nucleotides that he cut from the ribosomal RNAs of different microorganisms. It was mind-numbing, labor-intensive work that left him, he later said, "just completely dulled down." Yet it was fruitful work, for he gradually unraveled the relationships among some 60 different kinds of bacteria.

Then, in 1976, Woese's colleague Ralph Wolfe sent him some methanogens on which to work his magic. Methanogens were curious bacteria that came in a variety of shapes but all produced methane as a byproduct of their metabolism. No one knew where they fit in with the other bacteria. Woese set to work to find out. To his surprise, however, the methanogens lacked the characteristic sequences of ribosomal RNA that all other bacteria share. As he told Wolfe, the methanogens didn't seem to be bacteria at all.

It was dramatic news, for it meant that in addition to the eukaryotes, and the bacterial prokaryotes, we humans were sharing the planet with a completely different domain of organisms—the archaea—never before recognized. Yet, although the story hit the front page of the *New York Times,* the news faded away immediately. For most people, the news was abstract and meaningless. And most of Woese's own colleagues ignored his discovery. They simply didn't believe him. Although Woese provided the answers to many of the questions that most intrigued the eminent microbiologist Roger Stanier, Stanier, who has since died, never acknowledged Woese. For years, Woese says, he waited in vain for some friendly acknowledgment from Stanier.

R.G.E. Murray, the author of the leading microbiology textbook, didn't even include the archaea until 1986, and then only as a subgroup within the prokaryotes—even though Woese's work had demonstrated that the archaea were distinct from other bacteria. A handful of other re-

Figure A Carl Woese discovered the archaea in 1977. The scientific community did not fully accept his results until 1997. *(courtesy of University of Illinois at Urbana-Champaign News Bureau)*

searchers stood behind Woese, including Wolfe, arguing with their colleagues on his behalf. Yet, Woese's work was otherwise almost universally snubbed.

Woese merely retreated back to his laboratory, continuing his work, gathering new and better data, and publishing a series of landmark papers on the evolution and classification of bacteria. Today Woese is highly acclaimed. He has been elected to the prestigious National Academy of Sciences and awarded both a MacArthur Foundation "genius" grant and microbiology's top honor—the Leeuwenhoek medal.

Great discoveries in science frequently meet initial rejection. But as Wolfe told *Science* magazine in 1997, "We can say, 'Oh, that's just the way science works.' But it was a personal experience for him. He's the one who had to live through it."

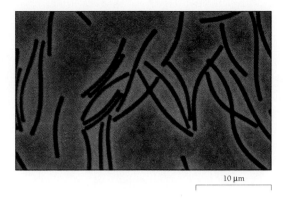

Figure 20-13 **Methanogens.** Methanogens, which *generate* the gas *methane*, include bacteria of all three standard shapes—bacilli, spirilla, and cocci. *(F. Widdel/Visuals Unlimited)*

How Do Biologists Classify the Eubacteria?

In spite of the difficulties of classifying prokaryotes, biologists do agree on some broad groupings. In one classification, the Eubacteria are divided into 13 phyla. For comparison, the Animalia include 33 phyla.

Just as zoologists classify animals according to whether they eat plants or other animals, microbiologists classify Eubacteria into broad categories according to the way they derive food. **Autotrophs** [Greek, *autos* = self + *trophos* = feeder] obtain carbon atoms directly from carbon dioxide, while **heterotrophs** [Greek, *heteros* = other + *trophos* = feeder], which include the great majority of prokaryotic species, derive carbon atoms (and also energy) from organic molecules such as glucose. Autotrophs may be divided into **photoautotrophs,** which use light energy to drive the synthesis of organic compounds from CO_2, and **chemoautotrophs,** which derive energy by oxidizing such inorganic substances as hydrogen sulfide (H_2S) or ammonia (NH_3).

Heterotrophs fall into two groups. **Photoheterotrophs** use light energy but also require organic compounds. **Chemoheterotrophs** use no light, relying exclusively on organic molecules for both energy and carbon atoms. Heterotrophic prokaryotes may be **decomposers,** which absorb nutrients from dead organisms, or **symbionts** [Greek, *syn* = together + *bios* = life], which absorb nutrients from living organisms. Symbionts that cause illness, such as *Salmonella newport,* are called pathogens. Others, such as *E. coli,* are usually harmless, or even helpful. Prokaryotic decomposers, like fungal decomposers, recycle the materials that were formerly part of other organisms.

Prokaryotes also vary in the way they use oxygen. Heterotrophs that require oxygen for respiration are called **obligate aerobes.** Those that can grow either with or without oxygen are called **facultative aerobes.** Still others, for which oxygen is a deadly poison, are called **obligate anaerobes.**

The Eubacteria include aerobic heterotrophs, anaerobic heterotrophs, and autotrophs. We will briefly describe examples of each of these three groups and the incredible diversity of environments in which these simple organisms thrive.

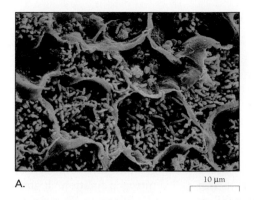

A.

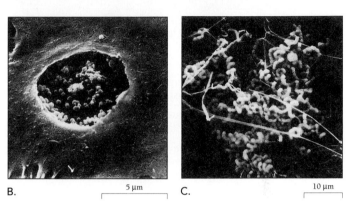

B. 5 μm C. 10 μm

Figure 20-14 Aerobic heterotrophs. A. Nitrogen-fixing bacteria growing in nodules on the roots of a plant. Some nitrogen-fixing bacteria live freely in soil and water, while others associate intimately with specific plants, living, for example, in tiny nodules on the roots of peas, beans, lupines, locusts, and other members of the pea family. The nitrogen-fixing bacteria provide the plant with a usable form of nitrogen, and the plant provides carbohydrates for energy. This relationship allows extraordinarily efficient fixation of nitrogen, as much as 100 times faster than that performed by free-living bacteria. B. *Chlamydia trachomatis.* Besides the enteric bacteria, the omnibacteria also include rickettsias and chlamydias, which live only as parasites inside animal cells and cause a variety of diseases. Shown here is *Chlamydia trachomatis,* which causes the most common sexually transmitted disease in the United States. C. *Streptomyces.* Actinobacteria form branching filaments like those of fungi. They are widely distributed in soils, where they decompose organic matter. *Streptomyces,* for example, produces the valuable antibiotic streptomycin and other substances that kill competing organisms. A few forms of actinobacteria cause human diseases such as tuberculosis and leprosy. *(A, C.P. Vance/Visuals Unlimited; B, David M. Phillips/Science Source/Photo Researchers; C, Frederick Mertz/Visuals Unlimited)*

The Aerobic Heterotrophs

The aerobic heterotrophs have tremendous biochemical diversity: some break down all kinds of organic compounds, while others can use nitrate (NO_3^-)—instead of oxygen—as the ultimate electron acceptor, and still others make powerful antibiotics that kill competitors. A few aerobic heterotrophs are the **nitrogen-fixing bacteria, pseudomonads, omnibacteria, actinobacteria,** and **myxobacteria** (Figure 20-14).

From the perspective of most other organisms on Earth, the most important prokaryotes may be the **nitrogen-fixing** aerobic bacteria (Figure 20-14A). Nitrogen-fixing bacteria provide nitrogen in forms such as ammonia and amino acids from which other organisms can make proteins. Although nitrogen gas (N_2) constitutes some 80 percent of our atmosphere, it is so chemically stable that most organisms are incapable of breaking the bond between the two nitrogen atoms to form complex organic compounds. As a result, all organisms rely on nitrogen-fixing bacteria to provide nitrogen in a usable form. The bacteria perform this service, called **nitrogen fixation,** at great cost, however, for the task requires a lot of energy. (Conversely, processes that yield nitrogen gas as a product release lots of energy. For this reason, many nitrogen-containing compounds, including commercial fertilizers, are explosives.)

The **pseudomonads,** known as "the weeds of the bacterial world," are a large and aggressive group of bacteria that seem to live everywhere—in soil, ponds, infected wounds, hot tubs, and even medicine bottles. Pseudomonads are typically straight or curved gram-negative rods with flagella. They require oxygen and produce CO_2. The extraordinary success of pseudomonads comes from their metabolic diversity. They can survive on carbohydrates, petroleum products, tough polysaccharides in natural fibers, and even antibiotics and pesticides.

Omnibacteria are the most common organisms on Earth (Figure 20-14B). These bacteria can all perform aerobic respiration and also have the extraordinary ability to use nitrate (NO_3^-) as an electron acceptor instead of oxygen. The most prolific and well-known are the common enteric bacteria [Greek, *enteron* = intestine], such as *E. coli,* which inhabit the intestines of animals. Omnibacteria may be rod-shaped, as is *E. coli,* or comma-shaped, as is the species that causes cholera.

The Anaerobic Heterotrophs

The anaerobic heterotrophs are mostly obligate anaerobes. They include, for example, the **fermenting bacteria,** which derive energy and carbon atoms from a variety of organic compounds in the absence of oxygen; and the **spirochetes,** which have a distinctive corkscrew shape (Figure 20-15).

The Autotrophs

The autotrophs consist of four phyla: the **chemoautotrophs,** which photosynthesize *without* oxygen, and three phyla of photoautotrophs, which photosynthesize *with* oxygen. The pho-

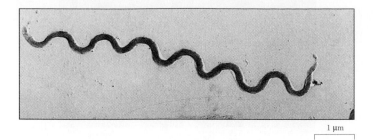

1 µm

Figure 20-15   **Swimming syphilis spirochete.** The most famous spirochete is *Treponema pallidum,* which causes the sexually transmitted disease syphilis. Spirochetes contain internal flagella enclosed between layers of the cell wall, by means of which they move differently from any other bacteria. Although some live freely and some are parasites, all spirochetes move by undulating quickly, even through the most viscous liquids. Most spirochetes are finicky about their diets and *T. pallidum* is no exception. Its requirements are so specific, in fact, that although it grows well inside the human body, biologists have been unable to grow virulent strains in the laboratory. *T. pallidum* is transmitted from one person to another by means of direct contact between moist mucous membranes. Until the introduction of penicillin in the 1940s, syphilis was incurable, progressive, and fatal. *(CDC/Science Source/Photo Researchers)*

toautotrophs consist of the **anaerobic photosynthetic bacteria,** the **cyanobacteria,** and the **chloroxybacteria.**

Biologists classify prokaryotes according to how they acquire energy and building blocks, whether they photosynthesize, and whether they use or tolerate oxygen. Biologists divide the Eubacteria into three main groups: the aerobic heterotrophs, anaerobic heterotrophs, and the autotrophs. The most ecologically important Eubacteria are (1) the cyanobacteria, the main photosynthetic prokaryotes, which fix carbon; and (2) the nitrogen-fixing bacteria, which supply the world's organisms with fixed nitrogen for making proteins.

In this chapter we have briefly surveyed the two domains of prokaryotes. Just as important, we learned how the biology of bacteria helps them acquire resistance to antibiotics. In the next chapter, we survey the first two kingdoms of eukaryotes, the Protista and the Fungi.

STUDY OUTLINE WITH KEY TERMS

Most **pathogenic** prokaryotes make us ill by releasing toxic chemicals that kill our own cells. These may be **endotoxins** or **exotoxins.** **Koch's postulates** are guidelines for determining whether a microorganism causes a specific disease, a topic of interest to **epidemiologists.** Resistance may allow a bacterium to survive exposure to an **antibiotic** or, alternatively, to actively destroy any antibiotic in its environment. Resistance genes accumulate in populations of bacteria,

first, through natural selection, and second, by moving from one strain of bacterium to another by means of plasmids and other mobile genes.

Prokaryotes tuck their DNA into a **nucleoid,** rather than a true nucleus. They also lack other membrane-bounded organelles. Most are single-celled organisms, though some aggregate to form colonies. They come in many shapes, but three forms dominate: the rod-shaped **bacilli,** the spiral-shaped **spirilla,** and the spherical **cocci.** Most have cell walls made of a **peptidoglycan.** Bacteriologists classify bacteria according to whether they are **gram-positive** or **gram-negative.** Some bacteria have specialized outer surfaces that form a **capsule,** a **slime layer,** or hairlike structures called pili.

Many prokaryotes are capable of directed movement or **taxis** (including **chemotaxis**), and many have **flagella,** whose structure and operation is different from those of eukaryotic flagella. Prokaryotes grow continuously and rapidly, and usually reproduce asexually.

Prokaryotes are biochemically more diverse than eukaryotes. They may be either **autotrophs** or **heterotrophs.** Some autotrophs (**photoautotrophs**) obtain energy from light, while others obtain energy from inorganic substances. Some heterotrophic prokaryotes derive energy from light (**photoheterotrophs**), but also require organic compounds. **Chemoheterotrophs** depend exclusively on organic compounds for both energy and carbon atoms. Heterotrophs may be **symbionts** or **decomposers.** Some heterotrophs require oxygen to live (**obligate aerobes**), others can live either in the presence or absence of oxygen (**facultative aerobes**), and some cannot live when oxygen is present (**obligate anaerobes**).

The metabolic versatility of prokaryotes allows them to live where eukaryotes cannot. Some prokaryotes provide benefits to humans, but others are pathogens, which cause disease. The identification of a pathogenic prokaryote is often an important step in understanding and curing a disease.

The metabolic abilities of prokaryotes provide a basis for their classification. Three groups of prokaryotes, together classified in the domain **Archaea,** differ dramatically from all other organisms. These include the **methanogens,** which make methane from carbon dioxide and hydrogen gas, the **halophiles,** which grow only in high concentrations of salts, and the **thermoacidophiles,** which derive energy by oxidizing sulfur compounds or methane.

The **Eubacteria** include about 10 phyla. Aerobic heterotrophs are biochemically diverse, but use the same metabolic pathways that are found in eukaryotes. These bacteria include **nitrogen-fixing bacteria, pseudomonads, omnibacteria, actinobacteria,** and **myxobacteria.**

Anaerobic heterotrophs include the **fermenting bacteria** (which often cannot live when oxygen is present) and the **spirochetes** (which contain internal flagella between layers of the cell wall).

Autotrophic prokaryotes include **chemoautotrophs** (which derive energy from a variety of inorganic compounds), **anaerobic photosynthetic bacteria, cyanobacteria,** and **chloroxybacteria.**

REVIEW AND THOUGHT QUESTIONS

Review Questions

1. What features of eukaryotes do prokaryotes lack?
2. What molecule makes up much of the cell wall of members of the Eubacteria but not of Archaea?
3. What is the difference between gram-positive and gram-negative bacteria? Which kind is susceptible to penicillin?
4. Describe two ways in which antibiotics work against bacteria. Why do these mechanisms not affect eukaryotic cells?
5. Describe the two major ways that antibiotic resistance is spread through populations of bacteria.
6. What features allow prokaryotes to reproduce so much more rapidly than eukaryotes? What advantages does this give bacteria?

7. What criteria do biologists use to classify bacteria?

Thought Questions

8. Are antibiotics helpful against a viral infection such as a cold or flu? Why or why not?
9. Animals breeders began giving animals antibiotics because studies showed that animals raised on antibiotics grew faster. Today, this effect is not as strong as it was 30 or 40 years ago. Why might antibiotics have this effect? Why might antibiotics no longer be so good at increasing growth rates?

SELECTED READINGS

Amábile-Cuevas, Carlos F., Maura Cárdenas-García, and Mauricio Ludgar, "Antibiotic Resistance," *American Scientist,* July–August, 1995.

Brock, T.D., *The Biology of Microorganisms,* 7th ed., Prentice-Hall, Englewood Cliffs, N.J., 1994.

Levy, Stuart B., *The Antibiotic Paradox, How Miracle Drugs Are Destroying the Miracle,* Plenum Press, New York, 1992. An engaging and authoritative discussion of the growing problem of antibiotic-resistant bacteria.

Margulis, L., and K.V. Schwartz, *Five Kingdoms: An Illustrated Guide to the Phyla of Life on Earth,* W.H. Freeman, New York, 1988.

Woese, C., "Archaebacteria," *Scientific American,* June 1988.

▶ On-line materials relating to this chapter are on the World Wide Web at http://www.saunderscollege.com/lifesci/
    Click on Tobin/Dusheck: *Asking About Life.*

The Fungus Fighters

In 1948, two government researchers set out to find a drug that would ultimately save thousands of human lives. In just two years, Elizabeth Lee Hazen and Rachel Brown succeeded in producing a drug that could safely cure deadly fungal infections in humans—a drug that is still a medical mainstay today. Both women had overcome great obstacles to obtain their scientific training. That they were highly educated, practicing scientists at all was something of a miracle. That they were able to make a major contribution to medicine was a testimony to their lifetime dedication to science and the respect they inspired in those around them.

Elizabeth Lee Hazen, born in August of 1885 to a Mississippi cotton farmer and his wife, was an orphan by the time she was two. Raised by relatives, Hazen grew up in a large family in Lula, Mississippi. After high school, she enrolled in the Mississippi Industrial Institute and College, the first state-supported college for women in the United States. Without such an institution she might never have gone to college, might never have studied physiology, anatomy, botany, zoology, and physics. Her grades were unremarkable. Like many people with great contributions to make, her strengths would not become obvious until she was much older.

After graduating from college, Hazen taught high school for six years. When she was 31, she enrolled as a graduate student at Columbia University, in New York, where one interviewer sized her up and then told her that she was so ignorant he doubted she had ever been to college. Nonetheless, she soon proved herself, and the following year Columbia awarded her a master of arts degree in biology. During World War I, she worked for the Defense Department diagnosing bacterial infections. After the war, she returned to Columbia, and, in 1927, at the age of 42, she received a Ph.D., becoming one of a small number of women with doctorates in the United States. She stayed at Columbia until 1931, when the New York State Department of Public Health lured her away to supervise its large medical pathology lab. But she maintained a connection with Columbia. In 1944, when the health department needed an expert on fungal infections, she returned to study mycology, at the same time continuing her full-time work in the pathology lab.

For most of us, fungi are no more than a mild curiosity. Mushrooms are a nice addition to a pizza. Mold is an annoyance in the shower. Some of us may have had unpleasant experiences with a fungus, a bout of athlete's foot, a yeast infection, nothing that a simple antifungal medication could not treat. In most people, the immune system's white blood cells quickly engulf and digest any fungal spores that make it into the body, and fungi are not a problem.

Dr. Paul A. Zahl/Photo Researchers

Under the right circumstances, however, a fungus can destroy a living human being as thoroughly as it can rot a dead stump in the forest. Diabetics, for example, have tissues that are unusually high in sugar. The warm, sugary environment of a diabetic's lung apparently creates a perfect environment for mucormycosis, the lush growth of a fungus in the order Mucorales. A common, usually harmless food mold, some species of Mucorales can grow through the lungs into the heart. Alternatively, this fungus may infect the sinuses, then run hyphae, long strands of the fungal body, throughout a person's head, destroying nerves and brain function. People with damaged immune systems—burn victims, transplant patients being treated with immunosuppressive drugs, and those with immune diseases such as AIDS—are all susceptible to fungal infections. For such people, fungi are a potentially life-threatening problem.

In the 1940s, when the first antibiotics came into use, medical experts discovered that patients treated with antibiotics became highly susceptible to fungal infections. It seemed that many of the bacteria killed by the antibiotics were important in keeping fungi at bay. Yet no antifungal drugs existed, and doctors were helpless to stop the spread of the often fatal infections.

Elizabeth Hazen was determined to find a drug that would help. By 1948, she had established an immense collection of pathogenic (disease-causing) fungi with which she and her army of technicians could diagnose fungal infections from people all over the eastern United States. Even more important, Hazen had for years been testing soil samples for bacteria that could kill infectious fungi. Now she had found two different bacteria that produced

The third substance was only mildly toxic and killed virtually any fungus it came into contact with. It looked like a miracle drug, another penicillin.

substances toxic to fungi. But what were the substances? Hazen needed the help of a talented chemist to isolate the active ingredient. One day in 1948, Hazen took a train from New York City to the health department's main office in Albany, New York, 150 miles away. There she had an appointment with senior researcher Gilbert Dalldorf, himself a talented bacteriologist and physician. Dalldorf listened to Hazen's proposal, then took her to see chemist Rachel Brown.

Like Hazen, Brown had succeeded in science against long odds. She was born in 1898 and raised in Springfield, Massachusetts, in a small, middle-class family. In 1912, however, when Rachel Brown was 14 years old, her father abandoned his family. Rachel's mother Annie then supported herself, her two children, and both her parents on a secretary's salary. Despite her poverty, Annie Brown was determined that her two children should go to the best schools she could afford. By Rachel's senior year in high school, however, her heart was set on Mount Holyoke College, a small, private college for women. But Rachel was turned down for a scholarship, and her mother could not possibly afford the tuition.

Miraculously, a wealthy friend of Rachel's grandmother offered to pay all of Rachel's expenses for a full four-year college education. Brown earned a B.A. in chemistry from Mount Holyoke and then put herself through graduate school at the University of Chicago. In 1926, she completed a Ph.D. dissertation in chemistry, and shortly after took a job in Albany with the New York State Department of Public Health, an institution that hired an unusual number of women scientists. There, she worked steadily for almost a quarter of a century, isolating proteins and other molecules from infectious bacteria and protists that could be used for diagnostic tests or for vaccines. When Elizabeth

Hazen and Gilbert Dalldorf stuck their heads in the door of her lab, it was the beginning of a fruitful collaboration and friendship that would last for more than 25 years.

Hazen explained that she had two strains of bacteria that could kill fungi, but that she needed to isolate the antifungal substances for further testing. In the months that followed, Brown and Hazen sent bulky package after bulky package between Albany and New York City. Hazen would send samples of bacteria, then Brown would extract different compounds for further testing by Hazen. The two researchers quickly isolated three fungicidal molecules from two kinds of bacteria. Two of these molecules were highly toxic to mice and therefore not useful as drugs. The third substance, however, was only mildly toxic and killed virtually any fungus it came into contact with; they called it fungicidin. It looked like a miracle drug, another penicillin.

At a meeting of the National Academy of Sciences in October 1950, Hazen and Brown presented their results. A reporter from the *New York Times* jumped all over the story, and within a day or two, the new life-saving drug was big news. Pharmaceutical companies began calling the state lab, demanding to be let in on the project.

Dalldorf and Brown nearly panicked. Funigicidin had been tested in animals, but never in humans. For clinical trials, huge quantitites of the drug had to be manufactured cheaply and quickly. Only a pharmaceutical company could do that. But Dalldorf and Brown knew, from previous experience with another drug, that they needed to patent fungicidin quickly before someone else did. Without a patent, to make clear who owned the rights to the drug, no pharmaceutical company would develop it. And Hazen, Brown, and Dalldorf wanted the drug to be used to save lives, the sooner the better.

By February, the researchers had filed a patent application with the United States Patent Office and a month after that licensed the patent to Squibb Pharmaceuticals. Fungicidin, it turned out, was a name already in use for another chemical, so Hazen and Brown renamed it nystatin (pronounced nye-stat-in), after New York State. Within five years, Squibb had found a way to produce nystatin cheaply and had successfully tested the drug in humans. As the first treatment for fungal infections in humans, nystatin was tremendously useful.

Squibb sold so much nystatin that between 1955 and 1979 alone, royalties from sales of the drug came to $13.4 million. Half of this went to a nonprofit foundation that funded scientific research. The other half went to the nonprofit Brown–Hazen Fund, which supported research and training in the medical and biological sciences and awarded grants to women scientists. Neither Hazen nor Brown ever accepted any of the enormous wealth that resulted from their discovery. Brown apparently helped put several young women through college, as she herself had been helped. Both Hazen and Brown continued to work into their 70s, content in their achievements, honored with medals, honorary degrees, and elections to prestigious societies.

KEY CONCEPTS

1. Protists are diverse, but all are eukaryotes.

2. Algae are protists that perform photosynthesis.

3. Protozoa are heterotrophic protists, most of which are motile.

4. Some protists resemble fungi in appearance.

5. True fungi lack flagella, have distinctive cell walls, and absorb rather than ingest food.

6. The classification of fungi is based on their life cycles and the way they grow.

7. Fungi play important ecological roles as decomposers, as partners in lichens, and in association with the roots of higher plants.

8. Fungi evolved from protists at least 430 million years ago.

In this chapter we will quickly survey two kingdoms of organisms: the fungi and the protists. Biologists used to define a fungus as a eukaryotic organism that is neither plant nor animal. However, that definition applies to the protists as well. One important difference between the fungi and the protists is their organization. Most fungi are multicellular, while most protists are single-celled. A few kinds of protists form colonies, but these are primitive collections of nearly identical cells. In contrast, fungi resemble plants and animals in having complex internal structures and specialized cells that serve different purposes.

Another difference between fungi and protists is how easily fungi can be classified. Most of the fungi are remarkably alike and make an easily recognized kingdom of organisms. But the protists are so diverse that some biologists believe that they should be divided into several kingdoms.

Protists, because they are mostly one-celled, are considered more primitive than the fungi. Certainly, the first protists

Figure 21-1 How many different protists live in a pond? A variety of protists live in both fresh water and salt water, including algae, dinoflagellates, diatoms. Some protists are motile, others drift or anchor themselves to some object. Different protists move by waving flagella or cilia, or by ameboid movements.

lived in the sea long before the first fungi evolved. However, protists are anything but simple, for they show extraordinary and elaborate specializations at the cellular level.

If we look at a drop of pond water under a microscope, we discover an amazing world of tiny, wriggling creatures whose existence we would not otherwise suspect. Many of these microscopic creatures are members of the kingdom Protista. The protists range from the simple amebas to the most complicated cells on our planet (Figure 21-1). Their variety raises many questions: Why are protists so diverse? How are the various protists related to one another? How are they different from

larger, more familiar organisms—such as mushrooms, mosses, or mammals? How are different from prokaryotes? And how do they derive energy?

WHAT ARE PROTISTS?

Protists [Greek, *protos* = first] include the algae (which photosynthesize like plants), the protozoans (which act more or less like animals), and some funguslike forms. Protists flourish wherever there is moisture. In some marine ecosystems, the brown algae and the red algae are responsible for trapping nearly all of the solar energy that eventually reaches other organisms. The abundance of protists in the fossil record suggests that protists have been successful for a long time.

Protists are eukaryotic. Protist cells have nuclei and other membrane-bounded organelles, and almost all have mitochondria. Most (but not all) also have chromosomes and undergo mitosis. All can reproduce asexually by fission, and some can perform meiosis and reproduce sexually.

At a biochemical level, protists are more or less identical to plants, animals, and fungi: all derive energy from glycolysis and respiration, and the photosynthetic protists all use water as a source of electrons. Protists lack the biochemical diversity of the prokaryotes, but they are the champions of cellular diversity. No other kingdom includes so many different kinds of cells.

Why Are Protists So Diverse?

Some protists are autotrophic and have chloroplasts. Others are heterotrophic and possess vesicles and lysosomes. Some are shapeless bags, while others have elaborately sculpted tests, or shells. Some are free-living, some parasitic. Some produce spores, others do not. Some are single-celled and others are multicellular like plants, animals, and fungi.

Some protists are **plankton** [Greek, *planktos* = wandering], organisms that float in the water, carried by currents (Figure 21-2). Others are **motile**, meaning that they move actively in search of food. Of the motile forms, some propel themselves with one or two long flagella (Figure 21-2). Some use hundreds

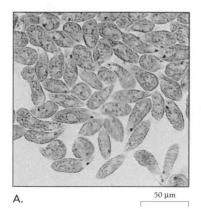

A. ⊢ 50 μm ⊣

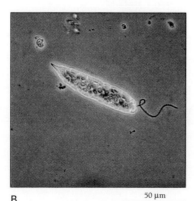

B. ⊢ 50 μm ⊣

Figure 21-2 Plant or animal? A. Euglena in sunlight are green with chloroplasts. B. Euglena in a dark environment use their flagella to go looking for light. *(A, Dwight Kuhn; B, Biological Photo Service)*

501

of cilia, and others use protrusions called **pseudopodia** [Greek, *pseudos* = false + *pous* = foot], temporary extensions of the cytoplasm.

With so much diversity, we may ask, What binds the protists together? The answer is, surprisingly little. What protists have in common is that they do not fit in with plants, animals, or fungi. When biologists regarded all organisms as either plants or animals, they classified all the multicelled protists as plants and all the single-celled protists as either **algae** (which, like plants, photosynthesize) or as **protozoa** (which, like animals, move under their own power).

The common pond-water protists of the genus *Euglena* illustrate why this classification scheme never worked very well (Figure 21-2). In the light, most species of *Euglena* develop chloroplasts and photosynthesize. In the dark, however, *Eu-*

glena lives a heterotrophic life, absorbing nutrients from its surroundings, and may even lose its chloroplasts. *Euglena,* then, is neither plant nor animal, but it is a fine protist.

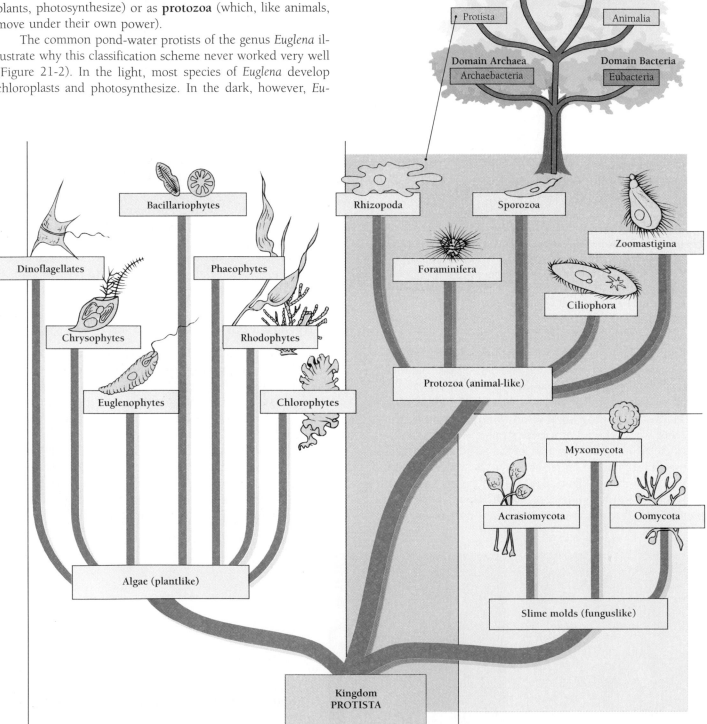

Figure 21-3 The diverse kingdom Protista.

The green alga *Chlamydomonas,* like *Euglena,* is a single-celled organism capable of both photosynthesis and directed movement. But *Chlamydomonas*'s close relative *Volvox* consists of colonies of hundreds to thousands of cells, each of which looks just like a free-living *Chlamydomonas.* Because the two algae are otherwise identical, it would be unreasonable to put one into the plant kingdom and one into the animal kingdom. As a result, biologists now put both groups into the Protista.

Protists are eukaryotic organisms that are defined not by what they have in common, but by not belonging to any of the other three eukaryotic kingdoms (Figure 21-3). In contrast, we define the other eukaryotic kingdoms, fungi, plants, and animals, partly by the way their cells function and develop. Plants and animals are all multicellular organisms that develop from embryos, organized collections of tissue that, through a highly predictable sequence of steps, develop into more-mature forms. Fungi are organisms, nearly always multicellular, that absorb nutrient molecules directly from their environments. The fungi do not photosynthesize and do not form embryos.

The diversity of the protists is both a joy and a headache. How can we classify them? Biologists do not agree on the number of phyla, or even on whether all the protists should be in the same kingdom: one protist researcher has even suggested that they really comprise some 20 separate kingdoms! Many biologists divide them according to their cellular organization and mode of nutrition, into plantlike organisms, animal-like organisms, and funguslike organisms. We will follow the classification scheme of Margulis and Schwartz, who divide the protists into 27 phyla, mostly on the basis of their cellular structures and life cycles. We will survey 16 of these phyla. For convenience, we group these into plantlike phyla (the algae), animal-like phyla (the protozoa), and the funguslike phyla.

Protists are metabolically more uniform than prokaryotes. However, partly because protists are a grab-bag kingdom (including every eukaryote that is not animal, plant, or fungus), they show enormous diversity at the cellular level.

How Do Protists Reproduce?

Every protist species goes through two phases in its life cycle—reproduction and dispersal. Dispersal may take many forms. Some free-living, motile protists swim or crawl. Others scatter seedlike **cysts,** enclosed structures that contain reproductive cells. Still others infect a host that moves, such as an insect, bird, or mammal.

Reproduction may be either asexual or sexual, depending on the species. Some protists are asexual and haploid throughout their life cycles and never undergo meiosis. Other protists are diploid for most of their life cycle and produce haploid gametes (Figure 21-4). Still other protists are haploid for almost all of their life cycles but produce a short-lived diploid zygote. Finally, some protists undergo **alternation of generations,** passing through both haploid and diploid phases.

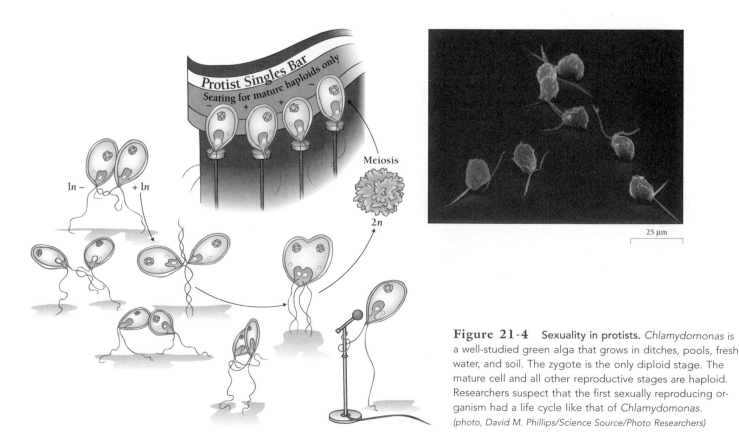

Figure 21-4 Sexuality in protists. *Chlamydomonas* is a well-studied green alga that grows in ditches, pools, fresh water, and soil. The zygote is the only diploid stage. The mature cell and all other reproductive stages are haploid. Researchers suspect that the first sexually reproducing organism had a life cycle like that of *Chlamydomonas.* (photo, David M. Phillips/Science Source/Photo Researchers)

ALGAE ARE PHOTOSYNTHETIC (PLANTLIKE) PROTISTS

Algae are relatively simple plantlike organisms that live in moist environments everywhere: in open water, ice, soil, as well as in puddles or pockets of water in trees or rocks. Most algae are *photoautotrophs,* organisms that supply themselves with energy by means of photosynthesis. Some, such as *Euglena,* can also live as heterotrophs. Just as plants do, algae use chlorophyll and other pigments to capture light for photosynthesis. However, whereas true plants photosynthesize with both chlorophyll *a* and chlorophyll *b,* most protists do not have any chlorophyll *b.* Algal protists also differ from true plants in the structure of their cell walls, the arrangement of flagella, and in their use of polysaccharides as food reserves rather than carbohydrates. In this section we will briefly survey 6 of the 12 phyla of algae: dinoflagellates, chrysophytes, euglenophytes, phaeophytes, rhodophytes, and chlorophytes.

Dinoflagellates

Most **dinoflagellates** [Greek, *dinos* = whirling + *flagellum* = whip] are single-celled organisms that float freely as plankton in warm oceans (Figure 21-5). A few can form colonies, and some live symbiotically with corals, sea anemones, and clams. The ultimate source of food in coral reef and other communities, in fact, is photosynthetic dinoflagellates. Dinoflagellates make up a major part of the phytoplankton, the microorganisms on which most aquatic ecosystems depend.

Of the thousands of species of dinoflagellates, the best known are those that grow rapidly to cause red tides, in which population explosions, or blooms, of dinoflagellates color the

Figure 21-6   Satellite photo of a red tide off the east coast of Italy. Dinoflagellate populations explode into "blooms," or "red tides," when conditions are good. Some species of dinoflagellates make poisons that clams, mussels, and other shellfish accumulate in their bodies, making these ordinarily edible shellfish deadly. Some dinoflagellates secrete toxins that kill whole schools of fish, whose flesh the dinoflagellates then feed on. During a bloom of such dinoflagellates, toxic vapors over the water can make boaters temporarily ill. *(Geospace/Science Photo Library/Photo Researchers)*

ocean red, brown, or other colors, depending on the species (Figure 21-6). Other red tides release deadly toxins. Because many dinoflagellates are bioluminescent, that is, they glow in the dark, they create dramatic nighttime displays.

Almost all dinoflagellates have a rigid wall, or **test,** made of cellulose and coated with silica. Around its circumference, the test contains two characteristic grooves, in which the two flagella are embedded at right angles to each other. Together the two flagella create the spinning that gives these organisms their name.

Dinoflagellates are very old. Because their hard tests are well preserved in the fossil record, we know that they have existed at least since the beginning of the Cambrian Period, some 590 million years ago. They are so ancient, in fact, that their chromosomes are organized differently from those of all other eukaryotes. When a dinoflagellate divides, its chromosomes segregate by attaching to the nuclear membrane rather than to a mitotic spindle. Dinoflagellates have no mitosis, no meiosis, and no sexual reproduction.

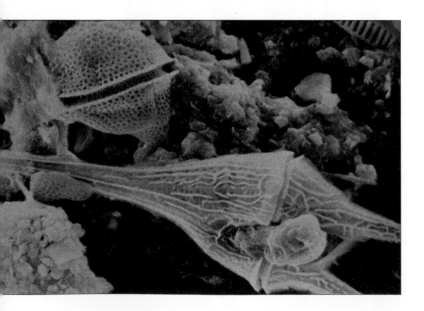

Figure 21-5   Dinoflagellates are renowned for their diversity and beauty. Shown here are the bioluminescent dinoflagellates *Ceratium* and *Peridineum.* *(Terry Hazen/Visuals Unlimited)*

Dinoflagellates are among the oldest and most primitive of the eukaryotes.

Chrysophytes (Golden Algae and Diatoms)

Golden-yellow pigments give the golden algae, or **chrysophytes** [Greek, *chrysos* = golden + *phyta* = plant], their name. As dinoflagellates do, many chrysophytes form hard tests containing silica and other minerals that help preserve them in the fossil record as far back as 500 million years ago. Each chrysophyte family contains both single-celled and colonial forms. The colonial forms may be spherical or loosely branched. Most chrysophytes are plankton, widespread in temperate lakes and ponds. Only one group lives in the oceans.

Chrysophytes reproduce asexually in one of two ways. A single cell, called a swarmer cell, may swim away from a colony to found another colony. Alternatively, the hundreds or thousands of cells of a mature colony may split into two groups, which then float away from each other to form two new colonies. Chrysophytes can also form cysts that resist cold temperatures and desiccation.

The diatoms are some of the most numerous and beautiful of the chrysophytes. Most of the 10,000 species of diatoms are single cells, although some form simple filaments or colonies. Each species has a unique and elaborate silica test (Figure 21-7). Most diatoms live as free-floating plankton in fresh or salt water. Some, however, can move on slime that they extrude from a slit between the two parts of the test. The tests of diatoms are well preserved in the fossil record, dating from the Cretaceous Period, between 65 and 135 million years ago.

Diatoms can reproduce sexually, forming gametes by meiosis. Usually, however, they reproduce asexually, dividing their chromosomes through ordinary mitosis. At cell division, each half of the test stays with one of the daughter cells, which then assembles a second half.

Chrysophytes (golden algae) have both single-celled and colonial species. Diatoms have elaborate tests and can reproduce sexually as well as asexually.

Euglenophytes

The **euglenophytes** [Greek, *eu* = true + *glene* = eyeball] derive their name from their light-detecting eye spots, which enable them to live a double life. In the dark, they live heterotrophically, by breaking down dissolved organic matter. When light becomes available, however, euglenophytes use their eyespots, and a single flagellum, to swim into the sunshine. Euglenophytes photosynthesize with both chlorophyll *b* (a pigment also used by higher plants), which gives them their grass-green color, and chlorophyll *a*. Except for the green algae, no other algae or any other protists make chlorophyll *b* (Table 21-1). Unlike plants and green algae, however, euglenophytes have no cell wall and store energy in a polysaccharide called paramylon, instead of in starch. Euglenophytes all reproduce asexually, using a primitive form of mitosis, with no clear mitotic spindle and no well-defined chromosomes. They also do not go through the stages of metaphase or anaphase.

Euglenophytes can live as either autotrophs or heterotrophs.

Phaeophytes (Brown Algae)

The **phaeophytes** [Greek, *phaios* = dusky brown], or brown algae, together with the rhodophytes (red algae), which we will discuss next, make up most of the organisms that we call seaweeds. Phaeophytes are all multicellular, and most live in the cool waters off temperate coasts. Phaeophytes include the biggest protists in the world—the giant kelps, which in deep water may be up to 100 m tall (Figure 21-8).

Most phaeophytes reproduce sexually. However, meiosis in phaeophytes produces **spores** rather than gametes. The spores germinate into new multicellular organisms in which all the

Figure 21-7 Diatoms may be even more beautiful than dinoflagellates. *(Robert Brons/Biological Photo Service)*

50 μm

Table 21-1 Which Protists Use the Same Chlorophylls as Plants?

Chlorophyll:	*a*	*b*	*c*	Storage Material
Plants	a	b		Starch
Chlorophytes	a	b		Starch
Euglenophytes	a	b		No starch
Dinoflagellates	a		c	Starch or oil
Phaeophytes	a		c	No starch
Chrysophytes	a		c	No starch
Rhodophytes	a			Starch

Figure 21-8 Kelp forests are home to a variety of other organisms. In shallow coastal waters, the giant kelp *Macrocystus* may grow at enormous rates, producing masses of brown algae that provide both food and shelter for fish, sea urchins, sea otters, and other organisms. Although the giant kelps are simpler than true plants, they have a few specialized structures. Floating photosynthetic blades capture light for photosynthesis. The blades are kept near the surface by gas-filled bladders and anchored to the sea floor by holdfasts.

cells are haploid. This haploid form is called a **gametophyte** because it then produces gametes. The diploid form is called a **sporophyte** because it produces spores. Such alternation of generations also occurs in green algae, red algae, fungi, and, of course, the true plants.

While the size and complexity of the phaeophytes far exceed those of most other protists, the individual cells of phaeophytes resemble those of other protists. In particular, the flagellated sperm of phaeophytes look just like the cells of chrysophytes. Phaeophytes make chlorophylls *a* and *c* and produce other pigments like those of chrysophytes (Table 21-1). But as distinct from higher plants, phaeophytes never produce chlorophyll *b* and do not synthesize starch.

Phaeophytes (brown algae), which are related to the chrysophytes, are the largest and most complex of the protists.

Rhodophytes (Red Algae)

Although the brown algae include the most spectacular seaweeds, most of the world's seaweeds are **rhodophytes** [Greek, *rodon* = rose], or red algae. Nearly all of the 4000 species of rhodophytes are marine, and most live in tropical seas, attached to rocks or other algae. Among the most beautiful of the rhodophytes are the coralline algae. These algae deposit calcium carbonate in their cell walls, and contribute to the building of coral reefs. Calcium-depositing rhodophytes have been preserved in the fossil record as far back as 500 million years.

The cell walls of red algae occasionally contain cellulose, but more often contain another polysaccharide, whose attached sulfate groups give the rhodophytes their characteristic slippery feel. Two such polysaccharides—agar and carrageenan—are useful to humans. Agar helps gel instant desserts, jellies, and baked goods, and is also the most widely used culture medium for growing bacteria in the laboratory. Carrageenan is used to stabilize oil and water emulsions in foods such as mayonnaise, as well as in paints and cosmetics.

All rhodophytes reproduce sexually and have complex life cycles. Meiosis can take place right after fertilization, with no sporophyte stage. Alternatively, the zygote can develop into a diploid organism, which later produces haploid spores that grow into male and female gametophytes. The female gametophyte produces a large ovum, while the male produces a much smaller sperm.

Rhodophytes (red algae) have distinctive pigments and complex sexual cycles.

Chlorophytes (Green Algae)

Chlorophytes [Greek, *chloros* = green], or green algae, are more like plants than any other group of protists. In fact, plants probably arose from a chlorophyte ancestor. Chlorophytes are

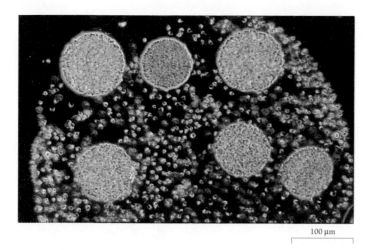

Figure 21-9 Colonial green alga—blobs of identical cells or multicellular individuals? Members of the colonial genus *Volvox* may contain from 500 to 60,000 cells arranged in a single-layered, hollow sphere. The cells in most colonies of green algae are identical. In *Volvox* and some other genera, however, there is some division of labor, suggesting a rudimentary form of multicellularity. For example, only some cells are able to reproduce. In addition, the cells may move smoothly through the water by coordinating the beating of their flagella, much as our heart cells coordinate their contractions. *(Alfred Owczarzak/Biological Photo Service)*

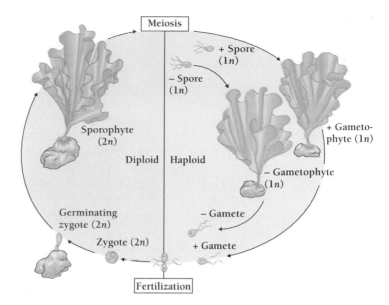

Figure 21-10 The life cycle of a sea lettuce. The marine Ulvophyceae take their name from *Ulva*, or sea lettuce, which grows in shallow seas throughout the temperate zones. Each ulva has a flat, leaflike thallus, which may be a meter or more in length, and a rootlike holdfast, which attaches the thallus to a rock or other object. *(photo, Andrew J. Martinez/Photo Researchers)*

unlike most of the other photosynthetic protists in having both chlorophylls *a* and *b* and in making starch, which they store within their chloroplasts as do plants. Among the rest of the protists, only the euglenophytes have both chlorophylls *a* and *b*, and only the dinoflagellates make starch. Although modern chlorophytes include several multicellular groups, biologists do not believe that any of them was the ancestor of the multicellular plants (Figure 21–9).

Chlorophytes come in a great array of colonial and multicellular forms. They range in size and complexity from the single-celled *Chlorella,* about 25 μm in diameter, to the seaweed-like *Codium magnum,* which may be 8 m long. Chlorophytes are mostly free living and aquatic, but some species live on the surface of snow, on tree branches, in the soil, or in close association with other organisms, including fungi, hydras, and other protists (Figure 21-10).

The simplest chlorophyte is *Chlorella,* which lacks flagella and is smaller than *Chlamydomonas. Chlorella* is among the most widespread of the green algae, living in fresh water, salt water, and soil. *Chlorella*-like species live as photosynthetic *symbionts* within protozoa as well as large numbers of invertebrate animals, including sponges, hydras, and flatworms. A close association between two organisms is called **symbiosis** [Greek, *sumbios* = living together]. Organisms living in a symbiotic relationship with one another are called symbionts.

Chlorophytes (green algae) are more like plants than other protists and also the most diverse of the plantlike protists.

PROTOZOA ARE HETEROTROPHIC (ANIMAL-LIKE) PROTISTS

Even though we no longer classify protists as plants or animals, we still use the old term **protozoa** [Greek, *protos* = first + *zoe* = life] to describe the protists that most resemble little animals. The protozoa include a wide variety of heterotrophs, most of which obtain their food by phagocytosis, ingesting particles or other cells.

Most protozoa are single-celled and motile. They are extraordinarily varied, and include eight phyla, of which we will survey just five: rhizopoda, foraminifera, sporozoa, ciliophora, and zoomastigina. Three of these—the rhizopoda, the ciliophora, and the zoomastigina—we distinguish by the way they move. Each of the other two has characteristic traits. We identify the sporozoa by their characteristic life cycles and the foraminifera by their distinctive shells.

Rhizopoda (Amebas)

Rhizopoda, or amebas, are common throughout the world, in oceans, in fresh water, in soil, and as parasites of animals. Hundreds of millions of people in the tropics (and about 10 million in the United States) harbor parasitic amebas that infect the lower intestinal tract. Fortunately, only about 20 percent of such infections produce the symptoms of amebic dysentery, which range from mild discomfort to severe diarrhea and death. In the most severe cases the ameba may cause abscesses that rupture the abdominal wall or penetrate the diaphragm and lungs.

An ameba continuously changes its shape, in an apparently chaotic way (hence the old name of one species—*Chaos chaos*). Amebas do this by means of **pseudopodia,** temporary extensions of membrane-enclosed cytoplasm. An ameba extends a pseudopod and then moves its cytoplasm into the pseudopod, so that the pseudopod becomes the body. The movement of amebas only appears chaotic, for they actually move in a directed manner toward sources of food. They then envelope the food in their pseudopodia.

All amebas reproduce through simple cell division. Some, such as the ones that cause amebic dysentery, can form cysts immune to desiccation or digestion by their hosts. A few construct distinctive shells (tests) by gluing together tiny grains of sand and other inorganic bits. Similar tests exist in the fossil record, with some dating back to the Precambrian, more than a half billion years ago.

Rhizopoda (amebas) use pseudopodia to move and ingest food.

Foraminifera

The **foraminifera** [Latin, *foramen* = little hole + *ferre* = to bear], or **forams,** are marine organisms whose tests are full of tiny holes (Figure 21-11A). The tests of forams are made of organic materials reinforced with grains of sand or minerals and range in size from about 20 μm to several centimeters. Most are tiny and live in the sand or attached to other organisms or stones. Two large groups are free-floating plankton and serve as food for clams, mussels, and other invertebrates that filter food from water.

Forams have complex sexual life cycles. Unlike most other protozoans, they have true alternation of generations, with extended haploid and diploid phases—just like plants. Some forams harbor within them photosynthetic protists—dinoflagellates, chrysophytes, and diatoms. Although forams are unicellular, they are subdivided into multichambered shells connected by cytoplasm.

The tests of forams are abundant in the fossil record, especially in the last 230 million years. Forams make up much of the chalky sediment on the bottoms of the seas. The famous white cliffs of Dover, England, are just such foramen sediments, uplifted by geologic processes (Figure 21-11B).

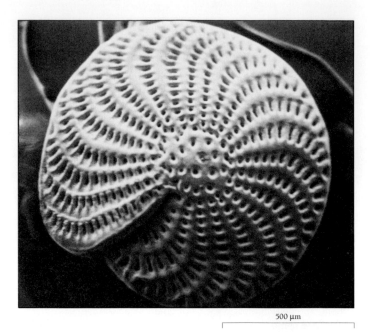

500 μm

Figure 21-11   The tests of foraminifera are rich in calcium. As these organisms die and drop to the sea floor, their tests form thick, white deposits.   *(Biophoto Associates)*

Foraminifera have complex sexual life cycles and tests studded with little pores.

Sporozoans

Sporozoans are all parasites. They reproduce by a combination of asexual and sexual processes, including the alternation of generations. Sporozoans have especially complex life cycles, usually involving two or more different hosts. Sporozoans are a scourge to their many animal hosts but a delight for puzzle-solving parasitologists. The members of one genus, *Plasmodium,* cause malaria, a disease that every year infects some 250 million people and kills 2 to 4 million of those infected. *Plasmodium* spreads among humans with the help of its other set of hosts—mosquitos of the genus *Anopheles* (Figure 21-12). The complex life cycles of sporozoans ensure continued success: at any moment, different phases of each species are present in several forms in several places in two or more hosts. They are virtually impossible to eradicate.

Sporozoans are nonmotile, spore-forming protozoa, which live as parasites in animals. They are known for their complex life cycles.

Ciliophora (Ciliates)

Ciliophora, or the **ciliates,** include about 8000 species of free-living, single-celled heterotrophs that live both in fresh water

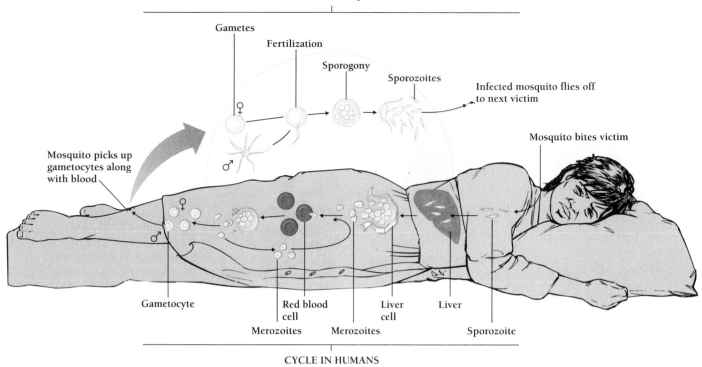

Gametes

Fertilization

Sporogony

Sporozoites

Infected mosquito flies off to next victim

♀

♂

Mosquito picks up gametocytes along with blood

Mosquito bites victim

Gametocyte

Red blood cell

Liver cell

Liver

Merozoites

Merozoites

Sporozoite

CYCLE IN HUMANS

Figure 21-12 How *Plasmodium* causes malaria. When a mosquito sucks blood from a human, it first injects a bit of saliva. The saliva contains an anticoagulant, a chemical that keeps the blood flowing freely and allows the mosquito to drink its fill. A mosquito infected with *Plasmodium* harbors a diploid form of the parasite—called a *sporozoite*—in the wall of its gut and in its salivary glands. When the mosquito injects its saliva, some of these sporozoites enter the human host's blood. In the human bloodstream, the sporozoites travel to the liver, where they divide several times by mitosis and develop into a new diploid phase, called a *merozoite*. The merozoites take up residence inside of red blood cells and divide rapidly. Eventually, the red cells rupture and release merozoites and toxins, which cause the awful fever and chills of malaria.

The merozoites infect more red cells or develop into *gametocytes*. The gametocytes, if sucked up by a second hungry mosquito, produce gametes, which form diploid zygotes, which embed themselves in the mosquito's intestinal wall. These develop into sturdy *oocysts,* which undergo mitosis to form thousands of sporozoites, some of which migrate to the salivary gland, where they can begin the cycle again.

The simultaneous presence of *Plasmodium* in the gut and salivary glands of mosquitos and in the blood and liver of humans has made it impossible to eradicate malaria. The draining of swamps and the use of insecticides such as DDT have destroyed breeding grounds for the mosquitos and reduced the incidence of malaria in some areas. Yet, worldwide, the number of new cases of malaria has more than doubled since the mid-1970s, for the mosquito has evolved resistance to insecticides such as DDT, and *Plasmodium* has evolved resistance to antimalarial drugs.

and in salt water. Their name [Latin, *cilium* = eyelash] reflects the presence, on their surface, of thousands of cilia, shorter and more numerous than eukaryotic flagella. The cilia, which may be arranged in bundles or in sheets, help ciliates move and feed (Figure 21-13). The best-known ciliate is probably *Paramecium,* a common resident of pond water, long used in teaching and research. Most ciliates lack shells, although a few marine ciliates form shells of sand held together by organic cements. The fossil record shows that such shell-forming ciliates have existed for about 100 million years.

Paramecia and other ciliophora (ciliates) are single-celled, free-living heterotrophs coated with cilia.

Zoomastigina (Flagellates)

The **zoomastigina** [Greek, *mastix* = whip], or **flagellates,** are a diverse group of single-celled heterotrophs. All zoomastigotes have at least one long flagellum, and some have thousands. Some are free-living in fresh or salt water, while others are parasites in animals. Reproduction of zoomastigotes is mostly asexual, though some species have a sexual cycle as well. Some convert from a flagellated to an ameboid form and back again, depending on the availability of food.

Zoomastigina (flagellates) are single-celled and use flagella to move.

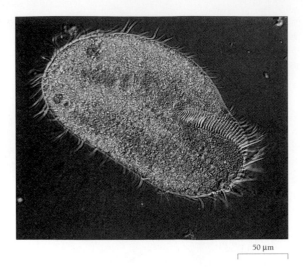

50 μm

Figure 21-13   The cilia of a *Paramecium*. The cilia surround the oral groove and form oarlike arrangements for movement. *(M. Abbey/Photo Researchers)*

FUNGUSLIKE PROTISTS

We have seen that some protists resemble plants and some resemble animals. The last group of protist phyla most resemble fungi. The names of these phyla reflect the fact that they were initially identified as fungi: all include the Greek root *mukes* (= fungus). We will describe three of the seven funguslike protist phyla: acrasiomycota, the cellular slime molds; myxomycota, the plasmodial slime molds; and oomycota, which include the water molds, mildews, and potato blight. While these organisms resemble fungi in their general appearance, they more closely resemble other protists in their cellular organization and modes of reproduction.

Acrasiomycota (Cellular Slime Molds)

Acrasiomycota [Greek, *acrasia* = confusing two things + *mukes* = fungus], or cellular slime molds, are peculiar organisms, animal-like at some stages of their lives and plantlike at others. They live in damp soil, in fresh water, on decaying vegetation, and especially on rotting logs. When nutrients are plentiful, they flourish as amebalike cells, consuming bacteria and other food particles by phagocytosis. When food is scarce, the amebas aggregate to form a multicellular stage that looks a bit like a miniature garden slug (Figure 21-14). The slug secretes a slime track and wanders about in search of food and light. When it finds food, it settles down and develops a mushroomlike *fruiting body* that produces spores. Each spore turns into an ameba. Reproduction is asexual; both the amebas and the slug are haploid.

Acrasiomycota (cellular slime molds) live as amebas and as multicellular slugs.

Myxomycota (Plasmodial Slime Molds)

Myxomycota [Greek, *muxa* = mucus + *mukes* = fungus]—the plasmodial, or acellular, slime molds—closely resemble the cellular slime molds in appearance and in life cycle. The migrating form of a plasmodial slime mold, however, differs greatly from that of a cellular slime mold. Instead of developing from an aggregation of individual hungry amebas, a plasmodial slime mold develops a **plasmodium**—a mass of cytoplasm with many nuclei but no boundaries between cells.

The plasmodial slime molds also differ from the cellular slime molds in having a sexual cycle. The amebas are haploid and two of them can form a diploid zygote. The nucleus of the zygote then divides repeatedly, but the cell merely grows larger and larger without dividing. The result is a plasmodium with many diploid nuclei. Unlike the cellular slime mold's slug, the plasmodium can both feed and move about. It propels itself by differential growth, growing more in one direction than another. When food or water becomes scarce, the slug forms a fruiting body, undergoes meiosis, and releases haploid spores.

Myxomycota (plasmodial slime molds) live as amebas or as a multinucleated plasmodium.

Oomycota

The **oomycota** most nearly resemble true fungi. They are either parasitic, deriving nourishment from living organisms, or *saprophytic,* deriving nourishment from dead organisms. Like true fungi, the oomycota extend threads into the tissues of their hosts, release digestive enzymes, and absorb the resulting organic molecules. The oomycota differ from the fungi, however, in having flagella on their motile spores. They reproduce sexually, and their life cycles include both haploid and diploid

500 μm

Figure 21-14   The slug of a cellular slime mold.   *(Cabisco/Visuals Unlimited)*

phases. The haploid gametes differ greatly in size; the large ovum is the source of this phylum's name, oomycetes [Greek, *oion* = egg + *mukes* = fungus], meaning "egg fungi."

Among the most common oomycotes are the water molds, which grow in fuzzy masses on the remains of algae or animals. In freshwater ecosystems, water molds decompose and recycle nutrients from dead organisms.

Some oomycotes cause diseases in animals or plants. Historically, the most important of these is *Phytophthora infestans,* which causes blight in potatoes. In 1845 and 1849, this oomycote infected and destroyed the Irish potato crop. A million Irish died of hunger, and another 1.5 million were forced to emigrate to the United States. A similar blight on the German potato crop late in World War I helped force an end to the war.

Oomycota resemble fungi, but their spores have flagella.

HOW DO TRUE FUNGI DIFFER FROM OTHER ORGANISMS?

We love fungi. We put fungi in our salads and fungi in our sauces. We use fungi to ferment wines and cheeses and to leaven bread. What is more, we need fungi. A walk through a wet wood demonstrates the enormous ecological importance of fungi: they are prodigious decomposers, recycling the nutrients of dead plants and animals. Without fungi, dead trees would fall and accumulate, and the soil, a little bit of which washes away in every storm, would never be replaced.

But fungi also plague us in countless ways. *Cryptococcus neoformans,* for example, is one of many unpleasant fungi that torment humans and other animals. Distributed worldwide, it infects first the lungs, then the membranes surrounding the brain, and sometimes the kidneys, bones, and skin. The first sign of infection is a cough. Usually, the symptoms that bring *C. neoformans*'s victims to a doctor, however, are headache, blurred vision, confusion, depression, or inappropriate speech or dress. Another infectious fungus attacks the sinuses, then the brain, producing intolerable headaches and ultimately consuming its victim alive.

Still other fungi destroy entire harvests of wheat, potatoes, and other crops. Fungi destroy in small ways as well as large. In the kitchen, fungi ruin bread, potatoes, jams, and sauces. In the bathroom, they attack our feet, producing athlete's foot, and grow on the walls of the shower and the porcelain surfaces of the toilet.

Fungi grow wherever there is sufficient warmth, moisture, and energy-rich organic molecules. Sometimes fungi find nourishment in surprising and bothersome places—the lining of a camera lens, clothing, shoes, the wooden planks of a ship. During the American Revolution, the British lost more ships to dry rot than to the fledgling American Navy. In World War II, fun-

gal infections of the skin took more men out of action in the South Pacific than did battle wounds.

Fungi are a little like plants and a little like animals. As long as biologists classified all organisms as either plants or animals, fungi posed a problem. Like animals, fungi are heterotrophic—they derive their energy from other organisms. But fungi enclose their cells in rigid walls as plants do. Fungi are distinct from both plants and animals, however: they take in food differently from animals, their cell walls are chemically different from those of plants, and they lack the embryonic stages of plants and animals. Everyone now agrees that the fungi deserve their own kingdom.

How Can We Characterize the Fungi?

With the exception of the single-celled yeasts and a few others, most fungi are **molds** composed of masses of threadlike filaments called **hyphae** [Greek, *hyphe* = web; singular, hypha]. The hyphae contain many nuclei (Figure 21-15). In some fungi, the nuclei all lie in a common cytoplasm, rather than in separate membrane-enclosed cells. Such fungi are said to be **coenocytic** [Greek, *koinos* = shared + *kytos* = hollow vessel]. In **dikaryotic** fungi (which have two nuclei per cell), walls called **septa** separate the nuclei within a filament. The septa, however, are perforated by large pores that allow the cytoplasm to flow along the hypha. Hyphae grow, branch, and intertwine to form a mass, called a **mycelium** [Greek, *myketos* = fungus]. Growth may be so rapid that in the time it takes you to read this page a single mycelium can add meters of hyphae.

Fungi's ability to grow rapidly helps compensate for another odd characteristic of their cells—lack of mobility. All fungal cells, even the gametes, lack flagella and cilia. Fungi can move in two ways only. The mycelium can move from place to place by growing, or the mycelium can produce spores that the wind scatters.

Another distinguishing characteristic of the fungi is their cell walls, which are made of a tough polysaccharide called

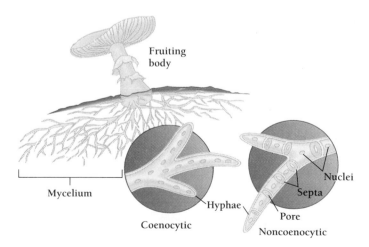

Figure 21-15 Two types of fungal hyphae. The nuclei within hyphae may all share the same cytoplasm (coenocytic) or they may be separated by septa.

BOX 21-1

Fungal infections today

In recent years, fungal infections have become increasingly common. Between 1991 and 1994, for example, fungal infections in one county in Southern California increased from 400 per year to 4500 per year. In fact, today nearly 40 percent of all deaths from hospital-acquired infections are due not to bacteria or viruses but to a handful of fungi.

Most hospital infections are due to *Candida albicans,* a virulent relative of baker's yeast. *Candida* infects 70 percent of AIDS patients. Even if science could eliminate *Candida,* however, other fungi would quickly take its place. *Cryptococcus, Histoplasma,* and *Aspergillus* are three of the next most common fungi that infect patients with impaired immune systems. Even ordinary baker's yeast, a normal inhabitant of the human mouth and the stuff we use to make bread rise, can grow out of control in individuals who are already very sick.

Healthy people also fall victim to some seemingly ordinary fungi. In California's San Joaquin Valley, for example, many people become infected by the soil fungus *Coccidioides,* which causes symptoms ranging from mild flulike illnesses to attacks in which the fungus spreads throughout the body (Figure A).

Against this onslaught of rot, health care workers have only a few broadly effective drugs. Nystatin and two other drugs derived from it are still the most effective way to treat fungal infections. One of the most powerful antifungals is amphotericin B, a drug with side effects so nasty that patients sometimes call it "amphoterrible." It can cause fever, chills, low blood pressure, headaches, vomiting, inflammation of blood vessels, and kidney damage.

Safe antifungal drugs with few side effects are much harder to find than safe antibiotics. This is because antibiotics, which

kill prokaryotes but not our own eukaryotic cells, work by exploiting any of a variety of differences in cell chemistry between prokaryotes and eukaryotes. But fungi are eukaryotes, like us. Their cells work like our own cells, and chemicals that are bad for fungi are usually bad for us too.

Nonetheless, the chemistry of fungal cells does differ from that of our own, and any differences can be exploited by researchers who design drugs. For example, the three drugs most commonly used all work by attacking ergosterol, a lipidlike solid that helps form the cell membranes of fungi. Our own cell membranes are made of cholesterol. Fungal cells also differ from ours in having a cell wall. Researchers' latest hope is to find a drug that attacks the cell walls of fungal cells without harming our own.

Figure A   Fungal spores are everywhere. The tiny haploid particles by which mushrooms and other fungi reproduce float in and out of our nostrils with each breath, land on bread and jams in our kitchens, and grow into dark streaks in our bathrooms. Outside our homes, a rain of fungal spores dusts trees and shrubs, lawns, dead leaves, and soil. Creeks, rivers, and lakes disperse fungal spores from one watershed to the next. And winds carry fungal spores from one end of the Earth to the other.   *(Biophoto Associates)*

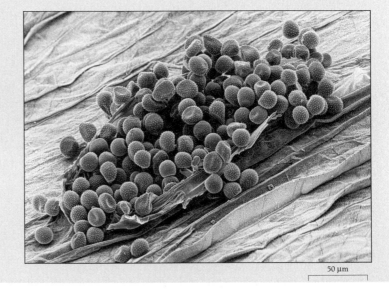

50 μm

chitin, the same substance in the hard shells of insects, spiders, and crustaceans. In plants, the major component of the cell wall is cellulose.

Recall that all fungi are heterotrophs, organisms whose food is organic molecules produced b other organisms. A fungus's food may be either dead or alive. Saprophytic fungi feed on the remains of dead organisms, while parasitic fungi feed on living organisms. Fungi obtain food, whether dead or alive,

by infiltrating the bodies of organisms and other food sources with long thin hyphae, usually just a few micrometers in diameter. The fungi extend their hyphae into the food, secrete digestive enzymes that break down the food, and absorb, rather than ingest, the resulting nutrients.

Fungi may reproduce asexually or sexually. Each cell (or nucleus) within a mycelium is usually capable of generating a new mycelium. Most fungi reproduce vegetatively (asexually).

Figure 21-16 A fairy ring. *(G. Carleton Ray/Photo Researchers)*

A single individual mycelium can produce separate individuals by means of simple cell division, by the breaking up of existing hyphae, or by budding. Fungi also reproduce by making and dispersing hardy spores capable of resisting heat, cold, drought, and lack of food. Spores are so small that the wind can scatter them all over the planet.

Spores, which are always haploid, may be made in a variety of intricate structures, called **sporophores.** Ordinary mushrooms are the sporophores of a large mycelium, which may be many meters in diameter (Figure 21-16). In general, asexual spores form when food is plentiful and the environment is otherwise welcoming. Fungi reproduce sexually when food is scarce or the environment has become unfavorable. Recall that sexual reproduction creates a greater diversity of genotypes in the next generation. This diversity increases the chances that a

few individuals will survive a hostile environment. Each haploid spore can germinate and produce a new haploid mycelium. Fungi differ from both plants and animals in that they do not go through any embryonic stages. Spores form mycelia directly.

Fungi are filamentous, lack flagella, have distinctive cell walls, absorb food, and reproduce by means of spores.

How Do Mycologists Classify Fungi?

Many mycologists still classify the funguslike protists as fungi. In this book, however, we consider the funguslike protists to be distinct from fungi (Figure 21-17). Fungi lack motility, and

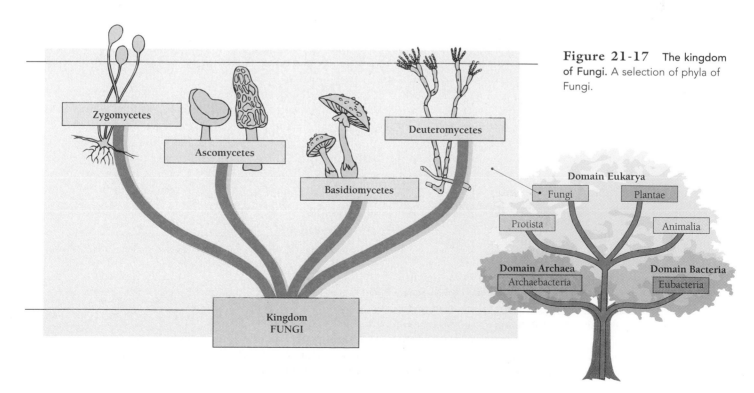

Figure 21-17 The kingdom of Fungi. A selection of phyla of Fungi.

Figure 21-18 Life cycle of black bread mold *Rhizopus stolonifer*.

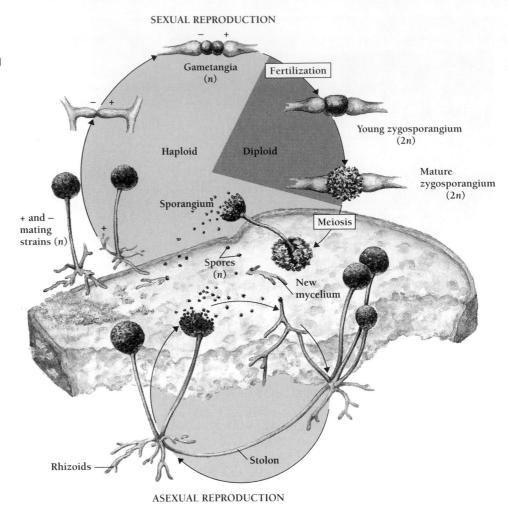

SEXUAL REPRODUCTION

Gametangia (*n*)

Fertilization

Haploid

Diploid

Young zygosporangium (2*n*)

Mature zygosporangium (2*n*)

+ and − mating strains (*n*)

Sporangium

Meiosis

Spores (*n*)

New mycelium

Rhizoids

Stolon

ASEXUAL REPRODUCTION

their cell walls differ from those of protists. Fungi also lack embryos and acquire nutrients differently from plants and animals. We will discuss examples of four phyla of fungi: zygomycetes, ascomycetes, basidiomycetes, and deuteromycetes.

Mycologists group fungi into these four phyla according to whether the hyphae have septa and how the fungi produce sexual spores. Most fungi are capable of reproducing both sexually and asexually. Asexual reproduction may occur vegetatively (as when parts of the mycelium divide to form new individuals) or through the production of spores by mitosis. Sexual reproduction always occurs by means of spores. The manner in which fungi produce their sexual spores is one of the two main characteristics that mycologists use to classify fungi.

Zygomycetes (Conjugation Fungi)

The mycelium of a **zygomycete** [zygote + *mukes* = fungus], such as bread mold, is coenocytic: it consists of hyphae without dividing septa and all the nuclei share a common cytoplasm. The only exceptions are two kinds of spore-forming structures, which form asexual or sexual spores. Most of the time, zygomycetes reproduce asexually by forming spores in spore-containing capsules called **sporangia**.

The common black bread mold, *Rhizopus stolonifer,* illustrates the sexual life cycle of a typical zygomycete (Figure 21-18). When zygomycetes reproduce sexually, they do so by forming a *zygosporangium,* a structure that produces haploid spores from a diploid zygote.

Whether zygomycetes reproduce sexually or asexually, they always use spores. Both kinds of spores travel on air currents and can, in the right environment, divide to produce a new mycelium. Because spores drift randomly, their arrival on a hospitable host (a piece of bread, for example) is entirely a matter of chance. A few zygomycetes, however, have evolved a means of actively launching themselves into likely new environments (Figure 21-19).

Bread molds and other zygomycetes are coenocytic and usually reproduce asexually by means of spores.

Ascomycetes (Sac Fungi)

Ascomycetes are a large and diverse group, with some 30,000 species (Figure 21-20). They include most yeasts, the powdery mildews, the molds, the cup fungi, the bread mold *Neurospora*

514

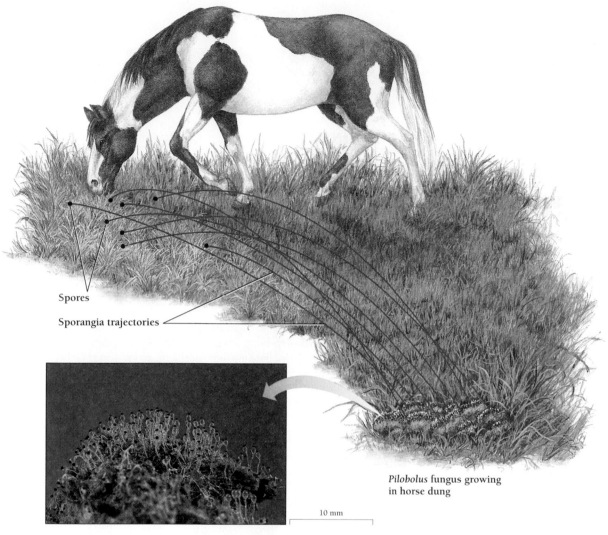

Spores

Sporangia trajectories

Pilobolus fungus growing
in horse dung

10 mm

Figure 21-19 Fungi disperse their spores effectively. *Pilobolus*, "the hat thrower,"
grows in the dung of horses and cattle. To disperse its spores to more dung, the hat thrower
relies on a light-sensing mechanism at the base of its sporangia. The fungus bends its
sporophores toward bright light, where grass is likely to grow, then fires its spores like can-
nonballs. The spores travel 2 m or more, with speeds of up to 50 km per hour. Spores that
land on blades of grass eaten by horses or cattle pass through the digestive tract with the
grass and emerge in the perfect environment—fresh dung. *(photo, courtesy Michael J. Wynne)*

(so useful to geneticists), and the truffles, prized by chefs for
their earthy flavor.

Ascomycetes [Greek, *askos* = wineskin + *myketos* = fun-
gus], or sac fungi, are so named because their sexual spores are
always produced in little sacs, called **asci** (singular, ascus). They
have no sporangia for making asexual spores. Their hyphae
usually have septa separating the nuclei. These septa, however,
are perforated. As a result, molecules and even organelles can
circulate throughout the cytoplasm.

The truffles, yeasts, mildews, and other ascomycetes have septa
dividing the nuclei of their hyphae and no sporangia. They
produce sexual spores in little sacs called asci.

Figure 21-20 The familiar cup fungi are ascomycetes. Bird's
nest fungi. *(Michael Fogden/DRK Photo)*

515

BOX 21-2

Ergotism

After *Penicillium* mold, the most famous fungus may be the ascomycete *Claviceps purpurea,* or ergot, which infects the flowers of rye and other grains. This peculiar sac fungus produces tiny spikes called sclerotia, or ergots, in the grain. The small, black or brown ergots—compact masses of hardened mycelium—replace a few of the seeds that would normally form in the seed head (Figure A).

Each ergot contains a witch's brew of deadly substances that can poison humans and other animals that eat the infected grain. In extreme cases, the symptoms of this poisoning, called ergotism, consist of hallucinations and fatal convulsions. In milder or chronic poisoning, sufferers may have headaches, muscle pains, numbness, and cold fingers and toes—all signs of constricted blood vessels that can result in gan-

grene, a usually fatal infection when not treated with antibiotics.

During the Middle Ages, ergotism was common in northern Europe and other areas where rye bread was popular. Also called St. Anthony's fire, ergotism could fell thousands of people in a season. In the year 994, for example, 40,000 people died from eating ergot-infected grain. Modern grains are cleaned of the dark masses of mycelium, and today ergotism is rare.

Ergot in small doses is medically useful. The main ingredient is an alkaloid—ergotamine tartrate. Ergotamine stimulates the smooth muscle of the peripheral and cranial blood vessels and also inhibits the neurotransmitter serotonin. As a result, ergotamine causes small arteries and smooth muscle fibers to contract. It is used to induce labor, to control bleeding from the uterus follow-

ing childbirth, to treat high blood pressure, and—in conjunction with caffeine—to relieve severe migraine headaches. Two other ergot alkaloids, bromocriptine and pergolide, are used in the treatment of Parkinson's disease. These two alkaloids help relieve the symptoms of Parkinson's disease by activating receptors for the neurotransmitter dopamine.

Ergot was also the original source of the psychoactive drug lysergic acid diethylamide (LSD). LSD induces mood changes —euphoria, anxiety, and depression—impaired judgment, and distorted perceptions of any sensations, including time, space, or self image. Despite LSD's reputation as a powerful hallucinogen, true hallucinations are apparently rare.

Figure A **Ergot.** At left, the small black ergot (or sclerotium) in the seed head of a rye plant. At right, a normal, uninfected seed head. *(Dennis Drenner)*

Basidiomycetes (Club Fungi)

The 25,000 species of **basidiomycetes,** or club fungi, include many of the most familiar fungi: the mushrooms, the bracket fungi, and the puffballs (Figure 21-21). These fungi take their name from the tiny, club-shaped **basidium** [Latin, = little pedestal; plural, basidia] that they use to produce spores. The basidia develop on the underside of a mushroom's cap, on the sides of the thin gills that radiate from the center like the pages of an open book (Figure 21-22). The mushroom itself, from

which the basidia develop, is the equivalent of a fruit and biologists call it a **basidiocarp** [Latin, *carpus* = fruit].

In the common field mushroom, the mycelium spreads underground and may be up to 30 m in diameter (Figure 21-16). The basidiocarps (mushrooms) develop at the periphery of the mycelium in "fairy rings"—often overnight. Folk legends once held that mushroom fairy rings appeared at the bidding of fairies who were tired of dancing. In reality, mushrooms are able to grow so quickly because a huge underground mycelium

Figure 21-21 **Basidiomycetes.** Bracket fungi and puffballs have the same sexual life cycle as supermarket mushrooms. *(J. Robert Stottlemyer/ Biological Photo Service; Michael Fogden/ DRK Photo)*

can quickly mobilize vast amounts of material. As the secondary mycelium exhausts the nutrients in the center of the ring, however, the ring grows outward in search of more nutrients. Because the mycelium typically grows at about 30 cm per year, we can estimate that a fairy ring 30 m in diameter is a century old.

Another group of basidiomycetes, less charming than mushrooms, but at least as important, are the rusts and smuts. Rusts and smuts ruin billions of dollars worth of wheat and other grains in the United States and elsewhere every year. As distinct from other basidiomycetes, however, rusts and smuts do not form basidiocarps. Instead, they are parasites with complex life cycles, usually involving hosts of two species.

Mushrooms, rusts, and other basidiomycetes are named for the club-shaped basidia that produce spores. They have septa and distinctive life cycles.

Deuteromycetes (Imperfect Fungi)

Mycologists classify fungi as zygomycetes, ascomycetes, or basidiomycetes, largely according to their mode of sexual spore formation. For some 25,000 named species of fungi, however, mycologists have so far discovered no sexual cycle at all. These are said to be "imperfect." Mycologists disagree, however, about whether the imperfect fungi form a separate phylum (or division).

Many imperfect fungi, for example, produce conidia (powdery asexual spores) and are probably ascomycetes. Mycolo-

gists therefore recognize the **deuteromycetes** as an artificial category. As the mode of reproduction of each species becomes known, mycologists are reclassifying them into one of the other three divisions of fungi.

Figure 21-22 **Like the pages of a book.** In basidiomycetes, the spore-producing basidia develop on the paperthin gills, which radiate from a central stalk.

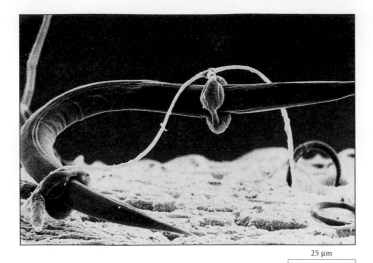

25 µm

Figure 21-23   A deuteromycete fungus lassoes a passing nematode worm.   *(Ken Barker)*

The deuteromycetes [Greek, *deuteros* = second + *mukes* = fungus] include many species that are important to humans. Some are parasites on the skin of humans and other mammals, causing ringworm, athlete's foot, and other skin diseases. These fungi are able to displace the bacteria that are normally abundant on skin because they are better able to derive nourishment from skin with their powerful digestive enzymes that break down tough skin proteins. These also compete with bacteria by inhibiting the growth of the bacteria with antibiotics. Recall that the first antibiotic used by humans came from a strain of the common deuteromycete mold *Penicillium.*

Some deuteromycetes are predaceous. Using sticky substances on their surface, these fungi trap protozoa and small animals. One species even snares its prey with quickly tightening hoops (Figure 21-23).

Penicillium and other deuteromycetes always reproduce asexually, which makes them hard to classify.

WHAT IMPORTANT ECOLOGICAL ROLES DO FUNGI PLAY?

Fungi have two important ecological roles, both of which depend on the power of the enzymes that fungi secrete. These extraordinarily potent enzymes can reduce substances as different as Wonder Bread and solid rock into small molecules that can be absorbed by fungi or other organisms. Using these enzymes, fungi decompose organic matter and make nutrients available to other organisms. Fungi, together with bacteria, recycle much of the world's organic matter from dead plants and animals, not to mention feces, leaves, and logs. Fungi also par-

ticipate in close, mutually beneficial symbiotic associations with other organisms. We will discuss two kinds of important fungal associations, lichens and mycorrhizae.

What Role Do Fungi Play in Lichens?

Each of the 25,000 recognized species of lichens looks like a moss or simple plant growing on a rock or a tree trunk. **Lichens** may be leafy, shrubby, or crusty. They come in a variety of colors, including black, white, red, orange, yellow, green, and brown. Lichens are not plants at all, however, but associations of fungi with photosynthetic partners (Figure 21-24). In all but about 20 lichen species, the fungal partner is an ascomycete. The photosynthetic partner in a lichen is either a green alga (a protist) or a cyanobacterium (a prokaryote).

The association between a fungus and its photosynthetic partner is not entirely mutual. The various fungi cannot live without the help of photosynthetic partners. The photosynthetic partners, however, are all capable of living without fungi. Most of the diversity of lichens comes from the fungi, for 90 percent of all photosynthetic partners in lichen are the same few organisms (from one of just three genera of green alga or of one genus of cyanobacteria). In contrast, the fungal partner is unique to each lichen. Nonetheless, the symbiotic association between fungi and their photosynthetic partners helps both parties.

The symbiotic association between lichens and their partners makes them especially tough. Fungi extract nutrients from rock and other harsh microenvironments. By doing so, fungi allow the algae or cyanobacteria to grow where they otherwise could not, while the photosynthetic algae provide a continuous source of energy to their fungal partners. Lichens therefore grow where no plant could survive—bare rock, mountaintops, the Arctic tundra, even the seemingly lifeless valleys of Antarctica (Figure 21-25).

Lichens usually reproduce asexually, either by fragmentation or by sporelike starters called soredia, which are clumps of algae enmeshed in a bit of mycelium. The fungal and photosynthetic components can also reproduce independently, either sexually or asexually.

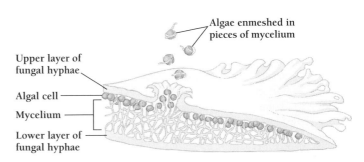

Algae enmeshed in pieces of mycelium

Upper layer of fungal hyphae

Algal cell

Mycelium

Lower layer of fungal hyphae

Figure 21-24   Lichen. Like a pie, a lichen is constructed with a protective crust of fungal hyphae above and below and a filling made of mycelium and algal cells.

Figure 21-25 Lichens are diverse. They are also able to absorb exquisitely tiny amounts of minerals and nutrients dissolved in rain and dew. While this ability is crucial to their survival in harsh settings, it makes lichens especially vulnerable to pollution. The reduced growth of lichens is a valuable indicator of deteriorating air quality. *(Larry Ulrich/DRK Photo)*

Lichens are mutually beneficial symbiotic associations of fungi and photosynthetic algae or cyanobacteria. In contrast to their photosynthetic partners, lichen fungi cannot live alone.

What Role Do Mycorrhizae Play in the Lives of Plants?

Fungi form close associations not only with photosynthetic bacteria and photosynthetic protists, but also with true plants. **Mycorrhizae** [Greek, *mukes* = fungus + *rhiza* = root] are symbiotic associations between fungi (usually a zygomycete or a basidiomycete) and the roots of plants (Figure 21-26). At least 80 percent of all plant species have mycorrhizae. Some biologists now think that even more plants depend on mycorrhizae. In such associations, the fungus's hyphae penetrate the plant roots and also extend into the surrounding soil.

What are these fungi doing? Although some fungi obtain sugar from the plant's roots, the mycorrhizae are not parasites, for they help the plant flourish by supplying phosphate and metal ions. Indeed, what the mycorrhizae do for plants is more obvious than what the plants do for the mycorrhizae. In one experiment, plants whose mycorrhizae had been removed grew more slowly than plants that had mycorrhizae. Plants with no mycorrhizae were then provided with fertilizer, however, and they grew as well as the plants with mycorrhizae. This experiment suggested that the mycorrhizae help plants pull nutrients from the soil.

Mycorrhizae are fungi on the roots of plants that supply minerals to the plant.

Where Did the Fungi Come From?

Because fungi have no hard parts, the fossil record of fungi is not rich. Nonetheless, fossil fungi can be found at least back to the Ordovician Period, 430 to 500 million years ago. The earliest fossils of land plants, which are about 400 million years old, contain petrified mycorrhizae, and some biologists now believe that plants could not have colonized the land without the aid of the fungi. Most mycologists think that the first fungi were zygomycetes, which have the simplest life cycle. Basidiomycetes probably derive from ascomycetes, which are in turn

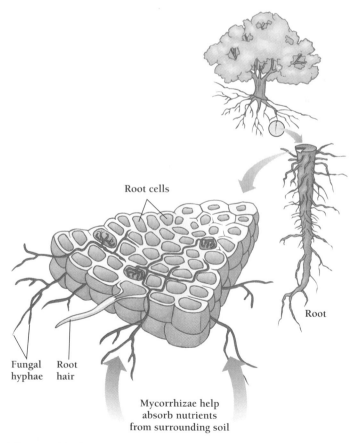

Root cells

Fungal hyphae   Root hair

Mycorrhizae help absorb nutrients from surrounding soil

Root

Figure 21-26 Mycorrhizae grow inside of the roots of trees and other plants.

descendants of the more primitive zygomycetes. Most imperfect fungi probably arose from ascomycetes that lost the sexual stages of their life cycles.

Fungi evolved from protists at least 430 million years ago. The fungi may have been responsible for the colonization of land by plants.

In this chapter we have surveyed the simplest eukaryotes, the Protista, as well as the Fungi, the first kingdom that consists entirely of multicellular organisms. In the next chapter, we will survey the plants, which along with the fungi are essential for humans and all other animals. On land, plants provide all the energy and simple organic materials that animals and fungi need to live. Plants and fungi are thus inextricably dependent on one another. Animals likewise absolutely depend on plants and fungi.

STUDY OUTLINE WITH KEY TERMS

Protists are eukaryotic organisms that are not fungi, plants, or animals. Protists include organisms that were once called **protozoa** (which move) and **algae** (which perform photosynthesis), as well as some organisms that resemble fungi. Most protists are single-celled organisms, but many are multicellular. Some protists are **plankton,** which drift. Others are self-propelled, or **motile.**

Protist life cycles vary greatly from species to species. Each species goes through two phases, reproduction and dispersal. Some are haploid throughout their life cycles and never undergo meiosis. Other species undergo **alternation of generations,** that is, they pass through both haploid and diploid phases. The diploid phase is the **sporophyte,** which produces **spores** (sometimes encased in **cysts**), and the haploid phase is the **gametophyte,** which produces the gametes.

Algae (plantlike protists) include 12 distinct phyla, of which this chapter surveys 7: **dinoflagellates, chrysophytes** (golden algae and diatoms), **euglenophytes, phaeophytes** (brown algae), **rhodophytes** (red algae), and **chlorophytes** (green algae). Dinoflagellates are among the oldest and most primitive of all eukaryotes and are marked by distinctive **tests.** Chrysophytes have both singled-celled and multicellular species, and diatoms can reproduce sexually as well as asexually. Euglenophytes can live as either autotrophs or heterotrophs. Phaeophytes are the largest and most complex of the protists. Rhodophytes have distinctive pigments, but no flagella. Chlorophytes are the most diverse of the algae.

Protozoa are heterotrophic protists, most of which are motile. They include eight phyla of which we survey five: **rhizopoda, foraminifera, sporozoa, ciliophora,** and **zoomastigina.** Rhizopods (amebas) move and ingest food with **pseudopodia.** Foraminifera (or **forams**) have tests (shells) studded with little pores. Sporozoans are nonmotile, spore-forming protozoa, which live as parasites in animals. Ciliophora (ciliates) are single cells, coated with cilia. Zoomastigina (**flagellates**) are single-celled and use flagella to move.

Many protists resemble fungi in appearance but other protists in cellular organization. These protists include seven phyla, of which we described three: **acrasiomycota, myxomycota,** and **oomycota.**

Acrasiomycota (cellular slime molds) live as amebas and as multicellular slugs, which can produce **fruiting bodies.** Myxomycota (plasmodial slime molds) live as amebas or as a multinucleated **plasmodium.** Oomycota resemble fungi and are saprophytic. Unlike fungi, however, their cells have flagella.

Fungi are heterotrophic (as are animals) and enclose their cells in rigid walls (as do plants). In contrast to both plants and animals, however, they do not undergo embryonic development. Fungi live mostly on land and, with a few exceptions, consist of a **mycelium,** a mass of intertwined filaments. Their cell walls are made from **chitin,** the same material used to build the external shells of insects. Fungi absorb rather than ingest nutrients, after predigesting their food with secreted enzymes.

Most fungi are **molds.** They may reproduce asexually or sexually. They also may reproduce and disperse as **spores** that resist harsh environmental conditions. Mycologists classify fungi by the way they produce spores (in **sporophores**) and by the appearance of their mycelia. In this chapter we surveyed four phyla of fungi: the zygomycetes, the ascomycetes, the basidiomycetes, and the deuteromycetes. **Zygomycetes** (conjugation fungi) are **coenocytic,** lacking **septa** between the nuclei in their **hyphae.** **Ascomycetes** (sac fungi) have septa (**dikaryotic**) but no **sporangia.** They produce spores in sacs called **asci.** **Basidiomycetes** (club fungi) have septa and distinctive life cycles. Each fruiting body of a basidiomycetes mushroom is called a **basidocarp.** On the thin gills beneath the cap lie the **basidia,** which produce spores. **Deuteromycetes** (imperfect fungi) reproduce asexually.

Fungi play important ecological roles, both as decomposers and as participants in **symbiotic** relationships. **Lichens** are mutually beneficial associations of fungi and photosynthetic algae or cyanobacteria. **Mycorrhizae** are mutually beneficial associations between fungi and the roots of plants. Fungi evolved from protists at least 430 million years ago. Some biologists believe that plants could not have colonized the land without the aid of mycorrhizae.

REVIEW AND THOUGHT QUESTIONS

Review Questions

1. Why is the kingdom Protista so much more diverse than the plants or animals?
2. How do the algae differ from the protozoa?

3. Which algae contain both chlorophylls *a* and *b*?
4. What features distinguish the slug of the plasmodial slime molds from that of the cellular slime molds?
5. What five features characterize the fungi?

6. To what phylum of fungi do common grocery store mushrooms belong? How is it that mushrooms can grow so quickly after a rain?
7. What are the two partners that compose a lichen? Which partner is more independent?
8. What do mycorrhizae do for plants?

Thought Questions

9. If you wanted to invent a new fungicide that would kill fungal cells but not human cells, what differences between the biology of fungal cells and biology of human cells might you exploit for this purpose?
10. Some paleontologists have speculated that vascular plants could not have evolved to the extent they have without the symbiotic relations with fungi they now depend on. How would you either test such a hypothesis or rephrase it in a way that would make it testable?

SELECTED READINGS

Arora, David, *Demystifying Mushrooms,* Ten Speed Press, Berkeley, California, 1988. An amusing yet expert discussion of the wonders of mushrooms and other fungi, combined with a well-illustrated field guide. Fun for anyone with even a passing interest in mushrooms, edible or otherwise.

Baldwin, Richard, *The Fungus Fighters: Two Women Scientists and Their Discovery,* Cornell University Press, Ithaca and London, 1981. A warm, somewhat formal account of the scientific and personal lives of Rachel Brown and Elizabeth Lee Hazen.

Bourke, Austin, *'The Visitation of God?' The Potato and the Great Irish Famine,* The Lilliput Press, Dublin, 1993. A history of the potato in Europe and a fascinating account of the science and politics of the great famine. It is well written, scholarly, and charmingly idiosyncratic. More detail than most readers will want.

Large, E.C., *The Advance of the Fungi,* Dover Publications, New York, 1962. A set of engaging essays describing crucial discoveries in 19th-century and early 20th-century mycology, including, for example, the fungus that caused the great Irish famine.

▶ On-line materials relating to this chapter are on the World Wide Web at http://www.saunderscollege.com/lifesci/
Click on Tobin/Dusheck: *Asking About Life.*

The Loves of Plants

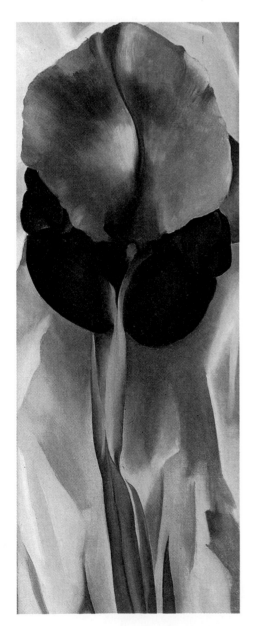

How to classify organisms is often a matter of dispute. Prokaryotes are commonly classified according to their metabolic pathways; protists, according to their morphology; and fungi, according to their style of reproduction. When 18th-century biologists first began to classify plants, they could not agree on what traits to use. By 1799, naturalists had suggested at least 52 different systems for classifying plants. One system grouped plants according to the form of the flower, the seed, and the seed coat. Other systems gave priority to the shape of the fruit.

It was Carolus Linnaeus's system that ultimately came into use, but not without heated arguments from both scientists and clergy. Linnaeus based his classification solely on the reproductive structures of flowers. Indeed, he was fascinated by the "marriages of plants," as he called them (Figure 22-1). The son of a country parson, Linnaeus equated sex with marriage and his descriptions of "vegetable coitus," as some called it, are simultaneously explicit and romantic.

As the historian of science Londa Scheibinger wrote,

> Linnaeus emphasized the "nuptials" of living plants as much as their sexual relations. Before their "lawful marriage," trees and shrubs donned "wedding gowns." Flower petals opened as "bridal beds" for a verdant groom and his cherished

bride, while the curtain of the corolla lent privacy to the amorous newlyweds.

Linnaeus divided plants first according to whether their marriages were public or private. Public marriages were those in which the flowers were visible to the naked eye; private marriages were those in which the flowers were too small to be seen. (The division corresponded to the public and private marriages then recognized in most of Europe. In England, the 1753 Marriage Act eliminated clandestine marriages by requiring a public proclamation before a marriage could be considered legal.)

Linnaeus further divided plants according to whether the male and female parts were together on the same flower, "in one bed," as he put it. Otherwise, if male and female parts were in separate flowers, they were in "two beds." Those in one bed, also called "hermaphrodites," were divided according to how many male parts (stamens) were present. These Linnaeus termed "husbands" to the single female pistil, or "bride." Flowers with one stamen were monandrian—having one man. (Linnaeus called his beloved wife Sara "my monandrian lily.")

As most flowers, however, have more than one stamen, indeed several, many naturalists and clergymen were horrified by Linnaeus's terminology. One St. Peters-

Dark Iris #1, 1927 by Georgia O'Keeffe

burg professor asked what God Almighty would introduce such "loathsome harlotry" into the plant kingdom. The bishop of Carlisle wrote to the Linnaean Society in 1808 that "a literal translation of the first principles of Linnaean botany is enough to shock female modesty."

The Englishman William Smellie, who edited the first edition of the Encyclopedia Britannica (1771), rejected the "alluring seductions" of plant sexuality. He was especially repulsed by Linnaeus's description of plants that had reproductive structures of both sexes as "hermaphrodites." Smellie compared Linnaeus's work to that of the most "obscene romance-writer."

Most naturalists avoided Linnaeus's more vivid imagry. One exception was Erasmus Darwin, Charles Darwin's maverick grandfather. His *Loves of the Plants* (1789) is a collection of poems depicting flowers as practitioners of free love and even incest.

Linnaeus's taxonomy had other drawbacks. For example, in his system the number of stamens determined the class to which a plant species belonged. But in some groups of plants, the number of stamens can vary from individual to individual. He also decided that the number of stamens (male parts) would determine class, while the number of pistils (female parts) would determine order, the next taxonomic group down. That is, he deemed the stamens more important than

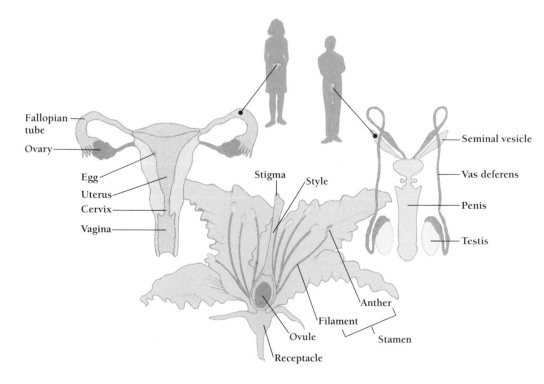

Figure 22-1 Linnaeus equated flower parts with the sexual parts of animals. He argued that the anthers are the testes; the filaments, the vas deferens; and the pollen, the seminal fluid. In the female parts, the stigma is the vulva; the style is the fallopian tube; the ovule is the ovary; and the seeds are the zygotes.

the pistils. This system neither corresponded to the phylogeny (relatedness) of plants, nor was the system arbitrary. It was simply Linnaeus's way of importing into science traditional views about the relative positions of males and females.

Although many of Linnaeus's species designations—genus and species—remain in use today, his classification of groups above the genus level have been abandoned in favor of systems that more accurately reflect evolutionary relatedness. As botanists have learned more about the lives of plants, they have acquired a better understanding of how plants live and how they have evolved.

In the rest of this chapter, we will see that plants are surprisingly diverse. Although the animal kingdom includes four times as many species as the plant kingdom, it is plants that first define a habitat—by their shape, size, and behavior. To a large extent, the animal inhabitants of grasslands differ from those of tropical forests because grasses differ structurally and chemically from trees. We can argue, then, that the plants lay the foundation for more habitat diversity than any other kingdom of organisms.

KEY CONCEPTS

1. A plant is a multicellular organism that performs photosynthesis and develops from an embryo.

2. Evidence for plant evolution comes from fossils and from comparisons of living species.

3. The nonvascular plants, or bryophytes (the mosses, liverworts, and hornworts), lack specialized elements for transporting water and nutrients.

4. The development of specialized vascular elements enabled plants to transport water and nutrients, to grow large, and to diversify.

5. Life on land offers both advantages and disadvantages to photosynthetic organisms.

6. Seeds allow plants to withstand dry environments.

7. Flowers and fruits of angiosperms have allowed their rapid diversification.

HOW DID PLANTS ADAPT TO LIFE ON LAND?

> The tree is an aerial garden, a botanical migration from the sea, from those earlier plants, the seaweeds; it has a purchase on crumbled rock, on ground. The human, standing, is only a different upsweep and articulation of cells. How treelike we are, how human the tree.
> —Gretel Ehrlich

The beauty of trees and flowers has inspired poets and gardeners alike. Plants delight the eye, tell us the season and even the time of day. They provide us with delicious foods, beautiful clothes, and strong building materials. Botanists like to say that plants provide "food, fiber, feed, fuel, and pharmaceuticals." But even this catchy list doesn't convey the importance of plants.

Through photosynthesis, plants (together with bacteria and protists) provide all of the energy we derive from food, whether that food is itself animal or plant. Plants living millions of years ago captured the energy that we now collect as coal, oil, or gas to heat our homes, to fuel our automobiles, and to drive our power plants. Plants provide the oxygen we breathe. Plants support the entire web of terrestrial life. Without plants, none of the other kinds of organisms could live on land.

A plant is a multicellular organism that performs photosynthesis and develops from an embryo. Almost all plants live on the land, and even those that live in water are descended from terrestrial plants. All plants, however, are descended from protists, and plants capture energy from the sun in much the same way as many of the marine protists. Nonetheless, the invasion of the land by plants nearly half a billion years ago required the evolution of structures different from anything found among plants' marine ancestors.

What Are the Advantages and Disadvantages of Life on Land?

The land offers three major advantages for photosynthetic organisms: more light (unfiltered by water), more carbon dioxide (which is more concentrated in air than in water), and a more reliable supply of minerals and other nutrients (which are more concentrated in soil than at the surface of a lake or sea).

Plants could not leave the water without solving a host of serious problems. To begin with, the seas provide a continuous supply of water. Water drives photosynthesis and provides a medium in which gametes and spores can disperse or meet for fertilization. Water also provides buoyancy for organisms.

Even a 30-meter tall brown alga can float. So aquatic plants, unlike terrestrial ones, require no skeleton or other structural support system.

To colonize the land, then, plants had to be able to obtain, conserve, and distribute water; to accomplish fertilization with little or no water; and to support their own weight. The evolution of plants depended principally on four adaptations:

1. A waxy **cuticle**, which coats the plant's exposed surfaces and reduces water loss through evaporation.
2. The ability to absorb water from dew, rainwater, or groundwater (using roots, for example).
3. Enclosed reproductive organs called **gametangia** in which gametes form.
4. Enclosed **sporangia** in which spores form.

All plants use some version of a waxy cuticle, an absorption system, and gametangia; plants share with protists and fungi life cycles characterized by "alternation of generations."

Land offers plants more light, more carbon dioxide, and more nutrients than the sea. But plants living on land need adaptations for absorbing and retaining water, for fertilization with little or no free-standing water, and, in the case of larger plants, some form of structural support.

What Is Alternation of Generations?

Recall that the cells of humans and most other animals are diploid, containing two of every chromosome, while their gametes (sperm and eggs) are haploid. The life cycle of every sexually reproducing organism includes a diploid phase and a haploid phase. In most animals, the haploid phase, consisting of single-celled sperm or ova, is short-lived or barely noticeable. Human sperm, for example, live only a few weeks. Human ova survive much longer, but are hidden away as individual cells.

In other kingdoms, however, the haploid stage is multicellular and constitutes a significant part of the life cycle. In plants, the haploid stage is a multicellular **gametophyte** that generates ova and sperm through mitosis, rather than meiosis. Plants are said to exhibit **alternation of generations,** a sexual life cycle in which the alternating haploid and diploid phases are both multicellular (Figure 22-2).

Plants exhibit alternation of generations, a sexual life cycle in which haploid and diploid phases are both multicellular.

How Are Vascular Plants Different from Nonvascular Plants?

Despite important similarities among all plants, plants may be divided into two groups—the tracheophytes (vascular plants)

and the bryophytes (nonvascular plants). Each group has different solutions for the problems of terrestrial life.

The **tracheophytes** [Latin, *trachea* = windpipe] include all the most familiar living plants, from fuschias to fir trees to ferns. All tracheophytes have pipelike tissues that conduct water and nutrients from one part of the plant to another. Because this system of conducting tissues resembles the vascular system of humans and other vertebrates, the tracheophytes are also referred to as the **vascular** plants. The tracheophytes' vascular system allows these plants to grow far larger than nonvascular plants, sometimes reaching enormous sizes.

The only true plants that lack such a vascular system are the **bryophytes** [Greek, *bryon* = moss], the mosses and their relatives. Because bryophytes lack a vascular system, they are much smaller than the tracheophytes. The bryophytes are also much less diverse, totaling only about 24,000 species, far fewer than the (approximately) 290,000 named species of tracheophytes.

Most botanists refer to the major groups of plants as divisions, equivalent to the phyla of animals that we will survey in the next chapter. However, the term phylum is becoming acceptable. Each division, like each phylum, contains classes, orders, families, genera, and species.

Some botanists divide the plants into only two divisions—Bryophyta and Tracheophyta. That is, they recognize plants with specialized vascular systems and those without. Most botanists, however, count 10 or 12 divisions of living plants, of which all but one (the bryophytes) are vascular plants. The divisions of vascular plants vary in their modes of reproduction and dispersal.

Table 22-1 lists 10 divisions. Although it gives both a formal Latin name and a common name for many of the divisions, we will use the common name if there is one (for example, "conifers" rather than the Coniferophyta).

Unlike Bryophytes, the single division of nonvascular plants, the 9 divisions of Tracheophytes are all vascular plants.

Table 22-1 The Divisions of Plants

Division	Number of Species
Bryophyta (mosses, liverworts, and hornworts)	24,000
Psilotophyta	small
Lycopodophyta	1,100
Equisetophyta (horsetails)	15
Pteridophyta (ferns)	13,000
Coniferophyta (conifers)	600
Cycadophyta (cycads)	100
Ginkgophyta (ginkgo)	1
Gnetophyta	70
Anthophyta (flowering plants)	275,000

Figure 22-2 The life cycle of a club moss. The club mosses are vascular plants, not true mosses (bryophytes). *(photo, Scott W. Smith/Earth Scenes)*

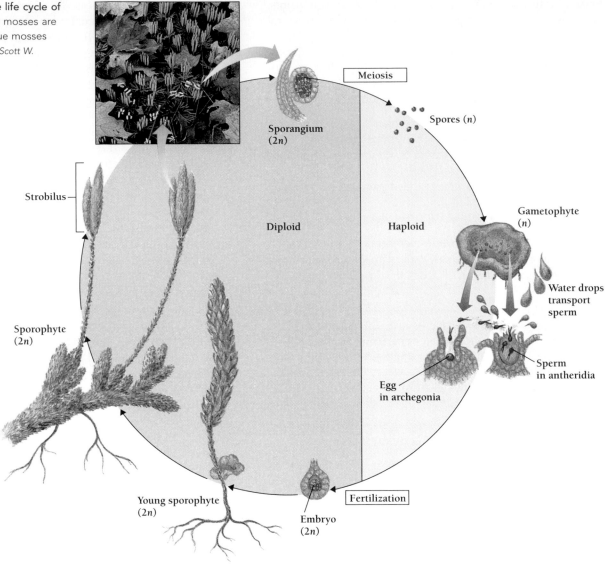

Meiosis

Sporangium
(2*n*)

Spores (*n*)

Diploid

Haploid

Gametophyte
(*n*)

Strobilus

Water drops
transport
sperm

Sporophyte
(2*n*)

Sperm
in antheridia

Egg
in archegonia

Young sporophyte
(2*n*)

Fertilization

Embryo
(2*n*)

How Do We Know About the Evolution of Plants?

The fossil record for plants, unlike that for protists, is excellent. Plant fossils are easier to find partly because plants are multicellular and therefore bigger, but also because the vascular plants, at least, have many hard parts such as bark, cones, pollen, and veins—all of which nicely survive the process of fossilization.

The oldest plant fossils date from the beginning of the Silurian Period, about 430 million years ago (Figure 22-3). These fossils are not only the oldest plant fossils, but the oldest terrestrial organisms. Until 430 million years ago, organisms apparently lived only in the seas. The first plants appeared on the land, probably in associations with fungi similar to present-day mycorrhizae. The successful invasion of the land by plants and fungi allowed the later evolution of terrestrial animals. Among these animals, the most populous and diverse were and continue to be the insects. As we will see, the evolution of the flow-

ering plants (angiosperms) went hand in hand with the evolution of insects and fungi.

We can deduce some of the characteristics of the first plants by noticing what all living plants have in common. All show alternation of haploid and diploid generations, all contain both chlorophylls *a* and *b*, all have cell walls made of cellulose, and all are multicellular. Almost all plants store energy as starch. All develop from embryos that are surrounded by nonreproductive (sterile) tissue. And, during cell division, all use a microtubular structure, the phragmoplast, to build a new cell wall between the daughter cells. Because plants share these striking characteristics with some green algae, most botanists believe that both nonvascular and vascular plants descended from the green algae.

True plants, however, have characteristics that set them apart from the green algae. In addition to a waxy cuticle, specialized structures for absorbing water, and specialized enclosed reproductive organs, plants also have **stomata** [Greek, *stoma* = mouth], tiny mouthlike pores that open and close to regulate

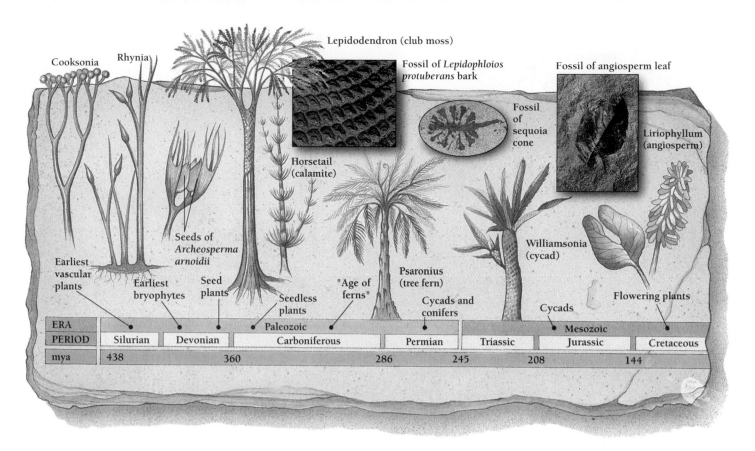

Figure 22-3 **The evolution of plants.** Notice the first appearance of vascular plants, bryophytes, seed plants, seedless plants, cycads and conifers, and flowering plants. *(Louise K. Broman/Photo Researchers; William E. Ferguson; McCutcheon/Visuals Unlimited)*

the flow of carbon dioxide and other gases in and out of the plant.

Evidence for plant evolution comes from fossils and from comparisons of living species. The first plants probably evolved from a common ancestor resembling a green alga.

What Does the Fossil Record Tell Us About the Evolution of Plants?

Although it is tempting to think of the nonvascular plants as more primitive than the vascular plants, the fossil record of vascular plants is actually older. The earliest bryophyte fossils date from the Devonian Period, about 400 million years ago, while the earliest vascular plants appear at the beginning of the Silurian Period, 430 million years ago (Figure 22-3).

By the time bryophytes first appeared in the early Devonian Period, several types of simple vascular plants were widespread, some of which are strikingly similar to modern species such as ferns and lycophytes. Botanists divide the vascular plants into those that produce seeds, such as the flowering plants and the conifers, and those that do not. To the seedless plants, the ferns and their allies, we owe our warm houses and speedy cars. These plants, buried deep beneath layers of sedi-

ment, have turned into the famous fossil fuels, mainly oil and coal, on which we are now so dependent. Seedless plants began flourishing in the Carboniferous Period, from 345 to 280 million years ago, when the world's climate was warm and wet. Shallow seas and dense swamps covered most of the land.

The growth of the earliest seedless plants, including ferns, horsetails, and lycophytes, was so luxuriant that a large proportion of these plants did not decay completely before being buried under more dead plants and other sediments. Over millenia, these buried, compressed plants became the major component of all fossil fuels—coal, oil, and natural gas. The large amounts of carbon in these fossils give the Carboniferous Period its name.

During the Carboniferous Period, plants grew taller and taller in an adaptive "light war." The tallest plants had access to the most sunlight and shaded the plants below. The tallest were 40-m lycophyte trees, as tall as large oaks. Far below this canopy of lycophytes spread vast stands of 18-m horsetails and 8-m (25-ft) ferns. Ferns, in fact, became so successful that the late Carboniferous Period is often called the "Age of Ferns."

Although a few Carboniferous plants had seeds, none had flowers. Seed plants, which first appeared in the late Devonian Period, were the progymnosperms [Greek, *pro* = first + *gymnos* = naked + *sperma* = seed], the ancestors of the modern gymnosperms. **Gymnosperms** are modern seed plants that do not have fruits or flowers, including, for example, the pine trees and other conifers, as well as palmlike cycads. Besides having

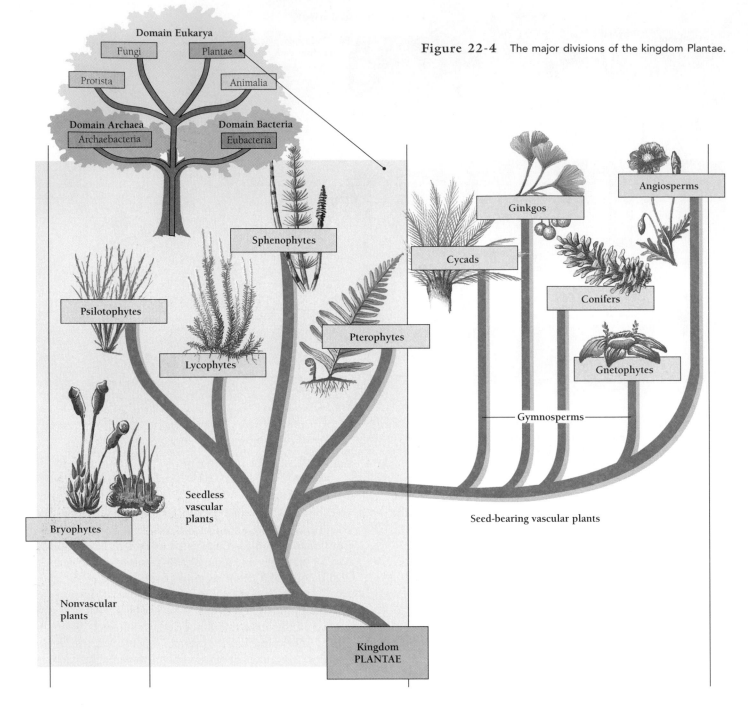

Figure 22-4 The major divisions of the kingdom Plantae.

Domain Eukarya
Fungi
Plantae
Protista
Animalia

Domain Archaea
Archaebacteria

Domain Bacteria
Eubacteria

Angiosperms

Ginkgos

Sphenophytes

Cycads

Psilotophytes

Pterophytes

Conifers

Lycophytes

Gnetophytes

Gymnosperms

Seedless vascular plants

Seed-bearing vascular plants

Bryophytes

Nonvascular plants

Kingdom PLANTAE

seeds, the progymnosperms had more elaborate branching and more complex vascular systems than the earlier vascular plants. The Carboniferous swamps contained conifers and palmlike cycads, as well as lycophytes, horsetails, and ferns. As the climate became drier, the giant lycophytes, horsetails, and ferns began to disappear, and the conifers and the cycads began to dominate the landscape.

At the end of the Paleozoic Era, about 225 million years ago, the Earth's climate changed dramatically, and for 100 million years the dominant plants were the conifers and great palmlike cycads. Then, about 130 million years ago, a new kind of plant appeared, one with flowers.

Flowering plants, which have both seeds and fruits, became the most abundant plants on Earth. Today we live in the Age of Flowers. Of the present 315,000 species of plants, including all bryophytes and all tracheophytes, an astounding 275,000 are flowering plants. All of the flowering plants fall into two groups. These are the **dicots,** or **broadleafed plants,** such as roses, poppies, oaks, and others; and the **monocots,** such as grasses, lilies, onions, daffodils, and cattails. Of the once dominant gymnosperms, only 529 species remain today.

Vascular plants predate nonvascular plants. To the first seedless plants of the humid Carboniferous Period we owe our warm houses. As the climate dried, first ferns, then conifers and cycads dominated the landscape.

How Did Angiosperms Evolve?

The flowering plants, or **angiosperms** [Greek, *angion* = vessel + *sperma* = seed], make up by far the largest division of plants (Figure 22-4). The angiosperms enclose their seeds in a vessel that we know as a fruit. A fresh peach is a fruit, and its pit is an enormous seed encased in a hard shell.

The first fossils of flowering plants date from the Cretaceous Period, about 125 million years ago. These early angiosperm fossils are leaves and pollen grains, rather than flowers.

In the absence of fossil flowers, evolutionary biologists who would like to know how angiosperms evolved from the gymnosperms have had to depend on comparisons among living plants. The magnolia, for example, is considered a very primitive plant with few advanced specializations. The magnolia has simple leaves, woody stems, and insect-pollinated flowers, features that probably characterized the first angiosperms.

In the 125 million years since the angiosperms first appeared, they have diversified both morphologically and biochemically. They have improved their transport system and developed *deciduous* leaves, which fall off seasonally, so that angiosperms are better able to withstand periods of drought than gymnosperms.

In addition, many modern angiosperm species have adaptations that prevent self-fertilization. Such species are more genetically diverse than gymnosperms and therefore better able to survive environmental changes. Angiosperm diversity and their dominance of the land attest to the success of their many evolutionary experiments.

Thanks to improved transport and deciduous leaves, many angiosperms have been able to colonize ever drier habitats. Adaptations preventing self-fertilization have contributed to angiosperms' enormous diversity and success.

In the rest of this chapter, we will review the ten divisions of plants, beginning with the ancient bryophytes and working our way up to the modern angiosperms, the flowering plants.

THE NONVASCULAR PLANTS, OR BRYOPHYTES, INCLUDE MOSSES, LIVERWORTS, AND HORNWORTS

The nonvascular plants, or bryophytes, include the mosses, which give the division its name, and two less common classes—the liverworts and the hornworts (Figure 22-5). Mosses and other bryophytes generally live in wet environments, some tropical and some temperate.

Because all bryophytes lack specialized vascular systems, they depend on free-standing water for both photosynthesis and fertilization by free-swimming sperm. Although bryophytes can live only where free-standing water occurs, they nonethe-

Figure 22-5 Sphagnum moss surrounding a skunk cabbage in the Quinalt Rain Forest, Olympic National Park, Washington. *(Pat O'Hara/DRK Photo)*

less disperse well and grow all over the world. One species of moss even grows high on an Antarctic mountain just a thousand miles from the South Pole.

The bryophytes play important but little understood roles in the world's ecosystems, recycling carbon and absorbing nutrients and toxic metals. Peat moss alone covers up to two percent of all land on Earth, an area the size of the United States. In some areas, bryophytes are the principal harvesters of solar energy, and all other organisms depend on them.

Bryophytes are also remarkable because the most obvious part of the bryophyte is not the diploid phase, but rather the haploid phase. When we look at a moss-covered stump, we are looking at the haploid phase, in which each cell has only one set of chromosomes.

How Do Bryophytes Absorb Water?

Although some bryophytes have elongated cells that conduct water, they lack the specialized transporting tissues of the vascular plants. They anchor themselves to surfaces by specialized structures called *rhizoids*. Unlike the roots of vascular plants, rhizoids are not specialized for absorbing water and nutrients. Instead, all parts of a bryophyte are absorbent, and the bryophytes consequently have a spongy feel. Bryophytes can grow on rocks, tree trunks, and other hard surfaces that resist the roots of vascular plants.

Bryophytes are small, usually less than 2 cm high and less than 20 cm long. In a moist environment, however, many plants may crowd together in a large mat. The closeness of the short plants allows them to retain water in the tiny spaces within the tangle. Because direct sunlight dries them out, bryophytes usually grow in the shade.

Figure 22-6 Life cycle of a moss.

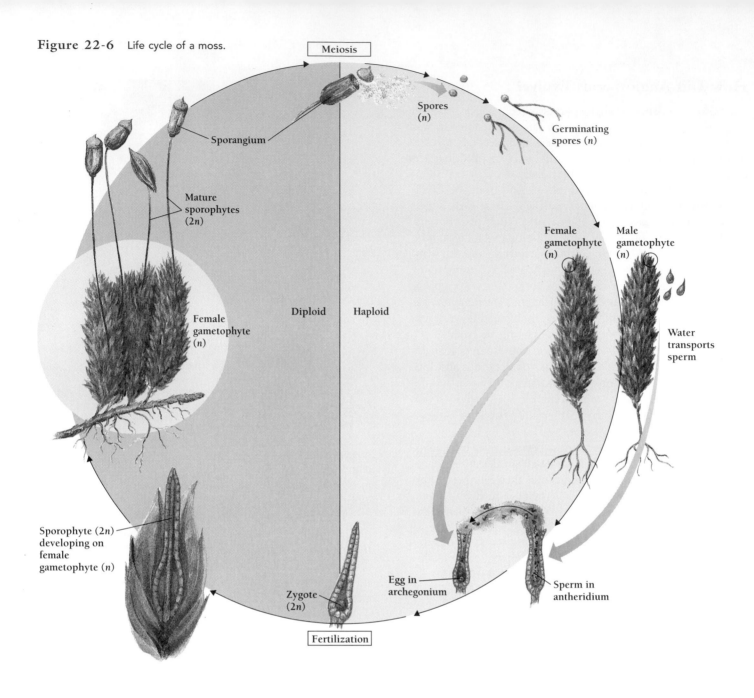

Meiosis

Sporangium

Spores
(n)

Germinating
spores (n)

Mature
sporophytes
(2n)

Female
gametophyte
(n)

Male
gametophyte
(n)

Diploid

Haploid

Female
gametophyte
(n)

Water
transports
sperm

Sporophyte (2n)
developing on
female
gametophyte (n)

Zygote
(2n)

Egg in
archegonium

Sperm in
antheridium

Fertilization

Although bryophytes lack a specialized transport system, all parts of their bodies are adapted to absorb water, which gives them a spongy feel.

How Do Bryophytes Reproduce?

Like the vascular plants, the fungi, and many protists, the bryophytes alternate generations. In contrast to the life cycles of the flowering plants, however, the most prominent phase of a bryophyte's life is the haploid **gametophyte,** which produces the gametes. Because the gametophyte is already haploid, it produces gametes—sperm or eggs—through mitosis, not meiosis. The gametes come together in the female gametophyte to form the diploid phase, called the **sporophyte** (Figure 22-6). The sporophyte produces haploid **spores** by means of meiosis,

just as animals and flowering plants do. But these haploid spores are not gametes. They are unlike any phase in our own life cycle, for once the haploid spore finds itself in a welcoming environment, its cells enter mitosis and the bryophyte grows directly into a leafy, green gametophyte.

The haploid gametophytes may be male or female. Each sex develops gametangia, structures in which the male or female gametes develop (Figure 22-6). The male gametangium, called an *antheridium,* produces sperm with two flagella. The female gametangium, called an *archegonium,* produces and houses the ova and is generally shaped like a flask with a long neck. Some bryophytes bear archegonia and antheridia on the same plant.

To reach the ovum in the archegonium, the sperm must either swim through water adhering to the spongy mat or be splashed into the vicinity of the neck of the archegonium (Figure 22-6). Sometimes insects carry moss sperm, suspended in water, from male to female plant. Once sperm are near an archegonium, chemical attractants draw them into the neck, and the sperm swim through a fluid in the archegonium's neck to reach the ovum at the center of the swollen base.

Fertilization occurs within the archegonium. Immediately after fertilization, the diploid zygote begins to divide by mitosis to produce the sporophyte. The sporophyte, which is usually not photosynthetic, consists of a stalk, as tall as 15 to 20 cm, at the top of which develops a fruitlike **sporangium** [Greek, *spora* = seed + *angeion* = vessel]. Cells inside the sporangium undergo meiosis to produce the haploid spores, which can develop into a new mossy gametophyte.

Because bryophyte spores need to germinate in an environment that will sustain them, dispersal is essential to bryophyte reproduction. Some bryophytes propel their spores from exploding capsules. Some depend on splashing rainwater. A few bryophyte species live on dung, and flies carry the spores to fresh dung. Once in a hospitable environment, the haploid spores germinate, divide by mitosis, and develop into mature, leafy, male or female gametophytes.

Bryophytes' life cycles differ from those of vascular plants significantly and from those of animals profoundly: in a bryophyte's life cycle, the haploid gametophyte phase dominates.

THE VASCULAR PLANTS, OR TRACHEOPHYTES, INCLUDE SEED PLANTS AND SEEDLESS PLANTS

The tracheophytes (vascular plants) are both more diverse and more numerous than the bryophytes. While the nonvascular plants consist of about 24,000 named species, the vascular plants include more than 290,000 species. Some taxonomists say that the number of vascular plant species runs into the millions. We need only look around us to appreciate the success of the vascular plants—wherever we see green, we nearly always encounter a vascular plant.

Of What Advantage Is a Transport System to a Plant?

The ability of vascular plants to transport water and nutrients allows them to grow to huge sizes and to develop specialized tissues. Vascular plants also have more division of labor than the bryophytes, with distinctive tissues and organs that include roots, leaves, and stems. Their transport systems contain two distinct kinds of conducting tubes: **xylem** [Greek, *xylon* = wood], which carries water and minerals from roots to the photosynthesizing leaves, and **phloem**, which distributes the sugars and other organic molecules made in the leaves to the rest of the plant (Figure 22-7). Phloem consists of elongated, living cells. Xylem consists of the stiff walls of dead cells. Vascu-

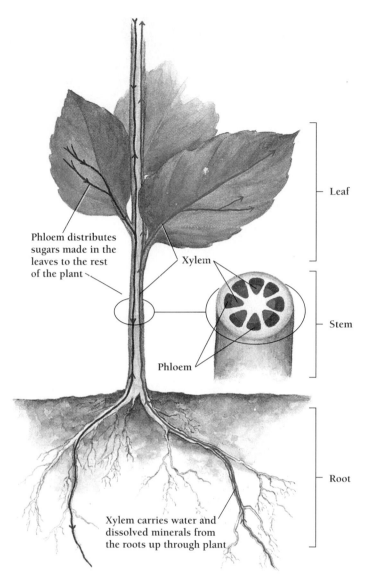

Figure 22-7   **Xylem and phloem.** Xylem conducts water and dissolved minerals from the roots to the leaves. Phloem conducts sugar from the leaves to the rest of the plant.

lar plants also produce a hard, supporting material called *lignin,* which usually lines the walls of the xylem. Lignin, a major component of wood, provides the extra support that allows trees to attain such enormous sizes.

Plants' ability to transport water and nutrients has allowed them to grow larger and develop specialized tissues.

How Do Vascular Plants Reproduce and Disperse?

Recall that in the bryophytes, the haploid gametophyte—the leafy, green part we call moss—is the more prominent part of the plant life cycle. In vascular plants, the diploid sporophytes

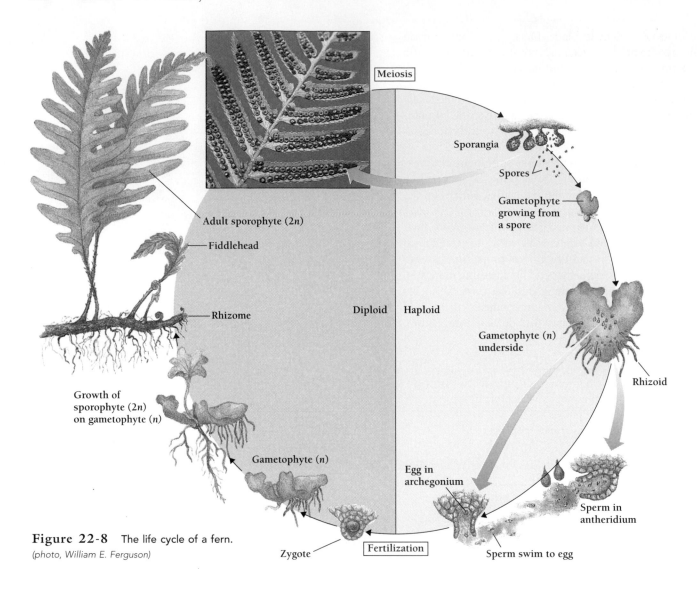

Figure 22-8 The life cycle of a fern.
(photo, William E. Ferguson)

Labels in figure: Meiosis; Sporangia; Spores; Gametophyte growing from a spore; Gametophyte (n) underside; Rhizoid; Sperm in antheridium; Sperm swim to egg; Fertilization; Egg in archegonium; Zygote; Gametophyte (n); Growth of sporophyte (2n) on gametophyte (n); Rhizome; Fiddlehead; Adult sporophyte (2n); Diploid; Haploid

are more prominent. The seed of a peach tree, for example, develops, in the right environment, into a mature sporophyte—a peach tree. In the spring, the peach tree blooms, producing flowers, which are reproductive structures. Parts of the flowers undergo meiosis to produce short-lived male and female gametophytes, which produce pollen and ova.

Vascular plants that produce seeds have at least one great advantage over the bryophytes. In colonizing new habitats, seed plants do not rely on the chance that a dispersed seed will fall into a benign and nourishing environment. Instead, the seed comes encapsulated in a tough coat that protects the embryo inside.

Within the safety of the plant's female reproductive organ, the plant embryo stops developing, and the mother plant packages the embryo, with a reserve of food, inside a protective seed coat. A seed is able to withstand long periods of cold and drought. When the seed finally encounters water and warmth, it germinates, and the embryo uses the stored food to finish developing into a photosynthetic seedling.

In vascular plants, the diploid phase dominates the life of the plant. The protective coat of a seed, and the food stored inside, allow a seed to wait indefinitely for suitable conditions for germination and growth.

Four Divisions of Tracheophytes Lack Seeds

Four divisions of living vascular plants lack seeds: the pterophytes (ferns), psilotophytes, lycophytes, and equisetophytes. (Another four divisions of seedless plants exist only in the fossil record.) We will discuss the ferns first, because they are the most familiar.

Pterophytes (ferns)

Some two-thirds of the 10,000 species of ferns live in the tropics, and the rest inhabit the temperate zones. Ferns range in size from a few centimeters to giant 30-m tree ferns, as tall as oak trees. Ferns are common in moist parts of shady forests. Their leaves (called *fronds*) often have a lacy appearance and consist of many *pinnae* [singular pinna; Latin, = feather], small leaflets that extend from a central stalk (Figure 22-8).

In the large ferns of the tropics, fronds grow from a central stem. In most ferns, however, fronds grow from a *rhizome*, a horizontal stem that spreads on or below the ground. Each frond develops from a characteristic coil called a fiddlehead. Most other seedless plants have **microphylls**—leaves with just one conducting vein. In contrast, fern fronds are **megaphylls**—large leaves with several veins that form complex networks of transport tubes.

Reproduction in ferns and other seedless plants resembles that in the bryophytes (Figure 22-8A). Fertilization can occur only if the released sperm can swim to the mature archegonia. Seedless plants therefore require a wet environment to provide at least a film of water in which the sperm can make their way. This need for water, together with the fragility of the gametophyte, limits the environments in which ferns and other seedless tracheophytes can grow.

As in other vascular plants, however, the sporophyte (which we recognize as the fern plant) is much more prominent than the gametophyte. The fern sporophyte has leaves, called **sporophylls**, which are specialized for reproduction. Clustered on the underside of the sporophylls are groups of sporangia called *sori* [singular sorus; Greek, *soros* = heap] (Figure 22-8B). The diploid sporangia produce haploid spores by means of meiosis. Some ferns catapult their ripe spores several meters into the air, where breezes carry them aloft. Such airborne spores may disperse over enormous distances.

On landing, each spore germinates to form a tiny nonvascular gametophyte, which is typically flat, thin, heart-shaped, and less than 1 cm wide. This gametophyte is green, photosynthetic, independent, and fragile. It usually bears both male and female reproductive structures.

Ferns have megaphylls, leaves with networks of transport tubes, and independent gametophytes.

The Psilotophytes

The psilotophytes contain just two genera, *Psilotum* and *Tmesipteris*. Both have photosynthetic leaflike scales (which lack conducting tissues) instead of true leaves. The common greenhouse weed *Psilotum*, or "whiskfern," is widespread in the tropics and subtropics. Unlike other vascular plants, *Psilotum* lacks true roots as well as leaves. Underground, it has a rhizome, with tiny branches called rhizoids. Above ground, its vertical stem forks several times, and bears sporangia and photosynthetic scales.

Psilotophytes, the simplest of the living vascular plants, lack true leaves.

Figure 22-9 Horsetails. Equisetophytes contain bits of silica that give them a rough feel. *(William E. Ferguson)*

Lycophytes

Lycophytes (also called lycopods) have true roots, stems, and leaves, as well as a simple arrangement of xylem and phloem. Lycophyte leaves are microphylls (with a single conducting vein). As in ferns, some of the leaves are sporophylls, with sporangia on their upper surfaces. The sporophylls may be green and look like ordinary leaves, or they may be distinct and grouped into spore-producing cones, called **strobili** [singular, strobilus; Greek, *strobilos* = cone]. The most common lycophytes are the club mosses, which form evergreen, mosslike mats beneath temperate forests. Their strobili resemble little clubs.

Lycophytes have true roots, stems, and simple leaves.

Equisetophytes

The living equisetophytes are all members of the single genus *Equisetum*, or horsetails (Figure 22-9). These plants grow in moist places along the banks of streams or at the edges of woods. Like the lycophytes, the equisetophytes have true roots, stems, and leaves. Their jointed stems contain many discrete strands of xylem and phloem. The scalelike leaves that grow from the joints are thought to be megaphylls (with branching vessels). The outer cell layer of the plant contains silica (the same material from which glass is made), which gives the plants a rough texture useful to campers and settlers for scouring pots. Indeed, another name for these plants is "scouring rush."

Equisetophytes have true roots, stems, and complex leaves. Their stems are jointed, and their outer cell walls are reinforced with silica.

How Do Seed Plants Reproduce?

Seeds are of great value in a dry, terrestrial environment because they liberate a plant from the need for free-standing water. **Seed plants** manage fertilization without water in the same way we do, through a form of internal fertilization. The gametophytes have no independent lives, but develop within the tissues of the sporophytes. The gametophytes of seed plants are tiny, even less prominent than in the seedless vascular plants.

Many seedless plants are *homosporous,* producing just one kind of spore and one kind of gametophyte. In contrast, all the seed plants are *heterosporous*—producing two kinds of gametophytes (male and female), from two kinds of spores on two kinds of sporangia. The sporophytes produce haploid microspores (male) and megaspores (female) in separate sporangia. In some species the male and female sporangia occur together on the same plant, while in others the two kinds of sporangia occur on male and female plants. In either case, the microspores and megaspores develop into distinct male and female gametophytes.

The female gametophyte stays within the megasporangium, which lies safe within the main body of the adult plant (the sporophyte), awaiting the arrival of the sperm. The megasporangium, also called a **nucellus** [Latin, *nucella* = a small nut], is in turn covered by one or two additional layers of tissue, called integuments, which will later develop into the protective seed coat. This combined structure—nucellus and integuments—is the **ovule.**

The small, male gametophytes are grains of pollen, released from the microsporangium, and carried, usually by wind or insect, to the female gametophyte. The male gametophyte then develops a slender pollen tube, which carries a sperm nucleus to the ovum produced by the female gametophyte. At fertilization, the sperm and ovum fuse to form a diploid zygote. The resulting zygote, still enclosed within the female gametophyte, begins to develop into an embryo.

After fertilization, the ovule and its contents become a **seed.** The seed includes (1) the new sporophyte embryo; (2) a female tissue that nourishes the developing embryo; and (3) the nucellus and integuments (now the seed coat) from the previous sporophyte generation. Inside the seed coat are the embryo and a source of food for its later development. The seed coat protects the young embryo from drying out or from being eaten. Once the seed germinates, the embryo resumes its development and growth.

In the following sections, we will discuss the five divisions of living seed plants. The 771 named species of the first four divisions produce seeds that are not enclosed in fruits. These plants are called **gymnosperms** [Greek, *gymnos* = naked + *sperma* = seed]. The 275,000 named species of the last division are **angiosperms** [Greek, *angion* = vessel], or flowering plants. Angiosperms produce seeds that are enclosed in a fruit.

Seeds, which enable plants to colonize relatively dry environments, consist of a diploid zygote and a source of food encased in a seed coat. Seed plants are heterosporous.

Gymnosperms: Four Divisions of Seed Plants Without Flowers

The four divisions of gymnosperms—conifers, cycads, ginkgos, and gnetae—are all vascular plants with seeds, but without flowers and fruits. They are not closely related to one another. We first discuss the most familiar division, the conifers.

Conifers

The conifers [Greek, *konos* = cone + *phero* = carry] take their name from their characteristic cones, which are male and female strobili. The conifers include many of the most common trees in the Northern Hemisphere—the pines, firs, spruces, and cedars—as well as the magnificent redwoods of the California and Oregon coasts. Conifers grow well in the short growing seasons of northern latitudes and high altitudes. They are mostly evergreen, keeping their leaves throughout the year. Individual leaves may last through two or three winters. (Bristlecone pines, some of which live to be more than 5000 years old, retain their needles for up to 45 years.) Conifer forests extend almost all the way across North America, Europe, and Asia.

Many conifers, including the familiar pines, have needle-like leaves that are well adapted for dry conditions, including, for example, northern winters, when the air is dry and the groundwater is frozen. A needle's small surface area and heavy cuticle minimize evaporation. Despite its simple shape, however, the needle is a megaphyll, usually with two vascular bundles. The stems of conifers are heavily reinforced with lignin, which gives trees the structural support to reach enormous sizes. The lignin in wood is also what makes it strong enough to support even the largest buildings.

Most conifers bear male and female cones on the same tree. The male, pollen-producing cones are generally small (1 to 2 cm long) and consist of hundreds of sporophylls. Wind carries the pollen grains, often in huge yellow clouds, to the female cones. The female, or ovulate, cone consists of tough scales arranged in a spiral. Figure 22-10 shows the life cycle of a typical conifer, a pine tree.

Conifers produce male and female gametophytes in cone-shaped strobili.

Cycads

The cycads are large-leafed plants that look like palms (Figure 22-11). Some cycads, called "sago palms," are used as ornamental plants in homes and gardens. But true palm trees are flowering plants, while cycads have neither flowers nor fruits: they bear naked seeds on the scales of cones near the top of each plant. Individual plants bear either male or female cones, but not both. Some people eat the seeds and underground stems of cycads, but cycads contain toxins that can cause gastrointestinal problems, degenerative brain disease, and even paralysis.

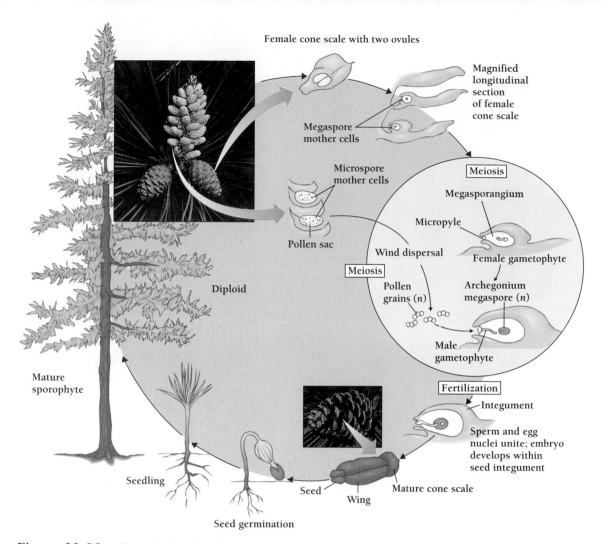

Female cone scale with two ovules

Magnified longitudinal section of female cone scale

Megaspore mother cells

Microspore mother cells

Meiosis

Megasporangium

Micropyle

Pollen sac

Wind dispersal

Female gametophyte

Meiosis

Diploid

Pollen grains (*n*)

Archegonium megaspore (*n*)

Male gametophyte

Mature sporophyte

Fertilization

Integument

Sperm and egg nuclei unite; embryo develops within seed integument

Seedling

Seed

Mature cone scale

Wing

Seed germination

Figure 22-10 Life cycle of a pine. Like other conifers, pines have two kinds of cones: pollen cones and seed cones. In the seed cones, which are the familiar "pine cones," each scale bears two ovules. Each ovule in turn contains a female sporangium, surrounded by a protective integument. A cell within the sporangium undergoes meiosis to produce four haploid cells, one of which undergoes mitosis to produce the female gametophyte. The gametophyte remains within the ovule and develops two or three archegonia, each of which in turn produces a single giant ovum, loaded with carbohydrates and proteins. In the integument surrounding the ovule is an opening for the entrance of pollen. In spring, a sticky fluid in the opening captures pollen grains from the wind. After a few months, the pollen gametophyte develops into a pollen tube, which makes its way to one of the archegonia in the ovule. One of the cells of the pollen tube divides to form two sperm nuclei, which then enter the archegonium. As much as a year after pollination, one of these nuclei finally fuses with the ovum to form a zygote, which develops within the ovule. When the new embryo has a root and several leaves, the seed is ready for dispersal. *(photos: top, Michael Giannechini/Photo Researchers; bottom, R.J. Erwin/Photo Researchers)*

The small male cones of cycads consist of microsporophyll scales, each of which contains many pollen-producing microsporangia. Female cones are much larger. The cone of one Australian cycad may measure 1 m in length and weigh 40 kg (90 pounds). As in conifers, pollen travels by wind to the ovule, enters through a micropyle, germinates to produce a pollen tube and sperm, and fertilizes the ovum.

***

Cycads produce male and female gametophytes in strobili that lie on separate plants.

***

Ginkgos

The ginkgo is the last living species in the Division Ginkgophyta, the rest of whose members went extinct as long as 280 million years ago. The ginkgo, or maidenhair tree, is a living fossil that has hardly changed in the last 80 million years. Yet this beautiful tree, which decorates many gardens and parks, is itself now extinct in the wild.

As in the cycads, ginkgo trees are either male and female. The male trees have cones (strobili), while the female carries fleshy (and smelly) seeds at the ends of short stalks. The seeds of the female tree have an odor so rancid that most people prefer to plant male trees only.

***

Ginkgos resemble cycads in their life cycle and conifers in their growth patterns.

***

535

Figure 22-11 *Zamia pumila* is the only species of cycad native to the United States. *(Walter H. Hodge/Peter Arnold, Inc.)*

Gnetae Have Characteristics of Both Gymnosperms and Angiosperms

The Gnetophyta are a small but diverse group of cone-bearing desert plants, which consist of three distinct genera—*Gnetum, Ephedra,* and *Welwitschia* (Figure 22-12). Of all the gymnosperms, the gnetae most resemble angiosperms. For example, some of their strobili look like flowers, and the vessels of their leaves and stems resemble those of angiosperms. But they have naked seeds and are certainly gymnosperms.

Figure 22-12 The gnetophyte *Ephedra viridas* (mormon tea) resembles an angiosperm. *(Noble Proctor/Photo Researchers)*

THE ANGIOSPERMS ARE SEED PLANTS WITH FLOWERS AND FRUITS

The diversity of the angiosperms dwarfs that of all other divisions of plants. Angiosperms, also called **Anthophyta** [Greek, *antho* = flower], range in size from the tiny duckweed *Wolffia,* less than 1 mm long, to *Eucalyptus* trees, taller than 100 m. With some 275,000 known species, angiosperms are the most widespread and familiar of plants. We see them daily in our lawns, gardens, parks, fields, and forests. They provide us with furniture, clothes, medicines, and dyes. We stock our kitchens with their seeds, fruits, leaves, stems, and roots. Angiosperms provide us with wine, beer, coffee, and tea. We use the more brilliant angiosperms to brighten our lives and enliven our loves.

Despite their diversity, all angiosperms share two important adaptations: flowers, which promote fertilization, usually by insects or other animals; and fruits, which promote the survival and dispersal of seeds.

Of What Use Are Flowers?

Flowers are reproductive structures that ensure the distribution of pollen, either by animals or by wind. Some flowering plants rely on wind alone to distribute pollen. Grasses, rushes, sedges, plantain, birch, poplar, and oaks, for example, have small inconspicuous flowers and no scent. Because wind pollination is so inefficient, wind-pollinated flowers release mind-boggling amounts of pollen. Individual plants of the weed *Mercurialis annua* may release one and a quarter billion grains of pollen.

Other flowers attract insects (especially bees, beetles, butterflies, and moths), birds (especially hummingbirds), and even bats and other small mammals. All these animals' visits are rewarded with a sugary fluid called nectar, a share of the high-protein pollen, or nutritious petals or other flower parts. Animals learn to associate these delicious foods with the seductive perfumes and bright colors of flowers. Not all of a flower's attractions are necessarily attractive to humans: the eastern skunk cabbage (*Symplocarpus foetidus*), for example, attracts pollinating flies by emitting an odor like that of dead fish.

Most flowers are specialized to ensure pollination by just one group of animals. Flowers pollinated by moths, for example, are usually large, white, and overwhelmingly fragrant at night, when moths fly. In contrast, flowers pollinated by butterflies are small and brightly colored, since butterflies fly during the day and rely on sight more than smell. One orchid species produces flowers that look so much like female bees that male bees come to mate with the flowers. As the male bee attempts to copulate with the flower, he is dusted with pollen, which he then carries to another flower in another futile attempt to mate.

Insects are as well-adapted to take advantage of flowers as flowers are adapted to take advantage of insects. One South African fly has a tongue nearly 3 in long, which allows it to take nectar that is out of reach for other insects. Insects and flowers provide a striking example of **coevolution,** the mutual adaptation of two separate evolutionary lines. Coevolution occurs when two species are significant factors in the lives of one another. In the case of insects and flowers, coevolution ensures a food supply for the insects and pollination for the plants.

Flowers attract animals that distribute pollen.

Of What Use Are Fruits?

A fruit is a mature ovary that encloses and protects seeds. Many fruits promote seed dispersal by animals or wind. The most familiar fruits are those that attract by appearance and taste, for example, peaches, apples, and berries. An animal usually digests the fleshy part of such fruits, but the seeds pass unharmed through the digestive system. (In fact, some seeds cannot germinate until they pass through the digestive tract of an animal.) Seeds excreted by the animal can end up far from the plant that produced them, together with a nourishing supply of fertilizer.

Other fruits use different stratagems to transport their seeds. Thistles and dandelion fruits, for example, soar like kites in the slightest breeze, and maple fruits whirl through the forest like free propellers. Burrs stick to the fur (or socks) of passing animals, while coconuts have spread to countless tropical islands by floating across the ocean. Touch-me-nots (*Impatiens*) propel their seeds from fruits that explode on touch.

A fruit is a ripened ovary that encloses and protects the seeds and usually enhances their dispersal.

How Do Angiosperms Reproduce?

In angiosperms, as in all the other seed plants, the diploid sporophyte generation is more prominent than the haploid gametophyte. In fact, the male and the female gametophytes of angiosperms are even tinier than those of the gymnosperms, and live their whole lives inside the sporophyte's flowers.

The flowers produce haploid microspores (male) and megaspores (female) in separate sporangia, from which the gametophytes derive. After meiosis the microspores divide once to give rise to two haploid nuclei—the tube nucleus and the generative nucleus (Figure 22-13). The male gametophyte resumes development only after it arrives on the **carpel,** the flower part that contains the ovule. The pollen tube now grows, and the generative nucleus divides into two sperm nuclei. By the time the pollen tube reaches the ovule, it contains two sperm nuclei.

Meanwhile, within the ovule, the female gametophyte has developed. Meiosis yields four haploid megaspores, only one of which survives. That megaspore divides three times by mitosis to produce the embryo sac, the mature female gameto-

phyte. The embryo sac has eight nuclei and seven cells. One cell (the central cell) contains two haploid nuclei (called the polar nuclei). Another cell becomes the haploid ovum.

Fertilization in angiosperms is different from that in any other organism. As we will discuss in more detail in Chapter 30, there are actually two separate fertilizations, together called a **double fertilization:** (1) one sperm nucleus fuses with the nucleus of the ovum to form the zygote, and (2) the other sperm nucleus fuses with the two nuclei of the central cell to form a triploid nucleus called the primary endosperm nucleus. The zygote then divides and develops into an embryo. The primary endosperm nucleus divides to form an **endosperm,** which will provide food for the developing embryo, and in some cases for the germinating seedling.

Flowers reproduce by means of double fertilization. Two sperm nuclei from the pollen grain fertilize two ova from the ovary, resulting in a diploid zygote and a triploid cell that forms the nutritious endosperm.

What Are the Parts of a Flower?

Like the gymnosperms, angiosperms are heterosporous. Separate sporangia produce male spores (microspores) and female spores (megaspores). As in the case of the gymnosperms, the sporangia lie on modified leaves, or sporophylls. In angiosperms, however, both male and female sporophylls are flower parts instead of cones. The male sporophylls are **stamens,** and the female sporophylls are **carpels.**

A flower is a short piece of stem that usually ends with four whorls, or circles, of modified leaves. These are the carpels, the stamens, the petals, and the sepals (Figure 22-13). The innermost parts are the carpels. Each carpel (or pistil) encloses one or more female sporangia (ovules) in a swollen base called the **ovary.** Pollen does not reach the ovules directly, but first falls on the stigma, the carpel's receptive area. The style, a slender tube through which the pollen tube enters the ovary, extends from the stigma to the top of the ovary. After fertilization and development, the ovule becomes the seed, and the ovary the fruit.

Just outside the carpels lies the next whorl of modified leaves, namely the stamens, or male sporophylls. Each stamen consists of an **anther,** a thick pollen-bearing structure, and a **filament,** a thin stalk that connects the anther to the base of the flower. Within the anther are four sporangia, called pollen sacs, where the pollen grains develop (Figure 22-13).

Outside the stamens is a whorl, or **corolla,** of **petals.** In flowers pollinated by insects or other animals, the petals are usually large and brightly colored. In flowers pollinated by wind, however, the petals are reduced or missing altogether. Sometimes the petals fuse, so that the corolla is an uninterrupted tube.

The fourth and outermost whorl is the **calyx,** a whorl of leaflike parts called **sepals,** which enclose the rest of the flower

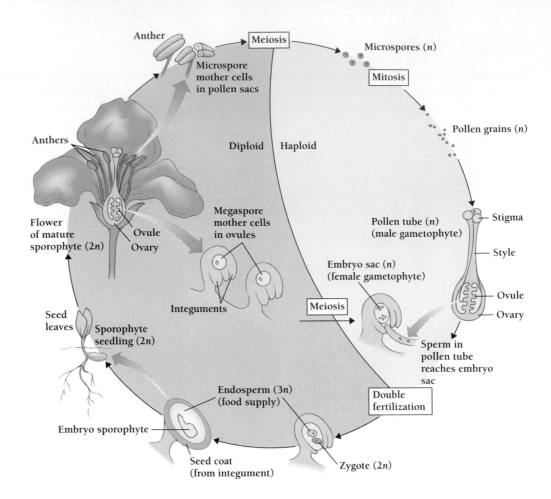

Figure 22-13 The life cycle of an angiosperm.

and protect it before it opens. The sepals are usually small and green, but in some flowers they may be large and colorful.

Flower structures vary widely. Those that lack one or more of the four whorls are said to be incomplete. Those that lack either stamens or carpels are said to be imperfect (as well as incomplete). Flowers may have radial symmetry, so that any cut along the axis of the flower will produce two identical halves. Or they may have bilateral symmetry, so that only one cut through the flower's axis will produce identical (or mirrored) halves.

Fruits are as diverse as the flowers that produce them. Some develop from a single carpel (with one or many seeds) or from several fused carpels with a single ovary. Others, such as raspberries or strawberries, derive from many separate carpels. And still others, like the pineapple, come from many flowers.

The male sporophyll is the stamen—including anther and filament—and the female sporophyll is the carpel (or pistil)—including style and ovary. The stamens and carpels are generally surrounded by a corolla of petals and a calyx of sepals.

The terrestrial environment is far more diverse than the marine environment. Extremes of temperature and rainfall, for example, demand a great range of adaptations. As a result, terrestrial organisms are themselves more diverse than their marine cousins. Animals, which we survey in the next two chapters, are especially diverse. But most animal species are insects, and the evolution of insects is closely tied to that of plants. Plants' colonization of the land 430 million years ago may have set the stage for the evolution of insects and other terrestrial animals.

STUDY OUTLINE WITH KEY TERMS

Plants are multicellular organisms that perform photosynthesis. They originated on the land, and most are terrestrial. The land offers three major advantages for photosynthetic organisms: more light, more carbon dioxide, and a more reliable supply of minerals and other nutrients. However, for plants to invade the land, they had to be able to obtain and conserve water.

The first plants probably evolved from green algae at least 430 million years ago. The first seed plants appeared by the late Devon-

ian Period. Fossils of the first flowering plants date from the Cretaceous Period, about 125 million years ago.

The evolution of plants depended principally on three adaptations: (1) a waxy **cuticle**, which coats the plant's exposed surfaces and reduces evaporation, (2) the ability to absorb water and minerals from dew, rainwater, or groundwater, and (3) jacketed reproductive organs. All plants also have **stomata** and life cycles characterized by **alternation of generations** between **gametophytes** and **sporophytes**.

The **gametangia** of the gametophytes produces gametes, and the **sporangium** of the sporophytes produces **spores.**

The **bryophytes** have no specialized conduction system for distributing water and nutrients and are therefore limited in the sizes to which they can grow. Bryophytes include mosses and two rarer classes, the liverworts and hornworts. The haploid (gametophyte) phase of bryophytes is more prominent than the diploid (sporophyte) phase.

Vascular plants are far more diverse and more numerous than the bryophytes. Vascular plants contain two distinct kinds of conducting tubes, the **xylem** (which carries water and minerals from roots to leaves) and the **phloem** (which distributes sugars made in the leaves to the rest of the plant). In vascular plants, the diploid phase of the life cycle is more prominent than the haploid phase.

Vascular plants disperse as diploid sporophytes. Most kinds of vascular plants disperse as **seeds,** which contain a diploid embryo, a food reserve, and a protective coat. Many **seed plants** have flowers and fruits, which aid in fertilization and in the dispersal of seeds.

Four divisions of vascular plants, or **tracheophytes,** lack seeds: pterophytes, psilotophytes, lycophytes, and sphenophytes. Pterophytes (ferns) have **megaphylls,** leaves with networks of transport tubes. Other tracheophytes have **microphylls,** leaves with just one vein. Psilotophytes are the simplest of the living vascular plants and do not have true leaves. Lycophytes are more complex than psilotophytes, but less so than ferns. Sphenophytes have jointed stems that are reinforced with silica.

Five divisions of tracheophytes bear seeds, but four of these, collectively called the **gymnosperms,** do not have flowers. A seed contains the embryo, the female gametophyte, the organ that produces the female gametophyte (the **nucellus**), and layers of tissue called the integuments from the previous sporophyte generation. The nucellus and integuments form the **ovule.**

Gymnosperms produce male and female gametophytes in **strobili.** We discussed the four divisions of gymnosperms: conifers, cycads, ginkgos, and gnetae. In conifers the strobili are separate cone-shaped structures on the same tree. Cycads produce male and female gametophytes in strobili that lie on separate plants. Ginkgos resemble cycads in their life cycle and conifers in their growth patterns. Gnetae have characteristics of both gymnosperms and flowering plants.

Anthophyta, or **angiosperms,** are the most diverse group of plants. The gametophytes of angiosperms are even more reduced than those of the gymnosperms. Both male and female gametophytes are part of the flowers of the **sporophyte,** as are the male and female **sporophylls.**

A flower is a short piece of stem that usually ends with four circles of modified leaves: the **carpels, stamens, petals,** and **sepals.** Each carpel—including stigma, style, and ovary—encloses the female sporangia. The stamens are the male reproductive structures. Each stamen consists of an **anther** and a **filament.** A **corolla** of petals, often brightly colored and scented, attracts insects that carry pollen grains (male gametophytes) from flower to flower. A **calyx** of sepals encloses and protects the rest of the flower before it opens.

Fertilization in angiosperms is different from that in any other organism and consists of two separate fertilizations—a **double fertilization.** The zygote then divides and develops into an embryo. The primary **endosperm** nucleus divides to form an endosperm, which will provide food for the developing embryo, and in some cases for the germinating seedling. Flowering plants are divided into **monocots** and **dicots** (or **broadleafed plants**), according to the form the embryo takes.

A fruit is a ripened, or mature, **ovary** that encloses and protects the seeds. Flowers and fruits allow angiosperms to interact with animals. All animals and plants have **coevolved.**

REVIEW AND THOUGHT QUESTIONS

Review Questions

1. Describe six traits that together set plants apart from members of the other kingdoms of organisms.
2. Name several advantages and disadvantages to life on land.
3. What traits enable plants to live on land?
4. Compare alternation of generations with the life cycle of an animal, such as ourselves.
5. How long ago do paleontologists find the first association between plants and mycorrhizae? What does this imply about the coevolution of these two groups of organisms?
6. What group of plants that flourished 300 million years ago provided the fossil fuels oil, coal, and natural gas? What is the name of that geologic period?
7. How do bryophytes absorb water? What important roles do mosses play in the world's ecosystems?

8. What is peculiar about the life cycle of a bryophyte?
9. What are xylem and phloem and what functions do they perform in vascular plants?
10. What is the sporophyte phase of a vascular plant?
11. What is the difference between a megaphyll and a microphyll?
12. What are the advantages of seeds to terrestrial plants?
13. What are the advantages of flowers and fruits to terrestrial plants? Name the parts of a flower.

Thought Question

14. Why did plant taxonomists abandon Linnaeus's system of classification? What characteristics of plants reflect their evolutionary relatedness? Why do these characteristics represent relatedness more than the number of stamens or the number of pistils?

SELECTED READINGS

Gensel, P.G., and Andrews, H.N., "The Evolution of Early Land Plants," *American Scientist,* 1987, 75, 478–489.

Niklas, K.J., "The Cellular Mechanics of Plants," *American Scientist,* 1989, 77, 344–349. How land plants support themselves against gravity.

Mauseth, James D., *Botany: An Introduction to Plant Biology,* 2nd ed., Saunders College Publishing, Philadelphia, 1995. A friendly introductory botany textbook.

▶ On-line materials relating to this chapter are on the World Wide Web at http://www.saunderscollege.com/lifesci/ Click on Tobin/Dusheck: *Asking About Life.*

Progress, Controversy, and the Burgess Shale

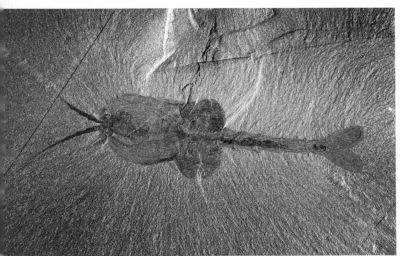

One day, about 530 million years ago, a mud slide the size of a city block suddenly collapsed onto the muddy bottom of a shallow sea, burying tens of thousands of small marine animals. Acres of mud covered the trapped organisms so abruptly that the creatures never had a chance to decay. Over the course of geologic time, the mud slide and all of its imprisoned organisms slowly turned to rock, perfectly preserving the animals within. Geologic processes lifted the rock and today it sits high in the Canadian Rockies and is called the Burgess Shale (Figure 23-1).

When the fossil remains of the tiny animals were finally discovered half a billion years later, they unleashed a bitter debate about the nature of evolution. Within the solid mass of the Burgess Shale, the perfectly preserved creatures revealed a remarkable diversity of animals, a diversity that suggested different things to different people.

The Burgess animals' first discoverer was paleontologist Charles Doolittle Walcott and his family. In 1909, Walcott, his wife Helena Walcott, and several of their children were collecting fossils in the Canadian Rockies when they came across an isolated slab of rock crammed with the small, perfectly preserved organisms. Walcott quickly found the source of this slab, higher up the slope of the mountain, and the following summer Charles, Helena, and their half-grown children returned and dug from the mountain more than 80,000 fossils.

Walcott was a distinguished leader of the early 20th century American scientific establishment. However, his role as top administrator at the Smithsonian, as well as his positions on the boards of other scientific organizations, left him little time for deep thought or serious research. As discoverer of the Burgess fossils, he was entitled to name and classify the new organisms. In his haste, however, he classified all of the Burgess animals as either trilobites (already well-known ancient organisms) or primitive members of still living groups (such as crustaceans, jellyfish, and polychaete worms). In short, Walcott mistakenly assumed that all his animals were members of already-well-known phyla (Figure 23-2).

Figure 23-1 Paleontologist Charles Doolittle Walcott and his wife Helena Walcott discovered a rich deposit of ancient organisms in the Canadian Rockies. *(Brown Brothers)*

Walcott well knew how little time he had given the Burgess animals. For the rest of his life, he dreamed of studying them in detail. In 1926, he announced his plans to retire so that he could spend his last years with the Cambrian animals. But he died just three months short of his planned retirement, and even though he was one of the leading scientists of his day, he never fully appreciated his own remarkable discovery.

Finally, in the 1970s, three British paleontologists—Harry Whittington, Derek Briggs, and Simon Conway Morris—undertook to reexamine the Burgess fossils. The fossils were not merely flat impressions, but fully formed stone animals, whose ancient bodies could be dissected and studied. Using dental drills, Whittington, Briggs, and Conway Morris carefully chipped away layers of stone tissues, revealing the anatomies of some of the most bizarre animals ever found (Figure 23-2). Altogether, Whittington, Briggs, and Conway Morris argued, the Burgess animals represented 25 basic body plans, of which only four survive in modern organisms.

In 1989, the paleontologist and popular writer Stephen Jay Gould wrote a book, called *Wonderful Life,* describing the animals of the Burgess Shale, as well as Walcott and the work of Whittington, Briggs, and Conway Morris. Gould argued that the diversity of life forms in the Cam-

---

Gould argued that evolution was not a progression from simple to complex, from few forms to many.

---

brian dwarfed anything we know today. The range of anatomical designs among the animals that died in the tiny block of mud 530 million years ago far exceeded that of all the creatures in today's oceans. The Burgess animals, wrote Gould, should have challenged the ablest classifiers of the time. Yet, Walcott had classified these bizarre organisms into well-known groups of organisms, such as jellyfish and shrimps.

Each of the body plans represented in the Burgess Shale, Gould said, easily could have been classified as a phylum. Only because many of these types are represented by just a single species had taxonomists hesitated to give each one a whole phylum to itself. Charles Walcott had failed to recognize the true diversity of these creatures, Gould argued, partly because Wal-

Figure 23-2   A few of the animals of the Burgess Shale. Floating in the upper left is a harp-shaped creature named *Marrella*. The large, green creature in the center is *Anomalocaris*, which grew up to a meter long. Just below is *Opabinia*—smaller, but with five eyes and a front-facing "nozzle." The small worm struggling in *Opabinia's* jaws is a *Burgessochaeta* worm. *Pikaia*, shown lower right, may be the oldest known Chordate—if not our direct ancestor, at least a relative of one. *Odontogriphus* (pink, ribbonlike animal) was a flat, swimming creature. *Hallucigenia* (the green creature walking on the sea floor) had seven pairs of spines and seven tentacles. The three reddish flame-shaped creatures are *Wiwaxia*, which probably crawled along the sea floor. What look like green plants on stems are actually *Dinomischus*, a sessile animal. *Vauxia*, shown behind *Anomalocaris*, was a sponge. In the lower left is *Aysheaia*, similar to a modern-day onychophoran. Shown in its burrow underneath the sea floor is *Ottoia*, which looked and behaved much like the priapid worms that still live along our coasts today.

cott had assumed that what he had found would fit into categories he already knew and partly because he assumed that evolution, by its nature, inevitably progresses from a few simple forms to a diversity of complex and modern forms (culminating in human beings). This idea—that evolution is progressive—is controversial, however, and depends on how the word "progressive" is defined.

Is evolution progressive? Does diversity increase? Do organisms become more complex? Gould argued that the diversity of the Burgess Shale provided evidence that evolution was *not* a progression from simple to complex, from few forms to many. He argued that life is less diverse today than during the Cambrian explosion. The Burgess Shale, he said, illustrated that multicellular life began in great

variety all at once. Since then, for reasons unknown, most of these forms became extinct. The organisms that survive today have, to some extent, been merely lucky survivors of various catastrophes in the history of the Earth.

Gould's arguments were not wholeheartedly accepted by other paleontologists and evolutionary biologists, however. Soon after the appearance of *Wonderful Life*, further studies of the Burgess fossils by Conway Morris and others suggested that the Burgess animals were not as novel as Conway Morris had initially believed. The animals could be comfortably grouped into smaller numbers of known phyla, although not necessarily the phyla that Walcott had selected.

Were the animals that lived 530 million years ago more diverse than those liv-

ing today? Probably not. While the Cambrian hosted some 11 phyla of animals, biologists have identifed about 35 phyla of animals living today. As far as is known, no whole phylum of animals has ever gone extinct. The dinosaurs, for example, were merely a group within the class Reptilia in the phylum Chordata, the phylum to which we humans belong. Certainly, the history of life suggests that diversity changes over time. Overall, however, diversity has increased since the first appearance of life 3.5 billion years ago. We can also say that organisms have become larger and more complex, on average, since the Cambrian and before.

In 1995, the philosopher Daniel Dennett attacked Gould, dissecting Gould's arguments about diversity and progress as aggressively as Gould had scrutinized

Walcott's taxonomy of the Burgess animals. Dennett insisted that evolution is, in fact, progressive. All living organisms, in all their current diversity, have descended, one generation at a time, from the same few ancestors. We know that most organisms share the same cell biochemistry and the same information technology in the form of DNA and RNA. Biologists therefore believe that modern organisms, whether complex or simple, are all descendants of a single ancestor. We may logically conclude that complex creatures such as ourselves have evolved, one generation at a time, from simpler organisms.

This increasing complexity may reasonably be described as progress, even though no goal is involved. Such progress does not, however, imply that evolution is *designed* to produce increasingly complex organisms. Nor does evolution *necessarily* produce increasing complexity. After all, we are surrounded by relatively simple organisms—whether bacteria, amebas, ferns, or flatworms—survivors, like us, of the ravages of both major catastrophes and the relatively minor ones that constitute natural selection. They are also descendants, like us, of a simple ancestor that lived more than 3.5 billion years ago.

KEY CONCEPTS

1. An animal is a multicellular, heterotrophic, eukaryotic organism that develops from an embryo.

2. Zoologists divide animals into two subkingdoms—the Parazoa (sponges) and Eumetazoa (all other animals). The members of the Eumetazoa are grouped according to their symmetry and internal body cavities.

3. The members of the subkingdom Parazoa are mostly sponges with neither symmetry nor organs.

4. Cnidarians and ctenophores are radially symmetrical animals without a true body cavity, or coelom.

5. Bilaterally symmetrical acoelomate animals are all solid worms. Some are free-living and some are parasites.

6. Pseudocoelomates have an internal body cavity between the endoderm and mesoderm.

7. Coelomate animals have many common properties, but two distinct patterns of development.

8. Mollusks are unsegmented animals that all share a similar body plan.

9. Annelids are segmented worms, consisting of series of identical or nearly identical sections.

10. Arthropods include the greatest number of species and the greatest numbers of individuals of all animal phyla.

WHAT IS AN ANIMAL?

As animals ourselves, we tend to see our own kingdom as the most interesting. Many of us have spent hours at zoos, observing the beautiful and the bizarre. Even the best zoos, however, barely hint at the variety of animals. A coral reef is a much better place to observe the tremendous diversity of animals. A coral reef swarms with fishes, shelled animals, worms, sponges, corals, sea anemones, and many other unfamiliar organisms. Most land dwellers are awed by the variety of shapes, domes, branches, fans, tubes, stars, as well as by the range of colors, from dark red to luminous blue.

What Features Characterize the Animals?

An animal is a multicellular, heterotrophic organism that develops from an embryo. All animals are eukaryotes and most reproduce sexually. A few, such as the corals, can reproduce by asexual processes, such as budding. Unlike plants, fungi, and protists, however, animals never show alternation of generations. With rare exceptions, the only haploid cells are the gametes (sperm and eggs).

Nearly all animals ingest their food—usually other organisms or their remains. Like fungi, a few parasitic animals absorb nutrients directly from their hosts. But, unlike fungi, these parasites develop from embryos, as other animals do. A few an-

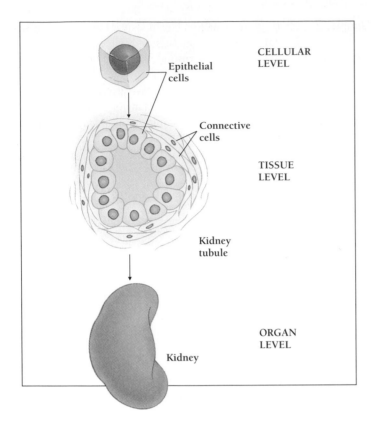

Figure 23-3   **The structure of animals.** A kidney is an organ, which is made up of two or more tissues. The two tissues in the kidney tubule, for example, are connective tissue and epithelial tissue, which are made of connective cells and epithelial cells, respectively.

imals—including many corals—appear to perform photosynthesis. Careful examination reveals, however, that these animals harbor symbiotic algae that are actually the ones performing photosynthesis.

Like plants and fungi, animals are highly structured. Almost all animals contain many different kinds of specialized cells. Groups of similar cells, along with the materials they secrete, called **matrix**, form **tissues**, units of structure and function. Two or more kinds of tissue in turn can form an **organ**, a structural unit with a distinctive function. For example, one class of cells can form an **epithelium**, a tissue that lines a surface (such as the outside of the body), while another kind of cell, along with its matrix, can form **connective tissue**, a network of loosely connected cells that may help support other body parts. A kidney, which contains both epithelial and connective tissues, is an example of an organ (Figure 23-3). The specialization of cells and the grouping of cells into tissues and organs allow animals to function efficiently.

Unlike plants, most animals can move about in search of food or mates. Some animals, such as the corals, are stationary but can move body parts to capture prey. Many have stages during development when they disperse. Almost all animals coordinate their movements in patterns that we can describe as behavior. For example, animals capture food, avoid predators, and breed. For all but the sponges, the coordination of such

behavior depends on the functioning of a network of nerve cells (neurons).

An animal is a multicellular, heterotrophic organism that develops from an embryo. All animals are eukaryotes and most reproduce sexually.

How Do Zoologists Classify the Animals?

Zoologists, biologists who study animals, estimate that living animal species number at least 4 million, and some estimates run as high as 30 million. Zoologists divide these millions of living animals into about 35 phyla (Figure 23-4). The 40,000 species of vertebrates (animals with backbones)—including fish, frogs, snakes, birds, and mammals—are just a subphylum of one of these 35 phyla. Vertebrates include a mere one percent of all animal species. Most animals have no backbones. The vast majority are insects (mostly beetles, in fact), snails, jellyfish, and worms.

Zoologists divide all these creatures into two subkingdoms. The subkingdom **Eumetazoa** [Greek, *eu* = true + *meta* = middle + *zoa* = animals] includes almost all of the 4 million named species of living animals. Excluded are the 5000 species of sponges and one other species, which make up the small subkingdom **Parazoa**. The sponges are the simplest animals, for they have neither regular symmetry nor organs.

How Are Animals Classified by the Way They Develop?

The Eumetazoa are divided into major groups of phyla according to how their embryos develop. After fertilization, an animal zygote divides by mitosis to produce a hollow ball of cells called a **blastula** (Figure 23-5). The blastula folds in on itself to form a **gastrula**, three cell layers surrounding a simple cavity called the **archenteron** [Greek, *arche* = beginning + *enteron* = gut] (Figure 23-6). The three cell layers of the gastrula develop into distinctly different tissues in the adult. The innermost layer, which surrounds the archenteron, is called the **endoderm** [Greek, *endon* = within + *derma* = skin] and gives rise to the intestines and other digestive organs. The outermost layer, called the **ectoderm** [Greek, *ecto* = outside], gives rise to skin, sense organs, and nervous system. The middle layer, the **mesoderm** [Greek, *mesos* = middle], gives rise to muscle, skeleton, and connective tissue. That all animals develop in the same general way provides further evidence that we share common ancestors.

In even the most complex animals the gastrula already contains the beginning of the adult body plan. We may view this plan as a tube within a tube. The inner tube is the digestive tract, with a mouth at one end and an anus at the other. The outer tube is the body wall. Between the two tubes (of most animals) is a fluid-filled space, within which the internal organs hang. In the most complex animals (including ourselves), a layer of mesoderm cells lines this space, and the lined cavity

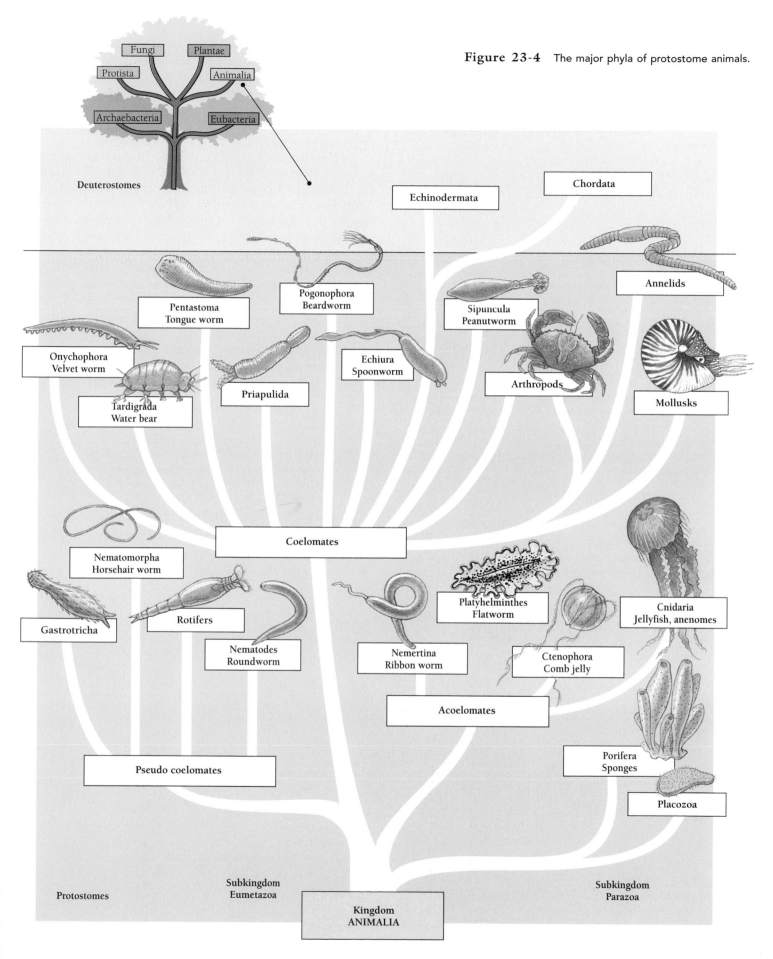

Figure 23-4 The major phyla of protostome animals.

Fungi

Plantae

Protista

Animalia

Archaebacteria

Eubacteria

Deuterostomes

Chordata

Echinodermata

Annelids

Pogonophora
Beardworm

Pentastoma
Tongue worm

Sipuncula
Peanutworm

Onychophora
Velvet worm

Echiura
Spoonworm

Arthropods

Mollusks

Priapulida

Tardigrada
Water bear

Coelomates

Nematomorpha
Horsehair worm

Platyhelminthes
Flatworm

Cnidaria
Jellyfish, anenomes

Gastrotricha

Rotifers

Nematodes
Roundworm

Nemertina
Ribbon worm

Ctenophora
Comb jelly

Acoelomates

Porifera
Sponges

Pseudo coelomates

Placozoa

Protostomes

Subkingdom
Eumetazoa

Subkingdom
Parazoa

Kingdom
ANIMALIA

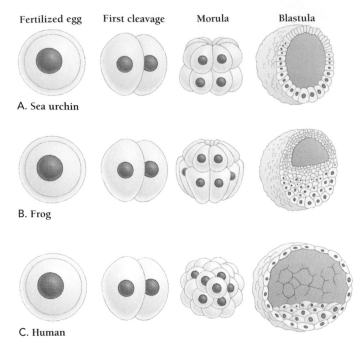

Fertilized egg   First cleavage   Morula   Blastula

A. Sea urchin

B. Frog

C. Human

Figure 23-5   All eumetazoan embryos begin development in the same fashion. Whether sea urchin, frog, or human, all true animals begin life by dividing to form a hollow ball of cells called a blastula. Although eggs and their zygotes may differ in size—from 1 mm for frogs to 0.1 mm for humans and sea urchins—all share a similar pattern of early development.

is called a **coelom** [Greek, *koilos* = hollow]. Animals with a coelom are called **coelomates**. These include, for example, snails, earthworms, insects, and all of the vertebrates.

The simplest animals, including the jellyfish and flatworms, have no cavity and are called **acoelomates** [Greek, *a* = with-

out]. In between are the **pseudocoelomates** [Greek, *pseudo* = false], which have an organ-containing cavity, but without the mesodermal lining of a true coelom (Figure 23-7). A pseudo-coel lies between the endoderm and the mesoderm and derives directly from the cavity in the blastula (Figure 23-5). Almost all the pseudocoelomates are tiny worms called nematodes.

The coelomates are further divided into two major groups according to another peculiarity of the early gastrula (Figure 23-6). At one end of the archenteron is an opening called the **blastopore.** In some animals the blastopore eventually develops into the adult mouth. In other groups of animals it develops into the anus, and the mouth develops later. Those in which the mouth develops first, from the blastopore, are called the **protostomes** [Greek, *protos* = first + *stoma* = mouth]. These include, for example, snails, earthworms, and all the arthropods—the spiders, insects, and their allies. In contrast, those in which the anus forms first, from the blastopore, and the mouth forms secondarily are called **deuterostomes** [Greek, *deuteros* = second + *stoma* = mouth]. Most coelomates are protostomes. We vertebrates and our cousins the echinoderms (sea stars and sea urchins, for example) are deuterostomes.

Zoologists divide the Eumetazoa according to the presence or absence of a coelom or pseudocoel and according to whether the blastopore develops into a mouth (protostomes) or an anus (deuterostomes).

How Are Animals Classified by Their Symmetry?

Animals are also divided according to their symmetry. The Parazoa (sponges) have no symmetry at all. Their lack of symmetry is one of the characteristics that sets them apart from the rest of the animal kingdom.

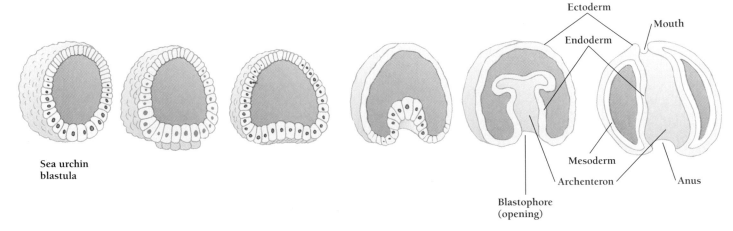

Ectoderm

Mouth

Endoderm

Mesoderm

Archenteron

Anus

Blastophore (opening)

Sea urchin blastula

Figure 23-6   The formation of the blastopore. The hollow blastula of all animal embryos folds in on itself to form the gastrula. The opening into which the cells flow during this process is called the blastopore. In some animals (protostomes), the blastopore becomes the mouth; in others (deuterostomes) it becomes the anus. In both kinds of animals, the hollow center (the archenteron) becomes "the tube within a tube," which later forms the gut.

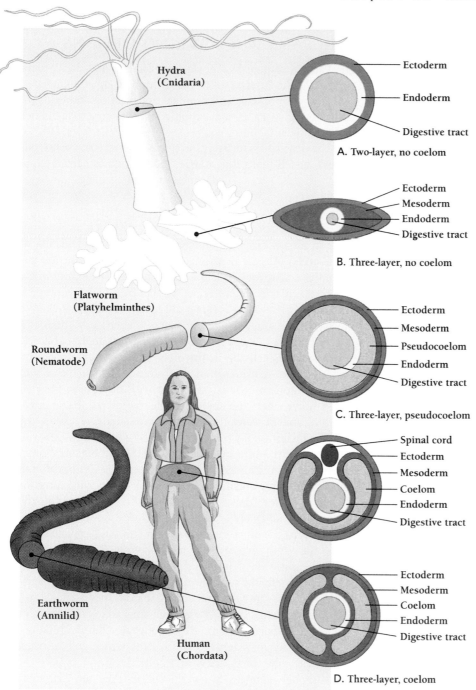

Figure 23-7 **Tissue layers.** A. A cnidarian has just two tissue layers and no coelom. B. The flatworms have three layers, but still no coelom. C. The roundworms have three layers and a pseudocoel. D. The true coelomates, including both earthworms and humans, have three layers and a true coelom, in which the gut is suspended.

The Eumetazoa have one of two kinds of symmetry. Animals with **radial symmetry**—the simple jellyfish and sea anemones, for example—can be rotated along their central axis without changing their appearance (Figure 23-8A). These animals have a top and bottom, but no front and back and no left and right. Many of these animals live most of their lives either floating passively in the water (jellyfish) or attached to a rock or some other support (sea anemones).

Most animals have **bilateral symmetry,** meaning that their left and right halves are (approximately) mirror images of one

another (Figure 23-8B). Such animals have a defined **anterior** (front) and **posterior** (back end). The *dorsal* surface is the back, or top, which, in most animals, faces the sky. The *ventral* surface is the part facing the Earth.

Bilateral symmetry implies some sort of head and, ordinarily, a preferred direction of movement—usually "head" first. Directed movement, in turn, demands a highly integrated system of muscles, nerves, and sense organs. For these reasons and others, bilaterally symmetrical animals are usually more complex than the radially symmetrical animals.

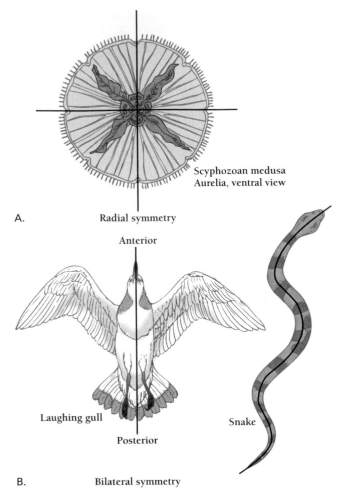

A.   **Radial symmetry**

Scyphozoan medusa
Aurelia, ventral view

Anterior

Laughing gull   Snake

Posterior

B.   **Bilateral symmetry**

Figure 23-8   Symmetry. A. Radial symmetry in a jellyfish. B. Bilateral symmetry in a laughing gull and a snake. All animals with bilateral symmetry have anterior and posterior ends and dorsal and ventral surfaces. Shown here are the gull's ventral surface and the snake's dorsal surface.

In general, the sense organs that detect prey and predators, as well as the neural networks that direct and coordinate movement, lie primarily near the front, the anterior end, of the animal (Figure 23-8B). Reproduction, digestion, and excretion tend to lie in the rear, the posterior end.

The Eumetazoa are further classified according to whether they have bilateral or radial symmetry. Bilaterally symmetrical animals, such as ourselves, are usually more complex than radially symmetrical animals.

In this chapter we will begin with the simplest animals, the sponges; move to the acoelomates, first those with radial symmetry, then those with bilateral symmetry; and then move to the pseudocoelomates. We will finish with an examination of the protostome coelomates. In the next chapter, we will survey the deuterostome coelomates.

THE PARAZOA HAVE NO ORGANS

The Parazoa contain two phyla, the Placozoa and the Porifera (sponges). The Parazoa differ greatly from all other animals, in showing no symmetry and minimal organization and cell specialization. The Parazoa have only simple connective tissues and no organs.

The Placozoa consist of a single species, *Trichoplax adhaerens,* which looks like a large, multicellular ameba. Unlike any protist, however, *T. adhaerens* produces gametes, develops as an embryo, and shows cell specialization. So it is a true animal.

The 5000 species of sponges (most of which are marine) come in many sizes, shapes, and colors, and live at all depths of the ocean. Almost all sponges are **sessile,** permanently anchored to rocks, logs, or coral. They range in size from a few millimeters to two meters. Some are shapeless blobs, while others resemble fans, cups, crusts, and tubes (Figure 23-9A). Like the synthetic "sponges" we use at the kitchen sink, the members of Porifera [Latin, *porifera* = hole-bearer] are full of holes.

A sponge's cells lie in three layers surrounding a central cavity, like the gastrula of all other animals. A sponge's outer, epithelial layer of cells is perforated with tiny holes, through which water enters the sponge. Each sponge pumps an enormous volume of water through its simple body. A sponge the size of a fingertip may pump 20 liters of water a day. This water, loaded with nutritious microorganisms and organic matter, enters through the pores, flows into the cavity, and exits through the sponge's large opening (or openings). The driving force for this flow comes from a layer of flagellated cells lining the sponge's central cavity (Figure 23-9B).

Between the outer epidermal layer and the flagellated cells lining the cavity is a layer of **mesenchyme** [Greek, *mesos* = middle + *enchyma* = infusion], loosely attached cells embedded within a jellylike substance. Within this simple tissue, amebalike cells capture food and shuttle it from the inside cells to the outer epithelial cells. These same amebalike cells also give rise to either sperm or eggs. Most sponges are hermaphrodites: a single organism is both male [like the Greek god Hermes] and female [like the goddess Aphrodite].

Although sponges are multicellular, they are extremely primitive. No nerve cells coordinate a sponge's responses, and each cell functions as an independent unit. Nonetheless, single sponge cells, separated from others by forcing a sponge through a piece of cheesecloth, will spontaneously reorganize themselves into a functioning sponge.

Sponges, whose fossils date from the early Cambrian Period, 530 million years ago, almost certainly evolved from flagellated protistan ancestors and diverged from all other animals early on.

The individual cells of parazoans resemble protists. But unlike protists, parazoans develop from embryos and produce sperm and eggs.

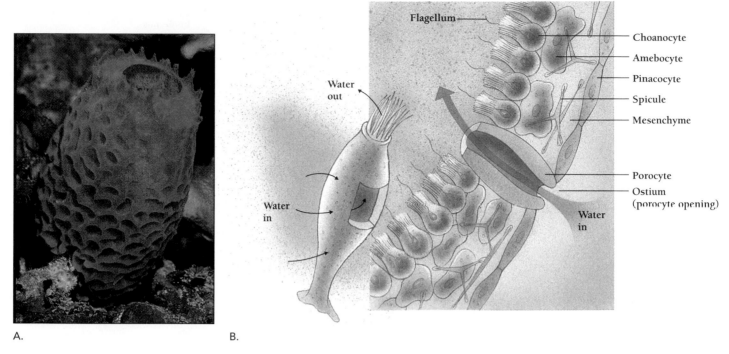

Figure 23-9   **Sponges.** A. An azure sponge from the Caribbean. B. Flagellated cells inside the sponge pump water through the pores (ostia) of the sponge. Other specialized cells digest microorganisms and organic matter filtered from the flowing water.   *(A, Brian Parker/ Tom Stack & Associates)*

RADIALLY SYMMETRICAL ACOELOMATES

The 9000 species of radially symmetrical acoelomates fall into two phyla, Cnidaria [pronounced ni-daria] and Ctenophora [pronounced ten-OFF-ora]. The cnidarians, which are simpler and far more numerous, are among the oldest animals in the fossil record, present well before the Cambrian explosion.

Cnidarians

Zoologists recognize four classes of **cnidarians**: Hydrozoa (hydroids), Scyphozoa (jellyfish), Anthozoa (corals and sea anemones), and the recently discovered Cubozoans. Almost all are marine (with the exception of *Hydra* and a few other freshwater hydrozoans). By far the largest class is Anthozoa, with about 6200 species. Hydrozoa include about 2700 species, and Scyphozoa only about 200.

Anthozoans [Greek, *anthos* = flower] take their name from their flowerlike appearance. They include sea anemones, sea pansies, sea fans, and sea whips. Each anthozoan has a cylindrical body with a crown of tentacles. Symbiotic algae enhance their plantlike appearance and also contribute to their nutrition, often allowing them to grow in water that is especially poor in nutrients. Especially impressive are the corals, whose secreted skeletons of calcium carbonate are the foundation of gigantic coral reefs.

The anthozoans have the most complex behaviors of the cnidarians. Some species of sea anemones, for example, feed on crabs, mussels, or even fish. A sea anemone can catch and consume prey with its tentacles. Some sea anemones even attack the tentacles of other sea anemones that intrude too closely.

All cnidarians contain two layers of cells that function as true tissues. But cnidaria have no organs, making them the least complex of the Eumetazoa. Cnidarians come in two basic body plans—**polyps**, which resemble cylinders, and **medusae**, which resemble bells (Figure 23-10). Both forms have a single mouthlike opening to a central cavity, and both use tentacles to capture food. Polyps are usually partly sessile, attached to rocks or other surfaces, with mouths and tentacles pointed upward. Medusae generally float free with the mouth pointed down and tentacles dangling, like the fringe on the edge of an umbrella. Some cnidarians are polyps throughout their life cycles, some are only medusae, and some cycle from one form to the other (Figure 23-11). Most cnidarians start out life as free-swimming larvae and become sessile as adults.

Both polyps and medusae have an internal digestive cavity, which is surrounded by an outer layer of epidermis and an inner layer of gastrodermis. Like higher animals, cnidarians perform most of their digestion extracellularly—that is, within the digestive cavity. Digestion in the cavity, however, is not complete, and the cells lining the cavity engulf fragments of the small animals that compose the cnidarian's diet.

Between the two cellular layers is the **mesoglea** [Greek, *mesos* = middle + *glia* = glue], a jellylike material that con-

Figure 23-10   The two body plans of the cnidarians. Jellyfish and other cnidarians have just two body plans: the polyp and the medusa. The polyp is the sessile form exemplified by sea anemones and hydras. The medusa, exemplified by the jellyfish, is basically a polyp turned over.

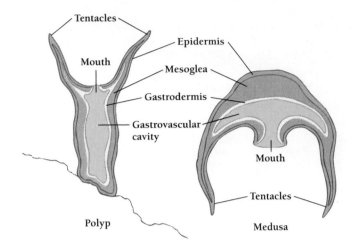

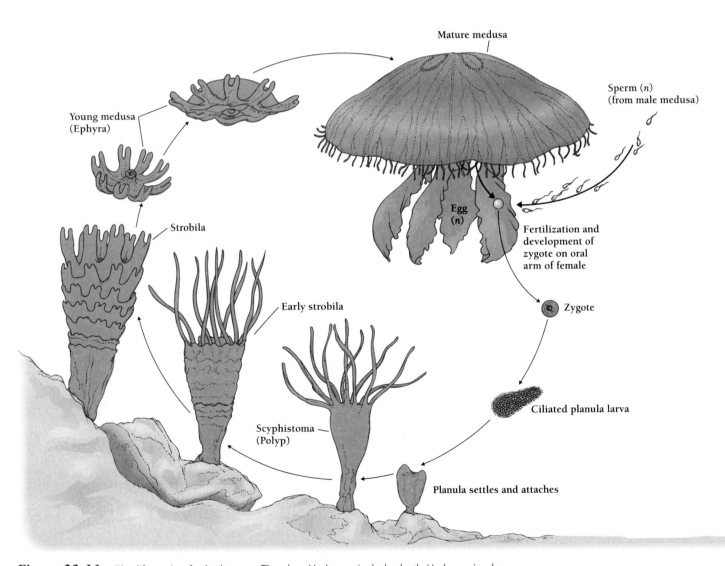

Figure 23-11   The life cycle of a hydrozoan. The class Hydrozoa includes both *Hydra*, a simple polyp often studied in biology classes, and the formidable Portuguese man-of-war (*Physalia*), capable of stinging to death an adult human. Most hydrozoans are jellyfish (also called jellies) that have both polyp and medusa stages. In the polyp stage, they are colonial, with many polyps budding around a branched central cavity. *Hydra* is an exceptional hydrozoan, for it lives as a solitary polyp, with no medusa stage. All hydrozoans shed sperm and eggs from testes and ovaries that project from the body wall.

tains few if any cells. In some medusae, such as the jellyfish, the mesoglea may be quite thick and gelatinous. In polyps, the mesoglea is usually thin, and in some cases (such as in the sea anemones) consists entirely of cells.

Cnidarians [Greek, *cnide* = nettle] take their name from specialized stinging cells, called **cnidocytes**, that lie on their tentacles. Each cnidocyte can use water pressure (some 140 times the pressure of the atmosphere) to fire a tiny barbed spear called a **nematocyst.** As whalers use harpoons threaded with rope to wound whales and drag the animals back to the whaling boat, cnidarians use nematocysts and their attached threads to disable and capture their prey.

Cnidarians work much more quickly than whalers, however. Only a few milliseconds elapse between the detection of the prey by the tickling of the cnidocyte's flagellumlike cnidocil and the firing of the nematocyst. After it has pierced its prey, the nematocyst can discharge a poisonous protein. The nematocysts of one cnidarian—the Portuguese man-of-war—produces a neurotoxin potent enough to kill a human swimmer.

Once the nematocysts have attached their tiny tethers, the tentacles draw the prey mouthward. Pulling and pushing the prey requires coordination among the tentacles, which is controlled by simple muscle cells and nerve cells. A cnidarian's nerve cells are unlike those in more complex organisms for they transmit impulses in both directions, rather than just one.

Cnidarians are radially symmetrical acoelomates with true tissues but no organs. Named for their specialized stinging cells, the cnidarians include the hydras, jellyfish, corals, and sea anemones.

Ctenophores (Comb Jellies)

The **ctenophores** [Greek, *ktenos* = comb + *phora* = motion], or comb jellies, are a small phylum of about 90 living species. Their name refers to the rows of comblike cilia that these beau-

Figure 23-12   A ctenophore, or comb jelly. These graceful animals are noted for their beauty.   *(James R. McCullagh/Visuals Unlimited)*

tiful animals carry in bands along their short bodies (Figure 23-12). Ctenophores seem to waltz through the water. By means of the coordinated action of their cilia, they move forward, mouth first, capturing prey with their sticky tentacles, while slowly rotating. They are hermaphrodites (simultaneously male and female) and usually shed both sperm and eggs into the open sea.

Because of the symmetrical arrangement of the combs and because one species of ctenophore has cnidocytes, zoologists once classified the ctenophores as cnidarians. Modern zoologists, however, argue that the lack of radial symmetry of the paired tentacles earns the comb jellies their own phylum. The ctenophores have existed separately for at least 400 million years.

Unlike the cnidarians, the ctenophores (comb jellies) are not radially symmetrical, for they possess paired tenacles. In addition, ctenophores move through the water by means of rows of cilia (the "combs").

BILATERALLY SYMMETRICAL ACOELOMATES

The cnidarians and ctenophores are radially symmetrical, or nearly so, but acoelomates may also have bilateral symmetry, as we do. Two phyla, Platyhelminthes (the flatworms) and Nemertea (the ribbon worms), are especially interesting. Flatworms include a host of parasites that infect hundreds of millions of people, as well as the planarian *Dugesia*, used in biology laboratories all over the world. The flatworms differ from more advanced animals in having only one opening to their digestive cavities. Mouth and anus are the same.

The ribbon worms, in contrast, are the simplest animals with two openings to their digestive tract. Unlike all other acoelomates, ribbon worms have a separate mouth and anus.

Platyhelminthes (Flatworms)

The 12,000 species of **Platyhelminthes** (flatworms) live in a wide range of environments. Most free-living species are marine, although some (such as *Planaria*) thrive in fresh water, and some are terrestrial. One species is specialized for living in bat guano. Zoologists divide the flatworms into three classes: Turbellaria (free-living flatworms), Trematoda (flukes), and Cestoda (tapeworms). All three classes share the same acoelomate body plan.

All flatworms have three body layers—an endoderm that lines the digestive cavity, an ectoderm on the outside, and a mesoderm in between. They have feeding, digestive, and reproductive organs, a brain, and a simple system for regulating water balance.

One thing flatworms do not have is a circulatory system. Some flatworms nonetheless grow to 60 cm, and tapeworms may be as long as 10 m (over 30 ft). How do such large animals distribute oxygen and nutrients to their cells? Flatworms survive without a circulatory system by keeping diffusion distances small. Their flattened shapes allow gases to diffuse rapidly to and from the outside environment, and the gut cavity of each flatworm is branched and extends throughout the body, so that no cell is far from the gut. The single opening to the flatworm's gut cavity serves as both mouth and anus. Likewise, the gut cavity itself serves both as a reservoir of nutrients and a collecting pool for wastes.

Turbellaria, the free-living flatworms, can reproduce either asexually or sexually. A free-living flatworm's simple nervous system allows it to move efficiently toward food and away

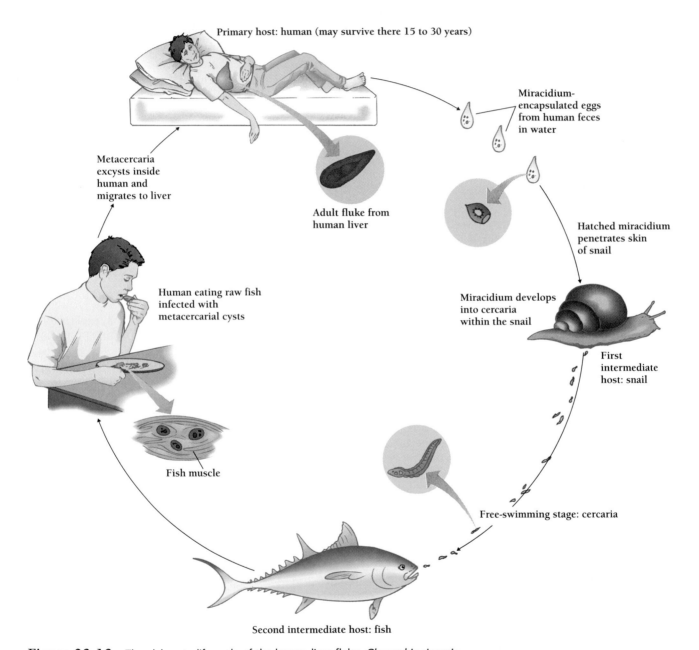

Figure 23-13   The elaborate life cycle of the human liver fluke, *Clonorchis sinensis.* The liver fluke lives its adult life as a 1- to 2-cm worm in the livers of humans, pigs, cats, and dogs. Each worm mates and reproduces, then deposits thousands of fertilized eggs. These develop into larvae, called miracidia, which leave the mammalian host in the feces. Development of the flukes then stops: the cycle continues only if the feces enter water and the larvae are eaten by a snail. Within the snail's digestive tract, each miracidium develops into a swimming larva, called a cercarium, which then reenters the water. The swimming larvae must burrow into the muscles of a fish within two days or they will die. In the fish's flesh, they form themselves into protected cysts, called metacercaria. Development resumes when a human (or other mammal) eats the raw infected fish.

from light. Some flatworms detect light with primitive paired eyespots on their heads. They can also detect odor. A good way to collect flatworms is to put a piece of meat in the bottom of a muddy pond. Chemicals diffusing from the meat stimulate receptors on the flatworm's head. When a flatworm smells the meat, it turns its body and moves toward its future meal.

The differences among the three classes of flatworms relate to differences in the way they eat and reproduce. The flukes and tapeworms are parasites. Their outer layer of cells resists the digestive enzymes of their hosts, and they lack eyespots and chemical receptors for detecting food. Flukes take in food through a specialized mouth, while tapeworms lack a digestive system altogether and simply absorb digested food from their hosts.

Parasites are specialized for rapid and prolific reproduction in their hosts, and they often evolve stripped down bodies, lacking many ancestral characteristics that are not useful for their specialized lives. Flukes and tapeworms illustrate many common adaptations of parasites: (1) rapid reproduction; (2) distinct stages that allow passage and dispersal through more than one host; (3) organs for attachment to their hosts; (4) specialized digestion (in the case of tapeworms, direct absorption); and (5) reduced sense organs.

As a result, the more complex free-living flatworms are probably more similar to the original flatworm ancestors than are the streamlined flukes and tapeworms.

Trematoda (flukes) range in size from less than a millimeter to more than 8 cm. All flukes (more than 8000 species) are parasites. Each attaches to the outside or the inside of its host by a hook or sucker. Like the free-living flatworms, flukes have a digestive system with a single opening.

Many flukes are hermaphrodites and generally reproduce by mutual copulation. They are far more prolific than free-living flatworms, producing 10,000 to 100,000 times as many eggs. Some flukes live their whole lives on a single host. Most, however, have more complicated lives, involving two or more hosts. The human liver fluke *Clonorchis sinensis,* for example, uses three distinct hosts to complete its life cycle—a mammal, a snail, and a fish (Figure 23-13).

Cestoda (tapeworms) are even more specialized for reproduction than are the flukes. An adult tapeworm has no digestive system at all. It lives in its host's gut, and its cells directly absorb passing nutrients.

A tapeworm maintains its easy life with a specialized attachment organ, called a **scolex,** at its anterior end. Behind the scolex lie a set of repeated segments, called **proglottids.** In the beef tapeworm, the proglottid chain may be as long as 10 m. Each proglottid is a complete reproductive unit, with both male and female organs. If two or more tapeworms live together, they fertilize one another. Otherwise, a tapeworm self-fertilizes. In either case, the proglottids toward the rear end of the tapeworm are filled with embryos, each within a separate shell, which pass, with the feces, to the outside world. If cattle eat vegetation carrying these embryos, they acquire tapeworms. In the United States, about one percent of all cattle have tapeworms, an important reason for eating only inspected beef.

Platyhelminthes (flatworms, flukes, and tapeworms) are the simplest animals with heads. They have three body layers, true organs and tissues, a brain, and an organ for regulating water balance. They do not have a one-way digestive system: they take in food and excrete wastes through the same structures. The parasitic members of this phylum possess fewer specializations than the free-living flatworms.

Nemertea (Ribbon Worms)

Ribbon worms are a group of about 750 species of free-living acoelomate animals, most of which live inconspicuously in the sea. A few species live in fresh water or on land in the tropics. Their flat, velvety bodies may be less than a millimeter long or as long as 30 meters (Figure 23-14). Each has a characteristic snout, or **proboscis**—a long, sensitive, retractable, and sometimes venomous, tube. A ribbon worm uses its proboscis to explore its environment, defend itself, and capture prey. Ribbon worms are abundant in tidal mudflats, where they spend their days hidden in the mud, emerging only at night to feed.

Ribbon worms are closely related to free-living flatworms, with similar nervous and reproductive systems (though each ribbon worm is usually either male or female, rather than hermaphroditic). Two important advances distinguish ribbon worms from flatworms: (1) a digestive tract with both a mouth and an anus, and (2) a circulatory system. The two-ended gut allows continuous eating, uninterrupted by the need to use the same opening for excretion. The second opening also permits more efficient extraction of nutrients from food, as it progresses in a single direction. The gut is like an assembly line; it performs different operations at each stage: the anterior end breaks down the food, the posterior processes wastes.

Figure 23-14   **Some ribbon worms grow up to 30 m (100 ft) long.** All have a two-ended digestive tract and circulatory system. *(Raymond Mendez/Animals Animals)*

BOX 23-1

The origin of feces

Did the evolution of the one-way digestive tract fuel the Cambrian explosion? In a controversial new theory, geochemists have suggested that the development of one-way digestive systems by metazoans provoked drastic changes in the deep ocean, changes that allowed metazoans to colonize the bottom of the sea in an unparalleled adaptive radiation.

Until the Cambrian explosion, some 540 million years ago (mya), life consisted almost exclusively of single-celled bacteria and algae. As photosynthetic algae proliferated, they pumped increasing amounts of oxygen into the atmosphere. To see how this change in the composition of the atmosphere affected ancient life, a group of geochemists led by Graham Logan measured the chemical signature of fossilized organic matter in ancient ocean deposits.

To their surprise, the researchers found that at the beginning of the Cambrian, a dramatic change in the proportion of different carbon isotopes occurred. Before 590 mya, the sea floor deposits were relatively high in ^{13}C, a heavy isotope of carbon. Sometime between 590 and 540 mya, however, the deposits became lower in ^{13}C and higher in ^{12}C. This switch in "isotope signature" coincided closely with the flowering of multicellular life known as the Cambrian explosion. The researchers could not help asking, Were the two events related?

It seemed possible. Organisms tend to have higher ratios of $^{12}C:^{13}C$ than the atmosphere. During photosynthesis, plants incorporate into sugars and other new compounds a lighter proportion of ^{12}C than ^{13}C. (This is because ^{12}C is lighter than ^{13}C and diffuses more quickly.) The result is that ^{12}C accumulates in photosynthetic organisms and all the organisms that eat them. Organic matter therefore has more ^{12}C than the atmosphere.

But nearly all organisms respire—a process that breaks down organic compounds and releases carbon dioxide. Respiration puts the ^{12}C back into the atmosphere. If photosynthesis and respiration are balanced, the ratio of ^{12}C to ^{13}C is also balanced. In that case, researchers would not see an excess accumulation of ^{12}C in ocean floor sediments.

But Logan and his colleagues had an idea that would account for the change in ratios. Before 590 mya, the researchers hypothesized, bacteria in the upper layers of the ocean decomposed nearly all organic waste, releasing organic ^{12}C back into the atmosphere. Early deep-sea sediments, then, were relatively high in ^{13}C.

Huge numbers of bacteria would have filled the upper layers of the ocean, and their respiration would have depleted ocean waters of oxygen. The surface would have remained oxygenated, but the water near the bottom would have contained very little oxygen. Few organisms could have survived there.

But what if the bacteria were suddenly prevented from decomposing organic matter? Logan and his colleagues suggested that the evolution of a one-way digestive tract made all the difference. Animals with a one-way gut can package wastes in concentrated packets (feces), which fall to the ocean floor before surface-dwelling decomposers can absorb and digest them. With less decomposition occurring in the upper layers of the ocean, the overall oxygen content of the water would have risen. Ultimately, argue the researchers, oxygen would have diffused into the deeper waters of the ocean—allowing colonization by oxygen-requiring metazoans. That, the researchers say, paved the way for the Cambrian explosion.

Did the first fecal pellets fuel the Cambrian explosion? Many researchers remain skeptical of this tentative train of hypotheses. But the idea that the origin of all the major animal phyla depended on the evolution of feces has a certain appeal.

The ribbon worm's circulatory system is relatively simple; in fact, some species have only two vessels. The blood is usually colorless, but may also be yellow, red, orange, or green. There is no heart to pump the blood in a particular direction: instead the vessels and the body wall contract irregularly, sloshing the blood through the circulatory system. These primitive adaptations make ribbon worms the most complex of the acoelomates.

Ribbon worms (Nemertea) have what flatworms have (nervous system and organs, for example). In addition, ribbon worms have a two-ended digestive system, a circulatory system, and a proboscis.

ASCHELMINTHS

Zoologists sometimes jokingly call the **Aschelminths** [Greek, *askos* = bag, *helminthos* = worm] the ash-can phylum, since it includes an assortment of unrelated animals that do not fit anywhere else. Like the Kingdom Protista, the Aschelminths are united more by what they do not have than by what they do. Because the members of this group are so different, the Aschelminths are not considered a single phylum, but an assemblage of eight or nine distinct phyla.

All of the bilateral animals we have seen so far have been acoelomates, which have solid bodies with no internal cavity (other than the gut). The Aschelminths also lack a coelom. Some Aschelminths, however, have a cavity called a **pseudo-**

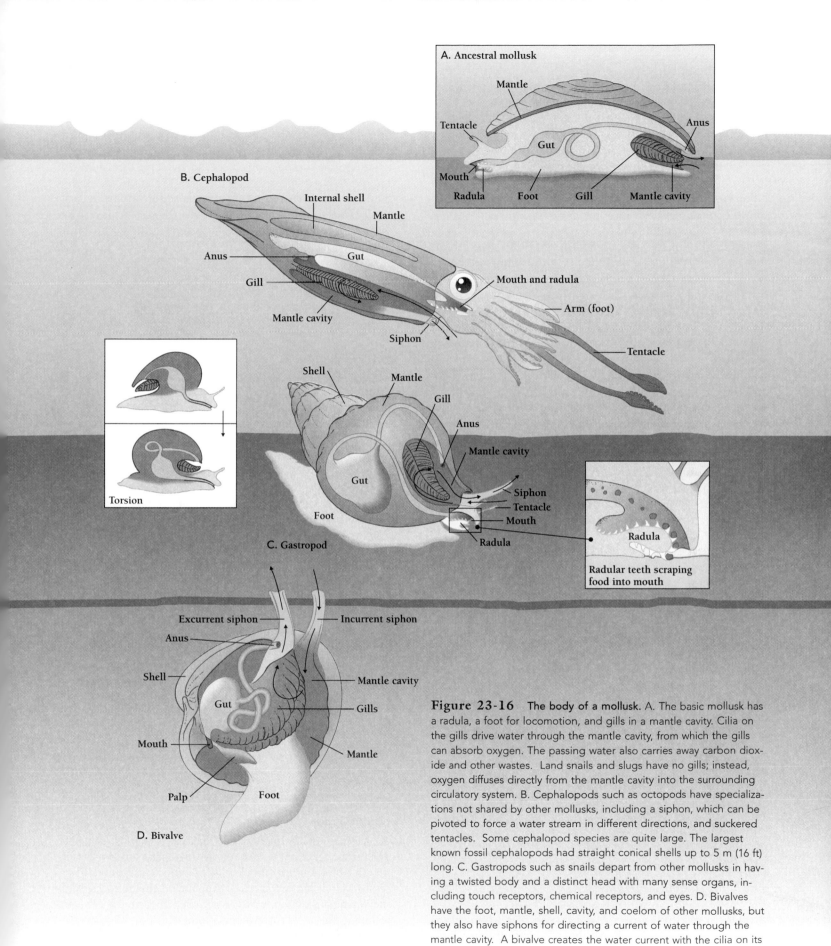

A. Ancestral mollusk

Mantle

Tentacle

Anus

Gut

Mouth

Radula   Foot   Gill   Mantle cavity

B. Cephalopod

Internal shell

Mantle

Anus

Gut

Gill

Mouth and radula

Mantle cavity

Arm (foot)

Siphon

Tentacle

Torsion

C. Gastropod

Shell

Mantle

Gill

Anus

Mantle cavity

Gut

Siphon

Tentacle

Mouth

Foot

Radula

Radula

Radular teeth scraping food into mouth

D. Bivalve

Excurrent siphon

Incurrent siphon

Anus

Shell

Mantle cavity

Gut

Gills

Mouth

Mantle

Palp

Foot

Figure 23-16 **The body of a mollusk.** A. The basic mollusk has a radula, a foot for locomotion, and gills in a mantle cavity. Cilia on the gills drive water through the mantle cavity, from which the gills can absorb oxygen. The passing water also carries away carbon dioxide and other wastes. Land snails and slugs have no gills; instead, oxygen diffuses directly from the mantle cavity into the surrounding circulatory system. B. Cephalopods such as octopods have specializations not shared by other mollusks, including a siphon, which can be pivoted to force a water stream in different directions, and suckered tentacles. Some cephalopod species are quite large. The largest known fossil cephalopods had straight conical shells up to 5 m (16 ft) long. C. Gastropods such as snails depart from other mollusks in having a twisted body and a distinct head with many sense organs, including touch receptors, chemical receptors, and eyes. D. Bivalves have the foot, mantle, shell, cavity, and coelom of other mollusks, but they also have siphons for directing a current of water through the mantle cavity. A bivalve creates the water current with the cilia on its gills and eats small organisms or food particles as they pass by its mouth and stick to the mucus on the gills.

557

Most mollusks have a **radula**, a rasping tongue covered with teeth made from chitin. Mollusks may use the radula to protect themselves, to capture prey, to scrape vegetation, or to tear food into tiny bits. A snail in an aquarium, for example, uses its radula to scrape algae from the aquarium's glass wall.

All mollusks have a specialized digestive tract, beginning with a mouth and ending with an anus. All have one or two kidneylike **nephridia**, tubular excretory organs that remove nitrogen-containing wastes from the coelomic fluid and regulate water and salt concentration.

All mollusks also have a circulatory system with a heart that receives oxygen-carrying blood from the gills and pumps it to other body tissues. Unlike humans, most mollusks have open circulatory systems, in which the heart pumps blood through spaces between the tissues, rather than exclusively through blood vessels. The only exceptions are the cephalopods, which have a closed, well-developed circulatory system.

Mollusks reproduce sexually. Usually, male and female gonads are in separate individuals, but some (such as land snails) are hermaphroditic (both male and female). Individuals of some species, including sea slugs and oysters, change sexes from male to female and back several times during a mating season.

Cross-fertilization is the rule, even among the hermaphrodites. Self-fertilization, however, does occur—an important adaptation for animals that sometimes move too slowly to find a mate. Among the land snails and slugs, fertilization is internal. In other mollusks, gametes are shed externally.

Mollusks are soft-bodied coelomates with sophisticated organs for respiration, circulation, digestion, and excretion. All the mollusks share a similar body plan, including a foot, a mantle cavity, and a visceral mass. Most have shells and a rasping tongue called a radula. They reproduce sexually, and their embryos develop into a characteristic larval stage.

Zoologists divide the mollusks into seven or eight classes of which we will briefly discuss three: the bivalves (which include clams, oysters, and mussels), the gastropods (which include snails, slugs, and abalones), and the cephalopods (which include squid, octopods, and the chambered nautilus).

In addition, the mollusks include *aplacophora* [Greek, *a* = without + *plak* = flat plate], shell-less, wormlike animals that live in the deep seas; *monoplacophora* [Greek, *mono* = one], which have a single dorsal shell; the chitons, or *polyplacophora* [Greek, *poly* = many], which have eight plates on their backs; and the tusk or tooth shells, which have tubular shells open at both ends.

Bivalves

Bivalves [Latin, *bi* = two + *valva* = part of a folding door] include clams, oysters, scallops, mussels, and other mollusks with two shells—a right shell and a left shell. A ligament hinges the two shells (or valves) and a muscular "foot" protrudes from between them (Figure 23-16D). Two large muscles, called adductor muscles, pull the shells together. Although most bivalves lead sedentary lives attached to rocks, some move about a little more. A clam, for example, uses its foot to burrow into the sandy or muddy bottom of the sea or river. The most active bivalves are the scallops, which "swim" by using their adductor muscles to clap their shells together, simulating a human swimmer's "frog kick." A scallop's enormous adductor muscles, tender and delicately flavored, are familiar to seafood lovers (although some restaurants give the name "scallops" to circular plugs of shark fin).

Bivalves such as clams, oysters, scallops, and mussels have two shells, and usually lead sedentary lives.

Gastropods

The **gastropods** are the most diverse class of mollusks, with some 80,000 named species, including snails, whelks, conches, limpets, abalone, and slugs. They have a distinctive twisted body and, except for the shell-less slugs, assymmetical spiral shell.

Gastropods differ from all other mollusks in having a mysterious 180-degree counterclockwise twist of the body (viewed from above) that occurs during embryonic development. One result is that both the anus and the mantle cavity end up at the front end (anterior end) of the body, not far from the mouth (Figure 23-16C). Torsion may occur very rapidly, sometimes over the course of just a few minutes. Biologists have many hypotheses about what advantages torsion might confer on gastropods, but they have not yet agreed on an answer.

The shells of gastropods show great variation in form and color. Most gastropods are marine, but many inhabit fresh water or land. Unlike the quiet bivalves, gastropods are active creatures, equipped with a well-developed head and a sensitive nervous system. Gastropods are free-living and consume plants, small animals, or decaying organic matter. A few species live as parasites.

Gastropods such as slugs, snails and abalone are active animals with twisted bodies, single shells (or no shell), and well-developed heads.

Cephalopods

The **cephalopods**, which include nautiluses, squids, cuttlefish, and octopods, are the most active and intelligent mollusks (Figure 23-16B). Although thousands of extinct cephalopods possessed well-developed shells, among the 600 living species only the *Nautilus* possesses an external shell (Figure 23-17). Surrounding the mouth are a set of tentacles that have evolved from the foot. Each species has a characteristic number of tentacles (or arms)—a squid has 10, an octopus 8 (hence its name), and a nautilus 80 or 90 tentacles. Cephalopods, which are all

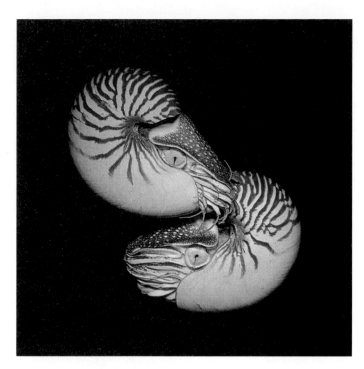

Figure 23-17 **The chambered nautilus.** The nautilus shell consists of a series of chambers arranged in a spiral. The nautilus occupies only the outermost chamber, and as the animal grows, it builds a new, larger chamber and moves in. The nautilus uses the remaining chambers to float or sink, at will. In the same way that a balloonist makes a balloon rise by adding more hot air to the balloon, the nautilus injects gas into its shell to float higher in the water. To sink, the animal siphons off the gas. Today, only a few species of nautilus exist, all living in the western Pacific Ocean. But relatives of the nautilus, called ammonites, flourished for hundreds of millions of years, from the Devonian Period to the end of the Cretaceous Period. *(Douglas Faulkner/Photo Researchers)*

Figure 23-18 A male squid caresses his mate with his many arms. *(Randy Morse/Tom Stack & Associates)*

carnivorous, use their tentacles to capture prey, to pull themselves along the sea bottom, and to steer themselves in open water.

Considering the intellectual limitations of their relatives, the bivalves and the gastropods, cephalopods are amazingly intelligent. In a laboratory aquarium, an octopus (plural, octopods) can learn to distinguish objects of different shapes and even to find its way through a maze. Octopods are very good at escaping, and they think nothing of disassembling an aquarium to do so. Presented with stones or small bricks, an octopus can build a protected enclosure, hide behind it, and bob up and down to look around. An octopus also shows much planning and patience in stalking and luring its prey. Cephalopod mating rituals are charmingly complicated (Figure 23-18).

An octopus's interesting behavior is controlled by the most sophisticated brain of any invertebrate. Information about the outside world comes from touch receptors on the tentacles, as well as from large and complex eyes. When we consider that octopus eyes evolved independently of vertebrate eyes, it is remarkable how similar the eye of an octopus is to our own (Figure 23-19). It is a classic example of convergent evolution.

Cephalopods respond to their environments not only with directed movements toward prey and away from predators, but also with a complex set of color changes. Most cephalopods have specialized skin cells, usually with three different pigments. The brain controls these cells in immediate response to the environment. An alarmed octopus, for example, may suddenly acquire dark stripes or spots, which presumably confuse potential predators. Yet another escape trick is to squirt a trailing cloud of dark ink. Some cephalopods, especially squids, can move amazingly fast by jet propulsion, squirting water from their mantle cavities.

Cephalopods, in which the foot has evolved into tentacles, are predatory, intelligent, and highly active.

What Do All Annelids Have in Common?

Earthworms and other annelids are common animals, although not nearly as diverse as the mollusks. The 12,000 species of annelids live in salt water, in fresh water, and on land, as both predators and scavengers. They range in size from less than 1 mm to over 3 m (10 ft). They may be red, pink, green, brown, or purple; they may be striped or spotted.

Annelids have long, segmented bodies consisting of identical (or nearly identical) sections called **metameres**. Segmentation (which also occurs in arthropods and vertebrates) allows animals to achieve larger sizes by repeating an already successful organizational plan. Injuries to individual metameres are less likely to be lethal, since other segments perform the same functions. In addition, the segments can move independently, which gives the animal more flexibility.

Like mollusks, annelids have specialized excretory organs called *nephridia*. Like us, annelids have a closed circulatory system, with blood contained entirely within vessels. Several vessels carry blood from one end of the body to the other, with

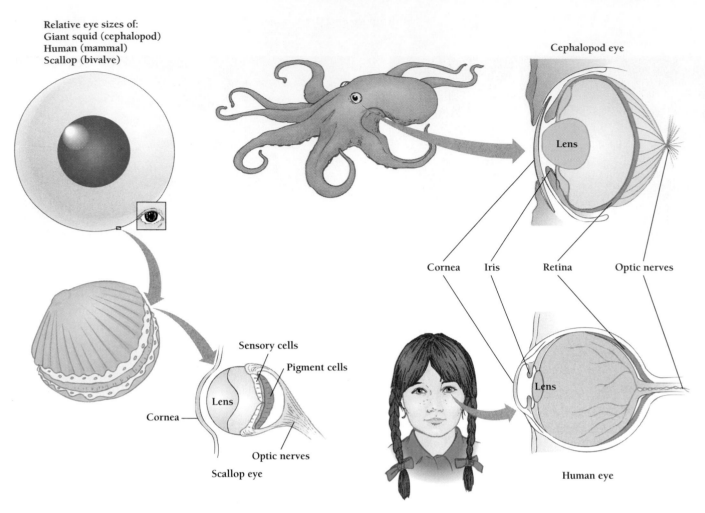

Relative eye sizes of:
Giant squid (cephalopod)
Human (mammal)
Scallop (bivalve)

Cephalopod eye

Lens

Cornea   Iris   Retina   Optic nerves

Sensory cells
Pigment cells
Lens
Cornea
Optic nerves
Scallop eye

Lens
Human eye

Figure 23-19 The cephalopod eye has long astounded biologists for its resemblance to the vertebrate eye. Cephalopod eyes can be enormous. In 1933, a squid 21 m long washed up on a New Zealand beach; its eyes measured 40 cm (16 in) across! Scallops have simple eyes that cannot focus. However, they have many of these eyes, which are probably useful for detecting the shadows of starfish and other predators.

many fine capillaries within each segment. Some of the larger vessels serve as hearts and pump the blood along its complicated circuits. Annelids have no gills or lungs, however. Gases move directly through the skin to the blood.

Zoologists divide annelids into three classes: (1) polychaetes are free-living bristle worms, most of which are marine; (2) oligochaetes, which include the earthworms and related forms, live in fresh water or salt water or on land; and (3) hirundinea, or leeches, are mainly freshwater parasites that suck blood.

Annelids have segmented bodies, consisting of identical or nearly identical sections, and a closed circulatory system.

Polychaetes

Most zoologists think that the earliest annelids were marine polychaetes (Figure 23-20). The polychaetes include many unusual and arresting species, including clamworms, plumed worms, scale worms, peacock worms, and sea mice. Each polychaete segment contains a pair of leglike paddles, called **parapodia,** which the worm uses to swim, crawl, or burrow, as well as to respire. On each parapodium is a bundle of bristles called **setae,** which give the polychaetes their name [Greek, *polukhaites* = many hairs, or bristle]. A polychaete worm's head is well developed and contains a variety of sense organs, including two to four pairs of eyes. Unlike the oligochaetes (earthworms), polychaetes have separate male and female sexes.

Polychaetes are primitive marine annelids. Each of a polychaete's segments has a pair of leglike parapodia with setae.

Oligochaetes

Oligochaetes include the common earthworms, the most familiar of the annelids. Each oligochaete segment, in contrast to that of the polychaetes, lacks parapodia altogether and contains just four pairs of bundled setae, leading to the class's name [Greek, *oligo* = few].

Figure 23-20 **Tube worm.** Many species of polychaetes—the tube worms—live in hidden places on the shallow sea bottom—under rocks, in burrows, or inside sponges, mollusks, or other animals. Some use their own secretions to build tubes where they live their entire lives. *(Joyce Burek/Animals Animals)*

Oligochaetes are masterpieces of segmentation. An earthworm, for example, consists of 100 to 175 nearly identical segments, with a few altered segments at the front and the rear. The forward segments contain specialized structures for eating, coordination, circulation, and reproduction, and the most posterior segment contains the anus (Figure 23-21).

Oligochaetes are mostly hermaphrodites and, typically, two worms will mutually cross-fertilize, with their heads pointing in opposite directions and their undersides touching. Each transfers sperm to specialized pores in its mate.

Oligochaete segments lack parapodia and have many fewer setae than polychaete segments.

Hirudinea

Because most Hirudinea, or leeches, are parasites, they are more specialized than other annelids. Their bodies are flattened rather than cylindrical, and they lack segments and setae. Many species have a sucker at each end of their bodies. Such leeches move by attaching first one sucker to a surface, drawing together the two ends of its body, and then shifting to the other sucker. Like the oligochaetes, from which leeches are thought to have evolved, leeches are hermaphrodites that cross-fertilize.

Metamere —
Setae —
Intestine —
(secretion, digestion)
Gizzard (grinding) Crop (storage) Pharynx (pump) Mouth (food intake)
Esophagus

Figure 23-21 **A single square meter of meadow soil may contain thousands of earthworms.** As an earthworm moves through soil, it swallows particles of soil. The soil passes through the digestive tract, where any organic matter is ground finely and absorbed. Undigested clay, sand, and other matter pass through the digestive tract and out the anus as small cylindrical "castings." A single worm eats its own weight in soil every day. Because earthworms are so common, they serve to keep soil mixed, which greatly benefits other soil animals and plants.

Leeches range in size from about 1 cm to as much as 30 cm. Most are carnivorous and live in fresh water. Some species—including the medicinal leech, *Hirudo medicinalis*—live on the blood of humans and other mammals. Until the 19th century, leeches were used to treat such conditions as asthma, rheumatism, and drunkenness. Physicians (or barbers, who traditionally performed leech therapy as well as cutting hair) applied leeches, who sucked up to three pints of blood. When the patient fainted from blood loss, the physician removed the leeches. Today, physicians use leeches in a more restrained manner.

A leech's impressive capacity to consume blood depends on powerful muscles of the pharynx (the anterior end of the digestive tract), on sharp teeth, and on a chemical that prevents blood clotting, called an anticoagulant. Physicians use leech anticoagulant to treat heart attack, stroke, and cancer.

Why Are Arthropods So Successful?

Arthropods, which include the insects, spiders, scorpions, shrimps, and many others, are the most successful phylum of animals on the planet. Arthropod species number at least a million, and zoologists estimate that between 2 and 30 million species remain to be described and named. The greatest number of species live in the tropics, but a student could probably collect hundreds (or even thousands) of species on any college campus.

Arthropods are more than just diverse, however. Zoologists estimate that the world today contains some 10^{18} (a billion billion) individual arthropods, hundreds of millions for every human being (a figure perhaps not shocking to any mosquito-bitten hiker). In a temperate climate, one square kilometer may contain 20 million arthropods. In the tropics, there are even more.

The enormous success of arthropods depends on both external and internal adaptations. All arthropods share two major external adaptations—an exoskeleton secreted by the cells of the epidermis and jointed limbs.

Jointed limbs, which give the arthropods their name [Greek, *arthron* = joint + *podus* = foot], allow them to move agilely, despite the rigidity of their exoskeletons. Like the annelids, arthropods have segmented bodies. Indeed, both phyla are probably descended from the same segmented ancestor. Some, like the centipedes and millipedes, have legs on nearly every segment. But many of the legs on an arthropod have become adapted for other functions. Over millions of years of evolution, legs have changed into mouthparts, antennae, gills, pincers, claws, and egg depositors.

These specialized appendages often develop from fused segments called **tagmata** [singular, tagma]. For example, the anterior segments of all arthropods are fused into a highly organized head. Other tagmata include the middle portion of an arthropod, called the **thorax,** and the rear portion, called the **abdomen.**

The **exoskeleton,** a hard external supporting framework, consists of layers of chitin and protein. It helps to protect the animals from predators and parasites and provides mechanical support. The exoskeleton also slows water loss, a feature that has allowed the arthropods to flourish away from water and dominate animal life on land since the end of the Devonian Period, more than 350 million years ago.

An exoskeleton has drawbacks, however. Because a hard exoskeleton prevents an animal from growing, many arthropods (spiders, for example) periodically must shed their exoskeletons and secrete new, larger ones in a process called **molting.** The process of molting is dangerous. After an arthro-

pod sheds its old armor and the new one is still hardening, the animal is highly vulnerable to predators. (Soft-shelled crab is a delicacy in many an East Coast restaurant.)

One solution is for an arthropod to do all its growing as a soft larva (as a caterpillar, for example) that lacks an exoskeleton. Once the animal has reached its adult size, it develops into the adult form in a process called **metamorphosis.** Some insects undergo gradual metamorphosis in stages that are not that different from one another (Figure 23-22). Others, such as flies, butterflies, beetles, and bees, undergo a complete transformation from a grub, caterpillar, or maggot to a flying insect. Metamorphosis and molting divide the life cycle of an insect into many distinct stages, each of which may have different food and lifestyle. In butterflies, for example, the caterpillar generally eats plants, while the adult butterflies drink nectar from flowers. In some arthropods, adults do not feed at all; their only role is to mate and die.

The internal anatomy of an arthropod is similar to that of an annelid. Both arthropods and annelids have a tubular gut that stretches from mouth to anus. The arthropod coelom is less prominent than that of an annelid and consists mostly of a cavity that encloses the reproductive organs.

The circulatory systems of arthropods are open. The heart pumps blood through the spaces of the body, and the blood returns through a series of one-way valves. The respiratory systems of arthropods vary according to taxonomic class and source of oxygen. Aquatic arthropods may have gills, with which they extract oxygen from the water in which they live. Spiders and some other air-breathing arthropods use stacks of modified gills, collectively called a **book lung,** to pull oxygen directly from the air.

Most terrestrial arthropods, including the insects, however, have neither lungs nor gills. Instead, they take in air through regulated openings in the body wall called **spiracles.** Air passes from the spiracles to special ducts called **tracheae.** Each trachea branches into tiny tracheoles, which deliver oxygen to individual cells throughout the body.

Reproduction in arthropods is almost always sexual. Male and female are usually separate animals, though some species are hermaphroditic. For terrestrial arthropods, fertilization must be internal, since there is no external water through which sperm can swim. Millipedes, for example, accomplish fertilization in a multilegged embrace. Elaborate courtship rituals help males and females of the same species recognize one another. Male fruit flies for example, do a little dance, during which the female decides if she wants to mate with him. In an experiment where fruit flies were kept in the dark for several generations, however, the males abandoned the dance—which the females could not see anyway.

Because arthropods that are predaceous do not always distinguish between mates and meals, copulation among predaceous arthropods can be a dangerous matter. Such animals may have especially protracted courtship rituals, which not only help them to recognize one another but also may serve to prevent one partner from eating the other.

A.   Eggs   Larva   Pupa (inside cocoon)   Adult

B.   Eggs   Young nymph   Nymph   Adult

Figure 23-22 **Complete and incomplete metamorphosis.** A. During complete metamorphosis an insect, such as this isabella moth, hatches from an egg and develops as a legless larva that looks more like a worm than a mature insect. During development a larva may
go through many successive stages, called *instars*, before settling into a usually inactive
stage called a *pupa*, or *chrysalis*. The pupa is usually encased in a protective cover, called a
cocoon. The metamorphosis from larva to pupa and pupa to adult often involves dramatic
changes in both external and internal structures. B. In incomplete metamorphosis, these
changes occur more gradually. In the harlequin bug (shown here) early stages of development, called nymphs, look much like the adult, and molting functions mostly to increase size
rather than to change form. Even in these metamorphoses, however, the juvenile forms may
lack wings.

Excretion in most terrestrial arthropods depends on unique organs called **Malpighian tubules.** These tubules absorb fluid from the blood and convert the nitrogen-containing waste molecules into insoluble crystals of uric acid or guanine. The Malpighian tubules also reabsorb and recycle salts and water, so that little water is wasted.

The nervous systems of arthropods are often complex. Many arthropods can move, run, and fly quickly, and many have complicated sexual and social behaviors. The nervous system consists of a double chain of ganglia that runs along the lower surface of the body. At the head, the chain curls upwards to form a brain—three pairs of fused ganglia. The sense organs—especially the eyes—of arthropods are more sophisticated than those of annelids. While some arthropods (such as the spiders) have simple eyes, most arthropod species have **compound eyes,** made up of numerous simple light-detecting units, called **ommatidia** [singular, ommatidium].

Arthropods include the greatest number of species of all animal phyla. External adaptations of the arthropods include the exoskeleton and jointed limbs. Internal adaptations of arthropods include circulatory, respiratory, reproductive, excretory, and nervous systems.

How Do Biologists Classify Arthropods?

The specialization of appendages, especially in the first few segments, is the basis of classification of living arthropods into three subphyla. (Extinct arthropods such as the trilobites may have their own subphyla.) In one subphylum—the **Chelicerates**—the first pair of appendages are mouthparts called chelicerae. The chelicerae serve as pincers or fangs, often associated with poison glands (Figure 23-23A).

In the other two subphyla, **Uniramia** and **Crustacea,** the first pair (or the first few pairs) of appendages are antennae, and the next pair are jaws, or mandibles (Figure 23-23B and C). These jaws differ from ours, which move up and down. Instead, an arthropod's mandibles crush and grind as they move from side to side. The Crustacea, which include water fleas, shrimps, lobsters, and crabs, are almost entirely aquatic, while the Uniramia, which include the insects, the centipedes, and the millipedes, are mostly terrestrial. The crustaceans all have branched (or biramous) appendages, while the Uniramia all have unbranched (or uniramous) appendages.

Biologists classify living arthropods into three subphyla according to the specialization of the appendages, especially the first two.

Figure 23-23 **The mouthparts of arthropods.** A. Chelicerae and pedipalps characterize chelicerates such as spiders and scorpions. B. Crustaceans such as this ghost crab have antennae, mandibles, and biramous appendages. C. The uniramia, including insects, millipedes, and centipedes, have antennae, mandibles, and uniramous appendages. *(A, John Cancalosi/Tom Stack & Associates; B, S.E. Georgia/ Animals Animals; C, Tom McHugh/Photo Researchers)*

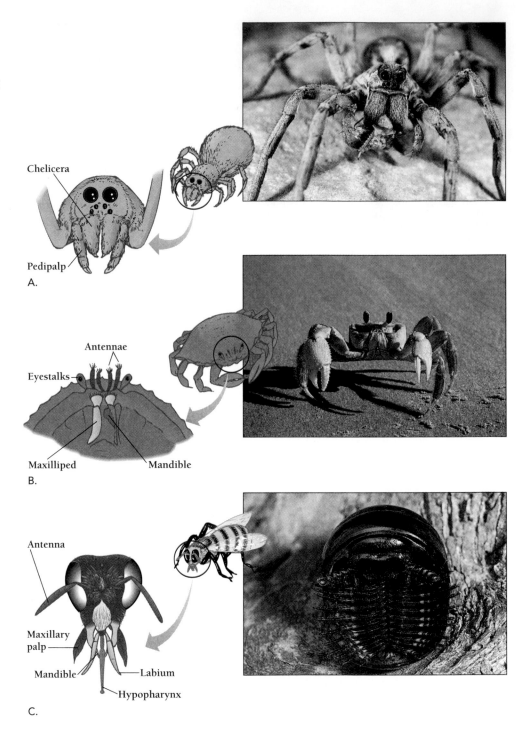

Chelicera

Pedipalp
A.

Antennae
Eyestalks
Maxilliped   Mandible
B.

Antenna
Maxillary palp
Mandible   Labium
Hypopharynx
C.

Chelicerates

The chelicerates are the only arthropods without antennae or jaws. The most anterior pair of appendages are the pincerlike chelicerae, and the second are the pedipalps, which perform different functions in the different classes (Figure 23-23A). The most familiar chelicerates are the horseshoe crabs and the spiders, of which there are some 35,000 named species. The spiders are just one of 11 orders of arachnids; arachnids are, in turn, just one of three classes of the chelicerate subphylum.

Spiders, mites, and other arachnids have six pairs of appendages (four pairs of legs, plus chelicerae and pedipalps, but no antennae). Like most other arachnids, spiders are predatory, capturing and eating insects and other small animals. Some spiders hunt and pursue their prey. Others trap their prey in elaborate silk webs. One spider genus, *Mastophora*, "fishes" for its prey with a silk line, tipped with a bit of glue. Spiders also use silk to make balloons (for making long trips), droplines (for making a quick exit), egg cases (for protecting their offspring), shrouds (for storing dead prey), and gifts from males (for luring attractive female spiders).

Spiders and other arachnids have no jaws and cannot chew. They digest only liquified food. Once a spider has trapped its prey, it injects a paralyzing poison through its chelicerae, tears and grinds a bit of its prey with chelicerae and pedipalps, then regurgitates a potion of digestive enzymes into the body of its prey, which liquifies the prey. The spider then sucks its meal into its gut, where it completes digestion.

The scorpions are the oldest order of arachnids, extending back at least to the Silurian Period, about 425 million years ago. Scorpions differ from most chelicerates in having clearly segmented abdomens. This allows the scorpion to curl its abdomen, holding its stinger aloft. Other arachnids include the distinctive daddy longlegs (or harvestmen) and the mites, of which there are some 30,000 named species and perhaps a million more as yet unnamed. Mites, including the ticks, are usually less than 1 mm long, although the largest are 2 cm long. They live almost everywhere, eating plants, fungi, and animals.

Chelicerates—such as spiders, mites, and horseshoe crabs—have no antennae and no jaws.

Crustaceans

The crustaceans include some 35,000 species of lobsters, crabs, shrimps, barnacles, and crayfish, as well as less conspicuous species such as water fleas, fairy shrimp, pillbugs, and copepods (Figure 23-24). Crustaceans usually have three pairs of chewing appendages (including the mandibles), many pairs of legs, and, unlike other arthropods, two pairs of antennae. Their legs are biramous, or split into two parts at the tip.

The most numerous crustaceans are the copepods. Indeed, next to the nematodes, the copepods are the most abundant animals on Earth. These tiny crustaceans, each a few millimeters long, are a major component of plankton, the microscopic plants and animals that float near the surface of the oceans. Copepods eat marine algae and are a major food in the diet of whales, nurse sharks, and other filter-feeding marine animals.

The largest crustaceans are the decapods, which include lobsters and crabs. Some lobsters may grow to be 60 cm long, and the Japanese spider crab is more than 3 m across. The decapods [Greek, *deka* = ten + *pous* = foot] have five pairs of legs.

The oddest of the crustaceans are the barnacles. One wit described these creatures as "nothing more than a little shrimp-like animal, standing on its head in a limestone house and kicking food in its mouth." Unlike other crustaceans, which are active, a barnacle lives its adult life in a single place, submerged in salt water and attached (by its head) to a submerged rock, piling, boat bottom, or even a whale. Barnacles use their appendages to direct passing food particles into their mouths. They secrete calcium-containing plates, which are firmly cemented to the underlying surface.

Crustaceans—such as lobsters, crabs, and shrimps—have jaws, biramous appendages, and, usually, two pair of antennae.

Figure 23-24 Krill. Small crustaceans such as these copepods are the major component of krill, the marine soup on which the great whales feed. *(William E. Ferguson)*

Uniramia

The Uniramia are the most diverse subphylum of arthropods, including 750,000 insects, 10,000 species of plant-eating millipedes, and 2500 species of carnivorous centipedes. Almost all Uniramia take in air through tracheae, use Malpighian tubules to rid themselves of nitrogenous waste, and have unbranched (Uniramous) appendages.

The myriapods [Greek, *myrioi* = countless + *pous* = foot], which include the millipedes [Latin, *mille* = thousand], the centipedes [Latin, *centum* = hundred], and two less familiar classes, are the most clearly segmented arthropods.

In insects, segmentation is clearest in the larvae, but adults are segmented as well. The body consists of three tagmata (fused segments): the head, the thorax, and the abdomen. The segments of the head are completely fused. The thorax consists of three segments, each of which carries a pair of legs, and, in many insects, one or two pairs of wings. The abdomen consists of 12 or fewer segments, none of which carries legs or wings.

Entomologists (scientists who study insects) divide the class Insecta into about 30 orders. These orders differ both in adult morphology and in their pattern of development. Some insects (for example, lice, fleas, and silverfish) have no wings, some (flies and mosquitos) have one pair of wings, some (dragonflies) have two similar pairs of wings, and some (beetles) have two pairs, each with a distinctive structure. The front wings of beetles are hard shells that serve to protect the delicate hind wings, which alone are responsible for flight.

Uniramia have one pair of antennae and one pair of mandibles.

Zoologists divided the animal kingdom into the Parazoa (mostly sponges) and the Eumetazoa (all the rest). The eumetazoans fall into three groups of phyla: the acoelomates (with no coelom), the aschelminths (with a pseudocoel or no coelom), and the coelomates (with a true coelom). We have briefly surveyed all four groups. These animals range from the simple sponges and jellyfish to animals as complex as insects or as alert as octopods. In the next chapter we will see that the deuterostome coelomates are also diverse, ranging in complexity and intelligence from sand dollars to humans.

STUDY OUTLINE WITH KEY TERMS

An animal is a multicellular, heterotrophic, eukaryotic organism that develops from an embryo. Most animals reproduce sexually. The developing embryo goes through a stage called a **blastula,** which is a hollow sphere made up of identical cells. The blastula folds into a three-layered **gastrula** surrounding the **archenteron.** The three cell layers of the gastrula are the **endoderm, ectoderm,** and **mesoderm.** These three layers develop into different types of **tissues** and **organs.** Tissues such as the **epithelial** and **connective** tissues are composed of specialized cells and their **matrix.** Organs are composed of two or more types of tissues.

Zoologists divide animals into two subkingdoms, the **Parazoa** and **Eumetazoa.** The Parazoa, which have no organs, include two phyla, the Placazoa and the Porifera. There is only one known species of Placazoa, but there are about 10,000 species of Porifera, or sponges. Sponges are usually **sessile.** Their cells lie in three layers (epidermal cells, **mesenchyme** cells, and flagellated cells) surrounding a central cavity.

Zoologists divided the eumetazoans according to the presence or absence of a **coelom** or **pseudocoel** and also according to whether they have **radial symmetry** or **bilateral symmetry.** Animals with bilateral symmetry generally have distinctive **anterior** and **posterior** ends. Animals with a coelom are called **coelomates.** Those with no coelom are called **acoelomates.** In between are the **pseudocoelomates.**

Almost all coelomate animals have digestive tracts with both a mouth and anus. The coelomates are divided into the **protostomes** and the **deuterostomes,** according to whether the embryonic **blastopore** develops into the mouth (protostomes) or the anus (deuterostomes). The majority of coelomates are protostomes, including the annelid worms, mollusks, and arthropods. However, two major phyla of coelomates are deuterostomes—the echinoderms and the chordates.

Cnidarians and **ctenophores** are radially symmetrical acoelomates. Cnidarians, which include jellyfish, hydras, corals, and sea anemones, have tissues but no organs. They have two layers of tissues enclosing a layer of **mesoglea.** They may have two body plans: **polyps** (which resemble cylinders) and **medusae** (which resemble bells). Both types have specialized stinging cells, called **cnidocytes,** which fire tiny barbs called **nematocysts.** Ctenophores (comb jellies) resemble the cnidarians, but also have distinctive features, including bands of cilia.

Platyhelminthes and **Nemertea** are bilaterally symmetrical acoelomates. Platyhelminthes (flatworms) are the simplest animals that have heads. They consist of three layers of cells. They have no circulatory systems, but their flat shapes allow gases and molecules to diffuse rapidly. There are three classes of flatworms: the **Turbellaria, Trematoda,** and **Cestoda.** The first class consists of the free-living flatworms, while the other two classes include only parasites. Parasites are specialized for rapid and prolific reproduction in their hosts. The tapeworm consists of a **scolex** and many **proglottids.** Nemertea (ribbon worms) have a two-ended digestive system, a circulatory system, and a **proboscis.**

The **Aschelminths** are a group of eight phyla, all bilaterally symmetrical animals with either no coelom or a pseudocoel. **Nematodes** (roundworms) are cylindrical and unsegmented, and have a one-way digestive system with both a mouth and an anus. The prominent pseudocoel of nematodes acts as a distribution route for nutrients, gases, and wastes; it also serves as a hydrostatic skeleton.

Mollusks (which include snails, slugs, clams, oysters, squids, and octopods) are unsegmented animals that typically have shells. Mollusks all have a **visceral mass,** a muscular **foot,** a **mantle,** a **mantle cavity,** a **radula,** and one or two kidneylike **nephridia.** Many shelled mollusks have an **operculum.** There are seven classes of mollusks, of which we discussed three: **gastropods, bivalves,** and

cephalopods. Gastropods have single shells, well-developed heads, and twisted bodies. Bivalves have two shells and most lead sedentary lives. Cephalopods have forward-pointing tentacles instead of a foot.

Annelids (which include earthworms, segmented marine worms, and leeches) have segmented bodies that consist of identical or nearly identical **metameres.** Annelids have closed circulatory systems, with blood contained entirely within vessels. Annelids are divided into three classes: polychaetes, oligochaetes, and hirundinea. Each segment of a polychaete (or bristle worm) contains a pair of paddlelike projections, called **parapodia,** each with a bundle of bristles, or **setae.** Oligochaete segments lack parapodia. Hirundinea (leeches) are more specialized than other annelids, and lack obvious segmentation.

Arthropods include the greatest number of species and the greatest numbers of individuals of all animal phyla. External adaptations of the arthropods include an **exoskeleton** and jointed limbs. Arthropods have many specialized appendages that develop from **tagmata,** fused segments that form the head, the **thorax,** and the **abdomen.** The specialization of appendages, especially in the first few segments, is the basis of classification of arthropods into three subphyla: **chelicerates, crustacea,** and **uniramia.**

Because an exoskeleton prevents an animal from growing, many arthropods periodically **molt.** Other arthropods do all their growing in larval form, then change into adults during **metamorphosis.**

Internal adaptations include circulatory, respiratory, reproductive, excretory, and nervous systems. Arthropods breathe by means of a **book lung,** or **spiracles** and **tracheae** and regulate excretion by means of **Malpighian tubules.** Some arthropods have simple eyes, but most have **compound eyes,** made up of numerous **ommatidia.**

REVIEW AND THOUGHT QUESTIONS

Review Questions

1. How is an animal different from a plant or a protist? Why are sponges grouped with animals? What animal features do sponges and cnidarians lack?
2. What criteria do zoologists use to classify the Eumetazoa?
3. What is a coelom and how is it different from a pseudocoel?
4. What specializations set parasites apart from other members of their phyla?
5. What is the difference between a protostome and a deuterostome?
6. Members of which class of mollusks would make the most interesting pets? Why?
7. Which phylum of protostomes includes the greatest number of species of any phylum of organisms?
8. What specific characteristics of a grasshopper tell you that it belongs in the class Insecta?
9. How do millipedes differ from centipedes? Which class includes more species?

Thought Questions

10. What is progress? In what sense is evolution progressive?
11. Zoologists use the following traits (among others) to classify the animals: number of embryonic cell layers, presence or absence of a true coelom, and symmetry. Do you think that such characteristics truly reflect relatedness between different groups of organisms? If you were going to classify the animals, what criteria would you consider using?

SELECTED READINGS

Crompton, John, *Ways of the Ant,* Nick Lyons Books, New York, 1954. An entertaining description of the lives of several kinds of ants. Great drama and amusing anthropomorphisms that don't go too far.

Dennett, Daniel C., *Darwin's Dangerous Idea: The Evolution and the Meanings of Life,* Simon & Schuster, New York, 1995. With the enthusiasm of a convert, Dennett, a philosopher at Tufts University, explores the implications of evolution.

Fabre, J. Henri, *The Life of the Spider,* Dodd, Mead, and Company, New York, 1919. An older classic. An entertaining and beautifully written description of the habits of several kinds of spiders. The book also contains beautiful engravings.

Gould, Stephen J., *Wonderful Life: The Burgess Shale and the Nature of History,* Norton, New York, 1989. Stephen Jay Gould rails against the idea of progress in evolution in another of his many charming books.

Holldöbler, Bert, and Wilson, Edward O., *Journey to the Ants: A Story of Scientific Exploration,* Harvard University Press, Cambridge, MA, 1994. A popular book about the lives of ants, with dramatic color illustrations.

Von Frisch, Karl, *The Dance Language and Orientation of Bees,* Harvard University Press, Cambridge, MA, 1993. A classic work in which von Frisch describes his experiments with honey bees. This book is somewhat technical, yet accessible and rewarding for a motivated lay reader.

▶ On-line materials relating to this chapter are on the World Wide Web at http://www.saunderscollege.com/lifesci/
Click on Tobin/Dusheck: *Asking About Life.*

DEUTEROSTOME ANIMALS: ECHINODERMS AND CHORDATES

Are We Upside Down?

In 1830, two French anatomists staged a public debate before the prestigious French Academy of Sciences on a topic so fundamental that historians of science have returned to it again and again. On one side was the great French anatomist and paleontologist Baron Georges Cuvier (1769-1832), who argued that all animals should be divided into four anatomical types. Cuvier said that the members of each group were anatomically similar to each other but completely unrelated to those in the other groups.

Cuvier's opponent was Étienne Geoffroy Saint-Hilaire (1772-1844), who argued that most familiar animals were of just one type, not four. Geoffroy Saint-Hilaire had maintained for years that mammals, birds, reptiles, and other vertebrates all had the same basic body plan, a stance that Cuvier and other biologists accepted.

In 1830, however, Geoffroy Saint-Hilaire took his argument one step further, proposing that all animals had the same body plan as the vertebrates. The main difference, he said, was that we vertebrates are turned upside down: where, for example, a lobster's nerve cord runs along its belly, a human's nerve cord runs along its back (Figure 24-1). As one modern biologist has put it, "If you lay down on your back and waved your arms, you would be doing what insects do when they walk."

For Geoffroy Saint-Hilaire, *unité de plan* (one plan, or singleness of plan) was a serious scheme, based on his dissections of a great number of animals. *Unité de plan* was provocative, he knew, not only because it was in direct opposition to Cuvier's theory of separate, unrelated groups. It also implied evolution. Cuvier believed in the fixity of species and had publicly ridiculed Lamarck and other proponents of evolution. According to Geoffroy Saint-Hilaire's hypothesis, however, humans and other vertebrates were a simple variation of the invertebrate body plan. His idea seemed to imply that vertebrates had evolved from invertebrates. Cuvier, never one to turn away from an argument, made *unité de plan* look like a joke. During the

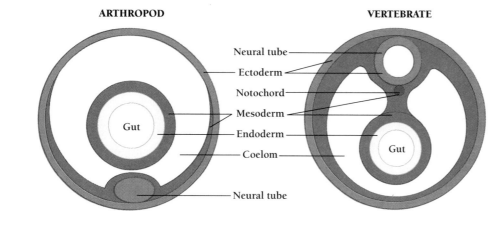

ARTHROPOD

VERTEBRATE

Neural tube

Ectoderm

Notochord

Mesoderm

Endoderm

Coelom

Gut

Gut

Neural tube

Figure 24-1 **Opposites?** The vertebrate body plan features a dorsal nerve and a ventral heart—an upside down version of the arthropod body plan, which has a ventral nerve cord and a dorsal heart.

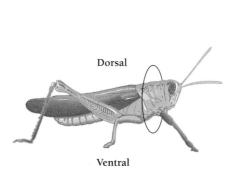

Dorsal

Ventral

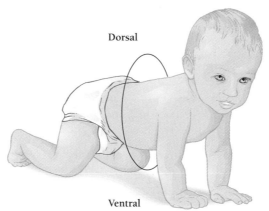

Dorsal

Ventral

debate, he listed, one by one, all of the differences between a duck and a squid.

Historians of science still argue over who won. But, regardless of who won, the reality was that Cuvier's scientific and political influence was so profound that for 164 years no biologist successfully resurrected *unité de plan* for serious discussion.

In fact, a researcher risked his or her reputation even to introduce the idea.

So the matter might well have rested had not modern biologists working on fruit flies and frogs recently begun to unravel the matrix of genes that underlie the development of the overall body plan. In the last decade, molecular biologists have

discovered a pair of genes that influence body plan in flies and a similar pair of genes in frogs.

In fruit flies, the gene *dpp* codes for a protein that somehow activates other genes necessary for the formation of **dorsal** structures—all the body parts that lie along the back of an animal. In addition, *dpp* seems to suppress the development of **ventral** structures—all the body parts that lie along the abdominal side of an animal—including nerve cells. A second fruit fly gene, *sog,* counteracts the effects of *dpp.* *Sog,* which is expressed in the ventral regions of the developing fruit fly, overrides *dpp,* allowing nerve cells and other ventral structures to develop.

site areas of frog and fly. Detlev Arendt and Katharina Nübler-Jung suggested that "the longitudinal nerve cords of insects and vertebrates derive from one and the same centralized nervous system in their common ancestor." In short, they had revived Geoffroy Saint-Hilaire's *unité de plan.*

They were not ridiculed, however. Instead, researchers in California immedi-

In a series of dramatic experiments, De Robertis and several other researchers in California and Wisconsin demonstrated that the fly genes *dpp* and *sog* work in frogs, and the frog genes *bmp-4* and *chordin* work in flies. *Sog* mRNA injected into frog embryos, for example, promoted the development of dorsal structures. Likewise, *chordin* mRNA injected into flies induced ventral development, as *sog* normally does. Further, when researchers sequenced *chordin* and *sog,* they discovered that the two genes had similar sequences.

> *"If you lay down on your back and waved your arms, you would be doing what insects do when they walk."*

When researchers looked for a gene similar to *dpp* in frogs, they found *bmp-4.* But although *bmp-4* acts much like *dpp,* it operates in the ventral region instead of the dorsal region. That is, *bmp-4* stimulates the formation of ventral structures in the frog embryo, and suppresses the formation of a nervous system.

No one made an explicit connection between the three genes, however, until September 1994, when two German researchers pointed out that *dpp* and *bmp-4* do the same thing, only acting in oppo-

ately uncovered a fourth gene. Edward De Robertis and his colleagues found that the gene *chordin* stimulates the organized development of nerve cells in the dorsal region of a frog embryo, just as *sog* does in a fly.

If Geoffroy Saint-Hilaire and his modern supporters were right, then *dpp* and *bmp-4* were essentially the same gene, just expressed in different parts of the embryo. Likewise, *sog* and *chordin* had to be the same gene, separated by half a billion years of evolution.

Has molecular biology redeemed *unité de plan?* Was Geoffroy Saint-Hilaire right, Cuvier wrong? Most biologists still hesitate to pronounce the matter settled, but nearly all are intrigued. If organisms as different as lobsters and linebackers share genes that help determine the development of the basic body plan, these organisms certainly share a common ancestor.

De Robertis, for one, is eagerly seeking more genes shared by vertebrates and invertebrates that might shed light on what that ancient ancestor was like. Until he and his colleagues succeed, however, even modern molecular biologists will hesitate a bit, reluctant to embrace an idea so long ridiculed.

KEY CONCEPTS

1. The deuterostomes consist of both invertebrates (echinoderms, arrow worms, and acorn worms) and vertebrates (fishes, amphibians, reptiles, birds, and mammals).

2. Echinoderms are radially symmetrical as adults and bilaterally symmetrical as larvae.

3. Arrow worms and acorn worms may resemble the ancestors of the chordates.

4. Chordates are deuterostomes that, at some time during their development, have a notochord, a dorsal hollow nerve cord, pharyngeal gill slits, and a postanal tail.

5. Vertebrates have a segmented spinal column and a distinct head with a skull and a brain.

6. Terrestrial vertebrates need special adaptations for life on land. These include adaptations for conserving water (kidneys and water-resistant skin), adaptations for breathing air (lungs), and adaptations for moving on land or through the air (strong legs or wings).

7. The evolution of the amniotic egg allows reptiles, birds, and mammals to reproduce on dry land.

DEUTEROSTOME COELOMATES

As we discussed in the last chapter, the early development of the deuterostome coelomates differs from that of the protostomes in several ways. In all deuterostomes, including ourselves, the tiny hole in the embryo called the **blastopore** develops into an anus instead of a mouth. The mouth develops later (or second), hence the term deuterostome [Greek, *deuteros* = second + *stoma* = mouth].

The deuterostomes consist of four phyla (Figure 24-2): **Echinodermata** ("spiny-skinned" animals such as sea stars), **Chaetognatha** (arrow worms), **Hemichordata** (acorn worms), and **Chordata**. The Chordata include the vertebrates—the fish,

Figure 24-2   The deuterostome phyla of the kingdom Animalia.

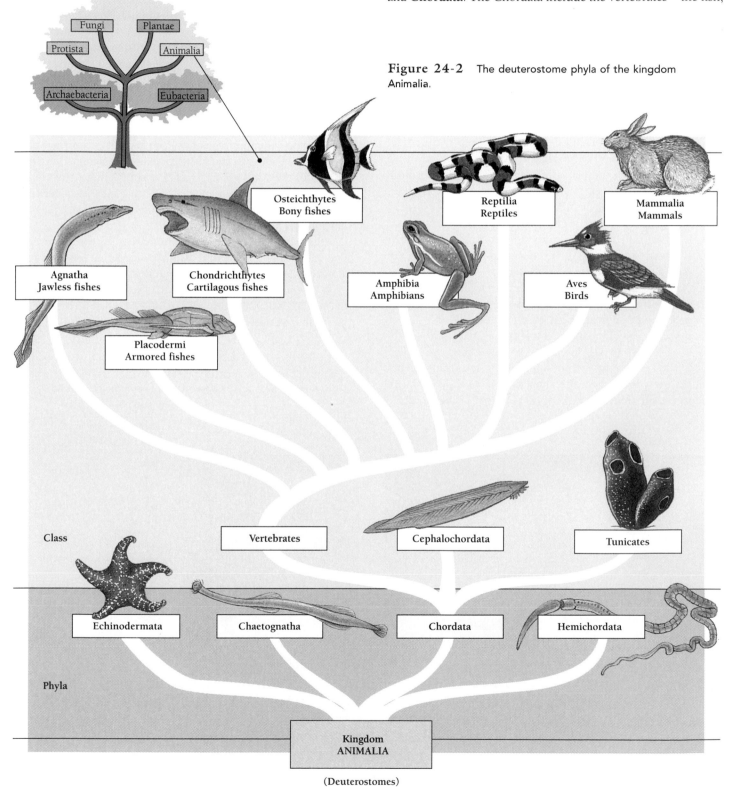

Fungi
Plantae
Protista
Animalia
Archaebacteria
Eubacteria

Osteichthytes
Bony fishes

Reptilia
Reptiles

Mammalia
Mammals

Agnatha
Jawless fishes

Chondrichthytes
Cartilagous fishes

Amphibia
Amphibians

Aves
Birds

Placodermi
Armored fishes

Class

Vertebrates

Cephalochordata

Tunicates

Echinodermata

Chaetognatha

Chordata

Hemichordata

Phyla

Kingdom
ANIMALIA

(Deuterostomes)

the amphibians, the reptiles, the birds, and the mammals—as well as the less well-known sea squirts and lancelets. While sea urchins and humans differ radically as adults, their embryos share many basic features.

Echinoderms

The 6000 species of echinoderms fall into 6 classes of marine animals. Among these are two of the most familiar inhabitants of tidepools—sea stars and sea urchins. The other four living classes of echinoderms are the sea lilies and feather stars, the brittle stars, the sea daisies, and the sea cucumbers. Another 20 classes of echinoderms are now extinct. The echinoderms are an ancient lineage. Fossil sea stars, sea cucumbers, and sea lilies from the Burgess Shale suggest that the echinoderms had diverged from the protostomes some 530 million years ago.

Echinoderms are plentiful on the bottoms of intertidal zones and deep oceans. The largest is a sea star one meter across. The smallest are just a few millimeters. Most echinoderms have calcium-rich spines that project from their skin, giving the phylum its name [Greek, *echinos* = hedgehog, a small European mammal covered with spines, like a porcupine].

As adults, all echinoderms are radially symmetrical, often with five nearly identical parts arranged around a central axis. Like other deuterostomes, however, the larvae of echinoderms have bilateral symmetry. Biologists believe that the echinoderms are descended from bilateral ancestors (Figure 24-3). Neither larvae nor adults show any sign of segmentation.

The sexes of echinoderms are separate, with two sets of gonads (testes or ovaries) in each arm. Males shed sperm and females shed eggs into the seawater, where fertilization takes place. In the open ocean, the chances that an individual egg will meet a sperm of the right species is small. Therefore, re-productive success depends on sheer numbers of gametes. A single female may produce more than 2 million eggs in a breeding season.

Sea Stars

A typical echinoderm is the sea star, or starfish. A sea star's body consists of a central region, with arms radiating out from the center. Under the center is the mouth, which opens into a digestive tract. Sea stars feed on all sorts of other animals, including sponges, corals, mollusks, crustaceans, worms, and even fish and oysters. However, a sea star prepares and eats an oyster rather differently from the way we would. A sea star's arms can bend and twist, allowing the sea star to move along the bottom, to grasp prey, and to right itself. Within the coelom is a unique set of canals, called the water vascular system, that helps some echinoderms control the movement of their arms (Figure 24-4). A primitive skeleton made of calcium-rich plates reinforces the whole body. Over this supporting structure is a delicate skin, with thousands of cells that are exquisitely sensitive to touch. Tiny, suction cup–like tube feet cover the underside of a starfish.

Starfish, and some other echinoderms, can regenerate arms that are damaged or broken away. In fact, some sea stars actually reproduce asexually: the animal breaks into two parts, and each regenerates the missing arms. Usually, however, echinoderms reproduce sexually, with males and females releasing gametes into the sea.

Echinoderms are ancient, spiny-skinned animals that are radially symmetrical as adults and bilaterally symmetrical as larvae.

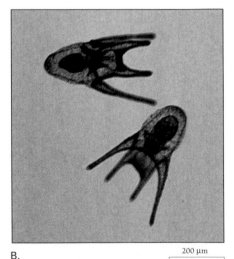

A.   B.   200 µm

Figure 24-3   A sea urchin adult and its larva. Although an adult echinoderm appears to have radial symmetry, a trait regarded by zoologists as somewhat backward, the larvae have the same bilateral symmetry as other deuterostomes.   *(A, Dave B. Fleetham Visuals Unlimited; B, Cabisco/Visuals Unlimited)*

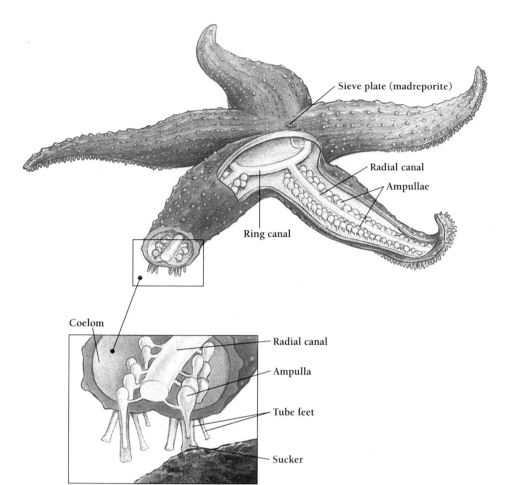

Figure 24-4   The sea star's water vascular system. Perhaps the most remarkable feature of sea stars and all other echinoderms is the water vascular system, which serves as both a circulatory system and an elaborate hydrostatic skeleton that allows complex movements of the tube feet. Water enters the system through a perforated sieve plate on the dorsal surface, which filters the water as it enters a tube. The filtered water flows down a tube into the ring canal, which surrounds the mouth (ventral) and from there into the radial canals that extend into each of the arms. In the arms, the radial canals branch to connect to long rows of tube feet. Each tube foot is also connected to a muscular fluid-filled sac, called an ampulla. In sea stars and other echinoderms, each tube foot is also attached to a sucker. The ampulla regulates fluid pressure, so that each tube foot may extend or retract, hang on or let go. Using the tube feet, a starfish can slowly pull itself along the bottom of the sea.

Arrow Worms and Acorn Worms

The 50 to 70 species of arrow worms are common marine organisms that range in length from 5 mm to 15 cm. Arrow worms feed on a variety of small marine animals, which they grasp with moveable hooks. Fossils of arrow worms date from late Cambrian times. Unlike echinoderm larvae or acorn worms, the body of an adult arrow worm extends beyond the anus, forming a tail. The only other deuterostomes with true tails are chordates, the phylum that includes the vertebrates. Because arrow worms have other properties in common with vertebrates, many evolutionary biologists think vertebrates are descended from ancestors that resembled arrow worms.

Still more similar—though not at first glance—to animals of our own phylum are the acorn worms. Most of the 65 species of acorn worms live in U-shaped burrows on the bottom of the sea. They range in length from 2.5 cm to 2.5 m (nearly 8 ft long). Zoologists classify these soft-bodied marine animals as hemichordates ("half chordates") because they resemble chordates in two respects: (1) they have **gill slits**, holes that directly connect the throat to the outside; and (2) in addition to a ventral nerve cord, they have the beginnings of a dorsal nerve cord (like our own spinal cord). As we will discuss below, both these features are fundamental characteristics of chordates.

Arrow worms and acorn worms may resemble the ancestors of the vertebrates because they possess a postanal tail. The acorn worms, in addition, have gill slits and a dorsal nerve cord like that of the chordates.

What Traits Characterize the Chordates?

The chordates include members of three subphyla. Members of two subphyla have no backbones: the tunicates and the cephalochordates. Members of the third subphylum, the vertebrates, have backbones. Members of all three subphyla exhibit four hallmark characteristics at least at some time in their lives: (1) a flexible rod running along the back (dorsal surface), called the **notochord**, which gives the phylum its name; (2) a **dorsal hollow nerve cord**, running between the notochord and the surface of the back; (3) **pharyngeal slits** (sometimes called gill slits), holes in the sides of the body that run from the inside of the intestinal tract to the outside surface of the animal; and (4) a segmented body and a postanal tail, one that extends beyond the anus (Figure 24-5).

The notochord is a long column of fluid-filled cells, with the character of a stiff, but flexible, sausage. In the tunicates

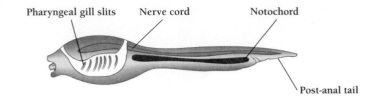
Tunicate larva (urochordate)

Pharyngeal gill slits   Nerve cord   Notochord

Post-anal tail

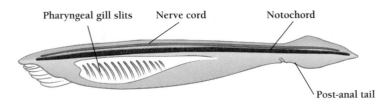
Amphioxus (cephalochordate)

Pharyngeal gill slits   Nerve cord   Notochord

Post-anal tail

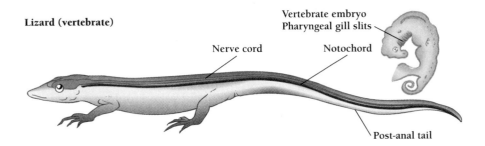
Lizard (vertebrate)

Vertebrate embryo
Pharyngeal gill slits

Nerve cord   Notochord

Post-anal tail

Figure 24-5   The four features of chordates.
The three chordate subphyla—the tunicates (sea squirts), the cephalochordates (amphioxus), and the vertebrates—all share four distinguishing features. These are pharyngeal gill slits, a dorsal nerve cord, a notochord, and a postanal tail.

and cephalochordates, as well as in some fish and larval amphibians, the notochord provides back support for the adult animal. In many vertebrates, however, the notochord disappears during embryonic development to be replaced by a bony or cartilaginous backbone. Cartilaginous fish have both a notochord and a surrounding vertebral column, while reptiles, birds, and mammals have only a vertebral column.

The dorsal nerve cord of embryos and its adult counterparts (the spinal cord and the brain) are partly hollow, and their inner spaces contain fluid. In contrast, the ventral nerve cords of annelids and arthropods are solid. The pharyngeal slits are present in the embryos of all chordates (including humans), but not in air-breathing adults. In early chordates, the slits may have served to allow filter feeding (as they do in tunicates), but fish, tadpoles, and other aquatic chordates use them primarily for respiration.

All chordates have a notochord, a dorsal hollow nerve cord, pharyngeal slits, and a postanal tail at some time in their lives.

What Have We in Common with Tunicates?

Adult tunicates do not look at all like other chordates (Figure 24-6). They are sessile marine animals that attach to rocks, boats, or the ocean floor. Their name derives from a tough outer coating, or tunic, made, amazingly, of cellulose. (Cellulose is common in the cell walls of plants, but rare in animals.) While

adult tunicates have pharyngeal gill slits, they have little else to recommend them as chordates. A tunicate larva, however, has a notochord, a dorsal hollow nerve cord, pharyngeal gill slits, and a postanal tail. It seems likely that the common ancestor of the chordates resembled a tunicate larva.

As larvae, tunicates have a notochord, a dorsal hollow nerve cord, pharyngeal slits, and a postanal tail, all characteristics of chordates.

What Have We in Common with Cephalochordates?

Cephalochordates [Greek, *cephalo* = head], or lancelets, include about 45 species of small, segmented, fishlike animals. Most belong to the single genus *Branchiostoma,* whose common name is amphioxus. In these animals, the notochord, which is present in adults as well as in embryos, extends all the way through the head, hence their name. Sets of chevron-shaped (<<<) muscles pull against this internal skeleton. The segmentation of a lancelet is superficially similar to that of annelid worms and achieves the same advantages. Cephalochordates have all four chordate characteristics.

Cephalochordates have notochords both as larvae and as adults.

Figure 24-6 Our chordate cousin the tunicate. To the average person, a tunicate might look more like a sponge than a close relative of the vertebrates. Like sponges, adult tunicates feed by filter feeding with a U-shaped digestive tract. Cilia in the pharynx pull water through an incurrent siphon into the mouth, past the pharyngeal slits, and out through an excurrent siphon. Sticky mucus traps food particles and carries them down into the digestive tract. The anus empties into the excurrent siphon. *(Dave Fleetham/Tom Stack & Associates)*

What Distinguishes the Vertebrates from Other Chordates?

Vertebrates have many features that distinguish them from other chordates. Two of the most important are (1) a segmented vertebral column made of units called vertebrae and (2) a distinct head, with a cranium (or skull) and a brain. Arches of the vertebrae encircle and protect the dorsal nerve cord, which becomes the adult vertebrate's spinal cord (Figure 24-7).

Vertebrates also have closed circulatory systems and characteristic internal organs—livers, kidneys, and hormone-secreting (endocrine) organs. The heart pumps blood through the circulatory system, through a system of fine capillaries that helps transport gases and nutrients to nearly every cell of the body. The same system carries metabolic wastes away from cells. Liver, kidneys, and other organs closely regulate the chemical composition of the blood, ensuring a nearly constant internal environment for all cells. (Part VII of this book deals with these organs and their operation in some detail.)

Most modern vertebrates have bony skeletons. **Bone** consists of fibers of the protein collagen, embedded primarily with crystals of calcium phosphate. Within this extracellular material are bone-making cells, as well as blood vessels and nerves. Bone is strong enough to support huge vertebrate bodies. Un-

like the exoskeletons of arthropods, bone can grow, allowing vertebrates to reach sizes far greater than those of the arthropods.

The embryos of all vertebrates have skeletons made of **cartilage,** also built from collagen, but softer and more elastic than bone, without calcium phosphate. Some parts of the adult skeleton retain cartilage, such as our ears and noses. But, except in the sharks and other cartilaginous fishes, bone mostly replaces cartilage in adults.

Vertebrate bone is so distinctive and so well preserved in the fossil record that biologists know more about the evolution of the vertebrates than about any other group. In this chapter, we survey eight classes of vertebrates. Four of these are groups of fishes and four are four-footed animals, or **tetrapods** [Greek, *tetra* = four + *pous* = foot]. The four classes of fishes include Agnatha (the jawless fishes), Placodermi (a class of extinct armored fishes), Chondrichthyes (fishes with skeletons of cartilage rather than bone), and Osteichthyes (bony fish). The tetrapods consist of the Amphibia (amphibians), Reptilia (reptiles), Aves (birds), and Mammalia (mammals).

Vertebrates have vertebral columns, skulls, and a well-developed brain. Most vertebrate skeletons are made of bone.

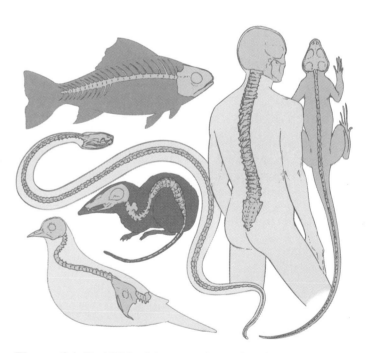

Figure 24-7 Which of these vertebrates has the strongest back? All vertebrates share a vertebral column composed of many vertebrae, as well as a thick skull that protects the brain. In most vertebrates, the vertebral column, skull, and skeleton consist largely of bone. In a few, such as the sharks and rays, these elements are flexible cartilage instead. The hero shrew (*center*), of Africa, is reputed to have the strongest backbone of any vertebrate. According to legend, an adult man can stand on this tiny insectivore's back without injuring the animal.

Agnatha Have No Jaws

Most jawless fishes are extinct. These extinct fishes, comprising some 60 families, are called **ostracoderms** [Greek, *ostrakon* = shell + *derma* = skin] because their skin contained a shell-like skeleton of bony plates. Ostracoderm fossils have been found in Cambrian deposits, showing that vertebrates have been around as long as the invertebrates (Figure 24-8). Although, traditionally, the agnathes have been lumped together, ongoing research suggests that they represent several separate lines (or clades).

The living jawless fishes include only about 60 species of cyclostomes [Greek, *kyklos* = circle + *stoma* = mouth], soft-bodied fish with round, jawless mouths. The cyclostomes, none of which have bony plates, include the lampreys and the hagfish. Lampreys and hagfish are hardly glamorous animals. Hagfish survive by eating polychaete worms and pieces of dead fish, scavenged from the muddy sea bottom. Lampreys are nearly all parasitic. A lamprey lives by attaching itself to another fish with its circular jawless mouth, then tearing a hole in its living victim with its rasping tongue. The lamprey then sucks whatever fluids flow from the wound.

A lamprey larva, like a lancelet, has a notochord and segmented muscle groups. It differs from a lancelet, however, in possessing a brain, a liver, and a kidney. Most zoologists think that the first vertebrate resembled a lamprey larva.

Lampreys and hagfish (the living members of the Agnatha) are the only vertebrates that lack jaws.

Placoderms, Now Extinct, Were Armored Fish with Jaws

At the beginning of the Devonian Period, some 405 million years ago, freshwater fish developed the first vertebrate jaws. These fish rapidly spread into the oceans, diversified, and multiplied. Paleontologists sometimes call the Devonian the "Age of Fishes." Among the prominent jawed fishes in Devonian seas were a class of armored fishes called **placoderms** [Greek, *plak* = plate + *derma* = skin]. The placoderms had jaws, paired fins, and much less armor than the ostracoderms.

Today, all vertebrates except the Agnatha have hinged jaws, with which they can seize, bite, and sometimes chew. Jaws permit a much more diverse diet than that of lampreys or hagfish. Jaws probably enabled placoderms and other jawed fishes to replace the ancient ostracoderms. By the end of the Devonian Period, armored placoderms had disappeared, replaced by the Chondrichthyes and Osteichthyes. These two classes of fishes still dominate the aquatic regions of the Earth's surface.

Placoderms were the first vertebrates with jaws.

Chondrichthyes Are Fish with Cartilaginous Skeletons

In most vertebrates, the soft cartilaginous skeleton of embryos is gradually replaced with bone. In sharks, skates, and rays, however, bone does not replace cartilage during development. Adult Chondrichthyes have lightweight cartilaginous skeletons, giving the class its name [Greek, *chondros* = cartilage + *ichthys* = fish]. Chondrichthyes, with only about 700 species, are much less diverse than the 30,000 species of bony fish (Osteichthyes).

Chondrichthyes also lack a swim bladder, a balloonlike organ that helps the bony fish float. Instead, sharks and their allies increase their buoyancy by storing large amounts of oil in their huge livers. Nonetheless, the Chondrichthyes are heavier than water and sink to the bottom when they are not moving (Figure 24-9).

Although sharks and rays lack the bony scales of the placoderms and ostracoderms, their skin is covered with small toothlike scales (denticles) that give it the feel of sandpaper. In addition, they have teeth (derived from their scales), which they use, along with their jaws and their ability to swim rapidly, to live as successful predators (Figure 24-9).

A.

Figure 24-8   The jawless fishes. A. Extinct ostracoderms. B. A modern lamprey. These fish have neither working jaws nor paired fins. (B, Berthoule-Scott/Jacana/Photo Researchers)

B.

A.   B.

Figure 24-9 The Chondrichthyes. A. Rays and skates have evolved for a life on the ocean floor, where they feed mostly on mollusks and crustaceans. Their pectoral fins are enormous, giving these fish their characteristic appearance. B. Sharks are mostly fast-moving predators with streamlined bodies, tapered at both ends. Lacking a buoyant swim bladder, a shark keeps from sinking by swimming constantly. (Some sharks take breaks on the sea bottom or in a sea cave.) Paired fins (pectoral and pelvic) on the underside help keep the moving shark afloat.   *(A, Jeff Rotman/Tony Stone Images; B, Doug Perrine/DRK Photo)*

Sharks have particularly well-developed senses that help them find their prey. Among these senses are keen vision and smell. The lateral line system, a row of tiny sense organs along each side of the body, detects changes in water pressure. Other receptors, originally derived from the lateral line system, detect tiny electrical currents, including those generated by muscular contractions of potential prey.

The Chondrichthyes have cartilaginous skeletons and no swim bladder.

Osteichthyes Are Fish with Bony Skeletons

The bony fish, Osteichthyes [Greek, *osteon* = bone + *ichthys* = fish], are the most diverse of all the vertebrate classes. More than 24,000 species fill oceans and seas, lakes and streams all over the world. They range in size (as adults) from about 1 cm to more than 6 m. The bony fish comprise more than 95 percent of all fish and half of all species of vertebrates.

Like the cartilaginous fishes, the bony fishes first appeared in the Devonian Period, but in fresh water rather than the oceans. Bony fish differ from the Chondrichthyes in having bony skeletons and thin, bony scales. Most bony fish also have a **swim bladder,** an air-filled sac that helps them control their buoyancy (Figure 24-10).

Zoologists divided the living Osteichthyes into two subclasses: the ray-finned fishes, Actinopterygii [Greek, *aktin* = ray + *pterygon* = wing], and the lobe-finned fishes, Sarcopterygii [Greek, *sarkodes* = fleshy]. Nearly all modern fishes are ray-finned: they have paired fins supported by thin, bony rays, originally derived from bony scales. Ray-finned fishes are some of the best swimmers in the world (Figure 24-11A and B). They are extraordinarily diverse and live virtually wherever there is water.

Unlike the fins of ray-finned fishes, the fins of lobe-finned fishes contain muscle as well as bone. Although once abundant, lobe-finned fishes are now unusual, including only four genera of modern fishes (Figure 24-11C). The lungfishes can use their sturdy fins to walk on land, and biologists believe that the first amphibians must have resembled lobe-finned fishes whose fins gradually evolved into limbs. Modern lobe-finned fishes, such as lungfish, live in stagnant freshwater ponds, using their lungs to extract oxygen they get by gulping air from the surface. Lungfishes can radically alter their metabolism to survive long periods of drought by burying themselves in the mud at the bottom of a pond.

Osteichthyes have bony skeletons, swim bladders, and paired fins.

VERTEBRATES NEED SPECIAL ADAPTATIONS FOR LIFE ON LAND

By the end of the Devonian Period, fish had appeared in every conceivable aquatic niche in seas, streams, and ponds. Except for a few lungfish adventurers, however, vertebrates had not found a way to live on the land. Plants and insects had already done so, however, and the land stood all-inviting, with great, lush forests and a diversity of tasty insects and other arthro-

Figure 24-10   The body plan of a bony fish. Fish share many features with terrestrial vertebrates, including a brain, a liver, and true kidneys. In addition, a bony fish has a gas-filled swim bladder from which evolved the terrestrial lung. By regulating gases in the swim bladder, a fish can adjust its buoyancy, allowing it to float at the level it wants without exertion. Bony fishes also possess a bony operculum over the gills that allows them to pump water across the gills, facilitating the exchange of oxygen and carbon dioxide.

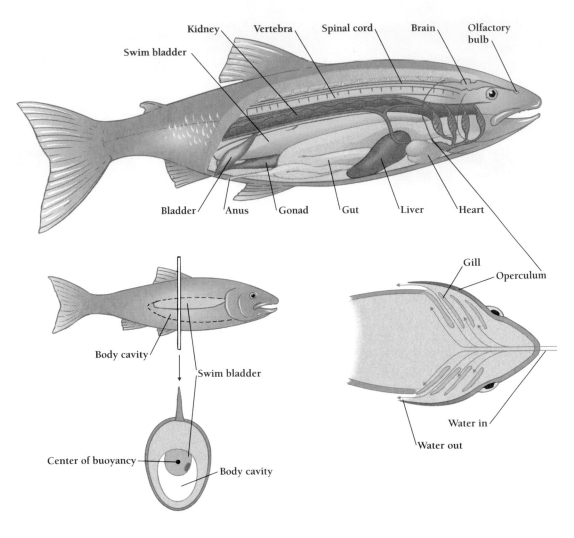

A.

B.

C.

Figure 24-11   The two subclasses of bony fishes: the ray-finned fishes and the lobe-finned fishes. A. The fast-swimming barracuda typifies the highly evolved ray-finned fishes. B. A pike can achieve accelerations of 12 to 24 g (times gravity), reaching top speeds of 6 meters per second (equal to a 4-minute mile). To find out how fish are able to swim so fast, researchers at MIT have built a robotic pike (*Esox lucius*), named "Robopike," which swims by itself. C. The 350-million-year-old marine coelacanth *Latimeria* is a lobe-finned fish that was thought to be extinct—until a fisherman pulled one from the sea in 1938.   *(A, Rondi/Tami Church/Photo Researchers; B, © Sam Ogden; C, Sally Bensusen/Science Photo Library/Photo Researchers)*

pods. The terrestrial environment offered good meals and few predators.

The movement from water to land was a dramatic event, accompanied by immense changes in vertebrate anatomy. To colonize the land, vertebrates had to solve several problems:

1. They needed to obtain and conserve water in a comparatively dry environment.
2. They needed to extract oxygen from air rather than water.
3. They needed strong skeletons to support their bodies—they could no longer depend for support on the buoyancy of water.
4. They needed to control fluctuations in body temperature, which can be much greater in air than in water.
5. They still needed a watery environment for fertilization and early development.

The terrestrial vertebrates—amphibians, reptiles, birds, and mammals—evolved adaptations to solve all of these and many other problems of land life.

Amphibians Are Terrestrial Animals That Begin Their Lives in Water

During the Devonian Period the freshwater lakes and streams dried up and flooded regularly, leaving many aquatic organisms stranded in foul ponds and muddy stream beds. Only fish able to extract oxygen from the air, with a primitive lung, were able to survive. Indeed, the Devonian Period saw the development of the two most important terrestrial adaptations: lungs and limbs.

Nearly all the freshwater fishes that survived the Devonian Period had a kind of lung, a simple sac enhanced with a rich supply of blood. The Devonian fish developed **double circulation**, in which blood circulates between the heart and lungs and between the heart and the rest of the body. All terrestrial vertebrates have double circulation.

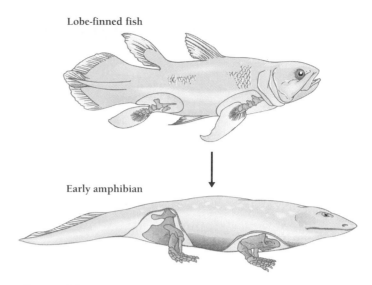

Lobe-finned fish

Early amphibian

Figure 24-12 The first amphibians are believed to have evolved from lobe-finned fishes at the end of the Devonian Period.

The Devonian Period freshwater vertebrates also developed legs, using their well-developed fins to actually walk across the bottoms of pools and possibly across land from one deep pool to another. The early amphibians were the first vertebrates to live most of their lives on land. The reptiles, birds, and mammals are all descendants of the earliest amphibians. The first amphibians somewhat resembled modern salamanders and lobe-finned fishes, and dominated the land for about 100 million years, until the rise of the reptiles in the Triassic (Figure 24-12). Some amphibians grew to lengths of three or four meters and were more terrestrial than modern amphibians. Modern salamanders, for example, have evolved several adaptations for moving about in shallow water, including limbs, strong tails, and a flat shape, that earlier amphibians did not possess.

Today, three orders of Amphibia exist (Figure 24-13): Urodela (salamanders), Anura (frogs and toads), and Gymnophiona (caecilians). The anurans are by far the most diverse amphibian order, including about 90 percent of the 3000

A.

B.

C.

Figure 24-13 Three orders of amphibians. A. Jordan's salamander, from North Carolina. B. A river frog from Florida. Frogs differ from toads in the quality of their skins: frog skin is smooth and moist, while toad skin is bumpy and dry. C. Gray caecilian. The caecilians are legless amphibians that live in the tropics. Like earthworms, they move by means of a hydrostatic skeleton—even though they have a backbone. (A, David M. Dennis/Tom Stack & Associates; B, Joe McDonald/DRK Photo; C, Michael Fogden/DRK Photo)

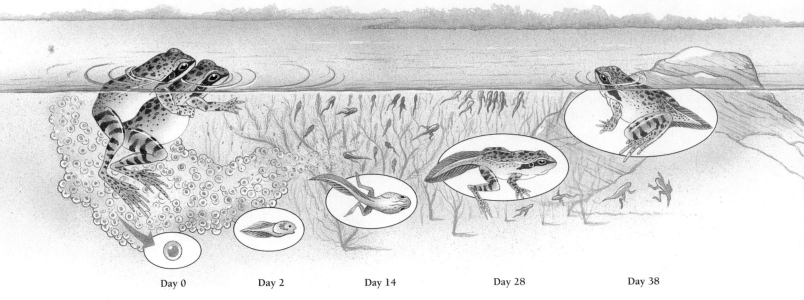

Day 0   Day 2   Day 14   Day 28   Day 38

Figure 24-14 **The development of an amphibian.** Most amphibians lay their eggs in water. The eggs develop into tadpoles, which eat and grow underwater. When the time is right, the tadpoles metamorphose into adult amphibians.

amphibian species. Anurans lack tails; urodeles have them. The names Anura [Greek, *an* = without + *oura* = tail] and Urodela [Greek, *oura* = tail + *delos* = visible] describe this simple distinction. The Gymnophiona [Greek, *gymnos* = naked + *ophioneos* = snakelike], which lack feet and resemble snakes or earthworms, are the least familiar amphibians. They are tropical amphibians that live underground and live like moles, pushing through the soil and searching for small invertebrates to eat. About 160 species are known.

Adult toads and frogs have powerful leg muscles for jumping and well-developed ears and vocal cords for communicating through air. They have extendible, sticky tongues with which they can expertly capture flying morsels. Amphibians can gulp air into their lungs, but they also absorb oxygen through their skins, which are moist and thin. In fact, some salamanders breathe entirely through their skins and the linings of their mouths.

An amphibian's life cycle generally consists of two stages, the first of which is aquatic and the second terrestrial—hence their name [Greek, *amphi* = double + *bios* = life]. Amphibians are only partly adapted to life on land. Their need for water dominates their lives. Unlike us, however, amphibians do not drink water. Instead, they must absorb it through their thin, damp skins. Such skin is not good at retaining water, and amphibians must confine themselves to very moist environments or dry out.

Most amphibians start life in water, when the male and female shed sperm and eggs together (Figure 24-14). Because amphibian eggs lack shells and special membranes to prevent water loss, their embryos almost always must develop in a very wet environment.

The fertilized egg (zygote) develops into an aquatic larva, or tadpole, which obtains oxygen through gills and uses its long tail to swim. Later, the tadpole becomes a terrestrial animal by losing its gills and developing lungs and limbs, in a process called **metamorphosis** (Figure 24-14). If it is an anuran, it also loses its tail. The adult amphibian is now capable of surviving on land.

A few species of frogs, salamanders, and caecilians have evolved direct development. The zygote develops directly into the terrestrial form, bypassing the larval form. Other species avoid the need for free-standing water for their larvae in other ways. They lay their eggs in moist places on land, in pockets of water in leaves, or in wet mossy places. Several species carry the developing eggs in special pouches on their backs or in the mouth. The South American frog (*Rhinoderma*), for example, lays its eggs on moist ground, after which groups of adults guard the developing embryos. When an embryo begins to move, a frog quickly takes it into its mouth (Figure 24-15).

Figure 24-15 **Some amphibians care for their young.** Here the reticulated poison dart frog carries two tadpoles to water. *(Michael Fogden/DRK Photo)*

BOX 24-1

Why are amphibians disappearing?

Costa Rica's spectacular golden toad (*Bufo periglenes*) has not been seen since 1989, Australia's gastric brooding frog (*Rheobatrachus silus*) was last seen in 1979, populations of the common toad (*Bufo bufo*) have declined on Norwegian coastal islands, and many species of frogs in California have vanished from most of their historic ranges (Figure A). Throughout the world, amphibian species are declining in number and, in some cases, going extinct locally.

Is some unknown phenomenon exterminating toads and frogs?

The question first arose at the First International Herpetological Congress, held in 1989 in Canterbury, England. At that meeting, herpetologists recounted what each thought was a personal sorrow: the disappearance of an amphibian species from a study site. So many scientists reported the same kinds of losses, however, that David Wake, a leading herpetologist in the United States, organized a 1990 meeting in Irvine, California, to discuss whether there was a global decline in amphibian populations. At that meeting and at a follow-up symposium, the herpetological community shared information about declines in amphibian populations in Australia, Canada, the western United States, the southeastern United States, Central America, the Amazon Basin, and the Andes.

For most amphibian populations, scientists do not have long-term data on population cycles. Without information about natural population cycles, it is impossible to determine whether, for any given species, a dip in numbers is natural variation or a decrease induced by human activities. Many frog species have boom-and-bust population cycles—populations explode one year and drop to almost nothing in another year. Such natural variation makes it difficult to interpret declines in some species.

Nonetheless, herpetologists agree that an ongoing, worldwide decrease in amphibian populations is occurring. They further agree that the primary cause is the fragmentation and destruction of habitat, as is unfortunately true for many other organisms.

Herpetologists do not yet agree what other reasons might account for the worldwide disappearance of amphibians. Some researchers say amphibian declines are just

Figure A   Golden toads. *(Michael Fogden/DRK Photo)*

part of the worldwide loss of biodiversity caused by destruction of forests, wetlands, and other habitats. The disappearance of amphibian species, they argue, is no different from the extinctions of thousands of other species.

Other researchers insist that amphibians—with their permeable skin, permeable eggs, and complex life cycles—are particularly vulnerable to certain environmental changes. In particular, amphibians seem to be especially sensitive to increases in UV radiation and chemical pollutants. Decreases in numbers of the North American tiger salamander (*Ambystoma tigrinum*) have been linked to pollution, especially acid rain. The Tarahumara frog (*Rana tarahumarae*) is extinct in the United States, and researchers have implicated high levels of the toxic metal cadmium. The massive dieoffs of fertilized eggs of the Cascades frog (*Rana cascadae*) and the Western toad (*Bufo boreas*) may be due to increases in UV radiation stemming from the destruction of ozone in the stratosphere; the eggs of both of these latter species are low in *photolyase,* an enzyme that helps repair UV-damaged DNA.

A host of nonnative predators have been accused of decimating amphibian populations. In the western United States, bullfrogs (*Rana catesbeiana*), nonnative trout, bass, and sunfish, all predators of amphibians, have been introduced into countless waterways. Losses of the Yavapai leopard frog (*Rana yavapaiensis*), the mountain yellow-legged frog (*Rana muscosa*), and the Chiricahua leopard frog (*Rana chiricahuaensis*) have all been attributed to such introduced predators.

Some researchers argue that an accumulation of environmental stresses, rather than a single cause, may reduce or destroy a population. For example, a population of frogs may become fragmented and reduced by the loss of wetland habitat. If a subsequent drought kills many more individuals, the population may never recover. Or stress from increased UV exposure and pollution may make individuals more susceptible to disease—one explanation for the extinction of 11 distinct populations of boreal toads (*Bufo boreas*) in western Colorado.

Frogs are showing signs of going the way of the dinosaurs. Species in habitats as different as Costa Rican cloud forest, Sonoran desert, and montane pools in the Sierra Nevada are succumbing. Unfortunately, beyond the obvious effects of habitat destruction on all species worldwide—destruction that proceeds as rapidly as ever—there is probably no one-size-fits-all explanation for the loss of amphibian species.

When the embryo finishes developing, the frog yawns, and a tiny froglet jumps out of his mouth.

A few salamander species do not complete metamorphosis. The mudpuppy and the axolotl, for example, retain their gills and continue to live exclusively in water even after they are capable of reproducing. The axolotl (*Amystoma*) can mature if conditions are right into an adult form with lungs instead of gills. But the mudpuppy (*Necturus*) never loses it gills or develops lungs.

Amphibians live a dual life. The larvae are usually aquatic and the adults are usually terrestrial.

Reptiles Can Live Their Entire Lives on Land

Reptiles [Latin, *reptare* = to crawl], unlike amphibians, are entirely adapted to life on land. Although some reptiles, such as the crocodiles and sea turtles, are adapted to aquatic life, they are descended from terrestrial reptiles and must breath air. They do not have gills.

Most of the 6000 species of living reptiles fall into three orders (Figure 24-16A–C): Chelonia (turtles and tortoises), Crocodilia (crocodiles and alligators), and Squamata (lizards and snakes). Of these, the Squamata are by far the most successful, with about 3300 species of lizards and 2300 species of

snakes. A suborder of the squamata are the Amphisbaenia, which include some 140 species of legless reptiles that resemble the amphibian caecelians. Like moles, the amphisbaenians have lost the use of their eyes and burrow through soil hunting invertebrates (Figure 24-16D). A fourth order of living reptiles includes only a single species, the tuatara, a lizardlike animal that survives only on a few islands off the New Zealand coast (Figure 24-16E).

Some of the most fascinating reptiles are now extinct. Although many of us think of dinosaurs when we think of extinct reptiles, only a fraction of the animals living in the Mesozoic Era, the Age of Reptiles, were actually dinosaurs. Others were swimming ichthyosaurs, and some were flying pterosaurs, for example (Figure 24-17). The long and glorious history of the reptiles began some 300 million years ago in the late Carboniferous Period. During the Permian and Triassic Periods the reptiles radiated into about 17 orders.

Reptiles were the first vertebrates to free themselves of the water. As a result, reptiles share with their descendants, the birds and mammals, some common adaptations to life on land. Their dry, watertight skin helps conserve water, serving the same function as the cuticles of insects and plants and our own dry skin. Because reptiles cannot breathe through such tough skins, they must depend entirely on their lungs for oxygen.

Reptiles and their descendants have also evolved more efficient kidneys for conserving water and legs better suited to walking on land. Among reptiles' many adaptations to land life, however, the most dramatic advance was the evolution of the

A.

B.

C.

Figure 24-16 The five orders of living reptiles. A. Galapagos tortoise. The Chelonia comprise the turtles and tortoises. B. Nile crocodile washing and releasing baby. The Crocodilia comprise the crocodiles, alligators, and caimans. C. Crested dragon lizard, Malaysia. The Squamata comprise the lizards and snakes. D. Slow worm, England. The Amphisbaenia comprise the ground-dwelling lizard worms. E. The lone species in the order Sphenodonta is the tuatara of New Zealand. *(A, Frans Lanting, Photo Researchers; B, Roger de la Harpe/Animals Animals; C, Mike Bacon/Tom Stack & Associates; D, Stephen Dalton/Photo Researchers; E, John Cancalosi/DRK Photo)*

D.

E.

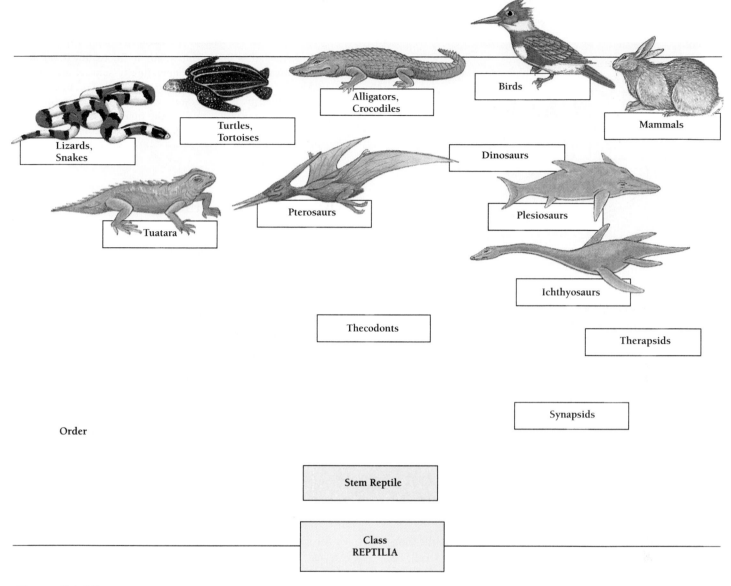

Figure 24-17 **Reptiles all.** The familiar lizards, snakes, turtles, and crocodiles are not the only animals in the reptile family album.

amniotic egg (Figure 24-18). The amniotic egg is so well adapted to a dry environment that even crocodiles and turtles, which live primarily in the water, return to the land to lay their eggs. The **amnion** is a membrane that encloses the developing embryo in its own little pond, eliminating the need for a separate aquatic stage. The amnion, in turn, lies within a porous shell that allows the exchange of gases with the surrounding air. Just beneath the shell is another membrane, called the **chorion** [Greek, = skin]. Also within the shell is the **yolk**, a rich food supply surrounded by a membrane called the **yolk sac.** A fourth membrane, the **allantois,** functions in both respiration and excretion.

All reptiles and their descendants have (or had) an amniotic egg. These include all birds and mammals. The amniotic egg is one of the characteristics that distinguishes all of these groups from other animals and shows that we are related.

In both reptiles and birds, a shell forms around the amniotic egg before it emerges from the female, and so fertilization

must occur before the shell forms. That is, fertilization must be internal. All male reptiles except the tuatara accomplish fertilization with a penislike copulatory organ. Crocodilians and turtles have a single penis, while snakes and lizards have a pair of "hemipenes." In all reptiles, both sexes possess a **cloaca,** a common entrance and exit chamber for the digestive, urinary, and reproductive systems. In the female, the cloaca serves as an entrance for sperm from the male and an exit for the fertilized eggs. The eggs are usually laid in soil, sand, or leaf mold, although a few reptiles give birth to live young. In either case, the zygote develops into a tiny young reptile that fully resembles its parents.

The evolution of the amniotic egg enabled reptiles to dominate the Earth for millions of years. Reptiles' descendants, the birds and the mammals, also have amniotic eggs.

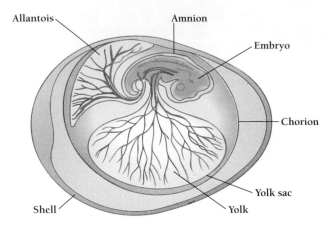

Figure 24-18 **The amniotic egg.** The most obvious feature of an egg is the yolk sac, which serves as a pantry, a storage area for fats, proteins, and other nutrients for the developing animal. But fish eggs and amphibian eggs also have a yolk sac. What distinguishes the amniotic egg are three other membranes: the amnion encloses the embryo in water; the allantois serves as a sort of outhouse, a separate area for wastes; and the chorion, the outermost membrane, encloses embryo, yolk, and allantois. The chorion helps regulate the exchange of oxygen and carbon dioxide between the embryo inside and the world outside.

How Do Terrestrial Vertebrates Regulate Body Temperature?

As we discussed in Part I of this book, life depends on the thousands of biochemical reactions that take place in every cell. Because the rate of every one of these reactions depends on temperature, the temperature of an organism is crucial to the way it lives. In aquatic animals, the temperature of the water around them usually establishes a fairly constant body temperature.

On land, however, air and ground temperatures fluctuate enormously. Animals that have left the water must therefore find a way to regulate temperature. Most reptiles do so by moving into and out of the sun. Because reptiles (as well as amphibians) depend on external sources of heat, they are called **ectotherms** [Greek, *ektos* = outside + *thermos* = heat].

In contrast, mammals and birds warm their bodies by capturing the heat released by metabolism and are called **endotherms** [Greek, *endo* = within]. Endotherms keep nearly constant body temperatures by regulating both the heat produced by biochemical reactions and the heat lost by evaporation. Notice that we have avoided calling reptiles "cold-blooded." They are not; they merely get their heat in a different way from endotherms.

The boundaries between ectotherms and endotherms are not always clear. Certain fast-swimming fish keep their muscles warmer than the surrounding water, to increase efficiency and speed. And some scientists speculate that many dinosaurs were endothermic.

Ectotherms allow body temperature to fluctuate with air temperature; they regulate temperature to some extent by basking in the sun or moving into the shade. Endotherms maintain their body temperature within a few degrees of a constant temperature most of the time. They regulate temperature by means of metabolic processes.

What Distinguishes the Birds from the Reptiles?

Birds are wonderfully diverse, with some 8700 species (Figure 24-19). Most species are distinguished by their distinctively colored plumage or their strange and haunting songs and calls. Yet birds are also remarkably uniform. They have three features that clearly distinguish them from the reptiles: feathers, flight, and endothermy. The ability to maintain a constant temperature in cold environments allows birds to live where no reptile could. The Arctic tern, for example, spends half its year north of the Arctic Circle, then flies halfway around the world to spend the rest of the year in Antarctica. Keeping a high body temperature and flying both require enormous amounts of energy. In return for this enormous energy expenditure, birds have access to vegetation, insects, and sea food that reptiles can never reach. Flight also provides terrific protection from most predators.

Most birds find food and shelter on land, though many birds depend on streams, ponds, and seas for their food. Like the reptiles, birds produce amniotic eggs and do not need water for fertilization or early development. Fertilization is always internal, but unlike the reptiles, only a few species possess a penis. Ducks and flightless birds such as ostriches, as well as herons, flamingos, chickens, and turkeys all have rudimentary penises. Most other birds, both male and female, have a cloaca. Birds accomplish fertilization by bringing together their cloacas. As in the tuatara, the sperm from the male's cloaca passes into the female's cloaca. Such intimacy often requires a long courtship process, and many bird species have elaborate mating rituals.

Feathers are made of keratin, the same protein that forms a lizard's scales and our own fingernails and hair. Flight feathers are an amazing adaptation to flight, but down, which insulates the bodies of birds, is also important. Many biologists now believe that feathers were originally an adaptation for warmth. Flight came later, as did a host of adaptations that lightened the body and made birds still better flyers. These included the loss of teeth (which lessens the weight of the head), the hollowing of the bones (which lightens the whole skeleton), and the reshaping of the breastbone (which forms a keel to which flight muscles attach) (Figure 24-20).

Feathers, flight, and endothermy distinguish birds from reptiles.

A.

B.

C.

Figure 24-19 **Birds are both distinctive and diverse.** Shown here are a great blue heron, a blue-throated hummingbird, and an emperor penguin and its young. *(A, Jim Zipp/Photo Researchers; B, Russel C. Hansen/Peter Arnold, Inc.; C, Barbara Cushman Rowell/DRK Photo)*

What Distinguishes the Mammals?

Mammals [Latin, *mammae* = breasts] take their name from their **mammary glands**, milk-producing organs in the female that characterize the mammals. No other vertebrates have mammary glands. Mother mammals nurse their young with warm milk, a rich mixture of fats, sugars, proteins, minerals, and vitamins, as well as antibodies that help defend newborns against infections.

Two other traits distinguish the mammals from other animals. Only mammals have hair and two sets of teeth ("baby" teeth and adult teeth). A mammal's hair, like a bird's feathers, helps conserve heat, a trait that is useful for an endotherm. Mammalian teeth, far more specialized than those of other ver-

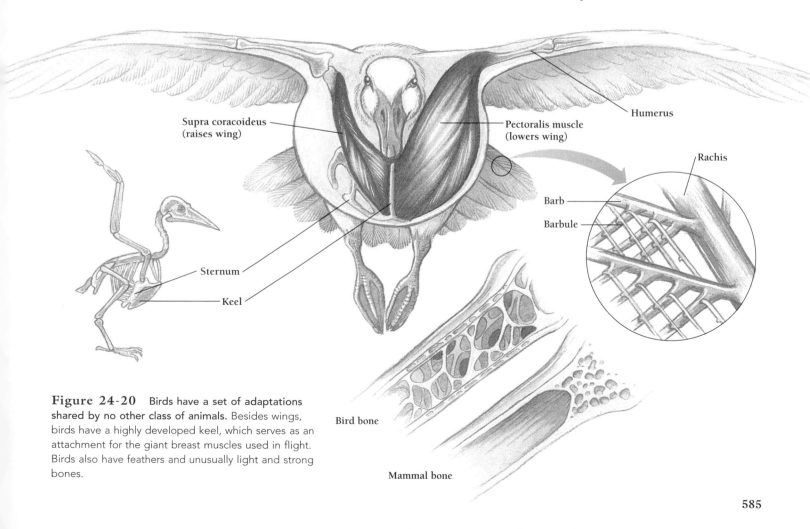

Figure 24-20 **Birds have a set of adaptations shared by no other class of animals.** Besides wings, birds have a highly developed keel, which serves as an attachment for the giant breast muscles used in flight. Birds also have feathers and unusually light and strong bones.

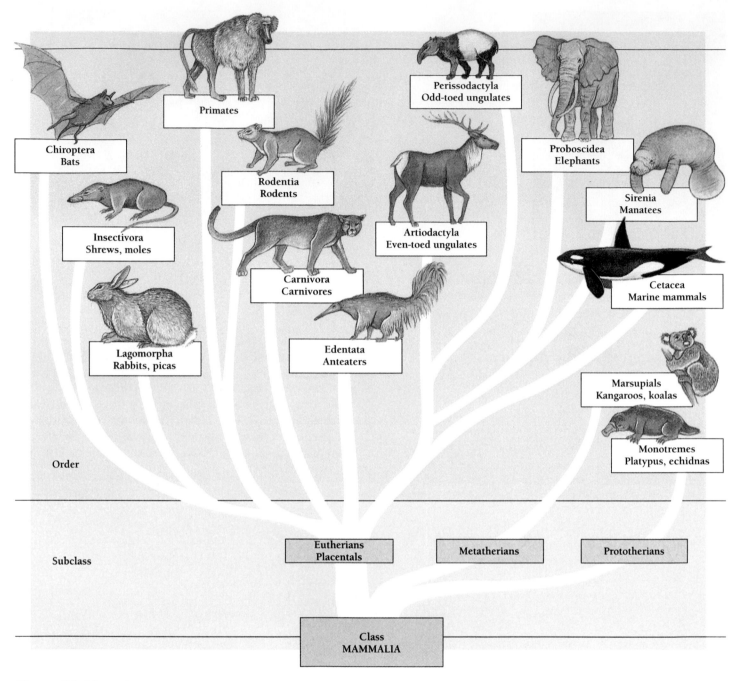

Figure 24-21 A few representative orders of the class Mammalia.

tebrates, help mammals get the tremendous amounts of energy they need to maintain a constant body temperature. Mammalian teeth may be specialized, for example, for grinding plants (horses and other herbivores), for shearing flesh (lions and wolves), for breaking bones (hyenas), or for seizing fish (dolphins and whales).

Mammalian teeth are extremely well preserved in the fossil record. Since a single tooth tells volumes about its owner, paleontologists know more about the evolution of mammals than is possible for other classes. In fact, some paleontologists now believe that the efficiency of the mammalian molar is at least partly responsible for the evolutionary success of the mammals since the end of the Cretaceous Period. When the

ruling reptiles died off, mammals radiated into about 30 orders, of which about two-thirds still have living representatives (Figure 24-21).

Mammalogists divide the living mammals into three subclasses—prototherians, metatherians, and eutherians. These subclasses have distinctive skulls and teeth that show that they first diverged from each other during the Mesozoic Era, even before the dinosaurs disappeared. These three subclasses differ markedly in the ways they reproduce.

Most mammals are **eutherians** [Greek, *therion* = wild beast] or **placental** mammals. In humans, elephants, rats, rabbits, and every other kind of eutherian mammal, embryonic development takes place within the uterus of the mother. In

Figure 24-22   A red kangaroo with its offspring. All marsupials carry their young in a special pouch that contains milk-secreting nipples. A female kangaroo may have a youngster at her side, another in the pouch, and a third in her womb.   *(Dave Watts/Tom Stack & Associates)*

Figure 24-23   Duck-billed platypus. Unlike other mammals, monotremes lay eggs. They are considered the most primitive of all the living mammals.   *(Alan Root/Okapia/Photo Researchers)*

fact, the word placental refers to the **placenta** [Latin, = flat cake], the flat structure that joins the lining of the uterus with the membranes of the fetus.

Metatherians, or **marsupials** [Latin, *marsupium* = pouch], such as the kangaroo and koala, reproduce differently from eutherians. Newborn marsupials are not babies at all, but undeveloped fetuses. A newborn red kangaroo, for example, is no bigger than a bumblebee (Figure 24-22). Emerging from the uterus only 33 days after fertilization, the tiny creature painstakingly pulls itself along its mother's fur until it finds a nipple. The mother's nipples are in a furry pouch, where the young kangaroo sleeps and nurses for about six months. Even after it begins to wander a bit on its own, the young kangaroo regularly returns to the pouch until it is almost eight months old. Meanwhile, the mother may have another, younger baby in the pouch, as well as an embryo in her uterus.

The **prototherians**, or **monotremes**, are the strangest mammals (Figure 24-23). They are endotherms with hair and mammary glands, like other mammals. Like the reptiles from which they and all other mammals are descended, however, monotremes lay eggs that are similar to those of reptiles and

birds. In addition, their digestive and reproductive systems empty into a cloaca, hence their name [Greek, *monos* = single + *trema* = hole]. The only living monotremes are the platypus and the spiny anteaters (or echidnas), which live in Australia and New Guinea. They are believed to be extremely primitive mammals, but exactly how they are related to other mammals is unknown. Mammals, both modern and ancient, are generally classified according to their tooth structure, and adult monotremes have no teeth.

Three characteristics set mammals apart from other vertebrates— mammary glands, hair, and two sets of teeth.

In this chapter we have seen how creatures as seemingly different as sea urchins and rabbits can share traits that show they are related. This chapter concludes our survey of the six kingdoms of organisms. In the next section, we see how ecological forces shape these many living organisms.

STUDY OUTLINE WITH KEY TERMS

Research in molecular biology supports the hypothesis, first advanced in the 19th century, that chordates have an inverted version of the same body plan as most protostome coelomates. Structures that are **dorsal** in frogs and other vertebrates are **ventral** in arthropods. Deuterostomes, however, differ from protostomes in several ways, including, for example, the fate of the embryo's **blastopore.** Deuterostomes include four phyla: **Echinodermata, Chaetognatha, Hemichordata,** and **Chordata.**

Echinoderms, such as starfish, are radially symmetrical as adults and bilaterally symmetrical as larvae. Echinoderms diverged from protostomes more than 500 million years ago. Chaetognatha (arrow worms) and Hemichordata (acorn worms) may resemble the ancestors of the chordates. Hemichordates have pharyngeal slits and a dorsal nerve cord, both of which are also chordate characteristics. They do not have, however, a notochord or a postanal tail, as do the true chordates.

Chordates share three main features at least sometime during their lives: a stiff flexible rod, called a **notochord,** running along the back, a **dorsal hollow nerve cord,** and **pharyngeal slits** (or **gill slits**). Chordates include the vertebrates and two other subphyla, the tunicates and the cephalochordates. Tunicates resemble other chordates as larvae, but not as adults. Cephalochordates have notochords both as larvae and as adults.

Vertebrates have heads and segmented spinal columns. The embryos of all vertebrates have skeletons made of **cartilage.** Most vertebrates replace the cartilage with **bone** during development. Vertebrates have closed circulatory systems and characteristic internal organs.

Four classes of vertebrates are fishes—Agnatha, Placodermi, Chondrichthyes, and Osteichthyes, and four classes are **tetrapods,** so-called four-footed animals—amphibians, reptiles, birds, and mammals. Agnatha (the extinct **ostracoderms** and the living lampreys and hagfish) are the only vertebrates that lack jaws. **Placoderms,** which are now all extinct, were the first vertebrates with jaws. In the Chondrichthyes (sharks and rays), bone never replaces cartilage. The Osteichthyes (bony fish) have bony skeletons and a **swim bladder.** One type of bony fish, with lobe-fins and primitive lungs, resembles the probable ancestors of the tetrapods.

Tetrapods have many special adaptations for land life. They have some or all of the following: **double circulation** and lungs; efficient kidneys; water-resistant skin; strong skeletons and jointed limbs; **ectothermy** and **endothermy;** and the **amniotic egg.**

Amphibians are semiterrestrial animals that always begin their lives in water but later develop lungs and other adaptations to live on land, in the process of **metamorphosis.** Reptiles and their descendants, the birds and mammals can all reproduce away from water as a result of the development of the **amniotic egg.** The different parts of the amniotic egg—the **amnion,** the **chorion,** the **yolk** and **yolk sac,** as well as the **allantois**—all contribute to the egg's ability to regulate water balance on dry land. Reptiles also developed watertight skins and strong skeletons. Reptiles and birds have a common exit for digestive, urinary, and reproductive systems, called the **cloaca.** Birds added feathers, flight, and endothermy, while mammals developed hair, two sets of teeth, and **mammary glands** for nursing their young. The mammals are divided into the **eutherians,** or **placental** mammals; the **metatherians,** or **marsupials;** and the **prototherians,** or **monotremes.**

REVIEW AND THOUGHT QUESTIONS

Review Questions

1. Describe three characteristics that distinguish the deuterostomes from the protostomes.
2. What characteristics do humans share with starfish (sea stars)?
3. What three characteristics define the chordates? Can you locate all of these features in your own body? For each feature, explain where it is or where it once was.
4. Rank the following organisms according to how closely related they are to humans: tunicates, lamprey larvae, and arrow worms.
5. Is the vertebral column, or backbone, actually made of bone in all vertebrates? What is bone?
6. Of what value are jaws?
7. Why must amphibians return to water to reproduce?
8. Name the layers of the amniotic egg. What classes of animals have amniotic eggs?

9. What is thermoregulation? Why is it important? In what two ways do vertebrates accomplish thermoregulation?
10. What three mammalian characteristics distinguish a human from a turkey?
11. What is the function of a placenta?

Thought Questions

12. Compare the vertebrate body plan with the arthropod body plan. What advantage(s) might there be to flipping over the arthropod body plan? How would you test your hypothesis?
13. How would the day-to-day lives of humans be different if we were marsupials instead of placental mammals?

SELECTED READINGS

Blaustein, Andrew R., and David B. Wake, "The Puzzle of Declining Amphibian Populations," *Scientific American,* Vol. 272, April: 52–57, 1995.

Diamond, Jared, *The Third Chimpanzee: The Evolution and Future of the Human Animal,* HarperPerennial, New York, 1993. Diamond asks how humans resemble and differ from the other primates. What traits seem to haunt us? What is our "natural" state?

Heinrich, Bernd, *Ravens in Winter: A Zoological Detective Story,* Summit Books, New York, 1989. A vivid account of a Maine winter spent watching the doings of ravens. Zoologist Heinrich brings to his story both humor and insight.

Montgomery, Sy, *Walking with the Great Apes, Jane Goodall, Dian Fossey, Biruté Galdikas,* Houghton Mifflin, Boston, 1991. A highly readable account of three women who pioneered an in-depth and personal style of studying primates and other animals in the wild. Jane Goodall (chimpanzees), Dian Fossey (gorillas), and Biruté Galdikas (orangatans) have left an indelible mark on how field research is conducted.

Owens, Mark and Delia, *Cry of the Kalahari: Seven Years in Africa's Last Great Wilderness,* Houghton Mifflin, Boston, 1984. A wonderful book by two young Americans who went to Africa's stark Kalahari Desert to study lions, hyenas, and other animals. The Owens survive drought, storms, and many hair-raising adventures. The book is funny, beautiful, and romantic.

▶ On-line materials relating to this chapter are on the World Wide Web at http://www.saunderscollege.com/lifesci/ Click on Tobin/Dusheck: *Asking About Life.*

PART V

Ecology

Mass of starfish, Rialto Beach, Olympic National Park, Washington.
(© Darrell Gulin/Natural Selection)

Gaia: Is the Biosphere an Organic Entity?

In the early 1960s, the National Aeronautics and Space Administration (NASA) began gearing up for an exploration of the solar system that would occupy most of the following two decades. The space agency wanted to search for signs of life on the other planets, especially Mars. NASA planned, in particular, to examine Martian soil for chemicals on which microorganisms could grow, or that might be products of life.

Analyzing the Martian soil would require new instruments specially designed for the project. The person who best knew how to design such instruments was James E. Lovelock, an eccentric English chemist. In 1958, Lovelock had conceived an idea for an ingenious device called the electron capture detector that could detect tiny amounts of chemicals. The device quickly became an invaluable tool for analyzing samples for trace quantities of pollutants, and led directly to the ban of the pesticide DDT. It was only natural, then, that NASA should invite Lovelock to the Jet Propulsion Laboratory, in Pasadena, California, to help design the instruments for detecting life on Mars.

Although NASA had asked Lovelock only to help design some very specific instruments, the chemist's fascination with the search for extraterrestrial life and his insatiable curiosity soon led him to reconsider the whole question of life on Mars. He wondered, for example, whether NASA was approaching the problem in the right way. What if Martian or other life didn't look or act anything like terrestrial life? Could it still be detected? If so, how? What, he wondered, is a universal description of life that would allow scientists to recognize life even in an alien form?

Lovelock's background in chemistry led him to approach these questions by considering the concept of entropy. Entropy is a measure of disorder in "systems." A **system** is an assemblage of interacting parts or objects. The solar system is a system. A household is a system. If any system is left alone, its entropy, or disorder, tends to increase, according to the Second Law of Thermodynamics. For example, a frozen Popsicle left in a warm room will gradually melt into a sticky pud-

dle as it reaches room temperature. The Popsicle's loss of its characteristic shape and temperature is a loss of order that represents an increase in entropy.

Organisms, by contrast, *decrease* local entropy, creating order. For example, plants convert carbon dioxide and water into highly structured leaves, stems, and flowers; animals convert amino acids and simple fats into muscle and brain tissue. This decrease in local entropy occurs only at the expense of energy from the sun, however. The decrease is local only. In the universe as a whole, entropy still increases.

A reduction in local entropy, Lovelock argued, was a sure sign of life. But how could one recognize a reduction in entropy on Mars, or anywhere else? Lovelock's colleagues at the Jet Propulsion Lab found his suggestion profoundly unhelpful. Even Lovelock acknowledged that NASA needed something tangible to search for.

Nonetheless, Lovelock couldn't stop thinking about how one could spot life in a strange place. Asking himself how an alien being might recognize the presence of life on Earth, he realized that an alien would probably notice the unlikely chemical composition of Earth's atmosphere. The atmosphere contains both large amounts of oxygen and a small amount of methane. These two gases readily react, forming carbon dioxide and water. If Earth

were a dead planet, without a continuous supply of fresh oxygen and methane, the two gases would interact until the methane was eliminated. Yet Earth's atmosphere always contains both gases in abundance—only because living organisms constantly provide new methane. Some comes from bacteria in marshes and rice paddies. But a surprisingly large amount of methane comes from the flatulence of cattle.

Lovelock realized that the Earth's chemically unlikely atmosphere could be seen as an extension of the planet's **biosphere**—the system of living things that covers the Earth. Then Lovelock extended

this idea still further and proposed a radical view. He envisioned the biosphere and the atmosphere, as well as the oceans and the soils, as constituting a single, immense entity—one that acts to preserve itself. Lovelock named this entity Gaia after the Greek Earth goddess Ge, from whose name the words "geology" and "geography" are derived.

Just as a warm-blooded animal regulates its blood temperature and chemistry, the biosphere, he reasoned, actively regulates the temperature and chemistry of the atmosphere, a process called **homeostasis.** One sign of Gaian homeostasis is an apparent limit on the ocean's saltiness. Geologists know that the rate at which eroding soils and rocks fill the oceans with salts should, theoretically, have made the ocean 40 times saltier than it is. Yet, somehow, the oceans have remained less than 10 percent saturated with salt for hundreds of millions of years.

Gaia supporters argue that living organisms keep the oceans from becoming

Lovelock envisioned the biosphere and the atmosphere, as well as the oceans and the soils, as constituting a single, immense entity—one that acts to preserve itself.

too salty. In areas of shallow water evaporation tends to concentrate salt. There, Gaia's defenders say, dense colonies of bacteria specially adapted to live in these high-salt environments remove salt.

The Gaia hypothesis, first published in the 1960s, amused and perplexed the scientific community. The public loved it. TV anchors glibly announced that the

591

Earth was one big organism. The concept of Earth as a superorganism carried enormous appeal, especially for nonscientists. Environmentalists seized on Gaia as a cultural metaphor, stressing the need to consider the biosphere and its environment—the atmosphere, the oceans, and the soils—as a whole. NASA's first breathtaking photographs of Earth as a cloud-swathed blue ball floating in space intensified Gaia's already enormous appeal (Figure 25-1).

Scientists found the idea of Gaia entertaining, but also embarrassing. Despite Lovelock's impressive credentials as an inventor, many scientists concluded he was a crank. Most academic ecologists were at pains to distance themselves from Lovelock. They had always distinguished their careful scientific work from free-wheeling political activity and undisciplined thinking of radical environmental groups. But Lovelock's credentials as a scientist seemed to blur the distinction, in the media at least, between the science of ecology and the politics of environmentalism. And although Lovelock denied that he intended Gaia to be understood as an actual organism, he at times described Gaia in almost human terms, as though "she" had intentions and goals. He referred, for example, to "Gaia's intervention" and "Gaian impatience."

The blur between Gaia and formal science increased when a biologist of national standing added her arguments to Lovelock's. In 1974, Lynn Margulis, a distinguished professor of botany at the

Figure 25-1   **Earth.** NASA photos persuaded most people that Earth's biosphere is precious, irreplaceable, and limited.   *(NASA)*

Amherst campus of the University of Massachusetts, began enthusiastically promoting her own slightly modified version of Gaia—to the horror of her scientific colleagues. Her provocative, take-no-prisoners style of argument alternately amused and infuriated fellow biologists. Yet despite Margulis's advocacy, few other reputable scientists have openly accepted the Gaia hypothesis today.

However one regards Gaia, it provides an excellent metaphor for the biosphere. In Margulis's view. Gaia is not a superorganism but a giant ecosystem, the highest level of organization of life. And the philosophical differences that fueled the debate about Gaia are, we will see, old issues in the science of ecology.

KEY CONCEPTS

1. Ecology is the study of the interactions of organisms with each other and with their physical environments.

2. An ecosystem is an interacting group of many species, together with nonliving environmental components.

3. The sum of all ecosystems is the biosphere.

4. Organisms in an ecosystem can be classed as producers or consumers.

5. Organisms dissipate energy. Nearly all the energy that supplies life on Earth comes from the sun. Autotrophs (producers) capture energy from the sun, in the form of sunlight. But only a small fraction of the energy captured by autotrophs is available to heterotrophs (consumers and decomposers). The rest of the energy is lost, usually as heat.

6. The biosphere recycles materials even as it dissipates energy.

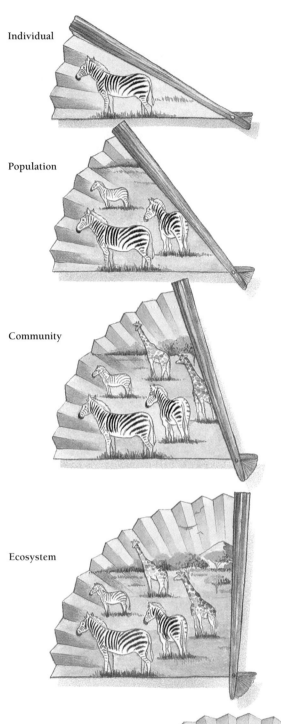

Individual

Population

Community

Ecosystem

Biosphere

WHAT IS AN ECOSYSTEM?

The word "ecology" is popularly associated with recycling soft-drink cans, planting saplings in city parks, and campaigning for stricter air and water pollution standards. Yet the science of ecology includes none of these activities. **Ecology** [Greek, *oikos* = home] is the study of the interactions of organisms with one another and with their physical environments. Ecology is concerned with understanding the relationship of organisms to their homes, in the broadest sense.

An individual organism is a member of a number of systems. For example, an individual is a member of a **population,** a breeding group of individuals of the same species that inhabit a common area; a member of a **community,** an interacting group of many species that inhabit a common area; and a member of an **ecosystem,** a community of organisms together with the nonliving parts of the community's environment (Figure 25-2).

An ecosystem consists of both **biotic** (living) and **abiotic** (nonliving) parts. The abiotic components of an ecosystem include forces (such as wind or gravity), conditions (such as light intensity, temperature, salinity, or humidity), and chemical substances in soil, air, and water. Chemical substances include both inorganic substances, such as nitrogen and water, and organic compounds that are the remains or products of living things.

An ecosystem has many of the properties of living organisms that we listed in Chapter 1. An ecosystem obtains energy from its environment, transforms chemicals, changes with time, and responds to environmental changes. And ecosystems, like

Figure 25-2 **Levels of organization.** An individual is a member of a population, of a community, of an ecosystem, and of the biosphere.

individual organisms, use energy to maintain a stable state. However, ecosystems differ from organisms in important ways. Unlike organisms, ecosystems do not evolve in the Darwinian sense, since they are not subject to natural selection. Further, an ecosystem does something that no organism does. It recycles. As Lynn Margulis has written, "no single organism supports its growth solely by eating its own waste and entirely cycling the carbon, hydrogen, sulfur and so forth needed for its body."

The living and nonliving components of an ecosystem interact with one another in such a way that it makes sense to consider them all as a unit. The physical environment—the atmosphere, the water, and the soil—both permits and limits life within an ecosystem. A freshwater lake, for example, provides all of the conditions necessary for certain fish and aquatic plants to flourish. Yet, the same lake would be inhospitable to desert tortoises and cacti.

Just as the physical environment affects organisms, organisms affect their physical environment. Trees, for example, block sunlight, change the characteristics of soil, and produce oxygen.

An ecosystem's boundaries may be well-defined: a pond is enclosed by its shores, a grassy meadow by the surrounding forest. More often, an ecosystem's boundaries are ill-defined: the pond blends little by little into marsh, and then into meadow.

In addition, each ecosystem, community, or population is part of a larger one. A pond ecosystem may be part of a meadow ecosystem, the meadow part of a forest, and the forest part of a two-thousand-mile drainage system such as the Mississippi River Valley. We may therefore speak of the frog population of the pond, of the meadow, of the forest, or of the drainage. To some degree, then, biologists may choose the boundaries of the ecosystem they wish to study, always keeping in mind the larger picture. Each ecosystem interacts with other ecosystems, because seeds and spores disperse, animals migrate, and flowing water and air carry organisms—and their products and remains—from one place to another.

All ecosystems taken together make up a much larger ecosystem, the biosphere. The biosphere differs from all other ecosystems in having fixed boundaries. Extending over the whole of the Earth's surface, the biosphere begins three kilometers underground and extends into the highest reaches of the atmosphere. It ends where the atmosphere ends and space begins, high above the layer of ozone (O_3) that protects organisms from ultraviolet light.

The concept of a biosphere is not unlike Margulis's Gaia. The biosphere is Gaia, but without Lovelock's personification. Gaia, as a metaphor, helps us remember that ecosystems, at times, function in complex and surprising ways.

Further, ecosystems, like other systems, are more than the sum of their parts. The properties of systems that cannot be predicted from a knowledge of the individual parts that compose them are called **emergent properties.**

The concept of emergent properties stands at the heart of ecology. For example, species richness, the number of species that a habitat will support, varies from place to place. Tropical rain forests are famous for their species richness. By comparison, redwood forests are species poor. The properties of tropical rain forests that enable them to support so many species are only partly understood. Climate plays a role, as do the evolutionary histories of such areas. The "patchiness" of the environment plays a role, as well. Mountainous areas can be divided into rich, moist valleys and drier, rockier hillsides. Whereas plateaus and savannas tend to be more uniform, offering fewer different habitats within the same amount of land. Ecologists have been arguing for a very long time about what determines species richness. One thing is certain: the answers cannot come by studying individual species or even individual ecosystems.

While cell biologists and molecular biologists study organisms by breaking them down into their component parts, ecologists do the opposite. They look at whole organisms in the context in which they live. Ecologists study the relations among organisms as much as they study the organisms themselves.

We can use Lovelock's Gaia metaphor to compare the way that a biologist studies an ecosystem with the way we would study an individual organism. In trying to understand the working of a single organism, we start by looking at the behavior of the intact organism. Next we might cut it up and study the arrangement and workings of its various parts. And finally, we might try to determine how each part contributes to the processes that make up the life of the organism.

We can similarly examine the parts of an ecosystem. In this chapter we will examine ecosystems as entities, and then see how the parts of communities interact. In succeeding chapters we will see how ecosystems and communities change over time, how human intervention affects ecosystems, how individual populations behave, and, finally, how individual organisms behave.

Ecology is the study of the interactions of organisms with each other and with their physical environment. An ecosystem is a functional unit consisting of both biotic parts and abiotic parts, including air, water, and soil.

HOW DOES ENERGY FLOW THROUGH ECOSYSTEMS?

The biotic part of an ecosystem consists of three kinds of organisms: **producers,** such as plants, which harvest energy directly from sunlight or, rarely, from inorganic molecules; **consumers,** typically animals, which obtain energy by eating producers or other consumers; and **decomposers,** such as bacteria and fungi, which live on the energy in the complex molecules of dead organisms. Particles of dead organic matter are called **detritus.**

What Are Trophic Levels?

Producers are all **autotrophs** [Greek, *auto* = self, same + *trophe* = to nourish], organisms that produce their own food. The vast majority are green plants, which obtain energy from the sun through photosynthesis. Consumers and decomposers are **heterotrophs** (Greek, *hetero* = other + *trophe* = to nourish), organisms that obtain food from other organisms.

Consumers may eat either producers or other consumers. Herbivores, such as deer, quail, and caterpillars, eat plants directly. Carnivores, such as mountain lions, hawks, and wasps, eat other animals—either herbivores or other carnivores. Often, several levels of consumers exist: hawks eat snakes, which eat ground squirrels, which eat lizards, which eat predaceous insects, which eat caterpillars, which eat plants (Figure 25-3). Such a sequence is called a **food chain.**

Each level of a food chain is called a **trophic level.** For example, in the food chain above, plants are in the first trophic level, caterpillars are in the second, spiders are in the third, birds are in the fourth, and hawks are in the fifth. Another way of designating trophic level is to call herbivores "primary consumers," to call carnivores that eat herbivores "secondary consumers," to call animals that eat secondary consumers "tertiary consumers," and so on.

Many animals, including humans, bears, and crows, eat from several levels of a food chain. Such animals are called **omnivores** [Latin, *omnis* = all + *vorus* = devouring] because they eat plants, herbivores, and other carnivores. Humans, for example, are known to eat a wide variety of organisms. We eat plants. We eat almost any herbivore, from cows to caterpillars. We also eat carnivores, such as sharks, frogs, snakes, and birds. And we eat other omnivores, such as pigs and bears. Nor do humans draw the line at scavengers and decomposers, for many people relish nothing more than lobster and mushrooms.

Among the one-celled organisms, the bacteria may be either producers or decomposers, and the protists may be producers, consumers, or decomposers. Most decomposers are bacteria or fungi. A few plants, called **saprophytes** [Greek, *sapro* = putrid + *phyte* = plant] also decompose dead material. And some species of millipedes, earthworms, termites, flies, lobsters, clams, catfish, and other animals also feed on dead organic matter, either exclusively or occasionally. Many of these are called **scavengers,** because they feed from whole carcasses.

We can see that a food chain may have many trophic levels. And every ecosystem has numerous food chains, each with

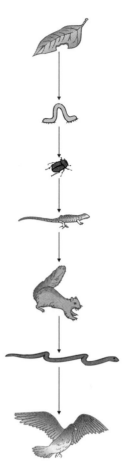

Figure 25-3   A food chain. In a food chain, the hawk eats the gopher snake, which eats the ground squirrel, and so on.

different numbers of trophic levels. Furthermore, a single organism may occupy different trophic levels in different food chains. The collection of all the food chains of an ecosystem is called the **food web,** a term that stresses the interconnectedness of the various food chains (Figure 25-4).

Describing an Ecosystem

Ecosystems have so many components, in so many intricate relations, that ecologists cannot ever completely describe a natural ecosystem. Further, each ecosystem is unique, whether it is a tiny pond or a vast area, such as the Serengeti Plain of Africa. Still, ecologists can attempt to describe the bare outlines of an ecosystem. One approach is to estimate the **biomass,** or aggregate dry weight, of producers, consumers, and decomposers.

For example, suppose an ecologist wanted to see if the ratio of the biomass of producers to the biomass of consumers

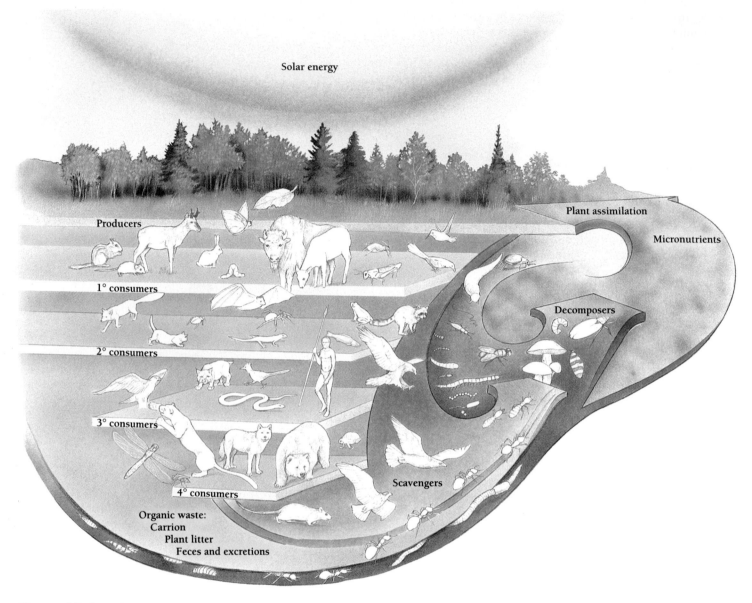

Solar energy

Producers

1° consumers

2° consumers

3° consumers

4° consumers

Organic waste:
Carrion
Plant litter
Feces and excretions

Scavengers

Decomposers

Plant assimilation

Micronutrients

Figure 25-4   A food web. A food web shows that the interactions among organisms are more complex. Organisms at each level may feed from several trophic levels, not just adjacent ones. In addition, decomposers return nutrients back to primary producers.

in a typical woodland pond was always the same. The ecologist would begin by comparing the ecosystem structures of different ponds, measuring, in particular, the biomass of all the species in each pond. The first step would be to collect samples of organisms that inhabit the different depths of the pond. To do this, an ecologist would use a net to catch fish, instruments to trap suspended and floating organisms within a column of water, and a mechanical shovel to dig up bottom-dwellers (Box 25-1).

The next step would be to sort, categorize, name, and count the different species of collected organisms. Filters or fine nets can separate organisms by size. The smallest organisms, many of them visible only with a microscope, include the plankton

(tiny, free-floating protists, plants, and animals), as well as similarly small organisms that live on the bottom. These organisms include both primary producers and the tiny protists and animals that feed on them.

To measure biomass, the ecologist would also weigh the organisms of each species. In the case of the producers, however, ecologists often measure the total amount of chlorophyll instead, since it is generally proportional to biomass. Chlorophyll, which is responsible for harvesting light energy for photosynthesis, is easy to measure with a spectrophotometer. Finally, a detailed study of the diets of each organism could show which species occupied which trophic level—in other words, who is eating whom and in what proportions.

BOX 25-1

Measuring photosynthesis in a pond

Measuring the rate of photosynthesis in a pond is easy. A researcher takes two water samples from the same depth and puts one into a clear bottle and the other into a "dark" bottle that light cannot enter. The researcher can then hang the two bottles back in the pond at the end of a rope. Inside each bottle, the many organisms carry on with life. Those in the clear bottle photosynthesize and respire. However, without light, those in the dark bottle can only respire.

Over time, the amount of oxygen released in the two bottles will be different. In the clear bottle, the amount of oxygen released will equal the amount produced through photosynthesis by the producers minus the amount consumed through respiration by all organisms. In the dark bottle, by contrast, the same kinds of organisms can only respire and therefore only consume oxygen. In the dark bottle, then, the amount of oxygen will drop steadily.

We can assume that the organisms in the clear bottle consume about the same amount of oxygen as those in the dark bottle. Therefore, the total amount of oxygen produced by photosynthesis in the clear bottle equals the amount we know is consumed in the dark bottle plus the "extra" amount of oxygen produced in the clear bottle. The total amount of oxygen produced in the clear bottle is a measure of the rate of photosynthesis in the pond as a whole.

Using the results from the two-bottle measurement, researchers can also estimate the *net primary productivity,* the amount of energy converted into organic molecules by the producers in a give time. On average, a plant stores about 3.5 kcal of energy (in glucose and other organic molecules) for each gram of oxygen released. The net primary productivity of the producers in the pond is the number of grams of oxygen released × 3.5 kcal/g of oxygen.

Suppose that when the results are summarized, however, the ecologist discovers that the biomass of the producers at each pond is the same, while the biomass of the consumers is different. In other words, the ratio of the biomass of consumers to producers is different at each pond. From this the ecologist would have to conclude that the biomass of the producers does not determine the biomass of the consumers.

At first this conclusion might be surprising. If we looked at a population of deer, for example, we would expect to see a connection between the number of deer—an approximate measure of deer biomass—and the amount of forage available to them. In fact, in terrestrial ecosystems, plant mass usually exceeds herbivore mass, which exceeds carnivore mass. As a result, a graphic presentation of the total mass of organisms at each trophic level of a forest ecosystem looks like a pyramid, a **pyramid of biomass** (Figure 25-5). A graph of the total numbers of organisms at each trophic level would have a similar shape, called a **pyramid of numbers.**

Ecologists describe ecosystems as pyramids of numbers and pyramids of biomass.

Energy Is Continually Dissipated in an Ecosystem

A pyramid of numbers may not always have a pyramidal shape. A single oak tree, for example, can support thousands of caterpillars, beetles, and other insect herbivores. A single dog can support thousands of fleas, ticks, and internal parasites (Figure 25-6). Neither does a pyramid of biomass always have a pryamidal shape. In aquatic ecosystems, for example, the biomass of primary consumers may exceed that of the photosynthetic plankton producers (Figure 25-7).

Apparently, the biomass of the consumers need not depend directly on the biomass of the producers. But why not? We know that consumers cannot consume more than is produced. What rules, then, govern the relative sizes of trophic levels? The answer lies in the flow of energy from one trophic level to

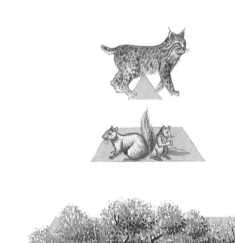

Figure 25-5 **A pyramid of biomass.** In a forest, the biomass of the primary producers is greater than that of herbivores and carnivores.

Figure 25-6 A pyramid of numbers in a canine community. A pyramid of numbers may look like an upright pyramid or it may be inverted. Because fleas and ticks are so small compared to a dog, a single dog can support large numbers of such parasites.

Figure 25-7 Inverted pyramid of biomass in an aquatic community. In aquatic communities, the primary producers may be eaten as fast as they multiply. So, while the accumulated biomass of the consumers is large, that of the producers may be small.

the next. Once the ecologist compares the energy flowing through the first trophic level with the energy flowing through the succeeding levels, it is clear that the producers always process more energy than the consumers. In fact, the ecologist discovers, this relationship holds in all ecosystems.

Beginning in the 1950s, ecologists became increasingly aware that one of the most important factors determining the structure of an ecosystem was the availability of energy. When they measured the energy used in an ecosystem, they discovered that energy becomes less and less available at successively higher trophic levels.

The energy that drives the Earth's biosphere comes from the sun. Producers use the energy of sunlight to make carbohydrates and other molecules. However, because every chemical and mechanical process expends considerable energy as heat, the producers can convert only part of the light energy into chemical energy. The lost energy is a consequence of the Second Law of Thermodynamics, as mentioned at the beginning of the chapter (Chapter 5). Primary consumers, in turn, use the chemical energy of the producers to make their own molecules. But, again, most of the energy is dissipated as heat. The amount of energy used at each trophic level in a specified time (usually a year) looks like a pyramid (Figure 25-8).

Such a **pyramid of energy** is always upright. The producers always use far more energy than the primary consumers, which always process far more energy than secondary, tertiary, and higher level consumers. Even when an ecosystem's pyramid of numbers and biomass is inverted, the pyramid of energy remains upright.

For example, at a given moment, the biomass of algae in a pond may be less than the biomass of fish. Yet, during the course of a year, the amount of sunlight energy converted into algae is actually greater than the amount converted into fish. The algae actually grow faster than the fish. If, over the course of a year, the total biomass of the algae were collected and

stored (instead of being consumed by the fish), it would be greater than the total biomass of the fish. But the fish eat the algae as fast as it grows, and convert only a fraction of it into fish. The rest is used to power movement and metabolism or is lost as heat.

Almost all energy for ecosystems comes from the sun. The rare exceptions are deepwater ecosystems fueled by high-energy sulfur compounds. As much as 99 percent of all the sunlight that bathes the Earth is absorbed by the atmosphere or reflected back into space by clouds, water, and rocks. The amount of sunlight reaching an ecosystem is easy to measure with a light meter, similar to the ones found on automatic cameras. Measurements reveal that the amount of sunlight reaching each square meter of land is about 1 to 2 million kcal per

Figure 25-8 Pyramid of energy. The pyramid of energy is always upright, no matter whether for a forest, a dog, or a pond.

year, the equivalent of the chemical energy in 30 to 60 gallons of gasoline, or the 2800 kcal of food each of us eats daily.

Plants absorb only a tiny fraction of this energy—3 percent or less. We cannot directly measure how much of this energy organisms capture. We can, however, estimate the total rate of photosynthesis and the total rate of respiration for all the organisms in a particular trophic level of an ecosystem. From that, we can estimate, in turn, the net energy stored. In a pond, for example, we can estimate the total rate of photosynthesis and respiration for all the microorganisms.

Calculations such as those in Box 25-1 reveal that plants convert 1 to 5 percent of the energy in sunlight to chemical energy. Some 40 percent of this energy is lost in plant respiration, leaving only 9000 kcal of stored energy per square meter per year for primary consumers. Not all of even this diminished energy is available to primary consumers. For example, few primary consumers eat tree trunks and root systems. Herbivores typically consume less than 20 percent of the energy that primary producers make available. In sum, of the 1 to 2 million kcal that strike a square meter of land each year, only about 2000 kcal/yr is actually available to primary consumers such as deer and cattle. Of this, herbivores convert only about 10 percent to tissues. Similar measurements of oxygen consumption by successive levels of consumers give estimates of energy conversion for each trophic level.

An energy pyramid expressing these values shows how much sunlight energy enters the system, how much is stored in the biomass of each trophic level, and how much flows between trophic levels. Such a pyramid is also called a **pyramid of productivity.** The productivity of a trophic level is the energy captured in the chemical bonds of new molecules each year for each square meter. The **primary productivity** is thus the productivity of the first trophic level, the producers. Primary productivity varies enormously from ecosystem to ecosystem (Table 25-1). A marsh, for example, is 35 times as productive as a desert and nearly 4 times as productive as a wheat field.

Each step in the pyramid is only about 10 percent as wide as the one below. That is, the organisms of any trophic level provide the next higher trophic level with only 10 percent of the energy that they have assimilated from the lower trophic level. The rest is lost as heat. This generalization is called the **ten percent law.**

The ten percent law helps us understand why ecosystems have so few trophic levels and so few individuals at the highest trophic level. If on a square meter of land, the primary consumers take in 2000 kcal/yr, secondary consumers (herbivores) will have only about 200 kcal/yr to live on, and tertiary consumers (herbivore-eating carnivores) will have only 20 kcal/yr to live on. We can immediately see that a food chain cannot have many trophic levels, usually not more than three or four, because at higher trophic levels, the amount of energy available becomes too small to support many individuals. In addition, to find enough food, top carnivores such as mountain lions and hawks must range farther and farther, which consumes more energy.

Table 25-1 Net Primary Productivity of Terrestrial Biomes

Ecosystems vary in their productivity. Estimates of net primary productivity are approximate only. Ecologists subtract respiration rates from the photosynthetic rates of different plant tissues, then extrapolate to the community level. Ecologists use the net production per gram of biomass of each species in a community. Unfortunately, we do not yet have accurate measurements of photosynthesis and respiration rates for all the species in any community. In addition, each plant species' net primary productivity varies, depending on its environment.

Ecosystem	Average Net Primary Productivity (grams dry matter/m^2/year)	Biomass (kg dry matter/m^2)
Swamp and marsh	3000	15
Tropical rain forest	2200	45
Temperate forest:		
evergreen	1300	35
deciduous	1200	30
Savanna	900	4
Boreal forest	800	20
Woodland and shrubland	700	6
Cultivated land	650	1
Temperate grassland	600	1.6
Lake and stream	400	0.02
Tundra and alpine	140	0.6
Desert and semidesert scrub	90	0.7
Extreme desert: rock, sand, ice	3	0.02

After R.H. Whittaker and G.E. Likens, 1975.

We can calculate the amount of area needed to support a single human being. Since the average man needs about 2800 kcal/day, the amount of energy that hits a square meter of land in a year, a single man—if he were able to photosynthesize—would need one square meter for each day of the year. That is, he would need 365 square meters of land to live on.

But men are not producers, they are consumers. If this hypothetical man were a strict vegetarian, he could, by the ten percent law, get by on about 3650 square meters, less than an acre. If he ate nothing but herbivores, he would need ten times more land—36,500 square meters, or 9 acres. If he ate any carnivores or omnivores, he would need even more land.

Extending this calculation, we can conclude that if the people of New York State subsisted entirely on hamburgers (without the bun), they would need all of the cows produced on 250,000 square miles of land, an area more than five times bigger than New York State. But if they switched to soy burgers, they could live off the food produced in just half the area of New York State (Figure 25-9).

Energy pyramids show that carnivores in general, and top carnivores especially, must find their food over wide areas.

Land needs to feed
the people of
New York State:

on soyburgers

on hamburgers

Figure 25-9   Ten percent law for hamburgers. Meat eaters consume the primary productivity of ten times as much land as do plant eaters.

Mountain lions, for example, roam over hundreds of square miles. Top carnivores are therefore the most mobile animals, including, for example, sharks, killer whales, and eagles. The total energy available in all the organisms at the top trophic level is too small to sustain another trophic level: that is why these carnivores are "top."

In every ecosystem, the pyramid of energy is narrower at the top. Energy is continuously dissipated in an ecosystem, with only 10 percent of the energy available to each trophic level passing to the next trophic level.

HOW DO ECOSYSTEMS RECYCLE MATERIALS?

We can see that at each trophic level energy becomes less available. However, although organisms constantly dissipate energy, they recycle materials. Within an ecosystem, materials such as carbon, nitrogen, and phosphorus constantly change form, but, unlike energy, these elements are rarely lost. Sometimes they are part of organisms, other times they are not. The carbon atoms in your fingernail have been, at different times, part of

an apple, part of a bicarbonate ion in the ocean, and part of a lump of coal. Carbon passes through many forms—both biotic and abiotic—in a system called a **biogeochemical cycle.**

Other materials pass through biogeochemical cycles as well. Oxygen, which makes up about 20 percent of the atmosphere, constantly cycles between the atmosphere, consumers, and producers. Animals and other consumers take it directly from the atmosphere to burn carbon compounds, releasing oxygen in the form of carbon dioxide. Plants take up carbon dioxide, remove the carbon, and release oxygen.

The biogeochemical cycles of some materials—such as water, carbon, nitrogen, and oxygen—involve the whole biosphere. A carbon atom in carbon dioxide or bicarbonate may travel all over the world on currents of wind or ocean. In fact, atoms such as oxygen and nitrogen become so widely dispersed that one scientist has calculated that each time one of us takes a breath, we breathe in a few atoms from Julius Caesar's dying breath. On the other hand, materials such as calcium, potassium, and phosphorus tend to stay put and cycle within smaller, more local ecosystems. We will discuss four biogeochemical cycles: those of water, carbon nitrogen, and phosphorus.

Organisms recycle materials even as they dissipate energy.

What Drives the Water Cycle?

All life is intimately connected to water. Flowing water refreshes and flushes ecosystems by bringing in minerals, organic substances, and even organisms and by carrying away wastes. Animals depend on water, which they get either by drinking or by eating plants or other animals. Plants process about 500 grams of water for every gram of biomass they produce. Plants pull water from the soil and release it as vapor through tiny pores (stomata) in their leaves, a process called **transpiration.** On a summer day, an average maple tree transpires more than 50 gallons of water every hour—enough to fill a large bathtub.

The water cycle involves both abiotic and biotic processes (Figure 25-10). At any moment, 97 percent of the water on our planet is in the oceans, which cover some 70 percent of the Earth's surface. Sunlight evaporates large quantities of water from the oceans, seas, lakes, ponds, rivers, and streams into the atmosphere. And plants transpire vast amounts of water. In fact, transpiration may make up as much as 90 percent of all the water that evaporates from the continents.

Yet the atmosphere is so vast that in spite of all this evaporation, the atmosphere as a whole contains only about one one-thousandth of one percent water vapor at any given moment. However, water vapor easily condenses into rain, which falls back to the Earth's surface. Then, more water evaporates and rains in a rapid cycle that drops nearly 3 feet of rain over the entire surface of the Earth each year.

Because the continents have mountains and extreme temperatures, proportionately more rain drops on the continents

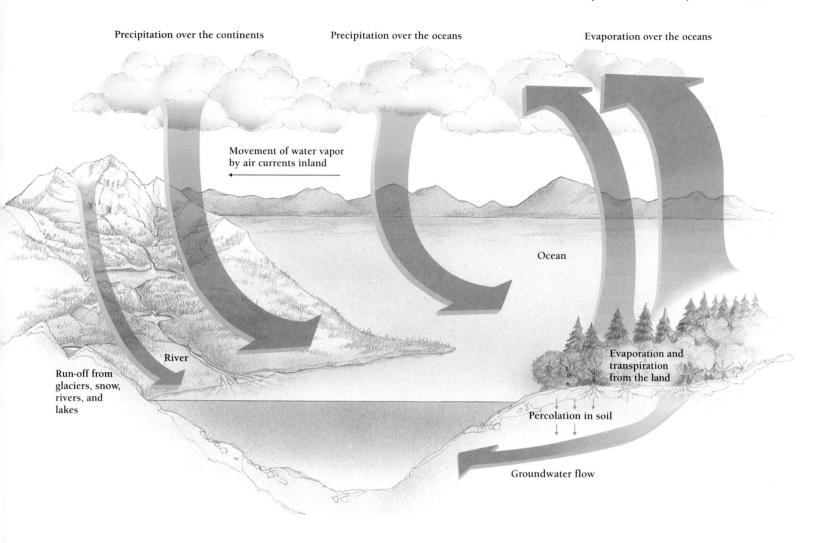

Precipitation over the continents   Precipitation over the oceans   Evaporation over the oceans

Movement of water vapor by air currents inland

Ocean

River

Evaporation and transpiration from the land

Run-off from glaciers, snow, rivers, and lakes

Percolation in soil

Groundwater flow

Figure 25-10 **The water cycle.** Much water that evaporates from the oceans drifts over the land and rains down, forming streams, rivers, and extensive groundwater flow.

than on the oceans. Of the water that evaporates into the atmosphere, 83 percent comes from the oceans. However, only 75 percent of rain falls on the oceans, so more rain falls on the land than comes from the land. Another way of expressing this is to say that, overall, water vapor moves from the oceans to the continents. The excess water returns to the oceans as run-off in streams and rivers.

Sunlight drives the evaporation of water from the oceans, the largest reservoir within the water cycle.

Carbon Cycles Through Earth, Air, Organisms, and Oceans

Much of the carbon in the biosphere is in the oceans, as bicarbonate (HCO_3^-) ions, and in the atmosphere, as carbon dioxide. The oceans contain about 30,000 billion metric tons

of carbon, and the atmosphere contains another 640 billion metric ions. Dissolved carbonates tend to precipitate to the bottoms of oceans and lakes, where they form thick sediments that later turn into sedimentary rocks. These geologic reserves hold some 18 million billion metric tons of carbon.

Organisms contain as much as 3000 billion metric tons of carbon, about 10 percent of what is dissolved in all of the oceans. Each year, terrestrial producers convert some 12 percent of atmospheric carbon dioxide into organic molecules. This carbon then flows through a worldwide food web (Figure 25-11). Organisms eventually return it to the oceans and atmosphere, in the form of carbon dioxide, through the process of respiration. Herbivores, carnivores, and decomposers all contribute to this carbon dioxide production. Even the producers themselves respire carbon dioxide.

The largest reserves of carbon, however, are in the fossil fuels—coal, oil, and methane (natural gas). These carbon-containing materials result from incomplete decomposition of ancient organisms. During some periods—the Carboniferous, es-

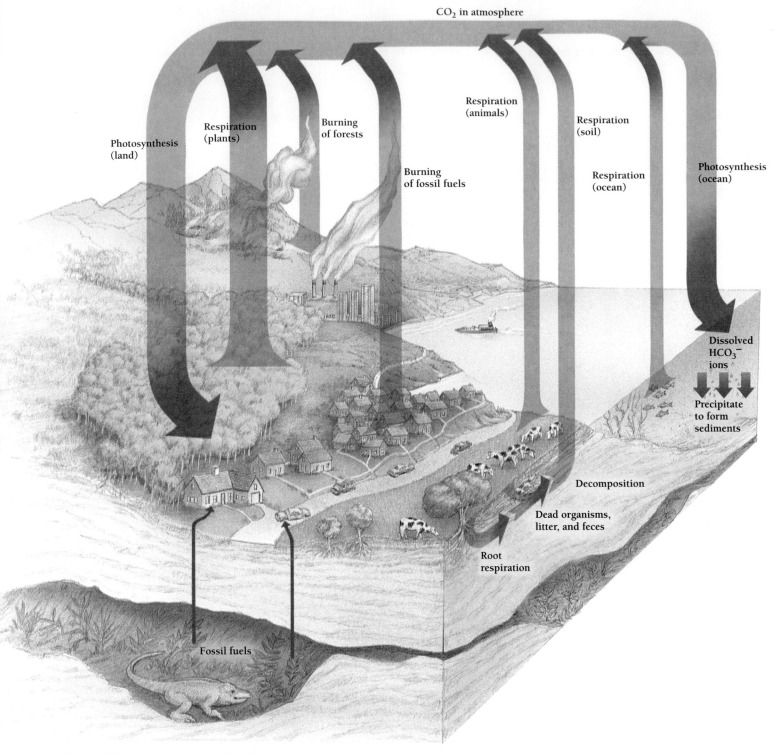

CO$_2$ in atmosphere

Respiration (plants)

Burning of forests

Respiration (animals)

Respiration (soil)

Photosynthesis (land)

Burning of fossil fuels

Respiration (ocean)

Photosynthesis (ocean)

Dissolved HCO$_3^-$ ions

Precipitate to form sediments

Decomposition

Dead organisms, litter, and feces

Root respiration

Fossil fuels

Figure 25-11 **The carbon cycle.** Carbon atoms move from the oceans to sedimentary rocks, from the atmosphere to plants and other organisms and back into the atmosphere, and from fossil fuels (ancient plants) back to the atmosphere.

pecially—decomposers did not convert all the organic molecules of these organisms back into carbon dioxide. Instead, the carbon accumulated. Geologists estimate that fossil fuels contain about 25 million billion (2.5 × 10^{16}) metric tons of carbon.

In the past two centuries, humans have begun returning these vast stores of fossil carbon to the carbon cycle by burning coal, oil, and gas. By reconverting fossil fuels to carbon dioxide—as well as by burning vast stretches of the world's forests—human societies have greatly increased the Earth's annual carbon dioxide production. Higher levels of carbon dioxide, combined with higher levels of other gases that tend to trap heat, have led to the greenhouse effect, which we discuss in Chapter 29.

Carbon atoms cycle through oceans and atmosphere, with large reservoirs in living organisms and in fossil fuels.

Why Is Nitrogen Both Common and in Short Supply?

Molecular nitrogen (N_2) is the most abundant element in the atmosphere, comprising 78 percent of the air we breathe. Nitrogen is also an indispensable part of the proteins and nucleic acids that all organisms need. Yet, ironically, the huge quantities of atmospheric nitrogen are not available to most organisms, and usable nitrogen is often in short supply (Figure 25-12).

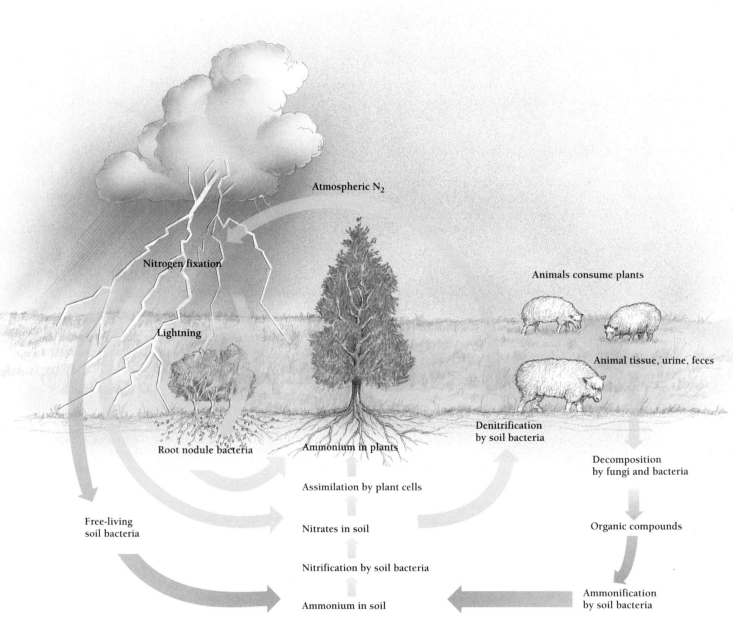

Figure 25-12   **The nitrogen cycle.** Nitrogen atoms in the atmosphere are fixed by bacteria and lightning and assimilated by plants. Nitrogen continuously cycles through the atmosphere, soils, oceans, and organisms.

One reason that atmospheric nitrogen is unavailable to most organisms is that N_2 is chemically so stable. The two atoms are held together by a powerful triple bond (Chapter 2). (This stability is the reason that nitrogen-containing compounds such as TNT are so explosive. They release lots of energy in their conversion to N_2.)

Because it takes a lot of energy to break the N_2 triple bond, most plants can take in nitrogen only as **ammonia** (NH_3) or nitrate (NO_3). Such nitrogen is called **fixed nitrogen,** in contrast to molecular nitrogen. Consumers and decomposers both obtain most of their nitrogen in the form of proteins from animal and plant tissues.

When animals and other organisms break down proteins during respiration, they convert them to ammonia. Because ammonia is toxic when concentrated, aquatic organisms such as fish dilute the ammonia with large quantities of water and flush it from the body. Terrestrial organisms, with less water available to them, first convert ammonia into something less harmful: **urea** in mammals such as ourselves, **uric acid** in birds, reptiles, and insects. Decomposers obtain energy from urea and uric acid by converting them back into ammonia, which can then be used once again by plants. This completes the cycle of fixed nitrogen among plants, animals, and decomposers.

Despite the rapid recycling of nitrogen between plants and animals, many ecosystems lose fixed nitrogen. For example, some soil bacteria obtain energy by converting ammonia to nitrite (NO_2) and nitrate (NO_3) in a process called *nitrification.* Nitrate is highly soluble, and heavy rains easily wash it out of soils and into lakes and streams. Eventually, it ends up in the oceans. Because nitrogen is often in short supply, plants soak up nitrate very easily and try to conserve it by removing it from dying plants, such as fallen leaves, and by storing it in roots and other living parts. However, such conservation measures are often thwarted. For example, when squirrels and caterpillars cut living leaves, the nitrogen-filled leaves fall to the forest floor, where their nitrogen may wash from the ecosystem.

Another important avenue of nitrogen loss involves certain bacteria—called *denitrifying bacteria*—which convert nitrate to atmospheric nitrogen. Denitrifying bacteria obtain energy by using nitrates to oxidize organic compounds.

We know that plants (and animals) depend on a steady supply of ammonia and nitrates. What counterbalances the steady losses of fixed nitrogen from an ecosystem? One important source is lightning, which fixes atmospheric nitrogen during storms. Rain washes the fixed nitrogen to the Earth.

However, the most important source of fixed nitrogen is **nitrifying bacteria.** Nitrifying bacteria in soil or water are specialized to expend the enormous energy needed to convert atmospheric nitrogen (N_2) into ammonia or nitrate. Some of these are free-living bacteria (such as cyanobacteria) that convert nitrogen to ammonia. Others live in mutualistic relationships with certain plants. For example, legumes (peanuts, peas, and beans) have bacteria of the genus *Rhizobium* living within tiny growths on their roots called root nodules. *Rhizobium* bacteria convert

atmospheric nitrogen into a form that plants can use, while the plants provide the bacteria with energy-rich sugars, which the bacteria need to do their work.

Nitrogen atoms move through the atmosphere, soil, oceans, and organisms.

Why Do Organisms Cling to Phosphorus Atoms?

Phosphorus is an indispensable part of both DNA and RNA, as well as ATP and the other energy-rich molecules of metabolism. It is also an important component of bone. Organisms need very little phosphorus compared to carbon and nitrogen. Yet this essential mineral is so rare—in usable form—that organisms have to work hard to get what little they need. For example, on average, plants contain about 3 percent phosphorus. But they must obtain it from water in the soil that contains only about 0.000003 percent usable phosphorus.

Phosphorus is available to plants when it is in the form of orthophosphate ($H_2PO_4^-$). In water, orthophosphate releases phosphate ion (PO_4^{3-}), which is also available to plants. Plants take up phosphate ion from soils, lakes, or seas extraordinarily rapidly—in minutes in the top layers of lakes. However, most forms of phosphorus are insoluble in water. These fall as solid particles to the bottoms of lakes and oceans. Lake and ocean sediments thus tend to accumulate high concentrations of unusable phosphorus salts that have literally fallen out of the ecosystem (Figure 25-13).

The phosphorus that drops out of aquatic ecosystems is replaced very slowly by phosphates that have washed into streams from terrestrial ecosystems. Terrestrial ecosystems recycle phosphorus even more efficiently than do aquatic ones. Within a forest ecosystem, phosphate moves rapidly from soil to plants, from plant to plant, and from plant to animal. The efficient uptake of phosphorus depends to a large extent on mycorrhizae, the fungal associations in plant roots. Nonetheless, terrestrial ecosystems steadily lose small amounts of phosphorus.

Once phosphorus has washed from terrestrial ecosystems and washed into the oceans, its return to the land is slow. Some phosphorus comes to shore within the wastes and bodies of animals that feed on fish and other sea life. Some returns when geologic processes lift sediments into mountains, which then gradually release phosphorus through erosion.

Because ecosystems are sensitive to phosphorus, a small increase in this essential mineral can make a big difference. In fact, when humans dump fertilizers and detergents into aquatic ecosystems, the extra phosphorus upsets the normal balance. Excess phosphorus causes dramatic blooms of photosynthetic

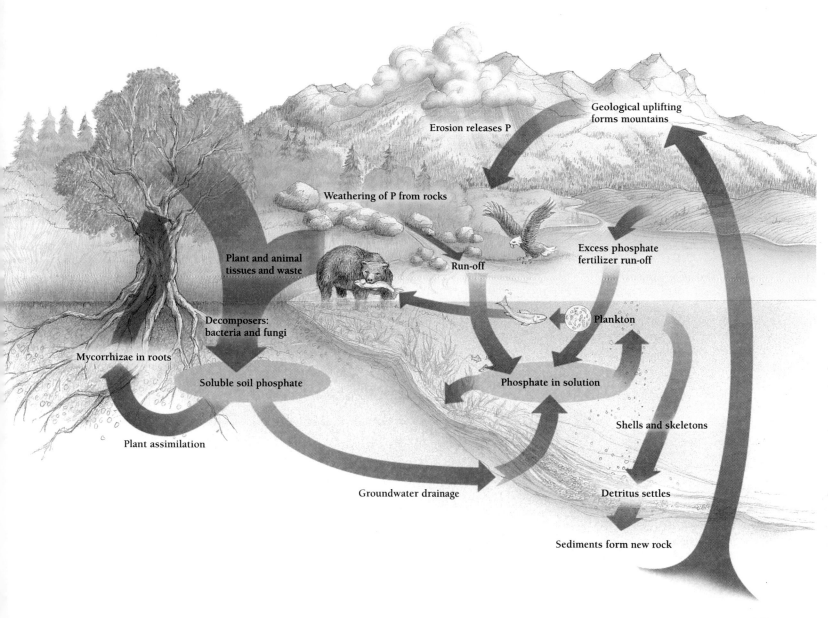

Figure 25-13 **The phosphorus cycle.** Phosphorus is both rare and indispensable to living organisms. It leaves ecosystems easily and returns slowly. When this scarce mineral is available, organisms take it up rapidly and efficiently.

algae and cyanobacteria. These clog the surface and cause declines in the oxygen of the water, killing fish and other organisms.

Phosphorus washes or precipitates out of ecosystems all too easily and returns only slowly. Consequently, when this scarce mineral is available, organisms take it up rapidly.

Until this point, we have limited our discussion to the flow of energy and the cycle of materials within ecosystems. However, ecosystems are made up of individual species. So we should also consider the characteristic roles each species plays in the flow of energy and the use of materials. In the next chapter, we will see how species interact and how ecosystems change in time. We will also examine the role of evolutionary history and current competition in establishing a community's structure.

STUDY OUTLINE WITH KEY TERMS

In the 1960s James E. Lovelock proposed that the Earth's atmosphere, oceans, and soils be considered integral with all living organisms, in a single, immense, self-regulating (**homeostatic**) superorganism called Gaia. The idea has not been accepted by most scientists, but remnants of his idea persist in the current understanding of what the biosphere is and how it works.

Ecology is the science concerned with understanding the relations among organisms and their environment. Ecologists study **populations, communities,** and **ecosystems.** A **system** is an assemblage of interacting parts or objects. An ecosystem is the living (**biotic**) community together with the nonliving (**abiotic**) components of the community's environment, including soil, water, air, and weather. The **biosphere** is the sum of all ecosystems. An ecosystem shares some properties of an organism, but not others. An ecosystem is a functional unit with **emergent properties.**

The biotic part of an ecosystem consists of **producers, consumers,** and **decomposers.** The producers are all **autotrophs,** which belong to the first **trophic level** in every **food chain.** Consumers and decomposers are **heterotrophs.** They occupy different trophic levels, from the second level up, depending on what they eat. Many animals, including humans, bears, and crows, eat from several levels of a food chain. Such animals are called **omnivores.** Decomposers such as **saprophytes** and **scavengers** break down whole organisms and **dentritus.**

Food chains are usually interwoven into a complex **food web.** One way of thinking about communities is in terms of a **pyramid of numbers** of organisms, a **pyramid of biomass** (dry weight), or a **pyramid of energy** (also called a **pyramid of productivity**). Only the pyramid of energy is always an upright pyramid.

Almost all the energy that drives ecosystems comes from the sun. Producers capture this sunlight energy, and organisms at higher trophic levels obtain what they can. Each trophic level has access to only about 10 percent of the **primary productivity** available to the one beneath it. This is called the **ten percent law,** and is another way of describing the upright energy pyramid.

Ecosystems recycle water, carbon, nitrogen, phosphorus, and many other materials in **biogeochemical cycles.** Water enters the atmosphere from the oceans through evaporation and from the continents and islands through evaporation and **transpiration** from plants. Atmospheric water condenses into clouds, then precipitates as rain, sleet, or snow. Rain and snowmelt form streams and rivers that carry fresh water to the oceans. Along the way, fresh water carries minerals, organic materials, organisms, and wastes into and out of ecosystems.

Plants and other producers absorb carbon dioxide and turn it into carbohydrates through photosynthesis. Producers, consumers, and decomposers all produce carbon dioxide by respiring—oxidizing carbohydrates back to carbon dioxide. In aquatic ecosystems, carbonates tend to precipitate to the bottom to form thick sediments. These sediments form rocks that are eventually lifted up and eroded. Large reserves of carbon also exist in the form of fossil fuels.

The Earth's atmosphere is 78 percent nitrogen. But this nitrogen is not available to most organisms until lightning or **nitrifying bacteria** transform it into **ammonia** or nitrate, which plants can use to make proteins. Consumers and decomposers get such **fixed nitrogen** by eating plant proteins, or each other. Most aquatic organisms eliminate waste nitrogen as ammonia. Terrestrial organisms eliminate it in the form of **urea** or **uric acid.** Ecosystems tend to conserve nitrogen, but some is lost when rain washes it from fallen organic matter and carries it to the sea.

Phosphorus washed into aquatic systems from terrestrial ecosystems is rapidly absorbed. Since phosphorus is relatively insoluble in water, the rest is lost in the form of precipitates that fall to the bottom. Phosphorus returns to terrestrial ecosystems extremely slowly by way of uplifted sediments and the guano of seabirds.

REVIEW AND THOUGHT QUESTIONS

Review Questions

1. What is ecology?
2. What distinguishes an ecosystem from a community?
3. What properties do ecosystems share with organisms? How are ecosystems different from organisms?
4. Distinguish between autotrophs and heterotrophs. Give an example of each.
5. Distinguish between a food chain and a food web.
6. Explain why the pyramid of energy is always narrowest at the top, but the pyramids of both numbers and biomass are not necessarily narrowest at the top.
7. Explain how the biomass of aquatic producers can be smaller than the biomass of the consumers.
8. Where does the carbon dioxide in the atmosphere come from?
9. In what three forms do organisms eliminate broken down pro-

teins? What form do aquatic organisms use, and why don't terrestrial organisms use the same form?

Thought Questions

10. Describe the steps of the nitrogen cycle. How is nitrogen lost from ecosystems? How is it replaced? (Mention two ways.) Why does it need to be "fixed"?
11. Suppose a strawberry farmer treated his fields with a fumigant (poison gas) that killed all of the bacteria in the soil. Describe two different things the farmer could do to ensure that the strawberries received an adequate supply of nitrogen.
12. Suppose that all of the primary producers were removed cleanly and instantaneously from the biosphere. List ten immediate consequences (e.g., declines in oxygen levels) and ten indirect consequences.

SELECTED READINGS

Botkin, Daniel B., *Discordant Harmonies: A New Ecology for the Twenty-First Century,* Oxford University Press, New York, 1990. An ecologist examines our relationship with nature.

Carson, Rachel, *Silent Spring,* Fawcett Crest, New York, 1962. A scholarly and beautifully written argument against the overuse of pesticides by a government biologist and prize-winning writer.

Leopold, Aldo, *A Sand County Almanac,* Oxford University Press, New York, 1966. A set of often lyrical discussions of the relations between human activity and the environment, by an eminent ecologist.

Lovelock, J.E., *Gaia: A New Look at Life on Earth,* Oxford University Press, New York, 1979. Lovelock's highly readable argument in favor of Gaia.

Worster, Donald, *The Wealth of Nature,* Oxford University Press, 1993. A collection of 16 essays on environmental history.

▶ On-line materials relating to this chapter are on the World Wide Web at http://www.saunderscollege.com/lifesci/
Click on Tobin/Dusheck: *Asking About Life.*

Volcano: An Ecosystem Is Destroyed

At 8:32 a.m., on May 18, 1980, a medium-sized earthquake shook Mount St. Helens in Washington State (Figure 26-1). In just 20 seconds, the entire north side of the mountain roared and slid away. It was the largest landslide in recorded history. Explosions from the volcanic eruption tore through the 2.5 cubic kilometers (0.6 cubic miles) of sliding debris, spewing rocks, ash, gas, and steam across adjacent valleys at velocities approaching the speed of sound. The series of blasts, 500 times as powerful as the bomb that flattened Hiroshima, destroyed an area the size of the city of Chicago—a forest with enough timber to build 300,000 houses. Every tree within 25 kilometers (15 miles) was either blown to bits, burnt to a crisp by 600°C gases, buried under tons of volcanic ash, or blown down. From the north face of the mountain poured a river of melted snow and ice, mud, boulders, and ash. In less than 10 minutes, this colossal river of rocks filled a valley the size of Manhattan to an average depth of 45 meters (deep enough to bury a 15-story building). A black plume of pumice and ash rose 25 kilometers into the air, and 520 million tons of ash began drifting eastward across the United States.

The death toll included an estimated 11 million fish, 27,000 grouse, 11,000 hares, 6000 black-tailed deer, 5200 elk, 1400 coyotes, 300 bobcats, 200 black bears, 57 human beings, and 15 mountain lions. Three more explosions in May, June, and July dropped thick blankets of pumice and ash. In the end, every exposed slope within 10 kilometers (6 miles) of the crater was stripped of trees and other vegetation and buried under 2 meters of ash and rock. In places, the mud, ash, and rock piled up 200 meters (an eighth of a mile) thick. The scene was as gray as a moonscape. For another 16 kilometers (10 miles) from the blast zone, thousands of full-grown fir and hemlock trunks lay flattened in the ash. It was hard to imagine how anything could have survived.

Yet, an enormous amount of life did survive (Figure 26-2). Along snow-covered ridges that faced away from the blast, saplings still bent by the weight of 1 or 2

▲ **Figure 26-2** **Life returns.** Ecologists predict that Mount St. Helen's ecosystem will return to mature coniferous forest in about 200 years. *(J. Lotter Gurling/Tom Stack & Associates)*

◀ **Figure 26-1** **Destruction of an ecosystem.** When Mount St. Helens erupted in 1980, it destroyed a 230-square-mile ecosystem, about the size of the city of Chicago. What was destroyed? *(David Weintraub/Photo Researchers)*

meters of winter snow, as well as seedlings and shrubs buried under the snow, came through unscathed. Shrubs that were obliterated at the surface, but whose roots survived underground, sprouted and grew anew. From underground, burrowing animals such as pocket gophers and a variety of insects and other invertebrates emerged. Within weeks, plants began growing on the naked slopes of the mountain, wherever rain had washed away the sterile ash to expose the old soil beneath.

Within 10 years, the slopes of Mount St. Helens were covered with dogwood, elder, huckleberry, and other shrubs, as well as seedlings of fir, hemlock, and other conifer trees. Ecologists say that by 2030, the area will begin to look like a young coniferous forest. In 200 years, the area should resemble an old-growth forest. The Mount St. Helens ecosystem—almost completely destroyed—will rebuild itself.

The process by which Mount St. Helens renews itself is called **succession,** which is the predictable change in numbers and kinds of organisms in a community over time. How and why succession occurs has been one of the burning questions in ecology for more than a hundred years.

KEY CONCEPTS

1. Ecologists view the groups of species that compose communities and ecosystems as both integrated and individual.

2. The interactions among species in an ecosystem include symbiotic, competitive, and predatory relations.

3. The relationships among species in an ecosystem depend on both current and historical interactions.

4. Over time, no two species can occupy the exact same niche in the same place.

5. The number of species that an ecosystem sustains depends partly on abiotic characteristics of the habitat.

6. Over time, communities undergo predictable changes, a process called succession.

7. Succession occurs because pioneer or successional species improve environmental conditions for subsequent species.

ARE COMMUNITIES REAL?

In 1898, one of America's first ecologists, Henry C. Cowles (1869–1933), published a groundbreaking paper on plant succession. As a graduate student at the University of Chicago, Cowles studied the succession of plants on sand dunes beside Lake Michigan. The sand dunes are created partly by the gradual retreat of Lake Michigan over thousands of years and partly by fierce winds that blow the sand from the former lake bottom into giant dunes. Cowles found that the dunes were first colonized by a distinct set of plants with extensive roots. Once these colonizers had stabilized a dune, other groups of plants then grew in amongst them and eventually replaced them.

Sand dunes first arise as the wind deposits sand around an obstacle, such as beach, or marram grass (Figure 26-3). More grasses then colonize the dune and stabilize it, allowing the germination of shrubs, such as bearberry and juniper and, later,

jack pine. Jack pine forest reduces the light intensity and increases moisture retention in the sand and in the accumulating soil, conditions that favor the germination of black oak trees. In some locations, black oak forest gives way to forests of beech and maple.

Animal life also changes during the succession. For example, the principal mammal present in the marram grass stage is the prairie deer mouse. Red squirrels and eastern chipmunks characterize the jack pine stage. The oak stage hosts fox squirrels and eastern chipmunks. Finally, with the establishment of the beech–maple stage, come red fox squirrels, gray squirrels, eastern chipmunks, and two kinds of shrews.

Although Cowles published no other groundbreaking papers, he continued to teach at the University of Chicago, inspiring hundreds of young ecology students. A warm, funny man of infectious high spirits, he was much beloved by his students, many of whom went on to do important work.

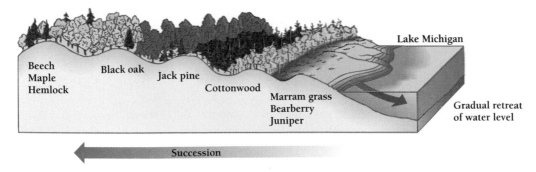

Figure 26-3 **Sand dune succession.** Henry C. Cowles showed that as the water level in Lake Michigan drops, the sand dunes left behind are gradually colonized by a succession of different plant communities. Beach grasses stabilize sand dunes, allowing the growth of shrubs and the formation of soil. Jack pines grow up through the shrubs, shading the soil. In the cool moist soil, oak trees germinate and grow, eventually shading out the jack pines. Oak forest may, in places, give way to beech and maple forest.

But plant succession's most important booster was Frederic Edward Clements (1874–1945). Clements was Cowles's opposite—arrogant, distant, and ascetic, and a prolific researcher and writer all of his life. Clements profoundly influenced 20th-century American ecology. He studied and described every major ecosystem in North America and published, in 1916, the definitive book on plant succession. Clements argued that in every geographic area, defined by its climate and soil, succession always tended toward a single, predictable community of plants, the **climax community.**

Clements argued that the climax community was an organism. Succession, he argued, was a developmental process, like the embryonic development of an organism. Clements wrote, "As an organism the [climax] formation arises, grows, matures, and dies . . . each climax formation is able to reproduce itself, repeating with essential fidelity the stages of its development." According to this view, a plant community is **integrated.**

An integrated community is one that consists of characteristic assemblages of species that interact with each other in predictable ways. According to this view, all maple tree forests have approximately the same species of plants and animals. Further, all these species interact in similar ways.

Ecologists found the integrated view compelling. Anyone can see that certain species tend to live together in particular physical environments. Deserts contain cacti and snakes; ponds contain water lilies, frogs, fish, and mosquitos. Every community is an integrated assemblage of interdependent species, and community boundaries are real, not mere abstract inventions of ecologists.

Clements's holistic view of plant communities, though hotly contested, is one of the most important philosophical themes in ecology. But its simplicity was also its weakness, and it invited criticism.

The first attack on Clements's view came in 1926, when a younger American ecologist named Henry Allan Gleason published a paper frankly challenging Clements. Gleason argued that communities are not integrated, but composed of discrete populations that merely happen to occupy the same habitat. Gleason wrote in part, "In conclusion, it may be said that every species of plant is a law unto itself, the distribution of which in space depends upon its individual peculiarities of migration and environmental requirements . . . [A plant] association is not an organism, scarcely even a vegetational unit, but merely a coincidence." In Gleason's view, communities were **individualistic.**

Gleason's attack on the much-revered Clements so embarrassed other ecologists that most refused to even discuss Gleason's heretical ideas. As Gleason later recalled, "Not one [ecologist] believed my ideas; not one would even argue the matter. . . . For ten years, . . . I was an ecological outlaw. . . ."

In time, though, more and more ecologists began to take Gleason's ideas seriously. A healthy and productive scientific conflict ensued between Clements's holistic ecologists and Gleason's individualistic ones. The main question was whether the assemblages of species that make up communities are "inte-

grated" or "individualistic." Are communities real functioning units that exist outside of human definitions?

Gleason and other individualists argued that every species has an independent distribution, and that, in effect, every community is unique. Many studies support this view. For example, a survey of Wisconsin forests showed that no two tree species consistently share the same range (Figure 26-4A). Maple trees may be associated with beech trees in one place, elms in another. In the individualistic view, a beech–maple forest is not an organic entity or community, but a mere association of two species that happen to share the same moisture requirements. In the individualistic view, communities are not necessarily discrete entities with clear boundaries. Species occupying the same community merely happen to share the same abiotic requirements. The individualistic view garners support from the many careful studies of species distribution patterns that show no obvious clusterings of species.

On the other hand, the integrated view of ecological communities rests firmly on thousands of studies of interactions between species, a few of which we will examine in this chapter. For example, the presence of sugar maples appears to limit the distribution of basswood trees, perhaps by competition for water or shading (Figure 26-4B). So, even though the unity of a community may not be as well defined as Clements envisioned, ecologists still think of communities as functioning units whose member species interact in complex ways.

The integrated and the individualistic views are not mutually exclusive ideas, but the extreme ends of a continuum. Contemporary ecologists recognize that both views are valid: a community is both an integrated set of interacting species and a group of organisms that occur together because their ranges happen to overlap. In the next section, we will examine a few of the ways in which different organisms in a community interact.

The individualistic and integrationist views represent extreme ends of a continuum of thought.

HOW DO SPECIES IN A COMMUNITY INTERACT?

Charles Darwin suggested that the flowering of clover in the English countryside depended on the local residents' affection for cats. Clover seed production, Darwin argued, depended on the presence of bumblebees. The number of bumblebees in turn depended on the number of field mice, which destroy the bees' nests. And the number of field mice depended on the number of cats in the neighborhood. Darwin was only guessing about the relation between cats and clover seed, but real examples of the complexity of species relations abound.

In fact, some ecologists argue that every species ultimately affects every other species. The American biologist Barry Commoner stressed this interaction and named it the "First Law of Ecology."

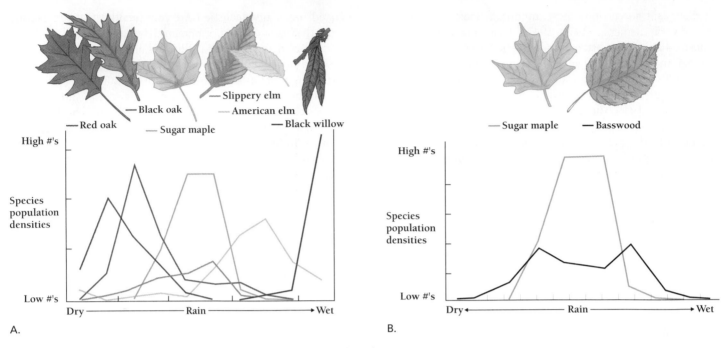

Figure 26-4 Are communities individualistic or integrated? Studies of trees in Wisconsin support both ideas. A. Oaks, maples, and other trees are distributed according to the dampness of the soil, independently of other trees. B. But basswood grows best where sugar maple is absent, even though it is capable of growing in the same soils.

The general attributes of a community depend on abiotic factors such as temperature, moisture, sunlight, and nutrients. Annual rainfall, for example, largely determines whether an area is a forest or a desert. However, abiotic factors alone do not define a community. Within each community, organisms interact with one another in many ways.

Interactions include those between the consumers and the consumed, between competitors, and between various species that associate with one another. Interactions between the consumers and the consumed include predator–prey interactions, herbivore–plant relations, and parasitism. For example, the wolf hunts the caribou, and the caribou eats the lichen. Species may also compete—for territory, prey, or other resources—or they may form intimate associations, wherein one species benefits from another or where two species help each other. The lichen, for example, turns out to be not one organism, but two, an association of a fungus and an alga.

Who Eats Whom?

We usually think of consumers as being larger and stronger than the consumed. Hawks, for example, generally prey on mice, squirrels, and other animals smaller than themselves. Most predators are indeed larger than their prey, although carnivores that hunt in groups, such as wolves, regularly bring down prey far larger than themselves. But herbivores are often smaller than the individuals they consume. Hordes of tiny oak moth caterpillars can strip all the leaves from a 60-foot oak tree. And parasitic consumers, such as ticks, fleas, and tapeworms, are generally smaller than their hosts.

Consumers, then, come in many forms. The main ones (besides the decomposers and scavengers) are herbivores, predators, and parasites. A **parasite,** by definition, consumes only part of the blood or tissues of its host and does not necessarily kill the host. In contrast, a **predator** usually (but not always) kills its prey and consumes most of its prey's body.

In some cases, the differences between predators and parasites may blur. For example, parasites do occasionally kill their host outright, and parasites frequently weaken the host so severely that the host dies of other causes. And some predators consume only part of their prey. For example, some predaceous fish bite chunks of flesh from other fish, without necessarily killing the victim.

Herbivores at first seem to fall outside these definitions, but they actually fit rather nicely. Leaf-eating caterpillars and deer, for example, browse on living plants, just like parasites. And seed-eating birds, which kill and consume an entire plant embryo, may be considered predators. In reality, ecologists rarely stress the parasite–predator distinction in herbivores.

How Do Plants and Animals Defend Themselves?

With notable exceptions, consumers' adaptations for finding and eating their favorite food exhibit a certain sameness: in the case of predators, swiftness, intelligence, acute senses, and sharp teeth; in the case of herbivores, patience and a good digestive system. In contrast, organisms have an arsenal of different weapons for defending themselves against potential consumers.

Figure 26-5 Camouflage. Many animals avoid being eaten by blending in with their background. *(Michael Fogden/DRK Photo)*

Figure 26-6 Chemical warfare. Everyone knows that skunks fight back with noxious chemicals. So do many plants and insects, such as this bombardier beetle. *(Thomas Eisner and Daniel Aneshansley/Cornell University)*

Camouflage

The subtlest defense against being eaten is to remain invisible. The rabbit, for example, simply dives down a hole or freezes, so that it blends into a background of shrubs or grass. But many organisms go to far greater lengths to **camouflage** themselves by mimicking twigs, leaves, stones, bark, and other materials in their environment (Figure 26-5). The walking stick, for example, looks just like a twig. Some species even sway slightly, as though an errant breeze were blowing.

Chemical Defense

Plants defend themselves against consumers in a variety of ways. Trees protect themselves from termites and other wood-eating creatures with thick layers of bark. Roses and cactuses fend off herbivores with thorns. The single most important defense that plants use, however, is toxic chemicals. The unforgettable wallop of hot chili pepper, the sharp tang of raw broccoli, and the deadly poison in poison hemlock all result from certain kinds of **secondary plant compounds**—chemicals that are not essential to a plant's normal metabolism, but which often serve a defensive purpose.

Chemical defense is not limited to plants. Bees, ants, and wasps inject a powerful acid into attackers. The bombardier beetle and the skunk both spray would-be predators with noxious chemicals (Figure 26-6). The famous poison-dart frog (*Phyllobates*) of South America and the skin and feathers of the pitohui bird of New Guinea contain the same deadly toxin (Figure 26-7).

Warning Coloration

If a skunk looked like a rabbit, it would be attacked over and over again, in spite of its potent defense. But because a skunk has a striking and memorable black-and-white pattern, predators quickly learn to avoid the animal. Among all animals, chemical deterrents work best if accompanied by a bright, memorable

design called **warning coloration.** Warning coloration, such as the yellow-and-black stripes of the aggressive, stinging yellow jacket, helps predators remember which prey to avoid. Predators will avoid any animal with the colors and patterns they associate with pain, illness, or other unpleasant experiences. Most

Figure 26-7 Warning coloration in a pitohui bird? In 1992, researchers discovered three species of birds in the genus *Pitohui*, whose skin and plumage contained the same toxin as that found in poison dart frogs of South America. Like wasps, skunks, and poison-dart frogs, pitohui birds bear striking colors that may function to warn predators. Research has shown that inconspicuously colored birds taste better than flashy ones. *(John Anderton)*

Figure 26-8 Müllerian mimicry. A. The monarch butterfly is so noxious that birds such as this jay quickly cough them up. B. Viceroy butterflies are unrelated but also toxic. Monarchs (left) and viceroys (right) have evolved amazingly similar color patterns. Each butterfly's message reinforces the other's: "Don't touch!"   *(A, Lincoln Brower; B, Thomas C. Emmel)*

A.

B.

warning colors are vivid shades of red, yellow, orange, and white combined with a contrasting shade of blue or black.

Because perdators avoid animals with warning coloration, other organisms copy the warning coloration of others. A situation in which one animal *mimics* the color pattern of another is called **mimicry.** Some perfectly harmless and quite edible flies, for example, mimic the black-and-yellow stripes of yellow jackets and bees. A situation in which a harmless animal imitates a dangerous one is called **Batesian mimicry.**

Ecologist Jane Brower demonstrated the effectiveness of Batesian mimicry by painting mealworms with bright paints and presenting both unpainted and painted mealworms to starlings. She dipped a few of the painted worms in a badtasting liquid. The rest were perfectly edible. The starlings quickly learned to avoid the painted mealworms, especially when more than about 60 percent of them tasted bad.

In a second form of mimicry, called **Müllerian mimicry,** two or more dangerous species evolve similar colors. The similar colors signal a similar hazard to any predators the two mimics share in common. For example, the monarch butterfly is toxic to birds. The monarch caterpillar feeds on milkweed, a plant that contains a toxic secondary plant compound. This toxin accumulates in the monarch and makes it toxic to birds. A bird that eats a monarch quickly becomes ill (Figure 26-8A). After a few such experiences, the bird avoids monarchs, and the monarch's striking warning colors help the bird remember.

A second butterfly, the viceroy, is unrelated to the monarch but also contains a toxin that makes birds sick. The viceroy looks remarkably like a monarch, and biologists believe these two butterflies have evolved to resemble one another (Figure 26-8B). The advantage to each butterfly is that birds that have learned to avoid the monarch also avoid the viceroy and birds that avoid the viceroy also avoid the monarch.

Müllerian mimicry helps predators learn the same lesson from a variety of prey species. Bees and wasps both use a similar pattern of yellow and black stripes. Anyone who has been stung by a bee also learns to avoid wasps and hornets. Such

encounters lead to greater safety for bees, wasps, and hornets. The more species that use these kinds of patterns and colors, the more this universal signal is reinforced.

Symbiosis: How Do Organisms Live Together?

A species may live in an intimate association with another species, an arrangement called **symbiosis,** meaning literally "living together." Symbiosis may be cooperative or antagonistic. Table 26-1 summarizes some of the possible interactions between any two species. Each species may affect the other positively, negatively, or not at all. In **parasitism,** one species benefits at another's expense. Tiny parasitic wasps, for example, lay their eggs inside of caterpillars. The wasp larvae literally eat the caterpillars alive as both animals mature.

Mutualism describes a symbiotic relationship between two species that mutually benefits each. The lichens that encrust rocks, for example, are mutualistic associations of fungi and algae. Another example is the bull's horn acacia (*Acacia cornig-*

Table 26-1 Interactions Among Organisms

Interaction	Effect on A	Effect on B
Competition between A and B	Harmful	Harmful
Predation by A on B	Beneficial	Harmful
Symbiosis		
Parasitism by A on B	Beneficial	Harmful
Commensalism of A with B	Beneficial	Harmless
Mutualism between A and B	Beneficial	Beneficial

Figure 26-9 **Commensalism.** The brightly colored anemone fish lives peacefully in the arms of the sea anemone, which normally eats fish this size. *(W. Gregory Brown/Animals Animals)*

era) of Central America, which has special enlarged thorns that house colonies of ants (*Pseudomyrmex ferruginea*). The acacia's nectar attracts the ants, which use it as their major food source. In return, the ants protect the acacia from being eaten by caterpillars and other herbivorous insects. This arrangement benefits both species. One study showed that acacias grown without ants for 10 months weighed less than one-tenth of those with intact ant colonies.

Another kind of symbiosis is **commensalism,** an intimate relationship between two species that helps one but neither helps nor harms the other. For example, certain species of fish live in amongst the poisonous tentacles of sea anemones, as commensals, benefiting from the protection of their host. For reasons unknown, the sea anemone, which catches and eats other species of fish, leaves the anemone fish alone (Figure 26-9).

Species interact as competitors, as predator and prey, and as symbionts.

Do Organisms Coevolve?

The mutualistic relationship between ants and bull's horn acacias demands special adaptations in each species. The acacia has evolved enlarged thorns with a tough woody exterior and a soft pithy center, easy for the ants to excavate for nests. The acacia also has "nectaries" at the bases and tips of its leaves to supply the ants with nectar. The ants exhibit special adaptations as well. While most species of ants are active only at night or only during the day, members of the species *Pseudomyrmex ferruginear* remain active and protect the acacia day and night.

These distinctive adaptations in the ant and the acacia appear to result from a long and mutual evolutionary history in which the needs of each species exert selective pressure on the other. The interdependent evolution of two or more species whose adaptations appear to be selected by mutual ecological interactions is called **coevolution.**

Biologists love to suggest likely examples of coevolution. For example, many pollinating insects specialize on flowers that seem, in turn, to be specially constructed to be pollinated by just one kind of insect. Mutualisms such as these present the most likely examples of coevolution. But demonstrating that coevolution has occurred in individual cases is difficult.

Further, all species that have ecological interactions may exert some selective pressure on one another. And virtually every species in a community interacts (however indirectly) with every other species. For example, the bacteria that break down organic matter to nutrients that plants can absorb through their roots are as essential to the herbivores and carnivores as to the plants. Coevolved relationships can be either very specific, as in the case of the acacias and the ants, or very diffuse. The concept of "diffuse coevolution" is another way of acknowledging that all organisms in a community affect one another.

Interactions between two species influence the evolution of both.

How Does Competition Affect Organisms?

Some species interact as consumers and consumed, and others as commensals or mutualists. Still others compete. In **competition** one organism uses a resource in a way that limits the availability of that resource to others. For example, if two species of deer both eat grass, and live in the same area, and the grass is in limited supply, the deer are in competition with each other.

What Is Interspecific Competition?

Competition in natural communities is often hard to detect. But laboratory studies show how competition occurs in simple situations and suggest questions for ecologists to ask about the role of competition in natural communities. The Russian ecologist G.F. Gause studied populations of two similar species of *Paramecium*—*P. caudatum* and *P. aurelia*. Each *Paramecium* species, cultured by itself, flourished on the bacterial food that Gause provided. But when Gause mixed the two species, *P. caudatum* disappeared from the cultured within two weeks. Gause provided only one environment and one kind of food for the two species, and *P. aurelia* was evidently better adapted to Gause's conditions than *P. caudatum.* One species thus replaced the other.

These and similar experiments have led to a generalization called the **competitive exclusion principle,** which says that when two species compete directly for exactly the same limiting resources, the more efficient species will eliminate the other. A **limiting resource** is one that is in short supply. For example, territories might be a limiting resource for a pride of lions, but air would not be. Similarly, water is often a limiting resource for terrestrial plants, but not for marine plants.

The competitive exclusion principle predicts that one species always wins, but the same species does not win under all conditions. For example, in a 40-year series of experiments at the University of Chicago, Thomas Park and his colleagues studied the effects of different environments on competition between two species of grain beetles (*Tribolium castaneum* and *Tribolium confusum*). These beetles were allowed to proliferate in containers of flour, the favorite food of the larvae. When grown separately, both species did best in warm, moist conditions. When grown together, *T. castaneum* drove *T. confusum* to extinction—but only when the container was warm and moist. In a cold, dry container, *T. confusum* prevailed.

According to the competitive exclusion principle, no two species can simultaneously exploit the same resources in the same place. The place in which an organism lives, along with the set of environmental conditions that characterize that place, is called its **habitat.** In contrast, the way an organism *uses* its environment is called its **niche.** The habitat of the desert tortoise, for example, is the desert. The niche that the tortoise occupies is that of burrowing herbivore. We may think of habitat as an organism's address and of niche as its occupation.

We can now restate the competitive exclusion principle: no two species can occupy the exact same niche in the same habitat indefinitely.

No two species can indefinitely occupy the exact same
niche in the same habitat.

Does Competitive Exclusion Occur In Natural Communities?

In the laboratory, species can be made to compete for the same limiting resources. But in nature, two species occupying the same habitat can, in practice, split up the niche in an infinite number of ways. For example, the ecologist Robert H. MacArthur (1930–1972) tested Gause's competitive exclusion principle in field studies of five species of North American warblers (Figure 26-10). All five bird species hunt for insects and nest in the same kind of spruce tree. The niches of these five

birds seem nearly identical, but MacArthur's studies showed that the five warblers had split up the spruce tree niche extremely finely. He showed that each species of bird hunts insects in a different part of the tree and nests at a separate time from the other warblers. As a result, the warblers' niches overlap without being identical. Ecologists call splitting the niche **resource partitioning.**

MacArthur's warblers avoided competition by splitting the niche. But other, more experimental research suggests that many organisms actively compete with one another. For example, two species of barnacles live in tidepools on the coast of Scotland. One species, *Balanus balanoides,* lives only on the lowest intertidal rocks, which are wet even during low tide. The other species, *Chthamalus stellatus,* lives higher up, in between the low-tide and high-tide lines.

To find out if the two species were competing, the ecologist Joseph H. Connell removed *Balanus* from some tidepools and *Chthamalus* from others. Connell found that where *Balanus* was removed, *Chthamalus* moved down to occupy the newly opened territory and survived there perfectly well. Likewise, where *Chthamalus* was removed, *Balanus* was able to survive on the higher, drier rocks.

Balanus dominated the lower levels because its heavy shell grows under the shell of *Chthamalus* and pries it up off of the rocks, preventing *Chthamalus* from growing on the lowest rocks. Higher up, however, *Chthamalus*'s higher tolerance for dry conditions allows it to prevail over the physically stronger *Balanus.*

Connell's two barnacle species illustrate the difference between a species's **fundamental niche,** its potential ability to utilize resources, and its **realized niche,** the resources that it actually uses in a particular community. The fundamental niche of *Chthamalus* included the lower rocks, but its realized niche did not. The fundamental niche of *Balanus* included the higher rocks, but its realized niche did not.

Similar experiments on rat parasites also illustrate competitive exclusion. Although the tapeworm and the spiny-headed worm both prefer the upper intestine of the rat, when the two worms are together, the spiny-headed worm takes the food-rich upper tract and displaces the tapeworm to the lower end of the intestine.

All these experiments, and many more like them, suggest that competition does occur in natural populations, and can even help determine species distributions.

Experiments show that organisms in natural environments
compete for limited resources.

Character Displacement: Does Competition Influence Evolution?

We have seen how competition between two species can lead to the exclusion of one species from a particular niche. Alternatively, however, species may evolve differences that reduce competition between them so that they can coexist. For example, two species of caterpillars that both eat members of the

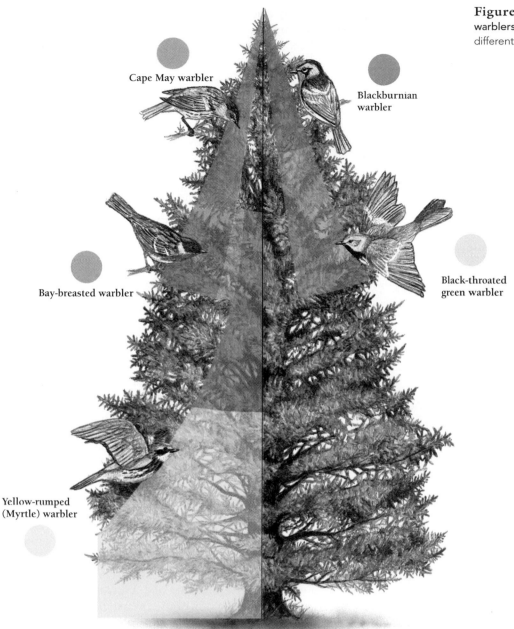

Figure 26-10   Resource partitioning among warblers. Five species of birds pick insects from different parts of a spruce tree.

Cape May warbler

Blackburnian warbler

Bay-breasted warbler

Black-throated green warbler

Yellow-rumped (Myrtle) warbler

rose family (which includes roses, apples, plums, and other species) might reduce competition by specializing, one on apple trees and one on roses. Any change in morphology, life history, or behavior that results from competition is called **character displacement.**

The finches of the Galápagos Islands provide some of the best-documented examples of character displacement. In the case of the Galápagos finches, character displacement has led to speciation. For example, the small ground finch *Geospiza fuliginosa* and the medium ground finch *G. fortis* occur both separately and together, depending on the island. On islands where the birds occur together, individuals of the species *G. fortis* have larger beaks and specialize on larger seeds than *G. fuliginosa*. On islands where the birds do not occur together,

however, *G. fuliginosa* has a larger beak and *G. fortis* has a smaller one. When separate, the two species eat many of the same seeds (Figure 26-11).

On islands where the two species occur together, they hybridize, producing offspring with intermediate sized beaks. When food is scarce (in times of drought), the hybrids reproduce poorly. When seeds are plentiful, however, the hybrids appear to reproduce *better* than their parent species.

Competition and adversity appear to drive these two species apart, while lack of competition and easy pickings drive them together. Currently, they are separate species. Their distinctive adaptations almost certainly have resulted from previous competition. As Joseph Connell put it, character displacement is "the ghost of competition past."

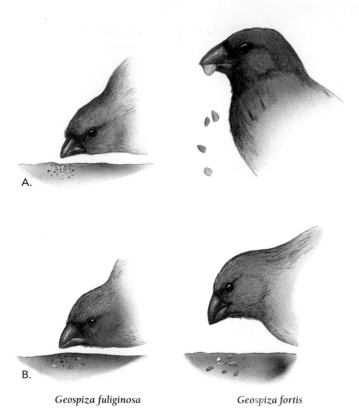

A.

B.

Geospiza fuliginosa   *Geospiza fortis*

Figure 26-11   Character displacement in Darwin's finches. A. On islands where the two species *Geospiza fuliginosa* and *Geospiza fortis* occur together, members of *G. fortis* have much larger beaks and eat larger seeds than do *G. fuliginosa*. B. On islands where the birds occur separately, individuals of each species have beaks of similar size.

Competition, over evolutionary time, causes character displacement through natural selection.

What Determines How Many Species an Ecosystem Can Sustain?

We have seen that if two competing species are to coexist, they must exploit different niches. Both resource partitioning and character displacement increase the number of species within an ecosystem.

Each species occupies a unique niche, and each niche has a unique set of properties, such as food supply, sunlight, and other resources. All the unique but overlapping niches fit together to form a kind of multidimensional jigsaw puzzle—the ecosystem. The number of niches is a measure of the richness and complexity of a community.

What Is Species Diversity?

The number of species in an ecosystem is called **species richness.** However, species richness is only a crude measure of diversity, since some species may occur in great numbers while others are rare. **Species diversity** is one name for a measure

of diversity that takes into account how common each species is. For example, because the city of San Diego has an unusually large and well-stocked zoo, the animal *species richness* of San Diego is rather high. However, only a few species, mostly humans, rats, mice, cockroaches, or ants, comprise the vast majority of organisms living in San Diego. Therefore the *species diversity* of animals is low.

How Many Species Can an Ecosystem Sustain?

Careful study of a particular ecosystem can reveal the number and kinds of species that compose it. But what determines the particular species makeup of a community? Do rainfall and other abiotic components determine the diversity of a biotic community? Or does the biological makeup of each community depend on its unique history—that is, on which plants and animals happen to arrive in what order? Are there any general rules that determine the diversity within a community?

Species diversity depends heavily on abiotic components. For example, ecologists have noticed that diversity generally increases from the poles to the equator, and from high elevations to low elevations. Physically varied areas (with hills, valleys, and streams) allow more species to have unique niches than flat, relatively featureless landscapes. Finally, the larger the size of an ecosystem, the more species it is likely to contain.

Biotic interactions such as character displacement may also increase the number of species in a community. In addition, predators can maintain diversity by preventing one species from successfully competing with another. When one prey species increases in numbers, for example, the predator is likely to prey preferentially on that species and keep it from overwhelming the rarer species.

Some ecologists have suggested that species diversity increases with productivity and stability. For example, a tropical rain forest, which has high diversity, also has enormous productivity and a stable, predictable climate. On the other hand, some relatively unproductive and unstable grasslands and deserts show very high diversity. Overall, there is little correlation between net primary productivity and diversity, and the relative importance of biotic and abiotic factors in ecosystem diversity is still a subject of ecological research.

One way that ecologists have approached this question is by studying the species diversity on islands. They have found that large islands, which have more habitat diversity and more niches, do indeed support more species than small islands. Surprisingly, however, the *number* of species on an island of a given size may remain stable—even as individual species flourish and disappear. For example, in 1968, the Channel Islands, off the coast of Southern California, had about the same number of bird species as they had had in 1917. But nearly a third of the species were different.

In the 1960s, two American ecologists, Robert MacArthur, of Princeton University, and Edward O. Wilson, of Harvard University, argued that the diversity of island species depends on a balance between the rates of immigration of new species and extinction of old species. According to the MacArthur–Wilson theory of island biogeography, a new island, devoid of already established species, should experience a high rate of immigration and a low rate of extinction, since there are fewer species to disappear. As more species arrive, the theory went, newer immigrants would have a harder time establishing themselves because of competition, and the rate of extinction would increase. MacArthur and Wilson also predicted that islands nearer the mainland should have a greater rate of immigration and, therefore, more species than more distant islands.

In 1969, Wilson and ecologist Daniel Simberloff devised an experimental test of the MacArthur–Wilson theory. They counted all the arthropod species they could find on several small mangrove islands (10 to 20 meters in diameter) in the Florida Keys, exterminated all the arthropods (mainly insects) with a chemical pesticide, and then monitored the gradual recolonization of the islands. Within 6 months, the islands had recovered their former arthropod diversity (Table 26-2). They had approximately the same number of species as before they were fumigated. However, the species that colonized the islands were different. This study suggested that the particular species that colonize an island may be a historical accident. But the number of species may well depend on the abiotic characteristics of the island.

Understanding the factors that govern species diversity on islands helps us understand species diversity on the continents. A forest surrounded by croplands or housing subdivisions stands as a kind of island, as does a meadow surrounded by forests or a mountain peak surrounded by valleys. For many species, a small island of habitat cannot support a population of sufficient size for animals to find mates or for animals at higher trophic levels to find sufficient prey. The stability of large ecosystems depends, in part, on having large enough "habitat islands" to replace declining species with new immigrants.

Species diversity, a characteristic of biological communities, is influenced by many abiotic and biotic factors.

Table 26-2 Diversity on Islands in the Florida Keys Before and After Defaunation

	Number of Species	
Trophic Level	Before	After
Herbivores	55	55
Scavengers	7	5
Detrivores	13	8
Wood borers	8	6
Ants	32	23
Predator	36	31
Parasite	12	9
Unknown	1	3
TOTAL	164	140

SUCCESSION: COMMUNITIES UNDERGO CHANGES OVER TIME

In the last section we saw that the characteristics of communities depend on the geography and form of the habitat as well as on the history of the community. In this section we will see how communities change over time.

What Is Succession?

After Mount St. Helens blew up in 1980, the near total destruction of the community of plants and animals that lived on its slopes initiated the gradual development of a new community. Windblown seeds from plants such as fireweed, pearly everlasting, and grasses first colonized the slopes of the volcano, rooting into the remaining soil wherever the thick volcanic ash had been washed away by erosion. Gradually these weedy plants covered the barren slopes of Mount St. Helens. In time, shrubs such as dogwood and huckleberry, and small trees such as fir and hemlock grew in amongst the first colonizers and replaced them. The trees would replace, or succeed, the shrubs, as the shrubs had begun to replace the weeds. Each set of plants represents a stage in **succession,** the change in numbers and kinds of organisms in a community over time. Ecologists distinguish between **primary succession,** the invasion of a completely new environment such as a sandbar or new volcanic island (Figure 26-1), and **secondary succession,** the sequence of stages in a community that has suffered serious damage. Examples of secondary succession include the reestablishment of life in a burnt or logged over forest and the return of natural communities to abandoned croplands—so-called old-field succession.

The first community in a succession is called a **pioneer community,** and the long-lived community at the end of a succession is called a **climax community.** Intermediate stages are sometimes called "successional" communities. Pioneer and successional plant communities are said to change over periods of from 1 to 500 years. These changes—in plant numbers and the mix of species—are cumulative and directional. Climax communities themselves change over time, but over periods of time greater than about 500 years. Thus, "climax community" is a relative term.

Pioneer communities and climax communities differ in a variety of other ways, summarized in Table 26-3. For example, pioneer communities consist of species that flourish in disturbed areas and breed rapidly. Climax communities are characterized by species that breed relatively slowly, but gradually take over in undisturbed areas. In general, the short-lived weedy species of pioneer communities (thistles, for example) are gradually replaced, or succeeded, by longer-lived climax organisms (Douglas fir trees, for example).

Most descriptions of succession refer to changes in vegetation because succession is a concept introduced by plant ecologists. However, other kinds of succession exist. For example, a deer's carcass, a rotting tree, or a pile of dung can support a succession of insects, bacteria, and fungi over periods of 2

Succession on a corpse

"The worms crawl in, the worms crawl out" goes the childhood ditty. But which "worms," and exactly when they crawl in and out, can tell special criminal investigators how long ago, and often where, a person died. By studying the crawling creatures collected from a corpse, forensic entomologists, scientists who apply their knowledge of insects to legal matters, can solve otherwise baffling cases.

The father of American forensic entomology, Bernard Greenberg of the University of Illinois at Chicago, likes to recount the case of a missing 9-year-old girl. After her partly skeletonized body was discovered in an abandoned apartment building, Greenberg collected fly larvae and pupae from the body and took them back to the laboratory. There he raised them at temperatures like those in the apartment building over the days preceding the girl's discovery. He soon learned that the flies took 21 days to mature under those conditions, which meant that the fly eggs had to have

been laid 21 days before the girl was found. The last time the little girl had been seen alive was 21 days earlier, near the apartment building in the company of a man in his 20s. As a result of Greenberg's testimony, this man was subsequently convicted of her murder.

Beetles, flies, and other arthropods colonize corpses in a predictable succession. "Flies are the best investigators of corpses," says Carl Olson, an entomologist at the University of Arizona. Metallic blue or green blowflies, drawn by odors that include a compound called cadaverine, can find an uncovered corpse within ten minutes of death. Blowflies and flesh flies lay eggs in wounds and moist natural body openings, such as eyes, ears, nose, and mouth. Inside the body, bacteria go to work decomposing the body, generating various gases in the process.

As the masses of maggots feed, their activity and that of the bacteria raises the temperature inside the corpse. "It's like when you walk into a room full of people—

notice how hot is?" Olson asks. "The maggots on a corpse create their own environment."

The maggots eventually puncture the skin, initiating what is called the decay stage and releasing what University of Hawaii researcher Lee Goff calls "strong, distinctly unpleasant odors." As the fly larvae complete their feast, they hop or crawl off to pupate in the soil. By that time, the latter part of the decay stage, the fly larvae have created a hospitable environment for the next wave of insects, primarily hide and rove beetles. It is a classic example of ecological succession. Any remaining larvae become food for the visiting beetles, which also finish off the flesh.

By the post-decay stage, only the skin and bones remain. A different set of beetles now arrive. These feed on dried skin and hair. They are the same insects that destroy fur coats and cashmere sweaters. When the body is reduced to bone alone, all of the carrion-feeding arthropods depart.

months or less. In such cases, there is no climax community, and succession occurs on a small scale in both time and space.

Succession in plant and other communities usually implies a progression through a series of transitional stages to a climax community.

Table 26-3 How Plant Community Traits Change in the Course of Succession

Trait	Early Stages	Later Stages
Biomass	Small	Large
Physiognomy	Simple	Complex
Role of detritus	Minor	Major
Net primary production	High	Low
Stability (absence of change)	Low	High
Plant species diversity	Low	High
Seed dispersal by	Wind	Animal
Seed longevity	Long	Short

After Barbour, 1980.

Why Does Succession Occur?

Clements argued that succession occurs because the organisms of nonclimax communities degrade their own environment. For example, in a succession that follows a forest fire, the first plants to invade are adapted to live in sunlight. But the shade created by these sun-loving plants prevents their own seedlings from growing and favors shade-tolerant plants.

While many of Clements's scientific heirs still contend that each stage in a succession prepares the way for the next, other ecologists argue that this is not always true. Some of the changes in a succession, they say, merely reflect differing rates of growth. Shrubs, for example, appear to come before trees even when both types of seedlings start at the same time, because the shrubs take less time to reach maturity. In fact, rather than preparing the way for new species, some species present in early stages of a succession may actually inhibit the establishment of the later species. For example, birch and aspen trees can prevent succession by white spruce when their wind-whipped branches knock the new growth off of young spruce trees.

We can see that species present at early successional stages can facilitate succession, tolerate succession, or inhibit succession. Especially in primary successions, however, pioneer organisms inevitably facilitate succession. This is only because no other organism can colonize a pile of sand or a piece of bare rock until the way has been broken. For example, the formation of soil from bare rock depends on the action of pioneer lichens and plants adapted to thin or nonexistent soils. Once these pioneer species have created a thin layer of soil, other species can move in and, often, take over.

Sand dunes are one of the best-studied examples of a primary succession. Henry Cowles's characterization of sand dune succession on the shores of Lake Michigan still stands as the classic example of succession.

Secondary succession in an abandoned field or even in a vacant urban lot, called "old-field succession," also has a stereotypical pattern. The first plants to arrive are grasses and herbs—nonwoody plants. The first herbs are **annuals,** which complete their life cycles, from seed to mature plant, in a single growing season and then die. Within a few years, however, **perennials,** which survive and produce seeds for 2 or more years, move in. Because perennials' roots are already grown at the beginning of the growing season, perennials can grow new foliage and produce new seeds faster than annuals. Thus, once perennials are established, they can often win the competition with annuals for light and water. Soon shrubs begin to appear, then sun-tolerant trees. As these trees grow and shade the smaller plants, shade-tolerant herbs and shrubs replace the earlier, sun-loving ones. After a very long time, shade-tolerant seedlings grow into shade-tolerant trees to produce a climax forest.

It is important to remember that in all cases where early successional species facilitate colonization by later species, the early species are not "preparing" the environment in any directed or purposeful sense. The changes some species make to their environment merely happen to benefit other species.

Clements argued that succession occurs because pioneer or successional species change their environment in such a way as to improve conditions for succeeding species.

What Determines the Character of a Climax Community?

The character of a climax community depends heavily on such abiotic factors as climate, soil, and terrain. In the dry, midwestern American prairies, for examples, grasses, not trees, dominate. In the moister Southeast, pines, oak, and hickory hardwoods dominate the forests. In fact, early ecologists believed that a given area had only one possible climax community. The climax community for the northeastern United States, for example, was believed to be beech–maple. Any other group of species growing in the Northeast was considered to be successional, no matter how long-lived.

Communities assemble themselves flexibly, however, and their particular structure depends on the specific history of the area. For example, giant sequoia forests in the southern Sierra Nevada include trees as old as 2000 years. These forests ap-

Figure 26-12 **A grove of giant sequoias.** Such rare groves depend on frequent forest fires to prevent succession to white fir and incense cedar. *(Barbara Gerlach/DRK Photo)*

pear to be a classic example of a long-lived climax community. However, the forests are maintained by periodic fires that sweep white fir and incense cedars from beneath the mammoth sequoias as often as every 7 or 8 years (Figure 26-12). When all fires are quelled, white fir and incense cedars grow up among the sequoias and compete for nutrients. Eventually, when a fire does come, the smaller trees pass fire up into the crowns of the giant sequoias, which kills them. Once that happens, the white fir and incense cedar take over. These results of human intervention challenge the idea that a sequoia forest is a climax community. It may instead be a long-lived successional community.

The character of a climax community depends on climate, terrain, and history.

What Determines the Long-Term Stability of an Ecosystem?

An ecologist who studies a pond today may well find it relatively unchanged a year from today. Individual fish may be replaced, but the number of fish will tend to be the same from one year to the next. We can say that the properties of an ecosystem are more stable than the individual organisms that compose the ecosystem.

At one time, ecologists believed that species diversity made ecosystems stable. They believed that the greater the diversity the more stable the ecosystem. Support for this idea came from the observation that long-lasting climax communities usually have more complex food webs and more species diversity than pioneer communities. Ecologists concluded that the apparent stability of climax ecosystems depended on their complexity. To take an extreme example, farmlands dominated by a single crop are so unstable that one year of bad weather or the invasion of a single pest species can destroy the entire crop. In contrast, a complex climax community, such as a temperate forest, will tolerate considerable damage from weather or pests.

The question of ecosystem stability is complicated, however. The first problem is that ecologists do not all agree what "stability" means. Stability can be defined as simply lack of change. In that case, the climax community would be considered the most stable, since, by definition, it changes the least over time.

Alternatively, stability can be defined as the speed with which an ecosystem returns to a particular form following a major disturbance, such as a fire. This kind of stability is also called **resilience.** In that case, climax communities would be the most fragile and the *least* stable, since they can require hundreds of years to return to the climax state.

Even the kind of stability defined as simple lack of change is not always associated with maximum diversity. At least in temperate zones, maximum diversity is often found in mid-successional stages, not in the climax community. Once a redwood forest matures, for example, the kinds of species and the numbers of individuals growing on the forest floor are reduced. In general, diversity, by itself, does not ensure stability.

Mathematical models of ecosystems likewise suggest that diversity does not guarantee ecosystem stability—just the opposite, in fact. A more complicated system is, in general, more likely than a simple system to break down. (A 15-speed racing bike is more likely to break down than a child's tricycle.)

Ecologists are especially interested to know what factors contribute to the resilience of communities because climax communities all over the world are being severely damaged or destroyed by human activities. The devastation caused by the explosion of Mount St. Helens pales in comparison to the destruction caused by humans. We need to know what aspects of a community are most important to the community's resistance to destruction, as well as its recovery.

Many ecologists now think that the relative long-term stability of climax communities comes not from diversity, but from the "patchiness" of the environment. An environment that varies from place to place supports more kinds of organisms than an environment that is uniform. A local population that goes extinct is quickly replaced by immigrants from an adjacent community. Even if the new population is of a different species, it can approximately fill the niche vacated by the extinct population and keep the food web intact.

Factors that contribute to an ecosystem's resilience include resistance to erosion and increased primary productivity. As long as the soil is intact, seeds will quickly sprout, and shrubs, ferns, herbs, and grass will hold the soil in place, preventing erosion. Partly damaged ecosystems often support greater plant growth than mature ones, since younger trees grow more vigorously than trees in a mature climax forest.

Some of the factors that contribute to ecosystem stability, the ability to recover from fires and other cataclysms, are patchiness of the physical environment, primary productivity, and resistance to erosion.

In this chapter we have seen how communities change over time and what factors, biotic and abiotic, influence the character of a community. In the next chapter, we will survey a selection of climax communities. Along the way, we will see how humans are damaging some of these climax communities.

STUDY OUTLINE WITH KEY TERMS

Henry Cowles did the first major American work on **succession,** and Frederic Clements published the first textbook describing succession. Clements also introduced the idea of a **climax community,** a single predictable community of plants that characterizes a particular area and climate. Clements believed that communities were **integrated**—real organic entities, with species playing the same roles in different communities. In contrast, the ecologist Henry Gleason argued that communities were **individualistic**—mere associations of species that happen to occupy the same area.

The species in communities interact in a variety of complex ways. They may compete or they may relate as **predator** and prey or as host and **parasite.** Organisms defend themselves from predators and herbivores by using **warning coloration, mimicry (Batesian** or **Müllerian), camouflage,** and **secondary plant compounds.** Species may share a **symbiotic** relationship, such as **parasitism, mutualism,** or **commensalism.** In evolutionary time, two or more species

may **coevolve** mutualisms and other special relations with one another.

Species may also **compete** for a **limiting resource.** The kind of community an organism occupies is its **habitat.** The way the organism uses the resources in its habitat is called its **niche.** The **competitive exclusion principle** says that if two species occupy the exact same niche, one species will eliminate the other through competition. Competition from other species may reduce the **fundamental niche** of an organism to a smaller **realized niche.** In **resource partitioning,** species break up a large niche into several smaller ones. Competition between species can cause them to evolve apart in a process called **character displacement.**

Communities vary in the number of species they contain (**species richness**), and in the relative proportions of each kind of species (**species diversity**). Ecologists are still trying to discover what determines the diversity of a community. Productivity, harshness of the environment, latitude, productivity, size of ecosystem, and historical accident each seem to play a role. In the 1960s Robert MacArthur and Edward O. Wilson proposed a **theory of island biogeography** to account for species diversity on islands of habitat. Tests of this theory suggest that the number of species on an island is a predictable consequence of the abiotic characteristics of the island, and not merely a historical accident. Habitat islands do not have to be conventional oceanic islands. They can be mountaintops or ponds in a forest, for example.

Primary succession occurs when organisms colonize a previously unoccupied area, such as a volcanic island or a sand dune. **Secondary succession** occurs when organisms recolonize an area that has suffered some catastrophe, but which still has its soil intact. In "old-field (secondary) succession," the first **pioneer** plant species in a **pioneer community** are **annuals,** followed by **perennials,** shrubs, and trees. The final stage in development is called the **climax community**. Intermediate stages are called **successional communities.**

Climax communities are either more stable than pioneer or successional communities or less stable, depending on the definition of stability that is used. Patchiness of the physical environment, primary productivity, and resistance to erosion are some factors that seem to contribute to ecosystem **resilience,** the ability of an ecosystem to recover from fires or other cataclysms.

REVIEW AND THOUGHT QUESTIONS

Review Questions

1. Contrast the integrationist view of communities with the individualistic one. Describe a piece of research that supports one view or the other.
2. Name a commensal and a parasite to humans.
3. Name a mammal, bird, reptile, amphibian, and insect that have warning coloration.
4. Define the competitive exclusion principle.
5. Give an example of a limiting resource for humans.
6. Define the terms niche and habitat. Give an example of each for the same organism.
7. What factors determine the number of species living on an island?
8. Give two definitions of ecosystem stability. What factors determine the long-term stability of an ecosystem?

Thought Questions

9. Describe a climax community somewhere near where you live. What criteria did you use to determine if it was a climax community? How would you go about testing your theory that it is a climax community?
10. Describe a series of small habitat islands that exist within 5 or 10 miles of your house and which might be subject to MacArthur and Wilson's theory of island biogeography.
11. You have discovered two butterflies that look nearly identical, but, from other evidence, clearly belong to different families. Assuming that one is mimicking the other, think of an experiment that would show which one was the mimic and which one was the "original."
12. If succession occurs because pioneer or successional species improve conditions for subsequent species, what can we say about climax species?

SELECTED READINGS

Clements, E.S., *Adventures in Ecology: Half a Million Miles . . . from Mud to Macadam,* Pageant Press, New York, 1960. Edith Clements's vivid account of her life exploring the American West at the turn of the century with her husband Frederic Clements and a score of other ecologists.

Hutchinson, G. Evelyn, *The Kindly Fruits of the Earth: Recollections of an Embryo Ecologist,* Yale University Press, New Haven and London, 1979. The autobiography of the great ecologist G.E. Hutchinson.

Owens, Mark and Delia, *Cry of the Kalahari: Seven Years in Africa's Last Great Wilderness,* Houghton Mifflin Co., Boston, 1984. A highly readable and personal account of two researchers' adventures as they studied the interactions among lions, hyenas, jackals, and other animals in the African desert.

Thoreau, Henry David, "Forest Succession in Trees," *The Natural History Essays,* Peregrine Smith, Salt Lake City, 1980.

▶ On-line materials relating to this chapter are on the World Wide Web at http://www.saunderscollege.com/lifesci/
Click on Tobin/Dusheck: *Asking About Life.*

How Many Biomes Can You See in a Day?

Death Valley, California, can be unbearably hot. With a record air temperature of 56.7°C (134°F), it is the hottest place in the world outside of Africa (where the hottest temperatures are only 1°C hotter). Nineteenth-century pioneers crossing this sunbaked desert in covered wagons found a flat, 150-mile-long valley, encrusted with salt deposits and nearly devoid of vegetation. Sunken along a geologic fault line and surrounded by jagged rocky mountains nearly as dry as the desert itself, Death Valley is the lowest place in North or South America.

Death Valley's lowest point is Badwater, 86 meters (282 feet) below sea level. Badwater is the last place in the world most people would want to go in the summer. Yet Badwater is the starting place for a punishing foot race that makes the Boston Marathon look like a jog around the block. Some 146 miles to the east and more than 4700 meters (15,000 feet) higher, lies the top of Mt. Whitney, the highest point in the contiguous United States, and the finish line for the grueling race. Each summer, a dozen runners attempt the 26-hour race from Badwater to Mt. Whitney (Figure 27-1).

Along the way, the runners pass through one of the most abruptly changing series of biological communities in the world, including all of the communities discussed in this chapter. The runners begin the race at 6 P.M. and run the first 50 miles at night. The flat floor of Death Valley itself supports little more vegetation than the moon. Daytime ground temperatures reach 182°C (200°F). Nighttime air temperatures may run to 77°C (95°F), and even in the dead of night the runners must stop for water every mile.

The runners climb out of Death Valley into the foothills and mountains of the Panamint Range. In the lower reaches, they pass through widely dispersed desert salt bush, then creosote bush. As in most **deserts,** only 10 percent of the ground has any vegetation whatever. The runners climb several thousand feet, and temperatures drop slightly. The creosote bush gives way to desert sage, a species that tolerates the higher altitude's cold winters. By the time the runners reach the top of Townes Pass (1500 meters), it is daylight. A blistering sun rises at their backs. Then they descend 900 meters into the temporarily cool shadow of the Panamints and cross a small valley and another range of mountains.

At noon the lead runners enter the Owens Valley, once one of the most fertile valleys in California. Today, desert has replaced grasslands and apricot orchards. And Owens Lake, 25 kilometers across,

Mickey Gibson/Animals Animals

climb up through the last of the desert and into the rich bottomland at the base of the Sierra Nevada. Here is a mountain stream winding through small meadows and dense groves of cottonwoods and aspens, the stream's banks overgrown with thickets of willow, its bottom lined with red metamorphic rocks and glistening fool's gold. The runners find relief in the odor of lush vegetation and cool water. But swarms of mosquitos soon begin to hum in their ears and cluster on their legs and arms, driving the overheated runners faster through the close, humid air of the canyon.

The runners have just 12 more miles to go. They have been running almost 24 hours, and it is evening again. Before them, rising above the lush valley, is a 6000-foot wall of red metamorphic rock, cut here and there by narrow, shrubby canyons. As the runners climb, they pass through dense thickets of shrubs, or mountain **chaparral**.

Gradually, as they gain altitude, the chaparral in the canyon bottom gives way to open **coniferous forests** of pine and fir (Figure 27-2). The air is cooler, but also thinner. Breathing becomes harder. The lead runners climb out of the canyon now and begin following a switchback trail up the side of a granite mountain. If they stop to look back, they can see the Panamint Range, looming darkly above Death Valley in the gathering dusk, and beyond, the purple Funeral Mountains.

As the runners continue to climb, the fir trees thin out, replaced by stunted white bark pines. Finally, above the timber line, the taiga gives way to alpine **tundra**, composed of grasses, nonwoody plants (herbs), and knee-high shrubs. Still higher, even these thin out, and the temperature drops rapidly. It is dark and almost freezing. An occasional low, windblown shrub or mountain wildflower dots

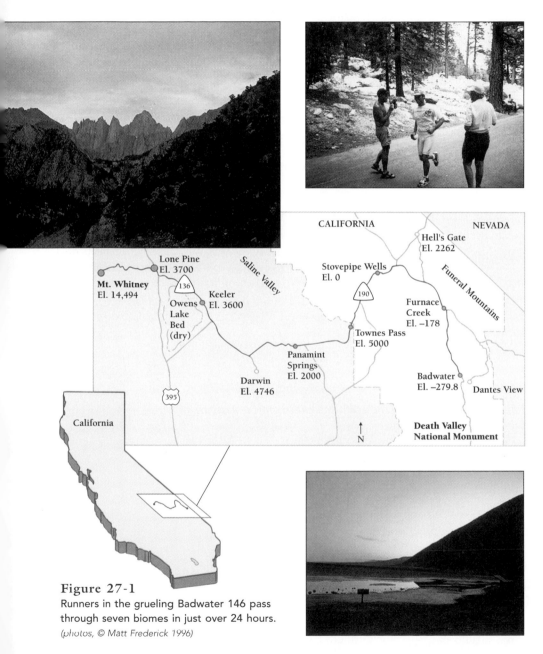

Figure 27-1
Runners in the grueling Badwater 146 pass through seven biomes in just over 24 hours.
(photos, © Matt Frederick 1996)

lies empty, thanks to the diversion of the Owens River to the millions of people living in the cities and subdivisions of Southern California.

The band of runners struggle north through the blistering heat to Lone Pine,

then turn west again. After crossing a narrow strip of fertile grassland, they begin to climb into the deep, narrow canyons of the eastern slope of the Sierra Nevada. There, they catch their first whiff of fresh, tumbling mountain water. Now they

Figure 27-2 Coniferous forest. At higher elevations and latitudes, cooler temperatures and moister soils support coniferous forests. *(Superstock)*

the bare slopes of the mountain. In the last, cold, dark mile, the runners hear only the crunch and chink of running shoes on gravel and rocks.

But even the summit of Mt. Whitney, 4418 meters (14,494 feet) high, is not quite bare rock. The summit is indeed too high, too cold, and too exposed to wind

and weather to support trees, shrubs, or herbs. But here and there lichens—symbiotic associations of fungi and photosynthetic algae—cling to the cold rocks.

The first runners to make it to the top collapse on the rocks, euphoric with accomplishment and oxygen deprivation. The rest of the runners are scattered across seven distinct climax communities, or

characteristics of their environment (including moisture, sunlight, and space) and on the specific history of the area (which species happened to migrate there at what time). Notwithstanding the importance of history, communities in similar climates look alike even when their similar species are unrelated. As a consequence, our planet contains a limited number of types of communities, or biomes. In this chapter we discuss several terrestrial biomes and their aquatic equivalents. (Ecologists do not use the word

---

Badwater is the starting place for a punishing foot race that makes the Boston Marathon look like a jog around the block.

---

biomes—desert, temperate grassland, temperate deciduous forest, chaparral, coniferous forest, krummholz, and tundra.

The distinctive makeup of these climax communities depends both on the

"biome" to refer to aquatic communities, reserving that term for land communities.) Along the way, we will see what effects humans have on each of these biomes and aquatic communities.

KEY CONCEPTS

1. Climate determines the character of an ecosystem.

2. Latitude and the rotation of the Earth, together with the presence or absence of mountain ranges, largely determine climate.

3. Areas with heavier rainfall support forests. Areas with less rainfall support grasslands.

4. The northern taiga and tundra have suffered the least human-inflicted damage.

5. Estuaries are productive and species rich, whereas ocean communities are, by comparison, unproductive and species poor.

WHAT DETERMINES THE CLIMATE OF A PARTICULAR REGION?

The diversity of species can be great even within a small community. For example, one ecologist found 163 species of beetles living in a single tree in the Panamanian rain forest. Another found 445 species of trees on just 2.5 acres of Brazilian rain forest. Fifteen of these trees species were new to science. The diversity of species worldwide is staggering. After all, although different communities have similar food webs and patterns of energy flow, every species has a unique niche.

Nonetheless, as we travel from place to place, we also see the similarities of widely spaced communities. Forests of spruce and other conifers stretch not only across Canada, Alaska, and the northern United States, but across Scandinavia and Russia (Figure 27-3). Similarly, grasses and large grazing animals dominate the plains of the North American Midwest, the Argentine pampas, Central Asia, the East African highlands, and much of Australia. Likewise, the deserts of the southwestern United States and northwestern Mexico, central Asia, Africa, and Australia all have plants with spiny or succulent leaves, and burrowing animals that emerge only at night.

A.

B.

Figure 27-3 Taiga here, taiga there. Whether in Canada, Scandinavia, or Russia, taiga is characterized by sparse, stunted coniferous trees. A. Siberian taiga. B. Spruce taiga in central Alaska.   *(A, © Roland Seitre/Peter Arnold, Inc; B, Tony Dawson/Words & Pictures/Tony Stone Images)*

What Is a Biome?

In each region, ecological succession leads to a characteristic climax community, or "biome." A **biome** is a geographical region characterized by a distinctive landscape, climate, and community of plants of animals. The existence of biomes reflects the comparable climates (especially temperature and moisture) of widely separated areas. But biomes differ in the particular species that they contain. For example, the grasslands of North America are dominated by deer, elk, bison, and other "ungulates," while the grasslands of Australia are dominated by different kinds of kangaroos. Differences in the individual species in a biome are partly the result of the unique evolutionary histories of each region and partly the result of differences in geography, such as the closeness of the oceans or the presence of mountain ranges.

How Do Sunshine and the Earth's Movements Determine Climate?

The most important determinants of biome type are temperature and moisture. These depend, in turn, on the intensity of sunlight and patterns of wind, rain, and ocean currents. All of these depend on the Earth's movements and the distribution of mountains and valleys.

The most critical factor is sunlight. Of some 10^{21} kilocalories of solar energy that enters the upper atmosphere each year, about half reaches the ground. The rest is reflected back into space or absorbed by the atmosphere. Plants capture less than 1 percent of the light that strikes the Earth's surface. The energy absorbed by the ground and by the atmosphere raises the temperature of the environment and is thus an important determinant of a region's climate.

The Earth's daily rotation constantly mixes the air, so that all the air along each latitude has a similar temperature. However, different latitudes have different climates, for two reasons: (1) the latitudes closest to the poles receive less solar energy than those near the equator, because the sun's rays arrive at a more oblique angle (Figure 27-4A); and (2) the latitudes closest to the poles experience more extreme seasons than those near the equator because the Earth's axis is tilted. As a result, the equatorial climate is both warmer and more unchanging than the temperate or polar climates (Figure 27-4B).

These differences in the temperature of the atmosphere cause air to move in currents that transfer heat away from the equator and toward the poles. Three consistent patterns, or rules, drive these currents: (1) hot air rises and cold air falls; (2) hot air holds more moisture than cold air; and (3) the rotation of the Earth twists the moving air.

These three rules lead to the pattern of air currents shown in Figure 27-5. At the equator, warm moist air rises, creating a region of low atmospheric pressure. Air from both hemispheres moves toward the equator, though not particularly forcefully. The movement of air from the tropics toward the equator creates the **tradewinds,** winds that blow from about 30° latitude, steadily toward the equator. In the northern hemisphere, tradewinds blow from the northeast, while in the southern hemisphere, they blow from the southeast. At the equator itself, not much wind blows: sailors refer to this lack of wind as the *equatorial doldrums* [Old English, *dol* = dull].

As the warm air from the equator rises and moves north and south, it cools and can no longer hold much water. Most of the surplus moisture falls as rain in the tropics, the regions between the equator and 30° north and south latitudes. So the climate of the tropics is warm and wet. By the time the equatorial air reaches 30° latitude, however, it has lost most of its moisture. Dry air is therefore typical of this latitude, which has most of the world's great deserts.

The now dry air drops back toward the Earth's surface as it cools. The cool, dry air moves along the Earth's surface toward the poles. The rotation of the Earth, however, twists these currents, so that the resulting winds, called the **westerlies,** come from the west. The westerlies are characteristic of lati-

Figure 27-4 Temperature and moisture are the two most important determinants of biome type. A. Average temperature partly depends on latitude. B. Because of the Earth's tilt, the equatorial latitudes are not only warmer than the polar latitudes, but also less variable.

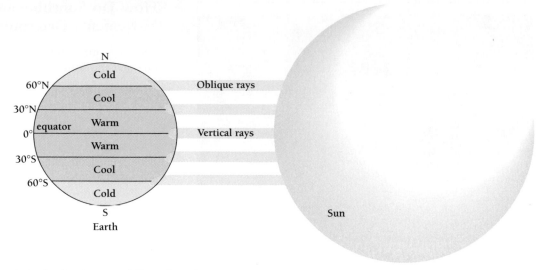

A. Angle of sun's rays at different latitudes

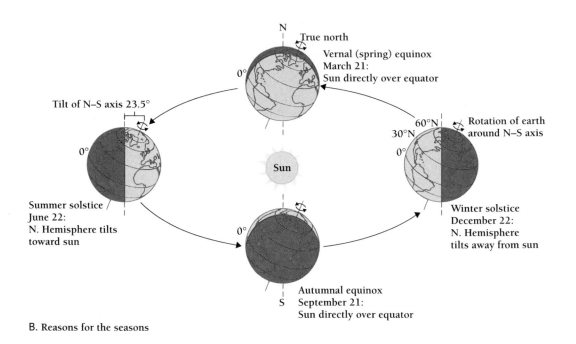

B. Reasons for the seasons

tudes between 30° and 60° in both hemispheres. As the air moves over the Earth's surface, it warms and picks up moisture again. At about 60° latitude, the air again rises to create the **polar fronts**, regions of low pressure, where the westerlies end and the polar easterlies begin (Figure 27-5).

Finally, near the poles, the air—again dry and cold—descends once more. The regions surrounding the poles are desertlike, while the regions around 60° north latitude harbor great coniferous forests. We would also expect such forests at 60° south latitude, but there are no land masses there.

The prevailing winds at each latitude, then, depend on the overall movement of warm air from the equator toward the

poles. In traveling toward the poles, air gains and loses both moisture and heat. The rotation of the Earth also changes the direction of the air movements, causing clockwise air currents in the Northern Hemisphere and counterclockwise currents in the Southern Hemisphere (Figure 27-5).

Physicists and meteorologists (scientists who study the weather) call the effect of rotation on movement the **Coriolis effect**, after Gaspard Gustave de Coriolis, a 19th-century French mathematician. The Coriolis effect is responsible for the clockwise twist of Northern Hemisphere wind currents, as well as of hurricanes, typhoons, and tornadoes. In the Southern Hemisphere, the Coriolis effect produces a counterclockwise

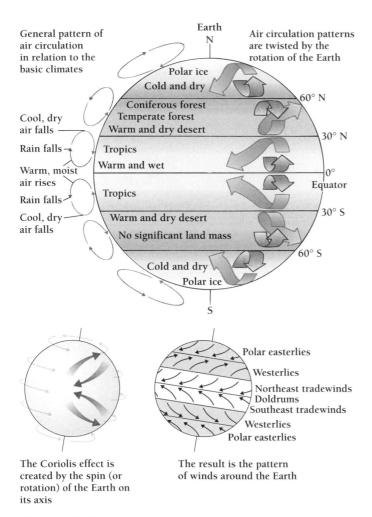

General pattern of
air circulation
in relation to the
basic climates

Earth
N

Air circulation patterns
are twisted by the
rotation of the Earth

Polar ice
Cold and dry

60° N

Coniferous forest
Temperate forest
Warm and dry desert

Cool, dry
air falls

Rain falls

30° N

Tropics
Warm and wet

0°
Equator

Warm, moist
air rises

Rain falls

Tropics

Cool, dry
air falls

Warm and dry desert

30° S

No significant land mass

60° S

Cold and dry
Polar ice

S

Polar easterlies

Westerlies

Northeast tradewinds
Doldrums
Southeast tradewinds

Westerlies

Polar easterlies

The Coriolis effect is
created by the spin (or
rotation) of the Earth on
its axis

The result is the pattern
of winds around the Earth

Figure 27-5 Air currents. A pattern of rising and falling air, combined with the Earth's rotation, causes air currents to flow east near the north and south poles, to flow west in the middle latitudes, and to flow east along the equator.

Figure 27-6 A rain shadow. A. Rain clouds driven inland from the Pacific Ocean hit California's Coast Range and ride up the slope, like skateboarders on a ramp. As the clouds rise, they cool and rain precipitates out, soaking the mountains. The clouds drop down the other side, warm up, and hold their remaining moisture. Over the Sierra Nevada, the clouds drop quantities of rain and snow. B. By the time the clouds reach the Great Basin, the moving air has lost nearly all its moisture and little rain falls in the desert beneath. *(A, Charlie Ott/Photo Researchers)*

twist to wind currents. Finally, the Coriolis effect shapes the currents of the oceans themselves, so that water also moves in great cycles.

Ocean currents also depend on the placement of the continents, so the cycles are generally more limited than those of the air. The Gulf Stream, for example, carries warm air from the tropics up the coast of the United States and then across the North Atlantic. The Gulf Stream is thus responsible for warming Britain, Ireland, and much of Northern Europe, which have much milder climates than would be expected for their latitudes.

Mountain ranges also influence the climate of the continents. First, mountains bring warm surface air to higher, colder altitudes, where moisture condenses into clouds and falls as rain or snow. The prevailing westerlies along the California coast, for example, drive warm, moist ocean air up and over first the Coast Range and then the Sierra Nevada (Figure 27-6A). As the air moves up the slopes of each range it cools and drops rain and snow. As the air moves east, down the farther side of each range, it again heats up. By the time the ocean air reaches the far side of the Sierra Nevada, it is very dry. East of the Sierra Nevada, all of the lands in the Owens Valley, Death Valley, most of Nevada, and parts of Oregon, Idaho, Utah, and New Mexico exist in a huge "rain shadow" called the Great Basin desert (Figure 27-6B). A **rain shadow** is the area adja-

A.

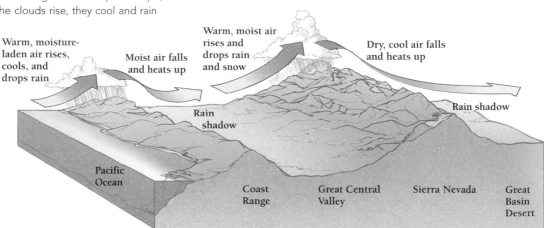

Warm, moisture-laden air rises, cools, and drops rain

Moist air falls and heats up

Warm, moist air rises and drops rain and snow

Dry, cool air falls and heats up

Rain shadow

Rain shadow

Pacific Ocean

Coast Range

Great Central Valley

Sierra Nevada

Great Basin Desert

B.

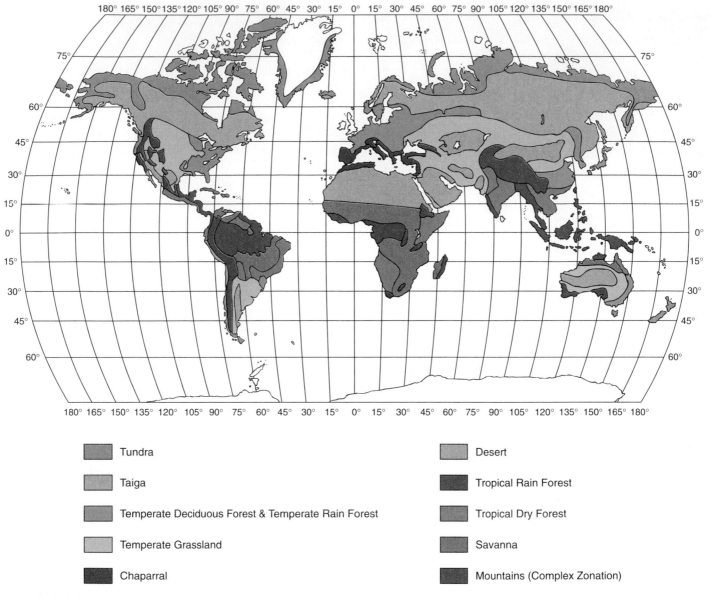

Tundra		Desert	
Taiga		Tropical Rain Forest	
Temperate Deciduous Forest & Temperate Rain Forest		Tropical Dry Forest	
Temperate Grassland		Savanna	
Chaparral		Mountains (Complex Zonation)	

Figure 27-7 **World distribution of ten biomes.** The sharp boundaries in this diagram are misleading, as biomes frequently grade together smoothly. In addition, each solid area actually represents an average of many vegetation types. Botanists divide California alone into 24 "floristic provinces."

cent to a mountain range, away from the prevailing winds, where little rain falls.

Tall mountains have other effects on climate as well. As we climb a mountain such as Mt. Whitney, the climate changes in the same way it does as when we travel from low latitudes to higher ones. The Appalachian Mountains, the Sierra Nevada, and the Rocky Mountains all contain extensive coniferous forests, which turn to alpine tundra above the **tree line**, the upper limit where subalpine trees grow.

Surprisingly, all of these geographic patterns of temperature and rainfall create just a few major types of climate. Within each climate similar adaptations are useful. If we view biomes in the individualist view, advocated by Gleason (Chapter 25),

we would say that similar biomes exist in different parts of the world because the same adaptations are of value in similar climates. The species from geographically separated communities differ greatly. The species of a desert biome are recognizably similar, however, whether the desert is in Asia, North America, Australia, or Africa. As a result, biomes contain many examples of convergent evolution. The most striking examples are the Australian marsupials illustrated in Figure 15-20. These animals have evolved the same adaptations to dozens of niches as placental mammals in South America, Africa, and elsewhere.

Biomes do not have fixed boundaries. They are land areas where the communities and their resident species look alike and have similar patterns of energy flow. In this chapter, we

BOX 27-1

How does acid rain alter biomes?

Who has not heard of acid rain and how it is destroying forests and freshwater lakes throughout Canada, the northeastern United States, and northern Europe? The dramatic damage to forests and freshwater ecosystems caused by acid rain has been well documented (Box 2-2). Whole forests of trees are dying. Fish, frogs, plants, and other organisms have disappeared from formerly productive lakes and streams. Yet, increasingly, ecologists are finding that acid rain also has the capacity to transform ecosystems in subtler ways.

Researchers have long known that acid rain increases the availability of nitrogen to plants. Indeed, some economists have argued that because nitrates from acid rain can stimulate plants to grow faster, plants will take up increasing amounts of carbon dioxide. According to this argument, the increased carbon uptake by plants should compensate for the carbon dioxide released through burning fossil fuels and therefore reduce global warming.

Although this is a reassuring thought, ecological research does not seem to support this hypothesis. In one study, researchers at the University of Minnesota applied nitrogen to plots of native Minnesota

tallgrass prairie for 12 years to imitate the excess nitrates that result from both acid rain and current agricultural practices. Because native grasses use nitrogen efficiently, they flourish in places where the nitrogen content of the soil is low. In the nitrogen-treated plots, however, exotic weeds flourished, outcompeting and displacing native Minnesota grasses.

But that was only half of the bad news. As prairie grasses die, their tissues break down very slowly, and over time much of their carbon, nitrogen, and other constituents remain stored in the soil. Such long-term storage is, in fact, the key to the famous, rich soils of the American midwest.

But in the Minnesota study, the weeds behaved differently from grasses. As the weeds died, their huge stores of nitrogen stimulated the growth of nitrogen-hungry soil microbes, which rapidly dismantled the exotic weeds into basic building blocks. The microbes released nitrates into the groundwater and carbon back into the air. The net effect was more carbon dioxide in the air—not less—and more nitrates in the soil, which stimulated the growth of more weeds.

Meanwhile, research on spruce forests in Germany's Fichtelgebirge mountains has

revealed that nitrates can pass through forest ecosystems without being incorporated into the trees. The excess nitrogen in acid rain supplies the trees with more nitrogen than they can use. Such excesses might be expected to stimulate their growth, but, unfortunately, it does not. The negatively charged nitrates raining down on forest soils bind with positively charged cations such as calcium and magnesium. Then, as the nitrates flow into lakes and streams, they take the positive cations with them. By one estimate, such acid leaching in some areas has removed in 40 or 50 years all of the calcium deposited in the soil for the last 500 years.

The prognosis for these calcium- and magnesium-poor forests is grim. Many trees are already dying and some ecologists predict that most trees will die. Other ecologists argue that trees deprived of calcium and magnesium habituate and simply grow quite slowly, like the tiny bonzai trees that gardeners grow in pots. The result is a miniature but ancient forest. Like the nitrogen-rich, weedy fields in Minnesota, however, these forests cannot be expected to store much carbon.

---

will describe eight major biomes: temperature deciduous forest, temperate grassland, chaparral, desert, savanna, tropical rain forest, taiga, and tundra (Figure 27-7).

Latitude and the Earth's rotation influence intensity and duration of solar radiation, wind, rain, and ocean currents. These factors, as well as terrain, help create distinct climates.

WHAT ARE BIOMES LIKE?

Why Do the Temperate Deciduous Forests Lack Top Predators?

Temperate deciduous forests receive from 80 to 140 centimeters of precipitation each year. However, annual rainfall

(and snowfall) is much more variable in the temperate zones than in the tropics or the polar regions. So communities within this biome may look rather different from one another. The particular type of forest vegetation depends on both temperature and rainfall. In the northern United States, for example, the dominant trees are maple, oak, hemlock, and birch, while in the southeastern states the dominant trees are pine, oak, and hickory.

Deciduous trees, which shed their leaves each autumn, characterize the temperate forest biome (Figure 27-8). The plants of a temperate forest occupy four vertical layers: trees, shrubs, herbs (nonwoody plants), and ground cover. The trees occupy the highest layer, their tops touching to form a nearly continuous canopy. In the shade beneath the trees are shrubs and bushes whose branches lie closer to the ground. The herb layer may be especially rich and beautiful in the spring. Finally, the ground layer consists of mosses and liverworts, often living among the accumulated leaf litter.

Figure 27-8 **Deciduous forest.** In the fall, aspens like these drop their leaves. *(A & L Sinibaldi/Tony Stone Images)*

Animal life is abundant in all the plant layers and in the soil beneath the litter (Figure 27-9). Trees and shrubs provide food, nest sites, and perches for many species of birds. Woodpeckers and chickadees, for example, live in cavities in the trees, while warblers and thrushes build nests on their branches. Mammals such as chipmunks and squirrels; reptiles such as snakes, lizards, and wood turtles; and amphibians such as salamanders and tree frogs also find food and homes in the forest.

People living in the eastern third of the United States, western Europe, Japan, Chile, and eastern China are most familiar with the temperate forest biome. The enormous human populations in all of these areas have greatly affected this biome. Humans have cut down much of the forest for lumber and firewood, and to make way for farms, towns, and cities. Most of the remaining forest is in small, isolated woodlots, often of 16 hectares (40 acres) or less.

One result of this breakup of the forests is the elimination of the enormous ranges needed to support carnivores at the top of the food chain. For example, a single mountain lion needs over 75,000 hectares (300 square miles). Even a single

family of gray foxes needs some 260 hectares (642 acres). It is not surprising, then, that the destruction of the temperate deciduous forests, together with intensive hunting, eliminated nearly all the top carnivores in the eastern American forests. Also gone, or nearly gone, are the red wolf, the bobcat, and the black bear.

Another result of breaking the forest into isolated patches is the creation of more forest "edge," areas where woods touch meadows or farmland. These edge communities provide opportunities for species that can take advantage of both environments, for example by nesting in the forest and foraging in the meadow. Among the beneficiaries of the increased forest edge are American robins and cardinals, woodchucks, and white-tailed deer. But squirrel foxes and many other species that compete poorly in an edge community have become ever rarer.

Temperate deciduous forests typically have warm,
rainy summers and cold winters.

Temperate Grasslands Have Been Replaced by Wheat Fields

Like temperate forests, **temperate grasslands** have well-defined seasons, with hot summers and cold winters (Figure 27-10). The low annual rainfall, usually 25 to 75 centimeters per year, is enough to keep grasslands from turning to deserts but not enough to sustain the growth of trees. In addition, grasslands usually have a dry period, during which fires are common. Wildfires and the lack of rain, together with heavy grazing by animals, keep trees from taking hold in this biome.

Grasslands extend over much of the Earth's surface, mostly in the interiors of the continents. They include the steppes of Russia, the veld of South Africa, the pampas of Argentina, the puszta of Hungary, and the prairies of the central United States and Canada. Before European settlement, more than a million square miles of grassland covered the middle of North America.

The types of grasses vary with the amount of rain and with season. In the eastern American prairie, for example, tall grasses, especially big bluestem, dominate; in the drier western plains—in the rain shadow of the Rocky Mountains—short grasses dominate.

Grasslands can support large numbers of mammals. For millions of years, African grasslands have supported huge herds of wild gazelles and zebras, while the Asian steppes have supported sheep and horses. All of these herbivores in turn supported populations of wolves, lions, humans, and other predators. Even the American prairie supported some 60 million bison and 40 million pronghorn antelope until 19th-century agriculture and sport hunting drove these populations to the brink of extinction.

In North America today, humans have virtually eliminated the temperate grassland biome. In place of hundreds of thou-

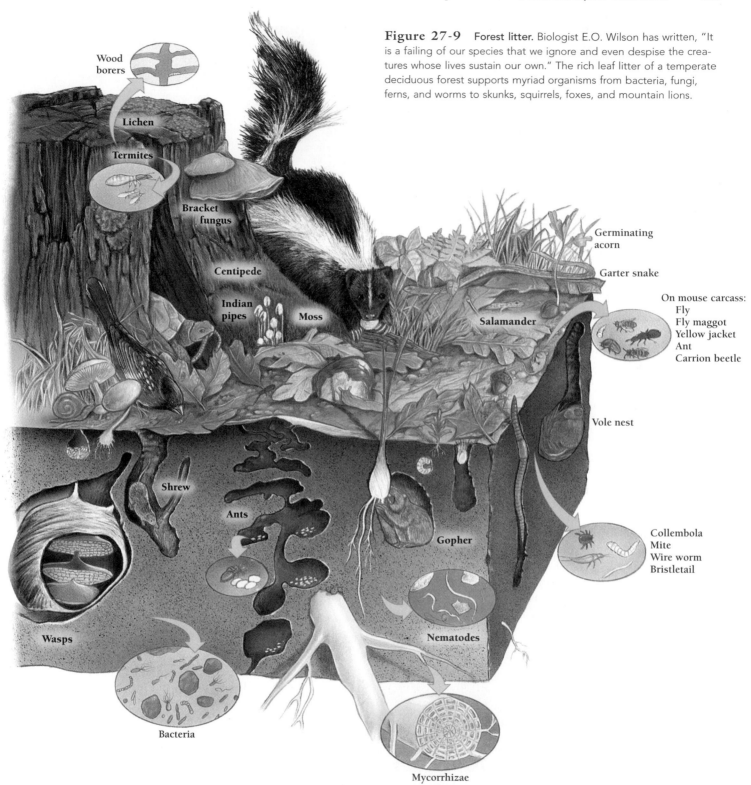

Figure 27-9 **Forest litter.** Biologist E.O. Wilson has written, "It is a failing of our species that we ignore and even despise the creatures whose lives sustain our own." The rich leaf litter of a temperate deciduous forest supports myriad organisms from bacteria, fungi, ferns, and worms to skunks, squirrels, foxes, and mountain lions.

sands of square miles of American and Canadian prairie we have created "the world's breadbasket," a feat of stupendous proportions. Cultivated corn has replaced big bluestem, and wheat has replaced the short grass. Mass-produced pigs, cows, and beef cattle have become the chief herbivores, humans the top carnivores. A few thousand bison still live in parks and reserves, and a few hundred thousand pronghorn antelope live

in the plains adjacent to the Rocky Mountains. Their natural predators, however, the Great Plains wolf and the grizzly bear, are gone.

Temperate grasslands have hot summers and cold winters, but less rainfall than temperate forests.

Figure 27-10   **Temperate grassland.** This wild prairie at the Living Prairie Museum in Winnipeg, Manitoba, has never been plowed. *(Tom McHugh/Photo Researchers)*

Figure 27-12   **Fire.** Brush fires in chaparral communities destroy million-dollar houses in minutes.   *(Reuters/Corbis-Bettmann)*

Chaparral Always Burns Eventually

Chaparral [Basque, *chabarro* = dwarf evergreen] dominates five widely separated temperate regions—California, central Chile, the shores of the Mediterranean Sea, southwestern Africa, and southwestern Australia. These regions, all on the west coasts of continents, share a "Mediterranean" climate, with mild, wet winters and hot, dry summers. In the summer, prevailing westerly winds blow cool moist sea air over the warm land. But the warmed air holds the moisture, so no rain falls.

Chaparral is characterized by dense shrubs and scattered, broad-leafed trees (Figure 27-11). The shrubs are typically spiny, thick, and dense. (A cowboy's leather "chaps," worn over his jeans, protect him as he rides through the chaparral.)

Typical mammals in California chaparral are bobcats, mule deer, brush rabbits, wood rats, and deer mice. Birds include rodent-eating hawks, and numerous ground-dwelling birds, such as towhees, sparrows, and quail.

Each of the five chaparral regions has its own species, but all chaparral species share common adaptations to summer drought. In addition, California chaparral species have evolved a variety of adaptations to fire. There, periodic wildfires clear the way for fresh growth and keep trees from shading out the chaparral. The seeds of some herbs and shrubs actually require exposure to the intense heat of a brush fire before they can germinate. Many perennial shrubs are also adapted to fire. When fire burns away their dry branches, many species resprout from still-intact roots.

One of the greatest dangers to chaparral communities is overzealous fire prevention by humans. But putting out small fires, humans ensure the accumulation of "fuel," or flammable plant material. Inevitably a fire so big and hot that firefighters cannot put it out sweeps thousands of acres, incinerating the roots of the chaparral vegetation and destroying hundreds of suburban houses (Figure 27-12). Because of the way such superhot fires affect the soil, serious erosion may follow (Figure 27-13).

Chaparral communities have hot, dry summers and mild, wet winters.

Figure 27-11   **Chaparral.** Dense shrubs and scattered trees characterize Mediterranean climates.   *(Tom McHugh/Photo Researchers)*

Desert Ecosystems Are Surprisingly Delicate

Deserts are relatively barren regions, with lower productivity than most other biomes. Deserts may be hot or cold, but all are dry, receiving less than 25 centimeters of rain per year. In the United States and Mexico, desert extends from the Sierra Nevada to the Rockies, and south to Baja California and northwestern Mexico. Other deserts extend across northern and southern Africa, Arabia, central Asia, central Australia, and parts of South America.

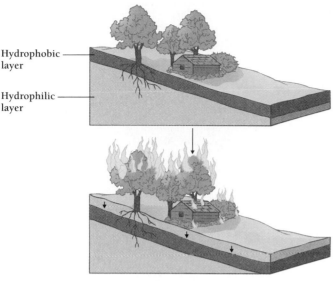

Hydrophobic layer

Hydrophilic layer

Fire burns the litter, vegetation, and structure. Intense heat drives hydrophobic layer down.

Hydrophilic layer

Hydrophobic layer

Rainwater accumulating in the upper hydrophilic layer may cause landslides.

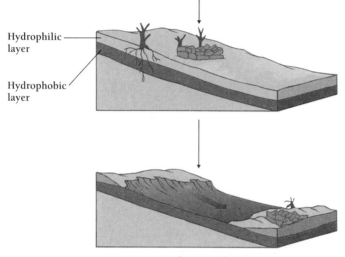

Figure 27-13 Building houses in fire-adapted communities is problematic. Controlling small brush fires to protect houses leads to a buildup of dead wood. When a fire finally burns out of control, it is hotter than usual. Such fires not only destroy houses, they also destroy trees and shrubs and damage soil. In chaparral and other communities, leaf litter and the soil just beneath it contain water-resistant plant compounds that prevent water from penetrating the soil. Heat from a hot fire forces this "hydrophobic layer" deeper into the soil. The uppermost layer is now hydrophilic and quickly absorbs all the rain that falls on it. After the fire, winter storms drench the upper layer of soil, leading to mudslides.

The dominant plants of a desert depend heavily on its temperature (Figure 27-14). Sagebrush, a bushy shrub, dominates the cold deserts of Nevada and Utah, while cacti and oily creosote bushes dominate the lower, warmer deserts of Arizona and Mexico.

Many desert plants have special adaptations to conserve water. Sagebrush, for example, has small leaves that minimize

Figure 27-14 Deserts. Creosote bushes punctuate the many deserts of the southwestern United States. *(Jim Steinberg/Photo Researchers)*

evaporation, while succulents (including the cacti) store water in fleshy tissue for future use. Cacti and other desert plants grow prickly spines that protect them from hungry or thirsty browsers.

Like desert plants, desert animals must also be able to tolerate dryness and hot summers. Many are active only during the cool hours of dawn and dusk. Lizards and snakes are especially successful, as are many birds, such as the cactus wren and the roadrunner, and rodents, such as kangaroo rats and mice. Many rodents **estivate**, retiring to underground burrows and entering a sleeplike state called *torpor* during the hottest and driest months of the summer. Round-tailed ground squirrels and several species of pocket mice estivate, as well as the cactus mouse (*Peromyscus eremicus*). Laboratory experiments by mammalogist Richard McMillan suggest that lack of food and lack of water can both trigger estivation in the cactus mouse.

A significant threat to American deserts is the off-road recreational vehicle, whose wheels destroy fragile desert plants and soils, as well as lizards, snakes, and tortoises. Perhaps the biggest threat to desert-adapted organisms is the conversion of desert habitat into suburban developments and farms. Heavy irrigation can destroy any soil, but desert soils are especially vulnerable. Unlike rainwater, irrigation water is loaded with dissolved minerals and salts (because it comes from rivers and wells). As irrigation water evaporates from hot soil, the minerals and salts are left behind as a deposit. In time, the soil becomes so caked with minerals and salts that farming is impossible. By then, even desert plants cannot grow in the poisoned soil.

Desert plants and animals are all adapted to conserve water. Animals also have adaptations for staying cool when temperatures soar.

Figure 27-15   Savanna. Grasslands and scattered trees in the Serengeti, in Tanzania.
(Art Wolfe)

Savannas Host Some of the Most Spectacular Grazing Animals

Savannas are tropical or subtropical grasslands punctuated by solitary trees or small clumps of trees (Figure 27-15). Like temperate grasslands, savannas have a regular dry season and periodic wildfires, conditions that limit the number of trees. But savannas generally have more rain (usually 90 to 150 centimeters per year) and hotter climates than temperate grasslands. Savannas cover much of Australia, south and central Africa, central South America, and parts of Southeast Asia.

Savanna grasses are tall, 2 meters or more in Africa, 1.5 meters in South America. The trees tend to be short, usually less than 10 meters tall. Deep, dense underground networks of roots allow grasses to obtain water where trees cannot and so survive the dry seasons.

Animal life in the savannas varies greatly from continent to continent. The best-known animals are those of the African plains. There, herds of zebras, wildebeest, and gazelles graze while elephants and giraffes browse on the trees. Among these herbivores, lions, cheetahs, hyenas, and other carnivores hunt and scavenge. The savanna also supports large, flightless, grazing birds—the ostrich in Africa, the emu in Australia, and the rhea in South America. Termites and their 7-meter-tall mounds, numbering 10 to 150 mounds per hectare (25 to 370 per acre), are also a striking presence on the savanna. Termites process up to a third of the plant litter, and serve as a major food source for birds and mammals.

Hunters have driven a few species to the brink of extinction. In East Africa, the seemingly infinite herds of grazing animals are irresistible to rapidly growing, protein-starved human populations. Elephants have become particular targets for poachers, not because of their protein or calories, but because of their ivory tusks. Their slaughter continues despite worldwide efforts to outlaw trading in ivory. Nonetheless, the real threat to African savanna is the continuing conversion of savanna to agriculture to support burgeoning human populations. While wildlife sanctuaries protect some of the African and South American savannas, the end of the savanna biome seems inevitable.

Tropical savannas have abundant rain and hot climates, with regular dry seasons.

Tropical Rain Forests Are the Most Productive and Diverse Communities in the World

Most **tropical rain forests** are within 10° of the equator—in the Amazon basin of South America, in central and west Africa, and in Malaysia, Indonesia, and New Guinea. The weather in the tropical rain forests is more or less constant throughout the year, with average temperatures of about 27°C (80°F) every month of the year. Every day has about the same pattern of temperature variation, often punctuated by afternoon thunderstorms. Rainfall is heavy, ranging from 2 to 4.5 meters per year.

Because warmth and moisture are plentiful in a tropical rain forest, the limiting abiotic factor is usually light. Tropical plants are so successful at capturing light that the forest floor is usually rather dim, with relatively little vegetation. It is quite unlike the usual image of the dense jungles, an image more characteristic of earlier stages in a tropical succession. Visitors often compare the tropical forest to a cathedral, with an open floor, high vaulted ceilings, and soft filtered light.

The soils of rain forests are surprisingly poor in nutrients. The heavy rainfall leaches nutrients from deep in the soil. The only nutrients lie just beneath the leaf litter at the very surface. As a result, trees must have extensive but extremely shallow roots. These vast networks of roots quickly take up and recy-

Figure 27-16 Tropical rain forest. Trees in the mist in a Costa Rican rain forest. *(Stephen J. Krasemann/DRK Photo)*

cle organic compounds and minerals from fallen leaves and dead organisms.

The tropical rain forests are nonetheless the most productive and diverse communities in the world. At least half, and perhaps as many as 90 percent, of all the Earth's species live in this biome. In a hectare of temperate forest, we may find two or three dominant species of trees with one or a few trees of another 10 to 20 species. In contrast, a hectare of tropical forest is home to 40 to 100 different species of trees.

Most tropical rainforest plants are broad-leaved evergreen trees (Figure 27-16). In contrast to a temperate forest, in which many plants are shrubs and herbs, more than two-thirds of the plants in a tropical forest are trees. Other plants common in tropical forests but rare in temperate forests are **lianas**, vines that are rooted in soil but climb into the canopy, and **epiphytes**, plants such as orchids that grow entirely on other plants. The common houseplant philodendron, for example, is a tropical liana.

As in a temperate forest, the vegetation forms discrete layers. Most trees form a continuous layer, called the **canopy**, 30 to 40 meters above the ground. Rising above the canopy are occasionally tall, isolated **emergents**, trees with umbrella-shaped crowns extending to a height of 50 meters or more. Below the canopy, a third layer of shorter trees 10 to 25 meters tall catch whatever light they can from gaps in the canopy (Figure 27-17). Finally, a short shrub layer consists of dwarf palms and giant herbs with especially large leaves, adapted to capture the faint light that filters through upper stories. Plants from this layer make excellent house plants, as they are adapted to grow in dim light.

The animals of tropical rain forests, like the plants, are extraordinarily diverse. Many species of birds, insects, snakes, bats, monkeys, and squirrels all live permanently in the canopy. The year-round supply of fruits and nectar are partly responsible for the specialization and diversity of birds, insects, and predators.

Emergent layer
to 50 m

Canopy layer
30–40 m

Understory
15–25 m

Shrub layer
3–15 m

Ground layer
0–3 m

Figure 27-17 The layers of the forest. Tropical rain forests are noted for their extensive layering. Shown here are the protruding emergent trees, the canopy, the understory, the shrubs, and the ground-level plants.

The forest floor hosts still more animals. Some live on underground plant parts, others on fruits and nectar. Besides the impressive array of ants, termites, other insects, and numerous invertebrates, the forest floor has an incredible variety of larger animals, higher in the food chain—anteaters and armadillos, pigs and their relatives, as well as poisonous and constricting snakes. The top carnivores of the forest floors used to be such large cats as the jaguar in South America and the tiger in Asia. However, these magnificent animals have all but succumbed to trophy and fur hunters and to the destruction of their habitat by humans.

To raise cash to support expanding urban populations, humans are each year logging an area of tropical rain forest equal to the size of Illinois. Tropical nations export timber and plywood from rainforest trees; metals mined from rainforest lands; and beef from cattle raised on clear-cut rain forest.

In Latin America, much of the land formerly occupied by rain forest now serves as cattle ranches to supply hamburger meat to American and European fast-food chains. Once cleared, however, former rain forest land does not serve farmers or ranchers well. Its soils, never rich to begin with, deteriorate rapidly. The average cattle ranch in Central America lasts only 6 to 10 years, after which the soil is too depleted of minerals to support grass for grazing cattle.

Future textbooks may list the tropical rain forest as a biome that has disappeared, since human activity threatens to eliminate it by about 2030. Accompanying the destruction of the rain forest will be the disappearance of millions of species of plants, insects, and other animals found nowhere else. Most of these species have not yet even been counted or named. Their value, in their contributions to the biosphere or in their possible uses as foods, fuels, or medicines, will probably never be known. Nor can we predict how the destruction of the tropical rain forests will affect the global climate.

Tropical rain forests are hot and wet, with little variation from season to season.

Taiga Is the Spruce–Moose Biome

Taiga, a Russian word, refers to the broad band of coniferous forest that extends across Canada, Alaska, Scandinavia, and Siberia (Figure 27-3). In the taiga, the winters are bitterly cold, and the summers are short.

The trees grow only 5 to 10 meters high, and the forests are less diverse than even the temperate forests, with only one or two species (often spruce and pine) dominating. Shrubs are relatively rare in the climax taiga forests, except in the early stages of succession, when blueberries, gooseberries, and other shrubs flourish. The herb layer, consisting of ferns and mosses, is also sparse in the climax taiga forest.

Moose are common herbivores in the taiga. Indeed, someone has called the taiga the "spruce–moose" biome. Smaller herbivores include mice, voles, squirrels, porcupines, and snowshoe hares. Predators include wolves, grizzly bears, lynx, and wolverines. Birds are plentiful and diverse, especially in the summer. Reptiles, however, are absent, with the exception of the garter snake. Insects—especially mosquitos—abound.

Because the taiga is so inhospitable, human populations are rare and have had relatively little impact. Nonetheless, logging has eliminated many of the original forests. In these areas deer have moved in, replacing the moose. A common parasite of the deer—the nematode brain worm—is fatal to moose and, in many areas, is accelerating the disappearance of the moose.

One of the greatest threats to the taiga comes from acid rain. Industry, electrical power plants, and automobiles produce oxides of sulfur and nitrogen that rise and travel hundreds or thousands of miles downwind. Eventually these oxides dissolve in rainwater or snow, making the precipitation acidic. Because prevailing winds carry the oxides northward, much of the acid rain falls on the taiga, where it acidifies groundwater, lakes, and ponds. The result has been the wholesale death of fish and forests in many regions of the Northern Hemisphere, including Scandinavia, eastern Canada, and the Adirondacks in New York State.

Taiga has cold winters and short, mild summers.

Tundra Has a Short Growing Season

The **tundra** [Finnish, *tunturi* = arctic hill], a vast, open land dominated by grasses and low shrubs, covers more than a fifth of the Earth's surface (Figure 27-18). The climate is cold and dry. Freezing temperatures can occur throughout the year, but the cool summers (no warmer than 10°C on average) usually provide a growing season of one or two months. **Permafrost,** permanently frozen ground less than a meter from the surface, underlies the soil during even the warmest summers.

Precipitation is low, less than 25 centimeters per year, and falls mostly as summer or autumn rains. Snow is rather light, with an average depth of only 10 to 20 centimeters. Despite the low rainfall, the soil is wet, both because the permafrost prevents drainage and because the cold air curbs evaporation. Bogs, ponds, and lakes punctuate the landscape, like puddles on a sheet of ice.

The arctic tundra lies beyond the northern boundary of the taiga, in northern Canada, Alaska, Scandinavia, and the former Soviet Union, as well as in Iceland and the shores of Green-

Figure 27-18   **Alaskan tundra.** Lupine and ferns on the Pribilof Islands, Alaska.   *(John Shaw/Tom Stack & Associates)*

land. Still nearer to the pole, and in the interior of Greenland, is an ice desert, utterly devoid of plant life. As we saw earlier in this chapter, alpine tundra lies on high mountains or plateaus in temperate regions.

All tundra vegetation is short, mostly less than 20 centimeters, and consists of grasses, shrubs, mosses, and lichens. The few small shrubs and trees grow on the banks of lakes or streams. All tundra plants have extensive underground roots, allowing them to propagate quickly during the brief summers.

The most numerous animals of the tundra are small rodent herbivores called lemmings, famous for their huge population explosions and subsequent crashes. Lemmings keep warm in winter by huddling together under the snow in nests of fur and grass. In contrast, the caribou (or reindeer) keep warm by simply being big. Their large size (and low surface-to-volume ratio) helps maintain a high body temperature. Other mammalian species of the tundra include polar bears (near the coasts), weasels, and foxes. In the summer, large numbers of birds arrive from the southern latitudes to nest and raise young. Swarms of flies and mosquitos fill the air. Especially in summer, the tundra supports more life than we might expect, but diversity is limited in comparison to that of the warmer biomes.

As is the case for the taiga, humans have had only limited impact on the tundra. Inuits of Alaska, Greenland, and Canada have long used the tundra as a hunting ground, while the Lapps (Sami) of northern Scandinavia have herded reindeer on the tundra and taiga. In the past 20 years, however, the energy needs of population centers have finally impinged on life in even this remote biome. Many people worry that the Alaska pipeline, which carries oil from the North Slope of Alaska across thousands of miles of tundra, will melt the permafrost or otherwise damage the fragile tundra.

Tundra has a cold, dry climate and a short growing season.

WHAT ARE THE AQUATIC COMMUNITIES LIKE?

Most of the Earth's surface is covered with water—about 70 percent by saltwater seas and oceans, and about 2 percent by freshwater lakes, ponds, rivers, and streams. Like terrestrial communities, aquatic communities in different geographical regions have plants and animals that resemble one another yet fall into just a few distinct types. Ocean communities, for example, differ from one another according to temperature, depth, and distance from shore. Ecologists do not use the word "biome" to refer to aquatic communities, reserving that term for land communities. As on land, however, the similarities among unrelated aquatic species illustrate convergent evolution—similar adaptations to similar environmental conditions.

Ocean Communities Have Many Subdivisions

Although the oceans cover two-thirds of the Earth's surface, they probably contain only about 10 percent of the planet's species. The total number of marine species is only about 160,000, far smaller than the millions of species—mostly insects and flowering plants—on land. All the oceans are continuous with one another, not broken up into habitat islands that allow geographical isolation and speciation (Chapter 17). Consequently, the oceans provide only limited opportunity for adaptive radiation (in coral reefs, for example).

Nonetheless, the oldest, largest, and most stable communities are those of the oceans. The oceans and their connecting seas extend over most of the Earth's surface to regions of widely varying air temperatures. But ocean waters resist changes in temperature both because of the high-heat retaining capacity of water and because of the continuous mixing of water by tides, winds, and ocean currents (Figure 27-5). As a result, the oceans are much more homogeneous from place to place on the globe than are the continents. In fact, because ocean temperatures are so stable, the oceans have a moderating effect on global climate.

In the last few decades, **oceanography,** the study of the seas and their ecosystems, has shown us that the seas are highly complex. The ocean bottom, for example, contains vast mountain ranges and rifts, as well as a shallow continental shelf bordering each continent. The average depth of the oceans is about 3 kilometers, but the Marianas Trench of the western Pacific Ocean dips to 11 kilometers below sea level—deeper than Mt. Everest is high. Like the continents, the ocean floor is slowly but constantly changing, as segments of the Earth's crust—called tectonic plates—move apart and collide.

Biologists distinguish **ocean communities** according to ocean depth and distance from shore. The **benthic division** consists of all organisms that live on the ocean bottom, including those that live in the **intertidal zone,** which lies between high tide and low tide, and those that live at the bottoms of deep canyons in the **abyssal zone** (Figure 27-19). The **pelagic division** consists of organisms that live in open water, above the bottom. In the pelagic division, microorganisms called **plankton** drift in the water, moving only where currents take them. Free-swimming organisms, such as fish and whales, feed on plankton or on each other and are sometimes called *nekton.*

The pelagic division is further divided into the **neritic zone,** which is farther from shore, over continental shelf, and the **oceanic zone,** which is out beyond the continental shelf over the deepest water.

The intertidal zone is submerged at high tide and exposed to air and pounding waves at low tide. Thus, the organisms living in the intertidal zone must be able to withstand drying, hot sun, and crashing waves. Intertidal organisms anchor themselves to rocks or even to grains of sand, and depend on a diversity of adaptations to protect them from drying out when

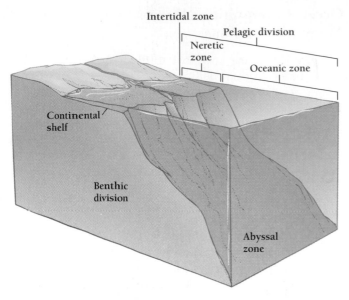

Figure 27-19 Dividing the ocean into zones. Oceanographers divide the ocean into divisions and zones according to the depth of the water and the distance from shore. The benthic division includes intertidal organisms and others that live on the bottom. The pelagic division includes the neritic zone (over the continental shelf) and the oceanic zone (deep water).

the waves recede. Tidepools contain a wonderful variety of intertidal species (Figure 27-20). Typical tidepool animals stay more or less in the same place, despite their ability to move.

The neritic zone consists of the coastal areas of the oceans, adjacent to islands and continents, where the water is shallower than in the open ocean. In the neritic zone, waters are shallower and algae, rather than microorganisms, are the dominant primary producers. While the total area of the neritic zone is much smaller than that of the oceanic zone, it contains many more ecological niches and species.

The oceanic zone is immense. In the oceanic zone, most primary production is carried out by photosynthetic cyanobacteria, diatoms, and dinoflagellates. The primary production of the oceanic zone is extremely low, comparable to that of a desert. The reason for this low productivity is that mineral nutrients, especially phosphorus and iron, become depleted as organisms die and their remains fall from the surface layers. Once these nutrients drop to the bottom of the ocean, there is generally nothing to return them to the surface.

A few exceptions exist. Off the coast of Peru, the Humboldt Current draws nutrients up from the bottom in a process called **upwelling.** Upwelling creates a rich soup that supports a rate of photosynthesis six times that found in other parts of the open ocean.

Where such currents bring nutrients, grazing protozoa and small animals feed on blooms of photosynthetic microorganisms. Small fish eat the tiny invertebrates, the larger fish eat the smaller fish. A marine food chain typically contains four or five trophic levels. (In contrast, terrestrial food chains seldom have more than three trophic levels.) This enormous productivity

Figure 27-20 The intertidal zone. The intertidal zone is characterized by organisms that cling to rocks as the waves crash around them. Shown here is a Pacific Coast tidepool. *(Nancy Sefton/Photo Researchers)*

supports, in turn, huge populations of fish-eating birds, as well as a large human fishing industry.

Every year, around Christmas, a natural experiment proves the importance of the Humboldt Current. In midwinter the current reverses itself. Warm water flows down the coast, preventing the nutrient-rich cold waters from rising to the surface. Usually these changes, which local fishermen call "El Niño" [Spanish, for "the Christ child"], are benign. However, every 25 years or so, the current reversal lasts long enough to produce catastrophic effects. In 1982 and 1983, for example, the failure of the Humboldt Current to provide its usual supply of nutrients led to a failure of the sardine industry and the wholesale starvation of fish-eating seabirds in the South Pacific.

Ocean communities are more homogeneous than land communities.

Estuaries Are Among the Most Productive Communities in the World

Estuaries are partly enclosed waters where freshwater streams or rivers meet the ocean. The water is less salty than the oceans, and its salinity is constantly changing. The flow of salt water and fresh water depends on the tides, which gives the estuaries their name [Latin, *aestus* = tide].

Estuaries are among the most productive of all natural communities. Tides and rivers provide a constant flow of nutrients. And, like intertidal zones, estuaries have a wide variety of specially adapted species. The primary producers in estuaries include plankton, algae, and large plants—eelgrasses, marsh grasses, and mangrove trees. The primary production in an estuary may be 20 times that of the open ocean, rivaling that of a tropical rain forest (Figure 27-21).

Figure 27-21 **Estuary.** Salt marsh in Baja California, Mexico. Estuaries are the most productive biomes in the world. *(William E. Ferguson)*

Because of this high productivity, estuaries provide breeding and nursery grounds for many species of fish and shellfish. Coasts with extensive estuaries support rich fisheries. Countries that destroy their wetlands ultimately destroy their fisheries. Estuaries also provide food and breeding grounds for many species of birds, amphibians, reptiles, and mammals.

Estuaries are especially vulnerable to human destructiveness. They are the first to receive pollution from streams and rivers, and they are prime targets for oceanfront development. Farms, condominium communities, football stadiums, industrial parks, and marinas sit atop filled-in wetlands the world over.

In the United States, the destruction of wetlands has slowed in the past 20 years. Cities and states have ceased to regard marshes as mere sewers and breeding grounds for mosquitos. Instead, people are beginning to realize that estuarine ecosystems are not only places of beauty but also irreplaceable contributors to the world food supply.

Estuaries are rich communities that support a diversity of species.

Why Are Lakes, Ponds, Rivers, and Streams Susceptible to Pollution?

Limnology is the study of freshwater lakes, ponds, rivers, and streams. Lakes and ponds are depressions in the ground that fill with water. They arise in many ways, among them the movement of glaciers, the changing courses of rivers, the action of volcanoes, and the dams of humans or beavers.

Many rivers and streams flow into lakes, bringing both water and dissolved nutrients. This material, both organic and inorganic, derives from the surrounding land, so the character of lakes and ponds varies from region to region. At the fringes of a lake, marshes and swamps form transition zones to the adjacent terrestrial communities.

As in ocean communities, primary production in a lake depends on the penetration of light. Both microorganisms and water plants use photosynthesis to capture light energy. Heterotrophic protists and small animals such as rotifers and crustaceans graze on the photosynthesizing microorganisms, while birds, fish, and invertebrates feed on the larger plants. Fish also eat one another. Fish of different sizes (sometimes of the same species) occupy different places in the food chain, as primary or secondary (or even tertiary) consumers. Despite the efficient aquatic food web in lakes, most of their organic matter, nonetheless, falls to the bottom as detritus, where it is consumed by bacteria and invertebrates. Far more energy flows through the detritus food chains than through the chains involving grazing herbivores and carnivorous fish.

In temperate regions, a lake's surface temperature varies greatly from season to season. In the summer, the warm air heats a layer of warm water as much as 20 meters thick (Figure 27-22). This surface layer, called the **epilimnion,** lies over the deeper, cooler waters, called the **hypolimnion,** whose temperature stays about 4°C year round. As winter approaches, the temperature of the epilimnion falls until it is the same as the hypolimnion. At that point, the waters of epilimnion and hypolimnion mix—an event called the **fall turnover.**

A similar turnover occurs in the spring, as the winter ice melts. The two turnovers bring dissolved nutrients from the bottom sediments into the upper waters, where photosynthesis occurs. The result is a seasonal growth of algae. The fall and spring turnovers also move oxygen from the surface down to the bottom-dwelling heterotrophs, which allows them to metabolize settled organic material.

A **eutrophic** lake is one that has abundant minerals and organic matter. The surfaces of eutrophic lakes tend to be clogged with algae and cyanobacteria. As these organisms die and drop to the bottom, they stimulate the lake-bottom decomposers to multiply. During the summer, the decomposers so deplete the hypolimnion of oxygen that the lake bottom can no longer support fish and other animals. As a result, the diversity of a eutrophic lake is low. In contrast, **oligotrophic** lakes have a limited supply of nutrients. These clear lakes have no permanent algal blooms, and they therefore contain more oxygen and support a more diverse community of organisms, including fish.

While eutrophication happens naturally over thousands of years, humans often accelerate the process by dumping nutrient-rich sewage and other wastes. Streams bring to lakes fertilizers from surrounding farms and phosphate detergents from nearby cities. As nutrients and minerals accumulate, a lake becomes eutrophic. Productivity increases, but diversity declines, and eventually the lake dies.

Freshwater ecosystems are unusually susceptible to pollution.

Figure 27-22 Deep temperate lakes turn over twice a year. Both turnovers bring nutrients to the surface of the lake, resulting in blooms of algae. Water's highest density occurs at 4°C. In spring, when sun-warmed ice (0°C) melts and reaches 4°C, it sinks below the colder layers, carrying oxygen downward and driving nutrients upward.

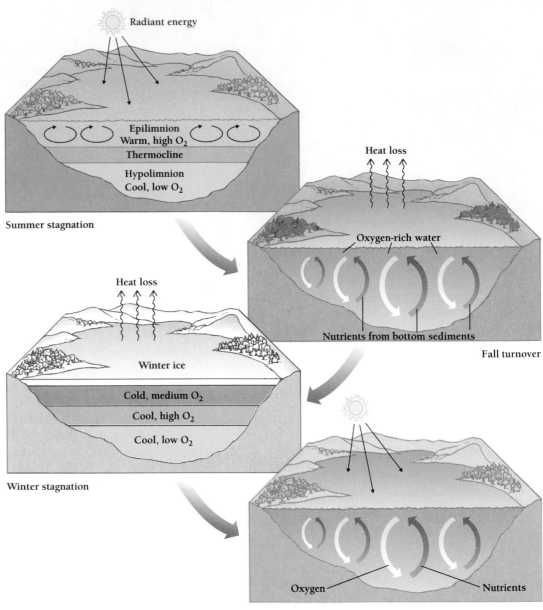

STUDY OUTLINE WITH KEY TERMS

Our planet contains a limited number of types of climax communities. These are called **biomes** if they are terrestrial, and simply aquatic ecosystems if they are aquatic. The character of these ecosystems is determined largely by climate, especially temperature and moisture. The average temperature, the total annual precipitation, and the variability in each of these values help determine the character of a biome.

The climate of a region is determined by the total solar energy absorbed and by wind and ocean currents. These depend on latitude and the rotation of the Earth. The Earth's rotation creates the **Coriolis effect,** which causes winds moving north or south to rotate, clockwise in the Northern Hemisphere, and counterclockwise in the Southern Hemisphere.

At the equator, hot, moist air rises and flows north and south over the tropics, dropping rain. **Tradewinds** blow cooler air from the tropics toward the equator. The **westerlies** are cool, dry winds that cross the Earth's surface between 30° and 60° latitude. These winds

warm and pick up moisture as they go, and, on reaching about 60° latitude, rise to create **polar fronts,** regions of low pressure.

Mountain ranges create **rain shadows** by forcing warm moist air to drop most of its moisture on one side of the range. Biomes at high elevations resemble those at high latitudes. At high altitudes, **coniferous forest** gives way to **krummholz** at the **tree line.**

Temperate deciduous forests, which receive 80 to 140 centimeters of precipitation each year, are characterized by a heavy cover of deciduous trees and a limited understory. Temperate deciduous forests, found in eastern North America, eastern China, and Europe, are commonly associated with dense populations of humans. These populations have cleared these forests repeatedly and extensively, and hunted to extinction or near extinction all of the major predators, including bears, wolves, and lions or tigers.

Temperate grasslands, which receive 25 to 75 centimeters of precipitation each year, are characterized by tall grasses in wet areas and short grasses in drier areas. Fire, insufficient water, and grazing animals all keep trees from becoming established in this biome. Much of the world's grassland, midwestern North America, for example, has been converted to agriculture, thus eliminating entire food chains.

Chaparral regions, which have mild, wet winters and hot, dry summers, are characterized by dense shrubs and scattered, broad-leafed trees, mostly drought and fire tolerant. Humans build homes in the chaparral, then suppress natural brush fires so long that when fire finally comes, fuel loads create a firestorm. Houses, chaparral, and valuable soil are all damaged or destroyed.

Deserts, which receive less than 25 centimeters of precipitation each year, are characterized by widely spaced drought tolerant plants, such as sage brush (in cold regions) and cacti (in hot regions). Some animals **estivate** during the hot, dry summers. The greatest human threats to deserts are off-road vehicles and agriculture.

Tropical **savannas,** which receive 90 to 150 centimeters of rain each year, are characterized by tall grasses and short, widely spaced trees. Savannas are hotter and wetter than temperate grasslands.

Tropical rain forests, which receive 200 to 450 centimeters of precipitation each year, are characterized by a dense **canopy** with a tall understory and little vegetation on the dim forest floor. **Lianas** and **epiphytes** grow on the trees, and **emergents** rise above the canopy. Tropical countries are cutting their rain forests down at a rate that will eliminate this biome within a few decades.

The cold **taiga** is characterized by short homogeneous pine or spruce forests. This biome has suffered less human-caused damage than most other biomes because it has so little to offer economically.

The cold, dry **tundra,** which receives less than 25 centimeters of precipitation each year, is characterized by short, open grassland, punctuated by bogs, ponds, and lakes. **Permafrost** underlies the soil all year long.

Oceanography has revealed the depth and character of the oceans and the organisms that live in them. **Ocean communities** can be divided into **benthic** and **pelagic divisions.** The benthic division includes the **intertidal** and the **abyssal zones.** The pelagic division includes the **neritic** and **oceanic zones. Plankton** float freely and travel passively only. Nekton, such as fish, propel themselves. Ocean food chains can have more trophic levels than terrestrial food chains, but the total number of species living in the oceans is far less than the number that live on land. Primary productivity is low, except in areas of **upwelling.**

Estuaries are marshes with partly fresh, partly salt waters. Estuaries are extremely productive and provide breeding grounds for many oceanic species. Humans have a long history of draining marshes and estuaries for farms or developments.

The study of freshwater lakes, ponds, rivers, and streams is called **limnology,** one of the oldest branches of ecology. In temperate regions, lakes divide into an **epilimnion** and a **hypolimnion.** During the **fall turnover,** these two layers mix, moving oxygen down to the bottom of the lake. Lakes that are rich in organic matter are **eutrophic.** Those lower in nutrients support a more diverse community and are said to be **oligotrophic.** Humans frequently pollute fresh water with sewage and industrial wastes, as well as the acid rain that results from burning coal and oil.

REVIEW AND THOUGHT QUESTIONS

Review Questions

1. Explain why the polar regions experience more extreme seasons than the equatorial regions.
2. What kind of biome exists in a rain shadow?
3. How does the division of a forest or other habitat into small sections interfere with top predators?
4. What keeps grasslands from changing, through succession, to forests?
5. Explain the difference between a liana and an epiphyte. Why do tropical plants often make good house plants?
6. What percentage of the Earth's surface is covered by water?

7. What organisms are the primary producers in the oceanic zone? What organisms are the primary producers in the neritic zone?
8. What causes fall turnover, and what benefits does turnover have for freshwater organisms?
9. What causes a lake or pond to become eutrophic?

Thought Questions

10. Explain why the soils of tropical rain forests are poor, even though tropical rain forests are so productive.
11. Why do tropical rain forests have so many species?
12. Why are the oceans species poor?

SELECTED READINGS

Austin, Mary, *The Land of Little Rain,* University of New Mexico Press, Albuquerque, 1995. Lyrical essays on the deserts of the American West.

Carson, Rachel L., *The Sea Around Us,* Oxford University Press, New York, 1961. An outstanding book describing the world's oceans, including sea life, geology, and weather.

Dietrich, William, *The Final Forest: The Battle for the Last Great Trees of the Pacific Northwest,* Simon & Schuster, New York, 1992. An accessible and well-written book about deforestation in the Northwest and the people on both sides of the battle over the fate of the last old trees.

Parkman, Francis, *The Oregon Trail,* Riverside Press, Houghton, Mifflin & Co, Cambridge, Massachusetts, 1847. Parkman's vivid account of his experiences during an 1846 trip through the prairies of Kansas, Nebraska, and Wyoming.

Wilson, Edward O., *The Diversity of Life,* The Belknap Press of Harvard University Press, Cambridge, Massachusetts, 1992. A popular discussion of (mainly tropical) ecosystems, including why there are so many species, why species diversity is important, what causes major extinctions, and how humans are contributing to modern extinctions.

▶ On-line materials relating to this chapter are on the World Wide Web at http://www.saunderscollege.com/lifesci/
Click on Tobin/Dusheck: *Asking About Life.*

POPULATIONS AND THE HUMAN PLACE
IN THE BIOSPHERE

Can a Population Grow Too Much?

In 1944, 29 reindeer were introduced to a lichen-covered rock island in the Aleutian Islands, west of Alaska. The reindeers' favorite food was lichen, and there was plenty of it on Saint Matthew Island. The reindeer ate and multiplied. By 1963, the 29 reindeer had burgeoned into a huge population. Six thousand animals, some weighing as much as 600 pounds, now roamed the island, pulling the last bits of lichen from the cold rocks.

By the end of the year, the slow-growing lichen had nearly vanished from 500-square-kilometer Saint Matthew Island. Then an especially long and harsh winter struck. By 1964, nearly all the reindeer had succumbed to hunger and cold. When spring arrived, only 41 females and one sterile male remained. Although these animals would scrape by on the remaining lichen for several more years, they could not reproduce. One by one they died, and the island returned to the lichen (Figure 28-1).

Saint Matthew Island doomed the reindeer in two ways. First, when the lichen began to disappear, the reindeer could not leave the island to browse elsewhere and allow the lichen to recover. Second, the island harbored no predators that might control the reindeer numbers. The reindeer simply multiplied until all the

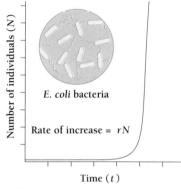

A. Exponential growth curve:
   Unrestrained growth of *E. coli*

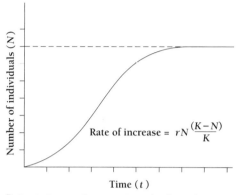

B. Logistic growth curve: Growth of *E. coli* restricted
   by waste accumulation and food and space
   availability

Figure 28-1   **How populations grow.** A. Exponential growth, or "J" curve. The rate of increase equals the growth rate times the size of the population at any given time. B. Logistic growth or "S" curve. The rate of increase is limited by *K*, the carrying capacity. C. The human population has increased exponentially. World politicians argue about the value of *K*, but not about whether there is a limit to the number of people the world can support.

food was gone. Their population had escaped from any restraints on its numbers—but only temporarily.

All natural populations are able to multiply as rapidly as the reindeer did. Indeed, biologists have provided a vivid example of this explosive population growth. They have calculated that if *all* of the descendants of just 100 California sea stars survived and reproduced for 15 generations, the number of sea stars would exceed the number of electrons in the visible universe.

In nature we might see such unrestrained growth when a species has just invaded a new territory, as, for example, when the protozoan *Giardia lamblia* finds an uninhabited human gut, or as in this story of a few reindeer being introduced to a formerly unoccupied island. But eventually, the population size will level off and very likely crash to a small number of individuals.

Fortunately, natural populations never achieve their maximum rate of natural increase. In fact, we find that although natural populations may cycle up and down a bit, they are surprisingly stable. An important question in population ecology, therefore, is, What limits the growth of natural populations?

The size of most populations within a community fluctuates from season to season and from year to year. Changes in abiotic conditions (such as rainfall) or in biotic interactions (such as the introduction of a new species) can dramatically influence the size of populations of different species. Natural populations usually remain rather stable over the long term. But while the population sizes of some species remain approximately constant year in and year out, others fluctuate wildly. In this chapter we will examine the mechanisms that determine population size.

One by one they died, and the island returned to the lichen.

Along the way, we will see that different species use different strategies to leave many offspring. Some species have a few offspring that they watch over carefully. Other species produce dozens or hundreds—even millions—of offspring, all of which must fend for themselves. We will see that the human population is not exempt from the rules that govern the growth and decline of populations. Human populations have been multiplying as rapidly as the reindeer. And although we are not confined to a small island in a nearly arctic ocean, we are confined to a small planet out of reach of any sanctuary. At the end of the chapter, we will see what specific factors limit the growth of our own population.

645

KEY CONCEPTS

1. Natural populations have the potential to grow exponentially. However, those that increase exponentially for long periods typically either reach a plateau or crash.

2. Both density-dependent and density-independent factors affect the population growth of a species.

3. Human population growth has accelerated in the last century.

4. Most humans live in heavily subsidized ecosystems.

5. The human population cannot continue its expansion indefinitely. It must reach a plateau or crash.

6. Hope for the future of our species and the Earth's biosphere depends on our ability to understand and to act.

HOW DO POPULATIONS GROW?

What Is Exponential Growth?

Under the most favorable conditions, the number of *E. coli* in a rich nutrient broth may double every 20 minutes. After the first 20 minutes, one bacterium forms two; after another 20 minutes, the two have formed four; after another 20 minutes, there are eight; and so on. We say that the bacteria have a **doubling time** of 20 minutes.

We can calculate the number of bacteria (N) from the number of times the population has doubled since the time at the beginning of the experiment (t) using the following equation:

$$N = 2^t$$

For example, if 60 minutes had elapsed, the population of *E. coli* would have doubled three times. So t would equal 3, and

$$N = 2^3 = 2 \times 2 \times 2 = 8$$

In this equation, t is an "exponent," hence we call the bacteria's growth **exponential growth,** a growth pattern in which a population doubles in some constant period of time.

Exponential growth means that the greater the population size, the faster the population increases. For example, we know that 16 *E. coli* cells would double in 20 minutes to 32 cells. Thus the population would increase by 16 cells. But a population of 16 million cells would also double in 20 minutes. That population would increase by 16 million cells over exactly the same time period. The large population and the small one would have the same doubling time.

A graph of the increasing number of bacteria looks like the letter **J** and is therefore called a **J-shaped curve** (Figure 28-1A). We can see that at the bottom of the J-shaped curve, the population size (N) increases rather slowly. Then it accelerates. When there are no limits to population increase, N increases faster and faster. If a population continued along this curve, with no limits to its growth, eventually N would be increasing almost infinitely fast.

We can describe this ever-increasing growth as follows: During unrestrained exponential growth, the rate of change in the number of organisms is proportional to the number of organisms present in the population. That is, the bigger the population, the faster it grows.

The growth rate, called **r,** is the population growth in any unit of time (for example, per minute). The growth rate r is also the difference between the average birth rate and the average death rate. We can see that the population will grow only if the birth rate exceeds the death rate. When a population's birth rate is at its maximum and the death rate is at its minimum, r is as large as it can be. This is called r_{max}, or the **intrinsic rate of natural increase.** Table 28-1 shows some values of r for various species.

The larger a population is, the faster it can grow. All populations are capable of growing exponentially.

Table 28-1   Intrinsic Rate of Increase and Generation Times

Organism	Birth Rate (per capita per day)	Generation Time (days)
E. coli	60.0	0.014
Flour beetle	0.12	80
Rat	0.015	150
Mouse	0.013	171
Dog	0.009	1000
17-year cicada	0.001	6.050

After Campbell, 2nd ed., p. 1081, after E.R. Pianka, *Evolutionary Ecology*, 3rd ed., Harper and Row, New York, 1983.

What Limits the Size of a Population?

Even in the laboratory, exponential growth cannot continue for long. We may see a J-shaped population curve when a bacterium has just colonized a glass flask containing a nutrient-rich medium. But even *E. coli* bacteria growing under nearly ideal laboratory conditions soon find limits to their growth. The bacteria run out of food and space, and their own wastes begin to poison them. The concentration of bacteria gradually reaches a plateau. If we look at the whole growth curve, we see that it consists of three parts—an initial "J" phase, in which the growth rate is accelerating; then a deceleration phase; and finally a gradual leveling off. The whole curve now looks more like a tilted and stretched out "S" (Figure 28-1B).

We can modify the growth equation to describe the decelerating growth that we see in natural populations. The equation should express the deceleration in terms of population size. That is, the larger the population size, or *N,* is, the faster growth should decelerate. We express this by taking the plateau value of the S-shaped curve of population growth and defining it as the **carrying capacity** of the environment for a particular species. The carrying capacity, also called **K,** is the maximum population density the environment can support indefinitely. *K* is also used to refer to the actual number of organisms that a habitat might support indefinitely.

When the environment is not yet filled to its capacity, there remains a fraction of *K* that is left to be filled. For example, if the carrying capacity for houseflies in Australia were 100 flies per square meter, and the population density *N* were 70 flies per square meter, then the fraction of *K* left would be 30 percent. Population ecologists take that fraction of *K* and multiply it by *rN* to get a measure of the change in the population growth over time. The equation they use, called the **logistic growth equation,** seems formidable to the nonmathematical:

$$\frac{dN}{dt} = rN\,\frac{(K - N)}{K}$$

But it merely describes the *restricted* growth of many natural populations. The rate of change of the growth of a population (the change in *N* divided by the change in time) equals the growth rate times the size of the population.

If the equation is plotted as a curve, the curve looks just like the S-shaped curve of *E. coli* bacteria. It starts out like the J-shaped curve, with an ever-increasing rate of population growth. That is, when the population at a given time (*N*) is very small, then the unfilled fraction of the carrying capacity is about 100 percent, or 1. Under those circumstances, the restricted growth is nearly equivalent to unrestricted exponential growth.

As the population increases, however, the fraction of unfilled *K* decreases and the growth rate slows. When population size reaches the carrying capacity, the environment cannot support any more growth. Then the fraction of unfilled *K* equals 0, and the population stops growing. The population is then said to have reached **zero population growth.**

What Determines Whether a Population Grows?

The carrying capacity of the environment for a particular species depends on the needs of the species and, therefore, on a multitude of factors—food supply, territories, predators, and competitors, to name just a few. Bees, for example, need nectar, pollen, and a safe place for a hive. Even if a given habitat has plenty of flowers, the carrying capacity for bees may be low if there are only a few good places for a hive within flying distance of the flowers.

Changes in a population size result from the combination of four processes: birth, death, **emigration** (movement out of the population), and **immigration** (movement into the population). The net change in population size is the sum of births and immigration, minus deaths and emigration.

Density-Dependent Factors

The rate of each of these four processes may depend on the population density. As a population's density approaches the carrying capacity, its members must increasingly compete with one another for limited resources. For example, one study of a population of 25 pairs of tawny owls at carrying capacity suggested intense competition. The 50 owls, theoretically capable of producing as many as 100 offspring per year, successfully fledged only 18 owlets. Because of limitations in the food supply—in this case, rodents—some owls were unable to feed their offspring, others did not bother to incubate their eggs, and still others did not even breed. Thus competition for food depressed the birth rate. But even 18 owls was too many. Since only 11 adults died, only 11 territories opened up. Just 11 of the 18 fledglings could find territories on which to breed. The other seven young owls had to emigrate.

We can see that high population density can drive the birth rate down. Animals run out of food or territories. Plants run out of water or light. High population density can also encourage emigration and discourage immigration (Figure 28-2). In addition, high population density can increase the death rate from predation, infectious diseases, and parasites.

Predation, parasitism, disease, competition (both intraspecific and interspecific), and emigration all limit growth in proportion to the density of the population. The denser the population, the more slowly it grows. Such factors are said to be **density-dependent.**

Density-Independent Factors

However, some populations are limited by **density-independent** factors, such as fire, drought, storms, tornadoes, volcanic eruptions, and other natural disasters. For example, severe

Figure 28-2   A plague of locusts. When the population density of certain species of locusts begins to rise, the individuals develop into longer winged, more gregarious types that like to fly together. When the proportion of these gregarious types reaches a certain threshold, they rise in great masses and emigrate to a new area, consuming everything in their path—a classic plague of locusts.   *(Gianni Tortoli/Photo Researchers)*

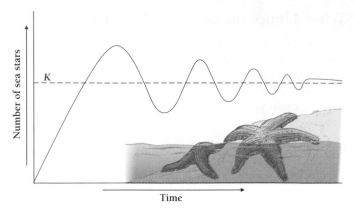

Figure 28-3   Starfish approach K. This starfish population grew rapidly, then fluctuated around K.

Natural populations do not follow the logistic curve exactly.

---

storms may all but eliminate a population of butterflies long before the caterpillar larvae run out of plants to eat. The changes in population density that result from density-independent limitations do not conform to the logistic growth equation.

Both density-dependent and density-independent factors limit population growth.

Do Natural Populations Actually Grow Exponentially?

The logistic curve gives an excellent description of the growth of bacteria in the laboratory. Populations in the wild also follow the logistic curve, but with more fluctuations. Often population density will temporarily **overshoot** the carrying capacity of the environment. For example, a population of sheep may exceed the carrying capacity, then drop far below the carrying capacity, and then cycle back up to K again.

When a population exceeds the carrying capacity of its environment, it must decrease. The population may then increase again, fluctuating up and down indefinitely. Many populations fluctuate up and down for a while, then gradually level off at the carrying capacity K (Figure 28-3). Whether the fluctuations continue or eventually disappear depends partly on the extent of the original overshoot. In turn, the overshoot depends partly on the intrinsic rate of natural increase r_{max}. A species capable of growing rapidly, such as a sea star, is much more likely to overshoot its carrying capacity than a species such as an elephant, whose intrinsic rate of growth is low.

Why Do Populations Undergo Regular Cycles?

Many populations show extremely regular fluctuations. In Alaska, for example, lemmings cycle every 3 to 6 years from a density of less than one animal per hectare (2.5 acres) to a high of more than 25 per hectare. At peak population densities, the lemmings devastate the food supply and attract predators. Reproduction all but ceases, the population plummets, and the cycle then repeats itself.

Many other mammals, birds, and insects show similarly regular cycles. A classic illustration of population cycles comes from the 200-year-old records of the Hudson Bay Company of Canada, which collected and sold pelts of both the snowshoe hare and its major predator, the Canadian lynx (Figure 28-4). Assuming that the number of pelts collected each year reflects the total population size, we see that both the hares and the lynxes have population cycles of about 10 years.

At one time, ecologists interpreted the parallel cycling of lynxes and hares as evidence that a predator population can control the size of the prey population. They reasoned that the greater the hare population, the greater the food supply for the lynx population, and, therefore, the greater the size of the lynx population. Then, went the argument, the large number of lynxes preying on the hares had the effect of increasing the death rate among hares, until there were too few hares to support the large population of lynxes.

If this were the case, however, we would expect the lynx curve to lag somewhat behind that of the hare: the lynx population should have declined only as their prey became rare, and the hare population should have recovered only after the lynx population decreased. But, as Figure 28-4 shows, lynx

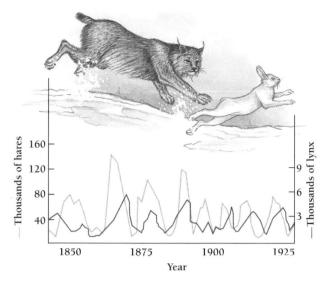

Figure 28-4 Population cycles of the snowshoe hare and the lynx. The two populations fluctuate, sometimes together, sometimes independently.

Figure 28-5 The Dust Bowl. Removal of native grasses, combined with a severe drought, destroyed nearly all plant life in some areas. In Oklahoma, the rich, prairie topsoil dried up and blew away. (Corbis-Bettmann)

populations do not consistently lag behind. Sometimes, in fact, the lynx populations begin to decline before the hares.

Further, snowshoe hare populations are known to cycle even on islands where no lynxes live. Ecologists think that the hares cycle independently of the lynxes, possibly following cycles in their own food supply. For example, trees and shrubs respond to heavy browsing by increasing the production of compounds that make them unpalatable.

Just as prey do not necessarily regulate the numbers of predators, however, plant growth does not necessarily regulate numbers of herbivores. A study by Batzli and Pitelka in the early 1970s showed that two populations of meadow voles, relatives of lemmings, cycled similarly even though one population had ten times as much food as the other population.

Cycling in natural populations remains hard to explain. Predators can and do play a role in regulating herbivore populations. For example, the Australian ladybird beetle introduced to California citrus groves in 1888 successfully controlled an outbreak of scale insects. However, other factors—including the supply of plant food, defensive behavior by plants, and self-regulation—often play even more important roles.

Density-independent factors such as weather can also cause irregular population fluctuations. During the Great Drought of the 1930s in the American Great Plains, for example, the average rainfall decreased from about 23 inches a year to about 16 inches. The crops withered and failed. American farmers had destroyed the prairie grasses that held the soil in place by plowing them under or by allowing cattle and sheep to graze the grass down to the roots. When the crops failed, nothing was left but bare ground, and the topsoil began to blow away, burying intact crops, fences, and farmhouses and devastating

large areas of Kansas, Oklahoma, Texas, and other states (Figure 28-5). Thousands of farm families abandoned their land, as vividly chronicled in John Steinbeck's *The Grapes of Wrath*.

The Great Drought was not an isolated incident, however, but a normal fluctuation in the weather. During the drought, the populations of some species declined (including big bluestem grasses and humans), while other species increased (including western wheatgrass, which grows mostly in the spring when water is most plentiful). When rainfall again increased in the early 1940s, the population of big bluestem grass and humans increased as well.

Some population cycles depend on specific interactions between species.

WHAT FACTORS INFLUENCE THE VALUES OF *r* AND *K*?

How Do *r*-Strategists and *K*-Strategists Differ?

Species differ greatly in their intrinsic rates of increase, r_{max}. Birds that lay six eggs per clutch, for example, have higher values of r_{max} than those that lay only two. An organism that reproduces while young has a greater r_{max} than one that delays reproduction. Longer periods of reproductive activity also favor higher rates of natural increase, although many species with high r_{max} may reproduce only once, but with a large number of young.

The r-Strategists

Some species are especially well-adapted to have high values of r_{max}. These species have evolved to reproduce as fast as they can whenever possible. A particularly striking reproductive strategy is that of the gall midge. When food is plentiful, the midge reproduces parthenogenetically, dispensing with fertilization by males. The mother produces eggs, which develop inside her. When the eggs hatch, the larvae sustain themselves by devouring their mother from the inside. They then quickly produce their own eggs. This behavior works until the energy available in the mother's tissues runs out. Then, the surviving midges must revert to a more conventional life cycle, eating fungi and reproducing sexually.

The gall midge illustrates an extreme example of an **r-selected species,** a species that has adaptations that support high values of r_{max}. Species that are r-selected tend to produce many small, quickly maturing offspring and expend little energy rearing them. In general, r-selected species are small and relatively short-lived. They are often pioneer species, adapted to exploit ephemeral, disturbed, or unpredictable habitats. Their numbers are likely to be controlled by density-independent factors, such as weather or local catastrophes. Sea stars release millions of eggs each year, but they invest no energy in caring for the resulting larvae. Normally, most of the larvae die. However, when conditions are right, sea stars and other r-selected species can flood a habitat with millions of offspring. Classic r-selected species also include dandelions, thistles, scotch broom, and other weeds that quickly multiply wherever soil is disturbed.

The reproductive strategy of r-selected species has been called "prodigal" and "opportunistic." These terms have the ring of moral chastisement, but these species can take advantage of new opportunities and changing conditions much more efficiently than species with low values of r_{max}.

The K-Strategists

Unlike r-selected species, many species have adaptations that increase their ability to maintain populations as close to the carrying capacity (K) as possible. These **K-selected species** tend to live long lives, reproduce late in life, produce few and large offspring, and provide extended parental care to each offspring. They often produce offspring sequentially in a series of small litters or clutches. K-strategists characterize climax communities, leading "prudent," rather than "prodigal," lives. Table 28-2 lists some typical characteristics of r-strategists and K-strategists.

Large mammals are frequently K-selected. For example, African elephants, which live 60 to 70 years, are classic K-strategists. The gestation period for the African elephant is 21 months, and the mother does not wean the single calf until it is 5 years old. Even after nearly 7 years, the calf continues to need some attention and will not be sexually mature until it is about 15 years old. K-selected plants include buckeyes and av-

Table 28-2	Alternate Life Strategies in Plants	
Trait	K-selection	r-selection
Climate	Predictable, fairly constant	Variable, uncertain
Mortality	Density dependent	Density independent
Survivorship	Type I and type II	Type III
Population size	Constant over time; at or near carrying capacity, K; no recolonization each year	Variable over time, recolonization each year
Competition	Heavy	Mild to none
Life span	Long	Short, less than one year
Reproduction	Slow development, delayed reproduction	Rapid development, early reproduction

After Barbour, Burk, and Pitts, after E.R. Pianka, "On r- and K-selection," *American Naturalist* 104:592–597, 1970.

ocados, both of which are trees that produce a relatively few enormous seeds.

While some species (such as gall midges) are clearly adapted to maximize r_{max}, and others (such as elephants) are adapted to live close to K, most species fall somewhere between the two extremes. For example, all trees are K-selected compared to dandelions, but the black willow, which lives about 70 years, might be considered r-selected compared to the giant sequoia, which lives 2500 years, or the bristlecone pine, which lives up to 5500 years. Sugar maples and sea anemones also illustrate the difficulty in defining distinct boundaries between r-selection and K-selection. Like r-selected species, both organisms produce millions of progeny, few of which survive to maturity; this large production allows them to flood new niches with their descendants. However, like K-selected species, those offspring that survive hold on tenaciously and reproduce for a long time.

Populations of K-selected species grow more slowly than populations of r-selected species. However, in a stable environment, populations of K-selected species persist.

How Do Life Strategies Affect Survivorship and Age Distributions?

Elephants and other K-strategists invest energy in the production and care of a few large offspring. As a result, the offspring have a good shot at living long enough to reproduce, usually many times. In contrast, most of the young of r-strategists die early. Only a small fraction have the opportunity to reproduce.

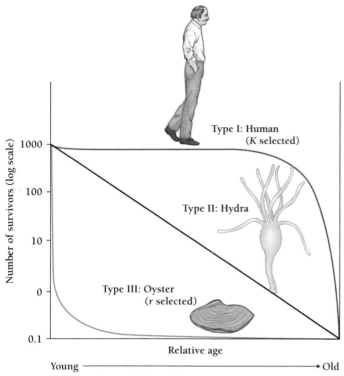

Survivorship curves: Types I, II, and III

Figure 28-6 Survivorship. Some organisms display a different pattern. Among oysters and other *r*-selected species, great numbers of young die at the beginning of life. Those that survive tend to live a long time (type III). In the type II pattern, exemplified by hydras, mortality is fairly steady throughout most of the life of the organism.

One way of representing these two contrasting life histories is a **survivorship curve,** a graph that shows the fraction of a population that is alive at successive ages. The opposite of survivorship is mortality, the fraction of a population that dies at a given age. Figure 28-6 shows survivorship for a *K*-selected population, humans in the United States. Humans in developed countries have a **convex** (or **type I**) survivorship curve, in which survivorship starts out high and decreases slowly with age until a certain point (in humans, about age 60), when survivorship begins to decrease more rapidly. Convex survivorship curves are characteristic of *K*-selected species. Such organisms have a good chance of surviving until they have finished reproducing and caring for their young. Then the chance of dying in a given year dramatically increases.

There are also species, such as hydras, for which the chances of dying do not change, no matter what their age. These organisms are as likely to die in midlife as when young or old. Hydras and other such organisms show a **diagonal** (or **type II**) survivorship curve—a straight, declining line (Figure 28-6). Finally, many species, especially those with high r_{max} values (oysters, for example), have the greatest chances of dying early in life. Their survivorship curves are called **concave** (or **type III**).

Intermediate survivorship curve:

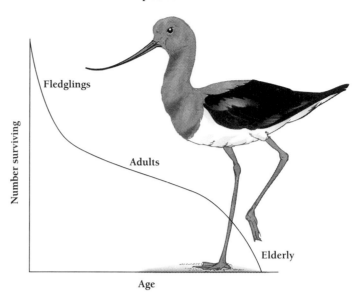

Figure 28-7 Intermediate survivorship. Survivorship need not look exactly like a type-I, -II, or -III curve. In avocets and other birds, it may be composed of parts of each.

Most real survivorship curves are mixtures of all three types of curves. They have phases that are concave, diagonal, and convex. Most songbirds, for example, have high mortality as fledglings. During the middle parts of their lives, when their chances of being caught by a predator or meeting with an accident are independent of their actual age, they have diagonal survivorship curves. At the end of their lives mortality increases as they age (Figure 28-7). Many plants have a similar pattern: high mortality of seeds and seedlings (a concave segment of the survivorship curve); constant mortality as saplings (a diagonal segment); then a long period of low mortality followed by rapid aging (a convex segment).

The *r*-strategists and *K*-strategists have different life expectancies.

The **age structure** of a population is the fraction of individuals of various ages. The distribution of ages in a population is closely related to the survivorship curve. Oysters, for example, have a concave survivorship curve, and most do not live very long. In contrast, humans have convex survivorship curves, and a greater fraction of the population lives for a relatively long time.

We can represent the age structure of a population by a graph such as those in Figure 28-8A, which show the number of people in each 5-year age group in the United States in 1997 and projected for 2025. Each such group forms a **cohort,** the set of individuals that enter the population, or are born, at the same time. In this case, we show males and females separately, since their age distributions are slightly different.

Figure 28-8 Age distributions. An age distribution chart is a survivorship curve turned on its side. A. In the United States, a declining birth rate, combined with an increasing life span, is transforming the age structure of the population from a pyramid to a rectangle. B. In Mexico and other countries with high birth rates, the age structure is more pyramidal because the proportion of young people is high compared to the number of older people. C. Hypothetical age distributions of expanding, stable, and declining populations. Mexico, India, and Nigeria have expanding populations; Sweden and Italy have stable populations; and Bulgaria, Hungary, and Russia have declining populations.

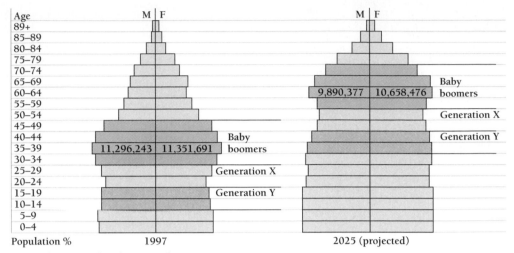

A. Population age distribution in the U.S.

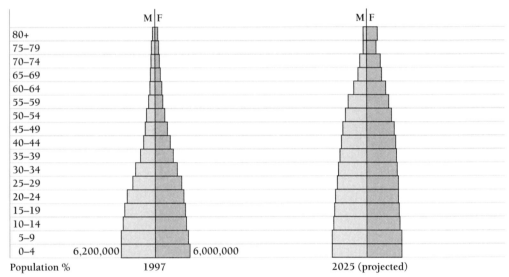

B. Population age distribution in Mexico

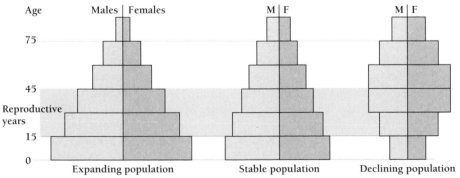

C. General patterns of age distribution

BOX 28-1

Why are age distributions of males and females different?

Age distributions such as those in Figure 28-8 traditionally show males and females separately. We usually think of males and females as being represented in about equal proportions, but in many sexually reproducing organisms large differences exist in the ratio of males to females.

In certain species of wasps, for example, most offspring are female. All fertilized eggs become females and all unfertilized eggs become males. Consequently, a shortage of males means that many eggs will go unfertilized and more males will hatch. On the other hand, an excess of males means that most of the eggs will be fertilized and will therefore be female. Other organisms use still more complicated negative feedback mechanisms to maintain their sex ratios.

Ants, known for their cooperative behavior, are less cooperative when the subject is sex ratios. A queen ant is equally related to her female and male offspring and bears equal numbers of each. But the males, which develop from an unfertilized egg, are haploid, and the female workers, which develop from fertilized eggs, are diploid. As a result, female workers who share the same father are more related to one another than they are to their brothers.

When Finnish researchers studied the sex ratios of ants from 59 colonies of wood ants, they discovered that in colonies where all the workers share the same father, the males die off disproportionately. Even though the queen lays equal numbers of male and female eggs (the researchers determined the sex of 3000 ant eggs), the adult females far outnumber the adult males. The researchers suspected that female workers either neglect the male larvae or kill them outright. Indeed, in the laboratory, eggs of males added to a nest of closely related females disappeared without a trace.

The situation was very different in colonies in which the queen had mated with two or more males. In such colonies, the females are less related to one another and the ratio of male and female adult ants remains about equal.

What about humans? Are our sex ratios 50:50? Most people know that women live longer than men, on average. As a result, among older people, women outnumber men. Not everyone knows, however, that in all countries, more baby boys are born each day than baby girls. The difference is not great. In North America, for example, 105 boys are born for every 100 girls.

By the time each cohort reaches reproductive age in the early twenties, however, the sex ratio is nearly exactly 50:50. This adjustment occurs gradually. Baby boys are slightly more likely than baby girls to die of infections and other health problems. Older boys are slightly more likely than girls to die in accidents.

Finally, adolescent boys are dramatically more likely to die than adolescent girls. The leading cause of death for all Americans between 5 and 27 years of age is motor vehicle accidents. But an adolescent boy driving a car is more than three times as likely to die in a car accident as an adolescent girl driving a car. Among 21- to 24-year-olds, the ratio approaches 4 to 1.

The physiological mechanisms that cause a sex ratio of 105 to 100 are not yet known. We can hypothesize, however, that this skewed ratio has some adaptive value. How would you test such a hypothesis in humans or in some other organism?

The age distribution in a population is not just a function of survivorship: it also depends critically on the birth rate. A rapidly reproducing population, such as that of Mexico, has a greater fraction of young people than does a population reproducing more slowly, such as the United States (Figure 28-8B). The age structure also depends on the ages of individuals who move into and out of the population (immigrants and emigrants). As a result, the age structure of the human population differs from country to country, often dramatically (Figure 28-8C).

We can predict the future age structure of a given population if we have the following information: (1) the present age structure; (2) the **mortality** of each cohort (the chances that individuals of a given age will die each year); (3) the age structure of immigrants and emigrants; and (4) the **fertility** of each cohort (the number of offspring each individual in a cohort is likely to produce).

Differences in survivorship curves lead to differences in age distributions.

Demography: How Do Scientists Measure Populations?

Studying the growth and changes in the structure of human populations is the concern of **demography** [Greek, *demos* = people + *graphos* = measurement], the statistical study of populations. Such calculations are important for economists, health planners, and insurance companies, as well as for biologists. Demographers use a variety of statistical measures to describe populations. Mortality is expressed in terms of the **death rate**, the number of individuals per 1000 who die each year. Similarly, fertility can be expressed in terms of the **birth rate**, the number of individuals per 1000 who are born each year. For example, in the United States in 1989, 3.9 million babies were born in a population of nearly 250 million, for a birth rate of 15.6 per 1000. In the same year, 2.1 million Americans died (8.4 per 1000). Net immigration was 700,000. So the total increase in the population was 2.5 million.

$$3.9 \text{ million} - 2.1 \text{ million} + 0.7 \text{ million} = 2.5 \text{ million}$$

Demographers use several measures to express population growth. One is the **annual rate of increase,** the actual percentage by which a population increases each year. For example, the 2.5-million-person increase in the 1989 population of the United States came from a population of 250 million, giving an annual rate of increase of about 1 percent. The annual rate of increase for the entire world in 1990 was about 1.8 percent.

Another measure of population growth is **completed family size,** the average number of children that reach reproductive age born to each family. For example, the average completed family size in the United States in 1992 was 2.0 children. **Replacement reproduction** is the family size at which each couple is replaced by just two descendants. At the replacement level, a population neither grows nor shrinks. Replacement level in the United States, for example, is 2.1 children per family. The additional 0.1 makes up for the small numbers of children who do not make it to reproductive age. We can see that the population of the United States is not quite replacing itself each generation. But we can also see that the total population of the United States is still growing, at the rate of 1 percent per year. This paradoxical growth is not due only to immigration, as people sometimes argue. Without immigration, the annual rate of increase would still be about 0.7 percent. In the next section, we will see how population growth can continue despite a reduction in average family size.

THE HUMAN POPULATION IS GROWING LOGARITHMICALLY

Before the end of the last ice age, about 10,000 years ago, the world population of humans consisted of about 5 million people, who supported themselves by gathering fruits, roots, and leaves and by hunting animals. Every region probably contained a fairly constant population, with deaths balancing births.

Then, at the end of the ice age, some 8000 to 10,000 years ago, humans invented agriculture. Agriculture increased the food supply, which permitted humans to reproduce at a far greater rate. The human population began a growth spurt that continues today.

In the 6000 to 8000 years following the invention of agriculture, the human population doubled more than four times to some 100 million people. In the 2000 years since, the world population has doubled nearly six more times. Today the world population stands at over 5.3 billion people. Another 100 million will be born next year.

Until the beginning of the Industrial Revolution, in the late 18th century, the human population grew exponentially, but the yearly absolute increase in numbers was small by today's standards. Then, in the 19th and 20th centuries, closed sewers, chlorinated water, refrigeration, vaccines, and antibiotics reduced the death rate dramatically. The death rate fell first in the industrialized countries of western Europe, then, country by country, in other industrialized areas of the world. After World War II, industrialized nations brought to developing countries sewer and water systems, vaccines and antibiotics, and death rates began to decline there as well.

After death rates declined in industrialized countries, birth rates soon followed. But in the rest of the world, birth rates have hardly dropped at all, and the net rate of population growth is now greater than ever before.

Between 1800 and 1930, the world population of humans grew from 1 billion to 2 billion, a doubling time of 130 years. The next doubling, to 4 billion, took only 45 years, ending in 1975. The doubling time is now estimated to be 39 years for the world as a whole, with higher growth rates in the less-developed countries. Kenya, for example, had a doubling time of 17 years in 1985. (The average completed family size was eight.)

Although the population curve is still exponential, the world population's annual rate of increase has declined from 2.1 percent in 1975 to 1.7 percent in 1987, rising, since then, to 1.8 percent. Demographers have estimated that even if the decline had continued, the global population would still not have leveled off—at 10.5 billion people—until sometime before the year 2100.

The human population has been growing exponentially for thousands of years. We are now so numerous that a doubling time of 40 years has enormous consequences.

Why Is Population Momentum Important?

As we saw earlier, the population of the United States continues to grow even though completed family size is below replacement level. In fact, even if tomorrow morning human couples began having no more than two children each, the world population would continue to grow for another 50 years, doubling to 10 billion by the mid-2000s. This paradox is called **population momentum.**

The reason for population momentum is simple: in a rapidly growing population, a large proportion of individuals will be young. In fact, about 20 percent of the people in the world today are younger than 15 years old. Even if they only replace themselves, which is highly unlikely, they will increase the world population by 20 percent (more than a billion individuals). Meanwhile they will continue to live alongside their children and grandchildren, and not begin to die until they are in their 60s, some 50 years from now (Figure 28-9).

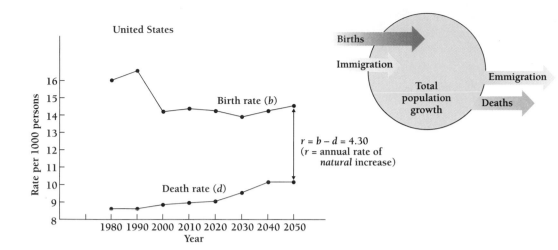

Figure 28-9 Population momentum. The age structure of current populations allows population experts, or demographers, to predict changes in birth rates and death rates. In the United States, the death rate will increase (as baby boomers age and die) and the birth rate will level off.

Even if human couples today began having no more than two children each, the world population would continue to grow for another 50 years.

MOST HUMANS LIVE IN HEAVILY SUBSIDIZED ECOSYSTEMS

Most humans live in cities, towns, and farms. Because these ecosystems consume more energy than they generate by photosynthesis, all require energy from other sources. Much of this energy ultimately comes from the fossil fuels—coal, oil, and natural gas—whose energy comes from sunlight captured by organisms living during the Carboniferous Era, 65 million years ago. Although experts disagree on exactly how long the supply of fossil fuels will last, everyone realizes that the supply is finite. The day will come when we have burned up all the coal, oil, and natural gas. Fossil fuels will be as much a thing of the past as dinosaurs are now.

Cities and Farms Depend on Energy Subsidies

Humans are not primary producers. In fact, we are both primary consumers (herbivores) and secondary consumers (carnivores). Sometimes we are even tertiary or quaternary consumers, as when we eat large fish. Because urban populations consume so much food energy, their principal sources of food are farms and ranches, outside cities and suburbs. As a result, agriculture **subsidizes,** that is, provides additional energy to, cities and suburbs.

Modern agriculture is itself subsidized (Figure 28-10). Its productivity depends not just on sunlight, but on chemical fertilizers that allow crops and livestock to grow at much higher densities than occurs in natural ecosystems. These chemical fertilizers are made with fossil fuels. An oil-subsidized farm thus has about 10 times the productivity of an unsubsidized, natural ecosystem.

Of course, the subsidized agricultural areas that supply cities and towns do not provide all the energy that urban areas consume. In addition to the million kilocalories each of us eats annually, we use 100 to 1000 times more energy for industry, transportation, heating, lighting, and other activities. The overall energy flow into an urban area may be 1000 times more than that in a pond or meadow.

Cities depend on energy subsidies and agriculture, which also depends on energy subsidies.

Ecosystems Altered by Humans Are Vulnerable to Pests

Human "ecosystems" are far less diverse than the natural communities they replace. Towns and cities consist mostly of humans, together with their gardens, parks, pets, pests, and parasites.

Farms are even more homogeneous, as they are usually devoted to a single species of plant or animal, or, at most, a few species. The mechanization of agriculture and the development of irrigation has allowed most farms to engage in **monoculture,** the raising of a single crop. Monoculture allows farmers to coordinate more efficiently each step from planting to harvest. In some countries, the entire economy may depend on the culture of a few crops.

The concentration of a few species in a small area makes cities and farms more attractive to pests and parasites. These range from rats, lice, and disease-causing bacteria to crop-destroying insects and rusts. In a natural ecosystem, a parasite must find its host species sparsely distributed among a wide variety of other species. But in a human-altered ecosystem, a parasite can find large numbers of hosts concentrated in a single place.

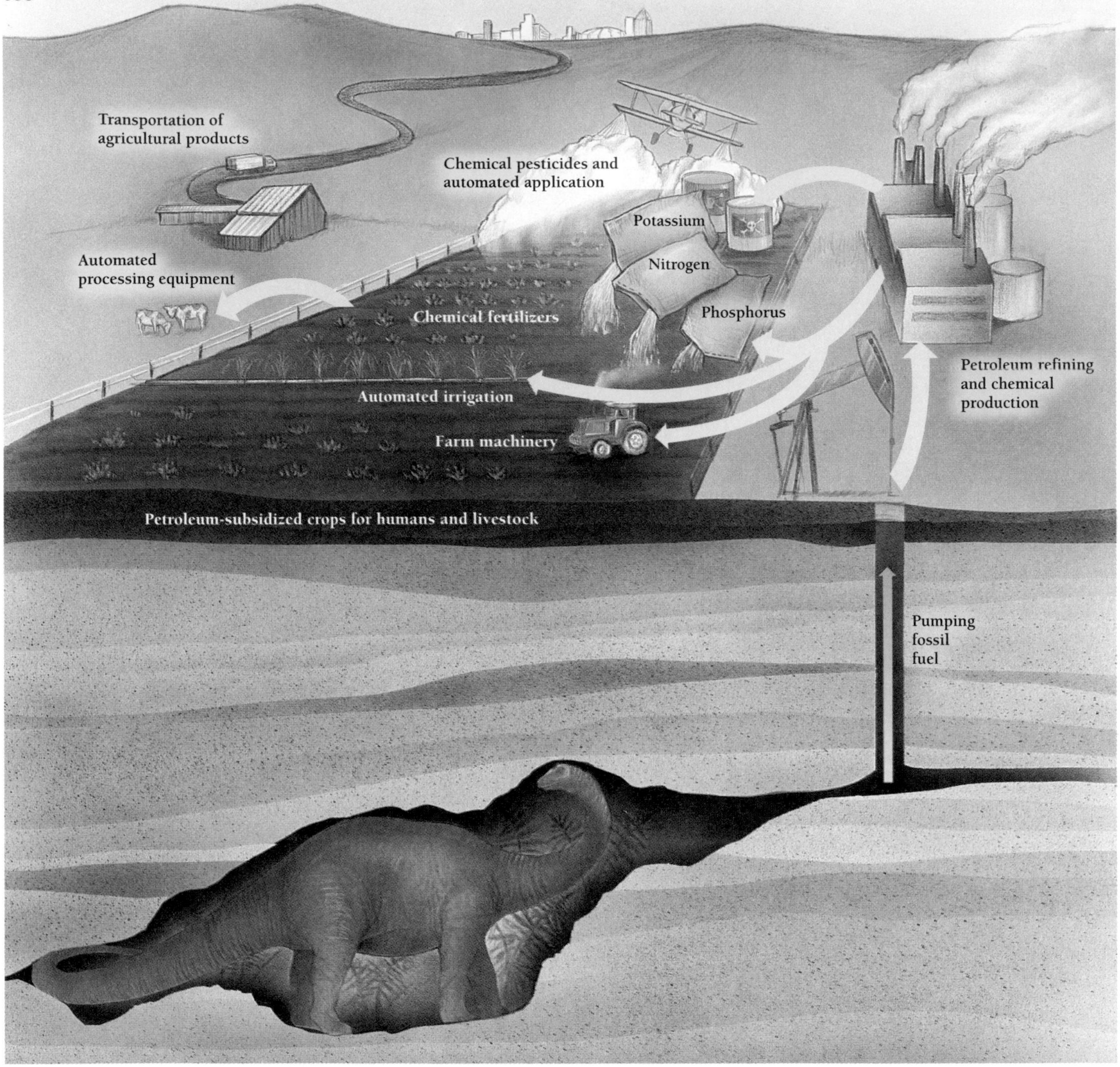

Figure 28-10 **Subsidized agriculture.** Agricultural subsidies and fossil fuel subsidies are the lifeblood of modern cities and suburbs. Modern agriculture, in turn, depends on energy subsidies from ancient stores of fossil fuels—carbon compounds whose energy came from the sun 65 million years ago. Oil-subsidized agriculture can have up to ten times the productivity of a natural ecosystem. However, the supply of fossil fuels is finite.

Modern agricultural crops are especially susceptible to devastating invasions by insects and other pests because intensive artificial selection for high yield has resulted in genetically identical crop varieties.

What Was the Green Revolution?

The Green Revolution, which began in the late 1960s, was an attempt to increase food production in developing countries by using agricultural methods similar to those used in the United States and other developed countries. The central effort of the Green Revolution was to develop, through intensive breeding, varieties of rice and wheat that produced up to ten times as much food as older varieties. Mexico, India, and China particularly benefited, as crop increases temporarily outstripped human population growth.

The new crop varieties use more of their photosynthetic energy to make edible seeds and less to make stems and roots.

However, plants can afford to do this only when nutrients are in such rich supply that they do not need extensive stem and root systems. Consequently, the Green Revolution's "miracle crops" depend heavily on fossil-fuel-based chemical fertilizers and irrigation. As the American ecologist H.T. Odum wrote, the increased food supply is "partly made of oil."

Although the Green Revolution improved crop yields in some places, it only slightly increased the world's food supply. Moreover, it reduced overall genetic diversity in agricultural crops. For example, some older strains of wheat are no longer available. These strains could be grown in some areas without chemical fertilizer or irrigation. But, as energy costs have risen, many farmers cannot afford to buy the fertilizer they need to grow newer varieties.

The Green Revolution was an energetically expensive attempt to increase the Earth's carrying capacity for the human population.

Why Do Modern Agriculture and Industry Use So Much Energy?

Agriculture requires not only the solar energy that drives photosynthesis, but also the energy expended by humans, so-called **cultural energy.** Cultural energy is the energy needed for cultivation, irrigation, transportation, and, in modern societies, the production of fertilizers, pesticides, and farm tools. In primitive societies, the cultural energy needed to produce each kilocalorie of energy (of grain) is about 0.2 kilocalories. More advanced societies produce greater **yields**, the amount of grain per acre, with less cultural energy by using metal hoes and sometimes animal labor to cultivate their crops. These groups expend only about 0.1 to 0.025 kilocalorie for each kilocalorie of food.

Modern mechanized agriculture, however, requires *more* cultural energy than primitive agriculture. For United States agriculture as a whole (in 1970), each kilocalorie of food energy required about 2 kilocalories of cultural energy.

Some energy can be saved by reduced dependence on machines. For example, some Nebraska farms are owned by Old Order Amish who do not use fuel-driven machinery. The energy costs of food produced on these farms are only about 35 percent that of their mechanized neighbors. The yield from the Amish farms, however, is only about half that on the non-Amish farms. This is the trade-off between energy costs and yields.

Cities use much more energy than farms, and the industrialized countries use the most energy by far. Each human city-dweller uses on average 100 to 1000 times more energy than he or she eats. This energy produces buildings, roads, dams, bridges, airports, and the multitudes of goods and services that keep our economies running. The more energy spent, the richer the economy; the richer the economy, the more energy its populace can afford. Conversely, poor countries are also energy poor. In 1981, the richest 25 percent of the world's population used about 80 percent of the total energy supply, while the poorest half used only 12 percent.

Mechanized agriculture and industry increase the energy costs of the human population.

ONE RESULT OF OVERPOPULATION IS POLLUTION OF THE BIOSPHERE

At the time of this writing, most experts agree that the world's farms produce enough to feed the current population. Delivering that food to all who need it, however, has proved to be a staggering problem. Each year about 10 million children and 10 to 15 million adults starve to death—enough people to repopulate the state of California. To put it another way, 70,000 people starve to death every day. At least another half a billion people are malnourished, with still more living on inadequate diets that leave them susceptible to disease and premature death. In Bangladesh, for example, life expectancy is 48 years; in the United States and Canada, it is 75 years.

The situation will not improve. Between 1950 and 1984, global grain production increased enough to stay ahead of population increases. But by 1986, production had begun to level off. In 1988, severe drought and crop failure in North America, the former Soviet Union, and China reduced world grain production by 5 percent. Grain stores from previous good years supplied most of the world's population during the drought. But the steadily rising demand for food is making it increasingly difficult to stockpile reserves for future worldwide crop failures.

Humans are not only using up limited resources. We are also ruining renewable resources, such as air, water, and soil, by dumping industrial waste, garbage, sewage, and other waste products. The more of us there are, the more garbage and pollution we create.

Pollution Has Degraded the Water and Soil

The problems of pollution are most apparent in freshwater communities, although marine ecosystems suffer from pollution as well. The time-honored method for disposing of wastes and sewage has been to dump them into a large body of water, with the expectation that the waste would be diluted enough to render it harmless. This is how pollution has started in freshwater streams, rivers, ponds, and lakes. Because polluted water threatens public health, considerable public attention and legislation has focused on cleaning up bodies of water.

Thermal pollution occurs when water is heated, as when, for example, a stream is used to cool a power plant. Cooling the plant inevitably warms the stream. The warm water kills some populations and causes others to multiply exponentially, throwing the ecosystem out of balance. As a result, thermal pollution destroys existing aquatic communities.

Sediments are particles suspended in water that settle out and drop to the bottoms of lakes, streams, rivers, and other bodies of water. Sediments accumulate in water as a result of erosion. They damage aquatic ecosystems by blocking light, burying organisms that live on the bottom, and filling in bodies of water. Lack of light prevents photosynthesis, limiting ecosystem productivity. Heavy sedimentation can bury the organisms living at the bottom of streams, and inundate coral reefs and shellfish beds.

The erosion that releases sediments into water is a natural process. Natural erosion has cut the Grand Canyon a mile into the Earth, for example. But human activities, such as agriculture, logging, overgrazing, strip mining, and construction, enormously accelerate erosion.

Biodegradable pollutants include carbon dioxide, sewage, fertilizers, and other wastes that organisms can consume. For example, nitrogen and phosphorus in fertilizers used in agriculture and home gardening wash into rivers and lakes and promote eutrophication. Raw sewage can have the same effect for the same reasons. In addition, raw sewage frequently carries bacteria and other organisms that cause diseases such as typhoid, cholera, and amebic dysentery. A particularly large producer of biodegradable pollution is the food industry. The unsaleable parts of harvested crops or slaughtered animals end up as wastes that are dumped. Proper biological treatment of these wastes could, however, produce useful organic fertilizers.

Nondegradable pollutants are another matter. Aluminum cans, plastics, glass, and hundreds of other human products now accumulate as solid wastes—usually mixed with biodegradable solid wastes in landfill. Some of these materials—aluminum soft drink cans are a prime example—require especially large amounts of energy to produce. As energy and raw materials become scarcer, there will be more economic incentive to separate the biodegradable wastes from recyclable metals and plastics. Many states, for example, now require deposits on soft drink containers, a practice that encourages recycling.

The hardest pollutants to deal with are poisons. These include heavy metals such as mercury and lead, reactive gases from smog, radioactive substances, pesticides, and an unknown but growing number of toxic chemicals used in industry and agriculture. Studies have documented the effects of some of these chemicals, but little is known about the effects of thousands of others. Still less is known about the interactions among these chemicals. For example, some substances can be relatively harmless by themselves, but toxic in combination with other compounds.

The effects of many toxic chemicals are amplified at higher levels in the food chain, a phenomenon called **biomagnification.** Organic compounds such as DDT accumulate in tissues, especially in the fats of animals (Figure 28-11). A herbivore may thus build up concentrations of a toxic substance to many times that in the groundwater or lake water. A carnivore that eats many such herbivores concentrates the substance still more. The highest concentrations are found in the top carnivores. For example, the trout in the Great Lakes have concentrations of pesticides and mercury that range from 1000 to 14 million times that in the water.

Poisons can affect many more organisms than their intended targets. Pesticides change whole communities, with major effects on the higher levels of the food chain. In addition, when such substances enter human food, they may cause acute reactions. More often, however, they accumulate and cause insidious, long-term damage. The consensus is that many of these substances increase people's susceptibility to cancer and other diseases and lead to long-term changes in brain function.

Water and soil pollution come in the form of thermal pollution, biodegradable pollutants, nondegradable pollutants, and poisons.

Pollution Has Also Degraded the Air

Heavy industrialization has often meant a darkening of the air and surrounding landscape, as particles from incompletely burned coal and oil spewed from factories and power plants. Industrial melanism, discussed in Chapter 16, p. 403, illustrates particularly well the effects of industrial activity on organisms.

Smoke and soot are only the most visible products of burning fossil fuels, and the most easily removed. The invisible chemical products of combustion, such as carbon dioxide, nitrogen oxides, and sulfur oxides, as well as the industrially important fluorocarbons, are responsible for a different set of problems.

Carbon Dioxide

Fossil fuels are mostly hydrocarbons, so the major products of combustion are carbon dioxide and water. As we use coal and oil to generate power and heat, we increase the amount of carbon dioxide in the atmosphere (Figure 28-12). Before the Industrial Revolution the concentration of atmospheric carbon dioxide was about 270 parts per million (ppm), that is, 270 microliters of carbon dioxide for every liter of air. In 1958, the average concentration of carbon dioxide in the atmosphere was about 315 ppm. By 1983, it had risen to 340 ppm.

The accelerating increase in atmospheric carbon dioxide may well lead to major changes in the biosphere. For example, carbon dioxide causes the atmosphere to warm. This is called the **greenhouse effect** because of the way it works. The glass windows of a greenhouse let light in. But once the light is inside, it degrades to heat. The heat, trapped inside by the glass, warms the greenhouse. Carbon dioxide in the Earth's atmosphere traps heat just like greenhouse glass, warming the Earth's surface. Scientists expect that, in the next 75 years, increased carbon dioxide concentration will cause the average temperature of the Earth's surface to increase 3° or 4°C. Such an increase is likely to melt the polar ice caps, raise sea levels worldwide, inundate coastal cities, and change the Earth's climate. No one knows precisely what impact the greenhouse effect will have on the Earth's carrying capacity for our species, but most projections suggest a decrease in the world food supply.

Figure 28-11   Biomagnification. Toxins not broken down or excreted tend to become more concentrated as they ascend the food chain—as shown by increasing concentration of dots. Plants and microorganisms begin the process by taking in contaminated water. They excrete the water but retain the toxin.

Figure 28-12   The greenhouse effect. Carbon dioxide and other gases trap heat in the same way as the glass walls of a greenhouse. Light passes through the atmosphere to be absorbed and reradiated as heat. But carbon dioxide (like glass) reflects the heat back inward, heating the atmosphere. As carbon dioxide levels rise, the temperature of the atmosphere increases.

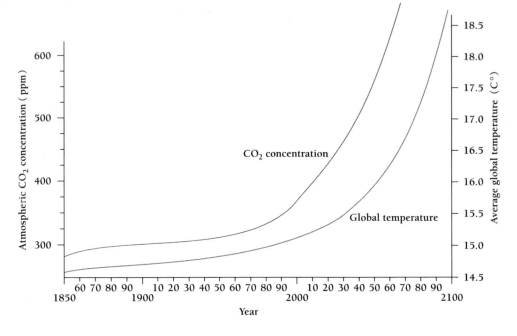

Sulfur and Nitrogen Compounds

Besides carbon dioxide, the most hazardous products of fossil fuel combustion are sulfur and nitrogen compounds. As the number of automobiles has increased, nitrogen oxides have entered the atmosphere at greater rates. The sulfur dioxide in the atmosphere, by contrast, comes mostly from the burning of coal and the smelting of metals.

Nitrogen and sulfur oxides may travel thousands of miles with air currents before dissolving in falling rain. The resulting so-called acid rain may thus devastate ecosystems a continent away from the source of the pollution. In addition, the energy of sunlight can convert nitrogen oxides in the atmosphere to the still more reactive compounds of photochemical

smog. In the short term, smog strains the lungs and heart. A smoggy day in Los Angeles prompts weather forecasters to advise older people to stay indoors and children not to play outdoors (Figure 28-13). In the long term, smog may cause lung and heart disease, and raise the incidence of cancer. Happily, the attention given to clean air has resulted in many fewer smoggy days than 20 years ago.

Fluorocarbons and the Ozone Layer

A group of industry-associated atmospheric chemicals that has raised much concern in recent years is the fluorocarbons, which are generally believed to have created the growing "ozone hole" over Antarctica. Fluorocarbons are small molecules derived

Figure 28-13   Los Angeles, California. Serious air pollution such as this not only affects human health, but that of plants and other organisms as well. Strict air pollution control laws have greatly reduced pollution in the Los Angeles basin. *(Frank S. Balthis Photography)*

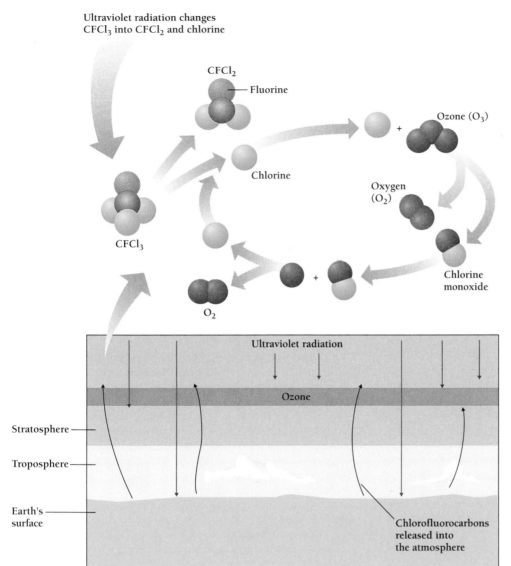

Ultraviolet radiation changes
CFCl₃ into CFCl₂ and chlorine

Figure 28-14   How do fluorocarbons break down the ozone layer? Ultraviolet light from the sun breaks down CFCl₃ into CFCl₂ and chlorine. The chlorine reacts with ozone to form oxygen and chlorine monoxide, thus depleting ozone in the upper atmosphere.

from hydrocarbons, in which fluorine and chlorine replace some of the hydrogen atoms. The most common fluorocarbon, **freon,** is widely used as a coolant in refrigerators and air conditioners. Some fluorocarbons serve in the electronics industry as solvents to clean circuit boards. Others are used to insulate homes or as propellants in spray cans. Products and services that depend on fluorocarbons total about $28 billion in sales a year in the United States.

Ozone is a highly reactive molecule made of three oxygen atoms, instead of the two in the oxygen we breathe. The layer of ozone gas 20 to 50 kilometers above the Earth, known as the **ozone layer,** absorbs most of the ultraviolet light from the sun, shielding living things from potentially dangerous amounts of ultraviolet light (Figure 28-14). Where the ozone shield has thinned, ultraviolet light reaching the Earth's surface causes increased mutations in DNA, destroys protein molecules, and increases the number of chemically reactive molecules in the atmosphere. For humans, the most obvious effect of increased ultraviolet light is higher rates of skin cancer.

Almost 20 years ago, atmospheric scientists began to document a decrease in the ozone layer. They traced the cause of the disappearing ozone to the presence of fluorocarbons in the atmosphere. In 1988, the evidence for the connection between increased fluorocarbons and decreased ozone became so clear that fluorocarbon manufacturers agreed to work toward the elimination of these compounds. At the time of this writing, industrial chemists are actively seeking safer substitutes, and atmospheric scientists are carefully watching the continuing expansion of a hole within the ozone layer over Antarctica. Even over North America, the ozone layer has already thinned some 2 or 3 percent.

Air pollution includes (1) visible particles of soot and other carcinogens; (2) invisible products of combustion and agriculture, which are raising the temperature of the Earth; and (3) fluorocarbons, which are damaging the ozone layer.

The future of the biosphere seems bleak indeed, and the material in this chapter is not comforting. There is reason for hope, however. Humans can reason, look ahead, and act. And there are now important signs of new awareness and responsibility. In 1988, for example, world leaders concluded international agreements to eliminate fluorocarbons and to dispose of a whole class of long-range nuclear weapons. These accomplishments suggest that national governments are beginning to act as if all people lived in a single world. In the 1950s, statesman and diplomat Adlai Stevenson wrote, "We travel together, passengers on a little spaceship."

STUDY OUTLINE WITH KEY TERMS

All living organisms are capable of **exponential growth.** Bacteria have a **doubling time** of as little as 20 minutes. Humans' doubling time is about 40 years. Unrestrained exponential growth results in a rapid and limitless increase in total population size over time. A graph of this increase is sometimes called a **J-shaped curve.** The maximum possible rate of increase for a species is r_{max}, also called the **intrinsic rate of natural increase.**

The growth of natural populations is limited by **density-dependent** factors such as space, resources, and predators, and by **density-independent** factors, such as weather and catastrophes. Density-dependent factors impose a limit to the growth of natural populations called the **carrying capacity** of the habitat, or **K.** When populations approach the carrying capacity of their habitat, the size of the population levels off, forming an **S-shaped curve.** Such restricted growth can be described by the **logistic growth equation.** When a population stops growing, it is said to have **zero population growth.**

Populations increase or decrease as a function of births, deaths, **immigration,** and **emigration.** When populations **overshoot** their carrying capacity they may crash, then rise and fall until the carrying capacity is reached or passed once more. Some populations cycle regularly and somewhat mysteriously. Ecologists used to think that predators drove such cycles. However, prey may cycle as a result of fluctuations in their food supply, and predators may cycle as a result of cycling in the prey. Also, a population may be somewhat self-regulating.

Ecologists classify the reproductive strategies into two general categories, **r-selected species** and **K-selected species.** K-strategists, which reproduce slowly, have **convex,** or **type I, survivorship** curves, while r-strategists have **concave,** or **type III,** survivorship curves. In between is the **diagonal,** or **type II,** survivorship curve typical of songbirds. If we know the **age structure** of a population, which represents the size of each age **cohort** in a population, the age structure of immigrants and emigrants, and the **fertility** and **mortality** of each cohort, we can predict the future age structure of the population. **Demography,** the study of human populations, expresses mortality as **death rate** and fertility as **birth, rate, replacement reproduction,** or **completed family size.** Demographers express the rate at which a human population increases as the **annual rate of increase.**

The global population of humans has been increasing exponentially for thousands of years. **Population momentum** means that even if humans immediately and permanently reduced their annual rate of increase to zero, the world population would continue to grow for another 50 years. Human agriculture is heavily **subsidized** with energy from fossil fuels and other forms of **cultural energy.** Cultural energy increases agricultural **yields.** However, when fossil fuels run out, energy subsidies will have to decline, and agricultural yield (and other kinds of productivity) will decline precipitously. Human-style **monocultures** are unusually vulnerable to crop-destroying insects, bacteria, fungi, and other pests. Increases in global food production are already falling behind increases in global population size.

Human overpopulation increases water and air pollution, with such attendant problems as **biomagnification,** the destruction of the **ozone layer,** and the **greenhouse effect,** as well as the chances of massive epidemics of disease. The threat of nuclear war, though now reduced, still stands as a major hazard to the biosphere.

REVIEW AND THOUGHT QUESTIONS

Review Questions

1. In your garage is a pair of healthy, mature Norway rats. Assume these animals can produce ten young every 3 weeks (actually Norway rats can produce 11). Assume also that each litter consists of five females and five males. Females are able to bear young at 12 weeks of age. If none of the adults or offspring die, and if conditions in your garage are so wonderful that none emigrate, how may rats will occupy your garage at the end of one year?
2. What is the intrinsic rate of natural increase for this pair of rats, expressed as the rate of increase per year?
3. What factors limit the carrying capacity of a habitat?
4. Give three examples of density-dependent limits to human population growth and three examples of density-independent limits.
5. What is replacement reproduction and why is replacement reproduction for humans 2.1 instead of just 2.0?
6. If the 1990 annual rate of increase persisted, what would the global population of humans be in 2000?
7. Why is subsidizing urban populations necessary?
8. What is the current advantage to subsidizing agricultural systems?
9. Approximately how many people starve to death each year?

Thought Questions

10. If a way could be found for the Earth to support 25 billion people, five times the current density, what other organisms and ecosystems would probably have to go to make room for these people? What could be retained? How would the world be better or worse than it is with a population of 5 billion?
11. Make a list of factors, both density-dependent and density-independent, that might limit the population of rats in your garage.

SELECTED READINGS

Diamond, Jared, *The Third Chimpanzee: The Evolution and Future of the Human Animal,* HarperCollins, New York, 1992. A series of related essays describing the evolution of humans and what kind of animal we are.

Ehrlich, Paul R., and Anne H. Ehrlich, *The Population Explosion,* Simon and Schuster, New York, 1990. A short, readable description of the human overpopulation problem.

Moss, Cynthia, *Elephant Memories: Thirteen Years in the Life of an Elephant Family,* William Morrow and Co., New York, 1988. How a family of elephants survived or succumbed to drought, poachers, disease, researchers, and more. This book beautifully illustrates two things: the individual events in the lives of animals that, added together, make populations grow or shrink; and the life of a field biologist.

▶ On-line materials relating to this chapter are on the World Wide Web at http://www.saunderscollege.com/lifesci/ Click on Tobin/Dusheck: *Asking About Life.*

Why Is Sociobiology Controversial?

In 1979, a retiring and bookish expert on the arcane habits of ants and other social insects was delivering a talk in Washington, D.C., when a group of angry students charged the podium and doused him with water. E.O. Wilson, an eminent Harvard professor, had unwittingly placed himself at the center of a furious controversy by writing an ambitious textbook on animal behavior. The 1975 publication of *Sociobiology: The New Synthesis* almost overnight transformed Wilson into a notorious and, to some people, repugnant political figure. Wilson became the target of vituperative public attacks from other scientists as well as students and nonscientists. The firestorm of criticism left the lay public with the impression that the field of sociobiology provided support for racism, sexism, classism, and even Nazism. **Sociobiology,** the study of behavior from an evolutionary perspective, got such a bad name, in fact, that biologists renamed it **behavioral ecology.**

The political maelstrom into which Wilson had plunged included several separate but intermingling currents, each with its own long history. These included the ongoing political debate over the role of political structure in shaping human behavior, the related "nature–nurture" debate, and a long-running battle between two groups of scientists.

Until about 1972, two separate disciplines for the study of animal behavior existed—the European science of **ethology,** the systematic study of animal behavior from a biological point of view, and the primarily American science of **experimental psychology.** Both ethologists and psychologists were interested in animal behavior, but their approaches, emphases, and philosophies differed markedly.

Experimental psychologists had, since the beginning of the 20th century, focused on *learned* behavior. They were primarily concerned with immediate mechanisms—what psychological processes controlled behavior in animals. For example, the Russian physiologist Ivan Pavlov discovered that he could teach a dog to salivate at the sound of a bell. First, Pavlov would ring a bell each time he fed the dog, when it naturally salivated. In time, the bell alone would cause the dog to salivate, a behavior Pavlov termed a **conditional reflex,** because the salivating, called a "reflex," could be elicited by the ringing of the bell.

Experimental psychologists tended to think of behavior as being mostly learned. The adaptive value of a given behavior was

Oxford Scientific Films/Animals Animals

of little interest, and all animals were considered more or less equivalent. As the famous experimental psychologist B.F. Skinner wrote in 1959, "Pigeon, rat, monkey, which is which? It doesn't matter."

Ethologists, by contrast, became deeply interested in how behavior helps animals survive. At the end of World War II, the ethologists Konrad Lorenz and Niko Tinbergen together began studying the natural behavior of animals in the wild, instead of studying animals' responses to artificial situations as experimental psychologists did. Tinbergen and Lorenz assumed that most animal behavior was adaptive and emphasized **innate** behavior—behavior that is not learned. For example, the tendency of newborn babies to suck is considered innate. Even though practice improves a baby's ability to suck, the *tendency* is inborn.

Psychologists, who were primarily interested in human behavior, were suspicious of ethology for two reasons. First, they believed that human behavior was so malleable that evolution was not relevant. Second, ethology reminded them of "Social Darwinism." This 19th-century idea held that white males occupied superior positions in Western society because they were intrinsically (genetically) superior to nonwhites, Jews, and women of any race. In the 20th century, this idea was renamed "biological determinism" and used to justify a variety of evils, including Hitler's Aryan breeding program and the Nazis' mass murder of Jews, Gypsies, and others, both in Germany and in the countries it occupied during World War II.

Today the debate over Social Darwinism is usually characterized as the "nature–nurture" debate. Put simply, the nature–nurture debate asks whether genes *or* environment controls the behavior of individuals. At one extreme, are people who argue that human behavior is almost entirely the product of environmental influences, so-called "nurture." They believe that behavior is learned, and that genes play only a minor role. At the other extreme are those who assert that most human behavior is under the control of the genes, so-called "nature," and that environmental influences are trivial.

The two camps battle especially bitterly over the relative importance of genetics and environment in criminality and intelligence. The biological determinists argue that the children of criminals will likely be criminals regardless of what society does for them, and the children of the ignorant and uneducated are mostly uneducable. The debate is more political than scientific, however, as few people now believe that behavior is entirely controlled by genes or entirely controlled by the environment.

When E.O. Wilson's book came out in 1975, it brought these conflicts to a head. In his book, Wilson argued that by the year 2000 both ethology and experimental psychology would wither away, to be replaced by sociobiology and neuroscience, the study of the nervous system. Psychologists couldn't help but be offended. For them, acknowledging the value of the evolutionary viewpoint was one thing. Retiring to the sidelines while neuroscientists and ethologists (turned sociobiologists) sorted out the problems of behavior was quite another.

Second, Wilson speculated freely—some said recklessly—about the possible genetic basis of various aspects of human social behavior. This speculation played directly into a hot nature–nurture debate of the 1970s. Some politicians argued that President Johnson's so-called "war on poverty"—which included any social program providing special education or economic support for the poor and the unemployed—was a waste of money because the poor were genetically inferior to everyone else. Many people felt that Wilson's book had provided ammunition that would be used to deprive the poor and the uneducated, especially minorities and women, of the assistance they needed to improve their lot in life.

The political maelstrom into which Wilson had plunged included several separate but intermingling currents, each with its own long history.

Because Wilson was (for the 1970s) somewhat conservative, some people were automatically suspicious of his ideas. A handful of biologists, including some self-described Marxists, unfairly accused Wilson of being a "biological determinist," an accusation that Wilson denied. Politicians and others continue to bend the work of biologists to their own ends. For biologists, however, the nature–nurture debate has lost much of its heat. Most biologists do not argue about whether behavior is controlled by genes or by environment; they agree that all behavior is clearly influenced by both.

KEY CONCEPTS

1. Although biologists sometimes refer to "learned" and "innate" behavior as if they were separate kinds of behavior, all behavior develops under the influence of both genetic inheritance and environmental experience.

2. Like other attributes, individual patterns of behavior affect an individual's ability to survive in a particular environment.

3. The genetic component of any pattern of behavior is therefore subject to natural selection.

4. Ecology, the relationship of an individual to its environment, determines the adaptive value of behavior.

5. Kin selection seems to explain many cases of altruism, but not all, and the phenomenon of altruism continues to puzzle and fascinate biologists.

6. Human behavior is a variety of animal behavior.

HOW IS BEHAVIOR ACQUIRED?

Behavior, like other characteristics of organisms, develops from the interplay of genes and environment. Because the behavioral phenotype of an animal develops over time under the influence of a continuously changing environment, genes and environment constantly interact in unique ways. As a result, even identical twins, who share the same genes, look different from each other.

A biologist asks the same kinds of questions about a pattern of behavior—such as feeding behavior—as about a morphological feature: How do genes and experience combine to establish and regulate the behavior pattern? How does a particular pattern of behavior contribute to the reproductive success of the individual?

An animal's behavior contributes to its ability to survive and reproduce in the same way as do its skeletal, cellular, and molecular characteristics. Since genes can shape behavior, natural selection must operate on genetically based behavioral variants, just as it does on genetically based structural variants. The adaptiveness of a given behavioral trait depends, to a large extent, on ecology, the relationship of an individual to its environment. In this chapter we will examine the ecological significance of behavior, as well as the role of natural selection in shaping behavior.

What Is Innate Behavior?

The human ability to walk is innate. Provided that we are physically able, we all begin walking at about the same age. Toddlers who have been unable to practice using their legs because their legs have been in casts during infancy, still begin to walk on schedule (Figure 29-1A). **Innate** behavior is defined as behavior that an animal engages in regardless of previous experience. The tendency for a dog to bark at strangers, whether canine, human, or otherwise, is innate. Most breeds bark at

least occasionally. A dog's upbringing and psychological well-being may influence how much it barks and at whom, so that environment plays a role as well, but the dog's tendency to bark is inborn.

By contrast, a dog *learns* who is a stranger and who is familiar. **Learning** is a change in an animal's behavior in response

A.

B.

Figure 29-1   Programmed behavior. A. Humans all begin walking at about the same age. However, the style of walking, whether short steps or long, for example, is learned. B. A herring gull chick begins begging for food soon after it hatches, a behavior that is partly innate and partly learned.   (A, Andy Cox/Tony Stone Images; B, Thomas W. Martin/Photo Researchers)

to a specific previous experience. Mammals may be the most sophisticated learners, but essentially all animals—from flatworms to fish—can learn.

Different species of animals are able to learn different things. A grizzly bear is good at learning how to catch salmon in a stream, while a killer whale is good at learning how to catch salmon in the ocean. Every animal has innate tendencies to learn different things. We can see, then, that learned and innate behavior are not completely separable.

Just as learning has an innate or genetic component, innate behavior often has a learned component. In humans, for example, styles of walking vary from culture to culture. Even innate behavior can be "learned." For example, the chick of a herring gull will peck at its mother's beak to beg for food the first time it encounters a parent (Figure 29-1B). Pecking is therefore regarded as innate.

The pioneering ethologist Niko Tinbergen reasoned that since a herring gull chick always pecks at a red spot on its parent's bill, the red spot may be what provokes the pecking behavior. To test this theory, Tinbergen presented chicks with models of gull bills with different colored spots. He found that the red spot elicited the pecking behavior better than any other colored spot, regardless of the shape of the model. Yet later research by biologist Jack Hailman showed that a newly hatched gull chick pecks at anything that looks remotely like a beak, including the beaks of its siblings. However, over the course of several days, the chick learns to recognize both its parents and food, so that it pecks in response to a much more limited range of stimuli.

Any behavior is likely to include both
innate and learned aspects.

Specific Stimuli Can Evoke Specific Responses

A highly stereotyped innate behavior is called a **modal action pattern.** Examples of modal action patterns abound. All dogs scratch their ears in the same way, lifting a rear leg above a foreleg. This behavior is so stereotyped that if a human scratches a dog's ear the same way that the dog's foot would scratch, the dog's hind leg often will rise above the foreleg and move up and down in midair as if scratching. Parrots and many other birds also scratch their heads by bringing a foot over the wing. This mode of head scratching is not necessarily required by their anatomy, however. A parrot is perfectly capable of bringing a foot under the wing, and does so to clean its bill.

A modal action pattern involves the coordinated activity of many muscles and nerves. The pattern is relatively inflexible, so that, for example, the contraction of one muscle always follows the contraction of another, as in the case of swallowing. However, a modal action pattern is more flexible than a reflex. In some cases, the pattern is partly fixed and partly flexible. For example, a greylag goose can roll a wayward egg back into her nest by rolling it with the underside of her bill (Figure 29-2). A bill-tucking movement allows her to pull a loose egg back toward her, but she also must adjust to the sideways wobble of the egg by moving her head from side to side. If an experimenter removes the egg after the goose has already begun tucking the egg, the goose will continue tucking. But the bird's adjustments to the side-to-side movements of the egg will cease. The tucking movement is fixed, or constrained, but the sideways adjustments are flexible.

Modal action patterns serve a purpose only when performed at the proper time and place. Some signal in the environment must therefore stimulate each modal action pattern. For example, a loose egg stimulates a greylag goose's chin tucking, and a fly provokes the characteristic flick of a frog's tongue. In each case, the animal's senses detect the stimulus and its nervous system issues the order to act. The key aspect of an object that triggers a modal action pattern is called a **sign stimulus.**

Not all cues are visual. Touch, sound, and smell can also evoke specific behavior patterns. For example, **pheromones,** chemicals made by one organism that influence the behavior of another organism of the same species, can also activate modal action patterns. The female silk moth emits a potent pheromone, called bombykol, that attracts males from a distance of over a mile. When the male detects as few as 200 molecules of bombykol, he flies upwind until he loses the scent, flies in a zigzag pattern until he detects the bombykol again, and continues in this pattern until he finds the female.

How does an animal distinguish between a proper sign stimulus and an inappropriate one? How, for example, does a herring gull chick know which object to peck? To answer this question ethologists have experimented with counterfeit stimuli.

These experiments show that an animal reacts to only a few aspects of its environment. Even with our own extraordinary brains and sensory systems, we simplify the complexity of the world. As a result, we can easily recognize public figures from simple line drawing caricatures. Animals with less developed nervous systems than our own also respond to caricatures. In fact, Tinbergen found that gull chicks pecked more

Figure 29-2 A modal action pattern. A greylag goose rolls a wayward egg back to her nest. If the egg is removed in the middle of this procedure, the goose continues tucking movements until she is back on the nest.

Figure 29-3   A supernormal stimulus. An oystercatcher prefers to brood a giant egg to either its own small egg or even the intermediate herring gull's egg. The giant egg is a supernormal stimulus.

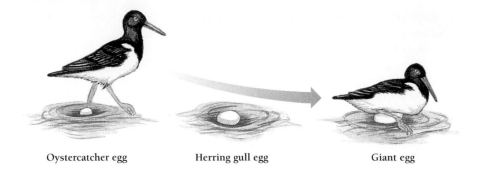

Oystercatcher egg   Herring gull egg   Giant egg

at a stick with three red spots than at a real gull's bill. For a herring gull chick, three red spots constitute a **supernormal stimulus,** a stimulus even more stimulating than anything normally encountered in nature (Figure 29-3).

Through such experimentation, ethologists have discovered the sign stimuli for many stereotyped behaviors. Consider the feeding behavior of a tick, for example. The female needs a blood meal to produce her eggs, but she is relatively immobile and so must wait for a meal to come to her. She can wait, if she has to, for as long as 18 years. She sits unmoving at the end of a twig, waiting for the chance to drop onto an appropriate host (Figure 29-4). At the exact moment that a warm mammal passes beneath the twig, she awakes, lets go of the twig, and drops. If she lands on the mammal, she inserts her feeding organ into the skin and sucks herself full of blood. If she misses and lands on the ground, she climbs up onto a nearby plant and renews her vigil.

Research from the 1930s showed that the tick's behavior was a response to three independent sign stimuli. The tick moves to the end of a twig in response to light. She lets go of the twig when she detects butyric acid (a component of sweat). And she begins her feeding in response to her host's warmth. Any warm surface will stimulate feeding—even a balloon filled with warm water. Light, butyric acid, and warmth all act as sign stimuli for components of the tick's behavior. More recently, researchers have shown that ticks will also drop off the twig in response to carbon dioxide, which animals exhale. Thus butyric acid and carbon dioxide seem to stimulate the same modal action pattern.

A modal action pattern is a stereotyped behavior triggered by a sign stimulus. The sign stimulus can be seen, heard, felt, or smelled.

How Do Experience and Heredity Interact to Influence Behavior?

In an early episode of the television show *Sesame Street*, Ernie is astonished that Bert has been able to teach Bernice the pigeon to play checkers. "Why a pigeon that can play checkers!" Ernie exclaims. "That must be the smartest pigeon in the whole world!" "She's really not that smart," demurs Bert. "In the ten games we've played? She's only beaten me twice."

Much Behavior Is Clearly Learned

Pigeons may or may not be able to beat Muppets at checkers, but they are surprisingly capable learners. Pigeons can learn to distinguish between an underwater scene with fish and one without. They can even be taught abstract ideas. Researchers can teach pigeons to choose one of two disks according to whether it matches the color of a third disk.

Every animal can learn something. Crows, upon seeing another crow sicken or die after eating poisoned bait, will shun the bait. Rats can be taught not only to run a simple maze accurately and consistently, but to run a mirror image of the maze, switching back and forth on cue. Even planarians, among the simplest animals, can be trained (Figure 29-5).

Animals can learn from one another, as well. For example, experiments have shown that blue jays prefer to eat what they have seen other jays eating. One of the most striking examples of culturally acquired learning in natural populations of animals has occurred in at least 11 species of small English birds, including blue tits, titmice, and others. Starting in around 1920,

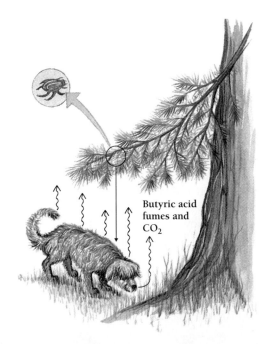

Butyric acid fumes and CO_2

Figure 29-4   Tick sign stimuli. A ticks waits at the end of a branch until it detects either butyric acid or carbon dioxide, then drops onto whatever animal is passing beneath.

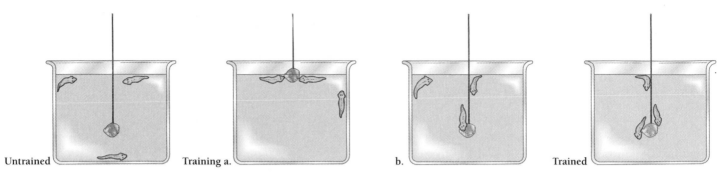

Untrained   Training a.   b.   Trained

Figure 29-5   Learning in flatworms. The simple planarian, with barely a brain to call its own, learns to find food suspended from a string. Untrained worms search the surface or the bottom of a container for food and they miss the meat hanging in the middle. During training, a piece of meat is suspended at the surface and then gradually lowered into the water. The flatworms learn to locate the string at the surface and then to follow it down to the meat. Trained worms quickly find food suspended in the water.

increasing numbers of birds learned to open the caps of milk bottles delivered to the front steps of English houses in order to drink the rich cream just under the cap. The cream is so attractive that the birds attack the bottles as soon as the delivery person is gone, and, in some areas, small flocks of birds follow the milk truck down the streets. If the bottles are capped with cardboard, the birds pull the whole cap off or peel it off in layers of paper. If the bottle is capped with foil, they poke a hole in the foil, then peel the foil off in strips. This behavior first appeared in a few limited areas, then, over a period of years, gradually spread, suggesting that the birds learned from one another.

But as we have seen, different animals are disposed to learn different kinds of things. Rats are good at mazes; birds and primates are good at visual recognition. When two species differ in their behavioral tendencies, we may suspect that these differences result from the genetic differences between the species. The apparent selective advantage of many species-specific behaviors is obvious. Yet because genes and environment interplay in complex ways to influence behavior, teasing out the exact role of genes in influencing behavior is extremely difficult, if not impossible, in most cases. Biologists ask, Which elements of behavior are under the influence of genes and are therefore subject to natural selection?

Genes Influence Behavior

Even in animals as learning-oriented as humans, some behaviors are unquestionably affected by genes. For example, research on certain human mutations suggests that a simple biochemical defect can lead to complex changes in behavior. The inability to produce the enzyme hypoxanthine phosphoribosyl transferase (HPRT), which is involved in certain aspects of DNA synthesis, leads to a disorder called Lesch–Nyhan syndrome, in which people compulsively mutilate themselves. Another genetic change called Tourette's syndrome (whose bio-

chemical basis is still unknown) leads to tics and unpredictable outbursts of loud obscene language and disruptive behavior.

The HPRT gene unquestionably affects human behavior and shows that genetics plays a role in human behavior. However, Lesch–Nyhan and Tourette's syndromes are abnormal. Examples of genetic differences (polymorphisms) that lead to alternate forms of normal behavior are unknown in humans and rare in other animals.

One interesting example, however, comes from two races of West Coast garter snakes (*Thamnopis elegans*). Garter snakes that live along the coast of California are known to relish slugs, while inland snakes will not touch them. To see if this difference in behavior was learned or genetically controlled, researchers hatched baby snakes in the laboratory so the snakes would have no opportunity to learn about slugs, then offered the young snakes a menu that included slugs. Despite their inexperience, snakes from the coastal population readily ate the slugs, while those from inland populations refused them. This result suggested that genes might well control the snakes' taste for slugs. In a separate experiment, the researchers crossbred the two kinds of snakes. Among the hybrids, more young snakes refused slugs than accepted them, and additional analysis suggested that a dominant gene was somehow responsible for slug refusal.

One reason that the influence of genes on behavior is so hard to find is that behavior, like most other species traits, does not usually depend on just one or two genes. Indeed, biologists have found that only about 2 percent of all recognizable traits are under the influence of just one gene. The vast majority of traits are polygenic. Two or more—possibly hundreds— of genes influence the expression of almost every morphological trait. Behavioral traits are no different. The complex genetics of phenotypic traits—including behavioral traits— means that genetics cannot easily determine the exact degree to which a given behavior pattern is genetically or environmentally influenced. Furthermore, as we shall see in the next section, just because a trait is under the influence of genes does not mean that the trait is fixed or unchangeable.

Examples of behaviors that seem to be controlled by genes alone are rare. Even those behavioral traits that are strongly influenced by the genes are, like most other traits, probably polygenic.

Behavior Depends on Both Genes and Environment

The nesting behavior of two small African parrots—the peach-faced lovebird and Fischer's lovebird—offers another example of how genes affect behavior. In this case, however, learning modifies a genetically influenced behavior. In both species of lovebird the female cuts long strips of bark or leaves for nest building. A peach-faced lovebird can carry up to six strips at a time by tucking them into her rump feathers, while a Fischer's lovebird carries these strips to the nest one at a time in her beak.

A hybrid of the two species tries both methods. When she builds her first nest, she starts out by tucking strips into her feathers, like her peach-faced parent. But she is inept, and the strips fall out before she can get to the nest. With time, she learns to carry just one strip at a time in her beak. After about three years, she gives up tucking strips into her feathers. However, just before she flies off with a strip in her beak, she still makes little sideways movements of her head, as if she were about to tuck a strip into her rump feathers.

Many behaviors may be strongly influenced by genes but still open to modification through learning.

Some Forms of Learning Occur Only at Particular Times

Just as humans learn languages most easily and best when they are young, many birds learn to sing their species song only when they are young. For many kinds of learning, timing is critical. A chaffinch raised in isolation, so that it cannot hear adult birds singing, never learns to sing an adult song. If a chaffinch hears the song as an adult, he cannot master the song. However, if the chaffinch hears the correct song during the first few months of life, he will learn to sing a song that resembles the chaffinch song, even if he cannot hear the correct song when he is actually practicing. The period of time during which an animal can learn a particular behavior pattern is called the **sensitive period,** similar to critical periods in morphological development.

The process by which an animal learns such behavior during the sensitive period is called **imprinting.** Konrad Lorenz provided the most famous example. Geese learn to recognize their mother by "imprinting on" the first moving object they see after hatching. Usually this object is the mother goose, who will ordinarily care for her brood. But newborn geese will also

Figure 29-6   Imprinting. Cygnets learn to recognize their mother and follow her soon after they hatch.   *(Gerard Lacz/Peter Arnold, Inc.)*

follow a matchbox on a string, a rubber ball, a flashing light, or even an ethologist (Figure 29-6).

A special form of imprinting, called **sexual imprinting,** helps birds recognize potential mates. Young birds learn to recognize what sort of creature to choose for a mate by imprinting on their siblings.

Sensitive periods exist for a variety of different kinds of behavior. For example, children born blinded by cataracts can learn to see if the cataracts are surgically removed before about age 10. If the cataracts are removed later, however, the children only see a random patchwork of colored shapes, which they cannot interpret. As infants we learn to correctly interpret complex patterns of shading and color as a three-dimensional world. But for newly sighted people, the patterns are meaningless.

Many complex behaviors can develop normally only during certain sensitive periods in development.

The Acquisition of Normal Behavior Can Have a Sensitive Period

Just as the development of normal vision depends on visual experience during a critical period, the development of normal social behavior depends on experience at sensitive periods. A young monkey isolated from other monkeys during the first year and a half of life does not develop social interactions with other monkeys when later given the chance. Instead, the deprived monkey crouches in the corner of its cage, rocking back and forth like a severely disturbed human child. A comparable 6-month period of isolation for an adult monkey, however, does not produce such problems.

Harry and Margaret Harlow, psychologists at the University of Wisconsin, attempted to find out what aspects of a young monkey's life prevented such mental disorders. A cloth-covered wooden dummy, they found, could partly prevent the symptoms of isolation. Such a "surrogate mother" would trigger a clinging response by the isolated monkeys.

The Harlows could fully prevent the symptoms of isolation by allowing the isolated monkey to spend as little as 15 minutes a day with a normally socialized monkey that spent the rest of its time with the colony. Most of the time this limited social experience produced full socialization only if it occurred during the sensitive period in the first year and a half of life. In some cases, however, even later exposure to persistently friendly monkeys could reverse the effects of social isolation. At least in monkeys, appropriate later experience can in some cases overcome deficits in early experience.

Even complex social behavior can have a sensitive period.

We have seen that genes play a critical role in shaping animal behavior. The slug-loving and slug-hating western garter snakes demonstrate that food preferences, for example, can be genetically determined. If natural selection can operate on such behavioral variants, then we must ask what factors determine the adaptiveness of various kinds of behavior. The answer lies in the ecology of behavior. Animals' interactions with each other and with their environment determine the adaptive value of behavior.

HOW DOES ECOLOGY INFLUENCE BEHAVIOR?

Economists have debated whether people make rational economic decisions. For example, if food costs twice as much at one store as another, most people go out of their way to shop at the cheaper store. On the other hand, they will not spend more money getting to the cheaper store than they will save. To a remarkable degree, then, people behave rationally. Animals, with far less intellectual capacity than humans, also behave surprisingly rationally about such decisions.

Animals Optimize Foraging

Crows open whelks, as well as other marine mollusks, by picking up the largest whelks, carrying them to a height of 5 meters, and dropping them on the rocky shore. If the whelk breaks, the crow feasts. If the whelk stays intact, the crow carries it aloft and drops it again, repeating the procedure until the whelk's shell breaks.

Ecologists studying this feeding strategy have shown that the crows maximize the amount of food they get *for the energy expended.* That is, the crows tend to choose the ideal, or optimal, solution to a feeding problem. Such energy-efficient feeding is called **optimal foraging.**

To see if the crows' strategy was optimal, researchers dropped whelks of different sizes from different heights. They found that the largest whelks were the most likely to break and that whelks broke consistently only if dropped from a height of 5 meters or more. The crows' strategy of choosing only the largest whelks and carrying them to about 5 meters was indeed optimal. Further experiments revealed that if a whelk didn't break on the first try, persistence with a single whelk was energetically more efficient than finding a new whelk and starting over. So, a crow's decision to drop the same whelk over and over proved equally sensible. Overall, the crows' feeding behavior is an efficient way of spending energy to get more energy.

When animals forage, they may efficiently maximize the amount of food they get for the energy expended. Ecologists call this process optimal foraging.

Interactions Between Species Affect Foraging

Maximizing energy intake is not the only consideration that guides feeding behavior, however. When predators are present, an animal must adjust its behavior to increase its own chances of surviving. One study of chickadee feeding, for example, revealed that the chickadees would eat seeds at a researcher's feeding tray only if the tray seemed safe. When researchers mounted a model of a hawk nearby, the chickadees took the seeds away to a safer place. Taking the seeds away cost the birds time and energy, but given the apparent presence of a hawk, moving the seeds was the only way to be sure of eating and surviving, too.

The presence of competing species can also change feeding behavior. For example, North American mink introduced into Sweden in the 1920s competed directly with European otters for food and caused otter populations to decline. However, the two animals continued to coexist. The mink ate a variety of foods, including fish, muskrats, rabbits, birds, and small burrowing mammals, while the otters carved out a specialized niche, feeding on crayfish, crabs, and other aquatic invertebrates only. The otters apparently defended this narrow niche aggressively, for they sometimes killed the mink.

We can say, then, that at least some aspects of feeding behavior depend on continuing interactions within individual communities. Competition, predation, mutualism, and other species interactions influence all kinds of animal behavior.

Foraging behavior depends, in part, on how an individual interacts with members of other species.

Individuals of the Same Species Compete for Resources or Mates

Animals of the same species may cooperate or compete with one another, according to the demands of their current environment and the constraints of their evolutionary histories. For example, in many species, reproduction requires close cooperation between the male and the female. Such animals exhibit stylized, species-specific courtship behaviors, including the well-known mating rituals of birds and fish (Figure 29-7).

Mating rituals allow females to assess the health and vigor of potential mates. Also, because mating season is often a time of heightened aggression, mating rituals allow both sexes to minimize the chances of being injured or killed by their partner. We can see that when such behavior patterns are influenced by the genes, natural selection must operate powerfully—increasing the frequency of those behavioral types that reproduce most successfully.

Agonistic Behavior

Reproduction and survival necessitate competition as well as cooperation. Among the most dramatic competitive behaviors are the ritualized fights between pairs of male deer. The outcome of such combat determines who will get to mate with a female. The clash of antlers is mostly symbolic: usually the fighters do not seriously injure each other. Other species use other symbolic contests—who can roar the loudest, stand the tallest, or stare the longest (Figure 29-8).

These contests are examples of **agonistic behavior** [Greek, *agonistes* = champion]—all the aspects of competitive behavior within a species, including aggression (outright attacks and fighting), aggressive displays, appeasement, and retreat. Ethologists use the term "agonistic behavior" instead of the word "aggression" to acknowledge that animals never exhibit pure aggression or pure fear. What most animals show (humans included) is a mix of fearful behavior (including appeasement and retreat) and aggression, sometimes more of one, sometimes more of the other.

Some animals use agonistic interactions to establish a **dominance hierarchy**, a ranking of individuals that fixes which animals have first access to resources or mates. Dominance hierarchies establish rules that allow animals to live in the same territory with limited resources. The classic example of a dominance hierarchy occurs in a flock of hens. Such a flock quickly establishes a "pecking order," which specifies who gets to peck whom without being pecked in return.

Animals compete with others of their own species for mates, resources, and rank through agonistic encounters.

Sexual Selection

The long-term result of competition for mates is **sexual selection**, the differential ability of individuals with different genotypes to acquire mates (see also Chapter 16). Some biologists distinguish sexual selection from other forms of natural selection, arguing that natural selection always results in the accumulation of adaptive structures and behaviors, while sexual se-

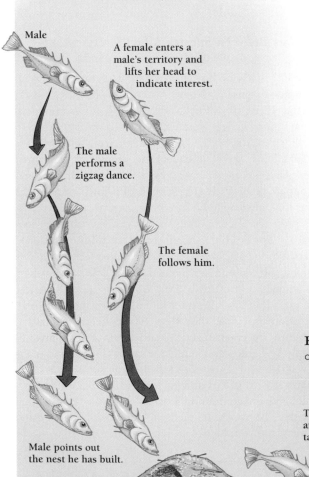

Male

A female enters a male's territory and lifts her head to indicate interest.

The male performs a zigzag dance.

The female follows him.

Male points out the nest he has built.

The female enters the nest, and the male touches her tail with trembling motions.

The female lays her eggs.

Female leaves the nest.

Male enters nest and fertilizes eggs.

Figure 29-7   Courtship in the stickleback. Mating rituals allow animals to recognize and evaluate one another.

Figure 29-8 **Agonistic behavior in rabbits.** Fighting between males is not confined to large animals such as deer or lions. Here two rabbits fight to establish dominance. Agonistic behavior also includes aggressive behavior and appeasement displays. *(Art Wolfe/Tony Stone Images)*

A.

Figure 29-9 The tsungara frog of Panama judges a potential mate by the sound of his voice. A. A tsungara frog. B. Males with deepest voices have the best chance of attracting a female and, unfortunately, the best chance of being eaten by a frog-eating bat. *(A, M.J. Ryan, University of Texas; B, Merlin D. Tuttle, Bat Conservation International/Photo Researchers)*

B.

lection often results in "maladaptive" structures and behaviors. But all adaptations to the various demands of survival and reproduction are a constant balancing act for organisms. The specific limitations of sexual reproduction are not fundamentally different from the specific limitations of gravity and chemistry.

Natural selection can act both on inherited variations in structures that are important in mating behavior, such as a buck's antlers or the peacock's tail, and on inherited mating behaviors. For example, in the tsungara frog (*Physalaemus*), which lives in Panama, female frogs consistently prefer larger males to smaller ones. The females choose the larger males, however, not by comparing male physiques but by listening to the male frogs' mating calls. The song consists of a whine, followed by one or more low-pitched "chucks." Bigger frogs have lower "chucks," so female frogs prefer frogs with deep voices (Figure 29-9A). The lower the "chuck," the more the female frog is attracted, even when the "chuck" comes from a tape recorder.

Unfortunately, the male frog's deep "chucks" also attract the attentions of a frog-eating bat, which, like the female frog, prefers a big frog to a little one (Figure 29-9B). The bat presents the male frog with a serious dilemma. If he chucks frequently to attract females, he risks his life by attracting bats. If he does not chuck enough, he will not be able to reproduce. To maximize his chances of reproducing while minimizing his chances of being eaten, the frog must optimize the number of chucks he emits. To whatever extent that variations in pitch are under genetic influence, sexual selection will tend to favor the evolution of low-pitched chucks—but maybe not too low.

Competition for mates over evolutionary time results in sexual selection, the differential ability of individuals with different genotypes to acquire mates.

Territoriality

In many species, an animal engages in competitive behavior to establish and defend a **territory,** an area occupied by an individual (or group of individuals), from which other individuals of that species are excluded. Some species, a few birds and fish, for example, use territories exclusively for mating. More commonly, species use territories for foraging or for raising young as well.

Establishing and defending a territory necessarily involves agonistic behavior. Animals use a wide variety of signals to mark their territories, including chemical scents, brightly colored plumage, or patrolling behavior. But actual aggressive encounters are relatively rare.

Defending a territory requires energy expenditure. Ecologists have studied the "economics" of territoriality and have shown that some bird species defend smaller territories when food is abundant than when food is scarce. When food is abundant, less territory is necessary to raise the same number of offspring. So, to conserve energy, birds defend a smaller territory, an example of optimal foraging. A few bird species even adjust the size of their territories on a daily basis.

Animals defend territories for raising young, mating, or foraging—as long as defending the territory will increase their reproductive potential.

Figure 29-10 **Zebras in flight.** Some biologists have theorized that the zebra's stripes make it hard for predators to see where one animal ends and the next begins. In the heart of the chase, a little confusion can reduce a predator's chance of catching its dinner. *(Jeremy Woodhouse/DRK Photo)*

SOME ANIMALS LIVE IN SOCIAL GROUPS

The tiger lives a solitary life, meeting up with other tigers rarely and not for much longer than it takes to mate. Its close relative the lion, on the other hand, lives in a tightly knit group called a pride. The members of a pride of lions are never far apart. Why do some animals live socially, while others live apart?

Social Behavior Has Advantages and Disadvantages

Living in groups affords distinct advantages (Figure 29-10). The greatest advantage to group living is that large groups of animals more quickly spot predators and other dangers than small groups of animals or solitary ones. For example, researchers have found that the more pigeons that are in a flock, the sooner the pigeons will notice a hawk and the less likely that the hawk will catch one of the pigeons.

Sometimes groups of animals can cooperatively defend themselves. When threatened, a herd of musk oxen will form a circle with their young in the middle and their horns pointed outward. When hundreds of honeybees simultaneously attack on animal bent on emptying the hive of its honey, the bees stand a better chance of succeeding than does a lone bee. Group living can be an advantage in finding food as well. Predators that hunt cooperatively, such as lions and wolves, can kill prey far larger than themselves (Figure 29-11).

A.

B.

Figure 29-11 **Social animals and loners.** A. Lions hunt animals much larger than themselves. B. Tigers hunt alone and kill smaller animals. *(A, Anup & Manoj Shah/DRK Photo; B, Tom Brakefield/DRK Photo)*

Social living has disadvantages as well, however. Animals living in groups must compete with each other for food, mates, and breeding areas. Individuals waste energy fighting or threatening one another. And individuals waste time and energy by searching for food in the same places. Further, groups of animals transmit diseases to one another and attract predators.

Whether animals live in groups depends on the ecological costs and benefits. Many songbirds stake out territories in the spring and actively repel all but their own mates and offspring. Yet when autumn arrives, and food becomes harder to find, the birds abandon their territories and roam the countryside in large flocks.

Sociality allows animals to spot both prey and predators more easily. But groups of animals are themselves more visible to predators, and individuals in groups must constantly compete for food, mates, and breeding areas.

Honeybees Are an Extraordinary Example of Social Behavior

The coordination of individuals within a colony of social insects rivals that of the cells within an individual. We can even view an insect colony as a superorganism, in which thousands or even millions of individuals function as a single unit. Social insects form enormous colonies. An African termite mound, for example, may contain 2 million workers. But the number of individuals alone is not what makes the social insects special. It is social organization that allows social species to take advantage of environmental resources unavailable to more solitary animals.

The familiar honeybee provides an excellent example. Humans long ago domesticated the honeybee to harvest the bees' useful honey and wax. Scientists since Aristotle have studied the workings of beehives and published tens of thousands of papers on bees. The fascination with bees has extended beyond the practical to the almost mystical. Humans admire the fact that the "busy bee" works so hard, and loyally and ferociously defends its queen (long thought to be a king). One writer even suggested that honeybees have "voluntarily escaped from the garden of Eden with poor fallen man for the purpose of sweetening his bitter lot."

A honeybee hive consists of double-layered wax combs that contain thousands of hexagonal cells. In these wax cells the bees store honey and pollen or house developing larvae. Each bee grows from an egg deposited in a separate cell. The egg develops into a larva, which then develops into a pupa and finally into an adult.

A hive may contain 20,000 to 80,000 workers, each engaged in a succession of tasks. At the beginning of a worker's life she brings nectar and pollen to the larvae. Later, she makes wax to enlarge the comb, removes the dead and the dying, guards the hive against intruders, and scouts the surrounding territory for food and nesting sites. Finally, toward the end of her short, 6-week life, she undertakes the most hazardous job of all, foraging for honey and pollen among the flowers.

All the workers in the hive are female, and all are sisters. None reproduce, however. Only a single female—the queen—lays eggs. Most of these eggs develop into sterile workers, but a few may become queens. Workers can create queens by feeding a few, select larvae a special rich diet. Normally, the hive's queen produces a "queen substance," a pheromone that prevents the workers from building special wax cells to raise new queens. In the spring, however, the queen's production of this chemical falls off, and the workers build the royal cells. Workers soon push the old queen out of the hive, and she flies off with a large swarm of workers to start a new hive (Figure 29-12).

About 8 days after the old queen leaves the hive, the new queens begin hatching. Each one can form a swarm and fly off to start a new colony. If a queen elects to stay in the old hive, however, she first seeks and destroys rival queen larvae. She identifies these competitors by making quacking sounds outside the queen cells. If another queen responds to this challenge, a fight to the death ensues. There can be only one queen in each colony.

Once a queen has disposed of her rivals and established her queenship, she mates repeatedly in a series of "nuptial" flights. She leaves the hive, approaches a group of males, and releases small amounts of queen substance. The pheromone attracts males, one of whom will mate with her. The male releases his sperm into the queen's genital chamber and quickly

Figure 29-12 Swarming honeybees. A swarm of bees will stay in one place for days at a time, while scouts look for a suitable site for a new hive. The scouts return to the swarm and dance the location of a hole in a tree or cave. Each scout, however, is dancing a different location. When all the scouts agree, the swarm sets off to build a new hive. *(Gilbert Grant/Photo Researchers)*

dies. The queen may mate with as many as 17 males over a 4-day period. In this way, she obtains enough sperm for a lifetime. She stores the sperm and uses it to fertilize her eggs, which she deposits one at a time into separate wax cells. A queen may live and lay eggs for more than 5 years.

All of the social insects, including termites, ants, wasps, and honeybees, have three traits in common: (1) they cooperate in raising their young; (2) only a few individuals reproduce, while sterile workers assist and defend the fertile queen; and (3) generations overlap, so that offspring help their mother.

Honeybees display extraordinarily complex social structure.

Is Altruism Adaptive?

In Chapter 16 we learned that natural selection favors traits that increase the likelihood that an organism's offspring will survive. Why, then, we must ask, do the sterile workers in a beehive persist in a behavior that can have no adaptive value for them individually? The workers increase the reproductive output of the queen, but at the expense of their own reproduction. Biologists call such an arrangement **altruism,** behavior that benefits others at the expense of the animal that performs the behavior.

People sometimes jump into frozen lakes to rescue others. We know that this behavior is risky, because many of these would-be rescuers themselves die. Such behavior is clearly altruistic. Similarly, ground squirrels risk their own lives to warn one another of an approaching predator. And the young of many species of birds, for example, help their parents rear the next generation. The social insects, however, demonstrate the most extreme example of altruism.

If altruistic behavior has a genetic basis, then biologists must ask how it could possibly evolve. We would expect natural selection to favor sacrifice by parents to help their offspring, but to act against the sacrifice of offspring to help their parents. The social insects, in which daughter workers sacrifice themselves for their mother queens, therefore provide a challenge to a simple model of natural selection.

Does Altruism Arise by Natural Selection?

How could evolution produce workers, however industrious, if they leave no offspring? For a while, Darwin himself thought that this paradox was fatal to his whole theory of evolution by natural selection. To save his theory Darwin argued that a family (the queen and her offspring, for example) must be the unit of selection, rather than the individual. By their self-sacrifice the workers perpetuate genes that they share with their sisters, one of whom will become a queen. What is selected is the ability to be altruistic. The altruistic family is successful in passing on its genes.

In the 1960s, the British biologist W.D. Hamilton suggested the theory of **kin selection,** a more general answer to the question of altruism in the social insects. He reasoned that an individual increases its reproductive output by helping relatives, which share its genes, to reproduce. Sisters (and brothers) are just as related to one another as they are to their parents. They share one-half of their genes. Thus, a worker bee helps ensure the continuation of her genes within her sisters and their descendants (recall that some of the sisters become queens).

Hamilton's work established the concept of "inclusive fitness." *Fitness* is a measure of selective advantage defined as the contribution to the next generation of one genotype in a population relative to the contribution of other genotypes. **Inclusive fitness** is the sum of an individual's genetic fitness (which includes that of its own direct descendants) plus all its influence on the fitness of its other relatives. Inclusive fitness, then, is the total success in passing alleles to the next generation, either by the efforts of a parent or by the altruistic acts of a relative.

One explanation for the adaptive value of altruism is inclusive fitness, which includes the concept of kin selection.

Why Might Intricate Societies Have Evolved Among Certain Insects?

Intricate societies evolved at least 11 separate times among the bees, ants, and other members of the order Hymenoptera, yet only once (in the termites) among all other insects. Why the difference? One reason may be that the hymenopterous social insect species have an unusual kind of sex determination, called **haplodiploidy.** In haplodiploidy males develop from unfertilized eggs and are therefore haploid, while females are diploid. Other than providing sperm in a single mating event, males contribute nothing to the economy of the hive. Haplodiploidy dramatically alters the usual genetic relationships among parents and siblings. It means that a worker's inclusive fitness is greater if she devotes her energies to her siblings than if she reproduces herself. Let's see how this works.

First consider the usual case, say in humans. Half your alleles come from your mother, half from your father. Each of your brothers and sisters also share one-half of their alleles with their mother.

What, then, is the chance of sharing an allele with your sister? Half of all your alleles came from your mother, and the chance that you and your sister inherited the same allele is 0.5. So the chance that you each inherited a given allele from your mother is $0.5 \times 0.5 = 0.25$. Similarly the chance that you each inherited the same allele from your father is 0.25. So the chance that you and your sister share any allele is $0.25 + 0.25 = 0.5$, exactly the same as the chance that you share an allele with your mother or father. A similar calculation shows that you share one-eighth of your alleles with your first cousins. The British evolutionary biologist, J.B.S. Haldane, summarized these relationships by saying, "I would gladly lay down my life for two of my brothers or eight of my first cousins."

As in the case of ordinary sex determination, the queen passes half her alleles to her female offspring. However, since males are haploid, a male passes on all his alleles. That is, his daughters inherit all of his genes. As a result, the (female) workers share 75 percent (rather than 50 percent) of their alleles with one another. Because the queen mates with many different males, not all of her offspring share the same father. However, the average relatedness will be greater than 50 percent.

If the workers were to have their own offspring, they would pass on 50 percent of their alleles. They can therefore pass on more of their genes by putting their energies into making sisters than they can by reproducing and making daughters. To put it another way, the workers more efficiently perpetuate their genes by helping their mother, the queen, produce more offspring (including new queens) than by producing eggs themselves.

Haplodiploidy in some social insects probably facilitates kin selection.

Does Inclusive Fitness Always Explain Altruistic Behavior?

The haplodiploid social insects are not the only animals that help one another. Grazing animals, birds, and other animals that live in groups commonly help protect one another. In as many as 300 species of birds, "helpers" may assist a breeding pair feed and protect the nestlings. Whether such behavior can likely be explained by kin selection, however, depends on how closely the altruists are related to one another.

A few studies suggest kin selection as an explanation for altruism. For example, when a pair of Florida scrub jays breeds, as many as six other jays may help feed and protect the nestlings. These helpers are usually the breeding jays' offspring from previous clutches, offspring that have yet to mate themselves. Because territories suitable for breeding are in short supply, these helpers may not be able to breed effectively. Like the honeybee workers, the young jays may be better off helping their siblings than trying to breed themselves.

An even more persuasive example comes from Belding ground squirrel populations high in the Sierra Nevada, in California. In the summer, female squirrels defend territories where they raise their offspring. The two biggest dangers to Belding ground squirrels, besides winter snows, are predators and other squirrels. Unrelated or distantly related squirrels regularly kill other ground squirrels, killing up to 8 percent of all pups. At the approach of a coyote or strange ground squirrel, some female squirrels stand up tall and sound a shrill alarm call that alerts other nearby squirrels (Figure 29-13). Other individuals who first sight a coyote or a stranger, make no alarm and simply dive into the nearest burrow silently, leaving the other squirrels to fend for themselves. The squirrels that sound the alarm increase their own chance of being attacked, so their behavior is altruistic.

Paul Sherman, a biologist now at Cornell University, asked why these squirrels risk their own lives to help others, and why some squirrels behave altruistically, while others do not. Sher-

Figure 29-13   How kinship affects behavior. A. At Tioga Pass, California, students watch individual Belding ground squirrels marked with numbers written with hair dye. These ground squirrels live in colonies consisting of related females and their offspring. B. Older females sound a loud alarm when predators approach. Males take no such chances and scurry silently away.   *(A, B, George D. Lepp)*

A.   B.

BOX 29-1

Tit for tat and prisoner's dilemma

The classic game theory analysis of the risks and benefits of cooperation and altruism is presented in the "game" *the prisoner's dilemma*. Imagine that you and a friend are arrested for committing a crime, and you are held in separate jail cells unable to communicate. The police immediately commence interrogating the two of you.

Suppose you claim innocence, but your friend confesses and incriminates you. You'll be in prison for a long time, but your friend will get off scot-free. On the other hand, you could get off yourself if you incriminate your friend and your friend pleads innocent. If you each confess (and incriminate each other) the state's job of convicting you will be easy, and your sentences will be light. But if you both plead innocent, the state will have little evidence against you, and your sentences will be even lighter.

What to do? Regardless of what your friend does, you are better off confessing, which is true for your friend too. On the other hand, if both of you confess, you will be worse off than if both of you pleaded innocent. The safest solution is to confess. Even though pleading innocent might get you off scot-free, it's the riskiest solution to the problem.

The game of prisoner's dilemma seems to show that selfish behavior is safer than cooperative behavior. In 1981, however, a political scientist named Robert Axelrod and a biologist named W.D. Hamilton combined forces to explore what happens when the game is played over and over. They reasoned that in real life, people and animals interact repeatedly. Perhaps cooperation might pay off in the long run.

To make the endeavor more interesting, Axelrod invited game theorists from sociology, political science, economics, mathematics, and biology to design strategies to compete in an ongoing computer tournament. Predictably, players that always cooperated where taken advantage of. But strategies of constant betrayal also did badly.

The winning long-term scheme was a simple strategy called *tit for tat*. In tit for tat, the player always cooperates in the first round (pleads innocent). After that, the player does whatever the opponent did in the last round. In short, you reward your friend for cooperating and punish your friend for ratting on you. Over time, tit for tat always won.

In the early 1990s, an even more successful strategy emerged from Axelrod's tournament. "Modified tit for tat" occasionally forgives other players for a betrayal, just often enough to nudge the game in the direction of cooperation and reciprocity. Although the prisoner's dilemma is a simple game, it suggests that reciprocity is a possible explanation for altruism in some situations.

---

man discovered that most of the squirrels that sound the alarm are yearling and older females, who are warning their female relatives—sisters, mothers, and daughters, as well as juvenile sons. However, males and childless females rarely warned other squirrels. The older female incurs a risk that is counterbalanced by the chance that she will save the life of a close relative. Her altruistic behavior may contribute to her inclusive fitness.

In the past 20 years many biologists, anthropologists, and other enthusiasts have tried to extend the concept of kin selection to virtually all altruistic behavior. But not all examples of altruistic behavior conform to the kin selection model. Dwarf mongooses, for example, help breeding pairs care for their young, but not all of the helpers are relatives. Kin selection does not account for such altruism.

There are other explanations for why animals behave altruistically, but one of the most interesting was first suggested by ethologist Robert Trivers. Trivers argued that animals capable of recognizing one another might engage in altruism on the understanding that the recipient would return the favor sometime in the future. Trivers called this idea **reciprocal altruism**, since the recipient reciprocates. Such a system at first seems too susceptible to cheaters to work. Biologists wondered, How can an animal ensure that the recipient of an altruistic act will return the favor? But research in the area of "game theory," a branch of mathematical logic applied to business, military sociobiological, and other problems, suggests that altruism can pay off in spite of cheaters.

Kin selection may explain altruism in other animals besides the social insects. But other explanations, such as reciprocal altruism, are possible, as well.

Understanding Human Behavior Poses Special Problems

Because much human behavior closely resembles that of other animals, we have every reason to assume that genes influence our behavior (Figure 29-14). However, our behavior, like that of other animals, is heavily influenced by learning. We call the extravagant complex of behaviors that we learn from one another "culture."

BOX 29-2

The nature–nurture question in humans

In nonhuman animals, biologists can demonstrate the role of environment in determining certain behavior by experimentally controlling the environment of an animal. Estimating the relative contributions of genetics and environment to behavior is difficult but sometimes possible.

In humans, however, true experiments are not possible. We cannot experimentally control the developmental environment of whole groups of individuals, or of even one. As a result, we can only *estimate* the role of genetics in any particular kind of behavior.

Two kinds of studies substitute for true experiments—"twin studies" and the related "adoption studies." In twin studies, researchers find identical twins who have

been raised apart and compare their behavior. For example, to see how important a role genetics plays in alcoholism, researchers might study how often twins either both become (or do not become) alcoholics compared to the number of cases where one twin becomes an alcoholic and the other does not. Their behavior might also be compared to that of twins raised together or to that of nonidentical twins or ordinary siblings raised apart or together.

Such studies are often revealing and intriguing. But adoptions are not controlled experiments, and therefore the results of twin studies are hardly definitive. For example, social workers try very hard to match adopted children with parents who

are similar to the biological parents. As a result, the environments of adopted siblings are likely to be very similar. On the other hand, even identical twins raised in the same family grow up to look and behave differently.

Twin studies and other kinds of adoption studies have shown, for example, that family environment and genetic endowment each play an essential role in determining IQ, a sometimes useful measure of academic intelligence. Yet, despite more than 30 different studies, researchers still do not know exactly how large a role genes play in determining IQ. Estimates for the heritability of IQ scores range from as low as 30 percent to as high as 70 percent.

Culture Evolves Rapidly

Culturally acquired behavior—whether it is how to get cream from a milk bottle or how to order cream from the milkman—is of adaptive importance and evolutionary significance in all

Figure 29-14 **Behavior is partly genetic.** Individual behaviors, such as smiling, are often shared by groups of related species. *(Tim Davis/Photo Researchers)*

species. Cultural practices are subject to a kind of selection at two levels. First, useful ideas tend to prevail more than harmful ones. Adaptive radiation and selection among variants occur with culture and ideas as well as with genes. Indeed, a new cultural practice can originate and spread far more rapidly than a new gene or set of genes. For example, cream-stealing behavior among birds spread many miles in just a few generations. Second, cultural learning can interact with natural selection. For example, individuals capable of quickly learning and adopting useful new ideas might tend to enjoy a selective advantage.

The advantages of various cultural practices, from styles of food gathering to social organizations, are the same whether the basis is genetic or cultural. In both cases, we can tell the same kind of adaptive fables. When a human parent gives a toddler fresh milk instead of bad milk, that is clearly a kind of behavior that has adaptive value. The parent's behavior reduces the chance that the child will get food poisoning. Checking the date of the milk carton, closing the carton immediately after use, and keeping the milk refrigerated are all behavior patterns with adaptive value. They are also behavior patterns that never appear in other species of animals. So the fresh-milk behavior pattern is unique to humans and clearly adaptive. But those two facts do not mean that the behavior is genetically determined. We might note that many groups of humans do not engage in this behavior. Is it because they lack the "fresh-milk gene"? Or is it because they have not learned the behavior?

A major problem for the biologist, then, is teasing out the relative influences of genes and culture on different behaviors, whether in humans of other animals. But too often both biologists and nonbiologists have made arguments that one or another human behavior is genetically controlled, and therefore more "natural," meaning good or right. In the absence of any data showing genetic polymorphism in social traits, such arguments are unfounded.

Much of human behavior is culturally acquired. Cultural behavior evolves faster than behavior that is under genetic control.

Is Human Behavior Influenced by Genes?

Because we are more interested in the behavior of humans than of any other animal, we often interpret animal behavior according to concepts that we know operate in human societies, such as calculating the economic costs of a particular feeding strategy. Conversely, biologists are especially interested in patterns of human behavior that resemble those of other animals. One of the most long-standing and controversial questions concerning human behavior is whether humans have instincts similar to those of animals.

Humans resemble other animals in having innately determined modal action patterns, but these patterns are subject to modification through experience. A baby's smile provides a wonderful example. Universal among all human cultures, a smile requires a stereotyped sequence of muscle movements.

Like modal action patterns in other species, characteristic releasers trigger a smile. Another human face is an especially powerful releaser, even for a young baby. Following the same approach that revealed the importance of the red spot on a gull's bill, students of human behavior discovered that for a 6-week old baby, the sign stimulus for smiling was the presence of more than one contrasting element in a face—the eyes (Figure 29-15).

At 6 weeks an appropriate dot pattern is a better releaser than a real face. As the baby matures, however, he or she learns to make subtler distinctions and comes to prefer a real face. By about 6 months, an unfamiliar face is more likely to evoke tears than a smile. For a blind baby, touch, smell, sounds, or other releasers may stimulate the smile. But the smile is the same, including the turning of the eyes toward the releaser. It seems unlikely, then, that babies learn to smile, though they do learn to refine their releasers and to modify the original muscular movements as they mature.

The extraordinary learning ability of humans often makes it hard to determine even the existence of modal action patterns or to study the genetic basis of behavior. However, biologists believe that many behaviors are part of our evolutionary heritage, even if both the behavior pattern and its releaser are modified by experience. For example, the adrenal gland in hu-

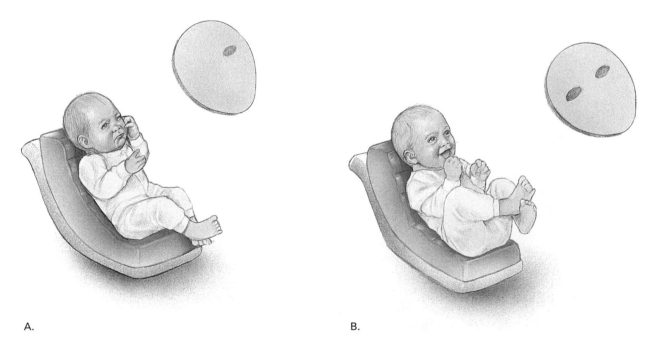

A.   B.

Figure 29-15 Releasing a smile. A flat, face-sized mask with one eye spot will not make a young baby smile. The same mask with two eye spots triggers a happy smile. Older babies learn to prefer familiar human faces.

mans (and in other vertebrates) produces the hormone epinephrine (adrenaline) when the brain perceives a threatening situation. This hormone coordinates a set of responses, called the "fight or flight" reaction, in the heart, lungs, circulation, and brain. Like other animals, we are likely to respond to the presence of, say, a grizzly bear, with a "fight or flight" response. But many people will respond just as intensely and just as automatically to a rude gas station attendant. We can conclude that much of human behavior, like that of other animals, is partly reflexive and partly learned.

Humans resemble lower animals in having innately determined modal action patterns, but these patterns are subject to modification through experience.

STUDY OUTLINE WITH KEY TERMS

The systematic study of natural (wild) behavior is called **ethology.** In contrast, the laboratory study of behaviors common to many animals is called **experimental psychology.** Social and other forms of behavior viewed in the context of evolution and ecology is the concern of the field of **sociobiology,** now called **behavioral ecology.**

Animal behavior is partly influenced by genes, partly by environment. Humans and other animals learn from one another as well as from their environment. Discussions of animal behavior can quickly become politicized.

Biologists are very interested in the relative importance of genes and environment in a variety of animal behaviors. If genes help to shape behavior, and natural selection operates on behavioral variants, then biologists must ask what ecological factors determine the adaptiveness of various kinds of behavior.

Innate behavior includes a baby's first smile and the salivating of a dog over his dinner bowl (a **conditional reflex**), but all behavior in all animals can be modified through **learning.** A **modal action pattern** is highly stereotyped innate behavior performed in response to a **sign stimulus.** An exaggerated sign stimulus is a **supernormal stimulus.** Stimuli may be sights, sounds, tactile experiences, or smells in the form of **pheromones.** Some forms of learning, called **imprinting,** occur only at special times in development, called **sensitive periods. Sexual imprinting** is one.

Animals seem to accurately weigh the costs and benefits of different foraging strategies in a process called **optimal foraging.** When animals compete for mates or resources they engage in **agonistic behavior,** which includes aggression, appeasement, and retreat. Animals such as chickens, dogs, and primates use agonistic behaviors to establish **dominance hierarchies.** Competition for mates results in **sexual selection.** Many animals defend a **territory,** which may encompass space and resources for mating, foraging, or raising young.

Social groups allow animals to spot both prey and predators more easily. However, groups of animals are themselves more visible to predators than are solitary animals. Also, the individuals in a group must constantly compete with one another for food, mates, and breeding areas.

The social insects display complex social structure and **altruism.** One explanation for the adaptive value of altruism is **inclusive fitness,** which includes the concept of **kin selection.** All of the social insects except the termites are **haplodiploid,** which means that sisters are more closely related to one another than to their own offspring. Such extreme relatedness facilitates kin selection and altruism. Another explanation for altruism is **reciprocal altruism.**

REVIEW AND THOUGHT QUESTIONS

Review Questions

1. In the past, what distinguished ethology from experimental psychology?
2. Define a modal action pattern and give an example.
3. What is a pheromone?
4. Give an example of a sensitive period.
5. Define and give an example of optimal foraging.
6. Distinguish between aggression and agonistic behavior.
7. Give an example of territorial behavior in dogs.
8. Why, according to the theory of inclusive fitness, are honeybee workers better off not reproducing?
9. Give an example of an innate response modified by learning in an adult human.
10. For how many second cousins should you lay down your life?

Thought Questions

10. Three pizza parlors offer specials on identical giant pizzas. One parlor will deliver the giant pizza for $25. The other two don't deliver. One is just around the corner, and its pizza is $20. The other is 8 miles away and charges $12. It costs you 50¢ a mile to drive your 1980 Buick Skylark, and the car needs a new battery. If you break down in the neighborhood where the $12 pizza is, you might get mugged. What is the optimal solution to this foraging problem?
11. If humans were haplodiploid, how would human societies be different?

SELECTED READINGS

Crompton, John, *A Hives of Bees,* Nick Lyons Books, 1987. An anthropomorphic but lively account of life in a beehive, by an amateur naturalist.

Crompton, John, *Ways of the Ant,* Nick Lyons Books, 1988. A vivid account of the natural history of various kinds of ants.

de Waal, Franz, *Peacemaking Among Primates,* Harvard University Press, Cambridge, Massachusetts, 1989. A prize-winning book about reconciliation and aggression in chimpanzees, humans, and other primates.

Fabre, Henri J., *The Life of the Spider,* Dodd, Mead, and Company, New York, 1912. A classic piece of 19th-century natural history, beautifully translated by A.T. de Mattos.

Heinrich, Bernd, *Ravens in Winter: A Zoological Detective Story,* Summit Books, New York, 1989. An exciting description of a zoologist's observations and experiments illuminating the social behavior of crows. The author's vivid prose and natural storytelling abilities bring this piece of behavioral ecology to life.

Montgomery, Sy, *Walking with the Great Apes,* Houghton Mifflin Co., Boston, 1991. A fascinating study of three women primatologists—Jane Goodall, Dian Fossey, and Birute Galdikas.

▶ On-line materials relating to this chapter are on the World Wide Web at http://www.saunderscollege.com/lifesci/
Click on Tobin/Dusheck: *Asking About Life.*

Structural and Physiological Adaptations of Flowering Plants

Conifers in Gunnison National Forest, Colorado. *(© Carr Clifton/Minden Pictures)*

Should We Eat Our Vegetables?

In 1983, a biochemist at the University of California, Berkeley, caused a stir in both the scientific community and in the health food industry when he asserted that enormous numbers of ordinary vegetables were loaded with mutagens, chemicals that cause DNA to mutate. Under the right circumstances, mutagens are capable of causing cancer, and natural chemicals, the biochemist argued, were as carcinogenic as anything manufactured by American industry.

Bruce Ames was already well known for his invention of the Ames test, an efficient method for testing the mutagenicity of chemicals. In the Ames test, plates of bacteria are treated with a chemical and the bacteria are then monitored for changes that indicate that their DNA has mutated. Because Ames's work had pointed up the hazards of various environmental toxins and mutagens, Ames was much admired by members of the environmental movement. In 1977, he had warned against the use of the pesticide ethylene dibromide, which he described as a "potent carcinogen." He also pointed out the hazards of a chemical used as a flame retardant in children's clothing and warned, more generally, against inadequately tested industrial chemicals.

But many people saw his 1983 paper as an about-face, in which he seemed to assert that pesticides, air pollution, and other environmental hazards were noth-ing compared to the toxins found in ordinary vegetables and meat. Indeed, Ames reviewed the scientific literature on mutagenic and toxic compounds in food and concluded, "the human dietary intake of 'nature's pesticides' is likely to be several grams per day—probably at least 10,000 times higher than the dietary intake of man-made pesticides." Ames listed scores of toxic chemicals found naturally in ordinary foods, including burned or browned meats and other foods, most fats—especially burned or rancid fats—and great numbers of compounds made naturally by plants.

Herb teas made from plants in the family Euphorbiaceae, for example, contain potent promoters of carcinogenesis. Wilted carrots contain a mutagen. Alfalfa sprouts contain a highly toxic analog of the amino acid arginine, called canavanine, which causes a lupuslike immune disease in monkeys fed alfalfa sprouts. The metabolism of the alcohol in beer, wine, and spirits generates a compound that is both a mutagen and a teratogen, a compound that causes birth defects. Many plants respond to any kind of damage by secreting high levels of toxins, many of which are mutagenic, teratogenic, or carcinogenic. Ames's list of plant and other food toxins went on for four pages in the journal *Science*.

Every plant that we eat has something in it that is bad for us, argued Ames. If we add all these compounds together, they far outweigh the traces of pesticides typically found on fruits and vegetables or in drink-

Figure 30-1 **Eat your carrots.** Carrots are full of fiber, potassium, and β-carotene, which your body converts to vitamin A, a crucial element in the functioning of the eye. Beta-carotene is also an antioxidant that helps protect against cancer, and may even help lower blood cholesterol. Avoid damaged or wilted carrots, which produce mutagens that promote cancer. *(Aaron Haupt/Photo Researchers)*

play a role in the incidence of both heart disease and certain kinds of cancer, particularly colon cancer and breast cancer, the two most common forms of cancer after lung cancer, which is caused by smoking. But that role can be both positive and negative. For example, one study showed that smokers who did not eat fruit regularly had a 30 percent higher incidence of lung cancer than those who ate fruit.

A large part of Ames's 1983 paper had been devoted to natural plant compounds, such as vitamin E, vitamin C, and β-carotene, that seem to prevent cancer. These compounds, called antioxidants, work by preventing the formation of free radicals, compounds that damage DNA. Free radicals are formed during normal metabolic processes in both animals and plants. Because chlorophyll-mediated photosynthesis tends to generate destructive free

heart disease, cataracts, and cancer. Lack of antioxidant-containing fruits and vegetables in the diet, argued Ames, was a major cause for these degenerative diseases.

We can summarize Ames's argument by saying that if we don't smoke or drink too much, stay away from obvious hazards such as asbestos mines, and eat our fruits and vegetables, we need not worry very much about trace amounts of pesticides or other contaminants in our food and drinking water (Figure 30-1).

On the other hand, the high levels of toxins that many workers are exposed to—in farming and in the processing of metals, for example—pose serious risks. Likewise, the high levels of contaminants found in toxic waste dumps pose distinct hazards to people and other organisms. Indeed, water and air pollution are frequently more threatening to other organisms than to us.

Although Ames continues to be the center of other controversies, his views on the risks and benefits of plant compounds have largely been accepted. Government agencies now regularly advise Americans to eat more fruits and vegetables, not only for the vitamins, but also because of the high levels of antioxidants (Table 30-1).

ing water. This is not to say that pesticides and other toxins are not dangerous at high levels. People who regularly expose themselves to toxic chemicals—field workers who spray pesticides, for example—are at great risk. But for the average person, Ames argued, the amounts of industrial chemicals we ingest are minuscule compared to what we get naturally from plants.

Ames's fellow researchers were appalled, accusing Ames of ignoring the well-known hazards of smoking, of backing off from his stance against both the use of dangerous pesticides and the chemical pollution of air and water. They even accused Ames of aiding corporations that manufacture and use dangerous chemicals. If nothing else, they asserted, he was confusing people about what constituted a real hazard and what did not.

Ames reacted defensively, arguing that he still believed in the strict testing of industrial chemicals. He insisted that the natural components of a diet could well

---

Every plant that we eat has something in it that is bad for us, argued Ames.

---

radicals, all plants manufacture β-carotene or other similar antioxidants to protect their DNA and other important molecules.

In subsequent years, Ames increasingly focused on the production of free radicals during normal metabolism as the ultimate cause for all the degenerative diseases associated with aging—including

In the rest of this chapter we discuss plant structure and the chemical compounds found in plants. Most of our discussion will concern the angiosperms (flowering plants). Despite their diversity, flowering plants have evolved common organizational features in response to common environmental demands.

KEY CONCEPTS

1. Flowering plants are largely terrestrial organisms that rely on underground roots and aboveground shoots to obtain water, minerals, carbon dioxide, and sunlight.

2. Most of the parts of a plant—roots, stems, and leaves—are composed of just three types of tissues.

3. The vascular system consists of xylem, which carries water and minerals from the roots, and phloem, which carries sugars and other nutrients from the leaves.

4. Plants produce secondary plant compounds, which protect plants against herbivores, pathogens, and other plants; provide structural support; and, sometimes, attract animal pollinators and seed dispersers.

WHAT ARE THE PHYSICAL CONSTRAINTS ON PLANTS?

Plants are, by definition, photosynthetic. They live by turning carbon dioxide into carbohydrates. To do this, plants need ample supplies of carbon dioxide, water, and sunlight. Recall from our discussion of photosynthesis in Chapter 7 that plants combine the hydrogen from water molecules (H_2O) with carbon dioxide (CO_2) to build the simple sugar glucose. The energy to accomplish this transformation comes from sunlight.

Until about 500 million years ago, photosynthetic organisms were largely aquatic, as was most life on Earth. Then, about 440 million years ago, late in the Ordovician, the first photosynthetic organisms invaded the land. Paleontologists have found traces of these little-known organisms in the form of spores, bits of cuticle, and tubes similar to the vessels that carry water in modern vascular plants.

Biologists suspect these early plants may have been descended from green algae, or chlorophytes, which can live in fresh water and even on land. Modern chlorophytes associate with fungi as land-dwelling lichens. Some researchers have suggested that plants evolved from "inside-out" lichens, in which the photosynthetic chlorophyte became the dominant element.

From some such primitive form, recognizable plants evolved in the Silurian and diversified throughout the 100 million years of the Devonian and Carboniferous. Plants have become the dominant feature of land life. Especially successful are the flowering plants, with more than 275,000 species now populating our planet. Plants constitute over 90 percent of the world's biomass. They are the major primary producers of terrestrial ecosystems and the ultimate source of food, for humans and all other terrestrial organisms.

How Did Terrestrial Plants Evolve?

For the aquatic ancestors of terrestrial plants, the land offered distinct advantages. Aquatic plants are limited to shallow water, where nutrients and sunlight combine in a narrow plane at the surface. In deeper waters, nutrients sink down to levels that are too dark for photosynthesis.

The land offered plenty of light and nutrients and far more carbon dioxide than was available in the sea. Compared to marine and freshwater environments, however, the land could provide little water, and the carbon dioxide was in a gaseous state, not dissolved in water. In addition, whereas water tends to buoy plants and other organisms against gravity, a terrestrial environment exerts a crushing force on these same organisms.

Photosynthetic organisms therefore needed new mechanisms for obtaining water and carbon dioxide and for supporting their bodies against the pull of gravity. Whereas aquatic plants absorb water and dissolved carbon dioxide from their surroundings, terrestrial plants have evolved roots that pull water from moist soil, as well as tiny pores in their leaves that collect carbon dioxide from the air. These pores, called **stomata** [Greek, *stoma* = mouth], open and close like tiny mouths.

Table 30-1	Foods That Are Good Sources of Three Antioxidants		
	Vitamin A (% RDA)	Vitamin C (% RDA)	Vitamin E (% RDA)
Apples (1)	—	10%	—
Apricots (3)	50%	—	—
Blueberries (1 cup)	—	33%	—
Strawberries (1/2 cup)	—	100%	—
Cantaloupe (1/2)	>100%	>100%	—
Orange (1)	—	>100%	—
Broccoli	90%	200%	—
Carrot (1)	400%	—	—
Red pepper (1)	84%	250%	—
Winter squash (1 cup)	150%	—	—
Sweet potato	500%	—	—
Almonds (1/2 cup)	—	—	170%
Mayonnaise (1 T)	—	—	80%
Peanuts (1/2 cup)	—	—	60%
Safflower oil (1 T)	—	—	45%

In addition, terrestrial plants harvest energy from sunlight using their leaves. Leaves are solar collectors: they are often broad and thin, which maximizes their exposure to the sun, and loaded with chlorophyll, the green pigment that plants use to capture the energy of sunlight.

Besides roots and leaves, most terrestrial plants rely on stems. Stems support leaves, flowers, and fruits and also provide an avenue for the transport of materials between the ground and the air. The leaves use water from the roots to support photosynthesis. The roots, in turn, use sugars manufactured in the leaves to supply energy to the ionic pumps that pull water from the surrounding soil.

Not all terrestrial plants have stems and transport tissues. The first vascular plants did not evolve until about 435 million years ago. And some 17,000 species of mosses and other nonvascular plants flourish today. The vast majority of terrestrial plants, however, are vascular plants, with roots, stems and leaves. The basic form of vascular plants has remained unchanged for hundreds of millions of years. This form, which solves all of the major problems of terrestrial life, is highly successful. Vascular plants live virtually everywhere on the planet where there is sunlight, carbon dioxide, and any amount of water at all. Only the polar ice caps lack plants altogether.

Terrestrial plants have roots that absorb water and nutrients from soil; broad, thin leaves that collect solar energy; and stomata that collect carbon dioxide and release water and oxygen.

The Shoot and the Root

While the defining characteristic of plants is that they photosynthesize, plants are not mere bags of photosynthetic protoplasm. Plants are, like animals and fungi, highly structured multicellular organisms. The forms that plants take represent solutions to easily defined problems.

The anatomy of a flowering plant reflects its need to derive sustenance from both ground and air. Most plants consist of an aboveground **shoot system**—stems, photosynthetic leaves, and flowers and other organs of reproduction—and an underground **root system**, which anchors the plant and absorbs water and minerals from the soil. Shoots and roots depend on one another utterly: roots could not exist without the energy provided by photosynthesis in the shoot, and shoots could not exist without the water and minerals absorbed by the roots.

Materials travel between shoot and root in the plant's **vascular system** [Latin, *vas* = vessel], the network of vessels in which fluids move through the plant. The vascular system consists of two tissues, the **xylem**, which transports water and minerals from roots to shoots, and the **phloem**, which transports the energy-rich products of photosynthesis. The phloem transports sugars from leaves to growing regions, reproductive structures, and root and also from storage areas (such as those of sugar beets, potatoes, or carrots) to the rest of the plant. Xylem and phloem tissues associate together in **vascular bundles.** In roots and stems, the vascular bundles frequently run in a central cylinder called the **stele.**

Terrestrial plants must collect water, carbon dioxide, and light, but they must also support their own bodies against the considerable pull of gravity and, finally, they must reproduce. Every plant is a compromise between design constraints imposed by the different functions of the plant. A plant that is well adapted for extracting water from the soil, for example, might not be the best at intercepting sunlight.

The more plants spread their leaves and branches to maximize their exposure to sunlight, for example, the stronger the stems and branches have to be to support the weight. If you take this textbook in your hand, you will notice that it is harder to hold it straight out to the side or in front of you than to hold it down at your side or even straight up above your head. Plants that spread their leaves in the sun face the same problem. Their ability to do so depended on the evolution of materials that are both enormously strong and remarkably lightweight.

The single plant structure that is most important to supporting a plant is the cell wall. Like us, plants are eukaryotes, whose cells have a nucleus, ribosomes, and other organelles. Unlike our cells, however, plant cells also have a **cell wall,** which consists of **cellulose,** other carbohydrates, and specialized proteins. For its weight, cellulose is the strongest material known. The nutshell of the Australian macadamia nut is stronger than commercial-grade aluminum, concrete, glass, or brick; yet the density of the nutshell is less than half that of these other materials.

All plant cells have a **primary cell wall** outside the plasma membrane. Some cells have a thicker **secondary cell wall,** which lies between the cell membrane and the primary cell wall (Figure 30-2). Cells with a secondary cell wall are frequently cells specialized to provide the plant with extra mechanical support. A polymer called **lignin** gives additional strength to the secondary wall, and in some cases to the primary wall, as well. Lignin has some of the same qualities as a hard, stiff plastic.

The cytoplasm of most plant cells fills only part of the space within the cell wall, often as little as 10 percent. The rest of the internal space consists of a large central vacuole, filled with a watery fluid called the cell sap. Water pressure in the central vacuole pushes a cell's contents hard against the cell wall. The pressure of the fluid contents of the vacuole against the cell wall, called turgor pressure, helps support plants against the pull of gravity. We have all seen that plants under water stress wilt. This is because they have lost turgor pressure.

Some plants are entirely soft and green and depend mainly on turgor pressure for support. Such herbaceous plants, or sometimes just herbs, are highly susceptible to wilting. Herbs include all plants that have no woody aboveground parts. Perennial herbs may have woody roots. (To a botanist, the term "herb" does not mean familiar kitchen herbs, many of which are, in any case, woody.) Woody plants, such as trees, shrubs, and other perennial plants, have woody parts that provide support independent of turgor pressure.

Figure 30-2 Primary and secondary cell walls. All plant cells have a primary cell wall, which is composed of cellulose, pectin (used to gel jellies), and a gluelike substance called hemicellulose. A secondary cell wall, if present, is usually similar but thicker than the primary cell wall and lies inside it. *(Biophoto Associates)*

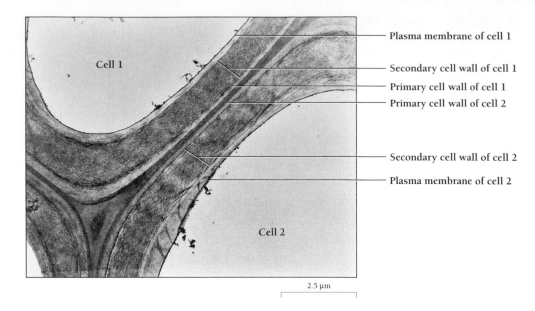

Cell 1

Plasma membrane of cell 1

Secondary cell wall of cell 1

Primary cell wall of cell 1

Primary cell wall of cell 2

Secondary cell wall of cell 2

Plasma membrane of cell 2

Cell 2

2.5 μm

Terrestrial plants include a shoot, with stems and leaves, and a root. A plant is supported against the pull of gravity by turgor pressure within its cells and by structural materials, such as cellulose and lignin, within the cell walls of support cells.

HOW ARE FLOWERING PLANTS STRUCTURED?

The cells of plants are organized into tissues, which are organized into tissue systems. Three tissue systems occur in all organs of a plant. These are the dermal, vascular, and ground tissue systems. The dermal tissue system makes up the outer, protective covering of the plant, the equivalent of our skin. The vascular tissue system comprises all of the tissues responsible for conducting. The vascular tissue system is embedded in the ground tissue system. The important differences in structure in root, stem, and leaf derive from differences in the way the vascular tissue system is distributed in the ground tissue. Tissues that consist of more than one cell type are called *complex tissues*. Tissues that consist of one cell type only are called *simple tissues*.

Primary Growth and Secondary Growth

All three tissue systems develop from regions of actively dividing cells called **meristems** [Greek, *meristos* = divided]. Meristems at the tips of roots and shoots are called **apical meristems** [Latin, *apex* = top]. Growth from apical meristem tissue is termed **primary growth.** Many plants also possess **lateral meristems,** which are cylinders of actively dividing cells within the roots and stems. Growth at the lateral meristems, called **secondary growth,** increases the thickness of a shoot or a root. Secondary growth is commonly (although not always) accompanied by the development of wood and cork (bark),

which replaces the dermal system as a protective covering. We discuss secondary growth (and wood) in more detail in Chapter 32.

The Dermal Tissue System

The **dermal tissue system** is the protective covering of the plant. It consists of the **epidermis,** which is primary tissue, and the **periderm,** which is secondary tissue (Chapter 32). The epidermis of the shoot secretes a **cuticle,** a waxy covering that keeps the aboveground parts of the plants from losing water. At first glance, the epidermis appears to be a single layer of identical cells. Closer examination, however, reveals that the epidermis is a complex tissue, meaning that it consists of more than one cell type, each with a distinct functional role. For example, the epidermis includes generalized epidermal cells, as well as the guard cells and the subsidiary cells of stomata. In addition, different plants may have glands, trichomes (hairs), secretory cells, and other specialized cells.

The Ground Tissue System

The **ground tissue system** consists of three main types of tissue—parenchyma, collenchyma, and sclerenchyma. Parenchyma and collenchyma are both simple tissues made of just one kind of cell, distinguished by the character of their cell walls. The cells of **parenchyma** [Greek, *para* = beside] are living cells that have only a thin primary cell wall. They are usually soft and succulent, with large vacuoles. Parenchyma cells make up most of the softer tissues of plants and they are frequently full of starch (Figure 30-3A).

The cells of **collenchyma** [Greek, *kolla* = glue] have a thick primary cell wall and often provide mechanical support, usually (but not only) in the growing regions of stems and leaves. Collenchyma cells often form bands or sheets just inside the epidermis. In celery, these bands are easy to find; they are the long strings just under the surface. In some collenchyma cells,

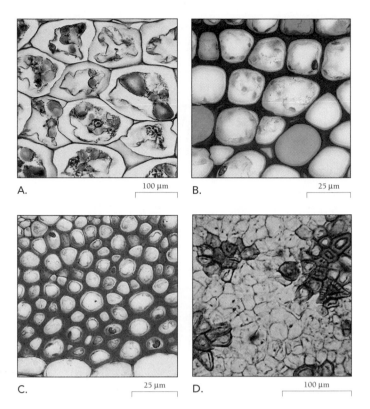

Figure 30-3 **Cell types in plants.** Most parts of a plant are made of just a few cell types. A. Delicate parenchyma cells make up most ground tissue. B. Tough-walled collenchyma cells strengthen stems and leaves. C. Fibrous sclerenchyma is tougher than collenchyma. D. Hard sclereids give pears their rough feel. The ripening of fruits often requires the breakdown of the lignin in sclereids. *(A, George Wilder/Visuals Unlimited; B, Ed Reschke/Peter Arnold, Inc.; C, Biophoto Associates/Science Source/Photo Researchers; D, Larry Mellichamp/Visuals Unlimited)*

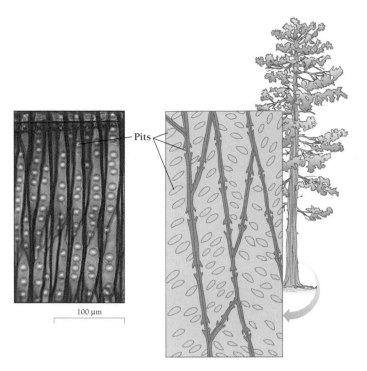

Figure 30-4 **Tracheid pit connections in a conifer.** While water flows smoothly from one vessel element to another, water in the tracheids must pass from one tracheid to another by way of holes or pits in the cell wall. Flowering plants have both tracheids and vessel members, but gymnosperms such as this pine have tracheids only. *(photo, John D. Cunningham/Visuals Unlimited)*

lignin accumulates irregularly in the secondary cell walls, particularly in the corners (Figure 30-3B).

Sclerenchyma [Greek, *skleros* = hard] consists of one of two cell types, fibers or sclereids. Fibers and sclereids, each of which exist in several varieties, have both primary and secondary walls and furnish mechanical support to the plant. Fibers and sclereids are usually not alive at maturity. Fibers consist of long, thin cells that join together, while sclereids occur as individual cells or in small groups (Figure 30-3C and D). Linen, which is woven from strands of sclerenchyma from the flax plant, illustrates the strength of sclerenchyma fibers.

The Vascular Tissue System

The **vascular tissue system** consists of the xylem, phloem, and associated tissues. Both xylem and phloem are themselves complex tissues. Xylem transports water and dissolved minerals and also provides mechanical support. Phloem transports sugars and other organic substances and provides some mechanical support.

Xylem Carries Fluids in Two Kinds of Nonliving Cells

Xylem carries water and its dissolved substances in two kinds of cells, tracheids and vessel elements. Both kinds of cells have lignin-reinforced secondary cell walls inside of primary cell walls. Although tracheids and vessel elements are the cells that conduct fluids, the complex xylem tissue also includes parenchyma cells, fibers, fiber-tracheids, and secretory cells.

Tracheids are long, thin, spindle-shaped cells. Water and dissolved substances move from cell to cell through **pits,** regions that lack secondary cell walls and watertight lignin (Figure 30-4). Pits may occur anywhere along a tracheid, but most form at the tapered ends of the tracheids. Pits form as pairs between adjacent cells, and water flows smoothly through the pits, from one hollow tracheid to the next.

In a **simple pit pair,** a region of secondary wall is absent (Figure 30-5A). While the tracheid is alive, plasmodesmata connect adjacent cells through the pit. As the tracheids mature, they lose their nucleus and cytoplasm. The primary walls remain, however, and form the walls of short channels between tracheids and parenchyma cells.

Because simple pits lack secondary cell walls, their presence in the xylem would reduce its mechanical strength. Plants therefore connect tracheids with another kind of pit pair, the

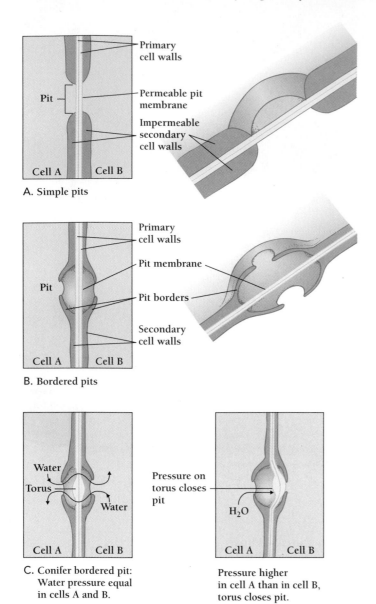

A. Simple pits

B. Bordered pits

C. Conifer bordered pit:
Water pressure equal
in cells A and B.

Pressure higher
in cell A than in cell B,
torus closes pit.

Figure 30-5   Pit pairs in living plant cells are simply openings in the cell wall. A. Two pits in adjacent cells form a pair of simple pits, also called a simple pit pair. B. In cells with a secondary cell wall, a bordered pit forms, which is stronger than a simple pit pair. C. In conifers, a bordered pit functions as a valve. When more water is flowing on one side of the valve than on the other, a buttonlike structure called the torus shuts the pit, preventing water loss.

bordered pit pair. In these, adjacent secondary wall overarches the pit membrane and reinforces the wall of the tracheid (Figure 30-5B). In conifers, bordered pits have an additional feature that allows each pit to function as a valve. A thickened region of the cell wall, called the *torus*, forms a buttonlike thickening in the middle of the pit (Figure 30-5C). If the hydrostatic pressure from one cell is greater than that on the other side of the pit, the torus presses against the pit opening and prevents water flow. During the growth of a shoot or root, these valves reduce water movement away from the growing regions.

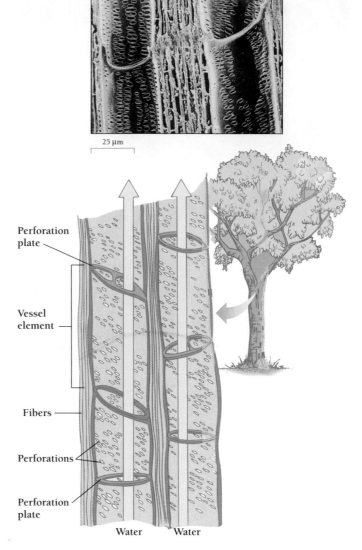

Figure 30-6   Vessel elements. Water passes directly through the perforation plate by way of holes that lack both primary and secondary cell walls. Vessel elements conduct water more efficiently than tracheids, but do not prevent the passage of air bubbles.   *(photo, G. Shih-R. Kessel/Visuals Unlimited)*

In addition, the valves guide water away from tracheids that have been blocked by bubbles of air.

Tracheids are the only conducting cells in conifers and other nonflowering plants. But a larger, more efficient kind of conduction cell, called a vessel element, evolved in flowering plants. **Vessel elements** are shorter, broader, and less tapered than tracheids. Vessel elements, which are open at each end, connect end to end to form long open channels (Figure 30-6). They are found almost exclusively in flowering plants. The end wall of each vessel element, called a **perforation plate**, contains one or more holes, through which water flows more easily than through the much smaller pits. While a tracheid pit lacks only the secondary wall, a vessel perforation lacks both primary and secondary walls.

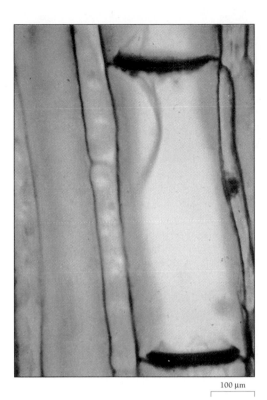

100 μm

Figure 30-7 **The living cells of phloem.** Phloem differs dramatically from xylem in consisting of living cells. Like xylem cells, phloem cells come in two types. Sieve cells are long and narrow with pointed ends, like tracheids. Sieve tube members are short and wide with flat ends, like vessel members. Similarly, only flowering plants have sieve tube members. Modified plasmodesmata between adjacent sieve tube members allow sugar to pass from cell to cell. Nearby companion cells load sugars in and out of the sieve tube members and probably supply the sieve tube members, which lack nuclei, with energy and useful molecules.   *(Randy Moore/Visuals Unlimited)*

The Conducting Elements of Phloem Are Living Cells

Unlike water transport in the xylem, the movement of materials through the phloem depends on the action of living cells. The cells that actually conduct fluid are the **sieve tube members,** so-named because stacks of them form the pipelike channels called **sieve tubes** (Figure 30-7). The complex phloem tissue also includes companion cells, parenchyma cells, sclerenchyma fibers, sclereids, and other cell types depending on species. The cell walls of the sieve tube members do not contain lignin.

The top and bottom end wall of each sieve tube member is called a **sieve plate.** A sieve plate has more and bigger pores than the sides of the sieve tubes, permitting more fluid flow through a tube than in and out of it. Even the pores in the sieve plate, however, resist water flow more than the much larger perforations in the vessel elements of the xylem. In addition, the sieve tubes are much narrower than the conducting tubes of the xylem. The sieve plate contains many sieve pores, open channels that develop from modified plasmodesmata.

Although sieve tube members are living cells, they are highly specialized and lack nuclei, ribosomes, and vacuoles.

Without nuclei and ribosomes, they are incapable of synthesizing the proteins they need to live. Next to each sieve tube member is a **companion cell**—a long, nucleated, fully functional parenchyma cell that supplies the sieve tube member with both proteins and energy-rich molecules. The companion cell is connected to the sieve tube member by plasmodesmata, through which materials flow to the sieve tube member.

Xylem consists of two kinds of nonliving cells: narrow tracheids and wider vessel elements. Phloem is made of living cells called sieve tube members, which accommodate the flow of materials through the phloem.

Monocots and Dicots: The Two Groups of Flowering Plants

When a seed first germinates, it has a root and a shoot. The very first leaves are quite simple in form and are called **cotyledons,** or seed leaves (Figure 30-8A). All flowering plants are classified into one of two categories according to whether their seeds have one cotyledon or two.

Monocotyledons, or **monocots,** have one cotyledon, while **dicotyledons,** or **dicots,** have two. Monocots consist of about 65,000 species, including all of the grasses, lilies, palms, and orchids. Most monocots are herbaceous. Dicots consist of about 170,000 species, including almost all trees and shrubs (but not including conifers and other gymnosperms, which are not flowering plants), as well as many herbs.

Other traits that distinguish monocots from dicots are the number of flower parts; the arrangement of vascular bundles; and the structure of the root system. The flower parts of a dicot usually come in multiples of four or five, while those of a monocot come in multiples of three (Figure 30-8C). The leaves of moncots and dicots have characteristic patterns of veins (Figure 30-8B): in monocots the veins generally run parallel to one another (as the leaves of corn and other grasses), while those of a dicot usually form a complex netlike pattern (as in a tomato leaf or a maple leaf).

Inside a dicot's stem, the vascular bundles lie in the form of a ring, while those of a monocot are scattered throughout the stem (Figure 30-8C). Inside the stele of a dicot's root, xylem and phloem often form a star shape (in cross section) that runs the length of the root. In many monocot roots, however, the stele has a central nonconducting core, called the **pith.** Bundles of xylem and phloem lie in distinct bundles around the outside of the central pith. Monocot and dicot roots are not consistently different, however. Some dicots have a central pith, and some monocot roots have a central xylem.

Monocots and dicots differ also in the structure of their root systems (Figure 30-8E). Dicot roots have lateral meristems and secondary growth, while monocots have no lateral meristems and are incapable of secondary growth. For this reason, monocots generally have a delicate system of many small roots branching from the stem. These tiny ("adventitious") roots

A. Embryos

Monocot
(One cotyledon)

Dicot
(Two cotyledons)

Parallel   Netlike

B. Leaf venation

Stems

Roots

0.5 mm   1.0 mm

0.5 mm   1.0 mm

C. Stems and roots

D. Flowers

Fibrous   Taproot

E. Roots

Figure 30-8 Characteristics of monocots and dicots. A. The monocot embryo (corn) has a single cotyledon, while the dicot (bean) has two cotyledons. B. Monocots have parallel veins in their leaves, while dicots have a more netlike arrangement of veins. C. Cross sections through the stem and root of a corn plant (a monocot) and a buttercup (a dicot). In monocots, vascular bundles in the stem are scattered throughout the ground tissue of the stem, but form a tight circle in the roots. In dicots, the vascular bundles cluster near the outer surface of the stem, but crowd to the very center in the roots. D. The flower parts of monocots, such as this iris, as well as lilies, corn, and other grasses, come in groups of 3. The flower parts of dicots, such as this buttercup, more often come in groups of 4 or 5. E. Monocots have masses of small roots. A dicot may have a single, large taproot.   *(A, left, Barry L. Runk/Grant Heilman Photography; right, Runk/Schoenberger from Grant Heilman; C, monocot stem, © Dwight Kuhn; others, Runk/Schoenberger from Grant Heilman; D, left, John Gerlach/Visuals Unlimited; right, John Gerlach/Tom Stack & Associates)*

often form a thin, fibrous mat, familiar to anyone who has tried to dig in a grassy lawn or field. In contrast, most dicots produce one or several large roots, with smaller roots branching off of larger roots. In many dicots, the entire root system connects to a single vertical root, called a taproot. A carrot, for example, is a taproot that has enlarged to serve as a storage organ.

Flowering plants fall into two classes that have characteristic differences in their embryos, flowers, leaves, roots, and vascular systems. These two classes are monocots and dicots.

HOW DO TERRESTRIAL PLANTS OBTAIN WATER?

Water and dissolved nutrients flow from the soil into the vascular system of the roots. They then move up through the stem to the plant's leaves and other organs. Some of the water contributes hydrogen atoms for the synthesis of sugars and other molecules. Ultimately, however, nearly all of the water that enters a plant through the roots subsequently departs by way of the stomata in the leaves as water vapor, which drifts away into the atmosphere.

Plants need other materials besides water from the soil and carbon dioxide from the air. To make amino acids and nucleotides, for example, plants need large amounts of nitrogen, sulfur, and phosphorus. Plants also need fairly high levels (about 1 milligram per gram of soil) of such elements as magnesium, calcium, and potassium. Finally plants need small amounts (less than 100 micrograms per gram of soil) of micronutrients such as molybdenum, copper, zinc, manganese, boron, iron, and chlorine. Plants obtain these nutrients almost entirely as ions dissolved in the water absorbed by their roots, so roots are the major suppliers of essential nutrients as well as of water.

Roots serve other functions as well: they anchor the plant to the ground, store carbohydrates, and produce several hormones that regulate growth. Finally, roots are important in vegetative reproduction.

How Do Roots Carry Water?

Roots vary greatly in their external form, but they have a common internal organization. A cross section of a primary root shows three concentric rings: the epidermis on the outside, the loose cells of the **cortex** beneath the epidermis, and the stele (the root's vascular system) in the center.

Most of the water that enters a root comes through **root hairs,** tiny projections, each of which is the extension of a single epidermal cell. Numerous root hairs lie just behind the root tips. Root hairs greatly increase the surface area of the epidermis. A single rye plant, for example, may have 10 billion root hairs and have a surface area of more than 400 square meters, as much as the walls of a small house. The epidermis and the root hairs secrete a slimy substance called **mucigel,** which enhances the absorption of water and minerals and lubricates the root tips as they force their way between soil particles.

Most of a root's cortex consists of parenchymal cells, large cells with large vacuoles. In many plants, these cells contain stored starch, most dramatically in food plants such as potatoes and carrots. The innermost layer of the cortex differs from the rest of the cortex and is called the **endodermis** [Greek, *endon* = within + *derma* = skin]. Water flows from the soil to the stele by two paths (Figure 30-9). In both cases, the water flows from the epidermis through the cortex to the endodermis, where the two paths converge.

The endodermis includes a structure that is key to roots' ability both to absorb water and to regulate its flow. Each cell of the endodermis is framed in a waxy, fatty substance called *suberin* that is impermeable to water (Figure 30-9). The boxy endodermal cells deposit suberin on only four of their six surfaces. The other two surfaces have normal cell walls and membranes that regulate the flow of water and minerals into the cells of the en-

dodermis. But the presence of this waxy wall, called the **Casparian strip,** prevents water from going between the cells. For water to move from the cortex to the vascular system, it must pass through the membrane and the cytoplasm of the cells of the endodermis. These cells can consequently regulate the amount of water and nutrients that enter the vascular system.

The endodermis is a selective barrier to the entrance of water and dissolved substances from the soil into the stele. Like other cell membranes, the endodermal membrane blocks the passage of some substances, allows other substances to diffuse passively into the cell, and provides special transport mechanisms for still others. The endodermis thus serves as a gatekeeper, regulating the contents of the fluid that enter the vascular system.

Within the cylinder formed by the endodermis is the stele, which contains the root's vascular system (Figure 30-9). At the center of the stele are the xylem and the phloem. Surrounding them is a sheath of parenchyma cells, called the **pericycle** [Greek, *peri* = around + *kykos* = circle].

> Water from the soil enters the root by way of root hairs, travels through the loose parenchymal cells of the cortex, and then moves through the endodermis into the vascular tissue.

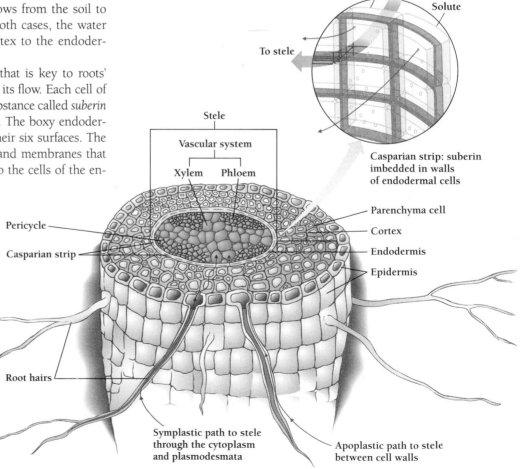

Figure 30-9 The two paths of water through the roots. The first path, the *symplast,* runs through the cytoplasm of the parenchyma cells of the cortex. Water and minerals pass from cell to cell via plasmodesmata. The second path, called the *apoplast,* lies in the material of the cell walls of the cortex. Water and dissolved minerals soak into the cell wall material of epidermis and cortex much as they soak into the cellulose fibers of paper towels. At the surface of the stele is a wall of cells called the Casparian strip. The cells of the Casparian strip are partly encased in waterproof wax. Water and minerals between cells in the cortex (the apoplast) cannot enter the stele unless the cells of the Casparian strip transport them inside.

Stems Carry Water and Minerals from Roots to Leaves and Nutrients from Leaves to Roots

We usually think of stems as long and thin, with various arrangements of attached leaves. But stems may have many forms and functions. Stems support the production of leaves, flowers, seeds, and fruits; elevate these organs toward the sun, pollinators, and seed dispersers; store nutrients and water; and preserve perennial plants through the stress of winter or summer. Some stems are photosynthetic. Our discussion here, however, focuses on the stem's role in transporting water and nutrients.

Like primary roots, primary stems consist of dermal, ground, and vascular tissues. As in the root, the dermal tissue (epidermis) is on the outside with the vascular system embedded within the ground tissue.

The stem's epidermis differs from that of a root in two respects—the absence of root hairs and the presence of a thick cuticle. The cuticle is a waterproof layer made of polyester and wax that encases the stem and prevents water loss.

Inside the stem's epidermis is a ring of cortex, surrounding a core of vascular tissue and pith. The stem's cortex is often thin, sometimes no more than a few millimeters. In some cacti, however, the cortex can be 20 to 30 cm thick. The outer parenchyma cells of the cortex, particularly of young stems, are sometimes green and photosynthetic. The stems of most plants do not have an endodermis.

In dicots, the vascular system of stems consists of vascular bundles of xylem and phloem surrounding the pith, which almost always consists entirely of parenchyma cells. By examining vascular bundles in successive sections of stem, researchers have followed their course as they run through the stem. In the simplest arrangement, for example, in the saguaro cactus, each bundle runs straight up the stem, branching to connect to the leaves. A branch that goes to a leaf is called a leaf trace. In other dicots, the vascular bundles branch frequently and may rejoin to form a network of interconnected channels. Each bundle thus provides a path between many roots and many leaves, and the plant may easily compensate for local damage.

The arrangement of vascular bundles differs between monocots and dicots. In a cross section of a monocot stem, the vascular bundles appear to be scattered. By following their courses, however, we see a defined pattern for each bundle. As a bundle ascends a stem, it gradually moves toward the center. It periodically branches outward to connect to other bundles or to leaves.

Leaves Are the Major Organs of Photosynthesis

Besides their major role in photosynthesis, leaves may have a variety of other functions—protecting buds and flowers and storing food for the embryo and young plant (Figure 30-10).

Leaves also help move water from the roots through the xylem, as we will see in Chapter 31.

Most leaves have a thin, flat part, called the **blade,** and a stalk, called a **petiole,** which connects the leaf to the stem (Figure 30-11). The region of the stem to which a petiole attaches is a **node,** and the region between nodes is an **internode.** Plants are modular in their organization, so that a leaf-node-internode unit can be repeated over and over to create a very large plant. The arrangement of leaves on their stems, as well as the sizes and shapes of leaves, influences their ability to capture light. Many plants position their leaves by movements of the petioles, so that leaf blades move out of the shade of other leaves.

The arrangement of leaves on a stem is characteristic of each species, with three principal patterns. In the most common arrangement, the **alternate** or spiral, the bases of the petioles form a spiral up the stem. In other plants, two or more petioles extend away from the stem from the same node. When there are only two leaves in each group, the pattern is called **opposite;** when there are three or more at each node, the pattern is called **whorled** (Figure 30-11).

Like stems and roots, leaves consist of dermal, ground, and vascular tissue systems. The leaf's epidermis is usually a single layer of cells covered with a layer of cutin and waxes. Many

Figure 30-10   **A diversity of leaves.** Leaves may be cotyledons, sepals, bud scales, or floral bracts. As organs of photosynthesis, they may be large or small, flat or round, waxy or hairy. Shown here are the flowerlike leaves of the pointsettia, the camouflaged leaves of the stoneflower, pine needles, and the bud scales, sepals, and new leaves of a horse chestnut.   *(Lefever/Grushow from Grant Heilman, © Kjell B. Sandved/Photo Researchers; © David Sieren/Visuals Unlimited; W. Ormerod/Visuals Unlimited)*

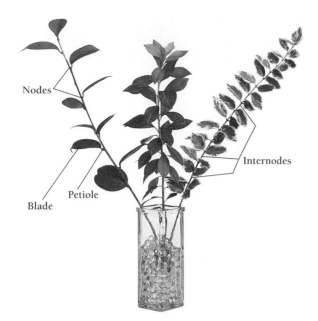

Nodes

Internodes

Petiole

Blade

Figure 30-11 **Leaf arrangements.** Taxonomists have invented dozens of ways to describe the arrangements, shapes, and textures of leaves. Here are just three arrangements (from left to right): alternate, whorled, and opposite. *(Paraskevas Photography)*

leaves have epidermal extensions, called leaf hairs, or trichomes. Leaf hairs have a variety of forms and functions, including shading, retardation of water loss, and defense against herbivores.

Both the upper and lower surfaces of a leaf contain specialized pores, called **stomata** [singular, stoma; Greek, = mouth], through which carbon dioxide, oxygen, and water vapor enter and exit the interior of the leaf (Figure 30-12). While the lower surface of a leaf may contain tens of thousands of stomata per square centimeter, the upper surface usually has few or none.

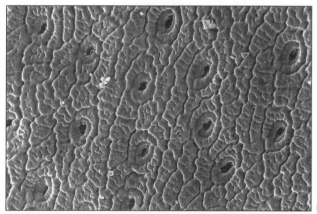

50 μm

A.

Figure 30-12 **The path of water in the leaf.** Leaves consist of two layers of epidermis and a central layer of mesophyll, through which run the vascular bundles, or veins. A. Stomata in the surface of the leaf open and close to regulate the flow of water vapor and carbon dioxide in and out of the leaf. B. Inside the leaf, water passes from the xylem into the parenchyma and then to the stomata. Carbon dioxide in the atmosphere enters the plant through the stomata and is taken up by photosynthetic parenchyma cells. *(A, U. Eggli, Municipal Succulent Collection, Zurich)*

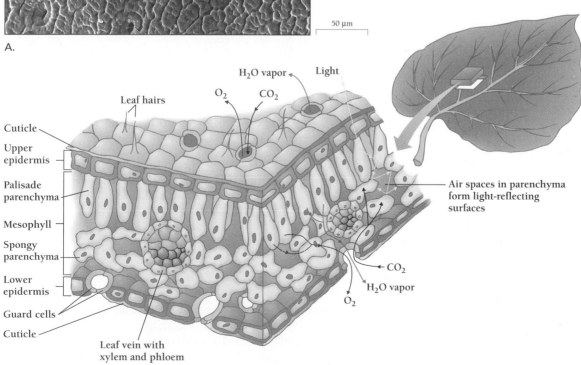

H_2O vapor   Light

O_2 CO_2

Leaf hairs

Cuticle

Upper epidermis

Palisade parenchyma

Mesophyll

Spongy parenchyma

Lower epidermis

Guard cells

Cuticle

Leaf vein with xylem and phloem

Air spaces in parenchyma form light-reflecting surfaces

CO_2

H_2O vapor

O_2

B.

BOX 30-1

The distinctive anatomy of C₄ plants

In Chapter 7, we described the C₄ pathway for the assimilation of carbon dioxide. Recall that in C₄ plants, the first reaction of carbon dioxide produces a 4-carbon compound, oxaloacetate, instead of the 3-carbon molecule phosphoglycerate. Oxaloacetate is converted to another compound that moves into the cells that make carbohydrates. There it breaks down to release carbon dioxide (Figure A).

In addition to a distinctive chemistry, C₄ plants also have a distinctive leaf organization. Surrounding each vascular bundle in the leaf is a sheath of cells, one or two layers thick, with especially large numbers of chloroplasts (Figure B). These bundle sheath cells are the principle sites of photosynthesis and glucose formation. Outside the bundle sheaths are mesophyll cells, which are primarily responsible for the uptake of carbon dioxide and its conversion into oxaloacetate. The mesophyll cells deliver the absorbed carbon dioxide, in the form of an organic acid, to the bundle sheath cells, which regenerate and use the carbon dioxide. Only the bundle sheath cells contain ribulose bisphosphate carboxylase (Rubisco), the first enzyme in the pathway of sugar synthesis. The distinctive anatomy that almost always accompanies the C₄ pathway is called kranz anatomy [German, *Kranz* = wreath].

The C₄ pathway and kranz anatomy concentrate carbon dioxide in the bundle sheath cells. These cells thus have a much higher concentration of carbon dioxide than the corresponding cells of C₃ plants.

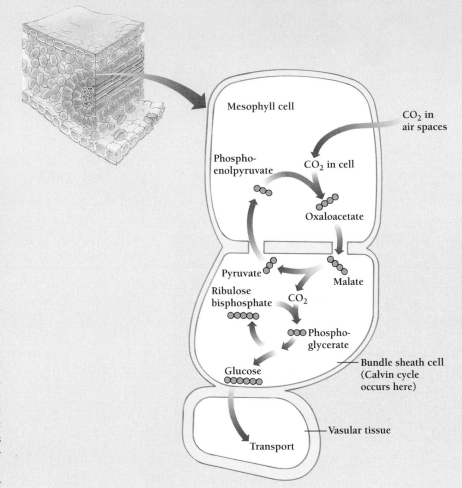

Figure A   The C₄ pathway is an adaptation to a dry environment. C₄ plants can keep their stomata closed, to minimize water loss, yet still obtain carbon for photosynthesis from malate stored in the bundle sheath cells.

Surrounding each stoma are two guard cells, specialized epidermal cells that serve as a valve, opening and closing the stoma according to environmental conditions inside and outside the leaf. Stomata open, for example, when the plant lacks carbon dioxide and yet can afford to lose water vapor through its leaves. Stomata close when the plant cannot afford to lose water. Guard cells have chloroplasts and respond to light. Other epidermal cells, however, generally do not contain chloroplasts.

The ground tissue of the leaf consists of **mesophyll** [Greek, *mesos* = middle + *phyllon* = leaf], green parenchyma cells that

are responsible for most of a plant's photosynthesis. Most leaves have two kinds of mesophyll—**palisade parenchyma** and **spongy parenchyma** (Figure 30-12). The cells of the palisade parenchyma, which lie just under the leaf's upper epidermis, are elongated and packed with chloroplasts. Below them are the cells of the spongy parenchyma, which are irregular but rounded in shape and separated by numerous air spaces. These also contain numerous chloroplasts.

Like the air-filled pockets in our lungs, the air spaces in the spongy parenchyma provide a route by which oxygen and carbon dioxide can diffuse to and from the outside air, via the

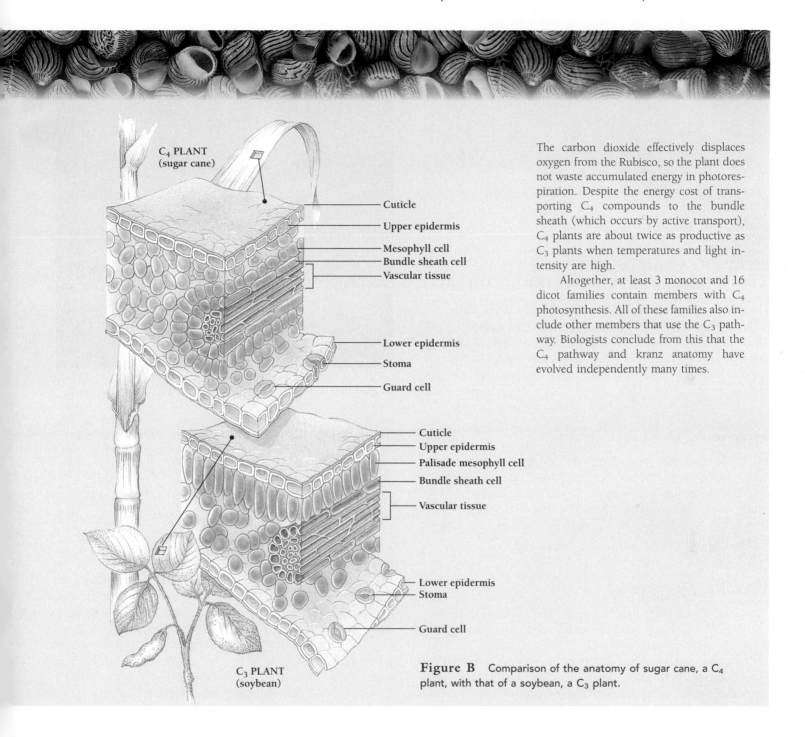

C$_4$ PLANT
(sugar cane)

— Cuticle
— Upper epidermis
— Mesophyll cell
— Bundle sheath cell
— Vascular tissue

— Lower epidermis
— Stoma
— Guard cell

— Cuticle
— Upper epidermis
— Palisade mesophyll cell
— Bundle sheath cell
— Vascular tissue

— Lower epidermis
— Stoma
— Guard cell

C$_3$ PLANT
(soybean)

Figure B Comparison of the anatomy of sugar cane, a C$_4$ plant, with that of a soybean, a C$_3$ plant.

The carbon dioxide effectively displaces oxygen from the Rubisco, so the plant does not waste accumulated energy in photorespiration. Despite the energy cost of transporting C$_4$ compounds to the bundle sheath (which occurs by active transport), C$_4$ plants are about twice as productive as C$_3$ plants when temperatures and light intensity are high.

Altogether, at least 3 monocot and 16 dicot families contain members with C$_4$ photosynthesis. All of these families also include other members that use the C$_3$ pathway. Biologists conclude from this that the C$_4$ pathway and kranz anatomy have evolved independently many times.

stomata. Water vapor follows the same route out of the plant (Figure 30-12). The internal air spaces also create many bright, reflecting surfaces within the leaf. Sunlight bounces back and forth within this tiny hall of mirrors, providing the photosynthetic pigments with many opportunities to absorb a photon.

Because cells near the top of a leaf shade those near the bottom, the lower cells have less light for photosynthesis. To minimize this effect, most leaves are thin, so that light can penetrate to the bottom-most parenchyma cells. In general, the tops of leaves are more specialized for light absorption and the bottoms for the exchange of gases and water.

One of the most obvious structures in a leaf is its vascular system. With the unaided eye, we can see the leaf's veins, which are its larger vascular bundles. The vascular bundles of a vein are continuous with those of the petiole and with the leaf traces of the stem. Inside the leaf, the veins branch into finer and finer vascular bundles. These may join to one another to form a continuous circular path, or they may end in special cells called terminal tracheids.

In monocots, leaf veins usually run parallel to one another. In dicots, the leaf usually has only one or a few large veins. But these branch into fine networks, so that every part of the leaf is no more than a few cells away from the vascular system. If the vascular bundles of a single dicot leaf were placed end to end, they would extend over two city blocks. The pattern of veins varies greatly among species.

Leaves, the primary photosynthetic organs of plants, consist of an outer layer of epidermis and a central region of mesophyll through which run the vascular bundles, or veins. The epidermis is dotted with stomata.

HOW DO PLANTS DEFEND THEMSELVES AGAINST HERBIVORES?

Plants are good to eat and they can't get away. Since the evolution of the first herbivores, plants have been fighting to stay alive. Bacteria eat plants, protists eat plants, fungi eat plants, and animals eat plants. Even other plants parasitize plants. How do plants defend themselves against this onslaught of hungry organisms?

Mechanical Defenses

Plants have numerous ways of preventing other organisms from taking a bite. One defense is the plant's waxy cuticle, which excludes both bacteria and fungi fairly well. Hairy leaves or very sticky leaves discourage caterpillars and other tiny herbivores (Figure 30-13). Spines, thorns, and prickles fend off larger mammalian herbivores.

Chemical Defenses

Plants' most effective defenses are probably not physical but chemical. Who has not regretted an afternoon jaunt through a patch of poison ivy or poison oak? Over hundreds of millions of years plants have evolved a pharmacopoeia of chemical defenses against herbivores. Most of these compounds are **secondary plant compounds. Primary compounds** include glucose, amino acids, ATP, and DNA, compounds that are essential for plants' day-to-day functioning. In contrast, secondary plant compounds are diverse chemicals that function in very specific ways. Many are particular to small groups of plants or even to a single species.

Just as plants have evolved thousands of ways of defending themselves against herbivores and pathogens, herbivores and pathogens have evolved as many ways of overcoming these defenses. For every shield there is a stronger sword, and for every bullet-proof vest, a more piercing bullet. The coevolution of plants and their attackers has led to a balanced ecosystem in which plants have so far managed to stay ahead. Cut-

Figure 30-13   **Leaf hairs.** Leaf hairs protect delicate plant epidermis from excessive sun and hungry insects, as well as from drying out and other dangers. Some leaf hairs secrete tarlike substances that make the leaf too sticky for caterpillars to walk on and too bitter for larger herbivores to eat.   *(© 1981 by Darwin Dale/Photo Reseachers)*

worm larvae raised on an artificial diet, for example, grow six times as large as those raised on any natural diet of leaves to which they are adapted. Without plant defenses, the world would be inundated with insects and other herbivores.

Plants manufacture thousands of these compounds, but most secondary plant compounds fall into three groups: terpenes, phenolic compounds, and alkaloids. **Terpenes** are lipids and the largest class of secondary compounds. Many are actually primary compounds that play important roles in plant metabolism. But vast numbers of terpenes seem to function solely to drive away or poison herbivores. Pyrethroids from *Chrysanthemum* leaves, for example, are insecticides. Peppermint, lemon, basil, and sage all derive their aromas from mixtures of terpenes called **essential oils,** which repel insects. Many plants store these repellent and toxic oils in glandular hairs on the surface of the epidermis, where the oils advertise the plant's toxicity to arriving insects.

Some terpenes, such as those found in sunflowers and sagebrush, repel mammals as well as insects and taste bitter to us. Certain varieties of alfalfa contain saponins, which kill cattle and other grazing animals by rupturing the membranes of red blood cells. Cotton plants derive much of their insect, bacterial, and fungal resistance from the terpene **gossypol,** which also happens to work as a contraceptive in human males.

Phenolic compounds are aromatic substances that play a variety of roles in plants. Aromatic chemicals are those whose molecular structure includes a flat carbon ring (Chapter 3). Some phenolics repel herbivores and pathogens, and some attract pollinators or fruit dispersers.

Others, such as **lignin,** both repel herbivores and provide mechanical support. Lignin is a major constituent of the walls of cells specialized to provide mechanical support or transport water. But lignin also makes plants relatively indigestible to both animals and microscopic pathogens. Surprisingly, lignin also has been isolated from the cell walls of algae, where it is believed to protect the algae from attack by bacteria and other microbes. Biologists now believe that lignin is an ancient compound, whose original function, to repel pathogens, has expanded to include mechanical support for terrestrial plants.

Some plants secrete phenolic compounds that poison any plants growing nearby. Desert plants, especially, resort to this unneighborly practice to eliminate competition for precious water. Other phenolics secreted into the soil regulate gene expression in nitrogen-fixing bacteria. Still others rapidly accumulate in the plant in response to bacterial or fungal infections. These phenolics form in a matter of hours around the infection site and kill a broad spectrum of fungal and bacterial plant pathogens. **Tannins,** commonly found in woody plants, are, like lignin, phenolic polymers that reduce growth and survivorship in many herbivores. Cattle, deer, humans, and apes avoid eating unripe fruits and other plant parts with high levels of tannins because of the tannins' sharp, astringent taste. This effect is caused by tannins' binding to enzymes in the saliva. Yet, humans like a bit of tannin in their tea, apples, blackberries, and red wine. Finally, the pigments responsible for most of the red, pink, purple, and blue colors of flowers, fruits, leaves, and other plant parts, called **anthocyanins,** are also phenolic compounds.

Alkaloids are nitrogen-containing secondary plant compounds found in 20 to 30 percent of all vascular plants. They include many of the most powerful drugs known to humans, including nicotine (the highly addictive active ingredient in tobacco), atropine, cocaine, morphine (from which heroin is made), the hallucinogen psilocybin, and the toxin strychnine.

The alkaloids in lupines, larkspur, poison hemlock, and other plants are highly toxic to herbivores, killing livestock and humans ignorant enough to eat them. (The Greek philosopher Socrates carried out his own execution with a tea made from poison hemlock.) These compounds can also be highly teratogenic. Milk from cattle and goats that have grazed on lupine can induce high rates of birth defects in pregnant humans or other animals who drink the contaminated milk. At lower doses, alkaloids such as caffeine, nicotine, cocaine, and morphine are used widely as stimulants and sedatives, for both medical and nonmedical purposes.

Other secondary plant compounds besides the terpenes, phenolics, and alkaloids include the **mustard oil glycosides.** These compounds give cabbage, broccoli, radishes, and other plants in the mustard family their characteristic pungent odor and flavor. Many insects and mammals are repelled by mustard oil glycosides. But others have learned to cope with these compounds, and some insect herbivores specialize on plants in the mustard family (Figure 30-14). For these insects, the mustard oil glycosides act as an attractant. Herbivores that specialize on noxious plants can often avoid competition with other herbivores. Many compounds that originally evolved as repellents to herbivores now act as attractants for certain groups of animals.

Secondary plant compounds are chemicals not needed for the major metabolic pathways of a plant. Many secondary plant compounds repel or poison insects and other herbivores.

In this chapter we have seen that the standard root-and-shoot form of a terrestrial plant allows the plant to collect water and minerals from the soil and to collect solar energy and carbon dioxide from the air. We have briefly discussed how leaves pull water from the roots and move sugar to the roots and how plants use secondary plant compounds to strengthen their bodies against the pull of gravity and to protect themselves from herbivores and pathogens. In the next chapter, we discuss in more detail how plants raise water from the roots to the tops of trees that may be hundreds of feet tall.

Figure 30-14   Attractive insect repellent. Most insects avoid plants in the mustard family because the leaves contain unpleasant compounds. The cabbage butterfly, however, prefers cabbages and other mustards. *(Holt Studios International/Photo Researchers)*

STUDY OUTLINE WITH KEY TERMS

Despite the enormous diversity of flowering plants, they share characteristic organizational features. Most consist of an underground **root system** and an aboveground **shoot system.** The shoot usually contains a stem, leaves, and flowers.

Materials travel between shoot and root through the plant's **vascular system,** a network of two kinds of transport tissues. The **xylem** transports water and minerals from root to shoot, and the **phloem** transports the energy-rich products of photosynthesis from leaves or storage areas to the rest of the plant.

Flowering plants fall into two classes—the **monocotyledons,** or **monocots,** and the **dicotyledons,** or **dicots**—that have characteristic differences in their embryos, flowers, leaves, roots, and vascular systems. The embryos of monocots have one **cotyledon** and the embryos of dicots have two. In both classes of plants, roots, stems, and leaves consist of three types of tissue systems—dermal, vascular, and ground. The **dermal tissue system** (or **epidermis** and **periderm**) is the protective outer covering, which includes the **cuticle.** The **vascular tissue system** lies inside, surrounded by **ground tissues.** Roots, stems, and leaves differ in the arrangement of vascular system and ground tissue. Tissues may consist of **parenchyma, collenchyma, sclerenchyma,** and other cell types.

The **primary growth,** or lengthening, of a plant occurs only in specialized regions of dividing cells, called **apical meristems,** at the tips of roots and shoots. **Secondary growth** is an increase in diameter of shoot or root that occurs in a cylinder of dividing cells called a **lateral meristem.**

Plant cells have **cell walls** made of **cellulose,** other carbohydrates, and proteins. All plant cells have a **primary cell wall** outside the plasma membrane. Some have a thicker **secondary cell wall** between the primary wall and the membrane. The polymer **lignin** strengthens many plant cell walls. Channels called **plasmodesmata** often extend through the cell walls and connect the cytoplasms of adjacent cells.

Stems carry water and minerals from roots to leaves and nutrients from leaves to roots. Like roots, stems consist of dermal, ground, and vascular tissue systems. In dicots, the ground tissue of a stem consists of two regions, the **cortex** and the **pith. Xylem** and **phloem** lie in **vascular bundles,** which lie between the pith and the cortex. The xylem lies toward the center and the phloem toward the outside of each bundle. The arrangement of vascular bundles differs between monocots and dicots.

Leaves are the major organs of photosynthesis. Each leaf consists of a (usually) flat **blade** and a stalk, called a **petiole,** that attaches the leaf to the stem. The petiole attaches to a **node** on the stem. In between each pair of nodes is the **internode.** The arrangement of leaves on a stem is characteristic of each species, with three principal patterns: **alternate, opposite,** and **whorled.** The surfaces of a leaf usually contain specialized pores, called **stomata,** through which carbon dioxide, oxygen, and water vapor enter and leave. The interior of a leaf is composed of **mesophyll,** which is made of **palisade parenchyma** and **spongy parenchyma.**

Xylem consists of two types of conducting cells—long, tapered cells, called **tracheids,** and shorter, flat-ended cells, called **vessel elements.** The xylem functions only after the cells die, leaving behind empty cell walls. Water (as well as dissolved substances) moves through the tracheids from dead cell to dead cell by way of pairs of **pits,** either **simple pit pairs** or **bordered pit pairs.** Water moves through vessel elements through **perforation plates** at either end.

The **sieve tubes** of phloem are made of living **sieve tube members** and their **companion cells.** Each end of a sieve tube member is a perforated **sieve plate** that allows materials to pass from cell to cell.

Water moves from soil to air by way of the xylem. Absorption and transport begin in the roots. Roots consist of three concentric cylinders of tissue. At the center is the **stele** (vascular tissue), which is surrounded by **cortex** (ground tissue), which is surrounded by epidermis. Water flows from the soil to the vascular system by way of the **mugicel**-covered **root hairs** and cortex. The innermost layer of the cortex is the **endodermis.** Within the endodermis is a layer of waxy, waterproof cell walls called the **Casparian strip,** which prevents water from flowing *between* the endodermal cells into the **pericycle** and stele. Water must flow instead *through* the living cells of the endodermis. The membranes regulate the movement of both ions and water into the stele.

Secondary plant compounds are chemicals not needed for the major metabolic pathways of a plant. Many secondary plant compounds repel or poison insects and other herbivores. **Primary compounds,** which include glucose, amino acids, ATP, and DNA, are compounds that are essential for plants' day-to-day functioning. Plants manufacture thousands of secondary plant compounds, but most fall into three groups: **terpenes, phenolic compounds,** and **alkaloids.** The terpenes, the largest class, include the pyrethroid insecticides, and the **essential oils** in peppermint, lemon, basil, and other strong-smelling plants. The terpene **gossypol** gives cotton plants much of their insect, bacterial, and fungal resistance.

Phenolic compounds include **lignin, tannins,** and **anthocyanins. Alkaloids,** such as nicotine and cocaine, are nitrogen-containing secondary plant compounds found in 20 to 30 percent of all vascular plants. Other secondary plant compounds besides terpenes, phenolics, and alkaloids include the **mustard oil glycosides.**

REVIEW AND THOUGHT QUESTIONS

Review Questions

1. What adaptations do terrestrial plants have that aquatic plants do not?
2. How is the root necessary for photosynthesis? How is the shoot necessary for photosynthesis?
3. How do monocots differ from dicots?
4. Is bread made from plants that are monocots or dicots? Name six other plants you know that are monocots and six that are dicots.
5. Name three general roles that secondary plant compounds can play in the life of a plant.

6. What is the difference between primary growth and secondary growth?
7. What does the word "meristem" mean literally and how does that meaning relate to what it is and does?
8. Draw a diagram of the endodermis, including the Casparian strip, and explain how these structures regulate the flow of water and minerals into the root xylem.

Thought Questions

9. Certain "prostrate" plants spread out over the ground without growing very tall. These plants can maximize their exposure to sunlight. Why don't all plants conserve energy and adopt this form? In other words, what are the advantages of growing tall?
10. When Bruce Ames first published his papers on the carcinogenicity of secondary plant compounds, many people felt that he was behaving irresponsibly. They worried that the public would misunderstand and (1) not bother to eat fruits and vegetables, and (2) not bother to avoid toxins that are genuinely dangerous. Some of his critics thought that Ames was endangering the lives of innocent lay people. Do you think Ames's behavior was irresponsible? Are scientists more obligated to publish what they know is true or more obligated to limit what they publish, according to how the public will likely respond to such information?

SELECTED READINGS

Berg, Linda R., *Introductory Botany: Plants, People, and the Environment,* Saunders College Publishing, Philadelphia, 1997. This book covers the basics of plant science in a friendly and unintimidating style.

Mauseth, James D., *Botany: An Introduction to Plant Biology,* 2nd ed., Saunders College Publishing, Philadelphia, 1995. An authoritative introduction to the biology of plants.

Moore, Randy, W. Dennis Clark, Kingsley R. Stern, and Darrell Vodopich, *Botany,* Wm. C. Brown Publishers, Dubuque, 1995. A friendly, well-written textbook on the science of plants.

Niklas, Karl J., *Plant Biomechanics: An Engineering Approach to Plant Form and Function,* Chicago, University of Chicago Press, 1992. This book is wonderfully written and highly accessible where Niklas is first introducing a topic or concluding one—that is, at the beginning and ending of each chapter. Much of this book, however, demands a familiarity with either botany or engineering beyond that of the average reader.

▶ On-line materials relating to this chapter are on the World Wide Web at http://www.saunderscollege.com/lifesci/ Click on Tobin/Dusheck: *Asking About Life.*

The Devastating Dry Leaf Creature

In the late 1980s, disaster threatened the burgeoning California wine industry. University of California scientists had for years advised grape growers to plant grapes known to be vulnerable to a devastating insect pest. By the early 1980s, nearly three hundred thousand acres of fertile farmland had been planted with a single kind of grape—all of it susceptible to *Phylloxera vitifoliae* [Greek, *phyllos* = leaf + *xeros* = dry], a tiny sucking insect that feeds on roots and leaves.

It was only a matter of time before the tiny relatives of aphids found the vines and destroyed them. Indeed, by the late 1980s, *Phylloxera* was destroying whole vineyards. The pattern was always the same. The leaves of a few plants would turn red and fall from the vines in midsummer. Within months, the branches and roots were dead, and plants nearby had begun to lose their leaves also. Yet scientists continued to recommend the susceptible grape, and grape growers continued to plant it.

It wasn't as if California grape growers hadn't heard of *Phylloxera*. In the 19th century, the infamous pest had devastated the Eu-ropean wine industry. In France alone, the insect utterly destroyed some two and a half million acres of vineyards. It had been a national disaster. Nor was the invasion confined to Europe. The pest infested vineyards all over the world. Indeed, the arrival of *Phylloxera* in California in 1880 from their native eastern United States prompted the State Legislature to mandate that the University of California create a department of viticulture ["grape-growing"], specifically to study *Phylloxera*.

The solution to the great wine blight, as it came to be called, came from a combination of American science and French technology. The American entomologist C.V. Riley worked out the biology of *Phylloxera*, demonstrating not only its complex life cycle, but showing that it was native to North America and lived in harmony with native North American grape species. The French hailed Riley as a savior, for he had provided the key to saving the international wine industry (Figure 31-1). The wild American grapes were useless for making wine, but they flourished despite the *Phylloxera* growing on their roots. By grafting high-quality French wine grapes onto wild American roots, or "rootstock," French viticulturists solved the *Phylloxera* problem.

How does *Phylloxera* kill grape vines? And how do some species resist the at-

Figure 31-1 Memorial at the School of Agriculture, Montpellier, France (1911). A sculptor has represented the rescuing of the French wine industry with American rootstock as a young American rescuing the ailing, old French vine. *(Don Ellis)*

tack? The insects cluster on the roots (and the leaves, as well, in the eastern United States). Each insect pierces the outer layers of the roots and injects saliva containing a cocktail of a plant hormone and enzymes into the tissues. The plant growth hormone auxin induces the roots to form a gall, a tumorlike ball of tissue, around the insect. Here the insect feeds, digesting the tissues of the roots.

The insects themselves do not appear to kill the vine, however, as they do not take enough energy and materials from the vine to hurt it much. Instead, the millions of insects cutting into the roots apparently make the plant susceptible to infections by fungi and bacteria. These secondary infections invade and destroy the vascular system in the roots. Unable to transport water and nutrients, the vine dies.

Viticulturists know comparatively little about how *Phylloxera* kill plants be-cause they have focused their attention on perfecting resistant rootstocks. In general, resistant rootstocks secrete tannins and other secondary plant substances that wall off the root galls and so prevent infections. In resistant rootstocks, the insects continue to feed without harming the plant. Susceptible roots cannot wall off the vascular system from infection.

Nineteenth-century grape growers developed and tested thousands of different rootstocks. Some did well in the cool, wet climate of northern Europe. Some grew better in warm, dry soil. Some produced heavy loads of fruit, but were susceptible to various diseases. Some could tolerate *Phylloxera* infestations, some could not. One rootstock, named AXR#1, produced copious amounts of fruit. But it was not highly resistant to *Phylloxera*.

At first, the insects mostly left AXR#1 alone; they seemed not to care for it. But in Sicily, and then in France and Africa, grapes grown on AXR flourished for a decade or two, before *Phylloxera* attacked them, and they withered and died. French viticulturists concluded that the tiny *Phylloxera* insects could adapt to and grow on any grape rootstock, given time, even AXR. The difference among rootstocks was that some could tolerate *Phylloxera* infestations and still live long, productive lives, while other rootstocks died under the attack. AXR was the kind that died. French viticulturists found that although *Phylloxera* did not at first attack AXR, the insects could adapt to AXR within 20 to 40 years. By the 1920s the French advised against the use of AXR.

But in the 1920s, Americans were in the midst of Prohibition, the 1919 law that prohibited the manufacture and sale of all alcoholic beverages. California grape growers dug up their sprawling vineyards, planted apricots, peaches, and plums, and subdivided the fertile land for housing developments. Without the wine grapes, *Phylloxera* gradually disappeared from the state. Not until after Prohibition ended, in 1933, did Californians begin replanting their vineyards. The new generation of growers knew little about *Phylloxera* except that vines planted on the right rootstock would be fine.

In the 1950s, Lloyd Lider, a researcher at the University of California, Davis, tested 18 rootstocks and found little difference in the quality of the grapes but a huge difference in the amount of fruit. He concluded that AXR produced the most fruit, and that it seemed to be "the nearest approach to an all-purpose stock for the coastal counties of California." Although AXR's resistance to *Phyl-*

The French hailed Riley as a savior, for he had provided the key to saving the international wine industry.

loxera was "not high," he wrote, *Phylloxera* would probably never be a serious problem in northern California's fertile valleys.

In the 1970s, American wine connoisseurs recognized for the first time that California wines were surprisingly good and far cheaper than French wines of the same quality. The California wine industry bloomed. Soon everyone, from retired dentists to multinational corporations, was planting vineyards, almost all on AXR rootstock. AXR now constituted a vast monoculture. If the *Phylloxera* could adapt to that one rootstock, the insects would eventually feast on hundreds of thousands of acres of grapes.

In the summer of 1982, Austin Goheen, a plant pathologist (an expert on plant diseases) from the University of California, Davis, recognized the telltale yellow colonies of the tiny *Phylloxera* insects on the roots of some dying grape vines. John Baritelle, the vineyard's owner, was horrified. Goheen's diagnosis might as well have been smallpox. Baritelle knew little about this historical disease except that it was disastrous.

Lider quickly reassured Baritelle and other worried vineyard owners, however, insisting that the infected rootstock was not AXR. Baritelle had undoubtedly been sold the wrong rootstock in some cases, Lider said, and only the non-AXR rootstock in the vineyard was infected with *Phylloxera*. Everything planted on AXR would be fine.

Jeffrey Granett, an entomologist, also from the UC Davis campus, collected some of Baritelle's *Phylloxera* and took it back to the laboratory. One of his assistants noticed that she could grow *Phylloxera* just as well on AXR as on any other kind of susceptible rootstock. The assistant's discovery should have been a wakeup call. The *Phylloxera* had either adapted to AXR (through natural selection) or a new strain of AXR-adapted *Phylloxera* had entered the state. Either way, the forecast was grim.

But Lider argued that, no matter how thoroughly *Phylloxera* might destroy AXR in the lab, the *Phylloxera* growing in Baritelle's vineyard was not on AXR. *Phylloxera* would not grow on AXR, he said.

In May 1984, Granett decided to grow the insects on known AXR rootstock in John Baritelle's vineyard. Within one year, the *Phylloxera*-infested AXR grapes were visibly sick. In two years, they were dead. By 1986, there was no longer any question that *Phylloxera* could kill vines growing on AXR.

Not until December 1989, however, did the university issue a press release warning growers to stop using AXR. The university had hesitated long enough to be certain the data were right, but also long enough to ensure that California's grapestock nurseries had time to stock alternatives, and so escape financial ruin. The grapestock nurseries were a major industry, integral to the entire wine-making industry. The state had an interest in protecting both the nurseries and the grape growers, but losing a few small growers would be easier on the state's economy than losing the nurseries. The grape growers and wine makers all depended on the nurseries.

Nonetheless, the growers felt angry and betrayed. Some assumed that scientific research came with an implicit guarantee. How, they asked themselves, could university researchers have made so serious an error, and why had they waited so long to reveal it to the growers?

Nonscientists sometimes expect scientists to be both objective and consistently right. Although science as a whole does remarkably well in living up to these enormous expectations, all individual scientists make mistakes in their work. In the long run, however, all scientific mistakes come to light.

Lider had watched AXR rootstocks work perfectly for nearly 30 years. For him to dismiss the idea that AXR was suddenly facing some calamity seems only natural. Why should everything change all at once? And how does any person gracefully face the fact that a basic tenet of his profession is a mistake?

In 1990, California grape growers began tearing out infected vines and replacing them with different rootstocks. No one could agree which single kind was best. Neighboring vineyards chose different rootstocks. A grower might plant four or five different kinds within a single vineyard, just to be safe. For the growers, the change was, and continues to be, enormously expensive. Many have been forced to sell their vineyards.

Phylloxera continue to infect northern California's vineyards, but newer rootstocks are able to wall off the tissue-piercing insects, and water and minerals from the soil flow smoothly up to the leaves and grapes in the hot sun above. The flow of water is as necessary to a plant's life as the flow of blood is to each of us. The flow of water both sustains photosynthesis and carries minerals and nutrients from the roots to leaves and from leaves to roots. Our own circulation system depends on the steady beat of our heart. Plants, however, do not have hearts. What, then, pumps water in plants? In the first part of this chapter we will see what forces drive water and minerals from the roots upward. Later, we will see how sugars manufactured in the leaves are transported down to the roots and other parts of the plant.

KEY CONCEPTS

1. Water moves from soil to air through the vascular system.
2. Water moves from solutions with higher potential energy to solutions of lower potential energy.
3. The driving force for the ascent of water is transpiration, with upward transport depending on water's cohesive properties.
4. Stomata regulate the rate of transpiration by opening and closing.
5. Water moving through the xylem also carries minerals from the soil.
6. The osmotic flow of water drives the transport of photosynthetic products in the phloem.

WHAT DRIVES WATER UP?

What Is the Route of Water from Soil to Air?

We know that water moves up from roots to leaves through the tracheids and vessels of the xylem: if we place a plant cutting in water that contains a dye, we can then trace the path of the dyed water throughout the plant. If the stems were transparent, we would be able to see the water move, for during the day it ascends rapidly, at about 10 to 100 cm/minute.

As we saw in the last chapter, water enters the roots via the root hairs, moves through the cortex, crosses the endodermis, and enters the xylem. Once in the xylem, water ascends from the roots to the leaves through parallel and usually interconnected xylem vessels in the stem. These vessels branch into leaf traces, and water flows through petioles into the leaves.

The water then enters cells of the mesophyll and evaporates at the cell surfaces. The water vapor diffuses through the air spaces of the mesophyll and exits the leaf by way of the stomata. The passing of water vapor from leaf to air is called **transpiration.**

The rate of transpiration is extraordinary: a leaf can transpire its own weight in water in less than an hour. A tree may transpire 50 gallons of water in an hour, up to 30,000 gallons during a growing season; a 40-acre cornfield uses 15 million gallons before harvest. In a plant's lifetime, it typically loses a hundred grams of water through transpiration for every gram of dry material it accumulates.

These tremendous quantities of water must sometimes move great distances. The tallest trees in the world are the California redwoods, *Sequoia sempervirens.* These trees are so large that entire churches have been built from a single tree. One tree, with a diameter of 20 feet and a height of over 200 feet, contained enough lumber to build 22 houses. The tallest redwood of all is 367 feet, the same height as a 29-story building, and nearly a third the height of the Empire State Building.

To pump water to such a height would require a pump exerting 150 pounds per square inch (psi) of pressure. But plants have no central pump. The cells of the xylem are themselves dead, so we can assume that these empty hulls exert no force. Plant scientists early in this century proposed three possible mechanisms for water transport: (1) water is pushed up from the roots; (2) capillary action in the xylem pulls water in the same way that a paper towel pulls water out of a glass; or (3) evaporation at the leaves creates a suction that pulls water up the xylem. But which? Many generations of plant scientists puzzled over this question before arriving at the currently accepted view.

Root Pressure: Can the Roots Push Water to the Tops of Tall Trees?

Recall from Chapter 4 that water tends to move from areas of purer water to areas with high concentrations of solutes. Thus, sugary raisins and cytoplasm-filled red blood cells both fill with water when placed in pure water. The tendency for water to move across a membrane in response to a difference in concentration of solutes is called **osmosis** [Greek, *osmos* = push, thrust].

When a plant cell is in a solution of water purer than the cytoplasm inside the cell membrane, water flows into the cell and causes the cell to swell. Most of the water accumulates in a central vacuole. As the vacuole swells, it exerts physical pressure against its cell wall. The pressure exerted by osmosis is called **osmotic pressure.**

The cells of the roots usually maintain higher concentrations of solute molecules than the soil; water from the soil therefore flows into the roots. The cells of the endodermis then actively pump ions into the xylem, a process that requires energy. The Casparian strip channels water *through* the cells of the endodermis, preventing water from flowing *between* the cells (Figure 30-9). The endodermal cell membranes regulate the concentration of ions in the xylem.

As the endodermis pumps ions into the xylem, the fluid inside the xylem accumulates a high concentration of ions and other solutes, and water from the cells of the cortex then flow in after them. The osmotic pressure in the xylem is called **root pressure.** As long as the roots continue to pump solutes into the xylem, water will continue to flow into the xylem. This water pushes water already in the xylem up the roots and into the stem. In small plants, root pressure can push water all the way up the stem and out of tiny holes at the margins of the leaves, in a process called **guttation** [Latin, *gutta* = a drop].

Early in the morning gardeners sometimes see the leaves of grass, tomatoes, strawberries, and other plants rimmed with tiny drops of water (Figure 31-2). These drops, which are not dew, are the result of root pressure. Guttation occurs on cool, humid mornings, when the air is too saturated for water to evaporate from the leaves.

Figure 31-2   Guttation demonstrates the action of root pressure. The droplets at the margins of these strawberry leaves are not condensation, or dew, but water forced from tiny holes.   *(Angelina Lax/Photo Researchers)*

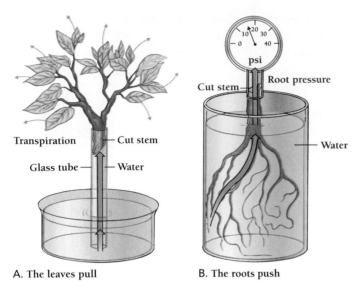

A. The leaves pull

B. The roots push

Figure 31-3   Root pressure. A. We've all seen cut flowers pull water up through their stems. B. Roots can also push water up through a cut stem. The osmotic pressure generated by the roots is a physical pressure that can be measured with a pressure gauge similar to that used to measure tire pressure.

How far can root pressure push a column of water? Researchers can directly measure root pressure by attaching a pressure-measuring device to a cut stem (Figure 31-3). Root pressures measured in this way are usually less than 1 atmosphere, which can support a column of water about 10 meters (30 feet) high—impressive, but not nearly enough to move water to the top of a redwood or eucalyptus tree. In any case, many plants (including redwoods and other conifers) do not develop any root pressure at all. Another clue that showed plant physiologists that root pressure cannot drive water to the tops of tall trees is that the water in the xylem is not usually under pressure. If we poke the xylem of a stem with a needle, water does not come spewing out. Instead, air is sucked up into the xylem, indicating that the water in the xylem is being sucked up by a vacuum.

Finally, the rate of water flow due to root pressure is too slow to account for the more rapid transport of materials from roots to leaves. The driving force of water transport must therefore come from some source other than root pressure.

Roots use ionic pumps to draw water from the soil. The most pressure that roots can generate is enough to raise water some 30 feet, but not the 300 feet or more of the tallest trees.

Can Capillary Action Raise Water to the Tops of Tall Trees?

The movement of water also depends on its adhesive and cohesive properties. Water's **adhesiveness** is its tendency to cling (by hydrogen bonds) to the surfaces of carbohydrates and other polar substances. Water's adhesive properties are what cause it

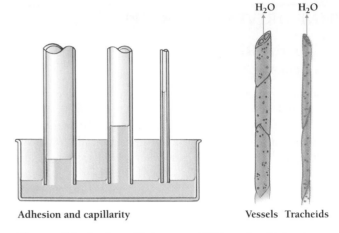

Adhesion and capillarity

Vessels Tracheids

Figure 31-4   Is capillarity enough? Water rises higher in narrower tubes than in wide ones. In xylem vessels and tracheids, water rises by capillarity about 0.5 meters, not nearly enough to get it to the top of a tall tree.

to soak into the cellulose of paper towels or cotton towels. Water also moves up the inside of any narrow tube, or **capillary**, including glass tubes and xylem (Figure 31-4). Because water is also **cohesive**, it sticks to itself very well and columns of water do not break as they move up narrow tubes, or capillaries. Could such capillary action drive water to the tops of tall trees? As it turns out, capillary action can drive water up only about half a meter (less than 20 inches), not nearly enough.

Although root pressure and capillary action are not enough to drive water to the tops of tall trees, both effects are nonetheless important to water movement in plants, as we will now see.

Water's adhesiveness and cohesiveness enable it to move up narrow tubes called capillaries.

The Driving Force for the Ascent of Water Is Transpiration

If neither root pressure nor adhesion of water to the inside of the xylem is enough to drive water up, then what does? The now accepted explanation for the ascent of water is called the **transpiration-cohesion theory**, first propounded in 1914 by the Irish botanist Henry Dixon. The transpiration-cohesion theory states that transpiration in the leaves pulls water up the stem in continuous columns (Figure 31-5A). For decades after Dixon's proposal, many plant physiologists doubted that the theory was correct. Acceptance of Dixon's hypothesis rested on answering two questions. First, was the osmotic pressure exerted by air enough to pull water up hundreds of feet from the soil? And, second, were long columns of water strong enough to be pulled up a tree like pieces of rope? Wouldn't the narrow columns of water in the xylem break under such tension?

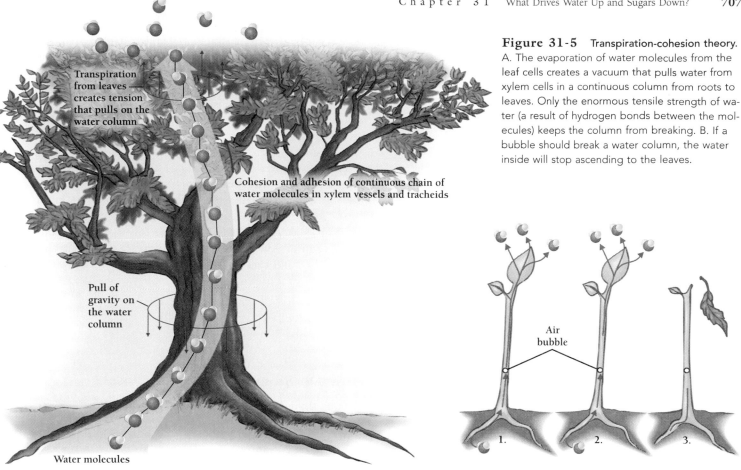

Transpiration from leaves creates tension that pulls on the water column

Cohesion and adhesion of continuous chain of water molecules in xylem vessels and tracheids

Pull of gravity on the water column

Water molecules

A. Transpiration-cohesion theory

Air bubble

1. 2. 3.

B. Breaking water column

Figure 31-5 Transpiration-cohesion theory. A. The evaporation of water molecules from the leaf cells creates a vacuum that pulls water from xylem cells in a continuous column from roots to leaves. Only the enormous tensile strength of water (a result of hydrogen bonds between the molecules) keeps the column from breaking. B. If a bubble should break a water column, the water inside will stop ascending to the leaves.

Henry Dixon's transpiration-cohesion theory raised two questions—one about the osmotic pressure exerted by air and one about the tensile strength of water.

Is the Water Potential of Air Enough To Draw Water to the Tops of Tall Trees?

Just as water tends to move into areas of high solute concentration, it also tends to move into areas with very few water molecules. Water molecules are pulled into dry air as effectively as they are pulled into cells with high concentrations of solutes.

Inside the mesophyll of a leaf, water is evaporating from virtually every cell surface, saturating the air spaces in the mesophyll with water vapor. If the air spaces remained saturated, water would stop moving out of the mesophyll cells and transpiration would cease. Outside the stomata, however, is the whole atmosphere, which is almost never saturated.

Air can be dry, as in the Mojave Desert, or very humid, as in much of the eastern United States. Even at 98 percent relative humidity, however, the tendency of water to evaporate is very high. The osmotic pressure that results from the difference in water concentration between the humid air and the liquid water inside a plant is enough to support a 300-meter column of water, nearly three times the height of the tallest trees.

In principle, then, the transpiration of water from the air spaces of the leaves to the atmosphere could provide the driving force for the ascent of water.

As water moves from the air spaces in the mesophyll to the atmosphere, more water moves into the air spaces by evaporation from the mesophyll cells. As water leaves these cells, water flows into them from the xylem. The result is the movement of water from soil to roots to xylem to leaves to air. The driving force for all this movement is the tendency for liquid water to evaporate into the air. This potential energy of water is also called **water potential.**

The water potential of even humid air is enough to raise water nearly three times as high as the tallest trees. The water potential of dry air is even greater.

Is the Flow of Water Through the Stem Coupled to Transpiration?

It is one thing for a process to be theoretically possible and another for it to actually occur and contribute to the workings of an organism. Plant scientists therefore asked whether water flow in the xylem increases as transpiration increases and decreases as transpiration decreases. In other words, are the two processes really linked? To address this question, researchers heated small amounts of water in the stem and then noted the time required for the heated water to arrive at a higher point. They found that fluid moves more quickly as transpiration increases during the day.

Is Water Cohesive Enough To Sustain the Transpiration Pull in a Tall Plant?

The transpiration-cohesion theory depends on the water columns in the xylem hanging together. If they should break apart, the top part of the column will be pulled up, but the bottom section will be left behind (Figure 31-5B). Opposing forces pull the column apart—transpiration pulling upward as gravity pulls downward. This tug-of-war extends throughout the plant, with each water molecule pulling at its neighbors via hydrogen bonds.

The great cohesiveness of water allows columns to hold together under tremendous pull. If an air bubble should interrupt a water column, however, the column breaks. In fact, air bubbles frequently arise in the xylem. In winter, when the air and the plant are cold, gases stay in solution and bubbles do not form easily. With the warmer temperatures of spring, however, gases in the xylem fluid form bubbles, disrupting the flow of water and minerals when it is most needed.

Woody plants have evolved a solution to this springtime hazard. Each spring, a burst of secondary growth creates new xylem routes. In many trees, most transport occurs in the newly formed xylem, in the outermost growth ring.

The tensile strength of water is sufficient to keep it from breaking under the tension generated by transpiration.

Is Water in the Xylem Really Under Tension?

The greater the rate of transpiration, the greater the tension in the xylem. The tension in the xylem of a big tree is strong enough to actually contract the walls of the xylem. Careful measurements show that during times of maximum transpiration tree trunks actually become smaller in diameter. When transpiration slows, the trunk relaxes to a larger diameter.

STOMATA REGULATE TRANSPIRATION BY OPENING AND CLOSING

Stomata are the gates through which carbon dioxide enters and water (and oxygen) leave. Far more water escapes than carbon dioxide enters, however. In many environments, plants can lose

more water by way of transpiration than is available in the soil. Plants therefore have various structural adaptations for preventing excess water loss.

All plants closely regulate water loss by opening and closing their stomata in response to environmental changes, such as light or water availability. Closing the stomata reduces water loss, but it also prevents the plant from collecting carbon dioxide and therefore limits photosynthesis. The trade-off between saving water and maintaining photosynthetic productivity is called the **transpiration-photosynthesis compromise.** The role of the stomata is to "provide food while preventing thirst."

To balance these needs, stomata usually open in response to light and to low levels of carbon dioxide. When the air is hot and dry, and excessive transpiration threatens to wilt the plant, the stomata close. Without supplies of carbon dioxide from the outside air, however, most plants drastically reduce the rate of photosynthesis. This is one reason that crop yields are low during a drought.

How Do Stomata Open and Close?

The tiny stomata occupy only about 1 percent of the total surface of a leaf. Each stoma lies over an internal air space, which in turn connects to the rambling air spaces within the mesophyll. The size and distribution of stomata permit the efficient diffusion of carbon dioxide into the air spaces and of water vapor from the air spaces into the atmosphere (Figure 31-6A).

Two guard cells surround each stoma. These cells can change shape, opening and closing the stoma. As the guard cells gain water, they become turgid and bend outward, opening the stoma. When water exits the guard cells, they become flaccid and close the gap between them (Figure 31-6B).

Guard cells regulate the flow of water by actively pumping potassium ions (K^+) from adjacent cells of the epidermis (Figure 31-6C). As potassium ions flow into the guard cells, chloride ions follow them inwards. Because of osmosis, water follows rapidly on the heels of these ions, the cells become turgid, and the stoma opens.

Several factors control the pumping of potassium ions and the opening of stomata, including light levels and water availability. For example, the stomata close in response to water stress as a result of the action of the plant hormone **abscisic acid.** As water becomes less available, the concentration of abscisic acid increases. Abscisic acid stimulates the flow of potassium ions out of the guard cells and the closing of the stomata.

Guard cells open in response to light and low carbon dioxide by pumping potassium ions from adjacent cells, which draws water into the guard cells and opens the stoma. In the presence of abscisic acid, guard cells lose potassium ions, water leaves the cells, and they relax, closing the stoma.

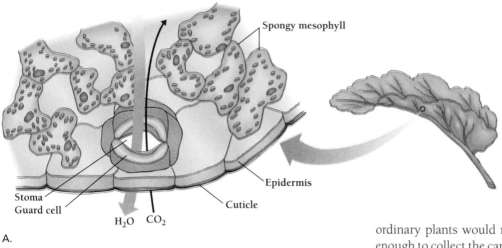

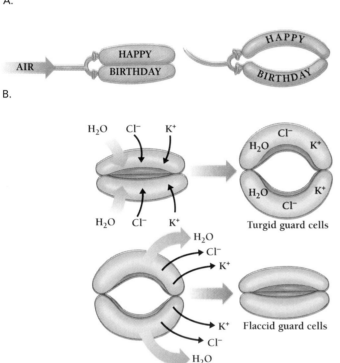

C. Increased concentration of abscisic acid

Figure 31-6 What causes stomata to open and close?
A. When the stomata are open, water molecules from the inside of the leaf diffuse into the surrounding atmosphere and carbon dioxide gas diffuses into the interior of the leaf. B. Like a pair of balloons being inflated, stomata fill with water and bend away from one another. C. Each stoma has a pair of guard cells, which actively pump in potassium ions. Chloride ions, then water molecules follow passively, filling the guard cells with water and causing the stoma to open. When abscisic acid causes potassium ions to flow out of the guard cells, the chloride ions and water molecules soon follow, closing the stoma.

How Can Plants Save Water Without Limiting Photosynthesis?

The transpiration-photosynthesis compromise prevents most plants from living in the driest deserts. In such places, most ordinary plants would not be able to open their stomata long enough to collect the carbon they need for photosynthesis. They would wilt before they got enough carbon dioxide to fuel photosynthesis.

Plants such as the cacti, however, have evolved a biochemical adaptation that gets them around this dilemma. Unlike other plants, **crassulacean acid metabolism (CAM)** plants can collect the carbon dioxide they need by opening their stomata only at night, when temperatures are cool and transpiration is low (Figure 31-7).

But photosynthesis cannot occur in the dark. The problem, then, is to store carbon dioxide until daylight. CAM plants have evolved a biochemical trick for storing carbon dioxide collected at night for use in photosynthesis during the day. In CAM plants, carbon dioxide enters the stomata at night and combines with a breakdown product of starch to form an organic acid (Figure 31-7). During the day, this acid, which is stored in the vacuole, gradually diffuses back into the cytoplasm, where it releases carbon dioxide for photosynthesis. One can actually taste this process: the leaves are sour in the early morning, from the accumulated acid, and sweet at the end of the day, after the acid disappears.

Crassulacean acid metabolism gives some plants another way of saving water.

WATER IN THE XYLEM CARRIES MINERALS FROM THE SOIL

Plant scientists have determined the elements needed for plant growth by growing plants with their roots in water instead of soil, a technique called hydroponic culture. By varying the minerals added to the water, researchers can determine which elements, and how much of each, are required for normal growth. These experiments have led to the identification of the elements essential for growth and completion of the life cycle. The same experiments have revealed the symptoms that result from deficiencies of particular nutrients (Table 31-1).

Figure 31-7   Dealing with dryness. CAM (crassulacean acid metabolism) plants such as cacti have evolved a unique biochemical pathway that allows them to store carbon dioxide. These desert-adapted plants open their stomata to collect carbon dioxide at night, when temperatures are cool and water loss is minimal. CAM plants store the carbon dioxide as malic acid until morning. During the day, CAM plants close their stomata (thus saving water) and use the carbon dioxide in malic acid to photosynthesize.

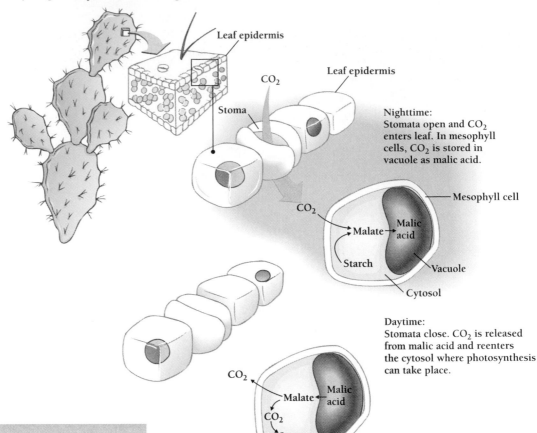

Leaf epidermis

Leaf epidermis

CO_2

Stoma

Nighttime: Stomata open and CO_2 enters leaf. In mesophyll cells, CO_2 is stored in vacuole as malic acid.

Mesophyll cell

CO_2

Malate → Malic acid

Starch

Vacuole

Cytosol

Daytime: Stomata close. CO_2 is released from malic acid and reenters the cytosol where photosynthesis can take place.

CO_2

Malate ← Malic acid

CO_2

Sugar

Table 31-1   Nutrients from the Soil

Nutrient	Concentration*	Some Symptoms of Deficiency
Nitrogen	1–4%	Uniform loss of color in leaves, first on oldest leaves
Potassium	0.5–6%	Yellowing at margins of leaves
Calcium	0.2–3.5%	Terminal bud dies, new leaves hooked at tip. Tips and margins withered
Phosphorus	0.1–0.8%	Plants stunted, leaves abnormally dark green
Magnesium	0.1–0.8%	Yellow between veins of leaf
Sulfur	0.05–1%	Leaves pale green with dead spots
Chlorine	100–10,000 ppm	
Iron	25–300 ppm	In young leaves, large veins, green; rest of leaf yellow
Manganese	15–800 ppm	Dead spots scattered over leaf surface; only veins and veinlets remain green
Zinc	14–100 ppm	
Copper	4–30 ppm	
Boron	5–75 ppm	Petioles and stems brittle; bases of young leaves break down
Molybdenum	0.1–5.0 ppm	

*Typical concentrations in healthy plants in % weight or parts per million (ppm)

Most Plants Obtain Both Water and Essential Minerals from the Soil

Both water and minerals usually come from the soil surrounding a plant's roots. Soils themselves are the product of both living and nonliving processes. They consist of particles of weathered rocks, and partially decayed organic matter.

The best soils for plant growth hold enough water to provide a continuing supply after a rainfall, but not so much that the roots cannot get the oxygen they need for respiration. The water capacity of soils depends in part on the sizes of the particles of weathered rock. Soils consisting only of small particles (less than 2 μm) are called **clays;** soils of medium sized particles (from 2 to 20 μm) are called **silts;** and soils of large particles (from 20 to 200 μm) are called **sands. Loams** consist of mixtures of all three particle sizes. The ideal loams contain about 20 percent clay, 40 percent silt, and 40 percent sand.

In most soils, earthworms, insects, bacteria, and fungi quickly convert dead organisms into large, organic molecules. The residue of such decay is a black or brown material, called **humus,** that decays much more slowly. Good soil has a high capacity for positively charged ions, which bind to negatively charged particles of clay and humus. Humus does not provide the plant with organic matter (as was once thought), but it does

BOX 31-1

Adaptations to limited water supply

All plants need ample supplies of water, so dry climates require special adaptations. Biologists sometimes divide plant adaptations to dry environments into four strategies—escape, resist, avoid, and endure. Desert annuals, for example, escape from drought (Figure A). Their seeds lie dormant, indifferent to any drought, until enough rain falls to wet the soil. Then, in a matter of days, they germinate, grow to maturity, and produce new seeds. Such seeds can survive a single dry season or 20 years of drought.

Cacti and other succulents resist drought by storing water (Figure B). CAM photosynthesis allows them to save water by keeping their stomata closed during the day. Shallow root systems quickly absorb water from the soil after each rainfall, and these plants store this water for future use.

Other plants such as desert palms and mesquite avoid water stress by growing only where rare water supplies exist. Palms grow only at oases, where their roots have access to relatively large amounts of groundwater. Mesquite bushes thrive by growing roots deep into the ground to water tables as much as 50 meters down.

Finally, some plants survive by enduring tremendous water losses. For example, while most plants die after losing 25 to 50 percent of their weight in water, the creosote bush can lose 70 percent of its weight in water and still survive (Figure C). How these plants survive such punishment is still unknown.

Even plants that grow in regions that ordinarily supply plenty of water have adaptations that allow them to survive water stress. The most common responses to drought are the slowing of cell growth and the reduction of transpiration by closing the stomata. Most plants can recover from temporary dry spells, although they may not make up for the lost period of growth.

In temperate climates, freezing temperatures make water unavailable for part of the year, and many species of plants cannot survive prolonged frosts. Surprisingly, the danger from frost turns out not to be from the ice itself. Ice crystals form only outside of cells and do not necessarily damage cells. Instead the presence of ice reduces the water potential, causing cells to lose water and therefore turgor.

Many plants can tolerate freezing temperatures, in some cases −25°C or lower. Winter rye and winter wheat, crocuses, tulips, and daffodils all grow slowly but steadily during the winter months. These frost-resistant plants prevent water loss by producing their own "antifreeze," which in some cases just consists of a concentrated solution of an amino acid. When spring comes, these plants are able to grow more rapidly, taking advantage of unfiltered sunlight before trees and taller plants leaf out.

Plants that are resistant to water stress—winter wheat, for example—are said to be acclimated or hardy. As plant biologists breed hardier plants, they greatly increase world food production. For example, if winter wheat and winter rye could withstand temperatures just 2°C colder than they now can, they could replace large tracts of spring wheat and rye in the United States, Canada, and Russia. Because winter crops make better use of the abundant spring rains, such a replacement could increase grain harvests by an astounding 25 to 40 percent.

Figure A   Desert annuals mostly exist as dormant seeds. When rain comes, the seeds germinate, grow into mature plants, bloom, and go to seed—all in a period of days or weeks. *(S. Krasemann/Photo Researchers)*

Figure B   Cacti and other succulents resist drought by storing water. *(M.C. Chamberlain/DRK Photo)*

Figure C   The creosote bush endures punishing dehydration. It can lose up to 70 percent of its water and still live. *(M. Patterson/Photo Researchers)*

provide a good reservoir of water and bound ions while also permitting oxygen to diffuse easily.

Most soils are composed of some combination of clay, silt, sand, and humus.

Why Do Roots Need Oxygen from the Soil?

Groundwater is full of dissolved minerals, usually in the form of ions. The concentrations of these ions in groundwater, however, are much less than their concentrations in plants. This is because plants actively transport ions into their root cells and concentrate them there, especially in the xylem. Because such active transport requires energy, root cells use large amounts of ATP (Chapter 6). They obtain ATP by oxidizing the products of photosynthesis. Thus, roots depend on the presence of oxygen in the soil. Most plants drown in water-logged soils.

Because roots actively transport ions from the soil, roots need oxygen to generate ATP by means of respiration.

Some Plants Obtain Minerals from Animals

Many plants live in areas where minerals, especially nitrogen, are in short supply. In many swamps and wetlands, for example, highly acidic water interferes with plants' ability to take up dissolved nitrates. Plants such as the Venus flytrap solve this problem by trapping and digesting insects for their nitrogen compounds, phosphates, and other ions (Figure 31-8).

Figure 31-8   **Hungry plants.** Some plants, such as this Venus fly-trap, live in environments that have insufficient supplies of nitrogen. Such plants may evolve techniques for trapping and digesting insects and other small animals. Some pitcher plants grow large enough to hold a regulation football (rugby).   *(William E. Ferguson)*

HOW DO PLANTS TRANSPORT THE PRODUCTS OF PHOTOSYNTHESIS?

We have seen that roots depend on the products of photosynthesis, but cannot themselves photosynthesize. How do roots and other plant parts obtain the energy and materials for building and maintaining cells and their myriad processes? To find out, modern biologists have tracked the movement of the products of photosynthesis.

Translocation Occurs in the Phloem

By exposing leaves to carbon dioxide containing radioactive ^{14}C, researchers can study the distribution of the radioactive carbon. Such studies show that the products of photosynthesis move entirely within the sieve tubes of the phloem and move from the leaves to the roots or to other areas where sugar is needed or stored.

In woody stems, functional sieve tubes are confined to the inner bark (Figure 31-9A). Removing a strip of bark around a tree trunk, a technique called **girdling**, interrupts the phloem and blocks the transport of photosynthetic products (Figure 31-9B). Girdling always kills the tree. Early colonists in North America girdled trees in the spring and felled them in the fall to harvest partially dried firewood.

In the early 18th century, two distinguished scientists—the Italian anatomist Marcello Malpighi and the English physiologist Stephen Hales—studied how girdling kills trees. They noted that the bark above the girdle initially remained healthy, while the bark below the girdle shriveled and died. This demonstrated the importance of downward transport from the leaves.

But a girdled tree usually fails to leaf out in the following spring, and before long, the whole tree dies. (A similar problem follows ground fires in young woods, where the trees may continue growth in the first year, but die in the second.) We must ask, then, how girdling kills the rest of the tree. The answer is that once the roots die, they no longer deliver water and minerals to the leaves.

Sugars are said to move from **source,** the site of production, to **sink,** the site of storage or consumption. A source may be either a site of photosynthesis, such as the leaves, or a storage site (such as the root) that releases sugars by breaking down starch. Sinks include the growing tips of roots and shoots; young leaves that are not yet active producers; fruits; and storage organs that are storing rather than releasing energy-rich molecules. The same organ—a potato tuber (an underground stem), for example—may be a sink in the summer, as it accumulates carbohydrates, and a source during the following spring, when it provides the plant with energy for the new growing season.

Sugars in plants move from sources, usually leaves or storage areas, to sinks, usually roots or storage areas. Sugars move by way of the phloem, a cylinder of living cells just under the outer bark, or cork, of trees and shrubs.

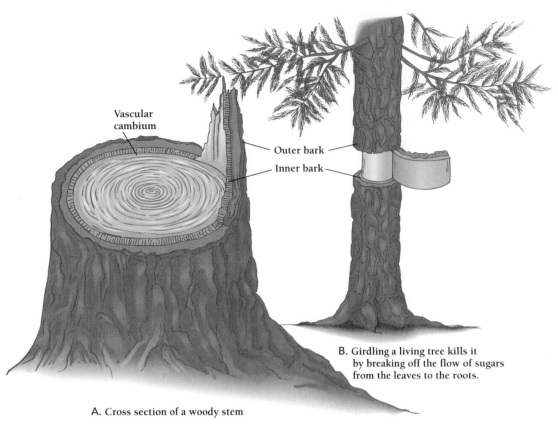

Vascular cambium

Outer bark

Inner bark

B. Girdling a living tree kills it by breaking off the flow of sugars from the leaves to the roots.

A. Cross section of a woody stem

Figure 31-9   Sugar on the outside. A. In woody plants, working sieve tube elements are confined to the inside of the bark. B. Stripping the bark from a tree will kill it if the strip goes all the way around. Such girdling destroys the sieve tubes and prevents the products of photosynthesis from reaching the roots.

What Is Inside a Sieve Tube?

Analyzing the contents of phloem was long a difficult problem for plant physiologists because the sieve tubes are small and collapse easily. Amazingly, the most useful method for analyzing phloem contents came from a 1953 suggestion of two insect physiologists.

These researchers were studying the nutrition of aphids, which derive their food from sap, the fluid contained in phloem (Figure 31-10). An aphid has a specialized mouth part, called a *stylet,* that pierces a sieve tube. Sap then flows from the phloem, through the stylet, directly into the aphid's body. The aphid extracts what it can, and the rest appears as "honeydew" at its rear end.

By cutting off the aphid and leaving the aphid's stylet embedded in the sieve tube, plant physiologists have been able to sample and to study the contents of the phloem. Botanists have found that the phloem contains a concentrated solution of sugars. Sucrose (table sugar) may make up as much as 30 percent of the phloem contents. Phloem sap also contains a few minerals, especially potassium ions, and carbohydrate derivatives, which vary considerably from plant to plant. Within the rose family (which includes apple, cherry, apricot, and pear trees), for example, sap contains little sucrose, but lots of the sugar sorbitol. As we will see in Chapter 33, sap also contains plant hormones.

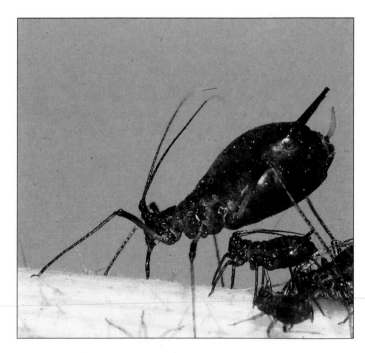

Figure 31-10   An aphid in action. Aphids unintentionally help researchers study the contents of phloem. An aphid passes its stylet into the phloem for a drink of sugar water. To collect the contents of the phloem, a researcher removes the aphid from its stylet.   *(Jerome Wexler/National Audubon Society/Photo Researchers)*

The Osmotic Flow of Water Drives Translocation in the Phloem

By using aphid stylets to follow the movement of radioactive sugars through the phloem, researchers found that translocation occurs much faster than could be explained by diffusion alone. They concluded that the contents of the phloem undergo some kind of transport. The explanation for how this transport occurs is called the **pressure flow hypothesis**, first advanced in 1926 by Ernst Munch, a German plant physiologist and artist.

Munch's insight was that the concentration of sugar is higher in the phloem near a source than near a sink. More water therefore flows into the sieve tubes near a source and pushes the contents in the direction of a sink. The pressure flow hypothesis predicts that the hydrostatic pressure within the sieve tubes is greater near a source than near a sink. Measuring the pressure within phloem tubes has been difficult, but several experiments have succeeded and have confirmed the prediction (Figure 31-11).

Ernst Munch formulated the pressure flow hypothesis to explain how sugars move once they are in the phloem.

What Draws Water and Sugar into the Phloem?

Once the concentrated sugar is in the phloem it makes sense that it should flow away from the areas of highest concentration. But what prevents the sugar (made in the mesophyll) from flowing back into the mesophyll? And what forces act to concentrate the sugar in the phloem? The answer to both questions is, once again, active transport.

The cells of the mesophyll are connected to one another by plasmodesmata and sugars flow freely from cell to cell. There are no plasmodesmata between the mesophyll cells and the cells of the phloem, however, and sugar must therefore always pass through two cell membranes, that of a mesophyll cell and that of a phloem cell (Figure 31-12A). Both passages require energy and are selective. The two cell membranes transport some molecules (such as sucrose) but not others (such as glucose). The sucrose concentration in phloem may be 2 to 3 times that in the mesophyll.

Sugar does not, however, flow directly from the mesophyll cells into the sieve tube members. Instead, sugar flows into **companion cells**, specialized cells adjacent to the sieve elements (Figure 31-12B). Unlike sieve tube members, companion cells possess nuclei and ribosomes and can make polypeptides.

The active pumping of sucrose is coupled to a pH gradient across the cell membrane (Figure 31-12C). The cell creates this gradient by spending ATP to pump out hydrogen ions, as in the case of the pH gradients across chloroplast and mitochondrial membranes. Hydrogen ions flow back into the companion cell via a transport protein that also carries sucrose.

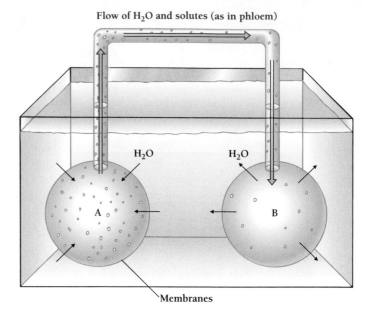

Flow of H_2O and solutes (as in phloem)

H_2O H_2O

A   B

Membranes

PRESSURE FLOW HYPOTHESIS

Figure 31-11   Munch's model for the pressure flow hypothesis. To test his hypothesis, Munch built a laboratory model that consisted of two selectively permeable membranes connected to each other by a tube: In this simplified diagram, the first membrane (labeled A) represents the source, and the second membrane (labeled B) represents the sink. Inside the first membrane, A, is a solution more concentrated than the solution outside of it. Water consequently flows into A and then up into the connecting tube. The solution inside of B is less concentrated than the solution outside. Osmotic pressure therefore drives water from the connecting tube into B. In accordance with Munch's hypothesis, the flow of water carries along any dissolved solutes at the same rate.

Once the sieve tube contains a high concentration of sucrose, osmotic pressure forces water into the phloem from the surrounding cells, driving the sugar away. Sap moves through the sieve tubes with a velocity of about 1 meter per hour, about the speed of the tip of the minute hand on a large classroom clock. (In contrast, water flows much more rapidly through the xylem, at a rate near that of the clock's second hand.) At the sink end of a phloem vessel, the phloem unloads its sucrose (aided by companion cells), and the water then enters nearby xylem and flows back up to the mesophyll.

The pressure flow model for phloem transport works in a single direction—from a source, with a high concentration of sucrose, to a sink, with a low concentration. One of the reasons that plant physiologists were slow to accept this hypothesis is that phloem transport can often occur both up and down the stem. Most researchers now think, however, that the flow within a single sieve tube goes in a single direction at a given time. Unlike the circulation of blood, then, the "circulation" of fluids in plants does not depend on a mechanical pump. The movement of fluids, both in xylem and in phloem, depends instead on active transport and osmotic pressure. The moving parts of a plant's pumps are all at the molecular level.

Solar energy

Figure 31-12 **Loading up on sugar.** A. Photosynthesis in the mesophyll cells generates sucrose sugar. B. The sucrose is actively transported through mesophyll cells, to a companion cell, and finally into a sieve tube member. C. The companion cells pump protons (hydrogen ions) out of the sieve tubes, creating a proton gradient. The cell membrane uses the proton gradient to pump sucrose into the phloem.

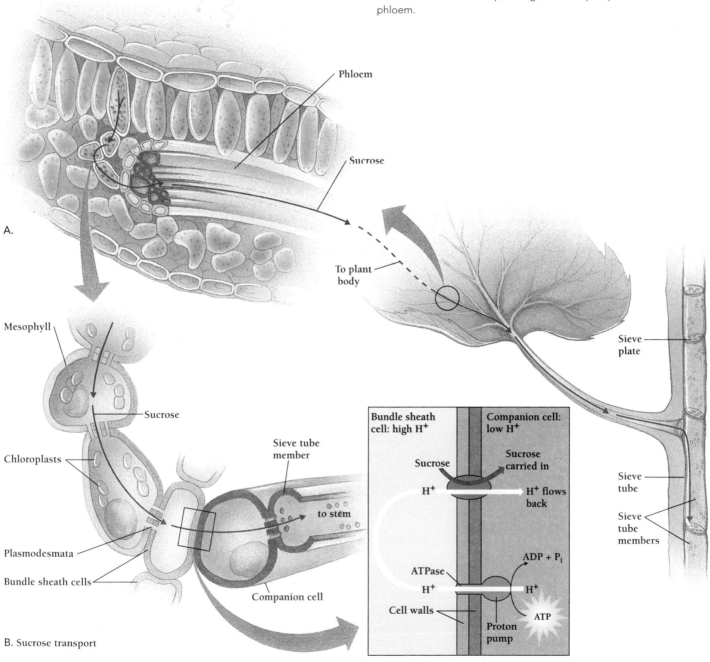

Phloem

Sucrose

To plant body

A.

Mesophyll

Sucrose

Chloroplasts

Sieve tube member

to stem

Plasmodesmata

Bundle sheath cells

Companion cell

B. Sucrose transport

Sieve plate

Sieve tube

Sieve tube members

Bundle sheath cell: high H$^+$

Companion cell: low H$^+$

Sucrose

Sucrose carried in

H$^+$

H$^+$ flows back

ADP + P$_i$

ATPase

H$^+$

H$^+$

Cell walls

Proton pump

ATP

C. Active transport

Cells in the mesophyll pump sugars by active transport from cell to cell and then into the phloem.

In this chapter we have seen how plants exploit the water potential of air to draw water and minerals from the ground up to the leaves; what processes open and close stomata; how sugars synthesized in the leaves are actively pumped into the phloem; and, finally, how those sugars move from areas of high sugar concentration to areas of low sugar concentration. In the next chapter, we discuss how plants generate a body plan and form the elements of root and shoot.

STUDY OUTLINE WITH KEY TERMS

Water moves into and out of living cells by **osmosis,** the flow of water through a selectively permeable membrane in response to differences in the concentrations of solutes. Water moves from areas of low solute concentration to areas of high solute concentration, and from areas of high water concentration to areas of low water concentration. Such differences in concentration are responsible for **water potential,** the potential energy of water.

In plant cells, the mechanical pressure of the cell wall opposes the inward flow of water from osmosis, or **osmotic pressure.** Turgor pressure, the pressure of a cell's contents against its cell walls, results from osmotic flow and provides mechanical support to nonwoody plants.

Solutes in the xylem cause water to flow, resulting in **root pressure.** Root pressure is beautifully demonstrated in **guttation,** the expression of droplets of water from the edges of leaves on cool, damp mornings.

Because of water's **adhesiveness** and **cohesiveness,** it tends to move up narrow tubes, or **capillaries.** But neither root pressure nor capillarity is enough to account for the movement of water to the tops of tall trees.

The major driving force for the ascent of water in the xylem is **transpiration.** And the driving force for transpiration is the low water potential of the air. According to the **transpiration-cohesion theory,** water vapor passes from the leaves into the atmosphere through open stomata. The evaporation of water from the leaves creates a water potential that pulls water up the xylem in continuous columns. Experimental measurements have established that water in the xylem is under tension, like a dangling rope with a weight at the bottom. The integrity of the water columns in the xylem is critical and depends on the cohesive properties of water.

Plants regulate transpiration and photosynthetic activity by opening and closing the stomata, a process mediated by the hormone **abscisic acid.** During hot, dry periods, plants must balance the rate of transpiration against the rate of photosynthesis, a trade-off called the **transpiration-photosynthesis compromise.**

Many plants have special biochemical and anatomical adaptations that reduce water loss while permitting the entry of enough carbon dioxide to sustain photosynthesis. **Crassulacean acid metabolism (CAM)** plants collect carbon dioxide by opening their stomata at night, when temperatures are cool and transpiration is low.

Most plants obtain minerals as well as water from the soil. Roots actively transport ions from the soil through cells of the cortex into the xylem. **Loams** are different combinations of variously sized particles of **sand, silt,** and **clay.** Decayed organic matter in soils is called **humus.** Insectivorous plants obtain nutrients from animals.

Many parts of a plant do not perform photosynthesis and require energy-rich molecules transported from the leaves. The movement of such molecules, mostly in the form of sucrose, occurs in the phloem. **Girdling** a tree or shrub interrupts the phloem and blocks the transport of photosynthetic products.

The osmotic flow of water, described in the **pressure flow hypothesis,** drives this movement. **Companion cells,** adjacent to the sieve elements, actively transport sugars into the phloem, and water follows by osmotic flow. The resulting pressure carries the contents of the phloem to other parts of the plants. Sugars move from a **source** to a **sink.** The circulation of nutrients in the phloem, like the circulation of water and minerals in the xylem, is best understood in terms of differences in water potential.

REVIEW AND THOUGHT QUESTIONS

Review Questions

1. Define osmotic pressure. In what different ways do plants use osmotic pressure to move materials from place to place?
2. Which is higher—the water potential of air or the water potential of a lake? Which is higher—the water potential of the inside of a raisin or the water potential of a pot of water? In which direction would water move in each of these comparisons?
3. Describe root pressure and guttation. How do the roots create root pressure?
4. Explain why water's natural cohesiveness is an essential element of the transpiration-cohesion theory.
5. How would you test the transpiration-cohesion theory?
6. What problem do CAM plants solve?

7. Name the three main particle sizes in soil. Which is biggest and which is smallest? The most fertile soils, or loams, are generally made up of what combination of these three kinds of particles?
8. Describe the pressure flow hypothesis. Is a potato a source or a sink?

Thought Questions

9. Why should xylem be composed of dead cells while phloem must be composed of living cells? What differences in the way these two parts of the vascular system function could explain this difference?
10. Do you feel that any of the characters, whether scientists or grape growers, described in the story about *Phylloxera* did anything wrong or unethical? Explain why or why not.

SELECTED READINGS

Berg, Linda R., *Introductory Botany: Plants, People, and the Environment,* Saunders College Publishing, Philadelphia, 1997. This book covers the basics of plant science in a friendly and unintimidating style.

Mauseth, James. D., *Botany: An Introduction to Plant Biology*, 2nd ed., Saunders College Publishing, Philadelphia, 1995. An authoritative introduction to the biology of plants.

Moore, Randy, W. Dennis Clark, Kingsley R. Stern, and Darrell Vodopich, *Botany,* Wm. C. Brown Publishers, Dubuque, 1995. A friendly, well-written textbook on the science of plants.

Niklas, Karl J., *Plant Biomechanics: An Engineering Approach to Plant Form and Function,* Chicago, University of Chicago Press, 1992. This book is wonderfully written and highly accessible where Niklas is first introducing a topic or concluding one—that is, at the beginning and ending of each chapter. Much of this book, however, demands a familiarity with either botany or engineering beyond that of the average reader.

▶ On-line materials relating to this chapter are on the World Wide Web at http://www.saunderscollege.com/lifesci/
Click on Tobin/Dusheck: *Asking About Life.*

The Origin of Corn

When Columbus left Spain in 1492, he was searching for a shortcut to China, Japan, and the rest of Asia. There he hoped to trade European goods for spices and silk. But when he reached the New World, one of his most valuable discoveries was not spices or silk but corn. Corn, also called maize, was good to eat, easy to grow, and bountiful.

Indeed, when Portuguese sailors finally reached the eastern coast of Asia in 1516, they brought, not European delicacies, but American corn. Within 50 years of Columbus's arrival in the West Indies, corn grew in every corner of the world. Wherever corn was introduced, it replaced native crops as an indispensable part of the diet. Corn became so thoroughly adopted in every tiny village where it was introduced that peoples in Africa, India, China, and even Turkey, came to believe that maize was native to their region.

Where did corn come from? Certainly it is American. Archaeologists have found corn (*Zea mays*) that is thousands of years old throughout North and South America. The oldest corn, from a site in Panama, is 7000 years old. Most botanists and paleontologists believe that corn originated in Mexico. But what we know as corn is the product of intense artificial selection. Its unparalleled yields allowed communities to grow far beyond what hunting and gathering or primitive farming could feed. The Western Hemisphere alone has thousands of varieties grouped into some three hundred races. Yet corn is almost unable to reproduce by itself; it is entirely domesticated (Figure 32-1). From what wild plant, then, is corn descended?

In the 1930s, the Harvard botanist Paul Manglesdorf backcrossed various corn hybrids to produce what he suggested was a model for wild-type corn. This approach would be similar to crossing several different breeds of dog to produce something that looked like a wolf. Manglesdorf found that as he backcrossed the corn, the cobs became smaller and the number of kernels inside the husk decreased until each corn kernel, or seed, was wrapped in its own husk, in the same manner as wheat. The ancestor of corn, he predicted, would be found to be an extinct, pod-bearing popcorn, which he called *pod corn*.

In the summer of 1948, two Harvard graduate students, in anthropology and botany, took a bus to Bat Cave, New Mexico. The cave had been occupied 3000 years earlier by people living on the shore of an ancient lake. As the two young men dug, they found hundreds of pieces of corn. The deeper they dug, the smaller and more primitive the corn cobs became. When they reached the bottom, they found tiny cobs of popcorn in which each kernel was enclosed in its own husk—in other words, Manglesdorf's pod corn.

Manglesdorf, who had given the two students $150 for bus fare and sleeping

Paraskevas Photography

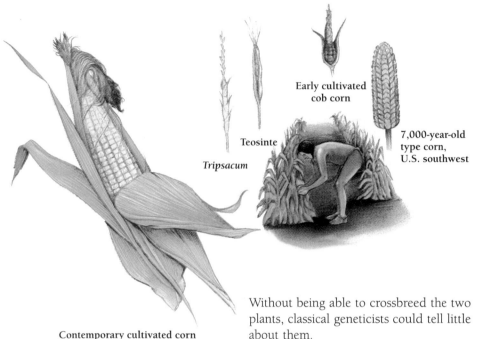

Tripsacum

Teosinte

Early cultivated
cob corn

7,000-year-old
type corn,
U.S. southwest

Contemporary cultivated corn

Figure 32-1   Corn kernels are bound tight. The tight wrappings around corn kernels make it impossible for the seeds to germinate. Even if a corn cob dropped to the ground to be buried in damp soil, the kernels could not germinate. Modern corn can only reproduce if humans (or other animals) remove the husk and plant the seeds individually.

bags, was delighted. He later wrote, "Seldom in Harvard's history has so small an investment paid so large a return." Manglesdorf took a few kernels of the ancient Bat Cave corn, dropped them into a pan of hot oil, and they popped!

But while Manglesdorf's own students and other archaeologists found the pod corn hypothesis persuasive, geneticists had their own ideas. The geneticist George Beadle had proposed, in 1939, that corn was descended from a Mexican grass called teosinte (tay-o SIN-tay). Nineteenth-century naturalists had first suggested that teosinte might be the ancestor of corn, and Beadle found that kernels of teosinte popped in hot oil were "indistinguishable from popped corn." Teosinte, however, has 40 chromosomes whereas corn has only 20, a difference that would prevent the two plants from interbreeding.

Without being able to crossbreed the two plants, classical geneticists could tell little about them.

Manglesdorf dismissed Beadle's teosinte hypothesis with a counter theory. The ancestor of corn is pod corn, he argued, and teosinte is a hybrid of ancient pod corn and another grass called *Tripsacum*. Modern corn, he suggested, is a mix of these three related grasses. *Tripsacum* did not have 20 chromosomes either, however, and the question of corn's

ancestry remained unresolved. For 40 more years, Manglesdorf and Beadle, as well as their intellectual descendants, picked at each others' theories. By the 1970s, however, many botanists thought that an extinct teosinte would prove to be the ancestor of corn.

In 1978, Hugh Iltis, a professor of botany at the University of Wisconsin, sent a Christmas card to a colleague at the University of Guadalajara, in Mexico. The card bore a drawing of a fanciful extinct teosinte, something that might resemble the ancestor of corn. Amused, the Mexican botanist challenged her students to find such a plant in the wild. Young Rafael Guzman took the challenge and spent his Christmas vacation searching for the hypothetical plant. In the mountains near Jalisco, Guzman found a new species of teosinte, seemingly the model for Iltis's drawing.

Guzman sent Iltis some seeds, and Iltis found, to his delight, that the plants had half the number of chromosomes that teosinte has. Like corn, *Zea diploperennis* has 20 chromosomes. The search for corn's ancestor, Iltis declared, was over. In-

Peoples in Africa, India, China, and even Turkey, came to believe that maize was native to their region.

deed, in the 1980s, genetic research at the University of Minnesota showed that approximately five genes control the traits that distinguish corn from *Zea diploperennis*. Once these genes are cloned, researchers hope to investigate how they are expressed.

719

Is the corn debate over? Most people now agree that corn is about 10,000 years old and probably descended from some form of teosinte from central Mexico. A likely ancestor for corn seems close at hand, yet paleobotanists and geneticists continue their comfortable bickering. As Richard Schultes of Harvard's Botanical Museum has said, "The origin of corn is a mess."

For most of us, it is a mess we can safely overlook. In comparison, the changes in corn's morphology under thousands of years of selective pressure are dramatic and revealing. Corn kernels, each of which is actually a fruit, are so tightly wrapped in the husk that the seeds can never disperse. Even if the whole husk is buried in moist earth, the resulting seedlings are so crowded that most of them die. Corn's artificially induced anatomy has made it entirely dependent on humans for reproduction. It is nearly alone in this odd defect. Nearly every other species of plant on Earth can reproduce and grow with no help at all.

In this chapter we will see how plants reproduce, how the seed forms, and how the embryo in the seed develops into a mature plant.

KEY CONCEPTS

1. Flowering plants undergo alternation of generations, with diploid sporophytes producing haploid gametophytes, which then produce haploid gametes. The haploid gametes then fuse to give rise to diploid sporophytes once more.

2. Plant embryos establish a body plan soon after fertilization and pause after seed production.

3. The primary growth and development of plants depend on the repeated generation of standard modules at the ends of the shoot and the root.

4. Specific genes control the development of flower parts.

5. Secondary growth depends on two kinds of lateral meristems.

6. Meristems may produce whole plants as well as primary and secondary growth.

HOW DO PLANT EMBRYOS ESTABLISH A BODY PLAN?

The life cycle of a flowering plant consists of two phases: the diploid **sporophyte,** which produces haploid spores by meiosis; and the haploid **gametophyte,** which produces haploid gametes by mitosis (Chapter 22). The cycling of the multicellular sporophyte and gametophyte phases is called **alternation of generations** (Figure 32-2). Most of our discussion here will concern the diploid, sporophyte generation, the dominant phase of the life of a flowering plant. (In Bryophytes such as liverworts and mosses, the gametophyte is often the dominant phase in the life cycle, but here we do not discuss these plants at all.) We begin by describing the development of the gametophytes, the production of gametes, and their union to produce a diploid zygote.

How Do Plants Make Pollen Grains?

The anther is the pollen-producing part of the stamen, attached to the flower's base with a filament (Figure 32-3A). Within the anther are four pollen sacs. Inside these sacs, diploid cells undergo meiosis to form haploid **microspores** [Greek, *mikros* = small], each of which develops into a male gametophyte.

Each microspore undergoes mitosis to produce two haploid cells, one *generative cell* and one *vegetative cell.* (A few flowering plants produce two generative cells and one vegetative cell.) The vegetative cell completely surrounds the generative cell. The two cells—generative and vegetative—are together enclosed within a common wall (that of the original microspore) to form a **pollen grain.** The outside of the pollen grain is often elaborately and distinctively sculpted (Figure 32-3B). Indeed, botanists often can recognize plants, extinct or living, by their pollen alone.

The anthers release the pollen, often immense quantities of it, and insects, wind, or other agents then carry some of it to the (female) stigma of a flower. Most pollen never reaches the stigma, however; it ends up, instead, in streams and lakes, on the ground, in the hives of bees, or even packed into small jars in health food stores.

Stamens produce haploid microspores, which develop into immature male gametophytes known as pollen grains.

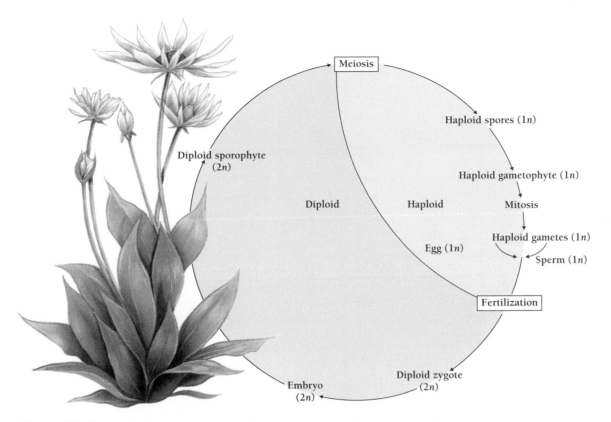

Figure 32-2 *Alternation of generations in flowering plants.* In flowering plants, the diploid sporophyte, which produces haploid spores, alternates with the haploid gametophytes, which produce haploid gametes. Both the sporophyte and the gametophyte are multicellular. The sporophyte is the conspicuous shoot and root, including most parts of the flowers. Inside the flowers are the gametophytes, which make the sperm and the eggs. Some plants have male and female gametophytes on separate plants; some on the same plant, but in different flowers; and some plants have both male and female gametophytes within the same flower.

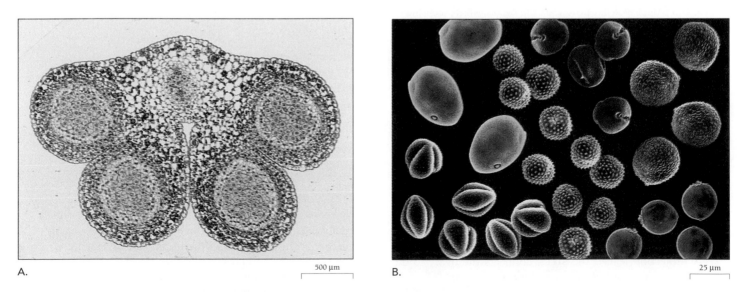

A.   500 μm

B.   25 μm

Figure 32-3 *How do flowers make pollen?* Anther of a lily. A. Each anther includes four pollen sacs, inside of which haploid microspores are formed. Each microspore divides and develops into a male gametophyte, a multicelled pollen grain. B. Pollen grains have distinctive textures and patterns that experts recognize. Shown here are pollen from ragweed, timothy brush, alder, and poplar.   *(A, Cabisco/Visuals Unlimited; B, Dr. Dennis Kunkel/Phototake, NYC)*

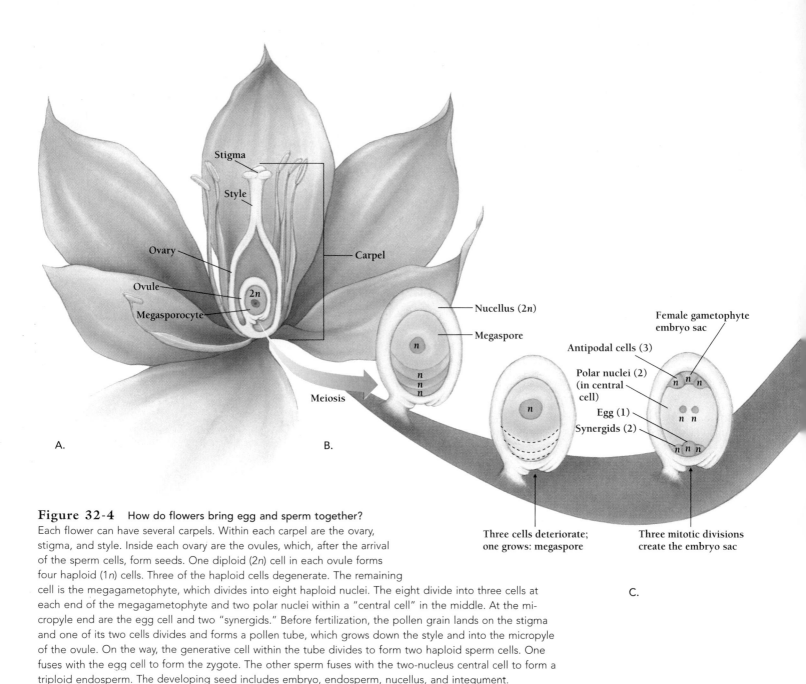

Figure 32-4 How do flowers bring egg and sperm together? Each flower can have several carpels. Within each carpel are the ovary, stigma, and style. Inside each ovary are the ovules, which, after the arrival of the sperm cells, form seeds. One diploid (2*n*) cell in each ovule forms four haploid (1*n*) cells. Three of the haploid cells degenerate. The remaining cell is the megagametophyte, which divides into eight haploid nuclei. The eight divide into three cells at each end of the megagametophyte and two polar nuclei within a "central cell" in the middle. At the micropyle end are the egg cell and two "synergids." Before fertilization, the pollen grain lands on the stigma and one of its two cells divides and forms a pollen tube, which grows down the style and into the micropyle of the ovule. On the way, the generative cell within the tube divides to form two haploid sperm cells. One fuses with the egg cell to form the zygote. The other sperm fuses with the two-nucleus central cell to form a triploid endosperm. The developing seed includes embryo, endosperm, nucellus, and integument.

How Do Plants Make Egg Cells?

Ovaries [Latin, *ovum* = egg], which produce the female gametophytes, are composed of one or more carpels (Figure 32-4A). A projection, called the **style**, connects the ovary to the **stigma**, the structure that receives the pollen grains.

The wall of the ovary produces one or more **ovules**, structures that, after fertilization, form a seed. One diploid cell in each ovule undergoes meiosis to form four haploid cells. One

of the four cells, called the **megaspore** [Greek, *megas* = large] develops into the **embryo sac**, which is the mature female gametophyte, while the other three haploid cells degenerate. The diploid tissue surrounding the megaspore is called the **nucellus** (Figure 32-4B and C). The megaspore nucleus undergoes three rounds of mitosis to form eight haploid nuclei. These eight nuclei form only seven cells, however, for one cell, the *central cell,* contains two nuclei. One of the other six cells of the embryo sac is the egg cell.

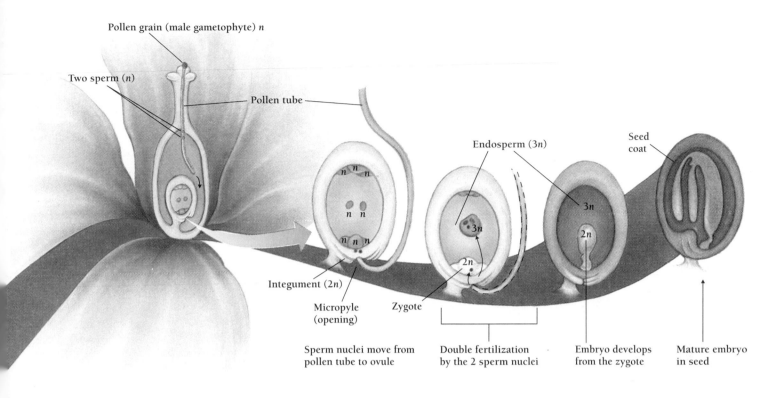

Pollen grain (male gametophyte) *n*

Two sperm (*n*)

Pollen tube

Endosperm (3*n*)

Seed coat

Integument (2*n*)

Micropyle (opening)

Zygote

Sperm nuclei move from pollen tube to ovule

Double fertilization by the 2 sperm nuclei

Embryo develops from the zygote

Mature embryo in seed

D.

E.

While the haploid gametophyte is developing, the surrounding diploid cells of the ovule are also dividing. The diploid tissue of the ovule forms the **integuments**, folds of tissue enclosing the gametophyte. The integuments eventually form the seed coat.

Carpels produce haploid megaspores, which develop into female gametophytes called embryo sacs. Each embryo sac has one egg.

How Does Fertilization in Plants Differ from That in Animals?

After a pollen grain lands on the stigma, the vegetative cell forms a **pollen tube**, into which the sperm cells pass. The pollen tube grows down the style and through a pore within the integument of the ovule to reach the embryo sac (Figure 32-4B). When the pollen tube is part way down the style, the generative cell divides to form two sperm cells, one of which may fuse with the egg cell. When the pollen tube arrives at the embryo sac within the ovule, the sperm cells are discharged.

Flowering plants differ from all other organisms in undergoing **double fertilization**, the simultaneous fusion of the two sperm cells with two cells of the embryo sac (Figure 32-4D). One sperm fuses with the egg to form the diploid zygote, which develops into an embryo and eventually into a plant. The other sperm fuses with the central cell (which, recall, has two nuclei), forming a cell with a triploid nucleus. The triploid central cell gives rise to the seed's **endosperm**, a triploid tissue that provides nourishment and hormones for the growing embryo. Shortly after fertilization, both the zygote and the endosperm nucleus undergo repeated rounds of mitosis. The developing seed consists of an embryo, the endosperm, the nucellus, and the integument. In most species, the endosperm is used up during development, and the mature, or ripe, seed has only the embryo, the nucellus, and the integument.

Flowering plants undergo double fertilization, the fusion of two sperm nuclei with two cells of the embryo sac to form (1) a diploid zygote and (2) a triploid endosperm.

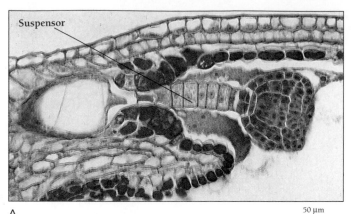

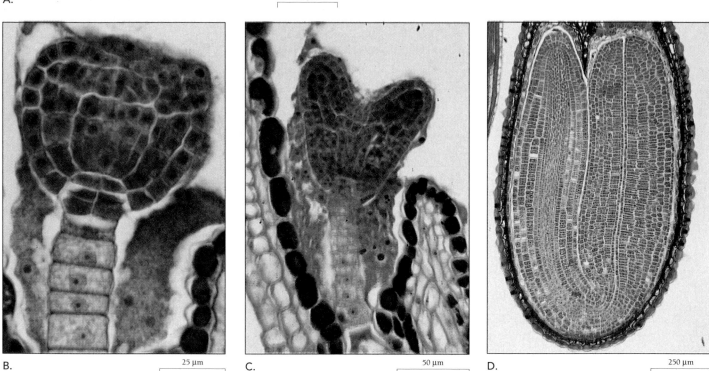

Figure 32-5   Stages in the development of a plant embryo.
A. Suspensor stage: like an umbilical cord, the suspensor connects the embryo to surrounding tissues. B. Globular stage: the embryo as a sphere. C. As the cotyledons form the embryo's shape becomes more heartlike. D. As the cotyledons grow out and then curve back toward the embryo, the embryo takes on a "torpedo" shape.   *(A,B, John D. Cunningham/Visuals Unlimited; C, D, Jack M. Bostrack/Visuals Unlimited)*

How Does a Plant Form a Seed?

The pattern of cell divisions in the early embryo decides the positions of its cells and the overall shape of the embryo. A typical dicot embryo, for example, changes its shape from a sphere to a heart to a torpedo.

The first cell division in a dicot embryo divides the zygote into two cells of unequal sizes. The larger, basal cell divides several times to form a narrow column of cells called a *suspensor,* which, like an umbilical cord, attaches the embryo to the surrounding tissue and nourishes the embryo (Figure 32-5A).

The smaller cell divides to form a spherical embryo, one end of which forms shoot apical meristem and one end of which

forms root apical meristem (Figure 32-5B). The fundamental shoot-root polarity of the embryo is thus established early in plant development.

Continuing division of the spherical embryo leads to the formation of three regions: the tissues that later form the epidermis, the ground tissue system, and the vascular system.

Cells of the shoot apical meristem soon begin to divide more rapidly, producing an embryo with a heartlike shape (Figure 32-5C). The continued division of cells in the shoot end produces "seed leaves," or **cotyledons.** The number of cotyledons divides flowering plants into two huge classes, the dicots and the monocots. **Dicots**—such as roses, soybeans, and most flowering trees—have two cotyledons, while **monocots**—such as palm trees, corn, and other grasses—have only one. The cotyledons absorb nutrients from the endosperm and later distribute them to the developing plant.

Above the attachment point of the cotyledons, the axis is called the *epicotyl* [Greek, *epi* = above], and below it is called the *hypocotyl* [Greek, *hypo* = under]. In the soybean and in many (but not all) other plants, the epicotyl also contains the plumule [Latin, *plumula* = a small feather], which develops into the first true leaves. At the end of the hypocotyl is the **radicle,** a primordial root that contains the root's apical meristem. As the cotyledons grow longer, they may curve back on the embryo to form a stage that in many species resembles a torpedo (Figure 32-5D).

At this stage, most of the embryo's cells resemble the large, differentiated cells of a mature plant, except that their plastids do not yet contain chlorophyll. The meristem cells remain small and undifferentiated. The shoot apical meristem is nestled between the cotyledons, just above their points of attachment, while the root apical meristem lies near the point of attachment of the suspensor. In most plants, by the time the embryo is mature, the endosperm is fully consumed. Some monocots retain the endosperm in the mature seed. In dicot plants such as peas, the developing plant derives nourishment from the cotyledons. In still other dicots, such as squash, the cotyledons nourish the seedling by photosynthesizing.

After the "torpedo" stage, the seed dries, and the seed coat hardens around the embryo. The embryo then enters a **dormant** [French, *dormir* = to sleep] stage, in which growth and development are suspended. The resumption of growth is called **germination.**

A seed contains a dormant embryo, with one or two cotyledons, surrounded by a tough seed coat.

How Does the Embryo Resume Development?

The first stage of germination is **imbibition** [Latin, *imbibere* = drink], the swelling of the seed with water. During imbibition, the seed may double or triple in size. Once the dormant embryo is hydrated, it awakes, with a burst of metabolic activity, cell division, and growth.

The first part of the embryo to break out of the seed coat is the radicle and the attached hypocotyl (Figure 32-6). The radicle immediately turns downward into the soil and forms a functioning root system that provides the growing plant with water and minerals. In the soybean and many other dicots, the top of the hypocotyl forms a hook that pushes up through the soil toward the sun, pulling the cotyledons and epicotyl behind (Figure 32-6). Once aboveground, the arching hypocotyl hook straightens, the green cotyledons spread out in the sun, and the epicotyl points upward and resumes its growth. In these

plants, most of the stem comes from the epicotyl, with only the first centimeter or so (below the cotyledons) deriving from the hypocotyl.

At germination, the seed swells with water, and the radicle breaks out of the seed coat and grows down into the soil. Development resumes with a burst of metabolic activity.

How Does the Embryo Obtain Nourishment?

Cells in the dormant embryo are not dead; they merely maintain a low level of metabolism. Germination, however, triggers a huge increase in biochemical activity, with energy and building blocks derived from stored starches, oils, and proteins. The presence of these stores also means that seeds are rich foods for a wide variety of animals, including ourselves (Figure 32-7).

The breakdown of these stores requires the action of enzymes that are not present (or are present in much lower amounts) in the ungerminated seed. In cereals such as corn, barley, and wheat, these enzymes are secreted by the **aleurone layer,** an outer layer of cells that surrounds the endosperm. After germination begins, the aleurone layer secretes enzymes that digest the starches and other storage molecules of the endosperm. One of these enzymes is **α-amylase,** which helps to break down starches stored in the endosperm.

The embryo itself must stimulate the aleurone layer to make digestive enzymes. If researchers remove the embryo, the aleurone layer secretes no enzymes and the endosperm sits undigested. The chemical signal that causes the aleurone layer to make digestive enzymes is the plant hormone gibberellin. As little as 10^{-11} grams (one hundred-billionth of a gram) of gibberellin induces the expression of the gene that encodes α-amylase.

At germination in cereals, the aleurone layer secretes enzymes such as α-amylase that release nutrients to the growing embryo.

What Environmental Cues Trigger Germination?

We can see the obvious effect of environment on plant growth by keeping a germinating seedling in the dark. Instead of straightening, the epicotyl arch keeps growing upward. The plant displays a condition called *etiolation* [French, *etioler* = to blanch or whiten], with a thin, spindly appearance, poor leaf development, and little or no chlorophyll production (Figure 32-8). Environmental cues also determine when a seed germinates.

As a seed, a plant embryo can suspend growth for days, months, years, decades, even centuries, depending on the species. Radioactive dating of lotus seeds from a former swamp bed in Manchuria, China, showed that they were about 1000

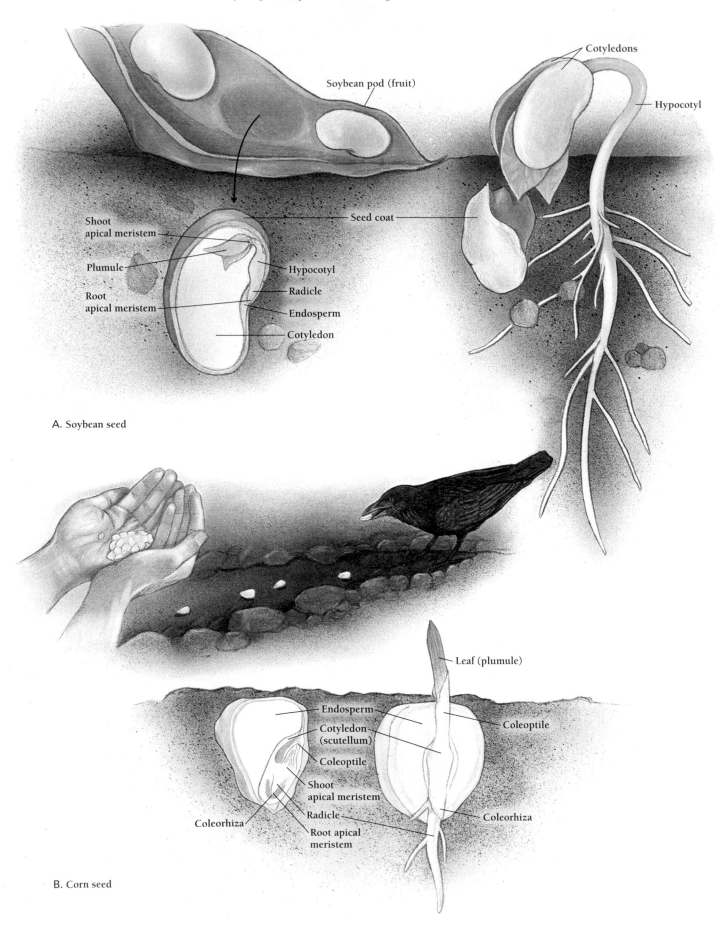

Soybean pod (fruit)

Cotyledons

Hypocotyl

Shoot apical meristem

Seed coat

Plumule

Hypocotyl

Root apical meristem

Radicle

Endosperm

Cotyledon

A. Soybean seed

Leaf (plumule)

Endosperm

Coleoptile

Cotyledon (scutellum)

Coleoptile

Shoot apical meristem

Coleorhiza

Radicle

Coleorhiza

Root apical meristem

B. Corn seed

◀ **Figure 32-6** **Germination in a dicot and a monocot.** A. When a soybean germinates, the radicle breaks out of the seed coat and turns down into the ground. The bent hypocotyl straightens and the two cotyledons unfold. B. In corn, the shoot grows straight out of the kernel through the tube of the coleoptile and the root grows down through the tube of the coleorhiza.

Figure 32-8 **Etiolation.** Corn plants grown in the dark are long and spindly compared to those grown in normal light. *(Blanche C. Haning)*

years old. Yet they still germinated. Still more amazingly, researchers have succeeded in growing plants from lupine seeds that were found in 10,000-year-old frozen sediments in Alaska. Botanists have concluded that seeds do not have a built-in clock that determines the time of germination, but instead depend on cues from their environment.

Plants use a wide variety of mechanisms to prevent premature germination. In many legumes, for example, a hard seed coat prevents the entrance of water or oxygen needed for germination. In these and other species, germination often requires *scarification,* harsh treatment to break the seed coat barrier. In nature, scarification conditions include fire, acid, and enzymes in an animal's digestive system; scraping against sand or rocks; successive freezing and thawing; or the action of a fungus. In the laboratory, researchers can scarify seeds with sandpaper, alcohol, or even sulfuric acid.

Other species prevent premature germination with chemical inhibitors, such as sodium chloride, cyanide, mustard oils, and other organic compounds. Rainfall then stimulates germination by washing away the inhibitors.

Light, water, temperature changes, stomach acid, and other environmental cues can all trigger germination.

Figure 32-7 **The size of the endosperm and the resulting seed varies widely among species.** Lettuce and tomato seeds are about 1 to 2 mm in diameter, while peas, beans and corn are about 10 times larger. The smallest known seeds are the microscopic seeds of orchids, and the largest are the "double coconuts" of the Seychelles nut palm. In general, the smaller the seeds, the more of them plants produce. Plants that produce very large seeds have only energy enough to make a few. These few large seeds, however, generally have a good chance of surviving. *(D. Cavagnaro/Visuals Unlimited)*

HOW DO PLANTS GENERATE A SHOOT AND A ROOT?

The most important organizing principle of a plant is the polarity between the shoot, the part of the plant above the ground, and the root, the part below the ground. The shoot-root polarity, roughly parallel to the head-tail polarity of an animal, is established early in the plant embryo.

Just behind the tip of both shoot and root are **apical meristems,** groups of *undifferentiated* cells that generate the specialized cells of the shoot and root (Figure 32-9). Undifferentiated cells are cells that resemble those in the embryo. Meristem cells can either divide, to give rise to more meristem cells, or differentiate.

An apical meristem can repeatedly generate the same set of plant parts, which serves as a kind of construction module. Each module, similar to segments in animals, may be repeated any number of times, depending on the conditions of plant growth.

A shoot apical meristem can generate a stem and a leaf. The region of the stem to which the leaf attaches is called a **node,** and the region between nodes is an **internode** (Figure 32-9). If we look from node to node, we see repeating modules, each consisting of a segment of stem (the internode), a leaf, and a meristem. In both shoot and root, the action of the apical meristems is responsible for the plant's primary growth.

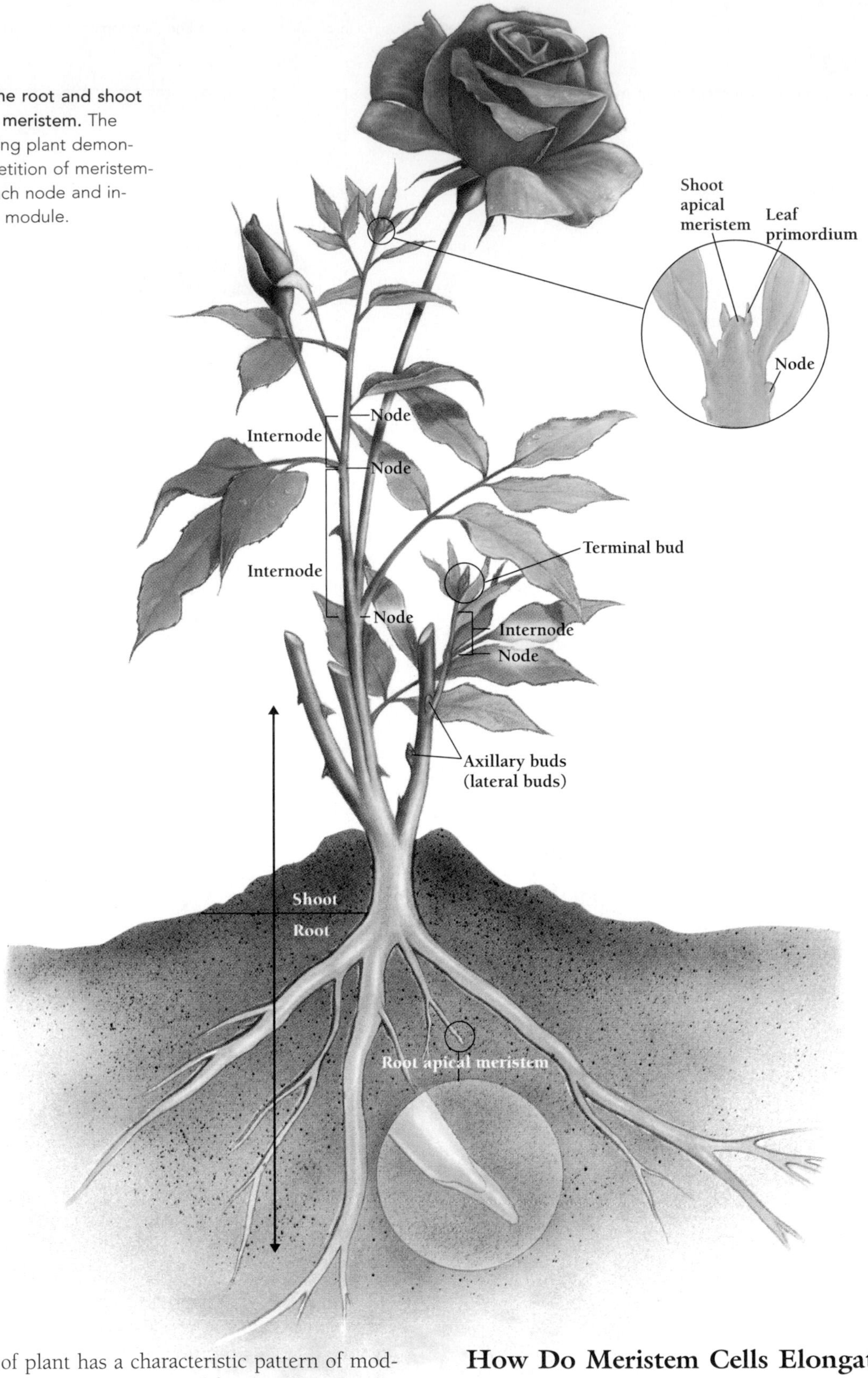

Figure 32-9 **The root and shoot each have an apical meristem.** The anatomy of a flowering plant demonstrates a regular repetition of meristem-derived modules. Each node and internode constitute a module.

Shoot apical meristem
Leaf primordium
Node

Node
Internode
Node

Terminal bud

Internode

Internode
Node

Node

Axillary buds (lateral buds)

Shoot
Root

Root apical meristem

Each species of plant has a characteristic pattern of modular repetition. Modular organization and fixed repetition rules underlie the branching patterns of stems, as well as the pleasing symmetry of many flowers and fruits.

Plants repeatedly generate modular structures at the apical meristems at the ends of the shoot and root.

How Do Meristem Cells Elongate and Differentiate?

In both shoot and root modules, undifferentiated meristem cells themselves are relatively small, with thin walls, prominent nuclei, and small vacuoles. Away from the meristem, these cells stop dividing and begin to differentiate into any of a variety of cell types. Most dramatically, they grow—mainly by taking water into a large central vacuole. The elongated, differentiated cells, already arranged in columns, are up to 50 times larger than those of the meristem.

Unlike animal cells, plant cells cannot move around and form new associations during development. After division ceases, they can only expand and sometimes bend. Plant development therefore depends heavily on orderly cell division and cell expansion. Every cell has its place. As cells divide within the meristem, the planes of cell division and rates of expansion establish future spatial relationships.

Cell expansion depends mainly on the osmotic influx of water into the central vacuole. Cellulose in the cell walls is laid down in such a manner that the cell wall tends to be weaker in one direction than in others. As a result, as the vacuole swells, the cells tend to elongate, rather than grow bigger in all directions. The exact orientation of cellulose fibers, which determines the direction of plant growth, depends on a number of plant hormones and many processes that are poorly understood.

Because of their effects on the patterns of cell division and expansion, plant hormones can dramatically influence the growth of an entire plant. The hormone ethylene, for example, causes a longitudinal orientation of the cellulose fibers. The parallel cellulose fibrils constrain cell expansion to the horizontal direction, leading to the production of short, fat seedlings. In contrast, gibberellin, another plant hormone, stimulates the transverse orientation of cellulose fibrils, leading to tall, thin shoots.

Although most of this discussion has centered on the growth and the differentiation of cells in the shoot, the same general considerations apply to the growth of the root. Nonetheless, the growth of roots and shoots have several distinctive characteristics.

The development of form in plants depends on oriented cell division and cell expansion.

Root Tips Grow and Mature in Three Zones

Primary growth of roots occurs almost exclusively near their tips. The growth and maturation of root tissues is complex. In general, however, the growth zone has three overlapping regions—the regions of cell division, cell elongation, and cell differentiation (Figure 32-10).

At the very tip of the root is the **root cap,** whose cells are full of Golgi complexes and vesicles. These organelles contribute to the synthesis of **mucigel,** a polysaccharide slime that lubricates the path of the growing root through the soil. The root cap also serves to protect the apical meristem, which lies just behind it.

We can demonstrate that the apical meristem contains most of the dividing cells in the root. If we allow growing roots to take up a radioactive precursor of DNA (^{3}H-thymidine), almost all of the labeled DNA shows up in the meristem, where mitosis is occurring.

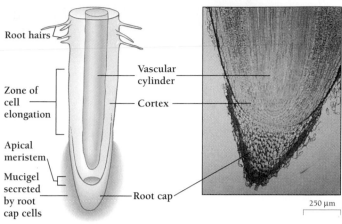

Longitudinal section of a root tip

Figure 32-10   **Growth in a root.** In the root, plant biologists distinguish three overlapping regions of growth—the root cap, the apical meristem, and a zone of elongation.   *(photo, Kevin and Betty Collins/Visuals Unlimited)*

The cells within the meristem have a small uniform size, but the pattern of cell division produces files of cells whose organization already resembles that of the mature root. Farther back along the root's axis is the **elongation zone.** The cells in each file are increasingly large. Cells farther from the tip are longer than those nearest the meristem, which have arisen more recently. Cells that are farther away from the meristem in the elongation zone are more mature and, therefore, more differentiated. The appearance of abundant root hairs marks the beginning of the differentiation zone. Root cells with root hairs are fully mature.

Growing root tips include three zones: the slimy root cap, the apical meristem, and the elongation zone.

Primary Growth of Shoots Is More Complicated Than That of Roots

In the shoot, the apical meristem must produce stem, leaves, branches, and flowers, so primary growth is more complicated than in the root. The region of primary growth may extend back as much as 10 to 15 cm behind the shoot's tip and may include several nodes (Figure 32-9). As in the root, the stem apical meristem contains the dividing cells. Cells produced in the meristem and arranged in concentric circles form files of cells that grow longer, pushing the meristem upward. The cells that are farthest away from the meristem are the most mature and the most differentiated.

Each leaf originates as a **leaf primordium,** a tiny extension of the apical meristem. Each primordium contains meristem cells, which together divide, expand, and differentiate to form a leaf. A **bud** consists of several leaf primordia, separated by internodes whose cells have not elongated (Figure 32-10B).

BOX 32-1

Why is *Arabidopsis thaliana* well-suited to genetic analysis?

Arabidopsis thaliana, a small, common weed, also called "wall cress," is so well-suited to genetic studies that it has been called the "botanical *Drosophila*." The major advantage *Arabidopsis* has over most plants is that it can be grown from a seed to a mature plant in as little as 5 weeks. In addition, *Arabidopsis* can self-fertilize, making it much easier to isolate homozygous strains than in *Drosophila. Arabidopsis* is small, usually less than 30 cm tall, and spindly enough that many plants can be grown easily and inexpensively in a small area. Its seeds are so tiny that 35,000 seeds occupy about 1 milliliter.

Geneticists have already identified more than 100 mutations in *Arabidopsis.* Some of these mutations lead to specific enzyme deficiencies, others to alterations in response to plant hormones, and still others to alterations in pattern. For example, one mutation leads to a missing shoot, another to a missing stem, another to a missing root. Mutations affect mature plants as well as seedlings. One mutant completely lacks a shoot meristem, both in the intact plant and in tissue culture. Other *Arabidopsis* mutations affect the formation of specific types of tissue, while still others

prevent the proper establishment of the embryo shoot-root axis.

Once a mutation has been found, it is a relatively straightforward—if time-consuming—task to identify and sequence the gene in which the mutation lies. The function of the gene can then be studied by putting it into normal or mutant *Arabidopsis.* Researchers have performed such studies for a relatively small number of genes but have already learned much about *Arabidopsis* development.

In addition to the **terminal buds,** which lie at the tip of a shoot, most plants have **axillary buds,** which form just above the points where leaves join the stem and which can develop into branches. Like the terminal buds, the axillary buds contain an apical meristem.

Many species show **apical dominance,** meaning that the presence of terminal buds suppresses the development of axillary buds. The terminal buds do this by releasing a plant hormone. The result is that the plant tends to grow taller than it is wide. Axillary buds near the tip of a plant are more inhibited than those farther away. That is why branches near the bottom of a conifer tree, for example, are much longer than those at the top. Removing the terminal bud allows the axillary buds to grow, and many gardeners routinely produce bushier plants by pinching off the terminal bud.

Each leaf starts as a tiny extension of the apical meristem. Terminal buds suppress the growth of axillary buds, an effect called apical dominance.

How Does a Flower Derive from a Specialized Bud?

Some buds are specialized to form flowers or a flowering shoot. In most species, flowers contain both **carpels,** female reproductive structures, and **stamens,** male reproductive structures. A flower consists of up to four sets of parts arranged in **whorls,** concentric circles of flower parts (Figure 32-11A). The outer two whorls, #1 and #2 in Figure 32-11A, are the **sepals,** the

individual green parts at the base of a flower, and the **petals,** the showy, usually colored, parts of the flower. Whorl #3 includes the male **stamens,** and whorl #4, the innermost whorl, includes the female **carpels.** The two inner whorls are said to be fertile, because they ultimately produce gametes—egg cells in whorl #4 and sperm cells in whorl #3.

Botanists have long noted that not all flowers have all four whorls. A flower that contains all four whorls is said to be a *complete* flower, while an *incomplete* flower is missing one or more whorls. A flower that contains both carpels and stamens (and produces gametes of both sexes) is said to be *perfect.* A flower missing either stamens or carpels is said to be *imperfect.* The primroses in Figure 32-11A are complete and therefore perfect flowers.

In contrast, the corn flowers in Figure 32-11B and C are incomplete and imperfect. The tassel-like corn flowers, at the ends of stems, have stamens but no carpels. The axillary flowers, at the end of "ears," have carpels but no stamens. Corn is an example of a *monoecious* [Greek, *monos* = single + *oikos* = house] species, in which both male and female flowers appear on the same plant. Willow trees and date palms are *dioecious* [Greek, *di* = two], with male and female flowers on different plants. (If we were to use this terminology for animals, we would say that humans are dioecious and snails are monoecious.)

The generalized flower develops from four circles, or whorls, of primordia. These form, from outside to inside, sepals, petals, stamens, and carpels. Many flowers are missing one or more of these parts.

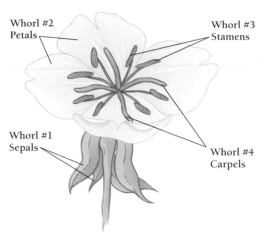

Whorl #2
Petals

Whorl #3
Stamens

Whorl #1
Sepals

Whorl #4
Carpels

Figure 32-11   The four whorls of flowers. A. A flower consists of up to four whorls of modified leaves. These are the sepals, the petals, the stamens, and the carpels. A flower with both carpels and stamens is perfect. A flower with all four whorls is complete. The evening primrose in A is complete and also perfect. B and C. Corn flowers possess two kinds of flowers on each plant, neither of which is complete or perfect. One set of flowers, the tassels, lack carpels and the other set, the silk, lack stamens.   *(B, William E. Ferguson)*

A. The four whorls of a complete, perfect flower—the evening primrose

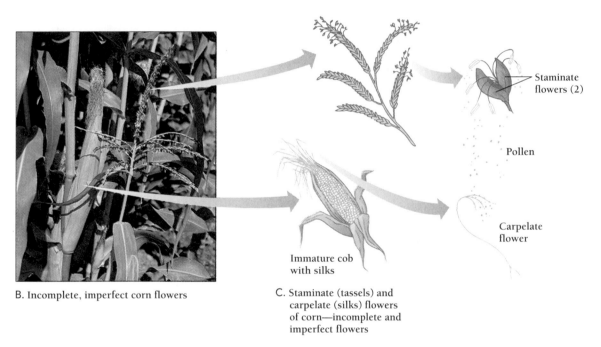

Staminate
flowers (2)

Pollen

Carpelate
flower

Immature cob
with silks

B. Incomplete, imperfect corn flowers

C. Staminate (tassels) and carpelate (silks) flowers of corn—incomplete and imperfect flowers

How Do Genes Regulate the Development of Flowers?

Among the most pressing questions being addressed by contemporary plant biologists are those dealing with the control of flower development: what molecules and cellular processes govern the development of complete versus incomplete flowers, or of male versus female flowers? In the last few years, major insights have come from the genetic analysis of plant mutants, particularly of the humble weed *Arabidopsis thaliana.*

Just as the genetic analysis of *Drosophila* and of *Caenorhabditis elegans* has revolutionized the understanding of animal development, so, it seems, that similar studies of *Arabidopsis* will lead to a new view of plant development. Many of the cellular decisions in plant development surprisingly resemble corresponding decisions in animal development, depending, for example, on regulation of gene expression by specific transcription factors.

Figure 32-12 shows abnormal *Arabidopsis* flowers that result from homeotic mutations, mutations in genes that influence the development of body plan. **Homeotic selector genes** are striking because the same sequences of DNA seem to regulate the development of a great diversity of organisms. Homeotic selector genes in fruit flies, for example, also regulate the development of the body in vertebrates. And genes with similar sequences and functions have been found in plants, vertebrates, and yeast cells. Despite great differences between plants and animals, both share common strategies and even common molecular designs for regulating the development of form.

In the *Arabidopsis* homeotic mutation *agamous*, the parts that would usually develop into stamens develop into petals and carpels (Figure 32-12B). Similarly, the *apetala3* mutation converts petals into sepals and stamens into carpels, and *apetala2* converts sepals into carpels and petals into stamens. Each of these homeotic selector genes changes the fate of one or more whorls.

Figure 32-12   Developmental mutations in the flower *Arabidopsis thaliana.* A. A normal flower with all four whorls: sepals, petals, stamens, and carpels. B. The *apetala2* mutation converts sepals into carpels and petals into stamens.   *(A, B, Leslie Sieburth, McGill University, Montreal)*

A.

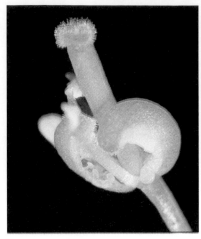

B.

Plant homeotic selector genes probably serve the same roles in all or most plants. Researchers have already found that similar genes serve the same roles in flower development in both *Arabidopsis* and snapdragons.

All of the homeotic selector genes for flower development discovered so far fall into one of three classes. To understand the action of these genes, researchers created a "triple mutant." This plant does not express any of the proteins known to participate in the development of flower parts. The result is a "flower" made entirely of leaves. If any of the three homeotic selector genes is expressed, however, the whorls then develop into flower parts. Researchers interpret this result to mean that meristems develop into leaves unless regulatory genes cause them to form something else. They say that leaf development is the "default pathway" of meristem development.

The homeotic selector genes responsible for flower development provide striking examples of the influence of genes on plant development. The next questions for researchers in this area concern the targets of the proteins specified by homeotic selector genes, with a view to understanding the molecular mechanisms of flower development.

Homeotic selector genes help regulate the development of flower parts in *Arabidopsis.* Mutations in these genes can result in extra petals, but no stamens, for example.

Secondary Growth Depends on Two Kinds of Lateral Meristems

Secondary growth refers to thickening, rather than the lengthening, of stems, branches, and roots. Secondary growth occurs not in the apical meristems but in two kinds of lateral meristems, or **cambia** [singular, cambium]. These are the **vascular cambium,** which produces the secondary xylem and phloem, and the **cork cambium,** which gives rise to *cork,* the waterproof, insectproof covering on woody stems and roots (Figure 32-13). The cells of both kinds of cambia lie in open-ended

cylinders that extend almost the entire length of woody stems and roots. Only the tips of woody stems and branches lack cambia.

The vascular cambium first appears between the xylem and the phloem. The vascular cambium adds secondary xylem to the outside of the primary xylem and secondary phloem to the inside of the primary phloem. In a woody stem, such as that of a tree, this process continues year after year. Each year the vascular cambium lays down new xylem outside the old and new phloem inside the old. The xylem accumulates from year to year as wood (Figure 32-13B). Among the most useful products of secondary growth are lumber, pulp for paper, cork, and root vegetables such as carrots, beets, turnips, and radishes.

As the trunk of a tree expands, it tears and destroys the mature epidermis. Just inside the epidermis, however, is the cork cambium, a layer of meristem cells that produce rows of protective cork cells. These tissues grow both outward, toward the bark of the tree, and inward, toward the phloem. The cells that are produced toward the outside deposit waxy, fatty suberin and other secondary compounds in their cell walls. The cells then die and the resulting bark continues to protect the tree from attacks by herbivores and pathogens. As the tree grows, the cork itself, or bark, stretches and tears. New inner layers repeatedly provide a continuous sheath of living and dying cells. We can identify different tree species by the characteristic pattern of tears in the bark—including the smooth strips of the *Eucalyptus,* the rough bark of the oak, and the regular cracks in the bark of ponderosa pine.

Secondary growth changes from season to season. In temperate zones, the vascular cambium is most active in the spring, producing large tracheids and vessel elements. In the summer, activity slows, and tracheids and vessel elements are much smaller. As a result, summer wood is much denser and darker than spring wood. The difference in appearance between the dark summer wood and the lighter spring wood creates a distinct layer of added vascular tissue called a **growth ring.** Each ring represents one season, or year, and the number of rings in a tree trunk reveals the age of the tree.

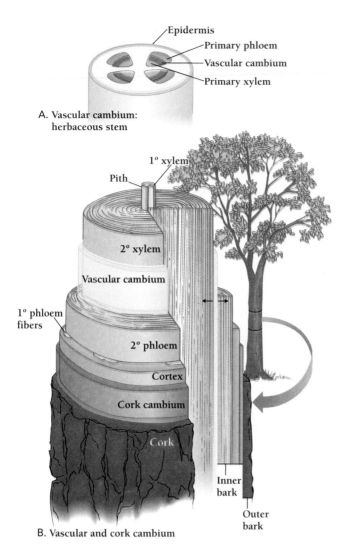

Epidermis
Primary phloem
Vascular cambium
Primary xylem

A. Vascular cambium: herbaceous stem

1° xylem
Pith

2° xylem

Vascular cambium

1° phloem fibers

2° phloem

Cortex

Cork cambium

Cork

Inner bark

Outer bark

B. Vascular and cork cambium

Figure 32-13   Secondary growth. A. In a herbaceous stem the cylinder of vascular cambium produces primary xylem (to the inside) and primary phloem (to the outside). B. In a woody stem, the vascular cambium produces secondary xylem (to the inside) and secondary phloem (to the outside). Everything outside of the vascular cambium is bark—including the secondary phloem, the cork cambium, and the periderm, which is produced by the cork cambium.

Secondary growth, which depends on vascular cambium, changes from season to season, resulting in characteristic growth rings in trees and other woody perennials.

MERISTEMS MAY PRODUCE WHOLE PLANTS

The same meristems that produce primary and secondary growth can also produce whole new plants. In a process called **vegetative reproduction,** plants reproduce nonsexually, with-

out fertilization or seeds. New strawberry plants form from runners, for example, and the houseplant *Kalanchoë diagremontiana* forms tiny new plants on its leaves.

Many people deliberately propagate their favorite house plants by means of vegetative reproduction. In some species, a stem cut from the main plant and placed in water generates new roots. Some species that generally do not form such "adventitious" roots, do so in the presence of the plant hormone auxin. Amateur and professional gardeners often dip cuttings into a solution of auxin to enhance root formation. When the rooted cutting is planted, it grows into a clone—a mature plant that is genetically identical to the parent plant.

One of the oldest and most commonly used means of artificial vegetative propagation is grafting. Horticulturists propagate several crops almost exclusively by grafting, including oranges, grapefruit, lemons, and limes; avocados, apples, and plums; roses; and grapes. In **grafting,** a horticulturist grafts a cutting of one woody plant onto the root of another plant. The stem cutting is called the *scion* and the root is called the *rootstock* (or simply the *stock*). Recall from Chapter 31 that French grape varieties are universally grafted onto American rootstocks. Rather than try to breed a single hybrid with the best grapes and the best roots, grape growers simply graft a variety that makes good grapes onto a rootstock that resists pests or supplies nutrients well. Most commercial fruit trees and roses are the products of such grafts.

The trick in grafting is to bind scion and stock so that their cambia come together. Secondary growth establishes continuity both within the xylem and the phloem of scion and stock. The buds, leaves, flowers, and fruits all derive from the scion, while the root system derives entirely from the stock.

Another method of propagating plants vegetatively is through tissue culture. Tissue culture has become increasingly important with the advent of genetically engineered plants. Carrots, cotton, tomatoes, and Douglas fir trees have all been propagated this way. The starting material is a *callus* (undifferentiated tissue) derived from the cut surface of a plant or from a single cell (which may or may not be genetically engineered.

In the proper chemical environment, a callus develops into a plantlet, a tiny plant with both shoot and root. This plantlet can develop into a whole plant, able to reproduce itself either sexually or vegetatively. Because the callus does not photosynthesize, the "proper chemical environment" must include a supply of energy, usually sucrose (table sugar), nutrients, and suitable plant hormones. In Chapter 33, we discuss how these hormones coordinate other responses of plants to their environments.

Meristems allow plants to reproduce themselves nonsexually in a process called vegetative reproduction that does not involve fertilization or seed production. Cuttings, grafting, and tissue culture allow the artificial propagation of desirable plant varieties.

STUDY OUTLINE WITH KEY TERMS

In **alternation of generations** in plants, the diploid **sporophyte** produces haploid spores, and the haploid **gametophyte** produces haploid gametes. A flower contains both male and female reproductive structures. **Stamens** produce haploid **microspores,** which develop into male gametophytes, inside pollen grains. **Carpels** produce haploid **megaspores,** which develop into female gametophytes, the **embryo sacs.** The male and female gametophytes in turn produce male and female gametes. Each **carpel** connects to a **style,** which connects to the **stigma.** Each **ovary** (made of one or more carpels) produces one or more **ovules,** which form the embryo sac, the **nucellus,** and the **integuments** of the developing seed.

Upon pollination, a **pollen grain** develops into a **pollen tube** with two sperm cells. Flowering plants undergo **double fertilization,** in which the two sperm nuclei fuse with two cells of the embryo sac within the carpel. One fertilization event produces the zygote, which grows into an embryo. The other gives rise to the **endosperm,** a triploid tissue that nourishes the growing embryo.

The zygote divides to form an embryo that first resembles a sphere, then a heart, then a torpedo. Development stops after the seed matures, when the embryo loses most of its water and becomes **dormant.** A seed consists of a dormant embryo and its adjacent endosperm, surrounded by a seed coat.

The resumption of growth is called **germination,** a process that begins with **imbibition,** the taking in of water. The first part of the embryo to emerge from the seed is the **radicle,** which contains the root's apical meristem and quickly forms a functioning root system. The shoot apical meristem lies above the point of attachment of the **cotyledons,** or seed leaves. Plants are divided into two classes—the **dicots** and the **monocots**—based on the number of **cotyledons.**

A seed may maintain its dormant embryo for many years, until the proper environmental changes bring about germination. Germination triggers a huge increase in biochemical activity, for which the growing plants derive energy from stored starches, oils, and proteins.

In grains, **α-amylase** and other enzymes secreted by the **aleurone layer** break down starches in the endosperm.

The primary growth of both roots and shoots depends on **apical** meristems. The growing root tip has three distinct zones, the **root cap,** which secretes **mucigel,** the apical meristem, and an **elongation zone.** The growing tip of a stem, however, must be able to produce stem, leaves, branches, and flowers. Each leaf originates as an extension of the apical meristem, called the **leaf primordium.** At each **node,** one or more **buds** each contain several leaf primordia. In between the nodes are the **internodes.** A bud may lie at the tip of a shoot, in which case it is called a **terminal bud,** or just above the points where leaves join the stem, in which case it is called an **axillary bud.** Many species show **apical dominance,** meaning that terminal buds suppress the development of axillary buds.

The four **whorls** of a flower may form carpels, stamens, **sepals,** and **petals.** Mutations in several **homeotic selector genes** in *Arabidopsis* can result in interesting deformities such as extra petals, but no stamens. Similar genes are found in vertebrates and yeasts.

Secondary growth depends on two kinds of lateral meristems, the **vascular cambium,** which produces secondary xylem and phloem, and the **cork cambium,** which produces cork, the outer protective covering of woody stems and roots. The rate of xylem growth changes from season to season, resulting in characteristic **growth rings.**

In **vegetative reproduction,** the same meristems that produce primary and secondary growth produce whole new plants. Many plants propagate in nature by producing adventitious roots or stems. Farmers, horticulturists, and researchers also propagate desirable varieties by stimulating adventitious root formation with auxin, by grafting a cutting of one plant together with the root or stem of another, or by allowing tiny plants to develop from undifferentiated tissue in artificial culture.

REVIEW AND THOUGHT QUESTIONS

Review Questions

1. What are the two products of double fertilization?
2. Describe the relationship between the embryo, aleurone layer, and starchy endosperm in promoting growth and germination in cereals.
3. Compare the following terms:
   (a) meristem versus node versus internode
   (b) terminal bud versus axillary bud
   (c) epicotyl versus hypocotyl versus radicle
4. Describe the three zones of growth in a root.

5. Explain how a single gene mutation can dramatically alter flower morphology.
6. Why is leaf development considered the default pathway of meristem development? How was this demonstrated?

Thought Questions

7. What interesting roses might be produced by applying our knowledge of *Arabidopsis* regulatory genes?
8. Why would yeasts, plants, and vertebrates share similar genes for regulating development?

SELECTED READINGS

Berg, Linda R., *Introductory Botany: Plants, People, and the Environment,* Saunders College Publishing, Philadelphia, 1997. This book covers the basics of plant science in a friendly and unintimidating style.

Fussel, Betty, *The Story of Corn,* Alfred A. Knopf, New York, 1992. An engaging account of the history, biology, and sociology of corn.

Mauseth, James. D., *Botany: An Introduction to Plant Biology,* 2nd ed., Saunders College Publishing, Philadelphia, 1995. An authoritative introduction to the biology of plants.

Moore, Randy, W. Dennis Clark, Kingsley R. Stern, and Darrell Vodopich, *Botany,* Wm. C. Brown Publishers, Dubuque, 1995. A friendly, well-written textbook on the science of plants.

Niklas, Karl J., *Plant Biomechanics: An Engineering Approach to Plant Form and Function.* Chicago, University of Chicago Press. 1992. This book is wonderfully written and highly accessible where Niklas is first introducing a topic or concluding one—that is, at the beginning and ending of each chapter. Much of this book, however, demands a familiarity with either botany or engineering beyond that of the average reader.

▶ On-line materials relating to this chapter are on the World Wide Web at http://www.saunderscollege.com/lifesci/
Click on Tobin/Dusheck: *Asking About Life.*

A Life Lived Full

In 1928, a young English chemist found himself on the job market in the middle of the worst economic depression in modern times. Unable to find an academic position in England, 25-year-old Kenneth Thimann landed a job as an instructor in biochemistry at the new California Institute of Technology, in southern California. In 1929, Thimann and his artist wife, Ann, sailed for America on their first wedding anniversary. It was the beginning of a new life, a life that would profoundly influence American agriculture and, even, the course of the Vietnam War.

Like many a biologist's career, Thimann's rests on an intriguing observation by Charles Darwin. In the 1880s, Charles Darwin and his son Francis performed a series of experiments designed to explain **phototropism,** the tendency of plants to bend toward light (Figure 33-1A). The Darwins showed that something in the tip of an oat seedling causes the seedling to bend toward light. When Charles and Francis Darwin masked the tips of seedlings with a lightproof foil, the seedlings no longer bent toward the light (Figure 33-1B). Later researchers found that the same mask applied in the growth region just below the tip had no effect on bending; the seedlings grew toward the light in a normal fashion.

Forty-five years later, the Dutch plant physiologist Frits Went and his father demonstrated that the "something" was a chemical substance in the tip of the plant. In 1926, the younger Went was serving his compulsory military service during the day and spending his evenings and nights working as a graduate student in his father's plant physiology laboratory at the University of Utrecht. Frits Went set out to isolate the growth-promoting substance from the tip of the oat seedling, using the strategy illustrated in Figure 33-2. On a tiny cube of gelatin, he placed as many plant tips as he could fit. After an hour, he removed the tips, reasoning that the hypothetical substance would have entered the cube of gelatin. He then placed the block on one side of a seedling stump that was growing in the dark and waited to see what would happen.

If the gelatin contained a substance that could alter growth, then the stumps should bend toward or away from the side with the cube of gelatin. At first nothing happened. By 3 a.m., however, the seedling had begun to curve away from the gelatin cube. Excitedly, Went ran home, burst into his parents' bedroom, and declared, "Father, come and see, I've got the growth substance!" Sleepily, his father suggested he repeat the experiment during the day. "If it is any good," the elder Went said, "it will work again, and then I can see it." Indeed, it worked again, and Frits Went named the mysterious growth substance **auxin** [Latin, *augmentum* = increase].

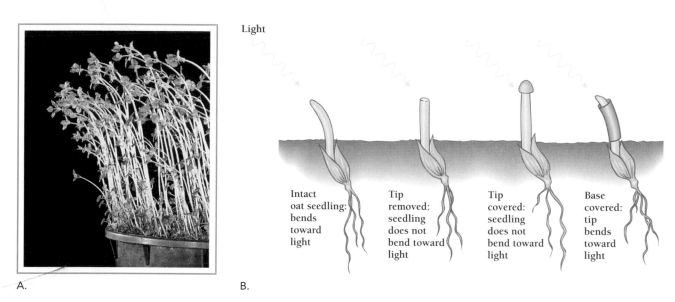

Light

Intact
oat seedling:
bends
toward
light

Tip
removed:
seedling
does not
bend toward
light

Tip
covered:
seedling
does not
bend toward
light

Base
covered:
tip
bends
toward
light

A.

B.

Figure 33-1 What makes plants bend toward light? A. Seedlings bending toward light. B. Charles and Francis Darwin showed that the part of the seedling that responds to light is the tip. When the tip is covered, the plant will not bend. Later research showed that covering the growth region, below the tip, has no affect on bending. *(A, Jerome Wexler/ Photo Researchers)*

When Kenneth Thimann arrived in California 3 years later, one of the first people he encountered was fellow instructor Herman Dolk. Dolk had been at the University of Utrecht with Frits Went, and he recruited Thimann to help him identify the chemical nature of auxin. Thimann and Dolk worked doggedly for 2 years, struggling to isolate the growth substance and solidifying a friendship.

Thimann's life seemed to couple tragedy with triumph, however, for in 1931, Dolk was killed in a car accident. Thimann was left to finish their work by himself (Figure 33-3). "It was kind of a lost feeling," he recalled almost 60 years later. "When Herman was killed, I was so deep into it that there was no choice but to go on."

Thimann went on to isolate the first known plant hormone, indole-3-acetic acid (IAA), which he showed to be the

compound responsible for all the actions of auxin. Then, in an inspired move, he determined that by substituting other chemical structures for the indole in natural auxin, he could make synthetic auxins with the same properties as the natural hormone (Table 33-1). But these synthetic auxins differed from the natural ones in one important respect. Plants did not possess enzymes for breaking down the synthetic versions. The synthetic auxins, therefore, persist in the plant much longer than natural ones.

Thimann and others began a series of experiments to see how auxin affects plant growth. Auxin, they found, promotes growth by stimulating rapid cell division in the apical meristems where it is produced. Auxin also travels down the stem,

"We look on nature as a book—every now and then we can turn a page. It's a great thrill."

stimulating cells behind the apical meristem to lengthen.

In 1934, Thimann and Frits Went, who had by then left Holland to join him at Caltech, discovered that auxin also stimulates plants to grow roots. It was a tremendously valuable discovery. Today, virtually every nursery in the United States and Europe uses artificial auxin to induce the formation of roots in cuttings. With auxin, desirable crops and flowers can be cloned indefinitely.

In the same year, Thimann and Folke Skoog, at the University of Wisconsin, showed that auxin could inhibit growth as well as promote it. Auxin is responsible for apical dominance, the tendency of the apical meristem to suppress growth in lateral meristems. In addition, auxin inhibits leaves, fruits, and flowers from falling. Today, many fruit growers spray their orchards with synthetic auxin to prevent the fruit from falling before all of the fruit in the orchard is ready for harvest.

But it was Thimann's next discovery that influenced agriculture most dramatically. Thimann had attracted the interest of Harvard University, which invited him to join its faculty. At Harvard, he noticed that although a little auxin may either promote growth or inhibit growth, large amounts of auxin are fatal to plants. Too much auxin induces such rapid cell division that the phloem becomes plugged with nonconducting cells, and the plant soon dies. Synthetic auxins are especially deadly because plants lack enzymes necessary to break down the synthetic versions.

During World War II, researchers at the University of Chicago developed Thimann's synthetic auxins for use as herbicides. The plan was to use the chemicals to defoliate areas hiding Japanese soldiers. In fact, the herbicides were never used against the Japanese; the atomic bomb made further ground battle unnecessary.

The herbicides nonetheless came into heavy domestic use immediately after the war. For the first time, road crews did not need to mow the edges of highways. They sprayed herbicide instead. Farmers no longer needed to plow under acres of weeds before planting a crop. Herbicides were easier, and not plowing reduced erosion of valuable topsoil as well. By using herbicides, corn and wheat farmers increased yields by 30 percent.

The most famous herbicides are the synthetic auxins 2,4-D and 2,4,5-T (Table 33-1). In the 1960s, American military strategists fighting to win the war against the North Vietnamese revived the idea of

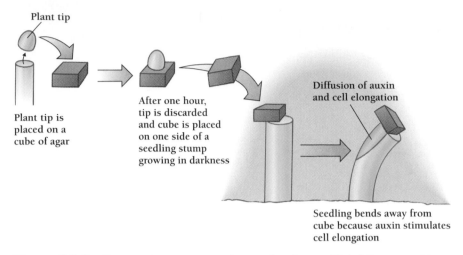

Figure 33-2 **Does a substance cause plants to bend toward light?** To answer this question, Frits Went placed the tips of seedlings on blocks of agar, in the hope that if a chemical substance were responsible, it would diffuse into the blocks of agar. He then balanced the blocks on the stumps of seedlings so that any substance in the agar would diffuse into only one side of the stem. The results confirmed his suspicions. Some substance causes the seedlings to bend. In addition, he learned that the seedling bends *away* from the substance.

using herbicides in warfare. They sprayed Vietnamese forests with heavy doses of herbicides to expose enemy positions. (They also destroyed crops with the same herbicides.) The trees dropped their leaves, revealing soldiers and others to American and South Vietnamese aircraft flying overhead.

The U.S. government experimented with a variety of herbicides, but the most commonly used herbicide was "Agent Orange," a mixture of 2,4-D and 2,4,5-T. Chemical companies, under pressure to supply huge amounts of these synthetic auxins to the government, produced bulk quantities that turned out to be heavily contaminated with the poison dioxin. In later years, American veterans who had been exposed to the dioxin attributed a variety of illnesses to Agent Orange. In the 1980s, 27,000 Vietnam veterans sued five chemical companies for illnesses they suffered after returning from Vietnam. The chemical companies settled out of court for $180 million.

Thimann took credit for the many uses to which his discoveries were put, but with reservations. He insisted that his research was always basic research, not applied research. "Improvements that are based on pure research are more fundamental than research based on finding applications," he said. He also argued that

2,4,5-T was not dangerous, pointing to a fire in 1976 at an herbicide factory in Seveso, Italy. "A lot of people got spattered with 2,4,5-T that contained dioxin." Exposed workers developed chloracne, a severe but temporary pimpling that is characteristic of dioxin exposure. "But that's not cancer," said Thimann. "It heals by itself after a time, just like any other acne. They've done studies galore, studied their babies and everything. In the years since, nobody's shown any serious effects, and researchers know the workers were really dosed."

Thimann spent nearly 30 years at Harvard University, devoting himself to teaching and mentoring hundreds of students. When he was 58, he left Harvard to return to California. At the brand new University of California, Santa Cruz, he assumed the task of building a first-class science faculty and creating an environment in which students could come into close contact with their professors. His secretary once recalled that the two of them sometimes worked until midnight to get all their work done, but Thimann never refused to see a student. Neither his research nor the work of building a new university campus ever prevented him from putting students first. Even after he retired from teaching at the age of 68, he continued to supervise undergraduate research for

Figure 33-3 **Kenneth Thimann.** When asked for some advice to give to students, Thimann said, "Never give up." *(UCSC Photo Lab)*

more than 15 years. In his 90s, he still went to work every day.

In 1983, the Italian government awarded Thimann the Balzan Foundation prize, one of the highest awards in biology. Great Britain's Royal Society and France's Academy of Science each elected him as a member. When Thimann's wife, Ann, died in 1991, he moved to Pennsylvania to be near his grown daughters. The University of Pennsylvania immediately provided him with an office and he continued his involvement in research and academic life until he died in January of 1997. He spent his last years studying senescence in plants. He never lost his love of science or of plants, once remarking, "We look on nature as a book—every now and then we can turn a page. It's a great thrill."

Table 33-1 Natural and Synthetic Plant Hormones and Their Effects

Hormone	Effects	Structure
Auxins	Stimulate cell elongation, root growth, differentiation and branching, apical dominance, development of fruit; acts in phototropism and gravitropism	Indoleacetic acid (IAA)
2,4-D	Synthetic auxin used as herbicide	
Cytokinins	Stimulate cell division and growth, germination, and flowering; delay senescence; affect root growth and differentiation	Zeatin
Gibberellins	Promote seed and bud germination, stem elongation, leaf growth, flowering and development of fruit; affect root growth and differentiation	Gibberellic acid (GA_3)
Abscisic acid	Inhibits growth; closes stomata during water stress; counteracts breaking of dormancy	
Ethylene	Promotes fruit ripening; promotes or inhibits growth and development of roots, leaves, flowers, depending on species; opposes some auxin effects	

KEY CONCEPTS

1. Hormones direct development and coordinate cellular and biochemical activities in many cells throughout the plant.

2. Auxin coordinates phototropism and has many effects on plant growth and development.

3. Gibberellins promote germination, trigger the mobilization of food reserves in germinating seeds, and stimulate growth in mature plants.

4. Cytokinins, in conjunction with auxin, stimulate cell division and growth.

5. Ethylene promotes leaf senescence and fruit ripening and inhibits elongation of stems and roots.

6. Abscisic acid promotes dormancy in buds and seeds.

7. In gravitropism, cells in the roots somehow detect the direction of gravity. If the root is horizontal, Ca^{2+} ions and auxin seem to facilitate the turning of the root downward.

8. Phototropism and solar tracking depend on a pigment other than chlorophyll.

9. Many responses to seasonal changes depend on a pigment called phytochrome.

HORMONES DIRECT DEVELOPMENT AND COORDINATE ACTIVITIES IN CELLS THROUGHOUT THE PLANT

A **hormone** is an organic compound produced in a tissue or organ and transported to another tissue or organ, called the **target,** where it produces one or more specific effects. Like animal hormones, plant hormones move from source to target both by diffusion from cell to cell and by transport in the vascular tissues. Plant hormones also exert their effects at very low concentrations, in doses of as little as a billionth of a gram per plant. Unlike animal hormones, however, plant hormones not only affect distant target cells but also influence the very cells that make them.

Plant physiologists have identified five major plant hormones: auxin, gibberellin, cytokinin, ethylene, and abscisic acid. In fact, auxin, gibberellin, and cytokinin are not single compounds, but each is a small family of related compounds. There are three naturally occurring auxins, at least 60 gibberellins, and several cytokinins. Other related compounds can trigger the same biological responses, but are not made by plants. These include compounds isolated from animals, fungi, and bacteria, as well as the products of laboratory synthesis.

We may ask how a few types of hormones can control complex sequences of development and respond to an ever-changing environment. The answer is that

1. The same hormones affect different tissues differently.
2. Hormones have different effects at different concentrations.
3. Hormones interact with one another in complex ways.

Auxin Coordinates Many Aspects of Plant Growth and Development

Frits Went's early experiments with auxin suggested that the hormone could mediate phototropism if light somehow caused more auxin to be concentrated on the side of the tip away from the light. In the early 1960s, Winslow Briggs, a plant physiologist at Stanford University, tested this prediction (Figure 33-4). Briggs first asked whether light affected the total amount of auxin in the growing tip of an oat seedling. After placing tips on blocks of agar, he illuminated some tips and kept others in the dark. He then used the degree of curvature to estimate the amount of auxin in each tip. Briggs found no difference in the amounts of bending or, therefore, auxin in the two sets of tips. Phototropism, he concluded, must depend not on the total amount of auxin in an illuminated tip, but on the distribution within the tip.

Briggs then looked for differences in auxin concentrations between the side of the tip nearer the light and the side away from the light. He inserted a thin piece of mica between the two sides. The mica completely separated the lighted side from the shaded side, and Briggs found no difference in the auxin levels of the two sides. When Briggs left a gap of half a millimeter near the very end of the tip, however, the auxin concentration on the shaded side rose to nearly twice that on the illuminated side. The light somehow caused the transport of auxin from the lighted to the shaded side of the tip. The higher concentration of auxin on the shaded side then caused that side to elongate more rapidly, leading to a curve in the direction of the light.

BOX 33-1

What is a bioassay?

A fixed amount of auxin gives a reproducible amount of curvature. Went's experiment established a *bioassay,* a method that estimates the concentration of a substance by measuring its biological activity. Went's bioassay allowed plant physiologists to estimate auxin concentrations by measuring seedling curvature with a protractor (Figure A). Another, newer approach is to measure the concentration of auxin and other plant hormones using chromatography, which depends on chemical rather than biological properties. However, bioassays continue to be useful for the identification of hormones and other functionally important molecules not only in plants, but in other types of organisms.

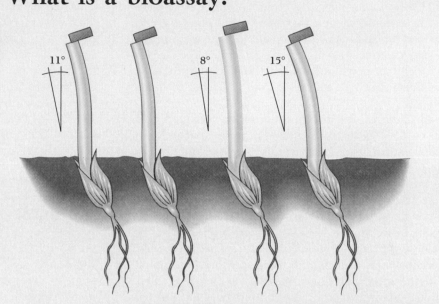

Figure A   In Winslow Briggs's experiment on how auxin is redistributed, he used degrees of bending in a seedling as a measure of auxin concentration. Nowadays, Briggs would probably use high-pressure liquid chromatography.

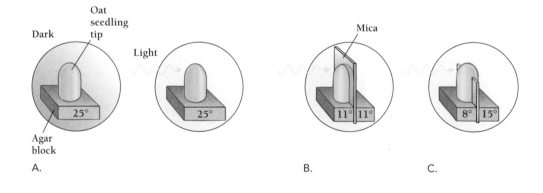

A.   B.   C.

Figure 33-4 Does auxin move from one side of the seedling to the other? Or does the seedling synthesize extra auxin on one side? Winslow Briggs's experiment demonstrated that auxin is redistributed in response to light. Briggs put oat seedling tips on blocks of agar then measured the auxin content of the blocks under different conditions. The numbers on the blocks of agar are the approximate degrees of bending the blocks caused in oat seedlings. A. The auxin concentration from a seedling in the dark is the same as the concentration when the seedling is exposed to light. Therefore, the seedling tip does not synthesize more auxin in response to light. B. The oat seedling tip is exposed to light on one side only and divided in half by an auxin-proof barrier. The resulting auxin concentrations are the same on each side. Therefore, extra auxin is not synthesized on the shaded side. C. If the barrier is incomplete, so that auxin can move from one side to the other, the auxin concentration becomes much higher on the shaded side. Therefore, auxin moves from the lighted side to the shaded side of the seedling tip.

Briggs's experiments, like those of Went, suggested that auxin acts at the level of individual cells, stimulating cell elongation in both shoots and roots. One way that auxin stimulates elongation is by regulating gene expression, thus stimulating the synthesis of several mRNAs (and therefore proteins). In addition to its effects on cell elongation, auxin also affects the gene expression in target cells. Some of these changes occur rapidly, within 10 to 20 minutes after exposure to auxin.

One area of influence is probably the cell wall. In elongation, bonds in the cell's rigid wall loosen, allowing the cell contents to expand. Turgor pressure exerted by the vacuole against the cell wall pushes and stretches the wall, and the cell wall then reforms around the expanded cell. The primary event in cell elongation, then, is the selective breaking of the bonds that make the wall rigid.

Besides controlling cell elongation and stimulating the formation of roots, auxin affects the growth of many other plant tissues. For example, the relative concentrations of auxin and gibberellin in vascular tissue help establish the developmental fate of individual cells—whether the cells will develop into xylem, which transports water and minerals from the roots, or phloem, which transports sucrose and other materials from the leaves.

Auxin also affects the course of development of plant cells in tissue culture. For example, a solution of auxin with 2 percent sucrose stimulates a callus culture (a mass of undifferentiated cells) to develop xylem, while auxin with 4 percent sucrose stimulates the development of phloem. Auxin also stimulates the development of fruit, sometimes even in the absence of fertilization. Unfertilized tomato flowers treated with auxin, for example, develop into seedless tomatoes.

Besides its growth-promoting activity, auxin also inhibits the development of axillary buds. In plants that show apical dominance, axillary buds do not develop into branches unless the terminal bud is removed. Even after removing the terminal bud, however, axillary buds do not develop if one places auxin on the cut surface.

Finally, as we have seen already, high concentrations of auxin are effective as herbicides. Synthetic auxins are especially effective because plants cannot break down these substances, and so they persist for a long time. Three synthetic auxins— 2,4-D (2,4-dichlorophenoxyacetic acid), 2,4,5-T (2,4,5-trichlorophenoxyacetic acid), and MCPA (methylchlorophenoxyacetic acid)—have been particularly popular because they are toxic to dicots and not to monocots. Grasses, corn, and wheat are monocots, so farmers can spray an entire field of corn or wheat, for example, killing all the dicot weeds without harming the crop. Similarly, ranchers can kill sagebrush and mesquite, both dicots, in grassy pastures or rangelands, and lawnkeepers can kill every dicot in a lawn of grass.

Phototropism results from the movement of auxin from the lighted side of the growing tip of a plant to the shaded side. Auxin stimulates cell elongation, the formation of roots, and the development of fruit. In conjunction with gibberellin, auxin helps determine whether cells differentiate into xylem or phloem. Auxin also inhibits the growth and development of axillary buds. Finally, in large concentrations, auxin and its synthetic analogs are potent herbicides.

How Do Gibberellins Act on Target Cells?

Plants have more kinds of **gibberellins** than of any other hormone. All the gibberellins are complicated organic molecules with 19 or 20 carbon atoms grouped into 4- or 5-ring struc-

tures (Table 33-1). Most plant species have several different kinds of gibberellin, some containing as many as 15. Some researchers think that most of the different forms are precursors or degradation products of a few active forms. Others think that different forms may have different biological effects.

Gibberellins generally affect the overall growth of intact plants far more than the other plant hormones. In fact, their discovery arose from studies of a disease of rice plants caused by a fungus called *Gibberella fujikoroi*. The disease, first described in Japan in the 1890s, causes rice plants to grow so tall that they cannot support their own weight. The plants topple over into the water and rot. Japanese farmers called the disease *bakanae,* "foolish seedling." In the 1930s, two Japanese researchers identified the chemical compound made by the fungus that is responsible for the extravagant growth of the rice seedlings. The active compound is the gibberellin GA_3. Later research in the 1950s showed that gibberellins are also present in uninfected plants and are likely to serve a role in the normal regulation of plant growth.

Like auxin, gibberellins seem to promote cell elongation by loosening cell walls, but with a more dramatic effect (Figure 33-5). If a segment of oat stem is treated with gibberellin and sucrose (to provide energy to the non-photosynthesizing tissue) the oat stem will grow 15 times as long as untreated oat stems. Unlike auxin, however, gibberellin's effects are chiefly confined to stems.

Gibberellins also stimulate the breakdown of starches and sucrose to monosaccharides (simple sugars). The increased glucose and fructose concentrations not only increase the availability of energy but also increase the solute concentration in-

Figure 33-5 The plant hormone gibberellin causes stems to lengthen dramatically. This flower "bolts" when levels of gibberellic acid rise. *(Courtesy of B.O. Phinney, University of California, Los Angeles)*

side the cell. More water then flows into cells and contributes to their elongation.

In some cases, gibberellins can stimulate cell division—for example, in the apical meristem of the shoot. Finally, as we already mentioned, gibberellins can act to change the pattern of expression of specific genes, as in the case of α-amylase induction in the germinating barley seed.

Some plant mutants—of peas, corn, beans, rice, and other crops—lack gibberellins altogether and are much shorter than their normal counterparts. When these dwarf mutants are treated with gibberellins, in some cases with as little as a billionth of a gram per plant, they grow to normal size.

Gibberellins appear to be important in the normal growth of shoots, the germination of seeds, the flowering of plants, and the mobilization of food reserves from the endosperm of cereals and cotyledons of other plants. Because gibberellins are present in most tissues of a plant and because they are active at such low concentrations, however, their precise role has been difficult to understand.

Gibberellins are complex organic molecules that promote cell elongation, mainly in stems; the breakdown of starches and sucrose; cell division; and gene expression.

Cytokinins Stimulate Cell Division and Growth

Since the early years of this century, researchers have known that some substances, now called **cytokinins,** stimulate cytokinesis (cell division) in plant tissue culture. Pieces of vascular tissue and extracts of coconut milk can produce this stimulation. In all cases, however, these materials stimulated cell division only in the presence of appropriate levels of auxin. Thus, cytokinin and auxin *together* affect cell division. As we will see, such interactions are the rule in plants, not the exception.

In addition to stimulating cell division, cytokinin and auxin together help determine the developmental fate of plant tissue. The relative levels of cytokinin and auxin establish, for example, whether a callus culture will continue to divide without differentiation or will differentiate into shoots or roots.

Although cytokinin and auxin work together, they also work in opposition to one another. The two hormones have contrasting effects on the growth of axillary buds. Auxin, applied to a cut stem, suppresses the growth of axillary buds, while cytokinin, applied directly to an axillary bud, stimulates cell division and growth. A plant bacterial infection called "witches' broom disease," provides an extreme example of the effects of cytokinin. The disease-causing bacteria secrete cytokinin, which stimulates the increased lateral branching. Although experimentally applied cytokinin and auxin affect the pattern of terminal and lateral branching, researchers are still not certain of the precise roles of these hormones in the normal regulation of lateral branching.

When a leaf is removed from a plant, it usually begins to lose chlorophyll and to turn yellow. This yellowing is part of the process of **senescence,** the breakdown of cellular components leading to cell death. Application of cytokinin to a cut leaf, however, delays senescence.

During the normal life cycle of a plant, cytokinin transported to leaves by the xylem appears to prevent senescence from occurring prematurely. Several fungi, and even two species of caterpillars, create "green islands" on otherwise yellowing leaves by secreting cytokinins.

Naturally occurring cytokinins are present at exceedingly low levels. The first cytokinin isolated from plant tissue, for example, required the extraction of 60 kg of corn kernels to obtain 1 mg of pure cytokinin. Researchers have found only a few cytokinins in plant tissues, but a number of other compounds, including natural products of several fungi and bacteria, have cytokinin activity.

Modern techniques have made it possible to determine the amounts of specific cytokinins at levels as low as a billionth of a gram, a nanogram (ng). This sensitivity has allowed the measurement of cytokinin concentrations in many plant tissues. At this time, however, researchers have been unable to demonstrate changes in cytokinin concentrations that can account for the response of plants to changed environmental cues. Some plant scientists therefore think that it is incorrect to call cytokinins hormones. Nor does anyone yet know exactly how cytokinins act on plant cells, although it seems likely that they affect the pattern of gene expression.

Together with auxin, cytokinins stimulate cell division, delay senescence, and determine how cells differentiate. Unlike auxin, cytokinins stimulate cell division and growth in axillary buds. Cytokinins function at extremely low concentrations.

How Did Researchers Identify Ethylene as a Plant Hormone?

Ethylene is a much simpler molecule than the other plant hormones, consisting of just two carbon atoms and four hydrogen atoms (Table 33-1). Unlike the other plant hormones, ethylene is a gas and is not transported in the vascular system. And unlike other hormones, plant or animal, ethylene is produced in reactions that have nothing to do with plants—in smoke from fires, for example.

The most dramatic, delicious, and economically important effect of ethylene is on fruit ripening. The ripening of a fleshy fruit (such as an apple or a tomato) is a complex process. Among the chemical changes that occur during ripening are the breaking down of chlorophyll, changing the fruit's color. Ripening also involves the breakdown of starches, increasing the fruit's sweetness, and the breakdown of cellulose, making it softer. Ethylene starts or speeds up these changes.

Ethylene also triggers the increased production of more ethylene, so that once ripening starts it spreads rapidly, both within a single fruit and from fruit to fruit. This positive feedback acceleration of ripening has practical consequences for fruit lovers. One can hasten the ripening of green fruit by keeping them in an enclosed bag to trap the ethylene and by adding one piece of fruit that is quite ripe.

Ethylene's effects also explain why "one rotten apple spoils the barrel." A rotten apple releases large amounts of ethylene and speeds up the ripening of those around it. These apples, in turn, release ethylene and soon the whole barrel is filled with rapidly overripening apples.

Fruit growers and shippers knew how to hasten ripening long before they knew the chemical cause. The ancient Chinese ripened fruit in a room with burning incense, and Puerto Rican growers once built bonfires near their crops to ripen their pineapples. Both these procedures work because the incense and wood fires produce ethylene.

The identification of the plant hormone ethylene came from a practical problem first recognized in Germany in 1864. Before the invention of electric power, city streets were lit with gas lights. Gas lines carrying "illuminating gas" to each street light often developed leaks. When nearby shade trees began to lose their leaves, workers discovered that the gas was the cause. The demonstration that the active component of illuminating gas was ethylene came, however, only in 1901. A Russian graduate student named Dimitry Neljubov showed that illuminating gas stimulated pea plants to grow horizontally and that the active component of the gas was ethylene.

Ethylene is present almost everywhere in a plant. Under the right conditions, it causes plants to drop both leaves and fruit, a process called **abscission**. In addition, it participates in many other responses to environmental changes. For example, ethylene causes a "triple response" in the stems of tomato seedlings—stems thicken, stop elongating, and begin to grow horizontally (Figure 33-6). Tomato seedlings grow in a similar way if they encounter a barrier as they emerge from the soil. The production of ethylene, primarily in the hook of the seedling's epicotyl, seems to coordinate the plant's ability to grow around objects. When a seedling finds its way to light, the amount of ethylene decreases, and the seedling resumes its upward growth. Tomato mutants that are unresponsive to ethylene do not show any aspect of the triple response (Figure 33-6).

Ethylene is a gas that stimulates fruit ripening and the release of more ethylene. Ethylene also triggers the abscission of fruit and leaves, and the triple response in seedlings.

Abscisic Acid Promotes Dormancy in Buds and Seeds

Abscisic acid [Latin, *abscissus* = to cut off] takes its name from its ability to stimulate the abscission (dropping) of leaves and fruit (Figure 33-7). Despite its name, abscisic acid plays only a minor role in abscission, but it does play a major role in the suspension of development in buds and seeds.

A.

B.

Figure 33-6   The triple response to ethylene. In seedlings, the stem thickens, stops elongating, and begins to grow horizontally. A. Triple response of normal tomato seedlings. B. Response of mutant tomato seedlings that are insensitive to ethylene.   *(Yen, Lee, Tanksley, Klee, and Giovanni,* Plant Physiology *(1195) 107:1343–1353)*

Unlike the other identified plant hormones, abscisic acid is primarily an inhibitor, not a stimulator. It antagonizes the effect of gibberellins and auxins. Abscisic acid slows the growth of oat seedlings, inhibits the germination of wheat embryos, and prevents the synthesis of α-amylase by the aleurone layer of barley seeds. At a cellular level, it inhibits the synthesis of both proteins and RNA. Abscisic acid also coordinates plant responses to a variety of environmental stresses, including drought, excess salt, waterlogging, cold, and mineral deficiency.

Abscisic acid suspends development in buds and seeds and helps stimulate abscission, thus antagonizing the effects of gibberellins and auxins.

Are There Other Plant Hormones?

Research in the past decade has shown that a number of other chemicals can coordinate cellular activities in different parts of a plant. For example, wounding a plant leaf can evoke both lo-

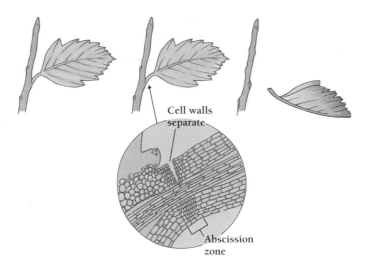

Figure 33-7 **Abscission.** Abscisic acid stimulates cells in the abscission zone at the base of a petiole to shrink and separate from one another. Finally, the attached fruit or leaf drops to the ground.

cal and long-distance effects, suggesting the involvement of one or more plant hormones. When caterpillars chew on a leaf, many plants produce *systemin,* an enzyme that inhibits the action of protein-digesting enzymes. The result is a bad case of indigestion for the caterpillar. Plants can also develop resistance to specific viruses, bacteria, or fungi, not only at the site of infection but also far away. In both cases, plant biologists have identified signaling molecules that may mediate both local and long-range effects.

Some researchers have suggested that three molecules—salicylic acid, jasmonic acid, and systemin—are plant hormones. Salicylic acid derives its name from the willow tree [Latin = *salix*], whose bark was used for centuries, by American Indians and ancient Greeks, to treat aches and fevers. Remarkably, salicylic acid, which we know as aspirin (acetylsalicylic acid), not only helps cure human ills, but helps prevent plant infections. In plants, salicylic acid promotes flowering and inhibits the production of ethylene.

One of salicylic acid's most striking actions is the flowers of a number of heat-generating (thermogenic) plants, such as those of skunk cabbage and the voodoo lily. In these flowers, salicylic acid stimulates the production of a mitochondrial electron carrier that produces heat instead of ATP. The heat production is so dramatic that, on the day of flowering, the flower's temperature increases by 14°C (25°F). The high temperature stimulates the vaporization of foul-smelling chemicals that attract pollinating insects.

In view of salicylic acid and other such signaling molecules, many plant scientists argue that plant hormones should be defined differently from animal hormones. Plant hormones can act both at long and short distances. They can also act on the same cells that produce them. One researcher has proposed that a plant hormone should be defined simply as "an endogenous plant substance that acts at low concentration to affect physiological processes."

The discovery of other plant hormones—such as systemin, salicylic acid, and jasmonic acid—suggested that plant hormones may act differently than animal hormones.

HOW DO PLANTS DETECT ENVIRONMENTAL CHANGES AND HOW DO THEY COORDINATE THEIR RESPONSES TO THOSE CHANGES?

Animals move to find new supplies of both energy and nutrients. To this end, they have evolved dozens of styles of locomotion, including swimming, running, and flying. In contrast, plants derive their energy from sunlight and their carbon from the air around them—an arrangement that makes large-scale movements unnecessary. Further, because most plants obtain water by means of roots that are permanently embedded in soil, they cannot move rapidly from place to place.

Plants are nonetheless active and responsive to their environments. Unlike animals, which have a permanent mature shape, plants change shape throughout their lives. We may compare the life of an animal to a play: it has a script with a beginning, a middle, and an end. In contrast, the life of a plant resembles an improvisation, whose form, content, and length depend on cues from the audience.

A plant shaded by other plants grows toward any available light. A low-growing plant may spread along the ground, re-rooting itself as it goes. A vine may grow up toward light or out across the ground. The roots of a plant growing in dry soil grow downward until they reach water. A plant suddenly exposed to more light than it is used to, grows new leaves that are less sensitive to light. A plant exposed to high winds grows stems or a trunk of increasing girth and strength.

Not all plant activity is slow or accomplished by means of gradual growth. Plants continuously adjust the position of their leaves or flowers to track the sun from hour to hour as it moves across the sky. Flowers open and close, according to the time of day and other variables. Some plants forcefully eject seeds from their fruits. Species of *Mimosa* (called "sensitive plants") rapidly fold their leaves in response to a light touch.

Despite such activity, plants are limited in their ability to seek desirable environments or avoid undesirable ones. They must make the most of what is available, and environmental cues dictate the pattern of growth. In response to light and gravity, for example, the shoot grows up and the root grows down into the earth. In response to changes in day length, temperature, or soil humidity, plants may drop their leaves, go dormant, or burst into flower. Depending on environmental cues, the same cells that give rise to leaves and stems may also give rise to flowers or roots.

Plants' ability to respond to changes in their environment seems remarkable, since their cells, which are enclosed in a cellulose wall, are more rigid and more tightly glued to one another than those of animals. When a leaf turns toward the morning sun, for example, cells on one side of the petiole must lengthen, while those on the other side remain the same size. For the leaf to grow toward the sun, cell expansion must be perfectly coordinated.

In the rest of this chapter, we see some of the molecular and cellular mechanisms for coordinating the responses of plants to environmental cues.

Like other living organisms, plants are highly responsive to changes in their environment. Over time, a plant may alter its shape, its structure, and even the positions of its leaves and flowers in ways that increase its ability to survive.

How Do Plants Respond to the Pull of Gravity?

Turning toward or away from a stimulus is called a *tropism* [Greek, *trope* = turning], so plants' response to gravity is called **gravitropism** (Figure 33-8A). In response to gravity, roots, especially primary roots, grow down, and shoots, especially primary shoots, grow up. Gravitropism allows plants to respond vigorously to their environment; if we turn a potted plant on its side, the root turns and grows downward and the shoot turns and grows upward. How exactly does a plant "know" which way is up?

A simple experiment demonstrated the location of a root's gravity detector: removal of the root cap abolishes the root's gravitropism. The response of the root must depend on the cells of the root cap. But which component of the root cap cells detects gravity?

Large cells in the root cap contain organelles called **amyloplasts,** each of which contains several starch grains (Figure 33-8B). The amyloplasts are denser than other organelles, and they consistently settle to the bottom of the cell. Plant physiologists long suspected that the amyloplasts are the root's **statoliths** [Greek, *statos* = standing + *lithos* = stone], or gravity detectors.

To test this hypothesis, researchers treated roots with gibberellin and cytokinin, which induce the digestion of the starch grains. With no amyloplasts, the root no longer exhibited gravitropism. Later, the same roots made new starch grains and recovered their gravitropism.

But this was not the end of the story. More recent research has not supported the hypothesis that amyloplasts allow plants to sense gravity. Timothy Caspar and his coworkers at Michigan State University genetically engineered *Arabidopsis thaliana* so that the plants could manufacture no starch and therefore no amyloplasts. These plants nonetheless demonstrated gravitropism. However, the response was not as strong as in plants with starch grains. Other research suggests that pressure between the cell membrane and cell wall stimulates gravitropism. Perhaps proteins between the cell membrane and cell wall detect such pressure.

Biologists do not yet know how plants sense, or detect, gravity, but they do understand how plants respond to gravity. As in phototropism, a change in the position of a shoot or root causes large differences in the rate of elongation of cells on the top and bottom. For example, the cells on the bottom of a horizontal stem may grow 10 times faster than those on the top. This difference is at least partly responsible for the upward bending of the stem.

Since this response is so similar to the phototropic response, we should not be surprised to learn that auxin plays a role in gravitropism as well. But differences in auxin concentrations between top and bottom are not great enough to account for the difference in the growth response. Other substances also participate. Among the suspected signals are ethylene and abscisic acid. The initial trigger, however, seems to be not a hormone, but calcium ions.

When a root is oriented horizontally, Ca^{2+} accumulates along the lower surface of the root and triggers a similar accumulation of auxin, which, in turn, inhibits cell elongation. As a result, the cells on the upper surface of the root grow more rapidly and force the root tip downward. When the root tip is oriented vertically once again, the asymmetrical distribution of calcium and auxin disappears, and the root grows straight.

The mechanisms for gravitropism is an area of active research, as plant physiologists are finally close to answering a question that has vexed them for 50 years. The research even

Figure 33-8 How do plants know which way is up? A. A fallen tree often sends new shoots upward, away from the pull of gravity. B. Because amyloplasts drop to the bottoms of cells, for many years biologists believed that the starchy grains helped plant cells detect the force of gravity. So far the evidence has not supported this theory. *(A, Tim Davis/Photo Researchers; B, Ed Reschke/Peter Arnold, Inc.)*

A.

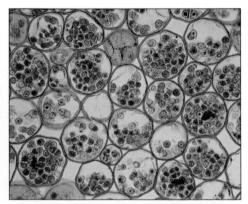

B.   50 μm

has a glamorous side, as interplanetary travel may ultimately depend on the ability to raise crops in the weightless environment of space.

Until recently, biologists believed that gravitropism was facilitated by starchy grains called amyloplasts. Contrary evidence suggests that pressure between the cell membrane and the cell wall may perhaps stimulate the release of Ca^{2+} and auxin that mediate the bending of roots and shoots in response to the pull of gravity.

What Pigment Facilitates Phototropism?

A plant's survival depends on its ability to get enough light to support photosynthesis, and plants therefore respond to light (or its absence) in many ways. Among these are phototropism and solar tracking, the turning of leaves to follow the sun, as well as nighttime sleep movements (Figure 33-9).

As in the case of gravitropism, we would like to know the nature of the light detector responsible for phototropism and solar tracking. One way of approaching this question is to study the effect of light of different wavelengths (which we perceive as different colors).

Every light-dependent process has a characteristic *action spectrum,* the effectiveness of different wavelengths in triggering a specific response (Chapter 7). Because the effect of light on a process depends on its absorption by some molecule, the action spectrum should correspond to that molecule's *absorption spectrum,* the relative absorption of different wavelengths of light (Figure 7-12). If chlorophyll were the detector for phototropism and solar movements, then the action spectrum for these processes should correspond to the absorption spectrum of chlorophyll. But this is not the case. The light-absorbing molecule responsible for phototropism and solar movements is therefore not chlorophyll, but some other compound that absorbs blue, but not red or yellow light.

Somehow, this light-absorbing compound stimulates the transport of auxin from the illuminated to the shaded side of

Figure 33-9 **Solar tracking.** Flowers and leaves of many plants turn to face the sun as it moves across the sky. *(D. Newman/Visuals Unlimited)*

a growing shoot tip. The absorbing pigment, however, is not necessarily in the tip itself.

Phototropism depends on a pigment other than chlorophyll.

How Do Plants Detect Changes in Season or Daylight?

Every gardener, farmer, or hiker knows that different plants flower in different seasons and at different times of day. The flowers of crocuses and daffodils appear in the spring, and those of carnations and black-eyed susans in the summer. Appropriate timing of flower and seed production determines how much sun and water will be available to new plants and which other organisms will be around to compete, to consume, or to pollinate, for example. In this section, we ask, How do plants detect and respond to seasonal changes?

In some species (such as cucumbers, peas, and tomatoes), the time of flowering depends only on maturity or size. Flowering may occur after a plant has grown for only a few weeks, or flowering may take months, or even years. In other species, flowering always happens around a certain date, even if the plant reaches maturity much earlier.

Understanding the seasonal control of flowering is a practical as well as a scientific issue. Soybean farmers, for example, once tried to stagger their harvests by planting fields at 2-week intervals. But they found that all the flowers appeared at once, late in the summer, and the whole harvest was ready at the same time regardless of the planting time.

For any species, the date of flowering varies with latitude and altitude. Plants in Massachusetts generally flower later than those in Georgia, for example. From this observation, biologists conclude that flowering does not result from the operation of an internal annual clock, in the way that our own sleep depends on an internal daily clock. Flowering response turns out to be an example of **photoperiodism**, the response to the relative lengths of day and night in their daily cycle. *Short-day plants* flower when days are short and nights are long, in the late summer, fall, or even winter. Examples include chrysanthemum, cocklebur, and corn (Figure 33-10).

In contrast, *long-day plants* require long days and short nights. These generally flower in the late spring or early summer and include spinach, sugar beets, and black-eyed susans. Finally, in *day-neutral plants,* such as tomatoes and dandelions, flowering does not depend on day length. Plants from tropical areas are often day-neutral, while those from regions closer to the poles are usually long-day plants.

One of the first questions that researchers asked was whether flowering depended on the length of the day or the length of the night. To find out, they grew short-day plants—cocklebur in one laboratory, soybeans in another—in artificial cycles of light and dark. The plants flowered whenever "days" were shorter than the "nights." It did not matter how long the "day" was, as long as it was shorter than the "night."

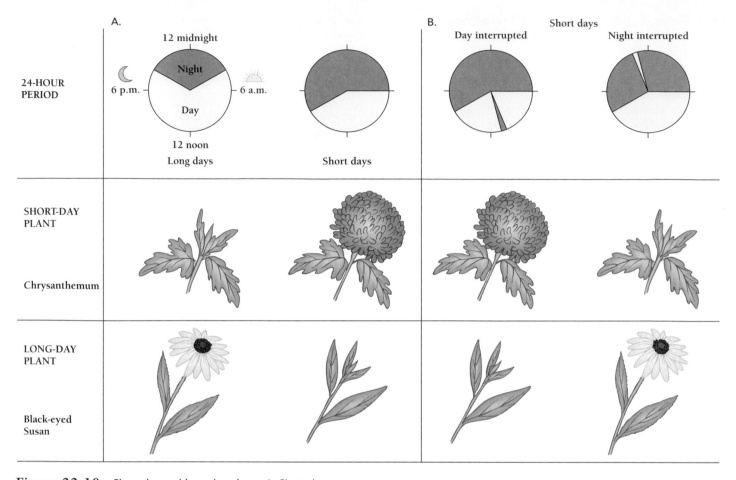

Figure 33-10   Short-day and long-day plants. A. Short-day plants such as chrysanthemums flower when days are short and nights are long. Long-day plants such as black-eyed susans flower when days are long and nights are short. B. The plant physiologists Hammer and Bonner asked if plants were sensitive to day length or night length. When they interrupted the day with a period of darkness, it made no difference in whether either kind of plant flowered. But when they interrupted the long nights with flashes of light, the short-day (long-night) chrysanthemum failed to flower, and the long-day (short-night) black-eyed susan flowered. The two researchers concluded that in both kinds of plants, it is the *length of the night* that determines flowering.

In 1938, Karl Hamner and James Bonner, plant physiologists at the University of Chicago, tried a different experiment. They interrupted the light period with a period of darkness or the dark period with a period of light (Figure 33-10). For both short-day and long-day plants, interrupting the artificial day made no difference in flowering. For short-day plants, however, breaking up the night prevented flowering, and for long-day plants, breaking up the night promoted flowering. The interruption did not need to be large—an ordinary light bulb turned on for half a minute was enough to prevent flowering. The researchers concluded that both short-day and long-day plants measure the length of the night, not the day. Short-day plants should really be called "long-night" plants, and long-day plants should really be called "short-night" plants.

Other work on the basis of photoperiodism was going on at the U.S. Department of Agriculture Laboratory in Beltsville, Maryland, where photoperiodism had first been discovered in the 1920s. The early work had immediately helped agriculture: farmers were better able to control when their crops flowered and set seed, breeders could test new varieties under controlled light-dark conditions, and florists could provide year-round supplies of flowers that had once been seasonal. But still no one understood the chemical basis of photoperiodism.

In 1940, however, researchers at Beltsville realized that they could learn about darkness detection by determining the action spectrum of light that most effectively interrupted the long nights required for soybeans to flower. Determining the action spectrum for such a complex process, however, was no easy matter. The researchers first demonstrated that the light effect was in the leaves and that illuminating a single leaf was enough to suppress flowering in the whole plant.

They built a large instrument that would light individual leaves in separate groups of plants with light of different colors. This experiment showed that red light and violet light were most effective in promoting flowering. The researchers concluded that the light detector must be a pigment that absorbs red and violet light, in other words, a green or blue pigment, perhaps even chlorophyll. Further measurements revealed, however, that the pigment was a new one, which they named

phytochrome [Greek, *phyton* = plant]. Phytochrome regulates a variety of processes that depend on the timing of dark and light, including flowering, germination, and leaf formation.

The Beltsville group discovered that they could reverse the effect of red light with a light flash of longer wavelength. The most effective wavelength for the red light effect was about 660 nm, and the most effective wavelength for the reversing flash was about 730 nm, in the "far-red" region of the spectrum. Sterling Hendricks, the plant physiologist who had built the instrument to determine the action spectrum, made a radical hypothesis to explain the far-red reversal. Phytochrome must exist in two interconvertible forms, one absorbing red light and one absorbing far-red light. He called the red absorbing form P_r and the far-red absorbing form P_{fr}. When P_r absorbs red light, it converts to P_{fr}, and far-red light converts P_{fr} back to P_r.

Hendricks and his colleagues proposed the following explanation for the effect of light and dark on flowering: (1) Sunlight contains more energy in the red than in the far-red part of the spectrum; (2) during the day, the red component in sunlight converts P_r to P_{fr}; and (3) at night, P_{fr} slowly converts back to P_r.

P_{fr} promotes flowering of long-day plants and inhibits flowering of short-day plants. That is, in short-day plants, flowering occurs only when P_{fr} levels are sufficiently low (Figure 33-11). Red light supplied at night decreases P_r and increases P_{fr}, suppressing flowering. High levels of P_{fr} thus inhibit flowering in short-day plants and promote flowering in long-day plants. Other plant physiologists were initially skeptical about the existence of phytochrome. One doubter waggishly termed it "a pigment of the imagination."

Hendricks' hypothesis, however, made a strong prediction: purified phytochrome should be a single compound whose absorption spectrum can switch from one form to another. This prediction stimulated a biochemical search that was long and difficult for two reasons: (1) phytochrome acts catalytically, so plants do not contain great amounts; and (2) P_{fr} is unstable, so researchers had to work in the dark to try to keep phytochrome in the P_r form.

After many years of work, several research groups isolated pure phytochrome. It is a protein linked to a much smaller organic compound that absorbs light. Purified phytochrome, however, behaves like the pigment imagined by Hendricks and his colleagues almost 40 years earlier. (Under natural sunlight, which contains both red and far-red light, both forms of phytochrome exist, with about 40 percent P_r and about 60 percent P_{fr}. In darkness, the P_{fr} level declines, either by reversion to P_r or by destruction by enzymes.

The action spectrum for other light-regulated changes suggested that phytochrome is important in many developmental processes. These processes vary from species to species but include seed germination, the elongation of new seedlings, the beginning of chlorophyll synthesis, and the production of enzymes needed for photosynthesis. For example, in mustard seedlings that have been grown in the dark and transferred to the light, phytochrome controls the following changes: opening of the hypocotyl hook, development of primary leaves, dif-

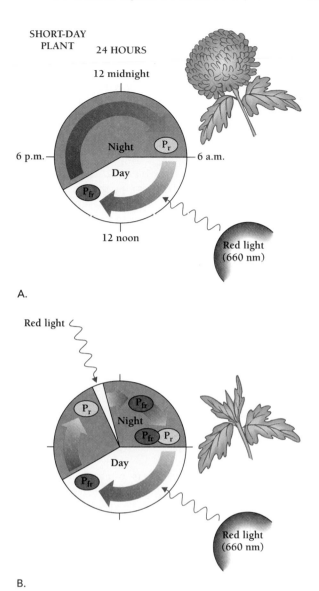

Figure 33-11   How long is the night? The flowering of long-day and short-day plants depends on the conversion of phytochrome from P_r to P_{fr} and back again. A. During the day, red light in sunlight converts P_r to P_{fr}. During the night, P_{fr} slowly converts back to P_r. When P_{fr} levels are sufficiently low, short-day plants such as chrysanthemums can flower. B. But a flash of light in the middle of the night resets the phytochrome clock, converting the P_r back to P_{fr}. The P_{fr} inhibits flowering.

ferentiation of xylem, degradation of storage protein, chlorophyll synthesis, protein synthesis, and RNA synthesis.

In almost every case, the developmental change controlled by phytochrome depends on the accumulation of P_{fr}. Red light (or daylight) thus promotes these processes and far-red light inhibits them. Phytochrome permits the sun, or the plant physiologist, to turn a molecular switch on or off. Some of the effects of phytochrome may depend on induced changes in the synthesis, degradation, or transport of plant hormones. Some effects, however, depend on a more direct influence on the production of particular mRNAs.

Plants detect the passing of the seasons by measuring the length of the night by means of a pair of pigments, called phytochromes. During the day, red light from sunlight converts P_r to P_{fr}. At night, in the absence of red light, P_{fr} gradually reverts to P_r. The longer the night, the lower the levels of P_{fr} become. High levels of P_{fr} inhibit flowering in some plants and promote flowering in others.

In this last chapter of our unit on plant anatomy and physiology we have examined the role of plant hormones in plants' growth and development and in their responses to their environment. We have reviewed both classic experiments and some of the latest research in these areas. In the last section of *Asking About Life,* we study the anatomy and physiology of animals, with an emphasis on vertebrates.

S T U D Y   O U T L I N E   W I T H   K E Y   T E R M S

Plant physiologists have identified five plant **hormones.** Each of these is an organic compound produced in one tissue or organ and transported to another tissue or organ, called the **target,** where it produces one or more specific effects.

The first plant hormone to be recognized was **auxin,** which (among other effects) coordinates **phototropism,** the growth of a plant toward light. Researchers can measure the amount of active hormones either with a *bioassay,* which estimates the amounts of hormones from their biological effects, or with chemical measurements. Illumination of one side of the growing tip of an oat seedling causes the transport of auxin away from the lighted side. The higher concentration in the shaded side causes it to grow more rapidly, leading the plant to curve toward the light.

Auxin triggers target cells to secrete hydrogen ions. This secretion makes the extracellular space more acidic and activates enzymes that break bonds within cell walls. Auxin also affects gene expression in target cells.

Other plant hormones have effects distinct from those of auxin. **Gibberellins** promote germination, trigger the mobilization of food reserves in germinating seeds, and stimulate growth in mature plants. **Cytokinins** stimulate cell division and growth and prevent leaf **senescence. Ethylene** promotes leaf senescence and fruit ripening and inhibits elongation of stems and roots. And **abscisic acid** promotes **abscission** and dormancy in buds and seeds.

A plant's response to environmental cues depends on its ability to detect such cues. Plants employ different detection mechanisms for different kinds of stimuli. Large cells in the root cap called **amyloplasts** seem to detect gravity, an example of **gravitropism.** Amyloplasts function as **statoliths.** One kind of unidentified pigment detects light and brings about phototropism. Another kind of pigment, called **phytochrome,** serves as a molecular switch for the **photoperiodism** of many seasonal processes, including flowering and seed production.

Absorption of red light converts phytochrome into a form that no longer absorbs red light, but only far-red light. Similarly, absorption of far-red light switches phytochrome back to a form that absorbs red light. Most of the processes controlled by phytochrome depend on the accumulation of enough of the far-red absorbing form.

R E V I E W   A N D   T H O U G H T   Q U E S T I O N S

Review Questions

1. What are the primary effects of auxin?
2. How is the process of senescence affected by cytokinins?
3. What is the triple response induced by ethylene? What is the experimental evidence suggesting that ethylene alters gene expression?
4. Why might you expect the level of abscisic acid to increase in a plant subjected to environmental stress?
5. Explain why botanists are beginning to doubt that amyloplasts help plants detect the pull of gravity.

6. Why can phytochrome be thought of as a molecular switch?

Thought Questions

7. Why do you think that some bacteria and fungi produce plant hormones? What advantage accrues to *Gibberella fujikoroi* in forcing seedlings to grow so fast?
8. What is a bioassay? Try to think of an example of a bioassay that could be used in animals.

S E L E C T E D   R E A D I N G S

Berg, Linda R., *Introductory Botany: Plants, People, and the Environment,* Saunders College Publishing, Philadelphia, 1997. This book covers the basics of plant science in a friendly and unintimidating style.

Mauseth, James. D., *Botany: An Introduction to Plant Biology,* 2nd ed., Saunders College Publishing, Philadelphia, 1995. An authoritative introduction to the biology of plants.

Moore, Randy, W. Dennis Clark, Kingsley R. Stern, and Darrell Vodopich, *Botany,* Wm. C. Brown Publishers, Dubuque, 1995. A friendly, well-written textbook on the science of plants.

Niklas, Karl J., *Plant Biomechanics: An Engineering Approach to Plant Form and Function,* Chicago, University of Chicago Press, 1992. This book is wonderfully written and highly accessible where Niklas is first introducing a topic or concluding one—that is, at the beginning and ending of each chapter. Much of this book, however, demands a familiarity with either botany or engineering beyond that of the average reader.

▶ On-line materials relating to this chapter are on the World Wide Web at http://www.saunderscollege.com/lifesci/
Click on Tobin/Dusheck: *Asking About Life.*

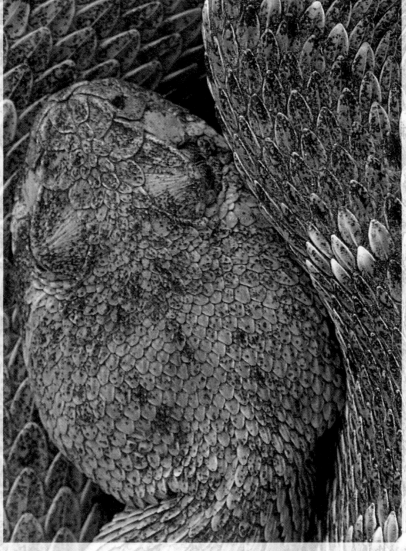

Western diamondback rattlesnake.   *(© John Gerlach/DRK Photo)*

Structural and Physiological Adaptations of Animals

Tyrannosaurus Wrecks: *T. rex* Treks, But Slowly

One of the high points of the movie classic *Jurassic Park* comes when an agile but stupid *Tyrannosaurus rex* pursues a Jeep at breakneck speed through a driving rain. Just when the *T. rex* seems ready to snap up actor Jeff Goldblum and down him like a dog biscuit, the gargantuan reptile runs out of steam and the three actors in the Jeep escape to (temporary) safety.

Could *T. rex* have run that fast? Zoologists and paleontologists say "no way." When an adult human runs, the force on each foot is hundreds of pounds—equal to the mass of the body times the acceleration of the foot against the ground. A 6.5-ton *T. rex* running at 45 miles per hour, say biologists, would exert so much force on its legs that they would snap like toothpicks.

Part of the drama of *Jurassic Park* stems from moviegoers' expectations that dinosaurs will be sluggish and clumsy. Indeed, the original reconstructions and paintings of dinosaurs from the early part of this century usually show dinosaurs moving very slowly or not at all. Early 20th-century paleontologists believed that dinosaurs were incapable of the kind of sprints at which warm-blooded animals excel. Accordingly, dinosaurs were nearly always depicted with all four feet on the ground (Figure 34-1). Yet, the velociraptors and other predatory dinosaurs in *Jurassic Park* and its sequel *The Lost World* leap, run, whorl, cock their heads like birds, and attack, swiftly and often intelligently. How can we tell which image is correct?

Because dinosaur skeletons share many features with modern reptiles—similar jaws, for example—early paleontologists assumed that dinosaurs resembled modern reptiles in other ways as well. They assumed that dinosaurs laid eggs instead of bearing live young, that they were solitary rather than social, and that they were ectotherms—cold-blooded animals whose activities depended on air temperature. In the last 20 years, all of these assumptions have been called into question.

In the 1970s, the maverick American paleontologist Robert Bakker electrified the imaginations of paleontologists, scien-

© John Sibbick

752

Figure 34-1 **Could a dinosaur outrun a human being?** Early researchers assumed that because dinosaurs were so big, they were as slow moving as elephants. *(Transparency #3883(3), Photo by K. Perkins/J. Beckett, courtesy Dept. of Library Sciences, American Museum of Natural History)*

tific illustrators, and small children by suggesting that dinosaurs might have been warm-blooded animals, more like birds than modern reptiles. In particular, Bakker examined *T. rex*'s leg bones and concluded that they were long enough to have allowed the monster to run up to 45 miles per hour. But Bakker's quick analysis left plenty of room for other researchers.

British zoologist R. McNeill Alexander, an expert on the physics of animal movement, jumped into the argument with a more thorough analysis of dinosaur leg bones. Alexander concluded that although many dinosaurs had the long legs that would have enabled them to run fast, their legs were not strong enough to carry them at high speeds. Alexander calculated the strength of a bone by comparing the size of a cross section of a bone with its length and the weight of the animal.

The femur (or thigh bone) of an African elephant, for example, has a fairly low "strength indicator" of about 7. A running elephant moves at little more than 11 miles per hours. (Charging elephants only

seem to be moving much faster.) In comparison, the femur of an African buffalo, which is capable of running 30 to 35 miles per hour, has a strength indicator of about 22. Alexander calculated that the femur of *T. rex* has a strength of about 9, a little stronger, perhaps, than that of an elephant, but nowhere near that of a buffalo. At best, Alexander estimated, a *T. rex* could run no more than about 18 miles per hour.

Alexander's analysis seemed conclusive until 1995, when a paleontologist and a physicist at Indiana University took the whole argument one heavy step further. What would happen, asked James Farlow and John Robinson, if a 6.5-ton *T. rex* pursuing a Jeep at 45 miles per hour happened to trip and fall? The answer, derived from simple physics, was decisive.

In a running position, *T. rex*'s head would be about 11 feet above the ground.

about 40 tons—enough to pulverize the creature's skull and excavate a crater 8 inches deep. Then, the colossal corpse would have skidded some 50 feet until the mangled body finally overtook the head, breaking the monster's neck. Not a pretty sight.

Thus, Farlow and Robinson confirmed Alexander's calculations, estimating that *T. rex*'s maximum safe speed was probably 18 to 22 miles per hour. A Jeep would easily leave the dinosaur in the dust. As for Jeff Goldblum and the rest of us, it is a good thing the problem of outrunning *T. rex* and its 6-inch teeth is purely theoretical.

How can biologists know the anatomy, physiology, and habits of long-dead animals? Based on zoologists' extensive knowledge of the anatomy and habits of living animals, biologists can draw detailed conclusions from the skeletons and teeth of extinct animals. Before R. McNeill Alexander estimated the movements of dinosaurs, he spent years studying how living animals run and jump.

As we ask how animals move in the rest of this chapter, we will focus on general ideas that account for why their skeletons and their muscles look and act the way they do. We will see that the shape of a structure determines how it works.

What would happen if a 6.5-ton **T. rex** *pursuing a Jeep at 45 miles per hour happened to trip and fall?*

At 45 miles per hour, the 5-foot-long head of a falling *T. rex,* with another 45 feet of its 6-ton body coming from behind, would hit the ground with a deceleration equal to 16 times the pull of Earth's gravity—

Zoologists sum up this idea by saying, "Form follows function."

The idea that form and function are related helps tie together the last eleven chapters of *Asking About Life*. In this last

753

section, we will review some of the major ideas in animal anatomy and physiology. The first chapters in this section cover the "housekeeping" functions of the body: movement, digestion, circulation, respiration, and excretion. These all serve to keep the body operating and provide many examples of the connection between form and function. We will also see how hormones coordinate and regulate all of these processes. Next we cover defenses, which

also maintain the body, and the nervous system, which integrates all of the processes so far discussed.

Most of the chapters in this last section deal with how the body is organized and how its parts work together, or **physiology.** One theme that ties all of physiology together is the idea of **homeostasis,** the tendency of living organisms to maintain their internal environment. Mammals and birds, for example, usually maintain a constant

body temperature, or nearly so. Similarly, the blood of all vertebrates is maintained at a constant acidity. All of the housekeeping functions of organisms maintain them in a condition in which they are most suited to reproduction and development—two aspects of life that are not considered housekeeping functions. As we will see, unlike many physiological functions, which are homeostatic, reproduction and development have everything to do with change.

KEY CONCEPTS

1. The shape of a species evolves in response to the uses to which that organism puts its body.

2. Large animals must have proportionately thicker bones than small animals. Likewise, fast runners must have proportionately stronger leg bones than sedentary animals.

3. Animals increase surface area in some parts of the body, including, for example, the lungs, the intestines, and the capillaries.

4. Countercurrent systems are a way for organisms to establish permanent gradients in temperature or concentration.

5. Vertebrates move by means of skeletal muscles acting across jointed bones.

6. Muscle contracts when actin and myosin fibers slide past one another—a movement powered by ATP.

FORM FOLLOWS FUNCTION

Recall from Chapter 24 that deuterostomes and their embryos share a common body plan—a tube within a tube. The inner tube is the intestinal tract, which extends from mouth to anus. The intestines and associated organs (such as the liver and pancreas) lie within a cavity, the **coelom,** which separates the inner and outer tubes. In vertebrates, the coelom has two distinct parts: the **thoracic cavity,** which encloses the heart and lungs, and the **abdominal cavity,** which encloses most of the length of the intestinal tract, as well as the liver, pancreas, and kidneys.

Running along the back is the backbone or spinal column. At the front (*anterior*) end of the spinal column is the skull; at the far (*posterior*) end is the tail. The bones of the tail are continuous with the bones of the backbone. Just behind the skull, the pectoral girdle supports the forelimbs—fins in fishes, arms in humans. Behind the pectoral girdle are the ribs, also attached to the backbone, and finally, the pelvic girdle, which supports the pectoral fins in fish and the hind limbs of all other vertebrates. Despite these commonalities, vertebrates differ enormously in size, shape, and physiology.

We can look at form at many levels. We can begin by looking at the overall shape of an animal. Fish, for example, are

streamlined, which eases their passage through water. We can look at form at the level of the skeleton, the muscles, or the individual organs of the body. We can look at the structure of the heart to see how it pumps, or we can look at the microstructure of heart muscle to see how it differs from that of skeletal muscle. We can find out how hormones influence blood pressure and how blood pressure influences the beating of the heart. We can look at the details of cell structure. Even the shapes of molecules suggest how they work.

Every organ of the body is shaped in a way that both increases its effectiveness and reveals its function. The shape of a single tooth tells volumes about the owner's way of life. The large shearing teeth of lions reveal that they survive by tearing flesh from other animals. In contrast, horses and elephants, which eat plant material, have broad, flat grinding teeth.

The stomachs of lions and horses are just as different. Because meat is a rich source of protein and other nutrients, animals that eat meat generally need to process their meals very little. As a result, their stomachs tend to be simple, small, and relatively inefficient (Figure 34-2).

In contrast, animals that eat grass, which is high in cellulose and hard to digest, must eat far more of it and digest it more thoroughly to gain the same amount of protein and calories. As a result, grass-eating herbivores often have immense

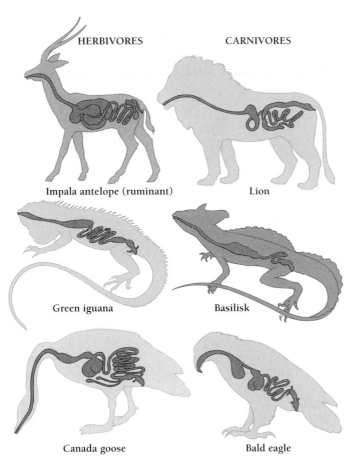

HERBIVORES   CARNIVORES

Impala antelope (ruminant)   Lion

Green iguana   Basilisk

Canada goose   Bald eagle

Figure 34-2 Grass eaters have big stomachs. Because plants, especially grasses, usually contain large amounts of cellulose and other indigestible compounds, an herbivore requires a large digestive system. The digestive system is large because herbivores must eat more material than carnivores to obtain the same quantity of nutrients. And the digestive system is large because, for full digestion to occur, the low-quality food that an herbivore eats must remain in the digestive tract for longer periods.

digestive tracts, sometimes with several stomachs, devoted to extracting every last nutrient from great quantities of low-quality food.

As we discuss the shapes of skeletons and muscles in this chapter, we must bear in mind that the forms of organisms result from successive adaptations over long periods of time. Our hands are adapted from paws, which evolved from the lobed fins of some long-extinct fishes. Likewise, the bones in our ears evolved from the jawbones of fishes. Every organ bears within it clues both to its function and to its evolution.

HOW DO SIZE, LOCOMOTION, AND SURFACE AREA AFFECT SHAPE?

The most obvious differences between animals are differences of size and surface area. Mice, for example, are small and round, while giraffes are large and leggy. Because a mouse is round, it has a small surface area for size. In contrast, a giraffe has a large surface area for its size. We will see, however, that as a

rule large animals have smaller *relative* surface areas than small animals.

Size

A mouse could not be as large as a mammoth, and a polar bear could not be as small as a cat. For every type of animal, there is an appropriate range of sizes, and the evolution of a bigger or smaller size demands a change in form, as well.

One reason is gravity. As the linear dimensions of an animal (or any object) double, its area increases fourfold. In other words, both the surface area of an animal and the cross-sectional area of its bones increase four times. In contrast, the volume and weight increase eight times. The result is that the animal's bones and muscles bear a proportionately heavier load.

In general, we can say that the heavier the animal, the thicker its bones must be to support its weight. For the graceful little gazelle to become as big as a rhinoceros, for example, it would have to develop thick, heavy legs, so that every pound of weight had the same cross-sectional area of bone to support it. A little gazelle cannot simply become a big gazelle. It must become something different. Its whole life-style must change, for it can no longer prance about and turn on a dime to escape a swift predator. Being big, however, the animal can defend itself against many predators if it is sufficiently aggressive, which, of course, the rhinoceros is.

Humans are no exception to this rule of size. Very tall humans, like very tall dogs, tend to suffer hip problems, a result of the enormous stresses generated by too much weight on too little bone.

Gravity affects the lives of large and small animals in other ways as well. Gravity will cause an ant (or any other object) dropped from an immense height to accelerate as it drops. Because force equals mass times acceleration, the force with which an object hits the ground is equal to its mass times the acceleration (which is 1 g, or gravity, on Earth). But an ant's mass is so small that the force with which it hits the floor is small—not enough to hurt it. As a result, an ant dropped from the top of a ten-story elevator shaft will likely walk away unscathed.

The same is true of mice, which can fall from great heights without injury. But larger animals are another story. Occasionally, a small child miraculously survives a fall from a second- or third-story window. More often, the child is seriously hurt. An adult human could not possibly survive such a fall without broken bones or worse.

But small animals' lack of mass is only part of the reason they survive falls better than large animals. Ants and other small creatures are also saved from falls by their relatively large surface areas. Wind resistance, which is proportional to surface area, slows any falling object until the object reaches a "terminal velocity." Like a parachute, an ant's relatively large surface area quickly slows the falling ant and its terminal velocity is therefore low. The ant hits the ground lightly and slowly, rights itself, and walks away. In contrast, an elephant, whose relative surface area is small, would take a very long time to reach its terminal velocity—far longer than it takes to fall ten stories.

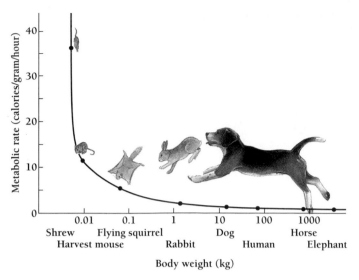

Figure 34-3   **Tiny animals burn a lot of fuel.** The smallest animals have the highest metabolic rates.

Figure 34-4   A mouse with the metabolic rate of a steer would need an eight-inch-thick fur coat to maintain a normal body temperature.

We can draw interesting conclusions from this kind of discussion. We can, for example, see that small animals are more suited to life in trees and other high places. We are not surprised to learn that while most primates live in trees, the larger primates, such as humans, gorillas, baboons, and chimpanzees, are more likely to be found safely on the ground. Likewise, large herbivores that eat the leaves of trees, such as elephants and giraffes, do not climb trees. Instead, they have evolved necks or trunks for reaching up into tall trees while keeping four feet firmly on the ground. As we saw at the beginning of this chapter, large animals must be careful to avoid not only heights, but even running too fast.

Because of gravity, large animals must have proportionately stronger and thicker bones than small animals. Large animals also fall much harder than small animals.

Metabolic Rate

As gravity influences size, so size influences metabolic rate. Mice, shrews, hummingbirds, and other small, warm-blooded animals have much greater metabolic rates than larger animals (Figure 34-3). Such small animals need both more calories and more oxygen to fuel this higher metabolism. To survive, a mouse must eat about one-quarter its own weight in food every day.

Why large animals have slower metabolisms is not entirely understood. One factor may be that larger animals have proportionately smaller surface areas. As a result, large animals lose a smaller proportion of their metabolic heat. One physiologist calculated that a mouse with the low metabolic rate of a steer would need an eight-inch-thick fur coat to keep it warm (Figure 34-4). Likewise, a steer with the high metabolic rate of a

mouse would produce so much heat that it could only maintain a stable internal body temperature if its hide were well above the boiling point of water.

But surface area is not the sole determining factor of metabolic rate. Even among cold-blooded animals (heterotherms) and one-celled organisms, which do not regulate their temperature, the larger organisms have the lowest metabolic rates. How and why size affects metabolic rate in sea stars and other heterotherms is one of the unanswered questions of biology.

Metabolic rate is partly a function of size.

Surface Area

Animals of the same size can vary enormously in shape. The webbed flippers of a sea lion look nothing like our own hands and feet. And the broad, flat grinding teeth of a horse appear to have little in common with the sharp teeth of a lion. One of the many changes that can occur in an organ over the course of evolution is an increase or decrease in surface area. Animals may increase or decrease the surface area of limbs or teeth, as well as the lungs, the digestive tract, and tiny vessels of the circulatory system.

Webbed feet are a way to increase surface area without increasing overall size. Aquatic animals from frogs to ducks need to be able to push large volumes of water in order to propel themselves through the water. Likewise, animals that fly must be able to push large volumes of air in order to keep themselves aloft. The wings of birds and bats are limbs whose surface area has increased enormously, whether by means of extended bones, long feathers, or wide membranes.

Another use for increased surface area is the absorption of nutrients. As we will see in later chapters, the lungs, the digestive tract, and the circulatory system all have enormous surface areas for moving nutrients and wastes to and from the cells of the body.

Figure 34-5 Countercurrent heat exchange. Whales and other marine mammals use countercurrent heat exchange to minimize heat loss from the flippers, whose large surface area is wonderful for swimming but a hazard in Arctic waters. As warm, arterial blood from the heart flows out into the flippers, it passes in close proximity to the returning venous blood, which is cold. The arterial blood becomes cooler and cooler as it enters the flippers, while the venous blood, warmed by the arterial blood, becomes warmer and warmer. In this way, body heat stays in the interior and the flippers remain cool.

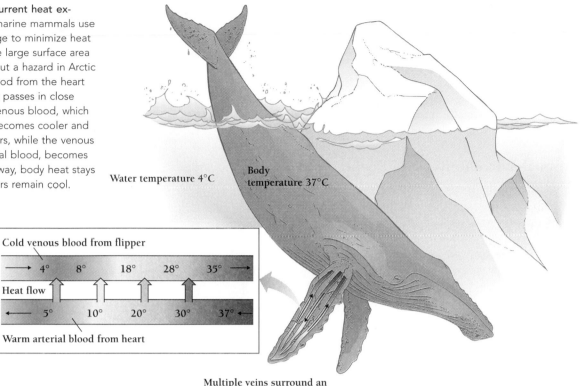

Water temperature 4°C

Body temperature 37°C

Cold venous blood from flipper

→ 4° 8° 18° 28° 35° →

Heat flow

← 5° 10° 20° 30° 37° ←

Warm arterial blood from heart

Multiple veins surround an artery to absorb its body heat, thereby minimizing energy loss to the cold ocean water

---

Many organs of the body rely on enormous surface areas to function. These include webbed feet, the lungs, the intestines, and the circulatory system.

Countercurrent Systems

Animals find all sorts of ways to use surface area for warming or cooling different parts of the body. Inside the noses of many vertebrates are elaborately folded **turbinate** bones, which increase the surface area of the inside of the nose. Since the mucous membranes inside the nose are moist, evaporation from this large surface area tends to cool the inside of the nose. Dogs use their cool noses to cool themselves off on a hot day. A dog inhales air through its wet nose, which cools the air before it enters the lungs. In the lungs, the cool air absorbs heat from the lungs and is then exhaled out through the mouth. As a result, the dog continually inhales cool air and exhales warm air. Because so much water evaporates inside the nose, it is important for dogs to drink lots of water on a hot day.

Desert animals, which conserve water by letting their body temperature rise on hot days, use the cool nose even more efficiently. The brain, unlike the other organs of the body, cannot function properly at high temperatures. Many of us have experienced the delirium that sometimes accompanies a high fever, for example. To keep the whole body cool would require

more water than is available to most desert animals. The solution is to allow the body to heat up while keeping the head cool. To do this, gazelles and antelopes circulate warm blood from the heart into a network of hundreds of small arteries. These arteries pass through a large sinus filled with cool venous blood from the nose. In the oryx, such an arrangement keeps the brain nearly 3°C cooler than the central arteries of the body.

Running the warm arterial blood past the cold venous blood is an example of a **countercurrent system,** a system in which fluids or gases are run past each other so that they can exchange something. The end result of a countercurrent system is a steep gradient—often in temperature. But the same system can be used to create pressure gradients—or concentration gradients. As we will see in Chapter 39, such gradients in the kidneys direct the movements of fluids.

Dolphins and whales use countercurrent systems to keep the cold blood from their flippers from chilling the rest of the body (Figure 34-5). The body of a whale that swims in the freezing waters of the Arctic Ocean is well insulated against the cold. But its flukes and flippers, which are thin and flat, become extremely cold. To prevent blood that has cooled in the flukes from chilling the body, the veins that carry the cold blood back to the body run adjacent to and surround the arteries that carry warm blood from the heart to the flippers. Thus, the warm arterial blood heats the cold venous blood before it enters the body. Likewise, the venous blood cools the arterial blood before it enters the flippers. This system, called a *countercurrent heat exchanger,* occurs also in the legs of wading birds, for the same reason, and in the testes of mammals (sperm develop only when the testes are kept cooler than the rest of the body).

Countercurrent systems enable animals to establish steep gradients of temperature or concentration.

Animal Locomotion and Shape

An animal's style of locomotion determines shape just as size does. In order to propel themselves forward, animals must push against whatever is available. Aquatic animals push against water, terrestrial animals push against the ground or other objects, and flying animals push against the air. Fast swimmers and flyers both tend to be streamlined. Slower movers are less so. But whereas most fish and dolphins use their tails for propulsion and their appendages for guidance, birds use their appendages for propulsion and their tails for guidance.

Terrestrial animals can slither, crawl, walk, run, or jump. Each style of movement uses the limbs in different ways. A deer and a kangaroo are similar in size and habits: both are medium-sized herbivores that escape from predators by running. But the kangaroo is adapted to hop on two legs and the deer is adapted to bound on four. These different styles give the two kinds of animals very different anatomies.

Style of locomotion partly determines the overall shape of an animal.

Swimmers

Animals exclusively adapted to swim rapidly in water all have the same oblong shape, also seen in fast cars, torpedoes, and the fuselages of airplanes. Vertebrate swimmers such as fish, seals, and eels propel themselves by undulating through the water. Invertebrate swimmers, such as shrimp and squid, have a similar shape. The strongest swimmers—dolphins, tuna, and mackerel, for example—all propel themselves by sweeping their tails back and forth. A swimmer's tail is broad and flat, a shape that forces water to flow against the tail. The tail exerts forces against the water both toward the sides and the rear. The forces to the side cancel each other out, however, and the net movement is forward.

Vertebrates less well-adapted for fast swimming rely on various sorts of webbed flippers and feet to move themselves through the water. Sea lions, penguins, and platypuses, for example, all push themselves through water with short strokes of their front appendages, often with the aid of webbed rear feet that undulate like a fish's tail.

Most swimmers undulate through the water.

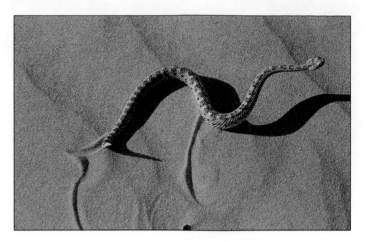

Figure 34-6 **Swimming on land.** The sidewinder of the North American deserts rolls along, throwing successive coils into the air. In this way, the snake minimizes its contact with hot sand. *(Wayne Lynch/DRK Photo)*

Snakes

Snakes usually undulate across the ground in the same way that eels move through water. Each curve of a snake's body pushes against small objects and rough surfaces on the ground. As with the side-to-side sweep of a fish's tail, the lateral forces mostly cancel each other out, while the force toward the rear pushes the snake forward.

Snakes on extremely smooth surfaces cannot undulate easily, as there is nothing for them to push against. In these circumstances, many snakes, especially short ones, resort to **rectilinear movement,** in which the snake progressively bunches up and relaxes its ventral scales, or scutes. The snake uses its scutes in the same way that a caterpillar uses its feet.

Another technique is **sidewinding,** in which the snake moves across loose or sandy soil by throwing successive coils sideways. The snake's body is in contact with the ground at two or three places at a time. The parts of its body in contact with the ground are momentarily stationary, whereas the loops that span the tracks are moving and are held clear of the ground (Figure 34-6). Most snakes are capable of this peculiar movement. Desert species such as the sidewinder rattlesnake, however, are especially good at sidewinding, which has the advantage of keeping much of the snake's body off the hot ground.

Snakes generally undulate like fishes, but many also move by rectilinear movement and sidewinding.

Flyers and Gliders

Flying offers enormous advantages to both invertebrates such as butterflies, locusts, and beetles and to vertebrates such as birds and bats. Flying offers a quick escape route from ground-dwelling predators. Flyers can wander great distances in search of mates, nest sites, and food; and flyers can migrate seasonally, so that they can enjoy summer and spring all year round.

Flyers can feed from the flowers at the tops of trees, as well as from any kind of food that is dispersed over a wide area. Some flyers even feed on other flyers. Bats and many birds eat insects, while some hawks specialize on other birds. Finally, good flyers can disperse widely, so that these species often colonize whole continents or hemispheres, unlike their earthbound cousins.

True flying demands extreme specializations that can limit flyers when they are on the ground. The most obvious adaptation is wings. All vertebrate flyers, including bats and birds, have adapted the forelimbs for use as wings. Wings are virtually useless for anything but flying. As a result, most vertebrate flyers walk on two legs instead of four. Because birds have no forelimbs, they use their hind limbs for grasping food and twigs in the same way that most other animals use the forelimbs.

But flyers need other adaptations as well. They are extremely light for their size. Birds have hollow bones and their skeletons are much lighter than those of ground dwellers. Birds also lack teeth, which tend to be heavy. Most flyers have reduced legs compared with their ancestors, which had larger, more developed legs. A few flyers, such as swifts, have legs so small and weak that they are used only for perching and roosting. Flyers reduce weight wherever possible. The reproductive organs of flyers often shrink when the animal is not actually breeding. Because herbivores must have large, heavy digestive tracts, few flyers eat leaves or grass. Most birds eat insects, flesh, or seeds, which, like meat, are high in protein and fat and low in cellulose. The birds that do eat leaves—geese and quail, for example—are large-bodied birds that take flight reluctantly and with difficulty.

Another requirement of flying is that the body be compact and stiff. Bicyclists know that a bicycle frame must be extremely stiff, so that the force from the pedals is transmitted to the wheels and not lost in flexion of the frame. The skeleton of a flyer must be similarly rigid, so that the power in the breast muscles is transmitted to the wings. Much of a bird's backbone and pelvis is fused into a solid mass of bone (Figure 34-7). At the front, the ribs are fused to a second segment of fused backbone and to the sternum, or breastbone, as well. In birds the sternum is enlarged into a keel, which serves as an attachment site for the huge breast muscles. The resulting skeleton is a light, inflexible cage of bone.

Flyers also have a large, well-developed cerebellum, a part of the brain that helps animals orient in three-dimensional space, and, in birds, well-developed eyes.

Birds can fly enormous distances and achieve amazing speeds. Thousands of golden plovers fly nonstop from the Aleutian Islands to the Hawaiian Islands every fall, neither eating nor sleeping for days. The total distance is some 2500 miles. Each year the Arctic tern flies 12,000 miles from the Arctic Circle to Antarctica, and then back. One pilot estimated the speed of a flock of migrating sandpipers at 110 miles per hour. A peregrine falcon in a vertical dive, or stoop, was clocked at a speed of 125 miles per hour. And wind tunnel experiments with Lagger falcons suggest that this species may reach a terminal velocity of 225 miles per hour.

Figure 34-7 **Built for flight.** The skeleton of a flying bird is rigid—like a good bicycle frame—to allow for the efficient transfer of force from the breast muscles to the wings. The backbone is fused to the pelvis and the ribs are fused to the backbone and sternum. The sternum is enlarged into a keel to which the huge breast muscles attach.

Flying animals have many specializations, including powerful wings, a light, rigid skeleton, hollow bones, reduced jaws and legs, and reproductive structures that shrink when not needed.

Walkers, Runners, and Jumpers

Most terrestrial vertebrates get about by walking, running, or jumping. Their speed depends on two factors—the length of each stride and the number of strides per unit of time A galloping horse covers about 23 feet (7 meters) at a stride. A racehorse can manage well over 150 strides per minute, so that it can run up to 70 km per hour (45 miles per hour). The much smaller cheetah has the same stride length as a horse, but it can sprint one and a half times as fast as a racehorse because it moves its legs one and a half times as fast. How can the cheetah have a stride as long as that of a racehorse? And how does the cheetah manage to move its legs so much faster than the horse?

Stride length. All fast-running animals attain great speed by increasing stride length and stride rate. One way that stride length increases over the course of evolution is when the length of the leg increases in proportion to other parts of the body.

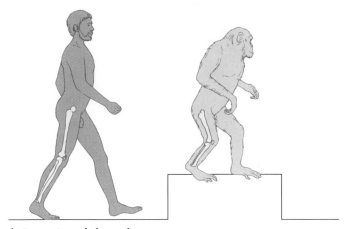

A. Proportionately longer legs

B. Flexible backbone

Figure 34-8   **Increasing stride length.** A. Humans are better adapted for running than chimps. Compared with the rest of our bodies, our legs are much longer than those of chimps. In addition, our toes are reduced, a trait typical of other running animals. B. Runners frequently have flexible spines, which increases the effective length of the stride.

The parts of the leg farthest from the body usually lengthen the most (Figure 34-8A).

Another way to increase stride length is to make the shoulder highly flexible. Cheetahs and other fast runners have a reduced shoulder girdle that allows them to move their forelimbs as far as possible. The clavicle is a mere vestige. In addition, the shoulder blades, or scapulas, may be turned 90° so that they are extensions of the leg. Effectively, this lengthens the leg still more. Some biologists theorize that dinosaurs' failure to evolve such modified shoulders forced the runners among them to evolve bipedalism. Cheetahs, dogs, weasels, horses, and other fast runners also extend their stride length by alternately flexing and extending their spines (Figure 34-8B).

Stride rate.   One might think that animal runners would move their legs faster simply by contracting their muscles faster. But, in fact, larger animals, which include most of the fastest runners, contract their muscles more slowly than small animals. One way that animals increase the rate of stride is by "gearing up" in the same way that bicyclists do.

Muscles that attach far from a joint, or lever, have more leverage than those that attach close to the joint. The armadillo uses its powerful limbs to dig through soil. As a result, its muscles tend to attach to the bone far from the joint. Its forelimbs move more slowly than those of a cheetah but with more power.

In contrast, muscles that attach close to the joint, or lever, move much faster, but are less powerful. Runners' muscles tend to attach very close to the joint.

Runners have other ways of increasing stride rate, but one important way is to keep the legs light. It is easier to move light legs than heavy ones. Consequently, antelopes, cheetahs, and other fast runners are lightly built compared with slower moving animals. A racehorse has lighter, longer legs than a draft horse. The bones of the outer parts of their limbs tend to be reduced and are often fused together for increased strength.

Animal runners increase overall speed by increasing stride length and stride rate. Long legs, specialized hips and shoulders, and a flexible spine all increase stride length. Light legs and muscle attachments close to the joint increase stride rate.

HOW DO VERTEBRATES MOVE?

Up to now, we have focused on how size and style of locomotion influence the overall shape of an animal. In the rest of this chapter, we will see how muscles and bone work together to move the body. We will also see how muscles actually contract.

As mentioned previously, every vertebrate possesses an internal skeleton, a solid supporting framework visible when all soft tissues are removed (Figure 34-9). The defining characteristic of vertebrates is the **backbone,** a column of hollow bony segments called **vertebrae** [singular, vertebra; Latin, *vertebratus* = jointed]. The **skull,** the bony case that encloses and protects the brain, lies at the forward (or, for humans, the top) end of a vertebrate's skeleton.

The point of attachment between any two bones is called a **joint.** Three kinds of connective tissue connect the bones and muscles of joints. First, joints may be bound together by a band of connective tissue called a **ligament.** Another type of connective tissue, called **cartilage,** cushions the joints. The tips of our noses and the outer ears, for example, are made of cartilage, as are the ends of many long bones such as the tibia and femur of the leg. The skeletons of fetal mammals, including humans, and the skeletons of sharks and rays are entirely cartilage, including no bone at all. A third form of connective tissue, a **tendon,** attaches the muscles to the bones.

Most bones can move relative to each other at the joints. The structure of a joint determines just how the adjacent bones can move. A knee joint, for example, enables the lower leg to swing back and forth, but not from side to side (Figure 34-10). In contrast, the joint between the thigh and the hip enables the thigh to swing in any direction, provided the connecting ligaments are sufficiently loose.

A few bones are immovable. The joints between the bones of the skull, for example, grow together at **sutures,** immovable joints that fuse separate bones into a rigid and protective helmet for the brain (Figure 34-9).

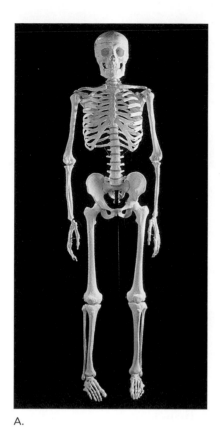

A.

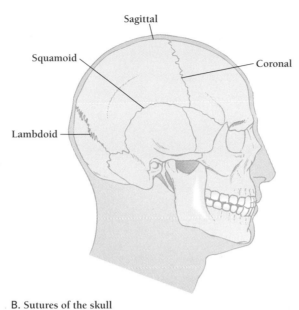

B. Sutures of the skull

Figure 34-9 A human skeleton. A. The bones of the "axial system" include the skull, rib cage, and spinal column. The bones of the "appendicular system" include the shoulder, or pectoral girdle (clavicle and scapula), and arms, as well as the pelvic girdle and legs. B. Not all joints move. During development, separate bones in the skull fuse at "sutures" to form a nearly impenetrable helmet of bone around the brain. *(A, SIU/Visuals Unlimited)*

Exoskeletons and Hydraulic Skeletons

Like vertebrates, invertebrates also move by contracting muscles against a rigid framework. An earthworm, for example, coordinates muscle action against a **hydrostatic skeleton,** a rigid fluid-filled space (the coelom), as described in Chapter 23 (Figure 34-11). Even a sea anemone (a cnidarian) has antagonistic muscle pairs. It can extend its body by contracting muscles that encircle the body cavity. Contraction of antagonist longitudinal muscles restores the animal to a squat form.

The skeletons of arthropods (most familiar in insects and edible crustaceans) are more similar to those of vertebrates, but they are outside, rather than inside, the body and are therefore called **exoskeletons** [Greek, *exo* = on the outside]. An arthropod's exoskeleton in some ways resembles a hollow cylindrical tube. As structural engineers discovered much later than arthropods, a hollow tube can support much more weight than can a solid rod of the same weight. Gram for gram, then, the exoskeleton of an invertebrate can support more body weight than the endoskeleton of a vertebrate. The major disadvantage of an exoskeleton, however, is that, unlike the bony endoskeleton, it cannot grow.

The flexion and extension of an insect's leg involves a pair of antagonist muscles, just as the flexion and extension of a vertebrate's forelimbs employs biceps and triceps. Having a skeleton outside, however, requires a different arrangement of muscles, rods, and joints. The same mechanical principles apply to invertebrates and vertebrates, and, in both cases, muscles can pull but not push.

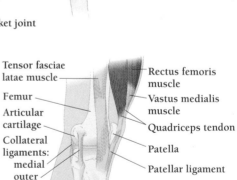

A. Hip: a ball-and-socket joint

B. Knee: a hinge joint

Figure 34-10 How do joints move? A. The thigh bone (femur) attaches to the pelvic girdle by means of a ball-and-socket joint. Like a joy stick, the femur moves freely both from side to side and up and down. It is limited only by the flexibility of the surrounding ligaments, tendons, and muscles. B. In contrast, the knee is more like the hinge on a door. It moves in one plane only. The lower leg can swing from behind the thigh, down to the ground, but no further, for the patella (knee cap) prevents the lower leg from swinging forward. An array of flexible tendons and ligaments both limit motion in the knee and allow slight twisting.

Figure 34-11 Hydrostatic skeleton. The earthworm's fluid-filled coelom provides a rigid hydrostatic skeleton, against which the muscles can push or pull. The earthworm uses a combination of circular and longitudinal muscles to inch through the soil.

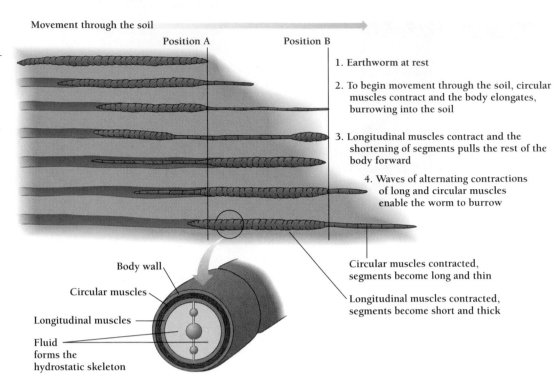

Movement through the soil

Position A   Position B

1. Earthworm at rest

2. To begin movement through the soil, circular muscles contract and the body elongates, burrowing into the soil

3. Longitudinal muscles contract and the shortening of segments pulls the rest of the body forward

4. Waves of alternating contractions of long and circular muscles enable the worm to burrow

Body wall

Circular muscles

Longitudinal muscles

Fluid forms the hydrostatic skeleton

Circular muscles contracted, segments become long and thin

Longitudinal muscles contracted, segments become short and thick

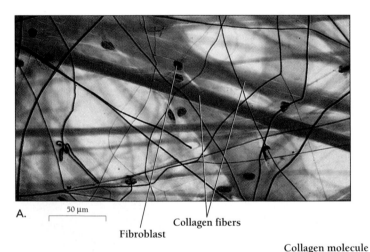

A.   50 μm

Collagen fibers

Fibroblast

What Forces Act on Bones and Connective Tissue?

Bones and the three kinds of connective tissue each must withstand very different sorts of forces. Ligaments and tendons must resist **tension,** the pulling action of two opposing forces. When we play tug-o-war, we exert tension on the rope and tension on our arms. Bones must resist tension and also two other kinds of stress—**compression,** the pushing action of two opposing forces, and **shear,** the twisting action created by forces that are not opposite one another. Tornadoes twist because opposing winds slide past one another and create intense shear forces.

Figure 34-12 Why is connective tissue so strong? A. The connective tissue of vertebrates is composed of fibroblasts and collagen fibers. B. A collagen fiber is composed, like a muscle fiber, of successive groups of fibers. Collagen molecules entwine to form collagen fibrils, bundles of which make up each collagen fiber. In connective tissue, these bundles often form perpendicular layers. Like plywood, which is made in the same way, these layers are together resistant to shear forces. (A, Ed Reschke/Peter Arnold, Inc.)

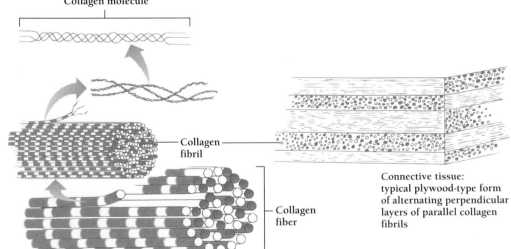

Collagen molecule

Collagen fibril

Collagen fiber

Connective tissue: typical plywood-type form of alternating perpendicular layers of parallel collagen fibrils

B.

We feel shear forces in our legs and hips when we suddenly while running, and we feel compression in our legs when we jump up and down.

Bones and connective tissue are enormously strong: a tendon half an inch in diameter could support a two-ton car. What makes connective tissue so strong? The microscope reveals that connective tissue consists largely of flat, irregularly shaped cells called **fibroblasts** embedded in a network of proteins and polysaccharides called the **extracellular matrix** (Figure 34-12A).

The fibroblasts secrete a fibrous protein called **collagen** and other substances into the extracellular matrix. The tough extracellular matrix is responsible for the diverse properties of connective tissues, which can vary from the soft transparency of the eye to the rocklike properties of bone. Collagen, which strengthens cartilage, ligaments, tendons, and bones, is by far the most abundant protein in the extracellular matrix and amounts to 25 percent or more of all animal protein. Collagen molecules form triple-stranded helices, which assemble into cross-linked cables, called **fibrils**. Fibrils in turn form large, interconnected **fibers.**

The organization of collagen helps connective tissues resist both tension and twisting shear forces. The fibrils resist tension by assembling in parallel along lines of stress, like the fibers of a rope. The fibrils resist shear forces by assembling in perpendicular layers, much like the alternating layers found in plywood (Figure 34-12B).

Bone's resistance to tension depends primarily on the combined strength of parallel collagen fibrils. But bone derives additional strength from needlelike crystals of calcium phosphate embedded in the extracellular matrix. The calcium phosphate crystals enable bones to resist both compression and shear forces. If a bone should crack, for example, the calcium phosphate crystals usually keep the crack from spreading far.

Tough collagen fibers arranged in parallel, like ropes, or in layers, like plywood, give bones and connective tissue much of their strength.

How Does Bone Form?

Adult bone consists of three major tissue types: **spongy bone tissue,** which generally lies at the ends and inside the bones; **red bone marrow,** a collection of "stem cells," which are undifferentiated cells embedded in the spongy bone, from which mature blood cells develop; and **compact bone tissue,** the hard, dense bone that surrounds the spongy interior (Figure 34-13).

We tend to think of our skeletons as both solid and unchanging. Certainly bones are solid, but unchanging they are not. Bones not only grow and repair themselves against continual wear and tear, they also change shape according to how they are used. The building and remodeling of bone is the task of two kinds of cells, the osteoblasts and the osteoclasts. The **osteoblasts** [Greek, *osteon* = bone + *blastos* = bud] are spe-

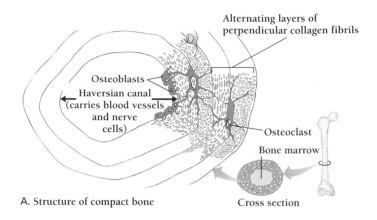

A. Structure of compact bone

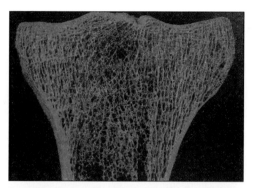

B. Spongy bone, red marrow, compact bone

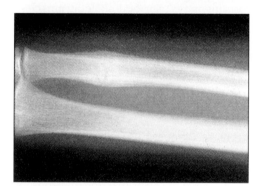

C. Broken bone in early stages of repair

Figure 34-13   The structure of bone. A. Compact bone consists of dense layers of perpendicular fibers of collagen. Embedded in the collagen layers are crystals of calcium and phosphate, which give compact bone its hardness. Within the bone are Haversian canals, thin tubes that run parallel to the length of a bone and contain blood vessels and nerve cells. B. Spongy bone, which lies in the interior to the compact bone and near the ends of long bones, consists of a lattice of collagen and calcium phosphate, similar to that of compact bone but more open. At the very center of some of the larger bones of the body is bone marrow, where blood cells develop. C. A bone in the early stages of repair. Initially, a broken bone mends with a mass of bone tissue, built up by bone-building cells called osteoblasts. Under the right circumstances, the interaction between the bone-destroying cells, the osteoclasts, and the osteoblasts results in a resculpted and functional bone.   *(B, Don W. Fawcett/Visuals Unlimited; C, Bates, M.D./Custom Medical Stock Photo)*

cialized cells that resemble fibroblasts and manufacture the bone matrix. The **osteoclasts** [Greek, *osteon* = bone + *klastos* = broken] are similar cells that digest collagen and bone matrix. In all living bone, the osteoclasts continually break down and absorb bone, while the osteoblasts continually form new bone.

Normally, the rate at which bone is created is equal to the rate at which it is absorbed. In bones that are growing, or in bones that are repairing damage, the osteoblasts produce more bone than the osteoclasts destroy (Figure 34-13C). But bones adjust their size and density according to the amount of compression they experience. The bones of people who exercise become thick and strong, while those of couch potatoes (and astronauts living at zero gravity) tend to become thin, brittle, and decalcified. If a person favors an injured leg, the injured leg becomes thin and decalcified, while the other leg bone becomes heavier and stronger than it was before. For this reason, physicians now encourage people with healing bone fractures to begin using the limb as soon as possible.

Osteoclasts and osteoblasts continually shape and reshape bones as we grow, heal, and change our habits.

How Do Skeletal Muscles Move Bones?

The human body contains more than 600 **skeletal muscles,** those muscles that are attached to bones (Figure 34-14). Skeletal muscles, together with the nerves that coordinate them, are responsible for our ability to move. Together, the skeletal muscles are the largest tissue in the vertebrate body, making up more than 40 percent of its weight. Some individual muscles are small, with only a few hundred muscle cells; others contain hundreds of thousands of muscle cells. The size, organization, and arrangement of our muscles establish which movements we can and cannot perform.

Muscles can only exert force in one direction. They can pull but not push. Most body movement therefore depends on the arrangement of **antagonistic pairs** of muscles, which pull in opposite directions. The contraction of the biceps muscle, for example, flexes the forearm, while the contraction of its antagonist, the triceps muscle, extends the forearm.

In the human body, some 600 skeletal muscles, arranged in antagonistic pairs, work our bodies by pulling against bone and each other.

How Do Smooth and Cardiac Muscle Differ from Skeletal Muscle?

Vertebrates have other muscles besides those that move the skeleton. These include **cardiac muscles,** which pump blood through the heart, and **smooth muscles,** which line the walls of hollow internal organs such as the intestines, the uterus, and the blood vessels. Smooth muscles also contract and relax the iris of the eye, and they form "goose bumps" to make hair stand on end when we are frightened or cold.

Because humans and other animals have voluntary control over most of their body movements, skeletal muscles are often called **voluntary muscles.** In contrast, smooth muscle and cardiac muscle are **involuntary muscles.** We cannot control the rate at which our intestines move or whether our heart beats.

The difference between voluntary and involuntary muscles lies in the types of nerves that control them. Skeletal muscles are generally under the control of a system of nerves called the **motor axis.** In contrast, the smooth muscles and the cardiac muscles are controlled by a parallel system of nerves called the **autonomic nervous system.**

The autonomic nervous system consists of two distinct sets of nerves—the *sympathetic* and the *parasympathetic* nerves. The **sympathetic nervous system** brings about the responses of the "fight or flight" reaction, slowing movements of the digestive system and speeding up the heart in preparation for vigorous action, for example. In contrast, the **parasympathetic nervous system** acts to save resources, by slowing the heart, for example, and increasing the action of the intestines. In general, we cannot consciously control the contractions of smooth muscles.

Smooth muscle, as its name suggests, is smoother than either cardiac muscle or skeletal muscle, both of which are **striated,** or striped. The **striations** consist of cross stripes spaced every 2 to 3 μm and reflect the regular arrangement of the protein filaments that are responsible for muscle contraction (Figure 34-15).

Smooth muscles and cardiac muscles are not attached to bones. Because they are under the control of the autonomic nervous system, cardiac and smooth muscles are not under conscious control.

Why Are Marathon Runners So Wiry?

Every skeletal muscle consists of many parallel **muscle fibers,** giant cells up to 4 cm long. Each muscle cell, or fiber, contains up to 100 nuclei, as well as the proteins responsible for muscle contraction.

Muscle fibers are of two kinds. Thick fibers, called **glycolytic fibers,** derive most of their energy from glycolysis and have few mitochondria. Thinner fibers, called **oxidative fibers,** derive most of their energy from respiration and have many mitochondria. The thin oxidative fibers power muscles over long periods, while the thick glycolytic fibers power muscles in short bursts.

Oxidative fibers depend on a continuing supply of oxygen, and therefore they are well supplied with blood capillaries. They also contain high levels of **myoglobin,** an iron-containing muscle protein that pulls oxygen from the blood. Both myoglobin and the rich supply of blood give oxidative fibers, and the muscles that contain them, a distinctive red color. The

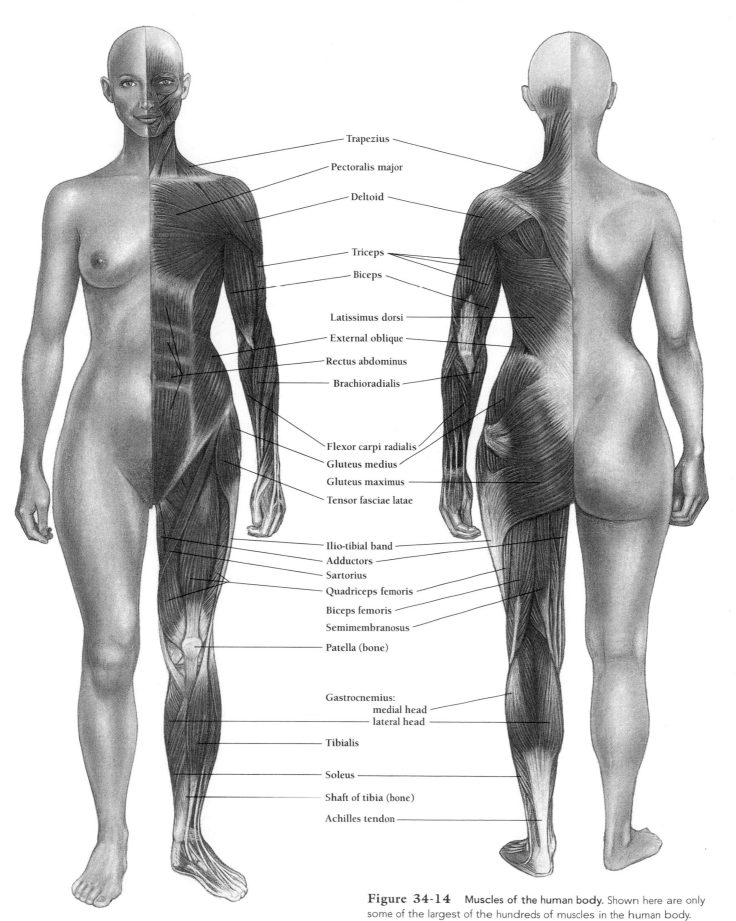

Trapezius

Pectoralis major

Deltoid

Triceps

Biceps

Latissimus dorsi

External oblique

Rectus abdominus

Brachioradialis

Flexor carpi radialis

Gluteus medius

Gluteus maximus

Tensor fasciae latae

Ilio-tibial band

Adductors

Sartorius

Quadriceps femoris

Biceps femoris

Semimembranosus

Patella (bone)

Gastrocnemius:
   medial head
   lateral head

Tibialis

Soleus

Shaft of tibia (bone)

Achilles tendon

Figure 34-14   **Muscles of the human body.** Shown here are only some of the largest of the hundreds of muscles in the human body.

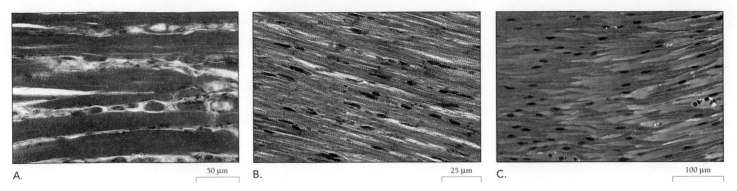

Figure 34-15   **The three kinds of muscle.** A. The striations that enable skeletal muscle to contract are distinctive in micrographs such as this. B. Cardiac muscle is also striated. C. Lacking striations, the muscles of the digestive tract and urogenital organs are called "smooth muscle."   *(A, David M. Phillips/Visuals Unlimited; B, Biophoto Associates/Science Source/Photo Researchers; C, M.I. Walker/Science Source/Photo Researchers)*

"red meat" in chicken and turkey legs is red because these muscles have a large proportion of oxidative fibers.

In contrast, the "white meat" of a chicken breast consists mainly of thick, white glycolytic fibers. Glycolytic fibers are also called "fast fibers" because they can deliver a lot of power in a short time. Fast fibers can only be used for short bursts of activity, however, and after a minute or so their energy supplies are exhausted.

Like bones, muscles grow or atrophy according to how much they are used. Brief, high intensity exercise—such as weight lifting—increases the diameter of the glycolytic fibers. This is why the muscles of a trained weight lifter bulge. The effect of aerobic exercise—such as long-distance swimming or running—is quite different. The muscle fibers do not increase in diameter, and the muscles do not bulge. Instead, invisibly, the oxidative fibers acquire more mitochondria, and the circulatory system's capacity to deliver oxygen to the muscles increases.

Sprinters and weight lifters therefore tend to develop big muscles with many fast fibers, while marathon runners tend to develop muscles with many thin, oxidative fibers, which accounts for their wiry builds.

Skeletal muscles consist of parallel fibers of two kinds. Glycolytic fibers make the big muscles of a weight lifter bulge and the breast of a chicken tender. Oxidative fibers power the long-distance runner's wiry legs.

How Are Contractile Proteins Arranged?

When early biologists first examined skeletal muscle under the light microscope, its striated appearance suggested that muscle was highly organized at the molecular level. As a result, in the 1950s, when biologists began to use the transmission electron microscope to study subcellular structure, skeletal muscle was among the very first tissues examined.

Such microscopic studies showed that a muscle fiber from a skeletal muscle consists of many **myofibrils** [Greek, *myos* = muscle + Latin, *fibrilla* = little fiber], threads 1 to 2 μm in diameter that run the whole length of the fiber (Figure 34-16). Each myofibril thread consists of a string of smaller units, called **sarcomeres** [Greek, *sarx* = flesh + *meros* = part], small cylinders each about 2.5 μm long. A 2.5-cm-long myofibril, then, consists of about 10,000 sarcomeres laid end to end.

Each sarcomere is marked by light and dark stripes of varying width. At one end of each sarcomere is a thin dark stripe called a Z band. The bands of each sarcomere are perfectly aligned with the bands of all the other sarcomeres within a myofibril. This alignment explains why muscle fibers appear striated. Between the two ends of the sarcomere, marked by the Z bands, are five more bands, which vary in thickness according to how much they contract.

The striped appearance of striated muscle derives from the ordered arrangement of striped sarcomeres. The light and dark bands of each sarcomere are aligned with those on adjacent myofibrils.

How Do Striated Muscles Contract?

When a muscle fiber contracts, the sarcomeres contract. The distance between the Z bands changes in direct proportion to the length of a muscle. For example, if the muscle shortens by 20 percent, each sarcomere shortens by 20 percent. The question "How does a muscle contract?" therefore becomes "How does a sarcomere contract?"

In long sections of a muscle fiber, the electron microscope reveals the basis of the banding pattern. Each sarcomere contains hundreds of tiny, parallel filaments. These filaments come in two sizes—**thin filaments,** made of **actin,** and **thick filaments,** made of **myosin.** The actin filaments come in pairs that extend from the Z bands inward toward each other (Figure 34-

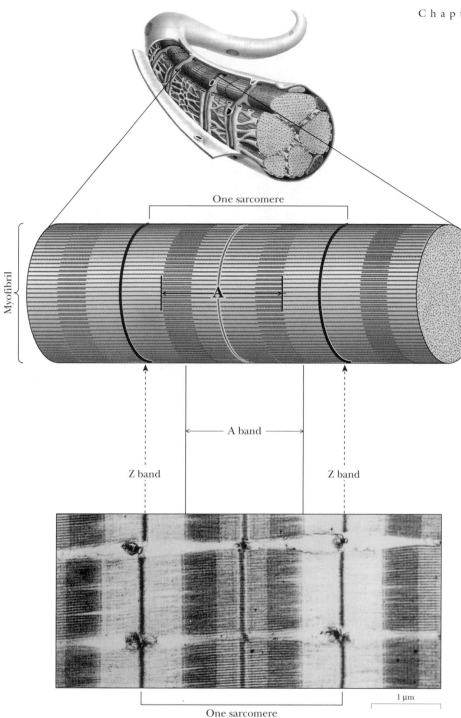

One sarcomere

Myofibril

A

A band

Z band   Z band

1 μm

One sarcomere

17). Parallel to the pairs of thin actin filaments are thick myosin filaments. During contraction, each pair of actin filaments slides together past the myosin filaments. When a muscle relaxes, the pairs of actin filaments slide apart again (Figure 34-17A). This model of muscle contraction is called the **sliding filament model.**

We can imagine how the filaments slide past each other during contraction, but what force pulls the actin molecules together and what role do the myosin filaments play? The answer came from particularly good electron microscope photographs. These showed that the thick and thin filaments were linked to each other at regular intervals by tiny **cross bridges.** The cross bridges are actually extensions of the myosin mole-

cules that make up the thick filaments (Figure 34-17B). Researchers guessed that the cross bridges perform the work of muscle contraction by pulling the thin and thick filaments over one another.

The hypothesis that the cross bridges provide the force for contraction suggested a testable prediction: the force generated by a muscle should depend on the degree of overlap between the thick and thin filaments. That is, if a muscle is sufficiently stretched, the thick and thin filaments have little overlap, and few cross bridges are available to move the filaments. Researchers tested this prediction by stretching muscles to different lengths and then measuring the amount of force the muscles could exert (by having the muscle pull against a spring).

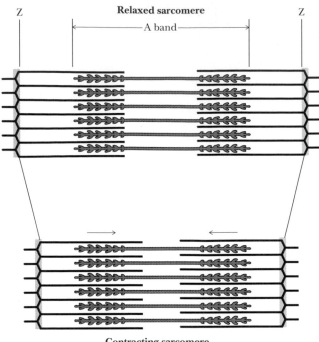

Relaxed sarcomere

A band

Z   Z

Contracting sarcomere

A.

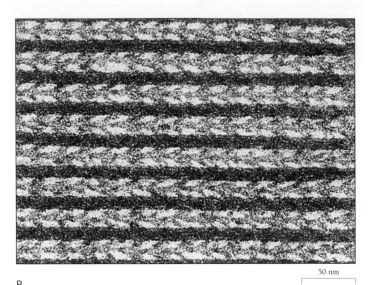

B.

50 nm

Figure 34-17 Muscle contraction. A. Thin filaments attached to the Z bands and thick filaments (the A band) slide past one another during the contraction and relaxation of the muscle. B. Cross bridges between the thin filaments and the thick filaments, such as those in this micrograph, provide the force for contraction. *(B, Mary Reedy, Duke University Medical Center)*

The results were exactly as predicted by the sliding filament model.

The light microscope and the electron microscope have revealed the organization of contractile proteins in skeletal muscle fibers and have suggested a model for their action. Cross bridges between thick and thin filaments are responsible for muscle contraction.

In this chapter we have seen some examples of how the function of an organ determines its form; what physical constraints affect form and physiology; and how animals use skeleton and muscle to move. In the next chapter we will find out how animals derive nourishment from food. We will examine both digestion, the breakdown of food particles to their constituent building blocks, and absorption of those building blocks by the gut. In the remaining chapters of this last section of *Asking About Life,* we will see many more specific examples of how form follows function.

STUDY OUTLINE WITH KEY TERMS

In this last section of *Asking About Life,* we look at the **physiology** of the body—how it is organized and how its parts work together. An overriding theme in all physiology is **homeostasis,** the tendency of living organisms to maintain their internal environment.

To understand function, we must understand structure. In all coelomate animals, the body is a tube (the digestive tract) within a tube (the body wall). In the space between the two tubes, the **coelom,** lie many of the main organs of the body. In vertebrates, the coelom consists of the **throacic cavity** and the **abdominal cavity.**

The shape of an animal depends on its size. Because of gravity, large animals must have proportionately stronger and thicker bones than small animals. Size also influences metabolic rate. Webbed feet, the lungs, the intestines, and the circulatory system are all organs with expanded surface areas. **Countercurrent systems** enable animals to establish steep gradients of temperature or concentration. Inside the noses of many vertebrates are elaborately folded **turbinate** bones, which increase the surface area.

Style of locomotion also determines shape. Swimmers, which undulate through the water, have a streamlined shape. Snakes undulate like fish, but many also move by **rectilinear movement** and **sidewinding.** Flying animals have many specializations, including powerful wings, a light, rigid skeleton, hollow bones, reduced jaws and legs, and reproductive structures that shrink when not needed. Runners increase overall speed by increasing stride length and stride rate. Long legs, specialized hips and shoulders, and a flexible spine all increase stride length. Light legs and muscle attachments close to the joint increase stride rate.

The defining characteristics of vertebrates are the **backbone,** a column of hollow bony segments called **vertebrae,** and the **skull,** the bony case that encloses and protects the brain. The point of attachment between any two bones is called a **joint.** Immovable joints among the bones in the head are called **sutures.**

All animals have some kind of skeletal system. Earthworms and some other invertebrates coordinate muscle action using a rigid fluid-filled space called a **hydrostatic skeleton.** Arthropods rely on an external **exoskeleton.**

Three kinds of connective tissue connect the bones and muscles of joints: **ligaments, tendons,** and **cartilage.** Ligaments and tendons must resist **tension.** Bones must resist tension and also two other kinds of stress—**compression,** the pushing action of two opposing forces, and **shear,** the twisting action created by forces that are not opposite one another.

Both bone and other kinds of connective tissue consist of flat, irregularly shaped cells called **fibroblasts** embedded in a network of proteins and polysaccharides called the **extracellular matrix.** The matrix consists of fibrous **collagen** and other substances secreted by the fibroblasts. Collagen molecules form triple-stranded helices, which assemble into cross-linked cables, called **fibrils.** Fibrils in turn form large, interconnected **fibers.** The fibers, which are arranged in parallel, like ropes, or in layers, like plywood, give bones and connective tissue much of their strength.

Adult bone consists of three major tissue types: **spongy bone tissue, red bone marrow,** and **compact bone tissue.** Specialized **osteoblasts,** similar to fibroblasts, manufacture the bone matrix. **Osteoclasts** continuously digest collagen and bone matrix. Osteoclasts and osteoblasts continuously shape and reshape bones as we grow, heal, and change our habits.

In the human body, some 600 **skeletal muscles**—arranged in **antagonistic pairs**—move our bodies by pulling against bone and one another. Besides skeletal muscle, vertebrates also have **cardiac muscles,** which pump blood through the heart, and **smooth muscles,** which line the walls of hollow internal organs such as the intestines, the uterus, and the blood vessels. Smooth muscle and cardiac muscle are not attached to bones.

Because we have voluntary control over most skeletal muscles, they are often called **voluntary muscles.** In contrast, smooth muscle and cardiac muscle are **involuntary muscles.** Skeletal muscles are generally under the control of a system of nerves called the **motor axis,** while the smooth and cardiac muscles are controlled by a parallel system of nerves called the **autonomic nervous system.** The autonomic nervous system consists of the **sympathetic nervous system,** which brings about the "fight or flight" responses, and the **parasympathetic nervous system,** which slows the heart and increases the action of the intestines.

Unlike smooth muscle, cardiac and skeletal muscle are **striated,** or striped. The **striations** consist of cross stripes spaced every 2 to 3 μm and reflect the regular arrangement of the protein filaments that are responsible for muscle contraction Every skeletal muscle consists of many parallel **muscle fibers,** giant cells up to 4 cm long. Each muscle cell, or fiber, contains up to 100 nuclei, as well as proteins responsible for muscle contraction.

Muscle fibers are of two kinds. The thick **glycolytic fibers** derive most of their energy from glycolysis and have few mitochondria. The thinner **oxidative fibers** derive most of their energy from respiration and have many mitochondria. Because oxidative fibers depend on a continuing supply of oxygen, they contain high levels of **myoglobin.**

Striated muscle consists of **muscle fibers,** which consist in turn of bundles of **myofibrils.** Each myofibril consists of bundles of **thick filaments** and **thin filaments.** The functional unit of a myofibril is a **sarcomere,** a short cylinder of thick filaments, which are made of **actin,** and thin filaments, which are made of **myosin.** At either end of each sarcomere is a thin, dark stripe called a Z band. According to the **sliding filament model,** thin filaments attach to the Z bands and thick filaments slide past during the contraction and relaxation of the muscle. **Cross bridges** between the thin filaments and the thick filaments provide the force for contraction.

REVIEW AND THOUGHT QUESTIONS

Review Questions

1. Why are an elephant's legs thick and straight?
2. Explain how countercurrent exchange systems work. Give an example.
3. Why are fish, birds, and ships streamlined?
4. Fibroblasts exist in which tissues of the body? What function do they serve?
5. What are the three kinds of bone found in adults?
6. Name the two partners in an antagonistic pair of muscles. What is the value of these two muscles working against each other?
7. What is the difference between voluntary muscles and involuntary ones?

8. Explain in detail why long-distance runners' muscles are so much smaller than those of body builders.

Thought Questions

9. Many people can consciously slow or speed up the rate at which their hearts beat. What mechanisms can you imagine that would allow voluntary control over this autonomic function?
10. Imagine a giant, 12 feet tall. For his legs to be as strong as a normal man's, how thick would his femurs need to be?

SELECTED READINGS

Alexander, R. McNeill, *Functional Design in Fishes,* Hutchinson University Library, London, 1974. An interesting, half-technical, half-popular account of how fishes work.

Gans, Carl, *Biomechanics: An Approach to Vertebrate Biology,* J.P. Lippincott, Philadelphia, 1974. A smorgasbord for anyone interested in biomechanics. Everything from how snakes manage to swallow and digest large eggs to gas flow in the respiratory and vocalization cavities of frogs.

Schmidt-Nielsen, Knut, *How Animals Work,* Cambridge University Press, Cambridge, 1972. A short book on some of the more interesting aspects of animal form and function. This well-written book is a classic.

Terres, John K., *How Birds Fly,* Harper & Row, New York, 1987. A charming, popular account of how birds fly and why.

▶ On-line materials relating to this chapter are on the World Wide Web at http://www.saunderscollege.com/lifesci/
Click on Tobin/Dusheck: *Asking About Life.*

Another American Shot Heard Round the World

American physiology began with a "bang." In the summer of 1822, a shotgun accidentally discharged into the belly of a 19-year-old French Canadian named Alexis St. Martin. His misfortune added immeasurably to the scientific understanding of the process of digestion.

St. Martin worked for the American Fur Company as a "voyageur," transporting goods and supplies between the company's remote outposts. That summer, St. Martin was one of a crowd of soldiers, Indians, and trappers who had jammed into

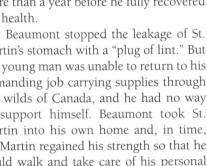

a trading post on Mackinac Island, between Lake Huron and Lake Michigan, to buy and sell pelts, coats, and moccasins and to trade stories after a long and lonely winter (Figure 35-1). St. Martin was less than three feet from the shotgun's muzzle, and the hole it made was bigger than the palm of his hand.

As the horrified crowd stood around the fallen young man, one of the store's clerks ran to fetch Dr. William Beaumont, the surgeon at nearby Fort Mackinac. When Beaumont arrived minutes later, he found a rib and part of a lung protruding from the wound, as well as a portion of St. Martin's stomach. The protruding stomach had a puncture, which, Beaumont later wrote, was "large enough to re-

ceive my forefinger, and through which a portion of his food that he had taken for breakfast had come out and lodged among his apparel. . . . I considered any attempt to save his life entirely useless." Beaumont nonetheless cleaned and dressed the wound, "not believing it possible for him to survive twenty minutes."

In fact, St. Martin recovered amazingly. By December, St. Martin's health was "daily improving, his spirits good, his appetite regular, his sleep refreshing." But St. Martin was to live the rest of his life with

a hole that led from the outside world directly into his stomach, and it would be more than a year before he fully recovered his health.

Beaumont stopped the leakage of St. Martin's stomach with a "plug of lint." But the young man was unable to return to his demanding job carrying supplies through the wilds of Canada, and he had no way to support himself. Beaumont took St. Martin into his own home and, in time, St. Martin regained his strength so that he could walk and take care of his personal needs. But he was still unable to return to work as a voyageur.

Then, one winter, almost two years later, Beaumont realized that St. Martin's protruding and perforated stomach pro-

Paraskevas Photography

Figure 35-1 William Beaumont and Alexis St. Martin. *(The Granger Collection, New York)*

vided "an excellent opportunity for experimenting upon the gastric fluids and the process of digestion." Beaumont began his first series of such experiments in May 1825, in Mackinac. A month later, the doctor was ordered to Fort Niagara, New York, and he took St. Martin with him in order to continue his experiments. Beaumont had come to realize that the unique experiments he could do with St. Martin would bring him great prestige in America's infant scientific community. In 1833, Beaumont published a full description of his experiments and his conclusions in a volume called *Experiments and Observations on the Gastric Juice and the Physiology of Digestion.*

Beaumont's work is a model of careful observation and experiment. In one study, for example, Beaumont showed that a piece of beef suspended into St. Martin's stomach on a string was digested over a period of several hours. Virtually the same visible changes occurred if a similar piece of beef was placed in a glass vial containing the "gastric juice" obtained from St. Martin's stomach with a glass tube. The

gastric juice in the vial, Beaumont showed, worked much better at body temperature, however, than at room temperature.

Beaumont later showed that the gastric juice contained hydrochloric acid. Yet, hydrochloric acid alone would not digest the meat. We now understand Beaumont's

results as demonstrating the action of the stomach enzyme **pepsin**, which hydrolyzes proteins into smaller fragments each containing a few amino acids. In the intestine, other enzymes break these fragments into individual amino acids. Pepsin catalyzes hydrolysis far more powerfully than hydrochloric acid. Pepsin works best in an acid environment, however, and much more rapidly at body temperature than at room temperature.

At the end of his volume, Beaumont listed 51 conclusions derived from his experiments and observations. This list was as important for the questions it raised as it was for the knowledge it imparted. For example, Beaumont noted that the secretion of gastric juice depended on the presence of food. From a 20th-century perspective, we want to know what chemical signals in food stimulate the secretion of gastric juice. How are the signals delivered? And what mechanisms trigger the production of hydrochloric acid and pepsin?

The questions are even more compelling if we ask them not only about humans but also about other animals. For

---

St. Martin was less than three feet from the shotgun's muzzle, and the hole it made was bigger than the palm of his hand.

---

example, Beaumont noted that how finely St. Martin chewed his food affected the way the food was digested in the stomach. We can then ask, How do birds, which lack teeth, and reptiles, which have teeth for biting but not grinding, digest their food efficiently?

KEY CONCEPTS

1. Animals depend on food for energy and raw materials.

2. An animal hydrolyzes macromolecules and lipids into their component building blocks and absorbs the resulting small molecules.

3. In vertebrates, digestion and absorption occur in a digestive tract that starts at the mouth and ends at the anus. The parts of the tract are specialized for different functions, including movement of the tract's contents and secretion of enzymes and mucus.

4. Nerve cells and hormones coordinate the actions of the different parts of the digestive system.

WHY MUST ANIMALS EAT?

The task of the digestive system is to derive both energy and raw materials from food. Food comes in many forms, and diets vary enormously among species, but digestive systems perform the same basic task: they break down the complex molecules of food, especially macromolecules and lipids, into simpler molecules that can be absorbed and incorporated in the animal's own metabolism.

Energy comes from complex molecules—sugars, polysaccharides, proteins, fats, and oils. Each molecule undergoes a series of enzyme-catalyzed conversions to simple molecules, such as carbon dioxide and water. In the case of glucose, the reactions of glycolysis, respiration, and oxidative phosphorylation produce the ATP molecules that power most of an animal's activities. Other simple molecules produced by digestion, such as amino acids and lipids, can undergo biochemical conversion into compounds that can also enter one of these same energy-producing pathways.

Although an animal can make many of the building blocks needed to produce its own large molecules, its food also provides a ready-made supply, including some that the animal cannot make itself. Most animals, for example, can make 10 of the 20 amino acids needed to make proteins, but they must obtain the other 10 from their food. Animals need other kinds of raw materials as well: minerals, inorganic elements, vitamins, and organic compounds that the animal cannot itself synthesize.

Individual species of animals derive their food from differing sources. **Herbivores** [Latin, *herba* = vegetation + *vorare* = to swallow], such as caterpillars, cows, and sea turtles, eat only plants. **Carnivores** [Latin, *carn* = flesh], such as spiders and seals, eat only animals, with many species specializing on particular parts, such as the blood or the skin. Still other animals only consume detritus, the remains of plants and animals, broken down into smaller fragments by the action of microorganisms. Humans are **omnivores** [Latin, *omnis* = all], and our diets range widely (Figure 35-2). Still, the processing of the food is always the same.

The process of deriving energy and raw materials from food is always the same.

How Much Must an Animal Eat?

Food must provide energy for all of an animal's activities. Physiologists usually estimate energy use by measuring oxygen consumption in animals performing various activities. Each liter of oxygen used corresponds to the expenditure of just under 5 kilocalories of energy. (Recall that one kilocalorie is 1000 calories, where one calorie is the energy required to raise the temperature of one gram of water by 1°C; tables in books and articles on diet and nutrition usually use "Calorie" with a capital C, which is the same as a kilocalorie.)

Animals differ in their energy requirements, according to their size, species, and activity levels. Energy consumption is roughly proportional to weight. For a moderately active college student, for example, daily energy use (in kcal) is about 15 to 20 times the weight in pounds. So a 120-pound woman may require 1800 to 2400 kcal per day, while a 160-pound man may require 2400 to 3200 kcal per day.

The consumption of energy depends on particular activities: running or swimming requires more energy than standing or floating. But every animal requires a certain amount of energy just to stay alive and awake. This minimal energy requirement is called the **basal metabolic rate.** For humans, the basal metabolic rate is about 1400 kcal per day for a 115-pound woman and about 1700 kcal per day for a 160-pound man. This is about the energy needed to keep a 75-watt bulb burning all day and night or to drive a mile in a small car. For humans and other animals, the basal metabolic rate is typically slightly more than half the energy used in ordinary activities.

Individual animal species have widely varying basal metabolic rates, even after accounting for their different sizes. For example, the basal metabolic rate, per gram, of an adult human corresponds to about 1 calorie per hour, whereas the cor-

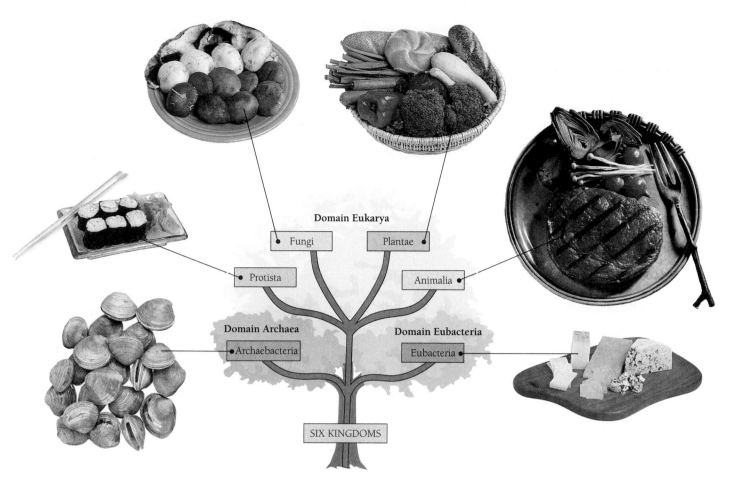

Figure 35-2 **Humans eat from nearly every kingdom.** Shown here are cheeses (eubacteria), mammal (animal), vegetables and bread (plants), mushrooms (fungi), and sushi wrapped in seaweed (protist). We have also included clams (animals) because they are known to harbor archaebacteria. *(Clams, sushi, mushrooms, vegetables, cheese, Paraskevas Photography; meat, Envision © Rudy Muller)*

responding basal metabolic rate for a shrew is 35 times higher. A 5-gram shrew must therefore consume about 4 kcal per day just to stay alive and another 4 kcal per day (a total of 8 kcal/day) to lead a normal shrew life, searching for beetle larvae in underground tunnels. A diet of pure protein or carbohydrate contains about 4 kcal per gram, so the shrew could sustain itself with only 2 grams per day on a diet that consists only of protein and carbohydrate. But a wild shrew is unlikely to find so refined a diet, and it has to eat an amount each day corresponding to its own body weight. To minimize the amount of food that must pass through its gut, a shrew must seek a high-quality diet, with little indigestible material. The best diet would be rich in fat—insect larvae, for example—since fat contains 9 kcal per gram, compared with 4 kcal per gram in carbohydrates and proteins.

Food must provide enough energy to supply an animal with its basic metabolic needs and to support its various activities.

What Are the Consequences of Too Much or Too Little Food?

If an animal takes in more food energy than it uses, the excess is stored as glycogen or as fat. This situation—called overnourishment—is common in human industrialized societies. Many people have more stored energy than they will probably ever need, causing strain on the heart and circulation.

If, on the other hand, an animal takes in less food energy than it needs, it must derive the extra energy from its own body. When stored energy runs out, animals begin to break down their muscles and other tissues. Undernourishment—taking in too few calories—occurs in modern human populations mostly in times of drought, war, or other catastrophe. In contrast, **malnourishment,** a deficiency in one or more essential nutrients, is unfortunately all too common: it affects at least 500 million people, 10 percent of the world's population. Malnourishment is an increasing problem even in wealthy countries such as the United States, where not everyone has equal access to a balanced diet or good nutritional advice.

Animals can usually adjust their food intake to match their activities. For example, when researchers diluted ordinary rat food with nonnutritive materials, the rats increased the amount they ate so that they acquired the needed amount of food energy. Damage to the hypothalamus, which is the part of the brain that controls appetite, however, can abolish the ability to adjust the diet. A rat with a damaged hypothalamus may feed without stopping, while another, with a different pattern of damage, may starve itself to death.

Animals usually regulate their food intake to match their activities, but this regulation can go awry either because of an inadequate food supply or because of a failure in normal regulation.

What Are the Causes of Malnourishment?

Malnourishment may result from a deficiency in protein, in minerals, or in vitamins. Each type of deficiency leads to a recognizable set of symptoms, many of which were first described in sailors long at sea.

Protein deficiency is the principal cause of human malnourishment, which affects some 100 million people worldwide. Protein deficiency disease is also called **kwashiorkor**, a term that originated in Ghana. Kwashiorkor is characterized by lethargy, severe anemia, change in hair color, inflammation of the skin, and a pot belly. Kwashiorkor is especially prevalent in children just after weaning, when their diet switches from milk to a single kind of starch—most often rice, corn, or cassava.

One of the tasks of the digestive system is to break down the proteins in food into individual amino acids, which are then transported to cells that make proteins. Those cells make ten or more amino acids from other molecules, but all the cells of an animal depend on dietary protein to provide the **essential amino acids**, those which in an animal cannot produce itself. Table 35-1 lists the essential amino acids in humans. Protein deficiency may result either from insufficient protein in the diet or from an improper ratio of essential amino acids.

Because animals do not store amino acids between feedings (as, for example, they store glucose in the form of glycogen), each meal must provide the proper proportions of essential amino acids needed to make proteins in that animal. For carnivores and for human populations that eat animal proteins—milk, eggs, or meat—the balance of amino acids is usually about right. But for herbivores and human vegetarians, the problem of amino acid balance may dominate all other health issues.

Many human societies depend on a single crop—one that grows well in the local climate, is inexpensive to grow, and is resistant to local pests. But the proportions of amino acids in a single plant species are never the same as those in an animal's own proteins. Proper balance requires mixing protein sources (Figure 35-3). An appropriate mixture—one that provides

Table 35-1	Essential Amino Acids in Humans	
Amino Acid	**Poor Sources**	**Good Sources**
Isoleucine	—	Essentially any good protein source: legumes, cereal grains, nuts, dairy, meat, eggs
Leucine	—	Essentially any protein source
Lysine	Peanuts, corn, oats, rice, wheat	Soybeans, other beans, milk, eggs, meat
Methionine	Beans, lentils, peanuts	Cereal grains, milk, meat, eggs
Phenylalanine	—	Essentially any good protein source
Threonine	—	Essentially any good protein source
Tryptophan	Beans, peanuts, corn, almonds	Cereal grains other than corn, wheat germ, milk, meat, cheese, eggs
Valine	—	Essentially any protein source

enough of each of the essential amino acids to sustain protein synthesis—is said to contain *complementary proteins*. Where enough food is available, traditional human societies have developed diets that contain complementary proteins—corn and beans, wheat and chick peas, rice and lentils. Corn, for exam-

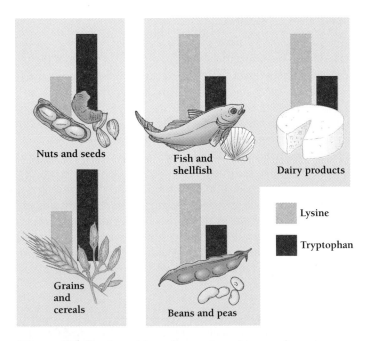

Figure 35-3 A nutritious diet must provide complementary proteins. Some foods are poor in lysine, while others are poor in tryptophan.

Table 35-2   Major Minerals Important in the Human Diet

Mineral	Good Sources	Symptoms of Deficiency	Functions
Sodium	Normal diet; table salt	Muscle cramps	Water balance; operation of muscles and nerves
Potassium	Fruits, vegetables, grains	Irregular heartbeat, fatigue, muscle cramps	Operation of muscles and nerves; acid-base balance
Chlorine	Table salt	Unlikely	Acid formation in stomach; nerve signaling
Calcium	Dairy, bony fish, leafy green vegetables, dried legumes	Osteoporosis	Formation of bone and teeth; clotting; nerve signaling
Phosphorus	Dairy, meat, cereals	Bone loss, weakness, lack of appetite	Formation of bone and teeth; energy metabolism (e.g., ATP); membrane phospholipids
Magnesium	Nuts, greens, whole grains	Nausea, vomiting, weakness	Enzyme action; nerve signaling
Sulfur	Protein foods	None, when protein needs are met	Proteins

ple, is low in lysine but high in tryptophan, while beans are low in tryptophan but high in lysine. In many parts of the world, however, only one protein source may be affordable, and protein deficiency disease is rampant.

Deficiencies in minerals can also cause malnourishment. Animals require inorganic ions for a wide variety of structures and processes. Some—calcium, phosphorus, potassium, sulfur, sodium, chlorine, and magnesium—are required in relatively large amounts. Others—iron, manganese, and iodine—are needed in smaller amounts, while still others—copper, zinc, molybdenum, cobalt, and selenium—are needed only in trace amounts. Finally, some minerals—including vanadium, silicon, and nickel—are needed by some animals but not necessarily by humans. Altogether, 26 of the 92 naturally occurring elements are known to be necessary for rats and chicks. Table 35-2 lists the essential minerals for humans, along with the symptoms of insufficiency for each mineral.

As is the case for the consumption of energy-rich molecules and of amino acids, animals appear to have some mechanisms for insuring that they take in the proper minerals. For example, herbivores seek salt licks to supplement the meager sodium present in plant tissues (Figure 35-4).

Vitamin deficiency can cause disease and the blockage of biochemical pathways. The first vitamin deficiency disease to be understood, *beriberi*, was initially recognized in Dutch troops stationed in Indonesia. The symptoms of this disease included the degeneration of muscles and nerves, spasms or rigidity of the legs, and mental confusion. In the late 19th century the Dutch government sent a team of researchers to find the cause of beriberi, suspecting that it resulted from a bacterial infection. The team worked for two years but they found no infectious agents.

Much of the troops' diet, however, consisted of "polished" rice, from which the husks had been removed to prolong storage. One of the researchers, Christian Eijkman, noticed that chickens that fed from the food from the kitchen and the mess hall developed symptoms like those of the stricken soldiers, whereas chickens fed elsewhere did not. Eijkman showed that

he could cure both the sick chickens and the sick men by feeding them unpolished rice. He later showed that the antiberiberi agent was a water-soluble organic compound, vitamin B_1, or thiamine, now known to be a cofactor in carbohydrate metabolism.

A similar story surrounds the discovery of another vitamin, ascorbic acid, or vitamin C. Ascorbic acid is the specific substance needed to cure *scurvy*, a disease characterized by bleeding gums, loosening teeth, and slow wound healing. Scurvy was common on sailing ships until the 18th century, when the British Navy began to supply lemons to their crews. When, in 1865, the Navy substituted limes for lemons, the word "limey" came to mean a British sailor.

The structure and biochemical function of ascorbic acid were not determined until the early 20th century, almost 200 years after its use to prevent scurvy. Ascorbic acid participates in the reactions that modify and strengthen collagen, the principal structural protein of connective tissue. Ascorbic acid also

Figure 35-4   Animals often obtain essential minerals at a salt lick.   *(Frank S. Balthis Photography)*

Table 35-3   Essential Vitamins in the Human Diet

Vitamin	Good Sources	Functions
Fat-Soluble		
A (retinol)	Leafy green and yellow vegetables, liver, eggs, fortified milk	Visual pigment; development of bone and teeth; development and maintenance of skin, gut, and other epithelia
D	Synthesized in skin exposed to sunlight; eggs, fortified milk	Bone growth; calcium absorption
E	Meat, milk, vegetable oils, whole grains	Prevents oxidative damage to cells; maintains levels of vitamin C
K	Green leafy vegetables; production by intestinal bacteria	Clotting; electron transport
Water-Soluble		
B_1 (thiamine)	Meat, milk, eggs, grains, legumes	Coenzyme in production of nucleic acids; development of connective tissues
B_2 (riboflavin)	Milk, meat, grains	Coenzyme in energy metabolism (e.g., FAD)
Niacin (nicotinic acid)	Meat, bread, potatoes	Coenzyme in energy metabolism and biosynthesis (e.g., NAD and NADP)
B_5 (pantothenic acid)	Milk, meat, eggs, yeast	Coenzyme in synthesis of fatty acids and steroids
B_6 (pyridoxine)	Meat, potatoes, spinach	Coenzyme in metabolism of amino acids
Folic acid (folate)	Vegetables, cereals, eggs, meat; also produced by bacteria in the gut	Coenzyme in metabolism of amino acids and nucleic acids
B_{12} (cobalamine)	Milk, meat	Coenzyme in metabolism of nucleic acids
Biotin	Legumes, nuts, eggs, liver	Coenzyme in metabolism of fats, glycogen, and amino acids
C (ascorbic acid)	Citrus fruits, potatoes, green leafy vegetables	Coenzyme in carbohydrate metabolism and in formation of connective tissue

serves as an *antioxidant,* limiting the damage caused to cells by oxygen and its products.

Some other vitamins have been harder to identify, because they are present at high enough concentrations in the ordinary diet that deficiency diseases are rare. Despite these difficulties, we now have a list of 13 essential vitamins required in the human diet (Table 35-3). Nutritionists usually speak of the vitamins by their letter names (such as "vitamin C"). Now that their chemical identities are known, however, biologists prefer their chemical names (such as "ascorbic acid").

Chemically, the 13 vitamins fall into two broad groups: *water-soluble vitamins* and *fat-soluble vitamins.* The water-soluble group includes vitamin C (ascorbic acid) and the B complex vitamins, which have a common designation only because they usually appear in the same foods. Each of the water-soluble vitamins is a precursor of a particular *coenzyme,* a small organic molecule that binds to an enzyme and plays a role in catalysis. The fat-soluble vitamins, including vitamins A, D, E, and K, play diverse roles: vitamin A, for example, is the precursor of retinol, the light-sensitive molecule in the visual pigments, whereas vitamin K participates in blood clotting.

Nutritionists agree that the listed substances are essential for humans, but not for all animals. Vitamin C appears to be essential only for humans, monkeys, and guinea pigs; other species can produce it themselves. Similarly, the fat-soluble vitamins are essential only for vertebrates. Nutritionists and physicians disagree, however, on how much of each vitamin a healthy diet should contain.

Malnourishment in humans usually results from an imbalance of amino acids in dietary protein, but it may also result from deficiencies in minerals or vitamins.

HOW DO SINGLE CELLS AND INVERTEBRATES DIGEST?

Most animals feed on complex fare—usually the whole, the parts, or the remains of other organisms. Their food usually consists of intact tissues, often in large chunks. The most nourishing parts of the food consists of carbohydrates, proteins, and lipids. In order to obtain this nourishment, however, an animal must (1) **digest** the food, that is, convert it into smaller molecules, such as glucose, amino acids, fatty acids, and glycerol; and (2) **absorb** the smaller molecules, that is, take them up into cells and into the circulation. Digestion and absorption are the business of the digestive system.

Digestion is essentially a chemical process that involves splitting bonds between the building blocks of macromolecules and lipids—the sugars of polysaccharides, the amino acids of proteins, and the various components of lipids. Each splitting reaction is a hydrolysis, that is, each is accompanied by the addition of a water molecule.

None of these hydrolysis reactions requires energy: all are exergonic and occur spontaneously. For digestion to occur at a useful rate, however, specific enzymes must participate, with each enzyme catalyzing a particular kind of reaction: *proteases* accelerate the hydrolysis of proteins; *glycosidases,* the hydrolysis of polysaccharides; *lipases,* the hydrolysis of lipids; and *nucleases,* the hydrolysis of DNA and RNA. But the enzymes and food molecules must first come into contact—no easy task for molecules immobilized within an intact tissue.

During digestion, enzymes hydrolyze macromolecules and lipids into their component building blocks.

Harsh Conditions Help Disrupt Tissue Interactions

Freeing food molecules to interact with enzymes requires harsh conditions. Humans help this process by cooking food at a high temperature or by marinating it in an acid solution, such as vinegar or lemon juice. Like other animals, we help break up the tissues in our food by using our teeth to bite, puncture, tear, and grind the food in our mouths. We expose the food to strong acid and powerful enzymes in our stomachs, which begin to break up the large macromolecules—carbohydrates, proteins, and nucleic acids. Then, the small intestine adds detergents (bile salts) that disrupt the nonpolar interactions of fats and oils.

The harsh conditions of digestion are a two-edged sword. They suitably prepare the consumable tissue, but they may also destroy the tissues of the consumer. To avoid this "self" digestion, animals perform digestion extracellularly: a membrane separates the digestive enzymes and the digesting food molecules from most of the proteins that function within the cell.

Animals protect their own cells from digestive enzymes by making digestion an extracellular process.

Digestion by Individual Cells

The simplest example of the isolation of the digestive process is the ingestion of liquid and dissolved molecules by a single cell. *Pinocytosis* is a form of *endocytosis,* the process by which a cell traps extracellular materials into a vesicle (an *endosome*) by folding of the plasma membrane inwards (Figure 35-5A). The pouchlike endosome, typically about 0.1 μm in diameter, fuses with a *lysosome,* full of hydrolytic enzymes, and the resulting small molecules then move into the cytosol.

Another type of feeding and digestion at the single-cell level is *phagocytosis.* The distinction between phagocytosis and pinocytosis is that phagocytosis involves larger particles than pinocytosis and that the membrane expands to surround the particle. Animals usually employ phagocytosis not for feeding but for removing cellular debris.

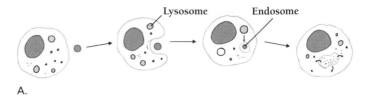

A.

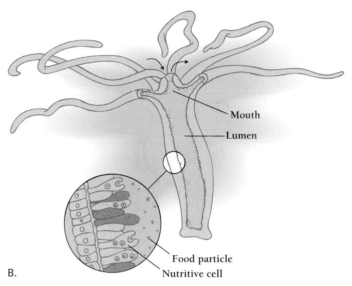

B.

Figure 35-5   Digestion in single cells and in simple organisms depends on endocytosis and phagocytosis. A. A cell takes in extracellular materials into an endosome. The endosome fuses to a lysosome, which is filled with digestive enzymes. After digestion, small molecules are released into the cytosol. B. In a hydra, initial digestion occurs outside the cells of the primitive digestive tract.

Digestion in Cavities with One Opening

How do animals deal with particles too large to enter a single cell? In such cases, animals rely on extracellular digestion chambers. The simplest digestion chambers, such as that of a hydra or a flatworm, have a single opening, which serves as both mouth and anus (Figure 35-5B). The hydra's arms move food into the chamber, and secreted enzymes then break down the food into smaller fragments, which then enter the surrounding cells by endocytosis. The hydra discharges the remnant of indigestible food through the opening. In flatworms, the digestive cavity ramifies through the body and also serves as a primitive circulatory system.

What Are the Advantages of a Two-Ended Digestive Tract?

Animals more complex than hydras and flatworms have digestive systems with two ends (Chapter 23). Food enters through the mouth and passes through an extended tube, variously called the **digestive tract,** the *gastrointestinal tract,* the *alimentary canal,* or the **gut** [Anglo-Saxon, = channel] (Figure 35-6).

Figure 35-6 The human digestive tract, with accessory organs.

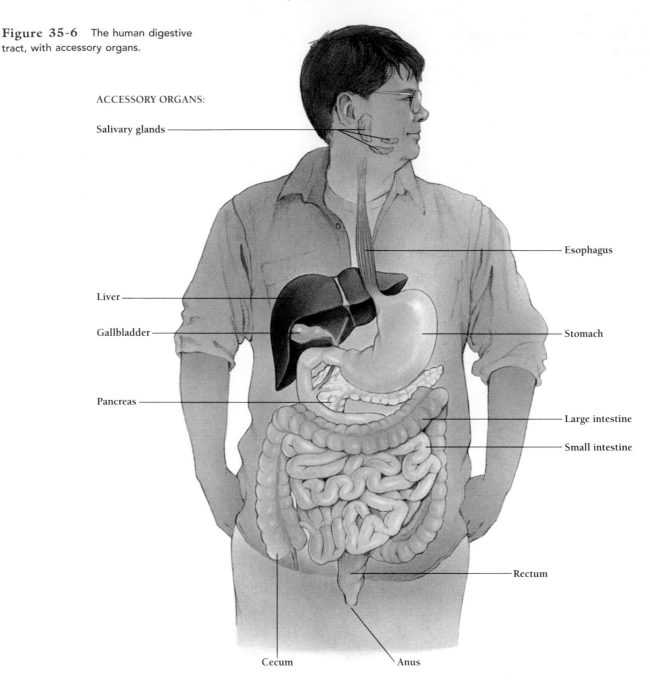

ACCESSORY ORGANS:

Salivary glands

Liver

Gallbladder

Pancreas

Esophagus

Stomach

Large intestine

Small intestine

Rectum

Cecum   Anus

The undigested and unabsorbed material leaves through the **anus,** the opening at the far end of the digestive tract. The remnants of the food, together with bacteria that inhabit the tract, form the **feces** [Latin, = dregs].

A digestive tract with two ends allows animals to process food more efficiently than a one-ended digestion vessel. Since the food passes each point in the tract only once, sequential processing can occur. Distinct regions of the tract may become specialized for different processes, each of which may require a different set of environmental conditions and a different set of enzymes.

The evolution of the digestive tract anticipated, by about half a billion years, the advantages of an industrial assembly line, in which workers, machines, and equipment are arranged so that the product passes consecutively from operation to operation until it is completed. As food passes down the digestive tract, it undergoes a series of successive reactions that ultimately allow the animal to absorb a variety of valuable molecules, while leaving indigestible materials behind. Indeed, we may think of the digestive tract as a "disassembly line" (Figure 35-7, page 780).

Behavioral adaptations also contribute to digestion. Some invertebrate carnivores, such as spiders and starfish, perform extracellular digestion by employing their victims' own bodies as a preliminary digestion vessel. A spider, for example, secretes digestive enzymes into the body of its prey. After the enzymes have done their hydrolytic job, the spider ingests its now liquefied prey for further processing.

In many invertebrates, the passage of food through the digestive tract is highly controlled by the animal's nervous system. In the lobster, for example, the muscles responsible for digestive movements are striated muscles like the voluntary muscles discussed in Chapter 34. The contraction of these muscles depends on stimulation by networks of nerve cells, which cause four different sets of digestive muscles to contract with different rhythms: (1) a rhythm responsible for the ingestion of food; (2) a rhythm responsible for the emptying of the sac where food is initially stored; (3) a rhythm that controls the calcified "teeth" that grind food within the lobster's stomach; and (4) a rhythm that controls a system of valves and sieves that allow the elimination of small particles and the retention of larger particles for further digestion.

A two-ended digestive tract allows the specialization of different regions for specific processes that take place in a fixed order.

HOW DOES THE VERTEBRATE DIGESTIVE TRACT FUNCTION AS A "DISASSEMBLY LINE"?

The digestive system actually performs four functions: (1) **movement** (or "motility")—agitating the food within specialized regions and pushing food through the system; (2) **secretion**—the production of lubricants, enzymes, and detergents; (3) **digestion**—the breaking down of large molecules to their component building blocks; and (4) **absorption**—taking up small molecules into cells or into the circulation. To visualize how the disassembly process occurs, we will follow the progress of a hamburger and lettuce on a bun, as it moves from mouth to anus. Before we do so, we will first describe the basic divisions of the tract.

How Is the Vertebrate Digestive Tract Organized?

Figure 35-7 shows the general organization of a vertebrate digestive system into five specialized divisions: (1) mouth and throat (headgut), (2) esophagus and stomach (foregut), (3) small intestine (midgut), (4) pancreas and biliary system, and (5) large intestine (hindgut). The **lumen** [Latin, = opening, light] of the gut, the inner space within the tract, is continuous—from mouth to throat to esophagus to stomach to small and large intestines. The pancreas and biliary system connect to the gut within the small intestine by means of separate ducts that empty into the gut.

The diameter of the gut's lumen varies widely from region to region, usually being widest in the stomach. Knowledge of the specialization of each division comes from anatomical stud-

ies. Microscopic studies have shown that in each division the lining of the gut lumen is an **epithelium**, a tightly connected sheet of cells.

Researchers have followed the movement of materials through the digestive tract using a number of methods. Beaumont and others inserted tubes into surgically prepared openings to sample the processed food at each stage. More recently, x-ray studies have allowed researchers and physicians to observe both the movement of food and the movement of the tract itself, without actually disturbing the process.

Food enters the digestive tract through the mouth and throat. The mouth captures the food and prepares it for entrance into the gut. All vertebrates other than lampreys and hagfish have jaws, and nearly all (except for birds, turtles, and a few other groups) have teeth. Jaws, beaks, teeth, tongue, and lips all participate in the grabbing and tearing of food.

Mammals are the only animals that can actually chew their food. Four sets of adaptations—jaws, tongues, cheeks, and teeth—allow mammals to pulverize food more finely than other vertebrates. Mammalian jaws and their attached muscles allow horizontal movement, needed for good grinding between the uneven surfaces of the large molars and premolars (Figure 35-8, page 782). Cheek and tongue muscles, which produce the suction needed for nursing, also allow the manipulation of food for grinding. Mammalian teeth are highly specialized as well: human teeth include *incisors,* whose chisel-like edges cut; *canines,* whose pointed crowns tear; and *premolars* and *molars,* whose flat, ridged surfaces grind and crush.

Mammals with different types of diets differ in the specialization of their teeth and other mouth parts. Carnivorous mammals have well-developed incisors and canine teeth, whereas herbivorous mammals may lack canines altogether and have incisors adapted for cutting vegetation (Figure 35-8B and C). Anteaters have no teeth and weak jaws, but they have specially adapted tongues and salivary glands that secrete quantities of viscous saliva (Figure 35-8D). The blue whale, the largest of all animals, has another mouth adaptation—an elaborate filtering apparatus that catches the small crustaceans (krill) that compose its diet (Figure 35-8E). In contrast to the array of mammalian mouth adaptations, a reptile's teeth appear to be unspecialized and are used only to grab and puncture rather than to cut and grind (Figure 35-8F).

The digestive system consists of five specialized divisions.

The Human Digestive System and Some Vertebrate Variations

To visualize the progress of a meal through the human digestive tract, imagine Lexie St. Martin, the great-great-great-great-great-great-great-great-granddaughter of Alexis St. Martin, whom we discussed in the beginning of the chapter, is a student at the University of New Hampshire, only 200 miles from Alexis's Montreal. Lexie is about to enjoy one of the burgers she has prepared for a sorority picnic.

Mouth, Pharynx, and Esophagus

Her first bite contains some bun, some lettuce, and a chunk of meat and fat. As she chews, her teeth tear and grind the bite into little bits.

The presence of food in the mouth stimulates the first secretions of the digestive tract. The **salivary glands** secrete both enzymes and **mucus**, a viscous, slippery substance that coats the food particles and lubricates their movements within the mouth and the digestive system. Mucus consists of water, salts, and *glycoproteins* (proteins attached to carbohydrates). The grinding action of the teeth exposes the food to the salivary enzyme **amylase**, which hydrolyzes the polysaccharides of the bun into shorter fragments. But the protein and lipid molecules within the bite are still intact.

Lexie's tongue shapes the chewed food and the lubricating mucus into a ball, called a **bolus**. Movement through the tract begins as the tongue and other muscles of the mouth push the bolus back into the throat, or **pharynx**. The pharynx is the common entryway both for food into the digestive tract and for air into the lungs. This being so, how does Lexie prevent the bolus from moving into her airways instead of her gut?

The bolus continues to move through the esophagus and into the stomach. Whether it will go into the esophagus or into the airway depends on the workings of *sphincters,* rings of muscles that surround a tube. Sphincters surround both the opening of the airway and the **esophagus**. Lexie's swallowing pushes the bolus into the esophagus. When she swallows, which takes about a second, breathing temporarily stops, and a flap, called the *epiglottis,* covers the closed sphincter at the entrance to the airway.

If a piece of food is too large to enter the esophagus, it may block the entrance to the *glottis,* through which air enters the lungs. The resultant choking can sometimes be fatal. The Heimlich maneuver can dislodge such a blockage by forcing the obstruction back into the mouth.

The initial movement of the bolus depends on swallowing. But how does Lexie control the complex movements of her

Figure 35-7 The digestive tract as a "disassembly line." Each region of the tract makes distinctive contributions to the process of digestion and absorption. Here simple sugars (monosaccharides) are shown as hexagons, amino acids as circles, and triglycerides as three-tooth combs.

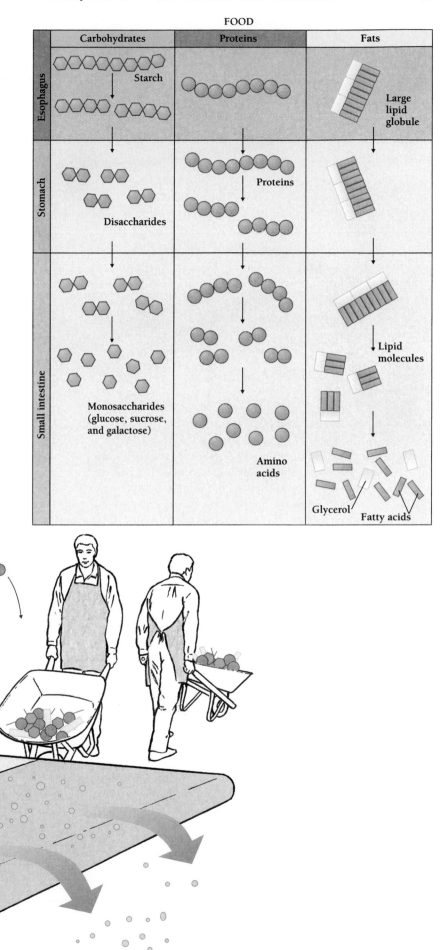

FOOD

	Carbohydrates	Proteins	Fats
Esophagus	Starch		Large lipid globule
Stomach	Disaccharides	Proteins	
Small intestine	Monosaccharides (glucose, sucrose, and galactose)	Amino acids	Lipid molecules Glycerol Fatty acids

DEPARTMENT OF ABSORPTION

Large intestine

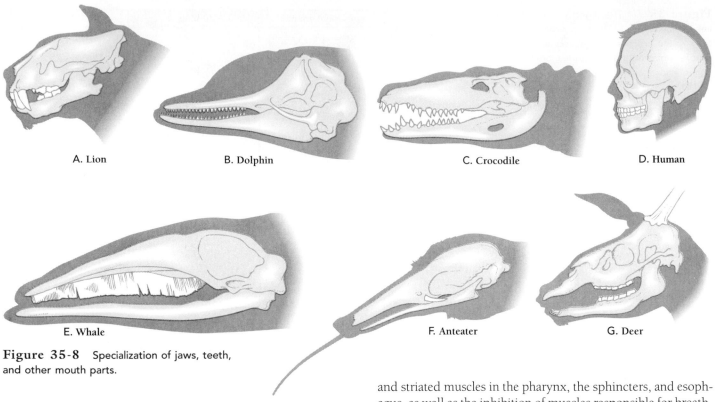

A. Lion   B. Dolphin   C. Crocodile   D. Human

E. Whale   F. Anteater   G. Deer

Figure 35-8   Specialization of jaws, teeth, and other mouth parts.

swallowing? The answer is that her control is only partly conscious: Animals can voluntarily initiate chewing and swallowing, but both processes are also involuntary reflexes, automatic responses to the presence of food in the mouth. Pressure receptors sense food against the gums, palate, teeth, and tongue, and chemical receptors on the tongue detect food by its taste. Swallowing involves the coordinated response of both smooth and striated muscles in the pharynx, the sphincters, and esophagus, as well as the inhibition of muscles responsible for breathing. A swallowing center in the lower part of the brain (in the medulla) manages the operation through a network of nerves that connect to the involved muscles.

Smooth muscles propel the bolus through the esophagus. **Peristaltic waves,** coordinated contractions of smooth muscles, move down the esophagus toward the stomach (Figure 35-9). During a wave, the muscles relax in front of the bolus

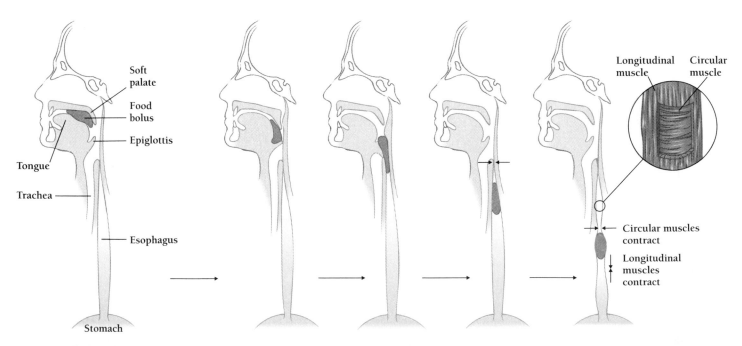

Soft palate

Food bolus

Epiglottis

Tongue

Trachea

Esophagus

Stomach

Longitudinal muscle   Circular muscle

Circular muscles contract

Longitudinal muscles contract

Figure 35-9   **Swallowing and the beginning of a bolus.** Peristaltic waves push the bolus down the esophagus and into the stomach.

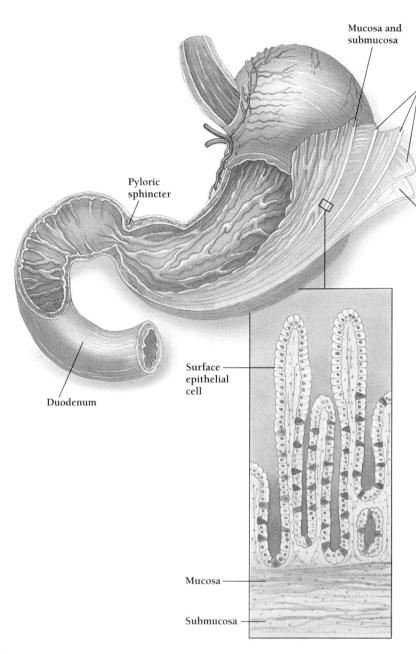

Mucosa and submucosa

Muscle layers

Pyloric sphincter

Serosa

Duodenum

Surface epithelial cell

Mucosa

Submucosa

Figure 35-10 The tissues and cells of a stomach.

Stomach

The esophagus empties directly into the **stomach** [Greek, *stoma* = mouth], the most dilated and most muscular section of the digestive tract (Figure 35-10). In humans, the empty stomach holds only about 50 ml (less than 2 ounces), but it can expand to contain an entire meal, which may occupy as much as 1.5 liters. As Beaumont's studies showed, the stomach accumulates food, disrupts it mechanically, and begins the process of protein hydrolysis. As Beaumont also observed, the stomach's lining has a velvetlike appearance, reflecting the folds of its epithelial surface.

In Lexie's case, bolus after bolus of Lexie's burger arrive in the stomach from the esophagus. After each bite, the sphincter at the stomach's entrance closes to prevent regurgitation of the contents back into the esophagus. Another sphincter at the stomach's exit prevents the contents from moving on prematurely. The presence of the food in the stomach stimulates the secretion of hydrochloric acid and pepsinogen. In the acid environment of the stomach, the pepsinogen immediately unfolds. The acid cuts away a part of the polypeptide chain to form **pepsin,** the stomach's protease, which begins to hydrolyze proteins from Lexie's burger. The acid also terminates the action of amylase, which entered the stomach along with the processed burger.

The smooth muscles surrounding the stomach churn its contents and the hydrochloric acid denatures the hamburger's proteins, increasing their susceptibility to pepsin digestion. The end result of the process is **chyme** [Greek, *khumos* = juice], a creamy, acidic liquid that passes into the small intestine. The fat from Lexie's burger remains completely undigested, and lipid molecules remain as suspended droplets within the chyme, giving the chyme the thick consistency of a fatty, burger-bun soup. The cellulose from the lettuce also remains undigested, and pieces of it remain dispersed within the chyme.

Peristaltic waves move the chyme to the stomach's exit. The exit sphincter opens briefly, allowing a small spurt of chyme to pass at a time.

In birds, the contents of the esophagus are less finely divided than in mammals, for birds have no teeth with which to

and contract behind it, pushing the bolus along toward the stomach. The path, about 25 cm long, goes straight down the chest cavity and through an opening in the diaphragm muscle, which is chiefly involved in breathing. Despite its downward course, gravity is not necessary for the movement of food into the stomach: Lexie could swallow her burger standing on her head, and astronauts can swallow when weightless.

Some animals have a saclike extension of the esophagus, called the **crop,** which can store food for later digestion. A bird, for example, can fill its crop with seeds or berries, which it can later regurgitate to feed its nestlings.

Digestion begins in the mouth with the division of the food, the start of enzyme hydrolysis, the formation of a bolus, and the movement of the bolus into the gut.

---

tear or grind their food. Food passes quickly from the esophagus (or crop) through an unspecialized stomach into the **gizzard,** a region specialized for the grinding of food. The inner surfaces of the gizzard are coated with a horny material that serves as an abrasive, along with small stones that the bird picks up and swallows. Seed eaters, such as the domestic chicken, have the most muscular gizzards. Gizzards contain pepsin and hydrochloric acid, so that the chyme that leaves a bird's gizzard is ready for the next steps of digestion.

The stomachs of cows—and those of moose, buffalo, camels, giraffes, and other ungulates—have special adaptations that enable them to extract energy efficiently from the hard-to-digest grasses that make up most of their diets (Figure 35-11). These animals are all **ruminants**—hoofed, horned, or antlered herbivores that can regurgitate partly digested food, called *cud,* for further chewing. A ruminant's stomach contains four distinctive chambers, which in a large cow may hold as much as 200 liters. Bacteria and protists in these chambers produce enzymes that break down cellulose, a feat that no animal can do by itself. So, in contrast to the chyme from Lexie's meal, the chyme entering a cow's intestine contains no undigested cellulose, but only energy-rich glucose molecules that result from the action of microbial enzymes.

The stomach mechanically disrupts food while hydrochloric acid and pepsin begin to hydrolyze proteins. Food leaves the stomach as semiliquid chyme.

Digestion and Absorption in the Small Intestine

Adaptations of the small intestine promote digestion and absorption. Chyme now enters the **small intestine,** the longest division of the digestive tract. The small intestine accomplishes two tasks: (1) it completes the digestion of macromolecules in the chyme and (2) it absorbs most of the useful products of this digestion.

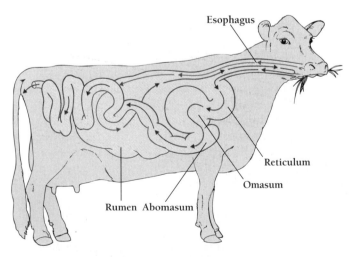

Figure 35-11   The four stomachs of a ruminant.

Complete digestion requires a relatively long time for enzyme action, and absorption requires a large surface area. In humans the small intestine is typically 3 to 5 cm in diameter and about 6 m long. The substantial length of the small intestine allows time for digestion and absorption.

In herbivores other than ruminants, undigested cellulose interferes with digestion and absorption. The small intestine is therefore proportionally much longer in these herbivores than in humans or other omnivores. In contrast, strict carnivores have a proportionally shorter small intestine: the carnivorous adult frog, for example, has a much shorter small intestine than a tadpole, which lives on a vegetarian diet.

The structure of the small intestine also reflects its specialization (Figure 35-12). As in the rest of the tract, an epithelium lines the lumen. Surrounding the epithelium is a layer of connective tissue, which contains blood vessels, nerves, and lymphatic ducts, which drain extracellular fluid into the circulatory system. Outside the connective tissue is a thin layer of smooth muscle. Together, the epithelium, the connective tissue, and the thin muscle layer are called the *mucosa.* Although the mucosa of the small intestine roughly resembles that in the other regions ·of the digestive tract, it is much more highly folded, greatly increasing the effective surface of the epithelium. Each fingerlike fold is called a **villus** [Latin, = shaggy hair; plural, **villi**], and each extends about 1 mm into the lumen.

Surrounding the mucosa is another layer of connective tissue and then two additional layers of smooth muscle. These are the muscles responsible for moving the chyme. Contractions of the inner, circular muscles squeeze the contents of the lumen; contractions of the outer, longitudinal muscles shorten the tube. Finally, another layer of connective tissue, the *serosa,* surrounds the entire tract, and the *mesenteries* connect the tract to the walls of the abdominal cavity.

Digestion within the **duodenum,** the first section of the intestine, and absorption within the rest of the small intestine depend on the continued mixing of the contents of the lumen. These movements depend on the action of the smooth muscles surrounding the small intestine, which are able to agitate the chyme in two distinct patterns. The first pattern, called *segmentation,* forces the contents back and forth, more or less in place. This action thoroughly mixes the chyme with the secretions of the liver and pancreas. It also allows prolonged absorption through the intestinal liming. Only after the intestine has absorbed most of the nutrients in a meal does the pattern of muscular contraction change. The second pattern of muscle contraction generates peristaltic waves. These push the remaining material toward the large intestine, the last region of the digestive tract.

The expansion of the epithelial surface extends to the cellular level, with surface membrane of each epithelial cell folded into **microvilli,** the membrane-covered extensions of epithelial cells (Figure 35-12). The epithelial cells are joined to each other through tight junctions, so no materials pass between the cells: all molecules and ions that move out of the lumen must pass through the epithelium. The ability to absorb materials therefore depends directly on the total surface area of the epithe-

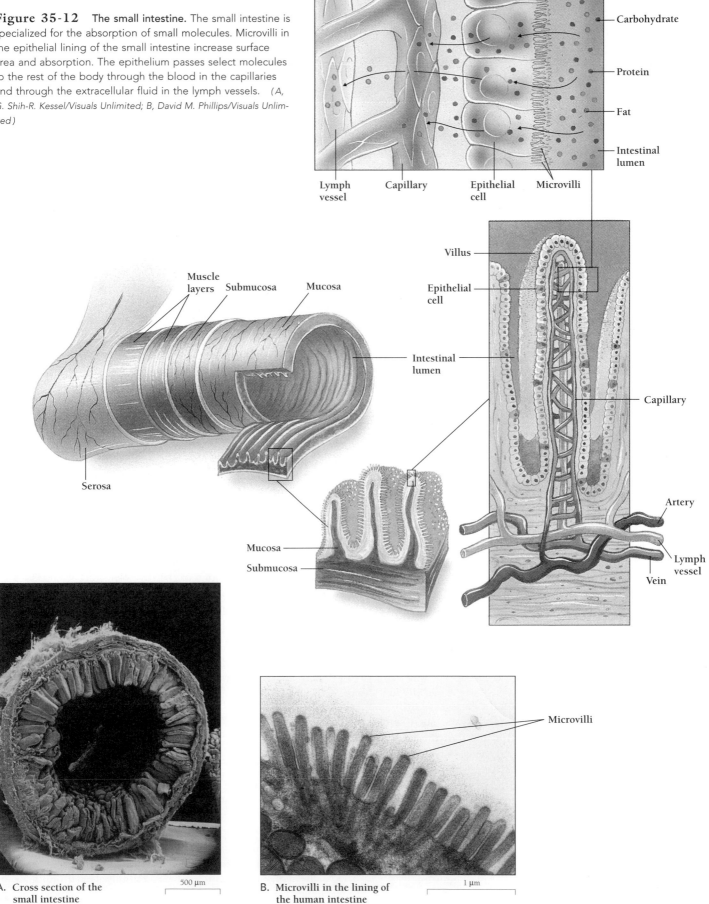

Figure 35-12 **The small intestine.** The small intestine is specialized for the absorption of small molecules. Microvilli in the epithelial lining of the small intestine increase surface area and absorption. The epithelium passes select molecules to the rest of the body through the blood in the capillaries and through the extracellular fluid in the lymph vessels. *(A, G. Shih-R. Kessel/Visuals Unlimited; B, David M. Phillips/Visuals Unlimited)*

Carbohydrate

Protein

Fat

Intestinal lumen

Lymph vessel   Capillary   Epithelial cell   Microvilli

Villus

Epithelial cell

Intestinal lumen

Capillary

Muscle layers   Submucosa   Mucosa

Serosa

Mucosa

Submucosa

Artery

Lymph vessel

Vein

A. Cross section of the small intestine

500 μm

B. Microvilli in the lining of the human intestine

1 μm

Microvilli

785

lium. Villi and microvilli greatly increase the total area of the small intestine. If the surface of the human small intestine were completely smooth, for example, its area would be about the size of a 28-inch television screen; instead, the total area through which it absorbs nutrients is almost 300 square meters, about the area of a tennis court.

The length of the small intestine allows sufficient time for digestion and absorption. Specializations that increase the surface area increase the efficiency of absorption. Smooth muscles agitate the contents of the lumen and propel them onward.

How Do the Pancreas, Liver, and Gallbladder Contribute to Digestion?

Digestion within the small intestine depends on secretions by the pancreas and liver. Most of the digestion within the intestine occurs in the duodenum, which, in humans, is about 25 cm long [about 12 "finger widths," hence its name, which comes from the Latin for "twelve']. The rest of the small intestine is specialized for absorption.

A common duct carries materials from the pancreas, the liver, and the gallbladder. The **pancreas** provides a panoply of digestive enzymes. The pancreatic enzymes, unlike pepsin, do not work at acidic pH, so the pancreas also secretes bicarbonate (HCO_3^-) ions, which neutralize the acidic chyme as it enters the duodenum.

Proteases from the pancreas carried into the lumen then chop the peptide fragments, produced by pepsin in the stomach, into lengths of a few amino acids. Other enzymes, called *peptidases,* on the surface of the microvilli then complete the hydrolysis of polypeptides into free amino acids, which can then be absorbed by the small intestine. By the time the remains of Lexie's meal have left the small intestine, essentially all the proteins have been hydrolyzed and all the resulting amino acids have been absorbed.

The digestion of polysaccharides (starch and glycogen) is also completed in the small intestine. Pancreatic amylase finishes the digestion of polysaccharides begun by salivary amylase. Maltose, sucrose, and other disaccharides accumulate in the small intestine. Again, enzymes on the surfaces of microvilli hydrolyze the disaccharides into simple sugars (such as glucose), ready for immediate absorption through the intestinal epithelium. Like the proteins, then, the starches in Lexie's chyme have been hydrolyzed and all the resulting sugars absorbed by the time the meal has left the small intestine. On the other hand, the cellulose from the bun and the lettuce remains intact, because Lexie has no enzyme that will catalyze its hydrolysis.

Nucleases produced by the pancreas hydrolyze the relatively small amounts of DNA and RNA in Lexie's burger, and their components are absorbed by the small intestine.

Processing the lipids is more difficult. Lipids are not soluble in water and so they usually aggregate into droplets whose oily interiors are not accessible to hydrolytic enzymes. As long as they remain intact, however, there is no hope of deriving nourishment. The liver secretes a detergent solution, called **bile,** however, which solves this problem. Bile is bitter, alkaline, and a distinctive green-yellow or brown-yellow color. Bile (also called "gall") moves from the liver to the gallbladder, which serves as a storage depot. A sphincter at the end of the bile duct opens only when food enters the intestine.

Just as dishwashing detergent breaks up the aggregations of fat in the food left on a dinner plate, so bile breaks up the fat droplets within the chyme. The movements of the small intestine mix the chyme with bile to form an *emulsion,* which contains tiny droplets of fat, about 1 μm in diameter, each coated with detergent molecules. Lipases, secreted by the pancreas into the duodenum, now hydrolyze the lipid molecules into fatty acids, glycerol, and other small molecules. All of these can be absorbed later by the small intestine.

Absorption of small molecules across the epithelium of the small intestine depends both on passive diffusion (for example, of free fatty acids across the plasma membrane) and on specific protein molecules on the plasma membrane. Some of these proteins are enzymes that finish the very last steps of digestion. Others are transport molecules. Some can carry molecules in either direction, but primarily they carry nutrient molecules from the lumen (where their concentration is high) into the cytosol (where their concentration is low). In some cases, the epithelial cells expend energy to ensure the effective transport of specific molecules, even when the concentration in the epithelium is higher than in the lumen.

The overall process of absorption is so efficient that, normally, by the time the chyme appears in the large intestine, no nourishing molecules remain—only water, dissolved ions, cellulose, and various unabsorbed molecules, together with mucus and other secretions of the digestive tract itself. On average, an adult human consumes about two liters of food and drink and secretes about seven liters of mucus, acid, and other fluids each day. The small intestine absorbs about 95 percent of this total, so that the total amount of liquid entering Lexie's large intestine on a typical day is 450 to 500 ml (about a pint).

Digestion depends upon enzymes produced by the pancreas and bile produced by the liver.

Large Intestine

The **large intestine** recovers water and dissolved ions. Most of the length of the **colon,** or large intestine, is specialized for the absorption of water and ions. Although the mucosa of the colon is not folded into villi like that of the small intestine, the epithelial cells that line the lumen have extensive microvilli. In humans, the junction between the small intestine and the colon is located near the bottom of the abdominal cavity, on the right side. The colon has three major sections: the *ascending colon,* which extends up the right side of the abdomen, the *transverse colon,* which crosses over to the left, and the *descending colon,* which extends downward on the left. The descending colon ends in a straight portion, called the **rectum** [Latin, = straight], in which feces are stored, until their elimination through the anus.

The large intestine removes most of the water that enters with the remains of the chyme, so that feces typically contain only about 100 ml of water and 50 g of solids per day. Both the small intestine and the large intestine remove water by pumping ions across the epithelium. Water follows, moving from a solution in the lumen with low solute concentration to a solution with high solute concentration.

If the colon is irritated, its contents may move too fast for efficient absorption, resulting in **diarrhea,** or watery feces. On the other hand, if the intestine's contents move too slowly, the epithelium absorbs too much water, resulting in **constipation,** the inability to move the bowels. Undigested cellulose provides *roughage,* also called *fiber* or *bulk,* which stimulates peristalsis. Insufficient roughage also leads to constipation.

The colon contains large amounts of bacteria, which make up about half the dry weight of the feces. These bacteria live on unabsorbed energy-rich molecules within the colon. The bacteria also produce vitamins—including many of the vitamin B complex—that can be absorbed and used.

Near the junction between the large and small intestine is the **cecum** [Latin, *caecus* = blind], a blind sac that diverts from the main route through the digestive tract. The human cecum is small and does not appear to serve any role in digestion or absorption. The *appendix,* a small, fingerlike projection at the top of the cecum, is now thought to have some role in the immune system. It is best known, however, for its tendency to become infected, usually prompting its surgical removal.

In contrast to the human cecum, the cecum of other mammals, especially herbivores, may be relatively large. In rabbits, for example, the cecum contains bacteria and protozoans that produce cellulose. Undigested material from the small intestine enters the cecum, where additional digestion can occur. The newly digested material, now containing energy-rich glucose molecules and other nutrients, leaves the cecum and travels down the large intestine. But the large intestine is not able to absorb these molecules, and they pass out in the feces. Rabbits, however, have evolved an effective (if unappetizing) recycling mechanism: they actually produce two types of feces, one of which consists of the energy-rich materials from the cecum. The rabbits then reingest these feces (but not the others) and extract the valuable molecules during the repassage through the gut.

The large intestine removes ions and water. Bacteria grow in both the colon and the cecum; in some species, bacteria in the cecum digest cellulose.

HOW DOES AN ANIMAL COORDINATE PROCESSES IN INDIVIDUAL ORGANS?

The digestive system must coordinate processes performed by cells that are far from one another. When food arrives in the stomach, for example, there must be acid and pepsin; and when chyme arrives in the intestine, there must be pancreatic enzymes, bile, and bicarbonate. How does this happen?

In multicellular organisms, chemical signals are responsible for almost all intercellular communication and are especially important in coordinating the responses of different organs to changes in the external or internal environment. In the case of the digestive system, both nerve cells and hormones use chemical signals to coordinate muscle activity, secretion, digestion, and absorption (Figure 35-13). After the intestine has absorbed the products of digestion, glucose serves as the principal energy carrier to the whole body. Hormones coordinate the absorption of blood glucose into cells throughout the animal.

Nerve Cells Help Control the Activities of the Digestive System

Part of the control of the digestive system depends on a complex network of nerve cells within the digestive tract itself. This network continuously receives information about the status of the tract from specialized receptors on the epithelial membranes. These receptors report on the contents of the lumen—how full it is, its total solute concentration, its pH, and the concentration of specific digestion products.

The nerve network processes this information and sends appropriate commands to cells within the tract. For example, when the small intestine is full (distended), the network increases the activity of its smooth muscles to increase mixing, digestion, and absorption. At the same time, other signals inhibit the smooth muscle activity in the stomach, preventing more material from entering the intestine.

The intestinal nerve network also receives information from the brain. Thus we respond to the sight, smell, and taste of food. The response includes activity of the salivary glands and the increased secretion of acid in the stomach. The Russian physiologist Pavlov discovered that the mere sounding of a bell would evoke acid secretion in dogs, provided that the dogs had learned to associate the bell's ringing with a forthcoming meal.

Hormones Coordinate the Activities of Distant Regions of the Digestive Tract

A **hormone** is a substance, made and released by cells in a well-defined organ or structure, that moves throughout the organism and exerts specific effects on specific cells in other organs or structures (Chapter 36). Both animals and plants use hormones to coordinate the actions of physically distinct tissues

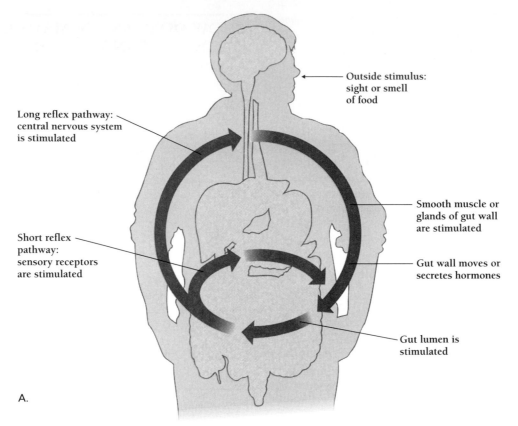

A.

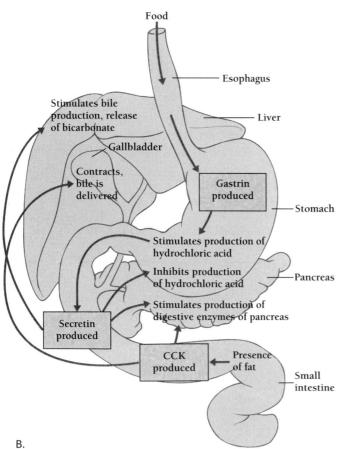

B.

Figure 35-13 Digestion is coordinated both by the nervous system and by hormones. A. The nervous system responds to the sight or smell of food. B. Hormone-secreting cells respond to food in a particular part of the intestinal tract. The secreted hormones regulate the activities of other parts of the gut. For example, when food moves into the stomach, hormone-producing cells release gastrin, which helps regulate the production of stomach acid. The endocrine cells are in the far end of the stomach, whereas the acid-producing cells are nearer to the esophagus end. Nerve stimulation can also lead to gastrin release. As acid moves from the esophagus end to the far end, gastrin release stops, and there is no further stimulation of acid production.

and organs. The structures upon which hormones act are called **target organs.** Target organs respond to hormones by changing their metabolism, the activity of their enzymes, or the expression of specific genes.

The digestive system uses hormones to coordinate many of its activities. Three small polypeptides—called secretin, gastrin, and cholecystokinin (CCK)—are known to be true hormones. All three are secreted by specialized cells; all travel through the circulation; and all act on specific targets to evoke specific effects. The digestive system also uses more than ten other small polypeptides as signaling molecules, but physiologists are unsure whether they arrive at their targets by means of the circulation or whether they act locally.

Polypeptide hormones, released in response to the contents
of the gut, regulate the secretion of hydrochloric acid,
bicarbonate, bile, and digestive enzymes.

STUDY OUTLINE WITH KEY TERMS

The digestive system allows animals to derive both energy and raw materials from food. Energy comes from the metabolism of small molecules, especially glucose, that result from the action of hydrolytic enzymes on the macromolecules in food. The amount of food that an animal consumes depends on its **basal metabolic rate,** the minimum energy needed to stay alive and awake, on its size, and on its activities. Failure to obtain enough of particular raw materials—including the **essential amino acids,** minerals, and vitamins—leads to **malnourishment** and disease. Protein deficiency disease, **kwashiorkor,** is disturbingly prevalent in the world population.

Individual animal species differ in their diets; **herbivores** eat only plants, and **carnivores** eat only animals. Humans are **omnivores,** eating plants, animals, and other organisms. In all cases, harsh conditions are necessary for hydrolytic enzymes to gain access to the molecules in food. The hydrolytic reactions always take place extracellularly, away from the cytosol, in lysosomes, in a digestive cavity, or in a **gut** or **digestive tract.**

The digestive tract of vertebrates consists of a continuous tube that leads from the mouth to the **anus.** Food enters through the mouth and undergoes a sequence of transformations as it passes through the **lumen** of the digestive tract. Enzyme action hydrolyzes macromolecules, and the resulting small molecules are absorbed in the small intestine. The remnants of food, together with secretions and bacteria from the large intestine, pass out of the anus as **feces.**

The digestive system has five specialized divisions: the mouth and **pharynx,** the **esophagus** and **stomach,** the **small intestine,** the **pancreas** and biliary system, and the **large intestine.** The digestive system accomplishes **movement, secretion, digestion,** and **absorption.**

The mouth is specialized for the capture and initial preparation of food. **Mucus,** secreted by the **salivary glands,** lubricates the movement of the food in the mouth, and **amylase** begins to hydrolyze starches and glycogen. The mouth shapes the food and mucus into a **bolus,** to be swallowed by the pharynx. Smooth muscle contractions within the esophagus form **peristaltic waves,** which propel the bolus into the stomach. The stomach agitates the food with more mucus and secretes hydrochloric acid and **pepsin,** a protease that works best in the acidic environment of the stomach. The result of the stomach's action is that the food is converted to **chyme,** a soupy liquid.

Individual groups of animals differ in the adaptations of their digestive tracts. Many birds have both a **crop,** an extension of the esophagus where food can be stored, and a **gizzard,** a specialized region beyond the stomach, where food can be ground against a hardened surface together with swallowed grit. **Ruminants** have highly specialized stomachs that allow the breakdown of cellulose, using enzymes produced by bacteria that live in the stomach.

In each region, the lumen is surrounded by an **epithelium.** The epithelium is specialized for different functions in different regions. In the small intestine, the epithelium and immediately surrounding tissue layers are folded into **villi,** and the epithelial cells themselves have extensively folded plasma membranes, or **microvilli.** Digestion occurs within the **duodenum,** using digestive enzymes from the pancreas and **bile** from the **gallbladder.** The resulting small molecules are absorbed by the rest of the small intestine, as the chyme continues its movement. The large intestine or **colon** removes water and ions as the remains of the food travel toward its last section, the **rectum.** Bacteria growing within the colon contribute about half the volume of the feces. Irritation of the colon can decrease absorption, leading to **diarrhea,** and insufficient bulk in the diet can lead to **constipation.** The **cecum** is a sac that extends just beyond the beginning of the colon and, in some species, is also the site of bacterial growth.

Nerves and chemical signals coordinate muscle activity, secretion, digestion, and absorption within the digestive system. **Hormones** are chemical signals that are released into the circulation by specialized cells or organs. Hormones evoke distinctive effects on specific **target organs.** In the digestive system, three polypeptide hormones regulate the secretion of hydrochloric acid, bicarbonate, bile, and digestive enzymes.

REVIEW AND THOUGHT QUESTIONS

Review Questions

1. Define the basal metabolic rate.
2. Why must vegetarians worry about balancing complementary proteins?
3. Give an example of a herbivore, a carnivore, and an omnivore. How do the teeth and digestive tracts of herbivores and carnivores differ?
4. Why is digestion always outside the cytosol? Describe the three major ways that organisms have for accomplishing this.
5. What is the function of the saliva?

6. How do hydrochloric acid and pepsin interact in the stomach?
7. Explain how peristaltic waves move food through the digestive tract. Draw a picture illustrating this process.
8. Explain what bile is, where it comes from, and how it functions in the digestive tract.

Thought Questions

9. Why must cows and other ruminants have such large stomachs?
10. If your pancreas no longer secreted its digestive enzymes, what food groups would you have to give up?

SELECTED READINGS

Beaumont, William, *Experiments and Observations on the Gastric Juice and the Physiology of Digestion,* Dover Publications, Mineloa, New York, 1996.

Stevens, C. Edward, and Ian D. Hume, *Comparative Physiology of the Vertebrate Digestive System,* 2nd ed., Cambridge University Press, Cambridge, 1995.

▶ On-line materials relating to this chapter are on the World Wide Web at http://www.saunderscollege.com/lifesci
Click on Tobin/Dusheck: *Asking About Life.*

Estrogens in the Environment

In the early 1990s, researchers announced a trend that grabbed headlines and the attention of men everywhere. Western men of the 1990s, smirked TV newscasters, make half the number of sperm that men did in the 1930s. As one reproductive physiologist quipped at a Congressional hearing in 1995, "Every man in this room is half the man his grandfather was."

The news media had a field day. "Sperm counts down? Penises shriveled? Hey, . . . don't blame it on the feminists," declared *Newsweek* magazine. Indeed, some researchers were blaming the downward trend in sperm counts on a new class of environmental pollutants called environmental estrogens, or *estrogenics,* natural and synthetic substances that mimic the physiological effects of the hormone estrogen. Were penises shriveling? Were men being doused in overwhelming quantities of the female hormone estrogen? Well, not actually.

Certainly sperm counts are down. But men generally produce far more sperm than they need, and men's fertility—as measured by the number of children they conceive and by other measures—is fine. And while it's true that a growing number of both natural and synthetic chemicals are turning out to have effects similar to those of estrogen, no one is sure what effects, if any, those chemicals actually have on our health. The idea that sperm counts are down because of environmental estrogens is no more than an intriguing hypothesis.

Still, the topic of environmental estrogens has excited a lively debate among researchers. On one side are those who argue that the rise in environmental estrogens very likely accounts for a host of problems, including a sharp rise in the rate of testicular cancer and a more gradual rise in the rate of breast cancer, not to mention the decline in sperm counts. On the other side are those who think the whole problem has been greatly exaggerated. Above all, the debate has raised more questions than researchers know how to answer.

Some questions researchers are asking are "What kinds of effects can environmental estrogens exert?" and "How do environmental estrogens compare with the real hormones they are said to mimic?"

A **hormone** is a substance secreted by the cells of a well-defined organ or tissue that moves throughout an organism and

© Stan Osolinski 1993/FPG

exerts specific effects on specific cells in other organs or structures. Hormones change metabolism, enzyme activity, or gene expression in **target cells** or **target organs.**

Estrogens are a group of hormones found in vertebrates that are responsible for the development of female secondary sexual characteristics and the functioning of the female reproductive organs. The most potent estrogen is *estradiol*. Estradiol is a steroid hormone, like progesterone or testosterone, secreted by the ovaries and the adrenal glands, which lie atop the kidneys. Estradiol maintains ovulation cycles, and large amounts of it are essential for a full-term pregnancy.

Environmental estrogenics include hundreds of different compounds. They are present at low levels in our environment and we ingest them in food and water. Most can bind to estrogen receptors on cells throughout the body, turn on estrogen-sensitive genes, and cause certain kinds of cells to proliferate. Some estrogenics have effects similar to natural estrogens made by the body in that they activate estrogen receptors. Some are antiestrogenic, meaning they counteract the effects of estrogen, preventing natural estrogens from binding to their receptors. Some are antiandrogenic, meaning they counteract the effects of testosterone and similar hormones, usually by blocking the receptors for testosterone.

Many estrogenics are *phytoestrogens,* compounds that mimic estrogens and are manufactured by plants. Other estrogenics are synthetic compounds manufactured by industry. The pesticides atrazine, chlordane, and DDT all have estrogenic ef-

fects. For example, a 1980 spill of DDT and related compounds in a Florida lake appears to have reduced the birthrate of alligators there by 90 percent. And the young male alligators that grew up in the DDT-laden lake had smaller-than-average penises.

The synthetic estrogen diethylstilbestrol (DES) was deliberately manufactured as a drug to treat pregnant women whose own estrogen levels were thought to be too low to sustain a normal pregnancy. Thousands of women were given this drug from 1948 until 1971, when the drug was banned for this purpose. Medical researchers had discovered that the daughters of women who had been given DES had a high rate of cancer of the reproductive organs, as well as sterility and other abnormalities.

have powerful effects. For example, PCBs (polychlorinated biphenyls) applied to turtle eggs individually had no effect on turtle development. When two different PCBs were combined, however, their estrogenic effects turn males into females. And research on yeast cells genetically engineered to express the human estrogen receptor showed that combinations of different pesticides had up to 1600 times the estrogenic potency of any of the chemicals individually. Yet researchers have not succeeded in replicating these results, and the significance of the experiments has come into doubt.

How can minuscule amounts of synthetic chemicals evoke powerful biological responses? The answer is probably that industrial chemists have accidentally made molecules with sizes, shapes,

---

The young male alligators that grew up in the DDT-laden lake had smaller-than-average penises.

DES differs from most estrogenics in that its estrogenic effects are even more powerful than those of estradiol. Most environmental estrogenics are very weak in their effects compared with natural estrogens. Because estrogenics are so weak and because individually they exist in low concentrations in our environment, many researchers have dismissed the idea that they could produce any dramatic effects in humans.

Preliminary research has shown that combinations of different compounds can

and charge distributions similar to the hormones and other molecules that animals use for internal signaling. Ironically, then, industrial chemists have unwittingly accomplished a major goal of many pharmaceutical chemists, whose very purpose may be to produce drugs that mimic natural chemical signals that may be lacking in some people. Understanding the biological effects of environmental chemicals requires knowledge of natural signaling molecules, the subject of this chapter.

KEY CONCEPTS

1. Nerve cells and hormones both contribute to homeostasis, the maintenance of a constant internal environment.

2. Hormones are chemical signals that are made in specific organs or cells and that move throughout an organism and affect processes in target cells and organs. Paracrine signals act as local chemical signals.

3. Hormones are chemically diverse but fall into two general classes—those that are soluble in lipids and those that are soluble in water.

4. Hormones and other signaling molecules bind to specific receptors in target cells and trigger characteristic responses.

5. Most water-soluble hormones act on the outside of a cell by stimulating the synthesis of an intracellular signal, called a "second messenger."

HOW DO CHEMICAL SIGNALS COORDINATE A RESPONSE TO A CHANGE IN THE INTERNAL ENVIRONMENT?

An animal, or any multicellular organism, depends on the co-ordinated activities of many cells and organs. The function of each cell in turn depends on the animal's ability to regulate its own **internal environment,** which means the external environment of an animal's cells (Figure 36-1). Most animal cells have only limited tolerance for environmental variation, and natural selection has produced remarkable adaptations that promote a stable chemical composition of the extracellular

space. The properties of every animal organ system—from the digestive system to the circulatory system to the respiratory system—reflect the need for **homeostasis** [Greek, *homos* = same + *stasis* = standstill], the maintenance of a constant internal environment.

In vertebrates, extracellular fluids are constantly renewed by the blood, which carries nutrients and oxygen to individual cells and waste products to the kidneys and lungs. The blood also carries chemical signals that mediate the responses to changes in the external or internal environment. In Chapter 35, for example, we saw how the presence of food in the stomach produced chemical signals—hormones—that stimulated the se-cretion of hydrochloric acid by the stomach, of hydrolytic en-zymes by the pancreas, and of bile by the liver. Animals also use the electrical activity of nerve cells to coordinate events in physically separated organs.

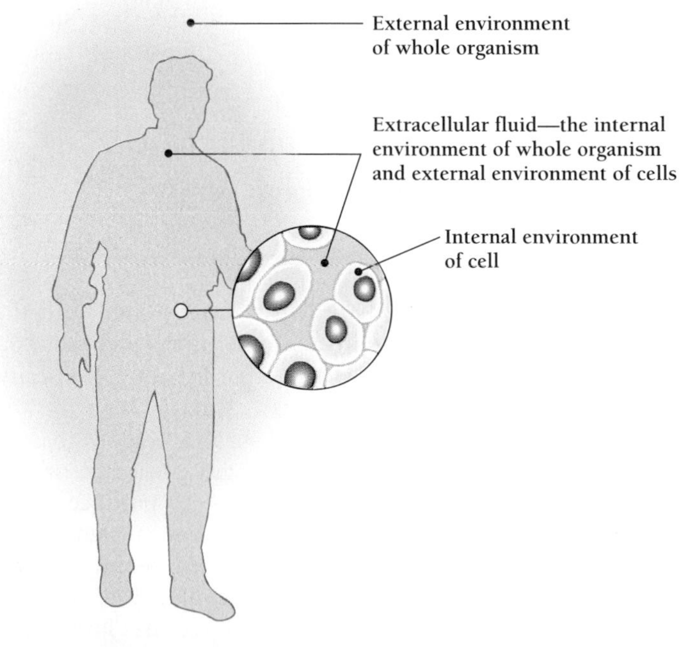

To maintain a constant internal environment, animals must coordinate the activities of many cells and organs.

How Do Organisms Send Chemical Signals?

Biologists distinguish among three types of chemical signals (Figure 36-2).

1. **Pheromones** are substances secreted by one organism that influence the behavior or physiology of another organism of the same species.
2. **Hormones** are chemical signals, made and released by a well-defined organ or structure, that travel long distances through the circulation and exert specific effects on specific cells in other organs or structures.
3. **Paracrine signals** [Greek, *para* = beside] are chemical sig-nals that affect only cells in the immediate vicinity of the signaling cells. Paracrine signals include **neurotransmitters,** which carry information between communicating nerve cells and from nerve cells to muscle cells; **prostaglandins,** which play an important role in blood clotting, inflamma-

Figure 36-1   An animal's "internal environment" is the extra-cellular environment of its individual cells.

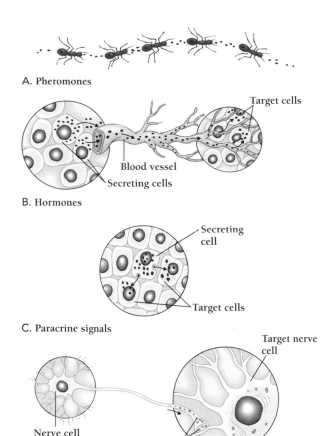

A. Pheromones

B. Hormones

Target cells

Blood vessel

Secreting cells

C. Paracrine signals

Secreting cell

Target cells

D. Neurotransmitters

Target nerve cell

Nerve cell

Neurotransmitters

Figure 36-2 **Types of chemical signaling.** A. Pheromones work between individuals. Ants lay down a trail of pheromones, which their nestmates follow and reenforce. B. Hormones travel long distances within the body. C. Paracrine signals travel short distances. D. Neuro-transmitters are specialized types of paracrine signals.

tion, and uterine contractions; and **growth factors,** which stimulate cell division and promote cell survival.

Since the beginning of the 20th century, physiologists have distinguished between signaling in two physiological systems—the *nervous system* and the **endocrine system** [Greek, *endo* = within + *krinein* = to separate]. At the chemical level, however, the two systems are not that different. In most cases of nerve signaling, the actual communication depends on signaling molecules that are identical or closely related to hormones.

Hormones are produced by endocrine cells or organs, which are specialized for secreting hormones into the general circulation. (In contrast, an *exocrine* [Greek, *exo* = outside] cell or organ makes products that are carried to specific targets by ducts, such as the duct that carries digestive enzymes into the small intestine.)

The major difference between hormonal and neural signaling is that hormones reach essentially all the tissues of the body, while nerves carry signals only to particular targets. A hormone signal resembles a broadcast message in that it may be received by any cell in the body, whereas a nerve signal can be received only by cells to which the nerve cell connects.

Another similarity of hormones and broadcast messages is that the same message can evoke different responses in different recipients. Consider, for example, the possible responses of people hearing a weather forecast that predicts a heavy snow-fall: skiers will react with delight, while farmers may fall into a funk. Similarly, the same hormone, secretin, inhibits acid secretion in the stomach, stimulates bile secretion in the liver, and triggers the release of digestive enzymes in the pancreas.

Hormones are chemical signals that travel through the entire body. They evoke characteristic responses in individual cells and organs.

How Does an Animal Keep Blood Glucose at a Nearly Constant Level?

In mammals, the homeostatic regulation of blood glucose levels depends on separate biochemical events in the liver, the pancreas, and the small intestine, as well as in organs, such as muscle, that use glucose. Shortly after a meal, cells in many organs begin to take up needed fuel and building blocks. The body is said to be in an **absorptive state,** in which cells take up glucose, make glycogen, and increase the synthesis of fats and proteins. The result of these processes is a drop in the blood concentration of glucose and of other fuel molecules. The body then goes into a **postabsorptive state,** in which liver cells produce more blood glucose from the breakdown of glycogen. Other cells return to deriving energy from internal sources of glycogen, fats, and proteins.

Two major hormones—insulin and glucagon—coordinate the switch between the absorptive and the postabsorptive states. **Insulin** is a protein hormone produced by endocrine cells of the pancreas (Figure 36-3). When the insulin-producing cells detect high levels of glucose, they increase the production of insulin. Insulin, which has been called the "hormone of plenty," stimulates cells throughout the body to take up glucose, thereby decreasing the glucose concentration in the blood. Insulin also stimulates biochemical processes in the liver, the muscles, and adipose (fat) tissue. The responses in the target organs are characteristic of the absorptive state—the increased synthesis of fats, proteins, and glycogen.

When no glucose remains in the small intestine, blood glucose decreases, and insulin secretion stops. Other cells of the endocrine pancreas begin to secrete **glucagon,** a protein hormone that, together with the low insulin concentration, signals the postabsorptive state: cells reduce their uptake of glucose and begin to break down glycogen, fats, and proteins. Whereas insulin promotes the synthesis of glycogen and inhibits its breakdown, glucagon inhibits glycogen synthesis and stimulates its breakdown into glucose (Figure 36-4).

In the disease diabetes, the failure to produce insulin or to respond to insulin keeps the body in a permanent postabsorptive state. The result is an excessive concentration of glu-

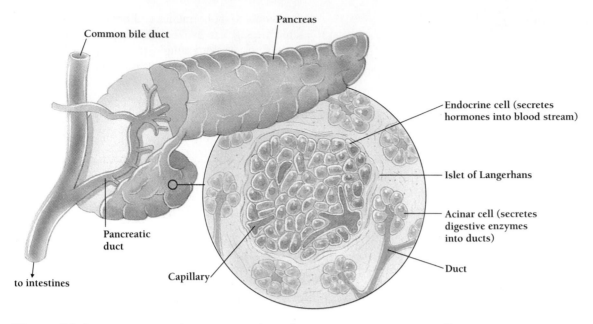

Figure 36-3   **The pancreas.** The pancreas produces digestive enzymes in exocrine cells (the acinar cells) and hormones in the endocrine cells.

cose in the blood, which leads to its presence in the urine and an excess production of urine. This condition, relatively common, is called **diabetes mellitus** [Greek, *diabetes* = siphon + Latin, *mellitus* = sweet], meaning the overproduction of sweet urine. The production of sugar-containing urine was in fact responsible for the discovery that the pancreas was an endocrine organ. In 1889, an observant animal caretaker noticed flies gathering around the urine produced by a dog whose pancreas had been removed. Without a pancreas, we now realize, the dog was unable to produce insulin, and glucose accumulated in the blood and urine, as it does in the blood and urine of diabetics.

About 10 to 20 percent of people with diabetes suffer from *insulin-dependent diabetes mellitus* (IDDM), also called *Type 1 diabetes* or *juvenile diabetes* because it generally begins early in life. People with IDDM fail to make any insulin because their immune systems have mistakenly attacked and destroyed the insulin-producing cells of the pancreas. Normal life depends on daily injections of insulin.

Most people with diabetes, however, do make insulin but their target cells fail to respond to it. This condition is called *Type 2 diabetes,* or *non-insulin-dependent diabetes mellitus* (NIDDM). NIDDM usually manifests itself in middle age, and it can usually be controlled by diet and by oral medicines.

Insulin, made when blood glucose is high, stimulates glucose absorption from the blood. Glucagon, made when blood glucose is low, promotes the production of more blood glucose from glycogen.

WHICH HUMAN ORGANS PRODUCE HORMONES?

Figure 36-5 highlights the known endocrine organs in humans. We have already encountered several of these in our discussion of digestion in Chapter 35—the endocrine pancreas as well as endocrine cells within the stomach and the small intestine.

Pancreas

The endocrine pancreas consists of 2 million clusters of cells that are physically distinct from the pancreas's exocrine cells (Figure 36-3). These clusters, called the *pancreatic islets* or *islets of Langerhans,* consist of at least three different cell types, each of which makes a different hormone. The α (alpha) *cells* produce glucagon, which stimulates the breakdown of glycogen to glucose in muscle cells; the β (beta) *cells* make insulin, which stimulates the uptake of glucose from the blood into muscle and other cells; and the δ (delta) *cells* make somatostatin, which decreases secretion, absorption, and smooth muscle action within the gut and inhibits the production of insulin and glucagon within the pancreatic islets themselves. Both α and β cells respond to the concentration of glucose in the blood: low glucose stimulates glucagon production by α cells, while high glucose stimulates insulin production by β cells.

Other parts of the digestive system also contain hormone-producing cells. Specialized cells of the stomach lining make *gastrin,* which stimulates hydrochloric acid production, and specialized cells in the small intestine make secretin and cholecystokinin (Chapter 35).

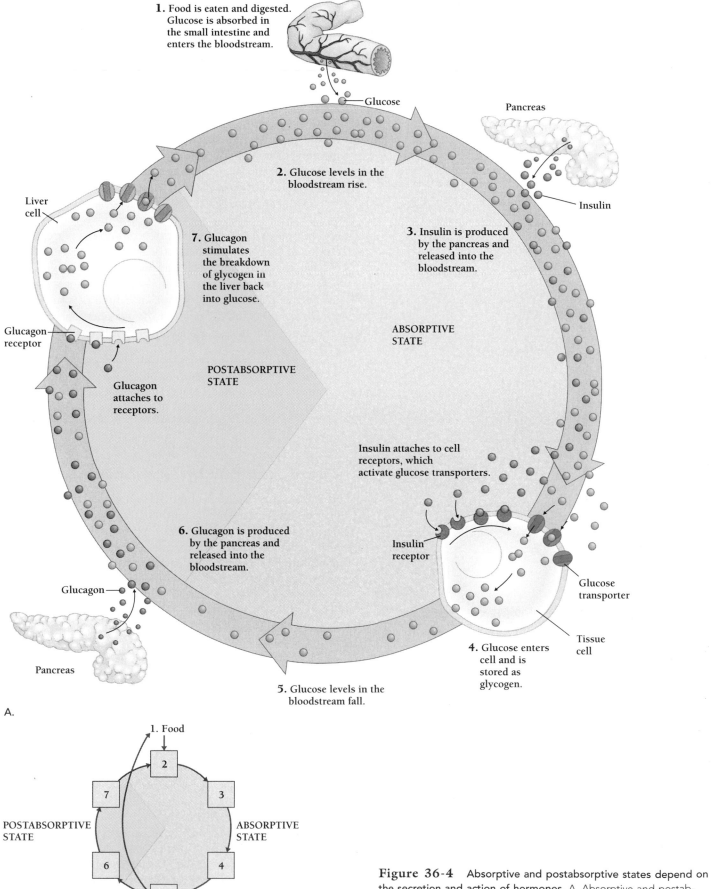

1. Food is eaten and digested. Glucose is absorbed in the small intestine and enters the bloodstream.

Glucose

Pancreas

Insulin

Liver cell

2. Glucose levels in the bloodstream rise.

3. Insulin is produced by the pancreas and released into the bloodstream.

7. Glucagon stimulates the breakdown of glycogen in the liver back into glucose.

ABSORPTIVE STATE

POSTABSORPTIVE STATE

Glucagon receptor

Glucagon attaches to receptors.

Insulin attaches to cell receptors, which activate glucose transporters.

6. Glucagon is produced by the pancreas and released into the bloodstream.

Insulin receptor

Glucose transporter

Glucagon

Tissue cell

4. Glucose enters cell and is stored as glycogen.

Pancreas

5. Glucose levels in the bloodstream fall.

A.

1. Food

2

7

3

POSTABSORPTIVE STATE

ABSORPTIVE STATE

6

4

5

B.

Figure 36-4 Absorptive and postabsorptive states depend on the secretion and action of hormones. A. Absorptive and postabsorptive states depend on the pancreatic hormones, insulin and glucagon. B. Simplified diagram of the process above.

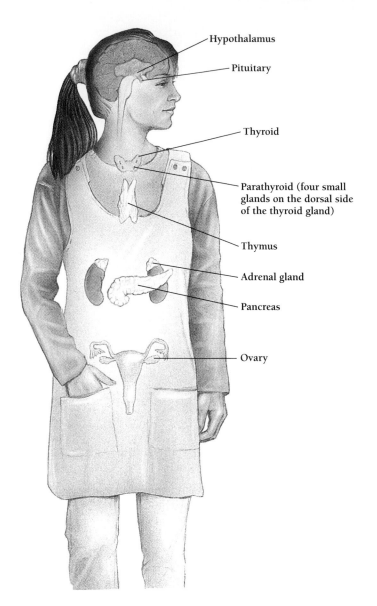

- Hypothalamus
- Pituitary
- Thyroid
- Parathyroid (four small glands on the dorsal side of the thyroid gland)
- Thymus
- Adrenal gland
- Pancreas
- Ovary

Figure 36-5   The major endocrine glands in a human.

Hormones produced in the digestive system help coordinate the absorption of food molecules both within the digestive system and in the rest of the body.

Adrenal Glands

The adrenal glands [Latin, *ad* = to + *renes* = kidneys] lie just next to the kidneys. Each adrenal consists of two parts—the outer **adrenal cortex** and the inner **adrenal medulla.** The adrenal cortex makes two hormones: *cortisol,* which mediates an animal's response to stress, stimulating the production of carbohydrate from protein and fat, and *aldosterone,* which helps regulate kidney function. In both cases, a particular region of the brain—the hypothalamus—controls the production of hormone.

The adrenal medulla, on the other hand, operates independently of the brain. Its primary hormonal output is the hormone **epinephrine** [Greek, *epi* = upon + *nephros* = kidney], also called *adrenaline* [from the adrenal gland], the mediator of the "fight or flight" reaction (Figure 36-6). Epinephrine speeds the heart, dilates the blood vessels, and increases the liver's production of glucose from glycogen. All these responses contribute to the "fight or flight" reaction that prepares an animal for immediate action and energy expenditure in the face of stress or danger. The adrenal medulla also produces substantial amounts of *norepinephrine,* a closely related compound.

Release of epinephrine into the blood readies the animal for immediate action and energy expenditure. The release of epinephrine and norepinephrine depends on the perception of danger or stress. Many kinds of stimuli can trigger this perception; the perceived danger can range from an approaching bear to an overbearing teacher. Transmitting a danger signal to the adrenal medulla involves both the central nervous system, which interprets the danger, and the peripheral nervous system, which transmits information to the adrenal medulla.

The adrenal cortex produces two hormones in response to hormone signals from the brain. The adrenal medulla produces epinephrine, the major mediator of the "fight or flight" reaction.

Thyroid and Parathyroid Glands

In humans, the two lobes of the **thyroid gland** are in the neck, just in front of the windpipe. In other vertebrates, the thyroid lies just below the windpipe. The thyroid produces *thyroid hormone,* which regulates metabolism and growth. Overproduction of thyroid hormone, called *hyperthyroidism,* results in uncontrolled, rapid metabolism, whereas underproduction, called *hypothyrodism,* may cause such symptoms as lethargy, weight gain, and intolerance of cold. Hypothyroidism during fetal life may result in abnormal development, mental retardation, and low metabolic rate. The thyroid gland also produces *calcitonin,* which stimulates tissues to remove calcium ions from the blood.

Adjacent to the thyroid gland is the **parathyroid gland.** This gland produces *parathyroid hormone,* whose effect is to increase the concentration of calcium ions in the blood—exactly the opposite effect from that of calcitonin. The antagonistic actions of parathyroid hormone and calcitonin are responsible for calcium homeostasis: between them, they keep blood calcium relatively constant. Low blood calcium triggers the parathyroid to release parathyroid hormone, and high blood calcium triggers the thyroid to release calcitonin.

Hormones from the thyroid gland regulate metabolism, growth, and calcium levels in the blood. Parathyroid hormone also contributes to calcium regulation.

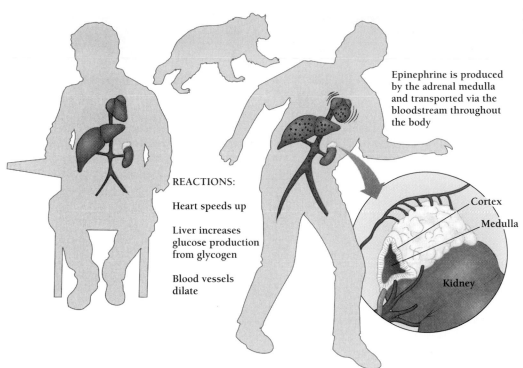

Figure 36-6 **Fight or flight.** Epinephrine (adrenaline) mediates the "fight or flight" reaction.

Epinephrine is produced by the adrenal medulla and transported via the bloodstream throughout the body

REACTIONS:

Heart speeds up

Liver increases glucose production from glycogen

Blood vessels dilate

Cortex

Medulla

Kidney

Figure 36-6 **Fight or flight.** Epinephrine (adrenaline) mediates the "fight or flight" reaction.

Testes and Ovaries

The gonads (testes and ovaries) produce the sex steroids: the **testes** produce *testosterone,* and the **ovaries** produce *estrogen* and *progesterone*. The sex steroids are responsible for the development and the maintenance of the secondary sexual characteristics as well as for the proper development of germ cells. Progesterone is especially important in maintaining pregnancy. We will have more to say about the role of the sex steroids in reproduction in Chapter 43.

The gonads secrete sex hormones.

The Hypothalamus and Pituitary Gland Regulate the Production of Hormones by Other Endocrine Organs

The **pituitary gland** is a small structure, about the size of a pea, at the base of the brain (Figure 36-7). Its name [Latin, *pituita* = slime] came from the mistaken belief that it produced the mucus of the nose. In humans, the pituitary has two main parts, called the *posterior pituitary* and the *anterior pituitary*.

The **posterior pituitary** is actually part of the brain itself; it consists of the terminals of nerve cells whose cell bodies lie in a part of the brain called the **hypothalamus** [Greek, *hypo* =

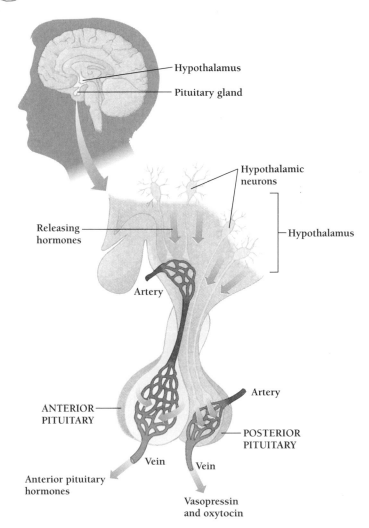

Hypothalamus

Pituitary gland

Hypothalamic neurons

Hypothalamus

Releasing hormones

Artery

Artery

ANTERIOR PITUITARY

POSTERIOR PITUITARY

Vein

Vein

Anterior pituitary hormones

Vasopressin and oxytocin

Figure 36-7 **The hypothalamus and pituitary and their hormones.** This diagram is stylized to emphasize the relationships of the hypothalamus and pituitary.

BOX 36-1

How do researchers identify hormones and glands?

Endocrinology, the study of hormones, has traditionally asked three questions about a suspected hormone: (1) what organ is responsible for its synthesis? (2) does the suspected hormone move through the circulation? and (3) what is its chemical identity?

To answer the first question, a researcher would determine whether a product of one organ affected the function of another. For example, a researcher would destroy a suspected endocrine organ and note the resulting changes. An often-cited example is the effect of removing the testes of a juvenile animal. The result is the failure to develop the secondary sexual characteristics that appear in males at sexual maturity. In roosters, a large red comb is a secondary sexual characteristic, and removing the testes results in the withering of the comb.

The second question—whether the suspected hormone travels through the circulation—distinguishes a hormone from a secretory (*exocrine*) product or a short-range *paracrine* signal. Researchers have addressed this question by restoring the suspected endocrine organ to another place in the body. In the case of a castrated rooster, transplanting testes into the abdominal cavity restores the comb. This result shows that the hormone responsible for the secondary sexual characteristics moves through the circulation rather than through a particular duct.

The third question—the chemical identity of the signaling molecule—is the hardest to answer. Researchers begin by demonstrating that an extract of the endocrine gland (for example, of a rooster testis) can produce the same effect as the intact gland. The next task is to isolate and identify the active compound and to determine its chemical structure. Finally, researchers synthesize the suspected hormone in the laboratory and show that the synthetic substance has the same effect as the suspected hormone. Again, in the case of the rooster's comb, pure testosterone can restore the comb just as well as an extract of rooster testes. The ability of a pure chemical, made in the laboratory, to mimic a hormone's effect establishes that hormone's chemical identity. The action must depend on that single hormone rather than some unknown contaminant in the original extract or on two or more hormones.

This straightforward approach has not always worked. Endocrine organs sometimes make more than one hormone, so that researchers must replace a number of substances in order to reverse the effects of organ removal. The pancreas, for example, makes both insulin and glucagon. And the pituitary gland releases at least nine hormones.

In vitro cultures of target cells or target organs provide another way that contemporary endocrinologists identify and study hormones, particularly in complex systems that involve many different signals and targets. Using both the traditional surgical methods with whole organisms and the more recent approaches with cultured cells and organs, endocrinologists have identified more than 50 hormones and more than 15 hormone-producing structures in vertebrates and similar numbers in invertebrates.

under + *thalmos* = inner chamber] (Figure 36-7). Some of the cells of the hypothalamus extend into the pituitary, and their terminals make up the posterior pituitary. These cells release several hormones, including *vasopressin,* or *antidiuretic hormone,* which stimulates water reabsorption in the kidney (Chapter 39). *Oxytocin,* another hormone released from the posterior pituitary, acts at childbirth to stimulate the uterus to contract and the mammary glands to eject milk.

The **anterior pituitary** is not part of the brain, but a separate gland. The anterior pituitary produces and releases eight peptide hormones: *corticotropin* (also called *adrenocorticotropic hormone,* or *ACTH*), which stimulates the production of steroids in the adrenal cortex; *endorphin,* a natural pain suppressor; *thyroid-stimulating hormone (TSH),* which stimulates the production of thyroxin in the thyroid gland; *growth hormone (GH),* which stimulates tissue and skeletal growth; *prolactin,* which stimulates milk production; *melanocyte-stimulating hormone (MSH),* which stimulates pigment production in specialized skin cells; *follicle-stimulating hormone (FSH),* which in females stimulates estrogen production by the ovaries and promotes the maturation of the follicle during the menstrual cycle and in males stimulates testosterone production; and *luteinizing hor-* *mone (LH),* which in females induces ovulation and stimulates estrogen production and in males increases testosterone production.

Distinct sets of cells within the anterior pituitary specialize in the production of each of these hormones. Some cells, for example, produce either growth hormone, ACTH, or TSH. Other cells produce FSH and LH, both of which act on gonadal tissue (ovaries and testes) and are therefore called *gonadotropins.*

Because the anterior pituitary stimulates so many target organs to produce other hormones, many endocrinologists have referred to it as the "master gland." In fact, however, the release of hormones by the anterior pituitary depends on other hormones produced by the hypothalamus.

The hypothalamus produces at least seven hormones that control the release of hormones by the anterior pituitary. Five of these are **releasing hormones,** each of which stimulates the release of a specific hormone, and two are **inhibiting hormones,** each of which inhibits hormone release. For example, *gonadotropin-releasing hormone (GnRH)* stimulates the secretion of FSH and LH by the anterior pituitary.

Hypothalamic hormones such as GnRH enter the general circulation through capillaries in a special structure called the median eminence (Figure 36-7). The median eminence also contains specialized veins that do not flow directly into the rest of the venous circulation. Instead, they flow into the anterior pituitary, where they divide and form a second capillary bed. This diversion gives the hypothalamic hormones more direct access to their targets in the anterior pituitary.

The regulation of hormone release often involves **negative feedback,** the use of information about the output of a system to reduce further output. In male mammals, for example, FSH and LH stimulate testosterone production in the testes. Testosterone from the testes enters the circulating blood and then acts on the hypothalamus to inhibit production of GnRH. The lowered GnRH in turn reduces the release of FSH and LH by the anterior pituitary, a state that reduces further testosterone production by the testes. The result is a negative feedback loop: increased testosterone levels lead to a reduction in testosterone synthesis.

In female mammals, the regulation of estrogen production depends on both negative feedback and positive feedback. High concentrations of estrogen inhibit GnRH release, ultimately leading to a lowered production of estrogen. But very high estrogen concentrations actually stimulate GnRH release, leading to a surge of LH. It is this surge of LH that triggers ovulation midway through the menstrual cycle.

The hypothalamus and anterior pituitary also participate in coordinating responses to neural information concerning the physical and emotional state of the body. Stress—produced, for example, by blood loss, fright, an argument, or even a biology final exam—increases the production of corticotropin-releasing hormone (CRH). Increased CRH leads to increases in ACTH and cortisol. High cortisol levels in turn heighten the body's reaction to stress: cortisol increases heart output, lung ventilation, blood flow to the muscles, and glycogen production in the liver. A small increase in CRH production (0.1 μg) leads to a larger increase in ACTH (1 μg), a still larger increase in cortisol (40 μg), and a still larger effect on glycogen synthesis (5600 μg). Unfortunately, in people (and other animals) with stressful lives, this amplifying cascade works all too well, leading to chronically high cortisol levels, sustained stimulation of the heart and circulation, and a higher incidence of cardiovascular disease.

In controlling the levels of its hormones, the hypothalamus exerts powerful control over hormone production in distant sites—not only in the adrenal cortex but also in the gonads, thryoid, breasts, and liver. In exerting this regulation, the hypothalamus integrates information both from the endocrine system and from the nervous system.

Information from the endocrine system also affects the functioning of the brain. Hormonal changes can lead to altered behavior. For example, a female dog in heat will accept the mating behavior (mounting) of a male. At other times, the female may respond to the same male behavior by just walking away. Experimenters can change the dog's response by injecting appropriate hormones. So the communication between the hormone system and the nervous system runs in both directions.

The hypothalamus is the major mediator of information between the brain and the endocrine system. Hormones made by the hypothalmus regulate hormone production elsewhere, especially in the anterior pituitary, which also produces hormones that regulate hormone production elsewhere.

HOW DO HORMONES FIND THEIR TARGETS?

Studies of the chemical identities of hormones and their modes of action have shown that hormones fall into two classes—those that are **lipid-soluble signals** (such as estrogen, testosterone, and other steroids), whose nonpolarity allows them to pass directly through the plasma membrane of the target cell, and **water-soluble signals** (polypeptides such as secretin and amines such as epinephrine), which cannot pass through the plasma membrane.

For both classes, action on a target cell begins when the signal molecule binds to a specific protein, called a **receptor.** Lipid-soluble signals bind to specific receptor molecules *inside* their target cells, while water-soluble signals bind to specific receptor molecules on the *surface* of their target cells (Figure 36-8).

Among the most important lipid-soluble signaling molecules are the **steroids,** nonpolar molecules that derive from cholesterol. *Glucocorticoids* (such as cortisol) and *mineralocorticoids* (such as aldosterone) are made in the adrenal cortex, while the *sex steroids* (estogen, progesterone, and testosterone) are made in the gonads. Chemically synthesized compounds that resemble estrogen and progesterone are widely used in birth control pills, and steroids that resemble testosterone have been abused by some athletes to build muscle mass. Use of these *androgens* [Greek, *andro-* = man; the generic term for testosteronelike steroids] increases the risks of sterility, cancer, and psychiatric disorders.

Water-soluble chemical signals include most of the **amines,** the simplest chemicals that act as hormones. A water-soluble

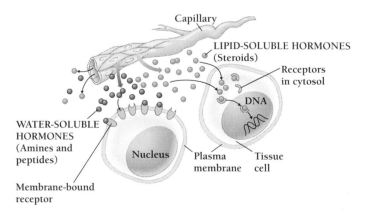

Figure 36-8   The differing actions of lipid-soluble and water-soluble hormones.

amine, such as epinephrine, cannot pass through the plasma membrane of its target cells and must act on the cell's surface. A few amines, however, such as the thyroid hormone thyroxin, however, are lipid-soluble. Thyroxin passes directly through the plasma membrane and acts in the same way as steroids and other lipid-soluble hormones.

Peptides, which are water-soluble, are by far the largest class of hormones. A peptide hormone may contain as few as 3 amino acids or well over 100. Among the shorter peptide hormones are those made by the hypothalamus, including the releasing factors, vasopressin, and oxytocin. The longer peptide hormones, which are classified as proteins, include insulin, glucagon, and the hormones of the anterior pituitary. Neither the small peptides nor the much larger proteins are lipid-soluble, and they do not cross the plasma membrane of their target cells. The action of these hormones depends entirely on their interactions with receptors on the surface of their target cells. These receptors in turn stimulate the synthesis or release of intracellular "second messengers" that influence the inner workings of their target cells.

For all kinds of chemical signals, signaling between cells involves six steps: (1) a cell must make the signaling molecule; (2) the cell must release the signaling molecule; (3) the molecule must travel to the target cells; (4) the target cells must detect the signal; (5) the target cells must respond to the signal; and, finally, (6) something must end the signaling process.

Hormones are chemically diverse, including steroids, amines, and peptides. Hormones are divided into those that are soluble in lipids and those that are soluble in water.

How Do Target Cells Detect Chemical Signals?

The response of a target cell to a particular signaling molecule depends on the presence of a specific receptor protein. The receptor binds to the signal and intitates the cell's response.

A receptor binds to a signaling molecule, such as a hormone, in much the same way that an enzyme binds to a substrate. Just as a substrate forms noncovalent bonds with amino acids at the *active site* of an enzyme, so a signaling molecule forms noncovalent bonds at the **binding site** of a receptor.

The affinity of a hormone receptor for a specific hormone is usually very high: The receptors on a cell may be fully saturated when hormone concentrations are one-millionth those at which most enzymes are saturated. The strong binding of hormones to their receptors helps explain why environmental contaminants that are present at extremely low concentrations can nonetheless interfere with natural signaling processes.

The response of a target cell to epinephrine can be mimicked by other synthetic compounds. Molecules that resemble a natural hormone can bind to the same receptor and initiate

the same response. For example, the synthetic compound *isoproterenol* binds to the same receptors as epinephrine, and either compound can be used to open the airways and ease the congestion caused by allergies and asthma.

Other molecules that are chemically similar may bind to the receptor and thereby prevent the binding of a natural signal. For example, *propranolol,* another synthetic compound that resembles epinephrine, prevents the action of epinephrine on heart muscle. By antagonizing the effect of epinephrine, propranolol prevents stressful increases of heart rate and blood pressure. For this reason, propranolol is widely used to treat people who have high blood pressure.

Signaling molecules, such as epinephrine, play an important part in the coordination of distinct organ systems. Studies of the action of hormones such as epinephrine have led to new therapies, not only for asthma and high blood pressure but for many other conditions as well.

Hormone action begins with binding to a receptor. Molecules that resemble a natural hormone can bind to the same receptor and initiate or prevent the same response.

Hormones May Target More Than One Kind of Receptor

Isoproterenol and propranolol are only two among hundreds of synthetic epinephrinelike compounds that researchers have studied. As these studies proceeded, researchers became increasingly aware that epinephrine could not be binding to only one kind of receptor molecule, since different tissues responded differently to the synthetic compounds. For example, although isoproterenol mimics the effects of epinephrine on the heart, isoproterenol does not inhibit smooth muscle contractions in the gut, as does epinephrine. One explanation for this difference was that there were several types of epinephrine receptors that differed in their ability to bind other agonists and antagonists. This explanation turned out to be exactly correct.

Biochemists were able to isolate the receptor molecules that bind to epinephrine and other hormones. Using molecular biological techniques, researchers have been able to identify and sequence the genes that encode each of these receptors. They found, as expected, that different tissues had different forms of epinephrine receptors. In fact, for virtually every type of signaling molecule researchers have found several types of receptors, which are deployed differently in different types of target cells. These findings begin to explain how different types of cells can respond so differently to the same signal.

Epinephrine binds to different receptor types in different tissues.

HOW DOES THE BINDING OF A WATER-SOLUBLE SIGNAL AT THE CELL SURFACE LEAD TO INTRACELLULAR RESPONSES IN THE TARGET CELL?

Water-soluble signaling molecules—including most amine and all peptide hormones—do not enter their target cells, but instead bind to receptors on the outer surface. The receptor does not itself change the inside of the cell. The receptor plays the role of a baton in a relay race in which the hormone or signaling molecule is just the first runner, or "first messenger." When the hormone binds to the outer surface, the receptor changes shape and stimulates the production, inside the target cell, of a **second messenger,** which relays to the cell's interior the information that a signal has arrived at the cell's surface. Researchers now know of four second messengers whose production is triggered by extracellular signaling molecules—cyclic AMP, calcium ions, diacyl glycerol, and inositol trisphosphate.

A particularly well studied part of the "fight or flight" reaction is an increase in the availability of energy, which is brought about by the increased breakdown of glycogen. Epinephrine increases the breakdown of glycogen by activating *glycogen phosphorylase,* the liver enzyme that hydrolyzes glycogen to glucose (Figure 36-9). Epinephrine itself acts only on the cell surface: the second messenger responsible for phosphorylase activation is **cyclic AMP (cAMP).** Addition of cAMP alone increases the activity of phosphorylase in liver extracts. Glucagon, another hormone that stimulates glycogen breakdown, also activates glycogen phosphorylase by means of cAMP.

The effects of cAMP (or of any second messenger) differ from cell to cell. For example, in the intestine, cAMP stimulates the pumping of sodium ions and water out of cells. People suffering from the infectious disease *cholera* have excess cAMP, which overstimulates the pumping of sodium ions and water, leading to diarrhea. The overproduction of cAMP results from the action of *cholera toxin,* a protein produced by *Vibrio cholerae* bacteria. Cholera toxin modifies the protein (called a G protein) that couples the receptor to the production of cAMP.

Figure 36-9 Some adrenergic receptors in the plasma membrane respond to epinephrine by stimulating the synthesis of cyclic AMP. Cyclic AMP activates a protein kinase, which activates another protein kinase (phosphorylase kinase), which stimulates the enzymatic activity of glycogen phosphorylase.

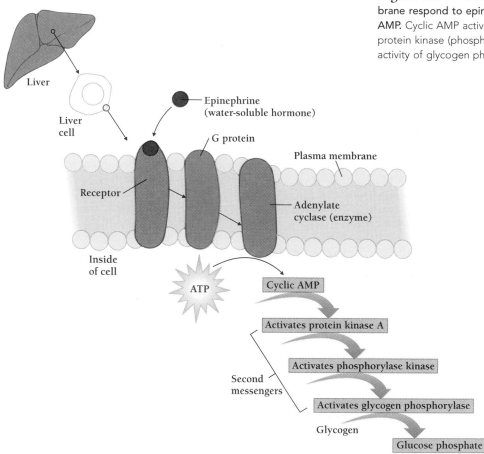

The coupling protein then becomes locked into its active form and cannot be switched off.

The binding of many water-soluble signals such as epinephrine to certain membrane receptors stimulates an increase of free calcium ions (Ca^{2+}) within the cytoplasm. Free Ca^{2+} ions then function as a second messenger to change the activity of specific enzymes and other proteins.

Water-soluble hormones act on a target cell's surface to stimulate the production of second messengers, such as cyclic AMP and calcium ions, which trigger intracellular events.

PARACRINE SIGNALS ACT OVER SHORT DISTANCES

Some chemical signals between cells do not travel through the blood, but act at much shorter distances. Such compounds are called *paracrine* signals (in contrast to *endocrine* signals, or hormones). Paracine signals affect only cells in the immediate vicinity of the signaling cell. Paracrine signals include *neurotrans-mitters* (signals by which nerve cells communicate), *prostaglandins* (signals by which other cells communicate, all derivatives of a 20-carbon unsaturated fatty acid), and *growth factors* (which stimulate cell division).

Some prostaglandins play an important role in the clotting of blood, others help bring about the inflammatory response, and still others induce smooth muscles to contract. For example, prostaglandins promote uterine contractions. The name "prostaglandin" (referring to the prostate gland in the male reproductive system) came from a practice, previously not understood, of using semen (which contains prostaglandins) to induce labor in pregnant women who had come to term.

A series of enzymes converts arachidonic acid, which is a minor component of membrane lipids, into various prostaglandins. Aspirin, ibuprofen, and acetaminophen all inhibit these pathways, so they act as both anti-inflammatory agents and as anticlotting agents. By reducing the formation of blood clots which can block the blood vessels that supply the heart, small amounts of aspirin significantly diminish the risk of a heart attack.

Prostaglandins are paracrine factors that exert their efforts through second messengers.

STUDY OUTLINE WITH KEY TERMS

To maintain a constant **internal environment,** an animal must coordinate the activities of many cells and organs. **Hormones,** produced in the organs of the **endocrine system** contribute to **homeostasis** by evoking characteristic responses within their **target cells** and **target organs.** Hormones are chemical signals that are carried by the circulation and act at long distances. **Paracrine signals** act only at short distances, while **neurotransmitters** act only over the short distances between nerve cells. **Pheromones,** on the other hand, influence the activities of other organisms of the same species.

Insulin and **glucagon** contribute to the regulation of blood glucose. Insulin establishes an **absorptive state,** in which cells take up glucose from the blood and make glycogen. Glucagon establishes a **postabsorptive state,** in which the liver produces more glucose and breaks down glycogen. The failure of insulin production or insulin action leads to excessive blood glucose, a condition called **diabetes mellitus.**

Endocrinologists have used surgical experimentation, chemical synthesis, and tissue culture to identify the hormones and organs that contribute to the endocrine system. Among the endocrine organs are the pancreas, stomach, small intestine, adrenal glands, **thyroid, parathyroid,** and gonads (**testes** and **ovaries**). The **adrenal cortex** produces hormones that are distinct from those of the **adrenal medulla.** Similarly, the **anterior pituitary** is distinct from the **posterior pituitary,** which is actually part of the **hypothalamus,** a part of the brain.

The hypothalamus and pituitary gland regulate the production of hormones by other endocrine organs. The hypothalamus produces

releasing hormones and **inhibiting hormones,** which travel within the **median eminence** and regulate the release of hormones by the anterior pituitary. The anterior pituitary produces hormones that regulate hormone production elsewhere in the body, in a manner often subject to **negative feedback.**

Hormones are chemically diverse and include **steroids, amines,** and **peptides.** Hormones are either soluble in lipids or soluble in water.

The binding of a hormone to a **receptor** initiates the response of the target cell. The hormone attaches to the receptor's **binding site.** Some compounds evoke the same response as the hormone itself, whereas other compounds prevent hormone binding and the subsequent target cell responses.

The receptors for **lipid-soluble signals** are usually intracellular proteins that regulate transcription. The receptors for **water-soluble signals** are usually membrane proteins with seven transmembrane segments. Signals can often bind to a family of slightly different receptors, which may lie on different target cells and which may evoke different cellular responses.

Most water-soluble signals (such as **epinephrine**) act by stimulating the production of a **second messenger,** which triggers the cell's responses. The most common second messenger is **cyclic AMP (cAMP).** Cyclic AMP binds to a protein kinase and stimulates the phosphorylation of specific protein targets. Several types of epinephrine receptor stimulate cAMP synthesis through a coupling factor. In the case of glycogen breakdown, cAMP amplifies the relatively small signal of epinephrine into the formation of thousands of glucose molecules.

Receptors may also use other second messengers, such as calcium ions. Like cAMP, calcium ions stimulate the phosphorylation of specific proteins. Some water-soluble hormones do not work through second messengers but directly stimulate protein phosphorylation by membrane-bound kinases.

Prostaglandins and **growth factors** are paracrine factors. Prostaglandins regulate blood clotting and smooth muscle contractions.

REVIEW AND THOUGHT QUESTIONS

Review Questions

1. How do hormones, paracrine signals, and pheromones differ from one another?
2. What is the difference between the nervous system and the endocrine system? What do these systems share in common?
3. What are the two major classes of hormones? How do they differ from each other?
4. What is a second messenger? Give an example of one.
5. How does diabetes mellitus (Type 1 diabetes) differ from Type 2 diabetes?

6. How is the anterior pituitary different from the posterior pituitary?
7. Why is the anterior pituitary not "the master gland"?
8. Name the eight peptide hormones released by the anterior pituitary and tell what each one does.

Thought Question

9. Suggest a mechanism by which DES, the synthetic estrogen given to pregnant women in the 1950s and 1960s, might cause cancers of the reproductive organs in the grown daughters of these women. How would you test your hypothesis?

SELECTED READINGS

Tobin, A., and R. Morel, *Asking About Cells,* Saunders College Publishing, Philadelphia, 1997, Chapter 19.

Sapolsky, Robert M., *Why Zebras Don't Get Ulcers,* W.H. Freeman, New York, 1994. A popular discussion of the role of hormones in stress, written by a first-rate biologist and writer. Sapolsky studies stress in wild baboons in Kenya.

▶On-line materials relating to this chapter are on the World Wide Web at http://www.saunderscollege.com/lifesci/ Click on Tobin/Dusheck: *Asking About Life.*

William Harvey, Miguel Serveto, and the Pulmonary Circulation

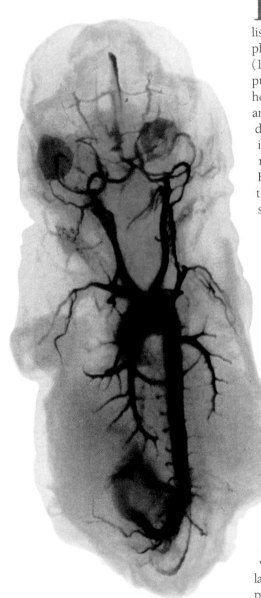

In 1628, an otherwise traditional English physician revolutionized medical physiology. In that year, William Harvey (1578–1657), physician to King Charles I, published the first accurate description of how blood circulates through the heart and body (Figure 37-1). Harvey's analysis depended on both a handful of radical ideas from different sources and on meticulous observation and experiment. His model of circulation overthrew orthodox views that had stood since the second century. His argument was logical and his evidence irrefutable, yet his model attracted angry attacks from physicians all over Europe. How could such a seemingly conventional doctor stir up so much controversy?

Harvey, one of ten children, was the son of a small-town businessman. Harvey was intelligent, polite, well-educated, and ambitious. In 1604, after he had finished medical school, he married the daughter of the personal physician to King James I. Despite intervention by Harvey's new father-in-law, Harvey's several applications to the College of Physicians were repeatedly rejected until 1607. Once accepted, however, he was able to apply for employment at St. Bartholomew's Hospital, one of two great London hospitals. In 1609, at age 31, Harvey was hired as an assistant, and when his supervisor died just a few months later, Harvey became the hospital's chief physician, a position he kept for 34 years.

Harvey became one of the most trusted doctors in all of England. Harvey adhered strictly to accepted medical practice, as taught by Aristotle nearly 2000 years earlier. His huge practice included Sir Francis Bacon and other eminent Englishmen. In 1618 he was appointed assistant court physician to King James I. A few years later, when the king's physician was out of the country and the king became ill, Harvey was chosen to oversee the dying king's care. After the old king's death, his son, the new king, was so grateful to Harvey that he adopted Harvey as his personal physician and friend. Throughout Harvey's life, he was an active member of the College of Physicians, and he accumulated great wealth before he died at age 80.

Yet beneath the breast of this orthodox English doctor beat the heart of an experimentalist. Harvey had learned anatomy in Italy under the guidance of one of the great anatomists of the 17th century, Hieronymus Fabricius. At the University of Padua, then the best medical school in Europe, Harvey had learned both to observe and to test his ideas by experiment. In his spare time, Harvey devoted himself utterly to the study of the circulatory system. From 1604 to 1642, he dissected insects, earthworms, reptiles, birds, and mammals, including his own deceased patients.

Unlike modern biologists, Harvey worked in comparative isolation. He could not afford to discuss his ideas with others

L. Li, B. Mercer, and E. Olson

Figure 37-1 **William Harvey.** The English physician William Harvey constructed the first accurate description of the circulatory system. *(The Granger Collection, New York)*

until he was certain that he was right. To be wrong would have been disastrous for his medical practice. As it was, many of his patients abandoned him when he published his classic work *de Motu Cordis* ["On the movement of the heart and blood"]. Seventeenth-century physicians looked on new ideas with suspicion. Harvey probably could propose such an extraordinary reinterpretation of human physiology because he was so established and otherwise so orthodox. Because he was a widely respected doctor and a close friend of the king of England, Harvey could not only say what he thought but he could also publish his ideas.

Just 75 years before Harvey published his discoveries, the Spanish theologian and physician Miguel Serveto (1509–1553) had published 1000 copies of a book that challenged basic ideas of both the Catholic Church and medicine. Serveto made some of the earliest arguments, for example, for the separation of

church and state, an idea that most of us now take for granted. An accomplished anatomist, he also noted, in the course of a discussion of the Holy Spirit, that the lungs do not deliver air to the heart, as everyone then believed. Air, he said, does not travel through the blood vessels: blood does. The heart, Serveto insisted, delivers blood to the lungs where air activates the blood before it returns to the heart. This circulation from heart to lungs to heart is now called **pulmonary circulation.**

As soon as Serveto's book was published, the Catholic Church arrested Serveto (as well as the man who had printed his book). Serveto was tried and convicted of heresy. Although all but a handful of his 1000 books were burned, Serveto himself escaped. He fled to Geneva, Switzerland, seemingly a haven of Protestantism. Serveto should have been safe in Geneva, for the Pope had far less power there. Yet, here, Serveto was again arrested and tried. The Protestant John Calvin, the theologian who with Martin Luther was most responsible for the Protestant Reformation, acted as a witness

against Serveto and repeatedly urged the court to execute him. The Geneva court condemned the physician to death, and in 1553, Miguel Serveto was burned alive.

The kind of persecution suffered by Serveto was continued in Harvey's time. Indeed, it was all too common. Serveto was in no way an exception, and 16th- and 17th-century scientists had every reason to watch what they said. Even Harvey, protected by his reputation and connections with the English court, must have hesitated to publish his life's work.

What Did William Harvey Discover?

Ancient scientists understood that the heart and blood were crucial for life. But how the heart and blood vessels were constructed and how these functioned were only partly understood. Before Harvey, the greatest contributor to the understanding of circulation was the Greek physician Galen (129–199). Galen, the highly educated son of a Greek architect, first described the anatomy of the heart and blood vessels based on his dissections of African monkeys.

In 157, Galen was appointed chief physician to the gladiators in the city of Pergamum. His close examination of the

How could such a seemingly conventional doctor stir up so much controversy?

horrendous wounds they suffered revealed much about the movement of blood in the body. In his later years, Galen became physician to the Roman emperor Commodus and performed dissections on living animals for the education and entertainment of onlookers. He demonstrated, for example, that the arteries carry blood rather than air and that nerves from the brain control movements in other parts of the body. Galen's extensive writings were preserved by medieval scholars and passed down to the anatomists of the 16th century. Fourteen hundred years after his death, Galen's views were endorsed by the Catholic Church, and he was then as much revered as Albert Einstein is today.

But Galen's understanding was only partly accurate. His ability to understand anatomy was limited by laws against the dissection of human bodies, and his ideas derived not only from his own observations but also from the ideas of still more ancient scholars, including Hippocrates and Aristotle. Galen thought that the movement of blood derived from the pulsing of the arteries. Galen viewed the heart itself not as a pump that pushed the blood through the vessels of the body, but as a sort of furnace that heated the blood.

Galen thought that blood originated in the liver and flowed to the right side of the heart, where it came in contact with air from the lungs. There the blood exuded some sort of toxic waste, which passed back to the lungs through the pulmonary veins. One could see these fumes, Galen said, as the clouds of breath that are visible on a cold day. The blood was then further refined as it passed through a fine filter between the right and left sides of the heart (Figure 37-2).

In many respects, Galen's understanding was accurate. For example, the blood does release toxic fumes into the lungs: this is carbon dioxide, the waste product of cellular respiration throughout the body. But Galen postulated microscopic holes in the heart that allowed the blood to flow from one side of the heart to the other. These holes do not exist, and 16th-century anatomists looked for them in vain. Most important, neither Galen nor other physicians until the 17th century had any idea that the blood might travel in a circle from heart to arteries to veins and back to the heart.

The presence of blood in the veins and arteries was obvious enough. And their mutual connection to the heart was clear. But no connection between the veins and the arteries was visible at all. Not only did it appear that veins and arteries were not connected, the blood in each system

appeared very different. Venous blood is dark, nearly purple, while arterial blood is bright red.

Why was the connection between arteries and veins invisible? The arteries that supply blood to our muscles and organs start from the heart with the diameter of a small garden hose. But these arteries split into smaller and smaller vessels. The smallest are the capillaries, which are so narrow that the red blood cells must pass through one cell at a time. After the capillaries pass through the tissues of the body, they merge together into larger vessels, which are veins. The tiny veins rejoin into larger and larger vessels, which ultimately return the blood to the heart. Capillaries are entirely invisible without a microscope. Neither Galen nor Harvey could know that the capillaries existed. Nothing appeared to connect the veins with the arteries.

In Galen's view, the blood in these two systems pulsed back and forth through the heart like ocean tides. The continual mixing inside the hot heart refined the blood in the same way a smelter refines metal in a furnace.

It seems amazing now that Galen's view dominated medicine and biology for so long. Yet, not until Serveto and other physicians began studying anatomy more thoughtfully did anyone improve on Galen's model. Harvey's mentor, Fabricius, described the one-way valves in the veins. Serveto and others noted the pulmonary circulation. It only remained for Harvey to recognize the true function of the heart and to discover the existence of the systemic circulation.

Despite the Church's efforts, the 17th century brought even greater enthusiasm for learning from direct experience. Among the factors contributing to this atmosphere were a decline in respect for political and religious authorities and an accelerating pace of technical invention. Among the new inventions were water pumps used by miners and firefighters. Some scholars have suggested that the water pump inspired Harvey to conceive of the heart as a pump.

Harvey recognized early that the heart functioned as some sort of pump, but its movements were so rapid that he initially despaired of ever understanding them. To-

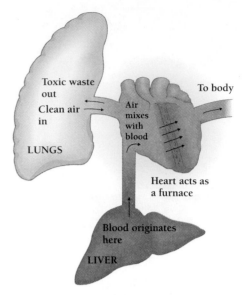

Figure 37-2   Galen's model of the circulatory system. Galen believed that blood originated in the liver and flowed to the heart. In the heart, the blood was refined by air from the lungs. The blood was further refined as it passed through a fine filter between the right and left sides of the heart.

day a scientist who wants to study a rapid process videotapes the process, then watches it in slow motion. Long before slow-motion cinematography, however, Harvey found a convenient way to slow down the motion of the heart. Instead of trying to observe the rapidly beating hearts of "warm blooded" mammals and birds, he began by studying the hearts of toads and snakes. By keeping these animals cool, he could slow down their hearts enough to observe the separate beats of the different parts of the heart.

With this understanding, Harvey performed dissections on living dogs and other animals so that he could see how the heart beat and where the blood went. As the animals died, their hearts beat more slowly, and Harvey could see the movements of each beat of the heart.

The heart of a mammal consists of four chambers—two ventricles and two atria. The atria fill with blood, which flows down into the more muscular ventricles through one-way valves. The valves shut behind the blood, and the ventricles then squeeze the blood forcefully out.

By cutting the tip off the left ventricle, Harvey was able to show that the muscular contraction of the ventricle squeezes

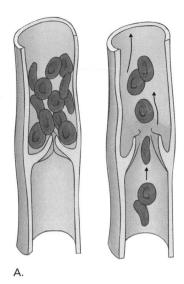

A.

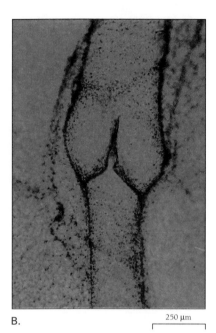

B.

250 μm

Figure 37-3 Veins contain one-way valves. A. Tiny flaps inside of veins prevent blood from flowing back toward the capillaries but allow blood to move toward the heart. B. A closed valve.
(B, John D. Cunningham/Visuals Unlimited)

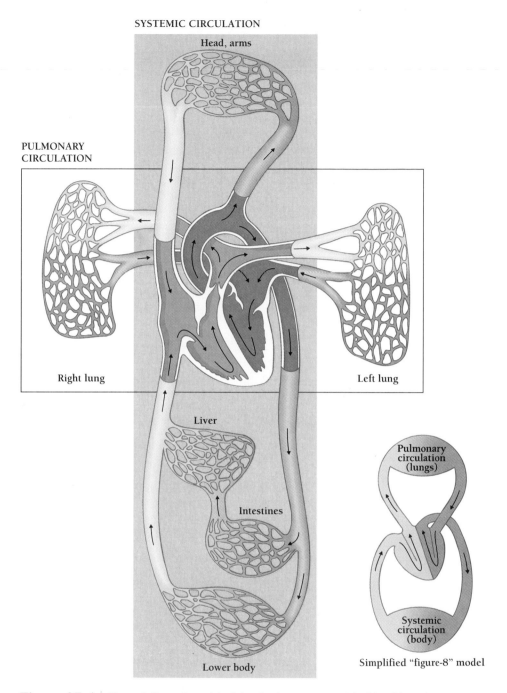

SYSTEMIC CIRCULATION

PULMONARY CIRCULATION

Head, arms

Right lung

Left lung

Liver

Intestines

Lower body

Pulmonary circulation (lungs)

Systemic circulation (body)

Simplified "figure-8" model

Figure 37-4 Harvey's figure-8 model of the circulatory system. The blood from the veins empties into the right side of the heart, which pumps it to the lungs. The blood returns from the lungs and then enters the left side of the heart, which pumps the oxygenated blood to the arteries.

the blood out of the heart. In contrast, earlier European anatomists had thought that the heart worked by sucking blood in. Harvey could see, however, that the heart pushed but did not pull. But if the heart cannot pull, he wondered, What forces the blood from the veins into the heart?

Harvey knew from his anatomical studies that the veins are filled with valves that allow blood to move toward the heart but not away from it (Figure 37-3). Like the valves in the heart itself, these flaps permit the passage of the blood in only one direction. The undeniable one-way flow of blood from the veins to the heart refuted Galen's hypothesis that blood moves back and forth like the tides. Harvey concluded that the veins deliver blood to the heart and the heart pumps the blood into the arteries. He was able to show that the pulse of the arteries was due to shots of blood pushed into them by the heart and was not something that the arteries did automatically.

How the blood got from the arteries back to the veins was a question he could not answer. But that the blood circulated—from heart to arteries to veins and back to heart—he was certain (Figure 37-4).

In addition, he now saw how Serveto's pulmonary circulation explained the complexity of the heart. The blood from the veins emptied into the right side of the heart, which pumped it to the lungs. The blood returning from the lungs entered the left side of the heart to be pumped into the arteries. In essence, the circulation consisted of two loops, one through the lungs and one through the body (Figure 37-4). Both loops joined at the heart, like the two parts of a figure "8."

Harvey bolstered his idea that the blood circulates with one final argument. He measured the amount of blood pumped with each heartbeat and showed that the amount of blood pumped each

hour far exceeds the total amount of blood in the whole body. This demonstrated that the same blood was passing through the heart over and over again. Somehow the blood was being recycled.

Harvey's model was not complete, however, because he could not show the connections between the outgoing arteries and the incoming veins. He simply had no way of seeing the capillaries. As Galen had theorized microscopic pores between the two halves of the heart, Harvey guessed that the arterial blood passed through the tissues and organs and back to the veins. Unlike Galen's theory, however, Harvey's was confirmed. Just 40 years later, the great Italian microscopist Marcello

Malpighi discovered the minute capillaries that complete the circulation.

Although Harvey's success is often attributed to him alone, it is clear that his achievement, like those of all other scientists, resulted partly from the work of those who came before him. His anatomical training in Italy, his exposure to Italian standards of observation, and his knowledge of the pulmonary circulation and the valves in the veins all prepared him to appreciate better than anyone the workings of the heart and circulation. Finally, his position in society and his relative safety in England enabled him to publish his theory without fear.

KEY CONCEPTS

1. The discovery of the circulation depended on both careful dissections and on well-designed experiments.

2. The circulating blood continually renews the fluids of the internal environment.

3. Small organic molecules and ions move through the walls of capillaries, the smallest blood vessels.

4. In mammals, the blood moves through two circles. One route (pulmonary) takes the blood from the heart to the lungs, where it acquires oxygen, and then back to the heart; the other (systemic) moves it from the heart to the tissues of the rest of the body, where it delivers oxygen, and then back to the heart.

5. The heart pumps blood through both circles.

6. The heart's pacemaker coordinates a cycle of muscle contractions in four separate chambers.

7. Nerves, hormones, and local influences determine the rate at which the heart beats and at which the blood flows to individual tissues.

8. Animals other than mammals have different patterns of blood circulation.

HOW DOES THE HEART PUMP BLOOD THROUGH THE BODY?

As Harvey showed, humans and other mammals have a double circulation: blood flows through two adjoining circles similar to a figure "8." In the first circuit, deoxygenated blood from the veins passes from the heart to the lungs, where oxygen attaches to the hemoglobin. The newly oxygenated blood then returns to the heart and passes into the second circuit, which delivers the oxygenated blood from the lungs to all the tissues of the body (Figure 37-5).

The blood, the heart, and the blood vessels together make up the **cardiovascular system.** The route of the blood is called the **circulation** [Latin, *circus* = circle]. The route through the lungs is called the **pulmonary** [Latin, *pulmo* = lung] **circulation,** and the route through the rest of the body is called the **systemic circulation.**

The muscular organ responsible for pushing the blood through the circulatory system is the **heart.** Birds and mammals have a single four-chambered heart, but other animals may have hearts with two or three chambers, and some animals have more than one heart. The earthworm, for example, has 10.

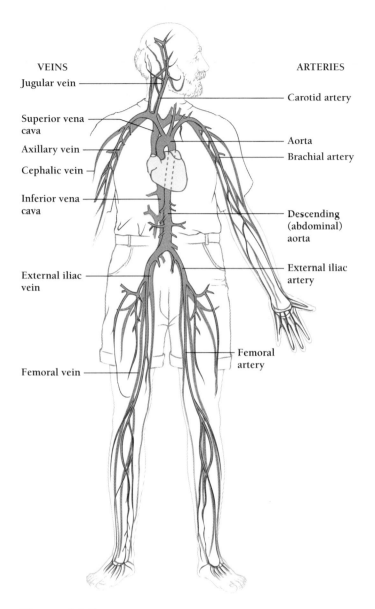

VEINS

Jugular vein

Superior vena cava

Axillary vein

Cephalic vein

Inferior vena cava

External iliac vein

Femoral vein

ARTERIES

Carotid artery

Aorta

Brachial artery

Descending (abdominal) aorta

External iliac artery

Femoral artery

Figure 37-5 The major vessels of the human circulatory system. In addition to the heart, the circulation includes arteries, arterioles, and capillaries, which carry oxygenated blood from the heart, and venules and veins, which carry deoxygenated blood back to the heart.

The Structure of the Heart

In humans, the heart is about the size of a clenched fist and is located in the chest just beneath the breastbone, or **sternum** [Greek, *sternon* = chest]. A fibrous sac, the **pericardium** [Greek, *peri* = around + *kardia* = heart], encloses the heart itself within a watery lubricating fluid. Four interconnected rings of connective tissue provide the frame for the heart's organization. Two rings form the openings between the atria and the ventricles, and two rings form the exit from the ventricles to the circulatory system. The rings are covered by the valves, which determine the direction of blood flow. The muscular

walls of the heart—the **myocardium** [Greek, *myos* = muscle]—attach to this fibrous skeleton to form the four chambers.

Several large arteries and veins attach directly to the heart, and it is sometimes hard to see where these vessels end and the heart itself begins (Figure 37-6). The largest vessel is the **aorta**, the artery that carries blood from the left ventricle to the rest of the body. The aorta arches over the top of the heart. The first branch of the aorta sends blood to the head. The two largest veins, which run up through the center of the body and carry deoxygenated blood from the body to the right atrium, are the superior and inferior **vena cava**. The **pulmonary artery** carries deoxygenated blood from the heart to the lungs, and the **pulmonary veins** carry oxygenated blood from the lungs to the heart.

The mammalian heart consists of two separate halves. The right half pumps oxygen-poor blood to the lungs; the left half pumps oxygen-rich blood to the head and the rest of the body. Each half of the heart contains two muscular sections: (1) a relatively thin-walled entrance chamber called an **atrium** [Latin, courtyard or entry] and (2) a thick-walled pumping chamber called a **ventricle** (Figure 37-6). The mammalian heart, then, consists of four chambers: the right and left ventricles and the right and left atria.

To review, oxygen-loaded blood from the lungs enters the left atrium and pours down into the left ventricle, the most muscular of the four chambers. The left ventricle then pumps the blood through the arteries to the capillaries, which supply and wash all the tissues of the body. The oxygen-depleted blood then returns from the capillaries to the veins, which converge and enter the right atrium. The right atrium pumps the blood into the right ventricle, which pumps the blood into the lungs.

In mammals, the heart pumps blood through two different circuits. One circuit passes through the lungs, while the other passes through all the other tissues of the body.

How Do Birds and Mammals Prevent Oxygen-Rich Blood from Mixing with Oxygen-Poor Blood?

In many animals, the freshly oxygenated blood mixes with deoxygenated blood from the tissues. This means that the tissues never receive blood with the highest possible oxygen content. For animals that move rapidly on a moment's notice, that arrangement isn't good enough. Birds and mammals need to maintain a constant body temperature, highly developed brains, and quick muscular responses, all of which require large supplies of concentrated oxygen.

The four-chambered hearts of mammals and birds prevent the oxygen-poor blood from the tissues from returning to the tissues before it has acquired oxygen in the lungs, always keeping the two kinds of blood separate. The organization of the

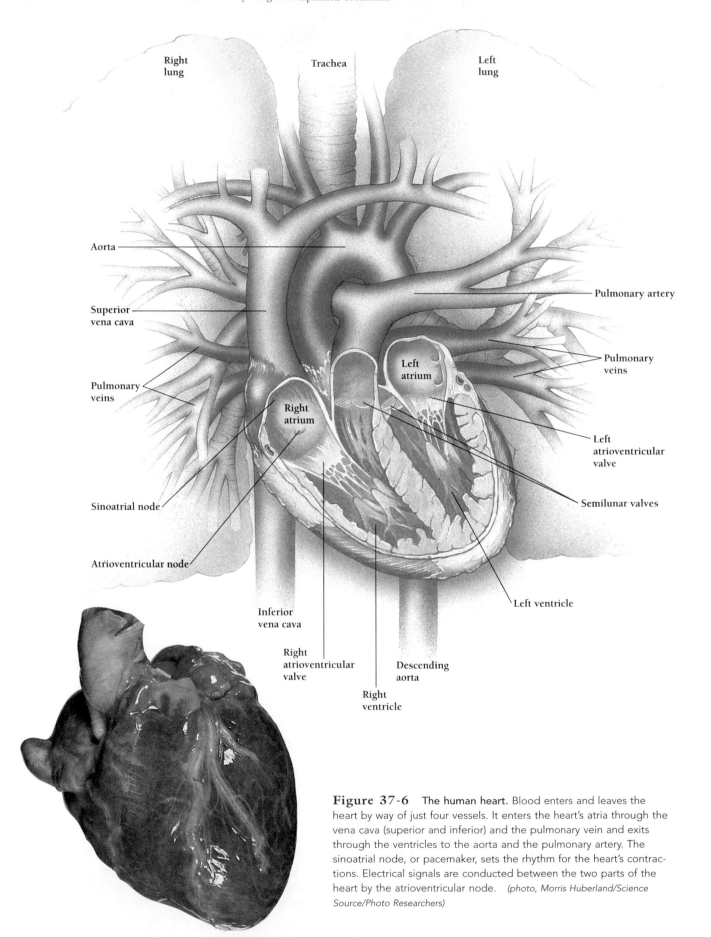

Right lung

Trachea

Left lung

Aorta

Superior vena cava

Pulmonary veins

Right atrium

Sinoatrial node

Atrioventricular node

Inferior vena cava

Right atrioventricular valve

Right ventricle

Descending aorta

Left atrium

Left ventricle

Pulmonary artery

Pulmonary veins

Left atrioventricular valve

Semilunar valves

Figure 37-6 **The human heart.** Blood enters and leaves the heart by way of just four vessels. It enters the heart's atria through the vena cava (superior and inferior) and the pulmonary vein and exits through the ventricles to the aorta and the pulmonary artery. The sinoatrial node, or pacemaker, sets the rhythm for the heart's contractions. Electrical signals are conducted between the two parts of the heart by the atrioventricular node. *(photo, Morris Huberland/Science Source/Photo Researchers)*

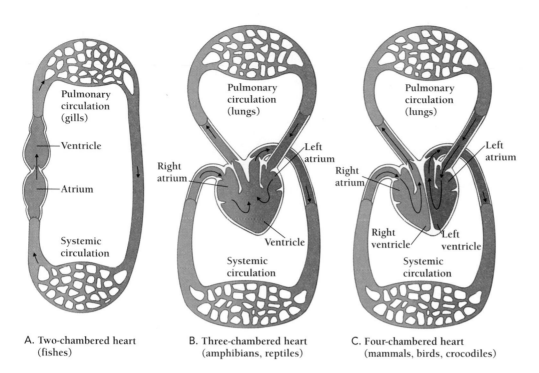

Figure 37-7 **Evolution of the heart.** A. Most fish have a two-chambered heart, which pumps a single stream of unoxygenated blood forward into the arteries that feed the gills. B. Three-chambered hearts, intermediate between those of fishes and mammals, are found in lungfish, amphibians, and reptiles. These hearts are not inefficient versions of the four-chambered heart, but more flexible versions. Blood passing through the turtle's three-chambered heart, for example, can bypass the lungs when the turtle is diving and oxygen supplies are low. C. The four-chambered hearts of mammals and birds pump two streams of blood, one oxygenated and one deoxygenated.

A. **Two-chambered heart** (fishes)

B. **Three-chambered heart** (amphibians, reptiles)

C. **Four-chambered heart** (mammals, birds, crocodiles)

circulatory system in these homeotherms maximizes the ability of the blood to deliver oxygen.

In contrast, the hearts of reptiles and amphibians contain only three chambers—two atria and one ventricle (Figure 37-7). Consequently, some mixing of oxygen-rich and oxygen-poor blood occurs. The ventricle's structure minimizes this mixing, however, so that the system works better than one might at first suspect.

Bony fishes get by with only two chambers, consisting of one atrium and one ventricle. The ventricle pumps blood to the gills, where it obtains oxygen in the gill capillaries. The oxygen-rich blood then flows directly from the gills to the tissue capillaries, where it provides oxygen and other nutrients.

The four-chambered hearts of birds and mammals keep deoxygenated blood from mixing with oxygenated blood. Such an arrangement helps guarantee that oxygen-rich blood flows to all of the tissues of the body.

WHY DO ANIMALS NEED BLOOD?

The "internal environment" of an animal is actually the external environment of its individual cells. The constancy of the internal environment is a necessary condition for life. The temperature and the concentration of individual molecules and ions must be such that they sustain but do not harm the cells of the body. Maintaining this environment requires a constant exchange of materials between the extracellular fluid and an animal's external environment.

Blood is the indispensable intermediary in this exchange. Blood delivers nutrients from the outside world to the extracellular fluid and transports wastes to organs that can dispose of them. Specifically, blood carries oxygen from the lungs and basic building blocks from the gut to the extracellular fluid while simultaneously moving carbon dioxide to the lungs and nitrogenous wastes to the kidneys. The blood also serves several other functions, some of which we discuss in other chapters: blood distributes hormones, it carries the molecules and the cells of the immune system, and it conducts heat to all parts of the body.

What Is the Connection Between Blood and the Extracellular Fluid?

The extracellular fluid contains many substances dissolved in water. Some, such as bicarbonate and phosphate ions, act as *buffers*, which prevent the fluid from becoming too acidic or too basic. These substances establish a stable environment. In addition, blood carries energy-rich molecules, building blocks, and waste products. From the gut to the extracellular fluid glucose, blood ferries amino acids, nucleosides, and other small molecules. From the lungs, the blood brings oxygen.

All the cells of the body are bathed in the extracellular fluid. The cell membranes serve as active gatekeepers that regulate the internal environments of the cells. The membranes usher in some components of the extracellular fluid, denying entry to others. The extracellular fluid would quickly become a cesspool of cell wastes if it were not for the blood, which carries these wastes to the liver for detoxification or to the kidneys for excretion.

Circulating blood maintains the extracellular fluid by
removing wastes and delivering nutrients.

What Cells Constitute the Blood of Vertebrates?

Blood carries all the substances that enter or leave the extracellular space. The fluid part of blood is called **plasma** [Greek, form or mold]. In invertebrates, blood consists only of plasma containing a balance of nutrients, wastes, and stabilizing buffers. Indeed, the chemical composition of plasma closely resembles that of the extracellular space. The major difference is that plasma contains about 7 percent protein, while the extracellular space contains only about 2 percent protein. The plasma of invertebrates contains huge protein molecules that do not leave the blood. The major proteins in invertebrate plasma are oxygen-binding proteins such as hemoglobin or hemocyanin.

In vertebrates, blood contains both plasma and vast numbers of cells. The blood carries three kinds of cells—red cells, or **erythrocytes** [Greek, *erythros* = red + *kytos* = receptacle]; white cells, or **leukocytes** [Greek, *leukos* = white]; and **platelets** [little plates] (Figure 37-8). By far the most numerous cells in the blood are the red cells. Each milliliter of blood contains about 5 billion erythrocytes, and a human adult contains some 25 trillion erythrocytes.

The most distinctive characteristic of blood, its bold red color, comes from the red blood cells themselves. And what makes red blood cells red is the oxygen-binding protein **hemoglobin** [Greek, *haima* = blood + Latin, *globus* = ball]. Each erythrocyte contains about 30 million hemoglobin molecules. The bold red color of hemoglobin comes from the iron-containing heme groups that help this molecule bind oxygen in

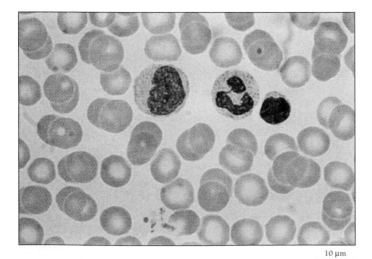

Figure 37-8   **Blood cells.** One milliliter (1 ml) of human blood normally contains about 5 million erythrocytes, as well as numerous leukocytes and platelets.   *(Biophoto Associates/Science Source/Photo Researchers)*

the lungs and then release it to other tissues. In contrast, the oxygen-carrying protein of many invertebrates is a beautiful blue color.

The white cells, or leukocytes, include several types of cells that contribute to the body's defenses against infection and tumors. The platelets are small cellular fragments in the blood that help blood coagulate. Because platelets are essential to prevent blood loss, we will discuss them with other aspects of the body's defense system in Chapter 40.

All these kinds of cells begin as true cells, with nuclei. In mammals, however, both the platelets and the red blood cells lose their nuclei, so that only the leukocytes are true cells. The precursors for all three kinds of cells arise in *stem cells* in the bone marrow. Stem cells are cells that have not differentiated into specific kinds of cells, such as blood cells or skin cells. In this respect, they resemble the cells of an early embryo.

As the erythrocytes mature, they lose their nuclei—in mammals, although not in other vertebrates. Some researchers postulate that the loss of the nucleus allows mammalian erythrocytes to pass more freely through the narrow capillaries of the circulatory system.

While invertebrate blood lacks cells, the blood plasma of
vertebrates contains large numbers of erythrocytes,
leukocytes, and platelets.

SMALL MOLECULES AND IONS MOVE BETWEEN THE BLOOD AND THE EXTRACELLULAR FLUID

In insects and other arthropods, and in most mollusks, the blood travels part of its circuit through open spaces, called **sinuses.** These invertebrate animals are said to have **open circulatory systems.** In an open system, the blood mixes freely with extracellular fluids and bathes the organs of the body. Vertebrates and some invertebrates (including earthworms), however, have **closed circulatory systems,** in which blood runs only within enclosed vessels. How then do materials move to and from the extracellular space? To answer this question, we must look at the structure of the blood vessels.

Arteries, Veins, and Capillaries

In most animals, blood leaves and returns to the heart via different routes, with different kinds of blood vessels. In vertebrates, the vessels that carry blood away from the heart are called **arteries** and those that carry blood back to the heart are called **veins.**

As the arteries leave the heart, they branch many times, and blood flows into smaller vessels called **arterioles.** The arterial blood then enters still finer vessels called **capillaries.** The network of microscopic capillaries is so extensive that no cell of the body is more than three or four cells away from a cap-

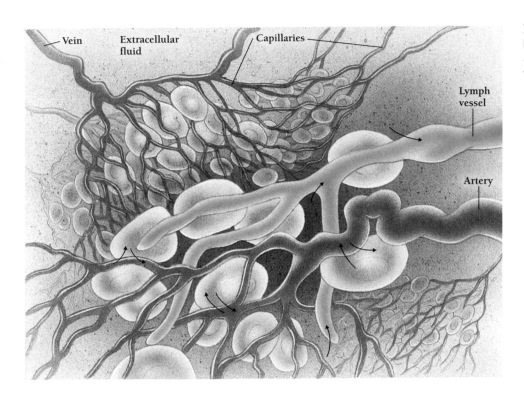

Vein
Extracellular fluid
Capillaries
Lymph vessel
Artery

Figure 37-9 Capillaries. Arterioles branch into capillaries, which are so fine and so numerous that no cell is more than four cells away from a capillary.

illary (Figure 37-9). If all of the blood vessels in your body were placed end to end, they would extend $2\frac{1}{2}$ times around Earth's equator.

The capillaries are the immediate source of the extracellular fluid. Although arteries and veins do not normally leak at all, capillaries leak constantly, seeping nutrients and oxygen into the surrounding space. The thin leaky walls of the capillaries also allow them to absorb cell wastes. The absorbed fluid, together with the suspended blood cells and the dissolved molecules and ions, flows into the smallest veins, called **venules.** The venules, in turn, converge to form larger veins for the final return to the heart.

Arteries carrying blood from the heart branch to form smaller arterioles, which branch to form extensive capillary beds, which converge to form venules, which converge into veins, which pour blood back into the heart.

The Lymphatic System

Not all of the extracellular fluid reenters the circulatory system through the veins. Some fluid reenters through the capillaries of the **lymphatic system** (Figure 37-10). The lymphatic system provides a secondary route for fluids from the extracellular space to the bloodstream. The lymphatic system also carries waste proteins and any large particles that cannot directly enter the capillaries. Without this function, we would die within 24 hours. Mammals, birds, reptiles, amphibians, and many fish all have a lymphatic system.

Like the blood in the circulatory system, the lymphatic tubes contain specialized white blood cells called **lymphocytes.** These immune cells are produced in the bone marrow and the thymus. As we will see in Chapter 40, lymphocytes respond to foreign proteins or kill microorganisms tagged by antibodies.

Extracellular fluid enters the lymphatic system through lymphatic veins and is funneled into two large thoracic ducts, which drain into the circulatory system by way of the veins of the neck. Fluid, dead or foreign cells, and unneeded proteins form the **lymph,** which moves through the lymphatic system passively. The normal movement of the body squeezes the lymph past one-way valves in the lymph vessels.

Lymphatic veins contain **lymph nodes,** regions where filterlike tissue separates cells and other detritus from the lymph. Inside the lymph nodes are phagocytic cells, which engulf material trapped in the lymph nodes. Dense concentrations of immune cells in the lymph nodes also monitor the lymph for signs of infection.

Dead and foreign cells and proteins in the extracellular fluid flow into the lymphatic system. In the lymph nodes, wastes are filtered from the lymph and injected by phagocytic white blood cells.

The Structure of Blood Vessels

The cellular organization of the blood vessels—arteries, capillaries, and veins—reflects their differing roles in the circulatory system. All three types of blood vessels are hollow tubes. Sur-

Figure 37-10   The lymphatic system. The lymphatic system carries the small amount of extracellular fluid that does not make its way back into the vascular capillaries. More important, the lymphatic system carries wastes from the extracellular fluid to the lymph nodes. Scattered throughout the body, the small lymph nodes filter dead cells, foreign cells, and other organic waste from the lymph. Lymphocytes ingest and digest the waste particles and the clean lymph exits the node. One-way valves keep the lymph draining in one direction, and eventually all the clean lymph drains into the "subclavian" veins (at the base of the neck) and reenters the circulatory system.

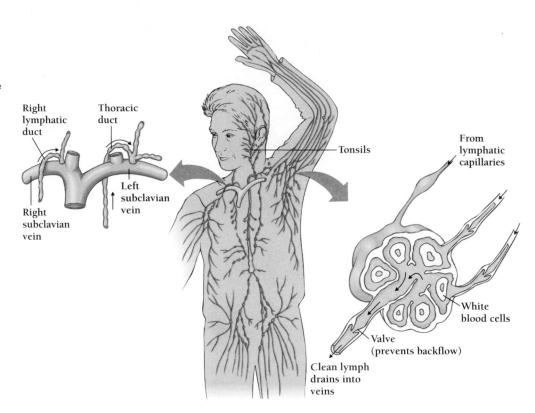

rounding the space inside, called the **lumen,** is a thin layer of cells called the **endothelium** (Figure 37-11).

Unlike the epithelial cells of the skin or gut, which are tough and tightly bound together, the endothelium of the capillaries is only one layer thick. The thin walls of capillaries permit the easy passage of substances from the lumen to the surrounding extracellular space.

The linings of arteries and veins are thick and relatively impermeable. In addition to their endothelial layers, the walls of arteries and veins contain layers of elastic fibers, collagen, and smooth muscles. Veins generally have thinner and less muscular walls than do arteries. The elastic walls of arteries and veins allow these vessels to narrow or expand—in response to changes in blood pressure or in response to signals from the autonomic nervous system to the smooth muscles.

But the vessels—especially the arteries in which blood pressure is highest—must not be too elastic. If they stretch too much, they can balloon out and flatten the capillaries in the surrounding tissues. To prevent such ballooning, collagen fibers like those found in connective tissue and bone strengthen and stiffen the vessel walls. If the collagen sheath of an artery fails and ballooning occurs, the result is an **aneurism.** An aneurism in the brain can cause tremendous damage. Sometimes aneurisms rupture a major artery, and death soon follows.

The endothelium of a capillary is only one cell thick. The walls of larger vessels are thicker and surrounded by elastic fibers, collagen, and smooth muscle.

How Does the Blood Acquire Oxygen?

Oxygen is one substance delivered by the blood that the body needs continuously. In the United States the single most common cause of death is the failure of the circulatory system to deliver oxygen to either the brain or the heart. A failure of the blood supply to the heart is called a **heart attack,** and a failure of the blood supply to the brain is called a **stroke.** A failure can result from damage to the blood vessels either by blockage or breakage.

In air-breathing vertebrates such as ourselves, oxygen enters the body and then the blood by way of the lungs. Blood from the heart flows into the lungs through the pulmonary arteries. As in other tissues, the arteries in the lungs divide into arterioles and then into a large network of capillaries.

The gases in the lungs, however, move in the opposite direction to those in the rest of the body (Figure 37-12). In most tissues, oxygen moves *out of* the capillaries to the body's tissues. In the lungs, however, oxygen moves from the lung tissue *into* the capillaries and binds to the hemoglobin molecules in the red cells. The blood in the lungs also releases carbon dioxide—the waste from cellular respiration—that has accumulated during the blood's passage through the rest of the body.

The refreshed blood—low in carbon dioxide and high in oxygen—then returns to the heart by way of the pulmonary veins. The pulmonary veins are the only veins in the body that carry red, oxygen-rich blood. From the heart, the oxygen-rich blood moves through the other loop of the circulatory system, releasing oxygen to respiring cells.

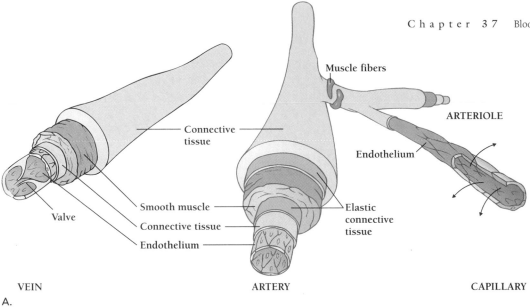

Muscle fibers

Connective tissue

Endothelium

ARTERIOLE

Elastic connective tissue

Smooth muscle

Connective tissue

Endothelium

Valve

VEIN

ARTERY

CAPILLARY

A.

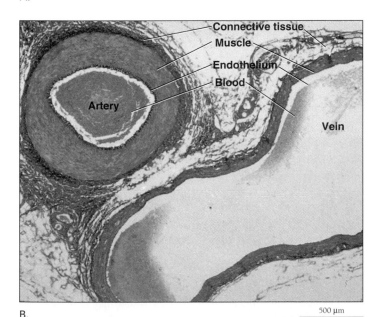

Connective tissue

Muscle

Endothelium

Blood

Artery

Vein

500 μm

B.

Figure 37-11   **Blood vessel layers.** A. Veins, arteries, and capillaries all include a thin layer of endothelial cells. Veins and arteries are further surrounded by connective tissue and smooth muscle, which contracts and relaxes to change the diameter of the vessel. Inside the lumen of a vein are one-way valves that force the blood back toward the heart. B. The walls of arteries are especially thick, the better to withstand the high blood pressures generated by the heart. C. Capillaries consist of a single layer of endothelial cells, which allows fluid and nutrients to pass into the extracellular space. Capillaries form dense networks that infiltrate all the tissues of the body. (*B, C, Biophoto Associates*)

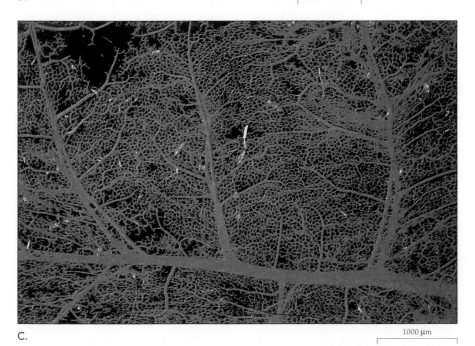

1000 μm

C.

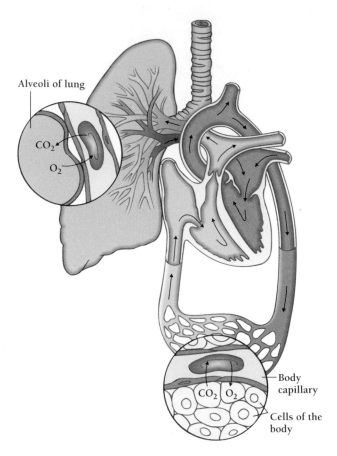

Figure 37-12   Oxygen movement. In the lungs, oxygen moves from the tissues of the lung into the capillaries. Elsewhere, oxygen moves from the capillaries into the tissues of the body.

Oxygen in the lungs moves into the blood in the capillaries, where it is transported to the heart in the pulmonary veins and then from the heart to the rest of the body.

WHAT MAKES THE HEART BEAT?

A healthy heart strictly coordinates the contractions of the atria and ventricles. In a resting human adult, the heart beats about 70 times per minute. Each beat consists of a cycle of contractions by the different chambers of the heart: the two ventricles contract simultaneously, as do the two atria.

Systole and Diastole

During half of each cycle, both the atria and the ventricles are relaxed. This period is called **diastole** [Greek, *dia* = between + *systellein* = to contract]. During the other half of the cycle, the atria contract, followed by the ventricles. The period of con-

traction lasts about 0.3 seconds and is called **systole** [Greek, *systellein* = to contract]. During systole, the blood pressure momentarily increases. This is why blood pressure is always expressed as two numbers. A normal reading for a person at rest is 120/80. The higher number is the systolic blood pressure and the lower number is the diastolic.

Within the heart, a system of valves ensures that the blood flows in one direction only. The valves between the atria and ventricles are called **atrioventricular valves;** those between the ventricles and the arteries are called **semilunar** [Latin, half-moon] **valves.** Each valve consists of flaps of connective tissue, which allow fluid to move in one direction but not the other. Blood from the ventricles cannot reenter the atria, and blood from the arteries cannot reenter the ventricles (Figure 37-6). Smaller valves in the veins ensure that blood flows only in a single direction through the circulatory system.

You can hear the heart working if you listen to your own heart with a stethoscope or to a friend's heart by putting your ear on his or her chest. You will hear that each heartbeat consists of two separate sounds—a low-pitched "lub" and a higher "dub." The first sound is that of the two ventricles contracting and the atrioventricular valves closing (Figure 37-13). The second is the closing of the semilunar valves behind the blood that the ventricles have expelled.

A trained person can detect some heart abnormalities just by listening. A defective valve, for example, can produce a "heart murmur," a low hissing sound in addition to the lub-dub. This sound comes from the movement of blood in the wrong direction, for example, from a ventricle back into the atrium. Such a defect may result from abnormal development of the valve or from an infection such as rheumatic fever.

The period when the muscles of the heart contract is called systole and the period in between, when they relax, is called diastole. One-way valves allow blood to flow from the atria to the ventricles but not back, and from the ventricles to the arteries but not back.

What Coordinates the Beating of the Heart Muscle?

The action of the ventricles in pumping blood to the lungs and the rest of the body is essential to life. To pump blood, the muscle fibers of the ventricles must all work together—contracting together, relaxing together, and pausing together. Occasionally, the muscles of the heart begin working against one another, so that many of the fibers are contracting when they should be relaxed. The result is **fibrillation,** continuous disorganized contractions.

Ventricular fibrillation, also known as cardiac arrest, accounts for one-quarter of all deaths in the United States. Fibrillation can affect the atria or the ventricles. If the atria fibril-

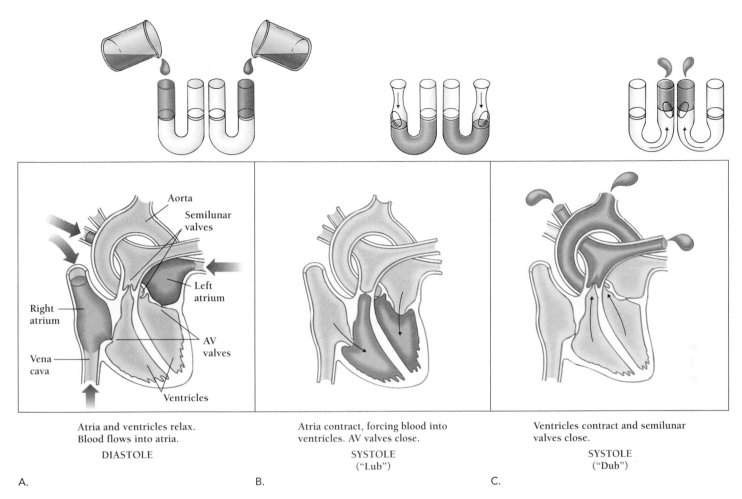

A.   B.   C.

| Atria and ventricles relax. Blood flows into atria. | Atria contract, forcing blood into ventricles. AV valves close. | Ventricles contract and semilunar valves close. |
| DIASTOLE | SYSTOLE ("Lub") | SYSTOLE ("Dub") |

Figure 37-13 **What's that sound?** The heart beats in two parts. A. During diastole, the atria and ventricles relax and blood fills the atria. The semilunar canals are still closed from the last beat. B. *Lub*. The atria contract, forcing the blood into the ventricles, and the atrioventricular valves close behind the blood, making a low-pitched "lub." C. *Dub*. During systole, the ventricles contract, forcing the blood into the aorta and the pulmonary artery. The semilunar valves close noisily behind the blood.

late, blood continues to flow into the ventricles, so that people with otherwise healthy hearts can live with fibrillating atria, and, in fact, it may go unnoticed. When the ventricles begin to fibrillate, however, death follows immediately and almost inevitably. During ventricular fibrillation, the different parts of the ventricles no longer contract simultaneously. Within a few moments they cannot pump any blood at all. Blood continues to fill the ventricles from the atria, and the ventricles become distended with blood. Because the ventricles are not pumping blood, however, their own blood supply, which comes from the coronary arteries, is cut off. Within 60 to 90 seconds, the ventricles become so weak from lack of oxygen that they cannot contract at all. A strong electric shock can sometimes stop the fibrillation and give the ventricles a chance to reestablish normal organized contractions.

The cells of the muscular myocardium perform the heart's work. Like skeletal muscles, heart muscles are striated, and contraction depends on sliding filaments of actin and myosin. The organization of heart muscle differs, however, from that of skeletal muscle. Whereas skeletal muscle fibers consist of single cells with multiple nuclei, heart muscle fibers consist of many cells, each with its own nucleus. The cell membranes separating these cells from one another allow extraordinarily rapid passage of ions and molecules, so that the cells behave nearly as one cell.

All the cells of the atria are interconnected in this way, as are all of the cells of the ventricles. If one muscle fiber in the ventricles is stimulated to contract, all of the other cells in the ventricles also contract, in a wave starting from the first fiber. The same is true of the atrial cells. But the atrial cells are physically separated from the ventricle cells by fibrous tissue, so stimulating the atrial cells to contract does not necessarily cause the ventricles to contract. The two groups of cells are, however, connected by a special tissue called the **atrioventricular (AV) node,** which conducts electrical signals from one part of the heart to the other.

As in the case of skeletal muscle, the immediate stimulus for contraction in cardiac muscle is an increase in calcium ion concentration. This increase depends on electrical changes across the cell membrane. In skeletal muscle, these electrical changes depend on stimulation by nerves. In cardiac muscle, however, the cells contract spontaneously.

About 1 percent of the cells of the myocardium are capable of rhythmic spontaneous contractions. One group of these cells forms a structure called the **pacemaker,** which contracts slightly more frequently than most myocardial cells. By initiating contraction before other cells in the heart, the pacemaker cells set the pace of contraction for the rest of the heart (Figure 37-6).

In the mammalian heart, the pacemaker lies near the top of the right atrium and is called the **sinoatrial (SA) node.** Electrical signals travel from the SA node to the AV node, which acts as a relay station and triggers contraction of the ventricles. The pacemaker thus determines the rhythm of the heart and coordinates the action of atria and ventricles.

The cells of the atria contract almost as one, as do the cells of the ventricles. Electrical signals for such contractions are conducted between the two parts of the heart by the atrioventricular (AV) node. A pacemaker, called the sinoatrial (SA) node, sets the rhythm for the heart's contractions.

Blood Pressure Depends on the Action of the Heart and the Properties of Blood Vessels

As William Harvey deduced, the pumping of the heart generates pressure that drives blood through the circulatory system. This pressure in the arteries changes from moment to moment as the heart contracts and relaxes. We can feel the fluctuations of pressure in the larger arteries, such as in the one that crosses the wrist, where we can feel our pulse. Each beat of the pulse corresponds to an expulsion of blood from the heart (systole). Medical workers usually measure blood pressure in the arteries of the upper arm.

The pressure within the vessels decreases as the arteries divide into finer and finer tubes. In the capillaries the pressure barely varies, and capillary blood flows smoothly. The capillary blood is still under pressure, however. This pressure drives some of the plasma from the capillaries into the surrounding extracellular space. Blood in the veins is under still less pressure and fluctuates hardly at all (Figure 37-14).

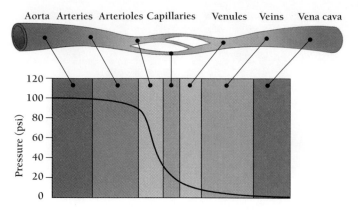

Figure 37-14   Pressure in the blood vessels. Blood pressure is highest in the aorta as the blood leaves the ventricles. In the arterioles and capillaries, the mean pressure drops from 100 to about 20 psi. Finally, as the blood flows into the veins, it is under virtually no pressure at all.

NERVES, GLANDS, AND OTHER TISSUES REGULATE THE FLOW OF BLOOD

The amount of blood pumped by the heart (the cardiac output) depends on two factors: (1) the **heart rate,** the number of beats per minute, and (2) the **stroke volume,** the volume of blood delivered by each ventricle. Because the blood travels in a circle, the volume that passes through the left ventricle must equal the volume that passes through the right ventricle. In a resting adult, the heart beats about 70 times a minute with a stroke volume of about 70 ml. So the volume of blood pumped each minute is about 70 × 70 ml = 4900 ml, or about 5 liters. This is approximately the total volume of blood. Each of us, while sitting still, pumps nearly all of our blood through the circulatory system once a minute.

When we exercise, both heart rate and stroke volume increase, and the amount of blood pumped by the heart may go as high as 35 liters (about 9 gallons) per minute. Our physical and emotional state affect both the heart rate and the stroke volume. Running upstairs will dramatically increase both heart rate and stroke volume. But subtler influences, such as a disturbing thought, can also influence the heart. Both nerves and hormones regulate the beating of the heart.

The heart varies the amount of blood it pumps depending on our activities and moods. The amount of blood pumped is a function of how fast the heart beats and how much blood it moves with each beat.

Regulation by Nerves and Hormones

Two systems of nerves influence heart rate—the **sympathetic** [Greek, *sym* = together + *pathos* = suffering] **nervous system**

SELECTED READINGS

Cantin, M., and Genest, J., "The Heart as an Endocrine Gland," *Scientific American,* February, 1986.

Eckert, R., Randall, D., and Augustine, G., *Animal Physiology: Mechanisms and Adaptations,* W.H. Freeman, New York, 3rd edition, 1988, Chapter 13.

Hochachka, P.W., "Brain, Lung, and Heart Functions During Diving and Recovery," *Science* 212, 509–514, 1981.

Jarvik, R.K., "The Total Artificial Heart," *Scientific American,* January, 1981.

Labarbera, M., and Vogel, S., "The Design of Fluid Transport Systems in Organisms," *American Scientist* 70, 54–60, 1982.

Schmidt-Nielsen, K., *Animal Physiology: Adaptation and Environment,* Cambridge University Press, New York, 3rd edition, 1983, Chapter 4.

Vander, A.J., Sherman, J.H., and Luciano, D.S., *Human Physiology: The Mechanisms of Body Function,* McGraw-Hill, New York, 4th edition, 1985, Chapter 11.

▶ On-line materials relating to this chapter are on the World Wide Web at http://www.saunderscollege.com/lifesci/
Click on Tobin/Dusheck: *Asking About Life.*

Stanton Glantz and the Tobacco Industry

In 1995, a Congressional committee astonished medical researchers by singling out a researcher from among thousands and canceling his research grant. For Congress to assail a particular scientist in this way was almost unheard of.

The federal government normally provides funds to agencies such as the National Institutes of Health, the National Science Foundation, and the National Cancer Institute (NCI) to distribute at the discretion of each agency. The NCI had awarded Stanton Glantz, a professor of medicine at the University of California, San Francisco, a three-year grant to study the effects of public policy on tobacco use.

Glantz had used the grant to study several aspects of tobacco and public policy. In particular, he had focused on how the tobacco industry works to keep Americans smoking. He then published a series of scientific papers in *JAMA* (the *Journal of the American Medical Association*) describing how the tobacco industry had for 30 years carefully concealed its knowledge that tobacco products are both deadly and addictive. Glantz's research also showed, for example, that elected officials who receive campaign contributions from tobacco companies are more likely to vote in ways that help the tobacco industry.

The government's House Appropriations Committee argued that because Glantz's work involved social science and political science the NCI could not continue to fund his work. But as Glantz put it, "You need to understand the vector of a disease in order to control it. The tobacco industry is the vector for lung cancer and heart disease." For Glantz, the cancellation of his grant was only the latest in a series of conflicts he had encountered in his attacks on the tobacco industry. Much of the turmoil had begun on May 12, 1994, when an unsolicited box of documents arrived at his office. These documents proved to be copies of internal memos and research reports from the tobacco company Brown and Williamson. Originally stolen from the company by a paralegal appalled by what he was reading, the 4000 pages of documents showed that Brown and Williamson and their legal advisors had understood the hazards of smoking since the 1960s.

Brown and Williamson immediately went to court for the return of the documents, demanded the names of any people who had read the papers, and, allegedly, or-

Vincent van Gogh, *Skull with cigarette*, 1885/Art Resource, New York

Figure 38-1 Actress Jean Harlow in 1932—wearing green and smoking Lucky Strikes cigarettes. When this advertisement and others like it failed to attract female smokers, American Tobacco hired promoter Edward Bernays to convince American women that green was the fashion color (and Lucky Strikes, the fashion cigarette). *(The Granger Collection, New York)*

Glantz's scientific papers analyzing the contents of the documents appeared in *JAMA*. Within weeks, the House Appropriations Committee canceled Glantz's grant.

Why Was the Tobacco Industry Fighting So Hard?

Until the late 19th century, tobacco smoking was viewed as a dirty habit, and only a few people smoked. Women did not, as a rule, smoke at all. Beginning in the 1880s, however, tobacco use began to increase. Tobacco manufacturers had discovered a way to process tobacco so that it could be inhaled, increasing smokers' exposure to nicotine. Addiction increased—especially during World War I, when soldiers were given cigarettes to calm their nerves and to pass the time during the long waits between battles. Not until the 1920s did women, emboldened by sophisticated advertising, begin smoking in significant numbers.

As early as the 1880s, one physician had already published a report suggesting that smoking caused cancer. Physicians

that wouldn't "irritate" the throat. Other ads represented Lucky Strikes as symbolizing freedom for women. In 1929, beautiful models marched in New York's Easter Parade, each brandishing a Lucky Strike cigarette—a "torch of liberty."

In the 1930s, the makers of Lucky Strikes found to their dismay that women were passing up the company's cigarettes. The company suspected that the problem might be the cigarette's new green packaging, which clashed with most women's clothes (Figure 38-1). Green just wasn't popular. What to do?

The American Tobacco Company turned to Edward Bernays, the world's most famous public relations man. Bernays's solution was to make green the most popular color in the United States. He held a "Green Ball" for well-connected women, and he enticed European fashion designers to create a collection of fashions, all in green. He held a Green Fashions Luncheon for fashion editors from major newspapers and magazines. He invented a Color Fashion Bureau that sent out press releases announcing that green was the

---

"If it were true that the [tobacco] companies steer clear of children, [as they say], the entire industry would collapse within a single generation."

---

dered a stakeout of the university's library, where the documents were stored.

But the cat was out of the bag. Similar documents had already arrived at several major newspapers and television networks. Although Brown and Williamson succeeded in forcing the library to keep the papers locked up for months, a judge eventually ruled that the papers were in the public domain.

On July 1, 1995, the University of California at San Francisco posted the Brown and Williamson documents on the Internet and they indeed became public documents. In the same month, several of

couldn't help noticing that nearly all of their patients with cancer of the mouth, throat, or lungs were smokers. But tobacco companies marketed cigarettes aggressively, and the government made no effort to intervene.

In the 1920s, the American Tobacco Company marketed Lucky Strikes cigarettes to women as a "healthy" cigarette

color of the season. In only six months, Bernays succeeded in making green the top fashion color of 1934. Women no longer hesitated to buy Lucky Strikes because of the color, and sales increased.

During World War II, cigarettes were again distributed to GIs along with K rations and letters from home. Smoking wasn't just accepted, it was encouraged

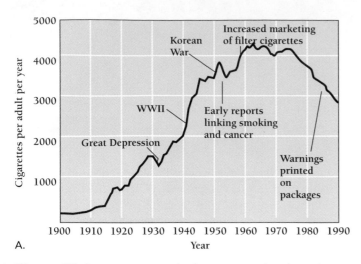
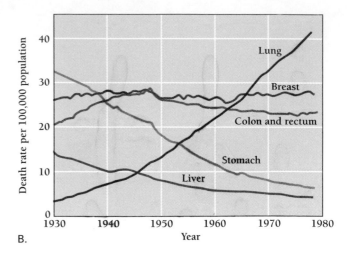

Figure 38-2   As cigarette sales have increased, so have the rates of lung cancer. Lung cancer among women climbed dramatically about 25 years after the increase in female smokers.

and expected. Hollywood movies showed virile actors and alluring actresses lighting each others' cigarettes. Smoking was glamorous, seductive, romantic.

Cigarette sales in the United States skyrocketed from 2 billion cigarettes per year in the 1880s to over 300 billion in the 1950s. Ten years later, sales had doubled again to nearly 600 billion per year (Figure 38-2).

As cigarette sales skyrocketed so, eventually, did lung cancer rates. Men had first begun getting lung cancer in large number in the 1930s, 25 years after the first major increase in cigarette sales. Women followed suit in the 1960s and 1970s, also 25 years after they began smoking in great numbers (Figure 38-2).

While deaths from nearly every other form of cancer declined or remained steady, the death rate from lung cancer soared. In 1930 the rate had been about three cases for every 100,000 people. By 1960, the lung cancer death rate was ten times that number.

In 1964, the Surgeon General of the United States issued a report stating that cigarette smoking was hazardous to health. For the first time in 75 years, cigarette sales leveled off. Two years later Congress passed a bill requiring cigarette manufacturers to label their product as hazardous. For the first time, cigarette sales actually dropped.

Tobacco companies were in disarray. The Brown and Williamson documents show that the company's own research

confirmed everything the government said. Indeed, tobacco industry research showing that cigarettes were both addictive and carcinogenic was better than the academic research on which the government had based its conclusions. Desperately, the tobacco industry began trying to develop the idea of a "safe" cigarette. In the meantime, they did what they could to keep people smoking. An entire industry, one that was central to the economies of a half-dozen states, was in serious danger.

Publicly, tobacco companies insisted that the scientific evidence was weak. They blitzed television viewers with advertisements showing healthy young people smoking cigarettes against scenic backdrops—green meadows, waterfalls, and soaring mountains. The Federal Communications Commission (FCC) countered with a new regulation stipulating that for every four cigarette commercials put on the air, broadcasters would have to air one public service announcement warning of tobacco's hazards. Soon television viewers were besieged with shocking ads from the American Cancer Society. Cigarette sales dropped precipitously. One-third of all smokers said the ads made them quit or seriously consider quitting.

The FCC ruling looked like a disaster for the tobacco industry. Beginning in 1970, cigarette companies confined their advertising to newspapers, magazines, billboards, and sporting events, and anti-smoking ads also disappeared from tele-

vision. The percentage of people smoking in the United States continued to decline, although not as rapidly as during the anti-smoking advertisements, from about 42 percent in 1965 to 32 percent in 1979.

In 1988, the Surgeon General again issued a report, this time stating that smoking was not only hazardous but also highly addictive. The nicotine in cigarettes, the report said, caused dependency in the same way that heroin and cocaine do. The tobacco industry vigorously attacked the report, arguing that the scientific evidence for such a conclusion was scanty or nonexistent. In fact, the Brown and Williamson documents show that the company had fully understood the addictive nature of tobacco for at least 25 years.

The "safe" cigarette the industry had sought never materialized. Cigarettes are inherently unsafe. Filters, low tar, and low nicotine cigarettes only force smokers to inhale more deeply or to smoke more cigarettes to get the same amount of nicotine. The tobacco industry's main concern over the years has been to insist that the evidence is not good. Yet, the three tobacco companies that own insurance companies charge smokers double the usual rate for life insurance. Indeed, the tobacco industry has spent millions of dollars on outside research in order to demonstrate that the issue is not yet settled and that "more research is needed." To date, some 50,000 scientific papers have been published documenting the health effects of tobacco smoking.

KEY CONCEPTS

1. Atmospheric oxygen must dissolve in water before it can participate in biochemical reactions.

2. Increased surface area speeds the exchange of molecules between gas and liquid.

3. Many aquatic animals use gills to obtain dissolved oxygen directly from the surrounding water. Most land animals use lungs to obtain oxygen from the air.

4. In mammals, the diaphragm and the muscles of the rib cage pull air into the airways that lead to oxygen-absorbing surfaces.

5. The molecular structure of hemoglobin allows it to take up oxygen in the lungs and release it to other tissues.

6. The respiratory system also allows animals to dispose of CO_2.

WHAT ARE THE HAZARDS OF SMOKING?

Smoking kills nearly half a million people in the United States every year, far more than the 40,000 deaths from AIDS and the 20,000 deaths from all illegal drugs. Each year, more people die from smoking than die from alcoholism, accidents (including all car accidents), AIDS, suicide, homicide, and all illegal drugs. Smoking is the leading preventable cause of early death in the Unites States. On average, smokers die seven years sooner than nonsmokers. One-third of smokers cut a full 21 years off their lives.

Tobacco kills in a variety of ways. It increases a person's risk of developing cancer of the mouth, throat, larynx, esophagus, cervix, bladder, kidneys, and pancreas. Tobacco smoke also causes *emphysema,* a disease of the lungs in which people gradually suffocate to death. But most deaths from smoking come in the form of cardiovascular disease or lung cancer. Of the million deaths from cardiovascular disease each year, one-third (about 330,000) are caused by smoking. Of the 146,000 deaths from lung cancer, 90 percent (or 130,000) are caused by smoking.

Although tobacco industry advertisements frequently suggest that the connection between tobacco smoke and premature death from cancer and heart disease is controversial, the connection is well established. Regularly smoking tobacco is much like playing Russian roulette with the last 20 years of one's life.

Do Smokers Hurt Society?

In an effort to garner support from nonsmokers, antismoking campaigns sometimes emphasize the effects of smoking on non-smokers. Indeed, an hour in a room filled with cigarette smoke exposes a nonsmoker to the same amount of toxins that a smoker inhales from one cigarette. Approximately 3000 non-smokers' deaths from lung cancer and several thousand more

from cardiovascular disease can be attributed to exposure to second-hand smoke at work or in the home.

Antismoking activists also point to the fact that smokers are sick more often than nonsmokers. Smokers spend 40 percent more work days at home sick than nonsmokers. Their absences from work and their excess medical care cost the United States some $65 billion a year.

But smokers are at far greater risk to themselves than to others. Their risk of a fatal illness is 30 to 40 times greater than that of nonsmokers. And although smokers cost society about 33 cents per pack of cigarettes they smoke, that figure is more than offset by the 52 cents per pack they pay in federal and state taxes. In short, it could be argued that nonsmokers reap nearly 20 cents a pack on smokers' self-destructive habit.

The main reason smokers are not a drain on society's coffers is that smokers—on average—die young and never collect social security benefits and Medicare benefits. They pay into the system when they are working, but it is the nonsmokers who collect at the end. The early deaths of the smokers save American taxpayers billions of dollars.

When Do People Begin Smoking?

Ninety percent of people who begin smoking after age 21 quit soon after. Lifelong smokers nearly always begin between the ages of 12 and 17. Indeed, 75 percent of smokers begin before the age of 18, often well before.

It is illegal, however, to sell tobacco products to anyone younger than 18, so tobacco companies are careful to state that they do not market their products to children and adolescents. Nonetheless, one study found that 90 percent of six-year-olds knew that Joe Camel represents cigarettes.

More important, if the tobacco companies did not market to those under 18, the market for their product would disappear. Who would take up smoking? Certainly not adults. In a book on the tobacco industry, *New York Times* reporter Philip Hilts wrote, "If it were true that the [tobacco] companies steer clear of children, as they say, the entire industry would collapse within a single generation." Adults rarely take up smoking. Children and teenagers do.

Why Do Smokers Continue To Smoke?

Young people who take up smoking find it difficult to quit because an important component of tobacco smoke is highly addictive. *Nicotine* is a narcotic similar to heroin and cocaine in both its effects and its addictiveness. Ninety-five percent of smokers are physically addicted, meaning that they experience physical *withdrawal* if they try to stop smoking. People who have been addicted to both heroin and cigarettes say that cigarettes are harder to give up. Unfortunately, cigarettes are also more deadly than heroin.

Some researchers have argued that a cigarette is nothing but a system for delivering nicotine to the body. It is an excellent delivery system: each puff on a cigarette delivers between 0.5 and 2.0 milligrams of nicotine to the bloodstream. Because the nicotine is inhaled, it reaches the brain within 6 to 7 seconds, twice as fast as injected heroin.

How Does Tobacco Smoke Damage the Cardiovascular System?

The three most dangerous substances in tobacco smoke are nicotine, carbon monoxide, and tar. Tar, a brown, oily substance, ruins the lungs, while nicotine and carbon monoxide work together to damage the heart and arteries.

Nicotine is a stimulant with a broad range of effects. By mimicking the neurotransmitter acetylcholine, it increases the heart rate and constricts the blood vessels, causing the blood pressure to increase. Nicotine also stimulates the release of antidiuretic hormone (vasopressin), a hormone that causes the kidneys to retain water, which can further raise blood pressure. In addition, nicotine increases the tendency of the blood platelets to form clots. As we saw in the last chapter, high blood pressure and abnormal clots are two of the factors that can trigger a heart attack or stroke.

Nicotine contributes to hardening of the arteries, or *atherosclerosis,* a major cause of heart attacks and strokes. By increasing blood pressure, nicotine causes damage to the arteries, which then become inflamed and scarred. Finally, nicotine increases levels of cholesterol in the blood, which, in response to the inflammation, forms deposits on the arterial walls (Figure 38-3). These deposits further narrow the arteries and increase the likelihood that a blood clot will plug an artery and cause a heart attack or stroke.

Smokers say nicotine makes them feel good. It relaxes the muscles and increases *alpha waves,* brain waves that characterize a pleasant, relaxed state. If you want to know what it feels like to increase alpha waves in the brain, sit down in front of the television for a few minutes.

While nicotine maintains the cigarette habit and raises blood pressure, carbon monoxide interferes with the body's

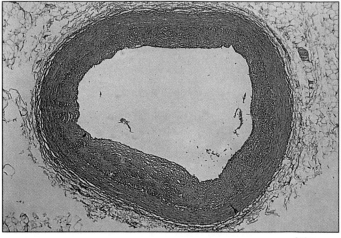

A. 250 μm

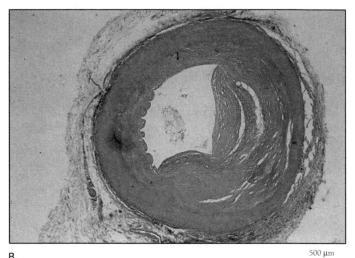

B. 500 μm

Figure 38-3 Normal and atherosclerotic blood vessels. A. A healthy, open artery. B. An artery scarred, hardened, and blocked by atherosclerotic deposits. *(A, Cabisco/Visuals Unlimited; B, Ober/ Visuals Unlimited)*

ability to supply oxygen to the cells of the body. Carbon monoxide is an odorless and toxic gas. Cigarettes are not the only source of carbon monoxide. Exhaust from automobiles and furnaces, for example, also release carbon monoxide. Government regulations outlaw levels higher than 10 parts per million (ppm) in homes or at work. Yet, cigarette smoke contains about 1600 ppm carbon monoxide.

In the blood, carbon monoxide binds to hemoglobin and displaces oxygen. Inhaling carbon monoxide reduces oxygen supplies to every cell in the body. The heart and the brain, however, are affected most. Oxygen deprivation in the brain impairs judgment, vision, and the ability to distinguish sounds. Oxygen deprivation in the heart reduces its ability to pump blood. But the high blood pressure induced by nicotine means that the heart needs to work harder than usual. The result can be muscle strain and long-term damage to the heart.

How Does Tobacco Smoke Damage the Lungs?

Tar is similar to the coating on the inside of a chimney. It consists of thousands of different substances. The tar in smoke interferes with lung function immediately. Because lungs need to be clean in order to collect oxygen most effectively, the lungs have three methods for self-cleaning. First, the lungs secrete slimy **mucus,** in which foreign particles become embedded. Second, tiny hairlike **cilia** sweep the mucus and particles upward into the throat, where the debris can be coughed out or swallowed. Finally, macrophages, large white blood cells, remove small particles, bacteria, and viruses from the inner surface of the lungs.

Within minutes of the first puff, however, tobacco smoke paralyzes both the macrophages and the cilia, so that they cannot clear the lungs. One cigarette paralyzes the cilia for an hour. Further smoking kills them outright. At the same time, the irritating smoke causes mucus to accumulate in the lungs, clogging the tiny passages through which air normally enters. The tar in the smoke settles throughout the lungs, further irritating the lungs and impairing the lungs' ability to absorb oxygen.

The only way to clear the lungs is for the smoker to cough constantly, the classic "smoker's cough." The lungs, clogged with mucus and tar and unable to rid themselves of bacteria and viruses, become infected and inflamed. Smokers have frequent respiratory infections, especially bronchitis and even pneumonia. These chronic infections, combined with continual coughing, further damage the lungs.

In **emphysema,** the progressive and fatal destruction of the lungs, the lungs break down and gradually become useless. The tiniest pockets of the lungs, the **alveoli,** provide the huge surface area on which the blood exchanges oxygen and carbon dioxide with the inhaled air. Without the alveoli, the lungs would be like two balloons, with a surface area of about one square foot. The alveoli, like the air pockets in a sponge, increase the inside surface area of the lungs 1300 times to that of a large house.

The constant coughing associated with chronic bronchitis, however, actually breaks the walls of the individual alveoli. Nitric acid and sulfuric acid produced by burning tobacco further weaken the walls of the alveoli, so that they break even more easily. As the walls of the alveoli break, the many small, bubblelike chambers become a few larger chambers, and the lung gradually loses its enormous surface area. In the end, the lung can no longer absorb enough oxygen to support life. In the last weeks or months of life, a person with emphysema must breathe from an oxygen tank. Finally, even that is not enough.

Tobacco smoke also causes cancer of the lungs. Cancer, recall, is the unregulated growth of the body's own cells. Tobacco smoke contains at least 50 different known *carcinogens,* chemical that cause cancer. Because tobacco smoke ruins the lungs' ability to cleanse themselves, these carcinogens permanently coat the lungs. Some of them damage genes that control cell division, so that the cells divide without restraint. Others damage enzymes that help regulate cell division. Still others enhance the effect of the other carcinogens. In time, small tumors develop, which eventually *metastasize,* or spread, to other areas of the body. Invasion of nearby nerves, for example, may cause partial paralysis.

Lung cancer is one of the least treatable forms of cancer. Untreated lung cancer patients live an average of 8 months. Ninety percent of those who are aggressively treated, with radiation and chemotherapy, die within 5 years. Tobacco smoke causes 90 percent of all cases of lung cancer. In societies that do not smoke, lung cancer is a rare disease.

Smoking tobacco grossly interferes with the healthy functioning of both the lungs and the cardiovascular system. The lungs and the heart and blood vessels are tightly interconnected. In the rest of this chapter we will see how healthy lungs (and other organs of respiration) work to absorb oxygen and how the circulation distributes the oxygen to the cells of the body.

WHY DO ANIMALS NEED OXYGEN?

Most animals get most of their energy from cellular respiration, a process that always requires oxygen. Because cellular respiration takes place in almost every cell of an animal's body, animal life requires the means both to acquire oxygen and to distribute it throughout the body. In this chapter, we discuss some of the ways that animals have solved the problem of taking oxygen from the environment and delivering it to individual cells—a process that is also called respiration.

The champion oxygen extractors are birds, some of which can fly at altitudes far beyond the abilities of the most ambitious human mountain climbers. Birds, mammals, reptiles, and amphibians obtain oxygen in the **lungs** and distribute it with the blood. Fish and other aquatic animals, however, extract oxygen with **gills,** while insects use tubes to distribute oxygen after extracting it from the air.

For oxygen to be useful, it must first dissolve in water, since only dissolved oxygen can participate in the chemical reactions of a cell. We begin, then, with a discussion of the way that oxygen and other gases dissolve in water.

To participate in biochemical reactions, oxygen must first dissolve in water.

Where Does Oxygen Come From?

Oxygen is far more plentiful in the air than in water. In air approximately 1 of every 5 molecules (or 21 percent) is oxygen, whereas in water, oxygen molecules make up fewer than 1 in 170,000 molecules (or 0.0006 percent).

Land animals—particularly insects and land vertebrates—have succeeded so fabulously because we are able to use oxygen in the atmosphere. But a land animal cannot use this abundant oxygen until it has dissolved in water. A land animal's first task, therefore, is to dissolve oxygen in water. Like land animals, aquatic animals also obtain oxygen from the atmosphere only after it is dissolved in water. In both cases, then, we want to know how oxygen dissolves in water.

Oxygen is far more plentiful in air than in water. Both land animals and aquatic animals depend on oxygen that has dissolved in water.

How Do Oxygen Molecules Move from Air into Water?

The average distance between molecules in a gas (such as air or even pure oxygen) is more than ten times the average distance between molecules in a liquid. The molecules of a gas have little interaction with one another. They move about at random, with an average speed that depends only on the temperature: the higher the temperature, the faster the molecules move.

When the molecules of a gas strike the surface of a liquid, most of them will bounce back into the space above the liquid (Figure 38-4). Sometimes, however, the solution will "capture" the gas molecule. Such a molecule enters the liquid: its free existence as a noninteracting gas molecule stops, as it sticks to the molecules of the liquid. Such dissolved gas molecules can themselves "escape" from the solution and become gas molecules again. Given enough time, gas molecules enter and leave

the solution at the same rate. Such molecules are said to have reached *equilibrium*.

Each time a gas molecule strikes a surface, it exerts a tiny impulse, which contributes to the pressure that a gas exerts on the surface. The total pressure depends both on the number of gas molecules that strike the surface (that is, the concentration) and on the speed with which each molecule strikes the surface (that is, the temperature).

At sea level, air exerts a total pressure that is enough to support a column of mercury about 30 inches (760 mm) high. We also say that the pressure is 760 torr, after Evangelista Torricelli, the 17th-century inventor of the mercury barometer (Figure 38-5). A barometer measures the total atmospheric pressure as it varies from day to day and place to place.

In air, about 21 percent of the molecules are oxygen (O_2), about 0.03 percent are CO_2, and almost all the rest are nitrogen (N_2). Because each molecule acts independently, each type of molecule independently contributes to the total pressure. Researchers conventionally give the concentration of a type of molecule in a gas as the **partial pressure,** the pressure exerted by that one type of molecule. The total pressure is the sum of all the partial pressures. In air, the partial pressure of O_2 is about 160 torr, that of N_2 is about 600 torr, and that of CO_2 is only about 0.2 torr.

Most molecules of a gas bounce off the surface of a solution, but some are captured and dissolved.

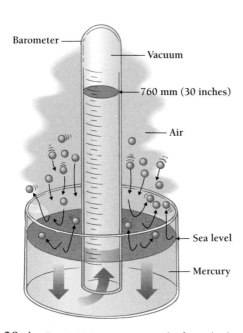

Figure 38-4   Barometric pressure results from the bouncing of gas molecules. Only a fraction of the gas molecules are "captured" by water and dissolve in it, so the concentration of any dissolved gas is low, making oxygen extraction a difficult problem.

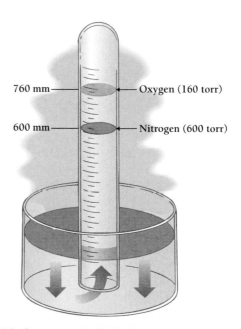

Figure 38-5   A mercury barometer. At sea level, the atmosphere can support of a column of mercury some 760 mm (30 inches) high. The oxygen molecules in the atmosphere contribute about 160 mm of this pressure. Most of the pressure comes from the bouncing of nitrogen molecules, with carbon dioxide molecules contributing only about 0.2 mm.

How Much Dissolved Gas Will a Solution Hold?

The equilibrium concentration of a type of molecule in a solution depends directly on its concentration in the gas phase, which is proportional to its partial pressure. For example, although 1 in 5 molecules in air is an oxygen molecule, dissolved oxygen molecules make up less than 1 in 100,000 of the molecules in a solution that has been exposed to air at room temperature.

The amount of dissolved gas also depends on temperature. Gas molecules are *less* soluble at higher temperatures. This is very different from most solids, which become *more* soluble at higher temperatures. For example, more sugar will dissolve in hot water than in cold. But heat a pot of water, and the gases come out of solution, forming bubbles.

Another familiar illustration is the behavior of a bottle of a "carbonated" drink. A carbonated drink contains dissolved carbon dioxide under pressure two to three times greater than atmospheric pressure. When you open a cold bottle of cola, the CO_2 comes out of solution and forms tiny bubbles. If the bottle is warm, however, more CO_2 comes out of solution. So many big bubbles form that opening a warm bottle often produces a mess of foamy overflow.

> The amount of dissolved gas in a solution depends on the temperature and the partial pressure of that gas.

How Does Surface Area Help Determine the Rate at Which a Gas Dissolves?

It is relatively easy to estimate the concentration of oxygen in a solution at a given temperature. It is harder to estimate how long the oxygen will take to reach equilibrium. The rate at which molecules of a gas can enter a solution depends on three factors: the temperature, the surface area, and the *concentration gradient,* or the difference between the concentration of the gas in air and in the solution.

Both the molecules of a gas and the molecules dissolved in a solution move by a process called *diffusion,* the spontaneous movement of a substance from a region of high concentration to a region of lower concentration. Diffusion results from the random movements of molecules.

We have already seen that higher temperatures reduce the amount of gas that can dissolve in a solution. However, the rate of diffusion *increases* with temperature. So even though the total oxygen dissolved in warm water is low, the rate at which the oxygen enters the water and reaches equilibrium is higher than in cold water.

The total amount of oxygen that moves across a boundary in a given time (per second, for example) is proportional to the difference in concentrations on the two sides of the boundary. That is why hospital patients are sometimes given high concentrations of oxygen to breathe; at high concentrations, the oxygen molecules more quickly dissolve in the fluid covering the lungs.

Large surface areas also speed the exchange of molecules. We understand intuitively that water in an open pan will evaporate more quickly than the same amount of water in a narrow-necked bottle. Similarly, oxygen exposed to water in an open pan will dissolve and equilibrate much more rapidly than oxygen exposed to the same amount of water in a narrow-necked bottle.

One reason that cells are as small as they are is that being small increases their surface area relative to their volume, the *surface-to-volume ratio* (Figure 38-6A). Each cell must have a large enough surface to permit enough oxygen to enter to support the needed level of respiration and ATP synthesis. Ordinarily, animal cells are no more than 20 to 30 μm in diameter.

Small multicelled animals such as rotifers or nematodes—up to about 1 mm in diameter—can obtain enough oxygen by diffusion alone. But larger animals require some special arrangement to increase the surface area through which dissolved oxygen can diffuse. Animals employ two general strategies to increase the total surface areas for gas exchange. They either fold a surface outward (*evagination*) or fold a surface inward (*invagination*) (Figure 38-6B). Evaginated breathing structures are called **gills**, and invaginated breathing structures are called **lungs**.

> The rate of gas movement into a cell or an organism depends on the surface area through which the gas can move. Evaginated breathing structures are called gills, and invaginated breathing structures are called lungs.

HOW DO AQUATIC ORGANISMS OBTAIN DISSOLVED OXYGEN?

Aquatic animals larger than about 1 mm obtain dissolved oxygen through gills. A sea star, for example, has tiny protuberances all over its surface, and a sea scallop has specialized gills within its shell. In each case, oxygen diffuses from the surrounding water across the expanded surface area into the animal's internal environment.

As oxygen diffuses from the water into the surface of the gills, however, the oxygen concentration in the surrounding water drops. To gather more oxygen, the animal must continually replace the oxygen-depleted water with new, oxygen-rich water. Some aquatic animals accomplish this by moving gills through the water, but this works well only if the gills are relatively small. Most organisms, therefore, move the water over the gills. In the scallop, for example, cilia on the gill surface move the water past the gill surfaces. In lobsters large paddles move water over the gills.

> Dissolved oxygen can diffuse into an organism through gills. Oxygen-depleted water must be replenished by movement.

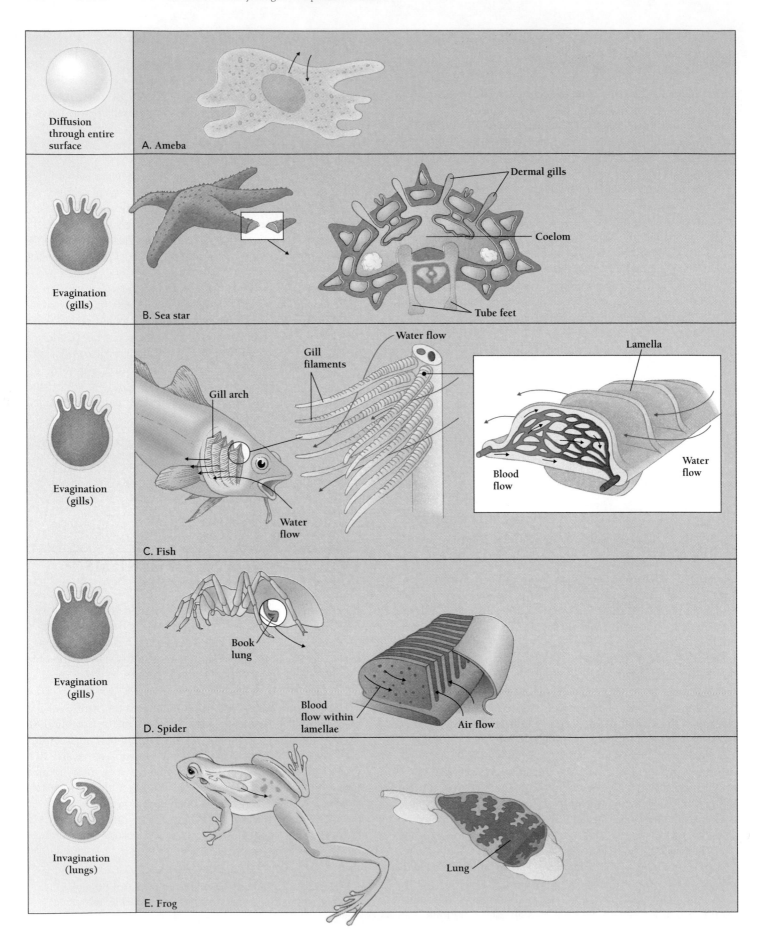

Diffusion through entire surface

A. Ameba

Evagination (gills)

B. Sea star

Dermal gills

Coelom

Tube feet

Evagination (gills)

C. Fish

Water flow

Gill filaments

Gill arch

Water flow

Lamella

Blood flow

Water flow

Evagination (gills)

D. Spider

Book lung

Blood flow within lamellae

Air flow

Invagination (lungs)

E. Frog

Lung

◀ **Figure 38-6** The ability of an animal to take in oxygen depends upon its surface area. A. Small organisms have much higher ratios of surface to volume than large ones, and organisms smaller than about 1 mm generally derive oxygen by diffusion alone. B. Gills increase the surface through which oxygen can diffuse. In a sea star, gills result from the evagination of the animal's outer surface. C. The gills of a fish are more extensive than those of invertebrates and are more efficient in extracting oxygen. D. Spiders develop internal gills, called "book lungs," which stay moist. E. Lungs arise by invagination (infolding) rather than evagination (outfolding).

How Do Gills Obtain Enough Oxygen To Power a Fast-Swimming Fish?

The oxygen requirements of a fast-swimming fish are greater than those of most invertebrates and other slower-moving fish. A mackerel, for example, has about 50 times as much gill area as a goosefish. Like scallops and lobsters, fish continuously move water over their gills. Fish may use muscles and valves in the mouth to pump water over the gills, or they may just hold their mouth partly open and swim constantly.

Fish adjust their behavior to maintain oxygen supplies. In one study, researchers kept mackerel in moving water that required them to swim at a constant speed. When the researchers reduced the oxygen content of the flowing water, each fish opened its mouth wider, allowing more water to pass through the gills and more oxygen to be extracted.

The gills of fish are even more highly organized than those of aquatic invertebrates. On each side of the head lie several gill arches, all covered by a protective bony flap called the operculum. Each arch carries two rows of gill filaments, and each filament carries many rows of parallel, platelike lamellae (Figure 38-6C). Water passes through the lamellae in a single direction, and a dense network of capillaries exposes the circulating blood to the oxygen in the water.

Fish move water over their gills by pumping or by swimming.

What Are the Disadvantages of Gills?

Why do fish die when they are taken from the water? Why can't the gills extract oxygen from the air?

In fact, some fish do not die when taken out of water. An eel, for example, survives well if kept cool and moist. Eels can even crawl over land—usually at night and in moist grass—from one body of water to another. But, in air, most of an eel's oxygen comes through the skin, not through the gills.

The problem with gills is that they collapse when exposed to air. They lack mechanical rigidity, and their surfaces therefore stick together so that they cannot provide an expanded area for the diffusion of gases.

In air, external gills quickly dry and lose the water into which oxygen must dissolve. Spiders solve this problem with moist internal gills (Figure 38-6D). Although this breathing ap-

paratus is called a *book lung* (because the outfolded gills resemble the pages of a book), they are actually gills, not lungs.

In air, gills do not have enough mechanical support to remain functional.

HOW DO AIR-BREATHING ORGANISMS OBTAIN OXYGEN?

Aside from spiders, air-breathing organisms nearly all obtain oxygen from infolded (invaginated) surfaces that do not collapse easily. Land snails, some fish, many amphibians, and all reptiles, mammals, and birds depend on **lungs**, localized organs of gas exchange that are always associated with the circulation. Insects, on the other hand, depend on **tracheae**, a complex set of tubes, which also arise by invagination, that carry air throughout the animal's body.

Lungs and tracheae are both extensions of tubes within an animal's body. They lie within the body cavity, where they are more protected than gills and less likely to collapse.

As in gills, oxygen that is to be absorbed by the lungs and tracheae first dissolves in a film of water. In both cases, the airways must be kept moist. Water is therefore essential to all respiration.

Almost all air-breathing vertebrates depend on moist lungs to acquire oxygen.

The Ability of a Lung To Extract Oxygen Depends on Surface Area and Blood Supply

Just as the surface area of gills limits the oxygen, and therefore the energy, available to fish of different species, the surface area of the lungs limits the oxygen-gathering ability of animals. Birds and mammals, which as endotherms require more energy and more oxygen, have elaborately folded lungs. Every cubic centimeter of a human lung has a total surface area of about 300 square centimeters, whereas the same volume of a frog lung has a surface area of only 20 square centimeters (Figure 38-6E).

The ability of a lung to extract oxygen depends not only on the surface area but also on the blood supply that carries oxygen away from the lungs. The more blood that flows through the lung, the more oxygen can be extracted. The simple lungs of several land snail species illustrate the relationship between blood supply and the ability to extract oxygen. Periwinkle snails that live just next to the ocean have primitive lungs and get most of their oxygen from seawater. Snails of a related species that live farther up on the beach have better developed lungs, with many more blood vessels.

More efficient oxygen extraction requires a large surface area and a rich blood supply.

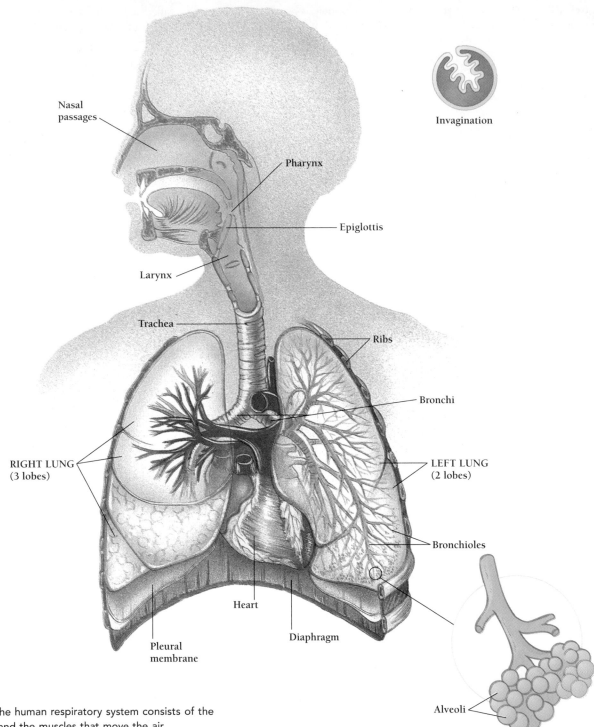

Figure 38-7   The human respiratory system consists of the lungs, the airways, and the muscles that move the air.

How Do Human Lungs Obtain Oxygen from Air?

Even in periwinkles, lungs do not function by themselves. They are part of a **respiratory system,** which consists of all the structures responsible for the exchange of gases between the blood and the external environment. In vertebrates, the respiratory system consists of the lungs, the airways, and the muscles that move the air (Figure 38-7).

Air enters the body through the mouth and nose. These two openings lead to the throat, or **pharynx,** a common passage for food and air. Food and air soon diverge into two separate branches, however. The *esophagus* leads to the stomach, while the *glottis* leads to the lungs. A flap called the *epiglottis* prevents food from entering the air tube during swallowing.

The most important challenge for the respiratory system is to bring air into contact with a large surface area that will allow oxygen to diffuse into the blood. But all the air enters the system through a single large tube. The first task of the respiratory system, then, is to spread the air over a large surface.

Before the air reaches the lung's working surface, the tube divides repeatedly. The first portion of the air tube is the glottis, which leads directly into the **larynx.** The larynx contains the vocal cords, folds of membrane that vibrate as air passes over them, producing sound. The larynx leads directly into a long tube, called the **trachea,** or windpipe. The trachea enters the chest, where it forks into two smaller tubes, the **bronchi,** which lead to the two lungs. Inside the lungs, the bronchi branch into smaller and smaller extensions, with thinner walls. The smallest of the bronchi are **bronchioles.**

No gas exchange takes place in the larynx, the bronchi, or the bronchioles: they are only passages to the working surfaces of the lungs. The surfaces on which oxygen dissolves are the **alveoli** [singular **alveolus;** from Latin, *alveus* = a hollow], hollow sacs that are richly supplied with blood. Here, hemoglobin molecules in the passing blood carry oxygen to the rest of the body. In mammals, the alveoli are dead ends. Air can leave the alveoli only the way it arrived, through a bronchiole. Breathing causes *tides* of air to move in and out of the alveoli.

The fine branching of the air passages and alveoli greatly increases the surface over which oxygen molecules come into contact with cell surfaces. If, for example, each lung were a smooth sphere with a volume of 1 liter, then the total surface area of the lungs would be about 0.1 square meter (about one square foot). Instead, the surface area of the alveoli of a healthy person is about 135 square meters, the area of a good-sized house and more than 80 times the outer surface area of the whole body.

Each lung of a healthy college student contains about 300 million alveoli. At rest, your lungs contain about 1.5 liters of air before you inhale and 2 liters of air after you inhale. The amount of air drawn in and then expelled in a single breath is called the **tidal volume.** For a healthy adult human, the tidal volume at rest is about half a liter (500 ml). During vigorous exercise, the tidal volume may increase to 3 liters or more.

The trachea, bronchi, and bronchioles together contain about 150 ml of air. The air in these conducting tubes represents the **dead space** of the airways, that is, the volume of air that does not come into contact with the surfaces of the alveoli. In a resting person, then, each breath brings into the lungs about 350 ml of fresh air (500 ml tidal volume −150 ml dead space). This air then mixes with the air remaining within the alveoli after the last breath.

The mammalian respiratory system consists of finely divided airways leading to hundreds of millions of blind sacs (alveoli), where oxygen is absorbed.

What Moves Air into the Lungs?

Mammals, birds, and most reptiles pull air into the lungs using muscles within the chest cavity. Amphibians and some rep-

tiles push, or gulp, air into the lungs using the muscles and valves of their mouths.

How do humans and other mammals generate the required suction? Our trick is to use our closed chest cavity like a pump. Using muscles in the chest, we expand the chest cavity and create a vacuum. The expansion reduces the pressure in the lungs compared with the pressure of the outside air, so that air flows passively inward.

Ventilation, the flow of air into and out of the alveoli, depends on the muscles of the chest cage, which surrounds the **thoracic cavity,** or chest cavity (Figure 38-7). Surrounding the thoracic cavity are the spinal column, 12 pairs of ribs, and the sternum, or breastbone. Muscles and connective tissue separate the top of the cavity from the head. The **diaphragm,** a sheet of muscle beneath the lungs, separates the thoracic cavity from the abdomen. Muscles and elastic connective tissue also run between the ribs and form the sides of the cavity. The lungs themselves lie within a fluid-filled sac attached to these connective tissues of the chest.

During **inspiration,** or inhalation, the diaphragm contracts and moves downward. This expands the chest cavity and decreases the pressure inside the lungs. At the same time, the rib muscles contract, which expands the chest. Since the pressure of the outside air now exceeds that inside the alveoli, air flows inward.

During **expiration,** or exhalation, the diaphragm relaxes into a steep dome on the cavity's floor. The ribs resume their relaxed positions, which further contributes to a decrease in the size of the cavity. As the volume decreases, the pressure increases, and air flows outward. At rest, we have only to relax our muscles and air flows outward.

During exercise, we increase the tidal volume by forcefully expelling (and then inhaling) more air. We do this by contracting special muscles that boost the lungs' internal pressure.

Mammals pull air into the lungs by expanding the thoracic cavity. They push air out of the lungs by relaxing the muscles of the thoracic cavity.

How Do the Cells of the Airways Stay Moist and Free of Debris?

Tightly connected epithelial cells line both the nonabsorbing surfaces of the airways and the absorbing surfaces of the alveoli (Figure 38-8). As we might expect from their differing roles, the cells of the airways differ from those of the alveoli.

Most of the cells that line the trachea, bronchi, and bronchioles have fringes of constantly beating cilia, while other cells in the airway linings secrete **mucus,** a protein solution like that secreted by the membranes of the nose. Mucus keeps the surface of the airways wet, which in turn keeps the air within the airways moist and prevents the surfaces of the alveoli from drying out.

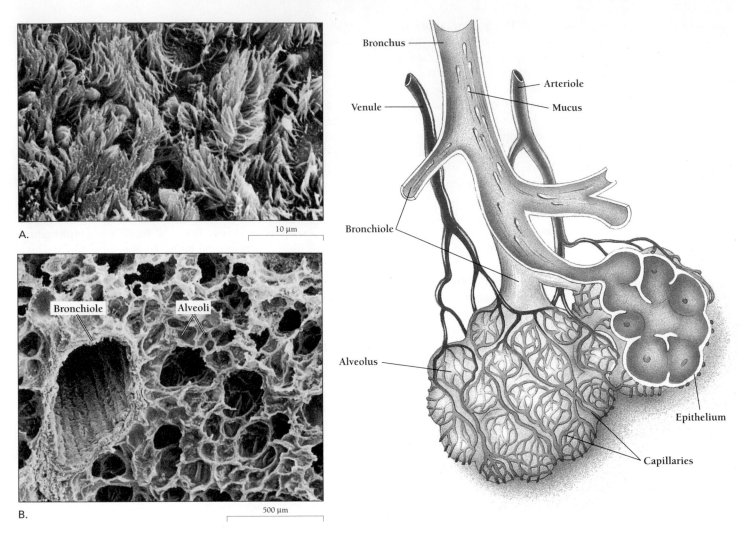

Figure 38-8   **Epithelia in the respiratory tract.** The epithelium that lines the airways does not absorb oxygen, whereas the epithelium that lines the alveoli does absorb oxygen. A. SEM of the inside of the trachea. B. SEM of a section of lung tissue, showing part of a bronchiole surrounded by spongy alveoli.   *(A, B, Biophoto Associates)*

Mucus also helps protect the lungs from becoming clogged with dust particles or infected with bacteria. Dust particles, often carrying bacterial passengers, stick to the mucus, and the beating cilia drive the dust-laden mucus out of the airways into the throat. Swallowing pushes the mucus into the digestive tract. The ciliated epithelium of the airways serves as an escalator for particles that would otherwise accumulate in the blind alveoli, where they could interfere with gas exchange and cause infection. The airways also contain *macrophages,* scavenger white blood cells that engulf the accumulated debris in the airways.

The respiratory tract traps dust and bacteria in mucus, which cilia propel towards the mouth. Macrophages also engulf debris in the airways.

How Do the Alveoli Move Oxygen into the Blood and Carbon Dioxide out of the Blood?

The primary function of the epithelial cells in the alveoli is to promote gas exchange. Unlike the cells of the airways, the cells of the alveoli have no cilia. They are also extremely thin, depending on a spongy mesh of connective tissue for mechanical support.

Fine capillaries run through the alveoli, carrying blood that binds oxygen and discharges carbon dioxide (Figure 38-9). The blood carries the oxygen into the pulmonary venules and veins and eventually to the heart.

As the oxygen in the inhaled air enters the alveoli, it dissolves in the liquid that coats the epithelium and diffuses across the epithelial layer. The oxygen molecules then move across

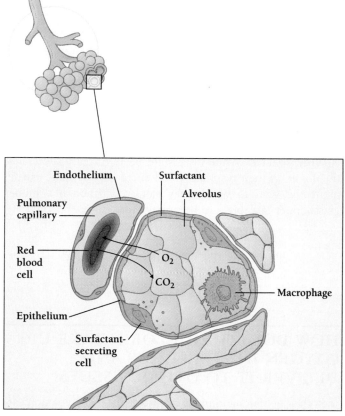

Figure 38-9   The cellular organization of alveoli and blood vessels. Oxygen molecules move from the inside of the alveolus, through the epithelium of the alveolus and the endothelium of the pulmonary capillary, into the oxygen-carrying red blood cell. CO_2 moves in the opposite direction—from the blood cells and plasma through the endothelium and the epithelium into the airway.

the **endothelium,** the epithelial cells that line the capillaries, and enter into the blood **plasma,** the liquid part of the blood. Oxygen molecules then diffuse from the plasma into the red blood cells where they can bind to hemoglobin. Carbon dioxide follows the reverse path, from red cells to blood plasma to endothelium to epithelium to the exhaled air.

The cellular structure of the alveoli promotes the diffusion of oxygen and carbon dioxide to and from the blood.

How Do Alveoli Overcome the Surface Tension That Resists Expansion?

Each inhaled breath pulls air into the alveoli, expanding the volume of each of these tiny sacs. The small size of each alveolus enables the lungs to have a huge area available for gas exchange. But the small size of each sac also increases the difficulty of expanding each alveolus, just as small balloons are harder to blow up than giant ones.

A liquid's resistance to an increase in surface area is called *surface tension.* To overcome this resistance, the alveoli produce a *surfactant,* a detergent that reduces surface tension and allows expansion. Infants born prematurely often lack surfactant. Such babies can breathe only with great difficulty and are said to be suffering from *hyaline membrane disease.* The prospects for survival are excellent, however, if the physician supplies a substitute surfactant.

Detergents within alveoli reduce surface tension and allow the lungs to expand more freely.

How Do Birds Extract Oxygen?

Among all the vertebrates, birds are the most efficient at extracting oxygen. Few mammals can live at altitudes above 6000 meters, where many birds routinely fly. The efficiency of bird lungs depends on the patterns of air passages and blood vessels, which differ from those of mammals.

When a bird takes a breath, the air passes not into the lungs but into a sac called the *posterior air sac,* which does not allow gas exchange (Figure 38-10). When the bird exhales, the contents of the posterior air sac move into the lungs. On

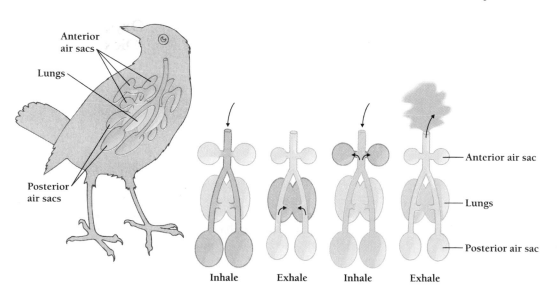

Figure 38-10   The two-cycle respiration of a bird is more efficient in extracting oxygen than the one-cycle respiration of a mammal. The route of the inhaled and exhaled air allows countercurrent exchange similar to that in a kidney or in a turbine engine.

the next breath, air moves out of the lungs into another sac, the *anterior air sac.* Finally, on the second exhalation, the contents pass back out into the atmosphere. The result of this two-cycle process is that air moves in a single direction through the lungs. In contrast to mammals then, birds have no dead space.

Meanwhile, the blood in the lungs flows in the opposite direction. The oxygen-poor blood from the heart first comes into contact with the air nearest the anterior air sac, which has the least oxygen. But the oxygen concentration of this partly depleted air is still much higher than in the depleted blood, so the net movement of oxygen is into the blood. As the blood acquires more oxygen, it continues to flow into lung regions that have still more oxygen. So even near the back of the lungs, where the blood has the most oxygen, the net flow of oxygen is into the blood.

Diffusion between two flows that move in opposite directions is called *countercurrent exchange.* Such an arrangement is especially efficient because it maintains the concentration difference even as materials move from one stream to the other.

In a bird's lungs, air flows in a single direction, allowing birds to extract oxygen more efficiently than do mammals.

How Do Insects Distribute Oxygen?

In all vertebrates and most invertebrates, the circulatory system distributes dissolved oxygen. Even in animals that absorb oxygen through the body surface, blood vessels just under the skin acquire dissolved oxygen and carry it to the rest of the body. In most invertebrates and in fish, oxygen moves through the gills to the circulation, whereas in most vertebrates, oxygen moves to the circulation through the surface of the lungs.

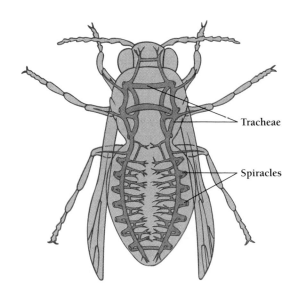

Figure 38-11   The respiratory system of an insect. Instead of distributing dissolved oxygen, insects distribute air itself through an extensive system of tracheae.

Insects, however, do not distribute dissolved oxygen at all. Instead, they distribute air itself through a system of tiny ventilation pipes called **tracheae** (Figure 38-11). Air enters the body through openings called **spiracles,** which lead directly to the tracheae. The advantage of this system is that it allows oxygen to diffuse as a gas, saving the energy of pumping it through the circulation. With this separate system to distribute oxygen, insects are able to tolerate the sluggishness of their open circulatory systems. While insects are phenomenally successful, the design of their respiratory and circulatory systems usually limits the sizes to which they can grow (though a lubber grasshopper is as big as a hummingbird).

In insects, oxygen diffuses to tissues through air tubes called tracheae.

HOW DOES HEMOGLOBIN TAKE UP OXYGEN IN THE LUNGS AND DELIVER IT TO OTHER TISSUES?

In both vertebrates and many invertebrates, the circulation carries oxygen to tissues throughout the body. If the blood consisted of only salts and water, as it does in many invertebrates, it would not be able to carry much oxygen, as very little oxygen can dissolve in water. Many invertebrates with little more than salt water for blood therefore take advantage of their circulatory systems simply to move oxygen faster than it could move by diffusion alone.

Other invertebrates and essentially all vertebrates, however, have special oxygen-binding proteins in the blood. These proteins are always colored pigments: red hemoglobin in vertebrates and some invertebrates, blue hemocyanin in mollusks and crustaceans, and occasionally other red or green proteins. The presence of these proteins increases the oxygen-carrying capacity of blood by more than 100 times.

Hemoglobin and the other pigments carry oxygen molecules in the blood much as a raft transports passengers down a flowing stream. In invertebrates, the oxygen carriers are dissolved directly in the blood. In vertebrates, however, tens of millions of hemoglobin molecules are packed into red blood cells. Packaging the hemoglobin into cells provides at least three advantages: (1) reduction in the osmotic pressure that would be generated by a high concentration of hemoglobin molecules; (2) the opportunity to control the chemical microenvironment of the oxygen-carrying molecules; and (3) a means for replacing hemoglobin molecules that do not work as well as when they were first made.

The mature red cell is a cell with limited function and with a limited lifetime. It does not have mitochondria, and in mammals it does not even have a nucleus. It chiefly serves as a container for the hemoglobin molecules.

In blood leaving the lungs, about 99 percent of the oxygen is in the form of **oxyhemoglobin,** hemoglobin that has

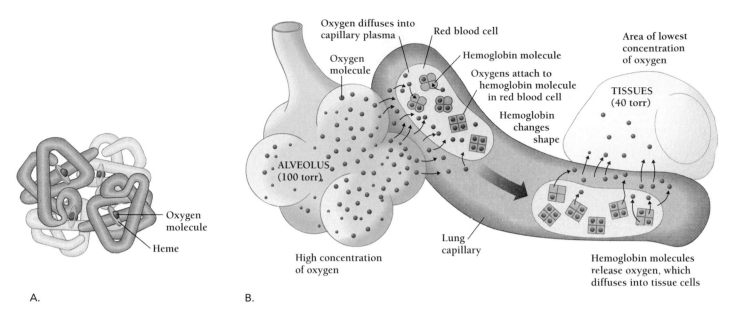

Figure 38-12 A hemoglobin molecule consists of four polypeptide chains, each of which is bound to a small, flat, iron-containing molecule called heme. A. Each heme can bind one molecule of oxygen, and each hemoglobin molecule can bind four molecules of oxygen. B. Hemoglobin molecules associate with oxygen in the lungs, where oxygen is present at high concentration (100 torr). Oxyhemoglobin, in red blood cells, moves through the circulation. In oxygen-consuming tissues, such as muscles, the oxygen concentration is lower (40 torr) and oxyhemoglobin dissociates into deoxyhemoglobin and oxygen. The released oxygen diffuses into the surrounding tissues.

combined with oxygen. For the oxygen to be useful in the tissues, oxyhemoglobin must dissociate back into oxygen plus **deoxyhemoglobin,** hemoglobin that has no bound oxygen.

Hemoglobin consists of four polypeptide chains, each of which contains a small organic molecule, called *heme* (Figure 38-12). In the middle of each heme is an iron atom, in its ferrous (Fe^{2+}) state. One oxygen molecule can bind to each heme, so that one hemoglobin molecule can bind four oxygen molecules.

In hemoglobin, oxygen binding is a cooperative act. As oxygen binds, the arrangement of the four polypeptides shifts. This change causes the hemoglobin molecule to a have a greater affinity for oxygen at higher oxygen levels—in the lungs—and a lower affinity at lower oxygen levels—in the tissues. The molecular properties of hemoglobin, then, contribute to its ability to deliver oxygen.

Oxygen-binding proteins greatly increase the oxygen-carrying capacity of blood.

HOW DOES THE BLOOD HELP DISPOSE OF CARBON DIOXIDE?

In addition to absorbing oxygen, the lungs also dispose of carbon dioxide (CO_2). As oxygen enters the blood, CO_2 leaves. CO_2 in the blood comes into equilibrium with CO_2 in the air,

just as oxygen in the blood comes to equilibrium with that in the air. The tissues of the body produce so much CO_2 that the lungs normally have much higher partial pressure of CO_2 than the outside air.

Once CO_2 dissolves in water, it can chemically combine with water to form carbonic acid:

$$CO_2 + H_2O \longrightarrow H_2CO_3$$

Carbonic acid in turn dissociates to give a hydrogen (H^+) ion and a bicarbonate (HCO_3^-) ion:

$$H_2CO_3 \longrightarrow H^+ + HCO_3^-$$

Most of the CO_2 carried by the blood is in the form of bicarbonate—both in red cells and in the plasma. About one-third of the CO_2, however, combines with hemoglobin (though not on the heme itself), so that the hemoglobin can ferry CO_2 as well as oxygen.

In the absence of a catalyst, the interconversions of dissolved CO_2 and bicarbonate are slow reactions. They happen too slowly to take up and release enough CO_2 during the brief passage of the blood through the capillaries of lungs and tissues. Even this simple reaction, then, requires a specific enzyme catalyst, called *carbonic anhydrase.* In red blood cells, carbonic anhydrase is the most abundant protein except for hemoglobin itself.

The blood carries carbon dioxide from tissues to the lungs.

Adult hemoglobin, fetal hemoglobin, and myoglobin

The concentration of dissolved oxygen is higher in the oxygen-providing lungs than in the oxygen-using tissues. The oxygen content of the air within the lungs approximates that of the atmosphere, but it is lower because of the dead volume of ventilation. Whereas the partial pressure of oxygen in the atmosphere is about 160 torr, that in the lungs is only 100 torr. In oxygen-using tissues, on the other hand, the partial pressure of oxygen is typically 40 torr.

By the time blood leaves the lungs, all the hemoglobin is in the oxyhemoglobin form. For hemoglobin to deliver oxygen, it must not only be able to take up oxygen but even more importantly it must let go. For example, myoglobin, the oxygen-binding protein of muscle cells, like hemoglobin, is fully saturated at 100 torr. But myoglobin is also fully saturated at 40 torr, and it is almost saturated at 20 torr. *Myoglobin,* therefore, could not possibly function as an oxygen carrier in the blood. Hemoglobin, on the other hand, is particularly well suited to serve as an oxygen carrier.

Researchers have been able to study oxygen binding to hemoglobin fairly easily, because oxyhemoglobin has a different color than deoxyhemoglobin. The oxygen-loaded blood of the arteries is bright red, while the oxygen-depleted blood of the veins is slightly blue-red. Through the use of a *spectrophotometer,* an instrument that measures the absorption of light of different wavelengths, researchers have determined how much oxygen is bound to hemoglobin at different partial pressures of oxygen. The fraction of oxyhemoglobin as a function of oxygen concentration is called an *oxygen dissociation curve.*

For the blood of adult humans, the fraction of oxyhemoglobin increases from a partial pressure of about 10 torr to about 60 torr. Above 60 torr, the fraction of oxyhemoglobin does not increase much: the hemoglobin is effectively saturated with oxygen. The shape of the whole curve vaguely resembles the letter "S," reflecting the cooperative interaction of the four polypeptides in hemoglobin.

Tissues that use oxygen have lower oxygen concentrations than the lungs. A resting muscle, for example, has an oxygen concentration of 40 torr. At this concentration, only about 75 percent of the hemoglobin is bound to oxygen. As the blood moves through the circulation from lungs to heart to resting muscle, it gives up about 25 percent of its oxygen.

Blood releases more oxygen to active tissues than to resting ones. During vigorous activity, the oxygen concentration in a muscle may be still lower, 10 to 20 torr. So, when oxygen and blood come to equilibrium, fewer hemoglobin molecules are bound to oxygen; that is, more of the oxyhemoglobin molecules let go of their passengers, and more oxygen becomes available to the active muscle.

Cells that are especially active may exceed their ordinary supply of oxygen. They then depend (temporarily) on glycolysis to provide ATP. The result is the accumulation of lactic acid, which decreases the pH. The lowered pH further increases the release of oxygen, since low pH decreases the bind-

HOW DO RED BLOOD CELLS, NERVES, AND HORMONES CONTRIBUTE TO OXYGEN HOMEOSTASIS?

Environmental changes (such as climbing a mountain) or internal traumas (such as severe blood loss) can interfere with the blood's ability to deliver oxygen. Happily, the body has adaptations that maintain oxygen delivery even under unusual conditions.

How Can Blood Deliver Oxygen at High Altitudes?

At high altitudes, the difference in the oxygen concentrations in the lungs and muscles is not as great as at sea level. When a person climbs a mountain, one of the first changes seen in the blood is an increase in the level of a three-carbon molecule called BPG (2,3-bisphosphoglycerate). BPG binds to hemoglo-

bin and increases its ability to deliver oxygen efficiently. Somewhat unexpectedly, BPG does this by *decreasing* hemoglobin's affinity for oxygen (Box 38-1).

On the other hand, animals such as llamas and vicuñas that have evolved adaptations for life at very high altitudes have hemoglobins with high oxygen affinity (Figure 38-13). This high affinity allows hemoglobins to pick up oxygen more efficiently in the lungs. Most animals, including humans, have hemoglobins that are adapted to life at lower altitudes.

The other adaptation to high altitude is an increase in the number of red blood cells. The increase in numbers of cells also increases the total amount of hemoglobin and the oxygen-carrying capacity of the blood. The stimulus for increased red cell production is a protein hormone, called *erythropoietin,* which is made in the kidney in response to low oxygen delivery.

In humans, low oxygen in the tissues triggers an increase in BPG, which increases the efficiency of oxygen delivery.

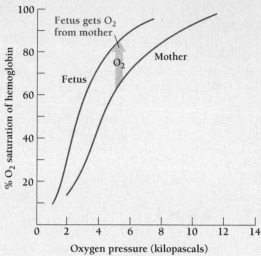

Figure A Oxygen dissociation curves for fetal and maternal hemoglobin. Because fetal hemoglobin has a higher affinity for oxgyen than does maternal hemoglobin, the maternal hemoglobin gives up oygen to the fetal hemoglobin.

ing of oxygen to hemoglobin. Again, the molecular properties of the hemoglobin molecule favor the unloading of oxygen in more active tissues.

Another adaptation of the hemoglobin molecules in many mammals helps increase oxygen delivery. These molecules bind to a relatively abundant compound in mammalian red blood cells, a three-carbon molecule called BPG (2,3-bisphosphoglycerate). The binding of BPG to hemoglobin decreases oxygen affinity, allowing it to release more oxygen in the tissues than would hemoglobin without bound BPG.

A fetus has a special problem in obtaining oxygen, since it depends not on its own lungs, which are not yet functional, but on its mother's. The blood of the fetus must somehow steal oxygen from the maternal blood, even as the maternal blood is supplying all the other tissues of the mother's body.

Maternal blood comes into close contact with fetal blood only in the *placenta*, a structure within the uterus. In order to move from the maternal to the fetal blood, the fetal hemoglobin must have a higher affinity for oxygen than the mother's hemoglobin.

The trick to making this happen is that the fetus makes a different hemoglobin, called *fetal hemoglobin,* or hemoglobin F (Figure A).

When researchers studied the oxygen-binding characteristics of hemoglobin F, however, they found that it bound oxygen with exactly the same affinity as adult hemoglobin, hemoglobin A. For decades, this lack of difference in oxygen affinity was totally confusing: how could fetal blood do its job?

The answer came from the discovery of the action of BPG. Although isolated hemoglobin F has the same oxygen affinity as hemoglobin A, fetal blood does have a greater affinity. In humans, the major molecular difference between hemoglobins A and F is that hemoglobin F does not bind BPG. The oxygen affinity of fetal blood is, therefore, much higher than that of hemoglobin A, which does bind BPG. The higher oxygen affinity means that the fetal hemoglobin can acquire oxygen from maternal blood and deliver it efficiently to the tissues of the developing fetus. The hemoglobin F can still deliver oxygen to fetal tissues, which have still lower oxygen concentrations and lower pH than in the placental circulation.

Figure 38-13 Animals that are adapted to life at high altitude, such as this llama, have hemoglobins with an especially high affinity for oxygen.

(François Gohier/Photo Researchers)

How Do Animals Regulate Breathing in Response to Changes in Oxygen and Carbon Dioxide Delivery?

The lungs and the blood, working together, accomplish both oxygen delivery and CO_2 removal. But how does an animal respond to lung failure, blood loss, or increased muscle activity, all of which may challenge the blood's ability to deliver sufficient oxygen?

As tissues use oxygen for respiration, they also produce CO_2 as a waste product. Increased respiration always leads to an increase in CO_2. In mammals, it is an increased CO_2 concentration in the blood that signals the need for deeper or more rapid breathing. Physiologists do not yet know exactly how this signal acts, but they do know that the coordination of breathing depends on the *breathing center,* a nerve complex, in a part of the brain called the *medulla,* that regulates the rate of breathing.

The firing of nerves in the breathing center triggers the contraction of the diaphragm and other muscles needed for inspiration. As the lungs expand, stretch receptors report back to the breathing center, the contraction signals to these muscles cease, and expiration occurs. High CO_2 stimulates the breathing center to coordinate faster and deeper breathing. The medulla also controls the circulatory system, changing the heart output according to the needs of the body. The maintenance of a constant internal environment means coordinating the actions of heart, blood vessels, and lungs to control the levels of oxygen and CO_2 as the status of the tissues changes.

STUDY OUTLINE WITH KEY TERMS

For oxygen to participate in biochemical reactions, it must first dissolve in water. The concentration of dissolved oxygen depends on its **partial pressure** and on the temperature. The rate at which oxygen dissolves depends on the surface area and the concentration gradient.

Both **lungs** and **gills** are folded surfaces with larger surface areas. Lungs are folded inward, gills are folded outward. An animal's **respiratory system** extracts oxygen from the environment. The circulatory system then distributes the oxygen throughout the body, often employing an oxygen-binding protein such as hemoglobin. Insects, however, distribute air directly to the tissues of the whole body from **spiracles** into a system of **tracheae.**

The respiratory system in mammals consists of the lungs, the airways, and the muscles that move air. Air moves into the lungs through the **pharynx,** the **larynx,** the **trachea,** the **bronchi,** and the **bronchioles.** The bronchioles lead to the **alveoli,** the sites of gas exchange.

Contractions of the **diaphragm** and the muscles of the chest are responsible for **ventilation,** the **inspiration** and **expiration** of air. Muscle movement expands the **thoracic cavity,** pulling air into the lungs. Each breath draws in and expels a volume of air called the **tidal volume.** Part of the new air stays within the **dead space** of the airways, while most of it enters the lungs.

The cells of the respiratory tract are specialized for several different functions. Some cells secrete **mucus,** which traps dust particles. Most cells of the upper tract have **cilia** that push dust-laden mucus out toward the pharynx. The cells of the alveoli form a specialized epithelium, well adapted for gas exchange. Oxygen and carbon dioxide diffuse through the epithelium of the alveoli and the **endothelium** of the capillaries into the blood **plasma.**

Hemoglobin binds and releases oxygen according to the oxygen concentration in the surrounding tissues. Oxygen combines with hemoglobin in the lungs to form **oxyhemoglobin.** In the muscles, oxyhemoglobin releases about 25 percent of its oxygen and re-forms **deoxyhemoglobin.** More active muscles have lower oxygen partial pressures, causing a greater release of oxygen. The low pH of active muscle also favors the release of oxygen.

The binding of oxygen to hemoglobin is cooperative. The binding of oxygen changes the three-dimensional structure of hemoglobin so that it is more likely to bind to oxygen molecules, up to a total of four per molecule, one per heme group and polypeptide chain.

The respiratory system also serves to dispose of CO_2. CO_2 in the tissues combines with water to form bicarbonate ions. Some CO_2 also binds directly to hemoglobin molecules. In the lungs, CO_2 passes through the alveoli and exits from the body. Accumulated CO_2 in the tissues stimulates the brain to command an increase in breathing rate.

REVIEW AND THOUGHT QUESTIONS

Review Questions

1. People who smoke cigarettes have increased rates of what three classes of disease?
2. Name the three most hazardous substances in tobacco smoke. What health effects does each have?
3. Define lungs and gills.
4. How do the lungs rid themselves of dust and microbes?
5. How does the diaphragm muscle expand and contract the chest cavity? Draw a diagram.
6. Define tidal volume and dead space.

7. How many oxygen molecules can a molecule of hemoglobin carry? Why?

Thought Questions

8. Why are scientific discoveries sometimes not immediately accepted by the general public?
9. Why should table salt and table sugar be more soluble in water than oxygen or CO_2?
10. Why does a drug that is inhaled reach the brain twice as fast as one that is injected?

SELECTED READINGS

Glantz, Stanton A., John Slade, Lisa A. Bero, Peter Hanauer, and Deborah E. Barnes, *The Cigarette Papers,* The University of California Press, 1996. A scholarly yet accessible discussion of the Brown and Williamson documents and the tobacco industry in general. This book is both encyclopedic and a great read.

▶ On-line materials relating to this chapter are on the World Wide Web at http://www.saunderscollege.com/lifesci/ Click on Tobin/Dusheck: *Asking About Life.*

Will Corn Chips Raise Your Blood Pressure?

In the spring of 1996, a team of researchers published a paper whose conclusions ran counter to decades of authoritative nutritional advice. Julian Midgley, Andrew Matthew, and their colleagues wrote that for most people an extremely low-salt diet would not lower blood pressure and in some people might actually cause a heart attack. At a time when salt consumption in the United States had fallen by 30 percent and supermarket aisles were jammed with low-salt crackers, low-salt chips, and low-salt soups, the researchers' stance dramatically contradicted accepted wisdom.

Americans love corn chips, potato chips, salty buttered popcorn, hot dogs, pizza, and greasy hamburgers. The more salt and fat, the better these foods seem to taste to us. Indeed, nutritionists uniformly agree that Americans consume far too much fat and far too much salt. Fat, especially animal and other highly saturated fat, contributes to both obesity and atherosclerosis, or hardening of the arteries. Both of these are major risk factors for heart attacks and strokes. Salt, we have been told, causes high blood pressure, a major risk factor for heart attacks, strokes, and kidney failure.

Even with recent declines in salt consumption, the average American consumes twice the recommended daily intake of salt. Nutritionists measure sodium intake, as opposed to sodium chloride (table salt) consumption. All simple salts containing sodium (actually sodium ions) contribute to our daily "salt" budget. The minimum sodium requirement for normal day-to-day health is 115 milligrams (mg). Yet most of us take in nearly 4000 mg each day. Government nutritionists, in an effort to recommend something reasonable, suggest a diet containing no more than 2400 mg per day.

The strongest evidence against sodium comes from research comparing high blood pressure in different countries. In countries where sodium consumption is high, blood pressure is also high, while in countries where sodium consumption is low, average blood pressure is also lower. One study of 32 countries found a consistent relationship between sodium consumption and high blood pressure, or *hypertension*.

But, according to Midgley and Matthew, a host of papers in the last ten years show that the connection between sodium and hypertension is indirect. To begin with, some researchers argue that a closer examination of people's behavior usually shows that those who consume a lot of sodium also drink more alcohol and tend to be overweight. Alcohol and obesity each cause hypertension, and these factors, rather than the salt itself, may be the cause of hypertension.

Researchers have long known that only a small percentage of people with

Paraskevas Photography

Table 39-1 Prevalence of Hypertension (≥140/90) in U.S. Adults, Ages 18–74

	White Men (%)	Black Men (%)	White Women (%)	Black Women (%)	All Adults (%)
All ages	33	38	25	39	30
18–24	16	11	2	10	9
25–34	21	23	6	15	14
35–44	26	44	17	37	24
45–54	43	55	36	67	41
55–64	51	66	50	74	53
65–74	59	67	66	83	64

Table 39-2 Researchers Estimate That 50 Million People in the United States Have Hypertension

Type	Range
High blood pressure (hypertension)	≥160/105
Mild hypertension	140–159/90–104
Borderline hypertension	120–140/85–89
Normal blood pressure	100–120/60–85

high blood pressure clearly benefit from a low-salt diet. In about half of all people with hypertension (just 5 to 10 percent of all Americans) the kidneys are unable to compensate for large amounts of sodium. In these "sodium-sensitive" individuals, a high-salt diet causes blood pressure to rise, while a low-salt diet causes blood pressure to drop. Sodium sensitivity gradually increases in those older than 45 or 50 years of age. Apparently, as we age, our kidneys lose some of their ability to regulate salt and fluid levels in the blood (Table 39-1).

In most people, however, the kidneys compensate almost magically. Thanks to kidneys, a typical 20-year-old can maintain normal blood pressure even after eating over the course of a day: five cafeteria pancakes, with a total of 4000 milligrams of sodium; two slices of pizza, with 2000 milligrams of sodium; and a can of spaghetti and meat balls, with 1000 milligrams of sodium. Nonetheless, considering the international data showing the strong association between sodium consumption and blood pressure, it was reasonable for health experts to recommend that Americans cut back on salt.

In January 1996, Midgley and Matthew decided to take a second look at all of the studies on blood pressure and sodium consumption in the previous several years. After discarding dozens of studies that they thought were poorly designed, they reanalyzed data from 56 clinical trials involving more than 3000 people. Midgley, Matthew, and their colleagues concluded that a low-sodium diet lowers blood pressure only in people older than 45 years of age who have high blood pressure. Even then, blood pressure drops only modestly—from 160/100, for example, to 156/99 (Table 39-2). And a low-sodium diet has virtually no effect on the blood pressure of people with normal blood pressure or young people with high blood pressure.

The researchers also questioned a basic assumption that any effect in older hypertensives would be beneficial. They argued that even in patients who had

Only a small percentage of people with high blood pressure clearly benefit from a low-salt diet.

lowered their blood pressure, there is no direct evidence that blood pressure lowered by means of a low-sodium diet actually prevents heart attacks and strokes or saves lives. Indeed, a super low-sodium diet may be dangerous. In one study, men with high blood pressure on extremely low-sodium diets had four times the risk of a heart attack as men consuming more sodium.

Should we eat as much salt as we can? Almost certainly not. Even for those of us whose kidneys are not sensitive to salt, too much salt burdens the kidneys and may possibly contribute to osteoporosis later in life. Will a high-salt diet raise blood pressure? It is beginning to appear that in most people, it does not, at least not directly.

We cannot resolve this complex controversy here. What we can do is equip ourselves with enough knowledge to follow the debate as it unfolds in the next few years. In particular, it will be useful to understand how kidneys remove salt from the blood, how different hormones regulate kidney function, and how sodium influences the release of such hormones.

KEY CONCEPTS

1. Organisms need to maintain a balance of water and solutes both inside each cell and in the extracellular fluid.

2. Organisms excrete nitrogenous wastes, usually diluted with water, by means of kidneys or other excretory organs.

3. The mammalian kidney draws all water and small molecules from the blood and then "puts back" essential ions and molecules.

4. Three hormones regulate the salt and water excretion of the mammalian kidney.

WHY DO ANIMALS NEED TO REGULATE WATER AND SALT?

Water is the major constituent of organisms: by weight it makes up more than 70 percent of the body. It is the medium in which cells are constantly bathed and is the major component of cells themselves. Most biochemical reactions occur in water, and water itself participates in many biochemical reactions. Cells and organisms therefore closely regulate the compositions of the solutions both inside and outside of cells. Even small changes in the concentrations of specific ions can dramatically change protein structures, reaction rates, and other cellular processes.

Cells and organisms spend enormous amounts of energy ensuring that they have the right amount of water (Figure 39-1). In terrestrial environments, keeping enough water in the body and its cells is a constant challenge. In freshwater environments, keeping water *out* is the challenge. Surprisingly, the marine environment is more like a terrestrial one than a freshwater one. Because the ocean is salty, marine fish tend to lose water through osmosis. Marine fish must expend energy to keep from losing water.

Having the right amount of water, however, is not enough. Cells also need the right amounts of ions and other molecules dissolved in that water. To maintain the correct concentrations of salts and other materials, organisms must somehow detect deviations from the norm. Then, when the deviation is large enough, the organism must take corrective action. This is, of course, the essence of homeostasis, the tendency of organisms to maintain a stable internal environment.

A.

B.

C.

Figure 39-1   Desert, sea, or fresh water. Healthy kidneys maintain correct blood pressure and water and salt balance regardless of the external environment. Thanks to adaptable kidneys, turtles and tortoises have evolved to live in environments in which the amount of salt and water vary enormously. A. An endangered desert tortoise (Mojave Desert). B. A green sea turtle (Hawaii). C. An eastern chicken turtle (North Carolina).   *(A, Robert Winslow/Tom Stack & Associates; B, William E. Ferguson; C, Jack Dermid/Photo Researchers)*

The principal engineers for regulating the watery environment of the vertebrate body are the **kidneys.** Kidneys regulate the volume and composition of the blood and the extracellular fluid. They maintain constant salt concentrations in the body fluids and regulate the disposal of water, wastes, and toxins. In this chapter, we will see how the kidneys and other excretory organs contribute to homeostasis.

Which Way Does Water Flow?

To understand the flow of water between an organism and its environment, we can think of an organism as a bag of watery fluid. For a single-celled organism, this description is literally true: the plasma membrane surrounds the cytoplasm, which contains a dilute solution of salts and organic molecules. The external environment of such an organism is always aqueous, with more or less salt.

Multicellular organisms contain a compartment that is neither cytoplasm nor external environment. This third compartment, the extracellular space, lies between the external environment and the membranes of individual cells. For organisms with circulatory systems, the extracellular space is continuous with the liquid part of the blood. The extracellular space contains a dilute solution of salts and organic molecules, whose concentrations differ from those in cytoplasm. For a human, the extracellular space is about 20 percent of the total volume of the body.

In general, water can flow freely between the extracellular space and the external environment or between the extracellular space and the cytoplasm. Which way, then, does water flow? In Chapter 3, we described the tendency of water to flow from dilute solutions into more concentrated solutions. When the concentrations of two such solutions become equal, net water flow then stops.

To talk about water flow we need a concentration. Animal physiologists use the term *osmolarity* (from the Greek *osmos* = impulse, thrust), which is roughly the sum of the concentrations of all the ions and molecules in a solution. When a pathway is open, water flows from a solution with lower osmolarity to a solution with higher osmolarity.

Between the external environment and the insides of the cells of multicellular organisms lies the extracellular space. The movement of water, ions, and molecules between the outside environment, the extracellular space, and the insides of the cells depends on the osmolarity of the fluids in each of these spaces.

Water and Salt Balance in Aquatic Animals

When the osmolarity of two solutions is the same, they are said to be *isotonic.* We can say, for example, that the body fluids of many marine invertebrates are isotonic with sea water. As a result, there is no net flow of water between such animals and their environment. These animals must still regulate their internal environment, however, for the concentrations of specific ions and molecules differ from those in sea water.

In contrast, freshwater animals are saltier than their environments, and their body fluids are said to be *hypertonic.* A hypertonic cell in a solution of fresh water tends to take in water until it bursts. Freshwater animals such as trout must therefore combat the relentless inflow of water with water-resistant skins. But having an impermeable skin limits a water dweller's ability to obtain oxygen, which must diffuse from the water into the gills. Consequently, freshwater animals tolerate some water influx and dispose of the excess in their urine.

The kidneys of freshwater animals excrete great volumes of urine that is more dilute than the body fluids. Freshwater animals produce such dilute, or *hypotonic,* urine by pumping ions out of the urine and back into the body. Producing a lot of urine, however, cannot solve all the problems of water and salt balance. Urine derives from the fluids of the extracellular space and contains dissolved salts and organic molecules. Excreting large amounts of urine, then, results in the loss of essential molecules and ions.

In contrast to freshwater fish, saltwater fish must counteract the tendency of water to flow outwards (Figure 39-2). So, while freshwater fish pump ions out of their urine to concentrate salts in their body fluids, saltwater fish actively pump salt ions out of their body fluids, into their urine, and back into the sea.

Nearly all vertebrates have body fluids whose salt concentration is much less than that of sea water. The jawless hagfish, among the most primitive of vertebrates, are the single exception: their body fluids have the same salinity as sea water.

All other saltwater fish, however, must continually compensate for water loss. Among other things, marine fish drink lots of sea water. (In contrast, freshwater fish do not drink at all.) This nearly constant drinking takes in salt as well as water, however, and marine fish must remove the extra salt, mostly through their gills. Sea turtles and sea birds, which share this problem, also have special organs that remove excess salt (Figure 39-3).

Freshwater animals rid themselves of excess water by excreting dilute urine. Marine and terrestrial animals tend to excrete salts and other wastes in minimal amounts of water.

How Do Land Animals Balance Water and Salt?

The major advantage of life on land is a rich supply of oxygen and a rich supply of plants to eat. The major problem, however, is getting and keeping water.

Land animals derive water from three sources: drinking, eating, and metabolism. Water comes from metabolism as a result of the breakdown of glucose during cellular respiration. Recall from Chapter 6 that the breakdown of one molecule of glucose produces six molecules of water.

Salt ions are removed from urine and returned to body

Large volume of hypotonic (dilute) urine is excreted

Small volume of hypertonic (salty) urine is excreted

Food and salt water

Salt ions are excreted

Food

Water and salt ions enter through gills

Saltwater fishes: body fluids are less salty than surrounding water

Freshwater fishes: body fluids are saltier than surrounding water

Figure 39-2   **Salt water and fresh water present different problems.** The bodies of salt-water fishes are less salty than the surrounding seawater, so they tend to lose water. These fishes must work to retain water while excreting salts and other ions and molecules. In contrast, the bodies of freshwater fishes are more salty than the surrounding water. Freshwater fishes must therefore excrete large amounts of water, while retaining previous salts. At the same time, the kidneys of both kinds of fishes need to excrete nitrogenous wastes.

Some land animals—such as frogs, salamanders, and earthworms—absorb water through their skins. Certain insects absorb moisture directly from the air. But most familiar animals obtain water by drinking or by eating.

The average adult human, for example, drinks about 1200 ml of fluids a day (a bit more than a quart) and obtains another 1000 ml in food. Metabolism yields another 350 ml of water, for a total input of about 2.5 liters per day.

Figure 39-3   **Salty tears.** Like many sea animals, this tortoise excretes excess salt in glands located near the eye.   *(M.P. Kahl/Photo Researchers)*

Normally, the body loses the same amount of water it takes in. Water leaves the bodies of land animals by three main routes: urine, feces, and evaporation. Evaporation takes place both from the outer surface of the body and from the surfaces of the lungs or, in invertebrates, other respiratory organs.

Most water leaves as urine. In humans, about 100 ml per day is lost in the feces, and about 50 ml is lost in sweat. Most of the rest leaves as evaporation from lungs and skin, as well as in various secretions (Figure 39-4).

Species differ in the total amount of water that they process each day. Desert animals, which have access to almost no water, conserve water meticulously. The kangaroo rat, a common desert rodent in the American Southwest, lives entirely on water derived from the metabolic breakdown of macromolecules in the seeds that it eats.

The kangaroo rat is much stingier with its water than nondesert animals. It lacks sweat glands entirely, its urine is highly concentrated, and it spends the daylight hours in cool underground burrows whose air is moister than that outside. It loses water mostly through evaporation from the lungs. Even this water loss, however, is less than it would be in other animals. The kangaroo rat has long nasal passages that cool and condense the water in exhaled air. The condensed water is then reabsorbed before the air leaves the kangaroo rat's body.

In all animals, water intake and loss may vary greatly from day to day. A human, for example, may drink more one day

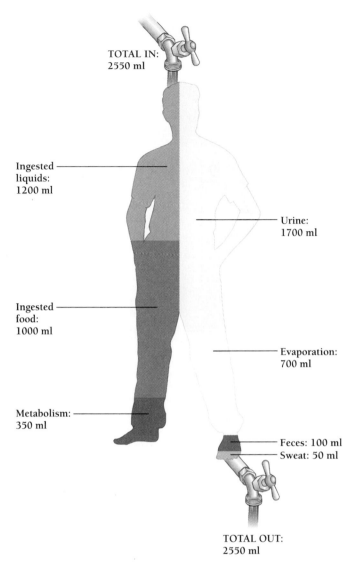

TOTAL IN:
2550 ml

Ingested
liquids:
1200 ml

Ingested
food:
1000 ml

Metabolism:
350 ml

Urine:
1700 ml

Evaporation:
700 ml

Feces: 100 ml
Sweat: 50 ml

TOTAL OUT:
2550 ml

Figure 39-4   **Water in, water out.** An average human ingests and excretes 2550 ml of water each day. Water comes into the body as food and drink and leaves in the urine, feces, and sweat and in the form of vapor from the lungs and other moist surfaces.

than the next, sweat profusely during vigorous exercise, or lose too much water because of diarrhea or vomiting. Humans and other animals must therefore regulate water flow. We do so mostly by varying our drinking and urine production. We are so successful that, under most circumstances, our water content varies less than 1 percent. The failure of such regulation leads either to **dehydration,** a physiological state in which too much water flows out of the body, or to water intoxication and **edema** (Greek, "swelling"), in which water accumulates in the tissues.

Land animals acquire water from water, from water in food, and from the metabolism of food. Animals lose water in urine, feces, and evaporation from moist surfaces such as the lungs.

HOW DO ANIMALS DISPOSE OF NITROGEN-CONTAINING WASTES?

In land animals, water and salt regulation is tightly connected to the disposal of a variety of wastes. Chief among these are the **nitrogenous** (nitrogen-containing) **wastes,** which are the products of protein breakdown. Nitrogenous wastes result from the metabolism of proteins and nucleic acids in food as well as from the normal breakdown of cellular components.

Cells and organisms rid themselves of nitrogenous wastes by means of **excretion,** a disposal process by which cells pass materials across a cell membrane. In contrast, the disposal of the remains of digested food is called **elimination.** Mammals excrete nitrogenous wastes in urine and eliminate digestive wastes in feces.

Fish and most aquatic animals accomplish excretion by the conversion of excess nitrogen to **ammonia** (NH_3). Ammonia-loaded urine flows rapidly into the watery environment, where it provides usable nitrogen to a variety of aquatic microorganisms and plants.

Ammonia, however, it toxic to animals. Aquatic animals can dilute the ammonia to harmless concentrations. But land animals do not have access to enough water to dilute the ammonia that much. Land mammals solve this problem by converting ammonia into **urea,** a relatively nontoxic organic compound, by means of enzymes secreted by the liver (Figure 39-5). The urea is less toxic, but it dissolves only in a fairly large amount of water. Animals must therefore secrete a minimum amount of water each day—in humans, about 500 ml. Mammals still must excrete enough water to dissolve urea, but less than if they had to excrete diluted ammonia.

Insects, land snails, most reptiles, and birds require even less water than mammals. Instead of disposing of nitrogenous wastes as urea, they produce a nearly insoluble organic compound called **uric acid** (Figure 39-5). Some insects, as well as the embryos of reptiles and birds, go one step further. They do not excrete the uric acid at all. They squeeze out as much water as they can and then deposit dry uric acid in their own bodies.

Cells and organisms rid themselves of nitrogenous wastes by means of excretion. Aquatic organisms excrete dilute ammonia; mammals excrete urea; and insects, land snails, birds, and most reptiles excrete uric acid.

HOW DOES THE MAMMALIAN KIDNEY FILTER BLOOD?

In vertebrates, the major organs of excretion are the kidneys. In addition to excreting nitrogenous wastes, the kidneys also regulate water balance, as well as the concentrations of salts and other organic molecules, including various wastes and tox-

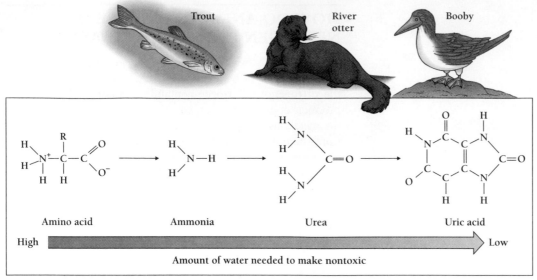

Trout   River otter   Booby

Amino acid   Ammonia   Urea   Uric acid

High   Low

Amount of water needed to make nontoxic

A.

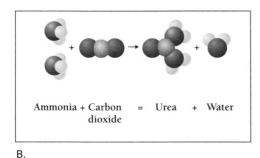

Ammonia + Carbon dioxide = Urea + Water

B.

Figure 39-5   Ammonia, urea, uric acid. Animals excrete nitrogenous wastes in different forms. A. Trout and other fish can excrete nitrogen as ammonia because they have access to plenty of water with which to dilute the toxin. Otters and most other mammals excrete urea, a much less toxic substance than ammonia. Many animals—including boobies and other birds, insects, land snails, most reptiles, and Dalmatian dogs—excrete nitrogen in the form of uric acid, which is a solid. B. Two molecules of toxic ammonia are transformed into urea and water through the addition of a molecule of carbon dioxide.

ins. By starting with the kidney's role in excretion, however, we can more easily see how it performs its other tasks.

Kidneys are paired structures that lie below the stomach and liver (Figure 39-6). Blood, carrying wastes from all over the body, arrives at each kidney by a **renal artery** [Latin, *renes* = kidney] and leaves by a **renal vein.** The two kidneys take wastes and water from the blood and produce urine that flows down into the **bladder** through a pair of tubes called the **ureters** [Greek, *ouron* = urine]. The bladder in turn drains to the outside of the body through the **urethra.**

In female mammals, the urethra empties directly to the outside. In male mammals, the urethra joins the reproductive tract and passes through the penis, a common exit for both sperm and urine. In most birds and amphibians, the ureter runs directly into a **cloaca,** a common exit chamber for the digestive, excretory, and reproductive systems.

How Does the Nephron Filter Blood?

The bland exterior shape of a kidney reveals little about how it works. The interior of a kidney, however, is intriguingly intricate (Figure 39-7). A section through a kidney shows an outer region, called the **cortex,** and an inner region, called the **medulla.**

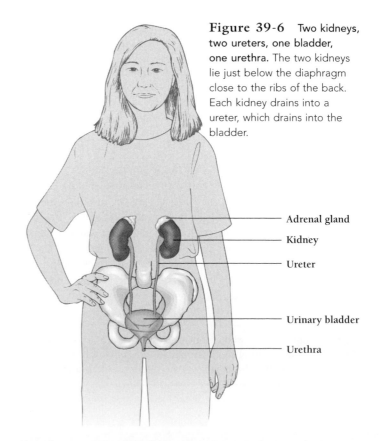

Figure 39-6   Two kidneys, two ureters, one bladder, one urethra. The two kidneys lie just below the diaphragm close to the ribs of the back. Each kidney drains into a ureter, which drains into the bladder.

Adrenal gland

Kidney

Ureter

Urinary bladder

Urethra

Figure 39-7 **Cross section through a kidney.** The outer cortex contains the renal tubules and Bowman's capsule for a million nephrons. The inner medulla contains all the loops of Henle.

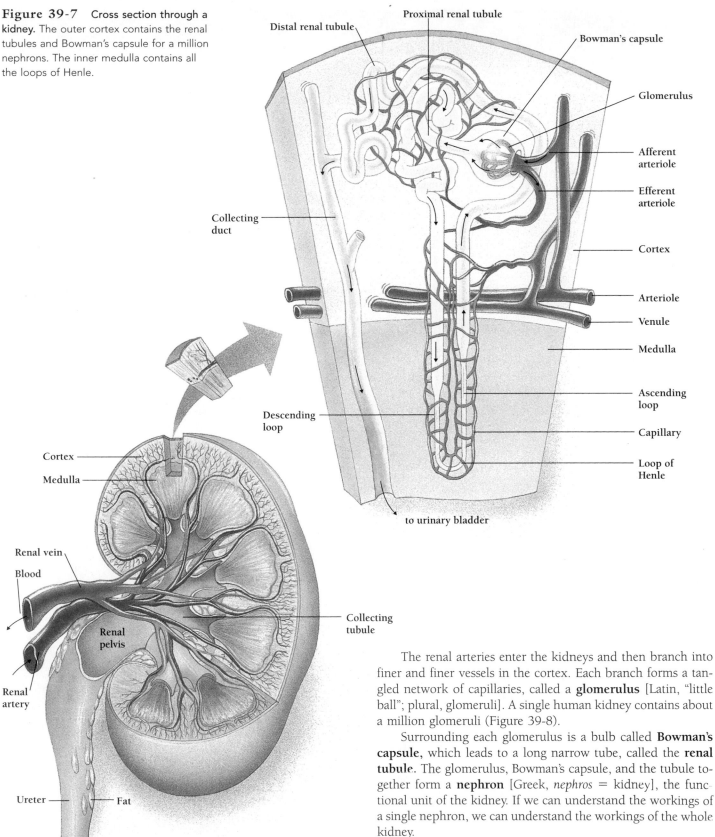

The renal arteries enter the kidneys and then branch into finer and finer vessels in the cortex. Each branch forms a tangled network of capillaries, called a **glomerulus** [Latin, "little ball"; plural, glomeruli]. A single human kidney contains about a million glomeruli (Figure 39-8).

Surrounding each glomerulus is a bulb called **Bowman's capsule**, which leads to a long narrow tube, called the **renal tubule**. The glomerulus, Bowman's capsule, and the tubule together form a **nephron** [Greek, *nephros* = kidney], the functional unit of the kidney. If we can understand the workings of a single nephron, we can understand the workings of the whole kidney.

The structure of the glomerulus and of Bowman's capsule provides a clue to how the kidney performs its first task—filtration. The capillaries of the glomerulus contain tiny pores

BOX 39-1

The artificial kidney

Infections, poisons, genetic diseases, tumors, and physical trauma can all severely damage the kidneys. In fact, more than 25,000 people die of kidney failure in the United States each year. When a person's kidneys begin to fail, urea and other substances build up in the blood. This condition is called uremia ("urine in the blood").

Many patients with kidney failure stay alive with the help of a device called a kidney dialysis machine. The patient's blood runs into a chamber containing a membrane that allows the passage of salt ions, as well as urea and other small molecules (Figure A). On the other side of the membrane is a large volume of a solution whose composition is the same as that of normal blood plasma. Sodium, chloride, and potassium ions, as well as glucose and other molecules, move passively across the artificial membrane and come to the same concentrations on each side. Urea and other small waste molecules diffuse across the membrane into the large volume, so that their concentration falls nearly to zero. The kidney dialysis machine thus restores the blood plasma to a nearly normal composition.

Patients who depend on dialysis machines cannot regulate their water balance with hormones, as the rest of us do. Instead, they must carefully balance their intake in food and drink to match losses from evaporation and elimination. Depending on the remaining abilities of their damaged kidneys, these patients produce little or no urine. Even the best dialysis machine, then, cannot approach the feats of the simplest kidney.

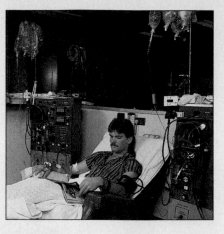

Figure A Kidney dialysis. A young man with damaged kidneys waits as a dialysis machine cleans urea and other small waste molecules from his blood.
(SIU/Photo Researchers)

Figure 39-8   A glomerulus. Every nephron includes a network of capillaries called the glomerulus. Blood pressure forces fluid out of the capillaries of the glomerulus and into an enclosed space called Bowman's capsule. The "filtrate" in Bowman's capsule flows into the proximal renal tubule for processing by the kidney.

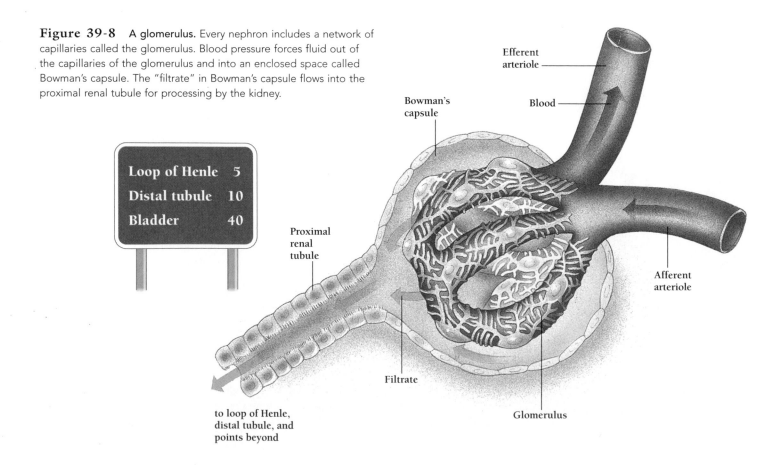

Loop of Henle   5
Distal tubule   10
Bladder   40

Efferent arteriole

Bowman's capsule

Blood

Proximal renal tubule

Afferent arteriole

Filtrate

to loop of Henle, distal tubule, and points beyond

Glomerulus

through which the liquid part of the blood (the plasma) can escape, leaving blood cells and macromolecules behind. Similarly, the cells of Bowman's capsule also contain sievelike openings through which the filtered plasma passes.

Blood from a renal artery enters each glomerulus and filters through the pores into Bowman's capsule. Blood cells and large molecules remain behind in the blood, but water, urea, and other small molecules in the blood freely pass into Bowman's capsule. The resulting liquid, called the **filtrate**, passes into the narrow renal tubule. The initial filtrate is similar in composition to the blood's plasma.

As blood enters a glomerulus, all of the constituents of the blood except the cells and large molecules filter into Bowman's capsule and the renal tubule.

How Do the Tubules Put Back Essential Molecules and Ions?

In humans, the total volume of filtrate is almost 200 liters per day (enough to fill a very large bathtub). But the nephrons must retain enough water to maintain the volume of the blood, as well as all the essential ions and molecules that form an indispensable part of the internal environment. The nephrons therefore recover more than 99 percent of the 200 liters of water and proteins. As a result, only 1.5 liters actually leaves the body as urine.

How can the kidney filter molecules so selectively? After the glomerulus and Bowman's capsule remove nearly everything from the blood, the filtrate flows down the length of the

tubule and the cells of the tubule selectively remove ions and molecules and return them to the blood. At the end of the tubule, the remaining filtrate exits via a collecting duct and empties into the ureter.

The renal tubules remove water and valuable ions and molecules from the filtrate and return them to the blood.

What Does the Structure of the Tubules Suggest About Their Workings?

Every tubule has a characteristic hairpin shape, with three main regions: (1) the wide **proximal tubule**, which lies just next to Bowman's capsule, in the kidney's outer cortex; (2) the narrow **loop of Henle** (pronounced HEN-lee), which first descends into the inner medulla and then ascends back into the cortex; and (3) the wide **distal tubule**, from which the urine flows into the collecting duct and then to the bladder (Figure 39-8).

We can see how a tubule works by examining the microscopic structure of its **epithelium**, the sheet of cells that lines the lumen of the tubule. The proximal tubule has great numbers of fingerlike *microvilli*, called *brush borders*. These fuzzy patches of microvilli help pump materials across the epithelium (Figure 39-9). Coupled with the brush borders are dense concentrations of mitochondria, which provide energy to power these pumps.

In contrast, the descending limb of the loop of Henle and the deepest part of the ascending limb have neither brush borders nor many mitochondria. We can guess from this difference that these parts of the loop of Henle do not actively pump materials out of the lumen.

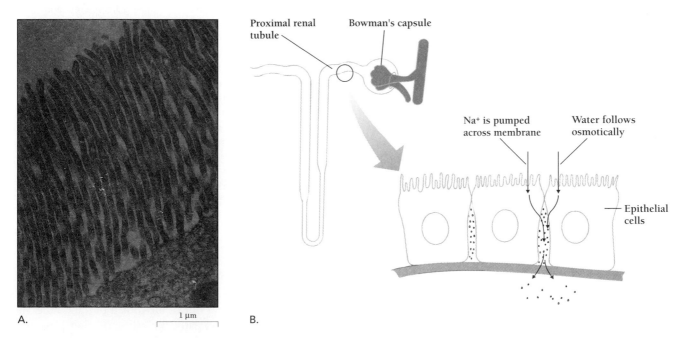

Figure 39-9 Brush borders. A. Microvilli, or brush borders, in the proximal tubules increase surface area and so promote the movement of ions. B. In the proximal tubules, brush border cells pull Na$^+$ ions from the filtrate. Water then follows the ions out of the tubule. *(A, Fred Hossler/Visuals Unlimited)*

Further along, in the upper portion of the ascending loop of Henle, in the distal tubule, and in the upper part of the collecting duct, we find that the epithelium again has brush borders and great numbers of mitochondria.

The three parts of a renal tubule differ in their structure and function. The proximal tubule is lined with microvilli, which pump ions across the epithelium. The descending limb of the loop of Henle is smooth and does not actively pump ions. The ascending limb of the loop of Henle and the distal tubule are lined with microvilli.

How Do the Tubules Regulate the Ion Content of the Lumen?

Researchers have been able to study the workings of the nephron by taking samples of the contents of different parts of the tubules with fine pipettes. When these samples were analyzed, researchers found that the concentrations of specific ions and molecules change along the length of the tubule. These concentration changes result both from the active pumping of ions across the membranes and from osmosis.

The changes in ion concentration begin in the epithelium of the proximal tubule, which actively pumps sodium ions out of the filtrate in the lumen. Other ions follow, and, because the epithelium is freely permeable to water, osmosis causes the water to follow the salt out of the lumen and into nearby capillaries.

By the time the filtrate in the lumen enters the loop of Henle, it has lost about 75 percent of its volume. The remaining fluid has far less water and salt, but it still has many small molecules, including urea. As a result, the fluid has the same concentration as the blood in the surrounding capillaries.

As we guessed from the absence of mitochondria and a brush border, neither the descending limb of the loop of Henle nor the deep part of the ascending limb actively transport ions (Figure 39-10). The deep part of the medulla, at the bottom of the loop of Henle, is very salty, however, and water in the lumen moves out through the walls of the tubule by osmosis, leaving the filtrate in the lumen even more concentrated.

Beyond the turn of the loop, the epithelium is impermeable to water and no more water escapes the lumen. The upper part of the ascending limb, however, actively pumps salt out of the lumen, and so does the upper section of the collecting duct (Figure 39-10). Because the tubule forms a loop, this fluid bathes the deep part of the loop of Henle. As a result, the intracellular fluid surrounding the loops of Henle becomes extremely salty. In contrast, the salt concentration in the outer part (the cortex) of the kidney is comparatively low.

Active pumping of ions by the cells in the wall of the renal tubules creates osmotic gradients that pull water from the filtrate.

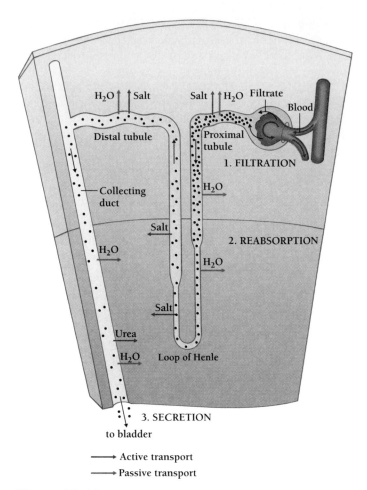

Figure 39-10   Countercurrent in a nephron. A nephron is a countercurrent system that extracts urea from the blood. Microvilli at the beginning of the renal tubule actively pump sodium ions out of the filtrate in the tubule. The filtrate, containing water, urea, and other molecules, descends into the loop of Henle, which is surrounded by the salty tissue of the medulla. Here, water passively diffuses from the tubule into a network of capillaries surrounding the loop of Henle. Beyond the turn of the loop, the epithelium is impermeable and water remains in the tubule. While the water is trapped inside, however, the same region of the tubule actively transports sodium and chloride (Na^+ and Cl^-) ions out. The fluid entering the collecting duct is consequently dilute. As this dilute fluid descends into the salty medulla, water flows out passively and the urea left in the duct becomes concentrated. Finally, as the water and urea pass through the saltiest portion of the medulla, even some urea flows out passively, further increasing the concentration of ions and molecules in the medulla. The remaining fluid, the urine, now flows from the collecting duct to the ureters and the bladder.

What Happens to the Urea?

Like salt, urea enters the tubule in the original filtrate. But unlike salt, urea cannot move out of the lumen of the tubule. As the filtrate moves along its path, the salt and water leave the lumen, leaving behind an increasingly concentrated solution of urea. Only when the urea-loaded filtrate reaches the far end of

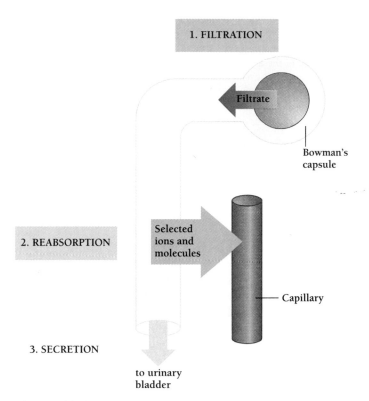

Figure 39-11   Filtration, reabsorption, and secretion. Each nephron *filters* cells and large molecules from the blood, *reabsorbs* salts and water, and *secretes* urea and water and waste materials.

the collecting duct can the urea diffuse out. As the concentrated urea exits the tubules, it raises the ion concentration of the fluid surrounding the deep part of the loop of Henle. The high urea content of this fluid, as well as its saltiness, is the driving force that pulls water and salt passively out of the lumen of the deepest part of the descending limb (Figure 39-10).

Urea stays in the renal tubule until it reaches the far end of the collecting duct. Here it accumulates in the surrounding tissues and draws water from the remaining fluid in the tubule.

How Does the Mammalian Kidney Form Urine?

The filtrate leaving the nephron is still dilute and not yet ready to be excreted. In the collecting ducts, the water content of the filtrate can remain dilute or become highly concentrated, depending on a host of variables we will explore at the end of this chapter.

As the collecting ducts pass through the deepest part of the medulla, where the concentrations of salt and urea are the highest, water and urea passively move out of the lumen. The urine that leaves the duct can therefore attain the same concentrations of sodium, chloride, and urea as the surrounding extracellular fluid in the deepest region of the loop of Henle.

For humans, this means a final maximum salt concentration almost four times that in blood plasma. In the kangaroo rat, the loop of Henle is longer, and the concentrating mechanism is still more effective. The salt concentration in the urine of a kangaroo rat can be 14 times that in the animal's blood plasma.

In addition to managing the flow of salt, urea, and water, the nephron also controls the concentrations of other substances in the blood and the urine. The nephron moves substances in both directions across the tubular epithelium. Transport of specific substances such as salt, water, and glucose from the filtrate to the blood is called **reabsorption** (Figure 39-11). Glucose reabsorption is so efficient, for example, that the urine of a healthy person normally contains no glucose at all.

Transport into the filtrate is called **secretion.** Secreted substances include potassium and hydrogen ions, ammonia, and organic acids and bases. In addition, the kidney secretes many foreign substances (antibiotics, for example, as well as other drugs and toxins).

A kidney nephron generates urine that contains high concentrations of urea, as well as various ions, ammonia, organic acids and bases, and foreign molecules. The nephron normally absorbs glucose.

HOW DOES A MAMMAL ADJUST WATER AND SALT IN RESPONSE TO CHANGING CONDITIONS?

Humans and other animals consume different amounts of water and salt from day to day, and they eliminate different amounts in their feces, sweat, and urine. Maintaining a constant level of water and salt concentration in the blood means that the kidneys must constantly adjust urine concentration and flow. Such homeostasis depends on a variety of complex mechanisms, including the control of blood flow through the kidneys and the control of fluid intake by the brain. Three hormones also regulate kidney function. The first, aldosterone, is a steroid hormone. The other two are polypeptides: antidiuretic hormone (ADH), or vasopressin, and atrial natriuretic factor (ANF).

Renal Blood Flow

Most organs of the body regulate blood flow independently of the flow supplied by the heart. The arterioles in a working muscle, for example, open up to increase the flow of blood throughout the muscle. Blood flow to the renal arteries of the kidneys is generally much more constant than blood flow to other areas of the body. Nonetheless, these arteries respond to variation in the concentration of salt and the end-products of protein metabolism in the blood. When the concentration of salt or protein wastes in the blood rises, blood flow in the renal arteries increases. The kidneys can then process more blood and more filtrate to remove excess salt, as after we have eaten a couple of slices of pizza.

An increase in the concentration of salt or protein wastes in the blood causes an increase in the blood flow in the renal arteries.

Thirst

Animals control water balance in the body by drinking as much water as they lose. The hypothalamus of the brain includes a "thirst center" that has *osmoreceptors,* receptors that detect changes in salt concentration. When these receptors are stimulated by an increase in salt concentration in the extracellular fluid, we experience the sensation of thirst, which lasts until we drink some water. Once we drink water, we do not feel thirsty for about 15 minutes. If the stomach is full, we may not feel thirst for up to 30 minutes. If, however, the salt concentration in the blood has not normalized, we become thirsty again until the proper concentration is achieved. A thirsty animal almost never drinks more water than the amount needed to relieve dehydration. We always drink almost exactly the right amount.

When salt concentrations in the blood increase, the brain's hypothalamus detects the increase and triggers a sensation of thirst.

Angiotensin and Renin

Angiotensin is the most powerful constrictor of blood vessels known. One ten-millionth of a gram (0.1 μg) of angiotensin can cause the blood vessels to constrict so much that blood pressure increases by as much as 20 points. Angiotensin works in conjunction with an enzyme secreted by the kidneys called **renin.**

Either a decrease in blood pressure or a decrease in the concentration of sodium ions in the blood causes the kidneys to secrete renin. Renin then acts to form angiotensin in the blood. Angiotensin constricts the arterioles and stiffens and constricts the veins, which reduces the total space in which the blood can move (Figure 39-12A). Angiotensin also constricts the renal arterioles, increasing blood pressure in the kidneys and causing them to retain both water and salt. This salt and water retention increases total blood volume. This increased volume of blood, confined in a circulatory system of limited size, is consequently under increased pressure. Angiotensin compensates for a decrease in sodium ions or blood pressure with a homeostatic build-up of body fluid and sodium ions.

A decrease in blood pressure or in sodium ions causes the kidneys to secrete renin, which forms angiotensin. Angiotensin constricts the arterioles and stiffens and constricts the veins, increasing blood pressure and causing the kidneys (and therefore the body) to retain both water and salt.

Aldosterone

The hormone **aldosterone** comes from the **adrenal glands** [Latin, *ad* = towards + *renes* = kidneys], which, as their name suggests, lie on top of the kidneys. Aldosterone helps keep the body from losing salt by stimulating sodium and potassium reabsorption in the distal tubules and collecting ducts of the kidneys, as well as in the sweat glands and in the intestines (Figure 39-12A). When the concentration of sodium in the blood decreases, the adrenal glands secrete more aldosterone.

Aldosterone regulates water loss as well as salt loss. People who cannot secrete aldosterone because their adrenal glands are missing lose large amounts of salt in the urine. Along with the salt goes a tremendous quantity of water. Most people can compensate by consuming salty foods and lots of water. If not, however, a person lacking aldosterone can become severely dehydrated. In that case, the volume of blood is so low that the heart cannot supply blood to the tissues. The result is severe shock and, often, death.

Aldosterone secretion depends not only on the adrenal glands themselves but also on chemical signals from the liver and electrical signals from the nervous system. This complex control system allows regulation of sodium ion reabsorption in response to changes in blood pressure, salt concentration, water loss, and disease.

Aldosterone prevents water loss and salt loss by stimulating the reabsorption of sodium and potassium ions in the kidneys, the sweat glands, and the intestines.

Antidiuretic Hormone (ADH), or Vasopressin

Like aldosterone, **antidiuretic hormone (ADH)** prevents water loss. ADH, also called **vasopressin,** resembles angiotensin in its ability to constrict the arterioles, although it does not constrict the veins. Antidiuretic hormone, as its name suggests, is the opposite of a *diuretic,* a substance such as caffeine or alcohol that stimulates water loss. ADH causes the body to retain water by increasing water permeability in the collecting ducts of the kidneys. Water then returns to the blood through the walls of the collecting ducts and the urine becomes more concentrated (Figure 39-12B). By this means ADH can raise blood pressure by as much as 40 points.

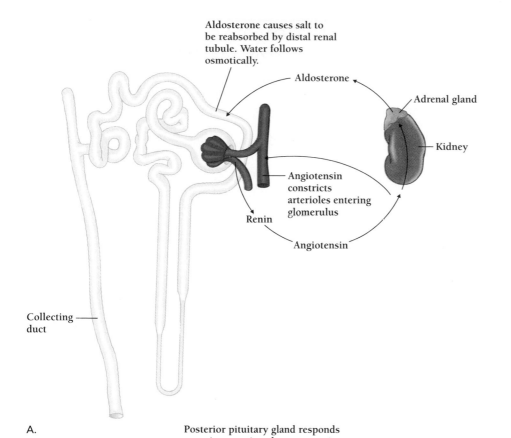

Aldosterone causes salt to be reabsorbed by distal renal tubule. Water follows osmotically.

Aldosterone

Adrenal gland

Kidney

Angiotensin constricts arterioles entering glomerulus

Renin

Angiotensin

Collecting duct

A.

Figure 39-12 Hormones and blood pressure. A. When blood pressure or sodium concentration drops, the kidneys secrete the enzyme renin, which acts to form angiotensin in the blood. Angiotensin counteracts a drop in blood pressure by constricting the arterioles and veins, which increases blood pressure and causes the kidneys to retain both water and salt. The hormone aldosterone also increases blood pressure by specifically stimulating re-absorption of salt (and, therefore, water) in the distal tubules, thus increasing total blood vol-ume. B. When the concentration of salt in the blood rises, the posterior pituitary gland se-cretes ADH (vasopressin), which increases wa-ter permeability in the collecting ducts of the kidneys. By causing the body to retain water, ADH can raise blood pressure by as much as 40 points. The heart limits such increases in blood pressure by secreting the hormone ANF. ANF inhibits the secretion of renin (and therefore angiotensin), aldosterone, and ADH (vasopressin). ANF also lowers blood pressure by relaxing smooth muscles, reducing thirst, and increasing the kidney's elimination of Na^+ and water by closing channels in the collect-ing ducts.

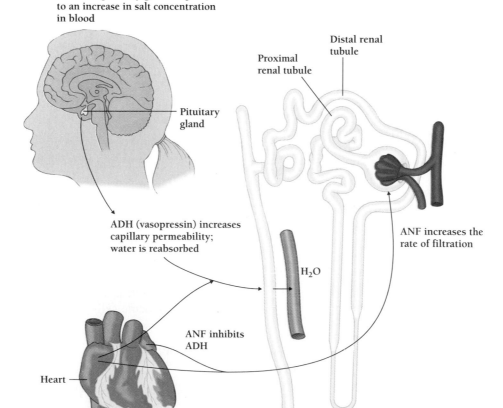

Posterior pituitary gland responds to an increase in salt concentration in blood

Distal renal tubule

Proximal renal tubule

Pituitary gland

ADH (vasopressin) increases capillary permeability; water is reabsorbed

H_2O

ANF increases the rate of filtration

ANF inhibits ADH

Heart

Atria of the heart respond to an increase in blood pressure by secreting ANF

Collecting duct

B.

The cells that make ADH lie in the **posterior pituitary,** a structure that lies within the brain (Chapter 36). The cells of the posterior pituitary respond to changes in the salt concentration of the blood. After a TV dinner, when the concentration of salt in the blood rises, the posterior pituitary secretes ADH into the blood. When the ADH reaches the kidneys, it stimulates the kidneys to reabsorb water. The urine becomes more concentrated, the blood becomes more dilute, and the salt concentration of the blood returns to normal. The total volume of the blood increases, however, and the result is an increase in blood pressure.

Other factors also influence ADH production and secretion. Alcohol, for example, inhibits the release of ADH, increasing urine flow. Alcohol is thus a diuretic. But alcohol also stimulates an appetite for salt. A beer drinker reaches for a bowl of salty pretzels, then for another beer. But while the salt stimulates the release of ADH, the alcohol inhibits its release. Pain, fear, cold, and stress can all affect ADH levels and the amount of urine production.

The same hormone systems that mediate changes in kidney operations also affect brain centers to stimulate thirst and appetite for salt. The body thus controls the composition of the blood and other fluids by regulating the transport of specific substances and altering drinking and eating behavior.

When the concentration of salt in the blood rises, the posterior pituitary secretes ADH, which increases water permeability in the collecting ducts of the kidneys. By causing the body to retain water, ADH can raise blood pressure by as much as 40 points.

Atrial Natriuretic Factor (ANF)

As we saw earlier, kidney function depends on blood pressure to provide the force for filtration and flow of the filtrate through the tubules. We have seen that an increase in ADH lowers urine volume and raises the volume of blood in the vessels. The result is an increase in blood pressure.

The heart itself limits such increases. In response to excessive blood pressure, cells in the smaller chambers of the heart (the atria) produce a hormone called **atrial natriuretic factor** [Latin, *atrium* = vestibule, referring to the heart chamber, + *natrium* = sodium + Greek, *ouron* = urine]. ANF inhibits the secretion of renin, aldosterone, and vasopressin, relaxes smooth muscles, reduces thirst, and increases the kidneys' elimination of sodium ions and water by closing channels in the collecting ducts. ANF thus increases water loss and lowers blood pressure (Figure 39-12B).

The heart produces a hormone that limits increases in water loss and lowers blood pressure.

EXCRETION, WATER, AND SALT IN NONMAMMALS

Animals in different environments must solve different problems of salt and water balance. Their solutions differ in ways that at least partly reflect their distinct evolutionary histories.

Vertebrates

Not surprisingly, the excretory systems of nonmammalian vertebrates differ from those of mammals. The kidneys of marine bony fish, for example, have nephrons without glomeruli and Bowman's capsules. Bony fish adjust the composition of the fluid in the tubules by secretion and reabsorption only. In contrast, freshwater bony fish have highly developed glomeruli, the better to rid the fish of excess water.

Both freshwater and saltwater fish have special cells in their gills that can pump chloride and other ions. The direction of the pumping depends on the environment. Freshwater fish pump salt ions into the body, whereas saltwater fish pump them outwards. Salmon and other fish that migrate from fresh water to salt water and back change the organization of these cells and the direction of salt pumping according to whether they are in fresh water or salt water.

The kidneys of most birds resemble those of mammals. Marine birds, which live on marine fish and drink salt water, however, have another means of disposing of salt. They possess special salt glands just above their eyes, which secrete a salty solution with twice the concentration of sea water. Some reptiles, including marine turtles, also possess salt glands. The salty tears produced by these salt glands appear to help these animals rid themselves of excess salt.

Flatworms

The simplest excretory organs are those of flatworms. These organs mainly excrete water, for most nitrogenous wastes are eliminated through the one-ended gut. The excretory organs of flatworms consist of two or more tubes that run the length of their bodies and empty through tiny pores. Leading into the tubes are many bell-shaped chambers, each formed from a single cell. Cilia, which line the interiors of these bell-shaped cells, propel liquids into the tubules. Under the microscope, the cilia appear to be flickering, an illusion which gives these cells their curious name, **flame cells.**

Earthworms

Although we think of the earthworm as a land animal, it spends most of its life in damp tunnels in the soil. Its body fluids therefore have a higher concentration of ions and molecules than the surrounding water, and the earthworm has the same problem as a freshwater animal—getting rid of excess water.

Like mammalian kidneys, earthworm **nephridia** (singular, nephridium) both filter and reabsorb. Each body segment con-

Figure 39-13 **Malpighian tubules.** In insects, the excretory organs are blind outpocketings of the gut.

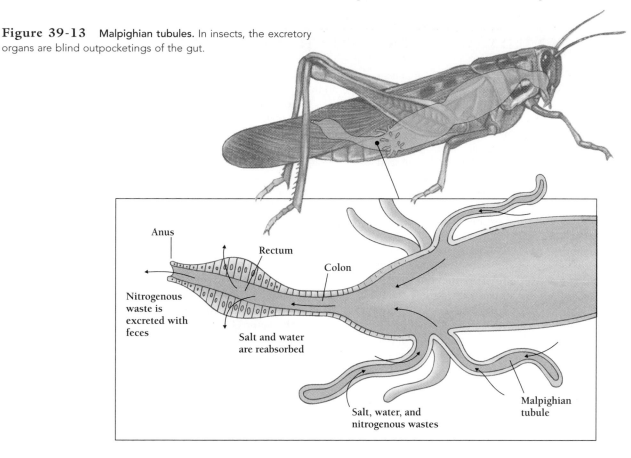

tains a pair of nephridia. Fluid in the earthworm's body cavity, or coelom, is under pressure. The pressurized fluid pushes through an opening in the tubular part of the nephridium. Inside the nephridium, ions and molecules re-enter the circulatory system. The dilute urine left behind passes from each nephridium through a pore in the body wall.

Marine Invertebrates

Crabs and most other marine invertebrates have body fluids with about the same concentration of salts as sea water. Many of these animals allow the concentrations of ions and molecules in their body fluids to fluctuate with that of their environment. In general, the disposal of excess water is not a problem for such animals.

Still, marine invertebrates must get rid of nitrogenous wastes and adjust the concentrations of specific ions. They accomplish these tasks with kidneys that, like those of vertebrates, work by filtration, absorption, and secretion. Like other aquatic animals, they eliminate excess nitrogen in the form of ammonia. Some invertebrates have arrangements of excretory organs very different from ours. Lobsters and other crustaceans, for example, perform excretion in their heads, through *green glands,* which function much like other invertebrate kidneys. And the abalone's two kidneys differ in function—the left kidney specializes in reabsorption and the right in secretion.

Insects

The excretory system of insects differs from all the other systems we have discussed. Insect excretory organs, called Malpighian tubules, are blind outpocketings of the gut. These tubules may number from two to several hundred. Their blind ends are bathed in the insect's body fluids (Figure 39-13).

Fluid flows into the tubules, but not because of a pressure difference (as in earthworms and marine invertebrates). Instead, the tubules actively pump potassium ions. Water, with dissolved nitrogen wastes, passively flows into the tubules by osmosis. The contents then flow into the gut, where the residues of digestion (the feces) mix with the products of excretion (the urine).

The insect's hindgut and rectum are especially efficient in removing water, so that the combination of feces and urine can be extremely dry—depending on the insect's diet. An insect that lives on fresh leaves has abundant water intake and excretes lots of liquid urine, while a termite, which dines on wood, excretes dry little pellets with the consistency of the compacted sawdust "logs" that city people buy at the supermarket for their fireplaces.

In this chapter we have seen why animals need kidneys and how they function to excrete water, salts, and nitrogenous wastes. In the next chapter we will see how animals defend themselves against infection by microorganisms, promote healing, and minimize blood loss—all important functions of the immune system.

STUDY OUTLINE WITH KEY TERMS

The "internal environment" of an animal is the external environment of individual cells. Animals must regulate the composition of the fluids in this extracellular space.

An animal's major problem in regulating its internal environment depends on the total salt concentration in its external environment. Freshwater animals tend to gain water and lose salts, while marine animals tend to lose water and gain salts. The major problem of land animals is to get and conserve water. Every animal has characteristic adaptations for regulating salt and water, for preventing **dehydration** or **edema.**

Animals must rid themselves of **nitrogenous wastes** from the proteins in their food. Marine invertebrates and fish excrete excess nitrogen in the form of **ammonia.** Mammals instead convert the waste nitrogen to **urea,** which is less toxic than ammonia. Birds, insects, and other animals excrete **uric acid. Excretion** and **elimination** are separate processes. Flatworms excrete nitrogenous wastes using **flame cells,** while earthworms use more complex **nephridia.**

Blood carries wastes to the kidneys through paired **renal arteries. Renal veins** carry the cleansed blood away from the kidneys. Urine leaves the kidneys in paired **ureters,** fills the **bladder,** and exits the body through a **urethra** or **cloaca.**

In vertebrates, the **kidney** is the major organ of excretion and of salt and water balance. The kidney regulates the internal environment by regulating the composition of the urine. This regulation depends on three processes—**filtration, reabsorption,** and **secretion.**

The **cortex** and **medulla** of the kidney consist of many separate functional units, called **nephrons.** Each nephron consists of a filtering unit—the **glomerulus** and **Bowman's capsule**—and **renal tubule** that ultimately drains into the ureter. As the **filtrate** passes through the tube, the cells of the surrounding **epithelium** adjust the concentrations of specific ions and molecules.

In mammals, the wide **proximal tubule** of the nephron bends into the narrow **loop of Henle** and exits the kidney through the wide **distal tubule.** This lies close to the end of the corresponding collecting duct. This arrangement allows the mammalian kidney to produce a concentrated urine.

Several hormones regulate the operation of the kidney. **Angiotensin** and **renin** cause the kidneys to retain water. **Aldosterone,** from the **adrenal glands,** controls the reabsorption of salt in the nephrons. **Antidiuretic hormone** (ADH), or **vasopressin,** and **atrial natriuretic factor** (ANF) regulate water reabsorption in the collecting ducts. ADH is secreted by the **posterior pituitary** and ANF is secreted by the heart.

REVIEW AND THOUGHT QUESTIONS

Review Questions

1. Why do marine fish drink continuously, but freshwater fish not at all? Why do freshwater amebas have a contractile vacuole, but marine amebas do not?
2. Sketch a nephron and collecting duct. How do the cells of the epithelium vary from segment to segment? How does the transport of ions vary from segment to segment? How does the epithelium's permeability to water vary?
3. Explain how differences in the transport of salt ions and in the permeability to water change the concentration of the filtrate as it passes down the tubule.
4. How does the cellular organization of the glomerulus and Bowman's capsule account for the passage of urea into the nephron? How does urea get to be more concentrated in the urine than in the blood? How does glucose get to be less concentrated in the urine than in the blood?
5. What are the water balance problems of flatworms, earthworms, lobsters, and insects? What organs does each use to solve these problems?
6. Why are people with kidney problems also likely to have high blood pressure as well?

7. Suppose you have an exceptionally spicy meal and drink two extra glasses of water at dinner. What effects do you expect the extra water to have on the secretion of hormones by the hypothalamus, adrenal glands, and heart?
8. Suppose you eat an exceptionally salty meal, without drinking any extra water. What effects do you expect the extra salt to have on the secretion of hormones by the hypothalamus, adrenal glands, and heart?

Thought Questions

9. The Dalmatian hound lacks the enzymes needed to produce urea. How do you think the operation of its kidneys differs from that in other dogs?
10. Why do sea birds and sea turtles have salt glands? Why can a sea gull survive by drinking sea water, but a shipwrecked sailor cannot?
11. Whales and seals do not have salt glands, but live well without fresh water. How do you think they might manage this?
12. What purpose might the highly salty tears of humans serve?
13. Why does alcohol act as a diuretic?

SELECTED READINGS

Beeuwkes, R., "Renal Countercurrent Mechanisms, or How to Get Something for (Almost) Nothing," in C.R. Taylor et al., (Eds.), *A Companion to Animal Physiology,* Cambridge University Press, New York, 1982.

Eckert, R., Randall, D., and Augustine, G., *Animal Physiology: Mechanisms and Adaptations,* W.H. Freeman, New York, 3rd edition, 1988, Chapters 4 and 12.

Hochachka, P., and Somero, G., *Biochemical Adaptations,* Princeton University Press, 1984.

Schmidt-Nielsen, K., *Animal Physiology: Adaptation and Environment,* Cambridge University Press, New York, 3rd edition, 1983.

Schmidt-Nielsen, K., "Salt Glands," *Scientific American,* 200, January, 1959.

Schmidt-Nielsen, K., and Schmidt-Nielsen, B., "The Desert Rat," *Scientific American,* 189, January, 1953.

Smith, H.W., *From Fish to Philosopher,* Little Brown, Boston, 1953; paperback edition by Doubleday Anchor Books, 1961. A classic discussion of the function and the evolution of the vertebrate kidney.

Vander, A.J., Sherman, J.H., and Luciano, D.S., *Human Physiology: The Mechanisms of Body Function,* McGraw-Hill, New York, 4th edition, 1985, Chapter 13.

▶ On-line materials relating to this chapter are on the World Wide Web at http://www.saunderscollege.com/lifesci/
Click on Tobin/Dusheck: *Asking About Life.*

The Danger Model

In the spring of 1996, researchers simultaneously published scientific papers that together challenged basic ideas in immunology that have stood for 50 years. Ever since a series of crucial experiments in the 1940s, biologists have thought that the immune system protects the body by distinguishing self from nonself and that the immune system actually learns which proteins and other molecules are self (and which are not) around the time of birth. This idea is called the *self-nonself model.*

Specifically, experiments done in the 1940s seemed to show that the immune systems of newborn mice worked very differently from those of adult mice. Newborn mice would tolerate injections of cells from other mice that adult mice would reject. Not only that, the mice exposed early in life tolerated the same foreign cells when they grew up, a sign that their immune systems had adopted these foreign cells as self.

But recently, Polly C.E. Matzinger, a biologist at the National Institutes of Health, has begun to argue that her research, and that of two other teams, shows that newborn immune systems do not differ much from those of adults. Furthermore, says Matzinger, the entire self-nonself model is wrong (Figure 40-1). The immune system does not distinguish between self and nonself at all, she says, but between cells and molecules that are dangerous and those that are not.

This new theory, which Matzinger calls the *danger model,* has provoked strong feelings in her fellow researchers. A few researchers are delighted with the danger model. Transplant biologists and surgeons, in particular, see great practical value in the danger model. If the model is correct, surgeons may be able to perform organ transplants far more safely and successfully than ever before. Later in this chapter we will see why.

The majority of biologists, however, dismiss the danger model as a nice idea with little supporting evidence. For Matzinger, the most frustrating criticisms come from a few biologists who insist that she isn't saying anything new at all, that her theory is just a different way of saying the same thing. Says Matzinger, "They haven't really jumped off the cliff with me."

Jumping Off the Cliff

When Polly Matzinger was in her early twenties, she could hardly have anticipated that her scientific work would be featured on the cover of one of America's most well-known scientific journals. Although she had studied biology in school, she abandoned science to try dog training, waitressing, and carpentry. She even did a stint as a Playboy bunny.

It was while serving drinks to two biologists at a restaurant in Davis, near the University of California, that she redis-

Figure 40-1 Biologist Polly C.E. Matzinger. Matzinger argues for a model of immune function called the danger model. *(Myriam I. Rosado)*

covered science, or science discovered her. The two biologists were arguing about an experiment, when Matzinger joined the conversation, pointing out some errors in their thinking. Astonished, they welcomed her into the conversation, and one of them ultimately persuaded her to return to school.

Today, Matzinger heads an immunology laboratory at the National Institutes of Health, near Washington, DC, (and still trains border collies). Her perpetual restlessness now expresses itself, however, in her intellectual outlook. Matzinger has never been one to assume that those who came before her necessarily knew best.

Early studies in immunology focused on the transplanting of organs or skin from one individual to another. Normally, such transplanted organs are rejected: the cells of the immune system literally attack and destroy the cells of the transplanted organ. Most heart, liver, and other transplants today are therefore done with the aid of drugs that suppress the immune system so that the transplanted organ is tolerated.

In the 1940s, researchers noticed that skin grafts between cattle twins were tolerated. In identical twins this might not have been surprising since such individuals share the same cell surface markers that seem to characterize self. But even fraternal twins, which have different cell surface markers, could accept skin grafts from each other, as long as they had shared a blood supply inside their mother. Researchers speculated that the twins learned to recognize each others' cells as self while they were in the womb together.

To test this theory, English biologist Peter Medawar showed that he could "teach" newborn mice to tolerate foreign cells. Medawar injected cells from a white mouse into brown fetuses while they were still developing inside their mother. When the brown mice grew up, he grafted onto them patches of skin from the same white

mouse. All of the brown mice that had been exposed to white mouse cells in the womb accepted these grafts and sported white patches of healthy fur. In contrast, mice that had not been exposed to the

white mouse's cells rejected skin grafts from that mouse.

Medawar's work confirmed the prediction of an Australian virologist named Frank MacFarlane Burnet, the researcher who had first formulated the self-nonself model of immunology. Specifically, Burnet had proposed that the immune system's main task is to distinguish self from nonself and that it learns this distinction just once—during fetal development. In 1960, Medawar and Burnet shared a Nobel Prize for their trailblazing work.

But the self-nonself model has always had a few problems, problems that had bothered Matzinger when she was just beginning to study immunology. The self-nonself model seemed to work, however, and she put her doubts aside.

Then, in 1989, Matzinger met a younger scientist named Ephraim Fuchs who had similar doubts. How, asked Fuchs, can the immune system learn to recognize self before birth when the self keeps changing throughout life? At puberty, for example, our bodies go through major changes and develop new kinds of cells. Later, a pregnant woman hosts fetal cells, which because they reflect the genetic makeup of the father cannot be recognized by the body as self. Yet, mothers'

If the danger model is correct, surgeons may be able to perform organ transplants far more safely and successfully than ever before.

---

immune systems only rarely attack their fetuses. How can the immune system of the fetal mother learn to recognize such foreign selves years before encountering them?

Also, many foreign cells and viruses take up residence in our bodies long after the immune system is supposed to have learned what self is. After birth, our intestines become loaded with bacteria, which our immune systems tolerate. And that is essential, for many of these bacteria are necessary to life. Some help us digest milk; some make vitamin K, which helps our blood clot; and others protect us from invasion by less friendly bacteria. The immune system tolerates bacteria in our mouths, throats, and noses. None of these organisms is self. How can the immune system learn to recognize all these foreign organisms long before it encounters them?

Matzinger recollected all her old doubts. She and Fuchs agreed that it was unlikely that the immune system actually recognized self. There were too many exceptions. But, in that case, what then did the immune system do? Matzinger's answer came from her knowledge of the way cells die. A normal cell that has lived a full life and died signals macrophages—big, cell-eating cells—to consume it. The dead cell is quietly consumed, and none of its contents are released into the extracellular environment. In contrast, a cell under attack by viruses eventually bursts open, releasing not only virus particles but also the cell's cytoplasm and organelles. It's a mess guaranteed to attract the attention of the immune system.

In fact, this scenario perfectly explains the behavior of **T cells,** special immune cells that recognize and destroy infected cells, cancerous cells, and also the cells of transplanted tissues. T cells apparently recognize the body's own cells by means of a set of highly individualized cell surface molecules, called MHC (major histocompatibility complex) proteins.

According to the self-nonself theory—as it has evolved over the years—T cells that bind to the body's own healthy cells automatically die unless a *second signal* indicates that the cells are infected or foreign. In this way, the immune system "learns" to leave the body's own healthy tissues alone. If the T cells detect the second signal, however, they become activated and migrate to a lymph node, where they activate other immune cells. According to the self-nonself model, the T cells

could not be activated by this second signal during fetal development. As a result, all the T cells that respond to anything in the fetal environment die. Medawar's mice tolerated the injection of foreign cells because the T cells that recognized them had died. There were therefore no T cells to recognize and destroy the injected cells.

No one knows, however, what the "second signal" is nor why it should fail to operate during fetal development. One theory is that the second signal comes from "dendritic cells." Like T cells, dendritic cells are scattered throughout the tissues of the body, and, like T cells, they migrate to the lymph nodes when activated (see page 860). Dendritic cells may collect molecules from foreign (or damaged) cells and display them to the T cells. These molecules, displayed as surface molecules, tell the T cells to multiply and launch an attack against cells bearing the telltale molecules.

But there is one hitch. Unlike T cells, dendritic cells cannot distinguish self from nonself. If dendritic cells carry the mysterious second signal and they cannot themselves distinguish self from nonself, then how do they avoid triggering constant immune attacks against the body's own tissues?

According to Matzinger and Fuchs, the dendritic cells are activated by the cytoplasm-spilling deaths of cells—regardless of whether the cells belong to the body or not. If cells are dying messily, that's a danger signal the dendritic cells can understand. If a measles virus-infected skin cell bursts open, spilling skin cell proteins and virus proteins, a nearby dendritic cell registers the disaster, drinks up the cell's contents, and heads for the nearest lymph node (where the T cells congregate) to sound the alarm (Figure 40-2). There, the dendritic cell displays some of these proteins on its cell surface, activating nearby T cells and marking which cells to attack.

While the self-nonself model says that the immune system learns self from nonself during the fetal period only, the danger model says that the immune system constantly redefines what is dangerous. If virus-infected skin cells are dangerous today, they may not be tomorrow. If no cells are being damaged, then the dendritic

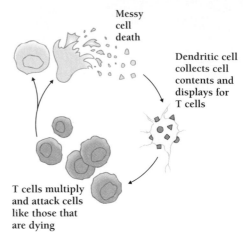

Figure 40-2   A simplified view of the danger model.

cells stop giving the second signal, and T cells that bind to skin cells die. As a result, the immune system can "quiet down," and become unreactive. In Matzinger's view we are constantly acquiring tolerance to our own proteins, to those of the bacteria that live with us, and to those of our own offspring. The self is redefined in a continuous process of adjustment.

Matzinger's theory explains why tumors and warts caused by viruses are often left alone by the immune system. The cells in these tissues are healthy and are not dying unnatural, messy deaths, so no second signal is given. After a while—without a second signal—the T cells that might react to them die off.

Matzinger speculates that organ transplants are rejected not because they are nonself but because they are full of dendritic cells activated by the trauma of being cut from the donor. As soon as the donated organ is transplanted, the donor's dendritic cells head for the lymph nodes of their new host and sound an alarm.

When Matzinger and her colleagues repeated Peter Medawar's mouse experiment, they were able to show that newborn mice are perfectly capable of rejecting foreign cells, so long as the cells are injected along with activated dendritic cells from the donor. The dendritic cells alert the T cells in the newborn mice.

Other researchers likewise found that infant mice, long thought to be incapable of fending off a certain virus, could mount a vigorous immune response, provided the dose of virus was lowered a thousand-fold from what other researchers had been giving the mice. They too concluded that the immune response of infant mice works in the same way as that of adult mice. The researchers suggested that in previous experiments the young mice, which have fewer T cells than adults, were simply overwhelmed by too many viruses.

Does the immune system recognize self and nonself, or does it recognize safe and dangerous? The research of Matzinger and others seems to pose a challenge to the currently accepted self-nonself model of immune response and offers some possible answers to long-standing difficulties with that model. The danger theory may eventually displace the current model if it offers more satisfactory answers and if the weight of future research supports its interpretation of the preliminary data. For now, its most important value may lie in reexamining long-accepted assumptions, raising new questions, and pointing new research in directions the self-nonself model could not. Nonetheless, in the rest of this chapter, we discuss the facts of immunology in the context of the self-nonself model, which most biologists still embrace.

KEY CONCEPTS

1. The skin provides the first barrier against invading bacteria, viruses, fungi, and other pathogens.

2. Inside the skin, inflammation, complement, interleukins, and interferon provide nonspecific internal defenses against infection.

3. The immune system includes two key components that recognize and eliminate foreign molecular shapes. These are circulating antibodies and cytotoxic cells.

4. The immune response demonstrates specificity, memory, diversity, and the ability to distinguish self from nonself.

5. The immune system's extraordinary capacity to recognize so many different kinds of molecules depends on the rearrangement of specific gene segments during the development of individual cells.

6. The descendants of each progenitor cell in the immune system express a single gene arrangement. These cells are subject to selection during development and to large expansion after exposure to specific challenges.

7. Cells interact with one another in producing an immune response, relying on cell surface molecules and secreted signaling molecules called cytokines.

THE CAST OF CHARACTERS

Our biosphere teems with pathogenic viruses, bacteria, fungi, protozoans, and parasitic worms that are all potentially harmful. Among vertebrates, at least, evolutionary success has depended heavily on our ability to prevent and combat invasions of these tiny pathogens. In this chapter we review three lines of defense used by mammals to protect themselves against pathogens.

The first line of defense is a barrier, the skin, that keeps out pathogens as effectively as a castle wall keeps out human invaders. However, any opening in such a barrier—whether by the mouth and eyes or a cut or puncture wound—allows pathogens free access to the body.

The next line of defense consists of **blood clots**, which plug any holes, and **inflammation**, which mobilizes white blood cells that phagocytose (engulf) bacteria, viruses, debris, and other elements.

The last line of defense is the **immune system**, a highly complex system involving roughly eight kinds of white blood cells. Some of these cells phagocytose pathogens, some punch holes in pathogens, and some secrete **antibodies**—molecular flags that mark certain cells for destruction by the other white cells. The immune system differs from the other lines of defense in its *specificity*. It can distinguish among individual pathogens, responding more vigorously to some than others and tolerating the body's own cells in some cases and not in others.

All of the cells engaged in the body's defense are white blood cells, or leukocytes [Greek, *leukos* = clear, white], which come from stem cells in the bone marrow. Stem cells are cells that can divide to produce identical cells or differentiate to produce one or more specialized cells. Four kinds of white blood cells—neutrophils, eosinophils, basophils, and mast cells—participate mainly in the inflammation response, while the two kinds of **lymphocytes** (B cells and T cells) and the **natural killer cells** (NK cells) participate in the immune responses. Finally, **macrophages**, giant cell-eating white cells, participate in both inflammation and immune responses (Table 40-1).

The body defends itself against invading pathogens with three lines of defense: the skin and the populations of microorganisms on the skin and in the gut, blood clots and inflammation, and the immune system.

Table 40-1   Types of White Blood Cells (Leukocytes)

NONSPECIFIC (INFLAMMATION RESPONSE)
  Neutrophils—phagocytic on bacteria and larger parasites
  Eosinophils—attack bacteria and larger parasites
  Basophils—reside in the bloodstream and release
    histamines, which trigger inflammation
  Mast cells—reside in the tissues and release histamines,
    which trigger inflammation and attract neutrophils. (IgG
    antibodies stimulate mast cells to release histamines.)

SPECIFIC
  T cells—kill infected cells, "altered self"
  B cells—make antibodies
  Natural killer cells (NK cells)—kill microorganisms and debris
    that are marked with antibodies
  Macrophages—phagocytic on microorganisms and debris
    that are marked with antibodies and release four kinds of
    cytokines, or intercellular signals

HOW DOES THE SKIN KEEP PATHOGENS OUT?

The skin is a flexible, stretchable barrier that repels the great majority of potential invaders. In humans, the skin makes up about 15 percent of the body's weight. Like the linings of the gut, mouth, and other internal surfaces of the body, the skin consists of epithelial cells, Unlike the thin, internal epithelia, however, the skin epithelium consists of a thick pile of cell layers, which together make up the **epidermis** [Greek, *epi* = on + *derma* = skin] (Figure. 40-3).

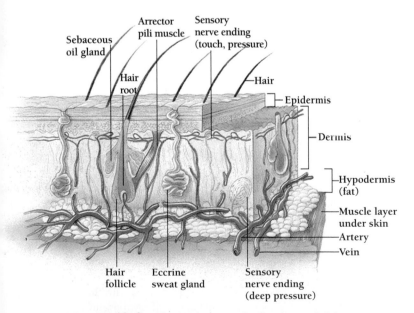

Figure 40-3   The epidermis is the first layer of defense against pathogenic microorganisms.

The outermost cells of the epidermis are nondividing cells, many of which are dead or dying, that protect the dividing cells beneath. These outer cells are continually exposed to scraping, drying, infection, and other injuries. On average, these cells live only about a month. As the outer cells die, still-dividing cells from the lower layers replace them.

Beneath the epidermis are the **dermis,** a thicker layer of living cells, and a layer of **subcutaneous tissue.** Embedded within these layers are hairs, nerve endings, muscles, sweat glands, oil glands, blood vessels, and sensors for heat, cold, and pressure.

At the top of the epidermal layers, the oil and sweat secreted by glands in the skin create a slightly acid environment, which is hostile to the growth of many microorganisms. The skin nonetheless swarms with microorganisms. Most are benign residents that cause little damage. Indeed, many actively compete with pathogenic organisms, reducing the chance that pathogens can enter the body.

The skin and its community of organisms are so effective a barrier that most microorganisms cannot enter the body unless they do so by way of the mouth, nose, or eyes. Even those that do enter on particles of food or dust usually die, because few survive exposure to **lysozyme,** an enzyme, found in saliva and tears, that digests the cell walls of many microorganisms.

A few microorganisms nonetheless reach the digestive tract or respiratory tract. In the respiratory tract, cilia sweep mucus and trapped microorganisms back toward the esophagus. This movement flushes most dust-borne microorganisms into the digestive tract, where they join a host of other microorganisms swallowed with food. These pass from the esophagus into the stomach.

The stomach's acid environment is enough to kill most microorganisms. Some—*Salmonella,* for example—survive and pass into the digestive tract. Vast numbers of microorganisms flourish in the digestive tract. Indeed, more bacterial cells inhabit the gut than there are cells in the body. As in the case of the microorganisms of the skin, however, most are benign and usually compete with pathogenic bacteria for space and resources. As a result, under normal conditions, most foreign microorganisms cannot survive in the gut.

The first barrier to invasion by microorganisms is the skin. Lysozyme in saliva and tears kill most foreign microorganisms that enter the eyes, mouth, or nose. Microorganisms that manage to enter the intestinal tract must compete with millions of others—almost all benign.

HOW DOES THE BODY DEFEND ITSELF WHEN THE SKIN IS BROKEN?

The skin provides the first line of defense against infection or invasion, and relatively few microorganisms breach its barrier.

Nonetheless, injuries, even minor ones, expose the inner body to infection. Animals continually suffer breaks in their skins. Some are major, life-threatening injuries, others minor. But even the tiniest wounds open the way for infection. Microorganisms that penetrate the skin, however, encounter a series of active defenses.

Any significant wound damages blood vessels as well as other tissues. Damaged blood vessels pose two dangers beyond the immediate infection of local tissues: blood loss and the rapid dispersal of microorganisms throughout the body by way of the circulating blood.

How Does the Body Control Blood Loss and Keep Pathogens Out of the Circulatory System?

Mammals have evolved two adaptations, each of which limits both bleeding and infection. In **wound healing**, normal cells—in both the epidermis and the dermis—divide when they lose contact with other cells and stop dividing once they have filled the gap. In **hemostasis** [Greek, *haima* = blood + *stasis* = standing], several mechanisms simultaneously hinder the flow of blood from the body and the passage of microorganisms into the body.

Hemostasis involves at least three interconnected mechanisms (Figure 40-4): (1) the contraction of smooth muscles in the damaged vessels, called **vasoconstriction**; (2) the clumping of small cellular elements in the blood, called **platelets**, at the site of injury; and (3) the formation of a **blood clot**, a mass of coagulated blood.

Injury immediately triggers the contraction of nearby blood vessels, which sometimes seals the leaky vessel completely but at least reduces the loss of blood. Injury to the wall of a blood vessel also exposes collagen, the major structural protein of connective tissue. Platelets in the blood adhere first to the collagen and then to each other, forming clumps of platelets that can completely plug tiny openings in the wall of the blood vessel. Such minute ruptures occur hundreds of times every day, only to be sealed by platelets. In addition, the platelets stimulate further vasoconstriction and initiate **clotting**, the mechanism by which the blood forms a gel-like lump that blocks blood flow.

Although platelets can help initiate clotting, clotting is strictly a property of the **plasma**, the liquid, noncellular part of the blood. Clotting can occur in the absence of any blood cells, even in a tube of plasma outside the body.

The body closes breaks in the skin by means of wound healing and breaks in the circulatory system by means of vasoconstriction, the aggregation of platelets, and blood clots. All of these processes prevent excessive blood loss and help prevent microorganisms from entering the circulatory system.

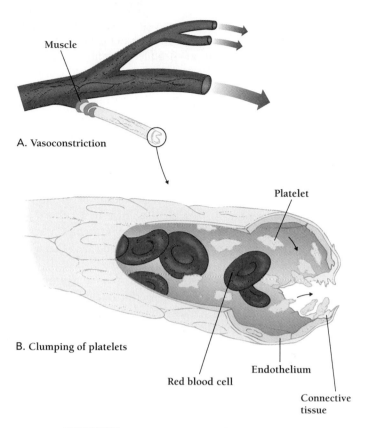

A. Vasoconstriction

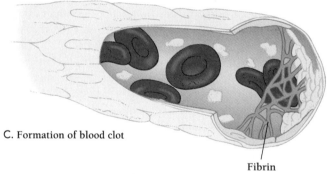

B. Clumping of platelets

C. Formation of blood clot

Figure 40-4 Hemostasis. The circulatory system has three ways of preventing blood loss. A. Vasoconstriction. B. Clumping of platelets. C. Formation of blood clots.

How Does the Body Prevent Damaging Blood Clots?

Although blood clots can keep an animal from bleeding to death, blood clots are themselves extremely dangerous. A single clot can block an arteriole in the brain (stroke), the heart (heart attack), or lungs (thrombosis), causing permanent damage or death. As a result, clots form only after a series of molecular checks.

Biochemists have studied the process of clotting intensively and identified several proteins responsible for clot formation. The clot itself consists of a network of fibers made of the protein **fibrin.** Fortunately, fibrin is not present in unclotted blood—otherwise clots would form constantly.

Fibrin derives from the precursor protein **fibrinogen**, which becomes fibrin only when the enzyme **thrombin** digests away part of the fibrinogen polypeptide. Thrombin itself is not normally in the blood either. Otherwise it too would cause the blood to clot all the time.

Instead of thrombin, blood plasma contains **prothrombin**, a precursor of thrombin. At the site of tissue damage, still another enzyme, called *Factor X* ["factor ten"], converts inactive prothrombin to active thrombin, initiating the clot. Factor X is itself normally inactive, but several other enzymes can convert it to an active form at the site of damage.

The various steps form a complex cascade of events that lead to clot formation. At each step, a single molecule of one enzyme activates many molecules for the next step. The result is a tremendous multiplication of the initial effect, so that a minor injury results in a large amount of fibrin.

The clotting system is so dangerous that it requires not only a series of checks but also a powerful antagonistic system that destroys fibrin. Like the clotting system, the anticlotting system consists of a multiplying cascade of enzymes. The last enzyme in the cascade is **plasmin**, which digests the bonds between fibrin molecules and dissolves the clot.

Plasmin, in turn, derives from **plasminogen**, a precursor protein in the plasma. The activation of plasminogen depends either on the action of plasma enzymes or on **tissue plasminogen activators** (tPA), which are enzymes that activate the anticlotting system. Some tissues have more tPAs than others. The uterus, for example, contains high levels of a tPA that prevents excessive clotting of menstrual blood.

Physicians treat heart attack victims with synthetic tPA, which, when given intravenously, stimulates the production of plasmin. The plasmin breaks up the clots that are blocking the arteries, and blood flow to the heart muscle resumes.

Besides tPAs, mammalian blood also contains various **anticoagulants**, substances that interfere with the action of the clotting cascade. Two such anticoagulants are **heparin** and **warfarin**. Heparin, which is made in specialized cells of the liver, lung, and gut is commonly used to prevent blood samples from clotting, as well as to prevent blood clots in the heart, brain, or lungs. Warfarin is used both as a drug, in people, and as a poison, in rats and mice, which die of internal bleeding if they consume the poison. Blood-sucking mosquitos and leeches also use anticoagulants to keep the blood of their victims flowing freely.

Because blood clots are dangerous, the synthesis of the active ingredient in clots, fibrin, is carefully regulated. Another system of anticlotting enzymes ensures that clots are rapidly dissolved.

How Do Plasma Proteins Contribute to Nonspecific Defense?

We have seen that a wound can trigger vasoconstriction, tiny platelet plugs, and blood clots. A fourth kind of defense is the **complement**, a set of blood proteins that attack microbiological invaders. Some of these proteins kill foreign organisms by boring holes through their membranes. Other complement proteins coat the microbes' surfaces to identify them as targets for phagocytic cells. Still other proteins stimulate special white blood cells called *mast cells*. Mast cells help trigger the next line of defense, which is **inflammation**, the redness, heat, swelling, and pain that result from local injuries.

Mast cells are large, round cells filled with small vesicles, or granules. Mast cells are distributed throughout the body's connective tissues (Figure 40-5). Any kind of damage to a tissue stimulates the mast cells to release their granules, which are loaded with **histamine,** the major stimulus for the inflammatory response. Allergy sufferers are especially familiar with histamine's effects, which are swelling and itchiness in the eyes, nose, throat, and other tissues. *Antihistamines* are drugs that block the effects of histamines.

Complement proteins and mast cells react nonspecifically to damage to body tissues. Mast cells release histamines, which help promote the inflammation response.

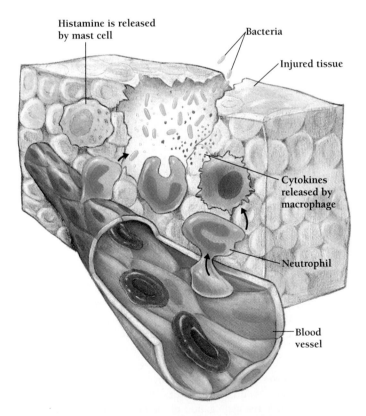

Histamine is released by mast cell

Bacteria

Injured tissue

Cytokines released by macrophage

Neutrophil

Blood vessel

Figure 40-5   Tissue damage such as a cut in the skin stimulates mast cells. Tissue damage stimulates the granules in a mast cell to release histamine. Mast cells also release cytokines, which attract migrating neutrophils and help initiate the (slower) immune response.

How Does Inflammation Help the Body Resist Infection?

Microscopic observations reveal a sequence of cellular events that contribute to the inflammatory response: (a) Capillaries *dilate,* or open, in the injured region. (b) Special white blood cells, called **neutrophils,** attracted by chemicals released by the mast cells, attach to the capillary walls. (c) The neutrophils move through the capillary wall into the damaged tissue. (d) The neutrophils phagocytose (engulf) microorganisms in the damaged tissue.

The open blood vessels increase the delivery of cells and plasma proteins, which contribute to the inflammatory and the immune responses. Among the specific cells that move into the inflamed region are macrophages, each of which counterattack invading organisms.

Macrophages [Greek, *makros* = large + *phagein* = to eat] are phagocytic cells—like neutrophils—but macrophages are larger and more widely distributed throughout the body (Figure 40-6). Immature macrophages (monocytes) make up about 5 percent of the cells in the blood. Mature macrophages are found in connective tissue in the liver, brain, spleen, lungs, and lymph nodes. Macrophages consume microorganisms and cellular debris and play a major role in initiating inflammation.

During inflammation, macrophages produce at least three small proteins called **cytokines.** Cytokines regulate cell division and protein synthesis in other white cells involved in both inflammation and the immune response. The three cytokines produced early in the inflammatory response are **tumor necrosis factor-α** (TNF-α), **interleukin-1** (IL-1), and **interleukin-6** (IL-6). The two interleukins derive their name from the fact that they act as messenger molecules between different kinds of white blood cells (leukocytes). By binding to specific receptors and initiating specific biochemical responses, interleukins initiate and regulate inflammation and immune response.

IL-1, for example, stimulates responses both in leukocytes and in various organs distant from the site of infection. It can cause a fever by stimulating cells in the *hypothalamus,* a region of the brain that regulates body temperature. A fever interferes with microbial growth and further promotes the body's immune defenses. IL-1 also attracts more phagocytic cells to the site of injury. And IL-1 stimulates the cells in the capillary walls so that passing neutrophils in the blood bind to the wall, then move through the wall and accumulate near the site of injury. Finally, IL-1 stimulates lymphocytes, the cells of the immune system.

Interleukin-6 stimulates stem cells in the bone marrow to produce more macrophages, while TNF-α acts locally on macrophages and neutrophils, stimulating them to increase their phagocytic activity.

Viral infection stimulates the infected cells to produce still another set of cytokines, called **interferons.** Interferons interfere with the replication of viruses by limiting protein synthesis in virus-infected cells. Even though the shutdown of protein synthesis may kill the cell, it prevents the release of more viruses. Interferons also stimulate phagocytosis and the immune response.

Inflammation helps the body resist infection by opening blood vessels, thus increasing the delivery of white blood cells, such as neutrophils and macrophages, as well as plasma proteins. Cytokines released by the macrophages regulate the inflammation response, stimulate the immune response, and send chemical signals to other parts of the body.

HOW DOES THE IMMUNE SYSTEM RECOGNIZE MOLECULES?

The most powerful and the most specific defenses against infection are those of the immune system. The immune system produces two distinct responses to invading organisms. The first is the production of **antibodies,** proteins that mark foreign molecules (often on the surfaces of foreign cells) by selectively binding to parts of the molecules that have just the right shape and charge. The second immune response is the production of cells that recognize and attack cells of the body that have become infected by viruses or microorganisms.

What Distinguishes the Immune System from Other Defenses?

Most of us have had contact with someone infected with chicken pox, a common childhood disease. If we have not had chicken pox, or been vaccinated against it, we are susceptible to infection. Once we have suffered its itchy ravages, however, we are no longer susceptible; we then say that we are immune to chicken pox. Four attributes characterize immunity: *specificity, memory, diversity,* and *self-nonself recognition.*

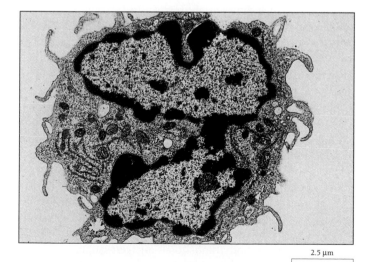

2.5 μm

Figure 40-6 **A macrophage.** Notice at left the endoplasmic reticulum and Golgi, which produce and secrete proteins such as interleukins. *(Don Fawcett/Visuals Unlimited)*

Immunity is **specific:** cells of the immune system recognize specific invaders and not others. If we are immune to chicken pox, we may still be susceptible to measles or other diseases. The specificity of the immune response distinguishes it from the **nonspecific** responses of clotting, complement proteins, inflammation, and interferons.

The immune system has **memory:** once the immune system has developed defenses against a particular invader, it can attack that molecule, organism, or virus again and again, whenever it appears, even decades later. The reappearance of the invader stimulates a **secondary immune response,** which is usually even stronger than the first response.

The immune system is capable of recognizing and attacking an incredible **diversity** of foreign invaders, including parasitic worms, fungi, bacteria, viruses, and pollen, as well as purified proteins and other molecules. Because the immune system responds powerfully to so many different kinds of cells and molecules, its greatest challenge is to refrain from attacking the cells of its own body. The immune system **tolerates** the other cells in the body, ignoring them unless they are infected by a pathogen. The immune system's apparent ability to distinguish the body's own components from others is called **self-nonself recognition,** in accordance with the model currently accepted by most biologists.

Until the 1950s, explaining these four properties of the immune response—specificity, memory, diversity, and self-nonself recognition—seemed an almost impossible challenge to biologists. Perhaps the most demanding aspect of the problem was the tremendous range of the immune response. The mammalian immune system can respond not only to proteins and other molecules created by any organism or virus, but even to molecules that are not natural products, ones that have been synthesized by organic chemists. How can an animal produce proteins and cells that specifically recognize substances that neither the animal itself nor any of its ancestors could ever have encountered in all of evolutionary history? And how can an animal have specific responses to so many different challenges?

The answer to all these questions was antibodies. The immune system produces a huge array of proteins, called *antibodies,* that bind to nonself molecules, which are called *antigens.* The array of antibodies is huge but finite. Each antibody recognizes a characteristic shape and charge distribution rather than a specific molecule. While the number of possible atomic arrangements of foreign molecules is impossibly large, there are, fortunately, a limited number of shapes: different arrangements of atoms look the same to the antibodies.

The immune response demonstrates specificity, memory, diversity, and the ability to distinguish self from nonself.

Which Cells Mediate the Immune Response?

Immune responses depend on cells. The immune cells are a special class of white blood cells called **lymphocytes,** which develop within the lymphoid tissues (including lymph nodes, spleen, thymus, and tonsils). A healthy human immune system contains about one trillion (10^{12}) lymphocytes. Under a light microscope, they all look alike, but they vary enormously in what they do.

Lymphocytes fall into three main classes: B lymphocytes, T lymphocytes, and natural killer cells. **B lymphocytes,** or **B cells,** make antibodies, which recognize foreign cells and molecules (Figure 40-7). In particular, antibodies recognize and bind to bacteria, fungi, and protists, as well as any other cells and molecules that do not belong.

Although antibodies are essential parts of the immune system, they are themselves mere markers and cannot kill bacteria or other cells. Antibodies are like the chalk marks the meter reader puts on the tires of cars parked in front of a parking meter. Chalk marks indicate when a car should be ticketed, or even towed. In the same way, antibodies mark cells for destruction by macrophages.

Antibodies have another important limitation. They do not recognize the body's own cells even when those cells harbor viruses or other pathogens. Because so many infections, such as AIDS, are intracellular (inside the cell), the body has a second set of lymphocytes, the **T lymphocytes,** or **T cells,** that recognize and destroy body cells that have become infected or damaged (Figure 40-7). In addition, **natural killer cells,** or **NK cells,** attack tumor cells and cells infected by a pathogen.

B lymphocytes, T lymphocytes, and natural killer cells are responsible for the immune response.

How Do Antibodies Recognize Cells and Molecules?

Although all the B cells look alike, they are not alike. Different cell lines produce different antibodies. In cell cultures, for example, each line of B cells makes just one kind of antibody. Researchers grow such lymphocytes, which ordinarily die in culture, by fusing them to cancerous lymphocytes. The result is a **hybridoma,** a cell line that is a hybrid of cells from a lymphoma (a cancer of lymphocytes) and normal B cells. The set of descendants from a single cell is a *clone.* Each clone makes only one kind of antibody, which is therefore called a **monoclonal antibody.** In contrast, the antibodies produced by a normal animal are made by the descendants of millions of different B-cell clones. Animals therefore produce millions of different antibodies.

The specificity of the immune system depends on having enough antibodies. **Antibodies** are proteins that selectively bind to molecules that have a particular shape and charge. Specifically, an antibody binds to an **antigen** [an *antibody generator*], any molecule that triggers an immune response. Among the most biologically important antigens are the coat proteins of viruses and the surface proteins, carbohydrates, and glycoproteins of bacteria.

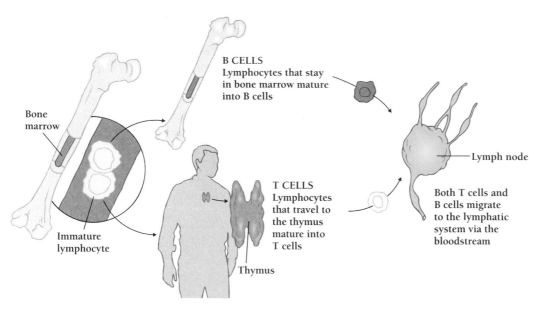

Figure 40-7 Cells involved in the immune response. Lymphocytes first develop in bone marrow. Those that migrate to the thymus mature into T cells (which recognize infected or cancerous body cells). Those that stay in the bone marrow mature into B cells (which make antibodies).

Every antigen contains one or more **epitopes** [Greek, *epi* = upon + *topos* = place], the specific shapes and charge distributions recognized by antibodies. Epitopes are often only small parts of molecules. A single epitope on a protein antigen, for example, typically consists of just five to eight amino acids (Figure 40-8). It is impossible to count the total number of possible epitopes, but biologists estimate that the immune system can make antibodies that recognize more than 100 million different epitopes. Because antibodies recognize shapes rather than exact atomic structure, an antibody may bind epitopes that are chemically different.

All antibodies are **immunoglobulins,** which are members of a group of globular proteins called globulins. There are five kinds of immunoglobulins—IgG, IgM, IgA, IgD, and IgE. These letters stand for "immunoglobulin G," "immunoglobulin M," and so on. All the immunoglobulins contain antigen-recognition sites made from two kinds of polypeptide chains, **light chains** and **heavy chains.** Immunoglobulins use thousands of different kinds of light and heavy chains. The five classes of immunoglobulins differ in their heavy chains. But, all the classes use the various light chains.

The most abundant class of immunoglobulins, **IgG,** consists of four polypeptide chains—two identical light chains and two identical heavy chains. Bonds between side chains connect the heavy chains to the light chains and the two heavy chains to each other. The resulting structure is shaped like a "Y," although some people have compared it to the head of the cartoon character Bullwinkle. Either way, antigens bind at the top of each of the two arms of the antibody, so that each IgG molecule contains two identical binding sites for antigens (Figure 40-9).

While most of the immunoglobulins have just two binding sites, IgA has four and IgM has ten. The IgM immunoglobulins, for example, have five identical IgG-like subunits linked together in a shape like a snowflake. IgM molecules consequently have ten identical binding sites.

The presence of two or more binding sites on each antibody explains the effectiveness of antibodies in combating infection. The surface of a microorganism may contain thousands of identical protein or carbohydrate molecules. An antibody with two binding sites can bind to the same site on two different bacteria. The result of hundreds of different antibodies binding to different sites on the bacteria is that each bacterium is covered with antibodies marking it for destruction. An antibody-covered bacterium is an easy target for phagocytic cells.

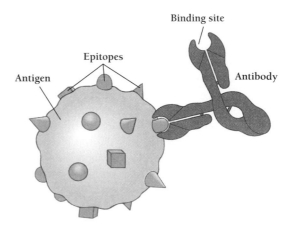

Figure 40-8 Structure of an antibody molecule.

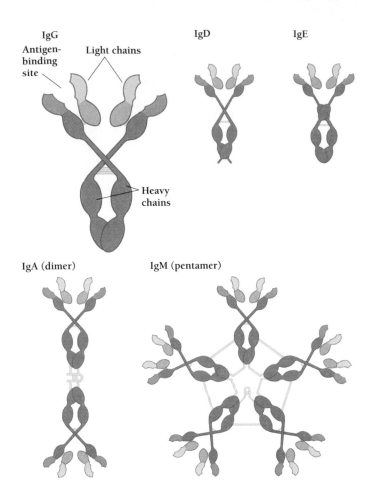

Figure 40-9   **The five classes of antibodies.** The five kinds of immunoglobulins differ in their heavy chains. Notice that IgG, IgE, and IgD have two binding sites each, while IgA has four and IgM has ten.

Each of the five classes of antibodies makes distinctive contributions to the immune response. IgG and IgM, for example, activate complement proteins, which destroy any cells marked with IgG or IgM. IgM molecules are the first to appear after exposure to antigen, while IgG molecules are made in greater amounts during a secondary response.

The remaining antibodies, IgA, IgD, and IgE, do not activate complement proteins. IgA molecules are especially abundant in tears, saliva, and milk. IgE antibodies are important to allergy sufferers, for they mobilize a distinct set of cellular and chemical responses. Certain antigens, such as ragweed pollen and dust mites, are especially effective at evoking the production of IgE. The IgE molecules circulate through the blood and attach themselves to mast cells, stimulating the release of histamine, which triggers the inflammatory response.

Researchers have suggested that the IgE response may defend against parasitic worms. Some people, however, mount this elaborate defense in response to antigens that are not dangerous at all. Such a response to a harmless antigen is called an *allergy*. Harmless antigens that evoke this response include substances in the feces of dust mites, cat dander (dandruff), dust, pollen, certain drugs, as well as foods such as nuts and

shellfish. When one of these antigens enters the respiratory passages of a sensitized person, histamine release stimulates increased mucus secretion and contraction of the airways. Typical allergy symptoms thus include congestion, sneezing, a runny nose, and difficulty in breathing. The most extreme response can result in death. In some cases, the response involves only the antibodies, macrophages, and other cells mobilized by inflammation and happens rapidly. In other cases—commonly with antigens in the leaves of poison ivy or poison oak—the reaction involves T cells and does not appear for days after exposure.

Antibody molecules recognize antigens by means of the specific shapes of their epitopes.

HOW DOES THE IMMUNE SYSTEM GENERATE SO MANY ANTIBODIES?

The diversity of antibodies depends on the immune system's ability to produce antibodies with many different three-dimensional structures. According to the **clonal selection theory,** different antibodies form because each has a unique amino acid sequence. Each amino acid sequence produces a distinct three-dimensional structure able to bind to a distinct epitope. The clonal selection theory requires that antibodies have an enormous number of different amino acid sequences—perhaps 100 million. From these, the immune system *selects* the set of antibodies it needs in response to the presence of a particular infection. This is the essence of the clonal selection theory.

If every different amino acid sequence requires a separate gene, however, the selection theory requires more genes than most vertebrates have in all their DNA. The human genome, for example, contains fewer than 100,000 genes. They cannot all be devoted to making antibodies. So how can thousands of genes carry the information to make millions of antibodies?

How Can Thousands of Genes Produce Millions of Antibodies?

Two facts make this problem less perplexing:

1. As we have seen already, every antibody is made from two different polypeptides, so that different polypeptides can be combined in ways that multiply their differences.
2. Two separate genes can be spliced together to specify one polypeptide, with surprising results. This is the reverse of the idea that one gene may specify more than one polypeptide (Chapter 11).

If lymphocytes could combine 10,000 light chains with 10,000 heavy chains, the cells could generate 100 million antibodies. But that would require 20,000 antibody genes, and most biologists doubt that the immune system makes up even that large a share of our genome.

Our genome, however, is just the genetic information passed from parent to child. What if, after we inherited our 2000 antibody genes, we let the genes in different cells become different from each other? In other words, what if antibody diversity was not inherited at conception but arose during the development of the individual?

In 1965, two researchers argued just that, suggesting that separate genes responsible for encoding different regions of immunoglobulin polypeptides somehow recombine during the development of each individual. Just as we can combine five shirts and two pairs of jeans to make ten unique outfits, so lymphocyte cells could combine 300 alleles for one polypeptide segment with four alleles for another to make 1200 different completed polypeptides.

The idea was intriguing, but for a long time, biologists had no way of knowing if this could really happen. Then, in 1976, biologist Susumu Tonegawa and his colleagues discovered that lymphocytes have genes that other cells do not have. Tonegawa showed that the novel gene in lymphocytes results from the joining of separate DNA fragments to make new genes. In a revolutionary discovery, he showed that two genes could join forces to make a single polypeptide.

We now know that every light chain is formed from three different gene segments. One segment has three alleles, one has four, and the third has approximately 300. The number of possible light chains is therefore about 3600 (= 3 × 4 × 300). Heavy-chain polypeptides, which form from a combination of four different gene segments, are even more diverse, with 130,000 types. The 130,000 heavy-chain polypeptides can be combined with the 3600 light-chain polypeptides to form almost half a billion different antibodies (Figure 40-10).

It now appears that mammals can make as many as a billion different kinds of antibody molecules. As if that were not enough, immunoglobulin genes mutate at high rates during the production of lymphocytes, increasing antibody diversity by a factor of ten. Most cells of the body quickly repair mutations in the DNA, but, by some still-unknown mechanism, lymphocytes tolerate mutations in the immunoglobulin genes.

Thousands of genes can produce millions of antibodies by (1) mixing thousands of polypeptides in different combinations and (2) by splicing genes into thousands of combinations not originally specified by the genome.

The Selection Theory Explains Memory and Self-Nonself Recognition

The generation of variation in immunoglobulin genes helps explain both the *specificity* and *diversity* of the immune response. To understand *memory* and *self-nonself recognition*, however, we need to understand the behavior of whole cells. First, we will look at the development of B cells, which make antibodies to cells and other elements that are clearly nonself. Then, we will discuss the roles of T cells, which recognize and destroy any of the body's own cells that have become a threat to the body.

Every B lymphocyte contains just two rearranged genes—one that constitutes an active heavy-chain gene and one that constitutes an active light-chain gene. As a result, every B cell and its descendants make only one kind of antibody. The particular gene arrangements within the B cell determine the structure of that antibody.

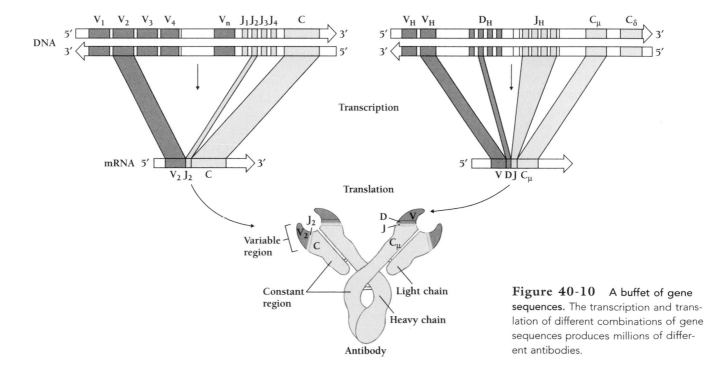

Figure 40-10   A buffet of gene sequences. The transcription and translation of different combinations of gene sequences produces millions of different antibodies.

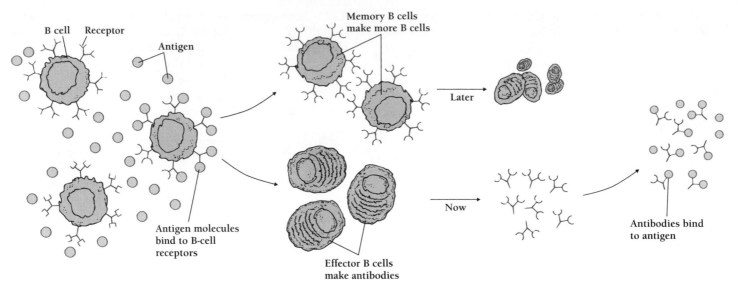

Figure 40-11   **The making of memory.** When B cells are exposed to antigen, they produce effector cells and memory cells. If the same antigen appears again, the many memory cells quickly produce large numbers of effector cells, mounting a much greater and faster immune response.

During development, precursors of B cells form a **lymphocyte library,** with each lymphocyte in the library containing a unique random arrangement of immunoglobulin genes. Each member of the B-lymphocyte library—and all its descendants—can make only one kind of antigen recognition site. But how do the right cells "know" when to spring into action, dividing and producing antibodies when their particular antigen appears?

B cells are able to respond to particular antigens because they have **B-cell receptors,** cell surface proteins with the same antigen recognition sites as the antibody that they produce. In fact, B-cell receptors are just membrane-bound forms of the antibodies (Figure 40-11).

When an antigen appears in the blood and binds to the B-cell receptors, the B cells begin dividing and differentiating along two separate pathways. One set of cells becomes **effector cells,** also called **plasma cells,** which actually make and secrete antibodies. The other set of B cells becomes **memory cells,** which can make more effector cells if the antigen ever appears again.

Memory cells are basically identical to unstimulated B cells. The only change is that there are many more of them after exposure to antigen. These cells have the same genetic rearrangements as the parent B cell, like their parent cells, they do not themselves make antibodies. When they next encounter the antigen, however, the memory cells rapidly produce effector cells. For this reason, the second exposure to an antigen often causes a much faster and stronger response than the first. The response may be so strong, in fact, that the body successfully fights off infection entirely. The multiplication of memory cells explains the lifelong immunity that children acquire after only one exposure to a disease such as chicken pox.

Despite the apparent specificity of antibodies, they are not perfectly specialized. Because lymphocytes respond to epitopes, which are very small parts of molecules, rather than to whole molecules, antigens whose epitopes are identical can precipitate an immune response for each other. In other words, antigens that mimic disease-causing bacteria or viruses are as effective at producing immunity as the pathogens themselves. A heat-killed polio virus, for example, produces immunity against polio, and an infection of cow pox, a mild disease, produces immunity against deadly smallpox.

When an antigen binds to the B cell receptors, the B cells divide and differentiate into effector cells, which actually make and secrete antibodies, or memory cells, which make more effector cells if the antigen ever appears again.

HOW DOES THE BODY DEFEND AGAINST GOOD CELLS GONE BAD?

The B cells produce antibodies that can recognize bacteria and bacterial products and target them for destruction. But because B cells do not respond to the body's own cells, B cells cannot respond to body cells that have become infected by viruses or other pathogens. Because infected cells are actively producing viruses or bacteria, it is important for the immune system to destroy them. We can think of them as good cells gone bad.

Cancer cells are also good cells gone bad. Recognizing body cells that have become infected or cells that are growing out of control is the task of the T cells (or T lymphocytes). T cells are also involved in rejecting organs and skin grafts surgically

BOX 40-1

Acquired immunodeficiency virus

In 1996, the number of Americans with AIDS, or acquired immunodeficiency syndrome, declined for the first time since the disease became known in 1980. Researchers were delighted and heralded the six-percent drop in new AIDS cases (from 60,620 cases to 56,730) as proof of the success of both preventive programs and treatments. Preventive programs have helped people avoid becoming infected with HIV. Treatment programs have helped those infected with HIV to avoid becoming ill and those who are ill to survive longer.

In the United States, 50,000 people die of AIDS each year. The majority of AIDS patients—about 69 percent—contract the disease either through homosexual relations or from intravenously injecting illegal drugs using contaminated needles. Safe-sex education programs have dramatically reduced the transmission of AIDS among homosexual men, and similar education programs have reduced the infection rate among intravenous drug users.

About 13 percent of AIDS patients acquire the virus through heterosexual contact, and that number is increasing. Heterosexuals are much less likely to practice safe sex than homosexuals and, as a result, the percentage of AIDS cases contracted through heterosexual contact is rising.

HIV is a retrovirus, an RNA virus that relies on the enzyme reverse transcriptase to transcribe its RNA into DNA inside the cells the virus infects. As you might recall from Chapter 12, different infectious viruses attack different kinds of cells in the body. Influenza virus attacks cells in the respiratory tract, for example, and poliovirus infects cells of the intestinal tract and, rarely, the nervous system. HIV infects cells of the immune system, specifically helper *T cells*. Helper T cells send cytokine messages that activate both B cells and cytotoxic T cells.

Different helper T cells have different receptors. About 60 percent of T cells have *CD4* receptors and another 20 to 30 percent have *CD8* receptors. Large numbers of $CD4^+$ T cells help activate other T and B lymphocytes. In contrast, $CD8^+$ T cells suppress the immune system.

Two of the main symptoms of AIDS are very low numbers of $CD4^+$ T cells and higher than average numbers of $CD8^+$ T cells. Healthy people have about 1000 $CD4^+$ T cells per ml of blood, while the sickest AIDS patients average less than 50 $CD4^+$ T cells per ml, with many individuals lacking these crucial immune cells altogether. The lack of helper T cells means that AIDS patients are unable to mount effective immune responses to infections or to certain kinds of cancers. As a result, they tend to die of pneumonia, fungal infections, and other diseases that most people's immune systems fend off.

Yet HIV continues to perplex medical researchers. As far as they can tell, HIV infects only one $CD4^+$ cell in every 400. Even the few cells that are infected seem relatively unaffected. And only one $CD4^+$ cell in 100,000 actually translates and assembles virus particles. Researchers wonder why the many uninfected cells don't multiply and protect AIDS patients from infections. Mysteriously, however, the $CD4^+$ cells just disappear.

Researchers suspect that the CD4 receptors are involved. Monocytes and macrophages also carry CD4 receptors on their surfaces, and they too are infected by HIV. They have fewer of these receptors, however, and perhaps for the same reason, fewer of them seem to die.

---

transplanted from other individuals. T cells not only recognize altered self, they also kill. T cells kill "bad" cells, a property called *cytotoxicity,* by boring holes in the cell membrane. Such T cells are called cytotoxic T cells.

We know the T cells are central to such defense because a child born without a thymus, where the T cells mature, lacks T cells and cannot mount an immune response to certain kinds of infections. Such a child has plenty of B cells and can withstand bacterial infections that multiply outside of cells. But the child is continually beset with infections by viruses, as well as by intracellular bacteria and fungi, all of which multiply inside of cells.

How Do T Cells Recognize Self?

T cells are of two kinds. **T cytotoxic (T_C)** cells kill cells recognized as altered self. **T helper (T_H)** cells send messages to nearby B cells and to T cytotoxic cells. Upon stimulation with a foreign antigen, some T_C cells differentiate into effector cells, called **cytotoxic T lymphocytes (CTL),** which actually kill the target cells. Both kinds of T cells recognize nonself antigens by means of T-cell receptors, but the two cell types differ biochemically and serve different defensive needs (Figure 40-12).

T_H cells produce chemical signals (cytokines) that activate both T and B cells. While some cytokines (such as IL-1, mentioned previously) can act throughout the body, most cytokines are *paracrine signals*, molecules that activate only those cells in the immediate vicinity. T helper (T_H) cells first bind the antigen and present it in a more effective form to the responding B cells.

The antigens recognized by T cells differ from those recognized by the antibodies of B cells in two respects: (1) antibodies recognize foreign antigens in solution as well as on cell surfaces, but T-cell receptors bind foreign antigens only on the

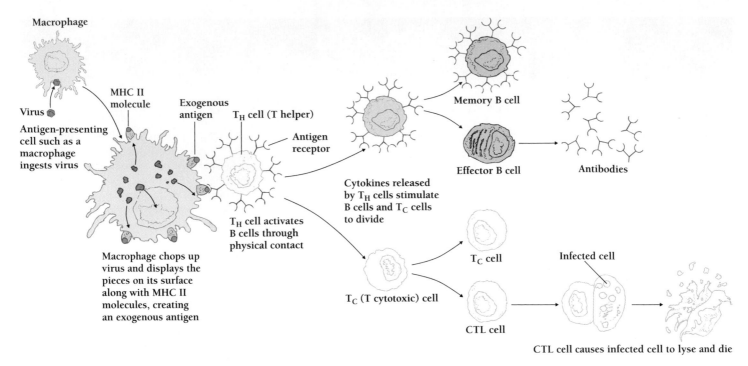

Figure 40-12   How do cells recognize self? The cells of the body display unique combinations of MHC proteins that identify the cells as belonging to the body. Phagocytic immune cells combine the MHC proteins of a dying cell with antigens from the virus or microorganism that infected the cell to form exogenous antigens. Exogenous antigens cause T_H cells to activate both B cells and T_C cells.

surfaces of cells that are recognizably self; and (2) antibodies recognize the three-dimensional shapes of proteins and carbohydrates, but T-cell receptors recognize only linear segments of polypeptide chains. These segments are short, usually only 20 amino acids long.

T cells recognize the cells of the body by means of the proteins encoded in a set of 40 to 50 genes called the **major histocompatibility complex,** or **MHC.** MHC proteins appear on the surface of essentially every cell in the body other than sperm cells.

MHC genes are extremely variable, with each gene existing in as many as 100 different alleles, so (excepting identical twins) every human's particular mix of MHC alleles is unique. It is on the basis of MHC proteins that transplant surgeons "match" donated organs. The MHCs of a donated organ must be similar to the MHCs of the recipient.

MHC proteins fall into three classes, two of which are important for our discussion. **MHC Class I** molecules are present on the surface of essentially every nucleated cell (except sperm). T cells recognize MHC Class I molecules as **endogenous antigens**—proteins made by the body's own cells. **MHC Class II** molecules are present on the surfaces of specialized antigen-presenting cells. These molecules alert the T cells to the presence of a virus or bacterium inside a cell. When one of these immune cells phagocytoses a virus or microorganism, the cell then combines antigens from the pathogen with MHC II molecules on its surface, creating **exogenous antigens**—pro-

teins that alert the T cells to the presence of agents that are infecting the cells of the body (Figure 40-12).

T cells only recognize altered-self antigens, specifically, antigens that are bound to MHC proteins. Unlike B cell receptors, which recognize the shapes of epitopes, T cell receptors recognize linear segments of polypeptide chains. MHC Class I and Class II genes specify membrane proteins that bind antigens and present them to T cells.

How Do T Cells Recognize Altered Self?

Unlike B cells, T cells do not recognize completely foreign cells. Human T cells would not, for example, recognize mosquito cells as invaders (although human B cells would). T cells recognize only those cells that are like others in the body but somehow different. Another way of stating this is to say that T cells recognize only nonself antigens in a self context, called **altered self.**

T cells recognize such altered-self antigens with the help of **T cell receptors**—molecules on the surfaces of T cells that recognize foreign antigens as altered self. The genes for T-cell receptors are related to immunoglobulin genes. As in the B cell immunoglobulin genes, rearrangements of T-cell receptor genes produce a T-lymphocyte library, each member of which con-

Autoimmunity: why does the immune system sometimes attack self?

Although the immune system is largely successful in distinguishing self from nonself, about 5 percent of the population suffers from some autoimmune disease, in which the immune system attacks the body's own tissues, causing tissue damage or malfunction.

In the disease myasthenia gravis, for example, antibodies block the muscle cell receptors that normally respond to acetylcholine, the chemical signal that triggers muscle contraction. When autoantibody levels are high, the muscles cannot be activated, resulting in weakness or even paralysis. One treatment for myasthenia gravis is to remove the antibodies from the patient's blood, which involves removing the blood plasma and then cycling it back into the body.

In insulin-dependent diabetes mellitus (juvenile diabetes), which affects about 0.5 percent of the United States population, T cells attack and destroy the insulin-producing cells of the pancreas. Unable to make their own insulin, patients must inject themselves with insulin every day.

Until the early 1970s, biologists believed that such autoimmune diseases resulted from a failure of a mechanism that should have eliminated all self-reactive lymphocytes early in life. Researchers have discovered, however, that all healthy adults contain some self-reactive lymphocytes. In most people and for almost all antigens, such self-reactive lymphocytes are held in check. In patients with autoimmune disease, something goes wrong. What goes wrong is only partly understood.

Some of the circumstances that bring about the multiplication of self-reactive lymphocytes help explain how it can happen. For example, the streptococcal bacteria that cause strep throat have surface antigens that closely resemble antigens on the cells in heart muscle, a happenstance called *molecular mimicry*. The result is that antibodies and T lymphocytes recognize both antigens and attack and damage not only the strep bacteria but the heart muscle cells as well. Such an attack on the heart is called *rheumatic fever*, a serious and often fatal disease that was quite common in the days before antibiotics quickly cured strep throat infections.

In the case of insulin-dependent diabetes, T cells that recognize an antigen of a common virus (Coxsackievirus) also recognize a peptide segment of an enzyme called GAD (glutamic acid decarboxylase) found in the insulin-producing cells. Researchers hypothesize that T cells that have proliferated in response to the Coxsackievirus infection attack pancreatic cells and so cause diabetes.

Like other exogenous antigens, mimicking antigens are presented to the immune system with MHC molecules. It is not surprising, then, that individuals with different MHC alleles differ in their susceptibility to particular autoimmune diseases. For example, individuals who express one group of MHC Class II genes are 100 times more likely to develop insulin-dependent diabetes than the population as a whole.

If autoimmunity to a mimicking antigen is indeed responsible for insulin-dependent diabetes and other autoimmune diseases, then the disease might be prevented with the same kinds of immunosuppressant drugs used to suppress the immune response when organs are transplanted.

Another possible approach would be to induce tolerance to the offending antigen—somehow reintroducing the antigen as self. This approach has worked for animal models of several autoimmune diseases, including insulin-dependent diabetes. A novel approach to inducing tolerance comes from the observation that most people are immunologically tolerant to the foods that they eat. By feeding the appropriate mimicking protein to people at risk for insulin-dependent diabetes and other autoimmune diseases, researchers hope to prevent a later autoimmune attack.

---

tains one antigen recognition site. T-cell receptors are even more diverse than immunoglobulins.

T-cell receptors recognize altered self.

How Does the Body Eliminate T Cells That Might Attack Healthy Cells?

Cells of the immune system undergo stringent selection as they develop. First, T cells that do not recognize self must be eliminated. During the maturation of T_C cells in the thymus, the main survivors are precursor cells that can recognize antigens within the individual's own MHC Class I molecules. Similarly, T_H precursor cells that can recognize antigens bound to the individual's own MHC Class II molecules survive.

Second, those T cells that do not distinguish between self and altered self must go. The T cell precursors that respond to self antigens (combined with self MHC receptors) also disappear, leaving only T cells that can recognize nonself antigens within self MHC molecules. Altogether, 90 to 99 percent of T cell precursors die within the thymus. The T cells that leave the thymus are highly selected to recognize altered-self antigens.

During maturation in the thymus, T cells undergo heavy selection, so that only T cells that recognize altered self survive.

Individuals Have Distinctive Antigens on Blood Cell Surfaces

The diversity and individuality of MHC proteins also allow a definitive identification of cells from one person and the tracing of paternity in genetic studies and legal disputes. Other surface antigens besides the MHC proteins, however, can direct an immune response to transplanted cells. Antigens on the surfaces of blood cells are especially important, because of the common use of blood transfusions to treat patients who have lost blood in an injury or in surgery. Individuals differ genetically in the kinds of antigens they produce, and each individual may be classified according to a number of "blood groups."

The most important classification is that of the ABO blood groups. Individuals of blood group A have an antigen called A on the surfaces of their red blood cells; those of blood group B have another antigen, called B; those of group AB have both A and B; those of blood group O have neither A nor B. An individual who does not make A antigen (that is, someone with type O or B blood) has antibodies to A antigen. They have these antibodies even if they have never been exposed to type A blood, probably because similar epitopes are present on the surfaces of common gut bacteria. Similarly, individuals who do not express the B antigen (people with blood type O or A) have antibodies against B.

Before physicians understood the nature of the antigenic differences, blood transfusions were dangerous procedures. If a person with type A blood received a transfusion of type B blood, their antibodies attacked the red cells, linked them together in big clumps, and clogged the blood vessels. Such an occurrence could easily be fatal.

Another antigen on the surface of red cells of some individuals is called the Rh factor—after the rhesus monkeys in which the factor was first identified. Unlike the situation with the A and B antigens, individuals who lack the Rh antigen do not generally make antibodies against it unless they are first exposed to Rh-positive blood. This would happen, for example, after transfusions of Rh-positive blood into an Rh-negative recipient—a situation that modern physicians avoid by testing the blood.

Another situation, far more common, is harder to control. Pregnancy frequently exposes Rh-negative women to Rh-positive blood. If a fetus receives the Rh factor gene from the father, the mother may develop antibodies against Rh factor. Such antibodies are not usually present at high concentrations during the first pregnancy with an Rh-positive fetus.

During the delivery of the baby, the mother is likely to be exposed to Rh antigens, triggering the creation of memory cells. Subsequent pregnancies with an Rh-positive fetus may then provoke a strong secondary response in the mother. The mother's antibodies can cross the placenta and attack the fetus's red cells. The fetus may die in utero or die at birth. Recent research has allowed physicians to prevent this immune response by treating the mother with anti-Rh antibodies, which remove memory cells for the Rh factor.

MHC proteins and other cell surface antigens make the cells of every individual unique.

In this chapter we have seen how the body defends itself against pathogenic viruses, bacteria, and other foreign invaders. The power of the immune system—its specificity, memory, diversity, and self-nonself recognition—comes from the ability of its cells to communicate with one another. In the next chapter we will find out how the cells of the nervous system communicate with one another and with the muscles and organs of the body.

STUDY OUTLINE WITH KEY TERMS

The body defends itself against invading pathogens with three lines of defense: the skin and the populations of microorganisms on the skin and in the gut, **blood clots** and **inflammation,** and the **immune system.** The skin consists of the **epidermis,** the **dermis,** and the **subcutaneous tissue.** The saliva and tears contain **lysozyme,** an enzyme that attacks bacteria.

The body closes breaks in the skin by means of **wound healing** and breaks in the circulatory system by means of **hemostasis,** which includes **vasoconstriction,** the aggregation of **platelets,** and **blood clots.** All of these processes prevent excessive blood loss and help prevent microorganisms from entering the circulatory system.

Because blood clots are dangerous, the synthesis of the **fibrin** in clots is carefully regulated. Fibrin derives from **fibrinogen** when the enzyme **thrombin** digests away part of the fibrinogen polypeptide. Blood **plasma** contains **prothrombin,** a precursor of thrombin. Another system of anticlotting enzymes destroys the fibrin in clots. **Plasminogen** in the blood plasma is turned into **plasmin,** which digests the bonds between fibrin molecules, dissolving the clot. **Tissue plasminogen activators** (tPA) activate the anticlotting system. Mammalian blood also contains **anticoagulants,** such as **heparin** and **warfarin.**

Complement proteins and **mast cells** react nonspecifically to damage to body tissues. Mast cells release **histamines,** which help promote the inflammation response. **Inflammation,** a set of responses to local injury, helps an animal to resist infection by increasing circulation, attracting and mobilizing phagocytic cells, and inducing signaling molecules. During inflammation, **neutrophils,** attracted by chemicals released by mast cells, attach to the capillary walls, move through the capillary wall into the damaged tissue, and phagocytose any microorganisms in the damaged tissue. **Macrophages** also consume microorganisms and cellular debris and play a major role in initiating inflammation by releasing three kinds of **cytokines: tumor necrosis factor-α** (TNF-α), **interleukin-1** (IL-1), and **interleukin-6** (IL-6). The interleukins act as messenger molecules between different kinds of white blood cells, initiating and regulating inflammation and immune responses. Interleukin-1 stimulates responses in phagocytic cells—in the endothelial cells of the capillaries and in the brain cells that control body temperature. Another set of cytokines, called interferons, interferes with the replication of viruses by limiting protein synthesis in virus-infected cells.

The immune system responds to invading organisms by producing **antibodies** and by producing lymphocytes—T cells and B cells—that recognize and attack cells of the body that have become infected by viruses or microorganisms. Four attributes characterize immunity: specificity, memory, diversity, and self-nonself recognition.

Cells of the immune system recognize **specific** invaders and not others. The immune system's **antibodies** recognize **antigens** by their **epitopes.** The specificity of the immune response distinguishes it from the **nonspecific** responses of clotting, complement proteins, inflammation, and interferons. The immune system also has **memory.** The reappearance of an invader stimulates a **secondary immune response,** usually even stronger than the first response.

The immune system can defend against a **diversity** of foreign invaders. But the immune system **tolerates** many cells in the body. In accordance with the model currently accepted by most biologists, the immune system's apparent ability to distinguish the body's own components from others is called **self-nonself recognition.**

Immune responses depend on three kinds of **lymphocytes,** which develop within the lymphoid tissues: **B lymphocytes,** or **B cells,** which make antibodies; **T lymphocytes,** or **T cells,** which recognize antigens on the surfaces of the body's own cells; and **natural killer cells,** or **NK cells,** which attack tumor cells and cells infected by a pathogen.

Biologists frequently make use of **hybridomas,** cloned cell lines that are hybrids of cells from a lymphoma and normal B cells, which make only one kind of antibody, or **monoclonal antibody.**

The simplest antibody molecules, called **immunoglobulin G** or **IgG,** consist of four polypeptide chains, two **heavy chains** and two **light chains.** Antibody diversity depends on the ability to combine many different light chains with many different heavy chains.

While the immune system can produce more than a billion antibodies, a single antibody-producing cell can make only one kind of antibody. Individual antibodies have different amino acid sequences. The immune response depends on an animal's ability to select among an existing repertoire of cells, each of which is each able to make a single type of antibody.

Both light and heavy chains of immunoglobulin are encoded in genes that undergo rearrangement. To form a functional gene, a light chain gene undergoes two rearrangements, and a heavy chain gene undergoes three rearrangements. Somatic mutations can further increase the diversity of immunoglobulin sequences.

During embryonic life, precursors of B cells form a **lymphocyte library,** with each lymphocyte containing a unique random rearrangement of immunoglobulin genes. B cells are able to respond to particular antigens because they have **B-cell receptors,** cell surface proteins with the same antigen recognition sites as the antibody that they produce. B cell receptors are membrane-bound forms of antibodies. According to the generally accepted **clonal selection theory,** an individual B lymphocyte proliferates when exposed to an antigen that binds to its particular antibody. Its descendants include **effector cells,** or **plasma cells,** which produce and export antibodies, and **memory cells,** which can later develop into effector cells. The clonal selection theory also suggests a mechanism for self-nonself recognition and immunological **tolerance.** Cells in the lymphocyte library that recognize self antigens are eliminated or suppressed.

T cells are of two kinds: **T cytotoxic (T_C)** cells, which kill cells recognized as **altered self,** and **T helper (T_H)** cells, which send messages to nearby B cells and to T cytotoxic cells. Upon stimulation with a foreign antigen, some T_C cells differentiate into effector cells, called **cytotoxic T lymphocytes (CTL),** which actually kill the target cells. Antigen recognition by T cells depends on **T-cell receptors** on the surface of T cells. T-cell receptors are even more diverse than antibodies.

T-cell receptors do not recognize free antigens, but only short polypeptides bound by the proteins of the **major histocompatibility complex (MHC).** Two classes of MHC proteins can present antigens to T cells: **MHC Class I** molecules, which are on the surface of all cells in the body, present **endogenous antigens** to T cytotoxic cells, and **MHC Class II** molecules, which present **exogenous antigens** on the surfaces of specialized antigen-presenting cells that interact with T helper cells.

REVIEW AND THOUGHT QUESTIONS

Review Questions

1. Name the body's three lines of defense against pathogens.
2. What is hemostasis?
3. What two systems prevent all of the blood from clotting at once? How do these two systems work?
4. Where do interleukins come from and what do they do?
5. How does the immune system make enough antibodies?
6. Lymphocytes come in three types. What are they and what function does each one serve?

7. Explain what a lymphocyte library is and why it is important to immune function.

Thought Questions

8. Why do you think that sperm are the only cells that do not have MHC Class I molecules on their cell surfaces?
9. Which facts presented in this chapter support the danger model and which support the self-nonself model?

SELECTED READINGS

Hall, Stephen S., *A Commotion in the Blood: Life, Death and the Immune System* (The Sloan Technology Series), Henry Holt & Co., New York, 1997. A popular history of immunotherapy as a treatment for cancer and a thoughtful critique of how medical stories are reported in the news.

Tizard, Ian R., *Immunology: An Introductuon*, 4th ed, Saunders College Publishing, Philadelphia, 1995. An academic introduction to immunology.

▶ On-line materials relating to this chapter are on the World Wide Web at http://www.saunderscollege.com/lifesci/
Click on Tobin/Dusheck: *Asking About Life.*

Rita Levi-Montalcini

When Nobel laureate Rita Levi-Montalcini (1909–) was a 21-year-old medical student, she asked her professor to suggest a suitable research project. Her professor, a neuroanatomist at the Turin School of Medicine, in Italy, suggested that she determine how the convolutions of the human brain were formed during development.

It was an impossible task. To begin with, it would have required that she study human fetuses, a rarity in a country where abortion is illegal. Beyond that, it was a mammoth project even for a well-established scientist with years in which to work on the problem. The question was just too big, too unfocused.

"It was a really stupid question, which I couldn't solve and no one could solve," Levi-Montalcini recalled at age 82. When the professor called her in a few months later to see how she was doing, he pronounced her attempts "real trash" and said she was not cut out for scientific research.

Yet the 22-year-old didn't let him discourage her. She quickly found a more manageable problem to work on and never looked back. Levi-Montalcini gradually taught the professor to respect her.

In the end, he proved a faithful friend and mentor until his death many years later. All her life, Levi-Montalcini has skillfully navigated obstacles that would stop the average person.

Born into an intensely patriarchal Italian-Jewish family, Levi-Montalcini made up her mind when she was still a child that she would not live under a husband's thumb, as her own mother did. She rebelled against her parents' assumption that she and her twin sister Paola would marry straight out of high school and begged, in vain, to be sent to a high school that would prepare her for college (Figure 41-1).

Nonetheless, after graduating from a high school for girls, which emphasized literature but not math or science, she found that she had no particular interests. Isolated and at loose ends, Levi-Montalcini read her way through her adolescence. She had ambition but neither goals nor occupation. She buried herself in books in which she was only mildly interested.

When Levi-Montalcini was barely 20, however, her beloved governess, Giovanna, died of stomach cancer, and Levi-Montalcini resolved to become a doctor. Levi-Montalcini knew that she had not had the education that she needed to go to medical school, so she decided to find a tutor to prepare her for the entrance exams. To make the endeavor more fun, she persuaded her cousin Eugenia to join her. Eight months later they both passed the entrance exams with flying colors.

Figure 41-1 As girls, Rita Levi-Montalcini (*shown here*) and her sisters, Paola and Nina, hoped to emulate the Brontë sisters, the famous 18th-century trio of English writers. As they grew older, Rita became a Nobel Prize–winning neuroscientist and Paola a famous painter. *(Washington University Archives)*

In 1930, they entered the Medical School at Turin, Italy. By 1936, they had earned degrees in medicine with top honors and had apprenticed themselves to a flamboyant Jewish neuroanatomist at Turin. After the false start with the human brain, Rita began work that fascinated her. "For the first time," she wrote in her autobiography, "I became passionate about research. . . ."

It was a passion that would be tested. In 1924, Mussolini had become dictator of Italy, and beginning in the 1930s his alliance with Hitler obliged him to take action against Italy's Jews. By the fall of 1938, a series of state decrees forbade all Jews from pursuing any kind of profession, including academic research and teaching. In 1939, Levi-Montalcini left the university at Turin, where she could no longer work without endangering both herself and her non-Jewish colleagues.

For a year, she lived quietly with her mother, her sister Paola, and her brother Gino. Her father had died several years earlier. She and Eugenia spent much of their time treating Jews who had fallen ill but could not be treated by non-Jewish doctors.

Then one day she received a visit from an old friend from medical school who demanded to know what sort of research she was doing. Under the wartime conditions, it hadn't occurred to her that she could do any research, and she said nothing. Distressed by her apathy, he scolded her, "One doesn't lose heart in the face of the first difficulties. Set up a small laboratory and take up your interrupted research. Remember Ramón y Cajal who in a poorly equipped institute, in the sleepy city that Valencia must have been in the middle of the last century, did the fundamental work that established the basis of all we know about the nervous system of vertebrates."

It was all the impetus Levi-Montalcini needed. She quickly set up a laboratory in her bedroom and decided to study chick embryos, as chicken eggs were relatively easy to obtain. Partly inspired by the thought of the Spanish anatomist Ramón y Cajal, she chose to study the development of the nervous system.

From 1941 until 1943, Levi-Montalcini studied the development of the nervous system in her bedroom. Always practical, she would carefully remove the tiny chick embryos from the eggs after she had finished an experiment and turn the eggs into omelets in the kitchen downstairs. Her brother Gino, after watching this operation once, categorically refused to eat any more of her omelets, although he had loved them until then.

Her research into how embryonic nerve tissue differentiates into specialized cells went splendidly. Wrote Levi-Montalcini:

> Now the nervous system appeared to me in a different light from its description in textbooks of neuroanatomy, where its structure is described as rigid and unchangeable. Only by following from hour to hour in different specimens, as in a cinematographic sequence, the development of nerve centers and circuits, did I come to realize how dynamic these processes are; how individual cells behave in a way similar to that of living beings; how plastic and malleable is the entire nervous system.

Levi-Montalcini made up her mind when she was still a child that she would not live under a husband's thumb, as her own mother did.

During development, the nerves of the peripheral nervous system appear to grow out from the spinal cord toward the limbs. The biologist Viktor Hamburger, at Washington University in St. Louis, had shown that embryos whose limb buds have been amputated do not develop nerves to supply the limb. He proposed that the limbs produce some special substance that tells the unspecialized nerve cells in the spinal cord what kind of cells they should turn into, as well as in what direction they should grow.

Levi-Montalcini's research suggested an alternative. She thought that the nerves in the embryonic spinal cord do specialize, or differentiate, but that they soon die without some special substance from the limb buds to sustain her growth. But her work was once more interrupted.

In 1943, German tanks approached Turin, and Levi-Montalcini and her family fled to Florence, abandoning the makeshift lab, their home, and their friends. In Florence, they took up lodging with a tolerant Catholic woman—assuring her that they were not Jewish. But Florence, too, was taken by the Germans, who mined the streets, blew up the bridges, and killed civilians. Finally, in September 1944, when the British liberated Florence, Levi-Montalcini went to work for the Red Cross as a practicing physician. Not until 1945 did she return to her research.

After the war, Hamburger, who had read her papers, invited her to come to the United States to work with him. Her theory that some substance sustains developing nerve cells turned out to be correct, and in the 1950s she and biochemist Stanley Cohen isolated the special substance, which they called *nerve growth factor*. Levi-Montalcini earned a place on the faculty at Washington University, and in 1986 she and Cohen shared the Nobel Prize for medicine and physiology for their work.

Their discovery has led to a more thorough understanding of the development of the nervous system and, more recently, to the possible use of nerve growth factor and similar proteins to stimulate the growth of damaged neurons, especially in the case of diseases such as Alzheimer's disease. Nerve growth factor itself not only stimulates the development of the nerves of the sensory system and the sympathetic system (which mobilizes the "fight or flight" response of the body) but it also acts on cells of the brain itself and on the immune system.

Levi-Montalcini, now in her 80s, lives in an apartment in Rome that she shares with her twin sister Paola, who is a well-known artist. Levi-Montalcini continues her scientific work, exploring the role of nerve growth factor on the adult immune and endocrine systems. As a Nobel laureate, she has been invited to serve on various scientific committees, a role that allows her to influence the direction of science both in Italy and internationally. She and her sister run a foundation to help teenagers find a path in life. Every week she talks to teenagers about their interests and helps them find work that interests them.

"You never know what is good, what is bad in life," Levi-Montalcini told the magazine *Scientific American* of the Nobel Prize–winning research done in her bedroom. "I mean, in my case, [working in partial isolation] was my good chance."

KEY CONCEPTS

1. The nervous system consists of neurons, which carry electrical signals, and glia, which provide metabolic and structural support to the neurons.

2. Neurons contain special proteins that allow them to carry electrical signals, called action potentials, over long distances.

3. Connections between neurons may be either electrical, mediated by gap junctions, or chemical, mediated by the release of signaling molecules called neurotransmitters.

4. Neurotransmitters act on target neurons either by directly changing the properties of ion channels or by altering the levels of second messengers such as cyclic AMP.

HOW ARE NERVE CELLS SPECIAL?

Even a generation after the acceptance of the cell theory, biologists were still not certain that the nervous system contained conventional cells. Until the late 19th century, anatomists studied the brain in the same ways that they studied other tissues, cutting, staining, and examining the tissues under a microscope. But when they examined the tissues of the brain, they did not see distinct cells, only dots, blurs, and dark spots that resembled nuclei (Figure 41-2A). The problem, we now know, was that the brain contains so many intertwined cells that the only way to see an individual cell is to stain one cell, while leaving its neighbors unstained. In the 1880s, Camillo Golgi developed just such a method, now called the Golgi method, for staining only a few cells in each sample (Figure 41-2B).

Shortly afterward, the great Spanish neuroanatomist, Santiago Ramón y Cajal, used Golgi's method to study the nervous system of a variety of species. Ramón y Cajal's careful work provided neuroscience with many deep insights into the organization and development of the nervous system. His most important contribution, however, was clear evidence that the nervous system consists of cells.

Nerve cells have exactly the same kinds of organelles as other cells (Figure 41-3). They come in a variety of shapes and sizes, but all have a distinctive look (Figure 41-4A). The most obvious difference between nerve cells and other cells of the

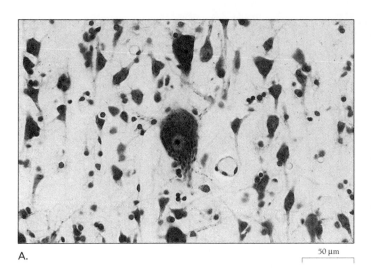

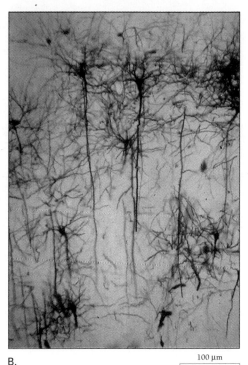

Figure 41-2 In the late 19th and early 20th century, anatomists debated whether the brain actually contains individual cells. A. With standard staining techniques, it is impossible to tell where one cell ends and another begins. B. The silver stain developed by Camillo Golgi allowed researchers to see the extended nature of individual cells. *(A, B, John D. Cunningham/Visuals Unlimited)*

body is their size. They can be immense. The nerve cells in the legs of a giraffe, for example, grow to several meters in length.

Nerve cells that carry information in the form of electrical and chemical signals are called **neurons.** But the nervous system also includes **glial cells,** which provide metabolic and structural support for the neurons. Glial cells do not themselves transmit electrical and chemical signals. Some glial cells serve as guides and scaffolding during neural development. Others, called *oligodendrocytes* and *Schwann cells,* wrap around nerve cell extensions in a *myelin sheath.* The myelin sheath speeds the transmission of signals through the nerves (Figure 41-4B).

Neurons exhibit many variations on a basic plan (Figure 41-4). Most have a distinct center, called the *cell body,* which is about the same size as most other cells. All neurons have long extensions from the cell body. Numerous short extensions, called **dendrites,** usually relay signals from other cells to the cell body. Longer, thicker extensions, called **axons,** usually carry signals away from the cell body and connect to other cells. Axons may be quite long, up to several meters. They may branch at their ends to make contact with target cells, but the number of axon branches is trivial compared with that in some dendrites. One type of cell in the brain's cerebellum, for example, has enough dendrites to relay information from more than 50,000 other cells. Between the axons and dendrites are highly specialized contacts, called **synapses.**

All of the specialized structures in neurons help them to carry information over long distances and make contacts with specific other neurons. Neurons also have particularly large numbers of mitochondria and ribosomes, suggesting that they are active users of energy and synthesizers of proteins. Studies of brain metabolism support this conclusion: the human brain, which makes up only 2 percent of the body's weight, uses nearly 20 percent of all the energy the body uses.

How Do Nerve Cells Carry Information?

Nerve cells carry innumerable different kinds of information. Whether we cut ourselves or catch sight of a loved one, the in-

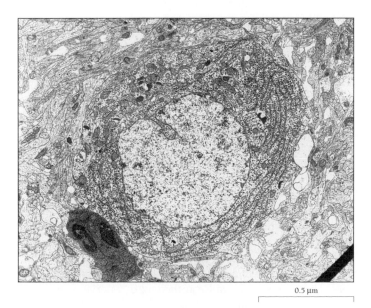

Figure 41-3 Electron microscopy reveals that nerve cells have exactly the same kinds of organelles as others cells—including mitochondria, ribosomes, and endoplasmic reticulum. *(Dr. Dennis Kunkel/Phototake)*

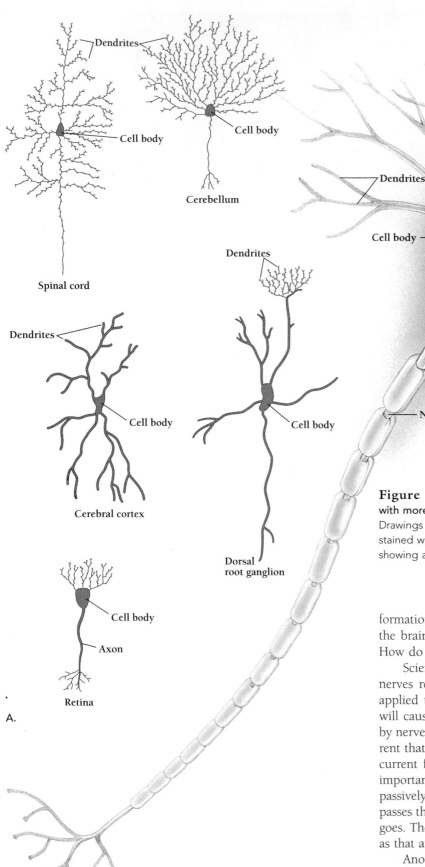

A.

Dendrites

Cell body

Spinal cord

Dendrites

Cell body

Cerebellum

Dendrites

Cell body

Cerebral cortex

Dendrites

Cell body

Dorsal
root ganglion

Cell body

Axon

Retina

B.

Dendrites

Cell body

Nucleus

Axon

Schwann
cell

Node of Ranvier

Figure 41-4   Nerve cells come in a variety of sizes and shapes, with more or fewer processes that connect them to other cells. A. Drawings of different types of nerve cells, taken from brain sections stained with Golgi's method. B. A schematic drawing of a nerve cell, showing axons enclosed in an insulating myelin sheath.

formation about the event is carried to the brain by nerves, and the brain's response is transmitted to our muscles by nerves. How do nerves do it?

Scientists have known since the late 18th century that nerves respond to electrical impulses. A small electric shock applied to the neurons that control a frog's leg, for example, will cause the leg to kick. The electrochemical pulses carried by nerves are entirely different, however, from the electric current that flows in the wires of your house. Household electric current flows much more rapidly than nerve impulses. More important, electric current gradually loses energy as it flows passively through a wire. In contrast, the electric pulse that passes through the body's nerves is constantly regenerated as it goes. The energy of the pulse at the end of a nerve is the same as that at the beginning.

Another important difference between electric current and nerve impulses is the way in which the energy is transmitted from cell to cell. Ordinary electricity can flow at a great rate or at a trickle, depending on how much electricity is originally supplied and on the resistance of the line. Nerve cells, however, always carry the same-sized pulses of energy. If a stimulus, such as a mild electric shock, is sufficient to trigger a nerve impulse, the impulse will be the same size no matter how strong or weak the stimulus is. On the other hand, if the stimulus is too weak to trigger an impulse, the nerve cell makes no response at all. The nerve cell has an all-or-none, or **threshold,** response. Any stimulus above a certain level triggers the same response.

We can ask ourselves, if the neuron has an all-or-none response, how can we detect subtle differences in temperature, light intensity, and other variable stimuli? Nerves convey differences in intensity in two ways. First, as the intensity of a stimulus increases, the frequency of pulses generated and transmitted also increases. Second, neighboring cells have different thresholds, so that as the intensity of a stimulus increases, the total number of responding cells increases. The result of these two processes is that as the intensity of a stimulus—say the temperature of your bathwater—increases, the number of impulses reaching your brain increases. In addition, the impulses come from a greater number of receptors if the bathwater is very hot than if it is just warm.

If nerve impulses are not simple electric currents, it is fair to ask what they are. In the rest of this chapter we will try to understand how neurons generate and transmit impulses and how these impulses are then passed from one neuron to the next.

Unlike electrical lines, nerves carry energy pulses that are always the same size and that do not dissipate during transmission. A nerve cell transmits an impulse only when stimulated by a signal that exceeds a minimum threshold.

How Does a Neuron Generate an Electrical Pulse?

Every cell has a chemical composition that is different from its surroundings. This difference results mostly from the activities of the cell membrane. As we saw in Chapter 4, cell membranes allow some molecules and ions to pass freely, while serving as impermeable boundaries to others. Membrane proteins that serve as **channels** allow the passage of specific molecules and ions. Some of these channel proteins have **gates,** which open and close the channels in response to environmental signals. Gates regulate the passive movement of ions or molecules through the membrane. Other membrane proteins actively **pump** ions or molecules through the membrane using the energy from ATP.

As a result of the operation of these channels, gates, and pumps, the inside of a cell has different concentrations of many molecules and ions than the outside. A difference in the concentration of molecules and ions on the two sides of a cell membrane is a **concentration gradient,** which is a source of po-

tential energy. If a concentration gradient also results in a net electrical charge on one side of the membrane relative to the other, then an electrical gradient also exists. Such a gradient is called **voltage,** or **electrical potential.**

Voltage is the driver of electric current. If we compare electricity to a waterfall, then the **current** is the amount of water, or the number of electrons, flowing per minute. The voltage is comparable to the height of the waterfall. The greater the voltage, the greater the work that can be performed and the greater the tendency for a reaction to happen spontaneously.

In a living cell, voltage does not result from height but from the difference in charge between one side of the cell membrane and the other. Researchers can measure the voltage, and, as we discuss later, they can even study the dependence of the voltage on the concentrations of particular ions. At rest, cells have a net negative charge that results in a potential (voltage) of about 70 millivolts across the plasma membrane (Figure 41-5A). During a nerve impulse, this net charge actually reverses itself for just a fraction of a second. This reversal, or pulse, passes along the length of the nerve cell membrane like a wave.

The -70-millivolt potential comes from a difference in the distribution of both negative and positive ions. But negative ions, such as chloride (Cl^-) and various proteins, generally do not move across the cell membrane to create the reversal in charge. Their concentrations are a constant during the generation of a nerve impulse. In contrast, the positive ions potassium (K^+) and sodium (Na^+) jump back and forth across the membrane like children playing hopscotch.

Thanks to an efficient sodium-potassium pump in the cell membrane, nerve cells contain a much lower concentration of sodium ions compared with that outside the cell. In contrast, the insides of cells have excess potassium ions. When a nerve cell is stimulated, sodium channels open and the cell membrane suddenly becomes permeable to sodium ions. As a result, the sodium ions rush across the cell membrane, attracted by the negatively charged cytoplasm, to the sodium-impoverished cell interior. As the sodium ions rush in, the inside of the cell gains a slight positive charge, at least near the membrane (Figure 41-5B). A few milliseconds later, the membrane loses its permeability to sodium ions, but it becomes permeable to potassium ions. Potassium ions now rush out of the cell, repelled by the cell's momentary positive charge. The cell then regains its customary -70-millivolt potential and it is ready to transmit another pulse.

The -70-millivolt potential of a nerve cell at rest is called the **resting potential.** Like the coil in a car engine (which stores the charge for the spark plugs), a nerve cell's resting potential represents a certain amount of potential energy that can be used to pass an impulse, under the right circumstances. In contrast, the sudden electrical change that occurs during an impulse is called the **action potential.**

Nerve cells maintain electrochemical gradients that give them a net charge leading to a potential of -70 millivolts.

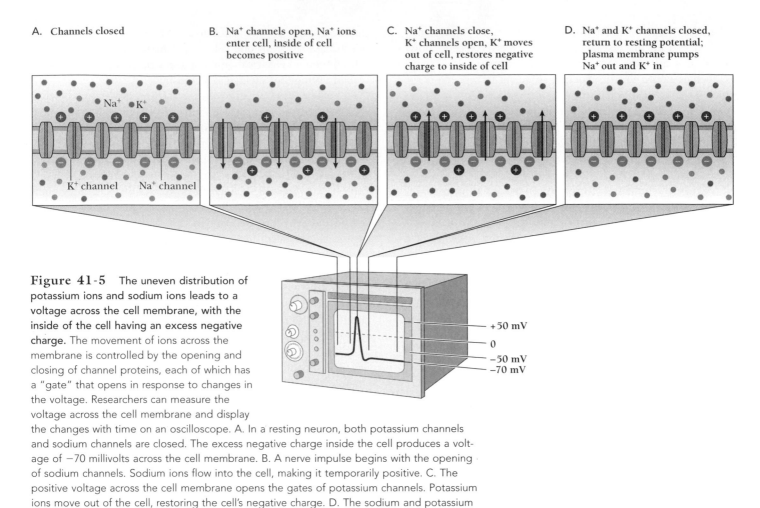

A. Channels closed

B. Na⁺ channels open, Na⁺ ions enter cell, inside of cell becomes positive

C. Na⁺ channels close, K⁺ channels open, K⁺ moves out of cell, restores negative charge to inside of cell

D. Na⁺ and K⁺ channels closed, return to resting potential; plasma membrane pumps Na⁺ out and K⁺ in

+50 mV
0
−50 mV
−70 mV

Figure 41-5 The uneven distribution of potassium ions and sodium ions leads to a voltage across the cell membrane, with the inside of the cell having an excess negative charge. The movement of ions across the membrane is controlled by the opening and closing of channel proteins, each of which has a "gate" that opens in response to changes in the voltage. Researchers can measure the voltage across the cell membrane and display the changes with time on an oscilloscope. A. In a resting neuron, both potassium channels and sodium channels are closed. The excess negative charge inside the cell produces a voltage of −70 millivolts across the cell membrane. B. A nerve impulse begins with the opening of sodium channels. Sodium ions flow into the cell, making it temporarily positive. C. The positive voltage across the cell membrane opens the gates of potassium channels. Potassium ions move out of the cell, restoring the cell's negative charge. D. The sodium and potassium channels are again closed, and the cell remains at its resting potential.

How Does a Neuron Maintain Its Resting Potential?

Measurements of resting potential across the membrane have shown that the inside of a neuron (or of any cell) has fewer positive ions than the outside. The main reasons there are so few positive ions inside the cell are (1) the membrane's sodium-potassium pump and (2) the membrane's selective permeability to sodium (Na⁺) and potassium (K⁺) ions. The sodium-potassium pump in a cell membrane uses the energy of ATP to push sodium ions out and pull potassium ions inside. For every ATP spent, three sodium ions leave and two potassium ions enter. The membrane then maintains the resulting uneven concentration of ions by restricting the passage of sodium and potassium.

The resting potential of a neuron arises because of the selectivity of ion pumps, which concentrate sodium ions outside the cell and potassium ions inside, and because of the selectivity of channels within the cell membrane, which allow different ions to pass at different rates.

How Does a Neuron Generate an Action Potential?

Electricians measure voltage with a voltmeter, a device that determines the tendency of electrons to flow between two contacts, called *electrodes*. The voltmeter measures this tendency in units called *volts*.

The voltages across cell membranes are (fortunately for curious biologists) much smaller than those in wall sockets. Even the small batteries used for toys and household appliances have voltages 1000 times greater than those of a cell. A standard radio battery, for example, generates about 1.5 volts, while most cells have voltages measured in millivolts. One millivolt is 1/1000 of a volt, so a radio battery generates 1500 millivolts. Neurobiologists measure these tiny voltages with an *oscilloscope,* a televisionlike screen that displays variations in voltage over time (Figure 41-5). Instead of standard metal electrodes, biologists use *microelectrodes,* tiny electrodes that contain salt solutions and are small enough to maintain contact with a single cell.

When a physiologist pokes a microelectrode into a nerve cell, the oscilloscope registers a negative voltage of about 70 millivolts, the resting potential. When stimulated, the nerve cell can suddenly and repeatedly change its electrical potential, an **action potential** that the electrodes also detect.

Understanding how nerve cells generate resting potentials and action potentials has come from experiments in which the internal voltages of cells could be accurately measured and also changed. Such experiments were impossible for a long time, however, because nerve cells are so small and standard electrodes are so big. Researchers were simply unable to put electrodes into cells. Doing the proper experiments required either a giant nerve cell or a very small electrode. Ultimately, biologists found the first and invented the second.

Most of our current appreciation of how a nerve cell fires—how it creates an action potential—came from experiments done on neurons from the squid. These neurons are responsible for the signals the squid uses to produce a jet of water, by which it can rapidly propel itself. The axons of these nerve cells are immense, up to 1 mm in diameter. They are so big, in fact, that until 1936, biologists thought that these nerve cells were blood vessels. With the discovery of their real nature, physiologists began collecting squid for work on neuron action potentials.

The most penetrating and influential work on squid action potentials was that of the English physiologists Alan Hodgkin and Andrew Huxley. In the 1930s and 1940s, Hodgkin, Huxley, and their collaborators took advantage of the size of the squid axon to measure voltage changes in the axon as the researchers varied ion concentrations inside and outside the cell. They found that the concentration of potassium ions determined the level of the resting potential. They also found a way to measure the ion concentrations within an axon by using a rubber roller to force the contents of the axon out into a dish.

Hodgkin and Huxley then showed that an action potential, unlike a resting potential, depends on the flow of sodium ions into the cell. In one experiment, for example, they removed all sodium ions from the extracellular fluid, replacing them with a larger ion, choline, that could not pass through a sodium channel in the membrane—even when the channel was open. The result was that the axon could no longer produce an action potential at all. This experiment underscored the conclusion that understanding the action potential requires an understanding of how the membrane temporarily changes its permeability to sodium ions.

The resting potential of a neuron depends on the difference in the concentration of potassium ions between the inside and outside of a cell. The action potential depends on the temporary opening of channels that admit sodium ions into the cell.

How Does a Membrane Change Its Permeability to Sodium Ions?

Hodgkin and Huxley discovered that a neuron opens its sodium channels after they developed a method called *voltage clamping,* which allowed them to control the voltage across a cell membrane. At the normal resting potential, they found, sodium channels are closed. When the cell is *depolarized,* or made less negative, however, the channels open briefly and admit a small horde of sodium ions. Because the channels' ability to admit sodium ions depends on the membrane potential, sodium channels are said to be **voltage-gated,** meaning that they open or close according to the voltage across the membrane.

Hodgkin and Huxley were able to explain the generation of action potentials in terms of the voltage dependence of sodium channels. When a nerve cell is sufficiently depolarized, the membrane reaches a threshold voltage, at which point sodium channels open and sodium ions flow inward. The flow stops within about 2 milliseconds, however, and the sodium channels become less likely to open than before. During this time, called the **refractory period,** the potassium ions flow out of the cell and restore the original resting potential.

During each action potential, then, some sodium ions move into the cell and some potassium ions move out of the cell. The flow of potassium ions outward increases during the action potential because the potassium channels are also voltage-gated. But the number of potassium ions lost is small compared with the total inside the cell, and the membrane's sodium-potassium pump soon restores the original distribution of both ions.

The time course of an action potential depends on the opening and closing of voltage-gated sodium and potassium channels.

How Do Voltage-Dependent Sodium Channels Open and Close?

Many poisons and stimulants affect the nervous system profoundly by binding to specific proteins important to neural function. For example, *tetrodotoxin,* which comes from the Japanese puffer fish ("*fugu*"), blocks the passage of sodium ions through the voltage-gated channels (Figure 41-6). Using

Figure 41-6   In Japan, the puffer (*fugu*) is a delicacy, but chefs are careful to remove the highly toxic skin and entrails before preparing the fish for the table. The active agent in the toxin is tetrodotoxin, which blocks sodium channels. Researchers used tetrodotoxin to isolate the sodium channel protein.   *(Jeffrey L. Rotman/Peter Arnold, Inc.)*

tetrodotoxin and similar toxins, biochemists have isolated the protein that forms the voltage-gated sodium channel. Electric eels are a particularly rich source of the sodium channel protein because their electric organs generate big voltages in the same way that neurons generate small voltages.

Soon after the biochemists isolated the channel protein, a group of Japanese scientists, led by S. Numa, isolated a gene encoding the channel protein, which they then sequenced. Knowing the amino acid sequence of the sodium channel allowed researchers to guess how it might work. Subsequent research confirmed that four polypeptide segments assemble to form a pore through which sodium ions can flow when the gate is open and that the gate opens only when the outside of the membrane has a positive charge, that is, when the cell is depolarized.

A sodium channel is a membrane protein that allows sodium ions to pass when the outside of the membrane has a positive charge.

How Does an Action Potential Move Down an Axon?

The properties of the voltage-gated sodium channel explain not only how action potentials occur but also how they can move for long distances along nerve cell membranes. We can list six consecutive events in the production of an action potential in a single spot in the membrane:

1. Something causes a local depolarization of the membrane, large enough to exceed the threshold.
2. Sodium channels open, and sodium ions flow into the cell.
3. As sodium ions flow inward, the inside of the membrane becomes locally positive.
4. The decreased polarization of the membrane causes more channels to open, increasing the positive charge of the membrane still further.
5. Finally (that is, after less than a millisecond), the sodium channels close (spontaneously).
6. Potassium ions then flow outward, restoring the membrane to the resting potential.

Notice that the pumping of sodium and potassium ions does not directly enter into the events of the action potential. This pumping serves only to maintain the distribution of ions responsible for the resting potential.

Now we can see how the action potential propagates itself. As the membrane becomes more positive at one spot, the charge spreads to the adjacent spot. That patch of membrane also becomes depolarized and opens its sodium channels, leading to further depolarization. In this way the action potential propagates itself down the axon, regenerating itself as it goes.

The action potential moves in just one direction, however. Recall that once the sodium channels have opened, they cannot immediately open again. As a result, an impulse cannot pass through again until after the refractory period. By that time the impulse has moved on. This prevents an impulse from moving backward. Each action potential passes through each membrane region only once.

An action potential propagates because a charge moves along the cell membrane, opening sodium channels and leading to further depolarization. Because each channel has a refractory period, the charge can move in only one direction.

How Do Nerves Speed the Transmission of Action Potentials?

Although the thick axons of the squid conduct impulses at speeds up to 10 meters per second, most nerve cells have much smaller axons that would normally conduct impulses much more slowly. Given the expected rate of conduction in the sensory nerves leading from our feet, for example, it would take a painfully long time for us to realize that we had stepped on a nail or come too close to a campfire. Most vertebrate nerves, however, are wrapped in **myelin,** a specialized, glistening sheath that allows a much more rapid conduction of nerve impulses.

The myelin sheath insulates most of the axon's membrane, preventing the passage of ions (Figure 41-7). Depolarization and ion flow are only possible in the **nodes of Ranvier,** gaps between the myelin wrappings, spaced about every millimeter along the axon. Instead of running down the axon continuously, then, the nerve impulse moves by **saltatory conduction,** jumping down a myelinated axon at rates up to 100 times faster than down an unmyelinated axon. An ordinary nerve conducts at about 1.2 meters per second, or about $2\frac{1}{2}$ miles per hour, the speed of a leisurely walk. In contrast, a myelinated nerve conducts at up to 120 meters per second, or about 250 miles per hour.

Nerves wrapped in myelin conduct action potentials rapidly because nerve impulses jump from node to node.

HOW DO NEURONS COMMUNICATE WITH ONE ANOTHER?

Ramón y Cajal's studies demonstrated that the nervous system consists of discrete cells. Since then, studies with the electron microscope have revealed that there is a distinct boundary, the **synapse,** between most communicating neurons (Figure 41-8). Nerve impulses flow in one direction only. The neuron sending information is the **presynaptic** neuron while the receiving neuron is the **postsynaptic** neuron. We have already seen that a neuron can send information over long distances down an axon. But we still need to understand how a presynaptic neuron (lying "upstream" from the synapse) influences the electrical activity of a postsynaptic neuron (to which it connects through a synapse).

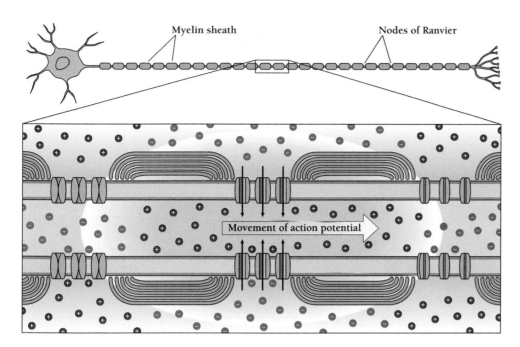

Myelin sheath   Nodes of Ranvier

Movement of action potential

Figure 41-7 A nerve impulse moves much more quickly down a myelinated axon because it jumps from node to node. Ion currents can only flow where the myelin sheath is interrupted, at the nodes of Ranvier.

The simplest synapse is the **electrical synapse,** which joins presynaptic and postsynaptic neurons through gap junctions, channels through the membranes of adjacent cells that allow ions and small molecules to pass freely from one cell to the next. An electrical synapse allows an action potential to continue traveling to the postsynaptic cell in the same manner and at about the same rate as it traveled down the presynaptic axon. All electrical synapses are **excitatory,** meaning that action potentials in the presynaptic cell stimulate action potentials in the postsynaptic cell.

The more common type of synapse is the **chemical synapse,** in which presynaptic and postsynaptic membranes do not join closely (Figure 41-8). Instead, the presynaptic and postsynaptic cells are separated by the **synaptic cleft,** a space of about 20 nm. Communication across such a gap requires the diffusion of one or more **neurotransmitters,** small signaling molecules made in the presynaptic cell that affect the electrical charge of the postsynaptic cell membrane. Chemical synapses may either excite or inhibit. So far, biologists have isolated at least 20 different neurotransmitters, with some estimates as high as 50. Among the small molecules that serve as neurotransmitters are several amino acids (glutamate and glycine) and amino acid derivatives (GABA, derived from glutamate; serotonin, derived from tryptophan; and dopamine, norepinephrine, and epinephrine, all derived from tyrosine). Other small-molecule neurotransmitters are acetylcholine and adenosine (Figure 41-9).

The electrical activity of the presynaptic neuron triggers the release of a neurotransmitter, which then diffuses across the synaptic cleft. A neurotransmitter can either excite or inhibit the activity of the postsynaptic neuron, but the effect is always delayed by the time needed for the neurotransmitter to diffuse across the gap, generally about 0.5 millisecond. No such delay

occurs in an electrical synapse. On the other hand, chemical synapses have two important advantages over electrical synapses. First, they may be either excitatory or inhibitory. Second, they can greatly amplify the signal of a small presynaptic neuron.

Most synapses in vertebrate central nervous systems are chemical rather than electrical. Electrical synapses occur in the vertebrate heart, where they coordinate the synchronous contraction of heart muscle cells. They also occur in a variety of invertebrate and vertebrate nerve circuits, where fast conduction between cells is advantageous and subtle modifications of a signal are unimportant. One such synapse is that responsible for mediating the escape reflex in a crayfish.

A synapse may be electrical or chemical. Electrical synapses are always excitatory, but chemical synapses may be excitatory or inhibitory.

How Does a Chemical Synapse Work?

Understanding how chemical synapses work is important not only to our understanding of the brain but also to medical and social problems. Essentially all the medications that affect the functioning of the brain act on synapses. Among these substances are sleeping pills, tranquilizers, antipsychotic medicines, and narcotics such as codeine, heroin, and cocaine. Each of these chemicals acts by mimicking or interfering with the production, release, or action of some neurotransmitter.

Let us look more closely at the structure of a chemical synapse. The best understood chemical synapse is the vertebrate *neuromuscular junction,* the synapse between a motor neu-

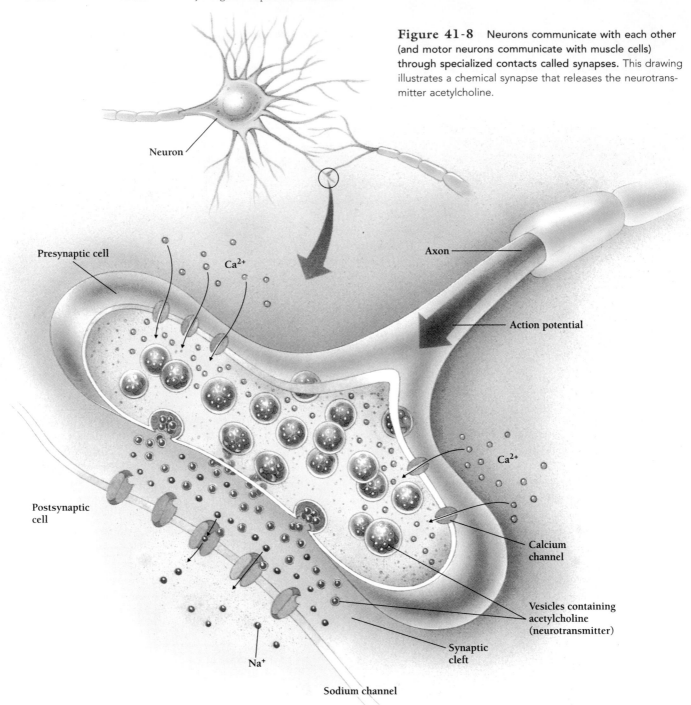

Figure 41-8 Neurons communicate with each other (and motor neurons communicate with muscle cells) through specialized contacts called synapses. This drawing illustrates a chemical synapse that releases the neurotransmitter acetylcholine.

ron and a voluntary muscle cell (Figure 41-10). These synapses are similar to those between neurons, but they are easier to study.

On the presynaptic side, the electron microscope reveals tens of thousands of tiny vesicles, each about 50 nm in diameter and enclosed by a membrane. These vesicles, which release their contents by exocytosis, are full of the neurotransmitter **acetylcholine.**

Knowing this much, we now want to know (1) what triggers the release of transmitter and (2) how the released neurotransmitter causes electrochemical changes in the postsynaptic cell.

The trigger for release of the acetylcholine appears to be an increase in intracellular calcium (Ca^{2+}) ions. The concentration of calcium ions in the cytoplasm is much lower than in the extracellular fluid. Calcium ions will always flow *into* a cell—if there is a route.

During an action potential, calcium ions flow into a presynaptic neuron, but they do so only locally—through the membrane next to the synapse and nowhere else (Figure 41-9). The calcium ions move through special calcium channels that open in response to the change in voltage during an action potential. Like sodium channels, calcium channels are voltage-gated.

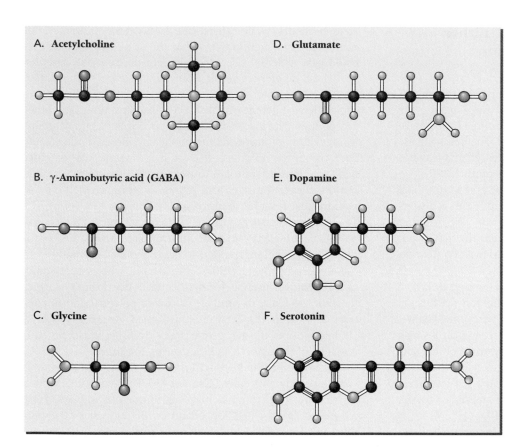

A. Acetylcholine

B. γ-Aminobutyric acid (GABA)

C. Glycine

D. Glutamate

E. Dopamine

F. Serotonin

Figure 41-9   Some of the smaller neurotransmitters—acetylcholine, GABA, glycine, glutamate, dopamine, serotonin.

A neuron has many fewer calcium channels than sodium channels. There are so few calcium ions in the cytoplasm, however, that even a small flow is enough to cause a big change in concentration and to trigger exocytosis and the release of acetylcholine.

At neuromuscular junctions, an increase in the concentration of calcium ions near the synapse triggers the release of the neurotransmitter acetylcholine.

What Does the Neurotransmitter Acetylcholine Do?

Once acetylcholine is released, it moves across the synapse of the neuromuscular junction to the postsynaptic membrane, where it binds to a receptor molecule. This acetylcholine receptor is itself an ion channel that allows the passage of sodium and potassium ions. Like the voltage-gated sodium channels and the voltage-gated calcium channels, this ion channel is gated. But it is acetylcholine rather than voltage that unlocks the gate. A molecule that specifically binds to another molecule is called a ligand, so the acetylcholine receptor is an example of a **ligand-gated channel.**

The acetylcholine receptor in muscle is a ligand-gated ion channel. When acetylcholine binds to the channel protein, it opens a pore that allows the passage of sodium and potassium ions.

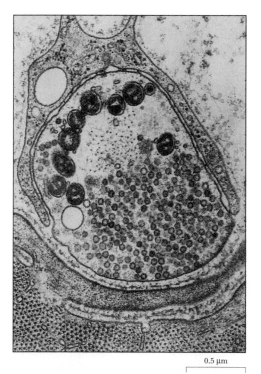

0.5 μm

Figure 41-10   The neuromuscular junction is a chemical synapse between a motor neuron and a muscle cell. In this electron micrograph, note the clustering of the small vesicles, which contain high concentrations of acetylcholine. Some of these vesicles are released by exocytosis when the motor neuron fires.   *(T. Reese & D.W. Fawcett/Visuals Unlimited)*

Neurotransmitters May Produce Either Excitatory or Inhibitory Effects in Postsynaptic Neurons

When sodium ions pass through the channel of the acetylcholine receptor, they depolarize the membrane, making the inside of the membrane less negative than before. This voltage change is called a **postsynaptic potential.** The postsynaptic potential triggered by acetylcholine excites the cell and is therefore called an **excitatory postsynaptic potential** (**EPSP**).

If an EPSP is sufficiently depolarizing, the postsynaptic cell fires an action potential. When the postsynaptic cell is a muscle cell, this action potential triggers contraction. On the other hand, when the postsynaptic cell is a neuron, the action potential travels through the cell body and axons to reach the next synapse.

A single EPSP may not depolarize a cell enough to fire an action potential. Whether the postsynaptic neuron reaches the threshold for firing often depends on more than one synapse, however. Many presynaptic neurons can converge on a single postsynaptic neuron. Some may be excitatory, while others may be inhibitory. Inhibitory neurotransmitters increase the negative charge on the inside of a membrane, making it less likely that it will fire. Inhibitory neurotransmitters are said to trigger an **inhibitory postsynaptic potential** (**IPSP**).

The most common inhibitory neurotransmitter in the brain is **gamma-aminobutyric acid** (**GABA**) (Figure 41-9B). Virtually every neuron in the brain can respond to GABA, and some 30 percent of all brain neurons make GABA. Most of these responses depend on the binding of GABA to the **GABA$_A$ receptor,** a ligand-gated chloride channel like the acetylcholine receptor of the neuromuscular junction.

Binding of GABA to GABA$_A$ receptors opens a channel that admits chloride ions, making the inside of the cell more negative. The negative charge of these ions counterbalances any positive sodium ions flowing into the cell and decreases the chance that the postsynaptic cell will fire an action potential.

The activity of a postsynaptic cell depends on the summing of the effects of excitatory and inhibitory potentials. A single postsynaptic neuron may integrate inputs from as many as 100,000 different presynaptic cells. Each neuron effectively serves as a microcomputer that determines whether to fire or not to fire according to the final balance of positive and negative ions coating its membrane.

The firing pattern of the postsynaptic cell depends on the summing of the responses to both excitatory and inhibitory inputs.

Neurons May Respond to the Same Neurotransmitters in Different Ways

By the mid-1970s, researchers had discovered and studied a handful of neurotransmitters. Some of these, like acetylcholine, were excitatory, while others, like GABA, were inhibitory. But neurobiologists soon realized that the same neurotransmitter could have different effects in different postsynaptic neurons.

These differences result from the existence of more than one type of receptor on each postsynaptic neuron. Different receptors have different responses to neurotransmitters and other compounds. The nicotine found in tobacco, for example, mimics the effects of acetylcholine in skeletal muscle and in some neurons but has no direct effect on heart muscle. On the other hand, the mushroom toxin muscarine stimulates heart muscle but not skeletal muscle. And muscarine stimulates only those neurons that respond to acetylcholine but not to nicotine.

Neurobiologists group acetylcholine receptors into two varieties, nicotinic receptors and muscarinic receptors. Cells with muscarinic receptors respond to acetylcholine differently from those with nicotinic receptors. For example, when smooth muscle and cardiac muscle receive acetylcholine from a neuron, they are less likely to produce an action potential and to contract than before the acetylcholine arrived. As a result, acetylcholine reduces the heart rate and slows the peristaltic contractions of the smooth muscles in the intestines.

Neurotransmitters work in the same way as hormones and other signaling molecules (Chapter 36). Postsynaptic receptors for all known neurotransmitters act in one of two ways. They are either ligand-gated ion channels, or they affect the production of a second messenger such as cyclic AMP. Neurotransmitters that affect cyclic AMP levels may act either to increase or to inhibit the synthesis of cyclic AMP.

Receptors for neurotransmitters respond either by opening ion channels or by triggering intracellular biochemical events, such as the production of cyclic AMP.

How Are Neurotransmitters Cleared from the Synapse?

We have seen how a neurotransmitter alters the postsynaptic neuron. But what terminates the action of the neurotransmitter?

Neurotransmitters must be cleared from the synaptic cleft to make way for the next set of signals. Termination usually involves specific **reuptake,** the pumping of transmitter from the synaptic cleft either into the presynaptic neuron or into surrounding glial cells. The recovered transmitter is then either recycled into vesicles or degraded by specific enzymes. In the case of GABA, for example, an enzyme called GABA transaminase converts GABA into an inactive compound. In the case of acetylcholine, an extracellular enzyme called **acetylcholinesterase** destroys the acetylcholine soon after it is released into the synapse. Both GABA and acetylcholine therefore have only brief periods in which to act on the postsynaptic cell.

The time during which a neurotransmitter acts is limited by its removal by enzymes and transporters.

BOX 41-1

What is the physical basis of addiction?

In spite of the social seriousness of drug abuse, research into the way that neuroactive drugs work has already led to important basic discoveries about the brain. One of the most startling of these discoveries occurred in the late 1970s, as researchers tried to understand why *opiates*—morphine, heroin, and related compounds—are such powerful drugs. The result of this research was the realization that the nervous system communicates information about pain using neurotransmitters that were previously unknown.

The route to this insight was as follows: (1) if opiates act powerfully on the brain, they must bind to specific receptor molecules; (2) the binding of radioactively labeled opiates to the brain should reveal the distribution of opiate receptors; and (3) the specific distribution and properties of these receptors suggest that they must ordinarily bind to some endogenous brain compound. In the 1970s, several research groups identified three classes of such natural opiates, called the *enkephalins,* the *endorphins,* and the *dynorphins*—polypeptides

that act as natural pain suppressers and regulators of mood.

In the early 1990s, after more than a decade of trying, molecular biologists were able to identify recombinant DNAs that specified the primary structures of opiate receptors. Understanding how these receptors work may allow us to understand the neural basis of pain and to develop ways of lessening it.

Many Psychoactive Drugs Act on Chemical Synapses

Not all the molecules that bind to neurotransmitter receptors are neurotransmitters A variety of substances—including, for example, nicotine, curare (a toxin long used in poison arrows), and α-bungarotoxin (from snake's venom)—all bind to the nicotinic acetylcholine receptor. While nicotine stimulates action potentials, however, the others block the receptor. Similarly, the smooth muscle–relaxing drug atropine inhibits the muscarinic acetylcholine receptor. All of these drugs are said to be psychoactive, meaning they alter mood and perception. And all work by either mimicking or disrupting the action of a neurotransmitter.

Many compounds that bind to neurotransmitter receptors have proved to be useful medications. For 2000 years, for example, physicians prescribed extracts of belladonna, a tall bushy herb called deadly nightshade (*Atropa belladonna*), to combat diarrhea. We now know that the extracts worked because a chemical within the belladonna, called atropine, blocks the effects of acetylcholine in the smooth muscles of the intestines.

Other naturally occurring or chemically synthesized compounds affect other neurotransmitter receptors. For example, *benzodiazepines,* which are commonly prescribed antianxiety drugs, bind to $GABA_A$ receptors and increase the effectiveness of GABA in opening the chloride (Cl^-) channel.

Still other substances influence the release, reuptake, or inactivation of specific neurotransmitters. Some of these are effective as painkillers, others are tranquilizers, still others are stimulants. Some drugs, such as cocaine, bring about euphoria and anesthesia; others, such as fluoxetine (Prozac), elevate mood and fight depression. Many drugs that act on the brain

are highly addictive and, therefore, have enormous social consequences.

HOW DO NEURAL NETWORKS MEDIATE BEHAVIOR?

Just as the connections within the sensory pathways establish what we extract from our sense organs, the connections within neural networks determine how an animal will respond to a stimulus.

The connections are often as direct and mechanical as those in household wiring. When we flip a switch the light comes on. In one experiment, a researcher "rewired" a spinal network in a frog. He cut the sensory nerve that led from the right hind limb of a frog to the spinal cord. He then reconnected the nerve to the other side of the spinal cord. Unpleasant stimulation of the right leg then caused the frog to move its left leg.

The behavior that follows the activation of a neural circuit may be quite complex—walking and breathing are two clear examples. It seems likely that many behaviors—including the modal action patterns that we discussed in Chapter 29—depend upon such neural connections. Once the appropriate connections develop, the proper stimulation of a "trigger" neuron can evoke a standard pattern of behavior. The trigger for action is called a "releaser."

How Do Neurons Mediate a Reflex?

Movements and behavior ultimately depend on coordinated signals from motor neurons. These **motor neurons** represent the final common pathway of the instructions that direct behavior.

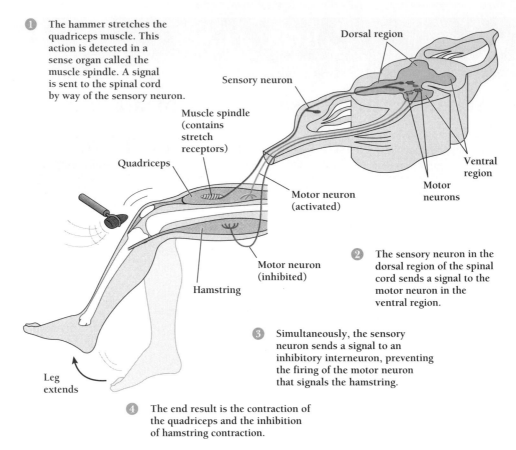

① The hammer stretches the quadriceps muscle. This action is detected in a sense organ called the muscle spindle. A signal is sent to the spinal cord by way of the sensory neuron.

Dorsal region

Sensory neuron

Muscle spindle (contains stretch receptors)

Quadriceps

Ventral region

Motor neurons

Motor neuron (activated)

Motor neuron (inhibited)

Hamstring

② The sensory neuron in the dorsal region of the spinal cord sends a signal to the motor neuron in the ventral region.

③ Simultaneously, the sensory neuron sends a signal to an inhibitory interneuron, preventing the firing of the motor neuron that signals the hamstring.

Leg extends

④ The end result is the contraction of the quadriceps and the inhibition of hamstring contraction.

Figure 41-11   **The knee jerk is an example of a monosynaptic reflex.** The physician's hammer stretches the quadriceps muscle. Stretch receptors in the quadriceps stimulate one set of motor neurons to fire, leading to the rapid contraction of the hamstring. The stimulated stretch receptors also activate a set of inhibitory interneurons, which temporarily prevent the firing of the motor neurons that stimulate the quadriceps.

The most complex behaviors—such as threading a needle, performing a laboratory experiment, or taking a midterm examination—originate in commands from the brain, particularly in the cerebral cortex. The least complex behaviors—such as withdrawing a hand from a flame—often do not depend on the brain at all but on signal processing in the spinal cord. Behavioral activity mediated by the spinal cord tends to be the most automatic, that is, the least influenced by thought or previous experience.

A **reflex** is the most automatic behavior pattern, with motor activity directly responding to a sensory stimulus. A reflex familiar to anyone who has had a medical examination is the jerking of the lower leg after a physician taps the knee with a little hammer. Let us see exactly how this happens.

The hammer hits a tendon, which stretches over the kneecap, or patella. This tendon connects a muscle in the thigh, the quadriceps, to a bone in the lower leg. The physician's hammer pulls the tendon and stretches the attached muscle. The muscle contains *stretch receptors*, modified muscle cells whose electrical output depends on the degree to which they are

stretched. Information from these receptors flows to sensory neurons in the dorsal part of the spinal cord, which in turn command the muscles to contract, which jerks the knee. The information passes through just one synapse, so the stretch reflex is said to be *monosynaptic*. The normal function of this reflex is to restore the position of the lower leg after a rapid movement in the other direction.

Information from the stretch receptors also travels to a second group of spinal cord neurons (Figure 41-11). These neurons inhibit the hamstrings muscles, which are "antagonistic" to the quadriceps muscles. By inhibiting the hamstrings from inhibiting the quadriceps, this inhibitory neuron allows the knee to jerk. Not only does the signal from the stretch receptor trigger one set of motor neurons to fire but it also inhibits another set. Yet another set of neurons in the spinal cord carries information about the status of the leg muscle to brain areas that coordinate movements and posture.

The connections involved in the stretch reflex form the simplest known neural circuit in vertebrates. Other circuits are more complicated, involving at least one interneuron. An **in-**

terneuron is a neuron that connects one or more neurons to other neurons with parallel functions (rather than to a sensory neuron or a motor neuron). An *inhibitory interneuron* prevents or slows the firing of the neuron with which it connects. Interneurons can also integrate information from many sources—from stretch receptors, from other somatic receptors, and from elsewhere in the nervous system, including the brain. Circuits in which such integration occurs are called *convergent*.

An interneuron can also participate in a *divergent* circuit, in which a single interneuron may coordinate the action of many motor neurons. Walking, for example, requires the cyclical activity of different groups of motor neurons. The needed coordination does not require the involvement of the brain, however; the interneurons of the spinal cord are enough. Input from the brain, however, can modify the activity of spinal cord motor neurons involved in walking.

In a reflex, input from a sensory receptor directly stimulates a motor neuron. More complicated neural circuits involve interneurons, which carry information among neurons with parallel functions.

HOW DOES EXPERIENCE MODIFY NEURAL NETWORKS?

Many neural networks are capable of **learning**, modification of neural activity and behavior as the result of experience. We humans are exquisitely aware of our learning, for example, as we master a new language.

Researchers distinguish between two types of learning—nonassociative learning and associative learning. In *nonassociative learning,* a person or other animal becomes more or less sensitive to a stimulus after repeated exposure. *Habituation,* for example, refers to the decrease in a behavioral response following repeated exposure to a stimulus: after we've lived in a new house for a while, we hardly notice the sound of the walls snapping at sunrise and the sounds of central heating going on and off. *Sensitization* refers to the increased behavioral response to a noxious stimulus: if a painfully loud sound is repeated, we are more likely to move away if we hear it again. The sound of the neighbor's dog barking is far more irritating after five hours than after five seconds.

Associative learning refers to the pairing of seemingly unrelated stimuli. The classic example of associative learning comes from the work of the Russian physiologist Pavlov, who trained dogs to salivate at the sound of a bell. Pavlov's trick was first to ring the bell (the *conditioned stimulus*) upon presenting the dog with dinner (the *unconditioned stimulus*). The dog would salivate at the sight and smell of the meat, which was accompanied by the ringing of the bell. Gradually the dog salivated

at the sound of the bell alone. But the sound of a can opener works just as well.

Researchers classify learning into nonassociative learning (habituation and sensitization) and associative learning (the pairing of stimuli).

Experience Can Modify the Gill Withdrawal Reflex of the Sea Hare, *Aplysia*

Experimenters have studied both nonassociative and associative learning in *Aplysia,* a type of mollusk also known as the sea hare. *Aplysia*'s simple nervous system consists of just eight paired and one unpaired clusters of nerve cells called **ganglia** [singular, *ganglion*] (Figure 41-12). Each ganglion consists of fewer than 2000 neurons whose large sizes (up to 1 mm in diameter) make them easy to identify.

Neurophysiologists also like to work with *Aplysia* because it reflexively withdraws its gills in response to a range of stimuli, from a simple touch on the siphon to a sharp blow or shock to the head or tail. *Aplysia* can learn to modify this reflex.

The neurons responsible for the gill withdrawal reflex lie in the abdominal ganglion. A sensory neuron from the siphon makes a single synaptic connection to the motor neuron that drives gill withdrawal, called L7. In addition, another neuron, called a facilitating interneuron, synapses with the axon terminal of the sensory neuron.

A single stimulation of the sensory neuron produces an action potential in L7. Repeated stimulation of the sensory neuron, however, leads to a *decrease* in the size of the action potential in L7. This decrease, called *synaptic depression,* underlies habituation. Depression results from a decrease in the amount of a released neurotransmitter called **serotonin.**

How Does an *Aplysia* Learn?

When researchers stimulate the facilitating interneuron, the synaptic depression reverses and the L7 motor neuron is said to be "facilitated." This synaptic facilitation explains sensitization. At Columbia University, Eric Kandel and his colleagues have determined the mechanism for synaptic facilitation. When the facilitating interneuron is stimulated it releases the neurotransmitter serotonin. Serotonin acts on receptors on the presynaptic axon terminals of the sensory neuron, stimulating the synthesis of cyclic AMP, which indirectly inactivates a potassium channel that helps repolarize the sensory neuron after firing. Serotonin has the effect of lengthening the period during which the axon terminal is depolarized. During the extended depolarization, more Ca^{2+} ions enter the axon terminal, which leads to a release of more neurotransmitter than usual and, consequently, a stronger response by the postsynaptic motor neuron.

A.

C.   **L7 neuron in abdominal ganglion**

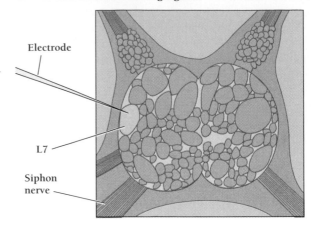

Electrode

L7

Siphon
nerve

B.   **Gill withdrawal reflex**

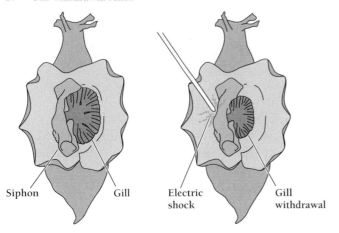

Siphon   Gill

Electric   Gill
shock   withdrawal

D.   **Circuit diagram**

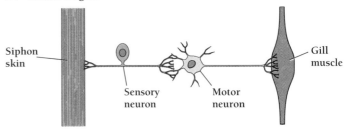

Siphon
skin

Sensory
neuron

Motor
neuron

Gill
muscle

Figure 41-12   The "sea hare" *Aplysia californica.* Despite its simple nervous system, this mollusk is still able to learn. A. *Aplysia.* B. Gill withdrawal, a reflex response to an electric shock. C. An L7 neuron in the abdominal ganglion. D. A circuit diagram of the gill withdrawal reflex.   *(A, Dan Gotshall/Visuals Unlimited)*

A single stimulus to the tail (or to the facilitating neuron) sensitizes the gill withdrawal reflex for several minutes as described above. Repeated stimulation of the tail, however, brings about a much longer period of sensitization, lasting days or even weeks. This long-term sensitization is a form of associative learning, not unlike that in Pavlov's dogs. Like short-term sensitization, long-term sensitization depends on the inactivation of a repolarizing potassium current. But, unlike short-term sensitization, long-term sensitization results from a change in gene transcription and protein synthesis. When researchers inhibit protein synthesis during the training period, the sea hare cannot learn.

The sea hare, *Aplysia,* has provided neuroscientists with the opportunity to study the cellular bases of both associative and nonassociative learning. Learning depends upon changes in synaptic signaling between individual neurons.

The insights from *Aplysia* have led to new approaches to studying the cellular basis of learning in other animals, including humans. In Chapter 42, we will see how the organization of the vertebrate brain—especially the human brain—allows behaviors far more complex than those of invertebrates.

STUDY OUTLINE WITH KEY TERMS

The operation of the nervous system depends on the properties of individual nerve cells and the way in which they connect with one another. The cells of the nervous system include **neurons,** which carry information in the form of chemical and electrical signals, and **glial cells,** which provide mechanical and metabolic support. A neuron's **axon** transmits a signal, while the **dendrites** connect to other cells.

Like all other cells, neurons have different concentrations of molecules and ions inside than outside. The selective action of **channels, gates,** and **pumps** results in a greater concentration (a **concentration gradient**) of positive ions outside the cell and an **electrical potential** (or **voltage**) of about −70 millivolts across the cell membrane. This **resting potential** can suddenly change when a neuron is stimulated, giving rise to an **action potential,** which can travel over large distances. Action potentials move more rapidly through **axons** that are surrounded by **myelin.**

The generation and propagation of action potential depend largely on the **voltage-gated** sodium channels in the neuronal membrane. These channels open when the voltage across the plasma membrane exceeds a **threshold,** and the electrical **current** into the cells increases. The **refractory period** of the channels ensures that action potentials move in a single direction down an axon. In myelin-enclosed axons, ion flow occurs only at the **nodes of Ranvier,** giving rise to rapid **saltatory conduction.**

Nerve cells communicate with one another at **synapses,** which may be either **electrical** or **chemical.** Electrical synapses allow action potentials to propagate in much the same way as they travel down axons. At a chemical synapse, the **presynaptic** cell responds to the arrival of an action potential by releasing a store of **neurotransmitter** (such as **acetylcholine,** glutamate, **GABA,** or **serotonin**) contained in synaptic vesicles.

The released neurotransmitter diffuses across the **synaptic cleft** to act on receptors on the **postsynaptic** neuron where its action may be either **excitatory** or **inhibitory.** These receptors may be **ligand-gated** ion channels, as in the case of the nicotinic acetylcholine receptors in muscle cells and **GABA$_A$ receptors** in neurons. Other neurotransmitter receptors resemble the membrane receptors for adrenaline and bring about their postsynaptic effects by stimulating or inhibiting the synthesis of second messengers, such as cyclic AMP. The action of neurotransmitters is terminated by **reuptake** or by degradative enzymes. Many psychoactive drugs act by mimicking or inhibiting the production, release, degradation, or action of neurotransmitters.

The simplest neural circuit is a **reflex,** which involves only one synapse. More complex circuits involve **inhibitory interneurons.** Even simple nerve circuits can mediate **learning,** as in the case of the sea hare, *Aplysia.* In this animal, the characteristics of the three neurons—a **sensory neuron,** a **motor neuron,** and a **facilitating interneuron**—are responsible for the modification of the gill withdrawal reflex. Short-term learning in this system depends on the effects of serotonin on potassium channels, while long-term learning depends on changes in protein synthesis.

REVIEW AND THOUGHT QUESTIONS

Review Questions

1. How are glia different from neurons? What role does each kind of cell play in the nervous system?
2. What is a resting potential? How does a neuron generate a resting potential?
3. What is an action potential? How does a neuron generate an action potential?
4. How does an electrical synapse differ from a chemical synapse?
5. What is a neurotransmitter?
6. Give an example of an excitatory neurotransmitter.

Thought Questions

7. What consequences—behavioral or otherwise—would you expect to see in an animal missing a gene that allows the synthesis of serotonin?
8. Eating carbohydrates causes an increase in blood sugar, which causes sleepiness. Suggest a hypothesis about neurotransmitters that would explain this phenomenon. How would you test your hypothesis?

SELECTED READINGS

Kandel, E.R., J.H. Schwartz, and T.M. Jessell, *Essentials of Neural Science and Behavior,* Appleton & Lange, Norwalk, Connecticut, 1995. An outstanding introduction to contemporary neuroscience, with especially clear diagram and explanations.

Snyder, S., *Drugs and the Brain,* Scientific American Books, 1986. A popular, although slightly outdated, examination of how psychoactive drugs imitate or interfere with neural signaling.

Tobin, A.J., and R. Morel, *Asking About Cells,* Saunders College Publishing, Philadelphia, 1997, Chapter 23, "The Cells of the Nervous System." A more detailed examination of the cellular and molecular structures responsible for the special properties of nerve cells.

► On-line materials relating to this chapter are on the World Wide Web at http://www.saunderscollege.com/lifesci
Click on Tobin/Dusheck: *Asking About Life.*

The Brain: A Single Mind or a Confederation of Minds?

"Does a falling tree make a sound if it falls when no one is there to hear it?" Philosophers and philosophically minded students have argued about this question for centuries. We now know that the answer is "no."

A falling tree makes no sound. Sound is a **perception,** the recognition and interpretation of an outside stimulus. In the case of a falling tree, the stimulus is a series of pressure waves, or vibrations, which arrive at the ear. It is the brain, however, that interprets these vibrations as sound, not the ear.

As humans, the air vibrations to which we have the strongest and most varied reactions are patterns that we perceive as words. Everyone knows that the vibration pattern we hear as "ma" conveys a particular meaning to an English speaker. Remarkably, people who speak many different languages also recognize slight variations of this particular vibration pattern as meaning "mother." But both the perception of the sound and the interpretation of its meaning depend on the experience of the hearer: a Chinese speaker, for example, hears tonal inflections in the vibration pattern that an English speaker may completely miss. To a Chinese speaker, "ma" may mean "mother" or "horse" or a fiber-producing plant.

We hear with our brains, however, not our ears. Even people with excellent hearing may be unable to understand language. Most English speakers, for example, cannot perceive words in vibration patterns that are obviously words to a Chinese speaker. Moreover, people who have suffered strokes on the left side of the brain are often unable—at least for a time—to recognize or to form words.

For the last 150 years, researchers have been trying to find out the physiological basis of word recognition and hearing in general. Researchers all knew enough to look beyond the ears to the brain itself, which was already known to be the site of perception and learning. But researchers could not agree about the role of the whole brain versus specific parts of the brain. The long debates about how the brain perceives words has contributed greatly to our understanding of the general question, "How do animals sense the outside world?" The brain is clearly important to perception. But is the whole brain important or do different parts of the brain perceive different things?

The notion that the brain is subdivided into specialized parts first gained prominence in the early 19th century from the work of Franz Joseph Gall, a German physician and neuroanatomist who worked in Vienna. Although the patterns

Art Wolfe/Tony Stone Images

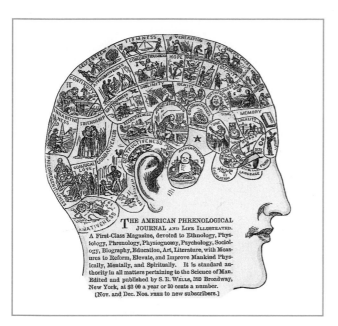

Figure 42-1 Phrenology. In the 19th century, scientists and the public began to realize that different parts of the human brain were responsible for different mental activities. Phrenologists imagined that this spatial specialization extended even into qualities of personality and spirituality. This idea became popular among nonscientists but was soon abandoned by the scientific community. *(The Granger Collection/New York)*

of folds are roughly the same in all human brains, individuals differ in the actual surface areas of different regions. Gall divided the brain into 35 areas and speculated that the sizes of each area varied with an individual's use of that area, just as the sizes of individual muscles vary with use (Figure 42-1). He then tried to establish a correlation between the size of an area and a particular emotional or intellectual capacity. Among the traits he assigned were such characteristics as benevolence, firmness, spirituality, combativeness, and amorousness.

Gall estimated these traits, moreover, not from the brain anatomy of people who had died but from the ridges and bumps on the skull of his living subjects. This prac-tice came to be known as *phrenology* [Greek, *phren* = heart, mind], an exercise that captured the imagination of many 19th-century intellectuals, including the writers George Eliot and Charles Dickens, in much the same way that astrology had. While phrenology was more of a social than a scientific phenomenon, it focused attention "not in our stars, but in our selves," and helped stimulate interest in the physical basis of the mind.

In the early 1820s, a young scientist subjected Gall's ideas to actual experiments. Pierre Flourens, who had recently graduated from medical school at age 19, developed methods for damaging specific parts of the brains of experimental animals. Flourens was able to demonstrate, for example, that destroying the *cerebellum,* a structure in the lower part of the brain, led to a loss of coordination. But when he removed slices of the *cerebral cortex,* the convoluted surface layer of the brain, he did not find any specific loss of particular traits, but rather a general loss of mental function.

In humans, the cerebral cortex is far larger than in other animals. Indeed, like our upright posture and our hairlessness, our immense cerebral cortex is one of the handful of traits that sets us apart from other primates. Flourens, therefore, drew a conclusion that supported his philosophical and religious views. Humans, he argued, had a single mind and a single soul. Despite the objections of Gall and others that Flourens was generalizing too freely, Flourens's conclusion found wide support among French intellectuals. Like his mentor Georges Cuvier, Flourens had, in his thirties and forties, a high standing both scientifically and socially, and he became the Perpetual Secretary of the prestigious Academie des Sciences. In addition, many scientists shared his philosophical opposition to the idea that the brain consisted of just so many independent machines.

Despite these philosophical objections, however, some neurologists continued to look for the relationship of specific functions—such as touching or hearing—with specific regions of the brain. For example, the British neurologist John Hughlings Jackson showed in 1870 that specific parts of the cerebral cortex were responsible for different types of sensation and movement.

The brain is clearly important to perception. But is the whole brain important or do different parts of the brain perceive different things?

Then, in 1876, a young German neurologist named Carl Wernicke described a patient who was able to speak, but who could not understand what he himself was saying. A telling illustration of the patient's problem came from a similar patient, interviewed a century later. In reply to the question, "Where do you live?" the patient answered, "I came there before here and returned there." After Wernicke's patient died, Wernicke dissected his brain and found that the cerebral cortex, the surface of the brain, had been damaged in a single place, probably by a stroke.

Some 15 years before Wernicke's report, a French neurologist named Pierre Paul Broca had described another patient who had difficulty with words, but it was of a different nature: Broca's patient could understand but not speak. After the patient's death, Broca found damage near the front, also on the left side of the cerebral cortex. Wernicke knew about the area Broca had described and its importance for language. He also knew that the cerebral cortex contained another area where damage led to a general failure of hearing.

The perception of a word is a relatively simple function compared with those in Gall's catalog (such as "conscientiousness" or "sublimity"). But Wernicke realized that word perception involved several different elementary processes, each of which occurred in a different region of the brain. His work became the basis for the now-accepted idea of *distributed processing*, in which different regions of the brain perform specific components of a complex process such as word perception. Even today, Wernicke's model of how the brain processes language still influences the thinking of neuroscientists.

Modern neuroscientists can now observe the involvement of different brain regions in word recognition. Using a technique called *positron emission tomography (PET)*, researchers can compare the metabolic activity of different regions of the cerebral cortex as human subjects are asked to perform various tasks (Figure 42-2). If a person has heard a single word read aloud, metabolism increases in the very areas predicted by Wernicke.

When a person reads the same word, however, metabolism increases in a totally different region. When he or she speaks the word, the researchers observed metabolic increases in still different regions, including Broca's area. And, when the person is asked to think about the word, a new pattern emerges that includes not only Wernicke's and Broca's areas but also other areas of the cortex.

Contemporary neuroscientists accept the view that individual regions of the brain perform specific functions and that complex functions of the brain require cooperation and information flow between several regions. In this chapter, we will focus on the central and peripheral nervous systems and on the roles of individual brain regions in **sensation**—the detection of external stimuli by vision, hearing, touch, taste, and smell. We will begin our discussion with an introduction to the nervous system.

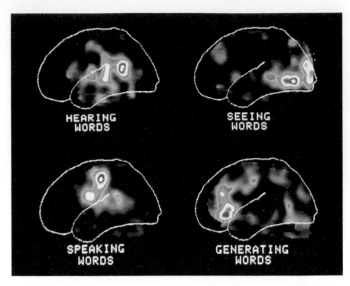

Figure 42-2   Different mental tasks involve different brain regions. Modern imaging techniques, such as these positron emission tomography (PET) scans, allow neuroscientists to observe the metabolic activity of individual regions of the brain. In these images, the hotter colors (red and yellow) show regions of high metabolic activity; cooler colors (blue and green) show regions of low metabolic activity. These images show metabolic activity just after a subject has heard words spoken, seen words written, spoken words, and thought words.   *(Marcus Raichle, Department of Neurology, Washington University School of Medicine)*

KEY CONCEPTS

1. Nervous systems are homeostatic organs that control muscles and regulate internal organs.

2. Nervous systems are also responsible for sensation, cognition, learning, emotion, mood, and consciousness.

3. Each type of sensation depends both on a sense organ and specific regions of the brain.

4. Each sensory system uses the same basic strategies to detect the quality, intensity, duration, and location of a stimulus.

5. Sense organs use specialized receptor molecules to detect pressure, light, or specific chemicals.

6. Networks of nerve cells represent only a limited number of features about each stimulus.

HOW DOES THE NERVOUS SYSTEM RESPOND TO ITS ENVIRONMENT?

One of the main functions of the nervous system is to maintain homeostasis. The nervous system as a whole receives reports on the state of the exterior world, as determined by vision, touch, hearing, smell, and taste. At the same time, it collects information about the interior world, as determined by receptors that monitor specific features of the internal environment, such as temperature, pH, oxygen supply, and levels of individual hormones.

The nervous system is the ultimate homeostatic organ, coordinating not only such internal regulatory mechanisms as our body temperature but also simple and complex behaviors such as shivering or knitting a sweater. A pain, a smell, a sound, or a visual image can send a signal that directs a homeostatic response, such as withdrawing a hand from a hot stove or running away from a charging elephant.

For the nervous system to maintain homeostasis, it needs to be able to (1) *detect changes* in its internal and external environments, (2) *interpret* the meanings of those changes, and (3) *respond* by sending appropriate signals to the muscles and organs of the body. Our eyes, nose, and other sense organs detect environmental changes, while our brain and spinal cord interpret signals from the sense organs and command the muscles and endocrine systems.

Connecting the sense organs and the brain are the nerves. Nerves carry signals from the sense organs to the brain or spinal cord, and from the brain or spinal cord to the muscles, organs, and endocrine system. The intricate connections between nerve cells—in the brain, spinal cord, and nerves—determine how animals respond to a variety of stimuli.

The nervous system is a cellular network that responds to changes in an animal's external and internal environment.

Do Invertebrate Animals Have Brains?

The simplest organisms that have nervous systems are jellyfish, hydra, and other cnidarians (Figure 42-3A and B). A hydra, for example, has a web of interconnecting neurons, called a **nerve network,** which carries information to all parts of its body. A stimulus anywhere on the hydra's body triggers reflexive muscle contractions. In jellyfish, nerve networks coordinate complex swimming movements. But, cnidarians have no brains, and, as far as anyone can tell, neither a hydra nor a jellyfish can learn. Their behavior does not appear to be influenced by experience.

The simplest animals capable of learning are the flatworms, and their nervous system resembles that of more complex animals (Figure 42-3C). The flatworm's nervous system consists of two parallel nerve cords, with nerves extending outward to the muscles. The two nerve cords converge near the front of the body in the simplest possible "brain." Like other bilaterally

A. Hydra

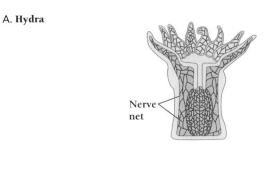

Nerve net

B. Jellyfish

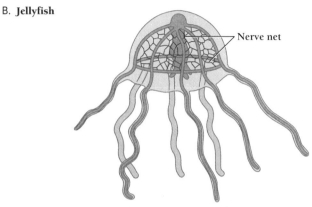

Nerve net

C. Planarian (flatworm)

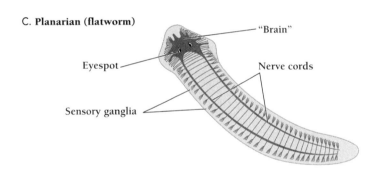

"Brain"

Eyespot

Nerve cords

Sensory ganglia

Figure 42-3   **Invertebrate nerve networks and nervous systems.** A. Nerve networks in a hydra. B. Nerve networks in a jellyfish. C. The nervous system of a flatworm, the simplest animal that is able to learn.

symmetric animals, flatworms have a well-defined direction of movement. They have eyespots at the front end, and they process sensory information in clusters of nerve cells called ganglia.

In even a simple brain, the convergence of two nerves in a ganglion provides many opportunities to generate complex patterns of neural activity and behavior. A simple brain also allows animals to modify such patterns by learning through association. Flatworms can, for example, be taught to move in a variety of patterns in order to find food (Figure 29-5). More complex animals have increasing degrees of *cephalization*, the concentration of sense organs and ganglia in the front (anterior) end.

Flatworms are the simplest animals that can learn. They have parallel nerve cords that converge near the front of the body.

How Is the Vertebrate Central Nervous System Organized?

The nervous systems of vertebrates show the highest degree of cephalization and the most sophisticated sense organs. Most neurons are concentrated in the **central nervous system (CNS),** which consists of the brain and the spinal cord (Figure 42-4). In vertebrates, the CNS lies *dorsal* to (above) the body, whereas in invertebrates the CNS is *ventral* (below).

In vertebrates, the central nervous system consists of the brain and spinal cord. The brain lies within a thick bony structure, called the *cranium,* and the spinal cord passes through a bony canal made by the stacks of vertebrae. Both the cranium

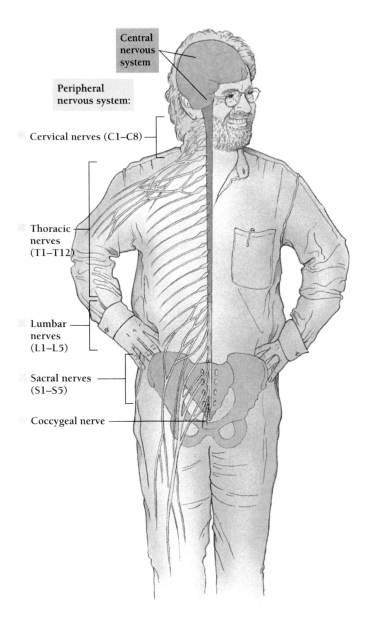

Figure 42-4 Nervous system of a human. The central nervous system consists of the brain and the spinal cord. The peripheral nervous system consists of the autonomic nervous system and the peripheral nerves.

and the vertebrae help protect the CNS from traumatic damage.

Also protecting the brain and spinal cord is a set of three membranes, which together are called the *meninges.* Bacterial or viral infection of the meninges—called *meningitis*—can be fatal. The central canal of the spinal cord and the connected *ventricles* of the brain are filled with a fluid, called the *cerebrospinal fluid* (or *CSF*). The CSF also surrounds the brain and spinal cord, filling the space between the outer surface and the meninges. The CSF cushions the brain and spinal cord and protects them against injury.

The CSF derives from the liquid part of the blood and, in many respects, resembles blood plasma in its chemical composition. But the capillaries that supply blood to the spinal cord and the brain are different from those that supply the rest of the body. The cells of the capillary walls are tightly connected to one another so that the passage of molecules is highly restricted, and the capillary walls are said to form a *blood-brain barrier* (or in the case of the spinal cord, a *blood-nerve barrier*).

Except for molecules with special transporters, only small, nonpolar molecules pass through the blood-brain barrier. Among the molecules that easily traverse the barrier, however, are alcohol, nicotine, heroin, and cocaine—partially accounting for the potency of these drugs of abuse. Pharmaceuticals that act on the brain or spinal cord must also traverse the blood-brain barrier, a major consideration in the design of medicines for neurological and psychiatric disorders.

The vertebrate central nervous system consists of the brain and spinal cord. Both are surrounded by bones, membranes, and cerebrospinal fluid.

The Spinal Cord

The **spinal cord,** a long bundle of specialized nerve fibers that runs from the brain to the tail, receives *sensory information* regarding the position of the limbs and sends *motor instructions* to the muscles of the limbs and trunk. The spinal cord also controls involuntary *autonomic functions,* such as the beating of the heart and the movements of the intestines. The spinal cord contains the sensory neurons and motor neurons that are responsible for stretch reflexes, such as the knee jerk, as we discussed in Chapter 41.

Nerve cells in the spinal cord called **motor neurons** directly stimulate muscle contraction. The nerve fibers from the motor cortex cross from left to right (or from right to left) as they pass through the medulla, so that the left side of the brain controls muscles on the right side of the body.

In cross section, the inside part of the spinal cord resembles a gray butterfly (Figure 42-5). This *gray matter* contains nerve cells, but no myelin. The lower parts of the double wings, which are called the *ventral horns* (or, in upright humans, the anterior horns) contain the motor neurons. The upper parts of the double wings, which are called the *dorsal horns* (or, in humans, the posterior horns) contain the bodies of the sensory

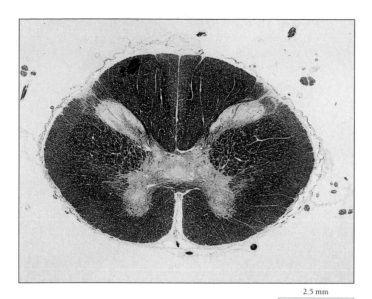

2.5 mm

Figure 42-5 The organization of the human spinal cord. In this cross section, the white matter (darkly stained) lies outside the gray matter (lightly stained), which surrounds the central canal in a shape that resembles a butterfly. The ventral horn contains the motor neurons that command muscle contraction, and the dorsal horn contains the cell bodies of sensory neurons. *(J.C. Revy/Phototake)*

nerves—including the stretch receptors. In the center of the gray matter is the fluid-filled *central canal,* which is the remnant of the interior of the *neural tube,* from which the entire central nervous system originally developed (Chapter 44).

The exterior part of the spinal cord consists of *white matter,* whose glistening appearance results from the myelin that encloses the nerve tracts that run up and down the spinal cord. These spinal tracts carry information from the brain to the motor neurons and from the sensory neurons back up to the brain.

The spinal cord can itself direct relatively complex behaviors even without a connection to the brain: a person or animal with a severed spinal cord can learn to move his or her legs in a coordinated rhythm and even to lift a leg over an obstacle. The ability of the spinal cord to direct coordinated movements without input from the brain is well known to people who have raised and slaughtered chickens: a beheaded chicken can run about a farmyard for some minutes after decapitation.

The spinal cord receives somatic sensory inputs concerning the position of body parts and sends motor instructions to voluntary muscles. It also coordinates many autonomic functions.

The Vertebrate Brain

Paleontologists' studies reveal that the brains of jawless fishes, almost half a billion years old, consisted of three major divisions: (1) a hindbrain, (2) a midbrain, and (3) a forebrain (Figure 42-6). The same divisions are apparent in today's fishes and in the embryos of contemporary vertebrates.

The hindbrain

The **hindbrain,** which forms a kind of cap on the spinal cord, consists of three parts, the **pons,** the **medulla oblongata,** and the **cerebellum.** Many of the nerves of the spinal cord form connections to the hindbrain, which integrates and processes both sensory information and motor instructions. Both the pons and the medulla are crossroads of information that flows between the brain and spinal cord. The medulla also contains centers that regulate breathing and blood circulation. The cerebellum, really an extension of the hindbrain, coordinates incoming sensory information and outgoing motor instructions so that the limbs move smoothly and the body maintains its upright posture. The cerebellum is also especially important for such tasks as eye-hand coordination.

The hindbrain integrates and processes both sensory information and motor instructions.

Midbrain

The **midbrain** is a major processor of visual information. In humans and other mammals, each side of the midbrain contains a region that processes visual information, especially in the reflexes that orient the head and neck toward a visual stimulus. Another region processes auditory information. Still another midbrain region, called the *substantia nigra,* is important in initiating movement. Parkinson's disease—which involves the destruction of many of the cells of the substantia nigra—severely limits a person's ability to move.

Together, the midbrain, the pons, and the medulla are called the *brainstem.* (Early neuroanatomists said that the brain consisted of three parts, the brainstem, the cerebellum and the cerebral hemispheres. This classification is based on appearance, however, not on developmental history of functional organization.) Running through the brain stem is a network of nerves, called the *reticular formation,* that receives information from all other parts of the brain. The reticular formation regulates arousal and attention, sleeping and dreaming.

The midbrain relays sensory information, helps organize movements, and directs attention.

Forebrain

In humans the **forebrain** is the locus of almost all conscious activity. It plays crucial roles in sensation, perception, movement, learning, memory, and mood. The forebrain consists of two divisions: the diencephalon and the cerebral hemispheres.

The **diencephalon** has two parts, the **thalamus,** the major integrator of sensory information about the *external* world, and the **hypothalamus,** the major integrator of information about the *internal* world. The hypothalamus is the master regulator of homeostasis. It regulates body temperature, eating and drinking, growth, response to stress or threats, mating, and pleasure.

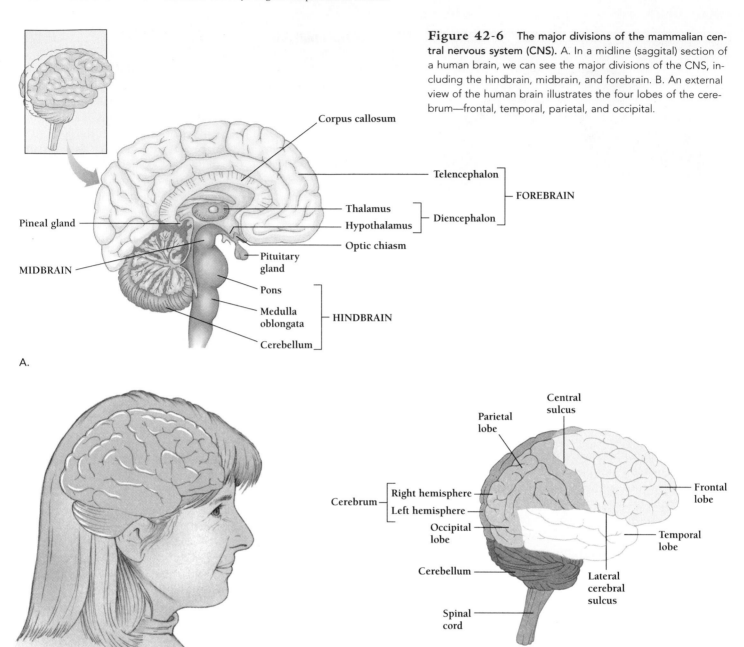

Figure 42-6 The major divisions of the mammalian central nervous system (CNS). A. In a midline (saggital) section of a human brain, we can see the major divisions of the CNS, including the hindbrain, midbrain, and forebrain. B. An external view of the human brain illustrates the four lobes of the cerebrum—frontal, temporal, parietal, and occipital.

The **cerebral hemispheres** consist of a thin surface layer, called the **cerebral cortex,** and a number of underlying structures. The gray matter—which consists of neurons, glia, and unmyelinated neuron extensions—is on the outside of the cerebral hemispheres, surrounding the white matter—myelinated axons. Recall that in the spinal cord, the white matter is outside and the gray matter is inside.

In humans the cerebral cortex has become so large that it is packed around the rest of the brain like a wadded-up bed sheet. The cerebral cortex completely dominates the appearance of the human brain. It consists of just six layers of cells and is only a few millimeters thick. In humans, the cortex contains more than 10^{11} neurons and has a total surface area of

about half a square meter. This huge surface fits into the brain case only because it is so highly convoluted, with characteristic folds (**gyri**) [singular, **gyrus**; Latin, = circle] and grooves, called **sulci** [singular, **sulcus**; Latin, = furrow] (Figure 42-6). The gyri and sulci provide landmarks for the division of the human cerebral cortex into four areas, called the *frontal, parietal, temporal,* and *occipital lobes.* The cerebral cortex, like the rest of the brain, consists of two almost symmetrical halves. The only direct connection between the two hemispheres is a thick bundle of nerve fibers called the *corpus callosum.*

Each of the four lobes of the cerebral cortex (on each side of the brain) shares the same cellular organization. The cortex consists of six cellular layers, with each layer containing neu-

BOX 42-1

How do neuroscientists study the living brain?

The human brain contains more than 100 billion neurons, each of which has some complicated pattern of ion currents, as we discussed in Chapter 41. As ions pass in and out of the membranes of these neurons, other ions also move in the extracellular space. Each of these currents is so small and there are so many neurons with independent inputs and outputs that it would seem unlikely that they would form any pattern. But they do. And we can study the voltage changes associated with these currents by placing electrodes on the surface of the scalp. Remarkably, the measured voltage changes reflect the state of the brain as a whole, in particular with regard to the state of arousal.

A record of the changes in voltage with time is called an *electroencephalogram*, or *EEG*. The EEG reveals the state of attention. The EEG pattern of a sleeping person is distinct from that of a waking person, and someone involved in mental activity has a different pattern than someone who is quietly vegetating. The pattern also changes during sleep. The alterations are associated with periods of dreaming and of rapid eye movements (or REM).

About 1 percent of the population suffers from epilepsy, a condition in which seizures disrupt normal activity. The seizures may cause massive muscle contractions (in *grand mal* epilepsy) or may bring about a brief but noticeable lapse in attention (in *petit mal* epilepsy). Such seizures result from electrical storms in the brain, in which a large number of neurons appear to fire action potentials in step with one another. The EEG during a seizure reflects such abnormal electrical activity, showing large periodic waves that are unusual in a waking person. Even between seizures, there may be occasional abnormalities in the EEG pattern of someone who suffers from epilepsy.

The passage of electrical currents through a neuron requires the restoration of the resting potential, a process that requires energy. Parts of the brain that are more active at a particular time use more energy and take up more glucose from the blood. Another way of visualizing the activity of the brain in a living person measures differences in glucose uptake.

The visualization of glucose uptake depends on the use of a special glucose derivative, called ^{18}F-fluoro-deoxyglucose (FDG). This compound enters cells along with glucose, but the cells cannot use it to produce energy. Instead, it accumulates in the cells. More active cells absorb more FDG than inactive cells, so the pattern of FDG uptake reflects the activity of the cells. A complicated and expensive procedure, called positron emission tomography (PET), can detect the presence of radioactive ^{18}F. A PET scan can produce a picture of the distribution of energy use in the brain, such that in Figure 42-2.

For example, the visual cortex of a person kept in the dark takes up little glucose. White light increases activity in the primary visual cortex, but a complex picture stimulates still more activity, not only in the primary visual area but in the association areas as well. Similarly, a subject who is listening to a Sherlock Holmes story shows greater activity in the parts of the brain known to be involved in hearing and in memory.

---

rons of characteristic shapes. Each lobe receives sensory signals (mostly via the thalamus), processes that information, and transmits instructions to some other part of the brain or spinal cord. The function of the rest of the cortex—called the **association cortex**—contributes to processing of both sensory input and motor output. More complex animals devote larger fractions of their cortex to association than do simpler animals. In a mouse, 95 percent of the cortex is sensory and motor and 5 percent associative. In humans, 95 percent is associative.

The frontal lobe deals principally with movement and smell, the parietal lobe with somatic sensation, the occipital lobe with vision, and the temporal lobe with hearing. The temporal lobe is particularly involved in the formation of memory, as is the underlying **hippocampus**, which is responsible for the first stages of memory storage.

Beneath the cerebral cortex lie other structures—the amygdala and a group of structures collectively called the basal ganglia. The **amygdala** coordinates autonomic (involuntary) responses to emotional states, especially anxiety. The **basal ganglia** help establish patterns of movement.

The forebrain plays important roles in sensation, perception, movement, learning, memory, and mood. The cerebral hemispheres consist of the cerebral cortex and underlying structures, which include the hippocampus, the amygdala, and the basal ganglia. In humans, the cerebral cortex is divided into four lobes, all of which share the same cellular organization.

How Do Humans Learn?

The association cortex, especially in the temporal lobe, is intimately involved in memory. **Memory** is the storage of knowledge about the world. So far, biologists have not been able to discover how or where memories form in the brain or how they are retained and recalled.

We do know that memories form in stages. **Short-term memories**—of experiences within the previous few seconds—can disappear after a blow to the head, electric shock, or drug treatment. The brain stabilizes **long-term memories**, however, so they usually remain after head traumas or other damage to the brain.

Memories may involve different degrees of consciousness. A *reflexive* memory is one that has a reflexlike quality—automatic and not dependent on awareness. In contrast, a *declarative* memory is a recalled thought or experience, a memory that a subject can summarize in a declarative sentence. Some skills—like those involved in driving a car—may be conscious, declarative memories at first, then later emerge as unconscious, reflexive memories.

Different brain structures may be involved in reflexive and declarative memories. Damage to a part of the cerebellum may destroy the ability to perform a particular task, while damage to the temporal lobe may interfere with long-term declarative memories.

The hippocampus has an important role in the formation of long-term memory from short-term memories and is also an important site in the generation and regulation of epileptic seizures. One of the most forceful illustrations of the role of the hippocampus came from a young man whose hippocampi were removed in an effort to stop persistent and otherwise untreatable seizures. After his operation, this man could still speak and understand language, he maintained memories from before his surgery, and he could remember for seconds or minutes. But he lost the capacity to form new long-term memories. Studies of both humans and experimental animals all point to the hippocampus as a crossroads of short-term and long-term memory.

One of the most amazing discoveries about memories came from tests performed during neurosurgery. Wilder Penfield, a Montreal brain surgeon, stimulated the temporal lobes of several of his patients. These patients vividly experienced past events. For example, one heard a specific melody each time that Penfield stimulated a particular spot in her brain. Despite the incredible amount of information gathered by physicians, psychologists, and neuroscientists, however, we are still far from understanding the basis of memory. As in all scientific disciplines, the early stages are, of necessity, more descriptive than theoretical.

Memories form in stages, with short-term memories including experiences of the preceding few seconds and long-term memories including experience from many years before. The hippocampus has an important role in the formation of long-term memories.

How Is the Vertebrate Peripheral Nervous System Organized?

The **peripheral nervous system,** which consists of all the nerve cells that are outside the central nervous system, connects the central nervous system (CNS) to the sense organs, muscles, and other organs. **Afferent** nerves carry *sensory* information *to* the CNS, while **efferent** nerves carry instructions for *motor* and other activity *from* the CNS. Efferent nerves carry instructions

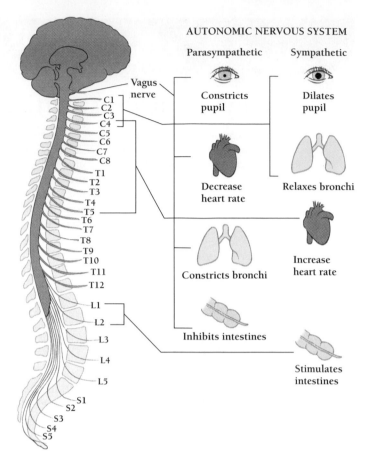

Figure 42-7 The human autonomic nervous system includes both the sympathetic and the parasympathetic nervous systems.

from the CNS to both muscles and other organs, including the endocrine glands (Figure 42-7).

The efferent nerves that carry information to the striated skeletal muscles make up the **somatic,** or voluntary, nervous system. Other efferent nerves, those of the **autonomic,** or involuntary, nervous system, carry instructions to the smooth muscles of the endocrine, digestive, reproductive, excretory, respiratory, and vascular systems, as well as to the cardiac muscles of the heart (Figure 42-7).

The autonomic nervous system consists of two distinct sets of nerves—the **sympathetic** and the **parasympathetic** nerves. These nerves usually have opposite effects. The sympathetic nervous system, along with the hormone epinephrine, triggers the "fight or flight" reaction. These responses involve the mobilization of energy stores as preparation for vigorous action. In contrast, the parasympathetic nervous system helps conserve energy by slowing the heart and increasing intestinal absorption.

The peripheral nervous system connects the CNS to sense organs and muscles. It includes nerves of both the somatic nervous system and the autonomic nervous system, which carries instructions to smooth muscles and to cardiac muscles.

How Does Information Move Through the Nervous System?

The nerves of the peripheral nervous system are further classified according to what kind of information they carry. The **auditory system,** for example, contains all the cells responsible for hearing, and the **visual system** contains all the cells responsible for seeing.

The receptor cells' information is in the form of **receptor potentials,** changes in the electrical voltages across their plasma membranes. Information from receptor cells moves to **relay neurons,** which convert information from one or more receptor cells into a series of electrical discharges, or action potentials, which, as we saw in Chapter 41, consist of temporary changes in ion movements across the plasma membrane. In every sense organ, relay neurons carry information (in the form of action potentials) to particular sets of neurons in the brain.

Whereas action potentials are all of the same intensity, or size, receptor potentials come in a range of intensities. But neurons can fire repeatedly, slowly or over and over in rapid succession. The graded signal of the receptor potentials, then, is interpreted as action potentials of different frequencies.

The electrical properties of nerve cells allow neurobiologists to probe individual cells and the interconnections of cells within the nervous system. Researchers can see how the brain responds to external stimuli by recording electrical currents from the surface of the brain, from within individual cells, and even from within minuscule patches of nerve cell membranes.

Researchers can find out which regions of the brain control which parts of the body by electrically stimulating the brain and then recording what happens. Such tests on humans are common during neurosurgery, when the patient is usually awake and aware. Since the brain has no pain receptors, such a procedure is not painful. Researchers can study both the final response, such as the movement of a limb, and also the intermediate steps along a neural pathway. In this way, researchers can begin to define the pathways of information transfer within the nervous system. These pathways involve communication between neurons, which are specialized for the conduction of electrical and chemical signals.

Neuroscientists can also map neural pathways in experimental animals by using special dyes. Nerve cells transport these dyes in the same direction in which they transmit information, so that when such a dye is injected into the motor cortex, for example, researchers can watch the dye move through the medulla and from there into the spinal cord.

Researchers have found that other dyes move in the direction opposite to that in which neural commands flow. These dyes allow neuroscientists to determine the sources of information processed by a particular set of nerve cells. Such studies show, for example, that cells of the motor cortex receive information from the spinal cord, from other regions of the cortex, and from brain structures that lie below the cortex. Each part of the nervous system and almost every nerve cell receives information from many sources. Ultimately, however, all new information comes from the sensory systems.

> Receptors detect characteristics of the external world and generate graded receptor potentials. Relay neurons convert the information encoded in receptor potentials into patterns of action potentials. Relays of neurons in each sensory pathway convey information to the brain.

HOW DO ANIMALS SENSE THEIR ENVIRONMENT?

Many individual parts of the brain contribute to *sensation,* the detection of external stimuli, and many more to *perception,* the conscious recognition and interpretation of those stimuli. Animals—human or otherwise—can respond to sensation, however, in ways other than conscious perception. Sensory stimuli can help control movements, regulate internal organs, and maintain arousal, without our being conscious of making any changes. For example, women regularly exposed to men's perspiration have more regular menstrual cycles than women not exposed to this constellation of chemicals.

What Senses Contribute to Sensation and Perception?

Humans have long recognized five senses—hearing, vision, smell, taste, and touch. Neuroscientists refer to the senses as five independent **modalities** (classes) of sensation: **audition,** or hearing; **vision,** or seeing; **gustation,** or tasting; **olfaction,** or smelling; and **somatic sensation.** Somatic sensation actually includes four independent modalities—touch (or pressure), pain, temperature, and sensing limb motion and position. Some animals detect additional qualities of the external world, such as electric fields (some fish), magnetic fields (some migrating birds), or infrared light (bees).

Sensation begins in **sensory receptors,** specialized cells that detect a particular property of the external world. Each receptor contributes to only one sense modality: receptors in the eye respond to light, those in the ear to vibration, and those in the taste buds and nose to particular chemicals. The sensory receptors for hearing, vision, taste, and smell are in specialized **sense organs**—the ears, eyes, taste buds, and the olfactory epithelium of the nose. Somatic sensation, on the other hand, depends on sensory receptors, dispersed in the skin and muscles, that report separately on pressure, pain, temperature, and limb position.

Every sensation except olfaction involves three stages:

1. **Transduction** translates the energy of a stimulus (such as light) into a change in the chemistry of the receptor cell.
2. **Transmission** carries a signal from the receptor cells to the central nervous system.
3. **Integration** combines the signals from many receptors to form a mental image of the outside world.

Each sensory system responds quantitatively to three characteristics of a stimulus—*intensity, duration,* and *location.* Sensory receptors within the ear, for example, produce larger receptor potentials in response to greater vibration intensities, which we perceive as louder sounds. The electrical activities of sensory receptors, relay neurons, and brain neurons indicate the beginnings and ends of each sound. Finally, nerve networks within the brain also allow an animal to coordinate information from both ears to locate a sound. Some species are much better than others at location: bats, dolphins, and even owls, for example, can precisely locate sounds that a human would be lucky to hear at all. On the other hand, humans' ability to locate visual stimuli is very good.

Sensory receptors also respond to other qualities of a stimulus. The ear, for example, distinguishes among different frequencies, which are perceived as high notes and low notes. Similarly, the eye distinguishes among different frequencies (or wavelengths) of light, which we perceive as different colors.

Sensation depends on signal transduction by the sensory receptor, the transmission of a signal to the central nervous system, and the integration of signals to form a mental image of the outside world.

What Path Does Sensory Information Take to the Brain?

Sensory receptors feed information to relay neurons. Some relay neurons simply transmit the information from a single sensory receptor to the next step of the neural chain. Other relay neurons do more: they integrate information from several sensory receptors or from several other relay neurons. When such integration occurs, the relay neuron extracts particular information from the combination of inputs. For example, some relay neurons in the eye detect bright spots surrounded by dark rings, while other relay neurons detect dark spots surrounded by bright rings.

Relay neurons usually send information first to the thalamus and then to particular regions of the cerebral cortex. Although all routes pass through the thalamus, they pass through different regions (*nuclei*) of the thalamus. And although all routes end in the cortex, they end in different regions of the cortex: visual information ends in the occipital lobe, auditory information in the temporal lobe, and somatic sensation in the parietal lobe.

Information from the olfactory system takes a different route, going first to the other parts of the brain before arriving at the thalamus and cortex. The specialness of the olfactory system probably reflects its ancient evolutionary origins and helps explain both our difficulty in naming odors and our intense emotional responses to certain smells such as the cologne or perfume of a parent or lover. The great 19th-century French novelist Marcel Proust provided an extraordinary account of his responses—including thoughts, memories, and fantasies—to the smells and tastes of tea and cake.

Each sensory receptor feeds information into a characteristic route. Each route is called a *labeled line* because the brain recognizes it as corresponding to a particular sense. Stimulation of a particular receptor, therefore, always evokes the same sensation, even when the stimulation is not entirely appropriate. For example, a person who suffers deafness because of a defect in the ear will "hear" tones if a researcher stimulates the relay neurons within the auditory pathway. Similarly, someone who receives a blow to the head may "see stars" because the blow stimulates receptors within the eye.

After the neural response to a sensation first arrives in the cortex, the primary sensory region further processes information and sends it to another region of the cortex. In the case of the visual system, there are at least 32 areas of the cortex that have representations of the visual world. Researchers do not yet know the significance of all these representations, but each is thought to extract different features of the visual image. One region, for example, appears to be highly specialized for the recognition of faces: people with damage to that area cannot identify faces but have otherwise normal memory.

The functional organization of sensory systems shows that the brain is an *active* agent in sensation; it is emphatically not a passive recipient of information from the sense organs. The brain extracts only certain features of the external world. The brain also communicates back to the sensory receptors and the relay neurons, actively modifying even the very beginnings of sensation.

Sensory information from each sensory modality passes separately through the thalamus and converges separately in different regions of the cerebral cortex.

HOW DO ANIMALS HEAR?

The perception of sound, like the detection of light or touch, depends on the ability to translate, or **transduce,** energy into electrochemical events in neurons. To understand how this transaction takes place, we must first understand the physical nature of a stimulus.

What we perceive as sound is actually pulsations of the molecules of the air. If we pluck the string of a guitar, it vibrates with a characteristic frequency. As it moves, it alternately pushes molecules together and apart, so that the surrounding air contains waves of pressure.

We can describe a wave in terms of the *frequency* of vibration, which we ultimately perceive as **pitch,** and of the *amplitudes* of the pressure differences, which we perceive as **loudness.** While this simple description works well for simple tones, such as those produced by a vibrating string, we will also want to understand the perception of more complicated sounds, such as those of a word. Fortunately, the same description applies, for any sound can be regarded as the sum of many tones of different frequencies and amplitudes. If we can understand how the ear transduces simple vibrations, then we will also understand how it transduces speech or music.

How Does the Mammalian Ear Detect Frequency and Amplitude?

The mammalian ear consists of three parts called the **outer ear,** the **middle ear,** and the **inner ear** (Figure 42-8). The outer ear collects sound the way a funnel collects liquid. It channels sound through the **auditory canal** to the **tympanic membrane,** or eardrum. The tympanic membrane moves and vibrates in response to variations in air pressure.

Behind the tympanic membrane is the middle ear, which, like the outer ear, is filled with air. This air comes into the ears by way of two tubes (one on each side), called the *Eustachian tubes,* from the throat. The Eustachian tubes are normally closed, so that the air in the middle ear is isolated from the air outside. Because the tubes are usually closed, the air in the middle ear can become pressurized relative to the outside air. A person who rapidly gains altitude in a plane or even an elevator sometimes experiences painful pressure on the eardrum. Fortunately, swallowing, chewing, and yawning can all open the Eustachian tubes and allow the middle ear to attain the same pressure as that outside.

Inside the middle ear, three small bones—called the *malleus* (hammer), *incus* (anvil), and *stapes* (stirrup)—transmit the movement of the tympanic membrane to the fluid-filled inner ear. The middle ear bones amplify the vibrations of the tympanic membrane, so that the fluid in the inner ear vibrates more forcefully than if it were directly linked to the eardrum.

Inside the inner ear is an organ that translates, or *transduces,* vibration into a signal that we understand as sound. This organ, called the **organ of Corti,** lies within the **cochlea** (Latin,

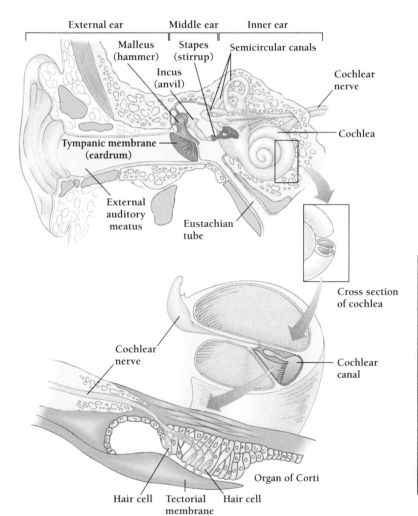

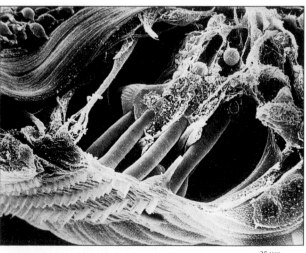

Figure 42-8   The human ear, showing the external ear, the middle ear, and the inner ear. The inner ear contains the organ of Corti. Hair cells within the organ of Corti are the detectors of sound waves.   *(photo, Dr. G. Bredberg/Science Photo Library/Photo Researchers)*

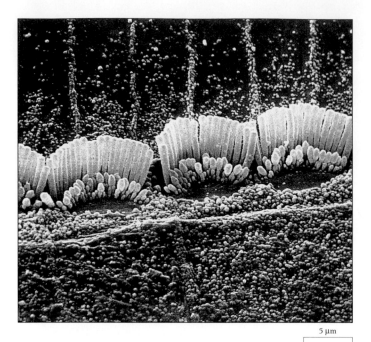

5 μm

Figure 42-9   Pressure-sensitive hair cells in the organ of Corti.
(Dr. G. Bredberg/Science Photo Library/Photo Researchers)

= snail), an enclosed tube that is coiled like a snail shell. When the fluid within the cochlea vibrates, the organ of Corti moves from side to side (Figure 42-8).

In the 1940s, Georg von Bekesy devised a way to watch these movements inside the cochlea. Von Bekesy removed a cochlea from a fresh human cadaver and sealed the lens of a microscope into the wall of the cochlea. Von Bekesy observed that low sound frequencies moved the organ of Corti towards the end of the cochlea, while higher frequencies moved it toward the middle ear.

The inner ear interprets this movement as sound by means of **hair cells,** stiff bundles of actin filaments simultaneously embedded in the membrane of the cochlea and the organ of Corti (Figure 42-9). When the organ of Corti moves, relative to the tectorial membrane, shear forces bend the hair cells. The bending of the hair cells alters the permeability of the membrane to ions.

Each frequency stimulates a different set of hair cells, and the signals pass from the organ of Corti to relay neurons in the rear of the brain. Relay neurons connect, in turn, to several other regions of the brain. One pathway leads to a part of the thalamus, then to the *primary auditory cortex* in the temporal lobe of the cerebral cortex.

As in the visual and the somatosensory systems, the brain simultaneously processes several separate "images" of sound. Each image provides different information. One set of cells, for example, appears to register differences in the timing of signals from the two ears; another set of cells registers differences in sound intensity. These two kinds of information are especially useful in locating the source of a sound.

Channels within the ear carry sound to the eardrum. The vibrations of the eardrum move the fluid in the inner ear, and hair cells in the organ of Corti detect the motion of the fluid. The brain interprets the patterns of signals from the organ of Corti as sounds.

How Does the Inner Ear Help Maintain Balance?

Hair cells are receptors that bend in response to the mechanical energy of air vibrations and transduce this energy into sound. But hair cells also bend in response to gravity. Consequently, many animals, including humans, use hair cells to determine the orientation of their bodies.

Fish and frogs also use hair cells to provide sensory information both about their own movements and those of potential predators and prey. In fish and frogs, the hair cells are part of the **lateral line system,** which consists of two canals that run just under the skin on each side of the body. The hair cells detect changes in water movements through the canals.

Within the mammalian inner ear are the **semicircular canals,** three fluid-filled tubes curved into half circles that lie at right angles to one another. As an animal moves its head, the fluid in the canals seeks its own level according to the pull of gravity. As the fluid flows, it pushes against hair cells in the canals.

The pattern of electrical activity from the hair cells tells the brain's cerebellum how the animal's head is oriented and in what direction it is moving. The cerebellum then integrates this information with other sensory information to determine how the muscles should move to maintain proper posture or to execute a particular movement.

Hair cells, within the semicircular canals of mammals and within the lateral line system of fishes and frogs, help establish the position and the movement of the body.

HOW DO ANIMALS SEE?

Animals have evolved many different kinds of light detectors. In all cases, the primary event in vision is the action of light energy on a pigment molecule that changes from one form to another. This change takes place within a specialized light-detecting cell called a **photoreceptor.**

Different kinds of eyes and different visual systems use different strategies to deliver light to the receptor pigment and to present the resulting information to the rest of the nervous system. Planarians, a type of flatworm, have photoreceptors within a sense organ called an *eye-cup.* The receptors detect a shadow at the cup's edge, thereby identifying the direction of a light source. But the planarian's eye does not form any image of the outside world and the flatworm lives in a world of light and shadow only.

In contrast, the eye of every vertebrate animal forms a distinct single image of the outer world. This image is projected against the back of the eye to a sheet of cells called the **retina** in much the same way that an image is projected against the film at the back of a camera. And just as photographic film records different frequencies and intensities of light from an image, photoreceptor cells in the retina contain pigment molecules that record the images before us.

In insects, the eye forms an image in a piecemeal fashion. The insect eye is really a collection of thousands of individual eyes, none of which forms a true image. Together, however, the parts of an insect's *compound eye* compose an accurate image of the outside world. Although the insect eye sees less detail than the best vertebrate eye, insects are extraordinarily good at detecting movement—especially of large objects.

The primary event in the detection of a visual image is the capturing of light energy by a light-sensitive pigment within a photoreceptor cell.

How Does a Vertebrate's Eye Form an Image of the Outside World?

The vertebrate eye is roughly spherical (Figure 42-10). Encasing and protecting the eye is a tough connective tissue, called the *sclera.* Light enters the front of the eye through a transparent region of the sclera, called the **cornea.** Just inside the sclera is a pigmented layer, the *choroid.* Behind the cornea, the choroid forms a muscular, donut-shaped disk called the **iris** that gives the eye its color. The iris surrounds the **pupil,** the dark, central opening through which light reaches the retina. The iris controls the size of the pupil in the same way that the iris of a camera controls the aperture of the lens. In dim light, the iris opens the pupil to let in more light. In bright light, the iris closes the pupil, limiting the amount of light that reaches the retina.

Just behind the pupil is a clear **lens,** which focuses images on the rear surface of the retina. The problem in both a camera and an eye is that the point in space where an image comes into focus depends on how close or far away an object is. Faraway objects form images immediately behind the lens, while nearby objects form images some distance further. A photographer deals with this problem by "focusing," moving the lens farther from the film for nearby objects and closer to the film for faraway objects.

The eyes of fish and amphibians work like a camera, moving the lens relative to the retina to focus objects of different distances. The mammalian eye, however, deals with the problem quite differently: it actually changes the shape of the lens according to whether objects are nearby or far away. Because a thicker lens bends light more, it forms the image closer to the

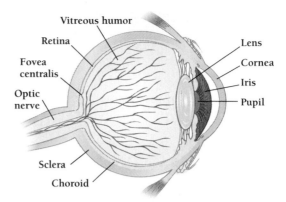

Figure 42-10 The human eye. The lens focuses images on the retina, which contains photoreceptor cells.

lens. When focused on nearby objects, the mammalian lens thickens to form a closer image; when focused on faraway objects, the lens thins. The focusing of objects at different distances is called **accommodation.**

In a vertebrate eye, the lens focuses an image on the surface of the retina.

Why Must Many of Us Wear Glasses?

A person who is **nearsighted** has eyes that are elongated from front to back. In spite of the thinning of the lens, the images of faraway objects come to a focus too soon in the space in front of the retina, so that the image on the retina is out of focus. A corrective concave lens spreads the light before it enters the eye, so that images converge on the retina (Figure 42-11).

A person who is **farsighted** has eyes that are too short. The result is that nearby images come to a focus beyond the retina (or would if they got that far), and images on the retina are out of focus. Even the thickening of the lens does not allow the formation of images of nearby objects. Correction of farsightedness requires a converging (convex) lens.

The lens changes its thickness and converging properties by means of muscles that lie in a structure called the *ciliary body.* Sometimes parts of the lens become cloudy, leading to the obscuring of vision and even to blindness. Such cloudy spots are called **cataracts.** An eye surgeon can remove a lens with a cataract and replace it with an artificial lens.

The eye's ability to focus also depends on the fluids through which light passes before and after it travels through the lens. The large space between the lens and the retina contains a transparent, jellylike substance, called the *vitreous humor.* A more watery substance called the *aqueous humor,* produced by the ciliary body, fills the space between the iris and the cornea. Too much pressure within the aqueous humor—caused by blockage of the ducts that drain the cavity—is called *glaucoma,* a condition that damages the eye and causes blindness.

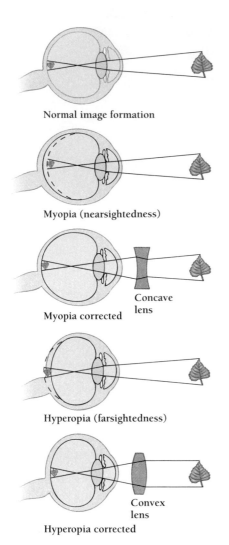

Normal image formation

Myopia (nearsightedness)

Concave lens

Myopia corrected

Hyperopia (farsightedness)

Convex lens

Hyperopia corrected

Figure 42-11 **Why do we have to wear glasses?** People with myopia have elongated eyes, so that images focus in front of the retina. People with hyperopia have shortened eyes, so that images focus behind the retina. In both cases, lenses can be used to correct vision.

Defects in the focusing apparatus cause eye disorders.

How Does the Retina Detect an Image?

The **retina** is a thin sheet of cells behind the vitreous humor. It consists of five types of cells arranged in distinct layers. The most numerous of these cells are the photoreceptors, which actually lie in the bottom layer of the retina, farthest from the light entering the eye. Light from the outside world must pass through the cornea, the lens, and the four other cell layers, including overlying nerve cells, before it reaches the photoreceptors. It is the photoreceptors, however, that convert the image into electrical signals that the brain can interpret.

The retina contains two kinds of photoreceptor cells: cylindrical **rod cells** and cone-shaped **cone cells** (Figure 42-12).

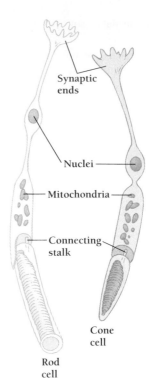

Synaptic ends

Nuclei

Mitochondria

Connecting stalk

Cone cell

Rod cell

Figure 42-12   Rods and cones. The two types of receptors in the vertebrate eye. Rods have only a single type of light-detecting pigment molecule, but cones may have one of three different pigments.

The rod cells are most dense around the edges of the retina and are phenomenally sensitive to light. When we are outside on a dark night and the pupil is opened up all the way, the 100 million rod cells in the human eye can detect even slight amounts of light and movement. The image that we see, however, is colorless and poorly defined. Because the rods are primarily around the periphery of the retina, we can actually detect faint objects more successfully by looking just to one side of the object, so that the image is projected not to the center of the retina, but to the edge. This phenomenon, useful to amateur astronomers, is called *averted vision.*

The cone cells are made for full-color vision in bright light. Unlike the rods, the 6 million cones of the human retina are concentrated in the central portion of the retina, called the **fovea.** As a result, even in bright light we see fully detailed images only by looking directly at an object. We perceive the whole visual world as colored only by constantly moving our eyes so that different images impinge on the fovea.

Both rods and cones contain special pigments called **rhodopsins** that absorb visible light and transmit a signal. Rods and cones differ, however, in the wavelengths of light to which they respond. Rods contain a single kind of rhodopsin, which is most sensitive to blue-green light. Cones come in three varieties, each of which has a distinct type of rhodopsin; one of these is most sensitive to blue light, one to yellow, and one to green. The four pigments are distinct proteins, called **opsins,** each of which binds the same small molecule, called **retinal.**

Each cone has just one kind of opsin. The brain compares the images produced by the different kinds of cones and interprets the differences as colors. Color blindness results from the lack of one or more type of cone opsin. Because the genes for the cone opsins are on the X chromosome, men, who have only one X chromosome, are more likely to lack one of these genes than women, who have two X chromosomes. Consequently, color blindness is more common in men than in women. About 5 percent of men lack the green-sensitive opsin and are red-green color blind.

Photoreceptor cells contain pigments that absorb visible light. The absorption of light alters the cell's electrical properties and initiates a series of events that transmits visual information to the brain.

How Does Information Travel from the Retina to the Brain?

We know a lot about the path that carries visual information from the photoreceptors to the brain. This information comes largely from the systematic examination of the electrical signals from individual cells in the pathway. It is clearly impossible to study each of the hundreds of millions of cells, but neuroscientists have managed to study thousands of them. Researchers find the same patterns of organization repeated throughout the visual system. In addition, the principles of information processing in the visual system seem to apply to other sensory systems.

At the beginning of this chapter we said that every sensory system consists of three stages: transduction, transmission, and integration. Sometimes, however, these three stages are not completely separate. In the visual system, the process of integration actually begins during transduction and transmission.

We have seen that transduction occurs in the rods and cones. These cells are located in the inner nuclear layer of the retina, and their absorbing pigments lie in membrane-bound compartments still deeper than their nuclei (Figure 42-13). The photoreceptor cells make contact with two other kinds of cells. These cells are called the **bipolar cells** and the **horizontal cells.** The bipolar cells are the main conduit for information from the photoreceptors. They connect to the **ganglion cells.** The ganglion cells in turn carry signals away from the retina to the brain itself. Also connected to the ganglion cells are **amacrine cells,** which lie in the same cell layer as the bipolar and horizontal cells.

The major pathway of transmission of visual information is from photoreceptors to bipolar cells to ganglion cells to the brain. There is not a point-to-point correspondence, however, as there is, for example, between the signal of a video camera and the image on a television screen. Instead, information from the more than 100 million receptors in the retina funnels into less than a million ganglion cells. This can happen because the

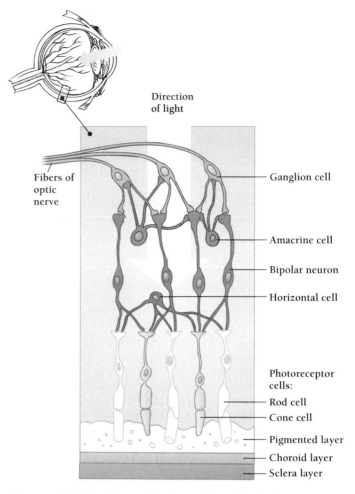

Direction of light

Fibers of optic nerve

Ganglion cell

Amacrine cell

Bipolar neuron

Horizontal cell

Photoreceptor cells:

Rod cell

Cone cell

Pigmented layer

Choroid layer

Sclera layer

Figure 42-13 **The organization of the retina.** Light must pass through the outer layers of the retina before encountering the photoreceptor cells, the rods and cones. Information passes from the photoreceptor cells to the bipolar cells to the ganglion cells. Horizontal cells and amacrine cells are interneurons.

ganglion cells receive a processed, or *integrated,* version of the visual world detected by the rods and cones.

The signals generated by each bipolar cell depend not only on the pattern of connections with photoreceptor cells but also on the action of the interconnecting horizontal cells. Similarly, each ganglion cell receives inputs both from bipolar cells and from amacrine cells. The horizontal and amacrine cells are examples of *interneurons,* which carry information between cells with parallel functions (Chapter 41). Interneurons often inhibit the action of neurons away from the immediate vicinity. In the retina, for example, the horizontal and amacrine cells inhibit nearby cells in a way that increases contrast in the image.

Because researchers can study the electrical activity of a single ganglion cell, they can discover what specific aspects of the external world a ganglion cell reports to the brain. Such studies have shown that every ganglion cell responds when light reaches a neighboring group of photoreceptors. Each ganglion then responds to a spot of light. Different ganglion cells respond to spots of different sizes. Cells near the center of the visual field respond to small spots and those near the side respond to larger spots. This difference partly explains why we can see with better resolution when we look directly at something.

Experiments on individual cells have revealed two types of ganglion cells: "on-center" cells, which increase their firing rate when light falls on the center of their fields, and "off-center" cells, which decrease their firing rate under the same circumstances. A ganglion cell seems to be specialized for detecting changes in light intensity between the center and the outside of its field of photoreceptors. A ganglion cell is relatively insensitive to the actual amount of light hitting the retina. There are at least three kinds of ganglion cells. Some respond to large groups of photoreceptors and probably detect large objects, such as an approaching bull elephant. Others respond to the fine details of an image, such as the down on the skin of a peach.

The photoreceptor cells of the retina transmit information to the bipolar cells, and the bipolar cells transmit to the ganglion cells. Connections among the photoreceptor cells, the bipolar cells, and horizontal cells and among the ganglion cells and amacrine cells are responsible for the integration of different aspects of the visual world.

There Is Much More to Vision Than Meets the Eye

The axons of all the ganglion cells of the retina together form a thick stalk called the **optic nerve.** The optic nerves from each eye converge in a structure called the *optic chiasma* (Figure 42-14). There, some of the fibers cross to the opposite side of the brain and some continue on the same side. The fibers are arranged so that the left side of the brain receives information about the right half of the visual field, from both eyes. Similarly, the right side of the brain receives information from both retinas about the left side of the visual field.

Most of the optic nerve fibers end in a region of the brain's thalamus called the *lateral geniculate nucleus.* There the cells lie in distinct layers, each of which receives input from the axons of one eye. Each lateral geniculate neuron receives input from several ganglion cells, and each neuron can integrate the information from these different ganglion cells. Because these neurons receive information from the ganglion cells, which responds to patches of light and dark, the neurons likewise respond to small spots or bars, not to overall changes in light level. Neighboring cells in the lateral geniculate nucleus connect to neighboring ganglion cells, so the overall organization of the image is maintained.

The signals that enter the lateral geniculate nucleus continue on to the *primary visual cortex,* a special region at the back of the brain (Figure 42-14). The cells of the visual cortex re-

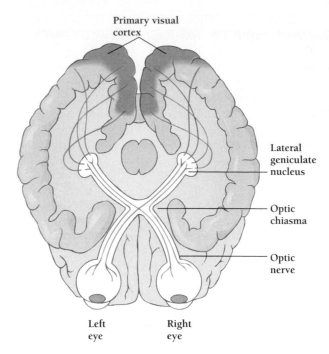

Figure 42-14   The pathway of visual information. Information from the retina passes by way of the optic nerve into the brain to the lateral geniculate nucleus of the thalamus. From there the information passes to the primary visual cortex on each side of the brain. Many other brain areas also receive visual information.

spond to still more integrated versions of the visual information gathered by the photoreceptors. Some cells respond to stationary bars or edges, others specifically respond to moving bars or edges. Still others respond not to bars but to stationary or moving corners. As in the case of the retina, connections between adjacent cells—both directly or through interneurons—increases the contrast in the image.

Information from the lateral geniculate and from the primary visual cortex also flows to other parts of the brain. In each case, the pattern of electrical activity in the neurons represents a "mental image" that reflects—although in distorted form—the spatial organization of the visual image. Each mental image represents the world slightly differently, emphasizing a particular aspect such as vertical bars, corners, color, three-dimensional form, or movement. We interpret these patterns of nerve activity as a visual representation of the world, even as we realize the limitations of our own perceptions (Figure 42-15). Ultimately, the interneural connections within the retina and brain establish not only what we see but also how we respond to it.

The axons of the ganglion cells form the optic nerve, which carries visual information to the brain. Optic nerve fibers end in the lateral geniculate nucleus of the thalamus, which serves both as an integrator of visual information and as a relay station.

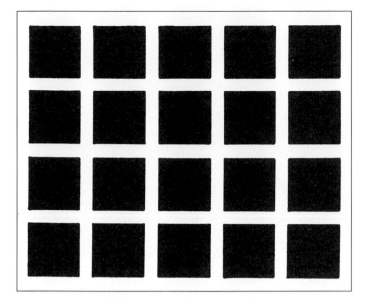

Figure 42-15 Visual illusions show that what we "see" depends on our expectations. Confusing cues can make it impossible to determine an unambiguous interpretation of some visual information. *(Corbis-Bettmann)*

OTHER SENSORY PATHWAYS

Our other sense organs present our brains with reports on other aspects of our external and internal worlds. In each case the corresponding mental image consists of a pattern of neural activity. Depending on which brain cells are active, we interpret these chemical changes and ion flows differently—as sound or smell, touch or taste.

Perception refers to the interpretation of these neural events as the conscious experience of objects and events in the external world. This definition raises a basic philosophical question: just who is doing the interpreting? Some people have envisioned the sensory pathways as television or telephone cables that carry electrical representations of pictures and sounds to some inner observer. The hard thing to understand about the brain is that there is no such inner observer sitting at a switchboard or video monitor. The brain does not convert the electrical impulses back into "real" images. Mental images are electrochemical patterns. The marvel is our ability to interpret these patterns and to respond to them. A particular mental image may thus stimulate a person to run, to talk, or just to feel good.

Somatic Sensations

The visual system processes information only from the external world. In contrast, the somatic sensory system reports on information from three sources—from the external world, as it impinges on the body's surface; from the position of body parts with respect to one another; and from the interior of the body. While the visual system responds to a single kind of stimulus, the somatic sensory system has four types of receptors—for touch (pressure), position, temperature, and pain.

Each kind of receptor converts information into electrical signals, which, like the signals of the visual system, travel to the central nervous system. For example, receptors in the skin transduce pressure into electrical activity. The receptors connect to neurons in the spinal cord or in the part of the brain called the medulla. The neurons in the spinal cord or medulla in turn connect to others in the thalamus, although in a different region from the lateral geniculate neurons that process visual information. Finally, the thalamus neurons connect to neurons of the cerebral cortex, in the region called the primary sensory area.

As in the visual system, the pathways of information roughly maintain the spatial organization of the receptors, so that the sensory cortex has a complete, although distorted, map of the whole body (Figure 42-16). Parallel pathways carry corresponding information from different kinds of receptors—for example, for touch or for position—to adjoining columns of cells within the sensory cortex.

Also, as in the case of the visual system, the brain contains more than one mental image of the touch information. The brain extracts different information from each of at least five parallel images of somatic sensation.

Sensory receptors throughout the body detect four kinds of somatic information—about touch, position, temperature, and pain. Somatosensory information passes to the spinal cord or to the medulla, through the thalamus, to the primary somatosensory area of the parietal lobe of the cerebral cortex.

How Do We Detect Smells and Tastes?

Perception of smell and taste depend on sensory receptors that respond to molecules from the external environment. Similar receptors are responsible for determining the chemical status of the internal environment. These initial receptor molecules are proteins that bind to the small molecules responsible for smell or taste. Like the receptors for hormones and neurotransmitters, binding initiates a cascade of cellular events that ultimately result in a signal being sent to the brain (Chapters 36 and 41).

Taste sensations depend on taste buds on the tongue and on the lining of the top of the digestive tract—the mouth, pharynx, larynx, and upper esophagus. A single taste bud consists of about 50 cells. All are modified epithelial cells. Some cells

Figure 42-16 Both sensory inputs from touch receptors to the cortex and motor outputs from the brain to individual body parts are highly organized. The motor "homunculus," shown here, illustrates the relationships and the relative sizes of the areas devoted to each body part. Face and hands are particularly well represented compared, say, with the back and legs.

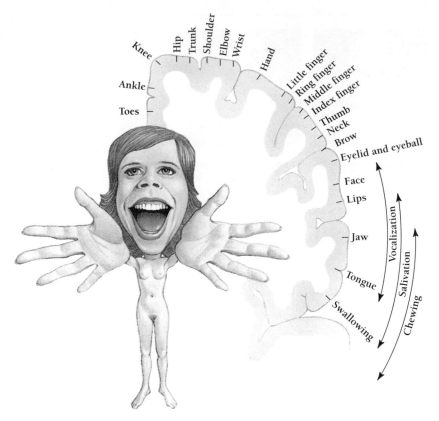

respond to dissolved molecules, while some provide support (Figure 42-17).

Biologists classify tastes into just four qualities—sweet, sour, bitter, and salty. A single taste bud can detect all four qualities, but taste buds at different locations tend to be more sensitive to some tastes than others. The human tongue, for example, detects sweet and salt near the tip, bitterness at the back, and sourness at the sides. Information from the taste buds flows to the medulla. Like sensory information from the eyes, ears, and skin, the mouth's contributions funnel through a distinct part of the thalamus before reaching the cerebral cortex. The pattern of electrical activity in individual nerve fibers differs according to the presented taste, so that some fibers seem to carry more information about sweetness and others about saltiness.

Four tastes are not, of course, enough to convey the subtle flavors of food. What we usually call the "taste" of a food is actually both taste and smell. The aromas of chocolate, vanilla, garlic, or lemon all come to our brains by way of our noses, not our tongues. That is why a stuffy nose can impair our ability to taste what we are eating. Smelling, or olfaction, is far

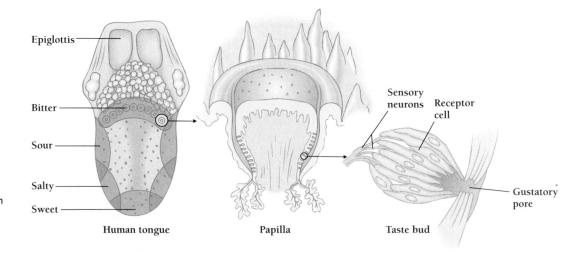

Figure 42-17 Distribution and structure of taste buds on the tongue.

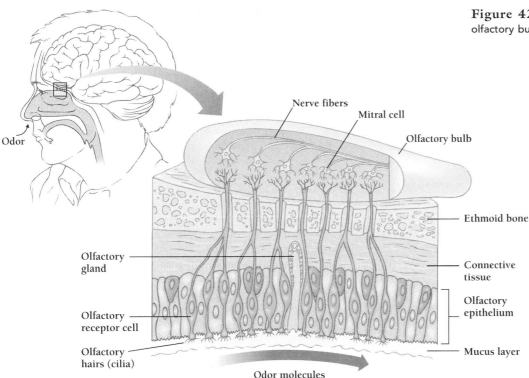

Figure 42-18 Olfactory receptors on the olfactory bulb.

more sensitive than taste. Devoted wine tasters, for example, can distinguish more than 100 different flavors in wine alone. Altogether a well-trained human can distinguish about 10,000 odors. Yet this is nothing compared with the olfactory powers of dogs, which may have 200 million receptors to our 10 million. Dogs can distinguish far more odors and at much fainter concentrations.

Researchers now suspect that humans have from 100 to 1000 distinct types of odor receptors. Genes for about two dozen of these have been found.

The smell receptors lie in the olfactory epithelium of the nose. Smell receptors, unlike taste receptors, are true neurons that connect directly to the brain's **olfactory bulb,** which extends into the space next to the nose. The pattern of electrochemical activity of the olfactory bulb reflects the type of odor detected by the olfactory epithelium (Figure 42-18). Unlike the relay stations for vision, hearing, and touch (all of which are in the thalamus), the olfactory bulb does not reflect the spatial organization of the outside world. As a result, our sense of smell provides no clues about the location of an odor.

Smell differs from the other senses in another way, as well. Electrical signals that represent smells do not reach the cerebral cortex directly, as do the signals for sound and sight. Instead, smell information flows to the parts of the brain that are most involved with emotions. These ancient regions of the brain—collectively called the **limbic system**—are thought to

have developed early in vertebrate evolution and are sometimes half-jokingly called our "reptilian brain." The distinctive path that olfactory information takes suggests that odors have long played an important role in the responses of organisms to their environment.

Sir John Eccles, who won the Nobel Prize for his studies of the synapse, once quipped, "The brain is so complicated that it staggers its own imagination." It seems unlikely that we will ever completely understand exactly how the brain accomplishes its uncounted activities. We cannot predict how much biology may reveal about the processing of sensory information, of the control of muscle movements, and of the modulation of mood, memory, personality, and mind and soul. One thing, however, it certain: intelligent wondering women and men will continue to use these marvelous organs to ask questions about ourselves and our fellows creatures.

The primary event in taste and smell is the binding of a small molecule to a specific protein. This event triggers changes in nerve activity in the brain. Information from taste receptors passes to the medulla, through the thalamus, to the cerebral cortex. Olfactory receptors, however, themselves lie within a brain structure, the olfactory bulb, and the signals that result from olfaction do not directly reach the cerebral cortex.

STUDY OUTLINE WITH KEY TERMS

The nervous system is a cellular network that responds to changes in an animal's external and internal environment. Whereas simpler animals have simple **nerve networks,** vertebrates have complicated **central nervous systems (CNS),** which are protected from physical damage by bones, membranes, and fluid.

In many cases, the outcome of nervous system processes is a command to contract a particular muscle. **Motor neurons,** the nerve cells that directly stimulate muscle cells, lie in the ventral horn of the **spinal cord.** Sensory information about touch, position, temperature, and pain moves from somatosensory receptors to the dorsal horn of the spinal cord and then up to the brain.

The vertebrate brain consists of the **hindbrain,** the **midbrain,** and the **forebrain.** The hindbrain, which consists of **pons, medulla oblongata,** and **cerebellum,** processes both sensory information going to the brain and motor instructions coming from the brain. The midbrain relays sensory information, helps organize movements, and directs attention.

The forebrain, which consists of the **diencephalon and cerebral hemispheres,** is the site of activities that include sensation, perception, movement, learning, memory, and mood. Within the diencephalon, the **thalamus** integrates information about the external world, while the **hypothalamus** is the major regulator of the internal world. The cerebral hemispheres consist of the **cerebral cortex** and underlying structures, including the **hippocampus, amygdala,** and the **basal ganglia.** In humans, the cerebral cortex is highly folded, with many bumps (**gyri**) and grooves (**sulci**). The cortex receives sensory information and also processes that information, largely in the **association cortex,** which is intimately involved in **memory. Short-term memories** are less stable than **long-term memories,** whose formation depends on the workings of the hippocampus.

The **peripheral nervous system** consists of both **afferent** nerves, which carry sensory information to the CNS, and **efferent** nerves, which carry motor instructions from the CNS. Efferent nerves include both **somatic** nerves that carry instructions to the voluntary muscles and the **autonomic** nerves that carry instructions to internal organs. The autonomic nervous system contains both the **sympathetic** and the **parasympathetic** nerves.

Sensation refers to the detection of a stimulus in the external world, and **perception** refers to its conscious recognition and interpretation. Both sensation and perception depend on **sensory receptors** and on the central nervous system. The receptors for four of the five traditional sensory **modalities—audition** (hearing), **vision** (seeing), **gustation** (tasting), and **olfaction** (smelling)—are in specialized **sense organs.** The fifth sensory modality—touch, or **somatic sensation**—includes touch, position and motion, pain, and temperature and depends on receptors that lie throughout the body. Sensation depends on **transduction** of an external signal into a neural signal, usually a **receptor potential;** the conversion of a receptor potential into a pattern of action potentials, by **relay neurons;** the **transmission** of the neural signal to cells of the CNS; and the **integration** of the signals from many receptors. Sensory information from each modality usually flows separately through the thalamus and converges separately in different regions of the cerebral cortex.

The **auditory system** detects and interprets differences in frequency, which we perceive as **pitch,** and differences in pressure, which we perceive as **loudness.** In mammals, sound passes from the **outer ear** through the **auditory canal** of the **middle ear** to the eardrum, or **tympanic membrane,** which transmits the vibrations to the fluid within the **inner ear.** Within the **cochlea** of the inner ear is the **organ of Corti,** whose **hair cells transduce** the vibrations into receptor potentials. Each frequency stimulates a different set of hair cells, and the brain interprets the patterns of stimulation as different patterns of sound. Hair cells within the **semicircular canals** of mammals or within the **lateral line system** of fish and amphibians detect pressure differences that arise from the body's position and motion.

Detection of light by the **visual system** starts with **photoreceptors,** light-detecting cells—**rods** and **cones**—in the **retina,** the thin sheet of cells at the back of the eye. Cones, which detect color, are concentrated in the central **fovea.** Both rods and cones contain **rhodopsins,** a light-absorbing pigment that consists of a protein (**opsin**) and **retinal.** Light enters the eye through the transparent **cornea** and passes through the **pupil,** whose opening depends on the action of muscles within the **iris.** The **lens** of the eye focuses images onto the retina. Misshapen lenses may result in **nearsightedness** or **farsightedness.** A cloudy spot on the lens is called a **cataract.** The ability to alter the shape of the lens to focus near or far is called **accomodation.**

The photoreceptor cells of the retina transmit information to the **bipolar cells,** and the bipolar cells transmit information to the **ganglion cells.** Connections among the photoreceptor cells, the bipolar cells, and **horizontal cells** and among the ganglion cells and **amacrine cells** are responsible for the integration of different aspects of the visual world. The axons of the ganglion cells form the **optic nerve,** which carries visual information to the brain. Optic nerve fibers end in the lateral geniculate nucleus of the thalamus, which serves both as an integrator of visual information and as a relay station.

The primary event in taste and smell is the binding of a small molecule to a specific protein. This event triggers changes in nerve activity in the brain. Information from taste receptors passes to the medulla, through the thalamus, and to the cerebral cortex. Olfactory receptors, however, themselves lie within a brain structure, the **olfactory bulb,** and the signals that result from olfaction do not directly reach the cerebral cortex. Olfactory information passes directly into the **limbic system,** the parts of the brain that are most involved with emotions.

REVIEW AND THOUGHT QUESTIONS

Review Questions

1. In what way is a nervous system a homeostatic organ?
2. What kind of cell detects and transduces the signal received by the sense organs?
3. How do sense organs relay signals to the brain or the spinal cord?
4. How does the output of a sensory receptor differ from the output of a neuron?
5. How do relay neurons integrate signals?
6. How does olfaction differ from other senses?
7. Distinguish between afferent and efferent neurons.

8. Draw a picture of the cochlea of the inner ear, including the organ of Corti, and explain how this structure registers sound as an electrochemical signal.

Thought Question

9. Explain in what ways the brain is integrated (a single mind) and in what ways the brain is a group of separately operating units (a confederacy of minds).

SELECTED READINGS

Kandal, E.R., J.H. Schwartz, and T.M. Jessell, *Essentials of Neural Science and Behavior,* Appleton and Lange, Norwalk, Connecticut, 1995. An outstanding introduction to neuroscience.

▶ On-line materials relating to this chapter are on the World Wide Web at
http://www.saunderscollege.com/lifesci/
Click on Tobin/Dusheck: *Asking About Life.*

Why Did Sex Evolve?

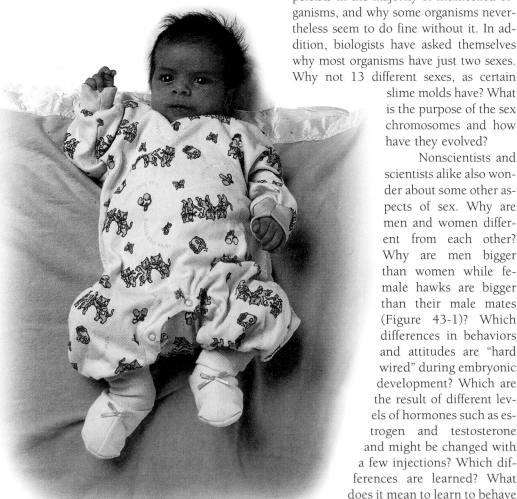

Textbooks tend to say that the purpose of sex is to recombine genes in novel ways, thereby increasing genetic diversity in each new generation. In truth, however, no one knows why sex evolved, why it persists in the majority of multicelled organisms, and why some organisms nevertheless seem to do fine without it. In addition, biologists have asked themselves why most organisms have just two sexes. Why not 13 different sexes, as certain slime molds have? What is the purpose of the sex chromosomes and how have they evolved?

Nonscientists and scientists alike also wonder about some other aspects of sex. Why are men and women different from each other? Why are men bigger than women while female hawks are bigger than their male mates (Figure 43-1)? Which differences in behaviors and attitudes are "hard wired" during embryonic development? Which are the result of different levels of hormones such as estrogen and testosterone and might be changed with a few injections? Which differences are learned? What does it mean to learn to behave a certain way? Are women inherently better drivers? Are men inherently better navigators? If so, why? How do homosexual men and women become homosexual? What attributes do they share with one another or with members of the opposite sex? And why? Discussions of all these questions nearly always lead to heated arguments.

Amazingly, biologists cannot confidently answer a single one of these questions. Instead, they themselves continue to argue about the most fundamental aspects of sex, wrangling as fiercely as any group of nonscientists. Nonetheless, biologists have some fascinating insights into a few of these questions.

The idea that sex evolved to increase genetic diversity is not universally accepted. For one thing, such increased diversity is not necessarily a good thing. Most organisms live in environments similar to those of their parents. If their parents survived and reproduced successfully in that environment, why shake up the genome on the off chance that something better will come up? Sex can erase adaptive traits as easily, probably a good deal more easily, than it can create them. If it ain't broke, why fix it?

One alternative theory comes from Richard E. Michod of the University of Arizona. Michod argues that sex evolved to repair damaged genes. Michod compares the genome to a vintage car that needs a steady supply of spare parts to keep it running. Any car collector knows

Paraskevas Photography

Figure 43-1 Sexual dimorphism in red-tailed hawks. In many species, the two sexes differ in size or other characteristics. In hawks, the female is larger than the male. *(Tom Bledsoe/DRK Photo)*

Figure 43-2 Chihuahua whiptail lizard. This unusual animal reproduces by parthenogenesis, the process by which eggs become activated and develop without any paternal genetic contribution. All specimens are females, and males are unknown. *(M.P. Kahl/Photo Researchers)*

that the best place to gets parts is from another car of the same model. Chances are good that what is not broken or worn out in one car will be usable in the other. From two broken cars, one whole one can be reconstructed.

In the same way, organisms can reconstruct a damaged genome using spare DNA. If a single strand is damaged, the adjoining strand can be used as a template for repair. But what if both strands are damaged? In that case, the correct sequence must be obtained from somewhere else.

Michod's research showed that bacteria can survive DNA damage (from mutagens such as ultraviolet light or chemicals) by swapping their own DNA with that of dead bacteria of the same species. In fact, Michod has found that only bacteria with damaged DNA actively scavenge for spare DNA, while those with healthy DNA do not bother. Bacteria best at recombining were most likely to survive.

As multicellular eukaryotes evolved from bacteria, says Michod, they used the same kind of recombination when making sperm and eggs. Then when the egg and sperm join, the two chromosomes, one from each parent, line up, duplicate, and swap DNA. As a result, each chromosome is a mix of genes from both parents. Recombinations, adds Michod, are most likely to occur in gaps in the DNA, often a sign of damage.

Such an exchange costs an individual, however. Thanks to sex, each individual passes on only half of its genes to each of its offspring. A female, for example, who could reproduce asexually could pass on all of her genes to every offspring, doubling her genetic representation in the next generation. Some females do reproduce this way. Plants often reproduce asexually, but so do whiptail lizards, aphids, and a variety of other organisms.

But offspring inheriting both sets of chromosomes from one parent run the risk of getting a double dose of damaged DNA. Sex keeps harmful recessive mutations masked, says Michod. Still, he does not explain why Chihuahua whiptail lizards, all of which are female, manage so well without spare parts from males (Figure 43-2). Michod's ideas are well-known, but they are not yet accepted.

The idea that sex evolved to increase genetic diversity is not universally accepted.

KEY CONCEPTS

1. Oogenesis and spermatogenesis are parallel processes in women and men that produce haploid oocytes and sperm.

2. Fertilization, syngamy, and conception all refer to the moment when the haploid sperm and haploid egg join to form a diploid zygote.

3. In mammals, male and female external genitalia develop from the same embryonic tissues, while internal genitalia develop from separate embryonic tissues.

4. Fertilization is facilitated by sexual intercourse, a process that, in humans, is divided into four phases: excitement, plateau, orgasm, and resolution.

5. A collection of related steroid hormones regulate the production of ova in females and the production of sperm in males. The anterior pituitary gland makes and releases hormones that control the secretion of these steroid hormones in both females and males.

HOW DO MAMMALS FORM GAMETES?

Few subjects occupy more attention, energy, and wonder than sexual reproduction. For species such as the mayfly, adult life consists only of mating and dying, sometimes without even the diversion of feeding. Like annual plants, such animals die before their young even appear. Humans and other mammals have a longer and more complex adult life with considerable energy devoted to the care of the next generation.

In mammals, successful reproduction involves many steps—production of eggs or sperm, mating and fertilization, nurturing the embryo within the uterus, delivery, nursing and cleaning, and continued care even after nursing has stopped. In both males and females, many organs contribute to these processes. All are influenced by hormones, organic compounds produced in one tissue or organ that produce specific effects in other tissues or organs. Several of the hormones that coordinate reproductive processes also regulate the development of the reproductive structures and behaviors in fetuses and, later, in adolescents.

In this chapter we examine the specific tissues and organs that allow sexual reproduction, focusing on human reproduction. We begin with the **gonads,** the paired organs where the gametes form and mature. Humans and nearly all other sexually reproducing organisms form just two kinds of gametes, eggs and sperm. Eggs, or **ova** [singular, *ovum*; Latin, egg], form in the **ovaries.** Sperm form in the **testes** [singular, *testis*; Latin]. In humans, the testes are called **testicles.**

Both eggs and sperm are haploid: they have just one set of chromosomes each. When an egg and a sperm come together at fertilization, or **syngamy,** the nuclei of the two gametes fuse and the resulting **zygote** is diploid: that is, it has two set of chromosomes. In humans, fertilization, syngamy, and **conception** all refer to the same event.

If all goes well, the zygote begins to divide and makes its way down the **oviducts** [Latin, *ovum* = egg + *ductus* = duct],

two long tubes to the **uterus** [Latin, womb]. The multicelled embryo implants itself in the uterus (Figure 43-3). When the embryo has developed the basic features of the organism it is to become, the bare outlines of all its organs, limbs, and so on, it is a **fetus.** In humans, the embryo is called a fetus nine weeks after fertilization.

How Do Female Mammals Produce Ova?

The female sex organs or ovaries are solid, egg-shaped organs. Each one is 4 cm long, flattened, and almond-shaped. The ovaries lie on the side walls of the lower abdominal cavity, or pelvis, one on each side, where they produce ova, or egg cells, and also sex hormones. Much of the process of **oogenesis,** the production of eggs, begins before birth.

In both females and males, mature germ cells (eggs or sperm) are descendants of **primordial germ cells,** the cells that represent the germ line in the early embryo. During the development of a female fetus, the primordial germ cells migrate to the outer surfaces of the ovaries. These cells divide thousands of times by mitosis to produce up to a million **oogonia,** diploid cells that, when they grow a little larger, are renamed **primary oocytes,** also diploid. The primary oocytes are each capable of dividing by meiosis to form a haploid ovum. They double their DNA and enter the first stage of meiosis (Figure 43-4). They will not complete meiosis, however, for 12 to 50 years.

A baby girl is born with an average of about 750,000 primary oocytes, each part way through meiosis. During her childhood, many of these cells die, so that by the time she reaches puberty she has about 200,000 primary oocytes. Each primary oocyte together with surrounding cells, called *granulosa* cells, forms a **follicle** [Latin, *folliculus* = small ball] (Figure 43-5). At puberty, the follicles begin developing, a few each month.

Several follicles fill with fluid each month, but usually only one of the several follicles in the two ovaries continues to ex-

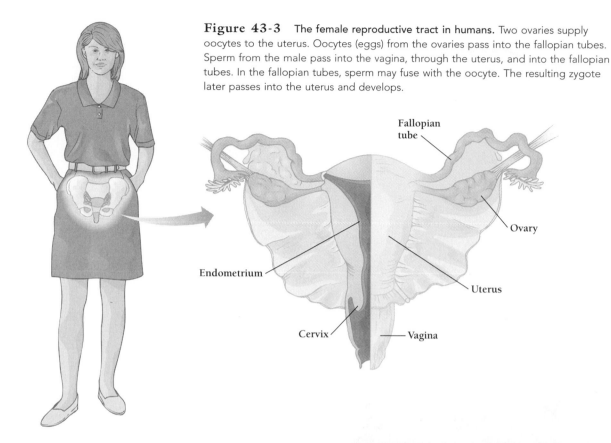

Figure 43-3 **The female reproductive tract in humans.** Two ovaries supply oocytes to the uterus. Oocytes (eggs) from the ovaries pass into the fallopian tubes. Sperm from the male pass into the vagina, through the uterus, and into the fallopian tubes. In the fallopian tubes, sperm may fuse with the oocyte. The resulting zygote later passes into the uterus and develops.

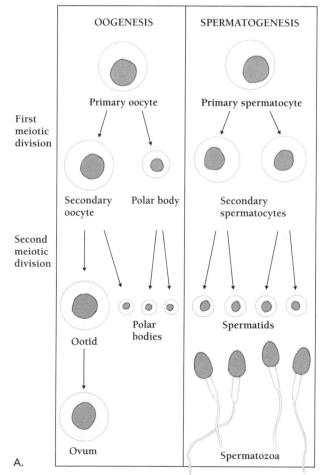

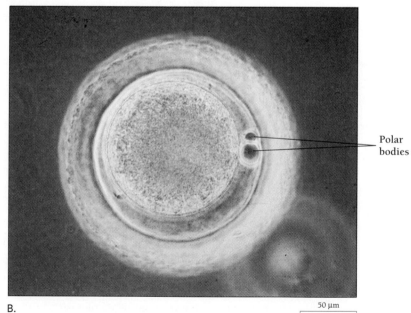

Figure 43-4 **Oogenesis and spermatogenesis.** A. The diploid primary oocyte and spermatocyte divide by meiosis (which is two divisions) to produce four haploid cells—called ootids and spermatids. One of the ootids contains virtually all of the cytoplasm from the original oocyte and becomes the ovum, or egg. The other three cells, called polar bodies, fall away. In contrast, all of the spermatids mature into sperm, developing a head, a midsection, and a tail, and losing their cytoplasm. B. In this photo of a rabbit zygote, two polar bodies still cling to the surface at about 3 o'clock. *(B, Biophoto Associates)*

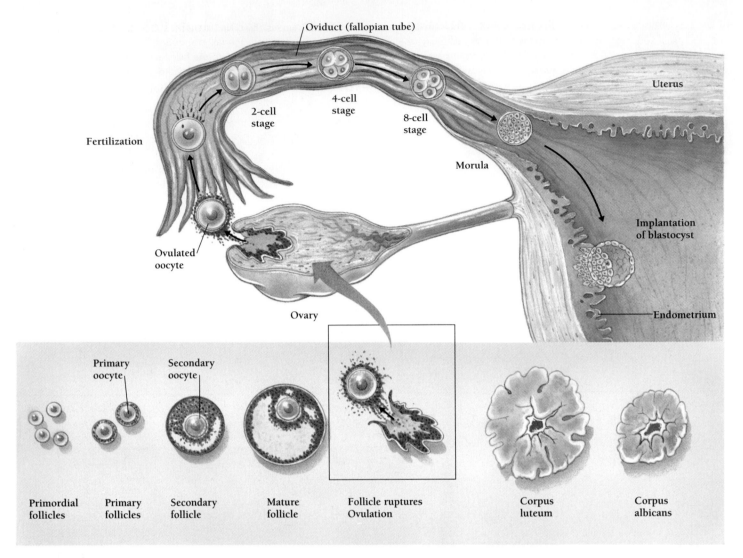

Figure 43-5 Development and maturation of primary oocytes and follicles. Inside the ovary, the primordial follicles (*bottom left*) enlarge to form secondary follicles. While inside the follicles, the primary oocytes mature into secondary oocytes. After the follicle ruptures (ovulation), the secondary oocyte migrates into the fallopian tubes. If the oocyte fuses with a sperm, it finishes meiosis II and divides to form a mature oocyte and a third polar body. The zygote begins dividing. After about a week, the blastocyst embeds itself in the wall of the uterus. In the ovary, the ruptured follicle meanwhile matures into a corpus luteum, then corpus albicans.

pand. The single maturing follicle expands so much, however, that it occupies as much as an eighth of the ovary's volume and bulges from the side of the ovary (Figure 43-5). At **ovulation,** the follicle bursts open, releasing the oocyte.

Before ovulation, the primary oocyte completes meiosis, producing two haploid cells—a large **secondary oocyte** and a small **polar body.** Polar bodies are tiny, spermlike cells that contain chromosomes but very little cytoplasm. Nearly all of the cytoplasm is reserved by the oocyte. The polar bodies are a mechanism for reducing the number of chromosomes and they do not participate in fertilization. The polar body may divide, but it cannot develop into an ovum.

Just before ovulation, the oocyte (secondary) enters meiosis II. By the end of meiosis II, the oocyte will divide again to form the ovum and another polar body, but in humans this

does not happen until after it unites with a sperm. The polar bodies may remain attached to the ovum until the ovum fuses with a sperm (Figure 43-4B).

The ovum itself contains a single set of chromosomes tightly packed into its nucleus and a large amount of cytoplasm, including all of the organelles and structures that most cells contain. In addition to mitochondria, endoplasmic reticulum, Golgi complex, and microtubules, for example, the egg also contains an abundance of high-fat yolk particles, which supply energy to the developing embryo.

Over a woman's lifetime, approximately one primary oocyte will mature into an ovum every 28 days or so. Of the 200,000 oocytes initially present in the ovaries, then, only 400 or 500 will complete meiosis between the onset of ovulation, at about age 13, and the end of ovulation, at around age 50.

All the rest of the oocytes degenerate, and women older than 50 have few if any oocytes. Of the 400 or so oocytes that mature, only a tiny fraction will meet a sperm and develop into a full-term baby.

A diploid oocyte produces one haploid ovum and two or three polar bodies.

Where Do Oocytes Go After They Leave the Ovary?

When the oocyte has erupted from the follicle in the ovary, it enters the abdominal cavity and then finds its way into the opening of one of the two oviducts that carry the egg into the uterus (Figure 43-5). In humans, the oviduct is called the **fallopian tube.** The end of each oviduct flares and partly encloses the adjacent ovary. Hairlike cilia around the opening of the oviduct draw fluid into the oviduct, sweeping the oocyte into the tube. The fallopian tubes do not fully enclose the ovary, however, and the oocyte often floats free in the abdominal cavity, at least for a time.

One might think that the oocytes would not enter the oviduct very reliably, but they do about 99 percent of the time. Even women who have had one ovary removed and the opposite fallopian tube removed can still conceive children with ease. For this to happen, an oocyte from the functioning ovary must cross the abdominal cavity to the opposite fallopian tube.

Inside the fallopian tube, more cilia propel the oocyte toward the uterus, the chamber in which—if the oocyte meets a sperm—the oocyte may develop into an embryo and then into a fetus. The uterus lies in the middle of the pelvis, above and behind the bladder. Both oviducts open into the uterus's hollow center, the womb (Figure 43-3).

The uterus is about the size and shape of an upside-down pear. It has thick muscular walls, bordered on the inside surface with a specialized lining called the **endometrium** [Greek, *endon* = within + *metro* = mother]. A rich supply of blood in the endometrium carries nutrients to the early embryo.

The narrower, lower part of the uterus is called the **cervix** [Latin, neck], which encloses a narrow passage from the uterus to the vagina. The cervix is a sphincter, a muscular ring that normally stays contracted. At the end of a pregnancy, however, when the fetus is **full-term,** ready to be born, the cervix opens to about 10 cm and the uterus contracts to push the fetus out through the cervix and the vagina and into the wide world.

The vagina connects the genital tract to the outside. During sexual intercourse, the vagina encloses the penis and receives the ejaculated sperm. At birth, the vagina expands to serve as the canal through which a baby emerges into the world.

The primary oocyte moves from the ovary, through the oviduct, into the uterus.

How Do Male Mammals Produce Sperm?

In mammals, the male gonads, or testes, are contained within the **scrotum** [Latin, bag], a pouch that lies outside the body (Figure 43-6). The temperature within the scrotum is usually a few degrees below body temperature. Sperm develop best at the lower temperature outside the body. Sometimes, in fact, hot baths or tight clothes can temporarily interfere with sperm maturation (although not reliably).

Each egg-shaped testis is 4 to 5 cm long, about the size of a golf ball, although not so heavy. A testis consists of hundreds of separate chambers filled with tightly coiled ducts, called **seminiferous tubules** [Latin, *semen* = seed + *ferre* = to bear]. The sperm mature in the seminiferous tubules (Figure 43-7). Each sperm cell, or **spermatozoan,** consists of a head, a midpiece, and a tail. Sperm cells are specialized for rapid movement. The head contains a dense nucleus, with a tightly packed haploid set of chromosomes. The **midpiece** contains a microtubule organizing center and dense concentrations of mitochondria—sources of ATP for the journey to the egg. Finally, the tail consists of a long flagellum, which, powered by the mitochondria, whips about and keeps the sperm cells moving (Figure 43-8).

Sperm, like ova, are descendants of primordial germ cells. In males, primordial germ cells undergo several mitotic divisions in the testes before beginning meiosis. Sperm derive from the dividing diploid cells called **spermatogonia** [Greek, *sperma* = sperm + *gonos* = offspring], which lie on the inner walls of the seminiferous tubules. When a spermatogonium divides through meiosis, the resulting haploid cells move into the lumen of the seminiferous tubule. The two meiotic divisions (meiosis I and II) produce four haploid cells, each of which is called a **spermatid** (Figure 43-4). These two divisions are analogous to the meiotic divisions that produce the secondary oocyte and its three polar bodies. The spermatids are like polar bodies. Each spermatid then undergoes further development to form a mature sperm with midsection and tail (Figure 43-8).

Each sperm becomes highly elongated and the nucleus becomes highly condensed, occupying only a fraction of the length of the mature sperm. Most of the length is taken up by a **flagellum,** which propels the sperm forward (Figure 43-8). The sperm also contains a special structure at its front—the **acrosome,** a lysosome whose enzymes will eventually allow the sperm to enter the egg.

The process of meiosis and development takes more than two and a half months. During this time, the developing sperm are surrounded by specialized cells, called **Sertoli cells,** which create a continuous lining between the outside of the tubule and its lumen. The Sertoli cells nourish the developing sperm, regulate the passage of nutrients from the blood, and secrete a fluid that fills the lumen (Figure 43-7). Among the seminiferous tubules is a matrix of **interstitial cells,** which synthesize the male sex hormone testosterone. The interstitial cells make up about 20 percent of the mass of the testes.

Spermatogonia undergo mitosis to form more spermatogonia, so the tissue continually renews itself and provides a continuous supply of sperm from puberty on. Each hour about

Figure 43-6   The male reproductive tract. Sperm form and mature in the testis, then pass into the epididymus until ejaculation, when they pass into the vas deferens and the penis. Fluids from the paired seminal vesicles and bulbourethral glands, as well as the single prostate gland, activate the sperm and speed them on their way.

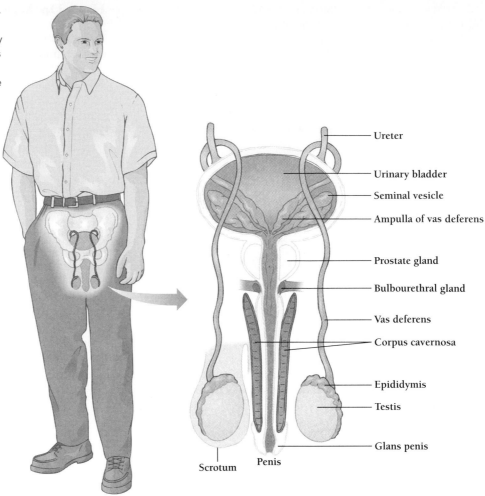

- Ureter
- Urinary bladder
- Seminal vesicle
- Ampulla of vas deferens
- Prostate gland
- Bulbourethral gland
- Vas deferens
- Corpus cavernosa
- Epididymis
- Testis
- Glans penis

Scrotum   Penis

200 million sperm mature, far more than can be used. Each ejaculation releases only about 200 million sperm, so unused sperm are either absorbed by other cells in the testes or passed in the urine.

Sperm cells form within the seminiferous tubules.

How Do Sperm Cells Travel from the Testes to the Penis?

Sperm maturation occurs in waves, so that all the spermatids in a given segment of a tubule are at about the same stage of development. As the spermatids mature, they separate from the Sertoli cells and pass into the lumen of the seminiferous tubules. From there, the sperm pass into the **epididymis,** an interconnected network of coiled ducts that leads to a single exit tube (Figure 43-7). The passage from the seminiferous tubules to the epididymis takes another 12 days, during which the sperm complete their maturation and the cells of the epididymis absorb the fluid that originally bathed the developing sperm.

Lining the inside of the epididymis are smooth muscles, which, during *ejaculation,* push the mass of mature sperm toward the **vas deferens** [Latin, *vas* = vessels + *deferre* = to carry down], a large, thick-walled duct. The two vas deferens, which are 45 cm (18 in.) long, head up into the abdominal cavity, loop around the bladder, and then empty into the urethra, the tube that carries urine from the bladder to the penis (Figure 43-6).

Mature sperm are stored in both the vas deferens and the lower portion of the epididymis. In many birds and some reptiles, sperm (or eggs in the female) and urine all pass through a single opening called the **cloaca** [Latin, sewer]. In all male mammals except the monotremes, and in some birds, some reptiles, and a variety of invertebrates, the penis delivers sperm directly into the female genital tract during sexual intercourse.

Just before the vas deferens ducts empty into the urethra, they pass a junction with one of the two **seminal vesicles,** which secrete sugars and other nutrients around the sperm. The two vas deferens ducts then join as they pass through the **prostate gland,** which secretes a thin, milky alkaline fluid into the lumen of the urethra. Next, the **bulbourethral glands** inject a small amount of mucus into the mix (Figure 43-6). The

Figure 43-7 **The development and maturation of sperm.** Inside the testis are hundreds of seminiferous tubules. On the inside wall of each tubule are diploid cells called spermatogonia, which divide, by meiosis, to form four spermatids. Over a period of two and a half months, the spermatids mature into sperm. Sertoli cells supply nutrients to the sperm and interstitial cells synthesize testosterone. Testosterone increases sex drive and aggressive behavior and suppresses the secretion of hormones that stimulate the interstitial cells to release testosterone.

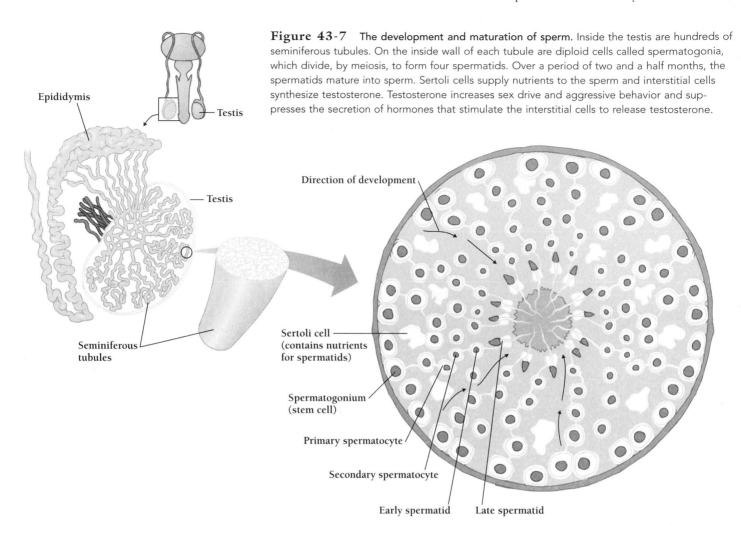

Epididymis

Testis

Testis

Seminiferous tubules

Direction of development

Sertoli cell (contains nutrients for spermatids)

Spermatogonium (stem cell)

Primary spermatocyte

Secondary spermatocyte

Early spermatid   Late spermatid

sperm cells, together with the fluid from the seminal vesicles, the prostate gland, and the bulbourethral glands, make up the **semen** [Latin, seed].

> Sperm cells move from the seminiferous tubules, through the male genital tract, and out through the penis.

SEXUAL INTERCOURSE: HOW DO THE EGG AND SPERM RENDEZVOUS?

Fertilization in mammals (and many other animals) occurs within the body of the female. Reproduction therefore requires that the male deliver mature sperm into the female genital tract.

When and Where Does Fertilization Occur?

For fertilization to occur, the secondary oocyte must encounter a sperm at a time when both the sperm and the oocyte are in

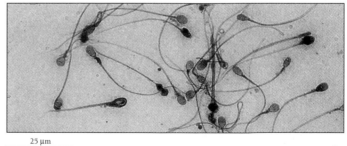

25 μm

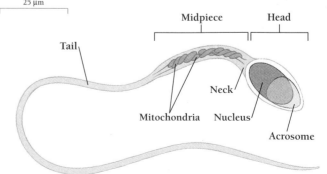

Tail

Midpiece   Head

Neck

Mitochondria   Nucleus

Acrosome

Figure 43-8 **Human sperm.** A mature sperm, or spermatozoan, consists of a chromosome-packed head, a mitochondrion-rich midpiece, and a long flagellum, or tail. *(photo, Ed Reschke/Peter Arnold, Inc.)*

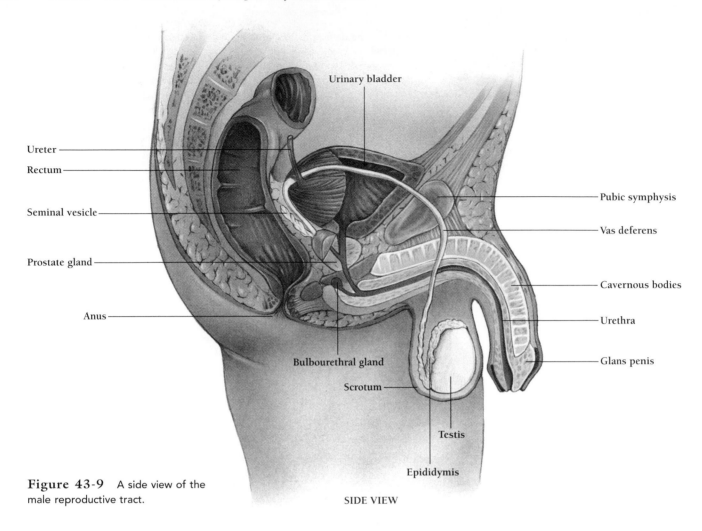

Figure 43-9   A side view of the
male reproductive tract.

SIDE VIEW

their prime. Oocytes survive only 24 hours after they are released from the ovaries. In contrast, sperm can survive in the
female reproductive tract for up to a week. As a result, the maximum period of fertility for a woman is eight days, the seven
days before ovulation plus just one day after ovulation.

In practice, however, sperm rarely retain any viability beyond three or four days. They may be alive and moving, but
they are incapable of fertilizing an egg. In addition, the viability of the egg declines rapidly after ovulation, so that the sperm
needs to encounter the egg within 12 hours after ovulation. In
reality, then, conception rarely happens unless intercourse occurs in the few days before ovulation or on the day of ovulation itself. Sperm deposited before ovulation are stored in the
cervix and released a few at a time.

Fertilization usually takes place in the oviduct, before the
secondary oocyte has arrived in the uterus. The oocyte moves
down the oviduct to meet the upward moving sperm. As we
will see in the next chapter, a single sperm fuses with the
oocyte's cell membrane and ejects its haploid nucleus into the
oocyte. Only then does the oocyte complete the second division of meiosis to form the mature ovum and a second (or

third) polar body. The chromosomes of the egg and sperm then
align, and the ovum becomes a zygote.

Usually, human sperm are viable for no more than 4 days after
ejaculation, and eggs are viable for no more than 12 hours after
ovulation. Consequently, conception rarely happens unless
intercourse occurs four or fewer days *before* ovulation.
Fertilization usually occurs in the oviduct.

How Do the External Genitalia Facilitate Sexual Intercourse?

The human penis contains a single exit tube, the **urethra,** which
ends in a slitlike opening. Surrounding the urethra is a spongy
cylindrical tissue, the **corpus spongiosum,** enlarged at each
end (Figure 43-9). At the base of the penis, this tissue is attached to the body, below the pelvis. At the far end of the penis, the tissue forms a smooth cap, called the **glans.** The glans
is full of sensory nerves that make the penis especially sensitive to mechanical stimulation. Above this tissue lie two ad-

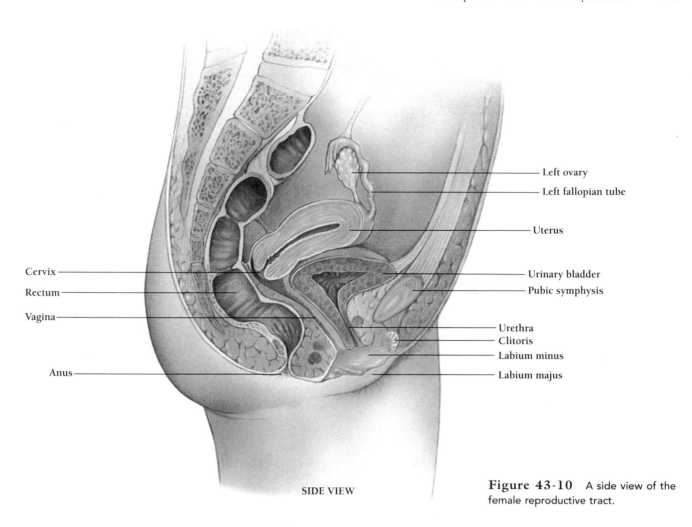

SIDE VIEW

Left ovary

Left fallopian tube

Uterus

Urinary bladder

Pubic symphysis

Urethra
Clitoris
Labium minus
Labium majus

Cervix

Rectum

Vagina

Anus

Figure 43-10   A side view of the female reproductive tract.

ditional spongy cylinders, each called a **corpus cavernosa**, which run along the length of the penis. These cylinders fill with blood during erection, as we will discuss shortly.

Fibrous tissue surrounds and separates the three cylinders. Blood vessels and nerves enter the penis within the fibrous tissue. Around the outside of the penis is a loose layer of skin, which ends in a flap, called the **foreskin**, that folds over the glans. The foreskin is one of the most sensitive parts of the penis, being at least as well supplied with nerves as the glans. In many cultures, it is surgically removed in the first days of life. Such removal, called **circumcision**, remains a common practice.

The external genitalia of the female are collectively called the **vulva** [Latin, *volvere* = to wrap] (Figure 43-10). Two thin skin folds, called the **labia minora** [Latin, small lips], surround the mouth of the vagina and the end of the urethra. These join at the front and form a hood over the **clitoris**, a diminutive penis. The clitoris includes two corpora cavernosa, a **bulb of the vestibule** that corresponds to the penis's corpus spongiosum and a glans that is especially sensitive to stimulation. On either side of the labia minora are two more skin folds, called

the **labia majora** [Latin, large lips], which enclose small amounts of fatty tissue. Bulbourethral glands secrete mucus at the mouth of the vagina.

Just inside the mouth of the vagina is a thin membrane of irregular ragged shape, called the **hymen.** The hymen often impedes penetration by the penis, but only at first. The first sexual intercourse nearly always tears a hymen that closes the vagina, so an intact hymen was long regarded as the only proof of virginity. Other events can also break the hymen, however, including disease, a fall, or vigorous exercise.

How Are the Male and Female Genitalia Alike?

The external male genitalia are most simply described as a female genitalia that have grown together and fused at the midline, with the clitoris enlarged to include the urethra. Testes and ovaries develop from the same embryonic structures. The labia majora and scrotum develop from the same structures and are supplied with the same sets of nerves. The surface of the labia minora likewise corresponds to the ventral surface of the penis and is similarly sensitive (Figure 43-11).

The internal genital tracts of males and females develop from separate embryonic structures. All embryos develop two kinds of ducts, Wolffian ducts and Mullerian ducts. In the male, however, the vas deferens and its attachments develop from the paired Wolffian ducts. In the female, the oviduct, uterus, and vagina develop from the paired Mullerian ducts. The ducts not used by each sex degenerate.

Many parts of the reproductive system develop from the same embryonic precursors in males and females.

The Male and Female Sexual Responses Each Consist of Four Stages

Studies of the human sexual response reveal a stereotyped cycle of responses in both males and females. These responses involve changes in blood flow and in the contractions of smooth and skeletal muscle. Sexual physiologists divide the sexual response cycle into four phases: excitement, plateau, orgasm, and resolution.

The first phase, **excitement**, is marked by increased blood flow to the clitoris, the labia minora, and the breasts in the female and to the penis and testes in the male. In each sex, the affected tissues become engorged with blood, causing *erection* of both penis and clitoris.

Erection results directly from an increase of blood flow and may happen quickly, sometimes within 5 to 10 seconds. Blood fills the tiny spaces within the spongy tissues of the clitoris and penis causing them to fill and assume a firm shape like that of a balloon filled with water or air. In females, erection simultaneously engorges the clitoris and tightens the tissues around the base of the vagina. During sexual intercourse, this tightening squeezes the penis.

The bulbourethral glands of the penis release a small amount of mucus, which lubricates the head of the penis and eases entry into the now tightened entry of the vagina. Most lubrication, however, comes from the female's **Bartholin's glands,** located beneath the labia minora and derived from the same structures as the bulbourethral glands in the male.

Erection in both females and males can occur as a result of physical stimulation of any part of the urogenital region, including both the sexual organs and the urinary tract. Stimulation of the glans, in particular, triggers a reflex response in which the blood vessels dilate.

Because nerve pathways from the brain control blood flow, visual stimuli, thoughts, or feelings can all cause erection and excitement even without physical stimulation. Just as the brain can stimulate erection and excitement, however, it can also inhibit excitement. Anxiety, fear, anger, or depression, for example, can all inhibit sexual response.

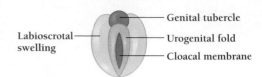

A. Indifferent stage

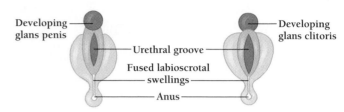

B. About 9 weeks development

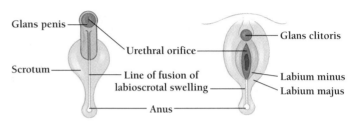

C. About 12 weeks development

Figure 43-11   The common origin of male and female external genitalia. A. From 4 to 7 weeks gestation, the genitals of male and female human embryos are indistinguishable. The "indifferent" genitals consist of a genital tubercle, a urogenital fold, and paired labioscrotal swellings on either side of the urogenital fold. B. By 9 weeks, the anus has formed and the labioscrotal swellings have begun to fuse along the midline at top and bottom. The structures of females and males begin to diverge, or differentiate. The labioscrotal swelling of males, for example, have fused more than those of females. C. By 12 weeks, the labioscrotal swellings have fused completely in the male. In females, the genital tubercle has shrunk, and the labioscrotal folds have completed fusion at top and bottom and developed into the labia majora and labia minora.

During the second, or **plateau**, phase, breathing and heart rate increase in response to continued stimulation. The end of the vagina dilates and the uterus pulls upward, creating a space near the cervix, where the semen can form a pool.

The third stage is **orgasm,** which refers to a whole complex of changes—smooth muscle contractions in the genital tract, skeletal muscle contractions throughout the body, and feelings of intense pleasure. Orgasm is typically brief, usually lasting 3 to 5 seconds.

In females orgasm increases the probability of conception, but it is not required. Conception can occur without orgasm. In males, however, sperm delivery requires **ejaculation,** the propulsion of sperm out of the penis. During ejaculation, contractions of the genital ducts and muscles in the penis force sperm out of the male's genital tract into that of the female. Ejaculation consists of two phases: (1) smooth muscle contractions in the prostate, vas deferens, and seminal vesicles force semen into the urethra in the penis and (2) smooth muscle contractions in the urethra force the semen out of the urethra. Some physiologists refer to the first set of contractions as orgasm, and they reserve the term ejaculation for the second set of contractions. A human male typically releases about 3.5 ml of semen, containing, on average, 200 million sperm (although these numbers are highly variable).

In the female, the smooth muscles in the uterus and the outer part of the vagina contract rhythmically and the cervix drops toward the pool of sperm. These contractions are, like those of the male, a reflex mediated by the spinal cord. Some research suggests that women who have orgasms less than a minute before the man, or up to 45 minutes after, retain most of the sperm in the reproductive tract, while those who have no orgasm, or have one more than a minute earlier, lose more sperm. It is also believed that orgasm-mediated contractions of the uterus and oviducts propel the sperm upward toward the descending ovum. In cows, the oviducts deliver sperm to the oocyte in five minutes, a rate of travel ten times as fast as that which the sperm can accomplish by beating their tails.

In the last phase, **resolution,** blood flow and muscle tension return to normal. In males, the penis ceases to be erect, and a second erection is not possible for a period of time that ranges from minutes to hours.

Sexual physiologists divide the sexual response cycle into four phases: excitement, plateau, orgasm, and resolution.

How Do the Sperm Reach the Egg and Penetrate Its Surface?

Once inside the vagina, the 200 million or so sperm will pile up against a mucus plug in the cervix, which keeps all but the healthiest, most active sperm out of the uterus. Those that cannot wiggle through the mucus plug end their days in the upper end of the vagina. The few hundred that make it through the cervix fan out into the uterus and up into the oviducts. In most instances of sexual intercourse, they find nary an oocyte. At the right time of the month, however, 400 to 500 sperm may arrive in one of the fallopian tubes and encounter a mature oocyte.

In many species of aquatic animals and other organisms the egg releases a chemical that attracts the sperm. Sea urchins, jellyfish, and seaweeds, for example, all use chemotaxis to attract sperm. Whether mammals use such mechanisms is unknown.

Even though the mammalian sperm may arrive near the oocyte within minutes of sexual intercourse, the sperm are incapable of fertilizing an egg until they have been in the reproductive tract for 5 or 6 hours. (Since the egg is only viable for 12 hours, intercourse must occur no later than 7 hours after ovulation for conception to result.) The head of each sperm is covered with a glycoprotein coat, which must be dissolved by enzymes in the female reproductive tract. Sperm not exposed to fluids from the female reproductive tract cannot fuse with an egg. After about 7 hours, the coating is dissolved, in a process called **capacitation.** The sperm become more motile and are now able to penetrate the outer layers of the egg.

Between the sperm's nucleus and the outer membrane is the acrosome, a lysosome packed with digestive enzymes. Under the influence of the female hormone progesterone, the outer membrane of the acrosome fuses with the membrane of the sperm head and releases the enzymes in the acrosome. These enzymes break down the outer layers of the egg, the corona radiata and the zona pellucida.

Once the sperm passes through the zona pellucida, this layer undergoes a reaction that makes it impenetrable to any other sperm. Fertilization by more than one sperm, or *polyspermy,* leads to an abnormal number of chromosomes, resulting in embryonic death or abnormal development. Eggs that are more than 12 hours old are unable to prevent polyspermy and the resulting embryos soon die.

After enzymes in the female reproductive tract capacitate the sperm, it swims to the egg. Progesterone causes the sperm's acrosome to release enzymes, which break down the outer layers of the egg, the corona radiata and the zona pellucida. Once the sperm has passed through the zona pellucida, it swells to prevent the entry of other sperm.

HOW DO HORMONES CONTROL GAMETE PRODUCTION?

A collection of hormones regulate both the production of sperm in males and the production of ova in females. Some hormones are unique to one sex or the other, but all are related. What controls the production of the steroid hormones by the gonads? How does the body ensure, for example, that estrogen and progesterone are made in the right amounts and at the proper times?

The answer is that another endocrine organ, the **anterior pituitary gland,** makes and releases two hormones that together control the secretion of the sex steroids—estrogen and progesterone in females and testosterone in males. The pituitary gland, embedded in the base of the brain, consists of two lobes, the anterior and the posterior. The pituitary gland lies directly beneath the **hypothalamus** [Greek, *hypo* = under + *thalamos* = inner room], a part of the brain that regulates the expression of many hormone systems (Chapter 36).

The anterior pituitary fulfills all the definitions of an endocrine gland. Removal of the anterior pituitary affects several target tissues, including the gonads, which shut down the production not only of the sex steroids but also of a variety of other hormones. Extracts of the anterior pituitary can restore all these functions. This restoration, however, depends not on

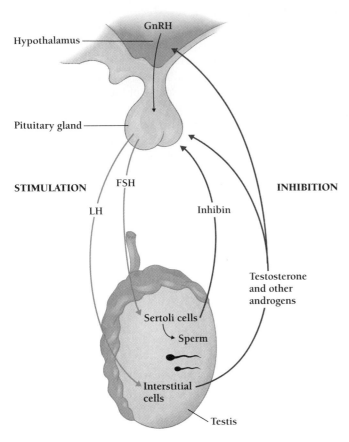

Figure 43-12   Hormonal regulation of sperm production in mammals. Here are just a few of the complex feedback systems that regulate male sperm production (and sexual behavior) in humans and other mammals. LH and FSH stimulate (*green arrows*) sperm production. Inhibin, testosterone, and other androgens inhibit (*red arrows*) sperm production.

a single compound but on six distinct molecules, all of them small proteins or polypeptides.

Two of these polypeptides, called **gonadotropins,** are hormones that act on the gonads in both males and females. The names of the gonadotropins, follicle-stimulating hormone (FSH) and luteinizing hormone (LH), come from their major effects in females. In males the concentration of FSH and LH are relatively constant after puberty. In females, however, the concentrations of FSH and LH change in a monthly cycle.

How Do Hormones Control Sperm Production?

The testes secrete testosterone, the most potent of a set of androgens, or male hormones. Like the ovaries, however, the testes also secrete estrogen, about one-fifth as much as is secreted by an adult female. We do not know, however, if estrogen plays a central role in male reproduction.

Some researchers suspect that it does. Like testosterone, estrogen can stimulate sexual behavior in a variety of male mammals, including primates, even after individuals have been

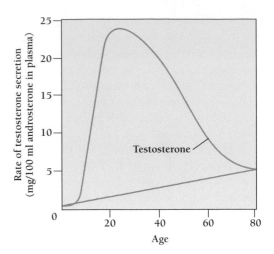

Figure 43-13   Increase and decrease in testosterone as a function of age. The concentration of testosterone in the blood of human males increases dramatically at puberty, then gradually drops beginning at about age 20.

castrated. In addition, the male brain seems to convert testosterone to estrogen, and estrogen levels in the brain rise and fall in synchrony with testosterone levels.

The role of testosterone in mediating sperm production and sex drive is clear. Once testosterone is released into the blood it circulates for only 15 to 30 minutes before it is bound by tissue receptors or degraded. As a result, testosterone levels are constantly renewed and closely regulated.

Hormones regulate the production of both sperm and testosterone by means of negative feedback (Figure 43-12). In the brain, the hypothalamus secretes **gonadotropin-releasing hormone, GnRH,** which stimulates the pituitary gland, also in the brain, to produce two hormones that males share with females—**luteinizing hormone (LH)** and **follicle-stimulating hormone (FSH).** LH stimulates the interstitial cells of the testes to produce testosterone and other androgens, while FSH stimulates the Sertoli cells to produce more sperm.

But testosterone and other androgens released by the interstitial cells into the circulation inhibit the production of both GnRH in the hypothalamus and LH in the pituitary. When LH production increases, androgen levels increase, shutting down the production of LH and, therefore, of androgens. Similarly, the Sertoli cells secrete the hormone *inhibin,* which enters the circulation and suppresses the secretion of FSH by the pituitary, shutting down sperm production.

Sperm production and testosterone production are thus intimately tied together. Not coincidentally, testosterone increases sex drive and aggressive behavior in general. When sperm are ejaculated, the hypothalamus releases LH and FSH, which together stimulate renewed production of sperm and testosterone.

Human males do not begin substantial testosterone production until they are about 10 years old. By 13, they are usually sexually mature and testosterone production increases rapidly until about age 20, when it begins a steady decline (Figure 43-13). By the late forties or fifties most men begin to experience a mild decrease in sexual function similar to that which women experience during menopause. This decrease is more gradual than in women, but, in some cases, it can include the hot flashes and sense of suffocation that can characterize menopause. Treatment with testosterone or estrogen can reduce such symptoms in men.

Hormones regulate the production of both sperm and testosterone by means of negative feedback. The hypothalamus secretes GnRH, which stimulates the pituitary gland to produce LH and FSH. LH stimulates the production of testosterone and other androgens, while FSH stimulates the Sertoli cells in the testes to produce sperm. But the Sertoli cells also secrete inhibin, which suppresses the secretion of FSH, shutting down sperm production.

What Are the Main Events of the Menstrual Cycle?

Beginning at puberty (between ages 10 and 17), human females undergo monthly reproductive cycles. Each cycle simultaneously produces a mature oocyte and prepares the wall of the uterus for **implantation,** the process in which a zygote burrows into the wall of the uterus. If a zygote does not implant, or implants but dies, the uterine wall is sloughed off in a process called *menstruation.* Each cycle is a highly coordinated series of events mediated by several hormones, many of the same ones that mediate sperm production in males. The cycle is divided into three phases, the menstrual phase, the preovulatory phase, and the postovulatory phase.

In 90 percent of healthy adult women, the entire menstrual cycle ranges from 21 to 38 days. The average is said to be about 28 days, sometimes called a *lunar month* because the phases of the moon also occur in a 28-day cycle. The menstrual cycles are *not,* however, coordinated with the phases of the moon.

Adolescents tend to have *irregular* cycles, which vary in length from one cycle to the next. As women mature, their cycles tend to become more regular. Women who live together tend to ovulate and menstruate at approximately the same time of the month. And women who live with men are likely to have more regular cycles than those who live alone. Menstruation and ovulation cease at menopause, which usually occurs in the late forties or early fifties.

Most variation in the length of the menstrual cycle occurs in the menstrual and preovulatory phases. The menstrual phase can range from 4 to 12 days, but it is usually 3 to 7 days. The period between ovulation and menstruation is nearly always 14 days, although 13 or 15 days are not impossible.

The menstrual cycle is divided into three phases, the menstrual phase, the preovulatory phase, and the postovulatory phase. The menstrual and preovulatory phases vary in length, but the postovulatory phase is nearly always 14 days.

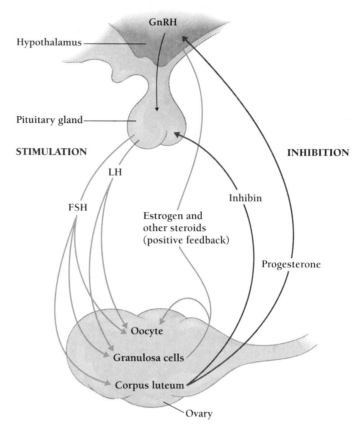

Figure 43-14 Hormonal regulation of oocyte production in mammals. Here are just a few of the complex feedback systems that regulate female egg production (and sexual behavior) in humans and other mammals. As in males, LH and FSH released by the pituitary stimulate (*green arrows*) oocyte production. Progesterone and inhibin inhibit (*red arrows*) the production of LH and FSH.

How Do Hormones Regulate the Ovarian Cycle?

Hormones regulate the **ovarian cycle**—the production of the oocytes and the regulation of the state of the endometrium (Figure 43-14). As in the male, the hypothalamus secretes gonadotropin-releasing hormone, GnRH, which stimulates the pituitary to produce LH (luteinizing hormone) and FSH (follicle-stimulating hormone), which together stimulate the ovary to develop mature oocytes and to produce estrogen and other steroid hormones.

During the menstrual phase, an increase in FSH (from the pituitary) stimulates 5 to 12 follicles in the ovaries to grow and mature (Figure 43-15). FSH causes growth and development of the oocyte and, together with LH, causes the follicle cells to release increasing amounts of estrogen, which promotes the growth of the follicle. This positive feedback cycle causes a build-up of estrogen in the blood in the days preceding ovulation. All but one of the follicles degenerates, leaving just one (usually) to develop a mature oocyte. The remaining follicle continues to secrete estrogen.

Figure 43-15 Hormonal regulation of the menstrual cycle in humans. Estrogen, LH, and FSH dominate ovulation, while progesterone and inhibin dominate the luteal (premenstrual) phase.

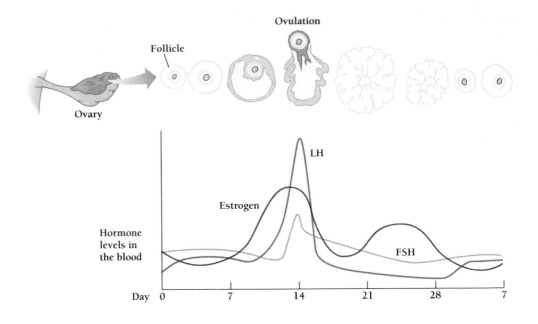

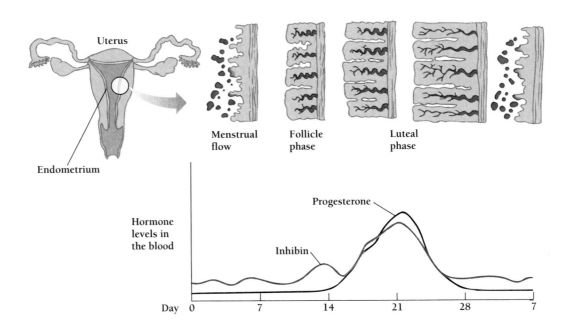

The increasing levels of estrogen trigger a second positive feedback cycle. Right around the time of ovulation, the high levels of estrogen stimulate the hypothalamus to release a surge of GnRH, which causes the release of a large surge of LH and FSH (Figures 43-14 and 43-15). The LH triggers ovulation. After the oocyte bursts from the follicle, much of the follicle is left behind in the ovary. There, stimulated by LH, it enlarges to form the **corpus luteum** [Latin, yellow body], a structure that secretes progesterone and a little estrogen. The progesterone stimulates the endometrium, the lining of the uterus, to thicken in preparation for implantation by the blastula. The progesterone also signals the hypothalamus to stop producing GnRH. As a result, the pituitary stops producing LH and FSH.

The corpus luteum also produces inhibin, which also inhibits the pituitary.

If the egg is not fertilized, the corpus luteum degenerates in 8 to 10 days. Inhibin and progesterone levels drop, and the pituitary begins producing FSH and LH once more. If the egg *is* fertilized, the zygote produces chorionic gonadotropin, a hormone similar to LH that maintains the corpus luteum. The presence of chorionic gonadotropin, or **HCG** (**human chorionic gonadotropin**), in blood or urine is the basis of most pregnancy tests. In that case, the corpus luteum expands and produces even more progesterone until about the fourth month of pregnancy.

Estrogen, then, participates both in negative feedback and—once each cycle—in positive feedback. In contrast, progesterone only promotes negative feedback. Together these two feedback systems bring about the regular cycles of ovaries and uterus.

In women, the hypothalamus secretes GnRH, which stimulates the secretion of LH and FSH, which together stimulate the ovaries to develop mature oocytes (and follicles) and to produce estrogen and other steroid hormones. The growth of the follicles causes a buildup of estrogen in the blood in the days preceding ovulation, which triggers the release of more GnRH, and, therefore, LH and FSH. LH triggers ovulation and development of the corpus luteum, whose progesterone and estrogen stimulate the thickening of the endometrium and the cessation of GnRH production.

How Do Hormones Regulate the Menstrual Cycle?

The lining of the uterus undergoes changes that follow the changes that occur in the ovaries. While the follicles are growing, the lining of the uterus, or endometrium, thickens to two or three times its thickness at the end of menstruation. After ovulation, the corpus luteum forms and its progesterone stimulates the endometrium, the lining of the uterus, to develop further. Blood vessels increase, and the endometrium secretes glycogen and other materials that facilitate implantation and nourish the zygote.

Estrogen and progesterone have other effects besides those on the endometrium, however. For example, estrogen increases the activity of cilia in the oviduct, while progesterone decreases such activity. Estrogen stimulates the cervix to secrete a clear, fluid mucus, which eases the passage of sperm into the uterus and oviduct. Progesterone, on the other hand, triggers the secretion of thick cervical mucus, which forms a mucus that prevents the entry of sperm or bacteria from the vagina into the uterus. When conception has occurred, the vaginal plug protects the embryo from infection.

If no fertilization has occurred, however, the richly prepared lining of the uterus disintegrates. The blood vessels weaken and break, and some 50 to 150 ml (about 3 to 10 tablespoons) of blood, mucus, vaginal secretions, and endometrial tissue washes out through the vagina. This bleeding, which may include bits of membranous tissue and blood clots, is called **menstruation** [Latin, *mens* = month]. Although light bleeding may occur after conception, several days of medium to heavy bleeding means that conception either has not occurred or it has occurred but the zygote has died, which happens in about 30 percent of conceptions.

After ovulation, the corpus luteum forms and its progesterone stimulates the endometrium, the lining of the uterus, to develop further. If no embryo implants, the endometrium breaks down and is washed away by a flow of blood, called menstruation.

HOW DO HUMANS CONTROL REPRODUCTION?

Humans, more than other animals, have sexual contact for reasons other than reproduction. Many students of human behavior have speculated on the evolutionary significance of sexual behavior that does not lead to new offspring. One hypothesis is that sex strengthens social bonding, increasing the chance that a couple will work together to nurture their young. In this way, sexuality may contribute to reproduction even when it does not lead to conception. Some biologists have speculated that this bonding is tied to a characteristic of our species unusual in other mammals: the human female is sexually receptive even at times when fertilization is impossible.

Humans do not necessarily want to bear as many children as possible, however, and the avoidance of conception is a major preoccupation for couples around the world (Figure 43-16). Although worldwide overpopulation is a real issue and of concern to millions of people, most individuals make decisions about how many children they want for personal, not ideological, reasons.

In certain primitive societies, mothers nurse their children until they are three years old. Because frequent nursing (at least eight times a day) inhibits ovulation, such mothers tend to bear children about four years apart. In modern societies, however, most women wean their babies much earlier. In the United States, for example, pediatricians regularly urge mothers to start giving babies solid food at six months, which often precipitates weaning. Many women nurse for only a few weeks, feeding their babies cow's milk, artificial milk (formula), and even solid food instead. In such cases, the mother begins to ovulate soon after giving birth and children may be spaced as little as 10 or 11 months apart. Such spacing quadruples the number of children that a woman can bear in her lifetime.

In areas of the world where infant mortality is high, families often have many children in the hopes that a few will survive to adulthood. Farm families often have several children, because the children can work and increase the productivity of the farm. But tradition also plays a role. Most societies are now largely urban, but city dwellers are the children or grandchildren of farmers, and the tradition of having many children continues.

Governments and religions also increase birthrates either by explicitly forbidding the use of birth control or by means of incentives (including tax deductions) that reward families with many children. In the past, societies ensured a steady supply of cheap labor and soldiers by discouraging the education of women and by banning information about contraception and other forms of birth control.

China, which has the largest population in the world, has dramatically lowered its birthrate over a period of decades by means of laws that punish people who have too many children. But such "top-down" policymaking doesn't always work. Italy has the lowest birthrate in the world, despite the fact that nearly all Italians are Catholic and the Catholic Church forbids both

World Birth-Control Use

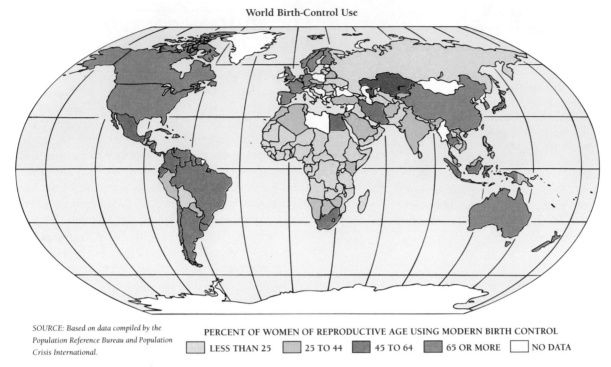

SOURCE: Based on data compiled by the Population Reference Bureau and Population Crisis International.

PERCENT OF WOMEN OF REPRODUCTIVE AGE USING MODERN BIRTH CONTROL

☐ LESS THAN 25   ☐ 25 TO 44   ▨ 45 TO 64   ■ 65 OR MORE   ☐ NO DATA

Figure 43-16 Birth control around the world. Different countries rely more or less heavily on birth control, depending on laws and education. *(Rodger Doyle © 1996)*

contraceptives and abortion. Some modern governments actively promote birth control.

In addition, in many cases, sexual activity precedes the beginning of reproductive life. Many societies, for example, expect couples to defer reproduction until long after sexual maturity, so that more young couples are sexually active than are reproductively active. **Birth control,** the conscious regulation of reproduction, is an important issue for many sexually mature humans. People have used four general approaches to birth control: abstinence, contraception, abortion, and sterilization. Each of these approaches is widely used and each has been the subject of intense social debate.

Abstinence

The most certain contraceptive method is to abstain from sexual intercourse altogether. **Abstinence** is common among adolescents. In the United States, about half of all teenagers younger than 19 abstain. Many adults in their twenties also continue to abstain, frequently until marriage. In addition many older adults abstain for a variety of reasons. Abstinence is favored, for example, by people who believe that the only reason for sexual intercourse is to reproduce (Table 43-1).

The **rhythm method** is a form of modified abstinence, in which a couple refrains from sexual intercourse at times in the monthly cycle when conception is likely to occur. The Catholic Church, which opposes most forms of birth control, allows the use of the rhythm method.

As we have seen, the mature oocyte is only fertilizable for 12 hours after ovulation, while viable sperm can wait around in the female reproductive tract for three or four days for the egg to burst from the ovary. If a couple abstains from four days before ovulation to one day after ovulation, no conception should occur. But because the preovulatory phase can vary enormously, the exact date of ovulation is hard to predict, and in many women, its timing may vary from month to month. In addition, some evidence suggests that women's interest in intercourse increases in the few days prior to ovulation. It is not surprising then, that the rhythm method does not provide reliable birth control. Of every 100 women practicing the rhythm method, as many as 20 become pregnant in one year (Table 43-1). For comparison, among sexually active women who practice no birth control, 85 out of 100 become pregnant in a year.

Several factors can increase the success of the rhythm method. If a woman has regular and predictable cycles, she has a better chance of correctly estimating the time of ovulation. If she knows when she ovulates each month, she can estimate when it will occur the following month. One clue is a one-half to one-degree increase in body temperature on the day of ovulation. In the days before ovulation, the quality of the mucus in the vagina becomes clear and also thinner and stringier than usual. In addition, the rupture of the follicle at ovulation sometimes causes pain, usually minor. Pain during ovulation is called **mittelschmerz** [German, *mittel* = middle + *schmerz* = pain, pain in the middle of the month]. Finally, a thorough knowledge of

Table 43-1   Contraceptive Methods and Their Effectiveness

Method	Chance of Pregnancy*	Risks	Protection from STDs?	Resulting Increased Deaths
Abstinence	0	None likely	Yes	0
Vasectomy	<1	Possible increase in prostate cancer, heart disease	No	1 in 300,000
Tubal ligation	<1	Infection, increase in heavy bleeding	No	1 in 67,000
Condom	3	None known	Latex, good	0
Birth control pill	<1	Heart disease, stroke especially in smokers	No	1 in 63,000† 1 in 16,000‡
Depro-Provera	<1	Possibly increase in osteoporosis	No	—
Norplant	<1	Infection at implantation site, abnormal vaginal bleeding	No	—
Morning-after pill	5 to 10	Nausea, cramps	No	—
Intrauterine device (IUD)	<2	Pelvic inflammatory disease	No	1 in 100,000
Diaphragm with spermicide	6	None known	Some	0
Cervical cap	9	None known	Some	0
Contraceptive sponge	9	None known	Some	0
Female condom	5	None known	Yes, good	0
Spermicide alone	6	None known	Some	0
Rhythm method (partial abstinence)	2	None likely	None	0
No method	85			0
Legal abortion in first 9 weeks				1 in 500,000
Pregnancy and childbirth				1 in 14,300

*When method is used correctly; expressed as percent per year.
†In nonsmokers.
‡In smokers.

female reproductive cycles can help. Some women use a modified rhythm method, using contraception during the week before ovulation.

To some people abstinence and periodic abstinence are the only acceptable methods of birth control.

Withdrawal

Another technique is **coitus interruptus** [Latin, *coitus* = sexual intercourse], or withdrawal, in which a man attempts to avoid conception by removing his penis from the vagina before ejaculation. Although this method is widely practiced in some parts of the world, it is one of the least reliable forms of birth control. One reason is that the secretions of the bulbourethral gland sweep small numbers of sperm into the vagina before ejaculation. In addition, withdrawal before orgasm requires the man to exert considerable resistance to the stereotyped pattern of the sexual response. In other words, he does not always remember to withdraw.

Barrier Methods Prevent the Union of Sperm and Ovum

Couples may use one or more forms of contraception to prevent conception while still having sexual intercourse. The eas-

iest of these methods to understand are those that physically prevent the union of sperm and ovum by imposing a **barrier** between them (Table 43-1).

Several contraceptive methods prevent the union of mature sperm and ova. A **condom** is a thin rubber sheath that covers the penis and prevents sperm from entering the vagina. By preventing actual physical contact between the male and female genital tracts, condoms prevent not only pregnancy but also sexually transmitted diseases (STDs), such as acquired immunodeficiency syndrome (AIDS), chlamydia, syphilis, and gonorrhea.

The major problems with condoms is that they break and sometimes fall off. In addition, if the condom is put on late, sperm may leak into the vagina before ejaculation, as in the withdrawal method. A condom must be put on as soon as the penis is erect, and it somewhat reduces the perceived stimulation of the penis and the vagina. It must be removed *immediately* after ejaculation, so that the sperm does not leak out from around the flaccid penis. Some couples prefer other methods of contraception, which, they feel, do not disrupt their sexual encounters.

A second method for preventing sperm from reaching the ovum is a **cervical cap,** a thimble-shaped rubber or plastic cap, about an inch in diameter, that fits tightly over the cervix. A **diaphragm** is larger than a cervical cap but works in the same way, serving as a barrier to the entrance of sperm. A **spermicide** [Latin, *cidere* = to kill], a cream or jelly that kills sperm, is an essential part of the effectiveness of both cervical caps and diaphragms. A woman may insert a diaphragm up to several hours before intercourse. A cervical cap may stay in place for several days. However, additional doses of spermicide must be inserted before each instance of intercourse.

When properly used, condoms, cervical caps, and diaphragms are effective contraceptives, with only two pregnancies per year per hundred women. With all three methods, however, improper usage and breakage lead to actual pregnancy rates of 10 to 13 per year per hundred women (Table 43-1).

The contraceptive sponge also provides a barrier to the entry of sperm into the uterus. Like the cervical cap and the diaphragm, the sponge fits over the cervix, but, unlike them, it requires no special fitting. Instead, the woman dips the sponge into water and inserts it over the opening of the cervix. The sponge, which may be left in place for as long as a day, not only blocks sperm entry but also releases a spermicide. Sponges by themselves give pregnancy rates of about 17 per hundred women per year and are therefore most effectively used in conjunction with another method, such as condoms.

Some couples attempt to prevent fertilization with spermicides alone. These include jellies, foams, creams, and suppositories. Each of these methods, by themselves, gives pregnancy rates of about 15 per hundred per year. Finally, some women attempt to prevent pregnancy with a **douche** [French, wash], the washing of the vagina immediately after intercourse. This method is ineffective, with a pregnancy rate of about 40 per hundred per year.

Barrier methods such as condoms and diaphragms prevent the sperm from reaching the oocyte.

Hormonal Contraceptives and IUDs

Researchers are now trying to develop chemical means of preventing sperm maturation. So far, however, no such "male pill" is yet available. In contrast, a great variety of *systemic* contraceptives, those which go into the bloodstream, exist for women.

Chemical birth control can prevent ovulation, fertilization, or implantation. One version of the female **birth control pill** consists of a combination of estrogen and progesterone. Together these prevent the anterior pituitary from secreting LH and thus stop ovulation. In one formulation, a woman takes the pills for 21 days and then stops for 7 days. The fall in steroids triggers menstrual flow, so that the result is a regular cycle without ovulation. Various versions of the *combination* birth control pill use different combinations of estrogen and progesterone (or similar compounds).

If used correctly, these formulations are almost entirely effective in preventing pregnancy, with pregnancy rates of less than 1 percent per year. The relatively high levels of hormone used, however, sometimes cause side effects, especially in the circulation, including high blood pressure and an increased risk of strokes and heart attacks. These risks are higher among smokers than among nonsmokers. The chances of fatal complications, however, are far lower than the mortality associated with pregnancy itself. For many couples the pill is the preferred method of birth control.

Another group of birth control pills contain only progesterone, with no estrogens. The pills (sometimes called "minipills") have few of the side effects of the combination pills, but they have a slightly higher pregnancy rate. They appear to work by inhibiting the movements of both ovum and sperm, as well as by interfering with implantation.

Another chemical method that interferes with ovulation involves the release of a substance from a capsule surgically placed under the skin. A single such implanted capsule, called Norplant, may last for up to 10 years.

Another chemical contraceptive is a progesterone derivative called DMPA [depo-medroxy progesterone acetate], or "**Depo-Provera.**" DMPA is taken as an injection, which appears both to suppress ovulation and to inhibit implantation for three months at a time. The risks associated with DMPA, however, are still the subject of debate. Some people argue that it is safer than the available birth control pills, while others point to serious problems, including depression and abnormal menstruation.

The **morning-after pill** is an oral contraceptive taken in higher than normal doses within 72 hours of unprotected intercourse. Like standard doses of birth control pills, morning-after pills effectively interfere with both conception and implantation. Side effects, such as nausea and cramps, are unpleasant but transient.

Two general methods are currently available for preventing implantation of the embryo in the uterine lining. The first of these, the **intrauterine device,** or **IUD,** is a small piece of plastic that is placed into the uterus, where it interferes with implantation. The idea of such a method of birth control appears to have come from an ancient practice of camel drivers, who would put pebbles into the uterus of a female camel before a long trip. The pebbles kept the camel from becoming pregnant during the trip. In a similar way, an IUD in the uterus of a human female appears to prevent pregnancy, although researchers do not know exactly how.

More than one hundred types of IUDs exist, some of which consist only of plastic, while others also contain copper or progesterone. One type of IUD that was sold in the United States between 1971 and 1974, the Dalkon Shield, produced serious side effects, including inflammations of the pelvis and spontaneous abortions, in a few cases resulting in death. Other types of approved IUDs are safer, although some women have infections, excessive bleeding, cramps, or other side effects. IUDs must be fitted and inserted (and eventually removed) by an experienced physician or other health professional.

Hormonal contraceptives suppress ovulation and inhibit implantation. The IUD also inhibits implantation.

Sterilization Provides Effective But Generally Irreversible Birth Control

Quick and relatively inexpensive surgical methods of male and female sterilization are now widely available. Although sterilization can sometimes be reversed, it is usually a permanent change, so individuals choosing it as a method of birth control must carefully consider such a choice.

Vasectomy, the cutting and tying of the vas deferens, prevents sperm from entering the urethra (Figure 43-17). Vasectomies are safe, do not require hospitalization, and are performed in less than an hour with local anesthesia. The procedure is usually permanent, however, and is suitable for men who are certain that they do not want to reproduce in the future.

Tubal ligation cuts or blocks the oviducts, or fallopian tubes, thus preventing eggs from reaching the uterus after ovulation (Figure 43-17). It is a more complicated operation than vasectomy. Because the oviducts lie within the abdomen, a tubal ligation requires one or more surgical incisions in the abdomen and a hospital-like setting. A surgeon can, however, perform a tubal ligation with only two small incisions in the woman's navel. The surgeon then inserts a laparoscope, a thin tube containing a viewing instrument, in one incision and the surgical instruments in the other incision. While modern techniques make this operation short and relatively uncomplicated, full recovery takes several weeks. Although reversals are occasionally possible, a tubal ligation should be considered permanent.

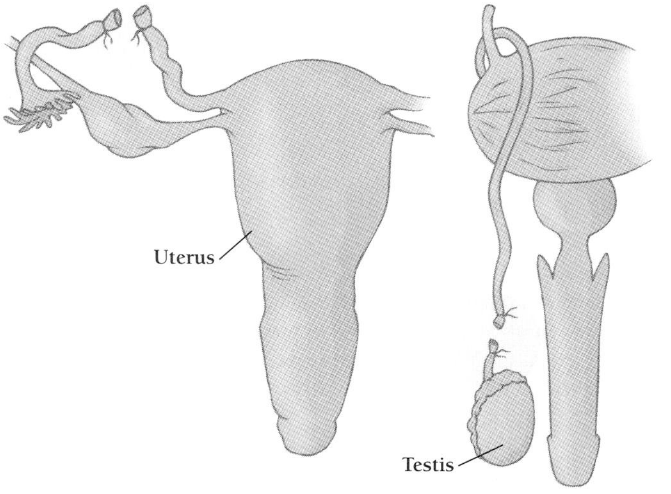

TUBAL LIGATION
Fallopian tubes are cut, tied, and cauterized

VASECTOMY
Vasa deferentia are cut and tied

Uterus

Testis

Figure 43-17 **Tubal ligation and vasectomy.** In tubal ligation, the fallopian tubes are either cut and tied, cauterized (burned), closed off with very tight rubber bands, which kill the tissue, or clamped with a tight clip. A tubal ligation requires abdominal surgery. Cutting and tying the vas deferens prevents sperm from reaching the penis. It is a much safer and simpler procedure than tubal ligation. Neither procedure should be considered reversible.

Women should elect this option only if they are certain that they will never again want to conceive.

Sterilization—vasectomy and tubal ligation—is highly effective and irreversible.

Induced Abortions Terminate Pregnancies After Implantation

Abortion ends pregnancy after the zygote has formed. **Spontaneous abortions,** or miscarriages, are extremely common: at least 30 percent of all conceptions end prematurely, usually in the first week or two. A large fraction of such spontaneously aborted embryos have chromosomal abnormalities.

An **induced abortion** is the deliberate removal of an embryo from the uterus. The ease of the procedure decreases with the age of the embryo. In the first trimester (the first three months), the procedure is brief and usually takes place in a clinic or doctor's office. The embryo is usually removed from the uterus by gentle suction between 3 and 10 weeks after conception. Early abortions, between 3 and 5 weeks after fertilization, are called **menstrual extractions.** Abortions between 5 and 10 weeks after fertilization are done by dilation and extraction (**D&E**). In some countries the drug **RU 486** has been used to induce abortions within the first nine weeks after conception. RU 486 has been used in France and China since the mid-1980s. A committee of the U.S. Food and Drug Administration (FDA) has recommended that RU 486 also be approved for use in the United States.

Induced abortions in the second trimester (the second three months) require different methods, which are performed only in a hospital. The most common method is to induce uterine contractions by injecting a solution of salt or of prostaglandin, a chemical that causes contractions, including those of the uterus. (RU 486 is also given in conjunction with prostaglandins.) Induced abortions in the third trimester are quite rare.

An induced abortion is the removal and destruction of the embryo, or fetus. A spontaneous abortion is the natural death of an embryo, often one with chromosomal defects.

Advances in Biology and Medicine Provide New Alternatives for Infertile Couples

About one couple in six in the United States is infertile, meaning that they do not conceive after a year of sexual relations without contraception. Many causes may underlie infertility, but about half of infertile couples are able to achieve a pregnancy. About a third of the time, infertility results from the man's failure to produce and deliver viable sperm; about a third of the time, the problem lies in the woman's inability to produce a fertilizable ovum or maintain a pregnancy once it has begun; and about a third of the time, combined factors are responsible.

A man's inability to produce mature sperm may result from infections (such as mumps), poor nutrition, or chemicals (including alcohol and other drugs) or radiation in the environment. Problems of sperm delivery may be due to blockage or scarring in the genital tract, which may result from sexually transmitted diseases, or from sexual problems such as impotence (inability to maintain an erection during intercourse) or premature ejaculation.

Infertility in a woman may result from infections of the genital tract (including sexually transmitted diseases), mechanical barriers (such as scarring) in the genital tract, scarring of the uterine wall or of the fallopian tubes, or from hormonal problems that interfere with ovulation or pregnancy.

Many infertile couples elect to adopt children. Others take advantage of an enormous industry of fertility specialists who sell sperm and eggs to infertile couples. If the woman is normal but the man cannot produce sperm, the woman can have a baby if someone places sperm from a donor, known or anonymous, in her vagina at the right time of the month. In rare cases, women who want to have a child without a male partner opt for the same procedure. In other cases of infertility, a man may produce normal sperm but not be able to deliver it in sexual intercourse. In that case, the sperm can be removed from the testes.

Women with normal ovaries and a normal uterus, but blocked oviducts (fallopian tubes) are candidates for a technique called *in vitro fertilization* (IVF), the fertilization of an ovum in the laboratory, which is followed by the implantation of the embryo back into the uterus.

A physician can use the woman's own ova (which must be removed surgically at the right time of month) or the ova of an anonymous egg donor. To accomplish IVF, a physician treats a woman (either the prospective mother or the donor) with gonadotropins to increase the number of mature follicles. Then, using ultrasound imaging to tell whether a follicle is mature, a surgeon removes the ova from the ovaries. The mature ova are placed in a dish and exposed to sperm.

After fertilization and some subsequent development, the physician transfers the embryos back into the uterus, where they can then develop normally. Many fertility clinics implant several embryos because most or all are likely to die after transfer to the uterus. Occasionally several will survive. And the previously infertile couple goes home with triplets or even quintuplets. In such cases, premature birth is extremely likely. Prematurity impairs the health of the baby and hospital costs are frequently astronomical. The cost of each *in vitro* baby averages $60,000 to $100,000. A single premature baby may cost more than a half million dollars.

In cases in which the woman is infertile, a couple may use the man's sperm to fertilize the ovum of a surrogate mother, a woman who will conceive and also carry the resulting embryo to term. Surrogate motherhood is a controversial procedure that has been the subject of much public debate because surrogate mothers sometimes ask for their children back. As a result, fertility clinics are increasingly using egg donors, healthy young women willing to donate eggs for $3000 to $5000.

Besides adoption, modern biotechnology offers a wealth of choices to infertile couples, from *in vitro* fertilization to surrogacy.

STUDY OUTLINE WITH KEY TERMS

Mammals form haploid gametes (**ova** and **spermatazoa**) for sexual reproduction from germ plasm in the **gonads,** which consist of the **testes** (**testicles**) in males and the **ovaries** in females. During **oogenesis,** the ovaries form diploid **oogonia,** which grow into **primary oocytes.** Each month several primary oocytes develop, with surrounding cells in the ovaries, into **follicles,** which swell with fluid. Just before **ovulation,** a primary oocyte completes meiosis I and forms a **secondary oocyte** and a **polar body,** which enters meiosis II but does not complete it.

In the male's **scrotum** are the testes, within which are the **seminiferous tubules,** which form the sperm. The seminiferous tubules are surrounded by **Sertoli cells,** which nourish the developing sperm, and **interstitial cells,** which secrete testosterone. On the inside walls of the seminiferous tubules, diploid **spermatogonia** divide through meiosis to form four haploid **spermatids,** which mature into sperm, with a head, a **midpiece,** and a tail-like **flagellum.** The front of the head contains an enzyme-packed lysozome called the **acrosome.**

The fusion of the egg and sperm (ovum and spermatozoan) is called fertilization, **syngamy,** or **conception.** The sperm moves into the **oviducts,** or **fallopian tubes,** where it encounters the **secondary oocyte** and fuses with it. The secondary oocyte and the polar body divide once more, forming a mature ovum and two or three polar bodies. The nucleus of the sperm and egg fuse, and the resulting diploid **zygote** moves down one of the two oviducts, or fallopian tubes, to the **uterus.**

The developing embryo then implants in the **endometrium** of the uterus. At about 9 weeks, all the major organs have formed, at least in general outline, and the embryo is called a fetus. When the fetus has matured at 9 months, it is said to be **full-term,** and is expelled by muscular contractions of the uterus.

Sperm leave the testes by way of the **epididymis,** the **vas deferens,** and the **urethra** of the penis. (In many birds and in some reptiles, the sperm leave by way of a **cloaca.**) Along the way, fluids from the **seminal vesicles,** the **prostate gland,** and the **bulbourethral glands** contribute to the formation of the **semen.**

The penis consists of the **corpus spongiosum,** paired **corpus cavernosa,** and the **glans,** all well supplied with blood vessels and nerves. A loose layer of skin, called the **foreskin,** covers the glans in males who have not been **circumcised.**

The female external genitalia consist of the vulva—including the labia majora and labia minora—and the clitoris. Like the penis, the clitoris includes a glans, paired corpus cavernosa, and a **bulb of the**

vestibule, like the corpus spongiosum. Inside the entrance to the vagina is a membrane called the **hymen** that is usually damaged during first intercourse.

In humans, sexual response consists of four stages—**excitement, plateau, orgasm,** and **resolution.** During the excitement stage, increased blood flow to clitoris and penis result in **erection** and tightening of the vaginal walls. In addition, the **Bartholin's glands,** in the female, and the bulborurethral glands, in the male, contribute lubrication. Enzymes in the female reproductive tract break down a coating on the sperm, a process called **capacitation.**

Gonadotropin-releasing hormone (GnRH) secreted by the **hypothalamus** regulates the secretion of **luteinizing hormone (LH)** and **follicle-stimulating hormone (FSH)** by the **anterior pituitary gland.** In males, hormones regulate the production of sperm and testosterone by means of negative feedback. LH stimulates the production of testosterone and other androgens, while FSH stimulates the Sertoli cells in the testes to produce sperm. But the Sertoli cells also secrete inhibin, which suppresses the secretion of FSH, shutting down sperm production.

In women, LH and FSH regulate the menstrual cycle and the **ovarian cycle.** As in men, the hypothalamus secretes GnRH, which stimulates the secretion of LH and FSH. These stimulate the ovaries to develop mature oocytes (and follicles) and to produce estrogen and other steroid hormones. The growth of the follicles causes a build-up of estrogen in the blood in the days preceding ovulation, which triggers the release of more GnRH, and therefore LH and FSH. LH triggers ovulation and development of the **corpus luteum.** Progesterone and estrogen from the corpus luteum stimulate the cessation of GnRH production and the development of the endometrium, the lining of the uterus. If no embryo implants, the endometrium breaks down and is washed away by a flow of blood, called **menstruation.** If fertilization occurs, the resulting zygote secretes **human chorionic gonadotropin (HCG),** which, like LH, maintains the corpus luteum.

Birth control measures include **sterilization** (**vasectomy** and **tubal ligation**), **abstinence,** the **rhythm method, coitus interruptus, intrauterine devices (IUDs),** and a variety of **barrier** methods. The rhythm method requires that the woman be alert to signals that indicate the pace of the menstrual cycle and the day of ovulation. These include the consistency of vaginal mucus, body temperature, and **mittelschmerz.** Barrier methods include two kinds of **condoms,** one to be worn by the man and one to be worn by the woman, the **cervical cap,** and the **diaphragm.** The efficacy of the cervical cap and the diaphragm are much improved by the use of **spermicides.** (A **douche** is so ineffective that it cannot be considered a method of birth control at all.)

Systemic contraceptives are synthetic hormones that prevent ovulation, fertilization, or implantation. These include the **birth control pill, Depo-Provera,** and the **morning-after pill.** All of these chemical contraceptives are for women only.

When contraceptives fail, some women arrange to have a doctor perform an **induced abortion,** the intentional destruction and removal of the embryo or fetus. An induced abortion is distinct from a **spontaneous abortion,** the death and expulsion of an embryo due to a chromosomal abnormality or other damage. Abortions may be induced chemically, using RU 486, for example, or physically. The most common two techniques are **menstrual extraction** and dilation and extraction (**D&E**). Both methods are performed in the first three months after fertilization.

REVIEW AND THOUGHT QUESTIONS

Review Questions

1. Compare oogenesis with spermatogenesis. What is the main difference between these two processes?
2. What is the difference between a primary oocyte and a secondary oocyte? Which fuses with the sperm and what happens then?
3. Draw a diagram with labels showing a cross section of a seminiferous tubule that illustrates how the sperm form and develop.
4. Describe the acrosome, where it is, and what its function is.
5. Can nonmotile sperm reach the oocyte? Why or why not?
6. What glands contribute fluids that make up the semen?
7. What gland secretes mucus during sexual intercourse?
8. What are the four stages of the human sexual response?

9. How do GnRH, LH, and FSH regulate the formation of mature oocytes and sperm? Draw a diagram from memory showing where these hormones are made and how they interact with the gonads. What do the initials stand for?

Thought Question

10. In mammals, the testes are external to the body cavity and sperm cannot form effectively inside the warmth of the body cavity. Why do you think this might be so? Did mammals evolve external testes and then evolve sperm that were temperature sensitive or did the temperature sensitivity evolve first? Why would either happen? Why are we different from birds and other vertebrates, which make healthy sperm inside of their bodies?

SELECTED READINGS

Carlson, Karen J., Stephanie A. Eisenstat, and Terra Ziporyn, *The Harvard Guide to Women's Health,* Harvard University Press, Cambridge, 1996. The encyclopedia format contains long entries on reproductive physiology, as well as on health in general. Many of the entries—on birth control, vasectomy, heart disease, and diabetes, for example—would be of interest to both men and women.

Guyton, Arthur C., *The Textbook of Medical Physiology,* 8th ed., W.B. Saunders Company, Philadelphia, 1990. This book is a surprise. It is an authoritative and encyclopedic reference book on human physiology, yet the writing is clear, concise, and even amusing.

▶ On-line materials relating to this chapter are on the World Wide Web at http://www.saunderscollege.com/lifesci/
Click on Tobin/Dusheck: *Asking About Life.*

How Do We Become Complex?

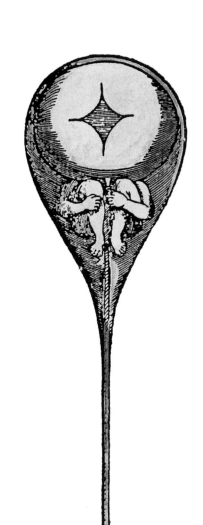

Most people know that reproduction requires two parents. But what exactly does each parent contribute? Until the 19th century, biologists were not sure. The Greek philosopher Aristotle argued that the female ovum provides formless substance, to which the male sperm gives shape, molding the egg into an individual as a potter shapes clay into a bowl. How sperm might accomplish such crafting, Aristotle did not say.

The question was—and remains—how does the complexity of a multicelled organism arise from the simplicity of a single egg? Eighteenth-century biologists answered this question by denying that simplicity ever existed. In this view, all the parts of the adult organism already exist at the earliest stages of life. The adult is *preformed* in the sperm or the egg.

The French biologist Charles Bonnet formalized this theory, called **preformation,** in 1745. He argued that each egg contains a complete embryo, and each embryo contains more complex embryos inside itself, and so on, like a set of Russian dolls. Many of Bonnet's contemporaries argued, however, that the complete embryo was in the sperm, not in the egg. The Dutch microscopist Antoine van Leewenhoek had already demonstrated the existence of sperm almost a hundred years before, and some microscopists even claimed to have seen a tiny creature, called a **homunculus,** curled up inside the sperm head, shown at left. Preformation implied that the whole organism resulted from the growth of a preformed miniature. At that time, no scientist had seen a mammalian egg or witnessed the coming together of an egg and sperm in either plants or animals.

The doctrine of preformation not only avoided the difficult problem of explaining development, it also provided a simple view of genetics: a child resembled its parents because it was already preformed in the ovum of the mother or the sperm of the father. The doctrine of preformation also presented a problem. If a child was already preformed—whether in egg or sperm—how would both parents contribute to the genetic makeup of the child?

Preformation left little room for evolution. Because all future generations must be contained within past generations, only a limited number of generations could be contained. Eventually, the tiniest embryos inside embryos would be too tiny to exist. Some biologists even tried to calculate how many embryos within embryos must have existed in Eve's ovaries.

Furthermore, preformation only worked as a theory as long as people believed that the Earth was only 4000 years old. As geologists discovered that life was millions of years old and chemists discovered that matter was not infinitely divisible, the underpinnings for the theory of preformation crumbled.

It was an Estonian embryologist, however, who provided the evidence that finally disproved preformation. In the 1820s, Karl Ernst von Baer described the gradual development of a mammalian zygote. He saw that a single-celled zygote divided and formed three layers of tissue, from which the embryonic organs and tissues then developed. It was a gradual process, like the one that Aristotle had suggested 2000 years before. The discovery and description of fertilization—the fusion of egg and sperm—in both plants and animals further undermined the theory of preformation.

By the middle of the 19th century, it was clear that Bonnet's theory of preformation was incorrect. Yet von Baer's work raised a new set of questions: if a single cell could divide and each descendant cell could specialize into different kinds of cells, how was this specialization accomplished? Did some cells get one set of genes while other cells got another set of genes? Or did every cell get a complete set of all the genes? In short, was each cell **totipotent**, capable of developing into any kind of cell? Or was each cell **differentiated** early on?

Nineteenth-century experimentalists were working at a disadvantage since they did not yet know that the genes were on the chromosomes or how genes were encoded in the DNA of the chromosomes. They only assumed that cells had genetic particles of some sort. In order to answer the question raised by von Baer's work, the German biologist Wilhelm Roux conceived an experiment that would tell whether or not embryonic cells were totipotent.

Roux reasoned that as the cells divided, each one got only some of the genes present in the zygote. If that were true, and if only half of the embryo were allowed to develop, the result would be an embryo with half its structures missing. In 1888, Roux took a two-celled frog embryo and destroyed one of the two cells by piercing it with a hot needle (Figure 44-1A). The remaining cell formed only half

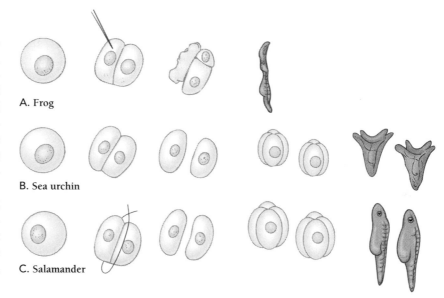

A. Frog

B. Sea urchin

C. Salamander

Figure 44-1 **Totipotency and preformation.** Is a single cell of an embryo as capable of forming a whole tadpole as all the cells in the embryo? A. Wilhelm Roux killed one cell of a two-cell frog embryo and got half a tadpole. B. Hans Driesch shook apart the two cells of a two-cell sea urchin embryo and got two complete sea urchins—the first known artificial clones of an animal. C. Hans Spemann repeated Roux's experiment. Instead of killing one cell, however, he separated the two cells of the salamander embryo with a fine hair. Both cells developed into normal salamander larvae.

an embryo. Roux concluded that embryonic cells are not totipotent. Instead, he argued that at each cell division the daughter cells receive half of the complexity in the dividing cell. As the zygote divides into more and more cells, these cells receive smaller and smaller shares of the genes in the zygote's original nucleus.

Within four years, however, Roux was challenged by another German biologist, named Hans Driesch. Driesch performed a similar experiment, but with dramatically different results (Figure 44-1B). Driesch used sea urchin embryos, and, instead of killing one cell and leaving it in place, he shook the two-celled sea urchin embryos until the pairs of cells fell apart.

As he watched, each of the two cells grew into a complete sea urchin. Driesch's result disproved the idea that each of the cells of an embryo had different information. An embryo, apparently, was not a

---

If each cell of an embryo could develop into a separate adult, then neither the egg nor the sperm could be said to exclusively harbor the individual.

---

mosaic of information. Instead, each cell may be totipotent, capable of forming all the structures of the adult.

Driesch's demonstration of totipotency in embryonic cells conclusively discredited the theory of preformation. For if each cell of an embryo could develop into a separate adult, then neither the egg nor the sperm could be said to exclusively harbor the individual.

But why had Roux's experiment suggested preformation? The German embryologist Hans Spemann wondered if Roux's embryos developed abnormally because the dead cell interfered with the development of the remaining embryo. Spemann repeated Roux's experiment, but instead of killing one cell, he used a fine hair (from a baby) to gently separate the cells of a two-cell salamander embryo (Figure 44-1C). Like Driesch, Spemann saw each cell develop into a whole, normal embryo. His result confirmed Driesch's view that each cell of a two-cell embryo can give rise to a whole adult.

The experiments of Driesch and of Spemann permanently laid to rest the theory of preformation and suggested that each cell inherits all of the genetic information contained in the zygote, not just part of it. But more questions remained. Where does complexity come from? How do cells become different from one another? How do organs and tissues form? How do the different parts of the embryo know how to arrange themselves?

Of all the wonders of life, none is so amazing as development. Each of us has developed from a single cell, barely visible to the naked eye. The central question of this chapter is how this development can occur. Much of our focus will be on the development of vertebrates, as they are familiar and illustrate the general problems of animal development. Along the way, however, we will also encounter some of the powerful insights revealed by research on flies and other invertebrates.

KEY CONCEPTS

1. During development, an animal changes from a single cell to a complex organized multicelled organism by means of cell division, cell movement, cell specialization, and pattern formation.

2. Sperm and egg cells are highly specialized to accomplish fertilization and to initiate development.

3. At the beginning of development, each cell contains all the genetic information necessary to produce an entire animal. As development proceeds, however, most cells become increasingly limited to a particular developmental pathway.

4. The pattern of gene expression of individual cells and their participation in the formation of organs and body parts depend on chemical signals, many of which act by regulating transcription of specific genes.

5. Virtually all animals appear to use the same mechanisms to achieve pattern formation and programmed cell death.

HOW DOES FERTILIZATION INITIATE EMBRYONIC DEVELOPMENT?

Fertilization, or syngamy, joins the haploid ovum and sperm to form a diploid zygote. Fertilization brings together genetic information from the maternal and paternal genomes and begins the life of a new individual.

Within a minute after a sea urchin sperm binds to the ovum, the ovum increases its rate of oxygen consumption. Within 10 minutes, the zygote begins intense metabolic activity, using energy at a much higher rate. Protein synthesis also increases dramatically, and the embryo begins to divide.

At one time, researchers guessed that the initial burst of protein synthesis in the sea urchin's zygote resulted from transcription (DNA to RNA) in the newly formed nucleus. But this hypothesis was wrong. Embryos that are treated with a drug that suppresses transcription display the same burst of protein synthesis as untreated embryos. But protein synthesis can only occur if mRNA is present. Where was the mRNA coming from if not from the nucleus of the zygote? Researchers were forced to conclude that the mRNA that directs early protein synthesis is mRNA that is already present in the egg cell. Such mRNA, called **maternal mRNA,** is produced in the egg cell—transcribed during oogenesis (egg formation) from the genes of the mother, not those of the zygote.

The early development of a zygote is entirely controlled by maternal proteins. In sea urchins, the zygote's nucleus is not only inactive during early development, it is unnecessary. Embryos lacking active nuclei still divide and develop into multicelled embryos.

Fertilization is both a genetic and a developmental event. Protein synthesis and development in early embryos is directed by maternal mRNA—mRNA transcribed from the genes of the mother.

HOW DO CELLS OF THE EMBRYO GIVE RISE TO CELLS OF THE ADULT?

Developmental biologists view the body plan of a vertebrate as "a tube within a tube" (Figure 44-2). The outermost tube, or **ectoderm** [Greek, *ecto* = outside + *derma* = skin], consists of the part of the animal that is in contact with the outside world—the epidermis, or outer skin layer, the nervous system, and the sense organs. The innermost tube, or **endoderm** [Greek, *endo* = inside + *derma* = skin], is the gastrointestinal tract, together with associated organs such as the pancreas and liver.

Between the ectoderm and the endoderm is the **mesoderm** [Greek, *mesos* = middle + *derma* = skin], which consists of connective tissues, such as bones, muscles, and tendons, and the cells of the blood. The mesoderm also includes a number of organs including the heart and kidneys.

All three layers of the adult vertebrate arise from a single-celled zygote. To see how this basic organization is accomplished we can study the early events of development and ask when the three-layer body plan first appears.

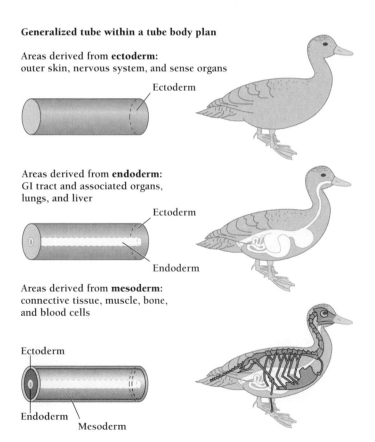

Generalized tube within a tube body plan

Areas derived from ectoderm:
outer skin, nervous system, and sense organs

Ectoderm

Areas derived from endoderm:
GI tract and associated organs, lungs, and liver

Ectoderm

Endoderm

Areas derived from mesoderm:
connective tissue, muscle, bone, and blood cells

Ectoderm

Endoderm

Mesoderm

Figure 44-2 **A tube within a tube.** An animal consists of a tube of endoderm (*yellow*) within a tube of ectoderm (*blue*). In between the two tubes the mesoderm (*red*) develops. In vertebrates such as this duck, the ectoderm forms the outer skin, nervous tissue, and sense organs; the mesoderm forms muscle, bone, blood cells, and connective tissue; and the endoderm forms the intestines, liver, and most other internal organs.

Biologists Separate Vertebrate Development into Eight Stages

Although the process of development is continuous, we can conveniently divide it into eight separate stages (Figure 44-3).

1. **Gamete formation**—the production of sperm and eggs, a process involving meiosis and cell specialization.
2. **Fertilization**—the fusion of the haploid egg and sperm to form a diploid zygote; fertilization initiates cleavage.
3. **Cleavage**—the division of the zygote into many smaller cells; in many vertebrates these cells form a hollow ball of cells called a **blastula;** in mammals the endpoint of cleavage is a **morula,** a solid ball of cells.
4. **Germ layer formation**—the movement of embryonic cells to form the three germ layers (ectoderm, mesoderm, and endoderm); the resulting structure is called a **gastrula** and the process is called **gastrulation.**
5. **Organ formation**—the movement and specialization of cells to form functioning organs such as the heart, kidneys, and nervous system.
6. **Growth**—the increase in size of an organism after the organs and body plan are established; growth involves both cell division and the production of extracellular materials (such as bone, cartilage, and hair) and converts the collection of beginning organs into an adult.
7. **Metamorphosis**—a series of changes in form in which the larval form (for example, caterpillar or tadpole) changes into an adult form (for example, butterfly and frog), which has a different morphology and lifestyle.
8. **Aging**—further development, which inevitably leads to death; once the adult form is established, cells die, extracellular tissues change, and the efficiency of organs decreases. Aging occurs at different rates in different species and in different individuals within a single species.

Why Does a Zygote Divide?

A zygote is much larger than the average body (somatic) cell, often containing large stores of yolk—a mixture of proteins, lipids, and carbohydrates that nourishes the embryo until it can feed itself. Such a large cell has two problems—too little surface area and too little nucleus for the amount of cytoplasm. Because of a zygote's small surface-to-volume ratio, its membrane can transport relatively fewer molecules—whether nutrients or wastes—than a smaller cell.

The zygote's nucleus presents a similar problem. The nucleus of a large cell and a small cell are the same size, containing the same amount of DNA and the same amounts of regulatory proteins. Yet the nucleus of a zygote has far more cytoplasm and membrane to regulate than a smaller cell. A single nucleus cannot meet the enormous demand for mRNA that an embryo has. In addition, the nucleus of a large cell cannot adequately control the distribution of mRNA (and, therefore, protein) in the cell.

Figure 44-3 Stages of verte-brate development. 1. Gamete formation. 2. Fertilization. 3. Cleavage. 4. Germ layer forma-tion. 5. Organ formation. 6. Growth and maturation. 7. Metamorphosis. 8. Aging.

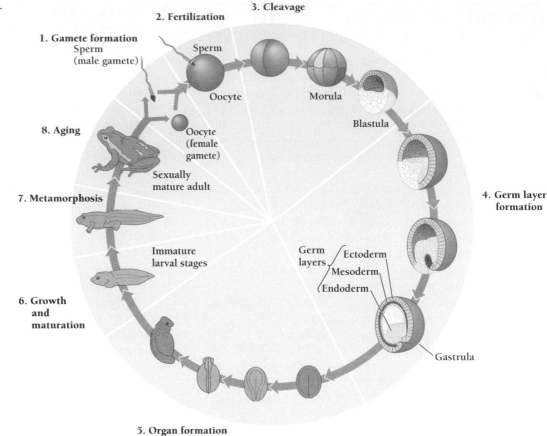

The division of the zygote into many smaller cells solves both of these problems (Figure 44-4). A zygote undergoes a series of rapid cell divisions, or cleavages. Because the cells do not grow between divisions, the resulting embryo is about the same size as the zygote. The subdivision of the zygote's cytoplasm into 100 or more cells increases both the surface area and the number of nuclei for the embryo's cytoplasm.

The specialization of cells in a multicellular organism requires that different cell types produce different kinds of mRNA.

It should not surprise us, then, that the first process undertaken by the embryo is rapid cell division. Cleavage not only increases surface area and potential mRNA production, it also allows each nucleus (and its cell) to specialize.

Cleavage converts the embryo from a single large cell to many smaller cells that can operate independently.

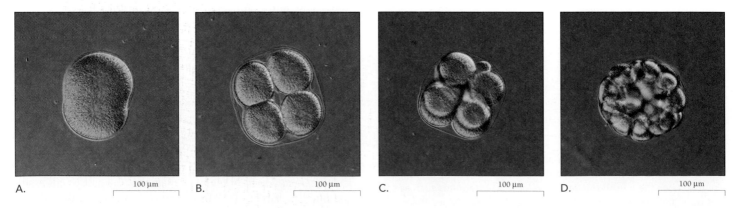

Figure 44-4 Cleavage in a sea urchin embryo. A. The beginning stages of the first cleavage. B. After two divisions, the embryo consists of four cells. C. 8–12 cell stage. D. 32 cell stage. *(David Fromson, California State University, Fullerton)*

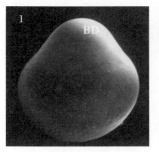
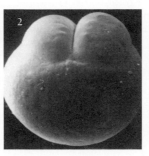
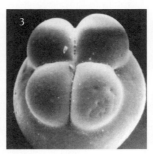
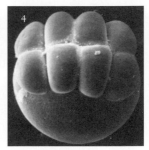
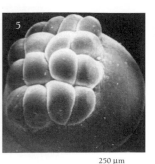

A. **Zebra fish**

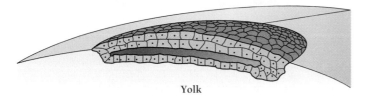

Yolk

B. **Bird**

Figure 44-5   Discoidal cleavage. A. In the yolky embryo of a zebra fish, cleavage is confined to a small disk on the outer surface of the yolk. B. The embryos of birds also divide by discoidal cleavage, which results in a flattened blastula.   *(A, from Beams and Kessel, 1976)*

How Do Zygotes Cleave?

Although the zygotes of all species of animals cleave, species differ dramatically in the pattern and pace of these divisions. The different cleavage patterns depend both on the orientation of the mitotic spindles in each division and on the distribution of yolk within the egg. Eggs that contain a lot of yolk either have relatively slow cleavage divisions or do not completely divide the yolk. Bird eggs and fish eggs, loaded with yolk, confine their cleavage divisions to a tiny disc of incompletely separated cells floating on the surface of the yolk (Figure 44-5).

Eggs with little yolk, such as those of the sea urchin, divide more or less equally and symmetrically. In the sea urchin, each division takes less than an hour. After seven such divisions the sea urchin embryo becomes a blastula, a hollow, fluid-filled ball of 128 cells (Figure 44-6A). The blastula consists of a single layer of cells, which continue to divide, eventually reaching 1000 to 2000 cells.

Cleavage in mammals occurs slowly, each division taking 12 to 24 hours. The orientation of the cleavage divisions is also unique, and, while the cells of a sea urchin embryo divide simultaneously, the individual cells of a mammalian embryo divide at different times.

After a mammalian embryo has divided to form about eight cells, the cells form a compact structure called a morula. The morula soon forms a **blastocyst** [Greek, *blastos* = germ + *cystos* = cavity], a modified blastula in which the cells enclose an internal cavity. The blastocyst contains two types of cells—the **trophoblast** [Greek, *trephein* = to nourish + *blastos* = germ], a prominent outer cell layer, and the **inner cell mass,** a small

group of cells that will eventually grow into the embryo itself (and subsequently into the adult). Cells derived from the trophoblast are responsible for the connections between the embryo and the mother—they will attach the embryo to the wall of the uterus and later become part of the **placenta,** a highly vascularized organ through which mother and fetus exchange nutrients and wastes.

During the cleavage stage of the amphibian embryo, cells near the top of the embryo, the **animal pole,** divide more rapidly than those near the bottom, the **vegetal pole.** The resulting blastula is several cell layers thick and more asymmetric than the sea urchin blastula. The cells near the vegetal pole are quite large and laden with yolk.

Embryos divide at different rates and in different ways depending on species and the amount of yolk in the egg. Sea urchins and amphibians divide to form a blastula. Mammals divide to form a blastocyst, consisting of a trophoblast and an inner cell mass.

How Does Gastrulation Set Up the Three-Layered Structure?

The embryo sets up the three germ layers and becomes "a tube within a tube" during the process of gastrulation. The most visible aspect of this process is the formation of the **archenteron** [Greek, *arche* = beginning + *enteron* = gut] or "primitive gut," the space inside the innermost tube. The archenteron is lined with endoderm and will become the digestive tract.

For an easy way to imagine the conversion of a blastula to a layered structure, think about pushing in one end of a tennis ball (Figure 44-6). Imagine pushing hard enough to create a sort of cup. The pushing-in process is called "invagination" and results in a two-layered cup. Now punch a hole in the closed end of the cup. (You will have to cut through two layers.) This exercise does not really explain how the layers form in real animals, but it does illustrate the geometry of gastrulation.

Recall that in sea urchins, vertebrates, and all other deuterostomes, the digestive tract has two openings—the mouth and the anus. The anus is formed from the original site of invagination. The mouth develops later, from the opening of the other end of the primitive gut—the hole we imagined putting in the two-layered end of the invaginated tennis ball.

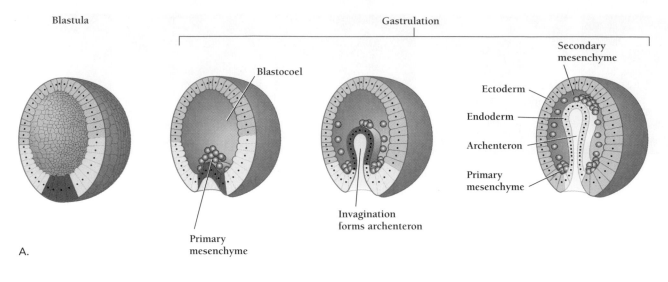

Blastula

Gastrulation

Blastocoel

Secondary mesenchyme

Ectoderm

Endoderm

Archenteron

Primary mesenchyme

Invagination forms archenteron

Primary mesenchyme

A.

B.   25 μm

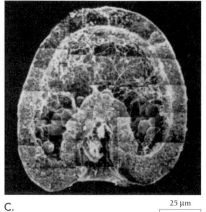

C.   25 μm

Figure 44-6   Gastrulation in a sea urchin. A. The sea urchin blastula is a hollow ball of cells. Cells from the bottom of the blastula, called primary mesenchyme, move into the blastocoel and become mesoderm. Endodermal cells invaginate smoothly in through the blastopore, to form the archenteron. Because sea urchins are deuterostomes, like vertebrates, the blastopore will eventually develop into the anus of the mature gut. As invagination proceeds, the cells at the tip of the archenteron detach to form the secondary mesenchyme. These cells form contacts with the wall of the blastocoel and actually pull the archenteron up to the wall. The cells of the secondary mesenchyme disperse into the blastocoel and proliferate to form most of the mesoderm. B. Scanning electron micrograph of early sea urchin gastrula C. Cross section of sea urchin gastrula.   *(B,C, from Morrill and Santos, 1985)*

Sea urchins and mammals gastrulate differently from one another. Sea urchin gastrulation resembles the tennis ball analogy: most of the cells of the blastula move smoothly in a tightly connected sheet. Mammals, however, gastrulate one cell at a time: each cell moves independently of the others.

The sea urchin blastula starts out as an almost symmetrical sphere. Soon, however, one side flattens, and a few cells move—as single cells—into the **blastocoel,** the hollow interior of the blastula (Figure 44-6). These independently moving cells, called the primary **mesenchyme** [Greek, *mesos* = middle + *enchyma* = infusion], attach to the interior wall of the blastocoel and eventually produce the sea urchin's calcium-containing skeleton.

Gastrulation is driven by the changing properties of individual cells. These changing properties include alterations in the attachments of cells to one another and transformations in cell shape, brought about by the cytoskeleton (Figure 44-7).

During gastrulation, a blastula (or blastocyst) invaginates, forming a two-layered cup. The inside of the cup is the archenteron, the inside of the future gut.

How Do Amphibians Gastrulate?

Gastrulation in amphibian embryos also depends on the coordinated behavior of sheets of cells and on the interactions of individual cells with extracellular matrices. Amphibians are favorite experimental animals for developmental biologists, partly because of the relatively large size of the egg (up to 1 mm in diameter). In addition, amphibians are the only vertebrates that can develop in a simple salt solution (rather than within an eggshell or uterus). Amphibian gastrulation, however, is more complicated than that of sea urchins, because of the presence of yolk.

Because the yolk-filled cells near the vegetal pole can move only sluggishly, amphibian cells gastrulate at the embryo's equator, rather than at the vegetal pole (Figure 44-8). Amphibian gastrulation begins when an indentation, called the **blastopore,** forms on the side of the blastula. Invagination at the blastopore depends on striking shape changes in cells at the blastopore lip, called bottle cells. Shortly after the initial invagination, adjacent cells follow the bottle cells into the invagination. The result is the formation of a hollow archenteron that is lined with endodermal (gut) cells from the surface of the blastula. As

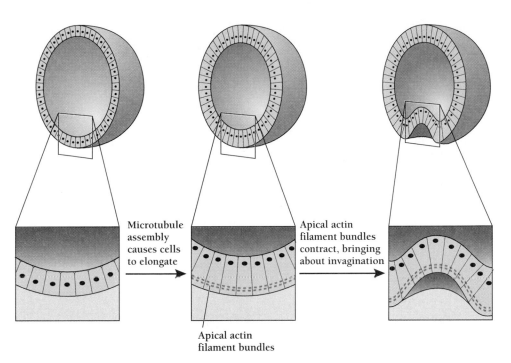

Figure 44-7 **Changes in cell shape underlie cell movement.** During gastrulation in a sea urchin, the cytoskeletons of the cells at the bottom of the embryo elongate the cells and then constrict them at their bases. The result is an arch of cells that begins invagination.

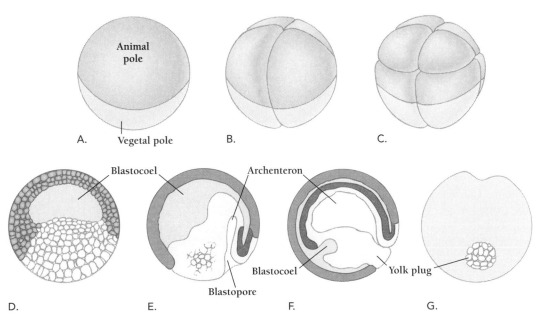

Figure 44-8 **Early development in a frog.** A. Zygote, showing animal pole and vegetal pole. B and C. Early cleavage stages. The cells at the animal pole divide into many smaller cells than those in the yolky vegetal pole. D. A longitudinal section of a blastula. E and F. Longitudinal sections through gastrulating embryos. As the endoderm (*yellow*) invaginates, a collection of mesodermal cells (*red*) forms at the blastopore (the future anus) and spreads across the embryo's back. This streak of red mesoderm will go on to form the notochord, which takes part in the formation of the spinal cord and spinal column. G. End view of late gastrula showing yolk plug and an indentation at top that will become the neural fold.

in the sea urchin, a few cells detach from the invaginating layers to form the mesoderm.

The cells of the future mesoderm help power gastrulation. These cells, which are among the first invaginating cells, crawl over the extracellular matrix on the roof of the blastocoel. Two proteins enable the mesodermal cells to attach to the blastocoel roof—*fibronectin*, in the extracellular matrix, and *integrin*, on the cell surfaces. Experiments have shown that when fibronectin and integrin are prevented from binding to each other the embryo cannot gastrulate.

Individual cells arrange themselves into tissues. In one experiment, researchers broke up ectoderm (skin and nerve) and endoderm (gut) tissues into individual cells and mixed the two kinds of cells together. The cells spontaneously formed new tissues: all the ectodermal cells with other ectodermal cells and all the endodermal cells with other endodermal cells.

Even more strikingly, when researchers mixed ectoderm, mesoderm, and endoderm cells, the cell types not only aggregated into three distinct germ layers, the cells arranged themselves with endodermal cells on the inside, ectodermal cells on the outside, and mesodermal cells in between. These experiments suggest that signaling molecules on the surfaces of cells may direct cells to the correct position in the embryo.

Amphibians invaginate smoothly through the blastopore. The cells of amphibian gastrula can spontaneously organize themselves into three layers of tissue.

How Do Mammals Gastrulate?

The early development of a mammalian embryo follows a pattern more similar to that in birds and reptiles than that in sea urchins or amphibians. In reptiles and their descendants, the birds and mammals, germ layer formation depends on the separation of parallel sheets of cells followed by movements of individual cells.

Recall that a mammalian blastocyst consists of a trophoblast and an inner cell mass. The first separation of cells within the inner cell mass produces two layers, called the **hypoblast** and the **epiblast** (Figure 44-9). The cells of the hypoblast enclose what will become the primitive gut, but these cells are later replaced by endoderm cells that are derived from the epiblast. The hypoblast stays just below the epiblast (separated from it by the blastocoel) to form the blastodisc, the equivalent of a blastula. All the structures of the adult derive from the blastodisc.

The formation of the germ layers now takes place. Cells of the blastodisc move toward a central line, leading to the formation of the **primitive streak,** the functional equivalent of the blastopore (Figure 44-10). Individual cells now move through a groove in the streak, into the blastocoel. These cells advance independently into the blastocoel, where they later separate into endoderm and mesoderm.

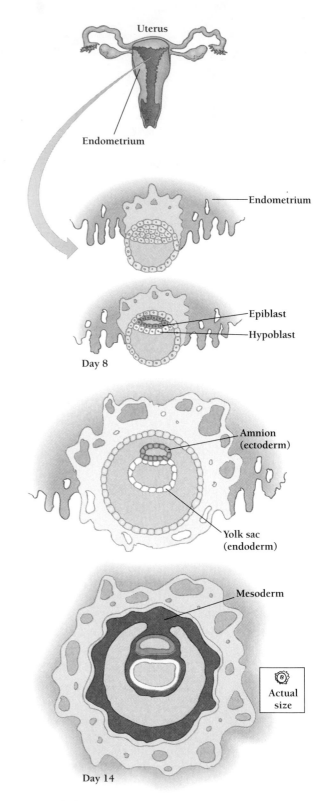

Figure 44-9   A developing human. In mammals, the inner cell mass splits into the epiblast and hypoblast. The hypoblast forms the blastodisc. These form the amnion and the embryonic disc.

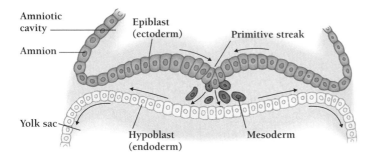
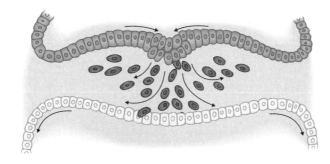
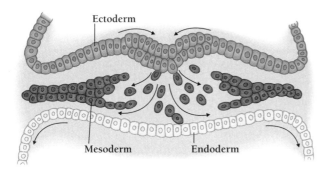

Figure 44-10   Gastrulation in mammals. In humans and other mammals, the epiblast cells invaginate along the primitive streak forming the mesoderm. The epiblast cells may also enter the hypoblast to form the endoderm.

The mammalian pattern of gastrulation, also found in birds and reptiles, is well suited to the yolky eggs of birds and reptiles, since only the relatively yolk-free cells of the epiblast move. This pattern appears to have evolved early in the history of the land-dwelling vertebrates, when the yolky, eggshell-encased eggs made reptiles and their descendants independent of water. The common pattern of embryonic development in reptiles, birds, and mammals supports the view that mammals evolved from an ancestor with an enclosed, yolky egg.

Even though the tiny eggs of mammals have little yolk, the resulting embryos gastrulate like those of their relatives, the birds and reptiles, whose eggs are heavy with yolk. The formation of the mammalian embryo takes place in a flattened disk, and the cells invaginate individually through the primitive streak, forming the gastrula.

How Does the Nervous System Form?

Once an embryo has established three germ layers, it next begins to form organs. An organ may derive from a single germ layer or may come from two different germ layers. In each case, organ formation consists of two major processes: **morphogenesis**, the creation of form, and **differentiation**, the specialization of cells. The formation of the brain in higher vertebrates illustrates many of the main features of organ formation.

At the end of gastrulation, mesoderm has separated from ectoderm and endoderm. In vertebrates, part of the mesoderm forms the **notochord**, a supportive cord that runs from head to tail, beneath the dorsal (back) surface. The notochord serves both as an internal skeleton for all vertebrate embryos and as an organizer for further embryonic development.

Just above the notochord, the ectoderm rearranges to form the nervous system. The cells along the central line of the embryo thicken to form the **neural plate**, a flat plate above the notochord (Figure 44-11). Then, along the embryo's long axis, the neural plate curls up, as individual cells in the neural plate constrict at one end. The elongated invaginated structure then pinches away from the surface to form a hollow tube, called the **neural tube**. The neural tube is the precursor for the entire central nervous system.

The cells left on the surface, above the neural tube, become skin ectoderm (epidermis), while the cells that had connected the tube to the surface—called **neural crest cells**—migrate away. Descendants of the neural crest cells develop into different types of cells, depending on where they go after they leave the neural tube. Many of them form nerve tissues. But others form pigment cells in the skin, for example. The entire process of forming the neural tube and neural crest cells is called **neurulation.**

The neural tube itself now folds to form the beginnings of the different regions of the brain and spinal cord. The front of the tube enlarges and then bulges on each side to start the development of the cerebral hemispheres. Just behind, a pair of bulges appear. These are the optic vesicles, which will become the eyes.

The notochord induces the formation of neural plate tissue, which folds into a hollow neural tube. The neural tube bends to form the different regions of the brain and the spinal cord. Neural crest cells from the surface of the neural tube migrate into distant parts of the embryo to form nerve and other tissues.

How Does the Eye Form?

The eye forms as a result of a simple set of cell movements (Figure 44-12). When the optic vesicle comes into contact with the overlying skin ectoderm, the ectoderm thickens to form a plate that will develop into the lens. Once the lens thickening occurs, the optic vesicle folds back on itself (or invaginates) to form a double-walled optic cup and the lens plate curls and

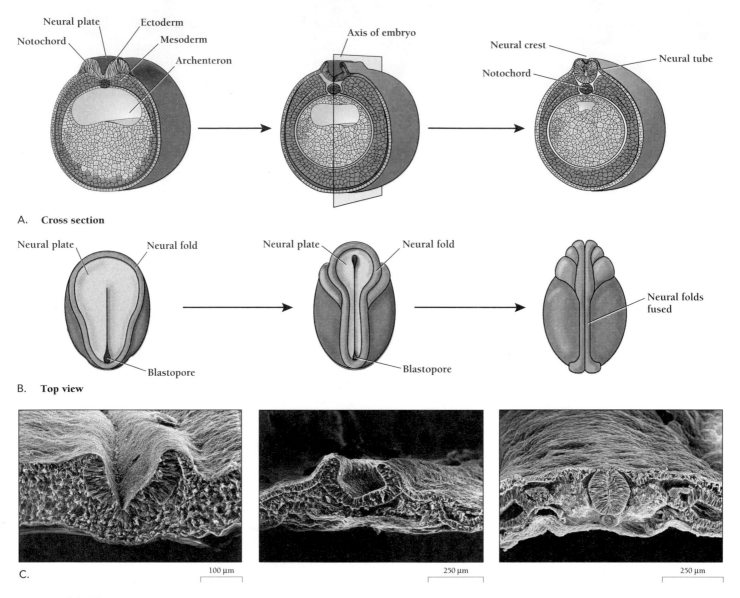

Figure 44-11   Formation of the neural tube. A. Cross section. B. Top view. C. Micrographs showing the neural folds coming together and fusing at the top.   *(C, K.W. Tosney)*

pinches away from the skin ectoderm to form the lens vesicle. The stalk that attaches the cup to the rest of the brain becomes the optic nerve, which will eventually carry visual information from the retina to the brain.

After this complex set of thickenings, foldings, pouchings, and pinchings, the cells of the lens, the retina, and the overlying skin ectoderm begin to differentiate. The cells of the lens make specialized proteins, called crystallins, that give the lens its optical properties. The cells overlying the lens change character to become the cornea, the transparent covering of the eye. Finally, the cells of the retina develop into light-detecting rod and cone cells. Through these events a seemingly uniform layer of cells in the neural tube develops into an eye, our most prized contact with our surroundings.

The development of the eye illustrates several generalizations about organ formation: (1) Organ formation precedes dif-

ferentiation: only after cells have taken their places do they start to specialize. (2) Organ formation often involves complex folding of sheets of cells. (3) Organ formation depends on interactions among cells that are brought together by movements of tissues and cells.

The optic vesicles of the neural tube interact with overlying ectoderm to form the retina, lens, and cornea of the eye.

How Do Organs Acquire the Right Shape and Size?

Sometime, relatively early in development, every organ appears as a recognizable **rudiment** [Latin, *rudimentum* = beginning], or initial stage, from which the final form will develop. In hu-

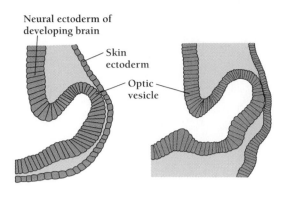
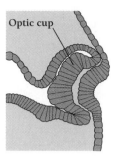
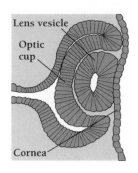

250 μm

Figure 44-12 Formation of the eye. A sheet of neural ectoderm folds to form the optic vesicle and optic cup. The skin ectoderm responds by forming the lens, as shown in the micrograph on the right. *(photo, courtesy of K.W. Tosney)*

mans, all these organs are in their proper places by the end of 9 weeks (Figure 44-13). The embryo is then said to be a fetus.

But the fetus is far from fully formed. The final form of each organ requires extensive cell specialization and growth. Each human hand and arm, for example, consists of 43 muscles, 29 bones, and hundreds of nerve pathways. None of these are present in the rudiment of an arm. In fact, except for the heart, the organs of an early fetus are largely nonfunctional. In addition, a 9-week human fetus is just a little over an inch long. Only with further growth and development does an early fetus become a baby.

Different parts of the body grow at different rates. A 9-week fetus, for example, has a head nearly the same size as all of the rest of its body. Although the head will never again be as proportionately large, it continues to grow and develop in advance of the other parts of the body. The brain, eyes, jaws, lungs, stomach, intestines, and kidneys—all structures that the baby will need when it is born—grow and develop most rapidly, while the legs and feet lag behind. Between 9 and 12 weeks the fetus doubles in length, the external genitalia appear and begin to develop, and the kidneys excrete their first urine.

Between 12 and 16 weeks, a human fetus doubles its length again. By 16 weeks, the ovaries are differentiated and the primary follicles contain oogonia. Between 17 and 20 weeks, the mother feels the first fetal movements. By 20 weeks, the fetus is about 10 inches long, and the testes have begun to descend.

By 22 weeks, the lungs, intestines, and kidneys are sufficiently developed that it is sometimes possible—with intensive medical care—for the fetus to survive outside the womb. Between 20 weeks and 26 weeks, the fetus begins to fill out. At 24 weeks, the lungs begin to secrete a *surfactant*, a detergent that allows the lungs to inflate with air (Chapter 38).

After 26 weeks, a fetus born prematurely has a good chance of surviving—because its respiratory system is mature and functioning. Between 26 weeks and 36 weeks, when most babies are born, the fetus gains weight in the form of fat. By 36 weeks, the circumference of the baby's abdomen equals the circumference of its head.

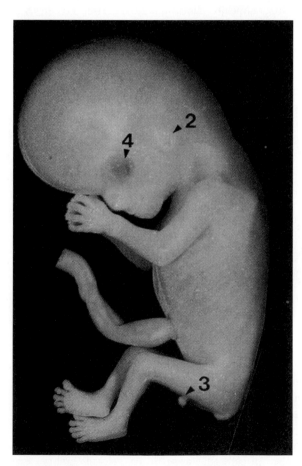

Figure 44-13 At 9 weeks a human embryo enters the fetal stage. All of its organs are formed and the rest of development consists of growth and the development of individual organs and limbs. The arm buds, for example, differentiate into bone, muscle, blood vessels, and nerves; the eyes become capable of vision; and the external genitalia take shape. While embryologists date embryos starting at conception, physicians add two weeks, dating the pregnancy from the first day of the mother's last menstrual period. A gynecologist would say that this embryo is 11 weeks old. *(From Color Atlas of Life Before Birth, Normal Fetal Development by Marjorie A. England, Mosby Yearbook Publishers, London, 1990)*

An embryo forms the bare outlines of all of its organs long before those organs are functional. At 9 weeks, a human embryo possesses rudiments of all the major organs and is said to be a fetus, but only the heart is functional.

Programmed Cell Death Contributes to Normal Development

Growth and development usually mean both the increase in size of individual cells and the increase in cell numbers due to cell division. Paradoxically, however, the death of cells is also a major part of development. In the last decade, researchers have increasingly realized that development requires extensive "programmed cell death," or **apoptosis**—the death and removal of cells. Cells undergoing apoptosis are distinctive both visibly and biochemically: their nuclei condense, their DNA is systematically digested, and they are methodically engulfed by other cells.

Apoptosis occurs in many developmental pathways in both vertebrates and invertebrates. It is particularly common among cells of the vertebrate immune and nervous systems, for example. Ordinarily, when a cell dies, as a result of attack by microbes, for example, the cell bursts, or lyses, and releases toxic cell products that can disrupt other developmental processes. Lysis can also provoke the immune system to initiate an immune attack on healthy cells. In apoptosis, cells die quickly. They are engulfed by phagocytes and digested without leaving a trace.

One of the best-studied examples in vertebrates is the programmed death of cells between the digits of the embryonic limbs (Figure 44-14). In the early embryo, a thin tissue lies between the digits (fingers and toes in humans). This tissue persists as the familiar webbing of a duck's hind limbs, but it is destroyed during development in chickens and other non-aquatic vertebrates (including humans). The process of making a hand or a foot resembles the work of a sculptor, who chisels away marble to create a statue.

Apoptosis is also well studied in the adult nematode worm *Caenorhabditis elegans* (*C. elegans*). The adult worm, which looks like a 1-mm-long transparent tube, consists of only about 1000 cells. Its simplicity and transparency have enabled researchers to track the precise pathway of every cell that arises during development.

A number of cells that arise in a *C. elegans* embryo undergo apoptosis. For example, the sisters of cells that develop into nerve cells of the adult worm normally undergo apoptosis. If a mutation prevents apoptosis, however, these nerve cells survive and grow and the worm's nervous system develops with too many neurons.

Apoptosis in vertebrates depends on gene action. Many of the genes that activate apoptosis are identical to genes that regulate the cell cycle. The product of one gene, called *bcl*-2, helps prevent apoptosis not only in mammals but also in *C. elegans*. Biologists can only conclude that the genes for cell death

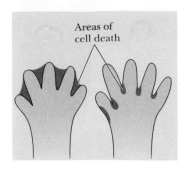

Figure 44-14 Sculpture and development. Like a sculptor chiseling stone, the programmed death of individual cells generates form in a developing animal. The human hand first forms as a finlike paddle that lacks both fingers and thumb. As cells die in selected areas, however, the general shape of the digits emerges. Later, the specific tissues—bones, muscles, nerves, and blood vessels—will emerge within each digit.

evolved at least 600 million years ago, when the common ancestor of mammals and nematodes first split into two lines.

Programmed cell death, apoptosis, plays a major role in embryonic development.

EMBRYONIC CELLS BECOME INCREASINGLY DIFFERENTIATED

Hans Driesch's experiment showed that each cell of a two-cell embryo has the ability to develop into a whole (rather than a half) sea urchin. With this and similar results in mind, embryologists have come to distinguish the **fate** of a cell—what it becomes during development—from its **potency**—what it could become if allowed to develop in another environment. In the case of the two-cell sea urchin embryos, the *fate* of each is normally to become the right or left half of the sea urchin, but each has the *potential* to become a whole organism.

What Kinds of Experiments Distinguish Potency and Fate?

At early stages of development and for relatively simple organisms, we may follow the fates of individual cells just by watching and photographing them. By observing the movements of cells in developing sea urchins, for example, biologists have determined which cells give rise to endoderm, which to mesoderm, and which to ectoderm.

In a frog, a bird, or a mammal, however, gastrulation is too complicated to follow without marking individual cells—with dyes, radioactive tracers, or particles of carbon. By following the movements of marked cells, embryologists have been able to construct "fate maps" that show what happens to each cell during development.

Defining the potential of a cell is more difficult. One way of determining the development potential of embryonic cells is to transplant them from one embryonic environment to another. In 1918 the German embryologist Hans Spemann performed a now-famous transplant experiment in two species of newts. One species was darkly pigmented and the other was pale.

Spemann transplanted pieces of the dark, pigmented embryo into the pale, unpigmented "host" embryo. Because of the pigment, he could distinguish easily between structures formed from the host cells and those formed from the transplanted cells. Having studied the fate maps of the newt, Spemann knew which embryonic cells should form neural plate and which should form skin. Cells known to form a particular tissue are described as "presumptive," as in "presumptive ectoderm cells" or "presumptive skin cells."

Spemann transplanted presumptive neural plate (nerve tissue) into an area of presumptive skin cells. If he transplanted the presumptive neural cells at the early gastrula stage, they formed skin instead of a neural plate. That is, transplanted cells developed according to their new environment. In short, their potency was greater than their fate.

If, on the other hand, Spemann waited until the late gastrula stage, the presumptive neural cells developed into neural plate, even though they were in the wrong environment. Spemann concluded that later in development, cells become *committed* to their normal developmental fate, gradually losing potency. Although presumptive neural plate cells look just like presumptive skin cells, they are already **determined**, meaning they can no longer develop according to their environment. In short, their fate is greater than their potency.

The fate of a group of cells is the tissue they will become during the course of normal development. The potency of a group of cells is the tissues that they could become under varying circumstances in different environments. Hans Spemann devised experiments for distinguishing a cell's fate from its potency.

Are Differentiated Animal Cells Totipotent?

Spemann's experiments raise two important questions: (1) How does a cell's environment influence its development? (2) How does a cell become determined, so that its environment no longer influences its development?

Modern biologists think of determination as the process that establishes which genes will be expressed and which will not, a process that is influenced by environmental signals. Spemann's transplantation experiment showed that determination in neural plate cells depends on environmental changes brought about by the movements of gastrulation. Further changes in the cellular environment bring about changes both in cell shape and in the pattern of gene expression that is responsible for neural development.

An easy explanation for determination would be that during development cells lose genes. For example, do presump-

tive neural cells lose the genes necessary to form skin proteins? In the 1950s, developmental biologists began to approach the question experimentally. Robert Briggs and Thomas King devised a method for nuclear transplantation, in which they removed the nucleus of a leopard frog egg and replaced it with a nucleus from a cell from another frog (Figure 44-15). The egg developed under the control of the transplanted nucleus. Briggs and King soon discovered that nuclei from a blastula were totipotent: many of them could direct the development of the whole frog.

Yet Briggs and King never were able to produce a swimming tadpole from a nucleus that had developed beyond the neurula stage. The researchers concluded that, in leopard frogs at least, nuclei lose the potential to direct complete development.

Later researchers found that nuclei from the intestines of the South African clawed toad tadpoles *could* generate a normal adult (Figure 44-16). Were fully differentiated animal cells totipotent? Until 1997, this pressing question remained unanswered.

Early transplantation experiments did not make clear whether differentiated animal cells were totipotent.

Can a Differentiated Nucleus Direct Development from Egg to Adult?

A singular consequence of Briggs and King's nuclear transplantation experiments was that, for the first time, biologists could produce a group of genetically identical animals. Because the nuclei from the blastula cells were identical (all resulted from the cleavage of a single zygote), the resulting leopard frogs were clones of one another.

For the first time, zoologists could clone an animal. Gardeners had been cloning plants for centuries, for every plant cell is totipotent and may develop into a complete adult. But no one had ever before cloned an animal. One important reason was that as animal cells differentiate, they seem to lose the ability to direct development. Why that should be so was not at all clear.

Animal breeders and biologists have long been motivated to clone animals. A farmer with a prize dairy cow would love to be able to clone such an animal. A whole herd of prize dairy cows would not only ensure a steady supply of milk for the life of the individual cows but it would so do indefinitely. More alluring still, a farmer with such a herd could sell the clones to other farmers with less productive animals.

Breeders would like to be able to clone adult animals with proven traits. No one can know an animal's worth until it is mature. But until recently, animal breeders never succeeded in cloning cells from adult animals. Biologists' efforts with mice had failed over and over. Animal breeders clone cells from embryos, but an embryo is basically an unknown: it is not a productive milk cow, a Derby-winning racehorse, or even a lab animal with a perfectly understood genotype and phenotype.

Figure 44-15   Do the nuclei of developing embryos remain totipotent as they differentiate? The experiments of Briggs and King suggested that as cells differentiate, their nuclei become less able to direct complete development. A nucleus from a blastula cell could direct the development of an egg into a normal leopard frog; a nucleus from a gastrula could direct development in some cases; but, a nucleus from a neurula cell could never direct complete development.

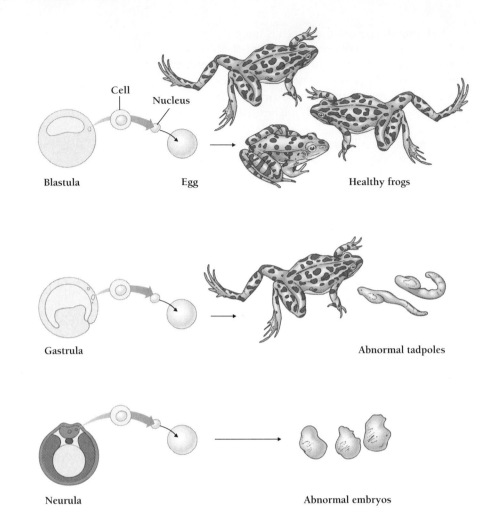

Blastula   Egg   Healthy frogs

Gastrula   Abnormal tadpoles

Neurula   Abnormal embryos

Then, in February of 1997, a team of biologists at the Roslin Institute, in Edinburgh, published a mind-bending result in the British journal *Nature*. The team, led by Ian Wilmut and Keith Campbell, had successfully reared a Scottish mountain sheep grown from an egg cell containing a nucleus transplanted from a mature cell from the udder of an adult sheep. In other words, the sheep, named Dolly, had been cloned from an adult animal. Dolly's impassive face appeared on the front pages of newspapers and magazines around the world (Figure 44-17). If biologists could clone sheep, commentators asked, why not people? Governments hastily passed laws banning the cloning of humans, although it was unclear if the cloning of humans would be possible.

Geneticists were equally excited. If they could clone lab animals, they realized, they could study the subtle effects of environment on whole colonies of genetically identical lab animals. They could study, for example, the genetics of development as well as genetic diseases. More lucratively, biotech-

Figure 44-16   Differentiated and totipotent? In the 1960s, John Gurdon repeated the experiments of Briggs and King in the African clawed toad. Gurdon transplanted nuclei from the intestines of swimming tadpoles to egg cells. Although most of the resulting zygotes did not develop normally, about 1 percent developed into normal adult frogs whose cells contained nuclei identical to those of the donor tadpole.

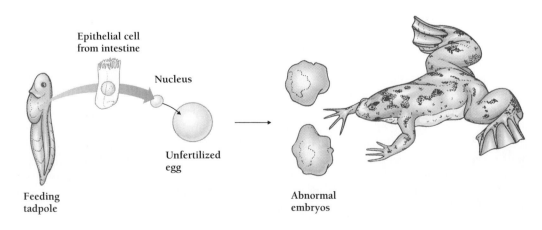

Epithelial cell from intestine

Nucleus

Unfertilized egg

Feeding tadpole

Abnormal embryos

Figure 44-17 **Hello, Dolly.** In 1996, Dolly the sheep was cloned from mammary cells taken from her clone (her genetic twin). *(AP/Wide World Photos)*

nologists could generate herds of genetically engineered animals, such as goats that secrete useful proteins in their milk.

But developmental biologists viewed the research in a different light. From their perspective, one of the most exciting aspects of Wilmut and Campbell's work was that it seemed to answer the question about totipotency. If it was true that Dolly's nuclei had come from a fully differentiated cell, then, at least in sheep, it was possible to say that differentiated cells remain totipotent. But do they? The answer lies in the details of the Edinburgh biologists' work.

The cloning of a Scottish mountain sheep in 1996 suggested that differentiated nuclei are totipotent.

How Did the Scottish Researchers Clone Dolly?

The first question that many people asked Wilmut and Campbell was, How did you do it? How had they prompted a differentiated nucleus to guide a full course of development starting in an egg cell? The answer was fairly simple, although the experiment had been repeated numerous times for one success (Figure 44-18).

The trick, Wilmut said, was to make the DNA of the donor cells behave more like the inactive DNA of an egg or sperm. They did this by growing the udder cells in culture and starving them of essential nutrients. Gradually, the cultured cells passed into a dormant state in which many genes shut down

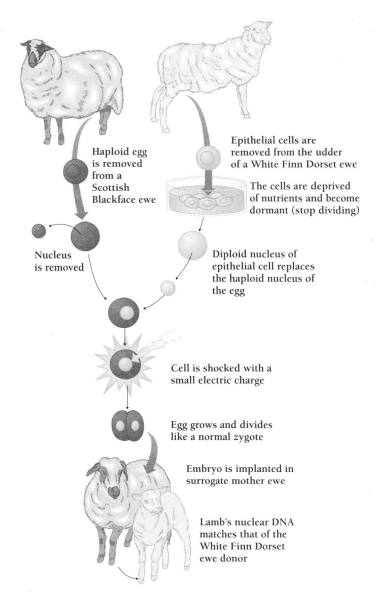

Figure 44-18 **Cloning Dolly.** The trick to cloning Dolly was to make differentiated cells less differentiated. By depriving the cultured udder cells of nutrients, the researchers induced the nuclei to enter a dormant state.

and replication was impossible. Once the DNA in the donor nuclei was dormant, the team of biologists began transferring the nuclei into egg cells. Painstakingly, they transplanted 277 nuclei, of which only one successfully developed.

During the first three divisions of the sheep embryo, the cells replicate their DNA without expressing any nuclear genes. All cell processes remain under the guidance of maternal proteins and RNA. The inactive DNA in the transplanted nucleus replicates but the cell transcribes no RNA from the DNA. Meanwhile, the nuclear DNA begins losing proteins that it brought from the udder cell, proteins that at first prevent the expression of the nuclear genes. By the third division, these proteins are replaced by proteins from the egg's cytosol. The maternal (egg) proteins then "reprogram" the nuclear DNA, and the embryo begins expressing its own genes.

Although maternal mRNA and maternal proteins are sufficient to guide cell division in sea urchins as far as the blastula stage, the nuclear DNA of most mammalian embryos comes into play much earlier. Mouse nuclei begin expressing their DNA after just one division. Human nuclei begin after two divisions.

Some biologists argue that the Scottish team succeeded where others had failed because the DNA of sheep takes so long to turn on. Three cleavage divisions perhaps allowed the egg cell proteins enough time to replace the udder proteins.

Wilmut, Campbell, and colleagues cloned Dolly the sheep by inducing cultured udder cells to enter a dormant state, in which gene expression was shut down. The nuclei from 277 such cells were then transplanted into sheep oocytes, which divided three times. Only one of the zygotes developed into an adult sheep.

Did Dolly Really Come from a Differentiated Cell?

Other biologists question whether Dolly is the product of a differentiated cell at all. Mammary cells, including cultured cells, typically contain not only differentiated cells for producing milk but a variety of other cells. They are a mixed population of cells. Some of these cells are **stem cells,** cells that can either produce more cells like themselves by cell division or undergo differentiation to one or more specialized cell types.

Adult vertebrates contain many kinds of precursor stem cells. Any cells that must regenerate during adult life—such as the cells of the skin, the linings of intestines, and the blood—come from a supply of stem cells. But stem cells are less differentiated than other cells. The developmental fate of stem cells—whether they proliferate to make more stem cells or differentiate to produce specialized cells—depends on cues from the environment.

Wilmut and Campbell do not know if the nucleus that gave rise to Dolly was a fully differentiated udder cell or an only partially differentiated stem cell. For the purposes of animal breeders, the difference is of minor importance. For developmental biologists, however, the question of whether a fully differentiated egg is totipotent is as interesting today as it was in 1950.

Because Dolly may have come from a stem cell nucleus, the totipotency of fully differentiated animal cell nuclei remains in question.

HOW DOES DEVELOPMENTAL FATE DEPEND ON CHEMICAL SIGNALING?

Since virtually every cell in an animal has the same genes, differences in cell fates must result from selective gene expression.

Differences in the patterns of gene expression must be influenced by cellular environment. The question then is, How do cues from the embryo influence the developmental fates of individual cells?

Can the Extraordinary Actions of the Primary Organizer in Amphibian Development Be Explained in Terms of Cells and Molecules?

Arguably the most spectacular experiment in the history of developmental biology was one performed by Hans Spemann's student Hilde Mangold. Recall that Spemann had established that cells become determined during gastrulation. Mangold and Spemann asked *how* cells become determined.

The two researchers correctly guessed that determination of the ectoderm into neural tissue or into skin depended on contact between the ectoderm and the underlying mesoderm, which derives from the cells that invaginate during gastrulation (Figure 44-8). Mangold asked what would happen if the invaginating cells—the dorsal lip of the blastopore—were transplanted to another region of the embryo. Amazingly, the transplanted dorsal lip not only invaginated it also induced the tissues around it to form a nearly complete second embryo (Figure 44-19).

Spemann termed the dorsal lip the **primary organizer,** meaning that it established the entire organization of the embryo. He was so astonished by Mangold's result that he com-

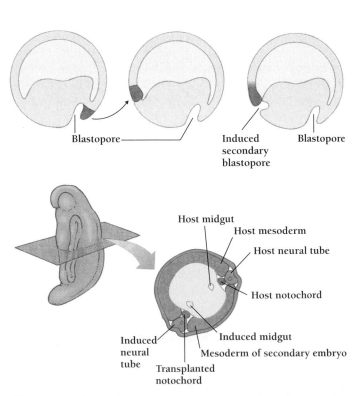

Figure 44-19   The primary organizer. Tissue from the lip of the blastopore, where gastrulation begins, can induce gastrulation and neurulation in an inappropriate place.

pared the action of the primary organizer to the workings of the mind—suggesting that it could not really be understood in chemical and physical terms.

Subsequent work, however, has demonstrated that the primary organizer's mechanism is chemical. For example, chemical extracts of the dorsal lip induce differentiation of ectoderm. Researchers do not yet fully understand the mechanism of embryonic induction. They have, however, identified a number of molecules that help to establish patterns of development within whole embryos and within individual organs.

One small molecule that is important in pattern formation is retinoic acid, which regulates the pattern of digit formation in the developing limb bud of the chick embryo and in regenerating limbs of newts. In the chick limb bud, added retinoic acid can induce the formation of extra digits, with the number of extra digits depending on the concentration of retinoic acid. In a regenerating newt limb, retinoic acid is present in a concentration gradient, with the highest concentration in the part of the limb farthest from the body. Retinoic acid is also present in relatively high concentrations in a chick embryo's primary organizer.

Hilde Mangold's transplantation experiments showed that the primary organizer establishes the overall orientation (axis) of an embryo.

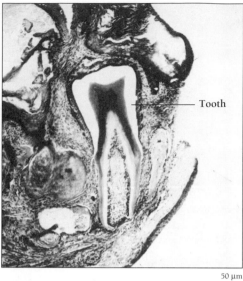

Tooth

50 μm

Figure 44-20 **Waking sleeping genes.** The power of tissues to induce differentiation in other tissues is enormous. Birds have not had teeth since they diverged from other reptiles millions of years ago. Yet when researchers transplanted mesoderm from the mouth of a mouse embryo under the epidermis of a chick embryo, the mouse cells induced the chick cells to make enamel in formations remarkably like teeth. *(From E.J. Kollar and C. Fisher, 1980, "Tooth induction in chick epithelium: expression of quiescent genes for enamel synthesis." Science 207:993-995. Copyright 1980 by the American Association for the Advancement of Science)*

How Do Cells "Know" Where They Are and What To Do About It?

Development requires both that individual cells express appropriate genes and that cells be organized into functioning tissues and organs. The development of a limb, for example, requires the differentiation of the cells of the muscles, skin, and bone into a characteristic pattern, distinct from the organization of the same cell types in a foot. How does the developing organism achieve this pattern?

Part of the answer to pattern formation lies in the interactions between layers of cells. For example, the skin ectoderm of a chicken can give rise to several different kinds of structures—fully tufted feathers, partly tufted feathers, scales, and claws (Figure 44-20). The developmental fate of the ectoderm depends upon the source of the underlying mesoderm. Thus, if wing ectoderm is placed over foot mesoderm, claws will develop instead of feathers.

The overall pattern of a limb, however, does not depend on short-range interactions between adjacent cell layers. Instead, special mechanisms specify positional information, chemical cues that establish the position of a cell in a pattern. The concentrations of particular substances, called **morphogens,** specify the contribution of each cell to a pattern, in the same way that a seat number can be used to specify the contribution of each member in a cheering section that shows the image of a team's mascot in a card display.

The formation of patterns during development depends on each cell's "knowing" *where* it is. It also depends on the interpretation of this information: each cell must also "know" *what* it is. At the tip of a developing leg bone, cells must develop into digits, while cells nearer the body must develop into the other bones, such as the femur and the tibia.

In vertebrates, the best-understood example of a morphogen is retinoic acid. In chicks, salamanders, and frogs, retinoic acid induces the formation of limbs—complete with humerus, radius, and ulna. Like a steroid, retinoic acid apparently directly influences gene expression by interacting with a transcription factor. Recall from Chapter 11 that steroid hormones such as progesterone and testosterone bind to receptor proteins inside a cell. The steroid-receptor complex then passes into the cell's nucleus and regulates the transcription of specific genes. Similar receptors bind to retinoic acid.

Which genes are expressed, however, differs from cell to cell. In the muscle cells of a male mammal, for example, testosterone stimulates cell division and growth, while in skin cells, testosterone stimulates the growth of hair. The cells' response to testosterone depends on already established differences among the cells.

Researchers have identified a number of genes responsible for pattern formation in *Drosophila*. The first of those identified were two similar genes that—when mutated—led to a couple of bizarre flies called bithorax, with two sets of identical wings rather than two different sets, and antennapedia, in

A.   B.   C.

Figure 44-21   Genes control pattern formation. A. A normal fruit fly, *Drosophila melanogaster*. B. A fly with the *antennapedia* allele has legs where its antennae should be. C. A *bithorax* mutant has two identical pairs of wings instead of two distinct pairs of wings. (A,B, Carolina Biological Supply Company/Phototake, NYC; C, David Scharf/Peter Arnold, Inc.)

which legs grew where antennae should have (Figure 44-21). Later studies revealed that the proteins encoded by bithorax and antennapedia act as transcription factors. Bithorax and antennapedia are two of a family of **homeotic selector genes,** genes whose expression affects overall body plan.

Another example of a transcription factor that directs development is a protein called bicoid protein. Normal development of the early embryo requires bicoid. When the *bicoid* gene mutates, the resulting embryo has no anterior structures. Researchers can produce exactly the same phenotype in a normal fly by removing cytoplasm from the anterior end of the embryo.

Similarly, researchers could "rescue" *bicoid* mutants by injecting them with the mRNA for the protein bicoid. The injected embryos developed normal anterior structures. These experiments suggested that the bicoid protein, normally present in the cytoplasm near the anterior pole, regulates the expression of other genes during development.

Pattern formation seems to result from the action of morphogens such as retinoic acid. Retinoic acid influences the transcription of genes and therefore the differentiation of cells.

Mammals and Flies Use Many of the Same Transcription Factors To Establish Patterns During Development

Information about the structure and action of genes that regulate *Drosophila* development led immediately to questions about the role of similar genes in mammalian development. Researchers asked whether mammalian genomes had counterparts of the amazing homeotic genes in *Drosophila*.

Mammalian counterparts, called **Hox** genes, were easy to find, for their sequences are remarkably similar to those in *Drosophila*. The mouse gene Hox 1.1, for example, shares 59 out of 60 amino acids with the *Drosophila* gene *antennapedia*. Even more amazingly, when researchers inserted two vertebrate Hox genes into flies, the flies suffered the same mutation in phenotype as if they had the fly mutations antennapedia and deformed.

The biggest surprise in the study of the Hox genes, however, has been that corresponding genes in flies and mice are arranged on the animals' chromosomes in the same order. And the order of genes on the chromosome matches the pattern of action within the embryo. Genes that affect head development lie at one end of the cluster, while those that affect tail development lie at the other. The same pattern holds for the arrangement of the Hox genes on the mouse chromosome and their pattern of expression in the primitive nervous system of the mouse embryo.

The most recent common ancestor of *Drosophila* and mice lived some 600 million years ago. We can conclude that during the last 600 million years flies and mice have used—with very few modifications—a system of specifying positional information that had already evolved during or before the Cambrian explosion. Although researchers have little information about the genetics of development in other arthropods and vertebrates besides mice and flies, it's reasonable to assume that if flies and mice share these genes then probably all arthropods and vertebrates and many other groups share these homeotic genes.

Animals as different as mice and fruit flies use the same gene products to establish the position and identifying of the parts of the body.

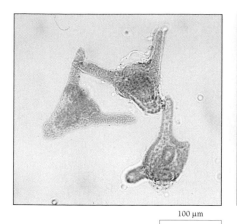

100 µm

Figure 44-22   Metamorphosis. Larval sea urchins and adult purple sea urchins.   *(left, www.biodisc.com; right, Tammy Peluso/Tom Stack & Associates)*

POST-EMBRYONIC DEVELOPMENT

Metamorphosis Converts a Larva into an Adult

A larva, the first stage at which an animal has an independent life, may not at all resemble the sexually mature adult. A frog embryo, for example, develops into a larva called a tadpole, which has a fishlike life—propelling itself with a large tail fin, getting dissolved oxygen through its gills, and subsisting on a vegetarian diet. The tadpole **metamorphoses** into an adult frog, which, in contrast, jumps instead of swims, breathes air with its lungs, and consumes flies and other invertebrates with its long tongue and big mouth.

Larvae are sometimes so different from the adult form that they appear to be different species. The axolotl, for example, inhabits lakes near Mexico City and has been considered a table delicacy since Mayan times. Only in 1920, when biologists induced metamorphosis in the laboratory, however, did they realize that the axolotl was actually a salamander.

Metamorphosis is more common among invertebrates than vertebrates. For example, the first independent form of a sea urchin is a larva, with an appearance and lifestyle totally different from the adult. The sea urchin larva or **pluteus** is mobile and elongated, moving easily on ocean currents (Figure 44-22). The adult sea urchin is spiny and more symmetrical, slowly grazing on the bottom of the sea.

The changing of a caterpillar into a butterfly is one of the most spectacular examples of metamorphosis. Many insects—including the beautiful butterflies, the more somber moths, and all the flies, beetles, and wasps—undergo **complete metamorphosis**, meaning that none of the tissues of the adult come from the larva (Figure 44-23). Instead, the adult, the *imago*, derives from 19 apparently unspecialized **imaginal discs**, groups

Figure 44-23   **Metamorphosis in insects.** Complete metamorphosis in the monarch butterfly. A caterpillar feeds on milkweed, then forms a pupa. Inside, the pupa metamorphoses into a butterfly.   *(1,3, Lior Rubin/Peter Arnold, Inc.; 2, Ed Reschke/Peter Arnold, Inc.; 4, Don Riepe/Peter Arnold, Inc.)*

of larval cells that are set aside for later development (Figure 44-24). In insects that undergo complete metamorphosis, the larval stages are specialized for feeding and growth, while the adults are adapted to dispersal and mating. An impressive example of this separation of tasks is the mayfly, whose adult form lives only one day. It flies about, mates, and dies.

During the development of insects that undergo complete metamorphosis, the embryo hatches into a wormlike larva (called a grub, a maggot, or a caterpillar). The larva may go through several stages of increasing size before entering an inactive phase, called a **pupa,** which may be enclosed in a cocoon. During the pupal stage, almost all the tissues of the larva are digested. The cells of the imaginal discs multiply, move about, and differentiate, to form the tissues and organs of the adult. Finally, the pupa digests away its cocoon and emerges as a sexually mature adult.

Not all metamorphoses are so complete, even in insects. Grasshoppers, cockroaches, and bugs, for example, undergo **gradual metamorphosis,** in which the embryo hatches into an immature form, called a **nymph,** which is more or less a miniature adult. The nymph then undergoes a series of molts, each time getting bigger and looking more like the adult (Figure 23-22).

Do humans and other mammals undergo metamorphosis? Scientists consider puberty a variation on the theme of metamorphosis. Before adolescence, mammals are dependent, sexually immature creatures, with size and body proportions different from those of adults. Adolescents undergo a burst of growth and develop a variety of sexual characteristics that enable them to reproduce. These changes, like those that occur during metamorphosis in insects and amphibians, are controlled by hormones.

Many amphibians and insects undergo complete metamorphosis. Some insects undergo gradual metamorphosis.

Aging Depends on the Action of Both Genes and Environment

Multicellular organisms lose their ability to reproduce with age. Bacteria and other unicellular organisms, in contrast, cannot be said to age. They go on dividing as long as they survive.

From the point of view of natural selection, organisms that can no longer reproduce are dispensable. It is not surprising, then, that for many multicellular organisms, including humans, the chance of death increases with time. The relatively long life span of humans probably derives from our long period of fertility and from the importance of adults in caring for the next generation.

Human aging is a major focus of medical concern. Many measures of vitality decline with age, including fertility, cardiac output, brain weight, lung capacity, and muscle mass. In ad-

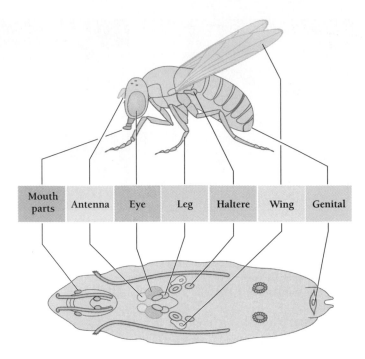

Figure 44-24   Imaginal discs. Imaginal discs in *Drosophila* are larval regions bound to develop into specific adult tissues.

dition, extracellular proteins of connective tissues lose their earlier resilience. Arteries harden, skin gets less elastic, and joints stiffen.

Aging also occurs at the cellular level. Cells taken from a young organism and grown in tissue culture can divide more times than can cells taken from an older organism. Cells appear to lose their capacity for proliferation. The speed with which wounds heal also decreases with age.

Part of the aging process must depend on genes. For example, different strains of laboratory mice differ in their average life spans. Environmental factors, such as disease, predators, parasites, or inadequate nutrition, also influence how rapidly we age. And genetically determined differences in aging may involve differences in susceptibility to environmental factors. Like other aspects of development, aging depends on the interaction of genes and environment. Aging—the decline in function of the different tissues and organs of the body—results partly from the way our genes function and partly from environmental factors.

In this final chapter of *Asking About Life,* we have surveyed the main events of developmental embryology. We have seen, for example, that although sea urchins, amphibians, and mammals may gastrulate differently, the result is the same—three germ layers that form skin, internal organs, and gut, respectively. We have also seen that the cells of early embryos are totipotent: each cell contains all the genetic information needed to construct a fully formed adult organism. The cloning of Dolly the sheep, along with other research, suggests that the nuclei of differentiated adult cells are very likely totipotent as well.

Does this mean that each cell in our bodies contains everything needed to make a new person? Could we scrape a cell from a woman's cheek and recreate her? The answer is no, for each of our cells and each of our nuclei is the product not only of the genes it carries and expresses but also of the environment in which it lives and develops.

The environment of a nucleus partly determines which genes are expressed and, therefore, the phenotype of the cell. The udder cell of a sheep may be driven into a state not unlike that of a stem cell, in which its nucleus can divide and re-differentiate to form the cloned sheep Dolly. But so far, this has occurred only when the nucleus has been transplanted into an egg cell. The cytoplasm of the original udder cell probably cannot sponsor embryological development.

In a similar way, we know the environment of a cell influences its fate. Ectoderm in contact with the notochord tends to become nervous tissue. Ectoderm in contact with the optic vesicle of the developing brain becomes lens tissue.

We can take another step back and see that the environment of a whole organism partly determines its phenotype as well. Wet soil produces a plant with a different phenotype than dry soil. Similarly, a calorie-rich, well-balanced diet produces an animal with a different phenotype than a diet deficient in either calories or specific nutrients.

Tongue in cheek, the British developmental and molecular biologist Sidney Brenner distinguished between two broad classes of developmental processes, those that occur by the "European Plan" and those that depend on the "American Plan." In the European Plan, Brenner argued, cells "do what their parents say" whereas in the American Plan, they "do what their neighbors say." In this chapter, we have learned that both plans operate—that development indeed requires genetic programs but that it also depends on environment and experience.

This lesson is a fitting place to end this book, since this conclusion leads to a slew of further questions: How does a set of genes specify a process that unfolds in time? Which features of the environment are important in affecting a particular process? How does a particular environmental trigger stimulate or inhibit the expression of the contributing genes? The answer to each question opens the door to still more questions.

Asking questions and seeking answers is one of the foremost ways of engaging with life. An unanswered question, like an unclimbed mountain or an unattained trophy, is always beckoning, calling us to do our best. We (the authors) hope that our book has provoked you (the readers) to ask your own questions about life.

STUDY OUTLINE WITH KEY TERMS

Animal development begins with a zygote that appears simple and homogeneous. Each species undergoes a characteristic sequence of developmental events, but all vertebrates pass through the same stages: **gamete formation, fertilization, cleavage, germ layer formation, organ formation, growth, metamorphosis** (in many but not all species), and **aging.**

According to the theory of **preformation,** each zygote contains a complete embryo, within which are more complete embryos. The embryo is sometimes called a **homunculus.** In some versions, these embryos within embryos come from the sperm. In other versions of preformation, the embryos are in the egg. When it became clear that a zygote creates form and differentiated tissues and cells where none (or nearly none) existed before, embryologists asked if each cell of a developing embryo was **totipotent** or if each cell was fully **differentiated,** meaning it had a specific, unchangeable fate. We now know that the cells of early embryos tend to be totipotent and the cells of even adults can sometimes be induced to develop into a fully formed adult.

The early development of a zygote is influenced by proteins translated from **maternal mRNA** transcribed during oogenesis from the genes of the mother.

Embryonic life starts with fertilization, after which cleavage divisions lead to the formation of a hollow ball of cells. In amphibians, cleavage results in an asymmetric ball of cells called a **blastula,** which has an **animal pole** that forms the embryo, and a yolky **vegetal pole.** Mammalian cleavage results in a solid ball of cells called a **morula,** which then forms a **blastocyst.** In mammals, the blastocyst consists of a **trophoblast,** which forms the tissues of the placenta, and an **inner cell mass,** which forms the embryo itself.

The hollow blastula or blastocyst then invaginates in a process called **gastrulation** that leads to the formation of a **gastrula,** with three germ layers—**ectoderm, mesoderm,** and **endoderm**—and a hollow space called the **archenteron,** which is the primitive gut.

In sea urchins, gastrulation begins when a few cells, called primary **mesenchyme,** attach to the **blastocoel** wall and form a calcium-containing skeleton. Amphibian gastrulation begins when an indentation, called the **blastopore,** forms on the side of the blastula. Mammals gastrulate when the inner cell mass separates into two layers, the **hypoblast** and the **epiblast.** The hypoblast forms a blastodisc, which then invaginates along the **primitive streak** to form the three germ layers—ectoderm, mesoderm, and endoderm.

In vertebrates, part of the mesoderm forms the **notochord,** a supportive cord that runs from head to tail that also serves as an organizer for further embryonic development. In particular, the notochord induces **neurulation.** The notochord induces the formation of **neural plate** tissue, which folds into a hollow **neural tube.** The neural tube bends to form the different regions of the brain and the spinal cord. **Neural crest cells** from the surface of the neural tube migrate into distant parts of the embryo to form nerve, skin, and other tissues.

The optic vesicles of the neural tube induce the overlying ectoderm to form the retina, lens, and cornea of the eye. Organ formation of this sort depends both on differentiation, the acquisition of specific proteins and subcellular structures, and on **morphogenesis,** the creation of form. In 9-week-old human embryos, all of the major organs appear as recognizable, although nonfunctional, **rudiments.** The embryo is then called a fetus. Programmed cell death, or **apoptosis,** contributes to normal development of the hands and feet and many other organs.

Embryologists distinguish a cell's **fate** from its **potency,** what it could become if it were given a different environment. Cells whose fate cannot be changed are said to be **determined.** Even in adults, **stem cells** remain totipotent and not determined.

In an embryo, cells differentiate according to their position in the embryo. Transplantation experiments have shown that, in amphibians, tissue from the dorsal lip of the blastopore, called the **primary organizer,** can establish the overall orientation of the embryo. Chemical signaling between different parts of the embryo provide positional information to individual cells, leading to the formation of harmonious patterns of organs and body parts. Substances whose concentration provides positional information are called **morphogens.**

Retinoic acid, for example, is a transcription factor, whose presence influences the transcription of genes.

Studies of mutants that affect the early development of *Drosophila* have revealed that the establishment of pattern depends on the production and distinctive distribution of a cascade of transcription factors. The genes for these transcription factors are called **homeotic selector genes.** Many of the proteins responsible for pattern formation in *Drosophila* are virtually identical to mammalian proteins, whose genes are called **Hox** genes.

Many animals **metamorphose** from a larval form such as a **pluteus** into an adult form such as a sea urchin. In **complete metamorphosis,** tadpoles turn into frogs or salamanders, caterpillars turn into butterflies or moths, and maggots turn into flies. In flies, the adult form derives from **imaginal discs** in the larvae. In **gradual metamorphosis,** a succession of **nymphs** gradually change from a larval form to an adult form.

REVIEW AND THOUGHT QUESTIONS

Review Questions

1. What are the eight stages of development in a vertebrate?
2. How did embryologists disprove preformation?
3. What does it mean for a cell to be totipotent? What does it mean for a cell to be determined?
4. In deuterostomes such as vertebrates and sea urchins, cleavage results in what basic form?
5. Draw a series of pictures of a mammalian embryo forming the blastocyst, trophoblast, and inner cell mass, and then gastrulating. Label as many tissues and spaces as you can.
6. Draw a picture of an embryo neurulating. What tissue induces neurulation?

7. How does the fate of a cell depend on morphogens? Give an example of a morphogen.

Thought Questions

8. If Dolly was cloned from a stem cell, rather than a fully differentiated udder cell, what are the consequences of that fact, both practical and theoretical?
9. Experiments in invertebrates suggest that the effects of maternal mRNA and other factors in the egg can influence phenotype for several generations. Design an experiment that would distinguish such maternal effects from those of the genotype. You may use any organism you like.

SELECTED READINGS

Browder, Leon W., Carol A. Erickson, and William R. Jeffery, *Developmental Biology,* 3e, Saunders College Publishing, Philadelphia, 1991. An authoritative and well-written introduction to embryology, with an emphasis on the historical and experimental aspects of the field.

Moore, Keith L., *The Developing Human: Clinically Oriented Embryology,* W.B. Saunders, Philadelphia, 1982. An in-depth look at both normal and abnormal development in humans. Fascinating reading, but not for those with weak stomachs.

Nilsson, Lennart, *A Child Is Born,* Delacorte Press, New York, 1929. This classic features now-famous photographs by Nilsson of living embryos from zygote to newborn. The text provides an accessible and gentle account of the events shown in the photographs.

▶ On-line materials relating to this chapter are on the World Wide Web at http://www.saunderscollege.com/lifesci/
Click on Tobin/Dusheck: *Asking About Life.*

PERIODIC TABLE OF THE ELEMENTS

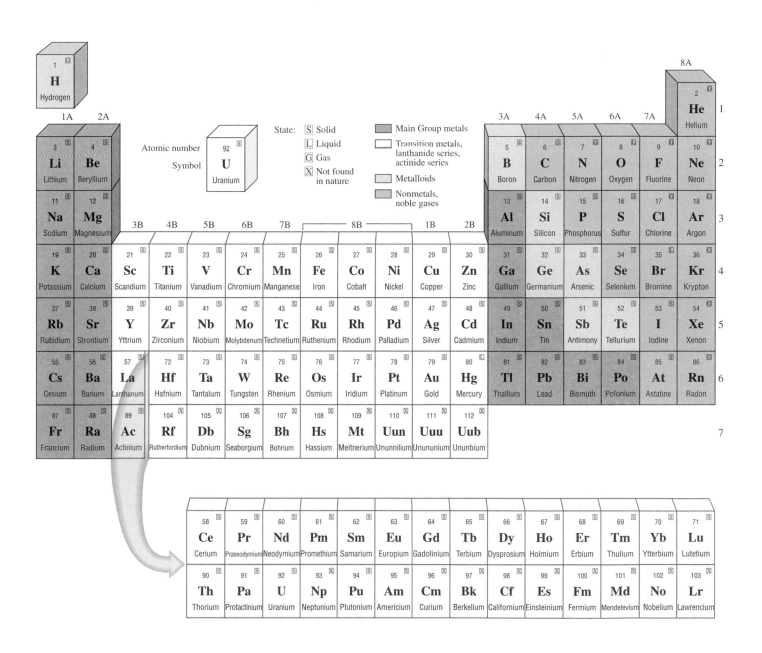

UNITS OF MEASURE: THE METRIC SYSTEM

Length

Standard Unit = Meter

1 meter (m) = 39.37 in

1 centimeter (cm) = 0.39 in   1 inch = 2.54 cm

1 kilometer (km) = 0.62 mi   1 mile = 1.61 km

Prefixes and Units of Length

Prefix	Meaning	Unit
kilo	thousand	kilometer (km) = 1000 m
centi	one-hundredth	centimeter (cm) = 0.01 m = 10^{-2} m
milli	one-thousandth	millimeter (mm) = 10^{-3} m
micro	one-millionth	micrometer (μm) = 10^{-6} m
nano	one-billionth	nanometer (nm) = 10^{-9} m

These prefixes are also used in units of volume and mass.

Volume

Standard Unit = Liter = 1000 cm³

1 liter (l) = 1.06 qt   1 qt = 0.94 l

1 milliliter (ml) = 0.03 fluid oz   1 fluid oz = 30 ml

1 l = 0.26 gal   1 gal = 3.79

Mass

Standard Unit = Kilogram = 1000 grams

1 kilogram (kg) = 2.21 lb   1 lb = 453.6 grams (g) = 0.45 kg

Temperature

Standard Unit = Centigrade

$°C = 5/9 (°F - 32)$ $°F = 9/5(°C) + 32$

Interval equivalents: 1 °Centigrade (°C) = 1.8 °F

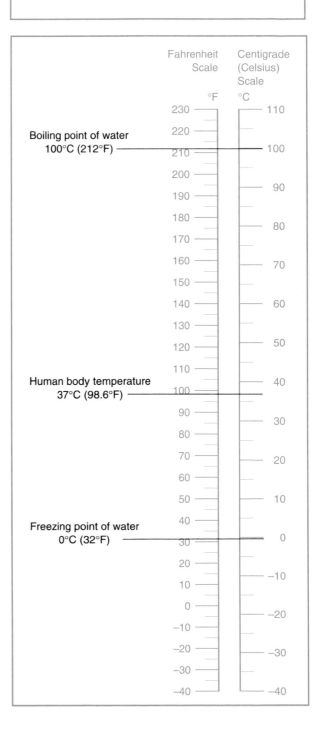

GLOSSARY

3′ end   The end of a polynucleotide at which the 3′ carbon of the nucleotide is not attached to another nucleotide but to a phosphate or a hydroxyl group.

5′ end   The end of a polynucleotide at which the 5′ carbon of the nucleotide is not attached to another nucleotide but to a phosphate or a hydroxyl group.

9 + 2 pattern   The pattern of microtubules in cilia and eukaryotic flagella; the pattern consists of nine doublet microtubules, each consisting of two microtubules fused along their length plus two singlet microtubules in the center.

α (alpha) carbon   In an amino acid, the carbon atom to which the carboxyl group and the amino group are both attached.

α (alpha) helix   A common secondary structure in proteins in which every carbonyl group of the polypeptide backbone is hydrogen bonded to the amino group four amino acids farther down the polypeptide chain.

A band (for *anisotropic,* meaning that it changes the direction of polarized light)   In muscle, the region of a sarcomere that contains myosin thick filaments.

abdomen   In a mammal, the part of the body between the chest (thorax) and the pelvis; the rear portion of an arthropod.

abdominal cavity   The part of the coelom that encloses most of the intestinal tract, as well as the liver, pancreas, and kidneys.

abiotic   Nonliving. Without life.

abscisic acid   The plant hormone that causes the abscission of leaves, dormancy in buds and seeds, and the closing of the stomata (by stimulating the flow of potassium ions out of the guard cells).

absolute age   The age in years of each of the layers of sedimentary rock.

absorb   To take up into cells and into the circulation.

absorption spectrum   The relative amounts of light of different wavelengths that a substance or solution absorbs.

absorptive state   The state of an animal shortly after a meal, during which cells in many organs take up glucose and other small molecules to make glycogen, fats, and proteins.

abstinence   Referring to birth control, refraining from sexual intercourse.

abyssal zone   Ocean waters deeper than 1000 meters.

acetyl CoA   A compound that consists of the two-carbon acetic acid (acetate) linked to a larger molecule called coenzyme A; acetyl CoA is the end point of glycolysis and a starting compound for the citric acid cycle.

acetylcholine   A neurotransmitter used in the neuromuscular junction to stimulate muscle contraction; acetylcholine is also used by parasympathetic nerves to slow the heart.

acetylcholinesterase   An extracellular enzyme that destroys acetylcholine soon after it is released into a synapse.

acid   A molecule (or part of a molecule) that can give up a hydrogen ion.

acoelomates   [Greek, *a* = without] The simplest animals, including the jellyfish and flatworms, that have no coelomic cavity.

acquired immunodeficiency syndrome (AIDS)   A disease characterized by a lack of CD4 T$_H$ cells; AIDS patients cannot mount effective immune responses to infections or to certain kinds of cancers.

acrasiomycota   [Greek, *acrasia* = confusing two things + *mukes* = fungus] Cellular slime molds—peculiar organisms that resemble animals at some stages of their lives and plants at others.

acridine dye   A chemical mutagen; it causes mutations by inserting itself into the backbone of the DNA double helix.

acrocentric   Refers to a chromosome in which the centromere is near one end; such a chromosome looks like a bent capital L.

acrosome   A specialized lysosome in a spermatozoan; it contains enzymes that allow the sperm to enter the egg.

actin   The protein that makes up actin filaments in nonmuscle cells and the thin filaments in muscle cells.

actin filaments   The most flexible elements of the cytoskeleton; formed by the association of actin molecules into long filaments about 7 nm in diameter.

action potential (also called a **nerve impulse**)   A rapid, transient, and self-propagating change of voltage across the membrane of a neuron or a muscle cell; action potentials allow long-distance signaling in the nervous system.

action spectrum   The relative effectiveness of different wavelengths in promoting a specific light-dependent process.

activation energy   The minimum energy needed to initiate a chemical reaction or other process.

active site   A groove or cleft on an enzyme's surface to which a substrate binds.

active transport   Movement of a substance across a membrane in a manner that does not occur spontaneously but requires an expenditure of energy.

adaptation   A genetically inherited structure or behavior acquired through the process of natural selection that contributes to the ability of an organism to live in its particular environment.

adaptive   Contributing to the fitness of a species.

adaptive radiation   The multiplication of a single species into many species, each with a separate ecological niche.

adenosine diphosphate (ADP)   [Greek, *di* = two] A nucleotide that consists of adenosine, ribose, and two phosphate groups; ADP is the product of the hydrolysis of a single phosphate from ATP.

adenosine triphosphate (ATP)   A nucleotide that consists of adenosine, ribose, and two phosphate groups; ATP is the universal energy currency, providing energy for many biochemical processes in all organisms.

adenylate cyclase   The enzyme that catalyzes the conversion of ATP into cyclic AMP.

adhering junction   A molecular assembly on the surface of an animal cell that connects the cell surface to actin filaments on the inner surface of the plasma membrane.

adhesiveness   The tendency to cling to a surface.

adrenal glands   [Latin, *ad* = towards + *renes* = kidneys] Endocrine glands that lie on top of the kidneys.

adrenal virilism   The development of male secondary sexual characteristics in women with excessive adrenal gland activity.

adrenergic receptor   [from Adrenaline (as in *adrenal* gland), the British trade name for **epinephrine**] One of a set of membrane proteins that binds to epinephrine and related molecules, thereby initiating a biological response.

aequorin   A protein, isolated from a luminescent jellyfish, that emits light when it binds Ca^{2+} ions.

aerobic   Requiring oxygen, usually for aerobic respiration.

afferent   Leading towards, as blood vessels entering an organ or nerve fibers carrying sensory information from peripheral nerves or sense organs to the central nervous system.

age structure   In a population, the fraction of individuals of various ages (or cohorts).

aging   Progressive developmental changes in an adult organism.

agonistic behavior   [Greek, *agonistes* = champion] All the aspects of competitive behavior within a species, including aggression, aggressive displays, appeasement, and retreat.

alchemy   An ancient study, one of whose aims was to turn "base" metals, such as lead or copper, into silver or gold.

aldosterone   A hormone, made by the adrenal glands, that stimulates sodium and potassium reabsorption in the distal tubules and collecting ducts of the kidneys.

aleurone layer   In a seed, a border of cells that surrounds the endosperm.

algae   Plantlike protists that perform photosynthesis using chlorophyll *a*.

alkaloid   A nitrogen-containing secondary plant compound.

allantois   One of the membranes that surrounds an amniotic embryo; the allantois functions in both respiration and excretion.

allele   One of several variant versions of the same gene.

allele frequency   The number of times the allele is present in the population divided by the total number of chromosomes on which the gene appears.

allergy   An inflammatory response to a harmless antigen; allergies involve the production of IgE antibodies and the release of histamine from mast cells.

allopatric speciation   [Greek, *allos* = other + *patra* = country] Species formation by geographical isolation.

allopolyploid   Description of the chromosome set that occurs when a hybrid doubles its chromosomes.

allosteric effector   (because it acts other than sterically)   A molecule or ion that changes the activity of an enzyme by binding to a site different from the active site.

altered self   In the immune response, nonself antigens presented in the context of an MHC molecule. These occur on the surface of a body cell when the cell is infected by a virus, bacterium, or other microorganism.

alternate   (also called **spiral**)   The arrangement of leaves on a stem in which the bases of the petioles form a spiral up the stem.

alternation of generations   A sexual life cycle in which haploid and diploid phases alternate.

altruism   Behavior that benefits others at the expense of the animal that performs the behavior.

alveoli   [singular, **alveolus**; Latin, *alveus* = a hollow] The expanded surfaces at the ends of the smallest passageways of the lungs; they provide the huge surface area for the exchange of oxygen and carbon dioxide.

Alzheimer's disease   A disease, associated with aging, in which patients suffer massive memory loss and disorientation.

ameboid movement   (also called **cell locomotion** or **cell crawling**)   The movement of an individual cell over a surface by extension of pseudopodia.

amine   A chemical compound that contains an amino group; many chemical signals are amines, derived from amino acids by the removal of a carboxyl group.

amino acid   A small molecule that contains both amino and carboxyl groups; the building block of polypeptides and proteins.

amino group   A functional group that consists of a nitrogen and two hydrogen atoms (NH_2).

aminoacyl tRNA synthase   A "decoding enzyme" that links a transfer RNA to its corresponding amino acid to form a "charged" (aminoacyl) tRNA.

aminoacyl-tRNA   A transfer RNA molecule linked at its 3′ end to the appropriate amino acid.

ammonia   A small molecule that consists of one nitrogen and three hydrogen atoms; it is secreted as a nitrogen-containing waste by many cells and organisms.

amniocentesis   [Greek, *centes* = puncture] Sampling of cells—usually with a syringe—in the amniotic fluid.

amnion   One of the membranes that surrounds a reptilian, avian, or mammalian embryo.

amniotic egg   An egg in which the embryo is surrounded by an amnion within a porous shell that allows the exchange of gases with the surrounding air.

amniotic fluid   [Greek, *amnion* = membrane around a fetus] The fluid that surrounds the fetus.

amphibians   Class of four-legged vertebrates that usually must reproduce in water.

amphipathic   [Greek, *amphi* = both + *pathos* = feeling] Having both a hydrophilic and hydrophobic region.

amylase   An enzyme found in saliva and pancreatic juice that hydrolyzes the polysaccharides of food into shorter fragments of maltose and glucose.

amyloplast   In a plant, a plastid that contains large granules of starches.

anabolism   [Greek, *ana* = up + *ballein* = to throw] The synthesis of complex molecules from smaller ones.

anaerobic   Without oxygen.

anaerobic photosynthetic bacteria   Photoautotrophic bacteria that live without oxygen.

anagenesis   (also called **phyletic evolution**)   Evolution within a single lineage.

analogous   Referring to structures that perform the same function though derived from different ancestral structures.

anaphase   [Greek, *ana* = up, again] The stage of mitosis during which sister chromatids separate.

anaphase A   The part of anaphase during which the chromosomes move.

anaphase B   The part of anaphase during which the poles of the spindle move apart.

anastral Having neither centrioles nor asters; a characteristic of mitosis in vascular plants.

anchoring junctions In animal cells, molecular assemblies that attach cells to each other or to extracellular matrix; anchoring junctions help provide strength to resist mechanical stress while not affecting the passage of molecules between them.

aneuploid Having an abnormal number of chromosomes.

aneurism Abnormal ballooning of the wall of a blood vessel.

angiosperms [Greek, *angion* = vessel + *sperma* = seed] The flowering plants.

angiotensin A hormone that stimulates the constriction of blood vessels.

Animalia The kingdom of animals—multicellular, heterotrophic organisms that undergo embryonic development.

anion (so called because it will move toward a positively charged electrode, or anode) A negatively charged ion.

annual rate of increase The actual percentage by which a population increases each year.

annuals Plants that complete their life cycles, from seed to mature plant, in a single growing season and then die.

antagonist A molecule that prevents a chemical signal from binding to its receptor and triggering its characteristic response in target cells.

antagonistic pair Muscles that pull in opposite directions.

antenna complex In a chloroplast, an association of chlorophyll and carotenoids that traps light and transfers the energy to the chlorophyll molecules that actually participate in photosynthesis.

anterior Toward the front of an animal.

anterior pituitary gland An endocrine organ that makes and releases a number of peptide hormones.

anther A thick pollen-bearing structure that is part of the stamen.

antheridium In plants and protists, the structure that produces sperm.

anthocyanins The phenolic pigments responsible for most of the red, pink, purple, and blue colors of flowers, fruits, leaves, and other plant parts.

Anthophyta [Greek, *antho* = flower] Angiosperms.

Anthozoa [Greek, *antho* = flower + *zoa* = animal] Sea anemones, corals, and related species.

anthropoid A monkey, ape, or human.

antibiotic A substance that kills (or interferes with the growth of) microorganisms.

antibody A blood protein that forms complexes with molecules (antigens), such as those on the surfaces of microorganisms.

anticodon A sequence of three nucleotides in transfer RNA that forms specific base pairs with the corresponding codon sequence in mRNA.

antidiuretic hormone (ADH) (also called **vasopressin**) A hormone, released by the posterior pituitary gland, that prevents water loss in the kidneys and constricts blood vessels.

antigen processing In the cellular immune response, degradation of an antigen into small peptides that associate with an MHC molecule.

antigen (an *antibody* generator) A molecule that stimulates the production of an antibody or of another immune response; a molecule that binds to an antibody or other recognition molecule within an animal's immune system.

antigenic determinant (also called an **epitope**) A specific shape and charge distribution to which antibodies or other recognition proteins can bind.

antioxidants Natural plant compounds, such as vitamin E, vitamin C, and β-carotene, that prevent the damaging effects of oxygen.

anus The opening at the far end of the digestive tract through which the undigested and unabsorbed material leaves.

aorta The artery that carries blood from the left ventricle to the rest of the body.

apical dominance In plant development, the suppression of lateral bud development by terminal buds.

apical meristem [Latin, *apex* = top + Greek, *meristos* = divided] In a plant, a self-renewing group of undifferentiated cells, just behind the tip of the shoot or root, that generates differentiated structures.

apoptosis [Greek, *apo* = away from + *ptosis* = fall] Programmed cell death.

aqueous Watery; dissolved in water.

Archaea (also called **Archaebacteria**) [Greek, *archein* = to begin] The kingdom (or "domain") of single-celled organisms that live under extreme environmental conditions and have distinctive biochemical features.

archegonium In ferns and mosses, the female reproductive organ that produces and houses the ova.

archenteron [Greek, *arche* = beginning + *enteron* = gut] The "primitive gut," the innermost tube of an animal embryo; it is lined with endoderm and will become the digestive tract.

aromatic Having a benzene-like planar ring of atoms.

artery One of the vessels that carry blood away from the heart.

arterioles Smaller vessels that branch off from arteries.

artifact A remnant or result of what the researcher does rather than of what exists in nature.

artificial selection The process by which animal breeders and farmers create new lines of plants and animals by selecting the most desirable individuals for breeding.

Aschelminths A group of eight phyla, all bilaterally symmetrical animals with either no coelom or a pseudocoel.

asci (singular, **ascus**) The little sacs in which ascomycetes produce their sexual spores.

ascomycetes (also called **sac fungi**) [Greek, *askos* = wineskin + *myketos* = fungus] So named because their sexual spores are always produced in little sacs.

asexual reproduction A process that produces offspring with genes from a single parent.

associative learning A type of learning in which the subject learns to respond to a stimulus (called the conditioned stimulus) that is not obviously related to the primary stimulus; the response to the conditioned stimulus arises because of its association with a stimulus that leads directly to a physiological response.

assortative mating Nonrandom mating.

aster [Latin, star] A starlike object visible in most dividing eukaryotic cells (other than those of vascular plants); the aster contains the microtubule organizing center.

astral Having an aster and centrioles that participate in mitosis; characteristic of mitosis in animals and in nonvascular plants.

astral microtubules Microtubules that extend from each pole of the mitotic spindle without attaching to any other visible structure.

asymmetric A carbon atom surrounded by four different groupings of atoms.

atherosclerosis Hardening of the arteries.

atom The smallest unit of matter that still has the properties of an element.

ATPase   An enzyme that hydrolyzes ATP into ADP and phosphate.

ATP synthase   An enzyme responsible for making ATP.

atrial natriuretic factor   [Latin, *atrium* = vestibule, referring to the heart chamber, + *natrium* = sodium + Greek, *ouron* = urine] A hormone produced in the smaller chambers of the heart (the atria).

atrioventricular (AV) node   A special tissue that conducts electrical signals from one part of the heart to the other.

atrioventricular valves   The valves between the atria and ventricles of the heart.

atrium   [Latin, courtyard or entry] One of the smaller thin-walled entrance chambers of the heart, through which blood enters the adjacent ventricle.

australopithecine   [Latin, *australis* = south + Greek, *pithekos* = ape] An early hominid.

autonomic nervous system   Involuntary nervous system; it coordinates the responses of smooth muscles, cardiac muscles, and other effector organs—including those of the endocrine, digestive, excretory, respiratory, and cardiovascular systems.

autonomous replication   Propagation of a virus or plasmid in which the replicating DNA remains separate from the DNA of the host cell.

autoradiography   A method that detects radioactive compounds by their ability to expose a photographic film or a photographic emulsion.

autosomes   The chromosomes that do not differ between males and females.

autotroph   [Greek, *autos* = self + *trophos* = feeder] An organism that can make its own organic molecules from simple inorganic compounds (such as carbon dioxide, water, and ammonia).

axillary bud (also called **lateral bud**)   [Greek, *axilla* = armpit] A bud that forms just above the point where a leaf joins the stem and which can develop into a branch.

axon   A type of neuronal process that usually carries signals away from the cell body and connects to other cells.

β (beta) structure   A common secondary structure in proteins in which the amino and carbonyl groups of the polypeptide chain are hydrogen bonded to the other polypeptide chains or to distant regions of the same chain folded back on itself.

B lymphocyte   Lymphocytes that when stimulated divide to form antibody-producing effector cells or memory cells.

bacilli   [singular, **bacillus**; Latin, little rod] Rod-shaped prokaryotic cells.

backbone   A column of hollow bony segments called vertebrae.

bacterial flagellum   A long structure, about 10 to 20 nm thick and up to 10 μm long, whose rotation propels a bacterium through a liquid.

bacteriophage (or **phage**)   [Greek, *bakterion* = little rod + *phagein* = to eat] A virus that infects bacteria.

balanced polymorphism   A balance of different alleles in a population.

basal body   The base of an axoneme; a cylinder about 500 nm long that resembles a centriole; the microtubule organizing center of a cilium or a eukaryotic flagellum.

basal metabolic rate   The minimum amount of energy required just to stay alive and awake.

base   A molecule (or part of a molecule) that can accept a hydrogen ion.

base substitution   A type of point mutation in which one base (nucleotide) is replaced by another.

basidiocarp   [Latin, *carpus* = fruit] The fruiting body of a mushroom.

basidiomycetes, or **club fungi**   Include many of the most familiar fungi: the mushrooms, the bracket fungi, and the puffballs.

basidium   [plural, **basidia**; Latin, little pedestal] A tiny, club-shaped structure that produces spores.

Batesian mimicry   A deception in which one species mimics another's warning coloration for protection.

behavioral ecology (formerly called **sociobiology**)   The study of behavior from an evolutionary perspective.

behavioral isolation   Reproductive isolation that results from different mating behaviors.

benthic division   All the organisms that live on the ocean bottom.

biceps   The major muscle on the front side of the upper arm.

bilateral symmetry   Structural symmetry such that one cut through the axis will produce identical (or mirrored) halves.

bile   A detergent solution, secreted by the liver; bitter, alkaline, and an ugly green-yellow or brown-yellow color.

binary fission   The process of cell division (in prokaryotes) in which a cell pinches in two, distributing its materials and molecular machinery more or less evenly to the two daughter cells.

binding site   The region of a protein molecule to which a substrate or a chemical signal binds.

binomial   A two-part name that includes both the genus and species.

bioassay   A method that estimates the concentration of a substance by measuring its biological activity.

biochemistry (also called **biological chemistry**)   The study of the structures and reactions that actually occur in living organisms.

biogeochemical cycle   The movement of a substance, such as carbon or nitrogen, through many forms—both biotic and abiotic.

biogeography   The study of the past and present distribution of plant and animal species.

biological species   The largest unit of a population of similar organisms in which gene flow is possible.

Biological Species Concept   A definition of species that says species are groups of actually or potentially interbreeding populations, which are reproductively isolated from other such groups.

biology   [Greek, *bios* = life, *logo* = word] The science of life.

biomagnification   Toxic chemicals are concentrated at higher levels in the food chain.

biomass   Aggregate dry weight of all organisms in a community or ecosystem.

biome   A geographical region characterized by a distinctive landscape, climate, and community of plants and animals.

biosphere   The system of living things that covers the Earth.

biotic   Living.

bipedal   Consistently walking on two rather than four feet.

birth control   The conscious regulation of reproduction.

birth control pill   A formulation of estrogen and progesterone that prevents the anterior pituitary from secreting LH, thereby stopping ovulation and preventing conception.

birth rate   The annual number of births per 1000 individuals in a population.

bivalent (also called a **tetrad**)   A chromosome pair visible during meiosis I.

bivalves   [Greek, *bi* = two + *valva* = part of a folding door] Includes clams, oysters, scallops, mussels, and other mollusks with two shells—a right shell and a left shell.

bladder   Where urine is stored until its elimination.

blade   The thin, flat part of a leaf.

blastocoel   In an animal embryo, the interior of the blastula.

blastocyst   [Greek, *blastos* = sprout + *cystos* = cavity] A modified blastula in which the cells do not lie within a single layer, but do enclose an internal cavity.

blastopore   An opening at the end of the archenteron.

blastula   [Greek, little sprout] A stage of an animal embryo that consists of a sphere with cells on the surface and fluid inside.

blending inheritance   The pattern of inheritance in which offspring appear to have characteristics intermediate between those of their parents.

bolus   [Greek, *bolos* = lump of earth] In digestion, a ball of macerated food and lubricating mucus.

bone   The hard connective tissue of a vertebrate; bone consists of fibers of collagen and crystals of calcium phosphate.

book lung   Stacks of modified gills that provide a surface of gas exchange in some animals, including spiders.

bordered pit pair   A region of the border between two plant cells in the xylem where the adjacent secondary wall overarches the pit membrane and reinforces the wall of a tracheid.

bottleneck effect   The restriction of genetic diversity of a population that results from the reduction of a large population to a few surviving individuals by a random disaster or harsh selection pressure.

Bowman's capsule   In a kidney, the bulb that surrounds each glomerulus.

BPG (2,3-bisphosphoglycerate)   A relatively abundant 3-carbon molecule in mammalian red blood cells. The binding of BPG to hemoglobin decreases oxygen affinity, allowing it to release more oxygen in the tissues than would hemoglobin without bound BPG.

branchial arches   In fishes and amphibians, the tissues between the gill slits.

breathing center   A nerve complex on which the coordination of breathing depends.

bronchi   In the mammalian respiratory system, the two tubes that branch from the tracheae and carry air to and from the lungs.

bronchioles   In the respiratory system, the smallest branches of the bronchi.

Brownian motion   (after Robert Brown, a 19th-century British surgeon and botanist) Jerky movements of small particles, reflecting the random thermal motion of molecules.

bryophytes   [Greek, *bryon* = moss] The mosses and their relatives.

bud   In a plant, the precursor of a leaf or a flower, consisting of several leaf primordia, separated by internodes whose cells have not elongated.

budding yeast   (*Saccharomyces cerevisiae*; also called baker's yeast and brewer's yeast)   A widely used experimental organism that divides by budding rather than by binary fission.

buffer   A molecule that easily converts between acidic and basic forms by donating or accepting one or more hydrogen ions.

bulb of the vestibule   One of the corpora cavernosa of the clitoris that corresponds to the penis's corpus spongiosum.

bulbourethral glands   The glands that inject a small amount of mucus into the semen.

callus   Undifferentiated tissue that forms at the cut surface of a plant.

calorie   A measure of energy; the amount of energy needed to raise the temperature of one gram of water 1°C.

calyx   The fourth and outermost whorl of leaflike parts called sepals.

Cambrian explosion   The burst of diversification—unmatched in the history of the world—in which all of the modern animal phyla that have fossilizable skeletons appeared.

camouflage   A defense against being eaten in which organisms mimic materials in their environment in an attempt to be invisible.

canopy   In a tropical rain forest, the continuous layer of trees 30 to 40 meters above the ground.

capacitation   The process in which the glycoprotein coat that covers the head of each sperm is dissolved by enzymes in the female reproductive tract.

capillary   One of the minute blood vessels that brings blood in closest contact with tissues; capillaries connect the finest arterioles with the finest venules.

capsid   A coat of protein on the outside surface of a virus.

capsule   A gelatinous layer of polysaccharides and proteins outside the cell wall; in disease-causing bacteria, the capsule protects from attack by the host's white blood cells, enabling the bacteria to overwhelm the body's immune system.

carbohydrate   A compound that contains the equivalent of one water molecule (one oxygen atom and two hydrogen atoms) for every carbon atom; includes sugars and polysaccharides.

carbonyl   A functional group that consists of a carbon atom attached to an oxygen atom by a double bond (C=O).

carboxyl   A functional group that contains one carbon atom and two oxygen atoms; the carbon atom forms a double bond with one oxygen atom and a single bond with the other.

carcinogen   A chemical that causes cancer.

cardiac muscle   The muscles that pump blood through the heart.

cardiovascular system   The blood, the heart, and the blood vessels together.

carnivore   [Latin, *carn* = flesh] An animal that eats only other animals.

carotenoid   A plant pigment, yellow or orange in color.

carpel   (also called a **pistil**)   In a flower, a female reproductive structure.

carrying capacity   The plateau value of the S-shaped curve of population growth. The maximum number of individuals of a species that a habitat can support.

cartilage   A type of connective tissue that cushions joints; cartilage consists of collagen, but without calcium phosphate.

Casparian strip   In plants, a waxy wall that extends around the walls of endodermal root cells; it restricts the movement of water and dissolved solutes.

catabolism   [Greek, *cata* = down + *ballein* = to throw] The reactions of metabolism that break down complex molecules, such as those in food.

catabolite activator protein (CAP)   A positive regulator of transcription in *E. coli*; CAP binds to the promoter region of the *lac* operon and increases transcription whenever cyclic AMP is present at sufficiently high levels.

catalyst   A substance that accelerates a chemical reaction and is not itself consumed in that reaction.

catarrhines   [Greek, *kata* = down + *rina* = nose] Monkeys that have downward pointing nostrils and often no tail at all.

catastrophism   The view that the discontinuities in the fossil record result from a series of catastrophes; some catastrophists argued that life arose anew after each catastrophe, rather than deriving from previously existing species.

cation (so-called because it will move toward a negatively charged electrode or cathode)   A positively charged ion.

cecum   [Latin, *caecus* = blind] A blind sac of the digestive tract, near the junction between the large and small intestine.

cell body   The center of a neuron, including its nucleus, as distinct from its axon or dendrites.

cell center (also called the **centrosome**)   The microtubule organizing center; a complex of proteins that in animal cells includes a pair of centrioles.

cell cortex   A meshwork of actin filaments just below the surface of a eukaryotic cell.

cell crawling (also called **cell locomotion** or **ameboid movement**)   The movement of cells over a surface.

cell cycle   The orderly sequence of events that accomplish cell reproduction.

cell division   The process by which a parent cell gives rise to two daughter cells that carry the same genetic information as the parent cell.

cell locomotion (also called **cell crawling** or **ameboid movement**)   The movement of cells over a surface.

cell plate   In a dividing plant cell, the precursor of a new cell wall between two daughter cells.

cell theory   The summary of the cellular basis of organisms: (1) All organisms are composed of one or more cells; (2) cells, themselves alive, are the basic living unit of organization of all organisms; and (3) all cells come from other cells.

cell wall   An external rigid structure surrounding all plant cells and most prokaryotes.

cellular oncogene (also called a **proto-oncogene**)   The cellular counterpart of a viral oncogene; cellular oncogenes participate in the normal control of growth and differentiation.

cellulose   A polysaccharide made by plants; the major structural material of wood, cotton, and paper.

central cell   A cell within the ovule of a flower; it contains two haploid nuclei, derived from the megaspore and develops into endosperm after fertilization.

central dogma   The frequently violated principle that "DNA specifies RNA, which specifies proteins."

central nervous system (CNS)   The brain and spinal cord.

central vacuole   A large membrane-enclosed space within a plant cell.

centrifugation   A method for separating macromolecules and subcellular structures according to size (and shape) by subjecting them to high gravitational fields in a spinning tube.

centriole   A pair of small cylindrical structures, each about 0.2 μm in diameter and 0.4 μm long, that lie at right angles to one another; centrioles are present at each pole of the mitotic spindle in animal cells and in some other eukaryotes.

centromere   The point at which the two chromatids of a single chromosome are joined.

centrosome (also called the **cell center**)   A microtubule organizing center that (in many cells) also contains a distinctive organelle called the centriole; the centrosome plays an important role in cell division.

cephalization   The concentration of sense organs and ganglia in the anterior end of an animal.

cephalochordates (also called lancelets)   [Greek, *cephalo* = head] About 45 species of small, segmented, fishlike animals. Most belong to the single genus *Branchiostoma*, whose common name is amphioxus.

cervical cap   A birth control device; a thimble-shaped rubber or plastic cap, about an inch in diameter, that fits tightly over the cervix and prevents fertilization.

cervix   [Latin, neck] The narrower, lower part of the uterus.

cestoda   Tapeworms.

chaetognatha   Arrow worms.

channel   A membrane protein that allows the passage of specific molecules or ions.

chaparral   [Basque, *chabarro* = dwarf evergreen] The dense growth of shrubs that is characteristic of the American Southwest and other regions, all on the west coasts of continents, that share a "Mediterranean" climate, with mild, wet winters and hot, dry summers.

chaperone   One of a number of special proteins that prevent promiscuous interactions by binding to unfolded polypeptides and catalyzing correct folding.

character displacement   In evolution, a change in morphology, life history, or behavior that results from competition.

chelicerae   The first pair of appendages in spiders and other arachnids; they may serve as pincers or as fangs that are associated with poison glands.

chemical equilibrium   [Latin, *aequus* = equal + *libra* = balance] A balance between forward and reverse chemical reactions.

chemical reaction   A transformation in which different forms of matter combine or break down.

chemical synapse   A distinct boundary between two neurons through which neurotransmitters diffuse.

chemical work   The energy-consuming formation or breakage of chemical bonds.

chemiosmosis   [Greek, *osmos* = to push] The linking of chemical and transport processes.

chemistry   The study of the properties and the transformations of matter.

chemoautotroph   An autotroph that derives energy by oxidizing such inorganic substances as hydrogen sulfide (H_2S) or ammonia (NH_3).

chemoheterotroph   A heterotroph that uses no light, relying exclusively on organic molecules for both energy and carbon atoms.

chemotaxis   The movement of a cell toward a higher (or, in some cases, a lower) concentration of a particular chemical.

chiasma   [plural, **chiasmata**; Greek, cross] The sites of exchange of DNA between homologous chromosomes during meiosis; chiasma is visible during prophase of meiosis I.

chitin   A tough, nitrogen-containing polysaccharide in the cell walls of insects, arachnids, and crustaceans.

chlorophyll   The green pigment of plants and protists; chlorophyll is responsible for absorbing light for photosynthesis.

chlorophytes   [Greek, *chloros* = green] Green algae.

chloroplast   A large, green, membrane-enclosed organelle that performs photosynthesis.

chloroxybacteria   Certain bacteria that live as photoautotrophs.

cholesterol   A small molecule that consists of four interconnected rings of carbon atoms; a component of cell membranes and the starting compound in the synthesis of steroids.

chondrichthyes   [Greek, *chondros* = cartilage + *ichthys* = fish] Sharks and rays—fish with cartilaginous skeletons.

Chordata   All vertebrate animals—fish, amphibians, reptiles, birds, and mammals—and animals that have notochords, including sea squirts and lancelets.

chorion   [Greek, skin] One of the membranes that surrounds an embryo; it lies just beneath the shell within an amniotic egg.

chorionic villus sampling   A procedure for sampling cells of the early embryo that uses fetal cells present in the placenta.

chromatid   One of the two separate but connected bodies that make up a chromosome at the beginning of mitosis, when the chromosomes first become visible; a chromatid contains a single long molecule of DNA.

chromatin   [Greek, *chroma* = color] A diffuse material within the nucleus of a nondividing eukaryotic cell; chromatin consists of DNA and proteins.

chromatography   A method for separating molecules according to their relative affinities for a stationary support, called the stationary phase, and a moving solution, called the mobile phase.

chromoplast   A plastid that contains the pigments of fruits and flowers.

chromosomal mutation   A change in a relatively large region of a chromosome.

chromosome   [Greek, *chroma* = color + *soma* = body; because it is stained by certain dyes] A discrete complex of DNA and proteins, visible with a light microscope within a dividing cell; originally used only for eukaryotic cells, but now also used to mean a single large molecule of DNA that contains the genes of a bacterium or a virus.

chromosome banding   The distinctive pattern, visible in a light microscope, that results from the selective binding of certain dyes to individual chromosomes.

chrysophytes   [Greek, *chrysos* = golden + *phyta* = plant] Golden algae and diatoms.

chyme   [Greek, *khumos* = juice] A creamy, acidic liquid that passes into the small intestine.

ciliophora   [plural, **ciliates**; Latin, *cilium* = eyelash] Some 8000 species of ciliated free-living, single-celled heterotrophs that live both in fresh water and salt water.

cilium   [plural, **cilia**; Latin, eyelid, from the hairlike appearance of a cilium]   A protein assembly, consisting of microtubules, that can move a cell through a liquid medium (or a liquid medium over a cellular surface); a single cell usually contains many cilia, often arranged in rows; cilia have the same organizational plan as eukaryotic flagella but cilia are much shorter.

circular   For DNA, forming a continuous loop; examples of circular DNAs include those of bacteria, energy-producing organelles, and many viruses.

circulation   (from "circle")   The route of the blood throughout the cardiovascular system.

circumcision   The surgical removal of foreskin.

citric acid cycle   (also called the **Krebs cycle** or the **tricarboxylic acid cycle**)   A set of reactions that converts the carbon atoms of acetyl CoA into carbon dioxide.

cladistics   [Greek, *clados* = branch] An approach to taxonomy whose first goal is to describe the groupings of organisms in a way that shows their phylogeny.

cladogenesis   Branching evolution.

class   A taxonomic group that consists of one or more orders; a subdivision of a phylum (or, for plants, of a division).

clay   A soil that consists only of small particles (less than 2 mm).

cleavage   A series of rapid cell divisions following fertilization in early animal embryos; cleavage divides the embryo without increasing its mass.

cleavage furrow   A groove formed from the cell membrane in a dividing cell as the contractile ring tightens.

climax community   The long-lived community at the end of a succession.

clitoris   A highly sensitive erectile tissue that is part of the female genitalia; it is homologous to the penis.

cloaca   A common entrance and exit chamber for the digestive, urinary, and reproductive systems in many vertebrates.

clonal selection theory   In immunology, a summary of the role of selective proliferation of responding lymphocytes, already programmed, into clones of cells that contain specific gene rearrangements and that recognize specific antigens.

clone   [Greek, *klon* = twig] A population of genetically identical individuals or cells descended from a single ancestor.

closed circulatory system   A circulatory system in which blood runs only within enclosed vessels.

cnidarians   [Greek, *cnide* = nettle] One of two phyla of radially symmetrical acoelomates; cnidarians include three classes: Hydrozoa (hydroids), Scyphozoa (jellyfish), and Anthozoa (corals and sea anemones).

cnidocytes   Specialized stinging cells that lie on the tentacles of cnidarians.

coacervate   [Latin, *coacervatus* = heaped up] Discrete tiny droplet into which proteins and polysaccharides can spontaneously concentrate.

cocci   [singular, **coccus**; Greek, *kokkos* = berry] Spherical-shaped prokaryotic cells.

codominant   Two alleles that each contribute to the phenotype of a heterozygote.

codon   A group of three nucleotides that specifies a single amino acid.

coelom   [Greek, *koilos* = hollow] A cavity lined by a layer of mesoderm cells.

coelomates   Animals with a coelom.

coenocytic   [Greek, *koinos* = shared + *kytos* = hollow vessel] Fungi in which the nuclei all lie in a common cytoplasm rather than in separate membrane-bound cells.

coenzyme   An organic molecule (but not a protein) that is a necessary participant in an enzyme reaction.

coevolution   The mutual adaptation of two separate evolutionary lines.

cohesion   An attraction between molecules of the same substance.

cohort   The set of individuals that enter the population (or are born) at the same time.

coitus interruptus   [Latin, *coitus* = sexual intercourse] A method of birth control, also called withdrawal, in which a man attempts to avoid conception by removing his penis from the vagina before ejaculation.

collagen   A fibrous protein, secreted by fibroblasts, that provides mechanical strength in cartilage, ligaments, tendons, and bones; collagen is by far the most abundant protein in the extracellular matrix.

collagen helix   A regular structure found principally in the structural protein collagen; it consists of three polypeptide chains wound around each other.

collenchyma   [Greek, *kolla* = glue] Plant cells that have a thick primary cell wall and often provide mechanical support; collenchyma are usually in the growing regions of stems and leaves.

colon   The lower end of the large intestine; it is specialized for the absorption of water and ions.

commensal   The species that benefits from the protection of its host in a commensalistic relationship.

commensalism   An association between individuals of two species in which one organism benefits without harming the other one.

communicating junctions   Membrane-associated structures that allow small molecules to pass freely between two adjacent cells.

community   An interacting group of species that inhabit a common area.

compact bone tissue   The hard, dense bone that surrounds the spongy interior.

companion cell   In plants, a long, nucleated, fully functional parenchyma cell that supplies a sieve tube member with proteins and energy-rich molecules.

comparative anatomy   The study of morphological similarities and differences among organisms; it often provides clues to evolutionary relationships.

competition   The interaction between organisms or species that depend on the same limited resource; competition occurs when one organism or species uses a resource in a way that limits the availability of that resource to others.

competitive exclusion principle   When two species compete directly for exactly the same limiting resources, the more efficient species will eliminate the other.

competitive inhibitor   A molecule whose inhibitory effects on an enzyme can be overcome by increased substrate concentration.

complement   A set of blood proteins that attack microbial invaders.

complete metamorphosis   The pattern of development in which there are distinct larval, pupal, and adult forms; this pattern is characteristic of many insects, including *Drosophila,* fleas, flies, beetles, wasps, moths, and butterflies.

completed family size   The average number of children that reach reproductive age born to each family.

complex tissue   Tissue consisting of more than one cell type, each with a distinct functional role.

complex transposon   A transposable element that contains other genes besides transposase, for example, a gene for antibiotic resistance.

compound   A substance produced, by a chemical reaction, from two or more different elements.

compound eye   An eye that consists of numerous simple light-detecting units; this type includes the eyes of most arthropod species.

compound microscope   A light microscope that contains several lenses.

compression   The pushing action of two opposing forces.

concave (or type III) survivorship curve   Description of a life history in which individuals have the greatest chances of dying early in life.

concentration   The number of molecules in a given volume.

concentration gradient (so named because the concentration changes in a graded manner)   A graded difference in the concentration of a substance.

conception   Fertilization or syngamy. The union of an egg and a sperm to form a zygote.

condensation (or dehydration condensation) reaction   The linking of two building blocks, accompanied by the removal of a water molecule.

conditional mutation   A change in DNA that alters a protein so that it can function under some conditions but not under others.

conditioned stimulus   In associative learning, the arbitrary stimulus that is paired with a stimulus (the "unconditioned stimulus") that directly evokes a physiological response.

condom   A thin rubber sheath that covers the penis and prevents sperm from entering the vagina; a condom is used both as a birth control device and to prevent the transmission of sexually transmitted diseases.

conformation   The three-dimensional arrangement of a polypeptide chain, the equivalent of its tertiary structure.

conjugation   A process in which two temporarily attached bacteria exchange genetic material.

connective tissue   A relatively sparse population of cells within a bed of extracellular matrix; includes bone and cartilage.

conservative   A possible pattern of DNA replication in which each parent molecule would remain intact and both strands of the descendant molecules would be newly assembled; actual DNA replication is not conservative, but semiconservative.

constant region   The part of an immunoglobulin light chain or heavy chain that is identical in all antibodies of a particular class.

consumer   An organism, usually an animal, that obtains energy by eating producers or other consumers.

contact inhibition of cell division   The ability of cells to stop dividing when neighboring cells touch each other.

contractile ring   A bundle of actin filaments that surrounds a dividing cell and pinches the cytoplasm in two during cytokinesis.

contrast   In microscopy, an object's ability to absorb more or less light than its surroundings.

control   In an experiment, the situation in which there has been no experimental manipulation; comparing measurements made in a control versus an experimental situation allows the evaluation of a hypothesis.

convergent evolution   The independent evolution of similar features in separate groups of organisms.

convex (or type I) survivorship curve   Description of a life history in which survivorship starts out high and decreases slowly with age until a certain point when survivorship begins to decrease more rapidly.

Coriolis effect   The twisting effect of rotation on movement, particularly of wind or of water.

corolla   In a flower, the whorl of petals.

corpus cavernosa   In the male reproductive system of mammals, two spongy cylinders that run along the length of the penis and fill with blood during erection.

corpus luteum   [Latin, yellow body] A structure within the ovary that develops from the ruptured follicle; it secretes progesterone and some estrogen.

corpus spongiosum   A spongy cylindrical tissue that surrounds the urethra.

cortex   The outer region of a cell, a tissue, or an organ.

cortical microtubules   Microtubules that encircle a dividing plant cell; they determine the placement of a new cell wall.

corticotropin (also called **adrenocorticotrophic hormone** or **ACTH**)   A peptide hormone, made in the anterior pituitary, that stimulates the production of corticosteroids in the adrenal cortex.

cotransport   The coupled transport of two substances across a membrane; cotransport depends on specific transmembrane protein molecules.

cotransporter   A transmembrane protein that binds to two or more molecules or ions and transports them across a membrane.

cotyledon (also called **seed leaves**)   [Greek, *kotyle* = a deep cup] The very first leaves that sprout from a seed; present in the embryos of seed plants, they contain stores of nutrients for use after germination.

countercurrent system A system in which heat, fluids, or gases run past each other in a manner that allows the exchange of heat or matter.

coupling factor (also called the F_0-F_1 **complex**) A protein complex in the mitochondrial membrane that uses the flow of protons to drive the synthesis of ATP.

covalent bond Shared arrangement of electrons that holds atoms together in molecules.

crassulacean acid metabolism (CAM) A variant of the C_4 pathway of the light-independent reactions of photosynthesis; CAM is used by succulent plants such as cacti.

creationism The view of the origin of life that says that the Earth was created very recently and that each species was created individually.

cristae Elaborate folds of the inner membrane of a mitochondrion.

Cro-Magnon An early species of *Homo sapiens*.

crop In the digestive tract of birds, a saclike extension of the esophagus; the crop can store food for later digestion.

cross bridges Extensions of myosin molecules that perform the work of muscle contraction by pulling the thin and thick filaments over one another.

crossbreeding (or **crossing**) The interbreeding of two genetically distinct organisms.

crossing over One type of genetic recombination; it involves the breakage and rejoining of single chromatids of homologous chromosomes.

crustacea A subphylum of arthropods in which the first pair (or the first few pairs) of appendages are antennae, and the next pair are jaws, or mandibles.

crystal A solid that is enclosed by geometrically regular faces; the starting material point for x-ray diffraction analysis.

ctenophores [Greek, *ktenos* = comb + *phora* = motion] Comb jellies, a small phylum of about 90 living species.

cultural energy Energy expended by humans in crop production.

current The movement of electrical charges.

cuticle A waxy covering that keeps the above-ground parts of plants from losing water.

cyanobacteria Photosynthetic bacteria; previously called "blue-green algae"; the most ancient photosynthetic organisms known and the probable ancestors of chloroplasts.

cyclic AMP (cAMP) Adenosine monophosphate in which the same phosphate group is linked to carbon atoms 3 and 5 of ribose; it is used as a "hunger" signal in bacteria and protists and as a second messenger in mammalian cells.

cyclic AMP phosphodiesterase The enzyme that hydrolyzes cyclic AMP to AMP.

cyclic AMP-dependent protein kinase (also called **A-kinase**) An enzyme that, when bound to cyclic AMP, transfers phosphate groups from ATP to sites in particular proteins.

cyclic photophosphorylation The production of ATP from light energy by a series of electron transfers that regenerate the absorbing chlorophyll; in plants, cyclic photophosphorylation depends on the flow of electrons from excited P_{700} in a cycle that regenerates P_{700}, which is then able to absorb another photon of light.

cyst An enclosed structure that contains reproductive cells.

cytochalasin A poison that interferes with the growth of an actin filament.

cytochrome [Greek, *kytos* = hollow vessel + *chroma* = color] One of a set of heme-containing electron carrier proteins that change color as they accept or donate electrons.

cytodifferentiation The development of a specialized cell.

cytokine One of a number of small proteins that regulate proliferation and protein synthesis in cells that participate in inflammation and in the immune response.

cytokinesis [Greek, *kytos* = hollow vessel + *kinesis* = movement] The division of the cytoplasm and formation of two separate plasma membranes.

cytokinin A plant hormone that influences the rate of division and differentiation.

cytoplasm [Greek, *cyto* = a hollow container, i.e., a cell + Latin, *plasma* = a thing molded or formed] In eukaryotes, that part of the cell outside the nucleus but inside the plasma membrane.

cytoplasmic streaming The movements of membrane-bounded organelles along actin filaments.

cytoskeleton A network of protein fibers that runs through the cytosol of eukaryotic cells; it consists of microtubules, actin filaments, and intermediate filaments, along with other associated proteins.

cytosol In eukaryotic cells, the part of the cytoplasm not contained in membrane-bounded organelles.

cytotoxic T lymphocyte (CTL) An effector cell, derived from a T_C cell, that actually kills a target cell.

cytotoxicity In the immune response, the ability of certain T lymphocytes to kill cells with altered surfaces.

daughter cells The cells produced from a single parent cell after cell division.

dead space The volume of air in the air passages that does not come into contact with the surfaces of the alveoli in the lungs.

death rate The number of individuals per 1000 who die each year.

deciduous Shedding leaves annually, as in many trees and shrubs.

decomposer An organism, such as a bacterium or fungus, that lives on the energy in the complex molecules of dead organisms.

degenerate Meaning that several codons are equivalent, that is, that they specify the same amino acid.

dehydration A physiological state in which the body lacks sufficient water.

deletion A type of mutation; the removal of one or more nucleotides.

demography [Greek, *demos* = people + *graphos* = measurement] The statistical study of populations.

denatured A protein that has lost its native, three-dimensional structure and its functional activity, while still having an unaltered primary structure.

dendrite A type of neuronal process that usually carries signals to the cell body.

dendritic cell In the immune response, a type of antigen-presenting cell; a highly extended cell present in many organs and tissues throughout the body.

denitrifying bacteria Bacteria that convert nitrate to atmospheric nitrogen.

density gradient A changing concentration of a dissolved substance—usually sucrose, glycerol, or cesium chloride—with the highest concentration at the bottom and the lowest at the top.

density-dependent Referring to population growth, limited in proportion to population density—the denser the population, the more slowly it grows.

density-independent Referring to population growth, limited by factors other than population density—such as fire, drought, and other natural disasters.

deoxyhemoglobin   Hemoglobin that has no bound oxygen.

deoxyribonucleic acid (DNA)   The long thin molecules in which organisms store and transmit genetic information.

Depo-Provera (DMPA, depo-medroxy progesterone acetate)   A birth-control pill; a progesterone derivative taken as an injection; it suppresses ovulation and inhibits implantation for three months at a time.

depolarizing   A voltage change across an excitable membrane that makes the inside of the cell less negative and more apt to produce an action potential.

depurination   The loss of a purine base from deoxyribose in DNA.

descent with modification   Evolution; change in organisms over time.

desert   A dry, relatively barren region, with lower productivity than most other biomes.

desmosome   An anchoring junction that consists of transmembrane proteins that attach to a cell's intermediate filaments.

detergent   An amphipathic molecule that interacts both with water and with hydrophobic molecules.

determined   In development, having a limited potency. A determined cell can no longer develop in accordance with new environmental signals.

detritus   Particles of dead organic matter.

deuteromycetes   Imperfect fungi.

deuterostomes   [Greek, *deuteros* = second + *stoma* = mouth] Those animals in which the anus forms first, from the blastopore, and the mouth forms secondarily.

diabetes mellitus   A disease characterized by the overproduction of sweet (sugar-containing) urine.

diagonal (or **type II**) **survivorship curve**   Description of a life history in which survivorship decreases in proportion to age, so that the curve is a straight, declining line.

diaphragm   (1) A sheet of muscle beneath the lungs that separates the thoracic cavity from the abdomen. (2) A birth control device. A membrane of latex that prevents sperm from entering the cervix.

diarrhea   [Greek, *diarrhein* = to flow through] Watery feces.

diastole   [Greek, *dia* = between + *systellein* = to contract] The half of the cycle of the heart in which both the atria and the ventricles are relaxed.

dicotyledons (also called **dicots**)   Flowering plants that have two cotyledons; there are about 170,000 species, including almost all trees and shrubs, but not the conifers and other gymnosperms, as well as many herbs.

differential gene expression   The production of differing amounts of individual proteins (and RNAs) in different types of cells and at different times of development.

differentiation   The process by which tissues and cells become specialized and different from one another.

diffraction   The scattering of electromagnetic waves by regular structures so that their interference produces a pattern of lines or spots; the diffraction of x rays by a crystal produces a pattern of spots that can be analyzed to reveal the structure of the molecules within the crystal.

diffusion   [Latin, *diffundere* = to pour out] Random movements of molecules that lead to a uniform distribution of molecules both within a solution and on the two sides of a membrane.

digestion   The process of hydrolyzing large molecules into smaller units such as glucose, amino acids, fatty acids, and glycerol.

digestive tract (also called the **gastrointestinal tract**, the **alimentary canal**, or the **gut**)   The tube, extending from mouth to anus, in which animals accomplish digestion, absorption, and elimination.

dikaryotic   Having two nuclei per cell, as in certain fungi.

dinoflagellates   [Greek, *dinos* = whirling + *flagellum* = whip] Single-celled organisms that float freely as plankton in warm oceans.

diploid   [Greek, *di* = double + *ploion* = vessel] Having two sets of chromosomes.

directional selection   In evolution, selection that shifts the frequency of one or more traits in a particular direction.

disjunction   In meiosis, the moving apart of two homologous chromosomes during anaphase I and of two sister chromatids in anaphase II.

dispersal   The spread of a species or group of species through a large area.

disruptive selection   In evolution, the opposite of stabilizing selection: it increases the frequency of extreme types in a population, at the expense of intermediate forms.

distal tubule   In the kidney, the wide tubule from which the urine flows into the collecting duct and then to the bladder.

disulfide   A covalent bond between two sulfhydryl groups; disulfides are a common cross link between two cysteine side groups within a single polypeptide or between two polypeptides.

diuretic   A substance, such as caffeine or alcohol, that stimulates water loss.

divergent evolution   The separation of one species of organisms into two (or more) species.

DNA polymerase   The enzyme that strings together nucleotides into DNA.

dominance hierarchy   A ranking of individuals that fixes (at least temporarily) who may dominate whom.

dominant   Refers to an allele that alone determines the phenotype of a heterozygote.

dormant   [French, *dormire* = to sleep] Referring to a stage in the development of a seed in which growth is suspended until restarted by environmental cues.

dorsal   Refers to the top, or back, of an animal.

dorsal hollow nerve cord   A flexible rod containing neurons and their processes that runs between the notochord and the surface of the back.

double circulation   A pattern of circulation in which blood circulates separately between the heart and lungs and between the heart and the rest of the body.

double covalent bond   A covalent bond in which two atoms share two pairs of electrons.

double fertilization   In flowering plants, the simultaneous fusion of two sperm nuclei from the pollen tube with two cells of the embryo sac.

doubling time   The time it takes a population to double.

Down syndrome (also called **Down's syndrome**)   A disorder, resulting from a particular chromosomal abnormality, that leads to mental retardation and the abnormal development of the face, heart, and other parts of the body; it is almost always caused by trisomy of chromosome 21.

downstream   In DNA or RNA, toward the 3′ end.

duodenum   [Latin, twelve; referring to its length of 12 "fingers"] The first section of the intestine, where most of digestion occurs.

dynamically unstable   Meaning that a microtubule or an actin filament loses tubulin or actin monomers and shrinks if it is not actively growing.

early cell plate In a plant cell, a small, flattened disc, formed in mitotic telophase, that is the beginning of a new cell wall.

Echinodermata "Spiny-skinned" animals such as sea stars.

ecological isolation A barrier to reproduction that results from differences in the ways that species live.

ecology [Greek, *oikos* = home] The study of the interactions of organisms with one another and with their physical environments.

ecosystem A community of organisms together with the nonliving parts of the community's environment.

ectoderm [Greek, *ektos* = outside + *derma* = skin] The outermost cell layer of an embryo, including, in vertebrates, the future skin, sense organs, and central nervous system.

ectotherm [Greek, *ektos* = outside + *thermos* = heat] An animal that depends on external sources of heat.

edema [Greek, "swelling"] A physiological state in which water accumulates in the tissues.

effector cell In the immune system, a cell that carries out a particular immune function.

efferent Leading away, as blood vessels leaving an organ or nerve fibers carrying information from the central nervous system to the muscles and other organs.

egg (also called an **ovum**) A female gamete.

ejaculation The propulsion of sperm out of the penis.

electrical potential (also called **voltage**) A measure of the potential energy that results from a charge separation.

electrical synapse A connection between two neurons through gap junctions.

electrical work Energy change caused by changing the separation of charges.

electrochemical gradient A double gradient composed of a chemical gradient (the difference in hydrogen ion concentration, or pH) and an electrical gradient (the difference in charge).

electrogenic Leading to the accumulation of charge on one side of a membrane.

electromagnetic radiation A form of energy—including light—that is transmitted through space as periodically changing electrical and magnetic forces.

electron A subatomic, negatively charged particle; its charge is exactly equal to that of a proton, but its mass is much smaller.

electron carrier One of the molecules that carry electrons from high-energy, reduced compounds (NADH, NADPH, and FADH$_2$) to oxygen.

electron microscope A powerful microscope that uses electrons instead of light waves to reveal structures; it has much higher resolution than a light microscope.

electron transport chain The pathway of electrons in oxidative phosphorylation or photophosphorylation.

electronegativity A measure of the tendency of an atom to gain electrons.

electrophoresis A method for separating charged molecules (such as proteins and nucleic acids) according to their ability to move in an electric field.

electrostatic interaction The attraction or repulsion of charges.

element A substance that cannot be reduced to simpler substances by chemical means.

elimination The disposal of the remains of digested food.

elongation The adding of additional nucleotides or amino acids to a growing polynucleotide or polypeptide.

elongation zone A region of the growth zone of a plant's root.

embryo [Greek, *en* = in + *bryein* = to be full of] A set of early developmental stages in which a plant or animal differs from its mature form.

emergent property A characteristic that arises only at complex levels of organization.

emergents In tropical forests, trees with umbrella-shaped crowns extending to a height of 50 meters or more.

emigration Movement out of the population.

emphysema A disease of the lungs characterized by irreversible enlargement of the air spaces.

endemic Found nowhere else in the world.

endergonic Refers to a process in which free energy increases.

endocrine [Greek, *endo* = within] Cells or organs that are specialized for secreting specific signaling molecules into the general circulation.

endocrine gland An organ that is specialized for secretion of a hormone into the general circulation.

endocrinology The study of hormones.

endocytosis [Greek, *endon* = within] The process of taking in materials from outside a cell in vesicles that arise by the inward folding ("invagination") of the plasma membrane.

endoderm [Greek, *endo* = inside + *derma* = skin] The innermost cell layer of an embryo, including the future gastrointestinal tract and associated organs such as the pancreas and liver.

endodermis [Greek, *endon* = within + *derma* = skin] In vascular plants, the innermost layer of the cortex in roots and stems.

endometrium [Greek, *endon* = within + *metro* = mother] The inner lining of the uterus.

endoplasmic reticulum (ER) [Greek, *endon* = within + *plasmein* = to mold + Latin, *reticulum* = network] An extensive and convoluted network of membranes within a eukaryotic cell.

endorphin One of three classes of polypeptides that act as natural pain suppressors.

endosperm In flowering plants, a storage tissue that develops during double fertilization.

endosymbiosis [Greek, *endon* = within + *syn* = together + *bios* = life] The close association of two organisms, one of which lives inside the other.

endosymbiotic theory The generally accepted view that present-day energy organelles are descended from prokaryotes that once lived within the early eukaryotic cells.

endothelium In the circulatory system, the epithelial cells that line the capillaries.

endotherm [Greek, *endon* = within] An animal that warms its body by capturing the heat released by metabolism.

endotoxin A toxic substance that is a component of a bacterial cell or other pathogen.

energy The capacity to perform work, to move an object against an opposing force.

enhancer (because it stimulates the transcription of neighboring DNA, without itself serving as a promoter; also called a **response element** or *cis*-**regulatory element**) A DNA sequence that regulates transcription by binding to a specific regulatory protein called a transcription factor.

entropy A formal measure of disorder; entropy has a high value when objects are disordered or distributed at random and a low value when they are ordered.

enzyme A large molecule, almost always a protein, that accelerates the rate of a specific chemical reaction.

enzyme-substrate complex The association of enzyme and substrate that forms in the course of catalysis.

eon   One of several very long periods of time into which geologists and paleontologists have divided the history of life; eons are much longer than eras; a billion years or more.

epicotyl   [Greek, *epi* = above] The axis of a developing plant above the attachment point of the cotyledons.

epidemiologist   [Greek, *epidemia* = how widespread a disease is] A researcher who studies the incidence and transmission of diseases in populations.

epidermis   In animals, the outer layer of the skin; in plants, the outermost layer of the embryo and the plant.

epididymis   In the male reproductive tract of mammals, an interconnected network of coiled ducts in which sperm are stored.

epilimnion   The warm surface water layer of a lake or pond.

epinephrine (also called **Adrenaline**)   A hormone, made in the adrenal medulla, that speeds the heart, dilates the blood vessels, and increases the liver's production of glucose from glycogen.

epiphyte   A plant that grows entirely on other plants.

episome   A virus or other genetic system that can propagate either autonomously or as an integrated part of the host's chromosome.

epistatic gene   A type of modifier gene that limits the expression of another gene.

epithelial tissue   [Greek, *epi* = upon + *thele* = nipple] Cells tightly linked together to form a sheet with little extracellular matrix.

epithelium   A tissue that lines a surface (such as the outside of the body or the inside of the lungs).

epoch   A relatively short period of geologic time; one of the subdivisions of the Tertiary and Quaternary Periods of the Cenozoic Era.

equilibrium   In a chemical reaction, the point at which no further net conversion of reactants and products takes place.

era   A period of time, ranging in length from 65 million to several billion years, into which paleontologists usually divide the history of life; eras are shorter than eons.

erection   The rigid state of the penis or clitoris, caused by the blood filling the tiny spaces within spongy tissues.

erythrocyte   [Greek, *erythros* = red + kytos = receptacle] A red blood cell.

Escherichia coli (E. coli)   A common eubacterial resident of the human gut; a favorite experimental organism for thousands of research and industrial biologists.

esophagus   The part of the digestive tract that connects the mouth and pharynx with the stomach.

essential amino acids   The amino acids that an animal cannot itself produce.

essentialism   The view, originally argued by Plato, that individuals, whether chairs, daisies, or people, are only distorted shadows of an ideal, or essential, form.

estivate   To pass time in a sleeplike state (torpor) during the hottest and driest months of the summer.

estuary   [Latin, *aestus* = tide] A partly enclosed body of water where a freshwater stream or river meets the ocean.

ethology   The study of animal behavior.

ethylene   A plant hormone; a two-carbon molecule containing a double bond.

etiolated   [French, *etioler* = to blanch or whiten] Having a thin, spindly appearance, poor leaf development, and no chlorophyll production.

Eubacteria   [Greek, *eu* = true] The commonly occurring prokaryotes that live in water and soil or within larger organisms.

euglenophytes   [Greek, *eu* = true + *glene* = eyeball] Protists that have distinctive light-detecting eye spots.

Eukarya   The domain that encompasses the Animalia, Plantae, Fungi, and Protista kingdoms.

eukaryotic   [Greek, *eu* = true + *karyon* = nucleus] Cells that contain a central nucleus and other membrane-bounded organelles.

eukaryotic flagellum   [plural, **flagella**; Latin, whip] A protein assembly, consisting of microtubules, that can move a cell through a liquid medium (or a liquid medium over a cellular surface).

Eumetazoa   [Greek, *eu* = true + *meta* = middle + *zoa* = animal] The larger of two subkingdoms of animals, including all of the animals except the sponges and the Placozoa.

euploid   Having the correct number of chromosomes.

eutherians (also called **placental mammals**)   [Greek, *therion* = wild beast] Animals whose embryonic development takes place within the uterus of the mother.

eutrophic   A lake that has excessive minerals and organic matter and insufficient oxygen.

evergreen   Keeping leaves throughout the year.

evolution   The process by which species arise and change over time. The idea that all organisms have descended from common ancestors.

excitatory   In the nervous system, leading to the production of action potentials.

excitement   A phase of sexual arousal that is characterized by increased blood flow to the clitoris, the labia minora, and the breasts (in the female) and to the penis and the testes (in the male).

excretion   A disposal process by which cells pass materials across a cell membrane.

exergonic   Refers to a process in which free energy decreases.

exocrine   [Greek, *exo* = outside] A cell or organ that makes products that are carried to specific targets by ducts, such as the duct that carries digestive enzymes into the small intestine.

exocrine organ   An organ whose products are carried to specific targets by ducts.

exocytosis   The export of molecules from a cell by a process that is approximately the reverse of pinocytosis; molecules to be exported are surrounded by membranes that move to the cell surface

exon   A segment of a gene (or of a pre-mRNA) that is also present in mature mRNA; most exon sequences encode polypeptide segments.

exoskeleton   [Greek, *exo* = on the outside] External skeleton.

exotoxin   A toxic substance secreted by bacteria.

exponential growth   A growth pattern in which a population repeatedly doubles in some constant period of time.

expiration   Exhalation.

extension   The unbending of a limb.

extracellular matrix   A network of proteins and polysaccharides found in connective tissue.

F-factor (for fertility factor)   A kind of episome in bacteria that can replicate either autonomously or in integrated form; it can move from one bacterium to another during conjugation.

F1 (or **first filial**) **generation**   [Latin, son] The initial progeny of a cross.

F2 (or **second filial**) **generation**   The progeny of the F1 generation.

facilitated diffusion   An increased rate of passive transport; it depends on the action of specific transporter molecules within the membrane.

facultative aerobe   A heterotroph that can grow either with or without oxygen.

facultative anaerobe   A microorganism that can live either anaerobically (by fermentation) or aerobically (using oxidative phosphorylation).

FAD (flavin adenine dinucleotide)   An electron acceptor in oxidative phosphorylation.

FADH$_2$ The reduced form of FAD.

fall turnover   In a lake, the annual mixing of the waters of epilimnion and hypolimnion.

fallopian tube   Human oviduct.

family   A taxonomic group that consists of one or more genera; a family is a subdivision of an order.

fat   A triacylglycerol that is solid at room temperature; the component fatty acids are usually saturated.

fate   What a cell or a tissue becomes during development.

fatty acid   A small molecule consisting of a hydrocarbon chain ending in a carboxyl group; a component of phospholipids and triacylglycerides.

feces   [Latin, "dregs"] Waste matter of digestion, discharged through the anus; feces consist of the remnants of food together with bacteria that inhabit the intestinal tract.

feedback inhibition   The inhibition of an enzyme reaction by the product of that reaction or of the end product of an entire pathway; a homeostatic mechanism.

female choice   A courtship pattern in which the female assesses and chooses the best male.

fermentation   [Latin, *fervere* = to boil] The anaerobic extraction of energy from organic compounds.

fermenting bacteria   Anaerobic heterotrophs that derive energy and carbon atoms from a variety of organic compounds in the absence of oxygen.

fertility   The number of offspring each individual in a cohort is likely to produce.

fertilization   The union of two haploid gametes to form a diploid cell or zygote.

fetal masculinization   The development of male characteristics in a fetus carried by a woman treated with male hormones during pregnancy.

fetus   In mammals, a stage of development in which all organs have formed.

fibril   A threadlike structure, made of smaller filaments; the term is used to refer to cross-linked cables of collagen molecules and to assemblies of actin and myosin filaments.

fibrillation   In the heart, continuous disorganized contractions.

fibroblast   A flat, irregularly shaped cell found in connective tissue.

fibronectin   A protein of the extracellular matrix.

fibrous protein   A protein with an elongated shape; fibrous proteins include most structural proteins.

filament   (1) A small protein fiber; (2) in plants, a thin stalk that connects the anther to the base of the flower.

filtrate   In the kidney, the water, urea, and other small molecules in the blood that freely pass into Bowman's capsule from the renal artery.

First Law of Thermodynamics   The statement that the total amount of energy stays constant in any process; that is, energy is neither lost nor gained—it only changes form.

fission yeast   *Schizosaccharomyces pombe,* a widely used experimental organism that divides by binary fission.

fitness   A measure of selective advantage; it is defined as the contribution to the next generation of one genotype in a population relative to the contribution of other genotypes.

fixation   A process that preserves cells for microscopic examination.

fixed nitrogen   Nitrogen in the form of ammonia (NH_3) or nitrate (NO_3).

flagellum   [plural, **flagella**; Latin, whip] In eukaryotes, a protein assembly, consisting of microtubules, that can move a cell through a liquid medium (or a liquid medium over a cellular surface); in prokaryotes, a protein assembly that moves like a propeller.

flexion   The bending of a limb.

fluid mosaic model   The accepted model of biological membrane structure; the model stresses that proteins and phospholipid molecules can move within each leaf of the lipid bilayer unless they are restricted by special interactions.

fluidity   A measure of the ability of substances to move within a membrane.

follicle   [Latin, *folliculus* = small ball] In the ovary, the granulosa cells surrounding a primary oocyte.

follicle-stimulating hormone (FSH)   A peptide hormone, made in the anterior pituitary that in females promotes the maturation of the follicle during the menstrual cycle and in males stimulates testosterone production.

food chain   The sequence in which consumers eat either producers or other consumers.

food web   The collection of all the interacting food chains of an ecosystem.

foot   In a mollusk, a muscular extension of the body that the mollusk uses for sensing, grabbing, creeping, digging, and holding on.

foraminifera (also called forams)   [Latin, *foramen* = little hole + *ferre* = to bear] Marine organisms whose tests are made of organic materials reinforced with grains of sand or minerals and full of tiny holes.

foreskin   A loose layer of skin around the outside of the penis, which ends in a flap and folds over the glans.

fossil   [Latin, *fossilis* = dug up] An object, usually found in the ground, that represents the remains or imprint of past life.

founder effect   The restriction of genetic diversity of a population that results from the founding of a new population by a small subset of a larger population.

frame-shift mutation   An insertion or deletion mutation that alters the groupings of nucleotides into codons.

free energy   A measure of available energy under the conditions of a biochemical reaction; the term is abbreviated G, after the American thermodynamicist J. Willard Gibbs.

free radical   A compound that is highly reactive because it contains an unpaired electron; it is biologically important because it can lead to DNA damage and to cell death; free radicals are formed during normal metabolic processes in both animals and plants.

free ribosomes   Ribosomes, present in the cytosol, that are responsible for the synthesis of soluble proteins.

fronds   Leaves of ferns that often have a lacy appearance.

fruiting body   A mushroomlike growth on cellular slime molds that produces spores.

functional group   A standard small grouping of atoms that contributes to the characteristics of an organic molecule.

fundamental niche   A species' potential ability to utilize resources.

Fungi   The kingdom that includes heterotrophic organisms, both multicellular and single-celled organisms.

G₀ The state of a cell that has withdrawn from the cell cycle.

G₁ The period of the cell cycle that represents the gap between the completion of mitosis and the beginning of DNA replication; it is also called the first growth phase.

G₂ The period of the cell cycle that represents the gap between the completion of DNA synthesis and the beginning of mitosis (of the next cell cycle).

gametangia In plants, the enclosed reproductive organs in which gametes form.

gamete [Greek, *gamos* = marriage] A specialized reproductive cell through which sexually reproducing parents pass chromosomes to their offspring; a sperm or an egg.

gamete formation The production of sperm and eggs.

gametic isolation Reproductive isolation that results from differences in the compatibility of gametes.

gametophyte The haploid form of a life cycle characterized by alternation of generations.

gap junctions Protein assemblies that form channels between adjacent animal cells.

gastrotricha Microscopic, bottle-shaped animals that live in freshwater or marine environments.

gastrula [Greek, little stomach] A stage of an animal embryo in which the three germ layers have just formed.

gastrulation The process of forming a gastrula.

gate The part of a channel protein that opens and closes the channel in response to environmental signals such as voltage or the binding of a neurotransmitter.

gene [Greek, *gen* = to produce] The unit of inheritance. A DNA sequence that is transcribed as a single unit and encodes a single polypeptide, a set of closely related polypeptides, ribosomal RNA, or transfer RNA.

gene rearrangement The cutting and splicing of segments of specific genes; it is known to occur within the immune system.

genetic code The relationship between nucleotide sequence in mRNA and amino acid sequence in polypeptides.

genetic linkage The tendency of two or more genes to segregate together.

genetic map (also called a **linkage map**) A summary of the genetic distances between genes.

genetic recombination Associations of genes that occur in offspring that did not exist in the parents.

genetics The study of inheritance.

genome The collection of all the DNA in an organism.

genotype The genes present in a particular organism or cell.

genus [plural, **genera**; Greek, *genos* = to be born] A group of species that share many morphological characteristics; a taxonomic grouping of one or more species; a subdivision of a family.

germ cells (or **germ line**) Gametes and the cells from which they arise.

germ layer One of the three tubes of the vertebrate embryo—ectoderm, mesoderm, and endoderm.

germ layer formation The movement of embryonic cells to give the three germ layers (ectoderm, mesoderm, and endoderm); gastrulation.

germ line theory The view that antibody diversity results from genetic information that is already present in the zygote.

germination [Latin, *germinare* = to sprout] The resumption of growth by a seed.

gibberellin A plant hormone.

gill slits Openings that directly connect the throat to the outside.

gills Evaginated breathing structures.

girdling Removing a strip of bark around a tree trunk.

gizzard A region of the intestinal tract of birds specialized for the grinding of food.

glans The smooth cap formed by the tissue at the far end of the penis.

glial cell A cell within the nervous system that does not itself transmit electrical and chemical signals, but which provides metabolic and structural support for neurons.

globular protein A relatively compact protein that is roughly spherical in shape.

glomerulus [plural, **glomeruli**; Latin, "little ball"] A tangled network of capillaries in the cortex of a kidney.

glucagon A protein hormone produced in the pancreas; a signal for the postabsorptive state; glucagon inhibits glycogen synthesis and stimulates its breakdown into glucose.

glucocorticoid A steroid hormone that stimulates the production of carbohydrate from protein.

glycerol A polar three-carbon molecule with three hydroxyl groups; a starting compound for triacylglycerides and phospholipids.

glycogen Polysaccharide used in animals for long-term energy storage.

glycogen phosphorylase The enzyme that breaks down glycogen into glucose.

glycolipid A molecule that consists of a lipid attached to carbohydrates.

glycolysis [Greek, *glykys* = sweet (referring to sugar) + *lyein* = to loosen] A set of ten chemical reactions that is the first stage in the metabolism of glucose.

glycolytic fiber A large muscle fiber that derives most of its energy from glycolysis and has few mitochondria.

glycoprotein [Greek, *glykys* = sweet] A protein that contains covalently attached carbohydrates.

glycosidase An enzyme that catalyzes the digestion of glycosidic bonds; it is used to digest starch or glycogen.

glycosidic bond The covalent bond between two sugar molecules in a polysaccharide or oligosaccharide.

glycosylation The process of adding sugars to a newly made protein; it takes place in the ER lumen and in the Golgi apparatus.

glyoxisome A specialized peroxisome present in the seeds of some plants; glyoxisomes provide energy for the growing plant embryo by breaking down stored fats.

goblet cells In the small intestine, specialized cells that produce mucus.

Golgi complex (also called the **Golgi apparatus**) In eukaryotic cells, a set of flattened discs of membrane, usually near the nucleus, involved in the processing and export of proteins.

gonad [Greek, *gonos* = seed] A gamete-producing organ; an ovary or testis.

gonadotrophin One of two hormones (FSH and LH), made in the anterior pituitary, that act on gonadal tissue.

gonadotropin releasing hormone (GnRH) A polypeptide hormone, made by the hypothalamus, that stimulates the secretion of FSH and LH by the anterior pituitary.

gossypol A terpene, made by cotton plants, that is responsible for resistance to insects, bacteria, and fungi; it also works as a contraceptive in human males.

gradualism The idea that species evolve gradually and continuously through the steady accumulation of changes, rather than through sudden changes.

granum [plural, **grana**; Latin, grain] A stack of thylakoids within a chloroplast, bounded by the thylakoid membrane.

gravitropism The response of a plant to gravity.

greenhouse effect The warming of the Earth as the result of the absorption of heat by carbon dioxide and other gases.

ground tissue In plants, the cells that occupy most of the interior of the embryo and the plant.

growth control A regulatory mechanism that prevents cell division by allowing the cell cycle to proceed under some conditions and to stop under others.

growth factor One of a number of protein paracrine signals that stimulate cell division and cell survival.

growth hormone (GH) A peptide hormone, made in the anterior pituitary, that stimulates tissue and skeletal growth, milk production, and other processes in humans, cows, and other mammals.

gut [Anglo-Saxon, "channel"] The gastrointestinal tract, the alimentary canal, or the digestive tract.

guttation [Latin, *gutta* = a drop] The process by which root pressure can push water all the way up the stem and out of tiny holes at the margins of the leaves.

gymnosperms [Greek, *gymnos* = naked + *sperma* = seed] Seed plants that do not have fruits or flowers, including, for example, pine trees and other conifers, as well as palmlike cycads.

habitat The place in which an organism lives, along with the set of environmental conditions that characterize that place.

habituation The decrease in a behavioral response following repeated exposure to a harmless stimulus.

half-life The time required for half the atoms of a radioactive isotope to decay.

halophiles The salt-loving bacteria, one of the three phyla of Archaea.

haplodiploidy An unusual kind of sex determination in which males develop from unfertilized eggs and are therefore haploid, while females are diploid.

haploid [Greek, *haploos* = single] Having a single set of chromosomes.

haplontic Having the haploid phase dominate the life cycle; in a haplontic life cycle, the only diploid cell is the zygote, which itself undergoes meiosis to produce haploid spores.

Hardy-Weinberg equilibrium A stable distribution of genotype frequencies maintained by a population from generation to generation.

heart The muscular organ responsible for pushing the blood through the circulatory system.

heart attack A failure of the blood supply to the heart.

heart rate The number of contractions (beats) per minute.

heat The form of kinetic energy contained in moving molecules.

heavy chain One of the polypeptides present in an immunoglobulin molecule.

heme An iron-containing organic molecule that gives hemoglobin and the cytochromes their red color; heme may donate or accept electrons, as its iron atom changes its charge between Fe^{2+} and Fe^{3+}.

Hemichordata Acorn worms.

hemizygous In male mammals, having one (instead of two) copies of a gene because the Y chromosome lacks the gene present on the X chromosome.

hemoglobin [Greek, *haima* = blood + Latin, *globus* = ball] The oxygen-binding protein that makes red blood cells red.

herbaceous plant (also called an **herb**) A plant that has no woody parts.

herbivore [Latin, *herba* = vegetation + *vorare* = to swallow] An animal that eats only plants.

heterokaryon [Greek, *heteros* = different + *karyon* = kernel, nucleus] A cell that has two nuclei from different sources.

heterosporous Producing two kinds of gametophytes (male and female) from two kinds of spores on two kinds of sporangia.

heterotroph [Greek, *heteros* = other + *trophe* = nourishment] An organism that cannot derive energy from sunlight or from inorganic chemicals but must obtain energy by degrading organic molecules.

heterozygous Having two different alleles for a single gene (in a diploid organism).

hexose [Greek, *hex* = six] A sugar that contains six carbon atoms; glucose is a hexose.

hierarchy An arrangement in which larger groups include smaller groups, which include still smaller groups.

high-energy bond A relatively unstable chemical bond that gives up energy as new, more stable, bonds form.

histamine The amino acid histidine minus the carboxyl group, a major stimulus for the inflammatory response; it is released by cells in damaged tissues; it dilates capillaries and increases their tendency to leak fluid.

histone [Greek, *histos* = web] One of a set of small, positively charged proteins that bind to DNA in eukaryotic cells.

homeostasis [Greek, *homeo* = like, similar + *stasis* = standing] The tendency of organisms to maintain a stable internal environment in the presence of a changing external environment.

homeotic mutation A mutation that causes the cells of an embryo to give rise to an inappropriate structure in the adult, for example, to legs instead of antennae.

homeotic selector gene In a plant, a gene that establishes the fate of one or more whorls of a developing flower.

hominid A member of the human family, Hominidae; the only living hominid is *Homo sapiens*.

hominoid [Latin, *homo* = human] Hominids and apes; all have large skulls and long arms and tend to walk at least partially erect.

homologous Arising from the same structures in a common ancestor.

homologous chromosomes The two matching chromosomes that align during meiosis I.

homosporous Producing just one kind of spore and one kind of gametophyte.

homozygous Having two copies of the same allele (in a diploid organism).

hormone [Greek, *horman*, = to urge on] A substance, made and released by cells in a well-defined organ or structure, that moves throughout the organism and exerts specific effects on specific cells in other organs or structures.

host range For a virus, the set of hosts that a particular virus can infect.

Hox gene Mammalian counterpart of a *Drosophila* homeo domain gene.

human chorionic gonadotropin (HCG) A polypeptide hormone made after conception; the basis of the most common tests for pregnancy.

human immunodeficiency virus (HIV) The retrovirus that causes AIDS; HIV infects and kills T_H lymphocytes by first binding to the CD4 protein on the cell surface.

humus   In soils, the residue of decayed dead organisms; a black or brown material that decays slowly.

hybrid   The progeny of a cross of two genetically distinct organisms.

hybrid breakdown   The weakness or sterility of the progeny of interspecies hybrids that are initially viable and fertile.

hybrid inviability   Chromosomal and genetic incompatibilities within hybrids.

hybrid sterility   The sterility of hybrids between two species.

hybridoma   A hybrid cell line derived from the fusion of a cancer cell (a lymphoma) to another cell, such as an antibody-producing cell.

hydrocarbon chain   A chain of connected carbon atoms, with hydrogen atoms sharing other available outer shell electrons.

hydrogen bond   A weak attraction between a hydrogen in one molecule that has a slight positive charge and a negatively charged atom in another molecule.

hydrogen ion (H^+)   The result of a dissociation of a water molecule; it is also called a proton, though it is actually a hydronium ion (H_3O^+).

hydrolysis   [Greek, *hydro* = water + *lysis* = breaking] Breaking the bond between two building blocks by adding a water molecule, reversing the dehydration-condensation reaction.

hydronium ion (H_3O^+)   A water molecule that has acquired an extra proton and a charge of +1; the proper name for a hydrogen ion.

hydrophilic   [Greek, *hydro* = water + *phili* = love] Water-loving; refers to a molecule (or a part of a molecule) that is soluble in water by virtue of its interactions with water molecules.

hydrophobic   [Greek, *hydro* = water + *phobos* = fear] Avoiding associations with water; nonpolar.

hydrophobic interaction   The association of nonpolar molecules.

hydrostatic skeleton   A rigid fluid-filled space (the coelom) that provides mechanical support in invertebrate coelomates.

hydroxide ion (OH^-)   A water molecule from which a hydrogen ion has dissociated; it has a charge of −1.

hydroxyl   The OH functional group, which allows molecules that contain it to form hydrogen bonds.

hymen   A thin membrane of irregular ragged shape just inside the mouth of the vagina.

hyperpolarizing   A voltage change (across the plasma membrane of a neuron) that makes the inside of the cell more negative and less apt to produce an action potential.

hypertension   High blood pressure

hypertonic   [Greek, *hyper* = above] Having a total concentration of solutes higher than that within a cell.

hypervariable region   The polypeptide segments, within the variable region of an immunoglobulin light chain or heavy chain, in which the most sequence variation occurs among antibodies; the hypervariable regions of an immunoglobulin molecule correspond exactly to the complementarity determining regions, where antigens bind.

hyphae   [singular, **hypha**; Greek, *hyphe* = web] The threadlike filaments of a fungus.

hypocotyl   [Greek, *hypo* = under] The axis of a developing plant below the attachment point of the cotyledons.

hypolimnion   In lakes and ponds, the layer of water beneath the epilimnion whose temperature stays about 4°C year-round.

hypothalamus   [Greek, *hypo* = under + *thalamos* = inner room] A part of the brain that regulates the expression of many hormone systems.

hypothesis   An informed guess—for example, about the way a process works or a structure is organized.

hypotonic   [Greek, *hypo* = under] Having a total concentration of solutes lower than that within a cell.

imaginal disc   In insect development, a group of larval cells from which adult structures later develop.

imago   Adult stage of an insect.

imbibition   The taking in of water at the time of seed germination.

immigration   Movement into a population.

immune response   The defense system by which animals resist microorganisms and other foreign tissues, including cancer.

immunoglobulin   An antibody; one of the members of a group of globular blood proteins called globulins.

impermeable junctions (also called **occluding junctions**)   Molecular assemblies that connect epithelial cells tightly together; they also prevent molecules from leaking between them.

implantation   In mammals, the process in which a zygote burrows into the wall of the uterus.

imprinting   The process by which an animal learns behavior during a sensitive period.

in vitro   [Latin, in glass] In a test tube.

in vivo   [Latin, in life] In a living organism.

inbreeding   Mating among close relatives, which greatly increases the number of homozygotes.

incipient species   Subspecies that do not interbreed.

inclusive fitness   The sum of an individual's genetic fitness (which includes that of its own direct descendants) plus all its influence on the fitness of its other relatives.

individualistic   The view that every species has an independent distribution, and that, in effect, every community is unique.

induced abortion   The deliberate removal of an embryo from the uterus.

induced fit   A change in the conformation of an enzyme brought about by the binding of the substrate.

induction   In embryonic development, the process by which one cell population influences the development of neighboring cells.

inflammation   A set of responses to local injury; characteristics of injured tissue include redness, heat, swelling, and pain.

inhibiting hormone   A peptide hormone, made by the hypothalamus, that inhibits the release of a specific hormone.

inhibitory   In the nervous system, tending to prevent the production of action potentials.

inhibitory interneuron   A neuron whose stimulation by one neuron prevents the firing of another neuron in the same area.

initiation   The start of synthesis.

initiation site   The first nucleotide actually transcribed from DNA into RNA.

innate   Behavior that an animal engages in regardless of previous experience.

inner cell mass   In a mammalian embryo, a small group of cells within a blastocyst that will eventually grow into the embryo itself and subsequently into the adult.

insertion   A type of mutation; the addition of one or more nucleotides.

insertion sequence (also called a **simple transposon**)   The simplest transposable element; a short length of DNA (up to a few thousand nucleotide pairs long) that can move from place to place in the genome.

inspiration   Inhalation.

insulin A protein hormone, produced by the pancreas, that regulates glucose uptake; a signal for the absorptive state; it promotes the synthesis of glycogen and inhibits its breakdown.

integral protein (also called a **transmembrane protein**) A membrane protein that spans the lipid bilayer.

integrated The view that a community consists of characteristic assemblages of species that interact with each other in predictable ways.

integrated replication A method for viral propagation in which the virus's DNA becomes integrated into the host cell's DNA and is replicated along with that of the host.

interferon A cytokine that interferes nonspecifically with the reproduction of viruses.

interleukin-1 (IL-1) A cytokine produced early during inflammation; it stimulates responses both in leukocytes and in organs distant from the site of infection.

intermediate filament In eukaryotic cells, a component of the cytoskeleton that consists of filaments 8 to 10 nm in diameter, thinner than microtubules but thicker than actin filaments.

internode In a plant, the region of a stem between nodes.

interphase The part of the cell cycle in which the chromosomes are not condensed and the cytoplasm is not dividing.

interstitial cells In the testes, the matrix amongst the seminiferous tubules that synthesizes the male sex hormone testosterone.

intertidal zone The zone that lies between high tide and low tide.

intrauterine device (IUD) A birth control device; a small piece of plastic or other material that is placed into the uterus, where it interferes with implantation.

intron (also called an **intervening sequence**) A segment of a gene (or of a pre-mRNA) that is transcribed into RNA but excised before the primary transcript matures into functional mRNA.

invagination The local folding of a cell layer to form an enclosed space with an opening to the outside.

inversion A mutation in which a segment of a chromosome is turned 180° from its normal orientation.

involuntary muscle A muscle that cannot be consciously controlled; smooth and cardiac muscles.

ion An atom (or a molecule) with a net electrical charge, the result of a different number of electrons and protons.

ionic bond A bond formed by ions with opposite charges.

iron-sulfur protein One of at least six proteins that participate in the electron transport chain; each contains an iron-sulfur center with two to four iron atoms and an equal number of sulfur atoms.

isomers Molecules that contain the same atoms arranged differently.

isotonic [Greek, *isos* = equal + *tonos* = tension] Having a total concentration of solutes that is the same as a cell's interior.

isotopes Forms of an element that have different numbers of neutrons.

J-shaped curve A graph of the exponential growth of a population; the curve resembles the letter J.

joint The point of attachment between two bones.

K-selected species Species with adaptations that increase their ability to maintain populations as close to the carrying capacity (K) as possible; these species tend to live long lives, reproduce late in life, produce few and large offspring, and provide extended parental care to each offspring.

karyotype [Greek, *karyon* = kernel or nucleus + *typos* = stamp] The chromosomal makeup of a cell.

kidney In vertebrates, the major organ of excretion and of salt and water balance; regulates the composition of the urine.

kin selection The tendency of individuals in some species to increase their reproductive output by helping relatives, which share their genes, to reproduce.

kinetic energy The energy of moving objects.

kinetics [Greek, *kinetikos* = moving] The study of the rates of reactions.

kinetochore A specialized disc-shaped structure that attaches the mitotic spindle to the centromere.

kinetochore microtubules A subset of polar microtubules that run from a pole of the mitotic spindle to a kinetochore.

Koch's postulates Rigorous criteria for identifying the pathogen for a given disease.

krummholz A forest of stunted trees that grow near timberline on a mountain.

kwashiorkor Protein deficiency disease.

labia majora [Latin, larger lips] In the external genitalia of human females, the skin folds on either side of the labia minora that enclose small amounts of fatty tissue.

labia minora [Latin, smaller lips] In the external genitalia of human females, the two thin skin folds that surround the mouth of the vagina and the end of the urethra.

lac repressor A protein that binds to the *lac* operator and prevents the expression of the *lac* operon.

lactose operon (or **lac operon**) A region of E. coli DNA that encodes three proteins used to derive energy from lactose: the enzyme β-galactosidase (which splits lactose into galactose and glucose) and two other proteins, called permease and acetylase.

lagging strand In replicating DNA, the newly made strand that is extended discontinuously.

larva A feeding form of an animal distinct from the later adult.

larynx The modified upper part of the trachea that contains the vocal cords, folds of membrane that vibrate as air passes over them, producing sound.

lateral bud (also called an **axillary bud**) A bud that forms just above the point where a leaf joins the stem and which can develop into a branch.

lateral meristem A cylinder of actively dividing cells within a root or stem.

leading strand In replicating DNA, the newly made strand that extends continuously, with DNA polymerase adding nucleotides to its 3′ end.

leaf primordium A tiny extension of the apical meristem; it grows into a leaf.

learning Modification of neural activity and behavior as the result of experience.

leukocyte [Greek, *leukos* = clear, white] White blood cell.

liana A vine that is rooted in soil but climbs into the canopy.

ligament A band of connective tissue by which joints may be joined together.

ligand [Latin, *ligare* = to bind] A molecule that binds to a specific binding site in a protein.

light chain One of the polypeptides present in an immunoglobulin molecule.

light source The part of a microscope that illuminates a specimen.

lignin A major constituent of the walls of cells specialized to provide mechanical support or transport water.

limiting resource A resource that is in short supply.

limnology   The study of freshwater ecosystems—lakes, ponds, rivers, and streams.

linkage group   A set of genes that do not assort independently because they are physically close to one another on the same chromosome.

linkage map   A summary of the genetic distances between genes; a genetic map.

lipid   A compound that is less soluble in water than in nonpolar solvents.

lipid-soluble signal   A chemical signal that can enter a target cell by passing directly through the plasma membrane.

liposome   An artificially produced vesicle that is surrounded by a phospholipid bilayer.

loam   A soil that consists of a mixture of clay, silt, and sand.

locus   [plural, **loci**; Latin, place] The position of a gene on a chromosome.

logistic growth equation   An equation that describes the growth of a population.

loop of Henle   In a kidney, the hairpin-shaped portion of the renal tubule.

lophophore   In three phyla of marine invertebrates, a filtering apparatus that catches food.

low-density lipoprotein (LDL)   A carrier protein in the blood that binds to cholesterol.

low-energy bonds   Relatively stable chemical bonds.

lumen   [Latin, light, an opening] An enclosed space, bounded either by membranes (as in the ER lumen) or by an epithelium (as in the lumen of the gut).

lungs   Invaginated breathing structures; localized organs of gas exchange that are always associated with the circulation.

luteinizing hormone (LH)   A peptide hormone, made in the anterior pituitary, that in females induces ovulation and stimulates estrogen production and in males increases testosterone production.

lymph   Fluid containing dead or foreign cells and waste proteins that passively moves through the lymphatic system.

lymph nodes   Regions in the lymphatic veins where filterlike tissue separates cells and other detritus from the lymph.

lymphatic system   The network of lymphatic vessels and nodes that provides a secondary route for fluids from the extracellular space to the bloodstream.

lymphocyte   One of a class of white blood cells that develop within the lymphoid tissues (including lymph nodes, spleen, thymus, and tonsils); the cells responsible for the immune response.

lymphocyte library   The collection of B lymphocytes, each containing a unique random rearrangement of immunoglobulin genes.

lysis   [Greek, *lysis* = loosening] The breaking down of a plasma membrane.

lysogenic cycle   [Greek, *lysis* = loosen + *genos* = offspring, because it can generate the destructive lytic cycle] The reproduction of a virus along with its host without causing the host cell to lyse.

lysogenic virus   A virus that reproduces either in a lytic cycle, in which it destroys its host, or along with the host; the virus can lie in a dormant (lysogenic) state and can be activated to enter a lytic cycle.

lysosome   A small membrane-bounded organelle that contains hydrolytic enzymes that break down proteins, nucleic acids, sugars, lipids, and other complex molecules.

lytic cycle   Viral reproduction that destroys the host.

lytic virus   A virus that destroys its host cell by lysing it.

macroevolution   The processes by which species and higher groupings (taxa) of organisms originate, change, and go extinct.

macromolecules   [Greek, *macro* = large] Large molecules formed by the polymerization of smaller building blocks.

macrophage   [Greek, *makros* = large + *phagein* = to eat] A large phagocytic cell, widely distributed throughout the body.

magnification   The ratio of the size of an image to the size of the object itself.

major histocompatibility complex (MHC)   A set of 40 to 50 closely linked genes, some of which encode proteins on the surface of every somatic cell; MHC proteins present antigens to the immune system.

male competition   Competition among males, usually for territory or access to females.

malnourishment   A deficiency in one or more essential nutrients.

Malpighian tubules   Excretory organs in arthropods.

mammary glands   Milk-producing organs in the female that characterize the mammals.

mantle   In mollusks, a specialized tissue that secretes a shell onto the dorsal surface of the visceral mass.

mantle cavity   In mollusks, an enclosed space, formed by folds in the mantle, that contains the mollusk's breathing organs.

marsupials (also called **metatherians**)   Mammals whose young are born in a fetal stage; marsupials carry the young in a pouch until they can fend for themselves.

mass   A measure of the amount of matter.

mass number   The total number of protons and neutrons in an atom's nucleus.

mast cell   A large round cell; mast cells are distributed throughout the connective tissues and are filled with small histamine-containing vesicles, which are released at the sites of tissue damage.

maternal effect genes   Genes that are expressed in the mother during oogenesis.

maternal mRNA   An mRNA that is already present in the egg before fertilization.

matrix   The intercellular substance within a tissue, or the interior substance of a mitochondrion.

matter   Any substance.

mechanical isolation   A reproductive barrier that derives from incompatibilities between male and female reproductive organs.

mechanical work   Energy change resulting from the movement of an object against a force.

median eminence   A structure that carries hypothalamic hormones to the anterior pituitary.

medulla   A part of the brain that regulates the rate of breathing; the inner region of a kidney or adrenal gland.

medusa   The bell-like body plan that represents one of the two basic body plans of the cnidarians.

megaphyll   A large leaf containing several veins that form complex networks of transport tubes.

megaspore   [Greek, *megas* = large] In a flower, the haploid cell within an ovule that develops into the embryo sac, the mature female gametophyte.

meiosis   [Greek, *meioun* = to make smaller] The process by which haploid gametes arise from diploid cells; meiosis distributes chromosomes so that each of four daughter cells receives one chromosome from each homologous pair.

meiosis I   The first of the two divisions of meiosis, during which homologous chromosomes pair and are distributed into two daughter cells.

meiosis II   The second of the two divisions of meiosis, during which sister chromatids are distributed to daughter cells.

membrane envelope   A membrane that surrounds a virus particle.

membrane-bound ribosomes   Ribosomes associated with a cell's internal membranes; they are responsible for the synthesis of secreted proteins as well as many membrane-associated proteins.

memory cell   In the immune system, a cell that can later be stimulated to produce effector cells with particular antigen specificity.

menstruation   [Latin, *mens* = month] The monthly shedding of blood, mucus, vaginal secretions, and endometrial tissue out through the vagina.

meristem   [Greek, *meristos* = divided] In plants, a region of undifferentiated, actively dividing cells.

mesenchyme   [Greek, *mesos* = middle + *enchyma* = infusion] Loosely attached cells embedded within a jellylike substance.

mesoderm   [Greek, *mesos* = middle + *derma* = skin] The middle layer of a vertebrate embryo, including the future connective tissues (bones, muscles, and tendons) and the cells of the blood.

mesoglea   [Greek, *mesos* = middle + *glia* = glue] In cnidarians, the jellylike material that lies between the two cellular layers.

mesophyll   [Greek, *mesos* = middle + *phyllon* = leaf] Green parenchymal cells that are responsible for most of a plant's photosynthesis.

messenger RNA (mRNA)   The RNA molecules that carry information from DNA to the ribosomes, where the mRNA is translated into a polypeptide.

metabolism   [Greek, *metabole* = change] All the chemical reactions occurring within an organism.

metacentric   The term referring to a chromosome in which the centromere is at or near the middle; such a chromosome looks like a V.

metamere   One of the identical (or nearly identical) sections that make up an annelid's long, segmented body.

metamorphosis   In many animal species, a series of dramatic changes in form leading from a larva to an adult.

metaphase   The stage of mitosis or meiosis during which chromosomes move halfway between the two poles of the spindle, where they accumulate in the metaphase plate.

metaphase plate   A disc formed during metaphase in which all of a cell's chromosomes lie in a single plane at right angles to the spindle fibers.

metastasize   Referring to cancer cells, to spread to other parts of the body.

metatherians (also called **marsupials**)   [Latin, *marsupium* = pouch] Animals whose newborns are relatively undeveloped fetuses.

methanogen   A methane-producing bacteria, one of the three phyla of Archaea.

MHC restriction   In the immune response, the inability of T cells to recognize foreign antigens except in cells with the same MHC type as the responding animal.

micelle   A cluster of amphipathic molecules.

microelectrode   An electrode that is small enough to enter or maintain contact with a single cell.

microevolution   Changes in the frequencies of alleles of genes in a population.

micronutrient   A substance required in minute amounts for the life of an organism; examples include molybdenum, copper, zinc, manganese, boron, iron, and chlorine.

microphyll   A leaf with just one conducting vein.

microsomes   Vesicles derived from fragments of the endoplasmic reticulum after experimental disruption of eukaryotic cells.

microsurgery   The physical manipulation of subcellular structures.

microtubule   In eukaryotic cells, the largest elements of the cytoskeleton; they consist of tubulin molecules assembled into hollow rods, about 25 nm in diameter and of variable lengths, up to several μm.

microtubule-associated protein (MAP)   A protein that binds to microtubules and influences their organization.

microtubule organizing center (MTOC)   The region of a eukaryotic cell from which microtubules emanate.

microvillus (plural, **microvilli**)   A membrane-covered extension of an epithelial cell.

midbody   In a dividing plant cell, the thin connection between daughter cells that persists until the end of cytokinesis; the midbody is packed with microtubules from the spindle apparatus.

midpiece   In a spermatozoan, a section that contains a microtubule organizing center and dense concentrations of mitochondria—sources of ATP for the journey to the egg.

mimicry   Pretending to be something else.

mineral   An inorganic substance, especially one required for the life of an organism.

minipill   A birth control pill that contains only progesterone, with no estrogens.

missense mutation   A change in DNA that alters the codon for one amino acid into a codon for another amino acid.

mitochondrion   [plural, **mitochondria**; Greek, *mitos* = thread + *chondrion* = a grain] In eukaryotes, the major energy-producing organelle; in cells that do not directly harvest sunlight, mitochondria produce nearly all of the ATP the cell needs to power the chemical reactions of the cell.

mitochondrial matrix   The compartment surrounded by the inner mitochondrial membrane.

mitosis   [Greek, *mitos* = thread] The process of the equal distribution of chromosomes during cell division.

mitotic spindle   An elongated structure that develops outside the nucleus during early mitosis; contains the microtubular machinery that moves the chromatids apart.

mittelschmerz   [German, *mittel* = middle + *schmerz* = pain] Pain during ovulation.

mobile gene   A gene whose chromosomal address changes.

modal action pattern   A highly stereotyped innate behavior.

model   A simplified view of how the components of a structure operate.

modern synthesis   The marriage of genetics and evolutionary theory that elucidated the genetic basis of variation and natural selection.

modifier gene   A gene that regulates the expression of another, separate gene.

molarity   The concentration of a dissolved substance in moles per liter of solution.

mold   A protist or fungus that grows as a downy coating on animal or vegetable matter.

mole   The amount of a substance in grams equal to its molecular mass relative to a hydrogen atom.

molecular genetics   The branch of genetics that studies how DNA carries genetic instructions and how cells carry out these instructions.

molecular weight   The mass of a molecule relative to that of a hydrogen atom.

molecule   A specific combination of individual atoms held together by covalent bonds.

mollusk   [Latin, *molluscus* = soft] A soft invertebrate inside a shell.

molting   The shedding of feathers, fur, or exoskeleton and the secretion of new ones.

Monera   The kingdom that, until recently, was considered to include all prokaryotes, including both the eubacteria and the archaea; when used now, the term refers only to the eubacteria.

monoclonal antibody   An antibody produced from a single hybridoma clone; each monoclonal antibody has a defined chemical structure and has the ability to recognize a single epitope.

monocotyledon (or **monocot**)   One of a large group of flowering plants that have one cotyledon. Monocotyledons consist of about 65,000 species, including all of the grasses, lilies, palms, and orchids. Most are herbaceous.

monoculture   The raising of a single crop.

monomer   [Greek, *mono* = single] Each of the component parts of a polymer.

monophyletic   A taxon that includes an ancestral species and all its descendants.

monosaccharide   [Greek, *mono* = one + *saccharine* = sugar] A simple sugar with three to nine carbon atoms; the building block for polysaccharides.

monosomy   The presence of only a single copy of a chromosome in a diploid cell.

monotreme (also called **prototherian**)   [Greek, *monos* = single + *trema* = hole] An egg-laying mammal; like other mammals, monotremes have hair and mammary glands; their digestive and reproductive systems empty into a cloaca, hence their name

morning-after pill   A birth control method that is taken after unprotected intercourse.

morphogen   A substance that specifies the position of a cell within a pattern.

morphogenesis   The development of form.

mortality   The probability that an individual of a given age will die each year.

morula   [Latin, mulberry] A solid ball of cells in an early mammalian embryo.

motor cortex   A region of the brain responsible for movements, with each part of the body specified by a distinct area.

motor neuron (also called a **motoneuron**)   A neuron that directly connects with a muscle and commands muscle contraction; the final common pathway of instructions that direct movement.

mucigel   In plants, the slimy substance that the epidermis and the root hairs secrete that enhances the absorption of water and minerals.

mucus   A viscous, slippery substance that coats the food particles and lubricates their movements within the mouth and the digestive system.

Müllerian mimicry   The resemblance of two or more equally dangerous species to each other; the similarities in color or form represent similar dangers to common predators, which therefore avoid all the mimicking species.

multicellular   Made of many cells.

muscle fiber   A giant muscle cell that contains the proteins responsible for contraction.

mustard oil glycosides   Compounds that give cabbage, broccoli, radishes, and other plants in the mustard family their characteristic pungent odor and flavor.

mutagen   An agent that increases the rate of mutation; a mutagen can be a chemical or a form of radiation.

mutation   A change in the nucleotide sequences of DNA.

mutualism   An association between organisms of two species from which both organisms benefit.

mycelium   [Greek, *mukes* = fungus] In a fungus, the mass into which hyphae grow, branch, and intertwine.

mycorrhizae   [Greek, *mukes* = fungus + *rhiza* = root] Symbiotic association between the root of a plant with the mycelium of a fungus.

myelin   An insulating structure around a nerve fiber; myelin consists of extensions of the plasma membrane of a Schwann cell or an oligodendrocyte.

myoblast   A cell that is a precursor of a muscle fiber.

myocardium   [Greek, *myos* = muscle] The muscular walls of the heart.

myofibril   [Greek, *myos* = muscle + Latin, *fibrilla* = little fiber] A thread, consisting of actin and myosin and about 1 to 2 μm in diameter, that runs the length of a muscle fiber.

myoglobin   An iron-containing muscle protein that pulls oxygen from the blood.

myosin   A long, two-headed protein that interacts with actin and generates movement in an ATP-dependent manner.

myxomycota   [Greek, *muxa* = mucus + *mukes* = fungus] The plasmodial, or acellular, slime molds; they closely resemble the cellular slime molds in appearance and in life cycle.

NAD$^+$ (nicotinamide adenine dinucleotide)   The major electron acceptor in oxidative phosphorylation.

NADH   The reduced form of NAD$^+$.

Na$^+$-K$^+$ATPase (also called the **sodium-potassium pump**)   An important active transporter that simultaneously transports sodium ions (Na$^+$) out of cells and potassium ions (K$^+$) into cells, using the energy of ATP.

natural selection   The differential survival and reproduction of individuals with certain inherited traits; the major mechanism for evolution.

Neanderthal   One of the early species of *Homo sapiens*.

negative assortative mating   A process in which individuals choose mates who differ somewhat from themselves.

negative feedback   The process of neutralizing external changes.

negative regulator   A protein that reduces transcription of a particular gene or operon.

nematocyst   In cnidarians, a tiny barbed spear fired with water pressure by cnidocytes.

nematodes   Roundworms; slender, cylindrical, and unsegmented.

nematomorpha   Horsehair worms.

nemertea   Ribbon worms.

nephridia (singular, **nephridium**)   Tubular excretory organs that remove nitrogen-containing wastes from the coelomic fluid and regulate water and salt concentration.

nephron   [Greek, *nephros* = kidney] The functional unit of the kidney formed by the glomerulus, Bowman's capsule, and the renal tubule together.

neritic zone   The zone of the ocean that is out from shore, over the continental shelf.

nerve growth factor (NGF)   A protein paracrine signal necessary for the growth and differentiation of specific nerve cells.

neural crest   A set of embryonic cells derived from the roof of the neural tube; neural crest cells migrate to different locations in the embryo and develop into a variety of different cell types in the adult.

neural plate   A flat plate above the notochord in a vertebrate embryo; the future nervous system.

neural tube   A hollow tube that forms from the neural plate, above the notochord in a vertebrate embryo; the precursor of the central nervous system.

neuromuscular junction   The chemical synapse between a motor neuron and a voluntary muscle cell.

neuron   A nerve cell specialized for the conduction of electrical and chemical signals.

neurotransmitter   A signaling molecule that transmits signals from a nerve cell either to another nerve cell or to a muscle or a gland.

neurulation   The process of forming the neural tube and neural crest.

neutrons   Neutral particles without electric charge found in the nucleus of an atom.

niche   The way an organism uses its environment.

nicotine   A naturally occurring compound (especially in tobacco) that acts as a stimulant and narcotic; like heroin and cocaine, it is highly addictive.

nicotinic acetylcholine receptor   A ligand-gated cation channel that responds to acetylcholine and nicotine; such receptors are found in the neuromuscular junction and elsewhere.

nitrification   The process by which energy is obtained by converting ammonia to nitrite (NO_2) and nitrate (NO_3).

nitrifying bacteria   Bacteria specialized to convert atmospheric nitrogen (N_2) into ammonia or nitrate.

nitrogen fixation   The conversion of atmospheric nitrogen gas into ammonia, which makes nitrogen available to organisms; nitrogen fixation is carried out only by specialized prokaryotes, some of which live within the roots of some plants.

nitrogenous base (so-called because it contains nitrogen and can accept hydrogen ions)   A small molecule that contains one or two aromatic rings of carbon and nitrogen, with attached hydrogen atoms; a component of a nucleotide.

nitrogenous wastes   Nitrogen-containing products of protein breakdown.

node   The region of the stem to which a petiole attaches.

node of Ranvier   A gap between myelin wrappings along a nerve axon.

nonassociative learning   Learning in which the subject changes sensitivity to a stimulus after repeated exposure.

noncompetitive inhibitor   A molecule whose inhibitory effects on an enzyme cannot be overcome by increased substrate concentration; a noncompetitive inhibitor usually binds to an enzyme at a location other than the active site.

noncovalent interaction   Attraction (or repulsion) between separate molecules or ions; noncovalent interactions include electrostatic interactions, hydrogen bonds, van der Waals interactions, and hydrophobic interactions.

noncyclic photophosphorylation   The production of ATP from light energy by a series of electron transfers that do not directly regenerate the absorbing chlorophyll; in plants, noncyclic photophosphorylation occurs within photosystem II, with electron flow ultimately depending on both photosystems.

nondisjunction   The failure of homologous chromosomes or sister chromatids to move apart during meiosis; nondisjunction results in a gamete having too many or too few chromosomes.

noninducible   Not made at higher levels in the presence of an inducer that would normally increase production.

nonpolar   Having an approximately uniform charge distribution.

nonsense   Not specifying any amino acid; a nonsense codon is also called a "stop" codon.

nonsense mutation   A change in DNA that changes the codon for an amino acid into a nonsense or termination codon and thereby leads to a truncated polypeptide.

norepinephrine   A neurotransmitter derived from tyrosine; norepinephrine is used by the central nervous system and the sympathetic nervous system.

notochord   A rod that runs from the front to the rear of a vertebrate embryo, beneath its back (dorsal) surface.

nucellus   [Latin, *nucella* = a small nut] In a flowering plant, the central portion of the ovule in which the embryo develops; the megasporangium.

nuclear envelope   The boundary of a nucleus; it consists of a double membrane separated by about 20 to 40 nm.

nuclear fission   The breakup of large nuclei, of uranium or plutonium, for example.

nuclear fusion   The fusion of two atomic nuclei to form a larger one—formation of a helium nucleus from two hydrogen nuclei, for example.

nuclear pore   An interruption of the nuclear envelope that forms channels between the contents of the nucleus and the cytosol.

nuclear reactions   The high-energy transformation of individual atoms; occurs in stars and in nuclear reactors.

nuclear transplantation   A technique for moving a nucleus from one cell to another.

nuclear winter   A world-wide darkening and cooling that is believed to be the devastating outcome of a large-scale nuclear war; the fine dust from nuclear bombs and from the fires they would ignite could create a dust cloud that would envelop the Earth and reduce sunlight (and photosynthesis) to a few percent of the present level; temperatures would plunge to below freezing in most parts of the world, triggering mass extinctions, similar to those of the Permian and Cretaceous Periods.

nuclease   An enzyme that catalyzes the hydrolysis of a nuclei acid.

nucleic acid   A macromolecule (DNA or RNA) formed by the polymerization of nucleotides.

nucleoid   In the prokaryotes, the restricted part of a cell that contains the cell's DNA; it is not surrounded by a membrane.

nucleolar organizer   A region of one or more chromosomes that contains the genes for ribosomal RNAs.

nucleolus   [plural, **nucleoli**; Latin, a small nucleus] A conspicuous structure within the nucleus, in which ribosomal RNAs are made.

nucleoplasm   The contents of the nucleus.

nucleoside   A nitrogenous base attached to a sugar by a covalent bond between a carbon atom of the sugar and a nitrogen atom of the base.

nucleosome   A DNA-histone complex, about 11 nm in diameter; each nucleosome contains a 146-nucleotide-long stretch of DNA and eight histone molecules (two each of H2A, H2B, H3, and H4).

nucleotide   A small molecule that consists of a nitrogen-containing aromatic ring compound, a sugar, and one or more phosphate groups.

nucleus   In eukaryotic cells, the membrane-enclosed structure that contains most of a cell's genetic information in the form of DNA.

objective   In a microscope, the lens nearest the sample.

obligate aerobes   Heterotrophs that require oxygen to live.

obligate anaerobe   A microorganism that can grow only in the absence of oxygen.

oceanic zone   The zone of the ocean that is beyond the continental shelf over the deepest water.

oceanography   The study of the seas and their ecosystems.

octet rule   The generalization that an atom is particularly stable and chemically unreactive when its outermost shell is full, meaning (usually) that it contains eight electrons.

ocular   The lens of a microscope nearest the eye.

oil   A triacylglycerol that is liquid at room temperature; the component fatty acids are usually unsaturated.

Okazaki fragment (named after its discoverer Reijii Okazaki)   In DNA synthesis, a stretch of DNA that is to be added to the lagging strand.

olfactory receptor   A protein, in the membrane of the olfactory epithelium, that binds to odorant molecules; binding is the first step in informing the brain about smell.

oligodendrocyte   A glial cell in the central nervous system; the oligodendrocyte's plasma membrane wraps around nerve cell extensions in a myelin sheet.

oligotrophic   Referring to a pond or a lake that lacks nutrients; oligotrophic lakes are clear and have no permanent algal blooms; they contain more oxygen and support a more diverse community of organisms than eutrophic lakes and ponds.

ommatidia (singular, **ommatidium**)   The simple light-detecting units that make up compound eyes.

omnibacteria   Bacteria that can perform aerobic respiration and can use nitrate (NO_3^-) as an electron acceptor; the most common organisms on Earth.

omnivores   [Latin, *omnis* = all + *vorus* = devouring] Animals that eat plants, herbivores, and other carnivores.

oncogene   [Greek, *onkos* = bulk, tumor] A gene whose product can change normal cells into cancerlike cells.

oocyte   A precursor of an egg cell.

oogenesis   The production of mature eggs.

oogonium   A precursor cell that is committed to forming female gametes (ova).

open circulatory systems   A circulatory system in which blood mixes freely with extracellular fluids and bathes the organs of the body.

open system   In thermodynamics, a region of space that exchanges both materials and heat with its surroundings.

operator   The DNA sequence in an operon to which a repressor protein binds.

operculum   In fish, a protective bony flap that covers several gill arches on each side of the head; in mollusks, a protective flap that covers the mantle cavity.

operon   In prokaryotes, a set of genes transcribed into a single mRNA.

opiates   Morphine, heroin, and related compounds.

opposite   The arrangement of leaves on a stem so that two petioles extend away from the stem from the same node.

optical isomers   Two molecules that differ only in the arrangement of four different groupings of atoms attached to a single carbon atom.

optimal foraging   Energy-efficient feeding.

orbital (so-called to distinguish from a planet's restricted "orbit")   A limited portion of an atom's space through which an electron moves.

order   A taxonomic group that consists of one or more families; a subdivision of a class.

organ   A structural unit with a distinctive function formed by two or more kinds of tissue.

organ formation   During embryonic development, the movement and specialization of tissues and cells to produce functioning organs such as heart, kidneys, and the nervous system.

organelle   A subcellular structure that performs a specialized task; in eukaryotic cells, many organelles are enclosed by membranes, which isolate the contents of the organelle from the rest of the cytoplasm.

organic   Carbon-containing; the term refers to all carbon-containing compounds, even when they have nothing to do with organisms.

organic chemistry   The study of the structures and reactions of carbon compounds.

organism   A living individual.

orgasm   A complex of changes that often accompany sexual intercourse—smooth muscle contractions in the genital tract, skeletal muscle contractions throughout the body, and feelings of intense pleasure.

origin of replication   A DNA sequence at which replication begins.

oscilloscope   A measuring instrument that displays voltages on a televisionlike screen, showing variations with time.

osmolarity   [Greek, *osmos* = push, thrust] The sum of the concentrations of all the ions and molecules in a solution.

osmosis   [Greek, *osmos* = push, thrust] The flow of water across a selectively permeable membrane as a result of concentration differences.

osmotic pressure   The pressure exerted by osmosis.

osteichthyes   [Greek, *osteon* = bone + *ichthys* = fish] Fish with bony skeletons.

osteoblasts   [Greek, *osteon* = bone + *blastos* = bud] Specialized cells that resemble fibroblasts and manufacture the bone matrix.

osteoclasts   [Greek, *osteon* = bone + *klastos* = broken] Specialized cells that digest collagen and bone matrix.

ostracoderms   [Greek, *ostrakon* = shell + *derma* = skin] Armor-plated jawless fishes, all of which are extinct.

outbreeders   Species with complex physical or behavioral adaptations that promote crossbreeding with individuals who are not closely related.

ova   [singular, **ovum**; Latin, egg] Eggs.

ovarian cycle   Refers to the events of the menstrual cycle within the ovaries—the production of the oocytes and the growth of the corpus luteum.

ovary   [Latin, *ovum* = egg] The organ (in an animal or a flower) that produces female germ cells (or gametophytes); a female gonad.

overlapping genes   A single stretch of DNA that contains the information for distinct polypeptides in different reading frames.

overshoot   Population growth that exceeds the carrying capacity.

oviduct   [Latin, *ovum* = egg + *ductus* = duct] One of two long tubes that lead to the uterus; in humans, it is called a fallopian tube.

ovulation   The release of the oocyte from the follicle.

ovule   [Latin, *ovulum* = little egg] The structure in a carpel that, after fertilization, forms a seed.

ovum   [plural, **ova**; Latin, egg] A female gamete; an egg.

oxaloacetate   A four-carbon compound that can combine with the acetyl group from acetyl CoA to produce citric acid, beginning the citric acid cycle.

oxidative fiber   A thin muscle fiber that derives most of its energy from respiration.

oxidative phosphorylation   The process that couples the oxidation of NADH and $FADH_2$ to the production of high-energy phosphate bonds in ATP.

oxidizing agent   The electron acceptor in a redox reaction.

oxygen evolving complex   In photosynthesis, a manganese-containing protein that directly participates in the electron transfers that convert coordinates of water to oxygen.

oxyhemoglobin   Hemoglobin that is bound to oxygen.

oxytocin   A peptide hormone, released by the posterior pituitary, that stimulates contractions of the uterus during childbirth.

ozone layer   The layer of ozone gas 20 to 50 kilometers above the Earth; ozone is a highly reactive molecule made of three oxygen atoms, instead of the two in atmospheric oxygen.

pacemaker   A group of the cells of the myocardium that are capable of rhythmic spontaneous contractions; they contract slightly more frequently than most myocardial cells; by initiating contraction before other cells in the heart, the pacemaker cells set the pace of contraction for the rest of the heart

paleontology   [Greek, *palaios* = ancient + *on* = being + *logos* = discourse] The study of ancient life.

palisade parenchyma   In plants, the mesophyll cells that lie just under the leaf's upper epidermis and are elongated and packed with chloroplasts.

paracrine signal   A chemical signal that acts only in the immediate region of its production.

parapatric speciation   A mechanism of speciation in which a species splits into two populations that are geographically separated, but still have some contact.

paraphyletic   Refers to a taxon that contains some but not all the descendants of the ancestral species.

parapodia   In polychaetes, a pair of leglike paddles in a single segment; parapodia are used in respiration and to swim, crawl, or burrow.

parasite   An organism that consumes parts of a larger organism; it does not necessarily kill the host.

parasitism   A symbiotic relationship in which one species benefits at another's expense.

parasympathetic nervous system   [Greek, *para* = beside, next to + sympathetic] A division of the autonomic nervous system; it generally acts to husband resources, for example by slowing the heart and increasing intestinal absorption.

Parazoa   The smaller of two subkingdoms of animals, which contains two phyla: the Placozoa and the Porifera (sponges); the parazoa differ greatly from all other animals in showing no symmetry and minimal organization and cell specialization—they possess only simple connective tissues and no organs.

parenchymal cells   Plant cells with thin cell walls, chloroplasts, large vacuoles, and the machinery to perform photosynthesis.

parental generation (also called **P generation**)   The original parents in a genetic cross.

passive transport   Movement of a substance across a membrane that occurs spontaneously, without the expenditure of energy.

patch clamping   A method that allows the measurement of currents across a tiny piece of membrane rather than across a whole cell membrane.

pathogens   Agents that cause disease.

pattern formation   The creation of a spatially organized structure during embryonic development.

pedigree   A family tree; pedigrees are often used to show the inheritance of a disease within a family.

pelagic division   The open waters of the ocean and the organisms within them.

pellet   The part of a sample that, during centrifugation, moves to the bottom of the centrifuge tube.

penetrance   The fraction of individuals with a particular genotype that show a corresponding phenotype.

pentose   [Greek, *pente* = five] A sugar that contains five carbon atoms; ribose and deoxyribose, which are components of nucleotides, are pentoses.

pepsin   The stomach enzyme that hydrolyzes proteins into smaller fragments each containing a few amino acids.

peptide   A molecule that consists of amino acids linked by peptide bonds; some peptides are water-soluble signals.

peptide bond   The covalent bond between two amino acid molecules in a polypeptide; these bonds are formed by the carboxyl group of one amino acid attaching to the amino group of another amino acid.

peptide bond formation   During protein synthesis, the transfer of an amino acid from aminoacyl-tRNA to the growing polypeptide chain.

peptidoglycan   The cell-wall material of many prokaryotes; peptidoglycan may consist of a single covalently linked molecule formed from polypeptides and polysaccharides.

perennial   Refers to a plant that lives and produces seeds for two or more years.

perforation plate   In the vascular system of a plant, the end wall of each vessel element; the perforation plate contains one or more holes through which water flows.

pericardium   [Greek, *peri* = around + *kardia* = heart] A fibrous sac that encloses the heart itself within a watery lubricating fluid.

pericycle   [Greek, *peri* = around + *kykos* = circle] In plant roots, a sheath of parenchyma cells just within the endodermis, but outside the vascular tissue.

periderm   In plants, the outer, secondary tissue; periderm is primarily cork.

period   In the geological time scale, a time interval of 30 to 75 million years that contains distinctive forms of life in its fossil record; a period is longer than an epoch, but shorter than an era.

peripheral nervous system (PNS)   Nerve cells outside the central nervous system (CNS); the PNS carries information from the CNS to muscles and organs and to the CNS from sense organs.

peripheral proteins   Membrane proteins located on the outer and inner surfaces of the plasma membrane.

peristaltic waves   [Greek, *peristellein* = to wrap around] Coordinated contractions of smooth muscles.

permafrost   Permanently frozen ground less than a meter from the surface that underlies the soil during even the warmest summers.

peroxisome   In eukaryotic cells, a kind of membrane-bounded vesicle; peroxisomes contain enzymes that use oxygen to break down molecules by pathways that produce hydrogen peroxide (H_2O_2).

petal   One of the showy, usually colored, parts of a flower.

petiole   The stalk that connects the leaf to the stem.

pH   The logarithm (to the base of 10) of the molar hydrogen ion concentration.

pH scale   Measures the concentration of H^+ and OH^- ions in a solution.

phaeophytes [Greek, *phaios* = dusky brown] Brown algae.

phagocytosis [Greek, *phagein* = to eat + *kytos* = hollow vessel] A type of cellular ingestion in which the cell's membrane surrounds a relatively large solid particle, such as a microorganism or cell debris.

pharyngeal slits (also called **gill slits**) Holes in the sides of the body that run from the inside of the pharynx to the outside surface of an animal.

pharynx The throat; a common entryway for food into the digestive tract, for air into the lungs, and for water into the gills.

phenetics A method of classification, based on all observable characteristics, that attempts to avoid the subjective choices of classical taxonomy.

phenolic compounds Aromatic substances that play a variety of roles in plants; some phenolics repel herbivores and pathogens, some attract pollinators or fruit dispersers.

phenotype [Greek, *phainein* = to show + *typos* = impression] The collection of all the properties of an individual organism.

phenotypic trait A single aspect of phenotype in which individuals may vary.

phenylketonuria (PKU) A human disease that results from the absence of the enzyme phenylalanine hydroxylase, which converts phenylalanine to tyrosine.

pheromone A substance secreted by one organism that influences the behavior or physiology of another organism of the same species; a pheromone can also activate modal action patterns.

phloem Conducting vessels that distribute the sugars and other organic molecules made in the leaves to the rest of the plant.

phosphate The ion (PO_4^{3-}) formed by the dissociation of hydrogen ions from phosphoric acid (H_3PO_4).

phosphodiester bond The links between nucleotides formed by the phosphate group of one nucleotide attaching to a carbon atom in the sugar component of another nucleotide.

phospholipid An amphipathic derivative of glycerol in which two hydroxyl groups attach to fatty acids and the third to a phosphate ester; phospholipids are principal components of biological membranes.

phosphorylase kinase An enzyme that transfers phosphate groups from ATP to glycogen phosphorylase.

photoautotroph An organism that supplies itself with energy by means of photosynthesis.

photochemical reaction center A complex of the chlorophyll molecules and proteins that convert captured light energy to chemical energy.

photoheterotroph A heterotroph that uses light energy but also requires organic compounds.

photon A package of energy; a light particle.

photoperiodism The response of a plant to the relative lengths of day and night.

photorespiration In plants, an oxygen-dependent process that does not produce ATP or NADPH; photorespiration converts ribulose bisphosphate into CO_2 and serine.

photosynthesis [Greek, *photo* = light + *syntithenai* = to put together] Process in which sugars are synthesized within the chloroplast and temporarily stored there.

photosystem I One of two distinct but interacting sets of electron transfer reactions responsible for storing light energy in high-energy chemical bonds; photosystem I best absorbs and uses light with wavelengths of about 700 nm.

photosystem II One of two distinct but interacting sets of electron transfer reactions responsible for storing light energy in high-energy chemical bonds; photosystem II best absorbs and uses light with wavelengths of about 680 nm.

phototropism The bending of a plant toward light.

phragmoplast [Greek, *phragmos* = fence + *plasma* = mold, form] A set of microtubules that extends between two dividing plant cells at right angles to the cell plate.

phylogeny [Greek, *phylon* = race, tribe + *geneia* = birth, origin] Evolutionary history.

physical anthropologist [Greek, *anthropo* = human] Researchers who study the evolution, anatomy, and behavior of humans.

physiology The study of how living organisms are organized and how their parts work together.

phytochrome [Greek, *phyton* = plant] A plant pigment involved in many processes that depend on the timing of dark and light, including flowering, germination, and leaf formation.

pigment A molecule that absorbs visible light and has color to human eyes.

pilus (plural, **pili**) In bacteria, a long appendage that serves as the means of attachment of conjugating bacteria and a conduit for the transfer of DNA.

pinnae [singular, **pinna**; Latin, feather] In plants, small leaflets that extend from a central stalk.

pinocytosis [Greek, *pinein* = to drink] A type of endocytosis; the nonspecific uptake of bits of liquid and dissolved molecules.

pioneer community The first community in a succession.

pistil (also called a **carpel**) In a flower, a female reproductive structure.

pith In plants, the unspecialized tissue within central nonconducting core of the vascular system.

pituitary gland A pea-sized structure at the base of the brain that releases at least nine hormones.

placenta [Latin, flat cake] Tissue that passes nourishment, oxygen, and wastes between the mother and the developing embryo in placental mammals.

placental mammals Mammals whose embryonic development takes place entirely within the uterus of the mother.

placoderms [Greek, *plak* = plate + *derma* = skin] Extinct armored fish with jaws.

plankton [Greek, *planktos* = wandering] Organisms that float in the water, carried by currents.

Plantae The kingdom that includes plants, autotrophic multicellular organisms that undergo embryonic development.

plasma [Greek, form or mold] The fluid part of blood.

plasma cell A cell specialized for the production of antibodies.

plasma membrane The membrane that surrounds a cell.

plasmid A circular DNA that can replicate autonomously.

plasmodesmata [singular, **plasmodesma**; Greek, *plassein* = to mold + *desmos* = to bond] Fine intercellular channels between plant cells, derived from vesicles trapped in the growing cell plate.

plasmodium A mass of cytoplasm with many nuclei but no boundaries between cells.

plasmolysis [Greek, *plasma* = form + *lysis* = loosening] In a plant cell, the flow of water out of the vacuole in a hypertonic solution and the resulting separation of the plasma membrane from the cell wall.

plastid A plant organelle surrounded by a double membrane; plastids include chloroplasts, chromoplasts, and amyloplasts.

platelet   A small, membrane-enclosed element that is a component of mammalian blood; it is formed as a cytoplasmic fragment of a precursor cell in the bone marrow; platelets contribute to clotting.

plate   In reference to the Earth's crust, a large block that moves with respect to other blocks at the rate of a few centimeters a year.

platyhelminthes   The phyla of the flatworms.

platyrrhines   [Greek, *platyus* = flat + *rine* = nose] Monkeys whose flat noses have widely separated nostrils that point sideways.

pleiotropy   [Greek, *pleios* = more + *trope* = turning] The capacity of a single gene to affect many aspects of phenotype.

point mutation (also called a **single-base mutation**)   A change in a single nucleotide pair of DNA; in some cases, a change in a few nucleotide pairs still qualifies as a point mutation.

polar   Having uneven distributions of electrical charge; having positive and negative ends (or poles).

polar body   The smaller daughter cell that results from the unequal meiotic division of an oocyte.

polar fronts   Regions of low pressure, where the westerlies end and polar easterlies begin.

polar microtubules   Microtubules that run from each pole of the mitotic spindle toward the equator.

pollen tube   An extension of a pollen grain, which carries the sperm nuclei into the ovule.

polyclonal   Referring to antibodies made in the descendants of many B cells, with each B-cell clone producing an antibody that recognizes a different epitope.

polygenic   Traits governed by many genes that vary smoothly and continuously within a population.

polymer   [Greek, *poly* = many + *meros* = part] A large molecule that consists of smaller identical (or nearly identical) subunits, called monomers.

polymerase chain reaction (PCR)   A method that specifically and repetitively copies a segment of DNA between two defined nucleotide sequences.

polymerization   [Greek, *polys* = many + *meros* = part] The assembly of many small molecules into a larger one.

polymorphic   [Greek, *poly* = many + *morph* = form] Having two forms of a given gene; the term may refer to a gene or to a population.

polynucleotide   A chain of nucleotides held together by phosphodiester bonds; DNA and RNA are polynucleotides.

polyp   A body plan that resembles a cylinder; one of the two basic body plans of the cnidarians.

polypeptide   A chain of amino acids held together by peptide bonds.

polyphyletic   Refers to a taxon that includes descendants of more than one ancestor.

polyploid   Having two or more complete sets of chromosomes, as in some plants.

polyribosome (also called a **polysome**)   Several ribosomes attached to a single mRNA.

polysaccharide   A carbohydrate macromolecule formed by the polymerization of simple sugars (monosaccharides).

polyspermy   Fertilization by more than one sperm.

polytene chromosomes   [Greek, *polys* = many + *tainia* = ribbon] Giant chromosomes, commonly studied in the larvae of *Drosophila* and other flies, that consist of about 1000 chromatids aligned in parallel.

polyunsaturated   Referring to a hydrocarbon chain (or fatty acid) with many double bonds.

Pongidae   Gorillas, chimpanzees, and orangutans.

population   A breeding group of individuals of the same species that inhabit a common area.

population genetics   The quantitative study of the processes by which variation is generated and passed on within populations.

population momentum   The continued growth of a population even though completed family size is below replacement level.

positional information   Chemical cues that establish the position of a cell in a pattern.

positioning   During protein synthesis, the binding of the next aminoacyl-tRNA to the A site in the ribosome-mRNA complex; the first step of elongation.

positive assortative mating   Choosing mates on the basis of similarities to the individual's own genotype.

positive regulator   A protein that increases the rate of transcription.

postabsorptive state   The state of an animal in which cells derive energy and building blocks by breaking down stored glycogen, fats, and proteins.

posterior   The back (tail-end) of an animal.

postsynaptic   Refers to the neuron that is "downstream" from a synapse; a postsynaptic neuron responds to the presynaptic neuron.

postsynaptic potential   A voltage change induced by the action of a neurotransmitter.

posttranscriptional processing   A series of chemical modifications that convert the primary transcript of a gene to a mature RNA.

posttranslational   Occurring after translation is completed.

postzygotic barrier   Barrier to gene flow which makes a zygote either inviable (certain to die) or sterile.

potency   What a cell or tissue could become during development if it were allowed to develop in another environment.

potential energy   A general term for energy that can ultimately be converted into kinetic energy.

pre-mRNA   The primary transcript of a protein-coding gene; the term is also used to refer to the partly modified, not yet mature mRNA precursor.

prebiotic evolution   The evolution of organisms from nonliving matter.

Precambrian   Referring to fossils that date from more than 590 million years ago, that is, before the beginning of the Cambrian Era.

predator   An organism that usually (but not always) kills its prey and consumes most of its prey's body.

preformation   The theory, now discredited, that all the parts of the adult organism already preexist at the earliest stages of life.

pressure flow hypothesis   The accepted explanation for the transport of sap through the phloem.

presumptive   Having a specified developmental fate.

presynaptic   Refers to the neuron that is "upstream" from a synapse; an action potential in the presynaptic neuron can trigger the release of neurotransmitter into the synapse.

prezygotic barrier   A barrier to gene flow that prevents the fusion of the sperm and egg to form a zygote (syngamy).

primary cell wall   The wall that lies outside the plasma membrane.

primary compounds   In plants, refers to compounds that are essential for day-to-day functioning; primary compounds include glucose, amino acids, ATP, and DNA.

primary growth   In a plant, growth from apical meristem tissue; the extension of length.

primary lysosome   A nearly spherical lysosome that has not yet fused with an endosome.

primary mesenchyme   [Greek, *mesos* = middle + *enchyma* = infusion] In an animal embryo, the first cells to move into the interior of a blastula.

primary oocyte   A precursor of an ovum; a diploid cell that is capable of dividing, by meiosis, to form a haploid ovum.

primary organizer   The term used by Spemann to describe the dorsal lip of the blastopore, meaning that its action established the organization of the entire early embryo.

primary productivity   The productivity of the first trophic level, the energy captured in the chemical bonds of new molecules each year for each square meter.

primary structure   The linear sequence of amino acids in each polypeptide chain of a protein.

primary succession   The invasion of a completely new environment such as a sandbar or new volcanic island.

primary transcript   The RNA first transcribed from a particular gene; the precursor of a mature RNA.

primates   The order of mammals that includes humans, apes, monkeys, and lemurs.

primer   An already existing polynucleotide to which additional nucleotides are added; all known DNA polymerases add nucleotides to a primer.

primordial germ cells   The precursors of both sperm and eggs, which arise early in development.

Principle of Independent Assortment (also called **Mendel's Second Law**)   The generalization that the alleles for one gene segregate independently of the alleles of another gene.

Principle of Segregation (also called **Mendel's First Law**)   The generalization that a sexually reproducing organism has two "determinants" (or genes, in modern terms) for each characteristic, and these two copies segregate (or separate) during the production of gametes.

probability   The chance that a given event will happen; it is calculated as the number of times an event has actually occurred divided by the number of opportunities it could have occurred; in the most familiar example, the probability of a tossed coin turning up "heads" is 1 (the number of sides that are heads) divided by 2 (the total number of sides).

proboscis   A long, sensitive, retractable, and sometimes venomous, tubelike snout characteristic of ribbon worms; a long, flexible snout or trunk characteristic of elephants. Insects often drink using a proboscis.

producers   Organisms, such as plants, that harvest energy directly from sunlight or, rarely, from inorganic molecules.

product rule   In the calculation of probability, the statement that the probability of two independent events taking place together is the product of their probabilities.

proglottid   One of the repeated segments of a tapeworm; a proglottid contains both male and female reproductive organs.

programmed cell death   The death of specific cells as a normal part of development.

prokaryotic   [Greek, *pro* = before] Cells that contain neither a nucleus nor other membrane-bounded organelles.

prolactin   A peptide hormone, made in the anterior pituitary, that stimulates milk production.

prometaphase   [Greek, *meta* = middle] Prometaphase was previously called early metaphase, the stage of mitosis during which the nuclear membrane disappears and the chromosomes attach to the spindle fibers.

promoter   A DNA sequence that specifies the starting point and direction of transcription; it is where RNA polymerase first binds as it begins transcription.

prophage   [Greek, *pro* = before] The dormant form of a bacteriophage in which the phage DNA has integrated into that of the host cell.

prophase   [Greek, *pro* = before] The first phase of mitosis, when the diffusely stained chromatin resolves into discrete chromosomes, each consisting of two chromatids joined together at the centromere.

proplastid   Common precursor of all plastids.

prostaglandin   One of 16 paracrine signals derived from a 20-carbon fatty acid called arachidonic acid.

prostate gland   In the reproductive tract of a male mammal, a gland that secretes a thin milky alkaline fluid into the lumen of the urethra.

protease (also called a **peptidase**)   A digestive enzyme that catalyzes the hydrolysis of peptide bonds; used to digest proteins.

protein   A macromolecule consisting of one or more polypeptides.

protein kinase   An enzyme that transfers the phosphate group from an ATP to a protein.

protein kinase A   An enzyme that, when stimulated by cyclic AMP, transfers a phosphate group from ATP to specific proteins.

protein kinase C   An enzyme that, when bound to diacylglycerol, transfers phosphate groups from ATP to sites in particular proteins.

proteinoids   Polypeptides thought to have been the first to appear in prebiotic evolution; proteinoids are almost certainly without specific sequences of amino acids.

Protista   [Greek, *protos* = first] The kingdom that consists mostly of single-celled organisms but that also contains some related multicellular species; protists include algae, water molds, slime molds, and protozoa.

proto-oncogene (also called a **cellular oncogene**)   The cellular counterpart of a viral oncogene; proto-oncogenes participate in the normal control of growth and differentiation.

protobiont   A primitive cell—thought to represent a stage of prebiotic evolution—that could concentrate organic molecules and begin to evolve the first metabolic pathway.

proton   A positively charged particle found in the nucleus of an atom; the word proton is commonly but inaccurately used to refer to a hydrogen ion (H^+, actually a hydronium ion H_3O^+), which comes from the dissociation of a water molecule.

proton channel   The route that hydrogen ions follow as they flow down an electrochemical gradient.

proton gradient   The difference in H^+ concentration across a membrane.

protoplasm   [Greek, *proto* = first + Latin, *plasma* = a thing molded or formed] A term used by nonbiologists to refer to the living material of organisms.

protostomes   [Greek, *protos* = first + *stoma* = mouth] Those animals in which the mouth develops first, from the blastopore; protostomes include mollusks, annelids, and arthropods.

prototherian (also called **monotreme**)   Egg-laying mammal; like other mammals, prototherians have hair and mammary glands, but their digestive and reproductive systems empty into a cloaca.

protozoa   [Greek, *protos* = first + *zoe* = life] The protists that most resemble little animals.

provirus   [Greek, *pro* = before] The dormant form of a virus in which the viral DNA has integrated into that of the host cell.

proximal tubule   The tubule that lies just next to Bowman's capsule, in the kidney's outer cortex.

pseudocoel   [Greek, *pseudo* = false + *koilos* = hollow] A cavity in certain invertebrates that lacks the mesodermal lining of a true coelom.

pseudocoelomates   [Greek, *pseudo* = false] Animals that have an organ-containing cavity that lacks the mesodermal lining of a true coelom.

Pseudomonads   A group of bacteria that are typically straight or curved gram-negative rods with flagella; they seem to live everywhere—in soil, ponds, infected wounds, hot tubs, and even medicine bottles.

pseudopodia   [Greek, *pseudos* = false + *pous* = foot] Temporary extensions of the cytoplasm.

pulmonary   [Latin, *pulmo* = lung] Relating to the lungs.

pulmonary artery   The artery that carries deoxygenated blood from the heart to the lungs.

pulmonary circulation   Circulation from heart to lungs to heart.

pulmonary veins   The veins that carry oxygenated blood from the lungs to the heart.

punctuated equilibrium   The idea that species change very little most of the time (stasis), and that most anatomical or other evolutionary change in individual species occurs during a geologically brief period at the time of speciation.

pupa   An inactive phase, following a larval stage, of insect development; a pupa may be enclosed in a cocoon.

purine   A nitrogenous base that contains a particular nine-membered double ring, with five carbon and four nitrogen atoms; adenine and guanine are purines.

pyramid of biomass   A graphic presentation of the total mass of organisms at each trophic level of an ecosystem.

pyramid of energy   A graphic presentation of the amount of energy used at each trophic level in a specified time (usually a year).

pyramid of numbers   A graphic presentation of the total numbers of organisms at each trophic level.

pyramid of productivity   An energy pyramid, which shows how much sunlight energy enters the system how much is stored in the biomass of each trophic level, and how much flows between trophic levels.

pyrimidine   A nitrogenous base that contains a particular six-membered ring, with four carbon and two nitrogen atoms; thymine, uracil, and cytosine are pyrimidines.

pyrophosphate   A molecule that consists of two phosphate groups linked together in a high-energy bond.

pyruvate   A three-carbon compound that is the end point of the first stage of glycolysis.

quantitative character   A phenotypic trait that may be described numerically within a range of values rather than as clear-cut alternatives.

quaternary structure   The relationship among separate polypeptide chains in a protein.

R group (also called a **side chain**)   Each amino acid's characteristic group of atoms.

r-selected species   A species that has adaptations that support high growth rate (r_{max}).

races (or **subspecies**)   Morphologically distinct subpopulations that can interbreed.

radial symmetry   Structural symmetry in which rotation along the central axis doesn't change the appearance.

radioactive isotope   A radioactive (unstable) isotope of an element; examples include ^{14}C and ^{3}H.

radula   In mollusks, a rasping tongue covered with teeth made from chitin.

rain shadow   The area adjacent to a mountain range, away from the prevailing winds, where little rain falls.

random drift   Changes in gene frequency not caused by selection, mutation, or immigration, but by random events.

reabsorption   In the kidney, the transport of specific substances such as salt, water, and glucose from the filtrate back into the blood.

reading frame   The grouping of nucleotides into codons that specify an amino acid sequence.

realized niche   The resources that a species uses in a particular community.

receptor   A protein to which a chemical signal first binds.

receptor-mediated endocytosis   A cell's uptake of specific substances that are recognized by receptor proteins on the plasma membrane.

recessive   The term referring to an allele that does not contribute to the phenotype of a heterozygote.

reciprocal altruism   An explanation for apparently selfless behavior; the hypothesis that animals engage in altruism with the expectation that the recipient would return the favor some time in the future.

recombinant   Containing a combination of genes not found in nature.

rectilinear   Movement in a straight line, especially in snakes.

rectum   [Latin, straight] The straight portion at the end of the descending colon in which feces are stored until their elimination through the anus.

reducing agent   The electron donor in a redox reaction.

reductionism   The effort to understand the whole in terms of the parts.

reflex   The most automatic behavior pattern, with motor activity directly responding to a sensory stimulus.

refractory period   In an axon, the period during which the further excitation does not result in channel opening and impulse propagation.

regulatory cascade   A set of reactions that amplify a signal from a relatively small number of molecules into a response that affects many more molecules.

regulatory mutation   A change in DNA that affects the amount (rather than the structure) of a polypeptide.

releasing factor (also called **releasing hormone**)   A peptide hormone, made by the hypothalamus, that regulates the release of specific hormones by the anterior pituitary.

renal artery   [Latin, *renes* = kidney] The artery that carries blood to each kidney.

renal tubule   In the kidney, a long narrow tube leading away from Bowman's capsule.

renal vein   The means by which blood leaves the kidneys.

renin   An enzyme secreted by the kidneys.

replacement reproduction   The family size at which each couple is replaced by just two descendants.

replication   [Latin, *replicare* = to fold back] Copying of a single DNA molecule (or a single set of DNA molecules) into two copies.

replication fork   A Y-shaped region of DNA where the two strands of the helix have come apart during DNA replication.

replication unit   In eukaryotic DNA replication cells, a group of 20 to 50 origins of replication that form replication forks at the same time.

reporter   A protein (or gene) whose activity serves as a measure of the ability of a test sequence to regulate expression in a particular type of cell.

reproductive unit   A species as the group of potential mates.

reproductively isolated   The term referring to populations that are unable to interbreed.

reptile   [Latin, *reptare* = to crawl] An air-breathing, egg-laying vertebrate with an external covering of scales or horny plates; reptiles include snakes, lizards, crocodiles, turtles, and dinosaurs.

resilience   One definition of ecosystem stability; the speed with which an ecosystem returns to a particular form following a major disturbance such as a fire.

resistance factor (also called an **R factor**)   A plasmid that contains a gene for an enzyme that inactivates an antibiotic.

resolution   (1) The minimum distance between two objects that allows them to form distinct images; (2) the part of the sexual response cycle in which blood flow and muscle tension return to normal after orgasm.

resource partitioning   Splitting an ecological niche.

respiration   The oxygen-dependent extraction of energy from food molecules.

respiratory chain   The pathway of electrons in oxidative phosphorylation.

respiratory system   All the structures responsible for the exchange of gases between the blood and the external environment.

response element (also called a *cis*-**regulatory element** or **enhancer**)   A DNA sequence that regulates transcription by binding to a specific protein, a transcription factor.

resting potential   The voltage across the cell membrane when a cell is not producing action potentials.

restriction enzyme   An enzyme that cuts DNA at a particular sequence.

retrovirus   An RNA virus that uses reverse transcriptase for viral replication.

reuptake   The pumping of transmitter from the synaptic cleft either into the presynaptic neuron or into surrounding glial cells.

reverse transcriptase   An enzyme that copies RNA into DNA; RNA-dependent DNA polymerase; reverse transcriptase is present in RNA tumor viruses and in HIV.

reversion   A mutation that restores the previous version of a gene sequence.

rhizoid   A thin, rootlike filament that enables mosses, liverworts, and fern gametophytes to anchor and to absorb nutrients.

rhizome   A horizontal stem that spreads on or below the ground from which fronds grow.

rhizopoda   Amebas.

rhodophytes   [Greek, *rodon* = rose] Red algae.

rhodopsin   The protein that detects light in the retina.

rhythm method   A birth control method; a form of modified abstinence, in which a couple refrains from sexual intercourse at times in the monthly cycle when conception is likely to occur.

ribbon model   A way of depicting the tertiary structure of a protein that stresses the secondary structures of a globular protein; in the ribbon model, an α helix is depicted as a coil, while β sheets are sets of arrows.

ribosomes   [Latin, *soma* = body + "*ribo*" because they contain *ribo*nucleic acid (RNA)] Complex assemblies of RNAs and proteins, about 15 to 30 nm in diameter, that are responsible for carrying out protein synthesis.

ribozyme   An artificial RNA enzyme.

RNA polymerase   The enzyme responsible for transcribing DNA into RNA.

root   The part of a plant below the ground.

root cap   A region of the growth zone of a plant's root; the root cap produces a polysaccharide slime that lubricates the path of the growing root through the soil and serves as a mechanical shield for the apical meristem, which lies just behind.

root hairs   Tiny projections, each of which is the extension of a single epidermal cell, through which most of the water that enters a root comes.

root pressure   Osmotic pressure generated by the difference in solute concentrations between root tissues and soil solution.

root system   An underground system that anchors the plant and absorbs water and minerals from the soil.

rotifers   Tiny aquatic aschelminths with saclike bodies.

rough ER   Endoplasmic reticulum that is associated with ribosomes.

RU 486   A drug used to induce abortions within the first nine weeks after conception.

rudiment   [Latin, *rudimentum* = beginning] Initial embryonic stage of an organ, from which the final form will develop.

ruminants   [Latin, *rumen* = throat] Hoofed, horned herbivores that can regurgitate partly digested food, called cud, for further chewing.

S (for DNA synthesis)   The period of the cell cycle during which DNA replicates.

salivary glands   Exocrine organs in the mouth that secrete both enzymes and mucus.

saltationism   The view, held by many 19th-century biologists, that evolution occurred by fits and starts.

saltatory conduction   The jumping movement of an action potential down a myelinated axon, from one node of Ranvier to the next; saltatory conduction occurs at rates up to 100 times faster than conduction down an unmyelinated axon.

sampling errors   Chance fluctuations that affect small samples but not large ones.

sands   Soils of large particles (from 20 mm to 200 mm).

saprophyte   [Greek, *sapro* = putrid + *phyte* = plant] A plant that decomposes dead material.

saprophytic   Deriving nourishment from dead organisms.

sarcomere   [Greek, *sarx* = flesh + *meros* = part] In muscle, a small repeating cylinder within a myofibril; each sarcomere is about 2.5 μm long; the contractile unit of a striated muscle.

sarcoplasmic reticulum   In a muscle fiber, a system of membrane-lined channels and sacs where Ca^{2+} ions are stored; a specialization of the endoplasmic reticulum.

saturated   A hydrocarbon chain (or fatty acid) in which all the bonds between carbon atoms are single bonds; the chain is called saturated because it contains the maximum number of hydrogen atoms.

savanna   A tropical or subtropical grassland punctuated by solitary trees or small clumps of trees.

scanning electron microscope (SEM) A type of electron microscope in which electrons are reflected from the surface of the observed object.

scavenger An animal that feeds from whole carcasses.

Schwann cell A glial cell in the peripheral nervous system; its plasma membrane wraps around nerve cell extensions in a myelin sheet.

scientific method A formal manner of formulating, testing, and eliminating hypotheses.

sclerenchyma [Greek, *skleros* = hard] A plant tissue that consists of cells, with thick cell walls; may be fibers or sclereids.

scolex A specialized attachment organ at the anterior end of a tapeworm.

scrotum [Latin, bag] A pouch that lies outside the body that holds the testes.

Scyphozoa The class of animals containing the jellyfish.

Second Law of Thermodynamics The statement that, while the total energy in the universe does not change, less and less energy remains available to do work.

second messenger An intracellular molecule whose synthesis and degradation depends on the action of an extracellular signal and which stimulates changes inside a target cell.

secondary cell wall The wall that lies between the cell membrane and the primary cell wall.

secondary growth Growth at the lateral meristems that increases the thickness of a shoot or a root.

secondary immune response A robust immune response triggered by the reappearance of a previously encountered antigen or microbe.

secondary lysosome An irregular lysosome that results from the fusion of a primary lysosome and an endosome; a secondary lysosome contains material that is in the process of being digested.

secondary mesenchyme In an animal embryo, cells that become mesoderm as gastrulation proceeds.

secondary oocyte A precursor of the ovum; a haploid cell produced from the primary oocyte by meiosis.

secondary plant compounds Chemicals that are not essential to a plant's normal metabolism, but which often serve a defensive purpose.

secondary sexual characteristics Hallmarks of sexual differentiation that appear at sexual maturity; in humans these include an increase in body size, a deepening of the voice, hair growth, and the development of sexual drive.

secondary structure Regular local structures resulting from regular hydrogen bonding within adjacent stretches of a polypeptide backbone; secondary structures include α helices and β structures.

secondary succession The sequence of stages in a community that has suffered serious damage.

secretion In the kidney, transport into the filtrate; secreted substances include potassium and hydrogen ions, ammonia, and organic acids and bases.

secretory vesicles In eukaryotic cells, small vesicles formed on *trans* side of the Golgi apparatus; they contain glycoproteins that are to be secreted.

section A thin slice of tissue prepared for microscopy.

sedimentary rock Rock formed by pressure from succeeding layers turning layers of mud, sand, and other sediment into rock.

segmentation gene A gene that regulates the development of segments within the early embryo.

segregation The separation of two homologous chromosomes during meiosis.

selective theory The view that the immune response depends on an animal's ability to select the set of antibodies to make in response to the presence of a particular infection.

selectively permeable Allowing the passage of some ions and molecules (especially of water) much more rapidly than others.

self-splicing Removal of intervening sequences from an RNA without the participation of other molecules.

selfish gene A sequence of DNA whose only role is to reproduce itself.

semen [Latin, seed] The sperm cells, together with the fluid from the seminal vesicles, the prostate gland, and the bulbourethral glands.

semiconservative The pattern of DNA replication in which half of each parent molecule (one strand) is present in each daughter molecule.

seminal vesicles In the male reproductive tract, organs that secrete sugars and other nutrients into the semen.

seminiferous tubules [Latin, *semen* = seed + *ferre* = to bear] The tightly coiled ducts in the testis in which the sperm matures.

senescence Aging; deterioration; in plant cells, the breakdown of cellular components leading to cell death.

sensitive period The period of time during which an animal can learn a particular behavior pattern.

sensitization The increased behavioral response to a noxious stimulus.

sepal One of the parts at the base of a flower; sepals are usually small and green, but in some flowers they may be large and colorful.

septa In a fungus, the walls that separate the nuclei within a filament.

Sertoli cells In the testes, specialized cells that surround the developing sperm; they nourish the developing sperm, regulate the passage of nutrients from the blood, and secrete a fluid that fills the lumen.

sessile Permanently anchored to rocks, logs, or coral.

seta (plural, **setae**) In annelids, a stiff, bristlelike projection.

sex chromosomes The chromosomes that differ between males and females; in humans, the 23rd pair of chromosomes.

sex steroid A steroid hormone that stimulates, maintains, and regulates reproductive organs and secondary sexual characteristics.

sex-determining region In male mammals, a gene on the Y chromosome that encodes a regulatory protein that promotes the sexual differentiation of the embryonic gonads into testes.

sex-linked gene A gene that has a different pattern of inheritance in males and females because the gene lies on a sex chromosome (usually the X chromosome in mammals).

sexual dimorphism A phenotypic difference between the sexes.

sexual imprinting A special form of imprinting that helps animals recognize potential mates.

sexual reproduction A process that produces offspring that have inherited genetic information from two parents rather than one; because the genes from each parent are likely to differ, sexual reproduction provides new combinations of genes.

sexual selection The differential ability of individuals with different genotypes to acquire mates.

shear The twisting action created by forces that are not opposite one another.

shell In an atom, a group of orbitals whose electrons have nearly equal energy.

shoot   The part of a plant above the ground.

shoot system   Stems, photosynthetic leaves, and flowers and other organs of reproduction.

sibling species   Two species that are extremely similar but reproductively isolated from one another.

sickle cell disease (also called **sickle cell anemia**)   A blood disorder that gets its name from the curled appearance of red blood cells in sickle cell patients; a genetic disease caused by a mutation in the gene encoding the β-polypeptide of hemoglobin.

sidewinding   Movement in which a snake moves across loose or sandy soil by throwing successive coils sideways.

sieve plate   In sieve tubes of a plant's phloem, the top and bottom end walls of each sieve tube member.

sieve tube members   In the phloem of a plant, the cells that actually conduct fluid.

sieve tubes   In the phloem of a plant, pipelike channels that carry organic matter from the leaves.

sign stimulus   The key aspect of an object that triggers a modal action pattern.

signal hypothesis   The statement that a hydrophobic signal peptide directs a growing polypeptide chain through the membrane of the endoplasmic reticulum into the ER lumen.

signal peptide   A polypeptide segment of 16 to 30 amino acids at the amino terminal end of the newly made protein; a signal peptide directs the newly made protein into the lumen of the endoplasmic reticulum.

silent mutation   A change in DNA that has no effect on phenotype.

silt   A soil of medium-sized particles (from 2 mm to 20 mm).

simple diffusion   [Latin, *diffundere* = to pour out] The random movement of like molecules or ions from an area of high concentration to an area of low concentration.

simple pit pair   A region between plant cells where the secondary wall is absent, allowing small molecules to pass.

simple transposon (also called an **insertion sequence**)   The simplest transposable element, a short length of DNA (up to a few thousand nucleotide pairs long) that can move from place to place in the genome.

sink   The site of storage or consumption of a substance.

sinoatrial (SA) node   The pacemaker, which lies near the top of the right atrium in the mammalian heart.

sinus   An open cavity; in insects and other arthropods and in most mollusks, it is a space through which blood travels.

sister chromatids   The two chromatids that make up a single chromosome; the sister chromatids are duplicate copies of the same genetic information.

site-directed mutagenesis   A technique that specifically alters a chosen nucleotide sequence.

skeletal muscles   The muscles that are attached to bones.

skull   The bony (or cartilaginous) case that encloses and protects the brain; it lies at the forward (or, for humans, the top) end of a vertebrate's skeleton.

sliding filament model   The accepted view of muscle movement, which holds that movement results from changing the relative positions of thick and thin filaments rather than from filament contraction.

slime layer   A tangled web of polysaccharides that, like a capsule, lies outside the cell wall.

smooth ER   Endoplasmic reticulum that is not associated with ribosomes.

smooth muscle   The involuntary muscles that line the walls of hollow internal organs such as intestines and blood vessels.

sociobiology (renamed **behavioral ecology**)   The study of behavior from an evolutionary perspective.

solute   A substance that dissolves within a solvent.

solvent   Any fluid in which other substances dissolve.

soma   The somatic cells of an organism, in contrast to germ cells or germ line.

somatic cells (also called the **soma**)   [Greek, *soma* = body] In a multicelled organism, the cells that do not give rise to germ cells or gametes.

somatic theory   The view that antibody diversity arises during development, rather than being inherited.

sori   [singular, **sorus**; Greek, *soros* = heap] In a fern, groups of sporangia clustered on the underside of the sporophylls.

source   The site of production (for example, of sugars in a plant).

speciation   The formation of a new species.

species diversity   A measure of diversity that takes into account how common individuals of each species are.

species richness   The number of species in an ecosystem.

sperm (also called a **spermatozoan**)   A male gamete.

spermatid   The four haploid cells produced during the two meiotic divisions (meiosis I and II).

spermatogenesis   The development of spermatogonia into mature sperm.

spermatogonium   [plural, **spermatogonia**; Greek, *sperma* = sperm + *gonos* = offspring] A diploid cell that is committed to forming sperm cells, but can still undergo mitosis.

spermatozoan (also called a **sperm**)   [Greek, *sperma* = seed + *zoos* = living] A fully differentiated male germ cell.

spermicide   [Latin, *cidere* = to kill] A cream or jelly that kills sperm; an essential part of the effectiveness of both cervical caps and diaphragms.

sphincters   Rings of muscles that surround a tube.

spiracle   In the insect respiratory system, an opening through which air enters the body; a spiracle leads directly to the tracheae.

spirilla   [singular, **spirillum**; Latin, *spira* = coil] Spiral-shaped prokaryotic cells.

spirochetes   Anaerobic heterotrophs that have a distinctive corkscrew shape.

spliceosome   A complex of pre-mRNA, proteins, and small nuclear RNAs that catalyzes the removal of introns and the splicing of exons.

spongy bone tissue   Tissue that generally lies at the ends and inside the bones.

spongy parenchyma   In a plant leaf, the mesophylls that lie just under the palisade parenchyma that are irregular but rounded in shape and separated by numerous air spaces.

spontaneous   A reaction that occurs without any external input of energy.

spontaneous abortion (also called **miscarriage**)   In mammalian development, embryonic or fetal death, usually resulting from congenital defects.

sporangium   [Greek, *spora* = seed + *angeion* = vessel] In plants and fungi, a structure within which cells undergo meiosis to produce the haploid spores.

spore   A cell that divides by mitosis to produce new individuals.

sporophyll   A leaf or leaflike structure that is specialized for the production of spores; in flowering plants, the term refers to carpels and stamens; in ferns sporophyll refers to a leaf that bears sporangia.

sporophyte The diploid form of a life cycle characterized by alternation of generations; a sporophyte produces haploid spores that give rise to haploid gametophytes by meiosis.

sporozoans Parasitic protists, including those that cause malaria; sporozoans undergo alternation of generations.

stabilizing selection Selection that tends to act against extremes in the phenotype, so that the average is favored.

stage On a microscope, a moveable platform that holds the observed specimen.

stain A dye that binds differently to cell components and thereby increases contrast for microscopy.

stamen In a flower, a male reproductive organ; the male sporophyll.

starch Polysaccharides used in plants for long-term energy storage.

Start (also called the **restriction (R) point**) The "point of no return" in the G_1 phase of the cell cycle; once a cell proceeds beyond Start, it proceeds through the rest of the cycle, including mitosis and cytokinesis.

stasis Stability over long periods.

statistics The science of collecting and analyzing numerical data.

statolith [Greek, *statos* = standing + *lithos* = stone] Gravity detector in a plant or animal.

stele In roots and stems, a central cylinder through which the vascular bundles frequently run.

stem cell A cell that can either produce more of itself by cell division or undergo differentiation to one or more specialized cell types.

stereoisomers Molecules with the same atoms and functional groups; two stereoisomers differ only in the spatial arrangements of their atoms.

steric inhibition Inhibition of an enzyme by a molecule whose shape resembles that of the substrate.

sternum [Greek, *sternon* = chest] The breastbone.

steroid A lipid-soluble molecule derived from cholesterol; many steroids are used as hormones.

stoma [plural, **stomata**; Greek, *stoma* = mouth] In a plant leaf, a tiny mouthlike pore that opens and closes to regulate the flow of carbon dioxide and other gases to the interior of the leaf.

stomach [Greek, *stoma* = mouth] The most dilated and most muscular section of the digestive tract.

stop codon A codon in mRNA that signals the end of a polypeptide chain.

striated Striped.

striations Cross stripes, 2 to 3 μm apart, that lie perpendicular to the long axis of a skeletal or cardiac muscle.

strobilus [plural, **strobili**; Greek, *strobilos* = cone] A conelike structure (such as a pine cone) that consists of overlapping sporophylls grouped around a central axis.

stroke Failure of the blood supply to the brain.

stroke volume In the action of the heart, the volume of blood delivered by a ventricle.

stroma [Latin, mattress] The region within a chloroplast bounded by the inner chloroplast membrane.

structural isomers Molecules that contain the same atoms, grouped in different ways to produce different functional groups.

structural mutation A change in DNA that leads to an alteration in the amino acid sequence of a polypeptide.

suberin The waxy substance, which is impermeable to water, that surrounds each cell of the endodermis.

subsidize Referring to the growth of crop plants, to provide additional energy.

substrate A reacting molecule in an enzyme reaction; a substrate is usually (but not always) much smaller than an enzyme.

succession Progressive, predictable change over time in kinds of species and numbers of species in a community.

sucrose Table sugar; a disaccharide consisting of two monosaccharides, glucose and fructose, linked together.

sugar A simple carbohydrate; a molecule that has the equivalent of one molecule of water (that is two hydrogen atoms and one oxygen atom) for every atom of carbon.

superfecundity A state of a population in which organisms are so fertile that their populations would increase exponentially if all offspring survived.

supernatant The part of a sample that, during centrifugation, remains in suspension.

supernormal stimulus A stimulus even more stimulating than anything normally encountered in nature.

suppressor mutation A mutation in one gene that prevents the phenotypic changes caused by a mutation in another gene.

surface tension The tendency of a substance to form a smooth round surface.

surface-to-volume ratio The amount of surface area for each bit of volume.

survivorship curve A graph that shows the fraction of a population that is alive at successive ages.

suture An immovable joint that fuses separate bones.

swim bladder An air-filled sac that helps fish control their buoyancy.

symbionts [Greek, *syn* = together + *bios* = life] Organisms living in a symbiotic relationship with one another.

symbiosis [Greek, *sumbios* = living together] A close association between two organisms.

sympathetic nervous system A division of the autonomic nervous system; the sympathetic nervous system initiates the "fight or flight" reaction.

sympatric speciation [Greek, *syn* = together + *patra* = country] The splitting of one species into two without geographical isolation.

synapse The distinct boundary between two communicating neurons.

synapsis [Greek, union] The pairing of homologous chromosomes in prophase I.

synaptic cleft A space of about 20 nm that separates the presynaptic and postsynaptic cells in a chemical synapse.

synchronous cell populations Cells that are all at the same stage of the cell cycle.

syngamy The coming together of an egg and a sperm at fertilization.

system An assemblage of interacting parts or objects.

systematics The scientific study of the kinds and diversity of organisms and of any and all relationships among them.

systemic circulation The route of the blood through the body, minus pulmonary circulation.

systole [Greek, *systellein* = to contract] The part of the cycle of the heart in which the atria contract, followed by the ventricles.

T cytotoxic cell (T$_C$ cell) A T lymphocyte that is responsible for the killing of cells recognized as nonself.

T helper cell (T$_H$ cell) A T lymphocyte that activates both the humoral and cellular immune responses.

T lymphocyte (because it matures in the *thymus* gland)   One of the cells responsible solely for cellular immunity or of the cells that participate in both humoral and cellular immunity.

T-cell receptor   A protein on the surface of a T lymphocyte that can bind to a complex of an antigen and MHC molecule.

tagmata (singular, **tagma**)   Fused segments.

taiga   The broad band of coniferous forest that extends across Canada, Alaska, Scandinavia, and Siberia.

tannin   A secondary compound commonly found in woody plants.

taproot   A single vertical root found in many dicots.

tar   An oily, viscous material, consisting of hydrocarbons, found in cigarette smoke and the insides of chimneys.

target organ   A structure upon which a hormone acts.

TATA box (after the first four nucleotides in the sequence)   A consensus sequence in eukaryotic promoters.

taxis   [Greek, to put in order] Directed movement.

taxol   A compound extracted from the bark of yew trees; taxol increases the stability of microtubules.

taxon   [plural, **taxa**; Greek, *taxis* = arrangement] A general term for any group of organisms at any level of the classification hierarchy.

taxonomist   A biologist who classifies organisms.

taxonomy   [Greek, *taxis* = arrangement] The naming and grouping of organisms.

telocentric   Refers to a chromosome in which the centromere is at one end.

telophase   [Greek, *telos* = end] The last phase of mitosis, during which the mitotic apparatus (including kinetochore, polar, and astral microtubules) disperses and the chromosomes lose their distinct identities.

temperate deciduous forests   Temperate forests that receive 80 to 140 centimeters of precipitation each year, composed of trees that lose their leaves in the winter.

temperate grasslands   Regions with well-defined seasons, with hot summers and cold winters; low annual rainfall keeps grasslands from turning to deserts but is not enough to sustain the growth of trees.

temperature-sensitive mutation   A change in DNA that alters a protein so that it functions normally at one temperature (called the permissive temperature, usually relatively low), but not at a second temperature (the restrictive temperature, generally higher than the permissive temperature).

template   A guide for the assembly of a complementary shape; in the context of DNA or RNA synthesis, one strand acts as a template for the assembly of a complementary sequence.

temporal isolation   Reproductive isolation that results from differences in the times at which two populations reproduce.

ten percent law   The generalization that the organisms of any trophic level provide the next higher trophic level with only 10 percent of the energy that they have assimilated from the lower trophic level.

tendon   A type of connective tissue that attaches the muscles to the bones.

tension   The pulling action of two opposing forces.

teratogen   A compound that causes birth defects.

terminal bud   A bud that lies at the tip of a shoot.

termination   The ending of chain growth.

termination signal   A DNA sequence that determines the end of transcription.

terpene   One of a class of unsaturated hydrocarbons that form the largest class of secondary compounds in plants.

terrestrial   Living on land.

territory   An area occupied by an individual (or group of individuals), from which other individuals of that species are excluded.

tertiary structure (or **conformation**)   The complete arrangement of all the atoms of a polypeptide.

testable   Refers to a hypothesis for which an experiment could be devised that would disprove the hypothesis if it were incorrect.

testcross   A cross between an individual that is homozygous for a recessive allele of a particular gene and an individual whose genotype (homozygote or heterozygote) for that gene is unknown; the phenotype of the progeny reveals the presence of the recessive allele in the parent.

testicles   Human testes.

testis (plural, **testes**)   A male gonad, which produces sperm.

tetrad, (also called a **bivalent**)   A united chromosome pair visible during meiosis I; a tetrad consists of four chromatids.

tetrapod   [Greek, *tetra* = four + *pous* = foot]   A four-footed animal.

tetrodotoxin   A poison, isolated from a species of puffer fish, that blocks the passage of Na^+ ions through voltage-gated channels.

theory   A system of statements and ideas that explains a group of facts or phenomena.

thermacidophiles   One of the three phyla of Archaea; the bacteria that inhabit hot sulfur springs such as those in Yellowstone National Park.

thermodynamics   [Greek, *thermo* = heat + *dynamis* = power, force]   The study of the transformations and relationships among different forms of energy.

thick filament   In striated muscle, a myosin filament, which is 14 nm in diameter.

thin filament   In striated muscle, an actin filament, which is 7 nm in diameter.

thoracic cavity   Chest cavity; that part of the coelom that encloses the heart and lungs.

thorax   The middle portion of an arthropod.

thylakoid   [Greek, *thylakos* = sac + *oides* = like] A flattened disc surrounded by the innermost membranes of a chloroplast.

thylakoid membrane   A membrane, within the stroma of a chloroplast, that delineates stacked vesicles, called grana.

thymine dimers   Two adjacent T nucleotides linked together in the same DNA strand; they are formed as a result of exposure to ultraviolet light.

thyroid stimulating hormone (TSH)   A peptide hormone, made in the anterior pituitary, that stimulates the production of thyroxin in the thyroid gland.

tidal volume   The amount of air drawn in and then expelled in a single breath.

tight junction   The main kind of impermeable junction in vertebrates; they are formed by the fusion of the membranes of adjacent cells.

tissue   A group of similar cells and associated intercellular material that performs one or more specific functions; tissues are often integrated with other tissues to form an organ.

tolerance   In the immune system, a state of induced unresponsiveness to proteins and other molecules; tolerance prevents immune attacks on the body's own components.

torus   A thickened region of the cell wall between two plant cells; a buttonlike thickening in the middle of a pit.

totipotent   Able to develop into a whole organism.

trachea   [plural, **tracheae**; from Greek, *arteria trakheia* = rough artery] A tube, which arises by invagination, that carries air within an animal's body; in vertebrates, the windpipe.

tracheids Long, thin, spindle-shaped cells in the xylem of plants.

tracheophytes [Latin, *trachea* = windpipe] Vascular plants; tracheophytes include all the most familiar living plants.

tradewinds Winds that blow from about 30° latitude, steadily toward the equator.

***trans*-acting factor** (also called a **transcription factor**) [Latin, *trans* = across, because they regulate the expression of other molecules] A protein that binds to DNA and regulates transcription.

transcription The production of RNA from DNA.

transcription factor (also called a ***trans*-acting factor**) A protein that binds to DNA and regulates transcription.

transfer RNA (tRNA) A small RNA molecule that serves as an adaptor in protein synthesis; each tRNA contains an anticodon that allows it to bind to a codon in mRNA and each becomes linked to a specific amino acid.

transformation The transfer of one or more genes from one organism to another.

transgenic Containing a particular recombinant DNA incorporated within the genetic material.

transition state A distorted form of a substrate that is intermediate between the starting reactant and the final product.

translation Protein synthesis; the conversion of information from an mRNA molecule into a polypeptide.

translational control Alterations in the rate at which specific mRNAs are translated.

translocation In chromosomes, a mutation in which part of one chromosome is moved to another chromosome; in protein synthesis, the movement of the growing polypeptide chain with its attached tRNA from the A site to the P site; the third step of the elongation cycle.

transmission electron microscope (TEM) A type of electron microscope in which the electron beam passes through a specimen.

transmission genetics The branch of genetics that deals with patterns of inheritance.

transpiration The process by which plants pull water from the soil and release it as vapor through stomata in their leaves.

transpiration-cohesion theory The accepted explanation for the movement of water from roots to leaves; transpiration in the leaves pulls water up the stem in continuous columns.

transpiration-photosynthesis compromise The trade-off between saving water and maintaining photosynthetic productivity.

transposable element A mobile gene that can replicate only in integrated form, as part of a host's chromosome.

transposase An enzyme that catalyzes insertion of a transposable element into new sites.

tree line The upper limit where subalpine trees grow.

trematoda Flukes.

triacylglycerol (also called a **triglyceride**) A nonpolar derivative of glycerol in which all three hydroxyl groups are attached to fatty acids.

triceps The major muscle on the back side of the upper arm.

triple covalent bond A covalent bond in which two atoms share three pairs of electrons.

trisomy [Greek, *tri* = three + *soma* = body] The presence of three, rather than two, copies of a chromosome.

trophic level Each level of a food chain.

trophoblast [Greek, *trephein* = to nourish + *blastos* = germ] In a mammalian embryo, a prominent outer cell layer in a blastocyst.

tropical rain forest A biome with relatively constant weather, with an average temperature of about 27°C (80°F) every month of the year and heavy rainfall, ranging from 2 to 4.5 meters per year.

tropism [Greek, *trope* = turning] The bending or curving toward or away from a stimulus.

tropomyosin In muscle, a rigid, rodlike protein, about 40 nm long, that stiffens the actin filament.

troponin In muscle, a large protein that binds both to tropomyosin and to free calcium ions.

true-breeding A breeding line in which offspring have the same phenotype, generation after generation.

tubal ligation An irreversible method of birth control; cutting or blocking the oviducts, or fallopian tubes, preventing eggs from reaching the uterus after ovulation.

tubulin One of two globular proteins, called α and β tubulin, from which microtubules are assembled.

tumor suppressor gene (also called an **antioncogene**) A gene that antagonizes the action of an oncogene; this gene is intimately involved in the normal regulation of cell growth.

tumor virus A virus that causes its host cell to lose its normal ability to regulate cell division.

tundra [Finnish, *tunturi* = arctic hill] A vast, open land, with a cold, dry climate; it is dominated by grasses and low shrubs and covers more than a fifth of the Earth's surface.

turbellaria The free-living flatworms; they reproduce either asexually or sexually.

turbinate Elaborately folded bones inside the noses of many vertebrates.

turgor pressure [Latin, *turgor* = swelling] In plant cells, osmotic pressure against the cell walls resulting from the higher concentration of dissolved molecules and ions in cytoplasm than in the surrounding fluid.

typological species concept The view that each species consists of individuals that are variants of a fixed underlying plan.

ultracentrifuge A high-speed centrifuge that can spin tubes at speeds up to 80,000 rpm, subjecting their contents to forces up to 500,000 × gravity.

ultrastructure Subcellular structures visible only with an electron microscope.

unconditioned stimulus In associative learning, the stimulus that directly leads to a physiological response.

unconventional gene A gene without a stable cellular address.

uniformitarianism The view, originated by Lyell, that the processes that now mold the Earth's surface—erosion, sedimentation, and upheaval—are the ones that have always molded it; geologic change is slow, gradual, and steady, not catastrophic.

Uniramia A subphyla of arthropods in which the first pair (or the first few pairs) of appendages are antennae and the next pair are jaws, or mandibles.

universal Used by all species.

unsaturated A hydrocarbon chain (or fatty acid) that contains at least one double bond between two carbon atoms; the chain is called unsaturated because it could accept two more hydrogen atoms per double bond.

upstream In a nucleic acid, toward the 5′ end.

upwelling The upward movement of water that draws nutrients from the bottom of the ocean.

urea The compound that mammals convert to ammonia to make it less harmful.

ureter   [Greek, *ouron* = urine] One of a pair of tubes through which urine flows from the kidneys to the bladder.

urethra   The means by which the bladder drains urine to the outside of the body; the single exit tube of the human urethra.

uric acid   A nearly insoluble organic compound produced by insects, land snails, most reptiles, and birds as nitrogenous waste.

uterus   [Latin, womb] The womb; in mammals, the portion of the female reproductive tract that is specialized to receive, nurture, and deliver a developing embryo.

vacuole   Within a eukaryotic cell, a membrane-bounded sac, without any obvious internal structure; a vacuole is larger than a vesicle.

van der Waals interactions   Weak attractive forces between molecules that are very close together.

variable region   In an immunoglobulin (antibody) molecule, the amino-terminal polypeptide segment, of about 110 amino acids, that contains the sequence variation responsible for antibody diversity.

vas deferens   [Latin, *vas* = vessel + *deferre* = to carry down] In the male reproductive tract, two large, thick-walled ducts that carry sperm from the testes.

vascular   Having a conducting system.

vascular system   [Latin, *vas* = vessel] In plants, the structures and cells that serve as a transport system; the vascular system forms the central core of the plant.

vasectomy   An irreversible method of birth control; the cutting and tying of the vas deferens, which prevents sperm from entering the urethra.

vasoconstriction   Contraction of the smooth muscles surrounding the arterioles.

vasodilation   A process causing arterioles to increase in size with consequent blood flow increase and reddening on the skin.

vasopressin (also called **antidiuretic hormone** or **ADH**)   A peptide hormone, released by the posterior pituitary, that stimulates water reabsorption by the kidneys.

vegetative reproduction   The process of producing a new plant without fertilization or seed production.

veins   In vertebrates, one of the vessels that carry blood back to the heart. In plants, the vascular tissue in the leaves.

velocity   In an enzymatic reaction, the rate of appearance of product.

vena cava (**superior** and **inferior**)   The two largest veins, which run up through the center of the body and carry deoxygenated blood from the body to the right atrium.

ventilation   The flow of air into and out of the alveoli.

ventral   The part of an animal facing the Earth.

ventricle   In the heart, a thick-walled pumping chamber.

ventricular fibrillation   Cardiac arrest.

venules   The smallest veins.

vertebrae   [singular, **vertebra**; Latin, *vertebratus* = jointed] The hollow, bony segments that make up the backbone.

vesicle   A membrane-bounded sac; some vesicles are present inside living eukaryotic cells and others form after a cell is broken open.

vessel elements   In flowering plants, structural elements of the phloem that are open at each end and connect end to end to form long open channels.

vestigial structure   [Latin, *vestigium* = footprint] A part of an organism, with little or no function, that reflects evolutionary history.

vicariance   The fragmentation of an already dispersed species or group of species.

villus   [plural, **villi**; Latin, shaggy hair] The highly folded mucosa of the small intestine that each extend about 1 mm into the lumen.

viral oncogene   A viral gene that affects growth control in a host cell.

virus   An assembly of nucleic acid (DNA or RNA) and proteins (and occasionally other components, such as lipids or carbohydrates) that can reproduce only within a living cell; viruses depend on cells to obtain energy and perform chemical reactions.

visceral mass   [Latin, *viscera* = internal organs] The main body of a mollusk; the visceral mass contains the intestinal tract, as well as the excretory and reproductive organs.

visible light   Electromagnetic radiation, with wavelengths from about 400 to about 750 nm, that can be perceived by human eyes and brains.

vitamin   An organic compound that an animal cannot itself synthesize but that is required in minute amounts for normal growth and metabolism; an essential component of the diet; many vitamins are cofactors in enzymatic reactions.

voltage clamping   A method that allows the control of the voltage across a cell membrane.

voltage (also called **electrical potential**)   A measure of the potential energy that results from a charge separation.

voltage-gated   Open or closed according to the voltage across the membrane.

voluntary muscles   Skeletal muscles; voluntary muscles enable voluntary control over body movements.

vulva   [Latin, *volvere* = to wrap] In animals, the collective name for the external genitalia of the female.

warning coloration   A bright, memorable design that helps the predator remember which prey to avoid.

water potential   The potential energy of water per gram (or kilogram); the difference between osmotic pressure and physical pressure on either side of a selectively permeable membrane.

water-soluble signal   A chemical signal that cannot pass through the plasma membrane and must act on its surface.

wavelength   The distance between the crests of two successive waves.

westerlies   Winds that come from the west caused by the rotation of the Earth, characteristic of latitudes between 30° and 60° in both hemispheres.

western blotting (after a similar blotting method, called "Southern blotting" because it was invented by Edward Southern)   A method for detecting a particular protein after electrophoresis by blotting the contents of the electrophoresis gel onto a paperlike support and incubating this "blot" with a specific antibody.

whorl   One of four concentric circles of flower parts at the end of a specialized stem.

whorled   The arrangement of leaves on a stem so that there are three or more petioles at each node.

wild type   An allele that is most highly represented in wild populations.

wobble   Nonstandard base pairing between the first base in the anticodon of transfer RNA and the third base of the codon in mRNA; wobble allows some tRNAs to recognize more than one codon for the same amino acid.

woody plants   Trees, shrubs, and other perennial plants that have woody parts that provide support independent of turgor pressure.

work The movement of an object against a force, or the conversion of energy into electrical energy, chemical energy, or concentration energy.

x-ray crystallography The analysis of the scattering, or "diffraction," of x rays by crystals.

xylem [Greek, *xylon* = wood] Conducting vessels that carry water and minerals from roots to the photosynthesizing leaves.

yield The amount of grain per acre.

yolk In an amniotic egg, a mixture of proteins, lipids, and carbohydrates that nourishes an embryo until it can feed itself.

yolk sac The membrane that surrounds the yolk in an amniotic egg.

Z band In muscle, the boundary of a sarcomere.

zero population growth The state of a population when its size reaches the carrying capacity and the environment cannot support any more growth; the state when a population reaches a stable size.

zoomastigina (also called **flagellates**) [Greek, *mastix* = whip] A diverse group of single-celled heterotrophs.

zygomycetes [Greek, *zygote* + *mukes* = fungus] Conjugation fungi.

zygosporangium A structure that produces haploid spores from a diploid zygote.

zygote [Greek, *zygon* = yoke] The first cell of an embryo, formed by the union of egg and sperm.

INDEX

Note: An italic page number indicates an illustration; "t" following a page number indicates a table.

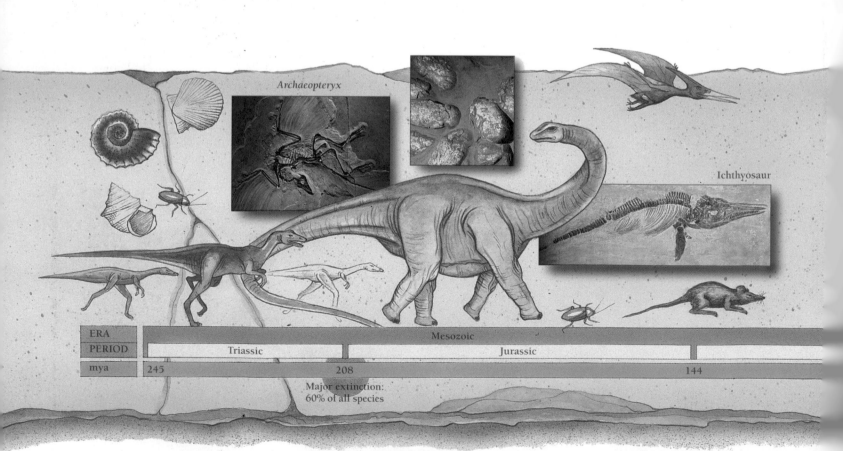

Archaeopteryx

Ichthyosaur

ERA		Mesozoic		
PERIOD		Triassic	Jurassic	
mya	245	208		144

Major extinction:
60% of all species